GARDNER'S CHEMICAL SYNONYMS AND TRADE NAMES

Gardner's Chemical Synonyms and Trade Names

Ninth Edition

EXECUTIVE EDITOR **Jill Pearce**

CONSULTANT **John Buckingham D Phil**

Gower Technical Press

Published by
Gower Technical Press Ltd,
Gower House,
Croft Road,
Aldershot,
Hants GU11 3HR,
England

Gower Publishing Company,
Old Post Road,
Brookfield,
Vermont 05036,
U.S.A.

British Library Cataloguing in Publication Data

Gardner, William

Gardner's chemical synonyms and trade names.—9th ed.
1. Chemicals—Dictionaries 2. Chemistry, Technical—Dictionaries
I. Title II. Pearce, Jill III. Gardner, William. Chemical synonyms and trade names
661' .003' 21 TP200

Library of Congress Cataloging-in-Publication Data

Gardner's chemical synonyms and trade names.

Includes index.
1. Chemicals—Dictionaries. 2. Chemicals—Trademarks. I. Pearce, Jill. II. Buckingham, John.
TP9.G286 1987 660'.03'21 86-29562

ISBN 0-291-39703-4

Preface

For more than 60 years and through a total of eight previous editions, *Gardner's* has been the best known and most widely used rapid source of information on trade names and synonyms for chemicals of commerce. In continuing this tradition with a new and greatly revised Ninth edition, the publishers have sought to maintain and enhance the prestige and value of this important reference work.

The degree of revision has probably been greater than ever previously undertaken and, for the first time, much material has been deleted, but with the addition of approximately 12,000 new entries, along with a comprehensive index, the result has been a larger volume containing at least 45 per cent new material. The trade entries have been obtained directly from chemical manufacturers worldwide, supplemented by a research programme into other secondary sources, including manufacturers' advertisements in the trade press.

The book is now in two parts. Part 1 is the alphabetical list of trade names and synonyms which in style is essentially unchanged from previous editions. Part 2, the coloured section, represents a considerable expansion of the very brief list of suppliers given as an appendix in the previous edition, and presents the names and addresses of all manufacturers who have cooperated in the production of this edition, together with a block listing of their products. Material contained in this volume comprises:

1 New trade names and synonyms derived from information supplied directly from manufacturers.
These are marked *

2 A proportion of entries retained from the 8th edition which are believed to be current but which have not been verified for this edition. These divide into two groups:
(i) trade names and synonyms which have not been verified by their designated manufacturers. This is either because there is insufficient information available concerning the manufacturer or because manufacturers have failed to supply information when requested.
(ii) trade names and synonyms without an accompanying manufacturer's affiliation.
These are marked †

3 New trade names deriving from the research programme described above.

Proprietary considerations

Every attempt has been made to ensure the accuracy of the information given in this new edition of *Gardner's*. However, the publishers cannot be held responsible for the accuracy of the information, and users are requested to bear in mind the following important considerations:

1. The reporting of a name in *Gardner's* cannot imply definitive legality in establishing proprietary usage. Questions of legal ownership of a particular name can only be resolved by due legal process.
2. A manufacturer may in some countries market its product under alternative names to those cited in *Gardner's*. Similarly, manufacture or marketing of a product may be licensed to a separate company in another country either under the same or a different name.
3. No indication is given in *Gardner's* whether or not trade names have been formally registered.

We are confident that users will find this new edition a wealth of useful information which is difficult to obtain from any other source. It is the intention to keep the database from which this book is produced regularly updated and to publish further revised editions at suitable intervals.

Companies wishing to submit material for inclusion in the 10th edition should write to: Jill Pearce, Gower Technical Press, Gower House, Croft Road, Aldershot, Hampshire GU11 3HR.

A

A-1 Thiocarbanilide. 1,3-diphenyl-2-thiourea; N,N-diphenyl-thiourea. Applications: vulcanization accelerator for fast-curing repair stocks; neoprene latex, natural rubber latex and cements; an activator for thiazole accelerators. (Monsanto Co).*

A.A.A. Ointment. A proprietary mercurial ointment containing salicylic acid, boric acid and zinc oxide, used to treat skin irritations. (Jenkins Laboratories Inc, Union, NJ).†

A.A.A. Spray. A proprietary preparation of benzocaine and cetalkonium chloride in an aerosol. A throat spray. (Armour Pharmaceutical Co).†

A-Acid. Anthraquinone-2-sulphonic acid. A dyestuff intermediate.†

A-acid. 2-Aminonaphthol,5-sulphonic acid, a dye intermediate.†

Aba-odo. An African term for a mixture of rubber latexes, probably those from *Funtumia elastica* and *Ficus Vogelii.*†

Abaca. Manila hemp, the inner fibre of *Musa textilis.*†

Abacid. An antacid preparation containing aluminium hydroxide gel and magnesium trisilicate. (Unichem).†

Abacid Plus. An antacid preparation containing aluminium hydroxide, magnesium trisilicate, dimethicone and dicyclomine hydrochloride. (Unichem).†

Abalyn. Methyl ester of rosin. It has a Gardner colour 6, Gardner Holt viscosity 21 and acid number 6. Applications: It is used in adhesives as a softening, plasticizing and tackifying resin and to increase open time. It is used in coatings to contribute gloss and fullness to lacquers. (Hercules Inc).†

Abanone. Magnesium phospho-tartrate.†

Abasin. A proprietary preparation of acetyl carbromal. A sedative. (Bayer & Co).*

Abatia. The leaves of *Abatia rugosa,* used as a black dye.†

Abavit B. Organo-mercury seed dressing. (Murphy Chemical Co).†

Abavit S. Organo-mercurial dip. (Murphy Chemical Co).†

Abbaflox. A proprietary preparation of flucloxacillin. Applications: Antibiotic. (Abbott Laboratories).*

Abbalgesic. A proprietary preparation of paracetemol with dextropropoxyphene. Applications: Analgesic. (Abbott Laboratories).*

Abbarome. Perfumery base. (Bush Boake Allen Ltd).

Abbavert. 2-Ethyl hexanal 1,2-ethane diol cyclic acetal. (Bush Boake Allen Ltd).

Abbcite No. 2. An explosive for coal mines containing ammonium nitrate, nitroglycerine and dinitrotoluene.†

Abbenclamide. A proprietary preparation of glibenclamide. Applications: Oral treatment of hypoglycaemia. (Abbott Laboratories).*

Abbloraz. A proprietary preparation of lorazepam. Applications: Anxiolytic. (Abbott Laboratories).*

Abbocillin-DC. Penicillin G Procaine. Applications: Antibacterial. (Abbott Laboratories).

Abbokinase. Urokinase. Applications: Plasminogen activator. (Abbott Laboratories).

Abbolactone. A proprietary preparation of spironolactone. Applications: Oedema in cirrhosis of the liver, nephrotic syndrome, congestive heart failure, potentiation of thiazide and loop diuretics, hypertension, Conn's syndrome. (Abbott Laboratories).*

Abbopramide. A proprietary preparation of metoclopramide. Applications: To promote gastric motility and emptying. (Abbott Laboratories).*

Abbopurin. A proprietary preparation of allopurinol. Applications: Gout prophylaxis, hyperuricaemia. (Abbott Laboratories).*

Abboxapam. A proprietary preparation of oxazepam. Applications: Anxiolytic. (Abbott Laboratories).*

Abboxide. A proprietary preparation of chlordiazepoxide. Applications: Anxiolytic. (Abbott Laboratories).*

ABC Liniment. Compound liniment of aconite (*Linimentum aconiti compositum BPC.*)†

Abeku Nuts. See NJAVE BUTTER.

Abelite. Blasting explosives containing ammonium nitrate and trinitrotoluene.†

Abel's Reagent. A 10 per cent solution of normal chromic acid. It is used in the micro-analysis of carbon steels, for etching.†

Abequito. Nimidane. Applications: Acaricide. (American Cyanamid Co).

Aberel. A proprietary preparation for treatment of acne. Retinoic acid. (McNeil Laboratories).†

Aberoid. A proprietary casein plastic material.†

Abicol. A proprietary preparation of reserpine and bendrofluazide used for control of hypertension. (Boots Company PLC).*

Abidec. A vitamin preparation containing Vitamin A, Vitamin D, thiamine, riboflavine, nicotinamide and ascorbic acid. (Parke-Davis).†

Abietic Anhydride. Synonymous with Colophony.†

Abisol. A 40 per cent solution of sodium hydrogen sulphite, $NaHSO_3$. A disinfectant and preservative.†

Abitol. Technical grade of hydroabietyl alcohol. It has a typical hydroxyl content of 4.8%. Applications: It is used as a tackifying resin and as a wetting aid for difficult to wet substrates, an intermediate in the production of stabilizers and as a synthetic resin for type E gravure inks. (Hercules Inc).†

Abocast. Adhesives, casting epoxies, solventless coatings. (Abatron Inc).

Abocrete. Concrete and masonry patching and resurfacing compound. (Abatron Inc).

Abocure. Catalysts, curing and hardening agents. (Abatron Inc).

Abol. Pesticides for garden use. (ICI PLC).*

Abol G. Garden aphicide. (ICI PLC, Plant Protection Div).

Abopon. A proprietary liquid inorganic resinous product which forms films in a few minutes on drying in air. It is recommended as an adhesive, a suspended medium for pigments and abrasives, and for sealing of surfaces to be lacquered or painted. It is stated to be a boro-phosphate.†

Aboseal. Sealants, caulks. (Abatron Inc).

Abracol. A registered trade mark for toluene sulphonic acid esters, used as plasticizers. Emulsifying agents. (Bush Boake Allen).*

Abradux. Extra tough and dense types of aluminium oxide. Applications: Used for abrasive industries, super refractories, sandblasting and for safes. (Lonza Limited).*

Abramant. Semi-friable aluminium oxide. Applications: Used for abrasive industries, super refractories, sandblasting and for safes. (Lonza Limited).*

Abramax. Pure white aluminium oxide. Applications: Used for abrasive industries, super refractories, sandblasting and for safes. (Lonza Limited).*

Abrarex. Special grades of pure aluminium oxide. Applications: Used for abrasive industries, super refractories, sandblasting and for safes. (Lonza Limited).*

Abrasit. Regular aluminium oxide, finely crystallized. Applications: Used for abrasive industries, super refractories, sandblasting and for safes. (Lonza Limited).*

Abrastol. (Asaprol, Asaprol-etrasol). Calcium - β - naphthol - α - sulphonate. It is used as a clarifier for wines.†

Abraum. A red ochre used to stain new mahogany.†

Abraum Salts. (Stripping Salt, Stassfurt Salts, Potash Salts). The names applied to the upper layers of mixed chlorides of magnesium, potassium, and dosium, overlying the beds of rock-salt at Stassfurt.†

Abril. Synthetic waxes. (Abril Industrial Waxes).*

Abrodil. A proprietary preparation of sodium iodomethanesulphonate.†

Abros. A heat-resisting alloy containing 88 per cent nickel, 10 per cent chromium, and 2 per cent manganese.†

A.B.S. Abbreviation for acrylonitrile butadiene styrene, an impact resistant moulding material.†

Absaglas. A registered trade name for flame retardant acrylonitrile- butadiene styrene.†

Absolute Acetic Acid. See ACETIC ACID, GLACIAL.

Abson A.B.S. 300. A grade of A.B.S. having medium impact resistance. Its good gloss makes it suitable for refrigeration applications. (BF Goodrich Canada, Kitchener, Ontario).†

Abson A.B.S. 213. A proprietary general-purpose grade of A.B.S. used in injection moulding and extrusion. (BF Goodrich Canada, Kitchener, Ontario).†

Abson A.B.S. 230. A proprietary general-purpose grade of A.B.S. possessing more toughness. Used in applications requiring a higher gloss. (BF Goodrich Canada, Kitchener, Ontario).†

Abson A.B.S. 500. A grade of A.B.S. having high-to-medium impact resistance and good strength over a wide temperature range. (BF Goodrich Canada, Kitchener, Ontario).†

Abyssinian. Gold (Talmi Gold, Cuivre Poli). A yellow alloy of copper and zinc. It usually consists of about 91 per cent copper and 8 per cent zinc, but sometimes contains 86 per cent copper, 12 per cent zinc, and 1 per cent tin. Employed in the manufacture of trinkets.†

Ac-Di-Sol. Croscarmellose sodium. Carboxymethylcellulose sodium that has been internally cross-linked. Applications: Pharmaceutic aid. (FMC Corporation).

Ac-zol. A mixture of ammonia, copper, zinc salts, and phenol. A wood preservative.†

Acacia Gum. See AMRITSAR GUM, and GUM ARABIC.

Acagine. A mixture of lead chromate and bleaching powder. Used to purify acetylene.†

Acajou Balsam. (Cardol). A material obtained from the fruits of *Anacardium occidentale* (mahogany nuts, elephant nuts)or *A orientale* by the extraction of the powdered nuts.It is employed in the preparation of indelible inks and colours for die-sinking work.†

Acaprin. Chemotherapeutic against piroplasmosis (babesiasis). Applications: Veterinary medicine. (Bayer & Co).*

Acarbodavyne. It is an aluminosilicate, sulphate and chloride of calcium and alkali metals; a variety free from CO_2.†

Acarin. Equivalent trade name for mitigan. (Agan Chemical Manufacturers Ltd).†

Accelemal. A proprietary rubber vulcanization accelerator. It is possibly thiocarbamide.†

Accelerase. Pancrelipase. A concentrate of pancreatic enzymes standardized for lipase content. Applications: Enzyme. (Organon Inc).

Accelerated Cement. A Portland cement containing a high proportion of lime.†

Accelerator A1. A proprietary thiocarbanilide rubber vulcanization accelerator.†

Accelerator A1010. A proprietary rubber vulcanization accelerator consisting of formaldehyde-aniline.†

Accelerator A17. A proprietary rubber vulcanization accelerator. It consists of methylene-*P*-toluidine.†

Accelerator A19. A proprietary rubber vulcanization accelerator. It is a modified ethylidene-aniline.†

Accelerator A20. A proprietary rubber vulcanization accelerator.†

Accelerator A22. A proprietary rubber vulcanization accelerator. It consists of di-*o*-tolyl-thiourea.†

Accelerator A32. A proprietary rubber vulcanization accelerator. It consists of condensation products of aldehydes and Schiff's base, *e.g.* butyraldehyde and butylidine aniline.†

Accelerator A50. A proprietary rubber vulcanization accelerator. It consists of aldehyde-amine.†

Accelerator A7. A proprietary rubber vulcanization accelerator made from 2 molecules of ethylidene aniline condensed with 1 molecule of acetaldehyde.†

Accelerator A77. Condensation product of acetaldehyde and aniline.†

Accelerator BB. A proprietary rubber vulcanization accelerator. It is butyraldehyde *p* - amino - dimethylaniline.†

Accelerator D. A proprietary rubber vulcanization accelerator. It is di-phenyl-guanidine.†

Accelerator DBA. A proprietary rubber vulcanization accelerator. It is di-benzyl-amine.†

Accelerator DT. A French proprietary rubber vulcanization accelerator. It is di-*o*-tolyl-guanidine.†

Accelerator E-A. - A proprietary rubber vulcanization accelerator. It consists of ethylidene-aniline.†

Accelerator G.M.F. A proprietary rubber vulcanizing accelerator. Quinone dioxime.†

Accelerator L. A French proprietary rubber vulcanization accelerator. It is thiocarbanilide.†

Accelerator Mercapto. A German rubber vulcanization accelerator. It is mercapto-benzthiazole.†

Accelerator PTX. A proprietary rubber vulcanizing accelerator. It is phenyl-tolyl-xylyl-guanidine.†

Accelerator R2. A proprietary rubber vulcanization accelerator. It is a condensation product from melthylene-dipiperidine and carbon disulphide.†

Accelerator R3. A proprietary rubber vulcanization accelerator. It is the zinc salt of a dithio-carbaminic acid.†

Accelerator R5. A proprietary rubber vulcanization accelerator. It is dithio-carbamate.†

Accelerator W29. A proprietary rubber vulcanization accelerator. It is a compound of diphenyl-guanidine with di-benzyl-dithio-carbaminic acid.†

Accelerator W80. A proprietary rubber vulcanization accelerator. It is the diphenyl-guanidine salt of mercapto-benzthiazole.†

Accelerator XLM. Rubber vulcanizing accelerator. Condensation product butyraldehyde - *P* - amino - dimethyl aniline.†

Accelerator XLO. Rubber vulcanizing accelerator. Two parts diphenyl-guanidine and 1 part magnesium oxide.†

Accelerator X28. A proprietary rubber vulcanization accelerator. It is stated to be impure tarry diphenylguanidine.†

Accelerator ZBX. A proprietary rubber vulcanization accelerator. It is zinc butyl xanthate.†

Accelerator ZPD. Proprietary rubber vulcanizing accelerator. Zinc penta methylene-dithio-carbamate.†

Accelerator Z88. A proprietary rubber vulcanization accelerator. It is the ammonia salt of mercapto-benzthiazole mixed with a softener.†

Accelerator 100. Aldehyde derivative of a Schiff's base, made from both butyraldehyde and acetaldehyde.†

Accelerator 108. A proprietary rubber vulcanizing accelerator. Tetramethyl thiyram disulphide 2/3, 2-merceapto-benxthiazole 1/3. (Naugatuck (US Rubber)).†

Accelerator 2P. A proprietary rubber vulcanizing accelerator. Piperidinium pentamethylene dithiocarbamate. (Anchor Chemical Co).*

Accelerator 4P. A proprietary rubber vulcanizing accelerator. Dipentamethylene thiuram disulphide. (Anchor Chemical Co).*

Accelerator 808. A proprietary rubber vulcanizing accelerator. Butyraldehyde aniline.†

Accelerators A11, 16E. Proprietary rubber vulcanization accelerators. They are aldehyde Schiff's bases.†

Accelerators A5-10. Proprietary formaldehyde-anilines rubber vulcanization accelerators.†

Accelerators F-A. Proprietary rubber vulcanization accelerators. The high melting one is formaldehyde-aniline, and the low melting one consists of methylene-dianilide.†

Accelerene V 1. A proprietary rubber vulcanization accelerator. It is stated to consist of equi-molecular proportions of *P*-nitroso-dimethylaniline and β-naphthol.†

Accomet. Paint pretreatment. (Albright & Wilson Ltd, Phosphates Div).

Accomet C. Metal pre-cleaning process. (Albright & Wilson Ltd, Phosphates Div).

Accosize. Modified alkenylsuccinic anhydride. Applications: The paper industry. (Cyanamid BV).*

Accostrength. Anionic polyacrylamide. Applications: The paper industry. (Cyanamid BV).*

Accostrength 72. Acrylamide copolymers (anionic). Applications: The paper industry. (Cyanamid BV).*

Accrotan. Self-basifying chrome tanning material. (British Chrome & Chemicals Ltd).

Accrox. Special refractory grades of chromic oxide. (British Chrome & Chemicals Ltd).

Accuglass. Spin-on Glass. Applications: Solutions of inorganic polymers (siloxanes or silicates) spun applied to silicon wafers to form insulating or passivating films used in the manufacture of integrated circuits. (Allied-Signal Planarization and Diffusion Products).*

Acculog. Standard volumetric concentrated solutions. Applications: Quantitative analytical chemistry. (Chemical Dynamics Corp).*

Accumulator Metal. An alloy called by this name contains 90 per cent lead, 9.25 per cent tin, and 0.75 per cent antimony.†

Accurac. Polymers based on acrylamide: can be described as non-ionic polyacrylamide, anionic polyacrylamide and cationic polyacrylamide. Applications: The paper industry. (Cyanamid BV).*

Accurac 33/35/41. Polyamine condensation products. Applications: The paper industry. (Cyanamid BV).*

Accurac 39. Polyacrylate. Applications: The paper industry. (Cyanamid BV).*

Accuspin. Spin-on Dopant. Applications: Solutions of impurity atoms (boron, phosphorus, arsenic or antimony) spun applied to silicon wafers used to dope silicon to form transistors and integrated circuits. (Allied-Signal Planarization and Diffusion Products).*

Accutane. Isotretinoin. Applications: Keratolytic. (Hoffmann-LaRoche Inc).

Ace-ite. A proprietary bituminous or asphalt composition.†

Ace-Sil. A proprietary trade name for microporous rubber. It is used for battery separators and filters.†

Aceito de Abeto. A turpentine of sharp acrid flavour, from the Mexican *Pinus religiosa.*†

Aceloid. A proprietary cellulose acetate material used as a moulding composition.†

Acelon. Cellulose acetate coated fabric. (May & Baker Ltd).*

Acelose. A proprietary cellulose acetate material.†

Aceplus. A proprietary cellulose acetate material.†

Acerado. (Fierroso). Names used for mercurial earths.†

Acerdol. Calcium permanganate, $CaMn_2O_8$. Used in gastro-enteritis and diarrhoea.†

Acertannin. The tannin of *Acer ginnula,* the Korean maple tree.†

Aceta. A proprietary brand of cellulose rayon. The name is also applied to a French nitro-cellulose lacquer.†

Acetadeps. Lanolin acetate. Applications: Oil-soluble emollient for cosmetics. (Westbrook Lanolin Co).*

Acetaloid. A proprietary cellulose acetate material in the form of rods, sheet, and moulding composition.†

Acetannin. (Diacetyl tannin). Acetyltannic acid.†

Acetargol. A mixture of formic and acetic acids.†

Acetazolamide. 5-acetamido-1,3,4-thiadiazole-2-sulphonamide. Diamox. (Lederle Laboratories).*

Acetest. A mixture of sodium nitroprusside, aminoacetic acid, disodium phosphate, and lactose, in tablet form, used to test for the presence of ketones in blood and urine. (Ames).†

Acetex. A proprietary safety-glass.†

Acetic Alcohol. A fixing agent used in microscopy. It consists of 6 parts absolute alcohol, 3 parts chloroform, and 1 part glacial acetic acid.†

Acetic Ester. (Acetic Ether). Ethyl acetate.†

Acetic Ether. See ACETIC ESTER.

Acetic Ethers. Alkyl acetates.†

Acetic-gelatin Cement. An adhesive obtained by dissolving gelatin in glacial acetic acid.†

Acetidin. Ethyl acetate†

Acetin Blue. Solutions of indulines in acetins (acetic esters of glycerol). See INDULINE, SPIRIT SOLUBLE.†

Acetinduline R. See INDULINE.

Acetine. A mixture of various acetyl derivatives of glycerol. Used in dyeing, either in a neutral form, or in one containing about 20 per cent acid.†

Aceto-carmine. A microscopic stain; solution of carmine in acetic acid.†

Aceto-caustin. Trichloroacetic acid. A caustic.†

Acetodin. See ASPIRIN.

Acetohexamide. Dimelor. An oral hypoglycaemic agent.†

Acetol. A proprietary cellulose acetate material in the form of flake.†

Acetonal. A 10 per cent solution of aluminium and sodium acetates, used in medicine.†

Acetone. See OIL OF ACETONE; PYROACETIC SPIRIT.

Acetone Alcohol. See GREENWOOD SPIRITS; MANHATTEN SPIRITS; STANDARD WOOD SPIRITS

Acetone Bromoform. See BROMETONE.

Acetone Chloroform. See CHLORETONE. The term is also used for chloroform prepared from acetone.

Acetonyl. A proprietary analgesic. It is the sodium salt of aspirin with potassium and sodium tartrates. (Upjohn Ltd).†

Acetophen. A proprietary preparation of aspirin, phenacetin and caffeine in capsule form. (Jenkins Laboratories Inc, Union, NJ).†

Acetophenetidin. An antipyretic and analgesic. Phenedidin. Ethoxyacetanilide.†

Acetophenone. See MITTEL HNA.

Acetopyrin. See ACOPYRIN.

Acetose. Acetyl-cellulose. Used in the manufacture of artificial silk.†

Acetosol. (Westrol). A trade name for tetrachloroethane.†

Acetosulphone. Promacetin. (Parke-Davis).†

AcetOxyl. Available in two strengths - AcetOxyl 2.5 and AcetOxyl 5. Contains benzoyl peroxide (either 2.5% or 5%) in an aqueous gel base. Applications: As an aid in the treatment of acne vulgaris. (Stiefel Laboratories (UK) Ltd).*

Acetryptin . A proprietary anti-hypersensitive agent. 5-Acetyltryptamine. (Warner-Chilcott, Morris Plains, NJ).†

Acetulan. Cetyl acetate and acetylated lanolin alcohol. Applications: Thin hydrophobic, spreading, emollient oil, plasticizer and cosolvent. Imparts velvety afterfeel, degreases waxy systems. (Amerchol Corporation).*

Acetyl. Anionic dyestuffs (level dyeing). Applications: Wool and wool blends. (Holliday Dyes & Chemicals Ltd).*

Acetyl Adalin. ACETYL CARBROMAL.†

Acetyl-amino-ethoxy-benzene. See PETONAL.

Acetyl-amino-salol. See PHENESAL.

Acetyl-cellulose. See ACETOSE.

Acetyl-choline Hydrochloride. See ACECHOLINE.

Acetyl-Digitoxin. α-digitoxin monoacetate. Acylanid.†

Acetyl-dinitro-butyl-xylene. See MUSK KETONE.

Acetylarsan. Dimethylamine acetarsol. (May & Baker Ltd).

Acetylcysteine. N -acetyl - l - cysteine. Airbron.†

Acetyldihydrocodeinone. Thebacon.†

Acetylene Black. (Shawinigan Black). A carbon black made by incomplete combustion of acetylene.†

Acetylene Blue 6B, BX, 3R. Dyestuffs. These colours dye cotton from a neutral bath containing sodium sulphate.†

Acetylene Stones. See ACETYLITH.

Acetylmethadol. Methadyl Acetate.†

Acetysal. See ASPIRIN.

Acheson's Deflocculated Graphite. (Dag). The trade mark for a lubricant obtained by macerating graphite with a solution of tannin for several weeks,forming a permanent emulsion. The graphite with water is called Aquadag. Oildag is prepared by pouring oil over the filtered dag, then freeing the material from moisture. See AQUADAG and OILDAG.†

Achilles Dipentene. Dipentene - commercial grade. Applications: Solvent for resins, waxes and oils, perfumery, wetting agent and antiskinning for paint. (Langley Smith & Co Ltd).*

Achilles Pine Oil. Pine oil - mixed isomers of terpene alcohol. Applications: Disinfectants/cleaners, plasticizer for epoxy resin, solvent and leveller in paint formulations, wetting agent for pigments. (Langley Smith & Co Ltd).*

Achilles Tall Oil Fatty Acid. Oleic acid/linoleic acid mixture with a rosin acid content. Applications: Alkyd resins, detergents, disinfectants, soaps and core oils. (Langley Smith & Co Ltd).*

Achirite. A name which has been used for dioptase.†

Achromycin. A proprietary preparation containing tetracycline hydrochloride. An antibiotic. (Lederle Laboratories).*

Achroodextrins. Intermediate products formed by the hydrolysis of starch or dextrin.†

Acid Alizarin. See CHROME.

Acid Alizarin Black. A dyestuff prepared from diazotized 6-nitro-2-amino phenol-4-sulphonic acid, and β-naphthol.†

Acid Alizarin Blue BB, GR. (Alizarin blue BB, GR, Alizarin Acid Blue BB). A dyestuff. It is the sodium salt of hexahydroxyanthraquinonedisulphonic acid. Dyes wool red from an acid bath, and blue from subsequent chroming. The mark GR gives duller shades.†

Acid Alizarin Brown B. An acid mordant dyestuff, which dyes wool from an acid bath.†

Acid Alizarin Green B, G. A dyestuff. It is the sodium salt of dimercapto-tetrahydroxy-anthraquinone-disulphonic acid. Dyes wool greenish-blue from an acid bath.†

Acid Black HA. A dyestuff. It is a British equivalent of Naphthol blue black.†

Acid Black N. A dyestuff. It is a British equivalent of Naphthylamine black D.†

Acid Blue B. A dyestuff for wool, of a greener shade than Acid Blue R.†

Acid Blue Black. A dyestuff. It is a British equivalent of Naphthol blue black.†

Acid Blue Black 428. A dyestuff. It is a British equivalent of Naphthol blue black.†

Acid Blue R. A dyestuff for wool, which it dyes from a bath of sodium sulphate containing a little sulphuric acid. Employed as a groundwork for logwood black.†

Acid Blue 6G. See CYANOL EXTRA.

Acid Bronze. Alloys containing from 82-88 per cent copper, 8-10 per cent tin, 2-8 per cent lead, and 0-2 per cent zinc. A metal containing 90 per cent copper and 10 per cent aluminium is also known as Acid Bronze.†

Acid Brown. (Dahl & Co.). See FAST BROWN G.†

Acid Brown. See FAST BROWN N.

Acid Brown G. An azo dyestuff, Dyes wool brown from an acid bath.†

Acid Brown R. (Vega brown R). A dyestuff. Dyes wool brown from an acid bath.†

Acid Carmoisine B. See FAST RED E.

Acid Cerise. See MAROON S.

Acid Clay, Japanese. See JAPANESE ACID CLAY.

Acid Eosin. Tetrabromofluorescein. The potassium salt forms soluble eosin, the sodium salt, Eosin C, and the ammonium salt, Eosin B. See EOSIN.†

Acid Fuchsine. See ACID MAGENTA.

Acid Green. See GUINEA GREEN B, FAST GREENS, and above.

Acid Green G. A dyestuff. It is a British equivalent of Guinea green B.†

Acid Haemotoxylin. See EHRLICH's ACID HAEMATOXYLIN.

Acid Levelling Red 2B. A dyestuff. It is a British equivalent of Azofuchsine B.†

Acid Magenta. (Acid Fuchsine, Fuchsine S, Magenta S, Cardinal S, Acid Roseine, Acid Rubine, Rubine S). A dyestuff. It is a mixture of the sodium or ammonium salts of the trisulphonic acids of rosaniline and pararosaniline. Sodium salts have the formulae $C_{19}H_{16}$. Dyes wool and silk red from an acid bath. Impure Acid magenta is sold as Acid maroon, Acid cerise, Maroon S, Grenadine S, and Cerise S.†

Acid Magenta II, Conc, L, N, N Extra. Dyestuffs. They are British equivalents of Acid magenta.†

Acid Maroon. See MAROON S.

Acid Mauve B. A sulphonated mauvaniline. It is an acid dyestuff, and dyes wool and silk mauve from an acid bath.†

Acid Metaborate. See BORAX.

Acid Milling Scarlet. A dyestuff. Dyes wool scarlet from an iron bath.†

Acid of Amber. Succinic acid, CH_2CH_2COOH.†

Acid of Lemons. Citric acid.†

Acid of Sugar. Oxalic acid.†

Acid Orange. See ORANGE II.

Acid Orange G. A dyestuff. It is a British equivalent of Orange II.†

Acid Red. A dyestuff. It is a British equivalent of Fast red.†

Acid Roseine. See ACID MAGENTA.

Acid Rubine. See ACID MAGENTA.

Acid Scarlet. A dyestuff. It is a British equivalent of Ponceau 2G.†

Acid Scarlet R. A dyestuff. It is a British equivalent of New Coccine.†

Acid Soaps. Soaps obtained by treating a dissolved soap with acid just insufficient to produce separation of the fatty acids. They are used in dyeing.†

Acid Tar. The waste acid from the washing of crude light oils of coal tar.†

Acid Violet 3BN. An acid colour, which dyes wool a less blue shade than 4BN.†

Acid Violet 4R. See FAST ACID VIOLET A2R.

Acid Violet 4RS. A dyestuff. It is a mixture of Acid violet with Fuchsine S. See RED VIOLET 4RS.†

Acid Violet 5RS. See RED VIOLET 5RS.

Acid Violet 6B. A dyestuff.Dyes wool bluish-violet from an acid bath.†

Acid Violet 6BN. A dyestuff. Dyes wool and silk bluish-violet.†

Acid Violet 7B. Dyes wool bluish-violet from an acid bath.†

Acid Violet 7BN. A dyestuff. Dyes wool and silk bluish-violet from an acid bath.†

Acid Violet 7BS, 5BNS, BNS. Dyestuffs which dye wool bluish-violet.†

Acid Vitriolated Tartar. Potassium bisulphate $KHSO_4$.†

Acid Wool Dyes. Colours which dye wool directly from from an acid bath.†

Acid Yellow. (Fast Yellow G, Fast Yellow GL, Acid Yellow G, Fast Yellow, Fast Yellow Extra, New Yellow L, Fast Yellow, Greenish, Fast Yellow S, Solid Yellow S). A dyestuff. Dyes wool and silk yellow from an acid bath.†

Acid Yellow D. See ORANGE IV.

Acid Yellow G. See ACID YELL0W.

Acid Yellow OO. See BRILLIANT YELLOW (ICI PLC).*

Acid Yellow R. See FAST YELLOW R.

Acid Yellow RS. See TROPAEOLINE O.

Acid Yellow S. See NAPHTHOL YELLOW S.

Acid Yellow 2G. See METANIL YELLOW S.

Acid Yellow 79210. A dyestuff. It is a British equivalent of Tartrazine.†

Acid-proof Ultramarine. See ULTRAMARINE, ACID-PROOF.

Acidax. A stearic acid for use in the rubber industry.†

Aciderm. Dyestuffs. Applications: Acid dyes for all kinds of leather. (Bayer & Co).*

Acidol Chromate Violet B. An acid chrome-developing dyestuff.†

Acidol Chrome Brilliant Blue BL. An acid chrome-developing dyestuff, giving navy blue shades on wool.†

Acidol Fast Violet BL. An acid colour suitable for wool printing.†

Acidol Grounding Yellow. An acid grounding dyestuff for leather.†

Acidol-Pepsin. A proprietary preparation of betaine hydrochloride, and pepsin, used in the treatment of achlorhydria gastritis. (Bayer & Co).*

Acid, Acetosalic. See ASPIRIN.

Acid, Chryseinic. See CAMPOBELLO YELLOW.

Acid, Disulphuric. See OLEUM.

Acid, F. See DELTA ACID.

Acid, Iso-anthraflavic. See ANTHRAFLAVIC ACID.

Acid, Laurent's Naphthalidinic. See LAURENT'S ACID.

Acid, Nordhausen. See FUMING SULPHURIC ACID.

Acid, NW. See NEVILE AND WINTHER'S ACID.

Acid, Pararosilic. See AURINE.

Acid, Polygallic. See STRUTHIIN.

Acid, Pyrosulphuric. See FUMING SULPHURIC ACID.

Acid, Rheic. See RHAPONTICIN.

Acid, Rhubarbaric. See RHAPONTICIN.

Acid, Rosamine A. (Violamine G). A dyestuff.†

Acid, Tannic. See GALLOTANNIC ACID.

Acid, Thymic. See THYME CAMPHOR.

Acid, Vitriolic. See OIL OF VITRIOL.

Acieral. An aluminium alloy.†

Acilan. Acid wool dyestuffs with good levelling power. (Bayer & Co).*

Acillin. Ampicillin. Applications: Antibacterial. (ICN Nutritional Biochemicals Corp).

Acinitrazole. 2 - Acetamido - 5 - nitrothiazole. Aminitrazole. Trichorad. Tritheon.†

Acintene. Turpentine products including alpha-pinene, beta-pinene middle boilers containing dipentene, technical grade anethole. Applications: *Alpha-pinene*: solvent, terpene resin monomer, intermediate for flavour and fragrance chemicals, intermediate for metal lubricant additive manufacture, camphor and camphene manufacture intermediate, petroleum extraction chemical. *Beta-pinene*: terpene resin monomer, intermediate for flavour and fragrance chemicals, solvent. *Middle-boilers containing dipentene*: pine oil extender, high kauri butanol value solvent. *Technical grade anethole*: anise flavour intermediate. (Arizona Chemical Company).*

Acintol D40LR, D30LR, D25LR. Tall oil fatty acid containing higher tall oil rosin content. Fatty acid range 59.1%-67.8%, rosin acids range 25%-45%, acid value range 185-188. Applications: Asphalt emulsions, cleaning formulations etc. (Arizona Chemical Company).*

Acintol FA-1, FA-1 Special, FA-2, FA-3, EPG, 746. Tall oil fatty acids in fatty acid range 92.8% to 99.0%, rosin acid range 0.5% to 4.5%, acid value range 194% to 198%. Applications: The fatty acids are used in the following industries: Agricultural, automotive, cleaning, cosmetic and medical supplies, housing and construction, leather, metal coatings, metal working, mineral flotation, petroleum, plastics and resins, rubber, surface coatings and textiles. (Arizona Chemical Company).*

Acintol R Type SFS. Formaldehyde-treated tall oil rosin. Applications: Pigment dispersing agent, fortified paper size, compounding cutting oils and buffing materials, surface coatings, printing ink vehicles and base stock for surfactants. (Arizona Chemical Company).*

Acintol R Type SM4. Maleated tall oil rosin. Applications: Pigment dispersing agent, fortified paper size, compounding cutting oils and buffing materials, surface coatings, printing ink vehicles and base stock for surfactants. (Arizona Chemical Company).*

Acintol R Type S, R Type SB, R Type 3A, R Type L03A, Tall Oil Rosin. Four grades of unmodified tall oil rosins. Colour range WG-3A. Acid value range 165-178, softening point range 74°C-78°C. Applications: Paper size, resinate emulsifiers, ester gum, synthetic resins, manufacture of rosin soaps, rubber compounding ingredients and many other rosin derivatives. (Arizona Chemical Company).*

Acintol, Liquaros. 60% rosin acids content solution in fatty acids. Applications: Pigment dispersing agent, fortified paper size, compounding cutting oils and buffing materials, surface coatings, printing ink vehicles and base stock for surfactants. (Arizona Chemical Company).*

Aciquel. Potassium Glucaldrate. Applications: Antacid. (McNeil Pharmaceuticals, McNEILAB Inc).

ACL. Chlorinated s-triazine triones. Applications: Machine dishwashing compounds, disinfection of public and private swimming pools, household and industrial cleaners, bleaching agents for both domestic and industrial washing, formulation of industrial bacteriocides. (Monsanto Co).*

ACL 56. Sodium dichloroisocyanurate dihydrate, 56% available chlorine product. Applications: Used in detergent applications, bleaches and pool disinfection. (Monsanto Co).*

ACL 59. Potassium dichloroisocyanurate, 59% chlorine product. Applications: Used for swimming pool disinfection and sanitizing; used as oxidizer in mechanical dishwashing compounds and other detergent applications. (Monsanto Co).*

ACL 60. Sodium dichloroisocyanurate, 62% chlorine product. Applications: Used in detergent applications, bleaches and pool disinfection. (Monsanto Co).*

ACL 90 Plus. Trichloroisocyanuric acid, 90% chlorine product. Applications: Used for pool disinfection in compacted forms, commercial bleaches and scouring powders. (Monsanto Co).*

Aclar. Range of protective thermoformable materials. Applications: Suitable for pharmaceutical packaging. (Allied Chemical Corporation).*

Aclon PCTFG. Homopolymer of chlorotrifluoroethylene. (Allied Chemical Corporation).*

Acme Rubber. A fine grade of American reclaimed rubber.†

Acme Yellow. See TROPAEOLINE O.

Acne-Aid Detergent Soap. A blend of high molecular weight fatty acids and selected detergents and contains sulphated surfactant blend. Applications: As an aid in the management of acne and any condition where greasy skin predominates. (Stiefel Laboratories (UK) Ltd).*

Acnidazil. A proprietary preparation of miconazole nitrate and benzoyl peroxide. Applications: For acne vulgaris. (Janssen Pharmaceutical Ltd).*

Acnil. A proprietary preparation of resorcinol, precipitated sulphur, and cetrimide. An acne treatment. (Fisons PLC).*

Aconite Liniment. See A.B.C. LINIMENT.

Acoomo Seeds. See OTE SEEDS.

Acopyrin. (Acetopyrin, Pyrosal, Phenazopirin). Antipyrine acetylsalicylate. An antipyretic.†

Acorga. Range of chemicals used in the extraction and treatment of metals and metalloids. (ICI PLC).*

Acorn Sugar. Quercitol, $C_6H_7(OH)_5$, found in acorns.†

Acpol. Polyester resin. (Freeman Chemical Corp, Subsidiary of H H Robertson Co).

Acquerite. A native alloy of silver and mercury, found in Chile, approximating to the formula $Ag_{12}Hg$.†

Acraconc. A range of printing concentrates. Applications: Dispersion thickeners for printing systems with a low white spirit content or without white spirit. (Bayer & Co).*

Acrafil. A registered trade name for flame retardant styrene-acrylonitrile.†

Acrafloc. Range of binders. Applications: For flocking of textiles. (Bayer & Co).*

Acraglas. Moulded acrylic synthetic resins.†

Acralen A. Aqueous dispersions of polyacrylic resins. Applications: Suitable for impregnated or coated substrates that must be fast to light. Especially suitable as binders for non-woven fabrics and padding materials. (Bayer & Co).*

Acramin. Pigment dyestuffs. Applications: For the pad dyeing of fabrics made of cellulosics, synthetic fibres and their blends. (Bayer & Co).*

Acramin. Pigment dyestuffs for the pad dyeing of fabrics from cellulosic fibres, synthetic fibres and their blends. (Bayer UK Ltd).

Acratamine Dyestuffs. Acid dyestuffs suitable for wool.†

Acrawax. A proprietary synthetic wax for lubrication and blending with other waxes to increase their melting-point.†

Acrawax C. N,N'-ethylenebis stearamide. A proprietary lubricant used in the manufacture of A.B.S., PVC and polystyrene. (Glyco Chemicals Inc).†

Acree-Rosenheim Reagent. A 1 : 6000 solution of commercial formalin.†

Acrex. Dinobuton. (Murphy Chemical Co).†

Acridine Orange. See PHOSPHINE.

Acridine Orange L, LP. Dyestuffs. They are British equivalents of Acridine orange.†

Acridine Orange NO. A dyestuff. It is the zinc double chloride of tetra-methyl-diamino-acridine. Dyes silk orange with greenish fluorescence, cotton mordanted with tannin, orange; also leather.†

Acridine Orange R Extra. A dyestuff. It is a salt of tetramethyl- diamino-phenyl-acridine. Dyes tannined cotton, orange red.†

Acridine Orange RS. A basic dyestuff for use in calico-printing.†

Acridine Red. A dyestuff. It is a mixture of Acridine orange and Pyronine.†

Acridine Red B, 2B, 3B. Dyestuffs. Dye silk or mordanted cotton, yellow shades of red.†

Acridine Scarlet. A dyestuff. It is a mixture of Acridine orange and Pyronine.†

Acridine Scarlet R, 2R, 3R. Dyestuffs. They are mixtures of Acridine orange with Pyronine.†

Acridine Yellow. A dyestuff. It is the hydrochloride of diamino-dimethyl-acridine, $C_{15}H_{16}N_3Cl$. Dyes silk greenish- yellow with green fluorescence and cotton mordanted with tannin, yellow. See PHOSPHINE.†

Acridine Yellow GR, R, 2R. Dyestuffs. They are British equivalents of Acridine yellow.†

Acriflex. An emollient, pale yellow vanishing cream containing chlorhexidine gluconate solution BP 1.25% v/w. Applications: Used to prevent infection in minor burns and scalds, scratches, cuts and abrasions, sunburn blisters and infected cracked skin. (Evans Medical).*

Acrilan. Acrylic fibre. Applications: For sweaters, handcraft yarns and carpets. (Monsanto Co).*

Acrilester. Two part polyester coating resin which is water-based and catalysed by the addition of equal parts of the two components. Applications: High gloss overcoat for a variety of surfaces which exhibits extremely high heat resistance, scuff resistance and smooth durable surface. (ADM Tronics Unlimited Inc).*

Acrilpact. Acrylic sealant. Applications: Sealant in building. (Siliconas Hispania SA).*

Acritamers. Acrylic, emulsifying, stabilizing and gel forming polymers. The thickening ability of these resins give formulations body while retaining the ability to be stirred, pumped or spread. (RITA Corporation).*

Acrodel. An insecticide preparation. (ICI PLC).*

Acrolite. A proprietary synthetic resin. It is a condensation product of glycerol and phenol or homologues of phenol used in varnishes and in the preparation of electrical insulating materials. Acrolite is also the trade mark for abrasive and refractory materials, the essential constituent of which is crystalline alumina.†

Acronal 14D. A plasticiser-free acrylic copolymer in the form of a 55 per cent. dispersion. (BASF United Kingdom Ltd).*

Acronal 160D. A plasticiser-free acrylic copolymer in the form of a 40 per cent. dispersion. (BASF United Kingdom Ltd).*

Acronal 21D, 27D and 30D. Plasticiser-free dispersions of thermosetting copolymers of various acrylic asters. (BASF United Kingdom Ltd).*

Acronal 350D. A plasticizer-free acrylic type copolymer in the form of a 45 per cent. dispersion. (BASF United Kingdom Ltd).*

Acronol Brilliant Blue. A dyestuff. It is a British equivalent of Setocyanine.†

Acronol Yellow T. A dyestuff. It is a British equivalent of Thioflavine T.†

Acrosyl. A saponified cresol disinfectant.†

Acry-Ace. Polymethyl methacrylate. A proprietary moulding powder. (Fuchow, Japan).†

Acrydur. Reactive acrylic resin systems. Applications: Seamless, industrial and commercial floorings. (Ulfcar International A/S).*

Acryl. Dyestuffs for acrylic fibres. (BASF United Kingdom Ltd).

Acrylafil G-40/20/FR and G-40/30/FR. Proprietary flame-retardant grades of styrene-acrylonitryle, reinforced with glass fibre. (Dart Industries Inc).†

Acrylamide. Acrylamide (2-propenamide). Applications: The mining industry. (Cyanamid BV).*

Acrylite. Acrylic sheet, and moulding and extrusion compounds. (Cyro Industries).

Acrylite H. Proprietary acrylic pellets used for injection moulding or extrusion. (Cyanamid of Great Britain).†

Acrylite M. Proprietary acrylic pellets providing medium heat resistance, used in injection moulding and extrusion where greater flow is required. (Cyanamid of Great Britain).†

Acryloid. A proprietary trade name for acrylic ester resins used for coatings.†

Acrylweld. Two-part acrylic adhesive. (Hardman Inc).*

Acrymul AM 123R. A proprietary trade name for a self cross-linking acrylic emulsion.†

Acrysol. Acrylic and urethane thickeners. Applications: Coatings. (Rohm and Haas Company).*

Acrysol LMW. Poly acrylic acid homo copolymer. Applications: Detergent, water treatment and textiles. (Rohm and Haas Company).*

Acrythane. Paint product - available in all colours for automotive/commercial vehicle use. Applications: Vehicle coating for original and refinishing also for high quality chemically resistant coating of machinery. (H Marcel Guest Ltd).*

Acsil. Sodium orthosilicate, sodium sesquisilicate. (Crosfield Chemicals).*

Acta. Throat drops. (Richardson-Vicks Inc).*

Actal. A proprietary preparation containing sodium polyhydroxy aluminium monocarbonate hexitol complex. An antacid. Activated aluminas. (Bayer & Co).*

Actellic. Insecticides containing pirimiphos-methyl. (ICI PLC).*

Actellifog. Liquid fogging insecticide formulation containing pirimiphos-methyl. (ICI PLC).*

Acterol. An irradiated ergosterol, a vitamin D concentrate.†

ACTH 40. Corticotropin. Applications: Hormone; glucocorticoid; diagnostic aid. (O'Neal, Jones & Feldman Pharmaceuticals).

Acthar. Corticotropin. Applications: Hormone; glucocorticoid; diagnostic aid. (Armour Pharmaceutical Co).

Acthar Gel. A proprietary preparation of corticotrophin gelatin for injection. (Armour Pharmaceutical Co).†

Acti-Dione. Cycloheximide. Applications: Antipsoriatic. (The Upjohn Co).

Acticarbone. Powdered and granulated activated carbon. Applications: Purification, decolourization, deodorization, separation and recovery in liquid of gas phase, in the chemical, petrochemical, pharmaceutical and food industries (Glucose factories, sugar refiners, oil refining, wine treatment). For the treatment of drinking and industrial water etc. Catalyst supports. (British Ceca Co Ltd).*

Acticrom. Dry powder reactive dyes. These dyes are of the triazynil type. Applications: Dyes used in the textile industry, mainly for dyeing cotton. (Multicrom SA).*

Acticulum. Conjugated glycopolypeptides. Applications: Cosmetic products to help stimulate cell metabolism. Strengthens skin against inflammation. (Active Organics).*

Actidil. Tablets and syrup. Proprietary formulations of triprolidine hydrochloride. Applications: For temporary relief of symptoms associated with allergic rhinitis. (The Wellcome Foundation Ltd).*

Actifed. Tablets, syrup, capsules and 12-hour capsules. Proprietary formulations of pseudoephedrine and triprolidine hydrochlorides. Applications: Temporary relief of upper respiratory symptoms, including nasal congestions, associated with allergy or common cold. (The Wellcome Foundation Ltd).*

Actifed with Codeine Cough Syrup. A proprietary formulation of pseudoephedrine and triprolidine hydrochlorides and codeine phosphate. Applications: Temporary relief of coughs and upper respiratory symptoms including nasal congestion, associated with allergy or the common cold. (The Wellcome Foundation Ltd).*

Actified Compound Linctus. A proprietary preparation containing Triprolidine hydrochloride, Pseudoephedrine hydrochloride and codeine phosphate. A cough linctus. (The Wellcome Foundation Ltd).*

Actigen. Live cell animal extraction. Applications: Cosmetic products - skin, body preparations and hair care products. (Active Organics).*

ActiMoist. Sodium hyaluronate (hyaluronic acid). Applications: Cosmetic products for high moisture retention/absorbtion. (Active Organics).*

Actiphyte. A plant extract in propylene glycol water. Applications: Cosmetic products - skin, body preparations and hair care products. (Active Organics).*

Actiplex. Actiblend - a combination of actiphytes. Applications: Cosmetic products - skin, body preparations and hair care products. (Active Organics).*

Actisize. Modified starch. Applications: Adhesives and paper. (Roquette (UK) Ltd).*

Activ 8. 1,10-phenanthroline (38%). Drier, accelerator and stabilizer for paint and coatings. Applications: Used in conjunction with manganese and/or cobalt in solvents and water thinned coatings. (Vanderbilt Chemical Corporation).*

Activated Alumina. Partially dehydrated aluminium trihydrate in the form of hard porous lumps. Has a strong affinity for water and will remove it from air.†

Activated Carbon. A carbon specially treated so that it is very highly adsorptive.†

Activated Sludge. A material obtained by allowing the growth of micro-organisms in the sludge deposited by sewage. it is used in the treatment of sewage †

Activator 736. Surface treated urea for easy dispersion in elastomers. Activator for thiazole, thiuram and dithiocarbamate accelerators. Odour reducer when used with nitrosoamine type blowing agents. (Uniroyal).*

Activax. A proprietary vaccine used in the treatment of fowl pox. (Coopers Animal Health, subsidiary of the Wellcome Foundation Ltd).*

Active Oxygen. See OXYGEN, ACTIVE.

Activex. Fungicides for garden use. (ICI PLC).*

Activit. A proprietary rubber vulcanization accelerator. It is thiocarbanilide.†

Activol. Gibberellic acid, a hormone growth regulator. (ICI PLC).*

Activox. Zinc oxides. (Durham Chemicals Ltd).*

Activox B. Colloidal zinc oxide. (Durham Chemicals Ltd).

Actol. Silver lactate.†

Actomol. A monoamine oxidase inhibitor being an antidepressant and containing mebanazine. (ICI PLC).*

Acton. Ethyl orthoformate, $CH(OC_2H_5)_3$.†

Actos 50. A proprietary preparation of prolonged action adrenocorticotrophic hormone. (Consolidated Chemicals).†

Actrapid. Insulin. Applications: Antidiabetic. (E R Squibb & Sons Inc).

Actrapid Human. Insulin Human. A protein that has the normal structure of the natural antidiabetic principle produced by the human pancreas. Applications: Antidiabetic. (E R Squibb & Sons Inc).

Actrel. Synthetic high boiling aromatic fluid. (Exxon Chemical International Inc).*

Actril. Selective weed killer. (May & Baker Ltd).*

Actrilawn. Selective weedkiller. (May & Baker Ltd).

Acturin. SEE DIUREXAN. (Degussa).*

Acupan. Nefopam hydrochloride. Applications: Analgesic. (Riker Laboratories Inc, Subsidiary of 3M Company).

Acupan Injection. Each ampoule contains nefopam hydrochloride 20 mg/ml. (Riker Laboratories/3M Health Care).†

Acupan Tablets. Each tablet contains nefopam hydrochloride 30 mg (Riker Laboratories/3M Health Care).†

Acylan. Acetylated lanolins. Applications: Used in the cosmetic and pharmaceutical industries. (Croda Chemicals Ltd).*

Acylan. Acetylated lanolin. (Croda Chemicals Ltd).

Acylanid. ACETYL DIGITOXIN. α-digitoxin monoacetate. (Sandoz).†

Ad-aluminium. A proprietary trade name for an alloy of 82 per cent copper, 15 per cent zinc, 2 per cent aluminium, and 1 per cent tin.†

Adabee. A proprietary preparation containing Vitamins A and C, thiamine, riboflavine, pryridoxine and nicotinamide. Vitamin supplement. (Robins).†

Adal. Alloying elements into aluminium alloys. (Foseco (FS) Ltd).*

Adalat. A cardioprotective coronary drug. (Bayer & Co).*

Adalin. Polyethylene emulsion nonionic. Applications: Softener for resin finishing. Improves sewability, abrasion resistance, crease angles and softness. (Henkel Chemicals Ltd).

Adalox. A proprietary trade name for coated abrasives for sanding metal or plastics.†

Adamac. Specially prepared tar for roads. (Thomas Ness Ltd).†

Adamant. Diamond or corundum.†

Adamantine Cement. See CEMENT, ADAMANTINE.

Adamite. A proprietary high-carbon nickel-chromium iron alloy used for dyes.†

Adamon. (1) Dibromo-dihydro-cinnamic acid bornyl ester. A sedative.†

Adamon. (2) The registered trade name for an antispasmodic. (AG Chemische Fabrik).†

Adamsite. (D.M.). 10-chloro-5 : 10-dihydrophenarsazine. A poison gas.†

Adanon. A proprietary trade name for Methadone.†

Adansonia Fibre. A fibre obtained from the bark of *Adansonia digitata*. Used for making rope and sacking, and also for special paper.†

Adaphax 758. A proprietary factice containing neither sulphur nor chlorine used as a processing aid in the manufacture of polyurethanes and PVC. It enables good electrical properties to be maintained in PVC and tackiness to be reduced. It can also be used in white nitrile mixes. (Hubron Rubber Chemicals, Failsworth, Manchester).†

Adapin. Doxepin hydrochloride. Applications: Antidepressant. (Pennwalt Corp).

Adaprin. A proprietary preparation of nicotinamide and acetomenaphthone, used to treat chilblains. (Ward, Blenkinsop & Co Ltd).†

Adaptinol. An anti-dazzle preparation. (Bayer & Co).*

Adarola. A proprietary casein plastic.†

ADC. Epoxy formulations used in tooling, adhesives, electronic encapsulants, coatings, sealants and ablatives. Applications: Ultrahitemperature laminating resins, fast setting hitemperature adhesives, pin hoe free RT laminating resin (requires no vacuum bag), chemical resistant coatings, thermally conductive and resistant coatings and encapsulants, construction sealants, binders and adhesives, soft tooling compounds, specialty coating for food industry and chemical industry. (ADC Resins).*

Adcora. A proprietary trade name for coating materials as follows:-

P6. A neoprene coating with good chemical and abrasion resistance and good flexibility.

H2. A hypalon (*qv*) coating for tanks etc. Possesses good resistance to halogens.

V. A Viton (fluoroelastomer) coating with very good heat stability resistant to temperatures up to 260°C.

A3. A hard coal tar epoxy coating for tanks and pipes for effluent and for resistance to fumes and weathering.†

Adcora SP. A solventless epoxide-based coating giving a hard finish resembling enamel.†

Adcortyl Cream. Triamcinolone acetonide in cream base. (E R Squibb & Sons Ltd).

Adcortyl in Orabase. Triamcinolone acetonide in ointment base. (E R Squibb & Sons Ltd).

Adcortyl Injection. Triamcinolone acetonide in aqueous vehicle. (E R Squibb & Sons Ltd).

Adcortyl Ointment. Triamcinolone acetonide in ointment base. (E R Squibb & Sons Ltd).

Adcortyl Spray. Traimcinolone acetonide in aerosol spray. (E R Squibb & Sons Ltd).

Adcortyl with Graneodin Cream. Triamcinolone acetonide, neomycin sulphate and gramicidin in cream base. (E R Squibb & Sons Ltd).

Adcortyl with Graneodin Ointment. Triamcinolone acetonide, neomycin sulphate and gramicidin in ointment base. (E R Squibb & Sons Ltd).

Adcortyl-A. A proprietary name for the acetonide of Triamcinolone.†

Adcote. Adhesives and primers for flexible packaging. (Williams (Hounslow) Ltd).

Add F. Agricultural chemicals. Liquid silage additive. (BP Chemicals Ltd).*

Add It To Oil. Alkylaryl poly (ethylene oxy). Applications: Natural oil emulsifier for use as an additive for conditioning oil for crop spray use. (Drexel Chemical Company).*

Add-add. The leaves of *Celastraceae serratus*, of Abysinnia. Used as an antiperiodic.†

Add-F. Liquid silage additive. (BP Chemicals Ltd).

Add-H. Liquid hay additive. (BP Chemicals Ltd).

Add-M. Liquid anti-moulding compound. (BP Chemicals Ltd).

Addabond. Substance for bonding new concrete to old. Applications: Particularly useful where it is impossible to dry out the substrate. (Addagrip Surface Treatments UK Ltd).*

Addacoat. Two part pack. Coloured material, giving a tile like finish. Applications: Applied to walls where cleaning is of paramount importance. Used in damp proofing of cellars and basements. (Addagrip Surface Treatments UK Ltd).*

Addacol. Organic and inorganic pigments. Applications: For colouring cement, concrete, bricks, etc. (Calder Colours (Ashby) Ltd).*

Addaflex. Flexible epoxy compound. Applications: For filling cracks and expansion joints in concrete floors. (Addagrip Surface Treatments UK Ltd).*

Addaflor. Solvent free coloured epoxy top coat. Oil, acid and chemical resistant. Strong enough to hold anti-slip chippings. Easily cleaned. Applications: Applied to concrete, brick, wood and steel. (Addagrip Surface Treatments UK Ltd).*

Addagrout. Three part pack. Epoxy based grouting material. Applications: For bedding machinery, filling anchor points and repairing cracks and depressions in floors. (Addagrip Surface Treatments UK Ltd).*

Addalevel. Levelling compound for laying between 3 and 6 mm thick. Applications: Applied to concrete and brick. (Addagrip Surface Treatments UK Ltd).*

Addamortar. Epoxy mortar - three part pack - fillers, hardener and resin. Applications: Filling cracks, holes and undulations, and for screeds. Can be power floated. (Addagrip Surface Treatments UK Ltd).*

Addapitch. Pitch epoxy material. Applications: Suitable for waterproofing and also for holding anti-slip chippings on ramps and loading bays etc. (Addagrip Surface Treatments UK Ltd).*

Addaprime. Two part pack - resin and hardener. Solvent free deep-penetrating epoxy primer. Can also be used as a coating. Strong enough to hold anti-slip chippings. Applications: Applied to concrete, brick, wood and steel. (Addagrip Surface Treatments UK Ltd).*

Addaseal. Polyurethane material of high quality (50% solids). Clear or coloured. Applications: For dust sealing concrete floors. Strong enough to hold small particles of anti-slip aggregate. (Addagrip Surface Treatments UK Ltd).*

Addasure. Solvent based epoxy coating. Two part pack. Applications: Used for coating walls and floors. (Addagrip Surface Treatments UK Ltd).*

Adder. An asbestos fabric grade of TUFNOL industrial laminates. (Tufnol).*

Additive-A. Modified calcium lignosulphonates. Applications: Clay conditioner for binding/dispersing clay products. (Reed Lignin Inc).*

Additol. Range of additives to improve properties such as flow and eliminate film defects etc. Applications: Paints and printing inks. (Resinous Chemicals Ltd).*

Adelan. Wool fat.†

Adelphane. A proprietary preparation of reserpine, and dihydrallazine sulphate. An antihypertensive. (Ciba Geigy PLC).†

Adeno. Adenosine phosphate. Applications: Nutrient. (O'Neal, Jones & Feldman Pharmaceuticals).

Adenotriphos. A proprietary preparation of adenosine triphosphoric acid sodium. (Rona Laboratories).†

Adenyl. A proprietary preparation of ADENOSINE mono-phosphoric acid used in the treatment of rheumatic diseases. (Rona Laboratories).†

Adepsine. See OZOKERINE.

Adepsine Oil. See PARAFFIN, LIQUID.

Adequan. Polysulphated glyco-saminoglycan for intra-articular injection. Applications: Treatment of lameness in horses. (Luitpold-Werk).*

Adexolin. Water miscible pale yellow, sugar free concentrate containing Vitamins A, C and D. Applications: Supplementation and prevention of deficiency of vitamins A, C and D which may occur during infancy and early childhood and in expectant and nursing mothers. (Evans Medical).*

Adflex. Compounded polymeric emulsion. Applications: Flexing agent for cement based adhesives. (Howlett Adhesives Ltd).*

Adheso. A proprietary trade name for a synthetic wax consisting of a modified polymerized terpene.†

Adigan. A trade mark for a digitalis preparation containing all the active principles except digitonin, this having been removed.†

Adimoll BO. Benzoyl octyl adipate. Applications: Monomeric plasticizer. (Bayer & Co).*

Adimoll DN. Diisononyl adipate. Applications: Monomeric plasticizer. (Bayer & Co).*

Adimoll DO. Dioctyl adipate. Applications: Monomeric plasticizer. (Bayer & Co).*

Adinol. A range of taurates, used as anionic surfactants. Applications: Used in the cosmetic industry as foaming agents. (Croda Chemicals Ltd).*

Adinol. Textile auxiliaries based on triethyl citrate. (Fine Dyestuffs & Chemicals Ltd).†

Adinol CT. Sodium N-methyl N-cocoyl taurate in white powder form. Applications: Anionic surfactant with high foaming and cleansing capacity, chemically stable, lime soap dispersant. Used in cosmetics and toiletries; pharmaceutical preparations. (Croda Chemicals Ltd).

Adinol T. Sodium N-methyl N-oleyl taurate in powder, paste, gel or liquid form. Applications: Biodegradeable anionic surfactant with detergent, wetting, emulsifying, dispersing and foaming properties. Used in the textile and dyeing industries; leather; paper. (Croda Chemicals Ltd).

Adipocere. A wax-like mass left when animal bodies decompose in the earth. It consists of the fatty acid salts of calcium and potassium.†

Adipon. Fatty alcohol sulphate, anionic. Applications: Detergent for scouring and milling of worsted fabrics. Softening effect. (Henkel Chemicals Ltd).

Adiprene. A trade mark for urethane rubbers, L-245, L-265, L-700, L-767, resistant to oil and radiation. (Du Pont (UK) Ltd).†

Adiprene. A range of polyether-based prepolymers offering high abrasion resistance, chemical resistance and electrical properties. (Uniroyal).*

Adiro. A platelet aggregation inhibitor, antithrombotic and anti-inflammatory drug. (Bayer & Co).*

Adirondackite. A rubber substitute made from sulphurized oils. Used in the proofing of cloth, and as an insulator.†

Adjab Fat. See NJAVE BUTTER.

Adjective Dyestuffs. (Mordant colours). Dyestuffs which require a mordant for fixing them are called adjective dyestuffs or mordant colours.†

Adju-Fluax. Influenza virus vaccine, bivalent types A and B with Adjuvant 65. Applications: Primary immunization and seasonal booster effect against influenza. (Merck Sharp & Dohme).*

Admerol. Modified oils. Applications: Paint. (NL Industries Inc).*

Admex. Polymeric plasticizers. Applications: plasticizing PVC and other polymers. (Nueodex Inc).*

Admiralty Antifriction Metal. Usually an alloy of 85 per cent tin and 15 per cent antimony.†

Admiralty Bell Metal. An alloy of 78 per cent copper and 22 per cent tin.†

Admiralty Brass. See BRASS.

Admiralty Brazing Metal. An alloy of 90 per cent copper and 10 per cent zinc.†

Admiralty Gun Metal. Alloys. Some contain from 87-90 per cent copper and from 10-13 per cent tin, whilst others consist of from 86-88 per cent copper, 6-10 per cent tin, and 2-6 per cent zinc.†

Admiralty Nickel. See ADNIC.

Admiralty White Metal. A bearing alloy containing 86 per cent tin, 8.5 per cent antimony, and 5.5 per cent copper.†

Admos Alloys. Brass alloys of varying composition containing small amounts of tin, nickel, lead, and iron.†

Admul. Mono and diglycerides and their acid derivatives. (PPF International Ltd).*

Admune. A proprietary proparation of inactivated influenza virus used to confer immunity to the disease. (Duncan, Flockhart).†

Adnic. (Admiralty Nickel). An alloy containing 70 per cent copper, 29 per cent nickel, and 1 per cent tin. It is resistant to corrosion and heat.†

Adol 62. Stearyl Alcohol. Applications: Pharmaceutic aid. (Sherex Chemicals Co Inc).

Adol 66. Isostearyl alcohol. Mixture of branched chain aliphatic 18 carbon alcohols, having an iodine value of 10 to 20, a saponification value of less than 3, and an acid value of less than 0.5. Applications: Pharmaceutic aid. (Sherex Chemicals Co Inc).

Adol 85. Oleyl alcohol. Applications: Pharmaceutic aid. (Sherex Chemicals Co Inc).

Adraganthin. See BASSORIN.

Adrenalin. Epinephrine. Applications: Adrenergic. (Parke-Davis, Div of Warner-Lambert Co).

Adrenocorticotrophic Hormone. Corticotrophin.†

Adrenocortrophin ACTH. Corticotrophin.†

Adrenoxyl. A proprietary preparation of adrenochrome monosemicarbazone dihydrate. (Horlicks).†

Adriamycin. A freeze-dried powder containing lactose and doxorubicin hydrochloride for injection. Applications: A prescription drug. (Hercules Inc).*

Adriblastina. Doxorubicin. Applications: Antineoplastic. (Farmitalia (Farmaceutici Italia)).

Adrin. A proprietary brand of Adrenalin (*qv*).†

Adroit. Cutting oils. (S & D Chemicals Ltd).†

Adronal. See CYCLOHEXANOL.

Adronal Acetate. (Hexalin Acetate). Cyclohexanol acetate, a resin solvent.†

Adroyd. A proprietary preparation of oxymetholone. An anabolic agent (Parke-Davis).†

Adrucil. Injectable fluorouracil. Applications: A prescription drug. (Hercules Inc).*

Adsol. See JAPANESE ACID CLAY.

Adsorbocarpine. Pilocarpine Hydrochloride. Applications: Cholinergic. (Alcon Laboratories Inc).

Adtac. Range of low molecular weight aliphatic hydrocarbon resins. Contributes a balance of tack and adhesive properties to elastomer systems. Typical softening point grades range from 10-25°C. Applications: Used in adhesives, in coatings and as a waterproofing agent. (Hercules Inc).†

Adurol. Monocloro and monobromo hydroquinones, $C_6H_3Cl(OH)_2$, and $C_6H_3Br(OH)_2$. Photographic developers.†

Aduvex. UV Light absorbers. (Ward Blenkinsop & Co Ltd).

Advagum. A proprietary plasticizer. A terpene resin.†

Advance. Selective herbicide. (ICI PLC, Plant Protection Div).

Advance Alloy. An alloy of copper with from 44-46 per cent nickel, and small quantities of iron and manganese.†

Advantage. Deposit inhibitors. Polymer-based descalers. Applications: Boiler deposit control, oil field steam generator deposit control. (Ashland Chemical Company).*

Advastab. PVC stabilizers for heat and light. (Morton Thiokol Inc, Carstab Div).

Advawax. Internal and external lubricants. (Morton Thiokol Inc, Carstab Div).

Advawax 280. A proprietary range of synthetic amide compositions used in paints, adhesives, asphalts and resins. (Cincinnati Milacron Chemicals Inc Reading, Ohio).†

Advil. Ibuprofen. Applications: Anti-inflammatory. (Whitehall Laboratories, Div of American Home Products Corp).

Advitacon. Oil soluble vitamins. (PPF International Ltd).*

Advitagel. Flour and confectionery emulsifier. (PPF International Ltd).*

Advitamix. Animal feed supplements. (PPF International Ltd).*

Advitaroma. Butter and meat flavouring. (PPF International Ltd).*

Advitrol. Anti-settling and thickening agents for paints, varnishes, lubricants, adhesives, coatings, putties and cosmetics. (Süd-Chemie AG).*

Advizor. Weedicide containing lenacil and chloridazon for use in sugar beet. (ICI PLC).*

Ad. A.M. A proprietary preparation of ephedrine hydrochloride and butethamate citrate. A bronchial anti-spasmodic. (Rybar).†

Aegerite. A trade name for Elaterite.†

Aeonite. A nickel silver containing 20 per cent nickel. A trade name for Elaterite.†

Aerial Cement. A term applied to cements which set in air, the setting being due to desiccation and carbonation.†

Aerialite. A proprietary synthetic resin.†

Aero. A proprietary trade name for rosin-glycerol varnish and lacquer resins.†

Aero Metal. An aluminium alloy, consisting mainly of aluminium with from 2.1-2.9 per cent magnesium, 0.3-1.3 per cent iron, and 0.2-0.6 per cent copper.†

Aero X. A proprietary rubber vulcanization accelerator.†

Aero 301 Xanthate. Sodium sec. butyl xanthate. Applications: The mining industry. (Cyanamid BV).*

Aero 303 Xanthate. Potassium ethyl xanthate. Applications: The mining industry. (Cyanamid BV).*

Aero 317 Xanthate. Sodium isobutyl xanthate. Applications: The mining industry. (Cyanamid BV).*

Aero 343 Xanthate. Sodium isopropyl xanthate. Applications: The mining industry. (Cyanamid BV).*

Aero 3477 Promoter. Sodium diisobutyl dithiophosphate. Applications: The mining industry. (Cyanamid BV).*

Aero 350 Xanthate. Potassium amyl xanthate. Applications: The mining industry. (Cyanamid BV).*

Aero 3501 Promoter. Sodium diisoamyldithiophosphate. Applications: The mining industry. (Cyanamid BV).*

Aerocol. Polyvinyl acetate adhesives. (Ciba-Geigy PLC).

Aerodri 100. Modified dioctyl-sulphosuccinate. Applications: The mining industry. (Cyanamid BV).*

Aerodri 104. Modified dioctyl-sulphosuccinate. Applications: The mining industry. (Cyanamid BV).*

Aerodri 200. Mixture of surfactants. Applications: The mining industry. (Cyanamid BV).*

Aerodux. Resorcinol/formaldehyde resins. (Ciba-Geigy PLC).

Aerofloat 208 Promoter. Sodium diethyl and sodium di-sec. butyl dithiophosphate mixture. Applications: The mining industry. (Cyanamid BV).*

Aerofloat 211 Promoter. Sodium diisopropyl dithiophosphate. Applications: The mining industry. (Cyanamid BV).*

Aerofloat 238 Promoter. Sodium di-sec. butyl dithiophosphate. Applications: The mining industry. (Cyanamid BV).*

Aerofonic. Compressed polyether acoustical foam. (ScotFoam Corp).

Aerofroth 65. Polypropylenglycol. Applications: The mining industry. (Cyanamid BV).*

Aerofroth 76. Mixture of higher alcohols. Applications: The mining industry. (Cyanamid BV).*

Aerofroth 88. 2 Ethylhexanol. Applications: The mining industry. (Cyanamid BV).*

Aerofroth 99. 2 Ethylhexanol tails. Applications: The mining industry. (Cyanamid BV).*

Aerolite. Urea/formaldehyde resins. (Ciba-Geigy PLC).

Aeromatt. Precipitated calcium carbonate for cosmetics. (John & E Sturge Ltd, Lifford).

Aeromin. An alloy of 91.6 per cent aluminium and 8.4 per cent magnesium.†

Aeron. See SELERON.

Aerophen. Phenol/formaldehyde resins. (Ciba-Geigy PLC).

Aerophine 3418A. Sodium diisobutyldithiophosphinate. Applications: The mining industry. (Cyanamid BV).*

Aeroplex. A proprietary safety-glass.†

Aeroseb-Dex. Dexamethasone. Applications: Glucocorticoid. (Herbert Laboratories, Dermatology Div of Allergan Pharmaceuticals Inc).

Aeroseb-HC. Hydrocortisone. Applications: Glucocorticoid. (Herbert Laboratories, Dermatology Div of Allergan Pharmaceuticals Inc).

Aerosil. A trade name for finely divided silicon dioxide. There are several grades the particle sizes of which range from 7 to 40 millimicrons. Used as a thixotropic agent for paints, polishes, epoxy resin, cosmetics and adhesives. (Bush Beach Ltd).†

Aerosil. Highly dispersed pyrogenic silica. Applications: Highly active filler for natural and synthetic rubber, especially for silicone rubber; as thickening agent for ointments, creams, toothpastes etc.; tabletting and dragée production auxiliary; thixotropizing agent for polyester resins, lacquers and printing inks; for maintenance of free-flowing characteristics of substances which tend to cake (free-flow agent); for the production of highest purity silicates; for blueprint papers; thickening agent for oils. (Degussa).*

Aerosil COK 84. A trade name for a mixture of Aerosil and alumina in 5:1 ratio. It is suited particularly for thickening aqueous and other polar systems. (Bush Beach Ltd).†

Aerosil Composition. A trade name for a mixture of aerosil with 15 per cent. starch, specially designed for tableting. (Bush Beach Ltd).†

Aerosil 130, 150, 200, 300, 380. A range of highly dispersed pyrogenic silica preparations. Applications: Highly active filler for natural and synthetic rubber; thickening agent; tabletting and dragée production auxiliary; thixotropizing agent for polyester resins; antisetting agent. (Degussa).*

Aerosol A-102. Ethoxylated alcohol half ester of sulphosuccinic acid in liquid form. Applications: Solubilizing, foaming, dispersing and emulsifying agent, non-dermatitic, lime soap dispersing, fluidizing, cleaning and surface tension reducing. Used in vinyl acetate and acrylate-based latexes especially cross-linkable types; agricultural products; germicides; cleaners; cosmetics and shampoos; foaming cement and wall-board. (Cyanamid of Great Britain Ltd, Chemicals Div).*

Aerosol A-103. Ethoxylated nonylphenol half ester of sulphosuccinic acid in liquid form. Applications: Solubilizing, foaming, dispersing and emulsifying agent, non-dermatitic, lime soap dispersing, fluidizing, cleaning and surface tension reducing. Used in acrylate-based systems; germicides; cosmetics; foaming cement and wall-board. (Cyanamid of Great Britain Ltd, Chemicals Div).

Aerosol A-196. Sodium dicyclohexyl sulphosuccinate in pellet form. Applications: Emulsion polymerization for styrene-butadiene based latexes. Adjuvant to various surfactant systems. (Cyanamid of Great Britain Ltd, Chemicals Div).*

Aerosol A-268. Disodium isodecyl sulphosuccinate in liquid form. Applications: Emulsifying, solubilizing and foaming agent for foamed coatings, used for emulsion and suspension polymerization of PVC. (Cyanamid of Great Britain Ltd, Chemicals Div).*

Aerosol AY. Sodium diamyl-sulphosuccinate as a liquid or waxy solid. Applications: Emulsion polymerization; leaching and electroplating. (Cyanamid of Great Britain Ltd, Chemicals Div).*

Aerosol C-61. Alkylamine-guanidine polyoxyethanol as a fluid paste. Applications: Cationic surfactant for wetting; dispersing and fixing agent for pigments, dyes and fillers; softener for textiles; lubricating; surface tension reduction; antistat agent; emulsifying agent alone or with Aerosol OT. (Cyanamid of Great Britain Ltd, Chemicals Div).*

Aerosol C61. Alkylamine-guanidine-polyoxyethanol. (Cyanamid BV).*

Aerosol IB45. Sodium diisobutyl sulphosuccinate. (Cyanamid BV).*

Aerosol MA-80. Sodium dihexyl sulphosuccinate in liquid form. Applications: Penetration, dispersing, emulsifying and solubilizing agent used in emulsion polymerization, batteries, electroplating and leaching. (Cyanamid of Great Britain Ltd, Chemicals Div).*

Aerosol OS. Anionic surfactant in powder form. Applications: Dispersing, wetting, solubilizing and emulsifying agent, used in emulsion polymerization; paints; plastics; hard surface and metal cleaners; agricultural powders. (Cyanamid of Great Britain Ltd, Chemicals Div).*

Aerosol OT and GPG. Range of anionic surfactants composed of sodium dioctyl sulphosuccinate. Liquid, powder or wax form. Applications: Wetting, dispersing, solubilizing and emulsifying agents used for suspension polymerization and emulsion polymerization; pigments, paints and printing inks; plastics; rust inhibitors; textiles; paper; petroleum; rubber; metal; cosmetics; agricultural chemicals; degreasers and lubricants; dry cleaning; dust laying; mold release for polymethyl methacrylate. (Cyanamid of Great Britain Ltd, Chemicals Div).*

Aerosol TR-70. Sodium bistridecyl sulphosuccinate in liquid form. Applications: Flushing, suspending, dispersing, emulsifying and solubilizing agent used in suspension polymerization, pigments, plastics, printing inks and rust inhibitors. (Cyanamid of Great Britain Ltd, Chemicals Div).*

Aerosol 18. Disodium N-octadecyl sulphosuccinamate in paste form. Applications: Foaming, suspending, lubricating and dispersing agent, used in emulsion polymerization, foamed insulation, cement, wallboard resins, cleaning, lubrication, cosmetics. (Cyanamid of Great Britain Ltd, Chemicals Div).

Aerosol 200. Disodium alkyl amidopolyethoxy sulphosuccinate. (Cyanamid BV).*

Aerosol 22. Tetrasodium N-(1,2 dicarboxyethyl) N-octadecyl sulphosuccinamate in liquid form. Applications: Solubilizing, foaming, dispersing and fluidizing agent used in all monomer systems, agricultural products, industrial and household cleaners, metal cleaners and cosmetics. (Cyanamid of Great Britain Ltd, Chemicals Div).*

Aerosporin Sterile Powder. A proprietary formulation of polymyxin B sulphate. Applications: Treatment of acute infections caused by susceptible strains of Pseudomonas aeruginosa. (The Wellcome Foundation Ltd).*

Aerothene. Chlorinated solvents used as vapour-pressure depressants and as a carrier solvent. (Dow Chemical Co Ltd).*

Aerotrol. A proprietary trade name for the hydrochloride of isoprenaline. An aerosol bronchodilator. (Abbott Laboratories).*

Aerox. Aero X.†

Aescorcin. See ESCORCIN.

Aeternol. A proprietary synthetic resin.†

Aethoxal. Fattening agent, oil components and dispersing agent. Applications: Shampoos; foam baths; non-foaming oil baths. (Henkel Chemicals Ltd).

Aethrol. A plastic of the pyroxylin-cellulose acetate type.†

Afalon. Herbicide for potatoes and carrots. (Hoechst UK Ltd).

A-Fax. A range of amorphous polypropylenes. Applications: Used in adhesives and sealants, used for asphalt modifications in construction and building industry, in carpet backing, in polyolefin modification, as sound deadening and rubber processing agents. (Hercules Inc).†

Afax. Continuous casting mould flux for all steel grades. (Foseco (FS) Ltd).*

Afcolene. A proprietary polystyrene. (Pechiney-St Gobain, France).†

Afenil. Calcium chloride urea.†

Afghan Yellow. A dyestuff produced by warming *p*-nitro-toluene-sulphonic acid with sodium hydroxide. It dyes unmordanted cotton in the presence of sodium chloride, and silk and wool from an acid bath.†

Aflaban. Feed preservative based on sorbic acid. Applications: Growth inhibitor for moulds, yeasts and bacteria in animal feeds. (Monsanto Co).*

Aflux. A range of fatty acid derivatives partly bound to highly active silica used as dispersing agents and internal lubricants in the rubber industry. Applications: Moulded and extruded technical articles. (Rhein-Chemie Rheinau).*

Afonic. Embossed sound absorbing foam. (ScotFoam Corp).

African Phosphates. Mineral phosphates found in Tunis and Algeria. They contain from 55-65 per cent calcium phosphate. Others found at Safaga and

Kosseir Contain 60-70 per cent calcium phosphate. Fertilizers.†

African Saffron. Carthamus.†

Afridi Wax. See ROGHAN.

Afridol Violet. A dyestuff. Produced by coupling diaminodiphenylurea with 2 molecules of H acid.†

Afrin. Oxymetazoline hydrochloride. Applications: Adrenergic. (Schering-Plough Corp).

Afrinol. Pseudoephedrine Sulphate. Applications: Adrenergic. (Schering-Plough Corp).

Afrisect. Insecticidal formulation. (Mitchell Cotts Chemicals Ltd).

Afrodit. A mineral, synonymous with Aphrodite.†

Afrol. Timber insecticide. (ICI PLC).*

Afrol. Timber insecticide. (Plant Protection (Subsidiary of ICI)).†

Afror Tyne Powder. A low-freezing explosive containing a nitrated mixture of glycerine and ethylene glycol and ammonium nitrate.†

A.F.S. A Canada balsam substitute made from aniline, formaldehyde and sulphur.†

Aftate. Tolnaftate. Applications: Antifungal. (Plough Inc).

Agalite. (Mineral pulp, Asbestine Pulp). A variety of talc (hydrated magnesium silicate). Used by papermakers.†

Agalma Black 10B. A dyestuff prepared from H-acid (8-amino-1-naphthol-3,6-disulphonic acid) by adding it to an equal amount of diazotized *p*-nitraniline, then coupling with diazotized aniline. It dyes wool and silk a greenish-black from an acid bath. It is the basis of many black dye mixtures.†

Agalyn. A proprietary material used for dentures.†

Agar-agar. (Bengal Gelatin or Isinglass, Ceylon Gelatin or Isinglass, Chinese Gelatin or Isinglass, Japanese Gelatin or Isinglass, Japan Agar, Layor Carang, Vegetable Glue). The material obtained from certain varieties of algae by boiling water. Used as a sizing for cloth, and as a culture medium for bacteria.†

Agarase. An agar-agar preparation.†

Agaricin. Agaric acid, a resin acid, obtained by extraction with alcohol of the fruit bodies of *Polyporus officinalis* and *Agaricus albus*. A febrifuge.†

Agarol. A proprietary preparation containing liquid paraffin, phenolphthalein and agar. A laxative. (Warner).†

Agar, Japan. See AGAR-AGAR.

Agate Ware. (Enamelled Iron-granite). Enamelled iron.†

Agate, White. See CHALCEDONY.

Agathine. (Salizone, Cosmin). Salicyl-α-methyl-phenyl-hydrazide, $C_6H_5(CH_3)$.N.NHCO(C_6H_4OH). Used in the treatment of neuralgia and rheumatism.†

Agatine. A proprietary phenol-formaldehyde resin in the form of sheet, rods, tubes, etc.†

Agavin. Thiosolucin-dihydrostreptomycin preparation for the veterinary field. (May & Baker Ltd).*

Age. (Axin). The fat of *Coccus Axin,* growing in Mexico. It consists of the glycerides of lauric and axinic acids.†

Age-Rite. A proprietary antioxidant for rubber. It is *N*-phenyl-β-naphthylamine. (Anchor Chemical Co).*

Age-Rite AK. A trade name for an antioxidant. - Polymethyl-dihydroquinoline. (Anchor Chemical Co).*

Age-Rite Alba. A proprietary trade name for a rubber antioxidant. It is para-benzyl-oxyphenol. (Anchor Chemical Co).*

Age-Rite Gel. A proprietary product. It is a combination or composition, consisting of ditolyamines and a selected petroleum wax. An antioxidant for rubber. (Anchor Chemical Co).*

Age-Rite Hipar. A proprietary trade name for a mixture of isopropoxy diphenylamine, diphenyl-para-phenylamine and phenyl-beta-naphthylamine. An antioxidant for rubber. (Anchor Chemical Co).*

Age-Rite HP. A proprietary trade name for a mixture of phenyl- betanaphtylamine and diphenyl-paraphenylene diamine. (Anchor Chemical Co).*

Age-Rite Resin D. A proprietary trade name for trimethyl-dihydro- quinoline in a polymerized state. (Anchor Chemical Co).*

Age-Rite Spar. Styrenated phenol. A proprietary anti-oxidant. (Anchor Chemical Co).*

Age-Rite Stalite. A proprietary trade name for polymerized trimethyl - dihydro-quinoline. An antioxidant. (Anchor Chemical Co).*

Age-Rite Stalite S. Alkylated diphenylamines. A proprietary antioxidant. (Anchor Chemical Co).*

Age-Rite White. A proprietary antioxidant for rubber It is di-b-naphthol-*p*-phenylene-diamine. (Anchor Chemical Co).*

Agene. A bleaching agent for flour. Nitrogen trichloride, NCl_3.†

Agerite. A full line of phenol and amine rubber antioxidants both primary and secondary. Applications: Used in all forms of rubber. (Vanderbilt Chemical Corporation).*

Agestan 68. Silver amalgam in tablet form. (Bayer & Co).*

Agfa-Gevaert. Imaging systems. Applications: Graphic and reprographic systems, x-ray, cinematography, office systems, photography and audio-video. (Agfa-Gevaert).*

Agidex. Glucoamylase for conversion of starch into dextrose. (Glaxo Pharmaceuticals Ltd).†

Agirite. See ACMITE.

Agma. Calcinated magnesite. (ICI PLC).*

Agnin. (Agnolin). Purified wool fat.†

Agnowax. Wool wax alcohols. (Croda Chemicals Ltd).*

Agotan. See ATOPHAN.

Agral. Wetting, spreading and emulsifying agent, for agricultural and horticultural pest control products. (ICI PLC).*

Agramm. Nitrogenous fertilizers. (ICI PLC).*

Agriben. Manure composter for processing liquid and solid manure for agriculture. (Süd-Chemie AG).*

Agricastrol. A range of lubricants and hydraulic fluids for use with farm machinery. Applications: Engine oils and hydraulic fluids for tractors of all kinds. (Burmah-Castrol Ltd).*

Agricol. A proprietary range of alginates used for root dipping. (Alginate Industries Ltd).†

Agridin 60. An emulsifiable insecticide. Contains active component Diazinone $C_{12}H_{21}O_3N_2SP$/, different emulgators and organic solvents. A light to dark brown liquid, 60% active component. Applications: Applied as an active insecticide against worms. (Chemical Combine).*

Agrilan. Anionic/nonionic emulsifiers. Applications: Agricultural toxicant dispersers. (Lankro Chemicals Ltd).†

Agrilite Alloy. A copper-lead alloy containing small amounts of tin.†

Agrimul. Anionic/nonionic emulsifiers. Applications: Agricultural toxicant dispersers. (Lankro Chemicals Ltd).†

Agriphlan 24. Trade name for an orange-red liquid containing 240g/litre trifluoraline. Applications: Herbicide against cotton-weeds, bean, tomato, pear, garlic and sunflower weeds. Very efficient against amaranth, bristle-grass, knap-weed etc. (Chemical Combine).*

Agrisol. Speciality co-solvents. (Lankro Chemicals Ltd).

Agrispon. Mineral and plant extracts in a water base containing cytokinin, B-vitamin, morphogenic and porphyrin activity to aid in increased plant metabolism and yield. Applications: For all agricultural, horticultural and forestry products. See also Turbo-Grass and Agro-Vita. (SN Corp/Appropriate Technology Ltd).*

Agritol. A proprietary ammonium dynamite explosive.†

Agritox. Selective weedkiller. (May & Baker Ltd).

Agro-Vita. Mineral and plant extracts in a water base containing cytokinin, B-vitamin, morphogenic and porphyrin activity to aid in increased plant metabolism and yield. Applications: For all agricultural, horticultural and forestry products. See also Agrispon and Turbo-Grass. (SN Corp/-Appropriate Technology Ltd).*

Agrocide. Insecticides. (ICI PLC).*

Agrosan. Organo mercury powder seed dressing. (ICI PLC).*

Agrosol. Liquid mercury seed dressing. (Plant Protection (Subsidiary of ICI)).†

Agrothion. Liquid insecticide. (ICI PLC).*

Agroxone. Selective weed killer. (ICI PLC).*

Agucarina. See SACCHARINE.

Ague Salt. Quinine sulphate.†

Agulin. Sheep vaccine. (Glaxo Pharmaceuticals Ltd).†

Agurin. Theobromine-sodium-acetate. A diuretic.†

Aguttan. Hydroxy-quinolinesulphonic acid.†

A-hydroCort. Hydrocortisone sodium succinate. Applications: Glucocorticoid. (Abbott Laboratories).

Ai Hao. A Chinese drug prepared from the leaves of *Artemisia vulgaris*. A remedy for haemorrhage and diarrhoea.†

Aiathesin. *o*-Hydroxy-benzoyl-alcohol (salicyl alcohol). Antirheumatic and antiseptic.†

Aicello. Photo-sensitive diazo film in screen printing technology. Applications: Stencil film for making photo-stencils in screen printing. (Aicello Chemical Co Ltd).*

Aich Metal. (Gedge's Metal, Sterro Metal). An alloy similar to Delta metal, except that it contains iron. It usually consists of 60 per cent copper, 38 per cent zinc, and 1.5-2 per cent iron, and is used for sheathing ships.†

AID. Mixture of polyols and salts. Used as a detergent. Applications: stabilization of motor fuels carburettor cleaning. (UOP Inc)*

Aimatolite. Synonym for Hematolite.†

Air Saltpetre. (Norwegian Saltpetre). A mixture of calcium nitrite and nitrate. It is produced by passing air over a series of high intensity alternating arcs, absorbing the gases produced by lime water, and evaporating.†

Air-hardening Steel. A trade term applied to a manganese tool steel containing some tungsten, which hardens when cooled in air.†

Airbron. A proprietary preparation of acetylcysteine. A broncial inhalant. (British Drug Houses).†

Airedale. A range of dyes of various classes. Applications: Dyeing of leather. (Yorkshire Chemicals Plc).*

Airets. Throat drops. (Richardson-Vicks Inc).*

Airex. PVC foam (soft and rigid), with a density from 50-400 kg/m^3. Applications: Used as protective padding and life jackets, as core material in sandwich construction used in boat building, automotive and aviation industries as well as in off-shore oil platforms, used in gymnastic mats, sealings and insulation. (Lonza Limited).*

Airglow. Bright nickel plating process. (Hanshaw Chemicals).†

Airol Roche. A proprietary preparation of tretinoin for treatment of acne. (Roche Products Ltd).*

Airstrip. Waterproof non-occlusive plaster. (Smith and Nephew).†

Air, Alkaline. See VOLATILE ALKALI.

Air, Dephlogisticated. See VITAL AIR.

Air, Nitrous. See NITROUS GAS OR AIR.

Air, Pure. See VITAL AIR.

AIT. Pigment dispersions. Applications: In-plant tinting of aqueous coatings. (Pacific Dispersions Inc).*

Aithalite. See ASBOLANE.

Aithesin. A proprietary preparation of ALPHAXALONE and ALPHADOLONE acetate used in the induction of anaesthesia. (Glaxo Pharmaceuticals Ltd).†

Aix Oil. (Var Oil, Riviera Oil, Bari Oil). Commercial varieties of edible olive oil.†

Ajax Alloy. A bearing metal. It contains 30-70 per cent iron, 25-50 per cent nickel, and 5-20 per cent copper.†

Ajax Phosphor Bronze. An alloy containing 81 per cent copper, 11 per cent tin, 7 per cent lead, and 0.4 per cent phosphorus.†

Ajax Plastic Bronze. See PLASTIC BRONZE.

Ajax Powder. An explosive consisting of potassium perchlorate, nitroglycerol, ammonium oxalate, wood meal, and small quantities of collodion cotton, and nitro-toluenes.†

Ajkaite. (Ajkite). A fossil resin found in Hungary.†

Ajkite. See AJKAITE.

Ajuin. Synonym for Haüyne.†

Ak Mudar. (Akanda, Akra Rui, Erukku Erukkam). The bark of *Calotropis gigantea* and *C. procera*. An important Indian drug.†

Akanda. See AK MUDAR.

Akarittom Fat. A solid fat from *Parinarium laurinum*. It melts at 49-50°C. and has an iodine value of 214.†

Akaustan. Flameproof finishing agents for textiles. (BASF United Kingdom Ltd).

Akbar. A rubber vulcanization accelerator. It is a condensation product of formaldehyde and *P*-toluidine.†

Akco Resins. A proprietary synthetic phenolic resin for varnish manufacture.†

Akee Oil. See OIL OF AKEE.

Akesol. Valeri-amido-quinine.†

Akineton. Biperiden. Applications: Anticholinergic; antiparkinsonian. (Knoll Pharmaceutical Co).

Akinetone. A trade mark for a benzyl compound, $C_6H_5CH_2NH_2.CO.C_6H_4.CO_2H$, an anti-spasmodic. (Knoll AG).†

Akra Rui. See AK MUDAR.

Akrinol. Acrisorcin. Applications: Antifungal. (Schering-Plough Corp).

Akrite. A tool alloy. It contains 38 per cent cobalt, 30 per cent chromium, 10 per cent nickel, 4 per cent molybdenum, and 2-5 per cent carbon.†

Akroflex CD. A proprietary trade name for an antioxidant composed of diphenyl-para-phenylene-diamine and neozene D.†

Akroflex DAZ. A proprietary amine blend used as a combined anti- oxidant and anti-ozonant. (Du Pont (UK) Ltd).†

Akrol. A proprietary synthetic resin.†

Akron. An alloy of 63 per cent copper, 36 per cent zinc, and 1 per cent tin.†

Akrotherm. Histamine, acetylcholine and cholesterol. Applications: Chilblains. (Napp Laboratories Ltd).*

Aktiplast. A range of zinc salts of unsaturated acids used as peptizing agents and dispersing agents and aromatic disulphides used as reclaiming agents in the rubber industry. Applications: Moulded and extruded articles, production of reclaims. (Rhein-Chemie Rheinau).*

Aktivin. See CHLORAMINE T.

Akulon. Nylon 6 and 66. (AKU Holland).†

Akulon K and M. Proprietary grades of Nylon 6. (Algemene Industriele).†

Akulon R2. A grade of Nylon 66. (Algemene Industriele).†

Akund. A vegetable down of the kapok class, from *Asclepias* species of South Africa.†

Akypoquat 129 and 130. Cationic surfactant in the form of a quaternary ester, supplied as a light yellow paste. Application: Biodegradable raw material for laundry softeners. (Chem-Y, Fabriek van Chemische Producten BV).

Akypoquat 131 and 8188R. Cationic surfactant in the form of a quaternary ester, supplied as a light yellow paste. Application: Biodegradable raw material for hair care products e.g. hair rinses and conditioning shampoos. (Chem-Y, Fabriek van Chemische Producten BV).

Akyporox NP105. Nonylphenol ethoxylate nonionic surfactant in liquid form. Applications: Emulsifier and wetting agent used in household and industrial detergents; institutional cleaners; textiles; pesticides; pulp and paper. (Chem-Y, Fabriek van Chemische Producten BV).

Akyporox NP150, NP200 and NP300. Nonylphenol ethoxylate nonionic surfactant in solid form. Applications: Emulsifier and wetting agent for light duty detergents eg cleaning specialities. (Chem-Y, Fabriek van Chemische Producten BV).

Akyporox NP40 . Nonylphenol ethoxylate nonionic surfactant in liquid form.

Applications: Emulsifier and wetting agent used for non-polar hydrocarbon solvents and oils.(Chem-Y, Fabriek van Chemische Producten BV).

Akyporox NP475, NP500, NP1000, NP1200 and NP1500. Nonylphenol ethoxylate nonionic surfactant in the form of a water-soluble solid. Applications: Emulsifier and wetting agent used in synthetic latices and polishes. (Chem-Y, Fabriek van Chemische Producten BV).

Akyporox OP250 and OP400V. Octylphenol ethoxylate nonionic surfactant in solid or liquid form. Applications: Emulsifier and wetting agent used in emulsification polymerization. (Chem-Y, Fabriek van Chemische Producten BV).

Akyporox OP40 and OP115. Octylphenol ethoxylate nonionic surfactant in liquid form. Applications: Emulsifier and wetting agent used in emulsification and stabilization of oil systems of the more hydrophobic compounds; emulsification in aqueous systems; selected products for cosmetics and toiletries. (Chem-Y, Fabriek van Chemische Producten BV).

Akyposal ALS. Ammonium lauryl sulphate as a light yellow liquid. Applications: Emulsifier and detergent used in shampoo and bubble bath preparations and in emulsion polymerization. (Chem-Y, Fabriek van Chemische Producten BV).

Akyposal BD and NPS. Sodium alkyl phenol ether sulphate in liquid form. Applications: Emulsifier for emulsion polymerization and emulsion and dispersion use. (Chem-Y, Fabriek van Chemische Producten BV).

Akyposal DE, DEG, LFS and LFS/G. Natural lauryl ether sulphate as a brown anhydrous liquid. Applications: Emulsifiers for pesticides, oilbaths, etc.; base for highly concentrated oil and cream bath formulations. (Chem-Y, Fabriek van Chemische Producten BV).

Akyposal DS and 23ST. Sodium lauryl ether sulphate in liquid form. Applications: Anionic surfactant with good foaming performance in hard water. Used in shampoos, foam baths, hand cleansers and fabric washing. (Chem-Y, Fabriek van Chemische Producten BV).

Akyposal EO and RLM. Sodium lauryl ether sulphate in liquid form. Applications: Anionic surfactant base material for foam baths, shampoos, hand cleansers, fabric washing. (Chem-Y, Fabriek van Chemische Producten BV).

Akyposal MGLS. Magnesium lauryl sulphate as a light yellow liquid. Applications: Detergent base for shampoos and toothpastes. (Chem-Y, Fabriek van Chemische Producten BV).

Akyposal MLES. Monoethanolamine lauryl ether sulphate as a slightly yellow opaque liquid. Applications: High foaming, low irritant surfactant raw material for high quality shampoos and foam baths. (Chem-Y, Fabriek van Chemische Producten BV).

Akyposal MLS. Monoethanolamine lauryl sulphate as a light yellow liquid. Applications: Emulsifier and detergent for shampoo and bubble bath preparations and emulsion polymerization. (Chem-Y, Fabriek van Chemische Producten BV).

Akyposal MS, PM and RO/E. Sodium lauryl ether sulphate in liquid form. Applications: Detergent used as a base for emulsion and pearl shampoos. PM is a pearlizing agent; RO/E is an egg shampoo base material. (Chem-Y, Fabriek van Chemische Producten BV).

Akyposal NLS and SDS. Sodium lauryl sulphate in the form of a light yellow liquid. Applications: Detergent and emulsifier for emulsion polymerization and shampoos. (Chem-Y, Fabriek van Chemische Producten BV).

Akyposal TLS. Triethanolamine lauryl sulphate in liquid or soft paste form. Applications: Detergent base for bubble bath and car shampoos. (Chem-Y, Fabriek van Chemische Producten BV).

Al Anodised. See ELOXAL.

AL terna GEL. Aluminium hydroxide, dried. Applications: Antacid. (Stuart Pharmaceuticals, Div of ICI Americas Inc).

Al-dur-ba. A patented alloy of 76 per cent zinc, copper, 22 per cent zinc, and 2 per cent aluminium.†

Al-kenna. The powdered roots and leaves of *Lawsonia inermis*. It is used in the East for dyeing the nails, teeth, and hair. See Alkanet.†

Alabaster. A form of gypsum, $CaSO_4$ $2H_2O$ used for ornamental carvings. Also see Onyx of Tecali.†

Alabaster, Oriental. A compact form of marble, $CaCO_3$.†

Alacetan. Aluminium aceto-lactate.†

Aladar. See ALUDUR.

Alalite. See DIOPSIDE.

Alamask. Industrial deodorants. (May & Baker Ltd).*

Alan-gilan. Cananga oil, a neutral oil from *Cananga odorata*.†

Alanex. Active ingredient: alachlor; 2-chlor-2',6'-diethyl-N-(methoxymethyl)-acetanilide. Applications: Pre-emergence and pre-plant incorporated herbicide for the control of most annual grasses and certain broadleaf weeds. (Agan Chemical Manufacturers Ltd).†

Alar. Growth regulator. (Murphy Chemical Ltd).

Alargan. An alloy of aluminium and silver, the surface having been dusted with platinum black and hammered or subjected to pressure. A platinum substitute.†

Alasil. A proriетary preparation of calcium acetyl-salicylate and colloidal aluminium hydroxide.†

Alathon. Polyethylene resins. (Du Pont (UK) Ltd).

Alazine. Active ingredients: alanex plus atranex. Applications: Ready formulated mixture of alachlor plus atrazine for use as a selective pre-emergence herbicide. (Agan Chemical Manufacturers Ltd).†

Albacar. Highly refined calcite. (Pfizer International).*

Albacer. A proprietary synthetic wax for increasing melting-point of waxes.†

Albalith. A white, light-resisting lithopone, used in the paint and rubber industries.†

Albamycin. A proprietary trade name for the calcium salt of Novobiocin.†

Albamycin Capsules. Novobiocin sodium. Applications: Antibacterial. (The Upjohn Co).

Albamycin G.U. A proprietary preparation of novobiocin calcium and sulphamethizole. Urinary antiseptic. (Upjohn Ltd).†

Albamycin T. A proprietary preparation containing Novobiocin and Tetracycline. An antibiotic. (Upjohn Ltd).†

Albanite. A bituminous material found in Albania.†

Albanose. A leucite rock.†

Albaphos Dental Na 211. A proprietary trade name for sodium monofluorophosphate. A fluorine component for toothpastes, the toxic effects of which are only 1/3 of those of sodium fluoride. (Hoechst Chemicals (UK) Ltd).†

Albarium. A lime obtained by burning marble. Used for stucco.†

Albata. A nickel-brass or low nickel-silver containing about 8 per cent nickel.†

Albatex. Dyeing and printing assistant. (Ciba-Geigy PLC).

Albatex OR. A proprietary trade name for a non-foaming polyvalent amide. Used as a levelling agent for vat dyes. (Ciba Geigy PLC).†

Albatra Metal. A nickel silver. It contains 57.5 per cent copper, 22.5 per cent zinc, I8.75 per cent nickel, and 1.25 per cent lead.†

Albegal. Dyeing and printing assistant. (Ciba-Geigy PLC).

Albegal CL. A proprietary trade name for an ester of sulphonated fat. It is used as a levelling agent in wool dyeing. (Ciba Geigy PLC).†

Albene. An acetate silk deadened with pigments in the spinning process.†

Albene. See VEGETABLE BUTTER.

Alberene. A blue-grey soapstone mined in Virginia.†

Alberit MF. A proprietary melamine formaldehyde thermosetting moulding compound used in the manufacture of tracking-resistant mouldings for the electrical industry. (Canadian Hoechst).†

Alberit MP. A proprietary melamine/phenol-formaldehyde thermosetting moulding compound. (Canadian Hoechst).†

Alberit PF. A proprietary phenol-formaldehyde resin thermosetting moulding compound. (Canadian Hoechst).†

Alberit VP. A proprietary unsaturated polyester thermosetting moulding compound used in the production of impact-resistant mouldings. (Canadian Hoechst).†

Albert. Basic slag for fertilizing purposes. (Fisons PLC).*

Albert Coal. See ALBERTITE.

Albertat. A proprietary range of chemical fillers, extenders and additives for products containing synthetic resins, such as additives for thickening and preventing setting in paints and varnish. (Chemische Werke, Albert).†

Albertol. Rosin modified phenolic resins. Applications: Printing inks, paints and varnishes. (Resinous Chemicals Ltd).*

Albertol IIIL. A phenol-resin condensation product melting at 106- 133°C. It has a saponification value of 15.8, is insoluble in alcohol, but soluble in linseed oil. It is stated to be a good substitute for kauri gum in the manufacture of oil varnishes. Rosin glycerine-diane-fromaldehyde condensate in the presence of alkali. (Chemische Werke, Albert).†

Albertol 142-R. Butyl phenol formaldehyde. (Chemische Werke, Albert).†

Albertol 175-A. The aluminium salt of unesterified Albertol IIIL (*qv*). (Chemische Werke, Albert).†

Albertol 237-R. Di-*iso* butylphenol-formaldehyde. (Chemische Werke, Albert).†

Albertol 326-R (387L). A rosin modified phenolic resin used in aircraft primers made from 1 part Diane (*qv*), 1 part rosin and 0.1 part paraformaldehyde. (Chemische Werke, Albert).†

Albertol 347Q. An ester gum-phenolic combination made from xylenol-formaldehyde rosin, pentaerythritol and glycerogen (*qv*). (Chemische Werke, Albert).†

Albertol 369-Q (209-L). An ester gum-phenolic combination made from phenol-formaldehyde, rosin, pentaerythritol and glycerogen (*qv*). (Chemische Werke, Albert).†

Albiogen. Tetramethylammonium Oxalate.†

Albion Metal. A sheet of metal containing tin and lead. It is formed by pressing together sheets of these metals.†

Alboferine. See ALBOFERRIN.

Alboferrin. (Alboferine). A phospho-albumen preparation of iron, a tonic.†

Albolene. See PARAFFIN, LIQUID.

Alboleum. Oil insecticide. (Plant Protection (Subsidiary of ICI)).†

Alboline. See PARAFFIN, LIQUID.

Albolineum. Oil insecticide. (ICI PLC).*

Albolit. A proprietary phenol-formaldehyde synthetic resin.†

Albondur. A Bondur alloy (see Bondur) coated on each side with pure aluminium to improve corrosion resistance.†

Albor Die Steel. A proprietary steel containing small amounts of chromium, molybdenum, and carbon.†

Alboresin. A proprietary urea-formaldehyde synthetic resin. Moulding composition.†

Albral. Flux for use with aluminium bronzes, silicon bronzes and high tensile brasses. (Foseco (FS) Ltd).*

Albrichrome. Textile dyestuffs. (Albright & Wilson Ltd, Phosphates Div).

Albricide. Fungicide. (Albright & Wilson Ltd, Phosphates Div).

Albrifloc. Flocculating agent. (Albright & Wilson Ltd, Phosphates Div).

Albrightex. Textile optical brighteners. (Albright & Wilson Ltd, Phosphates Div).

Albrilan. Textile dyestuffs. (Albright & Wilson Ltd, Phosphates Div).

Albrilene. Textile dyestuffs. (Albright & Wilson Ltd, Phosphates Div).

Albrilon. Textile dyestuffs. (Albright & Wilson Ltd, Phosphates Div).

Albrilube. Textile lubricant. (Albright & Wilson Ltd, Phosphates Div).

Albrinol. Textile dyestuffs. (Albright & Wilson Ltd, Phosphates Div).

Albrinyl. Textile dyestuffs. (Albright & Wilson Ltd, Phosphates Div).

Albriquest. Sequestering agent. (Albright & Wilson Ltd, Phosphates Div).

Albriscour. Textile scouring agent. (Albright & Wilson Ltd, Phosphates Div).

Albrisolve. Textile dyeing agent. (Albright & Wilson Ltd, Phosphates Div).

Albrisperse. Textile dispersing agent. (Albright & Wilson Ltd, Phosphates Div).

Albritone. Textile levelling agent. (Albright & Wilson Ltd, Phosphates Div).

Albrivap. Boiler scale inhibitor. (Albright & Wilson Ltd, Phosphates Div).

Albumaid Preparations. A group of proprietary preparations of beef protein hydrolysate, amino acids, carbohydrates, vitamins and minerals used in the treatment of aminoacidaeraias and mal-absorption syndromes. (Scientific Hospital Supplies).†

Albumen Paper. A photographic printing-out paper, the sensitive substance being silver chloride, and the ground, albumen.†

Albumen Powder. (Aleuronate). Dried and powdered gluten.†

Albuminar. Albumin human. Applications: Blood volume supporter. (Armour Pharmaceutical Co).

Albumotope I-131. Albumin, iodinated I 131 serum. Applications: Diagnostic aid; radioactive agent. (E R Squibb & Sons Inc).

Alburex. Vegetable proteins. Applications: Animal feedstuff. (Roquette (UK) Ltd).*

Albustix. A prepared test strip of tetra-bromphenol blue with a citrate buffer, used to detect protein in urine. (Ames).†

Albutannin. (Protan). Albumen tannate.†

Alcacement. See ALCEMENT.

Alcad. A duralumin coated with pure aluminium.†

Alcaine. Proparacaine Hydrochloride. Applications: Anaesthetic. (Alcon Laboratories Inc).

Alcalase. Proteolytic enzyme prepared by submerged fermentation of a selected strain of Bacillus licheniformis. Applications: Used in the detergent industry. (Novo Industri A/S).*

Alcan. A trade mark for alloys of aluminium coded as follows:

GB-99.8 per cent. Wrought, non heat-treatable high purity aluminium.

GB-1S. Wrought, non heat-treatable 99.5 per cent. aluminium.

GB-2S Wrought, non heat-treatable commercially pure aluminium.

GB-3S. Wrought, non heat-treatable aluminium. Stronger and harder than 2S.

GB-B53S, 54S, D54S, A56S, M57S. Wrought non heat-treatable aluminium alloys in which magnesium is the main additive.

GB-50S. Wrought heat-treatable aluminium alloy. Forms well in the "W" (solution heat-treated) condition.

GB-B51S, 65S. Wrought, heat-treatable medium strength alloys of aluminium.

GB-100. Cast, non heat-treatable commercially pure aluminium.

GB-160. Cast, non heat-treatable aluminium alloy.

GB-B320. Cast, non heat-treatable medium strength aluminium alloy.

GB-B116. Cast, heat-treatable alloy available in four conditions "M" (as cast), "P" (precipitation treated), "W" (solution treated) and "WP" (fully heat treated).

GB-350. Cast, heat-treatable, aluminium alloy specially impact resistant. Good resistance to marine conditions. (Hoechst Chemicals (UK) Ltd).†

Alcaphos 24. Strongly alkaline silicated solid. Applications: Formulated for heavy duty soak cleaning in steel fabricated metals. (Invequimica & CIA SCA).*

Alcement. (Alcacement). Fused cement prepared in the electric furnace from bauxite and lime. It contains approximately 40 per cent. CaO, 40 per cent. Al_2O_3, 10 per cent. SiO_2, and 10 per cent. Fe_2O_3.†

Alchemy. Metal carboxylates. Applications: Driers for paint or printing ink. (Manchem Ltd).*

Alcian. Dyestuffs, blues, greens and yellow dyes formed by introducing chloromethyl groups into phthalocyanin and its deriviatives by means of dichlorodimethyl either in pyridine containing aluminium chloride. (ICI PLC).*

Alcin. A proprietary preparation containing sodium magnesium aluminium silicate and basic magnesium aluminate. An antacid. (Reckitts).†

Alcoa 108. A proprietary aluminium alloy with 3 per cent silicon and 4 per cent copper.†

Alcoa 112. A proprietary aluminium alloy containing 7-8.5 per cent copper, 1-2 per cent zinc, and up to 1.7 per cent of other metals, mostly iron.†

Alcoa 122. A proprietary alloy of aluminium with 10 per cent copper, 0.2 per cent magnesium, and 1.2 per cent iron.†

Alcoa 145. A proprietary aluminium alloy containing 10 per cent zinc, 2.5 per cent copper, and 1.2 per cent iron.†

Alcoa 195. A proprietary alloy containing aluminium with 4 per cent copper.†

Alcoa 2-S. A proprietary trade name for a commercially pure aluminium.†

Alcoa 220-TA. A proprietary aluminium alloy containing aluminium with 10 per cent magnesium.†

Alcoa 24-S. A proprietary wrought aluminium alloy containing 93.7 per cent aluminium, 1.5 per cent magnesium, 4.2 per cent copper, 0.6 per cent manganese.†

Alcoa 3-S. A proprietary alloy of aluminium, containing small amounts of copper, iron, silicon, and zinc.†

Alcoa 32-S. A proprietary alloy of aluminium with 12 per cent silicon, 0.8 per cent nickel, 1 per cent magnesium, and 0.8 per cent copper.†

Alcoa 356. A proprietary alloy of aluminium containing 4 per cent silicon, 0.3 per cent magnesium.†

Alcoa 43. A proprietary alloy of aluminium and silicon. Contains 5 per cent silicon.†

Alcoa 47. A proprietary alloy of aluminium with 12.5 per cent silicon.†

Alcoa 515. A proprietary alloy containing aluminium with magnesium, silicon, and iron.†

Alcoa 535. A proprietary alloy of aluminium with 0.25 per cent chromium, 1.25 per cent magnesium, and 0.7 per cent silicon.†

Alcobon. A proprietary preparation of flucytosine. A systemic anti-fungal agent. (Roche Products Ltd).*

Alcobronze. An alloy of copper and aluminium of golden colour. Very hard.†

Alcodrill HPD-D. Dry free flowing powdered carboxylate copolymers. Applications: Deflocculant in water based drilling. (Alco Chemical Corporation).*

Alcodrill HPD-L. A polycarboxylate copolymer solution. Applications: Deflocculant for water based drilling. (Alco Chemical Corporation).*

Alcoform. A solution of formaldehyde in one of a variety of alcohols. Applications: Production of butylated resins and methylated resins. Methyl alcoforms are used for the production of ion-exchange resins. (Synthite Ltd).*

Alcogas. A trade mark for a mixture of alcohol (anhydrous) and hydrocarbons in varying proportions.†

Alcogum AN 10. High viscosity sodium polyacrylate thickener for latex systems. Applications: Carpet backing, latex foam, adhesives, dispersants. (Alco Chemical Corporation).*

Alcogum L-11. A high efficiency alkali-swellable acrylic emulsion thickener. Applications: Adhesives, latex thickening, paint thickening. (Alco Chemical Corporation).*

Alcogum L-27. A high efficiency alkali-swellable acrylic emulsion thickener. Applications: Adhesives, latex thickening, dispersion thickening. (Alco Chemical Corporation).*

Alcogum L-36. A reactive alkali activated acrylic emulsion polymer. Applications: Paper coating, latex compounding. (Alco Chemical Corporation).*

Alcogum L-52. A self-crosslinking alkali activated emulsion thickener. Applications: Adhesives, latex thickening. (Alco Chemical Corporation).*

Alcogum L-60. An alkali activated 'associative' emulsion thickener. Applications: Paint, adhesives, cleaners, wall joint compounds. (Alco Chemical Corporation).*

Alcogum 296-W. A high viscosity sodium polyacrylate thickener. Applications: Adhesives, paint, cement additive, protective colloid. (Alco Chemical Corporation).*

Alcogum 310. A high molecular weight inverse-emulsion copolymer (organic phase medical white oil). Applications: Adhesives, coatings. (Alco Chemical Corporation).*

Alcogum 9635. A high viscosity sodium polyacrylate thickener. Applications: Latex adhesive in the tufted carpet industry. (Alco Chemical Corporation).*

Alcogum 9661. A high viscosity sodium polyacrylate thickener. Applications: Textile coatings for upholstery. (Alco Chemical Corporation).*

Alcogum 9710. A sodium polyacrylate thickener. Applications: Coatings, packaging, adhesives, latex thickening. (Alco Chemical Corporation).*

Alcohol, Denatured. See METHYLATED SPIRIT.

Alcohol, Phenic. See PHENIC ACID.

Alcolec 532. A proprietary preparation of vinyl-based resin in bead form. A carboxylated vinyl copolymer used in the formulation of flexographic printing inks and paper lacquers. (Allied Colloids Ltd).†

Alcolite. A proprietary product used for denture purposes.†

Alcolube CL. A cationic polyethylene/-lanolin emulsion used as a substantive softener for polyamide and polyester fibres. (Allied Colloids Ltd).†

Alcon-Efrin. Phenylephrine Hydrochloride. Applications: Adrenergic. (Alcon Laboratories Inc).

Alconate L-80. Concentrated form of Petronate L, an anionic surfactant of the petroleum sulphonate type. Applications: Emulsifier and wetting agent, used when low oil content is required. (Witco Chemical Ltd).

Alcopal FA. A foaming agent for aqueous systems used for carpet backing.†

Alcopar. A proprietary preparation containing Bephenium Hydroxynaphthoate used as an anthelmintic. (The Wellcome Foundation Ltd).*

Alcopol AH New. Amine salt of sulphated higher fatty acid ester in the form of a brown fluid oil. Applications: Powerful wetting, penetrating and emulsifying agent used in pigments; paint; leather; emulsion polymerization; dry cleaning; cutting oils; agricultural chemicals. (Allied Colloids Ltd).

Alcopol FA. Anionic surfactant in which the anion is a long chain sulphosuccinamate, in the form of a fluid dispersion. Applications: Powerful surface active agent used in the production of low density latex foams with good wet stability. (Allied Colloids Ltd).

Alcopol O. Range of anionic surfactants in which the anion is di- octyl sulphosuccinate. Applications: Emulsifiers and powerful wetting agents with applications in industries such as paper, textiles, asbestos, plastics and photographic film, metals, pest control, detergents and degreasing, dust control, glass cleaning oils, lubricants, paints, pigments, printing inks and a wide range of proprietary products eg hand cleansers, cosmetics.(Allied Colloids Ltd).

Alcopol OB. Sodium di-iso-butyl sulphosuccinate as a water/alcohol solution. Applications: Powerful wetting agent in the presence of electrolytes. (Allied Colloids Ltd).

Alcopol OD. Sodium di-tri-decyl sulphosuccinate in water/alcohol solution. Applications: Emulsifier for oils, solvents, waxes and polymers. (Allied Colloids Ltd).

Alcopol OS. Sodium di-hexyl sulphosuccinate in water/alcohol solution. Applications: Emulsifier and powerful wetting agent in the presence of electrolytes, used for oils, solvents, waxes, polymers and windscreen wash concentrate. (Allied Colloids Ltd).

Alcopol T. Sodium salt of a sulphated higher fatty acid ester as a low viscosity pale yellow liquid. Applications: Powerful wetting, penetrating and emulsifying agent used in pigments; paints; leather; emulsion polymerization; dry cleaning; cutting oils, agricultural chemicals. (Allied Colloids Ltd).

Alcor 7. A clad combination of stainless steel and aluminium-containing magnetic stainless interlayer, used in the production of high quality cookware and magnetic induction heating stoves. (Pfizer International).*

Alcosperse. A range of speciality dispersing and levelling agents. They are used in the dyeing of synthetic fibres. (Allied Colloids Ltd).†

Alcosperse 104. A polycarboxylate solution polymer. Applications: Anti-redeposition agent, slurry stabilization. (Alco Chemical Corporation).*

Alcosperse 107-D. A dry powdered poly-carboxylate polymer. Applications: Dispersant in paper, board, coatings and paint. (Alco Chemical Corporation).*

Alcosperse 144. An acrylic solution polymer. Applications: Slurry preparation. (Alco Chemical Corporation).*

Alcosperse 149-C. An acrylate solution polymer. Applications: Paper coatings, pigment slurries. (Alco Chemical Corporation).*

Alcosperse 169. A sodium polyacrylate solution polymer. Applications: Dispersant in adhesives, paint, kaolin and calcium carbonate. (Alco Chemical Corporation).*

Alcosperse 175. A low molecular weight carboxylate copolymer solution. Applications: Laundry detergent, incrustation inhibitor. (Alco Chemical Corporation).*

Alcosperse 249. Ammonium polyacrylate solution polymer. Applications: Dispersant in paint, adhesives. (Alco Chemical Corporation).*

Alcosperse 602. Polycarboxylate solution polymer. Applications: Anti-redeposition agent. (Alco Chemical Corporation).*

Alcotreat PC 95. A high molecular weight cationic polymer solution. Applications: Water/oil clarification in oil field applications. (Alco Chemical Corporation).*

Alcotreat 182. A high molecular weight inverse-emulsion polymer. Applications: Drilling fluids, mineral extraction. (Alco Chemical Corporation).*

Alcovar. Fast dyes for spirit and cellulose varnishes. (Williams Div of Morton Thiokol Ltd). *

Alcryn. Thermoplastic elastomer. (Du Pont (UK) Ltd).

Alcumite. A proprietary corrosion-resisting alloy containing 87.5 per cent copper, 7.5 per cent aluminium, 3.5 per cent iron, and 1.5 per cent nickel.†

Alcuronium Chloride. Diallyldinortoxiferin dichloride. Alloferin.†

Aldactide. A proprietary preparation of spironolactone and hydroflumethiazide. A diuretic. (Searle, G D & Co Ltd).†

Aldactone. A proprietary preparation of spironolactone. A diuretic. (Searle, G D & Co Ltd).†

Aldamine. A proprietary trade name for acetaldehyde-ammonia $CH_3COCH.(OH).NH_2$. The material is used in the manufacture of plastics and as a pickling inhibitor for steel, and is a rubber vulcanizing accelerator.†

Aldehol A. An oxidized kerosene to be used in U.S.A. for denaturing methylated spirit.†

Aldehyde. Acetaldehyde, CH_3CHO.†

Aldehyde C14. A proprietary flaming material. Undecalactone.†

Aldehyde Green. (Aniline Green, Usébe Green). A diphenylmethane dyestuff. Dyes silk green from an acid bath.†

Alder Bark. The bark of *Alnus glutinosa.* Used for fixing yellow dyes and as a tanning material.†

Alderton's Solution. A solution of ammonium ichthosulphonate, in glycerol †

Aldiphen. See DINITRA.

Aldobond. Blend of synthetic or natural elastomers and resins in a solvent or aqueous medium. Applications: Adhesive compositions for industrial fabricating operations. (Aldo Products Co Inc).*

Aldocoat. Blend of synthetic or natural elastomers and resins in a solvent of aqueous medium. Applications: Industrial specialty coatings. (Aldo Products Co Inc).*

Aldocorten. A proprietaty preparation of aldosterone. (Ciba Geigy PLC).†

Aldoform. A formaldehyde preparation.†

Aldogen. A mixture of trioxymethylene and bleaching powder. An antiseptic.†

Aldol. 3-Hydroxybutanal, $H_3CCH(OH)CH_2CHO$. Formerly used as a hypnotic.†

Aldomet. Methyldopa. Applications: Antihypertensive. (Merck Sharp & Dohme).*

Aldomet Ester Hydrochloride. Methyldopate hydrochloride. Applications: Antihypertensive. (Merck Sharp & Dohme, Div of Merck & Co Inc).

Aldones. Flavour bases. (Bush Boake Allen).*

Aldoretic. Methyldopa, amiloride hydrochlorothiazide. Applications: Antihypertensive. (Merck Sharp & Dohme).*

Aldoril. Methyldopa, hydrochlorothiazide. Applications: Antihypertensive. (Merck Sharp & Dohme).*

Aldosperse. Emulsifiers. Applications: Food. (Jan Dekker BV).

Aldosterone. 11β,21-dihydroxy-3,20-dioxopregn-4-en-18-al. Aldocorten; Electrocortin.†

Aldrey. An aluminium alloy of Swiss origin used for electrical conductors. It contains 98.7 per cent aluminium, 0.6 per cent silicon, 0.4 per cent magnesium, and 0.3 per cent iron.†

Aldrin Dust. Insecticide. (Murphy Chemical Ltd).

Aldur. A urea-formaldehyde resin.†

Aldydale. A proprietary phenol-formaldehyde synthetic resin.†

Alecra. Chromium plating processes. (Albright & Wilson Ltd, Phosphates Div).

Alegar. Vinegar.†

Alepol. A proprietary preparation of selected sodium salts of hydrocarpus oil acids.†

Aletodin. See ASPIRIN.

Aleudrin. (Dichloramal). The carbamic acid ester of α-dichloro- isopropyl alcohol $(CH_2Cl)_2.CH.O.NH_2$. An analgesic and sedative. A proprietary trade name for the sulphate of Isoprenaline. A bronchodilator. (Lewis Laboratories).†

Aleuronate. See ALBUMEN POWDER.

Alevaire. A proprietary preparation containing tyloxapol. A mucolytic. (Bayer & Co).*

Alexan. A proprietary preparation of cytarabine. An antineoplastic agent. (Pfizer International).*

Alexander Green. See MOUNTAIN GREEN.

Alexandrian Laurel Oil. See LAUREL NUT OIL.

Alexipon. Ethylacetylsalicylate.†

Alexis. Aluminium etchant. (Albright & Wilson Ltd, Phosphates Div).

Alexite. A proprietary trade name for an aluminium oxide abrasive.†

Alfa. A variety of esparto grass used in the manufacture of paper. It is also the term for a synthetic tannin, a red-brown liquid containing 23 per cent tanning substance, 11 per cent non-tannins, 66 per cent water, and a trace of sulphuric acid.†

Alfacron. Poultry house insecticide. (Ciba-Geigy PLC).

Alfadex. Pyrethrin insecticide. (Ciba-Geigy PLC).

Alfavet. Alfaprostol. Applications: Prostaglandin. (Hoffmann-LaRoche Inc).

Alfenide. See NICKEL SILVERS.

Alferium. A proprietary alloy of aluminium with 2.5 per cent copper, 0.62 per cent magnesium, 0.5 per cent manganese, and 0.3 per cent silicon.†

Alferon. Alpha interferon - natural (injectable form). Applications: Used for the treatment of genital warts. (Interferon Sciences Inc).*

Alferric. Aluminium sulphate. (Laporte Industries Ltd).*

Alflorone. A proprietary trade name for the 21-acetate of Fludrocortisone†

Alfodex. Fungal-α-amylase. (Glaxo Pharmaceuticals Ltd).†

Alfol. A proprietary trade name for an aluminium foil in a crumpled condition used for heat insulation.†

Alforder. A proprietary synthetic resin.†

Alformin. A 16 per cent solution of basic aluminium formate, $Al_2(OH)_2(HCO_2)_4$. An astringent and antiseptic.†

Alfralat. A proprietary name for glyceryl phthalate resins. (Chemische Werke, Albert).†

Alfrax B301. A commercial grade of bubble aluminium oxide. (Carborundum Co).†

Alftalat. Alkyd resins (oil modified polyesters). Applications: Air drying decorative paints, air drying and stoving industrial finishes. (Resinous Chemicals Ltd).*

Algae Treat. 30% quatenary ammonia compound. Applications: Used for algae control in cooling water systems. (Delaware Chemical Corp).*

Algalex 104. A solution of the sodium salt of a chlorinated phenyl derivative of methan containing surface active and antifoam agents. A non- toxic non-corrosive bactericide for water systems. (Kinnis and Brown).†

Algalith. A proprietary algine plastic.†

Algarobilla. A vegetable tanning material. It consists of the pods of *Caesulpinias brevifolia* of Chile, and contains about 60 per cent tannin.†

Algarobillin. A dye product obtained from the carob tree, *Ceratonia siliqua*, found in the Argentine. It is employed for dyeing cloth, khaki.†

Algaroth, Powder of. See POWDER OF ALGAROTH.

Algarotti Powder. See POWDER OF ALGAROTH.

Algarovilla. A Columbian name for a copal resin obtained there.†

Alger Metal. An alloy of 90 per cent tin and 10 per cent antimony. A silvery-white alloy used in making jewellery.†

Algestone Acetonide. A progestational steroid. 16α, 17α- Isopropylidenedi-oxypregn-4-ene-3,20-dione.†

Algier's Metal. A jeweller's alloy.
(*a*) Consists of 90 per cent tin and 10 per cent antimony, and is used for the manufacture of forks and spoons; (*b*) contains 94.5 per cent tin, 5 per cent copper, and 0.5 per cent antimony. It is used for making hand bells.†

Algin. A gelatinous substance which is the residue from the water maceration of seaweed, during the process of obtaining iodine. It is used as a substitute for isinglass.†

Alginade. Alginate ice cream stabilizer. (Kelco/AIL International Ltd).

Alginoid Iron. See ALGIRON.

Algipan Balm. A proprietary preparation of methylnicotinate, glycol salicylate, histamine and capsicin. An embrocation. (Wyeth).†

Algiron. (Alginoid iron). An iron compound of alginic acid (from seaweed). It contains 11 per cent iron.†

Algistat. Preparations for water treatment. (BDH Chemicals Ltd).

Algodon. Cotton wool.†

Algodon de Seda. The fibre of *Calotropis gigantea* is known in Venezuela by this name.†

Algoflon. PTFE resins. (Montedison UK Ltd).*

Algol. A trade mark for a range of dyestuffs. (Casella AG).†

Algol Blue C. A dyestuff. It is equivalent to Indanthrene blue GC.†

Algol Blue CF. A dyestuff. it is the chloro derivative of indanthrene.†

Algol Blue K. A dyestuff. It is dimethyl-indanthrene.†

Algol Blue 3G. A dyestuff. it is a dihydroxyindanthrene.†

Algol Blue 3R. (Algol Brilliant Violet 2B). A dyestuff. It consists of dibenzoyldiaminoanthrarufin.†

Algol Bordeaux 3B. A dyestuff. It is an anthraquinoneimide.†

Algol Brilliant Orange FR. A dye stuff. It is benzoyl-1:2:4-triaminoanthraquinone.†

Algol Brilliant Red 2B. A dyestuff. It is 1:5-dibenzoyldiamino-4-hydroxy-anthraquinone.†

Algol Brilliant Violet 2B. See ALGOL BLUE 2B.

Algol Green B. A dyestuff. It consists of dibromodiaminoindanthrene.†

Algol Grey B. A dyestuff. It is 1:5-diamino-anthraquinone condensed with 1-chloroanthraquinone. The product is reduced with alkali sulphide, and then nitrated.†

Algol Olive R. A dyestuff. It is prepared by the action of chlorosulphonic acid upon dibenzoyldiaminoanthraquinone.†

Algol Orange R. A dyestuff. It is $\alpha\beta$-dianthraquinonylamine.†

Algol Pink. A dyestuff. It consists of benzoyl-4-amino-1-hydroxy-anthraquinone.†

Algol Red R Extra. A dyestuff. It is 1,5-dibenzamido-8-hydroxy-anthraquinone.†

Algol Red 5G. A dyestuff. It is 1,4-dibenzamidoanthraquinone.†

Algol Scarlet G. A dyestuff. It consists of 1-benzamido-4-methoxy-anthraquinone.†

Algol Violet B. A dyestuff. It is benzamido-4,5,8-trihydroxyanthraquinone.†

Algol Yellow R. (Hydranthrene Yellow ARNAG New, Indanthrene Yellow GK, Caledon Yellow 3G). A dyestuff. It is 1,5-dibenzamidoanthraquinone.†

Algol Yellow WG. (Hydranthrene Yellow AGR). A dyestuff. It is 1-benzamidoanthraquinone.†

Algol Yellow 3G. A dyestuff. It is 1-succinamidoanthraquinone.†

Algulose. A very pure cellulose used in paper-making. it is obtained from kelp.†

Ali-Clean. Acid based aluminium cleaner. Applications: Cleaning aluminium and magnesium alloy products especially car wheels. Removes dirt, brake dust and oxidation just by brushing on and washing off. (Hermetite Products Ltd).*

Alibated Iron. Iron coated with alumininum to form a protective covering.†

Alibi. Weedicide, containing bifenox and linuron. (ICI PLC).*

Alidine. The dihydrochloride of Anileridine.†

Aliette. Fungicide. (May & Baker Ltd).

Alimemazine. Trimeprazine.†

Alimet. Methionine hydroxy analog feed supplement. Applications: Liquid source of amino acid (methionine) activity for poultry and other animal feeds. (Monsanto Co).*

Alipal CO-128. The ammonium salt of ethoxylate sulphate. Applications: Air-entraining agent for concrete; foaming agent for the production of light-weight cements; frothing agent for gypsum wallboard, foaming agent for the petroleum industry. (GAF Corporation).*

Alipal CO-433. The sodium salt of sulphated nonylphenoxypoly-(ethyleneoxy)-ethanol. Applications: Base for scrub soaps, car washes, rug and hair shampoos, emulsifier for vinyl polymerization and petroleum waxes, antistatic agent for plastic materials and synthetic fibres. (GAF Corporation).*

Alipal CO-436. Ammonium salt of sulphated nonylphenoxypoly-(ethyleneoxy)-ethanol. Applications: Used as a detergent base for high-foaming dishwashing formulations, scrub soaps, car washes, rug and hair shampoos. Useful as surfactant base in pharmaceutical preparations. (GAF Corporation).*

Alipal EP-110, EP-115, EP-120. A range of increasingly hydrophilic ammonium salts of sulphated nonylphenoxypoly-(ethyleneoxy)-ethanol. Applications: Versatile primary emulsifiers and stabilizing agents. (GAF Corporation).*

Alipal HF-433. An ammonium salt of sulphated nonylphenoxypoly-(ethyleneoxy)-ethanol. Applications: High-foaming anionic surfactant. Excellent base for shampoos or bubble bath due to considerable mildness to the skin. (GAF Corporation).*

Alipal SE-463. A sodium salt of alkyaryl polyether sulphonate. Applications: Good foaming characteristics, coupled with detergency, rinsability and low level toxicity, make it an excellent base for cosmetic and pharmaceutical products. (GAF Corporation).*

Aliso. Aluminium isopropoxide. Applications: Cosmetics, pharmaceuticals. (Manchem Ltd).*

Alistell. Herbicide, containing linuron, 2,4-DB and MCPA. (ICI PLC).*

Alite. The chief constituent of Portland cement. It consists mainly of calcium ortho-silicate, $3CaO.SiO_2$, with calcium oxide, CaO, and a certam amount of calcium aluminate, and ferrite.†

Alival. 3-Iodo-1,2-propanediol, $CH_2I.CHOH.CH_2OH$.†

Alizanthrene Blue GC. A dyestuff. It is a British equivalent of Indanthrene blue GC.†

Alizanthrene Blue GCP. A dyestuff. It is a British equivalent of Indanthrene blue GCD.†

Alizanthrene Dark Blue 0. A dyestuff. It is a British equivalent of Indanthrene dark blue BO.†

Alizanthrene Olive G. A dyestuff. It is a British equivalent of Indanthrene olive G.†

Alizarin. Acid wool dyes. (Bayer & Co).*

Alizarin Acid Blue BB. See ACID ALIZARIN BLUE, BB, GR.

Alizarin Acid Blues. Dyestuffs prepared from anthrachrysone. They consist of polyoxyanthraquinonedisulphonic acids, and dye wool from an acid bath.†

Alizarin Acid Green. A dyestuff prepared by reducing dinitro-anthra-chrysone-disulphonic acid.†

Alizarin Astrol. A dyestuff. It is 1-methylamino-4-sulpho-toluidine-anthraquinone,†

Alizarin Black Blue. See ALIZARIN CYANINE BLACK G.

Alizarin Black P. A dyestuff. It is flavopurpurinquinoline. It dyes chromed wool violet-grey to black. Used also in calico printing.†

Alizarin Black S. (Alizarin black SW, SRW, WR, Alizarin Blue Black RW, Alizarin Blue Black SW, Naphthazarin S). A dyestuff. It dyes chromed wool black, and is used in printing.†

Alizarin Black S. A dyestuff. It is the sodium bisulphite compound of Alizarin Black P. It dyes chromed wool grey to black, and is also used in calico printing.†

Alizarin Black SRA. See ALIZARIN BLACK S.

Alizarin Black SW. See ALIZARIN BLACK S.

Alizarin Black WR. See ALIZARIN BLACK S.

Alizarin Black WX. A dyestuff. It is a tetrahydroxynaphthalene.†

Alizarin Blue. (Alizarin Blue A, AB,ABI, BSS, DNW, F, GW, R, RR). A dyestuff. Dyes mordanted fabrics blue. BSS is the sodium salt.†

Alizarin Blue BB, GR. See ACID ALIZARIN BLUE BB, GR.

Alizarin Blue Black RW. See ALIZARIN BLACK S.

Alizarin Blue Black SW. See ALIZARIN BLACK S.

Alizarin Blue Green. A dyestuff. It is the sulphonic acid of a trihydroxy-anthraquinoline-quinone.†

Alizarin Blue S. (Alizarin Blue ABS, Anthracene Blue S). A dyestuff. It is the bisulphite compound of Alizarin blue. Dyes chromed fabrics. Used in printing.†

Alizarin Blue SNG, SNW, SWN. Dyestuffs identical or isomeric with anthracene blues (*qv*).†

Alizarin Blue WA. See ALIZARIN BLUE XA.

Alizarin Blue XA. (Alizarin Blue WA). The soluble sodium salt of Alizarin blue (*qv*).†

Alizarin Bordeaux BA, BAY. Dyestuffs. They are British equivalents of Alizarin bordeaux B.†

Alizarin Bordeaux B, BD. (Alizarin Cyanine 3R, Quinalizarin). A dyestuff. It is hydroxyanthraquinone. It dyes wool mordanted with alumina, bordeaux, and with chrome, dark violet blue.†

Alizarin Bordeaux G, GG. Dyestuffs isomeric with Alizarin bordeaux B.†

Alizarin Brilliant Blue B. A dyestuff. It is a British equivalent of Alizarin saphirol B.†

Alizarin Brilliant Blue GS. An equivalent of Alizarin saphirol B.†

Alizarin Brilliant Violet R. A dyestuff. It is a British equivalent of Alizarin irisol R.†

Alizarin Brown. See ANTHRACENE BROWN.

Alizarin Brown. A dyestuff. It is nitroalizarin, and dyes wool yellow-brown with a chrome mordant.†

Alizarin Brown M. A dyestuff. It is a British equivalent of Metachrome brown B.†

Alizarin Cardinal. See ALIZARIN GARNET R.

Alizarin Carmine. See ALIZARIN RED S.

Alizarin Chestnut. A dyestuff. It is aminoalizarin, obtained by the reduction of Alizarin orange (β-nitroalizarin).†

Alizarin Claret. See ALIZARIN GARNET R.

Alizarin Cyanine AC. A dyestuff. It is a British equivalent of Alizarin cyanine R.†

Alizarin Cyanine Black G. An anthracene dyestuff. It dyes wool with chrome mordants.†

Alizarin Cyanine G. A dyestuff. It is the imide of tri- or tetrahydroxy-anthraquinone, and dyes wool mordanted with alumina, blue, and with chrome, bluish-green.†

Alizarin Cyanine Greens. (Quinizarin Greens, Alizarin Viridine). Dyestuffs obtained by treating quinizarin with excess of aniline and sulphonating. The sodium salt is Quinizarin green. They dye wool mordanted with chrome, green.†

Alizarin Cyanine 3G. A dyestuff. It is a sulphonic acid of a polyaminohydroxyanthraquinone. Dyes wool.†

Alizarin Cyanine 3R. See ALIZARIN BORDEAUX B, BD.

Alizarin Dark Blue. A mordant dyestuff. It gives indigo blue shades on chrome mordanted wool.†

Alizarin Dark Green. A dyestuff obtained by the treatment of naphthazarin melt with phenols. It dyes chromed wool grey-green to greenish-black.†

Alizarin Dark Red B,BB.G.5G. Acid dyestuffs suitable for dyeing wool.†

Alizarin DCA, JCA, CAF, YAR. Dyestuffs. They are British equivalents of Flavopurpurin.†

Alizarin Delphinol B. A dyestuff. It is a British equivalent of Alizarin saphirol B.†

Alizarin Dephinol. A blue anthracene dyestuff for wool or silk.†

Alizarin Direct Biue A3G. An acid dyestuff suitable for silk.†

Alizarin Direct Blue A. An acid dye stuff suitable for dress goods and curtains, also for silk and acetate silk.†

Alizarin Direct Brilliant Light Blue R. An acid dyestuff suitable for silk.†

Alizarin Emerald. An acid dyestuff for wool or silk.†

Alizarin FA. See FLAVOPURPURIN.

Alizarin Fast Blue, BHG. A mordant dyestuff for wool.†

Alizarin for Violet. See ALIZARIN.

Alizarin Garnet R. (Alizarin Cardinal, Alizarin Claret, Alizarin Granat R). A dyestuff. It is α-aminoalizarin. It dyes wool mordanted with alumina, bluish-red. It is also used in calico- printing.†

Alizarin GB. See FLAVOPURPURIN.

Alizarin GD. See ANTHRAPURPURIN.

Alizarin GI. See FLAVOPURPURIN.

Alizarin Green. See COEULEINE and COEULEINE S.

Alizarin Green B. A dyestuff. It is dihydroxynaphthoxazoniumsulphonate. It dyes chromed fabrics green.†

Alizarin Green S. A dyestuff. It is a mixture of the bisulphite compounds of tri- and tetrahydroxy-anthraquinone-quinolines and their sulphonic acids, chiefly. It dyes chromed wool bluish-green, and is also used for cotton printing.†

Alizarin Green S. A dyestuff. It gives green lakes with nickel or chromium mordants, and dyes chrome mordanted wool or cotton bluish-green. Employed in printing.†

Alizarin Heliotrope. A mordant dyestuff, which dyes unmordanted wool bluish-red from an acid bath, and chromed wool, bluish-violet. Used in calico-printing.†

Alizarin Indigo Blue. A dyestuff which consists of 1,2,5,7,8-pentahydroxy-anthraquinoline.†

Alizarin Indigo Blue S. A dyestuff. Dyes wool (chromed) indigo blue.†

Alizarin Irisol. (Solway purple). An anthracene dyestuff. Dyes wool bluish-violet from an acid bath.†

Alizarin Maroon. A mordant dyestuff. It consists of amino-alizarin mixed with amino-purpurins, and dyes wool mordanted with alumina, garnet-red, and with chrome, maroon. It is used in wool, silk, and cotton dyeing.†

Alizarin No.10CA. See FLAVOPURPURIN.

Alizarin No.6. See PURPURIN.

Alizarin OG. See ALIZARIN ORANGE.

Alizarin Oil. See TURKEY RED OILS.

Alizarin OK. See ALIZARIN ORANGE.

Alizarin OR. See ALIZARIN ORANGE.

Alizarin Orange. (Alizarin Orange A, AO, D, N, OG, OK, OR). A mordant dyestuff. It is a dihydroxy-nitro-anthraquinone. Dyes cotton mordanted with alumina, orange; with iron, reddish-violet; and with chrome, brown. Also employed for wool.†

Alizarin Orange AOP. See ALIZARIN ORANGE POWDER.

Alizarin Orange G. A dyestuff. Dyes alumina mordanted wool or cotton orange.†

Alizarin Orange Powder. (Alizarin Orange AOP). A dyestuff. It is the sodium salt of Alizarin orange.†

Alizarin Paste. See ALIZARIN.

Alizarin Powder. See ALIZARIN.

Alizarin Powder SA. See ALIZARIN RED S.

Alizarin Powder W. See ALIZARIN RED S.

Alizarin Puce. A colour obtained in calico-printing by using a mixture of aluminium and iron mordants with alizarin.†

Alizarin Pure Blue. See ALIZARIN SKY BLUE.

Alizarin Purple. See GALLOCYANINE DH, BS.

Alizarin Red S. (Alizarin S, 2S, 3S, SA, W, WS, Alizarin Carmine, Alizarin Powder SA, W). A mordant dyestuff. It is the sodium salt of alizarinmono-sulphonic acid. Dyes wool mordanted with alumina, scarlet, and with chrome, bordeaux red.†

Alizarin Red WGG. Flavopurpurin for wool.†

Alizarin Red 3WS. A dyestuff. it is the sodium salt of flavopurpurinsulphonic acid.†

Alizarin RF. See ANTHRAPURPURIN.

Alizarin RG. See FLAVOPURPURIN.

Alizarin RT. See ANTHRAPURPURIN.

Alizarin RX. See ANTHRAPURPURIN.

Alizarin SA. See ALIZARIN RED S.

Alizarin Saphirol B. (Solway blue, Durasol acid blue B, Alizarin brilliant blue GS, Alizarin brilliant blue B, Alizarin delphinol B, Alizurol sapphire). A dyestuff. It is the sodium salt of diaminoanthrarufindisulphonic acid. Dyes wool blue from an acid bath.†

Alizarin SAR. A dyestuff. It is a British equivalent of Anthrapurpurin.†

Alizarin SC. See ANTHRAPURPURIN.

Alizarin Sky Blue. (Alizarin Pure Blue). A dyestuff which is the mono-sulphonic acid of 1-amino-4-*p*-toluidino-2-bromoanthraquinone.†

Alizarin SSA. See ANTHRAPURPURIN.

Alizarin SSS. Sodium 1,2,6-trihydroxyanthraquinonesulphonate. An acid dyestuff.†

Alizarin SX. See ANTHRAPURPURIN.

Alizarin SX Extra. See ANTHRAPURPURIN.

Alizarin S,2S,3S. See ALIZARIN RED S.

Alizarin Violet. See GALLEINE. In calico printing and cotton dyeing, this term is applied to the colour produced by alizarin red with an iron mordant.

Alizarin Viridine. See ALIZARIN CYANINE GREENS.

Alizarin W. See ALIZARIN RED S.

Alizarin WG. See ANTHRAPURPURIN.

Alizarin WS. See ALIZARIN RED S.

Alizarin X. See FLAVOPURPURIN.

Alizarin YCA. See FLAVOPURPURIN.

Alizarin Yellow. See ANTHRACENE YELLOW.

Alizarin Yellow A. A mordant dyestuff. It is trihydroxybenzophenone. Dyes cotton mordanted with alumina and lime, a golden yellow. Used in printing.†

Alizarin Yellow C. A mordant dye-stuff. It is gallacetophenone (trihydroxyacetophenone). Dyes cotton mordanted with alumina, yellow; with chrome, brown; and with iron, black.†

Alizarin Yellow FS. A mordant and acid dyestuff. Dyes chromed wool yellow.†

Alizarin Yellow GG, Paste. (Alizarin Yellow GGW Powder). A mordant or acid dyestuff. It is *m*-nitrophenylazosalicylic acid. Dyes chromed wool yellow.†

Alizarin Yellow R. (Orange R, Terracotta R). A dyestuff. It is *p*-nitrophenylazosalicylic acid. Dyes chromed wool yellowish-brown.†

Alizarin Yellow RW. A dyestuff. It is the sodium salt of Alizarin yellow R.†

Alizarin Yellow SG. (Azoalizarin Yellow 6G, Tartrachromin GG). A dyestuff prepared from *p*-phenetidine and salicylic acid.†

Alizarin Yellow W. A dyestuff obtained by the condensation of β-naphthol with gallic acid in the presence of zinc chloride. It dyes chrome mordanted wool.†

Alizarin Yellow, Paste. A mordant dyestuff. It is ellagic acid. Dyes chromed wool sulphur yellow.†

Alizarine. Anionic dyestuffs (level dyeing). Applications: Wool and wool blends. (Holliday Dyes & Chemicals Ltd).*

Alizarine Cyanine R. A dyestuff consisting of 1,2,4,5,8-penta oxyanthraquinone, $C_{14}H_8O_7$. It dyes wool mordanted with alumina, violet, and with chrome, blue.†

Alizurol Sapphire. A dyestuff. It is a British equivalent of Alizarin saphirol B.†

Alk. See CHIAN TURPENTINE.

Alka. Range of alkali detergents for the food industry. Applications: Bottle washing and tank cleaning. (Harshaw Chemicals Ltd).*

Alka-Donna. a proprietary preparation containing magnesium trisilicate, aluminium hydroxide and belladonna extract. An antacid. (Carlton Laboratories).†

Alka-Donna P. A proprietary preparation of Alka-Donna with phenobarbitone. An antacid and sedative. (Carlton Laboratories).†

Alkagel. A trade name for alginates made for waterproofing.†

Alkali Blue. (Nicholson's Blue, Fast Blue, Soluble Blue, Guernsey Blue). An acid dyestuff. It consists of a mixture of the sodium salts of triphenylrosaniline monosulphonic acid, and triphenylpararosaniline monosulphonic acid. Dyes wool blue from a bath made alkaline with borax.†

Alkali Blue D. See METHYL ALKALI BLUE.

Alkali Blue XG. (Soluble Blue XG, Non-mordant Cotton Blue). A dyestuff consisting of the sulphonic acids of β-naphthylated rosaniline. Dyes cotton and silk from an acid bath.†

Alkali Blue 6B. See METHYL ALKALI BLUE.

Alkali Brown D. See INGRAIN BROWN.

Alkali Cellulose. (Hydrated Cellulose). The product of the reaction between cotton and caustic soda. When hydrolyzed by water it gives hydrated cellulose.†

Alkali Dark Brown G, V. (Alkali Red Brown 3R). A dyestuff which consists of mixed disazo compounds from benzidine, tolidine, or dianisidine, with one molecule of the bisulphite compound of nitroso-β-naphthol, and one molecule of α-aminonaphtholsulphonic acid. Dyes cotton and half-wool dark brown.†

Alkali Fast Red B, R. An acid colour which dyes wool bluish-red.†

Alkali Green. (Viridine). An acid dyestuff. Dyes wool and silk green from an acid bath.†

Alkali Reclaim. Rubber, reclaimed.†

Alkali Red Brown 3R. See ALKALI DARK BROWN G, V.

Alkali Violet. An acid dyestuff. Dyes wool from an alkaline, neutral, or acid bath, bluish-violet.†

Alkali Yellow. See ORIOL YELLOW.

Alkali Yellow R. A dyestuff. Dyes cotton yellow.†

Alkalin. See METHYL ALKALI BLUE.

Alkaline Air. See VOLATILE ALKALI.

Alkalit. A proprietary synthetic resin obtained by heating the sodium salt of phenolphthalein with toluoyl chloride.†

Alkalsite. An explosive containing 25-32 per cent potassium perchlorate, ammonium nitrate, trinitrotoluene, and other constituents.†

Alkanet. (Alkanna, Anchusin). Terms applied to two different plants, *Lawsonia inermis* and *Anchusa tintoria,* whose roots are the source of a red dye, anchusine (alkannin), $C_{15}H_{14}O_4$. The name is applied to the dye as well as to the plant.†

Alkanna. See ALKANET.

Alkannin. Anchusine, $C_{15}H_{14}O_4$. See ALKENET.†

Alkanol. Surface active agents. (Du Pont (UK) Ltd).

Alkanolamine. Dimethyl ethanolamine. Applications: Solubilizer of synthetic resins for water soluble paints, raw material for ion exchange resins and coagulants. (Yokkaichi Chemical Co Ltd).*

Alkasal. (Alkasol). A combination of aluminium salicylate and potassium acetate, used in medicine.†

Alkasit. A proprietary cellulose adhesive.†

Alkasol. See ALKASAL.

Alkasperse 25. A proprietary series of pigment dispersions based on a short oil xylol-thinned alkyd used in the colouring of medium to fast air- drying surface coatings. (Colloids Ltd).*

Alkastar 83. Organic brightener system. Applications: Alkaline non cyanide zinc electroplating. (Harshaw Chemicals Ltd).*

Alkathene. (Polyethylene, Polythene). Solid polymers of ethylene prepared by subjecting ethylene to extremely high pressures under carefully controlled conditions of temperature. (ICI PLC).*

Alkawet. Amphoteric surfactant. Applications: Used as a wetting agent. (Lonza Limited).*

Alkeran Tablets. A proprietary formulation of melphalan. Applications: For the palliative treatment of multiple myeloma and for the palliation of non-resectable epithelial carcinoma of the ovary. (The Wellcome Foundation Ltd).*

Alkermes. See KERMES.

Alkolite. A proprietary phenophthalein resin.†

Alkydal. Phthalate resins modified with oil or fatty acids. Applications: For use in the formulation of paints and varnishes. (Bayer & Co).*

Alkydal L4SU. 57 per cent glyceryl phthalate, 43 per cent linseed and synthetic fatty acids.†

Alkydal RD. Adipic acid non drying alkyd resin.†

Alkydal STK. Phthalic anhydride-tri-methyolethane and caster oil-synthetic fatty acids.†

Alkynol. Saturated polyesters free from oil and fatty acids. Applications: For use in the formulation of coil coatings and high-grade stoving finishes. (Bayer & Co).*

All-Mine Pig. See STAFFORDSHIRE ALL-MINE PIG.

All-O. A proprietary liquid soap for use as a rubber lubricant.†

Allactol. Aluminium lacto-tartrate.†

Allan Red Bronze. An alloy of 62.5 per cent copper, 30 per cent lead, and 7-5 per cent tin.†

Allan Red Metal. An alloy of 50 per cent copper, and 50 per cent lead.†

Allbee with C. A proprietary preparation of Vitamins B and C. (Robins).†

Allegheny Metal. A proprietary corrosion-resisting alloy. It contains iron with 17-20 per cent chromium and 7-10 per cent nickel.†

Allegheny 33, 44, 55, 66. Corrosion resisting alloys (formerly Ascoloy 33, 44, 55, 66). They contain iron and chromium : 33 contains 12-16 per cent chromium, and 55 contains 26-30 per cent chromium.†

Allegron. A proprietary preparation of nortiptyline hydrochloride. An anti-depressant. (Dista Products Ltd).*

Allenite. Synonym for Pentahydrite.†

Allenoy. A proprietary molybdenum steel.†

Allen's Metal. An alloy of 55.3 per cent copper, 44-6 per cent lead, and 0.1 per cent tin.†

Allergospasmin. Reproterolhydrochloride and DNCG. Aerosol. Applications: Prophylaxe and therapy of Asthma Bronchiale. (Degussa).*

Allethrin. A synthetic insecticide structurally similar to pyrethrin.†

Alletorphine. A. semisynthetic morphine analgesic.†

Alley Stone. See ALUMINITE.

Alliance Chrome Blue-black B. A dyestuff. It is a British equivalent of Chrome fast cyanine G.†

Alliance Chrome Blue-black R. A dyestuff. It is a British equivalent of Palatine chrome blue.†

Alliance Fast Red A Extra. A dyestuff. It is a British equivalent of Fast red.†

Alligator Wood. The wood of *Guarea grandifolia,* of West India.†

Allingite. A commercial variety of amber. It melts at 300° C.†

Alliols. Crude sulphur-bearing spirits obtained from the first distillation fraction of light tar oils. They contain impure sulphides of carbon.†

Allisan. Horticultural fungicide. (Boots Pure Drug Co).*

Alloferin. A proprietary preparation of alcuronium chloride. A muscle relaxant. (Roche Products Ltd).*

Alloprene. Chlorinated rubber. (ICI PLC).*

Allopurinol. 1H-Pyrazolo[3,4-d] - pyrimidin-4-ol. Zyloric. An uricosuric drug.†

Alloxan. Mesoxalylurea, $C_4H_4N_2O_5$.†

Alloy AMF. An alloy containing 50-60 per cent nickel for low temperature use.†

Alloy AM4-4. An aluminium-magnesium alloy. It contains magnesium with 4 per cent aluminium, 0-4 per cent manganese, and 0-15 per cent silicon.†

Alloy AM7-4. An alloy containing magnesium with 7 per cent aluminium, 0-4 per cent manganese, and 0-15 per cent silicon.†

Alloy AP33. An aluminium alloy containing 4.5 per cent copper and 0-4 per cent titanium.†

Alloy JL. An aluminium alloy containing 4-5 per cent copper, 0-41 per cent iron, and 0-35 per cent silicon.†

Alloy L10. An alloy containing aluminium with 10 per cent copper and 1 per cent tin.†

Alloy L11. An alloy containing aluminium with 7 per cent copper and 1 per cent tin.†

Alloy L24. See ALLOY Y.

Alloy L5. An aluminium alloy containing, in addition to aluminium, 13 per cent zinc and 2.8 per cent copper.†

Alloy L7. An alloy containing aluminium with 14 per cent copper, and 1 per cent manganese.†

Alloy L8. An alloy containing aluminium with 12 per cent copper.†

Alloy MG7. An alloy consisting mainly of aluminium with magnesium and manganese. It has mechanical properties similar to Duralmin, and is stated to be highly resistant to corrosion.†

Alloy N. An alloy of 91 per cent aluminium, 6 per cent copper, and 3 per cent manganese.†

Alloy NCT3. An alloy containing 44.5 per cent iron, 37.5 per cent chromium, 17-

5 per cent nickel, and 0.5 per cent manganese.†

Alloy Steel. The term applied to a steel containing one or more elements in addition to carbon. Also see STEELS V. and W.†

Alloy T. An alloy containing aluminium with 3.8 per cent magnesium, 0.5 per cent iron, 0.5 per cent silicon, and 0.1 per cent copper.†

Alloy W.7. 1 per cent. silicon, 4.5 per cent. manganese, balance nickel with controlled zirconium addition. (Wiggin Alloys Ltd).†

Alloy W.9. 1 per cent. silicon, 4-5 per cent. manganese, balance nickel with controlled zirconium addition. (Wiggin Alloys Ltd).†

Alloy Y. (Alloy 24). An alloy of aluminium with 4 per cent copper, 2 per cent nickel, and 1.5 per cent magnesium.†

Alloy 109. An alloy containing 88 per cent aluminium and 12 per cent copper.†

Alloy 122. An alloy containing 88.6 per cent aluminium, 10 per cent copper, 1.2 per cent iron, and 0.25 per cent magnesium.†

Alloy 142. An alloy containing 92.5 per cent aluminium, 4 per cent copper, 2 per cent nickel, and 1.5 per cent magnesium.†

Alloy 145. An aluminium alloy containing 10-11 per cent zinc, 2-3 per cent copper, and 1-1.5 per cent iron.†

Alloy 195. An aluminium alloy containing 4-5 per cent copper, and not more than 1.2 per cent silicon, 1.2 per cent iron, 0.35 per cent magnesium, and 0.35 per cent zinc.†

Alloy 2L5. An aluminium alloy with 12.5-14.5 per cent zinc and 2.5-3 per cent copper.†

Alloy 2L8. An aluminium alloy containing 11-13 per cent copper.†

Alloy 2129. A special nickel-iron alloy having fairly high permeability and excellent mechanical properties.†

Alloy 3L11. An aluminium alloy containing 6-8 per cent copper and tin may be added up to 1 per cent.†

Alloy 39. An alloy containing aluminium with 3.75-4.25 per cent copper, 1.2-1.7 per cent magnesium, and 1.8-2.3 per cent nickel.†

Alloys RR. A series of aluminium alloys containing aluminium with 0.5-5 per cent copper, 0.2-2.5 per cent nickel, 0.05-5 per cent magnesium, 0.6-1.5 per cent iron, 0.05-0.5 per cent titanium, and 0.2-5 per cent silicon.†

Alloys Wm. White-bearing metals. Wm5 contains 78.5 per cent lead, 15 per cent antimony, 5 per cent tin, and 1.5 per cent copper, with a specific gravity of 10.1. Wm10 consists of 73.5 per cent lead, 15 per cent antimony, 10 per cent tin, and 1.5 per cent copper, and has a gravity of 9-7. Wm42 contains 42 per cent tin, 41 per cent lead, 14 per cent antimony, and 3 per cent copper. Wm80 consists of 80 per cent tin, 10 per cent antimony, 10 per cent copper.†

Alluman. An alloy of aluminium with 10-20 per cent tin and 4-6 per cent copper.†

Ally. Suplhonyl weedkiller. (Du Pont (UK) Ltd).

Allyl Mustard Oil. See OIL OF MUSTARD.

Allyloestrenol. 17α-allyloestr-4-en-17β-ol. Gestanin.†

Allylprodine. 3-Allyl-1-methyl - 4 - phenyl-4-propionyloxypiperidine.†

Allzarin SDG. See FLAVOPURPURIN.

Almacarb. Tablets of aluminium hydroxide-magnesium carbonate co-dried gel, antacid. (British Drug Houses).†

Almadina. See POTATO GUM.

Almag. A proprietary aluminium alloy similar in composition to Alferium (*qv*).†

Almagel. A proprietary preparation of aluminium hydroxide gel and magnesium hydroxide. Applications: An antacid. (Ayrton Saunders plc).*

Almandine Ruby. See SPINEL.

Almasilium. An aluminium alloy containing 1 per cent magnesium and 2 per cent silicon.†

Almazine. Lorazepam. Applications: Pharmaceutical preparation for the treatment of anxiety. (M A Steinhard Ltd).*

Almeidina Gum. See POTATO GUM.

Almelec. A proprietary alloy of aluminium containing 0.7 per cent magnesium, 0.5 per cent silicon, and 0.3 per cent iron.†

Almen's Reagent. A solution containing 5 grams tannic acid in 240 cc of 50 per cent alcohol, to which has been added 10 cc of a 25 per cent solution of acetic acid. A precipitate is given with nucleoproteins.†

Almevax. A proprietary preparation of live rubella vaccine to immunize against Rubella. (The Wellcome Foundation Ltd).*

Alminate. Dihydroxyaluminium aminoacetate. Applications: Antacid. (Bristol-Myers Co).

Almo Steel. Proprietary chrome-molybdenum steels.†

Almocarpine. Pilocarpine Hydrochloride. Applications: Cholinergic. (Ayerst Laboratories, Div of American Home Products Corp).

Almora. Magnesium gluconate. Applications: Replenisher. (O'Neal, Jones & Feldman Pharmaceuticals).

Almstab. Stabilizers for use in PVC compositions. (Associated Lead Manufacturers Ltd).*

Alnagon. Aluminium sulphonaphthalate in glycerin and water.†

Alneon. An aluminium alloy for castings. It contains 10-20 per cent zinc, a small quantity of copper, and a little nickel.†

Alnovin. Fendosal. Applications: Anti-inflammatory. (Hoechst-Roussel Pharmaceuticals Inc).

Alnovol. A series of spirit soluble pure phenolic resins. (Chemische Werke, Albert).†

Alnovol. Non heat hardening phenol/formaldehyde resins. Applications: Printing inks and rubber reinforcement. (Resinous Chemicals Ltd).*

Alocrom. Chemical pretreatment for aluminium. (ICI PLC).*

Alomite. A trade name applied to a variety of sodalite used as an ornamental stone.†

Alon. A registered trade mark. Fumed alumina.†

Alophen. A proprietary preparation of aloin, phenolphthalein, ipecacuanha, strychnine and belladonna. A laxative. (Parke-Davis).†

Aloral. Tablets containing 100mg and 300mg allopurinol BP. Applications: For the treatment of gout: Primary hyperuricaemia. Secondary hyperuricaemia: Prophylaxis of uric acid and calcium oxalate stones. (Lagap Pharmaceuticals Ltd).*

Alox. A proprietary trade name for a series of methyl esters of higher alcohols.†

Aloxidone. 3-Allyl-5-methyl oxazolidine-2,4-dione. Allomethadione. Malidone.†

Aloxite. A trade mark for abrasive and refractory materials consisting essentially of alumina.†

Alpacca. (Alpakka). An alloy of 64 per cent copper, 19 per cent zinc, 14.5 per cent nickel, 2 per cent silver, 0.4 per cent iron, and 0.12 per cent tin. It is a nickel silver.†

Alpakka. See ALPACCA.

Alpax. See SILUMIN.

Alperox C. A registered trade mark for lauroyl peroxide. A catalyst for polymerization.†

Alpex. Cyclized rubber resins. Applications: Chemical resistant coatings. (Resinous Chemicals Ltd).*

Alpex 4505. Cyclized rubber used for surface coatings and printing inks.†

Alpha Chymar. Chymotrypsin. Applications: Enzyme. (Barnes-Hind Inc).

Alpha Daphnone. Pureα-isomethylionone. (Bush Boake Allen).*

Alpha-Cillin. Pivampicillin hydrochloride. Applications: Oral antibiotic with broad-spectrum activity against Gram-positive and Gram-negative bacteria. (Merck Sharp & Dohme).*

Alpha-Ruvite. Hydroxocobalamin. Applications: Vitamin. (Savage Laboratories, Div of Byk-Gulden Inc).

Alphachroic. Chrome dyestuffs. (JC Bottomley and Emerson).†

Alphadolone. An anaesthetic component present in althesin. 3α,21-Dihydroxy-5α-pregnane-11,20-dione.†

Alphadrol. Fluprednisolone. Applications: Glucocorticoid. (The Upjohn Co).

Alphalin. Vitamin A. Applications: Vitamin (anti-xerophthalmic). (Eli Lilly & Co).

Alphamine. Midodrine hydrochloride. Applications: Antihypertensive; vasoconstrictor. (Centerchem Products Inc).

Alphamint. Peppermint blend. (Bush Boake Allen).*

Alphanol. Medium chain length alcohols, forming a plasticizer. (ICI PLC).*

AlphaRedisol. Hydroxocobalamin. Applications: Vitamin. (Merck Sharp & Dohme, Div of Merck & Co Inc).

Alphasol OT. A proprietary trade name for the sodium salt of an alkyl ester of sulphosuccinic acid.†

Alphatrex. Betamethasone dipropionate USP Applications: Indicated for relief of the inflammatory and pruritic manifestations of corticosteroid responsive dermatoses. (Altana Inc).*

Alphenate. A series of phenolic-alkyd plasticized resins. (Chemische Werke, Albert).†

Alphide. A proprietary trade name for a cold moulded refractory ceramic.†

Alphogen. (Alphozone, Succinoxate). Succinyl peroxide, $(COOH.CH_2.CH_2CO)_2O_2$. An antiseptic.†

Alphol. α-Naphthol salicylate, $C_6H_4(OH)(COOC_{10}H_7)$. An antiseptic and antirheumatic.†

Alphosyl. A range of proprietary preparations containing allantoin and coal tar extract. Applications: For dermatological use - anti-psoriatic. (Stafford-Miller).*

Alphosyl HC.. Proprietary preparation containing allantoin, coal tar extract and hydrocortisone. Applications: Dermatological use - anti-psoriatic. (Stafford-Miller).*

Alphozone. See ALPHOGEN.

Alpine Blue. (Fast Wool Blue). A dyestuff. Dyes wool and silk blue from an acid bath.†

Alpivicin. Pivampicillin hydrochloride. Applications: Oral antibiotic with broad-spectrum activity against Gram-positive and Gram-negative bacteria. (Merck Sharp & Dohme).*

Alplate. A proprietary aluminium coated steel.†

Alpolit. Unsaturated polyester resin in styrene. Applications: Glass fibre reinforced laminate, casting and potting. (Resinous Chemicals Ltd).*

Alprenolol. A beta adrenergic receptor blocking agent. 1-(2-Allylphenoxy)-3-isopropylaminopropan-2-ol.†

Alprokyds. Proprietary drying oil and non-drying oil modified alkyd resins.†

Alquifon. (Black Lead Ore, Potter's Ore). A mineral. It consists of zinc sulphide, and is used in pottery to give a green glaze.†

Alquifou. See ALQUIFON.

Alreco. Aluminium and aluminium alloy ingot. Applications: Aluminium die cast industry. (Reynolds Metal Co).*

Alresat. Maleinized rosin esters. Applications: Nitrocellulose lacquers, printing inks and to improve gloss in air drying paints. (Resinous Chemicals Ltd).*

Alresat 313-C. Maleic anhydride-rosin-glycerine (m.p. 95° C). (Chemische Werke, Albert).†

Alresen. Alkyl phenol/formaldehyde resins, terpene phenolic resins. Applications: Oil varnishes and adhesives. (Resinous Chemicals Ltd).*

Alresen 260-R. Ortho cresol-turpentine condensate cooked with butylated diane (*qv*) condensate. (Chemische Werke, Albert).†

Alromin Ru 1000. Antistatic agent. (Ciba Geigy PLC).†

Alsace Green. See DINITROSO-RESORCIN.

Alsace Green J. See GAMBINE Y.

Alsace Grey. See NEW GREY.

Alsace Gum. See DEXTRIN.

Alsi. A pigment consisting of finely ground aluminium-silicon alloy used to give durable and rust-preventative paints.†

Alsibronz. Wet ground muscovite mica. (Franklin Mineral Products Co).

Alsica Alloys. Aluminium-silicon-copper alloys.†

Alsifer. A proprietary alloy of 40 per cent silicon, 40 per cent iron, and 20 per cent aluminium. A hardener alloy for adding silicon to aluminium alloys.†

Alsifilm. A material made from bentonite, used in place of mica.†

Alsimag. Ceramic materials. Used for insulation and for the dielectric of condensers. (3M).*

Alsimag 754. A beryllia ceramic. (3M).*

Alsimag 779. Leachable ceramic cores for precision metal castings. (3M).*

Alsimin. A similar alloy to Alsifer.†

Alsol. Aluminium acetotartrate, a germicide.†

Alstat. Very powerful antistatic agent. Applications: All stages of textile processing. (Altex Chemical Co Ltd).

Alstromed A 18 LV. N-octadecyl sodium sulphosuccinamate solution. Applications: Foaming agent, frothing agent and emulsifying agent. (Alco Chemical Corporation).*

Alsynates. Metal carboxylates based on C_8 - C_{10} branched chain synthetic aliphatic carboxylic acids. Applications: Driers for paint or printing ink catalyst for unsaturated polyester. (Manchem Ltd).*

Altacaps. Hydrotalcite plus dimethicone. Applications: Indigestion, peptic ulceration. (Roussel Laboratories Ltd).*

Altacite Plus. Proprietary product of hydrotalcite, dimethicone. Applications: Antacid, flatulence. (Roussel Laboratories Ltd).*

Altal. A proprietary trade name for triphenylphosphate.†

Altan. Veterinary laxative. (May & Baker Ltd).*

Altax. Benzothiazyl disulphide used as a rubber accelerator. Applications: For natural and synthetic rubbers, used as both primary and secondary accelerators in NR and SBR copolymers. Retard-plasticizer in G type neoprenes, core modifier in W types. (Vanderbilt Chemical Corporation).*

Altene DG. Trichloroethylene. Applications: Degreasing solvent. (Atochem Inc).†

Altolube. Excellent scrooping agent. Applications: Processing of warp knit nylon. (Altex Chemical Co Ltd).

Altowhite. Fillers calcined aluminium silicate. (Georgia Kaolin Co).

Alu-Cap. Each capsule contains dried aluminium hydroxide gel BP 475 mg as a white powder. (Riker Laboratories/3M Health Care).†

Alubarb. A proprietary preparation of aluminium hydroxide, phenobarbitone and belladonna. A gastrointestinal sedative. (Norton, H N Co).†

Aludrox. A proprietary preparation containing aluminium hydroxide gel. An antacid. (Wyeth).†

Aludrox CO. A proprietary preparation containing alumina-sucrose powder and magnesium hydroxide. An antacid. (Wyeth).†

Aludrox SA. A proprietary antacid preparation containing aluminium hydroxide gel, magnesium hydroxide, gel, magnesium hydroxide, *sec.* butobarbitone and ambutonium bromide. (Wyeth).†

Aludur (Aladar). An alloy of aluminium and silicon, containing from 5-20 per cent silicon. Also see SILUMIN.†

Alugan. A proprietary preparation of bromocyclen. A veterinary pesticide.†

Aluline. Allopurinol. Applications: Pharmaceutical preparation for the treatment of gout. (M A Steinhard Ltd).*

Alum. Potassium aluminium sulphate, $K_2SO_4Al_2(SO_4)_3 24H_2O$,ammonium aluminium sulphate $(NH_4)_2SO_4Al_2(SO_4)_3$ VbXV24H_2 O, and aluminium sulphate, $Al_2(SO_4)_3$ $9H_2O$, are all known by this name, but it is usually applied to the potassium salt.†

Alum-haematoxylin Solution. A microscopic stain. It consists of 1 gram haematoxylin dissolved in a saturated solution of ammonium alum (100 cc) and 300 cc distilled water. The solution is filtered after it has assumed a dark red colour, and 0.5 gram thymol added as preservative.†

Alumails. A range of inorganic smelted products for use in the surface coating of metals. Applications: Special purpose enamels for various uses as well as for electrophorectic or powder electrostatic application. (Bayer & Co).*

Aluman. An alloy of 88 per cent aluminium, 10 per cent zinc, and 2 per cent copper.†

Alumantine. A proprietary refractory containing 60-65 per cent. Al_2O_3.†

Alumbro. A trade mark for a 2 per cent. aluminium-brass alloy having good resistance to corrosion by sea water and marine atmospheres. (I M I (Kynoch) Ltd, Imperial Chemical Industries).†

Alumedia. Complex of alkyd resin and aluminium alkoxide. Applications: Preparation of high solids paint. (Manchem Ltd).*

Alumel. An electrical resistance alloy containing 94 per cent nickel, 1 per cent silicon, 2 per cent aluminium, 2.5 per cent manganese, and 0-5 per cent iron.†

Alumetised Steel. A steel which has been sprayed with aluminium and then heat-treated to give a surface of alloy.†

Alumilite. A proprietary trade name for chemical coatings applied to aluminium electrically.†

Alumin. Sodium aluminate, $Na_2Al_2O_4$.†

Alumina. Aluminium oxide, Al_2O_3.†

Aluminac. A similar alloy to Alpax (*qv*), for making die castings.†

Aluminate Cement. A cement made from bauxite. It usually contains about 40 per cent. Al_2O_3, 40 per cent lime, 15 per cent iron oxide, silica, and magnesia.†

Aluminium Brass. Alloys of from 59-70 per cent copper, 26-40 per cent zinc, 0.3-5.2 per cent aluminium, and sometimes a little iron.†

Aluminium Bronze. There are various alloys under this name. Those containing a high percentage of aluminium are termed light, and have from 83-89 per cent aluminium and 11-17 per cent copper. The other type is called heavy, and contains from 85-95 per cent copper and 5-15 per cent aluminium. The heavy alloys are sometimes used in the manufacture of jewellery. Also see CUPROR, CUPRO-ALUMINIUM, and MANHARDT'S ALUMINIUM BRONZE.†

Aluminium Cadmium. Alloys of 90-93 per cent aluminium, 2.5-3.5 per cent cadmium, and 4-6 per cent copper.†

Aluminium Cloflbrate. A pharmaceutical used in the treatment of arteriosclerosis. Di-[2-(4-chlorophenoxy)-2-methyl-propionato] hydroxyaluminium.†

Aluminium Iron. (Ferro-aluminium). An alloy of iron and aluminium. It is used for refining iron, also as a permanent ingredient for increasing the strength. A 15 per cent alloy has been used for crucibles exposed to high temperatures.†

Aluminium Iron Brass. An alloy containing 61.1 per cent copper, 35.3 per cent zinc, 1.1 per cent iron, and 2.3 per cent aluminium.†

Aluminium Iron Bronze. An alloy containing 85-89 per cent copper, 6-9 per cent aluminium, and 3-7 per cent iron.†

Aluminium Magnesium Bronze. An alloy of 89-94 per cent copper, 5-10 per cent aluminium, and 0.5 per cent magnesium.†

Aluminium Manganese. An alloy of aluminium with 2-3 per cent manganese.†

Aluminium Manganese Brass. An alloy consisting of 56-3 per cent copper,40 per cent zinc, 2.7 per cent manganese, and 1 per cent aluminium.†

Aluminium Manganese Bronze. An alloy of 89 per cent copper, 9.6 per cent aluminium, and 1-2 per cent manganese.†

Aluminium Nickel. An alloy containing varying amounts of nickel with aluminium. One alloy consists of 76.4 per cent nickel, and 23.6 per cent aluminium.†

Aluminium Nickel Bronze. An alloy containing 85 per cent copper, with from 5-10 per cent aluminium and 5-10 per cent nickel.†

Aluminium Nickel Zinc. An alloy consisting of 85 per cent aluminium, 10 per cent nickel, and 5 per cent zinc.†

Aluminium Silicon Alloy C. A British Chemical Standard alloy. It contains

12.74 per cent silicon, 0.34 per cent iron, 0.005 per cent manganese, 0.020 per cent zinc, 0.006 per cent titanium, and 0.010 per cent copper.†

Aluminium Silver. (Silver Metal). An alloy consisting of 57 per cent copper, 20 per cent nickel, 20 per cent zinc, and 3 per cent aluminium. The name is also applied to an alloy of 95 per cent aluminium, and 5 per cent silver.†

Aluminium Tin Bronze. An alloy of 85 per cent copper, 10 per cent tin, 2-5 per cent aluminium, and 2 per cent zinc.†

Aluminium-Nickel-Titanium. An alloy of 97.6 per cent aluminium, 2 per cent nickel, and 0.4 per cent titanium.†

Aluminium, Victoria. See PARTINIUM.

Alumino-Vanadium. An alloy of aluminium and vanadium, obtained by adding a mixture of vanadium pentoxide and powdered aluminium to liquid aluminium. Used as a deoxidizing agent.†

Aluminoferric. Consists of crude aluminium sulphate, and contains some iron sulphate. It is used as a precipitating agent in sewage and refuse liquids treatment, and also for removing suspended matter from boiler feed water. (Laporte Industries Ltd).*

Aluminoid. A trade mark for goods of the abrasive and refractory type, the essential constituent being crystalline alumina.†

Aluminon. Aurinetricarboxylic acid, a dyestuff which is used as a test for aluminium in solution.†

Aluminox. A trade mark for articles as abrasives and refactories. The essential constituent is crystalline alumina.†

Alum, Exsiccated. See BURNT ALUM.

Alum, Paperhanger's. See PAPER-HANGER'S ALUM.

Alundum. A registered trade mark for various types of goods, such as grinding wheels, abrasive and refractory grain, refractory articles and cement, porous plates, crucibles, and other articles made from crystalline alumina, or alumina which has been electrically fused and crystallized.†

Alunex. Chlorpheniramine maleate. Applications: Pharmaceutical preparation for the treatment of allergy. (M A Steinhard Ltd).*

Aluni. An aluminium-nickel alloy used as an anode for deposition of the alloy coating.†

Alupent. Metaproterenol sulphate. Applications: Bronchodilator. (Boehringer Ingelheim Pharmaceuticals Inc).

Alupent Expectorant. A proprietary preparation of orciprenaline sulphate and Bisolvon. A bronchial antispasmodic. (Boehringer Ingelheim Ltd).†

Alupent Obstetric. A proprietary preparation of ORCIPRENALINE used in the management of premature labour. (Boehringer Ingelheim Ltd).†

Alupent-Sed. A proprietary preparation of amylobarbitone and orciprenaline sulphate. A bronchial antispasmodic. (Boehringer Ingelheim Ltd).†

Aluphos. A proprietary preparation containing aluminium phosphate. An antacid. (Fisons PLC).*

Alupram. Diazepam. Applications: Pharmaceutical preparation for the treatment of depression. (M A Steinhard Ltd).*

Alusec. Aluminium organic complexes. Applications: Rheology modifiers for high solids, air drying paint systems. (Manchem Ltd).*

Alusil. Aluminium silicate. (Crosfield Chemicals).*

Alusil ET. Synthetic aluminium silicate of controlled particle size used an extender in emulsion paint. (Crosfield Chemicals).*

AluTninium Titanium Bronze. An alloy of from 89-90 per cent copper, 9-10 per cent aluminium, 1 per cent iron, and a trace of titanium.†

Aluzyme. A proprietary preparation of aneurine hydrochloride, riboflavin, pyridoxine, niacin, pantothenic acid and folic acid. Vitamin supplement. (Phillips Yeast Products).†

Alveograf. Durapatite. Applications: Prosthetic aid. (Sterling Drug Inc).

Alvex. Highly alkaline detergent. (Crosfield Chemicals).*

Alysine. Sodium Salicylate. Applications: Analgesic. (Merrell Dow Pharmaceuticals Inc, Subsidiary of Dow Chemical Co).

Alytol. Bitumen mastix. Applications: Roof repair and maintenance. (Vedag GmbH).*

Alzen. An alloy of 66 per cent aluminium, and 33 per cent zinc.†

Amalgam. A name applied to alloys of metals with mercury. It is also used as a term for a native alloy of mercury and silver with a formula varying between AgHg and Ag_2Hg_3.†

Amandol. A proprietary grade of benzaldehyde. (May & Baker Ltd).*

Amantadine. A drug used in the treatment of the Parkinsonian syndrome. 1-Adamantanamine. SYMMETREL is the hydrochloride.†

Amaranth. See FAST RED D, and SCARLET 6R.

Amaranth A. A dyestuff. It is a British equivalent of Fast red D.†

Amargosite. A trade name for a clay of the Bentonite type.†

Amarin. Triphenyldihydroglyoxaline, $(C_6H_5\ C\ NH)_2$:CH C_6H_5.†

Amatols. Mixtures of trinitrotoluene and ammonium nitrate. An 80/20 amatol contains 80 parts ammonium nitrate and 20 parts trinitrotoluene. They are high explosives.†

Amax - XLP. Low phosphorus copper. Oxygen free copper plus 0.001 - 0.005 per cent phosphorus. Conductivity 98 per cent IACS. Copper + silver - 99.95%, phosphorus - 0.001 - 0.005%. Applications: Ideal for applications where a low phosphorus content is beneficial and good conductivity with resistane to embrittlement must be ensured. (Amax Inc).*

Amax and Amax No 1. N-oxydiethylene 2-benzothiazole-sulphenamide - primary and secondary accelerators for rubber. Applications: Primary and secondary accelerators - safe at processing temperature and active over a wide curing range. Particularly advantageous in SBR tyres compounded with fine particle furnace blacks. (Vanderbilt Chemical Corporation).*

Amazonite. See AMAZON STONE.

Ambazyme. Amyloglucosidase. (ABM Chemicals Ltd).*

Amber. A fossil resin formed in certain beds of clay and sand, stated to be derived from *Pinites succinifer*. The following are varieties of amber: Succinite (m.pt. 25O-300° C.), Gedanite (m.pt. 150-180° C.), Glessite (m.pt. 250-300° C.), Beckerite, an opaque variety.†

Amber Oil. See OIL OF AMBER.

Amber Salt. See SALT OF AMBER.

Amber-Guaiacum Resin. A variety of guaiacum resin. It is not an amber.†

Amberdeen. See BAKELITE.

Amberglow. A proprietary phenol formaldehyde synthetic resin.†

Ambergris. A grey, wax-like product found in the sea. It occurs in certain conditions of the intestines of the sperm whale. The chief constituent is ambrein, $C_{23}H_{40}O$, and it is used in perfumery.†

Amberite. A smokeless powder consisting of 71 per cent nitro-cotton, 18.6 per cent barium nitrate, 1.3 per cent potassium nitrate, 1.4 per cent wood meal, and 5.8 per cent petroleum jelly. Also see BAKELITE.†

Amberlite. Ion-exchange resins. (Rohm & Haas (UK) Ltd).

Amberlite IRP-64. Polacrilin. A synthetic ion-exchange resin, supplied in the hydrogen or free acid form. Applications: Pharmaceutic aid. (Rohm & Haas Co).

Amberlite IRP-88. Polacrilin Potassium. Applications: Pharmaceutic aid. (Rohm & Haas Co).

Amberoid. See NICKEL SILVERS.

Amberol. A phenol-formaldehyde resin combined with rosin or other resin. Used in the varnish industry.†

Ambersil. Silicone emulsion for shell moulding and hot box processes. (Foseco (FS) Ltd).*

Amber, Artificial. See MELLITE.

Amber, Friable. See GEDANITE.

Amber, Pressed. See AMBROID.

Amber, Soft. See GEDANITE.

Ambiflo. Synthetic lubricants which are colourless, viscous liquids made by combining or polymerizing propylene oxide or propylene oxide and ethylene oxide. Applications: Used in equipment for metal working, heat transfer and as automotive brake fluids, internal combustion engines, gears and bearings. (Dow Chemical Co Ltd).*

Ambilhar. A proprietary preparation of niridazole. An antibilharzal agent. (Ciba Geigy PLC).†

Ambiteric. Ampholytic surfactant. (ABM Chemicals Ltd).*

Ambiteric D. High molecular weight substituted betaine as a creamy unctuous mass. Applications: Good alkali stability, wetting, foaming, detergency and solubilizing properties, used in industrial cleaner formulations; perfume solubilization; antistat. (ABM Chemicals Ltd).

Ambitrol. A series of formulated engine coolants made from glycols, deionized water and suitable inhibitors. Applications: Stationary engines operating for transmission of natural gas and petroleum products, electrical power generated systems, irrigation systems and drilling operations. (Dow Chemical Co Ltd).*

Ambodryl. A proprietary preparation of bromodiphenhydramine hydrochloride. (Parke-Davis).†

Amborate. Formoxy methyl isolongifolene. (Bush Boake Allen Ltd).

Amborol. Hydroxy methyl isolongifolene. (Bush Boake Allen Ltd).

Amboryl Acetate. Acetoxymethyl isolongifolene. (Bush Boake Allen Ltd).

Ambra. A phenol-formaldehyde synthetic resin.†

Ambrac Metal. A corrosion resistant nickel silver containing copper, nickel, zinc, and manganese.†

Ambraloys. A proprietary trade name for alloys of copper and aluminium with zincorion.†

Ambramycin. A proprietary trade name for tetracycline.†

Ambrasite. See BAKELITE.

Ambrene. (Musk Ambrette). An artificial musk perfume. It is dinitro-tert-butyl-*m*-cresol-methyl ether.†

Ambrite. A resin found in the lignite of Auckland, New Zealand.†

Ambroid. (Pressed Amber). A product consisting of small fragments of amber heated under pressure.†

Ambrol. A phenol-formaldehyde synthetic resin.†

Ambush. Broad spectrum insecticide containing permethrin. (ICI PLC).*

Amcide. Weed killer. (Albright & Wilson Ltd).*

Amcill. Ampicillin. Applications: Antibacterial. (Parke-Davis, Div of Warner-Lambert Co).

Amcron. Oxygen free copper plus 0.7 to 1.2 per cent chromium. Conductivity 82 per cent IACS. Chromium 0.7 to 1.2 %, copper + silver + chromium - 99.95%. Applications: Principal uses are based on its good compressive yield strength, creep resistance and thermal fatigue properties at moderately elevated temperatures. (Amax Inc).*

AME 4000. Modified epoxy resin with superior strength/weight characteristics. Applications: Fibreglass power boats and sailboats. (Ashland Chemical Company).*

Ameen. Long chain aliphatic amines. (Akzo Chemie UK Ltd).

Ameisine. Aluminium formate.†

Amenide. Quinfamide. Applications: Anti-amebic. (Sterling Drug Inc).

Amerchol BL. Multisterol absorption base of lanolin sterol esters and higher alcohols. Applications: Nonionic w/o emulsifier. (Amerchol Corporation).*

Amerchol C. Emollient absorption base containing fractions of surface active lanolin sterols and esters. (Amerchol Corporation).*

Amerchol CAB. Solid emollient multisterol extract of lanolin alcohols in petrolatum. Applications: W/o emulsifier activity, ideal for pharmaceutical vehicles well tolerated on dry and injured skin. (Amerchol Corporation).*

Amerchol H-9. Absorption base containing cholesterol esters and free sterols. Applications: Natural emollient and nonionic w/o emulsifier. (Amerchol Corporation).*

Amerchol L-101. Multisterol extract of lanolin alcohols in mineral oil. Applications: A natural moisturizer and emollient. (Amerchol Corporation).*

Amerchol L-99. Liquid absorption base, less fractionated than AMERCHOL L-101 with a broader range of lanolin and other alcohols for w/o emollient activity. (Amerchol Corporation).*

Amerchol Polysorbate. Emulsifying and solubilizing in foods. Defoamer. Applications: Vegetable oils, vitamins, beet sugar, yeast, cottage cheese. (D F Anstead Ltd).

Amerchol RC. Concentrated lipophilic lanolin alcohol fraction with lubricating, non-tacky, barrier properties, particularly suited for makeup systems. (Amerchol Corporation).*

Amerchol 400. Lipophilic fraction of lanolin and other alcohols and petrolatum. Applications: Moisturizing emollient with lubricating and barrier characteristics. (Amerchol Corporation).*

Amercor. Corrosion inhibitors. Blends of neutralizing amines, filming amines and/or oxygen scavengers. Applications: Steam/condensate line corrosion control, oil field line corrosion control. (Ashland Chemical Company).*

Amerfloc. Coagulants and flocculants. Applications: Liquid/solids separation. (Ashland Chemical Company).*

Amerfloc Plus. Fluid flocculants. Applications: Liquid/solids separation. (Ashland Chemical Company).*

Amergel. Foam control agents. Applications: Process systems - paper mills, pulp mills. (Ashland Chemical Company).*

Amergize. Deposit modifier/combustion improver. A unique blend of oil soluble organometallic compounds. (Ashland Chemical Company).*

Amergy. Fuel oil treatments. Blends of preburner treatments, slag modification agents and/or combustion catalyzing agents. Applications: Fuel, oil, waste fuel treatment. (Ashland Chemical Company).*

Ameri-Bond. Modified calcium lignosulphonates. Applications: Pelleting aids for animal feeds. (Reed Lignin Inc).*

Americaine. Benzocaine. Applications: Anaesthetic (American Critical Care, Div of American Hospital Supply Corp).

American Antifriction Metal. See ANTIFRICTION METAL.

American Blue. See PRUSSIAN BLUE.

American Chrome Yellow. See CHROME YELLOW.

American Cloth. (Leather Cloth). This material usually consists of a mixture of oxidized linseed oil, rosin, fillers, and colouring matter, spread on a cotton cloth.†

American Mace Butter. See OTOBA BUTTER.

American Sienna. See INDIAN RED.

American Silver. A nickel silver. It contains copper, zinc, nickel, iron, lead, tin, and sometimes small quanities of aluminium and phosphorus.†

American Vermilion. See CHROME RED.

Ameripol 1903. A trade mark for a high styrene synthetic rubber prepared from Ameripol 1511 cold SBR type latex and 85 per cent. bound styrene reinforcing resin. (Goodrich-Gulf Chemicals, Ohio).†

Amerite. A proprietary trade name for rubber derivatives and rubber-like resins in aqueous dispersion.†

Amerlate LFA. Lanolin fatty acids. Applications: Forms o/w soaps with monovalent neutralizers and w/o soaps with polyvalent cations. (Amerchol Corporation).*

Amerlate P. Isopropyl ester of lanolin fatty acids. Applications: Suitable for pigment dispersion in lipstick and makeup. Ideal plasticizer for waxy systems. (Amerchol Corporation).*

Amerlate W. Isopropyl ester of lanolin fatty acids. Applications: Lubricant, plasticizer and dispersing agent. (Amerchol Corporation).*

Amerol. The methyl ester of saccharin.†

Amerone. Perfumery base. (PPF International Ltd).†

Ameroyal. Evaporator treatment. A concentrated liquid blend of polyelectrolyte scale inhibitors and antifoam agents. Applications: Used to prevent scale deposition and foaming in conventional marine evaporators thereby minimizing the need for acid cleaning. (Ashland Chemical Company).*

Amerplex. Deposit and corrosion inhibitors. Blends of antiscalants and corrosion control agents. Applications: Preboiler, boiler and afterboiler deposit and corrosion control, oil field steam generator deposit and corrosion control. (Ashland Chemical Company).*

Amerscan MDP Kit. Medronate disodium. Applications: Pharmaceutic aid. (Amersham Corp).

Amerscreen. UV absorber. Applications: Sunscreen. (D F Anstead Ltd).

Amersite. Corrosion inhibitors. Oxygen scavenging agents. Applications: Preboiler and afterboiler corrosion inhibitors, oil field corrosion inhibitor. (Ashland Chemical Company).*

Amerstat. Liquid and solid microbial control agents used for control of slime in the papermaking process for bacterial control in the sugar process and for a preservative in aqueous systems. Applications: Paper mills, sugar mills, preservation of paints, coatings, adhesives, mineral slurries, drilling muds, animal glues, latent metal working fluids and paper coatings. (Ashland Chemical Company).*

Amerstat 252. Microbial control agents. Combination of 5-chloro-2-methyl-y-isothiazolin-3-one and 2-methyl-y-isothiazolin-3-one. Applications: Adhesives and mineral slurries. (Ashland Chemical Company).*

Amertrol. Deposit inhibitors. Deposit control agents. Applications: Boiler deposit control. (Ashland Chemical Company).*

Amerzine. Corrosion inhibitors. Organically catalyzed hydrazine. Applications: Preboiler and afterboiler corrosion control, oil field line corrosion control and chromate reduction. (Ashland Chemical Company).*

Amesec. A proprietary preparation of aminophylline, ephedrine hydrochloride and amylobarbitone. A bronchial antispasmodic. (Eli Lilly & Co).†

Amesite. A Chloritic mineral.†

"A" Metal. A nickel-iron-copper alloy containing 6-8 per cent copper. Used for making audio-frequency transformers.†

A-methaPred. Methylprednisolone sodium succinate. Applications: Glucocorticoid. (Abbott Laboratories).

Amethopterin. A proprietary trade name for methotrexate.†

Amethyst Violet. (Iris Violet). A dyestuff. It is tetra-ethyl-diamino-phenyl-phenazonium chloride. It is a basic dye, and dyes silk violet, with red fluorescence.†

Ametox. Specially purified and sterilized sodium thiosulphate for use in metallic poisoning. (May & Baker Ltd).*

Ametrex. Active ingredient: ametryne; 2-ethylamino-4-isopropylamino-6-methylthio-1,3,5-triazine. Applications: Selective pre-and post-emergence herbicide, also used as an aquatic herbicide and vine desiccant. (Agan Chemical Manufacturers Ltd).†

A.M.F. A Nickel-Iron Alloy. It contains nickel, manganese, and carbon.†

Amfac. A proprietary preparation of antimenorrhagic factor, used in the treatment of functional uterine haemorrhage. (Armour Pharmaceutical Co).†

Amfaid. Surface active agents and detergents. (ABM Chemicals Ltd).*

Amfecloral. An appetite suppressant. α - Methyl-N-(2,2,2-trichloro-ethylidene)-phenethylamine.†

Amfipen. A proprietary preparation of ampicillin. Applications: An antibiotic. (Brocades).*

Amfix. High speed fixer for photographic processing. (May & Baker Ltd).*

Amfonelic Acid. A stimulant for the central nervous system. 7-Benzyl-1-ethyl-4-oxo-1,8-naphthyridine-3-carboxylic acid.†

Amgard. Flame retardants. (Albright & Wilson Ltd, Phosphates Div).

Amianth. See AMIANTHUS.

Amianthoid. See ASBESTOS.

Amianthus. (Amianth, Mountain Flax). A white and satiny variety of asbestos.†

Amical Biocides. Mildewcide, algicide, yeasticide for protection of polymeric systems. Diiodomethyl-p-tolysulphone, both solid and liquid forms available. Applications: Protection of adhesives, latex paints, sealants, caulks, pigment dispersions and other polymeric systems. (Abbott Laboratories).*

Amicar. Amino caproic acid. Applications: Haemostatic. (Lederle Laboratories, Div of American Cyanamid Co).

Amicarbalide. An anti-protozoan for veterinary use. 3,3'-Diamidino-carbanilide. Diampron is the isethionate.†

Amichrome. Pre-metallized dyes. (ICI PLC).*

Amidate. Etomidate. Applications: Hypnotic. (Abbott Laboratories).

Amidated Cotton. See COTTON, AMIDATED.

Amide Powder. A blasting powder similar to gunpowder. It contains 40 per cent potassium nitrate, 38 per cent ammonium nitrate, and 22 per cent charcoal.†

Amidephrine. A vasoconstrictor and nasal decongestant. 3-(1-Hydroxy-2-methylaminoethyl)methanesulphonanilide. DRICOL is the mesylate.†

Amidin. A solution of starch in water.†

Amido-G-Acid. (Amido-G-Salt). β-Naphthylamine-6,8disulphonic acid. A dyestuff intermediate.†

Amido-G-Salt. See AMIDO-G-ACID.

Amido-R-Acid. (Amido-R-salt). 2-Naphthylamine-3:6-disulphonic acid.†

Amido-R-Salt. See AMIDO-R-ACID.

Amidogene. An explosive consisting of potassium nitrate, magnesium sulphate, wood charcoal, bran, and sulphur.†

Amidol. 2 :4-Diamino-phenol. Used as a photographic developer.†

Amidonaphthol Red GL. An acid dyestuff for dress materials.†

Amidone. A proprietary trade name for methadone.†

Amidox. Ethoxylated alkylolamides. Applications: Emulsifiers, detergents, wetting agents. (Stepan Company).*

Amiesite. A proprietary asphalt-rubber product used in road surfacing.†

Amigen. Protein Hydrolysate. Applications: Replenisher. (Travenol Laboratories Inc).

Amikapron. A proprietary trade name for tranexamic acid.†

Amikin. Amikacin sulphate. Applications: Antibacterial. (Bristol Laboratories, Div of Bristol-Myers Co).

Amilee. See PARAFFIN, LIQUID.

Amilorin. Amiloride hydrochloride. Applications: Potassium conserving agent, diuretic. (Merck Sharp & Dohme).*

Amine. Large range of cationic surfactants composed of primary amines, secondary amines or tertiary amines in liquid, solid or paste form. Applications: Widespread use in industry and in household and personal care formulations, though mostly in the form of derived quaternaries and various salts. (Keno Gard (UK) Ltd).

Amine D. Dehydroabietylamine. Applications: Used as asphalt additive, as cationic collectors for calcite, sylrite, mica, feldspar, vermicilulite and phosphate rock concentration operations. (Hercules Inc).†

Amine 0. Dewatering agent and corrosion inhibitor. (Ciba Geigy PLC).†

Aminess. Parenteral solution, tablets. Applications: Essential amino acid for the treatment of ureamia. (KabiVitrum AB).*

Aminic Acid. Formic acid, HCOOH.†

Amino Gluten MG. Maize gluten amino acids. Applications: Used in the food and cosmetic industries. (Croda Chemicals Ltd).*

Amino Plastics. A term applied to moulded materials made from urea-formaldehyde condensation products.†

Amino-G Acid. 2-Naphthylamine-6,8-disulphonic acid.†

Amino-PF. A proprietary preparation of amino acids. An amino-acid supplement. (Pfizer International).*

Amino-R Acid. 2-Naphthylamine-3,6-disulphonic acid.†

Aminofoam C. A range of coco fatty acids and collagen amino acids. Applications: Used in the cosmetic and pharmaceutical industries. (Croda Chemicals Ltd).*

Aminoform. See HEXAMINE.

Aminogen I. A rubber vulcanization accelerator. It is α-naphthylamine $C_{10}H_2N$.†

Aminogen II. A rubber vulcanization accelerator. It is *p*-phenylenediamine, $C_6H_4(NH_2)_2$.†

Aminogran. A proprietary PHENYL-ALANINE - free food used in the treatment of phenylketonuria. (Allen & Hanbury).†

Aminophyllin. Aminophylline. Applications: Relaxant. (G D Searle & Co).

Aminoplex. A proprietary preparation of amino-acids, sorbitol, ethanol, vitamins and electrolytes used for parenteral nutrition. (Geistlich Sons).†

Aminorex. An appetite suppressant. 2-Amino-5-phenyl-2-oxazoline. Apiquel is the fumarate.†

Aminosol. Protein Hydrolysate. Applications: Replenisher. (Abbott Laboratories).

Aminox. Reaction product of di-phenylamine and acetone. Gives protection against oxygen and heat deterioration. Used in tyre carcass, heels, soles, mechanicals, proofing sundries and wire insulation. Effective in natural and nitrile rubbers and nylon. (Uniroyal).*

Aminoxid. Foam stabilizer; wetting; auxiliary; polymerization accelerator. Applications: Shampoos, detergents; Aqueous dispersions; photography; electroplating; vulcanizing. (Th Goldschmidt Ltd).

Aminutrin. A proprietary preparation of aminoacids for oral nutrition. (Geistlich Sons).†

Amipaque. Metrizamide. Applications: Diagnostic aid. (Sterling Drug Inc).

Amiphenazole. 2,4-Diamino-5-phenylthiazole. Daptazole.†

Amisyn. A proprietary preparation of acetomenaphthone and nicotinamide, used in the treatment of chilblains. (Armour Pharmaceutical Co).†

Amitraz. A veterinary ascaricide. N-Methylbis-(2,4-xylyliminomethyl)-amine. Taktic; Triatrix; Triatox.†

Amitril. Amitriptyline hydrochloride. Applications: Antidepressant. (Parke-Davis, Div of Warner-Lambert Co).

Amitryptyline. 3-(3-Dimethyl-aminopropylideone)-1,2 : 4,5-di-benzocyclohepta-1,4-diene. Laroxyl, Saroten and Tryptizol are the hydrochloride.†

Ammon-Carbonite. An explosive. It contains ammonium nitrate, flour, nitro-glycerin, and collodion wool.†

Ammon-Dynamite. An explosive. It contains 40 per cent nitro-glycerin, 10 per cent wood meal, 10 per cent sodium nitrate, and 40 per cent ammonium nitrate.†

Ammon-Foerdite I. An explosive. It contains ammonium nitrate, flour, nitro-glycerin, collodion wool, glycerin, diphenylamine, and potassium chloride.†

Ammon-Gelatin-Dynamite. A blasting explosive. It consists of 50 per cent nitro-glycerin, 2.5 per cent collodion cotton, 45 per cent ammonium nitrate, and 2.5 per cent rye meal.†

Ammon-Halalit. An explosive containing nitro-glycerin, ammonium nitrate, vegetable meal, nitro-compounds, and potassium perchlorate.†

Ammonal. An explosive. It consists of 30 per cent trinitrotoluene, 47 per cent ammonium nitrate, 22 per cent aluminium powder, and 1 per cent charcoal.†

Ammonaldehyde. See HEXAMINE.

Ammondyne. A coal-mine explosive containing 9-11 per cent of nitro-glycerin, 45-51 per cent of ammonium nitrate, 8-10 per cent of sodium nitrate, 17-19 per cent of ammonium oxalate, and 11-13 per cent of wood meal.†

Ammonia Dynamites. Explosives. They usually contain nitro-glycerin, wood pulp, ammonium nitrate, sodium nitrate, and calcium or magnesium carbonate.†

Ammonia Gelignite. An explosive containing 29.3 per cent nitro-glycerin, 0.7 per cent nitro-cotton, and 70 per cent ammonium nitrate.†

Ammonia-Olein. The trade name for a form of sulphonated castor oil.†

Ammonia-superphosphate. See NITROPHOSPHATE.

Ammoniacal Turpethum. Hydrated dimercuri-ammonium sulphate, $(NHg_2)_2SO_4\ 2H_2O$.†

Ammoniated Mercury. See WHITE PRECIPITATE.

Ammonia, Copperized. See WILLESDEN FABRICS.

Ammonio-Mercuric Chloride. See MERCURAMMONIUM CHLORIDE.

Ammonioformaldehyde. See HEXAMINE.

Ammonit C (Anfo-explosives). Primed from the bottom of the borehole by Gelatine Donarit 1 and detonating fuse. (Dynamit Nobel Wien GmbH).*

Ammonite. An explosive. Contains ammonium nitrate, trinitrotoluene, and sodium chloride.†

Ammonium Iron Alum. Iron alum (ammonium-ferric sulphate).†

Ammonium Muriate. See SALAMMONIAC.

Ammonium Superphosphate. See NITROPHOSPHATE.

Ammonium-Syngenite. Synonym for Koktaite.†

Ammonocarbonous Acid. Hydrocyanic acid, HCN.†

Ammonyx. Wetting, foaming, foam stabilization; conditioner; wetting conditioner. Applications: Detergent/sanitizers; bubble baths; bath oils dishwash concentrates; hair colouring systems; softeners and cleansers. (Millmaster-Onyx UK).

Ammonyx CA, CA Special, 4 and 4B. Stearyl dimethyl benzyl ammonium chloride in paste form. Applications: Cationic surfactant used as a conditioner in cream and colour rinses and protective hand creams. (Millmaster-Onyx UK).

Ammonyx CETAC. Cetyl trimethyl ammonium chloride in liquid form. Applications: Cationic surfactant used in conditioner for cream rinses. (Millmaster-Onyx UK).

Ammonyx DME. Cetyl dimethyl ethyl ammonium bromide in paste form. Applications: Emulsifying and wetting agent used in pharmaceuticals. (Millmaster-Onyx UK).

Ammonyx KP. Oleyl dimethyl benzyl ammonium chloride in liquid form. Applications: Conditioner and antistatic agent for clear non-alcoholic cream rinses. (Millmaster-Onyx UK).

Ammonyx SKD. Complex di-fatty quaternary compound in paste form. Applications: Cationic surfactant softener for home and industrial use. (Millmaster-Onyx UK).

Ammonyx T. Cetyl dimethyl benzyl ammonium chloride in liquid form. Applications: Emulsifying and wetting agent which is acid stable, used in synthetic latex dispersions. (Millmaster-Onyx UK).

Ammonyx 27. Tallow trimethyl ammonium chloride in liquid form. Applications: Antistatic and demulsifying agent used in emulsion breaking; plastic and other surfaces. (Millmaster-Onyx UK).

Ammonyx 4002. Stearyl dimethyl benzyl ammonium chloride in powder form. Applications: Conditioner and antistatic agent in aerosol hair sprays and shampoos; emollient and conditioner in lipsticks, lotions and toiletries. (Millmaster-Onyx UK).

Ammonyx 4080. Formulated imidazoline sulphate in liquid form. Applications: Cationic surfactant softener for home and industrial use. Can be used with BTC 2125M to form SO/SAN softener sanitizer products. (Millmaster-Onyx UK).

Ammonyx 485 and 490. Stearyl dimethyl benzyl ammonium chloride in powder form. Applications: Cationic surfactant used in conditioner for cream rinses. (Millmaster-Onyx UK).

Ammonyx 781. Alkyl methyl isoquinolinium chloride in liquid form. Applications: Algaecide and germicide for low cost industrial water treatment eg in oil recovery, cooling towers and paper mills. (Millmaster-Onyx UK).

Amoloid HV. Ammonium alginates. Applications: Textile printing, ceramic binding, can sealant. (Kelco, Div of Merck & Co Inc).*

Amoloid LV. Ammonium alginates. Applications: Textile printing, ceramic binding, can sealant. (Kelco, Div of Merck & Co Inc).*

Amoxicap. A proprietary preparation of amoxicillin. An antibiotic. (Pfizer International).*

Amoxil. Amoxicillin. Applications: Antibacterial. (Beecham Laboratories).

Amoxisyrup. A proprietary preparation of amoxicillin. An antibiotic. (Pfizer International).*

Amoxycillin. An antibiotic. 6-[(-)- α-Amino-4-hydroxyphenylacetamido] penicillanic acid.†

AMP. Adenosine phosphate.†

Ampco. An aluminium bronze. It contains from 86-92 per cent copper, 7-11 per cent aluminium, and 1.3 per cent iron.†

Ampen. A proprietary preparation of ampicillin. An antibiotic. (Pfizer International).*

Amphenol. A proprietary trade name for polystyrene products.†

Amphionic. Ampholytic surfactant. (ABM Chemicals Ltd).*

Amphionic 25B. High molecular weight amino-acid derivative, supplied as a golden liquid. Applications: Alkaline cleaning and sanitizing formulations and biocidal soaps, being an efficient biocide with a broad spectrum of kill. Good stability in the presence of electrolytes; compatability with other types of surface agent; dispersant. (ABM Chemicals Ltd).

Amphobac. Bactericidal amphoterics. Applications: Used in shampoos and industrial cleaners. (Lonza Limited).*

Amphocerin. Mixture of higher molecular fatty alcohol and wax esters. Applications: Water-in-oil type ointments and creams with good spreading properties. (Henkel Chemicals Ltd).

Amphojel. Aluminium hydroxide. Applications: Antacid. (Wyeth Laboratories, Div of American Home Products Corp).

Ampholak QTE. Tallow amphoteric in liquid form. Applications: Non-irritant hair conditioning additive. (Zeta Euro-American Technical Services Ltd).

Ampholak XCE. Coco amphoteric in liquid form. Applications: Alkaline industrial cleaners. (Zeta Euro-American Technical Services Ltd).

Ampholak XCG. Coco amphoteric in liquid form. Applications: High viscosity and gel non-irritant shampoos. (Zeta Euro-American Technical Services Ltd).

Ampholak XCO. Coco imidazolinium derivative in liquid form. Applications: Non-irritant shampoos and toiletry products. (Zeta Euro-American Technical Services Ltd).

Ampholak XJO. A C8 imidazolinium derivative in liquid form. Applications: Industrial applications. (Zeta Euro-American Technical Services Ltd).

Ampholak XOO. Oleic imidazolinium derivative as a viscous liquid. Applications: Non-irritant additive for shampoos, toiletries and hand cleaners. (Zeta Euro-American Technical Services Ltd).

Ampholak XSA. Stearyl amphoteric in liquid form. Applications: Non-irritant hair conditioning additive. (Zeta Euro-American Technical Services Ltd).

Ampholak YCE. Coco amphoteric in liquid form. Applications: Alkaline industrial cleaners. (Zeta Euro-American Technical Services Ltd).

Ampholan. Complex amphoteric surfactants in high foam formulations. Applications: Froth flotation, fire fighting, toiletries. (Lankro Chemicals Ltd).†

Ampholan B171. Cocoamido propyl betaine in liquid form. Applications: Foaming and wetting agent, for toiletries, industrial cleaners, cement, gypsum & latex. (Diamond Shamrock Europe Corporation).

Ampholyte SKKP 70. Amphoteric surfactant with dispersing and corrosion inhibiting properties. Applications: Emulsion paints, pigment grinding. (Keno Gard (UK) Ltd).

Amphoram. Amines and derivatives. Primary, secondary, tertiary fatty mono-, di- and polyamines from C8 to C22. Quaternary ammonium salts. Amine oxides. Amine salts. Ethoxylated amines and polyamines. Amino-acids. Special nonionic derivatives. Additives. Applications: Surface active agents used as auxiliaries in road, mining, textile and fertilizer industries, in metallurgy, as lubricating agents, additives for fuel oils and soil stabilization. (British Ceca Co Ltd).*

Amphosol CA. Derivative of alkyl amido propyl N-dimethyl amino acetic acid, as a clear yellow liquid. Applications: Shampoos and bubble baths. (KWR Chemicals Ltd).

Amphosol DM and DMA. Acetyldimethyl alkylammonium chloride, sodium salt or in acid form. Clear yellow liquid. Applications: Bacterial detergent preparations. (KWR Chemicals Ltd).

Amphoterge. Substituted imidazoline amphoterics. Applications: Used in shampoos and industrial cleaners. (Lonza Limited).*

Amphotropin. Hexamethylene-tetramine - camphorate, $(C_6H_{12}N_4)_2.C_8H_{14}(COOH)_2$. A urinary antiseptic.†

Ampicillin. 6 - [D(-) - α-Amino-phenylacetamido] penicillanic acid. [D (-) α-Aminobenzyl]penicillin. Penbritin.†

Ampiclox. A proprietary preparation of ampicillin and cloxacillin. An antibiotic. (Beecham Research Laboratories).*

Ampilar. Capsules containing 250mg and 500mg ampicillin as ampicillin trihydrate syrup: Powder for re-constitution to contain 125mg/5ml and 250mg/5ml ampicillin. Applications: For treatment of infections of respiratory, urinary and digestive tract where infection may be due to more than one pathogen. Antibiotic active against shigella, E coli, proteus mirabilis, haemophilus influenza, bordetella pertussis and other Gram-negative bacteria, actinomyces muris ratti and other actinomycetes and penicillin sensitive Gram-positive pathogens. (Lagap Pharmaceuticals Ltd).*

Amplex. A proprietary preparation of chlorophyll. A deodorant. (Ashe Chemicals).*

Amprol. Amprolium. Applications: Coccidiostat. (Merck Sharp & Dohme, Div of Merck & Co Inc).

Amritsar Gum. See GUM, AMRITSAR.

Amron. A proprietary vinylite base plastic for coatings.†

Amsco Steel. A proprietary high manganese steel containing 12-13 per cent manganese and 1.2 per cent carbon.†

Amsidyl. Amsacrine. Applications: Antineoplastic. (Parke-Davis, Div of Warner-Lambert Co).

Amsil. Oxygen free copper plus 8 oz to 30 oz per ton silver. Conductivity 100 per cent IACS. C10400 - oxygen free copper plus 8 to 12 oz per ton silver. C10500 - oxygen free copper plus 10 to 15 oz per ton silver. C10700 - oxygen free copper plus 25 to 30 oz per ton silver. Applications: The principal uses are based on its good creep strength at elevated temperatures and its high softening point. (Amax Inc).*

Amstat. Tranexamic Acid. Applications: Hemostatic. (Lederle Laboratories, Div of American Cyanamid Co).

Amsulf. Sulphur copper alloy containing oxygen-free copper and 0.3 per cent. sulphur. Conductivity 96 per cent. IACS. (Amax Inc).*

Amtel. Tellurium-copper alloy containing oxygen-free copper and 0.5 per cent. tellurium. Conductivity 93 per cent. IACS. (Amax Inc).*

Amuno. Indomethacin. Applications: Anti-rheumatic. (Merck Sharp & Dohme).*

Amvis. An explosive containing ammonium nitrate.†

Amyl Ledate. Lead dimethyldithiocarbamate 50% in oil. Liquid dithiocarbamate rubber accelerator. Applications:

Recommended for improved dynamic properties in natural and polyisoprene rubbers. (Vanderbilt Chemical Corporation).*

Amyl Mustard Oils. Amyl thiocarbimides, $C_5H_{11}N.CS$.†

Amylase. Diastase, an enzyme which renders starch soluble, by converting it into maltose.†

Amylenol. Amyl salicylate, $C_6H_4(OH)(CO_2.C_5H_{11})$. Sometimes used in the place of sodium salicylate in cases of rheumatism.†

Amylit. A diamalt compound, which is an enzymic product. It is used for desizing in the textile industry.†

Amylmetacresol. An anti-infective present in STREPSILS. It is 6-pentyl-*m*-cresol.†

Amylodextrin. See SOLUBLE STARCH.

Amylogen. A soluble starch.†

Amyloid. Concentrated sulphuric acid dissolves cellulose, gradually converting it into dextrin, and ultimately into dextrose. If the solution as soon as it is made is diluted with water, a gelatinous hydrate is produced. This substance is known as amyloid.†

Amylon. See MALTOBIOSE.

Amylopsin. Pancreatic diastase.†

Amylozine Spansule. A proprietary preparation of trifluoperazine and amylobarbitone. A sedative. (Smith Kline and French Laboratories Ltd).*

Amylozyme. Amylase starch converting enzyme. (ABM Chemicals Ltd).*

Amylum. Starch $(C_{12}H_{20}O_{10})n$.†

Amysal. Arayl salicylate, $C_6H_4(OH)COOC_5H_{11}$.†

Amytal. Isoalmyl-ethyl-barbituric acid. An an aesthetic applied intravenously. A proprietary preparation of amylobarbitone. A hypnotic. (Eli Lilly & Co).†

Amytal Sodium. Amobarbital sodium. Applications: Hypnotic; sedative. (Eli Lilly & Co).

Amzirc. Zirconium copper alloys containing oxygen-free copper and 0.13 to 0.20 per cent. zirconium. (Amax Inc).*

Anabalm. A proprietary preparation of oleoresin capsicum, menthol, histamine hydrochloride, beta-chloroethyl salicylate and squalene. An embrocation. (Crookes Laboratories).†

Anabolex. A proprietary preparation of stanolone. Anabolic agent. (Lloyd, Hamol).†

Anacal. A proprietary preparation containing mucopolysaccharide, polysulphuric acid ester, prednisolone lauro-macrogol - 400 and hexachlorophene, used in the treatment of haemorrhoids. (Luitpold-Werk).*

Anacardic Acid. Obtained from *Anacardium occidenale* (cashew nut). The ammonium salt is used as a vermifuge.†

Anacobin. A proprietary trade name for cyanocobalamin.†

Anacotine. Narcotine.†

Anadonis Green. A pigment. It is hydrated chromium sesquioxide.†

Anadrol. Oxymetholone. Applications: Androgen. (Syntex Laboratories Inc).

Anaesthesin. (Benzocaine, Anaesthone). Ethyl-*p*-aminobenzoate, $NH_2C_6H_4.CO_2.C_2H_5$. Used as a substitute for cocaine as a local anaesthetic, also to check vomiting.†

Anaesthesin, Soluble. See SUBCUTIN.

Anaesthone. See ANAESTHESIN.

Anaesthyl. See ANESTILE.

Anaflex. A proprietary preparation of polynoxylin. (Geistlich Sons).†

Anafranil. A proprietary preparation of chlomipramine hydrochloride. An anti-depressant drug. (Ciba Geigy PLC).†

Anagenite. See CHROME OCHRE.

Anahaemin. A proprietary preparation of a solution of an active erythropoietic fraction of liver. A haematinic. (British Drug Houses).†

Analar. Laboratory reagents and chemicals. (British Drug Houses).†

Analgesic Balm (AS). A proprietary preparation of methyl salicylate, capsicum oleoresin, methyl nicotinate, menthol and camphor. Applications: Arubifacient. (Ayrton Saunders plc).*

Analgesine. See ANTIPYRINE.

Analgin. A proprietary preparation of aspirin, phenacetin, codeine, Caffeine and phenolphthalein. An analgesic. (Norton, H N Co).†

Analoam. Soil testing reagent. (Murphy Chemical Co).†

Analock. A proprietary preparation of mepirizole for the relief of pain and inflammation. (Pfizer International).*

Ananase. A proprietary preparation of bromelains, used as an anti-inflammatory agent. (Rorer).†

Anaphe Silk. See SILK, ANAPHE.

'A" Naphtha. See BENZINE.

Anapolon. A proprietary preparation of oxymetholone. An anabolic agent. (Syntex Laboratories Inc).†

Anaprox. Naproxen sodium. Applications: Anti-inflammatory; analgesic; antipyretic. (Syntex Laboratories Inc).

Anaroids. A proprietary preparation containing resorcinol, powdered gall, bismuth subgallate, titanium dioxide, zinc oxide, boric acid, balsam of peru and kaolin. An antipruritic. (Rybar).†

Anasite. An explosive. It consists of ammonium perchlorate and myrabolans, usually with some sodium or potassium nitrate, and a small quantity of agar-agar.†

Anaspalin. A form of wool fat.†

Anasthol. A mixture of ethyl and methyl chlorides. A local anaesthetic.†

Anatola. Vitamin A. Applications: Vitamin (anti-xerophthalmic). (Parke-Davis, Div of Warner-Lambert Co).

Anatomical Alloy. An alloy of 53.5 per cent bismuth, 19 per cent tin, 17 per cent lead, and 10.5 per cent mercury.†

Anatto. See ANNATTO.

Anavar. Oxandrolone. Applications: Androgen. (G D Searle & Co).

Anazol. Ethyl phthalate.†

Ancaflex. A proprietary trade name for a series of boron trifluoride based polymers. Used as curing agents for epoxy resins to give flexible products. (Anchor Chemical Co).*

Ancaflex 70 and 150. A proprietary range of polymeric hardeners based on boron trifluoride. (Pacific Anchor Chemical Corp).†

Ancamide 280, 400. A proprietary trade name for a fluid complex polyamide used as a curing agent for epoxy resins. (Anchor Chemical Co).*

Ancamine LO. An epoxy hardener for use at low temperatures. Free from phenolic odour and possessing a low irritation index. (Anchor Chemical Co).*

Ancamine LT. A proprietary trade name for an activated aromatic amine curing agent for epoxy resins. (Anchor Chemical Co).*

Ancamine MCA. A modified cyclo-aliphatic polyamine used on floorings and coatings. (Anchor Chemical Co).*

Ancamine XT. A faster version of LT above.†

Ancaris. Thenium compound. (The Wellcome Foundation Ltd).

Ancatax. A proprietary rubber accelerator. Dibenzthiazyl disulphide. (Anchor Chemical Co).*

Ancazate BU. A proprietary trade name for a self dispersible zinc butyl dithiocarbamate. Used as an accelerator for vulcanization and an anti- oxidant for rubbers. (Anchor Chemical Co).*

Ancazate EPH. Zinc ethyl phenyl dithiocarbamate. (Anchor Chemical Co).*

Ancazate ET. Zinc diethyl dithiocarbamate. (Anchor Chemical Co).*

Ancazate ME. A proprietary accelerator. Zinc dimethyl dithiocarbaroate. (Anchor Chemical Co).*

Ancazate Q. A proprietary complex of zinc dithiocarbamate used as an accelerator. (Anchor Chemical Co).*

Ancazate XX. A proprietary rubber accelerator. Butyl dithiocarbamate. (Anchor Chemical Co).*

Ancazide ET. A proprietary rubber accelerator. Tetraethyl thiuram disulphide. (Anchor Chemical Co).*

Ancazide IS. A proprietary rubber accelerator. Tetramethyl thiuram monosulphide. (Anchor Chemical Co).*

Ancazide ME. A proprietary rubber accelerator. Tetramethyl thiuram disulphide. (Anchor Chemical Co).*

Ancef. Cefazolin sodium. Applications: Antibacterial. (Smith Kline & French Laboratories).

Anchoic Acid. (Lepargylic Acid). Azelaic acid, $C_9H_{16}O_4$.†

Anchor. A proprietary trade name for a vanadium tool steel.†

Anchor 1040, 1115, 1170, 1171 and 1222. A proprietary group of boron- trifluoride epoxy hardener curing agents. They consist of modified amine complexes of boron trifluoride. (Pacific Anchor Chemical Corp).†

Anchoracel. A proprietary rubber vulcanization accelerator. It is thiocarbanilide.†

Anchorite. A light rubber product used in rubber mixings.†

Anchorlube G-771. Animal/vegetable oil in water emulsion. Applications: Metal cutting and tapping compound for stainless and other hard to work metals or operations on sensitive appliances. (Anchor Chemical Co).†

Anchred. A red oxide used in rubber mixings.†

Anchusin. See ALKANET.

Ancobon. Flucytosine. Applications: Antifungal. (Hoffmann-LaRoche Inc).

Ancofen. Tablets of meclozine hydrochloride, ergotamine tartrate and caffeine for relief of migraine. (British Drug Houses).†

Ancolan. A proprietary trade name for the dihydrochloride of Meclozine- antihistamine and anti-emetic. (British Drug Houses).†

Ancoloxin. Tablets of meclozine hydrochloride and pyridoxine hydrochloride taken to combat nausea and vomiting of pregnancy. (British Drug Houses).†

Ancotil. See ALCOBON. (Roche Products Ltd).*

Ancovert. Tablets of meclozine hydrochloride with nicotinic acid - vertigo and Meniere's disease. (British Drug Houses).†

Ancrack. Naptalam plus dinitro. Applications: Pre-emergent herbicide for use on peanuts and soybeans. (Drexel Chemical Company).*

Ancrod. An anti-coagulant. The active principle is obtained from the venom of the Malayan pit-viper *Agkistrodon rhodostoma*, acting specifically on fibrinogen. See ARVIN.†

Ancudite. Synonym for Kaolinite.†

Ancyte. Piposulfan. Applications: Antineoplastic. (Abbott Laboratories).

Andalusite. A mineral. It is a silicate of aluminium, Al_2SiO_5.†

Andeer's Solution. A 9 per cent aqueous solution of resorcinol.†

Anderol. Synthetic lubricants. Applications: Lubrication of compressors, crankcases and other industrial uses. (Nueodex Inc).*

Andersil. High flash point liquid polysilicate. Applications: Stable liquid polysilicate binders for coatings and precision casting moulds. (Anderson Development Company).*

Andersonite. A mineral. It is $6[Na_2CaVO_2(CO_3)_2\ 6H_2O]$.†

Anderson's Solution. A solution containing potassium hydroxide and pyrogallol used for absorption of oxygen.†

Andreolite. See HARMOTONE.

Andreson's Acid. 1-Naphthol-3,8-di-sulphonic acid.†

Andrez 8000. 85% bound styrene/butadiene. Applications: Used to improve the processing properties and increase the hardness and modulus of many rubber compounds. (Anderson Development Company).*

Andro LA 200. Testosterone Enanthate. Applications: Androgen. (O'Neal, Jones & Feldman Pharmaceuticals).

Andro 100. Testosterone. Applications: Androgen. (O'Neal, Jones & Feldman Pharmaceuticals).

Androcur. A proprietary preparation of CYPROTERONE acetate used in the control of hypersexual and deviant behaviour in males. (Schering Chemicals Ltd).†

Androgyn LA. Testosterone Enanthate. Applications: Androgen. (O'Neal, Jones & Feldman Pharmaceuticals).

Androx 3961. Shield. A water displacing fluid meeting DTD 900/4942.†

Andur. Urethane based prepolymers, water emulsion thermoplastics and coatings. Applications: For use in the manufacturing of urethane elastomers, varnishes and paints, adhesives, etc. (Anderson Development Company).*

Andursil. A proprietary preparation of aluminium hydroxide gel, magnesium hydroxide and carbonate, and simethicone. An antacid. (Ciba Geigy PLC).†

Anergan 25. Promethazine Hydrochloride. Applications: anti- emetic; antihistaminic. (O'Neal, Jones & Feldman Pharmaceuticals).

Anergan 50. Promethazine Hydrochloride. Applications: Anti- emetic; antihistaminic. (O'Neal, Jones & Feldman Pharmaceuticals).

Anesone. See ACETONE CHLOROFORM.

Anestacon. Lidocaine. Applications: Anaesthetic. (Alcon Laboratories Inc).

Anestan. Bronchial asthma preparation. (The Boots Company PLC).

Anesthol. See ANESTILE.

Anesthyl. See ANESTILE.

Anestile. (Anaesthyl, Anesthyl, Anesthol). A mixture of ethyl and methyl chlorides. An anaesthetic.†

Anethaine. An emollient, white, vanishing cream containing tetracaine, hydrochloride BP 1% w/w. Applications: It quickly and effectively relieves itching and stinging to check scratching. Gives prompt temporary relief from insect bites, stings, nettle rash, heat rash, skin itches, chapping and chaffing. (Evans Medical).*

Aneurin. Vitamin B_1. See THIAMINE.†

Angarite. A Russian cast basalt (*qv*) used as an electrical insulator.†

Angel Red. See INDIAN RED.

Angio-Conray. Iothalamate sodium. Applications: Diagnostic aid. (Mallinckrodt Inc).

Angioneurosin. Nitro-glycerine.†

Angiotensin Amide. A hypertensive. Val^5 -hypertensin II-Asp-β-amide. Asn-Arg-Val-Tyr-Val-His-Pro-Phe.†

Angiovist 282. Diatrizoate meglumine. Applications: Diagnostic aid. (Berlex Laboratories Inc, Subsidiary of Schering AG).

Angioxyl. A proprietary preparation of pancreatic extract insulin free.†

Angised. A proprietary preparation of Glyceryl trinitrate in a stabilized base. A vasodilator used in the treatment of angina pectoris. (The Wellcome Foundation Ltd). *

Anglopyrin. See ASPIRIN.

Anhydrite. Calcium sulphate. Applications: Concrete substitute and soil conditioner. (Pacific Chemical Industries Pty Ltd).†

Anhydrol Forte. Colourless evaporative solution containing 20% w/v aluminium chloride hexahydrate. Applications: For the topical treatment of hyperhidrosis specifically involving axillae, hands or feet. (Dermal Laboratories Ltd).*

Anhydron. Cyclothiazide. Applications: Diuretic; antihypertensive. (Eli Lilly & Co).

Anhydrone. A proprietary name for a perchlorate of magnesium, a drying agent for gases.†

Anidrisorb. Anhydro-sorbitols. Applications: Pharmaceutical encapsulation. (Roquette (UK) Ltd).*

Anilazone. See ANILIPYRIN.

Anileine. See MAUVEINE.

Aniline Black. (Aniline Black in Paste, Fine Black, Oxidation Black). A black dyestuff produced on the fibre by the oxidation of an aniline salt.†

Aniline Black in Paste. See ANILINE BLACK.

Aniline Blue, Spirit Soluble. (Rosaniline Blue, Opal Blue. Spirit Blue O, Gentian Blue 6B, Hessian Blue, Fine Blue, Lyons Blue, Light Blue, Parma, Bleu de Nuit, Bleu Lumidre). The hydrochloride, sulphate, or acetate of triphenyl-rosaniline and triphenyl-*p*-rosaniline. Dyes wool and silk greenish-blue.†

Aniline Brown. See BISMARCK BROWN.

Aniline Green. See ALDEHYDE GREEN.

Aniline Mauve. See MAUVEINE.

Aniline Orange. See VICTORIA YELLOW.

Aniline Pink. See SAFRANINE.

Aniline Purple. See MAUVEINE.

Aniline Red. See MAGENTA.

Aniline Rose. See SAFRANINE.

Aniline Salt. Aniline hydrochloride, $C_6H_5NH_2HCl$.†

Aniline Violet. See MAUVEINE.

Aniline Yellow. (Spirit Yellow). A dyestuff. It is the hydrochloride of aminoazobenzene. It is not used as a dye, but for the preparation of acid yellow and indulines.†

Anilipyrin. (Anilazone). A mixture of antipyrine and acetanilide.†

Anilotic Acid. Nitro-salicylic acid.†

Animal Charcoal. (Bone Charcoal). This term is used for all charcoal produced by the ignition of animal substances with exclusion of air, but more particularly to that obtained from bones. This material contains approximately 10 per cent carbon, and 90 per cent mineral matter, mainly calcium phosphate. Used for absorbing dyes and other purposes.†

Animal Glycerin. Neatsfoot oil.†

Animal Oil. See BONE OIL.

Animal Oil, Dippel's. See BONE OIL.

Animal Rouge. See CARMINE.

Animal Starch. Glycogen, found in the blood and liver of mammals.†

Anime Resin. (Gum Anime). A fossil copal resin from South America.†

Anime Resin, West Indian. See TACAMAHAC RESIN.

Anionyx 12EO. Disodium monolauryl ether sulphosuccinate in liquid form. Applications: Detergent and foaming agent used in hair shampoos, bubble baths, rug and upholstery shampoos. (Millmaster-Onyx UK).

Anionyx 12S. Disodium monolauryl-amidosulphosuccinate in liquid form. Applications: Non-irritating detergent and foaming agent used in hair shampoos and bubble baths, rug and upholstery shampoos. (Millmaster-Onyx UK).

Aniscol. Anti-scale paints to reduce oxidation losses and surface decarburization. (Foseco (FS) Ltd).*

Anisidine. 2-Methoxyaniline.†

Anisidine Ponceau. See ANISOL RED.

Anisidine Scarlet. See ANISOL RED.

Anisol Red. (Anisidine Ponceau, Anisidine Scarlet). An acid dyestuff. Dyes wool and silk red from an acid bath.†

Anisoline. See RHODAMINE 3B.

Anisotheobromine. An addition product of theobromine-sodium and sodium anisate.†

Anisotropine Methylbromide. Octatropine methylbromide.†

Anka Steel. (V2A Steel). Nickel-chromium steels containing from 15-16 per cent chromium and from 7-10 per cent nickel.†

Annaline. See GYPSUM.

Annatto. (Anatto, Annotto, Arnatto, Arnotto, Orleans, Rocou). A natural dyestuff, obtained from the fleshy covering of the seeds of the ruccu tree, *Bixa orellana*.†

Annealing Carbon. (Carbon of Cementation, Carbon of Normal Carbide). The names given to the carbon found combined in steel.†

Annotto. See ANNATTO.

Anobolex. A proprietary trade name for stanolone.†

Anode Mud. (Anode Slime). The material which falls to the bottom of the electrolysing vessel during the electrolytic refining of copper. It contains copper (10-25 per cent.) gold (0.7-2 per cent.) silver (5-40 per cent.) tellurium, antimony, and arsenic.†

Anode Slime. See ANODE MUD.

Anodesyn. A proprietary preparation containing ephedrine hydrochloride, lignocaine hydrochloride and allantoin. An antipruritic. (Boots Company PLC).*

Anodised Aluminium. See ELOXAL, ALUMINIUM ANODISED.

Anodyne. See ANTIPYRINE.

Anodynon. Ethyl chloride, C_2H_5Cl.†

Anogon. The mercury salt of 2,6-diiodophenol-4-sulphonic acid.†

Anol. *p*-Propenyl alcohol.†

Anonaid. Anionic surfactants and detergents. (ABM Chemicals Ltd).*

Anonaid TH. Sodium di-octyl sulphosuccinate in liquid form. Applications: Wetting and emulsifying agent for the textile, leather and paper industries and for emulsion polymerization. (ABM Chemicals Ltd).*

Anone. Cyclohexanol.†

Anorvit. Tablets of ferrous sulphate, ascorbic acid and acetomenaphthone - hypochromic anaemia. (British Drug Houses).†

Anotex. Dyes for anodized metal. (Pointing Ltd).†

Anovlar. A proprietary preparation of norethisterone acetate and ethinyl oestrodiol. Oral contraceptive. (Schering Chemicals Ltd).†

Anquil. A proprietary preparation of benperidol used in the control of deviant sexual behaviour. (Janssen Pharmaceutical Ltd).*

Ansa. Alkyl naphthaline sodium sulphonate. (Albright & Wilson Ltd).*

Ansaid. Flurbiprofen. Applications: Analgesic; anti-inflammatory. (The Upjohn Co).

Ansal. Phenazone salicylate.†

Ansax. Fertilizer prilled ammonium nitrate. (L & K Fertilisers Ltd).

Anscor. Lightly calcined sea water magnesia. (Steetley Refractories Ltd).

Anserine. A natural peptide from muscle. It is β-alanyl methyl histidine.†

Ansol A. A proprietary blend of anhydrous denatured ethyl alcohol with small percentages of esters, other alcohols, and hydrocarbons. A nitro-cellulose and resin solvent.†

Ansol B. A preparation similar to A, except that B contains a small amount of normal butyl alcohol in the place of amyl alcohol.†

Ansol E-121. A trade mark for ethylene glycol dimethyl ether. (Ansul Co).†

Ansol E-161. A trade mark for triethylene glycol dimethyl ether. TRIGLYME. (Ansul Co).†

Ansol E-181. A trade mark for tetra-ethylene glycol dimethyl ether. TETRAGLYME. (Ansul Co).†

Ansol M. A preparation similar to A but containing no higher alcohols.†

Ansol PR. A preparation similar to M but with a higher ester content.†

Ansolysen. Pentolinium tartrate. (May & Baker Ltd).

Anspor. Cephradine. Applications: Antibacterial. (Smith Kline & French Laboratories).

Ant Flip. A gel bait to control ants. Boric acid is the active ingredient. Applications: Household, hospital, restaurant use for controlling pharoah and other sweet eating ants. Workers carry back to nest to kill the colony. (Colonial Products Inc).*

Ant Killer. Insecticide. (Fisons PLC, Horticulture Div).

Ant Oil. See OIL OF ANTS.

Antabuse. Disulfiram. Applications: Alcohol deterrent. (Ayerst Laboratories, Div of American Home Products Corp).

Antacidin. See ANTACEDIN.

Antak. Alcohol. Applications: Contact tobacco sucker control material. (Drexel Chemical Company).*

Antara HR-719. Water soluble, slightly viscous, clear, biodegradable lubricant. Applications: Suitable for use in synthetic, semi-synthetic or aqueous systems. When neutralized with triethanolamine it exhibits rust inhibitory effects. (GAF Corporation).*

Antara LB-400. Oil and water soluble lubricity additive, rust inhibitor and emulsifier. Applications: Useful in lubricating and rolling oils, hydraulic, water-based and emulsifiable oil cutting fluids. (GAF Corporation).*

Antara LE-500. Oil and water soluble lubricant and emulsifier. Applications: Useful in oil and water based cutting fluids, hydraulic fluids and rolling fluids. (GAF Corporation).*

Antara LE-600. Oil and water soluble lubricant and emulsifier. Applications: Uses similar to ANTARA LE-500 only more hydrophilic. (GAF Corporation).*

Antara LE-700. Highly water soluble lubricant which has outstanding compatability with concentrated electrolyte solutions. Applications: Used as a lubricity additive in water based cutting fluids. (GAF Corporation).*

Antara LF-200. Oil and water soluble lubricant and emulsifier. Applications: Useful in oil and water cutting fluids, hydraulic fluids and rolling oils. (GAF Corporation).*

Antara LK-500. Linear alcohol based low-foaming lubricant. Applications: Used as a lubricity additive in water-based cutting fluids. (GAF Corporation).*

Antara LM-400. Oil soluble, water dispersible lubricant. Applications: Used as a rust inhibitor or in formulating aqueous cutting fluids. Also used in rolling oils, cutting oils and hydraulic fluids etc. (GAF Corporation).*

Antara LM-600. Oil and water soluble lubricant and emulsifier. Applications: Used in oil and water based cutting fluids, hydraulic fluids and rolling oils. (GAF Corporation).*

Antara LP-700. Non-foaming lubricant. Applications: Used as lubricity additive in water-based cutting fluids. (GAF Corporation).*

Antara LS-500. Oil and water soluble lubricant and emulsifier. Applications: Useful in oil and water based cutting fluids, hydraulic fluids and rolling oils. (GAF Corporation).*

Antarox BL-214. A modified linear aliphatic polyether. Applications: Surfactant used in ambient temperature metal cleaning and textile finishing operations. (GAF Corporation).*

Antarox BL-225. A modified linear aliphatic polyether. Applications: Used in textile wetting, cleaning and as a rinse aid in commercial dishwashing. (GAF Corporation).*

Antarox BL-236. A modified linear aliphatic polyether. Applications: Low foaming surfactant particularly suited for high temperature spray metal cleaning. Rinse aids for commercial and industrial use. (GAF Corporation).*

Antarox BL-240. Modified linear aliphatic polyether. Applications: Suitable for high temperature spray metal cleaning and textile wet operations and in formulating rinse aids. (GAF Corporation).*

Antarox BL-330. A modified linear aliphatic polyether. Applications: Useful derusting agent for powdered detergents. Used in laundry and dairy cleaners, bottle washing compounds, mechanical-dishwashing detergents and spray-metal and steam-cleaning systems. (GAF Corporation).*

Antarox CA. Series of nonionic surfactants of the octylphenol ethoxylate type in liquid, aqueous solution or wax form. Application: Emulsifiers, detergency, stabilizers and dyeing assistants. Used in detergent compounding; dry cleaning detergents; solvent emulsion cleaners; pesticide and floor polish emulsifiers; textile, paper and miscellaneous cleaners; solvent emulsification; steam and electrolytic cleaning; metal pickling; vinyl acetate and acrylate polymerization; synthetic latices. (GAF (Gt Britain) Ltd).

Antarox CO-210 and CO-430. Nonionic surfactants of the nonylphenol ethoxylate type in the form of an oil-soluble liquid. Application: Foam control; dispersants in petroleum oils. (GAF (Gt Britain) Ltd).

Antarox CO-520 and CO-530. Nonionic surfactant of the nonylphenol type in liquid form. Application: Surfactants which have oil-water solubility balance and are used as coupling agents, de-icing fluids and in dispersing applications. (GAF (Gt Britain) Ltd).

Antarox CO-610, CO-630, CO-660, CO-710, CO-720 and CO-730. Nonionic surfactants of the nonylphenol ethoxylate type in liquid form. Application: Detergency, wetting, spreading and penetrating agents used in all types of detergents; metal processing; textiles; food processing; pesticides; water-based paints. (GAF (Gt Britain) Ltd).

Antarox CO-850, CO-880 and CO-887. Nonionic surfactants of the nonylphenol type in wax or aqueous solution form. Application: Detergency, wetting, emulsification, demulsification agent, effective at high temperatures. Used in crude petroleum oil emulsions, synthetic latex, polyester resins and silicone emulsions. (GAF (Gt Britain) Ltd).

Antarox CO-890, CO-897, CO-970, CO-977, CO-980, CO-987, CO-990 and CO-997. Nonionic surfactants of the nonylphenol ethoxylate type in wax or aqueous solution form. Application: Highly water-soluble surfactants, effective at high temperatures and in concentrated electrolyte solutions. Used in synthetic latices, polishes and floor waxes. (GAF (Gt Britain) Ltd).

Antarox CTA-639. Nonionic surfactant consisting of modified alkylphenol ethoxylate in liquid form. Application: Water-soluble wetting and emulsifying agent used in latex paint; titanium dioxide; oil or alkyd additives. (GAF (Gt Britain) Ltd).

Antarox DM. Series of nonionic surfactants of the dialkylphenol ethoxylate type in liquid, paste, wax, flake or solid form. Application: Emulsifiers, detergency, dispersing and wetting agents, which range from oil-soluble to water-soluble. Used in agricultural chemicals; leather; emulsion polymerization; acid- based cleaners; aerosols; cosmetics; insecticides; wax emulsions; inks;.paint; lacquers. (GAF (Gt Britain) Ltd).

Antarox LF-222. Modified aromatic polyether. Applications: Ambient temperature spray metal cleaning agent. (GAF Corporation).*

Antarox LF-224. Modified aromatic polyether. Applications: Ambient temperature spray metal cleaning agent. (GAF Corporation).*

Antarox LF-330. A modified aliphatic polyether. Applications: Detergent and wetting agent. (GAF Corporation).*

Antarox LF-344. A modified aliphatic polyether. Applications: Low-foaming detergent and wetting agent. (GAF Corporation).*

Antarox RC-520. Nonionic surfactant of the dodecyl phenol ethoxylate type in liquid form. Application: Low-foam re-wetting agent for paper towels and tissues. (GAF (Gt Britain) Ltd).

Antarox RC-620 and RC-630. Nonionic surfactants of the dodecyl phenol ethoxylate type in liquid form. Application: Detergency, wetting, emulsifying and dispersing agents(RC-620 has higher foam properties). Used in acid cleaning; detergent sanitizing; rug shampoos; de-dusting; grease cutting; agricultural compounds. (GAF (Gt Britain) Ltd).

Antas. A proprietary preparation containing pepsin, aluminium hydroxide, magnesium trisilicate, belladonna, magnesium hydroxide. An antacid. (Consolidated Chemicals).†

Antasil. An antacid/deflatulent. (ICI PLC).*

Antec Farm Fluid S. A mixture of organic acids, high molecular weight phenols, low molecular phenols and surfactant. Applications: For the disinfection of all types of livestock buildings. (Antec International Ltd).*

Antec Longlife 250 S. A mixture of organic acids, high molecular weight phenols, synthetic boicide and surfactant. Applications: For the disinfection of all types of livestock buildings. (Antec International Ltd).*

Antec OO-Cide. Two part ammonia release system incorporating a biocide. Applications: Coccidiacide - For the treatment of all livestock buildings. (Antec International Ltd).*

Antelope. Acid sodium pyrophosphate. (Albright & Wilson Ltd).*

Antepan. Piperazine anthelmintic. (Coopers Animal Health, subsidiary of thr Wellcome Foundation Ltd).*

Antepar. Piperazine Citrate. Applications: Anthelmintic. (Burroughs Wellcome Co).

Antepsin. Sucralfate - basic aluminium salt of sucrose octa sulphate. Applications: Duodenal ulcer, gastric ulcer, chronic gastritis. (Ayerst Laboratories).†

Anthelvet. Tetrahydrozoline Hydrochloride. Applications: Adrenergic. (McNeil Pharmaceuticals, McNEILAB Inc).

Anthical. A proprietary preparation of mepyramine maleate and zinc oxide. An antihistamine skin cream. (May & Baker Ltd).*

Anthiomaline. Hexa lithium antimovtri mercaptosuccinate. (May & Baker Ltd).

Anthion. Potassium persulphate, used as a hypo eliminator in photography.†

Anthiphen. Dichlorophen. (May & Baker Ltd).

Anthisan. Hydrogen maleate of Mepyramine. (May & Baker Ltd).*

Anthium Dioxcide. A proprietary additive based on stabilized chlorine dioxide. A deodorant designed to remove odours caused by residual monomers in resins. It works by oxidation. (International Dioxide Inc, NV).†

Anthocyanins. The red, blue, and violet colouring matters of plants.†

Antholite. See ANTHOPHYLLITE.

Anthra-Derm. Anthralin. Applications: Antipsoriatic. (Dermik Laboratories Inc).

Anthracene Acid Black. A dyestuff obtained by coupling amino-salicylic acid with α-naphthylamine-sulphonic acids (1 : 6 and 1 : 7), the product being diazotized, and coupled withβ-naphthol-disulphonic acid R.†

Anthracene Blue. 1,2,4,5,6,8-Hexahydoxyanthraquinone.†

Anthracene Blue New WG. A dyestuff obtained by the long heating of anthracene blue with caustic soda. It dyes chromed wool greenish-blue.†

Anthracene Blue S. See ALIZARIN BLUE S.

Anthracene Brown R, G. (Anthragallol, Alizarin Brown). A dyestuff. It is trihydoxyanthraquinone. Dyes cotton mordanted with chromium, brown. Also see CHROME FAST BROWN FC.†

Anthracene Chrome Black F, FE, 5B. An acid azo dyestuff. It dyes with chromium mordants.†

Anthracene Chrome Yellow BN. A dyestuff. It is a British equivalent of Milling yellow.†

Anthracene Dark Blue W. A mordant dyestuff. It is similar to Anthracene blue WR.†

Anthracene Green. See COERULEINE AND COERULEINE S.

Anthracene Red. An acid dyestuff. It dyes wool fast red from an acid bath.†

Anthracene Violet. See GALLEINE.

Anthracene Yellow. (Alizarin Yellow). A dyestuff. It dyes chromed wool greenish-yellow.†

Anthracene Yellow BN. See MILLING YELLOW.

Anthracene Yellow C. (Fast Chrome Yellow C, Fast Mordant Yellow). An acid dyestuff. It dyes chromed wool, yellow.†

Anthracene Yellow Paste. A mordant dyestuff. It is dibromodihydoxymethyl-coumarin, and dyes chrome mordanted wool greenish-yellow.†

Anthrachrysone. 1,3,5,7-Tetrahydroxy-anthraquinone.†

Anthracite. A hard coal, containing 85-95 per cent carbon. It burns with little smoke.†

Anthracite Black B. (Phenylene Black). An acid dyestuff. It dyes wool black from an acid bath.†

Anthracyl Chrome Green. A dyestuff obtained from diazotized picramic acid and naphthionic acid.†

Anthraflavic Acid. 2,6-Dihydroxyanthra-quinone.†

Anthraflavone. A dyestuff which is prepared by treating 2-methylanthraquinone with condensing agents. It dyes cotton yellow shades.†

Anthragalanthranol. Trioxyanthranol, $C_{14}H_{10}O_4$.†

Anthragallol. See ANTHRACENE BROWN.

Anthrahydroquinone. Oxanthranol $C_{14}H_{10}O_2$.†

Anthranilic Acid. *o*-Aminobenzoic acid, $C_6H_4(NH_2).COOH$.†

Anthranol. Available in three strengths: 0.4, 1.0 and 2.0 ointments. A smooth soft ointment containing dithranol (in one of the three strengths indicated) w/w in a base containing cetyl alcohol, liquid paraffin, soft white paraffin and sodium sulphate with salicyclic acid. Applications: For the topical treatment of sub-acute and chronic psoriasis including psoriasis of the scalp. (Stiefel Laboratories (UK) Ltd).*

Anthrapurpuranthranol. Trihydroxy-anthranol, $C_{14}H_6(OH)_4$.†

Anthrapurpurin. (Isopurpurin, Alizarin GD, Alizarin RF, Alizarin RT, Alizarin RX, Alizarin SC, Alizarin SSA, Alizarin SX, Alizarin SX Extra, Alizarin WG). A mordant dyestuff. It is 1, 2, 7 trioxyanthraquinone, $C_{14}H_5O_2(OH)_3$, and dyes red shades on alumina mordanted wool. Also see ALIZARIN SAR.†

Anthraquinone. Discharging auxiliary for textiles. (BASF United Kingdom Ltd).

Anthraquinone Black. A dyestuff produced by the fusion of 1:5-dinitroanthraquinone with sodium polysulphide. It dyes cotton black from an alkaline or sulphide bath.†

Anthraquinone Blue. A dyestuff. It is 2,4,6,8-tetrabromo-1,5-diamino anthraquinone, heated with *p*-toluidine, and sulphonated.†

Anthraquinone Green GX. An anthracene dyestuff closely related to Alizarin viridine.†

Anthraquinone Violet. An anthracene dyestuff.†

Anthraquinone Violet 3R. A dyestuff. Anthraquinone violet.†

Anthrarobin. Dihydroxyanthranol, $C_{14}H_{10}O_2$. Used in the treatment of skin diseases.†

Anthrarufin. 1,5-Dihydroxy-anthraquinone.†

Anthrasol Brown IVD. A dyestuff. The sulphuric ester of the reduced form of the condensation product of 6,7-dichloro-5-methoxythioindoxyl and the 4,5-benzo derivative of thioindoxyl.†

Anthrasol Yellow 129. The di-o-sulphuric ester of 1, 5-di(*m*- trifluoro-benzamido)anthraquinone in the reduced form.†

Anthraxolite. A variety of anthracite.†

Anthropic Acid. A mixture of palmitic and stearic acids.†

Anti-Oxidant AH. A proprietary aldol-alpha-naphthylamine resin. (Bayer & Co).*

Anti-Oxidant AP. A proprietary aldol alpha-naphthylamine powder. (Bayer & Co).*

Anti-Oxidant DDA. A proprietary derivative of diphenylamine. (Bayer & Co).*

Anti-Oxidant DNP. Di-β-naphthyl-p-phenylenediamine. (Bayer & Co).*

Anti-Oxidant DOD. 4, 4'-Dioxydiphenyl. (Bayer & Co).*

Anti-Oxidant EM. A 30 per cent. aqueous emulsion of a diphenylamine derivative. (Bayer & Co).*

Anti-oxidant MB. Phenylenethiourea. A proprietary antioxidant. (3). 2-mercaptolbenzimidazole. (Bayer & Co).*

Anti-Oxidant PAN. Phenyl -α-naphthylamine. (Bayer & Co).*

Anti-Oxidant RR 10 N. A proprietary range of alkylated phenols. (Bayer & Co).*

Anti-Oxidant SP. A proprietary styrenated phenol. (Bayer & Co).*

Anti-oxidant 2246. A proprietary preparation of 2,3 - methylene - bis (4 - methyl - 6 tertiary - butyl) phenol. (Cyanamid of Great Britain).†

Anti-oxidant 4010. A proprietary preparation of N-phenyl-N'-cyclohexyl-*p*-phenylenediamine.†

Anti-oxidant 425. A proprietary preparation of 2,2-methylene-bis(4-ethyl-6-tertiary-butyl) phenol. (Cyanamid of Great Britain).†

Anti-white Lead. See OIL WHITE.

Antiar. The milky juice of the upas tree. Used as an arrow poison.†

Antibacterin. Chloramine T.†

Antiblu/Antiboror. Fungicide/insecticide. Applications: Freshly sawn timber. (Hickson & Welch Ltd).*

Anticorodal. An aluminium alloy containing 1 per cent silicon, 0.06 per cent magnesium, and 0.06 per cent manganese.†

Antidine. The phenyl ether of glycerol.†

Antidol. A proprietary preparation of ethosalamide, paracetamol and caffeine. An analgesic. (Lewis Laboratories).†

Antidust F. A proprietary group of surface-active agents in combination with polyvalent alcohols, used in a concentrated aqueous solution to prevent undesirable surface tackiness in sheets and extrudates of plastics and rubber materials. (Rhein-Chemie Rheinau).*

Antidust 2. A proprietary preparation similar to antidust F, but containing a particularly fine zinc stearate in aqueous dispersion. (Rhein-Chemie Rheinau).*

Antifoam ET. An emulsified form of Antifoam T. (Bayer & Co).*

Antifoam T. Tri-*n*-butyl phosphate. An antifoaming agent used in the manufacture of paper coatings. (Bayer & Co).*

Antifriction Metal. Also called White Metal (*qv*). There are several varieties which are used for bearings. (*a*) Used for rapid working, consists of 77 per cent antimony, 17 per cent tin, and 6 per cent copper; (*b*) Extra hard, containing 82 per cent antimony, 12 per cent tin, 2 per cent zinc, and 4 per cent copper; (*c*) Medium hard, consisting of 26 per cent antimony, 72 per cent tin, and 2 per cent copper; (*d*) American, containing 20 per cent antimony, 78 per cent lead, 1 per cent zinc, and 1 per cent iron; (*e*) Babbitt's, consisting of 7 per cent antimony, 89 per cent tin, and 4 per cent copper. Other Babbitt's bearing metals contain from 55-90 per cent zinc, 1-29 per cent tin, 3-8 per cent copper, and 1-3 per cent antimony, and still another class consists of from 10-80 per cent lead, 1-75 per cent tin, 8-20 per cent antimony, and 0-5 per cent copper.†

Antifungin. The trade name for magnesium borate, a fungicide.†

Antigrison. An explosive. It consists of 27 per cent nitro-glycerin, 1 per cent nitro-cotton, and 72 per cent ammonium nitrate.†

Antihypo. Potassium percarbonate, $K_2C_2O_6$. Used as a bleaching agent and hypo eliminator in photography.†

Antikamnia. See AMMONAL.

Antilirium. Physostigmine Salicylate. Applications: Cholinergic. (O'Neal, Jones & Feldman Pharmaceuticals).

Antiluetin. Potassium-ammonium-antimonyl-tartrate, $[SbO(C_4H_4O_6)_2DF]KNH_4]\ H_2O$. A trypanocide.†

Antilux. Blends of selected paraffins and micro waxes, which protect rubber articles from damage by the sun, ozone and weathering. Applications: Technical moulded and extruded articles, articles subjected to dynamic stress, tyres, conveyor belts, cable coverings, articles fit for foodstuffs quality. (Rhein-Chemie Rheinau).*

Antilux AOL. A proprietary blend of paraffinic hydrocarbons used as an anti-weathering agent in natural and synthetic rubbers. (Rhein-Chemie Rheinau).*

Antiminth. A proprietary preparation of pyrantel pamoate. An anthelmintic. (Pfizer International).*

Antimonial Lead. An alloy of 87 per cent lead and 13 per cent antimony.†

Antimonic Acid. Antimonic oxide, Sb_2O_5.†

Antimonin. See LACTOLIN.

Antimony Ash. Obtained by roasting the grey sulphide in air. It consists chiefly of Sb_2O_4, and is used for the preparation of antimony compounds.†

Antimony Blende. See RED ANTIMONY.

Antimony Cinnabar. See RED ANTIMONY.

Antimony Crocus. See SULPHURATED ANTIMONY.

Antimony Glass. Produced in the preparation of antimony from its ores. It is antimony oxysulphide, and is used for colouring glass yellow.†

Antimony Lead. See HARD LEAD.

Antimony Mordant. See TARTAR EMETIC POWDER.

Antimony Regulus. See REGULUS OF ANTIMONY.

Antimony Saffron. See SULPHURATED ANTIMONY.

Antimony Salt. Double salts of antimony fluoride, SbF_2, with alkali sulphates, or with alkali fluorides. Used as mordants in dyeing.†

Antimony Vermilion. A red pigment. It is antimony oxysulphide, $Sb_6S_6O_3$.†

Antimony, Tartrated. See TARTAR EMETIC.

Antinonnin. See VICTORIA YELLOW.

Antinutisin. Pyrogallol-mono-ethyl ester.†

Antioxidant BA. A proprietary antioxidant. It is aldol- α-naphthylamine powder.†

Antioxidant PBN. See also NEOZONE D. A proprietary antioxidant. It is phenyl-β-naphthylamine.

Antioxidant RES. A proprietary antioxidant. It is aldol-α-naphthylamine resin.†

Antioxidant 431. Non-discolouring and non-staining antioxidant for dry rubber and latex. (Uniroyal).*

Antioxidant 449. A phenolic phosphite antioxidant. A non-discolouring stabilizer for EDPM polymers. Also economical replacement for BHT in compound work. (Uniroyal).*

Antioxidant 451. Alkylated hydroquinone. For synthetic rubbers and plastics. Used as a stabilizer for synthetic rubbers, such as polybutadiene and as an antioxidant for uncured adhesives. It functions as an antioxidant in both black and non-black cured compounds and also latex compounds. (Uniroyal).*

Antioxygene A. A proprietary trade name for phenyl-alpha- naphthylamine. An antioxidant. (Soc Anon des matieres colorantes et produits chimiques de Saint-Denis Paris (Allied Colloids Bradford)).†

Antioxygene AFL. A proprietary trade name for a Ketone-amine reaction product. (Soc Anon des matieres colorantes et produits chimiques de Saint-Denis Paris (Allied Colloids Bradford)).†

Antioxygene AN. A proprietary trade name for aldol-alpha-naphthylamine paste. An antioxidant. (Soc Anon des matieres colorantes et produits chimiques de Saint-Denis Paris (Allied Colloids Bradford)).†

Antioxygene BN. A proprietary trade name for β-naphthol. An antioxidant. (Soc Anon des matieres colorantes et produits chimiques de Saint-Denis Paris (Allied Colloids Bradford)).†

Antioxygene CAS. A proprietary trade name for a mixture of phenyl-alpha-naphthylarnine and meta-toluylene-diamine. (Soc Anon des matieres colorantes et produits chimiques de Saint-Denis Paris (Allied Colloids Bradford)).†

Antioxygene INC. A proprietary trade name for aldol-alpha- naphthylamine in powder form. An antioxidant. (Soc Anon des matieres colorantes et produits chimiques de Saint-Denis Paris (Allied Colloids Bradford)).†

Antioxygene MC. A proprietary trade name for phenyl-beta- naphthylamine. An antioxidant. (Soc Anon des matieres colorantes et produits chimiques de Saint-Denis Paris (Allied Colloids Bradford)).†

Antioxygene RA. A proprietary trade name for aldol naphthylamines. Antioxidants. (166) .†

Antioxygene RES. A proprietary trade name for an aldol-alpha- naphthylamine resin. An antioxidant. (Soc Anon des matieres colorantes et produits chimiques de Saint-Denis Paris (Allied Colloids Bradford)).†

Antioxygene RM. A proprietary trade name for aldol naphthylamine. An antioxidant. (Soc Anon des matieres colorantes et produits chimiques de Saint-Denis Paris (Allied Colloids Bradford)).†

Antioxygene RO. A proprietary trade name for aldol naphthylammine. An antioxidant. (Soc Anon des matieres colorantes et produits chimiques de Saint-Denis Paris (Allied Colloids Bradford)).†

Antioxygene STN. A proprietary trade name for phenyl-alpha- naphthylamine mixed with meta-toluylene-diamine and stearic acid. An antioxidant. (Soc Anon des matieres colorantes et produits chimiques de Saint-Denis Paris (Allied Colloids Bradford)).†

Antioxygene WBC. A proprietary trade name for an antioxidant. (Soc Anon des matieres colorantes et produits chimiques de Saint-Denis Paris (Allied Colloids Bradford)).†

Antiozonant AFD. Non-staining antiozonant. (Bayer UK Ltd).

Antipynin. Sodium metaborate $NaBO_2.H_2O$. Antiseptic.†

Antipyoninum. Neutral sodium tetraborate, prepared by fusing together borax and boric acid.†

Antipyreticum. See ANTIPYRINE.

Antipyrine. (Analgesine, Anodynin, Antipyreticum, Methozin, Metozin, Paradyn, Phenazone, Phenylon, Pyrazine, Pyrazoline, Sedatine). 1-Phenyl-2,3-dimethyl-pyrazolone. A febrifuge and analgesic.†

Antiquax. Wax polishes. Applications: Furniture. (James Briggs & Sons Ltd).*

Antique Purple. A colouring matter. It is 6:6'-dibromoindigo.†

Antiquinol. Allyl-phenyl-cinchonic ester. An antiseptic and analgesic.†

Antisepsin. (Bromanilide). See AMMONAL.†

Antiseptin. A mixture of boric acid zinc iodide, and thymol. An antiseptic dusting powder.†

Antisettle. Thixotrope. (Cray Valley Products Ltd).

Antistatic 812 and 813. A proprietary range of antistatic compounds for plastics. They are 100 per cent. active phenolic ethoxylates, used in proportions of 5-7 per cent. (Farbwerke Hoechst, Frankfurt-am-Main).†

Antistatic 816. A proprietary antistatic compound for plastics, 95 per cent. active lauric imide, used in polyethylene, PVC and polystyrene. (Farbwerke Hoechst, Frankfurt-am-Main).†

Antistatin. Antistatic finishing agent for textiles. (BASF United Kingdom Ltd).

Antistin. A proprietary preparation of antazoline hydrochloride. (Ciba Geigy PLC).†

Antistitin-Privine. A proprietary preparation of ANTAZOLINE sulphate and NAPHAZOLINE nitrate used in the treatment of allergic rhinitis. (Ciba Geigy PLC).†

Antitan. A tannin remover. (S & D Chemicals Ltd).†

Antithrombin. Infusion substance. Applications: Cofactor of heparin and major guardian of the haemostatic balance. (KabiVitrum AB).*

Antitoxine. See AMMONAL.

Antitrem. Trihexyphenidyl Hydrochloride. Applications: Anticholinergic; antiparkinsonian. (J B Roerig, Div of Pfizer Laboratories).

Antivert. Meclizine hydrochloride. Applications: Anti-emetic. (J B Roerig, Div of Pfizer Laboratories).

Antoban. A veterinary anthelmintic. (Coopers Animal Health, subsidiary of the Wellcome Foundation Ltd).*

Antoin. A proprietary preparation of aspirin, calcium carbonate, citric acid, phenacetin, codeine phosphate and caffeine citrate. An analgesic. (Cox, A H & Co Ltd, Medical Specialities Divn).†

Anton N. A trade mark for a general-purpose antioxidant. It is an octylated diphenylamine used in many elastomers. (Du Pont (UK) Ltd).†

Antox N. Octylated Diphenylamine. A registered trade name for a general purpose antioxidant with only mild discolouring and staining characteristics. (Du Pont (UK) Ltd).†

Antoxyl. A preparation consisting mainly of sodium hyposulphite. A preservative for perfumed soaps.†

Antozite. Phenylene diamines - rubber antiozonants. Applications: Used in NR, IR, BR, SBR and CR rubbers as antiozonants. (Vanderbilt Chemical Corporation).*

Antracol. A wettable powder containing 70% propinals. Applications: Controls potato blight, hop downy mildew, apple scab, leafspot on celery, blackcurrants and gooseberries, downy mildew on grapes and suppression of yellow rust on winter wheat. (Bayer & Co).*

Antraderm. A proprietary preparation of dithranol. Applications: A dermatological product. (Brocades).*

Antrenyl Duplex. A proprietary preparation of oxyphenonium bromide. Gastrointestinal sedative. (Ciba Geigy PLC).†

Antromid-S. Clofibrate. Applications: Antihyperlipoproteinemic. (Ayerst Laboratories, Div of American Home Products Corp).

Anturan. A proprietary preparation of sulphinpyrazone. An antirheumatic. (Ciba Geigy PLC).†

Anturane. Sulphinpyrazone. Applications: Uricosuric. (Ciba-Geigy PLC).

Anugesic. A proprietary preparation of boric acid, balsam of Peru, bismuth oxide, bismuth subgallate, resorcinol, zinc oxide and pramoxine. Analgesic suppositories. (Warner).†

Anugesic-HC. A proprietary preparation of PRAMOXINE hydrochloride, HYDROCORTISONE acetate, zinc oxide, Peru balsam, benzyl benzoate, bismuth oxide and resorcinol, used in the treatment of haemorrhoids. (Warner).†

Anusol. A proprietary preparation containing boric acid, zinc oxide, bismuth subgallate, bismuth oxide, Peru balsam and resorcinol. An antiprunitic. (Warner).†

Anusol-HC. A proprietary preparation of benzyl benzoate, bismuth subgallate, bismuth oxide, resorcinol, Peru balsam, zinc oxide and hydrocortisone acetate, used in the treatment of haemorrhoids. (Warner).†

Anvil Brass. See BRASS.

Anxine. Tablets of dexamphetamine sulphate, cyclobarbitone and mephenesin. (Allen & Hanbury).†

Anzon. Flame retardant compositions. (Anzon Ltd).

Aolept. A neuroleptic. (Bayer & Co).*

AOR/GR. A proprietary name for technically-comminuted rubber. (Societ Indochine de Plantations d'Hvas (SIPH), Ivory Coast).†

Aosoft. Tetrafilcon A. Applications: Contact lens material. (American Optical Corp).

Apagallin. Tetraiodophenol-phthalein. An antiseptic and indicator.†

Apallagin. See NOSOPHENE.

Aparatine. See VEGETABLE GLUE.

Apatef. Injectable antibiotic, containing cefotetan. (ICI PLC).*

Aperione. Phenolphthalein.†

Apernyl. Socket pellets for prophylactic and therapeutic care of extraction wounds. (Bayer & Co).*

A-Petroleum Naphtha. (Cleaning Oil). That fraction of petroleum distilling at 120-150° C.†

Apex 400. A proprietary alloy of aluminium with silicon.†

Apexior. Heat-resistant organic coating for water-side corrosion protection of steam generating equipment and auxiliaries. Applications: Steam generating equipment, feed water heaters, evaporators, steam turbines, diesel cylinder liners, condensate tanks. (Dampney Company Inc.)*

Aphox. Aphicide containing pirimicarb. (ICI PLC).*

Aphrogene. Premetallized dyes. (ICI PLC).*

Aphthite. An alloy of 800 parts copper, 25 parts platinum, 10 parts tungsten, and 170 parts gold.†

Aphtite. Zinc and cadmium-containing nickel bronzes. Used for high-grade imitation silver products.†

Apiezon. A range of high quality oils, waxes and greases prepared by molecular distillation of low volatility hydrocarbon feedstocks. The range was originally developed for high vacuum applications but they find use in several other sectors of industry. (Shell Chemicals).†

Apigenine. A yellow dyestuff obtained by decomposing the glucoside apiine found in parsley.†

Apiol. (Apiol, White). The crystalline constituent of parsley oil. Liquid apiol is essential oil derived from an apiol-bearing variety of parsley. Sometimes used as a diuretic.†

Apisate. A proprietary preparation containing diethylpropion, aneurine hydro-chloride, riboflavine, pyridoxine hydro-chloride and nicotinamide. An anti-obesity agent. (Wyeth).†

APL. Gonadotropin, chorionic. Applications: Gonad-stimulating principle. (Ayerst Laboratories, Div of American Home Products Corp).

Aplisol. Tuberculin. Applications: Diagnostic aid (dermal reactivity indicator). (Parke-Davis, Div of Warner-Lambert Co).

Aplitest. Tuberculin. Applications: Diagnostic aid (dermal reactivity indicator). (Parke-Davis, Div of Warner-Lambert Co).

Apollo Red. (Archil Substitute Extra, Orchil Extract N Extra, Orchil Substitute N Extra). An acid dyestuff. It is the sodium salt of *p*-nitrophenylazo-α-naphthylaminedisulphonic acid. Dyes wool red from an acid bath.†

Apolloy. A proprietary copper-iron alloy containing 0.25 per cent copper and 0.08 per cent carbon.†

Apolomine. A proprietary preparation of hyoscine hydrobromide, benzocaine, riboflavine, pyridoxine hydrochloride, and nicotinamide. An antinauseant. (Bayer & Co).*

Aponal. Amyl carbamate, $NH_2COOC_5H_{11}$. A somnifacient.†

Aposafranine. A dyestuff. It is diazotized safranine boiled with alcohol.†

Aposet 707. A proprietary trade name for a ketone peroxide catalyst for polyesters.†

A.P.P. Stomach Powder. A proprietary preparation containing calcium carbonate, magnesium carbonate, magnesium trisilicate, bismuth carbonate, aluminium hydroxide gel, homatropine methylbroroide, papaverine and phenobarbitone. An antacid. (Consolidated Chemicals).†

Apperitive Saffron of Iron. Ferric subcarbonate.†

Appretan. Washfast finishing agents. (Hoechst UK Ltd).

Appretan Ant. A proprietary dispersant surfactant for finishing woven fabrics. An acrylate-based copolymer. (Alma Paint and Varnish Co, Ontario).†

Appretan CPF. A proprietary poly-vinyl acetate dispersion surfactant used for finishing woven, non-woven and knitted fabrics. (Alma Paint and Varnish Co, Ontario).†

Appretan GM. A proprietary poly-vinyl-acetate-based dispersion surfactant sed in the finishing of especially light-weight fabrics, knitted fabrics and non-wovens. (Alma Paint and Varnish Co, Ontario).†

Appretan TN. A proprietary dispersion surfactant for finishing woven, non-woven and knitted fabrics. A vinyl acetate copolymer. (Alma Paint and Varnish Co, Ontario).†

Appreteen. Water soluble size. (S & D Chemicals Ltd).†

Apresoline. A proprietary preparation of hydralazine hydrochloride. Anti-hypertensive. (Ciba Geigy PLC).†

Apresoline Hydrochloride. Hydralazine hydrochloride. Applications: Antihypertensive. (Ciba-Geigy Corp).

Aprinox. A proprietary preparation of bendrofluazide. A diuretic. (Boots Company PLC).*

Aptine. Alprenolol hydrochloride. Applications: Anti-adrenergic. (Astra Pharmaceutical Products Inc).

Apyonin. See PYOCTANIN.

Apyrogen. A proprietary preparation of pyrogen-free water for injection. (Searle, G D & Co Ltd).†

Aqua fortis. Nitric acid, HNO_3.†

Aqua Gro "G" Granular. Blended non-ionic soil wetting agent. 100% active Aqua-Gro 40% wt., (polyoxyethylene ester of cyclic acids - 47%, polyoxyethylene ether of alkylated phenols - 47%, silicone antifoam emulsion - 6%), vermiculite - 60% wt. Applications: Granular wetting agent to aid water penetration and drainage. Used by greenhouses, nurseries amd interior plantscapers in the manufacture of horticulture growing and potting media. (Aquatrols Corporation of America).*

Aqua Ivy, AP. Poison Ivy Extract, Alum Precipitated. Applications: Ivy poisoning counteractant. (Miles Pharmaceuticals, Div of Miles Laboratories Inc).

Aqua Magic. Silicones. Applications: Stain preventive for fabrics. (Adasco-Inc).*

Aqua Mer. A totally aqueous dry film photoresist, which comprises a three layer sandwich construction of polyolefin/photopolymer/polyester. (Hercules Inc).†

Aqua regia. A mixture of one volume of nitric acid and three volumes of hydrochloric acid.†

Aqua Thix. Modified polysaccharide thixotrope. Applications: Thickening agent and protective colloid for water-based coating compositions. (Nueodex Inc).*

Aqua-Gro "L" Liquid. Blended non-ionic soil wetting agent. Polyoxyethylene esters of cyclic acids - 47%, polyoxyethylene esters of alkylated phenols - 47%, silicone antifoam emulsion w- 6%. Applications: Increased water penetration into and out of soils and horticultural media. Used in manufacture of horticultural growing and potting media; on golf courses, sports turf, lawns and exterior landscapes; drainage of compacted soils and puddles; *poa annua* seedhead inhibition; spreader-activator; adjuvant; hydroseeding; dew and frost control. (Aquatrols Corporation of America).*

Aqua-Gro "S" Spreadable. Blended non-ionic soil wetting agent. 100% active Aqua-Gro -15% wt; (polyoxyethylene ester of cyclic acids - 47%, polyoxyethylene ether of alkylated phenols - 47%, silicone antifoam emulsion 6%) ground corn cobs - 85%. Applications: Granular soil wetting agent to aid water penetration and drainage. Used on golf courses, sports turf, lawns and exterior landscapes. (Aquatrols Corporation of America).*

Aquabase. Waterborne automotive paints. (ICI PLC).*

Aquacillin. Penicillin G Procaine. Applications: Antibacterial. (Schering-Plough Corp).

Aquaclene. Tablets containing chloramine. (British Drug Houses).†

Aquacoat. Water-based overlacquers. (The Scottish Adhesives Co Ltd).*

Aquadag. Colloidal graphite in water used as a lubricant for drawing tungsten and molybdenum filament wires, for metal forming operations such as extrusion, as an aid to cutting and for forming electrically conducting coatings. (Acheson Colloids).*

Aquadrome. Swimming pool water disinfectant. (Great Lakes Chemical (Europe) Ltd).

Aquaflex. Tetrafilcon A. Applications: Contact lens material. (UCO Optics Inc).

Aquafloc. Flocculants. (Dearborn Chemicals Ltd).

Aquaforte. Water-based primer/adhesive with low solids and containing no organic solvents. Applications: Extrusion primer for film to film, paper and foil, adhesion promoter of inks to foil. (ADM Tronics Unlimited Inc).*

Aquagel. A proprietary trade name for a colloidal bentonite. Also a proprietary hydrated silicate of alumina for water-proofing cement.†

Aqualac. A proprietary shellac.†

Aqualite. A proprietary phenol-formaldehyde synthetic resin laminated product and bearing material requiring only water as lubricant.†

Aqualon. An absorbent fibre which comprises an internally cross-linked form of sodium carboxymethylcellulose. (Hercules Inc).†

Aqualose. Alkoxylated lanolin and lanolin derivatives. Applications: Water-soluble emollients, o/w emulsifiers and wetting agents for cosmetics. (Westbrook Lanolin Co).*

Aqualube. A proprietary plasticizer for water soluble materials.†

Aquamephyton. Phytonadione. Applications: An aqueous colloidal solution of vitamin k_1 for parenteral administration. (Merck Sharp & Dohme).*

Aquamet M. A liquid formulation suitable for precipitating metal ions from solution. Applications: Waste water treatment. (Alco Chemical Corporation).*

Aquanol. Liquid blend of polymeric resins and modified silxcane in a petroleum solvent base. Applications: Used as an internal sealer and an external water repellant for concrete, brick, treated concrete block, stucco, wood and stone. (Secure Inc).*

Aquapel. Fatty acid ketene dimer sold in aqueous emulsions either alone or in combination with other chemicals. Applications: Neutral sizing agent for paper and paperboard. (Hercules Inc).*

Aquaperle. Textile auxiliary chemicals. (ICI PLC).*

Aquaphoril. SEE DIUREXAN. (Degussa).*

Aquaplex. A proprietary trade name for alkyd resin dispersed in an aqueous medium. An emulsion varnish and lacquer resin vehicle for stucco, etc.†

Aquapol. Hydrophilic urethane prepolymer. (Freeman Chemical Corp, Subsidiary of H H Robertson Co).

Aquaprint. Mouldable impression material, synthetic resins. (BP Chemicals Ltd).*

Aquaresin. A proprietary plasticizer. It is glyceryl boriborate.†

Aquarite. Water treatment chemicals. (Albright & Wilson Ltd, Phosphates Div).

Aquaseal. A proprietary trade name for a cold applied plastic bitumen for waterproofing and sealing.†

Aquasil. Window desiccants. (Laporte Industries Ltd).

Aquasoft. Water based textile ink. Applications: Direct silk screen for T-shirts, athletic garments, aprons, tote bags, draperies, tablecloths, caps, wallhangings. Adhesion is excellent to cotton, blends and most synthetic fabrics. Always test print before production run. (International Coatings).*

Aquasol A. Vitamin A. Applications: Vitamin (anti- xerophthalmic). (Armour Pharmaceutical Co).

Aquasol E. Vitamin E. Applications: Vitamin E supplement. (Armour Pharmaceutical Co).

Aquasperse. Colourant dispersions. Applications: Colouring of emulsion and other water-borne coating compositions. (Nueodex Inc).*

Aquastore. Cross-linked polyacrylamide. (Cyanamid BV).*

Aquasun. Sun protection products, to protect skin while promoting a tan. (Richardson-Vicks Inc).*

Aquatec. A proprietary trade name for a waterproofing material. It is a paraffin wax emulsion sometimes with aluminium acetate.†

Aquatreat AR-225-D. A free flowing low molecular weight sodium poly-methacrylate. Applications: Dispersant and desludger in water systems, cooling towers, boilers and heat exchangers. (Alco Chemical Corporation).*

Aquatreat AR-232. A low molecular weight sodium polymethacrylate solution. Applications: Dispersant and desludger in water systems, cooling towers, boilers and heat exchangers. (Alco Chemical Corporation).*

Aquatreat AR-626. A low molecular weight acrylate copolymer solution. Applications: Scale prevention in cooling towers. boilers, heat exchangers, oil field applications. (Alco Chemical Corporation).*

Aquatreat AR-648. A polycarboxylate copolymer solution. Applications: Scale prevention in cooling towers, boilers and heat exchangers. (Alco Chemical Corporation).*

Aquatreat AR-7-H. A high molecular weight acrylic acid polymer. Applications: Adhesives, lithographics, latex stabilization. (Alco Chemical Corporation).*

Aquatreat AR-900. A low molecular weight sodium polyacrylate. Applications: Scale prevention in cooling towers and heat exchangers. (Alco Chemical Corporation).*

Aquatreat DNM-30. A fungicide/bactericide composition of sodium dimethyldithio-carbamate and Nabam (EPA Reg No 31910-2). Applications: Controlling bacteria and fungi growth in industrial recirculating water cooling towers, pulp mills, paper mills and can sugar mills. (Alco Chemical Corporation).*

Aquazym. Alpha-amylase produced by submerged fermentation of a selected strain of Bacillus subtilis. Applications: Intended for use in the desizing of textiles. (Novo Industri A/S).*

Arabin. See GUM ARABIC.

Aracast. Heterocyclic epoxide resins. (Ciba-Geigy PLC).

Araldite. A proprietary trade name for a series of epoxy resins used for casting, encapsulating, laminating, surface coating and as an adhesive. (Ciba Geigy PLC).†

Aralen Hydrochloride. Chloroquine hydrochloride. Applications: Antiamebic; antimalarial. (Sterling Drug Inc).

Aralen Phosphate. Chloroquine phosphate. Applications: Antimalarial; antiamebic; suppressant. (Sterling Drug Inc).

Aramid Fibre. See KEVLAR.

Aramine. Metaraminol. Applications: Vasopressor agent for the treatment of acute shock. (Merck Sharp & Dohme).*

Aranox. p-(p-toluenesulphonylamido)-diphenylamine. Semi-non-discolouring antioxidant for natural, SBR and neoprene. It is a strong inhibitor of the pro-oxidant metals such as copper and manganese. Used in balloon fabrics, proofing, clothing, light-coloured sundries and wire insulation. (Uniroyal).*

Arazate. Zinc dibenzyldithiocarbamate. An ultra-accelerator, fast curing, low critical temperature, medium curing range up to 250°F (121°C). Non-discolouring and non-staining. Used in latex for adhesives, dipping and other compounds which require frequent replenishing with fresh material. Has exceptional resistance to precure in presence of ammonia. (Uniroyal).*

Arbacet. Arbaprostil. Applications: Antisecretory. (The Upjohn Co).

Arbeflex. Plasticisers. (Robinson Brothers Ltd).*

Arbeflex 489. A heat and light and secondary plasticizer for PVC. Oxirane oxygen 3.3 per cent. min. Iodine number 3 max; viscosity 98 cS/25° C. (Robinson Brothers Ltd).*

Arbeflex 550. A low priced moderate performance plasticizer for PVC. Sap. value 380 mg KOH/gm. Acid value 1 mg KOH/gm. (Robinson Brothers Ltd).*

Arbestab. Antioxidants and UV-stabilizers for polymers. Applications: Useful in a number of plastics materials but particularly effective in polyolefins. (Robinson Brothers Ltd).*

Arbo. Generic name for a range of putties, mastics and sealants. (Adshead Ratcliffe & Co Ltd).*

Arbocaulk. An acrylic, emulsion based sealant for gun application. Applications: Principal use for internal pointing applications. (Adshead Ratcliffe & Co Ltd).*

Arbocel. Wood cellulose. (ICI PLC).*

Arbocrylic. An acrylic, solvent based sealant for gun application. Applications: Principal use is for sealing external joints in building structures. (Adshead Ratcliffe & Co Ltd).*

Arboflex. A glazing compound based on a blend of vegetable oils, plasticisers and butyl rubber. Applications: Used for bead glazing aluminium and sealed timber window frames. (Adshead Ratcliffe & Co Ltd).*

Arbofoam. Polyurethane foam packed in an aerosol dispenser. Applications: Used to seal and insulate gaps around pipes and duct work and as a fixing, gap filling adhesive for doors and windows. (Adshead Ratcliffe & Co Ltd).*

Arbogard. Herbicides. (ICI PLC).*

Arbokol. A range of single and two component, polysulphide and epoxy/polysulphide sealants. Applications: Sealants in building joints, floor joints and double glaze unit construction. (Adshead Ratcliffe & Co Ltd).*

Arbolite. A putty composition based on a blend of vegetable oils. Applications: Used for face glazing steel window frames. (Adshead Ratcliffe & Co Ltd).*

Arbomast. A range of gun applied sealants based on vegetable oils or butyl rubber. Applications: Low cost, general purpose, sealants for a range of applications. (Adshead Ratcliffe & Co Ltd).*

Arborsan. Creosote and wood preservatives for timber. (Lancashire Tar Distillers Ltd).

Arboseal. A range of preformed mastics strips based on butyl rubber and polybutenes. Applications: Used for making watertight and dustproof seals between components, where the joint is under compression. (Adshead Ratcliffe & Co Ltd).*

Arbosil. A range of RTV, silicone based single component sealants.

Applications: Sealing a wide range of industrial and building applications between a variety of substrates. (Adshead Ratcliffe & Co Ltd).*

Arbostrip. Self adhesive foam strips based on plasticized PVC. Applications: Compression sealants for draught proofing and similar applications. (Adshead Ratcliffe & Co Ltd).*

Arbyl. Dispersing and levelling agent. Applications: Used for dyeing, a wetting agent and detergent for pretreatment, designing and dyeing in the textile industry. (Degussa).*

Arbylen. A wetting agent and detergent used for pretreatment, desizing and dyeing in the textile industry. (Degussa).*

Archiardite. Synonym for Dachiardite.†

Archibald's Stain. A microscopic stain. Solution A contains 0.5 gram thionin, 2.5 grams phenol, 1 cc formalin, and 100 cc distilled water, and solution B contains 0.5 gram methylene blue, 2.5 grams phenol, 1 cc formalin, and 100 cc distilled water. For use mix equal parts of A and B, and filter.†

Archil. (Orchil, Orseille, Persio, Cudbear, Orchellin). A natural colouring matter obtained from *Roccela tinctoria* and other lichens. The colouring principle is orcin, $C_6H_3(CH_3)(OH)_2$, which, in the presence of air and ammonia, oxidizes to a violet dye. orcein, $C_{28}H_{24}N_2O_7$. The alkali salts dye wool and silk. It is sold in three forms: (*a*) A pasty mass called archil; (*b*) a mass of drier character named persio; and (*c*) as a reddish powder known as cudbear or French indigo. Commercial brands of cudbear are Cudbear O, I, II, extra, fine, violet, red-violet, blue-violet and red. Orchellin is a product very rich in the dyestuff.†

Archil Brown. An azo dyestuff prepared by the action of diazotized amine azo-benzene-disulphonic acid upon α-naphthylamine, $HSO_3\ C_6H_4NH\ C_6H_3\ HSO_3\ N:N\ C_{10}H_6NH_2$.†

Archil Carmine. A dyestuff containing archil colouring matter in a state of great purity.†

Archil Purple. A dyestuff similar to archil carmine (*qv*).†

Archil Red A. An azo dyestuff, prepared by the action of diazotized amino-azo-xylene upon β-naphtholdisulphonic acid.†

Archil Substitute. See APOLLO RED

Archil Substitute N Extra. See APOLLO RED.

Archil Violet. See FRENCH PURPLE.

Archil, Earth. See EARTH ARCHIL.

Arcolloy. A non-magnetic alloy of iron with 12-16 per cent chromium, and less than 0.12 per cent carbon, 0.5 per cent manganese, 0.025 per cent phosphorus, 0.025 per cent sulphur, and 0.5 per cent silicon.†

Arcoloy. A proprietary copper-silicon casting alloy containing 97.25 per cent copper, 2.63 per cent silicon, 0.12 per cent iron, and 0.01 per cent phosphorus.†

Arctite Injection Mortar. Expanding cement. Applications: Brickwork - prevent rising damp (DPC). (Arcmann-Denmark A/S).*

Arctite Slurry 200 B. Waterproofing cement. Applications: Concrete - high water pressure. (Arcmann-Denmark A/S).*

Arctite Tanking Mortar 500. Cement based waterproofing compounds, non toxic. Applications: Concrete and brick structures. (Arcmann-Denmark A/S).*

Arcton. A range of fluorinated hydrocarbon refrigerants and aerosol propellants (fluorocarbons) (*qv*). (ICI PLC).*

Ardeer Powder. An explosive. It contains 31-34 per cent nitro-glycerin, 11-14 per cent kieselguhr, 47-51 per cent magnesium sulphate, 4-6 per cent potassium nitrate, and 0.5 per cent ammonium or calcium carbonate.†

Ardenite. A proprietary synthetic resin. A moulding composition.†

Ardinex. A proprietary preparation of guaiphenesin, methaqualone hydro-chloride and ephedrine hydrochloride. A bronchial antispasmodic. (Boots Company PLC).*

Ardmorite. A variety of Bentonite (*qv*), found in the Pierre shales at Ardmore, South Dakota.†

Ardux. Modified urea/formaldehyde resin. (Ciba-Geigy PLC).

Areca Red. A red colouring matter extracted from the areca nut.†

Arecaidine. N-methyl-Δ-3-tetrahydro-pyridine-3-carboxylic acid.†

Arecoline. The methyl ester of N-methyl-Δ-3-tetrahydropyridine-3-carboxylic acid.†

Arelon. Selective herbicide. (Hoechst UK Ltd).

Aremsol A. Coco amido betaine in liquid form. Applications: Foam booster for conventional shampoos, conditioning shampoos and conditioning rinses. (Ronsheim & Moore Ltd).

Arenka. A proprietary high-strength yarn manufactured from aramides (aromatic polyamides). (Enka Glanzstoff).†

Arenolite. An artificial siliceous-argillaceous-calcareous stone.†

Aresenid. Arsenic acid solution. Applications: Wood preservative. (Mechema).*

Aresin. Herbicide for potatoes and leeks. (Hoechst UK Ltd).

Areskap. A proprietary trade name for a butyl-phenyl-phenol sodium sulphonate.†

Aresket. A proprietary trade name for a wetting agent. It is stated to be a butyl-diphenyl sodium sulphonate.†

Aresklene. A proprietary trade name for a dibutyl-phenyl-phenol sulphonate. A mould lubricant and emulsifier.†

Aretan. A mercurial fungicide. (ICI PLC).*

Aretone 270. A proprietary trade name for a fine mica used as a pigment in paints.†

Arfonad. A proprietary preparation of the (+)-camsylate of trimetaphan. A vasodilator. (Roche Products Ltd).*

Argal. See ARGOL.

Argasoid. A low-grade nickel silver.†

Argentai. An alloy of 85 per cent copper, 10 per cent tin, and 5 per cent cobalt.†

Argentalium. An aluminium alloy containing antimony.†

Argentan. A nickel silver. It consists of 56 per cent copper, 26 per cent nickel, 18 per cent zinc, and 1 per cent iron, and is used as an electrical resistance alloy.†

Argentichthol. See ICHTHARGAN.

Argentum Metal. See METAL ARGENTUM.

Argilla. See CHINA CLAY.

Argipressin. An anti-diuretic hormone.†

Argiroide. See NICKEL SILVERS.

Argobase. W/o absorption bases containing lanolin and/or lanolin alcohols. Applications: Emulsifiers for cosmetics and ointments. (Westbrook Lanolin Co).*

Argol. (Argal). A crystalline crust deposited on the sides of the vat in which grape juice has been fermented. It contains 40-70 per cent tartaric acid, principally as potassium hydrogen tartrate.†

Argol, Red. See ARGOL.

Argol, White. See ARGOL.

Argonin L. A silver caseinate containing 10 per cent silver.†

Argonol. Liquid lanolin derivatives. Applications: Fluid emollients and moisturizers for cosmetics. (Westbrook Lanolin Co).*

Argotone. A proprietary preparation of ephedrine hydrochloride, and silver vitellin. Nasal drops. (Rona Laboratories).†

Argowax. Refined lanolin alcohols. Applications: W/o emulsifier in cosmetics and pharmaceuticals. (Westbrook Lanolin Co).*

Argozie. (Arguzoid, Argozoil). An alloy of from 54-56 per cent copper, 23-38 per cent zinc, 2-4 per cent tin, 2-3.5 per cent lead, and 13.5-14 per cent nickel.†

Arguzoid. An alloy of 56 per cent copper, 23 per cent zinc, 4 per cent tin, 3.5 per cent lead, and 13.5 per cent nickel.†

Arguzoil. See ARGOZIE.

Argyrolith. (China Silver, Electroplate). Names given to alloys containing 50-70 per cent copper, 10-20 per cent nickel, and 5-30 per cent zinc. Alfenide and Argentan are similar alloys. They are nickel silvers (German silvers).†

Ariabel. Inorganic and organic cosmetic pigments. Applications: Colouring of cosmetic products. (Williams Div of Morton Thiokol Ltd). *

Ariagran. High strength water soluble, granular food colours. Applications: Colouring of foodstuffs and pharmaceuticals. (Williams Div of Morton Thiokol Ltd). *

Arianor. Semi permanent, water soluble hair colours. Applications: Incorporated in hair products intended to colour hair. (Williams Div of Morton Thiokol Ltd).*

Ariavit. High strength water soluble powder food colours. Applications: Colouring of foodstuffs and pharmaceuticals. (Williams Div of Morton Thiokol Ltd).*

Aricel. Melamine-formaldehyde resins - usually 80% solids. Applications: Cross-linking polymer systems, textile finish, filter papers, non-woven binder. (Astro Industries Inc).*

Aricyl. Injectable deodorant and tonic. Applications: Veterinary medicine. (Bayer & Co).*

Aridex. A retexturing and reproofing aid. (Laporte Industries Ltd).*

Arigal PMP. A proprietary trade name for a solution of an organic mercuric compound which, when used together with Arigal C, imparts a mildew resistant and rot-proof finish on cellulosic fibres; fast to water, washing and dry cleaning. (Ciba Geigy PLC).†

Arigran. Granular food colours of guaranteed purity. (Williams (Hounslow) Ltd).

Arilvax. A proprietary vaccine against yellow fever. (The Wellcome Foundation Ltd).*

Aripol. Poly-electrolyte flocculating agent. (Steetley Chemicals Ltd).

Aristar. Ultrapure reagents & solvents. (BDH Chemicals Ltd).

Aristocort. Triamcinolone. Applications: Glucocorticoid. (Lederle Laboratories, Div of American Cyanamid Co).

Aristocort Acetonide. Triamcinolone Acetonide. Applications: Glucocorticoid. (Lederle Laboratories, Div of American Cyanamid Co).

Aristocort Forte Parenteral. Triamcinolone Diacetate. Applications: Glucocorticoid. (Lederle Laboratories, Div of American Cyanamid Co).

Aristocort Syrup. Triamcinolone Diacetate. Applications: Glucocorticoid. (Lederle Laboratories, Div of American Cyanamid Co).

Aristoflex. Hair lacquer resins. (Hoechst UK Ltd).

Aristol, Naphthol. See NAPHTHOL ARISTOL.

Aristospan. Triamcinolone Hexacetonide. Applications: Glucocorticoid. (Lederle Laboratories, Div of American Cyanamid Co).

Arizole Anethole Extra. Anethole. Applications: Anise flavouring material. (Arizona Chemical Company).*

Arizole Pine Oil. Pine oil. Applications: Disinfectants and cleaners, odourant, frothing agent in mineral flotation, solvent. (Arizona Chemical Company).*

Arizona DRS-40, DRS-42, DRS-43, DRS-50, DRS-51E Disproportionated Tall Oil Rosin Soaps. One sodium, four potassium soaps of disproportionated rosin. Total solids range 71%-84.5%, viscosity range 450-1700 cps at 75°C. Applications: Used as emulsifiers in SBR, ABS, neoprene and solution butadiene production. (Arizona Chemical Company).*

Arizona DR22 Disproportionated Tall Oil Rosin. Colour WW, softening point 66°C. Applications: Preparation of emulsifiers. Widely used in SBR production. (Arizona Chemical Company).*

Arizona DR24 Disproportionated Tall Oil Rosin. Colour WG-N, softening point 47°C. Applications: Preparation of emulsifiers. Widely used in SBR production. (Arizona Chemical Company).*

Arizona DR25 Disproportionated Tall Oil Rosin. Partically saponified anhydrous sodium soap, colour WG, softening point 75°C. Applications: Preparation of emulsifiers. Widely used in SBR production. (Arizona Chemical Company).*

Arizona Shellac. See SONORA GUM.

Arizona 208 Tall Oil Fatty Acid Ester. Isooctyl ester of tall oil fatty acid. Applications: As a low temperature plasticizer for chlorprene nitril and **Hypalon** elastomer systems. A replacement for higher-priced sebacates, oleates and adipates. (Arizona Chemical Company).*

Arizona 258 Tall Oil Fatty Acid Ester. 2-ethylhexyl ester of tall oil fatty acid. Applications: As a low temperature plasticizer for chlorprene nitril and **Hypalon** elastomer systems. A replacement for higher-priced sebacates, oleates and adipates. (Arizona Chemical Company).*

Arklone. Trichlorotrifluoroethane, a solvent for cleaning. (ICI PLC).*

Arko Metal. An alloy of 80 per cent copper, and 20 per cent zinc.†

Arkopal. Alkylphenol polyglycol detergent bases. (Hoechst UK Ltd).

Arkopal N. Range of nonionic surfactants of the nonylphenol ethoxylate type in liquid, paste or wax form. Application: Wetting, dispersing, foaming, emulsification, detergent and cleaning agent used in domestic and industrial cleaners, disinfectant cleaners, auxiliaries for textile, leather and fur dressing, metal working, rubber, electroplating, pesticides, plant protection, building industry, anti-dusting and many other uses. (Hoechst UK Ltd).

Arkopon T. Sodium oleoyl methyl tauride in powder form. Applications: Dispersing and wetting agent used in wettable powders, plant protection and pest control. (Hoechst UK Ltd).

Arlacel. Sorbitan or glycerine fatty acid esters. A non-ionic surfactant. (ICI PLC).*

Arlacel 80. Sorbitan Monooleate. Applications: Pharmaceutic aid - surfactant. (ICI Americas Inc).

Arlacel 83. Sorbitan Sesquioleate. Applications: Pharmaceutic aid - surfactant. (ICI Americas Inc).

Arlacel 85. Sorbitan Trioleate. Applications: Pharmaceutic aid - surfactant. (ICI Americas Inc).

Arlacide. Chlorhexidine salts. (ICI PLC).*

Arlagard. Bacteriastat. (ICI PLC).*

Arlamol. Blends of specific fatty acid ester A non-ionic surfactant. (ICI PLC).*

Arlamol E. Polyoxyproplene 15 Stearly Ether. Applications: Pharmaceutical aid. (ICI United States Inc).

Arlatone. Polyoxyethylene glycerine fatty acid or fatty acid esters. (ICI PLC).*

Arlatone 507. Padimate O. Applications: Ultraviolet screen. (ICI Americas Inc).

Arlef-100. A proprietary preparation of flufenamic acid. An antirheumatic. (Parke-Davis).†

Arlidin. Nylidrin hydrochloride. Applications: Vasodilator. (USV Pharmaceutical Corp).

Arlin. A proprietary name for polyethylene film. (Poly-Version Inc, Tulsa, Oklahoma).†

Arlinflex. Rigid vinyl. (Arlington Mills).

Arlix. Piretanide. Applications: Diuretic. (Hoechst-Roussel Pharmaceuticals Inc).

Armac. Series of cationic surfactants which consist of acetate salts of primary amines, and which are soluble. Applications: Pigment flushing, froth flotation and flocculation, particularly in mineral flotation; petroleum processing; leather; paper; pigments and surface coatings; ceramics. (Akzo Chemie UK Ltd).

Armco. A trade mark for ingot iron and stainless steel.†

Armco Ingot Iron. A trade mark for a very pure iron, 99.84 per cent pure.†

Armeen. Range of cationic surfactants composed mainly of primary amines, in liquid, paste or solid form. Applications: Degree of surface activity varies with composition. They have the ability to change mineral surfaces from hydrophobic to hydrophilic. Used in petroleum; road emulsions; plastics; rubber; textiles; leather; herbicides; fungicides; rodent repellents; mineral flotation; paper; pigments and surface coating; water and sewage treatment; wax; sealant formulations; cement curing. (Akzo Chemie UK Ltd).

rmeen. A range of coco, oleyl and stearyl amines. Applications: Used as chemical intermediates, anti-caking agents and in secondary oil recovery. (Harcros Industrial Chemicals).*

rmeen DM Series. Cationic surfactants of the tertiary amine type, in liquid form. Applications: Manufacture of amine salts eg acetates, quaternary ammonium salts, amine oxides, substituted betaines; catalysts in flexible foams; lubricating oils, especially as acid scavengers; DMHTD used in electrophoretic coatings; DMCD used as a polymerization inhibitor in butadiene plants and as corrosion inhibitor in monochloracetic acid plants. (Akzo Chemie UK Ltd).

rmenian Bole. See INDIAN RED.

rmenian Cement. A jeweller's cement containing gum mastic, isinglass, gum ammoniac, alcohol, and water. It is made by soaking the isinglass in water and mixing it with the spirit containing the gums.†

rmid. Long chain aliphatic amides. (Akzo Chemie UK Ltd).

rmide. See CAMITE.

rmids. Long chain aliphatic amides. (Armour Hess Chemicals).†

rmite. Vulcanized fibre. (Spaulding Fibre Co).

rmix 146. Formulated product. Applications: Speciality product designed to enhance the effectiveness of MSMA formulations. (DeSoto Inc).*

rmix 176. Formulated product. Applications: Tank mix which provides a combination of wetting, sticking, spreading and penetration. (DeSoto Inc).*

Armofilm. Long chain filming amine emulsion. (Akzo Chemie UK Ltd).

Armoflo. Conditioner - hygroscopic salts and fertilizers. (Akzo Chemie UK Ltd).

Armogard. Fuel oil additives. (Akzo Chemie UK Ltd).

Armogel. Thickening agent. (Akzo Chemie UK Ltd).

Armogloss. Cationic car wash additive. (Akzo Chemie UK Ltd).

Armohib. Acid inhibitors. (Akzo Chemie UK Ltd).

Armohib 25 and 28. Cationic surfactants consisting of aliphatic nitrogen derived materials, in the form of an amber liquid. Applications: Inhibitors used in acid pickling; plant cleaning; oil well acidizing. 25 developed for use with sulphuric, phosphoric, citric and sulphamic acids; 28 for use with hydrochloric acid. (Akzo Chemie UK Ltd).

Armor-ply. A proprietary trade name for metal bonded plywood.†

Armostat. Antistatic agent for plastics. (Akzo Chemie UK Ltd).

Armoteric LB. Amphoteric surfactant supplied as a yellowish liquid. Applications: Baby shampoos; bubble baths; strong acid and alkaline cleaning detergents. (Akzo Chemie Nederland BV).

Armoteric SB. Amphoteric surfactant supplied as a yellowish paste. Applications: Baby shampoos; bubble baths; strong acid and alkaline cleaning detergents. (Akzo Chemie Nederland BV).

Armourcote. Unreinforced fluorocarbon coatings and coatings reinforced with stainless steel, molybdenum and ceramic. Applications: Low friction and non-stick surfaces in baking and food processing, and general industrial applications. (Fothergill Tygaflor Ltd).*

Armowax. Synthetic waxes. (Armour Hess Chemicals).†

Armowax EBS. A proprietary trade name for N-N'ethylene bis-stearamide. (Armour Hess Chemicals).†

Armstrong Acid. See SCHAEFFER'S ACID

Armstrong and Wynne's Acid. 1-Naphthol-3-sulphonic acid. †

Armstrong's δ-Acid. 1,5-Naphthalene disulphonic acid.†

Armul 17. Formulated product. Applications: Emulsifier for paraffinic hydrocarbon crop oils. (DeSoto Inc).*

Armul 22, Armul 44, Armul 66, Armul 88. Blended emulsifiers. Applications: Emulsifiers which, when used in combinations, can be used to formulate a wide variety of pesticide products. (DeSoto Inc).*

Armyl. A proprietary trade name for Lymecycline.†

Arnatto. See ANNATTO.

Arnaudon Green. A green pigment prepared by stirring up 128 parts ammonium phosphate and 149 parts potassium bichromate with water to a paste, and heating the mixture to 170-180° C. See CHROMIUM GREEN.†

Arneel DN. A proprietary trade name for the dimerized product of octadecene and octadecadiene nitriles. A vinyl plasticizer. (Armour Pharmaceutical Co).†

Arneel HF. A proprietary trade name for the 18, 20 and 24 carbon atom fatty acid nitriles. Vinyl plasticizers. (Armour Pharmaceutical Co).†

Arneel S. A proprietary trade name for a derivative of octadecene and octadecadiene nitriles. A vinyl plasticizer. (Armour Pharmaceutical Co).†

Arneel TOD. A proprietary trade name for a derivative of octadecene and octa decadiene nitriles. A vinyl plasticizer. (Armour Pharmaceutical Co).†

Arnica Yellow. An azo dyestuff prepared by condensing *p*-nitrotoluenesulphonic acid with *p*-aminophenol, in the presence of aqueous caustic soda. Dyes cotton golden-yellow from a salt bath.†

Arnite A.K.U. A trade name for a polyethylene glycol terephthalate injection moulding material. (Algemene Industriele).†

Arnite G. Thermoplastic polyester grade for injection moulding and extrusion. (Algemene Industriele).†

Arnotto. See ANNATTO.

ARO. Asbestos and non-asbestos friction material. Applications: Friction material for brakes and clutches. (Caramba Chemie GmbH).*

Arochem. (Aroplax). A proprietary trade name for soft oil modified alkyds.†

Arofene. Formaldehyde polymers with phenol and substituted phenols usually supplied in solvent other than water. Supplied in drums and tank wagon. Applications: Paper impregnation: Air, oil and fuel filters. Fibre bonding: Non-wovens of all types. Laminates: Rolled and flat stock. Adhesives: High performance. (Ashland Chemical Company).*

Aroflat. Polyester/alkyd resins. Polymers of polyhydric alcohols and polybasic acids which may have fatty acid components. Applications: Primary component of paints and surface coatings. (Ashland Chemical Company).*

Aroflat. Alkyd resins. Applications: Paint. (NL Industries Inc).*

Aroflint. Polyester-epoxy resins. Applications: Paint. (NL Industries Inc).*

Arofoam. Unsaturated polyester - two component system. Applications: Structural applications such as acrylic tubes and showers with fibreglass reinforcement. (Ashland Chemical Company).*

Arolon. Emulsions, dispensions and water reducible polymers. Applications: Paints and surface coatings. (Ashland Chemical Company).*

Arolon. Water reducible resins. Applications: Paint. (NL Industries Inc).*

Aromabator PC-80. A tamed and stabilized non-toxic chlorine dioxide complex concentrate formulated for use as an additive deodorant without free chlorine release. It is formulated for use as an air-borne spray for industrial applications. Applications: Effectively arrests mal-odours caused by virae, fungi, bacteria and coliform densities, including escherichia coli, klebsiella aerogenes and enterobacter aerogenes when added to coolants, cutting oils, industrial sumps, sludge pits, cooling towers,waste water, marine holding stations and ship's bilge areas, animal housing and rest room surfaces. It is formulated for spray applications. (Punati Chemical Corp).*

Aromabator PC-88. A tamed and stabilized non-toxic chlorine dioxide complex concentrate formulated for use as an additive deodorant without free chlorine release. It is formulated for use as an air-borne spray in home and farm applications. Applications: For use in kitchens, toilets, outhouses, pet and animal housing and can be safely sprayed on fabric and non-fabric surfaces and also on air conditioning and humidifier filters. (Punati Chemical Corp).*

Aromaplas. Range of perfumes for plastics. (PPF International Ltd).†

Aromasol. Solvents. (ICI PLC).*

Aromex. Powdered perfumery compounds. (Bush Boake Allen).*

Aromix. Solvent emulsifier concentrates for pesticide formulations. (Plant Protection (Subsidiary of ICI)).†

Aromox. Long chain aliphatic amine oxides. (Akzo Chemie UK Ltd).

Aron Alpha. A proprietary cyano-acrylate adhesive. (Toagosie Chemical Co).†

Aroplax. See Arochem.

Aroplaz. Alkyd resins. Applications: Paint. (NL Industries Inc).*

Aroplaz. Polyester/alkyd resins.. Polymers of polyhydric alcohols and polybasic acids which may have fatty acid components. Applications: Primary component of paints and surface coatings. (Ashland Chemical Company).*

Aropol. Unsaturated polyester resins, including orthophthalic, isophthalic and other specialty polymer types. Applications: Fibreglass reinforced polyester applications in construction, transportation, gel coats, marine, consumer, electrical and corrosion resistant markets. (Ashland Chemical Company).*

Aropol Phase Alpha. Low profile unsaturated polyester resins. Applications: Sheet moulding compound for compression moulding into Class A exterior automotive panels (hoods, roofs, deck lids) and truck panels (hoods, tilt cabs). (Ashland Chemical Company).*

Aropol Phase II. Low profile unsaturated polyester resins. Applications: Sheet moulding compound for compression moulding into Class A exterior automotive panels (hoods, roofs, deck lids) and truck panels (hoods, tilt cabs). (Ashland Chemical Company).*

Aropol WEP. Unsaturated polyester resins which can form water emulsions. Applications: Casting of decorative art and other applications. (Ashland Chemical Company).*

Aroset. Pressure sensitive acrylic polymers in both solution and emulsion forms. Applications: Pressure sensitive tapes, labels, decals, transfer films, foams etc. (Ashland Chemical Company).*

Arotap (Phenolic). Formaldehyde polymers with phenol and other modifiers supplied in water and non aqueous solvents. Supplied in drums and tank wagons. Applications: Paper coating and impregnation: Air, oil and fuel filters. Fibre bonding: Non wovens of all types. Laminates: Rolled and flat stock. Adhesives: High performance. (Ashland Chemical Company).*

Arotech. Acrylamate polymer. Applications: Fibreglass reinforced parts requiring superior strength properties for automotive and other applications. (Ashland Chemical Company).*

Arotex. Growth regulator containing chlormequat for use on wheat, oats or rye. (ICI PLC).*

Arothix. Alkyd resins. Applications: Paint. (NL Industries Inc).*

Arova 16. 1,4-Dioxacyclohexadecane-5,16-dione. Applications: Musk perfume. (Huls AG).†

Arovit. See RO-A-VIT. (Roche Products Ltd).*

Arpocox. Arprinocid. Applications: Coccidiostat. (Merck Sharp & Dohme, Div of Merck & Co Inc).

Arpylene. Propylene compounds. (Norsk Hydro Polymers Ltd).

Arquad. Range of cationic surfactants composed of alkyl quaternary ammonium chlorides in mainly liquid form. Applications: Effective in killing micro-organisms at low concentrations. Used in sanitizing foodstuffs, catering, blanket sterilization; algal control;

mould inhibition; air conditioning; softening agents for textiles; laundry, dry cleaning; paper; corrosion inhibition; petroleum eg in drilling; emulsification eg in road making, metal cleaning, insecticides; antistatics eg plastics, rubber; latex foam rubber; cosmetics; leather; flocculants. (Akzo Chemie UK Ltd).

Arquad DMHTB-75 per cent. A proprietary trade name for hydrogenated tallow and benzyl dimethyl and distearyl benzyl quaternary ammonium chloride. An emollient and conditioning agent in toiletries and shampoos. A demulsifier and dispersing agent. (Armour Hess Chemicals).†

Arquel. Meclofenamic acid. Applications: Anti- inflammatory. (Parke-Davis, Div of Warner-Lambert Co).

Arquerite. See ACEUERITE.

Arrconox AHT, DNL and DNP. A proprietary range of anti-oxidants used in the manufacture or processing of rubber. (Rubber Regenerating Co).†

Arrconox GP. Proprietary name for a general-purpose staining anti-oxidant for rubber.†

Arrconox S.P. A non-staining antioxidant. (Rubber Regenerating Co).†

Arrcopep. A rubber-reclaiming agent.†

Arrcorez 16. A butyl rubber curing resin. (Rubber Regenerating Co).†

Arrcorez 17. A tackifying resin. (Rubber Regenerating Co).†

Arret. A proprietary preparation of loperamide. Applications: An antidiarrhoeal. (Janssen Pharmaceutical Ltd).*

Arrhenal. (Arsinyl, New Cacodyl). Sodium methylarsinate, $AsO(CH_3)(ONa)_2$ $6H_2O$.†

Arrhenalic Acid. Methylarsenious acid.†

Arrow Poison, Kombe's. See KOMBE'S ARROW POISON.

Arrow Tool Steels. Proprietary steels containing 0.9-1.02 per cent chromium, 0.16-0.20 per cent vanadium, 0.5-0.6per cent manganese, and 0.20-0.30 per cent carbon.†

Arrowroot, Queensland. See TOUSLES-MOIS STARCH.

Arsenic Acid. Arsenic oxide, As_2O_5.†

Arsenic Bronze. An alloy of 80 per cent copper, 10 per cent tin, 9.2 per cent lead and 0.8 per cent arsenic.†

Arsenic Glass, Red. See REALGAR.

Arsenic Greens. See SODA GREENS.

Arsenic Orange. See REALGAR.

Arsenic, Red. See REALGAR.

Arsenic, Ruby. See REALGAR.

Arsinette. Arsenate insecticides. (Plant Protection (Subsidiary of ICI)).†

Arsinyl. See ARRHENAL.

Art Bronze. An alloy of 80-90 per cent copper, and 5-8 per cent tin.†

Artane. Trihexyphenidyl Hydrochloride. Applications: Anticholinergic; antiparkinsonian. (Lederle Laboratories, Div of American Cyanamid Co).

Artane Sustets. A proprietary preparation of benzhexol hydrochloride, in a sustained release form, used in the treatment of Parkinson's disease. (Lederle Laboratories).*

Arthripax Cream. A proprietary preparation of benzyl salicylate, glycol salicylate, terebene, menthol, ephedrine hydrochloride and capsicin. An embrocation. (Nicholas).†

Arthrobid. Sulindac. Applications: Anti-rheumatic. (Merck Sharp & Dohme).*

Artic. A proprietary trade name for methyl chloride used in refrigeration.†

Artificial Amber. See MELLITE.

Artificial Chappe. See STAPLE FIBRE.

Artificial Gutta-Percha. See SOREL'S GUTTA-PERCHA SUBSTITUTE.

Artificial Musk. See MUSK, ARTIFICIAL.

Artificial Ochre. See MARS YELLOW and SIDERINE YELLOW.

Artificial Pine-needle Oil. See PINE-NEEDLE OIL, ARTIFICIAL.

Artificial Rubber. See LAKE'S INDIA-RUBBER COMPOUND, and THINOLINE.

Artificial Turpentine. See TURPENTINE, ARTIFICIAL.

Artificial Ultramarine. See ULTRAMARINE.

Artificial Wool. See STAPLE FIBRE

Artist's Linseed Oil. See LINSEED OIL, ARTIST S.

Arubren. Chlorinated paraffin plasticizer. (Bayer UK Ltd).

Arvetane. A proprietary adhesive containing polyurethane. (Arveta SA, Basel, Switzerland).†

Arvin. A proprietary preparation of ancrod. An anti-coagulant. (Berk Pharmaceuticals Ltd).*

Arylan. Alkaryl sulphonic acids, salts and blends with nonionics. Applications: For anionic detergency, emulsification, emulsion and polymerization. (Lankro Chemicals Ltd).†

Arylan CA. Anionic surfactant in liquid form. Applications: Emulsifier for solvents, herbicidal and pesticidal preparations. (Diamond Shamrock Process Chemicals Ltd).

Arylan PWS. Anionic surfactant in viscous liquid form. Applications: Emulsifier for mineral oils, kerosine, waxes and chlorinated solvents. (Diamond Shamrock Process Chemicals Ltd).

Arylan S. Anionic surfactant as pale cream flakes. Applications: Primary emulsifier and wetting agent for emulsion polymerization and wettable powders. (Diamond Shamrock Process Chemicals Ltd).

Arylan SBC. Anionic surfactant in acid form, supplied as a brown viscous liquid, based on broader cut alkylate. Applications: Emulsifier for phenolic systems eg cresylic or tar acids. (Diamond Shamrock Process Chemicals Ltd).

Arylan SC. Anionic surfactant in liquid form. Applications: Base for liquid detergents; primary emulsifier for emulsion polymerization. (Diamond Shamrock Process Chemicals Ltd).

Arylan SNS. Anionic surfactant supplied as a buff powder. Applications: Dispersing agent. (Diamond Shamrock Process Chemicals Ltd).

Arylan SP. Anionic surfactant in acid form, supplied as a brown viscous liquid. Low free oil and inorganic content. Applications: Biodegradable intermediate for detergents, especially liquids. (Diamond Shamrock Process Chemicals Ltd).

Arylan SX. Anionic surfactant in flake form. Applications: Base for detergents, wetting agent for detergent powders. (Diamond Shamrock Process Chemicals Ltd).

Arylan TE/C. Anionic surfactant in liquid form. Applications: Emulsifier for speciality waxes, chlorinated solvents. (Diamond Shamrock Process Chemicals Ltd).

AS. Nitrocellulose with nitrogen content of 11.3 to 11.7%. Applications: Coatings on cellophane and in converting operations for paper coatings. (Hercules Inc).†

ASA. Aspirin. Applications: Analgesic; antipyretic; antirheumatic. (Eli Lilly & Co).

Asa Dulcis. Benzoin.†

Asafoetida. A gum-resin. It is the dried juice of the roots of *Ferula narthex* and *F. Foetida.* A nerve stimulant and antispasmodic.†

Asagran. Acetylsalicylic acid BPC granules. (Monsanto PLC).

Asaprol-Etrasol. See ABRASTOL.

Asaprol,. See ABRASTOL.

Asarabacca Oil. See OIL OF ASARABACCA.

Asarol. (Asaronic Camphor, Asarum Camphor). Propenyltrimethoxybenzene, $C_6H_2(CH : CHCH_3)(OCH_3)_3$. An emetic and cathartic.†

Asaronic Camphor. See ASAROL.

Asarum Camphor. See ASAROL.

Asbestine Pulp. See AGALITE.

Asbesto-Wet. Blend of polyoxyethylene esters of mixed organic acids (47%) and polyoxyethylene ether of alkylated phenols (47%) containing a silicone defoamer (6%) for ease of handling. Applications: For dust control and wet removal of asbestos. (Aquatrols Corporation of America).*

Asbestos, Bastard. See BASTARD ASBESTOS.

Asbestos, Blue. See CROCIDOLITE.

Asbestos, Canadian. See CHRYSOTILE.

Asbestos, Elastic. See MOUNTAIN CORK.

Asbestos, Micro. See MICRO ASBESTOS.

Asbestos, Platinized. See PLATINIZED ASBESTOS.

Ascabiol. A proprietary preparation of benzyl benzoate, used in the treatment of scabies and pediculosis. (May & Baker Ltd).*

Aschtrekker. See TOURMALINE.

Ascinin P. Used in oil based paints and varnishes. (Bayer & Co).*

Ascinin R. Used in oil based paints and varnishes. (Bayer & Co).*

Ascinin Special. Used in oil based paints and varnishes. (Bayer & Co).*

Ascoloy. See ALLEGHENY METAL.

Ascon. A proprietary preparation containing dried aluminium hydroxide gel, magnesium trisilicate and hyoscyamine hydrobromide. An antacid. (Cox, A H & Co Ltd, Medical Specialities Divn).†

Ascorbic Acid. Vitamin C.†

Ascorbin. Sodium Ascorbate. Applications: Vitamin. (Merrell Dow Pharmaceuticals Inc, Subsidiary of Dow Chemical Co).

Asellacrin. Somatropin. Applications: Hormone. (Calbiochem-Behring Corp).

Aseptisil. An alkaline bottle washing detergent. (Staveley Chemicals Ltd).†

Aserbine. A proprietary preparation of malic acid ester of propylene glycol, malic acid, benzoic acid and salicylic acid used as a desloughing agent. (Horlicks).†

Ashberry Metal. (Ashbury Metal). An alloy of 80 per cent tin, 14 per cent antimony, 2 per cent copper, and 1 per cent zinc.†

Ashbury Metal. See ASHBERRY METAL.

Ashes, Blue. See CHESSYLITE and MOUNTAIN BLUE.

Ashland Hi-Sol 10. An aromatic hydrocarbon solvent. Applications: Has high solvency for paints, varnishes, resins and insecticides. (Ashland Chemical Company).*

Ashland Hi-Sol 15. An aromatic hydrocarbon solvent. Applications: Used in baked enamels and in chlorinated rubber finishes. (Ashland Chemical Company).*

Ashland Kwik-Dri. An aliphatic hydrocarbon solvent. Applications: Used in cleaning compounds, waxes, polishes and as a resin solvent. (Ashland Chemical Company).*

Ashland Lacolene. An aliphatic hydrocarbon solvent. Applications: Used as a diluent for lacquer. (Ashland Chemical Company).*

Ashlene. Polyamide (nylon, thermoplastic resins). Applications: Injection moulding and extrusion. (Ashley Polymers).*

Ash, Antimony. See ANTIMONY ASH.

Ash, Blue. See SMALT.

Ash, Bone. See BONE ASH.

Ash, Pearl. See PEARL ASH.

Ash, Soda. See SODA ASH.

Ash, Tin. See TIN ASH.

Ash, Ultramarine. See ULTRAMARINE ASH.

Asilone. A proprietary preparation containing polymethylsiloxane and aluminium hydroxide. An antacid. (Berk Pharmaceuticals Ltd).*

Asilone Paediatric. A proprietary preparation containing polymethylsiloxane and carob flower. An antacid. (Berk Pharmaceuticals Ltd).*

Asiphenin. Acetylsalicylic acid and phenacetin.†

Askure. Acid catalyst used in conjunction with Koolkat resin binders for the cold set bonding of sand moulds and cores. (Foseco (FS) Ltd).*

Asma-Vydrin. A proprietary preparation of adrenaline, atropine, papaverine, pituitary extract and chlorbutol. A bronchial inhalation. (Lewis Laboratories).†

Asmal. A proprietary preparation of theophylline, ephedrine and phenobarbitone. A bronchial antispasmodic. (Norton, H N Co).†

Asmapax. A proprietary preparation of ephedrine resinate, theophylline and bromvaletone. A bronchial antispasmodic. (Nicholas).†

Asmatane Mist. Epinephrine bitartrate. Applications: Adrenergic. (Riker Laboratories Inc, Subsidiary of 3M Company).

A-Sol. Vitamin A. Applications: Vitamin (anti-xerophthalmic). (The Purdue Frederick Co).

Asp. An asbestos paper grade of TUFNOL industrial laminates. (Tufnol).*

Aspartame. A Sweetening agent. 3-Amino-N-(α- methoxycarbonylphenethyl) succinamic acid. L-Aspartyl-L-phenylalanine methyl ester.†

Aspellin. A proprietary preparation containing menthol, camphor, aspirin, methyl salicylate, glycerin ammonia, citronella oil and methylated spirit. A liniment. (Radiol Chemicals).†

Aspergurn. A proprietary preparation of aspirin. phenacetin, codeine phosphate and caffeine. An analgesic. (Norton, H N Co).†

Asphalt. Species of natural bitumen, sometimes mixed with mineral matter.†

Asphalt Rock. See ROCK ASPHALT.

Asphaltenes. Constituents of bitumen insoluble in hexane, but soluble in carbon tetrachloride.†

Asphaltine. See PETROLINE.

Asphaltum Oil. See OIL OF ASPHALTUM.

Asphalt, Natural. See BITUMEN.

Asphalt, Retin. See RETINITE.

Asphalt, Syrian. See SYRIAN ASPHALT.

Asphalt, Trinidad. See TRINIDAD ASPHALT.

Aspiquinol. An antirheumatic. (Bayer & Co).*

Aspirin. (Acetodin, Acetysal, Acetosalic Acid, Helicon, Salacetin, Coxpyrin, Anglopyrin, Salaspin, Empirin, Regipyrin, Aspirgran, Asposal, Aspro, Atonin, Atylin, Nupyrin, Salantin). Trade names for acetylsalicylic acid. Used for rheumatism, neuralgia, and headache.†

Aspirol. An injectable analgesic. (Bayer & Co).*

Asplit. A range of acid and chemical resisting cements. (Prodorite).*

Asplosal. See ASPIRIN.

Aspon. Acid sodium orthophosphate for laundry use. (Albright & Wilson Ltd).*

Aspro. See ASPIRIN.

Aspulum. A mercury derivative of chlorophenol. A seed preservative.†

Asquirrol. Mercury dimethoxide, $(CH_3O)_2.Hg$.†

Assaf. Silicone foam control agent. (ABM Chemicals Ltd).*

Assam White. See GUTTA-SUSU.

Asset. Selective herbicide. (FBC Ltd).

Asterite. Acrylic dispersions. (ICI PLC).*

Asterol. Proprietary antifungal preparations containing 2-dimethylamino-6-(beta-diethylamino-ethoxy)-benzothiazole and its salts. A skin fungicide. (Roche Products Ltd).*

Asthmatussin. A proprietary preparation of guaiphenesin, ephedrine sulphate and phenobarbitone. A bronchial antispasmodic. (Robins).†

Asthmatussin-T. A proprietary preparation of guaiphenesin, ephedrine sulphate, phenobarbitone and theophylline. A bronchial antispasomodic. (Robins).†

Astick. Adhesive promoter for asphaltic shingle. Applications: Shingle tab adhesive additive. (Chemseco).*

Astingol. A proprietary preparation of dimethyl phthalate and diethyl toluamide. Applications: An insect repellant. (Ayrton Saunders plc).*

Aston RC. Special cationic surfactant in paste or liquid form. Applications: Antistat for rugs; reduces resoiling after shampooing. (Millmaster-Onyx UK).

Astra. Dyestuff. Applications: For the paper, printing ink, surface coatings and office supplies industries. (Bayer & Co).*

Astradur A and T. A registered trade mark for high impact PVC.†

Astraflex. Dyestuffs. Applications: For the printing ink industry. (Bayer & Co).*

Astragal. Retarders for dyeing polyacrylonitrile fibres. (Bayer & Co).*

Astral Oil. See KEROSENE.

Astralex. A proprietary range of chemicals used in the bright plating of nickel. (Albright & Wilson Ltd).*

Astralon. A registered trade mark for PVC polymers in sheet form.†

Astraphloxine FF. A sensitizer and dyestuff obtained from trimethyl indolenine methiodide and alkyl orthoformate in pyridine solution.†

Astrazon. Dyestuffs. Applications: For dyeing polyacrylonitrile fibres. (Bayer & Co).*

Astro Floctite. Acrylic and other polymer blends. Applications: Adhesive for flocking auto parts, carpets, mats, wall plaques, assorted items. (Astro Industries Inc).*

Astro Mel. Melamine-formaldehyde resins - usually 80% solids. Applications: Water resistant corrugated boxes, abrasive non-woven pads, textile finish. (Astro Industries Inc).*

Astrolith. A proprietary trade name for a special lithopone. A pigment.†

Astroplax. A hydraulic gypsum cement. Astryl. Sodium *p*- glycollylarsanilate. (May & Baker Ltd).*

Astroturf. Polyethylene. Applications: Doormats. (Monsanto Co).*

Astrowet 0-75. Dioctyl sodium sulphosuccinate solution. Applications: Wetting agent and emulsifier. (Alco Chemical Corporation).*

Astryl. Sodium p-glycollylarsanilate. (May & Baker Ltd).

Asulox. Selective weedkiller. (May & Baker Ltd).

Asuntol. Preparation for the control of ectoparasites, mange mites included, on all domestic animals. Applications: Veterinary medicine. (Bayer & Co).*

A.T. 10. A proprietary preparation of dihydrotacbysterol. Used in vitamin D deficiency. (Bayer & Co).*

Atabrine Hydrochloride. Quinacrine Hydrochloride. Applications: Anthelmintic; antimalarial. (Sterling Drug Inc).

Atar Phenol. A natural phenol derived by fractionation from crude tar acids. It is a colourless, crystalline solid at ambient temperatures, with a distinct cresylic odour. Applications: In the manufacture of phenol-formaldehyde resins and novolacs, in disinfectants, in selective weedkillers, as a preservative, and in epoxies eg Bisphenol A. (Sasolchem).*

Atarax. A proprietary preparation of hydroxyzine. An ataractic. (Pfizer International).*

Atasorb. A proprietary preparation of activated attapulgite. An anti-diarrhoeal. (Eli Lilly & Co).†

Atasorb-N. A proprietary preparation of activated attapulgite, neomycin sulphate, and pectin. An antidiarrhoeal. (Eli Lilly & Co).†

Atensine. A proprietary preparation of diazepam. A tranquillizer. (Berk Pharmaceuticals Ltd).*

Aterite. A nickel silver. It usually contains from 47-68 per cent copper, 17-38 per cent zinc, 10-14 per cent nickel, 1.5-1.9 per cent iron, and 0.16-2.2 per cent manganese.†

Atgard. A proprietary preparation of Dichlorvos. An insecticide.†

Athrombin-K. Warfarin Potassium. Applications: Anticoagulant. (The Purdue Frederick Co).

Atinosol. A proprietary thallium acetate solution.†

Atiran. A potato fungicide. (Plant Protection (Subsidiary of ICI)).†

Ativan. Lorazepam. Applications: Tranquilizer. (Wyeth Laboratories, Div of American Home Products Corp).

Atlac. Polyester and vinyl ester resins. (ICI PLC).*

Atlas Orange. See ORANGE II.

Atlas Red. A dyestuff. Dyes cotton terracotta-red from an alkaline bath.†

Atlas Red, Patent. See GERANINE.

Atlas Scarlet No. 3. An azo dyestuff. It is sulphoxyleneazo-β-naphtholdisulphonic acid. The sodium salt is the commercial product.†

Atlas Steel. A proprietary hot die steel containing 9-11 per cent tungsten, 3.25-

3.5 per cent chromium, and a little vanadium.†

Atlas 10 Bronze. A proprietary trade name for an aluminium bronze containing 9.0 per cent aluminium, 7.0 per cent lead with copper.†

Atlas 89. A proprietary trade name for an alloy of copper with 9.0 per cent aluminium and 1.0 per cent iron.†

Atlas 90. A proprietary trade name for an aluminium bronze containing 90.0 per cent copper with 10.0 per cent aluminium.†

Atlox. Blends of anionic and ionic surfactants. (ICI PLC).*

Atmido. A siliceous earth, used as a filtering medium, also as a rubber filler.†

Atmoid. A mineral filler for rubber goods. It consists of almost pure silica.†

Atmos. Glycerine fatty acid esters. (ICI PLC).*

Atoleine. See PARAFFIN, LIQUID.

Atolex ASL/C. Cationic surfactant in liquid form. Applications: Lubricating and antistatic agent. (Standard Chemical Company).

Atolex ASL/C100. Cationic surfactant in the form of a thick liquid. Applications: Antistatic lubricant. (Standard Chemical Company).

Atolex AST/3. Cationic surfactant in paste form. Applications: Antistatic agent. (Standard Chemical Company).

Atolex DA/25. Naphthalene sulphonate in liquid form. Applications: Anionic dispersing liquid and levelling assistant particularly for disperse or acid dyes on acrylics or polyester. (Standard Chemical Company).

Atolex Polythene Emulsions. Full range of cationic surfactants in liquid form. Applications: Softening and lubricating agents used in additives to resin finishes. (Standard Chemical Company).

Atolex QE. Cationic surfactant in liquid form. Applications: Retarding agent. (Standard Chemical Company).

Atoline. See PARAFFIN, LIQUID.

Atomite. Fine ground calcium carbonate. (Thompson, Weinman & Co).

Atomol. Nasal decongestant with prolonged action. (Allen & Hanbury).†

Atonin. See ASPIRIN.

Atophan. (Quinophan, Agotan). 2-Phenylquinoline-4-carboxylic acid. Used in the treatment of gout and sciatica.†

Atoquinol. Allylphenylcinchoninic ester. A powerful uric acid solvent and eliminator.†

Atosil. A phenothiazine derivative with sedative effect on autonomic nervous system and antihistaminic properties. (Bayer & Co).*

Atoxycocaine. Ethocaine hydrochloride.†

Atped 400. Polyethylene Glycol. Applications: Pharmaceutic aid - ointment base; suppository base; solvent; tablet excipient; tablet and/or capsule lubricant. (ICI Americas Inc).

Atped 600. Polyethylene Glycol. Applications: Pharmaceutic aid - ointment base; suppository base; solvent; tablet excipient; tablet and/or capsule lubricant. (ICI Americas Inc).

Atpeg 300. Polyethylene Glycol. Applications: Pharmaceutic aid - ointment base; suppository base; solvent; tablet excipient; tablet and/or capsule lubricant. (ICI Americas Inc).

Atpet. Range of surfactant properties including emulsifying, wetting, dispersing, anti-foam, anti-rust. Applications: Oil additives. (Atlas Chemical Industries (UK) Ltd).

Atraflow Plus. A liquid formulation containing 264g atrazine and 214g aminotriazole per litre as a suspension concentrate. Applications: It may be used where total weed control is required including industrial sites, paths, kerbs and channels, drives and hard tennis courts, hardstanding and storage areas. (Burts & Harvey).*

Atramentum Stone. A mixture of ferric and ferrous sulphates with ferric oxide. Used in the manufacture of inks.†

Atramet Combi. Active ingredients: atranex plus ametrex. Applications: Ready formulated mixture of atrazine plus ametryne for use as a selective pre- and post-emergence herbicide. (Agan Chemical Manufacturers Ltd).†

Atranex. Active ingredient: atrazine; 2-chloro-4-ethylamino-6-isopropylamino-

1,3,5-triazine. Applications: Pre- and post-emergence herbicide. (Agan Chemical Manufacturers Ltd).†

Atrixo. A silicone hand cream. (Smith and Nephew).†

Atromid-S. A proprietary preparation of clofibrate, used to reduce blood cholesterol levels. (ICI PLC).*

Atropisol. Atropine sulphate. Applications: Anticholinergic. (CooperVision Inc).

Atroscine. Optically inactive scopolamine (*dl*-hyoscine).†

Atrovent. Ipratropium bromide. Applications: An anticholinergic bronchodilator. (Boehringer Ingelheim Ltd).†

Atroxindol. The anhydride of *o*-amino-α-phenylpropionic acid, $C_6H_4(NH)(CH\ CH_3)CO$.†

A/T/S. Erythromycin. Applications: Antibacterial. (Hoechst-Roussel Pharmaceuticals Inc).

Attaclay. Fine particle size sorbent attapulgite (magnesium aluminium silicate), used as a chemical conditioning agent (free-flow agent). Applications: Prilled ammonium nitrate and urea fertilizers and other bulk granular chemicals. (Engelhard Corporation).*

Attacote. Fine particle size sorbent attapulgite (magnesium aluminium silicate), used as a chemical conditioning agent (free-flow agent). Applications: Fire extinguishing chemicals. (Engelhard Corporation).*

Attaflow. Liquid (slurry) attapulgite (hydrous magnesium aluminium silicate), used as a suspending agent. Applications: Liquid suspension fertilizers, flowable pesticides. (Engelhard Corporation).*

Attagel. Colloidal attapulgite (hydrous magnesium aluminium silicate), used as a thixotropic agent, gellant and suspending aid. Applications: Paints, coatings, adhesives, inks, caulks and liquid suspension fertilizers. (Engelhard Corporation).*

Attapulgus. Colloidal attapulgite (hydrous magnesium aluminium silicate), used as suspending agent. Applications: Oil well drilling clay, particularly salt water formations. (Engelhard Corporation).*

Attasorb. Fine particle size sorbent attapulgite (magnesium aluminium silicate), used as a chemical conditioning agent (free-flow agent). Applications: Powdered detergents, agricultural chemicals. (Engelhard Corporation).*

Attenuvax. Attenuated line of measles virus derived from Enders' attenuated Edmonston strain. Applications: Measles vaccine. (Merck Sharp & Dohme).*

Aturbane. A proprietary preparation of phenglutarimide hydrochloride, used in Parkinsonism. (Ciba Geigy PLC).†

ATV Steel. A complex nickel-chromium austenitic steel.†

Aubepine. Anisaldehyde. Used in perfumery.†

Audax. A proprietary preparation of choline salicylate, ethylene oxide polyoxy propylene condensate. Analgesic ear drops. (Napp Laboratories Ltd).*

Audicort. A proprietary preparation of triamcinolone acetonide, neomycin undecylenate and benzocaine used as ear-drops in the treatment of otitis externa. (Lederle Laboratories).*

Audrey. Automatic dielectric analyzer. (Tetrahedron Association Inc).

Auel Solder. An alloy of 63 per cent tin, 35 per cent zinc, 1.7 per cent copper, and 0.3 per cent aluminium.†

Augmentin. A proprietary preparation of amoxycillin with potassium clavulanate. Applications: An antibiotic. (Beecham Research Laboratories).*

Augsburg Metal. A brass containing 72 per cent copper, and 28 per cent zinc.†

Aunativ. Solution for injection. Applications: Human immunoglobulin anti-hepatitis B. (KabiVitrum AB).*

Auracet. A proprietary preparation of aluminium acetotartrate and lead subacetate. Ear drops. (Ward, Blenkinsop & Co Ltd).†

Auralgan. Glycerol containing phenazone and benzocaine. Applications: Ear drops to relieve pain and inflammation of acute otitis externa and facilitate the removal of ear wax. (Ayerst Laboratories).†

Auralgicin. A proprietary preparation of ephedrine, benzocaine, chlorbutol, potassium hydroxyquinoline sulphate, phenazone and glycerine. Ear drops. (Fisons PLC).*

Auramine. (Auramine O, Auramine I, II, and conc., Aureum, Pyoctanin). A diphenylamine basic dyestuff, now little used.†

Auramine G. A dyestuff. Dyes tannined cotton greenish-yellow.†

Auramine 0. See AURAMINE.

Auramine, Conc. See AURAMINE.

Auramine, Medicinal. See PYOCTANIN and AURAMINE.

Aurantia. (Imperial Yellow). An acid dyestuff. It is the ammonium salt of hexanitrodiphenylamine. Dyes wool and silk orange from an acid bath.†

Aurantiène. A residue containing terpenes left behind in the refining of orange oil. Used as a perfume for soaps.†

AurantiinW. (Hesperidine). Naringin, $C_{21}H_{26}O_{11}$.†

Aurantine. The trade name for Osage orange extract (from the bark of a shrub), used in the textile industry for tanning.†

Aureocort. A proprietary preparation of triamcinolone acetomide and chlortetracycline hydrochloride used in the treatment of skin disorders. (Lederle Laboratories).*

Aureolin. (Cobalt Yellow, Indian Yellow). A yellow pigment. It is a double nitrite of potassium and cobalt, $K_6Co_2(NO_2)_{12}$ $3H_2O$.†

Aureoline. See PRIMULINE.

Aureomycin. Chlortetracycline hydrochloride. Applications: Antibacterial; antiprotozoal. (Lederle Laboratories, Div of American Cyanamid Co).

Aureosin. A chlorinated fluoresceine.†

Aureum. See AURAMINE.

Aurine R. See PAEONINE.

Aurine Red. See PAEONINE.

Auro-Protasin. Colloidal gold.†

Aurochin. Quinine *p*-aminobenzoate.†

Aurocyanase. A colloidal gold and potassium double cyanide.†

Auromet 55. A proprietary trade name for an alloy containing 76-80 per cent copper, 10-12 per cent aluminium, 4-6 per cent iron, and 4-6 per cent nickel.†

Auronal Black. A dyestuff obtained by the fusion of dinitro-*p*-aminodiphenylamine with sodium polysulphide. It dyes cotton a blue-black from a sodium sulphide bath.†

Aurora Yellow. See CADMIUM YELLOW.

Aurotine. An acid or mordant dyestuff for wool. It is the sodium salt of tetranitrophenolphthalein.†

Aurum. Gold, Au.†

Austenite. A characteristic constituent of very highly carbonized steel, containing more than 1.1 per cent carbon.†

Australian Gold. A gold-silver alloy containing 8.33 per cent silver, used for coinage.†

Australian Sandarac. See SANDARAC RESIN.

Austrapol. Styrene - butadiene and polybutadiene polymers. Applications: Tyres, retread, general rubber goods and plastics modification. (Australian Synthetic Rubber Co).*

Austrian Cinnabar. See CHROME RED.

Austrostab. A full range of stabilizer systems for PVC, containing stabilisers (Pb, Pb/Ba, Cd, Ba/Cd, Ca/Zn), lubricants, pigments, fillers and modifiers. Applications: Completely ready-for-use formulations for applications such as PVC pipe, cable, profile. (Blielberger Bergwerke Union).*

Austrox. Melted lead oxide granules. Applications: Used as raw material for speciality glasses such as crystal glass and glass for electronic tubes. (Blielberger Bergwerke Union).*

AuSub. Gold substitute inorganic die attach pastes. Applications: Electronic applications in computer and for military and aerospace uses. (Johnson Matthey Chemicals Ltd).*

Autogal. A trade mark for a flux used for soldering and welding aluminium. It is a mixture of the halogen salts of the alkali metals.†

Automate. Liquid dyestuffs for petroleum. (Williams (Hounslow) Ltd).

Automolite. A mineral, ZnO Al_2O_3.†

Autumn Lawn Food. Lawn fertilizer. (Fisons PLC, Horticulture Div).

Avabond. Adhesives for packaging, woodworking, textiles etc. (Avalon Chemical Co Ltd).

Avadex. 2,3-Dichloroallyl-diisopropyl-thiocarbamate. Applications: Herbicide for wild oats. (Monsanto Co).*

Avadex BW, FAR-GO. Triallate herbicide. (Monsanto PLC).

Avamid. Polyimide pre-pregs. (Du Pont (UK) Ltd).

Avantine. A brand of isopropyl alcohol. An anaesthetic. (Laporte Industries Ltd).*

Avatec (as sodium). Lasalocid. Applications: Coccidiostat. (Hoffmann-LaRoche Inc).

Avazyme. Chymotrypsin. Applications: Enzyme. (Wallace Laboratories, Div of Carter-Wallace Inc).

Avecolite. A proprietary phenol-formaldehyde synthetic resin.†

Aventyl. A proprietary preparation of nortriptyline hydrochloride. An antidepressant. (Eli Lilly & Co).†

Aventyl Hydrochloride. Nortriptyline hydrochloride. Applications: Antidepressant. (Eli Lilly & Co).

Aveonal. An alloy of aluminium with 4 per cent copper, 0.05 per cent magnesium, 0.05 per cent manganese, and 0.05 per cent silicon.†

Aversin. Paraffin wax emulsion with zirconium salts. Water repellent finishing. Applications: Non-permanent water-repellent finish compatible with resins, suitable for all kinds of fibres. (Henkel Chemicals Ltd).

Avertin. A proprietary preparation of bromethol, used to control eclampsia in toxaemia of pregnancy. (Bayer & Co).*

Avgard. Anti-misting kerosene. (ICI PLC).*

Avialite. A proprietary trade name for an alloy of copper with about 9.0 per cent aluminium and 1.0 per cent iron.†

Aviamide-6. Policapram. Applications: Pharmaceutic aid. (Avicon Inc).

Aviashine. A blend of solvents, carriers, abrasives and waxes. Applications: Aircraft maintenance chemical. Provides effective cleaning for paintwork, chrome and other metal surfaces. Can be polished to give a durable bright and protective finish. (The Kent Chemical Company).*

Aviawash. A blend of detergents and surfactants with inhibitors. Applications: Aircraft maintenance chemical for aircraft exterior cleaning. For cleaning painted and unpainted external surfaces of aircraft and ground equipment. (The Kent Chemical Company).*

Avicel. Microcrystalline cellulose. Applications: Pharmaceutic aid. (FMC Corporation).

Avicell-RC. A proprietary trade name for a chemically pure colloidal cellulose. Forms thixotropic dispersions both mechanically and thermally stable. Edible and metabolically alert. Used as a thickening agent. (Honeywell and Stein).†

Aviester. Pegoterate. Applications: Pharmaceutic aid. (Avicon Inc).

Avilon. Metal complex dyes. (Ciba-Geigy PLC).

Avional D. An alloy of aluminium with 3.9 per cent copper, 0.5 per cent nickel, 0.5 per cent magnesium, 0.55 per cent silicon, and 0.3 per cent iron.†

Avirol. See MONOPOLE SOAP.

Avirol. Textile and leather auxiliary. (Hickson & Welch Ltd).

Avisol. Neutral soluble sulphathiazole for poultry. (May & Baker Ltd).*

Avistin. A range of basic fatty acid condensation products. Applications: Used for the manufacture of cationic preparing and finishing agents for textiles. (Huls AG).†

A-Vitan. Vitamin A. Applications: Vitamin (anti-xerophthalmic). (Janssen Pharmaceutica).

Avitex. Surface active agents. (Du Pont (UK) Ltd).

Avitex. A proprietary trade name for textile finishing materials. Consists of sulphated higher alcohols.†

Avitone. Dyeing assistants. (Du Pont (UK) Ltd).

Avitrol. 4-aminopyridine. Applications: Treated grain baits for pest bird control. Classified as a 'Restricted Use' pesticide. (Avitrol Corporation).*

Avivage. Combination of sulphated fats with special additives. Applications: Softener, stabilizer in bleaching liquors. (Henkel Chemicals Ltd).

Avivan. Finishing agent. (Ciba-Geigy PLC).

Avivan. (Tetrapol, Koloran, Flerhenol). Wetting out agents consisting of sulphonated oils with organic solvents.†

Avloclor. A proprietary preparation of Chloroquine phosphate. Anti-malarial. (ICI PLC).*

Avlosulfon. A proprietary preparation of dapsone, used in the treatment of leprosy. (ICI PLC).*

Avoca. Toughened silver nitrate. Applications: To be used only under medical supervision.(Avoca Pharmaceuticals).*

Avoilefin. Polipropene 25. Pharmaceutic aid. (Avicon Inc).

Avolan. Levelling agents for wool. Applications: Brightening agents to correct faulty dyeings on wool. (Bayer & Co).*

Avomine. A proprietary preparation of promethazine theoclate. (Promethazine 8-chlorotheophyllinate). An anti-emetic. (May & Baker Ltd).*

Avoparcin. A growth promoter. A glycopeptide antibiotic obtained from cultures of *Streptomyces candilus* †

Avotan. Avoparcin. Applications: Antibacterial. (Lederle Laboratories, Div of American Cyanamid Co).

Axall. Herbicide. (May & Baker Ltd).

Axelglo. Proprietary, polymeric based polishes to maintain and renew fibreglass and metal (painted and unpainted) surface lustre. Applications: Moulded fibreglass products (yachts, camper tops etc) and automotive. (Axel Plastics Research Laboratories Inc).*

Axerophthol. Vitamin A.†

Axetin. A proprietary preparation of hydroxizine hydrochloride, ephedrine sulphate and theophylline. An antiasthmatic. (Pfizer International).*

Axin. See AGE.

Axiquel. Valnoctamide. Applications: Tranquilizer. (McNeil Pharmaceuticals, McNEILAB Inc).

Axite. An explosive. It is a smokeless powder which contains guncotton, nitroglycerine, petroleum jelly, and a little potassium nitrate.†

Axoridin. Latamoxef. Applications: Semi-synthetic broad-spectrum beta-lactam antibiotic for parenteral administration. (Merck Sharp & Dohme).*

Aygestin. Norethindrone acetate. Applications: Progestin. (Ayerst Laboratories, Div of American Home Products Corp).

Ayrtol. A proprietary preparation of chloroxylenol. Applications: A disinfectant. (Ayrton Saunders plc).*

AZ. Toothpaste to help fight plaque and cavities. (Richardson-Vicks Inc).*

Azactam.. A proprietary preparation of aztreonam. Applications: An antibiotic (Squibb, E R & Sons Ltd).*

Azacyclonal. Diphenyl-4-piperidylmethanol.†

Azafran. See SAFFRON.

Azaleine. See MAGENTA.

Azalomycin. A mixture of related antibiotics produced by *Streptomyces hygroscopicus*, var. *asalomyceticus*.†

Azamine 4B. See BENZOPURPURIN 4B.

Azapen. Methicillin sodium. Applications: Antibacterial. (Pfizer Inc).

Azapetine. 1-Allyl-2,7-dihydro-3,4:5,6-dibenzazepine.†

Azapropazone. An analgesic and anti-inflammatory. 5-Dimethylamino-9-methyl-2-propyl-1*H*-pyrazolo-[1,2-*a*][1,2,4]-benzotriazine-1,3(2*H*)-dione.†

Azaribine. A preparation used in the treatment of psoriasis. 2-β-D-Ribofuranosyl-1,2,4-triazine-3,5-(2*H*, 4*H*)dione 2',3',5'-triacetate. 6-Azauridine 2',3',5',- triacetate.†

Azarine S. A mordant dyestuff. It is the ammonium bisulphite compound of dichlorophenolazo-β-naphthol. Used for pinks and reds in calico printing, and for dyeing silk.†

Azatadine. An anti-histamine. It is 5,6-dihydro-11-(1-methyl-4-piperidylidene) benzo(h)cyclohepta(b)-pyridine.†

Azathioprine. 6-(1-methyl-4-nitroimidazol-5-ylthio)purine.†

Azeosin. See AZO-EOSIN.

Azide Chrome Black Blue BL. A direct cotton blue suitable for artificial silk.†

Azidine Brilliant Yellow 6G Extra. A direct cotton colour suitable for artificial silk.†

Azidine Chrome Brown BB, BG, BR. Direct cotton browns after treatment with potassium bichromate and copper sulphate. Suitable for artificial silk.†

Azidine Fast Red F. See DIAMINE FAST RED.

Azidine Fast Scarlets. Dyestuffs obtained by passing carbonyl chloride into a mixture of *m*-toluylenediamine-4-sulphonic acid and J-acid.†

Azidine Fast Yellow G. A dyestuff obtained by warming *p*-nitrotoluenesulphonic acid with caustic soda. It gives golden shades on cotton in the presence of sodium chloride, and on wool and silk from an acid bath.†

Azidine Wool Blue R. A tetrazo dyestuff obtained by coupling diazotized tolidine with β-naphthol-8-sulphonic acid and 8-amino-α-naphthol-5-sulphonic acid.†

Azilex. A proprietary preparation of azulene. An antipruritic. (Ingasetter).†

Azindone Blue G. (Anzindone Blue R). Indigo blue dyestuffs closely allied to the naphthyl dyes.†

Azine Blue, Spirit Soluble. See INDULINE, SPIRIT SOLUBLE.

Azine Green. A safranine dyestuff, $C_{30}H_{25}N_4Cl$, the product of the reaction between nitrosodimethylaniline hydrochloride and 2:6-diphenylnaphthylenediamine.†

Azine Green GB. (Azine Green TO). A basic dyestuff. Dyes tannindid cotton dark green.†

Azine Green S. An acid dyestuff. It is the sodium salt of azine green sulphonic acid. Dyes wool bluish-green from an acid bath.†

Azine Green TO. See AZINE GREEN GB.

Azine Scarlet G. A dyestuff. Dyes yellow-scarlet.†

Aziplex. Blended metal chelate. (ABM Chemicals Ltd).*

Azlin. Azlocillin. Applications: Antibacterial. (Miles Pharmaceuticals, Div of Miles Laboratories Inc).

Azo Acid Black B, G, R, TL, 3BL. Acid dyestuffs employed in wool dyeing.†

Azo Acid Black TL Extra. A mixture of a blackish-blue, a green, a violet-blue, and an orange dye. It dyes wool black from an acid bath containing sodium sulphate.†

Azo Acid Blue B. A dyestuff belonging to the same class as Victoria violet BS. It dyes wool pure blue.†

Azo Acid Blue, 4B, 6B. Acid dyestuffs for wool.†

Azo Acid Brown. An acid colour for wool, which it dyes from an acid bath.†

Azo Acid Carmine. An acid dyestuff for wool.†

Azo Acid Magenta B, G. Acid dyestuffs for wool.†

Azo Acid Rubine. See AZO RUBINE S.

Azo Acid Rubine 2B. See FAST RED D.

Azo Acid Violet 4R. An acid colour, dyeing wool reddish-violet from an acid bath.†

Azo Acid Yellow. See CITRONINE B.

Azo Black. See BLUE BLACK B.

Azo Black Blue. (Azo Navy Blue). A dyestuff. Dyes cotton grey to dark violet from a boiling salt bath.†

Azo Black O. See BLUE BLACK B.

Azo Blue. A dyestuff. Dyes cotton greyish-violet.†

Azo Brown 0. See FAST BROWN N.

Azo Cardinal G. An acid dyestuff for wool.†

Azo Corinth. A dyestuff. Dyes cotton brown from a soap bath.†

Azo Galleine. A mordant dyestuff. Dyes chromed wool blackish-violet. It is used for calico printing, giving dark violet shades.†

Azo Green. A dyestuff. Dyes chromed wool green.†

Azo Mauve B. A dyestuff. Dyes cotton blackish-blue-violet from an alkaline bath.†

Azo Navy Blue. See AZO BLACK BLUE.

Azo Orange R. A dyestuff. Dyes cotton orange.†

Azo Orchil R. An acid dyestuff, giving brownish-red colour on wool.†

Azo Red A. A dyestuff allied to Palatine red (*qv*).†

Azo Violet. A dyestuff. Dyes cotton bluish-violet from a soap bath.†

Azo Yellow. See CITRONINE B.

Azo Yellow M. See CITRONINE B.

Azo Yellow 3G Extra Conc. See CITRONINE B.

Azo-Standard. Phenazopyridine Hydrochloride. Applications: Analgesic. (Alcon Laboratories Inc).

Azoalizarin Black. A dyestuff. Dyes chromed wool black.†

Azoalizarin Bordeaux W. A dyestuff. Dyes chromed wool bordeaux, and is used for wool printing.†

Azoalizarin Yellow 6G. See ALIZARIN YELLOW 5G.

Azobenzene Red. See BIEBRICH SCARLET.

Azocarmine B. (Rosinduline 2B). A dyestuff. Dyes wool bluish-red, and is used as an archil substitute.†

Azocarmine BX Powder and G Paste. Acid dyestuffs similar to azocarmine B and G.†

Azocarmine G. (Rosazine, Rosarin). A dyestuff. Dyes wool bluish-red. †

Azochromine. A mordant dyestuff. It is tetrahydroxyazobenzene. Dyes chromed wool and cotton dark brown. Used in cotton printing.†

Azococcin G. See TROPAEOLINE OOOO.

Azococcin 2R. (Double Scarlet R, Xylidine Scarlet, Jute Scarlet R). An acid dyestuff. Dyes wool red from an acid bath.†

Azococcin 7B. See CLOTH RED G.

Azocochineal. A dyestuff. Dyes wool red from an acid bath.†

Azocoralline. A dyestuff. It is the sodium salt of *p*-acetaminophenylazo-β-naphtholdisulphonic acid. Dyes wool from an acid bath.†

Azodiphenyl Blue. See INDULINE SPIRIT SOLUBLE.

Azoene. Fast red salt (ponceau fast L salt). A pinkish cream powder for use in automatic SGO-T assays. (British Drug Houses).†

Azoeosin. (Azeosin). An acid dyestuff. Dyes wool eosin red from an acid bath.†

Azofix. Azoic colours. (Fine Dyestuffs & Chemicals Ltd).†

Azoflavin. See CITRONINE B.

Azoflavin S or 2. See CITRONINE B.

Azoflavin 3R Extra Conc. See CITRONINE B or 2B.

Azoform. Monoacetylazoxytoluene.†

Azofuchsine B. (Acid Levelling Red 2B). An acid and acid mordant dyestuff. Dyes wool from an acid bath magenta red, becoming violet-black on chroming.†

Azofuchsine G. An acid and acid mordant colour. Dyes wool magenta red from an acid bath.†

Azofuchsine GN Extra. A dyestuff, similar to Azofuchsine G.†

Azogen R. A dyestuff similar to Azophor red PN.†

Azogrenadine S. See SORBINE RED.

Azoground. Azoic colours. (Fine Dyestuffs & Chemicals Ltd).†

Azoguard. A stabilizer for diazo compounds.Imperial Chemical Industries.

Azol. 4-aminophenol-hydrochloride in solution. (Johnsons of Hendon).†

Azolan. Active ingredient: aminotriazole; 1,2,4-triazol-3-ylamine. Applications: Weedkiller with good translocation characteristics for the control of perennial and annual weeds. (Agan Chemical Manufacturers Ltd).†

Azolid. Phenylbutazone. Applications: Antirheumatic. (USV Pharmaceutical Corp).

Azolith. A proprietary pigment containing 71.0 per cent. $BaSO_4$ and 29.0 per cent. ZnS.†

Azomagenta G. A dyestuff obtained by diazotizing sulphanilic acid, and treating the product with S acid.†

Azoman. A proprietary trade name for hexazole.†

Azonaphthol Red S. See FAST RED D.

Azone. Laurocapram. Applications: Pharmaceutic aid. (Nelson Research).

Azonigrin. An acid dyestuff. Dyes wool from an acid bath brownish black.†

Azophor Blue. See DIANISIDINE BLUE.

Azophor Red. See PARANITRANILINE RED.

Azophosphine GO. A basic dyestuff. Dyes cotton direct from an acid bath orange-red.†

Azoprint. Azoic printing colours. (Fine Dyestuffs & Chemicals Ltd).†

Azopurpurin 4B. A direct cotton dyestuff, which it dyes orange from a salt bath.†

Azorapid. Azoic printing colours. (Fine Dyestuffs & Chemicals Ltd).†

Azoresorcin. See DIAZORESORCIN.

Azorubine. See AZORUBINE S.

Azorubine A. See AZORUBINE S.

Azorubine S. (Azo acid rubine, Azorubine, Azorubine A, Brilliant carmoisine O, Brilliant crimson, Carmoisine, Chromotrope FB, Crimson B, Fast red C, Mars red, Nacarat, Rainbow Red L.LAS.S.WS). A dyestuff. It is the sodium salt of *p*-sulphonaphthaleneazo-α-naphthol-*p*-sulphonic acid. Dyes wool red from an acid bath.†

Azorubine 2S. An azo dyestuff prepared by the action of diazotized amino-azo-benzene-sulphonic acid upon α-naphthol-mono-sulphonic acid.†

Azosan. A fungicide. (May & Baker Ltd).

Azostix. A prepared test strip of urease, bromothymol blue and buffers, for the semi-quantitative determination of blood urea levels. (Ames).†

Azotox. Insecticides. (ICI PLC).*

Azovan Blue. The tetrasodium salt of 4,4'-di-[7-(1-amino-8-hydroxy-2,4-disulpho) naphthylazo]-3,3'-bitolyl. Evans blue.†

Azure Blue. See MOUNTAIN BLUE and COBALT BLUE.

Azuresin. Prepared from carbacrylic cation exchange resin and azure A dye (3-amino-7-dimethyl-amino-phenazathionium chloride). Diagrex Blue.†

B

BA 27. A non-heat-treatable alloy of aluminium containing 3.5 per cent magnesium and 0.4 per cent manganese.†

BAA. Battery acid. (Berk Spencer Acids Ltd).

Babbitt's Metals. White bearing metals, which usually contain 83-90 per cent tin, 7.3-11 per cent antimony, and 3.7-8.3 per cent copper. Hard alloys contain 83.3 per cent tin, 8.3 per cent antimony, and 5.5-8.3 per cent copper, whilst an ordinary Babbitt consists of 89 per cent tin, 7.3 per cent antimony, and 3.7 per cent copper. Lead often replaces tin in these alloys, and a marine Babbitt contains 72 per cent lead, 21 per cent tin, and 7 per cent antimony. A soft Babbitt metal contains 84 per cent tin, 7.4 per cent antimony, 5.6 per cent lead, and 3 per cent copper. Another alloy, also known as Babbitt, consists of 90 per cent tin and 10 per cent copper, and still another contains 69 per cent zinc, 19 per cent tin, 5 per cent lead, 4 per cent copper, and 3 per cent antimony.†

Babe's Solution. A mixture of equal parts of a concentrated alcoholic solution and an aqueous solution of safranine. Microscopic stain.†

Babul Bark. Acacia bark.†

Bacacil. A proprietary preparation of bacampicillin hydrochloride. An antibiotic. (Pfizer International).*

Bacdip. Preparation for the control of ectoparasites, especially of multi-host ticks on all domestic animals. Applications: Veterinary medicine. (Bayer & Co).*

Bachite. A specially prepared carbon made from anthracite.†

Baciguent. Bacitracin. Applications: Antibacterial. (The Upjohn Co).

Bacillarite. Synonym for Leverrierite.†

Bacitracin. An antibiotic produced by a strain of Bacillus subtilis.†

Bacnelo. A proprietary mixture of 3 per cent. sulphur, 2 per cent. coal tar solution, zinc oxide and calcium hydroxide, used in the treatment of skin disorders. (Ascher, Kansas City, Mo).†

Bacote. Paper coating insolubilizers. (Magnesium Elektron Ltd).

Bacterase. Bacterial enzyme or diastase. (ABM Chemicals Ltd).*

Bactipront. A proprietary preparation of co-trimoxazole. A urinary tract antibacterial. (Pfizer International).*

Bactiram. Amines and derivatives. Primary, secondary, tertiary fatty mono-, di- and polyamines from C8 to C22. Quaternary ammonium salts. Amine oxides. Amine salts. Ethoxylated amines and polyamines. Amino-acids. Special nonionic derivatives. Additives. Applications: Surface active agents used as auxiliaries in road, mining, textile and fertilizer industries, in metallurgy, as lubricating agents, additives for fuel oils and soil stabilization. (British Ceca Co Ltd).*

Bactocill. Oxacillin sodium. Applications: Antibacterial. (Beecham Laboratories).

Bactratycin. Tyrothricin. Applications: Antibacterial. (Wallace Laboratories, Div of Carter-Wallace Inc).

Bactrim. A proprietary preparation of trimethoprim and sulphamethoxazole. An antibiotic. (Roche Products Ltd).*

Bactrim Roche. Bactericide. (Roche Products Ltd).

Badam Kohee. See CHOOLI-KI-TEL.

Badin Metal. A name used for an alloy of iron with 8-10 per cent aluminium, 18-20 per cent silicon, and 4-6 per cent titanium. Used for adding silicon to steel.†

Bagasses Oil. See PYRENE OIL.

Bahia Arrowroot. Tapioca.†

Bahn Metal. A lead base bearing alloy containing copper, 0.7 per cent calcium, 0.6 per cent sodium, and 0.04 per cent lithium.†

Baicalite. Synonym for Baikalite.†

Baiculein. 5:6:7-trihydroxyflavone.†

Bailey Solder. An alloy containing 70 per cent tin, 16 per cent zinc, 10 per cent aluminium, and 4 per cent phosphor tin.†

Bakelite. Phenolics. (BP Chemicals Ltd).*

Bakelite. Phenolic moulding powders. (Sterling Moulding Materials).*

Bakelite A and B. (Resinite). Bakelite (a phenol-formaldehyde resin), soluble in certain solvents. (Bakelite Corporation).†

Bakelite C-9. A proprietary trade name for soft oil modified alkyds. (Bakelite Corporation).†

Bakelite Dilecto. A laminated product consisting of paper or fibre cemented together with a phenol-formaldehyde resin. (Bakelite Corporation).†

Bakelite DQD-3269. A proprietary trade name for an ethylene-vinyl acetate copolymer. A plastics material which retains its flexibility and toughness at low temperatures. (Bakelite Corporation).†

Bakelite Micarta. A similar product to BAKELITE DILECTO. (Bakelite Corporation).†

Baker P-2C. A proprietary trade name for cellosolve ricinoleate. A vinyl plasticizer.†

Baker P-8. A proprietary trade name for glyceryl triacetyl ricinoleate. A plasticizer for vinyl polymers and GR/S, neoprene GN, ethyl cellulose, and perbunan.†

Baker's Anaesthetic Ether. Diethyl ether or anaesthetic ether. (May & Baker Ltd).*

Baker's P and S Liquid. A proprietary mixture of 1 per cent. phenol, sodium chloride, liquid paraffin and water. (Baker Laboratories, Miami, Florida).†

Baker's P.8. A proprietary plasticizer for polyvinyl chloride and copolymers of PVC, GR/S, Neoprene GN, ethyl cellulose, and perbunan. It is glyceryl triaceto-ricinoleate.†

Baker's Salt. Ammonium carbonate, $(NH_4)_2CO_3$.†

Bakfil. Granular infill materials for use behind Garnex boards in continuous casting tundishes. (Foseco (FS) Ltd).*

Baking Soda. Sodium hydrogen carbonate, $NaHCO_3$.†

Bako Nuts. See NJAVE BUTTER.

Bakurol. See PARAFFIN, LIQUID.

B & L 70. Lidofilcon A. The material contains 70% of water. Applications: Contact lens material. (Bausch & Lomb, Professional Products Div).

B.A.L. A proprietary preparation of dimercaprol, used to treat heavy metal intoxication. (Boots Company PLC).*

BAL in Oil. Dimercaprol. Applications: Antidote to arsenic and gold and mercury poisoning. (Hynson Westcott & Dunning, Div of Becton Dickinson & Co).

Balanced Salt Solution. A proprietary solution of 0.49 per cent. sodium chloride. 0.075 per cent. potassium chloride and 0.036 per cent. calcium chloride. (Alcon Universal).†

Balata. The coagulated milky juice of *Mimusops globosa* (the balata or bullet tree) of South America. Used for electrical insulation, for transmission belts, and as soles for shoes.†

Baldwin's Phosphorus. Calcium nitrate, $Ca(NO_3)_2$.†

Bale Blue. A dyestuff prepared by the Condensation of nitroso-dimethyl-aniline with diphenyl-2:7-naphthylenediamine.†

all-bearing Steel. See CHROME STEELS.

allistite. A trade mark for a smokeless powder containing nitroglycerin and collodion cotton.†

alm Oil. See OIL OF BALM.

almex Medicated Lotion. A proprietary protective lotion of lanolin, hexachlorophene, allantoin, balsam Peru and silicone oil in a non-mineral oil base. (Macsil Inc, Philadelphia, Pa).†

al-nela. An alloy of 28 per cent nickel, 66 per cent copper, with 6 per cent manganese, silicon, iron, and zinc.†

alneol. Mineral oil. Applications: Laxative; pharmaceutic aid. (Rowell Laboratories Inc).

alnetar Liquid. A proprietary preparation used to treat skin disorders. It is a mixture of 2.5 per cent. crude coal tar, a lanolin derivative, mineral oil and a non-ionic emulsifier. (Westwood Pharmaceuticals, Subsidiary of Bristol-Myers Co).†

alsam of Capivi. See OLEO-RESIN COPAIBA.

alsam of Copaiva. See OLEO-RESIN COPAIBA.

alsam of Peru. An oleo-resin obtained from the bark of *Myroxylon pereinaeg.* It contains esters of cinnamic and benzoic acids. It is antiseptic, and is used in medicine and perfumery.†

alsam of Tolu. The product of the tree *Myroxylon toluifera.* Used in medicine and in perfumery.†

alsams. Resins containing benzoic or cinnamic acids.†

alsam, Carpathian. See RIGA BALSAM.

alsam, Copalm. See LIQUIDAMBAR.

alsan. Proprietary name for a specially purified preparation of balsam Peru. (Macsil Inc, Philadelphia, Pa).†

alsan-Katel. See MECCA BALSAM.

altane CF. 1,1,1 trichloroethane. Applications: Degreasing solvent. (Atochem Inc).†

altimore Chrome Yellow. See CHROME YELLOW.

Bamate. A proprietary name for MEPROBAMATE. A tranquillizer. (Century Pharmaceuticals Inc, Indiana).†

Bambuk Butter. See SHEA BUTTER.

Bamo. A proprietary name for MEPROBAMATE. A tranquillizer. (Misemar Pharmaceuticals Inc, Springfield, Mo).†

Banamine. Flunixin meglumine. Applications: Anti-inflammatory; analgesic. (Schering-Plough Corp).

Banflex. Orphenadrine citrate. Applications: Relaxant; antihistaminic. (O'Neal, Jones & Feldman Pharmaceuticals).

Banistyl. Dimethothiazine. (May & Baker Ltd).

Banlene Plus. Selective herbicide. (FBC Ltd).

Banminth. Pyrantel Tartrate. Applications: Anthelmintic. (Pfizer Inc).

Banocide. A proprietary preparation of diethylcarbamazine citrate. Applications: Used in the treatment of filariasis. (The Wellcome Foundation Ltd).*

Banox. Additive to de-icing salts. (Albright & Wilson Ltd, Phosphates Div).

Banthine. Methantheline bromide. Applications: Anticholinergic. (G D Searle & Co).

Banwee. Residual herbicide. (Fisons PLC, Horticulture Div).

Banweed. A suspension concentrate containing napropamide. Applications: Soil applied residual for control of annual grasses and annual broad-leaved weeds in field and container grown nursery stock. (Fisons PLC).*

Banweed-S. A suspension concentrate containing napropamide and simazine. Applications: Soil applied residual for control of annual grasses and annual broad-leaved weeds in field and container grown nursery stock. (Fisons PLC).*

Baquacil. Polymeric biguanide swimming pool sanitizer. (ICI PLC).*

Baquatop. Hydrogen peroxide for the treatment of swimming pools. (ICI PLC).*

B.A.R. Butyl acetyl ricinoleate. A vinyl plasticizer. (Lankro Chemicals Ltd).†

Barabar. Oxidation and corrosion inhibitors. (Exxon Chemical International Inc).*

Baragel. Rheological additives. Applications: Grease, underbody coatings. (NL Industries Inc).*

Barberite. A corrosion-resisting alloy containing 88.5 per cent copper, 5 per cent nickel, 5 per cent tin, and 1.5 per cent silicon. It is stated to have a high tensile strength in addition to good corrosion-resisting properties.†

Barbidex. A proprietary preparation of dexamphetamine resinate, and phenobarbitone. Antiobesity agent. (Nicholas).†

Barbivis. Phenobarbital. Applications: anticonvulsant; hypnotic; sedative. (Alcon Laboratories Inc).

Barbouze's Alloy. An alloy of aluminium with 10 per cent tin.†

Barbul Bark. (Barbura Bark). Acacia.†

Barbura Bark. See BARBUL BARK.

Bardac. Twin chain quaternary ammonium compounds. Applications: Germicidal applications in hospitals, institutions and industrial water treatment. (Lonza Limited).*

Bardase. A proprietary preparation containing phenobarbitone, hyoscyamine sulphate, hyoscine hydrobromide, atropine sulphate and aspergillus oryzae enzymes. Gastric sedative. (Parke-Davis).†

Bardew. Fungicide. (FBC Ltd).

Bardyne. Isdopher. Applications: A typical biocide with useful hospital and animal health applications. (Lonza Limited).*

Barfoed's Reagent. Copper acetate, 1 part, is dissolved in water, 15 parts. To 200 cc of this solution, 50 cc of 68 per cent acetic acid are added. Used to distinguish glucose and other monosaccharides from disaccharides, such as lactose and maltose. With glucose red cuprous oxide is produced.†

Bari Oil. See AIK OIL.

Bariform. Barium sulphate for X-ray diagnosis. (Laporte Industries Ltd).*

Bario Metal. An alloy containing 90 per cent nickel, 1.22 per cent tungsten, 0.29 per cent silicon, and 4.25 per cent chromium. It is stated to be acid and heat-resisting. A soft Bario metal is stated to contain 60 per cent cobalt, 20 per cent chromium, and 20 per cent tungsten, and a hard one 30 per cent cobalt, 30 per cent chromium, 25 per cent tungsten, 10 per cent manganese, and 5 per cent titanium.†

Bario-Muscovite. Synonym for Oellacherite.†

Bariostrontianite. Synonym for Stromnite.†

Baritite. Synonym for Baryte.†

Barium Chrome. See BARIUM YELLOW.

Barium Yellow. Barium chromate, $BaCrO_4$.†

Barium-Phosphoruranite. Synonym for Uranocircite.†

Bark. See QUERCITRON.

Bark Eleuthera. See SWEET BARK.

Barkite B. Di(dimethylcyclohexyl)oxalate. A plasticizer for cellulose lacquers (*qv*). (Laporte Industries Ltd).*

Barley Bloom. Design - clarifying additive. Applications: Beverage clarification. (NL Industries Inc).*

Barlox. A range of alkyl dimethyl amine oxides. Applications: Used for detergents, shampoos and textile processing applications. (Lonza Limited).*

Baros. A heat-resisting alloy containing 90 per cent nickel and 10 per cent chromium.†

Baros Camphor. See CAMPHOL.

Barosperse. Barium sulphate. Applications: Diagnostic aid. (Mallinckrodt Inc).

Barotrast. Barium sulphate. Applications: Diagnostic aid. (Armour Pharmaceutical Co).

Barquat. Alkyl dimethyl benzyl quaternary ammonium compounds. Applications: For germicidal applications in hospitals, institutions and industrial water treatment, also used as antistatic agents. (Lonza Limited).*

Barquinol HC. A proprietary preparation of hydrocortisone and clioquinol. An antibacterial dermatological agent. (Fisons PLC).*

Barras. A pine resin.†

Barroisite. A mineral. It is a variety of amphibole. Synonym for Carinthine.†

Barronia Metal. An alloy of from 80-83 per cent copper, 11-17 per cent zinc, 4-4.5 per cent tin, 0.4-1.15 per cent iron, and 0.5-0.65 per cent lead.†

Barseb. Hydrocortisone acetate. Applications: Glucocorticoid. (Barnes-Hind Inc).

Barseb HC. Hydrocortisone. Applications: Glucocorticoid. (Barnes-Hind Inc).

Barsilowsky's Base. Aminoditolyl *p*-toluquinonediimine.†

Barsowite. Synonym for Anorthite.†

Barth's Blue. Indigo sulphate.†

Barwood. See REDWOODS.

Baryta. Barium monoxide, BaO.†

Baryta, Caustic. See CAUSTIC BARYTA.

Baryta Green. See ROSENTHIEL'S GREEN.

Basacryl Salt. Levelling agents for textile dyeing. (BASF United Kingdom Ltd).

Basacryl/Bafixan. Dyestuffs for textile transfer printing. (BASF United Kingdom Ltd).

Bascal. Acid donor for textile dyeing. (BASF United Kingdom Ltd).

Base Oil. See BLOWN OILS.

Basex. A proprietary base-exchange material for water softening. It has an approximate composition $Na_2Al_2O_314SiO_2$. It is stated to have high softening capacity.†

Base, Millon's. See MILLON'S BASE.

Basic Chloride. See POWDER OF ALGAROTH.

Basic Lead Sulphate. See BLUE BASIC LEAD SULPHATE.

Basilen. Dyestuffs for cellulosic fibres. (BASF United Kingdom Ltd).

Basilex. A wettable powder containing tolclofos-methyl. Applications: Protective fungicide for use on all ornamentals and some edible crops against *Rhizoctonia*. (Fisons PLC).*

Basinetto Silk. See GALETTAME SILK.

Basle Blue. A safranine dyestuff.†

Basle Blue BB. A dyestuff. Dyes cotton mordanted with tannin and tartar emetic blue from a neutral bath.†

Basle Blue R. A dyestuff. Dyes cotton mordanted with tannin and tartar emetic, blue.†

Basle Blue RS. An acid dyestuff. It is a sulphonated derivative of Basle blue R.†

Basle Blue S. A dyestuff. It is the sodium salt of Basle blue sulphonic acid. Dyes wool and silk blue from an acid bath.†

Basofor. A proprietary trade name for a specially precipitated barium sulphate.†

Basolan DC. Wool finishing agent. (BASF United Kingdom Ltd).

Basopal. Wetting agents for textiles. (BASF United Kingdom Ltd).

Basosoft. Softening agents for textiles. (BASF United Kingdom Ltd).

Bassorin. (Adraganthin). Tragacanthin, the mucilage from gum tragacanth.†

Bastanet. A decolorising carbon.†

Bastard Saffron. See SAFFLOWER.

Bastose. The cellulose of jute.†

Basudin. Organophosphorus insecticides. (Ciba-Geigy PLC).

Batana Oil. See PATAVA OIL.

Batavian Dammar. See DAMMAR RESIN.

Batchite. An activated carbon produced from coconut shells.†

Bath Metal. A brass containing 83 per cent copper and 17 per cent zinc. Another alloy consists of 55 per cent copper and 45 per cent zinc.†

Battal. Fungicide, containing carbendazim. (ICI PLC).*

Batterium Metal. A high copper alloy containing 89 per cent copper, 9 per cent aluminium, and 2 per cent nickel, etc.†

Battery Copper. A brass containing 94 per cent copper and 6 per cent zinc.†

Battery Manganese. See MANGANESE BLACK.

Batu Gum. An East Indian Dammar.†

Baudische's Reagent. Cupferron (*qv*).†

Baudoin's Metal. Complex nickel silvers. One contains 72 per cent copper, 16.6 per cent nickel, 1.8 per cent cobalt, 2.25 per cent tin, and 7.1 per cent zinc, and another 75 per cent copper, 16 per cent nickel, 2.25 per cent zinc, 2.75 per cent tin, 2 per cent cobalt, 1.5 per cent iron, and 0.5 per cent aluminium.†

Baum's Acid. See SCHAEFFER'S ACID.

Baurach. Synonym for Borax.†

Bavarian Blue DBP. See METHYL BLUE.

Bavarian Blue DSF. (Methyl Blue, Water Soluble, Navy Blue B, Methyl Blue for Silk MLB, Marine Blue B). A dyestuff. It is the sodium salt of triphenyl-*p*-rosanilinedisulphonic acid., with some trisulphonic acid. Dyes silk blue from a soap bath.†

Bavarian Blue, Spirit Soluble. See DIPHENYLAMINE BLUE, SPIRIT SOLUBLE.

Bay Oil. See OIL OF BAY.

Bayberry Tallow. See LAUREL WAX.

Bayblend. Thermoplastic polymer blends. Applications: For the manufacture of injection mouldings for use in the electrical and automotive industries and for domestic appliances. (Bayer & Co).*

Bayblend. Polycarbonate/ABS blends, characterized by their cost/performance benefits. (Mobay Chemical Corp).*

Baybond. Hydrous polyurethane dispersion. Applications: For glass fibre sizes. (Bayer & Co).*

Baycast. Biomedical preparation. Applications: Polyurethane supportive dressing. (Bayer & Co).*

Baychrom A. A mineral tanning material containing chromic oxide. (Bayer & Co).*

Baycillin. A semisynthetic oral penicillin preparation. (Bayer & Co).*

Baycoll. Hydroxyl polyethers and hydroxyl polyesters. Applications: Used in conjunction with Desmodur in the production of reaction adhesives. (Bayer & Co).*

Baycryl. A range of acrylic resins. (Bayer & Co).*

Baycuten. An anti-eczematous preparation. (Bayer & Co).*

Bayderm A. Solutions of anionic dyestuffs. Applications: For spraying, curtain coating and printing as well as for shading the finish. (Bayer & Co).*

Bayderm KF. Solutions of cationic dyestuffs for brilliant spray dyeings and for shading finishing liquors. Applications: Leather industry. (Bayer & Co).*

Bayderm Lacquers Auxiliaries. Polyurethane-based products for leathers. (Bayer UK Ltd).

Bayer Base Plates Glass-Clear. Plastic plates. Applications: For the preparation of individual trays, for taking bites and trial fittings and for protective plates for surgical wounds. (Bayer & Co).*

Bayer CM. Chlorinated polyethylene. (Bayer UK Ltd).

Bayertitan. Titanium dioxide pigments. (Bayer UK Ltd).

Bayer's Tonic. A tonic and restorative. (Bayer & Co).*

Bayferon. Interferon inducer for cattle for prophylaxis and treatment of infections due to interferon-sensitive infectious agents and for paramunisation. Applications: Veterinary medicine. (Bayer & Co).*

Bayferrox. A range of iron oxides. Applications: Used as pigments in plastics and linoleum and in the rubber industry, used in electroceramics and in the manufacture of magnetic tapes. (Bayer & Co).*

Bayfidan. An emulsifiable concentrate containing 250g/l triadimenol. Applications: To control powdery mildew, rusts and rhychosporium in winter and spring crops of wheat, barley, oats and rye. (Bayer & Co).*

Bayfill. Polyols for the Bayfill system (semi-flexible foams). Applications: Used for the foam backing of flexible facings, e.g. in the automotive industry (instrument panels, arm rests etc.). (Bayer & Co).*

Bayfit. Polyols for the Bayfit system (flexible cold-curing moulded foams). Applications: Used in the upholstery sector (all-foam furniture), all-foam car seats etc. (Bayer & Co).*

Bayflex. Polyols for the Bayflex system (semi-flexible integral-skin foams).

Applications: Used in the manufacture of shoe soles and in the automotive industry (elastic bumpers, flexible body parts, steering wheel coverings, instrument panels etc). (Bayer & Co).*

Bayfolan. Foliar feed containing macro and micro nutrients. Applications: For all agricultural and horticultural crops to help recover from the effects of adverse conditions such as drought, low temperatures or waterlogging. (Bayer & Co).*

Baygal. A polyurethane based casting resin. Applications: For use in the electrical industry and for rock consolidation. (Bayer & Co).*

Baygen. Lacquers and auxilliaries based on polyurethane reactive lacquers. Applications: Leather industry. (Bayer & Co).*

Baygen Lacquers and Auxiliaries. Cold lacquer finishes based on polyurethane reactive lacquers. (Bayer UK Ltd).

Baygenal. Dyestuffs. Applications: For high-grade chrome upper leathers. (Bayer & Co).*

Bayhibit AM. 2-Phosphorobutane 1,2,4-tricarboxylic acid. Applications: Sequestering agent for calcium and magnesium ions (scale control), scale controlling additive for use in industrial cleansers, liquefaction of ceramic slips and slurries, wetting agent and liquefier for emulsion paints. (Bayer & Co).*

Baykanol AK, HLX, SL. Dyeing auxilliaries. High quality, almost neutral syntans for levelling aniline dyeing. Applications: The products facilitate the neutralization of chrome leathers and improve the fullness. (Bayer & Co).*

Baykanol Liquor TN. Light fast special liquor for suede and white leathers. Applications: Leather industry. (Bayer & Co).*

Bayleton. A wettable powder systemic fungicide containing 25% w/w triadimefon. Applications: May be applied to winter and spring crops of wheat, barley, oats or rye. For the control of powdery mildew, rhynchosporium, yellow rust, brown rust, crown rust (oats) and snow rot (winter barley). (Bayer & Co).*

Bayleton BM. A wettable powder systemic fungicide containing 12.5% w/w triadimenol and 25% w/w carbendazim. Applications: To control eyespot, mildew and early attacks of yellow and brown rust on winter wheat and winter barley, rhynchosporium on winter barley and eyespot and mildew on winter rye. (Bayer & Co).*

Bayleton CF. Wettable powder fungicide with contact and systemic properties containing 6.25% w/w triadimefon and 65% w/w captafol. Applications: To control powdery mildew, yellow and brown rust, leaf spot and glume blotch and to reduce the late-season ear disease complex on spring and winter wheat. (Bayer & Co).*

Bayleton 5. Wettable powder systemic fungicide containing 5% w/w triadimefon. Applications: To control powdery mildew on apples, hops, raspberries, strawberries and other cane fruits plus American gooseberry mildew on all varieties of blackcurrants and gooseberries. (Bayer & Co).*

Baylith. Techical grade oxides and zeolites. Applications: Uses include intensive driers for use in polyurethane systems, paints and varnishes, plastics and solvents. (Bayer & Co).*

Baylon. Low density polyethylene hompolymers and ethylene/vinyl acetate copolymers. Applications: For film, extrusion coating, cable and other applications. (Bayer & Co).*

Baylube. Polyether bases for synthetic lubricants. (Bayer UK Ltd).

Baymer. Raw materials to be foamed with Desmodur for the Baymer system (hard polyisocyanurate foams) used in the building industry. (Bayer & Co).*

Baymicin. A bactericidal broad spectrum antibiotic with specific action against gram-negative micro-organisms. (Bayer & Co).*

Baymid. PCD foam - polycarbodi-imide foam. (Bayer UK Ltd).

Baymidur. A polyurethane based casting resin. Applications: For use in the electrical industry, for rock consolidation and for the formulation of core sand binders. (Bayer & Co).*

Baymin. Flotation chemicals. (Bayer UK Ltd).

Baymix. Preparation for the control of gastrointestinal worms in cattle. Applications: Veterinary medicine. (Bayer & Co).*

Baymol A and D. Range on non-ionic tanning auxilliaries with grease emulsifying and degreasing effect. (Bayer & Co).*

Baynat. Polyols for the Baynat system (thermoformable rigid foams). Applications: For a variety of wall and ceiling coverings, also with decorative facings (automotive industry, furniture industry). (Bayer & Co).*

Bayolin. A percutaneous antirheumatic and anti-inflammatory agent. (Bayer & Co).*

Bayo-n-ox. Growth promoter for pigs, synthetically prepared, no antibiotic. Applications: Increases weight gain, improves feed conversion and reduces rearing losses. (Bayer & Co).*

Baypen. A broad spectrum penicillin. (Bayer & Co).*

Bayplast. Organic pigments. Applications: For the plastics' industry. (Bayer & Co).*

Baypreg. Polyols for the Baypreg system (filler and glass fibres containing resin compounds for the SMC technology). (Bayer & Co).*

Baypren. Polymers of 2-chlorobutadiene. Applications: They are used for rubber goods with excellent resistance to weathering and ageing, good flame-retardant behaviour and insensitivity to many chemicals. (Bayer & Co).*

Baypress. Nitrendipine. Applications: Antihypertensive. (Miles Pharmaceuticals, Div of Miles Laboratories Inc).

Baysical. A range of inorganic fillers. Applications: Uses include improving the whiteness and opacity of paper, preserving the free-flowing capacity of table salt, as a grinding aid in powders with the tendency to cake and as an extender in emulsion paints. (Bayer & Co).*

Baysilone. A range of silicone fluids. Applications: Used in the production of polishes and water repellants, transfer media, dampening fluids, hydraulic fluids, dielectrics, lubricants in the production of plastics and man-made fibres and in the metal industry. (Bayer & Co).*

Baysin. Compact colours for full grain/corrected grain side leather and splits. (Bayer UK Ltd).

Baysport. Polyols for the Baysport system (non-cellular polyurethane elastomers). Applications: Used for surfacing in sports areas, playgrounds and similar facilities. (Bayer & Co).*

Baystal. Butadiene-styrene copolymers with self-crosslinking groups and possible additional comonomers. Applications: For special applications in a variety of industries e.g. the textile, paper and rubber industries. (Bayer & Co).*

Baysynthol. Synthetic sizing agents. Applications: For the paper industry. (Bayer & Co).*

Baytan. Dry powder containing 25% w/w triadimenol and 3% w/w fuberidazole. Applications: A seed treatment for barley, wheat, oats and rye. To control the important seed and certain soil borne diseases including loose smut, covered smut, foot rot, leaf stripe, bunt and early attacks of mildew and rhynchosporium. (Bayer & Co).*

Baytec. Crosslinking agents to be processed with Desmodur for the Baytec system (casting type non-cellular polyurethane elastomers). Applications: For use in the manufacture of technical goods. (Bayer & Co).*

Baytherm. Polyols for the Baytherm system (rigid foams). Applications: Used in heat/cold insulation and in the building industry. (Bayer & Co).*

Bayticol. Preparation for the control of all resistant strains of one and multi-host ticks. Applications: Veterinary medicine. (Bayer & Co).*

BB Accelerator. A rubber vulcanization accelerator. It is a derivative of dimethyl-*p*-phenylene-diamine. (BF Goodrich Chemical Co).†

B.B.D.C. Standard Alloy. An alloy of 88.5 per cent copper, 10 per cent tin, 1 per cent nickel, 0.25 per cent lead, and 0.25 per cent phosphorus.†

BCF. Bromochlorodifluoromethane/Halon 1211 fire extinguishant. (ICI PLC).*

B.C. 500. A proprietary vitamin supplement containing thiamine, riboflavine, pyridoxine, calcium partothenate, cyanocobalamin and ascorbic acid. (Ayerst Laboratories).†

BC500 With Iron. Vitamin supplement containing ferrous fumarate, thiamine, riboflavine, nicotinamide, pyridoxine, calcium pantothenate, ascorbic acid. Applications: Treatment of iron deficiency when therapeutic doses of water-soluble vitamins are also required. (Ayerst Laboratories).†

Beale's Carmine. (Beale's Stain). A microscopic stain. It contains 1 gram carmine, 1.5 cc solution of ammonia, 80 cc glycerin, 25 cc water, and 120 cc alcohol.†

Bean Ore. A variety of Limonite.†

Bear. A cotton fabric grade of TUFNOL industrial laminates. (Tufnol).*

Bearing Bronze. See BRONZE BEARING METAL.

Bearing Metals. (Antifriction Metals, White Metals). There are many varieties of these alloys. Some of these have lead in the highest proportion with tin and antimony as the other main constituents. They contain from 42-86 per cent lead, 7-20 per cent antimony, 2-42 per cent tin, and 0-2 per cent copper. Some have tin as the predominating metal with copper and antimony as the other ingredients. These have tin from 38-91 per cent, antimony from 4-26 per cent, and copper from 2-21 per cent, another type have copper in high proportion with zinc, tin, and sometimes lead in addition. Such alloys contain 72-90 per cent copper, 0-22 per cent zinc, 2-20 per cent tin, and 0-15 per cent lead. Finally, a smaller class contain 47-90 per cent zinc, 1-38 per cent tin, 0.5-8 per cent copper, 0-4 per cent lead, and 0-6 per cent antimony.†

Bearium. Proprietary alloys of copper with 17.5-28 per cent lead and 10 per cent tin. Bearing metal.†

Beatin. A trade mark used in Germany for the sale of Famel Syrup (*qv*). (Optrex).†

Beaudouin's Reagent. 1 per cent furfural in alcohol.†

Beaver Steel. A proprietary nickel chromium steel containing 1.5 per cent nickel, 0.75 per cent chromium, 0.6 per cent manganese, and 0.55 per cent carbon.†

Bebate. A proprietary preparation containing betamethasone 17- benzoate. A Steroid skin preparation. (Warner).†

Becantyl. A proprietary preparation containing sodium dibunate. A cough syrup. (Horlicks).†

Bechilite. See PRICITE.

Beckol. A proprietary trade name for synthetic alkyd resin.†

Beckolin. A proprietary trade name for synthetic oil lacquer and varnish resins.†

Beckolloid. A proprietary synthetic resin plastic.†

Beckopol. A proprietary trade name for phenolated copal varnish and lacquer resins.†

Beckosol. A proprietary trade name for synthetic oil lacquer and varnish resins. Also a trade name for synthetic alkyd resin used in paints.†

Becksol. A proprietary trade name for alkyd synthetic resins.†

Beckton White. See LITHOPONE.

Beclovent Inhaler. Beclomethasone dipropionate. Applications: Glucocorticoid. (Glaxo Inc).

Becomel. A proprietary preparation containing nicotinamide, thiamine, pyridoxine and riboflavine. Vitamin supplement. (Crookes Laboratories).†

Beconase. A proprietary preparation of BECLOMETHASONE dipropionate as an aerosol, used in the treatment of allergic rhinitis. (Allen & Hanbury).†

Beconase Nasal Inhaler. Beclomethasone dipropionate. Applications: Glucocorticoid. (Glaxo Inc).

Becosal. A range of fire extinguishing powders which are also compatible with foam. (Degussa).*

Becosed. A proprietary preparation of phenobarbitone, aneurine, riboflavin and nicotinamide. (Norton, H N Co).†

Becosym. A proprietary preparation containing thiamine, riboflavine, nicotinamide and pyridoxine. A Vitamin B complex. (Roche Products Ltd).*

Becotide. A proprietary preparation of BECLOMETHASONE dipropionate as an aerosol. Used in asthma therapy. (Allen & Hanbury).†

Becxopox. Epoxide resins. Applications: Paints, adhesives, sealants and encapsulants. (Resinous Chemicals Ltd).*

Bedesol. Synthetic resins.†

Bedranol. Pink film coated tablets containing 10mg, 40mg, 80mg and 160mg propranolol hydrochloride BP. Applications: Bedranol is used in the treatment of hypertension, angina pectoris, cardiac dysrhythmias, tachycardia, anxiety, essential tremor, prophylaxis of migraine, adjunctive therapy in thyrotoxicosis. (Lagap Pharmaceuticals Ltd).*

Beechwood Sugar. See WOOD SUGAR.

Beehive Balsam. A proprietary preparation of glycerin, honey, lemon and ipecacuanha. Applications: A cough linctus. (Ayrton Saunders plc).*

Been Oil. See OIL OF BEEN.

Beesix. Pyridoxine Hydrochloride. Applications: Vitamin. (O'Neal, Jones & Feldman Pharmaceuticals).

Beet Molasses. See MOLASSES.

Beet Sugar. See SUCROSE.

Beetle. Foundry resins. (BIP Chemicals Ltd).

Beetle. Laminating resins. (BIP Chemicals Ltd).

Beetle. Polyester resins. (BIP Chemicals Ltd).

Beetle. Nylon 6 compounds. (BIP Chemicals Ltd).

Beetle. Nylon 6.6 compounds. (BIP Chemicals Ltd).

Beetle. Urea formaldehyde moulding compounds. (BIP Chemicals Ltd).

Beetle. Urea, melamine, alkyd and polyester resins. (BIP Chemicals Ltd).

Beetle. Synthetic resin, moulding powders, synthetic adhesives, lacquer, textile, paper and leather resins. (BIP Chemicals Ltd).

Beetle Resin BT 333. A cyclic reactant. Recommended for soft mechanical finishes on cellulosic fabrics, also soft crease resistant and shrink resistant finishes on cellulose/synthetic fibre blends. (BIP Chemicals Ltd).*

Beetle Resin BT 334. A cyclic reactant-modified melamine cross linking agent. Recommended for chlorine resistant finishes on cotton fabrics and easy care finishes on cellulose/synthetic fibre blends. (BIP Chemicals Ltd).*

Beetle Resin W69. A modified urea-formaldehyde resin. Recommended for glueing "Vac-Vac" treated timbers where difficulties may be experienced with standard adhesives. (BIP Chemicals Ltd).*

Beflavine Roche. A proprietary preparation of riboflavin (Vitamin B_2). Applications: Vitamin supplement. (Roche Products Ltd).*

Beflavit. Vitamin B2 (riboflavin) preparations. (Roche Products Ltd).

Behn Oil. See OIL OF BEEN.

Beilstein. Synonym for nephrite.†

Beinstein. Synonym for osteocolla.†

Bel. See BAEL.

Bela. See BAEL.

Belclene. Water treatment chemicals. (Ciba-Geigy PLC).

Belco. Cellulose car finish paints and resins. (ICI PLC).*

Beldex. ASS modifiers for PVC. (Borg Warner Chemicals).

Belgard. Water treatment chemicals. (Ciba-Geigy PLC).

Beligno Seeds. See OLIVES OF JAVA.

Belite. Antifoaming agents. (Ciba-Geigy PLC).

Bell Brass. See BRASS.

Bell Metal. Alloys usually consisting of from 78-80 per cent copper and 20-22 per cent tin, but sometimes lead and zinc are added. The alloy containing 78 per cent copper and 22 per cent tin has a specific gravity of 8.7, and melts at 890° C. Another bell metal consists of 83 per cent aluminium, 10 per cent manganese, and 7 per cent cadmium. Used for casting bells and other purposes.†

Bell Pepper. The fruit of *Capsicum grossum*.†

Belladenal. A proprietary preparation of phenobarbitone and belladonna leaf. A sedative. (Sandoz).†

Bellasol. Foam preventives. (Ciba-Geigy PLC).

Bellauxine. Water treatment chemicals. (Ciba-Geigy PLC).

Bellergal. A proprietary preparation of belladonna, ergotamine tartrate and phenobarbitone. A sedative. (Sandoz).†

Bellite. Explosives for coal mine. Consists of ammonium nitrate and dinitrobenzene with or without sodium chloride.†

Belloid. Dispersing agents. (Ciba-Geigy PLC).

Beloc. A proprietary preparation of metoprolol tartrate used for hypertension, cardiac arrythmias. (Pfizer International).*

Belpro. A water based acrylic co-polymer temporary coating and remover. (Also known as TEMPRO). (ICI PLC).*

Belro. Wood rosin derivative insoluble in water, soluble in organic solvents, fats and oils. It has a typical acid number of 114 and drop softening point of 90°C. Applications: Construction adhesive. (Hercules Inc).†

Belsoft. Blend of acid amides and fatty alcohol sulphates, anionic. Applications: Softener for cellulosic and synthetic fibres, especially suitable for terry cloths and textured polyester. (Henkel Chemicals Ltd).

Belsol. Wetting and penetrating agents. (JC Bottomley and Emerson).†

Beltherm. Fuel oil additive. (Ciba-Geigy PLC).

Belzak AC. Alpha sodium glucoheptonate dihydrate crystals sequestering agent. Applications: Metal cleaning and processing, various industrial cleaning compounds, bottle wash, set retarder in concrete and trace metals for agriculture. (Belzac Corporation).*

Belzak BL-50. Beta sodium glucoheptonate 50% liquid. Applications: Metal cleaning and processing, various industrial cleaning compounds, bottle wash, set retarder in concrete and trace metals for agriculture. (Belzac Corporation).*

Bemal. An alloy of 70 per cent copper, 29 per cent zinc, and traces of lead and iron.†

Ben Oil. See OIL OF BEEN.

Benacine. A proprietary preparation of diphenhydramine and hyoscine hydrobromide. Antinauseant. (Parke-Davis).†

Benadon. A proprietary preparation of pyridoxine. An anti nauseant. (Roche Products Ltd).*

Benadryl. Diphenhydramine hydrochloride. Applications: Antihistaminic. (Parke-Davis, Div of Warner-Lambert Co).

Benafed. A cough suppressant. A proprietary preparation of diphenhydramine hydrochloride. dextramethorphan hydrobromide, pseudoephedrine hydrochloride, ammonium chloride. sodium citrate, chloroform and menthol in a syrup. (Parke-Davis).†

Benalite. A proprietary trade name for a lignin plastic. Cured lignin sheets.†

Benaloid. A proprietary trade name for uncured lignin sheets.†

Benapen. Benethamine Penicillin. (Glaxo Pharmaceuticals Ltd).†

Benathix. Rheological additives. Applications: Unsaturated polyester laminating resins. (NL Industries Inc).*

Benatol. Pure benzyl alcohol. (Bush Boake Allen).*

Benazalox. Selective herbicide. (FBC Ltd).

Bendalite. A proprietary trade name for a cast styrene synthetic resin.†

Bendizon. An antihypertensive with a depot effect. (Bayer & Co).*

Bendogen. Tablets containing 10mg and 50mg bethanidine sulphate BP. Applications: Treatment of all grades of hypertension. (Lagap Pharmaceuticals Ltd).*

Bendopa. Levodopa. Applications: Antiparkinsonian. (ICN).

Bendroflumethiazide. Bendrofluazide.†

Benedict Plate. A nickel silver. It contains 57 per cent copper, 28 per cent zinc, and 15 per cent nickel.†

Benedict-Nash Reagent. For the ammonia content of the blood. Potassium carbonate, 8 grams, is placed in a 100 cc flask and 50 cc of distilled water added, then 12 grams of potassium oxalate crystals. The solution is boiled to about 30 ml and then diluted to 50 cc with ammonia-free distilled water. This is repeated and finally diluted to 80 ml.†

Benedict's Creatinine Solution. Creatinin, 100 mg, is placed in a 100 cc flask with some distilled water, 1 cc of concentrated hydrochloric acid added, and the whole made up to 100 cc with distilled water.†

Benedict's Molybdic Acid Reagent. For blood phosphorus. Pure molybdic acid, 20 grams, is placed in 25 cc of 20 per cent caustic soda solution and warmed to 50° C. Dilute to 200 cc with distilled water, mix, and filter. Add 200 cc concentrated sulphuric acid, keeping the solution cool.†

Benedict's Picric-picrate Solution. Normal solution of caustic soda, 125 ml, is added to 700 ml distilled water at 80° C., and then 36 grams of picric acid added and dissolved by shaking. It is finally made up to 1,000 cc†

Benedict's Reagent for Blood Phosphorus. Sodium hydrogen sulphite, 30 grams, is dissolved in 100 ml distilled water, 1 gram hydroquinone added, and the whole diluted with water to 200 ml.†

Benedict's Solution for Glucose (Qualitative). A solution of 173 grams sodium citrate and 100 grams anhydrous sodium carbonate is made using 600 cc water. The solution is diluted to 850 cc Copper sulphate, $CuSO_4$ $5H_1O$, 17.3 grams, is dissolved in 100 cc water, diluted to 150 ml, and added. The mixture is used to test for glucose by adding 5-10 drops of the glucose solution to 5 cc of the reagent and boiling. A red, yellow, or green precipitate is produced, the final colour depending upon the amount of glucose present.†

Benedict's Solution for Glucose (Quantitative). A solution is made by dissolving 200 grams of sodium carbonate, Na_2CO_3 $10H_2O$, or 100 grams Na_2CO_3, 200 grams sodium citrate, and 125 grams potassium thiocyanate in enough water to make 800 cc, and filtered. Copper sulphate, $CuCO_4$ $5H_2O$, 18 grams, is dissolved in 100 cc water and poured into the above solution, and 5 cc of a 5 per cent solution of potassium ferrocyanide are added, and the whole diluted to 1 litre. Used for the determination of glucose, 25 cc of the reagent = 0.05 gram of glucose.†

Benedict's Standard Glucose Solution. (*a*) For urine : pure glucose, 167 mg is dissolved in 400 cc water, 5 cc toluene added, and the whole mixed and diluted to 500 cc with water. (*b*) For blood : pure glucose, 0.1 gram, is dissolved in 400 cc of a 2 per cent picric acid in a 500 cc flask. It is mixed and diluted to the mark with the saturated picric acid solution. 3 ml = 0.6 mg glucose.†

Benedict's Standard Phenol Solution. Resorcinol, 11-62 mg in 100 cc 0.1 normal hydrochloric acid solution. 5 cc = 0.5 mg phenol.†

Benedict's Standard Phosphorus Solution. Potassium dihydrogen phosphate, 0.4394 gram, is dissolved in 800 ml distilled water, and made up to 1,000 ml with chloroform water (10 ml with 200 ml water). 5 ml contains 0.025 mg phosphorus.†

Benedict's Sulphur Reagent. Used to test for sulphur in urine. It consists of 200 grams of sulphur-free copper nitrate and 50 grams of sodium or potassium chlorate dissolved in distilled water to 1 litre.†

Benefex. Active ingredient: benfluralin; N-butyl-N-ethyl-2,6-dinitro-4-trifluoromethylaniline. Applications: Pre-emergence herbicide with a wide range of weed control both of annual grass weeds and broad-leaved weeds. (Agan Chemical Manufacturers Ltd).†

Benemid. Probenecid. Applications: Uricosuric agent for the treatment of chronic gout. (Merck Sharp & Dohme).*

Benerva. A proprietary preparation of thiamine hydrochloride (Vitamin B_1). (Roche Products Ltd).*

Benerva Compound. Vitamin B1, B2 and nicotinamide. (Roche Products Ltd).

Benexol Roche. A proprietary preparation of vitamins B1 and B6. (Roche Products Ltd).*

Bengal Blue. See INDULINE, SOLUBLE.

Bengal Blue G. See GENZOAZURINE G.

Bengal Catechu. See CUTCH.

Bengal Isinglass. See AGAR-AGAR.

Bengal Kino. See KINO, BENGAL.

Bengal Red. See ROSE BENGAL.

Bengaline. See INDOINE BLUE R.

Bengue's Balsam. 20% menthol, 20% methyl salicylate in a lanolin base. Applications: Rubefacient - for the symptomatic relief of muscular pain and stiffness (including rheumatic pains), also as a decongestant by application to the chest or by inhalation. (Bengue & Co Ltd).*

Beni Oil. See GINGELLY OIL.

Benical. A proprietary preparation of dextromethorphan, pseudoephedrine and chlorphenamine. Cough and cold preparation. (Roche Products Ltd).*

Beniseed Oil. See GINGELLY OIL.

Benjamin Gum. See GUM BENJAMIN.

Benjamin Oil. See OIL OF BENJAMIN.

Benlate. Fungicide for garden use (sold in UK on behalf of Du Pont). (ICI PLC).*

Bennatate. A proprietary trade name for a cellulose acetate plastic.†

Benné Oil. See GINGELLY OIL.

Benoquin. Monobenzone. Applications: Depigmentor. (Elder Pharmaceuticals Inc).

Benoral. A proprietary preparation of benorylate. An antirheumatic drug. (Sterling Research Laboratories).†

Benoxyl. A proprietary preparation of benzoyl peroxide used in dermatology as an antibacterial agent. (Stiefel Laboratories (UK) Ltd).*

Bensapol. A proprietary trade name for a wetting agent consisting of sulphonated oils and a solvent.†

Bensuccin. A proprietary preparation of benzyl succinate.†

Bensuldazic Acid. A fungicide. (5-Benzyl-6-thioxo-1,3,5-thiadiazin-yl) acetic acid. DEFUNGIT is the sodium salt.†

Bentalol. Specially pure benzyl alcohol. (Bush Boake Allen Ltd).

Benteine. Specially pure benzyl acetate. (Bush Boake Allen Ltd).

Bentene. Butyraldehyde - aniline. A proprietary rubber accelerator. (Naugatuck (US Rubber)).†

Bentokol. Foundry coal dust replacement additive. (Foseco (FS) Ltd).*

Bentone. Rheological additives. Applications: Paint, ink, grease, caulks, sealants, cosmetics and adhesives. (NL Industries Inc).*

Bentone Gel. Rheological additives. Applications: Pregelled bentone additive for cosmetics. (NL Industries Inc).*

Bentone SD. Super dispersing rheological additives. Applications: Paint, ink, caulks, sealants and adhesives. (NL Industries Inc).*

Bentonite. A variety of bedded clay which forms thixotropic aqueous suspensions. It is used especially in oil-well drilling.†

BentoPharm. Bentonite, pharmaceutical quality. Applications: Suspension and thickening agent. (Bromhead & Denison Ltd).*

Bentyl. Dicyclomine hydrochloride. Applications: Anticholinergic. (Merrell Dow Pharmaceuticals Inc, Subsidiary of Dow Chemical Co).

Benvic. Granulated PVC compound. (Laporte Industries Ltd).

Benylate. Benzyl benzoate. Applications: Pharmaceutic necessity for dimercaprol. (Sterling Drug Inc).

Benylin DM. Dextromethorphan hydrobromide. Applications: Antitussive. (Parke-Davis, Div of Warner-Lambert Co).

Benylin Expectorant. A proprietary preparation containing diphenhydramine hydrochloride, ammonium chloride, sodium citrate, chloroform and menthol. A cough linctus. (Parke-Davis).†

Benzac. Benzoyl peroxide. Applications: Keratolytic. (Owen Laboratories Div, Dermatological Products of Texas, Alcon Laboratories Inc).

Benzal Green. See MALACHITE GREEN.

Benzaldehyde Green. See MALACHITE GREEN.

Benzamin. Diazotizing dyes for cotton and other vegetable fibres, textiles from regenerated cellulose. (Bayer UK Ltd).

Benzanil. A range of direct dyes. Applications: Dyeing of cotton and viscose fibres. (Yorkshire Chemicals Plc).*

Benzanthronequinolines. Violet blue dyestuffs from α-aminoanthraquinone.†

Benzarene Blue 2B. A dyestuff. It is a British equivalent of Diamine blue 2B.†

Benzarene Brown G. A dyestuff. It is a British equivalent of Benzo brown G.†

Benzarene Green B. A dyestuff. It is a British equivalent of Diamine green B.†

Benzedrex. Proplhexedrine. Applications: Adrenergic. (Smith Kline & French Laboratories).

Benzets. A proprietary preparation of benzalkonium chloride and benzocaine. Throat lozenges. (Norton, H N Co).†

Benzex. A proprietary product. Benzyl cellulose.†

Benzhydrol. Diphenyl methanol, $(C_6H_5)_2.CH.OH$.†

Benzilan. Active ingredient: chlorobenzilate, ethyl-4,4'-dichlorobenzilate. Applications: Agricultural acaricide. (Makhteshim Chemical Works Ltd).†

Benzine. (Petroleum Naphtha, Benzoline, "A" Naphtha). That fraction obtained in the refining of petroleum which boils from 70-120° C., of specific gravity 0.725-0.737.†

Benzine, Heavy. See LIGROIN.

Benzine, Petroleum. See C-PETROLEUM NAPHTHA.

Benzine, Standard. See STANDARD BENZINE.

Benzo. Direct dyes. Applications: Suitable for cotton and rayon goods where no special demands are made on fastness properties. (Bayer & Co).*

Benzo Black. A direct cotton dyestuff.†

Benzo Black-blue G. A dyestuff. Dyes cotton black-blue from an alkaline bath.†

Benzo Black-blue R. A dyestuff. Dyes cotton dark bluish-violet from a soap bath.†

Benzo Black-blue 5G. A dyestuff. Dyes cotton greenish-black.†

Benzo Blue. (Diamine Blue, Benzo Cyanine, Congo Blue, Congo Cyanine, Chicago Blue, Columbia Blue). Tetrazo dyestuffs from toluidine or dianisidine and aminonaphtholsulphonic acids.†

Benzo Blue RW. A dyestuff identical with Chicago blue RW and Diamine blue RW. It is a similar dye to Chicago blue 2R.†

Benzo Brown B. (Orion Brown B, Enbico Direct Brown R). A dyestuff. Dyes cotton brown from a neutral salt bath.†

Benzo Brown BX and BR. Dyestuffs, which are derivatives of chrysoidine. They are identical with Cotton brown A and N.†

Benzo Brown G. (Benzarene Brown G, Orion Brown G, Enbico Direct Brown BB, Chlorazol Brown GM). A dyestuff. Dyes cotton yellowish-brown from a neutral salt bath.†

Benzo Brown SR. See INGRAIN BROWN.

Benzo Chrome-black B. A dyestuff similar to Benzo brown.†

Benzo Chrome-brown 3R. A polyazo dyestuff of the same class as Benzo indigo blue. Dyes unmordanted cotton.†

Benzo Cuprol. Direct dyestuffs whose wet fastness and light fastness are considerably improved by an aftertreatment with copper salts. (Bayer & Co).*

Benzo Dark Brown. A direct cotton dyestuff.†

Benzo Dark Green B. A polyazo dyestuff, similar to Diamine green G and Columbia green. Dyes cotton bluish and yellowish-olive.†

Benzo Fast Grey. A direct cotton dyestuff, giving bluish-grey shades from an alkaline bath.†

Benzo Fast Orange S. See BENZO FAST SCARLET GS.

Benzo Fast Pink 2BL. (Chlorazol Fast Pink BK). A dyestuff. Dyes cotton pink.†

Benzo Fast Red. See BENZO FAST SCARLET GS.

Benzo Fast Scarlet GS. (Benzo Fast Scarlet 4BS, 8BS, Benzo Fast Orange S, Benzo Fast Red, Chlorazol Fast Scarlet 4BS, Direct Fast Scarlet SE, Paramine Fast Scarlet 4BS, Direct Fast Orange SE, Paramine Fast Orange S, Chlorazol Fast Orange R). Cotton dyes.†

Benzo Fast Scarlet 4BS. See BENZO FAST SCARLET GS.

Benzo Fast Scarlet 8BS. See BENZO FAST SCARLET GS.

Benzo-furane Resin. See PARACOUMARONE RESIN.

Benzo Grey. A dyestuff. Dyes cotton grey.†

Benzo Indigo Blue. A dyestuff. Dyes cotton indigo blue.†

Benzo Olive. A dyestuff. Dyes cotton greenish-olive from a neutral salt bath.†

Benzo Orange R. (Paramine Orange G, R, Chlorazol Orange RN). A dyestuff. Dyes cotton orange from an alkaline bath, also chromed wool.†

Benzo Pure Blue. Identical with Diamine pure blue and Congo pure blue. See DIAMINE SKY BLUE.†

Benzo Pure Blue A. Identical with Diamine pure blue A and Congo pure blue A.†

Benzo Pure Blue 4B. Identical with Diamine blue 4B and Chicago blue 4B.†

Benzo Red Blue G, R. Dyestuffs identical with Columbia blue G, R, and Diamine blue LG, LR.†

Benzo Sky Blue. See DIAMINE SKY BLUE.

Benzoazurin G. (Benzoazurin 3G, Bengal Blue G). A dyestuff. Dyes cotton blue from an alkaline bath.†

Benzoazurin R. A dyestuff. It is a mixture of Benzoazurin G and Azo blue.†

Benzoazurin 3G. See BENZOAZURIN G.

Benzocaine. See ANAESTHESIN.

Benzocyanine 3B, B, R. Substantive cotton dyestuffs, identical with the corresponding marks of Congo cyanine and Diamine cyanine. Employed in calico printing.†

Benzodent. Analgesic denture ointment, for denture discomfort. (Richardson-Vicks Inc).*

Benzoflavine. A dyestuff. It is the hydrochloride of diaminophenyldimethylacridine. Dyes silk, wool, and mordanted cotton yellow.†

Benzoflex. Plasticizer. (Velsicol Chemical Corp).

Benzoflex 2-45. A proprietary trade name for diethylene glycol dibenzoate. (Tennessee Corporation).†

Benzofloc. Organic and inorganic specialty flocculant and coagulant polymers. Applications: Solids flocculation in industrial and municipal solids, dissolved or collodial form, colour, turbidity, algae removal. Specialty application. (Benzsay & Harrison Inc).*

Benzoin, Flowers of. See FLOWERS OF BENJAMIN.

Benzoin Yellow. A dyestuff obtained by the condensation of benzoin with gallic acid in the presence of cold sulphuric acid. Dyes chromed wool yellow.†

Benzonaphthol. Benzoyl-β-naphthol, $C_{10}H_7O\ COC_6H_5$. An antiseptic.†

Benzopurpurin B. An azo dyestuff. Dyes cotton red from an alkaline bath.†

Benzopurpurin 10B. (Cotton Red 10B). A dyestuff. It is the sodium salt of dimethoxydiphenyldisazobinaphthionic acid, $C_{34}H_{26}N_6O_8S_2Na_3$. Dyes cotton carmine red from an alkaline bath.†

Benzopurpurin 4B. (Cotton Red 4B, Sultan Red 4B, Erie Red B Conc., Imperial Red, Eclipse Red, Victoria Red, Fast Scarlet, Azamine 4B). A dyestuff. Dyes cotton red from an alkaline, and wool from a neutral bath.†

Benzopurpurin 6B. A dyestuff. Dyes cotton red from an alkaline bath.†

Benzopyrocatechol. Dihydroxybenzophenone.†

Benzoquinone. Quinone, $C_6H_4O_2$.†

Benzoresorcin. Dihydroxy-benzophenone.†

Benzoyl Green. See MALACHITE GREEN.

Benzoyl Pink. (Rose de Benzoyl). A dyestuff. Dyes unmordanted cotton pink.†

Benzoyl Superoxide. See LUCIDOL.

Benztrone. Oestradiol benzoate BP in ethyl oleate BP for injection. Applications: May be used for oestrogen replacement therapy associated with the menapause, female hypogonadiom and dysmenorrhoea. Also used for the treatment of malignant neoplasm of the breast of post menapausal women and of the prostate in men. (Paines & Byrne Ltd).*

Benzyl. Acid wool dyestuffs. (Clayton Aniline Co).†

Benzyl Blue. A dyestuff made from rosaniline, which dyes wool, silk, and cotton. The term is also used for a mixture of methyl violet and malachite green.†

Benzyl Green B. A dyestuff obtained by combining *o*-chlorobenzaldehyde with ethylbenzylanilinesulphonic acid, and oxidizing the product.†

Benzyl Violet. (Paris Violet 6B, Methyl Violet 6B, Methyl Violet 5B, Methyl Violet 6B Extra, Violet 5B, Violet 6B). A dyestuff, which is chiefly a mixture of the hydrochlorides of benzylpentamethyl-*p*-rosaniline and hexamethyl-*p*-rosaniline. It dyes silk and wool violet, and cotton after mordanting with tannin and tartar emetic.†

Benzyl Violet 5BN. A dyestuff. It is a British equivalent of Formyl violet S4B.†

Benzylidene Rubber. A product obtained by treating crepe rubber in carbon tetrachloride with benzyl chloride in the presence of aluminium chloride. It is almost insoluble in organic solvents.†

Benzyphos. Sodium dibenzylphosphate.†

Beogex. A proprietary preparation containing sodium acid phosphate and sodium bicarbonate. A laxative. (Pharmax).†

Bepadin. Bepridil hydrochloride. Applications: Vasodilator. (Wallace Laboratories, Div of Carter-Wallace Inc).

Bepanthen. A proprietary preparation of pantothenic alcohol or dexpanthenol, used in the treatment of paralytic ileus. (Roche Products Ltd).*

B-Petroleum Naphtha. (Ligroine, Naphtha, Cleaning Oil). That fraction of petroleum distilling at 100-120° C., Of specific gravity 0.707-0.772.†

Beplete. A proprietary preparation of phenobarbitone, thiamine hydrochloride, riboflavine, pyridoxine hydrochloride, nicotinamide and alcohol. (Wyeth).†

Beplex. A proprietary preparation of thiamine, riboflavine and nicotinamide. (Wyeth).†

Bercotox. Cattle dips and sprays. (The Wellcome Foundation Ltd).

Berelex. Gibberellic acid, a hormone growth regulator. (ICI PLC).*

Bergaminol. (Bergamiol). Artificial oil of bergamot. Linalyl acetate. Used in perfumery.†

Bergamiol. See BERGAMINOL.

Bergauf. Skin protective soap with 0-48-G skin protective agent. Medicated foot-spray with bactericidal and fungicidal properties. Eye cleansing cream for intensive and soothing cleansing of the area around eyes. Applications: For protection and care of the skin under environmental stress. (Dynamit Nobel Wien GmbH).*

Berger Colorizer - Full Gloss/Vinyl Matt/Vinyl Silk. Gloss - alkyd resin based thinned with white spirit. Matt - V.A. copolymer emulsion, water based. Silk - V.A. copolymer emulsion, water based. Applications: 450 colours available, high gloss external and internal application, matt and silk interior application. (Berger Jenson & Nicholson Ltd).*

Berger Cuprinol Woodpaints and Woodstains. Primer, Matt and Gloss: Acrylic copolymer emulsion, water thinned. Sheen finishes: Alkyd based - white spirit thinned. Applications: Exterior microporous wood paints and stains giving protection and decoration. (Berger Jenson & Nicholson Ltd).*

Berger Mixture. A mixture of zinc, potassium chlorate, a chlorinating agent, such as carbon tetrachloride, and a filler, such as kieselguhr. A smoke screen material.†

Berkatens. Verapamil. Applications: For the treatment of angina pectoris, supraventricular tachycardia, atrial fibrillation and flutter, hypertension. (Berk Pharmaceuticals Ltd).*

Berkazide. Bendrofluazide. (Berk Pharmaceuticals Ltd).*

Berkmycen. A proprietary preparation of oxytetracycline. An antibiotic. (Berk Pharmaceuticals Ltd).*

Berkolol. Propranolol. Applications: For the treatment of hypertension, angina pectoris, cardiac dysrhythmias and tachycardia. (Berk Pharmaceuticals Ltd).*

Berkozide. A proprietary preparation of bendrofluazide. A diuretic. (Berk Pharmaceuticals Ltd).*

Berlin Alloy. An alloy containing from 52-63 per cent copper, 26-31 per cent zinc, and 6-22 per cent nickel.†

Berlin Brown. A pigment produced by charring Prussian blue. It is a mixture of ferroso-ferric oxide and charcoal, and is used as an artist's colour.†

Berlin Red. See INDIAN RED.

Berlin White. See WHITE LEAD.

Bernit. A proprietary trade name for a cellulose acetate plastic.†

Bernthsen's Violet. (Isothionine). A dyestuff. It is β-aminodiphenylimide, and is isomeric with Lauth's violet. Dyes reddish-violet.†

Berocca. Proprietary preparation of vitamins B1, 2,5, 6 and 12, vitamin C, biotin and nicotinamide. (Roche Products Ltd).*

Berol. Block polymers. Applications: Machine dishwashing and rinse aids; emulsion polymerization. (Berol Kemi (UK) Ltd).

Berol 259. Nonionic surfactant consisting of nonylphenol ethoxylate in liquid form. Application: Foam depressor. (Berol Kemi (UK) Ltd).

Berol 26 and 02. Nonionic surfactant of the nonylphenol ethoxylate type, in liquid form. Application: Emulsifiers; solvent cleaners. (Berol Kemi (UK) Ltd).

Berol 267 and 09. Nonionic surfactant of the nonylphenol ethoxylate type, in liquid form. Application: Liquid cleaners; wetting agents; emulsifiers. (Berol Kemi (UK) Ltd).

Berol 269. Dinonylphenol ethoxylate in liquid form. Application: Nonionic surfactant used in emulsifiers. (Berol Kemi (UK) Ltd).

Berol 272 and 716. Dinonylphenol ethoxylate in liquid or paste form. Application: Nonionic low foaming detergents. (Berol Kemi (UK) Ltd).

Berol 278, 281, 282, 291 and 292. Nonionic surfactants of the nonylphenol ethoxylate type in liquid or wax form. Application: Emulsion polymerization. (Berol Kemi (UK) Ltd).

Berol 452 and 475. Sodium alkyl ether sulphate in liquid or paste form. Applications: Anionic surfactants used in shampoos and bath preparations and the textile industry. (Berol Kemi (UK) Ltd).

Berol 472. Sodium cetyl sulphate in liquid form. Applications: Anionic surfactant used in emulsion polymerization. (Berol Kemi (UK) Ltd).

Berol 474. Sodium lauryl sulphate in paste form. Applications: Anionic surfactant used in emulsion polymerization. (Berol Kemi (UK) Ltd).

Berol 480. Triethanolamine lauryl sulphate in liquid form. Applications: Anionic surfactant used in shampoos and bath preparations. (Berol Kemi (UK) Ltd).

Berol 484. Monoethanolamine lauryl sulphate in liquid form. Applications: Anionic surfactant used in shampoos and bath preparations. (Berol Kemi (UK) Ltd).

Berol 490. Anionic surfactant in liquid form. Applications: Emulsion polymerization. (Berol Kemi (UK) Ltd).

Berol 496. Anionic surfactant in the form of a paste. Applications: Powder and liquid detergents. (Berol Kemi (UK) Ltd).

Berol 513 and 525. Anionic surfactant in acid form, in which the anion is a phosphate ester. Supplied as a paste.

Application: Detergent auxiliary. (Berol Kemi (UK) Ltd).

Berol 518. Anionic surfactant in acid form in which the anion is a phosphate ester. Supplied as a wax. Application: Foam regulator. (Berol Kemi (UK) Ltd).

Berol 563. Alkyl polyglycol ether ammonium methyl sulphate in liquid form. Applications: Cationic surfactant used in alkaline cleaning eg vehicle cleaning. (Berol Kemi (UK) Ltd).

Berol 594. Hydroxy-ethyl 2 alkylimidazoline in liquid form. Application: Cationic surfactant used in water displacing acid cleaners in industrial cleaning; corrosion inhibition. (Berol Kemi (UK) Ltd).

Berol 733, 521 and 522. Potassium phosphate ester in liquid form. Application: Hydrotrope for non-ionics. (Berol Kemi (UK) Ltd).

Berol 822. Anionic surfactant in liquid form. (Berol Kemi (UK) Ltd).

Berotec. Fenoterol. Applications: Bronchodilator. (Boehringer Ingelheim Pharmaceuticals Inc).

Berpak. A 2 component cold curing polyurethane encapsulant. Sold in kits containing moulds and electrical accessories. Applications: Jointing of underground mains electrical cables and telecommunications cables. (Berger Elastomers).*

Berries, Yellow. See PERSIAN BERRIES.

Bersch Bearing Metal. An alloy of 93 per cent aluminium and 7 per cent nickel.†

Berthier's Alloy. A copper-nickel alloy containing approximately 32 per cent nickel.†

Berubigen. Cyanocobalamin. Applications: Vitamin. (The Upjohn Co).

Berylla. Beryllium oxide.†

Beryllium Bronze. An alloy of 97.5 per cent copper with 2.5 per cent benyllium.†

Berzeline. Synonym for Haüyne.†

Besconus. Textile processing aid. (Crosfield Chemicals).*

Beta-Air. Procaterol Hydrochloride. Applications: Bronchodilator. (Parke-Davis, Div of Warner-Lambert Co).

Beta-cardone. A proprietary preparation of sotalol hydrochloride used in the treatment of cardiac dysrhythmias. (Duncan, Flockhart).†

Beta-naphthol Orange. See ORANGE II.

Betacortril. A proprietary preparation of betamethasone. A corticosteroid. (Pfizer International).*

Betacortril Forte. A proprietary preparation of betamethasone. A corticosteroid. (Pfizer International).*

Betadine. Povidone-iodine. Applications: Infection, as a topical antiseptic. (Napp Laboratories Ltd).*

Betaine. Trimethyl-glycine, found in molasses.†

Betalin 12 Crystalline. Cyanocobalamin. Applications: Vitamin. (Eli Lilly & Co).

Betaloc. A proprietary preparation of METAPROBOL tartrate used in the treatment of angina pectoris. (Astra Chemicals Ltd).†

Betanal E. Selective herbicide. (FBC Ltd).

Betapen-VK. Penicillin V Potassium. Applications: Antibacterial. (Bristol Laboratories, Div of Bristol-Myers Co).

Betaprone. Propiolactone. Applications: Disinfectant. (O'Neal, Jones & Feldman Pharmaceuticals).

Betaserc. Betahistine. Applications: Counteracts disorders of the inner ear, Menière's disease. (Duphar BV).*

Betasol Ot-A. A proprietary trade name for a wetting agent. It is a sulphonated ester.†

Betathane. Solid Polyurethane elastomers. Applications: Engineering based. (Hallam Polymer Engineering Ltd).*

Betatrex. Betamethasone valerate USP. Applications: Indicated for the relief of the inflammatory and pruritic manifestations of corticosteroid responsive dermatoses. (Altana Inc).*

Betazole. Ametazole.†

Bethanidine. N-Benzyl-N'N'-dimethylguanidine. Esbatal is the sulphate.†

Betnelan. A proprietary preparation of betamethasone. (Glaxo Pharmaceuticals Ltd).†

Betnesol. A proprietary preparation of betamethasone disodium phosphate. (Glaxo Pharmaceuticals Ltd).†

Betnesol-N. A proprietary preparation of betamethasone sodium phosphate and neomycin sulphate. (Glaxo Pharmaceuticals Ltd).†

Betnovate. A proprietary preparation of betamethazone 17-valerate. A dermatological corticosteroid. (Glaxo Pharmaceuticals Ltd).†

Betnovate-C. A proprietary preparation of betamethasone valerate and clioquinol used in dermatology. (Glaxo Pharmaceuticals Ltd).†

Betnovate-N. A proprietary preparation of betamethasone valerate and neomycin sulphate used in dermatology. (Glaxo Pharmaceuticals Ltd).†

Betol. (Naphthosalol, Salinaphthol). β-Naphthyl salicylate.†

Betrox. Sodium chloride fertilizer. (ICI PLC).*

Betsolan. Betamethasone 21 phosphate veterinary preparations. (Glaxo Pharmaceuticals Ltd).†

Bettendorf's Reagent. A concentrated solution of stannous chloride in fuming hydrochloric acid. Used for the detection of arsenic.†

Beutene. Butyraldehyde-aniline condensation product. Fast-curıng accelerator most active above 250°F (121°C). Moderately nonscorchy at processing temperature; compatible with Channel and furnace blacks, reclaimed rubber and acidic softeners. Hard clays retard its activity. Fast cures in hard rubber, wire insulation and mechanical goods. (Uniroyal).*

Bevacid. Tall oil fatty acids and distilled tall oil. Surface coatings, synthetic resins, soaps and oils. (Bergvik Sales Ltd).*

Bevaloid. Emulsifier for natural oils; dispersant. Applications: Leather industry: non-aqueous systems. (Bevaloid Ltd).

Bevaloid DA 6805. Cationic surfactant in liquid form. Applications: Levelling agent used in dyeing polyamide with acid dyes. (Bevaloid Ltd).

Bevaloid 111. Sodium polycarboxylate in liquid form. Applications: Long term stability dispersant for paint. (Bevaloid Ltd).

Bevaloid 1299. Sodium dioctyl sulphosuccinate in liquid form. Applications: Powerful wetting agent used in textiles and detergent dry cleaning. (Bevaloid Ltd).

Bevaloid 211. Sodium polycarboxylate of low molecular weight, in liquid form. Applications: Dispersant for paper coating and paint, sometimes used in conjunction with polyphosphates. (Bevaloid Ltd).

Bevaloid 35 and 36. Anionic surfactant in powder form. Bevaloid 36 is a higher molecular weight version of Bevaloid 35. Applications: Dispersant for dyestuffs and pigments,and in concrete and resin technology. (Bevaloid Ltd).

Bevaloid 6423. Sodium di-isobutyl sulphosuccinate in flake form. Applications: Anionic surfactant having good stability to electrolytes, used in emulsion polymerization and electroplating. (Bevaloid Ltd).

Bevaloid 6522. Cationic surfactant in liquid form. Applications: Softener for acrylic fibres. (Bevaloid Ltd).

Bevaloid 6703. Low molecular weight polycarboxylate in liquid form. Applications: Dispersant for calcium carbonate, china clay. (Bevaloid Ltd).

Bevaloid 6744. Sodium polycarboxylate of low molecular weight, in liquid form. Applications: Dispersant for drilling muds. (Bevaloid Ltd).

Beviros. Tall oil resins. Applications: Surface coatings, synthetic resins and binders. (Bergvik Sales Ltd).*

Bevitack Resins. Esters and derivatives of distilled tall oil resin. Applications: Components of adhesives and coatings. (Bergvik Sales Ltd).*

Bewoid. Modified rosin emulsion. (Tenneco Malros Ltd).

Bewopac. Modified rosin emulsion. (Tenneco Malros Ltd).

Bex. Bakery compound. (Albright & Wilson Ltd, Phosphates Div).

Bexfilm. Acetate films, polyester films. (ICI PLC).*

Bexfilm 'A'. Cellulose diacetate. (ICI PLC, Petrochemicals & Plastics Div).

Bexfilm 'O'. Reprographic polyester film base. (ICI PLC, Petrochemicals & Plastics Div).

Bexfilm 'P'. Polyester. (ICI PLC, Petrochemicals & Plastics Div).

Bexfilm 'S'. Cellulose triacetate gel. (ICI PLC, Petrochemicals & Plastics Div).

Bexfilm 'T'. Cellulose triacetate. (ICI PLC, Petrochemicals & Plastics Div).

Bexloy. Automotive engineering resins. (Du Pont (UK) Ltd).

Bexphane. Polypropylene film. (Hercules Ltd).

Bextasol. A proprietary preparation of betamethasone valerate as an aerosol, used in asthma therapy. (Glaxo Pharmaceuticals Ltd).†

Bexton. Propachlor herbicide is used for pre-emergence grass and broadleaf weed control in corn and grain sorghum. (Dow Chemical Co Ltd).*

Beyrichite. See MILLERITE.

Bezalip. Bezafibrate. Applications: Antihyperlipoproteinemic. (Norwich Eaton Pharmaceuticals Inc).

BHC. Hexachlorocyclohexane. A powerful insecticide.†

Bhimsiam Camphor. See CAMPHOL.

BI Ammonium Phosphate. BI ammoniumphosphate tehnical grade. Applications: Other chemicals, water treatment, fire fighting, cosmetic products, food industry, pharmaceutical products. (Rhone-Poulenc NV/CdF Chimie AZF).*

Biactol. Antibacterial face wash to help clear acne blemishes. (Richardson-Vicks Inc).*

Biakmetals. A group of alloys some of which are zinc-copper alloys with small amounts of nickel or manganese or both, and others are aluminium-zinc-copper alloys, or zinc-aluminium alloys.†

Bial's Reagent. A reagent used for the detection of pentoses, giving a green colour. It consists of 1 gram orcinol in 500 cc of 30 per cent hydrochloric acid to which 30 drops of 10 per cent ferric chloride have been added.†

Biarsan. Proquazone. Applications: Anti-inflammatory. (Sandoz Pharmaceuticals, Div of Sandoz Inc).

Biasbeston. A synthetic varnish.†

Biavax II. Rubella and Mumps vaccine, live attenuated. Applications: For immunization against German Measles and Mumps. (Merck Sharp & Dohme).*

Biborate of Soda. See BORAX.

Bibra Alloy. An alloy of lead with smaller amounts of bismuth and tin.†

Bichromate. See BICHROME.

Bichrome. (Chrome, Bichromate). Terms used generally for potassium dichromate, but occasionally for sodium dichromate. Used as a mordant for wool dyeing.†

Bicillin. A proprietary preparation of penicillin and procaine penicillin. Applications: An antibiotic. (Brocades).*

Bicillin L-A. Penicillin G Benzathine. Applications: Antibacterial. (Wyeth Laboratories, Div of American Home Products Corp).

BiCNU. Carmustine. Applications: Antineoplastic. (Bristol Laboratories, Div of Bristol-Myers Co).

Bicor. Oriented and non-oriented polypropylene films. Applications: Packaging of food and non food products and special industrial applications. (Mobil Plastics Europe).*

Bidormal. A proprietary preparation of pentobarbitone sodium and buto-barbitone. A hypnotic. (Allen & Hanbury).†

Biebrich Acid Blue. A triphenyl methane dyestuff. It is an acid wool dye.†

Biebrich Acid Red 4B. See CHROMOTROPE 2R.

Biebrich Black AO, 4AN, 6AN, 3BG, 4BN, RO. Acid dyestuffs, producing shades of black on wool from an acid bath, sometimes with the addition of copper sulphate.†

Biebrich Patent Black BO. A dyestuff produced by coupling diazotized α-naphthylaminedisulphonic acid with α-naphthylaminesulphonic acids, diazotizing the product, and coupling it with β-naphtholdisulphonic acid R.†

Biebrich Scarlet. (Ponceau 3R or 3RB, Ponceau B, Fast Ponceau B, New Red

L, Imperial Scarlet, Azobenzene Red, Old Scarlet, Ponceau B Extra, Scarlet Red, Scarlet B, New Red, Scarlet EC). A dyestuff. Dyes wool scarlet from an acid bath.†

Biebrich Scarlet R, Medicinal. See SUDAN IV.

Bielzite. A resin-asphalt containing sulphur and nitrogen.†

Bife. A proprietary mixture of thiamine hydrochloride and ferrous sulphate, used in the treatment of iron deficiency. (Jenkins Laboratories Inc, Auburn, NY).†

Bifiteral. Lactulose. Applications: Treatment of obstipation and (pre-) hepatic coma. (Duphar BV).*

Bigall. Bismuth subgallate, $Bi(OH)_2 C_7H_5O_5$.†

Bigatren. Basic bismuth oxyquinoline sulphonate.†

Bigwood-Ladd Nitroprusside Reagent. For acetone. To 10 cc glacial acetic acid add 10 cc of a 10 per cent solution of sodium nitroprusside.†

BIK. Surface treated urea for easy dispersion in elastomers. Activator for thiazole, thiuram and dithiocarbamate accelerators. Odour reducer when used with nitrosoamine type blowing agents. (Uniroyal).*

Bikini Cream. A proprietary suntan cream. (Ayrton Saunders plc).*

Bikorit. White special fused corundum (crystalline aluminium oxide). Applications: Production of ramming mixes, shape bricks and crucibles for the lining of high temperature furnaces. Moulding material for precision casting moulds and the casting of aggressive special steels. Raw material for the electroslag remelting process, separating agent for the annealing processes. (Dynamit Nobel Wien GmbH).*

Bilarcil. An anti-schistosomal preparation. (Bayer & Co).*

Bilein. Sodium tauroglycocholate.†

Bilevon-Solution. Preparation against liver flukes (Fasciola hepactica and Fasciola gigantica), for subcutaneous injection in cattle. Applications: Veterinary medicine. (Bayer & Co).*

Bilgen Bronze. An alloy of 97 per cent copper, 1.9 per cent tin, 0.52 per cent iron, and 0.24 per cent lead.†

Biligrafin. A proprietary trade name for the bis-meglumine salt of Iodiparnide. (Schering Chemicals Ltd).†

Bilimiro, also Bilimiron. Iopronic acid. Applications: Diagnostic aid. (Bracco Industria Chimica SPA).

Bilivist. Ipodate sodium. Applications: Diagnostic aid. (Berlex Laboratories Inc, Subsidiary of Schering AG).

Bilopaque. Tyropanoate Sodium. Applications: Diagnostic aid, radiopaque medium, cholecystographic. (Sterling Drug Inc).

Biloptin. A proprietary trade name for Sodium Ipodate. (Schering Chemicals Ltd).†

Bilostat. A proprietary trade name for dehydrochloric acid.†

Bilston. A basic slag for fertilizing purposes. (Fisons PLC).*

Bilt-Plates. Clay. Applications: Used as filler when high brightness is not a prerequisite - paint and paper. (Vanderbilt Chemical Corporation).*

Biltricide. An anti-schistosomal preparation. (Bayer & Co).*

Bima's Redwood. See REDWOODS.

Binarite. See MARCASITE.

Bindschedler's Green. A dyestuff. It is a tetramethylindamine hydrochloride.†

Binotal. A broad-spectrum penicillin preparation. (Bayer & Co).*

Bioacid. A range of alkyl aryl sulphonic acids. Applications: Detergent for use in the manufacture of liquid and powdered cleaning compounds. (Harcros Industrial Chemicals).*

Bio-add. Fish/meat silage additive. (BP Chemicals Ltd).

Biocide. Animal feed preservative. (BP Chemicals Ltd).

Biogastrone. A proprietary preparation of carbenoxolone sodium. A gastro-intestinal sedative. (Berk Pharmaceuticals Ltd).*

Biogen. (Magnesium Perhydrol). A mixture of magnesium oxide and magnesium peroxide. It contains from 15-25 per cent magnesium peroxide, MgO_2, and is used as a bleaching agent and antiseptic. Pure magnesium peroxide is also called magnesium perhydrol, and is known by the names Novozone and Magnodat.†

Biomate. Microbiological growth control agents. (Dearborn Chemicals Ltd).

Biomydrin. Phenylephrine Hydrochloride. Applications: Adrenergic. (Parke-Davis, Div of Warner-Lambert Co).

Biopal NR-20. Nonylphenoxypoly(ethyleneoxy) ethanol iodine complex. Applications: Concentrated iodophor. Used for cleaning, sanitizing and disinfecting hospital, biological laboratory and dairy equipment. (GAF Corporation).*

Biopal NR-20 W. Nonylphenoxypoly(ethyleneoxy) ethanol iodine complex. Applications: Concentrated iodophor suitable for formulating "no rinse" sanitizing solutions. (GAF Corporation).*

Biopal VRO-20. Nonylphenoxypoly(ethyleneoxy) ethanol iodine complex. Applications: Concentrated iodophor. Used for cleaning, sanitizing and disinfecting hospital, biological laboratory and dairy equipment, breweries, multiple-use eating and drinking utensils in bars, restaurants, etc. (GAF Corporation).*

Biopar Forte. A proprietary preparation of Vitamin B_{12} and intrinsic factor. Haematinic. (Armour Pharmaceutical Co).†

Biopol. Poly B-hydroxybutyrate (PHB) biodegradable thermoplastic. (ICI PLC).*

Bioques. A series of enzyme producing biological cultures used to degrade industrial and municipal organically fouled waste water. Both dry and liquid compositions available. (Ques Industries).*

Bioques Q. A liquid chemical formulation which can liquify and digest complex fats, oils, grease, cellulose, proteins and starch. Applications: For use in toilets, drains and grease traps. (Ques Industries).*

Bioques Z. A blend of highly active, broad spectrum bioactive cultures that have the ability to digest and liquify organic wastes. Applications: Designed for use in all sewage systems, lagoons, sink drains and traps, and grease traps. (Ques Industries).*

Bioral. A proprietary preparation of carbenoxolone sodium. Used to treat peptic ulceration. (Berk Pharmaceuticals Ltd).*

Bio Soft. Alkylbenzene sulphonic acid and alkylbenzene sulphonate. Applications: Detergents. (Stepan Company).*

Biosoft C100. Anionic surfactant as an ivory paste. Applications: Detergent paste or liquids. (KWR Chemicals Ltd).

Biosoft D. Anionic surfactant in slurry form. Applications: Liquid detergents. (KWR Chemicals Ltd).

Biosoft N-300. Anionic surfactant, in liquid form. Applications: Liquid detergents. (KWR Chemicals Ltd).

Biosoft S and D-35X. Anionic surfactant in liquid form. Applications: liquid detergents. (KWR Chemicals Ltd).

Biosoft S-100 and JN. Anionic surfactant in acid form, supplied as a viscous liquid. Applications: Intermediate in detergent preparations. (KWR Chemicals Ltd).

Biosol. Neomycin sulphate. Applications: Antibacterial. (The Upjohn Co).

Biosol. Water soluble coolant for use in glass manufacturing. (Specialty Products Co).*

Biosone. A proprietary preparation of enoxolone. Antipruritic skin ointment.†

Biosperse. Microbiocidal agents used to control microbiological growth. Applications: Cooling towers, air washers, pasteurizers, cooling water systems, oil field water systems and metal working fluids. (Ashland Chemical Company).*

Biostat A.1. Oxytetracycline for fish preparation. (Pfizer International).*

Bio Terge. Alpha olefin sulphonate. Applications: Detergent, shampoo, bubble bath. (Stepan Company).*

Bio-terge AS-40 and AS-90F. Sodium alpha-olefine sulphonate in liquid or flake form. Applications: Liquid form used for dishwashing detergents and car washing; flake form used in powdered detergents. (KWR Chemicals Ltd).

Biotexin P. Novobiocin and penicillin veterinary preparation. (Glaxo Pharmaceuticals Ltd).†

Biotrase. A proprietary preparation of trypsin and bithional. Treatment of ulcers and wounds. (Lloyd, Hamol).†

Biotren. A proprietary preparation of glycine, zinc bacitracin, neomycin sulphate, *l* - cysteine and di-threonine. An anti-bacterial powder. (Carlton Laboratories).†

Birch Oil. See OIL OF BIRCH.

Bird Manure. See GUANO.

Birlane. Liquid seed dressing containing chlorfenvinphos for winter wheat. (ICI PLC).*

Birmabright. A corrosion-resisting alloy containing 96 per cent aluminium, 3.5 per cent magnesium, and 0.5 per cent manganese.†

Birmasil Alloy. A special alloy which is a nickel-aluminium-silicon alloy containing up to 3.5 per cent nickel and from 8.13 per cent silicon. It has high tensile strength.†

Birmidium. A proprietary trade name for an alloy of aluminium with smaller amounts of copper, nickel, and magnesium. It is similar to Y-alloy.†

Birmingham Nickel Silver. See NICKEL SILVER.

Birmingham Platina. A brass containing 47 per cent copper, 53 per cent zinc, and 0.25 per cent iron.†

Birmite. See BURMITE.

Birox. Resistor compositions. (Du Pont (UK) Ltd).

Bisformasal. See FORMASAL.

Bisiumina Suspension. A proprietary preparation containing bismuth aluminate. An antacid. (MCP Pharmaceuticals).†

Bislumina. A proprietary preparation containing bismuth aluminate and magnesium oxide. An antacid. (MCP Pharmaceuticals).†

Bismarck Brown. (Aniline Brown, Gold Brown, Manchester Brown, Phenylene Brown, Vesuvine, Leather Brown, Cinnamon Brown, English Brown). A dyestuff. Dyes wool, leather, and tannined cotton reddish-brown.†

Bismarck Brown R. See MANCHESTER BROWN EE.

Bismarck Brown T. See MANCHESTER BROWN EE.

Bismate. Bismuth dimethyldithio-carbamate - accelerator for rubber. Applications: Used in NR, 1R, BR and SBR for high temperature speed vulcanization. (Vanderbilt Chemical Corporation).*

Bismucyn. A proprietary preparation of bismuth sodium tartrate. Throat lozenges. (Rybar).†

Bismuth Brass. A nickel silver. It contains from 47-52 per cent copper, 30-31 per cent nickel, 12-21 per cent zinc, 0-5 per cent lead, 0-1 per cent tin, and 0.1-1 per cent bismuth.†

Bismuth Bronze. An alloy. (*a*) Consists of 1 part bismuth with 16 parts tin ; (*b*) contains 1 part bismuth, 63 parts copper, 21 parts spelter, and 9 parts nickel. Resists sea-water.†

Bismuth Magister. See MAGISTER OF BISMUTH.

Bismuth Solder. Alloys of from 25-50 per cent tin, 25-40 per cent bismuth, and 25-40 per cent lead. One containing 33 per cent bismuth, 33 per cent tin, and 33 per cent lead, melts at 284° F. A solder containing 40 per cent bismuth, 40 per cent lead, and 20 per cent tin melts at 111° C. Also See ROSE S, KRAFT'S, and HOMBERG'S METALS.†

Bismuth Subcarbonate. (Oxycarbonate of bismuth). Bismuth carbonate.†

Bismuth White. (Paint White, Pearl White, Spanish White, Fard's Spanish White). Basic bismuth nitrate, $BiO(NO_3)$, a pigment.†

Bismuth Yellow. Bismuth chromate, $3Bi_2O_3.2CrO_3$, a pigment.†

Bismutol. Sodium potassium bismuthyl-tartrate.†

Bisoflex. Plasticizers. (BP Chemicals Ltd).*

Bisoflex DNA. Dinonyl adipate. A vinyl plasticizer. (BP Chemicals Ltd).*

Bisoflex L79. A higher straight chain phthalate plasticizer. (BP Chemicals Ltd).*

Bisoflex L911. A higher straight chain phthalate plasticizer. (BP Chemicals Ltd).*

Bisoflex ODN. A trade mark for an adipic ester plasticizer. (BP Chemicals).†

Bisoflex 100. A trade mark for a plasticizer. Diisodecyl phthalate. (BP Chemicals).†

Bisoflex 1002. A trade mark for a polymeric plasticizer. (BP Chemicals).†

Bisoflex 1007. A trade mark for a polymeric plasticizer. (BP Chemicals).†

Bisoflex 102a and DOA. Dioctyl adipate. A vinyl plasticizer.†

Bisoflex 104. A trade mark for an adipic ester plasticizer. (BP Chemicals).†

Bisoflex 106. A primary low temperature plasticizer for PVC and synthetic rubber. (BP Chemicals Ltd).*

Bisoflex 130. A trade mark for ditridecyl phthalate. A plasticizer. (BP Chemicals).†

Bisoflex 610. A trade mark for a plasticizer. C_6 - C_{10} aliphatic alcohol phthalate. (BP Chemicals).†

Bisoflex 619. A trade mark for a plasticizer. C_6 - 6_{10} trimellitates. (BP Chemicals).†

Bisoflex 79A. The adipate of mixed C_7-C_9 alcohols. A vinyl plasticizer. (BP Chemicals Ltd).*

Bisoflex 791. Plasticizer. Dialkyl (C_7 - C_9) phthalate. (BP Chemicals Ltd).*

Bisoflex 799. A trimellitate plasticizer for higher temperature resistant PVC for cable use. (BP Chemicals Ltd).*

Bisoflex 8N. Condensation product of 2-ethyl hexyl urethane and formaldehyde. A vinyl plasticizer. (BP Chemicals Ltd).*

Bisoflex 81. Dioctyl phthalate. A vinyl plasticizer. (BP Chemicals Ltd).*

Bisoflex 810. A trade mark for a plasticizer. C_8-C_{10} aliphatic phthalate.†

Bisoflex 819. A trade mark for a plasticizer. C_8 - C_{10} trimellitates. (BP Chemicals).†

Bisoflex 82. Dioctyl phthalate. A vinyl plasticizer. (BP Chemicals Ltd).*

Bisoflex 88. The phthalate of a C_8 alcohol. A vinyl plasticizer. (BP Chemicals Ltd).*

Bisoivomycin. A proprietary preparation of Bisolvon and oxytetracycline hydrochloride. An antibiotic. (Boehringer Ingelheim Ltd).†

Bisol. Solvents, plasticizers, intermediates. (BP Chemicals Ltd).*

Bisolene. Proprietary liquid fuels. (BP Chemicals Ltd).*

Bisolite. Solid fuels, e.g. metaldehyde. (BP Chemicals Ltd).*

Bisolube. Oil additives. (BP Chemicals Ltd).*

Bisolvon. A proprietary preparation of bromhexine hydrochloride. A mucolytic agent. (Boehringer Ingelheim Ltd).†

Bisomer. Specialty chemicals for use as intermediates in the production of various products in industry, particularly surface coatings, adhesives and sealants. (BP Chemicals Ltd).*

Bisomer DALP. A trade name for a diallylphthalate plasticizer. (BP Chemicals).†

Bisomer DAM. A trade name for a plasticizer. Dialkyl maleate. (C_7 - C_9 alcohols). (BP Chemicals).†

Bisomer DBF. A trade name for a dibutyl fumarate plasticizer. (BP Chemicals).†

Bisomer DBM. A trade name for a dibutyl maleate plasticizer. (BP Chemicals).†

Bisomer DNM. A trade name for a dinonyl maleate plasticizer. (BP Chemicals).†

Bisomer DOM. A trade name for a diethyl hexyl maleate plasticizer. (BP Chemicals).†

Bisomer D10M. A trade name for a Diisooctyl maleate plasticizer. (BP Chemicals).†

Bisomer 2HEA. 2-Hydroxy-ethyl-acrylate. A monomer which permits the production of polymers with side chain hydroxyl groups suitable for further reaction (cross-linking) and the production of thermosetting acrylic adhesives. (BP Chemicals Ltd).*

Bisomer 2HEMA. 2-Hydroxy-ethyl methacrylate. A monomer which permits the production of polymers with side chain hydroxyl groups suitable for cross-linking and the production of thermosetting acrylic surface coating adhesives. (BP Chemicals Ltd).*

Bisomer 2HPMA. 2-Hydroxy-propyl-methacrylate. (BP Chemicals Ltd).*

Bisoprufe. Preparations for waterproofing cements. (BP Chemicals Ltd).*

Bisoxyl. Bismuth oxychloride in liquid suspension. An antisyphilitic. (British Drug Houses).†

Bistabillin. Penicillin preparation. (Boots Pure Drug Co).*

Bi-Tarco. Tar compounds for roads. (Thomas Ness, North Thames Gas Board).†

Bitran. Cationic surfactant range, composed of high molecular weight imidazoline compounds and their salts. Viscous liquids/ brown aqueous solutions. Applications: Adhesion aids, anti-corrosives, dispersants, flocculants, dewatering agents. Used in road maintenance; metal working; paints; emulsion cracking; effluent treatment; pigment dispersion. (ABM Chemicals Ltd).

Bitran H. A proprietary long-chain cyclic polyamine. (Glovers Chemicals Ltd).†

Bitrex. Denatonium benzoate used as a denaturant. Available as powder and in aqueous ethanol, methanol and N propanol. Applications: A denaturant for use in alcohols and aversive agents in other hazardous household fluids as a prevention to poisoning by ingestion. (MacFarlan Smith Ltd).*

Bitter Almond Oil Green. See MALACHITE GREEN.

Bitter Salt. See EPSOM SALTS.

Bittern. The mother liquor remaining after the crystallization of sodium chloride from sea-water. It is a source of magnesium, and also contains bromides and iodides.†

Bitumastic. A proprietary trade name for a spirit paint made from refined coal-tar pitch, etc.†

Bitumen. (Mineral Pitch, Natural Asphalt, Compact Bitumen, Jew's Pitch, Bitumen of Judaea, Naphthine). A hard pitchy material found at the surface of the Dead Sea, and in the pitch lake at Trinidad. The terms are applied to a number of mineral substances containing mainly hydrocarbons. Asphalt mineral, mountain or rock tallow, or hatchettine, and mineral caoutchouc or elaterite, are all bitumens. A mixture of asphalt with drying oil, used as a pigment, is also known as bitumen.†

Bitumen of Judaea. See BITUMEN.

Bitumen, Elastic. See ELATERITE.

Bitumuls. Asphalt products. (Chevron).*

Bitusize. Asphalt products. (Chevron).*

Bituvar. Anticorrosive paint. (JC Bottomley and Emerson).†

Biuret Reagent (Gies). A 10 per cent potassium hydroxide solution to which 25 cc of a 3 per cent copper sulphate solution per litre has been added. A purple-violet or pinkish-violet colour is obtained with $CONH_2$ groups, such as are contained in biuret and oxamide.†

Bivert. Amine salts of organic acids, aromatic acid, aromatic and aliphatic petroleum distillate. Applications: For use with all pesticides (herbicides, insecticides and fungicides) to control evaporation, increase plant coverage and control drift. (Stull Chemical Company).*

Black and White Bleaching Cream. Hydroquinone. Applications: Depigmentor. (Plough Inc).

Black Bronze. An alloy of 50 per cent lead, 40 per cent tin, and 10 per cent antimony. Used in the cutlery trade.†

Black Catechu. See CUTCH.

Black Copper. See BLISTER COPPER

Black Cummin. Nigella seeds.†

Black Diamond. See CARBONADO.

Black Drop. (*Acetum opii*). A vinegar.†

Black Gold. See MALDONITE.

Black Grip. Hard polyurethane elastomer resin, mboca, silica. Applications: Solid fork truck tyres, belting (conveyor), friction drive wheels, drum rotators. (Roayle Polymers Ltd).*

Black Hypo. See BURNT HYPO.

Black Iron. See PYROLIGNITE OF IRON.

Black Iron Liquor. See PYROLIGNITE OF IRON.

Black Lead. See GRAPHITE.

Black Lead Ore. See ALQUIFON.

Blackley Blue. A dyestuff. Used for dyeing paper pulp. See SOLUBLE BLUE.†

Black Liquor. See PYROLIGNITE OF IRON.

Black Manganese. See MANGANESE BLACK.

Black Mordant. See PYROLIGNITE OF IRON.

Black Out. Elastomeric dispersion. Applications: Finishing agent for decorative and protective flexible finishes on rubber products. (Vanderbilt Chemical Corporation).*

Blackox. Fe_3O_4. Applications: Foundry grade black iron oxide for core and mould use with particular emphasis on elimination of sub-surface porosity and carbon streaking when phenolic urethane sand binder systems are in use. (DCS Color & Supply Co Inc).*

Black Oxide of Iron. See MAGNETIC IRON ORE.

Black Paste. See FAST BLACK.

Black Powder. See GUNPOWDER.

Black Solder. Usually an alloy of 58 per cent zinc, 40 per cent copper, and 2 per cent tin.†

Black Varnish, Burma. See THITSI.

Bladder Green. See SAP GREEN.

Blagden Resins. Synthetic resins for the coatings industry. Applications: Industrial and decorative paints and printing inks. (Blagden Chemicals Ltd).*

Blagdenite. Unsaturated polyester resins. Applications: Reinforced plastics, marine, transport and industrial. (Blagden Chemicals Ltd).*

Blanc de Zinc. Zinc oxide, ZnO.†

Blanc D'Argent. See FLAKE WHITE.

Blanchite. See HYDROSULPHITE.

Blancol. Optical brightening agents. Applications: Whitening of textiles. (Holliday Dyes & Chemicals Ltd).*

Blancol N. The sodium salt of sulphonated naphthalene-formaldehyde condensate. Applications: In papermaking disperses pigments, clays and other solids, prevents pitchcoagulation, reduces two sideness, improves sizing. Bleaching, dispersing, levelling and neutralizing agent for leather. (GAF Corporation).*

Blancorol AC. Chrome aluminium mixed complex tanning materials. (Bayer & Co).*

Blandine. See PARAFFIN, LIQUID.

Blandlax. A laxative. (Boots Pure Drug Co).*

Blandofen CAZ. Complex polyalkyl amido imidazolinium sulphate as a white-yellow viscous liquid. Applications: Cold water dispersible cationic surfactant, stable to freeze thaw cycles. Used in softeners for fabrics and paper. (GAF (Gt Britain) Ltd).

Blandofen CT. Distearyl dimethyl ammonium chloride concentrate in isopropanol/water. Supplied as a white-yellow soft paste. Not cold water dispersible. Higher viscosity than other cationics in the range. Applications: Cationic surfactant used for fabric softeners. (GAF (Gt Britain) Ltd).

Blandofen FA. Complex polyalkyl amido imidazolinium sulphate as a yellow viscous liquid. Applications: Cationic surfactant, thixotropic at low temperatures, specially developed for bulk deliveries. Used in fabric softeners. (GAF (Gt Britain) Ltd).

Blandthax. A blackquarter/anthrax vaccine. (Boots Pure Drug Co).*

Blankit. Reductive bleaching agents for textiles. (BASF United Kingdom Ltd).

Blankit. See HYDROSULPHITE.

Blanko-Blech. An alloy of 80 per cent copper and 20 per cent nickel.†

Blankophor. Fluorescent whitening agents for textiles and detergents. Applications: Textile finishing agent. (Bayer & Co).*

Blanose. Sodium salt of carboxymethyl cellulose. Applications: stabilizer and thickener in aqueous systems for food and non-food uses. (Hercules Inc).*

Blanthax. Blackquarter-anthrax vaccine. (The Wellcome Foundation Ltd).

Blasting Gelatin. A blasting explosive. It is the jelly-like mass obtained when nitro-cotton is added to nitro-glycerin, and contains from 90-93 per cent nitro-glycerin, and 7-10 per cent dry nitro-cotton.†

Blatt Gold. A brass containing 77 per cent copper and 23 per cent zinc.†

Blatt Silver. An alloy of 91.1 per cent tin, 8.25 per cent zinc, 0.35 per cent lead, and 0.23 per cent iron.†

Blattanex. Blattanex 20 - emulsifiable concentrate containing 20% propoxur. Blattanex Residual Spray - an aerosol containing 2% propoxur and 0.5% dichlorros. Applications: Quick knock-down and lasting residual control of crawling and flying pests. (Bayer & Co).*

Blazon. Water soluble, biodegradable colourants. Applications: For spray pattern indicators for application of herbicides, pesticides in the lawn and turf, forestry and industrial weed control areas. (Milliken & Company).*

BLE. Reaction product of diphenylamine and acetone. Used in natural, IR, SBR, BR, neoprene and nitrile rubbers. Particularly recommended in tyre treads and carcass to combat the effects of heat and mechanical flexing. A general purpose antioxidant where discolouration and staining are not factors. Also available as a 75% active powder on silica. (Uniroyal).*

Bleach Liquor. (Bleaching Liquid, Liquid Chloride of Lime). Prepared by passing chlorine through milk of lime. It is a solution of chlorinated lime.†

Bleaching Liquid. See BLEACH LIQUOR.

Bleaching Powder. (Calcium oxymuriate, Chloride of lime, Chlorinated calcium oxide, Chlorinated lime, Oxymuriate of lime, Calcium hypochlorite). Prepared by saturating slaked lime with chlorine. An oxidizing and bleaching agent.†

Blenda. See OIL WHITE.

Blenderm. A proprietary preparation of polythene adhesive tape for surgical use. (3M).*

Blendex Modifiers. ABS polymers. Applications: Injection moulding, sheet extrusions. (Borg Warner Chemicals).*

Blende, Mercury. See CINNABAR.

Blenoxane. Bleomycin sulphate. Applications: Antineoplastic. (Bristol Laboratories, Div of Bristol-Myers Co).

Bleomycin. An anti-neoplastic antibiotic produced by *Streptomyces verticillus.*†

Bleph-10 Liquifilm. Sulphacetamide Sodium. Applications: Antibacterial. (Allergan Pharmaceuticals Inc).

Bleph-10 SOP. Sulphacetamide Sodium. Applications: Antibacterial. (Allergan Pharmaceuticals Inc).

Bleu de Lyon. See ANILINE BLUE, SPIRIT SOLUBLE.

Bleu de Nuit. See ANILINE BLUE, SPIRIT SOLUBLE.

Bleu de Paris. See METHYL BLUE.

Bleu de Saxe. See SMALT.

Bleu D'Azure. See SMALT.

Bleu Direct. See METHYL BLUE.

Bleu Lumière. See ANILINE BLUE, SPIRIT SOLUBLE.

Blex. Insecticide containing pirimiphos-methyl. (ICI PLC).*

Blister Copper. (Black Copper). A copper produced from copper matte by oxidation, then heating with coal in the furnace, when the metallic oxides are reduced. It contains 90-95 per cent copper, 1-2.5 per cent iron, and 0.5-2.5 per cent sulphur.†

Blister Gas. See MUSTARD GAS.

BLO. γ-butyrolactone. Applications: Solvent for polyacrylonitrile, polystyrene, fluorinated hydrocarbons, cellulose triacetate and shellac. Intermediate for aliphatic and cyclic compounds. Agricultural chemical. (GAF Corporation).*

Blocadren. Timolol maleate. Applications: Beta-blocker for the relief of hypertension, angina pectoris and migraine. (Merck Sharp & Dohme).*

Blocazide. Trimolol maleate and hydrochlorothiazide. Applications: For the relief of hypertension. (Merck Sharp & Dohme).*

Block Brass. See BRASS.

Blockain Hydrochloride. Propoxycaine Hydrochloride. Applications: Anaesthetic. (Sterling Drug Inc).

Blocuretic. Timolol maleate combined with hydrochlorothiazide and amiloride hydrochloride. Applications: For the relief of hypertension. (Merck Sharp & Dohme).*

Blood Meal. A nitrogenous fertilizer prepared by coagulating blood, drying and grinding the product. It contains on an average, from 11-14 per cent nitrogen, and 0.75 per cent phosphorus.†

Blood Red. See INDIAN RED.

Bloodit. A synthetic gum for process engraving. (Johnsons of Hendon).†

Blown Oils. (Oxidized Oils, polymerized Oils, Soluble Castor Oils, Base Oil). When semi-drying vegetable oils, marine animal oils, and liquid waxes are warmed at from 70-120° C., and a current of air blown through them, the oils oxidize to viscid fluids, miscible with mineral oils. Such oils are known as soluble castor oils. They are rich in triglycerides of the hydroxy-acids, and are used as lubricants. If the blowing is continued for a long time, thickened oils result.†

Blue Asbestos. See CROCIDOLITE.

Blue Ashes. See MOUNTAIN BLUE.

Blue Basic Lead Sulphate. A basic lead sulphate containing lead oxide, sulphite, sulphide, zinc oxide, and carbon, produced from lead ore by volatilization. It is used in rubber mixing and in priming paint.†

Blue-black B. (Azo Black O, Azo Black). A dyestuff. Dyes wool bluish-violet from an acid bath.†

Blue, Brunswick. See BRUNSWICK BLUE.

Blue CB. See INDULINE, SOLUBLE.

Blue CB, Spirit Soluble. See INDULINE, SPIRIT SOLUBLE

Blue Copper. See COVELLITE.

Blue Dot. A double-base-type formulation which minimizes charge weight and moisture absorption. Graphite glaze enables smooth granule flow. Applications: Ammunition; it is designed for use in magnum shotshell loads. It can also be used in magnum handgun loads. (Hercules Inc).†

Blue Flax. Flemish flax.†

Blue Gold. A jeweller's alloy, consisting of 75 per cent gold and 25 per cent iron.†

Blue Gum. Eucalyptus gum.†

Blue Mould. *Penicillium Glaucum,* a fungus. See PENICILLIN.†

Blue Powder. (Zinc Dust, Zinc Fume). A by-product in the smelting of zinc. It consists of a mixture of finely divided zinc and zinc oxide. It has the power of absorbing hydrogen, and is, therefore, used in the chemical industries. Employed to discharge locally the colour of dyed cotton goods. It is also used for the recovery of gold from the cyanide solution of the metal.†

Blue PRC. See CHROMOCYANINE V AND B.

Blue Pyoctanin,. See PYOCTANIN.

Blue Ridge. See Catalpo.

Blue Salt. See INDIGOSOL O.

Blue Silver. See NIELLO SILVER.

Blue 1900. (Deep Blue Extra R, Violet Moderne). Leucogallocyanines, giving blue to violet shades upon a chrome mordant.†

Bluish Eosin. See ERYTHROSIN.

B.M. Mixture. A mixture for producing smoke screens. It contains 35 per cent zinc, 42 per cent carbon tetrachloride, 9 per cent sodium chlorate, 5 per cent ammonium chloride, and 8 per cent magnesium carbonate.†

Bobierre Metal. A brass containing 58-66 per cent copper and 34-42 per cent zinc. Ships' sheathing metal.†

Bodenstein. Synonym for amber.†

Bodryl. A proprietary preparation of bromodiphenhydramine hydrochloride, aspirin, phenacetin, phenylephrine, caffeine and aluminium hydroxide dried gel. Cold remedy. (Parke-Davis).†

Bog Manganese. See WAD.

Bohemian Earth. (Veronese Earth, Tyrolean Earth, Seladon Green, Terre Verte, Stone Green). Green earths, which are products of the disintegration of minerals, chiefly of the hornblende type. Stone green is a mixture of ground green earth and white clay, and has been used for the manufacture of waterproofing paints.†

Bohemian Topaz. Synonym for Citrine.†

Bohnalite B. A proprietary trade name for an aluminium alloy containing 4.5 per cent copper and 0.3 per cent magnesium.†

Bohnalite J. A proprietary trade name for an alloy of aluminium with 10 per cent copper.†

Bohnalite S43. A proprietary trade name for an alloy of aluminium with about 5 per cent silicon.†

Bohnalite S51. A proprietary trade name for an alloy of aluminium with small amounts of magnesium, silicon, and iron.†

Bohnalite U. A proprietary trade name for an alloy of aluminium with 13 per cent silicon.†

Bohnalite Y. A proprietary aluminium alloy containing small quantities of copper, nickel, and magnesium.†

Bohrmittell Hoechst. Metal working lubricant and coolant. (Hoechst UK Ltd).

Boiled Oil. Linseed oil, which has been boiled with litharge to render the oil more "drying." The term is also applied to linseed oil which has been heated for some time. The tendency to oxidize is thereby increased.†

Boiler Plug Alloy. An alloy of 8 parts bismuth, 5-30 parts lead, and 3-24 parts tin.†

Boiler-Aid. Various liquid or powder blends of boiler water deposit and corrosion inhibitors. Applications: Steam and hot water boiler treatment. (Schaefer Chemical Products Company).*

Bolda. Fungicide, containing carbendazim, maneb and sulphur. (ICI PLC).*

Bole. See INDIAN RED.

Bole, Armenian. See INDIAN RED.

Bole, Red. See INDIAN RED.

Bole, Venetian. See INDIAN RED.

Bole, White. See CHINA CLAY.

Bolfo. Preparation for external use on dogs and cats against ecto parasite infestation. Applications: Veterinary medicine. (Bayer & Co).*

Bolster Silver. A nickel silver. It is an alloy of 65.5 per cent copper, 16 per cent zinc, 18 per cent nickel, and 0.5 per cent lead.†

Bombay Catechu. See CUTCH.

Bombay Gum. See EAST INDIA GUM.

Bond-A-Tint. Pigment dispersions. Applications: Colouration of bonded polyurethane carpet underlay. (Pacific Dispersions Inc).*

Bond-Plus. Adhesives, industrial coatings for packaging, laminating, paper converting, wood-working, labelling, foil laminating, paper cups, FDA approved adhesives and coatings. Applications: Packaging - case sealing, labelling, lap-gluing; Paper converting - foil laminating, paper to paper, paper to film laminations, paper cups, coatings; Wood-working - water resistant adhesives. (Industrial Adhesives Company).*

Bond-Plus HM. Adhesives, industrial coatings for packaging, laminating, paper converting, wood-working, labelling, foil laminating, paper cups, FDA approved adhesives and coatings. Applications: Packaging - case sealing, labelling, lap-gluing; Paper converting - foil laminating, paper to paper, paper to film laminations, paper cups, coatings; Wood-working - water resistant adhesives. (Industrial Adhesives Company).*

Bonderite 67. A trade mark for a corrosion inhibiting treatment for iron, steel, aluminium and zinc which converts the metal surface into a zinc phosphate coating. (The Pyrene Co).†

Bonding Agent M 3. A methylene donor. Used in combination with Bonding Agents R6 in natural, IR, BR, SBR, neoprene or nitrile rubbers to give improved adhesive bonds to fabrics such as cotton, nylon, polyester, rayon and glass, as well as to metals. (Uniroyal).*

Bonding Agent P 1. 4,4'-methylene-bis-(phenyl carbanilate). For pre-dip solutioning of polyester tyre and industrial cord prior to a secondary treatment with an RFL system. The two dip treatment provides superior adhesion to other known treatment systems. (Uniroyal).*

Bonding Agent R 6. A resorcinol donor. Used in combination with Bonding Agent M3 in natural rubber, IR, BR, SBR, neoprene or nitrile rubber to give improved adhesive bonds to fabrics such as cotton, rayon, nylon, polyester and glass, as well as to metals. (Uniroyal).*

Bondogen. A mixture of an oil soluble sulphonic acid of high molecular weight with a high boiling alcohol and a paraffin oil. Applications: Used as a peptizing agent and strong plasticizer for all elastomers. Functions as a scorch retarder. (Vanderbilt Chemical Corporation).*

Bondur. An aluminium alloy containing 4.2 per cent copper, 0.3-0.6 per cent manganese, and 0.5-0.9 per cent magnesium. It is a corrosion-resisting alloy.†

Bone Ash. (Bone Earth). The ash obtained by heating bones in air. It contains from 67-85 per cent basic calcium phosphate, 2-3 per cent magnesium phosphate, 3-10 per cent calcium carbonate, a little caustic lime, and a little calcium fluoride. Used as a cleaning and polishing material, as a manure, and for the manufacture of phosphorus.†

Bone, Dry. See CALAMINE.

Bone, Earth. See BONE ASH.

Bone Meal, Steamed,. See STEAMED BONE MEAL.

Bone Oil. (Animal Oil, Dippel's Oil, Oil of Hartshorn, Bone Tar). A dark brown oil, rich in pyridine bases, obtained by the distillation of bones. Used for denaturing spirits.†

Bone Tar. See BONE OIL.

Bone, Vitriolised. See VITRIOLISED BONES.

Bonine. Meclizine hydrochloride. Applications: Anti-emetic. (Pfizer Inc).

Bonjela. A proprietary preparation containing choline salicylate and cetyl-dimethyl benzyl-ammonium chloride. (Lloyd's Pharmaceuticals).†

Bonner L-894. A proprietary trade name for a polyester resin.†

Bonomycin. Sancycline. Applications: Antibacterial. (Pfizer Inc).

Bontex. Range of synthetic blended detergents. (Unilever).†

Bonzi. Growth regulator for use on ornamental plants. (ICI PLC).*

Boracic Acid. Boric acid, H_3BO_3. A preservative.†

Boracite, Crystalline. See BORACITE and STASSFURTITE.

Boral. Aluminium-boro-tartrate. An antiseptic and astringent prescribed in skin diseases, and in inflammation of the ear.†

Borascu. General non-selective weedkillers. (Borax Consolidated Ltd).

Borateem. Borates. (Borax Consolidated Ltd).

Borax. (Acid Metaborate, Biborate of Soda). Sodium dimetaborate, $Na_2B_4O_7$ $10H_2O$.†

Boraxo. Industrial hand cleaners. (Borax Consolidated Ltd).*

Boraxusta. Calcined borax.†

Borax, Methylene Blue. See SAHLI'S STAIN.

Borcher's Metal. Non-corrosive alloys. One alloy contains 64.6 per cent nickel, 32.3 per cent chromium, 1.8 per cent molybdenum, and 0.5 per cent silver. Others are stated to contain 30 per cent chromium, 34-35 per cent cobalt, and 34-35 per cent nickel; whilst alloys consisting of 60 per cent iron, 36 per cent chromium, 4 per cent molybdenum, and 65 per cent chromium and 35 per cent iron respectively, are also known as Borcher's metal.†

Bordeaux Cov. See CONGO VIOLET.

Bordeaux DH. See FAST RED D.

Bordeaux Extra. See CONGO VIOLET.

Bordeaux G. (Claret G). A dyestuff. Dyes wool red from an acid bath.†

Bordeaux G, BD, GG, GD, GDD. Dyestuffs allied to Fast red B.†

Bordeaux Mixture. A fungicide for plant diseases, especially for those which attack potatoes. (*a*) Consists of 5 lb copper sulphate, 5 lb lime, and 50 gallons water. (*b*) Contains 6 lb Copper sulphate, 4 lb lime, and 50 gallons water. It is made by the addition of the lime to the copper sulphate solution. Also see

PEACH BORDEAUX MIXTURE, SODA BORDEAUX MIXTURE, and POTASH BORDEAUX MIXTURE.†

Bordeaux R or B. (Claret Red). A dyestuff isomeric with Bordeaux G.†

Bordeaux Red. A dyestuff. It is azonaphthylnaphtholsulphonic acid.†

Bordeaux S. See FAST RED D.

Bordeaux Turpentine. See FRENCH TURPENTINE.

Borden. Foundry sand binders. (Borden (UK) Ltd).

Borderland Black. Active ingredient: Mesurol - 3,5-dimethyl-y-(methylthio) phenol methylcarbamate. Applications: Seed protectant to prevent sprout pulling by birds in newly planted corn. (Borderland Products Inc).*

Borester. Organic boron compounds. (Borax Consolidated Ltd).

Borester. Organic borates. (Manchem Ltd).

Borester A. Triaryl borate $C_{21}H_{21}O_3B$.†

Borester N. Tri-3,5,5-trimethylhexyl borate $C_{27}H_{57}O_3B$.†

Borester 2. Tri-*n*-butyl borate $C_{12}H_{27}O_3B$.†

Borester 25. Trimethoxyboroxine (methyl metaborate) $(CH_3OBO)_2$. This is used for preventing polymerization of liquid sulphur dioxide.†

Borester 7. Tri-2-methyl-2,4-pentadienyl-diborate $C_{18}H_{36}O_6B_2$.†

Borester 7. Tri-hexylene glycol diborate. Applications: Timber preservation. (Manchem Ltd).*

Borester 8. Tri-*m,p*-tolyl borate) $C_{21}H_{21}O_3B$.†

Boresters. A series of esters of boric acid. They are used as a convenient means of introducing boron into organic media such as paints and plastics.†

Borethyl. Triethylboron, $(C_2H_5)_3B$.†

Borfax. Shaped lightweight sideliner lines for steel ingot heading. (Foseco (FS) Ltd).*

Borite. A trade mark for goods used as abrasives and refractories. They consist essentially of crystalline alumina.†

Borium. A fused tungsten carbide diamond substitute formed by exposing tungsten at high temperatures to carbon monoxide or hydrocarbon gases.†

Borneol Camphol. (Baros Camphor, Bhimsiam Camphor, Borneo Camphor, Dryobalanops Camphor, Malay Camphor, Sumatra Camphor). Borneol, $C_{10}H_{17}OH$. a terpene from *Dryobalanops camphora*.Used in perfumery, in celluloid manufacture, and in medicine as an antiseptic and stimulant. The term Camphol has also been applied to oxanilide (diphenyl-oxamide), $CO(NHC_6H_5)(NHC_6H_5)CO$, a softening agent for cellulose esters.†

Borocil. Non-selective weedkiller. (Borax Consolidated Ltd).

Borofax Ointment. A proprietary formulation of boric acid and lanolin. Applications: An externally applied emollient for applications to burns, abrasions etc. (The Wellcome Foundation Ltd).*

Boroflux. A mixture of boron sub-oxide with boric anhydride and magnesia. Used to the extent of about 1 per cent to deoxidize copper during purification.†

Boroglyceride. (Boroglycerine). Glyceryl borate, $C_3H_5BO_3$. Used as a preservative for wines and fruits.†

Boroglycerine. See BOROGLYCERIDE.

Borolon. A trade mark for articles made for abrasive and refractory purposes. They consist essentially of crystalline alumina.†

Boron, Manganese. See MANGANESE BORON.

Borosalicylic Acid. A solution containing 4 per cent each boric and salicylic acids. Has been used as an antiseptic.†

Borosalyl. See BORSALYL.

Boroxo. A heavy-duty soap powder hand-cleanser, based on borax. (Borax Consolidated Ltd).*

Borsalyl. (Borosalyl). The sodium salt of borosalicylic acid, $B\ OH(O\ C_6H_4\ COOH)_2$. An antiseptic.†

Bortin "45". Brucella abortus vaccine for veterinary purposes. (Glaxo Pharmaceuticals Ltd).†

Borvicote. A proprietary range of emulsions of vinyl homopolymers and copolymers. (Borregaard, Norway).†

Bosch-bakelite. See BAKELITE.

Boselon. Packaging materials, plastic film made of low density polyethylene or other thermoplastic, containing a volatile corrosion inhibitor effective for rust protection of ferrous metals. Applications: Packaging for metal parts for the purpose of corrosion protection. (Aicello Chemical Co Ltd).*

Bostik. Wide variety of formulations based on natural and synthetic materials (rubbers, resins, polymers, etc.). Applications: Adhesives, sealants and coating compounds for a wide variety of bonding, sealing and productive coating applications in all types of industries; also a consumer range. (Bostik Ltd).**

Botany Bay Kino. See KINO, AUSTRALIAN.

Botrilex. A horticultural fungicide containing quintozene. (ICI PLC).*

Botryogen. (1) See RED VITRIOL.†

Bottle Green. See BRONZE GREEN.

Bouchardt's Reagent. A solution of 1 part iodine and 2 parts potassium iodide in 20 parts water. An alkaloid reagent giving a brown precipitate.†

Bourbonne's Metal. An alloy of 50.48 per cent tin, 48.8 per cent aluminium, 0.25 per cent copper, and 0.33 percent iron.†

Bourbouze Aluminium Solder. An alloy of 47 per cent zinc, 37 per cent tin, 10 per cent aluminium, and 5 per cent copper.†

Bourbouze Solder. An alloy of 83 per cent tin and 17 per cent aluminium.†

Bovatec (as sodium). Lasalocid. Applications: Coccidiostat. (Hoffmann-LaRoche Inc).

Bovinox. Cattle dip and spray. (The Wellcome Foundation Ltd).

Bowhill's Stain. A microscopic stain. It contains 15 cc of a saturated alcoholic solution of orcin, 10 cc of 20 per cent tannic acid solution, and 30 cc water.†

Bowl Spirit. See SCARLET SPIRIT.

Bowlingite. See STEATITE.

Bow-wire Brass. See BRASS.

Boxite. A trade mark for articles of the abrasive and refractory class. They consist essentially of crystalline alumina.†

Box Oil. See OIL OF BOX.

Boxolon. Herbicide, containing clopyralid, bromoxynil and mecoprop. (ICI PLC).*

Bozefloc. Ployacrylic flocculants. (Hoechst UK Ltd).

Bozetol. A proprietary trade name for a wetting agent. It is a sulphonated castor oil product.†

Bozzle. Container for agrochemicals, used with the 'Electrodyn' sprayer. (ICI PLC).*

BP LDPE. Low density polyethylene. (BP Chemicals Ltd).

BP Mycocide. Liquid preservative based on propionic acid. (BP Chemicals Ltd).

BP Polystyrene. Polystyrene. (BP Chemicals Ltd).

BR Destral. Heavy-duty non-selective herbicide. Applications: Weedkilling. (Borax Consolidated Ltd)*

Bradophen. Quaternary ammonium bacericide. (Ciba-Geigy PLC).

Bradosol. A proprietary preparation of domiphen bromide. Throat lozenges (Ciba Geigy PLC).†

Bradsyn. Polyethylene softeners for textile finishing. (Hickson & Welch Ltd).*

Braemer's Reagent. A tannin reagent. It consists of 1 gram sodium tungstate and 2 grams sodium acetate in 10 cc of water.†

Brakol. A water treatment alkali. (Laporte Industries Ltd).*

Branderz. Synonym for Idrialite.†

Brasivol. An abrasive cleaning paste in three grades, each containing synthetic aluminium oxide in a non-irritant soap-detergent base. Applications: For the treatment of acne vulgaris. (Stiefel Laboratories (UK) Ltd).*

Brasoran 50WP. Sutstituted urea herbicide. (Ciba-Geigy PLC).

Brass. A copper-zinc alloy of varying proportions. Usually it contains more than 18 per cent zinc, and lead is sometimes added to the extent of 1-2 per cent. Ordinary brass contains 67 per cent copper and 33 per cent zinc. This alloy has a specific gravity of 8.4 and melts at 940° C. The alloy containing 60 per cent copper and 40 per cent zinc has a specific heat of 0.0917 at from 20-100 C., and melts at 900° C., and that containing 72 per cent copper and 28 per cent zinc 0.094 at 14-98° C. The alloy consisting of 90 per cent copper and 10 per cent zinc melts at 1040° C., and that containing 70 per cent copper and 30 per cent zinc at 930° C. The specific resistance of brass at 0° C. varies from 7-8 micro-ohms per cm.^{3}.†

Bras-sicol. A contact fungicide containing 50% w/w quintozene, used for the control of diseases in turf. (Burts & Harvey).*

Brassicol. Fungicide for soil and turf treatment. (Hoechst UK Ltd).

Brassoline. See ZAPON VARNISH.

Brass, Iron. A brass containing from 1-9 per cent iron.†

Brass, Iserlohn. An alloy of 64 per cent copper, 33.5 per cent zinc, and 2.5 per cent tin.†

Brass, Japanese. See SIN-CHU.

Brass, Leaded. Alloys of from 71-79 per cent copper, 4.5-9.5 per cent lead, 8.5-23 per cent zinc, and traces to 3 per cent tin.†

Bravit. A proprietary preparation of thiamine hydrochloride, riboflavine, pyridoxine hydrochloride, nicotinamide and ascorbic acid. A vitamin supplement. (Galen).†

Braxo. Industrial hand cleanser. (Borax Consolidated Ltd).

Brazil Wood. See FUSTIC and REDWOODS.

Brazilian Elemi. Elemi from *Icica icariba*.†

Brazing Solder. Alloys of from 35-45 per cent copper, 45-57 per cent zinc, and 8-10 per cent nickel. One alloy, containing 50 per cent copper and 50 per cent zinc, is used for brazing, and for acetylene soldering an alloy of 90 per cent copper and 10 per cent zinc is used.†

Breaxit. Performance chemicals for oil field. (Exxon Chemical International Inc).*

Brebent. Bentonite clay for binding, suspending and emulsifying sands. (Laporte Industries Ltd).

Brebond. Bonding clay for foundry sands. (Laporte Industries Ltd).

Breecht's Double Salt. Potassium dimagnesium sulphate.†

Breedervac-I. Bursal disease vaccine, inactivated vaccine. Applications: immunization of poultry. (Intervet America Inc).*

Breedervac-II. Bursal disease, Newcastle disease, inactivated vaccine. Applications: immunization of poultry. (Intervet America Inc).*

Breedervac-III. Bursal disease, Newcastle bronchitis-reovirus vaccine, inactivated vaccine. Applications: immunization of poultry. (Intervet America Inc).*

Breedervac-IV. Bursal-Newcastle-bronchitis-reovirus disease, inactivated vaccine. Applications: immunization of poultry. (Intervet America Inc).*

Bregel. Drilling bentonite. (Laporte Industries Ltd).

Bremen Blue. See VERDITER BLUE.

Bremen Green. See VERDITER BLUE.

Brentan. A proprietary preparation containing miconazole nitrate. Applications: Antifungal. (Janssen Pharmaceutical Ltd).*

Breokinase. Urokinase. Applications: Plasminogen activator. (Sterling Drug Inc).

Breon. A trade mark. Vinyl materials, nitrile and acrylic rubbers. (British Geon).†

Breon. Synthetic rubber and latices. (BP Chemicals Ltd).*

Breon GA 301A. A proprietary high-quality insulation grade of polyvinyl chloride suitable for use at temperatures up to 105° C. (BP Chemicals Ltd).*

Breon GA 302A. A hard insulation grade of polyvinyl chloride suitable for use at 85° C. (BP Chemicals Ltd).*

Breon GA 304A. A soft insulation and sheathing grade of heat-resisting polyvinyl chloride, suitable for use at 85° C.†

Breon GA 314A. A high-quality soft insulation and sheathing grade of polyvinyl chloride able to withstand high temperatures. It is typically used to sheathe the insulated wiring in electric blankets. (BP Chemicals Ltd).*

Breox. Polyalkylene glycols and polyethylene glycols, brake fluids, lubricants. (BP Chemicals Ltd).*

Bresille Wood. See REDWOODS.

Bresin 2. Thermoplastic resins derived from natural materials extracted from pinewood. Applications: Used in construction adhesives and mastics. (Hercules Inc).*

Bresin 2E. Thermoplastic resins derived from natural materials extracted from pinewood. Applications: Used in construction adhesives and mastics. (Hercules Inc).*

Bretonite. Iodoacetone, $CH_3CO.CH_2I$.†

Bretylate. A proprietary preparation of bretylium tosylate, used in the treatment of cardiac arrhythmias. (The Wellcome Foundation Ltd).*

Bretylol. Bretylium tosylate. Applications: Anti-adrenergic; cardiac depressant. (American Critical Care, Div of American Hospital Supply Corp).

Brevibloc. Esmolol hydrochloride. Applications: Anti-adrenergic. (American Critical Care, Div of American Hospital Supply Corp).

Brevidil. A proprietary preparation of suxemethonium bromide, a muscle relaxant. (May & Baker Ltd).*

Brevidil E. Short acting muscle relaxant. (May & Baker Ltd).

Brevidil M. Suxamethoniom salts. (May & Baker Ltd).

Breviol. Dyeing auxiliary. Applications: Dispersing and levelling agent for dyeing blends of polyester/wool and polyester/acrylic; dispersing agent for cationic dyestuffs; dyeing in general. (Henkel Chemicals Ltd).

Brevital Sodium. Methohexital sodium. Applications: Anaesthetic. (Eli Lilly & Co).

Brexin. Chlorpheniramine maleate USP, pseudoephedrine hydrochloride USP. Applications: Indicated for the temporary relief of symptons of the common cold, allergic rhinitis and sinusitis. (Altana Inc).*

Bri-Nylon. Nylon yarns. (ICI PLC).*

Bricanyl. A proprietary preparation of terbutaline sulphate used in asthma therapy. (Astra Chemicals Ltd).†

Bricanyl Expectorant. A proprietary preparation of terbutaline sulphate and guaiphenesin. An expectorant cough medicine. (Astra Chemicals Ltd).†

Brietal. A proprietary preparation of methohexitone sodium. An intravenous barbiturate anaesthetic. (Eli Lilly & Co).†

Bright Red Oxide. See INDIAN RED.

Brightray Alloy B. A trade mark for an electrical resistance alloy of 59 per cent. nickel, 16 per cent. chromium, 0.3 per cent. silicon and the balance iron. (Wiggin Alloys Ltd).†

Brightray Alloy C. A trade mark for an electrical resistance alloy of 19.5 per cent. chromium, 1.5 per cent. silicon, 0.04 per cent. rare earth elements and the balance nickel. (Wiggin Alloys Ltd).†

Brightray Alloy F. A trade mark for an electrical resistance alloy of 37 per cent. nickel, 18 per cent. chromium, 2.2 per cent silicon and the balance iron. (Wiggin Alloys Ltd).†

Brightray Alloy H. A trade mark for an electrical resistance alloy of 19.5 per cent, chromium, 3.6 per cent. aluminium and the balance nickel. (Wiggin Alloys Ltd).†

Brightray Alloy S. A trade mark for an electrical resistance alloy of 20 per cent. chromium and the balance nickel. (Wiggin Alloys Ltd).†

Brij. Emulsifier polyoxyethylene alkyl ethers. (ICI PLC).*

Brij 96. Polyoxyl 10 Oleyl Ether. Applications: Pharmaceutic aid. (ICI Americas Inc).

Brij 97. Polyoxyl 10 Oleyl Ether. Applications: Pharmaceutic aid.(ICI Americas Inc).

Briklens. Laundry detergents. (Laporte Industries Ltd).*

Briline. Diazo compounds and coupling agents. (Bridge Chemicals).†

Brilliant Acid Bordeaux. A dyestuff. It is a British equivalent of Fast red D.†

Brilliant Alizarin Blue G and R. (Indochromine T). A dyestuff. Dyes chromed wool, cotton, and silk blue.†

Brilliant Alizarin Cyanine G. An analogous product to Anthracene blue WR.†

Brilliant Alizarin Cyanine 3G. An analogous product to Anthracene blue WR.†

Brilliant Archil C. A dyestuff. Dyes wool red.†

Brilliant Azofuchsine 2G. A dyestuff. It is a British equivalent of Brilliant Sulphone Red B.†

Brilliant Azurin B. A direct cotton dyestuff.†

Brilliant Azurin 5G. (Hexamine Azurine 5G). An azo dyestuff from dianisidine and dihydroxynaphthalenesulphonic acid. Dyes cotton blue from a salt bath, and wool from a neutral bath.†

Brilliant Benzo Blue 6B. See DIAMINE SKY BLUE.

Brilliant Benzo Green B. A direct cotton dyestuff.†

Brilliant Black 5G. A dyestuff. Dyes cotton bluish-black.†

Brilliant Blue. Naphthazine blue mixed with Naphthol black.†

Brilliant Bordeaux B, 2B. Dyestuffs. They are British equivalents of Fast Red D.†

Brilliant Bordeaux S. An acid dyestuff.†

Brilliant Carmoisine 0. See AZO RUBINE S.

Brilliant Chrome Red Paste. A mordant dyestuff used in calico printing.†

Brilliant Cochineal 2R, 4R. Acid colours belonging to the same group as Palatine scarlet. They dye wool and silk red from an acid bath.†

Brilliant Congo 2R. A dyestuff obtained from tetrazotized tolidine, amino R-salt, and F-acid.†

Brilliant Cotton Blue, Greenish. See METHYL BLUE.

Brilliant Cresyl Blue 9B. A dyestuff prepared from nitrosodimethyl-*m*-aminocresol and benzyl-*m*-aminodimethyl-*p*-toluidine. Dyes tannined cotton and silk.†

Brilliant Crimson. See AZO RUBINE S.

Brilliant Crimson 0. A dyestuff. It is a British equivalent of Fast Red D.†

Brilliant Croceine LBH. A dyestuff. It is a British equivalent of Brilliant Croceine M.†

Brilliant Croceine M. (Brilliant Croceine 3B, Brilliant Croceine, Bluish, Cotton Scarlet, Cotton Scarlet 3B Conc., Ponceau BO Extra, Paper Scarlet, Blue Shade, Croceine Scarlet 3B, 9187K, Brilliant Croceine LBH). A dyestuff. Dyes wool and silk red from an acid bath. Also used in paper staining.†

Brilliant Croceine 3B. See BRILLIANT CROCEINE M.

Brilliant Croceine, Bluish. See BRILLIANT CROCEINE M.

Brilliant Gallocyanine. See CHROMOCYANINE V AND B.

Brilliant Geranine. See GERANINE.

Brilliant Glacier Blue. See STETOCYANINE.

Brilliant Green. (Malachite Green G, New Victoria Green, Ethyl Green, Emerald Green, Fast Green J, Diamond Green C, Smaragdgreen, Solid Green J, Solid Green TTO, Fast Green S). A dyestuff, which is the sulphate or double zinc chloride of tetraethyldiaminotriphenylcarbinol. Dyes wool, silk, jute, leather, and cotton mordanted with tannin and tartar emetic, green.†

Brilliant Heliotrope 2R Conc. A dyestuff. It is a British equivalent of Methylene Violet 2RA.†

Brilliant Indigo. Dyestuffs for dyeing and printing of cellulosic fibres. (BASF United Kingdom Ltd).

Brilliant Indigo B. 5,6,7-Trihydroxyflavone.†

Brilliant Induline. A dyestuff, which is a sulphonic acid of an induline.†

Brilliant Milling Green B. A diphenylnaphthylmethane dyestuff.†

Brilliant Oil Crimson. A dyestuff. It is a British equivalent of Rosaniline.†

Brilliant Orange. See PONCEAU 4GB.

Brilliant Orange G. See PONCEAU 4GB.

Brilliant Orange GL. A dyestuff. It is a British equivalent of Ponceau 4GB.†

Brilliant Orange OL. A dyestuff. It is a British equivalent of Orange GT.†

Brilliant Orange R. See SCARLET GR.

Brilliant Orange 0. See ORANGE GT.

Brilliant Orchil C. An azo dyestuff, which dyes wool and silk.†

Brilliant Phosphine SG. An acridino dyestuff.†

Brilliant Ponceau G. See PONCEAU R.

Brilliant Ponceau GG. See PONCEAU 2G.

Brilliant Ponceau SR. A dyestuff from diazotized naphthionic acid and G-acid.†

Brilliant Ponceau SR. A dyestuff. It is a British brand of New Coccine.†

Brilliant Red. See FAST RED.

Brilliant Rhoduline Red. See RHODULINE REDS B and G.

Brilliant Scarlet. See IODINE RED.

Brilliant Scarlet 2R. A dyestuff similar to Palatine scarlet. It gives a light scarlet colour.†

Brilliant Scarlet 4R. A dyestuff similar to Palatine scarlet.†

Brilliant Scarlet 4R. A dyestuff. It is a British equivalent of New Coccine.†

Brilliant Sulphonazurine R. A direct cotton dyestuff.†

Brilliant Sulphone Red. (Brilliant Azo-fuchsine 2G, Vega Red S). A dyestuff of the same class as Fast Sulphone Violet 5BS.†

Brilliant Ultramarine. See ULTRAMARINE.

Brilliant Yellow. (4) (Naphthol Yellow RS). A dyestuff. It is the sodium salt of dinitro-α-naphtholsulphonic acid, $C_{10}H_5N_2O_8SNa$. Dyes wool and silk yellow from an acid bath.†

Brilliant Yellow. (1) A pigment. It is a mixture of cadmium sulphide and white lead.†

Brilliant Yellow. (2) (Acid Yellow OO). A dyestuff prepared from the sulphonic acids of mixed toluidines.†

Brilliant Yellow. (3) A dyestuff, $C_{26}H_{18}N_4O_8S_2Na_2$. Dyes cotton yellow from an acid bath. Used for paper-staining.†

Brilliant Yellow S. (Yellow WR, Brilliant Yellow, Curcumine). A dyestuff. It is the sodium salt of *p*-sulphophenylazo-diphenylaminedisulphonic acid, $C_{18}H_{13}N_3O_6S_2Na_2$. Dyes wool and silk yellow.†

Brimstone. Sulphur.†

Brimstone, Flowers of. See FLOWERS OF SULPHUR.

Briosil. A non-precious metal alloy. Applications: For dentistry and dental engineering. (Degussa).*

Briotril. Active ingredients: bromotril plus iotril. Applications: Ready formulated mixture of bromoxynil and ioxynil used for selective post-emergence weed control. (Agan Chemical Manufacturers Ltd).†

Briphos. Range of anionic surfactants composed of aliphatic or nonylphenol based ethoxylated phosphate esters (mixture of mono- and di-esters) in acid form. Viscous liquids or paste. Wide range of surfactant and physical properties depending an the molecular structure of the compound chosen from the range. Application: Agricultural chemicals; cleaners; cosmetics; toiletries; dry cleaning; inks; lubricants; surface coatings. (Albright & Wilson Ltd, Phosphates Div).

Briquest. Phosphonate derivatives. (Albright & Wilson Ltd, Phosphates Div).

Brisgo II. A thermoplastic resin derived from rosin acids. Applications: Hog carcass dehairing composition. (Hercules Inc).*

Bristagen. Gentamicin sulphate. A complex antibiotic substance, produced by Micromonospora purpurea nsp. It has three components, sulphates of gentamicin C_1, gentamicin C_2, and gentamicin C_{1A}. Applications: Antibacterial. (Bristol Laboratories, Div of Bristol-Myers Co).

Bristamycin. Erythromycin stearate. Applications: Antibacterial. (Bristol Laboratories, Div of Bristol-Myers Co).

Bristocycline. A proprietary preparation of tetracycline. An antibiotic. (Bristol-Myers Co Ltd).†

Bristol Brass. See BRASS.

Britannia Metal. An alloy of from 74-91 per cent tin, 5-24 per cent antimony, and 0.15-3.68 per cent copper, sometimes with small quantities of zinc, lead, and bismuth. A Britannia metal containing 90 per cent tin and 10 per cent antimony has a specific gravity of 7.9, and melts at 260° C. Formerly used in the manufacture of cheap tableware, such as teapots and spoons, also as an antifriction metal.†

Britesil. Hydrous sodium polysilicate. Applications: Used as an alkaline builder in powdered or granular laundry detergents, dishwashing detergents, household cleaners. (The PQ Corporation).*

Britesorb. Clarifying agent. Hydrous silica. Applications: Clarifying and chill proofing agent for beer, clarifying or fining agent for wines, fruit juices etc., selective absorbent of proteins and metals. (The PQ Corporation).*

British Barilla. See BARILLA.

British Gum. (Starch Gum, Vegetable Gum, Gommeline, Artificial Gum). Dextrin, $n(C_6H_{10}O_5)$, obtained by the action of diastase on starch paste, or by heating starch with a trace of acid. Used as an adhesive.†

Britonite. An explosive containing nitroglycerin, potassium nitrate, wood meal, and ammonium oxalate.†

Brittalite. See HYDROSULPHITE NF.

Brittox. Herbicide. (May & Baker Ltd).

Britulite. See HYDRALDITE C EXT.

Brix. Cupola flux. (Foseco (FS) Ltd).*

Brix Metal. Alloys of from 60-75 per cent nickel, 15-20 per cent chromium, 5 per cent copper, 1-4 per cent tungsten, 4 per cent silicon, 3 per cent titanium, 2 per cent aluminium, and 1 per cent bismuth. Stated to be non-corrosive.†

Brixil. Cupola flux and silicon additive. (Foseco (FS) Ltd).*

Briz. A scouring powder. (Unilever).†

Broad Salt. Ground rock salt.†

Brocade Colours. See BRONZE COLOURS.

Brocadopa. A proprietary preparation of levodopa used in the treatment of Parkinson's disease. (Brocades).*

Broenner's Acid. 2-Naphthylamine-6-sulphonic acid.†

Brolac Dualcote Acrylic Primer/Undercoat. Acrylic copolymer emulsion, water thinned. Applications: Interior primer for soft and hardwoods and an interior and exterior undercoat for Brolac Full Gloss. (Berger Jenson & Nicholson Ltd).*

Brolac Eggshell Low Odour. Alkyd resin based, white spirit thinned. Applications: Interior application, can be used in kitchens and bathrooms giving satin sheen finish. (Berger Jenson & Nicholson Ltd).*

Brolac Full Gloss. Alkyd resin based, white spirit thinned. Applications: High gloss protective finish for interior and exterior use by the professional decorator and specifier. (Berger Jenson & Nicholson Ltd).*

Brolac PEP Vinyl Matt & Vinyl Silk Emulsions. V.A. copolymer emulsion based, water thinned. Applications: Vinyl matt is for both exterior and interior application giving a matt finish. Vinyl silk is for interior application giving a silk sheen. (Berger Jenson & Nicholson Ltd).*

Brolac Primers, Sealers and Surface Preparation Products. Various alkyd/oleo resinous, white spirit thinned. Applications: Protective primers/sealers for painting surfaces. (Berger Jenson & Nicholson Ltd).*

Brolac Specialist Coatings. Various specialist resins. Applications: Maintenance and protection of steel, floors, concrete and other surfaces requiring extra resistance and performance. (Berger Jenson & Nicholson Ltd).*

Brolac Superflat Emulsion. Applications: V.A. copolymer emulsion, water thinned. Applications: Interior application, especially suited to new plasterwork. (Berger Jenson & Nicholson Ltd).*

Brolac Tartaruga. V.A. copolymer emulsion, water thinned. Applications: High build textured wall finish for interior and exterior use. (Berger Jenson & Nicholson Ltd).*

Brolac Undercoat. Alkyd resin based, white spirit thinned. Applications: Used in preparation for all types of Brolac finishes. (Berger Jenson & Nicholson Ltd).*

Brolac Varnishes. Alkyd resin base (some P.U. modified) - thinned white spirit. Applications: Surface sheen to soft and hardwoods. (Berger Jenson & Nicholson Ltd).*

Brolac Weathercoat No. 1 - finely textured. V.A. copolymer emulsion, water thinned. Applications: Masonry exterior coatings. (Berger Jenson & Nicholson Ltd).*

Brolac Weathercoat No. 2 - smooth. V.A. copolymer emulsion, water thinned. Applications: Exterior masonry, ideal for airless spray equipment. Suitable for buildings subjected to atmospheric pollution. (Berger Jenson & Nicholson Ltd).*

Brolac Weathercoat No. 3. Pliolite resin based, white spirit thinned. Applications: Smooth finish to masonry, can be applied in cold, wet and damp conditions with no separate sealer requirement. (Berger Jenson & Nicholson Ltd).*

Brolene. A proprietary preparation of propamidine isethionate. An ocular antiseptic. (May & Baker Ltd).*

Bromal. Tribromoacetaldehyde, CBr_3 CHO. Formerly used in medicine.†

Bromanil. Tetrabromoquinone, $C_6Br_4O_2$.†

Bromanilide. See ANTISEPSIN.

Brombutol. See BROMETONE.

Bromcresol Purple. Dibromo-*o*-cresolsulphonphthalein. An indicator.†

Bromeikon. Sodium tetrabromophenol-phthalein.†

Bromelains. A concentrate of proteolytic enzymes derived from Ananas comosus Merr. Ananase. Recommended as a substitute for pancreatin. Prepared from pineapple juice. A pale brown powder. (British Drug Houses).†

Bromelia. See NEROLIN II.

Brometone. (Acetone-bromoform, Brombutol). Tribromo-tert-butyl alcohol, CBr_3 $C(CH_3I_2OH)$. Formerly used in medicine as a sedative.†

Bromex. Active ingredient: naled; 1,2-dibromo-2,2-dichlorovinyl dimethyl phosphate. Applications: Fast-acting agricultural insecticide of short to moderate residual action. (Makhteshim Chemical Works Ltd).†

Bromicide. Cooling water biocide. (Great Lakes Chemical (Europe) Ltd).

Bromide Solution, Rice's. See RICE'S BROMIDE SOLUTION.

Bromidine. A dry mixture of sodium bisulphate with sodium or potassium bromide and bromate. A disinfectant.†

Bromindigo FB. (Ciba Blue 2B, Durindone Blue 4B). 5,7,5',7'-Tetrabrom-indigo.†

Bromindione. 2-(4-Bromophenyl)indane-1,3-dione. Circladin. An anti-coagulant.†

Bromine Salt. A mixture made by saturating caustic soda with bromine. draining off the mother liquor, and adding sodium bromate. Used in the extraction of gold ores.†

Bromo-Gas. Methyl bromide with chloropicrin. (Great Lakes Chemical (Europe) Ltd).

Bromo-purpurin. Bromotrihydroxy-anthraquinone.†

Bromochloral. Dichlorobromo-acetaldehyde, CCl_2.Br.CHO.†

Bromochloroform. Dichlorobromo-methane, $CHCl_2$ Br.†

Bromocoll. A registered trade mark currently awaiting re-allocation by its proprietors to cover a range of pharmaceutical products. (Casella AG).†

Bromocresol Green. Tetrabromo-*m*-cresolsulphonphthalein. An indicator.†

Bromocyclen. An insecticide and acaricide 5-Bromomethyl-1,2,3,4,7,7 hexachloronorborn-2-ene. ALUGAN.†

Bromodan. A brominating agent. (Fisons PLC).*

Bromodeine. A proprietary preparation containing bromoform, codeine hydrochloride, liquid extract of krameria, wild cherry and senega. A cough linctus. (Crookes Laboratories).†

Bromoform. Tribromomethane, $CHBr_3$.†

Bromoform. Bromoform. Applications: Gem and mineral testing. (Geoliquids).

Bromol. Tribromophenol, $C_6H_2(OH)Br_3$. Used as a caustic and disinfectant.†

Bromolaurionite. A mono-hydrated lead oxydibromide, $PbO\ PbBr_2\ H_2O$, obtained by heating a solution of lead acetate and sodium bromide for 12 hours.†

Bromonitroform. Bromotrinitromethane, $CBr(NO_2)_3$.†

Bromophenol Blue. Tetrabromophenol-sulphonphthalein. An indicator.†

Bromophenol Red. Dibromophenol-sulphonphthalein. An indicator.†

Bromopicrin. Tribromonitromethane, $CBr_3(NO_2)$.†

Bromopyrin. Monobromoantipyrine, $C_{11}H_{11}BrN_2O$.†

Bromotril. Active ingredients: bromoxynil octanoate; 2,6-dibromo-4-cyanophenyl octanoate. Applications: Selective post-emergence control of a wide range of annual broadleaf weeds in winter and spring cereals and in corn. (Agan Chemical Manufacturers Ltd).†

Bromowagnerite. A compound of calcium bromide and phosphate, $Ca_3(PO_4)_3\ CaBr_2$.†

Bromox. Potassium bromate dispersion in food grade filler. Applications: Flour additive. (Diaflex Ltd).*

Bromoxylenol Blue. An indicator, $C_{23}H_{20}O_5Br_2S$, prepared by adding bromine to xylenol blue dissolved in glacial acetic acid.†

Bromoxynil. 1,4-Dibromo-3-cyano phenol. A specific herbicide for use in cereal crops.†

Bromsulphalein. Sulphobromophthalein Sodium. Applications: Diagnostic aid. (Hynson Westcott & Dunning, Div of Becton Dickinson & Co).

Brom-tetragnost. Sodium tetrabromophenolphthalein.†

Bromthymol Blue. Dibromothymol-sulphonphthalein. An indicator.†

Bronchilator. A proprietary preparation of isoetharine methanesulphonate, phenylephrine hydrochloride and thenyldiamine hydrochloride. A bronchial antispasmodic. (Bayer & Co).*

Broncho-Binotal. A preparation for bacterial bronchopumlonary infections. (Bayer & Co).*

Bronchodil. SEE BRONCHOSPASMIN. (Degussa).*

Bronchopront. A proprietary preparation of ambroxol hydrochloride. A mucolytic. (Pfizer International).*

Bronchospasmin (Bronchodil). Reproterol. B_2-Mimetikum. Tablets, ampoules and aerosol. (Degussa).*

Bronco. Active ingredients are 2.6 pounds of 2-chloro-2'6'-diethyl-N-(methoxymethyl) acetanilide and 1.4 pounds of the isopropylamine salt of glyphosate. Applications: Herbicide for no-till farming. (Monsanto Co).*

Bronkaid Mist. Epinephrine. Applications: Adrenergic. (Sterling Drug Inc).

Bronkephrine. Ethylnorepinephrine hydrochloride. Applications: Bronchodilator. (Sterling Drug Inc).

Bronkometer. Isoetharine mesylate. Applications: Bronchodilator. (Sterling Drug Inc).

Bronkosol. Isoetharine hydrochloride. Applications: Bronchodilator. (Sterling Drug Inc).

Bronner's Acid. 2-Naphthylamine-6-sulphonic acid.†

Bron-Newcavac-M. Bronchitis-Newcastle disease, inactivated vaccine. Applications: immunization of poultry. (Intervet America Inc).*

Bronocot. Cotton seed dressings. (ICI PLC, Plant Protection Div).

Bronopol. 2-Bromo-2-nitropropane-1, 3-diol.†

Bronopol-Boots. Pharmaceutical and cosmetic preservative. (The Boots Company PLC).

Bronotabs. Milk testing preservative. (The Boots Company PLC).

Bronox. Selective herbicide. (FBC Ltd).

Brontyl. A proprietary preparation of proxyphylline. Bronchial antispasmodic. (Lloyd, Hamol).†

Bronze. Alloys usually consisting of copper and tin in varying proportions, often with zinc, and occasionally with lead. The copper varies from about 74-95 per cent. the tin from 1-20 per cent. zinc from 0-17 per cent. and the lead from 0-18 per cent. The alloys containing from 70-91 per cent copper and 9-30 per cent tin vary in melting-point from 750- 1030° C. and in specific gravity from 8.79-8.93.†

Bronze "A". A British chemical standard. It contains 85.5 per cent copper, 9.96 per cent tin, 1.86 per cent zinc, 1.83 per cent lead, 0.25 per cent phosphorus, 0.24 per cent antimony, 0.07 per cent iron, 0.04 per cent nickel, and 0.06 per cent arsenic.†

Bronze Acetate. A calcium acetate prepared from crude pyroligneous acid and lime. It contains from 60-70 per cent acetate. The name is also applied to an impure variety of lead acetate prepared from the same acid.†

Bronze, Acid. See ACID BRONZE.

Bronze, Ajax. See AJAX BRONZE.

Bronze, Allan Red. See ALLAN RED BRONZE.

Bronze, Aluminium. See ALUMINIUM BRONZE.

Bronze, Aluminium Iron. See ALUMINIUM IRON BRONZE.

Bronze, Aluminium Magnesium. See ALUMINIUM MAGNESIUM BRONZE.

Bronze, Aluminium Manganese. See ALUMINIUM MANGANESE BRONZE.

Bronze, Aluminium Tin. See ALUMINIUM TIN BRONZE.

Bronze, Aluminium Titanium. See ALUMINIUM TITANIUM BRONZE.

Bronze, Arsenic. See ARSENIC BRONZE.

Bronze, Artificial. See TOMBAC RED, and MANNHEIM GOLD.

Bronze, Atlas. See ATLAS BRONZE.

Bronze Bearing Metals. Very variable alloys. One type contains from 70-91 per cent tin, 7-26 per cent antimony, and 2-22 per cent copper; whilst another class contains from 70-86 per cent copper, 4-20 per cent tin, 0-30 per cent zinc, and 0-15 per cent lead. Also see KOCHLIN'S BEARING BRONZE.†

Bronze, Beryllium. See BERYLLIUM BRONZE.

Bronze, Bilgen. See BILGEN BRONZE.

Bronze, Black. See BLACK BRONZE.

Bronze Blues. Types of Prussian blue.†

Bronze Browns. Mixtures of unburnt umb[e]r and chrome yellow, toned with Cassel brown or chrome orange. Pigments.†

Bronze, Cadmium. See CADMIUM BRONZE.

Bronze, Calsun. See CALSUN BRONZE.

Bronze, Carbon. See CARBON BRONZE.

Bronze, Carloon. See CARLOON BRONZE.

Bronze, Caro. See CARO BRONZE.

Bronze, Chinese. See CHINESE BRONZE.

Bronze, Chromax. See CHROMAX BRONZE.

Bronze, Chromium. See CHROMIUM BRONZE.

Bronze, Cobalt. See COBALT BRONZE.

Bronze, Coin. See COINAGE BRONZE.

Bronze Colours. (Brocade Colours). Powdered metals or metallic alloys mixed with linseed oil varnish. Brocade colours are not so finely powdered.†

Bronze, Conductivity. See CONDUCTIVITY BRONZE.

Bronze, Cornish. See CORNISH BRONZE.

Bronze, Cowles' Aluminium. See COWLES' ALUMINIUM BRONZE.

Bronze, Damar. See DAMASCUS BRONZE.

Bronze, Dawson. See DAWSON BRONZE.

Bronze, Durbar. See DURBAR BRONZE.

Bronze, Eclipse. See ECLIPSE BRONZE.

Bronze, Eisler's. See EISLER'S BRONZE.

Bronze, Elephant. See ELEPHANT BRONZE.

Bronze, Emerald. See EMERALD BRONZE.

Bronze, File. See FILE BRONZE.

Bronze, Gold. See SAFFRON BRONZE.

Bronze, Graney. See GRANEY BRONZE.

Bronze Greens. (Bottle Green, Olive Green). Pigments. They usually consist of Brunswick greens mixed with ochre and umber, or of Chrome green with black and yellow colours.†

Bronze, Gurney. See GURNEY BRONZE.

Bronze, Harmonia. See HARMONIA BRONZE.

Bronze, Harrington. See HARRINGTON BRONZE.

Bronze, Helmet. See HELMET BRONZE.

Bronze, Herbohn. See HERBOHN BRONZE.

Bronze, Hercules. See HERCULES BRONZE.

Bronze, Holfos. See HOLFOS BRONZE.

Bronze, Hydraulic. See HYDRAULIC BRONZE.

Bronze, Instrument. See INSTRUMENT BRONZE.

Bronze, Japanese. See JAPANESE BRONZE.

Bronze, Kern's Hydraulic. See KERN'S HYDRAULIC BRONZE.

Bronze, Kochlin's Bearing. See KOCHLIN'S BEARING BRONZE.

Bronze, Kuhne's Phosphor. See KUHNE'S PHOSPHOR BRONZE.

Bronze, Lafond. See LAFOND BRONZE.

Bronze, Lead. See LEAD BRONZE.

Bronze, Leaded. See LEADED BRONZE.

Bronze, Liquid. See LIQUID BRONZES.

Bronze, Litnum. See LITNUM BRONZE.

Bronze, Lowroff Phosphor. See LOWROFF PHOSPHOR BRONZE.

Bronze, Lumen. See LUMEN BRONZE.

Bronze, Machine. See MACHINE BRONZE.

Bronze, Magenta. See MAGENTA BRONZE.

Bronze, Manganaluminium. See MANGANALUMINIUM BRONZE.

Bronze, Manganese. See MANGANESE BRONZE.

Bronze, McKechnie's. See McKECHNIE'S BRONZE.

Bronze, Medal. See MEDAL BRONZE.

Bronze, Mirror. See MIRROR BRONZE.

Bronze, Naval. See NAVAL BRONZE.

Bronze, Needle. See NEEDLE BRONZE.

Bronze, Nickel. See NICKEL BRONZE.

Bronze, Nickel Aluminium. See NICKEL ALUMINIUM BRONZE.

Bronze, Nickel Manganese. See NICKEL MANGANESE BRONZE.

Bronze, Olympic. See OLYMPIC BRONZE.

Bronze, Optical. See OPTICAL BRONZE.

Bronze, Oranium. See ORANIUM BRONZE.

Bronze, Phono. See PHONO BRONZE.

Bronze, Phosphor. See PHOSPHOR BRONZE.

Bronze, Plastic. See PLASTIC BRONZE.

Bronze, Platinum. See PLATINUM BRONZE.

Bronze, Reich's. See REICH'S BRONZE.

Bronze, Resistance. See RESISTANCE BRONZE.

Bronze, Roman. See ROMAN BRONZE.

Bronze, Saffron. See SAFFRON BRONZE.

Bronze, Screw. See SCREW BRONZE.

Bronze, Sea-Water. See SEA-WATER BRONZE.

Bronze, Sheathing. See SEA-WATER BRoNZE.

Bronze, Silicon. See SILICON BRONZE.

Bronze, Silver. See SILVER BRONZE.

Bronze, Silzin. See SILZIN BRONZE.

Bronze, Statuary. See STATUARY BRONZE.

Bronze, Steel. See UCHATIUS BRONZE.

Bronze, Stone's. See STONE'S BRONZE.

Bronze, Sun. See SUN BRONZE.

Bronze, Telegraph. See TELEGRAPH BRONZE.

Bronze, Tin. See TIN BRONZE.

Bronze, Tobin. See TOBIN BRONZE.

Bronze, Tungsten. See TUNGSTEN BRONZE.

Bronze, Turbadium. See TURBADIUM BRONZE.

Bronze, Turbiston. See TURBISTON BRONZE.

Bronze, Uchatius. See UCHATIUS BRONZE.

Bronze, Valve. See VALVE BRONZE.

Bronze, Vanadium. See VANADIUM BRONZE.

Bronze Varnishes. These are usually prepared by mixing suitable dyes with spirit varnishes. Used for colouring articles to give iridescence.†

Bronze, Vulcan. See VULCAN BRONZE.

Bronze Wire. An alloy of 98.75 per cent copper, 1.2 per cent tin, and 0.05 per cent phosphorus.†

Bronze, Wolfram. See TUNGSTEN BRONZE.

Bronze, Zinc. See ADMIRALTY GUN METAL.

Bronzing Liquids. Consist of volatile liquids which will hold up the metal and some material which will keep the metallic powder from rubbing off after it has been applied. The best one contains pyroxylin dissolved in amyl acetate to which the metallic powder is added. A cheaper form consists of rosin dissolved in an aromatic hydrocarbon with metallic powder added, and a very cheap one is a solution of sodium silicate with metal powder.†

Bronzing Powder. See MOSAIC GOLD.

Bronzing Solder. An alloy of 50 per cent zinc and 50 per cent copper.†

Bronzite. A pyroxene mineral. It is $16[(Mg,Fe)SiO_3]$. †

Brotopon. A proprietary preparation of haloperidol used for psychotic disorders. (Pfizer International).*

Brovon. Atropine methonitrate, adrenaline and papaverine. Applications: Bronchospasm. (Napp Laboratories Ltd).*

Brown Barberry Gum. See MOROCCO GUM.

Brown, Catechu. See CATECHU BROWN.

Brown Haematite. A hydrated oxide of iron, an iron ore, $2Fe_2O_3\ 3H_2O$.†

Brown Lead Ore. (Linnets). A mixture oflead phosphate and lead chloride, $3[Pb_3(PO_4)_2]+PbCl_2$.†

Brown Lead Oxide. Lead dioxide, PbO_2.†

Brown Madder. A pigment. It is a lake prepared from madder root.†

Brown Manganese Ore. See MANGANITE.

Brown NP and NPJ. A dyestuff from diazotized *p*-nitraniline and pyrogallol, $C_{12}H_9N_3O_5$. Dyes chrome mordanted wool.†

Brown Ochre. See YELLOW OCHRE.

Brown Ore. A variable mixture of hydrated oxides of iron, usually $2Fe_2O_33H_2O$.†

Brown Oxide of Iron. Peroxide of iron.†

Brown Oxide of Tungsten. Tungsten dioxide, WO_2.†

Brown Pink. (Stil de Grain). A pigment prepared from Turkish or Persian berries. It is a lake precipitated by alum from a decoction of the colouring matter.†

Brown PM. A dyestuff. It is the hydro chloride of *p*-aminophenylazo-*m*-phenylene-diamine, $C_{12}H_{14}N_5Cl$. Dyes tannined cotton dark brown.†

Brown Precipitate. Iodine dissolved in potassium iodide.†

Brown Red. See INDIAN RED.

Brown Red Antimony Sulphide. See KERMES MINERAL.

Brown Terre Verte. See BURNT SIENNA.

Brown Ultramarine. Ultramarine in which the sulphur constituent is replaced by selenium.†

Broxli. A proprietary preparation containing phenethicillin potassium. An antibiotic. (Beecham Research Laboratories).*

Brozgerite. See CLEVITE.

Brücke's Reagent. This reagent consists of 50 grams potassium iodide in 500 cc water saturated with mercuric iodide (120 grams) and made up to 1 litre. A precipitating agent for proteins.†

Brufen. A proprietary preparation containing ibuprofen. An anti-rheumatic drug. (Boots Pure Drug Co).*

Brugère Powder. An explosive. It is a priming composition, containing 54 per cent ammonium picrate, and 46 per cent potassium nitrate.†

Brulidine. A proprietary preparation of dibromopropamidine isethionate. Antiseptic skin cream. (May & Baker Ltd).*

Brunner's Salt. Obtained by dissolving vermilion in potassium mono-sulphide. It has the formula, $HgSK_2S\ 5H_20$.†

Brünnichite. See APOPHYLLITE.

Brunol. A proprietary preparation of *n*-butyl salicylate.†

Brunswick Black. Asphalt or pitch mixed with turpentine and linseed oil, and heated.†

Brunswick Blue. (Celestial Blue). A pigment produced by mixing 50-90 per cent barytes with Prussian blue.†

Brunswick Green, New. See CHROME GREENS.

Brush Wire. A brass wire containing 64.25 per cent copper, 35 per cent zinc, and 0.75 per cent tin.†

Brush-B-Gon. Brush killer. (Chevron).*

Brussels System. A systemaic insecticide. (Murphy Chemical Co).†

B.R.V. A coal-tar distillate consisting chiefly of high boilng constituents. It is used as a rubber softener.†

Bryrel. Piperazine Citrate. Applications: Anthelmintic. (Sterling Drug Inc).

B.S. Sea Water Alloy. An aluminium alloy containing 7.5-9.5 per cent magnesium, 0.2 per cent silicon, and 0.2-0.6 per cent manganese. It has a tensile strength of 45-55 kg per sq mm†

BTC. Range of cationic surfactants of the alkyl quaternary ammonium chloride type in liquid or powder form. Applications: Disinfectant, deodorant, germicide, algicide, slimicide and fungicide. Used in swimming pool and industrial water treatment; dairy and food processing equipment. (Millmaster-Onyx UK).

BTG Alloy. A heat-resisting alloy containing 60 per cent nickel, 12 per cent chromium, 1-4 per cent tungsten, and balance iron.†

Bu-Gas. Butane. (Chevron).*

Bu-White. Disproportionated resin derivatives. (Tenneco Malros Ltd).

Bubber Shet. Prilled urea contains minimum of 46% available nitrogen. Applications: Used as a fertilizer to supply nitrogen to the crops for better yield per acre. (Dawood Hercules Chemicals Ltd).*

Bubblefil. A proprietary trade name for regenerated cellulose.†

Bucarpolate. A pyrethrum synergist. (Bush Boake Allen).*

Buchu Camphor. Diosphenol, $C_{10}H_{15}O_2$, the chief constituent of the essential oil obtained from Buchu leaves from *Barosma beturima*. It is antiseptic.†

Buckland's Cement. A label cement consisting of 50 per cent gum arabic, 37.5 per cent starch, and 12.5 per cent white sugar. It is mixed with a little water for use.†

Buckroid. A very tough form of pure vulcanized rubber. Used for making mats, and for other purposes.†

Buckthorn Green. See SAP GREEN.

Bucrol. Carbutamide.†

Buctril. Selective weedkiller. (May & Baker Ltd).

Budale. A proprietary preparation of paracetamol, codeine phosphate and butobarbitone. A sedative. For veterinary use. (Dales Pharmaceuticals).*

Bufa. A proprietary trade name for dibutyl phthalate. (Koninklijke Nederlandsche Gist-En Spiritusfabriek).†

Bufferin. A proprietary preparation of aspirin, magnesium carbonate and aluminium glycinate. An analgesic. (Bristol-Myers Co Ltd).†

Bug Gun. Ready for use insecticide spray. (ICI PLC).*

Bug-Geta. Slug and snail pellets. (Chevron).*

Bull Metal. An alloy similar to Delta metal in composition.†

Bullet Brass. See BRASS.

Bullseye. A smokeless powder. The double-base-type formulation minimises charge weight and moisture absorption. Graphite glaze enables smooth granule flow. Applications: Ammunition. (Hercules Inc).*

Bullseye. Water soluble, biodegradable colourants. Applications: For spray pattern indicators for application of herbicides, pesticides in the lawn and turf, forestry and industrial weed control areas. (Milliken & Company).*

Bumal. Modified rosin emulsion. (Tenneco Malros Ltd).

Bumex. Bumetanide. Applications: Diuretic. (Hoffmann-LaRoche Inc).

Buminate. Albumin human. Applications: Blood volume supporter. (Hyland Therapeutics, Div of Travenol Laboratories Inc).

Buna AP. A range of ethylene/propylene rubbers. Applications: Used as modifiers for polyolefins, moulded articles and hoses, blend component for improving flowability and green strength. (Huls AG).†

Buna CB. Cis-1,4-polybutadiene polymers. Applications: These include tyres, conveyor belting, mountings and roll covers. (Bayer & Co).*

Buna EM. A range of styrene/butadiene rubbers; emulsion polymerization. Applications: Used for tyres, injection mouldings and industrial rubber goods. (Huls AG).†

Buna S. A proprietary trade name for a vulcanizable synthetic rubber. It is a butadiene-styrene co-polymer.†

Buna SL. A range of styrene/butadiene rubbers; solution polymerization. Applications: Used for tyre treads, industrial rubber goods, shoe soles and heels. (Huls AG).†

Buna SS. Contains a larger percentage of styrene. An important series of synthetic products is made by co-polymerizing butadiene with 10-30 per cent of another polymerizable substance such as styrene or acetonitrile, *e.g.* Buna N, etc.†

Buna Vi. A range of vinyl/butadiene rubbers. Applications: Used for industrial goods and tyres. (Huls AG).†

Bunatex. A range of styrene/butadiene rubber latex. Applications: Used as raw materials for foam backing for carpets and textiles, mattresses, upholstery materials for the furniture industry, shoe components, rubberized hair bitumen modifiers and friction linings. (Huls AG).†

Bunsenine. See WHITE TELLURIUM.

Bunte's Salt. Sodium-ethyl thiosulphate. (*qv*).†

Buprenex. Buprenorphine hydrochloride. Applications: Analgesic. (Norwich Eaton Pharmaceuticals Inc).

Burcop. Soda/Bordeaux fungicide. (McKechnie Chemicals Ltd).

Burez. Disproportionated resin derivates. (Tenneco Malros Ltd).

Burgess Solder. An alloy of 76 per cent tin 21 per cent zinc, and 3 per cent aluminium.†

Burgess-Hambuechen Solution. A solution containing 275 grams ferrous ammonium sulphate and 1000 cc water. Used for the electro-deposition of iron, using a current density of 6-10 amps. per sq ft. at 30° C.†

Burgundy Lake. A proprietary trade name for a red lake containing organic colours aluminium hydroxide, and Blanc fixé.†

Burgundy Mixture. For the prevention of blight or potato disease, and its cure. (*a* Copper sulphate dissolved in 5 gallons water, and made up to 35 gallons. (*b*) Contains 5 lb washing soda dissolved in 5 gallons water. (*b*) is added to (*a*) and stirred.†

Burgundy Pitch. The resinous exudation of the European silver fir. Artificial substitutes have been made.†

Burinex K. A proprietary preparation of bumetanide with potassium chloride in slow-release form. A diuretic. (Leo Laboratories).†

Burkeite. (Gauslinite). A double salt of sodium carbonate and sulphate.†

Burma Black Varnish. See THITSI.

Burmite. (Birmite). Burmese amber of a reddish-brown colour.†

Burnol Acriflavine Cream. A proprietary product. It is an emulsified cream containing neutral acriflavine for use in cases of burns, wounds, etc. (Boots Pure Drug Co).*

Burnt Alum. (Exsiccated Alum). Potash alum, which has been heated at low redness.†

Burnt Carmine. A pigment obtained by calcining carmine.†

Burnt Hypo. (Black Hypo, Eureka Compound). A mixture of lead thiosulphate and sulphide, and sulphur. Used in the vulcanization of rubber.†

Burnt Iron. Iron which has been heated to a high temperature for a long time. It is brittle.†

Burnt Lime. See LIME.

Burnt Magnesia. Magnesium oxide MgO.†

Burnt Nickel. A term used for a grey pulverulent nickel precipitated by too strong a current during its electro-deposition.†

Burnt Ochre. See INDIAN RED.

Burnt Pyrites. Pyrites which have been burnt until 70 per cent of ash is left. It consists of iron oxide, Fe_2O_3.†

Burnt Roman Ochre. An orange pigment obtained by calcining Roman ochre.†

Burnt Sienna. (Brown Terre Verte, Raw and Burnt Umber, Cappagh Brown, Mars Brown, American Sienna). Earths or ochres, raw or calcined, or artificial ochre. They are brown pigments containing iron or iron and manganese. Burnt sienna usually contains 65-75 per cent. Fe_2O_3, and 13-20 per cent. SiO_2. Also See INDIAN RED.†

Burnt Topaz. When Brazilian topaz is heated, it changes from a cherry-yellow to a rose-pink, being then known as burnt topaz.†

Burnt Umber. Umber which has been heated, whereby its colour is somewhat reddened. Raw umber is a brown earthy variety of ochre, coloured by oxides of iron. Also see BURNT SIENNA and UMBER.†

Buro-sol Concentrate. Aluminium acetate. Applications: Astringent. (Doak Pharmacal Co).

Burow's Solution. A 7.5-8 per cent solution of aluminium acetate. (*Liquor alumini acetatis, BPC*)†

Burr Brass. See BRASS.

Bursoline. Sulphonated oil for tanners. (Clayton Aniline Co).†

Burtolin. A tree growth inhibitor containing 185g/litre maleic hydrazide (as the potassium salt). Applications: It is used to control shoots on the trunk and suckers around the base of street trees. It also inhibits the development of buds on the trunk which remain dormant following treatment. (Burts & Harvey).*

Buscopan. A proprietary preparation containing hyoscine-N-butylbromide. An antispasmodic. (Boehringer Ingelheim Ltd).†

Bush Metal. An alloy of 72 per cent copper, 14 per cent tin. and 14 per cent yellow brass.†

Buspar. Busprione hydrochloride. Applications: Tranquilizer. (Mead Johnson & Co).

Butacite. Polyvinyl butyral. (Du Pont (UK) Ltd).

Butaclor. Polychloroprene rubber. (BP Chemicals Ltd).

Butacote. A proprietary preparation of phenyl butazone. An anti- rheumatic drug. (Ciba Geigy PLC).†

Butakon A2554. A proprietary butadiene copolymer rubber. (Revertex, Harlow, Essex).†

Butakon ML 577/1. A proprietary butadiene/methyl methacrylate latex. (Revertex, Harlow, Essex).†

Butane Tetrol. See TETRA-NITROL.

Butanex. Active ingredient: butachlor; N-(butoxymethyl)-2-chloro-2',6'-diethylacetanilide. Applications: Selective pre-emergence and early post-emergence weed control in transplanted, direct seeded and upland rice. (Agan Chemical Manufacturers Ltd).†

Butanox. Ethylmethylketone peroxide. (Akzo Chemie UK Ltd).

Butasan. A proprietary trade name for a rubber vulcanization ultra accelerator. It is the zinc salt of dibutyl dithiocarbamic acid.†

Butasan vulcanization Accelerator. Zinc dibutyldithiocarbamate. Applications: Functions as a non-discolouring stabilizer in non-curing applications and in butyl rubber. (Monsanto Co).*

Butazate. Zinc dibutyldithiocarbamate. An ultra accelerator, fast curing from 212°F (100°C) up. Low critical temperature, medium curing range below 250°F (121°C). Non-staining and non-discolouring. Active dithiocarbamate accelerator available for latex compounding. Has strong tendency to precure. (Uniroyal).*

Butazate 50D. A 50% active water dispersion of Butazate. Ready-to-use form for latex compounding. It is used in combination with Ethazate 50D and OXAF for greater economy and improved physical properties. (Uniroyal).*

Butazolidin. Phenylbutazone. Applications: Antirheumatic. (Ciba-Geigy Corp).

Butazolidin Alka. A proprietary preparation of aluminium hydroxide gel, magnesium trisilicate and Butazolidine. (Ciba Geigy PLC).†

Butazolidin with Xylocaine. A proprietary preparation of phenylbutazone and lignocaine. (Ciba Geigy PLC).†

Butea Gum. See KINO, BENGAL.

Butesin Picrate. Butamben picrate. Applications: Anaesthetic. (Abbott Laboratories).

Butex. (1) A proprietary trade name for a synthetic rubber.†

Butex. (2) Esters of 4 hydroxybenzoic acid. (Bush Boake Allen).*

Butisol. See BUTABARBITAL.

Butocresiol. See EUROPHENE.

Butofan D. Polybutadiene dispersion. (BASF United Kingdom Ltd).*

Butox. A proprietary preparation of polyisobutylene-isoprene. (Hardman Inc).*

Butoxone. A selective weed killer. (ICI PLC).*

Butoxyl. 1-Methoxybutyl acetate.†

Butoxyl. Solvent for lacquers. (Hoechst UK Ltd).

Butoxyne 497. A mixture of hydroxyethyl ethers of butynediol. Applications: Nickel brightener in electroplating. Pickling inhibitor used prior to plating copper. Corrosion inhibitor for specialty application such as in aerosol cans. (GAF Corporation).*

Butt Brass. See BURR BRASS.

Butter, Abjab. See NJAVE BUTTER.

Butter, American Mace. See OTOBA BUTTER.

Butter, Bambuk. See SHEA BUTTER.

Butter, Cacao. See COCOA BUTTER.

Butter, Coco-nut. See COCO-NUT BUTTER.

Butter, Cocoa. See COCOA BUTTER.

Butter Colour. See ANNATTO.

Buttercup Yellow. See ZINC YELLOW.

Butter, Galam. See SHEA BUTTER.

Butter, Gamboge. See GAMBOGE BUTTER.

Butter, Goa. See KOKUM BUTTER.

Butter, Kanga. See LAMY BUTTER.

Butter, Kanja. See LAMY BUTTER.

Butter, Kokum. See KOKUM BUTTER.

Butter, Lamy. See LAMY BUTTER.

Butter, Macaja. See MOCAYA OIL.

Butter, Macassar Nutmeg. See MACASSAR NUTMEG BUTTER.

Butter, Mace. See MACE BUTTER.

Butter, Mineral. See MINERAL BUTTERS.

Butter, Njave. See NJAVE BUTTER.

Butter, Nutmeg. See NUTMEG BUTTER.

Butter of Paraffin. Soft paraffin.†

Butter of Sulphur. Precipitated sulphur.†

Butter, Palm. See PALM BUTTER.

Butter, Papua Nutmeg. See MACASSAR NUTMEG BUTTER.

Butter, Para. See PARA PALM OIL.

Butter, Shea. See SHEA BUTTER.

Butter, Sierra Leone. See LAMY BUTTER.

Butter, Vegetable. See VEGETABLE BUTTER.

Butter, Wax. See WAX BUTTER.

Butter Yellow. (Oil Yellow, Butyro flavine). An azo dyestuff. It is dimethylaminoazobenzene, $C_6H_5\ N_2.C_6H_4\ N(CH_3)_2$. Formerly used for colouring butter and oils.†

Button Brass. See BRASS.

Button Lac. See LAC.

Button Metal. An alloy of 57 per cent zinc and 43 per cent copper.†

Button Solder. (White Solder). Usually contains 50 per cent tin, 30 per cent copper, and 20 per cent brass; or 33 per cent copper, 27 per cent brass, and 40 per cent zinc.†

Butvar. Polyvinyl butyral resins are white, free-flowing powders supplied in seven resin types or grade forms with various hydroxyl content whose solutions provide a wide range of viscosities. Applications: Recommended for upgrading the coating performance of thermosetting phenolics, ureas, melamines, epoxies, alkyds, polyurethanes and polyesters. (Monsanto Co).*

Butyl Carbitol. Diethylene glycol monobutyl ether. A lacquer solvent boiling at 240° C.†

Butyl Cellosolve. Ethylene glycol monobutyl ether. It is a nitrocellulose solvent, and is used in the manufacture of brushing lacquers.†

Butyl Namate. Water solution of sodium di-n-butyldithiocarbamate. Latex accelerator. Applications: Imparts fast precure and cure rates. Recommended where long compound life is not required. (Vanderbilt Chemical Corporation).*

Butyl Rubber. A proprietary trade name for a co-polymer of isobutylene with a small percentage of diene such as butadiene. An unsaturated synthetic rubber possessing the minimum unsaturation required for vulcanization.†

Butylated Hydroxytoluene. 2,6-Di-*t*-butyl-*p*-cresol.†

Butylite. A proprietary butyl rubber sealant. (Polymeric Systems Inc, Valley Forge, Pa).†

Butyrin. The glyceryl ester of butyric acid. It is found in butter.†

Butyroflavine. See BUTTER YELLOW.

Butyroin. $C_3H_7.CO.CH(OH)C_3H_7$. 5-Hydroxy-4-octanone†

Butyrone. Dipropyl ketone, $CO(CH_2\ CH_2\ CH_3)2$.†

Buxine. Berberine.†

B_1Vac. B_1 type, B_1 strain, Newcastle vaccine. Applications: immunization of poultry. (Intervet America Inc).*

BWF. Extruded acrylic and polycarbonate profiles. Applications: Shopfitting, sign work. (Cornelius Chemical Co Ltd).*

BX 310. A range of polypropylene films of different gauge thicknesses and widths. Applications: Packaging. (Hercules Inc).*

BXA. Diarylamine-ketone-aldehyde reaction product. For CR, NBR, SBR, non-blooming easily disperses. Protects against heat and oxygen. Brown discolouration in stocks exposed to light, may stain light coloured materials in contact with cured stocks. Used in tyre treads, carcass, inner tubes, insulated wire, soles, heels and mechanicals. (Uniroyal).*

B-X-A. A proprietary anti-oxidant. A product of the reaction diarylamine-ketone-aldehyde. (Rubber Regenerating Co).†

BXT. Polypropylene film. Applications: For wrapping tobacco products. (Hercules Inc).*

Byacin. Iodophor in liquid solution. Applications: Controlling storage rots in potatoes. (Wheatley Chemical Co Ltd).*

Byatran. Iodophor and TBZ as dry granules. Applications: Controlling soil borne diseases in growing potato crops. (Wheatley Chemical Co Ltd).*

Bygran F. Tecnazene and iodophor as dry granules. Applications: Controlling sprouting, dry rot and other storage diseases in potatoes. (Wheatley Chemical Co Ltd).*

Bygran S. Tecnazene as dry granules. Applications: Controlling sprouting and dry rot in stored potatoes. (Wheatley Chemical Co Ltd).*

Bynel. Co-extrudable adhesive resins. (Du Pont (UK) Ltd).

Bynin Amara. A proprietary preparation of iron phosphorus and nux vomica in Bynin liquid malt. Tonic. (Allen & Hanbury).†

C

4C. Stabilized and treated blood cells in an artificial plasma medium; used e.g. as 4C, 4C plus. Applications: Reference control for automated blood cell analysers. (Coulter Electronics Ltd).*

C Acid. 2-Naphthylamine-4,8-disulphonic acid.†

Cabbage Oil. See OIL OF CABBAGE.

Cabelec compounds. Compounds capable of conducting electricity. Applications: Compounds for use in the manufacture of electrically conductive and antistatic articles. (Cabot Plastics Ltd).*

Cabflex DIOZ. A trade mark. Diiso-octyl azelate.†

Cablinol. A speciality perfumery chemical. (PPF International Ltd).†

Cabol 100. A trade mark. A hydrocarbon oil type vinyl plasticizer.†

Cabuflx. An adhesive for cellulose acetate butyrate. (May & Baker Ltd).*

Cabulite. Cellulose acetate butyrate film and sheet. (May & Baker Ltd).*

Cacao Butter. See COCOA BUTTER.

Cacao Oil. See OIL OF COCOA.

Cachou. See CUTCH.

Cacodyl. (Alkarsin). Tetramethyldiarsine, $As_2(CH_3)_4$.†

Cacodylic Acid. Dimethylarsinic acid, $As(CH_3)_2O(OH)$.†

Cacodyl, New. See ARRHENAL.

Cadaverine. Pentamethylenediamine, $NH_2(CH_2)_5NH_2$. A base found in ergot and formed by bacterial decomposition.†

Cadmia. See CADMIUM YELLOW.

Cadmium Bronze. A copper alloy containing 0.5-1.2 per cent cadmium used for telephone and trolley wire.†

Cadmium Lithopone. (Cadmopone). A pigment analogous to lithopone, in which cadmium replaces zinc. It is made by the precipitation of cadmium sulphate solution with barium sulphide, and contains 38 per cent cadmium sulphide.†

Cadmium Orange. (Cadmium Red). Cadmium sulphide, CdS, with smaller amounts of cadmium selenide and heavy spar.†

Cadmium Red. See CADMIUM ORANGE.

Cadmium yellow. (Cadmia, Jaune Brilliant, Pale Cadmium, Orient Yellow. Radiant Yellow, Aurora Yellow, Daffodil). A pigment. It is cadmium sulphide, CdS.†

Cadmiumised Zinc. Zinc metal placed in a 2 per cent cadmium sulphate solution for five minutes, then well washed. A reducing agent.†

Cadmiumspat. Synonym for Otavite.†

Cadmium, Pale. See CADMIUM YELLOW.

Cadmopone. A registered trade mark. See CADMIUM LITHOPONE.†

Cadmopur. A range of cadmium pigments. Applications: For use especially in plastics such as polyethylene, polystyrene, polyamides, PVC, etc, as well as in rubber. (Bayer & Co).*

Cadon. Engineering thermoplastics, impact-modified styrene-maleic anhydride terpolymers. Applications: Used in automotive interior/exterior parts, business machines and appliance housings, electrical equipment, electronic parts, plumbing and industrial parts. (Monsanto Co).*

Cafadol. A proprietary preparation of paracetamol and caffeine citrate. An analgesic. (Typharm).†

Cafaspin. A restorative and analgesic. (Bayer & Co).*

Cafergot. A proprietary preparation of ergotamine tartrate and caffeine, used for migraine. (Sandoz).†

Cafformasol. See FORMASAL.

Caflon. Foam stabilizer. Applications: Liquid detergents; shampoos; bubble baths; hand cleaners. (Cargo Fleet Chemical Co Ltd).

Caflon MIS. Anionic surfactant as a pale amber clear liquid. Applications: Emulsifier for hand cleaning gels and degreasers. (Cargo Fleet Chemical Co Ltd).

Caflon MS33. Monoethanolamine alkyl sulphate in liquid form. Applications: Liquid detergents and shampoos. (Cargo Fleet Chemical Co Ltd).

Caflon NAS 25. Anionic surfactant as a pale yellow liquid. Applications: High foaming wetting agent for liquid detergents. (Cargo Fleet Chemical Co Ltd).

Caflon SA and SNA. Anionic surfactant in acid form, supplied as a dark brown liquid. Applications: Detergent intermediate. (Cargo Fleet Chemical Co Ltd).

Caflon SS28. Sodium alkyl sulphate in liquid form. Applications: Liquid detergents and shampoos. (Cargo Fleet Chemical Co Ltd).

Cake Lac. See LAC.

Cal Plus. Calcium chloride dihydrate, $CaCl_2.2H_2O$. Applications: Replenisher. (Mallinckrodt Inc).

Cal-C-Vita. Proprietary preparation of vitamin B6, vitamin C, vitamin D and calcium. (Roche Products Ltd).*

Cal-Grid. A calcium-aluminium alloy used in the production of calcium-containing long-life lead battery plates. (Pfizer International).*

Cal-Tint. Colourant dispersions. Applications: Colouring of aqueous and non-aqueous coating compositions. (Nueodex Inc).*

Calaba Oil. See LAUREL NUT OIL.

Calaband. A proprietary zinc paste and calamine bandage. Applications: Fairly acute and subacute eczema, erythema and dermatitis after plaster removal. (Seton Products Ltd).*

Calac. Calcium acetate. (Mechema).*

Caladryl. A proprietary preparation of diphenhydramine, camphor and calamine in an aerosol spray, for dermatological use. (Parke-Davis).†

Calamine. (Zinc Spar, Spathic Zinc Ore, Smithsonite). A term applied to both the silicate and carbonate of zinc, usually the carbonate, $ZnCO_3$, found native. In mineralogy, the name calamine is employed for the silicate, while the term smithsonite is given to the carbonate.†

Calamine, Cupriferous. See TYROLITE.

Calamine, Prepared. See PREPARED CALAMINE.

Calamine, White. See ZINC SULPHIDE GREY.

Calaton. A textile finishing agent. (ICI PLC).*

Calavite. A proprietary preparation containing vitamin A, ascorbic acid, CYANOCOBALAMIN, nicotinamide, thiamine and CALCIFEROL. A vitamin supplement. (Carlton Laboratories).†

Calbor. Borated animal bone and borate containing flux. (Borax Consolidated Ltd).*

Calbrite. Dentifrice grade of dicalcium phosphate. (Albright & Wilson Ltd, Phosphates Div).

Calbux. A proprietary ground limestone. (ICI PLC).*

Calc Sinter. See TRAVERTINE.

Calcareous Tufa. See TRAVERTINE.

Calcars. Calcium arsenate. (Mechema).*

Calcene. A trade name applied to a precipitated calcium carbonate with 2 per cent organic material. It is prepared for use as a rubber filler.†

Calcet. Contains calcium gluconate and calcium lactate. Applications: Replenisher. (Mission Pharmacal Co).

Calcetal. Calcium acetylsalicylate.†

Calcibind. Cellulose sodium phosphate, an insoluble, non-absorbable ion-exchange resin made by phosphorylation of cellulose. It contains approximately 34% of inorganic phosphate and approximately 11% of sodium. Applications: Antiurolithic. (Mission Pharmacal Co).

Calcibronat. A proprietary preparation of calcium bromido- lactobionate. A Sedative. (Sandoz).†

Calcichrome. Cyclo - tris - 7 - (1 azo - 8 - hydroxy - naphthalene - 3 : 6 disulphonic acid). A mauve crystalline powder. A sensitive and specific reagent for calcium giving a red colour with Cal^{2+} ions in alkaline solution. (British Drug Houses).†

Calcidine. Calcium iodide, CaI_2.†

Calciferol. Synthetic Vitamin D. Vitamin D_2.†

Calcimar. Calcitonin: a polypeptide hormone that lowers the calcium concentration in plasma. Applications: Regulator. (USV Pharmaceutical Corp).

Calcinite. See CARBORA.

Calcinol. Calcium iodate, $Ca(IO_3)_2$.†

Calciparine. Heparin calcium. Applications: Anticoagulant. (American Critical Care, Div of American Hospital Supply Corp).

Calcisorb. Each sachet contains sodium cellulose phosphate as a white to beige fibrous powder. (Riker Laboratories/3M Health Care).†

Calcitare. A proprietary preparation of porcine CALCITONIN used in the treatment of Paget's disease. (Armour Pharmaceutical Co).†

Calcitonin. A polypeptide hormone of ultimobranchial origin, extractable from the thyroid gland of mammalian species or the ultimobranchial gland of non-mammals. It lowers the calcium concentration in the plasma of mammals. See Thyrocalcitonin†

Calcium Chel 330. Pentetate Calcium Trisodium. Applications: Chelating agent. (Ciba-Geigy Corp).

Calcium Disodium Versenate. Edetate calcium disodium. Applications: Chelating agent. (Riker Laboratories Inc, Subsidiary of 3M Company).

Calcium Diuretin. Theobromine and calcium salicylate.†

Calcium Gummate. See GUM ARABIC.

Calcium Hydrochlorphosphate. A mixture of calcium chloride and phosphate.†

Calcium Lactophosphate. A mixture of calcium lactate, calcium acid lactate, and calcium acid phosphate. It contains about 2 per cent. P_2O_5.†

Calcium Oxymuriate. See BLEACHING POWDER.

Calcium Petronate 25H, 25C and 300. Calcium petroleum sulphonate. Applications: Detergents in lube oil additives, rust inhibitors, emulsifiers. (Witco Chemical Ltd).

Calcium Resonium. A proprietary preparation of calcium polystyrene sulphonate used in the treatment of hyperkalaemia. (Winthrop Laboratories).*

Calcreose. Calcium creosotate.†

Calcydic. Calcium monohydrogen phosphate and Vitamin D. (Allen & Hanbury).†

Calderol. Calcifediol. Applications: Regulator. (The Upjohn Co).

Caldiox. Calcium plumbate. (Associated Lead Manufacturers Ltd).*

Caldo Bordeles Valles. Bordeaux mixture plus adjuvants. Applications: Wettable powder used as protective fungicide for foliage application to ornamental and crop plants. (Industrias Quimicas Del Valles SA).*

Caldura. A high temperature (240° C) resistant resin, containing aromatic hydrocarbon groups linked by oxygen and methylene bridges. (Associated Electrical Industries (GEC)).†

Caledon. Vat dyes. (ICI PLC).*

Caledon Blue. A dyestuff synonymous with Indanthrene blue.†

Caledon Blue GC. A dyestuff. It is a British equivalent of Indanthrene blue GC.†

Caledon Blue GCD. A dyestuff. It is a British brand of Indanthrene blue GCD.†

Caledon Blue R. A British dyestuff equivalent to Indanthrene X.†

Caledon Brown. Indanthrene brown BB.†

Caledon Dark Blue O. A dyestuff. It is a British equivalent of Indanthrene dark blue BO.†

Caledon Green. Indanthrene green B.†

Caledon Jade Green. The dimethyl ether of 12,12'-dihydroxydibenzanthrone. A dyestuff giving blue-green shades.†

Caledon Pink. Indanthrene pink B.†

Caledon Purple. See INDANTHRENE DARK BLUE BO.

Caledon Red. Indanthrene red BN.†

Caledon Violet. Indanthrene violet R extra.†

Caledon Yellow. Indanthrene yellow G.†

Caledon Yellow 3G. A dyestuff. It is a British brand of Algol yellow R.†

Calgon. Water softening compound. (Albright & Wilson Ltd, Phosphates Div).

Calgonite. Dishwashing machine detergents. (Albright & Wilson Ltd, Phosphates Div).

Calibre. A family of polycarbonate resins used widely in automotive, electronics, business machines, medical and housewares applications. (Dow Chemical Co Ltd).*

Calibrite. A trade mark for series of aluminium pigments for plastics.†

Caliche. Crude nitrate of soda (Chile saltpetre). The soluble salts which cement together the sand, clay, and stones in the mixture known as caliche are mainly sodium nitrate and sodium chloride.†

Calido. An alloy of 64 per cent nickel with from 15-25 per cent iron, 8-12 per cent chromium, and 3-8 per cent manganese.†

Calido Brass. An alloy of 70 per cent copper, 30 per cent zinc, and traces of iron.†

Calido-elalco. A heat-resisting alloy containing 60 per cent nickel, 24 per cent iron, and 16 per cent chromium.†

Caligesic. Benzocaine, calamine and metacresol. Applications: Topical relief of inflammatory and pruritic conditions of the skin. (Merck Sharp & Dohme).*

Caliment. Foodgrade of dicalcium phosphates. (Albright & Wilson Ltd, Phosphates Div).

Calipharm. Pharmaceutical grade of dicalcium phosphate. (Albright & Wilson Ltd, Phosphates Div).

Calite. An alloy of 50 per cent iron, 35 per cent nickel, 10 per cent aluminium, and 5 per cent chromium. It is very resistant to heat, and melts at 2777° F.†

Calkleen. Industrial cleaning compound. (The Wellcome Foundation Ltd).

Callaica. Synonym for turquoise.†

Callusolve. Amber coloured paint containing 25% benzalkonium chloride bromine. Applications: For topical use in the treatment of warts, especially multiple or mosaic warts. (Dermal Laboratories Ltd).*

Calmitol. A proprietary preparation of chloral hydrate, camphor, menthol, iodine, hyoscyamine and zinc oxide in an ointment base. Antipruritic. (Horlicks).†

Calofil. Calcium carbonate. (John & E Sturge Ltd, Lifford).

Calofil A4 and Bl. A trade name for precipitated calcium carbonate fillers.See also CALOPAKE PC, CALOFORT-S HAKUENKA CCR, WINNOFIL S, VEDAR.†

Calofort S. Surface activated precipitated calcium carbonate. Applications: Sealants, adhesives, PVC plastisols, PVC extrusions and rubber compounds. (Sturge Lifford).*

Calogerasite. Synonym for Simpsonite.†

Calomel. (Calomel, Sublimed, Precipitated, Calomel). Mercurous chloride, Hg_2Cl_2.†

Calomel, Precipitated. See CALOMEL.

Calomel, Sublimed. See CALOMEL.

Calomic. A nickel-iron-chromium resistance alloy. It is used for electric heater elements operating up to 900° C.†

Calonutrin. A proprietary preparation of mono-, poly- and di- saccharides used in the tube feeding of invalids. (Geistlich Sons).†

Calopake. Ultrafine precipitated calcium carbonate. Applications: Paper, paint, printing ink, plastic and rubber industries. (Sturge Lifford).*

Caloreen. Glucose polymer. Applications: Nutritional supplement. (Roussel Laboratories Ltd).*

Caloride. A proprietary trade name for cakes of calcium chloride for drying purposes.†

Calorite. Alloys. One contains 65 per cent nickel, 15 per cent iron, 12 per cent chromium, and 8 per cent manganese; and another 65 per cent nickel, 23 per cent iron, and 12 per cent chromium.†

Caloxal CLP 45. A dispersion of high grade calcium oxide in chlorinated paraffin used as a desiccant for rubber and plastics. (Sturge Lifford).*

Caloxol. A range of dispersions of calcium oxide in oil wax or plasticizer. Applications: Dessicant for use in rubber and plastics industries. (Sturge Lifford).*

Caloxol CP2. A proprietary calcium oxide desiccant for rubber. (Sturge Lifford).*

Calpol. Paracetamol compounds. (The Wellcome Foundation Ltd).

Calprona K. Mould inhibitor. (BP Chemicals Ltd).

Calquat. Industrial cleaning compound. (The Wellcome Foundation Ltd).

Calsan. Sanitary disinfectant. (The Wellcome Foundation Ltd).

Calsept. A proprietary preparation of calamine, menthol, cetrimide, camphor and zinc oxide. Antiseptic skin cream. (Norton, H N Co).†

Calsil. Precipitated silica with a proportion of calcium oxide. Applications: Medium-active reinforcing filler with outstanding extrusion characteristics, good processibility for technical rubber articles. (Degussa).*

Calsolene Oil. General industrial wetting agent, emulsifier and emulsion breaker; biodegradable. Applications: Oil and petroleum industry; gas manufacture; control of foam generation. (ICI PLC, Organics Div).

Calstrip. Industrial cleaning compound. (The Wellcome Foundation Ltd).

Calsun Bronze. A proprietary trade name for an aluminium bronze containing copper with 2.5 per cent aluminium and 2 per cent: tin.†

Calsynar. A proprietary preparation of synthetic salmon CALCITONIN used in the treatment of Paget's disease. (Armour Pharmaceutical Co).†

Calthane. Two part solvent free, urethane elastomer systems. Applications: Moulds for making plastic, concrete and plaster parts. (Cal Polymers Inc).*

Calthor Suspension. Ciclacillin. Applications: Orally active broad spectrum amino-penicillin. (Ayerst Laboratories).†

Calthor Tablets. Ciclacillin. Applications: Orally active broad spectrum antibiotic (amino-penicillin). (Ayerst Laboratories).†

Calurea. A nitrogenous fertilizer. It is a mixture of urea and calcium nitrate containing 34 per cent nitrogen.†

Calx. Lime, CaO.†

Calyptol. A proprietary preparation of eucalyptol, terpineol and oil of pine needles, thyme and rosemary. A bronchial inhalant. (Smith and Nephew).†

C.A.M. A proprietary preparation of ephedrine hydrochloride and butethamate citrate. A bronchial antispasmodic. (Rybar).†

Camalox. Contains aluminium hydroxide gel, precipitated calcium carbonate and magnesium hydroxide. Applications: An antacid and laxative. (William H Rorer Inc).

Cambe Wood. See REDWOODS.

Cambilene. A selective weedkiller. (Fisons PLC).*

Cambison. A proprietary preparation of prednisolone, neomycin and quinolyl denrivatives used in dermatology as an antibacterial agent. (Hoechst Chemicals (UK) Ltd).†

Cambrelle. Melded fabrics. (ICI PLC).*

Cambrite. A smokeless powder, containing 22-24 per cent nitro-glycerin, 3-4.5 per cent barium nitrate, 26-29 per cent potassium nitrate, 32-35 per cent dried wood meal, 1 per cent calcium carbonate, and 7-9 per cent calcium chloride.†

Camelia Metal. An alloy of 70.2 per cent copper, 14.7 per cent lead, 10.2 per cent zinc, 4.2 per cent tin, and 0-5 per cent iron.†

Camenthol. A mixture of camphor and menthol for inhalation.†

Camite. (Dimondite, Haystellite, Armide, Straus metal, Purdurum, Phoran, Hartmetall). Proprietary trade names for tungsten carbide materials.†

Camoquin. A proprietary preparation containing amodiaquine and hydrochloride. An antimalarial drug. (Parke-Davis).†

Campaign. Post-emergence herbicide. (Murphy Chemical Ltd).

Campanuline. See MUSCARINE.

Campeachy Wood. See LOGWOOD.

Camphacol. See CAMPHACOLLASIS.

Camphorated Oil. (*Linimentum camphoroe B.P.*) It consists of 200 parts camphor, and 800 parts olive oil.†

Camphor, Asaronic. See ASAROL.

Camphor, Asarum. See ASAROL.

Camphor, Baros. See CAMPHOL.

Camphor, Bhimsiam. See CAMPHOL.

Camphor, Borneo. See CAMPHOL.

Camphor, Dryobalanops. See CAMPHOL.

Camphor, Laurel. See JAPAN CAMPHOR.

Camphor, Malay. See CAMPHOL.

Camphossil. See CAMPHOSAL.

Camphre de Persil. See PARSLEY CAMPHOR.

Campiodol. Iodized rape-seed Oil.†

Campobello Yellow. (French Yellow, Chryseinic Acid). A dyestuff. It is the sodium salt of 4-nitro-α-naphthol.†

Campolon. A proprietary preparation of liver extract with Vitamin B_{12} A haematinic. (Bayer & Co).*

Campovit. An injectable vitamin B complex with liver extract. (Bayer & Co).*

Camwood. (Cambe Wood). See REDWOODS.†

Camyna. A proprietary preparation of thioxolone. (Boehringer Ingelheim Ltd).†

Canacert. Food-stuffs colouring matter meeting Canadian regulations. (Pointing Ltd).†

Canadian Asbestos. See CHRYSOTILE.

Canadian Certicol. Food colours meeting Canadian specifications. (Williams (Hounslow) Ltd).

Canadium. An alloy of 1 part palladium, 2 parts platinum, and 6 parts nickel. Used as a substitute for platinum.†

Canadol. See GASOLINE.

Canagel 75. Semigelatin dynamite. It has good water resistance, high weight and volume energy and relatively high detonation velocity. Applications: Construction and building industry, mining, explosives. (Hercules Inc).*

Canarin. (Persulphocyanogen Yellow). A yellow colouring matter obtained by the action of bromine upon potassium or ammonium thiocyanate. Used in calico printing.†

Canary Yellow. See CHROME YELLOW.

Cancer Serum. See CANCROIN.

Candeptin. A proprietary preparation of candicidin in a petrolatum base. Antimonial vaginal ointment. (LRC International Ltd).†

Candeptin-N. A proprietary preparation of candicidin and neomycin, in a petrolatum base. Antibiotic skin ointment. (LRC International Ltd).†

Candle Tar. See STEARINE PITCH.

Cane Sugar. See SUCROSE.

Canilep. Vaccines for dogs. (Glaxo Pharmaceuticals Ltd).†

Canilep D.D. A combined double-dose vaccine for dogs. (Glaxo Pharmaceuticals Ltd).†

Canilin D. A canine distemper vaccine. (52O).†

Canopar. Thenium Closylate. Applications: Anthelmintic. (Burroughs Wellcome Co).

Cantabiline. Hymecromone. Applications: Choleretic. (Lipha SA).

Cantamega 1000. A fibre based vitamin and mineral supplement. Applications:

Dietary supplement. (Larkhall Laboratories Ltd).*

Cantamega 2000. A comprehensive vitamin and mineral dietary supplement. (Larkhall Laboratories Ltd).*

Cantil. Mepenzolate bromide. Applications: Anticholinergic. (Merrell Dow Pharmaceuticals Inc, Subsidiary of Dow Chemical Co).

Canzler Wire. An alloy of 98.8 per cent copper, 1 per cent silver, and 0.2 per cent phosphorus.†

Cao. Butylated paracresol. Applications: Antioxidants for food, plastic, rubbers and other general purpose requirements. (PMC Specialties Group Inc).*

Caoutchouc. Rubber.†

Caoutchouc, Mineral. See ELATERITE.

C-A-P. Cellulose acetate phthalate. Applications: Pharmaceutic aid. (Eastman Chemical Products Inc, Subsidiary of Eastman Kodak Co).

Cap Copper. An alloy of 95-97 per cent copper and 3-5 per cent zinc. It is used for deep drawing and stamping.†

Cap Gilding Brass. See BRASS.

Capa. Caprolactone and polymers based on it. (Interox Chemicals Ltd).

Capastat. Capreomycin sulphate, anti-tuberculant agent. (Dista Products Ltd).

Capastat Sulfate. Capreomycin sulphate, an antibiotic produced by streptomyces capreolus. Applications: Antibacterial. (Eli Lilly & Co).

Cape Blue. See CROCIDOLITE.

Capexco. A coal mine explosive consisting of 32-34 per cent nitro-glycerin, 0.5-1.5 per cent nitro-cotton, 24-25 per cent sodium nitrate, 30-32 per cent ammonium oxalate, and 8-10 per cent wood meal.†

Capitol. Clear aqueous gel containing 0.5% benzalkonium chloride. Applications: For the topical treatment of pityriasis capitis and other seborrhoeic scalp conditions where there is scaling and dandruff. (Dermal Laboratories Ltd).*

Capivi. See OLEO-RESIN COPAIBA.

Capivi Balsam. See OLEO-RESIN COPAIBA.

Capla. A proprietary preparation of mebutamate. An antihypertensive. (Horlicks).†

Caplenal. Allopurinol. Applications: For the treatment of gout, primary hyperuricaemia, secondary hyperuricaemia, prophylaxis of uric acid and calcium oxolate stones. (Berk Pharmaceuticals Ltd).*

Caposil. A trade mark for calcium silicate and asbestos based insulating materials. Caposil 1400 withstands 1400° F. (760° C.) and Caposil HT withstands 1850° F. (1000° C.). (Cape Insulation Cape Asbestos Co).†

Capoten. A proprietary preparation of captopril. Applications: An ACE inhibitor used in the treatment of cardiovascular disorders. (Squibb, E R & Sons Ltd).*

Capoten Tablets. Captopril. (E R Squibb & Sons Ltd).

Capozide. Contains captopril and hydroflumethiazide. Applications: Antihypertensive; enzyme inhibitor; diuretic. (E R Squibb & Sons Inc).

Cappagh Brown. (Euchrome, Mineral Brown). A natural pigment used in oil-painting, containing hydrated oxide of iron and 27 per cent manganese dioxide. See BURNT SIENNA.†

Capramin. Decongestant and antipyretic tablets. (Allen & Hanbury).†

Capreomycin. A proprietary antibiotic produced by *Streptomyces capreolus* present as the sulphate in capastat. (Dista Products Ltd).*

Capri Blue Gon. A dyestuff. It is the zinc double chloride of dimethyldiethyldiaminotoluphenazonium chloride, $C_{17}H_{20}N_3OCl$. Dyes tannined cotton greenish-blue.†

Caprinol. An antihypertensive preparation. (Bayer & Co).*

Capriton. A proprietary preparation of chlorpheniramine maleate, phenylephine hydrochloride, aspirin, caffeine. Decongestant and antipyretic tablets. (Allen & Hanbury).†

Capron. Range of type 6 nylon polymers including reinforced grades of nylon 6 and either glass fibres or a blend of glass fibres and selected minerals. Applications: Power tool housings, chain saw housings and other components, housings for lawn and garden equipment, automotive cooling fans and other engine compartment applications and exterior automotive body parts. (Allied Chemical Corporation).*

Capron Alpha 8200 C, 8202 C and 8203 C. A proprietary range of alpha type nylon moulding compounds used in the production of stiff mouldings. (Allied Chemical Corporation).*

Capron 8200. A proprietary nylon claiming superior mechanical properties. (Allied Chemical Corporation).*

Capron 8202. A proprietary nylon used for injection moulding. (Allied Chemical Corporation).*

Capron 8206 S. A proprietary nylon of good flexibility used in extrusion and moulding operations. (Allied Chemical Corporation).*

Capron 8230. 6% glass fibre. (Allied Chemical Corporation).*

Capron 8231 and 8233. A range of glass-filled nylons. (Allied Chemical Corporation).*

Capron 8250, 8251 and 8253. A proprietary range of copolymer grades of nylon possessing high impact strength. (Allied Chemical Corporation).*

Capron 8270. A proprietary modified nylon 6 compound used in blow moulding. (Allied Chemical Corporation).*

Caprosem. A proprietary preparation containing Testostorone, 17-chloral hemi-acetal acetate. Male sex hormone. (Leo Laboratories).†

Caprun. Polyamide 6. (Allied Chemical Corporation).*

Capsolin. A proprietary preparation of oleoresin capsicum, camphor, oil of turpentine and eucalyptus. A rubefacient. (Parke-Davis).†

Capsule Metal. An alloy of 92 per cent lead and 8 per cent tin.†

Captan. A foliage fungicide. (ICI PLC).*

Captan Granular. Captan as dry granules. Applications: Controlling soil borne fungal diseases in tomato, lettuce and strawberry. (Wheatley Chemical Co Ltd).*

Captan 83 WP. Fungicide - apple and pear scab. (ICI PLC, Plant Protection Div).

Captan-Col. A captan fungicide. (Plant Protection (Subsidiary of ICI)).†

Captan-50. A foliage fungicide. (Plant Protection (Subsidiary of ICI)).†

Captax. 2-mercaptobenzothiazole - primary rubber accelerator. Applications: Primary accelerator for both natural and synthetic rubbers. (Vanderbilt Chemical Corporation).*

Captax-disulphide. See ALTAX.

Caradate. A range of polyether polyols and isocyanates used for the production of flexible and rigid polyurethane foams. They may be used in furniture and automotive sealing, also in pipes and tanks, being effective insulators. They also have a use in domestic freezers and refrigerators. (Shell Chemicals).†

Caradol. A range of polyether polyols and isocyanates used for the production of flexible and rigid polyurethane foams. They may be used in furniture and automotive sealing, also in pipes and tanks, being effective insulators. They also have a use in domestic freezers and refrigerators. (Shell Chemicals).†

Carafate. Sucralfate. Applications: Anti-ulcerative. (Marion Laboratories Inc).

Caramba. Applications: Rust solvents, penetrating oils, anti-corrosives, preservative preparations, anti-corrosion lacquers, underbody coatings and cavity preservation compounds for vehicles. Cleaning, polishing, scouring and abrasive preparations for metals, motors and machine parts, for window panes of motor cars, antifreeze and de-icer. (Caramba Chemie GmbH).*

Caramba Felgenglanz. Detergent for cleansing wheel-rims. (Caramba Chemie GmbH).*

Caramba Felgenneu. Detergent for cleansing wheel-rims. (Caramba Chemie GmbH).*

Caramba Lackkrone. Applications: Cleans, polishes and seals in one operation paintwork and motor cars. (Caramba Chemie GmbH).*

Caramba Perlglanz. Highly effective and concentrated product for lacquer care. Applications: Sealant for motor cars. (Caramba Chemie GmbH).*

Carb-O-Sep. Carbarsone; N-Carbamoylarsanilic acid. Applications: Antiamebic. (Whitmoyer Laboratories Inc).

Carbagel. A proprietary trade name for a drying agent. It is an activated carbon containing calcium chloride.†

Carbamide. Urea, $CO (NH_2)_2$.†

Carbanilide. Diphenylurea, $C_6N_5NH.CO.NHC_6H_5$.†

Carbanillic Ether. Phenylurethane.†

Carbarsone. 4-Carbaminophenylarsonic acid.†

Carbaryl. An insecticide. 1-Naphthyl methylcarbamate.†

Carbathene. A proprietary trade name for an ethylene N-vinyl carbazole copolymer. A moulding compound form stable over 240° C. (Standard Telecommunications Laboratories).†

Carbazole Blue. A dyestuff, obtained by fusing carbazole with oxalic acid.†

Carbazole Violet. A dyestuff obtained by fusing 9-ethylcarbazole with oxalic acid.†

Carbazole Yellow. A dyestuff. Dyes cotton from a boiling alkaline bath.†

Carbergan. Carbon tetrachloride and promethazine in oil. (May & Baker Ltd).*

Carbetamex. Herbicide. (May & Baker Ltd).

Carbex. See CARBORA.

Carbicon. See CARBORA.

Carbide Black. A dyestuff. It is a British brand of Columbia black FF extra.†

Carbide Black BO. A polyazo dyestuff. It dyes cotton bluish-black in the presence of sodium chloride.†

Carbide Black D. A dyestuff. It is a British equivalent of Zambesi black D.†

Carbide Black RI. A dyestuff similar to Carbide black BO.†

Carbilys. Used to render starch insoluble in the manufacture of paper and cardboard, and also in the textile and spinning process industries. (Roquette (UK) Ltd).*

Carbitol. A proprietary solvent. It is diethylene glycol monoethyl ether, $C_2H_5O\ CH_2\ CH_2\ O.CH_2\ CH_2\ OH$. A solvent for cellulose nitrate, shellac, copal, rosin, etc. It is used in dopes for artificial leather and is added to brushing lacquers.†

Carbo Alumina. A trade mark for goods of the abrasive and refractory class. They consist mainly of crystalline alumina.†

Carbobrant. See CARBORA.

Carbocaine. Mepivacaine hydrochloride. Applications: Anaesthetic. (Sterling Drug Inc).

Carbo-corundum. A trade mark for articles made from crystalline alumina. They are refractories and abrasives.†

Carboform. Resin impregnated glass, carbon or aramid fabrics. Applications: Advanced composite moulded components for military and aerospace applications. (Fothergill Tygaflor Ltd).*

Carbofrax. A carborundum refractory containing more than 90 per cent silicon carbide.†

Carbo-gel. Silica gel impregnated with carbon, for use to absorb organic vapours.†

Carbogran. Silicon carbide. Applications: Used in abrasive industries. (Lonza Limited).*

Carbogran E. Silicon carbide. Applications: Used in the electrical industry. (Lonza Limited).*

Carbogran UF. Silicon carbide. Applications: For use in sintered ceramics and for abrasion resistant surfaces. (Lonza Limited).*

Carbokaylene. A proprietary preparation of vegetable charcoal with colloidal aluminium silicate.†

Carbol-gentian-violet Solution. A microscopic stain. It contains 10 cc of a saturated alcoholic solution of gentian violet and 90 cc of a 5 per cent phenol solution.†

Carbol Methyl Violet. A microscopic stain. It contains 10 parts of an alcoholic solution of methyl violet 6B with 90 parts of a 5 per cent aqueous solution of phenol.†

Carbol Xylene. A microscopic clearing solution containing 3 parts xylene and 1 part phenol.†

Carbolan. Super-milling acid dyes. (ICI PLC).*

Carbolfuchsine. (Ziehl's stain). A microscopic stain for bacteria. It contains 5 parts fuchsine, 25 parts phenol, 50 parts alcohol, and 500 parts water.†

Carbolic Acid. Phenol, C_6H_5OH. In trade, the term is used for pure phenol, the cresols and their mixtures with phenol, and also for crude tar oils.†

Carbolic Acid Crystals. Pure phenol.†

Carbolic Camphor. See PHENOL CAMPHOR.

Carbolic Oils. See MIDDLE OILS.

Carbolite. See KARBOLITE and CARBORA.

Carbolon. A trade mark for abrasive articles consisting essentially of silicon carbide.†

Carbolox. See CARBORA.

Carboloy. A trade mark for hard metal compounds consisting of tungsten carbide and cobalt for cutting glass and for high-speed tools.†

Carbomang. A proprietary trade name for a steel containing 1-1.25 per cent manganese, 0.45 per cent chromium, 0.5 per cent tungsten, and 1 per cent carbon.†

Carbomant. Silicon carbide. Applications: Used in abrasive industries and wire sawing. (Lonza Limited).*

Carbomer. A suspension agent. A polymer of acrylic acid cross-linked with allyl sucrose. CARBOPOL.†

Carbomucil. A proprietary preparation of charcoal, magnesium carbonate, and sterculia. A antidiarrhoeal. (Norgine).†

Carbonado. (Black Diamond). A black, compact variety of diamond, used in the steel crowns of rock-drills.†

Carbon Black. Modern carbon black is obtained by burning natural gas in a regulated air supply, and the black deposited on a metal surface.†

Carbon Bronze. An alloy containing 75.5 per cent copper, 9.75 per cent tin, and 14.5 per cent lead. A white metal used for bearings.†

Carbon of Cementation. See ANNEALING OF CARBON.

Carbon of Normal Carbide. See ANNEALING CARBON.

Carbon Steels. See STEEL A2, E, etc.

Carbon 4E. A liquid formulation containing 48% trichloropyrester. Applications: Controls woody weeds and perennial broad-leaved weeds in forestry and non-cropped areas. (Burts & Harvey).*

Carbondale Silver. An alloy of 66 per cent copper, 18 per cent nickel, and 16 per cent zinc. See NICKEL SILVERS.†

Carbonet. Non-adherent gauze dressing. (Smith and Nephew).†

Carbonin. Granular carbonaceous additive for molten iron and steel. (Foseco (FS) Ltd).*

Carbon, Silicized. See SILFRAX.

Carboplastic. A proprietary fireproof plastic cement, the main constituent of which is carborundum. It is suitable for repairing furnaces.†

Carbopol 910.. Carbomer 910. The viscosity of a neutralized preparation containing 2.5g of Carbomer 910 in 500ml water is not less than 2,500 centipoises and not more than 7000 centipoises. Carbomer 910 is a polymer of acrylic acid, cross-linked with a polyfunctional agent. Applications: Pharmaceutic aid. (B F Goodrich Chemical Group).

Carbopol 934. Carbomer 934. The viscosity of a neutralized preparation containing 2.5g of carbomer 934 in 500ml water is not less than 3,000 centipoises and not more than 4,000 centipoises. Carbomer 934 is a polymer of acrylic acid, cross-linked with a polyfunctional agent. Applications: Pharmaceutic aid. (B F Goodrich Chemical Group).

Carbopol 934P. Carbomer 934P. Polymer of acrylic acid, cross-linked with a polyfunctional agent. Applications: Pharmaceutic aid. (B F Goodrich Chemical Group).

Carbopol 940. Carbomer 940. Polymer of acrylic acid, cross-linked with a polyfunctional agent. Applications: Pharmaceutic aid. (B F Goodrich Chemical Group).

Carbopol 941. Carbomer 941. A neutralized preparation containing 2.5g of carbomer 941 in 500ml water is not less than 4,000 centipoises and not more than 11,000 centipoises. Carbomer 941 is a polymer of acrylic acid, cross- linked with a polyfunctional agent. Applications: Pharmaceutic aid. (B F Goodrich Chemical Group).

Carbo-Pulbit. Oral antidiarrhoeic. Applications: Veterinary medicine. (Bayer & Co).*

Carbora. (Carbonite, Carbolite, Carbolox Mimico, Carbobrant, Carbicon, Corex, Silexon, Crystolite, Sterbon, Storabon, Natalon, Natrundum, Lotens, Calcinite, Idilon, Carbex). Proprietary names for silicon carbide materials. Abrasives.†

Carboraffin. Powdered activated carbon. Applications: Used for decolourizing and improving the odour and taste of liquids mainly in the chemical and food industries. (Bayer & Co).*

Carborex. Silicon carbide - SiC. Applications: For abrasive, refractory, metallurgical and other usages. (Orkla Exolon A/S & Co).*

Carborite. See CARBORA.

Carborundum. The name applied to abrasives, the main constituent of which is silicon carbide, SiC.†

Carbosal. A range of multipurpose fire exinguishing powders for extinguishing fires of incandescent solid, liquid or gaseous materials. (Degussa).*

Carboset Resins. Acrylic copolymers: Alkaline water dispersions, dry, granular solid, polymer solutions in organic solvents and high viscosity dry liquid polymer. Applications: Paper coatings, electronics, industrial coatings, printing inks, adhesives and cosmetics. (BF Goodrich (UK) Ltd).†

Carbosorb. Soda synthetic silicate carbon dioxide absorbent. (BDH Chemicals Ltd).

Carbostyril. 2-Hydroxyquinoline.†

Carbotex. A brand of natural rubber and carbon black. Used in rubber mixings.†

Carbowax. Polyethylene glycols. (Union Carbide (UK) Ltd).†

Carbowax Sentry. Polyethylene Glycol. Applications: Pharmaceutic aid - ointment base; suppository base; solvent; tablet excipient; tablet and/or capsule lubricant. (Union Carbide Corp).

Carbox Metal. An alloy of 84 per cent lead, 14 per cent antimony, 1 per cent iron, and 1 per cent zinc.†

Carboxide. A proprietary fumigant containing 1 part ethylene oxide and 8 parts carbon dioxide.†

Carbrital. A proprietary preparation of carbromal and pentobarbitone sodium. A hypnotic. (Parke-Davis).†

Carburet of Iron. See GRAPHITE.

Carburetted Hydrogen, Heavy. See OLEFIANT GAS.

Carcanol G23. A proprietary trade name for a fraction of cashew nut shell liquid.†

Cardamist. A proprietary preparation of glyceryl trinitrate in aerosol form, used in angina pectoris. (Nicholas).†

Cardiacap. A proprietary preparation of pentaeythritol tetranitrate. A vasodilator used for angina pectoris. (Consolidated Chemicals).†

Cardiacap A. A proprietary preparation of pentaeythritol tetranitrate and amylobarbitone in a sustained release form, used in angina pectoris. (Consolidated Chemicals).†

Cardice. Solid carbon dioxide. (The Distillers Co (Carbon Dioxide) Ltd).

Cardilan. Isoxsuprine. Applications: Promotes blood flow rate. (Duphar BV).*

Cardilate. Oral/sublingual tablets. A proprietary formulation of erythrityl tetranitrate. Applications: For the prophylaxis and long-term treatment of frequent or recurrent anginal pain and reduced exercise tolerance associated with angina pectoris. (The Wellcome Foundation Ltd).*

Cardinal. A dyestuff. It is a mixture of chrysoidine and fuchsine. Used for dyeing cotton red.†

Cardinal Red J. A dyestuff. It is a British brand of Fast red.†

'ardinal Red S. See MAROON S.

ardinal Red 3B. A dyestuff. It is a British equivalent of Azorubine S.†

'ardinal S. See ACID MAGENTA.

'ardio-Green. Indocyanine green. Applications: Diagnostic aid. (Hynson Westcott & Dunning, Div of Becton Dickinson & Co).

'ardiuix. Benziodarone.†

'ardol. See ACAJOU BALSAM.

'ardolite. A proprietary cashew nut derivative of the phenol aldehyde polymer class. Also proprietary trade name for plasticizers, resins, rubber-like polymers, and solvents.†

'ardophylin. Aminophylin tablets, ampoules and suppositories. (Fisons PLC).*

'ardox. An explosive utilizing liquid carbon dioxide.†

'ardura. A proprietary trade nane for non-drying alkyd resins with excellent weathering properties used for surface coatings. (Shell Chemicals).†

'ardura E. This glycidyl ester of the synthetic fatty acid, Versatic 10, is a versatile intermediate for the production of stoving enamels, nitrocellulose lacquers and urethane paints. (Shell Chemicals).†

'ariflex. A proprietary trade name for a range of synthetic rubbers (styrene butadiene, cis-polyisoprene and polybutadiene). (Shell Chemicals).†

'ariflex Butadiene Rubber (BR). Polybutadienes are resilient and possess very good abrasion resistance. They are available in straight and oil-extended versions being used in car tyres and in the manufacture of high-impact polystyrenes as well as many industrial applications. (Shell Chemicals).†

'ariflex Isoprene Rubbers (IR). These are chemically very similar to natural rubber. They are available in straight and oil-extended versions being used in tyre carcasses, footwear and belting. (Shell Chemicals).†

'ariflex S. A proprietary styrene - butadiene rubber. (Shell International).†

Cariflex Styrene-Butadiene Rubbers (SBR). A family of general purpose synthetic rubbers available as straight, oil-extended, carbon-black masterbatch or resin/rubber masterbatch. They are widely used in car tyres, footwear, adhesives and a range of industrial and domestic products. (Shell Chemicals).†

Cariflex Thermoplastyic Rubbers (TR). These materials possess the inherent elasticity of rubbers yet can be processed as thermoplastics. They are used in the manufacture of adhesives or blended with thermoplastics for improved impact properties. When compounded they are used as carpet-backing or injection-moulded for footwear and a range of industrial and domestic products. (Shell Chemicals).†

Carin. See HEXAMINE.

Carina. A proprietary trade name for polyvinyl chloride (*qv*). (Shell Chemicals).†

Carindaden. A proprietary preparation of carindacillin sodium. An antibiotic used in the treatment of genito-urinary tract infections. (Pfizer International).*

Carinex. A proprietary trade name for a series of polystyrenes and toughened polystyrenes. (Shell Chemicals).†

Carinex SB41. A proprietary trade name for an easy processing polystyrene. (Shell Chemicals).†

Carinex SI 73. A proprietary high-impact grade of polystyrene possessing good flow properties in moulding. (Shell Chemicals, Ireland).†

Cariod. Papain. Applications: Enzyme. (Sterling Drug Inc).

Caritrol. Diethylcarbamazine citrate. (May & Baker Ltd).

Carletti's Indicator. Phenolphthalein which has been reduced by caustic soda and zinc dust.†

Carlona. A proprietary trade name for high and low density polyethylenes. (99). See also TELCOTHENE, ALKATHENE, RIGIDEX.†

Carlona LB 157. A proprietary low-density polyethylene used in the production of film. (Shell Chemicals).†

Carlona LF 456. A proprietary polyethylene having high impact strength, used in the production of film. (Shell Chemicals).†

Carlona LF 459. A proprietary low-density polyethylene possessing good optical and mechanical properties for the production of film. (Shell Chemicals).†

Carlona P. A proprietary trade name for polypropylene. *See also* PROPATHENE, NOBLEN.†

Carlona P PLZ 532. A proprietary talc-filled, heat-stabilised polyethylene used in the production of rigid mouldings. (Shell Chemicals).†

Carlona P PY 61. A proprietary polypropylene homopolymer used in the production of film, especially of fibrilla ed stretched tapes for use as plastic string. (Shell Chemicals).†

Carlona 460. A proprietary polyethylene used in the production of film. It has high impact strength but contains a slip additive. (Shell Chemicals).†

Carlona 462. A proprietary polyethylene containing slip additives, for optical and mechanical use. (Shell Chemicals).†

Carlona 463. A proprietary polyethylene containing slip additives for optical and mechanical film. (Shell Chemicals).†

Carlona 55-004. A proprietary high density polyethylene possessing high resistance to stress cracking, used in the blow-moulding of containers. (Shell Chemicals).†

Carlona 60-010. A proprietary high density polyethylene used in the extrusion blow-moulding of thin-walled bottles. (Shell Chemicals).†

Carlona 60-060. A proprietary high density polyethylene used in the injection-moulding of heavy-duty containers. (Shell Chemicals).†

Carlona 60-120. A proprietary high density polyethylene used in the injection-moulding of thin sections. (Shell Chemicals).†

Carloon Bronze. An alloy of 75.5 per cent copper, 14.5 per cent lead, and 10 per cent tin.†

Carlton Suspension N.K. A proprietary preparation containing neomycin sulphate and light kaolin. An anti-diarrhoeal. (Carlton Laboratories).†

Carmalum. A microscopic stain. It consists of carminic acid in aqueous alum.†

Carminaph. See SUDAN I.

Carminaph J. See SUDAN G.

Carminaphtha. See SUDAN I.

Carmine. A pigment. It is that preparation of cochineal which contains most colouring matter, and least aluminium base. See COCHINEAL.†

Carmine Base. See COCHINEAL.

Carmine Lake. (Lac Lake, Indian Lake). Cochineal carmine prepared by precipitating a decoction of cochineal with alum or stannic chloride, with addition of acid oxalate or tartrate of potassium. See COCHINEAL.†

Carmine Red. Obtained by boiling a dilute aqueous solution of carminic acid with a few drops of a mineral acid. It gives coloured lakes.†

Carminette Blue. A pigment. It is a red lea coloured with a bluish eosin.†

Carminette Blue-red. (Carminette Reddish yellow, Carminette Warm Red, Carminette Warm Dark Red). Pigments, which are similar to Carminette blue and yellow.†

Carminette Reddish-yellow. See CARMINETTE BLUE-RED.

Carminette Warm Dark Red. See CARMINETTE BLUE-RED.

Carminette Warm Red. See CARMINETTE BLUE-RED.

Carminette Yellow. A pigment. It consists of red lead coloured with eosin, extra yellowish.†

Carmine, Safflower. See SAFFLOWER.

Carmoisine. See AZORUBINE S.

Carmoisine L, LAS, S, WS. Dyestuffs. The are British brands of Azorubine S.†

Carnallite. A double chloride of potassium and magnesium, $MgCl_2$ KCl $6H_2O$, found in the Stassfurt deposits. It is a source of potassium chloride and potash manures.†

Carnauba Wax. A wax derived from the carnauba palm, *Corypha cerifera*. The wax is found on the young leaves. It is greyish, yellowish, or greenish in colour and when freshly purified melts at 85-86° C.†

Carnitine. See NOVAIN.

Carnoid. A proprietary trade name for a casein plastic.†

Carnotine. See PRIMULINE.

Carnot's Reagent. Basic bismuth nitrate (100 grams) is dissolved in hot, concentrated hydrochloric acid, and diluted to a litre with 92 per cent alcohol.†

Carnoy's Fluid. A mixture of absolute alcohol and glacial acetic acid. Used for fixing animal tissue.†

Caro Bronze. An alloy of 92 per cent copper, 8 per cent tin, and 0.25 per cent phosphorus. A bearing metal.†

Caroat. A peroxomonosulfate compound with about 4.5% active oxygen. Applications: Used for production of cleansers of all types e.g. denture cleansers, household and toilet cleaners, for detoxicating cyanidic waste water. (Degussa).*

Caromax. A range of closecut aromatic hydrocarbon solvents. (Carless Solvents Ltd).

Carophyll. Pigments for animal feedstuffs. (Roche Products Ltd).

Caro's Acid. Persulphuric acid, H_2SO_2†

Caro's Reagent. Obtained by dissolving ammonium or potassium persulphate in concentrated sulphuric acid. It is a pasty oxidizing agent. Used for testing alkaloids.†

Caroubier. An acid dyestuff, giving crimson shades on wool.†

Carovax. A proprietary pasteurella vaccine used in veterinary work. (Coopers Animal Health, subsidiary of The Wellcome Foundation Ltd).*

Carp. A cotton fabric grade of TUFNOL industrial laminates. (Tufnol).*

Carp Brand. Salt. (ICI PLC).*

Carpenters Wood Glue. PVA liquid. Applications: Fast grab wood glue. (Wessex Resins & Adhesives Ltd).*

Carrisorb. Attapulgite clay. Applications: An absorbent used in treatment and bleaching of oils (mainly mineral). (Bromhead & Denison Ltd).*

Carset. Hardener for cold-set sand moulds. (Foseco (FS) Ltd).*

Carsil. Binder for the CO^2 process. (Foseco (FS) Ltd).*

Carsilon. Silicon carbide. Applications: Used in special refractories. (Lonza Limited).*

Carsonol. Alcohol sulphates. Applications: Used for shampoos and bubble baths where high foaming and mildness are desired. (Lonza Limited).*

Carsonon. Nonionic sulphates and wetting agents. Applications: Used in industrial and household cleaning products where good detergency is required. (Lonza Limited).*

Carsoquat. Quaternary ammonium compounds. Applications: Used as creme rinses in personal care applications. (Lonza Limited).*

Carsosoft. Quaternary ammonium compounds. Applications: Used as fabric softeners. (Lonza Limited).*

Carterite. A proprietary trade name for resin-pulp moulding compound.†

Cartolac. General-purpose carton lacquers. (The Scottish Adhesives Co Ltd).*

Cartose. Dextrose. Applications: Replenisher. (Sterling Drug Inc).

Cartrol. Carteolol hydrochloride. Applications: Anti-adrenergic. (Abbott Laboratories).

Carvacrol. 2-Methyl-5-isopropyl-phenol. $C_{10}H_{14}O$.†

Carvene. See HESPERIDENE.

Carylderm. Carbaryl. Applications: Infestation by lice. (Napp Laboratories Ltd).*

Casabet. Coco amido betaine and coco amido sulpho betaine. Betaine type surfactants characterized by low toxicity and irritancy and high foam production properties. Applications: Used in the formulation of household cleaning products in industrial and institutional detergent products and in personal care products. (Thomas Swan & Co Ltd).*

Casabet 655. Coco amido sulpho betaine as a semi-viscous liquid. Applications: Foam booster, irritancy depressant, stable over wide pH range. Used in shampoos, bubble baths and liquid soaps. (Thomas Swan & Co Ltd).

Casahib. Amine function compounds. Applications: Corrosion inhibitors for crude oil and gas production. (Thomas Swan & Co Ltd).*

Casamer. Epoxy and urethane acrylates. Applications: Electron beam and ultraviolet curing systems. (Thomas Swan & Co Ltd).*

Casamid. Reactive and non-reactive polyamides, polyaminoamide/imine, aromatic amines, aliphatic and cycloaliphatic amines, amine adducts. Applications: Epoxy resin curing agents - used in adhesives, paints, flooring, surface coatings, powder coatings, civil engineering, electrical encapsulation and potting, filament winding, flexographic and gravure ink binders. (Thomas Swan & Co Ltd).*

Casamids. Polyamides for use as epoxy resin curing agents, adhesive and thixotropic agents. (Thomas Swan & Co Ltd).*

Casamine. Aminoethyl alkyl imidazolines and hydroxyethyl alkyl imidazolines. Applications: Anti-fungal agents, emulsifiers in agricultural sprays, ore flotation and acid detergent formulations. (Thomas Swan & Co Ltd).*

Casamox. Amine oxides. Applications: Non ionic detergent of versatile properties for the formulation of a wide range of cosmetic, detergent and industrial products. Lime soap dispersant, detergent thickener and foam booster. (Thomas Swan & Co Ltd).*

Casan Pink. See PYRONINE G.

Casaquat. Imidazoline derived quaternary ammonium compounds. Applications: Wide use in shampoos, hair conditioners, skin preparations, bacteriacidal and algaecidal application. Most important outlet is in the formulation of fabric softeners, both domestic and commercial. (Thomas Swan & Co Ltd).*

Casateric. Fatty acid imidazoline derived carboxylate amphoteric surfactants. Applications: Used to reduce irritancy of other surfactants used in shampoo, personal care and domestic applications. Also used in industrial cleaning, textile processing, ore benefaction, oil-gas production. (Thomas Swan & Co Ltd).*

Casathane. Prepolymers based on polyether and polyester diols. Applications: Casting systems for the engineering industry. (Thomas Swan & Co Ltd).*

Cascade. A nitroglycerin dynamite. Applications: Construction and building, explosives, mining. (Hercules Inc).*

Cascade. Photographic wetting agent. (May & Baker Ltd).

Cascamite. Precatalyzed urea formaldehyde powdered resin glue. Applications: General purpose wood glue - waterproof. (Wessex Resins & Adhesives Ltd).*

Cascara Sagrada. The dried bark of *Rhamnus Purshiana*. A laxative.†

Cascaras. Cascara tablets.†

Casco. Casein-based adhesives. (Borden (UK) Ltd).

Casco-resin. Urea-formaldehyde resins. (Borden (UK) Ltd).

Cascophen. Phenolic resins. (Borden (UK) Ltd).

Cascophen Resorcinol Resin RS216/RXS-8. Liquid resorcinol formaldehyde resin with powder catalyst. Applications: General purpose wood glue - weatherproof to BS1204WBP. (Wessex Resins & Adhesives Ltd).*

Casein Glue. Made by stirring casein with 25 per cent distilled water, and 1-4 per cent sodium bicarbonate, adding another 25 per cent distilled water, standing, and adding an antiseptic to prevent mould. It can be applied cold. Borax is sometimes used instead of the bicarbonate. A substitute for glue. Another preparation consists of 100 parts casein, 10 parts alum, and 3-5 parts soda ash, mixed with 500 cc water.†

Casein Magnesia. A preparation which consists of powdered casein, water, and magnesia. It will fix mineral pigments.†

Casein Paints. Paints formed by the addition of a powder containing casein and alkali, to water. The colouring matter and lime are added.†

asein Plastics. See under FANTASIT, GALA, LACRINITE, LACTOLOID, LUPINIT and PAPYRUS.

asein, Serum. See GLOBULIN.

asein Silk. Casein dissolved in alkali or zinc chloride solution, and spun into an acid bath.†

aseogum. A solution of casein in lime water. Used as an adhesive, or for impregnating linen and cotton fabrics.†

asilan. Calcium caseinate. (Glaxo Pharmaceuticals Ltd).†

asolithe. See GALLATITE.

asoron G. Dichlobenil. Applications: Weedkiller. Direct, selective weed killing action in orchards, vineyards, flower beds and parkland areas and along rail tracks, motorways and waterways. (Duphar BV).*

assava. A food product. It consists of the starch obtained from the roots of the Manioc, *Manihot utilissima*.†

assava Meal. Ground cassava root. Also see FARINE.†

assel Brown. See VANDYCK BROWN and UMBER.

assel Green. Rosenthiel's green (a manganate of barium). Used as a pigment.†

assel Yellow. See TURNER'S YELLOW.

assella's Acid. β-Naphtholsulphonic acid δ or F.†

asselmam's Green. A pigment made by mixing boiling solutions of copper sulphate and an alkaline acetate. It consists of $CuSO_4\ 3Cu(OH)_2 4H_2O$.†

assia Oil. See OIL OF CASSIA.

ast Brass. Brass which is not required to be spun, rolled, drawn or hammered. It is made by melting together the copper and zinc. This usually contains 66 per cent copper with zinc, and the lead varies from 1-3 per cent.†

ast Iron D2. A British chemical standard. It is a grey phosphoric cast iron in the form of fine turnings. It contains 1.31 per cent silicon, 1.07 per cent phosphorus, and 1.64 per cent manganese, also about 2.5 per cent graphitic carbon, 0.8 per cent combined carbon, and 0.03 per cent sulphur.†

Cast Yellow Brass. Usually an alloy of 67 per cent copper, 31 per cent zinc, and 2 per cent lead. It melts at 895° C.†

Castaldo. Natural rubber (CIS-1, 4-polyisoprene compound). Applications: Mould making for jewellery casting. (FE Knight Inc).*

Castaloy. A proprietary trade name for high carbon, high chromium steel.†

Castellanos Powder. A dynamite containing nitro-glycerin, nitro-benzene, fibrous material, and kieselguhr.†

Castethane. Two-component systems used in the manufacture of cast polyurethane elastomers. (Dow Chemical Co Ltd).*

Casting Copper. An American copper. It contains 98.5-99.75 per cent copper. Used for casting.†

Casto-Magic. A liquid oil derivitive applied to concrete forms to release forms from concrete and improve appearance of concrete. Applications: Concrete form release agent. (Rostine Mfg & Supply Co).*

Castomer. 2-component high duty polyurethane elastomer systems. (Baxenden Chemical Co Ltd).*

Castor Oil, Soluble. See BLOWN OILS.

Castrol GTX. High performance motor oil. Applications: For all types of motor car engines. (Burmah-Castrol Ltd).*

Castrol Turbomax. A heavy duty high performance mineral-based crankcase lubricant for commercial vehicles. Applications: Service, fill and top-up turbocharged and naturally aspirated diesel engines (except Detroit Diesel) and virtually all petrol units. (Burmah-Castrol Ltd).*

Catabond. A proprietary trade name for a phenol-formaldehyde liquid resin used for plywood manufacture where a waterproof bond is required.†

Cata-Chek. Applications: For catalysts for polymerization and adhesion of urethanes and other synthetic organic polymers, in Int Class 1. (Ferro Corporation).*

Cataflot. Amines and derivatives. Primary, secondary, tertiary fatty mono-, di- and polyamines from C8 to C22. Quaternary ammonium salts. Amine oxides. Amine salts. Ethoxylated amines and polyamines. Amino-acids. Special nonionic derivatives. Additives. Applications: Surface active agents used as auxiliaries in road, mining, textile and fertilizer industries, in metallurgy, as lubricating agents, additives for fuel oils and soil stabilization. (British Ceca Co Ltd).*

Catafor. Electrostatic paint and PU foam additive. (ABM Chemicals Ltd).*

Cataid. Surface active agents and detergents. (ABM Chemicals Ltd).*

Catalazuli. A proprietary trade name for a phenol-formaldehyde synthetic resin product.†

Catalex. A proprietary trade name for an expanded phenol formaldehyde plastic.†

Catalpo. A colloidal china clay, prepared from Cornish china clay, used as a fine pigment to take up colours. It is also used in rubber mixings. Other proprietary trade names for china clays are: Blue Ridge, Dixie, Devolite, Langford, Par, Stockalite, and Suprex. Dixie is a hard clay while Catalpo is soft.†

Catapol SR. A proprietary range of polyurethane elastomers ranging in hardness from 20-60 Durometer A. (Longfield Chemicals, Warrington).†

Catapres. A proprietary preparation of clonidine hydrochloride. An anti-hypertensive. (Boehringer Ingelheim Ltd).†

Catapres-TTS. Clonidine. Applications: Antihypertensive. (Boehringer Ingelheim Pharmaceuticals Inc).

Catarase. Chymotrypsin. Applications: Enzyme. (CooperVision Inc).

Catavar. A proprietary phenol-formaldehyde surface coating or laminating varnish.†

Catechol. See PYROCATECHIN.

Catechol, Methyl. See METHYL CATECHOL.

Catechu. See CUTCH.

Catechu, Bengal. See CUTCH.

Catechu, Black. See CUTCH.

Catechu, Bombay. See CUTCH.

Catechu, Brown. Prepared from Bismarck brown. A direct cotton dyestuff, $C_{30}H_{28}N_{14}$.†

Catechu, Cubical. See CUTCH.

Catechu, Gambier. See CUTCH.

Catechu, Pale. See CUTCH.

Catechu, Pegu. See CUTCH.

Catechu, Yellow. See CUTCH.

Catex. Imidazoline derived cationic and amphoteric surfactants. Applications: Textile softeners, antistats, foaming agents and solubilizing detergents. (Thomas Swan & Co Ltd).*

Catigene BR 80 B. Lauryl dimethyl benzyl ammonium bromide as a clear yellow liquid. Applications: Algicides, bactericides, fungicides, deodorants. (KWR Chemicals Ltd).

Catigene Brown N. (Cold Blacks B and R) Dyestuffs of the same class as thion blue B. They dye cotton in a cold bath, containing sodium sulphate and soap.†

Catigene DC/100. Myristyl dimethyl benzyl ammonium chloride as a white powder. Applications: Bactericides for pharmaceuticals. (KWR Chemicals Ltd).

Catigene Indigo R Extra. A sulphide dyestuff, recommended for cop-dyeing.†

Catigene Red-brown. A dyestuff obtained by the action of alkali sulphides and sulphur upon aminohydroxy-phenazines.†

Catigene SR. Quaternary alkylamine acetate as a white fluid paste. Applications: Cationic surfactant used in textile softeners. (KWR Chemicals Ltd).

Catigene T80. Alkyl dimethyl benzyl ammonium chloride as a clear yellow liquid. Applications: Bactericides, fungicides, algicides. (KWR Chemicals Ltd).

Catigene 4513. Mixture of the chlorides of alkyl dimethyl benzyl ammonium and alkyl dimethyl ethyl benzyl ammonium as a clear yellow liquid. Applications: Algicides, bactericides, fungicides, deodorants. (KWR Chemicals Ltd).

Catiomaster-C. 3-chloro-2-hydroxypropyl trimethylammonium chloride (50% aq solution). Applications: Quaternary cationic agent for starch and cellulose. (Yokkaichi Chemical Co Ltd).*

Cationic Softener X Concentrate. Alkyl-imidazoline derivative in paste form. Applications: Softener with good substantivity, used as glass fibre mordant and lubricant. (Millmaster-Onyx UK).

Catisol AO 100. Oleylamine acetate as a yellow soft wax. Applications: Cationic surfactant used as an antistatic agent for glass fibres. (KWR Chemicals Ltd).

Catosal. Organic phosphorus preparation. Applications: A stimulator of metabolism used in veterinary medicine. (Bayer & Co).*

Cat's Eye Dammar. See DAMMAR RESIN.

Cat's Gold. (Cat's Silver). Very finely powdered mica is sometimes called by these names. The term "cat's gold" is also used for mosaic gold (a tin sulphide).†

Cat's Silver. See CAT'S GOLD. Synonym for Mica.

Cat's-eye Resin. An East Indian dammara resin obtained from *Pinus dammara* or *Dammara alba*.†

Caucho Blanco. A rubber obtained from different species of the genus *Sapium* which belongs to the *Enphorboaceoe* family and found in the northern part of South America.†

Caust X. Sodium phosphates, silicates and precipitating agents which are compounded into solid bricks. Applications: Used as a descalant in the caustic removal sections of bottle washers. (Delaware Chemical Corp).*

Caustic Baryta. Barium hydroxide, $Ba(OH)_2$.†

Caustic Lime. See LIME.

Caustic Potash. Potassium hydroxide, KOH.†

Caustic Soda. Sodium hydroxide, NaOH.†

Caustic, Toughened. See TOUGHENED CAUSTIC.

Caustic, Vienna. See VIENNA CAUSTIC.

Caustic, White. See WHITE CAUSTIC.

Causul. A nickel-copper-chromium cast iron of marked acid and alkaline resistance. It is used in the manufacture of valves intended particularly for handling sulphuric acid and caustic soda.†

Caved-S. A proprietary preparation containing glycyrrhizinic acid as powdered block liquorice, bismuth sub-nitrate, aluminium hydroxide gel, light magnesium carbonate, sodium bicarbonate and powdered frangula bark. (AKU Holland).†

Caytur 21 & 22. A trade mark for a range of curing agents used in urethane elastomers. (Du Pont (UK) Ltd).†

Caytur 4. A proprietary partial complex of zinc chloride with benzothiazyl disulphide used as a cross-linking agent in sulphur-curable urethane elastomers. Formerly known as LD-55. (Du Pont (UK) Ltd).†

Cazin. An alloy of cadmium and zinc, containing 82.6 per cent cadmium.†

Ceanel. A proprietary preparation of phenylethyl alcohol, cetrimide and undecenoic acid. Applications: Scalp antiseptic used as shampoo. (Quinoderm Ltd).*

Ceara Wax. See CARNAUBA WAX.

Cecagel. Silica gel. Applications: Gas drying - catalyst support. (British Ceca Co Ltd).*

Cecaperl. Expanded perlite. Applications: Industrial and cryogenic insulation (tanks and methane tankers). (British Ceca Co Ltd).*

Cecarbon. Granular activated carbon. Applications: Purification, decolourization, deodorization, separation and recovery in liquid of gas phase, in the chemical, petrochemical, pharmaceutical and food industries (Glucose factories, sugar refiners, oil refining, wine treatment). For the treatment of drinking and industrial water etc. Catalyst supports. (British Ceca Co Ltd).*

Cecasil. Calcium silicate. Applications: Industrial fillers. (British Ceca Co Ltd).*

Ceclor. Cefaclor. Applications: Antibacterial. (Eli Lilly & Co).

Ce-Cobalin. Syrup containing vitamins B_{12} and C. Applications: Provides vitamin supplementation in deficiency states such as self imposed dietary restrictions as in strict vegetarianism or following therapeutic diets for gastro-intestinal ulceration. (Paines & Byrne).*

Cecolene 1. A proprietary trade name for trichlorethylene.†

Cecolene 2. A proprietary trade name for perchlorethylene.†

Cedarite. See CHEMAWINITE.

Cedilanid. A proprietary preparation of lanatoside C. (Sandoz).†

Cedilanid Ampoules. A proprietary preparation of deslanoside. (Sandoz).†

Cedocard. A proprietary preparation of sorbide nitrate used in the treatment of angina pectoris. (Tillots Laboratories).†

Cedrat Oil. See OIL OF CEDRAT.

Ceduran. A proprietary preparation of NITROFURANTOIN with deglyceryr-rhizinated liquorice. An antibiotic. (Tillots Laboratories).†

Ceeline. See COERULEUM.

CeeNU. Lomustine. Applications: Antineoplastic. (Bristol Laboratories, Div of Bristol-Myers Co).

Cefa-Lake. Cephapirin sodium. Applications: Antibacterial.(Bristol Laboratories, Div of Bristol-Myers Co).

Cefadyl. Cephapirin sodium. Applications: Antibacterial. (Bristol Laboratories, Div of Bristol-Myers Co).

Cefizox. Ceftizoxime sodium. Applications: Antibacterial. (Smith Kline & French Laboratories).

Cefmax. Cefmenoxime hydrochloride. Applications: Antibacterial. (TAP Pharmaceuticals).

Cefobid. A proprietary preparation of cefoperazone sodium. An antibiotic. (Pfizer International).*

Cefobine. A proprietary preparation of cefoperazone sodium. An antibiotic. (Pfizer International).*

Cefobis. A proprietary preparation of cefoperazone sodium. An antibiotic. (Pfizer International).*

Cefomonil. Cefsulodin sodium. Applications: Antibacterial. (TAP Pharmaceuticals).

Cekas. A heat-resisting alloy containing 59.7 per cent nickel, 11.2 per cent chromium, 28 per cent iron, and 2 per cent manganese.†

Celacol. Hydroxy propyl methyl/methyl cellulose. (Courtaulds Chemicals & Plastics Water Soluble Polymers Group).

Celafuse. A proprietary range of polyamid resins. Celafuse 100 is a terpolyamide with good melt flow properties (melting point 103-108° C.). Celafuse T is a terpolyamide with a higher viscosity (melting point 115- 125° C.). Celafuse CP has a melting point of 145-150° C. with good hydrophobic properties. Celafuse SG is a plasticized co-polyamide with a melting point of 110-120° C. (British Celanese).†

Celanese. (Lustron). Celanese is an Englis name for cellulose acetate silk. The san type of silk is known in America as Lustron.†

Celanese Nylon 6/6. Polyamide (PA). Applications: Coil bobbins, window winder mechanisms, lighter bodies and valve rocker covers. (Celanese Limited).†

Celanex. Thermoplastic polyester (PBT) polylutylentephthalate. Applications: Domestic equipment housings, plugs and sockets, switches. Telecommunications, keyswitches, switchboard components. Automotive, distribution housings, exterior door handles and wiper components. (Celanese Limited).†

Celanex 917 and J101. A proprietary rang of glass-reinforced polyester thermoplastics. (493) .†

Celastic. A proprietary trade name for a pyroxylin product.†

Celasyl. Fast dyestuffs for artificial fibres. (JC Bottomley and Emerson).†

Celatene Colours. Proprietary colours which are amino-anthraquinone derivatives.†

Celatom. Diatomaceous earth. Applications: Filter powder. (Flexibulk Ltd).*

Celbenin. A proprietary preparation containing methicillin sodium. An antibiotic. (Beecham Research Laboratories).*

Celeron. A proprietary trade name for a synthetic resin.†

Celescot. A proprietary trade name for a pyroxylin product.†

Celeste. See WILLOW BLUE.

Celestial Blue. See BRUNSWICK BLUE.

Celestine Blue B. See COREINE 2R.

Celestol. A proprietary trade name for an alkyd synthetic resin.†

Celestols. A proprietary trade name for polybasic acid-polyhydric alcohol fatty acid type synthetic resins.†

Celestone. Betamethasone. Applications: Glucocorticoid. (Schering-Plough Corp).

Celestron. A condensation product of phenol and formaldehyde with fillers. Used as an insulator.†

Celevac. A proprietary preparation containing oxyphenisatin diacetate. A laxative. (Parke-Davis).†

Celite. A registered trade mark for a material for separating impurities and other matters from fluids, and used as an aid in filtering, dehydrating, and demulsifying of liquids. It is also used as a filler in the plastics industry. The term is also applied to a constituent of Portland cement and clinker. It consists of a solution of dicalcium silicate in dicalcium aluminate.†

Cellacephate. A partial mixed acetate and hydrogen phthalate ester of cellulose, used as an enteric coating.†

Cellanite. A proprietary trade name for a synthetic resin paper product.†

Cellastine. A proprietary trade name for a cellulose-acetate product.†

Cellesta. A proprietary trade name for a pyroxylin product.†

Cellestren. Dyestuffs for cellulosic/polyester blends. (BASF United Kingdom Ltd).

Cellidor. Cellulose acetate, acetate butyrate and propionate compounds. Applications: For the manufacture by injection moulding and/or extrusion of spectacle frames, tool handles, seating furniture etc. and the production of fluidized bed coating powders. (Bayer & Co).*

Cellit. Cellulose acetate butyrate and propionate. Applications: For photographic films, electrical insulating films as well as block casting. (Bayer & Co).*

Celliton. Dyestuffs for acetate and triacetate. (BASF United Kingdom Ltd).

Cellobond. HEC water soluble polymer. (BP Chemicals Ltd).*

Cellocaps. A proprietary viscose product.†

Celloidin. (Photoxylin). A substance which is obtained from collodion by precipitating it from its solution in alcohol and ether. It consists of pure nitro-cellulose. Used in microscopy and in surgery.†

Cellolyn. A range of dibasic-acid-modified rosin esters. Applications: Tackifier for adhesives, used for type C gravure inks, used in nitrocellulose coatings. (Hercules Inc).*

Cellomold. A registered trade mark for a cellulose acetate moulding material and other plastics.†

Cellophane. (Registered Trade Mark). A brand of regenerated cellulose film produced from viscose by treatment with sulphuric acid and/or ammonium salts. The film is non-inflammable and is insoluble in water, alcohol, and oils and was formerly used extensively as a transparent wrapping material. See also VISCOSE and VISCOSE SILK.†

Cellophane. (Trade name for Benelux). Regenerated celulose film, plain and coloured. P - uncoated films; M - nitrocellulose-coated films; X - copolymer-coated films. Applications: Film to use on high speed automatic packaging machines and for incorporation into laminates. (UCB nv Film Sector).*

Cellosolve. Ethanediol ethers and ether esters. (Union Carbide (UK) Ltd).†

Celluclast. A cellulase preparation made by submerged fermentation of a selected strain of the fungus Trichoderma reesei. Applications: Can be used in any case where the aim is break-down of cellulosic matter for production of fermentable sugar, reduction of viscosity or increase in yield of valuable products of plant origin. (Novo Industri A/S).*

Cellucraft. A proprietary trade name for nitro-cellulose spray coating.†

Celluflex M179. A proprietary trade name for alkyl-aryl plasticizer.†

Cellufluor. Chemical equivalent for calcofluor white. Applications: Fungal stain for cytopathology. (Polysciences Inc).*

Cellulate. A proprietary trade name for a cellulose acetate plastic.†

Cellulith. A material made by grinding cellulose in water until a jelly is produced and boiling until hard. Used as a binding agent for carborundum wheels, also as a packing material.†

Celluloid. (Xylonite). Composed of a soluble nitro-cellulose mixed with camphor, obtained by gelatinizing nitro-cellulose by means of a solution of camphor in ethyl alcohol. It can be moulded, and was formerly used extensively for making toys, combs, and other articles.†

Celluloid-caoutchouc. A material prepared by dissolving rubber and celluloid in cyclohexanol and mixing the solutions. An elastic substance is produced.†

Cellulose Acetate Plastics. See under ERINOFORT, LANOPLAST, LANZOID, RHODIALITE, SERACELLE, TENITE, VITREOCOLLOID.

Cellulose Gum. Sodium carboxymethyl cellulose. (Hercules Ltd).

Cellulose Pitch. The residue obtained from the evaporation of the waste sulphite lye, from the treatment of wood in the sulphite process. Used for making briquettes.†

Cellulose Turpentine Oil. See SULPHITE TURPENTINE OIL.

Cellulose, Hydrated. See ALKALI CELLULOSE.

Cellulose, Starch. See FARINOSE.

Cellulosine. A proprietary trade name for a bleached celluloid.†

Celluplastic. A proprietary trade name for a plasticized cellulose used for containers.†

Cellushi. A proprietary viscose product use for packing.†

Celluvarno. A proprietary trade name for cellulose nitrate surfacing material.†

Celmar. A registered trade mark for polypropylene/glass fibre reinforced structures. (British Celanese).†

Celmontite. A coal mine explosive containing 65.5-68.5 per cent ammonium oxalate, 10.5-12.5 per cent trinitrotoluene, and 19.5-21.5 per cent sodium chloride.†

Celogen. A range of non-staining and non-discolouring, non-toxic and odourless nitrogen blowing agents for sponge rubber and expanded plastics. (Uniroyal).*

Celontin. Methsuximide. Applications: Anticonvulsant. (Parke-Davis, Div of Warner-Lambert Co).

Celontin Kapseals. A proprietary preparation of methsuximide. An anti-convulsant. (Parke-Davis).†

Celsit. An alloy similar in composition to Stellite.†

Celta. See LUFTSEIDE.

Celtex. Extruded cellular ceramic filters fo cast irons. (Foseco (FS) Ltd).*

Celtid. A proprietary trade name for a pyroxylin product.†

Celtite. An explosive. It consists of 56-59 per cent nitro-glycerin, 2-3.5 per cent nitro-cotton, 17-21 per cent potassium nitrate, 8-9 per cent wood meal, and 11 13 per cent ammonium oxalate.†

Celulon. Series of cellular polyurethane coating materials suitable for application by brush or spray. (Unitex Ltd).*

Cement, Adamantine. A mixture of powdered pumice and silver amalgam. Used in dentistry.†

Cement, American. A rubber cement made from 10 parts rubber, 6 parts chloroform, and 2 parts mastic.†

Cementation Copper. See CEMENT COPPER.

Cementation Steel. Obtained by heating bars of good malleable iron, packed wit nitrogenous matter, or wood charcoal.†

Cement Copper. (Cementation Copper, Copper Precipitate). Copper produced from copper liquors and mine liquors, by means of iron (pig iron, scrap iron, or spongy iron). It is usually contaminated with arsenic, antimony, and iron.†

Cement, Giant. See ACETIC-GELATIN CEMENT.

Cement, Hercules. See ACETIC-GELATIN CEMENT.

Cement, Iron. See EISEN-PORTLAND CEMENT.

Cementite. Triferrous carbide, Fe_3C, the hardest component of steel.†

Cementite, Independent. Cementite in rectilinear lamellae.†

Cementkote. See CERESIT.

Cement, Le Farge. See GRAPPIER CEMENT.

Cement, Magnesia. See SOREL CEMENT.

Cement Mortar. A mixture of natural slag or Portland cement, sand, and water. Lime is also added.†

Cement, Pozzolana. See EISEN-PORTLAND CEMENT.

Cement Prodor. A range of acid resisting cements. (Prodorite).*

Cement, Puzzolana. See EISEN-PORTLAND CEMENT.

Cement, Roman. See ROMAN CEMENT.

Cement, Slag. See EISEN-PORTLAND CEMENT.

Cemsave. Ground granulated blastfurnace slag (GGBFS) for blending with Portland cement in the concrete mixer to produce Portland Blastfurnace Cement concrete. Applications: All classes and types of concrete to give long term durability. Additionally when required will provide high chemical and sulphate resistance, low heat properties and reduce the risk of alkali aggregate reaction. (Frodingham Cement Co Ltd).*

Cemset. Accelerators for use in cement bonded sand moulds and cores. (Foseco (FS) Ltd).*

Cendevax. A proprietary rubella vaccine. (Smith Kline and French Laboratories Ltd).*

Ceneg. A proprietary trade name for dinonyl phthalate. A vinyl plasticizer (Koninklijke Nederlandsche Gist-En Spiritusfabriek).†

Centari. Acrylic enamels. (Du Pont (UK) Ltd).

Centrac. Profadol Hydrochloride. Applications: Analgesic. (Parke-Davis, Div of Warner-Lambert Co).

Centralite. Dimethyldiphenylurea. Used in explosive powders.†

Centrax. Prazepam. Applications: Sedative. (Parke-Davis, Div of Warner-Lambert Co).

Centrifugal Syrup. A selective weedkiller. (Marks, A H & Co Ltd).*

Centyl. A proprietary preparation of bendrofluazide. A diuretic. (Leo Laboratories).†

Centyl K. A proprietary preparation of bendrofluazide and potassium chloride in a slow release core. A diuretic. (Leo Laboratories).†

Cephreine. Pure citronellyl acetate. (Bush Boake Allen).*

Cephrol. Pure dextro citronellol. (Bush Boake Allen).*

Cephthalothin. 7-(2-Thienylacetamido)cephalosporanic acid.†

Cephulac. Lactulose. Applications: Treatment of obstipation and (pre-) hepatic coma. (Duphar BV).*

Ceplac. Erythrocine. Applications: An aid to the teaching of oral hygiene. (Berk Pharmaceuticals Ltd).*

Ceporex. A proprietary preparation containing cephalexin. An antibiotic. (Glaxo Pharmaceuticals Ltd).†

Ceporin. A proprietary preparation containing Cephaloridine. An antibiotic. (Glaxo Pharmaceuticals Ltd).†

C.E. Powders. Explosive powders, containing tetryl- or tetranitromethylaniline.†

Cepton. A range of medicated skin care products. (ICI PLC).*

Cerabrit. Synthetic waxes. (Abril Industrial Waxes).*

Ceracine. See FAST RED.

Ceracolor. Colours in tubes ready for use for painting on porcelain, bone china and earthenware. (Degussa).*

Ceralumin B. An aluminium alloy containing 1.5 per cent copper. 1.5 per cent nickel, 0.2 per cent magnesium, 0.7 per cent iron, 1.5 per cent silicon, and 0.1 per cent cerium. Used for pistons, cylinder heads, etc.†

Ceralumin "C.". A proprietary alloy of aluminium with 2.5 per cent copper, 1.5 per cent nickel, 1.2 per cent iron, 1.2 per cent silicon, 0.8 per cent magnesium, and 0.15 per cent cerium.†

Ceramitalc. Finely ground talc. Applications: Used in ceramic applications where higher fired strength is required. Will also be used as an auxilliary flux. (Vanderbilt Chemical Corporation).*

Ceramite. A solution of fluro-silicates, used as a disinfectant, as a preservative for wood, and for hardening cements.†

Ceramol. Refractory mould and core coatings. (Foseco (FS) Ltd).*

Cerasine Orange G. See SUDAN G.

Cerasine Red. See SUDAN III.

Cerasin(e). A registered trade mark currently awaiting re-allocation by its proprietors. (Casella AG).†

Ceratex. A trade mark for wax and rubber containing impregnating compounds for textiles and papers.†

Ceravase. A standardized solution of refined papain or purified papain concentrate. Applications: It is used in the brewing industry for the prevention of colloidal haze caused by repeated chilling and warming of beer. (Pfizer International).*

Cerazole. Sulphathiazole. Applications: Antibacterial. (Beecham Laboratories).

Cercobin. Fungicide. (May & Baker Ltd).

Cerebrose. Galactose, $C_6H_{12}O_2$.†

Cereclor. Series of secondary plasticizers manufactured from chlorinated waxes. The percentage of chlorine is indicated by the number after the name, e.g. Cereclor 70. (ICI PLC).*

Cereflo. A purified bacterial beta-glucanase preparation produced by submerged fermentation of a selected strain of Bacillus subtilis. Applications: Used as a supplementary glucanase preparation when masking malt or mixtures of malt and barley. (Novo Industri A/S).*

Cerelose. A commercial glucose.†

Ceremix. Contains the following enzyme activities: alpha-amylase, beta-glucanase and proteinase. Applications: Used in the brewing process when a proportion of the malt is replaced by barley and for the production of malt extract and barley syrups. (Novo Industri A/S).*

Cerere. Tricresylmercuroacetate. It is a mixture of mono- and diacetate derivatives of the three cresols, with about 75 per cent of the mono-derivative, and containing about 57 per cent mercury. It accelerates the germination of grain and affords protection against animal and vegetable parasites.†

Ceres. Fat-soluble dyestuffs. Applications: For shoe and floor polishes, office supplies, waxes, oils, fats, fuels, plastics, surface coatings and printing inks. (Bayer & Co).*

Ceresit. (Driwal, Impervite, Cementkote). Waterproofing compounds for cement, consisting mainly of calcium carbonate, alum, and calcium soap, sometimes with more or less free oil or fat. Sold in the form of powder to be mixed with dry cement, or as a paste to be mixed with water.†

Ceresol. Liquid mercury seed dressing for use on cereals. (ICI PLC).*

Cerespan. Papaverine Hydrochloride. Applications: Relaxant. (USV Pharmaceutical Corp).

Cerevax. Liquid seed dressings containing carboxin and thiabendazole. (ICI PLC).*

Cerex. A proprietary thermoplastic stated to be resistant to deformation at 100° C. It is a copolymer containing carbon, hydrogen, and nitrogen - probably of the acrylonitrile type.†

Cerfluorite. A compound, $(Ca_3Ce_2)F_6$, prepared artificially.†

Cergem. Gemeprost. Applications: Prostaglandin. (G D Searle & Co).

Ceridust. Micronized polyethylene wax. (Hoechst UK Ltd).

Cerin. See OZOKERITE.

Cerise. See MAGENTA.

Cerise B, BB, G, 2YS. Dyestuffs. They are British brands of Magenta.†

Cerit. An alloy similar in composition to Stellite.†

Cérite. A French synthetic resin of the phenol-formaldehyde type.†

Ceritone. An aromatic flavouring chemical. (PPF International Ltd).*

Cerium Copper. An alloy containing 10 per cent cerium metals and 90 per cent copper. Used as a deoxidizer in the preparation of non-ferrous metals.†

Cerone. Growth regulator containing 2-(chloroethyl) phosphonic acid for winter barley (sold under licence from Union Carbide). (ICI PLC).*

Cerosin. See OZOKERITE.

Cerotin. Ceryl cerotate, $C_{26}H_{53}COOC_{27}H_{55}$ Occurs in Chinese wax.†

Cerotine Orange C Extra. See CHRYSOIDINE R.

Cerotine Yellow R. A dyestuff obtained from diazotized aniline and resorcinol.†

Ceroxin GL. A proprietary trade name for 12-hydroxy stearic acid. A lubricant for plastics processing. (JH Little).†

Ceroxin GMO. A proprietary trade name for a partially esterified fatty acid made from naturally occurring saturated or unsaturated fatty acids and a polyhydric alcohol. A lubricant for plastics processing. (JH Little).†

Ceroxin GMR. A proprietary trade name for naturally occurring saturated or unsaturated fatty acids partially esterified with a polyhydric alcohol. A lubricant for pvc processing. (JH Little).†

Ceroxin GMSI. A proprietary trade name for a solid lubricant for pvc processing made from naturally occurring saturated or unsaturated fatty acids partially esterified with a polyhydric alcohol. (JH Little).†

Ceroxin TRI. A proprietary trade name for hydroxystearic acid glyceride. A lubricant for pvc which does not cause discoloration and which is particularly suitable for compounds for electrical purposes. (JH Little).†

Cerrobase. An alloy of lead and bismuth used as a pattern metal in foundry work.†

Cerrobend. An alloy of lead, bismuth, tin, and cadmium used for tube and section bending in foundry work.†

Cerromatrix. A proprietary alloy of bismuth, lead, tin, and antimony. It expands on cooling.†

Certi-fired. Resistor compositions. (Du Pont (UK) Ltd).

Certistain. High quality microscopy stains. (BDH Chemicals Ltd).

Certolake. Water insoluble, aluminium lake food colours. Applications: Colouring of foodstuffs and pharmaceuticals. (Williams Div of Morton Thiokol Ltd). *

Certrol. Selective weedkiller. (A H Marks & Co Ltd).

Cerubidin. Rubidomycin. (May & Baker Ltd).

Cerubidine. Daunorubicin hydrochloride. Applications: Antineoplastic. (Ives Laboratories Inc).

Cerulean Blue. See COERULEUM.

Cerumol. A proprietary preparation of *p*-dichlorbenzene, chlorbutol and oil of terebinth. Ear drops. (LAB Ltd).*

Ceruse. See WHITE LEAD.

Cervagem. A proprietary preparation of gemeprost. A prostaglandin analogue. Applications: Softening and dilitation of the cervix uteri in first trimester abortion. (May & Baker Ltd).*

Cervantal. A broad-spectrum penicillin combination with an extended range of action. Also known as Totocillin. (Bayer & Co).*

Cesol. Chlor-methyl-pyridine-β-carbonic acid methyl ester. A registered trade name. (E Merck, Darmstadt).†

Cestarsol. Arecoline acetarsol. (May & Baker Ltd).

Cetaceum. Spermaceti.†

Cetacourt. Hydrocortisone. Applications: Glucocorticoid. (Owen Laboratories

Div, Dermatological Products of Texas, Alcon Laboratories Inc).

Cetaffine. Cetyl alcohol. Applications: Raw material for cosmetics and pharmaceuticals. (Laserson & Sabetay).*

Cetal. An alloy of 87 per cent aluminium and 13 per cent silicon.†

Cetamide. Sulphacetamide Sodium. Applications: Antibacterial. (Alcon Laboratories Inc).

Cetaped. Veterinary antiseptic preparation. (ICI PLC).*

Cetavlex. A preparation of cetrimide in a cream base. Antiseptic skin cream. (Sold in UK by Care Laboratories Ltd). (ICI PLC).*

Cetavlon. A preparation of cetrimide. Skin disinfectant and wound cleanser. (Sold in UK by Care Laboratories Ltd). (ICI PLC).*

Cetec. A proprietary trade name for a cold moulding bituminous compound. (Non-refractory.)†

Ceteprin. A proprietary preparation of emepronium bromide, used in the control of frequency of micturition. (Goodrich-Gulf Chemicals, Ohio).†

Cetine. Cetyl palmitate, $C_{15}H_{31}COOC_{16}H_{33}$. Occurs in spermaceti.†

Cetiprin. A proprietary preparation of emepronium bromide, used in the control of frequency of micturition. (KabiVitrum AB).*

Cetiprin Novum. Tablets, solution for injection, syrup. Applications: Emepronium carragenate for the treatment of micturition disorders. (KabiVitrum AB).*

Cetomacrogol 1000. Polyethylene glycol 1000 monocetyl ether. Polyoxyethylene glycol 1000 monocetyl ether.†

Cetosan. A mixture of the higher alcohols of spermaceti, mainly cetyl and octodecyl alcohols, with petroleum jelly.†

Cevalin. Sodium Ascorbate. Applications: Vitamin. (Eli Lilly & Co).

Ceylon Isinglass or Gelatin. See AGAR-AGAR.

Ceyssatite. A white earth consisting of almost pure silica. An absorbent powder.†

CG-80. Dairy pipeline cleaner/sterilizer. (Ciba-Geigy PLC).

Chalcostibnite. A mineral, $CuSbS_2$.†

Chaldegal. Foam stabilizers, solubilizers. Applications: Detergent formulations; shampoos. (Hoechst UK Ltd).

Chalkone. 1,3-Diphenyl propenone, $C_6H_5.CH:CH.CO.C_6H_5$.†

Chalk, Prepared. See PREPARED CHALK.

Chalk, Red. See INDIAN RED.

Chalk, Tailor's. See TAILOR'S CHALK.

Chamotan. Leather dressing oils. (Marflee Refining Co).*

Charcoal, Bone. See ANIMAL CHARCOAL.

Chatelier Solder. An alloy of 70 per cent tin, 25 per cent zinc, 2 per cent aluminium, and 1.5 per cent phosphorus.†

Chatterton's Compound. Mixtures of tar, rosin, and gutta-percha. Used for cementing gutta-percha to wood and metals.†

Chaubert's Oil. Consists of 75 per cent oil of turpentine, and 25 per cent oil of hartshorn.†

Chaulmestrol. Consists of the ethyl esters of the fatty acids of chaulmoogra Oil.†

Chaulphosphate. Sodium dichaulmoogryl-β-glycerophosphate. Used in leprosy.†

Chaval Nickel Silver. See NICKEL SILVERS.

Chavicol. *p*-Allylphenol.†

Chavosol. (Chavosote). A dental antiseptic containing *p*-allylphenol.†

Chavosote. See CHAVOSOL.

Check Brass. See BRASS.

Checkmate. Herbicide. (May & Baker Ltd).

Checkmate. Sodium Fluoride. Applications: Dental caries prophylactic. (Oral-B Laboratories Inc).

Cheelox B-13. Mixed alkyldiaminepolyacetic acids as sodium salts and alkanolamines. Applications:

Organic sequestrant for calcium and magnesium. Clarifier for liquid soaps and shampoos, industrial cleaner solutions and herbicide formulations, water softener for liquid detergents, scale preventive in soaker-alkali formulations. (GAF Corporation).*

Cheelox BF Acid. Ethylene diaminetetracetic acid. Applications: Intermediate for the preparation of organic and inorganic salts, as well as metallic chelates. (GAF Corporation).*

Cheelox BF-12. Ethylene diaminetetracetic acid tetra sodium salt. Applications: Components of soaker-alkali cleaners to prevent and remove scale formation, water softeners for incorporation into cationic sanitizers and detergent sanitizers, used in pulp and paper industry to remove metallic ions. (GAF Corporation).*

Cheelox BF-13. Ethylene diaminetetracetic acid tetra sodium salt. Applications: Components of soaker-alkali cleaners to prevent and remove scale formation, water softeners for incorporation into cationic sanitizers and detergent sanitizers, used in pulp and paper industry to remove metallic ions. (GAF Corporation).*

Cheelox BF-78. Ethylene diaminetetracetic acid tetra sodium salt. Applications: Components of soaker-alkali cleaners to prevent and remove scale formation, water softeners for incorporation into cationic sanitizers and detergent sanitizers, used in pulp and paper industry to remove metallic ions. (GAF Corporation).*

Cheelox DTPA-14. Diethylenetriaminepentaacetic acid penta sodium salt. Applications: Used in pulp and paper industry in the peroxide bleaching stage to remove heavy metal ions known to cause destruction of peroxide. (GAF Corporation).*

Cheelox FE-12. Mixed alkyldiaminepolyacetic acids as sodium salts and alkanolamines. Applications: Contains an additive preferential ability to dielate iron ions in aqueous systems at any temperature and at a pH of 3.5 - 13.0. For use in controlling calcium. (GAF Corporation).*

Cheelox HE-24. N-(Hydroxyethyl)-ethylene diamine triacetate, trisodium salt. Applications: All-purpose chelating agent useful for iron in the pH range of 6.5 to 9.5. Also sequesters magnesium. (GAF Corporation).*

Cheelox NTA-Na_3. Nitroacetate tri sodium salt. Applications: General purpose sequestrants which complex iron in the acid pH range only. (GAF Corporation).*

Cheelox NTA-14. Nitroacetate tri sodium salt. Applications: General purpose sequestrants which complex iron in the acid pH range only. (GAF Corporation).*

Chel. Chelating agents. (Ciba Geigy PLC).†

Cheltenham Salts. A mixture of 34 parts sodium sulphate, 23 parts magnesium sulphate, and 50 parts sodium chloride.†

Chem-Rez. Phenolic and furfuryl alcohol-based resin systems cured through the application of acid catalyst or with heat. Applications: Production of foundry cores and moulds. (Ashland Chemical Company).*

Chemalog. Specialty chemical catalogue. Applications: Chemicals are sold through Chemalog to researchers for research and development applications. (Chemical Dynamics Corp).*

Chematex. Synthetic latex cement for bedding and jointing tiles and bricks. (Prodorite).*

Chemawinite. (Cedarite). A pale yellow Canadian amber.†

Chemdur. Liquid elastomeric urethane membrane for tanking rooms and pits. (Prodorite).*

Chemglaze. High performance single and two pack urethane coatings both moisture cure and catalyzed cure. Applications: Preserve, protect and beautify all types of substrates including rubber, plastic, concrete, metal and glass. (Lord Corporation (UK) Ltd).*

Chemical Red. See INDIAN RED.

Chemigum. A patented synthetic rubber derived from petroleum. It is a butadiene copolymer.†

Chemline. A range of anti-abrasive glass tiles and ceramic tiles for coal and coke bunkers and hoppers. (Prodorite).*

Chemlok. Rubber to metal bonding agents and adhesives. (Durham Chemicals Ltd).

Chemlube. A series of diester and polyolester synthetic based oils. Applications: Compressor oils, chain oils, gear oils, automotive engine oils, high temp oils, etc. (Ultrachem Inc).*

Chemol. A registered trade mark currently awaiting re-allocation by its proprietors. (Casella AG).†

Chemset. Artificial and synthetic resins all adapted for setting. Applications: Adhesives, chemical products for potting, impregnating, flooring, casting, moulding, laminating, building, sheathing, coating, enamelling, waterproofing and for the hardening of resins. (R F Bright Enterprises Ltd).*

Chenzinsky-Plehn's Solution. A microscopic stain. It contains 0.25 gram eosin, 50 grams of 70 per cent alcohol, 100 grams of a saturated solution of methylene blue, and 100 cc distilled water.†

Cheque. Mibolerone. Applications: Anabolic; androgen. (The Upjohn Co).

Chestnut. See MAGENTA and UMBER.

Chestnut Brown. See UMBER.

Chia Oil. An oil obtained from the Mexican plant *Salvia hispanica*. The raw oil dries slowly, but the boiled oil is a good drying oil.†

Chian Turpentine. (Alk, Chio Turpentine, Chios Turpentine, Cyprian Turpentine, Scian Turpentine). Names applied to the oleo-resin obtained from the bark of *Pistachi terebinthus*, a tree in the Mediterranean and Asia Minor.†

Chicago Acid. 2S Acid.†

Chicago Blue. See BENZO BLUE.

Chicle. A gum obtained from *Achras* species and others of Mexico, British Honduras, and Venezuela. It has been used in the manufacture of chewing gum, but has now been considerably superseded by other gums, e.g. Jelutong. Other names are Tunogum, Zapoto gum.†

Chierite. A proprietary moulding material of urea formaldehyde. (Butese, Italy).†

Chierol. A proprietary phenolic moulding material. (Butese, Italy).†

Chilcote. Dressings for metal chills. (Foseco (FS) Ltd).*

Chili Nitre. See CHILI SALTPETRE.

Chili Saltpetre. (Chili Nitre, Soda Salt petre, Peru Saltpetre, Cubic Saltpetre, Soda Nitre, Cubic Nitre, Nitratine). Sodium nitrate, $NaNO_3$, found as deposits in Chile and Peru.†

China Blue. See SOLUBLE BLUE.

China Clay. (Porcelain Clay, Cornish Clay, Kaolin, Argilla, White Bole, Pipeclay, Porcelain Earth, Catalpo). The purest form of clay (a silicate of aluminium, $Al_2O_3\ 2SiO_2\ 2H_2O$), found naturally. It is used in the manufacture of china, in distemper work, and for other purposes.†

China Green. See MALACHITE GREEN.

China Oil. Peru balsam.†

China Silver. See ARGYROLITH and NICKEL SILVERS.

Chinaldine. 2-Methylquinoline.†

Chinaspin. A preparation for the relief of colds and flu. (Bayer & Co).*

Chinese Bronze. Alloys of from 72.5-74 per cent copper, 15-18.5 per cent lead, 10-14 per cent zinc, and 1-5 per cent tin. One alloy, also known as Chinese bronze, contains 78 per cent copper and 22 per cent tin.†

Chinese Cinnamon Oil. See OIL OF CHINESE CINNAMON.

Chinese Glue. Shellac dissolved in alcohol. Used for joining wood, earthenware, or glass.†

Chinese Green. See SAP GREEN.

Chinese Isinglass. See AGAR-AGAR.

Chinese Jute. A commercial fibre derived from *Abutilon avicennoe* (Indian mallow). It is not a jute.†

Chinese Nickel Silver. See NICKEL.

Chinese Orange. A brownish-orange lake pigment made from an alizarin derivative precipitated on an aluminium base.†

Chinese Red. See CINNABAR and CHROME RED.

Chinese Scarlet. See CHROME RED.

Chinese Silver. (Peru Silver). A German silver, containing a little aluminium.†

Chinese Tallow. (Vegetable Tallow). A waxy substance obtained from the outer coating of the fruit of *Stillingia sebifera*, of China. It has a specific gravity of 0.918-0.922, a solidifying point of 24-34° C., a saponification value of 179-206, and an acid value of 2.4. See also STILLINGIA OIL.†

Chinese Vermilion. See SCARLET VERMILION.

Chinese Wax. (Vegetable Spermaceti, Tree Wax). Also wrongly called Japanese wax. Insect wax obtained from the insect *Coccus ceriferus*, or *C. pela*, which deposits wax on certain trees. The wax has a melting-point of 65° C., a saponification value of 92.9, an acid value of 13, and an iodine value of 15.2.†

Chinese White. See ZINC WHITE. The name is sometimes applied to barium sulphate.

Chinese White Copper. An alloy of 40 per cent copper, 31 per cent nickel, 25 per cent zinc, and 2 per cent iron.†

Chinese Wood Oil. (Wood Oil). Tung oil obtained from the seeds of *Aleurites* species in China and Japan. It has a specific gravity of 0.936-0.944 at 15° C., a refractive index 1.510-1.525 at 15° C., a saponification value of 188-197, and an iodine value 154-176. A drying oil which has been used in the manufacture of linoleum.†

Chinese Yellow. (Persian Yellow, Spanish Yellow, Royal Yellow). Arsenic Trisulphide. A pigment.†

Chinine. Quinine.†

Chinolin. See LEUCOLINE.

Chinosol. See QUINOSOL.

Chio Turpentine. See CHIAN TURPENTINE.

Chiorazol Brown B. A British dyestuff. It is equivalent to Diamine brown B.†

Chiorophenine Orange R and GO. Intermediate reduction products of curcuphenine.†

Chios Turpentine. See CHIAN TURPENTINE.

Chisso-rite. A proprietary synthetic resin made by condensing formaldehyde and acid oils from low temperature coal carbonization.†

Chizeuilite. See ANDALUSITE.

Chlor Cresol Green. Tetrachlor-*m*-cresolsulphonphthalein.†

Chloractil. A proprietary preparation of CHLORPROMAZINE hydrochloride. A tranquillizer. (DDSA).†

Chloral. Trichloroacetaldehyde,$Cl_3.CCHO$.†

Chloral Betaine. Chloral hydrate - betaine adduct. Somilan.†

Chloral Hydrate. The covalent hydrate of chloral, $Cl_3CCH(OH)_2$. A hypnotic.†

Chloral Iodine. A solution of chloral hydrate (50 grams in 20 cc water) saturated with iodine. Used for the detection of starch grains.†

Chloralamide. Chloral formamide, $CCl_3.CH(OH).NH.CHO$. A mild hypnotic and sedative.†

Chloralurethane. Carbochloral.†

Chloramide. See CHLORALAMIDE.

Chloramide of Mercury. See WHITE PRECIPITATE.

Chloramine B. Sodium benzenesulphochloroamide, $C_6H_5\ SO_2\ NNaCl+2H_2O$. Used like Chloramine T.†

Chloramine Blue 3G and HW. Polyazo dyestuffs which dye mixed fabrics. The brand 3G gives a greenish-blue on unmordanted cotton, and HW, a blackish- blue.†

Chloramine Orange. See MIKADO ORANGE G and 4R.

Chloramine T. (Tochlorine, Tolamine, Chloramine-Heyden, Pyrgos, Mianin, Aktivin). Sodium *p*-toluenesulpho-chloramide,$CH_3C_6H_4\ (SO_2NClNa)\ 3H_2O$. An antiseptic used in medicine. Solutions of Chloramine T are also used as detergents and bleaching agents. Zauberin, Mannolit, Gansil, Glekosa, Purus, and Washington Bleach are trade names for washing and bleaching materials containing Chloramine T as the active principle.†

Chloramine Yellow. (Chlorophenine Y, Diamine Yellow A, Diamine Fast

Yellows B, C, and FF, Oxyphenine, Oxyphenine Gold, Thiophosphine J, Columbia Yellow). Oxidation products of dehydrothiotoluidinesulphonic acid or the latter and primuline together. Dye cotton from a neutral bath, and silk from an acid bath, yellow.†

Chloramine-Heyden. See CHLORAMINE T.

Chloranil. Tetrachlorquinone, $C_6Cl_4O_2$.†

Chlorantine. Dyestuffs fast to light. (Clayton Aniline Co).†

Chlorarsine. Cacodyl, $As_2(CH_3)_4$.†

Chlorazoi Black FFH. A dyestuff. It is a British equivalent of Columbia black FF extra.†

Chlorazol. Direct cotton dyestuffs. (ICI PLC).*

Chlorazol Black BH. A dyestuff. It is a British brand of Diamine black BH.†

Chlorazol Black SD. A dyestuff. It is a British brand of Zambesi black D.†

Chlorazol Blue B. A dyestuff. It is a British brand of Diamine blue 2B.†

Chlorazol Blue R and 3G. A dyestuff. Dyes cotton blue from a salt bath.†

Chlorazol Blue RW. A dyestuff. It is a British equivalent of Chicago blue RW.†

Chlorazol Blue 2R. A dyestuff. It is a British brand of Diamine blue BX.†

Chlorazol Blue 3B. A British dyestuff. It is equivalent to Diamine blue 3B.†

Chlorazol Blue 6G. See DIAMINE SKY BLUE.

Chlorazol Brown GM. A dyestuff. It is a British equivalent of Benzo brown G.†

Chlorazol Brown LF. A dyestuff. It is a British equivalent of Trisulphone brown B.†

Chlorazol Brown M. A dyestuff. It is a British brand of Diamine brown M.†

Chlorazol Brown 2G. A dyestuff. It is a British brand of Trisulphone brown 2G.†

Chlorazol Deep Brown. A dyestuff identical with Trisulphone brown B, except that the end component is *m*-tolylene-diamine.†

Chlorazol Diazo Blue 2B. A dyestuff. It is a British equivalent of Diaminogen blue BB.†

Chlorazol Dyes. A registered trade mark applied to certain dyestuffs.†

Chlorazol Fast Orange D. A dyestuff. It is a British brand of Mikado orange.†

Chlorazol Fast Orange R. A British dyestuff. It is equivalent to Benzo fast orange S.†

Chlorazol Fast Pink BK. A dyestuff. It is a British equivalent of Benzo fast pink 2BL.†

Chlorazol Fast Red FG. A dyestuff. It is a British equivalent of Diamine fast red.†

Chlorazol Fast Scarlet 4BS. A dyestuff. It is a British brand of Benzo fast scarlet 4BS.†

Chlorazol Fast Yellow 5GK. A British equivalent of Cotton yellow (The Badische Co.)†

Chlorazol Green BN. A dyestuff. It is a British equivalent of Diamine green B.†

Chlorazol Green G. A dyestuff which is the same as Diamine green G (Cassella). Dyes unmordanted cotton, green.†

Chlorazol Orange RN. A British dyestuff. It is equivalent to Benzo orange R.†

Chlorazol Pink Y. A dyestuff. It is a British brand of Rosophenine 10B.†

Chlorazol Sky Blue FFS. A dyestuff identical with Benzo sky blue, Chicago blue 6B, and Diamine sky blue FF. See DIAMINE SKY BLUE.†

Chlorazol Violet N. A dyestuff. It is a British equivalent of Diamine violet N.†

Chlorazol Violet WBX. A dyestuff. It is a British equivalent of Trisulphone violet B.†

Chlorazol Yellow GX. A dyestuff. It is a British brand of Direct yellow R.†

Chlorazone. See CHLORAMINE T.

Chlorbutol. See CHLORETONE.

Chlorcahücit. A German explosive.†

Chlorcosan. A liquid chlorinated paraffin wax.The chlorine content varies from 27- 35 per cent.†

Chlordispel. Dishwashing detergent. (The Wellcome Foundation Ltd).

Chloresium. Chlorophyllin copper complex obtained from chlorophyll by replacing the methyl and phytyl ester groups with alkali and the magnesium with copper. Applications: Deodorant. (Rystan Company Inc).

Chloretone. Chlorobutanol. Applications: Pharmaceutic aid. (Parke-Davis, Div of Warner-Lambert Co).

Chlorez. Chlorinated paraffin, solid. Applications: Flame retardant in plastics, rubber, coatings etc. (Dover Chemical Corp).*

Chlorfenvinphos. An insecticide. 2-Chloro-1-(2,4-dichlorophenyl)-vinyl diethyl phosphate. BIRLANE.†

Chlorhydrol. Aluminium chlorohydrate. Applications: Anhidrotic. (Reheis Chemical Co).

Chloride of Lime. See BLEACHING POWDER.

Chloride of Lime, Liquid. See BLEACH LIQUOR.

Chlorin. See DINITROSO-RESORCIN.

Chlorina. Chloramine.†

Chlorinated Calcium Oxide. See BLEACHING POWDER

Chlorinated Lime. See BLEACHING POWDER.

Chlorinated Rubber. See DETEL, TORNESIT.

Chloritane. Sodium chlorite. (Interox Chemicals Ltd).

Chlormytol. A proprietary preparation of chloramphenicol and prednisolone used in dermatology as an antibacterial agent. (Parke-Davis).†

Chloroben. *o*-Dichlorobenzene. Used in sewage treatment.†

Chlorobromal. Chlorodibromo-acetaldehyde, $CCl.Br_2.CHO$.†

Chlorobromhydrin. α-Chloro-α-bromo isopropyl alcohol, $CH_2Br.CH\ (OH).CH_2Cl$.†

Chlorobromoform. Chlorodibromo methane, $CHCl.Br_2$.†

Chlorofin 42. Chlorinated paraffin extender for vinyl plastics containing 40 per cent. chlorine. (Hercules Inc).*

Chlorofin 70. A proprietary trade name for a chlorinated paraffin extender for vinyl plastics containing 70 per cent chlorine.†

Chloroform, Nitro. See CHLOROPICRIN.

Chloromycetin. Chloramphenicol. Applications: Antibacterial; antirickettsial. (Parke-Davis, Div of Warner-Lambert Co).

Chloromycetin Intramuscular. A proprietary preparation containing Chloramphenicol. An antibiotic. (Parke-Davis).†

Chloromycetin Kapseals. A proprietary preparation containing Chloramphenicol. An antibiotic. (Parke-Davis).†

Chloromycetin Palmitate Suspension. A proprietary preparation containing chloramphenicol. An antibiotic. (Parke-Davis).†

Chloromycetin Pure. A proprietary preparation containing chloramphenicol. An antibiotic. (Parke-Davis).†

Chloromycetin Succinate. A proprietary preparation containing chloramphenicol, (as the sodium salt of the monosuccinic ester). An antibiotic. (Parke-Davis).†

Chloromycetin Suppositories. A proprietary preparation containing chloramphenicol. An antibiotic. (Parke-Davis).†

Chloroneb Systemic Flowable Fungicide. 30% chloroneb. Applications: Seed treatment for control of *rhizoctonia solani*, *pythium* spp and *sclerotium rolfsii* on cotton, beans, soybeans and sugar beets. (Kincaid Enterprises Inc).*

Chloroneb 65W Fungicide. 65% chloroneb wettable powder. Applications: Seed treatment to suppress seeding blights, soreshin and pre- and post-emergence damp-off caused by *rhizoctonia solani*, *pythium* spp and *sclerotium rolfsii* on cotton, beans, soybeans and sugar beets. (Kincaid Enterprises Inc).*

Chlorophenine G. A dyestuff similar to Oxyphenine, Oxyphenine gold, and Thiophosphine J. Its properties and application are the same as Chloramine yellow.†

Chlorophenine Orange RR and RO. A dyestuff. Dyes cotton bright orange.†

Chlorophenine Y. See CHLORAMINE YELLOW.

Chlorophyll. (Leaf Green, Chromule). The green colouring matter of plants, leaves, and stalks. It is a magnesium compound, and has been used for colouring confectionery and liqueurs.†

Chloropicrin. (Nitrochloroform). Trichloronitromethane, $CCl_3(NO_2)$.†

Chloroprene. 1:3-chlor-2-butadiene. Used as a protective coating and in synthetic rubber manufacture by polymerization. The polymerized product bears the name of Neoprene.†

Chloropryl. See ISOPRAL.

Chloroptic. Chloramphenicol. Applications: Antibacterial; antirickettsial. (Allergan Pharmaceuticals Inc).

Chloropyramine. Halopyramine.†

Chloros. Disinfectant containing sodium hypochlorite. (ICI PLC).*

Chlorosoda. A proprietary form of solidified sodium hypochlorite for use as a bleaching agent. A small proportion of a saturated fatty acid, such as lauric acid, is incorporated.†

Chlorotex. Reagent for estimating residual chlorine in water. (BDH Chemicals Ltd).

Chlorothalonil. Tetrachloroisophthalonitrile. Applications: Raw material for fungicides. (SNIA (UK) Ltd).†

Chlorothene. A line of inhibited 1,1,1,-trichloroethane solvents used in industry. (Dow Chemical Co Ltd).*

Chlorothene (VG). An inhibited grade of 1-1-1-trichlorethane. Degreasing solvent. (Dow Chemical Co Ltd).*

Chlorovin. A registered trade mark for polyvinyl chloride.†

Chlorowax. A trade mark for chlorinated paraffins. (US Industrial Chemical Corp).†

Chlorowax LV. A chlorinated paraffin. A vinyl plasticizer. (US Industrial Chemical Corp).†

Chlorowax 40. Liquid chlorinated paraffin. A vinyl plasticizer. (US Industrial Chemical Corp).†

Chlorowax 50. Chlorinated paraffin (50 per cent. chlorine). A vinyl plasticizer. (US Industrial Chemical Corp).†

Chlorowax 70. Resinous chlorinated paraffin. A vinyl plasticizer. (US Industrial Chemical Corp).†

Chloroxethose. Hexachlorodivinyl ether $(CCl_2{:}CCl)_2O$.†

Chlorozone. A bleaching liquor prepared by passing chlorine into caustic soda.†

Chlorquinaldol. 5,7- Dichloro-8-hydroxy-2-methylquinoline. Steroxin.†

Chlor-Tabs. Effervescent chlorine tablets. (PPF International Ltd).*

Chlor-tetragnost. Sodium tetrachlorophenolphthalein.†

Chlor-Trimeton. Chlorpheniramine maleate. Applications: Antihistaminic. (Schering-Plough Corp).

Chloryl. Ethyl chloride, C_2H_5Cl, used as an anaesthetic.†

Chlorylen. Trichlorethylene.†

Chlotride. Chlorothiazide. Applications: For the relief of oedema, oedema accompanying premenstrual tension and control of hypertension. (Merck Sharp & Dohme).*

Chlumin. An aluminium alloy resistant to sea water. It contains chromium, a few per cent of magnesium, and iron.†

Chobile. A proprietary extract of ox bile with oxidized oxbile acids, used as a stimulant to bile secretion. (Mallinckrodt Inc).*

Cholan DH. Dehydrocholic acid. Applications: Choleretic. (Pennwalt Corp).

Cholebrine. Iocetamic acid. Applications: Diagnostic aid. (Mallinckrodt Inc).

Choledyl. Oxtriphylline. Applications: Bronchodilator. (Parke-Davis, Div of Warner-Lambert Co).

Cholesterol. Cholesterol. (Croda Chemicals Ltd).

Cholestrophane. Dimethylparabanic acid, $C_5N_6N_2O_3$.†

Cholestyramine. A Styryl-divinylbenzene copolymer (about 2 per cent. divinylbenzene) containing quaternary ammonium groups. Cuemid.†

Choletec. Mebrofenin. Applications: Diagnostic aid. (E R Squibb & Sons Inc).

Cholografin. Iodipamide. Applications: Pharmaceutic necessity for iodipamide meglumine injection. (E R Squibb & Sons Inc).

Cholografin Meglumine. Iodipamide meglumine. Applications: Diagnostic aid. (E R Squibb & Sons Inc).

Choloxin. Dextrothyroxine sodium. Applications: Antihyperlipoproteinemic. (Flint Laboratories, Div of Travenol Laboratories Inc).

Cholumbrin. A proprietary preparation of sodium tetraiodophenolphthalein.†

Choron. Gonadotropin, chorionic. Applications: Gonad-stimulating principle. (O'Neal, Jones & Feldman Pharmaceuticals).

CHP. *N*-Cyclohexyl-*s*-pyrrolidone. Applications: Solvent, reaction intermediate, textile auxilliary, cosmetic ingredient. (GAF Corporation).*

Chrismaline. See OZOKERINE.

Christofle. See NICKEL SILVERS.

Christolit. A proprietary plastic of the phenol-formaldehyde type.†

Christophite. Synonym for Marmatite.†

Chrogo U42. An alloy of 40 per cent gold, 45 per cent copper, 14 per cent nickel, 1 per cent chromium, and traces of platinum. A dental alloy.†

Chroma-Chem. Colourant dispersions. Applications: Colouring of industrial coating compositions. (Nueodex Inc).*

Chromacethin Blue. A dyestuff obtained by the condensation of gallo-cyanines with aromatic alkylated amines.†

Chromagan. Nickel-chromium steel for the manufacture of table ware.†

Chromagen. Ferrous fumarate USP, ascorbic acid USP, cyanocobalamin USP, desicated stomach substance. Phosphorous free vitamin and mineral dietary supplement. Applications: Indicated for the treatment of all anaemias responsive to oral iron therapy. Use during pregnancy and lactation. (Altana Inc).*

Chromalay. Materials for thin layer chromatography. (May & Baker Ltd).*

Chromalbin. Albumin, chromated Cr51 Serum. Applications: Radioactive agent. (E R Squibb & Sons Inc).

Chromaline. A chrome mordant made by reducing chromic acid with glycerin. Used for printing chrome colours on wool.†

Chromaloy. Nickel-chromium-iron alloys. The specific gravity varies from 8.15-8.35, and the melting-point 1360-1390° C.†

Chromaluminium. An alloy of aluminium, chromium, and other metals. It has a specific gravity of 2.9.†

Chroman B. A Rohn alloy containing 64 per cent nickel, 20 per cent iron, 15 per cent chromium, and 1 per cent manganese.†

Chroman Co. A Rohn alloy containing 79 per cent nickel, 20 per cent. chromium, and 1 per cent manganese.†

Chromanil Black BF, 2BF, 3BF, RF, 2RF. Direct cotton blacks obtained after treatment with potassium bi-chromate.†

Chromargans. Non-oxidizing steels, contain chromium. Non-magnetic, used for the manufacture of turbine blades, valves, cutlery, etc.†

Chromastral. Pigments for paints and plastics. (ICI PLC).*

Chromatised Gelatin. (Chrome cement, chrome glue). Made by adding 1 part potassium dichromate to 5 parts of a solution (5-10 per cent. of gelatin. A cement for glass.†

Chromatogram. Materials for thin layer chromatography. (Kodak Ltd).

Chromax. An electrical resistance alloy containing 75 per cent nickel and 25 per cent chromium.†

Chromax Bronze. An alloy of 15 per cent nickel, 67 per cent copper, 12 per cent zinc, 3 per cent aluminium, and 3 per cent chromium.†

Chromazone Blue R. A dyestuff. Gives blue shades on chromed wool.†

Chromazone Red A. A dyestuff. Dyes wool from an acid bath.†

Chromazurines. Dyestuffs similar to Delphine blue.†

Chromazurol S. A dyestuff. It is *o*-chlorobenzaldehydesulphonic acid, condensed with salicylic acid, and oxidized.†

Chrombral. Flux for copper/chromium alloys. (Foseco (FS) Ltd).*

Chrome. (Acid Alizarin, Acid Anthracene, Diamond Salicine). Acid chrome colours for wool.See BICHROME.†

Chrome Alum. Potassium chromium alum, $Cr_2(SO_4)_3$ K_2SO_4 $24H_2O$.†

Chrome Amalgam. An alloy of chromium and mercury obtained by the electrolysis of chromic chloride, using a mercury cathode.†

Chrome Black. Anhydrous copper chromate.†

Chrome Black A. A dyestuff. It is a British equivalent of Eriochrome black A.†

Chrome Bordeaux. A mordant dyestuff, used for printing with acetate of chromium mordants, giving claret-red shades.†

Chrome Bordeaux 6B Double. A dyestuff similar to Chrome bordeaux.†

Chrome Bronze. Crystalline chromium oxide obtained by heating potassium bichromate with sodium chloride or in a stream of hydrogen. It has a metallic sheen, a specific gravity of 5.61, and cuts glass.†

Chrome Brown. Manganese chromate.†

Chrome Brown P. A dyestuff prepared by the action of picramic acid upon *m*-aminophenol.†

Chrome Brown PA. A dyestuff. It is *p*-nitrophenylazopyrogallol.†

Chrome Brown RO. See FAST BROWN N.

Chrome Brown RR. A dyestuff. Dyes chrome wool brown, and cotton reddish-brown with a chrome mordant.†

Chrome Carmine. See CEROME RED.

Chrome Cinnabar. See CHROME RED.

Chrome Die Steel. See CHROME STEELS.

Chrome Emerald Green. See CHROMIUM GREEN.

Chrome Fast Black A. A dyestuff. It is a British brand of Eriochrome black A.†

Chrome Fast Black FW. A British dyestuff. It is equivalent to Diamond black F.†

Chrome Fast Brown FC. A dyestuff. It is a British equivalent of Anthracene brown R.†

Chrome Fast Cyanine B, BN. Dyestuffs. They are British brands of Palatime chrome blue.†

Chrome Fast Cyanine G. (Neochrome Blue-black B, Solochrome Black 6B, Fast Chrome Cyanine, Alliance Chrome Blue-black B, Eriochrome Blue-black B). A dyestuff prepared from diazotized 1-amino-2-naphtholsulphonic acid, and α-naphthol.†

Chrome Fast Yellow. See MILLING YELLOW.

Chrome Fast Yellow G. A dyestuff. Dyes Chromed wool yellow.†

Chrome Fast Yellow 0. A dyestuff. It is a British equivalent of Milling yellow.†

Chrome Garnet. See CHROME RED.

Chrome Green. Tetramethyldiaminotriphenylcarbinol-*m*-carboxylic acid, $C_{24}H_{25}N_2O_3$, is known as Chrome green. It dyes chromed wool green, also used in cotton printing. The name has also been used for various other pigments.†

Chrome IA. An alloy of 80 per cent nickel and 20 per cent chromium.†

Chrome Iron. (Ferrochrome). An alloy of iron and chromium, usually containing from 62-68 per cent chromium. Used in the manufacture of chrome steel.†

Chrome-nickel Steel, High. These alloys usually contain 70-75 per cent iron, 17-20 per cent chromium, and 8-10 per cent nickel.†

Chrome Orange. See CHROME RED.

Chrome Oxide Green. See CHROME GREENS.

Chrome Patent Blacks TG, TB, T, TR. Wool dyestuffs which are employed in an acid bath.†

Chrome Prune. A mordant dyestuff. It gives claret shades with chrome mordants.†

Chrome Red. (Austrian Cinnabar, Chinese Red, Persian Red, Victoria Red. Vienna Red, Derby Red, Chrome Cinnabar, Chrome Orange, American Vermilion, Chrome Garnet, Chrome Ruby, Chrome Carmine, Chinese Scarlet). Pigments consisting of basic lead chromate, $PbCrO_4$ $Pb(OH)_2$, or a mixture of this substance with chrome yellow.†

Chrome Red R. A mordant dyestuff, used in calico printing, with chromium acetate. It gives red shades.†

Chrome Ruby. See CHROME RED.

Chrome Steels. These steels represent a range of alloys containing from 0.2-2 per cent carbon and from 0.5-15 per cent chromium. Ball-bearing steel contains 1 per cent carbon and 1 per cent chromium, chrome die steel contains 2 per cent carbon and 12 per cent chromium, and stainless steel 0.2-0.4 per cent carbon and 11-13 per cent chromium.†

Chrome Steels, High. These usually contain from 64-81 per cent iron and 18-35 per cent chromium.†

Chrome-tin Pink. Obtained by calcining a mixture of stannic oxide and a small amount of chromic oxide.†

Chrome Violet. See MAUVEINE.

Chrome Violet. A dyestuff. Dyes chromed wool violet, and is used in calico printing.†

Chrome Violet. A dyestuff. It is the sodium salt of aurine-tricarboxylic acid, $C_{22}H_{13}O_{10}Na_2$. Gives a reddish-violet colour in calico printing with chrome mordants.†

Chrome Yellow. A pigment. It is normal lead chromate, $PbCrO_4$. The compounds, $PbCrO_4.PbSO_4$ and $PbCrO_4$ $2PbSO_4$, are also known as chrome yellow.†

Chrome Yellow D. See MILLING YELLOW.

Chrome Yellow GG. A dyestuff prepared from diazotized *o*-anisidine and salicylic acid.†

Chromeduol. Concentrated chrome sulphate powders. Applications: Tanning, mineral dyeing and electrolytic purposes. (Lancashire Chemical Works Ltd).*

Chromels. Nickel/chromium or nickel/chromium/iron alloys.They are used for heating elements, and as moulds for glass.†

Chrometan. Chrome tanning powders. (British Chrome & Chemicals Ltd).

Chrometrace. Chromic chloride hexahydrate. Applications: Supplement. (Armour Pharmaceutical Co).

Chromglaserite. A double salt, $3K_2CrO_4$ Na_2CrO_4.†

Chromidium. A nickel-chromium cast iron.†

Chromiform. A reagent used for the preservation of milk samples. It consists of pastilles containing 0.25 gram potassium dichromate and 0.25 gram trioxymethylene.†

Chromine G. See THIOFLAVINE S.

Chromitope Sodium. Sodium Chromate Cr 51. Applications: Diagnostic aid; radioactive agent. (E R Squibb & Sons Inc).

Chromium - cobalt - molybdenum Steel. See COBALT-CHROMIUM-MOLYBDENUM STEEL.

Chromium Bronze. The term applied to copper-zinc or copper-tin alloys to which chromium has been added up to 5 per cent.†

Chromium Copper. An alloy of copper and chromium, Containing 10 per cent chromium. Used in the manufacture of hard steels. Also added to increase elasticity.†

Chromium Green. (Guignet's Green, Pannetier Green, Arnaudon Green, Matthieu-Plessy Green, Chrome Emerald Green, Mittler's Green, Permanent Green, Emerald Green, Veridian, Chrome Green, Grüne's chromoxyd, French Veronese green). Green pigments consisting of hydrated sesquioxide of chromium, with phosphate or borate of chromium. Also see CHROME GREENS.†

Chromium Manganese. An alloy containing 30 per cent chromium and 70 per cent manganese. Used in the manufacture of hard steels. Also added to copper to increase elasticity.†

Chromium Mica. See FUCHSITE.

Chromium Molybdenum. An alloy of 50 per cent chromium and 50 per cent molybdenum. Used in the manufacture of hard steels.†

Chromium-molybdenum Steel. Alloys containing from 0.06-1.2 per cent carbon, traces to 15 per cent molybdenum, and traces to 6 per cent chromium.†

Chromium Nickel. An alloy of 10 per cent chromium and 90 per cent nickel, also 50 per cent chromium, and 50 per cent nickel. Used in the manufacturce of hard steels.†

Chromium Oxide, Green. See GREEN OXIDE OF CHROMIUM.

Chromium Oxide, Red. See RED OXIDE OF CHROMIUM.

Chromium Sandoz. A proprietary preparation of chromium sesquioxide. (Sandoz).†

Chromium Steels. See CHROME STEELS.

Chromium-vanadium Steel. An alloy of this type contains O.3-0.4 per cent carbon, 1-1.5 per cent chromium, and O.15-0.25 per cent vanadium.†

Chromium-vanadium-molybdenum Steels. Alloys of iron Containing 0.1-0.55 per cent carbon, 0.22-1.45 per cent molybdenum 0.8-1.5 per cent chromium, and 0.15-0.45 per cent vanadium.†

Chromocyanine V and B. (Blue PRC, Brilliant Gallocyanine). Sulphonic acids of leuco-gallocyanines. They dye violet to blue shades with chrome mordants. Employed in calico printing.†

Chromocyanines. See INDALIZARIN.

Chromoferrite. See CHROME IRON ORE.

Chromogen C, LL. See CHROMOTROPE ACID.

Chromogen I. See CHROMOTROPE ACID.

Chromol. Metachrome dyestuffs. (James Robinson & Co Ltd).†

Chromolay. Materials for thin layer chromatography. (May & Baker Ltd).

Chromo-nitric Acid. See PERENYI'S FLUID.

Chromosal B. A mineral tanning material containing chromic oxide. (Bayer & Co).*

Chromotrope Acid. (Chromogen I, Chromogen C, LL). Dihydroxy-naphthalene-disulphonic acid. An acid dye for wool.†

Chromotrope FB. See AZORUBINE S.

Chromotrope 2B. A dyestuff. It is the sodium salt of *p*-nitro-phenylazo-1,8-dihydroxynaphthalene-3,6-disulphonic acid. Dyes wool from an acid bath, bluish-red, becoming blue to black on chroming.†

Chromotrope 2R. (Biebrich Acid Red 4B, Lighthouse Chrome Blue B, XL Carmoisine 6R). A dyestuff. It is the sodium salt of phenylazo-1,8-dihydroxynaphthalene-3,6-disulphonic acid. Dyes wool bluish violet.†

Chromotrope 6B. A dyestuff. It is the sodium salt of *p*-acetaminophenylazo-1,8-dihydroxynaphthalenedisulphonic acid. Dyes wool violet-red from an acid bath.†

Chromovan Steel. A proprietary trade name for a non-sparking tool steel containing 12.5 per cent chromium, 0.8 per cent molybdenum, 1 per cent vanadium, and 1.6 per cent carbon.†

Chromoxan Colours. Dyestuffs. They are aldehydes of the naphthalene series, condensed with hydroxyacids of the benzene series, and oxidized.†

Chronin. An alloy of 84 per cent nickel and 15 per cent chromium.†

Chronite. Heat-resisting alloys containing 63-67 per cent nickel, 13-16 per cent chromium, 12-20 per cent iron, 0-0.4 per cent silicon, 0.1 per cent manganese, and 0-0.8 per cent aluminium.†

Chronne Blue. (Neochrome Blue-black R). A dyestuff. Dyes chromed wool blue, but is chiefly used for cotton printing.

Also the name for a chromium-silicon-phosphate, obtained by fusing a mixture of potassium bichromate, fluorspar, and silica.†

Chronogyn. Danazol. Applications: Anterior pituitary suppressant. (Sterling Drug Inc).

Chronulac. Lactulose. Applications: Laxative. (Merrell Dow Pharmaceuticals Inc, Subsidiary of Dow Chemical Co).

Chronulac. Lactulose. Applications: Treatment of obstipation and (pre-) hepatic coma. (Duphar BV).*

Chrysamin R. A dyestuff. Dyes cotton yellow from a soap bath.†

Chrysaureine. See ORANGE II.

Chrysazin. Dihydroxy anthraquinone.†

Chrysazol. Dihydroxyanthracene.†

Chryseinic Acid. See CAMPOBELLO YELLOW.

Chryseoline . See TROPAEOLINE 0.

Chryseoline Yellow. See TROPAEOLINE O.

Chrysin. 1, 3-Dihydroxyflavone. A pigment found in poplar buds.†

Chrysine. See OZOKERINE.

Chrysitin. See MASSICOT.

Chrysocale. A jeweller's alloy, containing 9 parts copper, 8 parts zinc, and 2 parts lead.†

Chrysochalk. (Gold-copper). An alloy containing 59-93 per cent copper, 8-39 per cent zinc, and 1.6-1.9 per cent lead. A jeweller's alloy.†

Chrysoform. Dibromodiiodohexa-methylenetetramine, $C_6H_8Br_2I_2N_4$. Antiseptic.†

Chrysoidine Crystal. A dyestuff. It consists of the hydrochloride of phenylazo-*m*-phenylenediamine, with some of the homologues from *o*- and *p*- toluidine. Dyes wool and silk orange.†

Chrysoidine R. (Cerotine Orange C Extra, Gold Orange for Cotton). A dyestuff. Dyes tannined cotton brown. It is the hydrochloride of phenylazo-*m*-tolylenediamine.†

Chrysoidine R, R Powder, R Crystals, Y Powder, Y Crystals, YRP, YL, Y, RL, Y, 1606, Pure Crystals, Supra Crystals. Dyestuffs. They are British equivalents of Chrysoidine.†

Chrysoine. See TROPAEOLINE O.

Chrysoine Extra. A dyestuff. It is a British equivalent of Tropaeoline O.†

Chrysoline. The sodium salt of benzyl fluoresceine. Dyes silk yellow. See TROPAEOLINE O.†

Chrysorin. An alloy of 66 per cent copper and 34 per cent zinc.†

Chrysotile. (Chrysotile Asbestos, Canadian Asbestos). An asbestos mineral, the average composition of which is 40.5 per cent silica, 41.5 per cent magnesium oxide, 14 per cent water, 2.6 per cent iron oxides, and 1.3 per cent aluminium oxide. It yields the best fibre for spinning, and is considered the best type of asbestos.†

Chrysotile Asbestos. See CHRYSOTILE.

Chymacort. A proprietary preparation of hydrocortisone acetate, neomycin palmitate and pancreatic enzymes, used in dermatology. (Armour Pharmaceutical Co).†

Chymar. A proprietary preparation of alpha-chymotrypsin. Treatment of wounds. (Armour Pharmaceutical Co).†

Chymar Ointment. A proprietary preparation of hydrocortamate hydrochloride, neomycin palmitate and pancreatic enzymes, used in dermatology. (Armour Pharmaceutical Co).†

Chymex. Bentiromide. Applications: Diagnostic aid. (Adria Laboratories Inc).

Chymocyclar. A proprietary preparation of tetracycline, trypsin and chymotrypsin. Antibiotic. (Armour Pharmaceutical Co).†

Chymoral. A proprietary preparation of trypsin and chymotrypsin. Digestive enzymes. (Armour Pharmaceutical Co).†

Chymosin. Pepsin.†

CI 336. Carbochloral.†

Cialit. The sodium salt of 2-ethyl mercury mercaptobenzoxazol-5-carboxylic acid. An antifouling pigment.†

Ciba. Vat dyes. (Ciba-Geigy PLC).

Ciba Blue B. A dyestuff. It is 5,7,5'-tribromindigo.†

Ciba Blue 2B. A dyestuff. It is tetrabromoindigo.†

Ciba Bordeaux. A dyestuff. It is 5,5'-dibromothioindigo.†

Ciba Green G. A dyestuff. It is dibromoβ-naphthindigo.†

Ciba Hellotrope B. A dyestuff. It is tetrabrom-indirubin.†

Ciba Red B. A dyestuff. It is 6, 6'-dichlorothioindigo.†

Ciba Red G. A dyestuff. It is dibromothioindigoscarlet.†

Ciba Scarlet G. A dyestuff obtained by the condensation of thioindoxyl with acenaphthenequinone.†

Ciba Violet A, B, and R. A dyestuff. It is halogenated thioindigo.†

Ciba Yellow G. A dyestuff. It is the dibromo derivative of Indigo yellow 3G.†

Ciba Yellow 5R. A dyestuff prepared from Ciba yellow G by reduction.†

Ciba 1906. A proprietary preparation containing thiambutosine. (Ciba Geigy PLC).†

Cibacet. Disperse dyes. (Ciba-Geigy PLC).

Cibacrolan. Reactive dyestuffs. (Clayton Aniline Co).†

Cibacron. Reactive dyes. (Ciba-Geigy PLC).

Cibalan. Wool dyestuffs. (Clayton Aniline Co).†

Cibalith-S. Lithium citrate tetrahydrate. Applications: Antimanic. (Ciba-Geigy Corp).

Cibamin. Modified melamine/formaldehyde resins. (Ciba-Geigy PLC).

Cibanite. A proprietary trade name for an aniline-formaldehyde resin resistant to water, oil, alkalis, and organic solvents.†

Cibanoid. A proprietary trade name for urea-formaldehyde synthetic resins.†

Cibanone. Vat dyes. (Ciba-Geigy PLC).

Cibanone Black B, 2B. A dyestuff prepared by fusing 2-methyl-benzanthrone with sulphur.†

Cibanone Blue 2G. A dyestuff prepared in a similar way to Cibanone black.†

Cibanone Brown. A dyestuff. It is amino-2-methylanthraquinone fused with sulphur.†

Cibanone Green. A dyestuff prepared as Cibanone black.†

Cibanone Orange R. A dyestuff prepared from a dichloromethylanthraquinone.†

Cibanone Yellow R. A dyestuff prepared from chloromethylanthraquinone.†

Cibaphasol 6042. A proprietary trade name for a fatty acid amide derivative giving improved quality and appearance to continuous dyeings. Used as a dyeing auxiliary for polyacrylonitrile fibres. (Ciba Geigy PLC).†

Cibatex PA. A proprietary trade name for a synthetic tanning agent of the phenol sulphonic acid type used as an improver of wet fastness in the dyeing of nylon. (Ciba Geigy PLC).†

Cibatex 248. A proprietary trade name for a phenol sulphonic acid derivative which gives improved wet fastness properties of dyeings. Used also as a synthetic tanning agent for dyes on nylon 66. (Ciba Geigy PLC).†

Cicatrin. A topical antibacterial. (The Wellcome Foundation Ltd).

Cicuta. Conium.†

Cidrase. Cider yeast.†

Cignolin. 1,8-Dihydroxyanthranol.†

Cilloral. Penicillin G Potassium. Applications: Antibacterial. (Bristol-Myers Co).

Cimet. Stainless alloys containing about 48 per cent chromium with iron.†

Cimmol. Cinnamyl hydride.†

Cimolite. See PURIFIED FULLER'S EARTH.

Cinchona Bark. See JESUITS' BARK, LOXA BARK, LUGO'S POWDER, PERUVIAN BARK, and TALBOR's POWDER.

Cincophen. (Phenylcinchoninic Acid). 2-Phenylquinoline-4-carboxylic acid, $C_{16}H_{11}NO_2$.†

Cindal. An aluminium alloy containing zinc and small amounts of magnesium and chromium.†

Cindol. Ciramadol hydrochloride. Applications: Analgesic. (Wyeth Laboratories, Div of American Home Products Corp).

Cindumix. Coaltar - bitumen mixture. Applications: Surface dressing material for roads. (Cindu Chemicals BV).*

Cinereine. An induline dyestuff obtained from azoxyaniline, aniline hydrochloride, and *p*-phenylenediamine.†

Cinnabar. (Chinese Red, Mercury Blend, Vermilion, Liver Ore, Patent Red, Cinnabarite). It is mercuric sulphide, HgS, used as a pigment.†

Cinnabar Green. See CHROME GREENS.

Cinnabar of Antimony. See RED ANTIMONY.

Cinnabar Red. See COTTON SCARLET.

Cinnabarite. See CINNABAR.

Cinnabar, Austrian. See CHROME RED.

Cinnabar, Chrome. See CHROME RED.

Cinnaloid. A proprietary preparation of rescinnamine. An antihypertensive. (Pfizer International).*

Cinnamein. A term applied to benzyl cinnamate. It is, however, also used for the mixture of ester and alcohol in the Balsams of Tolu and Peru, which are not extractable by alkalies from an ethereal solution.†

Cinnamene. See STYROL.

Cinnamol. See STYROL.

Cinnamon Brown. See BISMARCK BROWN.

Cinnamon Leaf Oil. See OIL OF CINNAMON LEAF.

Cinnamon Oil. See OIL OF CINNAMON.

Cinnar. A proprietary preparation containing cinnarizine. Applications: Anti-motion sickness. (Janssen Pharmaceutical Ltd).*

Cinobac. Cinoxacin. Applications: Antibacterial. (Eli Lilly & Co).

Cinopal. Fenbufen. Applications: Anti-inflammatory. (Lederle Laboratories, Div of American Cyanamid Co).

Cinquasia. High performance organic pigments. (Ciba-Geigy PLC).

Cin-Quin. Quinidine Sulphate. Applications: Cardiac depressant. (Rowell Laboratories Inc).

Cinquinolite. See WHITE COPPERAS.

Cipralan. Cibenzoline. Applications: Cardiac depressant. (Hoffmann-LaRoche Inc).

Circacid. Dairy hygiene compound. (The Wellcome Foundation Ltd).

Circadet. Surface active detergent. (Ciba-Geigy PLC).

Circaline MK 11. Hypochlorite pipe line cleaner/sterilizer. (Ciba-Geigy PLC).

Circanol. Ergoloid mesylates. Applications: Cognition adjuvant. (Riker Laboratories Inc, Subsidiary of 3M Company).

Circosan. Dairy hygiene detergent sterilizer. (The Wellcome Foundation Ltd).

Cirin. A proprietary preparation of aspirin and ascorbic acid 6-1. An analgesic. (Zemmer Co Inc, Oakmont, Pa).†

Cirrasol. A textile softening agent, fibre lubricant or antistatic agent. (ICI PLC).*

Cismollan BH. Highly bactericidal and fungicidal soaking agent for dried and salted raw hides, outstanding wetting action. (Bayer & Co).*

Cisplatin. Injection containing Cisplatin. Applications: Antineoplastic. (Lederle Laboratories).*

Citanest. Prilocaine Hydrochloride. Applications: Anaesthetic. (Astra Pharmaceutical Products Inc).

Citarin-L. Anthelmintic against lungworms and gastrointestinal nematodes in cattle, sheep, goats and pigs. Applications: Veterinary medicine. (Bayer & Co).*

Cithrol. Polyethylene-glycol esters, used as nonionic surfactants. (Croda Chemicals Ltd).*

Cithrol GMS A/S. Acid stable glyceryl monostearate, a nonionic surfactant. Applications: Used in the cosmetic, food and pharmaceutical industries. (Croda Chemicals Ltd).*

Citobaryum. A proprietary preparation of special barium sulphate.†

Citraclean. Citric acid metal cleaning process. (John & E Sturge Ltd, Selby).

Citra-Gran. A proprietary preparation containing tartaric acid sodium bicarbonate, citric acid, potassium bicarbonate, magnesium sulphate, lithium benzoate, anhydrous sodium phosphate and calcium lactophosphate. An antacid. (AKU Holland).†

Citralka. A proprietary preparation containing sodium acid citrate. (Parke-Davis).†

Citranova. Synthetic citrus oils.†

Citrene. See HESPERIDENE.

Citrest. A citric acid ester. (Cyclo Corporation).*

Citridic Acid. See EQUISETIC ACID.

Citrin. Vitarain P. See QUERCITIN.†

Citroflex A-2. Acetyl triethyl citrate. A vinyl plasticizer. (Pfizer International).*

Citroflex A-4. Acetyl tributyl citrate. A vinyl plasticizer. (Pfizer International).*

Citroflex A-8. Acetyl tri-2-ethyl hexyl citrate. A vinyl plasticizer. (Pfizer International).*

Citroflex A2. Plasticizer acetyl triethyl citrate. (Pfizer Ltd).

Citroflex 2. Triethyl citrate. A vinyl plasticizer. (Pfizer International).*

Citroflex 4. Tributyl citrate. A vinyl plasticizer. (Pfizer International).*

Citron Yellow. See ZINC CHROME.

Citronella Oil. See OIL OF CITRONELLA.

Citronine. See NAPHTHOL YELLOW S.

Citronine A. See NAPHTHOL YELLOW S.

Citronine B or 2B. (Azo Yellow, Azo Yellow M, Azo Yellow 3G Extra Conc., Azo Acid Yellow, Azoflavine S or 2, Azoflavine 3R Extra Conc., Jaune Yellow, Indian Yellow, Indian Yellow G, New Yellow, Citronine 2AEJ, Citronine NE, Helianthin, Jasmin). An azo dyestuff. Dyes wool and silk yellow from an acid bath.†

Citronine NE. See CITRONINE B.

Citronine 2AEJ. See CITRONINE B.

Citrosodine. Sodium citrate, C_6.†

Citrozone. Vanadium sodium citrochloride.†

Claforan. Cefotaxime sodium. Applications: Antibiotic. (Roussel Laboratories Ltd).*

Clairsol. Highly dearomatised aliphatic hydrocarbon solvents. (Carless Solvents Ltd).

Clampdown. Silage treatment. (May & Baker Ltd).

Claradin. A proprietary preparation of aspirin in an effervescent base. An analgesic. (Nicholas).†

Clar-Apel. A proprietary viscose plastic used as a packing material.†

Clarcel. Diatomaceous earth filter aid. Applications: Liquids filtration in chemical, pharmaceutical and food industries. (British Ceca Co Ltd).*

Clarcel Flo. Perlite based filter aid. Applications: Liquids filtration in chemical, pharmaceutical and food industries. (British Ceca Co Ltd).*

Claret G. See BORDEAUX G.

Claret Red. See BORDEAUX R or B.

Clarfina. Synthetic zeolite. Applications: Raw material for detergents. (SNIA (UK) Ltd).†

Clarifloc. Flocculants and coagulants. Applications: Clarification of water and waste water, sludge treatment, thickening and dewatering. (Allied Corp Water Treatment).*

Clarifloc. Carrageenan. (ABM Chemicals Ltd).*

Clarin. A proprietary preparation of hydroxizine hydrochloride, ephedrine sulphate and theophylline. An antiasthmatic. (Pfizer International).*

Clarit. A particularly pure calcium bentonite for the adsorptive stabilization of wines, fruit juices, vinegar etc. (Süd-Chemie AG).*

Clarite. Pretreatment agent. (Ciba-Geigy PLC).

Clark's Patent Alloy. An alloy of 75 per cent copper, 7.2 per cent zinc, 14.5 per cent nickel, 1.9 per cent tin and 1.9 per cent cobalt.†

Clar-O-Cel. Cellulose fibre filter aids. Applications: Liquids filtration in chemical, pharmaceutical and food industries. (British Ceca Co Ltd).*

Clarosan. Aquatic herbicide. (Ciba-Geigy PLC).

Clarstabil. Clarification of wines, beers, etc. (Minas de Gador SA).*

Clarus Metal. A light aluminium alloy used for sheets and tubes.†

Clarvin. Clarification of wines, beers, etc. (Minas de Gador SA).*

Claudilithe. See GALALITH.

Claussenite. See HYDRARGILITE.

Clay, Cornish. See CHINA CLAY.

'lay, Pipe. See PIPECLAY.

'lay, Porcelain. See CHINA CLAY.

'lay, Potter's. See PIPECLAY.

'lay, White. See CHINA CLAY.

'laymaster. Polystyrene boards of low density coloured pink. Applications: Compressible fill beneath concrete foundations. (Vencel Resil Ltd).*

'laysil. A silicate adhesive. (Crosfield Chemicals).*

'layton Black D. A dyestuff prepared from nitrosophenol. It dyes cotton black in a sodium sulphide bath.†

'layton Cloth Red. (Stanley Red). A dyestuff. Dyes wool and silk red from an acid bath.†

'layton Fast Blacks. (Clayton Fast Greys). Dyestuffs which are probably sulphides or thiosulphonic acids of aniline black and its analogues. They dye cotton from a sodium sulphide bath, and when applied with caustic soda upon glucose-prepared calico, give black prints.†

'layton Fast Greys. See CLAYTON FAST BLACKS.

'layton Wool Brown. An azo dyestuff obtained from aniline, sulphanilic acid, naphthionic acid, and *m*-phenylenediamine. Dyes wool and silk brown from an acid bath.†

'layton Yellow. (Turmerine, Thiazol Yellow, Mimosa, Thiazol Yellow G, Titan Yellow). A dyestuff. Dyes cotton greenish-yellow from a salt bath.†

'LD 2. Croscarmellose sodium. Carboxymethylcellulose sodium that has been internally cross-linked. Applications: Pharmaceutic aid. (Buckeye Cellulose Corp).

'lean Wiz-9. Proprietary solvent formulation to clean and strip build-up and residue of resin and mould release from metal and composite moulds. Applications: Fibreglass, polyurethane foam and rubber. (Axel Plastics Research Laboratories Inc).*

Clean-Up. General moss-killer and sterilizer for garden use. (ICI PLC).*

Clear by Design. Benzoyl penoxide. Applications: Keratoytic. (Herbert Laboratories, Dermatology Div of Allergan Pharmaceuticals Inc).

Clear Eyes. Naphazoline hydrochloride. Applications: Adrenergic. (Abbott Laboratories).

Clearam. A range of modified starches that require heating in aqueous solution to develop their viscosity. Applications: Food industry - to develop viscosity in a wide number of applications e.g. sauces, gravies, custards etc. (Roquette (UK) Ltd).*

Clearasil Adult Care. Medication for adult skin blemishes. (Richardson-Vicks Inc).*

Clearasil Super Strength. Acne treatments with 10% benzoyl peroxide (tinted or vanishing cream or lotion formula), to help clear acne blemishes and absorb excess skin oil. (Richardson-Vicks Inc).*

Clearine. Eye drops. (The Boots Company PLC).

Clearon. Sodium dichloroisocyanurate di-hydrate. (FBC Ltd).

Clearpol. Alginate for brewing. (Alginate Industries Ltd).†

Clearsite. A proprietary trade name for cellulose acetate used for transparent containers.†

Clearsol. Disinfectants. (Tenneco Organics Ltd).

Clebrium Alloys. Heat-resisting alloys. One contains 76.5 per cent iron, 13 per cent chromium, 3.6 per cent molybdenum, 2.6 per cent carbon, 2 per cent nickel, 1.5 per cent silicon, and 0.75 per cent manganese.†

Clelands Reagent. Dithiothreitol. A white powder, m.p. 37° C. A reagent for protecting -SH groups in certain biochemical systems. (British Drug Houses).†

Clemantine. See METHYLENE VIOLET 2RA.

Clemantine Girofle. See SAFRANINE MN.

Clenesco. A proprietary trade name for anhydrous sodium metasilicate.†

Cleocin. Clindamycin. Applications: Antibacterial. (The Upjohn Co).

Clerici Solution. A molecular mixture of thallium malonate and formate. Used for floating mineral specimens to determine the specific gravity.†

Clerit. A horticultural fungicide. (ICI PLC).*

Clerite. A horticultural fungicide. (Plant Protection (Subsidiary of ICI)).†

Cleroxide. A stabilizer for polyvinyl chloride. (Akzo Chemie GmbH, Düren).†

Cleve's Acid. β-Acid is α-naphthylamine-6-sulphonic acid, or γ-Acid is α-naphthylamine 3-sulphonic acid, θ-Acid or J-acid is α-naphthylamine 7-sulphonic acid and θ- or δ-acid is α-nitronaphthalene-7-sulphonic acid.†

Clevite. A registered trade mark. (Clevite Corporation of Cleveland, Ohio).†

Cliché Metal. An alloy of 33 per cent tin, 46 per cent lead, and 21 per cent cadmium. Another alloy called by this name contains 48 per cent tin, 32.5 per cent lead, 9 per cent bismuth, and 10.5 per cent antimony.†

Climacel. Protective moisturizer. (Richardson-Vicks Inc).*

Climatone. A proprietary preparation of ethinoloestrodiol and methyl testosterone used in the treatment of pre-menstrual menopausal symptoms. (Pharmax).†

Climax. A magnetic alloy containing 30 per cent nickel and 70 per cent iron. An alloy stated to contain 73 per cent iron, 24.4 per cent nickel, and 2.6 per cent manganese has also been called Climax.†

Climax 193. An alloy of 68 per cent iron, 28 per cent nickel, 2 per cent chromium, and 1 per cent manganese.†

Climeline. A sodium phosphate used as a water softener.†

Clinafarm. A proprietary preparation containing imazalil. Applications: Fumigation. (Janssen Pharmaceutical Ltd).*

Clindrol. Alkanolamides both diethanolamides and monoethanolamides of fatty acids. Applications: Surface active agents, detergents, thickeners, foam stabilizers for aqueous systems. Lubricants and rust inhibitors. (Clintwood Chemical Company).*

Clindrol EGDS. Glycol stearates. Applications: Impart pearlescence and opacity to aqueous solutions. (Clintwood Chemical Company).*

Clindrol SDG. Glycol stearates. Applications: Impart pearlescence and opacity to aqueous solutions. (Clintwood Chemical Company).*

Clindrol SEG. Glycol stearates. Applications: Impart pearlescence and opacity to aqueous solutions. (Clintwood Chemical Company).*

Clindrol SEG-S. Glycol stearates. Applications: Impart pearlescence and opacity to aqueous solutions. (Clintwood Chemical Company).*

Clinifeed Favour. Protein, carbohydrate, fat, vitamins and minerals. Applications: Liquid feed. (Roussel Laboratories Ltd).*

Clinifeed ISO. Proprietary preparation of protein, carbohydrate, fat, vitamins and minerals. Applications: Liquid feed. (Roussel Laboratories Ltd).*

Clinifeed Protein Rich. Protein, carbohydrate, fat, vitamins and minerals. Applications: Liquid feed. (Roussel Laboratories Ltd).*

Clinifeed 400. Protein, carbohydrate, fat, vitamins and minerals. Applications: Liquid feed. (Roussel Laboratories Ltd).*

Clinimycin. A proprietary preparation of OXYTETRACYCLINE. An antibiotic. (Glaxo Pharmaceuticals Ltd).†

Clinistix. A prepared test strip of glucose oxidase, peroxidase and *o*-toluidine, used for the detection of glucose in urine (Ames).†

Clinitest. A prepared test tablet containing copper sulphate, sodium hydroxide, citric acid, and sodium carbonate, used for the detection of reducing substances in urine. (Ames).†

Clinitetrin. A proprietary preparation of tetracycline hydrochloride. An antibiotic. (Glaxo Pharmaceuticals Ltd).†

Clinium. A proprietary preparation of lidoflazine. Applications: For angina pectoris. (Janssen Pharmaceutical Ltd).*

Clinoril. Sulindac. Applications: Anti-rheumatic. (Merck Sharp & Dohme).*

Clinostrengite. Synonym for Phosphosiderite.†

Clinovariscite. Synonym for Metavariscite.†

Clipper. Tree growth regulator. (ICI PLC).*

Cliradon. See KETOBEMIDONE.

Clistin. Carbinoxamine maleate. Applications: Antihistaminic. (McNeil Consumer Products Co).

Clistin-D. Contains acetaminophen, carbinoxamine maleate and phenylephrine hydrochloride. Applications: Analgesic; antipyretic; antihistaminic; adrenergic. (McNeil Consumer Products Co).

Cloderm. Clocortolone pivalate. Applications: Glucocorticoid. (Ortho Pharmaceutical Corp).

Clomid. Clomiphene citrate. Applications: Gonad-stimulating principle. (Merrell Dow Pharmaceuticals Inc, Subsidiary of Dow Chemical Co).

Clonevac D-78. Gumboro D-78 strain. Applications: immunization of poultry. (Intervet America Inc).*

Clonevac-30. B_1 type, cloned lasota strain, Newcastle vaccine. Applications: immunization of poultry. (Intervet America Inc).*

Clonevac-30T. B_1 type, cloned lasota strain, turkeys, Newcastle vaccine. Applications: immunization of poultry. (Intervet America Inc).*

Clonopin. Clonazepam. Applications: Anticonvulsant. (Hoffmann-LaRoche Inc).

Clont. A preparation active against T. vaginalis. (Bayer & Co).*

Cloparin. Range of chlorinated paraffins. (SNIA (UK) Ltd).†

Cloparin. Chlorinated paraffins. Applications: plasticizers for PVC, paints; additives for lubricating and cutting oils and flame resistant additives. (Caffaro Industrial).*

Cloparol 50. Chlorinated paraffins. (SNIA (UK) Ltd).†

Cloparten. Chlorosulphonated paraffins. Applications: Raw material for the preparation of emulsifiable synthetic greasing agents for hides and leather. (Caffaro Industrial).*

Cloparten Z. Chlorosulphonated paraffins. (SNIA (UK) Ltd).†

Clophen. Synthetic liquid insulants and coolants for the electrical industry. Applications: Used as high grade, flame-retardant impregnants for capacitors and as liquid coolants for transformers and rectifiers. (Bayer & Co).*

Clorafin. Chlorinated paraffin. It is non flammable, chemically resistant and has good plasticizing properties. Applications: plasticizer in inks, additive in cutting oils and drawing compounds, plasticizing resin in chlorinated rubber corrosion-resistant coatings, plasticizer for vinyl resins, waterproofer and flameproofer for textiles. (Hercules Inc).*

Cloran. A trade mark for a chlorinated anhydride thermal and chemical stablilizer for polymers. (UOP Chemicals).†

Clorpactin WCS-90. Oxychlorosene sodium. Applications: Anti- infective, topical. (Guardian Chemical, Div of United-Guardian Inc).

Clorpactin XCB. Oxychlorosene. The hypochlorous acid complex of a mixture of the phenyl sulphonate derivatives of aliphatic hydrocarbons. Applications: Anti-infective, topical. (Guardian Chemical, Div of United-Guardian Inc).

Clortex. Chlorinated rubber. Applications: Resin for the preparation of paints and varnishes. (Caffaro Industrial).*

Clortol. Chlorine releasing sterilant. (ABM Chemicals Ltd).*

Clostrin. Combined vaccine for sheep. (520).†

Cloth Brown G. A dyestuff. Dyes chromed wool brownish yellow.†

Cloth Brown R. A dyestuff. Dyes chromed wool brownish red.†

Cloth Oil. That fraction obtained from the residue of Russian petroleum by distillation, of specific gravity 0.875.†

Cloth Orange. A dyestuff. Dyes chromed wool brownish orange.†

Cloth Red B. A dyestuff. Dyes chromed wool red.†

Cloth Red BA. See CLOTH RED B.

Cloth Red B. . A dyestuff. Dyes chromed wool brownish-red from an acid bath. Other names for this colour are Cloth red O, Fast bordeaux O, Cloth red BA, and Fast milling red B.†

Cloth Red G Extra. See CLOTH RED G.

Cloth Red R. See CLOTH RED G.

Cloth Red 0. See CLOTH RED B.

Cloth Scarlet G. A dyestuff. Dyes wool red from an acid bath.†

Cloth, Leather. See AMERICAN CLOTH.

Cloustonite. A variety of asphalt.†

Clout. Herbicide. (May & Baker Ltd).

Clovean. A detergent for bakehouse usage. (Laporte Industries Ltd).*

Clovotox. Selective weedkiller. (May & Baker Ltd).

Clowes' Solution. A solution containing 160 grams potassium hydroxide in 200 cc water and 10 grams pyrogallol. Used in the Hempel pipette for the absorption of oxygen.†

Cloxapen. Cloxacillin sodium. Applications: Antibacterial. (Beecham Laboratories).

Cloxicap. A proprietary preparation of cloxacillin. An antibiotic. (Pfizer International).*

Cloxisyrup. A proprietary preparation of cloxacillin. An antibiotic. (Pfizer International).*

Clozan. A proprietary preparation of clotiazepam. An antianxiety agent. (Pfizer International).*

Clozaril. Clozapine. Applications: Sedative. (Sandoz Pharmaceuticals, Div of Sandoz Inc).

Clysar. Shrink film. (Du Pont (UK) Ltd).

Clysodrast. Bisacodyl tannex: water soluble complex of biscacodyl and tannic acid. Applications: Laxative. (Armour Pharmaceutical Co).

Coactabs. Amdinocillin Pivoxil. Applications: Antibacterial. (Hoffmann-LaRoche Inc).

Coactin. Amdinocillin. Applications: Antibacterial. (Hoffmann-LaRoche Inc).

Coal Black. See MINERAL BLACK.

Coal Oil. See KEROSENE.

Coalatex. A proprietary coagulating material for use in coagulating rubber latex. It primarily consists of oxalic aci[d] and oxalate, and is a good coagulating agent in 10 per cent solution.†

Coaldet. An electric blasting cap comprising copper-bronze alloy shell. Applications: Designed especially for initiating explosives in coal mining. (Hercules Inc).*

Coalite N.T.P. Non-toxic trixylenyl phosphate plasticizer. (Coalite Fuels & Chemicals Ltd).†

Coaltec. Wood preservative. (Coalite Fuels & Chemicals Ltd).

Coarse Metal. (Matte). An impure mixtur[e] of ferrous and cuprous sulphides produced in copper smelting. It usually contains from 20-75 per cent copper, 12-45 per cent iron, and 19-25 per cent sulphur. The impurities consist of arsenic, antimony, bismuth, manganese, nickel, cobalt, zinc, lead, selenium, tellurium, gold, and silver.†

Coarse Para. See SERNAMBY.

Coarse Solder. See PLUMBER'S SOLDER.

Cobac. Cobalt acetate. (Mechema).*

Cobadex Ointment. A proprietary preparation of hydrocortisone in a silicone base for dermatological use. (Cox, A H & Co Ltd, Medical Specialities Divn).†

Cobalamin. Vitamin B_{12}. See CYANOBALAMIN.†

Cobalin-H. Injectable hydroxobalamin BP. Applications: Main indications include Addisonian pernicious anaemia prophylaxis and treatment of other macrocytic anaemias due to B_{12} deficiency and tobacco amblyopia and Leber's atrophy. (Paines & Byrne).*

Cobalt Blue. (Cobalt Ultramarine, Leithner's Blue, Leyden Blue, Gahn's Ultramarine, Vienna Blue, Azure Blue, Hoxner's Blue, Hungary Blue, King's Blue). A blue pigment consisting of an aluminate of cobalt, prepared by calcining cobalt compounds with clay or alumina. Phosphoric acid and zinc oxide are often added. Used as an oil colour.†

balt Brass. Acid-resisting alloys of 52 per cent copper, 17-25 per cent zinc, and 22-30 per cent cobalt.†

balt Bronze. A phosphate of cobalt and ammonium.†

balt Brown. A pigment prepared by calcining a mixture of ammonium sulphate, cobalt sulphate, and ferrous sulphate.†

balt Green. (Rinmann's Green, Zinc Green). A pigment prepared by heating the precipitated oxides of cobalt and zinc. Also see GELLERT GREEN.†

balt Manganese Pink. A pigment obtained by mixing a thin paste of magnesium carbonate with cobalt nitrate solution, drying, and heating.†

balt Oxide, Prepared. See PREPARED COBALT OXIDE.

balt Pink. Cobalt phosphate, $Co_3(PO_4)_2$.†

balt Red. (Cobalt Rose). Cobalt oxide, Co_2O_3.†

balt Rose. See COBALT RED.

balt Steel. Alloys of steel with cobalt, used for certain parts of electrical machinery requiring a high permeability. An alloy containing 34.5 per cent cobalt, at high inductions, has a higher permeability than pure iron. Also see PERMANITE and KS MAGNET STEEL.†

obalt Ultramarine. See COBALT BLUE.

balt Violet. A pigment prepared by mixing solutions of cobalt sulphate and sodium phosphate. A precipitate of hydrated cobaltous phosphate is formed, which upon fusing produces a violet pigment.†

obalt Violet, Pale. See PALE COBALT VIOLET.

obalt Vitriol. Cobaltous sulphate, $CoSO_4$ $7H_2O$.†

obalt Yellow. See AUREOLIN.

obalt 254. A paste containing compounds of cobalt. Applications: Loss-of-dry inhibitor for non-aqueous coating compositions, inhibitor of gel-time drift in unsaturated polyester resins. (Nueodex Inc).*

Cobalt-chromium-molybdenum Steels. Usually an alloy of iron with 3.05 per cent molybdenum, 2.16 per cent chromium, 1.33 per cent cobalt and 0.65 per cent carbon.†

Cobaltron Steel Alloy. A proprietary alloy containing iron with about 11 per cent chromium, 2.25 per cent cobalt, 1.5 per cent carbon, 1.25 per cent molybdenum, and 0.25 per cent tungsten.†

Cobalt, Tin-white. See TIN-WHITE COBALT.

Coban. Monensin, produced by Streptomyces cinnamonensis, and used as the sodium salt. Applications: Antibacterial; antifungal. (Eli Lilly & Co).

Cobastab. A proprietary preparation of cyanocobalamin for injection. (Boots Company PLC).*

Cobatope-60. Cobaltous chloride Co 60. Applications: Radioactive agent. (E R Squibb & Sons Inc).

Cobex. Weedkiller containing dimitramine. (ICI PLC).*

Cobox. Cobalt oxide. (Mechema).*

Cobrol. A photographic developer. (May & Baker Ltd).*

Coccinin. See PHENETOLE RED.

Coccinine B. A dyestuff. Dyes wool red from an acid bath.†

Coccolite. See AUGUTE.

Cochineal. (Carmine, Carmine Lake). The dried bodies of the female shield louse, *Coccus Cacti*. It is the source of carmine, a red colouring matter. Used for dyeing scarlet on tin mordanted wool, but chiefly for the preparation of pigments. Carmine lake consists chiefly of the aluminium salt of carminic acid. Two qualities of cochineal are white or silver-grey, and black cochineal or Zaccatila.†

Cochineal Scarlet G. A dyestuff. It is the sodium salt of phenylazo-α-naphtholsulphonic acid. Dyes wool brick-red from an acid bath.†

Cochineal Scarlet PS. See PALATINE SCARLET.

Cochineal Scarlet 2R. A dyestuff. It is the sodium salt of tolueneazo- α-naphtholsulphonic acid. Dyes wool red from an acid bath.†

Cochineal Scarlet 4R. A dyestuff. It is the sodium salt of xyleneazo- α-naphtholsulphonic acid. Dyes wool red from an acid bath.†

Cochranite. Titanium dicyanide.†

Cochrome. An alloy of 60 per cent cobalt, 12 per cent chromium, 24 per cent iron, and 2 per cent manganese. It has been used in the place of Nichrome for the elements of electrical heating apparatus.†

Coclopet. Copper chloride (petroleum refining grade). (Mechema).*

Coclor. Industrial cupric chloride (35/36 per cent. Cu). (Mechema).*

Cocloran. Anhydrous copper chloride (46-47 per cent. Cu). (Mechema).*

Coco-Diazine. Sulphadiazine. Applications: Antibacterial. (Eli Lilly & Co).

Coco-nut Fibre. See COIR.

Coco-nut Milk. See COCONUT BUTTER.

Coco-nut Oil. See COCONUT BUTTER.

Coco-Quinine. Quinine Sulphate. Applications: Antimalarial. (Eli Lilly & Co).

Cocoa Butter. (Cacao Butter). The fat extracted from the seeds of *Theobroma cacao.* The seeds contain from 35-45 per cent of the butter. It has an acid value of 0.6-1.3 (oleic acid), a saponification value of 192-193, and melts at 33-34° C. It contains the glycerides of arachic, palmitic, oleic, stearic, and lauric acids.†

Cocoa Oil. See OIL OF COCOA.

Cocoaline. See VEGETABLE BUTTER.

Coconut Butter. (Coconut Oil, Coconut Milk, Copra). The sweet watery liquid contained in the coconut is called coconut milk. This disappears and gives place to a soft edible pulp, which hardens in the air, and is sold as copra. Copra contains from 60-70 per cent oil, which is extracted as coconut oil or butter.†

Cocose. See VEGETABLE BUTTER.

Codelcortone. Prednisolone. Applications: Certain endocrine and non-endocrine disorders responsive to corticosteroid therapy. (Merck Sharp & Dohme).*

Codelsol. Prednisolone phosphate. Applications: Parenteral corticosteroid for certain endocrine and non-endocrine disorders. (Merck Sharp & Dohme).*

Codelspray. Prednisolone phosphate. Applications: Topical spray for corticosteroid-responsive skin conditions. (Merck Sharp & Dohme).*

Co-Deltra. Prednisone. Applications: Certain endocrine and non-endocrine disorders responsive to corticosteroid therapy. (Merck Sharp & Dohme).*

Codiazine. Sulphadiazine. Applications: Antibacterial. (Beecham Laboratories).

Codidoxal. A proprietary preparation of doxycycline hyclate and codeine. An antibiotic and antitussive. (Pfizer International).*

Codinyl. A proprietary preparation of ephedrine hydrochloride, menthol, codeine phosphate, syrup of prunes and kola. Cough linctus. (A C Hatrick Chemicals Pty).†

Codiphen. A proprietary preparation of aspirin and codeine phosphate. An analgesic. (Pfizer International).*

Codis. A proprietary preparation of aspirin and codeine phosphate. An analgesic. (Reckitts).†

Codite. A proprietary trade name for a vulcanized fibre and pure cotton cellulose plastic tubing.†

Codoil. See RETINOL.

Codol. See RETINOL.

Codroxomin. Hydroxocobalamin. Applications: Vitamin. (O'Neal, Jones & Feldman Pharmaceuticals).

Codur. A proprietary trade name for a synthetic clear or coloured baking enamel.†

Co-Elorine Pulvules. A proprietary preparation of amylobarbitone and tricyclamol chloride. (Eli Lilly & Co).†

Coerulean Blue. See COERULEUM.

Coeruleine. (Alizarin Green, Anthracene Green, Coeruleine A). A dyestuff obtained by heating gallein with concentrated sulphuric acid, $C_{20}H_{10}O_6$. Dyes chromed wool, silk, or cotton, green. Used in cotton printing.†

Coeruleine A. See COERULEINE.

Coeruleine S. (Alizarin Green, Anthracene Green S, Coeruleine SW). The bisulphite compound of coeruleine. Dyes chromed wool, silk, or cotton, green. Used in calico printing.†

Coeruleine SW. See COERULEINE S.

Coeruleum. (Coeline, Sky Blue, Coerulean Blue). A blue pigment prepared by calcining a mixture of cobaltous sulphate, tin salt, and chalk, or by treating tin with nitric acid, adding a solution of cobalt nitrate, evaporating, and heating. Calcium sulphate is also usually found in the pigment.†

Co-Ferol. A proprietary preparation of ferrous fumarate and folic acid. A haematinic. (Cox, A H & Co Ltd, Medical Specialities Divn).†

Cofill. A finely ground mixture (50:50) of highly dispersed silica-resorcin. Applications: Adhesion powder for bonding rubber mixtures to treated and untreated synthetic textiles and metal fabrics. (Degussa).*

Cogentin. Benztropine mesylate. Applications: For the treatment of Parkinsonism and allied disorders of the nervous system. (Merck Sharp & Dohme).*

Cogesic. Prodilidine Hydrochloride. Applications: Analgesic. (Mead Johnson & Co).

Cogest. Pancreas substance, hog bile, papain-pepsin complex, diastase of malt. Papain 48.6 mg and pepsin 48.6 mg specially granulated and stabilized. Applications: As a digestive aid, for the relief of simple indigestion. (Pharmaceutical Basics Inc).*

Cognac Oil. See OIL OF COGNAC.

Co-Hydeltra. Prednisolone. Applications: Certain endocrine and non-endocrine disorders responsive to corticosteroid therapy. (Merck Sharp & Dohme).*

Cohedur. Bonding agents. (Bayer UK Ltd).

Cohydrol. A colloidal graphite solution.†

Coinage Bronze. An alloy of 95 per cent copper, 4 per cent tin, and 1 per cent zinc. It has a specific gravity of 8.9 and melts at 900° C.†

Cojene. Analgesic and antipyretic tablets. (Fisons PLC).*

Coke. The carbonaceous residue from the distillation of coal. It amounts to 70-80 per cent.†

Colac. Laxative tablets for relief of constipation. (Richardson-Vicks Inc).*

Colace. Dioctyl sulphosuccinate. A proprietary laxative. (Mead Johnson Laboratories).†

Colamine. Aminoethanol.†

Colanyl. Pigments for resin dispersions. (Hoechst UK Ltd).

Colasta. A proprietary trade name for a phenol-formaldehyde synthetic resin.†

Colastex. A patented mixture of Colas (cold asphalt) and rubber latex for improving and hardening roads.†

Colbenemid. Probenecid and colchicine. Applications: For the maintenance treatment of gout. (Merck Sharp & Dohme).*

Colcar D. A proprietary modified diethyl-phthalate range of carriers for the dying of triacetate fibres. (Allied Colloids Ltd).†

Colcolor. Carbon black/plastic concentrate. Applications: Used for pigmenting of paper and cement for the plastics processing industry, pigmenting of PE, PP, EVA, PVC, PS, SAN and ABS. (Degussa).*

Colcothar. See INDIAN RED.

Cold Blacks B, R. See CATIGENE BROWN N.

Cold Varnishes. Varnishes obtained by heating linseed oil to 105° C. for four and a half hours, adding manganese borate, linoleate, or resinate, and stirring the mass with compressed air.†

Coles Solder. An alloy of 82 per cent tin, 11 per cent aluminium, 5 per cent nickel, and 2 per cent manganese.†

Colestid. Colestipol hydrochloride, a basic anion-exchange resin, highly cross-linked and insoluble. Applications: Antihyperlipoproteinemic. (The Upjohn Co).

Col-Evac. Sodium Bicarbonate. Applications: Replenisher; alkalizer.

(O'Neal, Jones & Feldman Pharmaceuticals).

Colex 1000 FR. Fire retardent polyester gel coat. Applications: Special gel coat resin for manufacture of imitation fuel effects for electric and gas. (Colourex Ltd).*

Colfarit. A platelet aggregation inhibitor, antithrombotic and anti-inflammatory drug. (Bayer & Co).*

Colfite. A proprietary trade name for a graphited laminate bearing material.†

Colifoam. White odourless aerosol foam containing hydrocortisone acetate. Applications: Topical treatment for ulcerative colitis. (Stafford-Miller).*

Collagenase. Clostridiopeptidase A; 3.4.4.19. Digests native collagen at about physiological pH. (182). Collargol (Crede's Silver). Colloidal silver, has been used medicinally.†

Collasan. Colloidal kaolin.†

Collasol. Soluble collagen, a protein derivative. Applications: Used in the cosmetic industry. (Croda Chemicals Ltd).*

Collatex. Ammonium alginate. (Kelco/AIL International Ltd).

Collaurin. A form of colloidal gold.†

Collene. Colloidal silver.†

Collet Brass. See BRASS.

Collet Steel. A manganese steel with small amounts of chromium and 0.75 per cent carbon.†

Collidine. Trimethyl- and/or methylethylpyridines.†

Colliron. A proprietary preparation of ferric hydroxide colloid. A haematinic. (British Drug Houses).†

Collocal-D. A proprietary preparation of calcium oleate and stearate and calciferol. (Vitamin D.) (Crookes Laboratories).†

Collodion. Nitro-cotton, a weakly nitrated cellulose, usually regarded as dinitrocellulose, $C_6H_8(NO_2)_2O_5$. Its solution in alcohol-ether is also known as collodion. Employed in medicine and photography.†

Collodion Cotton. See GUNCOTTON.

Collodion Silk. An artificial silk made from collodion spun by precipitation in a liquid bath.†

Collokit. Technetium Tc 99m Sulphur Colloid. Applications: Radioactive agent. (Abbott Laboratories).

Collone. Emulsifying wax. (ABM Chemicals Ltd).*

Colloresin D. A proprietary product. It is a methyl cellulose soluble in cold water.†

Colloresin DK. A proprietary product. An alkyl ether of cellulose soluble in cold water insoluble in hot water. For use as a thickening agent in textile printing.†

Collosol Argentum. A proprietary preparation of colloidal silver, used as an ocular antiseptic. (Crookes Laboratories).†

Collosol Calamine. A proprietary preparation of colloidal calamine. A skin cream. (Crookes Laboratories).†

Collosol Manganese. A proprietary preparation of colloidal copper, chromium and manganese. Skin antiseptic. (Crookes Laboratories).†

Collotone. A proprietary preparation of iron and ammonium citrate, iron and manganese citrate, potassium glycerophosphate, sodium glycerophosphate, thiamine hydrochloride, tincture of nux vomica and caffeine citrate. A tonic. (28O).†

Collozets. A proprietary preparation of tyrothricin and cetylethyldimethylammonium ethyl sulphate, used for sore throat. (Crookes Laboratories).†

Collozine. Colloidal zinc hydroxide.†

Collubarb. A proprietary preparation of aluminium hydroxide gel, phenobarbitone and atropine sulphate. An antacid. (British Drug Houses).†

Collys. Native starches. Applications: Cardboard industry. (Roquette (UK) Ltd).*

Colmonoy. A proprietary trade name for a chromium boride, an abrasive.†

Colmonoy No. 6. A corrosion-resisting alloy with about 75 per cent nickel base. An essential constituent is chromium boride.†

Colofac. Mebeverine. Applications: Spasmolytic, counteracts intestinal spasms. (Duphar BV).*

Cologel. A proprietary preparation containing methylcellulose. A laxative. (Eli Lilly & Co).†

Cologne Brown. See VANDYCK BROWN and UMBER.

Cologne Yellow. See CHROME YELLOW.

Colombo Root. Calumba root.†

Colomycin. A proprietary preparation containing colistin sulphate. An antibiotic. (Pharmax).†

Colona Steel. A proprietary trade name for a nickel-chromium steel containing some manganese.†

Colophony. (Rosin, Resin). The residue which remains after the volatile oils have been removed by the distillation of crude turpentine. It consists of abietic acid and other diterpene acids. It is used in the manufacture of varnishes, and for a variety of other purposes.†

Color Seal. A two step application of sealers. Applications: It seals, hardens and dustproofs concrete, providing an attractive, coloured, very high gloss, tough abrasion and chemical resistant floor. (Secure Inc).*

Color-Max. Pigment/plasticizer dispersions. Applications: Pourable colours for plastisols and other systems. Offered custom matched to standard. (W J Ruscoe Co).*

Colorado Silver. An alloy of 57 per cent copper, 25 per cent nickel, and 18 per cent zinc. It is a nickel silver (German silver). Also see NICKEL SILVERS.†

Colortrend. Colourant dispersions. Applications: Volumetric machine colouring of coating compositions. (Nueodex Inc).*

Col-o-tex. A proprietary trade name for lacquer coated fabrics.†

Colour-Chem. Organic pigment powders. Applications: Paints, printing inks, plastics and rubber. (Colour-Chem Limited).*

Colours, Mordant. See ADJECTIVE DYESTUFFS.

Col-o-vin. A proprietary trade name for polyvinyl synthetic resin coated fabrics.†

Colrex Compound. Codeine phosphate. Applications: Antitussive; analgesic. (Rowell Laboratories Inc).

Colrex Expectorant. Guaifenesin. Applications: Expectorant. (Rowell Laboratories Inc).

Colsol. Fatty amide. Applications: Cotton and wool softener and lubricant. Finishing agent. (Scher Chemicals Inc).*

Coltapaste. Zinc paste and coal tar bandage. (Smith and Nephew).†

Coltrock. A proprietary trade name for a phenolic plastic.†

Colturiet. A wide range of epoxy coatings, coal tar, solvent free and solventless coatings. Also tank coatings. (Sigma Coatings).*

Coltwood. A proprietary trade name for a phenolic plastic.†

Colugel. A proprietary aluminium hydroxide gel used as an antacid. (Ulmer Pharmacal Co, Minneapolis).†

Coluitrin. See PITUITRIN.

Columbia Black B. A dyestuff. Dyes cotton black.†

Columbia Blacks 2BX, 2BW. Dye stuffs belonging to the same group as Columbia black B.†

Columbia Brown R. A direct cotton dyestuff.†

Columbia Chrome Black 2B. A direct cotton dyestuff.†

Columbia Fast Blue 2G. A dyestuff similar to Chicago blue 2R.†

Columbia Green. (Direct Green CO). A dyestuff. Dyes cotton green.†

Columbia Orange R. A direct cotton dyestuff.†

Columbia Red 8B. See GERANINE.

Columbia Resin. A proprietary trade name for a transparent thermosetting synthetic resin.†

Columbia Yellow. See CHLORAMINE YELLOW.

Coly-Mycin M Parenteral. Colistimethate sodium. Applications: Antibacterial. (Parke-Davis, Div of Warner-Lambert Co).

Coly-Mycin S. Colistin sulphate. Applications: Antibacterial. (Parke-Davis, Div of Warner-Lambert Co).

Comac. General agrochemicals. (McKechnie Chemicals Ltd).

Combantrin. A proprietary preparation of pyrantel pamoate. An anthelmintic. (Pfizer International).*

Combelen. Neuroplegic for use as tranquillizer in domestic animals. Applications: Veterinary medicine. (Bayer & Co).*

Combined Seed Dressing. Fungicide/insecticide. (Murphy Chemical Ltd).

Combovac-30. B_1 type, lasota strain, Newcastle with Massachuttes and Connecticut types. Applications: immunization of poultry. (Intervet America Inc).*

Comelian. SEE CORMELIAN. (Degussa).*

Comet. (2) Liquid soaps. (Unilever).†

Comet Metal. An alloy of 67 per cent iron, 30 per cent nickel, 2.2 per cent chromium, with small quantities of manganese and copper.†

Comfort Eye Drops. Naphazoline hydrochloride. Applications: Adrenergic. (Barnes-Hind Inc).

Comital/Comital L. Anti-convulsant preparations. (Bayer & Co).*

Comox. Cobalt and molybdenum oxides on alumina. (Laporte Industries Ltd).*

Compact Bitumen. See BITUMEN.

Compak. Composite kit of photographic processing chemicals. (May & Baker Ltd).

Compalox. Speciality aluminium oxide product. Applications: Used for the purification of waste water. (Lonza Limited).*

Compazine. Prochlorperazine. Applications: Anti-emetic. (Smith Kline & French Laboratories).

Compimide. High temperature bismalemide thermosetting resins. (Boots Company PLC).

Compitox. A selective weedkiller. (May & Baker Ltd).*

Complamex. A proprietary preparation of XANTHINOL NICOTINATE. A peripheral vasodilator. (Norton, H N Co).†

Complamin. Xanthinol Niacinate. Applications: Vasodilator (peripheral). (Riker Laboratories Inc, Subsidiary of 3M Company).

Complete. Denture cleanser toothpaste, to clean dentures. (Richardson-Vicks Inc).*

Comploment. Vitamin B6 in a controlled release tablet. Applications: Vitamin B6 deficiency. (Napp Laboratories Ltd).*

Compoglas. A proprietary trade name for a thermoplastic composition containing glass fibres.†

Compolon Forte. An injectable liver vitamin B_{12} preparation. (Bayer & Co).*

Compound, Lake's. See LAKE'S INDIARUBBER COMPOUND.

Compralgyl. An all-purpose analgesic. (Bayer & Co).*

Comprena. Food colour and flavour compounds. (PPF International Ltd).*

Compron. Veterinary compressed products. (May & Baker Ltd).*

Comtek. Sheetfed offset printing inks. Applications: Commercial printing. (Allied Signal Sinclair and Valentine).*

Conacure. Epoxy and polyurethane curing agents. (Conap Inc).*

Conadil. Sulthiame. Applications: Anticonvulsant. (Riker Laboratories Inc, Subsidiary of 3M Company).

Conapoxy. Epoxy resin systems. Applications: Potting and encapsulating tooling applications. (Conap Inc).*

Conathane. Polyurethane resin systems and elastomers. Applications: Tooling applications and potting and encapsulating applications. (Conap Inc).*

Concentrated Size. Powdered glue.†

Conceptrol. Nonoxynol 9. Applications: Spermaticide. (Ortho Pharmaceutical Corp).

Concordin. Protriptyline hydrochloride. Applications: Antidepressant with anergic properties. (Merck Sharp & Dohme).*

Concrete Oil of Mangosteen. See KOKUM BUTTER.

Concurat-L. Anthelmintic against gastrointestinal nematodes and lungworms in cattle, sheep, goats and pigs. Applications: Veterinary medicine. (Bayer & Co).*

Condens-Aid. Blends of amines. Applications: Steam and return line corrosion inhibitors. (Schaefer Chemical Products Company).*

Condenser Foil. An alloy of 90 per cent lead, 9.25 per cent tin, and 0.75 per cent antimony.†

Condensite. See BAKELITE.

Condensol. Catalysts for textile resin finishing. (BASF United Kingdom Ltd).

Conditioner Base. Blend of conditioning and stabilizing agents. (Croda Chemicals Ltd).

Condor. See OIL WHITE.

Conductivity Bronze. A bronze containing copper with 0.8 per cent cadmium and 0.6 per cent tin.†

Conducto-Lube. Silver powder and petroleum oil. Applications: Electrical lubricating. (ADC Resins).*

Conducto-Wrap. Copper foil tape with conductive pressure sensitive adhesive. Applications: EMI shielding and electrical splicing. (Custom Coating and Laminating Corporation).*

Condux. Electrically conductive elastomers. (Hardman Inc).*

Condy's Fluid. A solution of aluminium permanganate with some aluminium sulphate. It is an oxidizing agent, and is used as a disinfectant and deodorizer.†

Con-Fer. A proprietary preparation containing ethinyloestrodiol and norethisterone acetate with ferrous fumar-ate tablets. An oral contraceptive with an iron supplement. (Parke-Davis).†

Congo Blue 2B. (Benzo Blue 2B, Diamine Blue 2B). A dyestuff. Dyes cotton blue.†

Congo Brown G. A dyestuff. Dyes cotton brown.†

Congo Brown R. A dyestuff. Dyes cotton brown.†

Congo Corinth B. A dyestuff. Dyes cotton brownish-violet from a soap bath.†

Congo Corinth G. (Cotton Corinth G, Congo Corinth GW). A dyestuff. Dyes cotton brownish-violet from a soap bath.†

Congo Corinth GW. A dyestuff. It is a British brand of Congo Corinth G.†

Congo Cyanine. See BENZO BLUE.

Congo Fast Blue B. A dyestuff. Dyes cotton blue.†

Congo Fast Blue R. A dyestuff. Dyes cotton blue.†

Congo GR. A dyestuff. Dyes cotton red from a soap bath.†

Congo Pure Blue. The same as Diamine pure blue.†

Congo Red. (Congo Red Conc., L, R, Extra, W Conc.). A dyestuff. Dyes wool or cotton red from a neutral or alkaline bath.†

Congo Red Conc., L, R Extra, W Conc. Dyestuffs. They are British equivalents of Congo Red.†

Congo Red 4R. (Congo 4R). A dyestuff. Dyes cotton red from a soap bath.†

Congo Rubine. A dyestuff. Dyes cotton bluish-red.†

Congo Sky Blue. See DIAMINE SKY BLUE.

Congo Violet. (Bordeaux COV, Bordeaux Extra, Direct Violet). A dyestuff. Dyes wool bordeaux red from an acid bath, and cotton violet from a salt bath.†

Congo 4R. An azo dyestuff from diazotized tolidine, resorcinol, and naphthionic acid. See CONGO RED 4R.†

Conlex. Specialised latices based on acrylic-styrene co-polymers. Applications: Floor polishes. (Williams Div of Morton Thiokol Ltd). *

Conn's Stain. A microscopic stain. It contains 1 gram rose bengal, 5 grams phenol, and 100 cc distilled water.†

Conotrane. A proprietary preparation of penotrane and silicone cream. Antiseptic skin barrier cream. (Ward, Blenkinsop & Co Ltd).†

Conovid. A proprietary preparation of norethynodrel and mestranol. Oral contraceptive. (Searle, G D & Co Ltd).†

Conovid-E. A proprietary preparation of norethynodrel and mestranol. Oral contraceptive. (Searle, G D & Co Ltd).†

Conray 325, also Conray-400. Iothalamate sodium. Applications: Diagnostic aid. (Mallinckrodt Inc).

Conray 420. A proprietary preparation of sodium and/or meglumine iothalamate. Applications: Intravascular X-ray contrast medium. (May & Baker Ltd).*

Conray, also Conray 30 and Conray 43. Iothalamate meglumine. Applications: Diagnostic aid. (Mallinckrodt Inc).

Conrex. Acrylic levelling resin. (Williams (Hounslow) Ltd).

Constab. Specialized masterbatches for thermoplastics. (Cornelius Chemical Co Ltd).*

Constantan. An alloy of 60 per cent copper and 40 per cent nickel. It has a specific gravity of 8.9, and melts at 1290° C. Used as an electrical resistance. It has a specific resistance of 48 micro-ohms per $cm.^3$, and a specific heat of 0.098 calories per gram at 0° C.†

Constantin. Electrical resistance alloys. One contains 54 per cent copper and 46 per cent nickel, and another consists of 54 per cent copper, 44 per cent nickel, 1.3 per cent manganese, and 0.4 per cent iron.†

Constructal. An aluminium alloy containing 3 per cent of alloying elements, chiefly zinc.†

Contax. A proprietary preparation containing the diacetate of oxyphenisatin. A laxative. (Cox, A H & Co Ltd, Medical Specialities Divn).†

Contax. Weed killer. (Chevron).*

Contex. An antifluxing agent for gold- and silversmiths and in the jewellery industry. (Degussa).*

Continental Clay. Agriculture grade sedimentary kaolin. Applications: Used in suspensions and is non-abrasive and non alkaline. (Vanderbilt Chemical Corporation).*

Continex Carbon Black. Reinforcing carbon black made by the oil furnace process. Applications: Used in the tyre, rubber, plastics, ink and paint industries. (Continental Carbon Australia).*

Contracid. A corrosion-resisting alloy stable to nitric acid, hydrochloric acid, sulphuric acid, and other reagents. It contains from 50-60 per cent nickel, 15-20 per cent chromium, 0-20 per cent iron, and up to 10 per cent molybdenum or tungsten.†

Contradet. A laundry contra-flow detergent. (Laporte Industries Ltd).*

Contrapar. Gloxazone. Applications: Anaplasmodastat. (The Wellcome Foundation Ltd).

Contrastol W. For contrast dyeings. (Bayer UK Ltd).

Convol. Concentrated volumetric solutions. (BDH Chemicals Ltd).

Cooksons. White oxide of antimony. (Associated Lead Manufacturers Ltd).*

Cook's Alloys. One contains 68.5 per cent antimony and 31.5 per cent zinc, and another consists of 57 per cent antimony and 43 per cent zinc.†

Cool-Amp. A mixture of sodium chloride, calcium carbonate, silver chloride, potassium/bitartrate. Applications: Powder for silver plating electrical bus bars, connectors and the like. (The Cool-Amp Conducto-Lube Company).*

Coolanol. Series of formulated silicate ester fluids used for a wide range of heat transfer, dielectric and hydraulic fluid applications. Applications: Span an operating-temperature range of -90°F to 700°F. Used primarily in aerospace and advanced electronic hardware as a coolant/dielectric. (Monsanto Co).*

Cool-Treet. Various liquid blends of cooling water deposit and corrosion inhibitors. Applications: Cooling water treatment. (Schaefer Chemical Products Company).*

Coomassie. Milling and half milling acid dyes. (ICI PLC).*

Coomassie Blue-Black. A dyestuff. It is a British brand of Naphthol blue black.†

Coomassie Navy Blue. A dyestuff. Dyes wool navy blue.†

Coomassie Scarlet 9012K. A dyestuff. It is a British equivalent of Double scarlet extra S.†

Coopane. Piperazine adipate. (The Wellcome Foundation Ltd).

opaphene. Hexachlorophane amthelmintic. (The Wellcome Foundation Ltd).

opercote. Insecticidal varnish for paper, board etc. (The Wellcome Foundation Ltd).

operite. (1) An alloy of 80 per cent nickel, 14 per cent tungsten, and 6 per cent zirconium. Used for cutting tools. A modified alloy contains tantalum.†

opermatic. Automated insecticide dispenser. (The Wellcome Foundation Ltd).

ooper's Gold. An alloy of 19 per cent platinum and 81 per cent copper.†

ooper's Pen Metal. An alloy of 50 per cent platinum, 37.5 per cent silver, and 12.5 per cent copper.†

ooper's Speculum Metal. An alloy of 58 per cent copper, 27 per cent tin, 10 per cent platinum, 4 per cent zinc, and 1 per cent arsenic.†

opac. Metal salt complex. Applications: Accelerator for unsaturated polyesters. Drier for non-aqueous coating compositions. (Nueodex Inc).*

opaiva. (Copivi). Copaiba. See OLEO-RESIN COPAIBA.†

opaiva Balsam. See OLEO-RESIN COPAIBA.

opal Oils. Oils obtained by the dry distillation of copal. Used for the preparation of oil varnishes.†

opalm Balsam. See LIQUIDAMBAR.

opaloy. A proprietary platinum-tin-antimony alloy with a small percentage of copper. A bearing metal.†

opal, Pontianac. See MACASSAR COPAL.

opal, Singapore. See MACASSAR COPAL.

opel Alloy. An alloy of 55 per cent copper and 45 per cent nickel.†

opene. A proprietary trade name for a polyterpene copolymer resin used in lacquers, paints, and varnishes.†

opernick. An alloy similar to Hypernick. The permeability is constant over a wide range of flux densities.†

opisil. Colour developer for the production of carbonless copying papers, thermoreactive papers and chemical test papers. (Süd-Chemie AG).*

Copivi. See COPAIVA.

Copperas. See IRON VITRIOL.

Copperas, Green. See IRON VITRIOL.

Copper Blue. See MOUNTAIN BLUE.

Copper Blue B. A disazo dyestuff, which dyes wool reddish-blue shades in the presence of copper sulphate.†

Copper, Cementation. See CEMENT COPPER.

Copper Green. A term applied to the mineral Malachite.†

Copper, Gold. See CHRYSOCHALK.

Copperised Ammonia. See WILLESDEN FABRICS.

Copper-nickel. An alloy of 50 per cent copper and 50 per cent nickel, used in the manufacture of nickel copper alloys, is known by this name. See NICKEL COPPER ALLOYS. The term is also applied to the mineral Niccolite, NiAs.†

Copperone. See CUPFERRON.

Copper, Red. See VIOLET COPPER.

Copper, Ruby. See CUPRITE.

Copper Rust. See MALACHITE.

Copper Silumin. An alloy of 85.6 per cent aluminium, 13 per cent silicon, 0.8 per cent copper, and 0.6 per cent iron.†

Copper Soap. Copper resinate.†

Copper Solder. An alloy of 2 parts lead and 5 parts tin.†

Copper Steel. An alloy of steel with up to 1 per cent copper, usually 0.5 per cent. It resists corrosion.†

Coppertox. Liverstock dips and sprays. (The Wellcome Foundation Ltd).

Coppertrace. Cupric chloride dihydrate. Applications: Supplement, trace mineral. (Armour Pharmaceutical Co).

Copper Water. Iron sulphate, $FeSO_4$.†

Copper, White. See NICKEL SILVER and TOMBAC, WHITE.

Coppesan. A copper fungicide. (Boots Pure Drug Co).*

Copra. See COCO-NUT BUTTER.

Coprah Oil. Coconut Oil. See COCONUT BUTTER.†

Coprantex. Dyeing and printing assistant. (Ciba-Geigy PLC).

Coprantine. Cotton dyestuffs.†

Coprol. A proprietary trade name for dioctyl sodium sulphosuccinate.†

Coptal. Textile auxiliary chemicals. (ICI PLC).*

Co-Pyronil. A proprietary preparation containing pyrrobutamine, methapyriline and cyclopentamine hydrochloride. (Eli Lilly & Co).†

Corafilm. Amine treatment. Applications: Steam lines. (Western Chemical Co).*

Coralline. See PAEONINE.

Coralline Red. See PAEONINE.

Coralline Yellow. (Yellow coralline). The sodium salt of aurine (*qv*).†

Corangil. A proprietary preparation of glyceryl trinitrate, pentaerythritol tetranitrate, diprophylline and papaverine hydrochloride, used in angina pectoris. (British Drug Houses).†

Corasole. Carbon black/plastic concentrate. Applications: Used for pigmenting of paper and cement for the plastics processing industry, pigmenting of PE, PP, EVA, PVC, PS, SAN and ABS. (Degussa).*

Coravol. Original amine process. Applications: Steam lines. (Western Chemical Co).*

Corban. Corrosion inhibitors. Applications: used to reduce metal loss from oil and gas well equipment caused by hydrogen sulphide, carbon dioxide and organic acids. The inhibitor coats the metal surfaces with a thin protective film. (Dow Chemical Co Ltd).*

Corbrite. Disazo yellow and azo red pigments. (Horace Cory PLC).

Corcert. Food colours meeting EEC specifications. (Horace Cory PLC).

Cordetec 100. Phthalic anhydride catalyst made of vanadium pentoxide and titanium dioxide with some promoters supported on a uniform ceramic ring. Applications: The catalyst is used to produce phthalic anhydride from o-xylene reaching yields between 107-109 kg of phthalic anhydride from o-xylene (100%). The catalyst can be used in normal process and in low energy process. (Corporacion de Desarrollo Tecnologico CA).*

Cordex. A proprietary preparation of prednisolone and aspirin. Anti-inflammatory agent. (Upjohn Ltd).†

Cordilox. A proprietary preparation of verapamil hydrochloride. Applications: An antihypertensive. (Abbott Laboratories).*

Cordite. (Cordite MD, Maximite). An explosive used as powder and filaments. It is a mixture of 65 per cent guncotton, gelatinized by means of acetone, 30 per cent nitroglycerine, and 5 per cent petroleum jelly.†

Cordite MD. See CORDITE.

Cordran. A proprietary trade name for Flurandrenolone.†

Cordran. Flurandrenolide. Applications: Glucocorticoid. (Eli Lilly & Co).

Coreine AR, AB. A dyestuff obtained by heating Coreine 2R with aniline and sulphonating. Dyes chromed wool blue. Chiefly used for printing.†

Coreine 2R. (Celestine Blue B). A dyestuff. It is the amide of diethylgallo-cyanine, $C_{17}H_{18}N_3O_4Cl$. Dyes chromed wool bluish-violet.†

Corekal A. Sodium alkylnaphthalene sulphonate.†

Corephen 10. Phenol formaldehyde. Applications: For acid-proof stoving finishes. (Bayer & Co).*

Corex. See CARBORA.

Corexit. Oil slick dispersants. (Exxon Chemical Ltd).*

Corfast. Azo red and yellow pigments. (Horace Cory PLC).

Corfix. A range of adhesives for all types of resin, silicate and oil-bonded cores. (Foseco (FS) Ltd).*

Corgard. A proprietary preparation of nadolol. Applications: The treatment of cardiovascular disorders. (Squibb, E R & Sons Ltd).*

Corgard Tablets. Nadolol. (E R Squibb & Sons Ltd).

Corgaretic. A proprietary preparation of nadolol and bendrofluazide. Applications: For use in the treatment of hypertension. (Squibb, E R & Sons Ltd).*

orgran. Granular organic pigments, toners, food colours. (Horace Cory PLC).

ori Ester. Glucopyranose-1-monophosphate.†

oriacide. Leather dyes. (ICI PLC).*

oriban. Diamphenethide fasciolicide. (The Wellcome Foundation Ltd).

orichrome. Titanium lactates. Employed as mordants and "strikers" in the leather industry.†

oridine. *n*-Propyllutidine.†

orilene. An aqueous degreasing agent for leather. (ICI PLC).*

orinal. A trade mark for a synthetic tannin prepared by condensing heavy tar oils with formaldehyde and then making the aluminium salt. Also see ESCO EXTRACT and NERADOL. (National Lead Co).†

orindite. The trade name for an abrasive and refractory of the carborundum type. It is obtained from bauxite by heating it with anthracite, and contains 69 per cent. Al_2O_3, with SiO_2 and Fe_2O_3.†

orioflavines. Red or reddish-brown dyestuffs, used for leather.†

oripact. Mineral fibre thermal-insulating boards and strips together with bitumen sheet. Applications: For non ventilated roofs and for all roof coverings. (Vedag GmbH).*

oriphosphines. Dyestuffs. They are alkylated aminoacridines.†

orisol. A proprietary preparation of suprarenal gland hormone.†

oriumine. Leather dyes. (ICI PLC).*

Cork. The outer bark of a tree, *Quercus suber*.†

Corlake. Food colour lakes. (Horace Cory PLC).

Corlan. A proprietary preparation of hydrocortisone hemi-succinate sodium for mouth ulcers. (Glaxo Pharmaceuticals Ltd).†

β-Corlan. A proprietary preparation of betamethazone for use in local oral ulceration. (Glaxo Pharmaceuticals Ltd).†

Corlar. Epoxy finish. (Du Pont (UK) Ltd).

Cormelian (Comelian, Labitan). Dilazephydrochloride. Tablet form. Applications: Treatment of Ischemic Heart Disease. (Degussa).*

Cormix. Concrete auxiliaries. (Crosfield Chemicals).*

Cormul. Range of cationic bitumen emulsifiers in liquid form. Application: Used in road surfacing: surface dressing tack coats; stone coating; grouting; slurry sealing. (Thomas Swan & Co Ltd).

Cornalith. See GALLATITE.

Cornish Bronze. An alloy of 78 per cent copper, 9.5 per cent tin, and 12.5 per cent lead. A white bearing metal.†

Cornish Clay. See CHINA CLAY.

Cornite. A hard vulcanite.†

Cornox. A selective weedkiller. (Boots Pure Drug Co).*

Cornox Plus. Selective herbicide. (FBC Ltd).

Cornoxynil. Selective herbicide. (FBC Ltd).

Cornuite. A protein-like mineral found in the diatomaceous earth from Neu-Ohe. It is a gelatinous albuminous material with a 3 per cent dry residue.†

Cornutine. Impure ergotoxine.†

Corolox. A trade mark for abrasive and refractory materials. The essential constituent is crystalline alumina.†

Corona. Lanolin BP. Applications: Pharmaceutical preparations. (Croda Chemicals Ltd).*

Coronet. Refined lanolin. Applications: Used as emulsifier, emollient, dispersing, solubilizing and wetting agent. (Croda Chemicals Ltd).*

Coronium. An alloy containing 80 per cent copper, 15 per cent zinc, and 5 per cent tin.†

Coronium Bromide. Strontium bromide, $SrBr_2$.†

Corotox. A trade mark for goods made for abrasive and refractory purposes. The essential constituent is crystalline alumina.†

Corowalt. A trade mark for materials of the abrasive and refractory type, the essential constituent of which is crystalline alumina.†

Corox. A proprietary insulating material having a great thermal conductivity and high electrical insulating power. It consists essentially of magnesium oxide.†

Corozo. Vegetable ivory, the seeds of *Phytelephas macrocarpa.*†

Corronel Alloy 230. A trade mark for an alloy of 35 per cent. chromium and 65 per cent. nickel. It is resistant to nitric and nitric/hydrochloric acid mixtures. (Wiggin Alloys Ltd).†

Corronel 220. A trade mark for a nickel - molybdenum - vanadium alloy with good resistance to hydrochloric, sulphuric and phosphoric acids under reducing conditions. (Wiggin Alloys Ltd).†

Corronil. An alloy of 70 per cent nickel, 26 per cent copper, and 4 per cent manganese. It is a corrosion resisting alloy.†

Corrosalloy. A proprietary trade name for stainless steels.†

Corrosiron. An iron-silicon alloy containing 12 per cent silicon. It is stated to be very resistant to acids.†

Corrosist. A proprietary trade name for nickel base corrosion resistant alloys. Used for valves in chemical plant subject to contact with chlorine.†

Corrosive Sublimate. Mercuric chloride, $HgCl_2$.†

Corseal. Refractory mouldable products for repairing damaged cores or sealing core joints. (Foseco (FS) Ltd).*

Corsodyl. A proprietary preparation of chlorhexidine gluconate used in the treatment of gingivitis and dental hygiene generally. (ICI PLC).*

Cortacream. A proprietary preparation of hydrocortisone acetate and silicon fluid on a dressing, used for ecxematous skin disorders. (Smith and Nephew).†

Cortaid. Hydrocortisone acetate. Applications: Glucocorticoid. (The Upjohn Co).

Cort-Dome. Hydrocortisone. Applications: Glucocorticoid. (Miles Pharmaceuticals, Div of Miles Laboratories Inc).

Cortef. A proprietary preparation of hydrocortisone for local use on skin. (Upjohn Ltd).†

Cortef Acetate. Hydrocortisone acetate. Applications: Glucocorticoid. (The Upjohn Co).

Cortef Oral Suspension. Hydrocortisone Cypionate. Applications: Glucocorticoid. (The Upjohn Co).

Cortelan. A proprietary preparation of cortisone acetate. (Glaxo Pharmaceuticals Ltd).†

Corten. A proprietary trade name for a chromium steel containing 0.5 per cent chromium, 0.10 per cent carbon, 0.1 pe cent manganese, 0.5 per cent silicon, and 0.3 per cent copper.†

Cortenema. Aqueous suspension of hydrocortisone 100 mg per 60 ml. Applications: A steroid enema as adjunct in the treatment of idiopathic non-specific ulcerative colitis. (Bengue & Co Ltd).*

Cortico-Gel. A proprietary preparation of corticotrophin gelatin. (Crookes Laboratories).†

Cortifoam. A proprietary preparation of hydrocortisone in aerosol form for dermatological use. (Pfizer International).*

Cortipix. A proprietary preparation of coal tar and hydrocortisone acetate, used in the treatment of ecxema. (Fisons PLC).*

Cortisporin. Cream, otic solution, ophthalmic suspension and otic suspension. Proprietary formulations of polymyxin B sulphate, neomycin sulphate and hydrocortisone. Applications: Treatment of steroid-responsive inflammatory ocular conditions or dermatoses for which a corticosteroid is indicated, or of superficial bacterial infections of the external auditory canal caused by organisms susceptible to the action of antibiotics. (The Wellcome Foundation Ltd).*

Cortisporin Ointment. Ophthalmic Ointment. Proprietary formulations of polymyxin B sulphate, bacitracin zinc, neomycin sulphate and hydrocortisone. Applications: For the treatment of corticosteroid-responsive dermatoses with secondary infection or steroid-responsive inflammatory ocular conditions where the risk of infections exists. (The Wellcome Foundation Ltd).*

ortistab. A proprietary preparation of cortisone acetate. (Boots Company PLC).*

ortistan. Cortisone in a solution of 25 mg/10 cc (Standex Laboratories, Columbus, Ohio).†

ortitrane. A proprietary preparation of phenylmercuric dinaphthylmethane disulphonate and prednisolone. Antibiotic steroid skin cream. (Ward, Blenkinsop & Co Ltd).†

ortocaps. A proprietary preparation of hydrocortisone acetate and neomycin sulphate, used in eye infections. (Crookes Laboratories).†

ortoderm. A proprietary preparation of hydrocortisone acetate for dermatological use. (Crookes Laboratories).†

ortoderm-N. A proprietary preparation of hydrocortisone and neomycin used in dermatology as an antibacterial agent. (Crookes Laboratories).†

ortone. Organic pigment toners. (Horace Cory PLC).

ortone Acetate. Cortisone acetate. Applications: Glucocorticoid. (Merck Sharp & Dohme, Div of Merck & Co Inc).

ortril. A proprietary preparation of hydrocortisone acetate. A corticosteriod. (Pfizer International).*

ortril Acetate-AS. Hydrocortisone acetate. Applications: Glucocorticoid. (Pfizer Inc).

ortrophin Gel ACTH. Corticotropin, repository. Applications: Hormone; glucocorticoid; diagnostic aid. (Organon Inc).

ortrophin Zinc ACTH. Corticotropin zinc hydroxide. Applications: Hormone; glucocorticoid; diagnostic aid. (Organon Inc).

Cortrosyn. Cosyntropin. Applications: Hormone. (Organon Inc).

Cor-Tyzine. A proprietary preparation of tetrahydrozoline hydrochloride and prednisolone. A nasal decongestant and anti-inflammatory. (Pfizer International).*

Corubin. An artificial corundum Al_2O_3. It is the alumina which constitutes the slag formed in the reaction between aluminium and metallic oxides (Thermit). Used for polishing purposes, and in the manufacture of fireproof stones.†

Corundite. A trade mark used for abrasive and refractory materials the essential constituent of which is crystalline alumina.†

Corvaton. Molsidomine. Applications: Anti-anginal; vasodilator. (Hoechst-Roussel Pharmaceuticals Inc).

Corvic. PVC, emulsion polymers and copolymers. (ICI PLC).*

Cosbiol. Perhydrosqualene (squalane). Applications: High quality oil for cosmetics and pharmaceuticals. (Laserson & Sabetay).*

Coscopin. A proprietary preparation containing noscapine. A cough linctus. (British Drug Houses).†

Coscopin Paediatric. A proprietary preparation containing noscapine. A cough linctus. (British Drug Houses).†

Coslettised Steel. A steel whose surface has been rust-proofed by dipping in a solution of iron phosphate and phosphoric acid.†

Cosmegen. Dactinomycin. Applications: Cytotoxic, antineoplastic antibiotic. (Merck Sharp & Dohme).*

Cosmegin Lyovac. A proprietary preparation of actinomycin D. A cytotoxic drug. (Montedison UK Ltd).*

Cosmetic Mercury. See WHITE PRECIPITATE.

Cosmic. Emulsifying agents. (Abril Industrial Waxes).*

Cosmin. See AGATHIN.

Cosmocil. Preservatives for cosmetics. (ICI PLC).*

Cosmos Alloy. A proprietary lead base alloy with small amounts of tin and antimony. A bearing metal.†

Cosmowax. Nonionic self emulsifying wax base. (Croda Chemicals Ltd).*

Cosylan. A proprietary preparation containing tincture of cocillana, liquid extract of squill and senega, antimony potassium tartrate, cascarin, ethylmorphine hydrochloride, menthol and syrup. A cough linctus. (Parke-Davis).†

Cotane. 2-Hydroxy biphenyl. (Coalite Fuels & Chemicals Ltd).

Cotazym. Pancrelipase. A concentrate of pancreatic enzymes standardized for lipase content. Applications: Enzyme. (Organon Inc).

Coterpin. A proprietary preparation of codeine, terpin, menthol pine oil and eucalyptus. Applications: A cough linctus. (Ayrton Saunders plc).*

Cothias Metal. An alloy of 67 per cent copper and 33 per cent tin. Used as a hardener for zinc alloys.†

Cotinazin. Isoniazid. Applications: Antibacterial. (Pfizer Inc).

Cotinin. See YOUNG FUSTIC.

Cotnion-Ethyl. Active ingredient: azinphos-ethyl; S-(3,4-dihydro-4-oxobenzo organophosphorous[d]-[1,2,3]-triazin-ylmethyl)0,0-diethyl phosphorodithioate. Applications: Persistent agricultural organophosphorous insecticide. (Makhteshim Chemical Works Ltd).†

Cotnion-Ethyl-Methyl. Active ingredients: cotnion-ethyl plus cotnion-methyl. Applications: Persistent agricultural organophosphorous insecticide combination. (Makhteshim Chemical Works Ltd).†

Cotnion-methyl. Active ingredient: azinphos-methyl; S-(3,4-dihydro-4-oxobenzo[d]-[1,2,3]-triazin-3-ylmethyl) 0,0-dimethyl phosphorodithioate. Applications: Persistent agricultural organophosphorous insecticide. (Makhteshim Chemical Works Ltd).†

Cotolan Fast. Dyestuffs for the one-bath of union materials - wool/cellulosics. (Bayer UK Ltd).

Cotopa. A form of textile insulating material composed of acetylated cotton yarn.†

Cottestren. Dyestuffs for cellulosic/polyester blends. (BASF United Kingdom Ltd).

Cottoclarin. Wetting agent. Applications: Scouring, bleaching, dyeing and wetting processes. (Henkel Chemicals Ltd).

Cotton, Amidated. Cotton which has had amino groups introduced into its molecule. This is accomplished by treating immunized cotton (*qv*) with aqueous ammonia. Amidated cotton has a great affinity for acid dyestuffs.†

Cotton Black. A dyestuff obtained by the fusion of *o' p*-dinitrodiphenylamine-sulphonic acid with sodium polysulphides. Dyes cotton brownish-black.†

Cotton Blue. Many of the direct cotton blues, such as Diamine, Benzo, and Congo blues, are sold under this name. See SOLUBLE BLUE and METHYL BLUE.†

Cotton Blue R. See MELDOLA'S BLUE.

Cotton Bordeaux. A dyestuff from diaminodiphenyleneketoxime with naphthionic acid. Dyes cotton from a soap bath.†

Cotton Brown A, N. A tetrazo dyestuff. See BENZO BROWN BX, BR.†

Cotton Brown R. See INGRAIN BROWN.

Cotton Corinth G. See CONGO CORINTH G.

Cottonex. Active ingredient: fluometuron; 1,1-dimethyl-3-(α,α,α-trifluoro-m-tolyl)urea. Applications: Residual herbicide effective against a wide range of both annual broafleaf weeds and grasses. (Agan Chemical Manufacturers Ltd).†

Cotton Fast Blue 2B. See MELDOLA'S BLUE.

Cotton, Mineral. See SLAG WOOL.

Cotton, Nitro. See GUNCOTTON.

Cotton Orange R. A dyestuff. Dyes cotton orange from a boiling bath.†

Cotton Orange 6305. A dyestuff. It is equivalent to Ingrain brown.†

otton Ponceau. An azo dyestuff from diaminodixylylmethane and naphthol disulphonic acid R.†

otton Red. See BENZOPURPURIN 4B.

otton Red 10B. A dyestuff. It is a British brand of Benzopurpurin 10B.†

otton Red 4B. See BENZOPURPURIN 4B.

otton Rhodine BS. A dyestuff. Dyes tannined cotton violet-red.†

otton Scarlet. See BRILLIANT CROCEINE M.†

otton Scarlet. A dyestuff. Dyes cotton red from a boiling alkaline bath, and is used for preparing lakes. Another name is Cinnabar red.†

otton Scarlet 3B Conc. See BRILLIANT CROCEINE M.

otton, Sthenosised. See STHENOSISED COTTON.

otton Wax. The wax found in cotton fibre. It melts at 82-86° C., and appears similar to carnauba wax.†

otton Wool. Gossypium.†

otton Yellow G. A dyestuff. Dyes cotton orange-yellow.†

otton Yellow G.(Chlorazol. Fast Yellow 5GK). A dyestuff. Dyes cotton yellow from a boiling alkaline bath.†

otton Yellow R. See ORIOL YELLOW.

otton Yellow 6307. A British dyestuff which is an equivalent of Diphenyl citronine G.†

ouloscope (Coulometric coating thickness gauge). Coulometric electrolytes. Applications: Measurement of metallic coating thickness of coatings deposited over metallic and non metallic substrates, or of metal foils. (Fischer Instrumentation (GB) Ltd).*

oumadin. Warfarin Sodium. Applications: Anticoagulant. (Endo Laboratories Inc, Subsidiary of E I du Pont de Nemours & Co).

oumalic Acid. α-Pyrone-3-carboxylic acid.†

oumalux. Optical brightening agent. (Ward Blenkinsop & Co Ltd).

oumarone Resin. See PARACOUMARONE RESIN.

Coupier's Blue. See INDULINE, SPIRIT SOLUBLE.

Coupler. Photosensitive coupler. (ABM Chemicals Ltd).*

Courcel. Mixed cellulose ethers. (Courtaulds Chemicals & Plastics Water Soluble Polymers Group).

Courline. A proprietary trade name for polypropylene monofil yarns. (British Celanese).†

Courlose. Sodium carboxymethyl cellulose. (Courtaulds Chemicals & Plastics Water Soluble Polymers Group).

Cournova. A proprietary trade name for polypropylene oriented slit film for the manufacture of strings, twines and ropes. (British Celanese).†

Court Plaster. Isinglass dissolved in water, alcohol, and glycerin, and painted on taffeta.†

Covar. Aliphatic solvent containing emulsifier. (Exxon Chemical International Inc).*

Coveral. Aluminium alloy cleansing flux. (Foseco (FS) Ltd).*

Coverite. A wetting agent. (Murphy Chemical Co).†

Covermark. A proprietary preparation of titanium dioxide with different pigments in a cream base. A skin masking cream. (Stiefel Laboratories (UK) Ltd).*

Covexin. A proprietary combined sheep vaccine. (Coopers Animal Health, subsidiary of The Wellcome Foundation Ltd).*

Cowles' Aluminium Bronze. An alloy containing 89-98.75 per cent copper and 1.25-11 per cent aluminium. One alloy contains small amounts of iron and silicon.†

Coxistac. Salinomycin - an ionophorous antibiotic used as an anticoccidial agent in poultry. (Pfizer International).*

Coxpyrin. See ASPIRIN.

Coyden. Coccidiostat is mixed with chicken feed to prevent coccidiosis (diarrhoea) in poultry. (Dow Chemical Co Ltd).*

Cozirc. Chemically combined cobalt-zirconium carboxylate. Applications: Drier for lead and/or barium free paint or printing ink. (Manchem Ltd).*

CPB. Dibutylxanthogen disulphide. A non-discolouring and non-staining low temperature ultra-accelerator for natural, SBR, nitrile and neoprene rubbers. (Uniroyal).*

CPE. (Chlorinated polyethylene) resins. When added to other plastics, CPE can be calendered, injection moulded or extruded into tough, chemical resistant, weather resistant products. Typical end uses would be pond liners, automotive hose and chemical transfer hose. (Dow Chemical Co Ltd).*

C-Petroleum Naphtha. (Petroleum Benzine, Safety Oil). That fraction of petroleum distlling at 80.100° C., of specific gravity 0.667-0.707.†

C.P.D. A rubber vulcanization accelerator. It is cadmiumpentamethylenedithiocarbamate.†

C.P.R. Multi-component systems used in the manufacture of rigid foams. (Dow Chemical Co Ltd).*

C-Quens. A proprietary preparation of mestranol (14 yellow tablets) and mestranol and chlormadinone (7 pink tablets). Oral contraceptive. (Eli Lilly & Co).†

CR Resins. A proprietary trade name for allyl resins.†

Cradocap. Cetrimide. Applications: Cradle cap. (Napp Laboratories Ltd).*

Craig Gold. An alloy of 80 per cent copper, 10 per cent nickel, and 10 per cent zinc. It is a nickel silver (German silver). Also see NICKEL SILVERS.†

Cranco. Lacquers and enamels. (ICI PLC).*

Crasnitin. An enzyme with anti-leukaemic effect and tumour-specific site of action. (Bayer & Co).*

Crastine. Thermoplastic polyester moulding compounds. (Ciba-Geigy PLC).

Craymer. Dimer acids: dimmer derivates. (Cray Valley Products Ltd).

Crayvallac. Thixotropes. (Cray Valley Products Ltd).

Cream 45. Moisturizing cream for dermatitis. (The Boots Company PLC).

Creamaffin. Liquid paraffin, BP, magnesium hydroxide BP laxative. (The Boots Company PLC).

Cream, Lambkin's. See LAMBKIN'S CREAM.

Crede's Silver. See COLLARGOL.

Crelan. A binder and crosslinking agent the formulation of powder coatings. (Bayer & Co).*

Cremalgex. A proprietary preparationof methyl nicotinate, capsicum oleoresin, glycol salicylate and histamine hydrochloride. A rubefacient. (Norton H N Co).†

Cremalgin. A proprietary preparation of methyl nicotinate, capsicin and glycol salicylate. An embrocation. (Berk Pharmaceuticals Ltd).*

Cremalys. Blends of gelatinized and non-gelatinised starches. Applications: Fo industry to develop creamy textures. (Roquette (UK) Ltd).*

Cremba. Compounded lanolin derived sterols. Applications: Cosmetic and pharmaceutical industries. (Croda Chemicals Ltd).*

Cremnitz. See FLAKE WHITE.

Cremnitz White. See FLAKE WHITE.

Cremodex. Sulphadiazine/sulphamethazine. Applications: Bacteriostatic. (Merck Sharp & Dohme).*

Cremomycin. Succinylsulphathiazole and neomycin. Applications: For the treatment of mild bacterial diarrhoeas (Merck Sharp & Dohme).*

Cremosan. A proprietary preparation of zinc oxide, colophony cresot, formaldehyde and thymol. Applications: An antiseptic cream. (Ayrton Saunders plc).*

Cremostrep. Succinylsulphathiazole, streptomycin sulphate and kaolin. Applications: Relief of the mild bacterial diarrhoeas seen in general practice. (Merck Sharp & Dohme).*

Cremosuxidine. Succinylsulphathiazole. Applications: Treatment of mild bacterial diarrhoeas. (Merck Sharp & Dohme).*

Crems White. See FLAKE WHITE.

Crenette. Non-woven glass scrim fabric. Applications: Reinforcement of filler media, floorcoverings, needle-felt, kraft paper and plastic sheeting. (Fothergill Tygaflor Ltd).*

Creolin. A coaltar black disinfectant containing 20% coaltar acids conforming to B.S.2462 BA. Applications: A disinfectant for public health, industrial and institutional application. (William Pearson).*

Creosol. Homocatechol methyl ester.†

Creosote. A term used in reference to the mixed phenols obtained from wood tar, coal tar, and other sources. Used for wood preservation.†

Creosote-guaiacol. (Homoguaiacol). Creosol.†

Cresamol. See KRESAMINE.

Cresatin. *m*-Cresol acetate. External antiseptic and analgesic.†

Cresavon. Hospital antiseptic soap. (Laporte Industries Ltd).*

Crescormon. Somatropin. Applications: Hormone. (KabiVitrum AB).

Cresol. Crude cresol contains approximately 35 per cent *o*-, 40 per cent *m*-, and 25 per cent *p*-cresol, $C_6H_4(CH_3)OH$. Antiseptic.†

Cresol Purple. *m*-Cresolsulphonphthalein. An indicator.†

Cresol Red. *o*-Cresolsulphonphthalein. An indicator.†

Cresolox. Disinfectants. (Tenneco Organics Ltd).

Cresotic Acids. Cresol carboxylic acids.†

Cresotine Yellow G. A dyestuff. Dyes cotton yellow.†

Cresotine Yellow R. A dyestuff. Dyes cotton yellow.†

Cressylite. A mixture of picric acid and trinitrocresol. Used in explosives.†

Crestalan. Lanolin and isopropyl esters. Applications: Used in cosmetic and household products. (Croda Chemicals Ltd).*

Crestavin. A proprietary trade name for vinyl acetate resins.†

Crester KZ. Polyglycerol esters. Applications: Food emulsifier. (Croda Chemicals Ltd).*

Cresyl Blue BB, BBS. A dyestuff. Dyes tannined cotton blue.†

Cresyl Violet B, BB. A dyestuff closely related to Capri blue GN.†

Cresyntan. See NERADOL.

Creto. A cationic surface-active agent. (Croda Chemicals Ltd).*

Crex. Mild laundry alkali. (ICI PLC).*

Crexathix. Castor oil derivatives. Applications: Thixotropic agent for paints and printing inks. (Blagden Chemicals Ltd).*

Crill. Sorbitan esters. Applications: Used as nonionic surfactants and food emulsifiers. (Croda Chemicals Ltd).*

Crillet. Polyethoxylated sorbitan esters. Applications: Used as nonionic surfactants and food emulsifiers. (Croda Chemicals Ltd).*

Crillon. Detergent and foam stabilizing properties; antistatic agents; anti-corrosive properties; skin protecting agents. Applications: Shampoos, bubble bath formulations; detergent cleaners; hand cleansers; cutting fluids and soluble cutting oil compositions. (Croda Chemicals Ltd).

Crimidesa. Anhydrous sodium sulphate. Applications: Detergents, glass, dyestuffs, etc. (Bromhead & Denison Ltd).*

Crimson Lake. A cochineal lake containing aluminium salts.†

Cristal. See FLINT GLASS.

Cristalline. See ZAPON VARNISH.

Cristite. A proprietary trade name for a steel containing 10 per cent chromium, 17 per cent tungsten, 3.5 per cent carbon and 2.5 per cent molybdenum.†

Criterion. Blend of pyridinium compounds, alcohols and saccharin. Applications: Nickel electroplating additive. (Taskem Inc).*

C-Ron. Ferrous fumarate. Applications: Haematinic. (Rowell Laboratories Inc).

C-Ron Forte. Ferrous fumarate. Applications: Haematinic. (Rowell Laboratories Inc).

Croak. Fluometuron plus MSMA. Applications: A herbicide for post-emergence control of broadleaf and grass weeds in cotton. (Drexel Chemical Company).*

Crocein B. A dyestuff. Dyes wool red from an acid bath.†

Crocein 3B. A dyestuff. Dyes wool red from an acid bath.†

Croceine. Phenylazo-β-naphthol-6-sulphonic acid.†

Croceine. 2-Naphthol-8-sulphonic acid.†

Crocein Orange. A dyestuff. It is an equivalent of Orange GT.†

Crocein Orange. See PONCEAU 4GB.

Crocein Scarlet B. A mixture of Crocein Scarlet 3B with Orange 3B.†

Crocein Scarlet O Extra. A dyestuff. Dyes wool and silk scarlet.†

Crocein Scarlet R. A mixture of Crocein Scarlet 3B with Orange 5B.†

Crocein Scarlet 2B. A mixture of Crocein scarlet 3B with Orange 7B.†

Crocein Scarlet 3B. (Ponceau 4RB). A dyestuff. Dyes wool scarlet from an acid bath, and cotton from an alum bath.†

Crocein Scarlet 3BX. An azo dyestuff isomeric with Fast red E, prepared from diazotized naphthionic acid and β-naphthol-α-sulphonic acid. The commercial product is the sodium salt.†

Crocein Scarlet 3B, 9187K. Dyestuffs. They are British brands of Brilliant crocein M.†

Crocein Scarlet 7B. (Ponceau 6RB, Crocein Scarlet 8B). A dyestuff. Dyes wool red from an acid bath.†

Crocein Scarlet 8B. See CROCEIN SCARLET 7B.

Crocein Yellow. A dyestuff.†

Crocell. A strippable coating composition. (Croda Chemicals Ltd).*

Crocidolite. (Blue Asbestos). A fibrous mineral. It is a member of the group of minerals known as soda-amphiboles. Found in South Africa and Western Australia. It is a hydrated silicate of sodium and iron approximating to the formula $(Fe.Na_2)_4Si_4O_{12}.FeSiO_3$. A pale blue asbestos found in Bolivia and South Australia differing chemically and in quality, is also termed crocidolite. The name crocidolite is also incorrectly applied to a variety of quartz, known as Tiger's eye.†

Crocus. Saffron, a colouring matter from the dried and powdered flowers of the saffron plant, *Crocus sativus*, used for colouring confectionery, and ferric oxide having a bluish tint used for polishing metals, are both known as crocus. See SAFFRON and INDIAN RED.†

Crocus Martius. A hydrated ferric oxide. It is used as a pigment for pottery and other purposes.†

Crocus of Antimony. See SULPHURATED ANTIMONY.

Crocus Saturni. See RED LEAD.

Croda Bath Oil Disperant. Synergistic blend of nonionic surfactants. (Croda Chemicals Ltd).

Croda Fluid. Lanolin rust preventatives. (Croda Chemicals Ltd).*

Crodacol. A range of fatty alcohols, used as nonionic surfactants. Applications: Pigment dispersing agents for cosmetics, perfumes and toiletries. (Croda Chemicals Ltd).*

Crodafos. Range of surfactants composed of alkyl ether phosphates in acid form. Each is a mixture of mono- and di- esters and may be converted to salts by neutralization with e.g. alkanolamine or metal hydroxide. Application: Detergents and coupling agents (for non-ionics) with wetting and emulsifying properties, they give low and unstable foam and are compatible with alkaline builders. Used in heavy duty clothes washing; hard surface cleaners; floor and industrial cleaners and steam cleaning systems; dry cleaning; dispersible solvent cleaners; cutting and rolling oils, grinding fluids, lubricants; hydraulic fluids; as antistatics, lubricants and softeners in synthetic fibre and wool processing; emulsion polymerization; herbicides, pesticides, fertilizers; waxes, polishes and surface coatings; cosmetics and pharmaceuticals. (Croda Chemicals Ltd).*

Crodalan. Surface active emollient; emulsifier and solubilizer; super-fatting, wetting and spreading agent. Applications: Oil/water cosmetics; perfume and essential oils for skin perfumes bath essences; all skin and hai care preparations; baby products; germicidal skin cleaners. (Croda Chemicals Ltd).

Crodalan AWS. Acetylated ethoxylated lanolin alcohol. Applications: Used in the cosmetic and pharmaceutical industries. (Croda Chemicals Ltd).*

Crodalan C24. Polyethoxylated cholesterol derivatives. Applications: Used in the cosmetic industry. (Croda Chemicals Ltd).*

Crodalan IPL. Isopropyl lanolate. Applications: General purpose raw material for cosmetics. (Croda Chemicals Ltd).*

Crodalan LA. Acetylated lanolin alcohols. Applications: General purpose raw material for cosmetics. (Croda Chemicals Ltd).*

Crodalan 1PL. Isopropyl lanolate. (Croda Chemicals Ltd).

Crodamet. Polyethoxylated fatty amines. Applications: Buffing and grinding compounding aids for the engineering industry. (Croda Chemicals Ltd).*

Crodamine. Tertiary amines. (Croda Universal Ltd).

Crodamine 1. Series of cationic surfactants composed of primary amines RNH^2, where the fatty alkyl group, R, may be lauric, palmitic, stearic, coconut, soya, tallow or oleic in origin. Solid, liquid or paste forms. D in the code number denotes distilled grade for when superior colour and heat stability are required. Applications: Corrosion inhibition; emulsifiers for herbicides; mineral flotation; pigment dispersion; anti-caking agents; auxiliaries for leather, textiles, rubber, plastics and metal industries. (Croda Chemicals Ltd).

Crodamine 2.C, 2.S and 2.HT. Cationic surfactant consisting of secondary amines R_2NH, in solid form, where the fatty alkyl group, R, may be coconut, soya or hydrogenated tallow in origin respectively. Applications: Corrosion inhibition; emulsifiers for herbicides; mineral flotation; pigment dispersion; anti-caking agents; auxiliaries for leather, textiles, rubber, plastics and metal industries. (Croda Chemicals Ltd).

Crodamine 3A. Series of cationic surfactants of the tertiary amine type, where the fatty alkyl groups are lauric, palmitic, stearic and coconut in origin. Supplied in liquid or paste form. Applications: Corrosion inhibition; emulsifiers for herbicides and synthetic resins; mineral flotation; pigment dispersion; anti-caking agents eg for high analysis NPK fertilizers; auxiliaries for leather, textiles, rubber, plastics and metal industries; catalysts in the production of polyurethane foam. (Croda Chemicals Ltd).

Crodamine 3ABD. N,N-Dimethyl-behenylamine. Applications: An intermediate for the preparation of quaternary ammonium compounds, betaines and amine oxides. Other uses include pharmaceuticals, solvents and anti-corrosives. (Croda Chemicals Ltd).*

Crodamine 3AED. N,N-Dimethyl-ecrucylamine. Applications: Used as an intermediate, other uses include pharmaceuticals, solvents and anti-corrosives. (Croda Chemicals Ltd).*

Crodamine 3AHRD and 3ARD. N,N-Dimethyl 18-22C amines. Applications: Used as intermediates. Other uses include pharmaceuticals, solvents and anti-corrosives. (Croda Chemicals Ltd).*

Crodamine 3AOD. N,N-Dimethyl-oleylamine. Applications: An intermediate for the preparation of quaternary ammonium compounds, betaines and amine oxides. Other uses include pharmaceuticals, solvents and anti-corrosives. (Croda Chemicals Ltd).*

Crodamol. A range of straight and branched chain mono and dibasic esters. Applications: Used as cosmetic emollients. (Croda Chemicals Ltd).*

Crodapearl. Modified glycol ester, used as a nonionic surfactant. Applications: Pearling agent for cosmetics, perfumes and toiletries. (Croda Chemicals Ltd).*

Crodapur. A proprietary trade name for lanoline (pure). Croda is crude lanoline.†

Crodasinic. N-Acyl sarcosines and their sodium salts, used as anionic surfactants. Applications: Foaming agents for cosmetics, perfumes and toiletries. (Croda Chemicals Ltd).*

Crodasinic L and LS35. Anionic surfactants of the N-acyl sarcosine type. L is lauroyl sarcosine in acid form, as a white solid; LS35 is sodium lauroyl sarcosinate in the form of a clear liquid. Properties include mild detergency, high foaming, bacteriostatic activity, enzyme inhibition, corrosion inhibition, hard water tolerance and stability in mildly acid formulations. Applications: Cosmetics and toiletries - dentifrices and shampoos; carpet shampoos; emulsion polymerization; metal treatment; food and food packaging; textiles; fine fabric detergents. (Croda Chemicals Ltd).

Crodaterics. A range of imidazoline based amphoteric surfactants. In general they are 2-alkyl-1-(ethylbetaoxypropanoic acid)- imidazolines in which the parent fatty acid varies. Properties include dense foaming and high foam stability; stability in acids, alkalies and strong electrolyte solutions; compatability with soaps; good detergency; wetting and emulsifying properties; sequestering properties; hard water solubility; compatability with quaternary germicides; germicidal and fungicidal properties. Applications: Acid and alkali detergents; agricultural sprays; antistats; corrosion inhibitor; toiletries; dry cleaning; emulsions; polishes, polymers, textile; metal treatment; paints and inks; petroleum; paper. (Croda Chemicals Ltd).

Crodax. Rust preventatives. (Croda Chemicals Ltd).*

Crodesta. A range of sucrose esters of fatty acids, used as nonionic surfactants. Applications: Pigment dispersing agents for paint. (Croda Chemicals Ltd).*

Crodet. A range of polyethoxylated fatty acids, used as nonionic surfactants. Applications: Used as fibreglass size and as lubricants in the textile industry. (Croda Chemicals Ltd).*

Crodex. Range of emulsifying wax. Applications: Pharmaceuticals and cosmetics. (Croda Chemicals Ltd).

Crodinhib. Amine borates, used as lubricants and plasticizers. Applications: Used as corrosion inhibitors for paints and oils. (Croda Chemicals Ltd).*

Crodol. A barrier cream. (Croda Chemicals Ltd).*

Crodon. The trade name for a type of chromium plate.†

Croduret. Polyethoxylated hydrogenated castor oil, used as nonionic surfactant. Applications: Used as dispersing, wetting and spreading agent. (Croda Chemicals Ltd).*

Crodyne BY 19. Gelatin. Applications: Used in the food industry. (Croda Chemicals Ltd).*

Crolactil. Acyl lactylates, used as anionic surfactants. Applications: Used as a wetting agent for paint. (Croda Chemicals Ltd).*

Crolan. Lanolin fatty acid esters. Applications: Used in the cosmetic industry. (Croda Chemicals Ltd).*

Crolastin. Partially hydrolysed elastin. Applications: Used in the cosmetic industry. (Croda Chemicals Ltd).*

Crolax. A proprietary preparation containing dioctyl sodium sulphosuccinate and dihydroxyanthaquinine. A laxative. (Crookes Laboratories).†

Croloy. A proprietary trade name for high chromium steel containing molybdenum and vanadium.†

Cromal. An alloy of aluminium with 2-4 per cent chromium and smaller amounts of nickel and manganese. It melts at 700° C., and is specially suitable for castings.†

Cromalit. Concentrated chrome alums. Applications: Tanning and drilling mud additive. (Lancashire Chemical Works Ltd).*

Cromalit 150. Concentrated basic potash chrome alum powder. (Lancashire Chemical Works Ltd).

Cromaloy II. An alloy of 80 per cent nickel, 15 per cent chromium, and 5 per cent iron.†

Cromaloy III. An alloy of 85 per cent nickel and 15 per cent chromium.†

Cromaloy IV. An alloy of 80 per cent nickel and 20 per cent chromium.†

Cromeen. Substituted alkyl amine derivatives of various lanolin acids. Applications: Used in the cosmetic industry. (Croda Chemicals Ltd).*

Cromo Steel. A proprietary trade name for a chrome-molybdenum steel.†

Cromophtal C-20. Pigment preparations. (Ciba-Geigy PLC).

Cromophtals. High performance organic pigments for plastics. A registered trade name. (Ciba Geigy PLC).†

Cromphytal M-20. Pigment preparations. (Ciba-Geigy PLC).

Cronite. An alloy of nickel and chromium. No. 1 contains 85 per cent nickel, and 15 per cent chromium.†

Cropeptone. Hydrolysed animal protein in aqueous solution. Applications: Used in the cosmetic industry. (Croda Chemicals Ltd).*

Cropol 60. Sodium di-octyl sulphosuccinate as an ethanol solution. Applications: Emulsifier and powerful wetting agent, used in a wide range of manufacturing industries, especially textiles. (Croda Chemicals Ltd).

Croquat. Quaternized polypeptides. (Croda Chemicals Ltd).

Crosanol. Fibre lubricants. (Crosfield Chemicals).*

Croscolor. Dyeing auxiliaries. (Crosfield Chemicals).*

Croscour. Scouring agents. (Crosfield Chemicals).*

Crosdurn. Durable finishes. (Crosfield Chemicals).*

Crosfield. Chemicals. (Crosfield Chemicals).*

Crosfield EP. Catalysts. (Crosfield Chemicals).*

Crosfield HP. Matting agents. (Crosfield Chemicals).*

Crosfield SP. Silicas. (Crosfield Chemicals).*

Crosil. Textile processing system. (Crosfield Chemicals).*

Crosilk. Silk amino acids. (Croda Chemicals Ltd).

Croslube. Yarn lubricants. (Crosfield Chemicals).*

Crosoft. Softening agents. (Crosfield Chemicals).*

Cross Dye Black FNG. See KHAKI YELLOW C.

Cross Dye Blacks. (Sulpho Blacks, Cross Dye Navy). Dyestuffs produced by the fusion of a variety of amino compounds and phenols with sodium polysulphide. They dye cotton dark blue to black.†

Cross Dye Navy. See CROSS DYE BLACKS.

Crostat. Anti-static agents. (Crosfield Chemicals).*

Crotein A,C,. Hydrolysed collagen. (Croda Chemicals Ltd).

Crotein CAA. Collagen amino acids. (Croda Chemicals Ltd).

Crotein HKP. Keratin amino acids. (Croda Chemicals Ltd).

Crotein HWE. Hydrolysed whole eggs. (Croda Chemicals Ltd).

Crotein Q. Quaternized hydrolysed collagen. (Croda Chemicals Ltd).

Crotorite. A cupro-manganese alloy. It contains 68 per cent copper, 30 per cent manganese, and 2 per cent iron.†

Cro-tung. A proprietary trade name for a chromium-tungsten steel.†

Crow. A cotton fabric grade of TUFNOL industrial laminates. (Tufnol).*

Crow Chex. Active ingredient: Copper oxalate. Applications: Seed protectant to prevent sprout pulling by birds in newly planted corn. (Borderland Products Inc).*

Crown. Tyres. (Chevron).*

Crown Solder. An alloy of 63 per cent tin, 18 per cent zinc, 13 per cent aluminium, 1 per cent lead, 3 per cent copper, and 2 per cent antimony.†

Cruisemaster. Tyres, batteries and accessories. (Chevron).*

Crusader. Post-emergence herbicide. (Murphy Chemical Ltd).

Cruverlite. A proprietary trade name for a luminous plastic moulding powder.†

Crylene. A proprietary rubber vulcanization accelerator. It is an acetaldehyde-aniline condensation product.†

Cryoseal. Water-based cold seal adhesives. (The Scottish Adhesives Co Ltd).*

Cryptolin. Gonadorelin acetate. Luteinizing hormone-releasing factor acetate hydrate. Applications: Gonad-stimulating principle. (Hoechst-Roussel Pharmaceuticals Inc).

Cryptone. (Titanox, Duolith. Titanolith, Tidolith). Proprietary trade names for white pigments containing lithopone and TiO_2.†

Cryptopyrrole. 3-Ethyl-2,4-Dimethyl-pyrrole.†

Crysmalin. See PARAFFIN, LIQUID.

Crysta. A bright zinc process. (Hanshaw Chemicals).†

Crystal. Sodium silicates. (Crosfield Chemicals).*

Crystal Glass. A glass composed of lead and potassium silicates.†

Crystal Naphthylene Blue R. See MELDOLA'S BLUE.

Crystal Polystyrene. Polymerization product of styrene monomer, with additives. Applications: General commodity plastic for extrusion and injection. (Lin Pac Polymers).*

Crystal Violet. (Crystal Violet 5BO, Crystal Violet O, Violet C, Violet 7B Extra, Methyl Violet 10B). A dyestuff. It is the hydrochloride of hexamethyl-para-rosaniline, $C_{25}H_{30}N_3Cl$. Dyes silk and wool violet, also cotton mordanted with tannin and tartar emetic. It has also been used as an antiseptic.†

Crystal Violet SBO. See CRYSTAL VIOLET.

Crystal Violet 0. See CRYSTAL VIOLET.

Crystalex. A proprietary trade name for an acrylic denture base.†

Crystalite. A proprietary trade name for an acrylic moulding powder.†

Crystalite. Organic brighteners. Applications: Cyanide zinc electroplating. (Harshaw Chemicals Ltd).*

Crystalline Boracite. See STASSFURTITE and BORACITE.

Crystal, Tin. See TIN SALTS.

Crystamycin. A proprietary preparation containing Benzylpenicillin sodium and Streptomycin. An antibiotic. (Glaxo Pharmaceuticals Ltd).†

Crystamycin Forte. A proprietary preparation containing Benzylpenicillin Sodium and Streptonlycin. An antibiotic. (Glaxo Pharmaceuticals Ltd).†

Crystapen. A proprietary preparation containing Benzylpenicillin Sodium. Sodium penicillin G. An antibiotic. (Glaxo Pharmaceuticals Ltd).†

Crystapen G. A proprietary preparation containing Benzylpenicillin (as the potassium salt). An antibiotic. (Glaxo Pharmaceuticals Ltd).†

Crystapen V. A proprietary preparation containing Phenoxymethylpenicillin (as the calcium salt). An antibiotic. (Glaxo Pharmaceuticals Ltd).†

Crystex. A proprietary viscose product used for packing.†

Crystic. A range of unsaturated polyester resins. Applications: The manufacture of reinforced plastics articles and castings and coatings. (Scott Bader Co Ltd).*

Crysticillin. Penicillin G Procaine. Applications: Antibacterial. (E R Squibb & Sons Inc).

Crystodigin. Digitoxin. Applications: Cardiotonic. (Eli Lilly & Co).

Crystoids. Hexylresorcinol. Applications: Antihelmintic preparation. (Merck Sharp & Dohme).*

Crystolite. See CARBORA.

Crystolon. A registered trade name for various types of goods such as grinding wheels, abrasives, and refractory grain, etc., made from silicon carbide.†

Crystran. Crystals for optical transmission. (BDH Chemicals Ltd).

CSI. Crofilcon A. Applications: Contact lens material. (Syntex Ophthalmics Inc).

CT-680. A polyurethane elastomeric resin possessing good heat resistance.†

CT-690, CT-700. Polyurethane elastomeric resins possessing greater heat resistance than CT-680. (Unichem).†

CTW. Epoxy resin based cements, screeds and coatings. (Prodorite).*

Cuba Black. (Dianil Black). A dyestuff. Dyes cotton black from an alkaline bath.†

Cuba Orange. A dyestuff prepared by the action of sodium sulphite upon diazo-naphthalene-sulphonic acid. Dyes wool orange.†

Cuba Wood. See FUSTIC.

Cubanite. A mineral, $CuFe_2S_4$.†

Cube Ore. See TYROLITE.

Cubical Catechu. See CUTCH.

Cubic Nitre. See CHILE SALTPETRE.

Cubic Saltpetre. See CHILE SALTPETRE.

Cucar. Copper carbonate. (Mechema).*

Cuclat. Kerosene sweeting catalyst. (Mechema).*

Cudbear. See ARCHIL.

Cudgel. A suspension of microcapsules containing fonofos. Applications: An insecticide for the control of vine weevil, sciarid fly and cabbage root fly. (Fisons PLC).*

Cufenium. An alloy of 22 per cent nickel, 72 per cent copper, and 6 per cent iron.†

Cufor. Copper formate. (Mechema).*

Cuite. Natural silk freed from the silk gum or sericin (*qv*). Raw silk consists of about 66 per cent fibroin, forming the real silk substance, and silk gum or sericin. The silk gum is removed by treating with hot neutral soap solution.†

Cuivre Poli. See ABYSSINIAN GOLD.

Cullet. Broken glass, used in the manufacture of glass.†

Cultar. Growth regulator for use on fruit trees. (ICI PLC).*

Cumar. See PARACOUMARONE RESIN.

Cumar Gum. See PARACOUMARONE RESIN.

Cumar Resin. See PARACOUMARONE RESIN.

Cumate. A proprietary trade name for a vulcanizing accelerator containing copper.†

Cumidine Ponceau. See PONCEAU 3R.

Cumidine Red. See PONCEAU 3R.

Cuniloy. An alloy of nickel, manganese, copper, and small quantities of lead.†

Cunitex. Bird and animal repellant. (May & Baker Ltd).*

Cupac. Copper acetate. (Mechema).*

Cupalit. A softening agent and lubricant. Applications: Used by the textile industry for the after treatment of all types of fibres. (Degussa).*

Cupar. Copper arsenate. (Mechema).*

Cuperatin. Copper albuminate.†

Cupertine. Bordeaux mixture 84%, maneb 8%. Applications: Wettable powder used as protective fungicide for foliage application to ornamental and crop plants. (Industrias Quimicas Del Valles SA).*

Cupertine Folpet. Bordeaux mixture 80%, folpet 10%. Applications: Wettable powder used as protective fungicide for foliage application to ornamental and crop plants. (Industrias Quimicas Del Valles SA).*

Cupertine Super. Bordeaux mixture 90%, cymoxanil 3%. Applications: Wettable powder used as protective and curative fungicide for foliage application to ornamental and crop plants. (Industrias Quimicas Del Valles SA).*

Cupferron. (Copperone). Nitro-sophenyl-hydroxylamine. The ammonium salt is used as a precipitating agent for copper, in the determination of copper.†

Cupolloy. Ferro-alloy briquettes for cupola. (Foseco (FS) Ltd).*

Cuprammonium Silk. (Cuprate Silk). Artificial silks made by dissolving cotton in a cuprammonium solution (copper hydrate dissolved in ammonia), and precipitating it in a fine thread.†

Cuprammonium Solution. See CUPRAMMONIUM SILKS and WILLESDEN FABRICS.

Cupranil Brown B. A dyestuff. It is a British equivalent of Diamine brown B.†

Cupranium. A name for certain brass and bronze alloys. The bronze contains tantalum and vanadium.†

Cuprase. A form of colloidal copper. A proprietary trade name for a vulcanizing accelerator containing copper.†

Cuprate Silk. See CUPRAMMONIUM SILK.

Cuprex. Copper alloy general purpose flux. (Foseco (FS) Ltd).*

Cupriferous Calamine. See TYROLITE.

Cuprimine. Penicillamine. Applications: Wilson's disease, cystinuria. (Merck Sharp & Dohme).*

Cuprinol. A copper naphthenate or sodium pentachlorphenate preparation used as a preservative for timber and fabrics.†

Cuprit. Copper alloy cleansing flux. (Foseco (FS) Ltd).*

Cupro-aluminiums. Alloys of aluminium and copper containing 1-20 per cent aluminium. They are also wrongly called aluminium bronzes.†

Cuprocyan. A double cyanide of potassium and copper.†

Cuprodine. A medium for coating steel with copper. (ICI PLC).*

Cuproid. Cuprous chloride. (Mechema).*

Cupromagnesium. An alloy of 90 per cent copper and 10 per cent magnesium. It has a specific gravity of 8.4, melts at 1290°C., and is used as a deoxidizer.†

Cupron. A proprietary trade name for a copper nickel alloy with a low temperature coefficient of resistance.†

Cuprophenyl. An after coppering direct dye. (Ciba Geigy PLC).†

Cupror. An aluminium bronze. It contains 94.2 per cent copper and 5.8 per cent aluminium.†

Cupro-steel. An alloy of steel with copper up to 4 per cent. Occasionally used for printing rollers and projectiles.†

Cupro-titanium. Usually an alloy of copper with 10 per cent titanium. Used as a deoxidizer in making brass and bronze castings.†

Cuprotungstite. See CUPROSCHEELITE.

Cupro-vanadium. An alloy usually containing from 10-15 per cent vanadium, 60-70 per cent copper, 10-15 per cent aluminium, and 2-3 per cent nickel.†

Curaseal. Natural and synthetic rubber latex based adhesives and coatings - cohesive, pressure-sensitive and synthetic resin emulsions. Applications: Numerous adhesive bonding applications. (Testworth Laboratories Inc).*

Curasol. Soil erosion inhibitor. (Hoechst UK Ltd).

Curatin. Doxepin hydrochloride. Applications: Veterinary antipruritic. (Pfizer Inc).

Curcuma. See TURMERIC.

Curcumeine. See CITRONINE B and FAST YELLOW N.

Curcumine. Geigy. See BRILLIANT YELLOW S.†

Curene. Catalysts and light and heat stabilizers. Applications: Catalysts and chain extenders for urethane polymers. (Anderson Development Company).*

Curenox-50. Copper oxychloride formulated to 50% metallic copper contents. Applications: Wettable powder used as protective fungicide for foliage application to ornamental and crop plants. (Industrias Quimicas Del Valles SA).*

Curgon. A proprietary trade name for naphthenate driers.†

Curithane. Reactive polyamines used as a hardener and curing agent for polyisocyanurate foams. (Dow Chemical Co Ltd).*

Curodex. Rubber odorants and deodorants. (Bush Boake Allen).*

Curolac. Precatalysed and two-pack curing lacquers. (The Scottish Adhesives Co Ltd).*

Curzate. Fungicide. (Du Pont (UK) Ltd).

Cusamon. Copper sulphate monohydrate. (Mechema).*

Cusatrib. Tribasic copper sulphate. (Mechema).*

Cusiloy A. A proprietary trade name for an alloy containing 95.5 per cent copper, 3 per cent silicon, 1 per cent iron, and 0.5 per cent tin.†

Cusyd. Copper thiocyanate. (Mechema).*

Cut Aid. ASTM S-315 Oil, 1.1.1. trichlorethane, mask odour No 3. Applications: Excellent cutting aid for general machine operations. Exceptional for disc sanding, filing and punch press operations. (Doyle Specialties).*

Cutal. See CUTOL.

Cutch. (Kutch, Catechu, Katechu, Gambier, Japan earth, Cachou, Cutt). Natural dyestuff. It consists of a brown or reddish-brown amorphous extract, obtained by boiling with water the wood of *Acacia* spp., *Mimosa* spp., or *Uncaria* spp. Contains catechins†

Cutina. Self-emulsifying raw material. Applications: Creams and emulsions. (Henkel Chemicals Ltd).

Cutlanego. An alloy of 50 per cent bismuth and 50 per cent tin. Used for tempering steel tools.†

Cutlass. Growth regulator for hedges. (ICI PLC).*

Cutonic. Micronutrient foliar sprays. (McKechnie Chemicals Ltd).

Cutt. See CUTCH.

CW 79. Lidofilcon B. The material contains 79% of water. Applications: Contact lens material. (Bausch & Lomb, Professional Products Div).

Cyanacryl. A proprietary trade name for a synthetic rubber with excellent resistance to oils and high temperatures. (Anchor Chemical Co).*

Cyanamer P35 - P70. Acrylamide copolymers (anionic). (Cyanamid BV).*

Cyananthrene B Double. An anthracene dyestuff prepared by melting benzanthrone-quinoline with alkali hydroxides.†

Cyanine. (Leitch's Blue, Quinoline Blue). A blue pigment consisting of a mixture of Cobalt blue and Prussian blue. A quinoline dyestuff,is also known as Cyanine or Quinoline blue. It is used as a panchromatic sensitizer, and also dyes silk.†

Cyanine. Anionic dyestuffs (level dyeing). Applications: Wool and wool blends. (Holliday Dyes & Chemicals Ltd).*

Cyanine B. A dyestuff obtained by the oxidation of Patent blue with ferric salts or chromic acid. Dyes wool indigo blue.†

Cyanine Blue. See CYANINE.

Cyanine Fast. Acid dyes. (Holliday Dyes & Chemicals Ltd).

Cyanine Moderns. Dyestuffs, which are condensation products of gallo-cyanines and allyldiamines.†

Cyanite, Fireproof. See FIREPROOF CYANITE.

Cyanochalcite. See CHRYSOCOLLA.

Cyanol Extra. (Acid Blue 6G). A dyestuff. Dyes wool and silk blue from an acid bath.†

Cyanol FF. A dyestuff of the same class as Cyanol extra. It dyes wool and silk from an acid bath.†

Cyanol Green B. A triphenylmethane dyestuff, which dyes wool from a bath containing sodium sulphate and sulphuric acid.†

Cyanolime. Calcium cyanide. (Mechema).*

Cyanolit-Hitemp. A proprietary range of high-temperature cyanoacrylate adhesives. (Unichem).†

Cyanosin. See PHLOXINE.

Cyanosin A. (Cyanosin, Spirit Soluble). The alkali salt of tetrabromodichloro-fluoresceine methyl ether. Used in silk dyeing.†

Cyanosin B. A dyestuff. It is the sodium salt of tetrabromotetrachloro-fluoresceine ethyl ether. Dyes wool bluish-red.†

Cyanosin Spirit Soluble. See CYANOSIN A.

Cyanthrene. See INDANTHRENE DARK BLUE BO.

Cyasorb UV-531. A trade mark for an ultraviolet light absorbing additive for plastics. (56). See TINUPAL also.†

Cyasorb 5411. Octrizole. Applications: Ultraviolet screen. (American Cyanamid Co).

Cybis. Nalidixic acid. Applications: Antibacterial. (Sterling Drug Inc).

Cybond WD-4517 and WD-4521. Proprietary two-component polyurethane adhesives of the solvent type. (Unichem).†

Cyclaine. Hexylcaine hydrochloride. Applications: Topical anaesthesia of intact mucous membranes. (Merck Sharp & Dohme).*

Cyclamic Acid. N-Cyclohexylsulphamic acid. Hexamic acid.†

Cyclamine. A dyestuff, obtained by the bromination of the product of the action of sodium sulphide upon dichlorofluoresceine. It dyes wool and silk bluish-red from a neutral bath.†

Cyclamycin. A proprietary preparation of triaceto-oleandromycin. An antibiotic. (Wyeth).†

Cyclanon. Wetting agents for textiles. (BASF United Kingdom Ltd).

Cyclapen-W. Cyclacillin. Applications: Antibacterial. (Wyeth Laboratories, Div of American Home Products Corp).

Cyclatex. A proprietary cyclized rubber. (Hubron Rubber Chemicals, Failsworth, Manchester).†

Cyclimorph. A proprietary preparation of Cyclizine tartrate and morphine tartrate. An analgesic. (The Wellcome Foundation Ltd).*

Cyclite. A proprietary cyclized rubber. (Durham Raw Materials).†

Cyclo. Surfactant for cosmetics, toiletries, pharmaceutical, processing, agricultural and other industries. (Baxenden Chemical Co Ltd).*

Cyclobarbitone. 5-(Cyclohex-1-enyl)-5-ethylbarbituric acid. Phanodorm.†

Cyclochem. A fatty acid ester and emulsifying and wetting agent. (Cyclo Corporation).*

Cyclocort. Amcinonide. Applications: Glucocorticoid. (Lederle Laboratories, Div of American Cyanamid Co).

Cyclofor. A paint additive. (Cyclo Corporation).*

Cyclogol. Low foaming detergent, rinse aid. Applications: Machine dishwashing, soluble bath oils. (Witco Chemical Ltd).

Cyclogyl. Cyclopentolate hydrochloride. Applications: Anticholinergic. (Alcon Laboratories Inc).

Cyclolac. ABS polymers. Applications: Injection moulding, sheet extrusions. (Borg Warner Chemicals).*

Cyclomide. A fatty acid alkanolamide. (Cyclo Corporation).*

Cyclonal Sodium. A soluble hexabarbitone. (May & Baker Ltd).*

Cyclonette. Emulsifiable wax composition. (Cyclo Corporation).*

Cyclonox. Cyclohexanone peroxide. (Akzo Chemie UK Ltd).

Cyclopar. Tetracycline Hydrochloride. Applications: Anti-amebic; antibacterial; antirickettsial. (Parke-Davis, Div of Warner-Lambert Co).

Cyclophos. A phosphate ester. (Cyclo Corporation).*

Cyclopol. A detergent composition. (Cyclo Corporation).*

Cyclo-Prostin. Epoprostenol sodium. Applications: Inhibitor. (The Upjohn Co).

Cyclops Metal. A nickel-chromium-iron alloy, containing 18 per cent nickel, 18 per cent chromium, and the rest iron. It resists corrosion.†

Cyclorans. Wetting-out agents for textiles. They contain an alcohol of high boiling point emulsified with a potassium olein soap. Also see TERPURILE.†

Cyclo-Rubber. Cyclized rubber materials. (Hardman Inc).*

Cyclorubbers. Thermoplastic products made by heating a mixture of rubber sheet with about 10 per cent of its weight of an organic sulphonyl chloride or an organic sulphonic acid. These products resemble gutta-percha or can be made to resemble shellac according to treatment.†

Cycloryl. A liquid surface active agent. (Cyclo Corporation).*

Cycloryl 580 and 585N. Sodium lauryl sulphate, conforming to BP specification. Powder or needle form. Applications: emulsification polymerization; emulsification. (Witco Chemical Ltd).

Cyclosan. 4 per cent. Calomel dust. (May & Baker Ltd).*

Cycloserine Roche. Tuberculostatic drug. (Roche Products Ltd).

Cyclospasmol. A proprietary preparation cyclandelate. A vasodilator. (Brocades).*

Cycloteric. An amphoteric surface active agent. (Cyclo Corporation).*

Cycloton. A cationic emulsifying agent. (Cyclo Corporation).*

Cycogan. Active ingredient: chlormequat 2-chloroethyltrimethyl-ammonium chloride. Applications: Versatile plant growth regulant widely used for the prevention of lodging in wheat. (Agan Chemical Manufacturers Ltd).†

Cycolac. ABS thermoplastic compounds. (Borg Warner Chemicals).

Cycom. Resin impregnated glass, carbon or aramid fabrics. Applications: Advanced composite moulded components for military and aerospace applications. (Fothergill Tygaflor Ltd).*

Cyfloc 6000. Modified polyamine. (Cyanamid BV).*

Cyfol. A proprietary preparation of ferrous gluconate, folic acid and Vitamin B_{12}. A haematinic. (Rybar).†

Cygna. Self-emulsifiable mineral oil based and water soluble synlube based products. Applications: Fibre processing aids and lubricants. (Thomas Swan & Co Ltd).*

Cykelin. Bodied oils. Applications: Paint and ink. (NL Industries Inc).*

Cyklokapron. Solution for injection, tablets, oral solution. Applications: Antifibrinolytic agent. (KabiVitrum AB).*

Cylert. Pemoline. Applications: Stimulant. (Abbott Laboratories).

Cymag. A rodent, rabbit and insect exterminator containing hydrocyanic acid. (ICI PLC).*

Cymbal Brass. See BRASS.

Cymbal Metal. A brass containing 78 per cent copper and 22 per cent zinc.†

Cymbilide. Cyclamen aldehyde. (May & Baker Ltd).*

Cymbush. Insecticide containing cypermethrin. (ICI PLC).*

Cymogran. Phenylalanine-free casein hydrolysate. (Allen & Hanbury).†

Cymperator. Insecticide containing cypermethrin. (ICI PLC).*

Cymyl Orange. An indicator. It is an azo dye obtained by combining diazotized sulphonated amino-cymene with dimethyl-aniline.†

Cynorex. A bright cyanide copper plating process. (Hanshaw Chemicals).†

Cyomin. Cyanocobalamin. Applications: Vitamin. (O'Neal, Jones & Feldman Pharmaceuticals).

Cypan. Hydrolized polyacrylonitrile. (Cyanamid BV).*

Cyperkill. Insecticidal formulation. (Mitchell Cotts Chemicals Ltd).

Cypersect. Insectical formulation. (Mitchell Cotts Chemicals Ltd).

Cypress Pine Resin. See PINE GU.

Cyprian Turpentine. See CHIAN TURPENTINE.

Cypromin. Rolicyprine. Applications: Antidepressant. (Schering-Plough Corp).

Cyren-A. A synthetic oestrogen for implantation in carcinoma of the prostrate. (Bayer & Co).*

Cyrene. A resin used in adhesives, coatings, and moulding compounds.†

Cyrez 963. Hexamethoxymethylamine. (Cyanamid BV).*

Cyrez 963/4 Powders. Hexamethoxymethylamine on silica carrier. (Cyanamid BV).*

Cystamin. See HEXAMINE.

Cystemme. A proprietary preparation of sodium citrate as granules intended for dissolution in water before taking. Applications: To alleviate the symptons of cystitis. (Abbott Laboratories).*

Cysto-Conray. Iothalamate meglumine. Applications: Diagnostic aid. (Mallinckrodt Inc).

Cystogen. See HEXAMINE.

Cystografin. Diatrizoate meglumine. Applications: Diagnostic aid. (E R Squibb & Sons Inc).

Cystopurin. Urinary antiseptic tablets. (Fisons PLC, Pharmaceutical Div).

Cystorelin. Gonadorelin acetate. Luteinizing hormone-releasing factor acetate hydrate. Applications: Gonad-stimulating principle. (Abbott Laboratories).

Cystospaz. Hyoscyamine. Applications: Anticholinergic. (Alcon Laboratories Inc).

Cytacon. A proprietary preparation of cyanocobalamin. Oral Vitamin B_{12} preparations. (Glaxo Pharmaceuticals Ltd).†

Cytadren. Aminoglutethimide. Applications: Adrenocortical suppressant; antineoplastic. (Ciba-Geigy Corp).

Cytamen. A proprietary preparation of cyanocobalamin Vitamin B_{12}. (Glaxo Pharmaceuticals Ltd).†

Cytomel. Liothyronine sodium. Applications: Thyroid hormone. (Smith Kline & French Laboratories).

Cytosar. A proprietary preparation of CYTARABINE. A cytotoxic drug. (Upjohn Ltd).†

Cytosar-U. Cytarabine. Applications: Antineoplastic; antiviral. (The Upjohn Co).

Cytotec. Misoprostol. Applications: Anti-ulcerative. (G D Searle & Co).

Cytoxan. SEE ENDOXAN. (Degussa).

Cytrel. A proprietary tobacco substitute. (Celanese Corporation).*

D

AB. A proprietary intermediate for various high-temperature plastics, used to make polypyrones and polyquinoxalines. It is 3, 3' - diamino - benzidine. (Upjohn Ltd).†

abinese. A proprietary preparation of chlorpropamide. An oral hypoglycemic agent. (Pfizer International).*

acron. A polyethylene terephthalate fibre having high strength and low water absorption. See also Terylene. (Du Pont (UK) Ltd).†

actil. A proprietary preparation containing piperidolate hydrochloride. An antispasmodic. (MCP Pharmaceuticals).†

ag. Colloidal dispersions of graphite and other products. (Acheson Colloids).*

agenite. A proprietary trade name for a bituminous asbestos-filled thermoplastic for accumulator cases.†

ahlia. See HOFMANN'S VIOLET and METHYL VIOLET B. Also a mixture of methyl violet and fuchsine.

ahl's Acids. Acid II, β-naphthylamine-4, 6-disulphonic acid. Acid III,β-naphthylamine-4,7-disulphonic acid.†

ahmenite A. An explosive containing ammonium nitrate, naphthalene, and potassium bichromate.†

aintex. A proprietary trade name for a wetting agent. It contains miscible terpene alcohols.†

airos. A dairy detergent. (Laporte Industries Ltd).*

airozon. A dairy sterilization agent. (Laporte Industries Ltd).*

airy Fly Spray. Synergized pyrethrins insecticide. (Ciba-Geigy PLC).

Dakamballi starch. A starch prepared from the fruit of *Aldina insignis*, a tree of British Guiana.†

Dakin's Solution. A mixture of hypochlorite and perborate of sodium, with small amounts of hypochlorous and boric acids. Antiseptic.†

Daktacort. A proprietary preparation of miconazole nitrate and hydrocortisone. Applications: For inflamed and infected skin conditions. (Janssen Pharmaceutical Ltd).*

Daktarin. A proprietary preparation of miconazole nitrate used as an anti-fungal agent. (Janssen Pharmaceutical Ltd).*

Dalacin C. A proprietary preparation of CLINDAMYCIN. An antibiotic. (Upjohn Ltd).†

Dalcaine. Lidocaine hydrochloride. Applications: Anaesthetic. (O'Neal, Jones & Feldman Pharmaceuticals).

Dalfratex. Flexible silica textiles. (The Chemical & Insulating Co Ltd).

Dalgan. Dezocine. Applications: Analgesic. (Wyeth Laboratories, Div of American Home Products Corp).

Dalivit. Multivitamin preparations. Applications: Indications for the prevention and treatment of vitamin deficiency states. (Paines & Byrne).*

Dalmane. A proprietary preparation of Flurazepam. A benzodiazepine hypnotic. (Roche Products Ltd).*

Dalmatian Insect Powder. See INSECT POWDER.

Dalnate. Tolindate. Applications: Antifungal. (USV Pharmaceutical Corp).

Dalpad. Coalescing agent assists film formation of certain latexes, maintains that characteristic during formulation storage and resists tendency (of films) towards water sensitivity. (Dow Chemical Co Ltd).*

Daltocel. Polyesters or polyethers for flexible foams. (ICI PLC).*

Daltoflex. Polyurethane flexible foams. (ICI PLC).*

Daltogard. A polyurethane foam additive. (ICI PLC).*

Daltolac. Polyurethane rigid foams. (ICI PLC).*

Daltomold. Trade mark (512) for a range of plastic moulding compounds of which some examples follow.

135:-A thermoplastic polyurethane elastomer of the polyester type. It has a Shore D hardness of 35.

140:-Similar to Daltomold 135 but with a Shore D hardness of 40.

150:-Similar to Daltomold 135 but with a Shore D hardness of 50.

160:-Similar to Daltomold 135 but with a Shore D hardness of 60.

230:-A thermoplastic urethane elastomer of the polyether type. It has a Shore D hardness of 30.

238 and 338:-Thermoplastic polyurethane elastomers of the polyether type, with a Shore D hardness of 38.

245:-A thermoplastic polyurethane elastomer of the polyether type, with a Shore D hardness of 45.†

Daltoped. Polyurethane systems for shoe soiling. (ICI PLC).*

Daltorez. Polyester resins for adding to polyurethane foams and elastomers. (ICI PLC).*

Daltorol. Polyester for printers rollers. (ICI PLC).*

Damar Bronze. See DAMASCUS BRONZE.

Damascenised Steel. A steel made by repeatedly welding, drawing out, and doubling up a bar composed of a mixture of steel and iron, the surface of which has been treated with an acid. The steel is left with a black coating of carbon, and the iron retains its metallic lustre.†

Damascus Bronze. (Damar bronze). A alloy of 76 per cent copper, 10.5 pe cent tin, and 12.5 per cent lead. A bearing metal.†

Damiana. The dried leaves of a Mexic plant, *Turnera diffusa*. A tonic.†

Dammar Resin. A resin obtained from *Hopea, Shorea*, and *Balanocarpus* species, mainly from Malaysia. The melting point usually varies from 9 200° C., acid value from 33-72, an ash from 0.04-0.52 per cent. Rock dammar from Burma is derived fro *Hopea odorata*, and has a specific gravity of from 0.98-1.013, a meltin point of 90-115° C., a saponificatio value of 31-37, acid value 31, and a 0.55-0.68 per cent. A Borneo damm from *Retinodedron rassak*, melts at 130-150° C., has an acid value of 1 150, and saponification value 159-1 Perak dammar, obtained from *Balanocarpus heimii* of Malay Stat is pale yellow and amber, has an ac value of 34-37, and a melting point 80-100° C. Batavian dammar is a g variety, melting at 100° C., and hav an acid value of 35.5. Dammar Ma Kuching or Cat's-Eye dammar is a grade dammar from species of *Hop* and has a melting-point of 80-100° and an acid value of 21-24. Damma Sengai, from trees of *Busseraceae* species of Malay. It is dark brown i colour, has a melting point of 120-1 C., and a low acid value. Dammar S is obtained from *Shorea ridleyana* o Malay. It is hard and dark in colour has a melting point of 190-220° C., an acid value of 26. Dammar Tema from *Shorea crassifolia* of Malay. A pale resin with a melting point of 82 85° C. Has an acid value of 17-25. Dammar Hiroe, a type of Borneo dammar having an acid value of 13. saponification value of 57-61, meltin point 190-200 °C., and is soluble in benzene, turpentine, and chloroform is suitable for use in lacquers and varnishes. Dammar Hitaru, from *Balanocarpus penangianus* of Mala melts at 140° C., has an acid value 16, and a saponification value of 34. Dammar Kapur, from Kapur, the

Borneo camphor wood tree. It has an acid value of 52 and a saponification value of 78. Dammar Kedon-dong from Kedondong, a name for a species of *Canarium* of Malay. It has a low acid value. Dammar, Meranti Tembaga, from a tree of same name of Malay. It has a melting point of 180-210°C., and an acid value of 30-41. Dammar Meranti Jerit, from a tree of same name, a species of *Shorea* of Malay. It melts at 60-70° C. and has an acid value of 11. Dammar Minyak from *Ayathio alba* of Malay. A milky-white resinous liquid with acid value of 130. Other varieties of dammar resin mainly of low grade are : Dammar Hitaru, acid value 14.2, melting-point 140-170° C.; Dammar Kepong, acid value 10.1, melting-point 160-180° C. ; Dammar Saraya, acid value 24.2, melting-point 135-175° C. ; Dammar Batu, acid value 18.1, melting point 140-180° C. ; Dammar Daging, acid value 23.3, melting point 120-160° C. Dammar is soluble in amyl acetate, benzene, carbon tetrachloride, and chloroform, and partly soluble in acetone and alcohol. It is used in the manufacture of spirit varnishes.†

Damoil. Phytonomic oil. Applications: Dormant and summer spray oil for use as a contact insecticide. (Drexel Chemical Company).*

Danbar. A proprietary preparation of veratrum viride alkaloids, for scalp application. (Gerhardt Pharmaceuticals).*

Daneral. A proprietary preparation of pheniramine *p*- aminosalicylate. (Hoechst Chemicals (UK) Ltd).†

Daneral-SA. A proprietary preparation of pheniramine maleate. (Hoechst Chemicals (UK) Ltd).†

Danex. Pyrithione Zinc. Applications: Antibacterial; antifungal; antiseborrheic. (Herbert Laboratories, Dermatology Div of Allergan Pharmaceuticals Inc).

Danex. Active ingredient: trichlorphon; dimethyl-2,2,2-trichloro-1-hydroxyethylphosphorate. Applications: Organophosphate agricultural insecticide with a very broad range of activity. (Makhteshim Chemical Works Ltd).†

Danforth's Oil. See NAPHTHA.

Danocrine. Danazol. Applications: Anterior pituitary suppressant. (Sterling Drug Inc).

Danol. A proprietary preparation of danazol, used as a suppressant of gonadotrophins. (Winthrop Laboratories).*

Dantafur. Nitrofurantoin. Applications: Antibacterial. (Norwich Eaton Pharmaceuticals Inc).

Danthron. 1,8-Dihydroxyanthraquinone.†

Dantrium. Dantrolene sodium. Applications: Relaxant. (Norwich Eaton Pharmaceuticals Inc).

Dantyl. A proprietary preparation of *p*-aminosalicylic acid phenylester, *p*-aminosalicylic acid and sucrose. Anti-tuberculous agent. (Leo Laboratories).†

Dantyl-Inah. A proprietary preparation of phenyl-*p*-aminosalicylic acid, *p*-aminosalicylic acid, isoniazid and sucrose. Antituberculous drug. (Leo Laboratories).†

Daonil. A proprietary preparation of GLIBENCLAMIDE used in the treatment of late-onset diabetes. (Hoechst Chemicals (UK) Ltd).†

Daotan. Polyurethane modified alkyd resins. Applications: Air drying paints and varnishes. (Resinous Chemicals Ltd).*

Daphnetin. 7,8-Dihydroxycoumarin. $C_8H_6O_4$.†

Daphnin. See EOSIN BN.

Dapon M. A trade mark for diallyl isophthalate. A moulding material. (FMC Corporation).†

Dapon 35. A trade mark for diallyl phthalate. A moulding material. (FMC Corporation).†

Daponite Sheet. Diallyl phthalate resin. Applications: Decorative impregnated paper. (Sumitomo Bakelite, Japan).*

Dappol. Boiler and cooling water treatment. (Laporte Industries Ltd).

Daprisal. A proprietary preparation of dexamphetamine sulphate, amylobarbitone, aspirin and phenacetin.

(Smith Kline and French Laboratories Ltd).*

Dapsetyn. A chloramphenicol/dapsone veterinary preparation. (Allen & Hanbury).†

Dapsone. Di-(4-aminophenyl)sulphone. Diaphenylsulphone. Avlosulfon.†

Dapsyvet. A chloramphenicol/dapsone veterinary preparation. (Allen & Hanbury).†

Daptazole. A proprietary preparation of amiphenazole. A central nervous stimulant. (Nicholas).†

Daraclor. A proprietary antimalarial compound. (The Wellcome Foundation Ltd).*

Daranide. Dichlorphenamide. Applications: Potent carbonic anhydrase inhibitor for control of glaucoma. (Merck Sharp & Dohme).*

Daraprim Tablets. A proprietary formulation of pyrimethamine. Applications: For the chemoprophylaxis of malaria due to susceptible strains of plasmodia or conjoint with fast-acting schizonticides to initiate transmission control and suppressive cure. Also for the treatment of toxoplasmosis used conjointly with a sulphonamide. (The Wellcome Foundation Ltd).*

Daratac SP 1025. A PVC emulsion adhesive for sticking PVC film to a substrate. (Grace, W R & Co).†

Darbid. Isopropamide iodide. Applications: Anticholinergic. (Smith Kline & French Laboratories).

D'Arcet's Alloy. (*a*) Consists of 50 per cent bismuth, 25 per cent lead, and 25 per cent tin, melting-point 93° C. ; (*b*) contains 50 parts bismuth, 25 parts lead, 25 parts tin, and 250 parts mercury.†

Darcil. A proprietary preparation of potassium phenethicillin. An antibiotic. (Wyeth).†

Darco. A decolourizing and refining carbon. A substitute for bone char.†

Darco. Activated carbon, powdered and granular. Applications: Water treatment, glucose and sugar purification, chemical manufacture, gas and air treatment. (Norit).*

Daricol. A proprietary preparation of oxyphencyclimine. An antispasmodic. (Pfizer International).*

Daricon. A proprietary preparation of oxyphencyclimine. An antispasmodic. (Pfizer International).*

Daritran. A proprietary preparation of oxyphencyclimine and meprobromate. An antispasmodic and tranquillizer. (Pfizer International).*

Daritrax. A proprietary preparation of oxyphencyclimine and hydroxyzine. A antispasmodic and tranquillizer. (Pfize International).*

Dark Cylinder Oils. See CYLINDER OILS.

Dark Green. See DINITROSO-RESORCIN.

Dark Oxide of Iron. See INDIAN RED.

Dark Red Gold. An alloy of 50 per cent gold and 50 per cent copper.†

Darmex. High technology lubricant. Applications: All types of industrial/automotive equipment. (Darmex Corp).*

Darmex Plus. High technology lubricant. Applications: All types of industrial/automotive equipment. (Darmex Corp).*

Dartalan. A proprietary preparation of thiopropazate dihydrochloride. A sedative. (Searle, G D & Co Ltd).†

Darvan. Full line of dispensing agents. Applications: Used for dispensing materials in rubber, paint, ceramics, plastics, cosmetics, pharmaceuticals, agriculture and household products. (Vanderbilt Chemical Corporation).*

Darvisul. Sulphaquinoxaline/idveridine coccidiostat. (The Wellcome Foundation Ltd).

Darvon. Propoxyphene Hydrochloride. Applications: Analgesic. (Eli Lilly & Co).

Darvon-N. Propoxyphene Napsylate. Applications: Analgesic. (Eli Lilly & Co).

Darwins AR 655B. Similar to 654A but richer in molybdenum. It will resist attack up to boiling point.†

Dasag. See SUROPHOSPHATE. A fertilizer.

astar. Cholesterol. (Croda Chemicals Ltd).

atac. Adhesive. Applications: Industrial adhesives for packaging, paper-converting and woodworking. (Datac Adhesives Ltd).*

atagel. Activated datem gel. (Croda Chemicals Ltd).

atamuls. Increases volume, fermenting stability and gas retention; improved crumb structure; prolonged shelf-life; improved dough compatability with machines. Applications: Bread products. (Th Goldschmidt Ltd).

atem. Di-acetyl tartaric esters of monoglycerides. Edible emulsifiers for use in lipsticks and similar products. (Croda Chemicals Ltd).*

atril. Acetaminophen. Applications: Analgesic; antipyretic. (Bristol-Myers Co).

audelin Solder. An alloy of 65.6 per cent tin, 12.2 per cent zinc, 1 per cent aluminium, 17.4 per cent lead, 3.1 per cent copper, and 0.4 per cent phosphorus.†

avenol. A proprietary preparation containing carbinoxamine maleate, ephedrine hydrochloride and pholcodine. A cough linctus. (Wyeth).†

avey's Gray. A pigment prepared from siliceous earths. Used principally in mixtures with other colours to reduce tones.†

avis Metal. An alloy of 67 per cent copper, 29 per cent nickel, 2 per cent iron, and 1.5 per cent manganese.†

awson Bronze. An alloy of 83.9 per cent copper, 15.9 per cent tin, and traces of antimony, lead, iron, arsenic, and zinc.†

axid. A proprietary preparation of bacampicillin hydrochloride. An antibiotic. (Pfizer International).*

.B.A. See ACCELERATOR D.B.A.

BA Accelerator. Dibenzylamine and mono-benzylamine blend. A non-staining and non-discolouring activator for CPB, other xanthates or carbon disulphide. Used in natural, synthetic and nitrile rubbers. (Uniroyal).*

BP. Dibutyl phthalate. A plasticizer for vinyl and other plastics.†

DBPC. A proprietary anti-oxidant. Di-*tert.*-butyl-para-cresol. (Koppers Co Inc).†

DC Cristobalite. Cristobalite containing investment material for precious metals. Applications: Dental preparation. (Bayer & Co).*

DC 150. Chemical catalyst system. Applications: Decorative bright chromium plating (using low chromic acid concentrations). (Harshaw Chemicals Ltd).*

DCI-3. Corrosion inhibitor. (Du Pont (UK) Ltd).

DCP. Dicapryl phthalate. A plasticizer for vinyl plastics.†

DC700. Chemical catalyst system. Applications: Decorative bright chromium plating. (Harshaw Chemicals Ltd).*

DDAVP. Desmopressin acetate. Applications: Antidiuretic. (USV Pharmaceutical Corp).

D.D.D. An accelerator for rubber vulcanization. It is dimethylamine dimethyldithiocarbamate.

DDT. Dichlorodiphenyltrichlorethane. A powerful insecticide.†

De-Acidite. An anion exchange material. (Permutit-Boby Ltd).†

Dead Silver. (Frosted Silver). Silver, whitened by heating in air, and immersed in dilute sulphuric acid.†

Deanase. A proprietary preparation of desoxyribonuclease for treatment of ulcers, bruises and abscess. (Consolidated Chemicals).†

Deanase D.C. A proprietary preparation of delta chymotrypsin. Digestive enzyme. (Consolidated Chemicals).†

Deanol. 2-Dimethylaminoethanol.†

Deanox. Iron oxides. (Deanshanger Oxides Ltd).

Deapril-ST. Ergoloid mesylates. Applications: Cognition adjuvant. (Mead Johnson & Co).

Debenal. Geriatric drug for dogs and cats. Applications: Veterinary medicine. (Bayer & Co).*

Debendox. A proprietary preparation of dicyclomine hydrochloride, doxylamine succinate and pyridoxine hydrochloride. Antiemetic. (Richardson-Vicks Inc).*

Debron 711. A proprietary coating compound manufactured from polyphenylene sulphide. (de Beers Laboratories Inc, Broadview, Ill).†

Deca-Durabolin. Nandrolone decanoate. Applications: Androgen. (Organon Inc).

Deca-Indocid. Dexamethasone and indomethacin. Applications: The treatment of rheumatoid arthritis and certain non-articular musculoskeletal disorders. (Merck Sharp & Dohme).*

Decaderm. Dexamethasone. Applications: Glucocorticoid. (Merck Sharp & Dohme, Div of Merck & Co Inc).

Decadex. Water-based elastomeric weatherproofing compound applied by brush or spray. Applications: Roof refurbishment, anti-carbonation/all application external coating in range of colours. (Liquid Plastics Ltd).*

Decadron. Dexamethasone. Applications: Certain endocrine and non-endocrine disorders responsive to corticosteroid therapy. (Merck Sharp & Dohme).*

Decadronal. Dexamethasone acetate. Applications: Parenteral therapy for sustained relief of certain endocrine and non-endocrine disorders responsive to corticosteroid therapy. (Merck Sharp & Dohme).*

Decadron Duofase. Dexamethasone. Applications: Prompt and sustained corticosteroid activity for the treatment of certain endocrine and non-endocrine disorders. (Merck Sharp & Dohme).*

Decadron-LA. Dexamethasone acetate. Applications: Adrenocortical steroid. (Merck Sharp & Dohme, Div of Merck & Co Inc).

Decadron Shock-Pak. Dexamethasone. Applications: For the adjunctive treatment of shock. (Merck Sharp & Dohme).*

Decalex. A photographic developer. (May & Baker Ltd).*

Decalin. Decahydronaphthalene, $C_{10}H_{18}$, a paint and resin solvent. Used as turpentine substitute.†

Decamphorized Oil of Turpentine. (Oxidized Oil of Turpentine). The residue from the manufacture of camphor. It consists mainly of dipentene.†

Decapryn Succinate. Doxylamine succinate. Applications: Antihistaminic. (Merrell Dow Pharmaceuticals Inc, Subsidiary of Dow Chemical Co).

Decaspray. Dexamethasone and neomycin. Applications: A topical spray for corticosteroid-responsive skin and eye conditions. (Merck Sharp & Dohme).*

Deccox. Decoquinate. (May & Baker Ltd).

Decelox. Zinc oxide, modified to give slow rubber cure. (Durham Chemicals Ltd).*

Dechlorane A-O. A proprietary synergistic agent for fire-retardant plastics. It is antimony oxide and contains halogens. (Kingsley and Keith Chemical Corp).†

Dechlorane Plus 515, 25, 2520, 1000. Non-plasticizing, non-reactive and thermally stable flame retardant additives differing only in particle size. Used in polymer formulations where hydrolytic stability, excellent electrical properties and performance criteria such as UL 94, IEEE 383 and VW-1 are important. Dechlorane Plus 1000 is designed for applications where a finely ground, narrow particle size distribution product is required such as in thin film, coating and adhesive applications. (Occidental Chemical Corp).*

Decholin. Dehydrocholic acid. Applications: Choleretic. (Miles Pharmaceuticals, Div of Miles Laboratories Inc).

Deck Seal-PD. Polysiloxane and methylmethacrylate in an aromatic and aliphatic solvent vehicle system. Applications: Sealer for parking decks, bridge decks, ramps, stadiums etc. (Nova Chemical Inc).*

Declinax. A proprietary preparation of dibrisoquine sulphate. An antihypertensive. (Roche Products Ltd).*

Declomycin. Demeclocycline hydrochloride. Applications: Antibacterial. (Lederle Laboratories, Div of American Cyanamid Co).

Decoart. Phenolic resin. Applications: Decorative laminates. (Sumitomo Bakelite, Japan).*

Deco Board P. Polyester resin. Applications: Decorative laminates. (Sumitomo Bakelite, Japan).*

Decol. A range of alkyl aryl sulphonates. Applications: All purpose liquid detergent concentrates, detergent bases for powdered cleansers, foaming agents for plaster board production, concrete foaming agents, wetting agents in textile processing and raw wool scouring detergent. (Harcros Industrial Chemicals).*

Decola. Melamine resin. Applications: Decorative laminates. (Sumitomo Bakelite, Japan).*

Decola Back Sheet. Phenolic resin. Applications: Back Sheet. (Sumitomo Bakelite, Japan).*

Decola Excel. Melamine resin. Applications: Decorative laminates. (Sumitomo Bakelite, Japan).*

Decola F. Melamine resin. Applications: Decorative. (Sumitomo Bakelite, Japan).*

Decola FG. Melamine resin. Applications: Decorative laminates. (Sumitomo Bakelite, Japan).*

Decola MA. Melamine resin. Applications: Decorative laminates. (Sumitomo Bakelite, Japan).*

Decola MF. Melamine resin. Applications: Decorative laminates. (Sumitomo Bakelite, Japan).*

Decolamide. A range of alkanolamides. Applications: Used as foam boost additives in detergents and shampoos, super fatting agents, opacifiers, thickeners, demulsifiers and emulsifiers. (Harcros Industrial Chemicals).*

Decola New Marine. Melamine resin. Applications: Decorative laminates. (Sumitomo Bakelite, Japan).*

Decola PFC. Melamine resin. Applications: Postform. (Sumitomo Bakelite, Japan).*

Deconyl. A proprietary trade name for a weatherproof nylon coating. (Plastic Coatings Ltd).*

Deco Poly. Polyester resin. Applications: Decorative laminates. (Sumitomo Bakelite, Japan).*

Decopress. Antithrombotic stocking. (Bayer & Co).*

Decorpa. A proprietary preparation of guar gum granules used as an antiobesity agent. (Norgine).†

Decrolin. Textile discharing and stripping agents. (BASF United Kingdom Ltd).

Decroline. See HYDROSULPHITE.

Decrysil. 4, 6-Dinitro-*o*-cresol.†

De De Tane. DDT. products. (*qv*). (Murphy Chemical Co).†

Deeline. See PARAFFIN, LIQUID.

Deenax. A proprietary anti-oxidant. 2, 6 di - *tert.* - butyl 4 -methyl - phenol. (Exxon Chemical Co USA).†

Deep Blue Extra R. See BLUE 1900.

Deep Feed. Liquid fertilizer. (Fisons PLC, Horticulture Div).

Defencin. A proprietary preparation of ISOXUPRINE resinate. A peripheral vasodilator. (Bristol-Myers Co Ltd).†

Defirust. A proprietary trade name for a rustless iron containing 12-15 per cent chromium and 0.1 per cent carbon.†

Deflamene. Formocortal. Applications: Glucocorticoid. (Farmitalia (Farmaceutici Italia)).

Defol. Sodium chlorate. Applications: Cotton defoliant, dessicant for corn, grain sorghum, sunflowers, safflowers, rice, soybeans, chili peppers and guar beans. (Drexel Chemical Company).*

Defolia. Defoliant for hops. (Murphy Chemical Co).†

Degadur. Methacrylic resins. Applications: Self-levelling coatings, mortar, as floor and wall coverings, road construction, ship deck coatings. (Degussa).*

Degalan. Polymethyl methacrylate compounds for injection moulding and extrusion. Applications: Production of light- and weather-resistant sheets, tubes and specially shaped profiles as well as finished parts such as car light assemblies, graduated dials, writing and drawing equipment, household utensils, watch crystals, optical lenses, lamp covers. (Degussa).*

Degalan V. An acrylic modifier for pvc†

Degalex. Aqueous pure acrylic emulsions. Applications: Used as binder for the production of high grade emulsion paints and synthetic resin plasters. (Degussa).*

Degament. Cold-curing, low-viscosity methacrylate resins for precast elements (concrete/polymer/composite). Applications: Coverings on buildings, walk on elements, pictures, sanitary components, stationary sport installations, machine mountings, stone bondings and marble agglomerates. (Degussa).*

Degapas. Aqueous solutions of acrylic acid and methacrylic acid polymers (S series) and their sodium salts (N series). Applications: For the dispersion of organic solid materials; as a builder in phosphate free or low phosphate cleaners; as thickening agents; for the complexing of multivalent metal cations; for the sizing of polyamide fibres. (Degussa).*

Degaplast. Acrylic resins and foam resins. Applications: Production of cast arm and leg prostheses, apparatus, support corsets, night splints, bed casts; as filler resin for last tip extensions and corrections. (Degussa).*

Degaser. A degassing agent for aluminium alloys. (Foseco (FS) Ltd).*

Degest-2. Naphazoline hydrochloride. Applications: Adrenergic. (Barnes-Hind Inc).

Deglas. Extruded acrylic sheet. Applications: To cover lighting fixtures, for light domes, illuminated signs, wash basins and bathtubs, roof canopies etc. (Degussa).*

Degopol. Components for polyurethane shoe soling systems. (ICI PLC).*

Degopur. Polyurethane systems. (ICI PLC).*

Degubond. A precious metal alloy. Applications: For dentistry and dental engineering. (Degussa).*

Degucast. A precious metal alloy. Applications: For dentistry and dental engineering. (Degussa).*

Degudent. A range of precious metal alloy. Applications: For dentistry and dental engineering. (Degussa).*

Degulor. A range of precious metal alloys. Applications: For dentistry and dental engineering. (Degussa).*

Degusorb. A range of activated carbons. Applications: Used for decolourizing, cleaning and deoderising in the chemical, pharmaceutical and food industry. Also used in the beverage industry. (Degussa).*

D.E.H. Epoxy curing agents (hardeners) specifically designed to provide a desirable range of properties and handling characteristics for the hardening of epoxy resins. (Dow Chemical Co Ltd).*

De Haën Salt. A double salt of antimony trifluoride and ammonium sulphate, $SbF_3(NH_4)_2SO_4$. A mordant.†

Dehybor. Borax from which the water of crystallization has been removed by heat. (Anhydrous borax.) It is widely used in glass, vitreous enamel, ceramic glaze and metallurgical industries wher borax has to be melted. (Borax Consolidated Ltd).*

Dehydrite. A registered trade name for magnesium perchlorate trihydrate. A drying agent.†

Dehydrocholin. A proprietary preparation of dehydrocholic acid. A laxative. (British Drug Houses).†

Dehymuls. Mixture of higher molecular esters mainly mixester of penthaerithrit fatty acid ester. Applications: Water-in-0il type ointment and creams. (Henkel Chemicals Ltd).

Dehyquart A. Cetyl trimethyl ammonium chloride in liquid form. Applications: Cationic surfactant used in hair conditioner, deodorants and skin cream. (Henkel Chemicals Ltd).

Dehyquart C. Lauryl pyridinium chloride, in re-crystallized powder form. Applications: Cationic surfactant used in hair conditioner, especially antistatics. (Henkel Chemicals Ltd).

Dehyquart CDB. Cetyl dimethyl benzylammonium chloride in liquid form. Applications: Cationic emulsifier used in hair cosmetics and antistatics. (Henkel Chemicals Ltd).

Dehyquart DAM. Di-stearyl dimethyl ammonium chloride in paste form. Applications: Cationic surfactant used as a softening agent in hair care preparations. (Henkel Chemicals Ltd).

Dehyquart LDB. Lauryl dimethyl-benzyl ammonium chloride in liquid form. Applications: Cationic surfactant with germicidal effect, used in body lotion. (Henkel Chemicals Ltd).

Dehyquart LT. Lauryl trimethyl ammonium chloride in liquid form. Applications: Cationic surfactant used in shampoos and hair conditioners. (Henkel Chemicals Ltd).

Dehyquart SP. Oxyethyl alkyl ammonium phosphate in liquid form. Applications: Cationic surfactant used in hair care preparations eg hair packs, rinsing agents, hair conditioners, hair tonics. (Henkel Chemicals Ltd).

Dehyton AB-30. Fatty amine derivative with betaine structure, in liquid form. Applications: Liquid shampoos, especially baby and special shampoos. (Henkel Chemicals Ltd).

Dehyton K. Betaine in liquid form. Applications: Additive for shampoos. (Henkel Chemicals Ltd).

Dekol. Textile dyeing levelling agents. (BASF United Kingdom Ltd).

Dekrysil. A proprietary preparation of 4, 6-dinitrocresol.†

Delac MOR. N-oxydiethylene-benzothiazole-2-sulphenamide. The most delayed action accelerator offered by Uniroyal. Activated by thiurams, dithiocarbamates, BIK, guanidines and aldehyde-amines. Non-discolouring and non-staining to rubber stocks and materials in contact with them. Used in tyre treads, carcass, mechanicals and wire jackets. (Uniroyal).*

Delac NS. N-tert-butyl-o-benzothiazole-2-sulphenamide. A delayed action accelerator very safe at processing temperatures but produces high modulus stocks at curing temperatures. Activated by thiurams, dithiocarbamates, aldehyde amines, guanidines and BIK. Non-discolouring and non-staining to rubber stocks and materials in contact with them. Used in tyre treads, carcass, mechanicals and wire jackets. (Uniroyal).*

Delac S. N-cyclohexyl-2-benzothiazolesulphenamide. An all-purpose delayed action accelerator which combines superior scorch safety with shorter curing cycles. Used in tyre tread, carcass, camelback and mechanical goods. (Uniroyal).*

Delac-S. A proprietary rubber accelerator. It is 2-cyclohexyl- 2-benzthiazyl sulphenamide. (Naugatuck (US Rubber)).†

Delafield's Haematoxylin. A microscopic stain prepared by adding 4 grams haematoxylin dissolved in 25 cc absolute alcohol to 400 cc of a saturated aqueous solution of ammonium alum. This solution is exposed for several days, filtered, and 100 cc glycerin and 100 cc methyl alcohol added. Allowed to stand and filtered.†

Delalot's Alloy. An alloy containing 80 per cent copper, 18 per cent zinc, 2 per cent manganese, and 1 per cent calcium phosphate.†

Delalutin. Hydroxyprogesterone caproate. Applications: Progestin. (E R Squibb & Sons Inc).

Delan-Col. A fungicide containing dithianon. (ICI PLC).*

Delanium. A proprietary trade name for carbon and graphite materials highly resistant to all chemicals except some oxidizing agents. (Powell Duffryn Quarries Ltd).†

Delaphos. Zinc orthophosphate. Applications: Anticorrosive pigment for paints, particularly primer fomulations. (ISC Alloys Ltd).*

Delatestryl. Testosterone Enanthate. Applications: Androgen. (E R Squibb & Sons Inc).

Delatynite. An amber from Delatyn in the Galician Carpathians. Low in succinic acid and free from sulphur.†

Delaville. Zinc dust (minimum 95% metallic). Applications: Zinc rich anticorrosive paints and reducing agent in chemical processes. (ISC Alloys Ltd).*

Delax. Non-gripping laxative. (The Boots Company PLC).

Delchowyte. A decolourizing carbon, prepared from peat.†

Delegol. A strong disinfectant. (Bayer & Co).*

Delegol-T. Disinfectant on a phenol basis. Applications: Veterinary medicine. (Bayer & Co).*

Delestrogen. Estradiol valerate. Applications: Oestrogen. (E R Squibb & Sons Inc).

Delf HD Aerosol Adhesive. Synthetic rubber and resin. Applications: Universal heavy duty adhesive. (D L Forster Ltd)*

Delfloc. An aqueous solution of a cationic polymer. Applications: A retention aid for improved retention of fillers and fibres and improved drainage and drying in paper and paperboard manufacture. (Hercules Inc).*

Delf Silicone Aerosol. 5% silicone in inert lubricant. Applications: Universal release and lubricating agent, reduces friction. (D L Forster Ltd)*

Delft Blue. A pigment. It is a mixture of indigo and ultramarine.†

Delf 534 Aerosol Adhesive. Synthetic rubber and resin. Applications: Plastic and rubber foam, carpet tiles and carpet products. (D L Forster Ltd).*

Delhi Rustless Iron. A corrosion resisting alloy containing 18 per cent chromium, 1.5 per cent silicon, and not more than 0.08 per cent carbon.†

Delicron. Elastomeric precision impression material. Applications: For double-mix technique. (Bayer & Co).*

Delimon. A proprietary preparation of 1-phenyl-2,3-dimethyl-4-(2 - phenyl - 3 - methyl - hydroxazino - methyl) - pyrazolone-(5)-hydrochloride, paracetamol and salicylamide. An analgesic. (Consolidated Chemicals).†

Delinal. Propenzolate Hydrochloride. Applications: Anticholinergic. (Merrell Dow Pharmaceuticals Inc, Subsidiary of Dow Chemical Co).

Delmate. Delmadinone acetate. Applications: Progestin; anti- androgen; anti-oestrogen. (Syntex Laboratories Inc).

Delnet. A non-woven fabric made fom high density polyethylene, polypropylene and polypropylene copolymers. Applications: Non-adherent facing for surgical dressings,feminine hygiene products, disposable press cloth, breathable laminate, fusible adhesive, reinforcement for paper, plastic foam and other non-woven fabrics. (Hercules Inc).*

Delo. Lubricating oil. (Chevron).*

Delphine Blue. A dyestuff. Dyes chromed wool indigo blue. Employed in calico printing.†

Delrin. A proprietary trade name for a stiff strong engineering plastic of the acetal resin type. It has excellent fatigue resistance and is used as a replacement for die cast parts in gears. bearings and housings. (Dupont (UK) Ltd). Special grades are used, or possess characteristics, as follows: *Delrin 100*, for mouldings in which toughness is a prerequisite; *Delrin 150*, for general extrusions, blown bottles, tubing and rod; *Delrin 507* contains a light stabilizer; *Delrin 550*, in general purpose moulding and wire-coating applications; *Delrin 900*, a special purpose high-flow resin. See CELCON (Celanese Corporation) and HOSTAFORM C (Hoechst Chemicals (UK) Ltd).†

Delsene. Fungicide. (Du Pont (UK) Ltd).

Delta. Two-piece can inks. Applications: Two-piece cans, side wall printing of pilfer proof closures and fast cure sheet fed metal containers. (Coates Industrial Finishes Ltd).*

Delta Acid. (F-acid). β-Naphthylamine-7-sulphonic acid.†

Delta Cortef. A proprietary preparation of prednisone. (Upjohn Ltd).†

Delta Cortelan. A proprietary preparation for the acetate of prednisone. (Glaxo Pharmaceuticals Ltd).†

Deltacortone. Prednisone. Applications: Treatment of certain endocrine and non-endocrine disorders responsive to corticosteroid therapy. (Merck Sharp & Dohme).*

elta-Dome. Prednisone. Applications: Glucocorticoid. (Miles Pharmaceuticals, Div of Miles Laboratories Inc).

elta-Genacort. A proprietary preparation of prednisolone. An anti-inflammatory agent. (Fisons PLC).*

eltacortril. A proprietary preparation of prednisolone. A corticosteroid. (Pfizer International).*

eltalin. Ergocalciferol. Applications: Vitamin. (Eli Lilly & Co).

elta Metals. The registered trade mark for a variety of metals, metallic alloys and metal articles.†

eltasone. Prednisone. Applications: Glucocorticoid. (The Upjohn Co).

eltastab. Proprietary preparations of prednisolone, e.g. the acetate. (Boots Company PLC).*

eltyl. A proprietary trade name for a plasticizer. It is a fatty acid ester.†

elweve. Non-woven fabric made from polypropylene by the film extrusion-embossing-orientation process. Applications: Excellent backing material, skirt and cambric liners for furniture, reinforcement for films, plastic foam, needlepunched non-wovens and wall coverings, bale coverings. (Hercules Inc).*

ema. A proprietary brand of TETRACYCLINE. (USV Pharmaceuticals Corp).†

emavet. Dimethyl sulphoxide in aqueous solution. (Ciba-Geigy PLC).

emerol. Meperidine hydrochloride. Applications: Analgesic. (Sterling Drug Inc).

emetox. Insecticide containing demeton-s-methyl. (ICI PLC).*

emix 7730, 7740, 7750 Emulsifiers. Blends of disproportionated fatty acid/rosin acids in various ratios. Customer blended to customer needs. Applications: Used as the sole base stock for production of sodium and potassium soaps of fatty acid/rosin acid emulsifier systems used in making SBR. (Arizona Chemical Company).*

emodur RF. Bonding agent. (Bayer UK Ltd).

Demser. Metirosine. Applications: For the treatment of phaeochromocytoma. (Merck Sharp & Dohme).*

Demulen. A proprietary preparation of MESTRANOL and ETHYNODIOL diacetate. An oral contraceptive. (Searle, G D & Co Ltd).†

Demulen 50. A proprietary preparation of ethinyloestrodiol and ETHYNODIOL diacetate. An oral contraceptive. (Searle, G D & Co Ltd).†

D.E.N. Epoxy novolac resins and solutions designed to provide high temperature service for epoxy-type applications. Applications: Coatings and adhesives for abrasives. (Dow Chemical Co Ltd).*

DenClen. Denture cleanser (powder, tablets and liquid), for cleaning dentures. (Richardson-Vicks Inc).*

Dendrid. A proprietary preparation of IDOXURIDINE used as eye-drops. (Alcon Universal).†

De-Nol. A proprietary preparation of tri-potassium di-citratobismuthate used in the treatment of peptic ulcers. (Brocades).*

Denquel. Sensitive teeth toothpaste, to relieve the discomfort of sensitive teeth. (Richardson-Vicks Inc).*

Densil. Industrial biocides. (ICI PLC).*

Densites. Mining explosives. They contain ammonium nitrate, sodium or potassium nitrate, and trinitrotoluene.†

Dentplus Special. A proprietary trade name for dicalium phosphate dihydrate. Used as a thickening agent, cleaning agent, and carrier in toothpaste. (Hoechst Chemicals (UK) Ltd).†

Denver Clay. See BENTONITE.

Denver Mud. See BENTONITE.

Deoxidine. Metal treating compositions. (ICI PLC).*

Deoxidizing Tubes. Pure copper tubes containing a variety of deoxidizing agents for the treatment of copper and nickel alloys. (Foseco (FS) Ltd).*

Deoxiphos 600. Liquid, phosphoric acid based cleaner. Applications: deoxidizer for aluminium and steel. (Invequimica & CIA SCA).*

Deoxylyte. Metal treating compositions. (ICI PLC).*

Depakene. Valproic Acid. Applications: Anticonvulsant. (Abbott Laboratories).

Depakote. Divalproex sodium. Applications: Anticonvulsant. (Abbott Laboratories).

depAndro 100. Testosterone Cypionate. Applications: Androgen. (O'Neal, Jones & Feldman Pharmaceuticals).

depAndro 200. Testosterone Cypionate. Applications: Androgen. (O'Neal, Jones & Feldman Pharmaceuticals).

Depat. *O,O*-diethyl phosphoroamidothioate. Applications: An intermediate for phosphorus pesticides. (A/S Cheminova).*

Depen. SEE TROLOVOL. (Degussa).*

Depepsen. Sodium Amylosulphate. Applications: Enzyme inhibitor. (G D Searle & Co).

depGynogen. Estradiol cypionate. Applications: Oestrogen. (O'Neal, Jones & Feldman Pharmaceuticals).

Dephosphex. Highly basic oxidizing supplementary flux for phosphorus removal from arc melted steel. (Foseco (FS) Ltd).*

Depixol. A proprietary preparation of flupenthixol decanoate. A neuroleptic. (Pfizer International).*

Deplet. A proprietary preparation of potassium teclothiazide. A diuretic. (Nicholas).†

Depmedalone 40, also Depmedalone 80. Methylprednisolone acetate. Applications: Glucocorticoid. (O'Neal, Jones & Feldman Pharmaceuticals).

Depocillin. A proprietary preparation of procaine penicillin. Applications: An antibiotic. (Brocades).*

Depo-Medrol. Methylprednisolone acetate. Applications: Glucocorticoid. (The Upjohn Co).

Depomedrone. A proprietary preparation of methyl prednisolone acetate. (Upjohn Ltd).†

Depo-Penicillin. Penicillin G Procaine. Applications: Antibacterial. (The Upjohn Co).

Depo-Provera. A proprietary preparation containing medroxyprogesterone acetate. (Upjohn Ltd).†

Depostat. A proprietary preparation of gestronal hexanoate used to treat benign prostatic hypertrophy and endometrial carcinoma. (Schering Chemicals Ltd).†

Depot Glumorin. A proprietary preparation of kallikrein bound to a high molecular weight steroid. (FBA Pharmaceuticals).†

Depot-Impletol. Pharmaceutical preparation. Applications: Neural therapeutic and diagnostic agents. (Bayer & Co).*

Depot-Padutin. A vasoactive enzyme of prolonged action. (Bayer & Co).*

Depovirin. Testosterone Cypionate. Applications: Androgen. (Hoechst-Roussel Pharmaceuticals Inc).

depPredalone. Prednisolone Acetate. Applications: Glucocorticoid. (O'Neal, Jones & Feldman Pharmaceuticals).

Depronal S.A. A proprietary preparation of dextropropoxyphene hydrochloride. An analgesic. (Warner).†

Depsoline. Textile auxiliary chemicals. (ICI PLC).*

Dequacaine. Amber coloured, circular lozenges with plain convex face. Each lozenge contains decqualinium chloride BP (approx 0.25 mg) and benzocaine BP (approx 10 mg) in a sugar base flavoured with menthol, camphor and peppermint oil. Applications: Relief of severe sore throats. (Evans Medical).*

Dequadin. An orange coloured, circular, compressed lozenge. Each lozenge contains dequalinium chloride BP 0.25 mg in a flavoured sucrose base. Applications: Local therapy of most of the common infections of the mouth and throat. (Evans Medical).*

Dequalone. Dequadin and prednisolone. (Allen & Hanbury).†

Dequaspon. A proprietary preparation of gelatin sponge impregnated with dequadin. Dental packing for haemorrhage. (Allen & Hanbury).†

Dequest. Phosphonates. Applications: Used in scale prevention and corrosion inhibition for industrial water treatment and to control metal ions in aqueous systems. (Monsanto Co).*

D.E.R. Epoxy resins. A range of solid, liquid, flexible and brominated thermosetting polymers. Characteristics: adhesion, hardness, flexibility, toughness, dimensional stability, clarity and chemical resistance. (Dow Chemical Co Ltd).*

Deracyn. Adinazolam mesylate. Applications: Antidepressant. (The Upjohn Co).

Derakane. Vinyl ester resins. Formulated for the reinforced plastics industry. Applications: To make articles by moulding, spray-up and filament winding, in sheet moulding compounds and in coatings. (Dow Chemical Co Ltd).*

Derakane (470-45). A proprietary polyvinyl ester resin possessing good chemical resistance to chlorinated solvents. (Dow Chemical Co Ltd).*

Derakane (510-40). A proprietary polyvinyl ester resin used in fire-retardant laminates. It contains 20 per cent. bromine. (Dow Chemical Co Ltd).*

Deraspan. Insulating panels use **Styrofoam** brand plastic foam as the core material and a variety of outside layers such as aluminium, plywood, etc. Applications: Cold storage warehouses, dairy coolers, cold rooms for food processing, etc. (Dow Chemical Co Ltd).*

Derby Red. See CHROME RED.

Derbyshire Spar. See FLUORSPAR.

Dericin. See FLORICIN.

Derifil. Chlorophyllin copper complex, obtained by replacing the methyl and phytyl ester groups with alkali and magnesium with copper. Applications: Deodorant. (Rystan Company Inc).

Derizine. See FLORICIN.

Dermacort. Hydrocortisone. Applications: Glucocorticoid. (Rowell Laboratories Inc).

Dermaffine. Oleyl alcohol. Applications: Raw material for cosmetics and pharmaceuticals. (Laserson & Sabetay).*

Dermafill. Impregnating agents in aqueous solution. Applications: Leather. (Colour-Chem Limited).*

Dermalac. Nitrocellulose lacquers and emulsions. Applications: Leather. (Colour-Chem Limited).*

Dermalex. A proprietary preparation of squalene, hexachlorophene and allantoin. A protective skin lotion. (Dermalex).†

Dermasulph. A proprietary preparation of polythionates. Sulphur skin ointment. (Crookes Laboratories).†

Dermatol. A basic bismuth salt of gallic acid, $C_6H_2(OH)_3.CO.Bi(OH)_2$. Used medicinally for the treatment of wounds and skin diseases, also as a remedy for perspiring feet.†

Dermogen. See EKTOGAN.

Dermogesic. Benzocaine, calamine and metacresol. Applications: Topical relief of inflammatory and pruritic conditions of the skin. (Merck Sharp & Dohme).*

Dermol. Bismuth chrysophanate, Bi $(C_{15}H_9O_4)_2Bi_2O_3$. Used in the treatment of skin diseases as a 5-20 per cent ointment.†

Dermonistat. A proprietary preparation of miconazole nitrate. An anti-fungal agent. (Janssen Pharmaceutical Ltd).*

Dermoplast. A proprietary preparation of benzocaine, benzethonium chloride, menthol, hydroxyquinalone benzoate and methyl paraben in the form of an aerosol, used as a soothing skin spray. (Ayerst Laboratories).†

Dermovate. A proprietary preparation of CLOBETASOL propionate used in the treatment of eczema and psoriasis. (Glaxo Pharmaceuticals Ltd).†

De Rossi's Stain. A microscopic stain. It consists of two solutions: (*a*) Tannic acid 25 grams, distilled water 100 cc ; (*b*) fuchsin 0.25 gram, phenol 5 grams, alcohol 10 grams, and distilled water 100 grams.†

Derris Dust. Garden insecticide. (ICI PLC, Plant Protection Div).

Derussole. Carbon black dispersions. Applications: For simple and dust-free dyeing of paints, lacquers, paper, cardboard, plastics, synthetic fibres, printing inks and mineral binders. (Degussa).*

Desavin. Plasticizer for use in coating materials based on unsaponifiable binders as well as in plastics dispersions and film coatings. (Bayer & Co).*

Descale. Acid Detergent for milkstone removal. (Ciba-Geigy PLC).

Deseril. A proprietary preparation of methysergide. Treatment of migraine. (Sandoz).†

Desferal. A proprietary preparation of desferrioxamine mesylate used to treat haemochromatosis and acute iron poisoning. (Ciba Geigy PLC).†

Desibyl Kapseals. A proprietary preparation of dried whole bile. (Parke-Davis).†

Desicchlora. A proprietary name for a perchlorate of barium, a drying agent to replace calcium chloride, sulphuric acid, and potassium hydroxide. It absorbs 20 per cent of its weight of water.†

Desimpal. An active oxygen compound. Applications: Used in the textile industry during peroxide bleaching in order to prevent precipitation of hardening silicate when stabilizing with water glass. (Degussa).*

Desmalkyd. Oil-modified polyurethanes. Applications: For the formulation of paints and varnishes as well as printing inks. (Bayer & Co).*

Desmocap. Masked polymeric isocyanates. Applications: For the formulation of coatings and sealants. (Bayer & Co).*

Desmocoll. Hydroxyl polyurethane. Applications: For solution-based adhesives with particular suitability for high-grade footwear sole bonding and production of plastics laminates. (Bayer & Co).*

Desmoderm. One-component polyurethane finishes. Applications: For polyurethane coated materials. (Bayer & Co).*

Desmoderm Foil. Microporous polyurethane foil. (Bayer UK Ltd).

Desmodur. Crosslinking agent for adhesives based on Baycoll, Baypren and Desmocoll. (Bayer & Co).*

Desmodur R. Bonding agent. (Bayer UK Ltd).

Desmodur 1L. Isocyanate prepolymer. Used with Desmophen (*qv*) to provide improved polyurethane finishes. (Bayer & Co).*

Desmodurs. Isocyanates for foams and elastomers. (Bayer UK Ltd).

Desmoflex. Polyols for the Desmoflex system (cold-curing casting system). Applications: Used in the manufacture of pipe seals and as formwork matting for textured exposed concrete. (Bayer & Co).*

Desmolac. Polyurethane resins. Applications: For coating flexible substances. (Bayer & Co).*

Desmopan. Thermoplastic polyurethane elastomer with various levels of shore hardness. Applications: For processing by injection moulding and extrusion. (Bayer & Co).*

Desmophens. Polyester/polyethers for foams and elastomers. (Bayer UK Ltd).

Desmorapid. Catalyst for adhesives based on Baycoll, Desmocoll and Desmodur. (Bayer & Co).*

Desodora. Skin protective soap with substances against bacterial attack and fungal infections of the skin. Applications: For protection and cleansing of the skin under environmental stress. (Dynamit Nobel Wien GmbH).*

Desogen. A proprietary preparation of dodecanoyl N methylaminoethyl-(phenyl carbamyl methyl)dimethyl ammonium chloride. Throat lozenges. (Ciba Geigy PLC).†

Desomorphine. 7,8-Dihydro-6-deoxymorphine.†

DeSonate AOS. C_{14}-C_{16} sodium alpha olefin sulphonate. Applications: High foaming detergent component for personal care products, light duty detergents and industrial cleaners. (DeSoto Inc).*

DeSonate SA. Dodecylbenzene sulphonic acid. Applications: Component for manufacture of many detergents. (DeSoto Inc).*

DeSonate SA-H. Alkylbenzene sulphonic acid. Applications: Emulsifier, chemical intermediate (non biodegradable). (DeSoto Inc).*

DeSonate 50-S. Sodium dodecylbenzene sulphonate, 53% active. Applications: Detergent component. (DeSoto Inc).*

DeSonate 60-S. Sodium dodecylbenzene sulphonate, 60% active. Applications: Detergent component. (DeSoto Inc).*

DeSonic DA-4. Ethoxylated decyl alcohol. Applications: Wetting and re-wetting agent for industrial applications. (DeSoto Inc).*

DeSonic DA-6. Ethoxylated decyl alcohol. Applications: Wetting agent for built scour systems. (DeSoto Inc).*

DeSonic N Series (ie. 4N, 9N etc). Ethoxylated nonyl phenols, 1-100 moles ethylene oxide. Applications: Emulsifiers, wetting agents, detergents and dispersants. (DeSoto Inc).*

DeSonic S Series. Octylphenol ethoxylates, various degrees of ethoxylation. Applications: Emulsifiers, wetting agents, detergents and dispersants. (DeSoto Inc).*

DeSonic 30C. Ethoxylated castor oil. Applications: Emulsifier, lubricant, dye leveller, antistatic agent and dispersant for various textile applications and the pulp and paper industry. (DeSoto Inc).*

DeSonol A. Ammonium lauryl sulphate. Applications: Detergent and shampoo component. (DeSoto Inc).*

DeSonol AE. Ammonium Laureth (3) sulphate. Applications: Component for detergents, shampoos and personal care products. (DeSoto Inc).*

DeSonol S. Sodium lauryl sulphate. Applications: Detergent and shampoo component. (DeSoto Inc).*

DeSonol SE. Sodium laureth (3) sulphate. Applications: Component for detergents, shampoos and personal care products. (DeSoto Inc).*

DeSonol SE-2. Sodium Laureth (2) sulphate. Applications: Component for detergents, shampoos and personal care products. (DeSoto Inc).*

DeSonol T. Triethanolamine (TEA) lauryl sulphate. Applications: Mild shampoo component. (DeSoto Inc).*

DeSotan SMO. Sorbitan mono-oleate. Applications: Lipophilic emulsifier, fibre lubricant and softener. (DeSoto Inc).*

DeSotan SMO-20. Ethoxylated sorbitan mono-oleate. Applications: Hydrophilic emulsifier and wetting agent. (DeSoto Inc).*

DeSotan SMT. Sorbitan monotallate. Applications: Lipophilic emulsifier, fibre lubricant and softener. (DeSoto Inc).*

DeSotan SMT-20. Ethoxylated sorbitan monotallate. Applications: Hydrophilic emulsifier and wetting agent. (DeSoto Inc).*

Destral. Non-selective herbicides. Applications: Weedkilling. (Borax Consolidated Ltd).*

Destral BR. Non-selective herbicides. Applications: Weedkilling. (Borax Consolidated Ltd).*

Desulfex. Fluxes for removal of sulphur from molten iron or steel. (Foseco (FS) Ltd).*

Desyrel. Trazodone Hydrochloride. Applications: Antidepressant. (Mead Johnson & Co).

Deteclo. A proprietary preparation ot chlortetracycline, tetracycline and democlocycline hydrochlorides. An antibiotic. (Lederle Laboratories).*

Detergyl. Textile auxiliary chemicals. (ICI PLC).*

Dethlac. An insecticidal lacquer, containing dichlorvos. Applications: A slow release fly killer. (Gerhardt Pharmaceuticals).*

Dethmor. Warfarin. Applications: A rodenticide. (Gerhardt Pharmaceuticals).*

Detigon. An antitussive. (Bayer & Co).*

Detigon Linctus. A proprietary preparation containing chlophedianol citrate and potassium guaiacol sulphonate. A cough linctus. (FBA Pharmaceuticals).†

Dettol. A proprietary trade name for a germicide containing chloroxylenols and terpineol. It is very powerful in action and yet non-poisonous.†

Deva. A range of precious metal alloys. Applications: For dentistry and dental engineering. (Degussa).*

Devarda's Alloy. An alloy of 45 per cent aluminium, 50 per cent copper, and 5 per cent zinc.†

Deward Steel. A proprietary non shrinking steel containing 1.55 per cent manganese, 0.3 per cent molybdenum, and 0.9 per cent carbon.†

Dewrance Metal. See DURANCE'S METAL.

Dexa-Rhinaspray. A proprietary preparation of neomycin,

dexamethasone and tramazoline. (Boehringer Ingelheim Ltd).†

Dexacillin. Epicillin. Applications: Antibacterial. (E R Squibb & Sons Inc).

Dexamist Ear Spray. Non-pressurized pump action spray containing dexamthethasone, neomycin sulphate and glacial acetic acid. Applications: Treatment of otitis externa in dogs. (Stafford-Miller).*

Dexawin. Racephenicol. Applications: Antibacterial. (Sterling Drug Inc).

Dexedrine Tablets. Dexamphetamine sulphate. Applications: Used in the treatment of narcolepsy, also indicated in children with refractory hyperkinetic states. (Smith Kline and French Laboratories Ltd).*

Dexil. Core breakdown agent. (Foseco (FS) Ltd).*

Dexine 521. A polyisobutylene material with a good resistance to chemical attack from oxidizing liquors up to 110° C. Used for lining and covering metal tanks. (Dexine Rubber Co Ltd).†

Dexine 656. A natural rubber based compound with good abrasion resistance used for lining metal tanks. (Dexine Rubber Co Ltd).†

Dexine 687. A natural rubber based material with good resistance to chemicals especially sodium hypochlorite. (Dexine Rubber Co Ltd).†

Dexine 759. A Hypalon (*qv*) based lining and covering material with very good resistance to chemical attack. It can be used with sulphuric acid at concentrations up to 95 per cent. (Dexine Rubber Co Ltd).†

Dexine 779. A polyurethane lining and covering material with a very good resistance to abrasion. (Dexine Rubber Co Ltd).†

Dexlar. Staple cement. (Du Pont (UK) Ltd).

Dexon. Polyglycolic Acid. Applications: Surgical aid. (Davis & Geck).

Dexon. See POLYGLYCOLIC ACID.

Dexone. Dexamethasone. Applications: Glucocorticoid. (Rowell Laboratories Inc).

Dexonite. A trade name for a proprietary hard rubber moulded material for electrical insulation. A proprietary trad name for ebonite. (Dexine Rubber Co Ltd).†

Dexoplas. A proprietary trade name for a butadiene-styrene plastics material use for constructing corrosion resistant fittings. (Dexine Rubber Co Ltd).†

Dexten. A proprietary preparation of dexamphetamine resinate. (Nicholas).†

Dextran. A proprietary trade name forα1-6 polyglucose or polyanhydroglucose. (Fisons PLC).*

Dextranomer. Dextran cross-linked with epichlorohydrin. A promoter of wound healing.†

Dextraven. A proprietary preparation of dextrans in normal saline, used to resto blood volume. (Fisons PLC).*

Dextrin. See BRITISH GUM, and MAZAM.

Dextrin-Maltose. See GLUCOSE SYRUP.

Dextrinozole. See OZOLE.

Dextroform. A condensation product of dextrin and formaldehyde. An antiseptic.†

Dextrostix. A proprietary test strip impregnated with a buttered mixture of glucose oxidase, peroxidase, and a chromogen system, used to estimate blood glucose. (Ames).†

Dextrozyme. A balanced mixture of glucoamylase and pullulanase obtained from selected strains of Aspergillus niger and Bacillus acidopullulyticus by submerged fermentation. Applications: Used in the starch syrup industry for the saccharification of liquified starch (maltodextrins) in the production of high dextrose syrups. (Novo Industri A/S).*

Dexytal. A proprietary preparation of sodium amylobarbitone and dexamphetamine sulphate. An antidepressive agent. (Eli Lilly & Co).†

D.F. 118. A proprietary preparation of dihydrocodeine bitartrate. An analgesic. (British Drug Houses).†

Dhak gum. See KING, BENGAL.

abeta. Glyburide. Applications: Antidiabetic. (Hoechst-Roussel Pharmaceuticals Inc).

abetic Cough Mixture. A proprietary preparation containing codeine phosphate, pholcodine, butethamate citrate, ipecacuanha liquid extract and squill liquid extract. A cough mixture. (Rybar).†

abetin. See LEVULOSE.

abiformin. A proprietary preparation of chlorpropamide, metformin hydrochloride. An oral hypoglycemic agent. (Pfizer International).*

abinese. A proprietary preparation of chlorpropamide. An oral hypoglycemic agent. (Pfizer International).*

abiphage. A proprietary preparation of chlorpropamide, metformin hydrochloride. An oral hypoglycemic agent. (Pfizer International).*

acetin. Glycerol diacetate, $(CH_3COOCH_2)_2CHOH$.†

acetone Alcohol. 4-Hydroxy-4-methyl-2-pentanone. It is a solvent for cellulose acetate and nitrate.†

iacetyl Tannin. See ACETANNIN.

adavin. Cleaning and stain removing agent. Applications: Textile finishing. (Bayer & Co).*

iadem Chrome. Acid mordant wool dyestuffs. (Akzo Chemie GmbH, Düren).†

iadem Chrome Black A. A dyestuff. It is a British equivalent of Eriochrome black A.†

iadem Chrome Black F New. A British dyestuff. It is equivalent to Diamond black F.†

iadem Chrome Black PV. A dyestuff. It is a British brand of Diamond black PV.†

iadem Chrome Blue-black P6B. A dyestuff. It is equivalent to Palatine chrome blue.†

iadem Chrome Red L3B. A dyestuff. It is a British equivalent of Eriochrome red B.†

iadur. Zirconia-alloyed aluminium oxide. Applications: Used for abrasive industries, super refractories, sandblasting and for safes. (Lonza Limited).*

Diafen. Diphenylpyraline hydrochloride. Applications: Antihistaminic. (Riker Laboratories Inc, Subsidiary of 3M Company).

Diaflex. Intraperitoneal dialysis solutions. (The Boots Company PLC).

Diaginol. Sodium acetrizoate. (May & Baker Ltd).

Diak. Curing agent. (Du Pont (UK) Ltd).

Diakon. Acrylic moulding and extrusion powders. (ICI PLC).*

Dial. Diallylbarbituric acid. A powerful sedative and hypnotic.†

Dialin. Dihydronaphthalene, $C_{10}H_{10}$†

Dialose. Docusate potassium. Applications: Stool softener. (Stuart Pharmaceuticals, Div of ICI Americas Inc).

Dialose Plus. Contains casanthranol and docusate potassium. Casanthranol is a purified mixture of the anthranol glycosides derived from Cascara sagrada. Applications: Laxative and stool softener. (Stuart Pharmaceuticals, Div of ICI Americas Inc).

Dialpha. Fungal alpha amylaze dispersion in food grade filler. Applications: Flour additive. (Diaflex Ltd).*

Dialume. Aluminium hydroxide. Applications: Antacid. (Armour Pharmaceutical Co).

Dialuramide. (Murexan). Uramil, $C_4H_5N_3O_3$.†

Dialyl. Methylamine and lithium citrate.†

Dialysed Iron. A solution containing ferric oxide and acetic acid, made by dialysing ferric acetate. Used medicinally.†

Diamalt. Malt extract. (ABM Chemicals Ltd).*

Diamex. Malt extract. (ABM Chemicals Ltd).*

Diamin(e). A trade mark for a range of dyestuffs. (Casella AG).†

Diamine. Range of cationic surfactants of the diamine type, in liquid, solid or paste form. Applications: Widespread use in industry and in household and personal care formulations, though mostly in the form of derived quaternaries and various salts. (Keno Gard (UK) Ltd).

Diamine Azo Blue RR. A direct cotton dyestuff.†

Diamine Blue RW, RG. (Diamine New Blue R and G, Chicago Blues, Chicago Grey, Diazo Blue, Columbia Blue R and G). Analogous dyestuffs to Diamine sky blue. They dye cotton from a neutral salt bath.†

Diamine Bordeaux B. A dyestuff allied to Diamine scarlet B. It dyes cotton red from an alkaline salt bath.†

Diamine Bordeaux S. A dyestuff allied to Diamine scarlet B. It dyes cotton as above.†

Diamine Brilliant Blue. A dyestuff. Dyes cotton blue.†

Diamine Brilliant Blue G. See DIAMINE SKY BLUE.

Diamine Bronze G. A dyestuff. Dyes cotton yellowish brown of metallic appearance.†

Diamine Brown M. (Chlorazol Brown M, Paramine Brown M, Direct Brown M). A dyestuff. Dyes cotton brown.†

Diamine Brown V. A dyestuff. Dyes cotton dark violet-brown.†

Diamine Brown 3G. An azo dyestuff similar in composition and application to Diamine brown B.†

Diamine Catechin B and G. Direct cotton dyestuffs, used in conjunction with potassium bichromate.†

Diamine Deep Black. A disazo dyestuff derived from diazotized di-*p*-amino-diphenylamine, coupled with 1 molecule of aminonaphthol-sulphonic acid G, and 1 molecule of *m*-tolylenediamine.†

Diamine Deep Blue B and R. A direct cotton dyestuff.†

Diamine Fast Yellows, B, C, FF. See CHLORAMINE YELLOW.

Diamine Gold. (Diamine Golden Yellow). A dyestuff. Dyes cotton yellow from a salt bath.†

Diamine Golden Yellow. See DIAMINE GOLD.

Diamine Grey. An azo dyestuff dyeing mordanted cotton from an alkaline bath.†

Diamine Jet Black OO. An azo dyestuf dyeing cotton from an alkaline bath

Diamine Jet Black OR. A direct cotton dyestuff, after-treated with potassiu bichromate.†

Diamine Jet Black RB, SS. Azo dyestu having the same application as Dian jet black OO.†

Diamine New Blue G. An azo dyestuff, which dyes cotton from an alkaline bath, and wool from an acid bath.†

Diamine New Blue R. A dyestuff simila the G mark.†

Diamine Orange G and B. A tetrazo dyestuff, which dyes cotton reddish orange shades from an alkaline salt bath, and wool from a sodium sulph bath.†

Diamine Pink. See DIAMINE ROSE.

Diamine Pure Blue. See DIAMINE SK BLUE.

Diamine Pure Blue FF. See DIAMINE SKY BLUE.

Diamine Rose. (Diamine Pink). A dyest Dyes unmordanted cotton pink shad

Diamine Scarlet B. (Direct Red B). A dyestuff. Dyes wool and silk from an acid or neutral bath, and cotton fron alkaline bath.†

Diamine Scarlet 3B. A dyestuff belongi to the same group as Diamine scarle B.†

Diamine Yellow A. See CHLORAMIN YELLOW.

Diamine Yellow N. A dyestuff. Dyes cot yellow.†

Diaminogen. A registered trade mark currently awaiting re-allocation by it proprietors to cover a range of dyest (Casella AG).†

Diaminogen Blue G. A dyestuff containi β-naphthol-3: 6-disulphonic acid as t final component.†

Diaminogen Extra. A dyestuff similar tc Diaminogen.†

Diamite Epoxy Brushkote. Two-compon solvent-release, epoxy compound use as a heavy-duty coating for protectin floors, walls and equipment against attrition and chemical attack. (Metalcrete Mfg Co).*

Diamite Epoxy Flooring. 100% solids epoxy. Applications: Used for repairing and surfacing floors subject to heavy traffic. (Metalcrete Mfg Co).*

Diammonphos. Diammonium phosphate, $(NH_4)_2HPO_4$. A fertilizer.†

Diamol. 2,4-Diaminophenol hydrochloride.†

Diamond, Black. See CARBONADO.

Diamond Black PV. (Diadem Chrome Black PV). A dyestuff prepared from *o*-aminophenolsulphonic acid, and 1,5-dihydroxynaphthalene.†

Diamond Brown Paste. An acid mordant dyestuff, giving brown shades on wool mordanted with bichromate. Used in calico printing.†

Diamond Cement. A cement containing 8 parts isinglass, 1 part gum ammoniacum, 1 part galbanum, and 4 parts alcohol. Used for mending china and glass. Also see ACETIC-GELATIN CEMENT.†

Diamond Fibre. A proprietary trade name for a vulcanized fibre ; a laminated acid-treated cotton cellulose.†

Diamond Flavine G. A dyestuff. Dyes chromed wool yellow.†

Diamond Green B. See MALACHITE GREEN.

Diamond Green C. See BRILLIANT GREEN.

Diamond Grey. See ZINC GREY.

Diamondite. An alloy of 95.65 per cent tungsten with 3.91 per cent carbon.†

Diamond Magenta. See MAGENTA.

Diamond Orange. A dyestuff similar to Diamond yellow.†

Diamond Salicine. See CHROME.

Diamond White. See OIL WHITE.

Diamond Yellow G. A dyestuff. Dyes chrome mordanted wool reddish-yellow.†

Diamond Yellow R. A dyestuff. Dyes chrome mordanted wool reddish-yellow.†

Diamonine. Textile auxiliary chemicals. (ICI PLC).*

Diamox. A proprietary preparation of Acetazolamide. A diuretic. (Lederle Laboratories).*

Dianabol. A proprietary preparation of methandienone. An anabolic agent. (Ciba Geigy PLC).†

Diane. Diphenylolpropane used as the phenolic reactant in resin manufacture. (Chemische Werke, Albert).†

Dianil Black. See CUBA BLACK.

Dianil Blue B. A dyestuff. Dyes cotton blue from a salt bath.†

Dianil Yellow. A dyestuff obtained by coupling diazotized primuline with ethyl acetoacetate.†

Dianisidine Blue. (Azophor Blue). A dyestuff. It is a copper derivative giving a reddish-blue on cotton.†

Dianthine. See ST. DENIS RED.

Dianthine. A trade mark for a range of dyes. (National Lead Co).†

Dianthine B. See ERYTHROSIN.

Dianthine G. See ERYTHROSIN G.

Diaparene. Methylbenzethonium chloride. Applications: Anti- infective, topical. (Sterling Drug Inc).

Diapen. A proprietary preparation of penicillin G benzathine. An antibiotic. (Pfizer International).*

Diaphan Oil. A mixture of methylhexalin and sodium oleate. Used in the preparation of transparent soaps.†

Diapid. Lypressin. Applications: Antidiuretic; vasoconstrictor. (Sandoz Pharmaceuticals, Div of Sandoz Inc).

Diasone Sodium Enterab. Sulphoxone Sodium. Applications: Antibacterial. (Abbott Laboratories).

Diastase. See AMYLASE, DIASTOFOR, MALTINE, FRENCH, and TEXTASE.

Diastatin. A proprietary preparation of nystatin. An antifungal. (Pfizer International).*

Diastix. A proprietary test strip impregnated with glucose oxidase and peroxidase, plus potassium iodide, used to detect glycosuria. (Ames).†

Diatomaceous Earth. See INFUSORIAL EARTH.

Diavite. Vitamin mixture dispersed in food grade filler. Applications: Flour additive. (Diaflex Ltd).*

Diax. A proprietary product used as a diastatic ferment.†

Diazamine. Direct dyes. (ICI PLC).*

Diazamine Fast Yellow H. A dyestuff. It is a British equivalent of Mikado yellow.†

Diazine. Direct duestuffs for cotton and artificial silk. (JC Bottomley and Emerson).†

Diazine Black. A dyestuff. It is safranine-azo-phenol. Dyes tannined cotton black.†

Diazine Blue. See INDOINE BLUE R.

Diazine Blue BR. See INDOINE BLUE R.

Diazine Blue 2B. A British dyestuff. It is equivalent to Diamine blue 2B.†

Diazine Green. (Janus Green B and G). A dyestuff. It is the chloride of safranineazodimethylaniline. Dyes cotton dull bluish-green.†

Diazine Green B. A dyestuff. It is a British brand of Diamine green B.†

Diazine Yellow R. A dyestuff. It is a British equivalent of Direct yellow R.†

Diazinon. A proprietary preparation of DIMPYLATE. An insecticide.†

Diazinon Liquid. Insecticide. (Murphy Chemical Ltd).

Diazitol. Diazonon insecticide. (501). †

Diazitol Liquid. Organophosphorus insecticide. (Ciba-Geigy PLC).

Diazo Black. DHL. A dyestuff. It is a British equivalent of Diamine black BH.†

Diazo Black 3B, G, H, BHN, R. Dyestuffs similar to Diazo black B.†

Diazo Blue 3R. A dyestuff having a similar application to Diazo black B.†

Diazo Blue-black RS. Diamine Black blue B.†

Diazo Bordeaux. A dyestuff allied to primuline, having the same application as primuline.†

Diazo Brilliant Black R. A similar dyestuff to the B mark.†

Diazo Brown. A tetrazo dyestuff, dyeing cotton direct.†

Diazo Brown R Extra. A dyestuff similar to Diazo brown.†

Diazo Fast Black. (Diazo Fast Black H Direct cotton dyestuffs which are developed on the fibre.†

Diazo Indigo Blue. An analogous produ to Diamidogen (*qv*).†

Diazol. Direct dyes. (ICI PLC).*

Diazol. Active ingredient: diazinon; 0,0 diethyl-0-2-isopropyl-6-methylpyrimidin-4-yl-phosphorothioate. Applications: Organophosphorous agricultural insecticide with acaricidal properties (Makhteshim Chemical Works Ltd)

Diazone. Fast dyestuffs for cotton. (JC Bottomley and Emerson).†

Diazopon SS-837. Polyoxyethylated alk phenol (nonionic), a water-soluble surfactant with dispersing and solubilizing properties. Applications Anticrock agent for naphthol dyeing soaping agent for naphthol-dyed stoc yarn and fabrics; leuco-vat-ester dye assistant for use in the acid bath to inhibit pigment agglomeration on th fibre. (GAF Corporation).*

Diazoresorcin. (Azoresorcin, Resazoin). Resazurin, $C_{12}H_9NO_4$.†

Diazoxide. 7-Chloro-3-methyl-1,2,4-benzo-thiadiazine 1,1-dioxide.†

Diazurin G. A similar dyestuff to the B brand.†

Dibenyline. A proprietary preparation o phenoxybenzamine hydrochloride. Vascular antispasmodic. (Smith Klin and French Laboratories Ltd).*

Dibenyline Capsules. Phenoxybenzamine hydrochloride (α-adrenergic receptor antagonist). Applications: Short terr management of hypertensive episodes associated with phaeochromocytoma. (Smith Kline and French Laboratori Ltd).*

Dibenzo GMF. Dibenzoyl-p-quinone-dioxime. A non-sulphur vulcanizing agent for natural, SBR, butyl and EPDM rubbers used to impart heat resistance to tyre curing bags, gasket and wire insulation, Used as a coage with peroxide curatives. (Uniroyal).*

Dibenzyline. A proprietary preparation containing phenoxybenzamine hydrochloride. (Smith Kline and Fre Laboratories Ltd).*

Dibexin. A proprietary preparation of Vitamin B. (Parke-Davis).†

Dibistin. A proprietary preparation of antazoline hydrochloride and tripelennamine hydrochloride. Antihistamine. (Ciba Geigy PLC).†

Dibnal. Dibutylaceturethane.†

Dibotin. A proprietary preparation of phenformin hydrochloride. An oral hypoglycaemic agent. (Bayer & Co).*

Dibrogan. Dibromopropamidine isethionate/promethazine cream. (May & Baker Ltd).*

Dibrom. Naled insecticide. (Chevron).*

Dibromin. Dibromobarbituric acid. An antiseptic.†

Dicalite 14, 14B, and 14W. Proprietary trade names for diatomaceous silica fillers. Used for heat insulating and as a filler for plastics, etc.†

Dicarburetted Hydrogen. See OLEFIANT GAS.

Dicestal. Dichlorophen. (May & Baker Ltd).*

Dichevrol. Dielectric oil. (Chevron).*

Dichlofuanide. A proprietary paint fungicide. N-Dimethylamino-N'-phenyl-N'-(fluorodichloromethylthio) sulphamide. (Bayer & Co).*

Dichlone. Dichloronaphthoquinone. A fungicide used as a seed dressing.†

Dichlor-Stapenor. An antistaphylcoccal penicillin. (Bayer & Co).*

Dichloramal. See ALEUDRIN.

Dichloramine T. *N*, *N*-Dichloro-*p*-toluenesulfonamide. A disinfectant.†

Dichloramine-M. Methyldiphenylmethyl-dichloramine.†

Dichloraminet. See DICHLORAMINE T.

Dichloroditane. A proprietary tradename for *p*- dichlorodiphenylmethane. (Bakelite Corporation).†

Dichlosuric. Hydrochlorothiazide. Applications: For the treatment of oedema and hypertension. (Merck Sharp & Dohme).*

Dichlotride. A proprietary trade name for Hydrochlorothiazide.†

Dicköl Varnish. See STAND OIL.

Dicloxin. Dicloxacillin sodium. Applications: Antibacterial. (Bristol Laboratories, Div of Bristol-Myers Co).

Diconal. A proprietary preparation of dipipanone hydrochloride and cyclizine hydrochloride. An analgesic. (The Wellcome Foundation Ltd).*

Dicosal. A range of fire extinguishing powders to extinguish metal fires including burning uranium. (Degussa).*

Dicotox. A selective weedkiller. (May & Baker Ltd).*

Dicrodamine. Series of cationic surfactants in the form of fatty alkyl propylene diamines or diamine salts, in which the alkyl group is coconut, tallow, hydrogenated tallow or oleic in origin. Applications: Intermediates for di-quaternaries and polyethoxylates; dispersants for pigments; corrosion inhibitors; drawing aids for copper wire and tubing; flexible hardeners for epoxy resins; auxiliaries for oil and petroleum industries; bitumen emulsions. (Croda Chemicals Ltd).

Dicron 45Sc. Organophosphorus insecticide. (Ciba-Geigy PLC).

Dicrylan. Finishing agent. (Ciba-Geigy PLC).

Dicrylan 270. A proprietary trade name for an aqueous emulsion of an acrylic resin for increasing the abrasion resistance of crease-resistant fabrics. (Ciba Geigy PLC).†

Dicumoxane. A proprietary preparation of cumetharol. An anticoagulant. (MCP Pharmaceuticals).†

Di-Cup. A range of dicumyl peroxide preparations. Applications: vulcanizing and polymerization agent. (Hercules Inc).*

Dicurane. Substituted urea herbicide. (Ciba-Geigy PLC).

Dicurane Duo. Cereal herbicide. (Ciba-Geigy PLC).

Dicycloverine. Dicyclomine.†

Dicynene. A proprietary preparation of ethamsylate. An anti-haemorrhagic agent. (Delandale Laboratories).*

Didi-Col. An insecticide. (ICI PLC).*

Didigram. An oil insecticide. (Plant Protection (Subsidiary of ICI)).†

Didimac. An insecticide. (ICI PLC).*

Didrate. Hydrocodone bitartrate. Applications: Antitussive. (Penick Corp).

Didrex. A proprietary preparation of benzphetamine hydrochloride. An antiobesity agent. (Upjohn Ltd).†

Didronel. Etidronate disodium. Applications: Regulator; pharmaceutic aid. (Norwich Eaton Pharmaceuticals Inc).

Die-casting Alloys. These are usually zinc-base alloys containing 86 per cent zinc, 7-10 per cent tin, 4-7 per cent copper, 0.5-1 per cent aluminium. Some alloys have a tin base, and a typical one contains 90 per cent tin, 4 per cent copper, and 6 per cent antimony.†

Di-el. A material used as an insulating material for high voltage engineering.†

Dielan-Col. Dithianon fungicide. (ICI PLC, Plant Protection Div).

Dieline. Sym-dichlorethylene, CHCl = CHCl. It is a solvent for cellulose acetate, rubber, and oils.†

Dielmoth. A moth proofing agent for wool. (Shell Chemicals UK Ltd).†

Diene. Polybutadiene rubber. Applications: High impact polystyrene and tyres. (Firestone Synthetic).*

Dienol. Colloidal manganese.†

Diepoxy. An epoxy mould material. Applications: Dental moulds. (Kemtron International Inc).*

Dieselect. Lubes. (Chevron).*

Dieselmotive. Lubes. (Chevron).*

Dieselube. Lubes. (Chevron).*

Di-esterex N. A proprietary trade name for a rubber vulcanizing accelerator stated to be 60 per cent of the dinitrophenyl ester of mercaptobenzthiazole and 40 per cent of the acetate of diphenylguanidine.†

Diethoxol. Monoethyl ether of diethylene glycol. (ICI PLC, Petrochemicals & Plastics Div).

Diethylin. The diethyl ether of glycerol.†

Di-Farmon. Herbicide, containing mecoprop and dicamba. (ICI PLC).*

Difflam Cream. Cream containing 3% benzydamine hydrochloride. (Riker Laboratories/3M Health Care).†

Difflam Oral Rinse. Solution containing 0.15% benzydamine hydrochloride. (Riker Laboratories/3M Health Care).†

Difolatan. Captafol fungicide. (Chevron).*

Diformyl. Glyoxal, $(CHO)_2$.†

Difusor. Peritoneal dialysis solutions. (The Boots Company PLC).

Digallic Acid. Tannic acid.†

Diganox. A proprietary preparation of digoxin. (L Wilcox & Co Ltd).†

Digibind. A proprietary formulation of digoxin immune Fab (ovine). Applications: Treatment of potentially life threatening digoxin intoxication. (The Wellcome Foundation Ltd).*

Digifortis Kapseals. A proprietary preparation of digitalis. (Parke-Davis).†

Digiglusin. Digitalis. Applications: Cardiotonic. (Eli Lilly & Co).

Digilanid. A proprietary preparation of a glycosidal complex of lanatosides, used for treatment of heart failure. (Sandoz).†

Digitalis. The dried leaves of the flowering plants of *Digitalis purpurea*. A cardiac stimulant.†

Diglycolamine Agent. 2-(2-Aminoethyl)ethanol. Applications: Gas sweetening agent. (Texaco Chemical Co).*

Diglyme. The dimethylether of diethylene glycol. A solvent for polystyrene, PVC/PVA copolymer (*qv*) and polymethyl methacrylate. (ICI PLC).*

Dihalo. Swimming pool water disinfectant. (Great Lakes Chemical (Europe) Ltd).

Dihydrohydroxycodeinone. Oxycodone.†

Dihydrohydroxymorphinone. Oxy-morphone.†

Dihydroisophorone. 3,5,5- tri-ethyl-cyclo-hexanone. A high boiling point ketone solvent for surface coatings.†

Di-iodoform. Tetraiodoethylene.†

Dijex. A proprietary preparation containing aluminium hydroxide and magnesium carbonate as a co-dried gel. An antacid. (Boots Company PLC).*

Dilabil. Dehydrocholic acid. Applications: Choleretic. (Sterling Drug Inc).

Dilangio. A proprietary preparation of bencyclane fumarate. A vasodilator. (Pfizer International).*

Dilantin. Phenytoin. Applications: Anticonvulsant. (Parke-Davis, Div of Warner-Lambert Co).

Dilasoft. Softening and filling agent with hydrophilic properties. Applications: Synthetic fibres and blends with cellulosics. (Sandoz Products Ltd).

Dilatin NA Liquid. A proprietary preparation used in the dyeing of 100 per cent. polyester goods. It is based on unchlorinated aromatic hydro- carbons. (Sandoz).†

Dilaudid. Hydromorphone hydrochloride. Applications: Analgesic. (Knoll Pharmaceutical Co).

Dilectene. A proprietary trade name for an aniline-formaldehyde synthetic resin.†

Dilecto. A proprietary trade name for a phenol-formaldehyde synthetic resin laminated product.†

Dillex. Gripe mixture. (The Boots Company PLC).

Dilor Elixir. Dyphylline. Applications: Indicated for relief of acute bronchial asthma and for reversible bronchospasm associated with chronic bronchitis and emphysema. (Altana Inc).*

Dilor-G. Dyphylline, guaifenesin USP. Applications: Indicated as a bronchodilator-expectorant for treating bronchial asthma, emphysema, bronchitis, pneumonitis and other related bronchopulmonary insufficiency conditions. (Altana Inc).*

Dilosyn. A preparation of methdilazine hydrochloride. (British Drug Houses).†

Dilver. An alloy containing 42 per cent nickel. It is used in filament lamps, as it has the same coefficient of expansion as glass.†

Dimacide. Acid dyes. (ICI PLC).*

Dimagel. A proprietary preparation containing dimagnesium aluminium tri- silicate. An antacid. (Lewis Laboratories).†

Dimagel-Belladonna. A proprietary preparation containing dimagnesium trisilicate and belladonna alkaloids. An antacid. (Lewis Laboratories).†

Dimatos. See INFUSORIAL EARTH.

Dimazon. Diacetylaminoazotoluene, $C_{18}H_{19}N_3O_4$, a red dye used in ointment or as a dusting powder.†

Dimedon. Dimethyldihydroresorcinol.†

Dimelor. A proprietary preparation of acetohexamide. An oral hypoglycaemic agent. (Eli Lilly & Co).†

Dimetane. Bromopheniramine maleate. Applications: Antihistaminic. (A H Robins Co Inc).

Dimethicone. A polydimethylsiloxane.†

Dimethoate. *O,O*-dimethyl-*S*-(N-methylcarbamoylmethyl) phosphorodithioate. Applications: It is an insecticide and acaricide with contact and plant systemic activity suitable for protection against a broad range of insects and mites. (A/S Cheminova).*

Dimethylaniline Orange. See ORANGE III.

Dimethyl-phenylene-green. The tetramethyl derivative of Phenylene blue, $C_{16}H_{20}N_3Cl$. Dyes silk and other fabrics.†

Dimilin. Diflubenzuron. Applications: Insecticides to counteract a number of harmful organisms occurring in agricultural, horticultural and forestry circles, fungus growth, sub-tropical cultures (weevils, cotton worm, many varieties of fruit insects, etc.). Blocks development from the larvae to the adult insect stage. Does not harm the environment. (Duphar BV).*

Dimipressin. A proprietary preparation of IMIPRAMINE. An anti- depressant. (RP Drugs).†

Dimotane Expectorant. A proprietary preparation containing brompheniramine maleate, guaiphenesin, phenylephrine hydrochloride and phenylpropanolamine hydrochloride. A cough mixture. (Robins).†

Dimotane Expectorant DC. A proprietary preparation containing dihydrocodeinone bitartrate, brompheniramine maleate, guaiphenesin, phenylephrine hydrochloride and phenylpropanolamine hydrochloride. A cough mixture. (Robins).†

Dimotapp Elixir. A proprietary preparation containing brompheniramine maleate, phenylephrine hydrochloride and phenylpropanolamine hydrochloride. A cough linctus. (Robins).†

Dimundite. See CAMITE.

Dimycin. Streptomycin and dihydrostreptomycin for veterinary purposes. (Glaxo Pharmaceuticals Ltd).†

Dimyril. A proprietary preparation containing isoaminile citrate. A cough suppressant. (Fisons PLC).*

Dinacrin. Isoniazid. Applications: Antibacterial. (Sterling Drug Inc).

Dinamene. Selective weedkiller. (Murphy Chemical Co).†

Dindevan. A proprietary preparation of phenindione. An anticoagulant. (British Drug Houses).†

Dingler's Green. A pigment. It is a chromium phosphate.†

Dinitra. (Aldiphen, Dinitrenal, Nitraphen, Diphen). 2 : 4-Dinitrophenol.†

Dinitrenal. See DINITRA.

Dinitro-naphthol Yellow. See MARTIUS YELLOW.

Dinitrophenol Black. (Immedial Black N, Sulphur Black T Extra, Thiophenol Black T Extra, Thiol Black, Cross Dye Black BX). Dyestuffs obtained by heating 2 : 4-dinitrophenol with sodium sulphide and sulphur under reflux. They dye unmordanted cotton black.†

Dinitrosoresorcin. (Fast Green O, Dark Green, Chlorin, Russian Green, Fast Myrtle Green, Alsace Green, Resorcin Green, Resorcinol Green, Solid Green O). Dinitroso-resorcinol, $C_6H_4N_2O_4$. Dyes iron mordanted cotton, green, and iron mordanted wool, dark green.†

Dinoram. Amines and derivatives. Primary, secondary, tertiary fatty mono-, di- and polyamines from C8 to C22. Quaternary ammonium salts. Amine oxides. Amine salts. Ethoxylated amines and polyamines. Amino-acids. Special nonionic derivatives. Additives. Applications: Surface active agents used as auxiliaries in road, mining, textile and fertilizer industries, in metallurgy, as lubricating agents, additives for fuel oils and soil stabilization. (British Ceca Co Ltd).*

Dioctyl. A proprietary trade name for dioctyl sodium sulphosuccinate.†

Dioderm. White aqueous cream containi[ng] 0.1% hydrocortisone BP. Application[s:] Topical treatment of eczema, dermati[tis] and all types of inflammatory, pruriti[c] and allergic skin conditions. (Dermal Laboratories Ltd).*

Diodoquin. A proprietary preparation of [di]iodohydroxyquinoline. An anti-amoeb[ic] drug. (Searle, G D & Co Ltd).†

Diofan. See also KUROFAN. Polyvinylidene chloride. (BASF Unite[d] Kingdom Ltd).*

Diofan D. Polyvinylidene chloride dispersion. (BASF United Kingdom Ltd).*

Dioform. 1,2-Dichloroethylene.†

Diogyn. Estradiol. Applications: Oestrogen. (Pfizer Inc).

Diogyn E. Ethinyl estradiol. Application[s:] Oestrogen. (Pfizer Inc).

Diogynets. Estradiol. Applications: Oestrogen. (Pfizer Inc).

Diol. Glycol, $CH_2OH.CH_2OH$.†

Di-On. See Diurex. (Agan Chemical Manufacturers Ltd).†

Dionil. A range of fatty acid amide polyglycol ethers. Applications: Dioni[l] has special detergent, soil suspending/levelling and protective colloid action and some of the range have an additional superfatting effect. (Huls AG).†

Dionosil. Propyliodone suspensions or powders. (Glaxo Pharmaceuticals Ltd).†

Diosal. Sodium diiodosalicylate.†

Diotroxin. L-Thyroxine and L-triiodothyronine mixture. (Glaxo Pharmaceuticals Ltd).†

Diox DR 22. A trade name for a rutile (titanium dioxide) type white pigment with a blue tone and good dispersabili[ty.] (US Industrial Chemical Corp).†

Dioxatrine. Benzetimide hydrochloride. Applications: Anticholinergic. (Jansse[n] Pharmaceutica).

Dioxine. (Gambine R). A dyestuff. It is a nitrosodihydroxynaphthalene, $C_{10}H_7NO_3$. Dyes iron mordanted fabrics, green, and chrome mordanted materials, brown.†

Dioxitol. A colourless, slightly hygroscopic liquid with mild odour. It is used as a solvent in paints, lacquers, textile printing inks and stains. It is effective as a metal degreasing agent and is used in production of safety glass. It is also used as a coupling agent for cutting oils and emulsifiable oils and in production of plasticizers and also as an extraction agent for essences and perfumes. (Shell Chemicals).†

Dioxogen. A 3 per cent solution of hydrogen peroxide, H_2O_2. A disinfectant.†

Dioxyanthranol. Anthrarobine, $C_{14}H_{10}O_3$. Used externally for skin diseases.†

Dipar. A proprietary preparation of PHENFORMIN hydrochloride. A hypoglycaemic agent. (Hoechst Chemicals (UK) Ltd).†

Dipasic. A proprietary preparation of aminosalicylate of isonicotinic acid hydrazide. An antituberculous drug. (Geistlich Sons).†

Dipentek. A technical grade of dipentaerythritol.†

Dipentene No. 122. Terpene liquid. Applications: Solvent and antiskinning agent in paints. (Hercules Inc).*

Di-Petronate Series. Range of diluted sodium petroleum sulphonates. Applications: Emulsifiers, dispersing and wetting agents for use when lower viscosity and easier handling than the Petronate Series is required. (Witco Chemical Ltd).

Dipex. A proprietary molud lubricant for rubber. It is a water-soluble sodium sulphonate obtained from petroleum-acid sludges.†

Diphasol. Dyeing and printing assistant. (Ciba-Geigy PLC).

Diphen 60-B. A phenol-urea formaldehyde resin.†

Diphenal. The sodium salt of diaminodihydroxybiphenyl. A photographic developer.†

Diphentoin. Phenytoin Sodium. Applications: Anticonvulsant. (Beecham Laboratories).

Diphenyl. Direct dyes. (Ciba-Geigy PLC).

Diphenylamine Blue. See METHYL BLUE.

Diphenylamine Blue, Spirit Soluble. (Spirit Sky Blue, Diphenylamine Opal Blue, Bavarian Blue, Spirit Soluble, XL Opal Blue). The hydrochloride of triphenylpararosaniline $C_{37}H_{30}Cl.N_3$.†

Diphenylamine Opal Blue. See DIPHENYLAMINE BLUE, SPIRIT SOLUBLE.

Diphenylamine Orange. See ORANGE IV.

Diphenylamine Yellow. See ORANGE IV.

Diphenyl Chrysoine G. A dyestuff obtained by the ethylation of the product of the condensation of *p*-nitrotoluenesulphonic acid with *p*-aminophenol, in the presence of caustic soda. Dyes cotton golden yellow.†

Diphenyl Chrysoine RR. A dyestuff obtained by the diazotization of the alkaline condensation product of dinitrodibenzyldisulphonic acid and aniline, and the combination of the diazo compound with phenol, and ethylation. Dyes cotton reddish-orange.†

Diphenyl Citronine G. (Cotton Yellow 6370). A dyestuff obtained by the condensation of dinitrodibenzyldisulphonic acid with aniline, in the presence of caustic soda. Dyes cotton greenish yellow.†

Diphenyl Fast Yellow. A dyestuff obtained by the condensation of dinitrodibenzyldisulphonic acid with primuline, in the presence of caustic soda. Dyes cotton yellow.†

Diphenyl Orange. See ORANGE IV.

Diphone. A range of sulphone chemicals. Applications: For use in continuous tinplating processes, and as monomers for engineering plastics. (Yorkshire Chemicals Plc).*

Diphosgene. See SUPERPALITE.

Dipidolor. A proprietary preparation of piritramide. An analgesic. (Janssen Pharmaceutical Ltd).*

Diplosal. Salicyl-salicylic acid, $HO.C_6H_4\ CO\ O\ C_6H_4\ COOH$.Has been used in medicine in chronic and acute rheumatism.†

Diplovax. Poliovirus Vaccine Live Oral. Applications: immunizing agent. (Pfizer Inc).

Dipolymer. A coumarone-indene resin.†

Diponium Bromide. Diperine bromide.†

Dippel's Animal Oil. See BONE OIL.

Dippel's Oil. See BONE OIL.

Dipping Metal. A jeweller's alloy containing 48 parts of copper and 15 parts of zinc.†

Diprivan. Injectable anaesthetic. (ICI PLC).*

Diprofarn. Dipyrone. Applications: Analgesic; antipyretic. (Farmitalia (Farmaceutici Italia)).

Diprosone. Betamethasone dipropionate. Applications: Glucocorticoid. (Schering-Plough Corp).

Dipsal. Dipropylene glycol salicylate. Applications: UV absorber, sunscreen. (Scher Chemicals Inc).*

Dipterex 80. Organophosphorus insecticide as a soluble powder containing 80% w/w trichlorphon. Applications: To control mangolf fly on beet, cabbage white, leaf minor and other caterpillars on brassicas. (Bayer & Co).*

Dirame. Propiram Fumarate. Applications: Analgesic. (Miles Pharmaceuticals, Div of Miles Laboratories Inc).

Direct Blue R. A dyestuff. Dyes cotton black-violet.†

Direct Blue 2B, 2BL, 2B Supra. Dyestuffs. They are British equivalents of Diamine blue 2B.†

Direct Blue 3B. A dyestuff. It is a British brand of Diamine blue 3B.†

Direct Blue-black 2B. A polyazo dyestuff which dyes unmordanted cotton, in the presence of sodium sulphate, and some sodium carbonate.†

Direct Brown J. A dyestuff. Dyes cotton brown.†

Direct Brown M. A dyestuff. It is a British equivalent of Diamine brown M.†

Direct Brown R. See POLYCHROMINE B.

Direct Deep Black E. A polyazo dyestuff which dyes cotton with the addition of salt to the bath.†

Direct Deep Black RW. A polyazo dyestuff, which dyes cotton violet-black in a boiling bath containing 5-15 per cent salt. Also recommended for dyeing linen and jute.†

Direct Dyestuffs. Dyestuffs which are used for cotton, and dye the fibre direct by combination with it. The Congo colours are types.†

Direct Fast Orange D2G, D2R. British colours equivalent to Ingrain brown.†

Direct Fast Orange SE. A dyestuff. It is a British equivalent of Benzo fast orange S.†

Direct Fast Orange 4R, RL, 2RL. Dyestuffs. They are British brands of Mikado orange.†

Direct Fast Red F. A British equivalent of Diamine fast red.†

Direct Fast Scarlet SE. A dyestuff. It is a British equivalent of Benzo fast scarlet 4BS.†

Direct Fast Yellow GL, 2GL, 3GL, R, RL. Dyestuffs. They are British brands of Direct yellow R.†

Direct Fast Yellow 2GLO. A British equivalent of Mikado yellow.†

Direct Green BG, G. Dyestuffs. They are British brands of Diamine green G.†

Direct Green B, BL. Dyestuffs. They are British equivalents of Diamine green B

Direct Green CO. See COLUMBIA GREEN.

Direct Grey. See NEW GREY.

Direct Grey B. A dyestuff. Dyes cotton steel-grey to bluish-black.†

Direct Grey R. A dyestuff. Dyes cotton reddish-grey to bluish-black.†

Direct Jet Black R and T. Direct cotton dyestuffs.†

Direct Orange. A direct cotton dyestuff.†

Direct Orange G. See MIKADO ORANGE G and 4R.

Direct Orange 2R. See MIKADO ORANGE G AND 4R.

Direct Red. An azo dyestuff. Dyes cotton red from an alkaline salt bath.†

Direct Red B. See DIAMINE SCARLET B.

Direct Scarlet. See GERANINE.

Direct Sky Blue. See DIAMINE SKY BLUE.

Direct Sky-Blue. A British dyestuff. It is equivalent to Diamine sky blue.†

Direct Sky-Blue GS. A British brand of Diamine sky blue.†

Direct Violet. See METHYL VIOLET B.

Direct Violet. A dyestuff. It is a British equivalent of Congo violet.†

Direct Violet BB. A dyestuff. Dyes cotton bluish-violet.†

Direct Violet R. A dyestuff. Dyes cotton reddish-violet.†

Direct Yellow G. (Direct Yellow R, (Chlorazol Yellow GX, Diazine yellow R, Enbico Direct Yellow R, Direct Fast Yellow GL, 2GL, 3GL, R, RL, Paramine Yellow R, 2R, Y). A dyestuff. It is the sodium salt of the so-called dinitroso-stilbene-disulphonic acid. It dyes cotton yellow from a salt bath, and wool from an acid bath.†

Direct Yellow R. See DIRECT YELLOW G.

Direct Yellow 2G, 4G. See MIKADO YELLOW.

Dirigold. See ORANIUM BRONZE.†

Dirocide. Diethylcarbamazine citrate. Applications: Anthelmintic. (E R Squibb & Sons Inc).

Dirubin. Refined corundum and crystalline aluminium oxide. Applications: Production of ramming mixes, shape bricks and crucibles for the lining of high temperature furnaces. Moulding material for precision casting moulds and the casting of aggressive steels, raw material for the electroslag remelting process, separating agent for annealing processes. (Dynamit Nobel Wien GmbH).*

Disadine. Antiseptic for topical use, presented as a dry powder spray. (ICI PLC).*

Disalcid. Salsalate. Applications: Analgesic; anti-inflammatory. (Riker Laboratories Inc, Subsidiary of 3M Company).

Disalcid Capsules. Each capsule contains salsalate 500 mg (Riker Laboratories/3M Health Care).†

Disalol. Phenylsalicylsalicylate, $HO.C_6H_4.CO.O.C_6H_4COOC_6H_5$.†

Disamide. A proprietary trade name for disulphamide. An oral diuretic. (British Drug Houses).†

Discase. Chymopapain. Applications: Enzyme. (Travenol Laboratories Inc).

Discelite.. A proprietary trade name for a diatomaceous silica filler.†

Discharge Lake R and RR. See PARANITRANILINE RED.

Discolite. (Formopon, Hydrosulphite AW, Rongalite C, Hydros). NaH $SO_2.CH_2O.2H_2O$, a reducing agent for stripping in dyeing.†

Disfico.. A trade name tor a vulcanized fibre used for electrical insulation.†

Disflamoll DPK. Diphenyl cresyl phosphate. Applications: Monomeric plasticizer. (Bayer & Co).*

Disflamoll DPO. Diphenyl octyl phosphate. Applications: Monomeric plasticizer. (Bayer & Co).*

Disflamoll TCA. Trichloroethylphosphate. Applications: Monomeric plasticizer. (Bayer & Co).*

Disflamoll TKP. Triscresyl phosphate. Applications: Monomeric plasticizer. (Bayer & Co).*

Disflamoll TOF. Trioctylphosphate. Applications: Monomeric plasticizer. (Bayer & Co).*

Disflamoll TP. Triphenyl phosphate. Applications: Monomeric plasticizer. (Bayer & Co).*

Disinfection Oil. See SAPROL.

Disipal (as HCl). Orphenadrine citrate. Applications: Relaxant; antihistaminic. (Riker Laboratories Inc, Subsidiary of 3M Company).

Disodium Edetate.. A chelating agent. Disodium dihydrogen ethylenediamine-*NNN'N'*-tetra-acetate.†

Dispargen.. A form of colloidal mercury.†

Disparit B. Trichlorethylene, C_2HCl_3. A disinfecting cleaning compound.†

Dispello. A proprietary preparation of salicylic acid, zinc chloride and hypophosphorous acid. Applications: A corn cure. (Ayrton Saunders plc).*

Disperse-Ayd. Dispersing agents. Applications: Paints, inks etc. (Cornelius Chemical Co Ltd).*

Dispersite. A proprietary trade name for a dispersion of rubber, rubber-like and film-forming resins in water.†

Dispersol. Condensation product of formaldehyde and sodium naphthalene sulphonate. Applications: Dispersible powders and aqueous dispersions eg for dyestuffs, pigments, pest control. (ICI PLC, Organics Div).

Dispex A40 and N40. Ammonium or sodium salt of polymeric carboxylate as a pale yellow liquid. Applications: Non-foaming pigment dispersing agent used in emulsion paints, sometimes in combination with polyphosphate dispersants. (Allied Colloids Ltd).

Dispex G40 and GA40. Sodium salt of carboxylated polymer in the form of a pale yellow liquid. Applications: Pigment dispersant for emulsion paints, especially sheen and gloss water based paints. (Allied Colloids Ltd).

Dispray. Disinfectant used for rapid disinfections of the skin before operations, injections or venepuncture. (ICI PLC).*

Dissolvine. Sequestering agents. (Akzo Chemie UK Ltd).

Distaclor. Cefaclor. Applications: For the treatment of respiratory tract infections including pneumonia, bronchitis, exacerbations of chronic bronchitis, pharyngitis, tonsillitis and sinusitis, otitis media and urinary tract infections. (Dista Products Ltd).*

Distalgesic. A proprietary preparation of dextropropoxyphene hydrochloride and paracetamol. An oral analgesic. (Dista Products Ltd).*

Distamine. A proprietary preparation of penicillamine hydrochloride, used in the treatment of heavy metal poisoning and Wilson's disease. (Dista Products Ltd).*

Distaquaine V-K. A proprietary preparation containing phenoxymethyl-penicillin potassium. An antibiotic. (Dista Products Ltd).*

Distec. Fatty acids and glycerides. (Akz Chemie UK Ltd).

Distoline. A proprietary trade name for commercial oleic acid obtained from vegetable oils.†

Disulphine Blue. A preparation of sulpha blue used as a visual diagnostic agent for circulatory disorders. (ICI PLC).*

Disulphine Blue A. A dyestuff. It is a Bri equivalent of Patent blue A.†

Disulphine Green B. See NEPTUNE GREEN.

Disulpho Acid S. 1-Naphthylamine 4,8-disulphonic acid.†

Disulphuric Acid. See FUMING SULPHURIC ACID.

Disyston FE-10. Granular systemic insecticide containing 10% w/w disufloton on Fullers Earth granules. Applications: To control aphids and certain aphid-borne virus diseases on potatoes, carrots, celery, marrows, parsley, French and runner beans and Brussels sprouts. (Bayer & Co).*

Ditate-DS. Testosterone enanthate USP, extradiol valerate USP for injection. Applications: Indicated in the treatme of moderate to severe vasomotor symptoms associated with the menopause. (Altana Inc).*

Ditensamine C, O and S. Cationic surfactants in the form of alkyl propylene diamines in which the alkyl group is coconut, oleic and tallow respectively. Liquid or solid in form. Application: Synthesis intermediate, bitumen emulsions, corrosion inhibitio (Tensia SA).

Dithane. Protectant mancozeb fungicide. (Rohm & Haas (UK) Ltd).

Dithizone. Diphenylthiocarbazone. Used for the detection of heavy metals.†

Dithocream. Pale yellow aqueous cream containing dithranol BP. Applications Recommended for the topical treatme of sub-acute and chronic psoriasis including psoriasis of the scalp. (Derm Laboratories Ltd).*

Dithrolan. Stiff yellow ointment containin 0.5% dithranol BP, 0.5% salicylic acid BP, in equal quantities of hard and so paraffin. Applications: For the topical treatment of quiescent psoriasis. (Dermal Laboratories Ltd).*

Ditropan. Oxybutynin chloride. Applications: Anticholinergic. (Marion Laboratories Inc).

Diucardin. Hydroflumethiazide. Applications: Antihypertensive; diuretic. (Ayerst Laboratories, Div of American Home Products Corp).

Diulo. Metolazone. Applications: Diuretic; antihypertensive. (G D Searle & Co).

Diumide-K. Frusemide and controlled release potassium chloride. Applications: Oedema. (Napp Laboratories Ltd).*

Diuresal. Tablets containing 40mg frusemide BP. Injection containing 20mg/2ml frusemide BP and 50mg/5ml frusemide BP. Applications: Diuretic for the management of oedema of cardiac, renal or hepatic origin. Pulmonary oedema, toxaemia of pregnancy, mild or moderate hypertension. (Lagap Pharmaceuticals Ltd).*

Diuretic Salt. Potassium acetate.†

Diurex. SEE DIUREXAN. (Degussa).*

Diurex (also Di-On). Active ingredient: diuron; 3-(3,4-dichlorophenyl)-1,1-dimethylurea. Applications: Residual herbicide effective against a wide range of both broadleaf weeds and annual grasses. (Agan Chemical Manufacturers Ltd).†

Diurexan (Diurex, Aquaphoril). Xipamide. Tablet form. Applications: Diuretic/antihypertensive. (Degussa).*

Diuril. Chlorothiazide. Applications: Diuretic. (Merck Sharp & Dohme, Div of Merck & Co Inc).

Diurnal-Penicillin. Penicillin G Procaine. Applications: Antibacterial. (The Upjohn Co).

Diurol. Active ingredients: azolan plus diurex. Applications: Multipurpose herbicidal mixture which eradicates a wide spectrum of established weeds, while preventing further weed germination for extended periods. (Agan Chemical Manufacturers Ltd).†

Diuron. Total and selective herbicide. (Staveley Chemicals Ltd).

Diver's Liquid. A liquid formed by absorbing ammonia in solid ammonium nitrate. It is capable of dissolving ammonium nitrate.†

Divipan. Active ingredient: dichlorvos; 2,2-dichlorovinyl dimethyl phosphate. Applications: One of the most useful fast-acting agricultural insecticides-acaricides. (Makhteshim Chemical Works Ltd).†

Dixie. See CATALPO.

Dixie Clay. Hydrated aluminium silicate (hard clay). Applications: Mineral filler used as filler, extender or reinforcing agent for paint, paper, rubber, ceramics, plastics and specialities. (Vanderbilt Chemical Corporation).*

Dixie 5 and Dixie Special 102. Proprietary carbon-black pigments.†

DLG-10. Dispersible zinc stearate. Applications: Sanding sealer aid in lacquers. (Nueodex Inc).*

DLG-20. Dispersible zinc stearate. Applications: Sanding sealer aid in lacquers. (Nueodex Inc).*

DLPA 375. Tablets of DL-phenylamine 375mg in a natural basis. Applications: As a dietary suplement. (Larkhall Laboratories Ltd).*

D.M. 10-chloro-5:10-dihydro-phenarsazine.†

D.M. See ADAMSITE.

DM-2. Mould repair/potting compound and general purpose adhesive. Epoxide resin based mastic use as a heat resistant encapsulating material. Applications: Encapsulation of electric components where solder connections are affected by heat. Repair of mould porosity in rotational plastic moulds. (Dynamold Inc).*

DMP. Dimethyl phthalate. Used in insect repellants and as a plasticizer.†

D.N.T. Dinitrotoluene, $CH_3.C_6H_3(NO_2)_2$.†

Do Do. Bronchitis remedy. (Ciba-Geigy PLC).

DOA. Diisooctyl adipate. A vinyl plasticizer.†

Dobane (Detergent Alkylate). Linear alkylbenzenes which yield light coloured, biodegradable sulphonates.

The range may be used for general detergent applications, from household powders to light duty liquids. (Shell Chemicals).†

Dobanic Acids JN and 83. Dark, viscous liquids which on neutralization give light coloured sulphonates particularly suitable for the production of high performance, liquid detergents. (Shell Chemicals).†

Dobanol. Detergent alcohol. (Mitsubishi Petrochemical Co).*

Dobanol Ethoxylates. Intermediates for the production of shampoo components, toiletry products, dishwashing liquids and washing powders. (Shell Chemicals).†

Dobanol Ethoxysulphates. Aqueous or aqueous/ethanol solutions for various applications, such as components for toiletry products, light-duty liquid detergents and high-performance liquid detergents. (Shell Chemicals).†

Dobanols. The colourless liquids are of high purity and are used as base materials for the manufacture of alcohol sulphates, alcohol ethoxylates and alcohol ethoxysulphates. They are also needed for production of detergents, wetting agents, dispersants and emulsifiers. (Shell Chemicals) †

Dobanox. Surfactant/emulsifier. (Shell Chemicals UK Ltd).

Dobatex. An anionic detergent. (Shell Chemicals UK Ltd).†

Dobbin's Reagent. Prepared by adding mercuric chloride solution to a solution of potassium iodide until a permanent precipitate is obtained. The solution is filtered and 1 gram of ammonium chloride added, then dilute caustic soda until a precipitate is formed. Filtered and made up to 1 litre. Used for detecting traces of caustic alkalis in soap.†

Dobell Solution. An aqueous solution containing 1.5 per cent sodium borate, 1.5 per cent sodium bicarbonate, and 0.3 per cent phenol and glycerin. An alkaline antiseptic.†

Dobutrex. Dobutamine hydrochloride. Applications: Cardiotonic. (Eli Lilly & Co).

Doca Acetate. Desoxycorticosterone acetate. Applications: Adrenocortica steroid. (Organon Inc).

Docklene. Selective herbicide. (FBC Ltd

Doctor Metal. An alloy of 88 per cent copper, 9.5 per cent zinc, and 2.5 per cent tin.†

Dodigen. Range of cationic surfactants c the quaternary ammonium chloride type, in liquid, paste or solid form. Applications: Antistatic agents, fabri conditioner and softener, fibre finishe water-repellant agents and dewaterin agents, wetting agents for oils, dispersing agents for pigments,flushin agents, foaming and wetting agents, spinning bath and viscous additives, flotation chemicals and anti-caking agents for rendering salts free-flowing corrosion inhibitors, anchoring and wetting agents for tars and bitumen, surface coatings, lacquers, adhesives and dispersions, disinfectants, hair cosmetics and auxiliaries for leather, textiles, rubber and metal industries. (Hoechst UK Ltd).

Doff. Range of insecticides and herbicide Applications: For horticultural/household use. (Doff Portland Ltd).*

Dolan. Acrylic fibre. (Hoechst UK Ltd).

Dolanit. Acrylic fibre. (Hoechst UK Ltd

Dolasan. A proprietary preparation of DEXTROPROPOXYPHENE napsylate and aspirin. An analgesic. (Lilly & Co).†

Dolene. Propoxyphene Hydrochloride. Applications: Analgesic. (Lederle Laboratories, Div of American Cyanamid Co).

Doler Brass. A proprietary alloy. It is a silicon brass.†

Dolmatil. Sulpiride. (E R Squibb & Son Ltd).

Dolo-Adamon. The registered trade nam for a strong pain relieving agent. It is mixture of sodium phenyldimethyl pyrazolone methylaminomethane sulphonate(Noramidazophenum), Codeine phosphate and 5-ethyl-5-crot barbituric acid (Crotarbital) and Adamon (*qv*). (AG Chemische Fabrik).†

Dolobid. Diflunisal. Applications: For the relief of pain. (Merck Sharp & Dohme).*

Dolomol. A white insoluble powder, consisting mainly of magnesium stearate, with small amounts of magnesium oleate and palmitate. Used as a dusting powder for skin.†

Dolophine Hydrochloride. Methadone Hydrochloride. Applications: Analgesic. (Eli Lilly & Co).

Doloxene. A proprietary preparation of dextropropoxyphene hydrochloride. An analgesic. (Eli Lilly & Co).†

Doloxene Compound-65. A proprietary preparation of dextropropoxyphene hydrochloride, phenacetin, aspirin, and caffeine. An analgesic. (Eli Lilly & Co).†

Doloxytal. A proprietary preparation of dexopropoxyphene hydrochloride and amylobarbitone. An analgesic and sedative. (Eli Lilly & Co).†

Dolphin Blue. A dyestuff, obtained from gallocyanine and aniline, and sulphonation. A mordant dye used in calico printing.†

Dolviran. An all-purpose analgesic. (Bayer & Co).*

Domba Oil. See LAUREL NUT OIL.

Dome-Acne. A proprietary preparation of sulphur and resorcinol acetate used in the treatment of acne vulgaris. (Dome).†

Domeboro. Aluminium acetate. Applications: Astringent. (Miles Pharmaceuticals, Div of Miles Laboratories Inc).

Domestos. Stabilized sodium hypochlorite. (Unilever).†

Domical. A proprietary preparation of amitryptiline hydrochloride. An anti-depressant. (Berk Pharmaceuticals Ltd).*

Dominate. A wettable powder containing a mixed culture of micro-organisms (*Anthrobacter* sp., *Aspergillus terreus, Bacillis subtilis, Bacillis thuringiensis, Bacteroides* sp., *Nocardia* sp., and *Pseudomonas* sp.) used to suppress growth of pathogenic soil fungi. Applications: Applied to the soil, used for a wide variety of crops. (Westbridge Research Group).*

Dominit 18. A similar explosive to Donarit V.†

Domoso. Dimethyl sulphoxide. Applications: Anti-inflammatory. (Syntex Laboratories Inc).

Donarit 1, Donarit 2, Donarit 3. Powdery ammon dynamites with the addition of aromatic nitro-compounds and explosive oil. They are particularly suitable for medium-hard rock and are used in quarries, in agriculture and forestry under dry conditions. (Dynamit Nobel Wien GmbH).*

Donarite. An explosive containing 80 per cent ammonium nitrate, 12 per cent trinitrotoluene, 4 per cent flour, 3.8 per cent nitroglycerin, and 0.2 per cent collodion wool.†

Donnagel. A proprietary preparation containing kaolin, pectin, hyoscyamine sulphate, atropine sulphate and hyoscine hydrobromide. An anti- diarrhoeal. (Robins).†

Donnagel-PG. A proprietary preparation containing kaolin, pectin, hyoscyamine sulphate, atropine sulphate, hyoscyamine hydrobromide and opium. An anti-diarrhoeal. (Robins).†

Donnatal. A proprietary preparation containing hyoscyamine sulphate, atropine sulphate, hyoscine hydrobromide and phenobarbitone. An antacid. (Robins).†

Donnazyme. A proprietary preparation containing hyoscyamine sulphate, atropine sulphate, hyoscine hydrobromide, phenobarbitone, pepsin, pancreatin and bile salts. An antacid. (Robins).†

Dontalol. Pharmaceutical preparation. Applications: Mouth-wash concentration. (Bayer & Co).*

Doom. A microbial insecticide in powder form containing viable spores of Bacillus popilliae, a specific pathogen which infects and kills Japanese beetle grubs. Ready-to-use. Applications: Japanese beetles lay their eggs in lawns, the eggs hatch into grubs which are numerous enough to destroy the grass. Milky disease spore powder is applied to the lawn in spots. Grubs become infected and die releasing more spores to infect and kill succeeding generations of grubs. Only one application is needed as the living spores are self-perpetuating. See also JAPIDERMIC. (Fairfax Biological Laboratory Inc).*

Doom Bark. Sassy Bark.†

Door Plate Brass. See BRASS.

DOP. Di-2-ethyl hexyl phthalate. A vinyl plasticizer.†

Dopamet. A proprietary preparation of methyldopa. A drug used in the treatment of hypertension. (Berk Pharmaceuticals Ltd).*

Dopar. Levodopa. Applications: Antiparkinsonian. (Norwich Eaton Pharmaceuticals Inc).

Dopastat. Dopamine hydrochloride. Applications: Adrenergic. (Parke-Davis, Div of Warner-Lambert Co).

Dope. The name given to various solutions of cellulose or cellulose compounds in acetone, amyl alcohol, amyl acetate, and other solvents. Used for painting aeroplane wings, and other purposes. Also see EMAILLITE.†

Dopram. A proprietary preparation of DOXAPRAM hydrochloride. A respiratory stimulant. (Robins).†

Dorantamin. Pyrilamine Maleate. Applications: Antihistaminic. (Dorsey Laboratories, Div of Sandoz Inc).

Dorbane. Danthron. Applications: Laxative. (Riker Laboratories Inc, Subsidiary of 3M Company).

Dorbanex Capsules. Each capsule contains danthron BP 25 mg and poloxamer 188 200 mg (Riker Laboratories/3M Health Care).†

Dorbanex Forte. Each 5 ml spoonful contains danthron BP 75 mg and poloxamer 188 1000 mg (Riker Laboratories/3M Health Care).†

Dorbanex Liquid. Each 5 ml spoonful contains danthron BP 25 mg and poloxamer 188 200 mg (Riker Laboratories/3M Health Care).†

Doré Silver. A silver containing small amounts of gold.†

Doriden. A proprietary preparation of glutethimide. A hypnotic. (Ciba Geigy PLC).†

Dormakil. Fungicide. (ICI PLC, Plant Protection Div).

Dormate. Mebutamate. Applications: Antihypertensive. (Wallace Laboratories, Div of Carter-Wallace Inc).

Dormethan. Dextromethorphan hydrobromide. Applications: Antitussive. (Dorsey Laboratories, Div of Sandoz Inc).

Dormicum. See hypnovel. (Roche Products Ltd).*

Dormone. A selective weedkiller containing 465g/litre 2,4-D as the diethanolamine salt. Applications: It may be used for the control of broad leaved weeds on amenity areas, golf courses, playing fields etc. (Burts & Harvey).*

Dormonoct. Loprazolam. Applications: Short term treatment of insomnia and/or nocturnal waking. (Roussel Laboratories Ltd).*

Doroma. A soporific. (Bayer & Co).*

Dorsacaine. Benoxinate hydrochloride. Applications: Anaesthetic. (Dorsey Laboratories, Div of Sandoz Inc).

Dorsital. Pentobarbital. Applications: Sedative. (Dorsey Laboratories, Div of Sandoz Inc).

Dosaflo. Weedicide, containing metoxuron. (ICI PLC).*

Dosulphin. A proprietary preparation of sulphaproxyline and sulphamerazine. An antibiotic. (Ciba Geigy PLC).†

D.O.T.G. Di-*o*-tolylguanidine. A rubber vulcanization accelerator.†

D.O.T.T. Di-*o*-tolylthiourea.†

Double/Bubble. Package for two-part epoxies. (Hardman Inc).*

Double Green S.F. See METHYL GREEN.

Double Nickel Salt. Nickel ammonium sulphate, $Ni(NH_4)_2.(SO_4)_2.6H_2O$. Used in the plating trade.†

Double Scarlet. See FAST SCARLET.

Double Scarlet R. See AZOCOCCINE 2R.

Double Seidlitz Powder. See SEIDLITZ POWDER, DOUBLE.

Double Shield. Marine antifouling not containing any tin compounds.

Applications: As antifoulings for marine use. (Llewellyn Ryland Ltd).*

oublet. Herbicide. (May & Baker Ltd).

ouble Twitchell Reagent. The barium salt of the sulphonated mixture of naphthalene and fatty acid. Used in the decomposition of fats. See TWITCHELL REAGENT.†

ouble White. A proprietary trade name for a general purpose potassium silicate cement for acid conditions, e.g., as a bedding and jointing material for tiles. (Haworth (ARC) Ltd).†

oucil. Base exchange metal. (Crosfield Chemicals).*

owanol. A line of glycol ethers and glycol ether acetates used as solvents in a variety of unrelated industrial applications. (Dow Chemical Co Ltd).*

owclene. Industrial solvents, primarily for metal and suede. (Dow Chemical Co Ltd).*

owco 179. A proprietary preparation of Chlorpyrifos. An insecticide. (Dow Chemical Co Ltd).*

ow DBR. Dibenzoyl resorcinol. An ultraviolet absorber for plastics. (Dow Chemical Co Ltd).*

ow H Alloy. See S.A.E. No. 50 ALLOY.

ow plasticizer No. 5. Diphenyl mono ortho xenyl phosphate. A vinyl plasticizer. (Dow Chemical Co Ltd).*

ow plasticizer No. 55. Technical grade diphenyl mono ortho xenyl phosphate. A vinyl plasticizer. (Dow Chemical Co Ltd).*

ow V9. Alpha methyl styrene derivative. A vinyl plasticizer. (Dow Chemical Co Ltd).*

ow 276-V2. Alpha methyl styrene derivative. A vinyl plasticizer. (Dow Chemical Co Ltd).*

owetch Deadline. Magnesium photoengraving plate. Manufactured in different gauges and sizes to meet requirements of the newspaper and printing industry. (Dow Chemical Co Ltd).*

owex. Ion exchange resins. Applications: Water softening and recovering waste or undesirable materials from process streams. (Dow Chemical Co Ltd).*

Dowex Monosphere. Ion exchange resins. (Dow Chemical Co Ltd).*

Dowfax. A family of di-sulphonated anionic surfactants used in a variety of end-use applications such as cleaning products and bleaches, latex production and agricultural products. (Dow Chemical Co Ltd).*

Dowfax 2A1. Sodium dodecylated oxydibenzene disulphonate as a light coloured free-flowing powder. Applications: Anionic surfactant with high solubility, stability, coupling ability and surface activity in strong aqueous solutions of acids, alkalies and salts. Moderate sudsing agent. Used in metal cleaning including soak tank, steam and electrolytic systems; textiles; shampoos and cosmetics; emulsion polymerization; pulp and paper, mining and food processing industries. (Dow Chemical Company Ltd).

Dowfax 9N12, 9N14/15 and 9N12W. Nonionic surfactant of the nonyl phenol ethoxylate type in liquid or semi-solid form. Application: General detergency and wetting, involving elevated temperatures. Used in light duty liquid detergents; cleaning specialities; soak tank and metal cleaners. (Dow Chemical Company Ltd).

Dowfax 9N2, 9N3 and 9N4. Nonionic surfactants of the nonylphenol ethoxylate type in the form of an almost colourless liquid. Application: Emulsifier, wetting agent, dry cleaning soap, antifoam. Used in chemical intermediates; pesticides; metal cleaners; latex paints; dry cleaning formulations. (Dow Chemical Company Ltd).

Dowfax 9N5, 9N6 and 9N7. Nonionic surfactants of the nonylphenol ethoxylate type in the form of an almost colourless liquid. Application: Emulsifier, wetting agent and dry cleaning soap, used in pesticides; wax and polish; metal cleaners; metal working compounds; dry cleaning formulations. (Dow Chemical Company Ltd).

Dowfax 9N8, 9N9 and 9N10. Nonionic surfactant of the nonylphenol ethoxylate type in liquid form. Application:

Detergent, emulsifier, wetting agent and penetrant. Used in household, industrial and institutional cleaners and specialities; textiles; pesticides; latex paints; pulp and paper; leather. (Dow Chemical Company Ltd).

Dowfrost. Heat transfer fluid. Inhibited propylene glycol used as a coolant in the manufacture of beer, wine, milk and other liquids. It is also used to freeze poultry and fish. (Dow Chemical Co Ltd).*

Dowfroth. Flotation frothers. Used by the mining industry in the recovery of minerals for ores. They are low-viscosity water-soluble liquids which produce highly selective foams. (Dow Chemical Co Ltd).*

Dowgard. Coolant/antifreeze. Designed to protect automobile radiators against overheating in summer and freezing in winter and contains additives to protect against foaming and corrosion. (Dow Chemical Co Ltd).*

Dowicide. Phenolic-based antimicrobials used as active ingredients in disinfectant formulations, and also as preservatives in a variety of applications such as metal working fluids, adhesives and cosmetic preparations. (Dow Chemical Co Ltd).*

Dowicide A. A proprietary trade name for sodium-*o*-phenylphenate. An antiseptic and germicide.†

Dowicide B. A proprietary trade name for sodium 2 : 4 : 5-trichlorphenate. An antiseptic and germicide.†

Dowicide C. A proprietary trade name for sodium-chloro-*o*-phenylphenate. An antiseptic and germicide.†

Dowicide F. A proprietary trade name for sodium tetrachlorophenate. An antiseptic and germicide.†

Dowicide 1. A proprietary trade name for *o*-phenylphenol. An antiseptic and fungicide.†

Dowicide 2. A proprietary trade name for 2 : 4 : 5-trichlorphenol. An antiseptic and fungicide.†

Dowicide 3. A proprietary trade name for chloro-*o*-phenylphenol. An antiseptic and fungicide.†

Dowicide 5. A proprietary trade name for brom-*p*-phenylphenol. A germicide.†

Dowicide 6. A proprietary trade name f tetrachlorophenol. An antiseptic.†

Dowicide 7. A proprietary trade name f pentachlorophenol. An antiseptic an fungicide.†

Dowicil. Preservatives which provide microbial protection for various applications such as cosmetic and personal care formulations, househol products, paints, adhesives, metal working fluids and latex emulsions. (Dow Chemical Co Ltd).*

Dowlex. Linear low density polyethylen resins used for cast and blown films, injection moulding, blow moulding a extrusion coating. (Dow Chemical C Ltd).*

Dowmetal Alloys. Aircraft alloys containing magnesium with small amounts of aluminium and mangane sometimes with the addition of small quantities of copper, cadmium, and z They have low specific gravity.†

Downright. Latex additives used to imp low and high temperature performan and durability in asphalt concrete applications. (Dow Chemical Co Ltd

Dowper. Inhibited perchloroethylene. Applications: Solvent for dry cleanin (Dow Chemical Co Ltd).*

Dowpon. Herbicide. Applications: Controlling grass species. Used primarily in sugar cane, sugar beets, orchards and also in non-crop applications such as railroads and rubber plantations. (Dow Chemical C Ltd).*

Dowtherm A. A proprietary trade name a product consisting of a mixture of biphenyl and biphenyl ether. It is use for heating industrial machinery to h temperatures (e.g. 200° C.) in place steam.†

Dowtherm 209. Heat transfer agent use as a temperature controlling liquid (coolant) for diesel-powered vehicles. (Dow Chemical Co Ltd).*

Doxinate. Docusate sodium. Applicatio Stool softener; pharmaceutic aid. (Hoechst-Roussel Pharmaceuticals Inc).

Doxylar. Capsules containing doxycyclin hydrochloride BP equivalent to 100m doxycycline. Applications: Clinically useful in treatment of a variety of infections caused by susceptible strains of Gram-positive and Gram-negative bacteria. These include pneumonia and other respiratory tract infections including acute and chronic bronchitis, genito-urinary infections and soft tissue infections. (Lagap Pharmaceuticals Ltd).*

DPR. Depolymerized rubber and compounds made therefrom. (Hardman Inc).*

DP. 250. A polyester vinyl plasticizer.†

DP/4137-16. A proprietary fast-curing polyester resin with low viscosity. A fire-retardant. (Synthetic Resins Ltd).†

D.P.G. Diphenylguanidine. A rubber vulcanization accelerator.†

Dracyl. Toluene, $C_6H_5CH_3$.†

Dragendorf's Reagent. (Kraut's Re-agent). Potassium iodobismuthate. Used for testing alkaloids.†

Dragon. Insecticide. (ICI PLC).*

Dragon Gum. Gum tragacanth.†

Dragonmat. Insecticide vapourising device. (ICI PLC).*

Dragon's Blood. A red resin. The two varieties are Palm Dragon's Blood, obtained from the rattan palm, *Daemonorops draco*, of Sumatra and Borneo, and Socotra Dragon's Blood, from *Dracoena cinnabari*, of Socotra, and the West Indies. Employed as a pigment, for the preparation of red lakes, and varnishes.†

Dramamine. A proprietary preparation of dimenhydrinate. An antiemetic. (Searle, G D & Co Ltd).†

Drapex. A registered trade mark for an epoxy plasticizer. (Argus Chemical Corporation).†

Drapex 3.2. A registered trade mark for a vinyl plasticizer derived from an epoxide. (Argus Chemical Corporation).†

Drat. An oil formulation containing 2.5g/litre chlorophacinone, a powerful anticoagulent rodenticide. Applications: Controls black rats, brown rats, house mice, long-tailed field mice, voles and musk rats. (Burts & Harvey).*

Drawing Brass. See BRASS.

Draza. Pellet formulation containing 4% methiocarb. Applications: To control slugs and snails in any crop. Draza reduces populations of leatherjackets. There is some evidence that cutworms, earwigs and millipedes are controlled. (Bayer & Co).*

Dreadnought Powder. An explosive containing 73-77 per cent ammonium nitrate, 14-17 per cent sodium nitrate, 4-6 per cent ammonium chloride, and 3-5 per cent trinitrotoluene.†

Dreft. A proprietary trade name for a washing material consisting of sodium lauryl sulphate.†

Drenamist. A proprietary preparation of adrenaline hydrochloride in an aerosol form. Bronchial antispasmodic. (Nicholas).†

Drene. A proprietary trade name for a shampoo containing sodium lauryl sulphate.†

Drenison. A proprietary preparation of flurandrenalone for dermatological use. (Eli Lilly & Co).†

Drenusil. A proprietary preparation of polythiazide. A diuretic. (Pfizer International).*

Dresden Thick Oil. The trade name for a thick turpentine or oleo-resin. It is similar to Venice turpentine, and is used as a vehicle for colours for painting.†

Dresinate. Potassium and sodium soaps of rosin, modified rosin and tall oil derivatives. Applications: Improve latex stability in rubber latices, used in formulating soluble cutting oils and drawing compounds, used as modifiers of heavy duty metal cleaners and other industrial cleaners. (Hercules Inc).*

Dresinol. A range of resin dispersions. Applications: Used to modify polymer properties in adhesives, used in the production of pigments and resinated colours in the graphics and inks industry, used in paint polymer emulsions and to help wet paint pigments. (Hercules Inc).*

Drewclean. Maintenance chemicals. Blends of sequestering agents, surfactants, inhibitors and/or dispersants. Applications: Ion exchange resin and filter media cleaning agents. (Ashland Chemical Company).*

Drewcor. Corrosion inhibitors. Blends of neutralizing amines, filming amines and/or oxygen scavengers. Applications: Steam/condensate line corrosion control, preboiler/boiler oxygen corrosion control. (Ashland Chemical Company).*

Drewfax. Deposit control agent used to control deposits in the papermaking process. Applications: Paper mills, pulp mills. (Ashland Chemical Company).*

Drewfax 400 Series. Flow and levelling agent. z Blend of silicone derivatives and glycol ethers. Applications: Flow and levelling aid. (Ashland Chemical Company).*

Drewfax 600 Series. Blend of surfactants. Applications: Air release agent. (Ashland Chemical Company).*

Drewfax 800 Series. Blend of silicone copolymers and glycol ether. Applications: Anticratering agent. (Ashland Chemical Company).*

Drewfloc. Coagulants and flocculants. Applications: Liquid/solids separation. (Ashland Chemical Company).*

Drewplex. Boiler water treatments. Blends of natural and synthetic sludge conditioning agents. Applications: Boiler water sludge conditioning. (Ashland Chemical Company).*

Drewplus. Foam control agent. Blend of mineral oils, silica derivatives and emulsifiers. Applications: Foam control agents in latex paint, adhesives, ink, textile, industrial coatings, latex/rubber, paper coatings and agricultural products. (Ashland Chemical Company).*

Drewtrol. Deposit inhibitors. Blends of phosphate-based antiscalants and dispersing agents. Applications: Boiler deposit control agents. (Ashland Chemical Company).*

Drexar 530. MSMA. Applications: Selective herbicide for post emergent weed control on lawns and ornamental turf. (Drexel Chemical Company).*

Dri-Sil. Silicone water repellents. (Midland Silicones).†

Driclor. A colourless alcoholic solution containing aluminium chloride hexahydrate. Applications: For the treatment of hyper-hydrosis of the axillae, the hands and the feet. (Stiefel Laboratories (UK) Ltd).*

Dricol. A proprietary preparation of amidephrine mesylate. A nasal decongestant. (Bristol-Myers Co Ltd).†

Drierite. A proprietary trade name for anhydrous calcium sulphate used for drying gases.†

Driers. A trade term for those substances which are added during the process of boiling linseed oil, to accelerate its drying properties. Driers appears to absorb the oxygen from the air and transfer it to the oil, thereby aiding its oxidation. The term is used for the oxides of lead, manganese, and cobalt, which were formerly used as driers. More recently, the oxalate, acetate, and borate of manganese have been employed, and at present, the metallic salts of fatty acids, such as lead and manganese linoleates are much used. The metallic resinates are also employed as driers. Also see BOILED OIL, TEREBINE, and LEAD TUNGATE.†

Driftol. Drift retardant. (Chevron).*

Drikold. A trade mark for solid carbon dioxide. A refrigerating agent.†

Drikold. Solid carbon dioxide blocks or slices. (ICI PLC).*

Drill Rod Brass. See BRASS.

Drinamyl. A proprietary preparation of dexamphetamine sulphate and amylobarbitone. (Smith Kline and French Laboratories Ltd).*

Drisdol. Ergocalciferol. Applications: Vitamin. (Sterling Drug Inc).

Drisoy. Modified soybean oils. Applications: Paint. (NL Industries Inc).*

Dristan Inhaler. Propylhexedrine. Applications: Adrenergic. (Whitehall Laboratories, Div of American Home Products Corp).

Dristan Long Lasting Nasal Mist. Oxymetazoline hydrochloride. Applications: Adrenergic. (Whitehall Laboratories, Div of American Home Products Corp).

ittel Silver. An alloy of 67 per cent aluminium and 33 per cent silver.†

ivanil. A range of alkylene oxide addition products. Applications: Used as a base for synthetic lubricants, base component for brake fluids, heat transfer medium, and a viscosity modifier for hydraulic fluids containing no mineral oil. (Huls AG).†

riverit. Stabilized methylene chloride. Applications: Used as a solvent for metal degreasing, also suitable for the treatment of aluminium. (Huls AG).†

riverol MPL. A partial phosphate ester. Applications: Corrosion inhibitor for the mineral oil industry. (Huls AG).†

riverol OMM. An amide/anhydride mixture. Applications: Corrosion inhibitor. (Huls AG).†

riveron. Methyl tert-butyl ether. Applications: An anti-knock agent for motor fuels. (Huls AG).†

rivolan. A range of dodecanedioic acid esters. Applications: Base components for synthetic and semisynthetic lubricants. (Huls AG).†

riwal. See CERESIT.

rolban. Dromostanolone propionate. Applications: Antineoplastic. (Eli Lilly & Co).

roleptan. A proprietary preparation of droperidol. Applications: Tranquilizer, premedicant, anti-emetic. (Janssen Pharmaceutical Ltd).*

romoran. A proprietary preparation of levorphanol tartrate. An analgesic. (Roche Products Ltd).*

roncit. Tapeworm drug for dogs and cats, also against Echinococcus spp. Applications: Veterinary medicine. (Bayer & Co).*

rossa. Protective hand lotion. Applications: For practice and laboratory. (Bayer & Co).*

rott. A proprietary pyroxylin plastic.†

roxychrome. A photographic colour developer. (May & Baker Ltd).*

ry and Clear. Benzoyl Penoxide. Applications: Keratolytic. (Whitehall Laboratories, Div of American Home Products Corp).

Dry Ice. Solid carbon dioxide. Its specific gravity is 1.56.†

Dry Lightning. Inhibited sulphamic acid - dry. Applications: Removes water formed deposits. (Western Chemical Co).*

Dry Seed TRIGGRR. A dry powder containing trace minerals, used to enhance germination, maturation and crop yields. Applications: Applied to seed prior to planting. Used for a wide variety of crops including corn, sorghum, wheat and vegetables. (Westbridge Research Group).*

Dryobalanops Camphor. See CAMPHOL.

Dryptal. A proprietary preparation containing frusemide. A diuretic used for oedema of cardiac, hepatic or renal angina. (Berk Pharmaceuticals Ltd).*

Dryvax. Smallpox Vaccine. Applications: immunizing agent. (Wyeth Laboratories, Div of American Home Products Corp).

D S M. Preparation containing 50% demeton-S-methyl. Applications: Insecticide. (L W Vass (Agricultural) Ltd).*

D-S-S. Docusate sodium. Applications: Stool softener; pharmaceutic aid. (Parke-Davis, Div of Warner-Lambert Co).

D-S-S Plus. Contains casanthranol and docusate sodium. Casanthranol is a purified mixture of the anthranol glycosides derived from Casara sagrada. Applications: Laxative and stool-softener. (Parke-Davis, Div of Warner-Lambert Co).

D-steel. A steel containing 1.1-1.4 per cent manganese, 0.33 per cent carbon and 0.12 per cent silicon.†

DTIC-Dome. Dacarbazine. Applications: Antineoplastic. (Miles Pharmaceuticals, Div of Miles Laboratories Inc).

D.T.S. Dehydrothio-*p*-toluidinesulphonic acid.†

Du Pont Accelerators. Du Pont is a registered trade mark for proprietary products. Accelerator No. 1 is composed of *p*-nitroso-dimethylaniline; No. 4 consists of aniline; No. 5 is a formaldehyde-aniline product; No. 6 is methylene dianilide; No. 8 consists of anhydro-formaldehyde-*p*-toluidine; No. 11 is triphenylguanidine; No. 12 is diphenylguanidine; No. 15 is composed of thiocarbanilide; No. 17 is di-*o*-tolylthiourea; No. 18 is di-*o*-tolylguanidine; No. 19 is a synthetic resin produced by the condensation of aliphatic aldehydes with aniline. The above are all used in the vulcanization of rubber.†

Du Pont Permissible No. 1. A trade mark for an explosive containing nitroglycerin, ammonium nitrate, wood pulp, and sodium chloride.†

Du Pont Powder. A trade mark for an explosive containing nitrocellulose, nitroglycerin, and ammonium picrate.†

Du-Ter. Fungicide, containing fentin hydroxide for prevention of potato blight and disease control in sugar beet. (ICI PLC).*

Du-Ter. Fentinhydroxide. Applications: Counteracts potato, rice, sugarbeet, groundnut blight etc. (Duphar BV).*

Duallor. A range of precious metal alloys. Applications: For dentistry and dental engineering. (Degussa).*

Dubbin. Mixtures of waxes and tallow with colouring matter, sometimes with the addition of rosin. Used to render leather waterproof, and to preserve it.†

Duco. A proprietary trade name for pyroxylin lacquers, containing cellulose nitrate.†

Ducobee-Hy. Hydroxocobalamin. Applications: Vitamin. (Sterling Drug Inc).

Dudley Metal. An alloy of 98 per cent tin, 1.6 per cent copper, and 0.25 per cent lead.†

Dufox. A potato fungicide. (Murphy Chemical Co).†

Dugro. Ronidazole. Applications: Antiprotozoal. (Merck Sharp & Dohme, Div of Merck & Co Inc).

Duhnul-balasan. See MECCA BALSAM.

Duke's Metal. A heat-resisting alloy, containing 81 per cent iron, 12 per cent chromium, 4 per cent cobalt, 1.5 per cent carbon, and small quantities o manganese, tungsten, and silicon.†

Dulceta. Textile auxiliary chemicals. (PLC).*

Dulcine. (Sucrol, Sucrene). Mono-*p*-phenetol-carbamide, $NH_2.CO.NH.C_6H_4\ OC_2H_5$. A sweetening substance. It is 200 tim sweeter than cane sugar.†

Dulcodos. A proprietary preparation containing bisacodyl and dioctyl so sulphosuccinate. A laxative. (Boehringer Ingelheim Ltd).†

Dulcolax. A proprietary preparation containing bisacodyl. A laxative. (Boehringer Ingelheim Ltd).†

Dulenza. A proprietary viscose silk.†

Dullray. A heat-resisting alloy contain 60 per cent iron, 34 per cent nickel 5 per cent chromium.†

Dulux. A range of various interior/ext paints and paint-related products. (PLC).*

Dumacene C13, NP707, NP7710 and NPX10. Alkylphenol ethoxylate nonionic surfactant. Applications: Textile scouring; wool washing. (Te SA).

Dumet. A copper-clad nickel-iron alloy

Duncaine. A proprietary trade name fc Lignocaine.†

Dunclad CE. A proprietary trade name a laminate of Penton (*qv*) and a synthetic rubber. Offers a highly corrosion resistant lining at temperatures up to 125-130° C. (Dunlop Rubber).†

Dunclad VN. A laminate of unplasticiz PVC and synthetic rubber designed enable metal tanks to be lined using special adhesives, giving a highly corrosive resistant surface to preven attack from oxidizing and other aci up to 85° C.†

Dunlop Grade 6167. A first quality but rubber compound which can be use to 110° C. in corrosive conditions. (Dunlop Rubber).†

Dunlop PL. A laminate of polypropyle and synthetic rubber for tank lining (Dunlop Rubber).†

nlop 6593. A high grade neoprene compound giving high chemical and abrasion resistance at elevated temperatures. (Dunlop Rubber).†

nnite. An American explosive. The main constituent is picric acid.†

o-Autohaler. Breath-actuated pressurized aerosol containing a suspension of isoprenaline hydrochloride BP 8 mg/ml, phenylephrine bitartrate 12 mg/ml; delivers 400 metered doses (0.16 mg isoprenaline hydrochloride, 0.24 mg phenylephrine bitartrate per dose). (Riker Laboratories/3M Health Care).†

o-Decadrin. Dexamethasone acetate and dexamethasone phosphate. Applications: Prompt and sustained corticosteroid activity for the relief of certain endocrine and non-endocrine disorders. (Merck Sharp & Dohme).*

uofilm. Collodion based product containing salicyclic acid BP and lactic acid BP. Applications: Topical application for the treatment of plantar and mosaic warts. (Stiefel Laboratories (UK) Ltd).*

uogastrone. A proprietary preparation of carbenoxolone sodium in a delayed release capsule, for the treatment of duodenal ulcers. (Berk Pharmaceuticals Ltd).*

uolith. See CRYPTONE.

uomac. Series of cationic surfactants which consist of the acetate salts of alkyl propylene diamines, and which are soluble. Applications: Pigment flushing, froth flotation and flocculation, particularly in mineral flotation; petroleum processing; leather; paper; pigments and surface coatings; ceramics. (Akzo Chemie UK Ltd).

uomatic. Injector/drencher gun. (May & Baker Ltd).

uomeen. Range of cationic surfactants of the alkyl propylene diamine type, possessing both primary and secondary amine groups. They form strongly bonded films on the surfaces of metal, textiles, plastics, etc. Applications: Used in petroleum; road emulsions; plastics; rubber; textiles; leather; herbicides; fungicides; rodent repellents; mineral flotation; paper; pigments and surface coating; water and sewage treatment; wax; sealant formulations; cement curing; metal working; emulsion for car underseals; carbon paper and typewriter ribbon. (Akzo Chemie UK Ltd).

Duomeen. A range of coco, oleyl and stearyl diamines. Applications: Used as waterproofing agents, bitumen emulsifiers, bitumen adhesion additives, anticorrosives and as agricultural sprays. (Harcros Industrial Chemicals).*

Duoquad. Diamine quaternary ammonium chloride. (Akzo Chemie UK Ltd).

Duoteric. Surfactant blend. (ABM Chemicals Ltd).*

Duovac-C. B_1 type, B_1 strain, Newcastle and Connecticut type. Applications: immunization of poultry. (Intervet America Inc).*

Duovac-M. B_1 type, B_1 strain, Newcastle and mild Massachuttes type. Applications: immunization of poultry. (Intervet America Inc).*

Duphalac. Lactulose. Applications: Treatment of obstipation and (pre-) hepatic coma. (Duphar BV).*

Duphaston. Dydrogesterone. Applications: Progestative, counteracts complaints caused by hormonal disorders in women. (Duphar BV).*

Duponol. Surface active agent. (Du Pont (UK) Ltd).

Duponol. A proprietary trade name for a wetting and emulsifying agent containing the sodium salt of sulphated higher fatty alcohols.†

Duponol LS. A proprietary trade name for sodium oleyl sulphate, a wetting agent.†

Duponol ME. A proprietary trade name for sodium lauryl sulphate, a wetting agent.†

Duponol WA. A proprietary trade name for a mixture of sodium salt of sulphated lauryl alcohol and lauryl alcohol.†

Duprene. A proprietary trade name for a synthetic rubber made by the polymerization of chloroprene. Resistant to heat, oil, ozone, and most other chemicals.†

Durabolin. Nandrolone phenpropionate. Applications: Androgen. (Organon Inc).

Duracillin. A proprietary preparation of procaine penicillin. An antibiotic. (Eli Lilly & Co).†

Duracore. High strength aluminium honeycomb cores. Applications: Military and aerospace sandwich panel applications. (Fothergill Tygaflor Ltd).*

Duracreme. A proprietary preparation of propylene glycol, glycerin, sodium alginate, boric acid and hexyl resorcinol. A spermicidal jelly. (LRC International Ltd).†

Duradiene. Styrene/butadiene solution SBR. Applications: Tyres, moulded goods and adhesives. (Firestone Synthetic).*

Duraform. A proprietary trade name for asbestos reinforced thermoplastics. They have greater stiffness, lower coefficient of expansion, higher heat distortion point, lower creep, higher tensile and flexural strengths than the basic resins.†

Duraguard. Polyester epoxy and epoxy polyester-based coating powders. RAL colour range. Applications: Electrostatic spray applications for decorative finishes, including wireworking, sheet metal fabrication and domestic wirework. (Plascoat Systems Ltd).*

Dural. Macro-crystalline regular aluminium oxide. Applications: Used for abrasive industries, super refractories, sandblasting and for safes. (Lonza Limited).*

Dural. See DURALUMIN.

Duralac. Barium chromate jointing compound conforming to specification DTD 369A. Applications: Sealing of joints between dissimilar metals, protection of metals in contact with wood, synthetic resin compositions, leather, rubber, fabrics etc. (Llewellyn Ryland Ltd).*

Duralcon. Antistatic textile cohesive agent. (J C Thompson & Co (Duron) Ltd).

Duralium. An alloy of aluminium with from 3.5-5.5 per cent copper and small amounts of magnesium and manganese.†

Duralloy. Dental silver alloys and mercu[illegible] for mixing amalgams. Applications: [illegible] dentistry and dental engineering. (Degussa).*

Duralon. A proprietary trade name for a vinyl plasticizer. A furan resin. (US Stoneware Co).†

Duraloy. Alloyed thermoplastics. Applications: Vehicle bumpers, high impact applications. (Celanese Limited).†

Duralum. An alloy of 79 per cent aluminium, 10 per cent copper, zinc, and tin.†

Duralumin. Alloys of aluminium with fro[illegible] 3-5.5 per cent copper, 0.5-1 per cent manganese, 0.5 per cent magnesium, and small quantities of silicon and iro[illegible] The alloy, containing 95.5 per cent aluminium, 3 per cent copper, 1 per c[illegible] manganese, and 0.5 per cent magnesium, has a specific gravity of 2[illegible] and melts at 650° C. These alloys are resistant to sea water and dilute acids.

Duramycin. A proprietary preparation of diabekacin sulphate. An antibiotic. (Pfizer International).*

Durana Metal. An alloy of 65 per cent copper, 30 per cent zinc, 2 per cent tin[illegible] 1.5 per cent iron, and 1.5 per cent aluminium, of a golden yellow colour.[illegible]

Duranalium. An aluminium alloy similar [illegible] composition and properties to hydronalium and B.S. sea-water alloy.

Durance's Metal. (Dewrance Metal). A bearing metal, consisting of about 33 per cent tin, 23 per cent copper, and 4[illegible] per cent antimony.†

Durand's Metal. An alloy of 66.6 per cen[illegible] aluminium, and 33.3 per cent zinc.†

Duranest. Etidocaine. Applications: Anaesthetic. (Astra Pharmaceutical Products Inc).

Durango. See GUAYULE.

Duranic. An alloy containing aluminium with from 2-5 per cent nickel and 1.5-2.5 per cent manganese.†

Duranit. A range of styrene/butadiene copolymers. Applications: Used as reinforcing resins in rubber mixes and as reinforcing dispersion for natural an[illegible] synthetic latex. (Huls AG).†

uranite. A proprietary trade name for a fast-baking synthetic enamel.†

uranthrene Blue GCD. A dyestuff. It is a British equivalent of Indanthrene blue GCD.†

uranthrene Blue RD Extra. A British dyestuff. It is equivalent to Indanthrene X.†

uranthrene Dark Blue BO. A dyestuff. It is a British brand of Indanthrene dark blue BO.†

uranthrene Dyes. A registered trade name for certain British dyestuffs.†

uranthrene Gold Yellow Y. A British brand of Pyranthrone.†

uranthrene Olive GL. A dyestuff. It is a British equivalent of Indanthrene olive G.†

uraplex. A proprietary trade name for alkyd varnish and lacquer resins.†

uraplus. Polymer emulsion. Applications: High performance floor polishes. (Rohm and Haas Company).*

urapro. Oxaprozin. Applications: Anti-inflammatory. (Wyeth Laboratories, Div of American Home Products Corp).

uraquin. Quinidine Gluconate. Applications: Cardiac depressant. (Parke-Davis, Div of Warner-Lambert Co).

uraSoft. Phemfilcon A. Applications: Contact lens material. (Wesley-Jessen, Div of Schering Corp).

urasol Acid Blue B. A dyestuff. It is a British brand of Alizarin saphirol B.†

urastic. A bitumen compound stated to be composed of high grade bitumen freed from organic acids and used as a protective coating.†

uration. Oxymetazoline hydrochloride. Applications: Adrenergic. (Plough Inc).

urax. N-cyclohexyl-2-benzothiazolesulphenamide - primary delayed action accelerator for rubber. Applications: Used in both natural and synthetic rubbers. Safe at processing temperatures, fast at curing temperatures. (Vanderbilt Chemical Corporation).*

urazol. Direct dyestuffs. (ICI PLC).*

Durbar Bronze. A proprietary trade name for an alloy of copper with 24 per cent lead and 4 per cent tin.†

Durbar Hard Bronze. A proprietary trade name for an alloy of copper with 10 per cent tin and 20 per cent lead.†

Durcoton. A proprietary phenol-formaldehyde resin impregnated textile.†

Durecol. A proprietary glycero-phthalic synthetic resin.†

Durehete 1050. A proprietary trade name for a steel containing 1 per cent. chromium, 1 per cent. molybdenum and 3/4 per cent. vanadium for bolting materials capable of operating at metal temperatures up to 1050° F. (Samuel Fox & Co Ltd).†

Durehete 900. A proprietary trade name for steel containing 1 per cent. chromium and 1/2 per cent. molybdenum. It is suitable for studs and bolts for service at temperatures up to 900° F. (482° C.). (Samuel Fox & Co Ltd).†

Durehete 950. A proprietary trade name for a steel containing 1 per cent. chromium, 1/2 per cent. molybdenum and 1/4 per cent. vanadium for studs and bolts for service at temperatures up to 950° F. (510° C.). (Samuel Fox & Co Ltd).†

Durel. Polyarylate. Applications: Microwave cookers,vehicle light lens and traffic light lens. (Celanese Limited).†

Durena Metal. See DURANA METAL.

Durene. . 1,2,4,5-Tetramethylbenzene, $C_6H_2(CH_3)_4$.†

Duresco. See LITHOPONE.

Durethan. Polyamide 6 and 6.6. Applications: For the manufacture by injection moulding of engineering components with good impact resistance, high dynamic strength, good abrasion and wear resistance as well as for production by extrusion of profiles, hose, film, rod and pipe. (Bayer & Co).*

Durethan B. A proprietary brand of Nylon 6. (Bayer & Co).*

Durethan BKV. Glass-filled polyamide 6. (Bayer & Co).*

Durethane. A proprietary range of polyurethane thermoplastics used in injection moulding. (Bayer & Co).*

Durex. A proprietary trade name for an alloy of 83 per cent copper, 10 per cent tin, and 4-5 per cent carbon. Also a proprietary trade name for phenol-formaldehyde synthetic resin.†

Durex White. Barium carbonate, $BaCO_3$.†

Durez. A synthetic resin of the phenol-formaldehyde type. It is oil soluble. There are other Durez phenol-formaldehyde thermosetting synthetic resins and moulding powders. (Occidental Chemical Corp).*

Durez 18783. Glass fibre filled, high impact diallyl phthalate moulding compound. (Occidental Chemical Corp).*

Durferrit. Carburizing, nitriding, annealing, hardening, tempering and heat transfer salts. Applications: Heat treatment of metal in the machine, machine tool, motor vehicle, aircraft and other metal working industries. (Degussa).*

Durham's Stain. A microscopic stain. It contains a saturated solution of stannous chloride and a 15 per cent solution of tannic acid in equal parts, with a few drops of an alcoholic solution of methylene blue.†

Duricef. Cefadroxil. Applications: Antibacterial. (Mead Johnson & Co).

Durichlor. A hydrochloric acid resisting alloy containing 81 per cent iron, 14.5 per cent silicon, 3.5 per cent molybdenum, and 1 per cent nickel.†

Duridine. Tetramethylphenylamine, $C_6H(CH_3)_4.NH_2$.†

Durimet Alloys. Proprietary alloys for acid resistance. Alloy A is stated to contain iron with 25 per cent nickel, no chromium, and 5 per cent silicon; B contains iron with more nickel than A, and with chromium content about one-third of the nickel, and 5 per cent silicon; alloy D contains iron with 15 per cent nickel, and smaller amounts of chromium and silicon.†

Durindone Blue 4B. A dyestuff. It is a British equivalent of Bromindigo FB.†

Durindone Blue 6B. A dyestuff. It is a British brand of Indigo MLB/6B.†

Durindone Red B. A British equivalent Thioindigo red B.†

Durindone Red 3B. A dyestuff. It is equivalent to Helindone red 3B, and of British manufacture.†

Durine. A formalin preparation.†

Duriron. An acid resisting alloy, which silicon-iron alloy. It contains 15.5 pe cent silicon, 82 per cent iron, 0.66 p cent manganese, 0.83 per cent carbo and 0.57 per cent phosphorus.†

Durisol. A trade mark. See PERMALI.

Duro Cement. A cement used in the manufacture of acid towers. It conta 96 per cent silica and 4 per cent sodi silicate.†

Durochrome Black. A dyestuff. It is a British equivalent of Diamond black

Durocide. Biocides. (Durham Chemical Ltd).*

Durodi Steel. A proprietary trade name a nickel-chromium-molybdenum stee

Duroftal. A proprietary synthetic resin.†

Duroftal 293-E. See DUROPHEN 330

Duroglass. A proprietary borosilicate resistance glass for chemical use.†

Duroil. Self emulsifiable oil. (J C Thom & Co (Duron) Ltd).

Durol. Monochlorohydroquinone.†

Durolastik. Waterproofing, roofing materials. (Weatherguard/Marbleloi Products Inc).*

Durolube. Textile yarn lubricant. (J C Thompson & Co (Duron) Ltd).

Duromine. Each capsule contains phentermine 15 mg or 30 mg as ion-exchange resin complexes. (Riker Laboratories/3M Health Care).†

Duromorph. A proprietary preparation c morphine in a long acting form. A hypnotic. (LAB Ltd).*

Duronze. A proprietary trade name for high-silicon copper alloy.†

Durophen. A trade name for a series of plasticized phenolic resins widely use in baking finishes. (Chemische Werk Albert).†

Durophen 127-B. See KUNSTHARZ H or DUROPHEN 373U. An ammonia condensed phenol formaldehyde resin melting at 55° C. (Chemische Werke, Albert).†

urophen 170W. A butylated diane formaldehyde condensate cooked with trimethylene glycol maleate (65 per cent. solids). (Bush Beach Ltd).†

urophen 218V. A butylated diane formaldehyde castor oil resin. (Chemische Werke, Albert).†

urophen 287W. A butylated phenol urea formaldehyde resin sold at 58 per cent. solids. (Chemische Werke, Albert).†

urophen 309W. A butylated xylenol formaldehyde resin containing butyl glyceryl adipate. (Chemische Werke, Albert).†

urophen 330V. A butylated diane formaldehyde resin cooked with glyceryl phthalate and synthetic fatty acids. (Chemische Werke, Albert).†

urophen 373U. See KUNSTHARZ HM or DUROPHEN 127-B.

urophenine Brown. A dyestuff prepared by boiling nitrosophenol with dilute sulphuric acid. It dyes cotton dark violet-brown from a sodium sulphide bath.†

urophet. Each capsule contains laevo- and dextro-amphetamine (ratio 1:3) 7.5 mg or 20 mg as ion-exchange resin complexes. (Riker Laboratories/3M Health Care).†

uroplaz 610, 810, 911. Proprietary trade names for phthalate esters of straight chain alcohols.†

uroprene. A registered trade name for a product obtained by the exhaustive chlorination of natural rubber. It can be moulded, and is soluble in benzene, coal-tar naphtha, and carbon tetrachloride. It is resistant to chemical action and is used in paints and varnishes.†

uroseal. Modified aluminium stearate, for waterproofing. (Durham Chemicals Ltd).*

urosehl. Epoxy and polyurethane resin systems. Applications: Casting and laminated structures, engineering patterns and toolmaking, electrical encapsulations. (Solochart Ltd).*

urosil. Medium-active silica. Applications: Especially well suited for extrusion and calendering mixtures. (Degussa).*

Duroslip. Antistatic textile fibres lubricant. (J C Thompson & Co (Duron) Ltd).

Durosoft. Fibre softener. (J C Thompson & Co (Duron) Ltd).

Durosol. Self emulsifiable oil. (J C Thompson & Co (Duron) Ltd).

Durostabe. Stabilizers for PVC. (Durham Chemicals Ltd).*

Duroterm. Investment material. Applications: For casting precious metal alloys. (Bayer & Co).*

Durotex. Yarn strengthening agent. (J C Thompson & Co (Duron) Ltd).

Durotint. Textile fugitive tints. (J C Thompson & Co (Duron) Ltd).

Durowynd. Hydrophylic wetting agent. (J C Thompson & Co (Duron) Ltd).

Durox. A mullite (*qv*) made by fusing kyanite and alumina. A proprietary trade name for an ammonium dynamite. An explosive.†

Duroxyn. Epoxide resins esterified with fatty acids. Applications: Chemical resistant paints. (Resinous Chemicals Ltd).*

Durrax. Hydroxyzine hydrochloride. Applications: Tranquilizer. (Dermik Laboratories Inc).

Dursban. Broad-spectrum insecticide. (Murphy Chemical Ltd).

Dursban. Insecticides comprising chlorpyrifos as active ingredient . These products are very effective. Applications: Used to control ticks on cattle, mosquitoes and other insects. (Dow Chemical Co Ltd).*

Durundum. See METALITE.

Duspatal. Mebeverine. Applications: Spasmolytic, counteracts intestinal spasms. (Duphar BV).*

Duspatalin. Mebeverine. Applications: Spasmolytic, counteracts intestinal spasms. (Duphar BV).*

Dustallay.. Wetting system for reducing dust in mines. (Foseco (FS) Ltd).

Dutch Camphor. Obtained from the wood of the Japanese camphor laurel, *Cinnamomun camphora*.†

Dutch Metal. An alloy of 80 per cent copper and 20 per cent zinc.†

Dutch Pink. A pigment. It is a yellow colour made by absorbing quercitron on barytes or alumina.†

Dutch Varnish. A solution of rosin in turpentine.†

Dutch White. A pigment consisting of 1 part white lead with 3 parts heavy spar.†

Duthane. A proprietary polyester-based polyurethane elastomer cross-linked with diols. (151). VULKOLLAN.†

Dutral. A proprietary trade name for an ethylene-propylene synthetic rubber copolymer suitable for tank linings, seals, hose, cables. (Shell Chemicals).†

Dutral-Co. A proprietary range of ethylene-propylene copolymers. (Montedison UK Ltd).*

Dutral-Ter. A proprietary range of ethylene-propylene-diene terpolymers (EPDM). (Montedison UK Ltd).*

Dutrex. Hydrocarbon oils. (Shell Chemicals UK Ltd).

Dutrex Process and Extender Oils. A number of aromatic extracts are produced during the refining of lubricating oils. By suitable selection and blending, a range of Dutrex grades with varying viscosities can be obtained. These hydrocarbon mixtures possess good solvent characteristics and excellent polymer compatibility. (Shell Chemicals).†

Dutrex 20, 25. A proprietary trade name for an extender-plasticizer for vinyl resins. A petroleum derivative. (Shell Chemicals).†

Duty Oil. See OIL OF RHODIUM.

Duvadilan. Isoxsuprine. Applications: Promotes blood flow rate. (Duphar BV).*

Duvadilan Retard. Isoxsuprine. Applications: Promotes blood flow rate. (Duphar BV).*

Duvoid. Bethanechol chloride. Applications: Cholinergic. (Norwich Eaton Pharmaceuticals Inc).

Duxalid. A proprietary trade name for a synthetic resin.†

Duxite. An explosive containing nitro-glycerin, collodion cotton, sodium nitrate, wood meal, and ammonium oxalate. Duxite is also the name app to a resin from lignite.†

Duxol. A proprietary trade name for a synthetic resin.†

DV. Dienestrol. Applications: Oestroge (Merrell Dow Pharmaceuticals Inc, Subsidiary of Dow Chemical Co).

D.X.L. High boiling tar acids. (Coalite Fuels & Chemicals Ltd).†

Dy-Chek. Four step process for detectin surface flaws in metallic and non-metallic components. (Foseco (FS) Ltd).*

Dyamul. A range of dyebath auxiliaries, including detergents, buffers, solvent lubricants and combinations of these. Applications: Assistants in textile processing and dyeing. (Yorkshire Chemicals Plc).*

Dyapol. A range of dye dispersants. Applications: Dyebath dispersant for disperse dyes. (Yorkshire Chemicals Plc).*

Dyazide Tablets. Combination of triamterene and hydrochlorothiazide. Applications: Potassium conserving diuretic for use as antihypertensive. (Smith Kline and French Laboratori Ltd).*

Dycastal. Crack prevention in aluminiur diecastings. (Foseco (FS) Ltd).*

Dycill. Dicloxacillin sodium. Applicatio Antibacterial. (Beecham Laboratorie

Dyclone. Dyclonine hydrochloride. Applications: Anaesthetic. (Astra Pharmaceutical Products inc).

Dycote. Foundry die coatings. (Foseco (Ltd).*

Dycron. Crystalline aluminium oxide. Applications: For lapping of gearings and hydraulics, motors, electronics, glass etc. (Dynamit Nobel Wien GmbH).*

Dyer's Lac. See LAC DYE.

Dyflor L90. A proprietary polyvinyl fluoride. It is processed in dispersion form.†

Dyflor 2000. Polyvinylidenefluoride (PVDF). Injection-, extrusion- and blow moulding compound. Applications: Manufacture of chemic apparatus, mechanical engineering, cable industry, electronics, manufacture of medical equipment. (Dynamit Nobel Wien GmbH).*

Dylene. A proprietary polystyrene. (Koppers Co Inc).†

Dylon. Colloidal graphite in aqueous or solvent carriers. Applications: High temperature parting agent, release compounds, with good electrical and thermal conductivity. (Dylon Industries Inc).*

Dylon. Blue colourant, concentrated liquid for colouring spray solutions (especially pesticide sprays) so that operator can see the spray coverage. (Regal Chemical Company).*

Dylonite. Resin impregnated impervious graphite. Applications: Heat exchangers, tubes and pumps with excellent resistance to high temperature corrosive chemicals. (Dylon Industries Inc).*

Dymacryl. 100% acrylic copolymers, colour stable inorganic pigments and surface penetrating agents. Available in 10 colours plus clear. Applications: Beautify and protect above-grade masonry surfaces against damage caused by water absorption. Used on architectural concrete, precast and poured concrete, GFR concrete, brick, natural stone, stucco and unglazed tile. (Dampney Company Inc.)*

Dymax Multi-Care Structural Adhesives. Acrylic structural adhesives, ultraviolet curing structural adhesives. Applications: Magnet, metal, glass and fibre optics bonding, assembly adhesives and coatings for electronics. (Dymax Engineering Adhesives).*

Dymel. Dimethyl ether propellant. (Du Pont (UK) Ltd).

Dymelor. Acetohexamide. Applications: Antidiabetic. (Eli Lilly & Co).

Dymerex. Partially dimerized rosin acids. Applications: Used to reinforce specialty adhesive products, in specialty protective coatings. (Hercules Inc).*

Dymsol 38C. A proprietary anionic, bio-degradable, polymerization emulsifier for improving the processing characteristics of S.B.R., nitrile rubber and neoprene. (Diamond Shamrock Chemical Co, NJ).*

Dynacal. Fused calcium oxide. Applications: Additive for the melting of high purity metallurgical products. Raw material for the electroslag remelting process. Auxiliary material for metallurgical desulphurization processes. (Dynamit Nobel Wien GmbH).*

Dynacast. Fusion cased bricks. Applications: For the lining of slab pusher furnaces, billet pusher furnaces, ingot pusher furnaces, forging furnaces, rocker bar furnaces, roller hearth furnaces, soaking pit furnaces, tundishes of continuous casting plants, steel degassing plants, high temperature furnaces. (Dynamit Nobel Wien GmbH).*

Dynacerin. Jojoba oil substitute, liquid wax ester. Applications: Oily component for emulsions with emollient properties in pharmaceuticals and cosmetics. For ointments, creams, liquid emulsions, external suspension, skin lotions, bubble baths and shampoos. (Dynamit Nobel Wien GmbH).*

Dynacet. Acetylated mono-glycerides derived from edible fat raw materials. Applications: Dip coatings on various foods, such as sausage and meat products increase freshness and reduce weight loss. (Dynamit Nobel Wien GmbH).*

Dynacoll. Reactive polyesters. Applications: Adhesive raw material for the formulation of reactive hot melts. (Dynamit Nobel Wien GmbH).*

Dynaflock. Sodium aluminate. Applications: Caustic product for paper industry, water treatment, ceramic industry and building industry. (Dynamit Nobel Wien GmbH).*

Dynagrout. Sodium aluminate. Applications: Reactive agent for forming injection gels (grout), for the mining and building industry. (Dynamit Nobel Wien GmbH).*

Dynagunit. Alkali aluminate formulations. Applications: Liquid concrete setting accelerator for the mining and building industry. (Dynamit Nobel Wien GmbH).*

Dynamag. Electromagnesia (magnesium oxide). Applications: Raw material for ramming mixes and shape bricks for the lining of high temperature furnaces. Raw material for welding powder and the coating of electrodes. (Dynamit Nobel Wien GmbH).*

Dynamar Brand Specialities. A line of speciality chemicals used in the manufacture/processing of elastomer and plastics compositions. (3M).*

Dynamine. A rubber vulcanization accelerator. It is diphenylguanidine.†

Dynamite. See KIESELGUHR DYNAMITE.

Dynamite Acid. Concentrated nitric acid, used for making 96 per cent mixed acids (34 per cent nitric acid + 62 per cent sulphuric acid).†

Dynamite Glycerin. Glycerin of specific gravity 1.263, containing 98-99 per cent. It contains no lime, sulphuric acid, chlorine, or arsenic.†

Dynamites. Explosives first patented by Nobel, which consist of nitroglycerine rendered shock-stable by absorption onto Keiselgurhr or some other absorbent.†

Dynamullit. Fused mullite (aluminium silicate). Applications: Production of shape bricks for high-charged zones in heating and melting furnaces. Moulding material for precision casting moulds. (Dynamit Nobel Wien GmbH).*

Dynamutilin. Tiamulin Fumarate. Applications: Antibacterial. (E R Squibb & Sons Inc).

Dynamutilin Aqueous Solution. Tiamulin hydrogen fumarate in aqueous solution. (E R Squibb & Sons Ltd).

Dynamutilin Water Soluble Powder. Tiamulin hydrogen fumarate in powder base. (E R Squibb & Sons Ltd).

Dynamyte. Dinoseb. Applications: Herbicide for use on beans, small grains, forage, cereal crops. (Drexel Chemical Company).*

Dynamyxin. Sulphomyxin. Applications: Antibacterial. (Pfizer Inc).

Dynapen. Dicloxacillin sodium. Applications: Antibacterial. (Bristol Laboratories, Div of Bristol-Myers Co).

Dynapol H. Low-molecular, saturated polyesters. Applications: For the production of high-quality baking enamels (industrial enamels, enamels for household appliances, enamels for vehicles (metallic finishes)). (Dynamit Nobel Wien GmbH).*

Dynapol L. High molecular copolyesters. Applications: Raw material for coil coating, can coating, metal decorating. (Dynamit Nobel Wien GmbH).*

Dynapol LH. Medium and low-molecular, saturated polyesters. Applications: Particularly suitable for coatings coming in contact with foodstuffs. Raw material for coil coating, metal decorating and adhesives. (Dynamit Nobel Wien GmbH).*

Dynapol P. Thermoplastic copolymers. Applications: Where protection against corrosion and good resistance to aggresive chemicals are required. (Dynamit Nobel Wien GmbH).*

Dynapol S. Thermoplastic, linear, saturated copolyesters. Applications: Adhesive raw materials for hot melts. (Dynamit Nobel Wien GmbH).*

Dynapor. Phenolic foaming resins. Applications: Flame inhibiting and rotproof insulating material for ceilings in flat roof construction, for wall and facade panelling or pipe sheatings. (Dynamit Nobel Wien GmbH).*

Dynasan. Fatty acid esters. Applications: For use as lubricants and retarding agents in tablets, crystallization accelerator in suppositories and chocolate, consistency regulators in ointments, creams and lotions, basic material for fatty powders. (Dynamit Nobel Wien GmbH).*

Dynasil. Silicic acid esters. Applications: Binders for the foundry industry and for inorganic zinc dust coatings. (Dynamit Nobel Wien GmbH).*

Dynasperse. Modified sodium lignosulphonates. Applications: Dyestuff dispersants. (Reed Lignin Inc).*

Dynaspinell. Fused spinel (magnesium aluminate). Applications: Raw material for the production of pyrometer sheaths crucibles, refractory bricks, ramming mixes. (Dynamit Nobel Wien GmbH).*

Dynastite. An explosive containing 94 per cent potassium chlorate, and 6 per cent barium nitrate, dipped in nitro-toluene.†

Dynasylan. Organo-functional silane adhesion promoters, alkyl silanes. Applications: Bonding agent between inorganic surfaces and organic polymers. Binders for the foundry industry. (Dynamit Nobel Wien GmbH).*

Dynasylan BSM. Alkyl-alkoxy-silane mixtures. Applications: Impregnating material for buildings. (Dynamit Nobel Wien GmbH).*

Dynat W. A proprietary brand of mechanically comminuted rubber from Malaysia.†

Dynatherm. Electromagnesia. Applications: Fused magnesia for heating elements. (Dynamit Nobel Wien GmbH).*

Dynatred. Polyurethane sealant. Applications: Used for sealing horizontal joints in parking decks, plazas, warehouse floors or other areas subject to heavy foot and vehicular traffic, particularly where slope exceeds 1%. (Pecora Corporation).*

Dynatrol. Polyurethane sealant. Applications: Used in general construction for caulking vertical expansion joints in walls and sealing around door and window frames. (Pecora Corporation).*

Dynaweld. Urethane polymeric sealant adhesive. Applications: Used for adhering pre-formed sections of Urexpan NR-200 to concrete or epoxy substrates and to each other. (Pecora Corporation).*

Dynazirkon. Zirconium oxide. Applications: Raw material for the production of high refractory, slag resistant crucibles and mouldings for the melting of high temperature alloys. Raw material for high wear resisting casting nozzles in continuous casting plants. (Dynamit Nobel Wien GmbH).*

Dyne. Iodophor/acid pipe line cleaner and sterilizer. (Ciba-Geigy PLC).

Dynemate 200. Hypochlorite/acid pipe line cleaner and sterilizer. (Ciba-Geigy PLC).

Dynobel. An explosive containing potassium perchlorate, nitroglycerin, ammonium oxalate, wood meal, and a little collodion cotton.†

Dyphene. Alkyl phenolic resins. Applications: Resins used in adhesives, rubber, elastomers, bonding, abrasives and friction materials. (PMC Specialties Group Inc).*

Dyrenium. Triamterene. Applications: Diuretic. (Smith Kline & French Laboratories).

Dyslysin. Cholalic anhydride.†

Dyslytite. See SCHREIBERSITE.

Dysoid. A bearing bronze containing 62 per cent copper, 18 per cent lead, 10 per cent tin, and 10 per cent zinc.†

Dytac. A proprietary preparation of triamterene. A diuretic. (Smith Kline and French Laboratories Ltd).*

Dytel. Leak detective dye. (Du Pont (UK) Ltd).

Dytide. A proprietary preparation of triamterene and benzthiazide. A diuretic. (Smith Kline and French Laboratories Ltd).*

Dytransin. Ibufenac. Applications: Analgesic; anti-inflammatory. (The Boots Company PLC).

DZ910. Dissociated zircon consisting of spheres of free silica with zirconia dendrites embedded therein. Applications: Manufacture of ceramic pigments and refractory compositions. (Ferro Corporation).*

E

E107. See AVERTIN.

E45 Cream. A proprietary preparation of white soft paraffin, light liquid paraffin and wool fat used as a skin cream. (Boots Pure Drug Co).*

E Alloy. An alloy of 76 per cent aluminium, 20 per cent zinc, 2.5 per cent copper, 0.2 per cent iron, 0.5 per cent manganese, 0.5 per cent magnesium, and 0.2 per cent silicon.†

Earth Archil. Archil (*qv*), contaminated with mineral matter. Used for the preparation of litmus.†

Earth, Bone. See BONE ASH.

Earth, Cassel. See VANDYCK BROWN.

Earth, Diatomaceous. See INFUSORIAL EARTH.

Earth, Gold. An Ochre. See YELLOW OCHRE.†

Earth Green. See LIME GREEN.

Earth, Infusorial. See INFUSORIAL EARTH.

Earth, Japan. See CUTCH.

Earth, Lemnos. See LEMNIAN EARTH.

Earth, Porcelain. See CHINA CLAY.

Earth, Red. See INDIAN RED.

Earth, Santorin. See SANTORIN EARTH.

Earth Wax. See OZOKERITE.

Earth, Yellow. See OCHRE.

Easisperse. Easy dispersing pigments. Applications: Printing inks and paints. (Manox Ltd).†

Easprin. Aspirin. Applications: Analgesic; antipyretic; antirheumatic. (Parke-Davis, Div of Warner-Lambert Co).

East India Gum. (Bombay Gum). A variety of gum arabic, pale amber or pinkish in colour.†

East Indian Balsam of Copaiba. A name given to Gurjun balsam (the oleo-resin from the stems of *Diptero-carpus* species).†

Eastman Inhibitor DOBP. A proprietary trade name for 4-dodecyloxy-2-hydroxybenzophenone. A UV light inhibitor suitable for use in PVC, polyesters, polystyrene and butyrate acrylic coatings.†

Eastman Inhibitor OPS. A proprietary trade name for *p*-octylphenyl salicylate. A UV light inhibitor suitable for polyolefins.†

Eastman Inhibitor RMB. A proprietary trade name for a UV light inhibitor for polar resins (cellulosics). It is resorcinol monobenzoate.†

Eastman Yellow. A yellow colouring matter used as a corrective filter in photography. It is the sodium salt of glucosephenylosazone-*p*-carboxylic acid.†

Eastman 910. A proprietary trade name for a cyano-acrylate adhesive which sets with the application of pressure. Variants are: *910 EM* for vinyls; *910 FS* for quicker setting; *910 MHT* for applications involving high temperatures. (Eastman Chemical International AG).†

Easton's Syrup. (*Syrupus Ferri Phosphatis cum Quinina et Strychnina B.P.*) Quinine sulphate and strychnine are added to the solution obtained by dissolving iron in concentrated phosphoric acid, then the syrup is added.†

Eastozone 32. A proprietary anti-oxidant. N, N'-dimethyl-N, N'-di-(1 methyl-propyl) - p - phenylenediamine. (Eastman Chemical Products).†

Easy-FLo. Fluxes for silver alloy brazing. (Johnson Matthey Chemicals Ltd).

Eau de Brouts. Petitgrain water.†

Eau de Goudron. Tar water.†

Ebert and Merz's α-Acid. 2, 7-Naphthalene-disulphonic acid, $C_{10}H_6(SO_3H)_2$.†

Ebert and Merz's β-Acid. 2, 6-Naphthalenedisulphonic acid.†

Ebner's Fluid. A mixture of 2.5 cc hydrochloric acid, 2.5 grams sodium chloride, 100 c.c water, and 500 cc alcohol. Used for decolourizing in bacteriological work.†

Eboli Green. A polyazo dyestuff which is a similar product to Diamine green G.†

Ebonestos. A trade name for a series of proprietary moulded products for electrical and heat insulation.†

Ebonised Monel. A monel metal with a fine finish produced by an oxidizing process.†

Ebonite. (Vulcanite, Hardened Rubber). A material prepared by vulcanizing rubber with up to 75 per cent sulphur or metallic sulphides, with the addition of chalk, gypsum, or other filling and colouring substances. Used as an insulating material.†

Ebontex. A proprietary trade name for an emulsified asphalt used for waterproofing tanks.†

Ebony Black. A blackish-brown dyestuff mixed with a blue dyestuff. It is used for dyeing cotton from a bath containing sodium sulphate and sodium carbonate, and half-wool. Also see GAS BLACK.†

Eborex. A proprietary preparation containing about 65-70 per cent sodium fluosilicate. It is a light fluosilicate for use as an insecticide.†

Ebrok. A proprietary trade name for a bituminous plastic.†

E.C. See ELECTROLYTIC CHLOROGEN.

Eca. Oil additive packages. (Exxon Chemical Ltd).*

E.C.A. Cresylic acids. (Murphy Chemical Co).†

Eccobond Paste 99. A proprietary one part thixotropic epoxy adhesive of high thermal conductivity. It is used in heat-sink applications. (Emmerson and Cummings Inc).*

Eccobond SF40. A proprietary low density, two-component epoxy-based adhesive and rigid filler. (Emmerson and Cummings Inc).*

Eccobond 114. A proprietary one-part, filled epoxy adhesive. (Emmerson and Cummings Inc).*

Eccocoat SJB. A proprietary epoxy resin used as a coating for semiconductor junctions. (Emmerson and Cummings Inc).*

Eccofloat. PG23. A proprietary polyester-resin-bound syntactic foam used to fill voids in submarine hulls. (Emmerson and Cummings Inc).*

Eccofloat EG35. A proprietary epoxy resin-bound syntactic foam material used in deep-sea applications. (Emmerson and Cummings Inc).*

Eccofloat Encapsulant 1421. A proprietary epoxy-resin-bound encapsulant used to protect under-sea components. (Emmerson and Cummings Inc).*

Eccofloat HG452. A proprietary polyester-resin-bound low-density float material for use in deep-sea applications. (Emmerson and Cummings Inc).*

Eccofloat PC61. A proprietary polyester-resin-bound castable material for use in deep-sea applications. (Emmerson and Cummings Inc).*

Eccofloat PP22 and 24. Proprietary grades of polyester-bound syntactic foam which can be packed *in situ* to fill voids and to make buoys. (Emmerson and Cummings Inc).*

Eccofloat SP 12, 20. A proprietary polyester-bound low-density syntactic foam for use where buoyancy is required in harbour and off-shore applications. (Emmerson and Cummings Inc).*

Eccofloat SS40. A proprietary polyurethane rubber-bound material used in the making of deep-sea diving suits. (Emmerson and Cummings Inc).*

Eccofloat UG 36. A proprietary polyurethane-bound semi-flexible non-compressible material. (Emmerson and Cummings Inc).*

Eccofloat US 35. A proprietary polyurethane - bound material - flexible, compressible and usable down to about 1000 ft depth of water. (Emmerson and Cummings Inc).*

Eccofoam PP. A proprietary group of hydrocarbon resin closed-cell foams used in high-frequency electrical applications. (Emmerson and Cummings Inc).*

Eccosorb Coating 268E. A proprietary epoxy coating which is brushed on to surfaces to increase their electrical loss in the L-band of the high- frequency range. (Emmerson and Cummings Inc).*

Eccosorb MF. A proprietary range of magnetically-loaded epoxy resins. (Emmerson and Cummings Inc).*

Eccosorb 269E. A proprietary epoxy coating having properties similar to those of Eccosorb Coating 268E but for use in the S to the K bands inclusive. (Emmerson and Cummings Inc).*

Eccospheres. Trade name for small hollow glass or silica spheres of diameter ranging from 10-250 microns. Used as a loading material for plastics to impart lightness and reduced permittivity which it does by virtue of the large airspace. (Emmerson and Cummings Inc).*

Ecepox PB1 and PB2. Epoxidized esters used as plasticizers and stabilizers for PVC compounds. Registered trade names.†

Èchappe Silk. A name for floss or waste silk.†

Echicaoutchin. A low-grade gutta-like material from *Alstonia scholaris*.†

Echo. Thiadiazole derivatives used as crosslinking agents. Applications: Used in the vulcanization of halogen-containing polymers such as polyepichlorohydrins and chlorinated polyethylene. Typical applications are extruded and moulded hose and tubing. (Hercules Inc).*

Echurin. A mixture of picric acid and nitro-flavin.†

Eclabron. Guaithylline. Applications: Bronchodilator; expectorant. (US Ethicals Inc).

Eclipse. Pelleted organic based fertilizers comprising a composted organic base and chemical N, P & K to form the analysis. Applications: A steady release, lower nitrogen fertilizer for horticultural crops, parks and gardens. (Humber fertilizers plc).*

Eclipse Black. A sulphide dyestuff.†

Eclipse Blue. A dyestuff. The indophenol from dimethyl-*p*-phenylenediamine and phenol when treated with alkali sulphites gives a sulphonic acid. The sodium salt of this acid is heated with sulphur and sodium sulphide to give Eclipse Blue.†

Eclipse Bronze. A proprietary trade name for a nickel-bronze containing 60-65 per cent nickel, 24-27 per cent copper, 9-11 per cent tin, and small amounts of iron, silicon, and manganese.†

Eclipse Brown B. A dyestuff obtained by heating *m*-tolylenediamine and oxalic acid with polysulphides.†

Eclipse Green G. A sulphide dyestuff.†

Eclipse Red. See BENZOPURPURIN 4B. Also Geigy's Eclipse red, by heating aminohydroxyphenazines with alkali sulphides and sulphur.

Eclipse Yellow. A dyestuff obtained by heating diformyl-*m*-tolylenediamine and sulphur at 240° C.†

Ecolac. A proprietary trade name for an air drying lacquer and adhesive for plastics.†

Ecolo. Compounded plastic material. (Mitsubishi Petrochemical Co).*

Ecomytrin. A proprietary preparation of amphomycin and neomycin. Antibiotic skin cream. (Warner).†

Econacort. A proprietary preparation containing econazole nitrate and hydrocortisone. Applications: The treatment of fungal and inflammatory skin disorders. (Squibb, E R & Sons Ltd).*

Econochlor. Chloramphenicol. Applications: Antibacterial; antirickettsial. (Alcon Laboratories Inc).

Econopred. Prednisolone Acetate. Applications: Glucocorticoid. (Alcon Laboratories Inc).

Ecoro. A proprietary packaging material of polypropylene loaded with calcium carbonate to ease disposal by incineration. (Mitsubishi Petrochemical Co).*

Ecostatin. A proprietary preparation of econazole nitrate. Applications: An anti-fungal. (Squibb, E R & Sons Ltd).*

Ecostatin Cream. Econazole nitrate in cream base. (E R Squibb & Sons Ltd).

Ecostatin Lotion. Econazole nitrate in lotion base. (E R Squibb & Sons Ltd).

Ecostatin Pessaries. Econazole nitrate vaginal pessaries. (E R Squibb & Sons Ltd).

Ecostatin Powder Solution. Econazole nitrate in aerosol spray. (E R Squibb & Sons Ltd).

Ecosyl. Concentrated biological silage additive. (ICI PLC).*

Ecotrin. Aspirin. Applications: Analgesic; antipyretic; antirheumatic. (Menley & James Laboratories).

Ecru Silk. Silk which has lost about 3-4 per cent of its weight of sericin or silk gum.†

Ectimar. Broad spectrum antimycotic. Applications: Used for the treatment of dermatomycosis in veterinary medicine. (Bayer & Co).*

Eczederm. Calamine in a soft emolient base. Applications: For the treatment of eczema. (Quinoderm Ltd).*

Edasil. Natural ion exchanger for improvement and conditioning of the soil. (Süd-Chemie AG).*

Edecrin. Ethacrynic acid. Applications: Potent diuretic for the relief of certain oedemas. (Merck Sharp & Dohme).*

Edecrin Sodium. Ethacrynate sodium. Applications: Diuretic. (Merck Sharp & Dohme, Div of Merck & Co Inc).

Edelfeka. A nickel-containing silver-copper-cadmium alloy.†

Edelresanol. A proprietary synthetic resin.†

Edelweiss. See OIL WHITE.

Edelwit. Hydrated lime $Ca(OH)_2$. Applications: Chemical industries, drinking water treatment, wastewater treatment. (BV Nekami).*

Edenol. A proprietary trade name for a neutral ester of adipic acid and a special compound of synthetic alcohols.†

Edenol B316. A proprietary epoxy plasticizer made of an epoxidized linseed oil and used in rigid PVC (Hental and Cie, Düsseldorf).†

Edenol B35. A proprietary alkyl-epoxy stearate-type plasticizer for plastisols. (Hental and Cie, Düsseldorf).†

Edenol D72. A proprietary alkyl-epoxy stearate-type plasticizer used for plastisols. (Hental and Cie, Düsseldorf).†

Edenol D82. A proprietary epoxy plasticizer. An epoxidized soya bean oil, it is used in both rigid and soft PVC (Hental and Cie, Düsseldorf).†

Edenol HS 235. A proprietary alkyl-epoxy stearate-type plasticizer for use in plastisols. (Hental and Cie, Düsseldorf).†

Edenol 74. A proprietary alkyl-epoxy stearate-type plasticizer. (Hental and Cie, Düsseldorf).†

Eder's Solution. A solution of mercuric chloride and ammonium oxalate used in photometric determinations.†

Edetic Acid. Ethylenediamine - NNN'N'-tetra-acetic acid. Versene Acid.†

Edicol. Colouring matter for use in foodstuffs, pharmaceuticals and cosmetics. (Sold under licence from ICI). (ICI PLC).*

Edimet. Polymethyl methacrylate.†

Edinol. A photographic developer. It contains *p*-aminosaligenin, acetone sulphite, potassium hydroxide, and potassium bromide.†

Edistir. A proprietary range of polystyrene moulding granules. (Montedison UK Ltd).*

Edolan. Resist agent for wool/polyamide by the exhaust process. (Bayer & Co).*

Edosol. A proprietary food product containing fat, protein, lactose and minerals used in low-sodium diets. (Cow and Gate).†

Edrisal. A proprietary preparation of amphetamine sulphate, aspirin, and phenacetin. (Smith Kline and French Laboratories Ltd).*

Edunine. Textile auxiliary chemicals. (ICI PLC).*

Edward's Speculum. A zinc and arsenic bearing bronze containing 63.3 per cent copper, 32.2 per cent tin, 2.9 per cent zinc, and 1.6 per cent arsenic.†

Eel Antifriction Metal. An alloy of 75 per cent lead, 15 per cent antimony, 6 per cent tin, 1.5 per cent cadmium, 0.5 per cent arsenic, and 0.1 per cent phosphorus.†

EES. Erythromycin ethylsuccinate. Applications: Antibacterial. (Abbott Laboratories).

Efcortelan. A proprietary preparation of hydrocortisone for dermatological use. (Glaxo Pharmaceuticals Ltd).†

Efcortelan Soluble. A proprietary preparation of HYDROCORTISONE, for injection in cases of shock and adrenal crisis. (Glaxo Pharmaceuticals Ltd).†

Efcortelan-N. A proprietary preparation of hydrocortisone and neomycin used in dermatology as an antibacterial agent. (Glaxo Pharmaceuticals Ltd).†

Efcortesol. A proprietary preparation of hydrocortisone as the 21- disodium phosphate ester, used to treat adrenal insufficiency and shock. (Glaxo Pharmaceuticals Ltd).†

Effersyllium. Psyllium Husk. Applications: Laxative. (Stuart Pharmaceuticals, Div of ICI Americas Inc).

Effervescent Tartrated Soda Powder. See SEIDLITZ POWDER.

Effesay. Sulphonated alcohols and detergents. (ABM Chemicals Ltd).*

Effico. A proprietary preparation of thiamine hydrochloride, nicotinamide, tincture of nux vomica and caffeine hydrate. A tonic. (Pharmax).†

Efflorescent Pyrites. See WHITE PYRITES.

Efflox. Water treatment chemical. (Berk Spencer Acids Ltd).

Efudex. Fluorouracil. Applications: Antineoplastic. (Hoffmann-LaRoche Inc).

Efudix. A proprietary preparation of fluorouracil. Antineoplastic agent. (Roche Products Ltd).*

Efuranol. A proprietary preparation of imipramine hydrochloride. An antidepressant. (Pfizer International).*

Efweko. Carbon black chips in nitro-cellulose or plastics. Applications: Used for dyeing lacquers, gravure and flexographic printing inks. (Degussa).*

Egalex. Liquid dyeing assistant. Applications: Textiles. (Tensia SA).

Egalisal. A fibre protection agent. Applications: Used in the textile industry for dyeing wool. (Degussa).*

Egalon Colours, Egalon Auxiliaries, Egalon Thinners. Pigments and auxiliaries for nitrocellulose finishing. (Bayer UK Ltd).

Egg Oils. See WHITE OILS.

Eglantine. A name which has been applied to both isobutyl benzoate and to isobutyl-phenol acetate. Used in perfumery.†

Egyptian Blue. (Vestorian Blue). $CaO.CuO._4SiO_2$, is the formula which corresponds to this ancient blue, used by the Romans.†

Egyptian Fibre. See VULCANISED FIBRE.

Egyptianised Clay. Clay rendered more plastic by the addition of tannin.†

EHIDA Kit. Etifenin. Applications: Diagnostic aid. (Amersham Corp).

Ehrhard's Metal. An alloy of 89 per cent zinc, 4 per cent copper, 4 per cent tin, and 3 per cent lead.†

Ehrlich-biondi Stain. A microscopic stain containing 100 cc of a saturated solution of Orange G, 30 cc of a saturated solution of acid fuchsin, and 5o cc of a saturated solution of methyl green.†

Ehrlich's Acid Haematoxylin. A stain used in microscopy. It contains 100 cc water, 100 cc absolute alcohol, 100 cc glycerin, 10 cc glacial acetic acid, 2 grams haematoxylin, and alum in excess.†

Ehrlich's Diazo Reagent. For indole: 4 grams *p*-dimethylamino-benzaldehyde in 380 cc alcohol and 80 cc concentrated hydrochloric acid. One volume of the solution to be tested is used with 1 volume of the reagent, a positive colour being red.†

Ehrlich's Haematoxylin. A microscopic stain. It consists of 30 grains haematoxylin, 100 cc absolute alcohol, 100 cc glycerin, 30 grains ammonium alum, and 100 cc distilled water.†

Ehrlich's Triple Stain. A microscopic stain for blood corpuscles. It contains 135 parts of a saturated aqueous solution of Orange G, 100 parts of a saturated aqueous solution of Methyl green, 100 parts of a saturated solution of Acid fuchsine, 100 parts of glycerin, 200 parts of absolute alcohol, and 300 parts water.†

Eichrome Red B. A dyestuff obtained from 1-amino-2-naphthol-4-sulphonic acid and 1-phenyl-3-methyl-5-pyrazolone.†

Eisenepidot. Synonym for Ferriepidote.†

Eisenglimmer. Synonym for Vivianite.†

Eisenphyllit. Synonym for Vivianite.†

Eisen, Silicon. See SILICON-EISEN.

Eisler's Bronze. A bronze containing 5.9 per cent tin. Used for art castings.†

Ekaline. Aliphatic polyglycol ether. Applications: Versatile dispersing, levelling and scouring agent. (Sandoz Products Ltd).

Ekammon. A proprietary preparation of aspirin, and vitamins C and K. An analgesic. (Ward, Blenkinsop & Co Ltd).†

Ekanda Rubber. A rubber obtained from the shrub, *Raphionacme utilis* in Angola.†

Ekatin. Insecticide containing thiometon for aphid control. (ICI PLC).*

E-Kote 3042. A trade name for a silver filled air drying epoxy coating material soluble in isobutyl ketone. (Allied Products Corporation).†

Ektebin. An anti-tubercular preparation. (Bayer & Co).*

Ektogan. (Ektogen, Zinc Perhydrol, Zinconal). A preparation of zinc oxide, containing from 40-60 per cent zinc peroxide. An antiseptic used for dressing wounds and burns, also as an astyptic.†

Ektogen. See EKTOGAN.

Elaine. See OLEIN.

Elanone. Lenperone. Applications: Antipsycotic. (A H Robins Co Inc).

Elaol. A proprietary plasticizer. It is stated to be dibutyl phthalate.†

Elaol VI. Flame retardant hydraulic fluid. Applications: For use as a safety precaution (fire and consequential damage) in mining. (Bayer & Co).*

Elaol 1. A trade name for a plasticizer made from C_4 to C_6 paraffin fatty acids and hexanetriol.†

Elaol 2. A trade name for a plasticizer made from C_6 to C_9 paraffin fatty acids and hexanetriol.†

Elaol 3. A trade name for a plasticizer made from C_4 to C_6 paraffin fatty acids and pentaerythritol.†

Elaol 4. A trade name for a plasticizer made from C_6 to C_9 paraffin fatty acids and pentaerythritol.†

Elaqua XX. Urea. Applications: Diuretic. (Elder Pharmaceuticals Inc).

Elargol. A silver finish for mica and plastics. (Ward, Blenkinsop & Co Ltd).†

Elase. A proprietary preparation of fibrinolysin and desoxyribonuclease. Dermatological stimulant. (Parke-Davis).†

Elasti-glass. A proprietary trade name for a vinyl copolymer used for belts, braces, raincoats, tobacco pouches, etc.†

Elastic Asbestos. See MOUNTAIN CORK.

Elastic Bitumen. See ELATERITE.

Elastite. A sulphurized oil rubber substitute. Also a proprietary flooring block made from asphalt, fibre, and fillers.†

Elastoid 1300. High build elastomeric coatings based on multiphase synthetic rubber copolymers. Tough, flexible, corrosion and weather resistant, abrasion and impact resistant. Applications: Heavy-duty protection in aggressive chemical, industrial and marine environments. Extremely low water vapour transmission rate. Used on metal, concrete, foam insulation. Unaffected by water immersion. (Dampney Company Inc.)*

Elastolac. A proprietary trade name for a shellac derivative. It is water and alcohol soluble.†

Elastolith. A synthetic resin. See BAKELITE.†

Elastosil. RTV-1 silicone rubbers. Applications: Adhesives, sealing, coating, sealants in the electrical and electronics industry, seals in the automobile industry. (Wacker Chemie GmbH).*

Elaterite. (Elastic Bitumen, Mineral Caoutchouc, Helenite). A fossil resin, resembling asphaltum, found in some of the lead mines in Derbyshire. It contains 6-7 per cent mineral matter, and is slightly soluble in ether.†

Elaterite, Artificial. A proprietary product made from liquid bitumen and vegetable oils, then treatment with heat and pressure with sulphur chloride, saltpetre, and sulphur. Used for water proofing and insulation.†

Elaterium. The dried sediment from the juice of *Ecballium elaterium*. The active principle is elaterin.†

Elavil. Amitriptyline. Applications: Antidepressant with sedative properties. (Merck Sharp & Dohme).*

Elayl. See OLEFIANT GAS.

Elbasol. Solvent dyestuffs. Applications: For colouration of non aqueous solvents. (Holliday Dyes & Chemicals Ltd).*

Elbelan. 2:1 metal complex dyestuffs. Applications: For wool, nylon and blends containing one or both of these fibres. (Holliday Dyes & Chemicals Ltd).*

Elbelene. Dyes for polypropylene. (Holliday Dyes & Chemicals Ltd).

Elbenyl. Anionic dyestuffs. Applications: Dyes specially selected for their suitability for dyeing nylon. (Holliday Dyes & Chemicals Ltd).*

Elbestret. De-suplhurization of gas. (Holliday Dyes & Chemicals Ltd).

Elcema. Cellulose powder. Applications: Highly pure pharmaceutical auxiliary in the production of tablets and dragées. (Degussa).*

Elcomet. A proprietary trade name for a steel containing chromium, silicon, copper, and nickel.†

Eldecort. Hydrocortisone. Applications: Glucocorticoid. (Elder Pharmaceuticals Inc).

Elder Oil. See OIL OF CABBAGE.

Eldisine. Vindesine Sulphate. Applications: Antineoplastic. (Eli Lilly & Co).

Eldopaque. Hydroquinone. Applications: Depigmentor. (Elder Pharmaceuticals Inc).

Eldoquin. Hydroquinone. Applications: Depigmentor. (Elder Pharmaceuticals Inc).

Electran. Reagents for electrophoresis. (BDH Chemicals Ltd).

Electrathane. Anti static castable polyurethane. Applications: Passive static discharge, charged transfer roller - photo copiers. (PEI Precision Elastomers Inc).*

Electraurol. A form of colloidal gold.†

Electric Bronze. An alloy of 87 per cent copper, 7 per cent tin, 3 per cent zinc, and 3 per cent lead.†

Electric Metal. See TELEGRAPH BRONZE.

Electrical Castings Brass. See BRASS.

Electricidal. Electro-colloidal iridium.†

Electriridol. Colloidal iridium.†

Electrisil. A proprietary silicone rubber composition used for insulating conductors. (General Electric).†

Electrisil 758. A proprietary flame retardant silicone rubber compound used to insulate high-voltage cables. (General Electric).†

Electrisil 9025. A proprietary silicone rubber compound used in applications where radiation and high temperatures may be encountered. (General Electric).†

Electrit. A trade mark for goods of the abrasive and refractory class, the essential constituent of which is crystalline alumina.†

Electro-filtros. A diaphragm material. It consists of grains of pure crystalline silica cemented together with a fused siliceous binding substance. Used in electrolytic processes.†

Electro-fused Cement. See FUSED CEMENT.

Electro-granodized Iron and Steel. A process for forming a rust-preventing coat on iron and steel. An alternating current plates a continuous coating of zinc phosphate.†

Electrocuprol. A form of colloidal copper.†

Electrodag. Dispersions of conducting pigment in resin. Applications: Silk screen printable conducting inks for membrane switches. Shielding coatings. Conducting coatings. Heat generating coatings. (Acheson Colloids).*

Electrodyn. Sprayers. (ICI PLC).*

Electrofine. A registered trade name for chlorinated paraffins for use as extenders in plasticized materials.†

Electrolon. See CARBORA.

Electrolyilc Chlorogen (E.C.). A chlorinated soda prepared by the electrolysis of brine.†

Electromartiol. A form of colloidal iron.†

Electromercurol. A form of colloidal mercury.†

Electronite. A safety explosive containing 75 per cent ammonium nitrate, 5 per cent barium nitrate, with wood meal and starch.†

Electronite No. 2. An explosive consisting of 95 per cent ammonium nitrate, and 5 per cent wood meal and starch.†

Electropalladiol. A form of colloidal palladium.†

Electroplate. See ARGYROLITH and NICKEL SILVERS.

Electroplatinol. A form of colloidal platinum.†

Electrorhodiol. A form of colloidal rhodium.†

Electrorubin. A trade mark for abrasive and refractory materials. The essential constituent is crystalline alumina.†

Electrose. A proprietary trade name for a shellac plastic.†

Electroselenium. A form of colloidal selenium.†

Electrotype Metal. An alloy of 93 per cent lead, 4 per cent antimony, and 3 per cent tin.†

Electrox. Photoconductive zinc oxides. (Durham Chemicals Ltd).*

Electrozone. A similar preparation to Chloros (sodium hypochlorite solution). A disinfectant.†

Electrum. See NICKEL SILVERS.

Electrundum. A trade mark for materials of the abrasive type and consisting essentially of alumina. Elektron. A registered trade mark used in connection with certain magnesiuro alloys containing up to about 10 per cent of various alloying constituents, such as aluminium, zinc, and manganese. Its specific gravity is about 1.8. It is used in cast and wrought forms for aero engines and other purposes.†

Elektra. Organic brightener system. Applications: Acid copper electroplating (decorative). (Harshaw Chemicals Ltd).*

Elemi Resins. These are somewhat soft oleo-resins from species of *Canarium*, principally *Canarium luzonicum*. The chief elemi of commerce is Manila elemi. The fresh resin contains from 20-30 per cent of essential oil which is composed mainly of hydrocarbons, the main one being the terpene phellandrene. The soft resin contains 15-20 per cent volatile oil, an ash of 0.02-0.2 per cent. and an acid value of 17-25. The hard resin has 8-9 per cent volatile matter, 0.2-1 per cent ash, and an acid value of 15-28. Other elemi resins are Yucatan elemi from *Amyris plumieri*, Mexican elemi from *Amyris elemifera*, Brazilian elemi from *Protium heptaphyllum*, African elemi from *Boswellia freriana*, and East Indian elemi from *Canarium zephyrenum*.†

Elemite. A proprietary trade name for a wetting agent and detergent. It is a combination of sulphonated oils and solvents.†

Elephant Bronze. An alloy of 85 per cent copper, 10.5 per cent tin, 2.75 per cent zinc, 1.5 per cent lead, and 0.1-0.2 per cent phosphorus.†

Elephant Nuts. See ACAJOU BALSAM.

Elephant-S Bronze. An alloy of 80.5 per cent copper, 10.2 per cent tin, 9 per cent antimony, and 0.1-0.3 per cent phosphorus.†

Elestol. A proprietary preparation of chloroquine phosphate, prednisolone and aspirin. An anti-inflammatory agent. (FBA Pharmaceuticals).†

Eleudron-Solution. Sulphonamide. Applications: Used for coccidiosis in poultry in veterinary medicine. (Bayer & Co).*

Eleuthera Bark. See SWEET BARK.

Elexar. A proprietary range of thermoplastic rubbers designed for use in the cable industry. (Shell Chemicals).†

Elfan A432. Amphoteric surfactant supplied as a clear yellowish liquid. Applications: Baby shampoos; bubble baths; strong acid and alkaline cleaning detergents. (Akzo Chemie Nederland BV).

Elfan KT550. Anionic surfactant in which the cation is sodium and the anion is composed of 50% coconut/50% tallow fatty alcohol sulphate (C12-C18). Supplied as a white paste. Applications: Detergent raw material for heavy and light duty detergents, and washing pastes. (Akzo Chemie Nederland BV).

Elfan NS 243S. Anionic surfactant in liquid or paste form. Applications: Shampoos, bubble baths, dishwashing liquids, light duty liquids, washing pastes, car shampoos. (Akzo Chemie UK Ltd).

Elfan NS 682 KS. Anionic surfactant as a yellowish paste. Applications: Shampoos, bubble baths, dishwashing liquids, light duty liquids, washing pastes, car shampoos. (Akzo Chemie UK Ltd).

Elfan NS242. Anionic surfactant in liquid or paste form. Applications: Shampoos, bubble baths, dishwashing liquids, light duty liquids, washing pastes, car shampoos. (Akzo Chemie UK Ltd).

Elfan NS252 S. Anionic surfactant in liquid or paste form. Applications: Shampoos, bubble baths, dishwashing liquids, light duty liquids, washing pastes, car shampoos. (Akzo Chemie UK Ltd).

Elfan OS 46. Sodium alpha-olefine sulphonate (C14/C16) as a yellowish liquid. Applications: Anionic surfactant for shampoos, bubble baths, dishwashing detergents, liquid and paste-form cleaners. (Akzo Chemie UK Ltd).

Elfan WA Series. Dodecylbenzene sulphonate as triethanolamine or sodium salt or in acid form. Supplied as liquid, paste or powder. Applications: Anionic surfactants used in heavy and light duty, all-purpose and dishwashing detergents, scouring and other powder formulations. (Akzo Chemie Nederland BV).

Elfan 200. Sodium lauryl sulphate (C12) as a fine white powder. Applications: Anionic surfactant used in toothpastes. (Akzo Chemie Nederland BV).

Elfan 240 and 240S. Sodium lauryl sulphate (C12/C14), either natural (240) or based on a synthetic fatty alcohol (240S). Supplied as a transparent to white paste. Applications: Detergent and emulsifier used in shampoos, light duty detergents and cleaning pastes. (Akzo Chemie Nederland BV).

Elfan 240M and 240M/S. Monoethanolamine lauryl sulphate (C12/C14), either natural (240M) or based on a synthetic fatty alcohol (240M/S). Supplied as a clear, yellowish, medium viscous liquid. Applications: Detergent and emulsifier for shampoos and bubble baths. (Akzo Chemie Nederland BV).

Elfan 240T and 240T/S. Triethanolamine lauryl sulphate (C12/C14), either natural (240T) or based on a synthetic fatty alcohol (240T/S). Supplied as a yellowish clear liquid. Applications: Detergent for shampoos and bubble baths. (Akzo Chemie Nederland BV).

Elfan 280. Sodium coconut fatty alcohol sulphate (C12-C18) as a white powder or paste. Applications: Detergent raw material for light duty detergents, all-purpose washing agents and hand cleansers. (Akzo Chemie Nederland BV).

Elfan 680. Sodium oleyl-cetyl alcohol sulphate as a yellowish- brown paste. Applications: Detergent for heavy and light duty detergent powders, washing and cleaning pastes. (Akzo Chemie Nederland BV).

Elfanol 510. Sodium sulphosuccinic acid monoester of a fatty acid alkylolamide.

Creamy coloured paste. Applications: Washing and cleaning pastes; hand cleansers. (Akzo Chemie UK Ltd).

Elfanol 616. Sodium sulphosuccinic acid monoester of an ethoxylated fatty alcohol. Yellowish viscous liquid. Applications: Shampoos, bubble baths, baby baths, liquid hand cleansers. (Akzo Chemie UK Ltd).

Elfanol 850. Sodium sulphosuccinic acid monoester of an ethoxylated fatty acid. Yellow-brown, nearly clear liquid. Applications: Baby baths, shampoos, bubble baths, liquid hand cleansers. (Akzo Chemie UK Ltd).

Elfanol 883. Sodium dioctylester of sulphosuccinic acid. Colourless to slight yellow liquid. Applications: Wetting agent for technical processes. (Akzo Chemie UK Ltd).

Elfapur N50. Nonylphenol ethoxylate nonionic surfactant in the form of a clear, nearly colourless liquid. Applications: Dishwashing detergents for automatic machines; emulsifiers for fats and mineral oils. (Akzo Chemie Nederland BV).

Elfapur N70. Nonylphenol ethoxylate nonionic surfactant in the form of a clear, nearly colourless liquid. Applications: Low foaming dishwashing detergents for automatic machines, industrial and solvent cleansers, fat-dissolving pastes, all-purpose washing pastes. (Akzo Chemie Nederland BV).

Elfapur N90, N120 and N150. Nonylphenol ethoxylate nonionic surfactants in liquid or paste form. Applications: Wide range of detergents, cleansers, car shampoos, fat-dissolving pastes, all purpose washing pastes. (Akzo Chemie Nederland BV).

Elfugin. Wash-fast antistatic agent. Applications: Synthetic fibres. (Sandoz Products Ltd).

Elhuyarite. A red allophane mineral.†

Elianite I. An acid-resisting alloy, containing 82 per cent iron, 15 per cent silicon, and 0.6 per cent manganese.†

Elianite II. An acid-resisting alloy, consisting of 81 per cent iron, 15 per cent silicon, 0.5 per cent manganese, 2.2 per cent nickel, 0.8 per cent carbon, and 0.06 per cent phosphorus.†

Eliminal. Aluminium - removing flux. (Foseco (FS) Ltd).*

Elinvar. A nickel steel containing 36 per cent nickel, 46 per cent iron, 12 per cent chromium, 4 per cent tungsten, and 1-2 per cent manganese. It has a very low temperature coefficient of the elasticity modulus, and is used for the more delicate parts of watches.†

Elite Fast. Anionic dyestuffs (neutral dyeing). Applications: For dyeing shades of wool with good wash fastness. (Holliday Dyes & Chemicals Ltd).*

Elityran. A proprietary preparation of throid extract. (Bayer & Co).*

Elkem Microsilica. Raw and processed amorphous silica (condensed silica fume). Applications: Refractories, polymers, insulation, fluid cracking catalysts and a range of chemical and mineral uses. (Elkem Chemicals Inc).*

Elkonite. A copper-tungsten alloy used for making welding dies. It has a Brinnell hardness of 225, a compression strength ot 208,000 lb per sq in, and is not annealed at red heat.†

Elkosin. A proprietary trade name for Sulphasomidine.†

Ellagitannin. A variety of tannin found in divi-divi, knoppern, and myrobalans.†

Ellagite. A mineral. It is a variety of natrolite.†

Elmarid. An alloy of 89 per cent tungsten, 4.5 per cent cobalt, 5.9 per cent carbon, and 0.4 per cent iron.†

Elner's German Silver. A nickel silver containing 57.4 per cent copper, 26.6 per cent zinc, 13 per cent nickel, and 3 per cent iron.†

Elocril. Organophosphorus insecticide. (Ciba-Geigy PLC).

Elotex. Redispersible homo or copolymer powders (powdered emulsions). Applications: Building products. (Ebnother Group).*

Eloxal. A proprietary trade name for an anodized aluminium.†

Eloxyl. Benzoyl peroxide. Applications: Keratolytic. (Elder Pharmaceuticals Inc).

Elsner Green. A pigment. It is a mixture of Genteles green (copper stannate), with fustic decoction.†

Elsner's Reagent. A basic zinc chloride solution obtained by dissolving 500 grams zinc chloride and 20 grams zinc oxide in 425 cc water and warming. A solvent for silk.†

Elspar. Asparaginase. Applications: The treatment of acute lymphocytic leucaemia. (Merck Sharp & Dohme).*

Eltesol. Aromatic sulphonates for laundry detergents. (Albright & Wilson Ltd).*

Eltesol ACS 60. Alkylaryl sulphonate anionic. Pale yellow liquid. Applications: Manufacture of liquid detergent formulations. (Albright & Wilson Ltd, Detergents Div, Marchon).

Eltesol CA65 and CA96. Alkylaryl sulphonate anionic in acid form. Applications: Hardening in resin production; plasticer for nitrocellulose laquers; activator for nicotine insecticides; de- scaler for resin bound sand castings; catalyst for the preparation of esters, acetylation etc. (Albright & Wilson Ltd, Detergents Div, Marchon).

Eltesol PSA. Alkylaryl sulphonate anionic in acid form. Applications: Hardening in resin production: plasticer for nitrocellulose laquers; activator for nicotine insecticides; de- scaler for resin bound sand castings; catalysts in the preparation of esters, acetylation etc. (Albright & Wilson Ltd, Detergents Div, Marchon).

Eltesol PX. Alkylaryl sulphonate anionic. Off-white flakes, powders or pellets. Applications: Cloud point depressants. Viscosity reduction in finished product. Used in light duty liquid detergents. (Albright & Wilson Ltd, Detergents Div, Marchon).

Eltesol ST90, ST Pellets and PT90. Alkylaryl sulphonate anionic. Off-white powder, flakes or pellets. Applications: Reduction of slurry viscosity before spray drying in heavy duty detergent powder manufacture. (Albright & Wilson Ltd, Detergents Div, Marchon).

Eltesol SX30. Alkylaryl sulphonate anionic. Straw cloured liquid. Applications: Solubilizers or coupling agents, used in heavy duty liquid detergent formulations. (Albright & Wilson Ltd Detergents Div, Marchon).

Eltesol SX93 and SX Pellets. Alkylaryl sulphonate anionic. Off- white flakes,powder or pellets. Applications: Reduction of slurry viscosity before spray-drying, used in heavy duty detergent powder manufacture. (Albright & Wilson Ltd, Detergents Div, Marchon).

Eltesol TA, TA65 and TA96. Alkylaryl sulphonate anionic in acid form. Applications: Hardening in resin production; plasticizer for nitrocellulose lacquers; activator for nicotine insecticides; de- scaling for resin bound sand castings; catalysts for the preparation of esters, acetylation etc. TA65 is an industrial grade widely used in the foundry industy as catalyst for setting or hardening of phenolic resins. (Albright & Wilson Ltd, Detergents Div, Marchon).

Eltesol TPA. Alkylaryl sulphonate anionic. Applications: Additive for acidic electrolyte solutions for continuous tin plating of strip iron and steel. (Albright & Wilson Ltd, Detergents Div, Marchon).

Eltesol XA, XA65 and XA90. Alkylaryl sulphonate anionic in acid form. Applications: Hardening in resin production; plasticer for nitrocellulose lacquers; activator for nicotine insecticides; de-scaler for resin bound sand castings; catalysts for the preparation of esters, acetylation etc. (Albright & Wilson Ltd, Detergents Div, Marchon).

Eltesol 4009 and 4018. Alkylaryl sulphonate anionic in acid form. Applications: Used widely in the foundry industry for curing cold setting phenolic resins. (Albright & Wilson Ltd, Detergents Div, Marchon).

Eltesol 4402, 4403 and FDA 55/8. Alkylaryl sulphonate anionic in acid form. Applications: Used widely in the foundry industry for curing cold setting resins. (Albright & Wilson Ltd, Detergents Div, Marchon).

Eltex. High density polyethylene. (Laporte Industries Ltd).

Eltex P. Polypropylene. (Laporte Industries Ltd).

Eltroxin. A proprietary preparation of thyroxine sodium. Thyroid hormone. (Glaxo Pharmaceuticals Ltd).†

Eludril Mouthwash. Antibacterial/antifungal mouthwash. 0.1% chlorhexidine, 0.1% chlorbutol, 0.5% chloroform. Dilute 10 ml: half glass warm water. Applications: Gingivitis, apthous/dental ulcers and mouth and throat infection. Does not stain teeth or composites. (Concept Pharmaceuticals Ltd).*

Eludril Spray. Aerosol spray containing chlorhexidine 0.05% and amethecaine 0.015%. Antibacterial/anti-fungal. Applications: Apthous/dental ulceration and for mouth and throat infections. (Concept Pharmaceuticals Ltd).*

Elvacite. Acrylic resins. (Du Pont (UK) Ltd).

Elvaloy. Resin modifiers. (Du Pont (UK) Ltd).

Elvamide. Nylon multipolymer resins. (Du Pont (UK) Ltd).

Elvanol. A proprietary polyvinyl alcohol. (Du Pont (UK) Ltd).†

Elvaron. A wettable powder containing 50% w/w dichlofluanid. Applications: To control botrytis on strawberries, raspberries, loganberries, blackberries, blackcurrants, redcurrants, gooseberries, outdoor grapes, tomatoes under cover, tulips and paeonies. It also controls cane spot and stamen blight, reduces raspberry mildew and gives some reduction of spur blight on raspberries, mildew and blackspot on roses, downy mildew on cauliflower seedlings. It shows a side effect against mildew on strawberries, leaf spot on blackcurrants and gooseberries. (Bayer & Co).*

Elvax. Ethylene - vinyl acetate copolymer resins. (Du Pont (UK) Ltd).

Elvax D. Proprietary dispersions of ionomers and vinyl resins. (Du Pont (UK) Ltd).†

Elverite. A proprietary trade name for charcoal iron used for crushing mills.†

EMA. Ethylene-maleic anhydride coploymers resins. Applications: Dispersing agents, film formers and chemical intermediates for use in capsule walls, liquid detergents and drilling muds, thickening agents in textile print pastes and cosmetics. (Monsanto Co).*

Ema Resins. EMA resins. (Monsanto PLC).

Emaillit. Bituminous on a solvent base. Applications: Primer. (Vedag GmbH).*

Emaline. Conventional and high build bituminous coatings. (Sigma Coatings).*

Embacel. Kieselguhr for gas chromatography. (May & Baker Ltd).*

Embacide. Sheep dip. (May & Baker Ltd).

Embacoid. A cement for cine films. (May & Baker Ltd).*

Embadot. Photographic developer. (May & Baker Ltd).

Embafix. Photographic fixer. (May & Baker Ltd).

Embafume. Methyl bromide fumigant. (May & Baker Ltd).*

Embalith. Photographic developer. (May & Baker Ltd).

Embamix. Potassium iodide mixtures. (May & Baker Ltd).*

Embanox. A food grade antioxidant. (May & Baker Ltd).*

Embaphase. Stationary phases for gas chromatography. (May & Baker Ltd).

Embaspeed. Rapid working photographic developer. (May & Baker Ltd).

Embatex. A cellulose acetate coated fabric. (May & Baker Ltd).*

Embathion. An insecticide. (May & Baker Ltd).*

Embatype. Photographic developer. (May & Baker Ltd).

Embazin. A proprietary preparation of sulphaquinoxaline sodium. A veterinary coccidiostat. (May & Baker Ltd).*

Embedyne. Chlorodyne. (May & Baker Ltd).*

Embelic Acid. An acid obtained from the fruit of *Embelia ribes*. Used in medicine for worms.†

Embequin. Di-iodohydroxyquinoline. (May & Baker Ltd).

Embesafe. Plastic coated glass bottles. (May & Baker Ltd).

Embesol. Photographic developer. (May & Baker Ltd).

Embond 168. Acrylic emulsion. Applications: Flooring adhesive. (Marley Adhesives).*

Embond 212. Synthetic rubber latex. Applications: Flooring adhesive. (Marley Adhesives).*

Embrithite. A mineral, $Pb_3Sb_2S_6$.†

Embrol. Photographic developer. (May & Baker Ltd).

Embutox. A selective weedkiller. (May & Baker Ltd).*

Emcol. Surfactant for cosmetics, toiletries, pharmaceutical, processing, agricultural and other industries. (Baxenden Chemical Co Ltd).*

Emcol CC-55. Polypropox quaternary ammonium acetate. Application: General non-irritant quaternary cationic surfactant used especially where anionic compatability is required. (Witco Chemical Ltd).

Emcol CC-9, CC-36 and CC-42. Polypropoxyl diethyl methyl ammonium chloride. Application: General non-irritant quaternary cationic surfactants used especially where anionic compatability is desired. (Witco Chemical Ltd).

Emcol E-607. Lapyrium chloride. Applications: Pharmaceutic aid. (Witco Chemical Corp).

Emcol K8300. Fatty alkanol-amide sulphosuccinate in liquid form. Applications: Anionic surfactant used in emulsion polymerization. (Witco Chemical Ltd).

Emcol 4100M, 4150 and 4161-L. Sodium alkanol-amide sulphosuccinate in liquid or soft paste form. Applications: Hair and carpet shampoos; 4161-L has low irritancy and is suitable for baby products; 4150 is used in emulsion polymerization and ore flotation. (Wit Chemical Ltd).

Emcol 4300. Sodium fatty alcohol ethoxylate sulphosuccinate in liquid form. Applications: Hair shampoos; foam baths; emulsion polymerization; baby products. (Witco Chemical Ltd).

Emcol 4350. Triethanolamine fatty alcoh ethoxylate sulphosuccinate in liquid form. Applications: Low irritancy hai shampoos and foam baths. (Witco Chemical Ltd).

Emcol 4500. Sodium di-octyl sulphosuccinate in liquid form. Applications: Wetting agent; dry cleaning. (Witco Chemical Ltd).

Emcol 4600. Sodium bis-tri-decyl sulphosuccinate in liquid form. Applications: Dispersing agent; dry cleaning. (Witco Chemical Ltd).

Emcol 4776. Sodium di-hexyl sulphosuccinate in liquid form. Applications: Dry cleaning. (Witco Chemical Ltd).

Emcor. Ultra-short fibre reinforcement based upon attapulgite (magnesium aluminium silicate) used in non-asbest friction compounds. Applications: No asbestos disc pads, drum linings, truck and railroad blocks, friction papers, gasketing. (Engelhard Corporation).*

Emcyt. Estramustine phosphate sodium. Applications: Antineoplastic. (Hoffmann-LaRoche Inc).

Emdite. A proprietary trade name for a 5 per cent. w/w aqueous solution of ethy ammonium ethyl-dithio-carbonate, an alternative to hydrogen sulphide in qualitative inorganic analysis. (British Drug Houses).†

Emdithene. Basically a range of P.U. resir specifically formulated for the protection of light electrical and electronic components, but offering better retention of flexibility than normal P.U. types under thermal cycli conditions at higher temperatures. Applications: Potting, encapsulation, conformal coating of components, I.C.' P.C.B.'s. (Robnorganic Systems Ltd).*

Emerald Bronze. An alloy of 50 per cent copper, 49.7 per cent zinc, and 0.3 per cent aluminium.†

nerald Green. (Schweinfurth Green, Mitis Green, Vienna Green, Paris Green, Verdigris Green, Emperor Green, New Green, Mineral Green, Original Green, Vert Paul, Veronese Green, Parrot Green, Imperial Green, Kaiser Green, Meadow Green, English Green, Patent Green). Formerly the name Emerald green was applied to the hydrous oxide, $Cr_2(OH)_6$, but it is now given to cupric aceto-arsenite, $Cu(C_2H_3O_2)_3.3CuAs_2O_4$, made by mixing a hot solution of white arsenic in sodium carbonate, with the calculated quantity of copper sulphate solution. Also see BRILLIANT GREEN, CHROMIUM GREEN, CHROME GREEN, and OIL GREEN.†

nery's L-110. A proprietary form of azelaic acid. It is used as a softener for alkyd resins.†

nery's L-114. A proprietary mixture of low molecular weight aliphatic acids in which pelargonic acid, C_8H_7COOH, predominates. It is used in the oil modification of alkyd resins.†

neside. A proprietary preparation of ethosuximide. An anticonvulsant. (LAB Ltd).*

netic Tartar. See TARTAR EMETIC.

netrol. A proprietary preparation of laevulose, dextrose and orthophosphoric acid. An anti-emetic. (Pharmax).†

nge. Magnesium hyposulphite in ampoules and tablets.†

MI-24. A trade mark. 2-ethyl-4-methylimid-azole. A curing agent for epoxy resins used in low proportions thus improving chemical resistance.†

mmensite. An explosive consisting of 5 parts Emmens acid, 5 parts ammonium nitrate, and 6 parts picric acid.†

modin. Trihydroxymethylanthraquinone $C_{15}H_{10}O_5$. Used in medicine.†

mol Keleet. A purified fuller's earth.†

mperor Alloy. A nickel-chromium alloy. It will resist a temperature of 1750°-1800° F.†

mperor Brass. An aluminium bronze. It consists of 60 per cent copper, 20 per cent aluminium, and 20 per cent zinc.†

mperor Green. See EMERALD GREEN.

Emphos. Series of anionic surfactants of the phosphate ester type showing a range of properties eg surface tension lowering, wetting, foaming, according to composition. Applications: Antistats; dry cleaning; emulsion polymerization; industrial alkaline cleaners; papermaking; pesticides; textile processing; extreme pressure lubricants; corrosion inhibitors; release agents; moisture barrier agents; oil well fluids; pigment dispersion. (Witco Chemical Ltd).

Empicol. A wetting agent. It is an aliphatic alcohol sulphate and aliphatic alcohol alkyl ether sulphate. (Albright & Wilson Ltd).*

Empicol AL30/T. Ammonium lauryl alcohol sulphate as an amber liquid/paste. Applications: Detergency and foaming agent for shampoos including carpet types. (Albright & Wilson Ltd, Detergents Div, Marchon).

Empicol DLS. Diethanolamine lauryl sulphate as an amber clear liquid. Applications: Anionic surfactant used in shampoos and liquid detergents. (Albright & Wilson Ltd, Detergents Div, Marchon).

Empicol EAB. Ammonium lauryl alcohol ethoxy sulphate as a clear liquid. Applications: Liquid and lotion shampoos, bubble baths. (Albright & Wilson Ltd, Detergents Div, Marchon).

Empicol EGB. Magnesium lauryl alcohol ethoxy sulphate as a clear liquid. Applications: Liquid and lotion shampoos, bubble baths. (Albright & Wilson Ltd, Detergents Div, Marchon).

Empicol EL. Monoethanolamine lauryl sulphate as an amber viscous liquid whose viscosity doubles at 50% dilution. Applications: Anionic surfactant which also contains a nonionic. Used in liquid shampoos. (Albright & Wilson Ltd, Detergents Div, Marchon).

Empicol ESB and ESC. Sodium lauryl alcohol ethoxy sulphate in liquid form. Applications: Liquid and lotion shampoos, bubble baths. (Albright & Wilson Ltd, Detergents Div, Marchon).

Empicol ETB. Triethanolamine lauryl alcohol ethoxy sulphate as a clear liquid.

Applications: Liquid and lotion shampoos, bubble baths. (Albright & Wilson Ltd, Detergents Div, Marchon).

Empicol LM and LMV. Sodium lauryl alcohol sulphate in solid form. Applications: Shampoo manufacture; textiles; certain types of household detergent powders. (Albright & Wilson Ltd, Detergents Div, Marchon).

Empicol LQ. Monoethanolamine lauryl sulphate in liquid form. Applications: Detergency and foaming agent for high quality shampoos in liquid or lotion form. (Albright & Wilson Ltd, Detergents Div, Marchon).

Empicol LX and LXV. Sodium lauryl alcohol sulphate in the form of highly soluble powder, needles or paste. Applications: Carpet and upholstery shampoos; other outlets where high solubility is required. (Albright & Wilson Ltd, Detergents Div, Marchon).

Empicol LY28/S. Anionic surfactant in the form of a pale yellow liquid/paste. Applications: Detergency and foaming agent for carpet shampoos, plastics, rubber and foam rubber. (Albright & Wilson Ltd, Detergents Div, Marchon).

Empicol LZP. Sodium lauryl alcohol sulphate as a white powder. Applications: Anionic surfactant used for toothpaste and other dental preparations. (Albright & Wilson Ltd, Detergents Div, Marchon).

Empicol LZ, LZV,LZG and LZGV. Sodium lauryl alcohol sulphate in solid form. Applications: Shampoos and other industries requiring powerful wetting agents and low salt content eg rubber and plastics. (Albright & Wilson Ltd, Detergents Div, Marchon).

Empicol LZ/E, LZV/E, LZ/D and LZV/D. Sodium lauryl sulphate as white powder or needles. USP or BP grade. Applications: Wide variety of industrial uses requiring emulsifying, dispersing and wetting properties in a high grade product. (Albright & Wilson Ltd, Detergents Div, Marchon).

Empicol MD. Sodium lauryl ethoxy sulphate as a pale straw liquid. Applications: Foaming agent with good hard water performance and low viscosity and irritancy. Used in baby a medical shampoos, foam baths, hand cleaners and fine fabric washing. (Albright & Wilson Ltd, Detergents Div, Marchon).

Empicol ML26. Magnesium lauryl sulphate. Pale straw liquid. Applications: Anionic surfactant used shampoos, toothpastes etc and in the plastics industry. (Albright & Wilson Ltd, Detergents Div, Marchon).

Empicol SCC, SDD, SFF, SGG and STT. Range of anionic surfactants of the di sodium sulphosuccinate type. Liquid c paste form. Applications: Anionic surfactants with low irritation characteristics used in the toiletry industry as ingredients for shampoos and bubble bath formulations. (Albrig & Wilson Ltd, Detergents Div, Marchon).

Empicol TC30/T and TCR/T. Anionic surfactant in liquid form. TCR/T contains alkylolamide foam booster. Applications: Detergency and foaming agent used in shampoos and liquid detergents. (Albright & Wilson Ltd, Detergents Div, Marchon).

Empicol TDL. Anionic surfactant in soft paste form. Applications: Detergency, wetting and foaming agent for shampoos, bubble baths and general bath preparations. (Albright & Wilso Ltd, Detergents Div, Marchon).

Empicol TL40/T, TLP/T and TLR/T. Triethanolamine lauryl alcohol sulpha in liquid form. Applications: Anionic surfactant for liquid and lotion shampoos. (Albright & Wilson Ltd, Detergents Div, Marchon).

Empicol 0045. Sodium lauryl sulphate in white powder form. Applications: Detergent and foaming agent for toothpaste and dental preparations, shampoos and foam bath products. (Albright & Wilson Ltd, Detergents Div, Marchon).

Empicryl. Pigment dispersant. Applications: Wide variety of water based paints. (Albright & Wilson Ltd, Detergents Div).

Empigen. Teriary amine and imidazoline derivates. (Albright & Wilson Ltd, Detergents Div).

Empigen AB, AH, AM and AY. Alkyl dimethyl amines in liquid form or as a waxy solid. Applications: Intermediates for manufacture of amine salts, amine oxides, quaternaries and betaines. (Albright & Wilson Ltd, Detergents Div, Marchon).

Empigen AS and AT. Alkylamido dimethyl amine as an amber liquid/ paste. Applications: Corrosion inhibitors in cutting oils. (Albright & Wilson Ltd, Detergents Div, Marchon).

Empigen BAC. Benzalkonium chloride as a pale yellow aqueous solution. Applications: Disinfection in food processing, canteens, dairies, brewing and bottling plants; well-boring, cutting oils; lens systems. (Albright & Wilson Ltd, Detergents Div, Marchon).

Empigen BB. Alkylbetaine, formula R-N+(CH3)2-CH2COO-, where R is mainly C12/C14. An almost colourless liquid. Applications: Foam booster, thickening agent, wetting agent, stabilizer; has good acid and alkali stability and some cationic properties. Used in shampoos and detergents; co-active in baby shampoos; static control. (Albright & Wilson Ltd, Detergents Div, Marchon).

Empigen BT. Alkylamido betaine, as an amber liquid. Applications: Foaming agent, stabilizer and thickening agent, with good performance in presence of soap and hard water. Used in toiletry preparations, especially bath care products. (Albright & Wilson Ltd, Detergents Div, Marchon).

Empigen CDR. Coconut imidazoline betaine, as an amber liquid. Applications: Mild surfactant with foaming properties, compatible with anionic, nonionic and cationic surfactants. Used in mild, non-irritant toiletries such as baby, childrens' and family shampoos and bath care products. (Albright & Wilson Ltd, Detergents Div, Marchon).

Empigen CM. Alkyltrimethyl ammonium metho sulphate as a pale straw liquid. Applications: Cationic surfactant used as a conditioner and antistatic agent in hair conditioning creams; cationic emulsifier. (Albright & Wilson Ltd, Detergents Div, Marchon).

Empigen XDR. A coconut imidazoline betaine formulated with a sodium lauryl ethoxy sulphate. Viscous yellow liquid. Applications: Mild and non-irritant toiletries, especially baby shampoos. (Albright & Wilson Ltd, Detergents Div, Marchon).

Empilan. Fatty acid esters, emulsifiers and foam stabilizers. (Albright & Wilson Ltd).*

Empilan BD. Blend of quaternary ammonium germicide with nonionic detergent. Clear pale yellow stable liquid. Applications: Bactericidal detergent used in washing of dishes, pans, glassware, dishcloths, etc. (Albright & Wilson Ltd, Detergents Div, Marchon).

Empilan NP9. Nonionic surfactant of the nonylphenol ethoxylate type in the form of a pale straw soft paste or liquid. Application: An all round surfactant used in textiles and other industries. (Albright & Wilson Ltd, Detergents Div, Marchon).

Empimin. Alcohol ethoxy sulphates. (Albright & Wilson Ltd, Detergents Div).

Empimin KSN. Sodium lauryl alcohol ethoxy sulphate in liquid form. Applications: Liquid detergents. (Albright & Wilson Ltd, Detergents Div, Marchon).

Empimin LAM. Sodium alkyl ether sulphate as a pale yellow liquid. Applications: Foaming agent in the manufacture of plaster board. (Albright & Wilson Ltd, Detergents Div, Marchon).

Empimin LR28. Anionic surfactant in liquid form. Applications: Foaming agent and emulsifier used in latex compounding. (Albright & Wilson Ltd, Detergents Div, Marchon).

Empimin LSM. Sodium alkyl ether sulphate as a pale straw liquid. Applications: Foaming agent for detergent and alkali cleaners and plaster board and rubber latices. (Albright & Wilson Ltd, Detergents Div, Marchon).

Empimin LS30. Anionic surfactant in liquid form. Applications: Foaming agent and emulsifier used in disinfectant formulation. (Albright & Wilson Ltd, Detergents Div, Marchon).

Empimin MA. Sodium dimethyl amyl sulphosuccinate as a clear solution in water/alcohol. Applications: Emulsifier and wetting agent particularly in the presence of electrolytes, used for oils, solvents, monomers, waxes and polymers and for emulsion polymerization. (Albright & Wilson Ltd, Detergents Div, Marchon).

Empimin MHH. Disodium N-lauryl sulphosuccinamate as a pale cream liquid. Applications: Foaming agent for rubber latices eg in carpet manufacture. (Albright & Wilson Ltd, Detergents Div, Marchon).

Empimin MKK. Disodium N-stearyl sulphosuccinamate as a pale cream soft paste or spray dried powder. Applications: Foaming agent for rubber latices eg in carpet manufacture. (Albright & Wilson Ltd, Detergents Div, Marchon).

Empimin MSS. Diammonium N-lauryl sulphosuccinamate as a pale cream liquid. Applications: Foaming agent for rubber latices eg in carpet manufacture. (Albright & Wilson Ltd, Detergents Div, Marchon).

Empimin MTT. Disodium N-oleyl sulphosuccinamate as a pale cream liquid. Applications: Foaming agent for rubber latices eg in carpet manufacture. (Albright & Wilson Ltd, Detergents Div, Marchon).

Empimin OT. Sodium dioctyl sulphosuccinate as a clear solution in water/alcohol. Applications: Emulsifier and wetting agent particularly in electrolyte solution, used for oils, solvents, monomers, waxes and polymers and for emulsion polymerization. (Albright & Wilson Ltd, Detergents Div, Marchon).

Empiphos. A synthetic detergent comprising aliphatic phosphate. (Albright & Wilson Ltd).*

Empiquat. An alkyl dimethylbenzyl-ammonium chloride. (Albright & Wilson Ltd).*

Empirin. Tablets or with codeine tablets. Proprietary formulations containing aspirin or aspirin and codeine phosphate. Applications: A general analgesic and anti-inflammatory agent. (The Wellcome Foundation Ltd).*

Empiwax. Oil/water emulsifier. Applications: Pharmaceutical and toilet preparations, ointments; penicillin cream bases. (Albright & Wilson Ltd, Detergents Div).

Emplets Potassium Chloride. Potassium Chloride. Applications: Replenisher. (Parke-Davis, Div of Warner-Lambert Co).

Empracet Codeine Phosphate. A proprietary formulation of codeine phosphate and acetaminophen. Applications: Relief of mild to moderately severe pain. (The Wellcome Foundation Ltd).*

Emralon. Dispersions of PTFE in resin. Applications: Dry film lubricants and parting agents. (Acheson Colloids).*

EMS 209. Microsilica with a proprietary coating and treatment. Applications: Incorporated in extruded PVC pipe to increase elastic modulus and pipe stiffness, increase impact resistance and to increase volumetric output from the extruder. (Elkem Chemicals Inc).*

Emsac Concrete Additive. Microsilica (condensed silica fume) based dry or slurried products formulated with dispensing agents such as those commonly used in the concrete industry meeting the requirements of ASTM C494. Applications: Used to improve the strength and durability of ordinary portland cement concretes, grouts and mortars. (Elkem Chemicals Inc).*

Emsodur. Cubical shaped polyamide. Applications: Deflashing thermoset parts. (Emser Industries).*

Emsodur. Shot blast grit produced to close dimensional tolerances for the deflashing of thermoset compression mouldings. Applications: Removing of flashing from casts. (EMS-Grilon (UK) Ltd).†

Emtryl. A proprietary preparation of dimetridazole. A veterinary anti-protozoan. (May & Baker Ltd).*

Emulan. Emulsification. Applications: Impregnation, lubrication, polishing and cleaning; anti-corrosive emulsifier. (BASF United Kingdom Ltd).

Emulan PO. Nonionic surfactant of the alkylphenol ethoxylate type as a colourless to pale yellow clear oil. Application: Emulsification, mainly in combination with other emulsifiers. Used in cold cleaners and other solvent based cleaners; drilling oil. (BASF United Kingdom Ltd).

Emulgeen P. A proprietary trade name for potassium ricinoleate. (S & D Chemicals Ltd).†

Emulgeen S. A proprietary trade name for sodium ricinoleate. (S & D Chemicals Ltd).†

Emulgen. A jelly-like mass used for the rapid emulsification of oils and resins. It contains tragacanth, gum arabic, pittoporad, glycerin, alcohol, and water.†

Emulphogene BC-420. Polyoxyethylated(3) tridecyl alcohol. Applications: Intermediate in the manufacture of high-foaming anionic surfactants. Oil soluble detergent and dispersant for use in petroleum oils. (GAF Corporation).*

Emulphogene BC-610. Polyoxyethylated(6) tridecyl alcohol. Applications: Low foaming detergent and wetting agent for mechanical-dishwashing formulations and spray-type alkaline cleaners. Emulsifier for silicones and petroleum oils. (GAF Corporation).*

Emulphogene BC-720. Polyoxyethylated(9.75) tridecyl alcohol. Applications: Foam builder and detergent, solubilizer for alkylaryl sulphonates, component of light and heavy-duty high foaming detergent formulations. (GAF Corporation).*

Emulphogene BC-840. Polyoxyethylated(15) tridecyl alcohol. Applications: Foam builder and detergent, solubilizer for alkylaryl sulphonates, surfactant for high temperature detergency, emulsifier and stabilizer for synthetic latices. (GAF Corporation).*

Emulphogene DA-530. Polyoxyethylated(4) decyl alcohol. Applications: Used as an intermediate in the manufacture of esters for various textile and industrial applications. (GAF Corporation).*

Emulphogene DA-630. Polyoxyethylated(6) decyl alcohol. Applications: An effective wetting agent in built scours for removal of soils prior to dyeing of fabric and as an intermediate in the manufacture of esters for various textile and industrial applications. (GAF Corporation).*

Emulphogene DA-639. Polyoxyethylated(6) decyl alcohol. Applications: A wetting agent in built scours for removal of soils prior to dyeing of fabric. (GAF Corporation).*

Emulphogene LM-710. Polyoxyethylated alkyl thioether. Applications: Emulsifier of all types of grease and soils. Scouring agent for wool. (GAF Corporation).*

Emulphogene TB-970. Linear aliphatic ethoxylate. Applications: Nonionic surfactant, suitable for most dry blending operations in detergent formulations, used to control dissolution rate of solid or block-type hard surface cleaners. (GAF Corporation).*

Emulphor. A proprietary trade name for a condensation product of ethylene oxide and an organic acid.†

Emulphor EL-620. Polyoxyethylated(30) castor oil. Applications: Emulsifier, dyeing assistant, antistatic agent and lubricant for synthetic fibres, solubilizer for perfumes and emulsifier for cosmetic preparations. (GAF Corporation).*

Emulphor EL-719. Polyoxyethylated(40) castor oil. Applications: Used in emulsifying vitamins and other pharmaceuticals. (GAF Corporation).*

Emulphor EL-980. Polyoxyethylated(200) castor oil. Applications: Emulsifier for mineral oil, triglycerides and alkyl esters, used in textile compounding as antistat, synthetic fibre lubricant and dyeing assistant. (GAF Corporation).*

Emulphor LA-630. Polyoxyethylated(9) coconut fatty acid. Applications: Emulsifier and wetting agent with use as lubricant and softener. Used in cosmetic and textile industries. (GAF Corporation).*

Emulphor ON-870. Polyoxyethylated(20) oleyl alcohol. Applications: Used as an emulsifier stabilizer for rubber latex emulsions, wetting agent in metal cleaners, in acid-degreasing wool and as a dyeing assistant and a chrome-tanning assistant. (GAF Corporation).*

Emulphor VN-430. Polyoxyethylated(5) oleic acid. Applications: Oil soluble surfactant used in cutting oils, degreasing solvents and metal cleaners, emulsifier for mineral oils, liquid fatty oils and peanut oils, as a dyeing assistant and in degreasing pickled skins, improving fat liquoring and increasing the tensile strength of high grade chrome side leathers. (GAF Corporation).*

Emulphor VT-650. Polyoxyethylated(9) stearic acid. Applications: Used as self-emulsifying lubricant for textile scouring agents for synthetic fibres and industrial applications. (GAF Corporation).*

Emulpon. Emulsifiers, dispersants, solubilizers for oils, solvents and waxes. Applications: Textile, agriculture and cosmetic industries. (Witco Chemical Ltd).

Emulsamin. A proprietary trade name for menthol diurethane, a wetting agent and detergent.†

Emulsene. Emulsifying agents. (Bush Boake Allen).*

Emulsi-Phos. Cheese emulsifier. Applications: Food grade phosphate specially designed for use in process cheese, cheese food and cheese spreads. (Monsanto Co).*

Emulsiderm. Pale blue/green liquid emulsion containing 0.5% benzalkonium chloride BP, 25% liquid paraffin BP, 25% isopropyl myristate BPC. Applications: An aid in the treatment of dry skin conditions, especially those associated with eczema, ichthyosis or xeroderma. It permits rehydration of the keratin by replacing lost lipids and its antiseptic properties assist in overcoming secondary infection. (Dermal Laboratories Ltd).*

Emulsifier L.W. Cyclohexylammonum oleate. (Laporte Industries Ltd).*

Emulsil. Silicone products food grade. Applications: Antifoams and release agents. (Siliconas Hispania SA).*

Emulsin. A ferment. It decomposes the glucoside, amygdalin, into grape sugar, benzaldehyde, and hydrocyanic acid.†

Emulsogen. Emulsifiers for oils and waxes. (Hoechst UK Ltd).

Emultex. Liquid emulsifier. Applications: Sizing oils. (Tensia SA).

Emultex AC 431. A proprietary vinyl acetate/acrylic ester copolymer emulsion used in general-purpose emulsion paints. (Revertex, Harlow, Essex).†

Emultex 307 and 328. A proprietary range of unplasticized vinyl acetate homopolymer emulsions stabilized with polyvinyl alcohol, used in the manufacture of adhesives. (Revertex, Harlow, Essex).†

Emulvin. Emulsifying agents. (Bayer UK Ltd).

E-Mycin. Erythromycin. Applications: Antibacterial. (The Upjohn Co).

E-Mycin E. Erythromycin ethylsuccinate. Applications: Antibacterial. (The Upjohn Co).

En-Dur-Lon. Rubber base coating, contains no abrasive. Applications: Anti-slip coating. Decorative coating. (W J Ruscoe Co).*

Enadel. A proprietary preparation of cloxazolam. An antianxiety agent. (Pfizer International).*

Enamel White. See LITHOPONE.

Enamelled Iron-granite. See AGATE WARE.

Enanth. Nylon 7.†

Enavid 5mg. A proprietary preparation containing norethynodrel and mestranol. (Searle, G D & Co Ltd).†

Enavid-E. A proprietary preparation containing norethynodrel and mestranol. (Searle, G D & Co Ltd).†

Enbico Chrome Black F. A dyestuff. It is a British equivalent of Diamond black F.†

Enbico Direct Brown B. A British equivalent of Benzo brown G.†

nbico Direct Brown G. A dyestuff. It is a British brand of Benzo brown B.†

nbico Direct Fast Pink Y Conc. A dyestuff. It is equivalent to Salmon red.†

nbico Direct Green B, B Extra Conc. Dyestuffs. They are British equivalents of Diamine green B.†

nbico Direct Yellow G. A British brand of Chrysamine G.†

nbico Direct Yellow R. A dyestuff. It is a British equivalent of Direct yellow R.†

nbucrilate. A surgical tissue adhesive. Butyl 2-cyanoacrylate. HISTOACRYL.†

ncapsulation. Thermosetting material. Applications: Moulded products. (Sumitomo Bakelite, Japan).*

nceladite. Synonym for Warwickite.†

ncem Steel. A proprietary trade name for a nickel-chromium-molybdenum steel.†

ndanil. Nylon dyes and pigments for plastics. (ICI PLC).*

ndcor. A wide range of high performance, corrosion-resistant coating systems. Types include: acrylic, alkyd, chlorinated rubber, epoxy, pretreatments, primers, polyurethane, vinyl and zinc rich. Applications: General industrial maintenance for metal and masonry surfaces in a wide variety of exposure conditions, including weathering and chemical attack, immersion service. (Dampney Company Inc.)*

ndegal. Polyester dyes and pigments for plastics. (ICI PLC).*

ndermol. A compound ointment vehicle, containing hydrocarbons of the paraffin series, and stearic acid amide.†

ndobil. Iodoxamic acid. Applications: Diagnostic aid. (Bracco Industria Chimica SPA).

ndocaine. A proprietary trade name for Pyrrocaine.†

ndocrocine. The orange-yellow colouring matter isolated from *Nephioniopsis endocrocea*, a lichen growing in Japan. It is a hydroxyanthraquinone, $C_{16}H_{10}O_7$.†

ndojodin. An injectable iodine preparation. (Bayer & Co).*

Endomirabil. Iodoxamic acid. Applications: Diagnostic aid. (Bracco Industria Chimica SPA).

Endotryptase. A proteolytic enzyme.†

Endoxan (Cytoxan, Genoxal, Sendoxan, Endoxana). Cyclophosphamide. Tablets and vials (dry substance). Applications: Cystostatic. (Degussa).*

Endoxana. SEE ENDOXAN. (Degussa).*

Endrate. Edetate disodium. Applications: Chelating agent pharmaceutic aid. (Abbott Laboratories).

Endrine. A proprietary preparation of ephedrine, menthol, camphor, and eucalyptol. A nasal decongestant. (Wyeth).†

Enduro Alloys. Proprietary corrosion resisting alloys of iron with chromium, or with nickel and chromium. Enduro A contains iron with from 16.5-18.5 per cent chromium; Enduro KA2 contains iron with 17-20 per cent chromium and 7-10 per cent nickel; Enduro S has iron with 12.5-14.5 per cent chromium.†

Endurol. Vat dye colours. (James Robinson & Co Ltd).†

Enduron. A proprietary preparation of methylclothiazide. A diuretic. (Abbott Laboratories).*

Endyne. Indoprofen. Applications: Analgesic; anti-inflammatory. (Adria Laboratories Inc).

Eneril. A proprietary preparation of paracetamol. An analgesic. (Nicholas).†

Enervite. Cod liver oil BP veterinary. Applications: Dietary supplement with vitamins A and D. (Marfleet Refining Co).*

England, Salts of. See EPSOM SALTS.

English Bearing Metal. Anti-friction and fitting metal. It contains usually 53 per cent tin, 33 per cent lead, 10.5 per cent antimony, and 2.5 per cent copper.†

English Blue. See MOUNTAIN BLUE.

English Brown. See BISMARCK BROWN.

English Green. See EMERALD GREEN and GREEN VERMILION.

English Metal. A jeweller's alloy containing 88 parts tin, 2 parts copper, 2 parts brass, 2 parts nickel, 1 part bismuth, 8 parts antimony, and 2 parts tungsten.†

English Powder. See POWDER OF ALGAROTH.

English Red. See INDIAN RED.

English White. See CHALK.

English White Bearing Metal. An antifriction metal, containing 77 per cent tin, 15 per cent antimony, and 8 per cent copper.†

English Yellow. See TURNER'S YELLOW and VICTORIA YELLOW.

Engobe. A fusible mixture of clay, telspar, and silica. Used for the manufacture of glazes on pottery.†

Engraver's Brass. See BRASS.

Enisyl. Lysine hydrochloride. Applications: Amino acid. (Person & Covey Inc).

Enkade. Encainide hydrochloride. Applications: Cardiac depressant. (Mead Johnson & Co).

Enlax. A proprietary preparation. It is a phenolphthalein preparation.†

Enmag. Magnesium ammonium phosphate fertilizer. (Scottish Agricultural Industries PLC).

Ensecote S. A proprietary trade name for a modified epoxy resin coating material for high temperature stoving or spraying.†

Enso DTO 10 - 30. Distilled tall oils with a rosin acids content of 10-30%. Applications: Alkyd resins, liquid soaps, emulator, latex. (Enso-Gutzeit OY).*

Enso Rosin. Tall oil rosin. Applications: Paper size. (Enso-Gutzeit OY).*

Ensol 2. Tall oil fatty acids with 2% rosin acids. Applications: Alkyd resins, liquid soaps. (Enso-Gutzeit OY).*

Entacyl. A proprietary preparation of piperazine adipate. An antihelmintic. (British Drug Houses).†

Entair. Capsules and syrup containing theophylline and guaiphenesin. Reduces secretion of mucus in chronic bronchitis. (182) .†

Entair Expectorant. A preparation of diphenhydramine hydrochloride and guaiphenesin. (British Drug Houses).†

Entair-A. A proprietary preparation of theophylline, guaiphenisin and ephedrine hydrochloride. A bronchial antispasmodic. (British Drug Houses)

Entamide. A proprietary trade name fordiloxamide. An amoebicide. (Boot Pure Drug Co).*

Enterfram. A proprietary preparation containing framycetin sulphate and light kaolin. An antidiarrhoeal. (Fison PLC).*

Entericin. Aspirin. Applications: Analgesic; antipyretic; antirheumatic. (Bristol-Myers Co).

Enteromide. A proprietary trade name fc the calcium salt of Sulphaloxic Acid. An antibiotic. (Consolidated Chemicals).†

Enterosan. A proprietary preparation of iodohydroxyquinoline chlorodyne, tincture of belladonna and kaolin. An antidiarrhoeal. (Mayfair Chemicals).

Entramin. 2-Amino-5 nitrothiazole pre m (May & Baker Ltd).*

Entramin A. Acetamidonitrothiazole premix. (May & Baker Ltd).*

Entrosalyl. A proprietary preparation of sodium salicylate in an enteric coated tablet. Antirheumatic drug. (Cox, A H & Co Ltd, Medical Specialities Divn).

Entrosalyl Standard. A proprietary preparation of enteric coated sodium salicylate. (Cox, A H & Co Ltd, Medical Specialities Divn).†

Entrox. Zinc oxide, coated to improve incorporation into rubber. (Durham Chemicals Ltd).*

Enusin Colours. Pigment colours, based o casein binders, for aniline and semi-aniline finishing. (Bayer UK Ltd).

Envacar. A proprietary preparation of guanoxan sulphate. An antihypertensive. (Pfizer International).*

Envirez. Unsaturated polyester resins formulated for low smoke, reduced smoke toxicity and low volatility combined with fire retardancy. Applications: Transportation (subways aircraft etc.) and construction materia requiring both fire retardancy and reduced smoke hazards. (Ashland Chemical Company).*

nzactin. Triacetin. Applications: Antifungal. (Ayerst Laboratories, Div of American Home Products Corp).

nzeon. Chymotrypsin. Applications: Enzyme. (Sterling Drug Inc).

nzypan. A proprietary preparation of pepsin, pancreatin, and bile. Digestive enzyme supplement. (Norgine).†

nzytol. A 10 per cent aqueous solution of choline borate.†

olite. Synonym for Realgar.†

osin. (Eosin A, Eosin B, Eosin C, Eosin A Extra, Eosin DH, Eosin GGF, Eosin G Extra, Eosin 3J, Eosin JJS, Eosin G, Eosin KS, Eosin Yellowish, Water Soluble Eosin, Acid Eosin, Eosin 4J Extra). The alkali salts of tetrabromo-fluoresceine, $C_{20}H_6O_5Br_4Na_2$. Dyes wool and silk yellowish-red. Used as a microscopic stain and fluorescent tracer dye.†

osin A. See EOSIN.

osin A Extra. See EOSIN.

osin B. The ammonium salt of tetra-bromo-fluoresceine. Also see EOSIN and EOSIN BN.†

osin BB. See SPIRIT EOSIN.

osin, Bluish. See ERYTHROSIN.

osin BN. (Eosin B, Eosin BW, Eosin DVH, Eosin Scarlet, Eosin Scarlet BB, Eosin Scarlet B, Daphnin, Scarlet J, JJ, and V, Nopalin, Imperial Red, Methyl Eosin, Saffirosine, Lutecienne, Kaiser Red, Kaiserroth, Eosin BS). The potassium or sodium salt of dibromodinitrofluoresceine, $C_{20}H_6N_2O_9Br_2K_2$. Dyes silk and wool bluish-red.†

osin BS. A dyestuff. It is a British equivalent of Eosin BN.†

osin BW. See EOSIN BN.

osin C. See EOSIN.

osin DH. See EOSIN.

osin DVH. See EOSIN BN.

osin G. See EOSIN.

osin G Extra. See EOSIN.

osin GGF. See EOSIN.

osin J. See ERYTHROSIN.

Eosin JJS. See EOSIN.

Eosin KS. See EOSIN.

Eosin Orange. (Eosin 3G, Salmon Pink). Varying mixtures of di- and tetra-bromo-fluoresceine, used in dyeing.†

Eosin S. See SPIRIT EOSIN.

Eosin Scarlet. See EOSIN BN.

Eosin Scarlet B and BB. See EOSIN BN.

Eosin Soluble in Spirit. See ERYTHRINE and SPIRIT EOSIN.

Eosin, Water Soluble. See EOSIN.

Eosin, Yellowish. See EOSIN.

Eosin YS. A British brand of Eosin.†

Eosin 10B. See PHLOXINE.

Eosin 3G. See EOSIN ORANGE.

Eosin 3J and 4J Extra. See EOSIN.

Eosolate. Silver acetylguaiacoltri-sulphonate. An antiseptic.†

EP Lead. Specially-prepared lead naphthenate. Applications: Used in formulating extreme pressure gear oils and greases. (Nueodex Inc).*

EP-1. *O,O*-diethyl phosphorodithioic acid. Applications: Used as an intermediate for organophosphorus insecticides. (A/S Cheminova).*

EP-2. *O,O*-diethyl phosphorochlorodithioate. Applications: Mainly used in the production of organophosphorus insecticides. (A/S Cheminova).*

Epanutin. A proprietary preparation of phenytoin sodium. An anticonvulsant. (Parke-Davis).†

Ephedrine. An alkaloid obtained from the Chinese plant Ma Huang. It is α-phenyl-βmethylaminopropanol, $C_6H_5.CHOH.CH.CH_3NH.CH_3$. It is used as a mydriatic and as an atropine substitute in ophthalmology.†

Ephos. A basic phosphate, containing 60-65 per cent tricalcium phosphate.†

Ephynal. A proprietary preparation of Vitamin E. A vasodilator. (Roche Products Ltd).*

Epichlorhydrin. Chloromethyl oxirane.†

Epicure. A trade name for polyamide curing agents for epoxy resins. (Shell Chemicals).†

Epi-Cure 8515. A proprietary amidoamine curing agent for epoxy resins. (Alma Paint and Varnish Co, Ontario).†

Epiethylin. Ethylglycide ether.†

Epiflex. A proprietary trade name for an epoxy resin expansion jointing material. (Haworth (ARC) Ltd).†

Epifoam. Muco-adherent, white, odourless foam containing hydrocortisone acetate and pramoxine hydrochloride. Applications: Treatment of perineal trauma. (Stafford-Miller).*

Epiglaubite. An impure calcium phosphate.†

Epihydrin. Propylene oxide, C_3H_6O.†

Epikote. A proprietary trade name for a series of epoxy resins whose characteristics may be modified by hardeners and other additives. (Shell Chemicals).†

Epikote DX-209-B-80. A proprietary 80% solution of epoxy resin in methyl ethyl ketone. It is used with EPIKURE 3400 as a curing agent. (Shell Chemicals).†

Epikote DX-210-B-80. A proprietary 80% solution of epoxy resin in methyl ethyl ketone for use in work involving carbon fibres. It is cured with EPIKURE 3400. (Shell Chemicals).†

Epikote DX-231-B-91. A proprietary solution containing 91% epoxy resin in methyl ethyl ketone for use in work involving carbon film. It is cured with EPIKURE 34OO. (Shell Chemicals).†

Epikure 3400. A proprietary curing agent for epoxy resins. (Shell Chemicals).†

Epilim. A proprietary preparation of sodium valproate. An anti- convulsant. (Reckitts).†

Epilink. Epoxy curing agents. (Akzo Chemie GmbH, Düren).†

Epilok. Epoxy curing agents. (Akzo Chemie GmbH, Düren).†

Epilon. A proprietary trade name for an epoxy resin cement. (Haworth (ARC) Ltd).†

Epinal. Epinephryl borate. Applications: Adrenergic. (Alcon Laboratories Inc).

Epinalin. A proprietary preparation of adrenaline and ephedrine in solution.†

Epiphassol. A viscous oily liquid containing naphthene-sulphonic acids. It is a similar preparation to Kontakt, and is used for cleaning cotton fabrics.†

Epirez 501. Butyl glycidyl ether. A proprietary reactive diluent for epoxy resin systems. (Alma Paint and Varnish Co, Ontario).†

Epirez 502. A proprietary aliphatic diepoxide. (Alma Paint and Varnish Co, Ontario).†

Epirez 520C. A proprietary epoxy resin of the bisphenol A type. (Alma Paint and Varnish Co, Ontario).†

Episol. A proprietary preparation of chlorodiethylaminoethoxyphenylbenzt (Halethazole). An antifungal agent. (Crookes Laboratories).†

Epitar. An epoxy resin additive. (Midland Yorkshire Tar Distillers).†

Epitate. An epoxy resin additive. (Midland Yorkshire Tar Distillers).†

Epitrate. Epinephrine bitratrate. Applications: Adrenergic. (Ayerst Laboratories, Div of American Home Products Corp).

Epivax. A proprietary vaccine used against canine distemper. (Coopers Animal Health, subsidiary of The Wellcome Foundation Ltd).*

Epocap. Two-part epoxy compounds. (Hardman Inc).*

Epocast. 8408. An epoxy resin system giving tough rubber-like castings.†

Epocrete. Two-part epoxy concrete materials. (Hardman Inc).*

Epocure. Epoxy curing agents. (Hardman Inc).*

Epodur. High solids and solventless epoxy maintenance coatings for long-term corrosion protection under extreme exposure conditions. Applications: Buried pipelines, marine and offshore equipment, heavy construction, chemical process equipment, water and sewage works equipment, structural steel, tank linings. (Dampney Company Inc.)*

Epodyl. A proprietary preparation of triethyleneglycol diglycidyl ether. An anticancer agent. (ICI PLC).*

Epok. Elastomeric sealant coating. (BP Chemicals Ltd).*

Epolast. Two-part epoxy compounds. (Hardman Inc).*

Epomarine. Two-part epoxy compounds for splash zone and for underwater application. (Hardman Inc).*

Epon. A proprietary epoxy resin. (Shell International).†

Epon 8280. A proprietary liquid epoxy resin for use in filled compounds. (Shell Chemicals).†

Eponac. A range of epoxy resins. Applications: Varnish industry, electrotechnical field and building industry. (SPREA Spa).*

Epontol. Pharmaceutical preparation. Applications: For the induction of anaesthesia and for intravenous anaesthesia of short duration. (Bayer & Co).*

Epophen. Epoxide resins. (Borden (UK) Ltd).

Eposet. Two-part epoxy compounds. (Hardman Inc).*

Eposis. Epoxy resins.†

Eposolve. Solvent for epoxy clean-up. (Hardman Inc).*

Epotuf. A registered trade name for a hardener for epoxy resins. (Reichhold Chemicals Inc).†

Epoweld. Two-part epoxy compounds. (Hardman Inc).*

Epoxidized X-70 and X-75. Polysiloxane resin in an aliphatic vehicle system. Applications: Water repellant for masonry. (Nova Chemical Inc).*

Epoxol G-5. Linseed oil epoxy resin.†

Epoxol 7-4. Soybean oil epoxy resin.†

Epoxol 80, 130. A proprietary trade name for vinyl plasticizers manufactured from soya bean. (Swift and Co).†

Epox-S. A proprietary trade name for an epoxidized triester vinyl plasticizer. (Ruco Divn).†

Epoxy Putty Pack (EP-3/EHP-12). Epoxy putty resin and hardener. Applications: Filling and fairing - can be used underwater. (Wessex Resins & Adhesives Ltd).*

Eppy/N. Epinephryl borate. Applications: Adrenergic. (Barnes-Hind Inc).

Eprolin. Vitamin E. Applications: Vitamin E supplement. (Eli Lilly & Co).

Eprylac. Acrylic adhesives, sealants and coatings. (BP Chemicals Ltd).*

Epsikapron. A proprietary preparation of 6-aminocaproic acid, used in the treatment of fibrinolysis. (KabiVitrum AB).*

Epsilan-M. Vitamin E. Applications: Vitamin E supplement. (Adria Laboratories Inc).

Epsoline. Phenolphthalein.†

Epsom Salts. (Salts, Salts of England, Hair Salt, Bitter Salt). Magnesium sulphate, $MgSO_4.7H_2O$.†

Eptoin. Soluble phenytoin. (Boots Pure Drug Co).*

Epurite. A mixture of bleaching powder, iron sulphate, and copper sulphate. It is used for the production of oxygen, which gas is obtained by action of water.†

Equadiol. A proprietary preparation of meprobamate and ethinyloestradiol, used for control of menopause. (Wyeth).†

Equagesic. A proprietary preparation of ethoheptazine citrate, meprobamate, asprin, and calcium carbonate. An analgesic. (Wyeth).†

Equal. Aspartame. Applications: Sweetener. (G D Searle & Co).

Equalised Guano. Natural guanos, blended or mixed with ammonium salts, to obtain definite proportions of nitrogen and phosphorus. A fertilizer.†

Equanil. A proprietary preparation of meprobamate. A sedative. (Wyeth).†

Equaprin. A proprietary preparation containing meprobamate and aspirin. An analgesic. (Wyeth).†

Equatrate. A proprietary preparation of meprobamate and pentaerythrityl tetranitrate, used in treatment of angina pectoris. (Wyeth).†

Equiben. Cambendazole. Applications: Anthelmintic. (Merck Sharp & Dohme, Div of Merck & Co Inc).

Equionic. Sanitizer/detergent compounds. (Glover (Chemicals) Ltd).†

Equipoise. Boldenone undecylenate. Applications: Anabolic. (E R Squibb & Sons Inc).

Equipose. A proprietary preparation of hydroxyzine pamoate. A sedative. (Pfizer International).*

Equiproxen. Naproxen. Applications: Anti-inflammatory; analgesic; antipyretic. (Syntex Laboratories Inc).

Equisetic Acid. (Citridic Acid). Aconitic acid, $C_3H_3(COOH)_3$.†

Equivert. A proprietary preparation of buclizine hydrochloride, An antinauseant. (Pfizer International).*

Equivurm Plus. A proprietary preparation containing mebendazole. Applications: Veterinary antihelmintic (horses). (Janssen Pharmaceutical Ltd).*

Eqvalan. Ivermectin. A mixture of ivermectin component B1a and ivermectin component B1b. Applications: Antiparasitic. (Merck Sharp & Dohme, Div of Merck & Co Inc).

Era Chrome Dark Blue B. A dyestuff. It is a British equivalent of Palatine chrome blue.†

Era CR1. A proprietary trade name for a steel containing 0.06 per cent. carbon, 0.30 per cent. silicon, 0.04 per cent. sulphur, 0.04 per cent. phosphorus, 1.0 per cent. manganese, 18.50 per cent. chromium and 9.00 per cent. nickel. It is used in the manufacture of forged steel pressure vessels with good corrosion resistance. (Hadfield, George & Co).†

Era CR15 (CB). A proprietary trade name for a steel containing 0.06 per cent. carbon, 0.50 per cent. silicon, 0.04 per cent phosphrous, 1.00 per cent manganese, 19.00 per cent. chromium, 10.00 per cent. nickel and 0.6 per cent. niobium. (Hadfield, George & Co).†

Era Dyes. These are registered trade names for certain British dyestuffs.†

Era Metal. A steel containing 21 per cent chromium and 7 per cent nickel.†

Era 147. A proprietary trade name for a steel containing 0.22 per cent. carbon, 0.20 per cent. silicon, 0.04 per cent. sulphur, 0.04 per cent. phosphorus, 0.50 per cent. manganese, 5.00 per cent. chromium and 0.50 per cent. molybdenum. It is used for forging steel pressure vessels for service with hydrogen. (Hadfield, George & Co).†

Era 164. A proprietary trade name for a steel containing 0.20 per cent. carbon, 0.25 per cent. silicon, 0.04 per cent. sulphur, 0.04 per cent. phosphorus and 1.5 per cent. manganese. It is used in the manufacture of forged steel pressure vessels for use with intermediate pressures. (Hadfield, George & Co).†

Eraclene. A registered trade mark for low density polyethylene. (Anic Agricoltura Spa).†

Eradacin. Capsules containing acrosoxacin. Applications: An antibacterial agent used for the treatment of gonorrhoea. (Winthrop Laboratories).*

Eradite. Sodium hyposulphite, $Na_2S_2O_4$.†

Eranol. A form of colloidal iodine.†

Ercal. A proprietary name for ergotamine tartrate. (Blue Line Chemical Co, St Louis, Missouri).†

Ercerhinol. A colloidal silver.†

Ercusol. Aqueous acrylic polymer dispersion. (Bayer & Co).*

Erdmann's Reagent. Made by adding 40 cc of concentrated sulphuric acid to 20 drops of a solution containing 10 drops of nitric acid (specific gravity 1.153) and 20 cc of water. Used in testing for alkaloids.†

Erganol. Dibenzyl ether. A softening agent for cellulose esters.†

Ergodryl. A proprietary preparation of ergotamine tartrate, caffeine citrate and diphenhydramine hydrochloride, used for migraine. (Parke-Davis).†

Ergol. A proprietary plasticizer. It is stated to be benzyl benzoate, $C_6H_5.COO.CH_2.C_6H_5$ Used as a softening agent for cellulose esters.†

Ergomar. Ergotamine tartrate. Applications: Analgesic. (Fisons Corp).

Ergometrine. A water-soluble alkaloidal substance isolated from ergot.†

Ergostat. Ergotamine tartrate. Applications: Analgesic. (Parke-Davis, Div of Warner-Lambert Co).

Ergot. A dark coloured fungus, which attacks damp rye and other grasses, and when contained in flour, causes ergotism. It is a mixture of alkaloids, and is used in midwifery for causing contractile action of the pregnant uterus.†

rgot Oil. See OIL OF ERGOT.

rgotamine. *p*-Hydroxyphenylethylamine.†

rgotine. *Extracum ergotoe B.P.*†

rgotrate Maleate. Ergonovine maleate. Applications: Oxytocic. (Eli Lilly & Co).

ricol. Guaiacol acetate.†

ricon. A proprietary trade name for a phenol-formaldehyde synthetic resin.†

rie Blue GG. A dyestuff which dyes cotton or wool, afterwards treated with copper sulphate.†

rie Red B Conc. See BENZOPURPURIN 4B.

rika B. A dyestuff. It is the sodium salt of methylbenzenylaminothioxylenolazo-α-naphtholdisulphonic acid, $C_{26}H_{19}N_3O_7S_3Na_2$. Dyes unmordanted cotton rose pink.†

rika G. An azo dyestuff from dehydrothio-*m*-xylidine and β-naphthol-γ-sulphonic acid. It dyes in a similar way to Erika B.†

rinofort. A proprietary cellulose acetate plastic.†

rinoid. A proprietary trade name for casein-formaldehyde synthetic resin insulating material. (Mobil Chemical Co).†

rio. Acid dyes. (Ciba Geigy PLC).†

riochlorine A, B, CB, BB. Green dyestuffs similar to Erika B.†

riochrome. Mordant dyes. (Ciba-Geigy PLC).

riochrome Azurole B. A dyestuff obtained by the condensation of *o*-chlorobenzaldehyde and *o*-cresotinic acid, then oxidation.†

riochrome Black A. (Chrome Black A, Chrome Fast Black A, Diadem Chrome Black A). A dyestuff prepared from diazotized 8-nitro-1-amino-β-naphthol-4-sulphonic acid and β-naphthol.†

riochrome Black T. A dyestuff obtained by the diazotization of 8-nitro- 1-amino-β-naphthol-4-sulphonic acid andβ-naphthol.†

Eriochrome Blue-black B. A dyestuff. It is equivalent to Chrome fast cyanine G.†

Eriochrome Blue-black R. See PALATINE CHROME BLUE.

Eriochrome Cyanine. A dyestuff obtained by the condensation of benzaldehyde, *o*-sulphanilic acid, and *o*-cresotinic acid, and oxidation.†

Eriochrome Phosphine. A dyestuff prepared from diazotized *p*-nitraniline-*o*-sulphonic acid and salicylic acid.†

Eriochrome Red B. (Diadem Chrome Red L3B). A dyestuff prepared from diazotized 1-amino-β-naphtholsulphonic acid and phenylmethylpyrazolone.†

Eriochrome Verdone A. A dyestuff prepared by diazotizing sulphanilic acid, combining it with *m*-amino-*p*-cresol, diazotizing the product, and combining it withβ-naphthol. It dyes wool claret red shades from an acid bath, which upon chroming becomes blue-green.†

Erioclarite. Pretreatment agents. (Ciba-Geigy PLC).

Eriocyanine A. A dyestuff. It is the sodium salt of tetramethyldibenzylrosanilinedisulphonic acid. Dyes wool reddish-blue from an acid bath.†

Eriocyanine B. A similar dyestuff to the above.†

Erioglaucine A. A dyestuff. It is the acid ammonium salt of the trisulphonic acid of diethyldibenzyldiaminotriphenylcarbinol. Dyes wool and silk greenish-blue from an acid bath.†

Erioglaucine RB, BB, B, J, GB. Dyestuffs similar to above.†

Erional. Dyeing and printing assistant. (Ciba-Geigy PLC).

Erionyl. Acid dyes for polyamide fabrics. (Ciba Geigy PLC).†

Eriopon. Surface active agents. (Ciba Geigy PLC).†

Erkantol. Wetting agents and padding auxilliaries. Applications: Textile finishing. (Bayer & Co).*

Erlangen Blue. See PRUSSIAN BLUE.

Erlicki's Solution. A hardening agent used in microscopy. It consists of potassium dichromate 2.5 parts, calcium sulphate 1 part and water 100 parts.†

Ermite. A proprietary trade name for a synthetic resin.†

Ernite. Synonym for Grossular.†

Erthro. Erythromycin. Applications: Antibacterial. (Abbott Laboratories).

Ertilen. Antibiotic veterinary ethicals. (Ciba-Geigy PLC).

Erukku Erukkam. See AK MUDAR.

Ervamine. A proprietary trade name for melamine-formaldehyde.†

Ervamix. A proprietary trade name for a fibrous glass reinforced polyester moulding compound.†

Ervevax. RA27/3 live attenuated rubella virus vaccine. Applications: Vaccination of pre-pubertal children against rubella. (Smith Kline and French Laboratories Ltd).*

ERYC. Erythromycin. Applications: Antibacterial. (Parke-Davis, Div of Warner-Lambert Co).

EryDerm. Erythromycin. Applications: Antibacterial. (Abbott Laboratories).

Erypar. Erythromycin stearate. Applications: Antibacterial. (Parke-Davis, Div of Warner-Lambert Co).

Ery-Ped. Erythromycin ethylsuccinate. Applications: Antibacterial. (Abbott Laboratories).

Ery-Tab. Erythromycin. Applications: Antibacterial. (Abbott Laboratories).

Erysilin. An erysipelas vaccine for swine. (Glaxo Pharmaceuticals Ltd).†

Erythrin. (Spirit Eosin, Methyl Eosin, Primrose Soluble in Alcohol, Erythrin Methyl Eosin, Eosin Soluble in Spirit). A dyestuff. It is the potassium salt of tetrabromofluorescein methyl ether. Dyes silk bluish-red, with a reddish fluorescence.†

Erythrin Methyl Eosin. See ERYTHRIN.

Erythrin X. See PONCEAU 5R.

Erythrobenzine. See MAGENTA.

Erythrocin. A proprietary preparation containing erythromycin as the lactobionate. An antibiotic. (Abbott Laboratories).*

Erythrocin Lactobionate-IV. Erythromy lactobionate for injection. Applicatio Antibacterial. (Abbott Laboratories).

Erythroglucin. See ERYTHROMANNITE.

Erythrolar. Tablets containing erythromycin stearate BP equivalent 250mg erythromycin and 500mg erythromycin. Granules pro syrup. Each 5ml of reconstituted suspension contains erythomycin ethylsuccinate equivalent to 250mg of erythromycin. Applications: For the prophylaxis and treatment of infections caused by erythomycin-sensitive organisms. (Lagap Pharmaceuticals Ltd).*

Erythromid. A proprietary preparation containing erythromycin. An antibiot (Abbott Laboratories).*

Erythroped. A proprietary preparation containing erythromycin (as the ethyl succinate). An antibiotic. (Abbott Laboratories).*

Erythrosiderite. A mineral, $2KC1.FeC1_3.H_2O$.†

Erythrosin. (Erythrosin B, Erythrosin D, Pyrosin B, Iodeosin B, Eosin Bluish, Eosin J, Rose B, Dianthine B, Primro Soluble, Soluble Primrose). The sodiu or potassium salt of tetraiodofluoresce $C_{20}H_6O_5I_4Na_2$. Dyes silk and wool bluish- red. Used for paper staining, a a sensitizer of silver bromide in the photographic industry, and in the production of orthochromatic dry plates.†

Erythrosin A. The sodium or potassium s of triiodofluorescein.†

Erythrosin B. See ERYTHROSIN.

Erythrosin BB. See PHLOXINE P.

Erythrosin D. See ERYTHROSIN.

Erythrosin G. (Dianthine G, Pyrosin G, Iodeosin G). The sodium or potassium salt of diiodofluoresceine, $C_{20}H_8O_5I_2Na_2$. Dyes wool yellowish-red, with yellowish-red fluorescence.†

Erythroxyanthraquinone. 2-Hydroxy-anthraquinone.†

Esaflon. SF6 gas. (Montedison UK Ltd).

Esbatal. A proprietary preparation of benthanidine sulphate. An antihypertensive. (The Wellcome Foundation Ltd).*

sbenite. A material made from cellulose, powdered mica, and magnesium silicate.†

scaid. Specialty hydrocarbon for mineral applications. Metal extraction solvent. (Exxon Chemical International Inc).*

scalol 507. Padimate O. Applications: Ultraviolet screen. (Van Dyk & Co Inc).

scane. Detergent intermediates. (Exxon Chemical Ltd).*

schel. A fine-grained light coloured smalt (*qv*).†

schka Mixture. A mixture of 2 parts by weight pure calcined magnesia and 1 part pure anhydrous sodium carbonate. Used for the determination of sulphur in coal by heating the coal with the mixture, then adding hydrochloric acid and barium chloride, when barium sulphate is precipitated.†

sco Extract. A synthetic tannin prepared from sulphonated heavy tar oils, by condensation with formaldehyde, and then forming the chromium salt. Also see NERADOL.†

scomer. Polyethylene wax. (Exxon Chemical International Inc).*

scopol. Liquid reactive hydrocarbon resin. (Exxon Chemical Ltd).*

scor. Polyethylene copolymer. Low density polyethylene. (Exxon Chemical International Inc).*

scorene. Specialty plastics. Linear low density PE. High density PE. (Exxon Chemical International Inc).*

scorez. Hydrocarbon resins from petroleum feedstocks. (Exxon Chemical Ltd).*

scorpal. Phencarbamide. Applications: Anticholinergic. (Bayer AG).

sCort. Prednicarbate. Applications: Glucocorticoid. (Hoechst-Roussel Pharmaceuticals Inc).

scorto. A proprietary brand of artificial silk.†

scoweld. Liquid epoxy adhesives and grouts. (Exxon Chemical Ltd).*

scutcheon Pin Brass. See BRASS.

Esdeform. Stain removers. (S & D Chemicals Ltd).†

Esdesol. A paint stripper. (S & D Chemicals Ltd).†

Esdogen. Wetting agent and solvent soaps. (S & D Chemicals Ltd).†

Eserine. Physostigmine, $C_{15}H_{21}N_3O_2$. An alkaloid obtained from the calabar bean.†

Eshalit. A bakelite (phenol-formaldehyde resin).†

Esidrex. A proprietary preparation of hydrochlorothiazide (6-chloro-3,4-dihydro-7-sulphamyl-1,2,4-benzothiadiazine 1,1-dioxide). A diuretic. (Ciba Geigy PLC).†

Esidrex-K. A proprietary preparation of hydrochlorothiazide and potassium chloride in slow release core. A diuretic. (Ciba Geigy PLC).†

Esien Andradit. Synonym for Skiagite.†

Eskabarb. Phenobarbital. Applications: Anticonvulsant; hypnotic; sedative. (Smith Kline & French Laboratories).

Eskacef. A proprietary preparation of cephradine. An antibiotic. (Smith Kline and French Laboratories Ltd).*

Eskacillin. Penicillin G Procaine. Applications: Antibacterial. (Smith Kline & French Laboratories).

Eskacillin 100. A proprietary prepara- containing benzylpenicillin (as the potassium salt). An antibiotic. (Smith Kline and French Laboratories Ltd).*

Eskacillin 100 Sulpha. A proprietary preparation of benzylpenicillin potassium and sulphadimidine. An antibiotic. (Smith Kline and French Laboratories Ltd).*

Eskacillin 200. A proprietary preparation containing procaine penicillin. An antibiotic. (Smith Kline and French Laboratories Ltd).*

Eskacillin 200 Sulpha. A proprietary preparation of procaine penicillin and sulphadimidine. An antibiotic. (Smith Kline and French Laboratories Ltd).*

Eskadiazine. Sulphadiazine. Applications: Antibacterial. (Smith Kline & French Laboratories).

Eskalith. Lithium Carbonate. Applications: Antimanic. (Smith Kline & French Laboratories).

Eskamel Cream. Combination of resorcinol and sulphur. Applications: Treatment of acne. (Smith Kline and French Laboratories Ltd).*

Eskimo. Lubes. (Chevron).*

Eskornade. A proprietary preparation of isopropamide, diphenylpyraline and phenylpropanolamine hydro chloride. A nasal decongestant. (Smith Kline and French Laboratories Ltd).*

Eskornade Spansule Capsules. Combination of isopropamide iodide, phenylpropanolamine hydrochloride and diphenylpyraline hydrochloride. Applications: Oral nasal decongestant with antihistaminic and anticholinergic activity. (Smith Kline and French Laboratories Ltd).*

Esmaillite. A dope consisting of a mixture of cellulose acetate and volatile solvent, usually ethyl formate.†

Esoderm. Lindane. Applications: Infestation by lice. (Napp Laboratories Ltd).*

Esophotrast. Barium sulphate. Applications: Diagnostic aid. (Armour Pharmaceutical Co).

Esorb. Vitamin E. Applications: Vitamin E supplement. (Wyeth Laboratories, Div of American Home Products Corp).

Esperase. A proteolytic enzyme prepared by submerged fermentation of an alkalophilic strain of the genus Bacillus. Applications: Added to detergent formulas in order to assist in the removal of protein-based strains, eg from blood, mucus, sweat and various food products. (Novo Industri A/S).*

Esrakon. Tablets for use in start up of electro slag refining process for steel billets and slabs. (Foseco (FS) Ltd).*

Essar (W). A proprietary trade name for a general purpose acid and alkali resistant furane resin cement for bedding in acid resisting tiles. (15O).†

Essence of Bergamot. Oil of Bergamot.†

Essence of Bigarade. Oil of Bitter Orange Peel.†

Essence of Bitter Almonds. Benzaldehyde†

Essence of Mirbane. (Oil of Mirbane). Nitrobenzene, $C_6H_5NO_2$.Formerly us for scenting soap.†

Essence of Pineapple Oil. See ANANAS OIL.

Essence of Resin. See ROSIN SPIRIT.

Essence of Tar. Creosote.†

Essence of Turpentine. See OIL OF TURPENTINE.

Essential Oil of Bitter Almonds. Benzaldehyde, C_6H_5CHO.†

Essential Salt of Lemons. (Salt of sorrel) acid potassium oxalate.†

Essential Salt of Urine. See MICROCOSMIC SALT.

Essex Powder. An explosive, containing 24 per cent nitro-glycerin, 0.5-1.5 per cent collodion cotton, 33-35 per cent potassium nitrate, 33-35 per cent whe flour, and 5-7 per cent ammonium chloride.†

Esshete CML. A proprietary trade name for a steel containing 1 per cent. chromium and 1 per cent. molybdenu possessing high creep strength and corrosion resistance. It is suitable for operating up to 1000 °F. (Samuel Fo & Co Ltd).†

Esshete CRM2. A steel containing 2 1/4 per cent. chromium and 1 per cent. molybdenum and superior properties t CML. It can be used up to 1100 °F. (Samuel Fox & Co Ltd).†

Esshete CRM5. A steel containing 5 per cent. chromium and 1 per cent. molybdenum suitable for tubes expose to high temperature steam. (Samuel F & Co Ltd).†

Esshete 1250. A steel containing 16 per cent. chromium, 10 per cent. nickel an 6 per cent. manganese. An austenitic creep resisting steel. (Samuel Fox & C Ltd).†

Esskol. Modified oils. Applications: Pain (NL Industries Inc).*

Estabex 2307, 2349. Epoxidized soya bea oils.†

Estabex 2386. Epoxidized monoesters.†

Estane. A proprietary range of thermoplastic polyurethane moulding and extrusion compounds. (BF Goodrich Chemical Co).†

tasol. Dimethyl esters of adipic, glutaric and succinic acids. Applications: Low toxicity compound solvent. General purpose solvent coil coatings, foundry resins. (Chemoxy International Ltd).*

ter Copal. An ester gum obtained by the interaction between glycerin and copal.†

tergel. Isopropyl myristate. Applications: Pharmaceutic aid. (Merck Sharp & Dohme, Div of Merck & Co Inc).

terol. A proprietary brand of benzyl succinate. Also a proprietary trade name for alkyd synthetic varnish and lacquer resins.†

sterolane. Disperse dyes. (ICI PLC).*

sterox. A fast, air-drying, modified epoxy sealer-dustproofer. For protecting interior concrete floors. (Secure Inc).*

sterpol. A proprietary synthetic resin.†

stersil. Ethyl and propyl salicylglycollic esters. (Johnsons of Hendon).†

stigyn. Tablets of ethinyloestradiol-oestrogen. (British Drug Houses).†

stinyl. Ethinyl estradiol. Applications: Oestrogen. (Schering-Plough Corp).

stol. Textile lubricants. (Crosfield Chemicals).*

stolan. Polyester polyols. Applications: Suitable for textile lamination foams, packaging foams, microcellular formulated systems, adhesives and flexible coatings. (Lankro Chemicals Ltd).†

stone. See LENICET.

storal. Menthyl borate.†

strace. Estradiol. Applications: Oestrogen. (Mead Johnson & Co).

stracyt. A proprietary preparation of estramustine. Antineoplastic agent. (Roche Products Ltd).*

stragol. *p*-Propenylanisol, CH_2:CH. $CH_2.C_6H_4OCH_3$.†

stralite. See BAKELITE.

strovis. A proprietary preparation of QUINESTROL used for the suppression of lactation. (Warner).†

tard's Reagent. Anhydrous chromium oxychloride, an oxidizing agent.†

teleen. Trigallic acetal.†

EternaBrite. Leafing aluminium pigments. Applications: Metallic printing inks (silver and gold). (Silberling Mfg Co Inc).*

Eternite. A slate-like mass made from 6 parts Portland cement and 1 part asbestos fibre. Used for roofing.†

Ethacol. Pyrocatechol monoethyl ether.†

Ethafoam. Polyethylene foam used principally in cushioning and packaging applications. (Dow Chemical Co Ltd).*

Ethal. Cetyl alcohol, $C_{16}H_{33}OH$.†

Ethanite. A proprietary plastic made from ethylene dichloride and calcium polysulphide.†

Ethanol. See SPIRITS OF WINE.

Ethavan. A proprietary trade name for ethylvanillin.†

Ethavan. Ethyl vanillin. Applications: Used by flavour manufacturers to replace part of the vanillin to give bouquet to the finished flavour or fragrance. (Monsanto Co).*

Ethazate. Zinc diethyldithiocarbamate. Fast cures at low temperature; medium precure rate activated by OXAF; used for latex compounding; in latex foam dipped goods and fabric coatings. (Uniroyal).*

Ethazate 50D. A 50% active water dispersion of Ethazate. Ready to use form for latex compounding. (Uniroyal).*

Ethene. See OLEFIANT GAS.

Etheramine. Cationic surfactants of the ether amine type, in liquid form. Applications: Widespread use in industry and in household and personal care formulations, though mostly in the form of derived quaternaries and various salts. (Keno Gard (UK) Ltd).

Ethereal Oil of Bitter Almonds. Benzaldehyde, $C_6H_5.CHO$.†

Etherin. See OLEFIANT GAS.

Ether, Methylated. Ether prepared from methylated spirit.†

Ethibond. Suture, nonabsorbable surgical. Applications: Surgical aid. (Ethicon Inc).

Ethidium. Animal health insecticide. (FBC Ltd).

Ethidol. A proprietary preparation. It is the ethyl ester of iodo-ricinoleic acid. It is stated to be suitable for intraglandular injection.†

Ethiflex. Suture, nonabsorbable surgical. Applications: Surgical aid. (Ethicon Inc).

Ethiodan. Ethyl 4 - iodophenylundec-10-enoate. Iophendylate - X-ray contrast medium for myclography. (British Drug Houses).†

Ethiodol. Ethiodized oil for injection. Applications: Indicated for use in radiographic exploration for hysterosalpinography and lymphography. (Altana Inc).*

Ethion. *O,O,O',O'*-Tetraethyl-*S,S'*-methylene di(phosphorodithioate). Applications: It has both acaricidal and insecticidal properties. Its acaricidal action is widely used in the abatement of cattle ticks. As an insecticide it is used on citrus, deciduous fruits, tea, cotton and ornamental plants. (A/S Cheminova).*

Ethnine. A proprietary preparation containing pholcodine. A cough linctus. (Allen & Hanbury).†

Ethobral. A proprietary preparation of phenobarbitone., butobarbitone and quinalbarbitone. A hypnotic. (Wyeth).†

Ethocel. Thermoplastic cellulose ethers used primarily in coatings-type applications such as gel lacquers, lacquers, adhesives, binders and hot melts. (Dow Chemical Co Ltd).*

Ethoduomeen. A range of fatty diamine ethoxylates. Applications: Used as wax emulsions, polishes, fuel additives, anti-static additives, dye assistants, viscose spinning additives, algicides, bactericides and disinfectants. (Harcros Industrial Chemicals).*

Ethofat. Ethoxylated fatty acids. Applications: Mineral flotation; metal working; paper; leather; pigments and surface coatings; plastic foams; textiles. (Akzo Chemie UK Ltd).

Ethofoil. A proprietary trade name for ethyl cellulose film.†

Ethomeen. A range of fatty amine ethoxylates. Applications: Used as wax emulsions, polishes, fuel additives, an[tu]static additives, dye assistants, viscos[e] spinning additives, algicides, bactericides and disinfectants. (Harc[ros] Industrial Chemicals).*

Ethomid. A range of ethoxylated amides. Applications: Used as lubricants. (Harcros Industrial Chemicals).*

Ethomulsion. A proprietary trade name [for] ethyl cellulose lacquer emulsion.†

Ethoquad. Ethoxylated quaternary ammonium salts. Applications: Used [as] electroplating bath additives and as bacteriostats. (Harcros Industrial Chemicals).*

Ethotal. Emulsifier dispersant, solubilize[r]. Applications: Textile; agricultural; cosmetics. (Witco Chemical Ltd).

Ethoxol. A proprietary glycol ether. (ICI PLC).*

Ethrane. Enflurane. Applications: An inhalation anaesthetic. (Abbott Laboratories).*

Ethrel. Growth regulator containing ethephon. Sold in UK on behalf of Amchem Products Inc. (ICI PLC).*

Ethrel-E. Growth regulator. (A H Marks & Co Ltd).

Ethrel-R. Defoliant . (A H Marks & Co Ltd).

Ethril. Erythromycin stearate. Applications: Antibacterial. (E R Squibb & Sons Inc).

Ethrine. Linctus of pholcodine. (Allen & Hanbury).†

Ethulon. Ethyl cellulose film for tracing a[nd] industrial purposes. (May & Baker Ltd).*

Ethulose. A proprietary preparation of alcohol and laevulose. A parenteral source of calories. (Geistlich Sons).†

Ethyl Acid Violet S4B. A dyestuff. It is a[n] equivalent of Victoria violet 4BS.†

Ethyl Blue. A dyestuff obtained by heatin[g] ethyl-diphenylamine with oxalic acid, by the action of ethyl chloride upon Diphenylamine blue. Dyes silk blue.†

Ethyl Cadmate. Activated cadmium diethyldithiocarbamate - primary accelerator. Applications: For NR an[d] synthetic rubers. Used with a thiazol[e]

Gives heat resistance and low compression set properties. (Vanderbilt Chemical Corporation).*

Ethyl Chloride BP. Ethyl chloride BP. Applications: For local anaesthesia. (Bengue & Co Ltd).*

Ethyl Eosin. See SPIRIT EOSIN.

Ethyl Eosin Rose JB. A dyestuff. $C_{22}H_{11}O_5Br_4K$, obtained by the alkylation of eosin.†

Ethyl Green. (Methyl Green). A dyestuff. It is the zinc double chloride of ethylhexamethylpararosaniline bromide, $C_{27}H_{35}N_3Cl_3BrZn$. Dyes wool mordanted with sodium thiosulphate and sulphuric acid or zinc acetate, and silk and cotton mordanted with tannin, bluish-green. Also see BRILLIANT GREEN.†

Ethyl Mustard Oil. Ethyl is othiocyanate $C_2H_5.N.CS$.†

Ethyl Parathion. *O,O*-diethyl-*O*-(4-nitrophenyl)phosphorothioate. Applications: Used as an insecticide for the protection of field crops, vegetables and fruit. (A/S Cheminova).*

Ethyl Purple 6B. See ETHYL VIOLET.

Ethyl Red. A dyestuff. Used as a sensitizer for silver bromide gelatin plates.†

Ethyl Sulphonal. See TETRONAL.

Ethyl Tellurac. Tellurium diethyl-dithiocarbamate. Rubber accelerator. Applications: Used in NR, SBR, NBR, EPOM and 11R with thiazole modifiers. Produces high modulus vulcanization. (Vanderbilt Chemical Corporation).*

Ethyl Tuads. A proprietary accelerator. Tetraethyl thiuram disulphide. (K & K Greef Chemicals Ltd).†

Ethyl Tuex. Tetraethylthiuram disulphide. A non-discolouring and non-staining accelerator. Sharp curing range with normal to high sulphur. Used in natural, butyl and nitrile rubbers, steam hose and calendered air-cured sheeting. (Uniroyal).*

Ethyl Violet. (Ethyl Purple 6B). A dyestuff. It is the hydrochloride of hexaethylpararosaniline, $C_{31}H_{42}N_3Cl$. Dyes silk and wool bluish-violet, and cotton mordanted with tannin and tartar emetic.†

Ethyl Zimate. A proprietary accelerator. Zinc diethyl dithiocarbamate. (K & K Greef Chemicals Ltd).†

Ethylan. Non-ionic ethoxylates of fatty alcohols, acids, amines, phenols and alkylolamides. Applications: For detergency, low foam wetting, latex stabilization, emulsification, cosolvency and petroleum recovery. (Lankro Chemicals Ltd).†

Ethylan BAB 20. Modified nonionic surfactant in liquid form, biodegradable, low foam. Application: Mechanized dishwashing; low foam emulsifier for cutting oils etc.; metal cleaning. (Diamond Shamrock Process Chemicals Ltd).

Ethylan BCP and KEO. Nonionic surfactant of the nonylphenol ethoxylate type as a clear colourless liquid. Application: Detergency, wetting, emulsification and antistatic agent used in textile scouring. (Diamond Shamrock Process Chemicals Ltd).

Ethylan BV. Nonionic surfactant of the nonylphenol ethoxylate type, as a white soft paste. Application: Solubilizer for alkylaryl sulphonates and essential oils; emulsifier for pesticides and herbicides. (Diamond Shamrock Process Chemicals Ltd).

Ethylan BZA. Modified alkylphenol ethoxylate in liquid form. Application: Low-foam nonionic surfactant used in mechanical dishwashing and metal cleaning. (Diamond Shamrock Process Chemicals Ltd).

Ethylan DP. Nonionic surfactant of the nonylphenol ethoxylate type, as a slightly hazy viscous liquid. Application: Foam stabilizing and suspending agent for liquid detergents. (Diamond Shamrock Process Chemicals Ltd).

Ethylan ENTX. Nonionic surfactant of the alkylphenol ethoxylate type in the form of an oil-soluble liquid. Application: Emulsifier for mineral oil and hydrophobic wax; mastic plasticizer. (Diamond Shamrock Process Chemicals Ltd).

Ethylan GMF. Nonionic surfactant of the alkylphenol type in liquid form.

Application: Wetting agent and emulsifier for hydrocarbon solvents. (Diamond Shamrock Process Chemicals Ltd).

Ethylan HA. Nonionic surfactant of the nonylphenol ethoxylate type, in the form of a water soluble wax. Application: Emulsifier for emulsion polymerization; detergent and wetting agent for use at high temperatures and electrolyte concentrations. (Diamond Shamrock Process Chemicals Ltd).

Ethylan HP. Nonionic surfactant of the nonylphenol ethoxylate type, in the form of a water soluble wax. Application: stabilizer for latex; wetting agent for electrolyte solutions; emulsifier. (Diamond Shamrock Process Chemicals Ltd).

Ethylan NP1. Nonionic surfactant of the nonylphenol ethoxylate type. Application: Defoaming agent. (Diamond Shamrock Process Chemicals Ltd).

Ethylan N92. Nonionic surfactant of the nonylphenol ethoxylate type, in the form of a water soluble wax. Application: Primary emulsifier for emulsion polymerization. (Diamond Shamrock Process Chemicals Ltd).

Ethylan PQ. Nonionic surfactant consisting of a modified alkylphenol ethoxylate in liquid form. Application: Solubilizer in iodophor manufacture. (Diamond Shamrock Process Chemicals Ltd).

Ethylan 20. Nonionic surfactant of the nonylphenol ethoxylate type, in the form of a white waxy solid. Application: Detergent and wetting agent for use at high temperatures and electrolyte concentrations; stabilizer for latex. (Diamond Shamrock Process Chemicals Ltd).

Ethylan 44. Nonionic surfactant of the nonylphenolethoxylate type, as a clear straw liquid. Application: Emulsifier and coupling agent for mineral oil and aliphatic solvents. (Diamond Shamrock Process Chemicals Ltd).

Ethylan 77 and TU. Nonionic surfactant of the nonylphenol ethoxylate type in liquid form. Application: Detergent and emulsifier used in dry cleaning; wool scouring; oils, hydrocarbon solvents an waxes. (Diamond Shamrock Process Chemicals Ltd).

Ethylbutylcarbinol. *Sec*-Heptyl alcohol.†

Ethylcyanine. A dyestuff. It is lepidinquinolineethylcyanine bromide, similar dye to Cyanine, using ethyl bromide instead of amyl iodide. Used a substitute for Cyanine in dyeing.†

Ethylol. Essentially ethyl alcohol containing approximately 6% isopropy alcohol. The product is denatured wit 3f% butyl alcohol and 1f% lead-free petrol. Applications: Solvent in paint and lacquers, printing inks, foundries, dyes, industrial detergents, explosives, polishes, degreasers, rust removers, manufacture of xanthates and esters. (Sasolchem).*

Ethylphenacemide. Pheneturide.†

Ethylthiurad. Vulcanization accelerator, tetraethylthiuram disulphide. Applications: Sulphur bearing accelerator and vulcanizing agent. Cu modifier for neoprene. (Monsanto Co).

Etocas. Emulsifiers and solubilizers for oi solvents and waxes. Applications: Cosmetics; metal working fluids; textiles; insecticides, herbicides; household products. (Croda Chemicals Ltd).

E-Toplex. Vitamin E. Applications: Vitamin E supplement. (USV Pharmaceutical Corp).

Etronite. The trade name for a synthetic resin-paper product. It is used for electrical insulation.†

Etrynit. Propatyl Nitrate. Applications: Vasodilator. (Sterling Drug Inc).

Eubeco. Vitamin B complex for injection. (Allen & Hanbury).†

Eucalyptol. Cineol, $C_{10}H_{18}O$, a terpene alcohol.†

Eucalyptus Gum. (Red Gum, Eucalyptus kino). See KINO, AUSTRALIAN.†

Eucalyptus Kino. See KINO, AUSTRALIAN.

Euchrome. See CAPPAGH BROWN.

Eucopine. A pine disinfectant containing substituted phenolic germicides. (William Pearson).*

Eudemine. A proprietary preparation of DIAZOXIDE used in the treatment of hypertension and hypoglycaemia. (Allen & Hanbury).†

Euderm. Finely disperse acrylic resin. Applications: Used in the leather industry for tightening the grain and as a base coat in pigment finishing. (Bayer & Co).*

Eugenglanz. Synonym for Polybasite.†

Euglycin. A proprietary trade name for Metahexamide.†

Eugynon 30 and 50. Proprietary preparations of ethinyloestrodiol and dinogestrel. Oral contraceptives. (Schering Chemicals Ltd).†

Euka-drya. A Dutch cellulose rayon.†

Eukanol. Covering pigment dispersions with casein binder for finishing. Applications: Leather industry. (Bayer & Co).*

Eukanol Colours. Aqueous, casein-based pigments for all types of leather. (Bayer UK Ltd).

Eukinase. A powder obtained by the desiccation of the pancreatic juice of swine, and contains mainly trypsin.†

Eulan WA. Mothproofing agent. Applications: Auxiliary for the protection of goods made from wool or wool blends against moths and carpet beetles (Anthrenus) where demands made on fastness to washing are not exacting, e.g. carpets and furnishing fabrics. (Bayer & Co).*

Eulan 33. General purpose textile finishing agent, fast to washing and processing. Applications: For the permanent protection of wool, hair, feathers and brush filaments against moths, carpet beetles (Anthrenus) and black carpet beetles (Attagenus). (Bayer & Co).*

Eumydrin. Methylatropine nitrate. Applications: Anticholinergic. (Sterling Drug Inc).

Eunatrol. Pure sodium oleate, used medicinally.†

Euosmite. A resin found in the lignite in Bavaria.†

Eupad. An antiseptic consisting of bleaching powder and boric acid in equal parts.†

Euperlan. Mixture of fatty alcohol ether sulphates with special additives. Applications: Production of pearly sheen shampoos. (Henkel Chemicals Ltd).

Euphoramin. A proprietary preparation of meprobamate and methyl-amphetamine hydrochloride. (Rybar).†

Euphorbia Gum. See POTATO GUM.

Euphyllin. A registered trade mark for a mixture of equal parts of primary and secondary theophylline - ethylene - diamine, $C_2H_4(NH_2)_2.C_7H_8N_4O_2$-$+C_2H_4(NH_2)_2.2C_7H_8N_4O_2$. Used in medicine. (Byk-Gulden Lomberg Chemische Fabrik GmbH).†

Eupinal. A proprietary preparation of cabeine iodide.†

Eurax-Hydrocortisone. A proprietary preparation of hydrocortisone and crotamiton for dermatological use. (Ciba Geigy PLC).†

Eureka Alloy. A trade mark for a copper-nickel alloy used for electrical resistance wires.†

Eureka Compound. See BURNT HYPO.

Euresol. Resorcinol Monoacetate. Applications: Antiseborrheic; keratolytic. (Knoll Pharmaceutical Co).

Eurobin. Chrysarobin triacetate. †

Europolymer. Thermoplastic polyurethanes for injection moulding, and as adhesives and coatings. (Avalon Chemical Co Ltd).

Europrene. A registered trade mark for general purpose butadiene styrene copolymers coded as follows:
CIS. 1-4 cis-polybutadiene
SS. high styrene copolymers
N. butadiene acrylonitrile copolymers. (Anic Agricoltura Spa).†

Eurylon. Maize starch, high in amylose. Applications: Generally food, textiles, cardboard and paper. (Roquette (UK) Ltd).*

Eusin. Brilliant, transparent pigment dispersions based on casein binders for aniline and aniline effect finishing. Applications: Leather industry. (Bayer & Co).*

Eusol. A solution containing 0.54 per cent hypochlorous acid, 1.28 per cent calcium biborate, and 0.17 per cent calcium chloride. It is made by shaking Eupad (*qv*) in water, and filtering.†

Eusolvan. A proprietary solvent stated to consist of ethyl lactate.†

Eutannin. A mixture of gallic acid and milk sugar. An intestinal astringent.†

Eutectal. An aluminium alloy containing small amounts of copper, manganese, magnesium silicide, silicon, and titanium.†

Euthatal. Pentobarbitone sodium solution. (May & Baker Ltd).*

Euthroid. Liotrix. A mixture of liothyronine sodium and levothyroxine sodium in a ratio of 1:1 in terms of biological activity, or in a ratio of 1:4 in terms of weight. Applications: Thyroid hormone. (Parke-Davis, Div of Warner-Lambert Co).

Eutonyl. Pargyline Hydrochloride. Applications: Antihypertensive. (Abbott Laboratories).

Euvalerol B. A proprietary preparation of valerian root with phenobarbitone. A sedative. (Allen & Hanbury).†

Euvitol. A proprietary preparation of fencamfamin hydrochloride. (Allen & Hanbury).†

EVA. Ethylene vinyl acetate. A flexible polythene-like polymer for moulding and extrusion. Adheres well to metals.†

Evac-Q-Mag. Magnesium citrate. Applications: Laxative. (Adria Laboratories Inc).

Evac-Q-Tabs. Phenolphthalein. Applications: Laxative. (Adria Laboratories Inc).

Evadyne. A proprietary preparation of BUTRIPTYLINE hydrochloride. An antidepressant. (Ayerst Laboratories).†

Evans Blue. Azovan blue.†

Evans' Cement. A metallic cement made by adding cadmium amalgam (74 per cent mercury) to mercury.†

Evatane. Ethylene-vinyl acetate copolymer. Applications: Hot melt adhesives. (Atochem Inc).†

Evazote. Crosslinke foamed eva. (BXL Plastics Ltd).

Everbrite. A proprietary trade name for an alloy of 60 per cent copper, 30 per cent nickel, 3 per cent iron, 3 per cent silicon and 3 per cent chromium.†

Eveready Prestone. The trade marks applied to ethylene glycol antifreeze.†

Evergreen. Lawn fertilizer combined with selective weedkiller. (Fisons PLC, Horticulture Div).

Everitt's Salt. Potassium ferrous ferro-cyanide, $K_2Fe.Fe.(CN)_6$.†

Everlastic. A proprietary trade name for a textile incorporating Duprene.†

Everlube. Family of oven cured dry film lubricant coatings containing solid lubricants and suitable binder system. Applications: Fasteners, slides, pins, clips and numerous applications requiring dry lubrication and corrosion protection particularly under high loads. (E/M Corporation).*

Everseal. A bituminastic liquid applied as a corrosion-resisting material.†

Eversoft Plastex. A low-freezing explosive containing a nitrated mixture of glycerin and ethylene glycol, ammonium nitrate, sodium chloride, wood meal, trinitrotoluene, and nitro-cotton.†

Eversoft Sea Mex. A low-freezing explosive containing a nitrated mixture of glycerin and ethylene glycol, ammonium nitrate, sodium chloride, and wheat flour.†

Eversoft Tees Powder. A low-freezing explosive consisting of a nitrated mixture of glycerin and ethylene glycol, ammonium nitrate, wood meal, and sodium chloride.†

Eversun. Sun protection products, to protect skin while promoting a tan. (Richardson-Vicks Inc).*

Evidorm. A proprietary preparation of hexobarbitone and cyclobarbitone calcium. A hypnotic. (Bayer & Co).*

Evipal. Hexobarbital. Applications: Sedative. (Sterling Drug Inc).

Evipan Sodium. Pharmaceutical preparation. Applications: For intravenous anaesthesia. (Bayer & Co).*

Evo-stik 873 Super. A proprietary synthetic rubber latex adhesive having a high content of solids. (Evode Plastics Ltd).*

Evoprene. Thermoplastic elastomer compounds. Applications: Wide range of industrial components, by extrusion or moulding. Applications involve automotive and commercial transport, building, cable and a variety of component parts for general industrial use..(Evode Plastics Ltd).*

Evramycin. A proprietary preparation containing triacetyloeandomycin. An antibiotic. (Wyeth).†

Ewer and Pick's Acid. 1, 6-Naphthalene-disulphonic acid, $C_{10}H_6(SO_3H)_2$.†

Ewol. See NERADOL.

Exaltone. A trade name for muskone (cyclopentadecanone), the perfuming principle of natural musk.†

Excelite. See METALITE. Also an American trade name for a thermo-setting fibrous plastic.

Excellerex. A proprietary rubber vulcanization accelerator. It is an aniline derivative.†

Excello. An electrical resistance alloy containing 85 per cent nickel, 14 per cent chromium, 0.5 per cent iron, and 0.5 per cent manganese. It is also the name for a carbon black used in rubber mixings.†

Excelo. A proprietary trade name for a hot die steel containing 2.5 per cent tungsten, 1.5 per cent chromium, 0.35 per cent vanadium, and 0.55 per cent carbon.†

Excelon. A proprietary trade name for acrylic denture material.†

Excelsior. A proprietary trade name for carbon black.†

Exelderm. Topical anti-fungal preparation containing sulconazole nitrate. (ICI PLC).*

Exem. Anionic, non-ionic and special anionic and non-ionic blends of emulsifier. Applications: Emulsifier for agricultural pesticides. (Makhteshim Chemical Works Ltd).†

Exgraphite. Expandable acid treated graphite. (Foseco (FS) Ltd).

Exkin. Oxime anti-skinning agents. Applications: Prevention of skin formation in solvent-based coating compositions. (Nueodex Inc).*

Exl-die Steel. A proprietary trade name for a non-shrinking steel containing 1.15 per cent manganese, 0.5 per cent chromium, 0.5 per cent tungsten, and 0.9 per cent carbon.†

Exlax. A proprietary phenolphthalein preparation.†

Exna. Benzthiazide. Applications: Diuretic; antihypertensive. (A H Robins Co Inc).

Exoderil. Naftifine hydrochloride. Applications: Antifungal. (Sandoz Pharmaceuticals, Div of Sandoz Inc).

Exolan Cream. Pale yellow aqueous cream containing 1% 1:8:9 triacetoxy-anthracene. Applications: For the topical treatment of sub-acute and chronic psoriasis including psoriasis of the scalp. (Dermal Laboratories Ltd).*

Exolit. Ammonium polyphosphate. (Hoechst UK Ltd).

Exolon. A trade mark for abrasive articles consisting essentially of silicon carbide.†

Expanded Graphite. A substance prepared by covering fiake graphite with an oxidizing agent, to produce a film of graphitic acid, then heating strongly to cause the particles to become distended.†

Expanding Solder. An alloy of 37.5 per cent lead, 6.75 per cent bismuth, and 56.25 per cent antimony. It expands on cooling, and is used for fixing metal into holes.†

Expansyl Spansule. A proprietary preparation of trifluoperazine hydrochloride, diphenylpyraline hydrochloride and ephedrine sulphate. (Smith Kline and French Laboratories Ltd).*

Explosive D. Ammonium picrate.†

Explosive Gum. Nitroglycerin gelatinized with collodion cotton. It contains 96 per cent of the former, and 4 per cent of the latter compound.†

Exprol. Photographic developer. (May & Baker Ltd).

Expulin. A proprietary preparation of PHOLCODINE, EPHEDRINE hydrochloride, CHLORPEENIRAMINE maleate, glycerin and menthol. A cough medicine. (Galen).†

Exsel. Selenium Sulphide. Applications: Antifungal; antiseborrheic. (Herbert Laboratories, Dermatology Div of Allergan Pharmaceuticals Inc).

Exsiccated Alum. This is burnt alum (*qv*).†

Exterol Ear Drops. Viscous solution containing 5% w/w urea hydrogen peroxide. Applications: An aid in the removal of hardened ear wax. (Dermal Laboratories Ltd).*

Extil. A proprietary preparation containing carbinoxamine maleate and pseudoephedrine hydrochloride. A cough suppressant. (British Drug Houses).†

Extil Compound Linctus. A proprietary preparation containing noscapine, pseudoephedrine, and carbinoxamino maleate. A cough linctus. (British Drug Houses).†

Extir. A proprietary range of expandable polystyrene beads. (Montedison UK Ltd).*

Extol. A proprietary trade name for a sulphated compound with solvents used as a detergent.†

Extra Bond. PVA liquid. Applications: Wood and general builders glue. (Wessex Resins & Adhesives Ltd).*

Extra Gilder's White. A variety of whiting.†

Extra White Metal. An alloy of 50 per cent copper, 30 per cent nickel, and 20 per cent zinc. It is a nickel silver (German silver).†

Extract of Gamboge. A compound of gamboge and aluminium oxide. A yellow pigment.†

Extract of Vermilion. See SCARLET VERMILION.

Extract, Indigo. See INDIGO CARMIN

Extrox. Coated zinc oxide. (Durham Chemicals Ltd).

Extrusil. Precipitated silica with a proportion of calcium oxide. Applications: Medium-active reinforcing filler with outstanding extrusion characteristics, good processibility for technical rubber articles. (Degussa).*

Extrusion-Plus. Amber coloured liquid used by cold heading fastener industrie Non-chlorinated and non-sulphonated oil. Applications: Extensively used by automotive fastener and aerospace fastener manufacturers. (Rustlan Chemical Co).*

Exxate. Acetates of higher oxo-alcohols. (Exxon Chemical International Inc).*

Exxate. Oxo-acetate solvents for high soli paint formulations. (Exxon Chemical Ltd).*

Exxsol. High quality dearomatized aliphatic solvent. (Exxon Chemical International Inc).*

Exxtraflex. Polyolefin film. (Exxon Chemical International Inc).*

Exyphen. A proprietary preparation of BROMPHENIRAMINE maleate, guaiphenesin, phenylephrine hydrochloride and phenyl propanolamine hydrochloride. A cough medicine. (Norton, H N Co).†

Eyelet Brass. See BRASS.

E-Z Mix. Series of dry liquid concentrates Applications: Rubber and vinyl. (The C P Hall Company).*

F

F12. See FREON.

F 31. Basic lead silico chromate. Applications: Anti-corrosive pigment for coatings systems. (NL Industries Inc).*

F-310. Gasoline additive. (Chevron).*

F789. A bactericide. It is the hydroxyethyl derivative of hexamethylene tetramine.†

F790. A bactericide. It is the methiodide of hexamethylenetetramine.†

F-1000. Aluminium hydroxide. Applications: Antacid. (Reheis Chemical Co).

Fabahistin. Pharmaceutical preparation also known as Incidal. Applications: Day-time antihistamine. (Bayer & Co).*

Fabrethane. A proprietary one-component foamable polyurethane, 100% solids. (Grace, W R & Co).†

Fabrex. Preformed insulating refractory shapes for various applications. (Foseco (FS) Ltd).*

Fabrifil. A proprietary trade name for a macerated cotton fabric filler.†

Fabrikoid. A trade mark for a fabric coated with pyroxylin.†

Fabroil. A proprietary trade name for a synthetic resin.†

Fabrol. Mucolytic. (Ciba-Geigy PLC).

Fabrolite. A synthetic resin of the phenol-formaldehyde type. (Associated Electrical Industries (GEC)).†

F-acid. See DELTA ACID.

Facteka. A proprietary trade name for a rubber substitute.†

Factice. A polymerization product of natural fatty oils and sulphur or sulphur chloride, a processing promoter for an economical processing of rubber. Applications: Extruded articles. (Rhein-Chemie Rheinau).*

Factis. A term applied to rubber substitutes prepared from oils.†

Factoprene NS. A proprietary hard factice. (Hubron Rubber Chemicals, Failsworth, Manchester).†

Factoprene Z. A proprietary soft factice. (Hubron Rubber Chemicals, Failsworth, Manchester).†

Factorate. Antihaemophilic factor. Applications: Antihaemophilic. (Armour Pharmaceutical Co).

Factrel. Gonadorelin hydrochloride. Luteinizing hormone-releasing factor hydrochloride. The source of this compound may be sheep, pig or other species. Applications: Gonad-stimulating principle. (Ayerst Laboratories, Div of American Home Products Corp).

Faexin Extract. The fatty acids of yeast.†

Fagacid. A product derived from beech wood tar. It is an antiseptic agent used in the preparation of soaps and plasters.†

Fahlun Diamonds. (Tin Brilliants). Lead-tin alloys, containing about 40 per cent lead, used for theatre jewellery.†

Fahralloy. A proprietary trade name for chromium-nickel-iron alloys.†

Fairco. See CRISCO.

Fairey Metal. A proprietary trade name for an alloy of aluminium with copper and smaller amounts of magnesium.†

Fairprene. A proprietary trade name for a chloroprene polymer for fabric coating.†

Faktex. A proprietary trade name for a yellow rubber substitute which can be dispersed in water for addition to latex.†

Falapen. A proprietary preparation containing Benzylpenicillin. An antibiotic. (British Drug Houses).†

Falcodyl. A proprietary preparation of pholcodine and ephedrine hydrochloride. A bronchial antispasmodic. (Norton, H N Co).†

Falkaloid. A proprietary trade name for a soft oil modified alkyd resin.†

Falkyd. A proprietary trade name for a soft oil modified alkyd resin.†

Falmonox. Teclozan. Applications: Antiamebic. (Sterling Drug Inc).

Famel. Range of cough syrups. (The Boots Company PLC).

Famel Syrup. A trade mark for a syrup comprising purified beech wood creosote rendered water soluble by means of lactic acid in combination with calcium lactophosphate, aconite and codeine. Used for treatment of infections of the lungs. See Beatin. (Optrex).†

Famid. Carbamate insecticide. (Ciba-Geigy PLC).

Famosan. Agricultural disinfectant. (The Wellcome Foundation Ltd).

Fan Blade Brass. See BRASS.

Fanasil. A proprietary preparation of sulfadoxine. Sulphonamide antibiotic. (Roche Products Ltd).*

Fangerine. Tablets containing phenolphthalein, chocolate, and sucrose. A laxative.†

Fanghidi Sclofani. A yellow powder of volcanic origin, consisting chiefly of sulphur, with small quantities of iron, calcium, and manganese.†

Fansidar. Proprietary combination antimalarial product containing sulfadoxine and pyrimethamine. (Roche Products Ltd).*

Fantan. Phenylcinchonoylurethane.†

Fantasit. A proprietary casein plastic.†

Fanzil. Sulphadoxine. Applications: Antibacterial. (Hoffmann-LaRoche Inc).

FAR Mark I through X. Elastomeric coatings made up of asphalt, neoprene, acrylic and kraton rubber imbedded in polyester fabric. Applications: Cold applied liquid or fully adhered membrane type roof systems. (Flex-Shield).*

Far-Go/Avadex BW. Trichloroallyl diisopropylthiocarbamate. Applications: Herbicide for wild oats. (Monsanto Co).*

Fard's Spanish White. See BISMUTH WHITE.

Farina. Flour, or potato starch.†

Farine. A term applied in the West Indies to a product obtained by grating fresh cassava root, draining away the juice from the wet pulp, and then heating the residue. It is also known as Cassava meal.†

Faringets. A proprietary preparation of myristyl benzalkonium iodine chloride. (Bayer & Co).*

Farinose. Starch cellulose, or the outer covering of the starch granule.†

Farlite. A proprietary trade name for a phenol-formaldehyde synthetic resin laminated product.†

Farmacel. Growth regulator for wheat and oats containing chlormequat. (ICI PLC).*

Farmaneb. Fungicide for prevention of potato blight. (ICI PLC).*

Farmer's Reducer. A photographic reducer. It consists of 100 cc of hypo solution (1 : 4), with from 5-10 cc of a 10 per cent solution of potassium ferricyanide.†

Farmon. Range of liquid herbicides of different formulations. (ICI PLC).*

Farnesol. A sesquiterpene alcohol prepared from nerolidol (*qv*). A perfume.†

Farrant's Medium. A microscopic medium. It consists of a mixture of equal parts of glycerin and arsenious acid, to which is added powdered gum arabic, allowed to stand and filtered.†

Farronic. A heat-resisting nickel-copper alloy.†

Fascol. Hexachlorophane anthelmintic. (The Wellcome Foundation Ltd).

Fasigyn. A proprietary preparation of tinidazole. An antiprotozoal. (Pfizer International).*

'asinex. Cattle and sheep flukicide. (Ciba-Geigy PLC).

'ast Acid Blue B. (Intensive Blue). A dyestuff obtained from dimethyl-aniline and α-naphthylamine disulphonic acid. Intensive blue is chemically identical.†

'ast Acid Blue R. (Violamine 3B). A dyestuff. Dyes wool and silk blue.†

'ast Acid Cerise. A dyestuff. It is a British equivalent of Fast acid fuchsine B.†

'ast Acid Eosin G. A phthalein dyestuff. It dyes wool in the presence of 10 per cent sodium sulphate, and 4 per cent sulphuric acid. Fast phloxin A and Irisamine are similar dyestuffs.†

'ast Acid Fuchsine B. (Fast acid Magenta B, Fast Acid Cerise, Fast Acid Magenta BL). A dyestuff.†

'ast Acid Magenta B. See FAST ACID FUCHSINE B.

'ast Acid Magenta BL. A dyestuff. It is equivalent to Fast acid fuchsine B, and is of British manufacture.†

'ast Acid Orange S. A dyestuff obtained by treating J acid with phosgene gas.†

'ast Acid Phloxine A. See FAST ACID EOSIN G.

'ast Acid Red. An acid dyestuff.†

'ast Acid Violet A2R. (Violamine R, Acid Violet 4R). A dyestuff. Dyes silk and wool reddish-violet.†

Fast Acid Violet B. (Violamine B). A dyestuff. Dyes wool and silk reddish-violet.†

Fast Acid Violet 10B. A dyestuff. It is the sodium salt of benzylethyltetramethyl-pararosanilinedisulphonic acid. Dyes wool violet-blue from an acid bath.†

Fast Black. (Fast Blue-Black). A dyestuff obtained by the action of nitrosodimethylanilinehydrochloride upon *m*-hydroxy diphenylamine. Dyes tannined cotton blue-black.†

Fast Black B. A dyestuff obtained by the action of sodium sulphide in aqueous solution upon 1,8-dinitronaphthalene. Dyes cotton black from an alkaline bath.†

Fast Black BS. A dyestuff obtained by the action of alkalis upon Fast black B. Dyes cotton deep black.†

Fast Blue. See ALKALI BLUE and MELDOLA'S BLUE.

Fast Blue B, Spirit Soluble. See INDULINE, SPIRIT SOLUBLE.

Fast Blue R, Spirit Soluble. See INDULINE, SPIRIT SOLUBLE.

Fast Blue R, 2R, 3R, for Cotton. See MELDOLA'S BLUE and INDULINE, SOLUBLE.

Fast Blue 2B for Cotton. See NEW BLUE B or G.

Fast Blue 2B or R. See NEW BLUE.

Fast Blue 6B for Wool. A soluble induline. See INDULINE.†

Fast Blue-black. See FAST BLACK.

Fast Blue, Greenish. A soluble induline. See INDULINE.†

Fast Bordeaux 0. Cloth red B.†

Fast Brown. A dyestuff. Dyes wool brown from an acid bath.†

Fast Brown G. (Acid Brown). A dyestuff. Dyes wool brown from an acid bath.†

Fast Brown N. (Acid Brown, Naphthylamine Brown, Azo Brown O, Chrome Brown RO, Lighthouse Chrome Brown R, Acid Brown G, Acid Brown R.) A dyestuff. Dyes wool brown from an acid bath.†

Fast Brown ONT, Yellow Shade. A dyestuff. Dyes wool and silk brownish-red. Also used for lakes.†

Fast Chrome Cyanine. A dyestuff. It is equivalent to Chrome fast cyanine G.†

Fast Chrome Cyanine 2B. A dyestuff. It is equivalent to Palatine chrome blue.†

Fast Chrome Yellow C. See ANTHRACENE YELLOW C.

Fast Chrome Yellow GG. An azo dyestuff obtained frozn *o*-anisidine and salicylic acid.†

Fast Cotton Blue B. See NEW BLUE B or G.

Fast Cotton Blue R. See INDOINE BLUE R.

Fast Cotton Brown R. See POLY-CHROMINE B.

Fast Green. See MALACHITE GREEN.

Fast Green. (Fast Green Extra, Fast Green Extra Bluish). A dyestuff. Dyes wool bluish-green from an acid bath.†

Fast Green Extra. See FAST GREEN.

Fast Green Extra, Bluish. See FAST GREEN.

Fast Green FCF. An American dyestuff prepared by the condensation of 2 molecules of ethylbenzylaniline sulphonic acid with 1 molecule-*p*-hydroxy-benzaldehyde-*o*-sulphonic acid followed by oxidation with lead peroxide.†

Fast Green G. See GALLANILIC GREEN.

Fast Green J. See BRILLIANT GREEN.

Fast Green M. A dyestuff. Employed for printing on cotton in conjunction with tannin.†

Fast Green S. See BRILLIANT GREEN.

Fast Green 0. See DINITROSO-RESORCIN.

Fast Grey D and S. Sulphide dyestuffs. They dye cotton in a solution containing sodium sulphide, caustic soda, and sodium chloride.†

Fast Light Orange G. A dyestuff. It is equivalent to Orange G, and is of British manufacture.†

Fast Marine Blue. See MELDOLA'S BLUE.

Fast Marine Blue G, BM, BG. See NEW BLUE B AND G.

Fast Milling Red B. See CLOTH RED B.

Fast Mordant Yellow. See ANTHRACENE YELLOW C.

Fast Myrtle Green. See DINITROSO-RESORCIN.

Fast Navy Blue G, BM, GM. See NEW BLUE B OR G.

Fast Navy Blue R. See MELDOLA'S BLUE.

Fast Navy Blue RM, MM. See MELDOLA'S BLUE.

Fast Neutral Violet B. A dyestuff. It is dimethyldiethyldiaminophenazonium chloride. Dyes tannined cotton violet.†

Fast New Blue for Cotton. See PARAPHENYLENE BLUE R.

Fast Oil Brown S. A dyestuff. It is equivalent to Sudan brown.†

Fast Oil Orange I. A British brand of Sudan I.†

Fast Oil Orange II. A dyestuff. It is a Br equivalent of Sudan II.†

Fast Oil Scarlet III. A British equivalent Sudan III.†

Fast Pink for Silk. See MAGDALA RE

Fast Ponceau B. See BIEBRICH SCARLET.

Fast Ponceau 2B. (Ponceau S extra, Scar S). A dyestuff. Dyes wool scarlet from an acid bath.†

Fast Red. (Fast Red A, Fast Red HF, Fa Red O, Brilliant Red, Roccellin, Rauracienne, Rubidine, Orseilline No 3, Orseilline No. 4, Ceracine, Cardina Red J, Alliance Fast Red A Extra, Fa Red A New, AL, G Extra, KG, Acid Red). A dyestuff. It is the sodium salt of *p*-sulphonaphthalene-azo-*β*-naphthol, $C_{20}H_{13}N_2O_4SNa$. Dyes wool red from an acid bath.†

Fast Red. See FAST RED D.

Fast Red A. See FAST RED.

Fast Red A New, AL, G Extra, KG. Dyestuffs. They are British equivalent of Fast red.†

Fast Red C. See AZORUBINE S.

Fast Red D. (Bordeaux DH, Bordeaux S, Fast Red, Fast Red NS, Fast Red EB Naphthol Red O, Naphthol Red S, Amaranth, Azo Acid Rubine 2B, OEnanthin, Wool Red Extra, Victoria Rubine, Brilliant Acid Bordeaux, Brilliant Bordeaux B, 2B, Brilliant Crimson O). A dyestuff. It is the sodiu salt of *p*-sulphonaphthaleneazo-*β*-naphthol-disulphonic acid. $C_{20}H_{11}N_2O_{10}S_3Na_3$. Dyes wool red fro an acid bath.†

Fast Red E. (Fast Red, Fast Red S, Acid Carmoisine). A dyestuff. It is the sodi salt of *p*-sulphonaphthalene azo-*β*-naphtholmonosulphonic acid. Dyes wo red from an acid bath.†

Fast Red EB. See FAST RED D.

Fast Red HF. See FAST RED.

Fast Red NS. See FAST RED D.

Fast Red PR Extra. An acid wool dye.†

Fast Red RBE Base. A dyestuff. It is 6-benzamino-*m*-4-xylidine hydrochloride.†

Fast Red S. See FAST RED E.

'ast Red SX. A British brand of Fast red BT.†

'ast Red TR Base. The hydrochloride of 5-chloro-2-amino-*o*-toluene.†

'ast Red 0. See FAST RED.

'ast Red 7B. See CLOTH RED G.

'ast Scarlet. See BENZOPURPURIN 4B.

'ast Scarlet. (Double Scarlet). A dyestuff obtained by the action of diazotized aminoazobenzenesulphonic acid uponβ-naphthol.†

'ast Scarlet B. A dyestuff. Dyes wool scarlet from an acid bath.†

'ast Scarlet TR Base. The hydrochloride of 2-chloro-6-aminotoluene.†

'ast Sulphone Violet 4R. A dyestuff of the same class as Fast sulphone violet 5BS.†

'ast Sulphone Violet 5BS. A monoazo dyestuff, giving wool and silk bluish-violet shades from an acid bath.†

'ast Vat Blue R. A Bohme dyestuff. It contains logwood extract, methylene blue, and a chrome mordant.†

'ast Violet. See GALLOCYANINE DH and BS.

'ast Wool Blue. See ALPINE BLUE.

'ast Yellow. See ORANGE IV and FAST YELLOW R.

'ast Yellow. See ACID YELLOW, ORANGE IV.

'ast Yellow B. An azo dyestuff. It is aminoazotoluenedisulphonic acid.†

'ast Yellow Extra. See ACID YELLOW.

'ast Yellow G. See ACID YELLOW.

'ast Yellow GL. A dyestuff. It is a British brand of Acid yellow.†

'ast Yellow N. (Curcumeine, Jaune Solide N, Orange N, Yellow OO). A dyestuff. It is the sodium salt of sulpho-*p*-tolueneazodiphenylamine. Dyes wool orange from an acid bath.†

'ast Yellow R. (Fast Yellow, Acid Yellow R, Yellow W). A dyestuff. It is the sodium salt of aminoazotolueneisulphonic acid. Dyes wool reddish-yellow from an acid bath.†

'ast Yellow S. See ACID YELLOW.

'ast Yellow, Greenish. See ACID YELLOW.

Fasteeth. Denture adhesive powder, for securing dentures. (Richardson-Vicks Inc).*

Fasteeth Extra Hold. To secure hard-to-hold lower dentures. (Richardson-Vicks Inc).*

Fastex. A proprietary trade name for a specially stabilized and purified rubber latex supplied in concentrations of 40 and 60 per cent.†

Fastin. Phentermine Hydrochloride. Applications: Appetite suppressant. (Beecham Laboratories).

Fastocaine. A proprietary preparation of lignocaine hydrochloride and noradrenaline. A local anaesthetic. (Leo Laboratories).†

Fat, Chocolate. See VEGETABLE BUTTER.

Fat, Mineral. See PETROLEUM JELLY.

Fat, Otoba. See OTOBA BUTTER.

Fat Ponceau. See SUDAN IV.

Fatsco. Sodium arsenate 3%. Applications: Ant poison for sweet eating ants, kills roaches, moles, mice, woodchucks etc. (Fatsco).*

Fat, Tacamahac. See LAUREL NUT OIL.

Faturan. See BAKELITE.

Faunolen. Canine contagious hepatitis vaccine. (The Wellcome Foundation Ltd).*

Faversham Powder No. 2. An explosive containing ammonium nitrate, potassium nitrate, trinitrotoluene, and ammonium chloride.†

Favierite No. 1. An explosive consisting of 88 per cent ammonium nitrate and 12 per cent dinitro-naphthalene.†

Favierite No. 2. An explosive containing 90 per cent of No. 1 and 10 per cent ammonium chloride.†

Fax. Edible fats. (Marfleet Refining Co).*

Faxola. Cooking oil. (Marfleet Refining Co).*

FBC CMPP. Selective herbicide. (FBC Ltd).

FBC Fly Dip. Insecticide. (FBC Ltd).

FBC MCPA. Selective herbicide. (FBC Ltd).

FBC Pirimicarb 50. Insecticide. (FBC Ltd).

FBC Protectant Fungicide. Fungicide. (FBC Ltd).

FBC Slug Destroyer. Mulluscicide. (FBC Ltd).

FBC Winter Dip. Insecticide. (FBC Ltd).

Feathered Tin. Granulated tin.†

Febkol Elastomer 110 and 122. Proprietary brands of polysulphide sealant. (FEB (Great Britain), Manchester).†

Febkol Plastomer 555. A proprietary polysulphide/epoxy sealant and adhesive. (FEB (Great Britain), Manchester).†

Febplate. A proprietary range of epoxy mortars. (FEB (Great Britain), Manchester).†

Febrilix. A proprietary preparation of paracetamol. An analgesic. (Boots Company PLC).*

Febset. A proprietary range of epoxy mortars. (FEB (Great Britain), Manchester).†

Fecap. A proprietary preparation of ferrous fumarate and folic acid. An iron supplement. (Pfizer International).*

Fe-Cap. A proprietary preparation of ferrous glycine sulphate used in the treatment of anaemia caused by deficiency of iron. (MCP Pharmaceuticals).†

Fe-Cap C. A proprietary preparation of ferrous glycine sulphate and ascorbic acid used in the treatment of anaemia. (MCP Pharmaceuticals).†

Fe-Cap Folic. A proprietary preparation of ferrous glycine sulphate and folic acid used in the treatment of anaemia during pregnancy. (MCP Pharmaceuticals).†

Fecraloy. A proprietary trade name for an alloy of iron, chromium, aluminium and yttrium under development for use in sintered form as a catalyst to assist in the control of atmospheric pollution by reducing the emission of carbon monoxide and other fumes from automobile exhausts. (Atomic Energy Establishment, Harwell).†

Feculose. The name given to various commercial starch esters.†

Fedralite. A proprietary trade name for a Vinsol resin-treated laminated paper.†

Fedrazil Tablets. A proprietary formulati of pseudoephedrine and chlorcyclcizin hydrochlorides. Applications: For temporary relief of nasal and sinus congestion associated with colds and h fever. (The Wellcome Foundation Ltd).*

Feedercalc. Micro computer programs to facilitate rapid reliable calculation of optimum feed requirements for all typ of iron and steel castings. (Foseco (FS Ltd).*

Feedex. Mouldable exothermic feeding compound. (Foseco (FS) Ltd).*

Feedol. Feeding compounds for non-ferro metals. (Foseco (FS) Ltd).*

Fefol. A proprietary preparation of ferrou sulphate and folic acid. A haematinic. (Smith Kline and French Laboratories Ltd).*

Fefol Spansule Capsule. Ferrous sulphate and folic acid. Applications: Prophylaxis of iron and folic acid deficiency during pregnancy. (Smith Kline and French Laboratories Ltd).*

Fefol Z Spansule Capsule. Ferrous sulpha folic acid and zinc. Applications: Prophylaxis of iron and folic acid deficiency during pregnancy, when inadequate diet calls for supplementar zinc. (Smith Kline and French Laboratories Ltd).*

Fefol-Vit Spansule Capsule. Ferrous sulphate and folic acid with thiamine mononitrate, riboflavine, pyridoxine hydrochloride, nicotinamide, ascorbic acid. Applications: Prophylaxis of iron and folic acid deficiency during pregnancy particularly when inadequat diet calls for supplementary vitamins B and C. (Smith Kline and French Laboratories Ltd).*

Fehling's Solution. An alkaline solution of potassio-tartrate of copper. It is prepared in two solutions. (*a*) Consistir of copper sulphate, and (*b*) a solution Rochelle salt (potassium sodium tartrate) and caustic soda. Used for th identification and determination of sugars.†

Fehling's Solution, Neutral. A solution made by adding 25 cc of a solution containing 2 grams copper (7.86 grams

$CuSO_4.7H_2O$) per litre to 25 cc of a solution containing 3.292 grams sodium carbonate and 20 grams Rochelle salt per litre. It is stated to be a much more sensitive reagent for the detection of sugars.†

Felamine. A proprietary preparation of cholic acid and hexamine. (Sandoz).†

Felden. A proprietary preparation of piroxicam. An anti-inflammatory, analgesic. (Pfizer International).*

Feldene. A proprietary preparation of piroxicam. An anti-inflammatory, analgesic. (Pfizer International).*

Felixite. An explosive. It is a 42-grain powder, and contains metallic nitrates, nitro-hydrocarbons, and 3 per cent petroleum jelly.†

Fellozine. Promethazine Hydrochloride. Applications: Anti- emetic; antihistaminic. (O'Neal, Jones & Feldman Pharmaceuticals).

Felsinosima. A trade name for cultures of *Bacillus felsineus*. Used for flax retting.†

Felsite. See ORTHOCLASE.

Felspar. (Potassium Felspar, Orthoclase). A potassium-aluminium silicate, ($K_2O.3SiO_2$.) + ($A1_2O_3.3SiO_2$). Used in the manufacture of porcelain, as a building material, and as a fertilizer.†

Felspar, Blue. Chessylite.†

Felspar, Potash. See FELSPAR.

Femergin. A proprietary preparation of ergotamine tartrate, used for migraine. (Sandoz).†

Femerital. A proprietary preparation containing nifuratel. Contains ambucetamide. (MCP Pharmaceuticals).†

Femin-9. Vitamin, mineral and fatty acid preparation. Applications: For premenstrual syndrome. (Marfleet Refining Co).*

Feminal. See GYNOVAL.

Feminone. Ethinyl estradiol. Applications: Oestrogen. (The Upjohn Co).

Feminor Sequential. A proprietary preparation of mestranol (15 pink tablets) and mestranol and norethynodrel (5 white tablets). Oral contraceptive. (LRC International Ltd).†

Feminor 21. A proprietary preparation of mestranol (16 pink tablets) and mestranol and norethynodrel (5 white tablets). Oral contraceptive. (LRC International Ltd).†

Femipausin. A proprietary preparation containing methyltestosterone and ethinylcestradiol. (Hoechst Chemicals (UK) Ltd).†

Femulen. A proprietary preparation of ETHYNODIOL diacetate, used as an oral contraceptive. (Searle, G D & Co Ltd).†

Fenafix. A proprietary trade name for a series of modified vinyl- pyrrolidone resins. Used to modify the properties of other vinyl films and also to improve the adhesion of difficult surfaces. (Fine Dyestuffs & Chemicals Ltd).†

Fenamisal. Phenyl Aminosalicylate.†

Fenasprate. 4-Acetamidophenyl O-acetylsalicylate.†

Fenbid Spansule Capsul. Ibuprofen. Applications: Sustained release anti-inflammatory agent indicated in rheumatoid arthritis and other painful conditions. (Smith Kline and French Laboratories Ltd).*

Fenitrothion EC. Organophosphorus insecticde. (Ciba-Geigy PLC).

Fennite. A fungicide. (Fisons PLC).*

Fenocin. A proprietary preparation of penicillin V potassium. An antibiotic. (Pfizer International).*

Fenocin Forte. A proprietary preparation of penicillin V potassium. An antibiotic. (Pfizer International).*

Fenoil. Range of chemicals for exploration, production and hehanced recovery of oil and natural gas. (Bayer UK Ltd).

Fenolac. Cyclized rubber used in adhesives and bonding agents. THERMOPRENE.†

Fenolite. An Italian synthetic resin material for use in electrical insulation.†

Fenopon AC-78. Anionic surfactant in which the cation is sodium and the anion is a coconut ester of isethionate. Powder form. Applications: Low salt content detergent with good foaming, lathering and dispersing properties, used in detergent bars, dentifrices, shampoos, bubble baths and other cosmetics. (GAF (Gt Britain) Ltd).

Fenopon CD. Ammonium ethoxy sulphate in liquid form. Applications: High foaming surfactant for high electrolyte aqueous systems; air entraining properties used in the concrete industry; foaming agent for light weight cements; frothing agent for gypsum wallboard. (GAF (Gt Britain) Ltd).

Fenopon CN-42. Sodium N-cyclohexyl-N-palmitoyl taurate in paste form. Applications: Low foaming detergent, dispersing agent, stabilizer. Used in mechanical dishwashing detergents; industrial cleaners; synthetic rubber emulsions. (GAF (Gt Britain) Ltd).

Fenopon CO. Sodium or ammonium nonylphenol ethoxy sulphate in liquid form. Applications: High foaming detergent with wetting, dispersing, emulsifying, antistatic and lime soap dispersion properties. Used for scrub soaps; car washes; rug and hair shampoos; vinyl polymerization; petroleum wax; plastics and synthetic fibres. (GAF (Gt Britain) Ltd).

Fenopon EP. Ammonium nonylphenol ethoxy sulphate in liquid form. Applications: Versatile primary emulsifier and stabilizing agent used in emulsion co-polymers. (GAF (Gt . Britain) Ltd).

Fenopon SE. Sodium alkyl-aryl poly-ether sulphonate in liquid form. Applications: Detergent base with good foaming and rinsability, used for cosmetic and pharmaceutical products and emulsion polymerization. (GAF (Gt Britain) Ltd).

Fenopon T-33 and T-43. Sodium N-methyl N-oleyl taurate as a clear liquid or a slurry. Applications: Anionic surfactant used in textiles, rug shampoos, cleaning and detergent formulations. (GAF (Gt Britain) Ltd).

Fenopon T-51. Sodium N-methyl N-oleyl taurate in the form of a readily soluble gel. Applications: Textile processing; latex emulsion stabilizer. (GAF (Gt Britain) Ltd).

Fenopon T-77. Sodium N-methyl N-oleyl taurate in powder form. Applications: Wetting and dispersing agent used in dry blending; industrial and herbicidal wetting powders; textiles; rug shampoos; cleaning and detergent formulations. (GAF (Gt Britain) Ltd).

Fenopon TC-42. Sodium N-coconut acid N-methyl taurate in the form of a slurry. Applications: Chemically stable detergent with foaming, lathering and dispersing properties. Used in detergent bars; shampoos; bubble baths; cosmetics. (GAF (Gt Britain) Ltd).

Fenopon TK32. Sodium N-methyl N-tall oil acid taurate in liquid form. Applications: Dispersing and suspending agent. Used as a precipitation inhibitor for salts of barium, calcium and strontium, and for scale prevention in oil well tubing and flow lines. (GAF (Gt Britain) Ltd).

Fenopon TN-74. Sodium N-methyl N-palmitoyl taurate in powder form. Applications: Dispersing and suspending agent used for dry blending into detergent formulations and in herbicides and insecticides. (GAF (Gt Britain) Ltd).

Fenopron. A proprietary preparation of fenoprofen calcium. An anti-inflammatory drug. (Dista Products Ltd).*

Fenostil. A proprietary preparation of dimethindinehydrogen maleate. (Zyma (UK) Ltd).†

Fenostil Retard. A proprietary preparation of DIMETHINDENE maleate. An anti-pruritic drug. (Zyma (UK) Ltd).†

Fenotec. Ester-phenolic resin binders for cold-set bonding of sand cores. (Foseco (FS) Ltd).*

Fenoval. Phenoval.†

Fenox. A proprietary preparation of phenylephrine hydrochloride. A nasal spray. (Boots Company PLC).*

Fentachol. Tranquillizing and anti-cholinergic preparations. (Allen & Hanbury).†

Fentazin. A proprietary preparation of perphenazine. A sedative. (Allen & Hanbury).†

[F]enton's Metal. A bearing metal. It contains about 80 per cent zinc, 15 per cent tin, and 5 per cent copper.†

[F]enton's Reagent. Hydrogen peroxide and a ferrous salt. Used for the oxidation of polyhydric alcohols.†

[Fe]ntro. Insecticide. (Murphy Chemical Ltd).

[F]eosol. Ferrous sulphate. Applications: Haematinic. (Menley & James Laboratories).

[F]eospan. A proprietary preparation of dried ferrous sulphate. A haematinic. (Smith Kline and French Laboratories Ltd).*

[F]eospan Spansule Capsule. Ferrous sulphate. Applications: Prevention and treatment of iron deficiency. (Smith Kline and French Laboratories Ltd).*

[F]eospan Z Spansule Capsule. Ferrous sulphate and zinc. Applications: Prevention and treatment of iron deficiency and when supplementary zinc is needed. (Smith Kline and French Laboratories Ltd).*

[F]eostat. Ferrous fumarate. Applications: Haematinic. (O'Neal, Jones & Feldman Pharmaceuticals).

[F].E.P. A proprietary name for Teflon 100. (Du Pont (UK) Ltd).†

[F]er-In-Sol. Ferrous sulphate. Applications: Haematinic. (Mead Johnson & Co).

[F]erad. Bismuth and boron additives for malleable cast iron. (Foseco (FS) Ltd).*

[F]eraloy. A nickel-steel-chromium alloy, having a specific gravity of 8.15, and melting at 1480° C.†

[F]eravol. A proprietary preparation of ferrous gluconate and folic acid. A haematinic. (Carlton Laboratories).†

[F]erbelan. Preparations of B Vitamins with iron. (British Drug Houses).†

[F]ergapol. Artificial and synthetic resins. Applications: Adhesives, paints, mastics and sealants. (Ferguson & Menzies Ltd).*

[F]ergatac. Artificial and synthetic resins. Applications: Adhesives, paints, mastics and sealants. (Ferguson & Menzies Ltd).*

[F]ergon. A proprietary preparation of ferrous gluconate. A haematinic. (Bayer & Co).*

Ferlosa. A proprietary range of synthetic pulp for the paper and cement industry. (Montedison UK Ltd).*

Ferlucon. A proprietary preparation of ferrous gluconate and aneurine hydrochloride. A haematinic. (British Drug Houses).†

Fermentation Amyl Alcohol. See FUSEL OIL.

Fermenticide. An antiseptic compound. (Bush Boake Allen).*

Fermet Alloy. An alloy of 74.5 per cent iron, 18 per cent nickel, 2.2 per cent manganese, 0.7 per cent tungsten, 0.3 per cent copper, and 0.35 per cent carbon.†

Fermin. A proprietary preparation of cyancobalamin in an oral form. A haematinic. (The Albion Group).†

Fermine. A proprietary trade name for methyl phthalate.†

Ferna-Col. A fungicide. (ICI PLC).*

Fernambuco Wood. See REDWOODS.

Fernasan. A non-mercurial seed dressing. (ICI PLC).*

Fernasul. A lime and sulphur fungicide. (Plant Protection (Subsidiary of ICI)).†

Fernesta. A selective weed killer. (Plant Protection (Subsidiary of ICI)).†

Fernex. Insecticide containing pirimiphos-ethyl. (ICI PLC).*

Fernico Alloy. See KOVAR ALLOY.

Fernide. A foliage fungicide. (ICI PLC).*

Fernimine. A selective weed killer. (ICI PLC).*

Fernol. Textile flame retardants. Applications: Flame retardant textile finishing. (Yorkshire Chemicals Plc).*

Fernoxone. A selective weed killer. (ICI PLC).*

Fero-Gradumet. Ferrous sulphate. Applications: Haematinic. (Abbott Laboratories).

Ferox-Celotex. A proprietary trade name for Celotex (*qv*) which has to be treated to resist attack by fungi and termites.†

Ferozon. A disinfectant.†

Ferquatac. Artificial and synthetic resins, emulsions of artificial and synthetic resins. Applications: Tackifiers and modifiers for water-based adhesives, coatings, mastics and sealants. (Ferguson & Menzies Ltd).*

Ferrantigorite. Synonym for Ferro antigorite.†

Ferrax. Liquid non-mercurial seed dressing. (ICI PLC).*

Ferri-Darotin. A sodium iron (iii) ethylenediamine tetracetic acid trihydrate and ammonium iron ethylenediamine tetracetic acid composition. Applications: Used as a bleaching bath component for the development of colour pictures, as an agent to act against chlorosos (a sickness resulting from a deficiency in iron), in agriculture and gardening, in hair cosmetics and for organic synthesis. (Degussa).*

Ferrical. A proprietary preparation of ferrous iron, calcium, aneurine hydrochloride, and phenolphthalein. A tonic. (Norton, H N Co).†

Ferrikalite. An artificial potassium ferric sulphate.†

Ferriplex. Iron chelate. (ABM Chemicals Ltd).*

Ferriplus. Iron chelate. (ABM Chemicals Ltd).*

Ferrisul. A trade name for ferric sulphate.†

Ferrite. Nearly pure iron. Phosphorus and sulphur may be present in minute quantities, but the carbon content is not more than 0.05 per cent.†

Ferro-aluminium. See ALUMINIUM-IRON. It contains up to 20 per cent. aluminium with iron, and is used in the preparation of iron and steel.

Ferro-argentan. An alloy resembling silver and containing 70 per cent copper, 20 per cent nickel, 5.5 per cent zinc, and 4.5 per cent cadmium.†

Ferro-boron. An alloy of iron and boron, containing from 20-25 per cent boron. It is added to steel.†

Ferrocap. A proprietary preparation of ferrous fumarate and Vitamin B_1. A haematinic. (Consolidated Chemicals).†

Ferrocap F 350. A proprietary preparation of ferrous fumarate and folic acid used in the treatment of anaemia during pregnancy. (Consolidated Chemicals).

Ferro-carbon-titanium. An alloy of iron a titanium, containing carbon. Used for making steel.†

Ferro-cerium. See AUER METAL.

Ferrochlor. A mixture of ferric chloride a calcium hypochlorite. Used to clarify water.†

Ferro-chrome. See CHROME IRON.

Ferro-chromium. A British Chemical Standard Alloy. Low carbon No. 203 contains 69 per cent and over chromiu 0.08 per cent carbon, and 0.01 per cer sulphur, while the high carbon alloy contains 71.4 per cent and over chromium, 5.09 per cent carbon, and 0.02 per cent sulphur.†

Ferro-cobalt. An alloy of 70 per cent cob with iron. Used for adding cobalt to steel.†

Ferrocobaltite. A mineral. It is a cobaltit containing iron.†

Ferrocontin. Ferrous slycine sulphate in a controlled release tablet. Applications Iron deficiency anaemia. (Napp Laboratories Ltd).*

Ferrocrete. A brand of rapid hardening Portland cement of high strength.†

Ferro-cupralium. An alloy of 75-80.5 per cent copper, 11-12 per cent aluminiur and 2-13 per cent iron.†

Ferro-Cure. Applications: For rubber compounding additives, such as antioxidants, vulcanizing agents and t like in Int Class 1. (Ferro Corporation).*

Ferrodic. A proprietary preparation containing ferrous carbonate and ascorbic acid. A haematinic. (Allen & Hanbury).†

Ferrodur. A substance containing Nitroli (*qv*). It is used for case hardening and tempering iron and steel.†

Ferroferrite. See MAGNETIC IRON ORE.

Ferrogen. An iron degasser, scavenger an deoxidant. (Foseco (FS) Ltd).*

Ferrograd. A proprietary preparation of ferrous sulphate in a slow release form Applications: A haematinic. (Abbott Laboratories).*

Ferrograd C. A proprietary preparation containing ferrous sulphate and ascorbic acid. A haematinic. (Abbott Laboratories).*

Ferrograd Folic. A proprietary preparation of ferrous sulphate and folic acid used in the treatment of anaemia during pregnancy. (Abbott Laboratories).*

Ferro-magnesite. Obtained by burning magnesite with iron ore. It is used as a lining for furnaces.†

Ferro-manganese. An alloy of manganese with iron and carbon, usually made in the blast furnace. High-grade alloys contain about 78 per cent manganese and 8 per cent carbon. These alloys vary from 50-80 per cent manganese, 10-42 per cent iron, 2 per cent silicon, and 5-8 per cent carbon.†

Ferro-molybdenum. An alloy of iron with 80 per cent molybdenum. Used in the place of molybdenum in the manufacture of hard steel.†

Ferromyn. Ferrous succinate compounds. (The Wellcome Foundation Ltd).

Ferron. An alloy of 50 per cent iron, 35 per cent nickel, and 15 per cent chromium. Also a building material prepared from the pickling liquor from steel mills. It consists of precipitated iron oxide.†

Ferro-nickel. (Nickel Iron). An alloy of iron and nickel, usually containing 25 per cent nickel.†

Ferronite. A solid solution of about 0.27 per cent carbon inβ- iron.†

Ferro-phosphorus. An alloy of iron and phosphorus, used in steel making for thin castings.†

Ferrophos Pigment. It is a refractory ferro-alloy developed for use in high performance specialty coatings. Its primary use is in zinc-rich coatings where it works with zinc dust to provide good corrosion resistance and weldability characteristics. Applications: For use in weldable coil coatings and conductive paints for both EMI shielding and antistatic needs. (Occidental Chemical Corp).*

Ferrophytin. Neutral colloidal inositol hexaphosphate of iron containing about 7.5 per cent iron and 6 per cent phosphorus. A tonic and haematopoietic.†

Ferro-silicon. Alloys of iron and silicon made in the arc type electric furnace. They are graded upon the silicon content. The ordinary grades containing 25, 45-50, 75, and 95 per cent silicon are used in steel works. The quality containing 95 per cent silicon generates hydrogen by treatment with boiling water. Also see SILICON.†

Ferro-silicon-aluminium. An alloy of iron, silicon, and aluminium, containing up to 15 per cent silicon.†

Ferrostabil. A stabilized ferrous chloride preparation used in medicine in the form of tablets and suppositories.†

Ferro-titanium. An alloy of iron and titanium, usually containing from 15-18 per cent titanium. It, however, sometimes consists of 23-25 per cent titanium, 70-72 per cent iron, and 5 per cent aluminium. Used as a purifying agent for steels.†

Ferrotubes. An iron degassing and scavenging agent. (Foseco (FS) Ltd).*

Ferro-tungsten. An alloy of iron and tungsten. It usually contains from 65-85 per cent tungsten, and from 1-2 per cent carbon. Used in the steel industry.†

Ferro-uranium. Alloys of iron and uranium, containing from 30-50 per cent uranium. Used in steel making.†

Ferro-vanadium. An alloy of iron and vanadium, containing from 20-40 per cent vanadium. It is added to steel and iron.†

Ferro-vanadium (No. 205). A British Chemical Standard alloy. Vanadium 52.2 per cent. (standard). It also contains carbon 0.16 per cent. sulphur 0.03 per cent. (not standard), silicon 1.18 per cent. phosphorus 0.05 per cent.†

Ferrox. (Ferrite). Trade names for yellow iron oxides used as paint pigments. They consist of 98-99 per cent. $Fe(OH)_3$, with calcium sulphate.†

Ferroxide. Synthetic iron oxide pigments. Ferric oxide. Applications: Surface coatings, plastics and cement colouring. (Mercian Minerals & Colours Ltd).*

Ferroxyl Reagent. A gelatin or agar-agar jelly containing phenolphthalein and potassium ferricyanide. When a piece of iron is placed in the jelly, colours are formed at the ends of the metal after a time. Iron ions give a colour of Turnbull's blue with the potassium ferricyanide, and hydroxylions a pink colour with the phenolphthalein.†

Ferrozell. A proprietary trade name for a synthetic resin.†

Ferro-zirconium. A 20 per cent zirconium alloy with iron. Used to remove nitrogen and oxides from steel.†

Ferrozoid. (Vestalin). Alloys. They are usually 28 per cent nickel steels and are used as electrical resistances.†

Ferrozone. A saccharated iron and vanadium compound.†

Ferrugo. Ferric hydroxide, $Fe(OH)_3$.†

Ferrul. An alloy of 54.6 per cent copper, 40 per cent zinc, 5 per cent lead, and 0.4 per cent aluminium.†

Ferrum. The Latin name for iron.†

Ferrutope. Ferrous citrate Fe 59. Applications: Radioactive agent. (Mallinckrodt Inc).

Ferrux. Feeding compound for ferrous metals. (Foseco (FS) Ltd).*

Ferry Alloy. A trade mark for an electrical resistance alloy of 54 per cent. copper and the balance nickel, a material with a very low temperature coefficient of resistance. (Wiggin Alloys Ltd).†

Ferry Metal. Alloys used for electrical resistance and containing 40-45 per cent nickel and 55-60 per cent copper. The name appears to be also applied to a bearing alloy and solder containing lead with 2 per cent barium, 1 per cent copper, and 0.25 per cent mercury.†

Ferrybar. A proprietary preparation of iron and ammonium citrate, riboflavine, nicotinamide and aneurine hydrochloride, used to prevent iron deficiency anaemia. (Rybar).†

Fersaday. A proprietary preparation of ferrous fumarate. A haematinic. (Glaxo Pharmaceuticals Ltd).†

Fersamal. A proprietary preparation of ferrous fumarate. A haematinic. (Glaxo Pharmaceuticals Ltd).†

Fersolate. A proprietary preparation of ferrous, manganese and copper sulphates. A haematinic. (Glaxo Pharmaceuticals Ltd).†

Fesovit. A proprietary preparation of ferrous sulphate, ascorbic acid and B vitamins used in the treatment of anaemia. (Smith Kline and French Laboratories Ltd).*

Fesovit Spansule Capsule. Ferrous sulphate, thiamine mononitrate, riboflavine, pyridoxine hydrochloride, nicotinamide and ascorbic acid . Applications: Haematinic with added vitamins for the prevention and treatment of iron deficiency, particularly in the elderly patient where inadequate diet calls for supplementary vitamins B and C. (Smith Kline and French Laboratories Ltd).*

Fesovit Z Spansule Capsule. Ferrous sulphate, thiamine mononitrate, riboflavine, pyridoxine hydrochloride, nicotinamide, ascorbic acid and zinc. Applications: Oral iron and zinc preparation, with added vitamins, for the prevention of iron deficiency when inadequate diet calls for supplementary zinc and vitamins B and C. (Smith Kline and French Laboratories Ltd).*

Festoform. A solid preparation of formaldehyde, obtained by mixing an aqueous solution of formaldehyde with a soda soap solution. An antiseptic disinfectant and deodorizer.†

Fettel. Herbicide, containing triclopyr, dicamba and mecoprop. (ICI PLC).*

Feuille Morte. A jeweller's alloy, containing 70 per cent gold and 30 per cent silver.†

Fevarin 50. Fluvoxamine. Applications: Treatment of mental and physical complaints caused by disturbed moods and humours. (Duphar BV).*

Feximac. A proprietary preparation of Buteximac in a cream base, used in the treatment of eczema. (Nicholas).†

Fiberlac. A proprietary trade name for cellulose nitrate lacquer.†

Fiberloid. See VISCOLOID. A proprietary trade name for a cellulose nitrate plastic resistant to oils.

Fiberlon. A proprietary trade name for a phenol-formaldehyde synthetic resin resistant to oils.†

Fiberod. Long, continuous, parallel reinforcing fibres set in thermoplastic resin systems. It is available in rod, bar, tape and bidirectional fabric foam. Applications: Thermoplastic moulding compounds for compression, transfer and injection moulding. Application areas include automotive, appliances, equipment, sporting and others. (Polymer Composites Inc).*

Fiberoid. An American grade of vulcanized fibre used for electrical insulation. It is also the name for a celluloid.†

Fiber Pare. A short length olefin fibre. Applications: Asphalt reinforcement for highway paving, patching, seal coating, crack sealing, curb mix designs. (Hercules Inc).*

Fibervorm. Wheat-fibre derivative. Applications: Treatment of complaints arising from inflation of the small intestine. (Duphar BV).*

Fibestos. A proprietary trade name for a cellulose acetate plastic.†

Fibral. Polishing pads and lenses. (Carl Freudenberg).*

Fibralda. A proprietary trade name for untwisted cellulose acetate fibres.†

Fibravorma. Wheat-fibre derivative. Applications: Treatment of complaints arising from inflation of the small intestine. (Duphar BV).*

Fibre B. See KEVLAR 49.

Fibre, Egyptian. See VULCANISED FIBRE.

Fibre, Grey. See VULCANISED FIBRE.

Fibre, Horn. See VULCANISED FIBRE.

Fibre, Red. See VULCANISED FIBRE.

Fibrestos. A silica-asbestos product used as an insulator.†

Fibre, Vegetable. See VULCANISED FIBRE.

Fibre, Whalebone,. See VULCANISED FIBRE.

Fibrino-plastic Substance. See GLOBULIN.

Fibrinoplastin. See GLOBULIN.

Fibro. A proprietary artificial silk product. It is a staple fibre (*qv*) made by the viscose process.†

Fibroc. A laminated product consisting of fibre impregnated with a phenol-formaldehyde resin.†

Fibron. A trade mark for a surfacing material for resurfacing and treatment of floors resulting in a plastic finish.†

Fibrotex. A proprietary trade name for a roofing cement consisting of asbestos mixed with oil and gum.†

Fibrox. A silicon oxycarbide. It is a thermal insulator.†

Fibrredux. Fibre reinforced laminating resins and adhesives. (Ciba-Geigy PLC).

Ficam. Public health insecticide. (FBC Ltd).

Ficel. Blowing agents. (Fisons PLC).*

Fi-Chlor. Chlorcyanurate used for rendering wool shrink-resistant. (Fisons PLC).*

Fi-Clor. Chlorinated cyanuric acid. (FBC Ltd).

Fichtelite. A hydrocarbon, $C_{18}H_{32}$, found in fossil coniferous resins.†

Ficoid. A range of proprietary skin creams containing fluocortolone used in the treatment of eczema and related dermatoses. (Fisons PLC).*

Ficortril. A proprietary preparation of hydrocortisone acetate. A cortico-steroid. (Pfizer International).*

Ficote. Range of coated fertilizers. (Fisons PLC, Horticulture Div).

Fi-Cryl. Acrylic copolymer solutions and dispersions. (Fisons PLC).*

Fiddle Gum. Gum tragacanth.†

Field's Orange Vermilion. See SCARLET VERMILION.

Fierroso. See ACERADO.

Fi-Gard. Rubber and plastics additives. (Fisons PLC).*

Filamid. Polymer-soluble dyes for spin colouration of polyamide. (Ciba-Geigy PLC).

Filastic. A proprietary rubber textile yarn in which the rubber latex impregnation takes place during spinning. The yarns contain 50 per cent of rubber.†

Filcryl. A proprietary acrylic polymer for dental fillings.†

File Bronze. An alloy of 64.4 per cent copper, 18 per cent tin, 10 per cent zinc, and 7.6 per cent lead.†

Filester. Dyes for PET bottles. (Ciba-Geigy PLC).

Filex. An aqueous concentrate containing propamocarb hydrochloride. Applications: A protective fungicide for use on all ornamentals and some edible crops against *Pythium*, *Peronospora* and *Phytophthora*. (Fisons PLC).*

Filicic Acid. An acid obtained from male fern. It is a vermicide.†

Fi-Line. Dyeline chemicals for photocopying. (Fisons PLC).*

Filite. A smokeless explosive. It is Ballistite (*qv*), drawn out into cords with the aid of a solvent.†

Fillite. A proprietary inert silicate in the form of sphenes. Hard as glass, it is used as a filling material for plastics. (Fillite (Runcorn))†

Fillite Hollow Microspheres. Free flowing hollow alumina silica microspheres. Applications: Concretes and various refractory cements, ceramic fillers, thermoplastics, undercoating filler material and in moulded compounds. (Fillite USA Inc).*

Fillite Solid Microspheres - PFA. Solid alumina silica microspheres. Applications: Concretes and various refractory cements, paints and pigment industry, ceramic fillers and in other moulded compounds. (Horn's Crop Service Center).*

Fillpak. Polyurethane foams. Applications: Sealing, filler. (James Briggs & Sons Ltd).*

Filmarone Oil. A 10 per cent solution of filmarone (present in fern roots) in castor oil. Used as a specific against worms.†

Filmite. White oil preparations. (Murphy Chemical Co).†

Filofin. Pigment dispersions for polypropylene fibres. (Ciba-Geigy PLC).

Filon. Glass fibre reinforced polyester (GRP) sheeting. Applications: Roofing and cladding. (BIP Chemicals Ltd).*

Filpro. Protein hydrolysates. Applications: Any savoury application - soups, frozen meals, snacks, convenience meals, meat pies and sausages. (PPF International Ltd).*

Filt-char. A proprietary brand of bone charcoal, used as a filtering medium.†

Filter-cel. A proprietary preparation of infusorial earth, used in filtering.†

Filtram. Laminated filter drains. (ICI PLC).*

Filtrez. Rosin resin. Applications: Used in printing inks, coatings and adhesives. (Monsanto Co).*

Filtrol. A trade mark for a decolourizing substance consisting of fine silica with a little aluminium silicate.†

Filtros. Gum rosin. Applications: Used in paints, varnishes and lacquers. (Monsanto Co).*

Filtrosol'A'. Homosalate. Applications: Ultraviolet screen. (Norda Inc).

Finalgon. A proprietary preparation of nonylic acid vanillymide and butoxy-ethyl nicotinate. (Boehringer Ingelheim Ltd).†

Finaplix. Trenbolone Acetate. Applications: Anabolic (veterinary). (Roussel-UCLAF).

Fine Black. See ANILINE BLACK.

Fine Blue. See ANILINE BLUE, SPIRIT SOLUBLE.

Fine Gold. A jeweller's alloy, containing 75 per cent gold and 25 per cent silver.†

Fine Gold Solder. See GOLD SOLDERS.

Fine Silver. 99.9 per cent pure silver.†

Fine Solder. See PLUMBER'S SOLDER

Finesse. Sulphonyl weedkiller. (Du Pont (UK) Ltd).

Finestol. A phloroglucinol dye coupler. (Fisons PLC).*

Finings. The term applied to isinglass dissolved in an acid such as tartaric acid. Used to clarify beer.†

Finizym. A fungal beta-glucanase preparation produced by submerged fermentation of a selected strain of Aspergillus niger. Applications: Used during fermentation and storage of the beer to prevent filtration difficulties and to prevent precipitation of beta-glucans. (Novo Industri A/S).*

nnish Turpentine. See TURPENTINE.

nntitan. Titanium dioxide. Applications: Paints, inks, rubbers, cosmetics, ceramics etc. (Cornelius Chemical Co Ltd).*

ntex 572. Quaternary ammonium compound in liquid form. Applications: Cationic surfactant used in fabric conditioning. (Berol Kemi (UK) Ltd).

iolax. A trade name for a resistance glass.†

ire-armour. Heat-resisting alloys containing 60-61 per cent nickel, 18-20 per cent chromium, 10-20 per cent iron, 0-1.8 per cent manganese, and 0.5 per cent carbon.†

irebrake. A fire retardant. (Borax Consolidated Ltd).*

irecheck. Water-based elastomeric fire retardant protective coating, applied by brush or spray. Applications: Asbestos encapsulation, fire retardant weatherproofing, combustable substrates. (Liquid Plastics Ltd).*

ireclay. Clay containing a considerable amount of free silica.†

irecol. An alginate fire-fighting suspension. (Alginate Industries Ltd).†

irecrete. A proprietary trade name for a calcined high alumina clay used as a refractory in furnaces.†

ire-damp. A gas, mainly consisting of methane, often found in coal mines.†

ireproof Cyanite. A fireproof paint. It is a mixture of aluminium and sodium silicates.†

iresaife. Ammonium phosphate fire retardant. (Scottish Agricultural Industries PLC).

irit. A foundry refractory coating. (Foseco (FS) Ltd).*

irmadent. Denture adhesive (powder, cream and liquid), for securing dentures. (Richardson-Vicks Inc).*

irnagral. A mineral drying oil. Used to replace up to 30 per cent of the linseed oil in putty. It is stated to give the putty a harder and smoother surface.†

irnis. Linseed oil and driers.†

First Choice Electroless Palladium. Electroless plating solution for the autocatalytic chemical deposition of palladium metal. Applications: Plating of electronic components for its low contact resistance and high wear properties and for the metalization of ceramics, plastics and other non-metallics. (Callery Chemical Company).*

Firthite. A proprietary trade name for a material consisting of a mixture of tungsten and other carbides.†

Fir Wool Oil. The oil of Scotch fir leaves.†

Fischer-Langbein Solution. A solution containing 450 grams ferrous chloride, 500 grams calcium chloride, and 750 cc water. Used for the electro-deposition of iron, with a current density of up to 120 amps. per sq ft. A temperature of 60-70° C. is employed.†

Fischer's Reagent. A test solution for sugars. It consists of 2 parts phenyl-hydrazine hydrochloride and 3 parts sodium acetate in 20 parts of water.†

Fischer's Salt. Potassium cobaltic nitrite, $K_3Co(NO_2)_6$. Used for the detection and determination of potassium.†

Fischer's Yellow. See FISHER'S SALT.

Fisetin. Tetrahydroxymethyl-anthraquinone, $C_{15}H_{10}O_6$. A yellow colouring from the wood of *Quebracho colorado*, etc.†

Fish Berry. The fruit of *Cocculus indicus* and *Anamirata paniculata*. It acts as a powerful fish poison. The active principle is picrotoxin, $C_{30}H_{34}0_{13}$.†

Fish Gelatin. See ISINGLASS.

Fisons MCPB. A selective weedkiller. (Fisons PLC).*

Fisons P.C.P. A weedkiller. (Fisons PLC).*

Fisons 18-15. A selective weedkiller. (Fisons PLC).*

Fi-Vi. A lightweight expanded PVC. (Fisons PLC).*

Fixanal. Analytical chemicals accurately weighed and sealed, ready for rapid volumetric solution.†

Fixaplus. X-ray fixer. (May & Baker Ltd).

Fixat. Inorganic moulding sand binders for foundries, also in connection with lustrous carbon formers to obtain better castings. (Süd-Chemie AG).*

Fixatek. X-ray fixer. (May & Baker Ltd).

Fixed White. Commercial barium sulphate, $BaSO_4$.†

Fixegal. Dispersing and levelling agent. Applications: Used for dyeing in the textile industry (Degussa).*

Fixin. Aluminium lactate. †

Fixinvar. An alloy having the same properties as Elinvar, but having greater stability.†

Fixodent. Denture adhesive cream for securing dentures. (Richardson-Vicks Inc).*

Fixogene. Textile auxiliary chemicals. (ICI PLC).*

Fixol. Adhesives. (Associated Adhesives).†

Fixopone. See OIL WHITE.

Fix-Sol. A concentrated fixing and hardening solution. (Johnsons of Hendon).†

Flagyl. A proprietary preparation of metronidazole. Applications: Treatment of anaerobic and protozoal infections. (May & Baker Ltd).*

Flagyl Compak. A proprietary preparation comprising metronidazole, taken orally, and vaginal pessaries containing nystatin. Used in the treatment of vaginitis. (May & Baker Ltd).*

Flagyl IV. Metronidazole hydrochloride. Applications: Antibacterial. (G D Searle & Co).

Flake Lead. See WHITE LEAD.

Flake Litharge. Litharge made by the oxidation of lead.†

Flake White. (Cremnitz, Kremnitz, Crems White, Blanc D'argent, Silver White, London White, Nottingham White, Cremnitz White). Lead whites. They are all carbonates of lead, and contain varying quantities of hydrated oxide of lead. Flake white is a variety of chamber white lead, obtained in flaky pieces by heating lead plates. A basic bismuth is also known as Flake white. The term Flake or Pearl white is sometimes used for oxychloride of bismuth.†

Flamarret. Magnesium oxide for fire protection. (Steetley Refractories Ltd).

Flamco. Mould and core dressings. (Fose (FS) Ltd).*

Flamenol. A proprietary trade name for a polyvinyl chloride synthetic resin.†

Flaming. A decolourizing agent for sugar juices.†

Flammacerium. Silver cerium nitrate. Applications: Treatment of serious burns. (Duphar BV).*

Flammastik. Incombustible cable coating applied by spray guns or spatula. (Degussa).*

Flammazine. Silver sulphadiazine. Applications: Treatment of burns. (Duphar BV).*

Flammex. Flame retardant. (The Wellco Foundation Ltd).

Flammocite. A safety explosive. It contai 44 per cent ammonium nitrate, 16 per cent sodium chloride, 14 per cent sodi nitrate, 10 per cent trinitro-toluene, 6 per cent nitro-glycerin, 5 per cent ammonium sulphate, and 5 per cent cellulose.†

Flandrac. A decolourizing agent for suga juices.†

Flar. A proprietary preparation containin a lactic ferment resistant to antibiotics thiamine, riboflavine, pyridoxine, cyanocobalamin, nicotinamide, sodium pantothenate, inositol, folic acid and liver extract. It is used in the treatmen of antibiotic side-effects. (Consolidatec Chemicals).†

Flat-Ayd. Dispersed flatting agents. Applications: Paints, inks etc. (Cornelius Chemical Co Ltd).*

Flavaniline. A quinoline dyestuff. It is 2-*p* aminophenyl-4-methylquinoline, $C_{16}H_{14}N_2$.†

Flavaniline S. A sulphonated flavaniline.†

Flavanthrene. (Indanthrene Yellow). A dyestuff, $C_{28}H_{14}N_2O_2$, obtained by the oxidation of β-aminoanthraquinone. Th leuco-compound dyes cotton blue, becoming yellow upon air oxidation.†

Flavaurine. A yellow dyestuff allied to Alizarin yellow.†

Flavaxin. Riboflavin. Applications: Vitamin. (Sterling Drug Inc).

Flavazine S. (Kiton Yellow, Pyrazine Yellow S). A dyestuff similar to

Tartrazine. Employed on wool, in a bath containing sodium hydrogen sulphate.†

Flavazol. A dyestuff, $C_{13}H_{12}N_2O_2$, prepared from *p*-toluidine and salicylic acid. It dyes yellow on chrome mordanted wool.†

Flavelix. A proprietary preparation containing mepyramine maleate, ephedrine hydrochloride, ammonium chloride and sodium citrate. A cough linctus. (Pharmax).†

Flaveosine. A dyestuff obtained by condensing *m*-acetaminodimethyl-aniline with phthalic anhydride. It dyes tannined cotton and wool reddish yellow, and silk golden yellow.†

Flavinduline. (Induline Yellow). A dyestuff. Dyes tannined cotton yellow.†

Flavine. Three materials are known by this name: (*a*) Diaminobenzophenone; (*b*) a grade of quercitron bark extract; and (*c*) diaminomethylacridinium chloride.†

Flavocents. Concentrated flavour compositions. (PPF International Ltd).*

Flavoline. 2-Phenyl-4-methylquinoline.†

Flav-O-Lok. Free-flowing non-hygroscopic flavour powders. Applications: Used in formulations in food, beverages and pharmaceuticals where liquid flavours cause difficulties. (Hercules Inc).*

Flavomycin. Bambermycins. Antibiotic complex, containing mainly moenomycin A and C. Applications: Antibacterial. (Hoechst-Roussel Pharmaceuticals Inc).

Flavone. 2-Phenylbenzopyran-4-one†

Flavopurpurin. (Alizarin No. 10CA, Alizarin FA, Alizarin GB, Alizarin GI, Alizarin RG, Alizarin SDG, Alizarin X, Alizarin VCA, Alizarin CAF, DCA, JCA, VAR) Trihydroxy-anthraquinone, $C_{14}H_8O_5$. Dyes cotton mordanted with alumina, red.†

Flavotint. Food colour and flavour compounds. (PPF International Ltd).*

Flaxedil. A proprietary preparation of gallamine triethiodide. A muscle relaxant. (May & Baker Ltd).*

Flax, Mountain. See AMIANTHUS.

Flax Seed. Linseed.†

Flax Wax. A wax associated with flax fibre and with the cortical tissues. The air-dried cortex contains as much as 10 per cent by weight of wax. It is removed by extraction with a volatile solvent.†

Flea-B-Gon. Flea killer. (Chevron).*

Flea Flip. A line of flea control products. Applications: Insecticide for household, pet, outdoor use in the control of fleas, ticks and other insects. Includes aerosols, concentrates and ready to use liquids. (Colonial Products Inc).*

Flectol H. Polymerized 1,2-dihydro-2,2,4-trimethyl-quinoline. Antioxidant. Applications: Resists effect of heat deterioration and normal ageing, for dry rubber and latex. (Monsanto Co).*

Flectol ODP. Octylated diphenylamine, antioxodant. Applications: Protects rubber against heat deterioration and normal ageing; for neoprene. (Monsanto Co).*

Flectol Pastilles. Polymerized 1,2-dihydro-2,2,4-trimethyl-quinoline, antioxidant. Applications: Used in tyres, belts, hose retread rubber and general mechanicals. (Monsanto Co).*

Flerhenol. See AVIVAN.

Fletcher's Alloy. An alloy of 95.5 per cent aluminium, 3 per cent copper, 1 per cent tin, 0.5 per cent antimony, and 0.5 per cent phosphor tin.†

Fletcher's Bearing Alloys. Aluminium base alloys. One contains 92 per cent aluminium, 7.5 per cent copper, and O.25 per cent tin, and another 90 per cent aluminium, 7 per cent copper, and 1 per cent zinc.†

Fleur. Intensive decorating colours with flux mask for enamel and earthenware. (Degussa).*

Flex Carbon. A proprietary carbon black.†

Flex-Shield. Elastomeric coatings made up of asphalt, neoprene, acrylic and kraton rubber imbedded in polyester fabric. Applications: Roof coating, maintenance cleaning products, degreaser, waxes and wall coatings. (Flex-Shield).*

Flexade Regular. Lubricant, corrosion inhibitor; easily soluble in water; biodegradable. Applications: Metal treatment. (Carboxyl Chemicals Ltd).

Flexalyn. A proprietary trade name for a plasticizer. It is diethylene glycol diabietate.†

Flexamine. A superflexing antioxidant containing 35% JZF. For use in heavy service truck tread and SBR treads, camelback, wire insulation, neoprene belting and moulded soles. Offers protection against copper and manganese. (Uniroyal).*

Flexane. Two part urethane. Applications: Making flexible moulds, cast parts and non-marring holding fixtures. Forming abrasion resistant and noise reduction linings. Encapsulating parts. (Devcon Corporation).*

Flexcote. Rubber/vinyl lacquer. Applications: Decoration and protection of rubber articles. (W J Ruscoe Co).*

Flexcrete. Polymer- modified cementitious mortars (range) for repair and maintenance of (reinforced) concrete. Applications: High-rise housing, PRC housing, any damaged or defective concrete structures. (Liquid Plastics Ltd).*

Flexeril. Cyclobenzaprine hydrochloride. Applications: A muscle relaxant and tranquillizer. (Merck Sharp & Dohme).*

Flexin. Selected compound of silicic acid, cationic. Applications: Anti-slip agent for fabrics and knitted goods. (Henkel Chemicals Ltd).

Flexite. A reclaimed rubber.†

Flexocel. Twin component flexible polyurethane foam systems. (Baxenden Chemical Co Ltd).*

Flexol. A proprietary trade name for vinyl plasticisers.†

Flexol Plasticizer 3GH. A proprietary trade name for a plasticizer. It is triethylene glycol-di-2-ethyl butyrate.†

Flexol Plasticizer 3GO. A proprietary trade name for a plasticizer. It is triethylene glycol-di-2-ethyl hexoate.†

Flexonyl. Pigments for acqueous flexographic inks. (Hoechst UK Ltd).

Flexoresin. Proprietary brands of glycol and glyceryl phthalates. Also polymerized terpenes.†

Flexsol 43. Deltafilcon A. Applications: Contact lens material. (Alcon Laboratories Inc).

Flextron. A non toxic, non biodegradable component, cold curing polyurethane sealant. Applications: Expansion and construction joints in water retaining structures. Sold in 3 viscosity grades, suitable for pouring or pressure gunni (Berger Elastomers).*

Flexzone. A range of antiozonants which offer high protection against flexing, ozone, heat and oxygen in rubber products. (Uniroyal).*

Flexzone 3C. A proprietary anti-oxidant. is N-isopropyl-N'-phenyl- p-phenylene diamine. (Naugatuck (US Rubber)).†

Flexzone 6-H. A proprietary anti-oxidant It is N-phenyl-N'-cyclohexyl- p-phenylene diamine. (Naugatuck (US Rubber)).†

Flint. A form of silica, SiO_2.†

Flint Alloy. A heat and corrosion-resistin alloy containing 83 per cent iron, 12.5 per cent chromium, 3 per cent carbon and 0.5 per cent silicon.†

Flintcast. A white iron made in the electr furnace. It resists abrasion.†

Flint Glass. (Lead Glass, Cristal). A glas composed of lead and potassium silicates. Used for hollow-ware, superi bottles, and optical work. See POTAS LEAD GLASS.†

Flint Metal. A proprietary trade name fo an alloy of iron with 4-4.5 per cent nickel, 1.25-1.75 per cent chromium, and 3-3.5 per cent carbon.†

Flint SSD. Silver Sulfadiazine. Applications: Anti-infective, topical. (Flint Laboratories, Div of Travenol Laboratories Inc).

Flinty Zinc Ore. Flinty calamine.†

Fliselina. Non-woven textile for the appa industry. (Carl Freudenberg).*

Flit. A proprietary insecticide.†

Flixapret. Synthetic resins for textile finishing. (BASF United Kingdom Ltd

Float Tin. Cassiterite, occurring in the so and formed by the disintegration of ti rocks, is called float tin.†

Floatstone. Porous opal.†

ochel. Poly-electrolyte flocculating agent. (Steetley Chemicals Ltd).

o-Cillin. Penicillin G Procaine. Applications: Antibacterial. (Bristol-Myers Co).

ocklok. Urethane moisture curing and catalyzed curing flock adhesives. Applications: Bonding of flock to SBR and EPDM cured rubber for mainly channelling applications in the car industry. (Lord Corporation (UK) Ltd).*

oclean 103. A proprietary preparation of polycarboxylic, alkyl sulphonic and organic acids. Applications: For removal of metallic salts from cellulosic reverse osmosis membranes. (Pfizer International).*

oclean 106. A proprietary preparation of polycarboxylic, alkyl sulphonic and organic acids. Applications: For removal of organics, silt and other particulate matter from cellulosis reverse osmosis membranes. (Pfizer International).*

loclean 107. A proprietary preparation of polycarboxylic, alkyl sulphonic and organic acids. Applications: For removal of organics, silt and other particulates from cellulosic reverse osmosis membranes. (Pfizer International).*

loclean 108. A proprietary preparation of polycarboxylic, alkyl sulphonic and organic acids. Applications: For removal of organics, silt and other particulates from cellulosic reverse osmosis membranes. (Pfizer International).*

loclean 303. A proprietary preparation of phosphates and chelating agents. Applications: For removal of metallic inorganics from cellulosic reverse osmosis mebranes. (Pfizer International).*

loclean 307. A proprietary preparation of phosphates and chelating agents. Applications: For removal of organics and silt from cellulosic reverse osmosis membranes. (Pfizer International).*

'lo-Con. Textile chemical. (Albright & Wilson Ltd, Phosphates Div).

Flocon 100 - Antiscalent. 35% aqueous solution of a proprietary polyacrylate polymer. Applications: As a scale control agent in desalination. (Pfizer International).*

Floex. A proprietary trade name for a wetting agent for paint, etc. It is a condensation product of higher fatty alcohols.†

Flogel. Slurry explosives. Applications: For a wide range of surface blasting. (Hercules Inc).*

Flolan. Epoprostenol sodium. Applications: Inhibitor. (Burroughs Wellcome Co).

Flomac. Granular desulphurizing agent for blast furnace iron and steel. (Foseco (FS) Ltd).*

Flo-Mo DEL, Flo-Mo DEH. Formulated products. Applications: Matched pair of emulsifiers for a broad range of pesticides. (DeSoto Inc).*

Flo-Mo Lowfoam. Modified alcohol ethoxylate. Applications: Surfactant for the formulations of agricultural adjuvants. (DeSoto Inc).*

Flo-Mo Suspend. Formulated product. Applications: Compatibility agent for use in agricultural formulations. (DeSoto Inc).*

Flo-Mo 1082. Formulated product. Applications: Emulsifier for petroleum based agricultural sprays. (DeSoto Inc).*

Flo-Mo 1093. Formulated product. Applications: Emulsifier for vegetable oil based agricultural sprays. (DeSoto Inc).*

Flo-Mo 5BMP. Free acid of a complex organic phosphate ester. Applications: Emulsifier for phosphated and chlorinated pesticides. Coupling agent for agricultural and industrial applications. (DeSoto Inc).*

Flo-Mo 80/20. Modified alcohol ethoxylate. Applications: Agricultural adjuvant. (DeSoto Inc).*

Florafoam. Rigid foam (urethane and phenolic) for floral arrangements. (Baxenden Chemical Co Ltd).*

Floranit. Wetting agent. Applications: mercerizing and caustic lye padding. (Henkel Chemicals Ltd).

Floraquin. A proprietary preparation of di-iodohydroxyquinoline. Vaginal antiseptic tablets. (Searle, G D & Co Ltd).†

Floratex. Melded fabrics. (ICI PLC).*

Florence Lake. (Vienna Lake, Paris Lake). Lakes produced from cochineal, by precipitating alkaline solutions of cochineal with alum, or with a mixture of alum and tin salts.†

Florentine Brown. (Vandyck Red, Hatchette Brown). A pigment. It is copper ferrocyanide.†

Floricin. (Florizine, Derizine, Dericin). A substance produced by heating castor oil to 300° C., and distilling 10 per cent of it. There remains the product termed "floricin", which solidifies at -20° C. It is also made by heating castor oil with formaldehyde. It is miscible with ceresin and petroleum jelly and is used as a vehicle for menthol and oil of eucalyptus. Unlike castor oil, it is insoluble in alcohol.†

Florida Earth. See BLEACHING EARTH.

Florida Phosphates. Mineral phosphates. There are two types, hard rock phosphates, containing 80 per cent calcium phosphate, and soft clay phosphates, containing 40-60 per cent calcium phosphate. Fertilizers.†

Floridin. A form of fuller's earth found in Florida, U.S.A. It removes the colouring matter from oils and waxes, and is used for this purpose.†

Florinef. A proprietary preparation of fludrocortisone 21-acetate. (Squibb, E R & Sons Ltd).*

Florinef Acetate. Fludrocortisone acetate. Applications: Adrenocortical steroid. (E R Squibb & Sons Inc).

Florisil. Chromatography absorbent. (Hermadex Ltd).*

Florite. A proprietary trade name for a carefully prepared and screened bauxite.†

Florizine. See FLORICIN.

Florone. Diflorasone diacetate. Applications: Anti-inflammatory. (The Upjohn Co).

Floropryl. Isoflurophate. Applications: Cholinergic (Merck Sharp & Dohme Div of Merck & Co Inc).

Florosal. Concrete additive. (Steetley Chemicals Ltd).

Florox. Benzoyl peroxide dispersion in fo grade filler. Applications: Flour additive. (Diaflex Ltd).*

Flor Sherry. Dry yeast for wine producti (Ciba-Geigy PLC).

Flosol. A proprietary trade name for colloidal barium silico-fluoride for horticultural purposes.†

Floss, Akund. See AKUND.

Floto. See PINE OILS.

Flotox. Sulphur pesticide. (Chevron).*

Flour of Sulphur. Powdered sulphur. It h been powdered by grinding, but is not so finely powdered as flowers of sulphur.†

Flour, Fossil. See INFUSORIAL EART

Flour, Mountain. Infusorial earth.†

Flovan. Proofing agents. (Ciba-Geigy PLC).

Flowers of Antimony. Formed when antimony burns in air. It consists of antimony oxide, Sb_4O_6. Used in medicinal preparations, and as a whit pigment.†

Flowers of Benjamin. (Flowers of Benzoi Benzoic acid, C_6H_5COOH.†

Flowers of Benzoin. See FLOWERS OF BENJAMIN.

Flowers of Bismuth. Bismuth oxide, Bi_2O obtained by burning bismuth metal at red heat.†

Flowers of Brimstone. See FLOWERS O SULPHUR.

Flowers of Camphor. Camphor in crystalline form.†

Flowers of Copper. The oxide of copper, CuO, produced when copper burns.†

Flowers of Sulphur. (Flowers of brimston Consists of minute crystals of sulphur, obtained by chilling sulphur vapour.†

Flowers of Tin. Stannic oxide, SnO_2. A polishing powder.†

Flowers of Zinc. See PHILOSOPHER'S WOOL.

Flow Blue. See FLOW-POWDER.

wfusor. (Urological and topical irrigation solutions. (The Boots Company PLC).

w-powder. A mixture of white lead and salt which gives off chlorine on heating. It is used for the production of flow blues (cobalt blues) on ceramics. Cobalt chloride is formed, which, being volatile, gives blues of varying intensity.†

ox. A fibrous material made from cellulose sulphate pulp. It is mixed with cotton, wool, etc., and woven. It is an art wool.†

oxapen. Floxacillin. Applications: Antibacterial. (Beecham Research Laboratories).

oxifral. Fluvoxamine. Applications: Treatment of mental and physical complaints caused by disturbed moods and humours. (Duphar BV).*

uates. This term is used for fluosilicates. It is also the name for waterproofing compounds consisting of solutions of sodium silcate, or silicofluoride, and other silicofluorides, such as those of zinc, magnesium, and aluminium.†

uax. Anti-influenza vaccine. Applications: For protection against influenza. (Merck Sharp & Dohme).*

ubenol. A proprietary preparation containing flubendazole. Applications: Veterinary antihelmintic. (Janssen Pharmaceutical Ltd).*

ucsil. Activated silica. (Crosfield Chemicals).*

uderma. Formocortal. Applications: Glucocorticoid. (Farmitalia (Farmaceutici Italia)).

udor Solder. An aluminium solder containing 56.5 per cent tin, 40 per cent zinc, 3 per cent lead, 0.2 per cent antimony, and 0.1 per cent copper.†

ugène 113. A proprietary non-inflammable solvent of low toxicity for cleaning precision equipment. It is trifluorotrichloroethane. (Rhone-Poulenc NV/CdF Chimie AZF).*

uidil. Cyclothiazide. Applications: Diuretic; antihypertensive. (Adria Laboratories Inc).

uidiram. Amines and derivatives. Primary, secondary, tertiary fatty mono-, di- and polyamines from C8 to C22. Quaternary ammonium salts. Amine oxides. Amine salts. Ethoxylated amines and polyamines. Amino-acids. Special nonionic derivatives. Additives. Applications: Surface active agents used as auxiliaries in road, mining, textile and fertilizer industries, in metallurgy, as lubricating agents, additives for fuel oils and soil stabilization. (British Ceca Co Ltd).*

Fluilan. Liquid lanolin. (Croda Chemicals Ltd).

Fluinlan. Liquid lanolin. (Croda Chemicals Ltd).*

Fluisil S55K. A low temperature silicone lubricant. Applications: For refrigeration machinery. (Bayer & Co).*

Fluisol. A range of sulphated castor oils. Applications: Various applications in many industries. (Yorkshire Chemicals Plc).*

Fluitex. Thinned (or fluidised) starches. Applications: Textiles, to strengthen and finish. (Roquette (UK) Ltd).*

Flukiver. A proprietary preparation containing closantel. Applications: Veterinary flukicide. (Janssen Pharmaceutical Ltd).*

Fluoderm. A proprietary preparation of fluorometholone, clioquinol and chlorphenesin for dermatological use. (British Drug Houses).†

Fluogen. Influenza virus vaccine. Applications: immunizing agent. (Parke-Davis, Div of Warner-Lambert Co).

Fluolite. Textile auxiliaries. (ICI PLC).*

Fluon. A trade mark for polytetra-fluoroethylene (P.T.F.E.). A hard plastics material with a very water repellant surface which has a working temperature range from -200 to +280° C. It has a low high frequency power factor and permittivity and outstanding non-stick properties. It is used in heat resistant glands, packings and bearings. See also TEFLON. (ICI PLC).*

Fluon VG 15. A proprietary PTFE powder loaded as to 15% with glass.†

Fluon VX2. A bronze-filled PTFE used as a bearing material.†

Fluon VX3. A lead oxide filled PTFE used as a bearing material.†

Fluonid. Fluocinolone acetonide. Applications: Glucocorticoid. (Herbert Laboratories, Dermatology Div of Allergan Pharmaceuticals Inc).

Fluor. See FLUORSPAR.

Fluor-Adelite. Synonym for Tilasite.†

Fluor-Amps. Fluorescein sodium 5ml amps 10 and 20%. Applications: Diagnostic aid in the determination of circulation time, examination of opthalmic vasculature and differentiation of malignant and healthy tissue, visualisation of gall bladder and bile duct before surgery and for all localization of brain tumours. (SAS Pharmaceuticals Ltd).*

Fluor-I-Strip. Fluorescein sodium. Applications: Diagnostic aid. (Ayerst Laboratories, Div of American Home Products Corp).

Fluoral. Sodium Fluoride. Applications: Dental caries prophylactic. (Oral-B Laboratories Inc).

Fluoram. Ammonium bifluoride.†

Fluoranar. Coil coating compositions. (ICI PLC).*

Fluoraz (US, UK, France, Germany only). Modified structure of tetrafluoroethylene and propylene copolymers. Applications: Used in producing sealing materials for chemical, petrochemical and other applications where temperature and chemical resistance are of paramount importance. (Greene, Tweed).*

Fluorchrome. Chromium fluoride, $CrF_3.4H_2O$. A mordant.†

Fluorel. Family of heat resistant fluoroelastomers rated at 400° F. continuously and 600° F. for short periods. It is a vinylidene fluoride hexafluoropropylene copolymer or vunylidene fluoride, hexafluoropropylene, tetrafluorethylene terpolymer. (3M).*

Fluorel Fluoroelastomer. Range of elastomers composed of Vinylidene Fluoride/Hexafluoropropylene and/or Tetrafluoroethylene. Applications: High performance elastomer used for seals/gaskets for aerospace, automoti and chemical industries. (3M United Kingdom plc).†

Fluoremetic. Antimony sodium fluoride, $SbF_3.NaF$.†

Fluoresbrite. Fluorescent monodisperse carboxylated microspheres - polymer beads containing fluorescent dye. Applications: An identification tag ar a size reference for agglutination test flow cytometry, instrument calibratio gel filtration, light scattering and phagocytosis. (Polysciences Inc).*

Fluoresceine. Resorcinolphthalein. It is obtained by heating phthalic anhydri with resorcinol. Its alkali salts dye sil and wool yellow, as does fluoresceine itself, with green fluorescence. See URANINE.†

Fluorescent. Dyestuffs that fluoresce in daylight and/or ultra-violet light. Applications: Textiles, resin pigments and various solvents. (Holliday Dyes Chemicals Ltd).*

Fluorescent Blue. (Resorcin Blue, Iris Blue). A dyestuff. It is the ammoniun salt of tetrabromoresorufin. $C_{12}H_6Br_4N_2O_3$. Dyes silk and wool bl with brownish fluorescence.†

Fluorescent Red 5B. A proprietary organ fluorescent-red dye used for colouring polystyrene polymethyl-methacrylate and unplasticized PVC (Farbwerke Hoechst, Frankfurt-am-Main).†

Fluorescite. Fluorescein sodium. Applications: Diagnostic aid. (Alcon Laboratories Inc).

Fluorfolpet. A proprietary fungicide applied as a paint. It is *N*-(fluordichloromethylthio) phthalimid. (Bayer & Co).*

Fluorinse. Sodium Fluoride. Application Dental caries prophylactic. (Oral-B Laboratories Inc).

Fluorite. See FLUORSPAR.

Fluoroform. Trifluoromethane, CHF_3.†

Fluorol. Sodium fluoride, NaF.†

Fluorolene. A registered trade mark for polytetrafluoroethylene. (Liquid Nitrogen Processing Corp, Melven, pa).†

uorolubes. Polymers of trifluorovinyl chloride containing 49 per cent fluorine and 31 per cent chlorine. These products are produced as light oils, heavy oils and greases. Applications: Lubricant and sealant for plug cocks, valves and vacuum pumps. Impregnant for gaskets and packings. Fluid for hydraulic equipment, heat exchange and instrument damping. Miscellaneous applications such as low-temperature lubricant, high-density fluid etc. Also available as thickened 'GR' greases, lubricants that are non-reactive with corrosive liquids and stong oxidants. (Occidental Chemical Corp).*

uorolux. Coil coating compositions. (ICI PLC).*

uoromar. Fluroxene. Applications: Anaesthetic. (Anaquest, Div of BOC Inc).

uoroplex. Fluorouracil. Applications: Antineoplastic. (Herbert Laboratories, Dermatology Div of Allergan Pharmaceuticals Inc).

uorosint. A registered trade mark for polytetrafluorethylene with a ceramic-like texture suitable for machining.†

uorouracil Roche. Proprietary preparation of fluorouracil. Antineoplastic agent. (Roche Products Ltd).*

uorspar. (Fluor, Fluorite, Derbyshire Spar). A mineral. It is calcium fluoride, $4[CaF_2]$.†

uothane. A proprietary preparation of halothane. A general anaesthetic. (ICI PLC).*

uprim. Proprietary cough and cold remedy containing vitamin C, dextromethorphan, ephedrine, phenindamine, retinol and salicylamide. (Roche Products Ltd).*

urogestone Acetate. A proprietary progestin. It is 9-fluoro-11β, 17-dihydroxy-pregn-4-ene-3, 20-dione, 17-acetate. (Searle, G D & Co Ltd).†

urothyl. Di-(2,2,2-trifluoroethyl)-ether. Indoklon.†

uscorbin. A proprietary preparation of vitamin C, quinine dihydrochloride and sulphate, phenacetin and caHeine. A cold remedy. (Cox, A H & Co Ltd, Medical Specialities Divn).†

Flush Plate Brass. See BRASS.

Flutec. Range of extremely inert, temperature stable, non-toxic, non-inflammable liquids exhibiting excellent electrical insulating properties and good heat transfer characteristics. (ISC Chemicals Ltd).*

Flutec PP1. A fluorinated hydrocarbon C_6F_{14}. Perfluoro-*n*-hexane.†

Flutec PP2. A fluorinated hydrocarbon C_7F_{14}. Perfluoromethylcyclohexane.†

Flutec PP3. A fluorinated hydrocarbon C_8F_{16}. Perfluoro- 1,3-dimethyl-cyclohexane.†

Flutec PP9. A fluorinated hydrocarbon $C_{11}F_{20}$. Perfluoro-1-methyldecalin.†

Fluvermal. A proprietary preparation containing flubendazole. Applications: Antihelmintic. (Janssen Pharmaceutical Ltd).*

Fluvia. See JELUTONG.

Fluvirin. An inactivated influenza vaccine (surface antigen) containing highly purified haemaglutinin and neuraminidase antigens. Applications: For protection against influenza. (Evans Medical).*

Fluxol. A hardwood pitch prepared from the distillation of hardwood. It has a specific gravity of 1.16-1.19, and melts at 185-203° F. It is used as a rubber softener.†

Fluzone. Influenza virus vaccine. Applications: immunizing agent. (E R Squibb & Sons Inc).

FL7P. Liquid fertilizer. (Fisons PLC, Horticulture Div).

FML Liquifilm. Fluorometholone. Applications: Glucocorticoid. (Allergan Pharmaceuticals Inc).

Foam Spar. See APHRITE.

Foam Tannin. Tannin extracted from sumach, or galls, by means of ether.†

Foam Tint. Pigment dispersions. Applications: Colouration of polyurethane foam. (Pacific Dispersions Inc).*

Foamacure. Defoaming agents. Applications: Defoaming of aqueous coating compositions. (Nueodex Inc).*

Foamaster. Antifoams/defoamers. (Diamond Shamrock Process Chemicals Ltd).

Fob Metal. See BRASS.

Focal. Flowable fungicide. (FBC Ltd).

Foerdite. An explosive consisting of 25 per cent nitroglycerin, 1.5 per cent collodion wool, 5 per cent nitro-toluene, 4 per cent dextrin, 3 per cent glycerin, 37 per cent ammonium nitrate, and 24 per cent potassium chloride.†

Foetid Quartz. See STINK QUARTZ.

Foilcote. Overprint lacquers for aluminium, foil, carton boards, labels etc. (The Scottish Adhesives Co Ltd).*

Foilgrip. Aluminium foil to paper laminating adhesives. (The Scottish Adhesives Co Ltd).*

Foil Lead. An alloy of 86.5 per cent lead, 12.5 per cent tin, and 1 per cent copper.†

Folcovin. A proprietary preparation containing pholcodine. A cough linctus. (Rybar).†

Folex. 350. A proprietary preparation containing ferrous fumarate and folic acid used in the treatment of anaemia during pregnancy. (Rybar).†

Folia-Feed. Foliar nutrient. (Murphy Chemical Ltd).

Foliac Super Red. A proprietary trade name for a graphite jointing compound. (Graphite Products).†

Foliar TRIGGRR. A liquid containing plant growth regulators (cytokinin and gibberellic acid) and trace minerals, used to increase crop yields and quality. Applications: Applied to the foliage. Used for a wide variety of crops including cotton, soybeans, wheat, fruits and vegetables. (Westbridge Research Group).*

Folic Acid. Pteroylglutamic acid. Part of the Vitamin B complex. Deficiency produces anaemia in man.†

Folicet. Folic acid. Applications: Vitamin. (Mission Pharmacal Co).

Folicin. Prophylatic folic acid and iron preparations. Applications: Suitable for the prophylaxis and treatment of pregnancy anaemia. (Paines & Byrne).*

Folicote Transpiration minimizer. Refined wax, emulsifiers, preservatives, minimum 50 per cent solids. All Folicote components are FDA approv for use on edible crops. Applications Reduces water loss from plant foliage winter protection, transplanting and transporting plants, christmas trees, wreaths, agricultural crops such as potatoes, corn, tobacco, transplants, stone and citrus fruits. (Aquatrols Corporation of America).*

Folimat. A systemic insecticide as an emulsifiable concentrate containing 57.5% w/w omethoate. Applications: To control wheat bulb fly and opomy in winter cereals, frit fly in winter cereals, spring cereals and autumn so grass leys, and aphids on hops, plums and damsons. (Bayer & Co).*

Folin - McEllroy Sugar Reagents. Qualitative. 100 grams sodium pyrophosphate, $Na_4P_2O_7.10H_2O$, 30 grams crystalline disodium monohydrogen phosphate, Na_2HPO_4, and 50 grams dry sodium carbonate i 900 cc water. Dissolve and add 13 gr copper sulphate dissolved in 200 cc water. Quantitative. (*a*) Acidified copper sulphate solution containing 6 grams copper sulphate in 900 cc wate Dissolve, and add 5 cc concentrated sulphuric acid. Make up to 1,000 cc (Phosphate-carbonate-thiocyanate dry mixture prepared by mixing 100 gran dry sodium carbonate, $Na_2CO_3H_2O$, and 30 grams potassium thiocyanate mixed in a large mortar.†

Folin-Dennis Solution. A solution prepare by adding slowly 400 cc of a 0.7268 p cent solution of silver nitrate to a solution containing 10 grams mercuric cyanide and 180 grams caustic soda, i 1,200 cc water. Used for the determination of acetone.†

Folin-McEllroy Lactose Reagents. (*a*) A saturated picric acid solution (2 gram picric acid in 100 cc water). (*b*) A 20 per cent sodium carbonate solution.†

Folin's Uranium Acetate Mixture. Uric a reagent. It consists of 500 grams ammonium sulphate, 5 grams uranium acetate, and 6 cc glacial acetic acid dissolved in 650 cc water and made u to 1 litre.†

lin's Uric Acid Reagent. Add to 160 cc water, 50 cc syrupy phosphoric acid. Heat to 85° C., and add 100 grams sodium tungstate. Boil for 1 hour under reflux. Place 25 grams lithium carbonate in a beaker, add 50 cc syrupy phosphoric acid and 200 cc water. Boil 10 minutes, cool, and add first solution. Mix and dilute to 1,000 cc†

llicle Stimulating Hormone (**F.S.H.**). An extract of human post-menopausal urine containing primarily the follicle-stimulating hormone. PERGONAL.†

llotropin. A follicle-stimulating hormone.†

llutein. Gonadotropin, chorionic. Applications: Gonad-stimulating principle. (E R Squibb & Sons Inc).

losan. Horticultural fungicides. (Plant Protection (Subsidiary of ICI)).†

lpan. Active ingredient: folpet; N-(trichloromethylthio)phthalimide. Applications: Agricultural fungicide. (Makhteshim Chemical Works Ltd).†

olpet. A proprietary fungicide applied as a paint. N-(Trichlormethylthio) phthalimid. (Bayer & Co).*

olvite. Folic acid. Applications: Vitamin. (Lederle Laboratories, Div of American Cyanamid Co).

omac. A fungicide for rubber. (ICI PLC).*

omblin. A proprietary range of perfluoro-polyether fluids. (Montedison UK Ltd).*

ome-Cor. Light-weight, rigid board made of extruded polystyrene foam securely bonded between two layers of tough kraft linerboard. Applications: Used in manufactured housing, automotive and graphic arts. (Monsanto Co).*

omescol. Non-ionic surfactant. (ABM Chemicals Ltd).*

omitine. A liquid extract from the fungi *Fomes cinnamomeus*.†

omrez. Polyesters for flexible pu foam and pu elastomers. (Baxenden Chemical Co Ltd).*

ontaine's Powder. An explosive consisting of potassium picrate and potassium chlorate.†

Foots. Matter deposited by oils on standing.†

Foraflon. PVDF. Applications: Extruded pipe and monofilament. (Atochem Inc).*

Foral (AX, 85 and 105). A range of hydrogenated rosin and rosin esters. Applications: Tackifiers and polymer-modifying resins in adhesives and in hot-melt-applied decorative, pressure sensitive and heat-sealable coatings. (Hercules Inc).*

Forane. Isoflurane. Applications: An inhalation anaesthetic. (Abbott Laboratories).*

Forane. Chlorinated fluorocarbons. Applications: Aerosol propellant, foam blowing agent and refrigerant. (Pacific Chemical Industries Pty Ltd).†

Foraperle. Leather finishing agents. (ICI PLC).*

Forbes Metal,. An alloy of 53.5 per cent zinc, and 46.5 per cent copper.†

Force. A malt preparation.†

Forceval. A proprietary preparation of vitamins and minerals. (Unigreg).†

Forcite. A trade mark for various types of explosives.†

Fordath Resins. A proprietary range of phenolic and urea resins used in foundry work. (Fordath Engineering Co, West Bromwich).†

Forest Bark. Chipped, ground or composted bark for use as a soil conditioner, planting aid or mulch. (ICI PLC).*

Forex. PVC foam panels, slightly expanded with density 700 kg/m^3 and 450 kg/m^3. Applications: Possible applications include fair stands and exhibition booths, screen printing, signs and displays, lightweight construction in automotive and aviation industries and wall coverings for wet rooms. (Lonza Limited).*

Forging Brass. See BRASS.

Forhistal Maleate. Dimethindene maleate. Applications: Antihistaminic. (Ciba-Geigy Corp).

Forit. Oxypertine. Applications: Antidepressant. (Sterling Drug Inc).

Forlan C-24. Water and alcohol soluble cholesterol. Applications: Provides a convenient and economical source for cholesterol with emollient, solubilizing and emulsifying properties. (RITA Corporation).*

Forlan Series. Cholesterol absorption bases. Applications: Highly absorptive, will couple with more than 100% of weight in water and aqueous liquids. Produces highly emollient and skin hydrating cold creams, night creams, cleansing creams and lotions. (RITA Corporation).*

Forlan "LM'. Synthetic lanolin. Applications: An ingredient that qualitatively approximates the composition of lanolin. It enhances emulsion stability in both o/w and w/o systems, assists in the wetting and dispersion of pigments in facial makeup, lipstick and eye shadow preparations. (RITA Corporation).*

Forlan "L'. Synthetic lanolin. Applications: An ingredient that qualitatively approximates the composition of lanolin. It enhances emulsion stability in both o/w and w/o systems, assists in the wetting and dispersion of pigments in facial makeup, lipstick and eye shadow preparations. (RITA Corporation).*

Forlay. Selective weed killers. (Murphy Chemical Co).†

Formac 40. A solution polymer based on acrolein and formaldehyde with an active ingredient content of 40% by weight, used to control and eliminate algae, fungi and bacteria from industrial water. (Degussa).*

Formadermine. Guaiacol methylene ether.†

Formagen. A dental cement consisting of two parts : (*a*) A liquid containing creosote, phenol, olive oil, and alcoholic formalin, and (*b*) a powder consisting of aluminium silicate, magnesium and zinc carbonates, and lime.†

Formal. Formaldehyde. See METHYLAL.†

Formaldehyde. See ALDOFORM, FORMAL, FORMALIN, FORMALITH, FORMITROL, FORMOLYPTOL, FESTOFORM, METHYLALDEHYDE, and PRESERVALINE.

Formaldehyde-ammonia 6 : 4. See HEXAMINE.

Formalin. (Formol, Formol-chloral). A 4 per cent aqueous solution of formaldehyde. The commercial produ often contains from 12-15 per cent methyl alcohol, to prevent the separat of polymerized compounds.†

Formalite. A trade name for phenol-formaldehyde resin moulded material for use in electrical insulation.†

Formalith. A formaldehyde solution.†

Forman. Chloromethyl menthyl ether, $C_{10}H_{19}O.CH_2C1$.†

Formanek's Indicator. Alizarin green use as an indicator. It gives a violet colou with *p*H0.3, pink with *p*H1.0, yellow with *p*H12, and brown with *p*H14.†

Formaniline. A rubber vulcanization accelerator. It is anhydro-formaldehy aniline.†

Formapex. A proprietary phenol-formaldehyde resin varnish.†

Formasal. A condensation product of formaldehyde and salicylic acid.†

Format. Weedicide, containing clopyralid (ICI PLC).*

Formax. Sodium formate. (May & Baker Ltd).*

Formel - NF. A mixture of 'Isceon' chlorofluorocarbons and chloromethanes, which is totally non-flammable. Applications: Dielectric fluid for use in distribution transforme (ISC Chemicals Ltd).*

Formex. A proprietary trade name for an enamelled wire. The enamel is flexible and heat resisting up to about 185° C. It contains a mixture of polyvinyl acet and a thermosetting phenol-formaldehyde synthetic resin. See THERMEX.†

Formica. A proprietary trade name for phenolic and urea resins.†

Formin. See HEXAMINE.

Formit. See BAKELITE.

Formite. See BAKELITE.

Formitrol. A formaldehyde preparation.†

Formkote. Dry film lubricant coating containing graphite and suitable binde system. Applications: High temperature titanium metal forming lubricant. (E/M Corporation).*

Formo-gelatin. See GLUTOL.

Formodac. Chemical products for use as additives in the manufacture of concrete. (BP Chemicals Ltd).*

Formol. See FORMALIN.

Formol-chloral. See FORMALIN.

Formolide. An antiseptic consisting of an aqueous solution of 15 per cent alcohol, 2 per cent boric acid, 4 per cent sodium benzoate, and 1 per cent formaldehyde.†

Formolites. Phenol-formaldehyde resins. See BAKELITE, ALBERTOL, and ISSOLIN.†

Formolyptol. A formaldehyde preparation.†

Formon. Solder and braze compositions. (Du Pont (UK) Ltd).

Formopon. See DISCOLITE.

Formosa Camphor. Camphor from China.†

Formose. A mixture of sugars obtained from formaldehyde by polymerization.†

Formrez. A registered trade name for a polyester-based polyurethane elastomer cross-linked with a diamine. (Witco Chemical Corporation).†

Formrez. Polyesters for flexible pu foam and pu elastomers. (Baxenden Chemical Co Ltd).*

Formula 405. Pregnenolone Succinate. Applications: Non-hormonal sterol derivative. (Doak Pharmacal Co).

Formula 'S'. Squash court plaster. (Prodorite).*

Formusol. Sodium formaldehyde sulphoxylate. (RV Chemicals Ltd).

Formusol SA. Special acetaldehyde sulphoxylate. (RV Chemicals Ltd).

Formvar. Polyvinyl Formal Resins. Applications: Suitable for formulating structural adhesives, wash primers, can and drum linings and wood and knot sealers. (Monsanto Co).*

Fornax. Finishing agents. (Ciba-Geigy PLC).

Fornitrol. A reagent used for the estimation of nitric acid, nitro-compounds, and nitrates.†

Forociben Premix. Sulphonamide animal feed additive. (Ciba-Geigy PLC).

Foroid. A proprietary preparation of sodium tetraiodophenolphthalein.†

Forster Powder. See VON FORSTER POWDER.

Fortafix. A range of high temperature resistant adhesive cements and sealing compounds. Applications: Bonding, sealing or insulating inorganic materials where heat resistance up to 1600°C is required. (Fortafix Ltd).*

Fortagesic. A proprietary preparation of pentazocine and paracetamol. An analgesic. (Sterling Research Laboratories).†

Fortex. Flavours for confectionery, foodstuffs and beverages. (Bush Boake Allen).*

Fortiflex. High density polyethylene (HDPE). Applications: Packaging, especially blow moulded and injection moulded containers and film. Automotive and industrial parts. Pipe for potable water and natural gas transport. Rotationally moulded storage tanks. (Soltex Polymer Corporation).*

Fortilene. Polypropylene (PP). Applications: Fibre applications including non-wovens and slit-tape. Packaging, particularly injection moulded containers, caps and closures, and blow moulded bottles, film, industrial and automotive parts. (Soltex Polymer Corporation).*

Fortimax. Reinforcing fillers for synthetic elastomers. (ICI PLC).*

Fortisan. A proprietary trade name for a synthetic fibre made from regenerated cellulose.†

Fortral. A proprietary preparation of pentazocine hydrochloride. An analgesic. (Sterling Research Laboratories).†

Fortress. Heat-resistant glass-fibre tissues. (ICI PLC).*

Fortrex (2). An activated clay for reinforcing natural and synthetic rubber. (Croxton and Garry Ltd).*

Fortunan. Haloperidol. Applications: Pharmaceutical preparation for the treatment of schizophrenia. (M A Steinhard Ltd).*

Fosalsil. A proprietary trade name for a natural diatomaceous material made into bricks or used as a cement.†

Foset. Acid-catalysed phenolic resin systems for cold-set bonding of sand cores. (Foseco (FS) Ltd).*

Fosferno. Parathion insecticide. (ICI PLC).*

Fosfor. A proprietary preparation of phosphorylcolamine. A tonic. (Consolidated Chemicals).†

Fosfostilben. SEE HONVAN. (Degussa).*

Foshell. Liquid phenol-formaldehyde Novalac resins for coating sand by the warm process. (Foseco (FS) Ltd).*

Fosoil. Oil-based binders for cold-set bonding of sand cores. (Foseco (FS) Ltd).*

Fossil Flour. See INFUSORIAL EARTH.

Fossil Salt. Rock Salt.†

Fossil Wax. See OZOKERITE.

Fossiline. See OZOKERINE.

Fostap. Preformed highly insulating launder liners. (Foseco (FS) Ltd).*

Fotofax. Zinc oxide high purity. Applications: Varistors, reprographics. (Manchem Ltd).*

Fotosensin. A condensation product of phthalic acid and resorcinol containing small proportions of copper and iron. Small quantities increase the root and stem growth of plants.†

Fototar. Coal tar. Applications: Anti-eczematic. (Elder Pharmaceuticals Inc).

Fouadin. Pharmaceutical preparation comprising trivalent antimony. (Bayer & Co).*

Fouane. Benzthiazide.†

Foulon. Paper dyes. (ICI PLC).*

Founders' Type. See TYPE METAL.

Foundrox. Fe_2O_3. Applications: Foundry grade red iron oxide for core and mould use in eliminating veining and other casting expansion defects. (DCS Color & Supply Co Inc).*

Foundry Clay. A clay containing from 80-90 per cent silica and 15-18 per cent aluminia.†

Foundry Pattern Metal. An alloy of 75 per cent zinc and 25 per cent tin.†

Fouramine. Leather dyes. (ICI PLC).*

Fourdrinier Wire. A brass containing from 80-85 per cent copper, and 15-20 per cent zinc.†

Fovane. A proprietary preparation of benzthiazide. A diuretic. (Pfizer International).*

Foxglove Oil. See OIL OF FOXGLOVE.

FP-Vac. Fowl pox vaccine. Applications: immunization of poultry. (Intervet America Inc).*

FR-1360. A proprietary flame-retardant material comprising tribromoneopentyl alcohol used in the production of flexible polyurethane foam. (Dow Chemical Co Ltd).*

FR-2406. Tris (2,3-dibromopropyl) phosphate. A proprietary fire- retardant additive used in acrylics, epoxies, latices, phenolics, polyesters, polystyrenes, polyvinyl chloride, rayon celluloses and polyurethanes. (Dow Chemical Co Ltd).*

Fracton. A refractory dressing. (Foseco (FS) Ltd).*

Fractorite. An explosive containing 90 per cent ammonium nitrate, 4 per cent resin 4 per cent dextrin, and 2 per cent potassium dichromate.†

Fractorite B. An explosive consisting of 75 per cent ammonium nitrate, 2.8 per cent dinitro-naphthalene, 2.2 per cent ammonium oxalate, and 20 per cent ammonium chloride.†

Fragarol. (Fragasol). The butyl ether of β-naphthol. A synthetic perfume.†

Fragaroma. Perfumery products. (May & Baker Ltd).*

Fragasol. See FRAGAROL.

Fraissite. It is benzyl iodide, $C_6H_5.CH_2.1$.†

Framycort. A proprietary preparation of framycetin sulphate and hydrocortisone used in dermatology as an anti-bacterial agent. (Fisons PLC).*

Framygen. A proprietary preparation of framycetin sulphate. A systemic antibiotic. (Fisons PLC).*

Framyspray. An antibiotic aerosol. (Fisons PLC).*

Francolor. Dyes, pigments and chemical auxiliaries. (ICI PLC).*

Frankincense. Olibanum, a gum resin.†

Frankincense, Common. See GUM THUS.

Frankincense, Indian. See OLIBANUM.

Frankonite. (Silitonite, Tonsil). German bleaching earths obtained from deposits in Germany.†

Franocide. A proprietary preparation of diethylcarbamazine citrate used in veterinary work. (Coopers Animal Health, subsidiary of The Wellcome Foundation Ltd).*

Franol. A proprietary preparation of theophylline, phenobarbitone and ephedrine hydrochloride. A bronchial antispasmodic. (Bayer & Co).*

Franol Plus. A proprietary preparation of throphylline, phenobarbitone, ephedrine sulphate, and thenyldiamine hydrochloride used as a bronchospasm relaxant. (Winthrop Laboratories).*

Frantin. A proprietary anthelmintic given to unweaned lambs. (Coopers Animal Health, subsidiary of The Wellcome Foundation Ltd).*

Frary Metal. A calcium-barium-lead alloy containing up to 2 per cent barium and 1 per cent calcium. Used as a bearing metal. It melts at 445° C.†

Fraude's Reagent. Perchloric acid, $HC10_4$.†

Fre-Flex. Focofilcon A. 2-Hydroxyethyl methacrylate polymer with methacrylic acid. The material contains 55% water. Applications: Contact lens material. (Optech Inc).

Fredo. Calcium hydrosulphite.†

Freeflo. Industrial cleaning compound. (The Wellcome Foundation Ltd).

Freeman's Non-poisonous White Lead. A pigment. It is a mixture of white lead, zinc white, baryta white, and magnesium carbonate.†

Freeteem. Fluxes for uphill teemed killed steel ingots. (Foseco (FS) Ltd).

Freezine. See PRESERVALINE.

Frelen. Cross-link polyethylene foam. (Carl Freudenberg).*

Fremy's Salt. Potassium hydrogen fluoride, KF.HF.†

French Automobile Bearing Metal. See BEARING METALS.

French Blue. See ULTRAMARINE.

French Cement. A mucilage of gum arabic mixed with powdered starch. Used by naturalists, artificial flower makers, and confectioners.†

French Chalk. A variety of Steatite or Soapstone (*qv*). It is a hydrated silicate of magnesium, and is used for marking cloth, removing grease from silk, as a filler, and for other purposes.†

French Chrome Yellow. See CHROME YELLOW.

French Maltine. See EXTRACT OF MALT.

French Ochre. See YELLOW OCHRE.

French Pine Resin. See GUM THUS.

French Pollsh. A polish for wood. It consists of shellac dissolved in alcohol.†

French Purple. (Archil Violet, Red Indigo, Perseo). A dyestuff obtained from lichens.†

French Turpentine. (Bordeaux Turpentine). The oleo-resin from *Pinus maritima*. It contains from 15-20 per cent essential oil and 70-80 per cent rosin.†

French Ultramarine. See ULTRAMARINE.

French Verdigris. Basic copper acetate.†

French Veronese Green. See CHROMIUM GREEN.

French White. Powdered talc.†

French Yellow. See CAMPOBELLO YELLOW.

Frenokone. A selective weedkiller. (Plant Protection (Subsidiary of ICI)).†

Freon. Fluorocarbons. (Du Pont (UK) Ltd).

Fresh Pak. Extra-thin polyethylene HD-film. Applications: Packaging or incorporation into laminates. (UCB nv Film Sector).*

Freudenberg Megulastik. Mechanical dampeners. (Carl Freudenberg).*

Freund's Acid. 1-Naphthylamine-3: 6-disulphonic acid.†

Friable Amber. See GEDANITE.

Frick's Alloys. Nickel silvers containing from 50-69 per cent copper, 18-39 per cent zinc, and 5-31 per cent nickel.†

Friedelite. A mineral, Mn_2SiO_4.†

Friedländer's Stain. A microscopic stain. It contains 50 grams of a saturated alcoholic solution of gentian violet, 10 grams of glacial acetic acid, and 100 cc water.†

Frigen. Fluorinated hydrocarbon refrigerant. (Hoechst UK Ltd).

Frimulsion. Hydrocolloid blends used to improve stability, appearance, texture, and mouthfeel of food. Applications: Food and beverage industries. (Hercules Inc).*

Frishmuth's Aluminium Solder. Alloys. One contains 67 per cent tin, 27 per cent lead, and 3 per cent aluminium. Another consists of 94 per cent zinc, 4 per cent aluminium, and 2 per cent copper.†

Frishout Solder. An alloy of 46 per cent tin, 23 per cent zinc, 15 per cent aluminium, 8 per cent copper, and 9 per cent silver.†

Fritzsche's Reagent. Dinitroanthraquinone, $C_{14}H_6N_2O_6$.†

Froben. Potent non-steroidal anti-inflammatory and analgesic agent. (The Boots Company PLC).

Frohde's Reagent. An alkaloid reagent. It consists of 0.5 gram of sodium molybdate dissolved in 100 cc of concentrated sulphuric acid.†

Frosted Silver. See DEAD SILVER.

Frost-Off. Anti-freeze liquid. Applications: De-icing fluid for automobile windshields and door locks. (Merix Chemical Co).*

Frother 4171. A terpene-alcohol based flotation reagent. Produces a more brittle froth than typical pine oil. Applications: For flotation of non-metallic minerals. (Hercules Inc).*

Froth, Copper. See TYROLITE.

Fruit Sugar. See LEVULOSE.

Frumil. Frusemide and amiloride. Applications: Management of oedema, in congestive cardiac failure, nephrosis, corticosteroid therapy, oestrogen therapy, ascites associated with cirrhosis. (Berk Pharmaceuticals Ltd).*

Frusid. A proprietary preparation of FRUSEMIDE. A diuretic. (DDSA).†

Fubol. Potato/lettuce fungicide. (Ciba-Geigy PLC).

Fuchsia. See METHYLENE VIOLET 2RA. The diethyl-safranines, also the chlorides of α- and β-diamylsafranines, are known under this name.

Fuchsianite. See MAGENTA.

Fuchsine. See MAGENTA.

Fuchsine Extra Yellow. (Fuchsine Scarlet). A dyestuff. It consists of Fuchsine mixed with Auramine, to give a yellower shade than fuchsine alone.†

Fuchsine S. See ACID MAGENTA.

Fuchsine Scarlet. See FUCHSINE EXTRA YELLOW.

Fuchsone. Diphenylquinomethane.†

Fucidin. A proprietary preparation containing sodium fusidate. An antibiotic. (Leo Laboratories).†

Fucidin V.P. A proprietary preparation of sodium fusidate and phenoxymethyl penicillin. An antibiotic. (Leo Laboratories).†

Fucidin-H. A proprietary preparation of sodium fusidate and HYDRO-CORTISONE acetate used in the treatment of dermatosis. (Leo Laboratories).†

Fudow. A proprietary range of phenolic moulding materials. (Fuchow, Japan).†

Fudowlite U. A proprietary range of urea formaldehyde moulding materials. (Fuchow, Japan).†

Fugata. A proprietary preparation of ticonazole. An antifungal. (Pfizer International).*

Ful-Glo. Fluorescein sodium. Applications: Diagnostic aid. (Barnes-Hind Inc).

Fulacolor. Reactive clay - carbonless copy paper. (Laporte Industries Ltd).

Fulbent. Bentonite clay for binding, suspending and emulsifying. (Laporte Industries Ltd).*

Fulbond. A bonding agent for foundry sands. (Laporte Industries Ltd).*

Fulbond. Bentonite. Applications: Foundry bentonite. (Minas de Gador SA).*

Fulcat. Clay catalysts. (Laporte Industries Ltd).*

Fulcat Catalysts. Acid treated montmorillonite clays. Applications: Catalysts for a range of organic reactions notably the alkylation of phenols. (Laporte Industries Ltd).†

Fulcin. A proprietary trade name for a preparation of griseofulvin BP. An antifungal agent. (ICI PLC).*

Fulgurite. An explosive containing 60 per cent nitro-glycerin and 40 per cent wheaten flour and magnesium carbonate.†

Fuligo. Soot.†

Fullasorb. Absorbent earth granules. (Laporte Industries Ltd).

Fullerite. A proprietary trade name for a slate powder used as a rubber filler.†

Fuller's Earth. A term applied to a sandy loam or argillaceous earth found in Surrey and Kent. It consists of aluminium and magnesium hydrosilicates. A deodorizer. Used for clarifying oils, and used in cosmetics, etc.†

Fulmargin. A solution of colloidal silver.†

Fulmenit. An explosive consisting of 86.5 per cent ammonium nitrate, 5.5 per cent trinitrotoluene, 2.5 per cent paraffin oil, 1.5 per cent charcoal, and 4 per cent guncotton.†

Fulminating Gold. A compound having the formula $2AuN_2H_3.H_2O$, prepared by the action of concentrated ammonia on gold hydroxide. It is explosive.†

Fulminating Platinum. Explosive compounds formed by acting upon ammonium platinochloride with potassium hydroxide†

Fulminating Silver. A compound of nitrogen and silver prepared by the action of ammonia on precipitated silver oxide. It is explosive.†

Fulmont. Highly acetivated bleaching earth. (Laporte Industries Ltd).

Fulmont Activated Bleaching Earths. Acid activated montmorillonite clays. Applications: Colour and impurity removal from edible oils, re-refining or purification of mineral oils. (Laporte Industries Ltd).†

Fulton White. A proprietary brand of lithopone (*qv*).†

Fulvicin-P/G. Griseofulvin. Applications: Antifungal. (Schering-Plough Corp).

Fulvicin-U/F. Griseofulvin. Applications: Antifungal. (Schering-Plough Corp).

Fulvite. An artificial titanium monoxide.†

Fumagillin. An antibiotic produced by certain strains of Aspergillus fumigatus.†

Fumexol. Pretreatment agents. (Ciba-Geigy PLC).

Fuming Nitric Acid. Nitric acid containing some of the lower oxides of nitrogen.†

Fuming Oil of Vitriol. See FUMING SULPHURIC ACID.

Fuming Sulphuric Acid. (Oleum, Nordhausen Sulphuric Acid, Pyrosulphuric Acid). It consists of sulphur trioxide dissolved in sulphuric acid. The commonest fuming acid contains 55 per cent sulphuric acid, and 45 per cent sulphur trioxide. Nordhausen sulphuric acid has a specific gravity of from 1.86-1.90.†

Fumite. General purpose insecticide in smoke form for greenhouse use. (ICI PLC).*

Fumite Dicloran. Smoke fungicide (active ingredient - dicloran). Applications: For use in enclosed areas against botrytis and rhizoctonia on protected crops. (Octavius Hunt Ltd).*

Fumite Dicofol. Smoke acaricide (active ingredient - dicofol). Applications: For use in enclosed areas against red spider and other mites on protected crops. (Octavius Hunt Ltd).*

Fumite Lindane. Smoke insecticide (active ingredient - gamma HCH). Applications: Broad spectrum insecticide for use in (enclosed) poultry, mushroom farms, ships' holds, public buildings, protected crops, against a wide range of pests. (Octavius Hunt Ltd).*

Fumite Permethrin. Smoke insecticide (active ingredient - permethrin). Applications: For use in enclosed areas against whitefly and other pests of protected crops, cockroaches on stored produce, domestic insect pests. (Octavius Hunt Ltd).*

Fumite Pirimiphos Methyl. Smoke insecticide (active ingredient - pirimiphos methyl). Applications: For use in enclosed areas against whitefly and other pests of protected crops, pests in grain silos etc. (Octavius Hunt Ltd).*

Fumite propoxur. Smoke insecticide (active ingredient - propoxur). Applications: For use in enclosed areas against whitefly and other pests of protected crops. (Octavius Hunt Ltd).*

Fumite TCNB. Smoke fungicide (active ingredient - tecnazene). Applications: For use in enclosed areas against botrytis and mildew on protected crops. (Octavius Hunt Ltd).*

Fumite Tecnalin. Smoke insecticide/-fungicide (active ingredients - gamma HCH/tecnazene). Applications: For use in enclosed areas against botrytis, whitefly and other insect pests on protected crops. (Octavius Hunt Ltd).*

Funduscein. Fluorescein sodium. Applications: Diagnostic aid. (CooperVision Inc).

Fungamyl. A purified fungal alpha-amylase preparation produced from a selected strain of Aspergillus oryzae. Applications: Used in the following industries: starch processing, brewing, alcohol production and baking. (Novo Industri A/S).*

Fungex. A copper containing fungicide. (Murphy Chemical Co).†

Fungi-Fluor. Stain solution and counterstain in an eight ounce kit for rapid identification of fungi. Applications: Fungi identification in clinical cell cultures and tissue biopsy samples. (Polysciences Inc).*

Fungilin. A proprietary preparation containing amphotericin B. An antibiotic. (Squibb, E R & Sons Ltd).*

Fungilin Cream. Amphotericin in cream base. (E R Squibb & Sons Ltd).

Fungilin Lozenges. Amphotericin in lozenge form. (E R Squibb & Sons Ltd).

Fungilin Ointment. Amphotericin in ointment form. (E R Squibb & Sons Ltd).

Fungitex 656. A proprietary trade name for a solution of an organo- mercuric complex. Used as a durable mildew proofing agent for textiles. (Ciba Geigy PLC).†

Fungitrol. Fungicides. Applications: Non-aqueous coating compositions, caulking compounds and other substrates. (Nueodex Inc).*

Fungitrol Tinox. Bis(tri-n-butyl)oxide. Applications: Wood preservative and anti-fouling coating compositions. (Nueodex Inc).*

Fungizone. A proprietary preparation containing amphotericin B. An antibiotic. (Squibb, E R & Sons Ltd).*

Fungizone for Infusion. Amphotericin for intravenous infusion. (E R Squibb & Sons Ltd).

Fungus Fighter. Fungicide. (May & Baker Ltd).

Furac No. 3. A. proprietary rubber vulcanization accelerator. It is the lead salt of dithiofuroic acid.†

Furacin. Nitrofurazone. Applications: Anti-infective, topical. (Norwich Eaton Pharmaceuticals Inc).

Furacin. Acid and solvent resisting cement for bedding and jointing tiles and bricks. (Prodorite).*

Furadantin. Nitrofurantoin. Applications: Antibacterial. (Norwich Eaton Pharmaceuticals Inc).

Furamazone. Nifuraldezone. Applications: Antibacterial. (Norwich Eaton Pharmaceuticals Inc).

Furamide. A proprietary preparation of diloxanide 2-furoate. An antiamoebic. (Boots Company PLC).*

Furanace. Nifurpirinol. Applications: Antibacterial. (Dainippon Pharmaceutical Company).

Furanculine. Dried yeast.†

Furbac. A proprietary accelerator. It is n-cyclohexyl - 2 - benzthiazyl sulph-enamide. (Anchor Chemical Co).*

Furfural Resins. Artificial resins obtained by the condensation of furfuraldehyde with phenols, cresols, or other similar bodies.†

Furnace-calamine. Masses consisting mainly of zinc oxide, formed during the smelting of zinciferous iron ores.†

Furnacite. Synonym for Fornacite.†

Furol Green. A dyestuff obtained by condensing furfural with dimethylaniline in the presence of zinc chloride, then converting the leuco base produced into the dyestuff by oxidation.†

Furosemide. Frusemide.†

Furotec. Furane resin binders for cold-set bonding of sand cores. (Foseco (FS) Ltd).

Furoxone. Furazolidone. Applications: Anti-infective, topical; antiprotozoal. (Norwich Eaton Pharmaceuticals Inc).

Fursatil CS 12. A proprietary cold-setting resin based on urea/furane. (Fordath Engineering Co, West Bromwich).†

Fursatil CS15. A faster-setting, lower strength variant of FURSATIL CS12. (Fordath Engineering Co, West Bromwich).†

Fursatil CS25. A fast-setting, medium strength variant of FURSATIL CS12 having a low fume level. (Fordath Engineering Co, West Bromwich).†

Fursatil CS30. A proprietary phenol/furane cold-setting resin. (Fordath Engineering Co, West Bromwich).†

Fursatil CS40. A proprietary plasticized urea cold-setting resin. (Fordath Engineering Co, West Bromwich).†

Fursatil CS60. A proprietary phenol/urea cold-setting resin. (Fordath Engineering Co, West Bromwich).†

Fursatil CS65. See FURSATIL CS60. (Fordath Engineering Co, West Bromwich).†

Fursatil CS71. A proprietary modified phenolic cold-setting resin. (Fordath Engineering Co, West Bromwich).†

Fursatil CS81. A proprietary modified urea formaldehyde cold-setting resin. (Fordath Engineering Co, West Bromwich).†

Furunculin. A yeast preparation†

Fusafungine. An antibiotic produced by Fusarium lateritium 437.†

Fusarex. Fungicide containing tecnazene for use on potatoes. (ICI PLC).*

Fusariol. A mercury-formaldehyde preparation. A seed preservative.†

Fuscamine Brown. A dyestuff. It is *m*-aminophenol.†

Fuscanthrene. An anthracene dyestuff, prepared by heating formaldehyde and certain diamino-anthraquinones with potassium hydroxide. It is a yellow-brown vat dye.†

Fuscochlorin. A dark-green pigment from algae.†

Fuscorhodin. A dark-red pigment from algae.†

Fused Cement. (Electro-fused Cement). Terms applied to an aluminous cement with a high alumina content.†

Fusel Oil. (Fermentation Amyl Alcohol, Potato Oil, Grain Oil, Marc Brandy Oil). A by-product in alcoholic fermentation, especially in the preparation of potato spirit, and in the rectification of alcohol. It consists mainly of mixed C_2-C_5 alcohols.†

Fusible Salt. See MICROCOSMIC SALT.

Fusible Salt of Urine. This is microcosmic salt.†

Fusible White Precipitate. See MERCUR-AMMONIUM CHLORIDE.

Fusidic Acid. An antibiotic produced by a strain of Fusidium. Fucidin is the sodium salt.†

Fusilade. Herbicide containing fluazifop butyl. (ICI PLC).*

Fussolon. A proprietary range of fluoro carbon and similar resins. (Daikin Kogyo Co, Osaka).†

Fustic. (Old Fustic, Brazil Wood, Yellow Wood, Cuba Wood). The chips or extract from *Morus tinctoria.* It is a natural dyestuff, the dyeing principles being morin, and maclurin (pentahydroxy-benzophenone), $C_{13}H_{10}O_6$. Chiefly used for dyeing wool yellow.†

Fustin. The diazobenzene compound of maclurin (obtained from fustic).†

Futura Flex. High performance protective and waterproofing coatings for harsh environments. (Baxenden Chemical Co Ltd).*

Futura Thane. High performance protective and waterproofing coatings for harsh environments. (Baxenden Chemical Co Ltd).*

Futurit. A proprietary plastic of the phenol-formaldehyde type.†

Fybogel. A proprietary preparation of ispaghula husk with sodium bicarbonate and citric acid. A laxative. (Reckitts).†

Fybranta. A proprietary preparation of bran and calcium phosphate. A laxative. (Norgine).†

Fydulan. Dichlobenil. Applications: Weedkiller. Direct, selective weed killing action in orchards, vineyards, flower beds and parkland areas and along rail tracks, motorways and waterways. (Duphar BV).*

Fydumas. Dichlobenil. Applications: Weedkiller. Direct, selective weed killing action in orchards, vineyards, flower beds and parkland areas and along rail tracks, motorways and waterways. (Duphar BV).*

Fydusit. Dichlobenil. Applications: Weedkiller. Direct, selective weed killing action in orchards, vineyards, flower beds and parkland areas and along rail tracks, motorways and waterways. (Duphar BV).*

Fyfanon. *O,O*-dimethyl-*S*-(1,2-di(ethoxy-carbonyl)ethyl) phosphorodithioate. Applications: It is a low toxic insecticide. It is effective against insect pests on livestock, stored crops, agriculture, home and garden. (A/S Cheminova).*

Fyrol 6. A dihydroxy-terminated phosphonate ester. An additive for rigid polyurethane foam imparting self extinguishing properties.†

Fytospore. Fungicide for the control of potato blight. (ICI PLC).*

Fyzol 11E. Adjuvant oil for herbicides. (FBC Ltd).

G

G-G. Guaifenesin. Applications: Expectorant. (Merrell Dow Pharmaceuticals Inc, Subsidiary of Dow Chemical Co).

G.500. A proprietary preparation containing hexamine mandelate and methionine. A urinary antiseptic. (Ward, Blenkinsop & Co Ltd).†

G-3300. Alkylaryl sulphonate anionic surfactant in the form of a reddish-brown liquid. Applications: Emulsifier, dispersant, used for pigments in paints. (Atlas Chemical Industries (UK) Ltd).

Gabbett's Stain. A microscopic stain. It contains 2 grams methylene blue, 25 cc sulphuric acid, with water up to 100 cc†

Gabbro. A coarse, crystalline rock, composed mainly of lime-soda felspar.†

Gabian Oil. An inflammable mineral naphtha.†

Gabraster. Polyester resins.†

Gabrosa. Carboxymethyl cellulose for drilling muds. (Montedison UK Ltd).*

G-acid. 2-Naphthol-6 or 8-sulphonic acid.†

Gadorgel. Clarification of wines, beers, etc. (Minas de Gador SA).*

Gadorgel Ocma. Bentonite. Applications: Drilling fluid and civil engineering to achieve high quality tixotropic muds. (Minas de Gador SA).*

Gadose. A grease prepared from cod-liver oil and lanoline. Used as a basis for ointments.†

Gaduol. An extract containing the alcohol-soluble constituents of cod-liver oil.†

Gad's Cement. A mason's cement, consisting of 3 parts clay and 1 part ferric oxide.†

Gafac BG-510. Water soluble phosphate ester. Applications: Soak-tank metal cleaning, steam cleaning and household cleaners. (GAF Corporation).*

Gafac BH-650. Very hydrophilic surfactant with excellent caustic solubility and stability. Applications: Suitable for use in formulations requiring good alkali builder solubility. Also for use in cotton processing particularly in mercerizing. (GAF Corporation).*

Gafac BI-729 and BI-750. Very hydrophilic surfactants with 89% and 100% activity respectively. Applications: Suitable for use in formulations requiring good alkali builder solubility. Useful in formulations of liquid drain cleaners. (GAF Corporation).*

Gafac BP-769. Low foaming acid ester hydrotrope. Good alkali stability. Applications: Well suited for use in low foaming liquid industrial cleaning formulations in which electrolyte content is very high. (GAF Corporation).*

Gafac GB-520. Surfactant dispersable in the emulsification of mineral oils soluble in aromatic solvents. Applications: Used in the formulation of lubricants for wool and synthetic fibres. Imparts softening and antistatic effects. (GAF Corporation).*

Gafac LO-529. Viscous surfactant soluble in water and many polar and non-polar solvents. Compatible with electrolyte solutions. Applications: In detergent concentrates, imparts good hard-surface detergency, moderate foaming and retardation of rusting and corrosion. Also prolongs the life of wax and resin floor finishes without damage to floor coverings. Economical antistatic for acrylic carpet yarn. Dry cleaning agent. (GAF Corporation).*

Gafac MC-470. Emulsifier for mineral oils. Soluble in mineral oils and aromatic solvents. Applications: In textile finishing, as a softener, antistatic agent and constituent of lubricants for synthetic fibres, filament yarns and wool. (GAF Corporation).*

Gafac PE-510. Light-coloured, water dispersable acid ester wetting agent with good electrolyte tolerance. Applications: Dry cleaning detergent. Useful in waterless hand cleaners. As primary polymerization emulsifier promotes formation of polyvinyl acetate and acrylic films having clarity, heat and light stability, stable neutral pH and corrosion inhibition. (GAF Corporation).*

Gafac RA-600. Surfactant based on linear alcohol. Effective coupling agent for non-ionic surfactants in liquid alkali detergent systems. Applications: Excellent textile wetting properties and hard surface detergency. Moderate foamer. Widely used in a variety of household cleaners. (GAF Corporation).*

Gafac RB-400. Surfactant based on fatty alcohol. Excellent emulsifier. Applications: Soluble in naphthenic and paraffinic oils and aromatic solvents. Soluble in water as neutralized salt. (GAF Corporation).*

Gafac RD-510. Oil and water-dispersable emulsifier for creams, lotions and clear cosmetic gels. Applications: In textiles, antistatic lubricant for yarns and fibres. (GAF Corporation).*

Gafac RE. Series of anionic surfactants in the form of acid phosphate esters where the hydrophobe is aromatic. Liquid or paste form. Applications: Variety of uses eg emulsion polymerization; dedusting agent for alkaline powders; emulsifier and stabilizer used in polyvinyl acetate and vinyl/acrylic co-polymers; dispersant and dyeing assistant for oil-emulsion printing pastes. (GAF (Gt Britain) Ltd).

Gafac RE-410. Oil soluble surfactant, but also water-soluble when neutralized. Applications: Effective emulsifier, used in emulsion polymerization. (GAF Corporation).*

Gafac RE-610. Surfactant with good compatibility in strong electrolyte systems. Useful in heavy-duty, all-purpose liquid formulations. Applications: Dedusting agent for alkaline powders, drycleaning detergents and pesticide emulsifiers. In emulsion polymerization promotes formation of polyvinyl acetate and acrylic films having clarity, corrosion inhibition, stable neutral pH and stability to heat and light. (GAF Corporation).*

Gafac RE-877. Emulsifier and stabilizing agent. Applications: Used in the preparation of polyvinyl acetate and vinyl acetate/acrylic copolymers. (GAF Corporation).*

Gafac RE-960. Light-coloured, water soluble dispersant and dyeing assistant. Applications: Oil emulsion printing pastes and emulsion polymerization. (GAF Corporation).*

Gafac RK-500. Low-foaming, non-phenolic hydrotrope. (GAF Corporation).*

Gafac RL-210. Surfactant. Applications: Used as mould release agent. (GAF Corporation).*

Gafac RM. Series of anionic surfactants in the form of acid phosphate esters in which the hydrophobe is aromatic. Supplied in liquid form. Applications: Emulsifiers and rust inhibitors used in chlorinated and phosphated pesticides; polythene; industrial cleaners and dry cleaning. (GAF (Gt Britain) Ltd).

Gafac RM-410. Oil soluble surfactant. Good antiseptic properties. Applications: Effective textile wetting agent, used in alkaline scouring operations. Used as potassium or sodium salt in drycleaning detergents. (GAF Corporation).*

Gafac RM-510. Surfactant soluble in aromatic solvent and kerosene. Dispersable in water, spontaneously emulsifies chlorinated and phosphated pesticide concentrates and polyethylene.

Applications: Useful in industrial cleaners and in dry cleaning. (GAF Corporation).*

Gafac RM-710. Surfactant soluble in aromatic solvent and kerosene. Dispersable in water, spontaneously emulsifies chlorinated and phosphated pesticide concentrates and polyethylene. Water soluble. Applications: Useful in industrial cleaners and in dry cleaning. (GAF Corporation).*

Gafac RP-710. Low-foaming hydrotrope for nonionic surfactants used in liquid or powder alkaline cleaners. Applications: Particularly effective in solubilizing low-foaming non-ionic surfactants even at elevated temperatures. (GAF Corporation).*

Gafac RS. Series of liquid anionic surfactants in acid phosphate ester form, in which the hydrophobe is aliphatic. Applications: Emulsifiers, wetting, scouring and cleaning agents, used in textile wet-processing; pre-treatment of resin finished cotton; dry cleaning. (GAF (Gt Britain) Ltd).

Gafac RS-410. Oil soluble surfactant. Applications: Extremely effective paraffinic oil emulsifier as triethanolamine salt. Also contributes to rust-inhibiting properties. (GAF Corporation).*

Gafac RS-610. Surfactant with high electrolyte tolerance. Applications: Emulsifier, penetrant, antistatic agent and drycleaning detergent. Pesticide emulsifier with liquid fertilizer solutions. In cotton processing, used in caustic boil and peroxide bleach. Also used in detergents, cleaners and in the paper trade. (GAF Corporation).*

Gafac RS-710. Alkali-stable detergent, emulsifier, wetting agent and dispersant. Applications: Used in wet processing of cellulosic and synthetic fibre textiles. (GAF Corporation).*

Gafamide CDD-518. Coconut oil diethanolamine condensate. Applications: Adjuvant for nonionics, alkylaryl sulphonates and sulphated fatty alcohols. Used in liquid manual-dishwashing formulations, drycleaning detergents, heavy-duty household detergents and industrial cleaners. (GAF Corporation).*

Gafen LB-400, LE-500 and LS-500. Organic phosphate esters in free acid form. Supplied as liquids. Applications: Extreme pressure additives for metal working fluids, with lubricant, emulsifier and rust inhibition properties. Used in lubricating and rolling oils, cutting and hydraulic fluids. (GAF (Gt Britain) Ltd).

Gafen LE-700, LP-700 and LK-500. Organic phosphate esters in free acid form. Supplied as liquids. Applications: Lubricity additives used in water-based cutting fluids with high concentration of inorganic rust inhibitors. (GAF (Gt Britain) Ltd).

Gafen LM-400. Organic phosphate ester in free acid form, supplied as a liquid. Applications: Oil-soluble, water-dispersible lubricant for use as a rust inhibitor in aqueous cutting fluids, rolling oils and hydraulic fluids. (GAF (Gt Britain) Ltd).

Gafen LM-600. Organic phosphate ester in free acid form, supplied as a liquid. Applications: Oil-soluble emulsifier and metal lubricant, used in rolling oils, cutting fluids and hydraulic fluids. (GAF (Gt Britain) Ltd).

Gaffix VC-713. Vinylcaprolactam/vinyl-pyrrolidone/dimethylaminoethylmethacrylate terpolymer. Applications: Film-forming, fixative resin for use in mousses, gels, glazes, lotions and hairsprays. (GAF Corporation).*

Gafgard 233 and 233E. Clear amber liquid blends of reactive monomers. Applications: Radiation curable coatings formulated to provide abrasion resistant surfaces to plastics, paper, wood and other substrates. (GAF Corporation).*

Gafgard 238. A general purpose, non-yellowing aliphatic urethane based oligomer. Applications: Used in tough scuff resistant coatings for flexible substrates, including plastics, textiles and leather. (GAF Corporation).*

Gafgard 245. Clear amber liquid blend of aliphatic oligomers and monomers.

Applications: Radiation curable coatings designed for high speed application/curing. (GAF Corporation).*

Gafgard 277. Clear amber liquid blend of aliphatic oligomers and monomers. Applications: Radiation curable coatings designed for flooring use. (GAF Corporation).*

Gafgard 280. Clear amber liquid blend of aliphatic oligomers and monomers. Applications: Radiation curable coating designed for curtain coating. (GAF Corporation).*

Gafite. A range of thermoplastic polyester moulding compounds. Applications: Resins which can replace metals as well as other plastics in numerous applications, e.g. under-the-hood parts, electrical/electronic components, appliance parts, hardware, pumps and hydraulic controls. (GAF Corporation).*

Gafite LW. A range of thermoplastic polyester moulding compounds. Applications: Automotive exterior parts such as rear end panels, cowl vents, fender extensions and headlight housings. (GAF Corporation).*

Gaflex. Thermoplastic polyester elastomers. Applications: Uses include industrial tubing, fuel lines, hydraulic hoses, flexible couplings, fasteners, gaskets, seals, boots and bellows for mechanical drives, wire and cable jacketing, noise dampening devices and pump parts. (GAF Corporation).*

Gafoam AD. Ammonium salt of ethoxylate sulphate. Applications: Foaming agent primarily used as an air-drilling surfactant for oil and gas wells. Also used in well cleanout and as a mobility control agent for carbon dioxide. (GAF Corporation).*

Gafquat. A range of vinylpyrrolidone/-quaternized dimethylaminoethyl-methacrylate copolymers. Applications: Film-forming polymers for hair and skin care products. Also used in deodorants and antiperspirants, shaving preparations, antiseptics, toilet soaps, skin creams, sunburn remedies. (GAF Corporation).*

Gafstat AD-510 and AE-610. Free acids of complex phosphate esters, in liquid form. Anionic surfactants in which the hydrophobe is aromatic. Applications: Internal antistatic agents for plastics, with heat stability and low toxicity. Compatible with PVC, polyolefins, polystyrene and many other plastics. (GAF (Gt Britain) Ltd).

Gafstat AS-610 and AS-710. Free acids of complex phosophate esters, in liquid form. Anionic surfactants in which the hydrophobe is aliphatic. Applications: Internal antistatic agents for plastics, with heat stability and low toxicity. Compatible with PVC, polyolefins, polystyrene and many other plastics. (GAF (Gt Britain) Ltd).

Gaftuf. Impact modified polybutylene terephthalate compounds. Applications: Hand tool housings, shrouds subjected to severe abuse, gasoline and brake clips etc. (GAF Corporation).*

Gagat. A variety of soft coal.†

Gahn's Ultramarine. See COBALT BLUE.

Gala. A proprietary casein plastic.†

Galactan. (Gelose). A gum, $(C_6H_{10}O_5)n$, from agar-agar.†

Galactomin. Proprietary trade name for an artificial milk food with low lactose content. (Cow and Gate).†

Galag. Magnesium impregnated metallurgical coke for iron desulphurisation. (Foseco (FS) Ltd).*

Galagum. A mixture of modified polysaccharides. It gives a colloidal solution when boiled in water. It is a protective colloid, and is used in making baker's and flavouring emulsions, and in cosmetic and hair lotions.†

Galahad A. A proprietary trade name for a steel containing 0.10 per cent. carbon, 0.50 per cent. silicon, 0.04 per cent. sulphur, 0.04 per cent. phosphorus, 0.60 per cent. manganese, 13.00 per cent. chromium and sometimes 1.50 per cent. nickel and 0.25 per cent. molybdenum. It is used in the manufacture of forged steel pressure vessels with good corrosion resistance. (Hadfield, George & Co).†

Galalith. (Erinoid). The trade name for a polymeric material, obtained by the action of formaldehyde upon casein.†

Galam Butter. Shea butter.†

Galden. A proprietary range of perfluoropolyether fluids. (Montedison UK Ltd).*

Galena, Pseudo. See ZINC BLENDE.

Galenite. A pigment. It is a basic sulphate of lead. The name is sometimes applied to Galena.†

Galenomycin. A proprietary preparation of OXYTETRACYCLINE. An antibiotic. (Galen).†

Galettame Silk. (Ricotti Silk, Neri Silk, Basinetto Silk). The residue of the silk cocoon after reeling.†

Galicar. An anti-friction bearing metal containing 83 per cent tin.†

Galipot. The resin from *Pinus maritima*.†

Gallacetophenone. Trihydroxy-acetophenone, $C_8H_8O_4$. Used medicinally as an antiseptic in skin diseases.†

Gallal. Aluminium subgallate, $Al_4(C_7H_2O_5)_3$. An antiseptic and astringent.†

Gallamine Blue. A dyestuff. It is the amide of gallocyanine, $C_{15}H_{14}N_3O_4Cl$. Dyes chromed wool blue, also used in calico printing.†

Gallanilic Blue. (Gallanilic Indigo P and PS, Tannic Indigo). A dyestuff obtained by the action of aniline Gallanilic violets R and B. The mark PS is the sulphonic acid of the product, and it dyes silk and wool from an acid bath, or upon a chrome mordant. The brand P gives indigo blue upon a chrome mordant.†

Gallanilic Green. (Fast Green G, Solid Green G). A dyestuff obtained by the nitration of Gallanilic indigo PS. Dyes chromed wool green.†

Gallanilic Indigo Blue P and PS. See GALLANILIC BLUE.

Gallanilic Violets R and B. (Gallanilic Violet BS). Dyestuffs. They are the anilides of dimethyl- and diethyl-gallocyanines. BS is the bisulphite compound. They dye metallic mordanted wool or silk, reddish-violet, and give bluer shades from an acid bath.†

Gallant. Herbicides based primarily on haloxyfop. (Dow Chemical Co Ltd).*

Gallatite. (Lactite, Lactoform, Cornalith, Ingalite, Lactorite, Sicalite, Proteolite). Casein preparations of a similar type to Galalith (*qv*).†

Gallazine A. A dyestuff obtained by the condensation of Gallocyanine withβ-naphthol-sulphonic acid S. Dyes chromed wool indigo blue. It is also employed in printing.†

Galleine. (Galleine A, Galleine W, Alizarin Violet, Anthracene Violet). A dyestuff. It is pyrogallol-phthalein, $C_{20}H_{12}O_7$. Dyes chrome mordanted wool, silk, or cotton, violet.†

Galleine A. See GALLEINE.

Galleine W. See GALLEINE.

Gallicin. Gallic acid methyl ester, $C_6H_2(OH)_3CO_2CH_3$. Used by oculists as an antiseptic in conjunctivitis.†

Gallipoli Oil. An olive oil used in the textile industries.†

Gallisin. That portion of commercial starch syrup which resists fermentation.†

Gallobromol. Dibromogallic acid, $C_6Br_2(OH_3)COOH$.†

Gallocarmine Blue. A dyestuff allied to Gallocyanine. It is obtained from gallamic acid and nitrosodimethyl aniline.†

Gallocyanine DH and BS. (Fast Violet, Solid Violet, Alizarin, Purple, Gallocyanine RS, BS, and D). A dyestuff. BS is the bisulphite compound. Dyes chromed wool bluish-violet. Employed in printing.†

Gallocyanine RS, BS, and D. See GALLOCYANINE DH and BS.

Gallocyanine S. A dyestuff. It is Gallocyanine sulphonic acid. Dyes chromed wool blue.†

Galloflavine. A dyestuff, $C_{13}H_6O_9$, obtained by the moderate oxidation of gallic acid in aqueous or alcoholic alkaline solution, by means of air. Dyes chromed wool yellow. Used in printing.†

Gallotannic Acid. (Tannic Acid). Digallic acid, $C_6H_2.(OH)_3CO.O.C_6H_2(OH)_2\text{-}COOH$.†

Gallo Violet. A dyestuff. It is a leucopyrogallocyanine.†

Gallstone. A yellow pigment obtained from the gall gladder of oxen.†

Galorn. A proprietary trade name for a casein plastic.†

Galt Glass. Polyester resin bonded glass fibre mouldings with a special surface giving good resistance to weather and chemicals.†

Galvanized Iron. (Zinced Iron). Iron coated with metallic zinc.†

Galvanit. A plating powder consisting of a mixture of the salt of the metal to be deposited (silver for silver plating), and a more electro-positive metal.†

Galvano Lac. A mixture of celluloid varnish with powdered metal.†

Gamanase. A hemicellulase prepared by submerged fermentation of Aspergillus niger. Applications: Particularly suited for applications where a rapid break-down of galactomannans to less viscous products is required. (Novo Industri A/S).*

Gamastan. Globulin, Immune. Applications: immunizing agent. (Cutter Laboratories, Miles Laboratories Inc).

Gambier. See CUTCH.

Gambine B. See DIOXINE.

Gambine G. See GAMBINE Y.

Gambine R. See DIOXINE.

Gambine Y. (Alsace Green J, Gambine G, Steam Green S, Mulhouse Green). A dyestuff. It isα-nitroso-β-naphthol, $C_{10}H_7NO_2$. Dyes iron mordanted fabrics green.†

Gambine Yellow. A dyestuff. It dyes chromed wool yellow.†

Gamboge. (Gummi Gutta). A gum resin, the product of *Garcinia morella* of Siam. It is a yellow pigment used for water colours, also for colouring spirit and other varnishes.†

Gamboge Butter. A fat obtained from *Garcinia morella*..†

Gambria. See JELUTONG.

Gamimune. Globulin, Immune. Applications: immunizing agent. (Cutter Laboratories, Miles Laboratories Inc).

Gamma Acid. 2-Amino-8-naphthol-6-sulphonic acid.†

Gamma-BHC Dust. Insecticide. (Murphy Chemical Ltd).

Gamma-Col. Insecticides containing gamma-HCH. (ICI PLC).*

Gammagee. Globulin, Immune. Applications: immunizing agent. (Merck Sharp & Dohme, Div of Merck & Co Inc).

Gamma-HCH Dust. Garden insecticide. (Murphy Chemical Ltd).

Gammalex. An insecticidal seed dressing in both powder and liquid form. (ICI PLC).*

Gammalin. Insecticides. (ICI PLC).*

Gamman. A Tunisian dyestuff.†

Gammar. Globulin, Immune. Applications: immunizing agent. (Armour Pharmaceutical Co).

Gammasan. Gamma HCH liquid seed dressing. (ICI PLC).*

Gammatox. Livestock dips and sprays. (The Wellcome Foundation Ltd).

Gammatrol. Nucleonic instruments for process control and measurement. (ICI PLC).*

Gammexane. Insecticides. (ICI PLC).*

Gammonativ. Normal immunoglobulin for intravenous use. (KabiVitrum AB).*

Ganex P-904. Water soluble alkylated vinylpyrrolidone polymers. Applications: A dispersant in aqueous agricultural chemical formulations or pigmented skin care products. (GAF Corporation).*

Ganex V. A range of alkylated vinylpyrrolidone polymers, soluble in mineral oil, organic solvents and other polymers. Applications: Used in cosmetics and toiletries as moisture barriers, adhesives, protective colloids and microencapsulating resins, dispersants for pigments and solubilizer for dyes. (GAF Corporation).*

Gangue. The earthy portion of an ore which leaves the metal when reduced.†

Ganicin. Zinc-rich coatings. (Du Pont (UK) Ltd).

Ganister. A rock mineral with a composition corresponding to a pure silica with about 1/10 of its weight of clay. Used in the manufacture of siliceous fire-bricks, and for lining furnaces.†

Ganocide. Fungicides containing drazoxolon. (ICI PLC).*

Gansil. See CHLORAMINE T.

Gant. A barrier cream. (Croda Chemicals Ltd).*

Gantanol. A proprietary preparation of sulphamethoxazole. An antibiotic. (Roche Products Ltd).*

Gantrez. A range of poly(methylvinyl ether/maleic anhydride) copolymers. Applications: Gelling agents, thickeners, stabilizers, explosive stabilizers, anticorrosion coatings and suspending aids. (GAF Corporation).*

Gantrisin. See GANTANOL. (Roche Products Ltd).*

Garamycin. Gentamicin sulphate. A complex antibiotic substance, produced by Micromonospora purpurea nsp. It has three components, sulphates of gentamicin C_1, gentamicin C_2, and gentamicin C_{1A}. Applications: Antibacterial. (Schering-Plough Corp).

Garbritol. Textile auxiliary. (Hickson & Welch Ltd).

Garcinia Oil. See KOKUM BUTTER.

Garcrete. Refractory castable. (Foseco (FS) Ltd).*

Gardenal. Phenobarbitone. (May & Baker Ltd).

Gardenal Sodium. A proprietary preparation of phenobarbitone sodium. A hypnotic. (May & Baker Ltd).*

Gardeniol. Phenyl-methyl-carbinyl-acetate, $C_6H_5.CH(OOC.CH_3).CH_3$. A perfume.†

Gardinal. Textile auxiliary products. (Hickson & Welch Ltd).

Gardinol. A proprietary trade name for wetting agents. They are sodium salts of sulphated higher fatty alcohols.†

Gardlite. A proprietary trade name for a synthetic resin, produced from toluene sulphonamide and formaldehyde.†

Garganine. A madder extract.†

Garj. A bituminous sandstone. It contains from 6-17 per cent bitumen.†

Garlon. Herbicides based primarily on trichlorpyr. (Dow Chemical Co Ltd).*

Garlon 2. Selective herbicide. (ICI PLC, Plant Protection Div).

Garnet Brown. A dyestuff. It is the potassium or ammonium salt of isopurpuric acid, $C_8H_4N_5O_6K$. Dyes wool and silk from an acid bath.†

Garnet Jade. Synonym for Grossular.†

Garnet Lac. See LEMON LAC.

Garnet Red. A pigment. It is usually a red lead coloured with Crocein. Another quality consists of Orange lead coloured with Ponceau 2R and 3R.†

Garnex. Refractory insulating expendable tundish linings. (Foseco (FS) Ltd).*

Garoflam. Flame retardant blends. Applications: Plastics and rubber. (Croxton and Garry Ltd).*

Garomix. Magnesium oxide and zinc oxide blend. Applications: Rubber goods. (Croxton and Garry Ltd).*

Garosorb. Desiccants. Applications: Plastics and rubbers. (Croxton and Garry Ltd).*

Garospers. Dispersed rubber chemicals. Applications: Rubber goods. (Croxton and Garry Ltd).*

Garozinc. Zinc oxides. Applications: Rubber goods. (Croxton and Garry Ltd).*

Garpak. Rammable refractory. (Foseco (FS) Ltd).*

Garseal. Refractory air-setting mortar. (Foseco (FS) Ltd).*

Gartop. Refractory insulating boards for covering the steel in continuous casting tundishes. (Foseco (FS) Ltd).*

Gartube. Insulating shrouds fitted on the bottom of teeming ladles. (Foseco (FS) Ltd).*

Garvox 3G. Insecticide. (FBC Ltd).

Gas Black. (Satin Gloss Black, Hydrocarbon Black, Hydrocarbon Gas Black, Silicate of Carbon, Jet Black, Ebony Black). A carbon black made by the incomplete combustion of natural gas. Used in rubber mixings.†

Gas Blue. See PRUSSIAN BLUE AND SODA BLUE.

Gas Oil. The name applied to all mineral oils intended for the preparation of gas, such as the light oils of brown coal tar, and shale oil. The term is also used for a fraction of Russian petroleum distillation of specific gravity 0.865-0.885, and flashing at 90° C. A burning oil.†

Gas, Pintsch. See OIL GAS.

Gasbinda. Sodium silicate binders for sand cores and moulds using CO_2 hardening process. (Foseco (FS) Ltd).*

Gasil. Micromised silica gel. (Crosfield Chemicals).*

Gasil EBC and EBN. Proprietary compounds of silica used in the paint, resin and plastics industries, *e.g.* in the matting of electron-beam- cured coatings. (Crosfield Chemicals).*

Gaskoid. A proprietary trade name for a rubber jointing resistant to oil and petrol.†

Gastalar. A proprietary preparation containing aluminium hydroxide, magnesium carbonate and sorbitol. An antacid. (Armour Pharmaceutical Co).†

Gastex. A proprietary gas black used in rubber mixings.†

Gastrils. A proprietary preparation containing aluminium hydroxide and magnesium carbonate as a co-dried gel. An antacid. (Smith and Nephew).†

Gastro Caloreen. A proprietary preparation of a polyglucose polymer used as a high-calorie food supplement. (Scientific Hospital Supplies).†

Gastrocote. A proprietary preparation of alginic acid, dried aluminium hydroxide gel, magnesium trisilicate and sodium bicarbonate. An antacid. (MCP Pharmaceuticals).†

Gastrografin. Diatrizoate meglumine. Applications: Diagnostic aid. (E R Squibb & Sons Inc).

Gastrovite. A proprietary preparation of ferrous glycine sulphate, ascorbic acid and calcium gluconate. A haematinic. (MCP Pharmaceuticals).†

Gat 15. Rousselot special gelatine powder. Applications: Developed as a substitute for Gum Arabic. (Rousselot Ltd).*

Gauduin's Fluid. A mixture of finely powdered cryolite, and a solution of phosphoric acid in alcohol. A soldering fluid.†

Gauging Metal. An alloy similar to Delta metal, but containing iron.†

Gaultheria Oil. Oil of wintergreen.†

Gaultheriasalol. Methylsalicylo salicylate, $HO.C_6H_4.COO.C_6H_4.COOCH_3$.†

Gaultheric Acid. Methyl salicylate, $CH_3.C_7H_5O_3$.†

Gauslinite. See BURKEITE.

Gaviscon. A proprietary preparation of alginic acid, sodium alginate, magnesium trisilicate, aluminium hydroxide and sodium bicarbonate. An antacid. (Reckitts).†

Gazelle. Acid phosphates for the baking trade. (Albright & Wilson Ltd).*

GDL. Glucono delta lactone. (Pfizer International).*

GE. 2557. A proprietary trade name for a polyester vinyl plasticizer. (General Electric).†

Geax. A trade name for a synthetic resin-paper product used as an electrical insulation.†

Geblitol. Sodium hydrosulphite. Used as a disinfectant.†

Gechophen. A trade name for chlorphenesin B.P. (British Drug Houses).†

Gedanite. (Friable Amber, Soft Amber). A resin found on the shores of the Baltic. It is a variety of amber, and melts at 150°-180° C. It is also called Soft amber. A variety of amber low in succinic acid.†

Gedeflex. A registered trade mark for dibutyl, butylbenzyl and dioctyl phthalates.†

Gedelite. A registered trade mark for phenolic moulding powders and resins.†

Gedge's Metal. See AICH METAL.

Gedrite. A yellow resin found with Prussian amber. It is also the name for a mineral.†

Gefarnil. A proprietary preparation of gefarnate. A gastrointestinal sedative. (Crookes Laboratories).†

Geko. Inorganic moulding sand binders for foundries, also in connection with lustrous carbon formers to obtain better castings. (Süd-Chemie AG).*

Gel Flo. A slurry explosive. Applications: Used in iron ore mining. (Hercules Inc).*

Gel II. Sodium Fluoride. Applications: Dental caries prophylactic. (Oral-B Laboratories Inc).

Gel Power. Slurry explosives. Applications: Underground non coal-type mines. (Hercules Inc).*

Gel Rubber. A term used for the residue of rubber left undissolved when raw rubber is treated with a solvent.†

Gelamite D. A nitroglycerin dynamite. Applications: Construction and building, explosives, mining, petroleum and related industries. (Hercules Inc).*

Gelaprime F. A nitroglycerin dynamite. Applications: Construction and building, explosives, mining, petroleum and related industries. (Hercules Inc).*

Gelatase. Bacterial proteolytic enzymes. (ABM Chemicals Ltd).*

Gelatin. A colourless and odourless glue, obtained either from calves' heads, or the cartilage and skins of young animals. Used as a food, for making jellies, in photography, for preparing negatives, and for various other purposes.†

Gelatin Carbonite. An explosive consisting of 25 per cent nitroglycerin, 0.7 per cent collodion wool, 7 per cent gelatin, 25 per cent sodium chloride, and 42 per cent ammonium nitrate†

Gelatin Silk. See VANDURA SILK.

Gelatinastralite. An explosive consisting of ammonium nitrate, with some sodium nitrate, up to 20 per cent dinitrochlorhydrin, and maximum amounts of 5 per cent nitroglycerin, and 1 per cent collodion cotton.†

Gelatine Donarit S. An explosive for seismic prospecting which, on account of special sensitisers, can achieve best results even under high pressure. Gelatine Donarit S is supplied in plastic tubes which can be coupled by means of screw threads to form longer charging units. (Dynamit Nobel Wien GmbH).*

Gelatine Donarit 1, Gelatine Donarit 2, Gelatine Donarit 3. Gelatinous ammon dynamites (ammon gelatins, ammon gelignites). They are suitable for use at the surface as well as underground and have a very good water resistance. Their main ingredients are explosive-oil, ammonium nitrate, aromatic nitrocompounds and cellulose nitrate. (Dynamit Nobel Wien GmbH).*

Gelatine Donarit 2 E. Gelatinous explosive. The main ingredients are explosive-oil, ammonium nitrate, nitrocellulose etc. (Dynamit Nobel Wien GmbH).*

Gelatin, Bengal. See AGAR-AGAR.

Gelatin, Ceylon. See AGAR-AGAR.

Gelatin, Chinese. See AGAR-AGAR.

Gelatin, Fish. See ISINGLASS.

Gelatin, Fluid. See OIL PULP.

Gelatin, Japanese. See AGAR-AGAR.

Gelatin, Vegetable. Agar-agar.†

Gelato-glycerin. Glycerin jelly.†

Gelbin. See STEINBUHL YELLOW.

Gelcharg. High viscosity, low cost polymer for gelling water-gel explosives. Applications: Thickener, water-blocking agent. (Hercules Inc).*

Gelcotar. Tar BP coal tar solution in an aqueous gel base. Applications: Used to treat psoriasis and eczema. (Quinoderm Ltd).*

Gelflex. Dimefilcon A. 2-Hydroxyethyl methacrylate polymer with methyl methacrylate and ethylenebis dimethacrylate. Applications: Contact lens material. (Dow Corning Ophthalmics Inc).

Gelkyd. Thixotropic alkyds. (Cray Valley Products Ltd).

Gellert Green. A variety of cobalt green, obtained by roasting and igniting metallic cobalt with 4-5 parts of saltpetre and 8-10 parts of zinc oxide.†

Gelline. A colloidal iodine gel.†

Gelobel. A proprietary trade name for a gelatine dynamite. An explosive.†

Gelofusine. A proprietary preparation of gelatin, calcium chloride and sodium chloride used in the treatment of shock. (Consolidated Chemicals).†

Gelose. See GALACTAN.

Gelosine. A mucilaginous material extracted from a Japanese algae. It is soluble in alcohol and water.†

Geloxite. An explosive consisting of 54-64 parts nitroglycerin, 4-5 parts nitrocotton, 13-22 parts potassium nitrate, 12-15 parts ammonium oxalate, 0-1 part red ochre, and 4-7 parts wood meal.†

Gelusil. A proprietary preparation containing aluminium hydroxide and magnesium trisilicate. An antacid. (Warner).†

Gelva. Polyvinyl acetate resin emulsions/solutions. Applications: Used to make pressure sensitive coatings and laminating adhesives for difficult-to-bond surfaces, used to formulate surface coatings and for specialty paper coatings. (Monsanto Co).*

Gelva RA 737, 784, 788, 858. Acrylic multipolymer solutions for use in pressure sensitive adhesives.†

Gelvatol. Powdered resins. (Monsanto PLC).

Gemex. Surfactants. (BP Chemicals Ltd).*

Gemglo. A proprietary trade name for styrene and methacrylate.†

Geminimycin. A proprietary preparation of gentamycin sulphate. An antibiotic. (Pfizer International).*

Gemlite. A proprietary trade name for a urea formaldehyde synthetic resin.†

Gemme. Crude turpentine.†

Gemonil. Metharbital. Applications: Anticonvulsant. (Abbott Laboratories).

Gemstone. A proprietary trade name for a phenol-formaldehyde cast resin.†

Gemstone M.1.2. A proprietary trade name for a phenolic laminating resin.†

Gen-che. Zinc, manganese, copper, sulphur. Applications: Chelated micronutrient for vegetable, tree and field crops. (Drexel Chemical Company).*

Genacort. A proprietary preparation of hydrocortisone for dermatological use. (Fisons PLC).*

Genal P4300-CM. A proprietary heat resistant but asbestos-free phenolic moulding compound. (General Electric).†

Genamid (R). Polyamide resins, liquid reactable. (Cray Valley Products Ltd).

Genamin. Large range of cationic surfactants which may be primary, secondary or tertiary amines, diamines or amine salts. Solid, liquid or paste forms. Applications: Antistatic agents, fabric conditioner and softener, fibre finishers, water-repellant agents and dewatering agents, wetting agents for oils, dispersing agents for pigments, flushing agents, foaming and wetting agents, spinning bath and viscous additives, flotation chemicals and anti-caking agents for rendering salts free-flowing, corrosion inhibitors, anchoring and wetting agents for tars and bitume[n] surface coatings, lacquers, adhesives and dispersions, disinfectants, hair cosmetics and auxiliaries for leather, textiles, rubber and metal industries. (Hoechst UK Ltd).

Genamin KDM. Cationic surfactant of the quaternary ammonium chloride type, composed of eicosyl/docosyl trimethyl ammonium chloride in paste form. Applications: Range of uses as in Genamin above. (Hoechst UK Ltd).

Genapol. General surfactant properties, lo[w] foaming power and ability to reduce th[e] foam of other surfactants.Stable to acids, alkalies and most metal salts. Applications: Drain aid in machine dis[h] and bottle washing. (Hoechst UK Ltd).

Genapol LRO. Sodium alkyl diglycol ether sulphate in which the alkyl group is derived from natural sources. Supplied as pale yellow clear liquid or slightly yellowish mobile paste. Applications: Foaming and lime soap dispersing agen[t] for shampoos, bubble baths, shower preparations and hand cleansing. (Hoechst UK Ltd).

Genapol ZRO. Sodium alkyl triglycol ethe[r] sulphate in liquid or paste form. Applications: Foaming, wetting and lime soap dispersing agent for shampoos, foam bath and other body cleansing formulations. (Hoechst UK Ltd).

Genasco. A proprietary trade name for a bituminous softener.†

Genasprin. Acetylsalicylic acid tablets. (Fisons PLC, Pharmaceutical Div).

Genatosan Skin Bar. A soap free detergent bar. (Fisons PLC).*

Genclor. Chlorinated polyvinyl chloride - adhesive. (ICI PLC).*

Genelit. A spongy bronze-like bearing metal prepared from a very finely ground mixture of copper, tin, and graphite. When the mixture is heated the graphite burns away, the copper and tin melt together, leaving behind a porous mass which is able to absorb large quantities of lubricating oil.†

Generon. A unit used to separate components. (Dow Chemical Co Ltd).*

Genesolv A Solvent. Trichlorofluoromethane. Applications: Cleaning solvent for metal, plastic and glass. Extractant. (Allied Corp).*

Genesolv D Solvent. Trichlorotrifluoroethane. Applications: Cleaning solvent for electronic, electrical and other high value assemblies. Carrier for specialty coatings. Drying agent. Dielectric fluid. (Allied Corp).*

Genetron Dry Refrigerants. Chlorofluorocarbon refrigerant fluids. Applications: Refrigeration and air conditioning equipment, expansion agents for plastic foam applications, aerosol propellants. (Norplex).*

Geneva. Gin.†

Genexol. A proprietary preparation of triisopropylphenoxypolyethoxyethanol. A spermicide. (Rendell).*

Genisol. Medicated shampoo. (Fisons PLC, Pharmaceutical Div).

Genitron. Blowing agents. (Fisons PLC).*

Genklene. Industrial solvents. (ICI PLC).*

Genoa Oil. Fine olive oil.†

Genochrome. A stabilized photographic colour developer. (May & Baker Ltd).*

Genoform. Methyleneglycolsalicylic ester. It is used in medicine, and is said to split up in the intestines into formaldehyde, salicylic and acetic acids.†

Genomoll P. Trichlorethyl phosphate plasticizer. (Hoechst UK Ltd).

Genoptic Liquifilm, also Genoptic SOP. Gentamicin sulphate. A complex antibiotic substance, produced by micromonospora purpurea nsp. It has three components, sulphates of gentamicin C_1, gentamicin C_2, and gentamicin C_{1A}. Applications: Antibacterial. (Allergan Pharmaceuticals Inc).

Genotherm. Flexible PVC film. (Hoechst UK Ltd).

Genoxal. SEE ENDOXAN. (Degussa).*

Genoxide. A registered trade name for a special quality of hydrogen peroxide for medical purposes. (Laporte Industries Ltd).*

Gensil. Silicone anti-foaming agents. (Bevaloid Ltd).

Genster. A trade mark for a manufacture of carbon.†

Gentamicin. An antibiotic produced by Micromonospora purpurea. Garamycin. Genticin is the sulphate.†

Genteles Green. (Tin Green, Tin-Copper Green). Copper stannate, a pigment.†

Genthane SR. (GS 338). A proprietary trade name for a polyurethane based moulding compound with a temperature range from -60° C. to +160° C.†

Genthelvite. A mineral. It is $(Zn,Fe,Mn)_8Be_6Si_6O_{24}S_2$.†

Gentian. The dried rhizome and roots of *Gentiana lutea*, or of other Gentiana species.†

Gentian Blue 6B. See ANILINE BLUE, SPIRIT SOLUBLE.

Gentian Violet. A dyestuff. It is a mixture of penta and hexamethyl-*p*-rosaniline hydrochlorides. Used as a microscopic stain, and as a bactericidal agent.†

Gentiannie. The zinc double chloride of dimethyldiaminophenazthionium chloride, $C_{14}H_{14}N_3SCl$. Dyes mordanted cotton bluish-violet.†

Gentisin. The yellow pigment of *Gentiana lutea*. It is 1:3:7-trihydroxyxanthone-3-methyl ether.†

Gentisone HC. A proprietary preparation of GENTAMICIN sulphate and HYDROCORTISONEL acetate used in the treatment of infections and inflammations of the ear. (Nicholas).†

Gentran 40. Dextran 40. Applications: Blook flow adjuvant; plasma volume extender. (Travenol Laboratories Inc).

Gentran 75. Dextran 75. A polysaccharide produced by the action of Leuconostoc mesenteroides on sucrose. Average molecular weight: 75,000. Applications: Plasma volume extender. (Travenol Laboratories Inc).

Genu. A range of carrageenan and pectin powders. Applications: Emulsifiers, stabilizers, thickeners and gelling agents. (Hercules Inc).*

Genuzan. A proprietary gum blend. Applications: Thickener and suspender in food and beverage industries, for personal care products and cosmetics and in the pharmaceutical and medical industry. (Hercules Inc).*

Geocillin. A proprietary preparation of carindacillin sodium. An antibiotic used in the treatment of genito-urinary tract infections. (Pfizer International).*

Geoline. See OZOKERINE.

Geolith. Grout for combatting spontaneous combustion in coal mines. (Foseco (FS) Ltd).

Geon 100. A proprietary trade name for polyvinyl chloride.†

Geon 140X31. A proprietary vinyl powder used in the manufacture of fluidised-bed and electrostatic coatings. (BF Goodrich Chemical Co).†

Geon 200. A proprietary trade name for polyvinylidene chloride (Saran).†

Geon 450X23. A proprietary vinyl chloride/acrylic copolymer used in latex paints. (AKU Holland).†

Geon 460X6. A proprietary vinyl chloride copolymer incorporating a synthetic anionic emulsifier. (BF Goodrich Chemical Co).†

Geon 590X3. A proprietary dioctylphthalate plasticized vinyl-chloride copolymer. (BF Goodrich Chemical Co).†

Geon 590X4. A proprietary phosphate ester plasticized vinyl-chloride copolymer. (BF Goodrich Chemical Co).†

Geon 590X6. A proprietary phosphate ester plasticized vinyl chloride copolymer. (BF Goodrich Chemical Co).†

Geopen. A proprietary preparation of carindacillin sodium. An antibiotic used in the treatment of genito-urinary tract infections. (Pfizer International).*

Geoseal. Soil grouting resins. (Borden (UK) Ltd).

Geostone. Class IV dental stone. Applications: For high quality models and dyes. (Bayer & Co).*

Geostop. Accelerated anhydrite for mines roadways stoppings. (Foseco (FS) Ltd).

Geracryl. Patterned extruded acrylic sheet Applications: Glazing, shower screens etc. (Cornelius Chemical Co Ltd).*

Geranine. (Brilliant Geranine, Titan Rose, Patent Atlas Ted, Columbia Red 8B, Direct Scarlet, Thiazine Red, Thiazine Brown). Dyestuffs of a similar nature t Erika B, Diamine rose, and Sultan red.

Geranium. See MAGENTA.

Geranium Crystals. Diphenyl ether, $C_6H_5.O.C_6H_5$. Used in perfumery.†

Geranium Oil. See ROSÉ OIL.

Gerhardt's Caustic. This consists of litharg boiled with potassium hydroxide until i is dissolved, and water is added.†

Germ-i-Tol. Benzalkonium chloride. Applications: Pharmaceutic aid. (Hexcel Chemical Products).

Germalgene. Trichlorethylene, C_2HCl_3.†

German Brass. See BRASS.

German Green. See CHROME GREEN.

German Silver. See NICKEL SILVERS.

German Silver Solder. Usually consists of 5 parts German silver, and 4 parts zinc.†

German Turpentine. See TURPENTINE.

German Yeast. Dried yeast.†

Germanin. Pharmaceutical preparation. Applications: Antitrypanosomal preparation. (Bayer & Co).*

Geronol. Blend of nonionic and anionic surfactants. Applications: Pesticide emulsifiers. (Geronazzo S.p.A).*

Geropon. Sodium salt of polycarboxylic acid. Applications: Emulsion paint antisetting agent/dispersing agent; dispersant for pesticide wettable powders; dispersant for SBR/NBR/-ABS and fluidizing agent for concrete. (Geronazzo S.p.A).*

Gesagard. Triazine herbicide. (Ciba-Geigy PLC).

Gesaprim. Triazine herbicide. (Ciba-Geigy PLC).

esatop. Triazine herbicide. (Ciba-Geigy PLC).

esilit. safety explosives. No. 1 contains 30.75 per cent nitroglycerin jelly, 5.25 per cent dinitrotoluene, 7 per cent sodium chloride, 18 per cent sodium nitrate, and 39 per cent dextrin. No. 2 consists of 30.75 per cent nitroglycerin jelly, 5.25 per cent dinitrotoluene, 22 per cent ammonium nitrate, 21 per cent sodium chloride, and 21 per cent dextrin.†

esteins-tremonit V. An explosive containing ammonium nitrate, nitroglycerin, vegetable meal, potassium perchlorate, and nitro-compounds.†

esteins-Westfalit B and C. Explosives. They are Ammonals containing dinitrobenzene, and dinitrotoluene respectively.†

esterol 50. Progesterone. Applications: Progestin. (O'Neal, Jones & Feldman Pharmaceuticals).

estinal. Amniotic fluid substitute and lubricant. Applications: Veterinary medicine. (Bayer & Co).*

estone. Progesterone BP suitable for oral or injectable administration. Applications: Used for the treatment of premenstrual syndrome and puerperal depression. Also indicated for habitual and threatened abortion and dysfunctional uterine bleeding. (Paines & Byrne).*

etah Wax. See JAVA WAX.

etosedine. Pharmaceutical preparation. Applications: Gastrointestinal sedative and antispasmodic. (Bayer & Co).*

evodin. A proprietary preparation of famprofazone, paracetamol, isopropyl phenazone and caffeine. An analgesic. (Geistlich Sons).†

evral. A proprietary preparation of multivitamins and minerals. (Lederle Laboratories).*

H5. Granular fertilizer. (Fisons PLC, Horticulture Div).

iant Cement. See ACETIC-GELATIN CEMENT.

iant Powders. Explosives. No. 1 contains 40 per cent nitroglycerin, 40 per cent sodium nitrate, 6 per cent rosin, 6 per cent sulphur, and 8 per cent kieselguhr. No. 2 consists of 36 per cent nitroglycerin, 48 per cent potassium or sodium nitrate, 8 per cent sulphur, and 8 per cent rosin.†

Giemsa's Stain. A microscopic stain for white blood corpuscles. It contains eosin, glycerin, and methanol.†

Gies Biuret Reagent. See BIURET REAGENT.

Gilding Metal. A jeweller's alloy of 90 per cent copper, and 10 per cent zinc. Another alloy contains 70 per cent copper, 17.5 per cent brass, and 12.5 per cent tin.†

Gilding Solutions. These generally consist of solutions of gold chloride and potassium carbonate in water. Used for the electro-deposition of gold.†

Gilsonite. Naturally occurring bitumen. Applications: Oil well cements and drilling fluids, printing inks, foundry sand additive, explosives, asphalt pavement sealer, bituminous paints. (Mercian Minerals & Colours Ltd).*

Gilsonite and Design. Uintaite resin. (Chevron).*

Gina. A proprietary preparation of propatylnitrate. Used to treat angina pectoris. (Bayer & Co).*

Ginal. A purifier for sugars. It contains the sodium alginate (from seaweed).†

Gingelly Oil. (Gingili Oil, Teal Oil, Teel Oil, Til Oil, Beni Oil, Benne Oil, Beniseed Oil). Sesamé oil obtained from the seeds of *Sesamum indicum* and of S. *orientale.* Used in the manufacture of margarine and soap, and as a burning oil.†

Ginger Grass Oil. See ROSÉ OIL.

Gingicain. A proprietary preparation of amethocaine gentisate and chlorbutol in aerosol. A local anaesthetic. (Hoechst Chemicals (UK) Ltd).†

Gingili Oil. See GINGELLY OIL.

Ginseng. The root of *Panax quinquefolium.*†

Gin-shi-bui-chi. A Japanese alloy of 30-50 per cent silver with copper.†

Gippon. Antiseptic paint. (JC Bottomley and Emerson).†

Girard's Reagent. Betaine hydrazide hydrochloride.†

Giroflé. See METHYLENE VIOLET 2RA.

Gitalin. A purified digitalin.†

Githagin. See STRUTHIIN.

Giv Tan "F". It is 2-ethoxyethyl para-methoxy cinnamate. Used in suntan lotions.†

Givgard DXN. 6-acetoxy-2,4 dimethyl-m-dioxane non-formaldehyde bactericidal and fungicidal agent for industrial use. Applications: All kinds of industrial water based systems such as emulsions, suspensions and dispersions. (Givaudan SA).*

Givsorb UV-2. Formamidine based high performance UV-absorber. Applications: Polyurethanes, polystyrene, polyamide, PVC and polyolefines. (Givaudan SA).*

Glacialin. See BOROGLYCERIDE.

Glacial Phosphoric Acid. See PHOSPHORIC ACID, GLACIAL.

Glacier Blue. A dye. Dyes silk, wool, and tannined cotton, greenish-blue.†

Gladiolln 0, I, II. Direct cotton dyestuffs, similar to Brilliant congo G.†

Glagerite. A Bavarian white clay.†

Glance Green. See MOUNTAIN GREEN.

Glance Pitch. See MANJAK.

Glance, Iron. See HAEMATITE.

Glascol HA2. A proprietary aqueous solution of acrylic copolymer. The ammonium salt provides hard, water resistant films. (Allied Colloids Ltd).†

Glascol HA4. See GLASCOL HA2. (Allied Colloids Ltd).†

Glascol HN2. A proprietary aqueous acrylic copolymer, the sodium salt of which gives hard and brittle films soluble in water. (Allied Colloids Ltd).†

Glascol HN4. A proprietary aqueous acrylic copolymer the sodium salt of which gives soft and flexible films soluble in water. (Allied Colloids Ltd).†

Glascol PA6. A proprietary acrylic copolymer supplied in the form of low viscosity aqueous solutions. The ammonium salt gives tacky, pressure-sensitive films resistant to water. (Allie Colloids Ltd).†

Glascol PA8. A proprietary acrylic copolymer supplied in the form of low viscosity aqueous solutions. The ammonium salt gives soft, tacky, pressure sensitive films resistant to water. (Allied Colloids Ltd).†

Glascol PN 8. A proprietary acrylic copolymer supplied in the form of low viscosity aqueous solutions. The sodium salt gives soft, tacky, pressure sensitive films soluble in water. (Allied Colloids Ltd).†

Glaser's Salt. Potassium sulphate and sulphite.†

Glass, Fluid. See SOLUBLE GLASS.

Glass-gall. See SANDIVER.

Glasgro. Range of granular fertilizers. (Fisons PLC, Horticulture Div).

Glass Guard. Viscous liquid applied to gla and metal for protection against acid based cleaners. Applications: Historica glass, aluminium window castings etc. (Nova Chemical Inc).*

Glass, Lead. See POTASH-LEAD GLASS.

Glass Liquor. See SOLUBLE GLASS.

Glass-maker's Soap. Manganese dioxide, MnO_2.†

Glass of Antimony. See ANTIMONY GLASS.

Glass, Red Arsenic. See REALGAR.

Glass, Silica. See VITREOSIL.

Glass Silk. Glass wool.†

Glass, Soda. See SODA-LIME GLASS.

Glass Sponge. A patented sponge-like product obtained by mixing glass wool with salt, heating, then dissolving out the salt.†

Glass, Water. See SOLUBLE GLASS.

Glassite. Magnetic oxide of iron from precipitation of ferrous sulphate with caustic soda. Also called black rouge. It is used for buffing.†

Glassona. Cellulose bonded fibre glass bandage. (Smith and Nephew).†

Glauber's Salt. (Mirabilite). Sodium sulphate, $Na_2SO_4.10H_2O$.†

Glauconic Acids. Bluish-violet dyestuffs, obtained by the successive action of pyroracemic acid and formaldehyde on aromatic primary amines.†

Glaucosil. The siliceous residue obtained by extracting greensand with mineral acids. It consists of practically pure silica. It is used as an absorbent for gases.†

Glauramine. A proprietary solution of specially purified auramine. A powerful antiseptic.†

Glaurin. A proprietary trade name for a plasticizer. It is diethylene glycol monolaurate.†

Glazamine. Textile auxiliary chemicals. (ICI PLC).*

Glaze 'N Seal Concrete and Masonry Sealer. Solvent base sealer. Applications: A clear non-yellowing acrylic for application to exterior and interior concrete, stone and masonry type surfaces. (Glessner Corporation Inc (GGI Products) DBA).*

Glaze 'N Seal Waterbase Clear Concrete and Brick Sealer. Waterbase sealer. Applications: an acrylic emulsion for application to interior concrete, stone and masonry type surfaces. Non-flammable. (Glessner Corporation Inc (GGI Products) DBA).*

Glazier's Salt. Potassium sulphate, K_2SO_4.†

Glean. Sulphonyl weedkiller. (Du Pont (UK) Ltd).

Glekosa. See CHLORAMINE T.

Glendion. Polyether and polyester polyols. (Montedison UK Ltd).*

Gleptosil. Gleptoferron. Application: Haematinic. (Fisons PLC, Pharmaceutical Div).

Glessite. A variety of amber melting at 250-300° C.†

Gletvax. Porcine E coli vaccine. (The Wellcome Foundation Ltd).

Glibadone. Glibenclamide. Applications: Pharmaceutical preparation for the treatment of maturity onset diabetes. (M A Steinhard Ltd).*

Glibenese. A proprietary preparation of glipizide. An oral hypoglycemic agent. (Pfizer International).*

Glievor Bearing Metals. One alloy contains 76 per cent lead, 14 per cent antimony, 8 per cent tin, and 2 per cent iron, and another consists of 73 per cent zinc, 9 per cent antimony, 7 per cent tin, 5 per cent lead, and 4 per cent copper.†

Glimmer. Mica.†

Glist. Mica.†

Glitzi. Household pad. (Carl Freudenberg).*

Glizarin Binder. Textile finishing cross-linking agents. (BASF United Kingdom Ltd).

Globe Granite. A stone similar to Ward's stone (*qv*).†

Globulin. (Fibrino-plastic Substance, Paraglobin, Paraglobulin, Serum Casein). Blood protein.†

Glocure. Benzoin ether. (ABM Chemicals Ltd).*

Glofoam. Synthetic wax. (ABM Chemicals Ltd).*

Glokem. Synthetic wax. (ABM Chemicals Ltd).*

Glokill. Heterocyclic biocide. (ABM Chemicals Ltd).*

Glokill PQ. Polymeric quaternary ammonium compound in the form of a pale yellow aqueous solution. Applications: Slow acting non- foaming biocide for water treatment where foam is a problem. (ABM Chemicals Ltd).

Glokill 77. Non surface active biocide. (ABM Chemicals Ltd).

Glomeen. Chlorine releasing sterilant. (ABM Chemicals Ltd).*

Glonoine Oil. See NITRO-GLYCERIN.

Glopol LS6 and L6. Salts of poly-acrylates in aqueous solution. Applications: Low foam anionic surfactants used as dispersing agents for pigments and extenders in aqueous systems; emulsion paint; ceramics; water evaporators; photography. (ABM Chemicals Ltd).

Glopol 461. Polymeric quaternary ammonium salt as a pale yellow slightly viscous liquid. Applications: Cationic surfactant used as resin to impart electro-conductive properties to paper and ceramics. (ABM Chemicals Ltd).

Gloquat. Cationic surfactant, biocide. (ABM Chemicals Ltd).*

Gloquat 1032. Di-alkyl quaternary ammonium salt as a brown viscous liquid. Applications: Cationic surfactant with hydrophobe, anticorrosive, dispersant and electrostatic properties. Used in pigment dispersion; paints and coatings; car washes. (ABM Chemicals Ltd).

Glossite. An abrasive, the active material of which is stated to be black oxide of iron.†

Glover's Wool. See TANNER'S WOOL.

Glowtein. Poultry feed additive - colourant. (Mitchell Cotts Chemicals Ltd).

Gloy. A trade mark. An adhesive said to be a mixture of dextrin and starch, with magnesium chloride.†

Glucagon. A polypeptide hormone produced in the alpha cells of the islets of Langerhans in the pancreas. Proprietary preparation. (Eli Lilly & Co).†

Glucal. A reducing compound, $C_6H_{10}O_4$, obtained by the reduction ofβ-aceto-bromo-glucose with zinc dust and acetic acid.†

Glucalox. Glycalox.†

Glucanal. Proprietary preparations of silver and anthraquinone glucosides.†

Glucanex. A beta-glucanase preparation produced by a selected strain of Trichoderma. Applications: Can be used in all cases where the aim is to improve the clarification and the filtrability of wines made from botrytized grapes. (Novo Industri A/S).*

Glucase. Maltase, an enzyme which converts maltose into glucose.†

Glucidex. A range of dried glucose syrups or maltodextrins. Applications: A wide range of food applications to provide for example bulk with or without sweetness. (Roquette (UK) Ltd).*

Glucina. Beryllium oxide, BeO.†

Gluckauf. A German safety explosive containing ammonium nitrate, wood meal, dinitrobenzene, and copper oxalate.†

Glucodin. A proprietary preparation of glucose and ascorbic acid. A food supplement. (Farley).*

Glucophage. A proprietary preparation of metformin hydrochloride. An oral hypoglycaemic agent. (Rona Laboratories).†

Glucose D. A preparation of glucose with vitamin D and calcium glycero-phosphate.†

Glucose or Sugar Vinegar. A vinegar prepared by the conversion of starch substance into sugar, by the action of dilute acids, followed by fermentation and acetification.†

Glucose Syrup. (Dextrin-Maltose). A partially hydrolysed starch employed in brewing and in confectionery.†

Glucotannin. A tannin found in Chinese rhubarb. It is 1-galloyl-β-glucose.†

Glucox. Glucose oxidase. (John & E Sturge Ltd, Selby).

Glue, Vegetable. See VEGETABLE GLUE and AGAR-AGAR.

Gluferate. A proprietary preparation of ferrous gluconate and ascorbic acid. A haematinic. (Wyeth).†

Glumal. Aceglutamide aluminium. Applications: Anti-ulcerative. (Kyowa Hakko Kogyo).

Glumorin. A proprietary preparation of kallikrein. (FBA Pharmaceuticals).†

Glurenorm. Tablets containing gliquidone. Applications: Oral hypoglycaemic agent used for treatment of non-insulin dependent diabetes. (Winthrop Laboratories).*

Glurub. A proprietary rubber-glue compound for rubber stiffening.†

Glutarol. Colourless evaporative solution containing 10% w/v glutaraldehyde. Applications: For the topical treatment of warts, especially plantar warts. (Dermal Laboratories Ltd).*

Glutoform. See GLUTOL.

Glutoid. See GLUTOL.

Glutol. (Glutoform, Glutoid, Formo-Gelatin). Formaldehyde gelatin obtained by evaporating a solution of gelatin with one of formaldehyde. Employed as an antiseptic powder for wounds. Glutol is also a name applied to a proprietary synthetic resin.†

Glutolin. Methyl cellulose.†

lutril. A proprietary preparation of glibornuride used in the treatment of diabetes mellitus. (Roche Products Ltd).*

lutrin/Goulac. Modified calcium lignosulphonates. Applications: Binders for foundry and refractory brick manufacture. (Reed Lignin Inc).*

lyakol. Diglyceryl ether tetracetate.†

lybrom. Pyrabrom. Applications: Antihistaminic. (CooperVision Pharmaceuticals Inc).

lycamyl. See GLYCERIN OF STARCH.

lycarbin. Glyceryl carbonate.†

lycene. A proprietary trade name for an alkyd synthetic resin used for dentures.†

lyceria Wax. A wax formed in the stem of cane grass, *Glyceria ranirgera*, of Australia. It melts at 82° C.†

lycerin. Commercial glycerol, $C_3H_8O_3$.†

lycerin-formal. A condensation product of glycerol and formaldehyde.†

lycero. See ESTER GUMS.

lycero - piperaz. Basic piperazine glycerophosphate.†

lycero-ester. A trade name for ester gum.†

lycerogen. A mixture of polyhydric alcohols obtained by inversion of sugar to hexose, then reduction with hydrogen and vacuum distilled. The final product is 40 per cent glycerine, 40 per cent propylene glycol and 20 per cent hexyl alcohols.†

lycerox. Water soluble emollients. Applications: Cosmetics and toiletries. (Croda Chemicals Ltd).

lycidyl Ether. 2,3-epoxypropyl (n-butyl, allyl) ether. Applications: Epoxy resin reactive diluent, raw material for curing agents of epoxy resins. (Yokkaichi Chemical Co Ltd).*

Glyco. See PARAFFIN, LIQUID.

Glycobiarsol. Bismuth glycollylarsanilate.†

Glycobrom. The glyceryl ester of dibromo-hydrocinnamic acid.†

Glycoline. See PARAFFIN, LIQUID.

Glyco Metal. An alloy of 70-74 per cent lead, 14-16 per cent antimony, and 8-12 per cent tin.†

Glyconyl. A photographic developer, the active constituent of which is *p*-hydroxyphenylglycine.†

Glycophenol. See SACCHARIN.

Glycosal. The mono-salicylic ester of glycerol.†

Glycosin. See SACCHARIN.

Glycosterine. A proprietary preparation. It is a glycol glyceryl stearate. Used to replace beeswax in certain polishes.†

Glycothymoline. An antiseptic, usually containing thymol, eucalyptol, menthol, borates, bicarbonates, benzoates. and glycerine.†

Glycozone. A proprietary preparation claimed to contain 5 per cent glyceric acid, and 90 per cent glycerin. An antiseptic. Also see PEROXOL AND HYDROZONE.†

Glyecin. Ethylthiodiglycol. A solvent used in treatment of wool for dyeing.†

Glyezin. Textile printing auxiliaries. (BASF United Kingdom Ltd).

Glymol. See PARAFFIN, LIQUID.

Glyoxal 40 per cent. Glyoxal. (Cyanamid BV).*

Glyptal Resins. Resinous products obtained by the interaction of glycerol and organic acids. A resin of this type is made by reacting upon glycerol with oleic acid and phthalic anhydride.†

Glyptal 2557, 2559. Proprietary trade names for polyester plasticisers. (General Electric).†

Glyrol. Glycerin. Applications: Pharmaceutic aid. (CooperVision Pharmaceuticals Inc).

Glysal. The mono-glycol ester of salicylic acid.†

Glysennid. Sennosides. Applications: Laxative. (Sandoz Pharmaceuticals, Div of Sandoz Inc).

Glysobuzole. Isobusole.†

Glyvenol. Tribenoside. Applications: Sclerosing agent. (Ciba-Geigy Corp).

GMD. Pigment dispersions. Applications: Colouration of aqueous and solvent based trade sale paints. (Pacific Dispersions Inc).*

Gmelin's Blue. See TURNBULL'S BLUE.

GMF. p-Quinonedioxime. Applications: A rapid and economical vulcanizing agent when used in conjunction with red lead. Gives fast-curing high-modulus stocks. Recommended for use in butyl curing bags, wire insulation and where a fast-curing high-modulus stock is desired. (Uniroyal).*

GMS 263. A solution of an acrylic resin for use in compounding pressure sensitive adhesives.†

Go Pain. A proprietary analgesic preparation of salicylamide, potassium salicylate, calcium succinate, *p*-amino benzoic acid, vitamins B, and C, and aluminium hydroxide. (De Pree Co, Holland, Michigan).†

Goa Butter. See KOKUM BUTTER.

Gobapur Acide Pur. Sulphuric acid and oleum. Applications: Other chemicals. (Rhone-Poulenc NV/CdF Chimie AZF).*

Gohi Iron. A proprietary trade name for iron containing manganese, sulphur, phosphorus, copper (total less than 0.125 per cent.).†

Gold, Black. See MALDONITE.

Gold Bronze. See SAFFRON BRONZE.

Gold Brown. See BISMARCK BROWN.

Gold Button Brass. See BRASS.

Gold Coinage. See STANDARD GOLD.

Gold Earth. An ochre. See YELLOW OCHRE.†

Gold, Mock. See MOSAIC GOLD.

Gold Ochre, Transparent. See YELLOW OCHRE.

Gold Orange. See ORANGE II, III, IV.

Gold Purple. (Purple of Cassius). A flocculent purple precipitate, produced by the addition of stannous chloride to a solution of gold chloride. Used in the manufacture of artificial gems, and for colouring porcelain.†

Gold Size. Consists of a mixture of 1 part yellow ochre, 2 parts copal varnish, 3 parts linseed oil, 4 parts turpentine, and 5 parts boiled oil.†

Gold Solders. Various alloys of gold, silver, and copper, sometimes with zinc. An ordinary gold solder contains 43 per cent gold, 30 per cent silver, 20 per cent copper, and 7 per cent zinc. A fine gol solder consists of 66 per cent gold, 22 per cent copper, and 12 per cent silver and a hard gold solder contains 37.5 p cent. 18-carat gold, 21 per cent silver, and 21 per cent copper.†

Gold Solder, Fine. See GOLD SOLDERS

Gold Solder, Hard. See GOLD SOLDER

Gold, Sterling. See STANDARD GOLD.

Gold, Talmi. See ABYSSINIAN GOLD.

Gold Yellow. See TROPAEOLINE O, VICTORIA YELLOW, and MARTIUS YELLOW.

Gold-copper. See CHRYSOCHALK.

Golden Acorn. Nutmeg.†

Golden Antimony Sulphide. (Golden Sulphuret of Antimony, Golden Sulphide of Antimony). Antimony pentasulphide, Sb_2S_5.†

Golden Dawn. Refined lanolins of pharmaceutical/cosmetic quality. Applications: Emollients and w/o emulsifiers. (Westbrook Lanolin Co).*

Golden Fleece. Hypo-allergenic and super-refined lanolins. Applications: Emollients and w/o emulsifiers in cosmetics and pharmaceuticals. (Westbrook Lanolin Co).*

Golden Hermetite. Golden gel, non hardening, used as gasket jointing compound. Applications: A high technology gasket compound with excellent chemical and temperature resistance. High tack stabilizes gasket during assembly. Clean to use. (Hermetite Products Ltd).*

Golden Ochre. See YELLOW OCHRE.

Golden Seal. *Hydrastis Canadensis.* Used as a source of yellow dye.†

Golden Sulphide of Antimony. See GOLDEN ANTIMONY SULPHIDE.

Golden Sulphuret of Antimony. See GOLDEN ANTIMONY SULPHIDE.

Golden Syrup. (Drip Syrup). This is the product obtained when raw or brown sugar is dissolved and the solution clarified with animal charcoal and the white sugar crystallized from it.†

Golden Wax. Mould release liquid wax for plastic moulding. (Specialty Products Co).*

Golden Yellow. See MARTIUS YELLOW and TROPAEOLINE O.

Goltix. A water dispersible granular formulation containing 70 per cent w/w metamitron. Applications: To control annual weeds in sugar beet grown on mineral and organic soils and red beet, fodder beet and mangolds grown on mineral soils. (Bayer & Co).*

Gommeline. See BRITISH GUM.

Gon. Tablets for children. (Ward, Blenkinsop & Co Ltd).†

Gonacrine. Preparation of 2,8-diamino-10-methylacridinium chloride and di-amino acridine. (May & Baker Ltd).*

Gonadotraphon FSH. Freeze dried serum gonadotrophin. Applications: Used to treat sterility in males due to defective spermatogenesis and is used to treat secondary amenorrhoea and anovulatory sterility in females. (Paines & Byrne).*

Gonadotraphon LH. Sterile freeze-dried preparation of chorionic gonadotrophin. Applications: Used in males for the treatment of delayed puberty, undescended testes, ectopic testes requiring surgery, oligospermia or aspermia with inactive testes. Used in the female for the indication of ovulation, recurrent abortion, nenorrhazia due to persistent follicular phase and secondary amenorrhoea. (Paines & Byrne).*

Gonadotrophon. .A group of proprietary preparations of human gonadotrophins. (Pharmax).†

Gonal. Santalol.†

Gondang Wax. See JAVA WAX.

Gong Metal. A brass containing 78 per cent copper and 22 per cent zinc.†

Goniosol. Hydroxypropyl methylcellulose. Applications: Pharmaceutic aid. (CooperVision Inc).

Gooch and Eddy Reagent. A reagent used for precipitating magnesium as magnesium carbonate. It contains 180 parts concentrated ammonia, 800 parts water, and 900 parts absolute alcohol, the solution being saturated with ammonium carbonate.†

Good Gulf. Gasoline. (Chevron).*

Goodrite. A proprietary trade name for vinyl plasticizers coded as follows:
GP-223. Dioctyladipate.
GP-235. Octyldecyl adipate.
GP-236. Didecyl adipate.
GP-261. A phthalate.
GP-265. Octyl decyl phthalate.
GP-266. Didecyl phthalate. (BF Goodrich Chemical Co).†

Good-rite Polyacrylates. Homopolymers and copolymers of acrylic acid and their salts in both liqud and dry powder forms. Applications: Water treatment, soap and detergents and dispersants. (BF Goodrich (UK) Ltd).†

Gopmann Solder. An alloy of 49.1 per cent tin, 20.3 per cent zinc, and 26 per cent lead.†

Gordon Superflex D. A trade name for a graft of stereospecific rubber and polystyrene for high impact. Izod 1.35-1.65. Elongation 27 per cent. Tensile strength 3800 p.s.i. Modulus 291,000 p.s.i. (flex). (PBI - Gordon Corp).†

Goùlac. A proprietary trade name for a concentrated sulphite pulp process waste. A binder.†

Goulard Powder. Lead acetate.†

GP-II. Porofocon B. Applications: Contact lens material. (Barnes-Hind Inc).

GP66 Miracle Cleaner. Non-ionic surfactant, synthetic detergent, wetting agents, emulsifiers, builders etc. Applications: A USDA A-1 certified cleaning compound used to clean forklifts, diesel engines, conveyors, robots, concrete floors, ovens, whitewalls, vinyl tops, boats, rugs and upholstery. (GP66 Chemical Corporation).*

G-P-D. Colourant dispersions. Applications: Colouring of non-aqueous coating compositions. (Nueodex Inc).*

G.P.V. A proprietary preparation of penicillin V. (Galen).†

GR Acid. α-Naphtholdisulphonic acid.†

Grafene. A proprietary trade name for a lubricant containing graphite and oils.†

Grafita. A proprietary trade name for a lubricant containing graphite and grease.†

Grafitix (Anti-graffiti). Non inflammable solvent mixture manufactured in aerosol spray. Applications: Removing of paint, felt, applied on stone, cement, bricks, wood, metal and fabrics. (S F C).*

Grafitix (Bâtiment). Non inflammable solvent mixture water miscible, containing special wetting agent. Applications: Specially designed for building trade applications. It strips all surfaces covered with paint, rough coat and plastic facing coatings. (S F C).*

Grafitix (Ravalement). Aqueous product, basic reaction, low viscosity, completely soluble in water. Applications: Cleaning and stripping the outside surfaces of buildings mainly in stone or cement soiled by atmospheric pollution. (S F C).*

Grafix. Graphic arts fixer. (May & Baker Ltd).

Grahamite. An asphaltic substance found in Mexico and Cuba. It is usually associated with mineral matter. It has a specific gravity of 1.17, melts at 175-230° C., up to 45 per cent mineral matter, and is very soluble in carbon disulphide.†

Grain Alcohol. Ethyl alcohol, C_2H_5OH.†

Grain Oil. See FUSEL OIL.

Grain Store Smoke. Insecticide. (ICI PLC, Plant Protection Div).

Grains D'Ambrette. Musk seeds.†

Grains of Kermes. See KERMES.

Grains of Paradise. (Guinea Grains). Seeds of *Amomum melegueta*.†

Grains, Guinea. See GRAINS OF PARADISE.

Gramazine. Residual herbicide. (ICI PLC, Plant Protection Div).

Gramixel. Herbicide containing paraquat and diuron. (ICI PLC).*

Gramonol. Weedkiller containing paraquat and monolinuron. (ICI PLC).*

Gramoxone. Paraquat weedkiller preparations. (ICI PLC).*

Gramoxone 100. Selective herbicide. (FBC Ltd).

Gramp's Solder. An alloy of 60.4 per cent tin, 36.1 per cent zinc, 3 per cent copper, 0.25 per cent lead, and 0.18 per cent antimony. An aluminium solder.†

Gram's Iodine Stain. This consists of 1 gram iodine, 2 grams potassium iodide, and 300 cc distilled water.†

Gram's Stain. A microscopic stain. It contains gentian violet.†

Gramuron. Herbicide containing paraquat and diuron. (ICI PLC).*

Graneodin. A proprietary preparation of neomycin sulphate and gramicidin used in dermatology as an antibacterial agent. (Squibb, E R & Sons Ltd).*

Graneodin Ointment. Gramicidin and neomycin sulphate in ointment base. (E R Squibb & Sons Ltd).

Graney Bronze. An alloy of 76.5 per cent copper, 9.2 per cent tin, and 15.2 per cent lead.†

Granodine. Metal treating compositions. (ICI PLC).*

Granodised Steel. Steel which has been treated with zinc phosphate to give the surface resistance to corrosion.†

Granol. Carbonized granulated peat.†

Granolube. Metal treating compositions. (ICI PLC).*

Granstock. Animal feed additive. (ICI PLC).*

Granuform. Free-flowing, dust-free, small white beads of 90.5 ± 1% paraformaldehyde. Applications: In the plastics industry for production of phenolic, urea, melamine and coating resins; in the chemical and pharmaceutical industries for example for chloromethylation processes; as disinfectant. (Degussa).*

Granulose. The inner part of the starch granule is known by this name.†

Graphalloy, Silver. A trade mark for a moulded graphite impregnated with silver used in electrical brushes and similar appliances.†

Graphite. (Black Lead, Plumbago). A mineral. It is a form of carbon. Used for making electrodes and crucibles, and as a lubricant.†

Graphite Metal. An antifriction and fitting metal. It contains 15 per cent tin, 68 per cent lead, and 17 per cent antimony.†

Graphitic Carbon. The black shiny flakes of carbon present in pig iron.†

aphitic Temper Carbon. The black amorphous carbon present in certain varieties of iron.†

aphitites. Graphites which swell on moistening them with strong sulphuric acid, and then heating them to redness. They are not true graphites.†

appier Cements. Hydraulic limes are slaked and passed through sieves. The hard lumps left on the sieve consist of unchanged limestone and calcium silicates. These are finely ground, and are then known as grappier cements. Le Farge cement belongs to this class.†

raslam. Herbicide. (May & Baker Ltd).

rasselerator 101. A trade mark for a rubber vulcanization accelerator. It is aldehyde ammonia.†

rasselerator 102. A trade mark for a rubber vulcanization accelerator. It is hexamethylenetetramine.†

rasselerator 508. A proprietary rubber vulcanization accelerator. It is butylideneaniline.†

rasselerator 552. A trade mark for a rubber vulcanization accelerator.†

rasselerator 833. A trade mark for a liquid aldehyde amine condensation product. It is a low temperature rubber vulcanization accelerator, and has antioxidant properties.†

ravidox. Pyridoxine Hydrochloride. Applications: Vitamin. (Eli Lilly & Co).

ravulac. Gravure overlacquers. (The Scottish Adhesives Co Ltd).*

razon 90. Selective herbicide. (Murphy Chemical Ltd).

rease, Wakefieid. See YORKSHIRE GREASE.

reen Acid. (Green Sulphonate, Green Sulphonic Acid). Crude mixtures of sulphonic acids from refining petroleum sludge.†

reen Chrome Rouge. Chromium oxide, CrO, used for buffing steels.†

reen, Cinnabar. See CHROME GREEN.

reen Copperas. See IRON VITRIOL.

reen Dot. A smokeless powder. Double-base-type formulation minimises charge weight and moisture absorption. (Hercules Inc).*

Green Earth. (Veronese Green, Veronese Earth). A natural pigment. It contains ferrous iron, silica, magnesia, alumina, and lime. Used as an absorbent for basic dyestuffs.†

Green Gold. Alloys of 75 per cent gold, with from 11-25 per cent silver, and 4-12 per cent cadmium.†

Green Iodide of Mercury. Mercurous iodide, HgI.†

Green Magic. Various granular fertilizer blends. Applications: fertilizers for lawns, gardens and flowers. (Horn's Crop Service Center).*

Green Mordant. Sodium thiosulphate. Used as a mordant for fixing aniline greens on fibre.†

Green Ochre. Usually a mixture of silica, clay, and ferrous hydroxide. Used as a base for cheap lakes.†

Green Oils. A fraction of oil obtained from shale by treating the distillate with sulphuric acid and sodium hydroxide and then again distilling.†

Green Oxide of Chromium. Chromic oxide, Cr_2O_3.†

Green Powder. See METHYL GREEN.

Green Ramie. See RAMIE.

Green Spar. See MALACHITE.

Green Sulphonate. See GREEN ACID.

Green Sulphuric Acid. See GREEN ACID.

Green Ultramarine. A pigment produced by heating kaolin, silica, sodium carbonate, sulphur, coal, and rosin, washing and grinding. It is used in water-colours, and is known as Lime green. When heated again with sulphur, it gives blue ultramarine. Also see ULTRAMARINE.†

Green Verdigris. See VERDIGRIS.

Green Verditer. See MALACHITE and VERDITER BLUE.

Green Vermilion. (English Green, Mineral Green). Pigments. They are usually prepared by treating Prussian blue, rendered soluble by oxalic acid, with a solution of potassium bichromate, and then adding a solution of lead acetate. Varieties of Schweinfurth green pass under this name. Also see CHROME GREEN.†

Green Vitriol. See IRON VITRIOL.

Green Wood Spirits. See ACETONE ALCOHOL.

Green-blue Oxide. An artist's pigment made by calcining a mixture of chromic oxide, aluminium oxide, and a cobalt salt.†

Greenfly & Blackfly Killer. Systemic insecticide. (Fisons PLC, Horticulture Div).

Greenish Blue. A dyestuff. It is the hydrochloride of tri-*p*- tolylrosaniline.†

Greenkeeper. Range of turf fertilizers. (Fisons PLC, Horticulture Div).

Greenkeeper Mosskiller. Turf fertilizer with iron sulphate. (Fisons PLC, Horticulture Div).

Greenol. Liquid iron plant nutrient. (Chevron).*

Grefco. A proprietary trade name for a chrome ore cement used as a refractory.†

Greggio. Crude Sicilian sulphur. Also see RAFFINATE.†

Gregoderm. A proprietary preparation of NEOMYCIN sulphate, mystatin, POLYMYXIN and HYDROCORTISONE used in the treatment of skin disorders. (Unigreg).†

Grenacher's Alum Carmine. An aqueous solution containing 1-5 per cent common or ammonia alum, boiled with 0.5-1 per cent carmine, and filtered. A microscopic stain.†

Grenacher's Borax Carmine. A microscopic stain. It contains 3 grams carmine, 4 grams borax, and 100 cc distilled water. After dissolving by heat, 100 cc of 70 per cent alcohol added, and the solution filtered.†

Grenadine. A dyestuff. It is a British equivalent of Magenta.†

Grenadine S. See ACID MAGENTA.

Grenaldine. See MAGENTA.

Grenat S. See MAROON S.

Grenat Soluble. (Soluble Garnet). A dyestuff. It is the ammonium salt of isopurpuric acid.†

G Resin. A proprietary trade name for cumarone-indene resins.†

Grey Acetate. Crude acetate of lime, prepared with distilled pyroligneous acid. It contains from 80-82 per cent calcium acetate, and 20 per cent water.†

Grey Antimony. Trigonal antimony obtained by allowing molten antimony to cool in a crucible.†

Grey Cast Iron. A cast iron containing much of its carbon in the uncombined state. A typical one contains 94 per cent iron, 3.5 per cent carbon, and 2.5 per cent silicon. It has a specific gravity of 7.0 and melts at 1230° C.†

Grey Fibre. See VULCANISED FIBRE.

Grey Forge Pig. A pig iron usually containing less silicon than other grey irons.†

Grey Gold. An alloy of gold and iron, sometimes with silver. One alloy contains 86 per cent gold, 8.5 per cent silver, and 5.5 per cent iron, and an other 83 per cent gold and 17 per cent iron.†

Grey Mixture. A mixture of 7 parts meal powder, with 100 parts saltpetre and sulphur. Used in fireworks.†

Grey R and B. See INDULINE, SOLUBLE.

Grey Tin. A form of tin obtained by exposing the metal to low temperatures. It reverts back to the ordinary form when heated.†

Grey Ultramarine. A pigment prepared by replacing the sodium constituents of ultramarine with manganese.†

Grifa. Lithium acetyl salicylate.†

Griffith's White. See LITHOPONE.

Grifulvin V. Griseofulvin. Applications: Antifungal. (Ortho Pharmaceutical Corp).

Grignard's Reagent. Magnesium reacts with alkyl and aryl halides, in the presence of ether, forming compounds of the type R.MgX.†

Grilamid. Granules, dried and ready for processing. It is a polymerization product of laurinlactam, a derivative of crude oil. Applications: Extrusion types for the manufacture of sausage skins, films, semifinished products and cable sheathing. Also electrical/electronic, machinery, mining, automotive and building industry. (EMS-Grilon (UK) Ltd).†

Grilamid TR. Amorphous, transparent copolyamide and their modifications. Applications: Granules for extrusion and injection moulding of transparent parts. (EMS-Chemie AG).*

Grilbond. Capped isocyanate compounds. Applications: Adhesive for fibres and rubber in tyre and belt industry. (EMS-Chemie AG).*

Grilene Swiss Polyester. Polyethylene-terephthalate - fibres and mono-filaments. Applications: For garments, non-wovens and sieves. (EMS-Chemie AG).*

Grilesta. Saturated polyester resins in granular form. Applications: Polyester/epoxy and polyester TGIC powder coatings, particularly for short stoving time or low curing temperatures. (EMS-Grilon (UK) Ltd).†

Grilloten ZT40, ZT80, PSE 141G, LSE 87, LSE 87K. Sucrose ricinoleate. Applications: Griloten products are mild solvent-free surfactants. They are non-toxic, non-sensitizing and anti-irritant. Both water and oil soluble, Grillotens' moisturizing and emulsifying properties do not affect formulation. (RITA Corporation).*

Grilon. Granules, dried and ready for processing. Applications: Machinery, electrical/electronic, building, medical and automotive. Extrusion types for food packaging, film production, semi-finished products, monofilaments and cable sheathing. (EMS-Grilon (UK) Ltd).†

Grilon BT. Polyamide-blends. Applications: Granules for extrusion and injection moulding of technical parts. (EMS-Chemie AG).*

Grilon C. Copolyamides PA6/12. Applications: Granules for extrusion to layers of composite films for packaging. (EMS-Chemie AG).*

Grilonit. Solid and liquid, modified and unmodified epoxy resins. Applications: Manufacture of solvent-based and solvent-free coating systems, flooring systems, sealers, bonding layers, crack injection systems, structural adhesives. (EMS-Grilon (UK) Ltd).†

Grilpet. Polyethylene terephthalate and their modifications. Applications: Granules for extrusion and injection moulding of technical parts. (EMS-Chemie AG).*

Griltex. Copolyamide and copolyester granules and powders. For coating of fusible interlinings. Excellent resistance to laundering and dry-cleaning. Short fusing cycles at moderate temperatures. Applications: Automotive, textile and machinery industry. (EMS-Grilon (UK) Ltd).†

Grimm Aluminium Solder. An alloy of 69 per cent tin, 29 per cent lead, 1.5 per cent zinc, and 0.75 per cent silver.†

Grinding Oils. Drying oils such as linseed, etc., used for grinding pigments for paints.†

Grindstone. A sandstone consisting almost entirely of quartz.†

Grisactin. Griseofulvin. Applications: Antifungal. (Ayerst Laboratories, Div of American Home Products Corp).

Grisounites. Coal mine explosives. They contain nitroglycerin, collodion cotton, ammonium nitrate, and sometimes magnesium sulphate in the place of ammonium nitrate. Others contain ammonium nitrate and nitro-naphthalene.†

Grisoutite. An explosive consisting of 53 per cent nitroglycerin, 14.5 per cent kieselguhr, and 32.5 per cent magnesium sulphate.†

Grisovin. A proprietary preparation containing Griseofulvin. An antibiotic. (Glaxo Pharmaceuticals Ltd).†

Gris-PEG. Griseofulvin. Applications: Antifungal. (Dorsey Pharmaceuticals, Pharmaceutical Div of Sandoz Inc).

GRO-HY. Nitrogen, phosphorus, potash plus trace elements as a slow-release fertilizer tablet. Applications: fertilizer for trees, shrubs and bushes. (Envhy Ltd).*

Grodex. Seed germination indicator. (May & Baker Ltd).*

Grodex. A range of emulsifying waxes, used as nonionic surfactants. Applications: Used as emulsifiers, emollients, dispersing, solubilizing and wetting agents by the pharmaceutical and fine chemicals industries. (Croda Chemicals Ltd).*

Grommett Brass. See BRASS.

Gross Solder. An alloy of tin, zinc, aluminium, lead and phosphorus.†

Grossmann Reagent. An ammoniacal solution of dicyandiamidine sulphate, $(C_2H_6ON_4)_2.H_2SO_4$. A reagent for nickel.†

Grossman's Alloy. An alloy containing 87 per cent aluminium, 8 per cent copper, and 5 per cent tin.†

Ground-nut Oil. A non-drying oil obtained from the ground-nut. It consists chiefly of the glycerides of oleic and linolic acids. It is edible, and when hydrogenated is used in the manufacture of margarine. Lower qualities are employed in soap-making.†

Groundhog. Residual herbicide. (ICI PLC, Plant Protection Div).

Grout. A mixture of cement and water. Sometimes including sand.†

Gru-gru Fat. The fat from the seeds of *Acrocomia sclerocarpa*, of the palm family.†

Gruber Solder. An alloy of 60 per cent tin, 25 per cent zinc, 2 per cent aluminium, 10 per cent copper, and 3 per cent cadmium.†

Grudekok. (Lignite Char). Lignite from which about 30-40 per cent water has been expelled. Used in the manufacture of briquettes.†

Grüne's Chromoxyd. See CHROMIUM GREEN.

GT50A. A proprietary preparation of calciferol, aneurin, neostigmine, and carbachol, for use in osteoarthritis. (Geistlich Sons).†

Guaiacum Resin. A resin obtained from the wood of *Guaiacum officinale* and *Guaiacum sanctum*.†

Guaic. Guaiacol resin.†

Guaiol. Tiglic aldehyde, C_5H_8O.†

Guanimycin. A proprietary preparation of dihydrostreptomycin sulphate, sulfaguanidine and kaolin. An anti-diarrhoeal. (Allen & Hanbury).†

Guano. (Bird Manure). A fertilizer. It consists of deposits of excrements and skeletons of birds and animals.†

Guanor Expectorant. A proprietary preparation of ammonium chloride, DIPHENHYDRAMINE hydrochloride, sodium citrate, chloroform and menthol. (RP Drugs).†

Guara. The ground fruits of a species of *Caesalpinia*, from Central and South America. A tanning material.†

Guaranine. See THEINE.

Guardar. A liquid adjuvent for reducing pesticide use and for enhancing the "sticking" of foliar-applied materials to leaf surfaces and seeds. Contents: acetic acid, 2 per cent, petroleum distillates, 1 per cent, polybutenes, 1 per cent, inert ingredients, 96 per cent. Applications: Pesticide reduction. When applying an insecticide, fungicide or certain other pesticides the application rate can be reduced by 50 per cent or more when using only 1.0 to 1.5 oz/acre Guardar mixed in the tank allowing the same ra[te] of kill. (SN Corp/Appropriate Technology Ltd).*

Guayale. (Durango). A rubber obtained from *Parthenium argentatum*.†

Guernsey Blue. See SOLUBLE BLUE.

Guettier Metal. An alloy of 62 per cent copper, 32 per cent zinc, and 6 per cent tin.†

Guido's Balsam. Liniment of opium.†

Guignet's Green. See CHROMIUM GREEN.

Guillaume Alloy. A proprietary trade name for an alloy of 66 per cent iron and 34 per cent nickel. It has a low coefficient of expansion.†

Guillaume Metal. An alloy of 64 per cent copper and 36 per cent bismuth.†

Guinea Carmine B. A dyestuff. It dyes wo[ol] from an acid bath, red shades with a bluish tinge.†

Guinea Grains. See GRAINS OF PARADISE.

Guinea Green B. (Guinea Green BV, Acid Green, Acid Green G). A dyestuff. It i[s] the sodium salt of diethyldibenzyl-diaminotriphenylcarbinoldisulphonic acid. Dyes silk and wool green from an acid bath.†

Guinea Green BV. See GUINEA GREEN B.

Guinea Red. An acid dye for wool and silk.†

Gulf Lite. Charcoal starter, patio torch fuel. (Chevron).*

Gulf Lubcote. Lube oil and anticorrosive. (Chevron).*

Gulf No-Rust. Lube oil and anticorrosive. (Chevron).*

Gulfad-C. Chemical additive for cements used in oil and gas wells. (Chevron).*

Gulfco. Lube oils and greases. (Chevron).*

Gulfcrest. Greases and motor fuel. (Chevron).*

Gulfcrown. Greases. (Chevron).*

Gulfcut. Cutting fluids. (Chevron).*

Gulfgem. Greases. (Chevron).*

Gulfknit. Lube oils. (Chevron).*

Gulfleet. Lube. (Chevron).*

Gulflex. Greases. (Chevron).*

Gulflube. Motor oils. (Chevron).*

Gulfpride. Motor oils. (Chevron).*

Gulfspin. Lube oil. (Chevron).*

Gulftene. Alpha olefins. (Chevron).*

Gulftex. Lubes. (Chevron).*

Gulftow. Lubes. (Chevron).*

Gulftronic. Electrostatic precipitators. (Chevron).*

Gulfwax. Paraffin wax. (Chevron).*

Gum Acacia. See GUM ARABIC.

Gum Animi. See ANIMI RESIN.

Gum Arabic. (Arabin, Calcium Gummate). Gum acacia from *Acacia senegal* and other species. See GUMS AMRITSAR, EAST INDIAN, KHORDOFAN, MOROCCO, PICKED TURKEY, SENEGAL, and SUAKIN.†

Gum Benguela. A semi-fossil copal used in varnishes.†

Gum Benjamin. Benzoin.†

Gum D. A gelatin dynamite (French). It contains 69.5 per cent nitroglycerin.†

Gum Dragon. Gum tragacanth.†

Gum E. A gelatin dynamite (French). It contains 49 per cent nitro-glycerin.†

Gum Elemi. Manila elemi. See MANILA ELEMI.†

Gum Juniper. Gun sandarac.†

Gum Kino. See KINO.

Gum Lac. See LAC.

Gum Lini. A gum made from linseed by treatment with water, then treating the mass with 90 per cent alcohol. The gum is soluble in water and is used as a substitute for gum arabic.†

Gum MB. A gelatin dynamite (French). It contains 74 per cent nitroglycerin.†

Gum Thus. (French Pine Resin). Common frankincense, the crude turpentine from French pine-trees.†

Gum Tragacanth. A gum obtained from shrubs of the *Astragalus* family.†

Gumbo. See BENTONITE.

Gummeline. Dextrin.†

Gummi Gutta. See GAMBOGE.

Gum, Amritsar. An acacia gum from *Acacia modesta*. It is used in calico printing.†

Gum, Arabian. See KHORDOFAN GUM.

Gum, Artificial. See BRITISH GUM.

Gum, Bombay. See EAST INDIA GUM.

Gum, Brown Barberry. See MOROCCO GUM.

Gum, Butea. See KINO, BENGAL.

Gum, Catechu. Catechu. See Cutch.†

Gum, Cowrie. Gum dammar.†

Gum, Cumar. See PARACOUMARONE RESIN.

Gum, Dhak. See KINO, BENGAL.

Gum, Fiddle. Gum tragacanth.†

Gum, Karite. See GUTTA-SHEA.

Gum, Locust Bean. See INDUSTRIAL GUM.

Gum, Mesquite. See MESQUITE GUM.

Gum, Mogador. See MOROCCO GUM.

Gum, Muccocota. See OCOTA COCOTA GUM.

Gum, Para. See BRAZILIAN GUM and JUTAHYCICA RESIN.

Gum, Persian. See INDIA GUM.

Gum, Red. See EUCALYPTUS GUM.

Gum, Seiba. See TUNO GUM.

Gum, Sennaar. See SUAKIN GUM.

Gum, Silk. See CUITE AND SERICINE.

Gum, Starch. See BRITISH GUM.

Gum, Talca. See SUAKIN GUM.

Gum, Talha. See SUAKIN GUM.

Gum, Toonu. See TUNO GUM.

Gum, Touchpong. See POUCKPONG GUM.

Gum, Tunu. See TUNO GUM.

Gum, Turkey. See KHORDOFAN GUM.

Gum, White Sennaar. See PICKED TURKEY GUM.

Gum, Wood. See TREE GUM.

Gum, Zapoto. See CHICLE.

Gun. Metal. An alloy containing from 89-91 per cent copper, 8-11 per cent tin, and 1-2 per cent zinc. The alloy, containing 90 per cent copper, 8 per cent tin, and 2 per cent zinc, has a specific gravity of 8.8, and melts at 1010°C.†

Guncotton. (Nitro-cotton, Pyroxylin, Collodion Cotton). An explosive. It is nitro-cellulose, made by acting upon cotton with nitric and sulphuric acids.†

Gunning's Reagent. A 10 per cent iodine solution in alcohol. Used for the detection of acetone in urine.†

Gunpowder. (Black Powder). A mixture of saltpetre. carbon, and sulphur, in varying proportions. An average one contains 75 per cent potassium nitrate, 10 per cent sulphur, and 15 per cent carbon. An explosive.†

Guntapite. A monolithic refractory material used for lining and repairing linings of steelmaking vessels. (Pfizer International).*

Gurdynamite. See KIESELGUHR DYNAMITE.

Gurjun Balsam or Oil. (Wood Oil). The oleo-resin from the stems of *Dipterocarpus* species.†

Gurley's Metal. An alloy of 86.5 per cent copper, 5.4 per cent zinc, 5.4 per cent tin, and 2.7 per cent lead.†

Gurney's Bronze. An alloy of 76 per cent copper, 15 per cent lead, and 9 per cent tin.†

Gurr. Biological stains and reagents.(BDH Chemicals Ltd).

Gusathion MS. A wettable powder containing 25% w/w azinphos-methyl and 7.5% w/w demeton-S-methyl sulphone. Applications: To control a wide range of biting and sucking pests on agricultural and horticultural crops. (Bayer & Co).*

Guthrie's Eutectic Alloy. An alloy of 47 per cent bismuth, 20 per cent tin, 20 per cent lead, and 13 per cent cadmium.†

Gutta-percha. The coagulated latex of species of *Palaquium* and *Payena*, of Malay, Sumatra, and Borneo. The material is plastic when hot and can be moulded. It was formerly used extensively for electrical insulaters, chemical containers etc.†

Gutta-percha, Artificial. See SOREL'S SUBSTITUTE.

Gutta-percha, Indian. See PALA GUM.

Gutta-shea. (Karite Gum). A product resembling gutta-percha obtained from an African tree, *Bassia Parkii*. The gutta is separated from the fat (Shea butter).†

Gutta-Siak. A low grade gutta-percha gum from *Payena Leerii* and other *Payena* species, of Siak, Sumatra. It is a mixed product with a high resin content.†

Gutta-sundik. A gutta-percha from *Payena* species.†

Gutta-susu. (Assam White, Gutta Gerip, Gutta Singarip, Borneo Rubber). A wild rubber mainly obtained from *Willughbeia firma*, of Borneo.†

Guvacine. An alkaloid derived from pyridine. It is 1,2,5,6-tetrahydro-pyridine-3-carboxylic acid. Its methyl ester is also called Guvacine.†

G Varnish. A proprietary trade name for varnish and lacquer resins made from glycerol and phthalic anhydride.†

Gyle. The resulting liquid after wort is treated with yeast, and aerated in the preparation of vinegar. It contains 6-7 per cent alcohol. The liquid is treated with acetic acid bacteria.†

Gynaflex. A proprietary preparation of noxytiolin, and lignocaine hydrochloride, used in vaginitis. (Geistlich Sons).†

Gyne-Lotrimin. Clotrimazole. Applications: Antifungal. (Schering-Plough Corp).

Gyno-Daktarin. A proprietary preparation of miconazole nitrate used in the veterinary treatment of vaginal candidiasis. (Janssen Pharmaceutical Ltd).*

Gynocardia Oil. An oil similar to chaulmoogra oil. It is obtained from the seeds of *Taraktogenon kurzii*.†

Gynocardic Acid. A term used for the acids contained in the oil expressed from the seeds of *Gynocardia odorata*.†

Gynogen LA 10. Estradiol valerate. Applications: Oestrogen. (O'Neal, Jones & Feldman Pharmaceuticals).

Gynol. Nonoxynol 9. Applications: Spermaticide. (Ortho Pharmaceutical Corp).

Gypsite. A deposit consisting of small grains of gypsum, disseminated through an earthy mass. It is used for the production of wall plastics.†

Gypsona. Plaster of Paris bandage. (Smith and Nephew).†

Gypsum. (Alabaster, Selenite, Annaline, Terra Alba, Satinite, Mineral White, Light Spar, Satin Spar, Lenzit). A mineral. It is calcium sulphate, $CaSO_4.2H_2O$. Used in plasters.†

H

H.11. A proprietary preparation of a polypeptide from male urine, used in the treatment of cancer. (Standard Laboratories).†

HA 819. Polyoxyethylene nonyl phenol. A water soluble material of medium chain length. Used as a detergent. (Honeywell Atlas).†

Hachimycin. An antibiotic produced by *Streptomyces hachijoensis*, used in the treatment of trichomoniasis. TRICHOMYSIN.†

H-acid. 1,8-Diaminonaphthol-3, 6-disulphonic acid.†

Haelan. A proprietary preparation of flurandrenalone used in the treatment of skin diseases. (Dista Products Ltd).*

Haelan-C. Flurandrenolone with clioquinol BP. Applications: For the topical management of those dermatological disorders complicated by bacterial or fungal infection. (Dista Products Ltd).*

Haelan, Haelan-X. A proprietary preparation of flurandrenolone for dermatological use. (Dista Products Ltd).*

Haelan Tape. Flurandrenolone. Applications: Adjunctive therapy for chronic recalcitrant dermatoses that may respond to topical corticosteroids and particularly dry-scaling and localised lesions. (Dista Products Ltd).*

Haemachates. Agates marked with red jasper.†

Haemalum. A microscopic stain. One gram haematoxylin or its ammonium salt is dissolved in 50 cc of 90 per cent alcohol, added to a solution of 50 grams of alum in 1,000 cc water, and filtered.†

Haemate-P. Antihaemophilic factor. Applications: Antihaemophilic. (Hoechst-Roussel Pharmaceuticals Inc).

Haematin. Haematoxylin.†

Haematophan. A haemoglobin preparation, obtained from defibrinated blood.†

Haemoplastin. (Haemostatic Serum). A fluid preparation from blood serum. It consists chiefly of prothrombin and thrombokinase.†

Haemostatic Serum. See HAEMO-PLASTIN.

Haemostop. A proprietary preparation of naftazone. Used to stop capillary bleeding by injection. (Consolidated Chemicals).†

Hagafilm. Water treatment chemicals. (Albright & Wilson Ltd, Phosphates Div).

Hagatreat. Water treatment chemicals. (Albright & Wilson Ltd, Phosphates Div).

Hager's Reagent. Picric acid (1 gram) dissolved in 100 cc water.†

Hagevap. Treatment for sea water evaporators. (Albright & Wilson Ltd).*

Hahnmann's Mercury. Black oxide of mercury.†

Haine's Solution. Copper sulphate, 8.314 grams, is dissolved in 400 cc water, 40 cc glycerol, and 500 cc of 5 per cent potassium hydroxide added. A test for sugar.†

Hair Salt. See EPSOM SALTS.

Halar. A melt processable fluoropolymer comprising a 1:1 alternating copolymer of ethylene and chlorotrifluoroethylene.

Applications: Uses include pharmaceutical packaging, aerosol linings, wire and cable insulation and jacketing, flexible printed circuitry and flat cable. (Allied Chemical Corporation).*

alar E-CTFE. Copolymer of ethylene and chlorotrifluoroethylene. (Allied Chemical Corporation).*

alaurant. A proprietary preparation of halibut liver oil, orange juice and calciferol. Applications: A vitamin mixture. (Ayrton Saunders plc).*

alberland Metal. A brass containing 87 per cent copper and 13 per cent zinc.†

alciderm. A proprietary preparation of halcinonide used in the treatment of skin diseases. (Squibb, E R & Sons Ltd).*

alcion. Triazolam. Applications: Sedative, hypnotic. (The Upjohn Co).

aldol. A proprietary preparation containing heloperidol. Applications: Anti-psychotic, anti-emetic. (Janssen Pharmaceutical Ltd).*

aldol Decanoate. A proprietary preparation containing heloperidol decanoate. Applications: Anti-psychotic. (Janssen Pharmaceutical Ltd).*

aldrate. A proprietary trade name for the 21-acetate of Paramethasone. (Eli Lilly & Co).†

aldrone. Paramethasone Acetate. Applications: Glucocorticoid. (Eli Lilly & Co).

alf-stuff. Refined wood cellulose obtained as sulphite pulp and in the form of thick sheets. It is used either alone or mixed with esparto pulp in the manufacture of paper.†

alf-wool Black. See UNION BLACK B, R.

Halibol. A proprietary preparation of halibut-liver oil with irradiated ergosterol.†

Haliborange. Orange flavoured and coloured chewable sugar coated tablets containing Vitamin A 750 ug, Vitamin D 5 ug, Vitamin C 25 mg Applications: As a vitamin supplement for adults and children. (Evans Medical).*

Halite. See ROCK SALT.

Haliverol. A proprietary preparation of halibut-liver oil with irradiated ergosterol.†

Hallcomid. Series of dimethyl amides. Applications: Grinding aids, dispersing agents, insecticides, bio-agents. (The C P Hall Company).*

Haloflex. Vinyl acrylic copolymers. (ICI PLC).*

Halog. Halcinonide. Applications: Anti-inflammatory (E R Squibb & Sons Inc).

Halon. A proprietary polytetra fluoroethylene. See HOSTAFLON C2. (Allied Chemical Corporation).*

Halotestin. Fluoxymesterone. Applications: Androgen. (The Upjohn Co).

Halothane. 2-Bromo-2-chloro-1,1, 1-trifluoroethane. Fluothane.†

Halothane M & B. A proprietary preparation of halothane. Applications: Volatile anaesthetic. (May & Baker Ltd).*

Halowax 1014. A proprietary trade name for hexachlornaphthalene. (Bakelite Corporation).†

Halowax 4000 B-2. A proprietary trade name for a chlorinated hydrocarbon vinyl plasticizer. (Union Carbide (UK) Ltd).†

Haloxil. Haloxon/oxyclozanide anthelmintic. (The Wellcome Foundation Ltd).

Halphen Reagent. A solution of sulphur (1 per cent. in carbon disulphide. Used to test for cotton-seed oil. To 1 cc of oil add 1 cc of reagent and 1 cc of amyl alcohol, and heat. A red colour is given with cotton-seed oil.†

Halso AG-125. A clear, colourless liquid having a monochlorotoluene isomeric composition different to that of Halso 99. It is for use only as an agricultural solvent. (Occidental Chemical Corp).*

Halso 99. A solvent grade of monochlorotoluene. It is a clear, colourless liquid with the characteristic odour of ring chlorinated aromatic compounds. Applications: Used as a solvent in many industries including paints and plastics and as a dye carrier, sludge solvent, rubber solvent and in metal-cleaning formulations and carbon-removal compounds. (Occidental Chemical Corp).*

Halumin. An aluminium alloy with 1.48 per cent copper, 2 per cent nickel, 2.3 per cent manganese, 0.47 per cent iron, and 0.09 per cent silicon. It is specially resistant to corroding agents.†

Halycitrol. Vitamin emulsion. Halibut liver oil. Each ml is standardized to contain not less than 276 mcg (920 iu) Vitamin A and 1.9 mcg (76 iu) Vitamin D. (LAB Ltd).*

Hamameli Tannin. A tannin originally isolated from *Hamamelis virginica*.†

Hambergite. A mineral, $BeOH.BeBO_3$.†

Hamburg Blue. See MOUNTAIN BLUE and PRUSSIAN BLUE.

Hamburg White. A pigment consisting of 1 part white lead and 2 parts heavy spar.†

Hamilton Metal. A brass containing 67 per cent copper and 33 per cent zinc. An alloy of 90 per cent zinc, with small quantities of copper, lead, antimony, and phosphor tin, is also known as Hamilton metal.†

Hammer Slag. A basic silicate of iron produced and used in the puddling process for iron.†

Hammonia Metal. An alloy of 64.5 per cent tin, 32.2 per cent zinc, and 3.2 per cent copper.†

Hamonite. An activated carbon, made from peat.†

Hampden Steel. A proprietary trade name for a chromium tool steel containing 12.5 per cent chromium, 0.25 per cent nickel, 0.25 per cent manganese, and 2.1 per cent carbon.†

Hamposyl L, C and O. Anionic surfactants of the sarcosinate type, either as sodium salts or in acid form. L is lauroyl sarcosine, C is cocoyl sarcosine and O is oleoyl sarcosine. Liquid, powder or waxy solid forms. Applications: Foaming and wetting agents, mild to skin and eyes, imparting softness and lubricity. Compatible with quaternary ammonium germicides, tolerant to hard water, soluble in highly alkaline systems, stability in mildly acid formulations, viscosity building, low cloud point. Used in rug shampoo; delicate fabric detergents; dishwashing detergents; rinse aids; alkaline cleaners; lubricants; hair shampoos; dentifrices; mouth washes; hand soaps; corrosion inhibition; wool scouring; acid dyeing; emulsion polymerization; polymer films; leather; petroleum production and recovery. (W R Grace Ltd).

Handi-Wrap. Plastic film for use as a household wrapping material. (Dow Chemical Co Ltd).*

Hang-ge. A drug. It is the root nodule of *Pinellia tuberifera*. Used as an antemetic in Japan and China.†

Hansa Oil. A proprietary trade name for polymerized marine animal oil for soap manufacture.†

Hansa Yellow G. (Monolite Yellow, Pigment Fast Yellow Conc. New). A dyestuff obtained by coupling diazotized *m*-nitro-*P*-toluidine with acetoacetic anilide.†

Hanus' Iodine Bromide Solution. Iodine bromide (10 grams), dissolved in 500 cc of glacial acetic acid. Used for determining the iodine value of fats and oils.†

Hard Aluminium. An alloy of 77 per cent aluminium, 11 per cent zinc, and 11 per cent magnesium. Another alloy contains copper in the place of zinc.†

Hardcote. Dressing for cores and moulds. (Foseco (FS) Ltd).*

Hard Cure. Balanced blend of sodium, potassium and meta silicates combined with surface tension reducing agents. Applications: For application to freshly placed concrete following finishing. (Secure Inc).*

Hardened Rosins. Metallic resinates prepared by heating rosin with metallic oxides, usually calcium, magnesium oxide and zinc oxide.†

Hardened Rubber. See EBONITE.

Hardenite. A collective name for austenite and martensite of eutectoid composition.†

Hard-finish Plaster. Plaster made from oven-burnt gypsum dipped in alum solution, and again calcined.†

Hard Gold Solder. See GOLD SOLDERS.

Hard-head. The name by which the impurities obtained from the refining of tin are known.†

Hardite. Heat-resisting alloys containing 55-65 per cent nickel, 15- 18 per cent chromium, 1-4 per cent silicon, 1-2 per cent manganese, the balance being iron.†

Hardite X. A heat-resisting alloy. It contains 82-86 per cent nickel, 10-13 per cent chromium, and 2 per cent manganese.†

Hard Jatoba. A Brazilian copal resin.†

Hard Lac Resin. A hard resin obtained by extraction of shellac with alkaline reagent to remove the soft resin.†

Hard Lead. (Antimony Lead). An alloy of lead with 10-30 per cent antimony. It is also known as antimony lead, and is used as a type metal.†

Hard Metal. The name usually applied to a tin-copper alloy, containing 1 part tin, and 2 parts copper.†

Hard Platinum. Platinum containing from about 5-30 per cent iridium.†

Hardset. Heat-curing hard rubber compound. (Hardman Inc).*

Hard Silver Solder. See SILVER SOLDERS.

Hard Solder. An alloy of 86 per cent copper, 9 per cent zinc, and 4 per cent tin.†

Hard Zinc. An alloy of 92 per cent zinc, 5 per cent iron, and 3 per cent lead.†

Hardware Brass. See BRASS.

Hargus Steel. A proprietary trade name for a die steel containing 1.0 per cent manganese, 0.35 per cent nickel, and 1.0 per cent carbon.†

Harle's Solution. A solution of sodium arsenite.†

Harlington Bronze. See HARRINGTON BRONZE.

Harmaline. See MAGENTA.

Harmogen. A proprietary preparation of piperazine oestrone sulphate used in the treatment of menopausal symptoms. (Abbott Laboratories).*

Harmomang A and B. Iron alloys containing carbon, manganese and molybdenum. Proprietary alloys.†

Harmonia Bronze. An alloy of 57 per cent copper, 40 per cent zinc, 1.8 per cent iron, and 0.4 per cent lead.†

Harmony. Lube oil, hydraulic fluid. (Chevron).*

Harrier. Weedkiller containing mecoprop, 3,6-dichloropicolinic acid and ioxynil as potassium salts. (ICI PLC).*

Harrington Bronze. (Harlington Bronze). An alloy of 55.75 per cent copper, 42.5 per cent zinc, 1 per cent tin, and 0.75 per cent iron. A white bearing metal.†

Harringtonite. An Irish mineral. It contains 41.4 per cent. SiO_2, 30.2 per cent. Al_2O_3, 11.2 per cent. CaO, 5.2 per cent. Na_2O, and 12.5 per cent. H_2O.†

Harris' Haematoxyiin Stain. A microscopic stain. It contains 1 gram haematoxylin, 10 cc alcohol, 20 grams alum, 0.5 gram mercuric oxide, and 200 cc water.†

Harrison's Indicator. A small amount of starch boiled with a few c.cs. of water, adding to it 100 cc of a freshly prepared 10 per cent potassium iodide solution.†

Hartin. A white resin found in the lignite in Austria.†

Hartite. A similar resin to Hartin, and found with it.†

Hartmetall. See CAMITE.

Hartolan. Wool wax alcohols. (Croda Chemicals Ltd).*

Hartolite. Modified woolwax alcohols. (Croda Chemicals Ltd).

Hartshorn Oil. See BONE OIL.

Hartshorn Salt. See SALT OF HARTSHORN.

Hartshorn Spirit. See SPIRIT OF HARTSHORN and VOLATILE ALKALI.

Harvesan. Mercurial seed dressing. (Boots Pure Drug Co).*

Harvite. A proprietary trade name for a shellac compound.†

Hascrome. A proprietary alloy. It is a manganese-chromium-iron welding rod.†

H.A. Solvent. A proprietary solvent. It is cyclohexyl acetate. It is a solvent for cellulose acetate and nitrate, rosin, rubber, oils, and metallic resinates.†

Hastelloy Alloys. A registered trade mark for nickel base acid and resistant alloys.†

Hatchette Brown. See FLORENTINE BROWN.

Hatchettine. See BITUMEN.

Havapen. A proprietary preparation containing Penamecillin. An antibiotic. (Wyeth).†

Hayem's Solution. A solution of 5 grams sodium sulphate, 1 gram sodium chloride, and 0.5 gram mercuric chloride in 200 cc water. Used in the examination of blood corpuscles.†

Haylite No.1. An English explosive. It contains 25-27 per cent nitroglycerin, 0.5-1.5 per cent collodion cotton, 19-21 per cent potassium nitrate, 19-21 per cent barium nitrate, 12-14 per cent wood meal, 6-8 per cent mineral jelly, and 10-12 per cent ammonium oxalate.†

Haylite No. 3. An explosive. It consists of 9.5 per cent nitroglycerin, 60 per cent ammonium nitrate, 5 per cent wood meal, 19.5 per cent sodium chloride, and 5 per cent ammonium oxalate.†

Haynes Alloy No. 25. A proprietary trade name for an alloy of cobalt, nickel, tungsten and chromium. Possesses exceptional mechanical properties up to 1800° F†

Haynes Metals. Alloys of from 10-75 per cent iron, 20-30 per cent chromium, and 5-25 per cent cobalt. A harder alloy contains 45 per cent cobalt, 40 per cent tungsten, and 15 per cent chromium, and a softer one, 62 per cent cobalt, 28 per cent tungsten, and 10 per cent chromium. They are stated to be non-corrosive.†

Haynon. A proprietary preparation of CHLORPHENIRAMINE. An anti-allergic drug. (RP Drugs).†

Hayphryn. A proprietary preparation of phenylephrine hydrochloride and thenyldiamine hydrochloride. An anti-allergic nasal spray. (Winthrop Laboratories).*

Haysite. Thermoset polyester resin, fillers and glass. Applications: Electrical insulation, corrosion resistant. Sold as custom moulded parts, sheets, moulding compound SMC-BMC and pultruded shapes mil spec, nema and UL recognised materials. (Haysite Reinforced Plastics).*

Haystellite. See CAMITE.

Hazol. A proprietary preparation of oxymetazoline hydrochloride. A nasal decongestant. (Allen & Hanbury).†

HB-40. Partially hydrogenated terphenyl. Applications: plasticizer for vinyl sheeting, films and fabric or paper coatings and for vinyl protective coatings and adhesives. (Monsanto Co).*

HBR. A compressible closed-cell polyethylene foam rod. Applications: Placed in a joint before applying a sealant and to assist the sealant in assuming the proper configuration. Als serves as a bond breaker to prevent three-side adhesion of sealant to joint substance. (Hercules Inc).*

H-B-Vax. Hepatitis B surface antigen. Applications: A vaccine for protection against hepatitis B. (Merck Sharp & Dohme).*

Head and Shoulders. Pyrithione Zinc. Applications: Antibacterial; antifungal; antiseborrheic. (Procter & Gamble Co)

Heart Shape Indigestion Tablets. A proprietary preparation of sodium bicarbonate, magnesium carbonate and calcium carbonate. Applications: An antacid. (Ayrton Saunders plc).*

Heavithane. A proprietary polyester based polyurethane elastomer crosslinked by diols. (HMC Wheels).†

Heavy Benzine. See LIGROIN.

Heavy Carburetted Hydrogen. See OLEFIANT GAS.

Heavy Spar, Artificial. A pigment. It is barium white.†

Heberden's Ink. Mist. Ferri. Aromat.†

Heckel's Solution. A solution containing sodium sulphite, benzoic acid, and water.†

Hecla. A proprietary trade name for alloy steels with the following codings :

Hecla 35 contains 0.15 per cent. carbon, 0.25 per cent. silicon, 0.04 per cent. sulphur, 0.04 per cent. phosphorus and 0.70 per cent. manganese. Used to manufacture forged steel pressure vessels.

Hecla 37 contains 0-04 per cent. car bon, 0-25 per cent. silicon, 0.04 per cent. sulphur, 0.04 per cent. phosphorus and 0.70 per cent. manganese and is used for the manufacture of forged steel pressure vessels.

Hecla 115 contains 0.35 per cent. carbon, 0.25 per cent. silicon, 0.04 per cent. sulphur, 0.04 per cent. phosphorus, 0.7 per cent. manganese and 1.00 per cent. nickel and is used for manufacturing forged steel pressure vessels.

Hecla 135 contains 0.60 per cent. carbon, 0.30 per cent. silicon, 0.30 per cent. manganese, 2.00 per cent. chro mium, 2.00 per cent. nickel and 0.45 per cent. molybdenum. Used for the manufacture of forged high tensile steel pressure vessels.

Hecla 138 contains 0.30 per cent. carbon, 0.30 per cent. silicon, 0.30 per cent. manganese, 2.00 per cent. chromium, 2.00 per cent. nickel and 0.45 per cent. molybdenum. It is used for the manufacture of forged high tensile steel pressure vessels.

Hecla 138H contains 0.40 per cent. Carbon, 0.30 per cent. silicon, 0.04 per cent. sulphur, 0.04 per cent. phosphorus, 0.60 per cent. manganese, 0.70 per cent. chromium, 2.70 per cent. nickel, 0.50 per cent. molybdenum and 0.25 per cent. vanadium (optional). It is used in the manufacture of forged high tensile steel pressure vessels.

Hecla 155 contains 0.12 per cent. carbon, 0.20 per cent. silicon, 0.04 per cent. sulphur, 0.04 per cent. phosphorus, 0.40 per cent. manganese, 2.30 per cent. chromium, 1.00 per cent. molybdenum. It is used in the manufacture of forged steel pressure vessels for use at higher temperatures.

Hecla 174 contains 0.30 per cent. carbon, 0.30 per cent. silicon, 0.04 per cent. sulphur, 0.04 per cent. phosphorus, 0.50 per cent. manganese, 5.00 per cent. chromium, 1.30 per cent. molybdenum and 0.90 per cent. vanadium. It is used in the manufacture of forged steel pressure vessels for hydrogen service and high tensile purposes.

Hecla 180 contains 0.40 per cent. carbon, 0.30 per cent. silicon, 0.04 per cent. sulphur, 0.04 per cent. phosphorus, 0.60 per cent. manganese, 1.00 per cent. chromium, 3.2 per cent. nickel, 0.50 per cent molybdenum and 0.25 per cent. vanadium. It is used in the manufacture of forged high tensile steel pressure vessels.

Hecla 306 contains 0.30 per cent. carbon, 0.20 per cent. silicon, 0.04 per cent. sulphur, 0.04 per cent. phosphorus, 0.50 per cent. manganese, 3.00 per cent. chromium, 0.50 per cent. molybdenum and 0.20 per cent. vanadium. It is used in the manufacture of forged steel pressure vessels for hydrogen service.

Hecla 307 contains 0.18 per cent. carbon, 0.20 per cent. silicon, 0.04 per cent. sulphur, 0.04 per cent. phosphorus, 0.50 per cent. manganese, 3.00 per cent. chromium, 0.50 per cent. molybdenum and 0.20 per cent. vanadium. It is used in the manufacture of forged steel pressure vessels for hydrogen service. (Hadfield, George & Co).†

Hecla Powder. An explosive. It is similar in composition to Giant powder (*qv*).†

Hector Bases. Basic substances obtained by the oxidation of thioureas with hydrogen peroxide. They are stated to be good vulcanization accelerators.†

Hectorite Laponite. A sodium magnesium lithium fluoro silicate. A white, iron free suspending and gelling agent for aqueous systems.†

Hedgehog Crystals. Crystals of ammonium urate found in urinary deposits.†

Hedulin. Phenindione. Applications: Anticoagulant. (Merrell Dow Pharmaceuticals Inc, Subsidiary of Dow Chemical Co).

Hegolit 3. A patented product which is a preparation of higher aliphatic alcohols, with a melting-point of 50° C. A plasticizer.†

Hégor. Shampoo, to wash and condition hair. (Richardson-Vicks Inc).*

Heidenhain's Chrome Haematoxyiin. A microscopic stain. It is produced by staining the object in 0.33 per cent solution of haematoxylin in water, then soaking in 0.5 per cent solution of potassium chromate.†

Heidenhain's Iron Haematoxylin. (*a*) A solution of 4 grams iron alum in 100 cc distilled water. (*b*) A solution of 0.5 gram haematoxylin crystals in 100 cc distilled water. It is used as a microscopic stain by placing sections in (*a*) for 1 hour, washing, and then into (*b*) for 1/2 hour.†

Heiloy. A proprietary trade name for stainless steels used for dairy utensils.†

Helarion. A pelleted bait containing metaldehyde. Applications: Control of slugs and snails. (Fisons PLC).*

Heleco. A proprietary pyroxylin preparation.†

Helenite. See ELATERITE.

Heliane. Vat dyes. (ICI PLC).*

Helicon. See ASP1R1N.

Heligoland Blue 3B. A direct cotton blue.†

Heligoland Yellow. A dyestuff. Dyes cotton yellow.†

Helindone Blue B. A dyestuff prepared by the oxidation of 9-chloro-3(2H) thio hanthrenone.†

Helindone Blue 2B. A dyestuff. It is 5:5'-dibromoindigo.†

Helindone Fast Scarlet R. A dyestuff. It is 5:5'-dichloro-6: 6'-di-ethoxythioindigo.†

Helindone Grey BR. A dyestuff. It is dichloro-7, 7'-diaminothioindigo.†

Helindone Grey 2B. A dyestuff. It is 7, 7'-diaminothioindigo.†

Helindone Orange. D. A dyestuff. It is dibromo-6,6'-diaminothioindigo.†

Helindone Orange R. A dyestuff. It is 6,6'-diethoxythioindigo.†

Helindone Pink BN. A dyestuff. It is 6,6'-dibroromodimethylthioindigo.†

Helindone Red B.A. dyestuff. It is 5,5-dichlorothioindigo.†

Helindone Red R. A dyestuff prepared by the oxidation of thioindoxyl.†

Helindone Red 3B. (Durindone Red 3B). A dyestuff. It is 5,5'-dithloro-6:6'-dimethylthioindigo.†

Helindone Scarlet S. A dyestuff. It is 6,6'-dithioethylthioindigo.†

Helindone Violet 2B. A dyestuff. It is dichlorodimethyldimethoxythioindigo.†

Helindone Yellow 3GN. A dyestuff. It is a urea derivative ofβ-amino-anthraquinone.†

Helio. Pigments. Applications: For surface coatings, printing inks, wallpaper, coloured coated paper and lake producing industries. (Bayer & Co).*

Helio and Helio Fast, Levanox. Acqueous preparations of organic and inorganic pigments. (Bayer UK Ltd).

Helio Fast Red RL. (Pigment Fast Red HL, Monolite Fast Scarlet R, RN, Sitara Fast Red, Lithol Fast Scarlet R, Permanent Red 4R, Pigment Fast Scarlet 3L Extra). A dyestuff prepared from diazotized *m*-nitro-*p*- toluldine, andβ-naphthol.†

Helio Fast Yeliow RL. A dyestuff. It is the dibenzyl derivative of a mixture of 1,5 and 1,8-diaminoanthraquinones.†

Heliochrysin. (Sun Gold). The sodium salt of tetranitro-α-naphthol.†

Heliofil. Pigment dyestuffs in paste form for the dope dyeing of viscous continuous filaments and stable fibres. (Bayer & Co).*

Heliolac. A nitro-cellulose lacquer.†

Heliophan. Homosalate. Applications: Ultraviolet screen. (R W Greeff & Co Inc).

Helizarin. Dyestuffs for textile fibres. (BASF United Kingdom Ltd).

Helmet Brass. See BRASS.

Helmex. A proprietary preparation of pyrantel pamoate. An antihelmintic. (Pfizer International).*

Helmezlne. Preparations of salts of piperazine. (Allen & Hanbury).†

Hema-Combistix. A proprietary test strip comprised of four separate tests. (1) Methyl red and bromothymol blue for pH. (2) A buffered mixture of glucose oxidase, peroxidase, *o*-toluidine, and a red dye for glucose. (3) Buffered tetrabromophenol blue for glucose. (4) O-toluidine for blood. Used to test urine. (Ames).†

Hemastix. A proprietary test-strip containing O-tolidine and an organic peroxide, used to detect the presence of blood in urine. (Ames).†

Hematest. A proprietary test tablet of *o*-toluidine and an organic peroxide, used to detect blood in faeces and urine. (Ames).†

Hematine Paste and Powder. See LOGWOOD.

Hemi-celluloses. These are contained in plant cell walls. They are reserve celluloses and are readily converted into hexoses and pentoses by acids and by the enzyme cytase.†

Hemlock Oil. See OIL OF HEMLOCK.

Hemofil. Antihaemophilic factor. Applications: Antihaemophilic. (Hyland Therapeutics, Div of Travenol Laboratories Inc).

Hemoplex. Vitamin B complex injection with liver. (Paines & Byrne).*

Hemostatin. Adrenalin.†

Hemoterge. Rinse and reference solution. Applications: Cleaning and blanking of automatic haemoglobinometers. (Coulter Electronics Ltd).*

Hemoxone. Weedkiller containing dichlorprop and MCPA as potassium salts. (ICI PLC).*

Hempel's Solution. A solution made by mixing a solution of 120 grams potassium hydroxide in 80 cc water with a solution of 5 grams pyrogallol in 15 cc water. Used for the determination of oxygen.†

Hemrids. A proprietary preparation of phenylephrine hydrochloride, amethocaine hydrochloride, bismuth carbonate and tyloxapol. Anaesthetic suppositories. (Bayer & Co).*

Henna. Derived from the leaves and roots of *Lawsonia inermis* or *L. alba.* Used as a dye, and for staining the hair.†

HEP. *N*-(2-Hydroxyethyl)-2 pyrrolidone. Applications: Solvent reaction intermediate, textile auxilliaries, cosmetic ingredient. (GAF Corporation).*

Hepacon. A proprietary name for a group of preparations containing lever extract and/or vitamins of the B complex. (Consolidated Chemicals).†

Hepacort Plus. A proprietary preparation of heparin, hydrocortisone acetate, and methyl parahydroxybenzoate. Local anti-inflammatory cream. (Rona Laboratories).†

Heparin Lock Flush. Heparin sodium. Applications: Anticoagulant. (Abbott Laboratories).

Hepar Sulphur. Potassium sulphide.†

Hepatolite. Technetium Tc 99m Disofenin. Applications: Diagnostic aid; radioactive agent. (Dupont-NEN Medical Products).

Hep-B-Gammagee. Hepatitis B immune globin. Applications: Post exposure prophylaxis against hepatitis B. (Merck Sharp & Dohme).*

Hepicebrin. Hexavitamin capsules. Applications: Vitamins, combination. (Eli Lilly & Co).

Heptaline. 3-Methylcyclohexanol.†

Heptaminol. 6-Amino-2-methylheptan-2-ol.†

Heptanal. Heptoic aldehyde, $CH_3(CH_2)_5CHO$.†

Hepteen Base. Heptaldehyde-aniline reaction product. Fast-curing, high temperature accelerator with maximum processing safety. Works well with basic furnace blacks used in natural rubber pure gum stocks. (Uniroyal).*

Heptene. A proprietary rubber vulcanization accelerator. It is heptaldehydeaniline.†

Heptokill. Insectidal formulation. (Mitchell Cotts Chemicals Ltd).

Heptomer. Gleptoferron. Applications: Haematinic. (Fisons PLC, Pharmaceutical Div).

Heptuss. Cyproheptadine. Applications: For the treatment of coughs and colds. (Merck Sharp & Dohme).*

Her. Chemical resorcinol Di (beta-hydroxyethyl) ether. Applications: Curative for urethane polymers. (Anderson Development Company).*

Herapathite. Quinine sulphate per-iodide.†

Herapath's Salt. Quinine iodo-sulphate.†

Herbazin Plus. A wettable powder containing simazine and aminotriazole. Applications: A quick acting herbicide for control of existing weeds with long term persistence. (Fisons PLC).*

Herbazin Special. A wettable powder containing atrazine, 2,4-D and aminotriazole. Applications: A quick acting herbicide for control of well established and problem weeds. (Fisons PLC).*

Herbazin Total. A granule containing atrazine. Applications: A persistent herbicide for the control of grasses and many annual and perennial broad-leaved weeds. (Fisons PLC).*

Herbazin 50. A wettable powder containing simazine. Applications: Long term maintenance of weed-free pathways, bare ground and other areas requiring total weed control. (Fisons PLC).*

Herbohn Bronze. An alloy of 71 per cent copper, 26 per cent tin, and 3 per cent zinc.†

Herchlor (C, C85 and C110). A range of elastomeric copolymers of ethylene oxide-epichlorohydrin. Applications: Specialty elastomers. (Hercules Inc).*

Herclor. Epichlorhydrin elastomer. (Hercules Ltd).

Herco. A proprietary trade name for a pine oil (*qv*).†

Hercobind DS. An oleoresin emulsion. Resistant to rain and wind erosion. Binds and agglomerates fine minerals. Applications: For dust control. (Hercules Inc).*

Hercoflat. Paint texturing/flatting agents. (Hercules Ltd).

Hercoflav. Formulated flavours for foods. (Hercules Ltd).

Hercoflex 600. Non-volatile, low molecular weight plasticizers. (Hercules Inc).*

Hercoflex 707. Non-volatile, low molecular weight plasticizers. (Hercules Inc).*

Hercoflex 707A. Non-volatile, low molecular weight plasticizers. (Hercules Inc).*

Hercoflex 900. A polymeric plasticizer. Applications: Contributes to clarity, green tack, permanance and low-temperature flexibility in polyvinyl acetate adhesives. (Hercules Inc).*

Hercofloc. A series of high molecular weight synthetic water soluble polymers. Applications: Flocculant polymers with many uses in water management. (Hercules Inc).*

Hercofroth. A series of frothers. Range includes pine oil, modified terpene alcohols, aliphatic-terpene alcohol, polypropylene glycol water-soluble based frothers. Applications: In flotation, frothers stabilize the air bubbles containing the unwetted particles into a froth that is easily removed from the liquid. (Hercules Inc).*

Hercol 2. Nitroglycerin dynamite. Applications: Construction and building, explosives, mining, petroleum and related industries. (Hercules Inc).*

Hercol 2X. Nitroglycerin dynamite. Applications: Construction and building, explosives, mining, petroleum and related industries. (Hercules Inc).*

Hercolube. A range of synthetic esters (polyol type) used as lubricants. Applications: Functional fluids where exposure to both high and low temperature is a primary consideration. (Hercules Inc).*

Hercolyn. Hydrogenated methyl abietate. (Hercules Ltd).

Hercolyn D. The hydrogenated methyl ester of rosin. Applications: Tackifying resin for adhesives, plasticizing and softening agent for chewing gum, used in flexographic, type T gravure and screen-process inks, used as a plasticizing resin in cellulose-based coatings. (Hercules Inc).*

Hercomix. A blasting agent containing ammonium nitrate and fuel oil. Applications: Suitable for use under dry borehole conditions. (Hercules Inc).*

Hercon 2. Nitroglycerin dynamite. Applications: Construction and building, explosives, mining, petoleum and related industries. (Hercules Inc).*

Hercon 2X. Nitroglycerin dynamite. Applications: Construction and building, explosives, mining, petroleum and related industries. (Hercules Inc).*

Hercon 32. Cationic cellulose reactive sizing emulsions. (Hercules Inc).*

Hercon 40. Cationic cellulose reactive sizing emulsions. (Hercules Inc).*

Hercon 48. Cationic cellulose reactive sizing emulsions. (Hercules Inc).*

Herco-Prills. Ammonium nitrate prills containing a minimum of 33.5% nitrogen. Applications: Used alone or for bulk blending of mixed fertilizers. (Hercules Inc).*

Hercoprime. Adhesion promoters for powder coatings. (Hercules Ltd).

Hercopruf. Glycol based freeze conditioning agents. Applications: Prevents ice binding of wet minerals and ice build-up on conveyor belts. (Hercules Inc).*

Hercose AP. A trade mark for cellulose acetate propionate. Used in lacquer manufacture.†

Hercose C. A trade mark for cellulose acetobutyrate for use in lacquer manufacture.†

Hercosett. Polyamide epichlorohydrin resin. Applications: Shrink-proofing of wool. (Hercules Inc).*

Hercosett 125. A water-soluble cationic resin containing reactive polyamide epichlorohydrin. Applications: Imparts shrink resistance to wool. (Hercules Inc).*

Hercosol. A trade mark for a solvent made from pine oil.†

Hercosol TP-S. 65% solids solution of dark pine-tree-derived resin in a mixed-terpene hydrocarbon solvent. Applications: Designed for rubber reclaimers who need an additional solvent or a second reclaiming solvent in their process. (Hercules Inc).*

Hercosplit WR. A nitroglycerin dynamite. Applications: Construction and building industry, explosives, mining, petroleum and related industries. (Hercules Inc).*

Hercotac AD. A low molecular weight modified aromatic hydrocarbon resin. Applications: Used mainly in pressure sensitive systems containing natural rubber and in hot-melts based on ethylene vinyl acetate copolymer. (Hercules Inc).*

Hercotac LA. A low molecular weight modified aliphatic hydrocarbon resin. Applications: Used mainly in pressure sensitive systems containing natural rubber, and in hot melts based on ethylene vinyl acetate copolymer. (Hercules Inc).*

Hercotuf. Powdered polypropylene. (Hercules Ltd).

Hercules Cement. See ACETIC-GELATIN CEMENT.

Hercules Metal. A bronze containing 85.5 per cent copper, 10 per cent tin, 2.5 per cent aluminium, and 2 per cent zinc. Another alloy contains 54 per cent copper, 36 per cent zinc, 7.5 per cent iron, and 2.5 per cent aluminium. The term is also used for an alloy of copper, nickel, and aluminium.†

Herculine FR. Surface explosive. Applications: Surface charge for geophysical prospecting. (Hercules Inc).*

Herculite. Moulding plaster for moulds used in making aluminium alloy castings. (Foseco (FS) Ltd).*

Herculon. Polypropyleneolefin fibre. Applications: For floor coverings, furniture and fixtures and textiles. (Hercules Inc).*

Herculoy. A patented alloy. It is a silicon bronze containing tin. It has high tensile strength and resists corrosion. Herculoy 418 contains copper with about 3 per cent silicon and 0.5 per cent tin.†

Hercures. Aryl and alkyl-aryl resins. Applications: Adhesives and hot melts. (Hercules Inc).*

Heresite. A proprietary trade name for a phenol-formaldehyde synthetic resin moulding compound.†

Herkules. An alloy of 50 per cent silver and 50 per cent copper. Used for fuse wire.†

Hermann's Fluid. (Platino-Aceto-Osmic Acid). A fixing agent used in microscopy. It contains 15 parts of a 1 per cent platinum chloride solution, 4 parts of a 2 per cent osmium tetroxide solution, and 1 part of glacial acetic acid.†

Hermite Fluid. A disinfectant. It contains magnesium oxide and hypochlorous acid, with from 4-5 per cent available chlorine.†

Herolith. A svnthetic resin of the phenol-formaldehyde type. See BAKELITE.†

Heron. A paper based grade of TUFNOL industrial laminates. (Tufnol).*

Herplex Liquifilm. Idoxuridine. Application: Antiviral. (Allergan Pharmaceuticals Inc).

Herrifex DS. A liquid containing 587.5g per litre '58.75 per cent' of mecoprop as the potassium salt. Applications: To control cleavers, common chickweed and a wide range of other broadleaved weeds in cereals, sports turf, grass seed crops and apple and pear orchards. (Bayer & Co).*

Herrisol. A liquid containing 35.4% w/w dicamba (Banvel D), MCPA and mecoprop. Applications: To control common chickweed, cleavers, knotgrass, redshank, scentless mayweed, corn spurrey and a wide range of other broadleaved weeds in cereals, grass crops and orchards. (Bayer & Co).*

Herschel's Crystals. Hydrated calcium-tetrahydroxytrisulphide, $Ca_3(OH)_4.S_3.8H_2O$.†

Hespan. Hetastarch. A starch that is composed of more than 90 percent of amylopectin and that has been etherified to the extent that an average of 7 to 8 of the OH groups present in every 10 D-glucopyranose units of starch polymer have been converted into OCH2CH2OH groups. Applications: Plasma volume extender. (American Critical Care, Div of American Hospital Supply Corp).

Hesperidene. (Citrene, Carvene) Limonene, $C_{10}H_{16}$.†

Hesperidine. See AURANTIIN.

Hessian Blue. See ANILINE BLUE, SPIRIT SOLUBLE.

Hessian Brown MM. A dyestuff. Dyes cotton brown.†

Hessian Yellow. A dyestuff. Dyes cotton yellow from a neutral or acid bath.†

Hesthasulphid. A proprietary brand of sodium sulphide. Used in leather manufacture.†

Hetacin-K. Hetacillin potassium. Applications: Antibacterial. (Bristol Laboratories, Div of Bristol-Myers Co).

Hetron. Chlorendic, bisphenol, furan and vinyl ester resins for chemical resistant applications. Halogenated resins for fire retardant applications. Applications: Fibreglass reinforced parts for the chemical resistant building construction, electrical and transportation markets. (Ashland Chemical Company).*

Heusler Alloy. An alloy of 66-68 per cent copper, 18-22 per cent manganese, 10-11 per cent aluminium, and 0-4 per cent lead.†

Heveacrumb. A proprietary compressed rubber crumbled in oil. (RRI Malaya).†

Heveatex. The trade name to denote a series of preserved, concentrated, or processed Hevea rubber latexes.†

Hevikote. Two peak pitch epoxy protective coating. (Thomas Ness Ltd).

Hex. See HEXAMINE.

Hexa. A rubber vulcanization accelerator. It is hexamethylene-tetramine.†

Hexa-Betalin. Pyridoxine Hydrochloride. Applications: Vitamin. (Eli Lilly & Co).

Hexacal. Fire retardant additives for rigid foams. (ICI PLC).*

Hexacarb. Black dye for leather. (Pointing Ltd).†

Hexacert. U.S.A. certified foodstuff colourant. (Pointing Ltd).†

Hexacide. Insect and vermin killer. (Pointing Ltd).†

Hexacol. Foodstuff colourants of guaranteed specification. (Pointing Ltd).†

Hexaderm. Fast dyes for leather. (Pointing Ltd).†

Hexadrol. Dexamethasone. Applications: Glucocorticoid. (Organon Inc).

Hexafoam. Isocyanate foams. (ICI PLC).*

Hexalan. Fast to light colours for wool. (Pointing Ltd).†

Hexalin. Cyclohexanol.†

Hexalin Acetate. See ADRONAL ACETATE.

Hexallac. Dyes for cellulose lacquers. (Pointing Ltd).†

Hexamethylamine. Hexamethylene-tetramine, $C_6H_{12}N_4$.†

Hexamethyl-para-rosaniline. See CRYSTAL VIOLET.

xamic Acid. Cyclamic acid. Applications: Sweetener. (Abbott Laboratories).

xamine. (Cystamin, Cystogen, Metramine, Urotropine, Formin, Naphthamine, Xametrin, Vesaloin, Urisol, Uritone, Hex, H.M.T., Formaldehyde-ammonia 6:4, Carin, Ammonioformaldehyde, Vesalvine). Hexamethy-lene-tetramine, $C_6H_{12}N_4$.†

xamine Azurine 5G. A dyestuff. It is a British brand of Brilliant azurine 5G.†

xanhexol. Mannite, $C_6H_{14}O_6$.†

xanitrin. Mannitol-hexanitrate, $CH_2O(NO_2)$ $(CHO.NO_2)_4$ CH_2ONO_2.†

xapar. Detergent preparations. (Pointing Ltd).†

xaplas. Plasticizers. (ICI PLC, Petrochemicals & Plastics Div).

xaplus. Plasticizers. (ICI PLC).*

xaryl D60L. Triethanolamine alkylaryl sulphonate in liquid form. Applications: Anionic surfactant. (Witco Chemical Ltd).

xasol. Alcohol soluble dyes. (Pointing Ltd).†

xatype. Dyes for doubletone printing inks. (Pointing Ltd).†

xavibex. Pyridoxine Hydrochloride. Applications: Vitamin. (Parke-Davis, Div of Warner-Lambert Co).

xela. Cellulose acetate and nylon dyes. (Pointing Ltd).†

xil. Hexanitrodiphenylamine.†

xnitrol. Leather stains. (Pointing Ltd).†

xo. Industrial detergent. (Crosfield Chemicals).*

xoil. Oil and varnish dyes. (Pointing Ltd).†

xopal. A proprietary preparation of inositol nicotinate. A peripheral vasodilator. (Bayer & Co).*

xoran. A sodium alkylaryl sulphonate. Applications: Scouring, wetting-out level bleaching, dyeing etc on all fibres. (Roehm Ltd, Hexoran Div).

xoran A15. Sodium alkylaryl sulphonate. Anionic surfactant in liquid form. Applications: Scouring, wetting out, level bleaching, dyeing etc for all fibres. (Roehm Ltd, Hexoran Div).

Hexsotate. Hexamethylene-tetramine sodio-acetate.†

Hexyltan. Hexyresorcinal and tannic acid. Applications: For the treatment of burns. (Merck Sharp & Dohme).*

Heyn's Reagent. The double chloride of copper and ammonia. Used to reveal ferrite in the micro-analysis of carbon steels.†

HFC. Compounded plastic material. (Mitsubishi Petrochemical Co).*

HF(2). A proprietary histidine-free food used in the dietary treatment of histidinaemia. (Cow and Gate).†

Hibbo. A proprietary aluminium bronze containing iron.†

Hibiclens. Chlorhexidine gluconate. Applications: Antimicrobial. (Stuart Pharmaceuticals, Div of ICI Americas Inc).

Hibidil. An antiseptic and bactericidal solution containing chlorhexidine. (ICI PLC).*

Hibiscrub. Antiseptic skin cleanser containing chlorhexidine. (ICI PLC).*

Hibisol. Antiseptic hand rub containing chlorhexidine. (ICI PLC).*

Hibispray. Antiseptic spray containing chlorhexidine. (ICI PLC).*

Hibitane. A proprietary preparation of chlorhexidene digluconate, an antiseptic. (ICI PLC).*

Hibosol. High boiling point solvents. (BP Chemicals Ltd).*

Hibudine. A proprietary trade name for a synthetic rubber, probably derived from butadiene.†

Hi-Build. Paint filler. (ICI PLC).*

Hi-Cat. Cationic starch. Applications: Paper industry, wet end additive. (Roquette (UK) Ltd).*

Hickstor. Tecnazene potato treatment. (Hickson & Welch Ltd).

Hicore 90. A nickel-chromium-molybdenum case-hardening steel for heavy motor vehicles and other gears.†

Hicoseen. A proprietary preparation of diethylaminoethyl phenylbutyrate, codeine phosphate and guaicol albuminate. A cough linctus. (Hommel AG).*

Hiduminium. A registered trade name for a range of aluminium alloys. (High Duty Alloys).†

HiGel. Aluminium stearate. Applications: Non-aqueous coating compositions. (Nueodex Inc).*

High Brass. See BRASS.

Hi-Gloss I. Solvent base sealer. Applications: An acrylic sealer with high solids for application to exterior and interior concrete, stone and masonry type surfaces for commercial use only. (Glessner Corporation Inc (GGI Products) DBA).*

Hi-heet. An American synthetic resin moulded product for electrical insulation.†

Hi Sol. Aromatic hydrocarbon solvents. Applications: Used in coatings. (Ashland Chemical Company).*

Hi Temp EC-1000. Emulsion of aliphatic oils, salts and soaps with water. Applications: Die casting and mould release agent, drilling and tapping fluid/coolant. (Hi Temp Lubricants Inc).*

Hi Temp EC-4000. Emulsion of aliphatic oils, salts and soaps with water. Applications: Die casting mould release agent. (Hi Temp Lubricants Inc).*

Hi Temp EC-5000. Emulsion of aliphatic oils, salts and soaps with water. Applications: Die casting mould release agent, die casting plunger lubricant. (Hi Temp Lubricants Inc).*

Hi-Zex. A trade mark for Japanese high density polyethylene. See also Rigidex. (Mitsui Co Ltd).†

High Tensile Brass. An alloy of 76 per cent copper, 22 per cent zinc, and 2 per cent aluminium.†

Hightensite. A trade name for a proprietary hard rubber composition for use in electrical insulation.†

Hiirogane. A blood-red coloured metal prepared either by the treatment of copper with a solution of copper sulphate and verdigris, or by heating a copper alloy with a paste containing a salt of copper, borax, and water.†

Hills-McCanna Alloy No.45. An alloy containing 88 per cent copper, 10.5 pe cent aluminium, and 1.5 per cent iron.

Himaizol. A proprietary food product containing vegetable oils used in the treatment of hypercholesterolaemia. (Cow and Gate).†

Hiotrol. Applications: For commercial an residential agents for the control and reduction of odours associated with latrines, toilets, lavatories, locker and shower rooms, kennels, livestock pens, industrial malodours, sewage wastes, garbage, land fills and lagoons, in Clas 6 (Int Class 5). (Ferro Corporation).*

Hioxyl. A cream containing stabilized hydrogen peroxide 1.5%. Applications Used in the treatment of leg ulcers, minor wounds and infections. (Quinoderm Ltd).*

Hipernick. A registered trade mark for a alloy of nickel and iron in equal parts. Used for making cores of audio-frequency transformers.†

Hipersil. A registered trade mark for high permeability silicon steel. It is used in high-frequency communications equipment.†

Hipersolv. High performance solvents for HPLC. (BDH Chemicals Ltd).

Hippuran I 131. Iodohippurate sodium I 131. Applications: Diagnostic aid; radioactive agent. (Mallinckrodt Inc).

Hippuryl Amide. N-benzoyl-glycinamide. $C_6H_5.CO.NH.CH_2.CO.NH_2$. A white powder. A substrate in studies of papa action. (British Drug Houses).†

Hipputope. Iodohippurate sodium I 131. Applications: Diagnostic aid; radioactive agent. (E R Squibb & Son Inc).

Hiprex. Each tablet contains methenamin hippurate 1 g. (Riker Laboratories/3M Health Care).†

Hirathiol. A compound used as a substitu for Ichthyol (*qv*).†

Hirudoid. Heparidoid 0.3g% in cream and gel bases. Applications: Treatment of varicose ulcers and concomitant symptons of varicose veins, superficial soft tissue injuries, bruising and haematomas. (Luitpold-Werk).*

Hismanal. A proprietary preparation containing astemizole. Applications: Non sedative antihistamine, allergic rhinitis and conjunctivitis (hay fever), other allergic reactions. (Janssen Pharmaceutical Ltd).*

Hispor. Cereal fungicide. (Ciba-Geigy PLC).

Hispril. Diphenylpyraline hydrochloride. Applications: Antihistaminic. (Smith Kline & French Laboratories).

Histadyl E.C. A proprietary preparation containing codeine phosphate, ephedrine hydrochloride, thenylpyramine fumarate, ammonium chloride and chloroform. A cough linctus. (Eli Lilly & Co).†

Histalog. The hydrochloride of 3 - (2 - aminoethyl)pyrazole. (Ametazole hydrochloride). (Eli Lilly & Co).†

Histantin. A proprietary preparation of chlorcyclizine hydrochloride. (Antihistamine). (The Wellcome Foundation Ltd).*

Histazarin. 2:3-Dihydroxyanthraquinone.†

Histo-Acryl. A proprietary preparation of ENBUCRILATE. A surgical tissue adhesive.†

Histogenol. A registered trade mark for a combination of nucleic acid and sodium-methyl-arsenate. Used in the treatment of tuberculosis. (High Duty Alloys).†

Histryl. A proprietary preparation of diphenpyraline hydrochloride. (Smith Kline and French Laboratories Ltd).*

Histryl Spansule Capsule. Dipenylpyraline hydrochloride. Applications: Antihistamine for use in hay fever and other allergies. (Smith Kline and French Laboratories Ltd).*

Hitenso. A proprietary trade name for a cadmium bronze.†

HiTint. Pigment dispersions. Applications: Colouration of solvent based coatings. (Pacific Dispersions Inc).*

Hittorf's Phosphorus. See BLACK PHOSPHORUS.

H-K Mastitis. Hetacillin potassium. Applications: Antibacterial. (Bristol Laboratories, Div of Bristol-Myers Co).

HMS Liquifilm. Medrysone. Applications: Glucocorticoid. (Allergan Pharmaceuticals Inc).

H.M.T. See HEXAMINE.

H.M.T.D. Hexamethylenetriperoxydiamine. A detonating explosive.†

Hobane. Herbicide, containing bromoxynil and ioxynil. (ICI PLC).*

Hochst New Blue. A dyestuff. It is the calcium salt of the di- and trisulphonic acids of trimethyltriphenyl-para-rosanilinetrisulphonic acid. Dyes wool blue from an acid bath.†

Hoenle's Cement. A cement, consisting of 2 parts shellac, and 1 part Venice turpentine.†

Hofmann's Blue. A pigment, $KFe(Fe(CN)_6) + H_2O$.†

Hoffmann's Violet. (Dahlia, Primula, Violet 4RN, Red Violet 5R Extra. Violet 5R, Violet R, Violet RR). A dyestuff. It is a mixture of the hydrochlorides and acetates of the mono-, di-, or trimethyl (or ethyl) rosaniline, and pararosanilines. Dyes wool, silk, and mordanted cotton violet.†

Hoffner's Blue. See COBALT BLUE.

Holcote. Thixotropic ready-for-use water based coatings for sand moulds and cores. (Foseco (FS) Ltd).*

Holfos Bronze. An alloy of copper with 11-12 per cent tin, 0.25 per cent lead, and 0.1-0.2 per cent phosphorus.†

Holite. A proprietary trade name for a synthetic resin for moulding and laminating.†

Holocaine Hydrochloride. Phenacaine Hydrochloride. Applications: Anesthetic. (Abbott Laboratories).

Holoxan (Mitoxana, Ifomide). Ifosfamide. Vials (dry substance). Applications: Cytostatic. (Degussa).*

Homac. A proprietary synthetic resin.†

Homagenets Aoral. Vitamin A. Applications: Vitamin (anti- xerophthalmic). (Beecham Laboratories).

Homapin. Homatropine methylbromide. Applications: Anticholinergic. (Mission Pharmacal Co).

Homberg's Metal. A fusible alloy, containing 3 parts bismuth, 3 parts tin, and 3 parts lead. It has a melting point of 122° C.†

Homberg's Phosphorus. Anhydrous calcium chloride, $CaCl_2$, which, when fused, and exposed to the sun, becomes phosphorescent in the dark.†

Homberg's Salt. Boric acid, H_3BO_3.†

Home-Tet. Tetanus Immune Globulin. Applications: immunizing agent. (Savage Laboratories, Div of Byk-Gulden Inc).

Homobarbital. See PROPONAL.

Homoguaiacol. See CREOSOTE-GUAIACOL.

Homokol. A sensitizer for silver bromide plates. It is a mixture of Quinoline Red with an isocyanine dye.†

Homophan. See PARATOPHAN.

Homophosphine G. An acridine dyestuff.†

Hondurite. A proprietary moulded composition, with cotton and a vulcanized binder, for electrical insulation.†

Honestone. See WHETSTONE.

Honey Stone. See MELLITE.

Honeycat. Catalysts for the control of air pollution. Applications: Off-highway exhaust purification catalysts, industrial air pollution control catalysts. (Johnson Matthey Chemicals Ltd).*

Honey Wax. Mould release based wax for plastic moulding. (Specialty Products Co).*

Honvan (Honvol, ST 52, Fosfostilben). Fosfestrol. Tablets and ampoules. Applications: Cytostatic. (Degussa).*

Honvol. SEE HONVAN. (Degussa).*

Hoof Oil. See NEATSFOOT OIL.

Hooker Brass. See BRASS.

Hooker's Green. See CHROME GREENS.

Hopcalite I. A mixture of 50 per cent manganese dioxide, 30 per cent copper oxide, 15 per cent cobalt oxide, and 5 per cent silver oxide. Used as a catalyst to oxidize carbon monoxide.†

Hopkin's-Cole Reagent. (Benedict's modification). Powdered magnesium (10 grams) placed in a conical flask and shaken up with enough water to cover the magnesium; 250 cc of a cold saturated solution of oxalic acid are added slowly, and the whole shaken and poured on to a filter. The filtrate is acidified with acetic acid and made up to a litre. Used to detect proteins.†

Hopkin's-Cole Reagent. To a litre of a saturated solution of oxalic acid 60 grams of sodium amalgam are added, and the mixture allowed to stand. It is then filtered and diluted with from 2-3 volumes of water. Used for the detection of proteins.†

Hopkin's-Cole Tyrosine C Reagent. This contains mercuric sulphate dissolved in a solution of sulphuric acid.†

Hopkin's Lactic Acid Reagent. Thiophene.†

Hopp II. A modified aqueous hop extract; standardized to 35 per cent reduced iso-alpha acids. Applications: For addition to malt beverages after fermentation to standardize bitterness. (Pfizer International).*

Hopper Salt. Sodium chloride, which has been caused to crystallise in large hollow cubes which float when alum is added to the bath.†

Horco X. A proprietary trade name for a thermosetting polyvinyl butyral synthetic resin.†

Hordaflex. Chlorinated paraffins. (Hoechst UK Ltd).

Hordalub. Chlorinated paraffins. (Hoechst UK Ltd).

Hordamer. Polyethylene primary dispersions. (Hoechst UK Ltd).

Hormone Rooting Powder. Captan and hormone compound. (Murphy Chemical Ltd).

Hormonin. Oestriol 0.27 mg, oestradiol 0.6 mg and oestrone 1.40 mg Applications: Replacement therapy in oestrogen deficiency. (G W Carnrick Co Ltd).*

Horna. Lead chromate yellow and molybdate red pigments. (Ciba-Geigy PLC).

Horn O' Plenty. Various granular fertilizer blends. Applications: Farm and garden fertilizers. (Horn's Crop Service Center).*

Hortus. Fertilizers. (Scottish Agricultural Industries PLC).

Hostacain. A proprietary preparation of butanilicaine phosphate, procaine phosphate, and adrenaline. A local dental anaesthetic. (Hoechst Chemicals (UK) Ltd).†

Hostacor. Corrosion inhibitors. (Hoechst UK Ltd).

Hostadur. A trade name for partially crystalline thermoplastic polyester based on ethylene terephthalate used for construction of rigid components, e.g. gears. (Hoechst Chemicals (UK) Ltd).†

Hostaflam. Flame retardants for polymers. (Hoechst UK Ltd).

Hostaflex. Vinyl acetate modifying copolymer. (Hoechst UK Ltd).

Hostaflon. A trade mark for polytrifluormonochlorethylene. (Hoechst Chemicals (UK) Ltd).†

Hostaflon C2. A proprietary polychlorotrifluoroethylene. (Hoechst Chemicals (UK) Ltd).†

Hostaflon ET. A proprietary ethylene tetrafluoroethylene copolymer. (Hoechst Chemicals (UK) Ltd).†

Hostaflon TF. A trade mark for polytetrafluorethylene. (Hoechst Chemicals (UK) Ltd).†

Hostaform. Highly crystalline acetal copolymer. (Hoechst Chemicals (UK) Ltd).†

Hostaform C. A trade name for acetal plastics. (Hoechst Chemicals (UK) Ltd). See also DELRIN. (Dupont (UK) Ltd). CELCON. (Celanese Corporation).†

Hostalen. A trade mark for a high density polythene. (Hoechst Chemicals (UK) Ltd). See also RIGIDEX. (BP Chemicals Ltd).†

Hostalen G. High density polyethylene. (Hoechst UK Ltd).

Hostalen GM. A proprietary polyethylene resin used for covering pipes and wires. (Hoechst Chemicals (UK) Ltd).†

Hostalen GP. A proprietary polyethylene resin used for making film, monofil and blow-moulded containers. (Hoechst Chemicals (UK) Ltd).†

Hostalen OO. A trade mark for isotactic polypropylene. A thermoplastic moderately rigid moulding material. (Hoechst Chemicals (UK) Ltd).†

Hostalen PP. Crystalline polypropylene. (Hoechst Chemicals (UK) Ltd).†

Hostalit. A trade mark for PVC in sheet form.†

Hostalit Z. A trade mark for a blend of PVC and chlorinated polyolefin. A thermoplastic used for manufacturing pipes. (Hoechst Chemicals (UK) Ltd).†

Hostalub. Lubricants for polymers. (Hoechst UK Ltd).

Hostamid. A proprietary trade name for a group of nylons. (Hoechst Chemicals (UK) Ltd).†

Hostanox. Antioxidants for polymers. (Hoechst UK Ltd).

Hostaperm. Pigments powders for prints and inks. (Hoechst UK Ltd).

Hostaphan. Polyethylene terephthalate films. (Hoechst UK Ltd).

Hostaphane. A proprietary trade name for polythylene terephthalate film. (Hoechst Chemicals (UK) Ltd).†

Hostaphat. Phosphoric acid ester emulsifier. (Hoechst UK Ltd).

Hostapon CAS. Mixture of sulphation products of fatty alcohol derivatives and fatty acid condensation products, supplied as a clear yellowish liquid. Applications: Optimal foaming agent with very good skin compatability, used in cosmetics e.g. cleansing agents for intimate hygiene, baby shampoos and foam baths. (Hoechst UK Ltd).

Hostapon CT. Anionic surfactant in which the cation is sodium and the anion is composed of medium chain length saturated fatty acids condensed with N-methyl taurine. Soft white paste. Applications: Foaming agent with good washing action and lime soap dispersing effect. Used in cream and liquid shampoos, foam baths etc. (Hoechst UK Ltd).

Hostapon KA. Anionic surfactant in which the cation is sodium and the anion is composed of medium chain length fatty acids condensed with oxy-ethane sulphonic acid. Supplied as a white powder of low hygroscopicity. Applications: Foaming and cleansing agent for toothpaste and all powder-form cosmetics. (Hoechst UK Ltd).

Hostapon KTW new. Anionic surfactant in which the cation is sodium and the anion is composed of medium chain length saturated fatty acids condensed with taurine. White powder. Applications: Foaming agent with good washing action and lime soap dispersing effect. Used in toothpastes; hair care; cream shampoos. (Hoechst UK Ltd).

Hostapon STT. Anionic surfactant in which the cation is sodium and the anion is composed of high molecular weight saturated fatty acids condensed with methyl taurine. Supplied as a white, soft paste. Applications: Good lime soap dispersing and washing power, used in the modification of cream-like and semi-liquid shampoos. (Hoechst UK Ltd).

Hostapon TF. Anionic surfactant in which the cation is sodium and the anion is composed of unsaturated fatty acids condensed with methyl taurine. Supplied as a clear, yellowish viscous liquid. Applications: Foaming, washing and lime soap dispersing agent used in clear liquid and cream shampoos; foam baths; toothpastes. (Hoechst UK Ltd).

Hostapor. Polystyrene foaming grades. (Hoechst UK Ltd).

Hostaprint. Pigment preparations for plastics printing. (Hoechst UK Ltd).

Hostapur OS. Sodium olefin sulphonate with an alkylene radical (C15-C18) as a liquid or free-flowing powder. Applications: Powdered detergents of all kinds, light duty detergents and cleaning agents, textile and leather auxiliaries, and upholstery and carpets. (Hoechst UK Ltd).

Hostapur SAS. Series of three anionic surfactants of the alkane sulphonate type, where the chain length is C13-C18 and the cation is sodium. Liquid, paste or flake form. Applications: Wetting and foaming agent used in detergents, washing-up liquids, cleaning agents of all kinds, textiles and leather auxiliaries. (Hoechst UK Ltd).

Hostatint. Pigment preparations for tinting paint. (Hoechst UK Ltd).

Hostatron. Foaming agents for polymers. (Hoechst UK Ltd).

Hostavin. UV stabilizers for polymers. (Hoechst UK Ltd).

Hostyren. Polystyrene resins. (Hoechst UK Ltd).

Hotspur. Herbicide for broad leafed weeds in cereals. (ICI PLC).*

Houghite. See HYDROTALCITE.

Houillite. Anthracite.†

Houseplant Long Lasting Feed. Coated fertilizer. (Fisons PLC, Horticulture Div).

Howard's Silver. Mercury fulminate, used for percussion caps.†

Howflex. Plasticizer. (Laporte Industries Ltd).

Howsorb. Sorbitol syrup aqueous solution. (Laporte Industries Ltd).*

Howstik. Rubber/resin solutions in various solvents. Applications: Contact adhesives. (Howlett Adhesives Ltd).*

Howtex. Ready mixed ceramic tile adhesives - water based; cement based powder adhesives; cement based powder grout and 2 part epoxy grout. Applications: Building adhesives and grout. (Howlett Adhesives Ltd).*

Howtol. Cyclic alcohols. (Laporte Industries Ltd).*

Hoyle's Metals. Bearing metals usually containing about 46 per cent tin, 12 per cent antimony, and 42 per cent lead.†

Hoyt Metal. A proprietary trade name for an antimonial lead containing 6-10 per cent antimony.†

H P Acthar Gel. Corticotropin repository. Applications: Hormone glucocorticoid; diagnostic aid. (Armour Pharmaceutical Co).

HRF Ayerst Laboratories. Gonadorelin, leutinising hormone/follicle stimulating hormone releasing hormone. Applications: Diagnostic use for evaluating the functional capacity and response of the gonadotropes of the anterior pituitary. (Ayerst Laboratories).†

H-scale. A proprietary trade name for a synthetic pearl essence.†

H.T. Monocalcium phosphate monohydrate. Applications: Leavening agent. (Monsanto Co).*

H.T.S. A salt mixture used as a heat transfer medium. It contains approximately 40 per cent sodium nitrite, 7 per cent sodium nitrate, and 50 per cent potassium nitrate by weight. The temperature limits are 290-1000° F.†

ber's Reagent. A solution of ammonium molybdate and potassium ferrocyanide. Used for the detection of free mineral acids.†

bl's Reagent. (*a*) Iodine (50 grams), dissolved in 1 litre of 95 per cent alcohol. (*b*) Mercuric chloride (60 grams), dissolved in 1 litre of alcohol. Used for obtaining the iodine value of fats and oils.†

gel A. A proprietary solution of the sodium salt of an acrylic copolymer, used for thickening natural and synthetic rubber latices. (Hubron Rubber Chemicals, Failsworth, Manchester).†

ugel AH. A proprietary viscous solution of an acrylic copolymer. (Hubron Rubber Chemicals, Failsworth, Manchester).†

ugel B. A proprietary trade name for emulsions of acrylic copolymer containing free carboxyl groups. (Hubron Rubber Chemicals, Failsworth, Manchester).†

ugel BC 10. A proprietary viscous solution, colourless and odourless, of the sodium salt of an acrylic polymer. (Hubron Rubber Chemicals, Failsworth, Manchester).†

ugel CH14. The sodium salt of a proprietary acrylic polymer. It is very viscous, and takes the form of a short non-stringy yellow gel. (Hubron Rubber Chemicals, Failsworth, Manchester).†

ulot's Solder. An alloy of 37.5 per cent tin, 37.5 per cent lead, and 25 per cent zinc amalgam. Used for soldering aluminium bronze.†

umagel. A proprietary preparation of paromycin sulphate, kaolin and pectin. An antidiarrhoeal. (Parke-Davis).†

umatin. A proprietary preparation containing paromomycin (as the sulphate). An antibiotic. (Parke-Davis).†

umber. Pelleted organic based fertilizers comprising a composted organic base and chemical N, P & K to form the analysis. Applications: A steady release, lower nitrogen fertilizer for agricultural crops and grassland. (Humber fertilizers plc).*

Humifen. Sulphonated aliphatic polyesters. Wetting agent. Applications: Dyeing; yarn textile finishing; dry cleaning detergents; glass cleaners; wallpaper removers; battery separators. (GAF (Gt Britain) Ltd).

Humifen BA-77. Anionic surfactant in powder form. Applications: Wetting, dispersing and anti-static agent for paints, printing inks, latex stabilization, leather, textile processing and agricultural chemistry. (GAF (Gt Britain) Ltd).

Humifen BX-78. Anionic surfactant in powder form. Applications: Wetting, penetrating, dispersing and emulsification agent, used for cotton, rayon, dyestuffs, leather, agricultural chemicals, insecticides, paper, rubber latex and polymerization. (GAF (Gt Britain) Ltd).

Humorsol. Demecarium bromide. Applications: Cholinergic. (Merck Sharp & Dohme, Div of Merck & Co Inc).

Humulin. Insulin Human. A protein that has the normal structure of the natural antidiabetic principle produced by the human pancreas. Applications: Antidiabetic. (Eli Lilly & Co).

Hünefeld Solution. This contains 25 cc alcohol, 5 cc chloroform, 1.5 cc glacial acetic acid, and 15 cc turpentine. It is used for the detection of blood.†

Hungary Blue. See COBALT BLUE.

Hungary Green. See MALACHITE GREEN.

Huppert's Reagent. A 10 per cent aqueous solution of calcium chloride used for the detection of biliary pigments in urine.†

Huron. An aluminium alloy containing from 3.5-6.6 per cent copper and small amounts of manganese, magnesium, and chromium. The name is also used for a chromium steel containing 12.5 per cent chromium, 1.0 per cent vanadium, and 0.2 per cent carbon. It is used for dies.†

Hurr Nut. Myrobalans.†

Husman Metal. An alloy of 74 per cent tin, 11 per cent antimony, 10.6 per cent lead, 4 per cent copper, 0.22 per cent iron, and 0.18 per cent zinc.†

Hu-Tet. Tetanus Immune Globulin. Applications: immunizing agent. (Hyland Therapeutics, Div of Travenol Laboratories Inc).

HVA-2. NN' - *m* - phenylenedimaleimide. A free radical regulator used as an auxiliary in the curing of HYPALON. (Du Pont (UK) Ltd).†

HVP. Hydrolysed vegetable proteins. Applications: Natural flavours for all types of processed foods to improve or impart a characteristic flavour. (Hercules Inc).*

Hyalase. Hyaluronidase spreading agent. (Fisons PLC, Pharmaceutical Div).

Hyamine 10-X. Di-isobutyl cresoxyethoxy ethyl dimethyl benzyl ammonium chloride in crystal form. Applications: Bactericidal deodorant used in disinfectant sanitizers and antiseptics. (Rohm & Haas (UK) Ltd).

Hyamine 1622. Benzethonium chloride. Applications: A typical biocide with hospital and animal health applications. (Lonza Limited).*

Hyamine 3500 and 2389. A range of quaternary ammonium compounds. Applications: Germicidal applications in hospitals, institutions and industrial water treatment. (Lonza Limited).*

Hyb-lum. An alloy of aluminium in which the alloying elements, consisting of about 2 per cent. are mainly nickel and metals of the chromium group. Used for reflectors of therapeutic lamps.†

Hycal. A proprietary preparation of dextrose and related compounds, used as a high-calorie diet supplement. (Beecham Research Laboratories).*

Hycar ATBN. A proprietary amine-terminated butadiene-acrylonitrile liquid polymer used in the curing and modifying of epoxy resins. (BF Goodrich Chemical Co).†

Hycar EP. A proprietary butadiene-styrene copolymer containing 50 per cent styrene (cf. 25 per cent in GR-S). It is fawn in colour. Also known as Hycar 10. It is not oil resistant.†

Hycar OR. A proprietary trade name for oil-resisting synthetic rubber.†

Hycar Reactive Liquid Polymer (RLP). RLP are homopolymers of butadiene copolymers of butadiene/acrylonitrile. Reactive groups are in both terminal positions of the polymer chain and, optionally, may have additional reacti groups pendent on the chain. There a three types of reactive groups commercially available. They are carboxyl, acrylated vinyl and seconda amine. Applications: CT polymers ar presently in epoxy structural adhesive epoxy encapulants, epoxy coatings, epoxy composites, epoxy potting/-encapsulation compounds and solid propellant binder. AT polymers are presently in epoxy adhesives, epoxy maintenance coatings, epoxy encapsulation and geographical cable fillers and civil engineering/construction applications. VT polymers are used i polyester bulk moulding compound BMC and sheet moulding compound SMC, anarobic adhesives, radiation curing compositions and fibreglass reinforced plastic polyester compositions. (BF Goodrich (UK) Ltd).†

Hycar VTBN. A proprietary vinyl-terminated butadiene-acrylonitrile copolymer. It is a liquid polymer used for modifying polyesters, polystyrene, etc. (BF Goodrich Chemical Co).†

Hycar 1203X17. A proprietary 70 : 30 blend of HYCAR medium-high acrylonitrile rubber and GEON PVC resin. (BF Goodrich Chemical Co).†

Hycar 1204X5. A proprietary 100 : 70 : 120 pre-fluxed blend of HYCAR medium-high acrylonitrile rubber, GEON PVC resin and a phthalate plasticizer. (BF Goodrich Chemical Co).†

Hycar 1204X9. A proprietary 100 : 70 : 100 pre-fluxed blend of HYCAR medium-high acrylonitrile rubber, GEON PVC resin and a phthalate plasticizer, protected by non-staining stabilisers. (BF Goodrich Chemical Co).†

car 1205X3. A proprietary 50 : 50 : 60 pre-fluxed blend of HYCAR medium-high acrylonitrile rubber, GEON PVC resin and a phthalate plasticizer. (BF Goodrich Chemical Co).†

car 1273. A proprietary 70 : 30 blend of carboxy-modified HYCAR acrylonitrile rubber and GEON PVC resin, possessing good resistance to abrasion. (BF Goodrich Chemical Co).†

car 1402 H82. A proprietary acryl-onitrile-butadiene copolymer in powder form, having a medium-high content of acrylonitrile. (BF Goodrich (UK) Ltd).†

car 1402 H83. A proprietary acrylo-nitrile-butadiene copolymer in powder form, having a high content of acrylonitrile. (BF Goodrich (UK) Ltd).†

car 1403 H84. A proprietary acrylo-nitrile-butadiene copolymer having a medium acrylonitrile content. It is supplied in powdered form and used when good behaviour at low temperatures is required. (BF Goodrich (UK) Ltd).†

car 2100. A proprietary range of polyacrylic-solution polymers used as pressure-sensitive adhesives. (BF Goodrich (UK) Ltd).†

car 2550H33. A proprietary reinforced styrene-butadiene copolymer rubber used in the manufacture of foam rubber. (BF Goodrich (UK) Ltd).†

car 2550H5. A proprietary aqueous, anionic dispersion of a cold- polymerized styrene-butadiene copolymer. (BF Goodrich (UK) Ltd).†

car 2550H55. A proprietary aqueous, anionic dispersion of a styrene-butadiene copolymer used in the foambacking of carpets. (BF Goodrich (UK) Ltd).†

car 2570H28 and 2570H29. A proprietary group of aqueous anionic dispersions of self-reactive styrene-butadiene copolymers, used in the making of carpet backings. (BF Goodrich (UK) Ltd).†

car 2570X5. A proprietary aqueous, anionic dispersion of a carboxy-modified styrene-butadiene copolymer reactive to heat. It is used for leather finishes and adhesives. (BF Goodrich (UK) Ltd).†

Hycar 2671H49. A proprietary aqueous anionic dispersion of a heat- reactive carboxy-modified acrylic polymer used in the making of surgical rubber materials. (BF Goodrich (UK) Ltd).†

Hycar 4021. A proprietary copolymer of ethyl acrylate having a small percent age of 2-chloro-ethyl vinyl ether. (BF Goodrich Chemical Co).†

Hycar 4032. A proprietary polyacrylic rubber used to make rubber seals and gaskets. (BF Goodrich (UK) Ltd).†

Hycar 4043. A proprietary acrylic rubber which remains flexible at minus 40° C, but which also gives good resistance to oil at high temperatures. (BF Goodrich Chemical Co).†

Hycar 4201. A proprietary copolymer of ethyl acrylate having a small percent age of 2-chlorovinyl ether. (BF Goodrich Chemical Co).†

Hycathane. A proprietary polyurethane elastomer. (FPT Industries).†

Hycol. A coaltar black disinfectant composed of 30% coaltar acids, hydrocarbon oils and vegetable oil soap. (William Pearson).*

Hycolin. A general purpose hospital disinfectant containing 16% substituted phenolic germicides in a detergent base. (William Pearson).*

Hycon. Photographic developer. (May & Baker Ltd).*

Hycote. Spray paints. Applications: Matched colours and touch up paints. (James Briggs & Sons Ltd).*

Hydan. 1,3-Dichloro-5,5-dimethyl-hydantoin. (ABM Chemicals Ltd).*

Hydantil. A proprietary preparation of methoin and phenobarbitone. An anti-convulsant. (Sandoz).†

Hydelta. Prednisolone. Applications: Glucocorticoid. (Merck Sharp & Dohme, Div of Merck & Co Inc).

Hydeltra-TBA. Prednisolone Tebutate. Applications: Glucocorticoid. (Merck Sharp & Dohme, Div of Merck & Co Inc).

Hydeltrasol. Prednisolene. Applications: For the treatment of certain endocrine and non-endocrine disorders responsive to corticosteroid therapy. (Merck Sharp & Dohme).*

Hydergine. A proprietary preparation of mesylates of dihydroergocornine, dihydroergocristine and dihydroergokryptine. Peripheral vasodilator. (Sandoz).†

Hydex. A range of sorbitol solutions. Applications: For use as humectants, sweeteners, viscosity improvers in toothpaste, toiletries, candy, chewing gum and special dietary food. (Lonza Limited).*

Hydracetin. See PYRODINE.

Hydraffin. Granulated carbon. Applications: Used for the treatment of drinking, process and waste water. (Bayer & Co).*

Hydralin. Methylcyclohexane.†

Hydramyl. Pentane, C_5H_{12}.†

Hydranthrene Blue RS Paste. A dyestuff. It is a British equivalent of Indanthrene X.†

Hydranthrene Dark Blue. A dyestuff. It is a British brand of Indanthrene dark blue BO.†

Hydranthrene Olive R. A British equivalent of Indanthrene olive G.†

Hydranthrene Yellow AGR. A British dyestuff. It is equivalent to Algol yellow WG.†

Hydranthrene Yellow ARN, AG New. Dyestuffs. They are British brands of Algol yellow R.†

Hydraphthal. A preparation containing 90 per cent tetralin, 5 per cent ammonium oleate, and 3 per cent water. A wetting-out and scouring agent.†

Hydrated Cellulose. See ALKALI CELLULOSE.

Hydrated Zinc Oxide. Zinc hydroxide, $Zn(OH)_2$.†

Hydraulic Bronze. An alloy of 83 per cent copper, 5 per cent lead, 5 per cent zinc, 5 per cent tin, and 2 per cent nickel. Another alloy contains 83 per cent copper, 10.8 per cent tin, 6 per cent zinc, and 0.1 per cent lead. Also see KERN'S HYDRAULIC BRONZE.†

Hydraulic Cements or Mortars. These ar prepared by calcining mixtures of calcium carbonate with from 10-30 p cent clay. Tricalcium silicate, $3CaO.SiO_2$, and tricalcium aluminate, $3Ca$(Al_2O_3, are formed.†

Hydraulic Limes. Limes containing from 15-30 per cent clayey matter (aluminium silicate). They are made burning impure limestones at a low temperature. They slake in water, bu show hydraulic properties.†

Hydraulic Mortar. See HYDRAULIC CEMENTS.

Hydrazine Yellow. See TETRAZINE.

Hydrea. A proprietary preparation of hydroxyurea, a carcino-chemotherapeutic agent. (Squibb, E R & S Ltd).*

Hydrea Capsules. Hydrocortisone preparations. (The Boots Company PLC).

Hydrenox-M. A proprietary preparation hydroflumethiazide. A diuretic. (Boot Company PLC).*

Hydro. Fertilizers, chemicals, gases and plastics. (Norsk Hydro AS).*

Hydroblok. Cross-linked modified polyacrylamide. (Cyanamid BV).*

Hydrobol. Water repellent finishes. (Cib Geigy PLC).†

Hydrobuna. A name applied to hydrogenated rubber.†

Hydrocarbon Black. See GAS BLACK.

Hydrocarbon Cement. A cement made b mixing heated pitch or tar with from one to four times its volume of calciu or magnesium sulphate. Used for pavi or building purposes.†

Hydrocarbon Gas Black. See GAS BLACK.

Hydrocarbon Oil. See PARAFFIN, LIQUID.

Hydrocarbon-aldehyde Resins. Resins obtained by the interaction of hydrocarbons such as naphthalene wit formaldehyde in the presence of sulphuric acid.†

Hydrocortone. Hydrocortisone. Applications: For the treatment of certain endocrine and non-endocrine disorders responsive to corticosteroid therapy. (Merck Sharp & Dohme).*

ydrodarco. Activated carbon. Applications: Water purification, purification of chemicals/-pharmaceuticals, air purification and solvent recovery. (American Norit Company Inc).*

ydroderm. Hydrocortisone, neomycin and bacitracin. Applications: Topical corticosteroid with topical antibiotics for the treatment of skin conditions. (Merck Sharp & Dohme).*

ydrodiuril. Hydrochlorothiazide. Applications: Diuretic. (Merck Sharp & Dohme, Div of Merck & Co Inc).

ydrofol Acid 1655. Stearic Acid. Applications: Pharmaceutic aid. (Sherex Chemicals Co Inc).

ydrogen Peroxide, Solid. See HYPEROL.

ydrogen Rubeanide. The amide of dithiooxalic acid, $CS(NH_2)CS(NH_2)$.†

ydrogenite. A mixture of 5 parts ferro-silicon, 90-95 parts silicon, 12 parts sodium hydroxide, and 4 parts slaked lime. When ignited it yields hydrogen.†

ydrogen, Dicarburetted. See OLEFIANT GAS.

ydrogen, Phosphoretted. See PHOSPHINE.

ydrohaematite. See TURITE.

ydrolact. Lactase. (John & E Sturge Ltd, Selby).

ydrolete. Calcium hydride, CaH_2.†

ydrolith. A 90 per cent calcium hydride which yields hydrogen on contact with water.†

ydro-Marc. Etafilcon A. Applications: Contact lens material. (Vistakon Inc).

ydromet. Methyldopa and hydro-chlorothiazide. Applications: For the treatment of hypertension. (Merck Sharp & Dohme).*

ydromox. Quinethazone. Applications: Diuretic. (Lederle Laboratories, Div of American Cyanamid Co).

ydromycin-D. A proprietary preparation of prednisolone and neomycin used in dermatology as an antibacterial agent. (Boots Company PLC).*

Hydron Blue. A dark-blue vat dye, obtained by reducing nitroso-phenol with carbazole to form an indophenol, which is heated with sodium sulphide, and subsequently with sulphur. Also used as a trade mark for a range of other dyestuffs. (Casella AG).†

Hydronal. Pure (+)-hydroxy citronel-laldehyde. (Bush Boake Allen).*

Hydronalium. An aluminium alloy containing from 7-9 per cent magnesium and small amounts of silicon and manganese. It is resistant to sea-water, soap, and soda.†

Hydronaphthol. β-naphthol, $C_{10}H_7OH$.†

Hydrone. An alloy of 35 per cent sodium and 65 per cent lead, which generates hydrogen by action of water.†

Hydronyx. A proprietary trade name for sodium sulphoxylate.†

Hydropalat A. A registered trade name for diethylhydrophthalate, a solvent.†

Hydropalat B. A registered name for dibutylhydrophthalate, a solvent.†

Hydrophobol. Proofing agent. (Ciba-Geigy PLC).

Hydro-resin A. A proprietary trade name for a water soluble resin.†

Hydros. Hyposulphite of soda, specially prepared for use as a reducing agent in vat colour dyeing. Sodium hydro-sulphite. (Albright & Wilson Ltd).*

Hydrosaluric. Hydrochlorothiazide. Applications: For the relief of oedema and hypertension. (Merck Sharp & Dohme).*

Hydrosol. An aqueous colloidal silver solution.†

Hydrosoy 2000. Hydrolysed soy protein. (Croda Chemicals Ltd).

Hydrospray. Hydrocortisone, phenyl-propanolamine, phenylephrine and neomycin. Applications: Adjunctive local treatment of certain nasal disorders. (Merck Sharp & Dohme).*

Hydrosulphite. (Rongalite, Hyraldite, Decroline, Redo, Blanchite, Blankit). Hydrosulphites used in dyeing, and for decolourizing sugar syrups.†

Hydrosulphite A. A 10 per cent solution of hydrosulphite NF, or hyraldite. Used for testing dyed fabrics.†

Hydrosulphite A.W. See DISCOLITE.

Hydrosulphite B. Prepared by acidifying 200 cc of hydrosulphite A with 1 cc acetic acid. Used for testing dyed fibres.†

Hydrosulphite BASF. A 90 per cent sodium hydrosulphite, $Na_2S_2O_4$.†

Hydrosulphite NF. (Rongalite Conc., Brittalite, Formosul). A condensation product of formaldehyde and sodium hydrosulphite. It consists of a mixture of formaldehyde sodium bisulphite, $NaHSO_3.CH_2O$, and formaldehyde-sodium sulphoxylate, $NaHSO_2.CH_2O$. Used as a discharger in calico printing.†

Hydrosulphite NF Conc. See HYRALDITE C EXT.

Hydrotek. Pigment dispersions. Applications: Colouration of aqueous coatings. (Pacific Dispersions Inc).*

Hydrous Wool Fat. See LANOLIN.

Hydroxal. A proprietary oral antacid consisting of a suspension of aluminium hydroxide. (Blue Line Chemical Co, St Louis, Missouri).†

Hydroxine Yellow G, L, L Conc-Dyestuffs. They are British equivalents of Tartrazine.†

Hydroxymimetite. A basic lead arsenate made artificially.†

Hydrozets. Hydrocortisone, bacitracin, tyrotricin, neomycin and benzocaine. Applications: For the effective relief of minor mouth and throat irritations. (Merck Sharp & Dohme).*

Hydrozone. (Glycozone, Pyrozone). Trade names for hydrogen peroxide, H_2O_2, used as an antiseptic in dental practice.†

Hydryl. The product of the interaction between mercuric oxide and Orsudan (*qv*).†

Hyflux. Soldering fluxes. (ABM Chemicals Ltd).

Hyflux M. Hydrazine hydrobromide. (ABM Chemicals Ltd).*

Hy-glo Steel. A proprietary trade name for a stainless steel containing 17.0 per cent chromium and 0.6 per cent carbon.†

Hygrol. Colloidal mercury.†

Hygroton. A proprietary preparation of chlorthalidone. A diuretic. (Ciba Geigy PLC).†

Hygroton. Chlorthalidone. Applications: Diuretic. (Parke-Davis, Div of Warner Lambert Co).

Hygroton K. A proprietary preparation of CHLORTHALIDONE and potassium chloride. A diuretic with a potassium supplement. (Ciba Geigy PLC).†

Hyjet. Hydraulic fluid. (Chevron).*

Hylastic. A high-manganese steel containing 1.6-1.8 per cent manganese and 0.35 per cent carbon.†

Hylene. A proprietary trade name for an organic diisocyanate used in the manufacture of polyurethane foam having a range of rigidities. (Du Pont (UK) Ltd).†

Hylite Color-Max. Fluorescent pigment/-plasticizer dispersions. Applications: Pourable colours for plastisols and other systems. Offered custom matched to standard. (W J Ruscoe Co).*

Hylorel. Guanadrel sulphate. Applications: Antihypertensive. (The Upjohn Co).

Hymod. Copper powder for sintering iron. (Steetley Chemicals Ltd).

Hymono. Distilled monoglycerides. Applications: Various applications in the food industry e.g., margarines, shortenings, bakery and dairy. (PPF International Ltd).*

Hyonic. Wetting agents. (Diamond Shamrock Process Chemicals Ltd).

Hyoscine. Scopalamine, $C_{17}H_{21}NO_4$, an alkaloid.†

Hypacel. Bleaching agent for textiles. (RW Chemicals Ltd).

Hypalon. A proprietary trade name for a synthetic rubber (chlorosulphonated polyethylene). It has good resistance to heat, oils, oxidizing chemicals, sunlight and weathering. It is colour stable and ozone- proof. (Du Pont (UK) Ltd).†

Hypaque Meglumine. Diatrizoate meglumine. Applications: Diagnostic aid. (Sterling Drug Inc).

Hypax. Oxidized wax/printing inks. Applications: Anti-corrosives. (Chemoxy International Ltd).*

Hyperab. Rabies Immune Globulin. Applications: immunizing agent. (Cutter Laboratories, Miles Laboratories Inc).

percal. A proprietary preparation of rauwolfia alkaloids. An anti-hypertensive. (Carlton Laboratories).†

percal B. A proprietary preparation of rauwolfia alkaloids and amylobarbitone, used in the control of hypertension. (Carlton Laboratories).†

perdol. Reserpine and hydro-flumethiazide. An antihyper-tensive. (Boots Pure Drug Co).*

perHep. Hepatitis B Immune Globulin. Applications: immunizing agent. (Cutter Laboratories, Miles Laboratories Inc).

perit. This is the same as hyperol.†

pernick. See HIPERNICK.

perol. (Solid Hydrogen Peroxide, Ortizon, Perhydrate, Hyperit). Proprietary names for a compound of hydrogen peroxide and urea, $CO(NH_2)_2.H_2O_2$. It contains 35 per cent hydrogen peroxide, which is obtained by dissolving in water or ether. One gram in 10 cc = a 10-volume strength solution of hydrogen peroxide.†

yperstat. Diazoxide. Applications: Antihypertensive. (Schering-Plough Corp).

ypertensin. A proprietary preparation of angiotensin amide. A vasoconstrictor. (Ciba Geigy PLC).†

yper-Tet. Tetanus Immune Globulin. Applications: immunizing agent. (Cutter Laboratories, Miles Laboratories Inc).

ypertussis. Pertussis Immune Globulin. Applications: immunizing agent. (Cutter Laboratories, Miles Laboratories Inc).

ypnodil. A proprietary preparation containing metomidate hydrochloride. Applications: Veterinary hypnotic (pigs). (Janssen Pharmaceutical Ltd).*

ypnomidate. A proprietary preparation containing etomidate. Applications: Induction of anaesthesia. (Janssen Pharmaceutical Ltd).*

ypnomidate Concentrate. A proprietary preparation containing etomidate hydrochloride. Applications: Induction of anaesthesia. (Janssen Pharmaceutical Ltd).*

Hypnorm. A proprietary preparation containing fentanyl citrate and fluanisone. Applications: Veterinary anaesthetic (dogs, rabbits and guinea pigs). (Janssen Pharmaceutical Ltd).*

Hypnovel. Proprietary preparation of midazolam. Benzodiazepine with hypnotic and sedative properties. (Roche Products Ltd).*

Hypo. Sodium thiosulphate $Na_2S_2O_3$.†

Hypo, Black. See BURNT HYPO.

Hypon. Analgesic. (The Wellcome Foundation Ltd).

Hyporit. The trade name for calcium hypochlorite, containing 80 per cent available chlorine. An antiseptic.†

Hypovase. A proprietary preparation of prazosin hydrochloride for the treatment of hypertension, left ventricular failure and Raynaud's Disease. (Pfizer International).*

Hypromellose. A surface-active agent. It is a partial mixed methyl and hydroxypropyl ether of cellulose. ISOPTO.†

Hyraldite. See HYDROSULPHITE.

Hyraldite A. See HYDROSULPHITE NF.

Hyraldite C Ext. (Hydrosulphite NF Conc., Britulite). Sodium formaldehyde sulphoxylate, $NaHSO_2.CH_2O.2H_2O$. Used as a reducing agent in calico printing.†

Hysa. Hyvis derivative used in surfactant manufacture and as a corrosion inhibiting agent. (BP Chemicals Ltd).*

Hyskon. Dextran 70. A polysaccharide produced by the action of Leuconostoc mesenteroides on sucrose. Average molecular weight: 70,000. Applications: Plasma volume extender. (Pharmacia Laboratories, Div of Pharmacia Inc).

Hysol. A proprietary polyurethane elastomer. (Dexter Corporation).†

Hysol MBI-02. A proprietary epoxy resin moulding powder modified for use as a load-bearing material. (Dexter Corporation).†

Hysol XC7-W529. A proprietary flexible, one-component epoxy casting and potting compound possessing good thermal shock properties. (Dexter Corporation).†

Hystar. Hydrogenated starch. Applications: For use in toiletries and special dietary food. (Lonza Limited).*

Hytak. Hot melt adhesives for packaging and product assembly formulated from a variety of polymers, resins, waxes, plasticizers and antioxidants. Applications: Carton sealing, bottle and can labelling, manufacture of disposable sanitary products, assembling of components in automotive and general industries and woodworking. (Hytak Ltd).*

Hytakerol. Dihydrotachysterol. Applications: Regulator. (Sterling Drug Inc).

Hytane. Substituted urea herbicide. (Ciba-Geigy PLC).

Hytemco. A proprietary trade name for an iron-nickel resistance alloy.†

Hyten M Steel. A proprietary trade name for a nickel-chromium-molybdenum steel.†

Hy-ten-sl. A proprietary trade name for an alloy of 66 per cent copper, 19 per cent zinc, 10 per cent aluminium, and 5 per cent manganese.†

Hytex. Printing ink alkyds. (Croda Resins Ltd).

Hythane. Polyurethane alkyds. (Croda Resins Ltd).

Hytherm. Heat resistant dyes for plastics. (Williams (Hounslow) Ltd).

Hytone. Hydrocortisone. Applications: Glucocorticoid. (Dermik Laboratories Inc).

Hytox. Germicidal detergent. (Unilever).*

Hytrel. A trade mark for a group of polyester elastomers used as thermoplastic rubbers. They are graded for hardness as follows:- Hytrel 4055, 92A; Hytrel 5550, 55D; Hytrel 6350, 63D. (Du Pont (UK) Ltd).†

Hytrol. A mixture of amino triazole, 2-4D Na salt, divron and simazine. A powder dissolved in water as a spray. Applications: A total weedkiller. (Agrichem Ltd).*

Hyvar. Bromacil weedkiller. (Du Pont (UK) Ltd).

Hyvar X. Bromacil. A weed killer. Used as a broad spectrum weed control agent in raspberries. (Du Pont (UK) Ltd).†

Hy-Vin. Homo and copolymeric poly vinyl chloride (S-PVC/E-PVC) as compound. Applications: For production of plastic articles. (Norsk Hydro AS).*

Hyvis. Polybutenes. (BP Chemicals Ltd).*

I

chiol. Silver fluoride. AgF.†

-IA Alloy. An alloy of 60 per cent copper and 40 per cent nickel. Used for electrical resistances.†

ex. Acid calcium phosphate for aerating purposes. (Albright & Wilson Ltd).*

rin. Fibrinogen I 125. A preparation of human fibrinogen labelled with iodine-125. Applications: Diagnostic aid; radioactive agent. (Amersham Corp).

R/IVP/P13. Live vaccines. Applications: For active immunization against infectious bovine rhinotracheitis, infectious pustulous vulvovaginitis/balanoposthitis and parainfluenza-3 disease of cattle and for active interferonisation. (Bayer & Co).*

ular. Tablets containing 200mg and 400mg ibuprofen BP. Applications: Indicated for its anti-inflammatory and analgesic effect in the treatment of rheumatoid arthritis (including juvenile rheumatoid arthritis or Still's disease), ankylosing spondylitis, osteoarthrosis and other non-rheumatoid (sero-negative) arthropathies. (Lagap Pharmaceuticals Ltd).*

-VAC. Infectious bronchitis vaccine. Applications: immunization of poultry. (Intervet America Inc).*

-VAC-H. Massachuttes Holland type bronchitis vaccine. Applications: immunization of poultry. (Intervet America Inc).*

-VAC-M. Massachuttes type bronchitis vaccine. Applications: immunization of poultry. (Intervet America Inc).*

dal. Terephthalic acid resins. Applications: Insulation of heat-resistant electrical conductors. (Dynamit Nobel Wien GmbH).*

Ice Colours. Colours formed on the fibre by treating it with a phenol, and then with a diazotized amine, in the presence of ice. They are also known as Ingrain colours.†

Iceland Spar. Crystalline calcium carbonate.†

Iceline. See PRESERVALINE.

Ice Melt. Calcium chloride pellets. Applications: Snow and ice melting, dust control and tyre weighting. (Standard Tar Products Company, Inc).*

Ichden. See ICHTHYOL.

Ichtammon. See ICHTHYOL.

Ichthaband. A proprietary zinc paste and ICHTHYOL bandage used in the treatment of eczema. (Seton Products Ltd).*

Ichthadone. A proprietary preparation of ammonium ichtho-sulphonate.†

Ichthammon. See ICHTHYOL.

Ichthammonium. See ICHTHYOL.

Ichthamol. See ICHTHYOL.

Ichthopaste. Zinc paste and ichthamaol bandage. (Smith and Nephew).†

Ichthosan. See ICHTHYOL.

Ichthosulphol. See ICHTHYOL.

Ichthymall. Ichthammol. Applications: Anti-infective, topical. (Mallinckrodt Inc).

Ichthynal. See ICHTHYOL.

Ichthynat. See ICHTHYOL.

Ichthyocolla. See ISINGLASS.

Ichthyodine. See ICHTHYOL.

Ichthyol. Ichthammol. Applications: Anti-infective, topical. (Stiefel Laboratories Inc).

Ichtolithium. See ICHTHYOL.

Icipen 300. A proprietary preparation of phenoxymethylpenicillin. An antibiotic. (ICI PLC).*

ICR. Catalyst. (Chevron).*

Ictotest. A proprietary test tablet of *p*-nitrobenzene diazonium, *p*-toluene sulphonate, salicylsulphonic acid and sodium bicarbonate, used for the detection of bilirubin in urine. (Ames).†

Ideal Alloy. An alloy of 53.5 per cent copper, 45 per cent nickel, 0.66 per cent iron, and 0.45 per cent manganese.†

Idemin. A proprietary preparation of meprobamate and benactyzine hydrochloride. Antidepressant. (Horlicks).†

Idilon. See CARBORA.

Iditol. A proprietary shellac substitute.†

Idoklon. Flurothyl. Applications: Stimulant. (Anaquest, Div of BOC Inc).

Idryl. Fluoranthrene, $C_{16}H_{10}$.†

Ifex. Ifosfamide. Applications: Antineoplastic. (Mead Johnson & Co).

Ifomide. SEE HOLOXAN. (Degussa).*

Igasurine. Impure brucine.†

Igelit PCU. A proprietary manufacture of polyvinylchloride.†

Igepal. A wetting agent.†

Igepal CA-210, CA-420 and CA-520. A range of nonionic surfactants. Applications: Emulsifiers for nonpolar hydrocarbon solvents and oils, in solvent emulsion cleaners and drycleaning detergents. Widely used as pesticide and floor polish emulsifiers. Also used as solubilizers for hair dye preparations. (GAF Corporation).*

Igepal CA-620 and CA-630. A range of nonionic surfactants. Applications: May be used in all phases of detergent compounding and aqueous processing in the textile and paper industries, in industrial metal cleaners, acid cleaners and floor cleaners, detergent-sanitizers and waterless hand cleaners. (GAF Corporation).*

Igepal CA-720. A nonionic surfactant, used as a hard surface detergent with aqueous solubility at high temperatures. Applications: It is used in hot spray, soak and steam-cleaning systems, electrolytic cleaning and metal pickling operations. (GAF Corporation).*

Igepal CA-887, 890 and 897. A range of nonionic surfactants available as a wax or as aqueous solutions. Applications: Primary emulsifiers for vinyl acetate and acrylate polymerizations and post stabilizer for synthetic latices. Also, a dyeing assistant and an emulsifier for fats and waxes. (GAF Corporation).*

Igepal CO-210. An oil-soluble surfactant and intermediate. Applications: Used for foam control, in low concentrations, a foam stabilizer for high foaming detergents and in high concentrations a defoamer for low foaming detergents. Coemulsifier in nonionic surfactant blends. (GAF Corporation).*

Igepal CO-430. An oil soluble surfactant used as a coemulsifier with CO-850 and CO-880. Applications: plasticizer and antistatic agent for polyvinyl acetate; freeze-thaw stabilizer for latex emulsions, intermediate for the synthesis of high foaming, water soluble sulphate esters. (GAF Corporation).*

Igepal CO-520. An oil soluble surfactant and intermediate for anionic surfactants. Applications: De-icing fluid for jet aircraft fuels and automotive gasoline, added to home storage tanks to inhibit rusting, dispersant for petroleum oils. (GAF Corporation).*

Igepal CO-530. An oil soluble surfactant and intermediate for anionic surfactants. Applications: Uses are similar to those given for CO-520 and it is also used as an emulsifier for silicones, agricultural compounds and mineral oils. (GAF Corporation).*

Igepal CO-610. A water soluble surfactant. Applications: Widely used for detergency, wetting and emulsification applicable where low foaming is particularly important. (GAF Corporation).*

Igepal CO-620. Water soluble detergent, wetting agent and emulsifier.

Applications: Primarily designed for use in heavy-duty liquid detergent compounding. (GAF Corporation).*

[Ig]epal CO-630. Nonionic surfactants. Applications: Water soluble detergents, wetting agents and emulsifiers. (GAF Corporation).*

[Ig]epal CO-660. Nonionic surfactants. Applications: Water soluble detergents, wetting agents and emulsifiers. (GAF Corporation).*

[Ig]epal CO-710. Nonionic surfactants. Applications: Water soluble detergents, wetting agents and emulsifiers. (GAF Corporation).*

[Ig]epal CO-720. Nonionic surfactants. Applications: Water soluble detergents, wetting agents and emulsifiers. (GAF Corporation).*

[Ig]epal CO-730. Nonionic surfactants. Applications: Water soluble detergents, wetting agents and emulsifiers. (GAF Corporation).*

[Ig]epal CO-850. A water-soluble detergent and wetting agent. Applications: Emulsifier for fats, oils, waxes, solvents, demulsifier for crude petroleum oil emulsions. (GAF Corporation).*

[Ig]epal CO-880. A nonionic surfactant. Applications: Used as a detergent for the high temperature scouring of textiles in pressure equipment. (GAF Corporation).*

[Ig]epal CO-887. An aqueous solution of CO-880. Applications: Surfactant for emulsion polymerization and post-additive stabilization. (GAF Corporation).*

[Ig]epal CO-890. A highly water-soluble emulsifier and stabilizer. Applications: Useful in concentrated electrolyte solutions, synthetic latices, floor waxes and polishes. (GAF Corporation).*

[Ig]epal CO-970. A very highly soluble surfactant effective at high temperatures and in concentrated electrolyte solutions. Applications: stabilizer for synthetic latices, emulsifier and stabilizer in the preparation of floor waxes and polishes. (GAF Corporation).*

[Ig]epal CTA-639W. Water-soluble surfactant with exceptional wetting and emulsifying properties. Applications: Used in latex paints as a wetting agent, titanium oxide and as an emulsifier for oil or alkyd additives. (GAF Corporation).*

Igepal DM-430. An oil soluble emulsifier. Applications: Used as an emulsifier in agricultural chemicals, emulsion polymerization and in inverse emulsion polymerization fat liquoring leather. (GAF Corporation).*

Igepal DM-530. An emulsifier. Applications: Used in acid-based cleaners, aerosol, cosmetic, insecticide and wax emulsions, textile finishing oils, dry-cleaning soaps, inks, lacquers and paints. (GAF Corporation).*

Igepal DM-710. A water-soluble surfactant for detergency and emulsification, especially where low foaming is desired. Applications: Detergent for washing paper mill felts. (GAF Corporation).*

Igepal DM-730. Highly water-soluble surfactant. Applications: Used in emulsion polymerization, latex stabilization and emulsifiable pesticide formulations. (GAF Corporation).*

Igepal DM-970. Unique high-melting flake or solid nonionic surfactant. Applications: Used in household detergents and in industrial detergent formulations. (GAF Corporation).*

Igepal OD-410. A solvent. Applications: Solvent for various resins (vinyl, phenolic, polyester, alkyd, nitro-cellulose, cellulose acetate), ingredient of metal cleaners, paint strippers and other cleaning compounds and as an ink vehicle. (GAF Corporation).*

Igepal RC-520. A low foaming rewetting agent. Applications: Suitable for paper towels, tissues and semichemical corrugating media. (GAF Corporation).*

Igepal RC-620. An all-purpose detergent and wetting agent. Applications: Used as a detergent in acid cleaning, detergent-sanitizing and grease cutting formulations, as an emulsifier and wetter in agricultural compounds, rug shampoos and whitewall tyre cleaners. (GAF Corporation).*

Igepon A. A proprietary product. It is an oleyl derivative of isethionic acid (hydroxyethanesulphonic acid), $C_{17}H_{33}.COO.CH_2.CH_2.SO_3.Na$. A detergent.†

Igepon AC-78. Coconut acid ester of sodium isethionate. A surfactant with low salt content, having good foaming, lathering and dispersing properties. Applications: Used in detergent bars, dentifrices, shampoos, bubble baths and other cosmetic preparations. Low eye and skin irritation levels. (GAF Corporation).*

Igepon T. A proprietary preparation. It is an oleyl derivative of taurine (aminoethanesulphonic acid), $C_{17}H_{33}.CO.NH.CH_2.CH_2.SO_3.Na$.†

Igepon T-33. Sodium *N*-methyl-*N*-oleoyl taurate, a versatile surfactant for washing piece goods, raw stock and yarns at all temperatures and under all pH conditions. Applications: Used for soaping prints as well as vat and naphthol-dyed fibres. Assistant in the boiling, bleaching, dyeing, wetting and finishing of textiles. (GAF Corporation).*

Igepon T-43. A slurry of sodium *N*-methyl-*N*-oleoyl taurate with more than 33% activity. Applications: For compounding detergent products. (GAF Corporation).*

Igepon T-51. Readily soluble gel of sodium *N*-methyl-*N*-oleoyl taurate. Applications: Used extensively in textile processing, and as a latex emulsion stabilizer. (GAF Corporation).*

Igepon T-77. Soft, non-dusting flakes of sodium *N*-methyl-*N*-oleoyl taurate with more than 67% activity, suitable for dry blending. Applications: Used as a wetting and dispersing agent for insecticidal and herbicidal wettable powders. (GAF Corporation).*

Igepon TC-42. Sodium *N*-coconut acid-*N*-methyl taurate. It is a chemically stable surfactant with good foaming, lathering and dispersing properties. Applications: Used in shampoos, bubble baths and cosmetic preparations. (GAF Corporation).*

Igepon TK-32. Sodium *N*-methyl-*N*-tall oil acid taurate, a surfactant with good dispersing and suspending properties. Applications: Used in the petroleum industry to prevent scale formation in oil well tubing and surface flow lines. (GAF Corporation).*

Igepon TN-74. Sodium *N*-methyl-*N*-palmitoyl taurate. Applications: A surfactant with good dispersing and suspending properties. Applications: Reported use in dry blending into detergent formulations. Herbicide and insecticide dispersant. (GAF Corporation).*

Igetaleim MA. Melamine resin. Applications: Adhesives. (Sumitomo Bakelite, Japan).*

Igetaleim UA. Urea resin. Applications: Adhesives. (Sumitomo Bakelite, Japan).*

Igewsky's Reagent. A solution of 5 per cent picric acid in absolute alcohol. Used for etching in the micro-analysis of carbon steels.†

Iglodine. A proprietary preparation of phenol and iodine. Applications: An antiseptic. (Ayrton Saunders plc).*

Ignicide. Phosphate base. Applications: Flameproof plywood and hardboard. (Stanley Smith & Co Plastics Ltd).*

Iguafen. Dispersing and soaping agent; solubilizing properties; retarder; antiprecipitant. Applications: Textiles particularly acrylics and wool. (GAF (Gt Britain) Ltd).

IHSA I-125. Albumin, Iodinated I 125 serum. Applications: Diagnostic aid; radioactive agent. (Mallinckrodt Inc).

Ilcocillin. Antibiotic veterinary ethicals. (Ciba-Geigy PLC).

Ildamen. Oxyfedrin. Tablets, ampoules and drops. Applications: Therapy for Ischemic Heart Disease. (Degussa).*

Iletin. See INSULIN.

Ilexan E. Inhibited ethylene glycol. Applications: Leak indicating liquid, antifreeze agent and solar fluid. (Huls AG).†

Ilexan HT. Polyethylene glycol. Applications: Antifreeze agent stable at high temperatures for solar collectors. (Huls AG).†

Ilexan P. Inhibited 1,2-propylene glycol. Applications: Anti-freeze agent and solar fluid for heating drinking water. (Huls AG).†

Ilexan S. A mixture of alkylbenzenes. Applications: Heat transfer agent. (Huls AG).†

Ilinol. Conjugated linseed oil. (Unilever).†

Illium. An acid-resisting alloy containing 60.65 per cent nickel, 21.07 per cent chromium, 6.42 per cent copper, 4.67 per cent molybdenum, 2.13 per cent tungsten, 1.04 per cent silicon, 1.09 per cent aluminium, 0.98 per cent manganese, 0.76 per cent iron, and 1.19 per cent carbon and boron.†

Illosome. Oral antibiotic. (Dista Products Ltd).

Ilmenite. (Titaniferous Iron Ore, Titanic Iron, Titaniferous Iron). A mineral. It consists of about 52 per cent titanic oxide, TiO_2, and 48 per cent ferrous oxide, FeO.†

Ilonium. A proprietary preparation of colophony, larch turpentine, turpentine oil, phenol and thymol. Applications: Used in the treatment of skin disorders. (Ilon Laboratories).*

Ilopan. Dexpanthenol. Applications: Cholinergic. (Adria Laboratories Inc).

Ilosone. Erythromycin estolate. Applications: Antibacterial.(Eli Lilly & Co).

Ilosone. Erythromycin estolate BP. Applications: Antibiotic. (Dista Products Ltd).*

Ilotycin. A proprietary preparation containing erythromycin, ethyl carbonate. An antibiotic. (Eli Lilly & Co).†

Ilozyme. Pancrelipase. A concentrate of pancreatic enzymes standardized for lipase content. Applications: Enzyme. (Adria Laboratories Inc).

Imacol. Low foaming lubricant. Applications: Prevention of creases and abrasions when wet finishing piece goods of synthetic fibres and blends with natural fibres. (Sandoz Products Ltd).

Imadyl. Proprietary preparation of carprofen. Antirheumatic and analgesic. (Roche Products Ltd).*

Imavate. Imipramine hydrochloride. Applications: Antidepressant. (A H Robins Co Inc).

Imaverol. A proprietary preparation containing enilconazole. Applications: Veterinary dermatomycoses. (Janssen Pharmaceutical Ltd).*

Imbretil. Carbolonium bromide.†

Imbrilon. A proprietary preparation of indomethacin used in the treatment of arthritis. (Berk Pharmaceuticals Ltd).*

Imesatin. β-Iminoisatin, $C_8H_6ON_2$.†

Imferon. A proprietary preparation of iron dextran. A haematinic. (Fisons PLC).*

Imiodid. A substance obtained by heating *p*-ethoxyphenylsuccinimide with a solution containing potassium iodide and iodine. A powerful antiseptic.†

Imlar. Vinyl coating. (Du Pont (UK) Ltd).

Immadium. An alloy of manganese bronze containing aluminium.†

Immedial. A trade mark for a range of dyestuffs. (Casella AG).†

Immedial Black. A dyestuff obtained by fusing 1-chloro-2,4-dinitrobenzene and *p*-aminophenol with sodium polysulphide.†

Immedial Black N. A dyestuff produced by the sulphurisation of dinitrophenol. See DINITRO-PHENOL BLACK.†

Immedial Black V. A dyestuff produced by the fusion of dinitrohydroxydiphenylamine with sodium polysulphide. It dyes cotton black from a sodium sulphide bath. By the oxidation of the colour on the fibre by hydrogen peroxide, it is converted into Immedial blue (indigo blue).†

Immedial Blue. A dyestuff produced from hydroxydinitrodiphenylamine, and sulphurising at low temperature.†

Immedial Bordeaux G. A dyestuff prepared by the action of alkaline sulphides and sulphur on aminohydroxyphenazines.†

Immedial Bronze. A dyestuff prepared by fusing dinitrocresol with polysulphides.†

Immedial Brown B. A brown sulphide dyestuff, giving yellowish-brown shades on cotton. It is manufactured by boiling 4-hydroxy-4'-aminodiphenylamine with a solution of sodium hydroxide, and heating the product with sodium polysulphides.†

Immedial Cutch. (Immedial Direct Blue). A dyestuff belonging to the same class as Immedial indone R.†

Immedial Green. A dyestuff manufactured by dissolving 4-dimethylamino-4-hydroxydiphenylamine in a solution of sulphur in crystalline sodium sulphide, copper sulphate being added to the solution. The dye is precipitated with salt. It gives bluish-green shades on cotton.†

Immedial Indone R. A dyestuff obtained from the indophenol produced by oxidizing *o*-toluidine with *p*-aminophenol by heating it with sodium polysulphide in aqueous or alcoholic solution. It dyes cotton.†

Immedial Maroon B. (Thionine Red-brown B). A dyestuff prepared by the action of alkaline sulphides and sulphur upon aminohydroxyphenazine. See IMMEDIAL BORDEAUX G.†

Immedial Orange C. A dyestuff produced by fusing tolylene-2,4-diamine with sulphur. The product is fused with sodium sulphide. Dyes cotton orange-brown.†

Immedial Orange N. A sulphide dyestuff made by fusing together the same materials as in Immedial yellow D.†

Immedial Pure Blue. (Sky Blue). A dyestuff manufactured by oxidizing a mixture of reduced nitrosodimethylaniline, and an aqueous solution of phenol, to an indophenol. This is converted by reduction into 4-dimethylamino-4-hydroxydiphenylamine, which is added to a hot solution of sulphur and sodium sulphide, then heated.†

Immedial Sky Blue. A dyestuff produced by the fusion of dimethyl-*p*-amino-*p*-hydroxydiphenylamine with sodium polysulphides at 110-115° C. Dyes cotton blue from a sulphide bath.†

Immedial Yellow D. A dyestuff obtained from *m*-tolylenediamine and sulphur at 190° C.†

Immetal. Diiodoerucic acid isobutyl ester.†

Immu-Tetanus. Tetanus Immune Globulin. Applications: immunizing agent. (Parke-Davis, Div of Warner-Lambert Co).

Immuglobin. Globulin, Immune. Applications: immunizing agent. (Savage Laboratories, Div of Byk-Gulden Inc).

Immuno-bed. Plastic embedding kit. Applications: For light microscopy immunohisto chemistry procedures. (Polysciences Inc).*

Imodium. A proprietary preparation of loperamide. An anti-diarrhoeal. (Janssen Pharmaceutical Ltd).*

Imogen. Sodium diaminonaphthol-sulphonate.†

Imogen Sulphite. A photographic developer.†

Impact. Fungicide containing permethrin. (ICI PLC).*

Impad. High density impact pads for continuous casting tundishes. (Foseco (FS) Ltd).*

Imperacin. A proprietary preparation of OXYTETRACYCLINE dihydrate. An antibiotic. (ICI PLC).*

Imperial Green. See EMERALD GREEN

Imperial Metal. An alloy of 80 per cent copper and 20 per cent nickel.†

Imperial Red. See BENZOPURPURIN 4B, EOSIN BN, and INDIAN RED.

Imperial Scarlet. See BIEBRICH SCARLET.

Imperial Violet. See REGINA PURPLE.

Imperial Yellow. See AURANTIA and CHROME YELLOW.

Impervite. See CERESIT.

Impletol. Pharmaceutical preparation. Applications: Neural therapeutic and diagnostic agents. (Bayer & Co).*

Imposil. Veterinary iron-dextran complex. (Fisons PLC, Pharmaceutical Div).

Impra. Organic wood preservatives and wood stainers. Applications: Brushing, spraying and dipping. (Chemische Fabrik Weyl GmbH).*

Impra-biolan. Ecological wood stainer for indoor and outdoor use. Applications: Brushing and spraying. (Chemische Fabrik Weyl GmbH).*

Impra-color. Wood stainer against fungi and insect attack. For outdoor use.

Applications: Brushing, spraying and dipping. (Chemische Fabrik Weyl GmbH).*

Impra-elan. Fast-drying satin gloss joinery stainer. Applications: Brushing and spraying. (Chemische Fabrik Weyl GmbH).*

Impraleum. A coaltar distillate for fencing. Applications: Brushing, spraying and dipping. (Chemische Fabrik Weyl GmbH).*

Impralit. Wood preservative salts and fire retardant salts. Applications: Dipping, vacuum-pressure method. (Chemische Fabrik Weyl GmbH).*

Impranil. Polyurethane products. Applications: For coating and laminating textiles. Fast to washing and to solvents. (Bayer & Co).*

Imprez. Petroleum resins. (ICI PLC).*

Impsonite. An asphaltic substance found in Arkansas and other places. It is difficultly fusible, and is only slightly soluble in carbon disulphide.†

Imron. Urethane finish. (Du Pont (UK) Ltd).

Imsol. Solvents. (ICI PLC).*

Imwitor. Emulsifier. Applications: For the pharmaceutical industry, cosmetic industry and food industry. (Dynamit Nobel Wien GmbH).*

Inalium. An aluminium alloy containing 2 per cent cadmium, 0.8 per cent magnesium, and 0.4 per cent silicon.†

Inapassade, A. proprietary preparation of sodium para-aminosalicylic acid and isoniazid. Antituberculous drug. (Geistlich Sons).†

Inapsine. Droperidol. Applications: Antipsychotic. (Janssen Pharmaceutica).

Incidal. Pharmaceutical preparation also known as Fabahistin. Applications: Day-time antihistamine. (Bayer & Co).*

Inco Chrome Nickel. (Inconel). Proprietary nickel alloys containing 12-14 per cent chromium and 6-7 per cent iron. They resist corrosion by the organic acids met within foodstuffs. The alloy containing 80 per cent nickel, 14 per cent chromium, and 6 per cent iron melts at 1390° C.†

Incoloy Ailoy 901. A trade mark for an alloy of 12.5 per cent. chromium, 5.7 per cent. molybdenum, 2.9 per cent. titanium, 42 per cent. nickel and the balance iron. (Wiggin Alloys Ltd).†

Incoloy Alloy DS. A trade mark for an alloy of 18 per cent. chromium, 2.3 per cent. silicon, 37 per cent. nickel and the balance iron. (Wiggin Alloys Ltd).†

Incoloy Alloy 800. A trade mark for an alloy of 20 per cent. chromium, 32 per cent. nickel and 48 per cent. iron. It is resistant to hydrogen/hydrogen sulphide corrosion. (Wiggin Alloys Ltd).†

Incoloy Alloy 825. A trade mark for an alloy of 40 per cent. nickel, 21 per cent. chromium, 3 per cent. molybdenum, 2 per cent. copper, 1 per cent. titanium and the balance iron. It is resistant to corrosion in hot oxidizing acid conditions. (Wiggin Alloys Ltd).†

Incomate IDL. Isostearamidopropyl dimethylamine lactate. Applications: A cationic salt used in clear rinses, conditioners, conditioning shampoos, compatible in anionic systems. (Croda Chemicals Ltd).*

Inconel. (Inco Chrome Nickel). A proprietary trade name for an alloy resistant to corrosion and heat, containing 80 per cent nickel, 14 per cent chromium, and 6 per cent iron.†

Inconel Ailoy X-750. A trade mark for an alloy of I5 per cent. chromium, 2.5 per cent. titanium, 0.9 per cent. aluminium, 0.9 per cent. niobium, 7 per cent. iron and the balance nickel. (Wiggin Alloys Ltd).†

Inconel Alloy 600. A trade mark for an alloy of 16 per cent. chromium, 7 per cent. iron and 77 per cent. nickel. It has good oxidation resistance at high temperatures. (Wiggin Alloys Ltd).†

Inconel Alloy 700. A trade mark for an alloy of 15 per cent. chromium, 29 per cent. cobalt, 3 per cent. molybdenum, 2.25 per cent. titanium, 3.3 per cent. aluminium and the balance nickel. (Wiggin Alloys Ltd).†

Incromate CDL. Cocoamidopropyl dimethylamine lactate. Applications: A cationic salt used in clear rinses, conditioners, conditioning shampoos, compatible in anionic systems. (Croda Chemicals Ltd).*

Incromate CDP. Cocoamidopropyl dimethylamine propionate. Applications: A cationic salt used in clear rinses, conditioners, conditioning shampoos, compatible in anionic systems. (Croda Chemicals Ltd).*

Incromate ODL. Oleamidopropyl dimethylamine lactate. Applications: A cationic salt used in clear rinses, conditioners, conditioning shampoos, compatible in anionic systems. (Croda Chemicals Ltd).*

Incromate SDL. Stearamido-propyldimethylamine lactate. Applications: A cationic salt used in pearlescent and opaque cream rinses, conditioning shampoos, hand cleaners and lotions. (Croda Chemicals Ltd).*

Incromectant AMEA-70. Acetamide MEA. Applications: Clarifying and detangling agents for shampoos. (Croda Chemicals Ltd).*

Incromectant LMEA-70. Lactamide MEA. Applications: Conditioning agents in cream rinses, humectants in creams and lotions. (Croda Chemicals Ltd).*

Incromine BB. Behenamidopropyldi-methylamine. Applications: Used as intermediate for hair conditioners and shampoo rinses. (Croda Chemicals Ltd).*

Incromine CB. Cocoamidopropyldi-methylamine. Applications: Used as intermediate for hair conditioners and shampoo rinses. (Croda Chemicals Ltd).*

Incromine IB. Isostearamidopropyl-dimethylamine. Applications: Used as an intermediate for hair conditioners and shampoo rinses. (Croda Chemicals Ltd).*

Incromine OPB. Oleamidopropyl-dimethylamine. Applications: Intermediate for hair conditioners and shampoo rinses. (Croda Chemicals Ltd).*

Incromine Oxide B. Behenyl dimethylamine oxide. Applications: Used in a variety of cosmetic, household and janitorial products. (Croda Chemicals Ltd).*

Incromine Oxide C. Cocoamidopropyl-dimethylamine oxides. Applications: Used in a variety of cosmetic, household and janitorial products. (Croda Chemicals Ltd).*

Incromine Oxide I. Isostearamidopropyl-dimethylamine oxide. Applications: Used in a wide variety of cosmetic, household and janitorial products. (Croda Chemicals Ltd).*

Incromine Oxide L. Lauryl dimethylamine oxide. Applications: Used in a variety of cosmetic and household products. They are good wetting agents in concentrated solutions. (Croda Chemicals Ltd).*

Incromine Oxide M. Myristyl dimethylamine oxide. Applications: Used in cosmetic, household amd janitorial products. (Croda Chemicals Ltd).*

Incromine Oxide O. Oleamidopropyl-dimethylamine oxide. Applications: Used in a wide variety of cosmetic, household and janitorial products. (Croda Chemicals Ltd).*

Incromine Oxide OD-50. Oleyl dimethylamine oxide. Applications: Used in cosmetic, household and janitorial products. (Croda Chemicals Ltd).*

Incromine Oxide S. Stearyl dimethylamine oxide. Applications: Used in a variety of cosmetic, household and janitorial products, a good wetting agent. (Croda Chemicals Ltd).*

Incromine SB. Stearamidopropy-ldimethylamine. Applications: Intermediate for hair conditioners and shampoo rinses. (Croda Chemicals Ltd).*

Incronam B-40. Behenamidopropyl betaine. Applications: Used as a surfactant for bubble baths, shampoos, cleansing lotions, hand cleaners, skin care and household products. (Croda Chemicals Ltd).*

Incronam OP-30. Oleamidopropyl betaine. Applications: Used as a surfactant for bubble baths, shampoos, cleansing lotions, hand cleaners, skin care and household products. (Croda Chemicals Ltd).*

Incronam 1-30. Isostearamidopropyl betaine. Applications: A viscosity building high foaming mild surfactant compatible with a wide range of materials. Uses include shampoos, bubble baths, cleansing lotions, hand cleaners, skin care and household products. (Croda Chemicals Ltd).*

Incronam 30. Cocoamidopropyl betaine. Applications: Uses include shampoos, bubble baths, cleansing lotions, hand cleaners, skin care and household products. (Croda Chemicals Ltd).*

Incroquat S-85 and SDQ-25. Stearyl dimethyl benzyl ammonium chloride preparations. Applications: Used as cationic emulsifiers, applications include hair rinses, skin creams and lotions. (Croda Chemicals Ltd).*

Incrosoft CF1-75. A range of alkyl imidazolinium methosulphates. Applications: Fabric softeners for preparation of clear detergent liquid concentrates. (Croda Chemicals Ltd).*

Incrosoft S-75 and S-90. A range of tallow imidazolinium methosulphates. Applications: Fabric softeners and anti-static agents. (Croda Chemicals Ltd).*

Incrosoft T-75 and T-90. A range of ditallow diamido methosulphates. Applications: Fabric softeners and anti-static agents. (Croda Chemicals Ltd).*

Incrosperse. Polymeric dispersants. (Croda Chemicals Ltd).

Indalca. A thickening agent derived from natural gum. Applications: Thickener for fabric-printing dyes. (Hercules Inc).*

Indalizarin. A dyestuff formed by the action of sulphites upon gallocyanine sulphonic acid. It represents a series of dyes known as chromocyanines.†

Indalizarin Green. Nitrogallocyanine sulphonic acid. Dyes chromed wool green.†

Indalizarin R and J. A dyestuff produced by the action of sulphites upon gallocyaninesulphonic acid. Dyes blue upon a chrome mordant. Used in printing.†

Indamine Blue. See INDAMINE BLUE R and B.

Indamine Blue R and B. A dyestuff. Dyes tannined cotton bluish-violet.†

Indamine 3R. (Indamine 6R, Rubramine). A dyestuff prepared by the action of nitrosodimethylaniline hydrochloride upon *o*-toluidine, or upon a mixture of *o*-toluidine and *p*-toluidine. Dyes tannined cotton reddish to bluish violet.†

Indamine 6R. See INDAMINE 3R.

Indanthren. A range of vat dyes, having 'unsurpassed' overall fastness properties on cotton and other vegetable fibres as well as on textiles made from regenerated cellulose. (Bayer & Co).*

Indanthrene. A dyestuff produced by fusing *β*-aminoanthraquinone with potassium hydroxide.†

Indanthrene Black. A dyestuff. It consists of Indanthrene green chlorinated on the fibre.†

Indanthrene Blue GC. (Caledon Blue GC, Alizanthrene Blue GC). A dyestuff. It is dibromoindanthrene.†

Indanthrene Blue GCD. (Alizanthrene Blue GCP, Caledon Blue GCD, Duranthrene Blue GCD). A dyestuff. It is dichloro-indanthrene.†

Indanthrene Blue RC. A dyestuff. It is monobromoindanthrene.†

Indanthrene Blue RS. See INDAN-THRENE X.

Indanthrene Bordeaux B. A dyestuff allied to Indanthrene red G.†

Indanthrene C. A dyestuff. It is a mixture of di- and tribromoindanthrene.†

Indanthrene CD. A dyestuff. It is dichloroindanthrene. Used in calico printing.†

Indanthrene Dark Blue BO. (Cyanthrene, Alizanthrene Dark Blue O, Caledon Dark Blue O, Duranthrene Dark Blue BO, Hydranthrene Dark Blue). A dyestuff. It is prepared by fusing benzanthrone with alkali.†

Indanthrene Gold Orange G. See PYRANTHRONE.

Indanthrene Gold Orange R. A dyestuff. It is the halogen product of Indanthrene gold orange G.†

Indanthrene Green. A dyestuff. It is the nitro-derivative of Indanthrene dark blue BO.†

Indanthrene Grey B. A dyestuff prepared by the alkaline fusion of 1,5-diaminoanthraquinone.†

Indanthrene Maroon. A dyestuff obtained by the alkaline fusion of the formaldehyde compound of 1,5-diaminoanthraquinone.†

Idanthrene Olive G. (Alizanthrene Olive G, Duranthrene Olive GL, Hydranthrene Olive R). A dyestuff prepared by the fusion of anthracene with sulphur at 250° C.†

Indanthrene Red G. A dyestuff obtained by condensing 2,6-dichloro anthraquinone with α-aminoanthraquinone.†

Indanthrene Red Violet 2RN. A dyestuff formed by the oxidation of 5 chloro-4, 7-dimethylthioindoxyl.†

Indanthrene S. The leuco compound of indanthrene. Used in printing.†

Indanthrene Scarlet G. A dyestuff obtained by the halogenation of Indanthrene gold orange G.†

Indanthrene Violet RT. A dyestuff produced by the halogenation of Indanthrene dark blue BO.†

Indanthrene X. (Indanthrene Blue RS, Duranthrene Blue RD Extra, Hydranthrene Blue RS Paste). A dyestuff. It is anthraquinone azine, $C_{28}H_{14}N_2O_4$. Dyes cotton from a reduced vat, blue. Employed in printing.†

Indanthrene Yellow. See FLAVANTHRENE.†

Indanthrene Yellow GK. A dyestuff. It is equivalent to Algol yellow R.†

Indazin A. proprietary trade name for a range of products used in the dyeing of textiles. (Casella AG).†

Indazurine B. A dyestuff. Dyes cotton reddish-blue.†

Indazurine BB. A dyestuff. Dyes cotton blue.†

Indazurine GM. A dyestuff. Dyes cotton blue.†

Indazurine RM. A dyestuff. Dyes cotton reddish-blue.†

Indazurine TS. A dyestuff. Dyes cotton reddish-blue.†

Inderal. A proprietary preparation of propranolol hydrochloride. (ICI PLC).*

Inderetic. Combined antihypertensive and diuretic containing propranolol hydrochloride and bendrofluazide. (ICI PLC).*

India Gum. (Persian Gum). A gum resembling gum arabic. See also BASSORA GUM.†

Indian Fire. A mixture of 7 parts sulphur, 2 parts realgar, and 24 parts potassium nitrate. Used for signalling, and in the manufacture of fireworks.†

Indian Frankincense. See OLIBANUM.

Indian Gutta-percha. See PALA GUM

Indian Lake. See CARMINE LAKE.

Indian Ochre. A native ferric oxide of North America.†

Indian Purple. A pigment. It is prepared by precipitating cochineal extract with copper sulphate.†

Indian Red. (Venetian Red, Venetian Bole, Rouge, Colcothar, Red Bole, Bole, Armenian Bole, English Red, Angel Red, Chemical Red, Pompeian Red, Berlin Red, Iron Minium, Iron Red, Persian Red, Raddle, Reddle, Red Rudd, Red Ochre, Red Chalk, Red Earth, Terra di Sienna, Mineral Purple, Stone Red, Prussian Red, Italian Red, American Sienna, Mineral Rouge, Crocus, Blood Red, Brown Red, Pale Oxide of Iron, Dark Oxide of Iron, Violet Oxide of Iron, Iron Saffron, Imperial Red, Nuremberg Red, Scarlet Ochre, Prague Red, Red Oxide, Scarlet Red, Rubrica, Sinopis, Lemnos Earth, Vandyck Red, Spanish Oxide, Turkey Red Oxide, Turkey Red, Bright Red Oxide). Red pigments consisting mainly of ferric oxide, Fe_2O_3, with varying amounts of natural argillaceous compounds. Some are natural products (ochres or earths), burnt or unburnt. but the names are also applied to products obtained by heating ferrous sulphate, or as a red residue from the manufacture of fuming sulphuric acid. This consists chiefly of ferric oxide, together with some basic ferric sulphate. The burnt product comes on to the market as Stone Red. The following are some of the raw materials used for the production of these pigments : red iron ore, haematite, iron ochre, and limonite. These pigments contain from 12-95 per cent ferric oxide.†

Indian Saffron. See TURMERIC.

Indian Yellow. (Piuri, Purree, Pioury). A pigment used in India, obtained from the urine of cows fed on mango leaves. The colouring principal is the magnesium or calcium salt of euxanthic acid, $C_{19}H_{16}O_{11}Mg.5H_2O$. Used as a permanent water and oil colour. The name is also used for Citronine B. See CITRONINE B and AUREOLIN.†

Indian Yellow G. See CITRONINE B.

Indican. Potassim-indoxyl-sulphate.†

Indicator, Universal. A mixture of indicators, usually methyl red, α-naptholphthalein, phenolphthalein, bromothymol blue, and cresol red. The colour indicates the *P*H value when added to the Solution.†

Indicolite. A mineral from Brazil. It is a blue tourmaline.†

Indigen D and F. See INDULINE, SPIRIT SOLUBLE.

Indigo. Indigotine, $C_{16}H_{10}N_2O_2$. Natural indigo is obtained by steeping the leaves of indigo-bearing plants in water then oxidizing the extract. Synthetic indigo is prepared by several methods. Used for cotton, wool, silk, by steeping the material in a vat containing the leuco compound, then exposing to air.†

Indigo Blue. (Indigotine). Indigo, $C_{16}H_{10}N_2O_2$. A mixture of methyl violet and malachite green is also known as indigo.†

Indigo Blue N and SGN. Dyestuffs. They are mixtures of cyanol, with green and red dyes.†

Indigo Carmine. Indigotindisulphonate sodium. Applications: Diagnostic aid. (Hynson Westcott & Dunning, Div of Becton Dickinson & Co).

Indigo Purple. A dyestuff. It is the sodium salt of indigotinsulphonic acid. It blues bleached cotton, thread, and silk in a soap bath.†

Indigo RBN and RB. Dibromo-indigo.†

Indigo Red. See INDIPURPURIN.

Indigo, Red. See FRENCH PURPLE.

Indigo, Soluble. See INDIGO CARMINE.

Indigo TRG. Indigo.†

Indigo Vat. See INDIGO WHITE.

Indigo White. (Indigo Vat). Used for the preparation of indigo vats. Fibres such as cotton, wool, or silk, are immersed in an alkaline bath of this compound, and then exposed to air, to precipitate indigo within the fibres.†

Indigo Yellow 3G. Produced by the interaction of indigo and benzoyl chloride in the presence of copper powder, $C_{23}H_{14}N_2O_2$.†

Indigosol DH. The sodium salt of the acid disulphuric ester of indigo. Used in dyeing and printing.†

Indigosol-O. (Blue Salt). A colourless leucoindigo. It is suitable for dyeing wool and cotton, no reducing vat being necessary.†

Indigotine. See INDIGO CARMINE and INDIGO BLUE.

Indigotine P. A dyestuff. It is the sodium salt of indigotinetetrasulphonic acid. Dyes wool bluish-violet from an acid bath.†

Indilitans. An alloy of 36 per cent nickel, 0.06 per cent carbon, 0.68 per cent manganese, 0.09 per cent silicon, and remainder iron. The alloy has a low coefficient of thermal expansion.†

Indio. Decorating colours for vitreous enamel. (Degussa).*

Indirubin. Indigo red, $C_{16}H_{10}N_2O_2$, a red colouring matter associated with indigo in amount usually from 1-5 per cent. Java indigo often contains up to 10 per cent. It is produced from the decomposition of indican.†

Indisin. A registered trade mark currently awaiting re-allocation by its proprietors. (Casella AG).†

Indocarbon. A trade mark for a range of dyestuffs. (Casella AG).†

Indochromine T. See BRILLIANT ALIZARIN BLUE G and R.

Indochromogen S. A dyestuff. It is used in calico printing, giving greenish-blue shades.†

Indocid. Indomethacin. Applications: Analgesic and anti-inflammatory agent for the relief of rheumatic conditions. (Merck Sharp & Dohme).*

Indocid-R. Indomethacin. Applications: Analgesic and anti-inflammatory agent for the prompt and sustained relief of certain rheumatic conditions. (Merck Sharp & Dohme).*

Indocin. Indomethacin. Applications: Anti-inflammatory. (Merck Sharp & Dohme, Div of Merck & Co Inc).

Indocybin. A proprietary trade name for Psilocybin.†

Indoil CPD 142 and CPD 143. A proprietary trade name for petroleum type vinyl plasticisers. (Standard Oil Co).†

Indoine Blue R. (Janus Blue, Naphthindone, Naphthindone Blue BB, Benga line, Vac Blue, Fast Cotton Blue R, Indole Blue R, Diazine Blue, Diazine Blue BR). A dyestuff. It is safranine-azo-β-naphthol. It dyes unmordanted and tannined cotton indigo blue.†

Indoklon. A proprietary trade name for Flurethyl.†

Indolar. Suppositories containing 100mg indomethacin BP, capsules containing 25mg, 50mg, SR capsules containing 75mg indomethacin BP. Applications: Non-steroidal analgesic and anti-inflammatory agent indicated in active rheumatoid arthritis, osteoarthritis, ankylosing spondylitis, acute gout. Also indicated in periarticular disorders such as bursitis, tendinitis, synovitis, tenosynovitis and capsulitis. Also indicated in inflammation, pain and oedema following orthopaedic procedures. (Lagap Pharmaceuticals Ltd).*

Indole Blue R. See INDOINE BLUE R.

Indoline. A basic navy blue dye.†

Indomed. Indomethacin. Applications: Anti-inflammatory. (Rowell Laboratories Inc).

Indon. Phenindione. Applications: Anticoagulant. (Parke-Davis, Div of Warner-Lambert Co).

Indonex VG. A proprietary trade name for aromatic hydrocarbon vinyl plasticizer. (Standard Oil Co).†

Indophenine. See PARAPHENYLENE BLUE R.

Indophenine Extra. See INDULINE, SPIRIT SOLUBLE.

Indoptol. Indomethacin. Applications: Eye drops for the prevention of aphakic cystoid macular oedema. (Merck Sharp & Dohme).*

Indorm. A proprietary preparation of propiomazine. A hypnotic. (Wyeth).†

Induline. A mixture of aryl-amino-azines, made by heating together amino-azobenzene, aniline, and aniline hydrochloride. Used in making inks.†

Induline 3B. See INDULINE, SOLUBLE.

Induline 3B Opal and 6B Opal. See INDULINE, SPIRIT SOLUBLE.

Induline 3B, Spirit Soluble. See INDULINE, SPIRIT SOLUBLE.

Induline 6B. See INDULINE SOLUBLE.

Induline 6B, Spirit Soluble. See INDULINE, SPIRIT SOLUBLE.

Induline Black. A dyestuff. It consists of the sulphonic acids of induline, and dyes silk, wool, and cotton.†

Induline Opal. See INDULINE, SPIRIT SOLUBLE.

Induline R and B. See INDULINE SOLUBLE.

Induline Scarlet. A safranine dyestuff. Dyes tannined cotton scarlet. Chiefly used in printing.†

Induline, Soluble. (Induline R and B, Induline 3B, Induline 6B, Fast Blue R and 3R, Fast Blue 2R, B and 6B, Sloeline RS and BS, Fast Blue greenish, Nigrosine, soluble, Solid Blue 2R and B, Blue CB, Grey R and B, Bengal Blue Induline A, 2B, 5B, BL, 5B Crystals, L332). A dyestuff. It is a mixture of the sodium salts of the sulphonic acids of the various spirit soluble indulines. Dyes wool or silk blue, reddish-blue, or bluish violet, from an acid bath, Used in silk dyeing, and in the manufacture of inks.†

Induline, Spirit Soluble. (Induline Opal, Fast Blue R, spirit soluble, Induline 3B, spirit soluble, Induline 6B, spirit soluble,

Induline 3B Opal and 6B Opal, Fast Blue B, spirit soluble, Azine Blue, spirit soluble, Indigen D and F, Printing Blue, Acetin Blue, Nigrosine, spirit soluble, Coupier's Blue, Spirit Black, Oil Black, Indophenine Extra, Blue CB, spirit soluble, Pelican Blue, Sloeline, Azo Diphenyl Blue, Violanlline). Mixtures of dianilido, amino, trianilido, and tetra-anilido-phenyl phenazonium-chlorides. Employed for the preparation of the corresponding water soluble colours. Also used (mixed with chrysoidine), for the preparation of black Spirit varnishes and polishes. Employed in calico printing.†

nduline Yellow. See FLAVINDULINE.

ndur. A proprietary trade name for a phenol-formaldehyde synthetic resin used for moulding.†

ndurite. An explosive containing 40 per cent guncotton, freed from lower nitrates, and 60 per cent nitro-benzene. The name is also applied to a moulding powder of the phenolformaldehyde condensation product type.†

ndusoil. A proprietary product. It is a refined Talleol.†

ndustrial Dyne. Iodophor/acid pipeline cleaner and sterilizer. (Ciba-Geigy PLC).

ndustrial Dynemate. Hypochlorite pipeline cleaner/sterilizer. (Ciba-Geigy PLC).

ndustrial Gum. (Tragasol, Locust Bean Gum). Carob bean gum, an ingredient of mucilages. It is also used as a protective colloid.†

nertex. Protective powder for molten magnesium (Foseco (FS) Ltd).*

nfacare. Baby bath, additive for bath. (Richardson-Vicks Inc).*

Infasoft. Shampoo, to wash and condition hair. (Richardson-Vicks Inc).*

Infavina. Vitamin B complex. Applications: Tonic. (Merck Sharp & Dohme).*

Inflamase. Prednisolone Sodium Phosphate. Applications: Glucocorticoid. (CooperVision Inc).

Influvac. Influenza virus vaccine. Applications: Protection against virus infection. (Duphar BV).*

Infusible White Precipitate. Dimercuridi-ammonium chloride.†

Infusorial Earth. (Celite, Fossil Flour, Kieselguhr, Tripolite, Mountain Flour). The siliceous remains of diatoms. Used as a non-conducting material for boilers, an absorbent for liquids and liquid manures, in the preparation of dynamite, and as a filling material in soaps, dyes, and rubber goods. Also see FILTER-CEL.†

Ingalite. See GALLATITE.

Ingotol. Non-ferrous chill and ingot dressing. (Foseco (FS) Ltd).*

Ingrain. (Primuline Red). An azo dyestuff produced by dyeing the fabric with primuline, diazotizing its amino group, and developing in a bath ofβ-naphthol.†

Ingrain Brown. (Alkali Brown D, Benzo Brown 5R, Cotton Brown R, Terracotta G, Cotton Orange 6305, Direct Fast Orange D2G, D2R). A dyestuff. It is the sodium salt of primulineazo-phenylenediamine, being similar to Ingrain, a developing bath of phenylenediamine being used.†

Ingrain Orange. A dyestuff produced in a similar manner to Ingrain, using resorcinol instead ofβ-naphthol.†

INH. Isoniazid. Applications: Antibacterial. (Ciba-Geigy Corp).

Inhalit. A proprietary preparation of menthol, eucalyptol and pumilio pine oil. Applications: A nasal decongestant. (Ayrton Saunders plc).*

Inipol. Amines and derivatives. Primary, secondary, tertiary fatty mono-, di- and polyamines from C8 to C22. Quaternary ammonium salts. Amine oxides. Amine salts. Ethoxylated amines and polyamines. Amino-acids. Special nonionic derivatives. Additives. Applications: Surface active agents used as auxiliaries in road, mining, textile and fertilizer industries, in metallurgy, as lubricating agents, additives for fuel oils and soil stabilization. (British Ceca Co Ltd).*

Injacom. A proprietary vitamin injection for animals. (Roche Products Ltd).*

Injex. Iron/dextran injection. (The Wellcome Foundation Ltd).

Ink Blue. (Ink Blue 8671, 7567). Dyestuffs. They are British equivalents of Soluble blue.†

Inkovar 335. Modified hydrocarbon resins. Applications: It is designed specifically for compatibility with cellulosic polymers. (Hercules Inc).*

Inkovar 617. Modified hydrocarbon resins. Applications: It is designed for low energy heatset ink vehicles and flushed colours. (Hercules Inc).*

Innovace. Enalapril maleate. Applications: For the treatment of hypertension and congestive heart failure. (Merck Sharp & Dohme).*

Inochrome. Premetallised dyes. (ICI PLC).*

Inocor. Amrinone. Applications: Cardiotonic. (Sterling Drug Inc).

Inoculin. Inoculating compound for cast iron. (Foseco (FS) Ltd).*

Inoderme. Leather dyes. (ICI PLC).*

Inopak. Prepacked mould inoculants for use in production of grey and ductile iron castings. (Foseco (FS) Ltd).*

Inotab. Tabletted mould inoculants for use in production of grey and ductile iron castings. (Foseco (FS) Ltd).*

Insect Bite Cream (AS). A proprietary preparation of antazoline, benzocaine and cetrimide. Applications: An antihistamine cream. (Ayrton Saunders plc).*

Insect Spray for House Plants. Contact fertilizer aerosol. (Fisons PLC, Horticulture Div).

Insect Wax. See CHINESE WAX.

Insidon. A proprietary preparation of opipramol dihydrochloride. A sedative. (Ciba Geigy PLC).†

Insomnol. A proprietary preparation of methylpentynol. A hypnotic. (25O).†

Instant Gasket. R.T.V. silicone gasketing material. Applications: Replaces cork, felt paper, asbestos, rubber and metal gaskets. Gasket made in situ squeezed from the tube using unique applicator nozzle. (Hermetite Products Ltd).*

Instant Ocean. A dry, granular mixture of salts for preparation of synthetic seawater. Applications: Aquariums for marine animals, culture medium and corrosion testing. (Aquarium Systems Inc).*

Instoms. Indigestion tablets. (Fisons PLC).*

Instrument Bronze. An alloy of 82 per cent copper, 13 per cent tin, and 5 per cent zinc.†

Insulin. A hormone isolated from the pancreas. Used in the treatment of diabetes.†

Insullac. A copal spirit varnish with the resin acids neutralized. An insulating material.†

Insulmag. Magnesium oxide for electrical cable filling. (Steetley Refractories Ltd).

Insural. Insulating refractory shapes for use with aluminium alloys. (Foseco (FS) Ltd).*

Insurok. A proprietary trade name for a phenol-formaldehyde synthetic resin used for moulding compounds and laminated products.†

Intal Compound. A proprietary preparation of disodium cromoglycate and isoprenaline sulphate. A bronchial inhalant. (Fisons PLC).*

Integrin. A proprietary preparation of oxypertine. A tranquillizer. (Bayer & Co).*

Intense Blue. A pigment. It is indigo refined by precipitation.†

Intensive Blue. See FAST ACID BLUE B.

Interacton. A proprietary preparation of cocarboxylase, glutaminase, histaminase, Co-enzyme A, ascorbic and allyl sulphide. (FAIR Laboratories).†

Intercept. Nonoxynol 9. Applications: Spermaticide. (Ortho Pharmaceutical Corp).

Interferon. A protein formed by the interaction of animal cells with viruses. It is capable of conferring on animal cells resistance to virus infection.†

Internol. See PARAFFIN, LIQUID.

Interol. See PARAFFIN, LIQUID

Interox H48. Magnesium monoperoxyphthalate hexahydrate low temperature bleach. (Interox Chemicals Ltd).

terstab. Stabilizers/additives for plastics. (Akzo Chemie UK Ltd).

timate Contact. Silicone release film comprising super clear polyester coated with silicone release. Applications: Release film for optical applications. (Custom Coating and Laminating Corporation).*

tob. Mould inoculants for iron castings. (Foseco (FS) Ltd).

tradex. A proprietary preparation of dextran in saline or dextrose solution. An intravenous infusion used to restore blood volume. (Glaxo Pharmaceuticals Ltd).†

traflodex. A proprietary preparation of dextran in saline or dextrose, used to improve capillary circulation. (Glaxo Pharmaceuticals Ltd).†

tralgin Gel. Benzocaine BP 2% w/w and salicylamide 5% w/w in an alcoholic gel vehicle. (Riker Laboratories/3M Health Care).†

tralipid. Fat emulsion for parenteral use. Applications: High-energy source for intravenous nutrition. (KabiVitrum AB).*

traval. A proprietary preparation of thiopentone sodium. Used as a short duration anaesthetic or for induction of anaesthesia. (May & Baker Ltd).*

traval Sodium. Thiopentone sodium. (Akzo Chemie UK Ltd).

trex. Functional surfactant blend. (ABM Chemicals Ltd).*

trex Asa. Antisetting additive for paints. (ABM Chemicals Ltd).

trex DW81. Corrosion inhibitor. (ABM Chemicals Ltd).

trex HA70. Organic phosphoric acid derivative as an amber liquid. Application: Hydrotroping agent used in the solubilising of other surface active agents in highly alkaline cleaner bases; base for acid cleaners for metals, etc. (ABM Chemicals Ltd).

tron. Interferon. Protein formed by the interaction of animal cells with viruses capable of conferring on animal cells resistance to virus infection. Applications: Antineoplastic; antiviral. (Schering-Plough Corp).

Intropin. Dopamine hydrochloride. Applications: Adrenergic. (American Critical Care, Div of American Hospital Supply Corp).

Introsolvan HS. A mixture of isohexyl and isoheptyl alcohol.†

Invaderm. Dyeing auxiliaries. (Ciba-Geigy PLC).

Invaderm C9B. A sodium aryl disalphonate in powder form used as an anionic level dyeing assistant for leather. (Ciba Geigy PLC).†

Invadine. A proprietary trade name for a wetting agent containing a sodium alkylphenylene sulphonate.†

Invalon. Dyeing and printing assistant. (Ciba-Geigy PLC).

Invar. An alloy of 36 per cent nickel and 64 per cent steel (0.2 per cent carbon in steel). It has very little expansion on heating, and is used for delicate instruments for measuring.†

Invariant. An alloy of 47 per cent nickel and 53 per cent iron. Has similar properties to Permalloy.†

Invaro Steel. A proprietary trade name for a tool steel containing 1.15 per cent manganese, 0.5 per cent tungsten, and 0.9 per cent carbon.†

Invasol. Fatliquors. (Ciba-Geigy PLC).

Invenol. Carbutamide.†

Invephos 20. Liquid chemical containing calcium to be diluted with water. Applications: Used to produce a micro-crystalline, corrosion resisting, paint bording, zinc phosphate coating on iron, steel and zinc. (Invequimica & CIA SCA).*

Invephos 21C. Liquid chemical containing calcium to be diluted with water. Applications: Corrosion preventing chemical used in pretreatment for electropainting, produces microcrystalline coatings on iron and steel. (Invequimica & CIA SCA).*

Inveres EVH. Plyvinyl acetate homopolimer. Applications: Sizing agent, adhesives for wood and paper - available in a wide variety of viscosities. (Invequimica & CIA SCA).*

Inveres K-82. Highly plasticized homopolimer of polyvinyl acetate.

Applications: Binder for carpet backing. (Invequimica & CIA SCA).*

Inversal. A proprietary preparation of ambazone. (FBA Pharmaceuticals).†

Inversine. Mecamylamine hydrochloride. Applications: For the treatment of hypertension. (Merck Sharp & Dohme).*

Invert Sugar. A mixture of molecular proportions of glucose and fructose, obtained in the hydrolysis of cane sugar by acids. It is used to improve wines, also in the manufacture of liqueurs, fruit preserves, and honey substitutes.†

Invertase. See SUCRASE.

Invertin. See SUCRASE.

Iodal. A substance prepared by the action of iodine upon a mixture of alcohol and nitric acid.†

Iodanisol. *o*-Iodoanisole. An antiseptic.†

Iodantifebrin. Iodoacetanilide.†

Iodeosin B. See ERYTHROSIN.

Iodeosin G. See ERYTHROSIN G.

Iodesin. Tetraiodofluoresceine, $C_{20}H_8 I_4O_5$.†

Iodex. A proprietary ointment containing 4% organically-combined iodine in a petroleum jelly base. (Smith Kline and French Laboratories Ltd).*

Iodicyl. Diiodosalicylic acid methyl ester.†

Iodin. Iodized arachis oil.†

Iodine Green. (Night Green, Pomona Green, Light Green, Metternich Green). A dyestuff. It is the zinc double chloride of heptamethylrosaniline chloride. Dyes silk green.†

Iodine-potassium Iodide Solution. See LUGOL'S SOLUTION.

Iodine Red. (Iodine Scarlet, Brilliant Scarlet, Scarlet Red, Royal Scarlet). A pigment. It is mercuric iodide, HgI_2†

Iodine Scarlet. See IODINE RED.

Iodised Oil of Wintergreen. See IODOZEN.

Iodized Salt. Common salt, NaCl, containing a very small amount of sodium iodide. Goitre in certain districts is associated with lack of iodide in the water supply.†

Iodoeosin. Erythrosin or Pyrosin. Used as an indicator.†

Iodoform. Triiodomethane, CHI_3.†

Iodoglobin. Diiodotyrosine, $HO.C_6H_2I_2.CH_2.CH(COOH)NH_2$.†

Iodohemol. See HEMOL.

Iodol. Tetraiodopyrrole, C_4I_4NH. Used externally as a disinfectant.†

Iodosorb. Cadexomer iodine powder. (ICI PLC).*

Iodotope I-125. Sodium Iodide I 125. Applications: Diagnostic aid; radioactive agent. (E R Squibb & Son Inc).

Iodotope I-131. Sodium Iodide I 131. Applications: Antineoplastic; diagnosti aid; radioactive agent. (E R Squibb & Sons Inc).

Iodotope Therapeutic. Sodium Iodide I 13 Applications: Antieoplastic; diagnostic aid; radioactive agent. (E R Squibb & Sons Inc).

Iodoval. Monoiodoisovalerylurea, $(CH_3)_2CH.CHI.CO.NH.CO.NH_2$.†

Iodozol. Diiodo-*p*-phenolsulphonic acid, $C_6H_2I_2(OH)SO_3H.3H_2O$.†

Iodron. Bulk milk tank sanitizer. (The Wellcome Foundation Ltd).

Iohydrin. (Iothion). Diiodoisopropyl alcohol, $CH_2I.CH(OH).CH_2I$.†

Ionac ECP-88. A proprietary acrylic polymer used in the formulation of coatings applied by electrostatic spray. (Ionac Chemical Co, Birmingham, NJ).†

Ionamin. A proprietary preparation of phentermine in a resin complex, used i the treatment of obesity. (Pennwalt Corp).*

Ionex. Luboil additives. (Shell Chemicals UK Ltd).†

Ionil. Salicylic Acid. Applications: Keratolytic. (Owen Laboratories Div, Dermatological Products of Texas, Alcon Laboratories Inc).

Ionol. A proprietary trade name for a phenolic antioxidant possessing a symmetrical structure thus giving a low power loss at high frequencies when us in dielectrics. (Shell Chemicals).†

nol CP. Antioxidants. (Shell Chemicals UK Ltd).

nol CPA-Feed. Antioxidants. (Shell Chemicals UK Ltd).

nol J65. Antioxidants. (Shell Chemicals UK Ltd).

nomer Resins. A name given to thermoplastic resins containing both covalent and ionic bonds. Carboxyl groups are located along the polymer chain by copolymerization to provide the anionic portion of the ionic cross links. Metal ions constitute the cationic part of the links. Sodium, potassium, magnesium and zinc are examples of ions used. See also SURLYN a proprietary trade name. (Du Pont (UK) Ltd).†

nox. Antioxidants. (Shell Chemicals UK Ltd).

nox 220. 4,4'-methylene-bis-2,6-di-*tert*. butyl phenol. An antioxidant of 98 per cent. minimum purity. (Shell International).†

nox 330. Non-toxic antioxidants for polymeric systems, thermoplastic rubbers and fatty acid distillates. (Shell Chemicals).†

nox 901. An ultra violet light stabilizer for polymeric systems. (Shell Chemicals).†

nox 99. General antioxidants. (Shell Chemicals).†

propane. Iodopropanol.†

pydone. 3,5-Diiodo-4-pyridone.†

san CCT. Iodophor teat dip. (Ciba-Geigy PLC).

san D. Iodophor/acid. (Ciba-Geigy PLC).

san Super Dip. Iodophor and glycerine teat dip. (Ciba-Geigy PLC).

san Teat Dip. Iodophor teat dip. (Ciba-Geigy PLC).

san Udder Cream. Iodophor udder cream. (Ciba-Geigy PLC).

san 4. Iodophor/acid disinfectant. (Ciba-Geigy PLC).

sol. See IODISTOL.

tect. Iodine indicator. (May & Baker Ltd).

Iotex. Selective weedkiller. (May & Baker Ltd).

Iothion. See IOHYDRIN.

Iotril. Active ingredient: ioxynil octanoate; 4-cyano-2,6-di-iodophenyl octanoate. Applications: Selective post-emergence herbicide which controls a wide range of annual broadleaf weeds in cereals, onions, leeks and sugar cane. (Agan Chemical Manufacturers Ltd).†

Ipecac. (Ipecacuanha). The dried root of *Uragoga ipecacuanha*.†

Ipecacuanha. See IPECAC.

Ipecine. Emetine, an alkaloid.†

Ipesandrine. A proprietary preparation of alkaloids of opium and ipecacuanha, and ephedrine hydrochloride. A cough linctus. (Sandoz).†

Ipexon. A proprietary preparation of guaiphenesin, ephedrine hydrochloride, sucrose and glycerine. A cough linctus. (British Drug Houses).†

Iphaneine. Pure isobutyl phenyl acetate. (Bush Boake Allen).*

Ipomic Acid. Sebacic acid, $CO_2H(CH_2)_8CO_2H$.†

Iporka. Foamable urea resins. (BASF United Kingdom Ltd).*

Ipral. A proprietary preparation containing trimethoprim. Applications: An antibiotic. (Squibb, E R & Sons Ltd).*

Ipropran. Ipronidazole. Applications: Antiprotozoal. (Hoffmann-LaRoche Inc).

I.P.S. Isopropyl alcohol. (Foseco (FS) Ltd).*

Ipsilene. A disinfecting gas made by heating ethyl chloride and iodoform under pressure.†

Irabond. Primer and adhesive systems for irathane coatings and linings. Applications: For use on a variety of substrates to enhance the performance of irathane. (Irathane International Ltd).*

Iragcet. Solvent soluble dyes. (Ciba Geigy PLC).†

Irasolve. Solvents for use in cleaning/-dissolving irathane. Applications: Assistance in surface preparation and cleaning of equipment. (Irathane International Ltd).*

Irathane. Elastomeric polyurethane coating and lining materials. Applications: Protection from severe abrasion and corrosion problems in mines, power plants, processing and offshore. (Irathane International Ltd).*

Ircogel. A proprietary trade name for thixotropic agents for PVC plastosols and organosols for cold dipping and general use to prepare non-drip compounds. They are metallo-organic complexes high in calcium content, having a particle size of 0.01 micron.†

Iretol. 1,2,3-Trihydroxy-5-methoxybenzene, $C_7H_8O_4$, used in perfumery.†

Irgaclarol. Wetting and scouring agents. (Ciba Geigy PLC).†

Irgacure. UV curing agents. (Ciba-Geigy PLC).

Irgaderm. Dyes for leather finishing. (Ciba-Geigy PLC).

Irgaferm BC Champagne. Dry yeasts for wine production. (Ciba-Geigy PLC).

Irgafin. Predispersed pigments for plastics polymers. (Ciba Geigy PLC).†

Irgafiner. Predispersed pigments for polyolefines. (Ciba Geigy PLC).†

Irgafos. Costabilizers for plastics. (Ciba-Geigy PLC).

Irgalan. Metal complex dyes. (Ciba-Geigy PLC).

Irgalevone. Dyeing and printing assistants. (Ciba-Geigy PLC).

Irgalite. Pigment dispersions for emulsion paints and other organic pigments. (Ciba Geigy PLC).†

Irgalite Blue GST. Beta form copper phthalocyanine blue pigment (Cl Pigment Blue 15) with excellent texture, dispersibility, gloss and high strength for letterpress and lithographic inks. (Ciba Geigy PLC).†

Irgalite C-20. Pigment preparations. (Ciba-Geigy PLC).

Irgalite Dispersed. Pigment plasticizer dispersions. (Ciba Geigy PLC).†

Irgalite M-20. Pigment preparations. (Ciba-Geigy PLC).

Irgalite MPS. Multipurpose pigment stainers for paints. (Ciba Geigy PLC).

Irgalite PDS. Predispersed pigment powders for paints. (Ciba Geigy PLC)

Irgalite PR. Predispersed pigment powder for rotogravure. (Ciba Geigy PLC).†

Irgalite Yellow BGW. Metaxylidide *bis*-arylamide yellow pigment (Cl Pigment Yellow 13) with excellent dispersibility and improved gloss and transparency f letterpress and lithographic inks. (Ciba Geigy PLC).†

Irgalite Yellow F4G. A proprietary monoazo pigment of the arylamide typ It has a greenish hue which makes it suitable for use in letterpress, offset litho, flexographic and gravure printing inks. (Ciba-Geigy PLC).†

Irgalon. Pretreatment agent. (Ciba-Geigy PLC).

Irganol. Acid dyes for wool (Ciba-Geigy PLC).

Irganox. A trade mark for a range of speciality antioxidants of the hindered phenol type developed initially for the stabilisation of polyolefines for use at high frequencies. (Ciba Geigy PLC).†

Irgapadol. Dyeing and printing assistants. (Ciba-Geigy PLC).

Irgaphor. Rubber masterbatch pigments. (Ciba Geigy PLC).†

Irgaplastol M-20. Pigment preparations. (Ciba-Geigy PLC).

Irgapyrol. Flame proofing agents. (Ciba Geigy PLC).†

Irgarol. Paint additives. (Ciba-Geigy PLC).

Irgasan. Bacteriostats. (Ciba-Geigy PLC).

Irgasan DP 300. Triclosan. Applications: Disinfectant. (Ciba-Geigy Corp).

Irgasol. Dyeing and printing assistant. (Ciba-Geigy PLC).

Irgasperse-s. Pigment dispersions for non-aqueous decorative paints. (Ciba-Geigy PLC).

Irgastab. Stabilizers for polymers. (Ciba-Geigy PLC).

Irgatan. Synthetic tanning agents for leather. (Ciba-Geigy PLC).

Irgatron. Premetallised dyes for polyamides. (Ciba Geigy PLC).†

gawax. Plastics lubricants. (Ciba-Geigy PLC).

gazin. A trade mark for a range of organic pigments derived from iso- indolinone and dioxazine. Typical colours are Yellow 2GLT, Yellow 3RLT, Orange RLT, Red 2BJT and Violet BLT. (Ciba Geigy PLC).†

gazin C-20. Pigment preparations. (Ciba-Geigy PLC).

gazin M-20. Pigment preparations. (Ciba-Geigy PLC).

goferm CM Montrachet. Dry yeasts for wine production. (Ciba-Geigy PLC).

idin. (Irisin). The powdered extract of iris.†

idium. A term used for alloys containing 77-83 per cent zinc, 1.1-1.25 per cent copper, and 15-22 per cent tin. Also an element of the Platinum group.†

ridium Black. A pigment for china. It is an oxide of iridium.†

ridium Steel. A German steel containing 4 per cent cobalt, 16 per cent tungsten, 3.5 per cent chromium, 0.67 per cent vanadium, 0.8 per cent molybdenum, and 0.6 per cent carbon.†

ridosmine. (Osmiridium). An alloy of Iridium and osmium containing 40-77 per cent iridium, and 20-50 per cent osmium. If there is more iridium, the alloy is called Nevyanskite, and Siserskite if the content of osmium is high.†

rigenin. 5,7,3'-trihydroxy-6,4',5'-trimethoxyisoflavone.†

riphan. (Triphan). The strontium salt of atophan.†

ris Blue. See FLUORESCENT BLUE.

ris Green. See SAP GREEN.

ris Violet. See AMETHYST VIOLET.

risamine G. (Rhodine 3G). A dyestuff. It is the ethyl ether of unsymmetrical dimethylhomorhodamine, $C_{25}H_{25}N_2O_3Cl$. Dyes tannined cotton, silk and wool, red.†

rish Pearl Moss. Caragheen moss, a gelatinous seaweed, *Chondrus crispus*. It contains carrageenin allied to pectin. It is employed as a substitute for isinglass, as a size for thickening colours in calico printing, and for stiffening silk.†

Irish Peat Wax. (Montana Wax, Montanin Wax). Waxes extracted from Irish peat are sold under these names. They resemble Montan wax.†

Irisin. See IRIDIN.

Irisol. Spirit soluble dyestuffs. Applications: For the surface coatings, printing ink, office supplies and plastics industries. (Bayer & Co).*

Irisol Fast. Spirit-soluble dyestuffs. (Bayer UK Ltd).

Irofol C. A proprietary preparation of ferrous sulphate, folic acid, sodium ascorbate and vitamin C. A haematinic. (Abbott Laboratories).*

Iron A. A British chemical standard. It is a cast iron containing 0.734 per cent combined carbon, 1.989 per cent silicon, 0.047 per cent sulphur, 0.049 per cent phosphorus, 0.688 per cent manganese, 0.042 per cent arsenic, 0.052 per cent titanium, and 2.387 per cent graphitic carbon.†

Iron-Andradite. Synonym for Skiagite.†

Iron B. A British chemical standard. It is a cast iron containing 0.39 per cent combined carbon, 0.031 per cent sulphur, 0.026 per cent phosphorus, 0.031 per cent arsenic, 0.108 per cent titanium, and 2.67 per cent graphitic carbon.†

Iron Black. Antimony precipitated as a fine powder, by action of zinc upon an acid solution of an antimony salt.†

Iron Brass. See BRASS, IRON.

Iron Buff. (Nankin Yellow). A mineral colour. It is ferric hydroxide.†

Iron Chamois. See IRON BUFF.

Iron D. (Phosphoric D). A British chemical standard. It contains 11.8 per cent phosphorus.†

Iron D2. A grey phosphoric cast iron in the form of fine turnings. It contains 1.31 per cent silicon, 1.07 per cent phosphorus, and 1.64 per cent manganese. The approximate quantities of other elements are 2/5 per cent graphitic carbon, 0.8 per cent combined carbon, and 0.03 per cent sulphur. It is a British chemical standard.†

Iron Earth, Blue. See BLUE IRON EARTH.

Iron Flint. An opaque variety of quartz containing iron.†

Iron Froth. A spongy variety of haematite.†

Iron G. A standard cast iron containing 1.82 per cent graphite carbon, 0.86 per cent combined carbon, 1.3 per cent silicon, 0.41 per cent manganese, 0.125 per cent sulphur, and 0.45 per cent phosphorus.†

Iron Glycerophosphate. A ferric glycerophosphate containing approximately 15 per cent iron.†

Iron Gohi. See GOHI IRON.

Iron L. (Nickel-chromium-copper-Austenitic Iron L). A British Chemical Standard. It contains 3.06 per cent total carbon, 2.26 per cent silicon, 1.01 per cent manganese, 0.119 per cent phosphorus, 13.45 per cent nickel, 3.96 per cent chromium, 4.73 per cent copper, and 0.031 per cent sulphur, the remainder being iron.†

Iron-Leucite. A mineral. It is $KFe^{\cdots}Si_3O_8$.†

Iron Liquor. See PYROLIGNITE OF IRON.

Iron Man. Vitamin-fortified tonic. (Richardson-Vicks Inc).*

Iron Minium. See INDIAN RED.

Iron Ore A, Haematite Type. A standard iron ore containing 58.19 per cent iron, 0.056 per cent phosphorus, 8.14 per cent silica, and 0.066 per cent sulphur. Used as a standard for checking analyses of iron ore.†

Iron Ore, Blue. See BLUE IRON EARTH.

Iron Putty. A mixture of ferric oxide and boiled linseed oil. Used for joints in iron pipes.†

Iron Pyrites, White. See MARCASITE.

Iron Pyrolignite. See PYROLIGNITE OF IRON.

Iron Red. See INDIAN RED.

Iron Saffron. See INDIAN RED.

Iron Vitriol. (Green Vitriol, Copperas, Green Copperas). Ferrous sulphate, $FeSO_4.7H_2O$.†

Iron-ore Cement. Cements in which a large proportion of the alumina is replaced by ferric oxide. Also see SIDERO CEMENT.†

Iron-Orthoclase. Synonym for Ferri-orthoclase.†

Ironac. An acid-resisting alloy of iron a[nd] silicon. It contains 13 per cent silicon[,] 84 per cent iron, 0.77 per cent manganese, 1.08 per cent carbon, an[d] 0.78 per cent phosphorus.†

Iron, Alginoid. See ALGIRON.

Iron, Alibated. See ALIBATED IRON.

Iron, Aluminium. See ALUMINIUM IRON.

Iron, Armco. See ARMCO IRON.

Iron, Burnt. See BURNT IRON.

Iron, Chrome. See CHROME IRON.

Iron, Delhi Rustless. See DELHI RUSTLESS IRON.

Iron, Dialysed. See DIALYSED IRON.

Iron, Electro-granodised. See ELECTR[O-]GRANODISED IRON.

Iron, Galvanised. See GALVANISED IRON.

Iron, Manganese. See FERRO-MANGANESE.

Iron, Nickel. See FERRO-NICKEL.

Iron, Titanic. See ILMENITE.

Iron, Titaniferous. See ILMENITE.

Iron, Wrought. See MALLEABLE IRO[N].

Iron, Zinced. See GALVANISED IRO[N].

Irox. A proprietary trade name for a synthetic yellow iron oxide.†

Irribral. A proprietary preparation of bencyclane fumarate. A vasodilator. (Pfizer International).*

Isarit,. A bleaching earth.†

Isarol. See ICHTHYOL.

Isatin Yellow. A dyestuff. It is sodium isatinphenylhydrazone-*p*-sulphonate, $C_{14}H_{10}O_4N_3SNa$. Dyes wool and silk greenish-yellow.†

Isceon. A proprietary range of halogen derivations of aliphatic hydrocarbons used as refrigerants and propellants. (ISC Chemicals Ltd).*

Isinglass. Sodium silicate solution. Used the clearing of liquids, such as beer a[nd] wine. Also see THAO.†

Isinglass, Bengal. See AGAR-AGAR.

Isinglass, Ceylon. See AGAR-AGAR.

Isinglass, Chinese. See AGAR-AGAR and THAO.

Isinglass, Japan. See AGAR-AGAR and THAO.

Ismelin. A proprietary preparation of guanethidine. An antihypertensive. (Ciba Geigy PLC).†

Ismelin. Anti-hypertensive. (Ciba-Geigy PLC).

Ismelin-Navidrex K. A proprietary preparation of guanethidine sulphate, cyclopenthiazide and potassium chloride in a slow release core. (Ciba Geigy PLC).†

Ismotic. Isosorbide. Applications: Diuretic. (Alcon Laboratories Inc).

Isoanthraflavic acid. 2, 7-Dihydroxy-anthraquinone.†

Iso-Autohaler. Breath-actuated pressurized aerosol containing a suspension of isoprenaline sulphate BP 4 mg/ml; delivers 400 metered doses (0.08 mg per dose). (Riker Laboratories/3M Health Care).†

Isobu-M-AMD. N-(Iso-butoxymethyl)-acrylamide. (Cyanamid BV).*

Isobutad. A mineral rubber or bitumen.†

Isoclad. Water-based, elastic anti-corrosive cladding for all ferrous and non-ferrous metals. Applications: Any corrosive environment including industrially polluted areas. (Liquid Plastics Ltd).*

Isocon. Refined, semi-refined and polymeric isocyanates. Applications: Curing coponents for use with Propocon polyether systems. (Lankro Chemicals Ltd).†

Iso-Cornox. Selective weedkillers. (Boots Pure Drug Co).*

Iso-Cornox 57. Selective herbicide. (FBC Ltd).

Isocracking. Catalyst. (Chevron).*

Isocreme. Compounded lanolin derived sterols. (Croda Chemicals Ltd).

Isocure. Liquid resins and catalysts used as binders for foundry core and mould production. A tertiary amine vapour-cured phenolic urethane resin system. Applications: When mixed with sand, these resins and catalysts act as an adhesive for foundry sands. These resin-bonded sand articles are used in the production of ferrous and non-ferrous castings. (Ashland Chemical Company).*

Isoderm. Comprehensive range of nitrocellulose seasons for leather, aqueous or organic systems. Applications: Leather industry. (Bayer & Co).*

Isodiphenyl Black. A dyestuff. Dyes cotton black.†

Isodulcite. (Rhamnodulcite). Rhamnose.†

Isodurindine. Tetramethylaniline, $C_6H(CH_3)_4.NH_2$.†

Isofoam. Twin component rigid polyurethane foam systems. (Baxenden Chemical Co Ltd).*

Isoform Oxiosol. *p*-Iodoxyanisol.†

Isogel. A proprietary preparation derived from the husks of mucilaginous seeds. A laxative. (Allen & Hanbury).†

Isoject-Streptomycin Injection. Streptomycin Sulphate. Applications: Antibacterial. (Pfizer Inc).

Isol. An oil forming a permanent emulsion with hot or cold water. Used for oiling textiles.†

Isolantite. A proprietary trade name for a ceramic material made from steatite with binders.†

Isolan, Isolan K. 1:2 metal complex dyestuffs. Applications: For dyeing wool and polyamide fibres. (Bayer & Co).*

Isolene. Liquid cis -1,-4 polyisoprene rubber polymer. (Hardman Inc).*

Isoleucine. 2-Amino-3-methylbutanoic acid, a natural protein aminoacid.†

Isolevin,. A proprietary trade name for the hydrogen tartrate of isoprenaline.†

Isolierstahl. See BAKELITE.

Isolit. (Isolose). Cements for fastening porcelain insulators to metal supports.†

Isolose. See ISOLIT.

Isomist. A proprietary preparation of isoprenaline sulphate. A bronchial antispasmodic. (Nicholas).†

Isomol. Thixotropic ready-for-use spirit based coatings for all types of iron, steel and non-ferrous castings. (Foseco (FS) Ltd).*

Isonal. Shading dyestuffs complementing the isolan range. (Bayer UK Ltd).

Isonaphthol. β-Naphthol, $C_{10}H_7OH$.†

Isonate. Diphenylmethane diisocyanate and modified forms used in the manufacture of polyurethane and polyisocyanurate products. (Dow Chemical Co Ltd).*

Isonex. A proprietary preparation of isoniazid used for the treatment of tuberculosis. (Pfizer International).*

Isonex Forte. A proprietary preparation of isoniazid used for the treatment of tuberculosis. (Pfizer International).*

Isoniazid. Isonicotinoylhydrazine.†

Isonol. Polyols used in the manufacture of polyurethane products. (Dow Chemical Co Ltd).*

Isopar. High purity isoparaffinic solvent. (Exxon Chemical International Inc).*

Isopaste. Foundry core paste that cures fast without heat. Applications: Paste foundry cores and/or moulds together. (Ashland Chemical Company).*

Isophane. Insulin preparation. (The Boots Company PLC).

Isoplac. A proprietary insulation.†

Iso-Planotox. Selective weedkiller. (May & Baker Ltd).

Isopral. An internationally-registered trade mark. (Casella AG).†

Isopropylan 33. Mixture of lanolin oil and isopropyl palmitate. Applications: Cost effective emollient and moisturizer. (Amerchol Corporation).*

Isopropylan 50. Mixture of lanolin oil and isopropyl palmitate. Applications: Emollient and moisturizer. (Amerchol Corporation).*

Isoptin. Verapamil Hydrochloride. Applications: Anti-anginal; cardiac depressant (anti-arrhythmic). (Knoll Pharmaceutical Co).

Isopto Alkaline. A proprietary preparation of HYPROMELLOSE for ocular use. (Alcon Universal).†

Isopto Atropine. Atropine sulphate solution for ocular use. Applications: Anticholinergic. (Alcon Laboratories Inc).

Isopto Carbachol. Carbachol solution for ocular use. Applications: Cholinergic. (Alcon Laboratories Inc).

Isopto Carpine. A proprietary preparation of pilocarpine hydrochloride solution for ocular use. (Alcon Universal).†

Isopto Cetamide. Sulphacetamide Sodium solution. An ocular antiseptic. Applications: Antibacterial. (Alcon Laboratories Inc).

Isopto Eserine. A proprietary preparation of physostigmine salicylate solution for ocular use. (Alcon Universal).†

Isopto Frin. A proprietary preparation of phenylephrine hydrochloride solution for ocular use. (Alcon Universal).†

Isopto Homatropine. A proprietary preparation of homatropine hydrobromide solution for ocular use. (Alcon Universal).†

Isopto Hydrocortisone. A proprietary preparation of hydrocortisone for ocular use. (Alcon Universal).†

Isopto Hyoscine. A proprietary preparation of hyoscine hydrobromide solution for ocular use. (Alcon Universal).†

Isopto P-ES. A proprietary preparation of pilocarpine hydrochloride and physostigmine salicylate solution for ocular use. (Alcon Universal).†

Isopto Plain. A proprietary preparation of HYPROMELLOSE for ocular use. (Alcon Universal).†

Isopurpurin. See ANTHRAPURPURIN.

Isoquinollne Red. See QUINOLINE RED.

α-Isorcin. See CRESORCIN.

Isordil. Isosorbide dinitrate. Applications: Vasodilator. (Ives Laboratories Inc).

Isordil Tablets. Isosorbide dinitrate. Applications: Angina pectoris, congestive heart failure. (Ayerst Laboratories).†

Isordil Tembids Capsules. Isosorbide dinitrate. Applications: Prophylaxis of angina pectoris. (Ayerst Laboratories).†

Isorubine. See NEW MAGENTA.

Isoset. Resins and additives used in the production of bonded foundry sand for the production of cores and moulds utilizing free radical curing mechanisms. Applications: A sand binder system used in the production of ferrous and non-ferrous castings. (Ashland Chemical Company).*

Isoset WD3-A322 Emulsion Resin. Proprietary, reactive, water-based polymer for crosslinking with CX-hardener at ambient temperature. Applications: Structural wood laminations, Type 1 waterproof/no-creep performance. (Ashland Chemical Company).*

Isoset WD3-CM402 Emulsion Resin. Proprietary, reactive, water-based polymer for crosslinking with CX-hardener at ambient temperature. Applications: Structural sandwich composite, metal or plastic faces to porous i.e. wood cores. (Ashland Chemical Company).*

Iso Soap. A solid sulphonic derivative of castor oil, soluble in hot water. Used as a bleaching, washing, and dressing agent in the textile industries.†

Isotachiol. Silver silico-fluoride, Ag_2SiF_6.†

Isotense. A proprietary preparation of syrosingopine. An antihypertensive. (Nicholas).†

Isoterge. Detergent solutions; used e.g. as Isoterge, Isoterge III concentrate. Applications: Cleaning of automatic blood cell analysers. (Coulter Electronics Ltd).*

Isothan Q-75. Lauryl Isoquinolinium bromide. Applications: Anti- infective. (Onyx Chemical Co).

Isothionine. See BERNTHSEN'S VIOLET.

Isoton. Diluents based on normal saline, used e.g. as Isoton II, Isoton III. Applications: Blood cell counting and sizing. Analysis medium for electrical sensing zone particle size analysers. (Coulter Electronics Ltd).*

Isotonic Salt Solution. See PHYSIOLOGICAL SALT SOLUTION.

Isotox. Insecticide seed treater. (Chevron).*

Isovanat. A proprietary preparation of vanadium in isotonic and isobaric solution.†

Isovue. Iopamidol. Applications: Diagnostic aid. (E R Squibb & Sons Inc).

Issolin. A phenol-formaldehyde resin, which is soluble in alcohol.†

Issolith. See BAKELITE.

Istin. A proprietary preparation of cisplatin. An antineoplastic. (Pfizer International).*

Istizin. See ISTIN.

Istizin. Danthron. Applications: Laxative. (Sterling Drug Inc).

Isuprel Hydrochloride. Isoproterenol hydrochloride. Applications: Adrenergic. (Sterling Drug Inc).

Isurol. See ICHTHYOL.

Italian Earth. A pigment. It is a sienna.†

Italian Green. A sulphide dyestuff obtained by heating *p*-nitrophenol with copper sulphate in water mixed with caustic soda solution and sulphur. Dyes cotton dull green.†

Italian Pink. See YELLOW CARMINE.

Italian Red. See INDIAN RED.

Itrol. Silver citrate, $C_6H_5O_7Ag_3$.†

Itrumil. A proprietary trade name for the sodium derivation of Iodothiouracil.†

I T Talc. Hydrous magnesium calcium silicate - talc. Applications: Used as fillers and extenders for paint, ceramics, plastics, rubber etc. (Vanderbilt Chemical Corporation).*

Ivaleur. A proprietary trade name for pyroxylin (cellulose nitrate).†

Ivax. A proprietary preparation of neomycin sulphate, sulphaguanidine and light kaolin. An antidiarrhoeal. (Boots Company PLC).*

Iversal. Ambazone.†

Iversal/Iversal-A cum anaesthetico. Pharmaceutical preparation. Applications: For the prophylaxis and treatment of mouth and throat infections. (Bayer & Co).*

Ivomec. Ivermectin. A mixture of ivermectin component B1a and ivermectin component B1b. Applications: Antiparasitic. (Merck Sharp & Dohme, Div of Merck & Co Inc).

Ivoride. A casein product used as an electrical insulation.†

Ivorin-Profalon. Herbicide for beans and potatoes. (Hoechst UK Ltd).

Ivosit. Selective contact herbicide. (Hoechst UK Ltd).

Ixan. Polyvinylidene chloride. (Laporte Industries Ltd).

Ixol. Halogenated polyols. (Laporte Industries Ltd).

Ixolite. A resin found in Austria.†

Ixper. Inorganic peroxides. (Interox Chemicals Ltd).

Izal. A distillate from coke residues. It is a proprietary disinfectant.†

J

Jabclad. Moulded expanded polystyrene panels. Applications: Insulation panel applied externally on masonry walls for subsequent rendering. (Vencel Resil Ltd).*

Jabdec. Roof insulation board. (Vencel Resil Ltd).

Jabdie. Laminate of 12 mm fibreboard with expanded polystyrene. Applications: Insulation of flat roofs, used under felt and mastic asphalt weatherproofing. (Vencel Resil Ltd).*

Jablina Insulating Panels. Laminate of expanded polystyrene with aluminium/paper facings. Applications: Insulation lining for factory buildings. (Vencel Resil Ltd).*

Jablite. Expandable polystyrene. (Vencel Resil Ltd).*

Jablite Cavity. Expanded polystyrene boards. Applications: Used for partially filling the cavity in masonry wall construction. (Vencel Resil Ltd).*

Jablite Flooring. Expanded polystyrene boards. Applications: Underfloor thermal insulation. (Vencel Resil Ltd).*

Jablite Insulation Board. Expanded polystyrene boards. Applications: Insulation of walls and floors, flat roof insulation. (Vencel Resil Ltd).*

Jablite Thermacel. Expanded polystyrene beads. Applications: Insulation of masonry cavity walls, infill for 'bean bags'. (Vencel Resil Ltd).*

Jablite Thermoclik. Laminate of bitumen roofing felt with expanded polystyrene. Applications: Insulation of flat roofs, used under felt and mastic asphalt weatherproofing (Vencel Resil Ltd).*

Jaborandi. The native name for several drugs of a sudorific and salivating character, obtained from the leaves and twigs of various species of *Pilocarpus*.†

Jabroc. See PERMALI.

Jacana Metal. An alloy of 70 per cent lead, 20 per cent antimony, and 10 per cent tin.†

J-acid. 6-Amino-1-naphthol-3-sulphonic acid.†

Jackson's Button Brass. See BRASS.

Jacoby Metal. An alloy of 85 per cent tin, 10 per cent antimony, and 5 per cent copper.†

Jacquemart's Reagent. An aqueous solution of mercuric nitrate with nitric acid. Used as a test for ethyl alcohol.†

Jadit. A proprietary preparation of bulclosamide and salicylic acid. A skin fungicide. (Hoechst Chemicals (UK) Ltd).†

Jadit H. A proprietary preparation containing BUCLOSAMIDE, salicylic acid and HYDROCORTISONE used in the treatment of inflammatory and fungal skin diseases. (Hoechst Chemicals (UK) Ltd).†

Jalap. The roots and tubers of certain convolvulaceous plants which yield purgative resins.†

Jalcase. A steel with a high resistance to wear with a soft core. It has forging properties.†

Jamaica Wood. See LOGWOOD.

Janimine. Imipramine hydrochloride. Applications: Antidepressant. (Abbott Laboratories).

Janthone. A synthetic perfume obtained by condensing citral or lippial with mesityl oxide. It has a violet odour.†

Janus Blue. See INDOINE BLUE R.

Janus Green B and G. See DIAZINE GREEN.

Janus Grey. A dyestuff prepared by the action of phenol upon diazosafranine. It is identical with Diazine black (Kalle).†

Janus Red. (Janus Brown B). A dyestuff. Dyes cotton from an acid bath.†

Jaon. Potassium Gluconate. Applications: Replenisher. (Adria Laboratories Inc).

Japan Agar. See AGAR-AGAR.

Japan Camphor. (Laurel Camphor). Ordinary camphor, $C_{10}H_{16}O$, which separates from the essential oil of *Laurus camphora.*†

Japan Drier. See TEREBINE.

Japan Earth. See CUTCH.

Japanese Acid Clay. (Kambara Earth). A clay having the formula $Al_2O_3.6SiO_2.XH_2O$ (X is larger than 6). It has powerful adsorptive and decolourizing properties. The dried clay has strong dehydrating action. This action is used in a commercial product named Adsol.†

Japanese Bell Metal. An alloy of 60.5 per cent copper, 18.5 per cent tin, 12 per cent lead, 6 per cent zinc, and 3 per cent iron.†

Japanese Brass. See BRASS and SINCHU.

Japanese Bronze. An alloy of from 81-83 per cent copper, 10 per cent lead, 4.6 per cent tin, and 0-1.8 per cent zinc.†

Japanese Gelatin. See AGAR-AGAR.

Japanese Lac. See LAC, JAPANESE.

Japanese Silver. An alloy of 50 per cent aluminium and 50 per cent silver.†

Japanese Steel. See K.S. MAGNET STEEL.

Japanese Wax. See CHINESE WAX.

Japan Isinglass. See AGAR-AGAR.

Japan Lac. See LAC, JAPANESE.

Japan Sago. The starch from *Cycas revoluta.*†

Japan Tallow. (Sumach Wax, Vegetable Wax). Japan wax, derived from *Rhus succedanea, Rhus vernicifera,* and *Rh[illegible] sylvestric.*†

Japan Varnishes. These are obtained by blending asphalt varnishes with dark coloured copal or amber varnishes.†

Japan Wax. See JAPAN TALLOW.

Japidermic. A microbial insecticide in powder form containing viable spores c[illegible] Bacillus popilliae, a specific pathogen which infects and kills Japanese beetle grubs. Ready-to-use. Applications: Japanese beetles lay their eggs in lawn[illegible] the eggs hatch into grubs which are numerous enough to destroy the grass. Milky disease spore powder is applied to the lawn in spots. Grubs become infected and die releasing more spores to infect and kill succeeding generation[illegible] of grubs. Only one application is need[illegible] as the living spores are self-perpetuating. See also DOOM. (Fairfax Biological Laboratory Inc).*

Jara Jara. 2-Methoxynaphthalene, used in perfumery.†

Jargonelle Pear Essence. A solution of isoamyl acetate in ethyl alcohol. Used for flavouring confectionery.†

Jarische's Ointment. An ointment containing pyrogallic acid.†

Jasmacyclene. A perfumery speciality. (PPF International Ltd).†

Jasmal. That fraction of the essential oil o[illegible] jasmine flowers distilling at 100° C.†

Jasmin. See CITRONINE B.

Jasmolide. A perfumery chemical. (PPF International Ltd).†

Jasper, Opal. See OPAL JASPER.

Jatex. A proprietary brand of pure concentrated rubber latex, 60 per cent.†

Jatobá Duro. A hard copal obtained from Ceará and Northern Bahia, Brazil. Use[illegible] in varnishes.†

Jatobá Lagrima. (Trapocá Resin). A rathe[illegible] soft copal from the Jatobá tree in Braz[illegible] Used in spirit varnish.†

Jatobá Resin. Brazilian copals from *Hymenoea courbaril* and *Hymenoea parvifolia.* There are hard and soft qualities. The soft is called jatobá tean and Trapocá.†

Jaulingite. A mineral resin obtained from a species of lignite.†

Jaune Acide. See ACID YELLOW.

Jaune Anglais. See VICTORIA YELLOW.

Jaune Brilliant. See CADMIUM YELLOW.

Jaune de fer. See MARS YELLOW.

Jaune de mars. See MARS YELLOW.

Jaune d'or. See MARTIUS YELLOW.

Jaune N. See FAST YELLOW N.

Jaune Solide N. See FAST YELLOW N.

Jaune Yellow. See CITRONINE B.

Java Olives. See OLIVES OF JAVA.

Java Wax. (Sumatra Wax, Gondang Wax, Kondang Wax, Getah Wax). A wax obtained from the bark of the gondang (wild fig) tree, *Ficus variegata*.†

Jaydalene. A proprietary soldering paste consisting of *o*-phosphoric acid with a base which vaporizes without decomposition, e.g., aniline, etc.†

Jayflex. Plasticizers. (Exxon Chemical International Inc).*

JB-4. Plastic embedding kit. Applications: Light microscopy. (Polysciences Inc).*

Jectoflo. Lime based fluxes for desulphurisation of steel by deep ladle injection. (Foseco (FS) Ltd).*

Jectomag. Magnesium based powders for sulphur removal from blast furnace iron. (Foseco (FS) Ltd).*

Jectothane. A proprietary polyester-based polyurethane thermoplastic injection-moulding compound. (Anchor Chemical Co).*

Jeffamine. A series of polyoxyalkylene-derived di-and triamines. Applications: Epoxy curing agents, polymer flexibilizers and specialty polyamides. (Texaco Chemical Co).*

Jeffox. A series of poly(oxyethylene) glycols and poly(oxypropylene) glycols and triols. Applications: Flexibilizers, humectants and intermediates. (Texaco Chemical Co).*

Jefron. Polyferose. A chelate complex of iron and a polymerized derivative of sucrose. Applications: Hematinic. (Merrell Dow Pharmaceuticals Inc, Subsidiary of Dow Chemical Co).

Jelly, Lieberkuhn's. See LIEBERKUHN'S JELLY.

Jelly, Mineral. See PETROLEUM JELLY.

Jelly, Petroleum. See PETROLEUM JELLY.

Jelly Rock. See WILKINITE.

Jelly, Vegetable. See VEGETABLE JELLY.

Jelonet. Paraffin gauze dressing. (Smith and Nephew).†

Jelutong. (Pontianac, Fluvia, Gambia). A resinous latex yielded by *Dyera costulata*. It contains from 19-24 per cent of rubber and 75-80 per cent of resin, and is used for mixing with rubber and for other purposes.†

Jenner's Stain. A microscopic stain for white blood corpuscles. It consists of (*a*) a solution of water-soluble, yellowish eosin, 0.5 gram, in 100 cc methyl alcohol, and (*b*) a solution of methyl blue, 0.5 gram, in 100 cc methyl alcohol. For use 25 cc of (*a*) are mixed with 20 cc of (*b*).†

Jeppel's Oil. See BONE OIL.

Jersey Lily White. A pigment. It is a lithopone.†

Jesuit's Balsam. The oleo-resin copaiba.†

Jesuit's Bark. Cinchona bark.†

Jesuit's Tea. See MATÉ.

Jet. A mineral. It is a fossilized wood, and falls between lignite and coal. Used for ornaments.†

Jet Black. See GAS BLACK.

Jeunite. Copper fungicide. (Murphy Chemical Co).†

Jeweller's Brass. See BRASS.

Jeweller's Rouge. The finest calcined ferric oxide or haematite.†

Jew's Pitch. See BITUMEN.

Jeyes Disinfectant. A disinfectant containing creosote, rosin, caustic soda, and water. It forms emulsions with water.†

Jicwood. See PERMALI.

Jodomiron. Iodamide. Applications: Diagnostic aid. (Bracco Industria Chimica SPA).

Jothion. A proprietary preparation of iodopropanol.†

Journal Box Brass. See BRASS.

JR Surfacer. Water soluble epoxy. Applications: Tool/pattern making applications and filler. (JR Technology Ltd).*

JSR-10. A proprietary A.B.S. material possessing high impact strength. (Japan Synthetic Rubber Co, Tokyo).†

JSR-12. A proprietary A.B.S. material possessing high impact strength. (Japan Synthetic Rubber Co, Tokyo).†

JSR-21. A proprietary A.B.S. resin capable of giving good surface finish in moulding operations. (Japan Synthetic Rubber Co, Tokyo).†

Juchten Red. Fuchsine mixed with chrysoidine, to give a yellower shade.†

Juglone. 5-Hydroxynaphthoquinone.†

Julin's Chloride. Hexachlorobenzene, C_6Cl_6.†

Justite. A substance approximating to the formula, $(Ca.Mg.Fe.Zn.Mn)_3Si_2O_7$, found in furnace slag. The name was formerly applied to the mineral Koenite.†

Jutahy. See JUTAHYCICA.

Jutahycica. (Jutahy). A copal from Brazil. It is obtained from the roots of *Hymenoea courbaril* and *Hymenoea parvifolia*.†

Jutahycica Resins. (Paragum, Resina Animé). Brazilian copal resins from *Hymenoea courbaril* and *Hymenoea parvifolia*.†

Jute Scarlet R. A dyestuff. It is a British equivalent of Azococcin 2R.†

Juvelith. See BAKELITE.

JZF. N,N'-diphenyl-p-phenylenediamine. An antioxidant for use in rubber, polyethylene, petroleum and vegetable oils and animal fats and oils. In natural rubber it protects against copper and manganese and gives protection against outdoor flexing and static weather cracking. Protects against thermal oxidation in poly-ethylene, inhibits gum formation and degradation at elevated temperature in petroleum oils. (Uniroyal).*

K

K154. Aluminium based clear masonry waterproofing solution. Applications: Treats masonry, slates, tiles and all porous substrates. (Liquid Plastics Ltd).*

K285. Absorbable dusting powder. (Boots Pure Drug Co).*

K34. See HEXACHLOROPHANE.

K.A. Alloy. An aluminium alloy resembling duralumin.†

Kabaite. A mineral wax of the ozokerite type.†

Kabikinase. A proprietary preparation of streptokinase. A fibrinolytic agent. (KabiVitrum AB).*

Kachin. A photographic developer. Its active constituent is pyrocatechol.†

K-acid. 5-Amino-4-hydroxynaphthalene-1,7-disulphonic acid.†

Kadox. A very fine zinc oxide used as a rubber filler.†

Kaempferol. The colouring matter of the blue flowers of *Delphinium consolida.* It is a trihydroxyflavonol.†

Kafil. Insecticide containing permethrin. (ICI PLC).*

Kagoo Oil. See KORUNG OIL.

Kainite. A salt found in the Stassfurt deposits, consisting mainly of potassium magnesium sulphate and magnesium chloride, $K_2Mg(SO_4)_2.MgCl_2.6H_2O$. The crude material consists of a mixture of kainite and rock salt, and contains 23 per cent of potassium sulphate. Used as potash fertilizer, and as a source of potassium sulphate.†

Kairoline. *N*-Ethyltetrahydroquinoline.†

Kaiser Green. See EMERALD GREEN.

Kaiser Red. See EOSIN BN.

Kaiserling Solution. A solution used for preserving tissue. It contains 3 grams potassium acetate, 1 gram potassium nitrate, 75 cc water, and 30 cc formaldehyde.†

Kaiserroth. See EOSIN BN.

Kalammon. A fertilizer containing 17 per cent nitrogen and 30 per cent calcium carbonate.†

Kalar. Crosslinked butyl rubber. (Hardman Inc).*

Kalbord. Insulating feeder head liners supplied in the form of flexible boards. (Foseco (FS) Ltd).*

Kalcrete. Refractory castable for use in foundry ladles. (Foseco (FS) Ltd).*

Kalene. Elastomeric compounds. (Hardman Inc).*

Kaleoilris. A proprietary filling compound.†

Kalex. Two-part polyurethane elastomers. (Hardman Inc).*

Kaliammon Saltpetre. A potassium ammonium nitrate prepared by mixing equivalent molecular proportions of solid potassium chloride and ammonium nitrate in the presence of a little water. A fertilizer.†

Kalif. A proprietary copper-lead bearing alloy. It melts at 952° C., and has a tensile strength of 10,000 lb per sq in at 21° C.†

Kalipol. Polyphosphate solution. (Albright & Wilson Ltd, Phosphates Div).

Kalipol 18. A proprietary potassium polyphosphate solution used in the manufacture of liquid detergents. (Albright & Wilson Ltd).*

Kalite. A proprietary form of chalk prepared by a special process whereby it has a very small particle size, and the particles are coated with a calcium soap. A rubber filler.†

Kaliuzoto. A fertilizer containing nitrogen, potassium, and organic matter manufactured from residual molasses.†

Kalkammon. A fertilizer. It is a mixture of ammonium chloride with 30 per cent chalk.†

Kalkeisenolivin. Synonym for Iron monticellite.†

Kalkor. Disposable refractory plugs for plug-bottom ingot moulds. (Foseco (FS) Ltd).*

Kalkstickstoff. See NITROLIM.

Kalleonicit. Pharmaceutical preparation also known as Nico Padutin. Applications: Combination therapy for physiological vasodilation. (Bayer & Co).*

Kalle's Acid. 1-Naphthylamine-2,7-disulphonic acid.†

Kallodent. Registered trade name for a methyl methacrylate thermoplastic material used for moulding dentures.†

Kallodoc. Acrylic powder for artificial teeth and eyes. (ICI PLC).*

Kalmex. Thermit compound for feeding ingots and castings. Also used in continuous casting tundishes to facilitate start up of casting. (Foseco (FS) Ltd).*

Kalmin. Insulating riser sleeves and shapes. (Foseco (FS) Ltd).*

Kalminex. Exothermic sleeves with highly insulating residual structures for lining feeder heads on iron and steel castings. (Foseco (FS) Ltd).*

Kaloempang Beans. See OLIVES OF JAVA.

Kalorex. Lightweight exothermic sideliner tiles for steel ingots. (Foseco (FS) Ltd).*

Kalpack. Refractory ramming compound for securing nozzles in foundry ladles. (Foseco (FS) Ltd).*

Kalpad. Mouldable and preformed insulating materials to assist in feeding of steel and iron casting sections. (Foseco (FS) Ltd).*

Kalrez. Fluoroelastomer parts. (Du Pont (UK) Ltd).

Kalseal. Refractory air-setting mortar used in foundry ladles. (Foseco (FS) Ltd).*

Kalsert. System for applying feeder sleeves by insertion into pre-formed cavities in moulds of cores. (Foseco (FS) Ltd).*

Kaltas. Solvent based bituminous binder. Applications: Roads. (Vedag GmbH).*

Kaltek. Highly insulating disposable ladle lining. (Foseco (FS) Ltd).*

Kalten. Combined hypertensive and diuretic containing atenolol, hydrochloride thiazide and aniloride hydrochloride. (ICI PLC).*

Kaltop. Exothermic anti-piping compounds in board form. (Foseco (FS) Ltd).*

Kalvan. A proprietary trade name for calcium carbonate for use in rubber to give wear resistance. It has an ultra fine particle size.†

Kalzana. A proprietary calcium-sodium lactate.†

Kalzose. A casein preparation containing calcium.†

Kamala. See KAMELA.

Kambara Earth. See JAPANESE ACID CLAY.

Kambe Wood. See REDWOODS.

Kamela. (Kamala). A dyestuff obtained from the seeds or fruits of *Mallotus phillpenis* or *Rottlera tinctoria*. Used in India as medicine, and for dyeing silk orange.†

Kamillosan. Pharmaceutical preparation containing the active principle of matricaria chamomilla. (Degussa).*

Kamoran. Actaplanin: a complex of glycopeptide-type antibiotics. Applications: Growth stimulant. (Eli Lilly & Co).

Kane's Salt. A salt prepared by dissolving mercuric nitrate in a boiling solution of ammonium nitrate. †

Kanfotrex. A proprietary preparation of kanamycin, amphymycin and hydrocortisone used in dermatology as an antibacterial agent. (Bristol-Myers Co Ltd).†

nga Butter. See LAMY BUTTER.

nigen. Chemical nickel plate. (Albright & Wilson Ltd).*

nja Butter. See LAMY BUTTER.

nnasyn. A proprietary preparation containing kanamycin (as the sulphate). An antibiotic. (Bayer & Co).*

nten. A variety of agar-agar from red Tegusa Seaweed of Japan.†

nthal Alloy. A trade mark for an iron alloy containing aluminium, cobalt, and chromium, and having a high degree of resistance to heat, a low electrical conductivity, and good hot and cold working properties.†

nthosine J. See TOLUYLENE ORANGE G.

nthosine R. See TOLUYLENE ORANGE R.

ntmelt. Non-melting industrial grease. (Specialty Products Co).*

ntrex. A proprietary preparation containing kanamycin sulphate. An antibiotic. (Bristol-Myers Co Ltd).†

ntrexil. A proprietary preparation containing kanamycin sulphate, pectin, bismuth carbonate and activated attapulgite. An antidiarrhoeal. (Bristol-Myers Co Ltd).†

antrim. Kanamycin sulphate. Applications: Antibacterial. (Bristol Laboratories, Div of Bristol-Myers Co).

antStik. A broad line of plastic mould release agents. (Specialty Products Co).*

aochlor. A liquid potassium product sold only on prescription. (Hercules Inc).*

aodene. Preparation for the treatment of diarrhoea. (The Boots Company PLC).

aogel. A proprietary preparation containing kaolin and pectin. An anti-diarrhoeal. (Parke-Davis).†

aolin. See CHINA CLAY.

aolinase. A purified kaolin.†

aomycin. A proprietary preparation containing neomycin sulphate and kaolin. An antidiarrhoeal. (Upjohn Ltd).†

aon. A range of potassium products sold only on prescription. Applications: For pharmaceutical and medical uses. (Hercules Inc).*

Kaon-Cl. Potassium Chloride. Applications: Replenisher. (Adria Laboratories Inc).

Kaopectate. A proprietary preparation containing kaolin, bentonite and pectin. An antidiarrhoeal. (Upjohn Ltd).†

Kaovax. A proprietary preparation of succinylsulphathiazole, and kaolin. An antidiarrhoeal. (Norton, H N Co).†

Kapak. A material made from the mineral rubber Elaterite. Used in rubber mixings.†

Kapazang Oil. A fat obtained from the seeds of *Hodgsonia heteroclita*.†

Kapex. Exothermic shapes to cover the surface of blind risers for iron and steel castings. (Foseco (FS) Ltd).*

Kapithamia Piscum. See WOOD-APPLE GUM.

Kapok. A cotton-like down produced in the seed-pods of the kapok tree. Used in upholstery.†

Kapsol. A proprietary trade name for a methoxy-ethyl-oleate. Used as a plasticizer.†

Kapsovit. Multi-vitamin capsules. (Allen & Hanbury).†

Kapton. A proprietary trade name for polyimide resin in the form of film. (Du Pont (UK) Ltd).†

Kapur Kachri. See SANNA.

Karakane. An alloy of 62.5 per cent copper, 25 per cent tin, 9.4 per cent zinc, and 3.1 per cent iron.†

Karamate. Apple scab fungicide. (Rohm & Haas (UK) Ltd).

Karate. Pyrethroid insecticide. (ICI PLC).*

Karathane. Powdery mildew fungicide. (Rohm & Haas (UK) Ltd).

Karaya Paste. A proprietary preparation of sterculia in isopropyl alcohol, used as a dressing around colostomies. (Abbott Laboratories).*

Karbolite. (Carbolite). A Russian artificial resin made from phenols, formaldehyde, and naphtholsulphonic acid.†

Karbos. A carbonaceous decolourizer and filtering medium.†

Karetnja. A bituminous insulation. It consists mainly of asphalt, with an aluminium stearate binder.†

Karite Gum. See GUTTA-SHEA.

Karma Metal. An alloy of 80 per cent nickel and 20 per cent chromium. It is a heat-resisting alloy and melts at 1415° C.†

Karmarsch Metal. An alloy containing 88.8 per cent tin, 7.4 per cent antimony, and 3.7 per cent copper.†

Karmex. Diuron weedkiller. (Du Pont (UK) Ltd).

Karolith. A casein preparation similar to Galalith, used for the manufacture of buttons and other objects.†

Karvol. Decongestant inhalent. (The Boots Company PLC).

Kasil. Potassium silicates supplied in varying SiO_2:K_2O ratios. Applications: Alkaline builder for liquid detergents and cleaners, binder for phosphors in cathode ray tubes. (The PQ Corporation).*

Kasof. Docusate potassium. Applications: Stool softener. (Stuart Pharmaceuticals, Div of ICI Americas Inc).

Kastle-Meyer Reagent. A 2 per cent solution of phenolphthalein in 20 per cent aqueous caustic potash decolourized by boiling with zinc powder. It gives a pink colour with copper salts when 4 drops of the reagent are added to 10 cc of the solution to be tested, and 1 drop of hydrogen peroxide (5 vols.).†

Kastor. A proprietary preparation of senna and sodium potassium tartrate. Applications: A laxative. (Ayrton Saunders plc).*

Katadolon. Flupirtine. Capsules and suppositories. Applications: Analgesic. (Degussa).*

Katalabu Gum. A gum of Nigeria, from *Acacia sieberiana*. An adhesive.†

Katanol. A trade mark for a range of mordants used in the dyeing of textiles. (Casella AG).†

Katapol OA-910. Polyoxyethylated(30) oleyl amine. Applications: Hydrophilic emulsifier and textile dyeing assistant, antiprecipitant to prevent cross-staining, stripping agent and dye leveller for ac dyes. (GAF Corporation).*

Katapol PN-430. Polyoxyethylated tallow amine. Applications: Water-soluble emulsifier for mineral oils and agricultural chemicals. (GAF Corporation).*

Katapol PN-730. Polyoxyethylated(15) tallow amine. Applications: Antistat synthetic fibre processing, intermediat for formulation of fatty ammonium quaternaries for use in textile processi of natural fibres. (GAF Corporation).

Katapol PN-810. Polyoxyethylated(20) tallow amine. Applications: Antistat and lubricant for wool and synthetic fibre processing. Also effective as antistat and co-emulsifier in synthetic fibre spin finishes. (GAF Corporation)

Katapol VP-532. Polyoxyethylated alklylamine (cationic). Applications: Dispersant for fibreglass strands in the manufacture of fibre glass mats, retarding agent in the application of cationic dyes to acrylic fibres, antiprecipitant for acid and cationic dyes. (GAF Corporation).*

Katapone VV-328. Quaternary ammoniu chloride. Applications: Water soluble acid-corrosion inhibitor for steel, copp and aluminium. (GAF Corporation).*

Katarsit. A calcium sulphite pellet for us as a dechlorinating agent for water.†

Katavel Oil. The oil from *Hydnocarpus wightiana*.†

Katbél - ki - gond. See WOOD-APPLE GUM.

Katchung Oil. Peanut oil.†

Katechu. See CUTCH.

Katemul IG-70. Isostearamidopropyl dimethyl glycolate. Applications: A cationic emulsifier and conditioner for skin and hair preparations. (Scher Chemicals Inc).*

Katemul IGU-70. Isostearamidopropyl dimethyl gluconate. Applications: Giv substantivity and mildness to creams, lotions and bath preparations. (Scher Chemicals Inc).*

Katharin. A proprietary trade name for carbon tetrachloride used as a grease remover.†

Kathon. Biocide. Applications: Preservative for coatings, floor polish and adhesives. (Rohm and Haas Company).*

Kathon 886. 5-Chloro-2-methyl isothiazolones. Applications: Used as biocides and preservatives in a wide range of industrial applications. (Rohm and Haas Company).*

Kathon 893. N-Octyl isothiazolones. Applications: Used as biocides in paint, leather, textiles, paper and plastics. (Rohm and Haas Company).*

Kathro. Semi-refined cholesterol. (Croda Chemicals Ltd).*

Katigen. A registered trade mark currently awaiting re-allocation by its proprietors to cover a range of dyestuffs. (Casella AG).†

Katonium. A proprietary preparation of ammonium and potassium polystyrene sulphonate. A diuretic. (Bayer & Co).*

Katorin. A proprietary preparation of potassium gluconate. A potassium supplement. (Boots Company PLC).*

Katzenstein Bearing Metal. See BEARING METALS.

Kau Drega. See TALOTALO GUM.

Kauk Catalyst. A proprietary trade name for a spherical catalyst of 5 mm diameter consisting of potassium salts and vanadium on a porous silica carrier. V_2O_5 content 6.5 per cent. It is used for converting SO_2 into SO_3.†

Kava-kava Resin. A mixture of resins and resin acids from the dried roots of *Piper methysticum*.†

Kawasaki Hakkinko. A proprietary Japanese steel containing 0.19 per cent carbon, 1.8 per cent silicon, 1.0 per cent manganese, 17.0 per cent nickel, 25.0 per cent chromium, and 0.2 per cent molybdenum. Offers resistance to hydrogen embrittlement.†

Kay Ciel. Potassium Chloride. Applications: Replenisher. (Berlex Laboratories Inc, Subsidiary of Schering AG).

Kaydox. 1,4-dichlorobenzene paste. (Murphy Chemical Co).†

Kayexalate. Sodium Polystyrene Sulphonate. Applications: Ion-exchange resin. (Sterling Drug Inc).

Kaylene. A proprietary preparation of colloidal aluminium silicate.†

Kaylene-ol. A proprietary preparation of colloidal aluminium silicate with liquid paraffin.†

Kaynitro. Concentrated nitrogen/pot ash fertilizer. (ICI PLC).*

K-Bond. Aluminium tripolyphosphate dihydrate or metaphosphate. Hardener for waterglass. Applications: Inorganic coatings, interior and exterior heat resistance and non-flammable. (Bromhead & Denison Ltd).*

K-Contin. A proprietary preparation of potassium chloride in a controlled release tablet. It is used as a potassium supplement in diuretic therapy. (Napp Laboratories Ltd).*

K de Krizia. A fragrance from Milan's foremost name in fashion design. (Richardson-Vicks Inc).*

Keene's Alloy. See NICKEL SILVERS.

Keene's Cement. The name for a number of different plasters prepared by various manufacturers. It is usually obtained from plaster of Paris, dipped into a solution of alum or aluminium sulphate, and recalcining.†

Kefadol. Cefamardole nafate. Applications: Indicated in the treatment of infections of the respiratory tract, genito-urinary tract, bones and joints, bloodstream (septicaemia), skin and soft tissue, gall bladder and peritoneum and pelvic inflammatory disease in women when due to susceptible micro-organisms. (Dista Products Ltd).*

Keffekilite. A fuller's earth.†

Keffekill. Synonym for Sepiolite.†

Keflex. A proprietary preparation containing cephalexin monohydrate. An antibiotic. (Eli Lilly & Co).†

Keflin. A proprietary preparation of CEPHALOTHIN sodium. An antibiotic. (Eli Lilly & Co).†

Kefzol. A proprietary preparation of CEPHAZOLIN Sodium. An antibiotic. (Eli Lilly & Co).†

Kekuna Oil. Bakoly oil (candlenut oil)†

Kelburon. R-EMPP. Applications: An elastomer-modified polypropylene used for automotive applications (bumpers, dashboards). (DSM NV).*

Kelcoloid HV. Propylene glycol alginates. Applications: Emulsions and low pH systems. (Kelco, Div of Merck & Co Inc).*

Kelcoloid LV. Propylene glycol alginates. Applications: Emulsions and low pH systems. (Kelco, Div of Merck & Co Inc).*

Kelcosol. High viscosity sodium alginates. Applications: Welding rods, dry wall cement. (Kelco, Div of Merck & Co Inc).*

Keldax. Ethylene interpolymer resin. (Du Pont (UK) Ltd).

Keleastoi. A proprietary trade name for a ricinoleate type of vinyl plasticizer. (Spencer Kellogg and Sons).†

Kelecin. SoybeaNLecithin. Applications: Emulsifying and dispersion agent, paint, mastics, feed and rubber. (NL Industries Inc).*

Kelene. Ethyl chloride, C_2H_5C1.†

Kelenmethyl. A mixture of ethyl and methyl chlorides. An anaesthetic.†

Kel-F 81. Family of thermoplatic extrusion and moulding materials resistant to high temperatures. Polychloro-trifuoroethylene. (3M).*

Kel-F-Elastomer 3700. Polychloro-trifluoroethylene-vinylidene fluoride, 30 : 70. A proprietary synthetic rubber resistant to high temperatures. (3M).*

Kelferron. A proprietary preparation of ferrous glycine sulphate. A haematinic. (MCP Pharmaceuticals).†

Kelfizina. Sulphalene. Applications: Antibacterial. (Abbott Laboratories).

Kelfizine W. A proprietary preparation of sulphametopyrazine. An antibiotic. (Farmitalia).*

Kelfolate. A proprietary preparation of ferrous glycine sulphate, and folic acid. A haematinic. (MCP Pharma-ceuticals).†

Kelgin. Sodium alginate, food grade. (Kelco/AIL International Ltd).

Kelgin F. Sodium alginates. Applications: Paper coating and sizing, textile printing, dyeing and sizing, alkaline carpet dyeing and wallpaper adhesives. (Kelco, Div of Merck & Co Inc).*

Kelgin HV. Sodium alginates. Applications: Paper coating and sizing, textile printing, dyeing and sizing, alkaline carpet dyeing and wallpaper adhesives. (Kelco, Div of Merck & Co Inc).*

Kelgin LV. Sodium alginates. Applications: Paper coating and sizing, textile printing, dyeing and sizing, alkaline carpet dyeing and wallpaper adhesives. (Kelco, Div of Merck & Co Inc).*

Kelgin MV. Sodium alginates. Applications: Paper coating and sizing, textile printing, dyeing and sizing, alkaline carpet dyeing and wallpaper adhesives. (Kelco, Div of Merck & Co Inc).*

Kelgin QH. Sodium alginates. Applications: Dispersible products for paper coating and textile applications. (Kelco, Div of Merck & Co Inc).*

Kelgin QL. Sodium alginates. Applications: Dispersible products for paper coating and textile applications. (Kelco, Div of Merck & Co Inc).*

Kelgin QM. Sodium alginates. Applications: Dispersible products for paper coating and textile applications. (Kelco, Div of Merck & Co Inc).*

Kelgin XL. Sodium alginates. Applications: Paper coating and sizing, textile printing, dyeing and sizing, alkaline carpet dyeing and wallpaper adhesives. (Kelco, Div of Merck & Co Inc).*

Kelgo-Gel HV. Clarified sodium alginates. Applications: High clarity gels. (Kelco, Div of Merck & Co Inc).*

Kelgo-Gel LV. Clarified sodium alginates. Applications: High clarity gels. (Kelco, Div of Merck & Co Inc).*

Kelgum. A linseed oil rubber substitute.†

Kelig. Modified sodium lignosulphonates. Applications: Metal complexing agents for micronutrient formulations, industrial cleaners and water treatment formulations. (Reed Lignin Inc).*

Kellite. A proprietary trade name for a synthetic resin.†

Kellox. Oxidized marine oils. Applications Paint, levelling additive. (NL Industries Inc).*

elly's Paint. A benzoated collodion containing tincture of benzoin, glycerin, and collodion. Used for painting on abrasions of the skin.†

elmar. Potassium alginates. Applications: Welding rods, dry wall cement, industrial gels (toys). (Kelco, Div of Merck & Co Inc).*

elmer. Polymerized marine oils. Applications: Paint, caulks, mill whites and non-sagging additives. (NL Industries Inc).*

elo-form. A proprietary trade name for ethyl aminobenzoate.†

elocyanor. A proprietary preparation of cobalt tetracemate in a glucose solution. An antidote for cyanide poisoning. (Rona Laboratories).†

elp. (Varec). Seaweed or the ash from seaweed. A source of iodine.†

elpchar. A decolourizing carbon obtained from seaweed by carbonising in two stages and extracting with water and dilute hydrochloric acid.†

elpol. Alkyd copolymers. Applications: Paint. (NL Industries Inc).*

elpoxy. Elastomer modified epoxy resins. Applications: Adhesives, composites and coatings. (NL Industries Inc).*

elprox. Thermoplastic EPDM-rubber. Applications: Automotive, appliances, industrial worms. (DSM NV).*

elp Salt. A mixture of potassium chloride, with some alkaline sulphates and carbonates, formed in the preparation of potassium chloride from kelp.†

elrinal. CM rubber. Applications: Used as a high-grade vulcanized rubber in the wire and cable industry and in the automotive industry. (DSM NV).*

elsol. Water reducible resins. Applications: Paint. (NL Industries Inc).*

elstar. Pigment dispersions. Applications: Colouration of non-polyurethane plastics. (Pacific Dispersions Inc).*

eltan. Range of EPDM terpolymers having differing Mooney viscosities. Applications: Used in the automotive industry for wires and cables, in the building industry and in domestic appliances and technical products. (DSM NV).*

Keltan TP. Thermoplastic rubber. Applications: Used in the automotive, medical and pharmaceutical industries, for wires and cables, in household equipment and shoes. (DSM NV). *

Keltex. Sodium alginates. Applications: Textile printing, silver recovery. (Kelco, Div of Merck & Co Inc).*

Keltex S. Specially designed sodium alginate. Applications: Buffered product for textile printing. (Kelco, Div of Merck & Co Inc).*

Kelthane. Milicide insecticide. (Rohm & Haas (UK) Ltd).

Kelthix. Thixotropic resins. Applications: Paint. (NL Industries Inc).*

Keltose. Ammonium calcium alginates. Applications: Dry wall cement. (Kelco, Div of Merck & Co Inc).*

Keltrol. Oil copolymers. Applications: Paint. (NL Industries Inc).*

Keltrol. Xanthan gum, food grade. (Kelco/AIL International Ltd).

Kelzan. Xanthan Gum. Applications: Textured paint, carpet printing, cleaners and polishes, water-based lubricants, coatings, ceramic glazes, pigment and dye suspensions, agricultural products, metal working products, foam dyeing/printing, coal slurries and other systems requiring suspension. (Kelco, Div of Merck & Co Inc).*

Kelzan D. Xanthan Gum. Applications: Used where combinations with galactomannans, such as guar gum, are advantageous, such as in slurry explosives. (Kelco, Div of Merck & Co Inc).*

Kelzan S. Xanthan Gum. Applications: Same as for Kelzan except used where ready dispersion is necessary. (Kelco, Div of Merck & Co Inc).*

Kemadrin Tablets. A proprietary formulation of procyclidine hydrochloride. Applications: Treatment of parkinsonism including the postenchephalitic, arteriosclerotic and idiopathic types. (The Wellcome Foundation Ltd).*

Kematal. Acetal copolymer (POM) - polyoxymethylene. Applications:

Domestic equipment. i.e., jug kettle bodies, components for washing machines, food preparation equipment. Plumbing - taps, cistern valves and basins. (Celanese Limited).†

Kemflorseal. Epoxy die material and epoxy floor and wall product. Applications: Dental moulds, floor sealer, wall decorating and stain resistance. (Kemtron International Inc).*

Kemgo. A proprietary trade name for inks for use with heat.†

Kemick. Heat-resisting paint. (ICI PLC).*

Kemira Phlogopite Mica. Mica. Applications: Filler for plastics, surface coatings, sound-deadening compounds etc. (Cornelius Chemical Co Ltd).*

Kemite. A ceramic material. Labstone is the name given to the material when used for laboratory bench tops.†

Kemlet Metal. An alloy consisting mainly of zinc, with aluminium and copper.†

Kemmat. A range of glycol esters. Applications: Used as emulsifiers for cosmetic and pharmaceutical creams and lotions, thickeners for shampoos and emulsifiers for mineral oils. (Harcros Industrial Chemicals).*

Kemmest. A range of glycol esters. Applications: Used as emulsifiers in the cosmetic and pharmaceutical industry. (Harcros Industrial Chemicals).*

Kemonic. A range of lauryl alcohol ethoxylates. Applications: Used as biodegradable detergents and emulsifiers. (Harcros Industrial Chemicals).*

Kemopol. Propylene oxide condensates of ethylene oxide/glycol adducts. Applications: Used as a detergent base for machine dishwashing and cattle antibloat. (Harcros Industrial Chemicals).*

Kemotan. A range of polysorbates. Applications: Used as emulsifiers in the food and cosmetic industry. (Harcros Industrial Chemicals).*

Kemp. The shorter fibres of Mohair.†

Kempol. A proprietary trade name for vulcanizable vegetable oil polymers.†

Kempoxy. An epoxy flooring material. Applications: Floor resurfacer, chemical resistance floors. (Kemtron International Inc).*

Kempy Wool. Wool prepared from sheep badly fed, or subjected to exposure. It dyes badly.†

Kemtop. Epoxy die material and epoxy fl and wall product. Applications: Denta moulds, floor sealer, wall decorating a stain resistance. (Kemtron Internation Inc).*

Kemwax. Ethylene bis-stearamide. Applications: Used as plastics lubrican and slip agents in plastics. (Harcros Industrial Chemicals).*

Kenacort. Triamcinolone. Applications: Glucocorticoid. (E R Squibb & Sons Inc).

Kenacort Diacetate Syrup. Triamcinolone Diacetate. Applications: Glucocorticoid. (E R Squibb & Sons Inc).

Kenalog. A proprietary preparation of triamcinolone acetonide. (Squibb, E R & Sons Ltd).*

Kenalog Injection. Triamcinolone acetonid in suspension in aqueous vehicle. (E R Squibb & Sons Ltd).

Kendex OCTG. Petroleum - corrosion inhibitor. Applications: Specialty anti-corrosion formulation for ferrous meta for tubular goods and other machined parts. (Witco Corporation).*

Kendex 0220. Petroleum hydrocarbon crude petrolatum - wax (black). Applications: Used as a fuel in fire logs or for further refining to petrolatum. (Witco Corporation).*

Kendex 0834. Petroleum hydrocarbon heavy petroleum resins. Applications: As a viscosity modifier in lubrication applications, plasticizer extender. (Witco Corporation).*

Kendex 0842. Petroleum hydrocarbon - cylinder stock. Applications: A base oi for lubrication formulations or plasticizer/extender. (Witco Corporation).*

Kendex 0847. Petroleum hydrocarbon - 15 bright stock. Applications: Base oil in the formulation of motor oil or lubricants. A plasticizer/extender in the rubber industry. (Witco Corporation).*

Kendex 0866. Petroleum hydrocarbon - extract. Applications: A plasticizer/extender. Can be used in adhesives. (Witco Corporation).*

Kendex 0898. Petroleum hydrocarbon - intermediate petroleum resins. Applications: Viscosity modifier for lubricants. plasticizer/extender. Additive stock for quench oils. (Witco Corporation).*

Kendurit. Skin protective soap with granules. Applications: For cleansing of dirty hands. (Dynamit Nobel Wien GmbH).*

Kenflex. Vinyl plasticizer. The condensation products of alkyl naphthalenes. (Kenrich Petrochemicals Inc).*

Kenflex A. Vinyl plasticizer. A dimethyl naphthalene derivative. (Kenrich Petrochemicals Inc).*

Kenmag. A proprietary dispersion of magnesium oxide. (Kenrich Petrochemicals Inc).*

Kenplast. Distilled aromatic and cumylphenol derived plasticizers. Typical composition: Kenplast G - aromatic distillate, Kenplast ES2 - cumylphenyl acetate. Applications: Non-reactive and reactive diluents in epoxies, primary plasticizer in urethanes. (Kenrich Petrochemicals Inc).*

Ken-React. Titanate and zirconate coupling agents. Typical composition: LICA 01 - titanium IV neoalkenolato, tris neodecanolato-O, LZ 01 - zirconium IV neoalkenolato, tris neodecanolato-O. Applications: Coupling agents, adhesion promoters, composite improvers. (Kenrich Petrochemicals Inc).*

Kensol KM Metal Cleaner. Liquid metal polish - water based suspension. Contains ammonia, oxalic acid and methanol among other ingredients. Applications: Primarily as an all-metal cleaner for commercial metal maintenance on architectural and ornamental metalwork as well as household use on brass, chrome, stainless steel, etc. (Kensol Corporation).*

Kensol KV Rust Retarder. Hydrocarbon mixture - blend of petroleum products. Applications: Primarily used as a rust preventative on polished steel and stainless steel on bank security equipment. (Kensol Corporation).*

Kensol KX Oxide Resistor. Hydrocarbon mixture - blend of petroleum products. Applications: Primarily an all-metal preservative for commercial metal maintenance on architectural and ornamental metalwork as well as household use to protect against rust and tarnish. (Kensol Corporation).*

Kensol 10. Petroleum naptha. Applications: Light naptha used as fuel in sporting and camping equipment or as a solvent. (Witco Corporation).*

Kensol 13. Petroleum naptha. Applications: Light petroleum naptha used in oil well deparaffinizing and as a solvent in blending asphalt. (Witco Corporation).*

Kensol 30. Petroleum hydrocarbon - regular mineral spirits or Stoddard solvent. Applications: Wide range of applications as a commercial solvent and as a parts cleaner or diluent. (Witco Corporation).*

Kensol 48T. Petroleum hydrocarbon - low odour petroleum distillate fraction. Applications: Used as a solvent in various commercial formulations or as an oil to roll aluminium foil. (Witco Corporation).*

Kensol 50T. Petroleum hydrocarbon - low odour middle petroleum distillate. Applications: Oil used in aluminium rolling applications and as a diluent in commercial formulations. (Witco Corporation).*

Kensol 51. Petroleum hydrocarbon - middle distillate or a light mineral seal oil. Applications: Used as a lubricant in the rolling of aluminium and as a diluent in commercial formulations. (Witco Corporation).*

Kensol 53. Petroleum hydrocarbon - mineral seal oil. Applications: Formulation of commercial products, coke absorber oil, compounding base stock, cuting oil. (Witco Corporation).*

Kensol 61. Petroleum hydrocarbon - mineral seal oil. Applications: Oil for compounding various commercial products, hydraulic oil base stock, light oil for diluents. (Witco Corporation).*

Kensol 80. Petroleum hydrocarbon - V M & P type naphtha. Applications: Solvent in paint or as a solvent in commercial products. (Witco Corporation).*

Kentish Rag. A siliceous limestone, used as an adulterant of Portland cement.†

Kentite. An explosive. It contains from 32-35 per cent ammonium nitrate, 32-35 per cent potassium nitrate, 16-18 per cent ammonium chloride, and 14-16 per cent trinitrotoluene.†

Kephos. Non-aqueous phosphating metal pre-treatment. (ICI PLC).*

Keramine H. Liquid protein ampules, to strengthen hair. (Richardson-Vicks Inc).*

Keramyl. A solution of hydrofluosilicic acid, H_2SiF_6.†

Keraphen. Sodium tetraiodophenol-phthalein.†

Kerasol. Tetraiodophenolphthalein, $C_{20}H_{10}I_2O_4$.†

Keratite. A name applied to a vulcanite.†

Keratol. A cellulose waterproofing compound.†

Kerb. Soil active herbicide. (Rohm & Haas (UK) Ltd).

Kerecid. A proprietary preparation of idoxuridine. An ocular antiseptic. (Smith Kline and French Laboratories Ltd).*

Kericompost. Compost for houseplants. (ICI PLC).*

Kerigrow. Liquid fertilizer for houseplants. (ICI PLC).*

Keriguards. Pellets containing fertilizer and insecticide in combination. (ICI PLC).*

Kerimid 500. A proprietary polyamide-imide polymer in solution form. (Rhone-Poulenc NV/CdF Chimie AZF).*

Kerimid 501. A proprietary polyamide-imide polymer in film form. (Rhone-Poulenc NV/CdF Chimie AZF).*

Kerimid 502. A proprietary polyamide-imide polymer in the form of a green paste, comprising a thermosetting polymer and an aluminium powder filler. (Rhone-Poulenc NV/CdF Chimie AZF).*

Kerimid 503. A proprietary polyamide-imide polymer in film form, coloured green. (Rhone-Poulenc NV/CdF Chimie AZF).*

Kerimid 601. A proprietary thermosetting polyimide used in the manufacture of glass-fibre laminates. (Rhone-Poulenc NV/CdF Chimie AZF).*

Keriroot. Hormone rooting powder. (ICI PLC).*

Kerishine. Leaf glosser for houseplants. (ICI PLC).*

Kerispikes. Food spikes for houseplants. (ICI PLC).*

Kerispray. Pesticide spray for houseplants. (ICI PLC).*

Keristicks. Capilliary sticks for houseplants. (ICI PLC).*

Kerite. A rubber substitute containing vegetable oil, waxes, bitumen, coal tar, sulphur, and a little tannin.†

Kerman. A disinfectant containing 22.5 per cent flurosilicic acid.†

Kermesin Orange. See ORANGE T.

Kern's Hydraulic Bronze. An alloy of 78 per cent copper, 12 per cent tin, and 10 per cent zinc.†

Kerogen. The organic matter contained in shale. It amounts to from 20-27 per cent and gives on distillation water, ammonia, gas, and oil.†

Kerol. Disinfectant fluid. (The Wellcome Foundation Ltd).

Kerolite. Space heater fuel. (Chevron).*

Keromask. A proprietary cosmetic preparation containing titanium oxide and ochre pigments. (Innoxa).†

Keronyx. A proprietary trade name for a casein plastic material used for the manufacture of combs, etc.†

Kerosene. (Paraffin Oil, Astral Oil, Coal Oil). A refined distillate of petroleum, 150-300° C. An illuminating oil.†

Kessco. Esters. Applications: Emollients, lubricants, emulsifiers. (Stepan Company).*

Kest. A proprietary preparation of magnesium sulphate and phenolphthalein used in the treatment of constipation. (Berk Pharmaceuticals Ltd).*

Kester. Fatty acid esters. (Croda Chemicals Ltd).*

Ketaject. Ketamine hydrochloride. Applications: Anaesthetic. (Bristol Laboratories, Div of Bristol-Myers Co).

Ketalar. Ketamine hydrochloride. Applications: Anaesthetic. (Parke-Davis, Div of Warner-Lambert Co).

Ketaset. Ketamine hydrochloride. Applications: Anaesthetic. (Bristol Laboratories, Div of Bristol-Myers Co).

Ketavet. Ketamine hydrochloride. Applications: Anaesthetic. (Bristol Laboratories, Div of Bristol-Myers Co).

Keto Resins. Artificial resins obtained by the polymerization of aldehyde ketone condensation products.†

Keto-Diastix. A proprietary test-strip used to detect ketones and glucose in urine. (Ames).†

Ketone Base. Tetramethyldiaminobenzophenone. An intermediate for dyes.†

Ketone Blue G and R. Dyestuffs of the same type as Patent blue V, N.†

Ketone Blue 4BN. A dyestuff. It is the sulphonic acid of ethoxytrimethylphenyltriaminotriphenylcarbinol, $C_{30}H_{31}Cl.N_2O$. Dyes wool and silk blue.†

Ketone Musk. An artificial musk perfume. It is dinitro-tert-butylxylyl methyl ketone.†

Ketonone. A proprietary trade name for benzoic acid derivatives used as plasticisers for cellulose acetate and cellulose nitrate.†

Ketonone B. A proprietary trade name for butylbenzoyl benzoate. A plasticizer.†

Ketonone E. A proprietary trade name for ethyl *o*-benzoyl benzoate. A plasticizer.†

Ketonone M. A proprietary trade name for methyl *o*-benzoyl benzoate. A plasticizer.†

Ketonone M.O. A proprietary trade name for methylethyl benzoyl benzoate. A plasticizer.†

Ketostix. A proprietary test strip of buffered sodium nitroprusside and glycine, used for the detection of ketones in urine, serum or milk. (Ames).†

Ketovite. A complete vitamin supplement for restricted or synthetic diets. Applications: Ketovite tablets used in conjunction with ketovite liquids will provide a complete vitamin supplement for use in conditions such as phenylketonuria, disorders of carbohydrate or amino-acid metabolism. (Paines & Byrne).*

Ketrax. Anthelmintic containing levamisole. (ICI PLC).*

Ketrax. See LEVAMISOLE.

Kevadon. Thalidomide. Applications: Sedative; hypnotic. (Merrell Dow Pharmaceuticals Inc, Subsidiary of Dow Chemical Co).

Kevlar. Aramid fibre. (Du Pont (UK) Ltd).

Kevlar 49. A proprietary aromatic polyamide fibre of great strength. See FIBRE B and ARAMID FIBRE. (Du Pont (UK) Ltd).†

Key Alloy. A nickel-silver containing 60-65 per cent copper, 20-26 per cent zinc, 12 per cent nickel, 1-2 per cent lead, and 0-0.4 per cent iron.†

Keydime. Ketene dimer emulsion. (Tenneco Malros Ltd).

Keystone. Adhesives. (Associated Adhesives).†

Keytrol. A total weedkiller containing aminotriazole, atrazine and 2,4-D in a wettable formulation. Applications: The mixture provides broad spectrum control of grassy and broad-leaf weed species, including deep-rooted perennials. (Burts & Harvey).*

Khaki. A colouring matter produced on the fibre. The material is dipped in chrome alum, ferrous sulphate, and pyrolignite of iron, and then passed through a solution of sodium silicate.†

Khaki Brown C. See KHAKI YELLOW C.

Khaki, Mineral. See MINERAL KHAKI.

Khaki Yellow C. (Khaki Brown C, Cross Dye Black FNG). Sulphur dyestuffs.†

Khari Salt. A native salt of India consisting chiefly of sodium sulphate. Used for curing skins.†

Khordofan Gum. (Arabian Gum, Turkey Gum). Gum arabic.†

Kick Plate Brass. See BRASS.

Kidnamin. A proprietary preparation of essential aminoacids used as a dietary supplement. (Kabivitrum Ltd).†

Kidney Cotton. Peruvian cotton.†

Kiel Compound. An insulating material containing rubber, sulphur, and mineral oil. It sometimes contains pumice and beeswax.†

Kienmeyer's Amalgam. An amalgam consisting of 2 parts mercury, 1 part tin, and 1 part zinc. Used as a coating for frictional electrical machines.†

Kien Oil. Turpentine oil obtained by the dry distillation of resinous wood.†

Kieralon. Textile scouring agents. (BASF United Kingdom Ltd).

Kieselguhr. See INFUSORIAL EARTH.

Kieselguhr Dynamite. (Gurdynamite, Dynamite). Ordinary dynamite, consisting of nitroglycerin absorbed by kieselguhr.†

Kil. Insecticides. (Fisons PLC).*

Kilianite. A proprietary synthetic resin product.†

Killed Spirits. A solution of zinc chloride. Prepared by dissolving zinc in commercial hydrochloric acid until action ceases.†

Kilmet. Selective weedkillers. (May & Baker Ltd).*

Kilnet. Selective weedkiller. (May & Baker Ltd).

Kinel 5502. A proprietary polyimide casting, potting and encapsulating resin. (Rhone-Poulenc NV/CdF Chimie AZF).*

Kinel 5514. A proprietary polyimide moulding composition reinforced with glass fibre. (450) (Rhone-Poulenc NV/CdF Chimie AZF).*

Kinel 5517. A proprietary free-sintering self-lubricated, heat- resistant polyimide moulding powder. (450) (Rhone-Poulenc NV/CdF Chimie AZF).*

Kinetic. No. 12. See FREON.†

Kinetine. A combination of quinine and bectine.†

Kinetite. An explosive. It is a mixture of the jelly formed by dissolving guncotton in nitrobenzene, with potassium chlorate or potassium nitrate a sulphur.†

Kinevac. Sincalide. Applications: Choleretic. (E R Squibb & Sons Inc).

Kingston Bronze. An alloy of 85 per cent copper, 12 per cent zinc, 2.5 per cent tin, and 0.05 per cent iron.†

King's Blue. See SMALT and COBALT BLUE.

King's Green. See OIL GREEN.

King's Yellow. A pigment. It is arsenic sulphide, and occurs naturally as Orpiment (*qv*). Also see CHROME YELLOW.†

Kinite. A proprietary trade name for steel containing 12.5-14.5 per cent chromium, 1.5 per cent carbon, 1.1 cent molybdenum, 0.7 per cent cobalt, 0.55 per cent silicon, 0.5 per cent manganese, and 0.4 per cent nickel.†

Kino. (Kino Gum). The dried juice obtained from incisions in the trunk of *Pterocarpus marsupium.* It resembles catechu, and is used in dyeing and medicine.†

Kino, Australian. (Kino, Eucalyptus, Kino, Botany Bay). The dried exudation of *Eucalyptus* species.†

Kino, Bengal. (Kino, Madras, Dhak Gum). Butea gum, from *Butea frondosa.*†

Kino, Botany Bay. See KINO, AUSTRALIAN.

Kino, Cochin. (Kino, Malabar). *Kino B.P.*†

Kino, Eucalyptus. See KINO, AUSTRALIAN.

Kino Gum. See KINO.

Kino, Madras. See KINO, BENGAL.

Kino, Malabar. See KINO, COCHIN.

Kirchberger Green. A pigment. It has the same composition as Scheele's green.†

Kish. Crystalline graphite formed in blast furnace slag during iron smelting.†

Kite. A paper based grade of TUFNOL industrial laminates. (Tufnol).*

Kiton. Acid wool dyestuffs. (Clayton Aniline Co).†

Kiton Fast Orange G. A dyestuff. It is a British equivalent of Orange G.†

Kiton Yellows. British equivalents of Flavazine S.†

Kittool Fibre. A fibre obtained from the leaves of a Ceylon palm, *Caryota urens*. Used in the manufacture of brushes.†

Kival. Insecticide. (May & Baker Ltd).

Klavikordal. Nitroglycerin. Applications: Vasodilator. (U S Ethicals Inc).

Klebcil. Kanamycin sulphate. Applications: Antibacterial. (Beecham Laboratories).

Kleenite. Denture cleanser powder. Applications: For cleaning dentures. (Richardson-Vicks Inc).*

Kleenmold. Graphite lubricants for use in the glass industry, principally as mould releases. (Specialty Products Co).*

Kleenup. Weed and grass killer. (Chevron).*

Klee's Salt. Acid potassium oxalate, $KHC_2O_4.H_2O$.†

Kleinenberg's Fat Mixture. A solution of cacao butter and spermaceti in castor oil. Used as an embedding material in microscopy.†

Kleinenberg's Fixative. Used in microscopy. It consists of 100 cc of a saturated aqueous solution of picric acid, 3 cc of sulphuric acid, and 300 cc of water.†

Kleinenberg's Stain. A microscopic stain. It consists of a saturated solution of alum and calcium chloride in alcohol (70 per cent. diluted with six times its volume of alcohol (70 per cent.) to which is added an alcoholic solution of haematoxylin until the colour is violet blue.†

Klein's Reagent. A saturated solution of cadmium borotungstate, $2(Cd_2H_2W_8O).7(WO_3)B_2O_3.H_2O$. Used for the separation of minerals.†

Klenal. Industrial cleaner. (Specialty Products Co).*

Klerat. Rodenticide. (ICI PLC, Plant Protection Div).

K-Lor. Potassium Chloride. Applications: Replenisher. (Abbott Laboratories).

Klorax. A solution of Chloramine T. An antiseptic.†

Kloref. A proprietary preparation containing betaine hydrochloride and potassium bicarbonate. (Cox, A H & Co Ltd, Medical Specialities Divn).†

Kluberlubrication. Specialty lubricant and grease. (Carl Freudenberg).*

Klucel. Hydroxypropylcellulose. Applications: Emulsion stabilizer, emulsification aid, whipping aid, suspending agent, thermoplastic resin and thickener. (Hercules Inc).*

Kluchol. Anethol benzoate.†

Klucine. A proprietary waterproofing compound†

K-Lyte. Potassium Bicarbonate. Applications: Pharmaceutic necessity. (Mead Johnson & Co).

K-Lyte/C1. Potassium Chloride. Applications: Replenisher. (Mead Johnson & Co).

KM Ammonium Metavanadate. 77-78% V_2O. (Kerr-McGee Chemical Corp).*

KM Fly Ash. Damp bulk form. (Kerr-McGee Chemical Corp).*

KM Manganese Dioxide. AB and SB battery active grades - 90% minimum MnO_2 for use in Leclanche, alkaline and zinc chloride dry cell batteries. (Kerr-McGee Chemical Corp).*

KM Muriate of Potash. White agricultural grade. 62% minimum K_2O in granular, coarse and standard grades. (Kerr-McGee Chemical Corp).*

'K'-Monel. A proprietary alloy. It contains nickel and copper in approximately the same ratio as in monel metal with the addition of 4 per cent aluminium.†

KM Pebble Lime. 90% minimum available CaO, in mill run, coarse, medium, fine and crushed grades. (Kerr-McGee Chemical Corp).*

KM Phosphate Rock. 68, 70, 72 and 75% BPL (bone phosphate of lime expressed as $Ca_3(PO_4)_2$) grades. (Kerr-McGee Chemical Corp).*

KM Potassium Chloride. High purity white industrial grade. 98.3% KCI (62% minimum K_2O equivalent). (Kerr-McGee Chemical Corp).*

KM Potassium Perchlorate. 99.7% $KClO_4$, industrial and military grades. (Kerr-McGee Chemical Corp).*

KM Sodium Chlorate. Technical grade - 99.5% minimum in bulk. Drummed may contain 0.25-0.5% anti-caking agent. Can be supplied in dry bulk, solution and with salt added to custom specifications. (Kerr-McGee Chemical Corp).*

KM Sodium Perchlorate. 60-64% $NaClO_4$, aqueous solution. (Kerr-McGee Chemical Corp).*

KM Vanadium Pentoxide. Fused flake and fine granular, 98% minimum V_2O_5. (Kerr-McGee Chemical Corp).*

Knapp's Solution. Mercurous chloride (10.8 grains) are treated with potassium cyanide solution until the addition of caustic soda causes no precipitate. Caustic soda solution (100 cc of specific gravity 1.145), added, and the whole diluted to 1 litre. Used for the estimation of glucose.†

Knauerit S. An explosive plaster charge of best adhesive strength which can be formed by hand to fit the shape of the underground. Therefore it is advantageous to use Knauerit S for demolition work e.g. for cutting off iron constructions, bridge girders, rails etc. (Dynamit Nobel Wien GmbH).*

Knauerit 2. A plaster-shooting (mud-capping) explosive of greatest brisance and high velocity of detonation, developed for high performance. It is used without stemming for pop shots. From the cartridge of Knauerit 2 slices of adequate thickness are cut to be closely pressed against the boulder. (Dynamit Nobel Wien GmbH).*

Kneiss Alloy. An alloy of 42 per cent lead, 40 per cent zinc, 15 per cent tin, and 3 per cent copper. Used for machine bearings. Another alloy contains 50 per cent zinc, 25 per cent tin, and 25 per cent lead.†

Knight's Patent Zinc White. See LITHOPONE.

Knittex. Finishing agent. (Ciba-Geigy PLC).

Koate. Antihaemophilic factor. Applications: Antihaemophilic. (Cutter Laboratories, Miles Laboratories Inc).

Kochenite. A fossil resin resembling amber.†

Kochlin's Bearing Bronze. An alloy of 90 per cent copper and 10 per cent tin.†

Koch's Acid. 1-Naphthylamine-3, 6, 8 trisulphonic acid.†

Kodaflex DMP, DEP, etc. A trade mark for the following plasticisers.
DMP. Dimethyl phthalate.
DEP. Diethyl phthalate.
DBP. Dibutyl phthalate.
DIBP. Diisobutyl phthalate.
DMEP. Di-(2-methoxyethyl) phthalate.
DOP. Dioctyl phthalate.
OIDP. Octyl isodecyl phthalate.
DIDP. Diisodecyl phthalate.
DOA. Dioctyl adipate.
DIDA. Diisodecyl adipate.
DOZ. Dioctyl azelate.
DBS. Dibutyl sebacate.†

Kodaloid. A proprietary trade name for a cellulose nitrate. It is made in the form of sheets.†

Kodapak. A proprietary trade name for transparent cellulose acetate film. Used for making packets.†

Koenig Solder. An alloy of 60 per cent tin, 30 per cent aluminium, and 10 per cent antimony.†

Koerzit. An alloy for permanent magnets containing 1.1 per cent carbon, 3.5 per cent manganese, 36 per cent cobalt, 4.8 per cent chromium, the remainder being iron.†

Koerzit, I., II., III. Proprietary cobalt steels containing 10, 20, and 30 per cent cobalt respectively.†

Koka Seki. A variety of pumice stone found in the Nüjima Islands, near Tokio. It is used as a building material.†

Koken. A proprietary synthetic resin.†

Koko. The leaves of *Celastrus buxifolia.* Used in Natal as a sumac substitute for tanning.†

Kokowai. A variety of rouge used by the Maori.†

Kokum Butter. (Garcinia Oil, Concrete Oil of Mangosteen). A fat obtained from the seeds of *Garcinia indica* or G. *purpurea.* It is composed of stearine, myristicine, and oleine.†

Kola Nut. (Kola Seeds). The seeds of *Cola acuminata* and *C. vera.*†

Kola Seeds. See KOLA NUT.

Koladex. Pick-me-up tablets containing natural kola nut. (LAB Ltd).*

Kolanticon. A proprietary preparation of aluminium hydroxide, magnesium oxide, dicyclomine hydrochloride and dimethicone used as a gastro- intestinal sedative. (Richardson-Vicks Inc).*

Kolantyl. A proprietary preparation containing dicyclomine hydrochloride, aluminium hydroxide, magnesium hydroxide and methyl cellulose. An antacid. (Richardson-Vicks Inc).*

Kolantyl-NV. A proprietary preparation of dicyclomine hydrochloride, aluminium hydroxide, magnesium hydroxide and magnesium trisilicate. An ant- acid. (Richardson-Vicks Inc).*

Kolax. An explosive of the same type as Carbonite.†

Kolbeckine. Synonym for Herzenbergite.†

Kol-kol Gum. A gum of Nigeria, from *Acacia senegal*.†

Kollag. Oildag. See OILDAG AND AQUADAG.†

Kollargol. See COLLARGOL.

Kollercast. Mouldable synthetic resins. Applications: Industrial purpose e.g. flooring. (Scott Bader Co Ltd).*

Kolm. A variety of bituminous coal found in Sweden. The ash contains from 1-3 per cent of uranium oxide, U_3O_8.†

Koloran. See AVIVAN.

Kombé Arrow Poison. Strophanthus, the seed of *Strophanthus hispidus*.†

Kombé Seeds. See KOMBÉ ARROW POISON.

Komed HC. Hydrocortisone acetate. Applications: Glucocorticoid. (Barnes-Hind Inc).

Kommoid. A sulphurized corn oil rubber substitute.†

Kompak. Granular product for use on cast iron to produce a compacted graphite structure. (Foseco (FS) Ltd).*

Kompolite. Flooring materials. (Weatherguard/Marbleloid Products Inc).*

Konakion. Phytomenadione. A preparation of Vitamin K. (Roche Products Ltd).*

Kondang Wax. See JAVA WAX.

Kondremul. Mineral oil. Applications: Laxative; pharmaceutic aid. (Fisons Corp).

Konel. Proprietary nickel-cobalt-iron alloys containing about 2.5 per cent titanium. They are high temperature resisting alloys and possess high tensile strength at elevated temperatures.†

Konilite. A silica in powder form.†

Konnan Bark. Obtained from *Cassia fistula* of Southern India. A tanning material.†

Kon Oil. (Kusum Oil). Macassar oil obtained from the seeds of *Schleichera trijuga*. It has a saponification value of 215-230, an iodine value of 48-69, and an acid value of 6-35.†

Konstrastin. Basic zirconium basic acetate.†

Konstructal. An aluminium alloy containing 1 per cent copper or 8 per cent zinc.†

Kontakt. A purified form of the Twitchell reagent, used for the hydrolysis of fatty glycerides.†

Kontrastin. Zirconium oxide, ZrO_2.†

Konyne. Factor IX Complex. Applications: Haemostatic. (Cutter Laboratories, Miles Laboratories Inc).

Konzentrole. A term used for essential oils free from terpenes and sesquiterpenes. Used for flavouring.†

Koolkat. Furane resin cold-set binders for sand cores. (Foseco (FS) Ltd).*

Kopan. See BAKELITE.

Kopols. Commercial products consisting of esterified copal resins. Used in varnishes.†

Koppeschaar Solution. A bromine solution of N/10 strength.†

Korad A. Acrylic film. (Rohm and Haas Company).*

Koraton. A proprietary trade name for a synthetic resin.†

Koreon. A basic chromium sulphate, $Cr(OH)SO_4$. Used in the tanning industry.†

Korlan. Insecticide. The active ingredient is Ronnel. Applications: Used on cattle for the control of ticks, flies, maggots and lice. (Dow Chemical Co Ltd).*

Koro. A proprietary trade name for an alloy of 98 per cent copper and 2 per cent nickel.†

Korogel. A proprietary trade name for a soft Koroseal (*qv*).†

Korolac. A proprietary trade name for solutions of Koroseal (*qv*). It is used in acid-resisting tank linings.†

Koron. Refractory coatings for use on ingot moulds, bottom plates and slag pots. (Foseco (FS) Ltd).*

Koronit. A German explosive.†

Koronium Bromide. The trade name for strontium bromide, $SrBr_2.6H_2O$.†

Koroplate. A proprietary synthetic paint in which Koroseal (*qv*) is the base. It is extremely resistant to acid fumes.†

Koroseal. A proprietary trade name for a rubber-like thermoplastic varying in hardness, from soft jellies to hard rubber. It is detained by treating highly polymerized vinyl chloride with plasticisers at high temperatures and cooling. It can be worked like rubber when hot but requires rather higher temperatures. It is resistant to light, water, oils, and most other chemicals. It is used for impregnating and coating paper, fabrics, and metals for the manufacture of tubing for corrosive materials, and cable sheathing. Other proprietary names for this rubber substitute are Welvic, Telcovin. See also under POLYVINYL CHLORIDE.†

Korpad. Anti-splash pads to minimise splash defects during direct teeming of steel ingots. (Foseco (FS) Ltd).*

Korteite. Synonym for Koenenite.†

Korung Oil. (Kagoo Oil). Pongam oil, obtained from the fruits of *Pongamia glabra*.†

Kosmos Black, 3XB, BB, and F4. A proprietary gas black used in rubber mixings. Also used as a black pigment.†

Kostil. A proprietary range of styrene acrylonitrile moulding granules. (Montedison UK Ltd).*

Kourbatoff's Reagents. (*a*) A 4 per cent solution of nitric acid in isoamyl alcohol. (*b*) A 20 per cent solution of hydrochloric acid in isoamyl alcohol, with the addition of one-third of its volume of a saturated solution of nitraniline or nitrophenol in alcohol. (*c*) Consists of 4 per cent nitric acid in acetic anhydride, methyl alcohol, ethyl alcohol, and isoamyl alcohol. (*d*) Contains 3 parts of a saturated solution of nitrophenol, and 1 part of a 4 per cent solution of nitric acid in ethyl alcohol. Used as etching agents in the micro-analysis of carbon steels.†

Kovar Alloy. (Fernico Alloy). A registered trade mark for an alloy of iron with 23-30 per cent nickel, 17-30 per cent cobalt and 0.6-0.8 per cent manganese. Used for glass to metal seals.†

KP 201. A proprietary trade name for a vinyl plasticizer. Dicyclohexyl phthalate.†

KP 555. A proprietary trade name for a vinyl plasticizer. Bis(dimethylbenzyl) ether.†

KP 90. A proprietary trade name for a vinyl plasticizer of the epoxy type.†

KP-23. A proprietary trade name for a plasticizer consisting of butoxyethyl stearate.†

Kraft Paper. A paper produced by the sulphate pulp process.†

Kraft's Metal. A fusible alloy containing 5 parts bismuth, 3 parts lead, and 1 part tin. It melts at 104° C.†

Kramerite. Synonym for Proberite.†

Krantzite. A variety of Retinite.†

Kraton. A proprietary name for a range of thermoplastic rubbers used for foot wear and as adhesives. (Shell Chemicals).†

Kraut's Reagent. See DRAGENDORF'S REAGENT.

Kremnitz. See FLAKE WHITE.

Kremser White. The purest form of white lead. See WHITE LEAD.†

Krems White. See FLAKE WHITE.

Krenite. A 48% water soluble liquid formulation of fosamine ammonium. Applications: Applied to unwanted brush in late summer or autumn prevents bud break leading to death of treated plants the following spring. (Burts & Harvey).*

Krennerite. See WHITE TELLURIUM.

Kresamin. (Cresamol). A mixture of 25 per cent tricresol with ethylenediamine,

$H_2N.CH_2.CH_2.NH_2$. A powerful antiseptic.†

Kresatin. *m*-Cresol acetate.†

K-Resin Polymer KR01. Transparent and shatter resistant styrene-butadiene block copolymer containing at least 70 weight percent polymerized styrene. Applications: An injection moulding grade for use where higher stiffness and warpage resistance is required. Used for dust covers, point of purchase displays, moulded boxes and containers, lids and office supplies. (Philips 66 Company).*

K-Resin Polymer KR03. Transparent and shatter resistant styrene-butadiene block copolymer containing at least 70 weight percent polymerized styrene. Applications: An injection moulding grade for use where higher impact resistance is required. Used for overcaps, moulded boxes and containers, toys, medical devices and tool handles. (Philips 66 Company).*

K-Resin Polymer KR04. Transparent and shatter resistant styrene-butadiene block copolymer containing at least 70 weight percent polymerized styrene. Applications: A resin for blending with general purpose polystyrene for sheet extrusion and thermoforming applications such as disposable cups and containers, blister packages and portion packaging. (Philips 66 Company).*

K-Resin Polymer KR05. Transparent and shatter resistant styrene-butadiene block copolymer containing at least 70 weight percent polymerized styrene. Applications: A resin for non-blended sheet extrusion and thermoforming, for blow moulding (both extrusion and injection) and for profile extrusion. Uses include blister packages, bottles, jars, medical devices and extruded tubes and profiles. (Philips 66 Company).*

K-Resin Polymer KR10. Transparent and shatter resistant styrene-butadiene block copolymer containing at least 70 weight percent polymerized styrene. Applications: A resin for blown or cast film production. Uses include medical packaging, shrink wrap, overwrap, skin packaging, produce wrap, windows for envelopes and boxes and twist wrap. (Philips 66 Company).*

Kresival. A German preparation. It contains the water-soluble calcium salts of the sulphonic acids of the cresols.†

Kresol Red. See CRESOL RED.

Kresolin. See KRESOPOLIN.

Kresopolin. (Kresolin). Preparations of crude carbolic acid. Disinfectants.†

Kriegr-o-dip. A proprietary trade name for liquid dyes for plastics. S-standard chemical dye. A-for cellulose acetate. W-powder dye for use in hot water. V-for polystyrene.†

Kristalex. α-methylstyrene copolymer hydrocarbon resins. Applications: Used for hot-melt product assembly adhesives and light coloured caulking compounds. (Hercules Inc).*

Krist-o-kleer. A proprietary trade name for a plasticizer containing 50 per cent dextrose and 50 per cent levulose.†

Krokoloy. A proprietary trade name for a steel containing 14 per cent chromium with some cobalt.†

Kromaplast. Blended dye pigments. Applications: Used in plastics to pigment. (Ampacet Corporation).*

Kroma Red. A precipitated red iron oxide for colour pigment use. (Pfizer International).*

Kromax. An electrical resistance alloy of 80 per cent nickel and 20 per cent chromium.†

Kromore. An alloy of nickel with 15 per cent chromium. Used for the heating elements in wire-wound electric furnaces. It has a specific resistance of 98 micro-ohms cm at 0° C.†

Kromosperse. Pigment dispersing agent. Applications: Dispersing pigments for use in non-aqueous and aqueous coating compositions. (Nueodex Inc).*

Kronagold. Family of lubricating oils. Applications: Gear oils, hydraulic oils, compressor oils, etc. (E/M Corporation).*

Kronaplate. Family of lubricating greases. Applications: Bearings, gears, cams, slides etc. (E/M Corporation).*

Krona-Syn. Synthetic lubricating fluids. Applications: Compressor fluids, high temperature oven chain lubricants. (E/M Corporation).*

Kronds. Titanium dioxide pigments. Applications: Paint, paper, glass, ceramics, plastics and ink. (NL Industries Inc).*

Kronos. Titanium dioxide pigment, used to impart opacity and brightness. Applications: Paints, inks, plastics, paper etc. (NL Chemicals (UK) Ltd)*

Kruppin. An electrical resistance alloy containing 28 per cent nickel and the rest iron.†

Kryalith. A proprietary trade name for a synthetic cryolite.†

Krylene 606. A registered trade mark for a cold polymerized, alum coagulated non-staining butadiene-styrene rubber.†

Krylene 608. A registered trade mark for a cold polymerized styrene butadiene rubber. (Polymer Corporation).†

Krynac 27.50. A proprietary acrylonitrile rubber. (Polymer Corporation).†

Krynac 34.35. A proprietary cold-polymerised gel-free oil-resistant butadiene/acrylonitrile rubber. (Polymer Corporation).†

Krynac 34.50. A proprietary cold-polymerised gel-free oil-resistant butadiene/acrylonitrile rubber. (Polymer Corporation).†

Krynac 34.60 SP. A proprietary nitrile rubber capable of withstanding temperatures up to 135° C. (Polymer Corporation).†

Krynac 34.80. A proprietary cold-polymerised gel-free oil-resistant butadiene/acrylontrile rubber. (Polymer Corporation).†

Krynac 823X2. A registered trade mark for a copolymer of acrylonitrile and butadiene containing a medium level of bound acrylonitrile. (Polymer Corporation).†

Krynac 833. A proprietary isoprene acrylonitrile rubber containing 31.0% bound acrylonitrile. Its Mooney viscosity is 70. (Polymer Corporation).†

Krynac 843. A proprietary medium nitrile rubber containing 50% dioctyl phthalate as an extender. (Polymer Corporation).†

Krynac 850. A registered trade mark for a vinyl-modified nitrile rubber. (Polymer Corporation).†

Krynac 881 and 882. Proprietary names for synthetic rubbers of the ethylacrylate type. (Polymer Corporation).†

Krynac 882X1. A registered trade mark for a low temperature resistant acrylic rubber for oil seals. (Polymer Corporation).†

Kryogen Blacks, G, BG, B, N. Dyestuffs produced from the condensation of dinitro-*m*-dichlorobenzene with *p*-aminophenol and its sulphonic and carboxylic acids. On thionation they give black, greenish-black, and bluish-black shades respectively.†

Kryptocyanines. A series of dyestuffs obtained by dissolving lepidine ethyl iodide in boiling alcohol and adding a solution of sodium ethoxide and formaldehyde, with exclusion of air. They are purple-black in colour, and are used as photo-sensitizing dyes.†

Krystalex. Hydrogenated hydrocarbon resins. Applications: Adhesives and hot melts. (Hercules Inc).*

Krystallazurin. A fungicide consisting of ammoniacal copper sulphate.†

Krystallos. Quartz.†

Krytox. A proprietary name for a range of fluorinated greases used as lubricants in aircraft and missiles. (Du Pont (UK) Ltd).†

K-Slag. Potassium basic slag. (Fisons PLC).*

K.S. Magnet Steel. A cobalt steel containing 35 per cent cobalt. It is suitable for short magnets. Also see PERMANITE, COBALT STEEL, and JAPANESE STEEL.†

K.S. Powder. An explosive. It is a 42-grain powder.†

K.Tab. Potassium Chloride. Applications: Replenisher. (Abbott Laboratories).

Kuhne Phosphor Bronze. An alloy of 78 per cent copper, 10.6 per cent tin, 10.45 per cent lead, 0.57 per cent phosphorus, and 0.26 per cent nickel.†

Kukident. Denture cleanser for cleaning dentures. Also denture adhesive for securing dentures. (Richardson-Vicks Inc).*

kkersite. An oil shale of Esthonia, of specific gravity 1.2-1.4. It contains about 55 per cent volatile matter, and when distilled at 500° C. yields from 70-80 gallons per ton of oil of specific gravity 0.92-0.93.†

k-Seng. A Chinese drug. It is the dried root of *Sophora fiavescens*, and contains the alkaloid matrine.†

nstharz HW. An ammonia condensed phenol formaldehyde resin melting at 55° C. (003), (Chemische Werke, Albert).†

nststein. An artificial stone made from magnesite.†

oxam. A cellulose solvent prepared by dissolving 50 grammes of copper sulphate in 300 cc water and adding ammonia until all the copper hydroxide is precipitated. The precipitate after filtration is dissolved in 25 per cent ammonia solution.†

pferdermasan. A salicyl-copper soap preparation containing 2 per cent copper. A bactericide.†

racap. A proprietary accelerator. 2 - mercaptobenzthiazole + dibenzthiazole disulphide. (BTP Cocker Chemicals Ltd).†

rade. Accelerator for rubber. (Akzo Chemie GmbH, Düren).†

rchi. The root bark of *Holarrhena antidysenteriea*. A febrifuge.†

rofan. Polyvinylidene chloride. (BASF United Kingdom Ltd).*

rofan D. Polyvinylidene chloride dispersion. (BASF United Kingdom Ltd).*

rom 1. A jewellery alloy of copper with tin and cobalt.†

romoji Oil. An oil from *Lindera* species. It is used in Japan for perfuming soaps and oils and contains α-phellandrene, nerolidol, linaloöl, and geraniol.†

ron. Herbicide. Contains silvex as the active ingredient. It is used in ponds and other still water for the control of aquatic weeds, as well as control of brush on rangeland. It is also used industrially on railroads or under power lines for the control of weeds and brush. (Dow Chemical Co Ltd).*

Kurrodur. A proprietary trade name for an alloy of copper with 0.75 per cent nickel and 0.5 per cent silicon.†

Kurrol Salts. Alkaline metaphosphates insoluble in water, but soluble in pyrophosphate solutions. They are produced by heating sodium trimetaphosphate or ethyl sodium phosphate.†

Kusum Oil. See KON OIL.

Kutch. See CUTCH.

Küttner Silk. An artificial silk prepared by the viscose process.†

K-White. Aluminium tripolyphosphate. Non-toxic white anti-corrosive pigment. Applications: Anti-corrosive coatings. (Bromhead & Denison Ltd).*

Kwik Dri. Aliphatic hydrocarbon solvent. Applications: Used in coatings, in fabric drycleaning and in cold degreasing. (Ashland Chemical Company).*

Kwikfill. Polyester two part resin filler. Applications: Mending damaged body panels, filling dents etc. (Hermetite Products Ltd).*

Kwik-Green. Nitrogen, sulphur, iron and zinc. Applications: Used on turf, shrubs, trees and potted plants to promote deep rich green foliage. (Lawn & Garden Products Inc).*

Kymene. Cationic wet-strength resins, including polyamide, polyamine, epoxide and urea-formaldehyde resins. Applications: Impart strength to wet paper and paperboard. They are used primarily as internal additives in papermaking processes, but also as cationizing agents for starch added internally and as insolubilizing agents for starch in size press and pigmented coatings. (Hercules Inc).*

Kynar SL. Copolymer of vinylidene fluoride and tetrafluorethylene. Applications: A proprietary grade of a vinylidene fluoride based copolymer used primarily for coatings applications on metal, cellulosic and synthetic substrates. It has a low temperature bake finish and can be formulated as a solution or dispersion composition. (Pennwalt Corp).*

Kynar 500. Polyvinylidene fluoride resin. Applications: A proprietary grade of polyvinylidene resin used primarily for architectural coatings applications from solvent based dispersions. (Pennwalt Corp).*

Kynite. An explosive containing 24-26 per cent nitroglycerin, 2-3 per cent wood pulp, 32-321/2 per cent starch, 31-34 per cent barium nitrate, and 0-0.5 per cent calcium carbonate.†

Kynol. A highly cross-linked amorphous phenolic polymer. It resists temperatures up to 2500° C.†

Kypfarin. Warfarin. (Mechema).*

Kyrock. A rock asphalt consisting of san with about 7 per cent bitumen.†

Kysite. A proprietary trade name for a phenol-formaldehyde synthetic resin with a fibre filler.†

L

Labarraque's Solution. A solution of chlorinated soda, containing 2½ per cent of available chlorine.†

Labdanum. (Ladanum). A resinous substance obtained from various species of *Cistus*. A stimulant expectorant.†

Labitan. SEE CORMELIAN. (Degussa).*

Labiton. A proprietary preparation of thiamine hydrochloride, extract of Kola nuts, syrup and glycerophosphoric acid. (LAB Ltd).*

Labophylline. A proprietary preparation of theophylline and lysine. A respiratory and cardiac stimulant. (LAB Ltd).*

Laboprin. A proprietary preparation of aspirin and lysine. An analgesic. (LAB Ltd).*

Labosept. A proprietary preparation of dequalinium chloride. An antibacterial lozenge taken orally. (LAB Ltd).*

Labrocol. Tablets containing 100mg, 200mg and 400mg labetalol hydrochloride. Applications: For the treatment of all forms of hypertension, and all grades of hypertension (mild, moderate and severe) when oral antihypertensive therapy is desirable. (Lagap Pharmaceuticals Ltd).*

Labstix. A proprietary test-strip used for the detection of pH, protein, glucose, ketones and blood in urine. (Ames).†

Labstone. See KEMITE.

Lac. (Gum Lac, Lacca, Button Lac, Sheet Lac, Shellac). The resinous excretion of the lac insect, *Laccifer lacca*, cultivated in India, Burma, and Siam. The insects living on the twigs become surrounded with the lac, and in this form it is known as stick-lac. The stick-lac is made into seed-lac, by removing the twigs, insect bodies, etc., by crushing, winnowing, and washing. and is finally purified by heating it in a cloth bag and forcing the lac through the cloth by twisting the bag. This is shellac. Rosin and orpiment are often added to give a light appearance to the material. The best quality shellac contains no rosin or orpiment, and has from 3.5-4.5 per cent wax. Shellac has many applications, being used as a constituent of polishes, varnishes, and lacquers, in electrical insulation, and in the manufacture of gramophone records and sealing waxes. A variety of Japanese wax obtained from the wood of *Rhus vernicifera* of Japan is also called Lac. It is used in lacquer.†

L-Acid. See LAURENT'S ACID.

Lacanite. A proprietary trade name for a shellac compound.†

Lacca. See LAC.

Laccain. A phenol-formaldehyde resin made with the aid of hydroxy acids, such as tartaric acid.†

Lac-Dye. (Lack-lack, Dyer's Lac). The colouring matter derived from Lac (*qv*). Stick-lac contains 10 per cent of colouring matter. Used for dyeing wool mordanted with aluminium or tin salts.†

Lacimoid. A proprietary trade name for a synthetic resin. It is used in laminated form for walls, etc.†

Lacitin Red B. The calcium lake of Lithol Red R.†

Lacitin Red R. The barium lake of Lithol Red R.†

Lack-lack. See LAC-DYE.

Lackmoid. The blue colouring matter obtained by heating resorcinol with sodium nitrite. Used as an indicator in alkalimetry.†

Lac Lake. See CARMINE LAKE.

Lacmoid. See LACKMOID.

Lacmus. Litmus.†

Lacolene. Aliphatic hydrocarbon solvent. Applications: Used in coatings, adhesives and printing inks. (Ashland Chemical Company).*

Lacorene. A proprietary trade name for a polystyrene moulding resin.†

Lacqran. A proprietary trade name for an ABS (*qv*) moulding resin.†

Lacqrene E. Antistatic polystyrene. (Aquitaine-Organico).†

Lacqrene 550. A proprietary polystyrene used in extrusion and injection moulding. (Aquitaine-Organico).†

Lacqrene 635, 811, 835 and 836. Proprietary polystyrenes used to produce extrusions of differing tensile strengths. (Aquitaine-Organico).†

Lacqrene 740. An impact and heat resistant polystyrene suitable for use at 90° C. (Aquitaine-Organico).†

Lacqsan E. Antistatic styrene acrylonitrile.†

Lacqsan 125 and 125L. Proprietary copolymers of styrene and acrylonitrile used in extrusion and injection moulding. (Aquitaine-Organico).†

Lacqtene 1070 MN20. A proprietary low-density polyethylene used in injection moulding. (Aquitaine-Organico).†

Lacqtene 1200 MN26. A proprietary low-density polyethylene used in injection moulding. (Aquitaine-Organico).†

Lacquer. Shellac dissolved in alcohol, and coloured with saffron, annatto, or dragon's blood.†

Lacrinite. A proprietary casein-phenolformaldehyde product.†

Lac Sulphur. Precipitated sulphur. It is very light in colour.†

Lac Sulphuris. Milk of sulphur.†

Lactase. An enzyme which converts lactose into *d*-glucose and *d*-galactose.†

Lacteol. (Lactigen, Lactilloids, Lactobacilline, Lactone). Preparations of lactic acid bacilli.†

Lacticare. Contains lactic acid and sodium pyrrolidone carboxylate in an oil-in-water viscous lotion base. Applications: For the symptomatic relief of hyperkeratotic and other chronic dry skin conditions caused by low humidity or the use of detergents. (Stiefel Laboratories (UK) Ltd).*

Lacticol. Stabilizer for milk based systems. (Kelco/AIL International Ltd).

Lactigen. See LACTEOL.

Lactilloids. See LACTEOL.

Lactin. Lactose, a sugar.†

Lactine. See VEGETABLE BUTTER.

Lactinium. A German preparation of neutral aluminium lactate. An astringent and disinfectant.†

Lactite. See GALLATITE.

Lactitis. A casein preparation containing borax and lead acetate. It is an artificial ivory.†

Lactobacilline. See LACTEOL.

Lactoform. See GALLATITE.

Lactoid. A casein preparation.†

Lactol. (Lactonaphthol). The lactic acid ester of β-naphthol, $CH_3CHOH.COO.C_{10}H_7$. An intestinal astringent.†

Lactolin. (Antimonin). The double salts of antimony lactate with alkalis, alkaline earths, and zinc salts. A convenient means for the transport of lactic acid. Also used as a substitute for tartar emetic in dyeing.†

Lactolith. A casein preparation.†

Lactoloid. A proprietary casein product.†

Lactonaphthol. See LACTOL.

Lactone. See LACTEOL.

Lactoprene. A patented vulcanizable synthetic rubber made from emulsified methyl or ethyl acrylate copolymerized with small quantities of a polyfunctional monomer such as butadiene, isoprene or allyl maleate. The copolymer is compounded with sulphur and accelerator and cured.†

Lactorite. See GALLATITE.

ctosan. Dairy hygiene detergent sterilizer. (The Wellcome Foundation Ltd).

ctose Molasses. Molasses obtained from the preparation of milk sugar.†

ctozym. A beta-galactosidase (lactase) preparation produced by submerged fermentation of a selected strain of the yeast Kluyveromyces fragilis. Applications: Sweet milk products. Production of low lactose milk for persons suffering from lactose malabsorption . The milk can be consumed either directly or after condensing/drying. (Novo Industri A/S).*

ctulose. A sugar used in the treatment of hepatic coma and chronic constipation. It is 4-*0*-β-D-galacto-pyranosyl-D-fructose. DUPHALAC.†

ac, Garnet. See LEMON LAC.

ac, Japanese. The lac obtained from *Rhus vernicifera*. It is a natural varnish or lacquer, and contains 85 per cent, of urushic acid.†

ac, Oil. See OIL LACS.

ac, Orange. See LEMON LAC.

ac, White. See WHITE INSECT WAX.

adalrod. Aluminium and aluminium alloy rod. Applications: Used in alloying steel. (Reynolds Metal Co).*

adanum. See LABDANUM.

adelloy. Ferro-alloy ladle additions. (Foseco (FS) Ltd).*

adropen. Flucloxacillin. Applications: Treatment of infections caused by gram positive organisms. (Berk Pharmaceuticals Ltd).*

aevo-Glucose. See LEVULOSE.

aevo-Pinene. Terebenthene, a terpene.†

aevuflex. A proprietary preparation of laevulose used as a parenteral calorie supplement. (Geistlich Sons).†

aevuline Blue. A solution of induline in acetin.†

afond's Bearing Metal. See BEARING METALS.

afond's Bronzes. Alloys of 80-98 per cent copper, 2-18 per cent tin, 0-2 per cent zinc, and 0-0.5 per cent lead.†

Lafou's Reagent. A sulphuric acid solution of ammonium selenite. An alkaloidal reagent.†

Lagos Silk Rubber. A rubber obtained from *Funtumia elastica* of Africa.†

Lahkrostat. Polar and non-polar compounds. Applications: Antistatic agents for PVC, polystyrene, polyolefins and ABS. (Lankro Chemicals Ltd).†

Lake Copper. An American class of copper containing 99.8 per cent of the metal.†

Lake, Florentine. See FLORENCE LAKE.

Lake, Indian. See CARMINE LAKE.

Lake, Lac. See CARMINE LAKE.

Lakeland AMA. Monosodium salt of an alkylamine dicarboxylate, as a clear liqud. Applications: Traffic film removers and hard surface cleaners. (Lakeland Laboratories Ltd).

Lakeland AMA LF. Salt-free low-foam version of Lakeland AMA. Applications: Low foam cleaners. (Lakeland Laboratories Ltd).

Lakeland PP. Series of anionic surfactants in which the base is a nonylphenol ethoxylate. Applications: Wetting agents, detergents, emulsifiers and lubricants which are stable to acid, alkaline and electrolyte conditions. Used for heavy duty industrial cleaners; kier boiling of cottons; metal cleaning; various emulsification applications. (Lakeland Laboratories Ltd).

Lake, Paris. See FLORENCE LAKE.

Lake Red C. A dyestuff prepared from 5-amino-2-chlorotoluene-4-sulphonic acid.†

Lake-red Ciba B. An insoluble pigment, made by the interaction between indigo and phenylacetylchloride. The soluble dyestuff is obtained by sulphonating the product.†

Lake Red D. A dyestuff. It consists of diazotized anthranilic acid condensed with β-naphthol.†

Lake Red F. A dyestuff. It is an equivalent of Lake red P.†

Lake Red P. (Lake Red F, Monolite Red P). A dyestuff obtained by diazotizing *p*-nitraniline-*o*-sulphonic acid, and condensing it withβ-naphthol.†

Lakes. Compounds of inorganic bodies with organic colouring matters, usually aluminium oxide or other metallic oxide.†

Lake's Indiarubber Compound. A compound consisting of saponified resin and vulcanized oil, incorporated with indiarubber or gutta-percha.†

Lake, Venetian. See CRIMSON LAKE.

Lake, Vienna. See FLORENCE LAKE.

Lake, Yellow. See YELLOW CARMINE.

Lalicopharsol. Mild emollient and emulsifier. Applications: All fine cosmetics; pharmaceuticals; non-irritative-barrier creams. (Solaver SA).

Lalitecsol. Mild emollient and emulsifier. Applications: All fine cosmetics; non-irritative-barrier creams. (Solaver SA).

Lalona Bark. The bark of *Weinmannia bojeriana*. It contains 13.75 per cent tannin, and is used for tanning.†

Lamalgin. A thickening agent used for textile printing. (Degussa).*

Lambrex. Emulsion explosives (slurries) of the Lambrex series, in which none of the ingredients is classified as an explosive. It is by the mixing process that cap sensitivity is obtained. Lambrex is characterized by greatest handling safety, and can be supplied in cartridges or can be mixed in a pump truck on site and pumped into the boreholes. Lawinit 100 is a special explosive for blasting avalanches. (Dynamit Nobel Wien GmbH).*

Lambrit (Anfo-explosives). Primed from the bottom of the borehole by Gelatine Donarit 1 and detonating fuse. (Dynamit Nobel Wien GmbH).*

Lamefin. A softening agent and lubricant. Applications: Used by the textile industry for the after treatment of all types of fibres. (Degussa).*

Lamefix. Printing oils and fixation accelerators used for textile printing. (Degussa).*

Lamegum. A thickening agent used for textile printing. (Degussa).*

Lamephan. A softening agent and lubricant. Applications: Used by the textile industry for the after treatment of all types of fibres. (Degussa).*

Lamepon. Dispersing and levelling agent. Applications: Used for dyeing and stabilizer for peroxide bleaching in the textile industry. (Degussa).*

Lameprint. A thickening agent used for textile printing. (Degussa).*

Lamex 173/FR. A self extinguishing type of the above resins complying with BS 476 Part I (Class II). (Croda Resins Ltd).†

Lamex 185. A trade name for a flexible amine preaccelerated polyester resin used for motor car body repairs. (Croda Resins Ltd).†

Lamex 186. A trade name for a rigid amine preaccelerated polyester resin used for motor car body repairs. (Croda Resins Ltd).†

Lamicoid. A proprietary trade name for a phenol-formaldehyde synthetic resin with a mica filler used for laminated products.†

Laminac EPX-176. A trade mark. A self extinguishing (flame resistant) polyester resin. Suitable for manufacture of reinforced plastics in the transportation industry.†

Laminated Talc. See MICA.

Laminic. A nickel-iron alloy.†

Lamitex. A proprietary trade name for a hard vulcanized fibre.†

Lamotte Standard Indicators. These are prepared by mixing 0.5 cc of the prepared commercial solutions of the following indicators with 10 cc of the special buffer solution M5: Thymol Blue (acid range), Bromophenol Blue, Bromocresol Red, Bromocresol Purple, Bromothymol Blue, Phenol Red, Cresol Red, Thymol Blue (alkaline range).†

Lampadite. An earthy variety of manganese dioxide, containing copper oxide.†

Lampblack. This is carbon in a fine condition prepared by the incomplete combustion of tar, colophony, vegetable oils, pitch or heavy oils.†

Lampit. Pharmaceutical speciality. Applications: Used against Chagas' Disease. (Bayer & Co).*

Lamprene. A proprietary preparation of CLOFAZIMINE. An anti-leprotic. (Ciba Geigy PLC).†

Lampronol. Leather dyes and finishes, pigments for plastics, dyes for printing inks. (ICI PLC).*

Lamy Butter. (Kanja Butter, Kanga Butter, Sierra Leone Butter). The fat obtained from the seeds of *Pentadesma butyracea* of West Africa.†

Lana Batu. Citronella oil.†

Lanacron. Wool dyestuffs. (Clayton Aniline Co).†

Lanacyl Blue BB. (Indigo Substitute). An azo dyestuff. It is prepared by coupling aminonaphtholdisulphonic acid with 5-amino-1-naphthol. An acid wool dye.†

Lanacyl Blue R. A dyestuff belonging to the same group as Lanacyl Blue BB.†

Lanacyl Navy Blue B. A dyestuff belonging to the same group as Lanacyl Blue BB.†

Lanadin. An alcoholic soap solution containing 87 per cent trichlorethylene. A wetting-out and scouring agent.†

Lanafuchsine S.B. See SORBIN RED.

Lanain. See LANOLIN.

Lanaire. Scouring preparations. (Crosfield Chemicals).*

Lanalin. See LANOLIN.

Lanamar. See LANELLA.

Lanamine. Mixed isopropanolamines myristate soap. Applications: Offers mild cleansing action with high foaming powder. Reduces normal stripping in shampoos. (Amerchol Corporation).*

Lanapex. Textile auxiliary chemicals. (ICI PLC).*

Lanasol. Reactive dyes for wool and silk. (Clayton Aniline Co).†

Lanbritol. Self emulsifying waxes. (Hickson & Welch Ltd).*

Lancare. Household detergents and toiletries. (Lankro Chemicals Ltd).

Lancashire Brass. See BRASS.

Lancer. Herbicide, containing flamprop-methyl for wild oat control. (ICI PLC).*

Lancosol. Mild emollient and emulsifier. Applications: All fine cosmetics; non-irritative-barrier creams. (Solaver SA).

Landemul. Demulsifier components for oil fiel production chemicals. (Lankro Chemicals Ltd).

Land Plaster. Ground gypsum, $CaSO_4.2H_2O$, a soil dressing.†

Landromil. Ticlatone. Applications: Antibacterial; antifungal. (Dorsey Laboratories, Div of Sandoz Inc).

Landshoff and Meyer's Acid. 2-Naphthylaminedisulphonic acid.†

Lanella. (Lanamar). Woolly types of fibre obtained from a kind of seaweed, *Posidonia oceania,* found in the Pacific.†

Lanesin. See LANOLIN.

Lanesta. Isopropyl esters of lanolic acids. Applications: Emollients, gloss agents, binding agents for lipsticks and other cosmetics. (Westbrook Lanolin Co).*

Lanestren. Dyestuffs for wool/polyester fibre blends. (BASF United Kingdom Ltd).

Laneto Series. Lanolin. Applications: Available at various degrees of ethoxylation giving the formulator a wide range of water solubilities to select from. LANETO moisturizes and lubricates, acts as a solubilizer of oils and perfumes and lends emollience to shampoos, bath preparations, creams, lotions, soaps and detergents. They also impart a natural lustre or sheen to the hair. As they are nonionic in character, they are compatible with a wide range of ingredients. (RITA Corporation).*

Lanette. Self emulsifying waxes. (Hickson & Welch Ltd).*

Lanette Wax. A proprietary trade name for a mixture of cetyl and stearyl alcohols.†

Lanette Wax Ester. A proprietary trade name for palmitic acid ester of cetyl and stearyl alcohols used in emulsions.†

Lanette Wax SX. A proprietary trade name for an emulsified mixture of cetyl and stearyl alcohols.†

Lanexol. Soluble in ethanol and water; plasticizer emollient and lubricant; super-fatting agent. Applications: Aqueous and alcoholic systems; shampoos; after-shaves; hair sprays; dyes and conditioners; hair rinses. (Croda Chemicals Ltd).

Lange Solution. A colloidal gold solution.†

Langford. See CATALPO.

Langford Clay. Hydrated aluminium silicate (soft clay). Applications: Low cost reinforcer and inert filler for paint, paper, rubber, ceramics, plastics and specialilties. (Vanderbilt Chemical Corporation).*

Lanichol. See LANOLIN.

Laniol. See LANOLIN.

Lanital. A proprietary trade name for a casein textile fibre made by dissolving casein in a dilute alkaline solution and extruding the viscous compound in the form of thin filaments. These are treated with acid and rendered insoluble by means of formaldehyde.†

Lanitop. Medigoxin. Applications: Congestive heart failure, cardiac arrhythmias. (Roussel Laboratories Ltd).*

Lankroflex. Range of epoxidized fatty acid esters and oils. Applications: plasticizers for polymers. (Lankro Chemicals Ltd).†

Lankroflex ED3. A proprietary epoxy plasticizer. Octyl epoxy stearate. (Lankro Chemicals Ltd).†

Lankroflex ED6. A proprietary epoxy plasticizer. (Lankro Chemicals Ltd).†

Lankroflex GE. A proprietary epoxidized vegetable oil used as a plasticizer or as an extender. See ABRAC A. (Lankro Chemicals Ltd).†

Lankrol. Sulphated oils. (Lankro Chemicals Ltd).†

Lankrolan. Textile auxiliaries - shrink proofing agent for wool. (Lankro Chemicals Ltd).

Lankroline. Sulphated oils and pigment finishes. (Lankro Chemicals Ltd).†

Lankrolyte. Sequestering agents. (Lankro Chemicals Ltd).

Lankromark. Range of mixed metal carboxylates, organotin compounds and organophosphites. Applications: stabilizers and antioxidants for polymers. (Lankro Chemicals Ltd).†

Lankromul. Oil spill dispersants. (Lankro Chemicals Ltd).

Lankroplast. Lubricants in thermoplastic processing, antifogging agents for PVC film and viscosity modifiers for PVC pastes. (Lankro Chemicals Ltd).†

Lankropol. Sulphated and sulphonated esters and acids. Applications: Used as wetting agents, emulsifiers in textiles, paint and emulsion polymers. (Lankro Chemicals Ltd).†

Lankropol KO Special. Sodium di-isooctyl sulphosuccinate, as a clear pale straw liquid containing mineral oil. Applications: Emulsifier and water carrier in solvents, used in hydrocarbon solvents, dry cleaning detergent formulations and as a dewatering aid. (Diamond Shamrock Process Chemicals Ltd).

Lankropol KO2. Sodium di-isooctyl sulphosuccinate as a clear pale straw liquid containing ethanol. Applications: Wetting agent and emulsion polymerization aid. (Diamond Shamrock Process Chemicals Ltd).

Lankropol KSB 22. Monoester sulphosuccinate as a clear liquid. Applications: Primary emulsifier in latex production. (Diamond Shamrock Process Chemicals Ltd).

Lankropol KSG 72. Monoester sulphosuccinate as a clear liquid. Applications: Low irritancy toiletry intermediate for shampoos, bubble baths, etc. (Diamond Shamrock Process Chemicals Ltd).

Lankropol ODS. Disodium N-octadecyl sulphosuccinamate in liquid or paste form. Applications: Foamimg agent used for latex foam systems in carpet backing. (Diamond Shamrock Process Chemicals Ltd).

Lankrosol. Alkyl aryl sulphonates, modified nonionics, phosphate esters. Applications: Used as hydrotropes, wetters in electrolyte solutions. (Lankro Chemicals Ltd).†

Lankrosol SXS. Anionic surfactant as a clear, pale straw liquid. Applications: Solubilizing and viscosity modification agent used in high active liquid detergents and hard surface cleaners. Also used for dye levelling in nylon dyeing. (Diamond Shamrock Process Chemicals Ltd).

Lankrosperse. Dispersing agents. (Lankro Chemicals Ltd).

Lankrostat. Autistatic agents for plastics. (Lankro Chemicals Ltd).

Lankrothane. Polyurethane derivates. (Lankro Chemicals Ltd).

Lannate. Methomyl insecticde. (Du Pont (UK) Ltd).

Lanocerin. Lanolin wax. Amorphous waxy ester fraction of anhydrous lanolin. Applications: Provides textural smoothness in emulsions and increases stability. (Amerchol Corporation).*

Lanogel 21. Ethoxylated (27 mol) lanolin. 50% aqueous gel, dispersible in water. Applications: Nonionic surfactant with lubricating and conditioning properties for detergent systems. (Amerchol Corporation).*

Lanogel 31. Ethoxylated (40 mol) lanolin. Applications: Nonionic surfactant, more water soluble than LANOGEL 21. (Amerchol Corporation).*

Lanogel 41. Ethoxylated (75 mol) lanolin. 50% aqueous gel, water souble, suitable for aqueous systems. Applications: Conditioner in soaps and shampoos. (Amerchol Corporation).*

Lanogel 61. Ethoxylated (85 mol) lanolin. Nonionic 50% gel soluble in water and alcohol. Applications: Conditioner in hydro alcoholic systems. (Amerchol Corporation).*

Lanogene. Lanolin oil, oil soluble fraction of lanolin (dewaxed). Applications: Soluble in many oils, esters and hydrocarbons making it most suitable for incorporating lanolin properties into clear systems. (Amerchol Corporation).*

Lanoiac. Lanolin paint. (Croda Chemicals Ltd).*

Lanoid. See MULSOID.

Lanolic. Co-solvent, emollient. Applications: Skin and hair cosmetics; make-up; shaving foams and creams. (Croda Chemicals Ltd).

Lanolic Acid. Distilled lanolin fatty acids. (Croda Chemicals Ltd).

Lanolin. (Hydrous Wool Fat, Lanalin, Lanain, Laniol. Lanesin, Lanichol). Purified wool fat. Used as a basis for ointments and cosmetics.†

Lanoline. A formulation of lanolin. Applications: An emollient to soothe and soften the skin. (The Wellcome Foundation Ltd).*

Lanoplast. A proprietary cellulose acetate.†

Lanoresin. A resin obtained from the washing of wool.†

Lanosol. Lanolised mineral oil. (Croda Chemicals Ltd).*

Lanosterol. Lanosterol. (Croda Chemicals Ltd).

Lanoxicaps. Solution in capsules. A proprietary formulation of digoxin. Applications: Treatment of heart failure, atrial fibrillation, atrial flutter and paroxysmal atrial tachycardia. (The Wellcome Foundation Ltd).*

Lanoxin. Tablets, injections, elixir paediatric. Proprietary formulation of digoxin. Applications: Treatment of heart failure, atrial fibrillation, atrial flutter and paroxysmal atrial tachycardia. (The Wellcome Foundation Ltd).*

Lanoxine-PG. A proprietary preparation of digoxin. (The Wellcome Foundation Ltd).*

Lanpharsol. Mild emollient and emulsifier. Applications: All fine cosmetics; non-irritative-barrier creams. (Solaver SA).

Lanpol. Polyoxyethylated lanolin fatty acids. (Croda Chemicals Ltd).

Lanstar. Antistatic agents, emulsifiers. Applications: Polystyrene; synthetic fibres; textile auxiliaries. (Lancashire Tar Distillers Ltd).

Lanstar NP100/50. Nonylphenol ethoxylate in aqueous solution. Applications: Nonionic surfactant effective at high temperatures and in concentrated electrolyte solutions. Used in emulsion polymerization; latex stabilization; waxes and polishes. (Lanstar Chemicals).

Lanstar NP2 and NP4. Nonylphenol ethoxylate nonionic surfactant in the form of an oil-soluble liquid. Applications: Foam depressors; emulsifiers. (Lanstar Chemicals).

Lanstar NP40, NP50 and NP100. Nonylphenol ethoxylate nonionic surfactant in wax form. Applications:

Water-soluble surfactants effective at high temperatures and in concentrated electrolyte solutions. Used in emulsion polymerization; latex stabilization; waxes and polishes. (Lanstar Chemicals).

Lanstar PCH, PC2 and PCO. Calcium petroleum sulphonate in liquid form. Applications: Emulsifier, dispersing and wetting agent. Used in cutting oil, lube oil and fuel oil additives; rust preventatives; leather and textile industry. (Lanstar Chemicals).

Lanstar PS. Sodium petroleum sulphonate in liquid form. Applications: Emulsifier, dispersing and wetting agent for cutting oil, lube oil and fuel oil additives; rust preventatives; leather and textile industry. (Lanstar Chemicals).

Lanstar PSW. Sodium petroleum sulphonate in liquid form. Applications: Water soluble anionic surfactant used in ore flotation; building industry; demulsification. (Lanstar Chemicals).

Lantecsol. Mild emollient and emulsifier for technical uses; rust inhibitor; bind material. Applications: Printing ink; rust inhibitors; petroleum; rubber adjuvants; lubricants; soaps; leather protection; painting industry. (Solaver SA).

Lanthana. Lanthanum oxide.†

Lanthanol LAL. Sodium alkyl sulphoacetate in powder or flake form. Applications: Anionic surfactant used in shampoos and bubble baths. (KWR Chemicals Ltd).

Lantol. A name applied to colloidal rhodium, and also to lanthopine, an alkaloid of opium.†

Lanvis. A proprietary preparation of thioguanine used in the treatment of leukaemia. (The Wellcome Foundation Ltd).*

Lapis-Oliario. See STEATITE.

Lapis Smiridis. Emery. (A trade mark.)†

Lapix. A proprietary trade name for a flux used in steel moulding. It contains carbon and clay.†

Lapofloc. Water treatment chemicals. (Laporte Industries Ltd).

Laponite. Synthetic clay products resembling hectorite with thixotropic gelling properties. Applications: Used as gels in toothpastes and cosmetics. Thixotropy also enables use in surface coatings, sprays, conferring anti-static properties. Absorbancy means Laponite is used as a retentive medium. (Laporte Industries Ltd).†

Lapotan. Basic aluminium sulphate. (Laporte Industries Ltd).

Lapparentile. Synonym for Tamarugite.†

Lapudrine. A proprietary preparation containing chlorproguanil hydrochloride. An antimalarial drug. (ICI PLC).*

Laquanol. Water thinnable resins. (Croda Resins Ltd).

Laracor. Film coated tablets containing 20mg, 40mg, 80mg and 160mg oxprenolol BP. Applications: Indicated for the treatment of angina pectoris, all grades of hypertension and sympathetically induced cardiac arrythmias. Also anxiety tachycardia, functional heart disorders, hypertrophic obstructive cardiomyopathy, as an adjunct to the treatment of thyrotoxicosis and dysrhythmias associated with anaesthesia. (Lagap Pharmaceuticals Ltd).*

Laractone. Tablets containing 25mg and 100mg spironolactone BP. Applications: Used in congestive heart failure, hepatic cirrhosis with ascites and oedema, malignant ascites, nephrotic syndrome, diagnosis and treatment of primary aldosteronism, essential hypertension. (Lagap Pharmaceuticals Ltd).*

Laraflex. Tablets containing 250mg and 500mg naproxen BP. Applications: Treatment of rheumatoid arthritis, osteoarthritis, ankylosing spondylitis, juvenile rheumatoid arthritis, acute gout and acute musculoskeletal disorders (such as sprains and strains, direct trauma, lumbrosacral pain, cervical spondylitis, tenosynovitis and fibrositis). (Lagap Pharmaceuticals Ltd).*

Laratrim. Tablets containing 80mg trimethoprim BP, 400mg sulphamethoxazole BP and 160mg trimethoprim BP, 800mg sulphamethoxazole BP. Suspension containing 40mg trimethoprim, 200mg sulphamethoxazole BP in each 5ml of suspension and 80mg trimethoprim, 400mg sulphamethoxazole BP in each 5ml. Applications: Used as an antibacterial agent. Effective against a wide range of Gram-positive and Gram-negative organisms. Of value in the treatment of respiratory, genito-urinary, gastro-intestinal tracts and skin infections and other bacterial infections. (Lagap Pharmaceuticals Ltd).*

Larch Extract. An extract of the bark of *Pinus larix*. Used for tanning.†

Larch Sugar. Melezitose, a trisaccharide $C_{18}H_{32}O_{16}$.†

Larch Turpentine. See VENICE TURPENTINE.

Lard. Purified hog's fat.†

Lardine Oil. An oil prepared from cotton-seed oil.†

Largactil. A proprietary preparation of chlorpromazine hydrochloride. A sedative. (May & Baker Ltd).*

Largon. Propiomazine Hydrochloride. Applications: Sedative. (Wyeth Laboratories, Div of American Home Products Corp).

Lariam. Mefloquine hydrochloride. Applications: Antimalarial. (Hoffmann-LaRoche Inc).

Larixin. Larixinic acid, $C_{10}H_{10}O_5$, obtained from larch bark.†

Larnol. Rust preventative. (Croda Chemicals Ltd).*

Larocin. Proprietary preparation of amoxycillin. Antibacterial. (Roche Products Ltd).*

Larodopa. A proprietary preparation of levodopa used in the treatment of Parkinson's disease. (Roche Products Ltd).*

Larotid. Amoxicillin. Applications: Antibacterial. (Beecham Laboratories).

Laroxyl. A proprietary preparation of amitriptyline hydrochloride. An antidepressive agent. (Roche Products Ltd).*

Larvacide. A proprietary trade name for chloropicrin used as an insecticide.†

Larvex. A proprietary clothes-moth remedy. It is a solution of sodium fluosilicate.†

Lasan. Anthralin. Applications: Antipsoriatic. (Stiefel Laboratories Inc).

Lasilso. Sodium metasilicate. (Laporte Industries Ltd).*

Lasix. A proprietary preparation of furosemide. A diuretic. (Hoechst Chemicals (UK) Ltd).†

Lasix + K. A proprietary preparation of FRUSEMIDE and potassium chloride. A diuretic containing a potassium supplement. (Hoechst Chemicals (UK) Ltd).†

Lasonil. Pharmaceutical preparation. Applications: For the treatment of sprains, contusions, bruises and varicose conditions. (Bayer & Co).*

Lasso. 2-Chloro-2', 6'-diethyl-N-(methoxymethyl) acetanilide. Applications: Pre-emergence herbicide. (Monsanto Co).*

Lassolatite. A mineral. It is a variety of opal.†

Lastex. A proprietary product. It is rubber latex threads spun with fibre.†

Lastil. Fungicide for bonded cork. (BDH Chemicals Ltd).

Lastilac. A proprietary moulding compound.†

Lasurite. See CHESSYLITE.

Latene. A proprietary rubber vulcanization accelerator. It is a solution of trimene base in rubber latex.†

Latensol AP8. A non-ionic surfactant. (BASF United Kingdom Ltd).*

Laterite. A material consisting of hydroxides of iron and aluminium, with sand and clay. It is used in India as a building stone.†

Latex. The milky emulsion, containing minute suspensions of oil rubber, which flows from incisions in the bark of rubber trees, such as *Hevea brasiliensis*.†

Latex Foam. A proprietary trade name for a type of cellular sponge rubber made from latex by a special method.†

Latex, Reversible. See REVERTEX.

Lathanol. Sodium lauryl sulphoacetate. Applications: Powder bubble baths, synthetic detergent bars. (Stepan Company).*

Latialite. Synonym for Haüyne.†

Latibon. Calcium formate. Applications: Water-soluble calcium component for use in milk replacers and as a safety guard. (Bayer & Co).*

Laudanon. A mixture of opium alkaloids.†

Laudanum. Tincture of opium *(Tinctura Opii* B.P.).†

Laurel Camphor. See JAPAN CAMPHOR.

Laurel Nut Oil. (Domba Oil, Alexandrian Laurel Oil, Poonseed Oil, Tacamahac Oil, Njamplung Oil, Calaba Oil, Dilo Oil, N dilo oil, Pinnay Oil, U dilo oil). Calophyllum oil, from the nuts of *Calophyllum* species. Used for medicinal and illuminating purposes, not an edible oil, being poisonous.†

Laurel Oil. Bayberry oil, obtained from the berries of the laurel tree, *Laurus nobilis*.†

Laurel Wax. (Myrtle Wax, Bayberry Tallow). Myrtle berry wax, obtained from *Myrica cerifera*.†

Laurent's Acid. (L-acid). I-Naphthyl-amine-5-sulphonic acid.†

Laurent's Aluminium Solder. The hard solder contains 63-74 per cent zinc and 19-30 per cent tin. The soft solder contains 60-70 per cent zinc, 16-27 per cent tin, and 12 per cent lead.†

Laurent's Naphthalldinic Acid. See LAURENT'S ACID.

Laureol. See VEGETABLE BUTTER.

Laurex. Higher fatty alcohols. (Albright & Wilson Ltd, Detergents Div).

Laurex. A proprietary trade name for a series of primary fatty alcohols. (Marchon France SA).†

Lauridit. Foam stabilizers and skin protecting additives. Applications: Liquid and powder formulations. (Akzo Chemie Nederland BV).

Laurodin. Bactericidal preparations of laurolinium acetate. (Allen & Hanbury).†

Laurol. Laurene, $C_{11}H_{16}$.†

Laurydol. Lauroyl peroxide. (Akzo Chem UK Ltd).

Laurydol B-50. Paste of Laurydol in a phthalate plasticizer.†

Lausofan. A hexamethylene ketone used for destroying insects of the vermin type.†

Lautal. A proprietary trade name for an aluminium alloy containing 4 per cent copper and 2 per cent silicon.†

Lauth's Violet. (Thionine). Aminoimino-iminodiphenyl sulphide, $C_{12}H_9N_3S$. Used as a microscopic stain.†

Lavacol. Rubbing alcohol. Applications: Rubefacient. (Parke-Davis, Div of Warner-Lambert Co).

Lavalloy. A proprietary trade name for a ceramic product made from mullite a alumina.†

Lavarock. A heat-treated steatite.†

Lavasteril. *p*-Chlor-*m*-cresol and *p*-chlor-*m*-thymol.†

Lavasul. A mixture of 40 per cent coke d with sulphur. Used for tank linings.†

Lavema. Oxyphenisatin acetate, Applications: Laxative. (Sterling Drug Inc).

Lavender Drops. Compound tincture of lavender.†

Lavender, Red. See RED DROPS.

Lavendol. Impure linalol.†

Laveran's Stain. (a) Eosin I gram, distille water 1,000 cc (b) A few crystals of silver nitrate are dissolved in 60 cc wa and sodium hydroxide added in excess. Silver oxide is precipitated. This is collected and added to a saturated solution of methylene blue. For use 4 c of solution (a) are mixed with 6 cc wa and 1 cc solution (b) added.†

Lavite. A heated steatite product.†

Lawinit 100. Emulsion explosives (slurries of the Lambrex series, in which none o the ingredients is classified as an explosive. It is by the mixing process that cap sensitivity is obtained. Lambrex is characterized by greatest handling safety, and can be supplied in cartridges or can be mixed in a pump truck on site and pumped into the boreholes. Lawinit 100 is a special explosive for blasting avalanches. (Dynamit Nobel Wien GmbH).*

wn Food. Lawn fertilizer. (Fisons PLC, Horticulture Div).

wn Plus. Fertilizer and selective weedkiller. (ICI PLC, Plant Protection Div).

wn Spot Weeder. Selective weedkiller aerosol. (Fisons PLC, Horticulture Div).

wn Weedkiller. Selective weedkiller . (Murphy Chemical Ltd).

wn Weeds Killer. Selective weedkiller. (Fisons PLC, Horticulture Div).

wnsman. Range of lawn aids for garden use such as fertilisers, weedkillers and a spreader. (ICI PLC).*

wnsman Spring Feed. Spring/summer lawn food. (ICI PLC, Plant Protection Div).

wnsman Weed and Feed. Fertilizer combined with selective weedkiller. (ICI PLC, Plant Protection Div).

wnsman Winterizer. Autumn lawn feed. (ICI PLC, Plant Protection Div).

wrowit. Synonym for Lavrovite.†

wsone. . Hydroxynaphthoquinone, $C_{10}H_6O_3$, the colouring matter of henna leaves.†

X-1C. Aloin, P.E. cascara sagrada, P.E. belladonna (total alkaloids 0.04 mg), P.E. rhubarb, ginger. Applications: A laxative for temporary relief of simple constipation. (Pharmaceutical Basics Inc).*

X-2C. Aloin, P.E. cascara sagrada, P.E. belladonna (total alkaloids 0.08 mg), P.E. rhubarb, ginger. Applications: A laxative for temporary relief of simple constipation. (Pharmaceutical Basics Inc).*

X-3. Aloin, podophyllum resin, belladonna extract (total alkaloids 0.1 mg), strychnine sulphate. Applications: A laxative for temporary relief of simple constipation. (Pharmaceutical Basics Inc).*

X-4. White phenolphthalein, P.E. cascara sagrada, pepsin, aloin, podophyllum resin, diastase of malt, ginger. Applications: A laxative for temporary relief of simple constipation. (Pharmaceutical Basics Inc).*

LAX-42. Yellow phenolphthalein, P.E. cascara sagrada, pepsin, aloin, podophyllum resin, diastase of malt, ginger. Applications: A laxative for temporary relief of simple constipation. (Pharmaceutical Basics Inc).*

Laxans. See PURGEN.

Laxatin. See PURGEN.

Laxatol. See PURGEN.

Laxatoline. See PURGEN.

Laxen. See PURGEN.

Laxiconfect. See PURGEN.

Laxin. See PURGEN.

Laxoin. See PURGEN.

Laxophen. See PURGEN.

Laxothalen. A phenol phthalein preparation.†

Laybourn's Stain. A microscopic stain. It consists of two solutions: (a) Toluidine blue 0.15 gram, malachite green 0.2 gram, glacial acetic acid 1 cc, alcohol (95 per cent.) 2 cc, and water 100 cc (b) Iodine 2 grams, potassium iodide 3 grams, and water 300 cc†

Layor Carang. See AGAR-AGAR.

Laysa. Acoustic absorbers. (Carl Freudenberg).*

Laysa Plan. Acoustic absorbers. (Carl Freudenberg).*

Laytex. A proprietary insulating material derived from rubber latex.†

Lazialite. Synonym for Haüyne.†

Lead, Antimony. See HARD LEAD.

Lead Ashes. The skimmings formed during the melting of lead. It consists mainly of oxide.†

Lead, Black. See GRAPHITE.

Lead Bronze. An alloy of from 70-90 per cent copper, 6-16 per cent lead, and 4-13 per cent tin.†

Lead Chamber Crystals. See NITROSYL-SULPHURIC ACID.

Leaded Bronze. An alloy of 88.5 per cent copper, 10 per cent zinc, and 1.5 per cent lead. Another source gives the following proportions: 80 per cent copper, 10 per cent tin, and 10 per cent lead. It has a melting-point of 945° C.†

Leaded Gun Metal. An alloy of 85.5 per cent copper, 2 per cent zinc, 9.5 per cent tin, and 3 per cent lead.†

Leaded Zinc Oxides. Pigments containing 20-35 per cent lead sulphate and 60-80 per cent zinc oxide.†

Lead Flake. See WHITE LEAD.

Lead Glass. See POTASH-LEAD GLASS and FLINT GLASS.

Lead Ochre. Lead monoxide, PbO, found naturally.†

Lead Ore, Black. See ALQUIFON.

Leadoxe. Lead dioxide. Applications: Polysulphide sealant. (Mechema).*

Lead Oxide, Puce. See BROWN OXIDE OF LEAD.

Lead Oxide, Red. See RED LEAD.

Lead, Shot. See SHOT METAL.

Lead Soap. Lead resinate.†

Lead Solder. An alloy of 50 per cent lead and 50 per cent tin used for soldering lead.†

Lead, Tempered. An alloy. It is Noheet metal.†

Lead Tungate. A preparation obtained from lead acetate and tung oil. Used as a drier in the preparation of paints.†

Lead Vinegar. A basic lead acetate, $Pb(C_2H_3O_2)_2.PbO.2H_2O$.†

Lead Water. A 1 per cent solution of basic lead acetate.†

Lead Yellow. Lead chromate, $PbCrO_4$.†

Leaf Green. See CHROME GREEN and OIL GREEN.

Leak Detector. Aqueous special soap solutions. Applications: Bubble forming fluids that detect and indicate leaks in any pressurized gas piping. (Dylon Industries Inc).*

Lean Coal. Coal of the poorest quality.†

Leantin. A proprietary trade name for a bearing alloy of lead and tin.†

Leather Black CT. A colour used for dyeing leather by the chrome process.†

Leather Brown. A dyestuff. Dyes leather and jute brown. Also see BISMARCK BROWN and PHOSPHINE.†

Leather Cloth. See AMERICAN CLOTH.

Leatherlubric. A proprietary trade name for a sulphonated sperm oil.†

Leather Yellow. See PHOSPHINE.

Lebanon No. 34. A proprietary trade na for an alloy of 20 per cent chromium, 30 per cent nickel, 3.25 per cent molybdenum, 5 per cent copper, and 3.25 per cent silicon. It is stated to be resistant to hydrochloric and sulphuri acid.†

Lebanon No. 48. A proprietary trade na for an alloy of 30 per cent chromium, 30 per cent nickel, 0.4 per cent carbo and the remainder iron.†

Lebbin Salt. A proprietary mixture containing nitrite. Used for curing meat.†

Lecin. An iron albuminate, stated to be a easily assimilable form of iron. It contains 20 per cent albumin and 0.6 per cent iron.†

Lecipon. A 10 per cent lecithin in a solu form.†

Lecitase. A commercial preparation of phospholipase A-2 (phosphatide-2-acy hydrolase, E.C.3.1.1.4), manufacture from porcine pancreatic glands. Applications: Used for the hydrolysis lecithins of both animal and vegetable origin for improvement of the emulsifying power. (Novo Industri A/S).*

Lecithan. A trade name for lecithin.†

Lecithcerebrin. Lecithin prepared from brain substance.†

Lecithin. (Yolk Powder). A substance m from egg yolks. It is a compound of choline, glycerol, phosphoric, and various fatty acids. Used as a compon of invalid foods.†

Lecithmedullan. A lecithin preparation from bone marrow.†

Lecithol. An emulsion of brain lecithin containing 1.5 per cent lecithin.†

Leclo. Lead chloride. (Mechema).*

Lectricon. A proprietary trade name for alkyl ammonium compounds disperse in mixed aromatic/aliphatic oils with film strength agents. For electro statically spraying onto form work for the casting of concrete. (British Solve Oils).†

Lectro Cast. A proprietary trade name f an alloy of iron with 2.75 per cent nic and 0.7 per cent chromium.†

cutyl. A combination of lecithin and copper cinnamate, containing 1.5 per cent copper.†

dac. Lead acetate. (Mechema).*

dca. Lead carbonate. (Mechema).*

ddel Alloy. An alloy of 86 per cent zinc, 9.5 per cent antimony, and 4.5 per cent copper. Another alloy used for bearings consists of 87.5 per cent zinc, 6.25 per cent copper, and 6.25 per cent aluminium.†

deburite. Austenite-cementite eutectic.†

debur's Metal. Bearing metals. One contains 85 per cent zinc, 10 per cent antimony, and 5 per cent copper. Another consists of 77 per cent zinc, 17.5 per cent tin, and 5.5 per cent copper.†

edercillin VK. Penicillin V Potassium. Applications: Antibacterial. (Lederle Laboratories, Div of American Cyanamid Co).

edercort. A proprietary preparation of triamcinolone. A steroid. (Lederle Laboratories).*

ederfen. Proprietory preparation containing fenbufen. Applications: Non-steroidal anti-inflammatory. (Lederle Laboratories).*

edermix. Combination dental kit consisting of a paste and cement with democlocyline and triamcinolone acetonide and a hardener. Applications: Dental treatment combining antibiotic and anti-inflammatory agents. (Lederle Laboratories).*

edermycin. A proprietary preparation containing demethylchlortetracycline. An antibiotic. (Lederle Laboratories).*

ederplex. A proprietary preparation of Vitamin B complex. (Lederle Laboratories).*

ederspan. A proprietary preparation of triamcinolone hexacetonide. A steroid injection. (Lederle Laboratories).*

edfo. Lead formate. (Mechema).*

edmin LPC. A proprietary trade name for lauryl pyridinium chloride. An emulsifier for waxes giving bacteriostatic polishes. (Hoechst Chemicals (UK) Ltd).†

edni. Lead nitrate. (Mechema).*

Ledrite Brass. A proprietary trade name for a leaded brass containing 60-63 per cent copper and 2.5-3.7 per cent lead.†

Leefex. A hop defoliant. (Plant Protection (Subsidiary of ICI)).†

Leegen. A blend of a sulphonated petroleum product and selected mineral oil on an inert carier. Applications: An effective dry foam plasticizer and processing aid for NR, SBR, 1R and 11R rubbers. (Vanderbilt Chemical Corporation).*

Lees. Yeast and various suspended matters of the must produced during the fermentation of grape juice.†

Le Farge Cement. See GRAPPIER CEMENT.

Leffmann and Beam's Glycerol Re-agent. For Reichert-Meissl number. It is a mixture of 180 cc pure glycerol and 20 cc of 50 per cent sodium hydroxide solution.†

Legion. Lube oil. (Chevron).*

Legumex Extra. Selective herbicide. (FBC Ltd).

Legupren. A foam system based on Legural (unsaturated polyester resins) and a blowing agent for the system. Applications: For combination with light fillers to produce Legupren based lightweight concrete for use in building construction. (Bayer & Co).*

Leguval. Unsaturated polyester resins for the production of SMC, BMC and DMC etc. Applications: The manufacture of glass reinforced corrugated sheet, light domes and light dome supports, cable distribution boxes, covers, light housings, bulk containers, boats, swimming pools and vehicle bodies. (Bayer & Co).*

Lehner Silk. A similar silk to collodion silk.†

Leinsaat Oils. Linseed oil obtained by extracting the residues from the presses.†

Leipsic Yellow. See CHROME YELLOW.

Leitch'S Blue. See CYANINE.

Leithner'S Biue. See COBALT BLUE.

Lekutherm. A range of epoxy resins and hardeners. Applications: For the production of adhesives, for use in the construction of models, form plates and core boxes used in foundries, for jigs and tool construction. (Bayer & Co).*

Lemarquand's Alloy. An alloy made from 75 per cent copper, 14 per cent nickel, 2.0 per cent cobalt oxide, 1.8 per cent tin, and 7.2 per cent zinc. It is stated to be very resistant to oxidation.†

Lemberg's Solution. A solution of aluminium chloride and extract of log wood. It colours calcite mineral surfaces violet.†

Lembergite. An artificial hydrous aluminium-sodium silicate.†

Lembor. Lemon-scented ceram, light duty handcleanser. Applications: Handcleanser suitable for both factory and office washrooms. (Borax Consolidated Ltd).*

Lemco. A proprietary meat extract.†

Lemery Salt. See SALT OF LEMERY.

Lemery's White Precipitate. See WHITE PRECIPITATE.

Lemnian Earth. (Lemnos Earth). A red, yellow, or grey, earthy substance consisting of a hydrated silicate of aluminium, found at Lemnos. It is an ochre, and is used as a pigment.†

Lemnos Earth. See LEMNIAN EARTH.

Lemol. Lemon oil substitute for flavouring. (Bush Boake Allen).*

Lemolac. A proprietary preparation. It is a very light mercurous chloride.†

Lemonal. A name for linalyl acetate.†

Lemon Balm. Oil. See OIL OF BALM.†

Lemon Chrome. See BARIUM YELLOW.

Lemon Delph. Cleansing milk and skin freshener, for daily care of the skin. (Richardson-Vicks Inc).*

Lemon Essence. Oil of lemon.†

Lemongrass Oil. See OIL OF LEMONGRASS.

Lemon Lac. (Orange Lac, Garnet Lac). Terms referring to the colour of lac, determined to some extent upon the tree from which it is obtained. See LAC.†

Lemon Plus. Throat drops. (Richardson-Vicks Inc).*

Lemon, Salt of. See SALT OF SORREL.

Lemon Yellow. See CHROME YELLOW and ZINC YELLOW.

Lendorm. Brotizolam. Applications: Hypnotic. (Boehringer Ingelheim Pharmaceuticals Inc).

Leneta. Paint testing panels and opacity charts. Applications: Testing of paint and inks. (Cornelius Chemical Co Ltd).*

Lenetol. Textile auxiliary chemicals. (ICI PLC).*

Lenirobin. Chrysarobin tetracetate, used a substitute for chrysarobin.†

Lenium. A proprietary preparation of selenium sulphide and bithionol. A ski cleanser. (Bayer & Co).*

Lenka. Strongly alkaline detergent powde (Shell Chemicals UK Ltd).†

Lensine. Alkali detergent powders. (Shell Chemicals UK Ltd).†

Lentana. A red earthy ironstone of the ty commonly called Laterite.†

Lentard. Insulin zinc. Applications: Antidiabetic. (E R Squibb & Sons Inc

Lente Iletin. Insulin zinc. Applications: Antidiabetic. (Eli Lilly & Co).

Lente Insulin. Insulin zinc. Applications: Antidiabetic. (E R Squibb & Sons Inc

Lenticillin. Procaine penicillin/benzathine penicillin. (May & Baker Ltd).

Lentizol. A proprietary preparation of AMITRYPTYLINE hydrochloride. An anti-depressant. (Warner).†

Lentopen. Penicillin G Procaine. Applications: Antibacterial. (Wyeth Laboratories, Div of American Home Products Corp).

Lenzit. See GYPSUM.

Lenzol. A pale cedar-wood oil of known viscosity and refractive index. Used for oil immersion in microscopy.†

Leo K. A proprietary preparation of potassium chloride used in potassium replacement therapy. (Leo Laboratories).†

Leonil. See NEKAL.

Leophen. Wetting agents for textile treatments. (BASF United Kingdom Ltd).

Lepandin. Highly dispersed pyrogenic silica. Applications: Used for nonslip finishing. (Degussa).*

Lepargylic Acid. See ANCHOIC ACID.

Lepidine. 4-Methylquinoline.†

Lepidone Violet. A dyestuff obtained from methyl-lepidone and phosphorus pentachloride.†

Lepro. Lead peroxide. (Mechema).*

Leptinol. An alcoholic extract of *Leptotaenia dissecta*. It has a therapeutic value.†

Leptovax-Plux. Vaccine. (The Wellcome Foundation Ltd).

Lergoban. Each tablet contains diphenylpyraline hydrochloride BP 5 mg in a porous plastic matrix. (Riker Laboratories/3M Health Care).†

Leridine. The phosphate of Anileridine.†

Leromoll. Plasticizer with minimum volatility and good resistance to alkalis and acids. Applications: For use in combination with unsaponifiable binders in the formulation of chemical-resistant coatings. (Bayer & Co).*

Le Sage Cement. (Plaster Cement). A natural cement obtained from nodules found at Boulogne.†

Lethalbine. A lecithin albuminate, containing 20 per cent lecithin.†

Lethane. Trade mark for insecticide concentrates. Supplied in petroleum distillate. Used in industrial insecticide sprays and mosquito larvicides.†

Lethi. Lead thiosulphate. (Mechema).*

Lethidrone. A proprietary preparation of nalorphine hydrobromide, used to counteract overdosage of morphine and similar drugs. (Allen & Hanbury).†

Lettuce Opium. See LACTUCARIUM.

Leucaniline. Triamino-diphenyl-tolyl methane, $(H_2N.C_6H_4)_2.CH.C_7H_6.$-$NH_2$.†

Leucarsone. 4-Carbaminophenyl arsonic acid. (May & Baker Ltd).*

Leucaurin. Trihydroxytriphenylmethane, $CH(C_6H_4.OH)_3$.†

Leuchtol. A proprietary synthetic resin.†

Leucite-Ferrique. Synonym for Iron leucite.†

Leuco-Malachite Green. (4-Dimethyl-aminophenyl)phenylmethane, $C_6H_5CH(C_6H_4N(CH_3)_2)_2$.†

Leucoargilla. See LITHOMARGE.

Leucobenzaurin. Di-*p*-hydroxy-triphenylmethane, $CH.C_6H_5(C_6H_4OH)_2$.†

Leucogallotbionines. Dyestuffs obtained by condensing alkyldiaminothiosulphonic acids with gallic acid or derivatives.†

Leucogen. Sodium hydrogen sulphite, $NaHSO_3$, used in bleaching and paper making.†

Leucol. An impure quinoline.†

Leucoline. Quinoline.†

Leuconate. Triphenylmethane triisocyanate.†

Leuconine. (Leukonin). An antimony preparation containing 98 per cent of sodium metantimoniate. Recommended as a substitute for tin oxide for enamelling.†

Leucotrope O. Dimethylphenylbenzyl-ammonium chloride. An indigo discharger giving yellow shades on indigo dyed cloth.†

Leucotrope W. The calcium salt of disulphonated Leucotrope O. An indigo discharger, giving yellow colours soluble in alkalis.†

Leucovorin. A proprietary preparation of calcium folinate used as an antidote to methotrexate. (Lederle Laboratories).*

Leukarion. See OIL WHITE.

Leukeran Tablets. A proprietary formulation of chlorambucil. Applications: Treatment of chronic lymphatic (lymphocytic) leukaemia, malignant lymphomas including lymphosarcoma, giant follicular lymphona and Hodgkins' disease. (The Wellcome Foundation Ltd).*

Leukon. A proprietary synthetic resin moulding powder.†

Leukonin. See LEUCONINE.

Leukotrop W. Textile discharging auxiliary. (BASF United Kingdom Ltd).

Leuna Gas. A proprietary trade name for compressed propane.†

Leunaphos. A fertilizer. It is a mixture of phosphate, nitrate, and sulphate of ammonia.†

Leuna Saltpetre. A double salt of ammonium sulphate and nitrate. A fertilizer similar to Chilean nitrate in its action.†

Leutalux. Inorganic and organic fluorescent substances. Applications: Fluorescent screens for electron tubes and fluorescent lamps. (Leuchstoffwerk GmbH).*

Levafil. Dope dyes for polyamide fibres. (Bayer UK Ltd).

Levafix. Reactive dyes. Applications: For dyeing cotton as well as textiles made from regenerated cellulose. (Bayer & Co).*

Levaflex. Thermoplastic rubber. (Bayer UK Ltd).

Levaform. A silicone release agent. Applications: For the rubber, plastics, man-made fibres, metal, food and pharmaceutical industries. (Bayer & Co).*

Levalan N. 1:2 metal complex dyestuffs. Applications: For dyeing wool and polyamide with great colouring strength, good solubility and outstanding fastness properties. (Bayer & Co).*

Levalin. A range of auxilliaries for economical, continuous processes for dyeing wool, cellulosics and synthetic fibres. (Bayer & Co).*

Levamisole. An anthelmintic. (-)-2,3,5,6-Tetrahydro-6-phenylimidazo [2,1-*b*] thiazole. KETRAX.†

Levapon. Textile detergents. (Bayer & Co).*

Levapren. Ethylene/vinyl acetate copolymers. Applications: These include electric wire and cable insulations that must withstand high temperatures or present good flame-retardant behaviour. (Bayer & Co).*

Levasint. A fluidized-bed coating powder comprising saponified ethylene/-vinylacetate copolymer. Applications: Uses include maintenance coatings on metal parts, especially for fencing, cable riggs, traffic furniture and applications in water area. (Bayer & Co).*

Levcarb. Range of non toxic carburizing salt compounds. Applications: Heat treatment of steel. (Leverton-Clarke Ltd).*

Levegal. A range of levelling auxilliaries for dyeing synthetic and cellulosic fibres. Applications: Carriers in polyester dyeing. (Bayer & Co).*

Levelan. Dilutes nonionic surfactants. (Lankro Chemicals Ltd).

Levelan P148. Nonionic surfactant of the nonylphenol ethoxylate type in liquid form. Application: stabilizer for latex. (Diamond Shamrock Process Chemicals Ltd).

Levelan P208. Nonionic surfactant of the nonylphenol ethoxylate type in liquid form. Application: Wetting agent for use at high temperatures; stabilizer for latex; emulsion polymerization aid. (Diamond Shamrock Process Chemicals Ltd).

Levelan P307. Nonionic surfactant of the nonylphenol ethoxylate type in liquid form. Application: Wetting agent at high temperatures. (Diamond Shamrock Process Chemicals Ltd).

Levelan P357. Nonionic surfactant of the nonylphenol type in liquid form. Application: Primary emulsifiers for emulsion polymerization. (Diamond Shamrock Process Chemicals Ltd).

Levepox. Liquid epoxy resins for solvent free coatings. (Bayer & Co).*

Levesol. Solubilizants and fixation auxiliaries for phthalogen dyes. (Bayer UK Ltd).

Levius. A proprietary preparation of sustained-release aspirin. An analgesic. (Farmitalia).*

Levn-Lite. Sodium aluminium phosphate. Applications: Leavening agent. (Monsanto Co).*

Levo-Chrome-Cobalt-Assortment. Complete line of products for the production of Co-Cr-partial dentures (duplicating materials, investments, alloys). (Bayer & Co).*

Levo-Dromoran. Levorphanol tartrate. Applications: Analgesic. (Hoffmann-LaRoche Inc).

Levogen. Aftertreating agents. Applications: For improving the wet fastness properties of dyeings. (Bayer & Co).*

Levogen LF. Amphoteric dyeing auxiliary. Applications: Used in the leather industry to enhance the surface dyeing. (Bayer & Co).*

Levophed. A proprietary preparation of noradrenaline acid tartrate, used to raise blood pressure. (Bayer & Co).*

Levopress. Dental preparation. Applications: Autopolymerisate for dentures, orthodontic appliances, repairs, etc., colour stable. (Bayer & Co).*

Levoprome. Methotrimeprazine. Applications: Analgesic. (Lederle Laboratories, Div of American Cyanamid Co).

Levothroid. Levothyroxine sodium. Applications: Thyroid hormone. (USV Pharmaceutical Corp).

Levoxin. Hydrazine. Applications: Used for corrosion control in water/steam circuits. (Bayer & Co).*

Levuline. A proprietary preparation used in the textile industry for finishing.†

Levulose. (Fruit Sugar, Diabetin, Laevo glucose, Sucro-levulose). Fructose, $C_6H_{12}O_6$.†

Lewasorb. Powdered ion exchange resins regenerated for immediate use. Applications: Used for water purification. (Bayer & Co).*

Lewatit. Cation and anion exchange resins based on polymerization type synthetic resins of differing mesh size and/or macroporous structure and with active groups of various acidities and basicities or selective properties. Applications: Water treatment, waste water treatment, adsorption, decolourizing, separation, decontamination, catalysis, nutrient carrier for hydroponics. (Bayer & Co).*

Lewisite. A military poison gas. It is β-chlorovinyldichloroarsine, CHCl : $CH.AsCl_2$. The name Lewisite is also used for a mineral, $5CaO.2TiO_2.3SbO_3$.†

Lewis Metal. An alloy of 1 part tin and 1 part bismuth having the property of expanding when cooling. It has a melting-point of 138° C. and is used for sealing and holding die parts.†

Lewisol 28. Maleic-modified glycerol ester of rosin. Applications: Used for Type C gravure inks and for nitrocellulose-based paper and wood coatings. (Hercules Inc).*

Lexan. A registered trade name for polycarbonate resins. Used for electrical moulding purposes where good transparency is required. 92 per cent. transparency and a refractive index of 1.58 are obtained. Very high creep resistance is exhibited under load. (General Electric).†

Lexan. BPA Polycarbonate. Applications: Plastic components for automotive, electrical, electronics, lighting, medical, packaging, audio etc. (GE Plastics Ltd).*

Lexan LS. A proprietary polycarbonate resin used for moulding lenses. (Canadian General Electric Co, Ontario).†

Lexan RP 700. A proprietary polycarbonate resin used in the rotational moulding of containers and road lighting fittings. (Canadian General Electric Co, Ontario).†

Lexan 130. Proprietary name for a range of high-viscosity polycarbonates. (Canadian General Electric Co, Ontario).†

Lexan 140. A proprietary range of low viscosity polycarbonates used in the moulding of intricate parts. (Canadian General Electric Co, Ontario).†

Lexan 141R. A proprietary polycarbonate resin used in injection moulding. It possesses intrinsic properties which facilitate mould-release. (Canadian General Electric Co, Ontario).†

Lexan 145. A proprietary free-flowing granular polycarbonate used in the casting of films from solution and for dip-coating. (Canadian General Electric Co, Ontario).†

Lexan 3312. A proprietary polycarbonate resin containing 20% glass reinforcement. (Canadian General Electric Co, Ontario).†

Lexan 3314. A proprietary polycarbonate resin containing 40% glass reinforcement. (Canadian General Electric Co, Ontario).†

Lexan 700. A proprietary polycarbonate resin used in rotational moulding.

(Canadian General Electric Co, Ontario).†

Lexan 700 SEl and SE2. Self-extinguishing grades of LEXAN 700. (Canadian General Electric Co, Ontario).†

Lexel. A registered trade mark for a cellulose acetate-butyrate insulating tape. It is flame retardant, has low moisture absorption and has a high dielectric strength.†

Lexite Granular Carpet. All-solids epoxy fortified with quartz granules. Applications: For providing a tough wearing surface over industrial and commercial floors to make them both skidproof and reasonably easy to clean. (Metalcrete Mfg Co).*

Lexite 100. 100% solids epoxy. Applications: Used as a floor surfacer where housekeeping and long range freedom from replacement are primary requisites. Produces a tough, 16 mil film in a single application. (Metalcrete Mfg Co).*

Lexotan. Proprietary preparation of bromazepam. Benzodiazepine with anxiolytic properties. (Roche Products Ltd).*

Lextron. A proprietary preparation of liver extract, and ferric ammonium citrate. (Eli Lilly & Co).†

Ley-Cornox. Selective weedkillers. (Boots Pure Drug Co).*

Leyden Blue. See COBALT BLUE.

LG Wax. A montan wax derivative used in the production of non-ionic and ionic dry-bright emulsions for floor polishes and similar applications. (Bush Beach Ltd).†

Libavius' Fuming Spirit. A solution of stannic chloride, $SnCl_4$.†

Libigen. Gonadotropin, chorionic. Applications: Gonad-stimulating principle. (Savage Laboratories, Div of Byk-Gulden Inc).

Libollite. A variety of asphaltum.†

Librax. See LIBRAXIN. (Roche Products Ltd).*

Libraxin. A proprietary preparation of chlordiazepoxide, and clidinium bromide. A gastro-intestinal sedative. (Roche Products Ltd).*

Librium. A proprietary preparation of chlordiazepoxide hydrochloride. A sedative. (Roche Products Ltd).*

Licella Yarn. A product made from narrow strips of wood-pulp paper. A jute substitute.†

Lichen Blue. See LITMUS.

Lichen Starch. Lichenin, $C_6H_{10}O_5$. A carbohydrate derived from Iceland moss.†

Lichen Sugar. Erythritol, $C_4H_6(OH)_4$.†

Lichner's Blue. A variety of smalt. It is a silicate of cobalt and potassium.†

Lichtenberg's Metal. An alloy of 50 per cent bismuth, 30 per cent. lead, and 20 per cent tin, melting at 91.5°C.†

Licomer. Polyethylene primary dispersions. (Hoechst UK Ltd).

Liconite. A rubber substitute made from bitumen and oils.†

Licowet. Fluorinated surfactants. (Hoechst UK Ltd).

Licryl-55. Licryfilcon A. 2-Hydroxyethyl methacrylate polymer with ethylene dimethacrylate. The material contains 55% water as 0.9% saline. Applications: Contact lens material. (Liquid Crystal Lens Co).

Licryl-70. Licryfilcon B. 2-Hydroxyethyl methacrylate polymer with ethylene dimethacrylate. The material contains 70% water as 0.9% saline. Applications: Contact lens material. (Liquid Crystal Lens Co).

Licuado Instante. Nutritional additive to milk. (Richardson-Vicks Inc).*

Lida-Mantle. Lidocaine. Applications: Anaesthetic. (Miles Pharmaceuticals, Div of Miles Laboratories Inc).

Lidarral. Lidamidine hydrochloride. Applications: Antiperistaltic. (William H Rorer Inc).

Lidex. Fluocinonide. Applications: Glucocorticoid. (Syntex Laboratories Inc).

Lidocaine. A proprietary trade name for Lignocaine.†

Lieben Solution. A solution of iodine in potassium iodide.†

Lieberkuhn's Jelly. The jelly formed by mixing egg serum with about one third of its volume of twice normal sodium hydroxide.†

ebmann and Studer's Acid. 1-Naphthol-7-sulphonic acid (1:7).†

fe. Oil analysis services. (Chevron).*

gdynite. A coal mine explosive consisting of 25-27 per cent nitro- glycerin, 27-29 per cent sodium nitrate, 10-12 per cent sodium chloride, and 30-33 per cent wood meal.†

ght Blue. See METHYL BLUE and ANILINE BLUE, SPIRIT SOLUBLE.

ight Carburetted Hydrogen. See MARSH GAS.

ight Fast Yellow 3G. A dyestuff belonging to the same class as Tartrazine.†

ight Green. See METHYL GREEN and IODINE GREEN.

ightguard. A line of extruded, expanded polystyrene foam insulation products having a protective coating on one side. (Dow Chemical Co Ltd).*

ighthouse Chrome Blue B. A dyestuff. It is a British equivalent of Chromotrope 2R.†

ighthouse Chrome Brown R. A dyestuff. It is a British brand of Fast brown N.†

ighthouse Chrome Cyanine R. A British dyestuff. It is equivalent to Palatine chrome blue.†

ighthouse Chrome Yellow 0. A British equivalent of Milling yellow.†

ightning Powder. A double base smokeless rifle powder.†

ight Red. A pigment prepared by calcining Oxford or yellow ochre.†

ight Spar. See GYPSUM.

ight Water Foam. Fluorochemical surfactants in solution. Applications: Premium grade fire fighting foam proportioned in water, used for flammable liquid fire fighting and protection. Used in Expro, crash fire rescue and major risk industries. (3M United Kingdom plc).†

ight White. See OIL WHITE.

ignin Dynamites. Explosives. They consist of nitro-glycerin absorbed by a mixture of wood pulp, and a nitrate, usually sodium nitrate.†

Lignite. Brown coal. It is of a more recent date than true coal and is intermediate between coal and peat.†

Lignite Char. See GRUDEKOK.

Lignite Wax. (Mineral Wax). See MONTAN WAX.

Lignorosin. Calcium lignosulphonate, a by-product in the manufacture of paper pulp. It is a dark-brown syrup, and is used as an assistant in mordanting wool with chrome.†

Lignosol. Calcium and sodium lignosulphonates. Applications: Product line of specialty chemicals for use in a variety of binding, dispersing, complexing, stabilizing or copolymerizing applications. (Reed Lignin Inc).*

Lignostab. Lignocaine hydrochloride injection. (The Boots Company PLC).

Lignostone. See PERMALI.

Lignum Vitae. Guaiacum wood.†

Ligro. A proprietary trade name for a crude pine fatty acid mixture.†

Ligroin. A term rather loosely applied. It usually denotes a refined distillate of petroleum oil having a boiling-point of 120-135° C., of specific gravity 0.73. Used as a polishing oil, and as a turpentine substitute in varnishes.†

Ligulin. The dyestuff of privet berries.†

Ligurite. A mineral. It is a variety of sphene.†

Lilacin. (Lilicin, Terpilenol). Terpineol, $C_{10}H_{17}OH$, a terpene alcohol.†

Lilaflot. Formulated cationic flotation agents based on primary fatty amines, primary ether amines, the corresponding diamines and quaternary ammonium compounds. Applications: Mineral flotation for potash; feldspar; quartz; iron, sulphide and phosphate ores; scheelite. (Keno Gard (UK) Ltd).

Lilamin. Range of cationic surfactants in liquid form, developed as anti-caking agents. Applications: Used as anti-caking agents in fertilizers, either alone or with small amounts of powders as parting agents. (Keno Gard (UK) Ltd).

Lilaminox. Foam boosters. Applications: Detergent formulations; shampoos; bubble baths. (Keno Gard (UK) Ltd).

Lilaminox. Foam boosters. Applications: Detergent formulations; shampoos; bubble baths. (Lilachim SA).

Lilaminox 100. A proprietary trade name for a bis-(2- hydroxyethyl)alkylamine oxide of the general formula $R(C_2H_4OH)_2N[RAW]O$ where R is lauryl, myristyl or cetyl. Used similarly to Lilaminox 10. (Guest Industrials).†

Lilicin. See LILACIN.

Lilion 6. Nylon 6 monofilament/staple fibre. Applications: Sewing threads, carpets, fur fabrics. (SNIA (UK) Ltd).†

Lilion 66. Nylon 66 monofilament. Applications: Sewing threads, carpets, fur fabrics. (SNIA (UK) Ltd).†

Lillite. An earthy mineral resembling Glauconite, of Bohemia.†

Lily of Valley, Artificial. Terpineol, $C_{10}H_{17}OH$.†

Lilyolene. See PARAFFIN, LIQUID.

Limanol. A preparation of salt-marsh mud used for rheumatism.†

Lima Wood. See REDWOODS.

Limbachite. A Serpentine mineral.†

Limbaki. Stretch velour fabric. (ICI PLC).*

Limbitrol. A proprietary preparation of amitriptyline and chlordiazepoxide. An anti-depressant. (Roche Products Ltd).*

Limbux. A registered trade name for a form of mechanically slaked lime. It is used in agriculture, building and construction. metallurgical and chemical industries. (ICI PLC).*

Lime. (Burnt Lime, Quicklime, Caustic Lime). Calcium oxide, CaO, produced by calcining calcium carbonate. Sometimes a mixture of calcium and magnesium oxides is sold under this name.†

Lime-Arsenic Greens. See SODA GREENS.

Lime Blue. A pigment. It consists of a mixture of copper hydroxide and lime, and is prepared by precipitating a solution of copper sulphate with milk of lime, with the addition of ammonium chloride. The name Lime blue is also applied to the lime fast coal-tar colour lakes, and to ultra-marine cheapened by the addition of gypsum.†

Lime, Burnt. See LIME.

Lime, Caustic. See LIME.

Lime Chrome. Calcium chromate, $CaCrO_4$.†

Lime Green. (Earth Green). A pigment obtained by precipitating copper sulphate solution with chalk or milk of lime. Also see GREEN ULTRA-MARINE.†

Limeite. A cement containing rubber, tallow, and lime. The addition of vermilion causes the mixture to harden.†

Lime Mortar. Mixtures of slaked lime and sand.†

Lime Nitrate. (Lime Saltpetre). Calcium nitrate, $Ca(NO_3)_2.2H_2O$. A fertilizer.†

Lime Nitrogen. See NITROLIM.

Limeolivine. Calcium orthosilicate Ca_2SiO_4.†

Lime, Pyrolignite. See PYROLIGNITE OF LIME.

Lime Saltpetre. See LIME NITRATE.

Lime Soap. Calcium resinate.†

Lime-Sulphur Dips. Sheep dips for the treatment of scab. They contain flowers of sulphur, lime, and water.†

Lime Water. A saturated solution of calcium hydroxide, $Ca(OH)_2$, in water.†

Limex. Sodium phosphate and metal preservatives. Applications: Used in water using equipment where lime scale and corrosion are a problem. (Delaware Chemical Corp).*

Limex 'G'. Sodium hexameta phosphates. Applications: Used in water treatment to sequester calcium minerals. (Delaware Chemical Corp).*

Limit. N-(Acetylamino) methyl-2-chlor-N-2,6-diethyl-phenylacetamide. Applications: Turf grass regulator. (Monsanto Co).*

Limit 33. Defoamer for paper industry. Applications: Used at the size press to defoam starch and mixtures of starch. (Monsanto Co).*

Limo. (Sablon, Tartrate of Lime). The raw materials obtained by the precipitation of tartaric acid in tartar works or wine distilleries.†

Limpetite. Air cured liquid applied synthetic rubber. Applications:

Protection of items subject to extreme conditions of corrosion, erosion and electrolytic action. (Protective Rubber Coatings (Limpetite) Ltd).*

nalux. Gloss paint finishes for marine use. (ICI PLC).*

naqua. Water thinnable linseed oil. Applications: Paint. (NL Industries Inc).*

ncocin. Lincomycin hydrochloride. Applications: Antibacterial. (The Upjohn Co).

ncomycin. An antibiotic produced by Streptomyces lincolnensis var. lincolnensis. Lincocin and Mycivin are the hydrochloride.†

nctifed. Expectorant. (The Wellcome Foundation Ltd).

nctified Expectorant. A proprietary preparation of Triprolidine, Pseudoephrine, Codeine phosphate and Guaiphenesin. An expectorant. (The Wellcome Foundation Ltd).*

indenol. Pure alpha terpineol. (Bush Boake Allen).*

indesite. Synonym for Urbanite.†

inearil. A linear alkylate. (Montedison UK Ltd).*

inen. A fabric manufactured from flax fibre.†

inevol Phthalates. These are plasticizers based on straight chain Linevol alcohols. They are characterized by exceptionally low volatility and excellent low temperature properties. A range of three phthalates allows a versatile plasticizing performance. (Shell Chemicals).†

inevol 79. Blend of primary alcohols, the notable characteristic being a high straight chain alcohol content. It is readily esterified by conventional means to give a high quality plasticizer. (Shell Chemicals).†

inevol 911. Blend of primary alcohols, the notable characteristic being a high straight chain alcohol content. It is readily esterified by conventional means to give a high quality plasticizer. (Shell Chemicals).†

ingraine. A proprietary preparation of ergotamine tartrate, used for migraine. (Bayer & Co).*

Lining Metal. Alloys of 70-90 per cent lead, 2-20 per cent tin, and 5-20 per cent antimony. They are used for bearings.†

Linituss. A proprietary preparation of linseed, liquorice and chlorodyne. Applications: A cough linctus. (Ayrton Saunders plc).*

Linklon. Crosslinkable polyolefin compounds. (Mitsubishi Petrochemical Co).*

Linklon-X. Crosslinkable polyolefin compounds. (Mitsubishi Petrochemical Co).*

Linnaeite. See LINNALITE.

Linnalite. (Linnaeite). Native cobalt sulphide, Co_3S_4.†

Linnets. See BROWN LEAD ORE.

Linocin. A proprietary preparation containing lincomycin hydrochloride. An antibiotic. (Upjohn Ltd).†

LinoCure. An alkyd oil urethane no-bake binder system. Applications: utilized in the production of foundry cores and moulds as a resin binder system. (Ashland Chemical Company).*

Linotype Metal. An alloy of 13.5 per cent antimony, 2 per cent tin, and 84.5 per cent lead. See also TYPE METAL.†

Linoxyn. See SOLIDIFIED LINSEED OIL.

Linseed Meal. Ground oil-cake.†

Linseed Oil, Artist's. Raw linseed oil which has been allowed to stand for weeks, then treated with litharge, and finally bleached by exposure.†

Linseed Oil, Refined. Raw linseed oil which has been treated with a 1 per cent solution of sulphuric acid.†

Linseed Oil Soap. A potashsoap.†

Linsol. Dyestuff solutions in oleic acid. (Williams (Hounslow) Ltd).

Linter's Starch. A soluble starch prepared by mixing raw starch with 7.5 per cent hydrochloric acid in water, allowing it to stand for several days, with stirring. The solution is decanted, and the starch washed with water and dried.†

Lintex A10. A proprietary aqueous emulsion of a styrene copolymer internally plasticized with a copolymerized ester. It contains cross linkable groups and is used as a binder for emulsion paints. (Chemische Werke Hüls).†

Linurex. Active ingredient: linuron; 3-(3,4-dichlorophenyl)-1-methoxy-1-methylurea. Applications: Selective herbicide for both pre- and post-emergence application. (Agan Chemical Manufacturers Ltd).†

Lipal. A proprietary name for a range of polyoxyethylene esters and ethers. (PVO International Inc).†

Lipases. Enzymes which split up fats.†

Lipinol O. An octyl fatty acid ester. Applications: Viscosity depressant especially for rotational moulding plastisols. (Huls AG).†

Lipinol T. Dibenzyltoluene. Applications: Secondary plasticizer, lowers the viscosity of PVC plastisols to a lesser degree than LIPINOL O. (Huls AG).†

Lipiodul Ultra-Fluid. A proprietary preparation of iodized oil fluid injection. Applications: X-ray contrast medium. (May & Baker Ltd).*

Lipivas. Halofenate. Applications: Antihyperlipoproteinemic; uricosuric. (Merck Sharp & Dohme, Div of Merck & Co Inc).

Lipoclor. Chlorinated natural fats. Applications: Additives for lubricating and cutting oils and raw material for the preparation of greasings for leather. (Caffaro Industrial).*

Lipoclor S. Chorinated animal fats. (SNIA (UK) Ltd).†

Lipoflavonoid. A proprietary preparation of choline bitartrate, inositol, dimethionine, ascorbic acid, lemon bioflavnoid, thiamine, riboflavine, nicotinamide, pyridoxine, panthenol and cyanocobalamin. Used for neurosensory deafness. (Lewis Laboratories).†

Lipo Gantrisin. Sulphisoxazole Acetyl. Applications: Antibacterial. (Hoffmann-LaRoche Inc).

Lipo-Hepin. Heparin sodium. Applications: Anticoagulant. (Riker Laboratories Inc, Subsidiary of 3M Company).

Lipolan. Carboxylated styrene/butadiene copolymer dispersions. Applications: Binders for carpet backing, textile coating, fabric lamination, needle felt material, nonwoven fabrics, shoe toe-caps and other textile finishes imparting a firm consistency. (Huls AG).†

Lipo-Lutin. Progesterone. Applications: Progestin. (Parke-Davis, Div of Warner-Lambert Co).

Lipotriad. A proprietary preparation of tricholine citrate, inositol, DL-methionine, cyanocobalamin, thiamine hydrochloride, riboflavine, nicotin amide, pyridoxine hydrochloride and pantothenyl alcohol. (Lewis Laboratories).†

Lipowitz's Alloy. An alloy of 50 per cent bismuth, 27 per cent lead, 13 per cent tin, and 10 per cent cadmium. It has a specific heat of 0.0345 calories per gram at from 5-50° C., and melts at 65° C. It is used for automatic sprinklers and other purposes.†

Lipozyme. An experimental preparation of an *immobilized lipase*. The fungal lipase is produced by submerged fermentation of a selected strain of *Mucor miehei*. Applications: Lipozyme can be used for interesterification, alcoholysis of oils and fats and synthesis of esters. (Novo Industri A/S).*

Liquaemin sodium. Heparin sodium. Applications: Anticoagulant. (Organon Inc).

Liquamar. Phenprocoumon. Applications: Anticoagulant. (Organon Inc).

Liquamycin. Tetracycline. Applications: Anti- amebic; antibacterial; antirickettsial. (Pfizer Inc).

Liquapen. Penicillin G Potassium. Applications: Antibacterial. (Pfizer Inc).

Liquemin. Proprietary preparation of heparin. Anticoagulant. (Roche Products Ltd).*

Liqueur de Ferraile. See PYROLIGNITE OF IRON.

Liqueur de van Swieten. Consists of 1 part mercuric chloride, 100 parts alcohol, and 900 parts water.†

Liquibor. Organic borate. (Manchem Ltd)

Liquibor 169. Potassium glycol borate. Applications: Corrosion inhibitor for glycol based brake fluids. (Manchem Ltd).*

Liquibor 524. A borax-glycol condensation product which is used at a concentration of 1-2 per cent as a corrosion inhibitor in synthetic hydraulic fluids.†

Liqui-Cee. Sodium Ascorbate. Applications: Vitamin. (American Critical Care, Div of American Hospital Supply Corp).

Liquidambar. (Copalm Balsam). A balsam obtained from a large Mexican tree. It contains cinnamyl cinnamate and styrene. It is also erroneously called liquid storax.†

Liquid Bases. Compounded lanoline derived sterols. (Croda Chemicals Ltd).

Liquid, Bleaching. See BLEACH LIQUOR.

Liquid Bronzes. Varnishes in which bronze colours are suspended.†

Liquid Chloride of Lime. See BLEACH LIQUOR.

Liquid Copper Fungicide. Fungicide. (Murphy Chemical Ltd).

Liquid Drier. See TEREBINE. Also a name given to a concentrated solution of calcium chloride.

Liquid Gold. Contains about 10 per cent of gold (as chloride), resin, lavender oil, and bismuth. Used for painting china.†

Liquid Lightning. Inhibited sulphamic acid - liquid. Applications: Removes water formed deposits. (Western Chemical Co).*

Liquidow. Liquid calcium chloride. Applications: For applications on unpaved roads to control dust. (Dow Chemical Co Ltd).*

Liquid Pitch Oil. See OIL OF TAR.

Liquid Resins. (Polyterpene, Sulphate Resin, Talleol). Semi-resinous compounds obtained as by-products in the manufacture of wood pulp by the sulphite and sulphate processes of paper making.†

Liquid Storax. Storax, a balsam. Also see LIQUIDAMBAR.†

Liquinure. Liquid fertilizer. (Fisons PLC, Horticulture Div).

Liquor, Glass. See SOLUBLE GLASS.

Liquorice. The dried root of *Glycyrrhiza glandulifera*.†

Liquor, Iron. See PYROLIGNITE OF IRON.

Liskonum Tablets. Lithium carbonate. Applications: Controlled release tablet for treatment of acute episodes of mania or hypomania and for the prophylaxis of recurrent manic-depressive illness. (Smith Kline and French Laboratories Ltd).*

Lissamine. Acid and direct dyestuffs. (ICI PLC).*

Lissamine Green B. See WOOL GREEN S.

Lissanol. Leather finishes. (ICI PLC).*

Lissapol. Synthetic detergents. (ICI PLC).*

Listerine. A proprietary antiseptic containing boric acid, benzoic acid, thymol, and essential oils of eucalyptus, gaultheria, and others.†

Litalbin. Lecithin albuminate.†

Litefax. Low density insulating sideliner tiles for small to medium steel ingots. (Foseco (FS) Ltd).*

Litex. A range of copolymer emulsions and dispersions. Applications: Used for paints and anticorrosion coatings. (Huls AG).†

Lithane. A proprietary preparation of lithium carbonate. An antidepressant. (Pfizer International).*

Litharge. (Massicot). Pigments consisting of lead monoxide, PbO. Litharge is obtained in silver refining, and has a more reddish colour than Massicot, which is made by roasting lead.†

Lithargrite. A mixture of oxide of lead and calcined magnesia. Used as a rubber filler.†

Lithex. A proprietary rubber vulcanization accelerator. It is lead dithio-benzoate precipitated on an inert base such as clay.†

Lithic Acid. Uric acid, $C_5H_4N_4O_3$.†

Lithionpsiilomelane. Synonym for Lithiophorite.†

Lithiopiperazine. A preparation of piperazine and lithium salts. A solvent for uric acid.†

Lithobid. Lithium carbonate. Applications: Antimanic. (Ciba-Geigy Corp).

Litho-Carbon. A material resembling asphalt, found in Texas.†

Lithoclastite. An explosive. It is a dynamite.†

Lithocolla. See LITHOMARGE.

Lithoform. A pretreatment for zinc. (ICI PLC).*

Lithographer's Varnish. See STAND OIL.

Lithographic Stones. Limestones prepared and polished.†

Lithol. Ammonium ichtho-sulphonate.†

Lithol Fast Scarlet R. See HELIO FAST RED RL.

Lithol Rubine B. (Permanent Red 4B). A dyestuff obtained from *p*-toluidine-*o*-sulphonic acid andβ-hydroxynaphthoic acid.†

Lithomarge. (Lithocolla, Leucoargilla). A compact clay found in rock fissures.†

Lithonate. Lithium carbonate. Applications: Antimanic. (Rowell Laboratories Inc).

Litho-oil. (Stand oil, polymerized oil). Raw linseed oil heated in such a manner that practically no oxidation occurs, thus thickening the oil through polymerization alone.†

Lithophone. See LITHOPONE.

Lithopone. (Beckton White, Duresco, Enamel White, Fulton White, Griffith's White, Jersey Lily White, Knight's Patent Zinc White, Lithophone, Marbon White, Nevin, Orr's White, Oleum White, Pinolith, Porcelain White, Ross's White, Zinc Baryta White, Zincolith, Zinc Sulphide White, Sulphide White). A pigment. It is a mixture of barium sulphate and zinc sulphide, with some zinc oxide, containing from 11-42 per cent zinc sulphide.†

Lithopone, Titanium. See TI-TONE.

Lithostar. Reprographic chemical. (Ward Blenkinsop & Co Ltd).

Lithostat. Acetohydroxamic acid. Applications: Enzyme inhibitor. (Mission Pharmacal Co).

Lithotabs. Lithium carbonate. Applications: Antimanic. (Rowell Laboratories Inc).

Lithyol. See ICETHYOL.

Litmus. (Lichen Blue, Lacmus, Tournesol, Turnsole). The colouring matter derive from different species of lichens. Used as an indicator. Also see ARCHIL.†

Litnum Bronze. An alloy of from 80-85 pe cent copper, 10-15 per cent aluminium and 4 per cent iron.†

Lito-Silo. See SOREL CEMENT.

Litrison. A proprietary multivitamin preparation containing vitamins B1, B2 B6 and B12, methionine, choline tartrate, dexpanthenol, biotin, folic aci and vitamin E. Used in liver disease. (Roche Products Ltd).*

Littoral. A decolorising agent for sugar juices.†

Liver of Antimony. The name applied to t impure double sulphides of antimony, obtained by heating antimony sulphide, Sb_2S_3, with various metallic sulphides, more especially with the alkali and alkaline earth sulphides.†

Liver of Sulphur. See SULPHURATED POTASH.

Liveroid. A proprietary liver extract.†

Liver Ore. See CINNABAR.

Liver Sugar. Glycogen.†

Lloyd's Reagent. A hydrous aluminium silicate prepared from fuller's earth. It absorbs alkaloids.†

Inconel Alloy 718. A trade mark for an alloy of 19 per cent. chromium, 3 per cent. molybdenum, 0.8 per cent. titanium, 5 per cent. niobium, 53 per cent. nickel and the balance iron. (Wiggin Alloys Ltd).†

Lo-ex. An aluminium alloy containing 14 per cent silicon, 2 per cent nickel, 1 per cent copper, and 1 per cent magnesium

Loadstone. See MAGNETIC IRON ORE

Loalin. A proprietary trade name for a polystyrene moulding compound.†

Loams. Natural mixtures of clay and sand Used for the manufacture of bricks and tiles.†

Lobak. A proprietary preparation of chlormezanone, and paracetamol. (Bayer & Co).*

Lobosol. Oxygenated solvents. (BP Chemicals Ltd).*

ocan. A proprietary preparation of amethocaine, amylocaine, bismuth subnitrate and zinc oxide. Anaesthetic anal suppositories. (British Drug Houses).†

ocasol. A proprietary milk preparation with a low content of calcium. (Cow and Gate).†

ocke's Solution. A saline solution containing dextrose and used for injections.†

ockite. Polyolefine. Applications: Continuous extruded strip high slip fendering for docks and similar applications. (Stanley Smith & Co Plastics Ltd).*

ocobase Cream and Ointment. A proprietary cream and ointment base. Applications: A base for creams and ointments. (Brocades).*

ocoid. Hydrocortisone butyrate. Applications: Glucocorticoid. (Owen Laboratories Div, Dermatological Products of Texas, Alcon Laboratories Inc).

ocorten. A proprietary preparation of flumethasone pivalate for dermatological use. (Ciba Geigy PLC).†

ocorten-N. A proprietary preparation of flumethasone pivalate and neomycin sulphate used in dermatology as an antibacterial agent. (Ciba Geigy PLC).†

ocoum. A gum-like mass prepared from starch paste and sugar.†

ocron P. Aluminium chlorohydrate. Applications: Anhidrotic. (American Hoechst Corp).

octite. A trade mark. Single component structural adhesives and thread locking materials possessing the unique property of setting when air is excluded. Some are based on oxygenated methacrylic molecules of patented formulation.†

Locust Bean Gum. See INDUSTRIAL GUM.

Lodestone. See MAGNETIC IRON ORE.

Lodine. Etodolac 200 mg capsules. Applications: Acute or long-term use in rheumatoid arthritis. (Ayerst Laboratories).†

Lodol. Dephosphorising agent for ladle treatment of rimming steels. (Foseco (FS) Ltd).*

Lodosin. Carbidopa. Applications: A decarboxylase inhibitor for concurrent use with levodopa in the treatment of Parkinson's disease. (Merck Sharp & Dohme).*

Lodosyn. Carbidopa. Applications: A decarboxylase inhibitor for concurrent use with levodopa in the treatment of Parkinson's disease. (Merck Sharp & Dohme).*

Loestrin 20. A proprietary preparation of ethinyloestradiol and norethisterone acetate. An oral contraceptive. (Parke-Davis).†

Lofenalac. Proprietary name for an infant milk feed low in PHENYLALANINE, used in the dietary treatment of phenylketonuria. (Bristol-Myers Co Ltd).†

Löffler Stain for Flagella. To 10 cc of a 20 per cent solution of tannin are added 5 cc of a cold saturated solution of ferrous sulphate and 1 cc of a solution of fuchsine or methyl violet.†

Löffler's Methylene Blue. A stain for bacteria. It consists of 100 parts of a solution of sodium hydroxide (1 in 10,000) and 30 parts of a saturated alcoholic solution of methylene blue.†

Loftine. See MULSOID.

Lofton Merritt's Stain. (*a*) A solution of 2 grams of malachite green in 100 cc water. (*b*) A solution of 1 gram of basic fuchsine in 100 cc water. The test solution contains 1 part of (*a*) and 1 part of (*b*), and is used to distinguish between unbleached sulphate and unbleached sulphite fibres, the former giving a blue or blue-green colour, the latter a purple or lavender.†

Logas. For degassing copper base alloys. (Foseco (FS) Ltd).*

LoGel. Aluminium stearate. Applications: Non-aqueous coating compositions. (Nueodex Inc).*

Logwood. (Campeachy Wood, Jamaica Wood, Hematine Paste and Powder, Steam Black). A natural dyestuff from the wood of *Haematoxylon* *campechianum*. It is sold as chips or extract. The wood contains haematoxylin, $C_{16}H_{14}O_6.3H_2O$, the dyeing principle, which is converted into haematin, $C_{16}H_{12}O_6$, the dyestuff, by oxidation. Used for dyeing black with a chrome mordant, for wool; with an iron mordant for silk, and with a chrome, or an iron and aluminium mordant, for cotton.†

Lohys Steel. A mild steel having a high magnetic permeability.†

Lokain. Lokaonic acid.†

Lokandi. See VENTILAGO MADRA-SPANTA.

Lokas. See SAP GREEN.

Lomag. Magnesium removing flux. (Foseco (FS) Ltd).*

Loman Steel. An abbreviation of "low-manganese steel." It contains from 7-10 per cent manganese.†

LO/MIT-1. Proprietary. Applications: Silver coloured, non-thickness dependent, low emissivity coating. Light and heat reflection. High temperature tolerant coating. (Solec - Solar Energy Corporation).*

Lomodex. A proprietary preparation of dextrans in saline or dextrose solution, used to improve capillary circulation. (Fisons PLC).*

Lomotil. A proprietary preparation of DIPHENOXYLATE hydrochloride and atropine sulphate. An anti-diarrhoeal. (Searle, G D & Co Ltd).†

Lomotil with Neomycin. A proprietary preparation containing diphenoxylate hydrochloride, atropine sulphate and neomycin sulphate. An anti- diarrhoeal. (Searle, G D & Co Ltd).†

Lomudase. A proprietary preparation of chymotrypsin, and isoprenaline sulphate. Treatment of bronchitis. (Fisons PLC).*

Lomupren. A proprietary preparation of isoprenaline sulphate. A bronchial antispasmodic. (Fisons PLC).*

Lomusol. A proprietary preparation of sodium cromoglycate as a nasal spray in the treatment of allergic rhinitis. (Fisons PLC).*

Lonarit. An acetyl cellulose product. It can be moulded and coloured.†

Londal. A proprietary trade name for certain aluminium alloys.†

London Blue Extra. See SOLUBLE BLUE.

London Paste. A paste made by adding a third of the weight of water to a mixture of equal parts of sodium hydroxide and powdered lime. A caustic.†

London White. See FLAKE WHITE.

Longifene. The hydrochloride of buclizine.†

Longlife Turf Foods. Fertilizers containing longlasting nitrogen for sports grounds and parks. (Scottish Agricultural Industries PLC).

Long Oil Varnishes. A classification of varnishes. They contain about 1 part of the solid constituents to 1 1/2 parts of drying oil. Short oil varnishes contain 1 part of solid to 1/2 part oil.†

Loniten. Minoxidil. Applications: Antihypertensive. (The Upjohn Co).

Lonsicar. Silicon carbide. Applications: Used for anti-slip and wear resistant concrete or resin surfaces. (Lonza Limited).*

Lontrel. Weedkiller containing 3,6-dichloropicolinic acid, dichlorprop and MCPA as potassium salts. (ICI PLC).*

Lontrel. Herbicides based primarily on clopyralid. (Dow Chemical Co Ltd).*

Lontrel Plus. Selective herbicide. (ICI PLC, Plant Protection Div).

Lonzaine. Betaine amphoterics. Applications: Used in shampoos and industrial cleaners. (Lonza Limited).*

Lonzest. A range of sorbitan esters, polyoxyethylene sorbitan esters and polyethylene glycol esters. Applications: Used for emulsion formation and stabilization in foods, pharmaceuticals and cosmetics. (Lonza Limited).*

Lonzoid. A proprietary cellulose acetate.†

Lopid. Gemfibrozil. Applications: Antihyperlipoproteinemic. (Parke-Davis, Div of Warner-Lambert Co).

Lopox. A registered trade mark for epoxy resins.†

Lopresor. A proprietary preparation of METOPROLOL tartrate used in the treatment of angina pectoris. (Ciba Geigy PLC).†

Lopresoretic. Beta-blocker. (Ciba-Geigy PLC).

Lopressor. Metoprolol tartrate. Applications: Anti-adrenergic. (Ciba-Geigy Corp).

Lopressor SR. Beta-blocker. (Ciba-Geigy PLC).

Loprox. Ciclopirox olamine. Applications: Antifungal. (Hoechst-Roussel Pharmaceuticals Inc).

Lorelco. Prescription drug for the treatment of hyper-cholesterolemia. (Dow Chemical Co Ltd).*

Loretine. (Sulphiolinic Acid). Iodohydroxyquinolinesulphonic acid, $C_9H_4NI(OH)SO_3H$. A germicide.†

Lorexane. Preparations of gamma benzene hexachloride.γ An insecticide. Antiparasitic hair lotion. (ICI PLC).*

Lorfan. A proprietary preparation of levallorphan tartrate, used in cases of narcotics overdosage and in obstetrical anaesthesia. (Roche Products Ltd).*

Lorival. A proprietary electrical insulation made from a synthetic resin of the phenol-formaldehyde type.†

Lorol. A mixture of alcohols produced by the reduction of coco-nut oil. Sulphated fatty alcohols. (Hickson & Welch Ltd).*

Lorol DA. Diethanolamine lauryl sulphate in liquid form. Applications: Anionic surfactant used in shampoos, milder in action than other alkyl sulphates in the Lorol series. (Ronsheim & Moore Ltd).

Lorol MA and MR. Monoethanolamine lauryl sulphate in liquid form. Lorol MR is a built product. Applications: Anionic surfactant for shampoos. (Ronsheim & Moore Ltd).

Lorol NH. Ammonium lauryl sulphate in liquid form. Applications: Anionic surfactant having a mild action, used in good quality shampoos. (Ronsheim & Moore Ltd).

Lorol TA and TAR. Triethanolamine lauryl sulphate in liquid form. Lorol TAR is a built product containing alkylolamide foam booster. Applications: Anionic surfactant used in shampoos. (Ronsheim & Moore Ltd).

Lorol TN and TNR. Triethanolamine/ammonium lauryl sulphate in liquid form. Lorol TNR contains alkylolamides. Applications: Anionic surfactant used in shampoos. (Ronsheim & Moore Ltd).

Lorsban. Insecticides comprising chlorpyrifos. Applications: Used to control ticks on cattle, mosquitoes and other insects. (Dow Chemical Co Ltd).*

Losan. A proprietary preparation. It consists of quinophan and sodium bicarbonate.†

Losetic. A proprietary procaine hydrochloride and adrenalin preparation.†

Losilphos. Ferophosphorus briquettes. Applications: Used in steelmaking for their phosphoros content. (Monsanto Co).*

Losite. Synonym for Vishnevite.†

Losophane. Triiodo-*m*-cresol, $C_6H_3(I_3)OHCH_3$. Used externally in skin diseases, as an antiseptic and astringent.†

Lotens. See CARBORA.

Lotioblanc. White Lotion. Applications: Astringent; protectant (topical). (American Critical Care, Div of American Hospital Supply Corp).

Lotrimin. Clotrimazole. Applications: Antifungal. (Schering-Plough Corp).

Lotusate. Talbutal. Applications: Sedative. (Sterling Drug Inc).

Louse Powder. Insecticide. (FBC Ltd).

Lovol. A mixture of aliphatic alcohols formed by the high pressure hydrogenation of coconut oil.†

Low Brass. See BRASS.

Lowerite. Strippable coating compositions. (Croda Chemicals Ltd).*

Lowroff Phosphor Bronze. Alloys of 70-90 per cent copper, 4-13 per cent tin, 5-16 per cent lead, and 0.5-1 per cent phosphorus.†

Lox. Liquid oxygen. Used in mining explosives and in rocket propellants.†

Loxa Bark. Pale cinchona bark.†

Loxiol G-70, G-71, G-72 and G-73. Proprietary names for a range of multifunctional polyesters of high molecular weight, used as additives to PVC compounds to reduce surface tackiness. (Henkel Chemicals Ltd).†

Loxitane. Loxapine. Applications: Tranquilizer. (Lederle Laboratories, Div of American Cyanamid Co).

Loxon. A proprietary preparation of HALOXON. A veterinary anthelmintic.†

Lozol. Indapamide. Applications: Antihypertensive; diuretic. (USV Pharmaceutical Corp).

L.P.D. An ultra-accelerator for rubber vulcanization. It is the lead salt of piperidine-pentamethylene-dithiocarbamate.†

LPS Brake Cleaner. A solvent based cleaner for brakes and brake components. (Holt Lloyd Corporation).*

LPS Electro Contact Cleaner. A premium blend of solvents which is safe to use on rubber paint, plastics and to clean and degrease sensitive electronic equipment. Applications: Cleans electrical/-electronic components and delicate mechanisms. (Holt Lloyd Corporation).*

LPS Engine Degreaser. A petroleum base cleaner which will emulsify with water. Applications: Engine degreaser. (Holt Lloyd Corporation).*

LPS Heavy-Duty Silicone Lubricant. A water based dry film silicone lubricant. Applications: Lubricant, safe for use on almost any surface, approved by U.S.D.A. for use in Federally inspected meat and poultry plants, approved for use in areas with incidental food contact. (Holt Lloyd Corporation).*

LPS Instant Cold galvanize. A combination of solvents, resin binders and 95% pure zinc metal which provides a galvanic coating similar to hot dipped galvanize. Applications: Cold galvanizing compound that inhibits corrosion for up to three years, acts as a primer or finish coat and repairs damaged or worn hot-dipped galvanized coatings. (Holt Lloyd Corporation).*

LPS Instant Super Cleaner/Degreaser. A combination of chlorinated solvents to clean and degrease metal parts and electrical equipment. Applications: Cleaner/degreaser for equipment. (Holt Lloyd Corporation).*

LPS Paint Remover. A blend of solvents, waxes and stripping aids for removing all types of paint from metal surfaces. Applications: Removal of paint and varnishes. (Holt Lloyd Corporation).*

LPS Tap-All. An engineered blend of lubricant aids and corrosion inhibitors that can be used for the machining of all metals. Does not contain sulphur or chlorine. Applications: As a lubricant when tapping, drilling, grinding, milling, threading, reaming, turning, boring, sawing or engraving all metals (except magnesium). (Holt Lloyd Corporation).*

LPS 1 Greaseless Lubricant. A blend of solvents, lubricants, moisture displacers and corrosion inhibitors to provide greaseless lubrication of delicate mechanisms to dry out sensitive electrical and electronic circuits. Applications: Lubricant, penetrant, water displacer, cleans electrical equipment. (Holt Lloyd Corporation).*

LPS 2 General Purpose Lubricant. A blend of lubricants and corrosion inhibitors that provide general purpose lubrication, corrosion inhibition and penetration of seized mechanisms. Applications: Lubricant, penetrant and quick release, corrosion inhibitor. (Holt Lloyd Corporation).*

LPS 3 Heavy Duty Rust Inhibitor. A blend of waxes, solvents, corrosion inhibitors and oils to provide long term protection of metal parts. Applications: Corrosion inhibitors, chain lubricant and anti-seize compound. (Holt Lloyd Corporation).*

LPS 500 Plus. A blend of waxes, oils and corrosion inhibitors used to lubricate and prevent corrosion in severe conditions. Applications: As a lubricant corrosion inhibitor or anti-seize compound. (Holt Lloyd Corporation).*

LSD. Lysergic acid diethylamide.†

Lubafax. A proprietary formulation of propylene glycol and hydroxypropyl methylcellulose. Applications: A surgical lubricant for selected medical implements. (The Wellcome Foundation Ltd).*

Lubasin. Adhesives for textile screen printing. (BASF United Kingdom Ltd).

Lubestat. Water soluble, biodegradable lubricants. Applications: Fibre lubricants, textile softeners, yarn processing aids. (Milliken & Company).*

Lubestine. Talc, asbestine substitute. Applications: Extender for paints and as a general purpose filler. (Bromhead & Denison Ltd).*

Lubix. Die lubricant for semi-continuous casting. (Foseco (FS) Ltd).*

Lubolid. Dry powder lubricants. (Dow Corning Ltd).

Lubrico. A proprietary trade name for an alloy of 75 per cent copper, 20 per cent lead, and 5 per cent tin.†

Lubrisol. Natural oil. Applications: Textile lubricant. (Scher Chemicals Inc).*

Lubrite B33. Bleaching of mineral oils, greases and waxes, including re-refined/recycled lubricating oil. (Minas de Gador SA).*

Lubrol. A fatty alcohol ethoxylate with a neutral reaction. Used as a surfactant. (ICI PLC).*

Lubrol N13. Nonylphenol ethoxylate nonionic surfactant in semi- solid form. Application: Wetting and emulsifying agent used in pest control, oleines, metal treatment and degreasing. (ICI PLC, Organics Div).

Lubrol N5. Nonylphenol ethoxylate nonionic surfactant in viscous liquid form. Application: Conversion of premium paraffin to self-emulsifiable solvent. Used in industrial cleaning and degreasing including metal surfaces. (ICI PLC, Organics Div).

Lucca Oil. Olive Oil.†

Lucerno. An alloy containing 67.9 per cent nickel, 27.5 per cent copper, 2.4 per cent iron, and 2.2 per cent manganese.†

Lucidene. Range of aqueous acrylic and styrene-acrylic emulsions in printing inks and overprint lacquers. (Williams Div of Morton Thiokol Ltd). *

Lucidol. Benzoyl peroxide. (Akzo Chemie UK Ltd).

Lucidril. A proprietary trade name for the hydrochloride of Meclofenoxate. (Lloyd's Pharmaceuticals).†

Lucilite. Silica hydrogel. (ABM Chemicals Ltd).*

Lucipal. Benzoyl peroxide compositions. (Akzo Chemie UK Ltd).

Lucite. (Perspex). Proprietary trade names for a methyl methracrylate and acrylate synthetic resins. The material is more transparent than glass and has a very high refractive index. (Du Pont (UK) Ltd).†

Lucitone. See VERNONITE.

Lucobit. A polyethylene bitumen blend in granular form. Applications: Extruded sheet for roofing, waterproof membranes in civil engineering. (BASF United Kingdom Ltd).*

Lucofen S A. A proprietary preparation of chlorphentermine hydrochloride. Antiobesity agent. (Warner).†

Lucovyl. A proprietary polyvinyl chloride. (Pchiney-St Gobain, France).†

Ludenscheidt's Button Brass. See BRASS.

Ludigol. Auxiliary for textile printing. (BASF United Kingdom Ltd).

Ludigol. M-nitrobenzene sulphonic acid, sodium salt. Applications: Used to accelerate the stripping of nickel from steel, brass or copper; to control the reaction rate in pickling nickel-chomium-iron alloys and for dissolving nickel and copper; additive in phosphate coatings of metals. (GAF Corporation).*

Ludiomil. Maprotiline hydrochloride. Applications: antidepressant. (Ciba-Geigy PLC).

Ludlum Alloy. A heat-resisting alloy of iron with from 13-17 per cent chromium, 1 per cent silicon, 1 per cent molybdenum, and 0.4 per cent carbon.†

Ludlum No. 602 Steel. A proprietary trade name for a steel containing 1.7 per cent silicon, 0.7 per cent manganese, 0.4 per cent molybdenum, 0.12 per cent vanadium, and 0.48 per cent carbon.†

Ludox. Colloidal silica. (Du Pont (UK) Ltd).

Lufibrol. Auxiliary for pretreatment of textiles. (BASF United Kingdom Ltd).

Luftseide. (Celta, Soie Nouvelle, Tubulated Silk). Artificial silks of the rayon type made with hollow central spaces. They are formed by adding gas-evolving materials to the viscous solution.†

Lugacin. 2ml snap-off ampoule containing a 2ml solution of gentamicin sulphate equivalent to 80mg gentamicin base. Applications: Bactericidal antibiotic active against extremely broad spectrum of Gram-positive and Gran-negative pathogens including escherichia coli, klebsiella, proteus, pseudomonas aeruginosa and antibiotic-resistant strains of staphlococcus aureus. (Lagap Pharmaceuticals Ltd).*

Lugol's Solution. Iodine-potassium iodide solution. Iodine (5 grams) are triturated with 5 grams of potassium iodide, and 100 cc of water, and diluted to 1 litre.†

Lugo's Powder. Powdered cinchona bark.†

Lukens Bone Wax. Made from natural bees wax, salicylic acid and a natural oil. Applications: Applied by surgeon in order to produce hemostasis in bone during surgery in which bone is cut. (Lukens International Corp).*

Lulea Tar. A variety of Stockholm tar.†

Lumarith EC. A registered trade mark for an ethyl cellulose thermoplastic resistant to oils.†

Lumarith ER. A registered trade mark for ethyl rubber.†

Lumbang Oil. A drying oil obtained by pressing the seeds of *Aleurites moluccana*. Used in the manufacture of paints and soap.†

Lumen Alloy 11-C. A proprietary trade name for an alloy containing copper with 10 per cent aluminium and 1 per cent iron.†

Lumen Bronze. An alloy of 86 per cent zinc, 10 per cent copper, 4 per cent aluminium, and 0.1 per cent magnesium. A softer variety contains 88 per cent zinc, 8 per cent aluminium, and 4 per cent copper.†

Lumicon. Dental preparation. Applications: Fixing, filling and lining cement. (Bayer & Co).*

Lumicon Non Gamma 2. Silver alloy, practically corrosion free. Applications: Dental speciality. (Bayer & Co).*

Lumicon Silver Amalgam/Powder. Minimum corroding silver alloy of standardized grain size distribution, mixable in any commercial mixer. (Bayer & Co).*

Lumilux. Luminous pigments. (Hoechst UK Ltd).

Luminal. Pharmaceutical preparation. Applications: Anticonvulsant, hypnotic and sedative agents. (Bayer & Co).*

Luminal Sodium. Phenobarbital Sodium. Applications: Anticonvulsant; hypnotic; sedative. (Sterling Drug Inc).

Luminalettes. Pharmaceutical preparations. Applications: Anticonvulsant, hypnotic and sedative agents. (Bayer & Co).*

Lumite. An aluminium alloy containing 5.6 per cent nickel and 1 per cent iron.†

Lummer's Solution. Hydrogen platinochloride, H_2PtCl_6 (3 grams in 100 cc water), with 0.02 gram lead acetate. Used for coating platinum electrodes with finely divided platinum.†

Lumnite Cement. An alumina cement consisting of about 40 per cent alumina 40 per cent lime, 15 per cent iron oxide 5 per cent silica, and magnesia. The material is made from bauxite, and is stated to be stronger than Portland cement.†

Lump-Lac. See LAC.

Lunar Caustic. Silver nitrate, $AgNO_3$, fused and cast into sticks or rods. An energetic caustic for wounds and sores.

Lunosol. A proprietary preparation. It is a colloidal silver chloride.†

Luo-calcite. A name given to the calcium bicarbonate occurring in solution in natural waters.†

Luo-Chalybite. A name given to the ferro bicarbonate occurring in solution in natural waters.†

Luo-Diallogite. A name given to the manganese bicarbonate occurring in solution in natural waters.†

Luo-Magnesite. A name given to the magnesium bicarbonate occurring in solution in natural waters.†

Lupeose. (Mannotetrose). Stachyose, $C_{24}H_{12}O_{21}$, a tetrasaccharide. It occurs in the tubers of *Stachy tubifera*.†

Luperco. A registered trade mark for organic peroxide compounds for polymerization catalysis, drying accelerator, oxidizing agent and bleaching applications. A consists of benzoyl peroxide with an inorganic filler and AC consists of benzoyl peroxide with an organic filler.†

Luperco AFR. A trade mark for 50 per cent. benzoyl peroxide in plasticizer. (Wallace and Tiernan, NY).†

Lupersol DDM. A trade mark for 60 per cent. methyl ethyl ketone peroxide in dimethyl phthalate. (Wallace and Tiernan, NY).†

Lupersol DEL. A trade mark for 60 per cent. methyl ethyl ketone peroxide in dimethyl phthalate. (Wallace and Tiernan, NY).†

Lupersol DSW. A proprietary trade name for methyl ethyl ketone peroxide. A liquid fire resistant peroxide containing 11.5 per cent. active oxygen. (Kingsley and Keith Chemical Corp).†

Lupersol 227. A registered trade mark for a 50 per cent. solution of diisobutyryl peroxide in mineral spirits. An organic peroxide for polymerizing. (Wallace and Tiernan, NY).†

Lupetazin. Dimethylpiperazine, $(C_2H_3(CH_3)NH)_2$. A uric acid solvent.†

Lupinit. A proprietary casein product.†

Lupolen. Low, medium, high density polyethylene. Density from 0.918 to 〉 0.960 g/cm³. Applications: Product range covers grades for film extrusion, blow moulding (specially high molecular HDPE), injection and compression moulding. (BASF United Kingdom Ltd).*

Lupolen V-2524EX and V-3510K. Proprietary low-density polyethylene copolymers. (BASF United Kingdom Ltd).*

Lupolen 1800 H/M/S. A proprietary low-density polyethylene used in injection moulding. (BASF United Kingdom Ltd).*

Lupolen 1810E. A proprietary low density polyethylene used in the manufacture of milk containers. (BASF United Kingdom Ltd).*

Lupolen 1810H. A proprietary low-density polyethylene used in blow moulding. (BASF United Kingdom Ltd).*

Lupolen 1812D and 1812EH. Proprietary polyethylenes used in the manufacture of wires and cables. (BASF United Kingdom Ltd).*

Lupolen 1814E. A proprietary low density polyethylene used in the manufacture of milk containers. (BASF United Kingdom Ltd).*

Lupolen 1852E/H. A proprietary low density polyethylene used in the manufacture of pipes. (BASF United Kingdom Ltd).*

Lupolen 2040EX and 2410DX. Proprietary low-density polyethylenes used in the manufacture of bags. (BASF United Kingdom Ltd).*

Lupolen 2410S. A proprietary low density polyethylene used in injection moulding. (BASF United Kingdom Ltd).*

Lupolen 2424H and 2425K. Proprietary low-density polyethylenes used in the manufacture of transparent packaging materials. (BASF United Kingdom Ltd).*

Lupolen 2430H. A proprietary low density polyethylene used in blow moulding. (BASF United Kingdom Ltd).*

Lupolen 2452 E. A proprietary low density polyethylene used in the extrusion of pipes. (BASF United Kingdom Ltd).*

Lupolen 3010 S. A proprietary low density polyethylene used in injection moulding. (BASF United Kingdom Ltd).*

Lupolen 3020 D. A proprietary low density polyethylene used in blow moulding. (BASF United Kingdom Ltd).*

Lupolen 3020 KX and 3025 KX. Proprietary low-density polyethylenes used as over-wrappings. (BASF United Kingdom Ltd).*

Lupolen 4261 AX. A proprietary high density polyethylene of high molecular weight. (BASF United Kingdom Ltd).*

Lupolen 5011 K. A proprietary high density polyethylene used in injection moulding. (BASF United Kingdom Ltd).*

Lupolen 5052 C. A proprietary high density polyethylene used in the extrusion of pipes. (BASF United Kingdom Ltd).*

Lupolen 6011 K. A proprietary high density polyethylene used in injection moulding. (BASF United Kingdom Ltd).*

Lupolen 804H and 1814H. Proprietary low-density polyethylenes used in the manufacture of liners and barriers. (BASF United Kingdom Ltd).*

Luprimol. Auxiliaries for pigment printing. (BASF United Kingdom Ltd).

Luprintan ATP. Printing auxiliary for synthetics. (BASF United Kingdom Ltd).

Luprintol. Auxiliaries for emulsion printing. (BASF United Kingdom Ltd).

Lupron. Leuprolide acetate. Applications: Antineoplastic. (TAP Pharmaceuticals).

Luran. Styrene acrylonitrile copolymer in granule form. Applications: Mouldings for domestic appliances and sanitaryware. (BASF United Kingdom Ltd).*

Luran KR 2517. A proprietary acrylonitrile-styrene copolymer containing 35% glass fibre. (BASF United Kingdom Ltd).*

Luran S. Styrene acrylonitrile, modified with acrylic ester, in granule form. Applications: Mouldings for external garden furniture, surfboards, road signs, automotive parts. (BASF United Kingdom Ltd).*

Luran 378P. A proprietary styrene acrylonitrile copolymer used in injection moulding and extrusion. LURAN KR2551. (BASF United Kingdom Ltd).*

Luran 757R. A proprietary acrylonitrile styrene-acrylonitrile copolymer used in injection moulding. (BASF United Kingdom Ltd).*

Luran 776S. A proprietary acrylonitrile styrene-acrylonitrile used in extrusion and injection moulding. (BASF United Kingdom Ltd).*

Lurantin Supra. Dyestuffs for celluosic fibres. (BASF United Kingdom Ltd).

Lurgi Metal. An alloy of lead with 2 per cent barium.†

Lurotex A25. Textile dyeing and finishing auxiliary. (BASF United Kingdom Ltd).

Luscin. Coal dust replacement additives f[illegible] iron foundry sands. (Foseco (FS) Ltd).

Lusol. General purpose maintenance spra[illegible] Applications: Lubrication, penetrating seized parts, freeing rusted mechanism[illegible] driving moisture out of electrical parts, corrosion prevention on metal surfaces. (Hermetite Products Ltd).*

Lustilac. A proprietary moulding compound.†

Lustra-Cellulose. Artificial silk.†

Lustran. Styrene/acrylonitrile and acrylonitrile/styrene butadiene moulding compounds. (Monsanto PLC).

Lustran ABS. Acrylonitrile-butadiene-styrene thermoplastics. Applications: Moulding and extrusion grades, refrigerator inner liners, business machine and appliance housings, automotive parts and industrial components. (Monsanto Co).*

Lustran SAN. Styrene-acrylonitrile moulding resins. Applications: Clear rigid thermoplastics for moulding high quality articles requiring transparency, brilliant surface, chemical resistance and stiffness. (Monsanto Co).*

Lustran Ultra ABS. ABS moulding grades superior moulded part appearance and gloss. Applications: Telephones, appliances, video cassettes, power tool housings, office equipment, toys, sporting goods and lawn and garden equipment. (Monsanto Co).*

Lustranyl. Solvent soluble dyes. (Williams (Hounslow) Ltd).

Lustre. Protein binders based on modified casein dispersions. Applications: Leather. (Colour-Chem Limited).*

Lustrex. Polystyrene moulding and extrusion rsins. Applications: Packaging, toys, appliance, photographic, furniture, audio cassette, medical and houseware markets, thermoformed containers, trays, cups and lids. (Monsanto Co).*

Lustron. See CELANESE. Also a proprietary trade name for polystyrene moulding compounds.

Lustrose. A proprietary compound used in the textile industry for sizing.†

Lusynton. Detergents and cleaning agents. (BASF United Kingdom Ltd).

Lutalyse. Dinoprost tromethamine. Applications: Oxytocic; prostaglandin. (The Upjohn Co).

Lutate. A proprietary preparation of HYDROKYPROGESTERONE caproate. (Savage Laboratories, Missouri City, Texas).†

Lutecin. See NICKEL SILVERS.

Lutensit A-ES. Sodium alkyl-phenol ether sulphate as a yellow liquid. Applications: Wetting and dispersing agent with many applications, sometimes in combination with nonionics; emulsifying agent for methylene chloride. (BASF United Kingdom Ltd).

Lutensit A-LBA. Anionic surfactant in yellow liquid form. Applications: Wetting, dispersing, cleaning and emulsifying agent used in household rinsing and cleaning formulations, all-purpose cleaners, and various industrial emulsions. (BASF United Kingdom Ltd).

Lutensit A-PS. Sodium alkane sulphonate as a slightly yellow liquid. Applications: Alkaline industrial, household and metal cleaners, steam jet cleaners and pickling baths. (BASF United Kingdom Ltd).

Lutensit An 10. A surfactant. (BASF United Kingdom Ltd).*

Lutensit AS. Range of anionic surfactants which are clear, almost colourless to slightly yellowish liquids. Applications: Detergency, foaming and wetting agents with emulsifying and dispersing effect on perfume oils. Used in liquid or pasty detergents and cleaning agents with high foam formation; shampoos, foam baths etc where the high purity permits use of high concentrations. (BASF United Kingdom Ltd).

Lutensit K-LC. Dimethyl C12/C14 fatty alkylbenzyl ammonium chloride in almost colourless aqueous solution. Applications: Dispersing and wetting agent with bactericidal and fungicidal effects. Used in conjunction with nonionics to produce disinfectant cleaners for the food and beverage industries and for miscellaneous household and industrial uses. (BASF United Kingdom Ltd).

Lutensol. Low foaming surfactants. Applications: Domestic machine dishwashing detergents and rinse aids. Low foam industrial cleaners. (BASF United Kingdom Ltd).

Lutensol AP. Range of nonionic surfactants of the alkylphenyl (generally nonylphenyl) ethoxylate type in liquid or soft wax form. Application: Wetting, detergency, emulsifying, dispersing and low to medium foaming agent used in household and industrial cleaners; oil and solvent emulsification; leather, paper, pulp, paints, lacquers and building industries. (BASF United Kingdom Ltd).

Luteol. Hydroxychlorodiphenylquinoxaline. An indicator.†

Luteolin. See WELD. An orange-yellow dyestuff.

Lutetia. Pigments, for paints and inks. (ICI PLC).*

Lutexal. Synthetic thickening agents. (BASF United Kingdom Ltd).

Lutidine. Dimethylpyridine, C_7H_9N.†

Lutofan. Polyvinyl chloride. (BASF United Kingdom Ltd).*

Lutofan D. Polyvinyl chloride dispersion. (BASF United Kingdom Ltd).*

Lutol. See CUTOL.

Lutonal. Polyvinyl ether. (BASF United Kingdom Ltd).*

Lutonal D. Polyvinyl ether dispersion. (BASF United Kingdom Ltd).*

Lutonal LC. Polyvinyl isobutyl ether. (BASF United Kingdom Ltd).*

Lutrabond. Polyester non-wovens for roof covers. (Carl Freudenberg).*

Lutradur. Polyester non-woven. (Carl Freudenberg).*

Lutron. Waterless foundry moulding material. (Foseco (FS) Ltd).*

Luvican M170. Poly-vinyl carbazole. (BASF United Kingdom Ltd).*

Luxalloy. Dental silver alloys and mercury for mixing amalgams. Applications: For dentistry and dental engineering. (Degussa).*

Luxene. A proprietary synthetic resin of the phenol-formaldehyde class.†

Luxene 44. A proprietary denture compound made from a vinyl copolymer.†

Luxol. Solvent soluble dyes. (Williams (Hounslow) Ltd).

Luxor. A combination of hydrolysed vegetable proteins and other ingredients. Applications: Used as a natural flavour in vegetables, meat, chicken and seafood dishes. (Hercules Inc).*

Luzerne. A proprietary trade name for a hard rubber.†

Luzidol. Benzoyl peroxide, $C_{14}H_{10}O_4$.†

Lyargol. Silver proteinate.†

Lycal. Lightly calcined sea water magnesia. (Steetley Refractories Ltd).

Lycanol. A proprietary trade name for Glymidine.†

Lycasin. Hydrogenated glucose syrup. Applications: A non-carbohydrate sugar substitute, food (confectionery) and pharmaceutical (syrups and lozenges). (Roquette (UK) Ltd).*

Lycetol. (Tetradine). Dimethylpiperazine tartrate. A solvent for uric acid.†

Lycine. The base of *Lycium barbarum*. It is identical with betaine.†

Lycopodium. A pale yellow powder consisting of the spores of *Lycopodium clavatum*.†

Lycopon. (Vatrolite). A proprietary trade name for sodium hyposulphite.†

Lyddite. (Melinite, Pertite, Shimose). Picric acid, an explosive.†

Lydian Stone. (Lydite, Touchstone). A siliceous slate containing about 84 per cent silica, 5 per cent alumina, and 1 per cent ferric oxide. Used for testing gold by rubbing it upon the stone. and testing the streak of metal produced with acid.†

Lydin. A name given to Mauve or Aniline purple. See MAUVEINE.†

Lydite. See LYDIAN STONE.

Lye Glycerin. Glycerin obtained from soap liquor.†

Lynite. A trade mark for an alloy of 88 per cent aluminium, 10 per cent copper, 1.. per cent iron, and 0.25 per cent magnesium. It has a specific gravity of 2.95.†

Lynx. A cotton fabric grade of TUFNOL industrial laminates. (Tufnol).*

Lyocol. Low foaming scouring,levelling and dispersing agent. Applications: Combined scouring and dyeing of polyester fibres; beam dyeing and jet application. (Sandoz Products Ltd).

Lyofix. Finishing agent. (Ciba-Geigy PLC).

Lyofix F. A proprietary trade name for a melamine-formaldehyde resin precondensate used in finishing paper makers' felts. (Ciba Geigy PLC).†

Lyofix 363. A proprietary trade name for a modified urea formaldehyde resin precondensate used in the crease-resistance finishing of cellulosic textiles. (Ciba Geigy PLC).†

Lyogen. Levelling agent. Applications: Wool, nylon, polyurethane and polyamide fibres; vat dyestuffs; acid, milling, chrome and direct dyestuffs. (Sandoz Products Ltd).

Lyonore. A proprietary trade name for a steel containing 0.2 per cent copper with some chromium and nickel.†

Lyons Blue. See ANILINE BLUE, SPIRIT SOLUBLE, and SOLUBLE BLUE.

Lyons Sugar. Sucramine, the ammonium salt of saccharine.†

Lyopan. A proprietary preparation of phenazone theophyllin and urethane.†

Lyophrin. A proprietary preparation of adrenaline hydrogen tartrate with sodium sulphate as a preservative. Treatment of glaucoma. (Alcon Universal).†

Lyovac Sodium Edecrin. Ethacrynate sodium. Applications: Diuretic. (Merck Sharp & Dohme, Div of Merck & Co Inc).

Lypsyl. Lipsalve. (Ciba-Geigy PLC).

Lyracamine. Acrylic dyes. (ICI PLC).*

Lysamine. Protein derived from vegetable matter. Applications: Animal feedstuffs supplement. (Roquette (UK) Ltd).*

Lysargine. A colloidal silver, containing 60 per cent of silver. An antiseptic.†

Lysase. Enzyme preparations. Applications: Conversion of starch or starch hydrolysis products. (Roquette (UK) Ltd).*

Lyse S. Reagents which destroy red blood cells to leave white blood cells in suspension for analysis; used e.g. as Lyse S, Lyse S III, Lyse S III Diff. Applications: White blood cell and haemoglobin determination. (Coulter Electronics Ltd).*

Lysidine. (Piperazenyl). Tetrahydropyrazine.†

Lysinex. A proprietary preparation of L-lysine hydrochloride and stanolone. An anabolic agent. (Lloyd, Hamol).†

Lysitol. See LYSOL.

Lysivane. Ethopropazine hydrochloride. (May & Baker Ltd).

Lysochlor. Chloro-*m*-cresol, $C_6H_3Cl(CH_3)OH$. A disinfectant. It is a similar preparation to Eusapyl (*qv*).†

Lysodren. Mitotane. Applications: Antineoplastic. (Bristol Laboratories, Div of Bristol-Myers Co).

Lysoform. A solution of formaldehyde in alcoholic potash soap solution. A disinfectant much like Lysol.†

Lysol. (Lysitol). Consists of crude cresols mixed with a soft-soap solution. A disinfectant used in surgery, and for cleaning floors and walls. Also see NEOLYSOL and SAPOCARBOL.†

Lysophan. Triiodocresol, $C_6HI_3(CH_3)OH$. An antiseptic.†

Lysulfol. Lysol with 10 per cent of sulphur.†

Lythol Oil. A commercial phenolated oil used as a disinfectant.†

Lytrol. An alcoholic solution of potassium-β-naphthol. A disinfectant.†

Lytron. Range of aqueous styrene co-polymer emulsions with controlled particle size. Applications: For use in paper coatings and opacifying liquid detergents. (Williams Div of Morton Thiokol Ltd). *

Lytta. Cantharides.†

M

M33, MN3. Polyimide resins. (Rhone-Poulenc NV/CdF Chimie AZF).*

M50. Basic lead silico chromate. Applications: Anti-corrosive pigments for coatings. (NL Industries Inc).*

MA 20. Soap powder for launderettes. (Unilever).†

Maali Resin. A yellowish-white resin resembling elemi of Samoa.†

Maalox. A proprietary preparation containing magnesium aluminium hydroxide gel. An antacid. (Rorer).†

M & B 693. Sulphapyridine. (May & Baker Ltd).

Macadamite. An alloy of 72 per cent aluminium, 24 per cent zinc, and 4 per cent copper.†

Macaja Butter. Mocaya oil.†

Macaloid. Rheological additives. Applications: Paint, sealants, agricultural products, cosmetics and ceramics. (NL Industries Inc).*

Macassar Nutmeg Butter. (Papua Nutmeg Butter). The fat from *Myristica argentae*.†

Mace Butter. Nutmeg butter, obtained from the seeds of *Myristica officinalis,* of the East.†

Mace Butter, American. See OTOBA BUTTER.

Mace Oil. See OIL OF MACE.

MacFarland's Alloy. Heat-resisting alloys. (*a*) Contains 59 per cent nickel, 30 per cent chromium, and 11 per cent copper. (*b*) Contains 46 per cent nickel, 43 per cent chromium, and 11 per cent copper.†

Macgill Metal. An alloy containing 88 per cent copper, 7 per cent nickel, 4.5 per cent iron, and traces of tin and lead. It resists corrosion.†

Machacon Juice. An alkaline decoction of the juice of the root of a plant having the same name. It is used to coagulate rubber latex.†

Machete. Butachlor. Applications: Herbicide. (Monsanto Co).*

Machine Bronze. Variable alloys containing 50-90 per cent copper, 25 per cent nickel, 0-30 per cent tin, and 0.8 per cent lead. Some alloys contain no nickel and also contain zinc. Another alloy of this type consists of 83 per cent copper, 16 per cent zinc, and 1 per cent tin.†

Machinery Brass. See BRASS.

Mach's Metal. An alloy of aluminium with 2-10 per cent magnesium.†

Macht's Metal. An alloy of 57 per cent copper and 43 per cent zinc, used for castings.†

M-Acid. 1-Amino-5-naphthol-7-sulphonic acid†

Mackechnie. Copper sulphate, pentahydrate, monohydrate and anhydrous. (McKechnie Chemicals Ltd).

Mackechnie's Bronze. Usually an alloy of 57 per cent copper, 41 per cent tin, 1 per cent zinc, 1 per cent iron, and 0.5 per cent lead.†

Mackenzie's Amalgam. Bismuth (2 parts) and lead (4 parts) are melted separately in crucibles, and each poured into 1 part mercury. These amalgams are then rubbed together.†

Mackenzie's Metal. An alloy of 70 per cent lead, 17 per cent antimony, and 13 per cent tin; also 68 per cent lead, 16 per cent antimony, and 16 per cent bismuth. Electrotype metals.†

Mack's Cement. Prepared by adding calcined sodium sulphate or potassium sulphate to dehydrated gypsum.†

Macocyn. A proprietary preparation of oxytetracycline. An antibiotic. (Pfizer International).*

Macquer's Salt. Potassium arsenate, KH_2AsO_4.†

Macrodantin. Nitrofurantoin. Applications: Antibacterial. (Norwich Eaton Pharmaceuticals Inc).

Macrodex. Dextran 70. A polysaccharide produced by the action of Leuconostoc mesenteroides on sucrose. Average molecular weight: 70,000. Applications: Plasma volume extender. (Pharmacia Laboratories, Div of Pharmacia Inc).

Macrogol Stearate. Polyoxyl 40 stearate. Polyoxyl 8 stearate.†

Macrogols. Polyethylene glycols.†

Macrolex. Fast dyestuffs. Applications: For dyeing polystyrene, styrene copolymers, polymethacrylate rigid PVC and polycarbonate. (Bayer & Co).*

Macrospherical 95. Aluminium chlorohydrate. Applications: Anhidrotic. (Reheis Chemical Co).

Macrynal. Hydroxy acrylic resins for use with polyisocyanates. Applications: Automotive paints and industrial finishes. (Resinous Chemicals Ltd).*

Maculanin. Potassium amylate.†

Macuprax. Bordeaux/cufraneb fungicide. (McKechnie Chemicals Ltd).

Madanite. A product used as a binding material made from 2 parts petroleum jelly and 1 part rubber.†

Madar Fibre. A bast fibre, known in India by this name. obtained from *Calotropis procera* and *C. gigantea*.†

Madder. The powdered root of the plant, *Rubia tinctorum*. The chief constituent is ruberythric acid. This acid is a glucoside, and is split up into alizarin and a sugar by the action of acids.†

Madder Indian Red. (Tuscan Red). Red oxides of iron improved in colour by the addition of an alizarin lake. Pigments.†

Maddrell Salts. Alkaline meta-phosphates which are insoluble in water. They are made by heating mono-sodium phosphate at 245-250° C.†

Madecassol. A proprietary preparation extract of *Centella asiatica* in an ointment, used for skin protection. (Rona Laboratories).†

Madol Oil. An oil obtained from the seeds of *Garcinia echinocarpa*.†

Madopar. A proprietary preparation of levodopa and benserazide used in the treatment of Parkinson's disease. (Roche Products Ltd).*

Madras Blue. A mixture of gallocyanine and logwood extract.†

Madras Kino. See KINO, BENGAL.

Madribon. A proprietary preparation containing sulphadimethoxine. Applications: Used as an anti-infective. (Roche Products Ltd).*

Mafura Fat. A fat obtained from the seeds of *Mafureira oleifera*. It contains from 92-95 per cent fatty acids, has an acid value of 31-32, and a saponification value of 202-207. Used in the manufacture of soaps.†

Mag-40. Liquid chemical defoliant of cotton and a desiccant in silverskin onion. (Makhteshim Chemical Works Ltd).†

Magadi Soda. An East African soda. It contains sodium carbonate and sodium bicarbonate.†

Magan. A range of magnesium salicylate tablets sold only on prescription for medical and pharmaceutical uses. (Hercules Inc).*

Magcoke. Magnesium impregnated coke for the desulphurization and nodularization of cast iron. (Foseco (FS) Ltd).*

Magecol. A proprietary lampblack suitable for rubber goods.†

Magenta. (Roseine, Fuchsine, Aniline Red, Rubine, Magenta Red, Magenta Roseine, Rubesine, Rubianite, Fuchsianite, Magenta Crystals, Amethyst, Ponceau, Maroon, Grenaldine, Geranium, Cerise, Russian Red, Azaleine, Chestnut, Solferino, Erythro-benzine, Rubianin, Harmaline,

Grenadine, Russian Red G, 967, Cerise B, BB, G, 2YS). A dyestuff. It consists of a mixture of the hydrochlorides or acetates of pararosaniline (triamino-triphenyl-carbinol), and rosaniline (tri-amino-diphenyltolyl-carbinol). Hydrochlorides = $C_{19}H_{26}N_3ClO_4$, and $C_{20}H_{28}N_3ClO_4$. Acetates = $C_{21}H_{21}N_3O_2$, and $C_{22}H_{23}N_3O_2$. Dyes silk, wool, leather, and cotton mordanted with tartar emetic and tannin, red.†

Magenta Base. See ROSANILINE.

Magenta BB Powder, Crystals FF, CV Conc., L, Superfine Powder. Dyestuffs. They are British equivalents of New magenta.†

Magenta Bronze. (Violet Bronze). Tungsten potassium bronze, $K_2W_4O_{12}$, prepared by the addition of tungsten trioxide to fused potassium tungstate, then heating in a current of hydrogen, and digesting in water and acids. The residue is magenta bronze.†

Magenta Crystals. See MAGENTA.

Magenta Red. See MAGENTA.

Magenta Roseine. See MAGENTA.

Magenta S. See ACID MAGENTA.

Magicote Masonry Paint. Vinyl copolymer based - water thinned. Applications: Brickwork, cement, stone, concrete, pebbledash, internal and external. (Berger Jenson & Nicholson Ltd).*

Magicote Non Drip and Liquid Gloss. Alkyd resin based, white spirit thinned. Applications: For internal and external application in the DIY market. Non drip needs no undercoat, liquid requires Magicote undercoat. (Berger Jenson & Nicholson Ltd).*

Magicote Solid Emulsion. Vinyl copolymer based - water thinned. Applications: Available in vinyl matt and vinyl silk to apply by roller to interior walls and ceilings. Supplied ready for use in roller tray. (Berger Jenson & Nicholson Ltd).*

Magicote Vinyl Matt. Vinyl copolymer based - water thinned. Applications: Washable finish for interior walls and ceilings. (Berger Jenson & Nicholson Ltd).*

Magicote Vinyl Silk. Vinyl copolymer based - water thinned. Applications: Washable finish for interior walls and ceilings. (Berger Jenson & Nicholson Ltd).*

Magisal. Magnesium acetylsalicylate. Use for similar purposes as aspirin.†

Magister of Bismuth. Basic bismuth nitra $BiO.NO_3.H_2O$.†

Magister of Sulphur. Sulphur precipitated in an amorphous condition from solutions of hyposulphites or polysulphides, by acids.†

Magistery of Lead. See WHITE LEAD.

Maglite Y. A trade mark for magnesium oxide. (E Merck, Darmstadt).†

Maglstral. An impure copper sulphate containing ferric oxide, sodium sulphate, and sodium chloride.†

Magmet. Magnetic metal particles. Applications: Used in coatings for magnetic tapes and discs. (Hercules Inc).*

Magnacryl. Second generation acrylic adhesives and radiation curable (UV/EB) adhesives and coatings. Applications: Pressure sensitive and permanent bonding. Specialty overcoating applications. (Beacon Chemical Company Inc).*

Magnalite. (1) An alloy of 94.2 per cent aluminium, 2.5 per cent copper, 1.5 pe cent nickel, 1.3 per cent magnesium, and 0.5 per cent zinc. It has a specific gravity of 2.8.†

Magnalite. (2) A mineral. It is $Mg_4Al_6Si_{12}O_{37}.13H_2O$.†

Magnalium. Alloys of magnesium and aluminium. One contains from 1-2 per cent of magnesium. It is lighter than aluminium, and is as hard as brass. Another alloy contains 10 per cent magnesium, and is used for parts of machinery, for cooking utensils, and fo optical mirrors. One alloy containing 9 per cent aluminium and 10 per cent magnesium has a specific gravity of 2.8.†

Magnamite. Carbonized or graphitized polyacrylonitrile fibre. Applications: Used to reinforce thermoset and thermoplastic resins, and as a replacement for steel, aluminium, titanium and other metals. (Hercules Inc).*

agnapen. A proprietary preparation of ampicillin and flucloxacillin. An antibiotic. (Beecham Research Laboratories).*

agnaphoscal. Multiple phosphate (sodium, calcium, magnesium phosphate) granulated. Applications: For mineral feeds and mixed feeds. (Bayer & Co).*

agnaspheres. Magnesium hydroxide for water treatment. (Steetley Refractories Ltd).

agna Tac. Specialty adhesives for commercial and industrial applications. Pressure sensitives, two part epoxies, solvent and water based systems, hot melts and hat sizings. (Beacon Chemical Company Inc).*

agnesia. Magnesium oxide, MgO.†

agnesia Alba. A basic magnesium carbonate of variable composition.†

agnesia Alum. Synonym for Pickeringite.†

agnesia Bleaching Liquid. Magnesium oxychloride, $Mg(OCl)_2$. Used for bleaching.†

agnesia Cement. See SOREL CEMENT.

agnesia White. Both magnesium oxide and magnesium carbonate are known by this name.†

agnesia-Citrate Mixture. Citric acid (20 grams), is dissolved in 20 per cent ammonium hydrate, and mixed with 1 litre of magnesia mixture (*qv*). Used in the determination of phosphorus magnesia mixture. Magnesium sulphate (1 part), and ammonium chloride (1 part) are dissolved in 4 parts of water with the addition of 8 parts of ammonia (specific gravity 0.96), and filtered. Used for the determination of phosphorus .†

agnesium Base Alloys. The magnesium in these alloys usually varies from 90-95 per cent.†

agnesium Chalcanthite. Synonym for Pentahydrite.†

agnesium Superoxyl. Magnesium peroxide, MgO_2.†

Magnesium - Halotrichite. Synonym for Pickeringite.†

Magnesium-Monel. A proprietary trade name for an alloy of 50 per cent magnesium and 50 per cent monel metal.†

Magnesium-Pectolite. Synonym for Vivianite.†

Magnesium-Perhydrol. See BIOGEN.

Magnesol. Hydrated, amorphous, synthetic magnesium silicate. Applications: Absorption, catalyst support, edible oil and fat reclamation, anti-caking, deodorization. (Reagent Chemical & Research Inc).*

Magnet Steel. Alloys of iron with from 5-50 per cent cobalt, 5-18 per cent nickel, 0-7 per cent manganese, 1-12 per cent tungsten, and 0-12 per cent chromium.†

Magnetic Black. A proprietary trade name for a finely ground magnetic iron oxide for use as a pigment.†

Magnetic Iron Ore. (Loadstone, Lodestone, Magnetite, Black Oxide of Iron, Ferroferrite). A black ore of iron. It is a ferroso-ferric oxide, $FeO.Fe_2O_3$, and contains over 72 per cent iron.†

Magnetic Oxide of Iron. See MAGNETIC IRON ORE.

Magnetic Pyrites. (Pyrrhotine, Pyrrhotite, Magnetic Sulphide of Iron). A mineral. In composition it approximates to ferrous sulphide, FeS, or ferroso-ferric sulphide, Fe_3S_4, but often contains nickel.†

Magnetic Sulphide of Iron. Magnetic iron ore.†

Magnetite. See MAGNETIC IRON ORE.

Magnilor. Nifuratel. (The Wellcome Foundation Ltd).

Magniotriplite. Synonym for Talktriplite.†

Magnisal. Magnesium nitrate hexa-hydrate. Applications: Agricultural - for curing magnesium deficiency via irrigation system, direct soil application or foliar spray. Technical - in metal, textile, ceramic and other industries. (Haifa Chemicals Ltd).*

Magno. An electrical resistance alloy containing 95 per cent nickel and 5 per cent manganese.†

Magnocid. Magnesium oxychloride, Mg(0H)(OCI). It has bleaching properties.†

Magnodat. See BIOGEN.

Magnolax. A proprietary preparation of magnesium hydroxide with liquid paraffin.†

Magnolia Metal. A bearing metal consisting mainly of antimony amd lead, and sometimes tin, with small quantities of iron and bismuth. (*a*) Contains 83 1/2 per cent lead and 16 1/2 per cent antimony; (*b*) consists of 79 per cent lead and 21 per cent antimony. Other alloys contain 78-80 per cent lead, 15-16 per cent antimony, 5-6 per cent tin, and 0-0.25 per cent bismuth.†

Magnolium. An alloy of 90 per cent lead and 10 per cent antimony.†

Magnox. Magnetic iron oxides. Applications: Used in coatings for magnetic inks, discs and tapes. (Hercules Inc).*

Magnum. A family of acrylonitrile-butadiene-styrene resins (ABS) used in appliances, automotive, recreational, medicine and electronics applications. (Dow Chemical Co Ltd).*

Magnum. 60" membrane, spiral wound configuration. Both TFC and Roga available. Applications: Reverse osmosis water treatment. (Allied-Signal Fluid Systems).*

Magnuminium. A proprietary trade name for magnesium-aluminium alloys.†

Magnus Green Salt. An ammoniacol platinum compound, $Cl_2Pt.2NH_3$.†

Magrex. Fluxes for magnesium alloys. (Foseco (FS) Ltd).*

Magrods. Magnesium oxide sticks. (The Chemical & Insulating Co Ltd).

Magspa. Fertilizer additive. (ICI PLC).*

Magtran. Magnesium fluoride optical qualities. (BDH Chemicals Ltd).

Mahogany Acid. Crude mixtures of sulphonic acid for the refining of petroleum sludge.†

Mahogany Brown. A sienna which has been ignited, ground wet, and made up in the form of pieces, and dried.†

Mahogany Nuts. See ACAJOU BALSAM. Njave butter.

Ma Huang. A Chinese drug. It contains t alkaloid ephedrine.†

Mahura. Bael fruit.†

Maidene 300. A copolymer similar to MALDENE 285 save that it is a 25% solution of the partial octyl ester in toluene. (Unichem).†

Maillechort. A nickel silver containing copper, zinc, nickel, and iron.†

Mainstay. Cod liver oil BP. Applications Dietary supplement with vitamins A a D. (Marfleet Refining Co).*

Maizena. A proprietary trade mark for cc starch.†

Maizolith. A material resembling hard rubber is obtained by the alkaline treatment of corn-stalk or corn-cob.†

Majamin. Sodium-β-tetralin sulphonate. Majamin-kalium, the potassium salt, and Majammonium, the ammonium salt, are similar products. They are added to soap and soap powders to increase lathering power.†

Majammonium. Ammonium-β-tetralin-sulphonate.†

Majeptil. Thioproperazine. (May & Bake Ltd).

Majolica. A pottery enamelled with a tin oxide enamel.†

Majunga Noir. A rubber yielded by *Landolphia perrieri.*†

Makalot. A proprietary trade name for a phenol-formaldehyde synthetic resin, used for moulding.†

Makon. Nonyl phenol ethoxylates. Applications: Detergents, emulsifiers. (Stepan Company).*

Makon NP6, NP10 and 4,8,12,14 and 30. Nonylphenol ethoxylate nonionic surfactant in liquid form. Application: Detergents, wetting agents and emulsifiers. (KWR Chemicals Ltd).

Makon OP6 and OP9. Octylphenol ethoxylate nonionic surfactant in liquid form. Application: Detergents, wetting agents and emulsifiers. (KWR Chemicals Ltd).

Makroblend. PC/polyester blends, characterized by their chemical resistance and low temperature impact strength. (Mobay Chemical Corp).*

Makrofol. Polycarbonate based electrical insulating film with high dielectric strength and long-term thermal stability up to 130°C. Applications: For dial plates and signalling and warning indicators on motor vehicles. (Bayer & Co).*

Makrolon. Polycarbonate for the manufacture of injection mouldings primarily for use in the electrical industry and for industrial mechanical applications. (Bayer & Co).*

Makrolon GV. A trade mark. Glass filled polycarbonate resin.†

Malabar Kino. See KINO, COCHIN.

Malabar Tallow. A fat obtained from the seeds of *Vateria indica.* It melts at 37.5° C., has a saponification value of 188.7-189.3, and an iodine value of 37.8-39.6.†

Malacca Primers. Primers for bonding materials such as rubber, wood, fabric, and metals. They usually have a rubber base.†

Malachite. (Mountain Green, Green Verditer, Copper Rust, Green Spar). A mineral used as a pigment. It is a hydrated basic copper carbonate, $CuCO_3.Cu(OH)_2$, and when ground is sold as Mountain green and Mineral green.†

Malachite Green. (Malachite Green B, New Victoria Green, New Green, Fast Green, Vert Diamant, Bitter Almond Oil Green, Benzaldehyde Green, Benzal Green, Diamond Green B, Victoria Green, Solid Green, Solid Green Crystals, Solid Green O, Benzoyl Green, Dragon Green, China Green, Hungary Green). A dyestuff. It is the zinc double chloride, oxalate, or ferric double chloride of tetra-methyl-*p*-amino-triphenyl-carbinol. Dyes wool, silk, jute, and leather bluish green, also cotton mordanted with tannin and tartar emetic. The name Malachite Green is also used at times for Malachite (*qv*).†

Malachite Green B. See MALACHITE GREEN.

Malachite Green G. See BRILLIANT GREEN.

Malachite Green, Spirit Soluble. The picrate of tetramethyldi-*p*-aminotri-phenylcarbinol. Used for colouring spirit varnishes.†

Malapaho. Oil of Panao, collected from *Dipterocarpus vernicifluus.*†

Malatex. A proprietary preparation of propylene glycol, malic acid, benzoic acid and salicylic acid. A de-sloughing agent. (Norton, H N Co).†

Malathion. An insecticide. Diethyl 2-(dimethoxyphosphinothioyl-thio)succinate. CYTHION.†

Malathion Dust. Insecticide. (Murphy Chemical Ltd).

Malathion Liquid. Insecticide. (Murphy Chemical Ltd).

Malay Camphor. See BORNEOL CAMPHOL.

Malazide. Plant growth inhibitor. (Fisons PLC).*

Maldene. Bonding agent. (Borg Warner Chemicals).

Maldene 285. A proprietary copolymer of butadiene and maleic anhydride supplied as a 25% solution in acetone for use as an intermediate. (Unichem).†

Maldene 286. A copolymer similar to MALDENE 285 save that it is a 25% solution of the partial ammonium salt in water. (Unichem).†

Maldene 288. A copolymer similar to MALDENE 285 save that the solids content in the solution is 35%. (Unichem).†

Maldene 289. A proprietary copolymer of butadiene and maleic anhydride partially ethyl-esterified and dissolved to 25% in ethyl alcohol. (Unichem).†

Maldene 292. A proprietary copolymer of butadiene and maleic anhydride supplied as a partially-butyl-esterified 25% solution in butanol. (Unichem).†

Maldene 293. A copolymer similar to MALDENE 285 save that it is a 25% solution of the partial amide-ammonium salt in water. (Unichem).†

Maldene 631. A copolymer similar to MALDENE 285 save that it is an 18% solution of a zinc-ammonium-complexed form of MALDENE 286 in water. (Unichem).†

Male Fern. The rhizome of *Aspidium felix-mas.*†

Malenite. A material containing an antimony double salt, $SbF_3.Na_2SO_4$, in addition to sodium fluoride, and the sodium compound of dinitro-phenol or dinitro-*o*-cresol. Used for impregnating wood.†

Malethamer. Maleic anhydride-ethylene polymer.†

Malezafin LV-4. Emulsifiable concentrate of butoxyethanol ester of 2,4-D acid. Applications: Low volatile, broadleaf herbicide for corn crops and pasture land. (Invequimica & CIA SCA).*

Malezafin 55 Plus. Emulsifiable concentrate of ethyl ester of 2,4-D acid and butoxyethanol ester of 2,4-DP acid. Applications: Broadleaf herbicide for pasture land. (Invequimica & CIA SCA).*

Malezafin 57 LV. Emulsifiable concentrate of a mixture of butoxyethanol esters of 2,4-D and 2,4-DP acid. Applications: Low volatile, broad leaf herbicide, brushkiller for pasture land. (Invequimica & CIA SCA).*

Malidone. Aloxidone.†

Malix. Tablets containing 2.5mg and 5mg glibenclamide BP. Applications: Indicated for the treatment of maturity-onset diabetes which is not adequately controlled by dietary measures alone. (Lagap Pharmaceuticals Ltd).*

Malladrite. Sodium fluosilicate, Na_2 SiF_2.†

Malleable Iron. (Wrought Iron). Practically pure iron, through which are scattered particles of slag or oxide.†

Malleable Nickel. Nickel commercially refined, and treated with a deoxidizing agent such as magnesium, and cast into ingots. It is suitable for hot or cold working.†

Mallebrein. Aluminium chlorate. An antiseptic and astringent.†

Mallet Alloy. A brass containing 74.6 per cent zinc and 25.4 per cent copper.†

Mallet Bark. The bark of *Eucalyptus occidentalis.* It contains from 35-52 per cent tannin.†

Malloydium. An alloy of 61 per cent copper, 23 per cent nickel, 14 per cent zinc, and 1 per cent iron. It is stated to be acid-resisting.†

Malogen CYP 200. Testosterone Cypionate. Applications: Androgen. (O'Neal, Jones & Feldman Pharmaceuticals).

Malogen LA 200. Testosterone Enanthate Applications: Androgen. (O'Neal, Jones & Feldman Pharmaceuticals).

Malonoben. A pesticide. It is 2-(3,5 di-*ter* butyl-4-hydroxybenzylidene) malononitrile.†

Maloprim. A proprietary preparation of dapaone and pyrimethamine used in the prophylaxis of malaria. (The Wellcome Foundation Ltd).*

Maloran. Substituted urea herbicide. (Ciba-Geigy PLC).

Malotte's Alloy. An alloy of 46 per cent bismuth, 20 per cent lead, 34 per cent tin. Melting-point is 203° F.†

Malros. Saponified rosin size. (Tenneco Malros Ltd).

Malt Extract. See EXTRACT OF MALT

Malta Grey. See NEW GREY.

Maltase. See GLUCASE.

Maltha. A variety of mineral tallow or wax found in Finland. It is also the name applied to certain types of soft bitumen

Malthactite. A clay of the fuller's earth type.†

Malthenes. See PETROLENES.

Malthite. A name for viscous bitumens.†

Maltine. See EXTRACT OF MALT.

Maltine, French. Diastase. See AMYLASE.

Maltisorb. Polyhydric alcohol. Applications: A non-carbohydrate sugar substitute in powder form, food (confectionery) and pharmaceutical. (Roquette (UK) Ltd).*

Maltobiose. (Amylon). Maltose.†

Maltol. 3-Hydroxy-2-methyl-4-pyrone.†

Maltose. Malt sugar, an isomer of cellobiose.α-4-glucosidoglucose.†

Maltyl. Dry malt extract. It contains about 90 per cent soluble carbohydrates.†

Maltzyme. Diastase, an enzyme.†

Maluminium. An alloy of 87 per cent aluminium, 6.4 per cent copper, 4.8 per cent zinc, 1.4 per cent iron, and traces of manganese, silicon, and lead.†

Mammol. Bismuth subnitrate. Applications: Pharmaceutic necessity. (Abbott Laboratories).

Man Oil. See BONE OIL.

Manaca. The dried root of *Brunfelsia hopeana,* of Brazil. An extract is used as a diuretic and diaphoretic.†

Manal. Manganese acetate. (Mechema).*

Manalox. Aluminium organic compounds. Applications: Rheology modifiers, ink industry, water repellent for masonry and damp proofing and water repellent for timber. (Manchem Ltd).*

Manalox AG. A proprietary trade name for Glycalox.†

Manalox AS. A proprietary trade name for Sucralox.†

Manchem. Metal carboxylates. Applications: Drier for paint or printing ink, additives for fuel oil or diesel oil. (Manchem Ltd).*

Manchester Brown. See BISMARCK BROWN.

Manchester Brown EE. (Bismarck Brown R, Manchester Brown PS, Vesuvine B). A dyestuff. It is the hydrochloride of toluenedisazo-*m*-tolylenediamine, $C_{21}H_{26}N_8Cl_2$. Dyes wool, leather, and tannined cotton reddish-brown.†

Manchester Brown PS. See MANCHESTER BROWN EE.

Manchester Yellow. See MARTIUS YELLOW.

Mancobride Mancanese. Cobalt bromide solution. (Mechema).*

Mancopper. A pesticide. It is an ethylene bisdithiocarbamate-mixed metal complex containing about 13.7 per cent manganese and about 4% copper.†

Mandarin G. See ORANGE II.

Mandarin G Extra. See ORANGE II.

Mandarin GR. See ORANGE T.

Mandarin Orange. See ORANGE III.

Mandelamine. Methenamine mandelate. Applications: Antibacterial. (Parke-Davis, Div of Warner-Lambert Co).

Mandelin's Reagent. An alkaloidal reagent, consisting of 1 gram of ammonium vanadate dissolved in 200 cc of concentrated sulphuric acid.†

Manderite. A preparation of asbestos.†

Mandol. Cefamandole nafate. Applications: Antibacterial. (Eli Lilly & Co).

Mandurin. A proprietary preparation of hexamine mandelate. A urinary antiseptic. (Harker Stagg).†

Manfloc. Sodium aluminate composition. Applications: Retention aid for paper water treatment. (Manchem Ltd).*

Mangabeira Rubber. A rubber from the small tree, *Hancornia speciosa* cultivated in Paraguay and Venezuela.†

Mangal. Chemical complex of manganese and aluminium. Applications: Drier for cobalt free paint or printing ink. (Manchem Ltd).*

Mangaloy. An alloy of nickel containing iron and manganese.†

Mangan-Monticellite. Synonym for Glaucochroite.†

Mangan-Neusilber. A nickel silver containing manganese. It contains from 59-72 per cent copper, 5-20 per cent zinc, 10-18 per cent nickel, and 2-20 per cent manganese.†

Manganaluminium Bronze. An alloy containing from 9-10 per cent manganese, 85.5-86 per cent copper, and 4 1/2-5 per cent aluminium.†

Manganar. Manganese arsenate.†

Manganated Linseed Oil. Linseed oil which has been boiled with manganese dioxide to increase its drying properties.†

Manganese Bistre. Manganese dioxide pigment.†

Manganese Black. Manganese dioxide, MnO_2.†

Manganese Blende. Native manganese sulphide, MnS.†

Manganese Blue. A pigment obtained by calcining a mixture of china clay (2 parts), oxides of manganese (3 parts), and barium nitrite, (8 parts).†

Manganese Boron. An alloy of manganese and boron. It is used for making other alloys.†

Manganese Brass. Variable alloys, usually containing 51-69 per cent copper, 1-4 per cent manganese, 0-3 per cent iron, 29-40 per cent zinc, 0-2 per cent tin, 0-2 per cent nickel, and sometimes aluminium. Alloys containing higher amounts of manganese approximate to 50-84 per cent copper, 12-25 per cent manganese, and 4-15 per cent zinc.†

Manganese Bronze. An alloy made by adding ferro-manganese or manganese to bronze. It usually contains from 82-83.5 per cent copper, 8 per cent tin, 5 per cent zinc, 3 per cent lead, and 0.5-2 per cent manganese, but an alloy containing 59 per cent copper, 39 per cent zinc. 1.5 per cent manganese, and 0.5 per cent iron, melting at 900° C. and having a specific gravity of 8.6, is also known as manganese bronze. The term is also used for manganese copper. The name is sometimes applied to a colouring matter produced on the fibre by the oxidation of manganese hydroxide.†

Manganese Cupro Nickel. This is usually an alloy containing from 65-83 per cent copper, 15-30 per cent manganese, and 2-8 per cent nickel.†

Manganese German Silver. An alloy of 80 per cent copper, 15 per cent manganese, and 5 per cent zinc.†

Manganese Green. See ROSENTHIEL'S GREEN.

Manganese Iron. See FERRO-MANGANESE.

Manganese Nickel. An alloy of from 51-82 per cent copper, 14-31 per cent manganese, and 3-16 per cent nickel. An alloy containing 95 per cent nickel and 5 per cent manganese, and melting at 1420° C. is also known as manganese nickel.†

Manganese Nickel Brass. Alloys containing 51-65 per cent copper, 5-40 per cent zinc, 0-2.78 per cent iron, 1.5-3.24 per cent manganese, and 2-18 per cent nickel.†

Manganese Nickel-silver. Alloys containing 50-70 per cent copper, 9-40 per cent zinc, 1-20 per cent manganese, and 2-20 per cent nickel.†

Manganese Ore A. A standard manganese ore. It contains 51.3 per cent manganese, 14.3 per cent available oxygen, 6.5 per cent silica, 1.3 per cent iron, and 0.22 per cent phosphorus.†

Manganese Ore, Brown. See MANGANITE.

Manganese Silver. See MANGANESE GERMAN SILVER.

Manganese Steel. An alloy of manganese and steel containing up to 20 per cent manganese. Commercial manganese steel contains 11-14 per cent mangane 1-1.3 per cent carbon, 0.3 per cent silicon. 0.05-0.08 per cent sulphur, an 0.05-0.08 per cent phosphorus.†

Manganese Velvet Brown. See UMBER.

Manganese Violet. (Nuremberg Violet, Mineral Violet, Permanent Violet). A pigment. It is a manganic phosphate, and is used as a mineral colour.†

Manganese White. A pigment. It is manganous carbonate, $MnCO_3$.†

Manganese-Aluminium Brass. An alloy of 56 per cent copper, 40 per cent zinc, 3 per cent manganese, and 1 per cent aluminium.†

Manganese, Battery. See MANGANESE BLACK.

Manganese, Bog. See WAD.

Manganese, Silico. See SILICO-MANGANESE.

Manganese, Thermit. See THERMIT MANGANESE.

Manganic. Nickel containing a small percentage of manganese.†

Manganin. Alloys usually containing 70-8 per cent copper, 4-25 per cent manganese, and 2-12 per cent nickel. One of the best varieties contains 83.6 per cent copper, 13.6 per cent manganese, 2.5 per cent nickel, and 0.. per cent iron. Later alloys consist of 8 81 per cent copper, 17-18 per cent manganese, and 1.5-2 per cent nickel. Used for electrical resistances.†

Manganite. (Brown Manganese Ore). A mineral. It is a hydrated oxide of manganese, $Mn_2O_3.H_2O$. It is also a name for a war gas which consisted of a mixture of hydrocyanic acid and arsenic trichloride. Synonym, ACERDESE.†

Mangano Steel. A proprietary trade name for a non-shrinking steel containing 1. per cent manganese, 0.2 per cent chromium and 0.95 per cent carbon.†

Mangano-Titanium. This is usually an all of manganese with 30 per cent titaniu and is used as a deoxidizer in making bronze and brass castings.†

Mangatrace. Manganese chloride, tetrahydrate. Applications: Supplement. (Armour Pharmaceutical Co).

Mangnamite. Graphite fibre. Applications: Structural strength applications. (Hercules Inc).*

Mangol. A powder consisting of basic magnesium hypochlorite. Used for testing alkaloids.†

Mangonic. A manganese-nickel alloy containing about 3 per cent manganese.†

Mangoxe. Manganese dioxide. Applications: Polysulphide sealant. (Mechema).*

Manguinite Cyanogen. Chloride, CN.C1.†

Manhardt's Aluminium Bronze. An alloy of 83.3 per cent aluminium, 6.25 per cent copper, 10.13 per cent tin, 0.16 per cent antimony, 0.05 per cent magnesium, and 0.08 per cent phosphorus.†

Manhatton Spirits. See ACETONE ALCOHOL.

Manila Elemi. Elemi from *Canarum commune.*†

Manjak. (Glance Pitch). A bitumen found in Mexico, South America, and West Indies. Used as a paint, as a roofing material, and in connection with drilling for oil.†

Manna. A sugary exudation which occurs in the rising sap of *Fraxinus ornus*, and *F. rotundifolia*. The crude material contains from 12-13 per cent water, 10-15 per cent sugar, 32-42 per cent mannitol, 40-41 per cent mucilaginous substances, organic acids, and nitrogenous matter. Australian manna is obtained from *Myoporum platycorpum*, and contains as much as 90 per cent of mannitol.†

Mannheim Gold. A brass containing 80 per cent copper and 20 per cent zinc. A jeweller's alloy. The term is also applied to a bronze consisting of 83.7 per cent copper, 9.3 per cent zinc, 7 per cent tin, with a little phosphorus.†

Mannol. Ethylacetanilide. A febrifuge. It is also used as a softening agent for cellulose esters.†

Mannolit. See CHLORAMINE T.

Mannotetrose. See LUPEOSE.

Manoblend. Mixtures of rubber chemicals and fillers. Applications: Rubbers, fillers. (Manchem Ltd).*

Manobond. Cobalt boro acylate. Applications: Rubber to metal bonding agents. (Manchem Ltd).*

Manocat. Metal carboxylates. Applications: Catalysts for unsaturated polyesters and polyurethane foams. (Manchem Ltd).*

Manofast. Thiourea dioxide. Applications: Reducing agent for dyes. (Manchem Ltd).*

Manofil. Chemically treated fillers. Applications: Rubber and plastic manufacture. (Manchem Ltd).*

Manolene 5203. A hexene copolymer.†

Manomet. Metal carboxylates. Applications: Stabilisers for PVC, pigment dispersing aids and 'kicker' for PVC foams. (Manchem Ltd).*

Manosec. Range of metal carboxylates based on synthetic organic acids. Applications: Paint driers. (Manchem Ltd).*

Manosil. Combinations of silica or silicates with speciality chemicals. Applications: Rubber, plastics. (Manchem Ltd).*

Manosperse. Dispersed rubber chemicals (various). Applications: Rubber industry. (Manchem Ltd).*

Manox. Insoluble sulphur. Applications: Rubber vulcanizing. (Manchem Ltd).*

Manox. Iron blue pigments. Applications: Printing inks, paints and fungicides. (Manox Ltd).†

Manoxol. Sulphosuccinate surface active agents. Applications: Wetting agent, dispersing agents. (Manchem Ltd).*

Manoxol MA. Sodium di-(methyl amyl) sulphosuccinate in 60% water/alcohol solution. Applications: As for Manoxol OT. (Manchem Ltd).

Manoxol OT60. Sodium di-iso-octyl sulphosuccinate. Applications: Used as a wetting and dispersing agent. (Harcros Industrial Chemicals).*

Manoxol OT, OT/P and OT/B. Sodium dioctyl sulphosuccinate, available as solid, water/alcohol solution, powder

(85% with 15% sodium benzoate), or specially pure grade for pharmaceutical use. Applications: Powerful wetting agents which can act as aids to dispersion, detergency and emulsification under appropriate circumstances. Usage includes adhesives; asbestos; agriculture and horticulture; bleaching; clays; determination of anionic detergents; dry cleaning and laundry; dust control; electro-plating; emulsion polymerization; etching; germicides; leather; metal technology; oil additives; paint/printing ink; paper making; pharmaceutical; photography; resins; rubber; textiles; wax polishes; wallpaper removal. (Manchem Ltd).

Manoxolot. A proprietary trade name for dioctyl sodium sulphosuccinate.†

Manqueta. (Manquta). African names for a fossil gum resin, resembling copal.†

Manro. Foam booster and stabilizer. Applications: Powder and liquid detergents; hair shampoos; hand cleaning jellies and cleaners. (Manro Products Ltd).

Manro ALS. Ammonium primary alcohol sulphate as a clear, pale yellow liquid. Applications: Foaming and wetting agent for hair, carpet and upholstery shampoos. (Manro Products Ltd).

Manro BA and NA. Dodecyl benzene sulphonic acid. Anionic surfactant supplied as a brown viscous liquid, low in free oil and inorganic content. Applications: Raw material for the production of detergents and emulsifiers such as powder and liquid detergents, hand cleaning gels, machine degreasers and tank cleaners. (Manro Products Ltd).

Manro BA and SBS. Anionic surfactant. Supplied as a clear pale yellow liquid with low inorganic content and controlled levels of minor constituents. Applications: Detergent and emulsifier eg light duty liquid detergents, hard surface cleaners; scouring and wetting in textile industries; emulsifiable insecticides; plastics; rubber. (Manro Products Ltd).

Manro BES. Range of anionic surfactants in liquid form. Anion: sodium. Cation: synthetic alcohol ethoxy sulphate. Application: High foaming agents use in liquid detergent formulations, bubb baths and shampoos. Industrial applications include use as a drilling a (Manro Products Ltd).

Manro D Paste. Sodium cetyl/oleyl sulphate as a white/pale yellow stiff paste. Applications: Wetting, dispersi and emulsifying agent, used in textile scouring and kier boiling. (Manro Products Ltd).

Manro DL28. Sodium primary alcohol sulphate as a water white liquid. Applications: Foaming and wetting agent with good handling characteristics, used in latex processin and emulsion polymerization. (Manro Products Ltd).

Manro DS 35. Sodium primary alcohol sulphate as a pale yellow liquid. Applications: Wetting and emulsification agent used in metal cleaner formulations etc. (Manro Products Ltd).

Manro HA. Dodecyl benzene sulphonic acid, derived from propylene tetramer. Anionic surfactant supplied as a brow viscous liquid, low in free oil and inorganic content. Applications: Raw material for production of detergents and emulsifiers such as powder and liquid detergents, hand cleaning gels, machine degreasers and tank cleaners. (Manro Products Ltd).

Manro HCS. Anionic surfactant. Supplie as a pale amber sparkling viscous liqui Applications: Emulsification of a wide range of chemicals, especially in emulsifiable solvent degreasing eg for hand cleaning jellies or engine cleaners (Manro Products Ltd).

Manro KXS. Potassium xylene sulphonate Anionic surfactant in liquid form. Applications: Cloud point depressant and solubilization agent, especially for heavy duty detergent formulations. (Manro Products Ltd).

Manro MA 35. Disodium octadecyl sulphosuccinamate as a pale cream liquid. Applications: Foaming agent in the manufacture of latex foams. (Man Products Ltd).

anro ML33. Monoethanolamine primary alcohol sulphate in pale yellow liquid form. Applications: Foaming and wetting agent for high quality hair shampoos and other toiletry products. (Manro Products Ltd).

lanro NEC. Anionic surfactant as a colourless liquid. Cation: sodium. Anion: natural or Ziegler alcohol ethoxy sulphate. Application: Foaming and cleaning agent for liquid and lotion hair shampoos. (Manro Products Ltd).

lanro NP. Range of nonionic surfactants of the nonylphenol ethoxylate type in liquid, paste or wax form. Applications: General purpose nonionic detergent bases; emulsifiers for agrochemicals and pesticides; emulsion polymerization. (Manro Products Ltd).

lanro PTSA. Toluene sulphonic acid. Anionic surfactant in liquid form. Applications: Intermediate in detergent manufacture, also has descaling and catalyst properties. Used in metal industries for resin bound sand castings; manufacture of esters, acetylation and resin production; hydrotrope production for detergents. (Manro Products Ltd).

lanro SDBS. Anionic surfactant. Supplied as a white/pale yellow paste with low inorganic content and controlled levels of minor constituents. Applications: Used as a detergent and emulsifier eg light duty detergents, hard surface cleaners; scouring and wetting in textile industries; rubber; plastics; emulsifiable insecticides. (Manro Products Ltd).

lanro SLS28. Sodium primary alcohol sulphate as a very pale yellow liquid. Applications: Foaming and wetting agent for carpet and upholstery shampoos ; foam rubber; emulsion polymerization in plastics and rubber industries. (Manro Products Ltd).

lanro SLS45. Sodium primary alcohol sulphate as a white paste. Applications: Foaming and wetting agent for hair shampoos. (Manro Products Ltd).

lanro STS. Sodium toluene sulphonate. Anionic surfactant in liquid form. Applications: Reduction of slurry viscosity before spray drying in heavy duty detergent powder manufacture. (Manro Products Ltd).

Manro SXS. Sodium xylene sulphonate. Anionic surfactant in liquid form. Applications: Cloud point depressant, coupling and stabilization agent for light and heavy duty detergents. (Manro Products Ltd).

Manro S.I.O.S. Sodium iso-octyl sulphate as a pale yellow, clear mobile liquid. Applications: Wetting agent with good alkali stability used in metal cleaner formulations and similar products. (Manro Products Ltd).

Manro TDBS. Anionic surfactant, pale amber viscous liquid. Applications: A detergent, mild to the skin, used in bubble baths, car shampoos and other detergents. Also used as an emulsifier in emulsification polymerization: (Manro Products Ltd).

Manro TL40. Triethanolamine primary alcohol sulphate in the form of a clear, pale amber liquid. Applications: Foaming and wetting agent for high quality hair shampoos and other toiletry products. (Manro Products Ltd).

Manro XSA. Xylene sulphonic acid. Anionic surfactant in liquid form. Applications: Intermediate in detergent manufacture, also has descaling and catalyst properties. Used in metal industries for resin bound sand castings; manufacture of esters, acetylation and resin production; hydrotrope production for detergents. (Manro Products Ltd).

Mansil. A proprietary preparation of oxamniquine. An antihelmintic. (Pfizer International).*

Mansonil. Veterinary preparation. Applications: Used against tapeworm infestation in ruminants, dogs and cats. (Bayer & Co).*

Mansu. Manganese sulphate. (Mechema).*

Mantadil Cream. A proprietary formulation of chlorcyclizine hydrochloride hydrocortisone acetate. Applications: Treatment of pruritic skin eruptions and other dermatoses. (The Wellcome Foundation Ltd).*

Mantin. Organotin/dithiocarbamate. (Ciba-Geigy PLC).

Manucol. Sodium alginate, food grade. (Kelco/AIL International Ltd).

Manucol Ester. Propylene glycol alginate. (Kelco/AIL International Ltd).

Manucol Ester EX/LL. A proprietary trade name for propylene glycol alginate. A food grade emulsifying agent. (Alginate Industries Ltd).†

Manucrème. Hand cream, to soothe dry and rough hands. (Richardson-Vicks Inc).*

Manugel. Alginate gelling powder. (Kelco/AIL International Ltd).

Manutex. Sodium alginate, technical grade. (Kelco/AIL International Ltd).

Manzate. Maneb fungicide. (Du Pont (UK) Ltd).

Maolate. Chlorphenesin carbamate. Applications: Relaxant. (The Upjohn Co).

Mapé. A coarse starch obtained from the fruit of *Inocarous edulis*.†

Mapico Browns. Proprietary trade names for iron oxide browns.†

Maple Sugar Sand. A by-product in the manufacture of maple sugar. The sap from the maple is evaporated in pans, and a precipitate forms when the water content is about 35 per cent. This precipitate is maple sugar sand. The chief constituent is calcium malate (60-80 per cent.) and malic acid is easily prepared from it.†

Maprenal. Etherified melamine/formaldehyde resins. Applications: Stoving finishes and acid curing lacquers. (Resinous Chemicals Ltd).*

Maprofix. A proprietary trade name for a sulphonated fatty alcohol used as a wetting agent and detergent.†

Maprofix ES-2. Sodium lauryl 2EO sulphate in liquid form. Application: Foaming and dispersing agent for shampoos, bubble baths and general cosmetics. (Millmaster-Onyx UK).

Maprofix ESY. Sodium lauryl 1EO sulphate in liquid form. Application: Emulsification, foaming and wetting agent for emulsion polymerization, shampoos and gels. (Millmaster-Onyx UK).

Maprofix MG. Magnesium lauryl sulphate in liquid form. Applications: Wetting, emulsifying and dispersing agent with low cloud point, used in non-alkaline shampoos and rug shampoos. (Millmaster-Onyx UK).

Maprofix NH and NHL. Ammonium la sulphate in liquid form. Application Detergency and foaming agent with high buffering capacity pH 6-7. Usec non-alkaline shampoos, for general cosmetic use and in industrial foams. (Millmaster-Onyx UK).

Maprofix TAS. Sodium tallow alcohol sulphate in paste form. Applications Detergent and dispersing agent used high temperature detergents and ore flotation. (Millmaster-Onyx UK).

Maprofix TLS. Triethanolamine lauryl sulphate in liquid form. Applications Detergent with good foam stability a low cloud point, used in clear shampc bubble baths, fine fabric detergents, industrial and household cleaners, an rug and upholstery shampoos. (Millmaster-Onyx UK).

Maprofix WAC-LA and LCP. Sodium lauryl sulphate in liquid form. High purity, low salt content and viscosity, light in colour. Applications: Wettir emulsification and dispersing agent, polymerization grade. (Millmaster-Onyx UK).

Maprofix WA, WAC and WAQ. Sodium lauryl sulphate in liquid or paste forn Applications: Detergency, foaming ar dispersing agent for liquid and cream shampoos; general cosmetic uses; rug and upholstery shampoos; latices and paint pigments. (Millmaster-Onyx U

Maprofix 563 and LK.USP. Sodium laur sulphate in powder or granular form. Applications: Detergency, foaming an emulsification agent , food and dentifrice grade. Also used in emulsio polymerization. (Millmaster-Onyx UK).

Maprofix 60S and 60N. Sodium or ammonium lauryl 3EO sulphate in liquid form. Application: Foaming surfactant for light duty detergents. (Millmaster-Onyx UK).

Mapromin. A proprietary trade name for sulphated fatty alcohol used as a wett agent.†

Maprosyl 30. Sodium N-lauroyl sarcosinate in liquid form. Applications: Wetting, emulsifying and dispersing agent which lowers the cloud point of cosmetic formulations and is virtually non-toxic. Used in chemically modified soap and corrosion inhibition. (Millmaster-Onyx UK).

Marabond. Calcium lignosulphonate. Applications: Oil well cementing. (Reed Lignin Inc).*

Marabout Silk. A white silk which still contains its gum. It is dyed and used for the manufacture of imitation feathers.†

Maracarb. Fractioned (low molecular weight) sodium lignosulphonate. Applications: Dispersant, humectant in dyestuffs, industrial cleaners and agricultural chemicals. (Reed Lignin Inc).*

Maracell. Purified/partially desulphonated sodium lignosulphonate. Applications: Lead acid battery paste expander, agricultural chemical formulations, stabilization of 2,4-D amine solutions in hard water, scale inhibitor for internal treatment of boiler water. (Reed Lignin Inc).*

Maracon. Calcium and sodium lignosulphonates. Applications: Concrete admixtures. (Reed Lignin Inc).*

Maranta. Arrowroot starch.†

Maranyl. A proprietary trade name for nylon moulding and extrusion compounds. (ICI PLC).*

Marasperse. Calcium and sodium lignosulphonates. Applications: Product line of specialty chemicals for use in a variety of binding, dispersing, complexing, stabilizing or copolymerizing applications. (Reed Lignin Inc).*

Marax. A proprietary preparation of hydroxyzine hydrochloride, ephedrine sulphate and theophylline. An antiasthmatic. (Pfizer International).*

Marbleloid. Flooring materials. (Weatherguard/Marbleloid Products Inc).*

Marblette. A proprietary trade name for a phenol-formaldehyde cast resin.†

Marbo. A proprietary trade name for a chlorinated rubber.†

Marbolith. See BAKELITE.

Marbon B. A proprietary trade name for a cyclo-rubber.†

Marbon Latex. Marmix reactive S.B.R. ABS latices. (Borg-Warner Chemicals).†

Marbon Resins. ABS polymers. Applications: Injection moulding, sheet extrusions. (Borg Warner Chemicals).*

Marbon White. A proprietary brand of Lithopone (*qv*).†

Marc Brandy Oil. See FUSEL OIL.

Marcain. A proprietary preparation of bupivacaine and adrenaline. Long acting local anaesthetic. (British Drug Houses).†

Marcaine. Bupivacaine hydrochloride. Applications: Anaesthetic. (Sterling Drug Inc).

Marcasite. (White Iron Pyrites, Coxcomb Pyrites, Radiated Pyrites). A mineral. It is disulphide of iron, FeS_2. The term is also occasionally applied to bismuth.†

Marcasol. See MARKASOL.

Marchies. See MARGINES.

Marcoumar. A proprietary preparation of phenprocoumon. An anticoagulant. (Roche Products Ltd).*

Marcs. The name given to the residue from wine factories, consisting of the stems and skins of grapes. It is used for making verdigris.†

Mareepa. Kernels of the fruits of the cokerite palm of British Guiana.†

Maretin. Naftalofos. Applications: Anthelmintic. (Bayer AG).

Marevan. A proprietary preparation of warfarin sodium. An anticoagulant. (British Drug Houses).†

Marexine-CA. FC-126 strain, frozen turkey herpes virus marek's vaccine. Applications: immunization of poultry. (Intervet America Inc).*

Marezine. Tablets and injections. Proprietary formulations of cyclizine hydrochloride or cyclizine lactate. Applications: Treatment of nausea and vomiting of motion sickness. (The Wellcome Foundation Ltd).*

Marezine Injection. Cyclizine ractate. Applications: Antinauseant. (Burroughs Wellcome Co).

Marezzo Marble. An artificial marble from oxychloride cement. It is used for building.†

Marfanil-Prontalbin. Sulphonamide powder. Applications: Used against wound infections in veterinary medicine. (Bayer & Co).*

Margalite. A phenol-formaldehyde resin product. It is used in the manufacture of varnishes and insulators.†

Margarodite. A mica having an appearance similar to talc.†

Margines. (Marchies). The residues obtained from the manufacture of olive oil.†

Margol. A mixture of volatile fatty acids, used as a flavouring material for margarine, to give it the taste of butter.†

Margosa Bark. Indian azadirach.†

Margosan-O. Active ingredient is azadirachtin, a tetranortriterpinoid. Applications: Growth regulator on many insects in various life stages, due to hormonal disruption preventing normal metamorphosis. Repellancy and/or antifeedency through olfactory or gustatory rejection by various flying and crawling pests. Currently restricted to non-food crop use. (Vikwood Botanicals Inc).*

Margraff Alloy. An alloy consisting of 58 per cent copper, 28 per cent tin, and.14 per cent zinc.†

Maricol. Magnesium ricinoleate, $(HO.C_{17}H_{32}.CO_2)_2Mg$†

Marignac's Salt. Potassium-stanno-sulphate.†

Marine Acid. Hydrochloric acid, HC1.†

Marine Blue. (New Methylene Blue NX). Mixtures of Methylene blue 2B with Methyl violet. Also see SOLUBLE BLUE.†

Marine Blue B. See BAVARIAN BLUE DSF.

Marine Fibre. A fibre obtained by dredging the shallow water of a gulf in South Australia. It is a hydrated lignocellulose.†

Marine Oil. A mixture of blown rape oil and a mineral oil. Used for marine engines.†

Marine Salt. Sodium chloride, NaC1.†

Marine Soap. A soap made from coco-nut oil, which is soluble in fresh and sea water.†

Marinol. Dronabinol. Applications: Anti-emetic. (Unimed Inc).

Mark 1330. A proprietary sulphur-containing organotin stabilized for use in PVC for injection moulding and pipe extrusion. (Argus Chemical Corporation).†

Mark 1414. A proprietary organotin mercaptide stabilizer for PVC used inthe extrusion of rigid pipe. (Argus Chemical Corporation).†

Mark 80. Organic additive system (non coumarin). Applications: Semi bright nickel electroplating (Duplex). (Harshaw Chemicals Ltd).*

Markasol. (Marcasol). Bismuth boro-phenate. Used as a substitute for iodoform.†

Markus Alloy. See NICKEL SILVERS.

Marlamid. A range of fatty acid alkanolamides. Applications: They stabilize foam, increase soil-suspending power and have a superfatting effect, they are intermediates for textile dressing agents and softeners and starting materials for fabric softeners. (Huls AG).†

Marlate Methoxychlor Insecticide. 50% methoxychlor wettable powder. Applications: Primarily for home use fo flowers, gardens, trees and ornamentals. (Kincaid Enterprises Inc).*

Marlate 2-MR Emulsifiable Insecticide. 24% methoxychlor in an emulsifiable solvent. Applications: Control of insects on livestock, agricultural premises, forest and shade trees, agricultural crops, ornamentals and flowers and for mosquito control. (Kincaid Enterprises Inc).*

Marlate 300 Flowable. 30% methoxychlor - 3.0 lbs/gal. Applications: Seed treatment for insect infestations. (Kincaid Enterprises Inc).*

Marlate 400 Flowable Concentrate. 40.5% methoxychlor - 4.0 lbs/gal.

Applications: For elms, forage and field crops, vegetables and seed treatment. (Kincaid Enterprises Inc).*

Marlate 50 WP. 50% methoxychlor wettable powder. Applications: Control of insects in stored grain and for livestock, vegetables and fruits. (Kincaid Enterprises Inc).*

Marlazin. A range of fatty amine polyglycol ethers. Applications: Used in alkali resistant and acid resistant industrial cleaners and dyeing auxiliaries. (Huls AG).†

Marlex. A proprietary trade name for high density polyethylenes coded as follows:
TR.885 for injection moulding thin walled containers. It has a density of 0.965 gm/c.c. and a melt index of 30.
TR.610 is for use as a wire and cable insulation. (Pacific Petroleums (Quebec).†

Marlex. Proprietary name for a range of polyethylenes. Grades include the following:- *55250*. Used in the injection moulding of household goods, toys and containers with thin walls.
6003. High-density. Used in the blow moulding of articles requiring a high degree of stiffness such as dust bins and similar large containers.
BHB 5003. A high-performance, high-density polyethylene copolymer used in the blow-moulding of detergent bottles and similar containers.
BHB 5012. Used in the injection moulding of large articles such as petrol tanks, industrial containers, school seating and outdoor furniture.
BHP TR-201. Used for the coating of wires and cables.
BHP TR-203. Used in the insulation of power cables.
BHP TR-551. A resin treated to resist attack by rodents.
BHM 5002. A high-density copolymer used in the blow-moulding of detergent bottles.
BHM 5402. Used in the blow moulding of light-duty bottles for bleaches and detergents.
BHM 5603. A resin used in the blow-moulding of bottles and tanks.
BHV TR-204. Possessing good resistance to stress cracking, it is used in the coating by extrusion of wire and cables.
BHV TR-553. A compound used in the extrusion coating of wire and cable. Specially treated to give protection against attack by rodents.
BMB 5040, 5065 and 5095. Used in the injection moulding of household goods and toys.
BMN 5565. Combining superior toughness, high impact strength, good resistance to stress cracking and easy processing characteristics, it is used in the injection moulding of industrial containers, boxes and crates.
BMN TR-880. A narrow distribution of molecular weight gives high impact strength and good resistance to warping. Used in the injection moulding of large flat surfaces.
BMN TR-980. Used in the rotational moulding of air ducts, etc.
BMN TR-995. Good impact strength. Used in the rotational moulding of bins, etc.
BXM 43065. Used to make sheets for forming into trays and similar large, thin, rigid or semi-rigid products.
CL-100-35. Used in the rotational moulding of agricultural and chemical sewage tanks and small engine and automobile fuel tanks, it cross- links during moulding.
EHB 6002. Used in the extrusion of film having a high degree of mechanical performance.
EHH 6007. A high-density resin used in the blow-moulding of bottles for milk, fruit juices and soft drinks.
EHB 6009. Of high density. Used in the blow-moulding of pharmaceutical bottles and industrial containers.
EHB 6009-MT. A variant of EHB 6009 giving off little odour.
EHM 6001. Used in the manufacture of barrier film.
EHM 6006. Of high density. Used in the blow-moulding of milk and water bottles.
EMB 6035. Having good impact strength and rigidity and a fast processing cycle, it is used in the injection moulding of high-quality components and containers.
EMB 6050. Giving a high-gloss finish

and possessing easy processing characteristics, it is used in the injection moulding of large household articles and toys.

EMN 6065. Used in the injection moulding of large articles requiring a high degree of stiffness.

EMN TR-885. Used in the injection moulding of thin-walled containers when good product rigidity and a high production rate are required.

EMN TR-960. Used in the rotational moulding of air ducts and large bottles.

EXM 55035. Used to make sheet for the thermo-forming of large components requiring maximum stiffness.

GF-830. Glass-reinforced and of high density, it is used in the injection moulding of articles requiring rigidity, ability to resist heat, and good bearing qualities.

HGR-120-01. A flame-retardant grade of polypropylene used in the manufacture of radio components and electrical mouldings.

HHM 5003. Used in the thermoforming of intricate components requiring to be accurately reproduced from a moulding.

HHM 5202 and 5502. High density polyethylenes used in the blow-moulding of bleach and detergent bottles. (Pacific Petroleums (Quebec)†

Marlex BMN TR-880. A high density polyethylene used for injection moulding. Applications: Milk cases, tote boxes, automotive and industrial components and high quality housewares. (Philips 66 Company).*

Marlex BMN 55500. A high density polyethylene used for injection moulding. Applications: Thin wall containers, toys and overcaps. (Philips 66 Company).*

Marlex CL-100. A crosslinkable high density polyethylene used for rotational moulding. Applications: Trash containers, industrial and agricultural chemical storage tanks, small engine snowmobile and automotive fuel tanks. (Philips 66 Company).*

Marlex CL-50. A crosslinkable high density polyethylene used for rotational moulding. Applications: Trash containers, chemicals and sewage tanks seats, boats, camper tops. (Philips 66 Company).*

Marlex EHM 6003. A high density polyethylene used for blow moulding. Applications: Large chemical tanks and parts such as trash cans requiring high stiffness. (Philips 66 Company).*

Marlex EHM 6003. A high density polyethylene used for sheet extrusion and thermoforming. Applications: Sheet, tote boxes, deep draw thermoformed parts. (Philips 66 Company).*

Marlex EHM 6006. A high density polyethylene used for blow moulding. Applications: Containers for bottling products such as milk and distilled water, fruit juices, etc. (Philips 66 Company).*

Marlex EHM 6007. A high density polyethylene used for blow moulding. Applications: Lightweight containers for bottling products such as milk and distilled water. (Philips 66 Company).*

Marlex EMN TR-885. A high density polyethylene used for injection moulding. Applications: Thin walled containers where higher production rate and rigidity is required. (Philips 66 Company).*

Marlex ER9-0002. A reinforced high density polyethylene used for sheet extrusion and thermoforming. Applications: Structural automotive applications, housings, tote boxes. (Philips 66 Company).*

Marlex ER9-0020. A reinforced high density polyethylene used for sheet extrusion and thermoforming. Applications: Structural automotive applications, seating, shrouds, housings. (Philips 66 Company).*

Marlex HGH-050. A polypropylene homopolymer used for injection moulding. Applications: Appliance parts and chemical equipment. (Philips 66 Company).*

Marlex HGL-050-01. A polypropylene homopolymer used for extrusion. Applications: Soda straws. (Philips 66 Company).*

Marlex HGL-050-01 (Antistatic). A polypropylene homopolymer used for injection moulding. Applications: Food containers, housewares and toys. (Philips 66 Company).*

Marlex HGL-120-01 (Antistatic). A polypropylene homopolymer used for injection moulding. Applications: Thin wall containers, medical supplies and housewares. (Philips 66 Company).*

Marlex HGL-200 (Antistatic). A polypropylene homopolymer used for injection moulding. Applications: Thin wall containers, medical supplies and housewares. (Philips 66 Company).*

Marlex HGL-350 (Antistatic). Controlled rheology polypropylenes used for injection moulding. Applications: Thin wall containers, housewares and closures. (Philips 66 Company).*

Marlex HGN-020-01. A nucleated polypropylene used for extrusion blow moulding. Applications: Drugs and toiletries. (Philips 66 Company).*

Marlex HGN-020-01. A polypropylene homopolymer used for extrusion. Applications: Profiles, sheet and solid phase pressure forming. (Philips 66 Company).*

Marlex HGN-120-01 (Nucleated). A polypropylene homopolymer used for injection moulding. Applications: Closures and food containers. (Philips 66 Company).*

Marlex HGN-200 (Nucleated). A polypropylene homopolymer used for injection moulding. Applications: Thin wall containers, medical supplies and housewares. (Philips 66 Company).*

Marlex HGN-200A. A controlled rheology polypropylene homopolymer used for injection moulding. Applications: Pill vials and medical jars. (Philips 66 Company).*

Marlex HGN-350 (Nucleated). Controlled rheology polypropylenes used for injection moulding. Applications: Thin wall containers, housewares and closures. (Philips 66 Company).*

Marlex HGX-010. A polypropylene homopolymer used for extrusion, blow moulding and fibre extrusion. Applications: Drugs, toiletries and strapping. (Philips 66 Company).*

Marlex HGX-030. A polypropylene homopolymer used for general extrusion and slit film and monofilament extrusion. Applications: Woven carpet backing and bags, rope and cordage. (Philips 66 Company).*

Marlex HGX-040. A polypropylene homopolymer used for general extrusion and slit film and monofilament extrusion. Applications: Woven carpet backing and bags, rope and cordage. (Philips 66 Company).*

Marlex HGX-330 (Controlled Rheology). A polypropylene used for multifilament extrusion. Applications: Multifilament staple. (Philips 66 Company).*

Marlex HGZ-050-02. A polypropylene homopolymer used for injection moulding. Applications: Food containers, housewares and toys. (Philips 66 Company).*

Marlex HGZ-120-02. A polypropylene homopolymer used for injection moulding. Applications: Thin wall containers, medical supplies and housewares. (Philips 66 Company).*

Marlex HGZ-120-04. A polypropylene used for multifilament extrusion. Applications: Multifilament staple. (Philips 66 Company).*

Marlex HGZ-200. A polypropylene homopolymer used for injection moulding. Applications: Thin wall containers, medical supplies and housewares. (Philips 66 Company).*

Marlex HGZ-350. Controlled rheology polypropylenes used for injection moulding. Applications: Thin wall containers, housewares and closures. (Philips 66 Company).*

Marlex HHM TR-130. A medium density polyethylene used for film extrusion. Applications: Merchant bags, produce bags, trash bags and multiwall bag liners. (Philips 66 Company).*

Marlex HHM TR-140. A high density polyethylene used for film extrusion. Applications: Produce bags, merchant bags, multiwall bag liners and trash bags. (Philips 66 Company).*

Marlex HHM TR-144. A high density polyethylene used for film extrusion. Applications: Merchant bags, produce bags, multiwall bag liners and trash bags. (Philips 66 Company).*

Marlex HHM TR-210. A high density polyethylene used for wire and cable coating. Applications: Primary insulation for telephone conductors. (Philips 66 Company).*

Marlex HHM TR-226. A high density polyethylene used for wire and cable coating. Applications: Foam skin insulation on telephone singles. (Philips 66 Company).*

Marlex HHM TR-230 Black. A high density polyethylene used for wire and cable coating. Applications: Telephone cable jacketing. (Philips 66 Company).*

Marlex HHM TR-232 Black. A high density polyethylene used for wire and cable coating. Applications: Power cable jacketing. (Philips 66 Company).*

Marlex HHM TR-250 Black. A high density polyethylene used for wire and cable coating. Applications: Aerial cable jacketing of drop wire, line wire and tree wire. (Philips 66 Company).*

Marlex HHM TR-418 (Black, Orange). A polyethylene used for pipe extrusion and injection moulding of pipe fittings. Applications: Gas distribution pipe, potable water pipe, engineered pipe. (Philips 66 Company).*

Marlex HHM 4903. A high density polyethylene used for blow moulding. Applications: Large industrial containers, fuel tanks and automotive parts. (Philips 66 Company).*

Marlex HHM 4903. A high density polyethylene used for sheet extrusion and thermoforming. Applications: Tote boxes, housings, trays. (Philips 66 Company).*

Marlex HHM 5202. A high density polyethylene used for blow moulding. Applications: Bleach and detergent containers and chemical packaging. (Philips 66 Company).*

Marlex HHM 5202. A high density polyethylene used for sheet extrusion and thermoforming. Applications: Tote boxes, trays, industrial housings, shrouds. (Philips 66 Company).*

Marlex HHM 5502. A high density polyethylene used for blow moulding. Applications: Bottles for bleach and detergents, industrial parts, ice chests and coolers. (Philips 66 Company).*

Marlex HHM-4515. A high density polyethylene used for injection moulding. Applications: 5-gallon shipping containers, institutional seating, fuel tanks and closures. (Philips 66 Company).*

Marlex HLM-020. A polypropylene homopolymer used for injection blow moulding. Applications: Drugs, toiletries, cosmetics and spices. (Philips 66 Company).*

Marlex HLN-120-01. A polypropylene homopolymer used for injection moulding. Applications: Closures and food containers. (Philips 66 Company).*

Marlex HLN-200 (Antistatic, Nucleated). A polypropylene homopolymer used for injection moulding. Applications: Thin wall containers, medical supplies and housewares. (Philips 66 Company).*

Marlex HLN-350 (Antistatic, Nucleated). Controlled rheology polypropylenes used for injection moulding. Applications: Thin wall containers, housewares and closures. (Philips 66 Company).*

Marlex HMN TR-942. A high density polyethylene used for rotational moulding. Applications: Agricultural and industrial containers and food handling containers. (Philips 66 Company).*

Marlex HMN 5060. A high density polyethylene used for injection moulding. Applications: Industrial containers, fuel tanks, closures and feeder tubs. (Philips 66 Company).*

Marlex HMN 54140. A high density polyethylene used for injection moulding. Applications: Industrial containers, crates and boxes, large frozen food containers. (Philips 66 Company).*

Marlex HMN 5580. A high density polyethylene used for injection moulding. Applications: Pails, housewares, closures and crates requiring good toughness. (Philips 66 Company).*

Marlex HMN 6060. A high density polyethylene used for injection moulding. Applications: Trays, industrial parts, beverage crates and safety helmets. (Philips 66 Company).*

Marlex HMN-4550. A high density polyethylene used for injection moulding. Applications: Agricultural and industrial containers, food handling containers, seating and fuel tanks. (Philips 66 Company).*

Marlex HMN-938. A high density polyethylene used for rotational moulding. Applications: Agricultural and industrial containers, FDA approved drums and tanks, trash containers. (Philips 66 Company).*

Marlex HMX-020-01 (Lubricant). A polypropylene homopolymer used for injection blow moulding. Applications: Drugs, toiletries, cosmetics and spices. (Philips 66 Company).*

Marlex HXM 50100. A high density polyethylene used for blow moulding. Applications: 55-gallon shipping containers, gasoline and chemical tanks. (Philips 66 Company).*

Marlex HXM 50100. A high density polyethylene used for sheet extrusion and thermoforming. Applications: Large formed parts, cattle feeders, pallets, boats. (Philips 66 Company).*

Marlex RGX-020. A polypropylene random copolymer used for extrusion blow moulding. Applications: Syrup bottles, food containers and toiletries. (Philips 66 Company).*

Marlex RGX-020. A polypropylene random copolymer used for general extrusion. (Philips 66 Company).*

Marlex RGX-020 (Antistat). A polypropylene random copolymer used for extrusion blow moulding. Applications: Syrup bottles, food containers and toiletries. (Philips 66 Company).*

Marlex RMN-020C. A polypropylene random copolymer used for injection blow moulding. Applications: Drugs, toiletries, cosmetics and spices. (Philips 66 Company).*

Marlex RMN-020C. A polypropylene random copolymer used for injection moulding. Applications: Premium housewares, cassette cases and medical supplies. (Philips 66 Company).*

Marlex RMX-020. A polypropylene random copolymer used for injection blow moulding. Applications: Drugs, toiletries, cosmetics and spices. (Philips 66 Company).*

Marlex 1708. A proprietary trade name for a low density polyethylene suitable for the extrusion of heavy duty film. Type 1 Class A Grade 4 resin has a density of 0.917 and a melt index of 0.8. (Pacific Petroleums (Quebec)†

Marlican. Alkyl benzene preparations. Applications: Used as a starting material for alkylbenzene sulphonates, secondary plasticizer, viscosity depressant for PVC plastisols. (Huls AG).†

Marlie's Alloy. An alloy containing 10 per cent iron, 35 per cent nickel, 25 per cent brass, 20 per cent tin, and 10 per cent zinc, which has been quenched in a mixture of acids.†

Marlinat. A range of sodium salts of sulphosuccinic acid esters. Applications: Highly effective wetting agents for the textile, paint and paper industries, used for the preparation of cleaners for sensitive textiles. (Huls AG).†

Marlipal. A range of fatty alcohol polyglycol ethers. Applications: Non-ionic surfactants used as bases for detergents and dish-washing preparations, having dispersing, wetting, detergent, cleaning, soil-suspending and homogenizing properties. (Huls AG).†

Marlon. A range of alkylbenzenesulphonates. Applications: Used for the manufacture of detergents and cleaners. (Huls AG).†

Marlon AMX. Anionic surfactant as a clear, homogenous, brownish liquid above 25°C. Applications: Base material for the following: dry-cleaning detergents; degreasing agents for metal working industries; floor cleaning. (Huls (UK) Ltd).

Marlon AS3. Anionic surfactant in acid form, supplied as a liquid. Applications: Intermediate for the manufacture of detergent raw materials. (Huls (UK) Ltd).

Marlon A350. Anionic surfactant in liquid form. Applications: High quality products with low salt content, used in liquid detergents and cleaning materials for domestic and industrial use. (Huls (UK) Ltd).

Marlon A360, A365 and A375. Anionic surfactant in paste form. Applications: High quality products with low salt contents, used for paste detergents and cleaning materials for domestic and industrial use. (Huls (UK) Ltd).

Marlon A390, A396 and ARL. Anionic surfactant in solid form. Applications: High quality products with low salt contents, used for powder detergents and cleaning materials for domestic and industrial use. (Huls (UK) Ltd).

Marlophen. A range of nonylphenol polyglycol ethers with wetting, detergent, dispersing and homogenizing properties. Applications: Used as textile and paper auxiliaries, as wetting agents for coal, rock dusts and pigments and as an additive for concrete manufacture. (Huls AG).†

Marlophen DNP. Nonionic surfactant of the dinonylphenol ethoxylate type in paste form. Application: Low foam dispersing agent for textile and paper auxiliaries. (Huls (UK) Ltd).

Marlophen X. Alkyl phenol ethoxylate nonionic surfactant in the form of a brown liquid. Application: Low foam wetting agent for pigment wetting, binding of coal and stone dusts, concrete additive. (Huls (UK) Ltd).

Marlophen 80 Series and 800 Series. Nonylphenol ethoxylate nonionic surfactants in liquid, paste or wax form. The range includes oil-soluble and water-soluble surfactants and synergism is often obtained with anionic ingredients. Application: Liquid, paste-like and solid detergents and scouring agents; textile and leather auxiliaries; dry cleaning; cellulose and paper; ceramics and concrete; dust control in many industries. (Huls (UK) Ltd).

Marlophor. A range of partial phosphate esters in the form of acids and salts. Applications: Used as low foam wetting agents for alkaline solutions, starting materials for the manufacture of acid cleaners, water and oil-soluble wetting agents for the textile and paper industries, antistatic agents and drycleaning detergents. (Huls AG).†

Marlophor CS, DS and NP. Organic phosphate ester surfactants. Applications: Base components for acid cleaning materials, with corrosion protection and rust dissolving properties and which are inert to common paint films. Used for hard surface cleaning, eg metal; glass; ceramics. (Huls (UK) Ltd).

Marlophor FC. Sodium alkylpolyglycol ether phosphate in liquid form. Applications: Low foam wetting agent for detergents for dishwashers and industrial cleaning. (Huls (UK) Ltd).

Marlophor MD. Organic phosphate ester surfactant. Applications: Wetting agent (oil and water soluble) and antistat, used in paper; textiles; natural and synthetic fibres; dry cleaning detergents. (Huls (UK) Ltd).

Marlophor T10. Amino alkyl-polyglycol ether phosphate in liquid form. Applications: Special formulation for antistatic dry cleaning detergents. (Huls (UK) Ltd).

Marlopon. A range of alkylbenzenesulphonates. Applications: Used for the manufacture of cosmetic detergents and dishwashing agents. (Huls AG).†

Marlopon ADS50. Anionic surfactant in liquid form. Applications: Base material for liquid detergents containing phosphates. (Huls (UK) Ltd).

Marlopon AT50. Anionic surfactant in liquid form. Applications: Neutral detergent raw material for cosmetic detergents, dishwashers, car shampoos etc. (Huls (UK) Ltd).

Marlosol. A range of polyglycol esters of fatty acids. Applications: Starting materials for preparing agents used in the synthetic fibre industry. (Huls AG).†

Marlotherm. A range of benzyltoluenes. Applications: Suitable for use as heat transfer mediums. (Huls AG).†

Marlowet. A wide range of emulsifiers. Applications: Emulsifiers for mineral oils, hydrocarbons, waxes, oleic acids, solvents, pesticides, cold cleaners, spindle oils, textile lubricants, furniture polishes and leather dressings etc. (Huls AG).†

Marlox. A range of alkylene oxide addition products. Applications: Components of low-foaming detergents and cleaners particularly for low-foaming dishwashing agents and for low-foaming textile finishing agents and processing aids. (Huls AG).†

Marls. Natural mixtures of clay and chalk (aluminium silicate and calcium carbonate). Used in the manufacture of cements. The term is also applied to friable earths which are devoid of chalk, such as those of Staffordshire.†

Marme's Reagent. Consists of 10 parts cadmium iodide, CdI_2, and 20 parts potassium iodide, KI, dissolved in 80 parts water. Used for testing for alkaloids.†

Marmite. A yeast extract. It is a food preparation resembling meat extract.†

Marmo Bardiglio de Bergamo. Vulpinite, a variety of anhydrite, mixed with silica. Used for ornamental purposes.†

Maroon. See MAGENTA.

Maroon S. (Grenat S, Acid Cerise, Cardinal Red S, Acid Maroon). Impure qualities of Acid Magenta (*qv*).†

Marphos. Phosphoric acid. (Borg-Warner Chemicals).†

Marplan. A proprietary preparation of isocarboxazid. An anti-depressant. (Roche Products Ltd).*

Marquis's Reagent. A solution of formalin in sulphuric acid. An alkaloidal reagent.†

Mars Brown. See BURNT SIENNA.

Mars Orange. A pigment obtained by the calcination of an artificial ochre, such as Mars Yellow (*qv*).†

Mars Purple. A pigment produced by the calcination of an artificial ochre.†

Mars Red. (Rouge de Mars). A pigment. It is an artificial ochre, prepared by calcining an artificial ochre such as Mars yellow. The term is also applied to an acid red dye. See AZORUBINE S.†

Mars Violet. (Purple Ochre). A pigment. It is prepared by calcining an artificial ochre such as Mars Yellow.†

Mars Yellow. A pigment. It is an artificial preparation of yellow ochre and is obtained by dissolving ferrous sulphate in water, warming, and introducing strips of zinc. The bright yellow precipitate is Mars yellow. The average composition is from 51-53 per cent ferric hydroxide, 23-24 per cent calcium oxide, 18-18.5 per cent carbon dioxide, and 3.5 per cent water. The yellow variety contains calcium sulphate.†

Marseilles Soap. Olive oil soap.†

Marsh Gas. (Light Carburetted Hydrogen). Methane, CH_4.†

Marsilid. A proprietary preparation of iproniazid. An anti-depressant. (Roche Products Ltd).*

Marsipol. Leather finishes. (ICI PLC).*

Martensite. A solid solution of carbon in iron, and is a characteristic constituent of steel which has been tempered at a temperature a little above the transformation point.†

Martifin. Ground and finely precipitated aluminium hydroxide. Applications: Used as a filler and coating pigment for the paper and cardboard industry. (Lonza Limited).*

Martin Steel. (Open Hearth Steel). Steel obtained in the Martin process by melting from 75 per cent of cast iron in a reverberatory furnace with the necessary quantity of wrought iron to obtain the required amount of carbon.†

Martinal. Ground and finely precipitated aluminium hydroxide. Applications: Used as a flame retardant for plastics and rubber. (Lonza Limited).*

Martinite. A mineral, $Ca_5H_2(PO_4)_4$.†

Martino's Alloys. Alloys containing 17.25 per cent pig iron, 3-4.5 per cent ferro-manganese, 1.5-2 per cent chromium, 5.25-7.5 per cent tungsten, 1.25-2 per cent aluminium, 0.5-0.75 per cent nickel, 0.75-1 per cent copper, and 65-70 per cent wrought iron. Used for drilling and cutting tools.†

Martin's Cement. A similar cement to Keene's, except that potassium carbonate solution is used instead of alumina.†

Martipol. Speciality aluminium oxide product. Applications: Used as polishing aluminas for the ceramic industry. (Lonza Limited).*

Martisorb. Speciality aluminium oxide product. Applications: Used for the purification of water. (Lonza Limited).*

Martius Yellow. (Naphthol Yellow, Naphthylamine Yellow, Gold Yellow, Primrose, Jaune D'Or, Manchester Yellow, Naphthalene Yellow, Naphthaline Yellow). A dyestuff. It is the ammonium, sodium, or calcium salt of dinitro-α-naphthol. Dyes wool golden yellow from an acid bath.†

Martoxin. Speciality aluminium oxide product. Applications: Used as a coating pigment for carbonfree self-copying paper. (Lonza Limited).*

Marvylan. Polyvinylchloride. Applications: Used in the plastics processing industry, applications in building construction (tubes and pipes, profiles and cables); in the packaging industry (bottles); as synthetic leather (bags, wall covering, clothing); in hoses etc. (DSM NV).*

Marweld M-17. Pipe and flange thread sealant. Epoxy resin, liquid sealant, two component, flexible. Applications: For sealing threads in flanged ductile iron and cast iron water pipes, EPA approved. (RJ Manufacturing Inc).*

Marzine. Travel sickness remedy. (The Wellcome Foundation Ltd).

Masmoran. A proprietary preparation of hydroxyzine. An ataractic. (Pfizer International).*

Masol. Cement mixture for pump packing in mines roadways. (Foseco (FS) Ltd).

Masonry Stain and Seal. Methylmethacrylate acrylic, pigments and polysiloxane resins in an aliphatic and aromatic solvent vehicle system. Applications: Semi-transparent stain for masonry. (Nova Chemical Inc).*

Masoten. Veterinary preparation. Applications: For the control of ectoparasites in fish. (Bayer & Co).*

Massa Estarinum. Neutral hard fats based on mixtures of triglycerides. Applications: Preparation of suppositories. (Dynamit Nobel Wien GmbH).*

Massaranduba. (Balata Rans, Brittle Balata). A pseudo gutta-percha derived from the sap of the Brazilian cow tree, *Mimusops elata*.†

Massecuite. The boiled mass of beet sugar syrup. It is a semi-solid mass formed during the evaporation of the sugar juice, and consists of sugar crystals and a thick syrup. It contains from 3.5-7 per cent water.†

Massicot. See LITHARGE.

Masterblok. Solid phenolic ester in panels. Applications: Master patterns, N/C trials, models, fixtures and vac forming tools. (JR Technology Ltd)..*

Masterbond. Cellulose acetate laminating adhesives. (The Scottish Adhesives Co Ltd).*

Mastercarb. Mineral filled concentrates in polyolefines. Applications: Plastics mouldings, sheet and film. (Colloids Ltd).*

Mastercolor. Masterbatches of thermoplastics. Applications: Colour and additive concentrates used in plastics. (Ampacet Corporation).*

Masterflam. Halogenic and non-halogenic masterbatches for making self-extinguishing thermoplastics. Applications: 'Masterflam' in a given let-down ratio with the most important thermoplastics (VAMP produces a Masterflam-Grade for each thermoplastic) makes it flame retardant in compliance with the most important national and international standards. (VAMP srl).*

Masteril. A proprietary preparation of drostanolone propionate. Treatment of mammary carcinoma. (Syntex Laboratories Inc).†

terone. Dromostanolone propionate. pplications: Antineoplastic. (Syntex aboratories Inc).

terwood. Wood-filled thermoplastic olymers. Applications: Plastics ouldings and extrusions. (Colloids td).*

tic. The name for an important resin btained from *Pistachia lentiscus*, from arious parts of the Mediterranean oast. It is used in the manufacture of pirit varnishes. The term Mastic is also pplied to a mixture of asphalt rock and rinidad pitch, used for paving.†

ticillin C/Masticillin M. Preparations ontaining penicillin, streptomycin or enicillin sulphonamide. Applications: or the treatment of mastitis in eterinary medicine. (Bayer & Co).*

strite. Iodophor mastitis treatment. The Wellcome Foundation Ltd).

suron. A proprietary trade name for a ellulose acetate plastic.†

talex. X-ray developer system. (May & Baker Ltd).

tali. Rubber obtained from the roots of various species of *Apocynaceae*.†

té. (Jesuit's Tea, St. Bartholomew's Tea). Paraguay tea, the dried leaves and shoots of an evergreen, *Ilex paraguayensis*. It contains caffeine and theobromine.†

teflex. Polyethylene resin. Applications: Tennis courts. (Sumitomo Bakelite, Japan).*

texil. Textile auxiliary chemicals. (ICI PLC).*

thesius Metal. A proprietary trade name for an alloy of lead with calcium and strontium in small amounts.†

tico-camphor. A camphor from the Peruvian matico.†

tikus. Rodenticide containing brodifacoum. (ICI PLC).*

trikerb. Soil and leaf active herbicide. (Rohm & Haas (UK) Ltd).

trix Alloy. An alloy of 48 per cent bismuth. 28.5 per cent lead. 14.5 per cent tin, and 9 per cent antimony. It expands on cooling and is used to hold tools in position.†

Matrix Brass. See BRASS.

Matromycin. A proprietary preparation of troleandomycin. An antibiotic. (Pfizer International).*

Matromycin Tao. A proprietary preparation of oleandomycin phosphate. An antibiotic. (Pfizer International).*

Matromycin-T. A proprietary preparation of troleandomycin. An antibiotic. (Pfizer International).*

Matt Blues. Cobalt aluminate colours used in the ceramic industry. They are prepared by calcining cobalt oxide with ammonia alum, with varying amounts of zinc oxide.†

Matt Salt. Acid ammonium fluoride, $NH_4F.HF$.†

Matte. See COARSE METAL.

Matteucinol. 6, 8-dimethyl-5, 7-dihydroxy-4'-methoxyflavanone.†

Mattheylec. Conductive adhesives and coatings. Applications: Metallizing preparations for the electrical and electronic industries. (Johnson Matthey Chemicals Ltd).*

Matthieu-Plessy Green. See CHROMIUM GREEN.

Matulane. Procarbazine Hydrochloride. Applications: Antineoplastic. (Hoffmann-LaRoche Inc).

Maucherite. A mineral, Ni_3As_2.†

Mauritius Elemi. Elemi from *Canarum mauritianium*.†

Mauve. See MAUVEINE.

Mauve Dye. See MAUVEINE.

Mauveine. (Mauve, Mauve Dye, Chrome Violet, Aniline Violet, Aniline Mauve, Aniline Purple, Perkin's Violet, Perkin's Purple, Violine, Violeine, Rosolane, Indisin, Anileine, Phenamin, Phenamein Purpurin, Rosein, Tyraline, Lydin). A dyestuff prepared by the oxidation of a mixture of aniline and toluidine by potassium chromate. Dyes silk reddish-violet.†

Mawele. A millet of East Africa.†

MaxEPA. Fish oil concentrate. Applications: Triglyceride - lowering agent. (Marfleet Refining Co).*

Maxhete. A steel containing nickel, chromium, tungsten, copper, and silicon.†

Maxibolin. Ethylestrenol. Applications: Anabolic. (Organon Inc).

Maxicam. Isoxicam. Applications: Anti-inflammatory. (Parke-Davis, Div of Warner-Lambert Co).

Maxidex. A proprietary preparation of dexamethasone, benzalkonium chloride and phenylmercuric nitrate. Anti-inflammatory eye drops. (Alcon Universal).†

Maxidex Ointment. Dexamethasone sodium phosphate. Applications: Glucocorticoid. (Alcon Laboratories Inc).

Maxiflor. Diflorasone diacetate. Applications: Anti- inflammatory. (Herbert Laboratories, Dermatology Div of Allergan Pharmaceuticals Inc).

Maxigard. Diesel engine water treatment. A multicomponent nitrite borate cooling water treatment. Applications: Used to prevent corrosion and mineral deposits as well as corrosion due to cavitation in medium and high speed diesel engines. (Ashland Chemical Company).*

Maxilon. Modified basic dyes. (Ciba Geigy PLC).†

Maxilvry Steel. A proprietary high nickel-chromium steel containing copper. It is stated to be corrosion resisting, and particularly resistant to attack by cider.†

Maximate. Contains 71g/litre carbendazim and 457g/litre manels. Applications: Cereal fungicide. (Farmers Crop Chemicals Ltd).*

Maximite. An explosive. It is similar to cordite.†

Maxitrol. A proprietary preparation of DEXAMETHASONE, NEOMYCIN and POLYMYXIN B used as eye-drops. (Alcon Universal).†

Maxium Metal. Castings of magnesium metal.†

Maxolon. A proprietary preparation of metoclopramide hydrochloride. An anti-emetic. (Beecham Research Laboratories).*

Maxon. Polyglyconate. Applications: Surgical aid. (Davis & Geck).

May Green. (Mignonette Green). Pigments. They are mixtures of Chrome yellow and Paris blue, with yellow predominating. Also see CHROME GREEN.†

Mayari Iron. An iron made from Cuban ores. Small amounts of vanadium and titanium are present which give strengt to the metal obtained from these ores.†

Mayari Steel. A Cuban low nickel-chromium steel.†

Mayclene. Selective herbicide. (FBC Ltd).

Mayer's Albumen. A fixing agent used in microscopy. It consists of 50 cc white o egg, 50 cc glycerin, and 1 gram sodium salicylate.†

Mayer's Carmalum. See CARMALUM.

Mayer's Haemalum. See HAEMALUM.

Mayer's Paracarmine Stain. See PARACARMINE.

Mayer's Picrocarmine Stain. See PICROCARMINE.

Mayer's Solution. Mercuric iodide dissolved in aqueous potassium iodide. An alkaloidal reagent.†

Maypon. Salts of collagen polypeptide fatt acid condensate. Applications: Used as hair care additives. (Harcros Industrial Chemicals).*

Maytee. Fenugreek seeds.†

Mazam. A dextrin of high molecular weight.†

Mazanor. Mazindol. Applications: Anorexic. (Wyeth Laboratories, Div of American Home Products Corp).

Mazarine Blue. See WILLOW BLUE.

M.B. General chemicals and pharmaceuticals. (May & Baker Ltd).*

M.B.A. N,N'-Methylenebisacrylamide. (Cyanamid BV).*

MBS. A terpolymer of methylmethacrylate, butadiene and styrene. Its properties include rigidity, hardness, high impact strength, heat resistance and good clarity.†

M B T. 2-Mercaptobenzothiazole. A medium temperature general-purpose accelerator for natural and synthetic rubbers. Very active above 240°F (116°C); moderately low activation temperature. Used in tyre tread, carcas and wire insulation. (Uniroyal).*

M B T S. Benzothiazyl disulphide. A general-purpose accelerator similar to MBT but with a higher activation temperature for greater processing safety. Very active above 287°F (142°C). Used in both natural and synthetic rubbers. (Uniroyal).*

McCrorie's Stain. A microscopic stain. Solution A contains 1 gram tannic acid, 1 gram potash alum dissolved in 40 cc distilled water. Solution B consists of 0.5 gram night blue dissolved in 20 cc absolute alcohol. The solutions are mixed and filtered.†

McGill Metal. A proprietary trade name for a group of copper-aluminium-iron alloys, one of which contains 89 per cent copper, 9 per cent aluminium, and 2 per cent iron.†

McNamee Clay. See LANGFORD CLAY but this clay has a higher brightness. Applications: Low cost reinforcer and inert filler for paint, paper, rubber, ceramics, plastics and specialities. (Vanderbilt Chemical Corporation).*

MCPB. 3-phenoxybutyric acid. A systemic fungicide active against chocolate spot disease in broad beans.†

MD 50. Diatrizoate sodium. Applications: Diagnostic aid. (Mallinckrodt Inc).

MD 60. Diatrizoate meglumine. Applications: Diagnostic aid. (Mallinckrodt Inc).

Meadow Green. See EMERALD GREEN.

Mearl Film. Iridescent film. Applications: Packaging, laminating, visual effect. (Cornelius Chemical Co Ltd).*

Mearlin. Titanium dioxide coated mica pearl pigments. Applications: Plastics, paints, automotive finishes. (Cornelius Chemical Co Ltd).*

Mearlmaid. A proprietary trade name for a pearl essence.†

Measac. Ethanolamine sesquisulphite. (Albright & Wilson Ltd).*

Measurin. Aspirin. Applications: Analgesic; antipyretic; antirheumatic. (Sterling Drug Inc).

Meat-Sugar. Inositol, $C_6H_6(OH)_6$.†

Mebadin. A proprietary trade name for dehydroemetine.†

Mebaral. Mephobarbital. Applications: Anticonvulsant; sedative. (Sterling Drug Inc).

Mebatreat. A proprietary preparation containing mebandazole. Applications: Veterinary antihelminthic (cats and dogs). (Janssen Pharmaceutical Ltd).*

Mebenvet. A proprietary preparation containing mebendazole. Applications: Veterinary antihelminthic (game birds, poultry). (Janssen Pharmaceutical Ltd).*

Mebryl. A proprietary preparation of embramine hydrochloride. An anti-allergic. (Smith Kline and French Laboratories Ltd).*

Mecadox. Carbadox - an antibacterial used in swine. (Pfizer International).*

Mecca Balsam. An oleo-resin obtained from *Balsamodendron gileadense*, of Arabia. It is known in India as "Balsan-katel," and is imported there under the name of "Duhnul-balasan." Balm of Gilead is another name for Mecca balsam.†

Meclan Cream. Meclocycline sulphosalicylate. Applications: Antibacterial. (Ortho Pharmaceutical Corp).

Meclomen. Meclofenamate sodium. Applications: Anti-inflammatory. (Parke-Davis, Div of Warner-Lambert Co).

Meco. A cupro-nickel alloy.†

Meconium. Opium.†

Mecpa. Selective weedkillers. (Murphy Chemical Co).†

Mecufix. An adhesive for polyester film. (May & Baker Ltd).*

Meculon. Metallised polyester film. (May & Baker Ltd).*

Mecysteine. Methyl Cysteine.†

Medac Cream. Acne cream. (Fisons PLC, Pharmaceutical Div).

Medal Bronze. An alloy of from 92-97 per cent copper, 1-8 per cent tin, and 0-2 per cent zinc.†

Medal Metal. An alloy of 84 per cent copper and 16 per cent zinc.†

Medang Losoh Oil. An oil from the wood of *Cinnamonum parthenoxylon*. It consists mainly of safrole.†

MedGel. Aluminium stearate. Applications: Non-aqueous coating compositions. (Nueodex Inc).*

Medialan KA Conc. Anionic surfactant in which the cation is sodium and the anion is derived from medium chain length fatty acids condensed with sarcosine. Yellowish white paste. Applications: Cleansing and foaming agent used as an additive to high quality body lotions and shampoos. (Hoechst UK Ltd).

Medialan KF. Palm kernel oil fatty acid sarcoside as a pale yellow clear liquid. Applications: Anionic surfactant giving good cleansing, bubble foam and skin compatibility, used in toilet preparations. (Hoechst UK Ltd).

Medialan LD. Sodium lauroyl sarcoside as a clear, slightly yellowish liquid. Applications: Anionic surfactant, synergistic stabilizing foam, dermatological compatibility. Used in toothpastes. (Hoechst UK Ltd).

Medicaire. For relief of nasal congestion. (The Wellcome Foundation Ltd).

Medihaler Ergotamine. Ergotamine tartrate. Applications: Analgesic. (Riker Laboratories Inc, Subsidiary of 3M Company).

Medihaler Iso. Isoproterenol sulphate. Applications: Adrenergic. (Riker Laboratories Inc, Subsidiary of 3M Company).

Medihaler-Duo. Pressurized aerosol containing a suspension of isoprenaline hydrochloride BP 8 mg/ml, phenylephrine bitartrate 12 mg/ml; delivers 400 metered doses (0.16 mg isoprenaline hydrochloride, 0.24 mg phenylephrine bitartrate per dose). (Riker Laboratories/3M Health Care).†

Medihaler-epi. Pressurized aerosol containing a suspension of adrenaline acid tartrate BP 14 mg/ml; delivers 400 metered doses (0.28 mg per dose). (Riker Laboratories/3M Health Care).†

Medihaler-Ergotamine. Pressurized aerosol containing a suspension of ergotamine tartrate BP 9 mg/ml; delivers 75 metered doses (0.36 mg per dose). (Riker Laboratories/3M Health Care).†

Medihaler-iso. Pressurized aerosol containing a suspension of isoprenaline sulphate BP 4 mg/ml; delivers 400 metered doses (0.08 mg per dose). (Riker Laboratories/3M Health Care).†

Medihaler-iso Forte. Pressurized aerosol containing a suspension of isoprenaline sulphate BP 20 mg/ml; delivers 400 metered doses (0.4 mg per dose). (Rike Laboratories/3M Health Care).†

Medihaler-Tetracaine. Tetracaine. Applications: Anaesthetic. (Riker Laboratories Inc, Subsidiary of 3M Company).

Mediker. Shampoo, to wash and condition hair. (Richardson-Vicks Inc).*

Meditar. A proprietary preparation of coa tar. Applications: A dermatological product. (Brocades).*

Medium VS. A preparation consisting of dry sweet whey, sodium caseinate, sodium citrate and soluble growth factors derived from *Saccharomyces cerevisiae*. Applications: It is used to prepare culture medium for the growth of thermophilic lactic acid bacteria. (Pfizer International).*

Medium 10. A preparation consisting of d sweet whey, non-fat dry milk, disodium phosphate and soluble growth factors derived from *Saccharomyces cerevisiae* Applications: It is used to prepare culture medium for the growth of thermophilic lactic acid bacteria. (Pfize International).*

Medium 7. A preparation consisting of dry sweet whey, sodium caseinate, disodium phosphate and soluble growth factors derived from *Saccharomyces cerevisiae* Applications: It is used to prepare culture medium for the growth of thermophilic lactic acid bacteria. (Pfize International).*

Medol. A combination of cresols and iodin Recommended as an antiparasitic in skin troubles.†

Medolit. A phenol-formaldehyde condensation product. It is a resinous material, and is recommended as a shellac varnish substitute.†

Medomet. A proprietary preparation of METHYLDOPA. An anti-hypertensive. (DDSA).†

Medomin. A proprietary preparation of heptabarbitone. A hypnotic. (Ciba Geigy PLC).†

Medro - Cordex. A proprietary preparation of methylprednisolone and aspirin. An antirheumatic agent. (Upjohn Ltd).†

Medrol. Methylprednisolone. Applications: Glucocorticoid. (The Upjohn Co).

Medrol. A proprietary trade name for Methylprednisolone.†

Medrol ADT Pak, also Medrol Dosepak. Methylprednisolone. Applications: Glucocorticoid. (The Upjohn Co).

Medrol Enpak. Methylprednisolone acetate. Applications: Glucocorticoid. (The Upjohn Co).

Medrol Stabisol. Methylprednisolone sodium phosphate. Applications: Glucocorticoid. (The Upjohn Co).

Medrone. A proprietary preparation of methylprednisolone. (Upjohn Ltd).†

Medrone Medules. A proprietary preparation of methylprednisolone. (Upjohn Ltd).†

Medrone Veriderm. A proprietary preparation of methylprednisolone acetate for dermatological use. (Upjohn Ltd).†

Medusa. A waterproofing compound mixed with cement.†

Meehanite. A proprietary trade name for a close-grained, pearlitic, sorbitic iron with properties superior to cast iron. It has good casting and machining properties.†

Meena Harma. A name given to an opaque variety of bdellium gum-resin.†

Mefarol. Disinfectant based on a quaternary ammonium compound. Applications: Veterinary medicine. (Bayer & Co).*

Mefoxin. Cefoxitin. Applications: A beta-lactam antibiotic effective against a wide range of Gram-positive and Gram-negative bacteria. (Merck Sharp & Dohme).*

Megace. Megestrol acetate. Applications: Antineoplastic. (Mead Johnson & Co).

Megaclor. A proprietary preparation containing Clomocycline. An antibiotic. (Pharmax).†

Meganite. An explosive containing nitro-glycerin, and dinitro-cellulose, to which has been added a nitro mixture to ensure complete combustion.†

Megaperm 4510. A magnetic alloy containing 45 per cent nickel, 45 per cent iron, and 10 per cent manganese.†

Megaperm 6510. A magnetic alloy containing 65 per cent nickel, 25 per cent iron, and 10 per cent manganese.†

Megaphen. Pharmaceutical preparation. Applications: Classical broad-spectrum neuroleptic. (Bayer & Co).*

Megapren C 150. A proprietary chloroprene rubber used in the manufacture of mouldings and extrusions resistant to sunlight, weathering and ozone. (Rhein-Chemie Rheinau).*

Megapren Si 10, 20, 30 and 60. A proprietary range of materials based on silicone rubber. (Rhein-Chemie Rheinau).*

Megapren U225. A proprietary polyurethane rubber. (Rhein-Chemie Rheinau).*

Megasil. Silicone dielectric materials (Midland Silicones).†

Megilp. A mixture of linseed oil and mastic varnish. Used in artist's oil paints.†

Megimide. A proprietary preparation of bemegride. (Nicholas).†

Meglumine. N-Methylglucamine.†

Megomit. A mica product used as an electrical insulator.†

Megum. Bonding agent for the bonding of rubber to metals and other materials under vulcanizing conditions. (Chemetall GmbH).*

Mekad. Antiskinning agent for paints. (Ward, Blenkinsop & Co Ltd).†

Meketone. Methyl-ethyl-ketone. Applications: Solvent in paint and lacquer thinners, natural and synthetic resins, gums and rubbers, printing inks, PVC cloth manufacture, cleaning agent for metal surfaces, adhesives and cements. Refining and dewaxing of mineral and lubricating oils. (Sasolchem).*

Mekong Yellow G. A dyestuff. Dyes cotton greenish-yellow from a soap bath.†

Mekong Yellow R. A dyestuff. Dyes cotton yellow.†

Mekure T1. Copper chrome arsenate. Applications: Wood preservative. (Mechema).*

Mekure T2. Copper chrome arsenate. Applications: Wood preservative. (Mechema).*

Melacos. Bottle washing detergents. (ICI PLC).*

Melafix DM. A proprietary trade name for a melamine formaldehyde resin. Used as a shrink-resistant in wool. (Ciba Geigy PLC).†

Melalith. A steatite-porcelain product.†

Melamit 200. A proprietary melamine formaldehyde cellulose moulding powder. (Bush Beach Ltd).†

Melampyrite. Dulcitol, $CH_2(OH)(CH.OH)_4.CH_2OH$.†

Melanate. Tramadol Hydrochloride. Applications: Analgesic. (The Upjohn Co).

Melanex. A proprietary trade name for Metahexamide.†

Melaniline. Diphenyl-guanidine, $NH:C(NH.C_6H_5)_2$.†

Melanogen Black. See SULPHUR BLACK T.

Melanoid. A colloidal bituminous paint material. It is used as a preservative paint for metal which is in contact with corrosive gases.†

Melanthrene B. An anthracene dyestuff obtained by heating diamino-anthraquinone with caustic potash. Dyes cotton grey.†

Melanthrene BH. A dyestuff. It is a British equivalent of Diamine black BH.†

Melatix. A proprietary melamine formaldehyde moulding compound. (Nisshin Boseki).†

Melax. Proprietary, polymeric based conditioning and sealing agents for fibreglass moulds and other porous surfaces. Applications: Fibreglass tools. (Axel Plastics Research Laboratories Inc).*

Melco. A synthetic milk made from the peanut.†

Melcril 4079. A trade mark for tetra-hydrofurfuryl acrylate. (Danbert Chemical Co, Oak Brook, Ill).†

Melcril 4083. A trade mark for an ethylene glycol acrylate phthalate. (Danbert Chemical Co, Oak Brook, Ill).†

Melcril 4085. A trade mark for a benzyl acrylate. (Danbert Chemical Co, Oak Brook, Ill).†

Melcril 4087. A trade mark for a phenoxy ethyl acrylate. (Danbert Chemical Co, Oak Brook, Ill).†

Melcril 5919. A trade mark for a melamine acrylate. (Danbert Chemical Co, Oak Brook, Ill).†

Meldola's Blue. (New Blue; Naphthylene Blue R, in Crystals; Fast Blue R, 2R, 3R, for Cotton, in Crystals; Cotton Blue R; Fast Navy Blue R, RM, MM; Naphthol Blue R, D; Cotton Fast Blue 2B; β-Naphthol Violet; New Fast Blue, for Cotton; New Fast Blue R Crystals, for Cotton; New Blue R; Fast Marine Blue; Phenylene Blue). A dyestuff. It is the zinc double chloride of dimethyl-amino-naphthophenoxazonium-chloride, $C_{18}H_{15}N_2ClO$. Dyes cotton mordanted with tannin and tartar emetic indigo blue. Used as a substitute for indigo.†

Melhi N, NS and NLM. Acidic, medium-hard, dark amber-coloured resins obtained as coproduct during processing of rosin to modified forms. Applications:Melhi resin is used in mastic adhesives where colour is not important, Melhi N, NS and NLM are chemically related to Melhi but have different properties and are used in low-cost hot-melt adhesives and construction mastic adhesives. Melhi resin is used in low-cost Type A gravure inks. (Hercules Inc).*

Melibiase. An enzyme which splits melibiose into glucose and galactose.†

Meligrin. A condensation product of dimethyl-hydroxyquinoline and methyl-phenylacetamide.†

Melilot. *p*-Methylacetophenone, $C_6H_4(CH_3).CO.CH_3$. It imparts the honey-like fragrance of sweet clover, and is used for perfuming soap.†

Melilot, Methyl. See METHYL MELILOT.

Melimax. For acetonaemia in cattle. (The Wellcome Foundation Ltd).

Melinar. PET polymer. (ICI PLC).*

Melinex. A proprietary trade name for polyethylene terephthalate film. An extremely tough material used for cable lapping, motor insulation and capacitors, valve diaphragms and conveyor belting. (ICI PLC See also MYLAR Dupont (UK) Ltd).*

Melinite. The French name for Lyddite (*qv*). It consists of 70 per cent picric acid and 30 per cent collodion cotton.†

Melinose. A mineral. It is wulfenite.†

Melioform. A ruby-red liquid containing 25 per cent formaldehyde and 15 per cent aluminium acetate. Used as a disinfectant.†

Melioran F6. Alcohol sulphate in paste form. Applications: Powerful detergent with wetting-out, dispersing, level dyeing and fibre protection properties. Used in the processing of wool, cotton, linen, silk, artificial and synthetic fibres. (REWO Chemicals Ltd).

Meliose. High fructose corn syrup. Applications: Food, for example soft drinks and confectionery. (Roquette (UK) Ltd).*

Melit. Melamine resins.†

Mellaril. Thioridazine Hydrochloride. Applications: Antipsychotic; sedative. (Sandoz Pharmaceuticals, Div of Sandoz Inc).

Mellaril-S. Thioridazine. Applications: Antipsychotic; sedative. (Sandoz Pharmaceuticals, Div of Sandoz Inc).

Mellavax. Calf salmonellosis vaccine. (The Wellcome Foundation Ltd).

Melleril. A proprietary preparation of thioridazine. A sedative. (Sandoz).†

Mellinese. A proprietary preparation of chlorpropamide. An oral hypoglycemic agent. (Pfizer International).*

Mellite. Stabilizers for PVC. (Lankro Chemicals Ltd).

Melmac. A proprietary trade name for a melamine-formaldehyde synthetic resin and adhesive.†

Melmex. Melamine formaldehyde compounds. Applications: Moulding. (BIP Chemicals Ltd).*

Melocol. A proprietary trade name for a polyamide-formaldehyde product.†

Meloids. Throat lozenges. (The Boots Company PLC).

Melolam. Melamine/formaldehyde resins. (Ciba-Geigy PLC).

Melopas. Melamine/formaldehyde resins/moulding compounds. (Ciba-Geigy PLC).

Melsed. A proprietary preparation of methaqualone. A hypnotic. (Boots Company PLC).*

Melsedin. A proprietary preparation of methaqualone hydrochloride. A hypnotic. (Boots Company PLC).*

Melsprea. A range of melamine moulding compounds. Applications: Tableware and ashtrays. (SPREA Spa).*

Melurac. A proprietary trade name for a melamine-urea-formaldehyde laminating synthetic resin.†

Membrettes. Progesterone. Applications: Progestin. (Wyeth Laboratories, Div of American Home Products Corp).

Memilite. See RANDANITE.

Menest. Oestrogens, esterified. Applications: Oestrogens. (Beecham Laboratories).

Menformom A. Estrone. Applications: Oestrogen. (Savage Laboratories, Div of Byk-Gulden Inc).

Meningovax-C. Meningococcal vaccine. Applications: For immunization against infection caused by strains of Group C meningococci. (Merck Sharp & Dohme).*

Menispermin. An extract of Canadian moon-seed.†

Menopax. A proprietary preparation of ethinyloestradiol, carbomal and bromvaletone used in the treatment of menopausal symptoms. (Nicholas).†

Menopax Forte. A proprietary preparation of ethinyloestradiol, methyltestosterone, CARBROMAL, BROMVALETIN and MEPHENESIN used in the treatment of menopausal disorders. (Nicholas).†

Mephaneine. Pure methyl phenylacetate. (Bush Boake Allen).*

Mephenytoin. Methoin.†

Mephetol. Herbicides. (ICI PLC).*

Mephine. A proprietary preparation of mephentermine sulphate. Used to raise blood pressure. (Wyeth).†

Mephosol. A proprietary preparation of mephenesin and salicylamide. An antirheumatic. (Crookes Laboratories).†

Mephyton. Phytonadione. Applications: For the treatment of certain coagulation disorders. (Merck Sharp & Dohme).*

Mepilin. A proprietary preparation of ethinylcestradiol and methyltestosterone. Treatment of senile and menopausal conditions. (British Drug Houses).†

Mepred. Methylprednisolone acetate. Applications: Glucocorticoid. (Savage Laboratories, Div of Byk-Gulden Inc).

Meprobase-200. Tridihexethyl chloride and meprobamate. Applications: Possibly effective as adjunctive therapy in peptic ulcer and in the irritable bowel syndrome, especially when accompanied by anxiety or tension. (Pharmaceutical Basics Inc).*

Meprobase-400. Tridihexethyl chloride and meprobamate. Applications: Possibly effective as adjunctive therapy in peptic ulcer and in the irritable bowel syndrome, especially when accompanied by anxiety or tension. (Pharmaceutical Basics Inc).*

Meprospan. Meprobamate. Applications: Sedative. (Wallace Laboratories, Div of Carter-Wallace Inc).

Meraneine. Pure geranyl acetate. (Bush Boake Allen).*

Merantine. Level dyeing wool dyestuffs. (Holliday Dyes & Chemicals Ltd).

Merbaphen. See NOVASUROL.

Merbentyl. A proprietary preparation containing dicyclomine hydrochloride. An antispasmodic. (Richardson-Vicks Inc).*

Mercerized Cotton. Cotton which has been immersed in a solution of sodium hydroxide. It has a lustrous appearance.†

Mercer's Liquor. A solution containing potassium ferricyanide. Used for etching.†

Merchloran. Chlormerodrin.†

Merclor D. A proprietary trade name for a solution of sodium hypochlorite NaOCl A bleaching agent.†

Mercolloid. A colloidal mercury sulphide.†

Mercoloy. A proprietary trade name for a nickel bronze containing 60 per cent copper, 25 per cent nickel, 10 per cent zinc, 1 per cent tin, 2 per cent lead, and 2 per cent iron.†

Mercresin. A proprietary trade name for a solution of 0.1 per cent *o*-hydroxy-phenyl-mercuric chloride and 0.1 per cent sec-amyl-tricresol in 50 per cent alcohol, 10 per cent acetone, and water. A germicide.†

Mercuhydrin. A proprietary trade name for Meralluride.†

Mercurammonium Chloride. (Fusible White Precipitate, Ammonio-mercuric-chloride). Mercuri-diammonium chloride, $N_2H_6Hg.Cl_2$.†

Mercuric Iodide, Red. See RED IODIDE OF MERCURY.

Mercuric Potassium Iodide. See MAYER'S SOLUTION.

Mercurichrome. A fluorescein derivative of mercury. A bactericide.†

Mercuricide. Lithio-mercuric-iodide, $3LiI+HgI_2$. A germicide.†

Mercuriocoleols. A double stearate of cholesterol and mercury.†

Mercuriol. A mercury amalgara with aluminium.†

Mercury Blende. See CINNABAR.

Mercury Iodide, Green. See GREEN IODIDE OF MERCURY.

Mercury Oxide, Red. See RED OXIDE OF MERCURY.

Mercury, Ammoniated. See WHITE PRECIPITATE.

Mercury, Cosmetic. See WHITE PRECIPITATE.

Meretestate. Testosterone. Applications: Androgen. (Sterling Drug Inc).

Merfusan. Horticultural product. (May & Baker Ltd).

Mergal. Fungicides and bactericide. (Hoechst UK Ltd).

Mergamma. Combined mercury and gamma HCH powder seed dressing. (ICI PLC).*

Merigraph. A range (18 varieties available) of viscous, UV light sensitive liquid photopolymers. Applications: They are used to make printing plates for letterpress, flexo and dry offset printing processes. (Hercules Inc).*

Merital. Nomifensine maleate. Applications: Antidepressant. (Hoechst-Roussel Pharmaceuticals Inc).

Meritol. A photographic developer. (Johnsons of Hendon).†

Merix. A line of aerospace, automotive, industrial, paper, photography, rubber, safety and sporting goods chemicals. (Merix Chemical Co).*

Merlon. Polycarbonate resins. These are characterized by their toughness, heat resistance, dimensional stability and clarity. (Mobay Chemical Corp).*

Merocets. A proprietary preparation of cetylpyridinium chloride. Throat lozenges. (Richardson-Vicks Inc).*

Merolan. Fungicide containing dithiocarbonate for use on potatoes. (ICI PLC).*

Merpafol. Active ingredient: captafol; 3a,4,7,7a-tetrahydro-2-[(1,1,2,2-tetrachlorotheyl)thio]1H-isoindole-1,3(2H)-dione. Applications: Nonsystemic agricultural fungicide. (Makhteshim Chemical Works Ltd).†

Merpan. Active ingredient: captan; N-[(trichloromethyl)thio]4-cyclohexene-1,2-dicarboximide. Applications: Broad-spectrum agricultural fungicide. (Makhteshim Chemical Works Ltd).†

Merpectogel. Phenylmercuric Nitrate. Applications: Pharmaceutic aid. (Poythress Laboratories Inc).

Merpentine. A proprietary trade name for a sodium alkyl naphthalene sulphonate product used as a wetting agent.†

Merpol. Surface active agent. (Du Pont (UK) Ltd).

Merpol DSR. A non-ionic softener for use in conjunction with acrylic type soil release agents in durable press finishes on textiles. (Du Pont (UK) Ltd).†

Merpol SH. A nonionic long chain alcohol-ethylene oxide condensate. A highly efficient biodegradable surfactant used in the preparation of textile fabrics for dyeing and finishing. (Du Pont (UK) Ltd).†

Merquinox. Fungicide for industrial applications. (Ward, Blenkinsop & Co Ltd).†

Mersalyl BDH. Injection and tablets. A proprietary diuretic (mercurial). (British Drug Houses).†

Mersil. Turf fungicide. (May & Baker Ltd).*

Mersilene. Suture, nonabsorbable surgical. Applications: Surgical aid. (Ethicon Inc).

Mersize. Paper sizing agent. Applications: Used in paper mill to reduce sizing costs, improve resistance to ink, water and other aqueous solutions. (Monsanto Co).*

Mersol. A proprietary trade name for a solvent.†

Mersolat. Sodium alkane sulphonate. Applications: Detergent for use in the production of special purpose materials, especially electrolyte-resistant wetting agents for textiles, leather auxiliaries, acid and alkaline cleansers. (Bayer & Co).*

Mertan. Polyurethane water emulsions, polyurethane solutions, granule form. Applications: Thermoplastic polyurethane (Heel tops, soles etc.). (Merquinsa).*

Mertec. A proprietary trade name for a chlorinated rubber-base paint.†

Merthioiate. A proprietary preparation of thiomersal. A skin antiseptic. Sodium ethyl-mercurithio-salicylate. (Eli Lilly & Co).†

Meruvax. Rubella Virus Vaccine Live. Applications: immunizing agent. (Merck Sharp & Dohme, Div of Merck & Co Inc).

Meruvax II. Rubella vaccine. Applications: immunization against German Measles. (Merck Sharp & Dohme).*

Mervan. Alclofenac. Applications: Anti-inflammatory. (Continental Pharma).

Mesamoll. Alkyl sulphonic acid ester of phenol. Applications: Monomeric plasticizer. (Bayer & Co).*

Mesamoll-Verdingrin. Paraffin sulphonated ester. A vinyl plasticizer. (Bayer & Co).*

Mesantoin. Mephenytoin. Applications: Anticonvulsant. (Sandoz Pharmaceuticals, Div of Sandoz Inc).

Mesgamma. Combined insecticide fungicide and seed dressing. (Plant Protection (Subsidiary of ICI)).†

Mesicerin. Tri-ω-hydroxymesitylene, $C_6H_3(CH_2OH)_3$.†

Mesidine. 2, 4, 6-trimethylaniline. A dyestuff intermediate.†

Mesitol. Range of aftertreating and resist agents. Applications: For polyamide, polyamide/cellulosic blends and half wool. (Bayer & Co).*

Mesontoin. A proprietary preparation of methoin. An anticonvulsant. (Sandoz).†

Mestinon. A proprietary preparation containing pyridostigmine bromide. Applications: Cholinestetase inhibitor. (Roche Products Ltd).*

Meta. Metaldehyde produced by the polymerization of acetaldehyde.†

Metabisulphite. Sodium metasulphite, $Na_2S_2O_5$.†

Metachrome. Chrome dyestuffs. (JC Bottomley and Emerson).†

Metachrome Brown B. (Alizarin Brown M, Metachrome Brown G). A dyestuff prepared from picramic acid and *m*-tolylene-diamine.†

Metachrome Brown G. A dyestuff. It is a British equivalent of Metachrome brown B.†

Meta Chrome Mordant. The name given to a mixture of potassium bichromate and ammonium acetate. Used as a mordant.†

Metacinnabarite. A mineral having the same composition as cinnabar, but black in colour.†

Metacon. Rust remover. (Croda Chemicals Ltd).*

Metacrylene. A proprietary trade name for a styrene-methyl methacrylate copolymer.†

Metacrylene BS. A proprietary trade name for an MBS terpolymer.†

Metaethyl. A mixture of methyl and ethyl chlorides. An anaesthetic.†

Metaferrin. An iron albumin compound containing 10 per cent iron. Used in the treatment of anaemia.†

Metaform. A packing material for packing stuffing-boxes, and consisting of powdered white metal, graphite, cylinder oil, and asbestos fibre.†

Meta Fuel. See META.

Metagon. Sodium hexametaphosphate. (Albright & Wilson Ltd, Phosphates Div).

Metahydrin. Trichlormethiazide. Applications: Diuretic; antihypertensive. (Merrell Dow Pharmaceuticals Inc, Subsidiary of Dow Chemical Co).

Metahydroboracite. Synonym for Inderborite.†

Metal Argentum. An alloy consisting of 85 1/2 per cent tin and 14 1/2 per cent antimony.†

Metal-furnace Slag. A slag formed and used in the preparation of copper metal. It is essentially a silicate of iron, and contains about 4 per cent copper.†

Metal Soaps. Salts of the heavy metals with fatty acids.†

Metalite. (Durundum, Idilite, Exelite, Orelite). Proprietary trade names for aluminium oxide abrasives.†

Metallic Sodium. Pharmaceutical processing. (Foseco (FS) Ltd).

Metallichrome. A finish for motor bodies. (ICI PLC).*

Metalline. An alloy consisting of 35 per cent cobalt, 30 per cent copper, 10 per cent iron, and 25 per cent aluminium. Used in jeweller's work.†

Metalset. Epoxy resin base adhesives and cements. (Smooth-On Inc).*

Metalyn 582. An ester of fatty acids designed for metal-rolling and working lubricants that require particular thermal and hydrolytic properties at elevated temperatures. Applications: It is used as an ingredient in wood preserving agents, as a secondary plasticizer for vinyl copolymer resin, as a plasticizer, softener and tackifier for rubber (nitrile) compounding and in various low cost adhesives, metal working lubricant. (Hercules Inc).*

Metamine Blue B and G. See NEW BLUE B or G.

Metamsustac. A proprietary preparation of methylamphetamine hydrochloride, in a slow release tablet. An antiobesity agent. (Pharmax).†

Metamucil. A proprietary preparation containing psyllium mucilloid, dextrose, benzyl benzoate, sodium bicarbonate, potassium biphosphate and citric acid. A laxative. (Searle, G D & Co Ltd).†

Metandren. Methyltestosterone. Applications: Androgen. (Ciba-Geigy Corp).

Metanil Orange I, II. Dyestuffs prepared by the action of diazotized *m*-amino-benzene-sulphonic acid uponα- and β-naphthol respectively. They are acid dyes.†

Metanil Yellow. (Orange MN, Tropaeoline G, Yellow GA, Victoria Yellow, Extra Conc., Orange MNO, Metanil Yellow Y). A dyestuff. It is the sodium salt of *m*-sulphobenzeneazo-diphenylamine, $C_{18}H_{14}N_3O_3SNa$. Dyes wool orange-yellow from an acid bath. Used in paper-staining, carpet printing, and for colouring varnishes. It is poisonous.†

Metanil Yellow S. (Acid Yellow 2G). The sulphonated product of Metanil Yellow (*qv*).†

Metanil Yellow Y. A dyestuff. It is a British equivalent of Metanil yellow.†

Metaniline Grey. See NEW GREY.

Metanium Ointment. Titanium dioxide 20%, titanium peroxide 5%, titanium salicylate 3%, titanium tannate 0.1% in a siliconised excipient. Applications: For the prevention and treatment of nappy rash and other macerated skin conditions. (Bengue & Co Ltd).*

Metaphlorone. See PHLORONE.

Metaprel. Metaproterenol sulphate. Applications: Bronchodilator. (Dorsey Pharmaceuticals, Pharmaceutical Div of Sandoz Inc).

Metaquest. Sequestering agents. (Fisons PLC).*

Metasal. Multipurpose fire extinguishing powders for extinguishing fires of incandescent solid, liquid and gaseous materials. (Degussa).*

Metasil AL. Metasil A reinforced with hydrate of aluminium. Applications: For the production of sparkling water e.g. mineral water manufacturers. (Stella Meta Filters).*

Metasil ALAG. Metasil AL carrying a surface coating of silver. Applications: For the production of sparkling water for drinking purposes from low quality water supplies. (Stella Meta Filters).*

Metasil A, A+. Extra A+, B, C, E. A pure diatomaceous silica, each powder having different retentive properties. Applications: For the coarse filtration of suspensions down to very fine filtrations giving brilliant filtrates on most liquids e.g. syrups, oils, chemicals, water, beer, vinegar, etc. (Stella Meta Filters).*

Metasil D. Polystyrene micro beads. Applications: Special barrier type precoat material. (Stella Meta Filters).*

Metasil DA. Carbon. Applications: decolourizing, dechlorinating and taste removal. (Stella Meta Filters).*

Metasil MQC. Hydrated silicate. Applications: For the adsorption of oxides in rolling oil coolants. (Stella Meta Filters).*

Metasil Purasil. Perlite (volcanic rock) type. Applications: For general clarification of liquids not too strongly acid or alkaline, e.g. swimming pool. (Stella Meta Filters).*

Metasil R. Chopped rayon fibre. Applications: Special barrier type precoat material. (Stella Meta Filters).*

Metasil SA, SB. Pure cellulose. Applications: For the filtration of products where silica is unsuitable e.g. miscella oils, removal of oil from condensate water. (Stella Meta Filters).*

Metasil W/2. Carbon. Applications: Filtration of strong alkali liquids. (Stella Meta Filters).*

Metaspirine. A proprietary preparation. It consists of acetyl- salicylic acid and caffeine.†

Metastrengite. Synonym for Phosphosiderite.†

Metastyrene. See POLYSTYRENE.

Metastyrol. See POLYSTYRENE.

Metasystox. Metasystox 55 is an emulsifiable concentrate containing demeton-S-methyl 580 grams per litre and Metasystox R is a liquid concentrate of oxydemeton-methyl 570 grams per litre. Applications: To control aphids, red spider mites and certain other pests on arable crops, fruit, market garden and glasshouse crops and ornamentals. Metasystox R will be the choice of the grower who prefers a product with a less penetrating odour. It may be used on all the crops for which Metasystox 55 is recommended apart from brassica crops. (Bayer & Co).*

Metatone. A proprietary preparation containing thiamine hydrochloride, and calcium, potassium, sodium, manganese, and strychnine glycerophosphates. A tonic. (Parke-Davis).†

Metco 450. A nickel/aluminium composite material for building up metal surfaces by spraying. Similar materials are Metco 451 (nickel-chromium), Metco 44 (nickel base/chromium) and Metco 51 (aluminium bronze).†

Meteorite. An alloy of aluminium with from 1-2 per cent zinc and 1-4 per cent phosphorus.†

Metformin. N'N'-Dimethyldiguanide.†

Meth-O-Gas. Ethylene dibromide formulation. (Great Lakes Chemical (Europe) Ltd).

Metha-Meridiazine. Trisulphapyrimidines (oral suspension). Mixture of Sulphadiazine, Sulphamerazine and Sulphamethazine. Applications: Antibacterial. (McNeil Pharmaceuticals, McNEILAB Inc).

Methacrol. Fabric and yarn lubricants. (Du Pont (UK) Ltd).

Methadose. Methadone hydrochloride. Applications: Analgesic. (Mallinckrodt Inc).

Methaplex. Methylmethacrylate acrylic in an aromatic and aliphatic solvent vehicle system. Applications: Water repellant and sealer for masonry. (Nova Chemical Inc).*

Metharbitone. 5,5-Diethyl-I-methyl barbituric acid.†

Methasan. Zinc dimethyldithiocarbamate. Applications: vulcanization accelerator. (Monsanto Co).*

Methasol. Dyes for printing inks. (ICI PLC).*

Methazate. A proprietary preparation of zinc dimethyl dithiocarbamate. (Naugatuck (US Rubber)).†

Methergin. A propetary preparation of methylergometrine maleate, used in obstetrics to stimulate uterine contraction. (Sandoz).†

Methergine. Methylergonovine maleate. Applications: Oxytocic. (Sandoz Pharmaceuticals, Div of Sandoz Inc).

Methethyl. A mixture of ethyl and methyl chlorides. A local anaesthetic.†

Methic. Basic dyestuffs. (ICI PLC).*

Methisul. A proprietary preparation of SULPHAMETHISOLE. An antibiotic. (RP Drugs).†

Methocel. Methylcellulose. Applications: Used in adhesives, cosmetics, foods, paints, pharmaceuticals and textiles. (Dow Chemical Co Ltd).*

Methocel A. Methylcellulose. Applications: Pharmaceutic aid. (Dow Chemical Company Ltd).

Methocel E,F,K. Hydroxypropyl methylcellulose. Applications: Pharmaceutic aid. (Dow Chemical Company Ltd).

Methocidin. A proprietary preparation of hydroxymethylgramicidin ephedrine and cetylpyridinium chloride. Throat lozenges. (Rona Laboratories).†

Methokill. Insecticidal formulation. (Mitchell Cotts Chemicals Ltd).

Methoklone. Stabilized methylene chloride (ICI PLC).*

Methotrexate. Preparations containing methotrexate. Applications:

Antineoplastic agent. (Lederle Laboratories).*

ethoxone. Selective weed killers. (ICI PLC).*

ethozin. See ANTIPYRINE.

ethral. Fluperolone acetate. Applications: Glucocorticoid. (Pfizer Inc).

ethyl Acetone. A crude fraction of wood distillation. Its principal constituents are acetone, methyl alcohol, and methyl acetate. Some samples contain from 70-80 per cent acetone. Used as a rubber solvent. Methyl ethyl ketone, $CH_3CO.CH_2CH_3$, is also known by this name.†

ethyl Alkali Blue. (Alkalin, Alkali Blue D, Alkali Blue 6B). A dyestuff. It is the sodium salt of triphenyl-*P*-rosaniline-monosulphonic acid, $C_{37}H_{30}N_3SO_4Na$. Dyes wool from an acid bath.†

ethyl and Ethyl Selenac. Selenium dimethyl or diethyl dithiocarbamate. Rubber accelerators. Applications: For NR, SBR and 11R. Also vulcanizing agents. Effective in low sulphur and sulphurless heat resistant compounds. (Vanderbilt Chemical Corporation).*

ethyl Aniline Green. See METHYL GREEN.

ethyl Aniline Violet. See METHYL VIOLET B.

ethyl-Atophan. See PARATOPHAN.

ethyl Benzene. (Retinaphtha, Phenyl Methane). Toluene. $C_6H_5CH_3$.†

ethyl Blue. (Brilliant Cotton Blue, greenish, XL Soluble Blue, Diphenylamine Blue, Bavarian Blue DBF, Soluble Blue 8B, 10B, Methyl Blue for Cotton, Light Blue, Pararosaniline Blue, Paris Blue, Night Blue). A dyestuff. It is the sodium salt of triphenylpararosanilinetrisulphonic acid, $C_{37}H_{26}N_3O_9S_3Na_3$. Dyes silk and mordanted cotton blue.†

ethyl Blue for Cotton. See METHYL BLUE.

ethyl Blue for Silk MLB. See BAVARIAN BLUE DSF.

ethyl Blue, Water Soluble. See BAVARIAN BLUE DSF.

Methyl Catechol. Guaiacol, $C_6H_4(OH)(OCH_3)$, the chief constituent of beech-wood creosote.†

Methyl Cellosolve. A proprietary name for the monomethyl ether of ethylene glycol. It is a colourless and nearly odourless liquid boiling at 124.5° C., and has the lowest boiling-point and greatest rate of evaporation of all available glycol ethers. It is a solvent for cellulose acetate, nitro-cellulose, and hydrocarbons.†

Methyl Cellulose. See TRYLOSE S.

Methyl Cumate. Copper dimethyldithiocarbamate - ultra accelerator. Applications: For SBR, 11R, EPDM as an ultra accelerator for high speed vulcanization. Not for use in NR or 1R. (Vanderbilt Chemical Corporation).*

Methyl Eosin. See ERYTHRIN.

Methyl-Erythrine. The methyl ester of tetrabromofluoresceine, $C_{20}H_7CH_3Br_4O_5$.†

Methyl- Glycocoll. (Sarcosine). Methyl amino-acetic acid, $CH_3.NH.CH_2.CO_2H$.†

Methyl Green. (Paris Green, Light Green, Double Green SF, Green Powder, Methyl Aniline Green). A dyestuff. It is the zinc double chloride of heptamethyl-pararosaniline chloride, $C_{26}H_{33}Cl_4N_3Zn$. Dyes silk green from a soap bath. Also see ETHYL GREEN.†

Methyl Indigo B. A dyestuff. It is *o*-methylindigotin, $C_{18}H_{14}N_2O_2$. Dyes cotton greenish-blue from a reduced vat.†

Methyl Indigo R. A dyestuff. It is *p*-methylindigotin, $C_{18}H_{14}N_2O_2$. Dyes cotton reddish-blue from a reduced vat.†

Methyl Isobutyl Niclate. Methyl nickel-di-isobutyldithiocarbamate - rubber antioxidants. Applications: Offers antiozonant and antioxidant protection in epichlorohydrin. Used for optimum aging and extremes. (Vanderbilt Chemical Corporation).*

Methyl Ledate. Lead dimethyldithiocarbamate - ultra accelerator. Applications: Recommended for NR, SBR, 11R, 1R,

BR rubbers. Used for ultra accleration, high speed, high temp vulcanization. (Vanderbilt Chemical Corporation).*

Methyl Melilot. Dimethylaceto-phenone, $C_6H_3(CH_3)_2CO.CH_3$. It has a similar odour to Melilot (*qv*).†

Methyl Mustard Oil. Methyl isothiocyanate, $CH_3.NCS$.†

Methyl Orange. See ORANGE III.

Methyl Parathion. *O*,*O*-Dimethyl-*O*-(4-nitrophenyl)phosphorothioate. Applications: Used extensively in cotton producing areas. (A/S Cheminova).*

Methyl Pulvate. Vulpic acid, $C_{19}H_{14}O_5$†

Methyl Tuads. A proprietary preparation of tetramethyl thiuram disulphide. (K & K Greef Chemicals Ltd).†

Methyl Violet B. (Paris Violet, Methyl Aniline Violet, Direct Violet, Dahlia, Pyoctanin, Violet 3B Extra). A dyestuff. It consists chiefly of the hydrochlorides of penta- and hexa-methyl-pararosaniline, $C_{24}H_{28}Cl.N_3$. Dyes silk and wool violet, and cotton mordanted with tannin and tartar emetic.†

Methyl Violet 5B. See BENZYL VIOLET.

Methyl Violet 6B. See BENZYL VIOLET.

Methyl Violet 6B Extra. See BENZYL VIOLET.

Methyl Water Blue. See BAVARIAN BLUE DSF.

Methyl Zimate. A proprietary preparation of zinc dimethyl dithiocarbamate. (K & K Greef Chemicals Ltd).†

Methylal. (Formal). Methylene-di-methyl ether $CH_2(O.CH_3)_2$.†

Methylanone. Methylcyclohexanone, a solvent for resins.†

Methylated Spirit. Ethyl alcohol adulterated with methyl alcohol, pyridines, dyestuffs, etc. to make it non-potable for industrial or laboratory use.†

Methylene Blue. A dyestuff. It is the methylo-chloride of trimethyl-amino-imino-imino-diphenyl-sulphide, $C_{16}H_{18}Cl.N_3S$. Used in dyeing, calico printing, and as a staining material in bacteriological work. It is used in medicine for malaria and neuralgis.†

Methylene Blue A Extra. See METHYLENE BLUE B and BG.

Methylene Blue B and BG. (Methylene Bl BB in Powder Extra D, Methylene Blu A Extra, Methylene Blue 2B, 2B Conc 2BZF, FZP, G, GS, R Conc, ZF, GSF Dyestuffs. The chloride (Methylene bl BG, BB in powder extra D, printing blue), or zinc double chloride (Methylene blue B, BB in powder extr dyeing blue), of tetra-methyl-diamino-phenazthionium. Chloride, $C_{16}H_{18}N_3SCl$; zinc double chloride, $(C_{16}H_{18}N_3SCl)_2 + ZnCl_2 + H_2O$. The dye cotton mordanted with tannin blue

Methylene Blue BB in Powder Extra. See METHYLENE BLUE B and BG.

Methylene Blue BB in Powder Extra D. S METHYLENE BLUE B and BG.

Methylene Blue T50. A dyestuff. It is a British equivalent of Toluidine blue O.

Methylene Blue 2B, 2B Conc., 2BZF, FZI G, GS, R Conc., ZF, GSF. Dyestuffs. They are British equivalents of Methylene blue B and BG.†

Methylene Blue, Löffler's. See LÖFFLER'S METHYLENE BLUE.

Methylene Green. A mono-nitro derivative of methylene blue. It dyes unmordante cotton bluish-green shades.†

Methylene Green G Conc., Extra Yellow Shade. Nitro-methylene blue, $C_{16}H_{17}Cl.N_4O_2S$. Dyes cotton bluish green.†

Methylene Grey. See NEW GREY.

Methylene Red. The methylo-chloride of dimethyl-amino-imino-phenyl-disulphide, $(C_8H_9N_2S_2Cl)_2$. A by-product in the manufacture of methylene blue, found in the mother liquor.†

Methylene Violet. A dyestuff. Dyes silk a cotton.†

Methylene Violet 2RA, 3RA. (Fuchsia, Safranine MN, Clemantine, Giroflé, Brilliant Heliotrope 2R Conc.). A dyestuff. It is dimethyl-diamino-pheny phenazonium chloride, $C_{20}H_{19}Cl.N_4$. It is employed in printing reddish-violet.†

Methylhexalin. A registered trade name f a mixture of three isomeric methylcyclohexanols. A solvent for fat resins, oils, and waxes.†

Methylindone. A dyestuff. It is an indoine blue prepared with amino-naphthol.†

Methylon. A range of substituted phenolic condensates. Applications: Adhesion promoters for polysulphide sealants, chemical-resistant drum and can linings. (Cornelius Chemical Co Ltd).*

Meti-Derm. Prednisolone. Applications: Glucocorticoid. (Schering-Plough Corp).

Meticortelone Acetate. Prednisolone Acetate. Applications: Glucocorticoid. (Schering-Plough Corp).

Meticorten. Prednisone. Applications: Glucocorticoid. (Schering-Plough Corp).

Metilar. A proprietary trade name for the 21-acetate of Paramethazone. (Syntex Laboratories Inc).†

Metillure. An acid-resisting alloy consisting of 17 per cent silicon, 81 per cent iron, 0.9 per cent manganese, 0.25 per cent aluminium, 0.6 per cent carbon, and 0.17 per cent phosphorus.†

Metiloil. A registered trade mark for methyl esters of ricinoleic acid and other fatty acids in castor oil. Used for tanning and in textiles and for the manufacture of sulphonated emulsifying agents. (Aquitaine-Organico).†

Metol. Mono-methyl-*p*-amino-*m*-cresol sulphate, $(C_6H_3(OH)CH_3(NH)CH_3)_2H_2SO_4$. A developing agent used in photography.†

Metol-Quinone. See METOLHYDROQUINONE.

Metolhydroquinone. (Metol-quinone). A photographic developer. It contains Metol (*qv*), hydroquinone, sodium phosphate, sodium sulphite, and potassium carbonate.†

Metopirone. A proprietary preparation of metyrapone, used to test pituitary gland function. (Ciba Geigy PLC).†

Metosyn. A proprietary preparation of FLUOCINONIDE. A steroid skin cream. (ICI PLC).*

Metox. Metoclopramide. Applications: Pharmaceutical preparation for the treatment of dyspepsia, flatulence and hiatus hernia. (M A Steinhard Ltd).*

Metozin. See ANTIPYRINE.

Metra. Phendimetrazine Tartrate. Applications: Appetite suppressant. (O'Neal, Jones & Feldman Pharmaceuticals).

Metramine. See HEXAMINE.

Metraspray. Tetracaine. Applications: Anaesthetic. (Riker Laboratories Inc, Subsidiary of 3M Company).

Metrax. Industrial detergent. (Crosfield Chemicals).*

Metrazol. Pentamethylene-tetrazole.†

Metreton. Prednisolone Sodium Phosphate. Applications: Glucocorticoid. (Schering-Plough Corp).

Metro IV. Metronidazole. Applications: Antiprotozoal. (American McGaw, American Hospital Supply Corp).

Metro Tiles. Industrial floor tiles for heavy abrasion and impact. (Prodorite).*

Metro-nite. A refined natural mineral composed of calcium carbonate, and carbonates and silicates of magnesium. Used in the paint industry as a pigment.†

Metrolyl. Tablets containing 200mg and 400mg metronidazole BP. Suppositories containing 500mg and 1g metronidazole BP. Injection 0.5 per cent w/v in 100ml bottles or minibags (500mg metronidazole per 100ml). Applications: Treatment of infections. Prevention of post-operative infections due to anaerobic bacteria. Treatment of acute ulcerative gingivitis and acute dental, pericoronitis and apical infections. Trichomonas infections, amoebiasis, giardiasis. (Lagap Pharmaceuticals Ltd).*

Metrotect. A proprietary bitumen paint for protective treatment of iron-work, etc.†

Metrotonin. A mixture stated to contain sparteine and acetyl-choline. Used in medicine.†

Metruien M.. A proprietary preparation containing ethynodiol and mestranol. (Searle, G D & Co Ltd).†

Metrulen. A proprietary preparation containing ethynodiol and mestranol. (Searle, G D & Co Ltd).†

Metso. Alkali silicates. Sodium metasilicate (anhydrous or pentahydrate) and sodium orthosilicate. Applications: Alkaline component of heavy duty household, institutional and industrial cleaning compounds, metal cleaning. (The PQ Corporation).*

Metso 22, 66, 99. Trade names for proprietary cleaning agents containing alkaline silicates.†

Metso 99. A patented crystalline sodium silicate. It contains 3 parts Na_2O, 2 parts SiO_2, and 11 parts H_2O on a molecular basis. A cleaning material.†

Metternich Green. See IODINE GREEN.

Metubine Iodide. Metocurine iodide. Applications: Blocking agent. (Eli Lilly & Co).

Mevasine. Mecamylamine hydrochloride. Applications: An oral antihypertensive agent. (Merck Sharp & Dohme).*

Mevilin-L. A freeze dried preparation of a living attenuated virus of the Schwarz strain which has been grown in chick embryo fibroblast-tissue cultures. Applications: Active immunization against Measles. (Evans Medical).*

Mexate. Methotrexate. Applications: Antineoplastic. (Bristol Laboratories, Div of Bristol-Myers Co).

Mexican Blue. The colouring matter of *Sericographis mohite*. Dyes wool and cotton purplish-blue direct.†

Mexican Eliemi. Elemi from *Amyris elemifera*.†

Mexican Onyx. A variety of calcite.†

Mexican Turpentine. See TURPENTINE.

Mexico Seeds. Castor oil seeds.†

Mexphalte. (Spramex, Shelspra). Trade marks for varieties of bitumen, used for road dressing and other purposes.†

Meyerhofferite. An artificially prepared mineral. It is calcium borate.†

Meyer's Solution. Mercury-potassium iodide solution, obtained by dissolving 13.35 grams mercuric chloride and 49.8 grams potassium iodide separately in water, mixing the solutions and diluting to 1 litre.†

Mezlin. Mezlocillin. Applications: Antibacterial. (Miles Pharmaceuticals, Div of Miles Laboratories Inc).

M.F.C. Materials for chromatography. (British Drug Houses).†

MGA. Melengestrol acetate. Applications: Antineoplastic; progestin. (The Upjohn Co).

Mgoa Rubber. Commercial name for the rubber from *Muscarenhasia elastica* of East Africa.†

MG2/MG4. Range of granular fertilizers. (Fisons PLC, Horticulture Div).

MHA. Methionine hydroxy analog. Applications: Calcium feed supplement for poultry and other animal feeds. (Monsanto Co).*

Mi-Col. A mildew fungicide. (Plant Protection (Subsidiary of ICI)).†

Mi-Gee Brand. Methylene iodide. Applications: Gem and mineral testing. (Geoliquids).*

Mianin. See CHLORAMINE T.

Miazine. Metadiazine.†

Mica. (Laminated Talc, Glimmer Glist). This material consists mainly of a double silicate of aluminium, and sodium or potassium. It also contains magnesium and iron silicates. Used for making fireproof window-panes and lamp-chimneys, also as an electrical insulating material.†

Micabond. A proprietary trade name for a material consisting of mica, shellac, and resin.†

Mica Cambric. See MICANITE CLOTH.

Micacoat. Chemically coupled muscovite mica, coarse and fine grinds: SiO_2, Al_2O_3, K_2O, Fe_2O_3, Na_2O, CaO, TiO_2, MnO_2, P, S. Applications: Polymer composites and high performance coatings. (NYCO).*

Micafil B. A proprietary synthetic resinvarnish-paper product used in electrical insulation.†

Micafil G. A proprietary electrical insulator made in the form of tubes from shellac, coated paper, and mica.†

Micafil S. A proprietary trade name for a shellac varnish-paper product used as an electrical insulation.†

Micafolium. A general name for electrical insulators made from mica splittings and paper.†

Mica-Kote. A proprietary trade name for a roofing felt made from asphalt impregnated felt.†

Micalex. See MYCALEX.

Micanite. A mica material built up of small plates of mica with an insulating material such as shellac, or on a foundation of paper or cloth. Used as an electrical insulating material.†

Micanite Cloth. (Mica Cambric, Toile Micanite). Products used for electrical insulation, made from mica splittings on a cotton-cambric backing.†

Micaphilit. Synonym for Andalusite.†

Micarta. A trade name for a range of varnished paper and fabric products using natural and synthetic resin varnishes. They are used for electrical insulation.†

Micarta Folium. A similar product to micafolium.†

Mica Silk. An electrical insulating tape made from mica splittings on a silk cloth.†

Micatin. Miconazole nitrate. Applications: Antifungal. (Ortho Pharmaceutical Corp).

Michler's Hydrol. *p-p* Bisdimethyl-aminobenzhydrol, HO . $CH[C_6H_4N(CH_3)_2]_2$.†

Michler's Ketone. Tetramethyldiamino-benzophenone, $[(CH_3)_2N.C_6H_4]_2CO$. Used for making dyestuffs.†

Micolette Micro-enema. Sodium lauryl, sulphoacetate, sodium citrate, glycerol, potassium sorbate, citric acid, sorbitol. Applications: Chronic or acute constipation in the rectum and sigmoid colon. (Ayerst Laboratories).†

Micoren. A proprietary preparation of cropropamide and crotethamide. A respiratory stimulant. (Ciba Geigy PLC).†

Micracet. Pigment preparations. (Ciba-Geigy PLC).

Micralax. A proprietary preparation containing sodium citrate, sodium alkyl sulphoacetate and sorbic acid. An enema. (Smith Kline and French Laboratories Ltd).*

Micralax Micro-enema. Combination of sodium citrate, sodium alkylsulphoacetate, sorbic acid together with glycerin, sorbitol and purified water. Applications: Micro-enema for relief of constipation. (Smith Kline and French Laboratories Ltd).*

Micro-asbestos. An Austrian asbestos of short fibre unsuitable for the ordinary uses of asbestos.†

Micro-Chek. Applications: For industrial, microbiological growth inhibitors and preservatives, namely, fungicides, fungistats, biocides, biostats, bactericides, bacteriostats, mildewcides and mildewstats in Class 5. (Ferro Corporation).*

Micro Dry. Aluminium chlorohydrate. Applications: Anhidrotic. (Reheis Chemical Co).

Micro-K. Potassium Chloride. Applications: Replenisher. (A H Robins Co Inc).

Microbator PC-78. A tamed and stabilized non-toxic chlorine dioxide complex concentrate formulated for use as an additive deodorant without free chlorine release. Applications: Effectively arrests mal-odours caused by virae, fungi, bacteria and coliform densities, including escherichia coli, klebsiella aerogenes and enterobacter aerogenes when added to coolants, cutting oils, industrial sumps, sludge pits, cooling towers,waste water, marine holding stations and ship's bilge areas, animal housing and rest room surfaces. (Punati Chemical Corp).*

Microbiotone. A peptic digest of beef tissue that is a water soluble granular product. Applications: It is used in various fermentations, veterinary biologicals and diagnostics as a nutrient for faster growth of various organisms. (American Laboratories Inc).*

Microcal. Calcium silicate. (Crosfield Chemicals).*

Microcal ET. Series of synthetic calcium silicates of controlled particle size used as extenders in emulsion paint. (Crosfield Chemicals).*

Microcillin. Pharmaceutical. Applications: Injectable penicillin especially against gram-negative "problem organisms'. (Bayer & Co).*

Microcosmic Salt. (Phosphorus Salt, Fusible Salt of Urine, Fusible Salt, Essential Salt of Urine). Sodium-ammonium-hydrogen phosphate, $Na(NH_4)H.PO_4.4H_2O$.†

Microdol (Extra). A trade name for ground Dolomite.†

Microgynon 30. A proprietary preparation of ethinyloestradiol and α- norgestrel. An oral contraceptive. (Schering Chemicals Ltd).†

Microlan. Powdered lanolin. (Croda Chemicals Ltd).*

Microlith-A. Pigment preparations. (Ciba-Geigy PLC).

Microlut. Norgestrel. Applications: Progestin. (Schering AG).

Micromeritol. An alcoholic solution of yerba buena.†

Micromet. A slowly soluble sodium metaphosphate. (Albright & Wilson Ltd).*

Micronase. Glyburide. Applications: Antidiabetic. (The Upjohn Co).

Micronor. Norethindrone. Applications: Progestin. (Ortho Pharmaceutical Corp).

Micropaque. X-ray grade barium sulphate. (Graesser Laboratories Ltd).

Micropore. A proprietary surgical tape of rayon with a hypoallergenic adhesive. (3M).*

Microporite. An insulating concrete made from ground silica and lime, hardened by treatment with steam.†

Microsan. A copper fungicide. (Mechema).*

Microseal. Process of deposition of thin dry solid film lubricant coating by impingement. Applications: Mould release for plastic moulds, bearings, sliding surfaces and vacuum lubrication. (E/M Corporation).*

Microsil. Rubber reinforcing agent. (Crosfield Chemicals).*

Microsperse. Range of carbon black dispersions in water. Applications: Paper, concrete, ink, paint etc industries. (Colloids Ltd).*

Microthene. Polyethylene powders. Applications: Rotational moulding, powder coating and specialty applications eg processing aids, surface improvers. (USI Chemicals).*

Microthene F. A registered trade name for ultra fine polyolefine powder dispersions for coating purposes.†

Microx. Zinc oxide, high surface area. (Durham Chemicals Ltd).*

Micryston. A proprietary name for a gro of hormone preparations used for replacement therapy. (LAB Ltd).*

Mictine. 1-Allyl-6-amino-3-ethylpymidin 2,4-dione.†

Mictral. Granules containing nalidixic ac sodium citrate, citric acid and sodium bicarbonate. Applications: Antibacterial agent for the treatment cystitis and lower urinary tract infections. (Winthrop Laboratories).*

Midamide. Amiloride hydrochloride. Applications: A potassium-conserving agent. (Merck Sharp & Dohme).*

Midamor. Amiloride hydrochloride. Applications: A potassium-conserving agent. (Merck Sharp & Dohme).*

Middle Oils. (Carbolic Oils). That fractic of coal tar distilling at 170-230° C.†

Midicel. A proprietary preparation containing sulphamethoxypyridazine. (Parke-Davis).†

Midrid. Isometheptene mucate 65 mg, dichloralphenazone 100 mg and paracetamol 325 mg Applications: Migraine and tension headache. (G W Carnrick Co Ltd).*

Midvale Alloys. Heat-resisting alloys AT alloy contains from 33-39 per cent nickel, 10-12 per cent chromium, 1.1-1.8 per cent manganese, and the balar iron. BTG alloy contains 60-62 per ce nickel, 10-11 per cent chromium, 1.2-1.5 per cent manganese, and the balan iron.†

Midvaloy H.R. A proprietary nickel-tungsten-chromium alloy. It resists corrosion.†

Migafar. Proofing agent. (Ciba-Geigy PLC).

Migafar AL. A proprietary trade name f an aqueous emulsion of fatty products suitable for the removal of chafe mark from dyed fabrics. (Ciba Geigy PLC).

Migatex. Finishing agents. (Ciba-Geigy PLC).

Migen. House dust mite vaccine. (Beecha Pharmaceuticals).

Might-Weld Multi-Care Structural Adhesives. Acrylic structural adhesives ultraviolet curing structural adhesives. Applications: Magnet, metal, glass and fibre optics bonding, assembly adhesives and coatings for electronics. (Dymax Engineering Adhesives).*

Migil. Migraine remedy. (The Wellcome Foundation Ltd).

Miglyol. Special oils/neutral oils , triglycerides. Applications: Production of oily suspensions, suppositories, ointments and creams. (Dynamit Nobel Wien GmbH).*

Mignonette Green. See MAY GREEN.

Mignonette-geranium Oil. An oil obtained by distilling geraniol over mignonette flowers.†

Migra Iron. A special pig iron for high quality castings obtained by a special heat treatment before casting, which results in a remarkably fine grain.†

Migraleve. A proprietary preparation of BUCLIZINE dihydrochloride, PARACETAMOL, CODEINE phosphate and dioctylsodium sulphosuccinate tablets with tablets of codeine, paracetamol and dioctyl sodium sulphosuccinate, used in the treatment of migraine. (International Chemical Co Ltd).†

Migrane-Dolviran. Pharmaceutical preparation. Applications: For the relief of migraine and migrainous headaches. (Bayer & Co).*

Migril. A proprietary preparation of Ergotamine tartrate, Cyclizine hydrocloride and cafeine, used in the treatment of migraine. (The Wellcome Foundation Ltd).*

Mikado Brown B, 3G0, M. A dyestuff obtained by the action of alkalis upon *p*-nitrotoluenesulphonic acid, in the presence of an oxidizing substance. Dyes cotton brown.†

Mikado Gold Yellow 2G, 4G, 6G, 8G. See MIKADO YELLOW.

Mikado Orange G, 4R. (Direct Orange 2R, Direct Orange G, Chloramine Orange, Stilbene Orange 4R, Paramine Fast Orange D, G, Direct Fast Orange 4R, RL, 2RL, Chlorazol Fast Orange D). A dyestuff obtained by the action of alkaline reducing agents upon Direct yellow. Dyes cotton yellow-orange to reddish-orange.†

Mikado Yellow. (Mikado Gold Yellow 2G, 4G, 6G, 8G; Direct Yellow 2G, 4G, Diazamine Fast Yellow H, Direct Fast Yellow 2GLO, Paramine Fast Yellow 3G, Stilbene Yellow 2G, 3G, 8G). A dyestuff obtained by the treatment of the condensation product of *p*-nitrotoluenesulphonic acid and sodium hydroxide with oxidizing agents. Dyes cotton yellow from a salt bath.†

Mikamycin. An antibiotic produced by Streptomyces mitakaensis. Mikamycin B is Ostreogrycin B.†

Mikolite. A proprietary material. It is a vermiculite which has been expanded by calcination giving a very fine product. It is used in paints.†

Mikrobin. Sodium-*p*-chlorobenzoate. A preservative for wines.†

Mila. Skin cream, milk cleanser and lotion, for daily care of the skin. (Richardson-Vicks Inc).*

Milanol. Basic bismuth trichloro-butyl-malonate.†

Milcap. Fungicide containing ethirimol and captafol for use on wheat. (ICI PLC).*

Mil-Col. A mildew fungicide and seed dressing. (ICI PLC).*

Milcurb. Fungicide containing dimethirimol. (ICI PLC).*

Mild Alkali. Sodium carbonate, Na_2CO_3.†

Mild Lime. Calcium carbonate (chalk), $CaCO_2$, is known in agriculture as mild lime.†

Mildvac-C. Connecticut type bronchitis vaccine. Applications: immunization of poultry. (Intervet America Inc).*

Mildvac-M. Mild Massachusetts type bronchitis vaccine. Applications: immunization of poultry. (Intervet America Inc).*

Milgard. Cleansing lotion, for babies and sensitive skin. (Richardson-Vicks Inc).*

Milgo. Fungicide containing ethirimol. (ICI PLC).*

Milid. See PROGLUMIDE.

Milk Glass. A soda or flint glass rendered opaque by the addition of a mineral phosphate.†

Milk of Asafoetida. A 4 per cent emulsion of asafoetida in water. Used as a sedative and carminative.†

Milk of Barium. A suspension of barium hydroxide, $Ba(OH)_2$.†

Milk of Lime. Slaked lime and water in a thin cream.†

Milk of Magnesia. A suspension of magnesium hydroxide, $Mg(OH)_2$.†

Milkstone. A mixture of milk salts and protein obtained from milk.†

Milk Sugar. Lactose,$C_{12}H_{22}O_{11}$†

Milk Sugar Rennet. See PEGNIN.

Milk Tree Wax. Cow tree wax.†

Mill Creek. Natural products, a line of hair and skin care products containing natural ingredients including keratin, aloe vera, jojoba, henna, elastin, apricot and chamomile. (Richardson-Vicks Inc).*

Millaloy. A proprietary trade name for a nickel-chromium steel containing 4 per cent nickel, 1.5 per cent chromium, and 0.4 per cent carbon.†

Millektrol. A German sodium carbohydrate preparation. A styptic.†

Millidet. An electric blasting cap with a bronze or aluminium shell. Applications: Used in quarries, open-pit mines, underground mining operations, coal stripping, shafts, tunnels and heavy construction projects. (Hercules Inc).*

Milling Blue. A dyestuff. It is the sodium salt of a sulphonic acid of diphenyldiaminophenylnaphthazonium chloride. Dyes chromed wool blue.†

Milling Green. An acid dyestuff, giving bluish-green shades on wool.†

Milling Green S. A quinone-oxime dyestuff. The iron salt of Gambine B is probably present in the dye.†

Milling Orange. A dyestuff. Dyes chromed wool orange-red from an acid bath.†

Milling Red B, FPG, G, FR, R. Acid mordant dyestuffs. They dye direct from an acid bath, or on chrome mordanted wool.†

Milling Silver. A nickel silver. It is an alloy of 56 per cent copper, 27.5-31 per cent zinc, 12-16 per cent nickel, and 0.5-1 per cent lead.†

Milling Yellow. (Chrome Yellow D, Anthracene Yellow BN, Mordant Yellow O, Chrome Fast Yellow, Monochrome Yellow Paste, Powder, Phorochrome Yellow Y, Chrome Fast Yellow O, Alizarin Yellow L, WS, Lighthouse Chrome Yellow O, Solochrome Yellow Y). A dyestuff. Dyes wool yellow from an acid bath.†

Millon's Base. Hydroxy-dimercuro ammonium hydroxide, $OH.Hg_2NH_2O$[illegible] Used for colouring porcelain.†

Millon's Reagent. Mercury dissolved in [illegible] equal weight of nitric acid (specific gravity 1.41), and the solution diluted to twice its volume. After standing, t[illegible] liquid is decanted from the precipitat[illegible] Used as a test for proteins.†

Millophyline. A proprietary preparation [illegible] diethylaminoethyltheophylline and camphorsulphonate, used as a cardia[illegible] and respiratory stimulant. Also as a bronchial antispasmodic. For veterin[illegible] use. (Dales Pharmaceuticals).*

Mills Plastic. A proprietary trade name [illegible] a vinylidene chloride synthetic resin.[illegible]

Milontin. A proprietary preparation of phensuximide. An anticonvulsant. (Parke-Davis).†

Milori Blue. See PRUSSIAN BLUE.

Milori Green. See CHROME GREENS

Milory Green. See CHROME GREEN[illegible]

Milowite. A proprietary amorphous silic[illegible] for paint, polishing, and chemical tr[illegible] It is very white in colour, and 90 per cent is below 0.01 mm particle size.[illegible]

Milstem. Liquid ethirimol seed dressing (ICI PLC).*

Milton. Sterilizing fluid and crystals, to sterilize baby bottles. (Richardson-Vicks Inc).*

Miltopan. Blend of alkylarylsulphonate and solvents, anionic. Applications: [illegible] purpose detergent. (Henkel Chemic[illegible] Ltd).

Miltown. Meprobamate. Applications: Sedative. (Wallace Laboratories, Di[illegible] Carter-Wallace Inc).

Milvan Steel. A proprietary trade name [illegible] a high speed tool steel containing 1[illegible] cent tungsten, 4 per cent chromium[illegible] 2 per cent vanadium.†

waloy. A proprietary trade name for a chromium-vanadium steel.†

mea. See MOMEA.

mico. See CARBORA.

mosa. See CLAYTON YELLOW and WATTLE BARK.

nadex. Orange flavoured, green syrup containing in each 5 ml the following vitamins and minerals: Vitamin A 650 iu, Vitamin D 65 iu, iron (as green ferric ammonium citrate) 12 mg, calcium glycerophosphate 11.25 mg, potassium glycerophosphate 1.125 mg, manganese sulphate 0.5 mg, copper sulphate 0.5 mg Applications: A vitamin and mineral supplement and appetite restorative for children and adults, particularly during and after illness. (Evans Medical).*

inafen. A proprietary artificial milk feed low in PHENYLALANINE used in the dietary treatment of phenylketonuria. (Cow and Gate).†

inamino. A proprietary preparation of vitamins, minerals and aminoacids used as a dietary supplement. (Consolidated Chemicals).†

inargent. An alloy of copper, nickel, and aluminium.†

inargentatum. An alloy of 56.82 per cent copper, 39.77 per cent nickel, 2.84 per cent tungsten, and 0.57 per cent aluminium.†

indererus's Spirit. A solution of ammonium acetate.†

ineral Acid. An inorganic acid.†

ineral Black. A pigment. It is a shale found naturally, and contains 70 per cent silica and 30 per cent carbonaceous matter.†

ineral Blue. See MOUNTAIN BLUE, PRUSSIAN BLUE, and VERDITER BLUE.

ineral Brown. See CAPPAGH BROWN and UMBER.

ineral Butters. A term formerly used for several of the metallic chlorides, such as those of antimony, arsenic, bismuth, and zinc.†

ineral Caoutchouc. See ELATERITE.

ineral Carbon. Anthracite.†

ineral Cotton. See SLAG WOOL.

Mineral Fat. See PETROLEUM JELLY.

Mineral Flour. A Florida clay used in rubber mixings.†

Mineral Giycerin. See PARAFFIN, LIQUID.

Mineral Green. See MALACHITE, SCHEELE'S GREEN, VERDITER BLUE, and GREEN VERMILION.

Mineral Grey. The ash from lapislazuli after the extraction of ultramarine. Also see SLATE GREY.†

Mineral Gum. See SOLUBLE GLASS.

Mineral Indigo. The blue oxide of molybdenum is known by this name.†

Mineral Jelly. See PETROLEUM JELLY.

Mineral Khaki. A mineral colour produced on the fibre by impregnating cotton with a mixture of ferrous and chromic acetates, drying, and then steaming. Mixtures of basic ferric and chromic acetates are formed on the material, which are fixed by passing the fibre through solutions of sodium carbonate and sodium hydroxide.†

Mineral Lake. A basic chromate of tin, prepared by adding potassium chromate solution to stannous chloride solution. Used for colouring paper, and in oil painting.†

Mineral Oils. Natural oils of the petroleum series.†

Mineral, Orange. See ORANGE LEAD.

Mineral Pitch. See BITUMEN.

Mineral Pulp. See AGALITE.

Mineral Purple. See INDIAN RED. Purple of Cassius is also sometimes called by this name.

Mineral Rouge. See INDIAN RED and JEWELLER'S ROUGE.

Mineral Rubber. Bitumens of the gilsonite type.†

Mineral Superphosphate. See SUPER-PHOSPHATE.

Mineral Syrup. See PARAFFIN, LIQUID.

Mineral Tallow. See BITUMEN.

Mineral, Turbith.. See TURPETH MINERAL.†

Mineral Umber. See UMBER.

Mineral Violet. See MANGANESE VIOLET.

Mineral Wax. See OZOKERITE and LIGNITE WAX.

Mineral White. See GYPSUM and BARIUM WHITE.

Mineral Wool. See SLAG WOOL.

Mineral Yeast. Torula, a yeast-like organism, used for fodder production.†

Mineral Yellow. See TURNER'S YELLOW and YELLOW OCHRE.

Minerallic 20 and 78. Filling compounds.†

Minflo. Mining flotation reagents. Applications: They are depressants that selectively separate gangue materials in mineral flotation. They increase the purity of the concentrate, increase efficiency of the collector and can act as filter aids to agglomerate fines and prevent filter plugging. (Hercules Inc).*

Mini Slugit Pellets. Metaldehyde slug killer. (Murphy Chemical Ltd).

Minihist. Pyrilamine Maleate. Applications: Antihistaminic. (Ives Laboratories Inc).

Minipress. A proprietary preparation of prazosin hydrochloride for the treatment of hypertension, left ventricular failure and Raynaud's Disease. (Pfizer International).*

Minite. A similar explosive to Kohlen carbonite (*qv*), without barium nitrate.†

Minium. See RED LEAD.

Minium, Iron. See INDIAN RED.

Minium Tego. A high dispersed red lead marketed in Germany.†

Minizide. A proprietary preparation of prazosin hydrochloride and polythiazide. An antihypertensive. (Pfizer International).*

Minlon. A proprietary brand of Nylon 66 reinforced with 40% mineral filler. (Du Pont (UK) Ltd).†

Minocin. Minocycline hydrochloride. Applications: Antibacterial. (Lederle Laboratories, Div of American Cyanamid Co).

Minofor. Alloys used by jewellers. They contain from 9-64 per cent antimony, 20-84 per cent tin, 2-10 per cent copper, and 1-10 per cent zinc.†

Minol. See PARAFFIN, LIQUID.

Minolite Antigrisouteuse. A Belgian explosive containing 72 per cent ammonium nitrate, 23 per cent so nitrate, 3 per cent trinitrotoluene, per cent trinitronaphthalene.†

Mint Camphor. Menthol.†

Mintacol Solubile. Pharmaceutical preparation. Applications: Miotic (Bayer & Co).*

Mintezol. Thiabendazole. Application anthelmintic. (Merck Sharp & Dohme).*

Mintite. A patent finish for rubber su It consists of powdered mica.†

Mint-O-Mag. Milk of magnesia. Applications: Antacid; laxative. (E Squibb & Sons Inc).

Min-U-Gel. Colloidal attapulgite clay Applications: For suspension syste (Hermadex Ltd).*

Min-U-Sil. Micronized natural silica. Applications: Silicon rubber, surfa coatings. (Cornelius Chemical Co Ltd).*

Minyak Kerung. An oleo-resin obtain from *Dipterocarpus* species of Ma It is obtained as a viscous liquid.†

Miochol. Acetylcholine chloride. Applications: Cardiac depressant; cholinergic; miotic; vasodilator. (CooperVision Pharmaceuticals In

Mio-Pressin. A proprietary preparatio rauwolfia serpentina, protoveratrin phenoxybenzamine hydrochloride. antihypertensive. (Smith Kline and French Laboratories Ltd).*

Miostat. Carbachol. Applications: Cholinergic. (Alcon Laboratories I

Mipolam. A proprietary trade name f polyvinyl chloride which, when plasticized, has properties resembli soft rubber. It is used in moulding for sheathing cables.†

Mirabilite. See GLAUBER'S SALT.

Miracil. D. A proprietary trade name the hydrochloride of Lucanthone.†

Miraculoy. A proprietary trade name steel containing 1.25 per cent nicke 0.65 per cent chromium, 0.4 per ce molybdenum, and 1.55 per cent manganese.†

radon. Anisindione.†

ralite. A light aluminium alloy which can be cast or rolled, and drawn into wire. It contains 12 per cent copper and 2 per cent tin.†

ramant. A cutting alloy of heat resisting metals with a definite fraction of stable and hard carbides, especially molybdenum and tungsten carbides in eutectic proportions.†

ra Metal. Acid-resisting alloys. One contains 74.7 per cent copper, 16.3 per cent lead, 6.8 per cent antimony, 0.91 per cent tin, 0.62 per cent zinc, 0.43 per cent iron, and 0.24 per cent nickel. Another consists of 75 per cent copper, 16 per cent lead, 8 per cent tin, and 1 per cent nickel.†

iranol. Liquid foaming agent with outstanding emulsifying properties. (Foseco (FS) Ltd).*

iranol CM. Imidazoline based amphoteric with whole coconut as fatty radical. Light amber liquid. Applications: Emulsifies grease, suspends particulate soil. (Venture Chemical Products Ltd).

iranol C2M. Imidazoline based amphoteric with whole coconut as fatty radical. Clear aqueous solution. Applications: Non- irritating surfactants, emulsifiers, solubilizers, stabilizers, used in shampoos, cleaners, pharmaceutical applications. (Venture Chemical Products Ltd).

iranol C2M-SF. Imidazoline based amphoteric with whole coconut as fatty radical. Clear amber liquid. Applications: Industrial cleaner formulations. (Venture Chemical Products Ltd).

iranol DM. Imidazoline based amphoteric with stearic fatty radical, in the form of a creamy white paste. Applications: Fibre softener, wool lubricant. (Venture Chemical Products Ltd).

iranol JEM. Imidazoline based amphoteric with caprylic and ethylhexoic fatty radicals. Clear aqueous solution. Applications: Used in bottle washing, wax stripping, degreasing etc. (Venture Chemical Products Ltd).

Miranol L2M-SF. Imidazoline based amphoteric with linoleic fatty radical, in aqueous solution form. Applications: Wax type polishes and floor finishes etc. (Venture Chemical Products Ltd).

Miranol SM. Imidazoline based amphoteric with capric fatty radical. Clear aqueous solution. Applications: Medicated, germicidal, rug and upholstery shampoos, hand soaps and surgical soaps. (Venture Chemical Products Ltd).

Mirasol. A proprietary trade name for alkyd varnish and lacquer resins.†

Miravon. Flow modifier. (ABM Chemicals Ltd).*

Mirbane Essence. See ESSENCE OF MIRBANE.

Mirbane Oil. See ESSENCE OF MIRBANE.

Mirion. A proprietary preparation. It is iodo-hexamine.†

Mirrolac. Calendering varnishes. (The Scottish Adhesives Co Ltd).*

Mirror Bronze. A copper and tin alloy, containing 28-35 per cent tin. It sometimes contains a little nickel.†

Mirvale. Potato sprout suppressant. (Ciba-Geigy PLC).

Mischzinn. A tin alloy. Theoretically it is an eutectic containing 63 per cent tin and 37 per cent lead, but in practice the tin is at least 55 per cent. antimony and copper must be more than 3.5 and 0.5 per cent respectively, and zinc is present in traces. It is prepared from metal scrap, chiefly bearing alloys.†

Miscible Carbon Disulphide. A mixture of carbon disulphide with castor oil, caustic potash, denatured alcohol, and water. An insecticide for destroying the Japanese beetle in the soil without serious damage to the plant.†

Misco. An alloy of 57.5 per cent iron, 15 per cent chromium, 25 per cent nickel, 1.5 per cent silicon, 0.5 per cent manganese, and 0.5 per cent carbon.†

Miscrome. A proprietary trade name for a corrosion-resisting alloy of iron with 28 per cent chromium.†

Mission Prenatal. Folic acid. Applications: Vitamin. (Mission Pharmacal Co).

Mistletoe Rubber. A rubber obtained from the fruit of certain *Loranthaceae* as parasites on the coffee tree.†

Mist-o-Matic. A liquid seed dressing. (Murphy Chemical Co).†

Mist-o-matic Ferrax. Fungicide seed treatment. (Murphy Chemical Ltd).

Mitaban. Amitraz. Applications: Scabicide. (The Upjohn Co).

Mitac 20. Insecticide/acaricide. (FBC Ltd).

Mitchalloy A. A proprietary trade name for an alloy of iron with 2.5 per cent nickel and 0.9 per cent chromium.†

Mithracin. A proprietary preparation of mithramycin. An antineoplastic. (Pfizer International).*

Mitigan. Active ingredient: dicofol; 2,2,2-trichloro-1,1-bis(chlorophenyl) ethanol. Applications: Specific miticide. (Agan Chemical Manufacturers Ltd).†

Mitigated Caustic. A fused mixture of 1 part silver nitrate with 2 parts potassium nitrate.†

Mitin. Moth proofing agents. (Ciba Geigy PLC).†

Mitine. A base for ointments prepared from an emulsion which is superfatted with a non-emulsifying fat. Wool fat is used as the fat, and milk as the serum-like liquid to the extent of 50 per cent.†

Mitis Green. See EMERALD GREEN.

Mitoxana. SEE HOLOXAN. (Degussa).*

Mitrelle. Polyester yarn. (ICI PLC).*

Mitschlich's Ammoniacal Salt. Probably a double compound of mercuroxy-ammonium nitrate, and mercuriammonium nitrate $(NH_2.Hg_2O)NO_3.(NH_2.Hg)NO_3.H_2O$.†

Mitsubishi Yuka-ECX. Electroconductive polymer. (Mitsubishi Petrochemical Co).*

Mitsubishi Yuka-SPX. Soft polyolefin. (Mitsubishi Petrochemical Co).*

Mittel AEP. A proprietary softening agent for cellulose esters. It is ethyl-*p*-toluenesulphonate.†

Mittel KP. Cresyl-*p*-toluenesulphonate. A softening agent for cellulose esters.†

Mittel L. A solvent resembling turpentine.†

Mittler's Green. See CHROMIUM GREEN.

Mix. Miscellaneous. Applications: stabilizing agent for liquid fertilizer-pesticide application. (Drexel Chemical Company).*

Mix-Kit. Package device for two-part compounds. (Hardman Inc).*

Mix Metal. Misch metal.†

Mixad. Sand conditioner. (Foseco (FS) Ltd).*

Mixed Acid. (Nitrating Acid). Any mixture of nitric and sulphuric acids used for nitrating. Usually it consists of 37 per cent nitric acid and 63 per cent sulphuric acid. Used for nitrating dyes and explosives.†

Mixed Ether. An ether containing two different alkyl radicals, as in ethyl-methyl ether. $C_2H_5.O.CH_3$.†

Mixed Metal. A term used for alloys of cerium, lanthanum, and praseodymium.†

Mixed Vitriol. (Salzburg Vitriol). Cupric-ferrous sulphate, $CuSO_4.3FeSO_4.28H_2O$.†

Mixol. A timber insecticide. (ICI PLC).*

MK Ammonium Perchlorate. 99.1-99.7% NH_4C2O_4, depending on grade. Manufactured in ordnance and industrial grades. (Kerr-McGee Chemical Corp).*

M-M-R. Measles, Mumps and Rubella vaccine. Applications: For combined protection against Measles, Mumps and German Measles. (Merck Sharp & Dohme).*

M-M-VAX. Measles and Mumps vaccine. Applications: Protection against Measles and Mumps infection. (Merck Sharp & Dohme).*

MN Powder. (Maxim-Nordenfelt Powder). An American guncotton powder gelatinized with ethyl acetate.†

M.N.T. Mononitrotoluene, $CH_3.C_6.H_4.NO_2$, an intermediate in the preparation of trinitrotoluene.†

Moac. Very finely divided mica.†

Moban. Molindone hydrochloride. Applications: Antipsychotic. (Endo

Laboratories Inc, Subsidiary of E I du Pont de Nemours & Co).

obilrap. Pallet wrap stretch films including one-side cling films, in thicknesses varying from 17 to 35 microns. Applications: Wrapping around pallets for increased product protection during transport and warehousing. (Mobil Plastics Europe).*

obilrapper. Pallet wrap with stretch film system, which can be handled by one person, offering a variety of stretch films. Applications: Wrapping pallets for increased product protection during transport and warehousing. (Mobil Plastics Europe).*

obilsol 44. Mobilsol 66. Registered trade marks for modifying diluents for epoxy resins. They give flexibility. (Mobil Chemical Co).†

ocap 10G. Nematicide. (Murphy Chemical Ltd).

ocasco Iron. A proprietary trade name for a nickel-chromium-molybdenum cast iron containing 1-1.35 per cent nickel, 0.25-0.3 per cent chromium, and 0.75 per cent molybdenum.†

ocaya Oil. (Macaja Butter). Paraguay palm oil obtained from the kernels of *Acrocomia scelerocarpa*.†

ocha-stone. Agates of white or brown chalcedony from India, with markings due to oxides of iron and manganese.†

lock Epsoms. Needle crystals of sodium sulphate, Na_2SO_4.†

lock Gold. See MOSAIC GOLD.

lock Lead. Both tungsten ore found in Cornwall and zinc blende are known by this name.†

lock Silver. An alloy of 84 per cent aluminium, 10 per cent tin, 5.5 per cent copper, and 0.1 per cent phosphorus.†

lock Turkey Red. Barwood red on cotton.†

lock Vermilion. Lead chromate.†

lodaflow. Resin modifier. Applications: Additive for improving flow, levelling and adhesion properties of surface coatings. (Monsanto Co).*

lodane. A danthron preparation. Applications: Laxative tablets. (Hercules Inc).*

Modane Soft. Docusate sodium. Applications: Stool softener; pharmaceutic aid - surfactant. (Adria Laboratories Inc).

Moddite. An explosive. It is a variety of Cordite, but is made with a nitro-cellulose partially soluble in ether alcohol.†

Modecate. A proprietary preparation of fluphenazine decanoate used in the treatment of psychotic disorders. (Squibb, E R & Sons Ltd).*

Modern Blue. See MODERN VIOLET.

Modern Blue CVI. (Modern Cyanines). Dyestuffs. They are derivatives of gallocyanine.†

Modern Heliotrope PH. A dyestuff obtained by condensing nitrosomono-ethyl-*o*-toluidine, with gallamide, and reducing.†

Modern Violet. (1900 Blue, Modern Violet R, Chromazol Violet). Dyestuffs. They are leuco derivatives of gallo-cyanines.†

Modern Violet N. A dyestuff. It is the leuco compound of a pyrogallo-cyanine.†

Modern Violet R. A dyestuff. It is a British equivalent of Modern violet.†

Modic. Adhesive polyolefin. (Mitsubishi Petrochemical Co).*

Modified Butacite. A propetary trade name for thermosetting polyvinyl butyral synthetic resin.†

Modified Soda. A mixture of sodium carbonate and bicarbonate used as a cleaning agent in laundries.†

Modified Vinylite X. A proprietary trade name for thermosetting polyvinyl butyral synthetic resin.†

Modinal: T. A proprietary trade name for a wetting agent consisting of a long chain alcohol sulphate.†

Moditen. A proprietary preparation of fluphenazine hydrochloride. A sedative. (Squibb, E R & Sons Ltd).*

Moditen Enanthate. A proprietary preparation of fluphenazine enanthate used in the treatment of psychotic disorders. (Squibb, E R & Sons Ltd).*

Moditen Tablets. Fluphenazine hydrochloride in tablet form. (E R Squibb & Sons Ltd).

Modrenal. Capsules containing trilostane. Applications: Used in the treatment of adrenal cortical hyperfunction. (Winthrop Laboratories).*

Moducren. Timolol maleate, hydrochlorothiazide and amiloride hydrochloride. Applications: For the treatment of hypertension. (Merck Sharp & Dohme).*

Modulan. Acetylated lanolin. Applications: Effective emollient barrier for skin care products. Improves lather and inhibits cracking in bar soaps. (Amerchol Corporation).*

Modulex. A trade mark for a carbon black for use as a pigment.†

Moduret 25. Hydrochlorothiazide and amiloride hydrochloride. Applications: For the treatment of hypertension. (Merck Sharp & Dohme).*

Moduretic. Amiloride hydrochloride and hydrochlorothiazide. Applications: Potassium-conserving diuretic for the treatment of oedema and hypertension. (Merck Sharp & Dohme).*

Moellon. See DÉGRAS, ARTIFICIAL.

Moerner's Reagent. This consists of 1 cc formalin with 45 cc distilled water and 55 cc of 50 per cent sulphuric acid.†

Mofix. Oxime/triazine herbicide. (Ciba-Geigy PLC).

Mogadon. A proprietary preparation of nitrazepam. A hypnotic. (Roche Products Ltd).*

Mogador Gum. See MOROCCO GUM.

Mohair. A material made from the hair of the Angora goat.†

Mohawk Steel. A proprietary trade name for a hot die steel containing about 14 per cent tungsten, 3.5 per cent chromium, 0.7 per cent vanadium, and 0.45 per cent carbon.†

Mohr's Salt. Ferrous ammonium sulphate, $FeSO_4.(NH_4)_2.SO_4.6H_2O$.†

Molaschar. A decolourizing carbon used for sugar juices.†

Molascuit. A cattle food. It is the fine fibre of the sugar cane or begasse, with cane molasses absorbed by it.†

Molasocarb. A decolourizing black made from molasses.†

Molasses. The non-crystallizable residu from sugar. Cane molasses contain per cent sugar, 20 per cent water, a 9 per cent ash, whilst beet molasses consists of 50 per cent sugar, 10 per salts, 20 per cent water, 10 per cent nitrogenous matter, and 10 per cent nitrogenous matter. Used as a cattle food.†

Molassine Meal. A mixture of molasses peat moss. A cattle food.†

Molco. Spirit based mould and core coatings. (Foseco (FS) Ltd).*

Mold Wiz Ext. A series of non-silicone non-wax, non-stearate polymeric bas mould release agents, solvent or wat based. Applications: Reinforced pla composites, injection moulding, polyurethane foam and natural and synthetic rubber. (Axel Plastics Research Laboratories Inc).*

Mold Wiz Int. A series of polymeric ba additive lubricants/release agents us as processing aids. Applications: Reinforced plastic composites, melamine/phenolic/urea laminates a overlays, thermoplastic injection moulding, natural and synthetic rubl and urethane elastomers. (Axel Plas Research Laboratories Inc).*

Moldabaster/Moldabaster S. Plaster of Paris adjusted to work with Moldan normal and quick setting. Applicati Dental speciality. (Bayer & Co).*

Moldano. Blue hard plaster of high Bri hardness. Applications: Dental speciality. (Bayer & Co).*

Moldaroc. Yellow, extra hard plaster, f especially resistant models. Applications: Dental speciality. (Bay & Co).*

Moldcote. Self drying spirit based moul and core coatings. (Foseco (FS) Ltd)

Moldesite. A range of phenolic mouldin compounds. Applications: Tool machines, motor industry, pottery, electrotechnical industry, sanitary fie and electrical field. (SPREA Spa).*

Moldex. Cross-linkable polyethylene res Applications: Moulding compounds. (Sumitomo Bakelite, Japan).*

Moler. A Danish diatomaceous earth containing 82.6 per cent silica, 5.33

cent aluminium oxide, a small proportion of ferric oxide, and organic matter. It is very light in weight, and is used in the manufacture of heat-insulating materials.†

Molera. A heat insulator obtained by mixing fine clay with cork dust, and firing.†

Molipaxin. Trazodone hydrochloride. Applications: Antidepressive. (Roussel Laboratories Ltd).*

Mol-Iron. Ferrous sulphate. Applications: Haematinic. (Schering-Plough Corp).

Molivate. A proprietary preparation of CLOBETASONE butyrate used in the treatment of eczema. (Glaxo Pharmaceuticals Ltd).†

Mollifex. Embedding aid in microscopy. (BDH Chemicals Ltd).

Mollin. A base for ointments. It is a soft soap containing 17 per cent of uncombined fat.†

Mollit. A proprietary trade name for a polystyrene synthetic resin.†

Mollit B. A proprietary trade name for glyceryl tribenzoate.†

Mollit I. Diethyldiphenylurea, a softening agent for cellulose esters.†

Mollite. See CHESSYLITE.

Mollphorus. A glycerin substitute consisting of raw and invert sugar.†

Moloie. A proprietary trade name for a manganese-molybdenum steel.†

Molybdate Red. A lead chromate pigment consisting of mixed crystals of lead chromate, lead sulphate, and a small proportion of lead molybdate. Its colour varies from reddish-orange to scarlet.†

Molybdenum Blue. A blue pigment, $4MoO_3+MoO_2$, a product of the reduction of molybdic acid.†

Molybdenum Indigo. A molybdenum oxide, Mo_5O_7, used for colouring rubber.†

Molybdenum Nickel. An alloy of 75 per cent molybdenum and 25 per cent nickel. Used in the manufacture of saws.†

Molybdenum Permalloy. A proprietary trade name for an alloy of 81 per cent nickel, 17 per cent iron, and 2 per cent molybdenum. It has a higher permeability than Standard Permalloy (*qv*).†

Molybdenum Steel. A variable alloy. It usually contains from 0.06-1.73 per cent carbon and 0.23-15 per cent molybdenum.†

Molybdenum, Chromiuum. See CHROMIUM MOLYBDENUM.

Molybdosodalite. A variety of sodalite containing nearly 3 per cent. MoO_3. It is green in colour.†

Molydag. Dispersions of molybdenum disulphide in resin/solvent systems. Applications: Dry film lubrication. (Acheson Colloids).*

Molykote. Speciality lubricants. (Dow Corning Ltd).

Molyte. A trade name for a patented mixture of calcium and molybdenum oxides with a flux.†

Molyvan. Friction reducers. Applications: Used as antiwear and extra fine pressure agents. Can be used as antioxidants. (Vanderbilt Chemical Corporation).*

Momea. (Mimea). A hemp preparation made in Tibet.†

Momordicine. Elaterin, $C_{20}H_{28}O_5$.†

Monacetin. Glyceryl monoacetate, $CH_2OH.CHOH.CH_2OOCCH_3$.†

Monachit. An explosive containing 12 per cent trinitro-xylene, 1 per cent charcoal, and 1 per cent collodion cotton.†

Monacrin. Aminacrine hydrochloride. Applications: Anti-infective, topical. (Sterling Drug Inc).

Monafax. Range of surfactants, each of which is a mixture of mono- and di-phosphate esters derived from ethylene oxide based surfactants. Mainly viscous liquids which are the acid form of the compound. These can be converted to salts of alkali, metal, amine or ammonia by simply mixing with the desired base. Application: emulsion polymerization; industrial cleaners e.g. soak tank cleaners, all-purpose and steam cleaners; herbicide and insecticide emulsifiers; metal working lubricants; dry cleaning detergents. (D F Anstead Ltd).

Monamate CPA. Sulphosuccinate half ester of an alkanolamide in liquid form.

Applications: High foaming agent which produces a dense rich lather and imparts a soft and silky feel to the skin and hair and has low irritancy. Used in mild shampoos, bubble baths, skin cleansers; rug shampoos; detergent formulations. (D F Anstead Ltd).

Monamate OPA. Sulphosuccinate half ester of an alkylolamide as a light yellow liquid. Applications: Produces flash foam and a rich dense lather with good rinsing and viscosity control. Used in shampoos eg baby and family types, gel face cleaners. (D F Anstead Ltd).

Monamine. Foam booster and stabilizers, emulsifiers, detergents, wetting agents, corrosion inhibitors, viscosity builders, lubricants, dispersants. Applications: Cosmetics; shampoos, dry cleaning detergents; metal cleaners, toiletries; rust inhibitors, metal cutting fluids, emulsifiable waxes; agricultual sprays, fuel oil additives, leather and fur preparations. (D F Anstead Ltd).

Monase. A proprietary trade name for the acetate of Etryptamine.†

Monaspor. Anti-bacterial. (Ciba-Geigy PLC).

Monastral. Insoluble phthalocyamine pigments. (ICI PLC).*

Monastral Colours. A proprietary trade name for copper phthalo-cyanine or its derivatives.†

Monaterge. Excellent detergency and wetting; low to moderate foaming; alkaline stability. Applications: Wide range of alkaline cleaners eg floor, wall, drain; liquid laundry detergents. (D F Anstead Ltd).

Monateric CA-35%. Amphoteric surfactant based on coconut imidazoline, in amber liquid form. Applications: Detergent, wetting, emulsifying and dispersing agent with good foam and lather. Used in cosmetics e.g. shampoos, bubble baths, skin cleaners, hair dye formulations, and in high foam floor and metal cleaners. (D F Anstead Ltd).

Monateric CAB. A coco-amido betaine as a clear yellow liquid containing 4.8% sodium chloride. Applications: Used in bubble bath and shampoo formulations (D F Anstead Ltd).

Monateric CDX38. An imidazoline derived dicarboxylic acid amphoteric in the form of a light amber viscous liquid. Applications: Good flash foam and lathering properties, used in baby shampoos, daily use shampoos and skin cleansers. (D F Anstead Ltd).

Monateric CEM-38%. Sodium salt of a dicarboxyethyl fatty acid derived from imidazoline, in liquid form, clear to hazy amber in colour. Applications: Detergent, wetting, emulsifying and solubilizing agent used in liquid medium duty all-purpose detergents, toilet bowl cleaners and aluminium brighteners. (D F Anstead Ltd).

Monateric CSH 32. Sodium salt of a dicarboxymethyl fatty acid derived from imidazoline, supplied as a clear yellow liquid. Applications: Mild detergent properties. Used in shampoo formulations, particularly baby and daily use types. (D F Anstead Ltd).

Monateric Cy Na-50%. An amber liquid based on a capryl imidazoline. Applications: Used in bottle washing, steam cleaners, wax strippers, all-purpose cleaners, de-greasers, Kier boiling, mercerization and acid-pickling. (D F Anstead Ltd).

Monateric ISA-35%. Amphoteric surfactant based on isostearic imidazoline. Supplied as an amber flowable gel. Applications: Used in shampoos such as protein, low pH and daily use types. (D F Anstead Ltd).

Monateric LF. An amber liquid, low foaming detergent with high acid and alkaline stability. Applications: Low or high temperature alkaline and acid cleaners, automatic car wash detergents steam cleaners and truck body and aircraft cleaners. (D F Anstead Ltd).

Monateric 1000. Amphoteric surfactant based on alkyl imidazolines. Light amber liquid. Applications: Detergent and wetting agent with corrosion inhibiting properties, suggested primary surfactant for metal cleaning, soak tank cleaners etc. (D F Anstead Ltd).

Monateric 811. Amphoteric surfactant based on alkyl imidazolines, in amber liquid form. Applications: Has corrosion inhibiting properties with detergent and surface active properties. Used in a broad range of non-corrosive cleaners and industrial detergents. (D F Anstead Ltd).

Monateric 85. An anionic modified amphoteric as a light amber viscous liquid. Applications: Shampoos, hair conditioning and skin-care products (D F Anstead Ltd).

Monawet MB-45. Sodium diisobutyl sulphosuccinate as a clear colourless liquid. Applications: Anionic surfactant for styrene/butadiene emulsion systems for rug backing. (D F Anstead Ltd).

Monawet MM-80. Sodium dihexyl sulphosuccinate as a clear colourless liquid. Applications: Anionic surfactant for pesticidal sprays, shampoos, detergents, latex paints, coatings and textile fibres. (D F Anstead Ltd).

Monawet MO. Series of anionic surfactants composed of sodium dioctyl sulphosuccinate in liquid form. Applications: Textiles eg cotton cloth desizing, printing and dyeing processes, wool carbonizing; agriculture eg liquid fertilizers, insecticides and fungicides; cosmetics eg creams and lotions, bath oils, shampoos. MO-70E used in latex paints, adhesives and sanitation. MO-70 and 70R used in dry cleaning. MO-84R2W used in non-aqueous pigment dispersion for printing inks. (D F Anstead Ltd).

Monawet MT. Anionic surfactants containing sodium ditridecyl sulphosuccinate in liquid form. Applications: Used for vinyl chloride suspensions and styrene emulsions for coatings, paints. MT-80H2W is used in non-aqueous pigment dispersions for printing inks. (D F Anstead Ltd).

Monawet SNO-35. Tetrasodium N-(1,2 dicarboxyethyl) N-alkyl(C18) sulphosuccinamate as a clear light amber liquid. Applications: Mild detergent with good wetting and calcium tolerance; emulsifier. Used in industrial detergents, cosmetic and textile products, speciality emulsions or dispersions for polymerization systems. (D F Anstead Ltd).

Monawet TD-30. Half ester of sulphosuccinic acid based on an ethoxylated fatty alcohol. Light yellow liquid. Applications: Wetting, foaming and emulsifying agent with low inorganic electrolyte content, used in emulsion polymerization, textile wet processing and a broad range of cosmetic and fine fabric detergent formulations. (D F Anstead Ltd).

Monazoline CY, C, O and T. 1-hydroxyethyl-2 alkylimidazoline where the alkyl portion is caprylic, coconut, oleic and tall oil respectively. Amber liquids. Applications: Cationic surfactants with wetting, emulsifying, detergency, thickening, corrosion inhibiting, antistatic, softening and bactericidal properties. Used in agricultural sprays, fungicides, herbicides; gravel to asphalt bonding; dairy cleaners; rinse aids; paints, sealants and inks; oil well recovery operations, sludge dispersion; corrosion control; plastics; metal treatment; textiles eg softening; ore flotation; leather processing. (D F Anstead Ltd).

Moncler Derma. Facial cream, gel, lotion and stick, to improve complexion. (Richardson-Vicks Inc).*

Mond Gas. A combustible gas produced by passing air and steam over heated coal or peat. It consists of a mixture of carbon monoxide, hydrogen, and nitrogen.†

Mond 70 Alloy. An alloy of 70 per cent nickel, 26 per cent copper, and 4 per cent manganese.†

Monel Alloy K-500. A trade mark for an alloy of 30 per cent. copper, 3 per cent. aluminium, 0.5 per cent. titanium and the balance nickel. (Wiggin Alloys Ltd).†

Monel Alloy 400. A trade mark for an alloy of 30 per cent. copper, 1 per cent. manganese, 2.5 per cent. (max) iron and the balance nickel. (Wiggin Alloys Ltd).†

Monel Alloy 414. As for alloy 400 but with high carbon content. (Wiggin Alloys Ltd).†

Monel, Ebonized. See EBONIZED MONEL.

Monel Metal. An alloy. The cast metal usually contains from 68-70 per cent nickel, 28 per cent copper, 2 per cent iron, 1 per cent silicon, and 0.25 per cent manganese; and the forged alloy consists of 68 per cent nickel, 28 per cent copper, 2 per cent iron, 1.5 per cent manganese, and 0.2 per cent silicon. There are sometimes traces of carbon, phosphorus, and zinc.†

Monel 400. A trade mark. Copper 30, Manganese 1, iron 2.5 max. nickel balance. A general engineering alloy with good resistance to corrosion by sea water, sulphuric, hydrochloric and phosphoric acids. (Wiggin Alloys Ltd).†

Monel 414. Monel 400 with a high carbon content. Improved machining properties. (Wiggin Alloys Ltd).†

Monesia. A South American vegetable extract, said to be obtained from the bark of *Lucuma glycyphloea*. An astringent used in diarrhoea.†

Monesin. A saponin-like substance extracted from the bark of the South African plant, *Chrysophyllum viridifolium*.†

Monetite. A calcium phosphate, $CaHPO_4$, found in guano.†

Monex. Tetramethylthiuram monosulphide. A non-staining and non-discolouring delayed action accelerator. Has a short sharp curing range with normal to high sulphur in natural rubber. Used in natural, SBR, butyl, nitrile and neoprene rubbers for wire insulation, druggist sundries, mechanicals, sponge and footwear. (Uniroyal).*

Monistat-Derm. Miconazole nitrate. Applications: Antifungal. (Ortho Pharmaceutical Corp).

Monit. Isosorbide mononitrate tablets for prophylaxis of angina pectoris. (ICI PLC).*

Monitan. Polysorbate 80. Applications: Pharmaceutic aid. (Ives Laboratories Inc).

Monite. A proprietary plastic.†

Monitor. Methamidophos insecticide. (Chevron).*

Monnex. A compound obtained by reacting urea with potassium carbonate. Used as a fire extinguisher. It is non-toxic. (ICI PLC).*

Mono Ammonium Phosphate. Mono ammonium phosphate technical grade. Applications: Other chemicals, water treatment, fire fighting, cosmetic products, food industry, pharmaceutical products. (Rhone-Poulenc NV/CdF Chimie AZF).*

Mono Ammonium Phosphate (Agricultural Grade). Mono ammonium phosphate. Applications: fertilizers, other chemicals. (Rhone-Poulenc NV/CdF Chimie AZF).*

Monobel. (A2 Monobel). A trade mark for a smokeless powder. It is a mixture of 9-11 parts nitro-glycerin, 56-61 parts ammonium nitrate, 8-10 parts wood meal, 0.5-1.5 parts magnesium carbonate, and 18.5-21.5 parts potassium chloride.†

Monocast. A trade mark for a cast nylon obtainable in large diameter cylinders.†

Monochrome Yellow Paste, Powder. Dyestuffs. They are British brands of Milling yellow.†

Monocid. Cefonicid sodium. Applications: Antibacterial. (Smith Kline & French Laboratories).

Mono-Coat. A non-transferring, semi-permanent mould release that gives multiple releases with no transfer to the moulded part. Applications: Mould release. (Chem-Trend).*

Monocortin. Paramethasone Acetate. Applications: Glucocorticoid. (Syntex Laboratories Inc).

Monocron. Active ingredient: monocrotophos; 0,,0-dimethyl-0-(2-methylcarbamoyl-1-methyl-vinyl)-phosphate. Applications: Contact and systemic agricultural insecticide belonging to the enolphosphates. (Makhteshim Chemical Works Ltd).†

Monodral. A proprietary preparation containing penthienate methobromide. An antispasmodic. (Richardson-Vicks Inc).*

Monoformin. The formyl derivative of glycerin, $C_3H_5(OH)_2(OCHO)$.†

Monogermane. Germanium tetrahydride, GeH_4.†

Monoglyme. Ethylene glycol dimethyl ether.†

Mono-Kay. Phytonadione. Applications: Vitamin. (Abbott Laboratories).

Monol. Calcium permanganate, $Ca(MnO_4)_2.4H_2O$. Used in the textile industry.†

Monolan. Ethylene and/or propylene oxide condensates. Applications: Used in low foam non-ionics, lubricants, dishwashing and cosolvency. (Lankro Chemicals Ltd).†

Monolastex Smooth. Fast drying water-based external or internal elastomeric coating - brush or roller applied. Applications: Outside/inside of housing, offices, prefabricated units. (Liquid Plastics Ltd).*

Mono-Line. A monolithic refractory material used for lining steelmaking vessels. (Pfizer International).*

Monolite. Insoluble lake colours and pigments for paints, inks and plastics. (ICI PLC).*

Monolite Dyestuffs. A registered trade name for certain dyestuffs.†

Monolite Fast Scarlet R. See HELIO FAST RED RL.

Monolite Fast Scarlet RN. A dyestuff. It is a British equivalent of Helio fast red RL.†

Monolite Red P. A British dyestuff. It is equivalent to Lake red P.†

Monolite Red R. See LITHOL RED B.

Monolite Yellow. See HANSA YELLOW G.

Mono-Lube. General purpose microemulsion metalworking fluid for machining and grinding ferrous and non-ferrous metals. Applications: Metalworking fluid. (Chem-Trend).*

Monomethyl Methylene Blue. Phenyl-trimethyl-thionine.†

Monopar. Stilbazium Iodide. Applications: Anthelmintic. (Burroughs Wellcome Co).

Monophane. Insulin preparation. (The Boots Company PLC).

Monoplas 279. A proprietary dialkyl (C_7-C_9) phthalate plasticizer.†

Monoplex. Specialty plasticizers. Applications: Vinyl and synthetic lubricants. (The C P Hall Company).*

Monoplex 5. A proprietary trade name for dibenzyl sebacate. A vinyl plasticizer.†

Monopol Oil. See TURKEY RED OILS.

Monopol Soap. (Avirol). Sulphonated oils similar to Turkey red oil. Used as wetting-out agents.†

Monopoxy. Single-component epoxy resin adhesive. (Hardman Inc).*

Monopyrate. See GALLOLS.

Monoresate. See EURESOL.

Monoset. Pre-packed ultra rapid hardening cementitious mortars and concretes. Applications: Raising of manhole covers and frames, repairs to motorways, repairs in tidal conditions and wherever a minimum downtime is needed. (Ronacrete Ltd).*

Monosulfiram. A parasiticide. Tetra-ethylthiuram monosulphide.†

Monosulph. Anionic surfactants. (Diamond Shamrock Process Chemicals Ltd).

Monotah. A name for guaiacol-methyl-glycollate.†

Monotard. Insulin zinc. Applications: Antidiabetic. (E R Squibb & Sons Inc).

Monotard Human. Insulin Human. A protein that has the normal structure of the natural antidiabetic principle produced by the human pancreas. Applications: Antidiabetic. (E R Squibb & Sons Inc).

Monotheamin and Amytal Pulvules. A proprietary preparation of theophylline monoethamolamine and amylo-barbitone. (Eli Lilly & Co).†

Monotheamin Pulvules. A proprietary preparation of theophylline mono-ethanolamine. (Eli Lilly & Co).†

Mono Thiurad. Tetramethylthiuram monosulphide. Applications: vulcanization accelerator. (Monsanto Co).*

Monotype Metal. See TYPE METAL.

Monox. A product containing mainly silicon monoxide, with some silicon, silicon dioxide, and small quantities of silicon carbide. It is obtained by heating sand with silicon, carborundum, or coke, in the electric furnace. A good thermal and electrical insulator. It decomposes water with the evolution of hydrogen.†

Monphytol. A proprietary preparation of boric acid, chlorbutol, methyl salicylate, salicylic acid and undecylenic esters. Antifungal skin powder. (LAB Ltd).*

Monsanto Salt. A proprietary trade name for *o*-chloro-*p*- toluene sodium sulphonate, $Cl.C_7H_6.SO_3.Na$.†

Monsell's Salt. Basic ferric sulphate, $Fe_4O(SO_4)_5$. Used in medicine.†

Montago. A German light alloy with a specific gravity lower than aluminium.†

Montan Pitch. The residue from the production of montan wax. The crude material gives an ash of 1.7 per cent, has an acid value of 3, and a saponification value of 6.†

Montana Gold. An alloy of 89 per cent copper, 10.5 per cent zinc, and 0.5 per cent aluminium.†

Montana Wax. See IRISH PEAT WAX.

Montanin Wax. See IRISH PEAT WAX.

Montanine. A liquid containing 31 per cent hydrofluosilicic acid. Recommended as a disinfectant for the walls of breweries and distilleries. It is obtained from by-products in the pottery industry.†

Montax. A proprietary trade name for a filler for rubber, etc. It is a mixture of hydrated magnesium carbonate and silica.†

Monteban. Narasin. Applications: Coccidiostat; growth stimulant. (Eli Lilly & Co).

Monterey Bayleton. Treademefon. Applications: A fungicide for the control of diseases in turf and the control of powdery mildew, rusts and blight of ornamental plants. (Lawn & Garden Products Inc).*

Monterey Foliar Nutrient 11-4-6. Nitrogen, phosphoric acid, potash and chelated zinc, iron and manganese. Applications: Used as a foliar nutrient on turf, ornamentals, trees, vegetables and house plants. (Lawn & Garden Products Inc).*

Monterey Herbicide Helper. Petroleum distillate and alkylphenoxy polyethoxy ethanols. Applications: A spreader - penetrant that makes weed killers work faster and better. (Lawn & Garden Products Inc).*

Monterey Iron Chelated 10%. Iron chelate. Applications: Correction of iron deficiencies in turf, ornamentals, fruit trees and vegetables, (Lawn & Garden Products Inc).*

Monterey Perc-O-Late Plus. Surfactants plus nitrogen, zinc, iron and manganese. Applications: Used for water penetration on turf, flower beds, potted plants, etc. (Lawn & Garden Products Inc).*

Monthier's Blue. A coloured compound obtained by the oxidation of the precipitate formed by the action of ammoniacal ferrous chloride upon potassium ferrocyanide, $(Fe_2)_2(Fe(CN)_6)_3.6NH_3.9H_2O$.†

Montigel. Bentonites with specific swelling properties for various technical purposes, such as for iron ore pelletisations and for sealing of sanitary land fill. (Süd-Chemie AG).*

Montmorillonite. A mineral, $Al_2O_3.SiO_2$. It is a colloidal clay similar to Bentonite, and is often combined with alkalies or alkaline earths. Used for decolourizing oils.†

Montothene G50. A proprietary trade name for an ethylene vinyl acetate copolymer. It is translucent, non-toxic and has good mechanical properties. It is used for film injection and blow moulded articles and extrusions. (Armour Pharmaceutical Co).†

Montpelier Yellow. See TURNER'S YELLOW.

Montreal Potash. Commercial potassium carbonate.†

Moogrol. A proprietary preparation. It is a mixture of the acids of the chaulmoogra series. Used as a therapeutic agent in leprosy.†

Moore Floc. Organic and inorganic flocculants, cationic, anionic, alum replacement and caustic replacement products. Applications: Flocculating fine dissolved or suspended solids in potable water plants, colour and collodial and turbidin solids, algae precipitating aid. (Benzsay & Harrison Inc).*

Moore's Ointment. Resin ointment.†

Moorland Tablets. For the relief of indigestion. (The Boots Company PLC).

Moplefan. A proprietary range of polypropylene films. (Montedison UK Ltd).*

Moplen. Polypropylene. A flexible hard, tough hydrocarbon thermoplastic used for moulding domestic ware and for electrical purposes. (Montedison UK Ltd).*

Morat White. (Moudan White). A white pigment. It is a clay found in Switzerland.†

Mordant Rouge. See RED LIQUOR.

Mordant Yellow O. See MILLING YELLOW.

Mordant, Green. See GREEN MORDANT.

Mordant, Red. See RED LIQUOR.

Moreau Marble. A marble prepared by immersing soft amorphous limestone in a bath of zinc sulphate, and drying.†

Morell's Solution. A disinfecting solution containing arsenious acid, caustic soda, and a small quantity of phenol, dissolved in water.†

Morestan. A wettable powder containing 25% w/w quinomethionate. Applications: To control red spider mites, including organophosphorus resistant strains, on apples, gooseberries, strawberries and marrows and American gooseberry mildew and leafspot on blackcurrants, American and European gooseberry mildew (with partial control of leafspot) on gooseberries, powdery mildew on marrows and willow anthracnose. (Bayer & Co).*

Morfast. Liquid dyes for surface coatings. (Williams (Hounslow) Ltd).

Morfax. 4-Morpholinyl-2-benzothiazole disulphide - rubber accelerator. Applications: For NR and synthetic rubbers. Provides good curing activity. Suggested for tyres and mechanical goods requiring maximum strength and quality. (Vanderbilt Chemical Corporation).*

Morflex P50. A proprietary n-alkyl phthalate plasticizer. (Pfizer International).*

Morflex 100. A proprietary trade name for diiso-octyl phthalate. A vinyl plasticizer.†

Morflex 125. A proprietary plasticizer. n-octyl n-decyl phthalate. (Pfizer International).*

Morflex 175. A proprietary trade name for octyl decyl phthalate. A vinyl plasticizer.†

Morflex 210. A proprietary plasticizer. Diethyl hexyl phthalate. (Pfizer International).*

Morflex 240. A proprietary plasticizer. Dibutyl phthalate. (Pfizer International).*

Morflex 310. A proprietary plasticizer. Di-2 ethyl hexyl adipate. (Pfizer International).*

Morflex 325. A proprietary plasticizer. n-octyl n-decyl adipate. (Pfizer International).*

Morflex 330. Di-decyl adipate. A vinyl plasticizer. (Pfizer International).*

Morflex 410. A proprietary plasticizer. Di-2-ethyl hexyl azelate. (Pfizer International).*

Morflex 510. A proprietary plasticizer. Tri-2-ethyl hexyl trimellitate. (Pfizer International).*

Morflex 525. A proprietary plasticizer. Tri (n-octyl n-decyl) trimellitate. (Pfizer International).*

Morflex 530. A proprietary plasticizer. Tri-isodecyl trimellitate. (Pfizer International).*

Morhal Resin. A resin obtained from *Vatica lanceoefolia*, of India.†

Morhulin. Cod liver oil and zinc oxide. Applications: Wounds, scalds and dermatitis. (Napp Laboratories Ltd).*

Morintannic Acid. See MACLURIN.

Morland's Salt. $(Cr(NH_3)_2(SCN)_4)HNH$. The guanidinium salt of the same complex as Reinecke's salt, formed as a by-product in the preparation of the latter salt.†

Mornidine. A proprietary trade name for Pipamazine.†

Moroccan Olive Oil. Argan oil from *Arganum sideroxylon.*†

Morocco Gum. (Mogador Gum, Brown Barberry Gum). A variety of gum acacia in the form of tears.†

Morocide. Fungicide for mildew. (Hoechst UK Ltd).

Moroline. Petrolatum, white. Applications: Pharmaceutic aid; protectant. (Plough Inc).

Moronal. Aluminium-formaldehyde-sulphite. An antiseptic and astringent.†

Morpan. Cationic surfactant, biocide. (ABM Chemicals Ltd).*

Morpan BC. Aqueous solution of benzalkonium chloride. Applications: Bactericide, algaecide and antistatic used in fields such as food; brewing; hospitals; farming; veterinary; general sterilizing.(ABM Chemicals Ltd).

Morpan NBB. A proprietary trade name for a 50 per cent active composition of octyldimethylbenzylammonium bromide. A low forming flocculating agent used as a filtering aid.†

Morphia. Morphine, $C_{17}H_{19}NO_3$.†

Morpholine. Tetrahydro-1 : 4-oxazine. It has a boiling-point of 128° C.†

Morpholinoethylnorpethidine. Morpheridine.†

Morsep. Cetrimide, Vitamin A and Vitamin D_2. Applications: Napkin rash. (Napp Laboratories Ltd).*

Mortar, Cement. See CEMENT MORTAR

Mortar, Hydraulic. See HYDRAULIC CEMENT.

Mortar, Lime. See LIME MORTAR

Mortha. A proprietary preparation of morphine hydrochloride and tetra-hydroaminacrine hydrochloride. An analgesic. (Ward, Blenkinsop & Co Ltd).†

Morto. Weedkiller and potato haulm destroyer. (Murphy Chemical Co).†

Morton's Fluid. A solution containing iodine, potassium iodide, and glycerin.†

Morwet EFW. Proprietary product. Applications: Wetting agent for various pesticide formulations. (DeSoto Inc).*

Mosaic Gold. (Mock Gold, Cat's Gold, Bronzing Powder, Tin Bronze). A flaky yellow form of disulphide of tin, SnS_2. It was formerly used for gilding and imitating bronze. A substance also called Mosaic gold is made from an amalgam of tin and mercury, with ammonium chloride and sulphur. Used as a bronzing material for plastics.†

Mosaic Gold. An alloy of 50 per cent copper, and 50 per cent zinc. One containing 65 per cent copper and 35 per cent zinc is also known as Mosaic gold.†

Mosaic Silver. An alloy of tin and bismuth

Moss Gun. Ready for use moss-killer in spray form. (ICI PLC).*

Moss Starch. Lichenin, $(C_6H_{10}O_5)x$.†

Mosskil. Lawn fertilizer with iron sulphate (Fisons PLC, Horticulture Div).

Mos-Tox. Moss eradicant. (May & Baker Ltd).*

Mota. Tablets of metaldehyde.†

Mother of Pearl Sulphur. (Nacreous Sulphur). A form of monoclinic sulphur obtained by heating sulphur with benzene at 140° C. It is unstable.†

Motilium. A proprietary preparation containing domperidone. Applications: Antiemetic. (Janssen Pharmaceutical Ltd).*

Motipress. A proprietary preparation of fluphenazine hydrochloride and nortriptyline hydrochloride. Applications: The treatment of depressive and anxiety states. (Squibb, E R & Sons Ltd).*

Motipress Motival Tablets. Fluphenazine hydrochloride and nortriptyline hydrochloride in tablet form. (E R Squibb & Sons Ltd).

Motival. A proprietary preparation of fluphenazine hydrochloride and nortriptyline hydrochloride. A tranquillizer. (Squibb, E R & Sons Ltd).*

Motopren. Colour pastes. Applications: For colouring ester and ether-based polyurethane. (Bayer & Co).*

Motrin. Ibuprofen. Applications: Anti-inflammatory. (The Upjohn Co).

Motrin-A. Ibuprofen aluminium. Applications: Anti-inflammatory. (The Upjohn Co).

Motung Steel. A proprietary trade name for a high speed steel containing 7.5-8.5 per cent molybdenum, 1.25-2 per cent tungsten, 3.5-4.5 per cent chromium, 0.9-1.5 per cent vanadium, 0.8 carbon, 0.2-0.4 per cent manganese, and 0.25-0.5 per cent silicon.†

Moudan White. See MORAT WHITE.

Mou-iéou. (Pi-yu). Chinese vegetable tallow.†

Mould Release Agent N 32. Brushable mould coating on a plastic base. (Chemetall GmbH).*

Mouldensite. A proprietary synthetic resin product used for electrical insulation.†

Mouldrite. Thermosetting moulding powders. (BIP Chemicals Ltd).*

Mountain Blue. (Azure Blue, Mineral Blue, Copper Blue, Hamburg Blue, English Blue). A basic copper carbonate. A pigment used by painters.†

Mountain Butter. A hydrated aluminium sulphate found in fibrous masses.†

Mountain Cork. (Elastic Asbestos). An asbestos which floats on water.†

Mountain Flax. See AMIANTHUS.

Mountain Flour. See INFUSORIAL EARTH.

Mountain Green. A pigment prepared by precipitating a boiling solution of alum and copper sulphate with a hot solution of sodium or potassium sulphate. Varieties of mountain green blended with white clay, or heavy spar, are known under the names of Alexander green, Glance green, Napoleon's green, and Neuwieder green. The mineral malachite was formerly mined under the name of mountain green, but modern mountain greens are the artificial copper pigments.†

Mountain Leather. (Mountain Paper). Thin, tough types of asbestos.†

Mountain Milk. An earth similar to infusorial earth, used as a rubber filler.†

Mountain Paper. See MOUNTAIN LEATHER.

Mountain Soap. See STEATITE.

Mountain Tallow. See BITUMEN.

Mountain Wood. A variety of asbestos.†

Mountford's Paint. A waterproof paint. It consists of asbestos, ground in water, potassium or sodium aluminate, and potassium or sodium silicate sometimes with oil, and zinc white.†

Mourey's Aluminium Solder. An alloy of 82 per cent zinc and 18 per cent aluminium.†

Mouse Killer. Rodenticide. (Murphy Chemical Ltd).

Mouser. Mouse killer. (ICI PLC).*

Moussett's Alloy. An alloy of 60 per cent copper, 27.5 per cent silver, 9.5 per cent zinc, and 3 per cent nickel.†

Movelat. A proprietary preparation of adrenocortical extract, salicylic acid and mucopolysaccharide polysulphuric acid ester used in the treatment of arthritis. (Luitpold-Werk).*

Mowchem. A grass growth regulator containing 240g/litre mefluidide. It suppresses most grasses for up to 8 weeks. Applications: Grassed areas not subject to heavy wear. (Burts & Harvey).*

Mowilith. Polyvinyl acetate dispersion and solids. (Hoechst UK Ltd).

Mowilith D. A trade mark for a 50 per cent dispersion of polyvinylacetate in water without plasticizer.†

Mowiol. Polyvinyl alcohol. (Hoechst UK Ltd).

Mowital. Polyvinyl butyl resin. (Hoechst UK Ltd).

Moxam. Moxalactam disodium. Applications: Anti-infective. (Eli Lilly & Co).

Moxisylyte. Thymoxamine.†

MP-1. *O*, *O*-dimethyl phosphorodithioic acid. Applications: Used as an intermediate for organophosphorus insecticides. (A/S Cheminova).*

MP-2. *O*,*O*-dimethyl phosphorochloridothioate. Applications: Mainly used in the production of organophosphorus insecticides. (A/S Cheminova).*

MPA. Rheological additive designed to improve pigment settling with little or no viscosity increase. Applications: Solvent based coatings, paint, ink, sealants and adhesives. (NL Chemicals (UK) Ltd)*

MPEM. *O,O*-dimethyl-*S*-methoxycarbonylmethyl phosphorodithioate. Applications: Used as an intermediate for phosphorus pesticides. (A/S Cheminova).*

MPI DMSA Kidney Reagent. Succimer. Applications: Diagnostic aid. (Medi-Physics Inc).

MPI Indium DTPa In III. Pentetate Indium Disodium In 111. Applications: Diagnostic aid; radioactive agent. (Medi-Physics Inc).

MPI Indium Oxine In 111. Indium In 111 Oxyquinoline. Applications: Radioactive agent; diagnostic aid. (Medi-Physics Inc).

MPI Krypton Kr 81m Gas Generator. Krypton, isotope of mass 81. Applications: Radioactive agent. (Medi-Physics Inc).

MPS 500. A chlorinated fatty acid ester. A vinyl plasticizer. (Occidental Chemical Corp).*

MR-1, MR-1A, MR-17A, MR-17B. Proprietary nomenclature for allyl resins.†

MRV 1000. Poly tetra fluorethylene polymer. Applications: Release agent for epoxy resins and any moulded plastics. Lubricants for drive belts, gears and most machinery. (Loes Enterprises Inc).*

M-R-VAX II. Measles and Rubella vaccine. Applications: Combined protection against Measles and German Measles. (Merck Sharp & Dohme).*

MS 4. A silicone electrical insulating compound. (Midland Silicones).†

MS Contin. Morphine sulphate. Applications: Analgesic. (The Purdue Frederick Co).

MS-26, DMS-4-828. Mouldable liquid shim materials. Epoxy resin based mastic, used as a spacer between engines or skin of aircraft or ships. Applications: Surface conforming structural epoxies used to fill gaps between metal parts and between metal structural members and composites such as graphite. (Dynamold Inc).*

MST. Morphine sulphate in a controlled release tablet. Applications: Chronic severe pain. (Napp Laboratories Ltd).*

Mucaine. A proprietary preparation of oxythazaine, aluminium hydroxide gel, and magnesium hydroxide. Treatment of oesophagitis. (Wyeth).†

Muccocota Gum. See OCOTA COCOTA GUM.

Mucicarmin. A solution of 2 parts carmine, 1 part aluminium chloride, and 4 parts water. A staining solution.†

Mucilage Oil. See OIL OF MUCILAGES.

Mucodyne. A proprietary preparation of carboxymethylcysteine. An expectorant. (Berk Pharmaceuticals Ltd).*

Mucomycin. A proprietary trade name for Lymecycline. (British Drug Houses).†

Mucomyst. Acetylcysteine. Applications: Mucolytic. (Mead Johnson & Co).

Mucopront. A proprietary preparation of carbocysteine. A mucolytic. (Pfizer International).*

Mucron. Decongestant. (Ciba-Geigy PLC).

Mudar Gum. A material obtained from *Calotropis giganteas*. It resembles gutta-percha, and contains about 20 per cent of a rubbery material.†

Mudge's Speculum Metal. An alloy of 69 per cent copper and 31 per cent tin.†

Mud, Lime. See CHANCE MUD.

M.U.F. A rubber antioxidant for white and lightly coloured rubber.†

Muga Silk. The product of the caterpillar, *Antheraca assama*, of Assam.†

Muhlhaus White. A pigment. It is lead sulphate, $PbSO_4$.†

Muldan. An orthoclase mineral.†

Mule Gum. A name sometimes applied to Ceara rubber.†

Mulgofen. Tallow amine. Water soluble emulsifier. Applications: Mineral oils; agricultural chemicals. (GAF (Gt Britain) Ltd).

Mulhouse Green. A dyestuff. It is Gambine Y.†

Muller's Fluid. A solution of phosphoric acid in alcohol. A soldering fluid for brass and copper. The term is also used for a hardening agent used in microscopy. It consists of potassium dichromate and sodium sulphate dissolved in distilled water.†

Mullex. A proprietary refractory material made from mixtures containing various proportions of clay and mullite.†

Mullfrax 301. A mullite-alumina product. (Carborundum Co).†

Mullicite. A mineral. It is blue iron earth.†

Mullite. A refractory material formed by heating sillimanite to a temperature of 1550° C. It is also made from the minerals andalusite. dumortierite, and Indian cyanite, and by the electric fusion of alumina and silica. Mullite has a melting-point of about 1800° C. and a low coefficient of expansion.†

Mulsivin. A proprietary preparation containing bromoform, codeine phosphate, tincture of aconite, belladonna alkaloids, ipecacuanha alkaloids, benzoin, balsam of tolu, storax and liquid paraffin. A cough linctus. (Rybar).†

Mulsoid. (Lanoid, Loftine). Trade names for Bentonite detergents, usually consisting of soap with from 25-50 per cent. Bentonite.†

Multaglut. A mixture of persulphate and calcium phosphate. Used to improve flour.†

Multibrol. An organic combination of bromine consisting primarily of the sodium derivative of brom-oleate with a bromine content of 16 per cent.†

Multicel. Dry powder dyes. Applications: Dyes used in the paper industry. (Multicrom SA).*

Multicet. Dry powder disperse dyes. Applications: Disperse dyes used in the textile industry for dyeing nylon and acetate. (Multicrom SA).*

Multicoild. Film cement. (May & Baker Ltd).

Multicrom. Dry powder inorganic and organic pigments. Applications: Used in the manufacture of inks, paints and as dispersions for textile printing. (Multicrom SA).*

Multicuer. Dry powder acid dyes. Applications: Dyes developed for the leather industry. (Multicrom SA).*

Multiflow. Resin modifier. Applications: Additive for nonaqueous industrial coatings; improves flow, reduces pinholes and craters and improves substrate wetting. (Monsanto Co).*

MultiGuard. Non-nitrite, molybdate formula. Applications: Closed system treatment. (Western Chemical Co).*

Multilind. A proprietary preparation of nystatin and zinc oxide. Applications: The treatment of fungal skin infections. (Squibb, E R & Sons Ltd).*

Multilind Ointment. Nystatin and zinc oxide in ointment base. (E R Squibb & Sons Ltd).

Multiluz. Dry powder direct dyes. Applications: Direct light fast dyes, used in the textile industry. (Multicrom SA).*

Multimet. Mixed metal carboxylates. Applications: Drier for paint or printing ink, PVC stabilisers. (Manchem Ltd).*

Multionic. Cooling water treatment. Applications: Corrosion and scale control. (Western Chemical Co).*

Multisil. Precipitated silica. Applications: Used in the rubber industry as a filler. (Multicrom SA).*

MultiSperse. Boiler treatment. Applications: Scale and sludge control. (Western Chemical Co).*

Multispray. Public health insecticide. (The Wellcome Foundation Ltd).

Multistix. A proprietary test strip for the detection of pH, protein, glucose, ketones, bilirubin, blood and urobilinogen in urine. (Ames).†

Multiter. Dry powder disperse dyes. Applications: Disperse dyes used in the textile industry for dyeing polyester fibres. (Multicrom SA).*

Multitherm IG-2. Paraffinic hydrocarbon. Applications: Heat transfer fluid. (Multitherm Corporation).*

Multitherm PG-1. Naphthenic hydrocarbon. Light white mineral oil USP. Applications: Heat transfer fluid. (Multitherm Corporation).*

Multivite. A proprietary preparation of Vitamins A, D, C and thiamine hydrochloride. Vitamin supplement. (British Drug Houses).†

Mulukilivary. A gum-resin obtained from *Balsamodedron berryi.* A myrrh substitute.†

Mumetal. A patented nickel-iron-copper alloy having the highest permeability of all known commercial materials. Its exceptional magnetic properties and low losses make it invaluable for cable loading, instrument transformers, relays, magnetic shields, etc.†

Mumpsvax. Mumps vaccine. Applications: For protection against Mumps. (Merck Sharp & Dohme).*

Mungo Fibres. Short fibres of shoddy (*qv*).†

Municol. Sodium alginate, food/pharmaceutical grade. (Kelco/AIL International Ltd).

Munjeet. The root of *Rubia munjista.* It contains purpurin, and is an important Indian dyestuff.†

Munjistin. (Purpuroxanthic Acid). Dihydroxyanthraquinonecarboxylic acid, $C_{15}H_8O_6$.†

Muntz Metal. A brass containing 60 per cent copper and 40 per cent zinc.†

Murac. An insulating material stated to be made by treating the latex of *Sapotaceae* species.†

Murald. Aldrin insecticides. (Murphy Chemical Co).†

Murcurite. A mercury fungicide. (Murphy Chemical Co).†

Murdiel. Dieldrin insecticides. (Murphy Chemical Co).†

Murex. A proprietary trade name for a manganese steel containing 3 per cent manganese, 1 per cent carbon, and 0.85 per cent nickel.†

Murexan. See DIALURAMIDE.

Murexide. (Naples Red). An obsolete red basic dyestuff, obtained by the action of nitric acid upon guano, and subsequently treating the product with ammonia.†

Murfite. Acaricide. (Murphy Chemical Ltd).

Murfixtan. A mercury fungicide. (Murphy Chemical Co).†

Murfly. An insecticide. (Murphy Chemical Co).†

Murfotox. Organo-phosphorus insecticides (Murphy Chemical Co).†

Murfume. Pesticidal smoke generators. (Murphy Chemical Co).†

Muriate of Ammonia. See SALAMMONIAC.

Muriate of Potash. Potassium chloride, KCl.†

Muriate of Soda. Sodium chloride, NaCl.†

Muriatic Acid. Hydrochloric acid, HCl.†

Murine Ear Drops. Carbamide peroxide, a compound of urea with hydrogen peroxide (1:1). Applications: Anti-infective, topical. (Abbott Laboratories).

Murine Plus. Tetrahydrozoline Hydrochloride. Applications: Adrenergic. (Abbott Laboratories).

Muripsin. A proprietary preparation of glutamic acid hydrochloride and pespi[illegible] (Norgine).†

Murlin Premium Ladle Wash. A blend of finely divided, inorganic solids disperse[illegible] in an aqueous medium. Applications: Used to coat and protect surfaces, too[illegible] and equipment coming in contact with molten aluminium. (Murlin Chemical Inc).*

Murman's Alloy. One contains 92 per cen[illegible] aluminium, 4.4 per cent zinc, and 3.6 per cent magnesium. Another consists of 72 per cent aluminium, 14.5 per ce[illegible] zinc, and 13.5 per cent magnesium.†

Murnil. Veterinary preparation. Applications: For activation of the ski[illegible] metabolism of dogs and cats. (Bayer [illegible] Co).*

Murphex. Disinfestation products. (Murphy Chemical Co).†

Murphicol. Pesticide suspensions. (Murp[illegible] Chemical Co).†

Murphos. Parathion insecticides. (Murph Chemical Co).†

Muscarine. (Campanuline). A dyestuff. [illegible] is the dihydroxy derivative of Meldol[illegible] blue.$C_{18}H_{15}N_2Cl.O_2$. Dyes cotton mordanted with tannin and tartar emetic blue. Employed for calico printing.†

Muscle Sugar. Inositol, $C_6H_6(OH)_6$.†

Mushet Steel. (Self-hardening Steel). Steels containing from 0.7-1.2 per cent carbon and 2-3 per cent tungsten. They require no quenching or tempering.†

Mushroom Sugar. Mannitol.†

Musiv Gold. An alloy of from 66-70 per cent copper and 30-34 per cent zinc.†

Musk. The dried animal secretion of the musk deer. It has been practically superseded by synthetic compounds.†

Musk Ambrette. See AMBRENE.

Musk, Artificial. See MUSK BAUR, KETONE MUSK, XYLENE MUSK and AMBRENE.

Musk B. See MUSK BAUR.

Musk Baur. (Musk B, Tonquinol). An artificial musk. It is 2 : 4 : 6-trinitro-*l*-methyl - 3 - tertiary - butyl - toluene, $C_6H(CH_3)(NO_2)_3.C(CH_3)_3$. Used for soap and toilet purposes.†

Musk C. See KETONE MUSK.

Musk, Ketone. See KETONE MUSK.

Musk, Xylene. See XYLENE MUSK.

Mussanin. An extract from *Albizzin anthelmintica*. A vermifuge.†

Must. Grape juice.†

Mustard Gas. (Yperite, Yellow Cross Gas). Dichloro-diethyl-sulphide. A military poison gas.†

Mustard Oil. Ethyl-isothiocyanate, C_2H_5NCS. Also see OIL OF MUSTARD.†

Mustard Oil, Methyl. See METHYL MUSTARD OIL.

Mustargen. Mechlorethamine hydrochloride. Applications: Antineoplastic. (Merck Sharp & Dohme, Div of Merck & Co Inc).

Mustine. NN-Di-(2-chloroethyl)methyl-amine. Chlormethine. A proprietary preparation of mustine hydrochloride. A carcino-chemotherapeutic agent. (Boots Company PLC).*

Mustone. A Japanese chloroprene polymer. See NEOPRENE.†

Mutamycin. Mitomycin. Applications: Antineoplastic. (Bristol Laboratories, Div of Bristol-Myers Co).

Muthmann's Liquid. Acetylene-tetra-bromide, $CHBr_2.CHBr_2$. A solvent.†

Muthol. See PARAFFIN, LIQUID. Also see MUTHMANN'S LIQUID.

M'Varavara. The roots of *Securidaca longipedunculata*. Employed medicinally.†

MVP. Methacrylate. Applications: Permanent structural adhesive for bonding metals, plastics and ceramics. (Devcon Corporation).*

MXM-7500. A proprietary epoxy putty used for filling in difficult radii and depressions in epoxy-fibre glass components. (Fiberite West Coast Corp, Orange, California).†

Myacide AS. Industrial biocide. (The Boots Company PLC).

Myambutol. A proprietary preparation containing ethambutol. An antituberculous agent. (Lederle Laboratories).*

Myanesin. A proprietary trade name for Mephenesin carbamate - muscle relaxant and tranquillizer. (British Drug Houses).†

Mybasan. A proprietary trade name for Isoniazid.†

My-B-Den. Adenosine Phosphate.†

Mycardol. A proprietary trade name for Pentaerythritol Tetranitrate.†

Mycelex. Clotrimazole. Applications: Antifungal. (Miles Pharmaceuticals, Div of Miles Laboratories Inc).

Mycifradin. A proprietary preparation of neomycin sulphate. An antibiotic. (Upjohn Ltd).†

Myciguent. A proprietary preparation of neomycin sulphate used in dermatology as an antibacterial agent. (Upjohn Ltd).†

Mycil. A proprietary trade name for Chlorphenesin. Preparations of chlorphenesin (= "Gecophen'). Anti-fungal. (British Drug Houses).†

Mycil Ointment. A smooth, white odourless water-in-oil cream containing chlorphenesin BP 1973 0.5% w/w. Applications: For the treatment and prevention of Athlete's Foot and other skin infections such as dhobie itch and prickly heat. (Evans Medical).*

Mycil Powder. A free flowing fine white odourless powder containing chlorphenesin BP 1973 1% w/w. Applications: For treatment and prevention of Athlete's Foot and other skin infections such as dhobie itch and prickly heat. (Evans Medical).*

Mycivin. A proprietary preparation containing lincomycin hydrochloride. An antibiotic. (Boots Company PLC).*

Mycocide. A slime control agent. (Great Lakes Chemical (Europe) Ltd).†

Mycodermine. A yeast preparation used in medicine.†

Mycolactine. A proprietary preparation of dried yeast, ox bile, lactic principles, frangula, aloes and belladonna. A laxative. (L Wilcox & Co Ltd).†

Mycose. Trehalose, $C_{12}H_{22}O_{11}$.†

Mycospor. Bifonazole. Applications: Antifungal. (Miles Pharmaceuticals, Div of Miles Laboratories Inc).

Mycostatin. Nystatin. Applications: Antifungal. (E R Squibb & Sons Inc).

Mycota. A proprietary preparation of undecenoic acid and zinc undecenoate. Treatment of fungal infections of the skin. (Boots Company PLC).*

Mycozol. A proprietary preparation of chlorbutol, salicylic acid, benzoic acid and malachite green. Treatment of fungal infections of the skin. (Smith and Nephew).†

Mydflex. A triethanolamine salicylate. Applications: Topical analgesic. (Hercules Inc).*

Mydfrin. Phenylephrine Hydrochloride. Applications: Adrenergic. (Alcon Laboratories Inc).

Mydochrome. Colour reversal processing system. (May & Baker Ltd).

Mydoneg. Colour film processing system. (May & Baker Ltd).

Mydoprint. Colour paper processing system. (May & Baker Ltd).

Mydriacyl. Tropicamide. Applications: Anticholinergic (ophthalmic). (Alcon Laboratories Inc).

Mydriasine. Atropine methyl-bromide, $C_{17}H_{23}NO_3.CH_3Br$. A mydriatic.†

Myelin. A white, fatty substance obtained from various animal and vegetable tissues.†

Myer's Naphthol Green. (Standard). For cholestrol determination. It consists of 50 mg naphthol green B dissolved in 1,000 cc water.†

Mykon. Wetting and detergent, dispersing and emulsifying properties. Applications: Textiles; general purpose detergent and wetting agent. (Warwick Chemical Ltd).

Mykon 817. Phosphated alcohol anionic surfactant as a clear liquid. Applications: Wetting and dispersing agent, scouring and bleach assistant, used primarily in the preparation and dyeing of cotton and synthetic fibres. (Warwick Chemical Ltd).

Mykonaid. Dispersing and levelling agent. Applications: Textiles especially in high temperature dyeing. (Warwick Chemical Ltd).

Mykroy/Mycalex. Machinable mouldable ceramic material made from glass and mica powder. The material is completely inorganic and is supplied in sheets, rods or parts moulded to specific specifications. It is machinable with standard shop tools and no after firing is required. Temperature range to +1300°F (700°C). It is a thermal and electrical insulator. Applications: For use in areas where dimensional stability is a requirement over a broad temperature range. High voltage switch gear, arc barriers, asbestos filled material replacement. Jigs and fixtures for induction or thermal heating, thermal barrier for moulding presses, vacuum furnace insulators, special printed circuit applications, thermal switch covers. Applications requiring close tolerances for either a machined or moulded part. (Mykroy/Mycalex).*

Mylanta. Antacid preparations. (ICI PLC).*

Mylar. A proprietary trade name for a polyester film used for electrical insulation, cable lapping, magnetic tape. It has a very high tensile strength. (10) See also MELINEX. (ICI PLC).*

Myleran. Busulphan. Applications: Antineoplastic. (Burroughs Wellcome Co).

Myleran Tablets. A proprietary formulation of busalfan. Applications: For the palliative treatment of chronic myelogenous (myeloid, myelocytic, granulocytic) leukaemia. (The Wellcome Foundation Ltd).*

Mylicon. Simethicone. Applications: Antiflatulent. (Stuart Pharmaceuticals, Div of ICI Americas Inc).

Mylocon. A proprietary preparation containing methyl polysiloxane in flavoured vehicle. (Parke-Davis).†

Mylol. An insect repellant. (Boots Pure Drug Co).*

Mylomide. A proprietary preparation of amylobarbitone and megemide. A sedative. (Nicholas).†

Mylosar. Azacitidine. Applications: Antineoplastic. (The Upjohn Co).

Mynah. A proprietary preparation of ethambutol and isoniazid used in the therapy of tuberculosis. (Lederle Laboratories).*

Myocardol. A proprietary preparation of pentaerythritol tetranitrate. A vasodilator used in angina pectoris. (Bayer & Co).*

Myochrysine. Gold sodium thiomalate. A mixture of the mono- and di- sodium salts of gold thiomalic acid. Applications: Antirheumatic. (Merck Sharp & Dohme, Div of Merck & Co Inc).

Myocrisin. A proprietary preparation of sodium aurothiomalate. (May & Baker Ltd).*

Myodil. Ethyl iodophenylundecylate. (Glaxo Pharmaceuticals Ltd).†

Myolgin. A proprietary preparation of aspirin, paracetamol, codeine phosphate, caffeine citrate and aceto-menaphthone. An analgesic. (Cox, A H & Co Ltd, Medical Specialities Divn).†

Myotonine. Bethanechol chloride in tablet form. Applications: A smooth muscle stimulant. Approved in the treatment of post operative urinary retention and the management of Reflux Oesophagitis. (Glenwood Laboratories Ltd).*

Myprozine. Natamycin. Applications: Antibacterial. (American Cyanamid Co).

Myrabola Oil. A German soap-making material. It is an oil stated to be a mixture of different fatty acids and their glycerides. It is obtained from fat waste.†

Myras. Internal gear pump. Applications: Light oil and viscous liquid pumping. (Sihi-Ryaland Pumps Ltd).*

Myrickite. A trade name for a chalcedony.†

Myringacaine Drops. Mercufenol chloride. Applications: Anti- infective. (The Upjohn Co).

Myristica. Nutmeg.†

Myritol. Skin compatible oil component and fattening agent. Applications: Cosmetic preparations. (Henkel Chemicals Ltd).

Myrj. Polyexyethylene alkylesters. (ICI PLC).*

MYRJ 45. A proprietary trade name for Polyoxyl 8 stearate.†

MYRJ 52. A proprietary trade name for Polyoxyl 40 stearate.†

Myrj 52. Polyoxyl 40 Stearate. Applications: Pharmaceutic aid. (ICI Americas Inc).

Myrj 53. Polyoxyl 50 Stearate. Applications: Pharmaceutic aid. (ICI Americas Inc).

Myrmekite. A mineral. It is quartz.†

Myrobalans. The fruit of *Terminalia chebula*, of India. This is the chief variety of this product, but there are at least five varieties of the commercial article which are named after the district where they are marketed. They contain from 24-39 per cent tannin and from 2.2-3.1 per cent ash. A solid extract is made containing from 50-60 per cent tannin.†

Myrrh. The true myrrh is that known as Herabol myrrh, a gum resin obtained from various species of *Balsamodedron* and *Cammiphora*. The acid value varies from 60-70; 24-40 per cent is soluble in alcohol, and 5-8 per cent in petroleum ether. Bisabol myrrh is derived principally from *Balsamea erythrea* and has an acid value of about 42. Myrrh is used mainly in toilet preparations, perfumery, and incense.†

Myrrh, Substitute. See MULUKILIVARY.

Myrtle Green. See DINITROSORESORCIN.

Myrtol. A refined myrtle oil. It is an essential oil containing myrtenol, pinene, and cineol. It is used in medicine for bronchial and pulmonary affections, and as an antiseptic.†

Mysoline. A proprietary preparation of primidone. An anticonvulsant. (ICI PLC).*

Mysoline. Primidone. Applications: Anticonvulsant. (Ayerst Laboratories, Div of American Home Products Corp).

Myspamol. Proquamezine fumarate. (May & Baker Ltd).

Mysteclin Capsules. Nystatin and tetracycline hydrochloride in capsule form. (E R Squibb & Sons Ltd).

Mysteclin Syrup. A proprietary preparation containing tetracycline and amphotericin. An antibiotic. (Squibb, E R & Sons Ltd).*

Mysteclin Tablets. A proprietary preparation containing tetracycline and nystatin. An antibiotic. (Squibb, E R & Sons Ltd).*

Mystery Gold. An alloy of 1 part platinum and 2 parts copper, with a little silver.†

Mystic Metal. An alloy of 88.7 per cent lead, 10.8 per cent antimony, 0.4 per cent iron, and 0.1 per cent bismuth.†

Mystin. A mixture of formaldehyde and sodium nitrite. A preservative.†

Mytolac. Acne lotion and cream, to clear acne blemishes. (Richardson-Vicks Inc).*

Mytrex. Neomycin sulphate and triamcinolone acetonide. Applications: Indicated in the treatment of corticosteroid-responsive dermatoses with secondary infection. (Altana Inc).*

Mytrex F. Nystatin and triamcinolone acetonide. Applications: Indicated for the treatment of cutaneous candidiasis. (Altana Inc).*

Myuizone. *p*-acetylaminobenzaldehyde thiosemicarbazone.†

N

N. Sodium silicate solution (3.22 SiO_2:Na_2O ratio; 41°Be density). Applications: Builder for laundry detergents and cleaners, adhesive for corrugated, spiral wound and laminated paper products, binder for foundry cores and moulds, manufacture of silica gels, zeolites, catalysts. (The PQ Corporation).*

N33. An alloy of cast iron with additions of nickel, copper, and chromium.†

Nacap. Sodium 2-mercaptobenzothiazole, 50% aqueous solution. Applications: Corrosion inhibitor, metal deactivator and alkaline buffer - petroleum lubricants. Also chemical intermediate. (Vanderbilt Chemical Corporation).*

Nacarat. See AZORUBINE S.

Nacconal. Sodium alkylbenzene sulphonate. Applications: Detergent, wetting agent. (Stepan Company).*

Nacconol. A proprietary trade name for a wetting agent containing a sodium alkyl aryl sulphonate or of similar constitution.†

Nacconol 35SL. Anionic surfactant in liquid form. Applications: Liquid detergents. (KWR Chemicals Ltd).

Nacconol 90F and 40F. Anionic surfactant in cream flake form. Applications: Powdered detergents. (KWR Chemicals Ltd).

Naclex. A proprietary preparation of hydroflumethiazide. A diuretic. (Glaxo Pharmaceuticals Ltd).†

Nacreous Sulphur. See MOTHER OF PEARL SULPHUR.

Nacton. Oral anticholinergic. (Beecham Pharmaceuticals).

Nacton. Poldine Methylsulphate. Applications: Anticholinergic. (McNeil Pharmaceuticals, McNEILAB Inc).

Nadavin. Agents to increase the wet strength of papers. (Bayer & Co).*

Nafcil. Nafcillin sodium. Applications: Antibacterial. (Bristol Laboratories, Div of Bristol-Myers Co).

Nafion. Perfluorosulphonic acid membrane. (Du Pont (UK) Ltd).

Naftocit. Vulcanizing accelerators for latex and rubber processing. (Chemetall GmbH).*

Naftolen. Plasticizers and extender oils for natural and synthetic rubber. (Chemetall GmbH).*

Naftolen R 100, 510, 530, 550, 570, X413, X414, X10, 134. Vinyl plasticisers. (Wilmington Chemical Co).†

Naftonox. Antioxidants for the production and processing of natural and synthetic rubbers and their latices, thermoplastics and adhesives. (Chemetall GmbH).*

Naftopast. Solid dispersions for improved and non-polluting processing in rubber compounds. (Chemetall GmbH).*

Naftozin. Stearic acids as processing auxiliaries in the production of compounds and as lubricants in the processing of thermoplastics. (Chemetall GmbH).*

Naganol. Veterinary preparation. Applications: For the control of trypanosomiasis of domestic animals. (Bayer & Co).*

Nageli's Solution. A solution containing a mixture of zinc chloride and iodide. A disinfectant.†

Nahcolite. A native sodium bicarbonate.†

Nairit P. MERCAPTAN-modified polychloroprene.†

Nalan. Water repellent. (Du Pont (UK) Ltd).

Nalcite. A proprietary trade name for a water softener containing an organic zeolit type exchanger materi.†

Nalcrom. Treatment for ulcerative colitis etc. (Fisons PLC, Pharmaceutical Div).

Nalfleet. Chemical treatments for the marine industry. (ICI PLC).*

Nalfloc. Water and effluent treatment chemicals. (ICI PLC).*

Nalfon. Fenoprofen calcium. Applications: Anti-inflammatory; analgesic. (Eli Lilly & Co).

Nalorex. Naltrexone. (Du Pont (UK) Ltd).

Nalutron. Progesterone. Applications: Progestin. (Sterling Drug Inc).

Nalzin. Zinc hydroxy phosphite. Applications: Non-toxic anti-corrosive pigment for paint. (NL Industries Inc).*

Nametal. Foil wrapped metallic sodium. (Foseco (FS) Ltd).*

Nancic Acid. Lactic acid, $C_3H_6O_3$†

Nandrobolic. Nandrolone phenpropionate. Applications: Androgen. (O'Neal, Jones & Feldman Pharmaceuticals).

Nandrobolic LA. Nandrolone decanoate. Applications: Androgen. (O'Neal, Jones & Feldman Pharmaceuticals).

Nangawhite. A non-staining antioxidant for rubber. (Rubber Regenerating Co).†

Nankin. See PHOSPHINE.

Nankin Yellow. See IRON BUFF and PHOSPHINE.

Nankor. A proprietary preparation of FENCHLORPHOS. An insecticide.†

Nansa. Alkylaryl sulphonates. (Albright & Wilson Ltd, Detergents Div).

Nansa AS40. Alkylaryl sulphonate anionic. Golden yellow opaque gel. Applications: Liquid detergents, car shampoos and general cleaning formulations. (Albright & Wilson Ltd, Detergents Div, Marchon).

Nansa BMC. Blended anionic nonionic foaming agent, in yellow liquid form. Applications: Air entraining agent for mortar and concrete. (Albright & Wilson Ltd, Detergents Div, Marchon).

Nansa HS. Alkylaryl sulphonate anionic. Cream powder. Applications: Spray-dried detergents; scouring powders; emulsion polymerization. Manufacture of styrene and butadiene polymers; copolymerization. Heavy duty detergents. Hard surface cleaners. Insecticide sprays; metal pickling, paper, textiles. (Albright & Wilson Ltd Detergents Div, Marchon).

Nansa LES42. Predominantly straight chain sodium dodecylbenzene sulphonate, lauryl ether sulphonate and magnesium xylene sulphonate blend giving an anionic in the form of a golde liquid. Applications: Dishwashing detergents; hard surface cleaners. (Albright & Wilson Ltd, Detergents Div, Marchon).

Nansa SL and SS. Alkylaryl sulphonate anionic. Supplied as liquid or cream paste, with low inorganic content. Applications: Liquid and powder detergents; emulsifiers and foamers in plastics and rubber industries; scourers and wetters in textiles. (Albright & Wilson Ltd, Detergents Div, Marchon).

Nansa SSA. Alkylaryl sulphonate anionic in acid form. Dark brown liquid, low inorganic content. Applications: Liquid and powder detergents. (Albright & Wilson Ltd, Detergents Div, Marchon).

Nansa TS60. Alkylaryl sulphonate anionic. Medium brown viscous liquid with low inorganic content. Applications: Dishwashing; bubble baths; car shampoos; emulsification de-sizing; pigment dispersion; emulsion polymerization. (Albright & Wilson Ltd, Detergents Div, Marchon).

Nansa UC. A range of detergent powders containing a sodium alkylbenzene sulphonate as the active ingredient.†

Nansa UCA/S and UCP/S. Spray-dried built powders based on straight chain alkylbenzene sulphonate. Blue, free-flowing powder, dust free. Applications: Anionic for built detergents. (Albright & Wilson Ltd, Detergents Div, Marchon).

Nansa YS94. Alkylaryl sulphonate anionic. Amber viscous liquid. Applications:

Hand cleaners; charge detergent systems. (Albright & Wilson Ltd, Detergents Div, Marchon).

Nansa 1042. Alkylaryl sulphonate anionic in acid form. Dark brown liquid. Applications: Detergent intermediate with low inorganic contents. (Albright & Wilson Ltd, Detergents Div, Marchon).

Nansa 1042/P. Alkylaryl sulphonate anionic in acid form. Dark brown liquid. Applications: Emulsifier for emulsion polymerization. (Albright & Wilson Ltd, Detergents Div, Marchon).

Nansa 1169/P. Alkylaryl sulphonate anionic. Supplied as a golden liquid. Applications: Surfactant for emulsion polymerization. (Albright & Wilson Ltd, Detergents Div, Marchon).

Nansen. A high-speed tungsten steel containing 18 per cent tungsten.†

Napalite. A dark-red wax which melts at 42° C. It is a hydrocarbon and occurs naturally.†

Napalm. An aluminium soap consisting of a mixture of oleic naphthenic and coconut fatty acids. Makes petrol thicken and gel. Used in flame throwers and fire bombs.†

Napelec. Non-draining impregnant (obtained by mixing Napvis with water). (BP Chemicals Ltd).*

Napeline. Benzaconine, $C_{32}H_{43}NO_{10}$.†

Napental. Pentobarbital Sodium. Applications: Hypnotic; sedative. (Beecham Laboratories).

Napgel. Anti-freeze. (BP Chemicals Ltd).*

Naphalane. A crude naphtha product containing soap. It is similar to Naphthalane (*qv*).†

Naphcon. Naphazoline hydrochloride. Applications: Adrenergic. (Alcon Laboratories Inc).

Naphtalin. See TAR CAMPHOR.

Naphtha. (Petroleum Naphtha). The less volatile portion obtained in redistilling Benzine (*qv*), boiling from about 95-100° C. The term is loosely applied, and is synonymous with mineral naphtha. Solvent naphtha is not a petroleum product. Danforth's oil, B.P. 80-110° C., is a similar material. Used as a solvent, and for burning. Also see B-PETROLEUM NAPHTHA†

Naphthalamine. See NAPHTHALIDAM.

Naphthalene Black 12B. A dyestuff. It is a British equivalent of Naphthol blue-black.†

Naphthalene Blue B. An acid dye for wool.†

Naphthalene Blue-black. A British brand of Naphthol blue-black.†

Naphthalene Green Conc. See NAPHTHALENE GREEN V.

Naphthalene Green V. (Naphthalene Green Conc.). A diphenyl-naphthyl-methane dyestuff, prepared by condensing naphthalenedisulphonic acid with tetramethyldiaminobenzhydrol, and oxidizing the leuco compound. Dyes wool from an acid bath.†

Naphthalene Red. See MAGDALA RED.

Naphthalene Rose. See MAGDALA RED.

Naphthalene Scarlet. See MAGDALA RED.

Naphthalene Yellow. See MARTIUS YELLOW.

Naphthalidam. (Naphthalidine).α-Naphthylamine, $C_{10}H_7.NH_2$.†

Naphthalidine. See NAPHTHALIDAM.

Naphthalin. Naphthalene, $C_{10}H_8$. See TAR CAMPHOR.†

Naphthaline Red. See MAGDALA RED.

Naphthaline Yellow. See MARTIUS YELLOW.

Naphthalol. (Betol). Naphthol salicylate.†

Naphthamine. See HEXAMINE.

Naphtharene Orange G. A dyestuff. It is a British equivalent of Orange G.†

Naphtharene Orange R. A British brand of Scarlet GR.†

Naphtharene Scarlet 2R. A British dyestuff. It is equivalent to Ponceau.†

Naphtha, Vinegar. See ACETIC ESTER.

Naphthazarin. 5, 6-dihydroxy-1, 4 naphthoquinone, $C_{10}H_4(OH)_2O_2$.†

Naphthazarin S. See ALIZARIN BLACK S.

Naphthazine Blue-black. 6B. A British equivalent of Naphthol blue black.†

Naphthazurin B, 2B, R. Basic dyes, giving navy blue shades on tannined cotton.†

Naphthindone. See INDOINE BLUE R.

Naphthindone, Blue BB. See INDOINE BLUE R.

Naphthionic Acid. 1-Naphthylamine 4-sulphonic acid.†

Naphthite. Trinitro-naphthalene. Employed in explosives.†

Naphthochrome. Wool dyestuffs. (Clayton Aniline Co).†

Naphthocyanine. Fast acid wool dyestuffs. (JC Bottomley and Emerson).†

Naphthocyanole. A homologue of Pinacyanol (a red sensitizer for silver bromide plates), prepared by the condensation of β-naphtho-quinaldine ethiodide with formaldehyde in the presence of alcoholic potash. Used as a red sensitizer for photographic plates.†

Naphthoformol. A product ofα-naphthol and formaldehyde. Used as a dusting powder.†

Naphthol Aristol. Iodo-naphthol, $C_{10}H_6I_2O_2$. Used as an antiseptic.†

Naphthol AS-BG. A dyestuff. It is the 2, 5-dimethoxyanilide of β-hydroxy-naphthoic acid.†

Naphthol AS-BS. A dyestuff. It isβ-hydroxynaphthoic-*m*- nitranilide.†

Naphthol AS-D. The *o*-toluidide ofβ-hydroxynaphthoic acid.†

Naphthol AS-G. A dyestuff. It is diacetoacetotolidide.†

Naphthol AS-OL. A dyestuff. It is the *o*-anisidide ofβ-hydroxynaphthoic acid.†

Naphthol AS-RL. A dyestuff. It isβ-hydroxynaphthoic-*p*-anisidide.†

Naphthol AS-TR. 5-Chloro-*o*-toluidide of β-hydroxynaphthoic acid.†

Naphthol Blue B. See NEW BLUE B or G.

Naphthol Blue G, R. Acid dyes giving navy blue shades on wool.†

Naphthol Blue R and D. See MELDOLA'S BLUE.

Naphthol Blue-Black, B, 10, L, S Conc. Dyestuffs. They are British equivalents of Naphthol blue-black.†

Naphthol Green B. A dyestuff. It is the ferrous sodium salt of nitroso-β-naphthol-β-monosulphonic acid, $C_{20}H_{10}N_2O_{10}S_2FeNa_2$. Dyes wool from an acid bath containing an iron salt.†

Naphthol Orange. See ORANGE II.

Naphthol Red S. See FAST RED D.

Naphthol Red 0. See FAST RED D.

Naphthol Scarlet 3R. An acid dyestuff.†

Naphthol Violet. See MELDOLA'S BLUE.

Naphthol Yellow. See MARTIUS YELLOW and NAPHTHOL YELLOW S.

Naphthol Yellow FY. A dyestuff. It is a British brand of Naphthol yellow S.†

Naphthol Yellow RS. See BRILLIANT YELLOW.

Naphthol Yellow S. (Naphthol Yellow, Sulphur Yellow S, Citronine A, Acid Yellow S, Naphthol Yellow FY). A dyestuff. It is the calcium, ammonium, sodium, or potassium salt of dinitro-α-naphthol-β-monosulphonic acid, $C_{10}H_4N_2O_8SNa_2$. Dyes wool and silk yellow from an acid bath.†

Naphtholite. A proprietary trade name for a light petroleum distillate used as a solvent, etc.†

Naphtholith. A bituminous shale.†

Naphthopone E. Dispersing agent. Applications: Used in naphthol AS developing baths for aftersoaping fast dyeings.. (Bayer & Co).*

Naphthoresorcin. 1, 3-Dihydroxy-naphthalene.†

Naphthoride. A proprietary preparation of naphthalene tetrachloride.†

Naphthosalol. See BETOL.

Naphthosultone. The anhydride of 1 - Naphthol-8-sulfonic acid.†

Naphthothiam Blue. A dyestuff obtained by warming a solution of nitronaphthalenesulphinic acid with zinc dust and potassium sulphite. A vat dye.†

Naphthyl Blue. A sulphonated milling blue. It dyes silk violet-blue with red fluorescence.†

Naphthyl Blue Black N. An acid dyestuff, giving navy blue shades on wool.†

Naphthylamine Yellow. See MARTIUS YELLOW.

Naphthylene Blue. See MELDOLA'S BLUE.

Naphthylene Blue R in Crystals. See MELDOLA'S BLUE.

Naphtopon E. Dispersing agent in Naphtol AS developing baths. (Bayer UK Ltd).

Napisan. Nappy sterilizer, to sanitize baby diapers. (Richardson-Vicks Inc).*

Naples Red. See MUREXIDE.

Naples Yellow. (Paris Yellow). A pigment. It is an antimonate of lead, $PbO.Sb_2O_5$. A mixture of this body with carbonate and chromate of lead is also sold under this name. Cadmium sulphide, CdS, and a pale yellow ochre have been called by this term.†

Naplithin. Lithium-β-hydroxynaphthalene-α-monosulphonate.†

Napliwi. A sand cemented together by the rubber latex from the roots of the Chondrilla plant in Russia. The roots are attacked by the larvae of certain insects, when the latex exudes and runs into the sandy soil where it coagulates. The sand usually contains about 2-2.5 per cent rubber and 10 per cent resin.†

Napoleon's Green. See MOUNTAIN GREEN.

Napolite. Synonym for Haüyne.†

Naprosyn. Naproxen. Applications: Anti-inflammatory; analgesic; antipyretic. (Syntex Laboratories Inc).

Napryl. Polypropylene. (BP Chemicals Ltd).*

Napsalgesic. Oral analgesic. (Dista Products Ltd).

Naptel. Polybutenes. (BP Chemicals Ltd).*

Napvis. Polybutenes. (BP Chemicals Ltd).*

Naqua. Trichlormethiazide. Applications: Diuretic; antihypertensive. (Schering-Plough Corp).

Narcan. Naloxone hydrochloride. Applications: Antagonist. (Abbott Laboratories).

Narceine. The sodium bisulphite compound of *p*-sulphobenzeneazo-β-naphthol, $C_{16}H_{12}N_2O_7S_2Na_2$. Used in calico printing.†

Narceol. A synthetic perfume. It is *p*-tolyl acetate, $CH_3.C_6H_4.O.COCH_3$.†

Narcodeon. Double salts of narcotine and codeine with di- and poly-basic acids.†

Narcophin. A double salt of morphine and narcotine.†

Narcotil. A mixture of ethyl and methyl chlorides. Used as an anaesthetic. Methylene chloride, CH_2Cl_2, used as a local anaesthetic, is also known by this name.†

Narcotine. A proprietary trade name for Noscapine.†

Narcyl. Ethylnarceine hydrochloilde.†

Nardil. A proprietary preparation of phenelzine dihydrogen sulphate. An antidepressant. (Warner).†

Narex. A proprietary preparation of PHENYLEPHRINE and CHLORBUTOL used as a nasal spray. (Norton, H N Co).†

Nargentol. A protein compound of silver, containing 24 per cent silver. An antiseptic.†

Nari Oil. See NJAVE BUTTER.

Narki. An acid-resisting silicon-iron alloy.†

Narphen. A proprietary preparation of phenazocine hydrobromide. An analgesic. (Smith and Nephew).†

Nasalcrom. Cromolyn sodium. Applications: Anti-asthmatic. (Fisons Corp).

Naseptin. A proprietary preparation of chlorhexidene hydrochloride and neomycin sulphate. (ICI PLC).*

Nasoflu. A proprietary live influenza vaccine. (Smith Kline and French Laboratories Ltd).*

Nata. Herbicide grass control. (Hoechst UK Ltd).

Natacyn. Natamycin. Applications: Antibacterial. (Alcon Laboratories Inc).

Natalon. See CARBORA.

Natamycin. An antibiotic produced by Streptomyces natalensis. Pimafucin. Pimaricin.†

Natene. A trade mark for Ziegler-type polyethylenes. (Pchiney-St Gobain, France).†

Natirose. A proprietary preparation of GLYCERYL trinitrate, ethylmorphine hydrochloride and hyoscyamine hydrobromide used in the treatment of angina pectoris. (L Wilcox & Co Ltd).†

Natisedine. A proprietary preparation of quinidine phenylethylbarbiturate used as a cardiac sedative. (L Wilcox & Co Ltd).†

Native Paraffin. See OZOKERITE.

Native Prussian Blue. Synonym for Vivianite.†

Natopherol. Vitamin E. Applications: Vitamin E supplement. (Abbott Laboratories).

Natritope Chloride. Sodium chloride Na 22. Applications: Radioactive agent. (E R Squibb & Sons Inc).

Natrium. Sodium, Na.†

Natrolith. A water-softening material said to consist of granulated clay which removes lime and magnesium salts from hard water when used as a filter.†

Natrosol. Hydroxyethylcellulose. Applications: Multipurpose nonionic cellulosic. Thickener, protective colloid, binder, stabilizer and suspending agent. (Hercules Inc).*

Natrundum. See CARBORA.

Natulan. A proprietary preparation of procarbazine. An anti-mitotic. (Roche Products Ltd).*

Natural Asphalt. See BITUMEN.

Natural Bone Ash, BCP 400 and BCP 600. Calcium hydroxyapatite. Applications: Used to coat and protect all surfaces contacted by molten non-ferrous metals. Used extensively in both aluminium and copper industries. (Murlin Chemical Inc).*

Naturetin. Bendroflumethiazide. Applications: Diuretic; antihypertensive. (E R Squibb & Sons Inc).

Naturvue. Hefilcon B. 2-Hydroxyethyl methacrylate polymer with 1- vinyl-2-pyrrolidinone and ethylene dimethacrylate. Applications: Contact lens material. (Milton Roy Co).

Naubuc. A nickel-silver. It contains 58 per cent copper, 25 per cent nickel, 16.25 per cent zinc, and 0.75 per cent iron.†

Nauganlite. A proprietary anti-oxidant. Alkylated phenol. (Rubber Regenerating Co).†

Nauganlite Powder. A proprietary name fo alkylated bis-phenol. (Rubber Regenerating Co).†

Naugawhite. An alkylated phenol type non-discolouring, non-staining, general purpose antioxidant. For dry rubber and latex. Recommended for latex foam non-staining tyre carcass, refrigerator gaskets, footwear, proofing, sundries, wire insulations such as SP-1 cord, automotive rubber and cloth backing compounds. Effective in natural rubber polyisoprene SBR, neoprene, butyl and nitrile rubbers. (Uniroyal).*

Naugex SD-1. 4,4'Dithiodimorpholine. A sulphur donor used as a partial or total replacement of sulphur for resistance to heat and ageing in NR, SBR, BR, NBR and EPDM. (Uniroyal).*

Nauguard BHT. 2,6-di-tertiary butylparacresol. For NR, SBR, CR, NBR. Non-staining antioxidant, full FDA approval for use in rubber stocks contacting food. Used as stabilizer for butadiene rubber and diene extended plastics. Available in technical and food grades. (Uniroyal).*

Nauguard H. Mixed alkylated diphenylamine. An antioxidant and stabilizer for SBR, BR, IR and NBR. (Uniroyal).*

Nauguard K. A bis-phenolic stabilizer. A low-cost, non-staining, non-discolouring liquid antioxidant recommended for the protection of SBR latex, carboxylated SBR latex and natural rubber latex products, offers protection against heat, sunlight and NO_2. (Uniroyal).*

Nauguard PAN. Phenyl alpha naphthylamine. General purpose antioxidant for natural rubber and neoprene. (Uniroyal).*

Nauguard Q. Polymerized 1,2-dihydro-2,2,4-trimethyl-quinoline. A general purpose antioxidant offering protection against heat and oxygen. Use in natural polyisoprene, SBR, BR, nitrile and EPDM in applications such as tyre carcasses, belts, hose, seals, mechanical goods, footwear and wire. Contact stains light yellow tan, very little migration stain at the 1.0 part level. (Uniroyal).*

Nauguard SP. Styrenated phenol. Antioxidant in dry rubber and latex compounds. A stabilizer for synthetic polymers. Non-discolouring and non-pigmenting. Non-staining by contact and migration. (Uniroyal).*

Nauguard T. Octylated phenol. Applications: General purpose non-staining and non-discolouring stabilizer for SBR. Provides good UV resistance. FDA approved. (Uniroyal).*

Nauguard 445. Substituted diphenylamine. Because of its unusual structure, 445 does not discolour during heat ageing. The combination of high amine-antioxidant activity and good colour make it an ideal anti-oxidant and heat stabilizer for rubbers and plastics. Especially effective in polyisoprene, neoprene and nitrile rubbers. (Uniroyal).*

Nauguard 475. A phenolic phosphite blend. Non-discolouring and non-staining SBR stabilizer. Functions as an antioxidant in natural, SBR, BR, nitrile and EPDM rubbers. (Uniroyal).*

Nauguard 477. Blend of amine antioxidants. An antioxidant for such products as belting, wire, hose, tyre treads and shoe soling. Used for natural rubber tread compounds in combination with a Flexzone antiozonant. (Uniroyal).*

Nauguard 495. Blend of alkylated mercaptobenzimidazole and an amine type antioxidant. Superior heat resistant antioxidant for nitrile rubber and EPDM. Very low extractability in oils and fuels. (Uniroyal).*

Nauli Gum. An oleo-resin from a tree found in the Solomon Islands. It contains 10 per cent volatile oil, 8 per cent resin, and 3 per cent water-soluble matter containing anisic acid.†

Navac. Vacuum processed metallic sodium. (Foseco (FS) Ltd).*

Naval Brass. See BRASS.

Naval Bronze. An alloy of 88.1 per cent copper, 9.74 per cent tin, and 2.04 per cent zinc.†

Navane. A proprietary preparation of thiothixene hydrochloride. A psychotherapeutic agent. (Pfizer International).*

Navane Hydrochloride. Thiothixene Hydrochloride. Applications: Antipsychotic. (Pfizer Inc).

Navidrex. A proprietary preparation of cyclopenthiazide and potassium chloride in slow release core. A diuretic. (Ciba Geigy PLC).†

Navidrex-K. A proprietary preparation of CYCLOPENTHIAZIDE and slow-release potassium chloride. A diuretic. (Ciba Geigy PLC).†

Navy Blue. See SOLUBLE BLUE.

Navy Blue B. See BAVARIAN BLUE DSF.

Navy Green Paint. A mixture of barium sulphate, lead chromate, and an organic blue.†

NBC. A proprietary name for nickel di-butyl dithiocarbamate. (Du Pont (UK) Ltd).†

N.C.T. Nitro-cellulose tutular, a pyro-collodion powder, made from a gelatinized nitro-cellulose, pressed in the form of rods.†

N.C.T.3 Alloy. See ALLOY N.C.T.3.

Ndilo Oil. See LAUREL NUT OIL.

Neacid. A pickling agent for gold- and silversmiths and in the jewellery industry. (Degussa).*

Nealpon. See PANTOPON.

Neapolitan Ointment. Mercury ointment.†

Neatsfoot Oil. A fixed oil obtained by boiling ox or cow's feet in water. It is also the name for a mixture of 1 part lard and 3 parts colza oil.†

Nebacortril. A proprietary preparation of neomycin sulphate, bacitracin and hydrocortisone. An antibiotic and anti-inflammatory. (Pfizer International).*

Nebasulf. A proprietary preparation of neomycin sulphate, bacitracin and sulphacetamide. A topical antibacterial. (Pfizer International).*

Nebcin. Tobramycin Sulphate (injection). Applications: Antibacterial. (Eli Lilly & Co).

Nebs. Acetaminophen. Applications: Analgesic; antipyretic. (Norwich Eaton Pharmaceuticals Inc).

Nebulin. Tecnazene in liquid fogging solution. Applications: Controlling sprouting and dry rot in stored potatoes. (Wheatley Chemical Co Ltd).*

Neburex. Active ingredient: neburon; 1-butyl-3-(3,4-dichlorophenyl)-1-methylurea. Applications: Selective herbicide for both pre- and post-emergence application. (Agan Chemical Manufacturers Ltd).†

Necol. Cellulose lacquers, organic enamels, adhesives and plastic wood. (ICI PLC).*

Nectandra Bark. Bebeera bark.†

Needle Bronze. An alloy of 84.5 per cent copper, 8 per cent tin, 5.5 per cent zinc, and 2 per cent lead.†

Needle Tin Ore. Acute pyramidal crystals of cassiterite.†

Neem Bark. Indian azadirach.†

Neem Oil. See VEEPA OIL.

Nefomolit. A proprietary trade name for a plastic made from mineral oil, formaldehyde, etc.†

Nefranutrin. A proprietary preparation of essential amino acids used in the dietary treatment of renal failure. (Geistlich Sons).†

Nefrolan. Chlorexolone. (May & Baker Ltd).

Negex. Chemical substances used in industry. Applications: Negative expanders for lead acid batteries. (Associated Lead Manufacturers Ltd).*

NegGram. Nalidixic acid. Applications: Antibacterial. (Sterling Drug Inc).

Negram. A proprietary preparation containing nalidixic acid. A urinary antiseptic. (Bayer & Co).*

Neguvon. Insecticide against cattle grubs, mange and worm infestation, in particular for control of ectoparasites in the poultry house. Applications: Veterinary medicine. (Bayer & Co).*

Neillite. A proprietary phenol-formaldehyde synthetic resin moulding compound.†

Neisser's Stain. A microscopic stain. (*a*) Solution contains 0.1 gram methylene blue, 2 cc alcohol, 5 cc glacial acetic acid, and 95 cc water. (*b*) Solution contains 0.2 gram bismarck brown in 100 cc boiling water.†

Nekal. (Leonil, Oranit, Neomerpin N). Wetting-out agents probably derived from alkylated naphthalene sulphonic acids. Used to assist the penetration of textiles by liquids. Also see TETRACARNIT and TETRAPOL.†

Nekal BA-77. Sodium alkylnaphthalene sulphonate. Applications: Dispersing agent in latex, paint and printing ink formulations, in plastic and synthetic latices. stabilizer in latex formulations, prevents coagulation in synthetic rubber. Used in textile processing as a wetting, dispersing and penetrating agent. (GAF Corporation).*

Nekal BX. Alkylaryl sulphonate anionic surfactant. Applications: Wetting, dispersing, emulsifying and foaming agents with widespread industrial applications eg in the pigment, paint and paper industries; in leakage control in pipes, fittings and valves. (BASF United Kingdom Ltd).

Nekal BX-78. Sodium alkyl naphthalene sulphonate. Applications: Wetting agent in the textile industry, leather industry, agricultural chemicals industry, paints and coatings industry, also used as a surfactant for rubber latex polymerization and emulsification. (GAF Corporation).*

Nekal NF. Sodium alkyl naphthalene sulphonate. Applications: Low-foaming, wetting and penetrating agent for continuous padding and long-liquor dyeing with vat, naphthol, sulphur and direct colours. Levelling agent for wool and mixed fibre dyeing. Dispersant for solids in oil. (GAF Corporation).*

Nekal SBS. Anionic surfactant in free acid form. Applications: Wetting, dispersing, emulsifying and foaming agent with widespread industrial uses eg in the paper, paint and pigment industries; in leakage control in pipes, fittings and valves. (BASF United Kingdom Ltd).

Nekal WS-25-1 and WS-25. Sulphonated aliphatic polyesters. Applications: Wetting agent and penetrant for the textile industry. (GAF Corporation).*

Nekal WT-27. Sulphonated aliphatic polyester. Applications: Wetting, rewetting and penetrating agent in drycleaning detergents, emulsion polymerization, glass cleaners, wallpaper removers and battery separators. (GAF Corporation).*

Nekanil. Textile dispersing and wetting agents. (BASF United Kingdom Ltd).

Nelco. A proprietary trade name for whiting.†

Nelio Resin. A proprietary trade name for a purified wood resin.†

Nema. Tetrachloroethylene. Applications: Anthelmintic. (Parke-Davis, Div of Warner-Lambert Co).

Nemacin. TBZ and iodophor in dry granular form. Applications: Controlling soil and seed borne diseases in onions. (Wheatley Chemical Co Ltd).*

Nembutal. A proprietary preparation of pentobarbitone sodium. A hypnotic. (Abbott Laboratories).*

Nembutal Sodium. Pentobarbital Sodium. Applications: Hypnotic; sedative. (Abbott Laboratories).

Neoamfo. Neomycin sulphate, thiostrepton and amphotericin in aqueous suspension. (Ciba-Geigy PLC).

Neoarsycodyl. Sodium methyl-arsenate.†

Neobacrin. A proprietary preparation of neomycin sulphate and zinc bacitracin. An antibiotic skin ointment. (Glaxo Pharmaceuticals Ltd).†

Neobiotic. Neomycin sulphate. Applications: Antibacterial. (Pfizer Inc).

Neobor. Disodium tetraborate pentahydrate. It is borax partially dehydrated to effect economy of transport and handling. (Borax Consolidated Ltd).*

Neocaine-surrénine. A proprietary preparation. It consists of ethocaine hydrochloride with adrenalin.†

Neo-Calglucon. Calcium glubionate. Applications: Replenisher. (Dorsey Pharmaceuticals, Pharmaceutical Div of Sandoz Inc).

Neo-Cantil. A proprietary preparation of MEPENZOLATE BROMIDE and NEOMYCIN sulphate. An anti-diarrhoeal. (MCP Pharmaceuticals).†

Neochrome Blue-black B. A dyestuff. It is equivalent to Chrome fast cyanine G.†

Neochrome Blue-black R. A dyestuff. It is equivalent to Chrome blue.†

Neocid. A proprietary preparation containing 5 per cent. DDT. (*qv*). An insecticide.†

Neocidol Veterinary Powder. Organophosphorous wound dressing. (Ciba-Geigy PLC).

Neocinchophen. Ethyl 6-methyl-2 phenylcinchonate.†

Neo-Cobefrin. Levonordefrin. Applications: Adrenergic. (Cook-Waite Laboratories Inc).

Neo Cortef. A proprietary preparation of hydrocortisone acetate and neomycin sulphate used in dermatology as an antibacterial agent. (Upjohn Ltd).†

Neocosal. An oil binder. Applications: Used for cleaning bodies of water and ground surfaces that have been contaminated with oil. (Degussa).*

NeoCryl. Thermoplastic acrylic resins in solid form, homo and copolymers of acryl-methacrylic and styrene. Applications: Used in various paints eg. marine paints, road paints and paints on plastic. Flexo, gravure and screen printing inks. Dry toner resins, aerosols and other applications. (Polyvinyl Chemie Holland NV).*

Neo-Cultol. Mineral oil. Applications: Laxative; pharmaceutic aid. (Fisons Corp).

Neo Cytamen. A proprietary preparation of hydroxycobalamin. (Glaxo Pharmaceuticals Ltd).†

Neo Duroterm L. Soldering investment material for the precious metal technique. Applications: Dental speciality. (Bayer & Co).*

Neo Duroterm 3, 5 and 7. Investment compound system for precious metal casts. Applications: Dental speciality. (Bayer & Co).*

Neoferrum. A proprietary preparation of ferric hydroxide. A haematinic. (Crookes Laboratories).†

Neogen. An alloy of 58 per cent copper, 12 per cent nickel, 27 per cent zinc, 2 per cent tin, and 0.5 per cent aluminium. It is a nickel silver (German silver). Also see NICKEL SILVERS.†

Neogest. A proprietary preparation of *dl*-norgestrel. An oral contraceptive. (Schering Chemicals Ltd).†

Neo-Hombreol. Testosterone Propionate. Applications: Androgen. (Organon Inc).

Neolan. Metal complex dyes. (Ciba-Geigy PLC).

Neoleptol. Triformyl-trimethylene-triamine.†

Neoleukorit. A proprietary synthetic resin of the phenol-formaldehyde class.†

Neoloid. Castor oil. Applications: Laxative. (Lederle Laboratories, Div of American Cyanamid Co).

Neolyn. Elastomeric alkyd type rosin-based resins with good grease resistance. Applications: It is used to impart flexibility and grease resistance to adhesives, is used for vinyl floor tiles and vinyl inks, in vinyl based and type T-gravure inks. (Hercules Inc).*

Neolysol. Lysol made with chlorocresol. An antiseptic.†

Neo-Medrone Acne Lotion. A proprietary preparation of sulphur, aluminium chlorhydroxide, methylpred nisolone acetate and neomycin sulphate used in dermatology for acne. (Upjohn Ltd).†

Neo-Naclex. A proprietary preparation of bendrofluazide. A diuretic. (Glaxo Pharmaceuticals Ltd).†

Neo-Naclex-K. A proprietary preparation of BENDROFLUAZIDE and potassium Chloride. A diuretic. (Glaxo Pharmaceuticals Ltd).†

Neo-protosil. A proprietary colloidal silver preparation. It contains about 20 per cent silver iodide, AgI, combined with a protein base, and is a germicide.†

Neo-Synephrine Hydrochloride. Phenylephrine Hydrochloride. Applications: Adrenergic. (Sterling Drug Inc).

Neomedrone Veriderm. A proprietary preparation of methylprednisolone and neolaycin used in dermatology as an antibacterial agent. (Upjohn Ltd).†

Neomerpin N. See NEKAL.

Neomin. A proprietary preparation containing neomycin sulphate. An antibiotic. (Glaxo Pharmaceuticals Ltd).†

Neomix. Neomycin sulphate. Applications: Antibacterial. (The Upjohn Co).

Neonalium. An alloy of aluminium with 6-14 per cent copper, 1 per cent nickel, and small amounts of other metals.†

Neonite. A 33-grain sporting rifle powder. It contains 10 per cent. barium or potassium nitrate, 6 per cent petroleum jelly, and insoluble nitrocellulose.†

Neopelline. An alkaloid, $C_{32}H_{45}NO_8$, obtained from *Aconitum napellus*.†

Neopen SS,. A proprietary trade name for a wetting agent containing sodium abietene sulphonate.†

Neophax. A proprietary trade name for brown factice-vulcanized vegetable oils.†

Neophax FA. A proprietary brown factice added to polychloroprene, nitrile rubber, HYPALON and polyurethane when maximum resistance to oil is required. (Hubron Rubber Chemicals, Failsworth, Manchester).†

Neophenoquin. A proprietary preparation of lithium phenyl-cinchoninate.†

Neophryn. A proprietary trade name for the hydrochloride of Phenylephrine.†

Neopine. Hydroxy-codeine, $C_{18}H_{20}NO_3(OH)$.†

Neoplen. Foamed polyethylene slabstock and mouldable beads. Applications: Cushion packaging of fragile items, anti-rattle stowage of tools in automotives. (BASF United Kingdom Ltd).*

Neopon. Surfactant for cosmetics, toiletries, pharmaceutical, processing, agricultural and other industries. (Baxenden Chemical Co Ltd).*

Neopon LAM. Ammonium lauryl sulphate, which may be based on natural or synthetic alcohols. Liquid form. Applications: Hair and carpet shampoos. (Witco Chemical Ltd).

Neopon LOA/F. Ammonium alcohol 2EO sulphate in liquid form. Application: Hair shampoos. (Witco Chemical Ltd).

Neopon LOS, LOS/F and LOS/NF. Sodium alcohol 2EO sulphate in liquid form. Application: Hair shampoos, foam baths, emulsion polymerization. (Witco Chemical Ltd).

pon LOT/F. Triethanolamine alcohol 2EO sulphate in liquid form. Applications: Hair shampoos, foam baths, liquid detergents. (Witco Chemical Ltd).

pon LS. Sodium lauryl sulphate, which may be based on natural or synthetic alcohols. Liquid form. Applications: Hair and carpet shampoos; emulsion polymerization. (Witco Chemical Ltd).

pon LT. Triethanolamine lauryl sulphate which may be based on natural or synthetic alcohols. Liquid form. Applications: Hair shampoos. (Witco Chemical Ltd).

pon 33. Sodium olefine sulphonate and sodium ether sulphate as a clear amber liquid. Applications: Foaming agent and detergent for cosmetic and household uses. (Witco Chemical Ltd).

opralac. Pigments for textile printing. (ICI PLC).*

oprene (GR-M). A generic term for synthetic rubbers made by polymerizing chloroprene, the latter being obtained from acetylene and hydrogen chloride. It has a greater resistance to oils and ozone than rubber. It is used particularly for belts in machinery where oil resistance is required.†

opurpurite. Synonym for Heterosite.†

opybuthrin. Synergized synthetic pyrethroids. (The Wellcome Foundation Ltd).

oqinine. Quinine glycerophosphate.†

oresit. A phenol-formaldehyde resin.†

oRez. Water and solvent based polyurethane resins. Applications: Paints, inks and special coatings on all substrates eg. wood, metal, plastic, concrete etc. (Polyvinyl Chemie Holland NV).*

osaccharin. See SACCHARIN.

osar. Cyclophosphamide. Applications: Antineoplastic; immunosuppressive. (Adria Laboratories Inc).

oscan. Gallium citrate Ga 67. Applications: Diagnostic aid; radioactive agent. (Medi-Physics Inc).

Neosorb. Sorbitol powders or syrups. Applications: Food, pharmaceutical and various industrial applications. (Roquette (UK) Ltd).*

Neosote. The phenoloids of blast-furnace tar. It contains a small quantity of phenol, and a large amount of cresols.†

Neospectra. A proprietary trade name for carbon black.†

Neosporin Products. Proprietary formulations of polymyxin B sulphate, neomycin sulphate and gramicidin and bacitracin zinc depending on the formulation. Applications: For short-term use in the treatment of topical infections (primary or secondary), due to susceptible organisms. (The Wellcome Foundation Ltd).*

Neostar. Organic brightener system. Applications: Acid zinc electroplating. (Harshaw Chemicals Ltd).*

Neosulphexine. Neohexal (hexamethylene - tetramine - salicyl - sulphonate).†

Neosyl. Amorphous silica (SiO_2) of very low bulk density and high absorption power, prepared by a patented process. It has a bulk density of 0.8-0.2 gm. per cc (Crosfield Chemicals).*

Neosyl C. A registered trade mark for a pure precipitated silica specially adapted for hard rubber.†

Neosyl MH. A registered trade mark for a modified form of precipitated silica containing about 5% of magnesium as oxide. A new white reinforcing agent for rubber.†

Neoteben. Pharmaceutical speciality. Applications: Antitubercular. (Bayer & Co).*

Neothane. A proprietary polyester based polyurethane elastomer cross linked with diamine. (Goodyear Tyre and Rubber).†

Neothene. Solvent. Applications: Gem and mineral testing. (Geoliquids).*

Neotrizine. Trisulphapyrimidines (oral suspension). Mixture of Sulphadiazine, Sulphamerazine and Sulphamethazine. Applications: Antibacterial. (Eli Lilly & Co).

Neotulle. A proprietary preparation of paraffin tulle with neomycin sulphate, zinc bacitracin, and polymixin B sulphate. Wound dressing. (Fisons PLC).*

NeoVac. Miscellaneous chemical compositions. Applications: Coatings and printing inks. (Polyvinyl Chemie Holland NV).*

Neovadine. Dyeing and printing assistant. (Ciba-Geigy PLC).

Neovax. A proprietary preparation of neomycin sulphate, succinylsulphathiazole and kaolin. An antidiarrhoeal. (Norton, H N Co).†

Neovit. A proprietary preparation of thiamine hydrochloride, riboflavine, pyridoxine hydrochloride, nicotinamide, manganese, sodium and potassium glycerophosphates. A tonic. (Rybar).†

Neozone A. A proprietary antioxidant for rubber. It is phenyl-α- naphthylanine.†

Neozone B. A proprietary trade name for meta-toluylene diamine. An antioxidant. (Imperial Chemical Industries), (Du Pont (UK) Ltd).†

Neozone C. A proprietary antioxidant for rubber. It resembles neozone standard, but contains less *m*-toluylene diamine.†

Neozone D. A proprietary antioxidant for rubber. It is pure phenyl-β-naphthylamine.†

Neozone E. A proprietary trade name for an antioxidant for rubber containing 75 parts of phenyl-beta-naphthylamine and 25 parts of meta toluylene diamine. (Imperial Chemical Industries), (Du Pont (UK) Ltd).†

NEP. *N*-Ethyl-2-pyrrolidone. Applications: Solvent, reaction intermediate, textile auxilliaries, cosmetic ingredient. (GAF Corporation).*

Nephril. A proprietary preparation of polythiazide. A diuretic. (Pfizer International).*

Nephroflow. Iodohippurate sodium I 123. Applications: Diagnostic aid; radioactive agent. (Medi-Physics Inc).

N.E. Powder. A 36-grain powder, containing metallic nitrates, and nitro hydrocarbons.†

Neptazane. Methazolamide. Applications: Carbonic anhydrase inhibitor. (Lederle Laboratories, Div of American Cyanamid Co).

Neptune Green. (Disulphine Green B). A dyestuff formed by the condensation of *o*-chlorobenzaldehyde and benzyl ethylaniline, oxidation and conversion into the sodium salt.†

Neptune Green S. An acid dyestuff, giving bluish-green shades on wool.†

Neradol. (Maxyntan, Paradol, Cresyntan, Ordoval G, 2G, Ewol). Synthetic tannins generally prepared by the condensation of phenol-sulphonic acids with formaldehyde, under conditions that only water-soluble products are formed. Also see ESCO EXTRACT and CORINAL.†

Neradol D. A condensed sulphonated cresol. A synthetic tannin.†

Neradol N. A synthetic tannin made from sulphonated naphthalene.†

Neral. β-Citral, $C_9H_{15}CHO$.†

Neramine. Wood stain dyes and pigments for plastics. (ICI PLC).*

Nerco. Solvents and detergents. (ABM Chemicals Ltd).*

Nercol. Soluble cutting oils. (ABM Chemicals Ltd).*

Nercolan. Blend of phenolic biocides. (ABM Chemicals Ltd).*

Nercosol. Viscosity depressant, anti-foaming. Applications: Vinyl plastisols. (ABM Chemicals Ltd).

Nerfinol. Textile finishing agent. (ABM Chemicals Ltd).*

Nergandin. An alloy containing 70 per cent copper, 28 per cent zinc, and 2 per cent lead. Used for condenser tubes.†

Neri Silk. See GALETTAME SILK.

Nerloate. Paint dryers. (ABM Chemicals Ltd).*

Nerogene D. A developer for Zambesi black.†

Neroli Oil. See OIL OF ORANGE FLOWERS.

Nerolidol. Methyl-vinyl-homo-geranyl carbinol. A perfume.†

Nerolin. (Yara-yara).β-Naphthol-methyl ester, $C_{10}H_7.O.CH_3$.β-Naphthol-ethyl ester is also known under this name.†

·olin II. (Bromelia).β-Naphthol ethyl ester, $C_{10}H_7.O.C_2H_5$. A synthetic perfume.†

·van. Wetting agents and finishing oils. (ABM Chemicals Ltd).*

·van CP. Anionic surfactant. (ABM Chemicals Ltd).*

·vanaid. Sequestering agent, metal chelate, filter aid powder. (ABM Chemicals Ltd).*

·vanase. Bacterial α-amylase. (ABM Chemicals Ltd).*

·ve Oil. See NEATSFOOT OIL.

·vin. An extract of meat.†

·sacaine. Chloroprocaine hydrochloride. Applications: Anaesthetic. (Astra Pharmaceutical Products Inc).

·sfield's Triple Tablets. A water steriliser. It consists of (*a*) a tablet containing an iodide and an iodate; (*b*) a tablet containing citric or tartaric acid; and (*c*) a tablet containing sodium sulphite. The addition of (*a*) to (*b*) liberates iodine from the iodate, and (*c*) removes free iodine.†

·ssler's Reagent. An alkaline solution of mercuric iodide in potassium iodide. Employed as a delicate test for ammonia.†

·ste Polyethylene. Polyethylene of low, medium and high density as well as linear low pressure polyethylene. Applications: Packaging, distribution, household and construction films, insulation and jacketing of wire and cable, production of pipes, moulded materials and paper coating. (Neste Oy Chemicals & Unifos Kemi AB).*

·stosyl Ointment. Benzocaine 2%, butyl aminobenzoate 2%, resorcin 2%, zinc oxide 10%, hexachlorophane 0.1% (in ointment form). Applications: Topical anaesthetic for the relief of local pain and irritation in lacerated skin conditions, also in haemorrhoids and anal pruritis. (Bengue & Co Ltd).*

·ethaprin Dospan. A proprietary preparation of etafedrine hydrochloride, bufylline, doxylamine succinate and phenylephrine hydrochloride used in the treatment of bronchospasm. (Richardson-Vicks Inc).*

Nethaprin Expectorant. A proprietary preparation of etafedrine hydrochloride, bufylline, doxylamine succinate and glyceryl guaiacolate. An expectorant. (Richardson-Vicks Inc).*

Netromycin. Netilmicin sulphate. Applications: Antibacterial. (Schering-Plough Corp).

Nettolin. Humus complex-fertilizer on peat basis for the culture of wine, hops, fruit and vegetables, for flowers and lawns. (Süd-Chemie AG).*

Neudorfite. A resinous hydrocarbon found in Bavarian coal pits.†

Neulactil. A proprietary preparation of pericyazine. A sedative. (May & Baker Ltd).*

Neuphor 100. An anionic rosin emulsion with 35% solids used as a sizing agent. Applications: Used in paper and paperboard to impart resistance to water and aqueous solutions. (Hercules Inc).*

Neuro-Phosphates. A proprietary preparation of calcium glycerophosphate, sodium glycerophosphate and strychnine. A tonic. (Smith Kline and French Laboratories Ltd).*

Neuro-Transentin. A proprietary preparation of adiphenine hydrochloride and phenobarbitone. (Ciba Geigy PLC).†

Neurocil. Pharmaceutical speciality. Applications: Neuroleptic. (Bayer & Co).*

Neutral Acriflavine. See EUFLAVINE.

Neutral Alum. A neutral basic alum obtained by the addition of sodium hydroxide to a solution of alum, until the precipitate produced is just re-dissolved.†

Neutral Blue. A dyestuff. It is dimethylaminophenylphenonaphthazonium chloride, $C_{24}H_{20}N_3Cl$. Dyes tannined cotton blue.†

Neutral Oils. The name given to the lightest lubricating oils from American petroleum. The term is also applied to refined coal-tar oils.†

Neutral Orange. (Pentley's Neutral Orange). A pigment. It is a compound of Cadmium yellow and Venetian red.†

Neutral Phosphate. A fertilizer prepared by digesting mineral phosphate, bonemeal, or a mixture of both, with small amounts of sulphuric acid. This renders the P_2O_5 more available. The product contains 20-25 per cent. P_2O_5, and is neutral.†

Neutral Red. (Toluylene Red). A dyestuff. It is the hydrochloride of dimethyldiaminotoluphenazine, $C_{15}H_{17}N_4Cl$. Dyes cotton mordanted with tannin and tartar emetic, bluish-red.†

Neutral Tartar. Potassium tartrate.†

Neutral Violet. A dyestuff. It is the hydrochloride of dimethyldiamino-phenazine, $C_{14}H_{15}N_4Cl$. Dyes cotton mordanted with tannin and tartar emetic, reddish-violet.†

Neutralaleisen. A Swedish silicon-iron alloy, which is stated to resist the action of acids.†

Neutralite. An asphaltic material made in Germany.†

Neutralol. See PARAFFIN LIQUID.

Neutraphylline. A proprietary preparation of diprophyline. A bronchial antispasmodic. (A Cox & Co).†

Neutrase. A bacterial proteinase made by submerged fermentation of a selected strain of Bacillus subtilis. Applications: Can be used in any case where the aim is breakdown of proteinaceous matter for production of peptides and amino acids. (Novo Industri A/S).*

Neutrichrome. Pemetallized dyes for wool and wool blends. (ICI PLC).*

Neutrogene. Azoic dyes. (ICI PLC).*

Neutrolactis. A proprietary preparation containing aluminium hydroxide gel, magnesium trisilicate, calcium carbonate and milk solids. An antacid. (Sandoz).†

Neutronynx S-60. Ammonium alkyl-phenol ether sulphate in liquid form. Application: High foaming dish washing agent; car wash; rug shampoos; glass cleaner. (Millmaster-Onyx UK).

Neutronyx 600 and 656. Nonionic surfactants of the nonylphenol ethoxylate type in liquid form. Applications: Detergency, wetting, emulsifying and dispersing agents. Used in household detergents; metal cleaning; degreasing; industrial cleaning; detergent sanitizers; insecticidal and herbicidal sprays; silicone emulsifiers. (Millmaster-Onyx UK).

Neuwied Blue. See LIME BLUE.

Neuwieder Green. See MOUNTAIN GREEN.

Nevada Silver. See NICKEL SILVERS.

Nevastain. An alloy of 86 per cent iron, 9.5 per cent chrondum, 4 per cent silicon, and 0.43 per cent carbon. A non-corrosive alloy.†

Nevastain R.A. A stainless steel alloy containing iron with approximately 16 per cent chromium, 1 per cent copper, 1 per cent silicon, 0.4 per cent manganese, 0.03 per cent phosphorus and sulphur, and 0.1 per cent carbon (maximum).†

Nevidene. A proprietary trade name for a coumarone resin.†

Nevile and Winther's Acid. α-Naphthol-4-sulphonic acid.†

Nevillac. A proprietary trade name for a phenol-indene-coumarone resin. Used in varnishes and paints.†

Neville. A proprietary trade name for coumarone-indene resins.†

Nevillite. A proprietary trade name for a hydrocarbon.†

Nevin. A proprietary brand of lithopone (*qv*).†

Nevindene. A proprietary trade name for coumarone-indene resins.†

Nevinol. A proprietary coumarone plasticising oil. A viscous liquid polyme[r] practically non-drying at room temperature.†

Nevraltein. A name for sodium-*p*-phenetidine-methane-sulphonate.†

Nevrosthénine. An alkaline solution of the glycero-phosphates of sodium, potassium, and magnesium.†

Nevyanskite. See IRIDOSMINE.

Newagit. A trade mark for abrasive and refractory materials consisting essentially of alumina.†

Newaloy. A proprietary trade name for a steel containing copper.†

ew Blue. See ULTRAMARINE, PRUSSIAN BLUE, and MELDOLA'S BLUE.

ew Blue B or G. (Fast Blue 2B for Cotton; Fast Cotton Blue B; Fast Navy Blue G, BM, GM; Naphthol Blue B; Fast Marine Blue G, BM, BG). A dyestuff. Dyes cotton mordanted with tannin and tartar emetic, blue.†

ew Blue R. See MELDOLA'S BLUE.

ew Brick. Blended organic and inorganic acids in combination with surfactants and wetting agents. Applications: Cleaner for brick in new construction (mortar smears, dirt etc). (Nova Chemical Inc).*

ew Brick (Heavy Duty). Blended organic and inorganic acids in combination with surfactants and wetting agents. Applications: Cleaner for brick in new construction (mortar smears, dirt etc). (Nova Chemical Inc).*

ew Brunswick Greens. See CHROME GREENS.

ew Cacodyl. See ARRHENAL.

ewcavac. Newcastle disease, inactivated vaccine. Applications: immunization of poultry. (Intervet America Inc).*

ewcavac-T. Newcastle disease, inactivated vaccine. Applications: immunization of poultry. (Intervet America Inc).*

ew Fast Blue F and H. A navy blue dyestuff for tannined cotton.†

ew Fast Blue for Cotton. See MELDOLA'S BLUE.

ew Fast Blue R Crystals for Cotton. See MELDOLA'S BLUE.

ew Fast Green 3B. See VICTORIA GREEN 3B.

ew Fast Grey. A basic dyestuff giving reddish-grey shades on tannined cotton.†

ew Fuchsine. See NEW MAGENTA.

ew Gold. A Bohme dyestuff. It contains quercitron extract, and a chrome mordant.†

ew Green. A dyestuff. It is dimethyldiaminonaphthyldiphenyl-carbinol hydrochloride. Used in calico printing. Also see MALACHITE GREEN and EMERALD GREEN.†

New Grey. (Nigrisine, Methylene Grey, New Methylene Grey, Malta Grey, Alsace Grey, Direct Grey, Nigramine, Metaniline Grey, Special Grey R). A dyestuff obtained by boiling nitrosodimethyl-aniline hydrochloride with water and alcohol, or by the oxidation of dimethyl-*p*-phenylenediamine. Dyes cotton mordanted with tannin silver-grey or blackish-grey.†

New Legumex. Selective weedkillers. (Fisons PLC).*

Newloy. An alloy of 64 per cent copper, 35 per cent nickel, and 1 per cent tin.†

New Magenta. (New Fuchsine, Isorubine, Magenta BB Powder, Crystals FF, CV Conc, L, Superfine Powder, New Magenta Crystals, New Roseine O). A dyestuff. It is the hydrochloride of triaminotritolylcarbinol, $C_{22}H_{24}N_3Cl$. Dyes wool, silk, leather, and tannined cotton, red.†

New Magenta Crystals. A dyestuff. It is a British equivalent of New magenta.†

New Methylene Blue. A greenish-blue dyestuff used for dyeing mordanted cotton and silk. It is obtained by condensing Meldola's blue with diphenylamine.†

New Methylene Blue GG. A dyestuff. It is tetramethyl-diamino-naphthophenoxazonium chloride, $C_{20}H_{20}N_3OCl$. Dyes tannined cotton greenish-blue. It dyes silk from a killed soap bath.†

New Methylene Blue N. A dyestuff. Dyes tannined cotton blue.†

New Methylene Blue NX. See MARINE BLUE.

New Methylene Grey. See NEW GREY.

New Murbetex. Pre-emergence herbicide. (Murphy Chemical Ltd).

New Patent Blue B, 4B. See PATENT BLUE V, N.

New Phosphine G. A dyestuff. It is dimethylaminotolueneazo-resorcinol, $C_{15}H_{17}N_3O_2$. Dyes leather and tannined cotton, yellow.†

New Pink. See PHLOXINE P.

New Printing Black SS, NR, NGR. Logwood extracts, which have an odour of acetic acid, mixed with sodium chlorate. Employed directly for printing cotton.†

New Red. See BIEBRICH SCARLET.

New Red L. See BIEBRICH SCARLET.

New Roseine O. A British brand of New magenta.†

New Solid Green. (New Victoria Green Extra). A dyestuff. It is analogous in composition to Malachite green, but is prepared with dichlorobenzaldehyde, instead of benzaldehyde. See MALACHITE GREEN and BRILLIANT GREEN.†

New Solid Green BB and 3B. See VICTORIA GREEN 3B.

Newton's Alloy. An alloy of 50 per cent bismuth, 31.2 per cent lead, and 18.7 per cent tin. It is a fusible alloy, and melts at 94.5° C.†

New Verdone. Selective weedkiller. (Plant Protection (Subsidiary of ICI)).†

New Victoria Green Extra. See NEW SOLID GREEN.

New White Lead. A sulphate of lead. †

New-wrap. A proprietary viscose packing material.†

New Yellow. See CHROME YELLOW and ORANGE IV.

New Yellow L. See ACID YELLOW.

New Zealand Dammar. A name given to Kauri copal resin, obtained from *Dammara Australis*.†

Nez. A proprietary preparation of phenylephrine hydrochloride, paracetamol, caffeine and ascorbic acid. Cold remedy. (Rybar).†

NFB. Phosphoric acids 80% and 85%. Applications: Used in chemical polishing of aluminium. (Monsanto Co).*

NFT fertilizer. Soluble fertilizer. (Fisons PLC, Horticulture Div).

Ngai Camphor. A camphor obtained from *Blumea balsamifera*. It is closely related to borneol.†

N'hangellite. An elastic bitumen.†

Niac. Niacin. Applications: Vitamin. (O'Neal, Jones & Feldman Pharmaceuticals).

Niacin. See NICOTINAMIDE.

Niagara Blue. This is the same as trypan blue.†

Niamid. A proprietary preparation of nialamide. An antidepressant. (Pfizer International).*

Niamide. A proprietary preparation of nialamide. An antidepressant. (Pfizer International).*

Nibiol. A proprietary preparation of nitroxoline. For veterinary use. (Dales Pharmaceuticals).*

Nibren. Flame-retardant impregnating, encapsulating and dipping waxes with high dielectric constants. Applications: For use in the manufacture of paper capacitors. (Bayer & Co).*

Nibren Wax. Chlorinated naphthalene.†

Nibrite. Barrel bright nickel plating process. (Hanshaw Chemicals).†

Nicametate. 2-Diethylaminoethyl nicotinate.†

Nicar. Nickel carbonate. (Mechema).*

Nicaragua Wood. See REDWOODS.

Nicat. Nickel catalysts and Raney-type nickel catalysts. (Crosfield Chemicals).*

Nicat NP/AC60PT. As NP/AC6oP but in the form of 1/8" pellets.†

Nicfo. Nickel formate. (Mechema).*

Ni-chillite. A proprietary trade name for a nickel-chromium-molybdenum cast iron.†

Nicholson's Blue. See ALKALI BLUE. Also a mixture of Alkali blue and Water blue.

Nichroloy. Electrical resistance alloys containing 23-75 per cent nickel, 7-20 per cent chromium, 7-50 per cent iron, and 1-3 per cent manganese. They are stated to resist corrosion by steam and dilute acids.†

Nichrome. A registered trade mark. Alloys of 54-80 per cent nickel, 10-20 per cent chromium, 7-27 per cent iron, 0-11 per cent copper, 0-5 per cent manganese, 0.3-4.6 per cent silicon, and sometimes 1 per cent molybdenum, and 0.25 per cent titanium. Nichrome I is stated to contain 60 per cent nickel, 25 per cent iron, 11 per cent chromium, and 2 per cent manganese, and Nichrome II 75 per cent nickel, 22 per cent iron, 11 per cent chromium, and 2 per cent manganese. Nichrome III-a heat-resisting alloy containing 85 per cent nickel and 15 per cent chromium. They are used as electrical resistance metals, and are stated to resist acids.†

Nichrosi. Alloys. (*a*) Contains 25-30 per cent chromium, 16-18 per cent silicon, balance nickel. (*b*) Contains 15-25 per cent chromium, 16-18 per cent silicon, balance nickel.†

Nickel 201. A grade of nickel containing 99.0 per cent. nickel (min) and 0.02 per cent. carbon (max). (Wiggin Alloys Ltd).†

Nickel. 204. A nickel containing 4 per cent. cobalt. (Wiggin Alloys Ltd).†

Nickel Aluminium Bronze. An alloy of 10-40 per cent nickel, 10-88 per cent copper, and 2-30 per cent aluminium. One alloy contains 20 per cent tin.†

Nickel Babbitt. A proprietary trade name for a tin-copper-nickel alloy used as a bearing metal for high speeds.†

Nickel Brass. A nickel silver. One alloy contains 55 per cent copper, 43 per cent zinc, and 2 per cent nickel, and another 50 per cent copper, 34 per cent zinc, 15 per cent nickel, and 0.1 per cent aluminium.†

Nickel Bronze. A nickel silver. It usually contains from 20-30 per cent nickel, 50-86 per cent copper, and 8-25 per cent tin, but other alloys contain 11-18 per cent zinc, and 0-18 per cent lead.†

Nickel Glance. A mineral, $Ni(AsS)_2$.†

Nickel Iron. See FERRO-NICKEL.

Nickel Manganese Bronze. An alloy containing 2.5 per cent nickel, 53.4 per cent copper, 39 per cent zinc, 1.7 per cent manganese, and 2.6 per cent tin, with small quantities of aluminium and lead.†

Nickel Oreide. A nickel silver. It is an alloy of from 63-87 per cent copper, 6-33 per cent zinc, and 2-7 per cent nickel.†

Nickel Silver Solder. An alloy of 35 per cent copper, 57 per cent tin, and 8 per cent nickel.†

Nickel Silvers. Ternary alloys of copper, nickel, and zinc, the standard of which is determined by the nickel content. Albatra, Alfenide, Alpacca, Amberoid, Ambrac, American silver, Aphit, Argentan, Argentin, Argiroide, Argentan solder, Argyroide, Argyrolith, Aterite, Benedict plate, Bismuth bronze, Bismuth brass, Bolster silver, Brazing solder, China silver, Carbondale silver, Charcoal nickel silvers, Christofle, Chromax bronze, Colorado silver, Craig gold, Electroplate, Electrum, Elner's German Silver, Frick's alloys, Keene's alloy, Key alloy, Lutecin, Maillechort, Markus alloy, Milling silver, Naubuc, Neogen, Neu-silver, Nevada silver, Nickel brass, Nickel bronze, Nickelin, Nickel oreide, Packfong, Packtong, Platinoid, Potosi silver, Rheotan, Ruolz alloys, Silverite, Silveroid, Spoon metal, Sterlin, Sterline, Suhler white copper, Toncas metal, Tuc-tur metal, Tungsten brass, Tutenag, Victoria silver, Victor metal, Virginia silver, Wessel's silver, White button alloy, White copper, White metal, White solder. Alfenide, Argyroide, and Christofle are plated German silvers. These alloys are used in the manufacture of tableware, and for other purposes.†

Nickel Steel. An alloy of nickel with steel, usually containing from 3-5 per cent of nickel, but sometimes a larger amount. One alloy contains 30 per cent of nickel, 1 per cent manganese, and 1 per cent chromium. Nickel steels are used for armour plates, ships' screws, boiler plates, cable wires, and gun barrels. Also see STEELS O and T.†

Nickel Yellow. A pigment obtained by treating nickel sulphate with sodium phosphate, and calcining the product.†

Nickel Zirconium. An alloy of 86.4 per cent nickel, 6 per cent aluminium, 6 per cent silicon, and 1.5 per cent zirconium. Used for cutting tools.†

Nickel 200. A grade of nickel containing 99.0 per cent. nickel. (Wiggin Alloys Ltd).†

Nickel 205. Nickel containing 99.0 per cent. min. nickel and low carbon. (Wiggin Alloys Ltd).†

Nickel 211. Nickel containing 5 per cent. manganese. (Wiggin Alloys Ltd).†

Nickel 212. Nickel containing 2 per cent. manganese. (Wiggin Alloys Ltd).†

Nickel 213. A nickel with improved machining properties. Nickel 96 per cent. min., Manganese 2 per cent. High carbon and silicon content. (Wiggin Alloys Ltd).†

Nickel 222. Nickel containing 99.5 per cent. min. nickel, 0.06-0.09 per cent. magnesium and very low impurity levels. (Wiggin Alloys Ltd).†

Nickel 223. Nickel containing 99.5 per cent. min. nickel, 0.035-0.065 per cent. magnesium and very low impurity levels. (Wiggin Alloys Ltd).†

Nickel 229. Nickel containing 97.5 per cent. min. nickel, 1.8-2.2 per cent. tungsten, 0.35-0.65 per cent. magnesium and 0.02-0.04 per cent. aluminium. (Wiggin Alloys Ltd).†

Nickel 270. Nickel 99.9 per cent. pure. (Wiggin Alloys Ltd).†

Nickeladium. A proprietary trade name for a nickel-vanadium cast steel.†

Nickelene. A name suggested for nickel silver (German silver) alloys.†

Nickelin. Electrical resistance allys of nickel and copper, usually with zinc. One alloy contains 55.3 per cent copper, 31 per cent nickel, 13 per cent zinc, 0.4 per cent iron, and 0.2 per cent lead; another 68 per cent copper and 32 per cent nickel; and a third 74.5 per cent copper, 25 per cent nickel, and 0.5 per cent iron.†

Nickeline. Synonym for Niccolite.†

Nickelite. See NICCOLITE.

Nickel-linnaeite. See POLYDYMITE.

Nickel-manganese-copper. An alloy of 73 per cent copper, 24 per cent manganese, and 3 per cent nickel. Used for electrical resistances.†

Nickel-molybdenum Steels. Alloys containing 0.13-0.54 per cent carbon, 0.12-4.4 per cent molybdenum, and 1.8 per cent nickel.†

Nickeloid. A proprietary trade name for a dual metal. It is zinc faced with nickel.†

Nickeloy. An alloy of 1.5 per cent nickel, 4 per cent copper, and 94 per cent aluminium.†

Niclad. A proprietary trade name for a duplex metal in which nickel or nickel alloy is deposited on steel or iron.†

Niclocide. Niclosamide. Applications: Anthelmintic. (Miles Pharmaceuticals, Div of Miles Laboratories Inc).

Nico. Nickel oxide. (Mechema).*

Nico Padutin. Pharmaceutical preparation also known as Kalleonicit. Application Combination therapy for physiological vasodilation. (Bayer & Co).*

Nico-400. Niacin. Applications: Vitamin. (Marion Laboratories Inc).

Nicobid. Niacin. Applications: Vitamin. (USV Pharmaceutical Corp).

Nicocap. Niacin. Applications: Vitamin. (ICN Nutritional Biochemicals Corp).

Nicolane. A proprietary trade name for Noxapine.†

Nicolar. Niacin. Applications: Vitamin. (USV Pharmaceutical Corp).

Nicolle's Carbol-thionin Blue. A microscopic stain. It consists of a mixture of 10 cc of a saturated solution of thionin blue in 50 per cent alcohol, and 100 cc of a 2 per cent carbolic acid solution.†

Nicon. An alloy of 70 per cent iron and 30 per cent nickel.†

Nicor. Mixture of nickel/cobalt oxide. (Mechema).*

Nicorette. Nicotine Polacrilex. Methacrylic acid polymer with divinyl benzene, complex with nicotine. Applications: Deterrent. (Merrell Dow Pharmaceuticals Inc, Subsidiary of Dow Chemical Co).

Nicoschwab. See UBA.

Nicotinamide. Nicotinic acid, niacin. One of the substances comprising the Vitamin B complex. Deficiency in the diet is partly the cause of pellagra, a skin disorder.†

Nicotine 40% Shreds. Insecticide smoke. (Murphy Chemical Ltd).

Nicotinyl Alcohol. 3-Pyridylmethanol.†

Nicral Alloys. Aluminium alloys containing varying percentages of nickel, chromium, and copper.†

Nicro-copper. An alloy of 98 per cent copper and 2 per cent nickel.†

Nicrolan. Wool grease-based compositions. Applications: Anti-corrosive coating materials for metal protection. (Westbrook Lanolin Co).*

Nicroman. A proprietary trade name for a tool steel containing 1 per cent chromium, 1.65 per cent nickel, 0.35 per cent copper, and 0.7 per cent carbon. An oil-hardening hob steel.†

Nicrosil. A proprietary trade name for an alloy of iron with 18 per cent nickel and 4-6 per cent silicon.†

Nicrosilal. An alloy of 71.2 per cent iron, 18 per cent nickel, 6 per cent silicon, 2 per cent chromium, 1.8 per cent carbon, and 1 per cent manganese. It is a non-magnetic grey cast iron which is resistant to staining and has great ductility.†

Nicu Steel,. A nickel steel containing 2.13 per cent nickel, 0.2 per cent copper, 0.51 per cent manganese, 0.03 per cent sulphur, 0.03 per cent silicon, and 0.006 per cent phosphorus.†

Nidazol. Metronidazole. Applications: Pharmaceutical preparation for the treatment of trichomonal infestation. (M A Steinhard Ltd).*

Nidrin. A proprietary preparation containing aluminium hydroxide and magnesium carbonate in a co-dried gel. An antacid. (Smith and Nephew).†

Niello Silver. (Russian Tula, Blue Silver). An alloy of silver, copper, lead, and bismuth with a bluish colour.†

Niferex. A proprietary preparation of a polysaccharide-iron complex used in the treatment of anaemia. (L Wilcox & Co Ltd).†

Niflor. Precision nickel/PTFE composite. Applications: Precision thickness coating for hard, self-lubricating and corrosion resistant surfaces on engineered components. (Fothergill Tygaflor Ltd).*

Nigagin. Methyl-*p*-hydroxybenzoate. A preservative.†

Night Blue. A dyestuff. Dyes silk and wool greenish-blue. Also see METHYL BLUE.†

Night Blue B. A dyestuff. Dyes wool and silk bluish-green from an acid bath.†

Night Green. See IODINE GREEN.

Night Green 2B. A dyestuff. Dyes wool and silk bluish-green from an acid bath.†

Night of Olay Nightcare. Cream. Applications: For nightly care of the skin. (Richardson-Vicks Inc).*

Nigramine. See NEW GREY.

Nigraniline. Aniline black, $C_{30}H_{25}N_5$.†

Nigre. The impure soap remaining after the good soap has been removed by running out. It contains iron soaps, caustic soda, and sodium chloride.†

Nigrin. Streptonigrin. Applications: Antineoplastic. (Pfizer Inc).

Nigrisine. See NEW GREY.

Nigrol. The residue obtained after the removal of kerosene, gasoline, and light oil from petroleum naphtha.†

Nigrosin Bases. For shoe polish, printing ink, office supplies and plastics industries. (Bayer UK Ltd).

Nigrosine ABKS, G, L, SG, SS, 1471, 7600. Dyestuffs. They are British equivalents of Water-soluble nigrosine.†

Nigrosine Crystals, B, BP, G, JB, R, W, WM, 79694. Dyestuffs. They are British brands of water-soluble nigrosine.†

Nigrosine, Soluble. See INDULINE, SOLUBLE.

Nigrosine, Spirit Soluble. See INDULINE, SPIRIT SOLUBLE.

Nigrosulphine. A sulphur black dye made from 2-hydroxy-*m*-phenylene-diamine and sulphur.†

Nigroth Metal. A proprietary trade name for a heat-resisting nickel-chromium cast iron.†

Nigrotic Acid. Dihydroxysulphonaphthoic acid.†

Ni-hard. Alloys of iron with from 4-5 per cent nickel, 1.5 per cent chromium, and varying amounts of silicon and carbon.†

Nihil. See NIL.

Nikoteen. An. American product containing 26 per cent nicotine. A fumigant.†

Nikro-trimmer Steel. A proprietary trade name for a nickel-chromium steel containing 0.3 per cent nickel, 0.55 per cent chromium, and 0.85 per cent carbon.†

Nikrome. Proprietary trade name for nickel-chromium steels. Nikrome M contains 2.25 per cent nickel, 1 per cent chromium, 0.45 per cent molybdenum, and 0.4 per cent carbon.†

Nil. (Nihil). Zinc oxide, ZnO.†

Nile Blue A. A dyestuff. It is diethyldiaminonaphthophenoxazonium sulphate, $(C_{20}H_{20}N_3O)_2SO_4$. Dyes tannined cotton blue.†

Nile Blue R. A dyestuff allied to Nile blue A and 2B, of a more reddish colour.†

Nile Blue 2B. A dyestuff. It is diethylbenzyldiaminophenoxazonium chloride, $C_{27}H_{26}N_2OCl$. Dyes tannined Cotton greenish-blue.†

Nilergex. A proprietary preparation of isothipendyl hydrochloride. (ICI PLC).*

Nilevar. A proprietary preparation of norethandrolone. An anabolic steroid. (Searle, G D & Co Ltd).†

Nilex. A proprietary 36 per cent nickel steel used for pendulums, etc., on account of its low coefficient of expansion.†

Nilo. A trade mark for controlled expansion alloys. Their compositions are coded as follows:

Alloy 36. 36 per cent. nickel, balance iron.

Alloy 42. 42 per cent. nickel, balance iron.

Alloy K45. 32 per cent. nickel, 13 per cent. cobalt, 54 per cent. iron.

Alloy P50. 50 per cent. nickel, 50 per cent. iron.

Alloy 475. 47 per cent. nickel, 5 per cent. chromium, balance iron.

Alloy 48. 48 per cent. nickel, balance iron.

Alloy 51. 51 per cent. nickel, balance iron.

Alloy K. 29 per cent. nickel, 17 per cent cobalt, balance iron. (Wiggin Alloys Ltd).†

Nilodin. A proprietary trade name for the hydrochloride of Lucanthrone.†

Nilomag. A trade mark for magnetic alloys made by a powder metallurgy process. Their compositions are as follows:

Alloy 471. 47 per cent. nickel, 50 per cent. iron, 3 per cent. molybdenum.

Alloy 475. 47 per cent. nickel, 5 per cent. chromium, balance iron.

Alloy 48. 48 per cent. nickel, balance iron.

Alloy K. 29 per cent. nickel, 17 per cent. cobalt, balance iron.

Alloy 51. 51 per cent. nickel, balance iron. (Wiggin Alloys Ltd).†

Nimocast Alloy 242. A patented alloy of 21 per cent. chromium, 10 per cent. cobalt, 10.5 per cent. molybdenum and the balance nickel. A trade mark.

Alloy 771. 77 per cent. nickel, 14 per cent. iron, 5 per cent. copper, 4 per cent. molybdenum.

Alloy 713. 13.4 per cent. chromium, 4.5 per cent. molybdenum, 1 per cent. titanium, 6.2 per cent. aluminium, 2.3 per cent. niobium and the balance nickel.

Alloy PE10. 20 per cent. chromium, 6 per cent. molybdenum, 6.5 per cent. niobium, 2.5 per cent. tungsten, and the balance nickel.

Alloy PK24. 10 per cent. chromium, 15.2 per cent. cobalt, 3 per cent. molybdenum, 5.2 per cent. titanium, 5.5 per cent. aluminium, and the balance nickel. (Wiggin Alloys Ltd).†

Nimol. An alloy of cast iron with 20 per cent monel metal and 2-4 per cent chromium. It is non-magnetic and has high resistance to corrosion by acid and sea water.†

Nimonic Alloy PE 11. A trade mark for an alloy of 18 per cent. chromium, 5.2 per cent. molybdenum, 2.3 per cent. titanium, 0.8 per cent. aluminium, 38 per cent. nickel, and the balance iron.

PE 13. 22 per cent. chromium, 1.5 per cent. cobalt, 9 per cent. molybdenum, 18.5 per cent. iron, 0.6 per cent. tungsten, and the balance nickel.

PK 31. 20 per cent. chromium, 14 per cent. cobalt, 4.5 per cent. molybdenum, 2.3 per cent. titanium, 0.4 per cent. aluminium, 5 per cent. niobium, and the balance nickel.

PK33. 19 per cent. chromium, 14 per cent. cobalt, 7 per cent. molybdenum, 2 per cent. titanium, 2.0 per cent. aluminium, and the balance nickel.

Alloy 75. 20 per cent. chromium, 0.4 per cent. titanium, and the balance nickel.

Alloy 80A. 20 per cent. chromium, 2.3 per cent. titanium, 1.3 per cent. aluminium, and the balance nickel.

Alloy 90. 20 per cent. chromium, 17 per cent. cobalt, 2.5 per cent. titanium, 1.5 per cent. alumininum. and the balance nickel.

Alloy 93 is the same as alloy 90 except that closer control is maintained.

Alloy 105. 15 per cent. chromium. 20 per cent. cobalt, 5 per cent. molybdenum, 1.2 per cent. titanium, 4.7 per cent. aluminium, and the balance nickel.

Alloy 108 has a closer compositional control.

Alloy 118. A fully vacuum-melted and cast version of alloy 115.

Alloy 115. 15 per cent. chromium, 15 per cent. cobalt, 3.5 per cent. molybdenum, 4 per cent. titanium, 5 per cent. aluminium, and the balance nickel. (Wiggin Alloys Ltd).†

Nimorazole. A drug used in the treatment of trichomoniasis. It is 4-[2-(5-nitroimidazol-1-yl) ethyl] morpholine. Nitrimidazine. NAXOGIN, NULOGYL.†

Nimotop. Nimodipine. Applications: Vasodilator. (Miles Pharmaceuticals, Div of Miles Laboratories Inc).

Nimox. Nickel molybdenum oxides on alumina. (Laporte Industries Ltd).*

Nimrod. Fungicide containing bupirimate for the control of powdery mildew. (ICI PLC).*

Nimrod T. Garden fungicide. (ICI PLC, Plant Protection Div).

Ninate. Alkyl benzene sulphonate. Applications: Emulsifier. (Stepan Company).*

Ninate 401. Anionic surfactant as a dark viscous liquid. Applications: Emulsifier, used in oil additives. (KWR Chemicals Ltd).

Ninate 411. Anionic surfactant as a light viscous liquid. Applications: Emulsifier for solvent degreasers and dry cleaning detergents. (KWR Chemicals Ltd).

Ninate 415. Anionic surfactant in light amber liquid form. Applications: Emulsifier for solvent degreasers and dry cleaning detergents. (KWR Chemicals Ltd).

Ninhydrin. (Triketol). Triketohydrindene-hydrate. A colorimetric reagent for aminoacids.†

Ninol. Alkylolamides. Applications: Thickeners, foam stabilizer, detergent. (Stepan Company).*

Ninox. Amine oxide. Applications: Foam enhancers and thickeners. (Stepan Company).*

Ninox. Foam stabilizer. Applications: Shampoos; foam baths; detergents. (KWR Chemicals Ltd).

Niobe Oil. See OIL OF NIOBE.

Niong. Nitroglycerin. Applications: Vasodilator (U S Ethicals Inc).

Nipa Salt. A material obtained by ignition of the plant *Nipa fructicans*. Used to coagulate rubber latex.†

Nipabenyzl. Benzyl 4-hydroxybenzoate. (Nipa Laboratories Ltd).*

Nipabutyl. Butyl 4-hydroxybenzoate. (Nipa Laboratories Ltd).*

Nipacombin. Compounded sodio-4-hydroxybenzoic acid. (Nipa Laboratories Ltd).

Nipagin. Methyl and/or ethyl 4-hydroxybenzoate. (Nipa Laboratories Ltd).

Nipagin A. Ethyl ester of *p*-hydroxybenzoic acid. A preservative. (Nipa Laboratories Ltd).*

Nipagin M. Methyl ester of *p*-hydroxybenzoic acid. A preservative. (Nipa Laboratories Ltd).*

Nipanox. Gallate synergent composition. (Nipa Laboratories Ltd).

Nipantiox. Butylated hydroxyanisole. (Nipa Laboratories Ltd).*

Nipasept. Compounded esters of 4-hydroxybenzoic acid. (Nipa Laboratories Ltd).*

Nipasol. Propyl-4-hydroxybenzoates. (Nipa Laboratories Ltd).*

Nipasol M. Propyl ester of *p*-hydroxybenzoic acid. A preservative. (Nipa Laboratories Ltd).*

Nipastat. Compounded esters of 4-hydroxybenzoic acid. Applications: Preservative. (Nipa Laboratories Ltd).*

Nipholite. Synonym for Khodnevite.†

Nipride. A proprietary preparation of sodium nitroprusside. Intravenous vasodilator. (Roche Products Ltd).*

Nipro (i). Ammonium sulphate. Applications: fertilizer. (Columbia Nitrogen Corporation).*

Nipro (ii). Caprolactam. Applications: Nylon manufacturing. (Columbia Nitrogen Corporation).*

Ni-resist. A corrosion and heat-resisting cast iron containing 12-15 per cent nickel, 5-7 per cent copper, and 1.5-4 per cent chromium.†

Nirex. A proprietary trade name for an alloy of 80 per cent nickel, 14 per cent chromium, and 6 per cent iron.†

Nirolex Expectorant Linctus. A proprietary preparation containing guaiphenesin, ephedrine sulphate and mepyramine maleate. A cough linctus. (Boots Company PLC).*

Nirostaguss. A non-rusting and heat-resisting 34 per cent chromium cast iron.†

Nisapas. A proprietary preparation of sodium para-aminosalicylic acid and isoniazid. An antituberculous agent. (Antigen International Ltd).†

Nisentil. Alphaprodine hydrochloride. Applications: Analgesic. (Hoffmann-LaRoche Inc).

Nispan Alloy C-902. A trade mark for an alloy of 42 per cent nickel, 5.2 per cent. chromium, 2.4 per cent titanium, 0.6 per cent aluminium, and the balance iron. (Wiggin Alloys Ltd).†

Nitolac. A proprietary polyurethane flooring material. (Tercol, Shifnal, Shropshire).†

Nitoman. A proprietary preparation of tetrabenazine. (Roche Products Ltd).*

Nitracc. Fertilizers. (ICI PLC).*

Nitrados. Nitrazepam. Applications: Relief of insomnia due to anxiety or stress. (Berk Pharmaceuticals Ltd).*

Nitral. A trade name for moist nitrous oxide used as a bactericide.†

Nitraline. Nitric acid for boiling water cleaning. (Ciba-Geigy PLC).

Nitralloy. A nitrided aluminium-chromium-molybdenum steel.†

Nitram. Ammonium nitrate fertilizer. (ICI PLC).*

Nitrammite. A native ammonium nitrate.†

Nitrammomkalk. A mixture of ammonium and calcium nitrates in a granular form of Norwegian manufacture. A fertilizer.†

Nitraniline N. Nitraniline mixed with sufficient sodium nitrite necessary for its diazotization.†

Nitraphen. See DINITRA.

Nitrapo. A product obtained from crude caliche by crystallization. It contains about 66 per cent sodium nitrate, 29 per cent potassium nitrate, and a little sodium chloride. It is used as a fertilizer.†

Nitrate of Tin. A mixture of stannous and stannic chlorides. Used by dyers.†

Nitrated Oils. Thick syrupy liquids obtained by treating castor oil or linseed oil with a mixture of concentrated sulphuric and nitric acids. They form homogenous mixtures with nitro-cellulose. Dissolved in acetone, these oils form varnishes, which are used for enamelling leather or similar material, and mixing paints.†

Nitrating Acid. See MIXED ACID.

Nitrazol. See PARANITRANILINE RED.

Nitre. See SALTPETRE.

Nitre Cake. A residue from the manufacture of nitric acid. It consists of a mixture of normal and acid, sodium sulphate.†

Nitre, Chile. See CHILE SALTPETRE.

Nitre, Cubic. See CHILE SALTPETRE.

Nitre, Soda. See CHILE SALTPETRE.

Nitre Spirit. See SPIRIT OF NITRE.

Nitre, Sweet. See SPIRIT OF SWEET NITRE.

Nitrex. A proprietary name for nitro furantoin. (Vale Chemical Co, Allentown, Pa).†

Nitric Ether. Ethyl nitrate, $C_2H_5NO_3$.†

Nitrided Steel. Steel which has been treated with ammonia gas at a temperature of 950° C., whereby nitrogen is absorbed on the surface giving a hard, non-brittle surface. Steels containing from 0.5-2 per cent aluminium and 0.5-4 per cent of other elements are used. See NITRALLOY.†

Nitrimidazine. See NIMORAZOLE.

Nitrite Rubber. A product obtained by treating latex with a nitrite and coagulating with an acid.†

Nitro-26. A nitrogeneous fertilizer. (Fisons PLC).*

Nitro Base. *p*-Nitrosodimethylaniline. An intermediate for dyes.†

Nitrocalcite. (Wall saltpetre). Calcium nitrate, $Ca(NO_3)_2$.†

Nitro-cellulose Varnishes. Varnishes used in the manufacture of artificial leather. They consist of nitro-celluloses dissolved in amyl acetate, and coloured. Used also for painting iron work.†

Nitro-chalk. A proprietary fertilizer consisting of an intimate mixture of chalk and ammonium nitrate in the form of a fine powder. (ICI PLC).*

Nitro-chloroform. See CHLORO-PICRIN.

Nitrocontin. Glyceryl trinitrate. Applications: Angina pectoris. (Napp Laboratories Ltd).*

Nitro-cotton. See GUNCOTTON.

Nitro-dextrin. A similar product to Nitro-starch. Used in explosives.†

Nitrodisc. Nitroglycerin. Applications: Vasodilator. (G D Searle & Co).

Nitrodracrylic Acid. *p*-Nitrobenzoic acid.†

Nitro-erythrite. Erythrol tetranitrate.†

Nitroferrite. An explosive. It contains 93 per cent ammonium nitrate, 2 per cent trinitronaphthalene, 2 per cent potassium ferricyanide, and 3 per cent sugar.†

Nitroform. Tetranitromethane.†

Nitrogen, Lime. See NITROLIM.

Nitroglycerin. (Glonoine Oil, Pyro-glycerin). Glycerol nitrates, usually trinitro-glycerol, $C_3H_5(O.NO_2)_3$.†

Nitrolac. A German nitro-cellulose lacquer.†

Nitrolignin. Wood which has been nitrated.†

Nitrolim. (Lime Nitrogen). A fertilizer. Commercial nitrolim contains 57-63 per cent calcium cyanamide, 20 per cent lime, 14 per cent graphite, and 7-8 per cent silica, iron oxide, and alumina.†

Nitrolite. An explosive. It contains nitro-glycerin.†

Nitromannite. Mannitol hexanitrate.†

Nitro-methylene Blue. Methylene green.†

Nitro-muriatic Acid. See AQUA REGIA.

Nitron. A base which forms a nitrate almost insoluble in water. It is used for the determination of nitric acid. It is also used as a rubber vulcanization accelerator. Also a proprietary trade name for a cellulose nitrate plastic.†

Nitronet. Nitroglycerin. Applications: Vasodilator. (U S Ethicals Inc).

Nitrophosphate. A fertilizer sometimes wrongly called ammonium superphosphate. It is prepared by mixing calcium superphosphate with ammonium sulphate. Some mixtures contain ammonium phosphate and calcium sulphate.†

Nitropress. Sodium Nitroprusside. Applications: Antihypertensive. (Abbott Laboratories).

Nitropropiol. Sodium-*o*-nitro-phenyl-propiolate.†

Nitrosamine Red. A dyestuff. It is diazotized nitraniline.†

Nitrosin Saltpetre. A proprietary preparation containing nitrite, used for curing meat.†

Nitroso Blue. See RESORCIN BLUE.

Nitrospan. Nitroglycerin. Applications: Vasodilator. (USV Pharmaceutical Corp).

Nitro-starch. (Xyloidin). A nitric ester of starch, probably the octonitrate $C_{12}H_{12}(NO_2)_8O_{10}$. Used for blasting explosives, either alone, or by mixing 10 per cent of it with a mixture of sodium nitrate and carbonaceous material.†

Nitrostat. Nitroglycerin. Applications: Vasodilator. (Parke-Davis, Div of Warner-Lambert Co).

Nitrosulphate. Ferric sulphate $Fe_2(SO_4)_3$. Sold as a mordant for dyeing.†

Nitrosyl Silver. Silver hyponitrite. $Ag_2N_2O_2$.†

Nitrosyl Sulphuric Acid. (Lead Chamber Crystals). Nitro-sulphonic acid, $NO_2(SO_2.OH)$.†

Nitrous Ether. Ethyl nitrite, $C_2H_5NO_2$.†

Nitrous Gas or Air. Nitrogen dioxide, NO_2.†

Nitrous Sulphuric Acid. A solution of nitrosyl-sulphuric acid (Weber's acid), in sulphuric acid.†

Nitroxan. A catalyst used for the conversion of ammonia into nitric acid. It is stated to be a compound of barium metaplumabate and barium manganate. The ammonia is oxidized to nitric acid, which is retained as barium nitrate.†

Nitroxoline. 8-Hydroxy-5-nitroquinoline. Nibiol.†

Nitto Nitoflon. Fluorocarbon plastic products. Tapes, films, tubes and mouldings. Used to modify non-adhesive fluoroplastic surfaces into bondable surfaces. Applications: Electric applications and electronics, food production, paper and pulp etc. (Nitto Electric Industrial Co Ltd).*

Nitto SPV. Surface protection adhesive films and sheets. Protects the surface of stainless steel, aluminium, decorated laminates etc from damage in transportation, storage and fabrication. Applications: Architectural structures, rolling stock, household articles. (Nitto Electric Industrial Co Ltd).*

Nivan. See OIL WHITE.

Nivaquine. A proprietary preparation of chloroquin sulphate. An anti-malarial. (May & Baker Ltd).*

Nivar. A proprietary trade name for Invar (*qv*).†

Nivebaxin. A proprietary preparation of neomycin sulphate, bacitracin zinc, polymyxin B sulphate, calcium phosphate, glycine and starch. (Boots Company PLC).*

Nivembine. Chloroquine/di-iodohydroxyquinoline tablets. (May & Baker Ltd).

Nivemycin. A proprietary preparation containing noemycin sulphate. An antibiotic. (Boots Company PLC).*

Nix Cream Rinse. A proprietary formulation of permethrin. Applications: Treatment of infestation with pediculus humanus variety capitis (the head louse) and its nits (eggs). (The Wellcome Foundation Ltd).*

Nixenoid. See VISCOLOID.

Nixon C/A. A proprietary brand of cellulose acetate.†

Nixon C/N. A proprietary brand of cellulose nitrate.†

Nixon E/C. A proprietary brand of ethyl cellulose.†

Nixonite. A proprietary trade name for a cellulose acetate plastic.†

Nixonoid. A proprietary trade name for a cellulose nitrate plastic.†

Nizin. Sulphanilate Zinc. Applications: Antibacterial. (Broemmel Pharmaceuticals).

Nizoral. A proprietary preparation containing ketoconazole. Applications: Anti-fungal. (Janssen Pharmaceutical Ltd).*

Njamplung Oil. See LAUREL NUT OIL.

Njatuo Tallow. A fat obtained from *Palaquium oblongifolium*.†

Njave Butter. (Nari Oil, Noumgou Oil, Adjab Fat). A fat obtained from the seeds of *Mimusops njave* or *djave*, also from *Bassia toxisperma* and *Bassia djave*. The nuts are known as Abeku, Bako, or Mahogany Nuts. The fat has a saponification value of 182-188 and an iodine value of 56-65.†

NLA-10. Dibutylphthalate. A vinyl plasticizer. (National Lead Co).†

NLA-20. Di-2-ethylhexylphthsiate. A vinyl plasticizer. (National Lead Co).†

NLA-30. Di-iso-decylphthalate. A vinyl plasticizer. (National Lead Co).†

NLA-40. Di-decylphthalate. (National Lead Co).†

N-Labstix. A proprietary test-strip used to detect pH, protein, glucose, ketones, blood and nitrite in urine. (Ames).†

NLF-32. A mixed adipate vinyl plasticizer. (National Lead Co).†

NLF-33. A modified adipate vinyl plasticizer. (National Lead Co).†

NM-AMD. N-Methylolacrylamide. (Cyanamid BV).*

Nobecutane. A proprietary preparation of acrylic resin dissolved in acetic esters used in aerosol form as a plastic wound dressing. (Astra Chemicals Ltd).†

Nobel Ardeer Powder. A dynamite containing 33 per cent nitro-glycerin, 49 per cent magnesium sulphate, 13 per cent kieselguhr, and 5 per cent potassium nitrate.†

Nobel Polarite. An explosive. It is a mixture of potassium perchlorate and ammonium nitrate, with trinitrotoluene, a little starch, and wood meal.†

Nobese. Phenylpropanolamine Hydrochloride. Applications: Adrenergic. (O'Neal, Jones & Feldman Pharmaceuticals).

Noblen. Polypropylene. See also PROPATHENE (2) and MOPLEN (Montedison UK Ltd).*

Nobrium. A proprietary preparation of medazepam. A tranquillizer. (Roche Products Ltd).*

Noctec. A proprietary preparation of chloral hydrate. Applications: A sedative. (Squibb, E R & Sons Ltd).*

Noctec Capsules. Chloral hydrate in soft gelatin capsule. (E R Squibb & Sons Ltd).

No-Del. A proprietary preparation of allyl isothiocyanate, ethyl nicotinate, methyl salicylate, eugenol, oil of turpentine, and cholesterol in a hydrophillic base. (Rybar).†

Nodulant. Magnesium/iron tablets for production of S.G. cast irons. (Foseco (FS) Ltd).*

Nogos. Organophosphorus insecticide. (Ciba-Geigy PLC).

Noheet Metal. (Tempered lead). An alloy of 98.4 per cent lead, 1.4 per cent sodium, 0.11 per cent antimony, and 0.08 per cent tin.†

Noil. An alloy of 80 per cent copper and 20 per cent tin.†

Noiret's Aluminium Solder. An alloy of 80 per cent zinc and 20 per cent tin.†

Nokol. Coal dust replacement additives for iron foundry green sands. (Foseco (FS) Ltd).*

Noltam. Tamoxifen citrate. Applications: Breast cancer and anovulatory infertility. (Lederle Laboratories).*

Noludar. A proprietary preparation of methyprylone. A hypnotic. (Roche Products Ltd).*

Nolvadex. A proprietary preparation of TAMOXIFEN used in the treatment of breast cancer. (ICI PLC).*

No-mag. A non-magnetic alloy of 77 per cent iron, 12 per cent nickel, 6 per cent manganese, 3 per cent carbon, and 2 per cent silicon. It has a high specific resistance, is close-grained, and of good mechanical properties.†

No-max. A proprietary trade name for a high speed molybdenum-tungsten steel for cutting tools.†

Nomaze. Aerosol nasal decongestant. (Fisons PLC, Pharmaceutical Div).

Nomex. A proprietary trade name for nylon fibre specially fabricated to withstand exposure to 500° F. It will not melt or drip. (Du Pont (UK) Ltd).†

Nonad Tulle. Paraffin gauze dressing. (Allen & Hanbury).†

Nonaid. Non-ionic surfactant. (ABM Chemicals Ltd).*

Nonanol. Plasticizer, containing 3:5:5-trimethyl hexanol. (ICI PLC).*

Noncorrodite. A proprietary trade name for a chromium steel.†

Nonex. Surfactants. (BP Chemicals Ltd).*

Nonflammable Decobest DA. Diallyl phthalate. Applications: Decorative laminates. (Sumitomo Bakelite, Japan).*

Nongo. A gum resembling tragacanth, obtained from *Albizzia brownei*, of Uganda.†

Non-gran Metal. A bronze containing 87 per cent copper, 11 per cent tin, and 2 per cent zinc.†

Nonidet. Ethylene oxide condensate. (Shell Chemicals UK Ltd).†

Non-mordant Cotton Blue. See ALKALI BLUE XG.

No-Nox. Motor fuel. (Chevron).*

Nonox. Group of anti-oxidants. (ICI PLC).*

Nonox NSN. Similar to Nonox NS.†

Nonox ZA. A proprietary anti-oxidant. 4-Isopropylamine diphenylamine. (ICI PLC).*

Non-pareil Metal. An alloy of 78 per cent lead, 17 per cent antimony, and 5 per cent tin.†

Nopalcol. Nonionic surfactants. (Diamond Shamrock Process Chemicals Ltd).

Nopalin. See EOSIN BN.

Nopco Foamaster. Antifoams/defoamers. (Diamond Shamrock Process Chemicals Ltd).

Nopco Worsted Oil 12. Textile processing aid. (Diamond Shamrock Process Chemicals Ltd).

Nopcocastor. Anionic surfactants. (Diamond Shamrock Process Chemicals Ltd).

Nopcochex. Metal processing aid. (Diamond Shamrock Process Chemicals Ltd).

Nopcofloc. Water treatment chemical. (Diamond Shamrock Process Chemicals Ltd).

Nopcogen. Speciality surfactant. (Diamond Shamrock Process Chemicals Ltd).

Nopcolene. Leather processing aid. (Diamond Shamrock Process Chemicals Ltd).

Nopcolube. Textile processing aid. (Diamond Shamrock Process Chemicals Ltd).

Nopcone. Textile processing aid. (Diamond Shamrock Process Chemicals Ltd).

Nopcosant. Dispersing agent with low foaming tendency. Applications: Highly effective with carbonates. (Diamond Shamrock Process Chemicals Ltd).

Nopcosize. Speciality sizes. (Diamond Shamrock Process Chemicals Ltd).

Nopcosulph. Anionic surfactant. (Diamond Shamrock Process Chemicals Ltd).

Nopcotan. Leather processing aid. (Diamond Shamrock Process Chemicals Ltd).

Nopcote. Paper processing aid. (Diamond Shamrock Process Chemicals Ltd).

Nopcotex. Textile softener lubricants. (Diamond Shamrock Process Chemicals Ltd).

Nopcowax. Synthetic waxes. (Diamond Shamrock Process Chemicals Ltd).

Nora. Rubber flooring and soling. (Carl Freudenberg).*

Noradran. A proprietary preparation of ephedrine hydrochloride, theophylline and papaverine hydrochloride. A bronchial antispasmodic. (Norma).†

Noradrenaline. (-)-2-Amino-1-(3,4-dihydroxyphenyl)ethanol. Levophed.†

Noraflor. Spun-bonded flooring for carpets (Carl Freudenberg).*

Noralastic. Acoustic flooring. (Carl Freudenberg).*

Noralen. Elastic roof cover material. (Carl Freudenberg).*

Noram. Amines and derivatives. Primary, secondary, tertiary fatty mono-, di- and polyamines from C8 to C22. Quaternary ammonium salts. Amine oxides. Amine salts. Ethoxylated amines and polyamines. Amino-acids. Special nonionic derivatives. Additives Applications: Surface active agents used as auxiliaries in road, mining, textile and fertilizer industries, in metallurgy, as lubricating agents, additives for fuel oils and soil stabilization. (British Ceca Co Ltd).*

Noramac. Amines and derivatives. Primary, secondary, tertiary fatty mono-, di- and polyamines from C8 to C22. Quaternary ammonium salts. Amine oxides. Amine salts. Ethoxylated amines and polyamines. Amino-acids. Special nonionic derivatives. Additives. Applications: Surface active agents used as auxiliaries in road, mining, textile and fertilizer industries, in metallurgy, as lubricating agents, additives for fuel oils and soil stabilization. (British Ceca Co Ltd).*

Norament. Special designed floors. (Carl Freudenberg).*

Noramid. Flooring material. (Carl Freudenberg).*

Noramium. Amines and derivatives. Primary, secondary, tertiary fatty mono-, di- and polyamines from C8 to C22. Quaternary ammonium salts. Amine oxides. Amine salts. Ethoxylated amines and polyamines. Amino-acids. Special nonionic derivatives. Additives. Applications: Surface active agents used as auxiliaries in road, mining, textile and fertilizer industries, in metallurgy, as lubricating agents, additives for fuel oils and soil stabilization. (British Ceca Co Ltd).*

oramox. Amines and derivatives. Primary, secondary, tertiary fatty mono-, di- and polyamines from C8 to C22. Quaternary ammonium salts. Amine oxides. Amine salts. Ethoxylated amines and polyamines. Amino-acids. Special nonionic derivatives. Additives. Applications: Surface active agents used as auxiliaries in road, mining, textile and fertilizer industries, in metallurgy, as lubricating agents, additives for fuel oils and soil stabilization. (British Ceca Co Ltd).*

oraplan. Flooring material. (Carl Freudenberg).*

oratex. A proprietary preparation of talc, kaolin, zinc oxide, cod liver oil and wool fat used as a protective skin cream. (Norton, H N Co).†

orbide. A proprietary trade name for an amorphous boron carbide. An abrasive.†

orbo. A proprietary trade name for phenolic resin.†

orcuron. Vecuronium Bromide. Applications: Blocking agent (neuromuscular). (Organon Inc).

ordel. A proprietary ethylene-propylene synthetic rubber. (Du Pont (UK) Ltd).†

ordel 2744. A trade mark for a fast curing EPDM hydrocarbon rubber possessing high green strength. (Du Pont (UK) Ltd).†

ordhausen Acid. See FUMING SULPHURIC ACID.

ordox. Cuprous oxide, paint grade, red, micro milled. Applications: Active ingredient in antifouling paints. (Nordox Industrier AS).*

oreplast. A proprietary trade name for laminated thermosetting, plastic thermosetting and thermoplastic synthetic resins.†

orepol. A proprietary trade name for vulcanizable vegetable polymers.†

oreseal. A proprietary trade name for a cork substitute made from low cost domestic raw materials. It is stated to be equal in strength to cork.†

orfer. A proprietary preparation of ferrous fumarate used in the treatment of anaemia. (Norton, H N Co).†

Norflex. Orphenadrine citrate. Applications: Relaxant; antihistaminic. (Riker Laboratories Inc, Subsidiary of 3M Company).

Norflex Injection. Each ampoule contains orphenadrine citrate BP 30 mg/ml. (Riker Laboratories/3M Health Care).†

Norflex Tablets. Each tablet contains orphenadrine citrate BP 100 mg in a slow release base. (Riker Laboratories/3M Health Care).†

Norfroth. Mixed propylene glycol iso-butyl ethers and surfactants. Applications: Ore flotation chemicals. (Chemoxy International Ltd).*

Norgesic. Each tablet contains orphenadrine citrate BP 35 mg and paracetamol BP 450 mg (Riker Laboratories/3M Health Care).†

Norge, Saltpetre. See AIR SALTPETRE.

Norgine. The sodium-ammonium salt of laminaric acid from seaweed, *Laminaria digitata* and *Saccharinus digitatus*. It is used in the treatment of textiles.†

Norgotin. A proprietary preparation of ephedrine, amethocaine, chlorohexidine acetate and propylene glycol used as eardrops. (Norgine).†

Norinyl-1, Norinyl-2. A proprietary preparation of norethisterone and mestranol. Oral contraceptive. (Syntex Laboratories Inc).†

Norisodrine Aerotrol. Isoproterenol hydrochloride. Applications: Adrenergic. (Abbott Laboratories).

Norisodrine Sulfate. Isoproterenol sulphate. Applications: Adrenergic. (Abbott Laboratories).

Norit. Activated carbon. Applications: Purification and decolourization of among others food products, pharmaceutical products, water (potable, process and waste), recovery of gold (CIP and CIL) and air purification (removal of H_2S, mercury, SO_2 etc). (Norit).*

Norit C. Activated carbon. Applications: Purification and decolourization of among others food products, pharmaceutical products, water (potable, process and waste), recovery of gold (CIP and CIL) and air purification (removal of H_2S, mercury, SO_2 etc). (Norit).*

Norit PK. Activated carbon. Applications: Purification and decolourization of among others food products, pharmaceutical products, water (potable, process and waste), recovery of gold (CIP and CIL) and air purification (removal of H_2S, mercury, SO_2 etc). (Norit).*

Norit R. Activated carbon. Applications: Purification and decolourization of among others food products, pharmaceutical products, water (potable, process and waste), recovery of gold (CIP and CIL) and air purification (removal of H_2S, mercury, SO_2 etc). (Norit).*

Norit RO. Activated carbon. Applications: Purification and decolourization of among others food products, pharmaceutical products, water (potable, process and waste), recovery of gold (CIP and CIL) and air purification (removal of H_2S, mercury, SO_2 etc). (Norit).*

Norithene. Air conditioning carbons. (Norit UK Ltd).

Norlestrin. A proprietary preparation of norethisterone acetate and ethinyl-oestradiol. Treatment of menstrual disorders. (Parke-Davis).†

Norlestrin 21. A proprietary preparation of norethisterone acetate and ethinyl oestradiol. Oral contraceptive. (Parke-Davis).†

Norleusactide. See PENTACOSACTRIDE.

Norlig. Calcium and sodium lignosulphonates. Applications: Product line of specialty chemicals for use in a variety of binding, dispersing, complexing, stabilizing or copolymerizing applications. (Reed Lignin Inc).*

Norlutate. Norethindrone acetate. Applications: Progestin. (Parke-Davis, Div of Warner-Lambert Co).

Norlutin. Norethindrone. Applications: Progestin. (Parke-Davis, Div of Warner-Lambert Co).

Norlutin "A". A proprietary preparation NORETHISTERONE acetate used i the control of menstrual irregularity. (Parke-Davis).†

Normacol. A proprietary preparation containing sterculia. A laxative. (Norgine).†

Normal Powder. A gelatinized gun-cotton powder.†

Normal Salt Solution. See PHYSIOLOGICAL SALT SOLUTION.

Normax. A proprietary preparation containing dioctyl sodium sulphosuccinate and 1-8-dihydroxyanthraquinone. A laxative. (Horlicks).†

Normodyne. Labetalol hydrochloride. Applications: Anti- adrenergic; (Schering-Plough Corp).

Norodur. ABS graft polymers and copolymers. Applications: For the manufacture of injection mouldings fo use in the household and consumer goods section, vehicle construction, the audio and furniture industries, the offi supplies and stationery sector, the photographic and film industries, electrical engineering and the toy and textile industries. (Bayer & Co).*

No-Roma. A proprietary preparation of paraformaldehyde, sodium carboxymethyl cellulose, sodium hydroxide an methylene blue. A medical deodorant. (Salt).†

Noroxin. Norfloxacin. Applications: Urinary antiseptic. (Merck Sharp & Dohme).*

Norpace. Disopyramide phosphate. Applications: Cardia depressant. (G D Searle & Co).

Norpar. High purity normal paraffin solvent. (Exxon Chemical Internationa Inc).*

Norplex laminates. Woven fibreglass fabr impregnated with thermoset resin, wit thin copper foil on one or two sides, pressed under high temperature and heat to form plastic laminated sheets. Applications: Substrate material for printed circuit boards. (Norplex).*

Norpramin. Desipramine hydrochloride. Applications: Antidepressant. (Merrell

Dow Pharmaceuticals Inc, Subsidiary of Dow Chemical Co).

orpramine. A proprietary preparation of IMIPRAMINE hydrochloride. An anti-depressant. (Norton, H N Co).†

or-Q D. Norethindrone. Applications: Progestin. (Syntex Laboratories Inc).

orsed. A proprietary preparation of cyclobarbitone and amylobarbitone. A hypnotic. (Norton, H N Co).†

orsolene. A registered trade mark for coumarone and polyindene resins. (CdF Chimie, France).†

orsomix. A registered trade mark for polyester compounds.†

orthovan. Sodium-*o*-vanadate.†

ortron. Selective herbicide. (FBC Ltd).

orust. Amines and derivatives. Primary, secondary, tertiary fatty mono-, di- and polyamines from C8 to C22. Quaternary ammonium salts. Amine oxides. Amine salts. Ethoxylated amines and polyamines. Amino-acids. Special nonionic derivatives. Additives. Applications: Surface active agents used as auxiliaries in road, mining, textile and fertilizer industries, in metallurgy, as lubricating agents, additives for fuel oils and soil stabilization. (British Ceca Co Ltd).*

orval. A proprietary preparation o-dioctyl-sodium sulphosuccinate in gelatine. A faecal softener. (Horlicks).†

o Vein Compound. A proprietory blend of red iron oxides. Applications: For use in the foundry industry to eliminate veining and other casting expansion defects. (DCS Color & Supply Co Inc).*

orvinyl. Homo and copolymeric poly vinyl chloride (S-PVC). Applications: As a main product or as an extender for production of plastic articles. (Norsk Hydro AS).*

orwegian Saltpetre. See AIR SALTPETRE.

oryl. Modified polyphenylene oxide. Applications: Plastic components for automotive, electrical, electronics, lighting, medical, packaging, audio etc. (GE Plastics Ltd).*

Noryl SE-100. A registered trade mark for self extinguishing ABS resins. (General Electric).†

Noryl SE-1000. Self-extinguishing acrylonitrile butadiene styrene plastic. (General Electric).†

Nosiheptide. A veterinary antibiotic. It is a peptide obtained from cultures of *Streptomyces actuosus 40037* or the same substance obtained by any other means.†

No-Swab. Mould coating containing solid lubricants and suitable binder system. Applications: Moulds for glass containers in I.S. machines for extension of mould swabbing cycles. (E/M Corporation).*

Notak. Products for non-destructive detection of surface flaws in iron and steel components. (Foseco (FS) Ltd).*

Nottingham White. See FLAKE WHITE.

Noumgou Oil. See NJAVE BUTTER.

Novacetoform. Aluminium acetate.†

Novaculite. A quartz rock used as an abrasive.†

Novadelox. Flour bleaching agent. (Akzo Chemie UK Ltd).

Novadine. Iodophor sanitizer for brewery plant. Applications: Sanitizing fermenting vessels, lager tanks, maturation tanks, bright beer tanks, fillers and mains. (Harshaw Chemicals Ltd).*

Novafed. Used primarily in the treatment of coughs, colds and upper respiratory conditions. (Dow Chemical Co Ltd).*

Novafil. Polybutester. Applications: Surgical aid. (Davis & Geck).

Novahistine. Used primarily in the treatment of coughs, colds and upper respiratory conditions. (Dow Chemical Co Ltd).*

Novain. Carnitine.†

Novalak Resins. Synthetic resins of the formaldehyde-phenol type.†

Novaldin. Dipyrone. Applications: Analgesic; antipyretic. (Sterling Drug Inc).

Novaloy 6521. A proprietary one component epoxy resin supplied in powder form for the coating of electronic components. (Rogers Corp, Rogers, Mass).†

Novantrone. Injection containing mitozantrone. Applications: Antineoplastic agent. (Lederle Laboratories).*

Novanyl. A range of acid dyes. Applications: Dyeing of polyamide fibres. (Yorkshire Chemicals Plc).*

Novapel. Leather finishes. (ICI PLC).*

Novaplaste. Leather dyes. (ICI PLC).*

Novathion. Dimethyl-KbKOKb±K-(3-methyl-4-nitrophenyl) phosphorothioate. Applications: It is an all round low toxic insecticide for forest protection, agriculture and public health. It is used especially where long term effect is desired. (A/S Cheminova).*

Novazole. Cambendazole. Applications: Anthelmintic. (Merck Sharp & Dohme, Div of Merck & Co Inc).

Noveloid. Proprietary products of cellulose esters and ethers.†

Novemol. A trade name for a wetting agent containing sulphonated terpene alcohols.†

Novester. Disperse dyes. (ICI PLC).*

Novex. Benzalbisdimethyldithio-carbamate. It has a specific gravity of 1.365, is a white crystalline powder, and melts at 175° C. It is used as a rubber vulcanization accelerator.†

Novidium. Homidium chloride. (May & Baker Ltd).

Novitane. A proprietary polyurethane elastomer. (BF Goodrich Chemical Co).†

Novite. A proprietary trade name for an alloy of iron with about 1.5 per cent nickel and 0.5 per cent chromium.†

Novobiocin. An antibiotic produced by Streptomyces niveus and Streptomyces spheroides. Streptonivicin. Albamycin is the calcium salt and Biotexin and Cathomycin are the sodium salts.†

Novocain. Procaine Hydrochloride. Applications: Anaesthetic. (Sterling Drug Inc).

Novocodine. Hexamethylenetetramine-di-iodide.†

Novodur. ABS thermoplastic. (Bayer UK Ltd).

Novofix. Photographic fixer. (May & Baker Ltd).

Novol. Cosmetic quality oleyl alcohol. (Croda Chemicals Ltd).

Novolac. A name applied by Baekeland to fusible and soluble phenol-formaldehyde resins.†

Novolen. A range of polypropylenes and polypropylene copolymers supplied in granular form. Applications: Injection moulding of technical parts and packaging, extrusion into special soft film, extrusion into fibres and tapes, blow moulded bottles. (BASF United Kingdom Ltd).*

Novoline. Photographic developer. (May & Baker Ltd).

Novolith. A photographic developer. (May & Baker Ltd).*

Novomatic. Photographic developer. (May & Baker Ltd).

Novonasco. A proprietary trade name for a wetting agent containing modified sodium alkylnaphthalenesulphonate.†

Novoplas. A proprietary trade name for an organic polysulphide synthetic elastic material.†

Novoprotin. A proprietary preparation of crystalline vegetable albumen.†

Novor. Crosslinker for rubber. (Durham Chemicals Ltd).*

Novotak. Photographic developer. (May & Baker Ltd).

Novotriad. Triple sulphonamides. (May & Baker Ltd).

Novozone. A German registered name for a magnesium peroxide, MgO_2, prepared for medical purposes. An antiseptic used internally, and externally as an ointment, for wounds and gatherings. Also see BIOGEN.†

Novozym. A proteolytic enzyme prepared by submerged fermentation of a selected strain of Bacillus licheniformis. Applications: Used in the manufacture of effective denture cleaners. (Novo Industri A/S).*

Novulatin. Ethynerone. Applications: Anti-fertility agent. (Merck Sharp & Dohme).*

Novulatum. Ethynerone. Applications: Anti-fertility agent. (Merck Sharp & Dohme).*

Noxyflex. A proprietary preparation of NOXYTHIOLIN used to irrigate the bladder. (Geistlich Sons).†

Noxythiolin. N-Hydroxymethyl-N'-methylthiourea.†

Nozinan. A proprietary preparation of methotrimeprazine. Applications: Sedative. (May & Baker Ltd).*

Nozolex. Free flowing refractory fillers used in sliding gate nozzles. (Foseco (FS) Ltd).*

NPH Iletin. Insulin, Isophane. Applications: Antidiabetic. (Eli Lilly & Co).

N.P.L. Alloy. An alloy of 94.5 per cent aluminium, 2.5 per cent nickel, and 1.5 per cent magnesium.†

NPP-1. *O*, *O*-di-n-propyl phosphorodithioic acid. Applications: Used as a high grade intermediate for organophosphorus insecticides and other products. (A/S Cheminova).*

NSAE Powder. Sodium alkyl-naphthalene sulphonate. Anionic surfactant in powder form. Applications: Wetting and dispersing agent with high electrolyte tolerance, eg agricultural dispersing agent for sulphur. (Millmaster-Onyx UK).

N.S.B. A proprietary preparation of noxytiolin, and vinyl pyrrolidone/vinyl acetate copolymer in an aerosol. Wound dressing. (Geistlich Sons).†

N-Serve. A line of nitrogen stabilizers based primarily on nitropyrin. (Dow Chemical Co Ltd).*

N.S. Fluid. A mixture of sodium chloride, aluminium chloride, and iron chloride.†

NTA. Sodium nitrilotriacetate powder and 40% solution. Applications: Domestic and industrial laundry detergents, hard surface cleaning formulations, boiler treatment, textile auxiliary. (Monsanto Co).*

Nuact. Loss-of-dry inhibitors. Applications: Prevention of loss of drying of non-aqueous coating compositions. (Nueodex Inc).*

Nuade. Slip and mar agents. Applications: Slip and mar aid for water-borne baking finishes. (Nueodex Inc).*

Nuba. A proprietary trade name for a thermoplastic coal-tar pitch and a cumarone-indene resin for paints.†

Nubain. Nalbuphine hydrochloride. Applications: Analgesic; antagonist. (Endo Laboratories Inc, Subsidiary of E I du Pont de Nemours & Co).

Nubex. Building remedial treatments. (Tenneco Organics Ltd).

Nubrite. A bright nickel plating process. (Hanshaw Chemicals).†

Nubun. A proprietary trade name for a synthetic rubber latex insulation for power and communication cables. It is made from a special modification of buna S synthetic rubber.†

Nuchar. A decolourizing agent of American origin.†

Nucleant. An aluminium alloy grain refiner. (Foseco (FS) Ltd).*

Nucol. Wide range of high build, conventional and modified chlorinated rubber and vinyl coatings. (Sigma Coatings).*

Nucoline. See VEGETABLE BUTTER.

Nucrel. Ethylene-methacrylic acid copolymer. (Du Pont (UK) Ltd).

Nuelin Liquid. Each 5 ml contains theophylline sodium glycinate 120 mg equivalent to theophylline hydrate BP 60 mg (Riker Laboratories/3M Health Care).†

Nuelin SA-250. Each tablet contains 250 mg anhydrous theophylline in a slow release formulation. (Riker Laboratories/3M Health Care).†

Nuelin Tablets. Each tablet contains theophylline BP 125 mg in microcrystalline form. (Riker Laboratories/3M Health Care).†

Nuelin-SA. Each tablet contains 175 mg anhydrous theophylline in a slow release formulation. (Riker Laboratories/3M Health Care).†

Nufol. Nitrogenous fertilizers. (ICI PLC).*

Nuglas. Tubes moulded with glass reinforcement. Applications: Electrical, structural and thermal applications, jigs, fixtures and tooling, lightweight support structures for antennae, lighting and surveillance. (Fothergill Tygaflor Ltd).*

Nujol. Mineral oil. Applications: Laxative; pharmaceutic aid. (Plough Inc).

Nulacin. A proprietary preparation containing milk, dextin, maltose base with magnesium trisilicate, calcium carbonate, magnesium oxide and oil of peppermint. An antacid. (Horlicks).†

Nulogyl. A proprietary preparation of NIMORAZOLE used in the treatment of trichomonas infections. (Bristol-Myers Co Ltd).†

Nuloid. See BAKELITE.

Nulomoline. A solution of partly inverted sugar. Used for some purposes as a substitute for glycerine.†

Numoquin. Ethyl-hydrocupreine.†

Numorphan. Oxymorphone hydrochloride. Applications: Analgesic. (Endo Laboratories Inc, Subsidiary of E I du Pont de Nemours & Co).

Numorphan Oral. A proprietary trade name for Hydromorphinol. Tablets of hydromorphinol-Analgesic. (British Drug Houses).†

Numotac. Each tablet contains isoetharine hydrochloride 10 mg in a porous plastic matrix. (Riker Laboratories/3M Health Care).†

Nuocure. Metal salts of organic acids. Applications: Driers for water-borne coating compositions, catalyst for polyurethane foam. (Nueodex Inc).*

Nuocure 28. Catalyst for polyurethane foams. (Durham Chemicals Ltd).

Nuodex. Metal salts of organic acids, fungicides and bactericides, bodying agents, dispersing agents. Applications: Driers for coating compositions, curing catalysts, biocides for coating compositions, industrial biocides, thickeners, pigment dispersants. (Nueodex Inc).*

Nuodex NA. A pigment dispersant. Applications: Dispersing pigments for use in vinyl and acrylic lacquer formulations. (Nueodex Inc).*

Nuodex 100. Compositions containing the dodecyldimethyl-benzylammonium salt of naphthenic acid. Applications: Antimicrobial preservatives for textiles, shoe linings and plasticized PVC compositions. (Nueodex Inc).*

Nuodex 321 Extra. Mercurial biocide. (Durham Chemicals Ltd).

Nuodex 84. 50% aqueous solution of the sodium salt of 2-mercaptobenzothiazole. Applications: Antimicrobial preservative for fibrous substrates, adhesives and aqueous emulsions. (Nueodex Inc).*

Nuodex 87. Biocidal wall washing compound. (Durham Chemicals Ltd).

Nuolate. Metal salts of tall oil fatty acids. Applications: Driers for non-aqueous coating compositions. (Nueodex Inc).*

Nuophene. Dihydroxy-dichloro-diphenylmethane. Applications: Fungicide for use on textiles, cordage and hair felt. (Nueodex Inc).*

Nuoplaz. Plasticizers. Applications: plasticizing PVC and other polymers. (Nueodex Inc).*

Nuosept. Antimicrobial preservatives. Applications: Preservation of aqueous coating compositions, pigment dispersions, inks, adhesives, caulks, metalworking fluids, paper coatings, drilling muds. (Nueodex Inc).*

Nuosperse. Dispersing agents. Applications: Pigment dispersants for coating compositions. (Nueodex Inc).*

Nuostabe. Stabilizers. Applications: Heat and light stabilization of PVC compositions. (Nueodex Inc).*

Nuostabe 1317. A proprietary trade name for a Ba/Cd/Zn stabilizer for PVC. (Durham Raw Materials).†

Nuostabe 1374. A proprietary trade name for a non sulphide staining stabilizer for PVC used in blown foams. (Durham Raw Materials).†

Nuostabe 1605. A stabilizer for PVC based on a barium, cadmium zinc complex. (Durham Raw Materials).†

Nuosyn. Metal salts of organic acids. Applications: Driers for non-aqueous coating compositions. (Nueodex Inc).*

Nupercainal. A proprietary preparation of cinchocaine hydrochloride. An anaesthetic ointment. (Ciba Geigy PLC).†

Nupercaine. A proprietary preparation of cinchocaine hydrochloride. Surface anaesthetic. (Ciba Geigy PLC).†

Nupercaine Hydrochloride. Dibucaine hydrochloride. Applications: Anaesthetic. (Ciba-Geigy Corp).

Nuprin. Ibuprofen. Applications: Anti-inflammatory. (The Upjohn Co).

Nuprin. A proprietary trade name for Sulphamoxole.†

Nupyrin. See ASPIRIN.

Nurac. A proprietary rubber vulcanization accelerator. It is stated to be diphenyl-guanidine. It has also been said to be thiocarbanilide.†

Nuram. Nitrogenous fertilizers. (ICI PLC).*

Nuremberg Gold. An alloy of 90 per cent copper, 7.5 per cent aluminium, and 2.5 per cent gold. A jeweller's alloy.†

Nuremberg Green. (Victoria Green, Permanent Green). Pigments. They are mixtures of chromium green with zinc yellow.†

Nuremberg Red. See INDIAN RED.

Nuremberg Violet. See MANGANESE VIOLET.

Nurofen. Ibuprofen. (The Boots Company PLC).

Nurolon. Suture, nonabsorbable surgical. Applications: Surgical aid. (Ethicon Inc).

Nusat. A satin finish nickel plating process. (Hanshaw Chemicals).†

Nu-Seals Aspirin. A proprietary preparation of aspirin in an enteric coated form. An analgesic. (Eli Lilly & Co).†

Nu-Seals Potassium Chloride. A proprietary preparation of potassium chloride in enteric coated tablets. (Eli Lilly & Co).†

Nu-Seals Sodium Salicylate. A proprietary preparation of sodium salicylate used as an analgesic and antipyretic. (Eli Lilly & Co).†

Nu-Set. Bristle-setting cement. (Hardman Inc).*

Nut, Abeku. See NJAVE BUTTER.

Nut, Chop. See ORDEAL BEAN.

Nut, Gooroo. See KOLA NUT.

Nut, Hurr. Myrobalans.†

Nut Oil. A term used in China for Tung oil (Chinese wood oil). The same name is also used for Walnut and Arachis oils.†

Nutmeg Butter. Expressed oil of nutmeg.†

Nutracort. Hydrocortisone. Applications: Glucocorticoid. (Owen Laboratories Div, Dermatological Products of Texas, Alcon Laboratories Inc).

Nutralys. Animal feedstuff. (Roquette (UK) Ltd).*

Nutramigen. A proprietary artificial lactose-free infant milk food. (Bristol-Myers Co Ltd).†

Nutramin. Flour improver - vitamin enrichment additive. (Akzo Chemie UK Ltd).

Nutramon. Aqueous ammonia. (ICI PLC).*

Nutranel. Protein, fat, carbohydrate, vitamins, minerals and trace elements Applications: Liquid feed. (Roussel Laboratories Ltd).*

Nutraphos. Nutrients for effluent treatment. (Scottish Agricultural Industries PLC).

NutraSweet. Aspartame. Applications: Sweetener. (G D Searle & Co).

Nutregen. A proprietary gluten-free wheat starch used in the dietary treatment of coeliac disease. (Energen).†

Nutrifos. Food grade tetrasodium pyrophosphate. Applications: Used in meat curing. (Monsanto Co).*

Nuts, Bako. See NJAVE BUTTER.

Nuts, Mahogany. See NJAVE BUTTER and ACAJOU BALSAM.

Nut, Split. See ORDEAL BEAN.

Nuvacon. A proprietary preparation of megestrol acetate and ethinyl oestradiol. Oral contraceptive. (British Drug Houses).†

Nuvamide. Triple sulphonamide with neomycin. (May & Baker Ltd).

Nuvan. Organophosphorus insecticide. (Ciba-Geigy PLC).

Nuvan Fly Spray. Organophosphorus/pyrethrin insecticide. (Ciba-Geigy PLC).

Nuvan Top Aerosol. Organophosphorus insecticde. (Ciba-Geigy PLC).

Nuvanol. Organophosphorus insecticide. (Ciba-Geigy PLC).

Nuvanol N. Organophosphorus insecticide. (Ciba-Geigy PLC).

Nuvis. Thixotropic bodying agents. Applications: Control of sag and flow of non-aqueous coating compositions. (Nueodex Inc).*

Nuxtra. Metal salts of synthetic organic acids. Applications: Driers for non-aqueous coating compositions. (Nueodex Inc).*

N.W. Acid. See NEVILLE'S AND WINTHER'S ACID.

Nyacol. Colloidal dispersions of silica (SiO_2), antimony pentoxide, alumina and other metal oxides. Applications: Binders for investment casting moulds and fibrous refractories, polishing agent for semiconductor wafers, rigid disks, etc., anti-slip coating for paper, textile fibres, etc., flame retardent for plastics (antimony oxide dispersion). (The PQ Corporation).*

Nyad Wollastonite. High aspect ratio and fine particle size calcium metasilicate minerals: CaO, SiO_2, Fe_2O_3, Al_2O_3, MnO, MgO, TiO_2. Applications: Plastics, coatings, refractories, fire resistant board, adhesives, rubber and elastomers and polymer concrete. (NYCO).*

Nycoat. Family of chemically coupled minerals; e.g. alumina trihydrate, barytes and celestite (varies depending on substrate). Applications: Speciality applications - plastics, high performance coatings, adhesives, rubber and elastomers. (NYCO).*

Nycor Barytes. 200 + 325 mesh barium sulphate: barium sulphate, silica, ferric oxide, manganese and lead. Applications: Coatings, rubber and elastomers, friction products, refractories and sound deadening compounds. (NYCO).*

Nycor Celestite. 200 + 325 mesh strontium sulphate: Strontium sulphate, barium sulphate, calcium carbonate and alumina. Applications: Coatings, rubber and elastomers, friction products, refractories and sound deadening compounds. (NYCO).*

Nyctal. See ADALIN.

Nydrane. A proprietary preparation of beclamide. An anticonvulsant. (Rona Laboratories).†

Nydrazin. Isoniazid. Applications: Antibacterial. (E R Squibb & Sons Inc

Nydur. Nylon 6. These impact modified nylon-6 grades offer low temperature impact, as well as abrasion, heat and chemical resistance. (Mobay Chemical Corp).*

Nyebar. A solution of a low surface energy fluorocarbon polymer in a fluorinated solvent. Applications: Provides a non-wettable surface which controls or prevents the migration or creep of lubricants or other fluids. (William F Nye Inc).*

Nyflake Muscovite Mica. Muscovite mica; coarse and fine grinds: SiO_2, Al_2O_3, K_2O, Fe_2O_3, Na_2O_3, CaO, TiO_2, MnO_2, P, S. Applications: Plastics, coatings, oil well drilling mud, adhesives. (NYCO).*

Nyglas. Chemically coupled ground glass; soda lime glass, platey, various particle sizes: SiO_2, Na_2O, CaO, MgO, Al_2O_3, K_2O, Fe_2O_3. Applications: Plastics and coatings. (NYCO).*

Nykon. Corrosion resistant rheological additive. Applications: Grease. (NL Industries Inc).*

Nylander's Reagent. An alkaline solution of bismuth subnitrate and Rochelle salt obtained by dissolving 40 grams Rochelle salt and 20 grams bismuth subnitrate in 1,000 cc of 8 per cent caustic soda. It is used for the detection of glucose in urine by boiling 5 parts of the glucose solution with 1 part of the reagent when reduction occurs and a black precipitate is produced.†

Nylatron. Nylon moulding compounds. (Polymer Corporation).†

Nylax. Red sugar coated tablets each containing thiamine hydrochloride BP 3 mg, phenolphthalein 60 mg, cascara dru extract BP 30 mg, aloin BPC 2 mg, powdered senna leaf 15 mg and bisacodyl BP 2 mg Applications: Nylax tablets are indicated for the short term therapy of functional constipation. (Evans Medical).*

Nylocrom. Dry powder acid dyes. Applications: Specially developed acid dyes, used in the textile industry for dyeing nylon. (Multicrom SA).*

Nylofanol. See ATOPHAN.

Nylofixan. Aromatic sulphonate, black-tanning agent. Applications: Polyamides. (Sandoz Products Ltd).

Nylomine. Dyestuffs for nylon and polyamide fibres. (ICI PLC).*

Nylon. A generic term for polyamide products prepared from adipic and related acids and hexamethylene and related diamines by condensation. It has a protein-like chemical structure. It is fabricated into bristles, fibres, and sheets.†

Nyloset Finish. Amide resin. Applications: Nylon builder and softener. (Scher Chemicals Inc).*

Nylox. Adhesive composition for nylon, etc. (Hardman Inc).*

Nyogel. An inorganically gelled series of greases based upon synthetic lubricating oils. Applications: Specialty lubrication of delicate machinery and engineereed components. (William F Nye Inc).*

Nypene. A proprietary trade name for a polyterpene hydrocarbon resin used in adhesives, paints, and varnishes.†

Nyrim. RIM nylon (Reaction Injection Moulding), prepared by in-mould polymerization of caprolactam (feedstock for nylon) to a nylon-6 block polymer with utilization of a catalyst 2.0 agents. Applications: Industry in general, in agriculture and automotive (body parts, for instance). (DSM NV).*

Nyspheres Hollow Glass Spheres. Fine particle hollow glass spheres, lightweight: Silica, alumina, iron oxides, calcium, magnesium and alkalis. Applications: Plastics, coatings, adhesives, cement products, refractories. (NYCO).*

Nystadermal. A proprietary preparation of nystatin and triamcinolone acetonide used in the treatment of fungal and eczematous skin disorders. (Squibb, E R & Sons Ltd).*

Nystan. A proprietary preparation containing nystatin. (Squibb, E R & Sons Ltd).*

Nystan Cream. Nystatin in cream base. (E R Squibb & Sons Ltd).

Nystatin. An antibiotic produced by *Streptomyces noursei*. MYCOSTATIN NITACIN, NYSTAM. Also present in LEDERSTATIN, MYSTECLIN, NYSTAN TA and SILTETRIN.†

Nystavescent. A proprietary preparation of nystatin in an effervescent pessary used in the treatment of vaginal candidiasis. (Squibb, E R & Sons Ltd).*

Nystavescent pessaries. Nystatin in effervescent vaginal pessary. (E R Squibb & Sons Ltd).

Nystex. Nystatin. Applications: Indicated in the treatment of cutaneous or mucocutaneous mycotic infections caused by candida albicans and other candida species. (Altana Inc).*

Nytal. Hydrous magnesium calcium silicate - talc. Applications: Used as fillers and extenders for paint, ceramics, plastics, rubber etc. (Vanderbilt Chemical Corporation).*

O

Oak Moss Resin. A resin obtained from lichens. Used as a fixative in perfumery.†

Oak Oils. Petroleum oils. Applications: Drawing, stamping, grinding and machining. (Oak International Chemical Inc).*

Oak Red. A colouring matter, phlobaphene, $C_{28}H_{22}O_{11}$, obtained by the hydrolysis of quercitannic acid.†

Oak Syncrolube. Synthetic (oil free) lubricant. Applications: Stamping and machining. (Oak International Chemical Inc).*

Obermayer's Reagent. A solution of ferric chloride in concentrated hydrochloric acid (4 gm. ferric chloride dissolved in 1 litre of concentrated hydrochloric acid). It is used for the detection of indoxyl in urine, indigo being formed if this substance is present.†

Obermine Black & Yellow. Phentermine Hydrochloride. Applications: Appetite suppressant. (O'Neal, Jones & Feldman Pharmaceuticals).

Obinese. A proprietary preparation of chlorpropamide, metformin hydrochloride. An oral hypoglycemic agent. (Pfizer International).*

Obsidene. A plastic residue from the distillation of petroleum.†

Obsidian. See PUMICE STONE.

Obturin. Soluble fluoresceine.†

Occidine. A fungicide. It contains copper sulphate, iron sulphate, sulphur, naphthalene, and calcium carbonate.†

Occlusin. Light cured dental filling composite. (ICI PLC).*

Occultest. A proprietary test tablet of *o*-toluidine, strontium peroxide, calcium acetate, tartaric acid and sodium bicarbonate, used for the detection of blood in urine. (Ames).†

Ocenol. The mixture of fatty alcohols derived from sperm oil. Also a proprietary trade name for technical oleic acid.†

Ochran. A yellow bole (or clay-earth).†

Ochre. (Yellow Ochre, Oxide Yellow, Chinese Yellow). A natural pigment consisting of hydrated oxides of iron and manganese, mixed with clay and sand. The term is frequently restricted to a pale, yellowish-brown variety.†

Ochre, Artificial. See MARS YELLOW and SIDERINE YELLOW.

Ochre, Brown. See YELLOW OCHRE.

Ochre, Burnt. See INDIAN RED.

Ochre, Cobalt. A mineral. It is cobalt bloom.†

Ochre, Golden. See YELLOW OCHRE.

Ochrematite. A mineral, $3(3Pb(AsO_4)_2.PbCl_2)4Pb_2MoO_5$.†

Ochre, Orange. See BURNT YELLOW OCHRE.

Ochre, Purple. See MARS VIOLET.

Ochre, Red. See INDIAN RED.

Ochre, Roman. See YELLOW OCHRE.

Ochre, Spruce. See YELLOW OCHRE.

Ochre, Transparent Gold. See YELLOW OCHRE.

Ocota Cocota Gum. (Muccocota Gum). Names applied in West Africa to varieties of copal.†

Ocre de Ru. Brown ochre.†

Octacosactrin. A corticotrophic peptide.†

Octaflex. A proprietary preparation of octaphen resin-acrylate and methacrylate polymers. Plastic wound dressing. (Ward, Blenkinsop & Co Ltd).†

Octamine. Octylated diphenylamine. A solid amine-type antioxidant, gives minimum discolouration with maximum protection. Effective in natural, SBR, BR, neoprene and nitrile rubbers. Used in tyre carcass, footwear, moulded heels and soles, proofing, sponge, automotive rubber, wire insulation and tiling. (Uniroyal).*

Octoate 2. Zinc di-2-ethylhexoate - rubber activator. Applications: Used in soluble cure systems in place of stearic acid and partial replacement of zinc oxide for natural and synthetic rubbers. (Vanderbilt Chemical Corporation).*

Octoil. A proprietary trade name for a plasticizer. Dioctyl phthalate.†

Octoil S. A proprietary trade name for a plasticizer. Dioctyl sebacate.†

Octonativ. Substance for injection. Applications: Factor VIII concentrate for the treatment of haemophilia. (KabiVitrum AB).*

Octopirox. Piroctone Olamine. Applications: Antiseborrheic. (Hoechst AG).

Octoran. Low foaming nonionic surfactant of the alkylphenol ethoxylate type in liquid form. Applications: Wetting, detergency, emulsification and antistatic agent. (Roehm Ltd, Hexoran Div).

Octorez. Chemical modified rosins. (Tenneco Malros Ltd).

Octosol. Modified rosin soap. (Tenneco Malros Ltd).

Octovit Tablets. Vitamin A, thiamine mononitrate, riboflavine, nicotinamide, pyridoxine, cyanocobalamin, ascorbic acid, cholecalciferol, tocopherol, calcium, ferrous sulphate, magnesium and zinc. Applications: A multivitamin/mineral product indicated where supplementation with vitamins and minerals may be of benefit. (Smith Kline and French Laboratories Ltd).*

Ocuba Wax. Ocuba fat, from *Myristica ocuba*.†

Ocusert. Pilocarpine. Applications: Antiglaucoma agent; cholinergic. (Ciba-Geigy Corp).

Ocusol. A proprietary preparation of sodium sulphacetamide, zinc sulphate and cetrimide used as eye-drops. (Boots Pure Drug Co).*

Odylen. Veterinary preparation. Applications: For external use against ectoparasites. (Bayer & Co).*

OEnanthin. See FAST RED D.

OEnolin. The red colouring matter of wine, precipitated by lime or basic lead acetate.†

OEstradin. A proprietary preparation of ethinyloestradiol, phenobarbitone, sodium bromide and glyceryl trinitrate. (Norton, H N Co).†

Oestroform. Injection, oestradiol benzoate. Tablets, oestradiol- oestrogenic hormone. (British Drug Houses).†

OFA. Fuel additive. (Chevron).*

OFHC Copper. A trade mark for oxygen-free, high conductivity copper.†

OFHC Copper. Oxygen-free, high conductivity copper. The composition per cent is: Nickel 0.0006, bismuth 0.001, cadmium 0.0001, lead 0.001, mercury 0.0001, oxygen 0.001 (max), phosphorus 0.003, selinium 0.001, sulphur 0.0018, tellurium 0.001, zinc 0.0001, iron 0.0005. 0.00 to 70 total max (tin, antimony, arsenic, bismuth, manganese, selenium, tellurium). Conductivity 101 per cent. (Amax Inc).*

Oftentral. A proprietary preparation containing miconazole nitrate, polymixin B sulphate. Applications: Eye infections in cats and dogs. (Janssen Pharmaceutical Ltd).*

OGA. Gasoline additive. (Chevron).*

Ogen. Estropipate. Applications: Oestrogen. (Abbott Laboratories).

Ogwin. A mixture of lime and starch used to increase the rate of sedimentation of solids in water.†

O-hi-o. A proprietary trade name for a die steel containing 12 per cent chromium, 1.55 per cent carbon, 0.85 per cent vanadium, 0.4 per cent cobalt. and 0.8 per cent manganese.†

Ohlan. Hydroxylated lanolin with three times the structural hydroxy content of lanolin. Applications: Superior nonionic w/o emulsifier. Pigment wetting agent and stabilizer for o/w systems. Adds tack, gloss and adhesion for makeup. Strengthens emollient barriers of skin care products and reduces cracking in bar soaps. (Amerchol Corporation).*

Ohm Oil. A mineral oil which has been treated or contains in solution an anti-oxidant, thereby stabilizing the oil and increasing its electrical resistivity.†

Ohmal. A resistance alloy similar in composition to manganin. It usually contains 87.5 per cent copper, 9 per cent manganese, and 3.5 per cent nickel.†

Ohmlac Kapak. A refined elaterite.†

Ohmoid. A proprietary trade nane for a phenol-formaldehyde synthetic resin laminated product used for electrical insulation.†

Oil, Alexandrian Laurel. See LAUREL-NUT OIL.

Oil, Animal. See BONE OIL.

Oil Asphalt. A thick fluid remaining after distilling crude petroleum. Used for roofing materials, and for paving when mixed with natural asphalt.†

Oilatum Application. A proprietary preparation of arachis oil and polyvinyl pyrrolidine. Used in dermatology. (Stiefel Laboratories (UK) Ltd).*

Oilatum Emollient. A proprietary preparation of liquid fraction of acetylated lanolin alcohols in a semi-dispersing emollient base. Used in dermatology. (Stiefel Laboratories (UK) Ltd).*

Oilatum Soap. Contains unsaponified arachis oil BP and high molecular weight fatty acid salts. Applications: As an aid in the management of dry, pruritic skin conditions such as senile pruritus, atopic dermatitis and in the drying stage of eczema. (Stiefel Laboratories (UK) Ltd).*

Oil Babulum. See NEATSFOOT OIL.

Oil, Bagasses. See PYRENE OIL.

Oil, Bari. See AIX OIL.

Oil, Batana. See PATAVA OIL.

Oil, Beniseed. See GINGELLY OIL.

Oil, Black. See INDULINE, SPIRIT SOLUBLE.

Oil Blue. A pigment prepared by introducing copper filings into boiling sulphur, and after cooling, boiling the mass with sodium hydroxide to remove excess of sulphur. It is applicable as an oil colour only. Also see PRUSSIAN BLUE.†

Oil Brown D. A dyestuff. It is equivalent to Sudan brown.†

Oil, Calaba. See LAUREL-NUT OIL.

Oil, Carbolic. See MIDDLE OILS.

Oil, Chinese. Peru balsam.†

Oil, Coal. See KEROSENE.

Oil, Coco-nut. See COCO-NUT BUTTER.

Oildag. Colloidal graphite in mineral oil used as a special lubricant and as an additive to lubricating oils and greases. (Acheson Colloids).*

Oil, Danforth's. See NAPHTHA.

Oil Die. A proprietary trade name for tool steel containing 1.6 per cent chromium, 0.45 per cent tungsten, and 0.9 per cent carbon.†

Oil, Dilo. See LAUREL-NUT OIL.

Oil, Dippel's. See BONE OIL.

Oil, Disinfection. See SAPROL.

Oil, Domba. See LAUREL-NUT OIL.

Oil, Duty. See OIL OF RHODIUM.

Oiled Silk. Thin silk fabric which has been impregnated with oil, usually linseed.†

Oilfos. Glassy sodium phosphate. Applications: Controls viscosity of oil well drilling muds. (Monsanto Co).*

Oil, Fousel. See FUSEL OIL.

Oil, Garcina. See KOKUM BUTTER.

Oil Gard. Blend of petroleum oil, viscosity index improvers. Applications: Reduces oil consumption in automotive vehicles, increases oils' viscosity index. (Gard Corporation).*

Oil, Ginger Grass. See ROSÉ OIL.

Oil, Gingili. See GINGELLY OIL.

Oil, Glonoine. See NITROGLYCERIN.

Oil, Grain. See FUSEL OIL.

Oil Green. (Maple Green, Leaf Green, Emerald Green). Pigments. They are mixtures of Chrome yellow and Paris blue. Also see CHROME GREEN.†

il Gutta-percha. A product obtained by heating together 100 parts gutta-percha, 10 parts olive oil, and 2 parts stearin. Used for moulding purposes.†

il, Hoof. See NEATSFOOT OIL.

il, Hydrocarbon. See PARAFFIN LIQUID.

ilite. Applied to both sintered bronze self lubricating bearings and all sintered ferrous parts and structural components. Applications: Oil impregnated self lubricating bronze bearing bushes supplied to a very wide range of industry. Ferrous components for motor industry - domestic appliance industry and domestic agricultural applications. (MB Powder Metal Group).*

il, Jeppel's. See BONE OIL.

il, Kagoo. See KORUNG OIL.

il, Kusum. See KON OIL.

il Lacs. These are obtained by adding to almost boiling oil varnish, fused copal or other resin, and diluting with oil of turpentine when using.†

il, Linseed. See LINSEED OIL ARTIST'S and LINSEED OIL REFINED.

il, Monopol. See TURKEY RED OILS.

il, Nari. See NJAVE BUTTER.

il, Neem. See VEEPA OIL.

il, Neroli. See OIL OF ORANGE FLOWERS.

il, Nerve. See NEATSFOOT OIL.

il, Nice. See AIX OIL.

il, Noumgou. See NJAVE BUTTER.

Dil of Absinthe. Wormwood oil.†

Dil of Ajava. Ajowan oil, the essential oil from the seeds of *Carum copticum.*†

Dil of Akee. A yellow non-drying butter-like fat from *Blighia sapida.*†

Dil of Allspice. Oil of Pimento.†

Dil of Aloes. The oil obtained from Socotrine aloes.†

Dil of Amber. An oil distilled from amber. The oils obtained from copal or dammar are also called by this name.†

Dil of Anthos. Rosemary oil.†

Dil of Ants. Olive oil, in which ants have been digested.†

Oil of Ants, Artificial. Furfural, $C_4H_3(CHO)O$.†

Oil of Apple. Amyl valerate, $C_4H_9COO C_5H_{11}$, used in the manufacture of fruit essences.†

Oil of Asarabacca. An oil obtained from the roots of *Asarum europoeum.*†

Oil of Asphaltum. The oil obtained from asphaltum.†

Oil of Aspic. See OIL OF SPIKE.

Oil of Balm. (Oil of Lemon Balm). The volatile oil from *Melissa officinalis.* Used as a diaphoretic.†

Oil of Bay. A volatile oil obtained from the leaves of *Myrcia acris.*†

Oil of Bay Berries. The oil expressed from the berries of *Laurus nobilis.*†

Oil of Been. (Oil of Behn, Oil of Ben). The oil expressed from the seeds of *Moringa aptera.*†

Oil of Behn. See OIL OF BEEN.

Oil of Benjamin. The oil obtained from benzoin, after the sublimation of benzoic acid.†

Oil of Benné. See GINGELLY OIL.

Oil of Bergamot, Artificial. See BERGAMINOL.

Oil of Birch. The volatile oil from *Betula lenta*, the sweet birch. It consists mainly of methyl salicylate. It is also the name for the oil from *Betula alba* the white birch.†

Oil of Bones. See BONE OIL.

Oil of Box. The oil obtained from box wood.†

Oil of Cabbage. (Oil of Elder). Olive oil in which elder leaves have been boiled.†

Oil of Cacao. See OIL OF COCOA.

Oil of Cassia. (Oil of Chinese Cinnamon). The oil distilled from the bark, leaves, and twigs of *Cinnamomum cassia*, of China and Java. It contains from 75-95 per cent cinnamic aldehyde, $C_6H_5CH : CH.CHO$.†

Oil of Cedrat. The oil obtained from citron peel.†

Oil of Chinese Cinnamon. See OIL OF CASSIA.

Oil of Cinnamon. The oil obtained from the bark of *Cinnamomum zeylanicum*, of

Ceylon. The yield varies from 0.5-1 per cent. It contains from 55-75 per cent cinnamic aldehyde, $C_6H_5.CH : CH.CHO$.†

Oil of Cinnamon Leaf. An oil distilled from the leaves of *Cinnamomum zeylanicum*. It contains from 75-90 per cent eugenol and safrole, and only traces of cinnamic aldehyde.†

Oil of Citronella. The oil obtained from *Andropogon nardus*. Its chief constituent is citronellal, $C_{10}H_{18}O$.†

Oil of Cocoa. (Oil of Cacao). Oil of theobroma (cacao butter, (*qv*).†

Oil of Cognac. A mixture of different esters. It consists partly of rectified wine fusel oil, and partly of an artificial grape juice. Artificial oil of cognac is made from coco-nut oil and alcohol, and from palargonium oil, and castor oil.†

Oil of Daget. Oil of birch tar.†

Oil of Dog-fish. Shark oil.†

Oil of Dolphin. Porpoise oil.†

Oil of Duty. See OIL OF RHODIUM.

Oil of Elder. See OIL OF CABBAGE.

Oil of Ergot. The residue left when an ethereal solution of tincture of ergot is evaporated.†

Oil of Exeter. Oil of elder, mixed with euphorbium and mustard.†

Oil of Flaxseed. Linseed oil.†

Oil of Foxglove. Olive oil, in which fresh leaves of the foxglove have been digested.†

Oil of Garlic. Diallyl sulphide, $(C_3H_5)_2S$.†

Oil of Geranium. See ROSÉ OIL.

Oil of Geranium, East Indian or Turkish. See ROSÉ OIL.

Oil of Gingelli. See GINGELLY OIL.

Oil of Ginger Grass. See ROSÉ OIL.

Oil of Gourd. Cucumber oil.†

Oil of Grain. See FUSEL OIL.

Oil of Hartshorn. See BONE OIL.

Oil of Hemlock. The volatile oil from *Pinus canadensis*, the hemlock spruce. The term is also applied to olive oil in which fresh leaves of *Conium maculatum* have been digested.†

Oil of Japanese Mint. This is not a true mint oil. It is derived from *Mentha arvensis* and is a partly dementholize oil.†

Oil of Lemon Balm. See OIL OF BALM

Oil of Lemon Grass. The oil obtained fro *Andropogon citratus*.†

Oil of Liquid Pitch. See OIL OF TAR.

Oil of Mace. The name erroneously give to expressed oil of nutmeg.†

Oil of Mirbane. See ESSENCE OF MIRBANE.

Oil of Mosoi Flowers. See YLANG-YLANG OIL.

Oil of Mucilages. Olive oil boiled with a decoction of marshmallow root, linsee and foenugreek seeds.†

Oil of Mustard. (Allyl Mustard Oil). Al isothiocyanate, $C_3H_5N : CS$.†

Oil of Neroli. See OIL OF ORANGE FLOWERS.

Oil of Nerves. See NEATSFOOT OIL.

Oil of Niobe. The methyl ester of benzoi acid. Used in the soap industry.†

Oil of Olay. Beauty fluid. Applications: care for the skin. (Richardson-Vicks Inc).*

Oil of Orange Flower. (Oil of Neroli). Th oil distilled from the flowers of the bit orange tree.†

Oil of Origanum. Oil of thyme. Also oil from *Origanum* species.†

Oil of Palma Christi. (Ricinus oil). Casto oil.†

Oil of Palmarosa. See ROSÉ OIL.

Oil of Partridge Berry. Oil of winter gree

Oil of Pear. See PEAR OIL.

Oil of Pelargonium. See ROSÉ OIL.

Oil of Pennyroyal. (Oil of Poley). Oil of pulegium. It consists chiefly of pulego $C_{10}H_{18}O$.†

Oil of Peter. (Oil of Petre). Rock oil, or mixture of 1 part of oil rosemary, 4 p turpentine, and 4 parts of Barbados ta

Oil of Petitgrain. The oil obtained from t leaves of the bitter orange tree.†

Oil of Petre. See OIL OF PETER.

Oil of Poley. See OIL OF PENNY-ROYAL.

Oil of Pompillon. An ointment of poplar seeds, also green elder ointment.†

Oil of Portugal. Oil of sweet orange peel.†

Oil of Ptychotis. Oil of ajowan.†

Oil of Rhodium. (Oil of Duty). The oil obtained from the root of *Genista canariensis*. Also a mixture of sandalwood oil, and otto of rose, or oil of rose geranium.†

Oil of Rose Geranium. See ROSÉ OIL.

Oil of Smoke. See CREOSOTE.

Oil of Spike. The volatile oil from *Lavandula spica*. It is also the name for a mixture of lavender oil, and oil of turpentine, coloured with alkanet.†

Oil of St. John Wort. The oil obtained by digesting the flowering tops of *Hypericum perforatum* in warm olive oil.†

Oil of Sweet Birch. Oil of betula from *Betula lenta*.†

Oil of Tar. (Oil of Liquid Pitch). Creosote.†

Oil of Tartar. Deliquescent potassium carbonate, K_2CO_3.†

Oil of Tea. The oil obtained from the seeds of *Camellia species*.†

Oil of Theobroma. See CACAO BUTTER.

Oil of Turpentine. (Spirits of Turpentine, Essence of Turpentine, Turps). Derived from the pine, *Pinus palustris* and *P. Taeda*, and from the Scotch fir, *Pinus sylvestris*.†

Oil of Turpentine, Oxidized. See DECAMPHORIZED OIL OF TURPENTINE.

Oil of Verbena. The oil obtained from *Verbena triphylla*. Also the name for oil of lemongrass.†

Oil of Vitriol. (O.V., Vitriolic Acid). Ordinary concentrated sulphuric acid, H_2SO_4.†

Oil of Wax. See WAX BUTTER.

Oil of Wheat. The oil obtained from bruised wheat.†

Oil of Wintergreen. Oil of gaultheria, obtained by distilling the leaves of *Gaultheria procumbens*, or from the bark of *Betula lenta*. The chief constituent is methyl salicylate, $C_6H_4(OH)COO.CH_3$.†

Oil Orange 0. A dyestuff. It is a British brand of Sudan G.†

Oil Orange, E. Dyestuffs. They are British equivalents of Sudan I.†

Oil, Oxidized. See BLOWN OILS.

Oil, Oxidized Linseed. See SOLIDIFIED LINSEED OIL.

Oil Paalsgaard. See SCHOU OIL.

Oil, Palmichristi. See OIL OF PALMA CHRISTI.

Oil, Paraffin. See KEROSENE.

Oil, Physic Nut. See PURGING NUT OIL.

Oil, Pinnay. See LAUREL-NUT OIL.

Oil, Poonseed. See LAUREL-NUT OIL.

Oil Pulp. (Thickener, Fluid Gelatin, Viscom). A gelatinous material made by heating aluminium oleate with mineral oil. It is sold to give increased viscosity to mineral oils.†

Oil Red Base. A British equivalent of Rosaniline.†

Oil Red S. An oil-soluble colour. It is toluene-azo-toluene-azo-β-naphthol.†

Oil, Ricinus. See OIL OF PALMA CHRISTI.

Oil, Riviera. See AIX OIL.

Oil, Roshé. See ROSÉ OIL.

Oil, Safety. See C-PETROLEUM NAPHTHA.

Oil, Scarlet. (Red B Oil, Soluble Extra Conc., Ponceau 3B). A dyestuff. It is toluene-azo-toluene-azo-β-naphthol, $C_{24}H_{20}N_4O$. Used for colouring oils and varnishes, also for cotton fibre.†

Oil, Scarlet AS. A dyestuff. It is a British brand of Sudan III.†

Oil Scarlet L, Y. Dyestuffs. They are British equivalents of Sudan II.†

Oil Shale. A sedimentary rock which yields from 12-60 gallons of shale oil per ton on distillation.†

Oil, Sherwood. See PETROLEUM ETHER.

Oil Skin. A waterproof material made by impregnating cotton or other fabric with hardening oils.†

Oilsol. Dyestuffs soluble in oils, waxes and plastics. (Williams (Hounslow) Ltd).

Oil, Soluble. See TURKEY RED OILS.

Oil, Soluble Castor. See BLOWN OILS.

Oil-soluble Resins. Synthetic resins of the phenol-formaldehyde type obtained by a fixing process, using rosin in the mixing. They dissolve in hydrocarbon solution and are no longer soluble in alcohol. Also see HYDROCARBON-ALDEHYDE RESINS.†

Oilstone. See WHETSTONE.

Oil-stone. See STINK-STONE.

Oil, Sulphated. See TURKEY RED OILS.

Oil, Sulphocarbon. See SANSE.

Oil, Sulphonated. See TURKEY RED OILS.

Oil, Sulphur Olive. See SANSE.

Oil, Teal. See GINGELLY OIL.

Oil, Teel. See GINGELLY OIL.

Oil, Theobroma. See COCOA BUTTER.

Oil, Thickened. See BLOWN OILS.

Oil, Til. See GINGELLY OIL.

Oil-Treet. Applications: Fuel oil conditioner and combustion catalyst. (Schaefer Chemical Products Company).*

Oil, Turkey Geranium. See ROSÉ OIL.

Oil Udilo. See LAUREL-NUT OIL.

Oil, Var. See AIX OIL.

Oil Varnishes. Solutions of resins in linseed oil.†

Oil, Veppam. See VEEPA OIL.

Oil Vermillion. A dyestuff. It is a British brand of Sudan R.†

Oil Vulcanized. See THINOLINE.

Oil Wax. See WAX BUTTER.

Oil White. (Light White, Leukarion, Albanol, Diamond White, Edelweiss, Snow White,. Anti-White Lead, Blenda, Condor, Fixopone, Nivan). White lead substitutes. They consist chiefly of lithopone mixed with white lead or zinc white, also with whiting, gypsum, magnesia, or silica.†

Oil Yellow. A dyestuff. It is a British equivalent of Sudan G.†

Oil Yellow. See BUTTER YELLOW.

Ointment, Blue. Mercury ointment.†

Oisanite. A mineral. It is anatase.†

Oiticica Oil. An oil obtained from the Brazilian plant *Conepia grandifolia*. It resembles Tung oil in its properties.†

Okerin. Antiozonant waxes, blends of petroleum waxes. Applications: For u in tyre sidewall compounds and gener rubber products. (Astor).*

Okol. A disinfectant consisting of an emulsion containing phenols. It is miscible with water.†

Okstan XO. A di-*n*-octyl tin mercaptide f stabilization of non toxic PVC compounds. (Croda Resins Ltd).†

Okstan X3. A trade name for a butyl tin mercaptide, a stabilizer for PVC with an exceptionally high heat stability. (Chemische Werke München Otto Bärklocker).†

Olay Beauty Bar. A cleansing bar to clea soften and smooth the skin. (Richardson-Vicks Inc).*

Olay Beauty Cleanser. Gentle, greaseless facial cleanser to leave skin soft and smooth. (Richardson-Vicks Inc).*

Olcotrop Leatber. A leather made from a species of shark skin.†

Old Fustic. See FUSTIC.

Old Plantation. Ammonium nitrate. Applications: fertilizer. (Columbia Nitrogen Corporation).*

Oleandocyn. A proprietary preparation of oleandomycin phosphate. An antibioti (Pfizer International).*

Oleandomycin. An antibiotic produced by certain strains of Streptomyces antibioticus.†

Olefiant Gas. (Heavy Carburetted Hydrogen, Dicarburetted Hydrogen, Elayl, Ethene, Etherin). Ethylene, C_2H_4.†

Olein. The tri-oleyl derivative of glycerol, $C_3H_5(OC_{18}H_{33}O)_3$. The term is applied commercially to any liquid oil obtained from solid fats by pressure, to crude oleic acid, and to the potassium, sodiur or ammonium salt of the sulphonate of oleic acid. Elaine is another name give to commercial oleic acid. Also see TURKEY RED OILS.†

Olein of Saponification. Commercial oleic acid prepared by the saponification of pure fats, and the separation from stearine, by pressing.†

Oleite. Sodium-sulpho-ricinoleate.†

Oleo. (Premier Jus). The oil expressed from beef-fat. Used in margarine manufacture.†

Oleobismuth. An oil suspension of bismuth oleate.†

Oleogen. See PAROGEN.

Oleoguaiacol. Guaiacol oleate, $CH_3O.C_6H_4.O.CO.C_{17}H_{23}$. Used as an antiseptic.†

Oleo-resin Copaiba. (Balsam of Copaiba, Balsam of Copaiva, Balsam of Capivi). Copaiba, an oleo-resin.†

Oleo-resins. Resins mixed with a volatile oil.†

Oleosol. Oildag. See OILDAG and AQUADAG.†

Oleum. See FUMING SULPHURIC ACID.

Oleum Spirits. A petroleum distillate.†

Oleum White. A pigment. It is a sulphide of zinc mixed with a small proportion of barium sulphate (a lithophone). Used as a rubber filler.†

Olex. Fat soluble powdered flavours. (Bush Boake Allen).*

Olibanum. (Indian Frankincense, Salaigugl). A gum-resin obtained from *Boswellia* species. It contains from 8-18 per cent of essential oil, 55- 57 per cent of resin, and 20-23 per cent gum (carbohydrates). It has a specific gravity of 1.2, an acid value of 45-88, and a saponification value of 65-120.†

Oligan. Shrink-resist additive. (Ciba-Geigy PLC).

Olinor. Blend of hydrocarbons and sulphated fatty alcohols. Applications: Softening agent for raising cotton goods. (Henkel Chemicals Ltd).

Oliolase. An iodized vegetable oil.†

Olive Green. See BRONZE GREEN.

Olive Green Oxide of Uranium. Uranoso-uranic oxide, U_3O_8.†

Olive Lake. A pigment. Originally it was prepared from green ebony, but is now exclusively a mixture.†

Olive Oil. See GALLIPOLI OIL, GENOA OIL, LUCCA OIL, PROVENCE OIL, RED OIL, SICILY OIL, and SPANISH OIL.

Olive Oils, Sulphur. See SANSE.

Olivenite. A mineral, $4CuO.As_2O_5$.†

Olives of Java. (Kaloempang Beans, Beligno Seeds, Sterculia Kernels). Seeds of *Sterculi foetida*, the source of Sterculia oil.†

Olivite. A substance having a rubber base. Used as an acid-proofing material in pumps.†

Olminal. A trade name for a commercial aluminium oleate. It contains 1.5 per cent aluminium.†

OLOA. Lubricating oil additive. (Chevron).*

Olobintin. A German preparation. It is a 10 per cent solution of a mixture of various rectified turpentine oils.†

Olutkombul. The gelatinous sap of the plant *Abroma angustum*. Used in India in dysmenorrhoea.†

Olympic Bronze. A proprietary trade name for an alloy of copper with 3 per cent silicon and 1 per cent zinc.†

Olympic Bronze G. A proprietary trade name for an alloy of copper with 22 per cent zinc and 1 per cent silicon.†

Olyntholite. Synonym for Grossular.†

Omal. Trichloro-phenol.†

Omarsan. A detergent and sterilizer. (PPF International Ltd).*

Omnilac. Non-nitrocellulose overlacquers. (The Scottish Adhesives Co Ltd).*

Omnilube. A series of synthetic oils and greases based on polyalphaolefins and various organic compounds. Applications: A variety of oils used for OEM applications in appliances, power tools, computers, automotive electrical motors and industrial maintenance applications. (Ultrachem Inc).*

Omnipaque. Iohexol. Applications: Diagnostic aid. (Sterling Drug Inc).

Omnipen. Ampicillin. Applications: Antibacterial. (Wyeth Laboratories, Div of American Home Products Corp).

Omnipen-N. Ampicillin. Applications: Antibacterial. (Wyeth Laboratories, Div of American Home Products Corp).

Omnivax. Anticlostridial vaccine. (The Wellcome Foundation Ltd).

Omnopon. A proprietary preparation of papavertum. An analgesic. (Roche Products Ltd).*

Omnopon-Scopolamine. Pre-anaesthetic medication. (Roche Products Ltd).

Omyastab. A secondary stabilizer for chlorinated polyester resins. It is a colourless cyclic organic compound containing ether linkages which are compatible with the polyester resin.†

Onadox-118. A proprietary preparation of aspirin and dihydrocodeine bitartrate. An analgesic. (British Drug Houses).†

Onamine RO. 1-(2-hydroxyethyl) 2 n-heptadecenyl-2 imidazoline in liquid form. Applications: Acid stable emulsifier used in corrosion inhibition, solvent degreasing, defoaming and demulsifying. (Millmaster-Onyx UK).

Onamine 12. Dodecyl dimethylamine in liquid form. Applications: Intermediate in the synthesis of surfactants, anti-oxidants, oil and grease additives. (Millmaster-Onyx UK).

Onamine 14. Tetradecyl dimethylamine in liquid form. Applications: Intermediate in the synthesis of surfactants, anti-oxidants, oil and grease additives. (Millmaster-Onyx UK).

Onamine 16. Hexadecyl dimethylamine in liquid form. Applications: Intermediate in the synthesis of surfactants, anti-oxidants, oil and grease additives. (Millmaster-Onyx UK).

Onamine 18. Stearyl dimethylamine in liquid form. Applications: Intermediate in the synthesis of surfactants, anti-oxidants, oil and grease additives. (Millmaster-Onyx UK).

Onamine 65, 835 and 1214. Alkyl dimethylamine in liquid form. Applications: Intermediate in the synthesis of surfactants, anti- oxidants, oil and grease additives. (Millmaster-Onyx UK).

Once. Hand cleanser. (Hardman Inc).*

Oncol. Insecticide containing benfuracarb. (ICI PLC).*

Oncor. Basic lead silico pigments. Applications: Anti-corrosive pigment for paint. (NL Industries Inc).*

Oncor Split Sphere. Design for Oncor products. Applications: Paint. (NL Industries Inc).*

Oncor 75. Flame retardant compositions. (Anzon Ltd).

Ondoita. A proprietary synthetic resin moulding powder.†

Ongard. Smoke suppressants. (Anzon Ltd

Onion's Alloy. A fusible alloy containing [illegible] per cent bismuth, 30 per cent lead and 20 per cent tin.†

Ontario Steel. A proprietary trade name for a non-shrinking steel containing 11 per cent chromium, 0.75 per cent molybdenum, 0.25 per cent vanadium, 0.35 per cent silicon, 0.30 per cent manganese, and 1.45 per cent carbon.†

O.N.V. A proprietary trade name for a rubber vulcanization accelerator. It isdiphenylcarbamyldimethyldithio-carbamate. †

Onyx. Consists chiefly of silica.†

Onyx Marble. A marble containing fossil shells.†

Onyx of Tecali. A variety of alabaster. Th colour varies from milk-white to pale yellow and pale green.†

Onyxide 172. Alkyl dimethyl ethyl benzyl ammonium cyclohexyl sulphamate in liquid form. Applications: Anti-fungal agent and preservative used in latex emulsions; paints; adhesives; coated fabrics; cutting oils. (Millmaster-Onyx UK).

Onyxide 3300. Alkyl dimethylbenzyl ammonium saccharinate in powder form. Applications: Germicide, conditioner and disinfectant with low skin and eye irritation, used in cosmeti and pharmaceuticals; hair preparations detergent-sanitizers; disinfectants. (Millmaster-Onyx UK).

Onyxide 75. Alkenyl(90% C18, 10% C16) dimethyl ethyl ammonium bromide in paste form. Applications: Cationic surfactant algaecide used in re-ciculating water systems; swimming pools; humidifiers. (Millmaster-Onyx UK).

Oolitic Limestone. A massive variety of calcium carbonate, used for building purposes.†

Opacite. Stannic chloride, $SnCl_4$.†

Opacode. An edible ink for pharmaceutica or food use. (Colorcon Ltd).*

Opacolor. Colouring system for sugar coated confectionery pieces. (Colorcon Ltd).*

Opadry. Complete film coating system in a dry form for reconstitution. Applications: Aqueous film coating, organic film coating and enteric coating. (Colorcon Ltd).*

Opal Blue. See ANILINE BLUE, SPIRIT SOLUBLE, and FINE BLUE.

Opal Blue XL. See DIPHENYLAMINE BLUE SPIRIT SOLUBLE.

Opal Blue 6B. See ANILINE BLUE, SPIRIT SOLUBLE.

Opal Jasper. It is silica, resembling jasper.†

Opal Violet. See SPIRIT PURPLE.

Opalite. A proprietary trade name for an amorphous silica.†

Opalon. A proprietary trade name for phenol-formaldehyde cast resins.†

Opalon 740. A trade name for a graft copolymer used as a semi rigid wire insulation.†

Opalux. Colouring system for sugar coated tablets. Applications: Sugar coating for pharmaceutical products. (Colorcon Ltd).*

Opalwax. Hydrogenated castor oil.†

Opaque Oxide of Chromium. A green pigment. It is sesquioxide of chromium.†

Opaspray. Colouring system for film coated tablets. Applications: Aqueous film coating colourant, organic film coating colourant. (Colorcon Ltd).*

Opatint. A multipurpose liquid colour dispersion formulated for the colouring of all types of food and confectionery. (Colorcon Ltd).*

Opazil. Acid and alkaline activated bentonites for the adsorption of detrimental substances in the papermaking process. (Süd-Chemie AG).*

Open Hearth Steel. See MARTIN STEEL.

Operidine. A proprietary preparation of phenoperidine hydrochloride. Analgesic supplement in anaesthesia. (Janssen Pharmaceutical Ltd).*

Ophorite. An ignition powder for projectiles. It consists of magnesium powder and potassium chlorate.†

Ophthaine. Proparacaine Hydrochloride. Applications: Anaesthetic. (E R Squibb & Sons Inc).

Ophthaine Solution. Proxymetacaine hydrochloride injection in aqueous solution. (Ciba-Geigy PLC).

Ophthalgan. Glycerin. Applications: Pharmaceutic aid. (Ayerst Laboratories, Div of American Home Products Corp).

Ophthalmadine. Idoxuridine 0.1% eye drops and 0.5% eye ointment. Applications: Treatment of Ocular Herpes Simplex. (SAS Pharmaceuticals Ltd).*

Ophthetic. Proparacaine Hydrochloride. Applications: Anaesthetic. (Allergan Pharmaceuticals Inc).

Opilon. A proprietary preparation of thymoxamine hydrochloride. A vasodilator. (Warner).†

Opium. The dried juice from the unripe capsules of *Papaver somniferum*. It contains morphine, codeine, narcotine, narceine, thebane, papaverine, and meconin.†

Opobyl (Pills). Desiccated liver 50 mg, sodium tauroglycocholate 50 mg, aqueous extract of boldo 10 mg, podophyllin 2 mg, alcoholic extract of euonymus 2 mg, aloes 20 mg per pill. Applications: Hepatic insufficiency, biliary stasis, habitual constipation caused by intestinal sluggishness. (Bengue & Co Ltd).*

Opogard. Mixed triazine herbicide. (Ciba-Geigy PLC).

Opol. Chemical polishers for plastics. (Laporte Industries Ltd).*

Opopanax. The dried juice from the roots of *Pastinaca opopanax*. It is used in perfumery, and medicinally as an antispasmodic.†

Oppalyte. Range of polypropylene films. Applications: Packaging of food and non food products and special industrial applications. (Mobil Plastics Europe).*

Oppanol. Polyisobutylene in the form of viscose liquids or rubbery solid crumb, depending on molecular weight. Applications: In sheet or film form for waterproofing or corrosion resistant coatings. To modify properties of polyolefine polymers and waxes used in film or impregnation. For sealants and adhesives. (BASF United Kingdom Ltd).*

Oppanol D. Polyisobutylene dispersion. (BASF United Kingdom Ltd).*

Opsan. Antibacterial eye drops. (Boots Pure Drug Co).*

Op-Sulfa 30. Sulphacetamide Sodium. Applications: Antibacterial. (Broemmel Pharmaceuticals).

Optannin. Basic calcium tannate.†

Optbas. Green, effervescent tablets which are dissolved in water to produce an antiseptic eye lotion. Each tablet contains adrenaline BP 0.6% w/w, phenylephrine hydrochloride BP 0.05% w/w and acriflovine BPC 1963 0.005% w/w. Applications: Relief of tired strained eyes, watering and bloodshot eyes, inflammation and catarrhal conjunctivitis. (Evans Medical).*

Optemet. Prefabricated disposable pouring trumpets for uphill teemed steel ingots. (Foseco (FS) Ltd).*

Opthaine. A proprietary preparation of proxymetacaine hydrochloride solution. An ocular anaesthetic. (Squibb, E R & Sons Ltd).*

Op-Thal-Zin. Zinc Sulphate. Applications: Astringent (ophthalmic). (Alcon Laboratories Inc).

Optical Bronze. An alloy of 89 per cent copper, 6.5 per cent zinc, and 4.5 per cent tin.†

Opticite. Polystyrene based packaging labelling material. (Dow Chemical Co Ltd).*

Opticorten. Steroid veterinary ethicals. (Ciba-Geigy PLC).

Opticrom. Eye drops. (Fisons PLC, Pharmaceutical Div).

Opticrom. Cromolyn sodium. Applications: Anti-asthmatic. (Fisons Corp).

Optimine. Azatadine maleate. Applications: Antihistaminic. (Schering-Plough Corp).

Optinol. A range of carriers or fibre-dilating agents. Applications: Dyebath assistants particularly for the dyeing of polyester. (Yorkshire Chemicals Plc).*

Optocillin. Broad-spectrum penicillin combination with penicillinase-fast component. (Bayer & Co).*

Optosil Liquid. Elastomeric isolating film on silicone basis. Applications: For dental techology. (Bayer & Co).*

Optosil Plus. Organic silicone polymer. Applications: For dental impressions. (Bayer & Co).*

Optosil/Optosil Hard. Elastomeric impression material and ancilliaries. Applications: Dental speciality for practice and laboratory. (Bayer & Co).*

Optox. Melted lead oxide granules of high purity. Applications: Basic raw material for optical glass. (Blielberger Bergwerke Union).*

Optran. High grade optical chemicals. (British Drug Houses).†

Optrex. Eye treatment preparations. (The Boots Company PLC).

Opulets. A proprietary preparation of atropine sulphate in gelatine, used as a long acting mydriatic as an aid in diagnosis, and in treatment of uveitis. (Pharmax).†

ORA. Refinery process stream additive. (Chevron).*

Orabase. A proprietary preparation containing sodium carboxy-methyl cellulose, pectin and gelatin in a liquid paraffin polyethylene base. (Squibb, E R & Sons Ltd).*

Oracet. Solvent soluble dyes. (Ciba-Geigy PLC).

Oragrafin Calcium. Ipodate calcium. Applications: Diagnostic aid. (E R Squibb & Sons Inc).

Oragrafin Sodium. Ipodate sodium. Applications: Diagnostic aid. (E R Squibb & Sons Inc).

Orahesive. A proprietary preparation containing sodium carboxy-methyl cellulose, pectin and gelatin. (Squibb, E R & Sons Ltd).*

Oralcer. A proprietary preparation of CLIOQUINOL and ascorbic acid used in the treatment of mouth ulcers. (Vitabiotics).†

Oraldene. A proprietary preparation of HEXETIDINE. An antiseptic mouth wash. (Warner).†

Oralith. Organic pigments.†

Oranabol. A proprietary trade name for oxymesterone.†

ange A. See ORANGE II.

ange Chrome. See CHROME RED.

ange ENL. See PONCEAU 4GB.

ange Essence. Oil of orange.†

ange Extra. See ORANGE II.

ange Flower Oil. See OIL OF ORANGE FLOWERS.

ange G. (Orange GG, Orange Yellow, Patent Orange, Kiton Fast Orange G, Fast Light Orange G, Naphtharene Orange G). A dyestuff. It is the sodium salt of benzeneazo-β-naphthol-disulphonic acid G, $C_{16}H_{10}N_2S_2O_7Na$. Dyes wool orange-yellow from an acid bath.†

range GG. See ORANGE G.

range GRX. See PONCEAU 4GB.

range GS. See ORANGE IV.

range GT. (Orange RN, Orange O, Orange N, Brilliant Orange O, Brilliant Orange OL, Crocein Orange (Brotherton & Co)). A dyestuff. It is the sodium salt of tolueneazo-β-naphthol-sulphonic acid, $C_{17}H_{13}N_2O_4SNa$. Dyes wool from an acid bath.†

range II. (Tropaeoline ooo No. 2, Mandarin G,β-Naphthol Orange, Mandarin G Extra, Chrysaureine, Gold Orange, Orange Extra, Atlas Orange, Orange A, Acid Orange, Orange P, Acid Orange G, Orange II Conc, Special). A dyestuff. It is the sodium salt of *p*-sulphobenzeneazo-β-naphthol, $C_{16}H_{11}N_2O_4SNa$. Dyes wool and silk orange from an acid bath.†

range II Conc., Special. Dyestuffs. They are British equivalents of Orange II.†

range III. (Orange No. 3, Methyl Orange, Porrier's Orange III, Dimethylaniline Orange, Helianthin, Tropaeoline D, Gold Orange, Mandarin Orange). A dyestuff. It is the sodium salt of *p*-sulpho-benzene-azo-dimethyl-aniline, $C_{14}H_{14}N_3SO_3Na$. Dyes wool orange from an acid bath. Used as an indicator in alkalimetry.

The sodium salt of *m*-nitrobenzene-azo-β-naphtholdi-sulphonic acid $C_{16}H_9N_3O_9S_2Na_2$, is also called Orange III. It dyes wool orange from an acid bath.†

Orange IV. (Diphenylamine Orange, Tropaeoline oo, Orange M, Diphenylamine Yellow, Fast Yellow, Orange W, Orange GS, New Yellow, Diphenyl Orange, Gold Orange, Orange N, Acid Yellow D). A dyestuff. It is the sodium salt of *p*-sulphobenzene-azodiphenylamine, $C_{18}H_{14}N_3O_3SNa$. Dyes wool orange-yellow from an acid bath.†

Orange L. See SCARLET GR.

Orange Lac. See LEMON LAC.

Orange Lead. (Orange Mineral, Orange Red, Sandix, Saturn Red). A red lead obtained by calcining powdered white lead. It is a better red lead than crystal minium.†

Orange M. See ORANGE IV.

Orange Mineral. See ORANGE LEAD.

Orange MNO. A British brand of Metanil yellow.†

Orange N. See ORANGE IV. See ORANGE GT. Also see FAST YELLOW N and SCARLET GR.

Orange No. 3. See ORANGE III.

Orange Ochre. See BURNT YELLOW OCHRE.

Orange Oxide of Uranium. (Yellow Sesquioxide of Uranium, Uranium Yellow). Sodium uranate, $Na_2U_2O_7$, has been called by these names.†

Orange P. See ORANGE II.

Orange R. See PONCEAU 2G, ORANGE T, ORANGE RR, and ALIZARIN YELLOW R.

Orange Red. See ORANGE LEAD.

Orange RL and RRL. Identical with Resorcin yellow.†

Orange RN. See ORANGE GT.

Orange RR. (Orange R). A dyestuff. It is the sodium salt of sulphoxylene-azo-β-naphthol, $C_{18}H_{15}N_2O_4SNa$. Dyes wool orange from an acid bath.†

Orange Russet. Rubens madder.†

Orange Spirit. See TIN SPIRITS.

Orange T. (Mandarin GR, Orange R, Kermesin Orange). A dyestuff. It is the sodium salt of sulpho-*o*-toluene-azo-β-naphthol, $C_{17}H_{13}N_2O_4SNa$. Dyes wool orange from an acid bath.†

Orange TA. A direct cotton dyestuff.†

Orange Tungsten. Saffron bronze, tungsten-sodium tungstate, $Na_2WO_4.W_2O_5$.†

Orange Vermilion. See SCARLET VERMILION.

Orange Vermilion, Field's. See SCARLET VERMILION.

Orange W. See ORANGE IV.

Orange Yellow. See ORANGE G.

Orange 0. See ORANGE GT.

Oranit. See NEKAL.

Oranium Bronze. (Dirigold). An aluminium bronze. It contains from 87-97 per cent copper and 3-11 per cent aluminum.†

Orap. Pimozide. Applications: Antipsychotic. (McNeil Pharmaceuticals, McNEILAB Inc).

Orasol. Solvent soluble dyes. (Ciba-Geigy PLC).

Orasone. Prednisone. Applications: Glucocorticoid. (Rowell Laboratories Inc).

Ora-Testryl. Fluoxymesterone. Applications: Androgen. (E R Squibb & Sons Inc).

Oratrast. Barium sulphate. Applications: Diagnostic aid. (Armour Pharmaceutical Co).

Oratrol. Dichlorphenamide. Applications: Carbonic anhydrase inhibitor. (Alcon Laboratories Inc).

Oravue. Iopronic acid. Applications: Diagnostic aid. (E R Squibb & Sons Inc).

Orbenin. A proprietary preparation containing cloxacillin. An antibiotic. (Beecham Research Laboratories).*

Orbicin. A proprietary preparation of diabekacin sulphate. An antibiotic. (Pfizer International).*

Orbigastril. A proprietary preparation of oxyphencyclimine and meprobromate. An antispasmodic and tranquillizer. (Pfizer International).*

Orbinamon. A proprietary preparation of thiothixene hydrochloride. A psychotherapeutic agent. (Pfizer International).*

Orbitol. Photographic developer. (May & Baker Ltd).

Orca. A French synthetic resin prepared from acrolein. It is used for electrical insulation.†

Orchidee. (Sanfoin). The isoamyl ester of salicylic acid or *o*-oxy-benzoic acid, $C_6H_4(OH)COOC_5H_{11}$. Used in perfumery.†

Orchidee. Resistant purple colours for porcelain, bone china and earthenware. (Degussa).*

Orchil. See ARCHIL.

Orchil Extract N Extra. See APOLLO RED.

Orchil Red A. (Union Fast Claret). A dyestuff. It is the sodium salt of xyleneazoxyleneazo-β-naphthol-disulphonic-acid, $C_{26}H_{22}N_4O_7S_2Na_2$. Dyes wool red from an acid bath.†

Orchil Substitute N Extra. See APOLLO RED.

Orchindone. Isobutyl salicylate. Used in perfumery.†

Orcin. Orcinol, $C_6H_3.CH_3(OH)_2$.†

β-Orcin. Betorcinol, $C_6H_2(CH_3)_2(OH)_2$.†

Ordeal Bark. The bark of *Erythrophloeum guincense*.†

Ordeal Bean. (Split Nut). Calabar bean, a source of eserine.†

Ordnance 204, 500. Oil based water soluble cutting oil (extreme pressure). Applications: Multi-purpose oil based soluble cutting oil for all metal removing operations where physical and chemical extreme pressure assistance is required to assist functional cooling properties on many metals where a soluble oil is preferred. (Sumner Oil Industries).*

Ordonezite. A mineral. It is $2[ZnSb_2O_6]$.†

Ordoval. A synthetic tannin made from sulphonated anthracene.†

Ordoval G.2G. See NERADOL.

Orea. See RESINS, ACROLEIN.

Ore, Flinty Zinc. Flinty Calamine.†

Ore-furnace Slag. A slag obtained when roasted copper sulphide ore is mixed with oxidized ores and slag and fused for the production of "coarse metal". It often contains unfused quartz and less than 1 per cent copper, and is mainly a silicate of iron.†

Oregon Balsam. The true oleo-resin is obtained from *Pseudotsuga mucronata*, but another product sold under the same name consists of a mixture of rosin and turpentine.†

Oreide. A yellow alloy resembling gold. It usually contains from 80-90 per cent copper, 10-14.5 per cent zinc, and 0-4.5 per cent tin.†

Orelite. See METALITE.

Ore, Liver. See CINNABAR.

Orellin. A yellow colouring matter found in annatto. It is probably an oxidation product of bixin, another colouring matter of annatto.†

Oreoselone. Oreoselin, $C_{14}H_{12}O_4$.†

Ore, Potter's. See ALQUIFON.

Oresmasin. Pigments for textiles. (Ciba-Geigy PLC).

Ore, Spathic Zinc. See CALAMINE.

Oretic. Hydrochlorothiazide. Applications: Diuretic. (Abbott Laboratories).

Ore, Titaniferous. See ILMENITE.

Oreton. Testosterone. Applications: Androgen. (Schering-Plough Corp).

Oreton Methyl. Methyltestosterone. Applications: Androgen. (Schering-Plough Corp).

Oreton Propionate. Testosterone Propionate. Applications: Androgen. (Schering-Plough Corp).

Ore, Yellow Copper. Copper pyrites.†

Orgamide R. A proprietary trade name for nylon 6.†

Organidin. Glycerol, iodinated. Iodinated dimers of glycerol. Applications: Expectorant. (Wallace Laboratories, Div of Carter-Wallace Inc).

Organol. Dyes for the petroleum industry. (ICI PLC).*

Organy. Pennyroyal, also Origanum.†

Orgatrax. Hydroxyzine hydrochloride. Applications: Tranquillizer. (Organon Inc).

Orglas. G.R.P. linings and coatings. (Prodorite).*

Orient Yellow. See CADMIUM YELLOW.

Oriental Blue. See ULTRAMARINE.

Oriental Powder. An explosive used in fireworks. It is a mixture of gamboge and potassium nitrate.†

Oriental Sweet Gum. See STORAX.

Origanum Oil. See OIL OF ORIGANUM.

Original Green. See EMERALD GREEN.

Orimeten. Treatment of breast cancer. (Ciba-Geigy PLC).

Orimune. Poliovirus Vaccine Live Oral. Applications: immunizing agent. (Lederle Laboratories, Div of American Cyanamid Co).

Orinase. Tolbutamide. Applications. Antidiabetic. (The Upjohn Co).

Orinase Diagnostic. Tolbutamide Sodium. Applications: Diagnostic aid (diabetes). (The Upjohn Co).

Oriodide-131. Sodium Iodide I 131. Applications: Antineoplastic; diagnostic aid; radioactive agent. (Abbott Laboratories).

Oriol Yellow. (Cotton Yellow R, Alkali Yellow). A dyestuff. It is the sodium salt of primulineazosalicylic acid. Dyes cotton yellow from an alkaline bath.†

Orion Blue R33. A dyestuff. It is a British brand of Trisulphone blue R.†

Orion Blue 2B. A dyestuff. It is a British equivalent of Diamine blue 2B.†

Orion Brown B. A dyestuff. It is a British equivalent of Benzo brown B.†

Orion Brown G. A British dyestuff. It is equivalent to Benzo brown G.†

Orion Green B. A dyestuff. It is a British brand of Diamine green B.†

Orion Green G. A British equivalent of Diamine green G.†

Orion Sky-blue. A dyestuff. It is a British equivalent of Diamine sky-blue.†

Orion Violet. A British dyestuff. It is equivalent to Trisulphone violet B.†

Orisol. A solution of berberine acid sulphate.†

Orisulf. A proprietary preparation of sulphaphenazole. An antibiotic. (Ciba Geigy PLC).†

Orlean. See ANNATTO.

Orlest 28. A proprietary preparation of ethinylcestradiol and

NORETHISTERONE acetate. An oral contraceptive. (Parke-Davis).†

Ormolu. One alloy contains 58 per cent copper, 25.3 per cent zinc, and 16 per cent tin, and another consists of 90.5 per cent copper, 3 per cent zinc, and 6.5 per cent tin.†

Ornalith. See BAKELITE.

Ornithine. α-δ-Diamino-valeric acid, $(H_2N)CH_2.CH_2.CH_2.CH(NH_2)COOH$.†

Ornithite. A tricalcic phosphate, $Ca_3P_2O_8.2H_2O$.†

Oroglas DR. An impact modified polymethacrylate. Applications: Injection moulding, extension. (Rohm and Haas Company).*

Oromid. Nylon granules. Applications: Engineering plastic. (SNIA (UK) Ltd).†

Oronite. Lubricating oil additive. (Chevron).*

Oropon. A bate for leather. It is composed of the enzymes of pancreas absorbed in sawdust or kieselguhr, and intimately mixed with ammonium chloride or boric acid.†

Orotan. Pigment dispersing agent. Applications: Coatings. (Rohm and Haas Company).*

Orovite. Oral vitamins B and C. (Beecham Pharmaceuticals).

Orpiment. (Yellow Sulphide of Arsenic). A mineral, As_2S_3. Also see KING'S YELLOW.†

Orpiment, Red. See REALGAR.

Orris Camphor. Essential oil of orris.†

Orris Root. The rhizome of *Iris florentina*.†

Orr's White. See LITHOPONE.

Orseilline. See ARCHIL.

Orseilline No. 3. See FAST RED.

Orseilline No. 4. See FAST RED.

Orseillne BB. A dyestuff. It is the sodium salt of sulphotolueneazo-tolueneazo-α-naphthol-*p*-sulphonic acid, $C_{24}H_{18}N_4O_7S_2Na_2$. Dyes wool red from an acid bath.†

Ortamine Brown. A dyestuff. It is *o*-dianisidine.†

Orthene. Acephate insecticide. (Chevron).*

Orthenex. Pesticide. (Chevron).*

Orthesin. See ANAESTHESIN, BENZOCAINE.

Ortho. Pesticides, fertilizers publications. (Chevron).*

Orthochrome T. *p*-Toluquinaldine *p*-toluquinoline-ethyl-cyanine bromide. A red sensitizer for silver bromide plates.

Ortho-Chrysotile. A mineral. It is $2[Mg_3Si_2O_5(OH)_4]$.†

Orthocide. Captan fungicide. (Chevron).*

Orthocoll. See THIOCOLL.

Ortho-Gro. Liquid plant food. (Chevron).*

Ortho-Klor. Insecticide killer. (Chevron).*

Ortholeum. Grease stabilizer and lubricant assistant. (Du Pont (UK) Ltd).

Orthomatic. Lawn sprayer. (Chevron).*

Orthorix. Lime-sulphur fungicide. (Chevron).*

Orthosil. A proprietary trade name for an anhydrous sodium orthosilicate. A detergent.†

Orthotrol. Drift retardant. (Chevron).*

Orthoxicol. A proprietary preparation containing methoxyphenamine hydrochloride, codeine phosphate and sodium citrate. A cough linctus. (Upjohn Ltd).†

Orthoxine. A proprietary preparation of methoxyphenamine. A bronchial antispasmodic. (Upjohn Ltd).†

Orth's Stain. A microscopic stain. It contains 1 gram lithium carbonate and 2.5 grams carmine in 100 cc water.†

Ortizon. See HYPEROL.

Ortol. Dyestuffs for wool and polyamide. (BASF United Kingdom Ltd).

Ortol. A photographic developer. Methyl-*o*-amino-phenol, $C_6H_4(OH)(NHCH_3)$ (2 mols), combined with hydroquinone (1 mol.), forms the basis of this developer.†

Ortolan. A proprietary insulation.†

Ortolan. Dyestuffs for wool and polyamide (BASF United Kingdom Ltd).

Ortosol. A mixture for dry cleaning. It consists of 10 per cent chlorobenzene, 88 per cent *o*- and *m*-dichlorobenzenes, and 2 per cent *p*-dichlorobenzene.†

Orudis. A proprietary preparation of ketoprofen used in the treatment of arthritis. (May & Baker Ltd).*

Oruvail. A proprietary preparation of ketoprofen in a pH sensitive controlled release system. Applications: Treatment of arthritis. (May & Baker Ltd).*

Orvus WA. A proprietary trade name for a wetting agent. It is sodium lauryl sulphate.†

Orzan. A range of ammonium and sodium lignin sulphonates. Applications: Used as cement grinding aids and as thinners for drilling muds. (Harcros Industrial Chemicals).*

Os Sepiae. (White Fish-Bone). The calcareous shell lying within the back of the cuttle fish. It consists mainly of calcium carbonate.†

Osage Orange. A material obtained from the bark of the Osage orange tree, containing 25 per cent of tannin. Used as a dyestuff.†

Osbil. Iobenzamic acid. (May & Baker Ltd).

Oscodal. A proprietary preparation of vitamin cod-liver oil product.†

Osimol. Dispersing and levelling agent. Applications: Used for dyeing. (Degussa).*

Osmal Black. Activated carbon which has absorbed ammonia. It can replace part of the red lead in rust-preventive pigments. It is used particularly for exposure to sewer gases.†

Osmic Acid. The name formerly used for osmium tetroxide, OsO_4.†

Osmiridium. See IRIDOSMINE.

Osmitrol. Mannitol. Applications: Diagnostic aid; diuretic. (Travenol Laboratories Inc).

Osmo-calamine. Colloidal calamine.†

Osmoglyn. Glycerin. Applications: Pharmaceutic aid. (Alcon Laboratories Inc).

Osmo-kaolin. A preparation of kaolin obtained by a patented electro-osmosis process. It has a high covering power and clinging properties, and is used in toilet powders.†

Osmondite. The stage in the transformation of austentite, at which the solution in dilute sulphuric acid reaches its maximum rapidity.†

Osmo-sil. A dye absorbent. It is a very pure form of silica, SiO_2.†

Osnol. A proprietary preparation of liquid paraffin with vitamin D.†

Ospolot. A proprietary preparation of sulthiame. An anticonvulsant. (FBA Pharmaceuticals).†

Ospolot/Ospolot Mite. Pharmaceutical preparation. Applications: Anticonvulsants. (Bayer & Co).*

Osram. An alloy of osmium and tungsten.†

Ossein. A variety of gelatin prepared from bones.†

Ossivite. A proprietary preparation of bonemeal and Vitamins A and D. (Wyeth).†

Ostamer. Polyurethane Foam. Applications: Prosthetic aid. (Merrell Dow Pharmaceuticals Inc, Subsidiary of Dow Chemical Co).

Ostan. A trade name for pure sodium and potassium hydroxides in disc form (5 mm diameter), suitable for analytical work and convenience in weighing.†

Ostelin. A proprietary preparation. It is a vitamin cod-liver oil product.†

Ostreocin. A proprietary trade name for a mixture of Ostreogrycins B and G.†

Ostreogrycin. Antimicrobal substances produced by Streptomyces ostreogriseus (specific substances are designated by a terminal letter; thus, Ostreogrycin B). Ostreocin is a mixture of Ostreogrycins B and G.†

Ostrilan. Sulphonamide quinoline hormone veterinary ethical. (Ciba-Geigy PLC).

Oswego. A proprietary trade mark for cornflour.†

Otamidyl. Antibacterial/antifungal ear drops. (May & Baker Ltd).

Ote Seeds. (Acoomo Seeds). Seeds of *Myristica angolensis*, of Nigeria. The seeds yield Kombe fat.†

Otita. A solution containing 1.8 grains quinine dihydrochloride per fluid ounce in glycerin and water.†

Otoba Butter. (Otoba Wax, American Mace Butter). Otoba fat, obtained from the fruit of *Myristica otoba*.†

Otoba Fat. See OTOBA BUTTER.

Otoba Wax. See OTOBA BUTTER.

Otopred. A proprietary preparation of CHLORAMPHENICOL, PREDNISOLONE and THIOMERSAL used as anti-infective ear-drops. (Typharm).†

Otoryl. Compound ear drops, veterinary. (May & Baker Ltd).

Otosporin. A proprietary preparation of polymyxin B sulphate, neomycin sulphate and hydrocortisone used as anti-infective ear-drops. (The Wellcome Foundation Ltd).*

Ototrips. A proprietary preparation of POLYMYXIN B, BACITRACIN and TRYPSIN used as anti-infective ear-drops. (Consolidated Chemicals).†

Otox. Ear drops. (The Boots Company PLC).

Otrivin Hydrochloride. Xylometazoline Hydrochloride. Applications: Adrenergic (vasoconstrictor). (Ciba-Geigy Corp).

Otrivine. Hay fever formula nasal decongestant CCP. (Ciba-Geigy PLC).

Otrivine-Antisitin. Nasal decongestant. (Ciba-Geigy PLC).

Ottasept. Applications: For chemical compositions having antiseptic, germicidal and fungicidal properties such as para chloro meta xylenol, in Class 18. (Ferro Corporation).*

Otto. Essential oil.†

Ouabaine Arnaud. A proprietary preparation of ouabaine used in cases of threatened heart failure. (L Wilcox & Co Ltd).†

Oulumer 70. Partially polymerized tall oil rosin. Applications: Adhesive tackifier. (Oulu Oy).*

Oulutac 105. Pentaerythritol ester of tall oil rosin. Applications: Adhesive tackifier. (Oulu Oy).*

Oulutac 20 EP. Triethylene glycol ester of tall oil rosin. Applications: Adhesive tackifier and plasticizer. (Oulu Oy).*

Oulutac 30. Ethylene glycol ester of tall oil rosin. Applications: Adhesive tackifier. (Oulu Oy).*

Oulutac 30 D. Aqueous emulsion of tall oil rosin based ethylene glycol ester. Applications: Adhesive tackifier. (Oulu Oy).*

Oulutac 80. Glycerol ester of tall oil rosin. Applications: Adhesive tackifier. (Oulu Oy).*

Oulutac 80 D. Aqueous dispersion of tall oil rosin based glycerol ester. Applications: Adhesive tackifier. (Oulu Oy).*

Oulutac 90. Pentaerythritol ester of tall oil rosin. Applications: Adhesive tackifier. (Oulu Oy).*

Oulutac 90 D. Aqueous dispersion of tall oil rosin based pentaerythritol ester. Applications: Adhesive tackifier. (Oulu Oy).*

Ounce Metal. A bronze consisting of 85 per cent copper, 5 per cent tin, 5 per cent zinc, and 5 per cent lead.†

Ouralpatti. See VENTILAGO MADRASPANTA.

Ourari. Curare.†

Outremer. See ULTRAMARINE.

Ouvarovite. A mineral. It is a lime-chrome-garnet.†

O.V. See OIL OF VITRIOL.

Ovaban. Megestrol acetate. Applications: Antineoplastic. (Schering-Plough Corp).

Ovac. A proprietary trade name for a rubber vulcanizing accelerator. It is a blend of the thiazole derivatives of two specially selected aldehyde-amines.†

Overnite. A slug killer. (Murphy Chemical Co).†

Ovicide. Miscible tar oil winter wash. (ICI PLC).*

Ovigest. For the weakly lamb. (The Wellcome Foundation Ltd).

Ovitelmin. A proprietary preparation containing mebendazole. Applications: Veterinary antihelmintic (sheep and goats). (Janssen Pharmaceutical Ltd).*

Ovol. A proprietary preparation of dicyclomine hydrochloride and dimethicone used in the treatment of infant colic (Carter-Wallace).†

Ovolecithin. See LECITHIN.

Ovran. A proprietary preparation of ethinyloestradiol and *d*- norgestrel. An oral contraceptive. (Wyeth).†

Ovranette. A proprietary preparation of ethinyl oestradiol and d-norgestrel. An oral contraceptive. (Wyeth).†

Ovrette. Norgestrel. Applications: Progestin. (Wyeth Laboratories, Div of American Home Products Corp).

O-V Statin. Nystatin. Applications: Antifungal. (E R Squibb & Sons Inc).

Ovulen 1 mg. A proprietary preparation of ethynodiol diacetate and mestranol. Oral contraceptive. (Searle, G D & Co Ltd).†

Ovulen 50. A proprietary preparation of MESTRANOL and ETHYNODIOL diacetate. An oral contraceptive. (Searle, G D & Co Ltd).†

Oxaf. A proprietary accelerator. Zinc mercaptobenzothiazole. (Naugatuck (US Rubber)).†

O X A F. Zinc-2-mercaptobenzo-thiazole. A medium temperature accelerator widely used in latex compounding. It is also used in proofing, wire and druggist sundries where fast curing and a minimum of odour are required. (Uniroyal).*

O X A F 50D. A 50% water active dispersion. Ready to use form for latex compounding. (Uniroyal).*

Oxalan. Oxaluramide, $C_3H_5N_3O_3$.†

Oxalantin. Leucoturic acid, $C_6H_6N_4O_6$.†

Oxalate Blasting Powder. A safty explosive powder containing 71 per cent nitre, 14 per cent charcoal, and 15 per cent ammonium oxalate.†

Oxalic Ether. Ethyl oxalate $(C_2H_5.COO)_2$.†

Oxalid. Oxyphenbutazone. Applications: Anti-inflammatory; antirheumatic. (USV Pharmaceutical Corp).

Oxalumina. An abrasive consisting of small crystals of alumina.†

Oxamine Black BR. A dyestuff. Dyes cotton black.†

Oxanid. Oxazepam. Applications: Pharmaceutical preparation for the treatment of schizophrenia. (M A Steinhard Ltd).*

Oxford Ochre. See YELLOW OCHRE.

Oxford Yellow. See YELLOW OCHRE.

Oxicebral. A proprietary preparation of vincamine. A cerebral vasodilator. (Pfizer International).*

Oxi-Chek. Applications: For antioxidants for plastics and elastomers. (Ferro Corporation).*

Oxidation Black. See ANILINE BLACK.

Oxide of Iron, Dark. See INDIAN RED.

Oxide of Iron, Magnetic. See MAGNETIC IRON ORE.

Oxide Of Iron, Pale. See INDIAN RED.

Oxide of Iron, Violet. See INDIAN RED.

Oxide of Lead, Red. See RED LEAD.

Oxide, Red. See INDIAN RED.

Oxide Yellow. See OCHRE.

Oxidized Linseed Oil. See SOLIDIFIED LINSEED OIL.

Oxidized Oil of Turpentine. See DECAMPHORIZED OIL OF TURPENTINE.

Oxidized Oils. See BLOWN OILS.

Oxilube. Polyoxyalkylene diols and derivatives. (Shell Chemicals UK Ltd).

Oxilubes. Clear liquids which are co-polymers of ethylene oxide and propylene oxide. The range is used as base components for compressor oils, greases, gear oils and aviation turbine lubricants. (Shell Chemicals).†

Oximony. A proprietary red oxide of iron used as a rubber pigment.†

Oxine. 8-Hydroxyquinoline. An analytical reagent for metal analysis.†

Oxiosol. See ISOFORM.

Oxi-tan. A proprietary tanning compound.†

Oxitex. Textile fibre lubricant. (Shell Chemicals UK Ltd).

Oxitol. A colourless, slightly hygroscopic liquid with mild odour. It is used as a solvent in paints, varnishes, inks and stains. It is also effective as an extraction agent for antibiotics and as a degreasing solvent. (Shell Chemicals).†

Oxolin. (Perchoid). A patented material made from oxidized oil, jute fibre, and sulphur. It is a rubber substitute.†

Oxone. (Oxolin). Compressed sodium peroxide, Na_2O_2. Used in washing powders.†

Oxonite. An explosive. It is made from 54 per cent of nitric acid (specific gravity 1.5), and 46 per cent of picric acid.†

Oxsoralen. Methoxsalen. Applications: Pigmentation agent. (Elder Pharmaceuticals Inc).

Oxucide. Piperazine Citrate. Applications: Anthelmintic. (Sterling Drug Inc).

Oxxef. Latamoxef. Applications: Semi-synthetic broad-spectrum beta-lactam antibiotic. (Merck Sharp & Dohme).*

Oxyalizarin. Purpurin, $C_{14}H_5O$.†

Oxyammonia. Hydroxylamine, NH_2OH.†

Oxyanthracene. Anthrol $C_{14}H_{10}O$.†

Oxybuprocaine. 2-Diethylaminoethyl 4-amino-3-butoxybenzoate. Novesine is the hydrochloride.†

Oxycamphor. Campholenic acid, $C_9H_{15}.CO_2H$.†

Oxycarbonate of Bismuth. See BISMUTH SUBCARBONATE.

Oxycel. A proprietary preparation of oxidized cellulose. Used to stop haemorrhage. (Parke-Davis).†

Oxychloride. A disinfectant. It is a solution of sodium hypochlorite, containing 10-12 per cent available chlorine.†

Oxychloride of Tin. See TIN SALTS.

Oxycholine. Muscarine, $C_5H_{15}O_3N$.†

Oxycinchophen. 3-Hydroxy-2-phenyl-cinchonic acid.†

Oxycymol. Carvacrol, $C_{10}H_{14}O$.†

Oxydase. An oxidizing enzyme.†

Oxydasine. Consists mainly of a 0.05 per cent solution of vanadic acid. An antiseptic recommended for wounds.†

Oxydiamine Orange G and R. Polyazo dyestuffs. They are substantive cotton dyes, but can also be employed on wool and silk.†

Oxydislin. A compound of silicon having the formula Si_2H_2O, prepared by treating calcium silicide with cold dilute alcoholic hydrochloric acid in the dark. It is a white solid spontaneously inflammable in air.†

Oxydon. A proprietary preparation containing Oxytetracycline. An antibiotic (RP Drugs).†

Oxydpech. Oxide pitch obtained as a residue from the distillation of the fatty acids obtained from the oxidation of paraffin.†

Oxygenated Oil. Olive oil through which chlorine has been passed for several days.†

Oxygenated Paraffin. See PAROGEN.

Oxygen, Active. Ozone, O_3.†

Oxygen Cubes. Made by mixing sodium peroxide and bleaching powder together and compressing into tablets. They contain 100 parts of bleaching powder (33-35 per cent available chlorine), and 39 parts of sodium peroxide. On contact with water, oxygen is evolved.†

Oxygenite. A mixture of perchlorates or nitrates with a combustible substance. When ignited the mixture produces oxygen, and the material is used for this purpose.†

Oxygen Powder. Sodium peroxide, Na_2O_2.†

Oxygen, Solid. See OXYGEN CUBES.

Oxyguard. A proprietary anti-oxidant. 2, 6-di-tertiary-butyl-*p*-cresol. (Naugatuck (US Rubber)).†

Oxyhaemoglobin. Haemoglobin.†

Oxyliquit. A blasting explosive. It is formed by rapidly mixing liquid air, rich in oxygen, with powdered charcoal, petroleum residues, or cotton wool.†

Oxylith. A compressed powder. It is a mixture of sodium peroxide and bleaching powder, which evolves oxygen on treatment with water. Also see OXYGEN CUBES.†

Oxylone. Fluorometholone. Applications: Glucocorticoid. (The Upjohn Co).

Oxymel. Clarified honey (80 per cent) mixed with acetic acid (10 per cent) and water (10 per cent).†

Oxymuriate. Chlorate.†

Oxymuriate of Lime. See BLEACHING POWDER.

Oxymuth Saca. A colloidal suspension of bismuth oxyhydrate.†

Oxymycin. Oxytetracycline. Applications: Antibacterial. (O'Neal, Jones & Feldman Pharmaceuticals).

Oxyneurine. Betaine, $C_5H_{11}NO_2$.†

Oxynone. A proprietary trade name for a rubber antioxidant. It is 2, 4-diamino-diphenylamine.†

Oxyper. Sodium carbonate peroxyhydrate. (Interox Chemicals Ltd).

Oxyphenine. Direct cotton dyestuffs. See CHLORAMINE YELLOW. (Clayton Aniline Co).†

Oxyphenine Gold. See SIMILARTO CHLORMINE YELLOW.

Oxytoluol. Cresol, $CH_3C_6H_4OH$.†

Oxytracyl. Antibiotic veterinary ethical. (Ciba-Geigy PLC).

Oxytri. Polymerized β-hydroxy-trimethylene sulphide.†

Oxytril. Selective weedkiller. (May & Baker Ltd).

Oxytril P. Selective herbicide. (Murphy Chemical Ltd).

Oxzone. Hydrogen peroxide. (ABM Chemicals Ltd).*

Ozamin. Benzopurpurin.†

Ozark White. A pigment used in mixed paints. It usually consists of approximately 60 per cent zinc oxide and 40 per cent lead sulphate.†

Ozogen. Hydrogen peroxide, H_2O_2.†

Ozokerine. (Adepsine, Chrismaline, Chrysine, Cosmoline, Fossiline, Geoline, Petrolina, Saxoline, Paraffin Jelly). Trade names applied to varieties of soft paraffin. Also see PETROLEUM JELLY.†

Ozokerite. (Mineral Wax, Earth Wax, Cerasin, Cerosin, Cerin, Fossil Wax). The solid residue left when petroleum evaporates, which occurs mixed with earth in most oil-bearing districts. It consists of hydrocarbons of the olefine series, and is chiefly used for the preparation of ceresine.†

Ozole. A term applied to volatile aromatic odours contained in certain dextrines and plant extracts. "Dextrinozole" is the name applied to the body giving the scent of commercial dextrin.†

P

P 13 N. A trade mark for a range of polyimide varnishes and coatings. (TRW Inc, Redondo Beach, California).†

P-2003-K and P-2020-T. Low-density polyethylenes, of Soviet origin.†

P-289. A proprietary name for a PVC plasticizer of the lead stearate type. (Haagen Chemie BV).†

P3. A proprietary degreasing material. It is a mixture of water-glass and tri-sodium phosphate in solid form. It is used for cleaning metals, glass, and textiles. It is stated to have no corrosive action on aluminium, aluminium alloys, tin, zinc, and brass.†

P-51. A proprietary name for dibasic lead stearate used as a heat stabilizer for PVC. (BF Goodrich (UK) Ltd).†

Paalsgaard Oil. See SCHOU OIL.

Pabagel. Aminobenzoic acid. Applications: Ultraviolet screen. (Owen Laboratories Div, Dermatological Products of Texas, Alcon Laboratories Inc).

Pabanol. Aminobenzoic acid. Applications: Ultraviolet screen. (Elder Pharmaceuticals Inc).

Pabracort. A proprietary preparation of HYDROCORTISONE acetate. A nasal decongestant spray. (Pharmax).†

Pabrinex. Vitamins B and C preparation. Applications: Used in the case of severe depletion or malabsorption of vitamins B and C particularly in alcoholism, after acute infections, post operatively and in psychiatric states. (Paines & Byrne).*

P.A.C. A proprietary manufacture of formaldehyde.†

Pacatal. A proprietary trade name for the hydrochloride or the acetate of Pecazine. A sedative. (Warner).†

Pacer. Tyres. (Chevron).*

Pacherite. A mineral, $BiVO_4$.†

Pacific Blue. A dyestuff, $C_{58}H_{49}N_6$. Dyes wool and cotton greenish-blue.†

Pacitron. L-tryptophan. Applications: Mild to moderate depressive illness. (Berk Pharmaceuticals Ltd).*

Packfong. A nickel silver. It contains from 26-44 per cent copper, 16-37 per cent zinc, and 32-41 per cent nickel. One alloy contains 40.4 per cent copper, 25.4 per cent zinc, 31.6 per cent nickel, and 2.6 per cent iron. See NICKEL SILVERS.†

Packtong. See NICKEL SILVERS.

Paco. (Pacos). A Peruvian term for ferruginous earth containing small quantities of metallic silver.†

Pacolol. A preparation of the lysol class.†

Pacos. See PACO.

Pacvac. Liquid ring vacuum pump, baseplate mounted complete with liquid/air separator. Applications: Vacuum applications where the customer requires a pump set where only electrical pipe connections have to be made. (Sihi-Ryaland Pumps Ltd).*

Pacwet. Applications: Dust suppressant (especially in coal mines both above and underground). (Pacific Chemical Industries Pty Ltd).†

Paddox. Herbicide, containing the sodium/potassium salts of MCPA, mecoprop and dicamba. (ICI PLC).*

Padutin. Enzyme with circulatory action. (Bayer & Co).*

Padutin 100. Enzyme. Applications: For the treatment of male infertility due to reduced number and mobility of spermatozoa. (Bayer & Co).*

Paedo Sed. A proprietary preparation of dichloralphenazone and paracetamol. A hypnotic. (Pharmax).†

Paeonine. (Coralline Red, Aurine Red, Rosophenoline, Aurine R, Coralline). A dyestuff. It is the sodium salt of the reaction product of alcoholic ammonia upon aurine. Dyes silk and wool.†

Pafra. Synthetic emulsion adhesives. Applications: Adhesives. (Pafra Ltd).*

Pagid. Asbestos and non-asbestos friction material. Applications: Drum brake linings for trucks and passenger cars, disc brake pads for passenger cars, clutch facings for trucks and passenger cars, rolled material for industrial application, slip bearings for tooling machines. Repair, servicing and maintenance of vehicle brakes. (Caramba Chemie GmbH).*

Painter's Naphtha. A petroleum distillate. It has a boiling-point of 105-200° C.†

Paint, Waterproof. See MOUNTFORD'S PAINT.

Paint, White. See BISMUTH WHITE.

Pakfong. See NICKEL SILVERS.

Pala Gum. (Indian Gutta-percha). A product from a Ceylon tree. The coagulated juice resembles gutta-percha.†

Paladac. A proprietary preparation containing Vitamin A palmitate, calciferol, thiamine, riboflavine, pyridoxine, nicotinamide and ascorbic acid. (Parke-Davis).†

Palaite. A mineral. It is a hydrated manganese phosphate.†

Palamoll 632. A polyester of adipic acid and propanediol. A plasticizer for PVC. (BASF United Kingdom Ltd).*

Palamoll 644 and 646. A polyester of adipic acid and butanediol. A plasticizer for PVC. (BASF United Kingdom Ltd).*

Palamoll 645 and 647. Propriety polyadipates having viscosities of 60 mPa.S and 10,000 mPa.S at 20° C respectively. They are used as polymeric plasticizers. (BASF United Kingdom Ltd).*

Palamoll 855. A proprietary polymeric plasticizer having a viscosity of 5000 mPa.S. (BASF United Kingdom Ltd).*

Palanil. Dyestuffs for polyester fibres. (BASF United Kingdom Ltd).

Palanil Carrier. Auxiliaries for dyeing polyester fibres. (BASF United Kingdom Ltd).

Palao Amarillo. A rubber obtained from the Mexican *Euphorbia fulva*. It has a high resin content.†

Palapent. Pentobarbital Sodium. Applications: Hypnotic; sedative. (Bristol-Myers Co).

Palatal. Unsaturated polyester resins or vinyl ester resins. Applications: Reinforced with glass fibre it is used in boat building, pipe manufacture, storage tank manufacture, automotive industry and the building industry. Unreinforced it is used to make body filler. (BASF United Kingdom Ltd).*

Palatal KR 1397. A proprietary unsaturated polyester based on chlorendic acid dissolved in monostyrene. It is used in the manufacture of articles made of glass-reinforced plastics. (Dexine Rubber Co Ltd).†

Palatal P5. A proprietary unsaturated polyester resin dissolved in monostyrene. It has low viscosity and is used in the manufacture of articles made of glass-reinforced plastics. (BASF United Kingdom Ltd).*

Palatal P8, P50T and P52TL. Proprietary polyester resins used in the manufacture of articles made of glass reinforced plastics. (BASF United Kingdom Ltd).*

Palatal S333. A proprietary polyester resin used in the manufacture of articles made of glass-reinforced polyesters. (BASF United Kingdom Ltd).*

Palatase. A fungal lipase produced by fermentation of a selected strain of Aspergillus niger. Applications: Used for production of certain Italian cheese types and other specialty cheeses in which a modest lipolysis is desired. (Novo Industri A/S).*

Palatin Fast. Dyestuffs for wool and polyamide. (BASF United Kingdom Ltd).

Palatine Chrome Black. A dyestuff. It is equivalent to Palatine chrome blue.†

Palatine Chrome Blue. (Eriochrome, Blue-black R, Chrome Fast Cyanine B, BN, Alliance Chrome Blue-black R, Fast Chrome Cyanine 2B, Palatine Chrome Black, Stellachrome Black L757, Diadem Chrome Blue-black P6B, Lighthouse Chrome Cyanine R, Era Chrome Dark Blue B). A dyestuff. It is obtained by combining diazotized 1-amino-2-naphthol-4-sulphonic acid with β-naphthol.†

Palatine Chrome Brown. A dyestuff prepared from diazotized *o*-amino-phenol-*p*-sulphonic acid, and *m*-phenylenediamine.†

Palatine Orange. A dyestuff. It is the ammonium salt of tetranitro-γ-diphenol. Dyes wool and silk from an acid bath.†

Palatine Scarlet. (Cochineal Scarlet PS, Brilliant Cochineal). A dyestuff. It is the sodium salt of *m*-xyleneazo-naphtholdisulphonic acid, $C_{18}H_{14}N_2S_2O_7Na_2$. Dyes wool scarlet from an acid bath.†

Palatinit. A mixture of sodium hydrosulphite (Blankit) and zinc dust. A bleaching agent.†

Palatinol A. A registered trade mark for plasticizer for cellulose lacquers. It is stated to be diethyl-phthalate, $C_6H_4(COO.C_2H_5)_2$. It dissolves cellulose nitrate, ester gum, coumarone, etc.†

Palatinol AH. Di-2-ethyl benzylphthalate and dioctylphthalate. PVC plasticizers for general application. (BASF United Kingdom Ltd).*

Palatinol C. Di-butyl phthalate. A plasticizer.†

Palatinol DN. Di-iso-nonyl phthalate. A plasticizer for PVC. (BASF United Kingdom Ltd).*

Palatinol D10. Di-iso-octylphthalate. A plasticizer for PVC. (BASF United Kingdom Ltd).*

Palatinol K. Di-butylglycol phthalate. A plasticizer for PVC. (BASF United Kingdom Ltd).*

Palatinol Z. Di-iso-decyl phthalate. A plasticizer for PVC. (BASF United Kingdom Ltd).*

Palatinol 1C. Di-isobutyl phthalate. A non volatile softening agent for cellulose esters. A trade mark.†

Palatlaol M. A non-volatile softening agen for cellulose esters. It is dimethyl phthalate. A trade mark.†

Palatone. A proprietary trade name for an acrylic denture material.†

Palau. A platinum substitute. It is an alloy of gold and palladium, usually 80 per cent gold and 20 per cent palladium. Another alloy termed Palau contains 60 per cent nickel, 20 per cent platinum, 10 per cent palladium, and 10 per cent vanadium.†

Pale Acid. Nitric acid containing less than 0.1 per cent nitrogen oxides.†

Pale Cadmium. See CADMIUM YELLOW.

Pale Catechu. See CUTCH.

Pale Cobalt Violet. A pigment. It is cobalt arsenite.†

Pale Lemon-yellow. A pigment. It is a chromate of barium.†

Pale Oils. A name applied to a distillate from the residue of petroleum which ha been treated with acid and soda, washe or filtered to a certain degree of refinin or colour. They have a light and mediu viscosity and are employed as lubricant for rapidly moving machinery.†

Pale Oxide of Iron. See INDIAN RED.

Pale Smalt. See SMALT.

Pale Yellow Gold. Alloys. One contains 91 per cent gold and 8.3 per cent silver, an another 91.6 per cent gold and 8.3 per cent iron.†

Paleva. A proprietary preparation of aspiri and paracetemol. An analgesic. (Concept Pharmaceuticals Ltd).*

Palfium. A proprietary preparation of dextromoramide. An analgesic. (MCP Pharmaceuticals).†

Palite. Chloromethyl chloroformate, $COC1.OCNH_2C1$. A military poison gas.†

alladium Asbestos. An asbestos coated with palladium used in gas analysis for the absorption of hydrogen.†

alladium Black. A finely divided palladium used as a catalyst in the hydrogenation of oils.†

alladium Gold. (White Gold). An alloy of 90 per cent gold and 10 per cent palladium. An alloy of 40 per cent copper, 31 per cent gold, 19 per cent silver, and 10 per cent palladium, is also known by this name.†

alladium Red. Ammonio-chloride of palladium. A red pigment.†

allgrip. Adhesives for use as an anti-slip agent on bags to be stacked. (Norsk Hydro AS).*

alliag. A range of precious metal alloys. Applications: For dentistry and dental engineering. (Degussa).*

allicid. Sodium tribismuthyl-tartrate.†

alm Butter. Palm oil.†

alm Oil, Para. See PARA PALM OIL.

alm Pitch. A pitch obtained by the treatment of palm oil with sulphuric acid.†

alm Wax. A yellow wax from *Ceroxylon andicola*. Used as a beeswax substitute.†

alma Christi Oil. See OIL OF PALMA CHRISTI.

almarosa Oil. See ROSÉ OIL.

almerite. A mineral. It is an aluminium-potassium phosphate.†

almine. See VEGETABLE BUTTER.

almitin. Commercial palmitic acid is incorrectly called by this name.†

alohex. Inositol niacinate. Applications: Vasodilator. (Sterling Drug Inc).

alorium. A platinum substitute. It is a white alloy of gold and platinum only distinguished from platinum with difficulty. It is a ductile, homogeneous alloy with a melting-point of 1310° C., and it remains stronger than platinum on heating.†

aludrine. A proprietary preparation of proguanil hydrochloride. An anti-malarial. (ICI PLC).*

alusol. Intumescent fireboard based on sodium silicate. Applications: Fire and smoke seals at fire resistant doors and service penetrations of fire walls. (BASF United Kingdom Ltd).*

α-Palygorskite. Synonym for Lassallite.†

Pamak. A range of tall oil fatty acids. Applications: Chemicals and chemical processing, construction and building, floor coverings, paints and coatings, rubber, soaps, detergents and household products. (Hercules Inc).*

Pamelor. Nortriptyline hydrochloride. Applications: Anti-depressant. (Sandoz Pharmaceuticals, Div of Sandoz Inc).

Pamergan. Promethazine/pethidine preparations. (May & Baker Ltd).

Pameton. Tablets containing paracetamol and DL-methionine. Applications: Analgesic for minor painful conditions. Particularly useful where the possibility of misuse or overdosage exists. (Winthrop Laboratories).*

Paminal. A proprietary preparation of hyoscine methobromide and phenobarbitone. (Upjohn Ltd).†

Pamine. A proprietary preparation of methscopolamine bromide. A gastro-intestinal sedative. (Upjohn Ltd).†

Pamn. The methonitrate of Pramine.†

Pamolyn. Tall oil fatty acids. (Hercules Ltd).

Pamolyn 100. Oleic acids from vegetable sources. Applications: Used as plasticizers, emulsifiers and textile and wet-processing aids, in the food and drink industry, in the manufacture of personal care products and cosmetics. (Hercules Inc).*

Pamolyn 100 FGK. Oleic acids from vegetable sources. Applications: Used as plasticizers, emulsifiers and textile and wet-processing aids, in the food and drink industry, in the manufacture of personal care products and cosmetics. (Hercules Inc).*

Pamolyn 100FG. Oleic acids from vegetable sources. Applications: Used as plasticizers, emulsifiers and textile and wet-processing aids, in the food and drink industry, in the manufacture of personal care products and cosmetics. (Hercules Inc).*

Pamolyn 125. Technical grade oleic acid derived from vegetable sources, contains

83% oleic acid and 14% linoleic acid. Applications: It is an anionic collector for oxide and non metallic mineral flotation, is used as a metal drawing aid, is used in the production of additives for oil and grease and to produce emulsion breakers. (Hercules Inc).*

Pamolyn 200. Technical grade linoleic acid derived from vegetable sources. Applications: Used in caulking and sealant compositions, used to produce oleoresinous printing ink vehicles, used for making epoxy resin ester coatings and pale colour-retentive fast drying alkyds. (Hercules Inc).*

Pamolyn 240. Technical grade linoleic acid derived from vegetable sources. Applications: Used in caulking and sealant compositions, used to produce oleoresinous printing ink vehicles, used for making epoxy resin ester coatings and pale colour-retentive fast drying alkyds. (Hercules Inc).*

Pamolyn 300. Conjugated linoleic acid derived fom vegetable sources. Applications: Used as epoxy ester resin intermediates for adhesives and sealants, used as chemical intermediates for conjugated double-bond reactions, used as modifiers of styrenated, vinylated and methacrylated alkyds. (Hercules Inc).*

Pamolyn 327B. A partially bodied, conjugated fatty acid. Applications: It is used as a replacement for G-H-viscosity dehydrated castor oil in short to medium oil alkyds and copolymer alkyd resins. (Hercules Inc).*

Pamolyn 380. Conjugated linoleic acid derived fom vegetable sources. Applications: Used as epoxy ester resin intermediates for adhesives and sealants, used as chemical intermediates for conjugated double-bond reactions, used as modifiers of styrenated, vinylated and methacrylated alkyds. (Hercules Inc).*

Pan Scale. The calcium sulphate, containing some sodium chloride, which settles out during the crystallization of salt from brine. It is sold as "salt lick" for cattle, also for manuring purposes.†

Panabath. Panacide preparation for water baths. (BDH Chemicals Ltd).

Panacide. Dichlorophen preparations. (BDH Chemicals Ltd).

Panaclean. Biocidal cleaner. (BDH Chemicals Ltd).

Panacryl. Cationic dyestuffs. Applicatior Dyeing acrylic fibres, modacrylic fibre and basic dyeable polyesters and nylor (Holliday Dyes & Chemicals Ltd).*

Panacur. Fenbendazole. Applications: Anthelmintic. (Hoechst-Roussel Pharmaceuticals Inc).

Panadeine CO. A proprietary preparation of paracetamol and codeine phosphate An analgesic. (Bayer & Co).*

Panadol. A proprietary preparation paracetamol. An analgesic. (Bayer & Co).*

Panadonin. A preparation of *Adonis davurica*, a Japanese herb. Advocated as a substitute for digitalis.†

Panama Bark. Quillaia bark.†

Panama Crimson. A colouring matter obtained from the leaves of a vine call "china."†

Panar. A proprietary preparation of containing pancreatin. (Armour Pharmaceutical Co).†

Panase. A combination of digestive enzymes of the pancreas, derived from the pancreatic glands of the pig.†

Panasorb. A proprietary preparation of paracetamol in a sorbitol base. An analgesic. (Bayer & Co).*

Panazyme. Fungal protease. (ABM Chemicals Ltd).*

Pancil T. Fruit tree canker paint. (Rohm & Haas (UK) Ltd).

Pancoxin (1). A proprietary preparation containing sulphaquinoxaline. A veterinary coccidiostat.†

Pancoxin (2). A proprietary preparation of AMPROLIUM. An anti-protozoan for veterinary use.†

Pancoxin (3). A proprietary preparation of ETHOPABATE. A veterinary anti-protozoal.†

Pancreas Diastase. Amylopsin, an enzyme.†

Pancrease. Pancrelipase. A concentrate of pancreatic enzymes standardized for lipase content. Applications: Enzyme.

(McNeil Pharmaceuticals, McNEILAB Inc).

Pancreatokinase. A mixture of Eukinase (*qv*), and pancreatin.†

Pancreol. A trypsin preparation, used for bating skins.†

Pancreozymin. A hormone obtained from duodenal mucosa.†

Pancrex. A pancreatic deficiency supplement. Applications: Used in the treatment of cystic fibrosis and chronic pancreatic steatorrhea. (Paines & Byrne).*

Pancrex V. A proprietary preparation of concentrated pancreatin used in the treatment of cystic fibrosis. (Paines & Byrne Ltd).*

Pandex. A selenium preparation for rubber vulcanization.†

Panelyte. A proprietary trade name for phenol-formaldehyde laminated products and paper, fabric, wood veneer, fibre glass, and asbestos base thermosetting plastics for structural work.†

Panets. Peadiatric paracetamol preparations. (The Boots Company PLC).

Panex. Diacetyltartaric ester of edible mono-diglycerides. Applications: Used as emulsifier/ antistaling additive for bread fats, mayonnaise and sauces. (Harcros Industrial Chemicals).*

Panilax. A registered trade name for materials made from aniline-formaldehyde synthetic resin. They are thermoplastic but have a softening point about 100° C.†

Panitrin. Papaverine nitrate in acetyl diethyl-amide.†

Pankreon. See PANCREON.

Panmycin. A proprietary preparation of tetracycline. An antibiotic. (Upjohn Ltd).†

Panmycin Hydrochloride. Tetracycline Hydrochloride. Applications: Anti-amebic; antibacterial; antirickettsial. (The Upjohn Co).

Panmycin Syrup. Tetracycline. Applications: Anti-amebic; antibacterial; antirickettsial. (The Upjohn Co).

Pannetier Green. See CHROMIUM GREEN.

Panok. A proprietary trade name for Paracetamol.†

Pan-O-Lite. Sodium aluminium phosphate. Applications: Food leavening agent. (Monsanto Co).*

Panolog. Triamcinolone acetonide, mycostatin, neomycin sulphate and thiostrepton in cream, ointment or lotion base. (Ciba-Geigy PLC).

PanOxyl Aquagel. White viscous gels in three grades containing 2.5%, 5% and 10% benzoyl peroxide. Applications: For use in the topical treatment of acne vulgaris. (Stiefel Laboratories (UK) Ltd).*

PanOxyl Wash. A viscous lotion containing benzoyl peroxide. Applications: An antibacterial cleanser for use in the topical treatment of acne vulgaris. (Stiefel Laboratories (UK) Ltd).*

Pansecretin. See DUODENIN.

Pantakaust. A fuel similar to meta.†

Pantal. An aluminium alloy containing 0.8-2 per cent magnesium, 0.4-1.4 per cent manganese, 0.5-1 per cent silicon, and 0.3 per cent titanium. It resists corrosion, and has a tensile strength of from 18-33 kg per sq mm.†

Pantarol. A proprietary product for the protection of metal. It is applied by brushing, spraying, and dipping. It resists the action of light, sea air, steam, acid fumes, but is destroyed by concentrated acids and alkalis.†

Pantene. Hair tonic, shampoo and conditioner to keep hair healthy and attractive. (Richardson-Vicks Inc).*

Pantene Grooming Lotion. Grooming lotion. (Roche Products Ltd).

Panteric. A proprietary preparation containing pancreatin (triple B.P. strength). (Parke-Davis).†

Panthoderm. Dexpanthenol. Applications: Cholinergic. (USV Pharmaceutical Corp).

Pantholin. Calcium pantothenate. Application: Vitamin. (Eli Lilly & Co).

Pantolit. A proprietary synthetic resin of the phenol-formaldehyde type.†

Pantopaque. Iophendylate. Applications: Diagnostic aid. (Alcon Laboratories Inc).

Pantopon. (Omnopon, Nealpon). Mixtures of the soluble hydrochlorides of opium alkaloids.†

Pantosept. A German antiseptic in which the active agent is hypochlorous acid.†

Pantothenyl Alcohol. Panthenol.†

Panturon. A proprietary preparation containing atropine sulphate, papaverine hydrochloride phenobarbitone, aluminium hydroxide gel, kaolin, magnesium carbonate, magnesium trisilicate and oil of peppermint. An antacid. (Norma).†

Panwarfin. Warfarin Sodium. Applications: Anticoagulant. (Abbott Laboratories).

Paoferro. The inner bark of the Brazilian ironwood tree. Used as an antidiabetic.†

Papain. See PAPAYOTIN.

Papaw. The seeds of *Asimina tribola.* An emetic.†

Papayotin. (Papain, Papoid). Papain, a vegetable digestive ferment obtained from the unripe fruit of the papaw tree. Has been used for dyspepsia.†

Paper Scarlet Blue Shade. See BRILLIANT CROCEINE M.

Paperad. Finely divided aluminium trihydrate paper pigment. Applications: Used in papers. (Reynolds Metal Co).*

Paper-clay. See BENTONITE.

Paperhanger's Alum. Aluminium sulphate, used for sizing paste.†

Paperine. A starch product used in paper manufacture.†

Paper, Mountain. See MOUNTAIN LEATHER.

Paper-spar. A mineral. It is a variety of calcite.†

Paper, Tetra-base. See TETRA-PAPER.

Paper, Vulcanized. See VULCANIZED FIBRE.

Papi. Polymeric diphenylmethane di-isocyanate products used in the manufacture of polyurethane products. (Dow Chemical Co Ltd).*

Papite. A tear gas. It is acrolein with stannic chloride.†

Papoid. See PAPAYOTIN.

Paposite. A mineral. It is a ferric sulphate

Pappenheim's Stain. (Pyronin Stain). A microscopic stain. It consists of 1 part of a concentrated solution of pyronin and 3 parts of a concentrated solution of methyl green.†

Paprika. Cayenne pepper.†

Papua Nutmeg Butter. See MACASSAR NUTMEG BUTTER.

Papyrus. A proprietary casein product.†

Par. See CATALPO.

Para Arrowroot. Tapioca.†

Parabar. Oxidation inhibitors. (Exxon Chemical Ltd).*

Parabis. Resin intermediate. A pure form of bisphenol A used as an intermediate in the production of polycarbonate and polysulphone resins. (Dow Chemical Cc Ltd).*

Para Blue. A basic induline dyestuff, obtained by heating Spirit blue with *p*-phenylene-diamine. Dyes tannined cotton greyish-blue.†

Parabolix 100. Anti-static and cleaning agent for electronics and light luminaires. Applications: Decreases electronic components, removes fingerprints and dirt from parabolic ligh fixtures. (Merix Chemical Co).*

Para Butter. See PARA PALM OIL.

Paracarmine. A microscopic stain. It contains 1 gram carminic acid, aluminium chloride, 4 grams calcium chloride, and 100 cc 70 per cent alcohol.†

Parachlor. Parachlormetacresol. *p*-chlor-*m*-cresol. A germicide.†

Paracodeine. Dihydro-codeine hydrochloride.†

Paracodol. A proprietary preparation of paracetamol and codeine phosphate. An analgesic. (Fisons PLC).*

Paracol. Wax and wax-rosin emulsions with excellent shelf, shear and chemical stability. Applications: Used in paper and paperboard and to impart resistanc to water and aqueous solutions. Also used in many types of wet or dry-forme building products. (Hercules Inc).*

ara-col. Paraquat based herbicide. (ICI PLC, Plant Protection Div).

aracolline. A rubber solution.†

aracon. A generic name for polyester elastomers. See PARAPLEX.†

aracort. Prednisone. Applications: Glucocorticoid. (Parke-Davis, Div of Warner-Lambert Co).

aracortol. Prednisolone. Applications: Glucocorticoid. (Parke-Davis, Div of Warner-Lambert Co).

aracoto Bark. Coto bark.†

aracoumarone Resin. (Cumar Resin, Cumar Gum, Coumarone Resin, Benzo-Furane Resin, Cumar). A synthetic resin produced from coal-tar distillates. Solvent naphtha distilling between 150 and 200° C. is used, and is polymerized with sulphuric acid, or aluminium chloride, tin chloride, ferric chloride and phosphoric acid. It contains *p*-coumarone, *p*-indene, and polymers of other hydrocarbons. The resin is used in the production of varnishes, polishes, artificial leather, and linoleum.†

aracril (NBR). A range of polymers offering heat, oil and abrasion resistance. (Uniroyal).*

aracure. Dispersions and emulsions of various chemical additives, primarily rubber latex antioxidants, accelerators and curatives. Applications: Chemical additives for rubber latex compounding. (Testworth Laboratories Inc).*

aradene. A proprietary trade name for coumarone-indene resins.†

aradione. Paramethadione. Applications: Anticonvulsant. (Abbott Laboratories).

Paradise Grains. See GRAINS OF PARADISE.

Paradol. See NERADOL.

Paradol. A range of fatliquors. Applications: Modification of leather handle. (Yorkshire Chemicals Plc).*

Paradow. Para-dichlorobenzene. Applications: Commercial and industrial solvents, deodorants and sanitary products. (Dow Chemical Co Ltd).*

Paradura. A proprietary trade name for a phenolic synthetic resin for varnish and lacquers.†

Paradyne. Fuel oil additives. (Exxon Chemical International Inc).*

Paradyne. Flow improvers. (Exxon Chemical Ltd).*

Parafecol. A paste containing phenol. A disinfectant.†

Paraffagar. A proprietary preparation of liquid paraffin and agar-agar.†

Paraffin Jelly. See OZOKERINE.

Paraffin Oil. See KEROSENE.

Paraffin Scale. Crude paraffin wax.†

Paraffin Wax. The wax obtained from petroleum, and from bituminous shales.†

Paraffinum Molle. See PETROLEUM JELLY.

Paraffin, Liquid. A mixture of liquid hydrocarbons obtained by the distillation of the liquid remaining after the lighter hydrocarbons have been removed from petroleum. It is decolourized and purified. The following names (proprietary and otherwise) are some of those used for liquid paraffin and similar preparations. Adepsine Oil, Albolene, Alboline, Amilee, Atoleine, Atolin, Bakurol, Blandine, Crysmalin, Deeline, Glycoline, Glyco, Glymol, Hydrocarbon oil, Interol, Internol, Lilyolene, Mineral glycerin, Mineral syrup, Minol, Muthol, Neutralol, Nujol, Paroleine, Paroline, Petralol, Petro, Petrolax, Petrolia, Petronol, Petrosio, Russol, Saxol, Seneprolin, Stanolax, Stanolind, Terraline, Terroline, Usoline.†

Paraffin, Native. See OZOKERITE.

Paraffin, Oxygenated. See PAROGEN.

Parafil. Fibre reinforced thermoplastic rope/cable. (ICI PLC).*

Parafilm. A proprietary trade name for a rubber composition.†

Parafix. Photographic fixer. (May & Baker Ltd).

Paraflex. Chlorzoxazone. Applications: Relaxant. (McNeil Pharmaceuticals, McNEILAB Inc).

Paraflow. Pour point depressants. (Exxon Chemical Ltd).*

Paraflow. A synthetic lubricating oil prepared by the condensation in the presence of anhydrous aluminium chloride of chlorinated wax with aromatic hydrocarbons.†

Paraflow. Dewaxing oils and pour depressants for lubricants and power transmission fluids. (Exxon Chemical International Inc).*

Paraform. Moulded insulation shapes. (The Chemical & Insulating Co Ltd).

Parafuchsine. (Para-magenta). A dyestuff. It is the hydrochloride of triaminotriphenylcarbinol, $C_{19}H_{26}N_3C1O_4$. Dyes wool, silk, and leather red, and cotton mordanted with tannin and tartar emetic, red.†

Paragearksutite. A mineral. It is $CaA1(F,OH)_5$. $3/4H_2O$.†

Paragite. A mineral. It is an apatite containing iron.†

Paraglas. Cast acrylic sheet. Applications: Used for production of light domes, illuminated signs, wash basins and bathtubs, safety coverings for machines, showcases, graduated dials, models etc. (Degussa).*

Paraglobin. See GLOBULIN.

Paraglobulin. See GLOBULIN.

Paragol. A rubber substitute made from oxidized oil.†

Paragon Steel. A proprietary trade name for a non-shrinking steel contains 1.55 per cent manganese, about 0.6 per cent chromium, and 0.25 per cent vanadium.†

Paragon-15. Polysiloxane resin in an aliphatic vehicle system. Applications: Water repellant for masonry. (Nova Chemical Inc).*

Paragrid. Synthetic reinforcement and support materials for coil engineering applications. (ICI PLC).*

Para Gum. See JUTAHYCICA RESINS.

Paragutta. A patented insulating compound for use in the manufacture of submarine telegraph and telephone cables. It is made from deproteinized rubber obtained by the heat treatment of rubber latex to hydrolyse the protein and washing. This rubber has reduced water-absorbing properties and is mixed with gutta-percha or balata and suitable waxes to produce paragutta.†

Parahypon. Analgesic. (The Wellcome Foundation Ltd).

Parake. A proprietary preparation of PARACETAMOL and codeine phosphate. An analgesic. (Galen).†

Paral. Paraldehyde. Applications: Hypnotic; sedative. (O'Neal, Jones & Feldman Pharmaceuticals).

Paralac. Solvated solutions of synthetic rubber and resins. Applications: Contact adhesives and coatings. (Testworth Laboratories Inc).*

Paralactic Acid. Sarco-lactic acid.†

Paralaudin. Diacetyldihydromorphine.†

Paralene. A range of synthetic tanning agents. Applications: Tanning of leather. (Yorkshire Chemicals Plc).*

Paralgin. A proprietary preparation of PARACETAMOL, cakeine and codeine phosphate. An analgesic. (Norton, H N Co).†

Paralink. Synthetic reinforcement and support materials for coil engineering applications. (ICI PLC).*

Paraloid EXL 2607. Acrylic and MBS additives for engineering plastics. Applications: Toughening polymers particularly for automotive and leisure segments. (Rohm and Haas Company).*

Paraloid K120N and KM-228. Acrylic modifiers for PVC. (Rohm and Haas Company).*

Paraloop. Synthetic reinforcement and support materials for coil engineering applications. (ICI PLC).*

Paralux. Lube oils. (Chevron).*

Param. Cyanoguanidine, $C_2H_4N_4$.†

Paramagenta. See PARAFUCHSINE.

Paramar. A mineral rubber used in rubber mixing.†

Paramax. A proprietary preparation of paracetamol with metoclopramide. Applications: For symptomatic treatment of migraine. (Beecham Research Laboratories).*

Paramel. A range of synthetic resins. Applications: Tanning of leather. (Yorkshire Chemicals Plc).*

Paramet Ester Gum. A proprietary trade name for rosin-glycerol synthetic resin for lacquer and varnish manufacture.†

Parametol. Emetine oleate solution in liquid paraffin.†

Paramid. Lube oils. (Chevron).*

Paramine Black BH. A dyestuff. It is a British equivalent of Diamine black BH.†

Paramine Black HW. A dyestuff. It is a British brand of Diamine black HW.†

Paramine Blue 2B New. A dyestuff. It is equivalent to Diamine blue 2B, and is of British manufacture.†

Paramine Brilliant Black FB, FR. Dyestuffs. They are British equivalents of Columbia black FF extra.†

Paramine Brown. A dyestuff. It consists of *p*-phenylenediamine oxidized on the fibre.†

Paramine Fast Brown B. A British equivalent of Diamine brown B.†

Paramine Fast Brown M. A dyestuff. It is a British brand of Diamine brown M.†

Paramine Fast Orange D, G. Dye stuffs. They are British equivalents of Mikado orange.†

Paramine Fast Orange S. A British dyestuff. It is equivalent to Benzo fast orange S.†

Paramine Fast Red F. A British brand of Diamine fast red.†

Paramine Fast Scarlet 4BS. An equivalent of Benzo fast scarlet 4BS.†

Paramine Fast Violet N. A dyestuff. It is a British equivalent of Diamine violet N.†

Paramine Fast Yellow 3G. A British brand of Mikado yellow.†

Paramine Green B, G. British equivalents of Diamine green G.†

Paramine Orange G, R. British brand of Benzo orange R.†

Paramine Yellow GG. A British dyestuff. It is equivalent to Nitrophenine.†

Paramine Yellow R, 2R, Y. Dyestuffs. They are equivalent to Direct yellow and are of British manufacture.†

Paramins. Additives for the petroleum industry. (Exxon Chemical Ltd).*

Paramol. *o*-Amino-*m*-hydroxy-benzy alcohol, $C_6H_3(OH)(CH_2OH)NH_2$. photographic developer.†

Paramol-118. A proprietary preparation of paracetamol and dihydrocodein bitartrate. An analgesic. (British Drug Houses).†

Paramontmorillonite. A mineral. It is a hydrated aluminium silicate.†

Paramorphan. Dihydro-morphine hydrochloride.†

Paramorphine. Thebaine, $C_{19}H_{21}O_3N$.†

Paramount. Lube oils. (Chevron).*

Paranitraniline Red. (Azophor Red, Para Red, Nitrazol, Discharge Lake R and RR). A dyestuff. It is *p*-nitro-benzeneazo-β-naphthol, $C_{16}H_{11}N_3O_3$. Used for dyeing cotton, and in the preparation of lakes for paper-staining.†

Paranol. A proprietary trade name for a phenol-formaldehyde synthetic resin.†

Paranolin. A soya-bean casein.†

Paranox. Detergents and dispersants. (Exxon Chemical International Inc).*

Parapak. Industrial oil additives. (Exxon Chemical Ltd).*

Para Palm Oil. (Para Butter). Pinot oil, a semi-drying oil from the seeds of *Euterpe oleracea*.†

Paraphenylene Blue G, R, and B. Dyestuffs formed by heating *p*-phenylenediamine with certain amino-azo compounds.†

Paraphenylene Blue R. (Fast New Blue for Cotton, Indophenine). A dyestuff obtained by the action of *p*-phenylene diamine upon the hydrochloride aminoazobenzene. Dyes cotton mordanted with tannin and tartar emetic, blue.†

Paraphthalein. A preparation of phenol-phthalein.†

Paraplex. Series of polymeric plasticizers. Applications: Flexible vinyl and ruber. (The C P Hall Company).*

Paraplex G.62. Epoxied soya bean oil. Applications: Vinyl plasticizer and vinyl stabilizer. (Rohm and Haas Company).*

Parapoid. Extreme pressure additives for gear lubricants. (Exxon Chemical International Inc).*

Paraquat. It is a herbicide which functions by stifling the chlorophyll uptake by plants. It is 1.1'-dimethyl-4-4'-dipyridium dichloride. Very toxic.†

Paraquat + Plus. Herbicide. (Chevron).*

Para Red. See PARANITRANILINE RED.

Pararosaniline. Triamino-triphenyl-carbinol. Pararosaniline chloride dyes wool and silk, purple-red, and cotton with mordants.†

Pararosaniline Base. An equivalent of Rosaniline.†

Pararosaniline Blue. See METHYL BLUE.

Parasiticine. A fungicide containing 57 per cent copper sulphate and sodium carbonate and bicarbonate.†

Parasulphurine S. See SULPHANIL YELLOW.

Paratac. Additive for lubricating oils and greases. (Exxon Chemical International Inc).*

Paratartaric Acid. Racemic tartaric acid, $C_4H_6O_6$.†

Paratect Bolus. Morantel tartrate - a cattle antihelmintic. (Pfizer International).*

Paratex. A proprietary preparation of thiamine hydrochloride, calcium and strychnine glycerophosphates, and sodium hypophosphate. A tonic. (The Albion Group).†

Parathion. O,O-diethyl O-*p*-nitrophenyl phosphorothioate. Diethoxy, nitro-phenoxy phosphorothioate. A powerful insecticide.†

Para-Thor-Mone. A proprietary preparation of parathyroid hormone. (Eli Lilly & Co).†

Paratie. Synthetic reinforcement and support materials for coil engineering applications. (ICI PLC).*

Paratol. Natural and synthetic rubber latex based adhesives and coatings - cohesive, pressure-sensitive and synthetic resin emulsions. Applications: Numerous adhesive bonding applications. (Testworth Laboratories Inc).*

Paratone. Viscosity index improver. (Exxon Chemical Ltd).*

Para Toner. Paranitraniline red. Used as a toner for lakes.†

Paratophan. (Homophan, Methyl-atophan). 6-Methyl-2-phenylquinoline-4-carboxyic acid.†

Paraweb. Fibre reinforced thermoplastic webbing for civil engineering applications and used as cargo slings, show fencing, windbreaks. (ICI PLC).*

Paraxin. Chloramphenicol.†

Parazol. Crude dinitrodichlorobenzene. It contains *m*-dinitro-*p*-dichlorobenzene, *o*-dinitro-*p*-dichlorobenzene, and *p*-dinitro-*p*-dichlorobenzene. It is used as a high explosive.†

Parazolidin. A proprietary preparation of PHENYLBUTAZONE and PARACETAMOL used in the treatment of arthritis. (Ciba Geigy PLC).†

Par Clay. Hydrated aluminium silicate (hard clay). Applications: Mineral filler used as filler, extender or reinforcing agent for paint, paper, rubber, ceramics, plastics and specialities. (Vanderbilt Chemical Corporation).*

Pardale. A proprietary preparation of paracetamol, codeine phosphate and caffeine. An analgesic. For veterinary use. (Dales Pharmaceuticals).*

Paredrine. Hydroxyamphetamine hydrobromide. Applications: Adrenergic. (Smith Kline & French Laboratories).

Pareira. The dried root of *Chondrodendron tomentosum*.†

Parel 58. Sulphur vulcanizable, elastomeric copolymer of polypropylene oxide and alkyl glycidyl ether. Applications: It is used for flexible wing seals in supersonic aeroplanes, it is used for motor mounts and other noise suppression applications requiring long term durability and is used as a blending rubber in tubes and tyres. (Hercules Inc).*

Parenamine. Protein Hydrolysate. Applications: Replenisher. (Sterling Drug Inc).

Parencillin. Penicillin G Procaine. Applications: Antibacterial. (Merrell Dow Pharmaceuticals Inc, Subsidiary of Dow Chemical Co).

Parenol. Consists of 65 per cent soft paraffin, 15 per cent wool fat, and 20 per cent distilled water.†

Parenol Liquid. Consists of 70 per cent liquid paraffin, 5 per cent white bees wax, and 25 per cent distilled water.†

Parentrovite. Parenteral vitamins B and C. (Beecham Pharmaceuticals).

Parenzyme. Trypsin, crystallized. Applications: Enzyme (proteolytic). (Merrell Dow Pharmaceuticals Inc, Subsidiary of Dow Chemical Co).

Parez Resins. Modified resins of melamine and formaldehyde. (Cyanamid BV).*

Parfenac. A proprietary preparation of bufexamac used as a skin cream. (Lederle Laboratories).*

Parian Cement. A cement which is similar to Keene's cement, except that a solution of borax is used instead of alum.†

Parianite. An asphaltum from the pitch lake at Trinidad.†

Parilene. A proprietary trade name for polyparaxylyene. A plastics material used for film manufacture for electrical purposes.†

Paris Blue. Finest Prussian blue (*qv*). Also see METHYL BLUE. The term is generally applied to a mixture of Prussian blue, Turnbull's blue, and Willow blue.†

Paris Green. See EMERALD GREEN and METHYL GREEN.

Paris Lake. See CARMINE LAKE.

Paris Red. See RED LEAD. A variety of rouge employed in polishing is also sold under this name.

Paris Salts. A disinfectant containing 50 parts zinc sulphate, 50 parts ammonia alum, 1 part potassium permanganate, and 1 part lime.†

Paris Violet. See METHYL VIOLET B.

Paris Violet 6B. See BENZYL VIOLET.

Paris Yellow. See CHROME YELLOW, TURNER'S YELLOW, and NAPLES YELLOW.

Parixistil. A proprietary preparation of hydroxyzine. An ataractic. (Pfizer International).*

Parkerized Steel. A patented process for the treatment of steel with iron and manganese phosphates to give the surface resistance to corrosion.†

Parker's Cement. See ROMAN CEMENT.

Parkesine. See CELLULOID.

Parlay. Growth regulator for growers of grass seed. (ICI PLC).*

Parlodel. Bromocriptine mesylate. Applications: Enzyme inhibitor. (Sandoz Pharmaceuticals, Div of Sandoz Inc).

Parlodion. A trade mark for a shredded form of pure collodion.†

Parlon. A proprietary trade name for a chlorinated rubber compound, rubber derivatives and rubber-like resins for use as a base for concrete paint and alkyd enamels.†

Parlon. Chlorinated rubber. It is available in six viscosity grades. Applications: Used as film-formers in adhesives, in corrosion resistant coatings for wood, metal and concrete, for wood floor finishes and sealers, in inks etc. (Hercules Inc).*

Parlon P. A range of chlorinated polypropylenes. It forms clear, hard, protective films that are resistant to chemicals, salt solutions and water. Applications: It is used as film formers in adhesives and sealants, the construction and building industry, for lumber and wood products, for floor coverings etc. (Hercules Inc).*

Parma. See ANILINE BLUE, SPIRIT SOLUBLE.

Parme R. See PRUNE, PURE.

Parmentine. A mixture of glycerin, gelatin, dextrine, sodium sulphite, and zinc sulphate. Used for sizing and finishing cotton, wool, and silk.†

Parmid. Tablets containing 10mg metoclopramide monohydrochloride BP. Syrup each 5ml containing 5mg metoclopramide monohydrochloride BP. Injection each 2ml ampoule containing 10mg metoclopramide monohydrochloride BP. Applications: Treatment of digestive disorders, nausea and vomiting, migraine, post-operative hypotonia, post-operative syndrome. (Lagap Pharmaceuticals Ltd).*

Parmr. The trade name for a blown bitumen residue. It is a mineral rubber for use in the rubber industry. Grade I melts as from 190-310° F., and Grade II at above 300° F.†

Parnate Tablets. Tranylcypromine sulphate (non-hydrazine monoamine oxidase inhibitor). Applications: Depressive illness particularly where phobic symptons are present. (Smith Kline and French Laboratories Ltd).*

Paroa-caxy Oil. The seed-oil of *Pentaclethra flamentosa.*†

Parodyne. See ANTIPYRINE.

Parogen. (Oleogen, Oxygenated Paraffin, Vasogen). Consists of 2 parts liquid paraffin, 2 parts oleic acid, and 1 part ammoniated alcohol (5 per cent.).†

Parogen, Thick. Consists of 6 parts hard paraffin, 24 parts liquid paraffin, 15 parts oleic acid, and 5 parts ammoniated alcohol (5 per cent.).†

Paroidin. Parathyroid. Applications: Regulator. (Parke-Davis, Div of Warner-Lambert Co).

Paroil. Chlorinated paraffin, liquid. Applications: Used in coating oils, industrial lubricant, flame retardant as well as plasticizer in plastics. (Dover Chemical Corp).*

Par-o-lac. An impregnating compound.†

Paroleine. See PARAFFIN, LIQUID.

Paroline. See PARAFFIN, LIQUID.

Paromomyicn. An antibiotic produced by Streptomyces rimosus forma paromomycinus.†

Paroven. A proprietary preparation of hydroxyethylrutosides used in the treatment of varicose veins. (Zyma (UK) Ltd).†

Parraynite. A trade name for a rubber compound which is used by X-ray operators to protect them from injury by exposure to the rays.†

Parrot Coal. Cannel coal.†

Parrot Green. See EMERALD GREEN and ZINC GREENS.

Parr's Alloys. Anti-corrosion alloys. One alloy contains 80 per cent nickel. 15 per cent chromium, and 5 per cent, copper. Another alloy contains 66.6 per cent nickel, 18 per cent chromium, 8.5 per cent copper, 3.3 per cent tungsten, 2 per cent alumimium, and 1 per cent manganese.†

Parsidol. Ethopropazine hydrochloride. Applications: Antiparkinsonian. (Parke-Davis, Div of Warner-Lambert Co).

Parsley Camphor. (Camphre de Persil). crystallized apiole.†

Parsolin. Organophosphorus insecticide. (Ciba-Geigy PLC).

Parson's Alloy. A proprietary trade name for an alloy of 56 per cent copper, 41.5 per cent zinc, 1.2 per cent iron, 0.7 per cent tin, 0.1 per cent manganese, and 0.46 per cent aluminium. It has a specific gravity of 8.4.†

Parstelin. A proprietary preparation of tranylcypromine, and trifluoperazine. (Smith Kline and French Laboratories Ltd).*

Parstelin Tablets. Combination of tranylcypromine sulphate and trifluoperazine hydrochloride. Applications: Treatment of depressive illness complicated by anxiety particularly where phobic symptons are present. (Smith Kline and French Laboratories Ltd).*

Partagon. A German preparation. It consists of rods containing silver chloride and - sodium-silver chloride.†

Partinium. (Victoria Aluminium). An aluminium alloy. It varies in composition, and often contains tungsten, copper, tin, zinc, and magnesium. One alloy contains 96 per cent aluminium, 2.4 per cent antimony, 0.8 per cent tungsten, 0.64 per cent copper, and 0.16 per cent tin. Another alloy consists of 88.5 per cent aluminium, 7.4 per cent copper, 1.7 per cent zinc, 1.3 per cent iron, and 1.1 per cent silicon.†

Partridge Berry Oil. See OIL OF PARTRIDGE BERRY.

Parvol. Contraceptive jelly, containing polyoxyethyleneothylcresol. (British Drug Houses).†

Parvol. A range of auxiliary products. Applications: For use as assistants in tanning processes. (Yorkshire Chemicals Plc).*

Parvoline. Dimethylethylpyridine, $C_9H_{13}N$.†

Parylene N. A plastic material used to make thin film membranes, 2-1000 Angstroms thick. (Union Carbide (UK) Ltd).†

Pasilex. Precipitated aluminium silicate. Applications: Used as filler for the paper industry. (Degussa).*

Pasonex. A proprietary preparation of aminosalicylate calcium, isoniazid, pyridoxine hydrochloride and menadione. An antitubercular. (Pfizer International).*

Passini's Solution. An aqueous solution of mercury and sodium chlorides and glycerine. Used to preserve animal tissue.†

Passow's Slag Cement. Prepared by blowing into liquid slag as it issues from the blast furnace, when it becomes granulated. It is then finely ground.†

Pastaccio. A residue from the manufacture of calcium citrate. It consists of vegetable cellulose with some hydrocarbons.†

Paste. See STRASS.

Paste Blue. See PRUSSIAN BLUE.

Patafol. Fungicide containing ofurace for the control of potato blight. (ICI PLC).*

Patafol Plus. Fungicide. (ICI PLC, Plant Protection Div).

Patava Oil. (Batana Oil). Coumou oil, a semi-drying oil obtained from the kernels of the Brazilian palm tree, *Oenocarpus batava*.†

Patchouli. An Indian herb, *Pogostemon patchouly*. Used in perfumery.†

Patent Atlas Red. See GERANINE.

Patent Bark. Commercial quercetin.†

Patent Black. An acid dyestuff. It is a substitute for logwood.†

Patent Blue A. (Disulphine Blue A). A dyestuff. Dyes wool greenish-blue.†

Patent Blue JOO. A dyestuff. It is a mixture of Patent blue and violet.†

Patent Blue V, N, Superfine, and Extra. (New Patent Blue B and 4B). Dyestuffs. The calcium, magnesium, or sodium salt of the disulphonic acid of *m* hydroxytetraallyldiaminotriphenylcarbinol. Dyes wool greenish-blue.†

Patent Fustin. (Wool Yellow). Dyestuffs. They are condensation products of fustic extracts and diazo compounds. They dye wool and cotton. The brand K is only partially oxidized, and is recommended for use with oxidizing mordants. The brand E is fully oxidized.†

Patent Green. See EMERALD GREEN.

Patent Greens 0 and V. Dyestuffs. They are mixtures of Patent blue and Acid green.†

Patent Orange. See ORANGE G.

Patent Phosphine. See PHOSPHINE.

Patent Red. See CINNABAR.

Patent Rock Scarlet. See ST. DENNIS RED.

Patent Yellow. See TURNER'S YELLOW.

Patent Zinc White. A pigment made by adding a soluble sulphide to a zinc chloride or zinc sulphate solution, filtering off the precipitate, drying it, and then calcining it. It has the composition, 5ZnS.ZnO.†

Patents. The small portion of very white flour obtained from wheat. It is poor in proteins, and is used for fancy breads.†

Path Gun. Ready for use herbicide spray. (ICI PLC).*

Path Weeds Killer. Total weedkiller. (Fisons PLC, Horticulture Div).

Pathclear. Long-acting weedkiller for paths, drives and patios. (ICI PLC).*

Pathilon. Tridihexethyl Chloride. Applications: Anticholinergic. (Lederle Laboratories, Div of American Cyanamid Co).

Pathocil. Dicloxacillin sodium. Applications: Antibacterial. (Wyeth Laboratories, Div of American Home Products Corp).

Pathozone. Cefoperazone sodium - an antibiotic used for mastitis in dairy cows. (Pfizer International).*

Patina. The green film which forms on copper and bronze mouldings. It consists of basic copper carbonate or other basic copper salts.†

Patoran. Substituted urea herbicide. (Ciba-Geigy PLC).

Pattern Metal. An alloy of 83 per cent copper, 10 per cent zinc, 4 per cent tin, and 3 per cent lead.†

Pattern Metal, Light. An alloy of 72 per cent aluminium, 16 per cent zinc, and 2 per cent copper.†

Pattinson's White Lead. A pigment. It is basic lead chloride, $PbCl_2.2Pb(OH)_2$.†

Pattonex. Active ingredient: metobromuron; 3-(4-bromophenyl)-1-methoxy-1-methylurea. Applications: Residual herbicide for use as a selective weedkiller. (Agan Chemical Manufacturers Ltd).†

Pattrex. Pattern plaster. (Foseco (FS) Ltd).*

Pattrit. Casting plaster. (Foseco (FS) Ltd).*

Pavabid. Papaverine Hydrochloride. Applications: Relaxant. (Marion Laboratories Inc).

Pavacol. A proprietary preparation containing pholcodine, papaverine hydrochloride, balsam of tolu, oil of clove, tincture of ginger, oil of aniseed, tincture of capsicum, oil of peppermint, glycerin, alcohol, chloroform and treacle. A cough linctus. (Ward, Blenkinsop & Co Ltd).†

Pavacol Diabetic Cough Syrup. A proprietary preparation containing agents as in Pavacol without carbohydrates. A cough linctus. (Ward, Blenkinsop & Co Ltd).†

Pavlin. Fraxin, $C_{16}H_{18}O_{10}$. A substance which occurs in the bark of the common ash.†

Pavulon. Pancuronium Bromide. Applications: Blocking agent. (Organon Inc).

Pavy's Solution. A modified Fehling's solution used for the determination of sugar. The solution consists of (*a*) 4.158 grams copper sulphate in 500 cc water, and (*b*) 20.4 grams sodium-potassium tartrate, 20.4 grams potassium hydroxide, and 300 cc strong ammonia, made up to 500 cc. Equal parts of (*a*) and (*b*) are used.†

Paxadon. Pyridoxine hydrochloride. Applications: Pharmaceutical preparation for the treatment of pregnancy sickness and peripheral neuritis. (M A Steinhard Ltd).*

Paxalgesic. Dextropropoxyphene with paracetamol. Applications: Pharmaceutical preparation for the treatment of mild to moderate pain. (M A Steinhard Ltd).*

Paxalol. Propranalol. Applications: Pharmaceutical preparation for the treatment of cardiac arrhythmias and prevention of myocardial reinfaction. (M A Steinhard Ltd).*

Paxane. Flurazepam. Applications: Pharmaceutical preparation for the treatment of insomnia. (M A Steinhard Ltd).*

Paxbestos. Asbestos products bonded with hydraulic cement and impregnated with bitumen. Used as insulating materials.†

Paxipam. Halazepam. Applications: Sedative. (Schering-Plough Corp).

Paxofen. Ibuprofen. Applications: Pharmaceutical preparation for the treatment of rheumatic pain. (M A Steinhard Ltd).*

Paxofenac. Diclofenac. Applications: Pharmaceutical preparation for the treatment of musculo-skeletal disorders and gout. (M A Steinhard Ltd).*

Paxolax. Bisacodyl. Applications: Pharmaceutical preparation for the treatment of constipation. (M A Steinhard Ltd).*

Paxolin. A synthetic resin bonded paper product used for insulating purposes.†

Payne's Grey. An oil and water colour prepared from black alizarin, madder, and indigo. ,†

Payne's Solution. A solution of sodium hypobromite.†

Payzone. A proprietary preparation of NITROVIN hydrochloride. A veterinary growth promoter.†

Pazo. An insecticide. (Murphy Chemical Co).†

P.B.N. Phenylbetanaphthylamine. A rubber antioxidant. See NEOZONE D.†

PC-1244. Defoamer. Applications: Suppress foaming tendencies in lubricating oil formulations. (Monsanto Co).*

PC-1344. Defoamer. Applications: Defoamer in nonaqueous hydrocarbon and solvent systems. (Monsanto Co).*

PCMX. Applications: For chemical compositions having antiseptic and germicidal properties, in Class 18. (Ferro Corporation).*

PDS. Polydioxanone. Applications: Surgical aid. (Ethicon Inc).

Pea-stone. See PISOLITE.

Peach Black. A variety of carbon black similar to lampblack.†

Peach Bordeaux Mixture. Consists of 3 lb copper sulphate, 9 lb lime, and 50 gallons water.†

Peach Wood. See REDWOODS.

Peacock Blue. A dyestuff. It is a mixture of Methyl violet and Malachite green.†

Peacock Blue Lake. The barium lake of Patent blue. Used in printing inks.†

Peacock Copper Ore. An iridescent copper pyrite, produced by the partial decomposition of the yellow mineral, $Cu_2S.Fe_2S_3$.†

Peanut Ore. A mineral. It is a variety of wolframite.†

Pear Oil. Isoamyl acetate, $CH_3.COO.C_5H_{11}$. Used in the manufacture of fruit essences for flavouring confectionery.†

Pearl. Kerosene. (Chevron).*

Pearl Alum. A specially prepared aluminium sulphate used in the paper industry.†

Pearl Ash. A variety of potassium carbonate, K_2CO_3.†

Pearl Dust. A registered trade name for a form of potassium carbonate, K_2CO_3, used as a filler.†

Pearl-hardening. Calcium sulphate, $CaSO_4$, used as a loading for paper.†

Pearl Powder. Bismuth oxychloride.†

Pearl Sinter. A variety of opal, SiO_2, found in volcanoes.†

Pearl Spar. A double carbonate of magnesium and calcium, $Mg.Ca(CO_3)_2$.†

Pearl White. (Flake White). Bismuth oxychloride, BiOCl, (blanc de perle). A basic bismuth nitrate, $Bi(OH)_2NO_3$, is also known as pearl white.

The term is sometimes used in connection with a white lead which has been tinted with Paris blue or indigo.

A preparation of mother of pearl is called by this name.

Also see BISMUTH WHITE and FLAKE WHITE.†

Pearlite. Iron carbide eutectoid, consisting of alternate masses of ferrite and cementite.†

Pearsol. Synthetic phenolic germicides in a terpeneol vegetable oil soap base. Applications: For general disinfection/antisepsis. A substitute for chloroxylenol solution. (William Pearson).*

Pearson's Cerate. Consists of 4 parts lead plaster, 1 part beeswax, and 3 parts almond oil.†

Pearson's Solution. A solution of dried sodium arsenate 1 per 100 to 1 per 1,000.†

Peat. The partially decayed remains of plants. Used as fuel.†

Peat Coal. An intermediate between peat and lignite.†

Pebax. Polyether block amides. Applications: Sparking applications, gears. (Atochem Inc).*

Pecan Oil. Oil obtained from the seed of the North American walnut, *Juglans niger*.†

Pechman Dyes. Coloured dehydration products of β-benzoyl-acrylic acid or its homologues.†

Pectamol. A proprietary preparation containing oxeladin citrate, in flavoured vehicle containing menthol, chloroform and glycerin. A cough linctus. (British Drug Houses).†

Pectase. A clotting enzyme, which produces vegetable jellies.†

Pectin. A polysaccharide substance soluble in water, which is a constituent of many fruits, such as apples, pears, and gooseberries, also of carrots and beetroot. Its aqueous solution gelatinizes on cooling. Used in jam and preserved fruit manufacture.†

Pectinex. A purified enzyme preparation produced from a selected strain of Aspergillus niger. Applications: Used in the case where the aim is breaking down of soluble and insoluble pectins with varying degrees of esterification, for reduction of viscosity, clarification, maceration of plant tissue and depectinization. (Novo Industri A/S).*

Pecutrin. Vitaminized mineral salt mixture. Applications: For individual dosing and as a feed additive for all animals kept for use. (Bayer & Co).*

PediaCare 1. Dextromethorphan hydrobromide. Applications: Antitussive. (McNeil Consumer Products Co).

Pediaflor. Sodium Fluoride. Applications: Dental caries prophylactic. (Ross Laboratories, Div of Abbott Laboratories).

Pediamycin. Erythromycin ethylsuccinate. Applications: Antibacterial. (Ross Laboratories, Div of Abbott Laboratories).

Peerless Alloy. A heat-resisting alloy containing 78.5 per cent nickel, 16.5 per cent chromium, 3 per cent iron, and 2 per cent manganese.†

Pegmatite. A felspathic rock, similar to Cornish stone.†

Pegnin. A preparation of lactose and rennet, which yields a finely divided curd from cows' milk. Used in infant food.†

Pegu. See CUTCH.

Pegu Brown. A direct cotton dyestuff.†

Pegu Catechu. See CUTCH.

Peka Glas. (PK Glas). A proprietary safety glass.†

Pekafix. Cold curing denture resin on the basis of methyl methacrylate. Applications: Dental preparation. (Bayer & Co).*

Pekatop. Heat-curing denture resin on the basis of methyl methacrylate. Applications: Dental preparation. (Bayer & Co).*

Pekatray. Cold curing plastic for the preparation of individual impression trays. Applications: Dental speciality. (Bayer & Co).*

Pelamag. Salt coated magnesium granules for desulphurizing blast furnace iron (sold under licence from Dow Chemical Company). (Foseco (FS) Ltd).*

Pelargone. A trade name for Nylon 9.†

Pelargonium Oil. See ROSÉ OIL.

Pelaspan. Expandable polystyrene resin used in manufacture of loose-fill packing. (Dow Chemical Co Ltd).*

Pelaspan GP. A trade mark. General purpose expandible polystyrene.†

Pelaspan Mold-a-Pac. For packaging products using PELASPAN-PAC loos fill coated with an adhesive to form a resilient moulded cushion. (Dow Chemical Co Ltd).*

Pelaspan PAC. A trade mark. Polystyren foam strands.†

Pelaspan 333FR. A flame retardant expandible polystyrene. (Dow Chemica Co Ltd).*

Pelaspan-Pac. Expanded polystyrene. Loose-fill packaging material. (Dow Chemical Co Ltd).*

Pelentan. A proprietary trade name for ethyl biscoumacetate.†

Pelican Blue. An induline dyestuff. See INDULINE, SPIRIT SOLUBLE.†

Peligot Blue. A pigment. It is a hydrated copper oxide.†

Peligotite. Synonym for Johannite.†

Pelionite. A coal of the cannel type.†

Pellethane. A wide range of polyurethane elastomers, 'rubber-plastics' used cured or uncured fabricated into various shapes and forms by conventional methods. (Dow Chemical Co Ltd).*

Pellitory Root. Pyrethrum root.†

Pellurin. Hexamethylene-tetramine hydrochloride.†

Pelonit D. Contains aluminium, has a strong shoving effect and is used where coarse fragmentation is required. (Dynamit Nobel Wien GmbH).*

Pelosine. Berberine.†

Pelouze's Green. (Prussian Green). Ferroso-ferric-ferricyanide, $Fe_3Fe_4(FeCy_6)_6$.†

Pembrite. See COLOGNE ROTTWEIL.

Pen A. A proprietary preparation of ampicillin. An antibiotic. (Pfizer International).*

Pen A/N. Ampicillin sodium. Applications: Antibacterial. (Pfizer Inc).

n Metal Brass. See BRASS.

nacolite G-1124 and G-1131. Proprietary trade names for resorcinol-formaldehyde glues.†

nak Dammar. See DAMMAR RESIN.

napar VK. Penicillin V Potassium. Applications: Antibacterial. (Parke-Davis, Div of Warner-Lambert Co).

naryl A. A proprietary trade name for a plasticizer. It is amyldiphenyl.†

naryl B. A proprietary trade name for a plasticizer. It is a diamyldiphenyl.†

nbritin. A proprietary preparation containing ampicillin. An antibiotic. (Beecham Research Laboratories).*

nchlor. A proprietary acid-proof cement made from cement powder and sodium silicate solution. Used for lining tanks.†

ndare. A name for Venezuelan chicle.†

ndecamaine. A surface-active agent present in TEGO-BETAINES. It is *NN*-dimethyl-(3-palmitamidopropyl)-glycine betaine.†

ndiomide. Azamethonium Bromide.†

ndramine. SEE TROLOVOL. (Degussa).*

netrol. Nasal decongestant. (The Boots Company PLC).

netrol. As insecticide against aphides. It is a sulphonated oxidation product of petroleum.†

netrol. A compound used as a textile detergent.†

enicals. A proprietary preparation containing Phenoxymethylpenicillin (as the calcium salt). An antibiotic (Leo Laboratories).†

enicals 333. A proprietary preparation containing Phenoxymethylpenicillin Calcium. An antibiotic. (Leo Laboratories).†

enicillin. The name given to the antibiotic principle of the mould *penicillium notatum*. The material is now prepared by special fermentation processes in large quantities.†

enicillin N. Adicillin.†

enicillin V Pulvules. A proprietary preparation containing Phenoxy-methylpenicillin. An antibiotic. (Eli Lilly & Co).†

Penicillin V Sulpha. A proprietary preparation of phenoxymethylpenicillin sulphadiazine, sulphamerazine and sulphadimidine. An antibiotic. (Eli Lilly & Co).†

Penicillinase. An enzyme obtained from cultures of Bacillus cereus which hydrolyses benzylpenicillin to penicilloic acid. Neutrapen.†

Penidural All Purpose. A proprietary preparation containing Benzathine Penicillin, Benzylpenicillin and Procaine Penicillin. An antibiotic. (Wyeth).†

Penisem. Penicillin G Potassium. Applications: Antibacterial. (Beecham Laboratories).

Penitriad. Trisulphonamide/potassium penicillin V preparations. (May & Baker Ltd).

Penjectin. Procaine penicillin intramammary injection. (May & Baker Ltd).

Pennettier's Green. See CHROMIUM GREEN.

Pennsalt TD-5032. Hexamethylditin. A contact and systemic insecticide.†

Pennyroyal Oil. See OIL OF PENNYROYAL.

Penotrane. A proprietary preparation containing hydrargaphen. (Ward, Blenkinsop & Co Ltd).†

Pensa's Rubber. A rubber substitute made from coal tar, petroleum tar, oil of turpentine, and boric or phosphoric acids.†

Penta G.P. 79. Rust preventatives. (Croda Chemicals Ltd).*

Penta-erythritol Tetrastearate (PET). A proprietary release agent used in injection-moulding processes. (Du Pont (UK) Ltd).†

Pentabor. Sodium pentaborate. (Borax Consolidated Ltd).*

Pentac Aquaflow. Acaricide. (Murphy Chemical Ltd).

Pentacizers. Proprietary trade names for plasticizers.†

Pentacosactride. A corticotrophic peptide.†

Pental. Trimethylethylene, C_5H_{10}.†

Pental A. A range of rosin esters. Applications: A tackifying agent for natural rubber and SBR based pressure sensitive adhesives. (Hercules Inc).*

Pental G. A range of rosin esters. Applications: Used in heat set lithographic inks. (Hercules Inc).*

Pental X. A range of rosin esters. Applications: Used in heat set lithographic inks. (Hercules Inc).*

Pental 28. A range of rosin esters. Applications: Used to modify nitrocellulose based coatings and lacquers. (Hercules Inc).*

Pental 8D. A range of rosin esters. Applications: To produce chewing gum base. (Hercules Inc).*

Pental 802A. A range of rosin esters. Applications: Used in heat set lithographic inks. (Hercules Inc).*

Pentalan. Pentaerythritol ester of woolwax fatty acids. (Croda Chemicals Ltd).

Pentaline. Pentachlorethane, $CHCl_2.CCl_3$.†

Pentalyn. Pentaerythritol esters of rosin, modified rosins, dibasic acid- and phenolic-modified rosins. Applications: Pentalyn C, H, K and 344 resins are used as tackifying and reinforcing resins for adhesives and sealants. Pentalyn 255, 261 and 856 resins are alkali-soluble resins used in emulsion floor polishes. Pentalyn A and H resins are used in the production of chewing gum base. Pentalyn synthetic resins are used in heat set offset inks, letterpress printing inks and flexographic inks. Pentolyn G, X, 802A, 802A Pale and 833 resins are used in overprint varnishes. (Hercules Inc).*

Pentamid CI2. Fatty acid polydiethanolamide. Applications: Metal-working corrosion inhibitor. (Pentagon Urethanes Ltd).*

Pentamid KH. Methyl N-octadecyl terephthalamate. Applications: Gelling agent for high-temperature greases. (Pentagon Urethanes Ltd).*

Pentamin BDMA etc. N,N-Dimethyl benzylamine. Applications: Polyurethane catalyst, epoxy curing agent. (Pentagon Chemicals Ltd).*

Pentaphane. A proprietary trade name for a film made from a chlorinated polyether (Polymerized 3,3-bis(chloromethyl)oxetane. (British Cellophane).†

Pentasol. A mixture of pure amyl alcohols It is stated to contain 75 per cent primary alcohol and 25 per cent secondary alcohol, and is obtained from pentane fraction of gasoline and is used as a varnish and lacquer solvent.†

Pentek. A proprietary trade name for a technical grade of pentaerithritol used in synthetic resins and in the paint and varnish industry.†

Pentelex. Photographic developer. (May & Baker Ltd).

Penthrane. A proprietary preparation of methoxyfluorane. Inhalation anaesthetic, used in obstetrics. (Abbott Laboratories).*

Penthrinit. An explosive. It is a plastic mixture of 80 per cent pentaerythritol-tetranitrate and 20 per cent nitroglycerine.†

Penthrit. Pentaerythritol tetranitrate.†

Pentids. Penicillin G Potassium. Applications: Antibacterial. (E R Squibb & Sons Inc).

Pentley's Neutral Orange. See NEUTRAL ORANGE.

Pentol. Timber fungicides. (Plant Protection (Subsidiary of ICI)).†

Pentonate DB. Fatty alcohol benzoate. Substitute for isopropyl myristate. Applications: Emollient. (Pentagon Chemicals Ltd).*

Pentonite. Sodium benzoate/sodium nitrite mixture. Applications: Corrosion inhibitor for aqueous systems. (Pentagon Urethanes Ltd).*

Pentonium 50, 80 etc. Alkyl dimethyl benzalkonium chlorides, alkyl trimethyl ammonium chlorides. Applications: Biocides. (Pentagon Chemicals Ltd).*

Pentostam. Injection of pentavalent sodium stibogluconate. (The Wellcome Foundation Ltd).

Pentothal. A proprietary preparation of thiopentone sodium. Intravenous anaesthetic. (Abbott Laboratories).*

Pentothal Sodium. Thiopental Sodium. Applications: Aneasthetic; anticonvulsant. (Abbott Laboratories).

ntovis. A proprietary preparation containing Quinestradol. (Warner).†

ntoxyl M. Perfumery speciality. (Bush Boake Allen Ltd).

ntrex. Phenolic-modified and maleic modified esters of rosin. Applications: Pentrex G and X rosin esters are used in heat set lithographic inks. Pentrex 28 is used in nitrocellulose-based coatings and lacquers. (Hercules Inc).*

ntrexyl. A proprietary preparation of AMPICILLIN. An antibiotic. (Bristol-Myers Co Ltd).†

ntritol. Pentaerythritol Tetranitrate. Applications: Vasodilator. (USV Pharmaceutical Corp).

ntrium. A proprietary preparation of chlordiazepoxide and pentaerythritol tetranitrate, used in the treatment of angina pectoris. (Roche Products Ltd).*

ntrone. Anionic surfactant. (ABM Chemicals Ltd).*

ntrone ON. A proprietary trade name for a 33 per cent. active composition of sodium-2-ethyl hexyl sulphate. A high purity, low fat content material used as a surface tension reducing agent for caustic soda solutions. (Glovers Chemicals Ltd).†

entrone ON. Sodium 2-ethyl-hexyl sulphate in liquid form. Applications: Anionic surfactant for dispersing and alkaline wetting, used in electroplating and lyepeeling. (ABM Chemicals Ltd).

entrone S. Range of anionic surfactants of the disodium mono- alkyl polyalkylene sulphosuccinate type. Supplied as liquids. Applications: Dispersing, foaming and emulsification agents with low toxicity, used in emulsion polymerization, surgical scrubs and shampoos. (ABM Chemicals Ltd).

entrosan. Dispersing and solubilizing agents. (ABM Chemicals Ltd).

entryate 80. Pentaerythritol Tetranitrate. Applications: Vasodilator. (O'Neal, Jones & Feldman Pharmaceuticals).

entyl. Amyl, C_5H_{11}.†

en-Vee. Penicillin V. Applications: Antibacterial. (Wyeth Laboratories, Div of American Home Products Corp).

Pen-Vee Drops. Penicillin V Benzathine. Applications: Antibacterial. (Wyeth Laboratories, Div of American Home Products Corp).

Pen Vee Dural. A proprietary preparation containing Phenoxymethylpenicillin Potassium and Benzathine Penicillin. An antibiotic. (Wyeth).†

Pen-Vee K. Penicillin V Potassium. Applications: Antibacterial. (Wyeth Laboratories, Div of American Home Products Corp).

Pen-Vee Suspension. Penicillin V Benzathine. Applications: Antibacterial. (Wyeth Laboratories, Div of American Home Products Corp).

Penyl. Polamide.†

Penzold's Reagent. A solution of diazo-benzosulphonic acid and potassium hydroxide. A reagent for sugar in urine.†

Peonine. See PAEONINE.

Pep Set. Resins and catalysts associated with the production of foundry cores and moulds. These resins and catalysts, when applied to foundry sand, provide a binder system useful in the production of foundry cores and moulds. These binder systems fall into the large category of no-bake binders used within the foundry industry and cure at room temperature. (Ashland Chemical Company).*

PEP-1. A polyethylene of Soviet origin.†

Pepcidine. Famotidine. Applications: For the treatment of gastric and duodenal ulcer and hypersecretory conditions such as Zollinger-Ellison syndrome. (Merck Sharp & Dohme).*

Peppermint Camphor. Menthol, $C_{10}H_{19}OH$. Used in medicine.†

Pepsin. A proteolytic enzyme of the mucous membrane of the stomach. It decomposes albuminous bodies into peptone.†

Peptacol- 10. A proprietary preparation of homatropine methylbromide and phenobarbitone. (Pharmax).†

Peptard. Each tablet contains 1-hyoscyamine sulphate 0.2 mg in a porous plastic matrix. (Riker Laboratories/3M Health Care).†

Peptavlon. Pentagastrin. Applications: Diagnostic aid. (Ayerst Laboratories, Div of American Home Products Corp).

Peptoil. Petroleum oil. Applications: Spray adjuvant for enhancing herbicide activity and defoliation performance with foam eliminator. (Drexel Chemical Company).*

Pepton 65. A proprietary trade name for zinc 2-benzamide thiophenate. M.p. 200-300 C. A peptizer for natural rubber. SBR and synthetic polyisoprene above 65° C. (Anchor Chemical Co).*

Peptorub. Pre-plasticized comminuted rubber.†

Peradinol. Textile dyeing auxiliaries. (Fine Dyestuffs & Chemicals Ltd).†

Peralvex. A proprietary preparation of anthraquinone glycosides and salicylic acid used in the treatment of mouth ulcers. (Norgine).†

Perapret. Additives for textile resin finishing. (BASF United Kingdom Ltd).

Perborax. Sodium perborate, $NaBO_3.4H_2O$, a washing and bleaching agent.†

Perborin. Sodium perborate, $NaBO_3$, a constituent of washing powders.†

Perborin M. A mixture of soap, soda, and sodium perborate. A dry soap.†

Perborol. See PERBORAX.

Percarbamid. A hydrogen peroxide preparation. Applications: Used in the cosmetics and pharmaceuticals industry. (Degussa).*

Perchlorethylene. Tetrachloroethylene, C_2Cl_4.†

Perchloron. A technical calcium hypochlorite containing 68.1 per cent available chlorine.†

Perchoid. See OXOLIN.

Per-clene. A proprietary trade name for perchlorethylene (*qv*).†

Percogesic. Analgesic, for enhanced pain relief. (Richardson-Vicks Inc).*

Percolaye. Attapulgite or Sepiolite clay. Applications: Refining of minerals and chemicals. (Bromhead & Denison Ltd).*

Percorten Acetate. Desoxycorticosterone acetate. Applications: Adrenocortical steroid. (Ciba-Geigy Corp).

Percorten M Crystules. Treatment of adrenocortical insufficiency. (Ciba-Geigy PLC).

Percorten Pivalate. Desoxycorticosterone pivalate. Applications: Adrenocortical steroid. (Ciba-Geigy Corp).

Percorten "M" Crystules. A proprietary preparation of deoxycortone pivalate. (Ciba Geigy PLC).†

Percresan. A mixture of cresols, soap, and water. Used as a disinfectant in 1-2 per cent solution.†

Percussion Cap Brass. See BRASS.

Perduren. An organic polysulphide synthetic rubber.†

Perdynamine. A compound of albumen and haemoglobin.†

Perecot. A copper fungicide. (ICI PLC).*

Peregal O. Polyoxyethylated fatty alcohol (nonionic). Applications: Dyeing assistant for use with basic and direct colours. Assistant for the dyeing, levelling and stripping of vat dyes. Levelling agent in acetate printing. (GAF Corporation).*

Peregal OK. Methylpolyethanol quaternary amine (cationic). Applications: Used as a vat dye retarder. (GAF Corporation).*

Peregal ST. Polyvinylpyrrolidone. Applications: Stripping assistant for cotton and rayon yarns or fabrics that have been dyed or printed with vat, sulphur or direct colours. Rag stripping assistant in high grade paper. (GAF Corporation).*

Pereiro Bark. The bark of *Geisso-spermum vellosii*. A Brazilian febrifuge.†

Pereman. A copper fungicide. (Plant Protection (Subsidiary of ICI)).†

Perenox. A copper fungicide. (ICI PLC).*

Perenyi's Fluid. (Chromo-nitric Acid). A fixing agent used in microscopy. It contains 3 parts of 92 per cent alcohol, 4 parts of 10 per cent nitric acid, and 3 parts of 0.5 per cent chromic acid. The objects are treated with alcohol after fixing.†

Perfix. High speed photographic fixer. (May & Baker Ltd).

Perflex. A proprietary trade name for unstretched vinylidene chloride (Saran *q.v.*).†

Perfumery Oil. Refined petroleum, of specific gravity 0.880-0.885. Used in perfumery.†

Perfusamine. Iofetamine hydrochloride. Applications: Diagnostic aid; radioactive agent. (Medi-Physics Inc).

Pergacid. Dyes for paper. (Ciba-Geigy PLC).

Pergalen. A proprietary preparation of sodium apolate, and benzyl nicotinate. Resolution of bruises and local trauma. (Hoechst Chemicals (UK) Ltd).†

Pergamin. (Pergamyn). A grease-proof paper made from cellulose pulp.†

Pergamyn. See PERGAMIN.

Pergantine. Pigment dispersions for paper. (Ciba-Geigy PLC).

Pergaprint. Crosslinking agents for starch. (Ciba-Geigy PLC).

Pergascript. Chemicals for carbonless paper. (Ciba-Geigy PLC).

Pergasol. Dyes for paper. (Ciba-Geigy PLC).

Pergenol. A mixture of sodium perborate and bitartrate. It gives hydrogen peroxide on the addition of water.†

Perglanz-Konzen-Trat B48 and B30. Blend of amphoterics and anionics, in the form of a white liquid. Applications: Sheen additive for shampoos, bath and shower preparations (Th Goldschmidt Ltd).

Perglow. A bright nickel plating process. (Hanshaw Chemicals).†

Perglycerol. An aqueous solution of sodium lactate. Used as a substitute for glycerol for medical and cosmetic purposes.†

Pergonal. A proprietary trade name for a follicle-stimulating hormone.†

Pergopak. Organic fillers. (Ciba-Geigy PLC).

Pergut. See ALLOPRENE. A proprietary trade name for chlorinated rubber.

Pergut. Chlorinated rubber. Applications: For the formulation of coatings with high water, chemical and low-temperature resistance as well as for use in printing inks. (Bayer & Co).*

Perhydrate. See HYPEROL.

Perhydrit. See HYPEROL.

Perhydrol. Hydrogen peroxide, H_2O_2, one volume of 30 per cent hydrogen peroxide giving 100 volumes of oxygen. Used for bleaching, also as a disinfectant.†

Perhydrol of Magnesia. See BIOGEN.

Peri Acid. 1-Naphthylamine-8-sulphonic acid.†

Periactin. Cyproheptadine hydrochloride. Applications: Relief of allergic and pruritic conditions and migraine headache. Also an appetite stimulant. (Merck Sharp & Dohme).*

Periactin-Vita. Cyproheptadine hydrochloride and multivitamins. Applications: An appetite stimulant which helps prevent hypovitaminosis which may be associated with poor eating habits or an inadequate diet. (Merck Sharp & Dohme).*

Perichthol. A proprietary preparation of ammonium ichthosulphonate.†

Peri-Colace. Contains casanthranol and docusate sodium. Casanthranol is a purified mixture of the anthranol glycosides derived from Casara sagrada. Applications: Laxative and stool-softener. (Mead Johnson & Co).

Perifenil. A proprietary preparation of pheneizine, and pentaerythritol tetranitrate. (Warner).†

Perigen. Formulations of permethrin. (The Wellcome Foundation Ltd).

Perikol. A sensitizer for silver bromide plates, prepared by treating the addition product of toluquinaldine and the ethyl ester of toluene-sulphonic acid with alcoholic potassium hydroxide.†

Perilla Oil. An oil obtained from the seed of the Asiatic mint, *Perilla ocymoides*. It is used as a drying oil in substitution for linseed oil.†

Periograf. Durapatite. Applications: Prosthetic aid. (Sterling Drug Inc).

Peripress. A proprietary preparation of prazosin hydrochloride for the treatment of hypertension, left ventricular failure and Raynaud's Disease. (Pfizer International).*

Peristaltin. A cascara preparation containing the water-soluble glucosides extracted from the bark of *Rhamnus* *purshiana*. It stimulates peristalsis without drastic purgative action.†

Peritrate. Pentaerythritol Tetranitrate. Applications: Vasodilator. (Parke-Davis, Div of Warner-Lambert Co).

Perium. Pentapiperium Methylsulphate. Applications: Anticholinergic. (William H Rorer Inc).

Perkadox. Organic peroxides. (Akzo Chemie UK Ltd).

Perkaglycerol. An aqueous solution of potassium lactate. Used as a substitute for glycerol for medical and cosmetic purposes.†

Perkin's Base. *p*-Tolylaminoditolyl-*p*-toluquinone diimine.†

Perkin's Purple. See MAUVEINE.

Perkin's Violet. See MAUVEINE.

Perklone. Perchloroethylene. A solvent for dry cleaning. (ICI PLC).*

Perlankrol. Anionic alcohol-, alcohol ether-amide ether- and phenol ether-sulphates. Applications: Used in toiletries, as foam boosters, emulsifiers, cement and gypsum foamers. (Lankro Chemicals Ltd).†

Perlankrol ADP-3. Anionic surfactant in which the cation is sodium and the anion is a synthetic primary alcohol ether sulphate. Applications: Base for the preparation of high foaming shampoos and toiletries. (Diamond Shamrock Process Chemicals Ltd).

Perlankrol ATL. Triethanolamine lauryl sulphate as a pale straw liquid. Applications: Base material in the preparation of high foaming shampoos and toiletries. (Diamond Shamrock Process Chemicals Ltd).

Perlankrol DAF. Anionic surfactant in liquid form. Applications: Foaming agent for toiletries and carpet shampoos. (Diamond Shamrock Process Chemicals Ltd).

Perlankrol DSA. Anionic surfactant in the form of a white slurry. Application: Detergent base for household and industrial cleaners; foaming agent for synthetic latex. (Diamond Shamrock Process Chemicals Ltd).

Perlankrol ESD. Anionic surfactant in liquid or gel form. Applications: High foam additive or booster for liquid household and industrial detergent formulations. (Diamond Shamrock Process Chemicals Ltd).

Perlankrol ESK-29. Sodium alkyl ether sulphate as a clear yellow liquid. Applications: Wetting and foaming agent for alkaline solutions, used for chlorphenolic disinfectants; liquid detergents; plaster board; concrete. (Diamond Shamrock Process Chemicals Ltd).

Perlankrol O. Sodium alkyl sulphate as a clear light amber liquid. Applications: Viscosity modifier and stabilizing agent for liquid detergents; wetting agent for use in high concentration electrolyte solutions. (Diamond Shamrock Process Chemicals Ltd).

Perlankrol PA, FF and FD. Ammonium alkyl-phenol ether sulphate as a hazy viscous amber liquid. Applications: Base, foam booster and stabilizer, used in liquid detergents; emulsifier used for cresylic acid and in emulsion polymerization. (Diamond Shamrock Process Chemicals Ltd).

Perlankrol SN and RN. Sodium alkyl-phenol ether sulphate in liquid form. Applications: Foam booster, stabilizer and emulsifier, used in industrial detergents and emulsion polymerization. (Diamond Shamrock Process Chemicals Ltd).

Perlankrol TM1. Fatty alkylolamide ether sulphate. Applications: Base for the preparation of high foaming shampoos and toiletries. (Diamond Shamrock Process Chemicals Ltd).

Perlate Salt. See SALT PERLATE.

Perlex. Bismuth oxychloride pearls. Applications: Decorative cosmetic products. (Williams Div of Morton Thiokol Ltd). *

Perlextra. Bismuth oxychloride pearls. Applications: Decorative cosmetic products. (Williams Div of Morton Thiokol Ltd). *

Perlit. Water repellents based on silicone, paraffin or fatty acid condensation products. Applications: Textile finishing. (Bayer & Co).*

Perlite. A eutectic product resulting from an alloy of ferrite and cementite in steel.†

Perlygel. A registered trade mark for benzoyl peroxide.†

Permabond. OPP laminating adhesives. (The Scottish Adhesives Co Ltd).*

Permador. A precious metal alloy. Applications: For dentistry and dental engineering. (Degussa).*

Permafuse. Modified phenolic adhesives. Applications: For bonding friction materials to their backing - brake shoes, clutch facings, transmission linings etc. (The Permafuse Corp).*

Permalba. A composite pigment consisting mainly of barium sulphate. An artist's colour.†

Perma-Leaf. Aluminium pigment paste. Applications: Protective coatings. (Reynolds Metal Co).*

Permalens. Perfilcon. Applications: Contact lens material. (CooperVision Inc).

Permali. (Jicwood, Jabroc, Durisol, Lignostone). A proprietary trade name for laminated products containing wood or paper impregnated with synthetic resin. Some are made from thin wood coated with synthetic resin solution and compressed under heat, others are impregnated under pressure, solvent removed and then compressed.†

Permalloy. A registered trade mark for alloys of nickel and iron containimg more than 30 per cent nickel. They are prepared by certain heat treatment and show unusual magnetic properties, giving a high initial permeability. One of the best alloys contains 78.5 per cent nickel and 21.5 per cent iron. Another alloy of this class contains 78.5 per cent nickel, 18 per cent iron, 3 per cent molybdenum, and 0.5 per cent manganese. A typical analysis gives 78.23 per cent nickel, 21.35 per cent iron, 0.04 per cent carbon, 0.03 per cent silicon, 0.035 per cent sulphur, 0.22 per cent manganese, 0.37 per cent cobalt, 0.1 per cent copper and traces of phosphorus.†

Permalon. A proprietary trade name for stretched vinylidene chloride (Saran *qv*).†

Permalose. Textile auxiliary chemicals. (ICI PLC).*

Permalux. Neoprene accelerator. (Du Pont (UK) Ltd).

Permalyn. Modified hydrocarbon resins. (Hercules Ltd).

Permanent Blue. See ULTRAMARINE.

Permanent Green. See CHROMIUM GREEN, NUREMBERG GREEN, and CHROME GREEN.

Permanent Orange R. A dyestuff prepared from 5-chloroaniline-2-sulphonic acid.†

Permanent Red 4B. See HELIO FAST RED RL and LITHOL RUBINE B.

Permanent Vermilion. A pigment. It is usually Orange mineral (*qv*), tinted with *p*-nitrailine.†

Permanent Violet. See MANGANESE VIOLET.

Permanent White. See ZINC WHITE.

Permanent Yeast. See ZYMIN.

Permanent Yellow. See BARIUM-YELLOW.

Permanite. A cobalt steel which has very high magnetic properties. Also see K.S. MAGNET STEEL and COBALT STEEL.†

Permapen. A proprietary preparation of penicillin G benzathine. An antibiotic. (Pfizer International).*

Permaplex. An ion exchange membrane. (Permutit-Boby Ltd).†

Permasect. Insecticidal formulation. (Mitchell Cotts Chemicals Ltd).

Perma Shield. An elastomeric coating. Applications: Suitable for application to concrete, brick, concrete block, stucco, metal, wood, sheetrock, masonite and plywood as a protective and decorative finish. (Secure Inc).*

Perma-Slik. Family of dry lubricant coatings containing solid lubricants and suitable binder system. Applications: Frictional surfaces requiring dry lubrication particularly on substrates which cannot tolerate oven cure. (E/M Corporation).*

Permatag. Insecticidal cattle eartag. (Mitchell Cotts Chemicals Ltd).

Permathin. Tetrafilcon A. Applications: Contact lens material. (UCO Optics Inc).

Permatol A. A proprietary trade name for a preservative for wood. It contains pentachlorphenol in oil.†

Permax. A nickel steel containing 76 per cent nickel, of French manufacture. It has magnetic properties.†

Permidan. Dimethylaminopyrazolone, $C_5H_9N_3O$.†

Perminal. Textile auxiliary chemicals. (ICI PLC).*

Perminvars. Proprietary alloys having exceptional magnetic properties. They are particularly suited for use in electrical communication circuits. One alloy contains 45 per cent nickel, 25 per cent cobalt, and 30 per cent iron, and has a high initial permeability.†

Permitil. Fluphenazine hydrochloride. Applications: Antipsycotic. (Schering-Plough Corp).

Permobel. Synthetic top coat for automobiles. (ICI PLC).*

Permonite. An explosive used in mines. It is a mixture of potassium perchlorate and ammonium nitrate, with trinitrotoluene, a little starch, and wood meal.†

Permutite. An artificially made zeolite, prepared by igniting together china clay (aluminium silicates), and (sometimes) quartz or sand, with alkali carbonates. Used for removing calcium and magnesium salts, sodium and potassium salts, and manganese and iron, from water.†

Permyl B-100. Applications: For stabilizers for halogenated hydrocarbon resins in US Class 6. (Ferro Corporation).*

Pernambuco Wood. See REDWOODS.

Pernax. An artificial gutta-percha made from rubber, wax, and rosin.†

Pernivit. A proprietary preparation of nicotinic acid and acetomenaphthone. (British Drug Houses).†

Pernomol. A proprietary preparation of chlorobutol, phenol, camphor, tannic acid and spirit Chilblain paint. (LAB Ltd).*

Peroheme 40. Occult blood test. (BDH Chemicals Ltd).

Perone. A proprietary trade name for pure hydrogen peroxide.†

Peronoid. A trade name for a mixture of copper sulphate and lime. A fungicide.†

Perox. Dyes for plastics. (Williams (Hounslow) Ltd).

Peroxal. A trade name for hydrogen peroxide, H_2O_2.†

Peroxide RH-2. A proprietary trade-name for a high melting, stable, aromatic organic peroxide used as a polymerization catalyst. (Compare LUPERCO, LUCIDOL.)†

Peroximon. A proprietary range of organic peroxides. (Montedison UK Ltd).*

Peroxol. (Pyrozone, Glycozone). Disinfectants. They are solutions containing hydrogen peroxide, sometimes mixed with other disinfectants.†

Peroxydol. Sodium perborate, $NaBO_3.4H_2O$. An antiseptic, deodorant, and bleaching agent.†

Peroxyl. Hydrogen peroxide. (May & Baker Ltd).*

Perpentol. A tetralin preparation used for cleaning wool.†

Persadox. Benzoyl peroxide. Applications: Keratolytic. (Owen Laboratories Div, Dermatological Products of Texas, Alcon Laboratories Inc).

Persantin. A proprietary preparation of dipyridamole. (Boehringer Ingelheim Ltd).†

Persantine. Dipyridamole. Applications: Vasodilator. (Boehringer Ingelheim Pharmaceuticals Inc).

Perseo. See FRENCH PURPLE.

Persian Balsam. Compound tincture of benzoin.†

Persian Berries. (Yellow Berries, Rhamnine). A natural dyestuff obtained from the dried unripe fruits of various species of *Rhamnus*. The dyeing principles are rhamnetin, or quercetin-mono-ethyl ether, $C_{16}H_{12}O_7$, rhamnazin or quercetin-dimethyl-ether, $C_{17}H_{14}O_7$, and quercetin, all as glucosides. Used for cotton printing with tin, chrome, or aluminium mordants, giving yellow to orange shades.†

ersian Berry Carmine. (Dutch Yellow). A pigment consisting of the aluminium and calcium lakes of the Persian berry colouring matters.†

ersian Gum. See INDIA GUM.

ersian Red. See INDIAN RED and CHROME RED.

ersian Yellow. A dyestuff. It is nitrotolueneazonitrosalicylic acid, $C_{14}H_{10}N_4O_7$. Dyes chromed wool yellow. Also used in cotton printing giving yellow shades with chromium acetate. Also see CHINESE YELLOW.†

ersiderm. Preparation for fashionable writing and lustre effects on suede. Applications: Leather industry. (Bayer & Co).*

ersiderm Black. Aniline pigment for the lustring of black suedes. (Bayer UK Ltd).

ersio. See ARCHIL and FRENCH PURPLE.

ersionin. An acetone extract of cud-bear.†

ersistol. Water-repellent agents for textiles. (BASF United Kingdom Ltd).

ersodine. A mixture of ammonium and potassium persulphates.†

ersoftal. Softeners for textiles. Selected brands simultaneously impart an antistatic effect. (Bayer & Co).*

ersoftal PE. Olignomer binder in polyester dyeing. (Bayer UK Ltd).

ersoftal PE Special. Oligomer binder. Applications: In polyester dyeing. (Bayer & Co).*

ersoz's Reagent. Zinc Oxide (2 grams), is added to a solution of zinc chloride (10 grams), in 10 cc water. It dissolves silk, and detects silk in the presence of wool.†

erspex. A trade mark for acrylic (methyl methacrylate) resins in sheet form. See also PLEXIGLAS, LUCITE, RESIN M. (ICI PLC).*

erstoff. (Diphosgene). A poison gas. Trichloromethyl-chloroformate, $ClCO.OCCl_3$.†

erstorp Phenolic Moulding Compound. Applications: Toilet seats, pan handles, meter cases, automotive, domestic accessories, law bowls, electrical accessories. (Perstorp Ferguson Ltd).*

Perstorp Urea Moulding Compound. Applications: Electrical accessories, toilet seats, closures. (Perstorp Ferguson Ltd).*

Persulphocyanogen Yellow. See CANARIN.

Perthane. Trade mark for an agricultural insecticide based on diethyldiphenyldichloroethane, supplied as a wettable powder or emulsifiable concentrate. Control insects on plants and livestock, also used as a moth protection for textiles.†

Pertinit. A proprietary synthetic resin of the urea-formaldehyde type.†

Pertite. An Italian explosive. The main constituent is picric acid. See LYDDITE.†

Pertofran. A proprietary trade name for the hydrochloride of desipramine.†

Pertofran. Antidepressant. (Ciba-Geigy PLC).

Pertofrane. Desipramine hydrochloride. Applications: Antidepressant. (USV Pharmaceutical Corp).

Pertscan-99m. Sodium Pertechnetate Tc 99m. Applications: Radioactive agent. (Abbott Laboratories).

Pertusa. A proprietary preparation containing ephedrine hydrochloride, tincture of belladonna, liquid extract of ipecacuanha, syrup of tolu, honey, citric acid and sodium benzoate. A cough linctus. (Boots Company PLC).*

Peru Balsam, Synthetic. See PERUGEN.

Peru Saltpetre. See CHILE SALTPETRE.

Peru Silver. See CHINESE SILVER.

Perugen. (Synthetic Peru Balsam). A synthetic Peru balsam made by mixing benzyl benzoate with storax, benzoic, and tolu balsams.†

Peruol. A registered trade mark currently awaiting re-allocation by its proprietors to cover a range of pharmaceuticals. (Casella AG).†

Peruscabin. Benzyl benzoate, $C_6H_5.CO_2.CH_2.C_6H_5$. It is the active constituent of Peru balsam, and is used in the same manner, and for the same purposes as Peruol.†

Peruvian Balsam. The oleo-resin of *Myroxylon Pereirae*, of Central America.†

Peruvian Bark. Cinchona bark.†

Peruvin. Cinnamyl alcohol, $C_6H_5CH : CHCH_2OH$.†

Pervon. Fully reacted polyurethane/pitch coating. (Sigma Coatings).*

Pescola Oil. An oil used in the tanning industry.†

Pest-B-Gon. Roach bait. (Chevron).*

Pestex. An insecticide. (Fisons PLC).*

Pestilizer. Phosphate esters. Applications: Compatibilizers for liquid fertilizers. (Stepan Company).*

Petameth. Racemethionine. Applications: Acidifier. (O'Neal, Jones & Feldman Pharmaceuticals).

Peter Oil. See OIL OF PETER.

Pethidine Roche. A proprietary preparation of pethidine. Narcotic analgesic. (Roche Products Ltd).*

Pethilorfan. A proprietary preparation of pethidine hydrochloride and levallorphan tartrate. An analgesic. (Roche Products Ltd).*

Petitgrain Oil. See OIL OF PETITGRAIN.

Petlon. Thermoplastic PET polyesters. Modified polyethylene terephthalate grades which offer good dimensional stability, high rigidity, high heat resistance, good electrical properties and chemical resistance. (Mobay Chemical Corp).*

Petra. Range of polyethylene terephthalate polymers, including glass reinforced compounds. Applications: Automotive ignition and carburettor components; electrical connectors, bobbins, relays and switches; intricate mechanical parts that maintain dimensional stability under load, or at temperatures above 150°C, or in the presence of moisture. (Allied Chemical Corporation).*

Petrac CP-11. Calcium stearate. Applications: Lubricant in processing of PVC. (DeSoto Inc).*

Petrac CP-12. Calcium stearate. Applications: Lubricant for PVC formulations, water repellent for concrete mixes. (DeSoto Inc).*

Petrac ZN-44. Zinc stearate. Applications: External lubricant (antitack) for rubber articles, internal lubricant in phenolic moulding resins and polystyrene. (DeSoto Inc).*

Petracin A and B. Dyestuffs obtained from petroleum by treatment with sulphuric acid, then with halogens in the presence of oxidizing agents. Upon neutralizing, the solution gives a precipitate of Petracin B, the solution containing Petracin A. A dyes silk and wool; B when heated with nitric acid, then treated with alkalis, dyes silk, cotton, and wool.†

Petralol. See PARAFFIN, LIQUID.

Petralon. A German name for a preparation of wood tar. An antiseptic.†

Petramin. Special disperse dyes for dyeing nickel-modified polypropylene fibres. (Bayer UK Ltd).

Petrasul.(Transite. A trade mark.) A synthetic stone material made from asbestos and Portland cement, which is sulphur-impregnated. The sulphur content varies from 15-35 per cent and the material can be coloured, and is suitable for counter-tops, or similar purposes.†

Petre. Potassium nitrate, KNO_3, used in pyrotechny.†

Petre Oil. See OIL OF PETER.

Petrex. A proprietary trade name for a polybasic acid used in synthetic resin manufacture, the essential constituent of which is 3-isopropyl-6-methyl-3 : 6 endo-ethylene - Δ_4 - tetrahydrophthalic anhydride.†

Petrex 7-75T. A solution of an alkyd-type resin derived from a terpene polybasic acid. Applications: Used in coatings for cellophane. (Hercules Inc).*

Petrified Asafoetida. An inferior type of asafoetida.†

Petrin. Pentrinitrol. Applications: Vasodilator. (Parke-Davis, Div of Warner-Lambert Co).

Petrinex. Propylene glycol mixtures. (ICI PLC).*

Petro. See PARAFFIN, LIQUID.

Petro AG Special. Alkyl naphthalene sulphonate. Applications: Water soluble anticaking agent for detergents, fertilizers and various salts. (DeSoto Inc).*

Petro BAF. Linear alkyl naphthalene sodium sulphonate. Applications: Hydrotrope and surfactant for detergent and cleaning formulations. (DeSoto Inc).*

Petro P. Alkyl napthalene sodium sulphonate. Applications: Wetting agent for pesticide and fertilizer formulations. (DeSoto Inc).*

Petro S. Alkyl napthalene sodium sulphonate. Applications: Soil conditioner. (DeSoto Inc).*

Petro ULF. Modified alkyl naphthalene sodium sulphonate. Applications: Low foam surfactant for various cleaning applications. (DeSoto Inc).*

Petro WP. Modified alkyl naphthalene sodium sulphonate. Applications: Wetting agent for industrial cleaning formulations and pesticide formulations. (DeSoto Inc).*

Petro 11. Alkyl naphthalene sodium sulphonate. Applications: Hydrotrope and surfactant. (DeSoto Inc).*

Petro 22. Modified alkyl napthalene sodium sulphonate. Applications: Low foam surfactant for various cleaning formulations. (DeSoto Inc).*

Petroacid. A proprietary trade name for a mixture of fatty acids obtained from petroleum distillates.†

Petrobenzol. A petroleum distillate. A solvent. It has a boiling-point of 61-96° C.†

Petroclastite. (Petroklastite). An explosive. It contains potassium nitrate, sulphur, coal tar pitch, and potassium dichromate.†

Petrofracteur. An explosive. It consists of 10 per cent nitrobenzene, 67 per cent potassium chlorate, 20 per cent potassium nitrate, and 3 per cent antimony pentasulphide.†

Petrogens. (Petroxolins). Similar to Vasogen (*qv*).†

Petroklastite. See PETROCLASTITE.

Petrol. A product of the distillation of petroleum. The term is synonymous with gasoline and petroleum spirit. Other names for the same product are naphtha, petroleum naphtha or mineral naphtha, benzoline, benzine, and carburine.†

Petrolagar. A proprietary emulsion of liquid paraffin and agar-agar.†

Petrolatum. See PETROLEUM JELLY.

Petrolatum Wax. A wax contained in the residual stock from the refining of petroleum waxes. It is one of them, the others being slop wax from the heavier wax distillate, and paraffin from the lighter wax distillate.†

Petrolax. See PARAFFIN, LIQUID.

Pétrole Hahn. Hair tonic, shampoo and conditioner to keep hair healthy and attractive. (Richardson-Vicks Inc).*

Petrolenes. (Malthenes). Constituents of bitumens which are soluble in hexane.†

Petroleum Benzine. See C-PETROLEUM NAPHTHA.

Petroleum Ether. (Gasoline, Solene). A distillate from petroleum oil, of boiling point 40-70° C., and specific gravity 0.64-0.65. It consists essentially of pentane and hexane, and is a solvent for resins. Petroleum oil of boiling point 70-90° C. (Sherwood oil), is called petroleum ether, also that fraction of boiling-point 120-140° C.†

Petroleum Jelly. (Mineral Jelly, Paraffinum Molle, Petrolatum). Varieties of soft paraffin. They consist of the yellow, semi-solid, purified residue left when petroleum is distilled, and contains several hydrocarbons. The mixture has a specific gravity of 0.87-0.90. An artificial product has been prepared by dissolving 1 part paraffin or cerasin in 4 parts liquid parafin. Also see OZOKERINE.†

Petroleum Naphtha. A term very loosely applied. It often denotes the first fraction of boiling-point up to 150° C., obtained from the distillation of crude petroleum oil, but is sometimes applied to any low boiling petroleum product. See BENZENE.†

Petroleum Pitch. Asphalt.†

Petroleum Spirit. (Light Petroleum). Both benzoline and naphtha are sold under this name. They are used as motor spirits, and for dry-cleaning cloths. See BENZINE and NAPHTHA, with all of which the terms are synonymous.†

Petroleum, Stockholm. Stockholm tar.†

Petrolia. See PARAFFIN, LIQUID.

Petrolig. Chrome and ferrochrome lignosulphonates. Applications: Oil well drilling mud conditioners. (Reed Lignin Inc).*

Petrolina. See OZOKERINE.

Petroline. A fraction of petroleum distillation of boiling-point 120-150° C., of specific gravity 0.722-0.737. Used for defatting, or cleaning. The term is also used for a volatile oil yielded by asphalt, when it is distilled with water.†

Petrolit. A German explosive containing potassium chlorate and mineral oil.†

Petromix 9. Balanced blend of sodium sulphonates, auxiliary fatty acid soaps and coupling agents with anti-foaming agents. Applications: Strong emulsification agent used in textile processing oils, cutting oils, metal degreasers. (Witco Chemical Ltd).

Petromor. Petroleum sulphonates. (Burmah-Castrol Ltd).*

Petronate L, HL, K, CR and S. Series of anionic surfactants in which the cation is sodium and the anion is petroleum sulphonate. Applications: Emulsifiers, dispersing and wetting agents. CR is rust preventative. Outlets include printing inks; dry cleaning soaps; leather oils; lubricating grease; metal working; mineral dressings; paint; petroleum additives; emulsion breaking; rubber; textile processing oils. (Witco Chemical Ltd).

Petronate RP. Blend of sodium and calcium petroleum sulphonates. Applications: Anionic surfactant used in rust preventative formulations. (Witco Chemical Ltd).

Petrone A4 and A6C. Amine salt of alkylaryl sulphonic acid. Brown viscous liquid. Applications: Solubilization, antistat, dewatering, formulation, pigment dispersion, emulsification. Corrosion inhibition properties. (ABM Chemicals Ltd).

Petronol. See PARAFFIN, LIQUID.

Petropul. A proprietary trade name for a synthetic resin.†

Petrosio. See PARAFFIN, LIQUID.

Petrostep. Petroleum sulphonates. Applications: Enhanced oil recovery. (Stepan Company).*

Petrostep A-70. Branched chain alkylate sulphonic acid, anionic surfactant properties, viscous amber liquid. Applications: An emulsifier intermediate for speciality products. (KWR Chemicals Ltd).

Petrosulfol. See THIOL.

Petrothene. Low, medium, linear low and high density polyethylene resins. Applications: Blown and cast film, injection moulding, blow moulding, extrusion coating, sheet and profile extrusion and wire and cable. (USI Chemicals).*

Petrothene XL. Crosslinkable polyethylen resins. Applications: Wire and cable. (USI Chemicals).*

Petrowet. Surface active agent. (Du Pont (UK) Ltd).

Petroxolins. See PETROGENS.

Peucedanin. Imperatorin. It occurs in the root of masterwort.†

Pevafix. An adhesive for polyvinyl alcohol film. (May & Baker Ltd).*

Pevalon. Polyvinyl alcohol film. (May & Baker Ltd).*

Pevidine. A proprietary preparation of povidone iodine used as a skin disinfectant. Marketed for veterinary use only. (Berk Pharmaceuticals Ltd).*

Pevikon. Homo and copolymeric poly viny chloride (E-PVC). Applications: As a main product for use in organosols and plastisols for production of plastic articles. (Norsk Hydro AS).*

Pewter. A variable alloy of from 73-89 pe cent tin, 1.6-6.7 antimony, 1-6.8 per cent copper, and 0-20.5 per cent lead, and sometimes zinc. The alloy containing 80 per cent tin and 20 per cent lead has a specific gravity of 10 a melts at 200° C. Another alloy of this class consists of 88 per cent tin. 7 per cent antimony, 3 per cent copper, and per cent zinc. A harder variety contains 75 per cent tin and 25 per cent lead.†

Pewter, Berthier's. An alloy of 72 per cent copper, 25 per cent zinc, 2 per cent lead, and 1 per cent tin.†

Pexalyn. Polar, acidic hydrocarbon-based synthetic resins. Applications: Pexalyn A500 and A600 are used as tackifier resins for solvent and emulsion adhesives and as modifier resins for ethylene/vinyl acetate copolymer and wax-based hot-melt coatings and adhesives. (Hercules Inc).*

Pexate. Metal resinates. (Hercules Ltd).

Pexid. Perhexiline Maleate. Applications: Vasodilator. (Merrell Dow Pharmaceuticals Inc, Subsidiary of Dow Chemical Co).

Pexid. A proprietary preparation of perhexiline maleate used in the treatment of angina pectoris. (Richardson-Vicks Inc).*

Pexite. Wood rosins. Applications: It is used in wax modification, as a chemical intermediate, in solder fluxes and in wax based coatings. (Hercules Inc).*

Pexol. Fortified rosin size for paper. (Hercules Inc).*

Peyron's Chloride. Diaminodichloroplatinum.†

Peyton Powder. An explosive. It is a nitro-cellulose and nitro-glycerin powder, containing 20 per cent ammonium picrate.†

Pfaudlon 301. A water suspension of Penton (*qv*).†

Pfeilringspalter. A catalyst used in the decomposition of fats. It is prepared by treating with sulphuric acid a mixture of hydrogenated ricinoleic acid and naphthalene.†

Pferrico. A cobalt treated iron oxide for magnetic media use. (Pfizer International).*

Pferrisperse. A high-solids iron oxide pigment slurry. (Pfizer International).*

Pferritan. A zinc ferrite compound for high temperature colour pigment use. (Pfizer International).*

Pferrocal. A steel-clad calcium wire effective in the production of high quality specialty steels. (Pfizer International).*

Pferromet. A metallic iron particle for magnetic tapes and disks. (Pfizer International).*

Pferrox. A gamma ferric oxide for magnetic tape and disk applications. (Pfizer International).*

Pfeufer's Green. A blue-green dye obtained from *Chlorospenium aeruginosum,* a fungus.†

Pfico$_2$-Hop. A non-isomerized carbon dioxide extract of hops. Applications: It is used by adding to the brew kettle during boiling in the manufacture of malt beverages, to add bitterness to this product. (Pfizer International).*

Pfico$_2$-Isohop. A modified aqueous hop extract produced from a liquid carbon dioxide base concentrate and standardized to 35% isomerized alpha acids. Applications: For addition to malt beverage after fermentation to standardize bitterness. (Pfizer International).*

Pfico$_2$-Redihop. A modified aqueous hop extract produced from a liquid carbon dioxide base concentrate and standardized to 35% reduced alpha acids. Applications: For addition to malt beverage after fermentation to standardize bitterness. (Pfizer International).*

Pfiklor. Potassium Chloride. Applications: Replenisher. (Pfizer Inc).

Pfinodal. A copper-nickel-tin alloy used in the fabrication of electronic connectors. (Pfizer International).*

Pfizercycline. A proprietary preparation of tetracycline. An antibiotic. (Pfizer International).*

Pfizer-E. A proprietary preparation of erythromycin stearate. An antibiotic. (Pfizer International).*

Pfizer-EM. A proprietary preparation of erythromycin stearate. An antibiotic. (Pfizer International).*

Pfizerpen. Penicillin G Potassium. Applications: Antibacterial. (Pfizer Inc).

Pfizerpen A. Ampicillin. Applications: Antibacterial. (Pfizer Inc).

Pfizerpen A. A proprietary preparation of ampicillin. An antibiotic. (Pfizer International).*

Pfizerpen Vk. Penicillin V Potassium. Applications: Antibacterial. (Pfizer Inc).

Pfizerpen-AS. Penicillin G Procaine. Applications: Antibacterial. (Pfizer Inc).

Pfizerpen-AS. A proprietary preparation of penicillin G procaine. An antibiotic. (Pfizer International).*

Pfizerquine. A proprietary preparation of chloroquine. An antimalarial. (Pfizer International).*

Phaltan. Fulpet fungicide. (Chevron).*

Phamosan. Skin protection ointments, lotions and soaps. Applications: For protection and care for the skin under environmental stress. (Dynamit Nobel Wien GmbH).*

Phanodorm. A proprietary preparation of cyclobarbitone. A hypnotic. (Bayer & Co).*

Phanteine. Pure linalyl acetate. (Bush Boake Allen).*

Phantol. Pure linalol. (Bush Boake Allen).*

Pharmagel. A proprietuy trade name for pure gelatin.†

Pharmakon. Zinc oxide BP. Applications: Pharmaceutical. (Manchem Ltd).*

Pharmasorb. Pharmaceutical grade attapulgite (hydrous magnesium aluminium silicate) used as an antidiarrhoeal agent and inert. Applications: Tablet or liquid antidiarrhoeal dosage forms, tableting aid, pharmaceutical carrier, inert, cosmetics. (Engelhard Corporation).*

Pharmatone. A hydrolysed pork tissue that is spray dried. It is over 90% protein and water soluble. Applications: Used in veterinary biologicals and food supplements, for its nutrient and high nitrogen content. (American Laboratories Inc).*

Phasal. White sustained-release tablets containing 300mg lithium carbonate BP. Applications: Treatment of acute manic or hypomanic episodes. Prophylaxis in manic-depressive disorders, recurrent depression. (Lagap Pharmaceuticals Ltd).*

Phaseomanite. Inositol, $C_6H_{12}O_6$.†

Phazyme. Pink, sugar coated tablet containing specially activated simethicone. Applications: Deflatulent. (Stafford-Miller).*

P.H.D. A proprietary trade name for a plasticizing oil.†

Phe-Mer-Nite. Aphenylmercuric Nitrate. Applications: Pharmaceutic aid. (Beecham Laboratories).

Phemerol Chloride. Benzethonium chloride. Applications: Anti- infective, topical; pharmaceutic aid. (Parke-Davis, Div of Warner-Lambert Co).

Phemox. Mercury fungicide. (Murphy Chemical Co).†

Phenac. A proprietary trade name for a phenolic synthetic resin for varnish and lacquer.†

Phenacemide. (Phenylacetyl)urea.†

Phenacetin. Acetyl-*p*-phenetidine, $C_6H_4(O.C_2H_5)\ NH.CO.CH_3$.†

Phenald Resins. A general term for phenol-formaldehyde resins.†

Phenaldine. A proprietary trade name for a rubber vulcanization accelerator. It is diphenylguanidine.†

Phenalein. Phenolphthalein.†

Phenalgin. See AMMONAL.

Phenalin. A proprietary trade name for a phenol-formaldehyde synthetic resin.†

Phenamein. See MAUVEINE.

Phenamin. See MAUVEINE.

Phenanthraquinone Red. A dyestuff. Dyes wool red from an acid bath.†

Phenaphen. Acetaminophen. Applications: Analgesic; antipyretic. (A H Robins Co Inc).

Phenazone. See ANTIPYRINE.

Phenazopirin. See ACOPRRIN.

Phenedin. See PHENACETIN.

Phenegol. The mercury-potassium salt of nitro-*p*-phenolsulphonic acid. A bactericide.†

Phenelzine. Phenethylhydrazine.†

Phenergan. A proprietary preparation of promethazine hydrochloride. (May & Baker Ltd).*

Phenester. A proprietary trade name for a synthetic resin of the coumaroneindene type.†

Phenethyl Alcohol. 2-Phenylethanol.†

Phenethyl Mustard Oil. *p*-Ethylphenyl-thiocarbimide, $SCN.C_6H_4.(C_2H_5)$.†

Phenetidine. Amino-phenyl ethyl ether, $C_6H_4(OC_2H_5)NH_2$. Antipyretic.†

Phenetidine Red. An azo dyestuff obtained from nitrophenetidine and β-naphthol.†

Phenetole. Phenyl ethyl ether, $C_6H_5.O.C_2H_5$.†

Phenetole Red. (Coccinin). An azo dyestuff, $C_2H_5.O.C_6H_4.N_2.C_{10}H_{14}(HSO_3)_2$. OH. It is homologous with Anisole red.†

Phenex. A proprietary trade name for a rubber vulcanization accelerator. An aldehyde-amine.†

Phenic Acid. (Phenic Alcohol). Phenol, C_6H_5OH.†

Phenic Alcohol. See PHENIC ACID.

Phenicienne. See PHENYL BROWN.

Phenicine. See PHENYL BROWN.

Phenin. A proprietary preparation of phenacetin.†

Phenindione. 2-Phenylindane-1,3-dione. Phenylindanedione. Dindevan.†

Phenistix. A proprietary preparation of ferric ammonium sulphate, magnesium sulphate and cyclohexylsulphamic acid impregnated on a test strip, used for the detection of phenylketonuria and ingestion of salicylates. (Ames).†

Phenitol. Phenylmercuric Nitrate. Applications: Pharmaceutic aid. (Alcon Laboratories Inc).

Phenmerzyl Nitrate. Phenylmercuric Nitrate. Applications: Pharmaceutic aid. (Merrell Dow Pharmaceuticals Inc, Subsidiary of Dow Chemical Co).

Phenobarbitone Spansule Capsules. Phenobarbitone in sustained release form. Applications: Anticonvulsant indicated in treatment of epilepsy. (Smith Kline and French Laboratories Ltd).*

Phenocitrain. A proprietary preparation of phenolic analogues, citral and LIGNOCAINE used in the treatment of varicose ulcers. (Whaley).†

Phenocyanine TC. A dyestuff. Dyes chromed wool blue.†

Phenocyanine TV. A dyestuff. Dyes chromed wool and silk blue. Employed in printing.†

Phenocyanine VS. A dyestuff. Used for printing blue upon chromed cotton.†

Phenocyanines. A series of dyestuffs produced by reacting upon gallocyanines with resorcin.†

Phenodip. Liquid, compounded esters of 4-hydroxybenzoic acid. Applications: Preservative. (Nipa Laboratories Ltd).*

Phenodorm. Pharmaceutical preparation. Applications: Soporific. (Bayer & Co).*

Phenodur. Heat hardening phenol/formaldehyde resins. Applications: Industrial paints, brake and clutch linings and abrasives. (Resinous Chemicals Ltd).*

Phenoflavine. A dyestuff. It is the sodium salt of *m*-sulphobenzeneazo-aminophenolsulphonic acid, $C_{12}H_9N_3S_2O_7Na_2$. Dyes wool yellow from an acid bath.†

Phenoform. See BAKELITE.

Phenolax. See PURGELLA.

Phenolax. Phenolphthalein. Applications: Laxative. (The Upjohn Co).

Phenol Blue. Dimethyl-amino-phenylimide, $C_{14}H_{14}N_2O$.†

Phenol Camphor. (Carbolic Camphor). Phenol with camphor *(Phenol cum camphorae B.P.)*.†

Phenolene and Phenolene Supra. Tin plating bath brighteners. (Ciba Geigy PLC).†

Phenol-Formaldehyde and Allied Synthetic Resins. See under BAKELITE, BECKACITE, CHRISTOLIT, DURCOTON, ERICON, FORMICA, FUTURITE, KILIANIT, LACRINITE, LORIVAL, LUXENE, NEOLEUKORITE, PANTOLIT, RAMOS, RESINOX, REVOLITE, ROCKITE.

Phenoline. A disinfectant identical with Lysol. It is a cresol made soluble in water by saponification.†

Phenolite. A proprietary trade name for a phenol-formaldehyde synthetic resin laminated product.†

Phenolphthalein Synthetic Resin. Alkalit.†

Phenolphthalein. The lactone of dioxy-triphenyl-carbinol-carboxylic acid, $C_{20}H_{14}O_4$. Used as an indicator in alkalimetry.†

Phenol Red. Phenol-sulphonphthalein. An indicator.†

Phenol Red. Tetrabrom-phenol-sulphon phthalein. An indicator.†

Phenol-sodium Sulphoricinate 25 per cent. This material contains 25 per cent phenol and 75 per cent of the sodium salt of sulphonated castor oil.†

Phenomauveine. A dyestuff prepared from nitrailine and diphenyl-*m*-phenylene-diamine.†

Phenomine. See TYRAMINE.

Phenonip. Broad-spectrum antimicrobial agent. (Nipa Laboratories Ltd).

Phenopreg. A proprietary trade name for phenolic impregnated fabrics and papers.†

Phenoro. Proprietary preparation of canthaxanthin and beta-carotene. Photoprotective agent. (Roche Products Ltd).*

Phenosafranine. (Safranine B Extra). A dyestuff. It is diaminophenyl phenazonium chloride, $C_{18}H_{15}N_4Cl$. Dyes cotton mordanted with tannin and tartar emetic, red.†

Phenosalyl. A mixture of phenol, salicylic acid, menthol, and lactic acid. An antiseptic.†

Phenosept. Mixture of propylene phenoxetol and para-chloro-meta-xylenol. Applications: Antiseptic. (Nipa Laboratories Ltd).*

Phenoweld. Phenolic adhesives for nylon etc. (Hardman Inc).*

Phenox. A proprietary trade name for phenylmercury hydroxide.†

Phenoxetol. Monoaryl ethers of aliphatic glycols. (Nipa Laboratories Ltd).*

Phenoxin. Carbon tetrachloride, CCl_4.†

Phenoxine. Chlorphenoxamine hydrochloride. Applications: Relaxant; anti-partkinsonian. (Merrell Dow Pharmaceuticals Inc, Subsidiary of Dow Chemical Co).

Phenoxylene Plus. A selective weed killer. (Fisons PLC).*

Phenoxylene 50. Selective herbicide. (FBC Ltd).

Phensedyi Linctus. A proprietary preparation containing promethazine hydrochloride, codeine phosphate and ephedrine hydrochloride. A cough linctus. (May & Baker Ltd).*

Phensedyl. Promethazine hydrochloride compound cough linctus. (May & Bake Ltd).

Phenurone. Phenacemide. Applications: Anticonvulsant. (Abbott Laboratories).

Phenyform. An antiseptic powder prepared from phenol and formaldehyde. Used as an indicator and for denaturing purposes.†

Phenylalanine. α-Amino-β-phenyl-pro pionic acid, $C_6H_5.CH_2CH(NH_2)COOH$.†

Phenylamine. Aniline, $C_6H_5NH_2$.†

Phenylene Black. See ANTHRACITE BLACK B.

Phenylene Blue. See MELDOLA'S BLUE.

Phenylene Brown. See BISMARCK BROWN.

Phenyl-Gamma Acid. 2-Phenylamino 8-naphthol-6-sulphonic acid.†

Phenylindanedione. Phenindone.†

Phenylon. See ANTIPYRINE.

Phenyl Methane. See METHYL BENZENE.

Phenyl-peri Acid. 1-Phenyl-naphthylamine-8-sulphonic acid.†

Phenyl Violet. See REGINA PURPLE.

Pheny-PAS-Tebamin. Phenyl Aminosalicylate. Applications: Antibacterial. (The Purdue Frederick Co).

Pheophytin. The brownish derivative obtained by treating chlorophyll with acid.†

Philadelphia Yellow G. See PHOSPHINE.

Philanized Cotton. Cotton material which has been treated with concentrated nitric acid to convert it into a wool-like fabric.†

Philosopher's Wool. (Flowers of Zinc). The zinc oxide produced in a flocculent condition by burning zinc.†

Phisomed. A proprietary preparation of chlorhexidine gluconate in a detergent base used as a skin cleanser. (Winthrop Laboratories).*

Phloba-tannins. Tannins which give the phlobaphene reaction.†

Phlobaphenes. Red or brown colouring matter from barks, usually oak bark.†

Phloroglucinol. 1, 3, 5-trihydroxy benzene, $C_6H_3(OH)_3$.†

Phlorol. *o*-Ethylphenol, $C_6H_4.C_2H_5$ (OH).†

Phlorone. (Metaphlorone). *p*-Xyloquinone, $C_6H_2(CH_3)_2O_2$.†

Phloxine. (Phloxine TA, Eosine 10B, Cyanosine). The sodium salt of tetra-bromotetrachlorofluorescein, $C_{20}H_2Cl_4Br_4O_5Na_2$. See PHLOXINE P.†

Phloxine P. (Phloxine, New Pink. Erythrosin BB). Dyestuffs. The alkali salts of tetrabromo-dichloro-fluoresceine $C_{20}H_4Cl_2Br_4O_5K_2$. Dyes wool bluish-red, and cotton mordanted with tin, alumina, or lead.†

Phloxine TA. See PHLOXINE.

Phobol. Finishing agent. (Ciba-Geigy PLC).

Phobotex FTN. A proprietary trade name for a fat modified melamine resin with outstandimg water-repellancy. It is fast to repeated washing in soap and soda and provides a durable water repellant finish for textiles. (Ciba Geigy PLC).†

Phobotone. Proofing agent. (Ciba-Geigy PLC).

Phocenic Acid. Isovaleric acid.†

Phoenicite. Basic lead chromate.†

Phoenix Alloy. An electrical resistance alloy containing 25 per cent nickel and 75 per cent iron.†

Phoenix Powder. An explosive contaning 28-31 per cent nitro-glycerin, 0-1 per cent nitro-cotton, 30-34 per cent potassium nitrate, and 33-37 per cent wood meal.†

Phoenixite. A proprietary pyroxylin product.†

Pholin's Alloy. An alloy of 77 per cent tin, 19 per cent bismuth, and 4 per cent copper.†

Pholtex. Each 5 ml spoonful contains pholcodine BP 15 mg and phenyltoloxamine 10 mg as ion-exchange resin complexes in a thixotropic vehicle. (Riker Laboratories/3M Health Care).†

Phono Bronze. A proprietary trade name for certain copper alloys containing about 1.25 per cent tin and small amounts of silicon and cadmium.†

Phoran. See CAMITE.

Phorochrome Yellow Y. A dyestuff. It is a British equivalent of Milling yellow.†

Phosbrite. Chemical brightening solutions. (Albright & Wilson Ltd, Phosphates Div).

Phos-Chek. Fire retardant, ammonium polyphosphate. Applications: Catalyst used in intumescent coatings for paints, used to help meet flammability requirements for rigid and flexible polyurethane foams. (Monsanto Co).*

Phosclene. Metal finishing cleaners. (Albright & Wilson Ltd, Phosphates Div).

Phosclere. Stabilizers and antioxidants. (Akzo Chemie GmbH, Düren).†

Phos-copper. A proprietary welding alloy composed essentially of copper with from 5-10 per cent phosphorus. It melts at 700° C., becoming extremely fluid at 750° C.†

Phosene. Synanthene, $C_{14}H_{10}$.†

Phosflex 300 and 400. Proprietary mixed triaryl phosphate ester plasticizers containing halogen and possessing better flame-retardant properties in PVC than tricresyl phosphate. (Stauffer Chemical).†

Phosforme. Monoethyl-phosphoric acid.†

Phosgene. Carbonyl chloride, $COCl_2$.†

Phosphaljel. Aluminium phosphate. Applications: Antacid. (Wyeth Laboratories, Div of American Home Products Corp).

Phospham. Phosphorus imidonitride, PN_2H.†

Phosphammite. Native ammonium phosphate.†

Phosphate, Vesta. See RHENANIA PHOSPHATE.

Phosphazote. A fertilizer. It is an intimate mixture of superphosphate and urea containing 4-11 per cent nitrogen and 10-14 per cent. P_2O_5.†

Phosphin. 3-Amino-9-*p*-aminophenylacridine. Used in the treatment of malaria.†

Phosphine. (Leather Yellow, Xanthine, Vitoline Yellow 5G, Patent Phosphine, Nankin, Philadelphia Yellow G, Chrysaniline, Acridine Yellow, Acridine Orange, Leather Brown, Phosphine II, N, P). The nitrate of chrysaniline (diaminophenylacridine, $C_{16}H_{26}N_4O_3$), and homologues. Dyes leather reddish-yellow.†

Phosphine II. See PHOSPHINE.

Phosphine N. See PHOSPHINE.

Phosphine P. See PHOSPHINE.

Phosphine, Patent. See PHOSPHINE.

Phospho-gélose. A German clarifier for sugar juice. It consists of 70 per cent phosphate of lime, and 30 per cent kieselguhr.†

Phosphocalcite. See PSEUDO-MALACHITE.

Phosphocol P32. Chromic phosphate P32. Applications: Radioactive agent. (Mallinckrodt Inc).

Phospholan. Alkyl aryl sulphonates, modified non-ionics, phosphate esters. Applications: Used as hydrotropes and wetters in electrolyte solutions. (Lankro Chemicals Ltd).†

Phospholan ALF-5. Phosphate ester in acid form, as an off-white solid. Applications: Foam control agent for detergent powders. (Diamond Shamrock Process Chemicals Ltd).

Phospholan KPE-4. Potassium phosphate ester as a pale yellow liquid. Applications: Hydrotrope used for solubilization of low foam surfactants into highly built liquids. (Diamond Shamrock Process Chemicals Ltd).

Phospholan PDB-3. Fatty alcohol ethoxylate in acid form, as a pale straw liquid. Applications: Wetting agent, detergent, emulsifier and lubricant which is stable in high concentrations of acid, alkali and electrolyte. Used for industrial and dry cleaning detergents, lubricants, hydrotropes, emulsifying agents and primary emulsifiers in the emulsion polymerization process. (Diamond Shamrock Process Chemicals Ltd).

Phospholan PNP-9. Nonylphenol ethoxylate in acid form as a dark amber viscous liquid. Applications: Wetting agent, detergent, emulsifier and lubricant which is stable in high concentrations of acid, alkali and electrolyte. Used for industrial and dry cleaning detergents, lubricants, hydrotropes, emulsifying agents and primary emulsifiers in the emulsion polymerization process. (Diamond Shamrock Process Chemicals Ltd).

Phospholine Iodide. Ecothiopate iodide. Applications: Chronic open-angle glaucoma, chronic angle-closure glaucoma after iridectomy, accommodative esotropia. (Ayerst Laboratories).†

Phospholutein. Lecithin.†

Phosphomort. Organo-phosphorous insecticides. (Murphy Chemical Co).†

Phosphor Bronze. A bearing metal. It is an alloy of from 70-97 per cent copper, 3-13 per cent tin, 0-16 per cent lead, and 0.1-1.0 per cent phosphorus and sometimes a little zinc. The alloys for casting usually contain 85-92 per cent copper, 7-13 per cent tin 0.3-1.0 per cent phosphorus, and traces of lead and zinc. For malleable phosphor bronze the alloys consist of 94-97 per cent copper, 3-5 per cent tin, and 0.1-0.35 per cent phosphorus. Also see AJAX BRONZE, KUHNE PHOSPHOR BRONZE, and LOWROFF PHOSPHOR BRONZE.†

Phosphor Copper. An alloy of copper with from 5-15 per cent of phosphorus. Used as an addition to other metals, and in the manufacture of phosphor bronze.†

Phosphor Resin. A resin prepared from triphenyl phosphamide by heating and passing carbon dioxide through the compound.†

Phosphor Steel. An alloy of steel with phosphorus.†

Phosphor Tin. An alloy of tin and phosphorus, containing up to 10 per cent phosphorus.†

Phosphoretted Hydrogen. See PHOSPHINE.

Phosphoric Acid. Phosphoric acid. Applications: Other chemicals. (Rhone-Poulenc NV/CdF Chimie AZF).*

Phosphoric Acid, Glacial. Metaphosphoric acid, HPO_3. Usually it contains some sodium phosphate.†

Phosphoric D. See IRON D.

Phosphoric Ether. Triethyl phosphate, $OP(OC_2H_5)_3$. Also see SULPHURIC ETHER.†

Phosphorous Ether. Triethyl phosphite, $(OC_2H_5)_3P$.†

Phosphorus Salt. See MICROCOSMIC SALT.

Phosphorus, Schenk's. See SCARLET PHOSPHORUS.

Phosphotope. Sodium Phosphate P 32. Applications: Antineoplastic; antipolycythemic; diagnostic aid; radioactive agent. (E R Squibb & Sons Inc).

Phosteem. Metal treating compositions. (ICI PLC).*

Phostin. Tin-plating process. (Albright & Wilson Ltd, Phosphates Div).

Photal. A photographic developer.†

Photine. Fluorescent whitening agents for paper textiles and detergents. (Hickson & Welch Ltd).*

Photoglaze. UV and EB cure coatings of varying viscosities and chemical composition. Applications: To protect and preserve the original appearance of a wide range of materials - vinyls, plastics, paper, wood, even metallized surfaces giving good resistance to abrasion, chemical stains and scuffing. (Lord Corporation (UK) Ltd).*

Photomer. Radiation curing chemicals. (Lankro Chemicals Ltd).

Photophor. A registered trade name for a calcium phosphide, Ca_2P_2. Used for signal fires.†

Photoxylin. See CELLOIDIN.

Phoxim-Methyl. A pesticide. O-α-cyanobenzylideneamino OO-dimethyl phosphorothioate.†

Phtalofix FN. Premordant. Applications: For PHTALOGEN K dyestuffs. (Bayer & Co).*

Phtalogen. Phthalocyanine dyes. Applications: To produce blue and green shades of outstanding fastness on cotton and regenerated cellulose. (Bayer & Co).*

Phtalogen K/Phtalogen N1. Heavy metal donor for phthalogen dyestuffs. Applications: In dyeing and textile printing. (Bayer & Co).*

Phtalotrop B. Resist agent. Applications: For phtalogen resist printing. (Bayer & Co).*

Phthalamaquin. Quinetolate. Applications: Relaxant. (Penick Corp).

Phthalofyne. See WHIPCIDE.

Phylatol. Biocidal preparation. (BDH Chemicals Ltd).

Phyldrox-G. A proprietary preparation of theophylline sodium glycinate, ephedrine hydrochloride and phenobarbitone. A bronchial antispasmodic. (Carlton Laboratories).†

Phyllocontin. A proprietary preparation of aminophylline in a controlled release tablet used in the treatment of bronchospasm. (Napp Laboratories Ltd).*

Phyllocontin. Aminophylline. Applications: Relaxant. (The Purdue Frederick Co).

Phyllol. Pure eugenol. (Bush Boake Allen).*

Phyomone. Growth promoting hormone. (ICI PLC).*

Physeptone. A proprietary preparation of methadone hydrochloride. An analgesic. (The Wellcome Foundation Ltd).*

Physeptone Linctus. A proprietary cough linctus containing methadone hydrochloride. A cough linctus. (The Wellcome Foundation Ltd).*

Physic Nut Oil. See PURGING NUT OIL.

Physiological Salt Solution. (Normal Salt Solution, Isotonic Salt Solution, Surgical Solution). This consists of 8.5 grams of sodium chloride in 1,000 cc of distilled water. It is sterilized and used for intravenous injection.†

Physostigmine. Eserine, $C_{15}H_{21}N_3O_2$, an alkaloid from the calabar bean.†

Physostol. A sterile solution of 1 per cent of an eserine salt in olive oil. Used in the treatment of certain eye diseases.†

Phytex. A proprietary preparation of boratannic complex in alcohol in ethylacetate solvent. Used in dermatology. (Pharmax).†

Phytic Acid. (Phytinic Acid). Inositol-hexaphosphoric acid, $C_6H_6(OPO_3H_2)_6$.†

Phytin. Calcium magnesium salt of inositolhexaphosphoric acid; contains about 22 per cent phosphorus, 12 per cent calcium, and 1.5 per cent magnesium. A powerful nerve and general tonic.†

Phytodermine. Antifungal preparations. (May & Baker Ltd).

Phytoforol. Vitamin E capsules. (British Drug Houses).†

Phytol. A primary alcohol, $C_{20}H_{39}OH$, obtained by the decomposition of chlorophyll, the colouring matter of plants.†

PIB. Isoprenaline and atropine. Applications: Bronchospasm. (Napp Laboratories Ltd).*

P.I.B. An abbreviation for polyiso-butylene.†

Pibiter. A proprietary range of moulding products based on saturated thermoplastic polyesters. (Montedison UK Ltd).*

Picamar. Propylpyrogallol dimethyl ether, $C_{11}H_{16}O_3$. Used in perfumery.†

Picco 5000. Aromatic hydrocarbon resin derived from petroleum. Applications: Used in higher-vinyl acetate content ethylene/vinyl acetate copolymer-based hot melt adhesives, used in caulking compounds, sealants and concrete curing resins, and in news ink and oil-based lithographic inks. (Hercules Inc).*

Picco 6000. Aromatic hydrocarbon resin derived from petroleum. Applications: Used in higher-vinyl acetate content ethylene/vinyl acetate copolymer-based hot melt adhesives, used in caulking compounds, sealants and concrete curing resins, and in news ink and oil-based lithographic inks. (Hercules Inc).*

Piccodiene. Aliphatic hydrocarbon resins, composed mainly of polydicyclo-pentadienes. Applications: Used in concrete curing compounds, in metallic paints and varnishes and as a compounding ingredient and process aid in many rubber goods. (Hercules Inc).*

Piccolastic. A range of styrene monomer hydrocarbon resins. Applications: Used for pressure sensitive, hot melt and solvent adhesives and as epoxy resin extenders. (Hercules Inc).*

Piccolastic A-5, A-25. Proprietary trade names for vinyl plasticizers based upon polymerized styrene and homologues. (Harwick Standard Chemical Co).†

Piccolyte. A proprietary trade name for thermoplastic terpene resins.†

Piccolyte A. Polyterpene hydrocarbon resins derived from α-pinene. Applications: Used in adhesives for tapes and labels, in construction adhesives, in packaging adhesives, as tackifiers for styrene-butadiene rubber and styrene-block-copolymer rubber. (Hercules Inc).*

Piccolyte C. A series of polyterpene hydrocarbon resins derived from *d*-limonene. Applications: Used as tackifiers for natural rubber based pressure sensitive adhesives and can sealants, used in the production of chewing gum base, used in packaging adhesives and used to promote gloss, adhesion and hardness in wax coatings. (Hercules Inc).*

Piccolyte HM110. A hydrocarbon-modified terpene resin. Applications: A tackifier resin of particular value in adhesives requiring high shear and high tack. (Hercules Inc).*

Piccolyte S. A range of polyterpene hydrocarbon resins derived from beta-pinene. Applications: Used in natural rubber-based pressure sensitive adhesives and styrene-butadiene rubber can sealants and in textile dry sizes. (Hercules Inc).*

Piccomer XX. A range of low molecular weight resins produced from petroleum derived monomers. Applications: Binder and plasticizer resins in putty and sealants, as felt saturants in felt-based sheet goods, in waterproofing treatments for paperboard, as softeners for elastomers in rubber compounding. (Hercules Inc).*

Picconol A100. A range of aliphatic hydrocarbon resin emulsions. Applications: Tackifiers in adhesives and in the production of waterproof finishes. (Hercules Inc).*

Picconol A200. Anionic terpene resin emulsions based largely on low molecular weight thermoplastic terpene resins produced from beta-pinene or **d**-limonene monomers. Applications: Used in water-based laminating, case-sealing adhesives, in laminating paper and as tackifiers for natural rubber latex. (Hercules Inc).*

Picconol A300. Anionic terpene resin emulsions based largely on low molecular weight thermoplastic terpene resins produced from beta-pinene or **d**-limonene monomers. Applications: Used in water-based laminating, case-sealing adhesives, in laminating paper and as tackifiers for natural rubber latex. (Hercules Inc).*

Picconol A400. A low molecular weight anionic pure monomer resin emulsion. Applications: Used with other aqueous thermoplastic and/or elastomeric systems to produce excellent adhesives and coatings. (Hercules Inc).*

Picconol A500. A range of aromatic hydrocarbon resin emulsions. Applications: Used as adhesives, laminants and tackifiers. (Hercules Inc).*

Picconol A600. A range of aromatic hydrocarbon resin emulsions. Applications: Used as adhesives, laminants and tackifiers. (Hercules Inc).*

Piccopale. Range of aliphatic hydrocarbon resins manufactured from petroleum-derived monomers. Applications: Used in high ethylene/vinyl acetate hot-melt and natural rubber-based adhesives, in can coatings for packaging, in varnishes to impart gloss and flowability, in paper saturation and as waterproofing agents for paper and textiles. (Hercules Inc).*

Piccopyn. A range of phenolic-modified terpene resins. Applications: Used as laminating agents and modifiers in ethylene/vinyl acetate copolymer-based packaging and tray forming adhesives, in specialty coatings, to modify specific polyurethanes, as tackifiers and process aids in specialty rubber compounding. (Hercules Inc).*

Piccotac. A range of aliphatic hydrocarbon resins produced from mixed monomers of petroleum origin. Applications: Developed especially for adhesives, particularly pressure sensitive and hot-melt types. (Hercules Inc).*

Piccotex. Vinyltoluene copolymer hydrocarbon resins. Applications: Used in hot-melt product assembly adhesives, for wax based coatings, in transparentizing paper, as dry sizing agents for textiles. (Hercules Inc).*

Piccotoner. Styrene acrylic copolymer hydrocarbon resins. Applications: Used as toner resins for dry reproduction inks. (Hercules Inc).*

Piccoumaron. A proprietary trade name of terpene varnish and lacquer resins.†

Piccoumarone Resins. Proprietary trade name for coumarone-indenes.†

Piccovar AB. Aliphatic hydrocarbon resin. Applications: Used as reinforcing agents in adhesives with high-temperature requirements, in heat set printing ink applications, in specialty coatings, in rubber compounding. (Hercules Inc).*

Piccovar AP. Aromatic hydrocarbon resins used as plasticizers, softeners and tackifiers. Applications: Suitable for use in various adhesive systems based on natural rubber, styrene-butadiene rubber and poly-chloroprene. (Hercules Inc).*

Piccovar L. Alkylated aromatic hydrocarbon resins. Applications: The resins are nonpolar, low molecular weight resins used as saturants, plasticizers and tackifiers in adhesives and hot melts. (Hercules Inc).*

Pichurim Camphor. A substance resembling laurel camphor obtained from pichurim beans.†

Picked Turkey Gum. (White Sennaar Gum). The best variety of gum acacia.†

Picket. Garden insecticide. (ICI PLC).*

Pickle Alum. Aluminium sulphate, $Al_2(SO_4)_3$. Used for packing and preserving.†

Pickle Green. A commercial variety of Scheele's green.†

Picoline. Methyl-pyridine, C_6H_7N.†

Pi-cone. A proprietary lithopone containing 15 per cent titanium oxide, 25 per cent zinc sulphide, and 60 per cent precipitated barium sulphate. It is stated to have a much higher covering power than ordinary lithopone.†

Picraalluminite. Synonym for Piero-allumogene.†

Picrasmin. Quassin, $C_{10}H_{12}O_3$.†

Picrate Powder. The name given to explosive powders, in which the main constituent is the potassium or ammonium salt of picric acid.†

Picric Acid. Trinitrophenol, $C_6H_2.OH(NO_2)_3$. Used for making explosives and dyes. A solution is employed in the treatment of burns, erysipelas, eczema. and gonorrhoea. Also see IGEWSKY'S REAGENT and PICRONTIC ACID.†

Picric Powder. An explosive consisting of ammonium picrate and potassium nitrate.†

Picro-aniline Blue. A stain used in microscopy. It is prepared by adding aniline blue to a saturated solution of picric acid in 92 per cent alcohol until the liquid becomes deep blue-green in colour.†

Picrocarmine. A microscopic stain obtained by mixing 1 gram carmine in 10 cc water and 3 cc strong ammonia solution, and adding the mixture to 200 cc of a saturated solution of picric acid.†

Picrocrichtonite. Synonym for Picroilmenite.†

Picronigrosine. An alcoholic solution of picric acid and nigrosin. A microscopic stain.†

Picrontric Acid. See PICRIC ACID.

Picro-sulphuric Acid. A liquid made by adding to 100 volumes water, 2 volumes sulphuric acid, and about 0.25 per cent picric acid. Used in microscopy as a fixing agent.†

Picryl Brown. A colouring matter obtained by sulphonating the nitro derivatives of secondary and tertiary aromatic amines. Dyes silk and wool yellow from an acid bath.†

Pictet Crystals. White crystals, $SO_2.XH_2O$ formed when liquid sulphur dioxide evaporates.†

Pictet's Fluid. Liquid carbon dioxide, used for freezing machines.†

Pictet's Liquid. A mixture of liquid carbon dioxide and sulphur dioxide. Used for producing low temperature.†

Pictolin. A mixture of liquid carbon dioxide and sulphur dioxide.†

Pictressin Tannate. Argipressin tannate. Applications: Antidiuretic. (Parke-Davis, Div of Warner-Lambert Co).

Pielanase. Enzyme. (ABM Chemicals Ltd).*

Pierrot Metal. An alloy consisting mainly of zinc, with smaller amounts of copper tin, antimony, and lead.†

Pif-Paf. Insecticide preparations. (The Wellcome Foundation Ltd).

Pig Iron. Cast iron in pieces of D section are called pigs. It is an impure form of iron, usually containing from 92-93 per cent iron, and at least 2.5 per cent carbon as graphitic carbon. It is classified into six classes, Nos. 1, 2, 3, 4, mottled, and white, according to the appearance of the fracture. No. 1 contains the largest flakes of gaphite, and white the smallest.†

Pig Iron, Grey Forge. See GREY FORGE PIG.

Pig Lead. Lead is obtained from galena by heating in a reverberatory furnace with a silica flux, then heated with coke, and sometimes lime in a cupola furnace. The lead drawn off is called pig lead.†

Pigeon Berry. Phytolacca.†

Pigmentar. A proprietary trade name for standardized pine tar prepared for use as a rubber softener in rubber compounding.†

Pigment Chrome Yellow L. A dyestuff made from diazotized *o*-toluidine, and 1-phenyl-3-methyl-5-pyrazolone.†

Pigment Fast Orange L Extra. A dyestuff. It is a British equivalent of Pigment orange R.†

Pigment Fast Red HL. See HELIO FAST RED RL.

Pigment Fast Scarlet 3L Extra. A dyestuff. It is a British brand of Helio fast red RL.†

Pigment Fast Yellow Conc., New. See HANSA YELLOW G.

Pigment Fast Yellow GRL Extra. A dyestuff obtained by coupling diazotized *m*-nitro-*p*-toluidine with aceto acetic toluidide.†

Pigment Orange R. (Pigment Fast Orange L Extra). A dyestuff obtained from daizotized *p*-nitro-*o*-oluidine coupled with β-naphthol.†

Pigment Purple. A dyestuff formed by diazotizing *o*-anisidine and coupling with β-naphthol.†

Pigment Scarlet 3B. A dyestuff obtained by diazotizing anthranilic acid and coupling it with R salt.†

Pigment 40-40-20. An American pigment. It contains 40 per cent zinc oxide, 40 per cent lithopone, 10 per cent silica, and 10 per cent asbestine.†

Pig, Silicon. See SILICON-EISEN.

Pilasonite. A mineral, Bi_3Te_2.†

Piliogrip Adhesive System for Styructural Bonding. 100% reactive urethane structural adhesives designed for bonding thermosets and thermoplastics and metals. Applications: SMC to SMC for automotive and truck body assemblies. SMC to metal for automotive body assemblies. (Ashland Chemical Company).*

Piliophen. Sodium tetraiodophenol phthalein.†

Pilocarp-phenol. Pilocarpine carbolate.†

Pilofrin. Pilocarpine Nitrate. Applications: Cholinergic. (Allergan Pharmaceuticals Inc).

Pilomiotin. Pilocarpine Hydrochloride. Applications: Cholinergic. (CooperVision Inc).

Pilot. Selective herbicide. (FBC Ltd).

Pimafucin. A proprietary preparation of natamycin used in the treatment of fungal infections. (Brocades).*

Pimaricin. A proprietary trade name for Natamycin.†

Pimeleine. Petrolatum.†

Pimelite. A mineral. It is meerschaum containing nickel.†

Pimple Metal. A term used for a type of copper metal produced from the "coarse metal" obtained from sulphide ores which have been fused with an excess of copper oxide in their purification.†

Pinachrom. *p*-Ethoxy-quinaldine-*p*-methoxy-quinoline-ethyl-cyanine-bromide. A red sensitizer for silver bromide plates.†

Pinacoline. Methyl-*tert*-butyl-ketone, $CH_3.CO.C(CH_3)_3$.†

Pinacyanol. A red sensitizer for silver bromide plates obtained by treating quinaldinium salts with formaldehyde followed by alkali.†

Pinaflavol. A basic dye used as a green sensitizer in photography.†

Pinakol. A pyrogallol photographic developer in which part of the alkali usually employed is replaced by sodium amino-acetate.†

Pinakol P. A photographic developer. Pyrogallol is the developing substance, but it also contains pinakol salt N.†

Pinakol Salt N. A 20 per cent solution of sodium aminoacetate, $CH_2.NH_2.COONa$. It replaces the alkali in organic developers.†

Pinakryptol. A green dye that is used as a photographic sensitizer.†

Pinaverdol. A green sensitizer for silver bromide plates.†

Pinchbeck. An alloy of from 83-93 per cent copper and 6-17 per cent zinc. A brass.†

Pincoffin. Commercial alizarin.†

Pine Gum. (Pine Resin, White Pine Resin, Cypress Pine Resin). Names applied to Australian sandarac resin, obtained from *Callitris quadrivalis* and *C. calcarata*. See SANDARAC RESIN.†

β-Pinene. Clear water white terpene liquid. Applications: Chemical intermediate. (Hercules Inc).†

Pine Oil. The name was originally applied to turpentine oils obtained from pine trees. The term is used in America to designate turpentine obtained by distilling pine wood. It is also used for the lighter oils of pine tar, a refined rosin oil, and a by-product in the manufacture of wood pulp by the sulphite process. The proprietary brands are obtained from waste wood and stumps of yellow pine by steam distillation. Used as an alcohol denaturant.†

Pine Resin. See PINE GUM.

Pine Resin, Cypress. See PINE GUM.

Pine Resin, French. See GUM THUS.

Pine Tar. See STOCKHOLM TAR.

Pine-needle Oil, Artificial. Bornyl acetate.†

Piney Tallow. Malabar tallow. an edible fat obtained from the seeds of *Vateria indica*, of East Indies.†

Pinguin. Alantol, $C_{10}H_{16}O$, obtained from the roots of *Inula elecampane.*†

Pink Salt. Ammonium-stannic-chloride $SnCl_4.2NH_4Cl$. Formerly used as a mordant for dyes.†

Pinna Silk. See BYSSUS SILK.

Pinnay Oil. See LAUREL-NUT OIL.

Pinolin. A name for rosin oil (the first distillate from rosin, boiling at from 78-250° C.).†

Pinolith. See LITHOPONE.

Pinwire Brass. An alloy of from 66-73 per cent copper and 27-34 per cent zinc.†

Pioneer. Single stage centrifugal glandless circulator. Applications: Domestic and industrial central heating circulator. (Sihi-Ryaland Pumps Ltd).*

Pioneer Alloy. An alloy of 20 per cent copper, 38 per cent nickel, 4 per cent silicon, 3 per cent molybdenum, 2 per cent tungsten, and the remainder iron.†

Pioury. See INDIAN YELLOW.

Pioxol. A proprietary trade name for Pemoline.†

Pipanol. A proprietary preparation of benzhexol hydrochloride, used in parkinsonism. (Bayer & Co).*

Pipeclay. Aluminium silicate. An abrasive. See CHINA CLAY.†

Piperazenyl. See LYSIDINE.

Piperazidine. Piperazine $C_4H_{10}N_2$.†

Piperazine Calcium Edetate. A chelate produced by reacting ethylene-diamine-NNN'N'-tetra-acetic acid with calcium carbonate and piperazine. Perin.†

Piportil. Pipotiazine Palmitate. Applications: Antipsychotic. (Ives Laboratories Inc).

Piportil Depot. A proprietary preparation of pipothiazine palmitate. Applications Depot tranquillizer. (May & Baker Ltd).*

Pip Pip. An abbreviated name for piperidinium - pentamethylene - dithiocarbamate. An accelerator for rubber vulcanization.†

Pipracil. Piperacillin Sodium. Applications: Antibacterial. (Lederle Laboratories, Div of American Cyanamid Co).

Piprelix. Worm expellent. (Boots Pure Drug Co).*

Pipricide. Veterinary anthelmintic. (The Wellcome Foundation Ltd).

Pipril. Injection containing piperacillin. Applications: Antibiotic. (Lederle Laboratories).*

Piptal. A proprietary preparation containing pipenzolate methobromide. A gastrointestinal sedative. (MCP Pharmaceuticals).†

Piptalin. A proprietary preparation of pipenzolate bromide and simethicone used as a gastro-intestinal sedative. (MCP Pharmaceuticals).†

Piral. Pyrogallol, $C_6H_3(OH)_3$.†

Pirazoline. Phenazone (antipyrin).†

Piria's Naphthionic Acid. α-Naphthylamine-4-sulphonic acid.†

Piriex. A proprietary preparation containing chlorpheniramine maleate, ammonium chloride and sodium citrate. A cough linctus. (Allen & Hanbury).†

Pirimor. Aphicide containing pirimicarb. (ICI PLC).*

Piriton. A proprietary preparation of chlorpheniramine maleate. (Allen & Hanbury).†

Pirsch-Baudoin's Alloy. An alloy of 71 per cent copper, 16.5 per cent nickel, 1.75 per cent cobalt oxide, 2.5 per cent tin, and 7 per cent zinc.†

Pistol. Formulated phenmedipham. (ABM Chemicals Ltd).*

Pitayin. Quinidine, $C_{20}H_{24}N_2O_2$. An alkaloid used in medicine.†

tch Barm. A cement made from casein, water-glass, and caustic lime.†

tch, Candle. See STEARIN PITCH.

tch, Glance. See MANJAK.

tch, Jew's. See BITUMEN.

tch Mineral. See BITUMEN.

tch Oil, Liquid. See OIL OF TAR.

t-ite No. 2. An explosive consisting of 23-25 per cent nitro-glycerin, 28-31 per cent potassium nitrate, 33-36 per cent wood meal, and 7-9 per cent ammonium oxalate.†

tocin. A proprietary preparation of oxytocin used to induce labour. (Parke-Davis).†

tralon. A German antiseptic. It is a wood tar derivative.†

tressin. Vasopressin (injection). Applications: Hormone (antidiuretic). (Parke-Davis, Div of Warner-Lambert Co).

tressin Tannate. A proprietary preparation of vasopressin tannate. Anti-diuretic hormone. (Parke-Davis).†

ttaccal. Eupittonic acid, $C_{25}H_{26}O_9$.†

itteliene. A mixture of coal tar and oil. An insecticide.†

itti. See VENTILAGO MADRASPANTA.

ittinite. A mineral. It is a variety of gummite.†

ittylen. A mixture of pine tar with formaldehyde. Used in skin diseases.†

iuri. See INDIAN YELLOW.

ivatil. Pivampicillin. Applications: A broad spectrum antibiotic ester of ampicillin less effected by the presence of food in the gut. (Merck Sharp & Dohme).*

ivofax. A dried yeast.†

ix. Pitch.†

ix Solubilis. (Soluble Pitch). A soluble modification of the tar obtained by sulphonating the tar obtained from peat.†

ixol. A form of wood tar soluble in water, made from tar and soap. A disinfectant.†

ixtonet. A trade name for a slate substiute. It is used as an electrical insulator.†

Pi-yu. See MOU-IÉON.

PJ1 Chain Lube, Blue Label. Applications: It keeps 'O' rings soft and pliable to maintain your chain's seal against wear causing dirt and moisture. (PJ1 Corporation).*

PJ1 Octane Plus. Applications: Together with racing fuel additives, it boosts the power of lower octane gasolines with complete safety. Gas stabilizers keep the gas in tank from going stale and prevent gum and varnish build-up in the carburettors and fuel system. (PJ1 Corporation).*

PJ1 Super Cleaner. Applications: Cleans and degreases all metal parts, also disperses water and works on all electrical parts that need to be cleaned or dried. (PJ1 Corporation).*

Placadol. A proprietary preparation of papaverine hydrochloride, homatropine methylbromide and codeine phosphate. (Ward, Blenkinsop & Co Ltd).†

Placet Alloy. An alloy of 60 per cent nickel, 20 per cent iron, 15 per cent chromium, and 5 per cent manganese. An electrical resistance.†

Placidyl. Ethchlorvynol. Applications: Sedative. (Abbott Laboratories).

Placodin. Contains Aloxiprin.†

Planavin. A trade mark for a nitrogenous weed-killer. (Shell Chemie GmbH, Germany).†

Planetol. Photographic preparations. (May & Baker Ltd).

Planidets. A proprietary preparation of dibromopropamidine embonate, chlorphenoctium amsonate and butyl aminobenzoate used in the treatment of mouth ulcers. (May & Baker Ltd).*

Planocaine. Procain hydrochloride. (May & Baker Ltd).

Planocaine. A proprietary preparation of ethocaine hydrochloride.†

Planochrome. Mercurochrome. (May & Baker Ltd).*

Planofix. Pre-harvest fruit drop inhibitor. (May & Baker Ltd).

Planotox. Selective weedkiller. (May & Baker Ltd).

Plant Indican. Indican, $C_{26}H_{31}NO_{17}$.†

Plantvax. Systemic fungicide containing oxycarboxin. (ICI PLC).*

Plaquenil. A proprietary preparation of hydroxychloroquin sulphate. An antimalarial. (Bayer & Co).*

Plasblak masterbatches. Granulated concentrates of carbon black in plastics. Applications: Colouring of thermoplastic resins and for imparting resistance to weathering degradation. (Cabot Plastics Ltd).*

Plasbumin. Albumin Human. Applications: Blood volume supporter. (Cutter Laboratories, Miles Laboratories Inc).

Plas-Chek. Applications: For plasticizers for vinyl halide resins, in Int Class 1. (Ferro Corporation).*

Plascoat Plasinter. Polyethylene coating powders. Applications: Coating domestic wirework, display stands, tools, clips, WRC approved, insulation of electrical components. (Plascoat Systems Ltd).*

Plascoat PPA. Polyolefin alloy. Thermoplastic coating powder. Applications: Coating wire dishwasher baskets, process valves, pipework and units, electrical switchgear, seat frames and the lining of fire extinguishers. (Plascoat Systems Ltd).*

Plascoat PPA 31 series. Resistant to aqueous chemicals at temperatures up to 100°C. (Plastic Coatings Ltd).*

Plasdone. Povidone. Applications: Pharmaceutic aid (dispersing and suspending agent). (GAF Corporation).

Plasdone C. A range of pyrogen free polyvinylpyrrolidones. Applications: Solubilizer, stabilizer, protective colloid for veterinary pharmaceuticals which minimizes toxic side effects and reduces irritation at site of infection. (GAF Corporation).*

Plasdone K. A range of polyvinyl-pyrrolidones. Applications: Tablet binder and coating agent, cohesive agent, stabilizer and protective colloid, detoxicant for many poisons and irritants, drug vehicle and retardant, film forming agent in medicinal aerosols. (GAF Corporation).*

Plasgon. A proprietary trade name for plastic gasket and joint cement.†

Plasmin. The proteolytic enzyme derived from the activation of plasminogen.†

Plasminogen. The specific substance derived from plasma which, when activated, has the property of lysing fibrinogen, fibrin and other proteins.†

Plasmosan. A proprietary trade name for Povidone.†

Plastacele. A proprietary trade name for a plasticized cellulose acetate compound.

Plastamid. Polyamides. (Croda Resins Ltd).

Plastammone. An explosive containing ammonium nitrate, glycerin, mononitro toluene, and nitro-semicellulose.†

Plastamol. Plastizers for PVC. (BASF United Kingdom Ltd).*

Plastazote. Crosslinked foamed polyethylene. (BXL Plastics Ltd).

Plaster Cement. Cements made from gypsum. See LE SAGE CEMENT.†

Plaster of Paris. A partially dehydrated gypsum, $2(CaSO_4).H_2O$. It is made from gypsum by heating the latter from 212-400° F., when 3 parts of the water of crystallization is given off.†

Plasteryl. An explosive. It is a mixture of 99.5 per cent trinitrotoluene and 0.5 per cent resin.†

Plast-E-Tint. Pigment dispersions. Applications: Colouration of plastisols and organisols. (Pacific Dispersions Inc).*

Plasthall. Series of monomeric and specialty plasticizers. Applications: Rubber and flexible vinyl. (The C P Hall Company).*

Plastibase. Ointment based containing polyethylene and mineral oil. (E R Squibb & Sons Ltd).

Plastic. Solvent dyestuffs. Applications: Colouration of plastics. (Holliday Dyes & Chemicals Ltd).*

Plastic A. A proprietary trade name for glyceryl tribenzoate.†

Plastic Bronze. An alloy of 64 per cent copper, 30 per cent lead, 5 per cent tin, and 1 per cent nickel.†

Plastic Magnet. Thermosetting material. Applications: Moulded products. (Sumitomo Bakelite, Japan).*

Plastic Metal. An alloy of 80.5 per cent tin, 9.5 per cent copper, 8.6 per cent antimony, and 1.4 per cent iron.†

Plastic Plant Product. See PLASTIC WOOD.

Plastic Steel. Paste containing 80% powdered steel and 20% epoxy resin capable of being hardened. (Devcon Corporation).*

Plastic Sulphur. Prepared by heating sulphur to 225° C. Used to a limited extent as a material for preparing moulds for electrotyping.†

Plastic Wood. (Plastic Plant Product). A material prepared by cooking vegetable matter with neutral salt solution followed by mechanical treatment to break down intercellular binding material. The name is also used to describe wood cellulose in solution with certain additional solvents such as ether or acetone. Used for filling up holes in wood or other materials.

"Plastic wood" is a registered trade mark for a brand of cellulose fibre filler.†

Plastic X. A proprietary plasticizer. It is tricresyl phosphate, and has a specific gravity of from 1.177-1.18.†

Plasticalk. A proprietary trade name for a plastic resin used as an adhesive and filler.†

Plasticede. A proprietary trade name for a plasticizer for clay. It contains tannins and lignin.†

Plasticizer E. A proprietary trade name for a chlorinated paraffin plasticizer.†

Plasticizer 13. The neutral esterification of *p*-oxybenzoic acid and 2-ethyl hexanol. A plasticizer for polyamides. (BASF United Kingdom Ltd).*

Plasticizer 28P. Dioctyl phthalate. A vinyl plasticizer.†

Plasticote. Vinyl/resin lacquer. Applications: Decoration of vinyl articles. (W J Ruscoe Co).*

Plastifix. Polychloroprene. (Bayer & Co).*

Plastikon. A proprietary trade name for a rubber putty for filling and adhesive purposes.†

Plastite. A name applied to a vulcanite.†

Plastitoy. Vinyl/resin lacquer. Applications: Decoration of vinyl toys. (W J Ruscoe Co).*

Plastitube. A proprietary trade name for cellulose acetate tubing.†

Plastogen. A mixture of an oil soluble sulphonic acid of high molecular weight with a paraffin oil. Applications: An effective plasticizer in all elastomers. Used in sponge and low dunometer stocks. (Vanderbilt Chemical Corporation).*

Plastokyd. Alkyd. (Croda Resins Ltd).

Plastokyd SC. Silicone resins. (Croda Resins Ltd).

Plastokyd 310. A proprietary trade name for a short oil linseed alkyd resin for rapid drying trichlorethylene paints. (Croda Resins Ltd).†

Plastolin I. A proprietary solvent. It is reported to be benzyl acetate, $C_6H_5.CH_2.OOC.CH_3$.†

Plastomenite. A German explosive powder made by incorporating 1 part nitro-lignin with 5 parts fused dinitrotoluene, and granulating. It may contain barium nitrate.†

Plastomoll BMB. N-Butylbenzene sulphonamide. A plasticizer for polyamides. (BASF United Kingdom Ltd).*

Plastomoll DMA. Adipic ester of hydrogenated higher cyclic alcohols. A vinyl plasticizer. (BASF United Kingdom Ltd).*

Plastomoll DOA. Di-2-ethylhexyl adipate. A plasticizer for PVC. (BASF United Kingdom Ltd).*

Plastomoll NA. Di-iso-nonyl adipate. A plasticizer for PVC. (BASF United Kingdom Ltd).*

Plastomoll TAH. Thiodibutyric acid ester of a synthetic octyl alcohol. A vinyl plasticizer. (BASF United Kingdom Ltd).*

Plastomoll WH. Ester of an aliphatic dicarboxylic acid. A vinyl plasticizer. (BASF United Kingdom Ltd).*

Plastomoll 34. A vinyl plasticizer. An aliphatic acid ester mixture. (BASF United Kingdom Ltd).*

Plastone. A proprietary rubber vulcanization accelerator. It is methylene diphenyl-diamine mixed with a small amount of stearic acid or with naphthalene or naphthalene oil.†

Plastone A. A proprietary trade name for a phenolic moulding compound containing cotton seed hull.†

Plastone B. A proprietary trade name for an inorganic and phenolic moulding compound.†

Plastopal 11. A modified urea-formaldehyde condensation product in the form of a 65 per cent. butanol-white spirit solution. (BASF United Kingdom Ltd).*

Plastoprene. Isomerized rubber resins. (Croda Resins Ltd).

Plastoprene No2/LV. A proprietary trade name for an isomerized rubber resin used for printing inks and chemical resistant coatings. (Croda Resins Ltd).†

Plastorit. Natural coalescance of quartz, chlorite and mica. Applications: Surface coatings including anti-corrosive paints, textured and powder. (Mercian Minerals & Colours Ltd).*

Plastose. Proprietary synthetic resin moulding powders.†

Plastosol. An antiseptic liquid plaster containing copper guaiacol-sulphonate, and penetrodine dissolved in a volatile organic solvent.†

Plastosperse. A range of pigments dispersed in plasticizers for use in plastics. (Colloids Ltd).*

Plastosperse 40. 40 per cent. carbon black dispersed in a plasticizer. (Colloids Ltd).*

Plastpak. Priming preparations and plasticizing additive. (ICI PLC).*

Plastplate. A proprietary trade name for moulded plastics plated with chromium, copper, gold, or nickel.†

Plastrotyl. An explosive. It is a plastic product prepared from trinitrotoluene, resin, collodion cotton, and crude dinitro-toluene. Sometimes larch turpentine is used.†

Plastules with Folic Acid. A proprietary preparation of ferrous sulphate, folic acid, yeast and liver extract. A haematinic. (Wyeth).†

Plastules with Liver. A proprietary preparation of ferrous sulphate, yeast and liver extract. A haematinic. (Wyeth).†

Plastyrol. Monomer modified alkyds. (Croda Resins Ltd).

Plastyrol E6X. A proprietary trade name for a styrenated epoxide ester resin. A rapid air drying resin. (Croda Resins Ltd).†

Plastyrol S-77X. A proprietary trade name for a styrenated alkyd resin. A very fast drying resin. (Croda Resins Ltd).†

Plastyrol S88X. A proprietary trade name for a styrenated alkyd resin for rapid drying finishes. (Croda Resins Ltd).†

Plasvita. Formaldehyde casein. Applications: Tablet disintegration agent. (Dynamit Nobel Wien GmbH).*

Plaswite Masterbatches. Granulated concentrates of titanium dioxide in plastics. Applications: Colouring of thermoplastic resins. (Cabot Plastics Ltd).*

Platalargan. An alloy of aluminium and silver with some platinum. It is similar to Alargan, except that it contains platinum. Used as a platinum substitute.†

Platamid. Polyamides for extrusion and injection moulding.†

Platamid. Polyamide copolymer, hot melt adhesives. Applications: Textile industry. (Atochem Inc).*

Plataril. Monofilament. Applications: Grass trimmer, fishing line. (Atochem Inc).*

Plate Black. A dyestuff of the same class as Diphenyl fast black.†

Plate Pewter. An alloy of 90 per cent tin, 6 per cent antimony, and 2 per cent each bismuth and copper.†

Plate Powder. Bone ash of which calcium phosphate, $Ca_3(PO_4)_2$, forms 80 per cent. is sold as a non-mercurial plate powder under the name of White rouge.†

Plate Sulphate. The double sulphate, $K_2SO_4.Na_2SO_4$, is called plate sulphate. It crystallizes from hot water, a flash of light accompanying the separation of each crystal.†

latina. An alloy of 53.5 per cent tin, and 46.5 per cent copper.†

latine. A brass containing 43 per cent copper and 57 per cent zinc.†

latine-autitre. A proprietary trade name for a platinum substitute containing 65-83 per cent silver and platinum.†

latinized Asbestos. Loosely fibred asbestos moistened with a concentrated solution of platinum chloride, dried, dipped into ammonium chloride solution, again dried, and brought to a red heat. It usually contains 8-8.5 per cent platinum, and is used in the manufacture of sulphuric anhydride in the contact process for sulphuric acid.†

latinite. A proprietary trade name for a nickel steel containing 46 per cent nickel, and 0.15 per cent carbon. It has a low coefficient of expansion, and can be sealed in glass.†

latino. An alloy of 11 per cent platinum and 80 per cent gold. It is resistant to fused potassium nitrate, and to alkalis.†

latino-aceto-osmic Acid. See HERMANN'S FLUID.

latinoid. An alloy of 60 per cent copper, 24 per cent zinc, 2 per cent tungsten, and 14 per cent nickel. It is used as the material which connects filaments with outside wires of electric lamps. It has the same coefficient of expansion as glass. Also employed as an electrical resistance.†

latinol. Cisplatin. Applications: Antineoplastic. (Bristol Laboratories, Div of Bristol-Myers Co).

latinor. An alloy of 2 parts platinum, 5 parts copper, 1 part silver, and 1 part nickel.†

latinum Black. Finely divided platinum metal.†

latinum Bronze. An alloy consisting of 90 per cent nickel, 9 per cent tin, and 1 per cent platinum.†

Platinum Gold. Variable alloy containing from 12-81 per cent copper, 9-58 per cent platinum, 0-70 per cent gold, 0-37 per cent silver, and 0-4 per cent zinc.†

Platinum Grey. See ZINC GREY.

Platinum Iridium. An alloy usually containing 90 per cent platinum and 10 per cent iridium.†

Platinum Silver. An alloy containing 66.6 per cent silver and 33.3 per cent platinum.†

Platinum Solder. An alloy usually consisting of 73 per cent silver and 27 per cent platinum.†

Platinum Substitute. An alloy of 72 per cent nickel 23.6 per cent aluminium, 3.7 per cent bismuth, and 0.7 per cent gold.†

Platinum Yellow. A barium chloroplatinate or other alkaline chloroplatinate. Used as a coating for fluorescent screens in X-ray work.†

Platinum, Soft. See SOFT PLATINUM.

Platnam. An alloy consisting of 56 per cent nickel, 31 per cent copper, 12 per cent lead. 0.48 per cent iron, and 0.32 per cent aluminium.†

Platnik. An alloy of nickel and platinum. A platinum substitute.†

Platol II. Spray on coating for arc furnace electrodes. (Foseco (FS) Ltd).*

Platone. A peptone powder that is water soluble. Applications: Used in electroplating industry. (American Laboratories Inc).*

Plegine. Phendimetrazine Tartrate. Applications: Appetite suppressant. (Ayerst Laboratories, Div of American Home Products Corp).

Plesmet. A proprietary preparation of ferrous glycine sulphate and folic acid used in the treatment of anaemia of pregnancy. (Napp Laboratories Ltd).*

Plessite. A German explosive containing potassium chlorate and mineral oil. It is also the name for the mineral - Gersdorffite (NiAsS).†

Plessy's Green. Chromic phosphate, $Cr(PO_3)_3$, a pigment.†

Plex-Hormone. A proprietary preparation of methyltestosterone, deoxycortone acetate, ethinyloestradiol and vitamin E used to treat male androgen deficiency. (Consolidated Chemicals).†

Plexar. Adhesive. Applications: For multilayer film and plate between polyolefins and polar plastics (polyamide, polyester) and metals. (DSM NV).*

Plexiglo. A proprietary trade name for a polish and cleaner for transparent plastics.†

Plexigum. Pure acrylic resins. Applications: Surface coatings, inks, firing lacquers, heat seal lacquers, road marking paints, cementitious coatings. (Cornelius Chemical Co Ltd).*

Pleximon. Polymeric/oligomeric methacrylate based resins. Applications: Thickeners, dental etc. (Cornelius Chemical Co Ltd).*

Plexisol. Acrylic resin solution. Applications: Paints, inks etc. (Cornelius Chemical Co Ltd).*

Plexol. Oil additives. (Rohm & Haas (UK) Ltd).

Plextol. Pure acrylic resin emulsions. Applications: Paints, concrete add mixtures, textile finishes, leather dressings, wood coatings etc. (Cornelius Chemical Co Ltd).*

Plialite. A proprietary product stated to be rubber resin.†

Plictran. Acaracide containing cyhexatin (sold in UK on behalf of Dow Chemical Co). (ICI PLC).*

Plictran. A line of miticides based primarily on cyhexatin. (Dow Chemical Co Ltd).*

Plimmer and Paine's Stain. A microscopic stain. It contains 10 grams tannic acid, 18 grams aluminium chloride, 18 grams zinc chloride, 1.5 grams rosaniline hydrochloride, and 40 cc 60 per cent alcohol.†

Plimmer's Salt. Sodium antimony tartrate.†

Plinthite. A red clay from Ireland.†

Pliobond. A broad line of solvent and water borne adhesives based on various polymer systems. Includes contact adhesives, heat reactive systems, pressure sensitives and elastomeric sealants. Applications: Used for bonding forms, metals, plastics, wood products, fibreglass, rubber in various construction, consumer, industrial and automotive applications. (Ashland Chemical Company).*

Pliobond Adhesives. Rubber/solvent base adhesives. Applications: Adhesive for bonding and sealing a variety of substrates (metal, rubber, wood, glass, ceramic, leather, cork, canvas, fibregla and aluminium). (Ruscoe, W R Co).†

Plio-Caulk. Butyl and acrylic caulking compounds. Supplied 1/10 gallon cartridges. Applications: Used to caulk and seal windows, doors, siding, stone and masonry. For interior and exterior use. (Ashland Chemical Company).*

Plioflex. A proprietary trade name for polyvinyl chloride.†

Plioform. A proprietary product. A type of rubber plastic obtained by the action of halogenated acids on rubber.†

Pliolite. A proprietary trade name for modified, isomerized rubber, rubber derivatives and rubber-like resins.†

Plio-Nail. High performance synthetic rubber based adhesives. Certified by PFS Corporation to meet the American Plywood Associations AFG-ol specification. Applications: Joints, sub floors, siding, decorative panels, gypsum board, fixtures and a wide variety of materials. (Ashland Chemical Company).*

Plio-Seam. Elastomeric rubber sealant. Provides a strong and durable seal for aluminium, steel, glass, wood, masonry, ceramics and fibreglass joints. Remains flexible and tough. May be painted. Available in clear, white, aluminium, black and architectural bronze. Applications: Used to install or repair downspouts, rain troughs, roof flashings, shower stalls and metal or wood structures. Weather seals glass in metal or wood storm windows and doors. (Ashland Chemical Company).*

Plio-Tac. Neoprene contact adhesive supplied in aerosol spray cans. Will not cavitate polystyrene foams. Also used to bond wood, carpet, vinyl fabric, metal, rubber and many other materials. Applications: Used for bonding insulation to various surfaces, craft assembly projects, rebonding carpet or vinyl flooring, counter or table top laminates, mounting signs and photos and attaching labels to various surfaces. (Ashland Chemical Company).*

o-Tac 38. Contact cement designed for DIY. Meets CPSA guidelines. Applications: For bonding decorative laminates to wood and metal surfaces. Also bonds leather, fabrics, unglazed ceramics, wall boards, cove base and carpets to themselves and each other. (Ashland Chemical Company).*

itex. A proprietary trade name for a wood and phenolic resin.†

omb. See BLEI.

ombit. A German acid-resisting material made from hard rubber, oleic acid, sulphuric acid, and sulphur.†

ondrel. Fungicide containing ditalimfos for the control of powdery mildew and scab (sold in UK on behalf of Dow Chemical Co). (ICI PLC).*

umbago. See GRAPHITE.

umbago Grease. A mixture of plumbago and tallow. Used for lubricating.†

umber's Solder. Usually a mixture of lead and tin, sometimes with a little antimony. Coarse solder contains 75 per cent lead and 25 per cent tin, and melts at 250° C. Ordinary solder (slicker solder) usually consists of 67 per cent lead and 33 per cent tin, and melts at 227° C. Plumber's fine solder, or soft solder, contains 50 per cent lead and 50 per cent tin, and melts at 188° C.†

umber's White Alloy. An alloy of from 54-58 per cent copper, 25-27 per cent zinc, 13-17 per cent nickel, 1-7 per cent lead, and sometimes 1 per cent tin and 1 per cent iron.†

lumbobinnite. Synonym for Dufrenoysite.†

lumbocalcite. See TARNOWITZITE.

lumboxan. A compound or solid solution of sodium manganate and sodium metaplumbate, of the composition, $Na_2MnO_4.Na_2PbO_3$. It gives up oxygen when treated with steam.†

lumbral. Copper/lead alloys flux. (Foseco (FS) Ltd).*

lumbrex. Flux for lead and alloys. (Foseco (FS) Ltd).*

lumbrit. Cleansing flux for lead alloys. (Foseco (FS) Ltd).*

lumose Mica. A variety of muscovite.†

Plurivite. Vitamin preparations. (The Boots Company PLC).

Pluronic F-68. Poloxamer 188. Waxy, nonionic surfactant of the poly(oxypropylene)poly(oxyethylene) copolymer type. Applications: Laxative. (BASF Wyandotte Corp).

Pluronic F68LF. Poloxamer 188 LF. Prilled solid nonionic surfactant of the poly(oxypropylene)poly(oxyethylene) copolymer type. Applications: Pharmaceutic aid. (BASF Wyandotte Corp).

Pluronic L-101. Poloxamer 331. Liquid nonionic surfactant. Applications: Food additive. (BASF Wyandotte Corp).

Pluronic L-62LF. Poloxamer 182Lf. Liquid nonionic surfactant of the poly(oxypropylene)poly(oxyethylene) copolymer type. Applications: Food additive; pharmaceutic aid. (BASF Wyandotte Corp).

Pluronic L62D. Poloxamer 182D. Liquid nonionic surfactant of the poly(oxypropylene)poly(oxyethylene) copolymer type. Applications: Pharmaceutic aid. (BASF Wyandotte Corp).

Plus. Range of fertilizers for garden use. (ICI PLC).*

Plus-Gas 'C'. Non corrosive cutting fluids. (Foseco (FS) Ltd).*

Plusbrite. A chemical for bright nickel plating. (Albright & Wilson Ltd).*

Pluviusin. A proprietary trade name for a synthetic resin of the urea type.†

Plymul 98-759. A range of polyvinyl acetate thermosetting emulsions used for bonding cellulose to cellulose. (Reichhold Chemicals Inc).†

Plyophen. A proprietary trade name for a phenolic laminating resin and varnish.†

Plyothene. A proprietary phenolic moulding material. (Reichhold Chemicals Inc).†

Plysolene (PIB). Polyisobutylene. Applications: For waterproofing insulation on pipelines, ventilation ducts, reservoirs etc. Also a roofing grade suitable for insulating roofs. (Plysolene Ltd).*

PMA 18. Solubilized form of phenylmercuric acetate. Applications:

Preservative and fungicide for aqueous paints. (Nueodex Inc).*

PMA 60. Powdered form of phenylmercuric acetate. Applications: Preservative and fungicide for aqueous paints. (Nueodex Inc).*

P.M.G. Metal. A proprietary trade name for an alloy of copper with 3-4 per cent silicon, 2 per cent iron, and 2 per cent zinc.†

P.M.T. Alloy. A proprietary alloy made as a substitute for Admiralty gun metal. It contains 88 per cent copper, 2 per cent zinc, and 10 per cent silicon, manganese, and iron.†

Pneulec Core Gum. A proprietary product. It is a linseed oil and wood extract material, and is used as a binder for the sand for cores in metal casting.†

Pneumatogen. A mixture of the peroxides of potassium and sodium.†

Pneumovax. Pneumococcal vaccine. Applications: Effective protection against certain pneumococcal infections. (Merck Sharp & Dohme).*

P.N.P. *p*-Nitro-phenol. Used as fungicide in the rubber industry.†

Pocan. Thermoplastic PBT polyesters. Polybutylene terephthalate grades which offer easy moulding combined with the high mechanical, thermal and electrical properties of polyesters. (Mobay Chemical Corp).*

Pocan. Thermoplastic polyester based on polybutylene terephthalate. Applications: For the manufacture of injection moulding with brilliant surface finish. Uses include household appliances, office machines and electrical components. (Bayer & Co).*

Pod Pepper. Capsicum.†

Podophyllum. The dried rhizome of *Podophyllum peltatum.* Used in medicine.†

Poilite. A trade name for an asbestos cement product used for building work.†

Poirrier's Orange III. See ORANGE III.

Poison Flour. Arsenious oxide, AS_4O_6.†

Poivrette. The ground stones of olive fruit. Used as an adulterant and toning agent in spices.†

Pokalon. Polycarbonate film. (Lonza Limited).*

Polaqua. Water based primer/adhesive with low solids and containing no organic solvents. Applications: Extrusion and lamination of film to paper and foil adhesion promoter to films, paper and foil. (ADM Tronics Unlimited Inc).*

Polar. Acid dyes for wool. (Ciba-Geigy PLC).

Polar Dynobel. (Polar Monobel No. 2, Polar Rex, Polar Saxonite, Polar Stomonal, Polar Super Clifite, Polar Thames Powder, Polar Viking). Proprietary low freezing explosives containing a mixture of nitrated glyc and polyglycerin or glycerin and ethylene glycol, ammonium nitrate, sodium chloride, wood meal, etc.†

Polaramine. Dexchlorpheniramine maleate. Applications: Antihistamin (Schering-Plough Corp).

Polaris. Boiler and cooling water treatm (Laporte Industries Ltd).

Polarwhite. A headless white paint. (JC Bottomley and Emerson).†

Polectron 430. Vinylpyrrolidone/styrene copolymer. Applications: A stable opacifier especially used in high pH systems, e.g. detergents and cold wav lotions. In surface coatings, provides adhesion, film toughness, colour receptivity and non-corrosiveness. (G Corporation).*

Poley Oil. See OIL OF PENNYROYA

Polidene. Emulsion polymers and copolymers based on vinylidene chloride. Applications: Paint, surfac coatings, textiles. (Scott Bader Co Ltd).*

Polidene 528F. A proprietary polyvinlyidene chloride emulsion. (A Stanley Manufacturing Co, Decatur, Ill).†

Poligen MMV. A proprietary aqueous dispersion of styrene/acrylic copolymers. (BASF United Kingdom Ltd).*

Poligen PE. An aqueous solvent-free dispersion of a polyethylene of averag molecular weight. (BASF United Kingdom Ltd).*

oliomyelitis Vaccine. Attenuated live poliomyelitis vaccine. Applications: immunization against poliomyelitis. (Smith Kline and French Laboratories Ltd).*

olisax. Polyethylene sacks. (ICI PLC).*

olish Turpentine. See TURPENTINE.

olishing Oil. A fraction of petroleum oil, having a boiling-point of 130-160° C., and a specific gravity of 0.74-0.77. Used as a turpentine substitute.†

olitarp. Polyethylene sheets. (ICI PLC).*

olitint. Proprietary dyes for plastics. No. 1 for methylmethacrylate. No. 2 for cellulose acetate, cellulose acetate butyrate, and ethyl cellulose.†

'ollack's Cement. A cement glycerin, litharge, and red lead.†

'ollinex. Hay fever pollen vaccine. (Beecham Pharmaceuticals).

'ollopas. Urea resin moulding compounds. Applications: Main fields of application; electrical engineering, sealing caps and sanitary equipments. (Dynamit Nobel Wien GmbH).*

'olmerite. Synonym for Taranakite.†

'olnac. A range of polyester resins. Applications: For G.F.R.P. building coverings, buttons, electrical industry, various applications in G.F.R.P., silos for fodder, varnishes for wood, pieces for industrial coachwork, prefab for the building industry, tanktrucks for raw material stockage, marble agglomerates for floorings and decoration for furniture etc. (SPREA Spa).*

'olocaine. Mepivacaine hydrochloride. Applications: Anaesthetic. (Astra Pharmaceutical Products Inc).

Polomyx. Modified acrylate based coarse particle size suspension. Applications: Architectural coating that is spray applied to gypsum board and other substrates and produces a textured, tone on tone or multi toned, seamless wallcovering. (Polomyx Industries Inc).*

Poloxalkol. A polymer of ethylene oxide, propylene oxide and propylene glycol.†

Poloxyl Lanolin. A polyoxyethylene condensation-product of anhydrous lanolin. Aqualose.†

Poly Check. Boiler water treatment. (Dearborn Chemicals Ltd).

Poly Pale. Polymerized rosin esters. (Hercules Ltd).

Poly-Chek. Applications: For additives for polymers, such as heat and light stabilizers, lubricants and the like, in Int Class 1. (Ferro Corporation).*

Poly-Eth. Plastic resins. (Chevron).*

Poly-Eth Hi-D. Plastic resins. (Chevron).*

Poly-Pale. A partially polymerized rosin. Applications: Hot melt coatings and adhesives. Poly-Pale Ester 10 is a tackifying resin for solvent and emulsion pressure sensitive adhesives and for hot melt packaging adhesives. (Hercules Inc).*

Poly-Pro. Plastic resins. (Chevron).*

Poly-zole AZDN. A trade mark. Azo diisobutyronitrile. A vinyl polymerization catalyst. Gives freedom from side reactions and is not readily poisoned. (National Polychemicals, Wilmington, Mass).†

Polyalk. A proprietary preparation of DIMETHICONE and aluminium hydroxide gel. An antacid. (Galen).†

Polybactrin. Antibacterial. (The Wellcome Foundation Ltd).

Polybead. Monodisperse latex polymer beads. Applications: An identification tag and a size reference for agglutination tests, flow cytometry, instrument calibration, gel filtration, light scattering and phagocytosis. (Polysciences Inc).*

Polyblack. Nitriles. (BP Chemicals Ltd).*

Polyblends. Nitrile/PVC blends. (BP Chemicals Ltd).

Polybor. Prepared product intermediate between borax and boric acid in its chemical properties. It is, however, more soluble in water than either of these and is therefore useful for obtaining sodium borate rich solutions which are almost neutral. (Borax Consolidated Ltd).*

Polybrene. A proprietary trade name for Hexadimethrine Bromide.†

Polycarbafil. A registered trade name for flame retardant polycarbonate materials.†

Polychol. Polyoxyethylated woolwax alcohols. (Croda Chemicals Ltd).

Polychrest Salt. Potassium sulphate, K_2SO_4. The term is also applied to Rochelle salt.†

Polychrome Blue of Unna. A dyestuff prepared by the action of potassium carbonate on Methylene blue. Used in microscopy.†

Polychrome, Methylene Blue. See TERRY'S STAIN.

Polychromine. See PRIMULINE.

Polychromine B. (Fast Cotton Brown R, Direct Brown R). A dyestuff obtained by boiling equal molecules of *p*-nitro-toluenesulphonic acid and *p*-phenylene-diamine with aqueous sodium hydroxide. Dyes cotton orange-brown.†

Polycillin. Ampicillin. Applications: Antibacterial. (Bristol Laboratories, Div of Bristol-Myers Co).

Polycillin-N. Ampicillin sodium. Applications: Antibacterial. (Bristol Laboratories, Div of Bristol-Myers Co).

Polycizer. A proprietary trade name for vinyl plasticizers as follows:
Coded 162-dioctyl phthalate.
Coded 332-dioctyl adipate. (Harwick Standard Chemical Co).†

Polyclar. Polyvinylpyrrolidones. Applications: Used for beverage clarification and stabilization. (GAF Corporation).*

Polyclear. Textile reduction clearing additive. (Diamond Shamrock Process Chemicals Ltd).

Polycon. Photographic developer. (May & Baker Ltd).

POLYCON II. Silafocon A. Applications: Contact lens material. (Syntex Ophthalmics Inc).

Polycroit. The colouring matter of saffron.†

Polycrol. A proprietary preparation containing methylpolysiloxane, aluminium hydroxide gel and magnesium hydroxide. An antacid. (Nicholas).†

Polycron. Disperse dyestuffs. Applications: Colouration of polyester fibres. (Holliday Dyes & Chemicals Ltd).*

Polycryl. Range of acrylic polymer emulsions. Applications: Textile finishing. (Williams Div of Morton Thiokol Ltd). *

Polycycline Hydrochloride. Tetracycline Hydrochloride. Applications: Anti-amebic; antibacterial; antirickettsial. (Bristol-Myers Co).

Polydur. Polyester resin moulding compounds. Applications: Electronic and electrical engineering. (Dynamit Nobel Wien GmbH).*

Polydymite. (Nickel-linnaeite). A miner[al] $(Ni.Co)_4S_5$.†

Polyeite. A proprietary polyester laminating resin. (Reichhold Chemic[als] Inc).†

Polyestradiol Phosphate. An oestrogen currently undergoing clinical trial as "Leo 114" and "Estradurin'. It is a polyester of oestra-1,3,5(10)-triene-3, 17β-diol and phosphoric acid.†

Polyethylene. See POLYTHENE, ALKATHENE.

Polyfax. Ointment for polymyxin B sulphate and bacitracin. (The Wellco[me] Foundation Ltd).

Polyfeed. Water soluble, chlorine free N-P-K fertilizer, with chelated micro-nutrients. Applications: For direct soi[l] application, via irrigation system or foliar spray. (Haifa Chemicals Ltd).*

Polyfilm. Polyethylene film. Applications: Used in packaging and industrial applications. (Dow Chemical Co Ltd)

Polyflex (1). A registered trade mark for flexible polystyrene sheet and fibre.†

Polyflex (2). A proprietary anti-oxidant. I[t] is 6-ethoxy-2,2,4- trimethyl-1,2-dihydroquinoline. (Naugatuck (US Rubber)).†

Polyflon. A proprietary brand of polytetrafluoroethylene (PTFE). (Daikin Kogyo Co, Osaka).†

Polyfon. A range of sodium lignin sulphonates. Applications: Used as concrete retard additives and solids dispersion additives. (Harcros Industri[al] Chemicals).*

Polyfusor Solutions. A range of sterile pyrogen free solutions. (The Boots Company PLC).

Polygalin. See STRUTHIIN.

lygallic Acid. See STRUTHIIN.

lygard. An alkylated aryl phosphite. Used as non-staining and non-discolouring stabilizer for SBR, BR, nitrile and EPDM polymers. Prevents resinification of the polymer during manufacture and storage. Protects the polymer against heat and oxygen degradation during processing. (Uniroyal).*

lygeline. A polymer of urea and polypeptides derived from denatured gelatin. Present in Haemacel.†

lyglactin. A synthetic suture capable of being absorbed by the patient's body. It is a mixture of lactic acid polyester with glycolic acid.†

lyglandin. A solution of the autacoid principles of the thyroid, parathyroid, ovary, testic, and pituitary gland substances.†

lygloss. Flexographic printing inks. Applications: Printing flexible film. (Allied Signal Sinclair and Valentine).*

lyglycolic Acid. A synthetic suture capable of being absorbed by the patient's body. Poly(oxy-carbonylmethylene). DEXON.†

lygon. Sodium tripolyphosphate. (Albright & Wilson Ltd, Phosphates Div).

olygrade. Degradable plastic materials. Applications: Degradable additive concentrates to cause degradation (after useful lifetime). (Ampacet Corporation).*

olyhalite. (Isobelite). A mineral which occurs in the Strassfurt deposits. It is a crystalline mixture of the sulphates of calcium, magnesium, and potassium, and is found with rock salt. It has the formula, $K_2SO_4.MgSO_4.2\text{-}CaSO_4.2H_2O$.†

olyhexanide. An anti-bacterial. Poly-(1 - hexamethylenebiguanide hydro-chloride). Present in TEATCOTE PLUS.†

olyhydrite. A mineral. It is a silicate of iron.†

olyisobutylene. (PIB, Isolene, Oppanol, Vistanex). Polymers of isobutylene. See under OPPANOL B. It is sold under the above trade names.†

Polykol. Poloxamer 188. Waxy, nonionic surfactant of the poly(oxy-propylene)poly(oxyethylene) copolymer type. Applications. Laxative. (The Upjohn Co).

Polylite. Alkylated diphenylamine. Closely related to octamine in chemical structure and may be considered a liquid form of Octamine. It has the same broad protective action with the same minimum of discolouration and staining. For use as an antioxidant and stabilizer in the manufacture of SBR and nitrile polymers and as an antioxidant in latex compounding. (Uniroyal).*

Polylite. (1) A mineral. It is a variety of pyroxene.†

Polylite. (2) A registered trade name for polyester resins for moulding purposes.†

Polylite (3). A proprietary anti-oxidant. Alkylated diphenylamine. (Reichhold Chemicals Inc).†

Polylithionite. A mineral. It is a variety of zinnwaldite containing lithium.†

Polymate. Cooling water treatment. (Dearborn Chemicals Ltd).

Polymeric Sealant Gun. Thermal extruder for rubber-like sealants. (Hardman Inc).*

Polymerized Oils. See BLOWN OILS.

Polymignite. A mineral, $4(Ca.Ce.Fe)O.(Ti.Zr)O_2.CaO.Nb_2O_5$.†

Polymon. Soluble dyes for plastics. (ICI PLC).*

Polymox. Amoxicillin. Applications: Antibacterial. (Bristol Laboratories, Div of Bristol-Myers Co).

Polymul. Range of polyethylene emulsions. Applications: Variety of protective and decorative coatings. (Diamond Shamrock Process Chemicals Ltd).

Polymyxin. Antimicrobial substances produced by Bacillus polymyxa. Specific substances are designated by a terminal letter, eg, Polymyxin B sulphate. AEROSPORIN is Polymyxin B sulphate. Present in DIPLOMYCIN, FRAMYSPRAY, POLYFAX and TRINACTRIC as the sulphate.†

Polynoxylin. Poly[methylenedi-

(hydroxymethyl)urea]. Anaflex Ponoxylan.†

Polyox. A range of water soluble, high molecular weight polymers of ethylene oxide. They are used in adhesives, binders, pharmaceuticals and lubricants.†

Polyoxyl 40 Stearate. Polyoxyethylene 40 stearate. Macrogol stearate 2000. Myrj 52 and 52S.†

Polyoxyl 8 Stearate. Polyoxyethylene 8 stearate. Macrogol stearate 400. Myrj 45.†

Polypel. Fertilizer. (Chevron).*

Polypentek. A proprietary trade name for polypentaerythritol.†

Polyphenyl Black. A polyazo dyestuff of the same class as Chloramine green B. It dyes cotton black.†

Polyphenyl Yellow R. A direct cotton dyestuff.†

Polyplasdone. Crospovidone. Applications: Pharmaceutic aid. (GAF Corporation).

Polyplasdone XL. Polyvinylpoly-pyrrolidones. Applications: Insoluble crosslinked polymer of *N*-vinyl-2-pyrrolidone used as a tablet disintegrant, complexing agent, detoxifier and anti-diarrhoea agent. Adsorbent in thin-layer chromatography. (GAF Corporation).*

Polyprene. Weber's name for rubber.†

Polypress. A proprietary preparation of prazosin hydrochloride and polythiazide. An antihypertensive. (Pfizer International).*

Polyquart H. Polyglycol polyamine condensation resin as a yellow viscous liquid. Applications: Pseudo cationic, used where a cationic surfactant is required, compatible with anionic surfactants. (Henkel Chemicals Ltd).

Polyrad. 5- and 11-mole ethylene oxide adducts of Amine D dehydro-abietylamine. Applications: Polyrad rosin is used as a corrosion inhibitor and detergent in petroleum-processing agents and for inhibiting hydrochloric acid used in industrial and household cleaners. (Hercules Inc).*

Polyram. Amines and derivatives. Primary, secondary, tertiary fatty mono-, di- and polyamines from C8 to C22. Quaternary ammonium salts. Amine oxides. Amine salts. Ethoxylated amines and polyamines. Amino-acids. Special nonionic derivatives. Additive Applications: Surface active agents us as auxiliaries in road, mining, textile and fertilizer industries, in metallurgy, as lubricating agents, additives for fue oils and soil stabilization. (British Cec Co Ltd).*

Polyseal. Polyurethane concrete sealer. Applications: Suitable for use on any concrete floor that is subject to heavy abrasive traffic or is exposed to chemicals or oils. (Secure Inc).*

Polysil. Silicone rubber gums and compounds. (Midland Silicones).†

Polysoft CA. Polyethylene. Applications: Antistatic agent, lubricant and softene compatible with resin finishes. (Scher Chemicals Inc).*

Polysolvan E. A proprietary trade name f a solvent mixture comprising the acetates of propyl, isobutyl and amyl alcohols.†

Polysolvan O. Coalescing solvent for adhesives. (Hoechst UK Ltd).

Polysolvan 0. A proprietary trade name f a solvent composed of the ester of isobutyl alcohol with glycollic and but glycollic acids.†

Polysolvan SHS. A proprietary trade nam for acetic acid esters of alcohols up to C_{11}.†

Polysorbate 20. Polyoxyethylene 20 sorbitan monolaurate. Sorbimacrogol laurate 300. Tween 20.†

Polysorbate 40. Polyoxyethylene 20 sorbitan monopalmitate. Sorbimacrogo palmitate 300. Tween 40.†

Polysorbate 60. Polyoxyethylene 20 sorbitan monostearate. Sorbimacrogol stearate 300. Tween 60.†

Polysorbate 65. Polyoxyethylene 20 sorbitan tristearate. Sorbimacrogol tristearate 300. Tween 65.†

Polysorbate 80. Polyoxyethylene 20 sorbitan mono-oleate. Sorbimacrogol oleate 300. Tween 80.†

Polysorbate 85. Polyoxyethylene 20 sorbitan trioleate. Sorbimacrogol trioleate 300. Tween 85.†

lysphaerite. A mineral, $(Pb.Ca)_3(PO_4)_2(Pb.Ca)_2Cl(PO_4)$.†

lysporin. Ointment and opthalmic ointment. A proprietary formulation of polymyxin B sulphate and bacitracin zinc. Applications: To prevent infection in minor cuts, burns and abrasions, treatment of superficial ocular infections involving the conjunctiva and or cornea due to susceptible organisms. (The Wellcome Foundation Ltd).*

lystab. Polymer additives. (Lankro Chemicals Ltd).

lystat. Anti-static agent. (Crosfield Chemicals).*

lystep. Various surfactants. Applications: Emulsion polymerization. (Stepan Company).*

lystyrene. A synthetic thermoplastic formed by the polymerization of monomeric styrene $C_6H_5CH : CH_2$. It is used in moulding parts for high frequency insulation and in the preparation of lacquers. Its high-frequency electrical properties are among the best.†

lystyrol. Plastic (polystyrene) granules for extrusion and injection moulding. Applications: Packaging - extrusion. Household appliances - injection moulding. (BASF United Kingdom Ltd).*

olystyrol KR 253 and KR 2538. Proprietary styrene-butadiene copolymers offering very high resistance to impact at low temperatures. (BASF United Kingdom Ltd).*

olystyrol KR 2536. A proprietary styrene/butadiene copolymer offering good resistance to impact and to deformation at high temperatures. (BASF United Kingdom Ltd).*

olystyrol 143E. A proprietary polystyrene having easy melt-flow and mechanical properties. (BASF United Kingdom Ltd).*

olystyrol 165H. A proprietary polystyrene similar to POLYSTYROL 143E but possessing greater mechanical strength. (BASF United Kingdom Ltd).*

olystyrol 168N. A proprietary polystyrene stabilized against ultra- violet light. (BASF United Kingdom Ltd).*

Polystyrol 427M. An impact-resistant polystyrene with good resistance to deformation at high temperatures. (BASF United Kingdom Ltd).*

Polystyrol 432F. A proprietary impact resistant styrene-butadiene copolymer. (BASF United Kingdom Ltd).*

Polystyrol 466 I. A proprietary styrene/butadiene copolymer with high impact resistance. (BASF United Kingdom Ltd).*

Polystyrol 472 D. A proprietary styrene/butadiene copolymer offering high impact resistance. (BASF United Kingdom Ltd).*

Polystyrol 473 E. A proprietary styrene-butadiene copolymer offering high impact resistance and easy flow properties. (BASF United Kingdom Ltd).*

Polystyrol 475 K. A proprietary styrene butadiene copolymer offering high resistance to impact. (BASF United Kingdom Ltd).*

Polytar. A proprietary preparation of liquid paraffin, tar cade oil, coal tar and arachid oil extract of coal tar used in the treatment of skin diseases. (Stiefel Laboratories (UK) Ltd).*

Polytelite. A mineral, $(Pb.-Ag_2)_4Sb_2S_7$, with $(Zn.Fe)_4Sb_2S_7$.†

Polyterpene. See LIQUID RESINS.

Polythene. The general term for a range of solid polymers of ethylene. See ALATHON, ALKATHENE, BAKELITE POLYETHYLENE, MONSANTO, PETROTHENE, TENITE POLYETHYLENE, HI-FAX, HOSTALEN, ROTENE, VESTOLEN, FORTIFLEX, FORTILENE, MARLEX, RIGIDEX and CARLONA.†

Polythiazide. 6-Chloro-3,4-dihydro-2-methyl-3-(2,2,2-trifluoroethylthio-methyl)benzo-1,2,4-thiadiazine-7-sulphonamide 1,1-dioxide. Nephril. Renese.†

Polytrend. Colourant dispersions, polyester extender and pigment vehicle. Applications: Colouring plastic compositions, extender for unsaturated polyester compositions. (Nueodex Inc).*

Polytrope. Rheological additives. Applications: Unsaturated polyester resins. (NL Industries Inc).*

Polyurax. Urethane intermediate polyols. (BP Chemicals Ltd).*

Polyvest C70. A polymeric chalk filler activator. Applications: Used for activating carbonate fillers in EPDM compounds. (Huls AG).†

Polyvest 25. A polymeric filler activator. Applications: Used for activating light-coloured silicate fillers in EPDM compounds. (Huls AG).†

Polyvidone. See POVIDONE. Polyvinyl-pyrrolidone.

Polyvinox,. Poly(butyl vinyl ether). Shostakovsky Balsam.†

Polyvinyl Chloride. PVC., Breon, Carina, Chlorovene, Corvic, Flamenol, Geon 100, Koron, Koroseal, Mipolam, Plioflex, Vinylite Q, and Welvic are trade names for polyvinyl chloride which may or may not be plasticized.†

Polyviol. Polyvinyl alcohol in a range of viscosities and degrees of saponification. Applications: Protective colloid for emulsions, dispersions and suspensions, textile auxiliaries (finishes, impregnating agents, sizes); thickening agent for glues and adhesives; release agent in the processing of polyester resins. (Wacker Chemie GmbH).*

Polyzote. A proprietary trade name for nitrogen-expanded synthetic resin plastics.†

Poly/Bed 812. Replacement for Shell's discontinued Epon 812. Applications: Embedding kit for light microscopy. (Polysciences Inc).*

Poly/Sep 47. Mixture of 47 buffers. Applications: Electrofocusing buffer for electrophoresis. Generates a stable linear pH in polyacrylamide gels in the pH range of 2.5-10. Avoids inconsistencies of conventional SCAM's. (Polysciences Inc).*

Pomace. The residue from the extraction of apple juice in cider manufacture. A cattle food.†

Pomade en Crème. Cold cream.†

Pombe. A beer made from Sorghum millet.†

Pomoloy. A proprietary trade name for a cast iron made by a special process.†

Pomona Green. See IODINE GREEN.

Pompeian Red. See INDIAN RED.

Pompey Red. Ferric oxide, Fe_2O_3.†

Pompholix. Zinc oxide, ZnO.†

Pompilion Oil. See OIL OF POMPILIO

Ponceau. Synonym for scarlet. See MAGENTA.†

Ponceau B. See BIEBRICH SCARLET.

Ponceau B Extra. See BIEBRICH SCARLET.

Ponceau BO Extra. See BRILLIANT CROCEINE M.

Ponceau G. See PONCEAU R.

Ponceau GL. A dyestuff. It is a British equivalent of Ponceau 2G.†

Ponceau GR. See PONCEAU R.

Ponceau GT. A dyestuff. It is toluene-azo-β-naphthol-disulphonic acid.†

Ponceau J. See PONCEAU R.

Ponceau JJ. See PONCEAU 2G.

Ponceau RT. A dyestuff. It is the sodium salt or toluene-azo-β-naphthol-disulphonic acid, $C_{17}H_{12}N_2O_7SNa$. Dyes wool orange-red from an acid ba It is isomeric with Ponceau GT from R salt.†

Ponceau R, 2R, G, GR. (Ponceau J, Xylidine Red, Xylidine Scarlet, Scarle G, Brilliant Ponceau G, Ponceau 2RX, Naphtharene Scarlet 2R, Rainbow Scarlet G, 2RS, Scarlet 2R, 2RL, R, 3R). A dyestuff. It is the sodium salt o xylene-azo-β-naphthol-disulphonic acid, $C_8H_9.N_2.C_{10}H_4$-$(HSO_3)_2ONa$.Dyes wool scarlet from a acid bath.†

Ponceau S Extra. See FAST PONCEAU 2B.

Ponceau 10RB. An acid dyestuff. It dyes wool or silk bluish-crimson from an aci bath.†

Ponceau 2G. (Orange R, Brilliant Poncea GG, Ponceau JJ, Acid Scarlet G, Ponceau GL). A dyestuff. It is the sodium salt of benzene-azo-β-naphthol-disulphonic acid R, $C_{16}H_{10}N_2O_7S_2Na_2$. Dyes wool reddish-orange from an acid bath.†

Ponceau 2R. See PONCEAU R.

Ponceau 2RX. A British brand of Ponceau.†

Ponceau 3B. See OIL SCARLET.

Ponceau 3G. (Scarlet 3G). A dyestuff. It is the sodium salt of sulpho-anisoil-azo-β-naphthol, $HSO_3.OCH_3.C_6H_3N$: $N.C_{10}H_6ONa$.†

Ponceau 3R. (Cumidine red). A dyestuff. It is the sodium salt of ethyl-dimethyl-benzene-azo-β-naphthol-disulphonic acid, $C_2H_5.(CH_3)_2 . C_6H_2 . N_2 . C_{10}H_4(HSO_3)_2ONa$. Dyes wool bluish-scarlet from an acid bath. See BIEBRICH SCARLET.†

Ponceau 3R. (Ponceau 4R, Cumidine Red, Cumidine Ponceau). A dyestuff. It is the sodium salt of ψ-cumene-azo-β-naphthol-disulphonic acid, $C_{19}H_{16}N_2O_7S_2Na_2$. Dyes wool bluish-scarlet from an acid bath.†

Ponceau 3RB. See BIEBRICH SCARLET.

Ponceau 4GB. (Croceine Orange, Orange ENL, Brilliant Orange, Brilliant Orange G, Orange GRX, Brilliant Orange GL, Rainbow Orange). A dyestuff. It is the sodium salt of benzene-azo-β-naphthol-β-sulphonic acid, $C_{16}H_{11}N_2O_4SNa$. Dyes wool orange-yellow from an acid bath.†

Ponceau 4R. See PONCEAU 3R.

Ponceau 4RB. See CROCEINE SCARLET 3B.

Ponceau 5R. (Erythrin X, Scarlet 5R). A dyestuff. It is the sodium salt of benzene -azo-benzene-azo-β-naphthol-trisulphonic acid, $C_{22}H_{13}N_4O_{10}S_3Na_3$. Dyes wool bluish-red from an acid bath.†

Ponceau 6R. See SCARLET 6R.

Ponceau 6RB. See CROCEINE SCARLET 7B.

Ponderax PA. A proprietary preparation of FENFLURAMINE hydrochloride in a sustained-release capsule. An appetite suppressant. (Servier).†

Pondermite. A mineral, $Ca_2B_6O_{11}. 4H_2O$. A source of boric acid.†

Ponder's Stain. A microscopic stain. It consists of 0.02 gram toluidine blue, 1 cc glacial acetic acid, and 2 cc absolute alcohol in 100 cc distilled water.†

Pondicherry Oil. (Nut Oil). Arachis oil.†

Pondimin. Fenfluramine hydrochloride. Applications: Anorexic. (A H Robins Co Inc).

Pondinil Roche. Proprietary preparation of mefenorex. Anorexic. (Roche Products Ltd).*

Ponite. A mineral. It is a variety of rhodocrosite containing iron.†

Ponolith. (Sunolith, Superlith). Lithopone pigments.†

Ponoxylan. Polynoxylin. Applications: Infective skin conditions including furuncolosis and pustular acne, burns. (Berk Pharmaceuticals Ltd).*

Ponsital. The registered trade name for a neuroleptic agent. It is 3-chloro-10 [LCB]γ - [N' - β' - (1'- methyl - 2" - oxo - imidazolidyl - 3') - ethyl - N - piperazinyl] -propyl[RCB]-phenothiazine dihydrochloride (AG Chemische Fabrik).†

Ponstan. A proprietary preparation of mefenamic acid. An analgesic. (Parke-Davis).†

Ponstel. Mefenamic acid. Applications: Anti-inflammatory; analgesic. (Parke-Davis, Div of Warner-Lambert Co).

Pontallor. A range of precious metal alloys. Applications: For dentistry and dental engineering. (Degussa).*

Pontianac. See JELUTONG.

Ponticin. See RHAPONTICUM.

Pontocaine. Tetracaine. Applications: Anaesthetic. (Sterling Drug Inc).

Pontocaine Hydrochloride. . Tetracaine Hydrochloride. Applications: Anaesthetic. (Sterling Drug Inc).

Pool-Chem (Chemical line for swimming pools). Varies with each product. Applications: Consumer product line for control of water chemistry in swimming pools. (Puma Chemical Co Inc).*

Poonac. Coconut cake, a cattle food. The term is also used for the residue from castor oil seeds after cold and hot pressing and solvent extraction. Used for caulking timber.†

Poonahlite. A mineral. It is a hydrated aluminium-calcium silicate.†

Poonseed Oil. See LAUREL-NUT OIL.

Pope's Solution. A solution of 1 part in 10,000 of a mixture of 10 parts 2 : 7-dimethyl-3 : 6-diamino-acridinium-methylo-chloride hydrochloride and 1 part crystal violet. An antiseptic for wounds.†

Poppy Capsules or Heads. The dried, immature fruit of *Papaver somniferum.*†

Populin. Benzoyl-salicin, $C_{20}H_{22}O_8$.†

Porcelain. A mixture of clay, quartz, and felspar. A normal mix consists of 50 per cent clay, 25 per cent quartz, and 25 per cent felspar.†

Porcelain Clay. Synonym for Kaolinite. See CHINA CLAY.†

Porcelain Earth. See CHINA CLAY.

Porcelain White. See LITHOPONE.

Porcelanite. A fused clay and shale found in burned coal seams.†

Porcelave. A proprietary trade name for a ceramic material.†

Porfiromycin. An antibiotic. 6-Amino-8 - carbamoyloxymethyl - 1, 1a, 2, 8, 8a, 8b-hexahydro-8a-methoxy-1,5-dimethylazirono [2', 3' : 3, 4] pyrrolo[1, 2-*a*]indole-4,7-dione.†

Porocel. A proprietary trade name for a carefully prepared and screened bauxite.†

Porofor. A range of chemical blowing agents for the production of cellular plastics primarily based on PVC, polyethylene, ABS and polystyrene. (Bayer & Co).*

Porosil-Clarcel. Diatomite fillers. Applications: Painting, fertilizers coating and defluoration. (British Ceca Co Ltd).*

Porous Alum. Sodium aluminium sulphate (soda alum), $Al_2(SO_4)_3.Na_2SO_4.24H_2O$.†

Porpezite. A native alloy of gold and palladium.†

Porphyry. A building stone having the same composition as felspar.†

Porporino. An alloy of mercury, tin, and sulphur. Used for decorating purposes.†

Porrier's Orange III. See ORANGE III.

Portagen. A proprietary artificial infant food containing medium chain triglycerides and non-lactose carbohydrates, for use in cases of intolerance of fat and lactose. (Bristol Myers Co Ltd).†

Portalac. Lactulose. Applications: Treatment of obstipation and (pre-) hepatic coma. (Duphar BV).*

Porter. A beverage. London porter conta[ins] 5.4 per cent alcohol.†

Portland Arrowroot. The starch from *Ar[um] maculatum.*†

Portland Cement. Made by heating an intimate mixture of argillaceous and calcareous substances, such as lime an[d] clay, and pounding the product. The material does not slake with water, an[d] has energetic hydraulic properties.†

Portland Stone. A limestone.†

Portsmouth Accelerator No. 3. A proprietary rubber vulcanization accelerator. It is phenyl-*o*-tolyl-guanidine.†

Portugal Oil. See OIL OF PORTUGAL.

Portugallo Oil. Essential oil of orange pe[el].

Portuguese Turpentine. See TURPENTINE.

Portyn. Benzilonium bromide.†

Portyn Kapseals. A proprietary preparati[on] containing benzinolinium bromide. An antispasmodic. (Parke-Davis).†

Porzite. Synonym for Mullite.†

Pos O Print. Ammonium hydroxide - 26·baume-29.4% concentrate, 24·baume-25.5% concentrate, 23·baume-23.5% concentrate, 20·baume-17.7% concentrate. Applications: Developing solution for blue prints in the engineering/drafting industries. Developing solution for microfilm in the micrographic industry (W D Service Company Inc).*

Posicor. Quazinone. Applications: Cardiotonic. (Hoffmann-LaRoche Inc).

Posistac. Salinomycin - an ionophorous antibiotic used as a growth stimulating nutritional aid in cattle and swine. (Pfizer International).*

Poskine. A depressant of the central nervous system. *0*-Propionylhyoscine. PROSCOPINE is the hydrobromide.†

skydal. Unsaturated polyester resins. Applications: Used for the formulation of furniture finishes with and without paraffin wax as well as for use in fillers. (Bayer & Co).*

st-Kite. Selective herbicide. (FBC Ltd).

st-4. Anti-settling additives. Applications: Paint. (NL Industries Inc).*

taba. Potassium para aminobenzoate in tablet, capsule and powder form. Applications: Approved in the management of Peyronie's Disease and Scleroderma. (Glenwood Laboratories Ltd).*

tarite. A mineral (Pd.Hg).†

tash. Potassium hydroxide, KOH. Potassium carbonate is also called potash. Another name for the carbonate is Salts of tartar.†

otash Alum. A double sulphate of potassium and aluminium, $Al_2(SO_4)_3.K_2SO_4.24H_2O$.†

otash Bordeaux Mixture. Contains 6 lb copper sulphate, 2 lb potassium hydroxide, and 50 gallons water.†

ot-ashes. Impure potassium carbonate.†

otash Felspar. See FELSPAR.

otash Glass. A glass containing silicate of potassium.†

otash-lead Glass. A glass usually containing from 40-50 per cent. SiO_2, 28-53 per cent. PbO, 8-11 per cent. K_2O, and 1 per cent. Al_2O_3 and Fe_2O_3.†

otash Lozenges. Potassium chlorate lozenges.†

otash Pellets. Compressed tablets of potassium chlorate. See POTASH LOZENGES.†

otash Salts. See ABRAUM SALTS.

otash Water-glass. A mixture of potassium silicates.†

otassalumite. A mineral. It is a potash alum.†

otassic Superphosphate. A manure made by combining calcium superphosphate with potash salts.†

otassium Amalgam. An alloy of potassium and mercury, formed by the combination of the elements.†

Potassium Cadmium Iodide. See MARME'S REAGENT.

Potassium Felspar. See FELSPAR.

Potassium Menaphthosulphate. Dipotassium 2-methyl-1,4-disulphatonaphthalene dihydrate. Vikastab.†

Potassium Muriate. See MURIATE OF POTASH.

Potassium Prussiate, Red. See RED PRUSSIATE OF POTASH.

Potassium Prussiate, Yellow. See YELLOW PRUSSIATE OF POTASH.

Potato Flour. Potato starch.†

Potato Gum. (Almadina, Euphorbia Gum). Almeidina gum, stated to be derived from *Euphorbia rhipsaloides*, of West Africa. The latex contains about 10 per cent rubber, 32 per cent water, 51 per cent resin, 1 per cent protein, 6 per cent insoluble matter, and gives an ash of 2.5 per cent. The dry material contains 14.3 per cent rubber and 75.8 per cent resin.†

Potato Oil or Spirit. The alcohol obtained from potato starch. Also see FUSEL OIL.†

Potato Rubber. See POTATO GUM.

Potazote. A French fertilizer containing 14 per cent nitrogen, as ammonium chloride, and 20 per cent potassium oxide as potassium chloride.†

Potentite. See TONITE.

Potenzol V. Alkyl aryl polyether alcohol as an emulsifiable compound. Applications: Wetting agent for herbicides (Invequimica & CIA SCA).*

P.O.T.G. Phenyl-*o*-tolyl-guanidine, a rubber vulcanization accelerator.†

Potin. An alloy of 72 per cent copper, 25 per cent zinc, 2 per cent lead, and 1 per cent tin.†

Potinjaune. See POTIN.

Pot Metal. An alloy of lead and copper.†

Potosi Silver. See NICKEL SILVERS.

Potstone. An impure steatite (*qv*).†

Potter's Clay. See PIPECLAY.

Potter's Ore. See ALQUIFON.

Pouckpong Gum. (Touchpong Gum). A rubber gum of British Guiana.†

Poudre B. (Vieille Powder). A French explosive. It is a smokeless powder made from a mixture of soluble and insoluble nitro-cellulose, thoroughly gelatinized with a mixture of ether and alcohol, rolled into sheets, and cut into strips.†

Poudre EF. A French explosive made from nitro-cellulose and binding material.†

Poudre J. A French explosive containing 83 per cent guncotton and 17 per cent potassium bichromate.†

Poudre Pyroxulée. A French sporting powder. It consists of insoluble nitro cellulose, with 35 per cent barium and potassium nitrates.†

Poudre Savory. Seidlitz powder.†

Poulenc 309. A proprietary preparation. It is sym-di-sodium-*m*- amino-benzoyl-*m*-amino-*p*-methyl-benzoyl-1-naphthyl-amino-4 : 6 : 8-trisulphonate-urea.†

Pounce. Powdered sandarac.†

Poutet's Reagent. Consists of 1 cc mercury dissolved in 12 cc nitric acid, specific gravity 1.42. Used for testing oils.†

P.O.V. Purified oil of vitriol (sulphuric acid containing 93-96 per cent. H_2SO_4).†

Povan. Pyrvinium Pamoate. Applications: Anthelmintic. (Parke-Davis, Div of Warner-Lambert Co).

Povidone. Poly(vinylpyrrolidone) . Plasmosan.†

Povidone-Iodine. A complex produced by reacting iodine with poly(vinyl-pyrrolidone). Betadine.†

Powax. Additive to hot rinse tanks used after acid pickling of steel sheet and rod. (Foseco (FS) Ltd).*

Powder of Algaroth. (Basic Chloride, English Powder, Powder of Algarotti). A mixture of antimony oxychloride, SbOC1, and antimony oxide, Sb_2O_3. Used in the preparation of tartar emetic.†

Powder of Algarotti. See POWDER OF ALGAROTH.

Powder 19/04/15H Black 904. A proprietary polyethylene used in rotational moulding and carpet-backing applications. It can be used in contact with foodstuffs.†

Powder 215 Natural. A proprietary 400-micron powder used in flame- retarda[nt] rotational mouldings.†

Powder 22/04/00A 400. A proprietary polyethylene powder of micron size having a low melting point. It is used for making interliners for fabrics and [a] carpet backing.†

Powder 26/04/00. A proprietary polyethylene powder possessing good rigidity, used in rotational moulding.†

Powdered Hydrocyanic Acid. The name applied to a calcium cyanide prepared from calcium carbide and hydrocyanic acid. It evolves hydrocyanic acid with moisture, hence the name. A fumigator.†

Powellite. A mineral. It is calcium molybdate, $CaMoO_4$.†

Power Gard. Proprietary blend of fuel additives. Applications: Improved fuel consumption, reduced carbon deposits, cleaner carburettor, reduced valve deposits. (Gard Corporation).*

Powers Terebine. Drier solution. Applications: For addition to certain paints and varnishes to speed drying. (Llewellyn Ryland Ltd).*

Powmet. Metal powders and premixes. (McKechnie Chemicals Ltd).

P.P.D. Piperidine-pentamethylene-dithio-carbamate. A rubber vulcanization accelerator.†

PPO. Polyphenylene oxide.†

P.P.S. A proprietary polyphenylene sulphide. A cross-linkable aromatic thermoplastic with a high modulus use[d] as a coating material capable of withstanding temperatures in the range 200°-260° C. (Liquid Nitrogen Processing Corp, Melven, pa).†

PP-Vac. Pigeon pox. Applications: immunization of poultry. (Intervet America Inc).*

Practolol. A beta adrenergic receptor blocking agent. It is 4-(2-hydroxy-3-isopropylaminopropoxy)acetanilide.†

Pradone Plus. Herbicide. (May & Baker Ltd).

Praenitrona. A proprietary trade name for Trolnitrate Phosphate.†

Prage Alizarin Yellow G. A dyestuff. It is *m*-nitro-benzene-azo-resorcylic acid, $C_{13}H_9N_3O_6$. Dyes chrome mordanted cotton pure yellow, and chromed wool brownish-yellow.†

Prage Alizarin Yellow R. A dyestuff. It is *p*-nitro-benzene-azo-resorcylic acid, $C_{13}H_9N_3O_6$. Dyes chromed wool and cotton orange-yellow.†

Pragmatar. A proprietary preparation of cetyl alcohol and coal tar distillate with sulphur and salicylic acid used in the treatment of dandruff. (Smith Kline and French Laboratories Ltd).*

Pragmatar Ointment. Combination of cetyl alcohol-coal tar distillate, precipitated sulphur and salicylic acid. Applications: Treatment of dandruff, seborrhoeic conditions and common scaly skin disorders where skin is unbroken. (Smith Kline and French Laboratories Ltd).*

Pragmoline. A proprietary preparation of acetyl-choline bromide.†

Prague Red. See INDIAN RED.

Praims. Cough drops. (Richardson-Vicks Inc).*

Prajmalium Bitartrate. A preparation used in treatment of arrhythmia of the heart. *N*-Propylajmalinium hydrogen tartrate.†

Pralidoxime. N-Methylpicolinaldoxime. Protopam is the iodide.†

Pramidex. A proprietary preparation of tolbutamide used in the treatment of diabetes. (Berk Pharmaceuticals Ltd).*

Pramiverine. 4, 4-Diphenyl-*N*-isopropylcyclohexylamine.†

Pramoxine. 4-(3-(4-Butoxyphenoxy) - propyl] morpholine.†

Prampine. O-Propionylatropine. PAMN is the methonitrate.†

Prantal. Diphemanil methylsulphate. Applications: Anticholinergic. (Schering-Plough Corp).

Prapagen WK, WKL and WKT. Cationic surfactants of the quaternary ammonium chloride type in liquid or paste form. Applications: Antistatic agents, fabric conditioner and softener, fibre finishers, water-repellant agents and dewatering agents, wetting agents for oils, dispersing agents for pigments,flushing agents, foaming and wetting agents, spinning bath and viscous additives, flotation chemicals and anti-caking agents for rendering salts free-flowing, corrosion inhibitors, anchoring and wetting agents for tars and bitumen, surface coatings, lacquers, adhesives and dispersions, disinfectants, hair cosmetics and auxiliaries for leather, textiles, rubber and metal industries. (Hoechst UK Ltd).

Prase. A mineral, SiO_2.†

Praseolite. Similar to Cardierite.†

Praxilene. Nafronyl oxalate. Applications: Vasodilator. (Lipha SA).

Prazepam. A muscle relaxant. 7-Chloro - 1 - (cyclopropylmethyl) - 1, 3 -dihydro - 5 - phenyl - 2*H* - 1, 4 - benzo - diazepin-2-one.†

Praziquantel. An anthelmintic currently undergoing clinical trial as "EMBAY 8440'. It is 2-cyclohexylcarbonyl-1, 3, 4, 6, 7, llb - hexahydro - 2*H* - pyra - zino(2, 1 - *a*]isoquinolin-4-one.†

Prazitone. An anti-depressant. 5-Phenyl - 5 - (2-piperidyl)methylbar-bituric acid.†

Prazosin. An anti-hypertensive. 1-(4-Amino -6, 7-dimethoxyquinazolin- 2-yl) -4-(2-furoyl)piperazine.†

PRC. Motor fuel additive. (Chevron).*

Prebane. Triazine herbicide. (Ciba-Geigy PLC).

Precef. Ceforanide. Applications: Antibacterial. (Bristol Laboratories, Div of Bristol-Myers Co).

Precipitated Calomel. See CALOMEL.

Precipitated Phosphate. Insoluble calcium phosphate.†

Precipitate, Red. See RED OXIDE OF MERCURY.

Precipitate, White Fusible. See MERCURAMMONIUM CHLORIDE.

Preconativ. Factor IX concentrate for the treatment of haemophilia B. (KabiVitrum AB).*

Predalone TBA. Prednisolone Tebutate. Applications: Glucocorticoid. (O'Neal, Jones & Feldman Pharmaceuticals).

Predazzite. A mineral. It is a mixture of calcite and brucite.†

Predef. Isoflupredone acetate. Applications: Anti-inflammatory. (The Upjohn Co).

Predef 2X. A proprietary preparation containing ISOFLUPREDONE 21-acetate. A veterinary steroid.†

Pred Forte. Prednisolone Acetate. Applications: Glucocorticoid. (Allergan Pharmaceuticals Inc).

Pred Mild. Prednisolone Acetate. Applications: Glucocorticoid. (Allergan Pharmaceuticals Inc).

Predne-Dome. Prednisolone. Applications: Glucocorticoid. (Miles Pharmaceuticals, Div of Miles Laboratories Inc).

Prednelan. A proprietary preparation of prednisolone. (Glaxo Pharmaceuticals Ltd).†

Prednesol. A proprietary preparation of PREDNISOLONE disodium phosphate. (Glaxo Pharmaceuticals Ltd).†

Prednis. Prednisolone. Applications: Glucocorticoid. (USV Pharmaceutical Corp).

Prednisolamate. Prenisolone 21-diethylaminoacetate.†

Prednisolone. 11β, 17α-21-Trihydroxy-pregna-1,4-diene-3,20- dione. 1,2-Dehydrocortisone. Metacortandralone. Codelcortone. Delta-Cortef. Delta-cortril. Delta-Stab. Di-Adreson-F. Precortisyl. Ultracorten-H. Prednelan is the acetate. Predsol is the disodium phosphate.†

Prednisone. 17α-21-Dihydroxypregna-1,4-diene - 3,11,20-trione-1,2-Dehydrocortisone. Metacortandracin. De cortisyl. Deltacortone. Di-Adreson. Ultracorten. Delta-Cortelan is the acetate.†

Prednylidene. 11β, 17α,21-Trihydroxy-16 - methylenepregna - 1,4 -diene - 3,20-dione. Dacortilene. Decortilen.†

Predsol. A proprietary preparation of prednisolone disodium phosphate. (Glaxo Pharmaceuticals Ltd).†

Predsol-N. A proprietary preparation of prednisolone sodium phosphate and neomycin sulphate. (Glaxo Pharmaceuticals Ltd).†

Pre-Empt. Selective herbicide. (FBC Ltd)

Prefin Liquifilm. Phenylephrine Hydrochloride. Applications: Adrenergic. (Allergan Pharmaceuticals Inc).

Prefix D. A granular formulation containing 6.75% dichlobenil. Applications: It provides season long weed control of both annual and perennial grasses and broad-leaved weeds. (Burts & Harvey).*

Pregaday. A proprietary preparation of ferrous fumarate and folic acid. Haematinic for use in pregnancy. (Glaxo Pharmaceuticals Ltd).†

Pregamal. A proprietary preparation of folic acid and ferrous fumarate. A haematinic. (Glaxo Pharmaceuticals Ltd).†

Pregfol. A proprietary preparation of folic acid and ferrous sulphate. A haematinic (Wyeth).†

Pregl's Solution. A solution of potassium iodide and sodium iodate with a little sodium chloride and bicarbonate.†

Pregnenolone. 3β-Hydroxypregn-5-en-20-one.†

Pregnyl. Gonadotropin, chorionic. Applications: Gonad-stimulating principle. (Organon Inc).

Pregolan. A chlorine compound giving 65-72 per cent available chlorine.†

Pregrattite. A mineral. It is a variety of paragonite.†

Pregwood. A proprietary synthetic resin impregnated wood, made by impregnating and then subjecting the wood to heat and pressure.†

Prehnite. (Jacksonite). A mineral. It is a silicate of aluminium and calcium, $2CaO.Al_2O_3.3SiO_2$.†

Pre-Kite. Selective herbicide. (FBC Ltd).

Preludin. Phenmetrazine Hydrochloride. Applications: Anorexic. (Boehringer Ingelheim Pharmaceuticals Inc).

Premaline. Herbicide. (May & Baker Ltd)

Premalox. Herbicide. (May & Baker Ltd).

Premarin Vaginal Cream. Conjugated Oestrogen (natural). Applications:

Atrophic vaginitis and postmenopausal atrophic urethritis. (Ayerst Laboratories).†

Premerge. 3 Herbicide. Applications: Used for the control of broadleaf weeds in peas, soybeans, potatoes and orchards. The active ingredient is dinoseb. (Dow Chemical Co Ltd).*

Premier Alloy. A heat-resisting alloy containing 61 per cent nickel, 11 per cent chromium, 25 per cent iron, and 3 per cent manganese.†

Premier Jus. See OLEO.

Premix. Masterbatches in thermoplastics. Applications: Pigmentation of thermoplastics. (Cornelius Chemical Co Ltd).*

Prempak. Natural conjugated oestrogen plus norgestrel. Applications: Menopausal and postmenopausal oestrogen replacement therapy and allied disorders such as postmenopausal osteoporosis, atrophic vaginitis, postmenopausal atrophic urethritis. (Ayerst Laboratories).†

Prempak-C. Natural conjugated oestrogen plus norgestrel. Applications: Menopausal and postmenopausal oestrogen replacement therapy for vasomotor symptoms (sweating, flushes), allied disorders such as postmenopausal osteoporosis, atrophic vaginitis, Kraurosis vulvae and atrophic urethritis. (Ayerst Laboratories).†

Prenite. A proprietary trade name for an asbestos sheet bonded with neoprene. A packing material.†

Prenomiser. A proprietary preparation of isoprenaline sulphate in aerosol form. A bronchial antispasmodic. (Fisons PLC).*

Prent. Pharmaceutical preparation. Applications: Cardioselective beta blocker. (Bayer & Co).*

Prenylamine. N - (3,3-Diphenylpropyl)-α-methyl phenethylamine. Segontin and Synadrin are the lactate.†

Prepagen. Textile softening agents. (Hoechst UK Ltd).

Prepagen WK. A proprietary trade name for 75 per cent. distearyl dimethyl ammonium chloride in isopropanol. Used as a softener in the laundry trade. (Hoechst Chemicals (UK) Ltd).†

Pre-Par. Ritodrine. Applications: Uterospasmolytic, prevents premature birth. (Duphar BV).*

Preparation 102. See LUARGOL.

Prepared Bark. See QUERCITRON.

Prepared Calamine. Obtained by calcining and powdering negative zinc carbonate or calamine, and freeing the product from gritty particles. It consists of zinc carbonate with some oxide of iron.†

Prepared Chalk. (*Creta proeparata B.P.*) Washed chalk or whiting.†

Prepared Cobalt Oxide. Cobalt oxide, CoO, obtained by heating the black oxide, Co_2O_3. Used in the ceramic industry.†

Preparing Salt. Sodium stannate, $Na_2SnO_3.3H_2O$. Used as a mordant in dyeing and calico printing.†

Prepcort. Hydrocortisone. Applications: Glucocorticoid. (Whitehall Laboratories, Div of American Home Products Corp).

Prepon. Casting and modelling wax, blue, green and ivory. Applications: Dental preparation. (Bayer & Co).*

Presamine. Imipramine hydrochloride. Applications: Antidepressant. (USV Pharmaceutical Corp).

Preservaline. (Freezine, Iceline). Names for formaldehyde used as a preservative for milk.†

Preservals. Parabens. Applications: Preservatives for cosmetics and pharmaceuticals. (Laserson & Sabetay).*

Preservative. Mixture of anti-microbials; preservative. Applications: Cosmetic preparations. (REWO Chemicals Ltd).

Preservol. Creosote, partially emulsified with pyroligneous acid. Used for preserving wood.†

Presidal. A proprietary trade name for Pentacynium Methylsulphate.†

Presinol. Pharmaceutical preparation. Applications: Antihypertensive. (Bayer & Co).*

Presinol Mite. Pharmaceutical preparation. Applications: Antihypertensive. (Bayer & Co).*

Presol. Dampproofer and wood preserver solvents. (Carless Solvents Ltd).

Presol W. Mercury fungicide solution. (Great Lakes Chemical (Europe) Ltd).†

Presomet. Black bitumous paint. (Thomas Ness Ltd).

Press N Seal. Caulking tape in roll form. Applications: Used to caulk windows, doors and other openings in construction buildings and homes. (Chemseco).*

Press-cake. The mill-cake formed by mixing the ingredients of gunpowder in the incorporating mill, is subjected to a high pressure to make press-cake.†

Pressed Amber. See AMBROID.

Pressimmune. A proprietary preparation of anti-human lymphocyte globulin used in tissue transplants to produce immuno-suppression. (Hoechst Chemicals (UK) Ltd).†

Pressolith. (Sillimanith). Earthenware porcelain products.†

Pressonex Bitartrate. Metaraminol bitartrate. Applications: Adrenergic. (Sterling Drug Inc).

Pressphan. A German name for press-boards made from wood pulp. Used as insulating materials.†

Presszell. A German synthetic resin varnish-paper product used as an electrical insulator.†

Prest-o-lite. A proprietary brand of acetylene gas compressed in cylinders.†

Presto Steel. A proprietary trade name for a steel containing 1.4 per cent chromium.†

Prestogen. Stabilizers for textile bleaching systems. (BASF United Kingdom Ltd).

Preston Salts. (Smelling Salts). Consist of acid ammonium carbonate, NH_4HCO_3.†

Prestone. The trade mark of National Carbon Company Inc. applied to ethylene glycol anti-freeze.†

Presuren. A proprietary trade name for Hydroxydione sodium succinate.†

Pretamazium Iodide. A preparation used in the treatment of enterobiasis. It is 4-(biphenyl-4-yl) - 3-ethyl - 2-[4-(pyrrolidin-1-yl) styryl] thiazolium iodide.†

Pretolone. Dyeing and printing assistant. (Ciba-Geigy PLC).

Preventol. A proprietary preparation. It is stated to be a chlorinated phenol dissolved in organic bases. It is used as a preservative for adhesives, etc, liable to attack by moulds or bacteria.†

Preventol A2. An organic inhibitor. Applications: For use in the formulation of fungicidal interior paints (emulsion and solvent based paints) excluding air-drying paints. (Bayer & Co).*

Preventol CL 5. An organic inhibitor. Applications: For use in cleansing acids and in the surface treatment of metals. (Bayer & Co).*

Prevex. PPE co-polymer. Applications: Injection moulding, sheet extrusions. (Borg Warner Chemicals).*

Priadel. A proprietary preparation of lithium carbonate used in the treatment of bipolar affective disorder. (Delandale Laboratories).*

Priamid. Lather booster and stabilizer, emuslifier; antistatic agents. Applications: Powdered detergents; toilet and shaving soap; shampoos; bubble baths; thermoplastics and synthetic fibres; liquid paste cleaners. (The Sales Organisation of Unichema International).

Pribramite. A mineral. It is a variety of sphalerite.†

Pricite. (Bechilite). A mineral. It is a calcium borate $3CaO.4BO_3.6H_2O$.†

Priderite. A mineral. It is $(K,Ba)_{1.3}(Ti,Fe^{\cdots})_8O_{16}$.†

Prilocaine. N-(α-Propylaminopropionyl)-*o*-toluidine. Citanest.†

Primacor. A family of adhesive polymers used for extrusion coating and layers in flexible packaging. (Dow Chemical Co Ltd).*

Primal. Acrylic emulsions. Applications: Decorative and industrial coatings, binders for textile and non-woven applications, floor polishes, leather, adhesives, cement modifiers. (Rohm and Haas Company).*

Primal. A solution of toluylene-diamine with neutral sulphite. Recommended as a hair dye. Also a proprietary trade name for a synthetic resin of the acrylate type used for leather finishes.†

rimalan. A proprietary preparation of mequitazine. Applications: Antihistamine. (May & Baker Ltd).*

rimapel. Soil retardent and leather chemicals. (Rohm & Haas (UK) Ltd).

rimaquine. 8-(4-Amino-1-methyl-butylamino)-6-methoxyquinoline.†

rimaquine Bayer. Pharmaceutical preparation. Applications: Antimalarial. (Bayer & Co).*

rimasol. Textile padding auxiliaries. (BASF United Kingdom Ltd).

rimatene Mist. Epinephrine. Applications: Adrenergic. (Whitehall Laboratories, Div of American Home Products Corp).

rimatol AA. Triazine amino triazole. (Ciba-Geigy PLC).

rimatol AD. Triazine amino triazole 2,4-D total herbicide. (Ciba-Geigy PLC).

rimatol AP. Triazine, pictoram total herbicide. (Ciba-Geigy PLC).

rimatol SE. Triazine amino triazole total herbicide. (Ciba-Geigy PLC).

rimax. Textile and leather chemicals. (Rohm & Haas (UK) Ltd).

rimazin. Dyestuffs for cellulosic fibres. (BASF United Kingdom Ltd).

rime Flavours. Meat and savoury flavours. Applications: Food products (soups, sauces etc) and processed meats. (Fries & Fries of Mallinckrodt Inc).†

rimenes. Amine stabilizers. (Rohm & Haas (UK) Ltd).

rimer Gilding Brass. See BRASS.

rimex. See CRISCO.

rimicid. Soil applied insecticide containing pirimiphos-ethyl. (ICI PLC).*

Primidone. 5 - Ethylhexahydro - 5 - phenylpyrimidine - 4,6 - dione. Mysoline.†

Primobolan. A proprietary preparation of methenolone acetate. An anabolic agent. (Schering Chemicals Ltd).†

Primobolan Depot. A proprietary preparation of methenolone enanthate. An anabolic agent. (Schering Chemicals Ltd).†

Primodos. A proprietary preparation of ethinyloestradiol and NORETHISTERONE acetate used in the treatment of secondary amenorrhoea. (Schering Chemicals Ltd).†

Primofax. Nosiheptide. Applications: Growth stimulant. (Rhone-Poulenc Industries).

Primogyn Depot. A proprietary preparation of oestradiol valerate used in the treatment of amenorrhoea and prostatic carcinoma. (Schering Chemicals Ltd).†

Primolut Depot. A proprietary trade name for the caproate of Hydroprogesterone. (Schering Chemicals Ltd).†

Primolut N. A proprietary trade name for Norethisterone. (Schering Chemicals Ltd).†

Primor. Lubricating and industrial oils. (Burmah-Castrol Ltd).*

Primotec. Seed dressing containing pirimiphos-ethyl. (ICI PLC).*

Primoteston Depot. A proprietary preparation of testosterone used in the treatment of male osteoporosis or sterility. (Schering Chemicals Ltd).†

Primperan. A proprietary preparation of metoclopramide hydrochloride. An anti-emetic. (Berk Pharmaceuticals Ltd).*

Primrose. See MARTIUS YELLOW and SPIRIT EOSIN.

Primrose Smokeless. A smokeless 42 grain powder.†

Primrose Soluble. See ERYTHROSIN.

Primrose Soluble in Alcohol. See ERYTHRIN.

Primula. See HOFMANN'S VIOLET. Also a mixture of Methyl violet and Fuchsine.

Primuline. (Polychromine, Carnotine, Thiochromogen, Aureoline, Sulphine, Primuline Extra). A dyestuff. It is the sodium salt of the mono-sulphonic acids of the dehydro-thionated condensation products of dehydro-thio-toluidine (mixed with some sodium-dehydro-thio-toluidine-sulphonate). It is chiefly $C_{28}H_{17}N_4O_3S_4Na$. Dyes cotton primrose yellow from an alkaline or neutral bath, and is employed for the production of ingrain colours. It also has application in photography.†

Primuline Base. *p*-Toluidine heated with sulphur.†

Primuline Bordeaux. Fabric dyed with primuline (*qv*), then passed through a solution of ethyl-β-naphthylamine.†

Primuline Brown. Fabric dyed with primuline (*qv*), then passed through a solution of *m*-phenylene-diamine.†

Primuline Extra. A British equivalent of Primuline.†

Primuline Orange. Fabric dyed with primuline (*qv*), then passed through a solution of resorcinol.†

Primuline Red. See INGRAIN.

Primus. Fibre-rich bar to help 'regularity'. (Richardson-Vicks Inc).*

Prince Rupert's Metal. See PRINCE'S METAL.

Prince's Blue. A mineral. It is a blue variety of Sodalite (*qv*). The slabs are polished for ornamental purposes.†

Prince's Metal. (Prince Rupert's Metal). An alloy. It is a variety of brass containing from 61-83 per cent copper and 17-39 per cent zinc. Another alloy, also called Prince's metal, consists of 84.75 per cent tin and 15.25 per cent antimony.†

Prince's Metallic. See PRINCE'S MINERAL.

Prince's Mineral. (Prince's Metallic). A clay containing about 40 per cent oxides of iron.†

Princillin. Ampicillin sodium. Applications: Antibacterial. (E R Squibb & Sons Inc).

Principen. Ampicillin. Applications: Antibacterial. (E R Squibb & Sons Inc).

Principen/N. Ampicillin sodium. Applications: Antibacterial. (E R Squibb & Sons Inc).

Printer's Acetate. Aluminium acetate, $Al(C_2H_3O_2)_3$.†

Printer's Iron Liquor. A deep black solution of ferrous acetate, containing some ferric acetate. It contains about 10 per cent iron.†

Printing Black for Wool. A dyestuff produced by the reduction of a mixture of 1 : 5- and 1 : 8-dinitro-naphthalene by means of glucose in alkaline solutio in the presence of sodium sulphite. Dy wool violet black from an acid bath. Employed in printing.†

Printing Blue. See INDULINE, SPIRIT SOLUBLE, and METHYLENE BLUE B and BG.

Printing Blue for Wool. A dyestuff obtain by the reduction of 1 : 8-dinitro-naphthalene with sodium sulphide in th presence of sodium sulphite and sodiun hydroxide. Dyes cotton violet-blue. Use in wool printing.†

Printing Green. See CHROME GREENS.

Printing Inks. Inks consisting of pigments incorporated with varnish made by heating linseed oil.†

Printogen. Printing oils and fixation accelerators used for textile printing. (Degussa).*

Printwash. Mixture of esters, alcohols, toluene. Applications: Cleaning of prin rollers. (Solrec Ltd).*

Prioderm. A proprietary preparation of malathion used in the treatment of infestation by lice. (Napp Laboratories Ltd).*

Priormatt. Synthetic emulsion paint. Applications: For decorative purposes on walls. For coating the interior of GRP boats. (Llewellyn Ryland Ltd).*

Priscol. A proprietary preparation of tolazoline hydrochloride. A vasodilator. (Ciba Geigy PLC).†

Priscoline Hydrochloride. Tolazoline Hydrochloride. Applications: Vasodilator (peripheral). (Ciba-Geigy Corp).

Prism. Gas separators. Applications: Separates industrial gases through use of gas permeable hollow fibre membranes. (Monsanto Co).*

Prismatic Emerald. See EUCLASE.

Prismatic Nitre. Potassium nitrate, KNO_3, so-called from the form of the crystals.†

Pristane. Iso-octadecane, $C_{18}H_{38}$.†

Pristinamycin. An antibiotic produced by Streptomyces pristina spiralis.†

Privine Hydrochloride. Naphazoline hydrochloride. Applications: Adrenergic. (Ciba-Geigy Corp).

Pro Seal. Silicone sealant, gasketing compounds. Applications: For making formed in place gaskets and for use within cut gaskets. (Novest Inc).*

Pro Weld. Two part cold welding compound. Applications: Bonding aluminium, brass and steel castings. (Novest Inc).*

Pro-Actidil. Prolonged action triprolidine hydrochloride tablets. (The Wellcome Foundation Ltd).

Proban. Textile flame retardant. (Albright & Wilson Ltd, Phosphates Div).

Pro-Banthine. Propantheline Bromide. Applications: Anticholinergic. (G D Searle & Co).

Probe. Wettable powder containing 75% methazole for pest emergence weed control. (ICI PLC).*

Probenecid. 4 - (Di-*n*-propylsulphamoyl)benzoic acid. Benemid.†

Proberite. A mineral. It is 2[Na,Ca, $B_8O_9.5H_2O$].†

Probilin. A preparation of phenol phthalein.†

Probimer. Photo-crosslinkable synthetic resin. (Ciba-Geigy PLC).

Probnal. Propyl-butyl-acet-urethane.†

Probucol. An agent used to control the increase of cholesterol in the blood beyond normal limits. It is 4,4'-(iso-propylidenedithio) bis - (2, 6- di-t-butylphenol.†

Procainamide. N-(4-Aminobenzoyl)-2-diethylaminoethamine. 4-Amylino-N-(2- diethylaminoethyl) benzamide. Pronestyl.†

Procal. Vacuum formed ceramic fibre insulation. (Foseco (FS) Ltd).

Procan. Procainamide Hydrochloride. Applications: Cardiac depressant. (Parke-Davis, Div of Warner-Lambert Co).

Procarbazine. N-4-Isopropyl-carbamoylbenzyl -N'-methylhydrazine. Natulan is the hydrochloride.†

Procardia. A proprietary preparation of nifedipine for the treatment of angina and hypertension. (Pfizer International).*

Procetyl AWS. Polypropoxylated cetyl alcohol. (Croda Chemicals Ltd).

Prochinor. Amines and derivatives. Primary, secondary, tertiary fatty mono-, di- and polyamines from C8 to C22. Quaternary ammonium salts. Amine oxides. Amine salts. Ethoxylated amines and polyamines. Amino-acids. Special nonionic derivatives. Additives. Applications: Surface active agents used as auxiliaries in road, mining, textile and fertilizer industries, in metallurgy, as lubricating agents, additives for fuel oils and soil stabilization. (British Ceca Co Ltd).*

Prochlorite. A mineral similar to chlorite.†

Prochlorperazine. 2-Chloro-10-[3-(4-methylpiperazin-1-yl)-propyl]pheno-thiazine. Compazine is the dimaleate or the edisylate.†

Procilene. Polyester/cotton dyes. (ICI PLC).*

Procinyl. Reactive disperse dyes. (ICI PLC).*

Procion. Reactive dyestuffs. (ICI PLC).*

Pro-cort. Hydrocortisone. Applications: Glucocorticoid. (Barnes-Hind Inc).

Proclonol. $\alpha\alpha$-Di-(4-chlorophenyl) - cyclopropylmethanol.†

Proctocort. Hydrocortisone. Applications: Glucocorticoid. (Rowell Laboratories Inc).

Proctofibe. Proprietary product of grain fibre, citrus fibre. Applications: Diverticular disease, irritable colon, constipation. (Roussel Laboratories Ltd).*

Proctofoam HC. Muco-adherent, white, odourless aerosol foam containing hydrocortisone acetate and pramoxine hydrochlorite. Applications: Topical anti-haemorrhoiddal. (Stafford-Miller).*

Procyclidine. 1-Cyclohexyl-1-phenyl - 3-pyrrolidinopropan-1-ol. Kemadrin is the hydrochloride.†

Procythol. A liver extract.†

Prodag. A trade mark for a colloidal form of graphite in water.†

Pro-Diaban. An oral anti-diabetic preparation. (Bayer & Co).*

Prodoraqua. Liquid waterproofer and hardener. (Prodorite).*

Prodorbond. A range of polymer modified heavy duty industrial floor finishes. (Prodorite).*

Prodorcrete GT. A range of urethane resin based floor toppings for acid, alkali and chemical resistance. (Prodorite).*

Prodorfilm. Series of special light stable wall coatings and potable linings for vessels and tanks. (Prodorite).*

Prodorflor. Acid and chemical resisting epoxy resin based floor finish. (Prodorite).*

Prodorglas. A range of stoved coatings for vessels and tanks to resist chemical corrosion. (Prodorite).*

Prodorglaze. A range of textured decorative multi-coloured wall coatings. (Prodorite).*

Prodorguard. Epoxy resin based floor coatings for application to concrete floors and walls. (Prodorite).*

Prodorite. An acid-resisting material. It is a concrete with a hardened pitch binder. The mineral part is carefully graded and mixed with the pitch. It is stated to be suitable for plants containing corrosive gases.†

Prodorlac. Heavy bituminous paint. (Prodorite).*

Prodorshield. Self-levelling epoxy resin based floor topping. (Prodorite).*

Prodox. Alkyl phenols. Applications: Intermediate chemicals for antidegradents, agricultural chemicals and resins. (PMC Specialties Group Inc).*

Prodoxol. A proprietary preparation of OXOLINIC ACID. A urinary antiseptic. (Warner).†

Product AAS 90. Anionic surfactant in the form of a free flowing powder. Applications: Powder type detergents and cleaning formulations where free flowing product is required. (Henkel Chemicals Ltd).

Product MB320. Cyclohexylamine lauryl sulphate in the form of a solid block which is water and oil soluble. Applications: Emulsifier for insecticides; printing ink manufature. (Ronsheim & Moore Ltd).

Pro-Etch. Patented glass etching system. Applications: Etching vehicle registration number on to glass car windows. (Hermetite Products Ltd).*

Profadol. 1-Methyl-3-propyl-3-(3-hydroxyphenyl)pyrrolidine.†

Profax. Insulating refractory sideliner tiles for killed steel ingots. (Foseco (FS) Ltd).*

Pro-Fax. Polypropylene, available as natural or coloured moulding pellets, coloured concentrates, stabilizer powder blend or base polymer. Applications: Suitable for a wide range of applications as can be extruded, blow-moulded, thermoformed, injection moulded etc. (Hercules Inc).*

Profenamine. Ethopropazine.†

Proferdex. Iron dextran. Applications: Haematinic. (Fisons Corp).

Proflavine. 3 : 6-Diamino-acridine-sulphate, $C_{13}H_{11}N_3.H_2SO_4.H_2O$. An antiseptic.†

Proflex. Analgesic and anti-inflammatory.(Ciba-Geigy PLC).

Progallin. Esters of gallic acid. (Nipa Laboratories Ltd).*

Proganol. See PROTARGENTUM.

Pro-Gas (Gas Disclaimed). Propanol. (Chevron).*

Progene. Liquid detergents. (Unilever).†

Progilite. A proprietary trade name for phenol-formaldehyde.†

Proglumide. An anti-gastrinic. 4-Benzamido *NN*- dipropylglutaramic acid. MILID.†

Proglycem. Diazoxide. Applications: Antihypertensive. (Schering-Plough Corp).

Progressite. An explosive containing 89 per cent ammonium nitrate, 4.7 per cent aniline hydrochloride, 6 per cent ammonium sulphate, and 0.2 per cent colouring matter.†

Progynova. A proprietary preparation of oeatradiol valerate used in the treatment of menopausal symptoms. (Schering Chemicals Ltd).†

Proheptazine. Hexahydro-1,3-dimethyl-4-phenyl-4-propionyloxy-azepine.†

Proidonite. A mineral, SiF_4.†

Proil. Rust preventatives. (Croda Chemicals Ltd).*

Proiodin. A combination of iodine with protein, containing 4.4 per cent iodine.†

Prokayvit Oral. Tablets of acetomenaphthone (vitamin K). (British Drug Houses).†

Proketazine. The dimaleate of Carphenazine.†

Proklar. Sulphamethizole. Applications: Antibacterial. (O'Neal, Jones & Feldman Pharmaceuticals).

Prokliman Ciba. A German product. It consists of tablets containing ovarial hormone, peristaltin, nitro-glycerin, pyramidone, and a sodium salt of caffeine.†

Proladone. A proprietary preparation of oxycodone pectinate. An analgesic. (Crookes Laboratories).†

Prolan. Preparations of gonadotropins. Applications: For the treatment of reproduction disorders. (Bayer & Co).*

Prolaurin. A proprietary trade name for propylene glycol mono-laurate.†

Prolein. A proprietary trade name for propylene glycol mono-oleate.†

Prolene. Suture, nonabsorbable surgical. Applications: Surgical aid. (Ethicon Inc).

Prolex. Active ingredient: propachlor; 2-chloro-N-isopropyl-acetanilide. Applications: Pre-emergence weed control of annual weeds. (Agan Chemical Manufacturers Ltd).†

Proline. α-Pyrolidine-carboxylic acid.†

Prolintane. 1-(α-Propylphenethyl)-pyrrolidine. 1-Phenyl-2-pyrrolidinopentane. Villescon.†

Prolit. Bleaching Earths (activated clays). Applications: For refining vegetable and mineral oils, vegetable and animal fats and in purifying solvents. As colour developing agent for carbonless copying papers. (Caffaro Industrial).*

Prolit. Activated bleaching earths. Applications: For the decolourizing of vegetable and mineral oils, animal and vegetable fats and solvents. (SNIA (UK) Ltd).†

Prolith. Photographic developer. (May & Baker Ltd).

Prolixin. Fluphenazine hydrochloride. Applications: Antipsycotic. (E R Squibb & Sons Inc).

Prolixin Enanthate. Fluphenazine enanthate. Applications: Antipsycotic. (E R Squibb & Sons Inc).

Proloid. Thyroglobulin. Applications: Thyroid hormone. (Parke-Davis, Div of Warner-Lambert Co).

Prolongal. Iron preparation. Applications: For the treatment of piglet anaemia. (Bayer & Co).*

Proloprim. Trimethoprim. Applications: Antibacterial. (Burroughs Wellcome Co).

Proloprim Tablets. A proprietary formulation of trimethorprim. Applications: Treatment of initial episodes of uncomplicated urinary tract infections due to susceptible organisms. (The Wellcome Foundation Ltd).*

Prolugen. Propylene glycol. (Ward, Blenkinsop & Co Ltd).†

Promacetin. Acetosulphone sodium. Applications: Antibacterial. (Parke-Davis, Div of Warner-Lambert Co).

Promapar. Chlorpromazine hydrochloride. Applications: Anti- emetic; antipsycotic. (Parke-Davis, Div of Warner-Lambert Co).

Promazine. 10-(3-Dimethylamino-propyl)phenothiazine. SPARINE is the embonate or the hydrochloride.†

Prometal. A variety of cast iron, used in the construction of furnace parts.†

Promethazine. 10-(2-Dimethylamino-*n*-propyl)phenothiazine. Phenergen is the hydrochloride.†

Promethazine Theoclate. Promethazine salt of 8-chlorotheophylline. Promethazine chlorotheophyllinate.†

Prométhée. See EXPLOSIF O3.

Promethoestrol. 3,4-Di-(4-hydroxy-3-methylphenyl) hexane. Methoestrol.†

Promethus. A blasting powder. It contains potassium chlorate, manganese dioxide, iron oxide, mono-nitro-benzene, turpentine oil, and naphtha.†

Prometrex. Active ingredient: prometryne; 2,4-bis-(isopropylamino)-6-methylthio-1,3,5-triazine. Applications: Selective pre- and post-emergence herbicide for the control of broadleaf and grass weeds in a variety of crops. (Agan Chemical Manufacturers Ltd),†

Promicrol. Ultra fine grain photographic developer. (May & Baker Ltd).

Prominal. A proprietary preparation of N-methyl-ethyl-phenyl-malonyl- urea. (methylphenobarbitone). A hypnotic. (Bayer & Co).*

Promintic. A proprietary preparation of METHYRIDINE. A veterinary anthelmintic.†

Promoloid. A fertilizer containing colloidal magnesium silicate. It is a Japanese product.†

Promoxolan. 4-Hydroxymethyl-2,2-di-isopropyl-1,3-dioxolan.†

Promozyme. A heat stable debranching enzyme obtained from a novel species of bacillus by submerged fermentation. It belongs to the group of debranching enzymes known as pullulanases. Applications: Used in the production of dextrose and maltose. (Novo Industri A/S).*

Pronalys. Analytical grade reagents. (May & Baker Ltd).*

Prondol. A proprietary preparation of iprindole. An antidepressant. (Wyeth).†

Pronel Capsules. A proprietary prparation of gelatin. Applications: Used in the treatment of flaking fingernails. (Bioglan Laboratories).*

Pronestyl. Procainamide Hydrochloride. Applications: Cardiac depressant. (E R Squibb & Sons Inc).

Pronestyl Solution. Procainamide hydrochloride in aqueous solution for injection. (E R Squibb & Sons Ltd).

Pronestyl Tablets. Procainamide hydrochloride in tablet form. (E R Squibb & Sons Ltd).

Pronethalol. 2-Isopropylamino-1-(2-naphthyl)ethanol. Alderlin is the hydrochloride.†

Proof Spirit. A term originally intended tc denote alcohol, that was just strong enough to ignite gunpowder, when bur upon it. It is alcohol containing 49.24 parts of alcohol to 50.76 parts of water by weight, or 100 volumes of alcohol tc 81.82 volumes of water. It has a specifi gravity of 0.920 at 15° C.†

Proof Vinegar. A vinegar containing 66 pe cent acetic acid.†

Proofite. A proprietary trade name for a product similar to aquatec.†

Propaderm. A proprietary preparation of beclomethasone dipropionate for dermatological use. (Allen & Hanbury).†

Propaderm A. A proprietary preparation o beclomethasone dipropionate and chlortetracycline hydrochloride for dermatological use. (Allen & Hanbury).†

Propaderm C. A proprietary preparation o beclomethasone dipropionate and clioquinol for dermatological use. (Alle & Hanbury).†

Propaderm N. A proprietary preparation o beclomethasone dipropionate and neomycin sulphate for dermatological use. (Allen & Hanbury).†

Propadrine. Phenylpropanolamine Hydrochloride. Applications: Adrenergic. (Merck Sharp & Dohme, Div of Merck & Co Inc).

Propaesin. (Propocaine). The propyl ester of *p*-amino-benzoic acid, $NH_2.C_6H_4.COO.C_3H_7$. A local anaesthetic.†

Propafilm. Balanced biaxially oriented polypropylene film. (ICI PLC, Petrochemicals & Plastics Div).

Propafoil. Metallised oriented polypropylene film. (ICI PLC).*

Propain. Paracetamol, codeine phosphate, diphenhydramine, caffeine in yellow scored tablets. Applications: Treatmen of headache, migraine, muscular pain, period pain and toothache. Also for the relief of symptons of influenza and feverish colds. (Luitpold-Werk).*

Propaklone. Industrial solvent. (ICI PLC).*

Propal. See PROPONAL.

Propalanin. Amino-butylic acid, $C_4H_9O_2N$.†

Propamidine. 1,3-Di(4-amidinophen-oxy)-propane.†

Propamine. Catalysts for polyurethane foams. (Lankro Chemicals Ltd).†

Propamine D. A liquid aliphatic tertiary amine catalyst miscible with water and organic liquids. It is tetramethyl ethylene diamine. (Lankro Chemicals Ltd).†

Propanal. Propionic aldehyde, CH_3CH_2CHO.†

Propanidid. Propyl 4-diethylcarbamoylmethoxy-3-methoxyphenyl-acetate. Epontol. FBA 1420.†

Propanol. Normal propyl alcohol, $CH_3.CH_2.CH_2OH$.†

Propanthelline Bromide. 2-Di-isopropylaminoethyl xanthen-9-carboxylate methobromide. Pro-Banthene Bromide.†

Proparacaine. A proprietary trade name for Proxymetacaine.†

Propathene. A trade mark for polypropylene. The lightest of the thermoplastics. It has good rigidity and tensile strength which are retained at elevated temperatures. It has excellent resistance to chemicals and has no tendency to environmental stress cracking. Used in plastics manufacture and as moulding and extrusion compounds. See also NOBLEN. (ICI PLC).*

Propatylnitrate. 1,1,1-Trisnitrato-methylpropane Etrynit. Gina.†

Propcorn. Chemical products for the treatment of corn. (BP Chemicals Ltd).*

Propenol. Allyl alcohol, CH_2 : $CH.CH_2OH$.†

Proper-Myl. A proprietary preparation of lyophilized yeasts, for injection. (Consolidated Chemicals).†

Properidine. A narcotic analgesic. Isopropyl 1-methyl-4-phenylpiperidine-4-carboxylate.†

Propetamphos. A pesticide. Z-0-2-iso-propoxycarbonyl-1-methylvinyl o-methyl ethyl phosphoramido = thioate.†

Propezite. A natural alloy of palladium and gold, containing 7 per cent gold.†

Propicillin. 6 -(α-Phenoxybutyramido)-penicillanic acid. (1-Phenoxypropyl)-penicillin. Brocillin and Ultrapen are the Potassium salt.†

Propine. Dipivefrin hydrochloride. Applications: Antiglaucoma agent. (Allergan Pharmaceuticals Inc).

Propinol. Propargyl alcohol, CH : $C.CH_2OH$.†

Propiofan. Polyvinyl propionate. (BASF United Kingdom Ltd).*

Propiofan D. Polyyinyl propionate dispersion. (BASF United Kingdom Ltd).*

Propiolactone. β-Propiolactone. Betaprone.†

Propiolic Acid. *o*-Nitro-phenyl-propiolic acid, $C_6H_4(NO_2)C : C.CO_2H$., is known commercially by this name. It is in the form of a thin paste.†

Propiomazine. 10-(2-Dimethylamino-propyl)-2-propionylphenothiazine. Dorevane Indorm Largon.†

Propione. Diethyl-ketone $(C_2H_5)_2.CO$. A hypnotic and anaesthetic.†

Propionylhyoscine. Poskine.†

Propiram Fumarate. N-(1-Methyl-2-piperidinoethyl)-N-(2-pyridyl)-propionamide fumarate. FBA 4503.†

Proplatinum. An alloy of 72 per cent nickel, 23.6 per cent silver, 3.7 per cent bismuth, and 0.7 per cent gold.†

Proplex. Factor IX Complex. Applications: Haemostatic. (Hyland Therapeutics, Div of Travenol Laboratories Inc).

Propocon. Formulated polyurethane systems for rigid, semi-rigid , microcellular foams. Applications: Used in insulation, construction, automotive and shoe soling. (Lankro Chemicals Ltd).†

Proponal. (Propal, Homobarbital). Dipropyl-malonyl urea or dipropyl-barbituric acid, $(C_3H_7)_2.C(CO_2)$-$.(NH)_2CO$. A soporific.†

Proponesin. A proprietary trade name for the hydrochloride of Tolpronine.†

Propoquad. Propoxylated quaternary ammonium salts. (Armour Hess Chemicals).†

Pro-Portion. Sodium Fluoride. Applications: Dental caries prophylactic. (Oral-B Laboratories Inc).

Proposote. Creosote-phenyl-propionate. A stimulating expectorant.†

Propranolol. 1-Isopropylamino-3-(1-naphthyloxy)propan-2-ol. Inderal is the hydrochloride.†

Propstearyl. Polypropoxylated stearyl alcohol. (Croda Chemicals Ltd).

Propyl Docetrizoate. Propyl 3-diacetylamino-2,4,6-tri-iodobenzoate. Pulmidol.†

Propyl Ether. Dipropyl-oxide, $(CH_2.CH_2.CH_3)_2O$.†

Propyl Zithate. Zinc isopropyl xanthate. Rubber accelerator. Applications: Used in natural amd synthetic cements and doughs. Non discolouring in presence of copper or iron. (Vanderbilt Chemical Corporation).*

Propylan. Polyether polyols for flexible, semi-rigid and rigid polyurethane foams. Applications: Used in upholstery, bedding, insulation, automotive, elastomers, coatings and as pigment dispersing media. (Lankro Chemicals Ltd).†

Propylan A350. A proprietary amine initiated polyether used in the manufacture of polyurethane foam. (Lankro Chemicals Ltd).†

Propylan G600. A proprietary polyoxypropylene trial of low molecular weight used in the production of rigid urethane foams and other urethane compositions, including elastomers. (Lankro Chemicals Ltd).†

Propylan RF55. A proprietary modified sorbitol-based polyether for making rigid, flame-proof urethane foams. (Lankro Chemicals Ltd).†

Propylhexedrine. 1-Cyclohexyl-2-methylaminopropane. Benzedrex. Eventin.†

Propyliodone. N-Propyl 3,5-di-iodo-4-pyridone-N-acetate. 3,5-Diodo-1-propoxycarbonylmethylpyrid-4-one. Dionosil.†

Propylol. Essentially normal propyl alcohol containing approximately 12% secondary butyl alcohol. Applications: Solvent in paints and lacquers, foundries, de-oiling waxes, grinding media for glass forming, manufacture of hair lacquers, floor polishes, latex rubber production, disinfectants, hand cleaners, degreasers, rust removers, printing inks, production of xanthates and other mining chemicals. (Sasolchem).*

Propyphenazone. An analgesic. 4-Isopropyl-2, 3-dimethyl-1-phenyl- 5-pyrazolone. Present in GEVODIN.†

Propyrin. Sodium thymol-benzoate.†

Propytal. Dipropyl-barbituric acid. See PROPONAL.

Proquamezine. 10-(2,3-Bisdimethyl-aminopropyl)phenothiazine. Myspamol.†

Proresid. A proprietary preparation of mitopopozide. An antimitotic. (Sandoz).†

Proscillaridin. 3β, 14β-Dihydroxybufa-4,20,22-trienolide 3-rhamnoside. Talusin.†

Proscopine. A proprietary trade name for the hydrochloride of Poskine.†

Prosobee. A proprietary artificial baby milk derived from soya, used in cases of intolerance of cows' milk. (Bristol-Myers Co Ltd).†

Prosol. A proprietary high protein food. (Cow and Gate).†

Prosopite. A mineral, $Ca(F.OH)_2.Al_2(F.OH)_3$.†

Prosparol. A proprietary preparation of ARACHIS OIL and water emulsion. A high calorie food. (Duncan, Flockhart).†

Prostaphlin. Oxacillin sodium. Applications: Antibacterial. (Bristol Laboratories, Div of Bristol-Myers Co).

Prostigmin. A proprietary preparation of neostigmine bromide, used in the treatment of myesthenia gravis and to counteract the effect of drugs resembling curare. (Roche Products Ltd).*

Prostin F2 Alpha. Dinoprost tromethamine Applications: Oxytocic; prostaglandin. (The Upjohn Co).

Prostin VR Pediatric. Alprostadil. Applications: Vasodilator. (The Upjohn Co).

Prostin/15M. Carboprost tromethamine. Applications: Oxytocic. (The Upjohn Co).

Protagon. Lecithin.†

Protan. See ALBUTANNIN.

Protargentum. (Protargin, Proganol). A compound of gelatin and silver. It contains 8 per cent silver, and is used in aqueous solution in medicine.†

Protargin. See PROTARGENTUM.

Protargolgranulat. A German product. It consists of 1 part protargol and 2 parts urea. It has the advantage of easy solubility.†

Protars. A dry arsenical fungicide prepared from talc, lime, and arsenic oxide.†

Protasan. Denture adhesive (powder, cream and liquid), for securing dentures. (Richardson-Vicks Inc).*

Protectoid. A proprietary trade name for a cellulose acetate plastic in the form of a non-inflammable film.†

Protectol. A brown, syrupy liquid used for the protection of animal products such as hair, wool, silk, skin, and leather from the action of alkaline liquids. It contains sodium lignin sulphonate.†

Protectyl. A solution containing 0.2 per cent mercury, 1 per cent salicylic acid, 3 per cent glycerin, and 95.8 per cent water. A disinfectant.†

Protegin X. A base containing petrolatum and oxycholesterin.†

Proteids. The same as albuminoids.†

Proteinase. Bacterial protease. (ABM Chemicals Ltd).*

Proteol. A combination of casein with formaldehyde. An antiseptic dusting powder.†

Proteryl. Emetine-bismuth iodide capsules.†

Protex. A proprietary safety glass.†

Protexulate. Loose-fill mineral powder for underground pipework. Applications: Underground pipework for heat insulation and as a water barrier to stop corrosion. (Croxton and Garry Ltd).*

Prothar. Factor IX Complex. Applications: Haemostatic. (Armour Pharmaceutical Co).

Protheite. A mineral. It is a variety of pyroxene.†

Protogest. Protein hydrolysate veterinary. (The Wellcome Foundation Ltd).

Protopam Chloride. Pralidoxime Chloride. Applications: Cholinesterase reactivator. (Ayerst Laboratories, Div of American Home Products Corp).

Protophane. Insulin, Isophane. Applications: Antidiabetic. (E R Squibb & Sons Inc).

Protormone. Veterinary injection of progesterone. (The Wellcome Foundation Ltd).

Protovit. Proprietary preparation containing the B complex vitamins, vitamins A, C, D and E, biotin and nicotinamide. (Roche Products Ltd).*

Protropin. Somatrem. Applications: Hormone. (Genentech Inc).

Provence Oil. The finest (Aix) olive oil.†

Proventil Inhaler. Albuterol. Applications: Bronchodilator. (Schering-Plough Corp).

Proventin 7. An active oxygen compound. Applications: Used in the textile industry as specific protection agent for polyamide fibres during peroxide bleaching and in the dyeing process. (Degussa).*

Provera. Medroxyprogesterone acetate. Applications: Progestin. (The Upjohn Co).

Provera Dosepak. Medroxyprogesterone acetate. Applications: Progestin. (The Upjohn Co).

Provocholine. Methacholine chloride. Applications: Cholinergic. (Hoffmann-LaRoche Inc).

Provol. A compound used for electrical insulation. It is a mixture of pitch, bitumen or similar materials, and mineral matter.†

Proxel. Industrial microbiocides. (ICI PLC).*

Proxitane. Peracetic acid. (Interox Chemicals Ltd).

Proxy. A hydrogen peroxide solution.†

Proxyl. A proprietary trade name for a pyroxylin denture material.†

Prozine. A proprietary preparation of meprobamate, and promazine hydrochloride. (Wyeth).†

Prozinex. Active ingredient: propazine; 2-chloro-4,6-bis-(isopropylamino)-1,3,5-triazine. Applications: Selective pre-emergence herbicide. (Agan Chemical Manufacturers Ltd).†

PR Spray. A proprietary preparaton of trichlorofluoromethane and dichlorodifluororoethane used as an analgesic spray. (Boots Pure Drug Co).*

PR Tablets. Analgesic. (The Boots Company PLC).

Prulet. Phenolphthalein. Applications: Laxative. (Mission Pharmacal Co).

Prune, Pure. (Parme R). A dyestuff. It is the methyl ether of gallocyanine (*qv*), $C_{16}H_{15}N_2O_5Cl$. Dyes tannined cotton, and wool bluish-violet. Used in calico printing.†

Prussian Black. A pigment prepared by calcining Prussian blue. It consists of carbon and oxide of iron.†

Prussian Brown. A pigment composed of sesquioxide of iron and alumina.†

Prussian Green. A pigment. It consists of Prussian blue and gamboge, with Prussian blue largely in excess. Also see PELOUZES GREEN.†

Prussian Red. See INDIAN RED.

Prussiate Black. A carbonaceous pigment obtained as a by-product in the manufacture of potassium ferrocyanide.†

Prussic Acid. Hydrocyanic acid HCN.†

Pruteen. Single-cell protein used as a feed additive. (ICI PLC).*

Prystal. A French proprietary rubber vulcanization accelerator. It is a formaldehyde-urea condensation product. Also a proprietary trade name for a cast, clear phenolic moulding.†

Prystaline. A proprietary moulding compound of the urea-formaldehyde type of synthetic resin.†

P.S.E. No. 15 Powder. An explosive. It is a mixture of ammonium perchlorate and rosin.†

Pseudo-alums. Double sulphates of aluminium and another metal containing a bivalent metal sulphate instead of a monovalent one. A type is $MnSO_4.Al_2(SO_4)_3.24H_2O$. They are not iso-morphous with the alums.†

Pseudo-galena. See ZINC BLENDE.

Pseudoephedrine. (+)-2-Methylamino-1-phenylpropan-1-ol (a stereoisomer of ephedrine).†

Pseudomorphine. Dehydro-morphine.†

Psoranide. Fluocinolone acetonide. Applications: Glucocorticoid. (Elder Pharmaceuticals Inc).

Psoriderm Bath Emulsion. Buff-coloured liquid emulsion containing 40% special coal tar extract. Applications: An aid in the treatment of sub-acute and chronic psoriasis. (Dermal Laboratories Ltd).*

Psoriderm Cream. Buff coloured aqueous cream containing 6% special coal tar extract, 0.4% lecithin. Applications: For the topical treatment of sub-acute and chronic psoriasis including psoriasis of the scalp and flexures. (Dermal Laboratories Ltd).*

Psoriderm Scalp Lotion. Amber coloured foaming liquid containing 2.5% special coal tar extract, 0.3% lecithin. Applications: For the topical treatment of psoriasis of the scalp. (Dermal Laboratories Ltd).*

Psorox. Dermatological lotion and ointment. (Fisons PLC, Pharmaceutical Div).

PS3/PS4/PS5. Range of granular fertilizers. (Fisons PLC, Pharmaceutical Div).

P.T.D. A rubber vulcanization accelerator. It is dipentamethylenethiuram disulphide.†

P.T.F.E. Polytetrafluoroethylene.†

P.T.M. A rubber vulcanization accelerator. It is dipentamethylene-thiuram-monosulphide.†

Ptychotis Oil. Ajowan oil.†

Puce Oxide of Lead. See BROWN LEAD OXIDE.

Pulluzyme. Pullulanase. (ABM Chemicals Ltd).*

Pulmadil Auto. Breath-actuated pressurized aerosol containing a suspension of rimiterol hydrobromide 10 mg/ml: delivers 300 metered doses (0.2 mg per dose). (Riker Laboratories/3M Health Care).†

Pulmadil Inhaler. Pressurized aerosol containing a suspension of rimiterol hydrobromide 10 mg/ml: delivers 300 metered doses (0.2 mg per dose). (Riker Laboratories/3M Health Care).†

Pulmo Bailly. Per 5 ml spoonful - guaiacol 75 mg, phosphoric acid concentrated 85 mg, codeine 7 mg Applications: Cough sedative and expectorant. (Bengue & Co Ltd).*

Pulmolite. Technetium Tc 99m Aggregated Albumin. Applications: Diagnostic aid; radioactive agent. (Dupont-NEN Medical Products).

Pulmovax. Pneumococcal vaccine. Applications: Effective protection against certain pneumococcal infections. (Merck Sharp & Dohme).*

Pulp, Asbestine. See AGALITE.

Pulpex E and P. Polypropylene pulps. Applications: They are special fibrous additives that can be mixed in any proportion with natural pulps and can be handled on conventional equipment, Very wide range of applications. (Hercules Inc).*

Pulp, Mineral. See AGALITE.

Pulse. A family of polycarbonate resin blends used for interior automotive applications. (Dow Chemical Co Ltd).*

Pulvatex. A trade mark for rubber (raw or partly prepared) for manufacture.†

Pumice Stone. (Obsidian). A volcanic mineral (lava froth), consisting mainly of aluminium silicate. Specific gravity 2.2-2.5. An abrasive.†

Pumiline. A proprietary preparation of pine oil.†

Pump Repair Putty. Ceramic filled epoxy resin. Applications: Repair, rebuild and protect pumps. (Devcon Corporation).*

Punctilious Ethyl Alcohol. Ethyl alcohol. Applications: Pharmaceuticals, cosmetics and personal care products, vinegar flavourings and foods. (USI Chemicals).*

Punicin. A colouring matter obtained from *Purpura capillus* and other shell fish.†

Purac. Decolourizing and absorptive activated carbon. (Lancashire Chemical Works Ltd).

Puratronic. High purity chemicals suitable for electronic materials. Applications: Electronic device materials, crystal growing, epitaxy. (Johnson Matthey Chemicals Ltd).*

Purbeck Stone. A limestone used in building.†

Purdox. A proprietary trade name for high purity recrystallized alumina. (Morgan Refractories Ltd).†

Purdurum. See CAMITE.

Pure Blue. See SOLUBLE BLUE.

Pure Chrysoidine RD, YD. British equivalents of Chrysoidine.†

Pure Scarlet. Mercuric iodide, HgI_2.†

Pure Soluble Blue. A British brand of Soluble Blue.†

Purex. A pigment. It is basic lead sulphate, and varies between $2PbSO_4.PbO$ and $3PbSO_4.PbO$.†

Purez. Polyester polyols based on adipic acid. (Avalon Chemical Co Ltd).

Purganol. A phenol-phthalein preparation.†

Purgatin. (Purgatol). Anthrapurpurin diacetate, $C_{14}H_5O_2(OH)(O.C_2H_3O)_2$. An aperient.†

Purgatol. Consists mainly of the anhydrides and lactones of fatty acids. Used for dressing hides. Also see PURGATIN.†

Purgen. (Laxans, Laxatin, Laxatol, Laxatoline, Laxen, Laxiconfect, Laxin, Laxoin, Laxophen, Phenolax, Purgella, Purgen, Purgo, Purgylum). Proprietary preparations of phenolphthalein, $C_6H_4(OC)C.(C_6H_4.OH)_2O$, sometimes with malic acid.†

Purging Nut Oil. (Physic Nut Oil). Curcas oil, from the seeds of *Jatropha Curcas*. Used in soap-making, and for lubricating.†

Purging Salt, Tasteless. Sodium dihydrogen phosphate, $NaH_2PO_4.H_2O$.†

Purgo. See PURGELLA.

Purgolade. See PURGELLA.

Purgylum. See PURGELLA.

Purifloc. Polyacrylamides used as flocculants. (Dow Chemical Co Ltd).*

Purinethol. A proprietary formulation of mercaptopurine. Applications: For remission induction, remission consolidation and maintenance therapy of acute leukaemias. (The Wellcome Foundation Ltd).*

Purochem. Organotin bactericides. (Akzo Chemie GmbH, Düren).†

Puromix. Food phosphate mixtures. (Albright & Wilson Ltd, Phosphates Div).

Puron. Food grade phosphates. (Albright & Wilson Ltd, Phosphates Div).

Purone. 2, 8-Dihydroxy-1, 4, 5, 6-tetra-hydro-purine.†

Purozone. A proprietary preparation. It consists of an alcoholic solution of the sodium salts and acids of wood tar or similar materials. A disinfectant.†

Purple Carmine. See MUREXIDE.

Purple Madder. A pigment. It is a lake formed by the precipitation of the colouring matter of madder in combination with a metallic oxide.†

Purple Ochre. See MARS VIOLET.

Purple of Cassius. See GOLD PURPLE.

Purpurin. (Alizarin No. 6). Trihydroxy anthraquinone, $C_{14}H_8O_5$. Dyes cotton mordanted with alumina, red, and with chromium, reddish-brown. Also see MAUVEINE.†

Purpuroxanthic Acid. See MUNJISTIN.

Purpuroxanthin. (Xanthopurpunn). *m*-Dihydroxy-anthraquinone, $C_{14}H_8O_4$.†

Purree. See INDIAN YELLOW.

Purrenone. (Purrone). Euxanthone, $C_{13}H_8O_4$, obtained from purree, a yellow dyestuff.†

Purrone. See PURRENONE.

Pursennid. A proprietary preparation containing sennosides A and B (as calcium salts). A laxative. (Sandoz).†

Purus. See CHLORAMINE T.

Pusher. Secondary oil recovery polymer is an additive to water injected into petroleum reservoirs. It decreases the mobility of the water flood in relation to the mobility of the crude. (Dow Chemical Co Ltd).*

Putrescine. Tetramethylene-diamine, $NH_2(CH_2)_4.NH_2$. A base found in ergot.†

Putty Powder. An impure stannic oxide, SnO_2, used for polishing glass. It is also used sometimes in rubber mixing specific gravity of 6.6.†

Puzzuolana. A volcanic material found in various parts of Italy, especially Puzzuoli. It is employed for the conversion of pure lime into a hydraulic lime.†

P.V.A. Polyvinyl alcohol.†

Pvacote. Compounded polymeric emulsion. Applications: Spray coating of asbestos stripped rooms to bind residue fibres. (Howlett Adhesives Ltd).*

PVC. Polyvinyl chloride.†

P V Carpine Liquifilm. Pilocarpine Nitrate. Applications: Cholinergic. (Allergan Pharmaceuticals Inc).

PVF2. A polyvinyl fluoride. See KYNAR.†

PVP-Iodine. Povidone-Iodine. Applications: Anti-infective. (GAF Corporation).

PX 104. Dibutyl phthalate.†

PX 108. Di-iso-octyl phthalate.†

PX 114. Decyl butyl phthalate.†

PX 120. Di-iso-decyl phthalate.†

PX 138. Dioctyl phthalate.†

PX 208. Di-iso-octyl adipate.†

PX 209. Dinonyl adipate.†

PX 404. Dibutyl sebacate.†

PX 408. Di-iso-octyl sebacate.†

PX 658. Tetrahydrofurfuryl oleate.†

PX 916. Triphenyl phosphate.†

Py-Ran. Anhydrous monocalcium phosphate. Applications: Used as acid component in leavening agents for self-raising flours, baking powders, cake mixes and corn meal. (Monsanto Co).*

Pybuthrin. Synergized pyrethrins. (The Wellcome Foundation Ltd).

Pycamisan. A proprietary preparation containing sodium aminosalicylate, isoniazid. An antituberculous agent. (Smith and Nephew).†

Pycasix. A proprietary preparation containing sodium aminosalicylate, isoniazid. An antituberculous agent. (Smith and Nephew).†

Pycazide. A proprietary preparation containing isoniazid. An anti-tuberculous agent. (Smith and Nephew).†

Pylkrome. A composite alloy steel. The base is mild steel, and it has a corrosion-resisting surface of high chrome or high chromium-nickel-iron alloys.†

Pylura. A proprietary preparation of adrenaline, benzyamine lactate and phenol used in the treatment of haemorrhoids. (Pharmax).†

Pymafed. Pyrilamine Maleate. Applications: Antihistaminic. (Hoechst-Roussel Pharmaceuticals Inc).

Pynosect. Insecticidal formulation. (Mitchell Cotts Chemicals Ltd).

Pyoctanin. (Pyoktanin). The name given to different coal-tar colours. (1) Yellow pyoctanin (auramine, *q.v.*), and (2) Blue pyoctanin (methyl violet, *q.v.*). Used in surgery as bactericides.†

Pyoctanin Blue. See PYOCTANIN.

Pyoctanin Yellow. See PYOCTANIN.

Pyoktanin. See PYOCTANIN.

Pyopen. A proprietary preparation containing carbenicillin (as the disodium salt). An antibiotic. (Beecham Research Laboratories).*

Pyracine. Phenazone.†

Pyradiolin. A proprietary synthetic plastic used as a dielectric material in wireless telegraphy, and for other purposes. It is a modified pyroxylin plastic.†

Pyralin. A registered trade mark for a celluloid product. Available in transparent, translucent, opaque coloured and colourless forms. Resistant to hydrocarbons and oils.†

Pyralin. (2) A registered trade name for polyimide high temperature resistant materials. (Du Pont (UK) Ltd).†

Pyraloxin. Oxidized pyrogallic acid, a dark brown powder obtained by oxidation with air and ammonia.†

Pyramid. Potassium silicates. (Crosfield Chemicals).*

Pyramidol Brown BG. A dyestuff. It is the sodium salt of diphenyldisazobiresorcinol, $C_{24}H_{18}N_4O_4Na$. Dyes cotton red.†

Pyramidol Brown T. A dyestuff. It is the sodium salt of ditolyldisazobi-resorcinol, $C_{26}H_{22}N_4O_4Na$. Dyes cotton brownish-red.†

Pyramine Orange R. A dyestuff. It is the sodium salt of disulphodiphenyldisazobinitro-*m*-phenylenediamine, $C_{24}H_{18}N_{10}O_{10}S_2Na_2$. Dyes cotton orange-red.†

Pyramine Orange 3G. A dyestuff. It is the sodium salt of diphenyldisazo-nitro-*m*-phenylenediamine-*m*-phenylenediaminedisulphonic acid, $C_{24}H_{19}N_9S_2O_8Na_2$. Dyes cotton yellowish orange.†

Pyramol. Silica sols for industry. (Crosfield Chemicals).*

Pyranet. Detergent sanitizer. (Crosfield Chemicals).*

Pyranil Black. A sulphide dyestuff.†

Pyranol. Sodium acetylsalicylate, $C_2H_3O.OC_6H_4COONa$.†

Pyranthrone. (Indanthrene gold orange G, Duranthrene Gold Yellow Y). A dyestuff prepared by treating 2 : 2'-dimethyl-1 : 1'-dianthraquinonyl with alkali or zinc chloride. Dyes orange tints.†

Pyrantimonite. See RED ANTIMONY.

Pyrasteel. A proprietary trade name for a heat-resisting alloy of iron with 25 per cent nickel, 14 per cent chromium, and 2.5-3.0 per cent silicon.†

Pyratex. 2-Vinylpyridine-styrene-butadiene terpolymer. Applications: Used to treat tyre cord and other textiles to improve their adhesion to rubber and enhance the compatability of the textile-rubber system. (Bayer & Co).*

Pyraton. A proprietary trade name for diacetone alcohol.†

Pyrax Talcs A and B. Qualities of pure white talc mineral from deposits in America. Used in rubber, textile, and ceramic industries.†

Pyrazinamide. Pyrazinoic acid amide. Zinamide.†

Pyrazine. See ANTIPYRINE.

Pyrazine Yellow S. A British equivalent of Flavazine S.†

Pyrazoline. See ANTIPYRINE.

Pyre-ML. Wire enamel and insulating varnish. (Du Pont (UK) Ltd).

Pyrene Oil. (Begasses Oil). An inferior olive oil.†

Pyrethrum Powder. An insecticide made from the powdered flowers of some species of pyrethrum plants. The active principle is rotenone (*qv*).†

Pyrgos. See CHLORAMINE T.

Pyribenzamine Citrate. Tripelennamine Citrate. Applications: Antihistaminic. (Ciba-Geigy Corp).

Pyribenzamine Hydrochloride. Tripelennamine Hydrochloride. Applications: Antihistaminic. (Ciba-Geigy Corp).

Pyricit. Sodium fluoroborate, $NaBF_4$.†

Pyricol. See ACOPYRIN.

Pyridium. Phenazophridine Hydrochloride. Applications: Analgesic. (Parke-Davis, Div of Warner-Lambert Co).

Pyrido Rubber. A polymerized acrolein-methyl-amine. A rubber-like material.†

Pyridoxine. Vitamin B_6.†

Pyrimithate. An insecticide. *o*-2-Dimethylamino-6-methylpyrimidin-4-yl *oo*-diethyl phosphorothioate. DIOTHYL.†

Pyrinex. Active ingredient: chlorpyrifos; 0,0-diethyl-0-3,5,6-trichloro-2-pyridyl-phosphorothioate. Applications: Organophosphorous agricultural insecticide effective against a broad range of insects. (Makhteshim Chemical Works Ltd).†

Pyrites, Cockscomb. See MARCASITE.

Pyrites, Coxcomb. See MARCASITE.

Pyrites, Efflorescent. See WHITE PYRITES.

Pyrites, Radiated. See MARCASITE.

Pyrites, White Iron. See MARCASITE.

Pyro. Pyrogallol, $C_6H_3(OH)_3$.†

Pyro-acetic Spirit. Acetone, $CH_3.CO.CH_3$.†

Pyro Alcohol. Methyl alcohol, CH_3OH.†

Pyro-bitumen. Bitumen which is insoluble in carbon tetrachloride.†

Pyrobrite. Copper plating processes. (Albright & Wilson Ltd, Phosphates Div).

Pyrocast. A proprietary trade name for a nickel-chromium cast iron.†

Pyrocatechin. (Catechol). Pyrocatechol, $C_6H_4(OH)_2$.†

Pyrocatechol Arsenic Acid. *o*-Hydroxyphenyl arsenate. A reagent for alkaloids.†

Pyro-Chek. Applications: For flame retardant additives for plastic resins, in Int Class 1. (Ferro Corporation).*

Pyrochlor. Fire-resistant and wood preservative. (Hickson & Welch Ltd).*

Pyrocollodion. A soluble nitrocellulose containing the highest practicable percentage of nitrogen, about 12.5 per cent.†

Pyro Cotton. A nitrated cellulose, not so fully nitrated as guncotton.†

Pyrodialite. An explosive. It contains 80-88 per cent potassium chlorate, 5-6 per cent charcoal, 10-18 per cent gas tar, and 3-4 per cent sodium and ammonium bicarbonates.†

Pyrodine. (Hydracetin).2-Acetyl-7-phenylhydrazine.†

Pyrofulmin. A yellow substance obtained by heating mercuric fulminate. It is probably a mixture of mercuric oxycyanide and oxide.†

Pyrogallic Acid. Pyrogallol $C_6H_3(OH)_3$. It absorbs oxygen, and is used in gas analysis.†

Pyrogastrone. Tablets or liquid containing carbenoxolone sodium with alginate and antacids. Applications: Used in the treatment of oesophagitis. (Winthrop Laboratories).*

Pyrogene Blacks and Blues. See SULPHUR BLACK T.

Pyrogene Blues and Greys. Dyestuffs produced by heating hydroxy-dinitro-diphenylamine and indophenols under pressure with polysulphides in alcoholic solution.†

Pyrogene Brown D. A dyestuff obtained by sulphurizing sawdust or bran.†

Pyrogene Dark Green. A dyestuff produced by sulphurizing *p*-aminophenol and its substitution derivatives with sodium sulphide and sulphur, in the presence of copper.†

Pyrogene Green. A sulphide dyestuff prepared by the fusion of various indophenols with polysulphides, in the presence of copper compounds.†

Pyrogene Indigo. A dyestuff obtained by heating indophenol, $C_2H_5.NH.C_6H_4N.C_6H_4O$, with polysulphides.†

Pyrogene Olive N. (Pyrogene Yellow M). Dyestuffs prepared by heating various methyl-amino and nitro-amino compounds with sodium sulphide, sulphur, and alkalis.†

Pyrogene Yellow M. See PYROGENE OLIVE N.

Pyroglycerin. See NITRO-GLYCERIN.

Pyrolignite of Iron. (Iron Liquor, Black Mordant, Black Liquor, Liqueur de Ferraile). Ferrous acetate, $Fe(C_2H_3O_2)_2$, prepared by the action of pyroligneous acid upon iron turnings. The solution also contains ferric acetate. Used in calico printing, and in dyeing, for the preparation of blue, violet, black, and brown colours.†

Pyrolignite of Lime. Calcium diacetate.†

Pyrolith. Fire retardant treatment for timber. Applications: Internal use only. (Hickson & Welch Ltd).*

Pyromic. A particularly pure form of induction melted nickel-chromium used for electric furnaces, heaters, ovens, etc.†

Pyromorphic Phosphorus. Prepared by heating red phosphorus with a trace of iodine at 280° C., or *in vacuo*.†

Pyronate. Alkyl aryl petroleum sulphonate, completely hydrophilic, water soluble. Dark in colour. Applications: Demulsification; froth flotation of non-metallic ores. (Witco Chemical Ltd).

Pyronil. A proprietary trade name for the phosphate of Pyrrobutamine. (Eli Lilly & Co).†

Pyronin Stain. See PAPPENHEIM'S STAIN.

Pyronine B. A dyestuff. It is tetraethyldiaminoxanthenylchloride, $C_{19}H_{23}Cl.N_2O$. Dyes cotton, wool, and silk, red.†

Pyronine G. (Casan Pink). A dyestuff. It is tetramethyldiaminoxanthenyl-chloride, $C_{17}H_{19}Cl.N_2O$. Dyes cotton, wool, and silk, red.†

Pyronite. See TETRYL.

Pyronium. A proprietary substitute for tin oxide in enamels. It is used in conjunction with tin oxide.†

Pyrophan. A combination of pyrogallol and dimethylamine.†

Pyroretin. A brown resin found in lignite.†

Pyros. A paramagnetic alloy consisting of nickel with 7 per cent chromium, 5 per cent tungsten, 3 per cent manganese, and 3 per cent iron. It is suitable for expansion pyrometers.†

Pyrosal. See ACOPYRIN.

Pyrosin R. A dyestuff. It is a mixture of Pyrosine B and G.†

Pyrosine B. See ERYTHROSIN.

Pyrosine G. See ERYTHROSIN G.

Pyrostibite. Synonym for Kermesite.†

Pyrostibnite. See RED ANTIMONY.

Pyrosulphuric Acid. See FUMING SULPHURIC ACID.

Pyrotin RRO. A dyestuff. It is the sodium salt of sulphonaphthaleneazo-α-naphthol monosulphonic acid, $C_{20}H_{12}N_2O_7S_2Na_2$. Dyes wool red from an acid bath.†

Pyrovatex. Proofing agent. (Ciba-Geigy PLC).

Pyroxylin Plastics. See AETHROL, ALCOLITE, DROTT, HELECO, PHOENIXITE, TRELIT, TROLIT F, VENITE.

Pyrozone. See PEROXOL.

Pyrrhol. (Pyrroline). Pyrrole, C_4H_5N.†

Pyrrhotine. See MAGNETIC PYRITES.

Pyrrhotite. See MAGNETIC PYRITES.

Pyrrodiazole. Triazole, $C_2H_3N_3$.†

Pyrrol Black. See SULPHUR BLACK T.

Pyrrole Blue A. An indigo-blue compound, $C_{24}H_{16}O_3N_4$, obtained by adding pyrrole to a solution of isatin in dilute sulphuric acid.†

Pyrrole Blue B. A compound, $C_{24}H_{16}O_2N_4$, obtained by adding pyrrole in glacial acetic acid to isatin in acetic and sulphuric acids cooled to 0° C.†

Pyrrole Red. A polymer of pyrrole, produced by hot hydrochloric acid.†

Pyrroline. See PYRRHOL.

Pyrvinium Pamoate. Viprynium Embonate.†

Pyxol. An emulsion of coal-tar acids with soap. A disinfectant.†

Q

Qaulineg. Colour film processing system. (May & Baker Ltd).

Qazul. Polyamide fibres for luggage. (ICI PLC).*

Q-Cast. A castable refractory monolithic material for high temperature applications. (Pfizer International).*

Q-Cel. Hollow microspheres. Silicate glass and ceramic microspheres produced in selected density and particle size ranges. Applications: Low density fillers for thermoset plastics, adhesives, sealants, sensitizers for emulsion and water gel explosives. (The PQ Corporation).*

Q-Crete. A monolithic refractory concrete for high temperature applications. (Pfizer International).*

Q-Gun. A gunnable refractory monolithic material for high temperature applications. (Pfizer International).*

Q-Therm. High temperature fluids with excellent thermal conductivity, temperature range and dielectric properties. Applications: For the design of new functional fluids and lubricants with low toxicity values, e.g., heat transfer fluids for process equipment, sealed electrical appliances. (Anderson Development Company).*

-Vibe. A vibratable refractory monolithic material for high temperature applications. (Pfizer International).*

uab 151. 2,3-Epoxypropyl trimethyl ammonium chloride. Applications: Cationic biopolymers (e.g. starch, cellulose, guar, gelatine, protein) and synthetic polymers (e.g. polyacrylic acid, acrylamide-acrylic acid copolymers, polyaminoamides, polyethylene imide). (Degussa).*

Quab 188. 3-Chloro-2-hydroxypropyl trimethyl ammonium chloride. Applications: quaternization of compounds with hydroxyl, amino and other functional groups, especially corresponding polymers, production of cationic polyelectrolytes. (Degussa).*

Quadrilan. Cationic surfactants. Applications: Used in fabric conditioning, biocides, industrial and vehicle cleaning. (Lankro Chemicals Ltd).†

Quadrilan AT. Quaternized fatty amine ethoxylate as a clear amber liquid. Applications: Antistatic agent for PVC and other polymers. (Diamond Shamrock Process Chemicals Ltd).

Quadrilan BC. Benzalkonium chloride (B.P. Grade) as a pale yellow liquid. Applications: Broad spectrum germicide for disinfectants and sanitizers. (Diamond Shamrock Process Chemicals Ltd).

Quadrilan MY 211. Specially developed cationic surfactant as a hazy amber liquid. Applications: Used in alkaline spray cleaning concentrates for vehicle and chassis cleaning and crate washing. (Diamond Shamrock Process Chemicals Ltd).

Quadrilan OD-75. Quaternary ammonium compounds as a stiff white paste. Applications: Softener and antistatic agent for textiles. (Diamond Shamrock Process Chemicals Ltd).

Qualidot. Machine lith developer system. (May & Baker Ltd).

Qualifix. Photographic fixer. (May & Baker Ltd).

Qualitol. Photographic developer. (May & Baker Ltd).

Quamilin. Formulations for permethrin. (The Wellcome Foundation Ltd).

Quantacure. Photo-curing compounds. (Ward Blenkinsop & Co Ltd).

Quantril. Benzquinamide. Applications: Antiemetic. (J B Roerig, Div of Pfizer Laboratories).

Quantrovanil. A name applied to ethyl-protocatechuic aldehyde.†

Quartz. A mineral, silica, SiO_2. In the main there are three types. (1) Crystalline, such as Tridymite and Cristobalite. (2) Crypto-crystalline, such as Chalcedony, and (3) hydrated silica or Opal.†

Quartz, Foetid. See STINK QUARTZ.

Quartz Glass. Fused silica glass.†

Quartzilite. A metallic carbide formed by action of silica and carbon at 2000-3000° C. It is suitable for electrical resistances.†

Quarzal. An alloy of aluminium with 15 per cent copper, 6 per cent manganese, and 0.5 per cent silicon. Used for the cylinders of internal combustion engines.†

Quasilan. Pure MDI prepolymers. Applications: Curing components for use with Propocon polyether systems. (Lankro Chemicals Ltd).†

Quaternary Steels. Steels containing two special elements in addition to the iron and carbon.†

Quatramine. A range of quaternary ammonium salts. Used in antiseptics, disinfectants, bactericides and algicides. (Harcros Industrial Chemicals).*

Quatrex. Epoxy resins used for the electronics industry. (Dow Chemical Co Ltd).*

Quatum. FCC catalysts. (Crosfield Chemicals).*

Quebrachite. (Quebrachitol). The monomethyl ether of laevo-inositol, $C_6H_6(OH)_5(OCH_3)$. It is found in the latex of *Hevea brasiliensis* (the rubber tree) to the extent of 1-2 per cent.†

Quebracho. *Loxopteryngium lorenzii.* The wood of this tree contains about 20 per cent of tannin, and is used in the form of an extract for tanning.†

Queensland Arrowroot. See TOUS-LESMOIS STARCH.

Queen's Metal. A jeweller's alloy. It is very variable in composition. It contains from 50-85 per cent tin, 7-16 per cent antimony, 0-16 per cent lead, 0-3.5 per cent copper, and 1-12 per cent zinc.†

Queen's Yellow. See TURPETH MINERAL.

Quelicin. Succinylcholine Chloride. Applications: Blocking agent. (Abbott Laboratories).

Quell Oil. Oil spill dispersants. (Lankro Chemicals Ltd).

Quellada. A proprietary preparation of gamma benzene hexachloride. Applications: For dermatological use - anti-parasitic. (Stafford-Miller).*

Quercitron. A colouring matter, sold as chips, or as a coarse powder, obtained by grinding the bark of *Quercus tinctoria* and *Q. nigra*. The dyeing principle is quercitin or flavin, $C_{15}H_{10}O_7$ which forms yellow lakes with aluminium and tin salts. Used for calico printing and wool dyeing, for yellows and browns. Flavin, Patent bark, and Prepared bark are commercial preparations.†

Querton. Range of cationic surfactants of the quaternary ammonium compound type, in solid, liquid or paste form. Applications: Cosmetics and detergents, eg hair conditioning, fabric softeners, disinfectants, detergent sanitizers; anti-caking agents for fertilizers; mineral flotation and extraction; anti-corrosion agents and additives in the petroleum industry; dispersing of paints, pigments and clays; antistatic and mould release agents in plastic/rubber; softeners, antistats etc in textiles; adhesion agents in road surfacing; use in pulp and paper sugar refining and pesticide industries. (Keno Gard (UK) Ltd).

Quesfloc. A series of organic polymers suitable for waste water clarification. It is available in both dry and liquid compositions. Applications: Industrial and municipal waste water (effluent) clarification. (Ques Industries).*

Quesfloc F11283-1. A high molecular weight anionic polymer which performs as a coagulant aid and sludge conditioning agent in a variety of solid-liquid separation processes. In its dry form, it is a white, dustless free-flowing granular powder. Applications: Used in conjunction with ferric chloride and alum for phosphorus removal. It can also be used as a component in a dual anionic-cationic system for sledge dewatering operations. (Ques Industries).*

Questran. Cholestyramine resin. Applications: Ion-exchange resin;antihyperlipoproteinemic. (Mead Johnson & Co).

Questran. A proprietary preparation of cholestyramine chloride used in the treatment of hypercholesterolaemia. (Bristol-Myers Co Ltd).†

Quevenne's Iron. Reduced iron.†

Quiacryl. Acrylic water emulsions. Applications: Leather finishes, textile, inks, floor polish, coatings. (Merquinsa).*

Quick-pach. A plastic fire-clay for making monolithic linings and quick repairs.†

Quickening Liquid. A solution of mercuric nitrate or cyanide. Used in electro-plating.†

Quickfloc. Ferrous sulphate - moist copperas. Applications: Waste and potable water, animal feed, iron oxide raw material. (NL Industries Inc).*

Quicklime. See LIME.

Quicksilver. Mercury, Hg. A pigment which consists of sulphide of mercury is also known under this name.†

Quicksol. Insulin preparation. (The Boots Company PLC).

Quide. Piperacetazine. Applications: Antipsychotic. (Merrell Dow Pharmaceuticals Inc, Subsidiary of Dow Chemical Co).

Quidur. Urethane prepolymers. Applications: Binder for cork-compositions, laminating adhesives. (Merquinsa).*

Quiess. Hydroxyzine hydrochloride. Applications: Tranquilizer. (O'Neal, Jones & Feldman Pharmaceuticals).

Quilastic. Polyurethyane water emulsions, polyurethane solutions, granule form. Applications: Leather finishes, synthetic leather, inks, shoe adhesives, textile. (Merquinsa).*

Quillaic Acid. An acid obtained from the inner bark of *Quillaja saponaria* (soap bark). An expectorant.†

Quinaband. A proprietary bandage coated with zinc paste, CALAMINE and IODOCHLOROHYDROXYEUINO LONE use in the treatment of leg ulcers. (Seton Products Ltd).*

Quinaglute. Quinidine Gluconate. Applications: Cardiac depressant. (Berlex Laboratories Inc, Subsidiary of Schering AG).

Quinaldine. 2-Methylquinoline.†

Quinalizarin. See ALIZARIN BORDEAUX B, BD.

Quinalspan. A proprietary preparation of quinalbarbitone sodium. A hypnotic. (Pharmax).†

Quinamin. Quinine Sulphate. Applications: Antimalarial. (Merrell Dow Pharmaceuticals Inc, Subsidiary of Dow Chemical Co).

Quindex. Fungicides containing copper 8-quinolinolate. Applications: Preservation of textiles, cordage, paper, adhesives and caulking compounds. (Nueodex Inc).*

Quindoxin. A growth promoter. Quinoxaline 1,4-dioxide. GROFAS.†

Quine. Quinine Sulphate. Applications: Antimalarial. (Rowell Laboratories Inc).

Quineine. A standard solution of *Cinchona succiruba,* containing 7 per cent of alkaloids, 5 per cent of which is quinine.†

Quinicardine. Quinidine sulphate, $(C_{20}H_{24}O_2N_2)_2.H_2SO_4$.†

Quinidex. Quinidine Sulphate. Applications: Cardiac depressant. (A H Robins Co Inc).

Quinizarine Blue. A dyestuff. It is the sodium salt of anilido-oxy-anthraquinone-sulphonic acid. Dyes wool reddish-blue from an acid bath, and chromed wool, a greenish-blue.†

Quinizarine Greens. See ALIZARIN CYANINE GREENS.

Quinn's Rubber. A patented rubber substitute made from rapeseed oil, petroleum, and chloride of sulphur.†

Quinocort. Hydroxyquinoline sulphate and hydrocortisone in a cream base. Applications: Inflamed dermatoses where infection is present. (Quinoderm Ltd).*

Quinoderm Cream. A proprietary preparation of benzoyl peroxide and potassium hydroxyquinoline sulphate in a cream base. Applications: Used in the treatment of acne and acneform eruptions. (Quinoderm Ltd).*

Quinoderm Cream with Hydrocortisone. Benzoyl peroxide and potassium hydroxyquinoline sulphate with hydrocortisone. Applications: Used in the treatment of inflamed acne. (Quinoderm Ltd).*

Quinoformine. A compound of quinic acid and hexamethylene-tetramine. Used as a solvent for uric acid.†

Quinol. Hydroquinone (*p*-dihydroxy-benzene), $C_6H_4(OH)_2$. Used as a photographic developer.†

Quinoline Blue,. See CYANINE.

Quinoline Green. A dyestuff prepared by the action of phosphorus oxychloride upon a mixture of quinoline and tetra-methyldiaminobenzophenone.†

Quinoline Red. (Isoquinoline Red). A dyestuff, $C_{26}H_{19}Cl.N_2$, obtained by the action of benzo-trichloride upon a mixture of quinaldine (methylquinoline), and isoquinoline. Dyes wool and silk rose-red. Also used for isochromatizing photographic plates.†

Quinoline Yellow. (Quinoline Yellow, Water Soluble). A dyestuff. It is the sodium salt of the sulphonic acids (chiefly the disulphonic acid), of quinophthalone, $C_{18}H_9NO_8S_2Na_2$. Dyes silk and wool greenish-yellow from an acid bath.†

Quinoline Yellow, Spirit Soluble. Quino-phthalone, $C_{18}H_{11}NO_2$. Used for colouring spirit varnishes and waxes. See SPIRIT SOLUBLE QUINOLINE YELLOW.†

Quinoline Yellow, Water Soluble. See QUINOLINE YELLOW.

Quinoped. A proprietary preparation of benzoyl peroxide and potassium hydroxyquinoline sulphate in a cream base. Applications: Used in the treatment of fungal foot infections. (Quinoderm Ltd).*

Quinophan. See ATOPHAN.

Quinosol. (Chinosol, Sunoxol). The potassium salt of oxy-quinoline sulphonic acid, $C_9H_6.NO.SO_3K$. An antiseptic.†

Quinovasugar. Quinovitol, $C_6H_9O(OH)_3$.†

Quintesse. Polyamide fibres for upholstery. (ICI PLC).*

Quintiofos. An insecticide. O-Ethyl O-8-quinolyl phenylphosphonothioate. BACDIP.†

Quisqueite. An asphaltum-like compound containing much sulphur. It is found in Peru.†

Quixalud. Halquinol in feed additive. (Ciba-Geigy PLC).

R

-Acid. 2-Naphthol-3, 6-disulphonic acid.†

R-acid. 7-Amino-1-naphthol-3, 6-disulphonic acid.†

-2 Crystals. A proprietary trade name for an ultra accelerator for latex, etc. It is the reaction product of carbon disulphide with methylenedipiperidine.†

-502, MR-502K, MR-502Y, MR-502P. Magnetic Rubber Inspection Material. Dimethyl poly siloxane combination used as a magnetic particle type of non destruct testing. Applications: Non destruct inspection of ferro magnetic parts, used for inspection of thread, gear roots, inaccessible areas, coated areas. Permanent record of the inspection is an added bonus. (Dynamold Inc).*

abalon. Thermoplastic elastomer. (Mitsubishi Petrochemical Co).*

abro. A proprietary preparation containing liquorice, bismuth subnitrate, magnesium carbonate, sodium bicarbonate, alder buckthorn bark and calamus rhizome. An antacid. (Rybar).†

achromate-51. Sodium Chromate Cr 51. Applications: Diagnostic aid; radioactive agent. (Abbott Laboratories).

ackarock. A blasting explosive, consisting of 79 per cent potassium chlorate and 21 per cent nitrobenzene, mixed sometimes with picric acid or sulphur.†

ackarock Special. An explosive similar to Rackarock, but containing 12-16 per cent picric acid.†

acumin. Racumin bait contains 0.0375% w/w coumatetralyl and is highly palatable and readily consumed by rats. Racumin Tracking Powder contains 0.75% w/w coumatetralyl. It is picked up by rats on their fur and feet and is then swallowed during their frequent pauses for grooming. Applications: For the control of rats. It is recommended that both tracking powder and bait be used together. (Bayer & Co).*

Radar. Broad spectrum fungicide containing propiconazole for use on cereals. (ICI PLC).*

Radarsan. See RAWSTOL.

Radauite. A mineral. It is a variety of labradorite.†

Raddle. See INDIAN RED.

Radex. Insulating powders for ladles and continuous casting tundishes. (Foseco (FS) Ltd).*

Radiant Yellow. See CADMIUM YELLOW.

Radicle Vinegar. Acetic acid, glacial.†

Radiocaps-131. Sodium Iodide I 131. Applications: Antineoplastic; diagnostic aid; radioactive agent. (Abbott Laboratories).

Radio-malt. A proprietary preparation consisting of malt extract with irradiated ergosterol (radiostol). It contains vitamins A, B, and D.†

Radiometal. A nickel-iron-copper alloy having high incremental permeability and low losses. It is largely used for radio transformers, relays, etc.†

Radiopaque. A special barium sulphate.†

Radiose. A French nitro-cellulose lacquer.†

Radiostol. A proprietary preparation of calciferol, vitamin D_2. (British Drug Houses).†

Radiostoleum. A proprietary preparation of vitamins A and D in oil.†

Radmolite. An electrical insulating material for supporting heating coils. It largely consists of diatomaceous earth, and having a slower rate of heat absorption, replaces fireclay.†

Radumine. A synthetic oxalic acid.†

R.A.E. 57 Alloy. An alloy of aluminium with 4 per cent copper, 2 per cent iron, and 0.5 per cent magnesium. Specific gravity 2.8.†

Raffinate. A refined Sicilian sulphur with impurities amounting to about 0.5 per cent. Also see GREGGIO†

Raffinose. Mellitose, $C_{12}H_{22}O_{11}$, a sugar†

Rainbow Custom Coloured Mortars. Blends of iron oxide colours and mortars. Applications: Preblended coloured mortar to produce more uniform colour in the mortar joints of a building. (DCS Color & Supply Co Inc).*

Rainbow Orange. A dyestuff. It is a British equivalent of Ponceau 4GB.†

Rainbow Red RB, NB. Dyestuffs. They are British brands of Azorubine S.†

Rainbow Scarlet G, 2RS. British equivalents of Ponceau.†

Rainbow Scarlet 4R. A dyestuff which is equivalent to New coccine. It is of British manufacture.†

Rainbow Ware. A proprietary synthetic resin of the urea-formaldehyde type.†

Rakel's Alloy. An aluminium bronze. It contains 87.5 per cent copper, 10.5 per cent aluminium, 1 per cent manganese, and 1 per cent lead or zinc.†

Ramasit. Textile softening and water-repelling agents. (BASF United Kingdom Ltd).

Rambufaside. A cardiac glucoside. 14-Hydroxy - 3 - (4 - *O* - methyl - α - L - rhamnopyranosyloxy) - 14β - bufa - 4, 20,22-trienolide. 4'- *o*-Methylproscillaridin.†

Ramenti Ferri. Iron filings.†

Ramet. A proprietary cutting material for steel alloys, cast iron, etc. It consists of tantalum carbide with nickel, and melts at 4100° C.†

Rametin. A proprietary preparation of NAPHTHALOPHOS. A veterinary anthelmintic.†

Ramie. (Rhea, Green Ramie). A fibre obtained from *Boehmeria tenacissima*..†

Ramie, Green. See RAMIE.

Ramix. A proprietary magnesite refractory.†

Ramos. A proprietary product. It is a phenol-formaldehyde resin.†

Ramrod. 2-Chloro-N-isopropyl-acetanilide, propachlor. Applications: Herbicide. (Monsanto Co).*

Ramtap. Fibrous tap hole plugs for electric arc and basic oxygen steelmaking furnaces. (Foseco (FS) Ltd).*

Ranciérite. Synonym for Ranciáte.†

Randanite. (Ceyssatite, Memilite). Varieties of hydrated silica or opal.†

Ranestol. Triclofenol Piperazine. Application: Anthelmintic. (Parke-Davis, Div of Warner-Lambert Co).

Raney Nickel. A form of finely divided nickel used in hydrogenating certain organic compounds.†

Ranide. Rafoxanide. Applications: Anthelmintic. (Merck Sharp & Dohme, Div of Merck & Co Inc).

Ranotex. Loop raising agent. (ABM Chemicals Ltd).*

Ransome's Stone. An artificial stone made by mixing sand with sodium silicate and a little chalk, or other similar material. The product is moulded to shape, and immersed in a solution of calcium chloride.†

Raolein 131. Triolein I 131. Applications: Radioactive agent. (Abbott Laboratories).

Rapadex. Photographic developer. (May & Baker Ltd).

Rapic Acid. A name which has been applied to the fatty acids of rape oil. It appears to be identical with oleic acid.†

pid. Garden insecticide. (ICI PLC).*

pid Fast. 3G. A dyestuff. It consists of the anilide ofβ-hydroxy-naphthoic acid mixed with *p*-chloro-*o*-nitro-aniline. Used in calico printing.†

pid Fast Red B. A dyestuff. It consists of the anilide of β-hydroxynaphthoic acid mixed with 5-nitro-*o*- anisidine. Used in calico printing.†

pid Fast Red BB. A dyestuff. It consists of a mixture of *p*-nitro-anilide of β-hydroxynaphthoic acid and the nitro samine salt of 5-nitro-*o*-anisidine.†

pid Fast Red GG. A dyestuff. It consists of the anilide of β-hydroxynaphthoic acid mixed with *p*-nitro aniline. Used in calico printing.†

pid Fast Red GL. A dyestuff. It consists of the anilide of β-hydroxynaphthoic acid mixed with *o*-nitro-*p*- toluidine. Used in calico printing.†

pid Fast Red 3GL. A dyestuff. It is a mixture of the aniline of β-hydroxy naphthoic acid and *p*-chloro-*o*-nitro aniline. Used in calico printing.†

-Pid-Gro. Plant food. (Chevron).*

pidogen. Dyestuffs. Applications: Used for dyeing cotton by the warp sizing process. (Bayer & Co).*

pidogen and Rapidogen N. For prints on vegetable fibres and on regenerated cellulose. (Bayer UK Ltd).

pidosept. A hand disinfectant. (Bayer & Co).*

pifen. A proprietary preparation containing alfentanil hydrochloride. Applications: Analgesic supplement in anaesthesia. (Janssen Pharmaceutical Ltd).*

pitard. A proprietary preparation of beef insulin crystals in a rapid acting solution of neutral pork insulin. (British Drug Houses).†

schit. A preservative for latex. It is *p*-chloro-*m*-cresol.†

sorite. Kernite (hydrated sodium borate), and in general boron ores and products. (Borax Consolidated Ltd).*

Rastinon. A proprietary preparation of tolbutamide. An oral hypoglycaemic agent. (Hoechst Chemicals (UK) Ltd).†

Rat Flip. Trade name for a line of rat and mouse bait products for both indoor and outdoor use. Applications: Anticoagulant active ingredients are safer around domestic animals and pets. Vitamin K is antidote. (Colonial Products Inc).*

Ratak. Rodenticide containing difenacoum. (ICI PLC).*

Ratox. Warfarin-based rodenticide. (The Wellcome Foundation Ltd).

Raudixin. A proprietary preparation of rauwolfia. An antihypertensive. (Squibb, E R & Sons Ltd).*

Rauracienne. See FAST RED.

Rau-Sed. Reserpine. Applications: Antihypertensive. (E R Squibb & Sons Inc).

Rauserfia. Rauwolfia Serpentina. Applications: Antihypertensive. (Penick Corp).

Rautrax Sine K. A proprietary preparation of rauwolfia and hydroflumethiazide used as an anti-hypertensive agent. (Squibb, E R & Sons Ltd).*

Rautrax Tablets. Rauwofia serpentina whole root, hydroflumethiazide and potassium chloride in tablet form. (E R Squibb & Sons Ltd).

Rauverid. Rauwolfia Serpentina. Applications: Antihypertensive. (O'Neal, Jones & Feldman Pharmaceuticals).

Rauwiloid. Each tablet contains selected alkaloid hydrochlorides of rauwolfia serpentina 2 mg (Riker Laboratories/3M Health Care).†

Rauxite. Proprietary trade name for urea-formaldehyde varnish and lacquer resins.†

Rauxone. A proprietary trade name for alkyd varnish and lacquer resins.†

Rauzene. A proprietary trade name for a phenolic varnish and lacquer resin.†

Rauzene Ester. A proprietary trade name for an ester gum.†

Ravocaine Hydrochloride. Propoxycaine Hydrochloride. Applications:

Anaesthetic. (Cook-Waite Laboratories Inc).

Ravoien. Plasticizer and rubber extender. (Burmah-Castrol Ltd).*

Ravolen 11(T). A proprietary trade name for a decolourized petroleum aromatic extract. (Burmah-Castrol Ltd).*

Raw Palmira Root Flour. See TALIPOT.

Raw Sienna. A yellow pigment. It is a native ferruginous earth.†

Raw Turkey Umber. See UMBER.

Raw Umber. See UMBER.

Rawstol. (Radarsan). Vermicides. They consist of solutions of fluosilicic acid.†

Raybar. A proprietary preparation of barium sulphate. (Fleet CB Co Inc, Lynchburg, Va).†

Rayo. An electrical resistance alloy consisting of 85 per cent nickel and 15 per cent chromium.†

Rayon. Artificial silk.†

Rayox. A proprietary trade name for titanium dioxide, TiO_2.†

Razoxin. Anti-cancer preparation containing razoxane. (ICI PLC).*

RBC. A proprietary preparation of phenylznercuric nitrate, iso-butyl paraamino benzoate, N-butyl para-amino benzoate, benzocaine, cholesterol and calaraine. Antipruritic. (Rybar).†

RD10. Roadway dust suppression agent. (Foseco (FS) Ltd).

RE-VAC II. Reovirus disease, inactivated vaccine. Applications: immunization of poultry. (Intervet America Inc).*

Reacrone. Acrylic resins. Applications: Paints, inks and metal decorating. (Resinas Sinteticas SA).*

Reactal. Alkyds. Applications: Paints, lacquers and varnishes, inks and metal decorating. (Resinas Sinteticas SA).*

Reactivan. A proprietary preparation of FENCAMFAMIN hydrochloride, thiamine, PYRIDOXINE, CYANOCOBALAMIN and ascorbic acid. A tonic. (E Merck, Darmstadt).†

Reacton. Rare earth metals and compounds. Applications: Electronic devices, phosphors, magnetic materials. (Johnson Matthey Chemicals Ltd).*

Reafor. Etherified amino resins. Applications: Paints, lacquers and varnishes, inks and metal decorating. (Resinas Sinteticas SA).*

Reafree. Saturated polyester. Applicatio Paints, inks and metal decorating. (Resinas Sinteticas SA).*

Realgar. (Ruby Sulphur, Red Orpiment, Red Arsenic, Red Arsenic Glass, Rub Arsenic, Arsenic Orange). Arsenic disulphide, As_2S_2. Realgarite. Synony for Realgar.†

Reamul. Aqueous dispersions (polyvinyl acetate, styrene. acrylic etc.). Applications: Paints and inks. (Resina Sinteticas SA).*

Réamur's Alloy. An alloy of 70 per cent antimony and 30 per cent iron.†

Reatane. Polyisocianates. Applications: Lacquers and varnishes. (Resinas Sinteticas SA).*

Reater. Modified polyester resins. Applications: Lacquer and varnishes. (Resinas Sinteticas SA).*

Reaumerite. A compound, $(Ca.Na_2)O.3SiO_2$, obtained by heating glass at its softening temperature.†

Réboulet's Solution. An aqueous solution calcium chloride, potassium nitrate, a alum. Used to preserve anatomical specimens.†

Recoil. A wettable powder containing 10 w/w oxadixyl and 56% w/w mancozet Applications: To control foliar and tu blight in potatoes. (Bayer & Co).*

Recoura's Sulphate. A chromium hexahydrated sulphate, $Cr_2(SO_4)_3.6H_2O$.†

Recovered Grease. The oil used to lubrica wool during spinning is recovered from the wash-water. It is used to manufacture a low-grade stearin.†

ReCovr (veterinary). Tripelennamine Hydrochloride. Applications: Antihistaminic. (E R Squibb & Sons Inc).

Recresal. A proprietary preparation of ac sodium phosphate in tablets.†

Rectified Spirit S.V.R. A specially rectifie ethyl alcohol 68-69 over proof, containing 96-97 per cent ethyl alcoho by volume. It is used in perfumery and in pharmaceutical extracts and tinctures.†

Recupex. Flux for recovering non-ferrous scrap. (Foseco (FS) Ltd).*

Red Acid. Nitric acid of 40° Bé, or stronger. It contains dissolved nitrogen oxides.†

Red Algar. Arsenic disulphide, As_2S_2.†

Redalloy. A proprietary trade name for brass containing 85 per cent copper, 14 per cent zinc, and 1 per cent tin.†

Red Antimony. (Antimony Blende, Pyrantimonite, Pyrostibnite, Antimony Cinnabar). A mineral. It is an oxysulphide of antimony, Sb_2O_3. $2Sb_2S$. Antimony cinnabar, or red antimony, is also obtained by treating antimony chloride with sodium thiosulphate in aqueous solution. Used as a pigment to replace ordinary cinnabar.†

Red Argol. See ARGOL.

Red Arsenic. See REALGAR.

Red Arsenic Glass. See REALGAR.

Red B. See SUDAN II.

Red B Oil-soiuble Extra Conc. See OIL SCARLET.

Red Bole. See INDIAN RED.

Red Brass. A brass containing 90 per cent copper and 10 per cent zinc. Also Tombac (*qv*), which has been pickled in acid.†

Red C. See SUDAN III.

Red Chalk. See INDIAN RED.

Red Charcoal. A wood charcoal made at low temperature. It contains hydrogen and oxygen.†

Red Chromate of Potash. Potassium bichromate, $K_2Cr_2O_7$.†

Red Chrome,. See CHROME RED.

Red Cobalt. A mineral. It is erythrite.†

Red Copper. See VIOLET COPPER.

Red Coralline. See PAEONINE.

Red Crocus. Ferric oxide, Fe_2O_3.†

Redd Citrus Specialties. A range of essential citrus oils, natural citrus aroma and natural citrus specialty products. Applications: Wide range of applications in the food and beverage industries, personal care products, cosmetics, soaps, detergents and household products. (Hercules Inc).*

Red Developer. β-Naphthol, used for developing red on fibre which has been treated with Primuline yellow.†

Reddingite. A mineral, 3(Mn.Fe)O. P_2O_5.†

Reddle. See INDIAN RED.

Red Dot. A smokeless powder. Applications: Designed for light and standard shotshell loads of all gauges. It can also be used in specific handgun cartridges. (Hercules Inc).*

Red Drops. (Red Lavender). Compound tincture of lavender.†

Red Earth. See INDIAN RED.

Redeptin Injection. Aqueous suspension of fluspirilene. Applications: Long acting intramuscular injection for treatment of schizophrenia. (Smith Kline and French Laboratories Ltd).*

Redeptin. A proprietary preparation of fluspiriline used in the treatment of schizophrenia. (Smith Kline and French Laboratories Ltd).*

Red Fibre. See VULCANISED FIBRE.

Red Gold. A jeweller's alloy containing 75 per cent gold and 25 per cent copper.†

Red Gum. See EUCALYPTUS GUM.

Red Hermetite. Red paste, semi hardening, used as gasket jointing compound. Applications: To supplement all gaskets flanged or threaded applications. Ensures leak free joints in most environments. (Hermetite Products Ltd).*

Rediclear. Hydrosulphite/dispersing agent blend. (RV Chemicals Ltd).

Redicote. Specialised cationic bitumen emulsifiers. (Armour Hess Chemicals).†

Red Indigo. See FRENCH PURPLE.

Red Iodide of Mercury. Mercuric iodide, HgI_2.†

Redisol. Cyanocobalamin. Applications: Vitamin. (Merck Sharp & Dohme, Div of Merck & Co Inc).

Red Lavender. See RED DROPS.

Red Lavender Spirit. See SPIRIT OF RED LAVENDER.

Red Lead. (Red Lead Oxide, Minium, Paris Red, Saturn Red). A pigment. It is oxide of lead, Pb_3O_4, made by heating litharge, PbO. There are several kinds on the market distinguished by their colour and amount of lead dioxide they contain. This varies between 18 and 34 per cent. A good quality for a paint contains 25 per cent lead dioxide.†

Red Lead Oxide. See RED LEAD.

Red Liquor. (Mordant Rouge). A solution corresponding to the formula, $Al_2(C_2H_3O_2)_6$, which appears to consist of a diacetate of aluminium, and acetic acid. Red liquor is largely used in dyeing and calico printing, especially for the production of red colours, for the manufacture of dense lakes, and for waterproofing woollen fabrics.†

Redmanol. See BAKELITE.

Red Metal. A term usually applied to an alloy of 90 per cent copper, and 10 per cent zinc.†

Red Mordant. See RED LIQUOR.

Red Nickel Ore. A mineral. It is niccolite or nickeline.†

Redo. See HYDROSULPHITE.

Red Ochre. See INDIAN RED.

Red Oil. Acid-treated distillates or residual oils of petroleum which are finished by a soda wash or by a clay treatment after the acid tar produced by sulphuric acid treatment has been allowed to settle. These oils are used for purposes of general lubrication. The term is also used for olive oil containing the red colouring matter of alkanet and for commercial oleic acid. Also see TURKEY RED OILS.†

Red Orpiment. See REALGAR.

Red Oxide. See INDIAN RED.

Red Oxide of Chromium. Chromium trioxide, CrO_3.†

Red Oxide of Lead. See RED LEAD.

Red Oxide of Mercury. (Red Precipitate). Mercuric oxide, HgO.†

Redoxon. Vitamin C preparations. (Roche Products Ltd).

Red Precipitate. See RED OXIDE OF MERCURY.

Red Prussiate of Potash. Potassium ferricyanide, $K_3Fe(CN)_6$.†

Redray. An electrical resistance alloy containing 85 per cent nickel and 15 per cent chromium.†

Red Rudd. See INDIAN RED.

Red Salts. Both crude sodium acetate and crude sodium carbonate, coloured red by ferric oxide, are known as red salts.

Red Saunderswood. See REDWOODS.

Red Soda. A solution of red ink containing a little gum arabic and sodium carbonate. It is used as a marking ink for " blue prints.'†

Red Star Powder. An explosive. It is a 33-grain smokeless powder containing metallic nitrates, nitro-hydrocarbons, and petroleum jelly.†

Red Storax. (Solid Storax). An artificial product obtained by mixing poor storax with sawdust, and pressing the mixture. Used for fumigating candles and powders.†

Reduced Turpentine. A mixture of turpentine oil with petroleum.†

Reducin. Triamino-resorcinol, $C_6H(NH_2)_3(OH)_2$. Used in photography as a developer.†

Redul. An oral hypoglycaemic agent. (Bayer & Co).*

Red Ultramarine. Obtained from blue ultramarine by the action of dry hydrochloric acid gas and oxygen, at 150-180° C. See ULTRAMARINE.†

Redurit. Electrically fused standard grade corundum. Applications: Production of abrasives, abrasive paper, discs and cloth. (Dynamit Nobel Wien GmbH).*

Reduxol Z. Soluble zinc formaldehyde sulphoxylate. (RV Chemicals Ltd).

Red Violet 4RS. (Acid Violet 4RS). A dyestuff. It is the sodium salt of dimethylrosanilinetrisulphonic acid, $C_{22}H_{22}N_3O_{10}S_3Na_3$. Dyes wool from an acid bath.†

Red Violet 5R Extra. See HOFMANN'S VIOLET.

Red Violet 5RS. (Acid Violet 5RS). A dyestuff. It is the sodium salt of ethyl rosanilinesulphonic acid, $C_{22}H_{22}N_3O_2S_3Na_3$. Dyes wool bluish-red from an acid bath.†

Red Vitriol. (Botryogen). A native ferroso-ferric sulphate from Sweden.†

Red Wash. A zinc sulphate solution containing red colouring matter.†

Red Water Bark. Sassy bark.†

ed Yacca Gum. See ACAROID BALSAM.

edwoods. (Red Dye Woods). These dye woods are divided into two classes: (1) Soluble, which comprise Brazil, Pernambuco or Fernambuco wood, Peach wood, Lima wood, Sapan wood, Bimas redwood, and Nicaragua wood. All of them contain the colouring principle brazilin, $C_{16}H_{14}O_5$, which, by oxidation, is converted into brazilein, $C_{16}H_{12}O_5$, which gives purple shades with chrome mordants, and crimson with alum.

2) Insoluble redwoods, consisting of Camwood or Cambe wood, Barwood, Saunderswood, Santalwood. Sandle wood or Sandelwood, Bresille wood, and Caliatur wood. The dyeing principle is santaline. These woods have a limited application for dyeing wool with alumina, chrome, tin, or iron mordants.†

eed Brass. See BRASS.

eese's Alloy. An alloy used in dental work. It contains 87 per cent tin, 8.6 per cent silver, and 4.4 per cent gold.†

ees's Thionin Stain. A microscopic stain. It consists of 1.5 grams thionin and 10 cc alcohol in 100 cc of 5 per cent solution of carbolic acid. It is used at the rate of 5 cc in 20 cc water.†

efagan. A proprietary preparation of salicylamide, phenacetin, caffeine and mebhydrolin napadisylate. (FBA Pharmaceuticals).†

efikite. A resin found in lignite.†

efined Silver. A silver usually contain ing from 99.7-99.9 per cent metal.†

efinex. Powdered attapulgite clay. Applications: Oil refining. (Hermadex Ltd).*

eflectafoam. Closed cell polyethylene backing with highly reflective aluminised polyester surface. Applications: Preventing heat loss behind radiators and insulating airing cupboards. (Piccadilly Products Ltd).*

Reflite. A proprietary synthetic resin moulding powder.†

Reflorit. Picric acid. It has been tried for the disinfection of seed-corn.†

Reform Phosphate. Rock phosphate which has been treated with small quantities of dilute acid to render it more porous, converting calcium carbonate into calcium hydrogen carbonate.†

Refrax. Bricks made from recrystallised silicon carbide. A refractory material.†

Regal Crown. Combination of growth stimulators to enhance plant growth by accelerating root growth. (Regal Chemical Company).*

Regalox. A sintered material comprising 88 per cent alumina.†

Regalstar. Herbicide to control crabgrass, goosegrass and other annual weeds. Applications: Applied in early spring to turfgrass and cultivated nursery fields prior to weed seed germination. (Regal Chemical Company).*

Regamycin. Porfiromycin. Applications: Antibacterial; antineoplastic. (The Upjohn Co).

Regenerated Turpentine. A product of synthetic camphor manufacture. It boils at 170° C.†

Regenex. Copper and nickel alloy flux. (Foseco (FS) Ltd).*

Regina Purple. (Regina Violet, Violet Imperial Rouge, Violet Phenylique, Phenyl Violet, Imperial Violet). A dyestuff. It is the acetate of *o*-tolyl-*p*-rosaniline, $C_{28}H_{27}N_3O_4$. Dyes wool reddish-violet.†

Regina Spirit Purple. See SPIRIT PURPLE.

Regina Violet. See REGINA PURPLE.

Regina Violet, Water Solubie. A dyestuff. It is the trisulphonic acid of Regina purple. Dyes wool reddish violet from an acid bath.†

Reginal. Proofing agent. (Ciba-Geigy PLC).

Regitine Mesylate. Phentolamine Mesylate. Applications: Anti- adrenergic. (Ciba-Geigy Corp).

Reglan. Metoclopramide hydrochloride. Applications: Anti-emetic. (A H Robins Co Inc).

Reglone. Desiccant and herbicide containing diquat. (ICI PLC).*

Reglox. Diquat herbicide. (ICI PLC, Plant Protection Div).

Regnis. Machine lubricant for glass forming machinery. (Specialty Products Co).*

Regonol. Pyridostigmine Bromide. Applications: Cholinergic. (Organon Inc).

Regu-mate. Altrenogest. Applications: Progestin. (Roussel-UCLAF).

Regulex. Gibberellic growth regulator. (ICI PLC).*

Regulus Metal. This is usually a 5-12 per cent antimony with lead.†

Regulus of Antimony. Produced by heating antimony ore, Sb_2S_3. It contains about 10 per cent of iron.†

Regulus of Venus. An alloy of copper and antimony, $SbCu_2$.†

Regutol. Docusate sodium. Applications: Stool softener; pharmaceutic aid. (Schering-Plough Corp).

Rehydrol. Aluminium chlorohydrex: co-ordination complex of basic aluminium chloride and propylene glycol or polyethylene glycol. Applications: Astringent. (Reheis Chemical Co).

Reich's Bronze. An aluminium bronze containing 85.2 per cent copper, 7.52 per cent iron, 6.6 per cent aluminium, 0.5 per cent manganese, and 0.15 per cent lead.†

Reicolit. A proprietary insulation.†

Reinecke's Salt. A metal chromammine, $Cr(NH_3)_2.(CNS)_4.NH_4$, produced when ammonium cyanate is melted and ammonium bichromate added.†

Reiset's First Base. Plato-diamine hydroxide, $Pt(NH_3.NH_3.OH)_2$.†

Reiset's First Chloride. Plato-diamine-chloride, $Pt(NH_3.NH_3.Cl)_2.H_2O$.†

Reith Alloy. An alloy of 75 per cent copper, 10 per cent tin, 10 per cent lead, and 5 per cent antimony.†

Rela. Carisoprodol. Applications: Relaxant. (Schering-Plough Corp).

Reldan 50. Insecticide. (Murphy Chemical Ltd).

Releasil. Silicone release agents. (Dow Corning Ltd).

Reloder 7. A smokeless powder. Applications: It is designed for rifle loads and 'benchrest' type reloads. (Hercule Inc).*

Relugan GT. An aldehyde tanning agent. is an aqueous solution of glutaraldehyde.†

Remafin. Pigment masterbatches for plastics. (Hoechst UK Ltd).

Remarcol. Sodium fluoride.†

Remazol. Reactive dyestuffs. (Hoechst U Ltd).

Remex. Desulphurizing flux for injection into steel melted in electric arc furnac (Foseco (FS) Ltd).*

Remiderm. A proprietary preparation triamcinolone acetonide and halquinol used in dermatology as an antibacteria agent. (Squibb, E R & Sons Ltd).*

Remiderm Cream. Triamcinolone acetonid and halquinol in cream base. (E R Squibb & Sons Ltd).

Remiderm Ointment. Triamcinolone acetonide and halquinol in ointment base.(E R Squibb & Sons Ltd).

Remiderm Spray. Triamcinolone acetonide and halquinol in aerosol spray. (E R Squibb & Sons Ltd).

Remnos. A proprietary preparation of NITRAZEPAM. A hypnotic. (DDSA).†

Remotic. A proprietary preparation of halquinol and triamcinolone acetonide used as a local antiinfective agent. (Squibb, E R & Sons Ltd).*

Remotic Capsules. Triamcinolone acetonide and halquinol in castor oil. (E R Squibb & Sons Ltd).

Remsed. Promethazine Hydrochloride. Applications: Anti-emetic; antihistaminic. (Endo Laboratories Inc, Subsidiary of E I du Pont de Nemours & Co).

Remtal SC. Selective herbicide. (FBC Ltd

Renacit 4. Non-staining antioxidant. (Bayer UK Ltd).

Renaglandin. An extract of the suprarenal gland.†

Renaleptine. Synthetic adrenalin.†

Renarcol. Tribromethyl alcohol.†

Renault Alloy. An aluminium alloy containing 88 per cent aluminium, 10 per cent zinc, and 2 per cent copper.†

Rencal. Phytate Sodium. Applications: Chelating agent. (E R Squibb & Sons Inc).

Rendells. A proprietary preparation of nonoxinol used in the form of contraceptive pessaries. (Rendell).*

Rendrock. An explosive. It is a modification of Lithofracteur, consisting of 40 per cent potassium nitrate, 40 per cent nitroglycerin, 13 per cent wood pulp, and 7.0 per cent paraffin or pitch.†

Renektan. Leather dyes. (ICI PLC).*

Renese. A proprietary preparation of polythiazide. A diuretic. (Pfizer International).*

Renex. Polyexyethylene alkyl or alkyl acryl esters. (ICI PLC).*

Rengasil. Pirprofen. Applications: Anti-inflammatory. (Ciba-Geigy Corp).

Rennet. A clotting enzyme, which coagulates milk by precipitating the casein. It is the aqueous or alcoholic infusion of the dried stomach of the calf.†

Rennet, Milk Sugar. See PEGNIN.

Rennilase. A milk clotting enzyme produced by a selected non-pathogenic strain of the fungus Mucor miehei. Applications: Used in cheese-making for coagulation as an alternative to calf rennet. (Novo Industri A/S).*

Rennin. A solid form of rennet.†

Reno-M. Diatrizoate meglumine. Applications: Diagnostic aid. (E R Squibb & Sons Inc).

Renoquid. Sulphacytine. Applications: Antibacterial. (Parke-Davis, Div of Warner-Lambert Co).

Renovue-DIP. Iodamide meglumine. Applications: Diagnostic aid. (E R Squibb & Sons Inc).

Renovue-65. Iodamide meglumine. Applications: Diagnostic aid. (E R Squibb & Sons Inc).

Renyx. A proprietary trade name for an alloy of aluminium with nickel, copper, and silicon.†

Reochlor (LF and 54). A trade mark for an extender-plasticizer for PVC. They are chlorinated paraffins. (Ciba Geigy PLC).†

Reoflam. 20, 40 and 60. A range of proprietary plasticizers used with PVC. (Ciba Geigy PLC).†

Reofos. A trade name for a range of synthetic organic phosphates. (Ciba-Geigy PLC).†

Reogen. See PLASTOGEN - more active ingredients. Applications: Effective in all elastomers. (Vanderbilt Chemical Corporation).*

Reolube. A trade name for a range of synthetic organic phosphates. (Ciba-Geigy PLC).†

Reolube FAD. A proprictary trade name for a long chain fatty acid mixture, the principal components being C_{14},C_{16}, and C_{18} acids. (Ciba Geigy PLC).†

Reomet. Metal treatment additives. (Ciba-Geigy PLC).

Reomol. Plasticizers. (Ciba-Geigy PLC).

Reomol. A trade mark (17) for vinyl plasticizers as follows: Reomol DBS dibutyl sebacate. DCP dicapryl phthalate. DOS di-2-ethyl-hexyl sebacate. D79S a mixture of heptyl and nonyl sebacates.†

Reomol BCF. A proprietary plasticizer. Butyl carbinol formal. (Ciba Geigy PLC).†

Reomol P. A proprietary trade name for dimethoxy ethyl phthalate. A chemical bonding agent for cellulose acetate staple fibre. (Ciba Geigy PLC).†

Reomol PBPS. A sebacic acid polyester and a small proportion of non polymeric ester. (Ciba Geigy PLC).†

Reomol TC9. A proprietary trade name for a chemical bonding agent for Terylene and cellulose triacetate fibres.†

Reomol 4PG. A proprietary plasticizer. Butyl phthalyl butyl glycollate. (Ciba Geigy PLC).†

Reoplast. Epoxy plasticizers/stabilizers. (Ciba-Geigy PLC).

Reoplex. Polyester plasticizers. (Ciba-Geigy PLC).

Reoplex 200, 220, 300. A trade mark. Vinyl plasticizers of the polyester type. (Ciba Geigy PLC).†

Reoplex 901. A trade mark for a plasticizer for PVC sheeting intended for manufacture of surgical and electrical tapes. (Ciba Geigy PLC).†

Reoplex 902. A plasticizer with good resistance to extraction by petroleum. (Ciba Geigy PLC).†

Reostene. A nickel-iron alloy.†

Repelit. A proprietary synthetic resin varnish-paper product used for electrical insulation.†

Repello DC. Resin-wax blend. Applications: Fabric water repellent. (Scher Chemicals Inc).*

Replicast CS. Patented process for making castings in which an expanded polystyrene pattern is coated with ceramic material then the pattern is burnt out. (Foseco (FS) Ltd).*

Replicast FM. Patented process for making castings using expanded polystyrene patterns in unbonded sand compacted under vacuum. (Foseco (FS) Ltd).*

Resacetophenone. 2, 4-Dihydroxy-acetophenone, $C_6H_3(OH)_2.CO.CH_3$.†

Resad. Polymer emulsions, polyvinyl acetate homopolymers and copolymers, acrylic and styrene acrylic polymers. Supplied in drums or bulk tankers. Applications: Used in adhesives, textile treatments and surface coatings. (Resadhesion Ltd).*

Resamine. A proprietary trade name for formaldehyde resins. (Chemische Werke, Albert).†

Resan. See BAKELITE.

Resarit. Acrylic moulding compound. Applications: Double and triple-walled sheets, rear lights, automotive parts, instrument covers, lampshades, condensor lenses, casings, covers for measuring instruments etc. (Resart-IHM AG).*

Resarix SF. Scratch resistant coating system for surface treatment. Applications: Optical industry (lenses, magnifying glasses, lenses for sunglasses, scales) and for head protection (visors for crash helemts and astronaut helmets). (Resart-IHM AG).*

Resart. A range of melamine moulding compounds. Applications: Moulded parts with tracking resistance, moulded parts with high-grade dimensional stability and electrical components such as switches and relays. (Resart-IHM AG).*

Resartglas GS. Cast acrylic sheets LDII. Applications: Roof windows, light domes, windscreens, caravan windows, displays, advertising gifts, furniture, for solarium equipment and solar beds - transparent to UVA light. (Resart-IHM AG).*

Resartherm. Glass fibre reinforced polyester. Applications: Electrical engineering household appliances, electric tools, car ignition systems. (Resart-IHM AG).*

Resart-PMMA XT. Standard extruded acrylic sheets 500 high impact. Applications: Advertising aids (letters, displays, advertising transparencies), engineering components (housings, machine covers etc.), lighting fittings (cover for long-field light fittings, exterior fittings etc.), roof hoods for caravans, drawing instruments, dome lights, door glazing, roof lights and caravan roof vents. (Resart-IHM AG).*

Resazoin. See DIAZORESORCIN.

Rescon. Disposable devices used for obtaining samples of molten steel. (Foseco (FS) Ltd).*

Resectisol. Mannitol. Applications: Diagnostic aid; diuretic. (American McGaw, American Hospital Supply Corp).

Reserpine. An alkaloid obtained from Rauwolfia serpentina Serpasil.†

Reserpoid. Reserpine. Applications: Antihypertensive. (The Upjohn Co).

Resibon. Phenolic resins. Applications: Metal decorating, adhesives, abrasives, thermal insulation refractories, interior can coatings. (Resinas Sinteticas SA).*

Resicart. Wet strength resins. (Ciba-Geigy PLC).

Residuren. Weedicide, containing chlorpropham. (ICI PLC).*

Resilia. A proprietary trade name for a silico-manganese spring steel.†

Resilita. Polyamide resins. Applications: Inks, paints and adhesives. (Resinas Sinteticas SA).*

Resilla. A proprietary trade name for a special silicon-manganese spring steel.†

Resilon. Silicon carbide. Applications: Used in the electrical industry. (Lonza Limited).*

Resimene. Melamine-formaldehyde resin. Applications: Used as binders in adhesives, moulded products and foundry cores, paint coating ingredients, printing ink product. (Monsanto Co).*

Resin. See COLOPHONY.

Resin Blende. Zinc blende, ZnS, of a yellow colour, is sometimes called by this name.†

Resin Essence. See ROSIN SPIRIT.

Resin Ether L. A proprietary synthetic resin for use as a cellulose-lacquer plasticizer. It is stated to be non drying, not susceptible to atmospheric oxidation, and to have a low acid value.†

Resin Lutea. (Acaroid Balsam). A name applied to yellow acaroid balsam, a yellow resin obtained from *Xanthorrhoea Hastile*.†

Resin M. See PERSPEX.

Resin M.S.2. Cyclohexanone condensation products. (Laporte Industries Ltd).*

Resin Oil. That fraction of the distillation of resin (colophony), which distils over from 300-400° C. It consists principally of terpineol, $C_{10}H_7OH$. Used as a lubricant.†

Resin Release N. Mould release agent for fibreglass reinforced hand layup moulding or casting. (Specialty Products Co).*

Resin Spirit. See ROSIN SPIRIT.

Resin WP. A proprietary trade name for a melamine based thermosetting resin used for crease-resisting finishes. (Ciba Geigy PLC).†

Resinette. A synthetic resin obtained from phenol and formaldehyde.†

Resinite. See BAKELITE.

Resinol. A varnish substitute obtained by the dehydrogenation of petroleum, distillation, and polymerization.†

Resinous Silica. A variety of hydrated silica or opal.†

Resinox. A proprietary synthetic resin moulding powder of the phenol-formaldehyde type.†

Resins, Acrolein. Resins obtained by the polymerization of acrolein by means of inorganic and organic bases or salts of iron and lead. Orea is a trade name for a resin of this type. Acrolein also condenses with phenols to form resins.†

Resins, Alkyd. Resins resulting from the interaction of a polyhydric alcohol and a polybasic acid. See BECKOSOL, ALKYDAL, and REZYL.†

Resin, Benzo-furane. See PARACOUMARONE RESIN.

Resin, Coumarone. See PARACOUMARONE RESIN.

Resin, Cumar. See PARACOUMARONE RESIN.

Resin, Cypress Pine. See PINE GUM.

Resin, Saliretin. See SALIRETINS.

Resin, Spiller's. See SPILLER'S RESIN.

Resin, Sulphate. See LIQUID RESINS.

Resin, Urea. See UREA and THIOUREA RESINS.

Resin, West Indian Anime. See TACAMAHAC RESIN.

Resin, White Pine. See PINE GUM.

Resiosol. See ANUSOL.

Resipol. Insaturated polyester. Applications: Paints, lacquers and varnishes, reinforced plastics. (Resinas Sinteticas SA).*

Resipol DL. A proprietary trade name for a plasticizer. It is glyceryl dilactate.†

Resipol ML. A proprietary trade name for a plasticizer. It is glyceryl monolactate.†

Resiren. Sublimable disperse dyestuffs. Applications: For heat transfer printing preferably on PES and other synthetics. (Bayer & Co).*

Resisco. A proprietary trade name for an alloy of 91 per cent copper, 7 per cent aluminium, and 2 per cent nickel.†

Resista. A glass similar in composition to Pyrex glass. It contains 70 per cent silica and 13.5 per cent boric oxide.†

Resista Steel. An alloy of iron, nickel, and manganese, which is ductile at low temperatures.†

Resistac. A proprietary trade name for an alloy of copper with 9 per cent aluminium and 1 per cent iron.†

Resistal. A heat-resisting alloy containing 63.5 per cent iron, 16.6 per cent nickel, 15 per cent chromium, 4.5 per cent silicon, and 0.3 per cent carbon.†

Resistance Bronze. A term for an alloy of from 84-86.5 per cent copper, 11.5-13.5 per cent manganese, and 2 per cent iron.†

Resistherm. Raw materials used for the formulation of wire enamels with high thermal stability. (Bayer & Co).*

Resistin. An electrical resistance alloy containing 84-86 per cent copper, 2 per cent iron, and 11-13 per cent manganese.†

Resistoflex. A proprietary trade name for polyvinyl alcohol synthetic resins.†

Resistolac. Heat resistant cigarette carton lacquers. (The Scottish Adhesives Co Ltd).*

Resistone. Cationic surfactant, biocide. (ABM Chemicals Ltd).*

Resistone QD. Alkylaryl quaternary ammonium salt in pale yellow aqueous solution. Applications: Bactericide and algaecide, useful in alkaline media, used in static suppression. (ABM Chemicals Ltd).

Resistopen. A penicillin combination especially active against 'problem organisms'. (Bayer & Co).*

Resistox. A proprietary antioxidant for rubber. It is an aldehyde-amine condensation product.†

Resithren. Combinations of disperse and vat dyes. Applications: For the one-bath dyeing of polyester/cellulosic blends. (Bayer & Co).*

Reslin. Synergized bioresmethrin. (The Wellcome Foundation Ltd).

Reslin S. Synergized pyrethroid/S-bioallethrin. (The Wellcome Foundation Ltd).

Resmax. A proprietary preparation of hydroxizine hydrochloride, ephedrine sulphate and theophylline. An antiasthmatic. (Pfizer International).*

Resochin. A range of antirheumatics. (Bayer & Co).*

Resocoton. Mixtures of disperse and reactive dyestuffs. Applications: For printing polyester/cotton blends. (Bayer & Co).*

Resoflavin. A dyestuff obtained by the oxidation of *m*-dihydroxy-benzoic acid in sulphuric acid solution, by means of ammonium persulphate. Dyes wool mordanted with chromium or alumina, yellow.†

Resoglass. See POLYSTYRENE.

Resoglaz. A proprietary trade name for a polymerized styrene.†

Resol. A disinfectant said to be made by saponifying wood tar with caustic potash, and adding wood spirit.†

Resolamin. Dyestuffs. Applications: For the one-bath dyeing of wool and polyester blends. (Bayer & Co).*

Resolin. Disperse dyestuffs. Applications: For polyester fibres. (Bayer & Co).*

Resolin P. A range of disperse dyestuffs. Applications: For the dyeing of polyamide fibres. (Bayer & Co).*

Resonium-A. A proprietary preparation of sodium polystyrene sulphonate, used in hyperkalaemia. (Bayer & Co).*

Resopol. A proprietary polyester laminating resin. (DSM Resines France SA).†

Resorbin. A registered trade mark currently awaiting re-allocation by its proprietors to cover a range of pharmaceuticals. (Casella AG).†

Resorcin. Resorcinol, $C_6H_4(OH)_2$.†

Resorcin Blue. Nitroso Blue. A dyestuff. It is the tannin compound of dimethyl aminophenoxazone, $C_{14}H_{12}N_2O_2$, and is produced on the fibre, giving indigo blue shades.†

Resorcin Blue. See FLUORESCENT BLUE. The name resorcin blue is also used for Lackmoid (*q.v.*).

Resorcin Brown. (Resorcin Brown A Conc., G, R, RBW, Resorcinol Brown). A dyestuff. It is the sodium salt of xyleneazoresorcinazobenzene-*p*-sulphonic acid, $C_{20}H_{17}N_4O_5SNa$. Dyes wool brown from an acid bath.†

Resorcin Brown, A Conc., G, R, RBW. British equivalents of Resorcin brown.†

Resorcin Green. See DINITROSO-RESORCIN.

Resorcin Yellow. See TROPAEOLINE O.

Resorcin Yellow 0 Extra. A British brand of Tropaeoline O.†

Resorcinal. A mixture of equal parts of resorcinol and iodoform. Used as an antiseptic dusting-powder.†

Resorcinol Brown. A dyestuff. It is a British equivalent of Resorcin brown.†

Resorcinol Green. See DINITROSO-RESORCIN

Resorufin. Hydroxyphenazone, $C_{12}H_7NO_3$.†

Resotren. An amoebacide. (Bayer & Co).*

Resovin. A proprietary synthetic resin of the vinyl type.†

Resovyl. Textile auxiliary chemicals. (ICI PLC).*

Respenyl. A proprietary preparation containing guaiaphenesin (guaiacol glyceryl ether). A cough suppressant. (Crookes Laboratories).†

Respiral. Cough drops. (Richardson-Vicks Inc).*

Respirot. Respiratory stimulant veterinary ethical. (Ciba-Geigy PLC).

Respumit. Antifoam. Applications: Dyeing and printing auxilliary. (Bayer & Co).*

Resticel. Expansible polystyrene.†

Restil. SAN copolymers.†

Restiran. ABS copolymers.†

Restirolo. Polystyrenes.†

Restor-E (Restoration Chemical Products). Varies with each product. Applications: Commercial line of products for professional and do-it-yourself restoration of a variety of surfaces (wood, floors, soft goods, odour control, corrosion control) following fire, floods, etc. (Puma Chemical Co Inc).*

Restoration Cleaner. Blended organic and inorganic acids in combination with surfactants and wetting agents. Applications: Heavier duty concentrated cleaner for building exteriors. (Nova Chemical Inc).*

Restoration Cleaner (Heavy Duty). Blended organic and inorganic acids in combination with surfactants and wetting agents. Applications: Heavier duty concentrated cleaner for building exteriors. (Nova Chemical Inc).*

Restoration Cleaner (Super Heavy Duty). Blended organic and inorganic acids in combination with surfactants and wetting agents. Applications: Heavier duty concentrated cleaner for building exteriors. (Nova Chemical Inc).*

Restoration Rinse. Blended organic and inorganic acids in combination with surfactants and wetting agents. Applications: Light duty cleaning for historical building exteriors. (Nova Chemical Inc).*

Restore-X Exterior Paint Remover. Green colour, heavy-bodied liquid, non-flammable, water soluble, sodium hydroxide remover. Applications: Surface preparation tool for removal of deteriorated, exterior paints and heavy-bodied stains. (Restech Industries Inc).*

Restore-X Weathered Wood Renewer. Blue colour, heavy-bodied liquid, non-flammable, water soluble, sodium hydroxide remover. Applications: Removes semi-transparent stain and the grey, weathered look from wood. (Restech Industries Inc).*

Resydrol. Water soluble synthetic resins. Applications: Paints and printing inks. (Resinous Chemicals Ltd).*

Retaminol. Paper auxilliaries. Applications: To increase the filler and pigment yields in paper manufacture, to improve drainage speed and for backwater clarification. (Bayer & Co).*

Retard. Potassium salt of maleic hydrazide. Applications: Growth retardant for trees, shrubs, ivy and grass. (Drexel Chemical Company).*

Retarder ESEN. Surface treated phthalic anhydride. A non-discolouring retarder of vulcanization in all stock with all accelerators at processing temperature with a minimum retarding action at curing tmperatures. (Uniroyal).*

Retargal. Levelling agent. Applications: Dyeing acrylic fibres with basic dyes. (Sandoz Products Ltd).

Retariox. A proprietary trade name for an aldehyde-amine condensation product.†

Retcin. A proprietary preparation of ERYTHROMYCIN. An antibiotic. (DDSA).†

Reten. High molecular weight synthetic water soluble polymers. Applications: Used as thickeners, flocculants, antistatic agents, film formers, adhesives, slip agents, solids-suspending agents and cross-linking agents. (Hercules Inc).*

Retene. Methylisopropylphenanthrene, $C_{18}H_{18}$.†

Retenema. A proprietary enema containing BETAMETHASONE valerate used in the treatment of proctitis. (Glaxo Pharmaceuticals Ltd).†

Reticusol. Soluble reticulin. (Croda Chemicals Ltd).

Retilox. A proprietary range of organic peroxides suitable for polymer crosslinking. (Montedison UK Ltd).*

Retin Asphalt. See RETINITE.

Retin-A. Tretinoin. Applications: Keratolytic. (Ortho Pharmaceutical Corp).

Retinaphtha. See METHYL BENZENE.

Retingan. Range of resin tanning materials. Applications: For the filling retannage of chrome upper leathers, particularly from cattle hides and sheepskins. (Bayer & Co).*

Retingan R6, R7, R4B. Resin tanning materials. (Bayer UK Ltd).

Retinite. (Retin Asphalt, Walchowite). A fossil resin found in brown coal. It occurs in Derbyshire and in Walchow. The material found near Walchow is a polymeric resin made up chiefly of sesquiterpenes.†

Retinol. (Codoil, Rosinol, Rosin Oil). A product obtained by the distillation of rosin. An antiseptic and a vehicle for ointments.†

Retinol. Vitamin A alcohol.†

Retnolite. A proprietary trade name for a phenol-formaldehyde synthetic resin.†

Retz Alloy. An alloy of 75 per cent copper, 10 per cent lead, 10 per cent tin, and 5 per cent antimony.†

Reuniol. See ROSEOL.

Reussinite. A reddish-brown resin found in certain coal deposits.†

Revalon. A proprietary trade name for an alloy of 76 per cent copper, 22 per cent zinc, and 2 per cent aluminium.†

Revatol. Reserving agent. Applications: Printing. (Sandoz Products Ltd).

Revatol S. A proprietary trade name for sodium *m*-nitrobenzenesulphonate.†

Reversible Latex. See REVERTEX.

Revertex. (Reversible Latex). A trade mar for a highly concentrated rubber latex produced by a patented process, in whic the latex is concentrated in the presenc of an alkaline protective colloid.†

Revlen. Esterifilcon A. Butyl methacrylate polymer with butyl acrylate and ethylene dimethacrylate. Applications: Contact lens material. (BioContacts Inc).

Revolite. A proprietary phenol-formaldehyde synthetic resin impregnated cloth.†

Revona. A proprietary trade name for a water-soluble aminoplast. A very effective pitch dispersant in paper making. (Ciba Geigy PLC).†

Revultex. A trade mark for vulcanized rubber latex of any concentration produced from revertex.†

Rewagit. Crystalline aluminium oxide as blasting corundum. Applications: Especially suited for descaling, derusting, roughening of work piece surfaces and blasting of austenitic steels (Dynamit Nobel Wien GmbH).*

Rewo-amid. Deodorant additive, perfume extender additive, versatile additive; foam booster - pearlizing agent, thickening agent, superfatting agent. Applications: Synthetic detergents, soaps, shampoos, bubble baths, cosmetics, toiletries. (REWO Chemicals Ltd).

Rewo-LAN/5. Woolwax alcohol sulphosuccinate in paste form. Applications: Reduces irritation and defatting properties of other surfactants. (REWO Chemicals Ltd).

Rewocor. Anti-corrosive additive. Applications: Soluble cutting oils etc. (REWO Chemicals Ltd).

Rewoderm S 1333. Unsaturated fatty acid alkylolamide sulphosuccinate in liquid form. Applications: Additive for improving dermatological properties of liquid detergents. (REWO Chemicals Ltd).

Rewominoxid. Foam booster; conditioner; fabric softener. Applications: Shampoos and foam baths. (REWO Chemicals Ltd).

Rewophos EAK 8190. Phosphate ester alkyl polyglycol ether in liquid form. Applications: Corrosion inhibitor, emulsifier, antistat for metal treatment agents. (REWO Chemicals Ltd).

Rewophos TD40. Phosphate ester alkyl polyglycol ether in liquid form. Applications: Low foaming surfactant for metal treatment and antistatic applications. (REWO Chemicals Ltd).

Rewophos TD70 and OP80. Phosphate ester alkykl polyglycol ether in liquid form. Applications: Surfactant/ hydrotrope used in metal treatment, antistatic applications and textile auxillary applications. (REWO Chemicals Ltd).

Rewopol B2003. Tetrasodium N-(1,2,dicarboxyethyl) N-octadecyl sulphosuccinamate in liquid form. Applications: Anti-gelling agent and cleaning agent, used in paper making felts; emulsifier and dispersing agent used in emulsion polymerization. (REWO Chemicals Ltd).

Rewopol DLS. Diethanolamine lauryl sulphate in liquid form. Applications: Raw material for shampoos, detergents etc. (REWO Chemicals Ltd).

Rewopol MLS. Monoethanolamine lauryl sulphate in liquid form. Applications: Raw material for shampoos, detergents etc. (REWO Chemicals Ltd).

Rewopol NL3. Sodium lauryl ether sulphate in liquid or gel form. Application: Surfactant used in the formation of shampoos, bubble baths and liquid detergents. (REWO Chemicals Ltd).

Rewopol SB FA/30. Fatty alcohol ether sulphosuccinate in liquid form. Applications: Extremely mild raw material for shampoos, bubble baths and baby products. (REWO Chemicals Ltd).

Rewopol SB IP. Fatty acid alkylolamide sulphosuccinate in liquid form. Applications: Raw material for shampoos, personal hygiene products, surgical scrubs. SB IPE is used as a catalyst for emulsion polymerization. (REWO Chemicals Ltd).

Rewopol SB-DO 70. Dioctyl sulphosuccinate in liquid form. Applications: Anionic surfactant used as a wetting and solubilizing agent. (REWO Chemicals Ltd).

Rewopol SBL 203. Fatty acid alkylolamide sulphosuccinate in liquid or powder form. Applications: Liquid is raw material for dry residue carpet shampoos; powder form is a non-hygroscopic, low- irritating raw material for carpet shampoos, bubble baths etc. (REWO Chemicals Ltd).

Rewopol SLS. Sodium lauryl sulphate in paste form. Applications: Raw material for shampoos, detergents etc. (REWO Chemicals Ltd).

Rewopol SMS. Alkyl disodium sulphosuccinamate as a clear liquid. Applications: Spreading and penetrating agent for latex emulsions. (REWO Chemicals Ltd).

Rewopol TLS. Triethanolamine lauryl sulphate in liquid form. Applications: Raw material for shampoos and bubble baths etc. (REWO Chemicals Ltd).

Rewopol TMS and ODS. Alkyl disodium sulphosuccinamate in liquid or paste form. Applications: Foaming agent for latex emulsions, emulsifier for emulsion polymerization. (REWO Chemicals Ltd).

Rewopon IM-BT. Fatty acid quaternary imidazoline in liquid/paste form. Applications: Cationic surfactant which improves adhesion of bitumens and other binding and coating agents. (REWO Chemicals Ltd).

Rewopon IM-OA. Fatty acid quaternary imidazoline in liquid/paste form. Applications: Rust inhibitor; antistat; emulsifier. (REWO Chemicals Ltd).

Rewoquat CPEM. Coco-penta-ethoxy methyl ammonium metho-sulphate in the form of a viscous liquid. Applications: Hair conditioner for shampoos; emulsifier for polymerization processes; antistatic agent. (REWO Chemicals Ltd).

Rewoquat CR 3099. Di-fatty acid ester dimethyl ammonium metho- sulphate as a viscous liquid. Applications: Softener, dry-cleaning agent, textile auxiliary, leather auxiliary. (REWO Chemicals Ltd).

Rewoquat W7500. Quaternary imidazoline in liquid form. Applications: Cationic surfactant used as a fabric softener. (REWO Chemicals Ltd).

Reworyl ACS. Ammonium cumene sulphonate. Anionic surfactant in liquid form. Applications: Hydrotrope. (REWO Chemicals Ltd).

Reworyl C. Cumene sulphonic acid. Anionic surfactant in liquid form. Applications: Catalyst for synthetic resins; hydrotropes. (REWO Chemicals Ltd).

Reworyl KXS. Potassium xylene sulphonate. Anionic surfactant in liquid form. Applications: Hydrotrope. (REWO Chemicals Ltd).

Reworyl NCS. Sodium cumene sulphonate. Anionic surfactant in liquid form. Applications: Hydrotrope. (REWO Chemicals Ltd).

Reworyl NTS. Sodium toluene sulphonate. Anionic surfactant in liquid form. Applications: Hydrotrope. (REWO Chemicals Ltd).

Reworyl NXS. Sodium xylene sulphonate. Anionic surfactant in liquid form. Applications: Hydrotrope. (REWO Chemicals Ltd).

Reworyl T. Para-Toluene sulphonic acid. Anionic surfactant in liquid form. Applications: Catalyst for foundry resins. (REWO Chemicals Ltd).

Reworyl X. Xylene sulphonic acid. Anionic surfactant in liquid form. Applications: Catalyst for synthetic resins. (REWO Chemicals Ltd).

Rewoteric AM-B13. Alkylamido-betaine as a slightly viscous liquid. Applications: Cosmetic, foam bath and shampoo formulations; foam booster. (REWO Chemicals Ltd).

Rewoteric AM-B13/T. Alkylamido-betaine as a slightly viscous liquid. Applications: Alkaline and acid cleanin[g] agents; car shampoos; metal cleaners. (REWO Chemicals Ltd).

Rewoteric AM-CA. Imidazoline-based ampholyte, in viscous liquid form. Applications: Mild foam bath and hair shampoos, baby toiletries. (REWO Chemicals Ltd).

Rewoteric AM-DML. Dimethyl-lauryl amino-betaine in liquid form. Applications: Baby shampoos; hard surface cleaners; steam-jet cleaners; pickling agents. (REWO Chemicals Ltd).

Rewoteric AM-G30. Imidazoline-based betaine, modified, in liquid form. Applications: Baby and child-care cosmetic formulations. (REWO Chemicals Ltd).

Rewoteric AM-KSF. Imidazoline-based ampholyte, salt-free, as a viscous liquid. Applications: Baby and child-care cosmetic products, personal hygiene toiletries. (REWO Chemicals Ltd).

Rewoteric AM-LL. Linear alkylamido amphoteric in liquid form. Applications: High skin compatible amphoteric surfactant for cosmetics. (REWO Chemicals Ltd).

Rewoteric AM-V. Imidazoline-based ampholyte in the form of a slightly viscous liquid. Applications: Cleaning agent for strongly alkaline and acidic media; corrosion inhibitor; wetting agent for pickling baths. (REWO Chemicals Ltd).

Rewoteric AM-2C. Imidazoline-based ampholyte as a highly viscous liquid. Applications: Mild foam bath and hair shampoos, personal hygiene toiletries. (REWO Chemicals Ltd).

Rewoteric AM-2L. Imidazoline-based ampholyte as a slightly viscous liquid. Applications: Mild foam baths, shampoos, personal hygiene toiletries. (REWO Chemicals Ltd).

Rex. A trade mark for abrasive goods consisting essentially of alumina.†

Rex 95. A proprietary trade name for a cobalt steel containing 5 per cent cobalt 14 per cent tungsten, and 4 per cent chromium, 2 per cent vanadium, and 0.5 per cent molybdenum.†

x-blak. A special preparation of carbon black, containing carbon, rubber, and glue, in varying proportions. Used in rubber mixings.†

xenite. A proprietary trade name for a cellulose acetate butyrate plastic.†

xhide. A rubber-glue stock material for use in rubber mixings.†

xite. A blasting explosive, containing 6.5-8.5 per cent nitro-glycerin, 64-68 per cent ammonium nitrate, 13-16 per cent sodium nitrate, 6.5- 8.5 per cent, trinitro-toluene, and 3-5 per cent wood meal.†

xol. An explosive. It is a mixture of ammonium perchlorate, potassium chlorate, rosin, zinc or aluminium, and mineral oil or wax.†

xoll Black. A sulphide dyestuff.†

xoteric XCE. Carboxyethylated coco amphoteric in liquid form. Applications: Alkaline industrial cleaners. (Jan Dekker BV).

xoteric XCG. Specially designed amphoteric in liquid form. Applications: Used in non-irritating conditioning shampoos, bath and toiletry products when high viscosity is required. (Jan Dekker BV).

exoteric XCO. Carboxymethylated coco imidazoline derivative in liquid form. Applications: High foaming, non irritating shampoos, bath & toiletry products. (Jan Dekker BV).

exoteric XJO. Carboxymethylated caprylic imidazoline derivative in liquid form. Applications: Detergent and wetting agent for low foaming alkaline and acid cleaners. (Jan Dekker BV).

exoteric XOO. Carboxymethylated oleic imidazoline derivative as a viscous liquid. Applications: Imparts softening properties and is used especially in high viscosity hand cleaners and detergent formulations. (Jan Dekker BV).

exoteric YCB. Carboxyethylated coco imidazoline derivative. Applications: High foaming detergent for use in all-purpose cleaners. Viscosity modulator, gel-builder and conditioner for low pH non-irritating shampoos, bath and toiletry products. (Jan Dekker BV).

Rexoteric YCE. Carboxyethylated coco amine derivative. Applications: High foaming light to heavy duty cleaners in acid to alkaline pH range. (Jan Dekker BV).

Rexoteric YJE. Carboxyethylated short chain amine derivative. Applications: Moderately foaming light to heavy duty cleaners, giving good surface tension reduction over the whole pH range. (Jan Dekker BV).

Rexoteric YOB. Carboxyethylated alkyl imidazoline derivative. Applications: Specially designed for use in corrosion inhibitors and as a softener. (Jan Dekker BV).

Rexoteric ZXCO. A blend of Rexoteric XCO and an anionic surfactant, in liquid form. Applications: Ready-made base for non-irritating shampoos, bath and toiletry products. (Jan Dekker BV).

Rextox. A proprietary trade name for a material consisting of copper with a layer of cuprous oxide forned on the surface of the metal at high temperatures.†

Rextrude. A proprietary trade name for a cellulose-acetate-butyrate plastic.†

Reyalite. Aluminium alloys containing lithium. (Reynolds Metal Co).*

Reyalith. Aluminium alloys containing lithium. (Reynolds Metal Co).*

Reycomp. Laminated sheets of aluminium having plastic inner cores. (Reynolds Metal Co).*

Reydox. Aluminium and aluminium alloy castings. (Reynolds Metal Co).*

Reynolds Wrap. Aluminium foil sheets and rolls. Applications: For consumer use in cooking, wrapping etc. (Reynolds Metal Co).*

Reynolon. Plastic film, laminated and unsupported. Applications: Consumer and food use. (Reynolds Metal Co).*

Reynolon. PVC with various plasticizers and stabilizers. Applications: Shrink film for presentation applications. (S Kempner Ltd).*

Rezex. Construction auxiliaries. (Crosfield Chemicals).*

Rezifilm. Thiram. Applications: Antifungal. (E R Squibb & Sons Inc).

Rezistal. Corrosion and heat-resisting steels consisting of iron with up to 0.4 per cent carbon, 1-5.5 per cent silicon, 8-26 per cent chromium, and 7-35 per cent nickel.†

Rezyl. A proprietary trade name for an alkyd synthetic resin.†

R G-Acid. α-Naphtholdisulphonic acid.†

R G-Salt. The sodium salt of R G-acid.†

RH Maneb 80. Protectant fungicide. (Rohm & Haas (UK) Ltd).

Rhamnetin. Quercetin-3-methyl-ether.†

Rhamnine. See PERSIAN BERRIES.

Rhamnodulcite. See ISODULCITE.

Rhaponticin. (Rheic Acid, Rheumin, Rhubarbaric Acid, Rhubarbarin). Chrysophanic acid, $C_{15}H_{10}O_4$, found in rhubarb root.†

Rhaponticum. (Rhapontin, Ponticin). The crystalline substance from the common English rhubarb.†

Rhapontin. See RHAPONTICUM.

Rhatany. The dried root of *Krameria triandra* and *K. argentea*.†

Rhea. See RAMIE.

Rheaform. Clioquinol. Applications: Anti-amebic; anti-infective, topical. (E R Squibb & Sons Inc).

Rheic Acid. See RHAPONTICIN.

Rhemattan. See ATROPHAN.

Rhenania Phosphate. (Vesta Phosphate). Prepared by sintering together in a furnace at 1200-1300° C., a mixture of raw phosphate, limestone, and alkali silicate. The resulting product approximates to the formula $Ca_2KNa(PO_4)_2$. The German phosphate contains 23-31 per cent soluble phosphoric acid and 40 per cent lime, and is made by treating raw phosphate with soda at high temperatures.†

Rhenanit V. A German explosive containing nitroglycerin, ammonium nitrate, vegetable meal, nitro-compounds, and potassium perchlorate.†

Rheniforming. Catalyst. (Chevron).*

Rhenish Dynamite. A solution of 75 per cent nitroglycerin in naphthalene, 2 per cent chalk or barium sulphate, and 23 per cent kieselguhr.†

Rhenoblend. Polymer blends. (Bayer UK Ltd).

Rhenocure. Accelerators for polymers. (Bayer UK Ltd).

Rhenodin. Separating agents. (Bayer UK Ltd).

Rhenodiv. A combination of surface active agents with film forming substances, partly enriched with corrosion inhibitors. Applications: It prevents the sticking together of raw rubber, uncured rubber compounds, blanks and extrudates. (Rhein-Chemie Rheinau).*

Rhenofit. Activators. (Bayer UK Ltd).

Rhenoflex. Chlorinated PVC. Applications: Lacquer industry, adhesives industry, pyrotechnics. (Dynamit Nobel Wien GmbH).*

Rhenogran. A range of polymer-bound rubber chemicals with an activity content of 80%. Applications: Technical moulded and extruded articles, tyres and cable coverings. (Rhein-Chemie Rheinau).*

Rhenomag. A range of magnesium oxide preparations of several qualities. Applications: Acid acceptor and vulcanization activator used in the rubber industry. (Rhein-Chemie Rheinau).*

Rhenomag. Magnesium oxide. (Bayer UK Ltd).

Rhenopor. Blowing agent. (Bayer UK Ltd).

Rhenopren. A polymeric processing promoter for the rubber industry. Applications: Technical moulded and extruded articles. (Rhein-Chemie Rheinau).*

Rhenosin. A range of thermoplastic hydrocarbon resins, used as softening resins and homogenizers in the rubber industry. Applications: For light and dark coloured compounds, e.g., tyres and conveyor belts. (Rhein-Chemie Rheinau).*

Rhenosorb. Very finely divided calcium oxide. It prevents porosity caused by moisture when cured continuously and without pressure. Applications: Rubber compounds of all kinds. (Rhein-Chemie Rheinau).*

Rhenosorb. Dessiccants. (Bayer UK Ltd).

eocin. Anti-settling and thickening agents for paints, varnishes, lubricants, adhesives, coatings, putties and cosmetics. (Süd-Chemie AG).*

eolate. Rheological additives. Applications: Water borne coatings, adhesives and paper coatings. (NL Industries Inc).*

eomacrodex. Dextran 40. A polysaccharide produced by the action of Leuconostoc mesenteroides on sucrose. Average molecular weight: 40,000. Applications: Blood flow adjuvant; plasma volume extender. (Pharmacia Laboratories, Div of Pharmacia Inc).

eonine. A dyestuff. It is the hydrochloride of tetramethyltriamino-phenylacridine, $C_{23}H_{24}N_4$. Dyes tannined cotton and leather brownish yellow.†

eostene. A nickel-iron alloy. It has a specific resistance of 77 microhms per cm.^{3} at 0° C.†

heotan I. An electrical resistance alloy containing 84 per cent copper, 12 per cent manganese, and 4 per cent zinc.†

heotan II. An alloy for electrical resistances. It contains 25 per cent nickel, 52 per cent copper, 5 per cent iron, and 18 per cent zinc. This alloy or a similar one has been called Rheostan.†

heox. Rheological additives. Applications: Paint. (NL Industries Inc).*

hesal. Glyceryl monosalicylate.†

hesonativ. Human immunoglobulin for prophylaxis of rhesus immunization. (KabiVitrum AB).*

heumajecta. A proprietary preparation of sulphurylsulphokinase, cholin- acetylase and catalase, used in the treatment of rheumatic diseases. (Enzypharm Biochemicals).†

heumin. See RHAPONTICIN.

hexite. See REXITE.

hine Metal. An alloy of 97 per cent tin and 3 per cent copper. It has a specific gravity of 7.35 and melts at 300° C.†

hodamine B. (Rhodamine O, Safraniline). A dyestuff. It is the hydrochloride of diethyl-*m*-aminophenol-phthalein, $C_{28}H_{31}Cl.N_2O_3$. Dyes wool and silk bluish-red with fluorescence, also tannined cotton, violet-red.†

Rhodamine G and G Extra. Dyestuffs, consisting chiefly of triethyl-rhodamine, $C_{26}H_{27}Cl.N_2O_3$. Dyes wool, silk, and tannined cotton, red.†

Rhodamine S. A dyestuff. It is the hydrochloride of dimethyl-*m*-amino-phenol-succineine, $C_{20}H_{23}Cl.N_2O_3$. Dyes cotton red, and is used for dyeing half silk goods, and for colouring paper pulp and wool.†

Rhodamine 0. See RHODAMINE B.

Rhodamine 12GM. A dyestuff. It is the ethyl ether of dimethyl-amino-ethoxy-rhodamine, $C_{25}H_{24}Cl.NO_4$. Dyes silk and tannined cotton yellowish-red.†

Rhodamine 3B. (Anisoline). A dyestuff. It is the ethyl ester of tetraethyl rhodamine, $C_{30}H_{35}Cl.N_2O_3$. Dyes wool, silk, and mordanted cotton, bluish-red.†

Rhodamine 6G. (Trianisoline). A dyestuff. It is the ethyl ester of symdiethyl-rhodamine, $C_{26}H_{27}Cl.N_2O_3$. Dyes silk and mordanted cotton red or pink shades.†

Rhodanate. Potassiura thiocyanate. Potassium sulphocyanate.†

Rhodazines. Dyestuffs prepared by the action of phenylhydrazine upon rosilic acid. They have no technical importance.†

Rhodeoretin. Jalapin, the chief constituent of Jalap resin.†

Rhodester. Polyester resins. (Rhone-Poulenc NV/CdF Chimie AZF).*

Rhodia-Phos. Sodium tripolyphosphate. Applications: Detergents , water treatment, paper industry, food industry, textile industry, animal food. (Rhone-Poulenc NV/CdF Chimie AZF).*

Rhodialite. A proprietary cellulose acetate.†

Rhodine 12GF. A dyestuff, $C_{49}H_{38}N_2O_8Cl_2$, obtained by the action of formaldehyde upon the etherified condensation product of dimethylaminobenzoylbenzoic acid and resorcinol. Dyes tannined cotton and silk yellowish-red, and is used for printing on cotton and silk.†

Rhodine 2G. A dyestuff. It is the ethyl ester of dimethylethylrhodamine, $C_{26}H_{27}Cl.N_2O_3$. Dyes silk, wool, and tannined cotton, red.†

Rhodine 3G. See IRISAMINE G.

Rhodinol. A terpene alcohol prepared from the oils of rose, geranium, and citronella. It is practically pure geraniol.†

Rhodione. See VIOLETTON.

Rhodite. See RHODIUM GOLD.

Rhodium Gold. (Rhodite). A native alloy of from 57-66 per cent gold and 34-43 per cent rhodium.†

Rhodizite. A borate of lime, $3CaO.4B_2O_3$, imported from the West Coast of Africa.†

Rhodoid. A proprietary preparation. It is a plastic material made from cellulose acetate.†

Rhodole. Intermediate products between fluoresceine-phthalein and the rhodamines.†

Rhodopas. Polyvinyl acetates. (Rhone-Poulenc NV/CdF Chimie AZF).*

Rhodopas AX. Polyvinyl acetate/polyvinyl chloride copolymers. (Rhone-Poulenc NV/CdF Chimie AZF).*

Rhodopas X. Polyvinyl chlorides. (Rhone-Poulenc NV/CdF Chimie AZF).*

Rhodoviol. Polyvinyl alcohol. (Rhone-Poulenc NV/CdF Chimie AZF).*

Rhoduline Reds G and B. (Rhoduline Violets, Brilliant Rhoduline Red). Alkylated safranines, which dye in a similar way to safranine.†

Rhoduline Violets. See RHODULINE REDS G and B.

Rhometal. A complex nickel-iron alloy having high electrical resistivity and retaining its permeability up to very high frequencies. It is used for television transformers, special radio transformers, HF alternators, etc.†

Rhoplex. A proprietary trade name for an acrylic resin for textile finishes.†

Rhotanium. A series of alloys, consisting mainly of gold (60-90 per cent.) and palladium, and in some cases with a small proportion of rhodium. They are said to be more resistant to hot concentrated sulphuric acid and fused caustic soda than lead.†

Rhovinal B. Polyvinyl butyrals. (Rhone-Poulenc NV/CdF Chimie AZF).*

Rhovinal F. Polyvinyl formals. (Rhone-Poulenc NV/CdF Chimie AZF).*

Rhubarbaric Acid. See RHAPONTICIN.

Rhubarbarin. See RHAPONTICIN.

Rhyolite. A volcanic rock. It usually contains lime and iron.†

Riamat. Thyroxine I 125. Applications: Radioactive agent. (Mallinckrodt Inc).

Riblene. A trade name for polyethylene. (ABCD Petrochimica).†

Riboflavin. Vitamin B_2. Lactoflavin.†

Rice Paper. A paper made from plant pith, particularly in China and Japan.†

Rice Rubber. An elastic cellulose product made from Japanese rice.†

Rice's Bromide Solution. A solution containing 125 parts bromine, 125 parts sodium bromide, and 1,000 parts water. Used in the determination of urea.†

Richardson's Speculum Metal. An alloy of 65.3 per cent copper, 30 per cent tin, 2 per cent silicon, 2 per cent arsenic, and 0.7 per cent zinc.†

Richard's Aluminium Solder. An alloy of 71.5 per cent tin, 25 per cent zinc, and 3.5 per cent aluminium.†

Riché Gas. A gas obtained during the dry distillation of wood. It contains, on an average, about 60 per cent carbon dioxide, 25 per cent carbon monoxide, 15 per cent methane, and a very small quantity of hydrogen.†

Ricin. The name given to the toxic constituents of castor beans.†

Ricinose. Castor oil.†

Ricinus Oil. See OIL OF PALMA CHRISTI.

Ricolite. See BAKELITE.

Ricotti Silk. See GALETTAME SILK.

Ridaura. Auranofin. Applications: Antirheumatic. (Smith Kline & French Laboratories).

Ridoline. Metal treating compositions. (IC PLC).*

Ridomil. Fungicide. (Ciba-Geigy PLC).

Ridosol. Metal treating compositions. (ICI PLC).*

Rifadin. Brand rifampin, an antibiotic used in the treatment of pulmonary tuberculosis. (Dow Chemical Co Ltd).*

Rifamate. A combination product consisting of brand rifampin and isoniazid. Applications: Used in the treatment of tuberculosis. (Dow Chemical Co Ltd).*

Rifamide. An antibiotic. RIFAMYCIN B diethylamide. RIFOCIN-M.†

Rifampicin. An antibiotic. 3-(4-Methyl piperazin - 1 - yliminomethyl)rifarmycin SV. RIFADIN, RIMACTANE.†

Rifamycin. Antibiotics isolated from a Streptomyces mediterranei.†

Rifleite. Explosive. It is a nitrocellulose gelatinized by acetone.†

Rifocin-M. See RIFAMIDE.

Rigidex. Polyolefins and derivatives - all for manufacture of plastics and plastic articles. (BP Chemicals Ltd).*

Rigidex X4RR. High density polyethylene having a density of 0.946. (BP Chemicals Ltd).*

Rigidex 3. High density polyethylene copolymer having a density of 0.946, melt index 0.3. (BP Chemicals Ltd).*

Rigidex 9. High density polyethylene with a density of 0.960 and a melt index of 0.9. (BP Chemicals Ltd).*

Rigidoil. Oil spill clean up products. (BP Chemicals Ltd).*

Rigilene. Polyethylene. Applications: Sheet and block for docks, defendering. (Stanley Smith & Co Plastics Ltd).*

Rigipore. Plastics - expandable polystyrene. (BP Chemicals Ltd).*

Rikospray Balsam. Benzoin topical protective in aerosol form containing benzoin BPC 12.5% w/w (equivalent to dissolved solids of benzoin 9% w/w), prepared storax BPC 2.5% w/w. (Riker Laboratories/3M Health Care).†

Rikospray Silicone. Topical protective in aerosol form containing aluminium dihydroxyallantoinate 0.5% w/w, cetylpyridinium chloride BP 0.02% w/w, dimethicone 1000 BPC to 100%. (Riker Laboratories/3M Health Care).†

Rilan Wax. Waxy acids and vegetable oils hardened by hydrogenation. It is obtained from the higher alcohols of the fat series.†

Rilata. Vulcanized bitumen linseed oil mixture containing 16 per cent sulphur.†

Rilsan. Nylon 11 and nylon 12, moulding goods, extrusion goods and fine powder coating for metal. Applications: Medical and sporting goods. (Atochem Inc).*

Rilsan 11 and 12. Nylon 11 and 12. (Atochem Inc).*

Rimactane. A proprietary preparation of rifampicin. An antibiotic and anti-tuberculous agent. (Ciba-Geigy Corp).

Rimactazid. A proprietary preparation of RIFAMPICIN and ISONIAZID used in the treatment of tuberculosis. (Ciba Geigy PLC).†

Rimadyl. Carprofen. Applications: Anti-inflammatory. (Hoffmann-LaRoche Inc).

Rimevax Injection. Live attenuated measles vaccine. Applications: immunization against measles. (Smith Kline and French Laboratories Ltd).*

Rimifon. Proprietary preparation of isoniazid. Antitubercular agent. (Roche Products Ltd).*

Rimthane. Multi-component systems used in the manufacture of polyurethane products via reaction injection moulding. (Dow Chemical Co Ltd).*

Ringer Solution. An isotonic solution containing 0.7 per cent sodium chloride, 0.03 per cent potassium chloride, and 0.025 per cent calcium chloride in water. Used in physiological experiments.†

Rinmann's Green. See COBALT GREEN.

Rinoxin. A rodenticide. (Gerhardt Pharmaceuticals).*

Rinsan. Dairy hygiene circulation cleaner. (The Wellcome Foundation Ltd).

Rintal. Broad-spectrum anthelmintic. Applications: Veterinary medicine. (Bayer & Co).*

Rintal Plus. Broad-spectrum anthelmintic plus boticide. Applications: For horses. (Bayer & Co).*

Rio Resin. A proprietary amorphous resin. (R T Vanderbilt Co Inc).†

Riopan. Magaldrate. Aluminium magnesium hydroxide sulphate, hydrate. Applications: Antacid. (Ayerst Laboratories, Div of American Home Products Corp).

Ripercol. A proprietary preparation containing levamisole hydrochloride. Applications: Veterinary antihelmintic (cattle). (Janssen Pharmaceutical Ltd).*

Ripping Ammonal. An explosive. It contains 84-87 per cent ammonium nitrate, 7-9 per cent aluminium, 2-3 per cent charcoal, and 3-4 per cent potassium dichromate.†

Rippite. An English explosive. It contains 56-63 per cent nitro-glycerin gelatinized with a small quantity of collodion cotton, potassium nitrate, wood meal, castor oil, ammonium oxalate, with the addition of calcium or magnesium carbonate, and a petroleum jelly.†

Risigallum. Synonym for Realgar.†

Riso. A material consisting of ammonium carbonate ground in a mixture of mineral and vegetable oils (Cycline oil). Used in the manufacture of sponge rubber.†

Risor. See PINE OILS.

Rissicol. A castor oil powder, containing 49 per cent castor oil and 36 per cent of inorganic matter, mainly magnesia.†

Ristin. A 25 per cent solution of ethylene-glycol monobenzol ester.†

Ristocetin. Anti-microbial substances produced by Nocardia lurida. Spontin contains Ristocetin A and Ristocetin B.†

Risunal. A proprietary preparation of β-diethylaminobutyric acid, aniline hydrochloride, isopropylphenazone, and ethyl and benzyl nicotinate. A rubefacient skin ointment. (Geistlich Sons).†

Ritacetyl. Acetylated lanolin. Applications: Fatting agent for soaps and shampoos. Forms a water-resistant film on the skin and hair and has excellent solubility in mineral oil. Recommended for hypo-allergenic formulations. (RITA Corporation).*

Ritachol. Liquid cholesterol emulsifier. Applications: Lends itself to a wide range of formulations. stabilizes emulsions, dispersions and suspensions. Serves as an epidermal moisturizer, lubricant and emollient. (RITA Corporation).*

Ritaderm. Petrolatum, lanolin, sodium PCA and polysorbate 85. Applications: Hydrophillic dual-action skin and hair moisturizer. Reinforces the lipid layer of the skin and contributes to the natural moisturizing factor. Used in hair conditioners, shampoos, soaps and high electrolyte preparations. Disperses and extends pigments in lipsticks, eye shadow and liquid makeups. (RITA Corporation).*

Ritalafa. Refined lanolin fatty acids. Applications: A mild, light coloured fatty acid that produces a water repellent film which resists rewetting in makeup preparations and enhances oil phase compatibility. Excellent valve lubricant for aerosols. (RITA Corporation).*

Ritalan. Lanolin oil. Applications: This liquid fraction of lanolin is a concentrated moisturizer, emollient and stabilizer, compatible with a broad range of ingredients, provides "slip" in lipstick, ameliorates irritancy in peroxide lotions, wave lotions, depilatories and hair straighteners. (RITA Corporation).*

Ritalan "C'. Isopropyl palmitate and lanolin oil. Applications: Helps disperse finely ground solids and pigments, epidermal penetrant, blending and wetting agent. Helps to stabilize o/w and w/o emulsions. (RITA Corporation).*

Ritalin Hydrochloride. Methylphenidate hydrochloride. Applications: Stimulant. (Ciba-Geigy Corp).

Ritasol. Isopropyl lanolate. Applications: Compatible with fatty acid esters, anhydrous alcohols, mineral and vegetable oils. Combines spreadability, lubricity and wetting characteristics for lipsticks, creams and lotions. (RITA Corporation).*

Ritawax. Lanolin alcohol. Applications: Purified nonsaponifiable fraction of lanolin. Used in creams, lotions, eye and makeup preparations and lipsticks. (RITA Corporation).*

Ritha. An alkaline deposit found on the land in India. Used as a soap substitute. Also see SAJJI.†

Rivalit P. A German explosive containing nitro-glycerin, ammonium nitrate, vegetable meal, nitro-compounds, and potassium perchlorate.†

Rivet Metal. An alloy of copper and tin, to which zinc is sometimes added.†

Riviera Oil. See AIX OIL.

Rivotril. A proprietary preparation of clonazepam. An anti-convulsant. (Roche Products Ltd).*

Rizolex. Fungicide. (FBC Ltd).

RJ-100. Styrene-allyl alcohol resins. Applications: Paint coating ingredient. (Monsanto Co).*

RMD. High molecular weight polymers based on sodium acrylate. (Cyanamid BV).*

RMR. High molecular weight polymers based on sodium acrylate. (Cyanamid BV).*

Roaccutane. Proprietary preparation of isotretinoin (13-cis-retinoic acid). Systemic treatment for acne. (Roche Products Ltd).*

Roachban. A cockroach insecticide. (Murphy Chemical Co).†

Roanoid. A proprietary urea-formaldehyde moulding compound.†

Roaster Slag. A slag produced in the purification of copper metal. It contains from 17-40 per cent copper as silicate and metal.†

Ro-A-Vit. Proprietary preparation of vitamin A (retinol). (Roche Products Ltd).*

Robac. Principally rubber accelerators and vulcanizing agents. Some activators and antioxidants. Applications: vulcanization and protection of natural and synthetic rubbers. (Robinson Brothers Ltd).*

Robac T.B.Z. A trade name for zinc thiobenzoate. A low temperature Peptizing agent for natural rubber.†

Robacure. Principally rubber accelerators and vulcanizing agents. Some activators and antioxidants. Applications: vulcanization and protection of natural and synthetic rubbers. (Robinson Brothers Ltd).*

Robalate. A proprietary preparation containing dihydroxy aluminium amino acetate. An antacid. (Robins).†

Robaxin. Methocarbamol. Applications: Relaxant. (A H Robins Co Inc).

Robengatope I-125. Rose Bengal Sodium I 125. Applications: Radioactive agent. (Mallinckrodt Inc).

Robengatope I-131. Rose Bengal Sodium I 131. Applications: Diagnostic aid; radioactive agent. (E R Squibb & Sons Inc).

Robertson Alloy. Consists of 1 part gold, 3 parts silver, and 2 parts tin. It is mixed with mercury for use as a dental filler.†

Robert's Reagent. For proteins. It consists of 1 volume of pure nitric acid with 5 volumes of a 40 per cent. (saturated) solution of magnesium sulphate.†

Robicillin Vk. Penicillin V Potassium. Applications: Antibacterial. (A H Robins Co Inc).

Ro-Bile. Pancrelipase. A concentrate of pancreatic enzymes standardized for lipase content. Applications: Enzyme. (Rowell Laboratories Inc).

Robimycin. Erythromycin. Applications: Antibacterial. (A H Robins Co Inc).

Robinul. Glycopyrrolate. Applications: Anticholinergic. (A H Robins Co Inc).

Robitet. Tetracycline. Applications: Anti-amebic; antibacterial; antirickettsial. (A H Robins Co Inc).

Robitussin. A proprietary preparation containing guaiphenesin. A cough linctus. (Robins).†

Robitussin A.C. A proprietary preparation containing codeine phosphate, guaiphenesin and pheniramine maleate. A cough linctus. (Robins).†

Roburite. An explosive used in mines. It consists of 86 per cent ammonium nitrate and 14 per cent chlorodinitrobenzene.†

Roburite I. An explosive containing 87.5 per cent ammonium nitrate, 7 per cent dinitro-benzene, 0.5 per cent potassium permanganate, and 5 per cent ammonium sulphate.†

Roburite III. An explosive containing 87 per cent ammonium nitrate, 11 per cent dinitro-benzene, and 2 per cent chloronaphthalene.†

Rocaltrol. Proprietary preparation of calcitriol. Vitamin D metabolite used in correction of abnormalities of calcium and phosphate metabolism. (Roche Products Ltd).*

Roccal. A proprietary preparation of benzalkonium chloride. A pre-operative skin cleanser. (Bayer & Co).*

Roccelline. See FAST RED.

Rocephin. Proprietary preparation of ceftriaxone. Broad spectrum cephalosporin antibiotic. (Roche Products Ltd).*

Rochdale Salt. See ROCHELLE SALT.

Rochelle Salt. (Rochdale Salt, Tartrated Soda, Seignette's Salt). Sodium potassium-tartrate, $C_4H_4O_6Na.K.4H_2O$. Used to reduce the silver salts in the silvering of mirrors, and in medicine as a mild aperient. It is the active constituent of seidlitz powders.†

Rock Ammonia. Ammonium carbonate $(NH_4)_2CO_3$.†

Rock Asphalt. Limestone or other material and found naturally, impregnated with bitumen.†

Rock Cork. A variety of asbestos. Also see ROCK WOOL.†

Rock Crystal. Transparent and colourless quartz.†

Rock Dammar. A variety of dammar resin derived from *Hopea odorata*, of Burma.†

Rock Salt. (Halite). Sodium chloride, NaCl, from sea water.†

Rock Scarlet BS. See ST. DENIS RED.

Rock Scarlet YS. A dyestuff. It is the sodium salt of azoxytoluenedisazo-β-naphthol-α-naphthol monosulphonic acid, $C_{34}H_{25}N_6O_6SNa$. Dyes wool scarlet from an acid bath.†

Rock Tallow. See BITUMEN.

Rock Wool. (Rock Cork). A furnace product made from self-fluxing siliceous and argillaceous dolomite in which the basic and acidic constituents are present in proportions that their fluxing action is nearly balanced. The molten rock at 2800-3000° F. is Atomized by a blast of steam under pressure. The wool treated with a binder is sold as rock cork. A heat insulator and corrosion-resisting packing.†

Rocksil. A proprietary trade name for rockwool insulating materials. It withstands 760° C. (Cape Insulation Cape Asbestos Co).†

Rocktex. A proprietary trade name for rock wool.†

Rocol P.R. A proprietary silicone-based spray used for mould release. (Rocol).†

Rocol R.S.7. A proprietary non-silicone-based wet film spray used for mould release. (Rocol).†

Rocou. See ANNATTO.

Rocryl. Speciality acrylic and methacrylic monomers. (Rohm & Haas (UK) Ltd).

Rocsol. Oxidized hydrocarbon waxes. (Croda Chemicals Ltd).

Rod Wax. A wax-like mass deposited on the drill rods in many petroleum oil wells.†

Rodea. Ricinoleic diethanolamide. Applications: Emulsifier, thickener, solvating agent, corrosion inhibitor. (Clintwood Chemical Company).*

Rodeo. Aquatic herbicide. (Monsanto Co).*

Rodinal. A photographic developer. The active constituent is *p*-aminophenol hydrochloride.†

Rodo. Blend of essential odours (deodorants). Applications: Neutralizes typical dry rubber odours and in emulsions act as deodorants in finished latex products. (Vanderbilt Chemical Corporation).*

Roebaryt. A barium sulphate prepared for use in X-ray work.†

Roesch's Aluminium Solder. An alloy of 50.2 per cent zinc, 50 per cent tin, 0.7

per cent antimony, and 0.2 per cent copper.†

Roferose. Dextrose monohydrate. Applications: Food, pharmaceutical and industrial applications. (Roquette (UK) Ltd).*

Roga. Cellulose acetate spiral wound membrane. Applications: Reverse osmosis water treatment. (Allied-Signal Fluid Systems).*

Rogé Cavaillès. Soap, bath additive and shampoo, for cleansing care of skin and hair. (Richardson-Vicks Inc).*

Roghan. (Afridi Wax). Obtained by boiling safflower oil for 2 hours, then putting it into vessels partly filled with water.†

Rogitine. A proprietary preparation of phentolamine mesylate. (Ciba Geigy PLC).†

Rogor E. Insecticide. (FBC Ltd).

Rohafloc. Methacrylate based ionic/non-ionic flocculants. Applications: Sludge dewatering, industrial waste treatment, paper sizing etc. (Cornelius Chemical Co Ltd).*

Rohn Alloys. Heat-resisting alloys containing nickel, chromium, and iron, sometimes with the addition of manganese and molybdenum. Also see CHROMAN.†

Rohrbach's Solution. A solution of barium and mercuric iodides (100 grams barium iodide and 130 grams mercuric iodide heated with 20 cc water to 150-200° C.). The solution is allowed to cool, when a double salt is deposited. The liquid is decanted. The salt is used in the separation of the heavy metals by gravity as it has a specific gravity of 3.58.†

Rohypnol. Proprietary preparation of flunitrazepam. A benzodiazepine with hypnotic properties. (Roche Products Ltd).*

Rokon. 2-Mercaptobenzothiazole. Applications: Corrosion inhibitor, metal deactivator and extreme pressure agent - petroleum lubricants. Also chemical intermediate. (Vanderbilt Chemical Corporation).*

Rol-man Steel. A proprietary trade name for a high-manganese steel containing 11-14 per cent manganese and 1-1.4 per cent carbon.†

Rolafix. Photographic fixer. (May & Baker Ltd).

Rolaids. Dihydroxyaluminium sodium carbonate. Applications; Antacid. (Parke-Davis, Div of Warner-Lambert Co).

Roll Sulphur. Sulphur which has been melted and poured into moulds.†

Rollit. Graphite powder. Applications: Used as mandrel bar lubricant. (Lonza Limited).*

Rollofix X100. One component expanding polyurethane foam comprising polyol, methyl di-isocyanate, freon and niax. Applications: Filling cavities (domestic and industrial). Bonding - sealing against noxious vapours and moisture. (Piccadilly Products Ltd).*

Rolox. Two-part epoxy compounds. (Hardman Inc).*

Roman Alum. Potash alum, $Al_2(SO_4)_3.K_2SO_4.24H_2O$.†

Roman Bronze. An alloy of 90 per cent copper and 10 per cent tin.†

Roman Cement. (Parker's Cement). A natural cement made by calcining the modules of argillaceous limestone mixed with calcareous spar, which occurs in London and other clays.†

Roman Ochre. See YELLOW OCHRE.

Roman Sepia. A brown pigment. It consists of sepia mixed with yellow browns.†

Roman Yellow. See YELLoW OCHRE.

Romanite. See ROUMANITE.

Romanium. An alloy of 97.43 per cent aluminium, 1.75 per cent nickel, 0.25 per cent copper, 0.25 per cent antimony, 0.17 per cent tungsten, and 0.15 per cent tin.†

Romanowsky's Stain. A microscopic stain. It consists of (A) methylene blue 2 grams, distilled water, 200 cc N/10 caustic potash 10 cc This is boiled, cooled, and 10 cc of N/10 sulphuric acid added. (B) Eosin 1 gram, distilled water 1,000 cc For use mix 1 cc of (A) with 6 cc of (B).†

Romilar. Proprietary preparation of dextromethorphan. Antitussive. (Roche Products Ltd).*

Romite. An explosive. It is ammonium nitrate mixed with a solid, melted hydrocarbon (paraffin or naphthalene), gelatinized with a liquid hydrocarbon (paraffin oil), and contains gelatinized potassium chlorate.†

Rompel's Alloy. An antifriction metal, containing 62 per cent copper, 10 per cent zinc, 10 per cent tin, and 18 per cent lead.†

Romperit G. A German explosive containing nitroglycerin, ammonium nitrate, vegetable meal, nitro-compounds, and potassium perchlorate.†

Rompun. Sedative, analgesic, anaesthetic and muscle relaxant. Applications: For use in cattle. (Bayer & Co).*

Ronafix. Styrene butadiene waterproof bonding additive. Applications: Used in concrete repair, the laying of thin screeds and floors and the fixing of building components. (Ronacrete Ltd).*

Ronase. Tolazamide. Applications: Antidiabetic. (Rowell Laboratories Inc).

Rondase. A proprietary preparation of hyaluronidase, used to facilitate absorption of injected fluids. (British Drug Houses).†

Rondec. Contains carbinoxamine maleate and pseudoephedrine hydrochloride. Applications: Antihistaminic and adrenergic. (Ross Laboratories, Div of Abbott Laboratories).

Rondec DM. Contains carbinoxamine maleate, dextromethorphan hydrobromide and pseudoephedrine hydrochloride. Applications: Antihistaminic; antitussive; adrenergic. (Ross Laboratories, Div of Abbott Laboratories).

Rondomycin. A proprietary preparation of methacycline hydrochloride. An antibiotic. (Pfizer International).*

Ronfalin. ABS - acrylonitril butadiene styrene. Applications: Plastic for general purposes and for special applications in the automotive and electrical industries: for toys, domestic appliances, telematic equipment and extruded products. Examples of applications are dashboard components and radiator grilles, switches, housings for hair driers and coffee makers, telephones, office machines, pipe and plate material. (DSM NV).*

Ronfaloy E. Blends/compounds with ABS (RONFALIN). Applications: For use in the 'open air'. (DSM NV).*

Ronfaloy V. Blends/compounds with ABS (RONFALIN). Applications: Telematic industry. (DSM NV).*

Ronfusil Steel. A proprietary trade name for a manganese-steel containing 12 per cent manganese.†

Rongalit. Textile discharging and reducing agents. (BASF United Kingdom Ltd).

Rongalite. See HYDROSULPHITE.

Rongalite C. A combination of the sodium salt of the unstable sulphoxylic acid and formaldehyde. A reducing agent used in the dye industry, and as a photographic developer. See DISCOLITE.†

Rongalite, Concentrated. See HYDROSULPHITE NF.

Ronia Metal. A brass containing small quantities of cobalt, manganese, and phosphorus.†

Roniacol. Nicotinyl alcohol. Applications: Vasodilator. (Hoffmann-LaRoche Inc).

Ronicol. A proprietary preparation of nicotinyl tartrate used in the treatment of circulatory disorders. (Roche Products Ltd).*

Ronnel. A proprrietary preparation of FENCHLORPHOS. An insecticide.†

Ronoxan. Antioxidant pastes for food. (Roche Products Ltd).

Ronstar. Herbicide. (May & Baker Ltd).

Root Guard. Diazinon insecticide. (Murphy Chemical Ltd).

Ropaque. Opaque polymer emulsion. Applications: Opacifying agent for decorative coatings, paper coatings etc. (Rohm and Haas Company).*

Rosamine. A dyestuff obtained from benzotrichloride and dimethyl-*m*-aminophenol, $C_{23}H_{23}Cl.N_2O$.†

Rosaniline. (Magenta Base, Rosaniline Base, O, SF, Pararosaniline Base, Oil Red Base, Brilliant Oil Crimson). Triamino-tolyl-diphenyl-carbinol.†

Rosaniline Base, 0, SF. Equivalents of Rosaniline.†

Rosaniline Blue. See ANILINE BLUE, SPIRIT SOLUBLE.

Rosanthrenes O, R, A, B, CB. Tetrazo dyestuffs, which produce shades similar to Turkey Red. They are prepared by coupling diazo compounds with *m*-amino-benzoyl-amino-naphthol-sulphonic acids, and dye from a bath of sodium sulphate, sodium carbonate, and soap.†

Rosarin. See-AZOCARMINE G.†

Rosaurine. Rosilic acid, $C_{21}H_{16}O_3$.†

Rosazine. See AZOCARMINE G.

Rose B. See ERYTHROSIN.

Rose Bengal. (Rose Bengal N, Rose Bengal AT, Rose Bengal G, Bengal Red). A dyestuff. It is the alkaline salt of tetraiododichlorofluoresceine, $C_{20}H_4Cl_2I_4K_2O_5$. Dyes wool bluish-red. Also see ROSE BENGAL 3B.†

Rose Bengal AT. See ROSE BENGAL.

Rose Bengal B. See ROSE BENGAL 3B.

Rose Bengal G. See ROSE BENGAL.

Rose Bengal N. See ROSE BENGAL.

Rose Bengal 3B. (Rose Bengal B, Rose Bengal). A dyestuff. It is the potassium salt of tetra-iodo-tetrachloro-fluoresceine, $C_{20}H_2Cl_4I_4K_2O_5$. Dyes wool bluish-red.†

Rose de Benzoyl. See BENZOYL PINK.

Rose Ester. A trade name for trichloro-methyl phenyl carbinyl acetate. M.Pt. 85-87° C. A rose scent used in the perfumery industry. (BTP Cocker Chemicals Ltd).†

Rose Food. Granular fertilizer. (Fisons PLC, Horticulture Div).

Rose JB. See SPIRIT EOSIN.

Rose JB, Alcohol Soluble. See SPIRIT EOSIN.

Rose Pink. A pigment prepared by dyeing whiting with Brazil wood. Used in paper-staining.†

Rose Plus. Granular fertilizer containing magnesium. (ICI PLC, Plant Protection Div).

Rose Quartz. A mineral. It is a variety of quartz, SiO_2, which is stated to owe its colour to manganese.†

Rose Vitriol. Cobalt sulphate.†

Rosé Oil. (Roshé Oil, Oil of Geranium, Oil of Rose-geranium. Oil of Pelargonium, Ginger Grass Oil, Turkish Geranium Oil, Oil of Palmarosa). Andropogon oils, obtained from a grass, *Andropogon nardus*. They contain geraniol, $C_{10}H_{18}O$.†

Rose-Geranium Oil. An oil obtained by distilling geraniol over rose flowers. Also see ROSE OIL.†

Roseclear. Combined insecticide and fungicide for garden use. (ICI PLC).*

Rosein. An alloy of 44.4 per cent nickel, 33.3 per cent aluminium, 11.1 per cent silver, and 11.1 per cent tin. Another alloy contains 40 per cent nickel, 30 per cent aluminium, 20 per cent tin, and 10 per cent silver. Used by jewellers.†

Roseine. See MAGENTA and MAUVEINE.

Roseine Crystals SF. A dyestuff. It is a British equivalent of New magenta.†

Roseline. Finishing agent. (Ciba-Geigy PLC).

Roselle Fibre. A Malay fibre similar to jute.†

Rosenöl. A synonym for rose otto.†

Rosenthiel's Green. (Baryta Green, Cassel Green, Manganese Green). A pigment. It is a manganate of barium, obtained by heating a mixture of oxides of manganese, barium nitrate, and heavy spar or kaolin.†

Roseol. (Reuniol). Names applied to citronellol, $C_9H_{17}.CH_2OH$, or to mixtures of this body with geraniol. Perfumes.†

Rosette Copper. This is obtained in thin films by throwing water on to the surface of molten copper and removing the crusts formed.†

Rose, Cobalt. See COBALT RED.

Rose's Metal. Fusible bismuth alloys. (1) Consists of 42 per cent bismuth, 42 per cent lead, and 16 per cent tin. It melts at 79° C. (2) Consists of 33.3 per cent bismuth, 33.3 per cent lead, and 33.3 per cent tin. It melts at 93° C. (3) Consists of 48.9 per cent bismuth, 27.5 per cent lead, and 23.6 per cent tin. It has a specific heat of 0.0552 cal per gm between 20-89° C.†

Roshé Oil. See ROSÉ OIL.

Rosin. See COLOPHONY.

Rosin Blende. See RESIN BLENDE.

Rosin Grease. A combination of rosin oil (*qv*) with lime.†

Rosin Oil. An oil obtained by the distillation of rosin from 300-400° C. Used as an adulterant for olive oil, also as a lubricant for iron bearings. See RETINOL.†

Rosin Pitch. Obtained by the distillation of rosin. It is the residue, and amounts to about 16 per cent.†

Rosin Soap. Sodium resinate.†

Rosin Spirit. (Essence of Resin, Resin Spirit). The first distillate from rosin, 78-250° C. It is a complex mixture of hydrocarbons somewhat resembling turpentine. Used as a substitute for turpentine.†

Rosin Tin. A yellow variety of the mineral Cassiterite or Tinstone.†

Rosinduline G. A dyestuff. It is the sodium salt of rosindone monosulphonic acid, $C_{22}H_{13}N_2Na.SO_4$. Dyes wool and silk scarlet, and is chiefly used for printing.†

Rosinduline 2B. See AZOCARMINE B.

Rosindulone. Rosindone.†

Rosindulone 2G. A dyestuff. It is the sodium salt of rosindone-β-mono-sulphonic acid, $C_{22}H_{13}N_2Na.SO_4$. Dyes silk and wool orange from an acid bath.†

Rosin, Hardened. See HARDENED ROSINS.

Rosinjack. See ZINC BLENDE.

Rosinol. See RETINOL.

Roskens. Hand conditioner. (Fisons PLC, Pharmaceutical Div).

Rosolane. See MAUVEINE.

Rosolane B, R, O, T. A dyestuff. Dyes silk violet-pink.†

Rosolene. A rosin oil obtained by the distillation of rosin.†

Rosoli. A liqueur.†

Rosophenine Geranine. A direct cotton colour.†

Rosophenine Pink. See ROSOPHENINE 10B.

Rosophenine 10B. (Rosophenine Pink, Thiazine Red R, Chlorazol Pink Y). A dyestuff. Dyes cotton pink to red shades.†

Rosophenine 4B. See ST. DENIS RED.

Rosophenoline. See PAEONINE.

Ross Alloy. A bronze containing 68 per cent copper and 32 per cent tin.†

Ross's White. See LITHOPONE.

Rota. PH indicator papers.(May & Baker Ltd).

Rotalin. Herbicide, containing linuron. (ICI PLC).*

wRotax. Industrial grade 2-mercaptobenzothiazole. Purified to reduce odour. Applications: Used as a corrosion inhibitor in automotive chemicals and industrial cleaners where a copper deactivator is required. It is also used to protect silver from sulphur blackening. (Vanderbilt Chemical Corporation).*

Rotenone. A crystalline material found in the roots of derris, a plant grown in the rubber plantations of the Malay Peninsula. It is also found in the South American "cube" plant. It occurs up to 5.5 per cent in derris and up to 7 per cent in "cube‘. It is used as an insecticide.†

Roter. A proprietary preparation containing bismuth subnitrate, magnesium carbonate, sodium bicarbonate and frangula. An antacid used in the treatment of peptic ulcers. (FAIR Laboratories).†

Rotercholon. A proprietary preparation of turmeric, ox-bile extract, peppermint oil, fennel oil, caraway oil and methyl salicylate used in the treatment of biliary disorders. (FAIR Laboratories).†

Rotersept. A proprietary aerosol spray containing CHLORHEXIUINE digluconate used ia the prophylaxis of puerperal mastitis. (FAIR Laboratories).†

Rotnickelkies. Synonym for Niccolite.†

Rotoval. Gravure printing inks. Applications: Package printing. (Allied Signal Sinclair and Valentine).*

otoxit. A resistant high-silicon-copper alloy. It is stated to be resistant to uric, fluosilicic, and fatty acids, to dilute hydrochloric, sulphuric, and acetic acids, to 30 per cent phosphoric acid, hydrogen peroxide, ammonia, lyes, and sulphates, but not to chromic, nitric, and lactic acids.†

otra Bark. The bark of *Rotra fotsy* and *R. meno*. The bark contains 12.6 per cent tannin.†

otten Stone. A soft, friable aluminium silicate, containing a little organic matter. It is used as a polishing material. The term is also sometimes applied to Tripoli (*qv*).†

ouen White. A pigment. It is a clay found near Rouen.†

ouge. Good qualities of rouge consist of very fine iron oxide, Fe_2O_3, and are used as abrasives. The finest rouge is prepared from safflower. See INDIAN RED.†

ouge de Mars. See MARS RED.

ouge, Mineral. See JEWELLER'S ROUGE.

ouge, Mordant. See RED LIQUOR.

ouge, Toilet. A mixture of carmine and chalk.†

ouge, Vegetable. See VEGETABLE ROUGE and SAFFLOWER.

ouge, White. See PLATE POWDER.

oumanite. (Romanite). The amber of Roumania, which much resembles the Prussian variety.†

oundup. Isopropylamine salt of glyphosphate. Applications: Post emergence herbicide. (Monsanto Co).*

ousselot Gelatine. Gelatine derived from animal tissue available in several mesh sizes. Applications: Photographic emulsions, pharmaceutical hard, soft and micro capsules, edible gelatines for confectionery, meat, dairy and dessert industries. (Rousselot Ltd).*

oussel's Solution. A solution of sodium phosphate.†

oussin's Black Salt. A compound, $Fe_3H_2N_4O_4S_5$, obtained by adding a solution of ferrous or ferric chloride slowly to the mixed solution of potassium nitrite and ammonium sulphide, and then boiling.†

Roussin's Red Salt. A salt, $Fe_2S_4N_2O_2Na_2.H_2O$, obtained by treating the sodium salt of Roussin's black salt with excess of acid after boiling, and then evaporating.†

Roussin's Salts. Salts of the type $KFe_4(NO)_7S_3$, obtained when nitric oxide is passed through a suspension of ferrous sulphide in a sulphide solution.†

Roux's Stain. A microscopic stain. It contains 0.5 gram gentian violet, 1.5 grams methyl green, and 200 cc distilled water.†

ROV. Rectified oil of vitriol (sulphuric acid containing 93-96 per cent. H_2SO_4).†

RR 53 Alloy. An aluminium alloy containing 91.85 per cent aluminium, 2.25 per cent copper, 1.3 per cent nickel, 1.5 per cent magnesium, 1.5 per cent iron, 1.5 per cent silicon, 0.1 per cent titanium. Used for die casting pistons.†

Rovamycin. Spyramycin. (May & Baker Ltd).

Rovel. Weatherable polymers used to manufacture spas, truck toppers, automotive trim. (Dow Chemical Co Ltd).*

Rovigon. Proprietary vitamin preparation containing vitamins A and E. (Roche Products Ltd).*

Rovimix. Vitamin supplements for animal feeds. (Roche Products Ltd).

Rovisol. Water miscible vitamins for animals. (Roche Products Ltd).

Rovral. Fungicide. (May & Baker Ltd).

Roxadyl. Rosozacin. Applications: Antibacterial. (Sterling Drug Inc).

Roxamine. A dyestuff. It is the sodium salt of dihydroxyazonaphthalene-sulphonic acid, $C_{20}H_{13}N_2Na.O_5S$.†

Roxarsone. An anti-protozoal and growth promoter. It is 4-hydroxy- 3-nitrophenylarsonic acid.†

Roxite. A synthetic resin product used for electrical insulation.†

Roxon. A proprietary synthetic resin of the phenol-formaldehyde type.†

Roxotit. A copper-silicon acid-resisting alloy.†

Royal Blue. (Saxon Blue, Azure Blue). Varieties of Smalt (*qv*).†

Royal Scarlet. See IODINE RED.

Royal Yellow. See CHROME YELLOW and CHINESE YELLOW.

Royalac 133. A modified dithiocarbamate. Produces excellent vulcanizates in EPDM polymers when used with normal sulphur levels. The processing safety is reduced as compared to the standard Monex/MBT acceleration system. The major feature of Royalac 133 is the high degree of tensile strength developed at short cure times while retaining equivalent compression set properties. (Uniroyal).*

Royalac 136. A proprietary accelerator. An ultra-accelerator developed for use in EPDM polymers. (Uniroyal).*

Royalac 140. Zn (hexadecyl-octadecyl mixture) isopropyl dithiocarbamate. To improve physical properties of EPDM diene blends. EPDM-NBR blends vulcanized with sulphur curing systems employing Royalac 139 and 140 compare favourably with CR compound in heat ageing, have essentially the same oil resistance level, exhibit better low temperature properties and are highly ozone resistant. (Uniroyal).*

Royalene (EPDM). A range of polymers used for ozone and weather protection. (Uniroyal).*

Roydalox. A proprietary trade name for alumina porcelain with good resistance to corrosion and thermal shock. It is used in ball mills and grinding balls. (Doulton).†

Roydazide. A proprietary trade name for silicon nitride. It is used for the manufacture of turbine blades and generally where a temperature of up to 1650° C. is present. (Doulton).†

RPM. Motor oil. (Chevron).*

RRV. A proprietary antioxidant for rubber. It is resorcylindene.†

R-Salt. The sodium salt of R-acid (*qv*).†

RS Nitrocellulose. Nitrocellulose. Applications: Used as a clear, tough, fast drying film former. (Hercules Inc).*

'R' Type Solvent. Adipate polyester plasticizer. Applications: Suitable for softening hot melt adhesive of packaging grade. (Nordson (UK) Ltd).*

Rubalt. A proprietary compound of rubbe bitumen, and benzene. It is a waterproof, rust-proof, and acid-resisting paint, and is stated to be high resistant to mineral acids, alkalis, chlorine, ammonia, and salt solution.†

Rubber. An elastic material contained in the latex of certain plants. The most important plant is *Hevea brasiliensis*, of South America, which yields the Para rubber of commerce. Other plants yielding similar latexes are *Castilloa elastica*, of Central America; *Castilloa ulei*, of Peru; and *Manihot glaziovii*, of Brazil. African rubbers are obtained from *Funtumia elastica* and *Landolphia owariensis*, Asiatic rubber from *Ficus elastica*, and Guayule rubb from *Parthenium argentatum*, of Texa and Mexico. Plantation rubber is obtained from cultivated *Hevea brasiliensis*. The latex is treated with a coagulating agent, usually acetic acid, and is then marketed in the form of smoked sheet, or pale crzepe. The latex itself is now shipped, coagulation being prevented by the addition of ammonia or sodium carbonate. It has been used in the preparation of paper, and for other purposes. Coagulated latex, as cr zepe, is used for soling shoes and other purposes, and the vulcanized material (with or without accelerators and fillers is used for tyres, mats, and many other purposes. Rubber as received contains resin in varying amounts, from 2 per cent for the finest pale crzepe to 15 per cent for Guayule and 70-80 per cent fo Jelutong.†

Rubber Cements. These are made by dissolving rubber in suitable solvents, such as coal-tar naphtha or carbon disulphide. Sometimes rosin or turpentine is added. Rubber cement is often a mixture of rubber and sulphur dissolved in oil.†

Rubber Formolite. A product obtained by the action of formaldehyde on a petroleum ether solution of a pale cr zepe rubber to which has been added concentrated sulphuric acid.†

Rubber Lead No. 4. A vulcanization accelerator for rubber. It contains organic and mineral constituents, and is manufactured by precipitating a substituted guanidine on a rubber lead base.†

Rubberite. An artificial rubber made from asphalt, oxidized oil, petroleum jelly, and sulphur.†

Rubberlene. A white refined petroleum product. It is used as a solvent, and can be substituted for carbon disulphide for dissolving rubber. It boils at 145-300° F.†

Rubber, Abba. A low-grade African rubber, probably from *Ficus vogelii*. It is a red rubber containing 55 per cent rubber and 45 per cent resin.†

Rubber, Frost. A name for sponge rubber.†

Rubber, Mineral. See ELATERITE.

Rubber, Silk. See CUITE, SERICINE, and SILK RUBBER.

Rubber-sulphur. Amorphous plastic sulphur, obtained from the Kobui sulphur mine, Japan.†

Rubbone. A patented composition stated to be rubber resin prepared by oxidizing rubber catalytically. It is used in paints, varnishes, etc., in electrical insulation, and in the impregnation of coils.†

Rubel Metal. An alloy of 55 per cent copper, 40 per cent zinc, and 5 per cent aluminium-iron-manganese-nickel alloy. Another alloy contains 51 per cent copper, 40 per cent zinc, and 5 per cent aluminium-iron-manganese-nickel alloy, and 4 per cent ferro-manganese.†

Rubelix. A proprietary preparation containing pholcodine and ephedrine hydrochloride. A cough linctus. (Pharmax).†

Rubellan. A mineral. It is a variety of mica.†

Rubelogen. Rubella Virus Vaccine Live. Applications: immunizing agent. (Parke-Davis, Div of Warner-Lambert Co).

Ruben's Brown. See VANDYCK BROWN.

Ruben's Madder. A pigment. It is a preparation of madder.†

Rubeosine. A nitrochlorofluorescein, obtained by the action of nitric acid upon aureosin.†

Ruberite. A name for red copper ore, Cu_2O.†

Rub-er-red. A proprietary pigment for rubber. It is a red iron oxide of fine particle size, which is acid and alkali free, and contains no soluble salts.†

Rub-erok. A propetary trade name for hard rubber for electrical insulation.†

Rubesine. See MAGENTA.

Rubianic Acid. Ruberythric acid, $C_{22}H_{28}O_4$.†

Rubianin. See MAGENTA.

Rubianite. See MAGENTA.

Rubidine. See FAST RED.

Rubine. See MAGENTA.

Rubine S. See ACID MAGENTA.

Rubini's Essence. A saturated solution of camphor in alcohol.†

Rubio Ore. A brown ore of iron, from Bilbao, in Spain.†

Rubmag. A proprietary trade name for a light magnesium carbonate used for rubber reinforcing.†

Rubox. Zinc oxide rubber grade. Applications: Rubber industry. (Manchem Ltd).*

Rubramin PC. Cyanocobalamin. Applications: Vitamin. (E R Squibb & Sons Inc).

Rubramine. See INDAMINE 3R.

Rubratope-57. Cyanocobalamin Co 57. Applications: Diagnostic aid; radioactive agent. (E R Squibb & Sons Inc).

Rubrax. A mineral rubber containing 98-99 per cent of material soluble in chloroform, with an ash less than 0.5 per cent.†

Rubrescin. An indicator prepared from resorcinol and chloral hydrate.†

Rubrica. A red pigment. It is a natural burnt ochre, containing varying quantities of iron. See INDIAN RED.†

Rubriment. A proprietary preparation of nicotinic acid benzyl ester and capsicin. An embrocation. (Horlicks).†

Rubrophen. Dihydroxytrimethoxy-triphenylmethane.†

Rub-tex. A proprietary trade name for a hard rubber.†

Ruby Arsenic. See REALGAR.

Ruby, Chrome. See CHROME RED.

Ruby Ore. Cuprous oxide, Cu_2O.†

Ruby Powder. A sporting 42-grain powder containing 50 per cent nitro-cellulose, metallic nitrate, 8 per cent nitro-hydrocarbon, and 6 per cent starch.†

Ruby Sulphur. See REALGAR.

Ruby Tin. A red variety of the mineral Cassiterite, or tinstone.†

Ruddle. See INDIAN RED.

Rufigallol. Hexahydroxyanthraquinone, $C_{14}H_8O_8$. Dyes chromed wool brown.†

Rufiopin. Tetrahydroxyanthraquinone, $C_{14}H_8O_6$.†

Rufocromomycin. An antibiotic produced by Streptomyces rufuchromogenus.†

Rufol. Dihydroxyanthraquinone.†

Ruge's Solution. A solution containing 1 cc glacial acetic acid, 2 cc formalin, and 100 cc distilled water.†

Rule Brass. See BRASS.

Rumensin. Monensin, produced by Streptomyces cinnamonensis, and used as the sodium salt. Applications: Antibacterial; antifungal. (Eli Lilly & Co).

Rumicin. Chrysophanic acid, $C_{15}H_{10}O_4$.†

Runa. Rutile type titanium dioxide. (Laporte Industries Ltd).*

Runge's Madder Orange. Rubiacin, a yellow, crystalline substance obtained from the madder root.†

Ruolz Alloys. See NICKEL SILVERS.

Rupel Alum. Roche alum.†

Ruselite. A proprietary trade name for an alloy of 94 per cent aluminium, 4 per cent copper, 2 per cent chromium, and 2 per cent molybdenum. It is stated to be resistant to corrosion.†

Rusma. A mixture of arsenic sulphide (orpiment), As_2S_3, with lime. It is made into a paste with water, and used for unhairing skins prior to tanning them.†

Ruspini's Solution. A styptic containing tannic acid, rose water, alcohol, and water.†

Russet Rubiate. A pigment. It is a preparation of madder.†

Russian Cast Brass. See BRASS.

Russian Green. See DINITROSORESORCIN.

Russian Red. Fuchsine mixed with safranine to give it a yellow shade. See MAGENTA.†

Russian Red G, 967. Dyestuffs. They are British equivalents of Magenta.†

Russian Tallow. A mixture of beef and mutton fat.†

Russian Tula. See NIELLO SILVER.

Russian Turpentine. The oleo-resin from *Pinus sylvestris* and *P. Ledebourii*.†

Russian White Lead. A white lead having the composition, $5PbCO_3.2Pb(OH)_2.PbO$.†

Russol. See PARAFFIN, LIQUID.

Rustban. Protective coatings. (Exxon Chemical Ltd).*

Rustlan Oil. Yellow to amber coloured liquid used as rust preventive and stamping oil. Applications: Wire mills and steel mills used to protect rust whi[le] shipping and inside storage. Metal stamping: used as a shallow stamping oil which need not be removed prior to spot welding like other oils. (Rustlan Chemical Co).*

Rustless Iron. A rustless steel containing about 0.1 per cent carbon. It is made i[n] the electric furnace by means of practically carbon-free ferro-chrome.†

Rustless Steel. See CHROME STEELS.

Rust-Tap. A light yellow non-flammable liquid used as 'tapping oil'. Residual coating acts as rust preventive. Applications: Used in machine shops fo[r] tapping operation of all metals except aluminium. Can be used as drilling lubricant. (Rustlan Chemical Co).*

Rutaform. Phenolic, melamine and granular polyester moulding powders. (Sterling Moulding Materials).*

Ruthenium Red. Ammoniated ruthenium oxychloride, $Ru_2(OH)_2Cl_4.7(NH_3).3H_2O$. Used as a microscopic stain, and as a reagent for pectin, plant mucin, and gum.†

Ruthmol. A proprietary preparation of potassium chloride, lactose and gluten-free starch. A sodium-free table salt. (Larkhall Laboratories Ltd).*

Rutiox. A rutile titanium dioxide pigment. (Laporte Industries Ltd).*

RVPaba Lipstick. Aminobenzoic acid. Applications: Ultraviolet screen. (Elder Pharmaceuticals Inc).

RX-56. Porofocon A. Applications: Contact lens material. (Rynco Scientific Corp).

RXXL. High boiling tar acids. (Coalite Fuels & Chemicals Ltd).†

Ryax C. Centre-line suspended process pump. Applications: Process pumping applications. (Sihi-Ryaland Pumps Ltd).*

Ryax F, O. Back pull out process pump. Applications: Process pumping application. (Sihi-Ryaland Pumps Ltd).*

Rybaferrin. A proprietary preparation of ferrous sulphate, manganese, copper, strychnine hydrochloride, thiamine, hydrochloride and nicotinic acid. A tonic. (Rybar).†

Rybarex. A proprietary preparation of chloroxylenol, papaverine hydrochloride, atropine methonitrate, methyl salicylate, menthol, benzocaine, pituitary extract, adrenaline, tri-iodophenol and a saline base. (Rybar).†

Rybarvin. A proprietary preparation of atropine methonitrate, ADRENALIN, papaverine hydrochloride, benzocaine and posterior pituitary extract, used in the treatment of asthma. (Rybar).†

Rybronsol. A proprietary preparation of phenazone iodoantipyrine, caffeine, and butethamate citrate. (Rybar).†

Ryflex. Belt driven single-stage centrifugal pump capable of being stacked one unit on top of another to a maximum of 2. Applications: Heating and cooling water applications where floor space is small and a stacked pump unit is required. (Sihi-Ryaland Pumps Ltd).*

Rylard. Marine finishes. Applications: For boats made from GRP (with correct Rylard GRP primer), wood, steel and aluminium. (Llewellyn Ryland Ltd).*

Rymel. A proprietary preparation containing ipecacuanha liquid extract, acetic acid, squill liquid extract, sodium citrate and glycerin. A cough linctus. (Rybar).†

Rynabond. A proprietary preparation containing phenylephrine tannate, pheniramine tannate and mepyramine tannate. (Fisons PLC).*

Rynacrom. A proprietary preparation of sodium cromoglycate used in the treatment of allergic rhinitis. (Fisons PLC).*

Rynite. Polyester resin. (Du Pont (UK) Ltd).

Ryotol. A proprietary preparation of phenoxyethanol, phenylmercuric nitrate and phenazone. (Rybar).†

Ryspray. A proprietary preparation of atropine methonitrate and isoprenaline hydrochloride in an aerosol. A bronchial antispasmodic. (Rybar).†

Rytherm. Condensate extraction pump set with low mounted tank. Applications: Several sizes of set suit various sizes of industrial central heating systems, pumps extract condensate at or near 100°C, thus reducing heat loss. (Sihi-Ryaland Pumps Ltd).*

Rythmatine. Meobentine sulphate. Applications: Cardiac depressant (Burroughs Wellcome Co).

Ryton. Polyphenylene sulphide (PPS). A thermoplastic moulding resin used for structural components and the encapsulation of semiconductors.†

Ryton A-100. High molecular weight polyphenylene sulphide with fibre glass. Applications: Injection moulded thick-walled heavy industrial parts such as mechanical components, pump housings, impellers and seals, (Philips 66 Company).*

Ryton R-4. Polyphenylene sulphide with 40% fibre glass reinforcement. Applications: Electrical electronic parts, automotive under-the-hood components, pump housings and impellers and valves, cams etc. (Philips 66 Company).*

Ryton R-7. Polyphenylene sulphide with fibre glass and mineral filler. Applications: Electronic components, structural electrical parts, automotive under-the-hood parts and replacement for thermosets. (Philips 66 Company).*

Ryvin. Single stage centrifugal pump vertical in-line. Applications: In-line circulating duties. (Sihi-Ryaland Pumps Ltd).*

S

S945. Monoclinic zirconium dioxide with a zirconia content (including hafnia) of 94.5%. Applications: Manufacture of ceramic pigments, welding fluxes and insulating material. (Ferro Corporation).*

S975. Monoclinic zirconium dioxide with a zirconia content (including hafnia) of 97.5%. Applications: Manufacture of pigments for ceramics and enamels and welding fluxes. (Ferro Corporation).*

S987. Monoclinic zirconium dioxide with a zirconia content (including hafnia) of 98.7%. Applications: Manufacture of ceramic pigments. (Ferro Corporation).*

S992. Monoclinic zirconium dioxide with a zirconia content (including hafnia) of 99.2%. Applications: Manufacture of lead-zirconate-titanate piezo electric ceramics, zirconates, zirconia technical ceramics, oxygen sensors, milling media and ceramic pigments. (Ferro Corporation).*

S994. Monoclinic zirconium dioxide with a zirconia content (including hafnia) of 99.4%. Applications: Manufacture of lead-zirconate-titanate piezo electric ceramics, zirconates, zirconia technical ceramics and oxygen sensors. (Ferro Corporation).*

Sabeco Metal. A proprietary trade name for copper with 21 per cent lead and 9 per cent tin.†

Sablon. See LIMO.

Sabulite. An explosive containing ammonium nitrate, charcoal, and calcium silicide.†

Sabutol. Essentially C_4 alcohols consisting of approximately 62% normal butyl alcohol, 20% isobutyl alcohol and 15% secondary amyl alcohol. Applications: Solvent in paints, printing inks, dyes, foundries, manufacture of butyl acetate and xanthate. (Sasolchem).*

Saccharase. See SUCRASE.

Sacchareines. Dyestuffs similar to rhodamines, obtained by condensing saccharin with dialkylated *m*-amino-phenol.†

Saccharin. (Glycophenol, Glycosin, Neosaccharin, Sycorin, Sykose). *o*-Anhydrosulphamidebenzoic acid (benzoic sulphimide), $C_6H_4(CO)$-(SO_2).NH. A sweetening substance which is 500 times sweeter than cane sugar.†

Saccharinol. See SACCHARIN.

Saccharinose. See SACCHARIN.

Saccharol. See SACCHARIN.

Sacholith. (Sachtolith). A pigment stated to be a specially prepared zinc sulphide.†

Sachsse's Solution. A solution containing 18 grams mercuric iodide, 25 grams potassium iodide, and 80 grams potassium hydroxide in a litre. Used for the determination of reducing sugars.†

Sachtolith. A proprietary trade name for zinc sulphide used as a pigment.†

S-acid. 1, 8-Amino-1-naphthol-5-sulphonic acid.†

2 S-acid 1 : 8-Dihydroxy-naphthalene 2 : 4-disulphonic acid.†

Sacred Bark. The bark of *Rhamnus purshianus*.†

SAE No 50 Alloy. (Dow H Alloy). A magnesium alloy containing 6 per cent aluminium, 3 per cent zinc, and 0.2 per cent manganese. An aircraft alloy.†

Safapryn. A proprietary preparation of aspirin and paracetamol. An analgesic. (Pfizer International).*

Safapryn-Co. A proprietary preparation of aspirin, paracetamol and Codeine phosphate. An analgesic. (Pfizer International).*

Safebond 3. Modified bitumen emulsion. Applications: Flooring adhesive. (Marley Adhesives).*

Safety Dynamite. An explosive consisting of 24 per cent nitro-glycerin, 1 per cent guncotton, and 75 per cent ammonium nitrate.†

Safety Nitro-powder. An explosive similar in composition to Giant powder.†

Safety Oil. See C-PETROLEUM NAPHTHA.

Safety-Cool. Heavy-duty cutting and grinding fluids specifically designed to cool, lubricate and protect metal surfaces in a variety of machining applications. Applications: Metalworking fluid. (Chem-Trend).*

Safex. A proprietary super-accelerator for rubber vulcanization. It is stated to be dinitro-phenyl-dimethyl-dithio-carbamate. It is used with zinc oxide.†

Saffil. High-temperature inorganic fibres. (ICI PLC).*

Safflor. See SAFFLOWER.

Safflower. (Bastard Saffron, Safflor, Safflower Extract). A natural dyestuff, it dyes silk and cotton red. It is sold under the names of Rouge végétable and Safflower Carmine, and is used as a cosmetic and pigment.†

Safflower Carmine. See SAFFLOWER.

Safflower Extract. See SAFFLOWER.

Saffron. A colouring matter obtained from the dried and powdered flowers of the saffron plant, *Crocus sativus*. Used for colouring confectionery. See SULPHURATED ANTIMONY.†

Saffron, Bastard. See SAFFLOWER.

Saffron Bronze. (Gold Bronze). Tungsten-sodium bronze, $Na_2W_3O_9$. Used as a pigment. The corresponding potassium salt is known as Violet bronze or Magenta bronze.†

Saffron, Dyer's. See SAFFLOWER.

Saffron, Indian. See TUMERIC.

Saffron, Iron. See INDIAN RED.

Saffron of Antimony. See SULPHURATED ANTIMONY.

Saffron Oil. Safflower oil from the seeds o *Carthamus tinctorius*. It has an acid value of 9.8 and a saponification value of 197.3.†

Saffron Substitute. See VICTORIA YELLOW.

Saffron Sugar. Crocase, $C_6H_{12}O_6$.†

Saffron Surrogate. See VICTORIA YELLOW.

Saflex. Polyvinyl butyral film. Applications: Interlayer for safety glass (Monsanto Co).*

Safraniline. See RHODAMINE B.

Safranine. (Safranine T, Safranine Extra G, Safranine S, Safranine FF Extra, Safranine Conc., Safranine GGS, Safranine AG, AGT, OOF, Safranine GOO, Aniline Rose, Aniline Pink). A dyestuff. It is a mixture of diamino-phenyl and tolyl-tolazonium-chlorides, $C_{21}H_{21}Cl.N_4$ and $C_{20}H_{19}Cl.N_4$. Dyes cotton mordanted with tannin and tarta emetic, red. Used in calico printing.†

Safranine. GOO. See SAFRANINE.†

Safranine AG, AGT, OOF. See SAFRANINE.

Safranine B. See PHENOSAFRANINE.

Safranine B Extra. See PHENOSAFRANINE.

Safranine Conc. See SAFRANINE.

Safranine Extra G. See SAFRANINE.

Safranine FF Extra. See SAFRANINE.

Safranine GGS. See SAFRANINE.

Safranine MN. See METHYLENE VIOLET 2RA.

Safranine RAE. (Tolusafranine). Dyes tannined cotton.†

Safranine S. See SAFRANINE.

Safranine Scarlet. A dyestuff. It is a mixture of auramine and safranine.†

Safranine T. See SAFRANINE.

Safranisol. Methoxy-safranine.†

frol. The methylene ether of allyl-pyrocatechol, $C_6H_3.C_3H_5.(O.OCH_2)$. It is found in oil of sassafras, and is obtained from red oil of camphor. Used in the place of oil of sassafras.†

frosine. See EOSIN BN.

gatal. Pentoparbitone sodium solution. (May & Baker Ltd).

hli's Reagent. A mixture of equal parts of a 48 per cent solution of potassium iodide and an 8 per cent solution of potassium iodate. Used to test for free hydrochloric acid in stomach contents.†

hli's Stain. A solution of borax and methylene blue in water. Used to stain nervous tissues and cell nuclei.†

t. Bartholomew's Tea. See MATÉ.

t. Denis Black. A dyestuff obtained by the fusion of *p*-phenylenediamine with sodium polysulphides. Dyes cotton greyish-blue to black.†

t. Denis Red. (Dianthine, Rosophenine 4B, Trona Red, Rock Scarlet BS, Patent Rock Scarlet). A dyestuff. It is the sodium salt of azoxytoluenedisazo-bi-α-naphtholsulphonic acid, $C_{34}H_{24}N_6Na_2O_9S_2$. Dyes cotton red from an alkaline bath.†

t. Helen's Powder. An explosive consisting of 92-95 per cent ammonium nitrate, 2-3 per cent aluminium powder, and 3-5 per cent trinitrotoluene.†

t. Ignatius Bean. The seed of *Strychnos ignatii*.†

t. John Wort Oil. See OIL OF ST. JOHN WORT.

aisan. Liquid drazoxolan seed dressing. (ICI PLC).*

ajji. An alkaline deposit found on the land in India. It is used as a soap substitute. Also see RITHA.†

akaloid. A synthetic resin. It is a polymerized sugar product obtained from sugar, dextrose, levulose, etc. It can be used for varnishes and lacquers and, when extruded, as an artificial silk.†

akoa Oil. An oil obtained from the seeds of *Sclerocarpa caffra*. It is a non-drying oil, and has a saponification value of 193.5.†

Sal Absinthii. Salt of wormwood, potassium carbonate, K_2CO_3.†

Salacetin. See ASPIRIN.

Sal Acetosella. Acid potassium oxalate.†

Salactol. Colourless evaporative paint containing 16.7% w/w salicylic acid BP, 16.7% lactic acid BP, 66.6% w/w flexible collodion BP. Applications: For the topical treatment of warts, especially plantar warts. (Dermal Laboratories Ltd).*

Sal Aeratus. Potassium bicarbonate, $KHCO_3$.†

Salaigugl. A local name for Olibanum (*qv*)†

Sal Alembroth. See SALT OF ALEMBROTH.

Sal Alkali Minerale. Sodium carbonate, Na_2CO_3.†

Sal Alkali Vegetable. (Sal Tartari). Potassium carbonate, K_2CO_3.†

Salamac. Compressed blocks of ammonium chloride. (ICI PLC).*

Sal Amarum. (Sal Catharticum, Sal Anglicum, Sal Seidlitense). Magnesium sulphate, $MgSO_4$.†

Sal Ammoniac. (Salmiak, Muriate of Ammonia). Ammonium chloride, NH_4Cl.†

Sal Anglicum. See SAL AMARUM.

Salantin. See ASPIRIN.

Salargyl. A protein-silver preparation.†

Salarmoniac. An ammonium chloride obtained from volcanoes of Central Asia.†

Salaspin. See ASPIRIN.

Sal Auri Philosophicum. Potassium bisulphate, $KHSO_4$.†

Salbulin Inhaler. Pressurized aerosol containing salbutamol BP, delivers 200 measured doses (100 mcg per dose). (Riker Laboratories/3M Health Care).†

Salbulin Syrup. Each 5 ml contains salbutamol BP 2 mg (Riker Laboratories/3M Health Care).†

Salbulin Tablets 2 mg. Each tablet contains salbutamol BP 2 mg (Riker Laboratories/3M Health Care).†

Salbulin Tablets 4 mg. Each tablet contains salbutamol BP 4 mg (Riker Laboratories/3M Health Care).†

Sal Chalybdis. Iron sulphate.†

Sal Commune. Sodium chloride, NaCl.†

Sal Cornu Cervi. Ammonium carbonate, $(NH_4)_2CO_3$.†

Sal Culinaris. Sodium chloride, NaCl.†

Salde Duobus. (Sal Polycrest). Potassium sulphate, K_2SO_4.†

Salde Uvas Picot. Effervescent antacid, for relief of digestive distress. (Richardson-Vicks Inc).*

Sal Digestnum Sylvii. Potassium chloride, KCl.†

Sal Diureticum. Potassium acetate, $(C_2H_3O_2)K$.†

Sal Duobus. See SAL DE DUOBUS.

Salenixon. Crude potassium sulphate, obtained in the manufacture of nitric acid.†

Saleratus. Sodium hydrogen carbonate, $NaHCO_3$.†

Salesthin. Methylene chloride, CH_2Cl_2.†

Sal-ethyl. Ethyl salicylate.†

Saletin. See ASPIRIN.

Salfuride. A proprietary preparation of NIFURSOL. A veterinary anti-protozoan.†

Salge Metal. An alloy of 4 per cent copper, 9.9 per cent tin, 1.1 per cent lead, and 85 per cent zinc.†

Salhar Gum. A gum-resin obtained from *Boswellia serrata*.†

Salicaine. See SALIGENIN.

Saliceral. Glyceryl monosalicylate.†

Salicine Red G, 2G. Dyestuffs similar to Salicine red B.†

Salicitrin. Citrosalic acid.†

Salicolen. Salene, a mixture of ethyl and methyl glycollic acid esters of salicylic acid.†

Salicylanilide. See SALIFEBRIN.

Salicylazosulphapyridine. Sulphasalazine.†

Salicylic Orange. A dyestuff obtained from sulpho-salicylic acid by bromination. Dyes wool or silk dark golden-yellow or orange.†

Salicylic Yellow. A dyestuff obtained from sulpho-salicylic acid by nitration. Dyes wool and silk yellow.†

Salidol. See LUCIDOL.

Salifebrin. Salicylanilide.†

Salinaphthol. See BETOL.

Sali-Presinol. Pharmaceutical preparation. Applications: Antihypertensive. (Bayer & Co).*

Saliretin Resins. See SALIRETINS.

Saliretins. (Saliretin Resins). Resins obtained from saligenin by either heating or treating it with formaldehyde. They are similar to phenol-formaldehyde resins.†

Salitre. Sodium nitrate, $NaNO_3$.†

Salizone. See AGATHINE.

Salkowski's Solution. A solution of phospho-tungstic acid. For albumen in urine.†

Sallit's Speculum Metal. An alloy of 64.6 per cent copper, 31.3 per cent tin, and 4.1 per cent nickel.†

Sally Nixon. Fused nitre cake (acid sodium sulphate).†

Sal Martis. Ferrous sulphate, $FeSO_4$.†

Salmiak. See SAL AMMONIAC.

Sal Mineral. Ferric oxide, Fe_2O_3.†

Sal Mirabil. Sodium sulphate, Na_2SO_4.†

Salmocid. A proprietary preparation of polynoxylin. (Geistlich Sons).†

Salmon Pink. See EOSIN ORANGE.

Salmon Red. A dyestuff. It is the sodium salt of diphenyl-urea-disazo-bi-naphthionic acid, $C_{33}H_{24}N_8Na_2O_7S_2$. Dyes cotton flesh colour to brownish-orange from a boiling alkaline bath.†

Salmon Red. A dyestuff. It is the sodium salt of diphenyl-thiourea-bi-naphthionic acid, $C_{33}H_{24}N_8Na_2O_6S_3$. Dyes cotton orange-red.†

Sal Nitre. Potassium nitrate, KNO_3.†

Salodine. A proprietary preparation. It is an iodized salt.†

Salol Red. A reddish-brown powder, $C_{19}H_{16}O_4$, obtained by heating salicyl-metaphosphoric acid with phenol. It dyes wool.†

Salol. Phenyl salicylate.†

Sal Polycrest. See SAL DE DUOBUS.

Sal Prunella. Potassium nitrate, KNO_3, in balls.†

Sal Rupellensis. Sodium-potassium tartrate.†

Sal Saturni. Lead acetate, $Pb(C_2H_3O_2)_2$.†

Sal Sedativus. Boric acid, H_3BO_3.†

Sal Siedlitense. See SAL AMARUM.

Sal Soda. Sodium carbonate, Na_2CO_3.†

Sal Succini. Succinic acid, $HOOC.CH_2.CH_2.COOH$.†

Salt, Abraum. See ABRAUM SALTS.

Salt, Amido-G. See AMIDO-G-ACID.

Salt, Amido-R. See AMIDO-R-ACID.

Salt, Bitter. See EPSOM SALTS.

Salt Cake. Crude sodium sulphate, Na_2SO_4, produced in the Leblanc soda process.†

Salt, Common. Sodium chloride, NaCl.†

Salt, Fossil. See ROCK SALT.

Salt, Fusible. See MICROCOSMIC SALT.

Salt, Hair. See EPSOM SALTS.

Salt Lick. See PAN SCALE.

Salt, Lister's. See LISTER'S ANTISEPTIC.

Salt, Mining. A mixture of sodium bromate and bromide. It was formerly used in the extraction of gold from its ores.†

Salt of Alembroth. (Salt of Wisdom, Sal-Alembroth). A compound of mercuric chloride and ammonium chloride, $2NH_4Cl.HgCl_2.H_2O$.†

Salt of Amber. Succinic acid, $C_2H_4(COOH)_2$.†

Salt of England. See EPSOM SALTS.

Salt of Hartshorn. Ammonium carbonate, $(NH_4)_2CO_3$.†

Salt of Lemery. Potassium sulphate, K_2SO_4.†

Salt of Lemon. See SALT OF SORREL.

Salt of Norton. Platinum tetrachloride, $PtCl_4.5H_2O$.†

Salt of Saturn. (Sugar of Saturn). Normal lead acetate, $Pb(C_2H_3O_2)_2$, was formerly known under these names.†

Salt of Soda. Sodium carbonate, Na_2CO_3.†

Salt of Sorrel. (Salts of Sonel, Salts of Lemon). The two acid salts of potassium oxalate, C_2O_4HK, and $C_2O_4KH.C_2H_4O_2.2H_2O$, are both sold under these names.†

Salt of Steel. Ferrous sulphate, $FeSO_4$.†

Salt of Tartar. See POTASH.

Salt of Tin. Stannous chloride, $SnCl_4$.†

Salt of Urine, Fusible. See MICROCOSMIC SALT.

Salt of Wisdom. See SALT OF ALEMBROTH.

Salt of Wormwood. (Sal Absinthii). Impure potassium carbonate, K_2CO_3, made from plant ash.†

Salt Perlate. Sodium phosphate, HNa_2PO_4.†

Salt, Phosphorus. See MICROCOSMIC SALT.

Salt, Pinakol. See PINAKOL SALT N.

Salt, Potash. See ABRAUM SALTS.

Salt, Roussin's. See ROUSSIN'S BLACK AND RED SALTS.

Salt, Seignette's. See ROCHELLE SALT.

Salt, Stassfurt. See ABRAUM SALTS.

Salt, Stripping. See ABRAUM SALTS.

Salt, Wonderful. Sodium sulphide, Na_2S.†

Sal Tartari. See SAL ALAKALI VEGETABLE.

Saltpetre. (Nitre). Potassium nitrate, KNO_3.†

Saltpetre, Cubic. See CHILE SALTPETRE.

Saltpetre Flour. Minute crystals of refined saltpetre, KNO_3, used in the manufacture of gunpowder.†

Saltpetre, Lime. See LIME NITRATE.

Saltpetre, Norwegian. See AIR SALTPETRE.

Saltpetre, Peru. See CHILE SALTPETRE.

Saltpetre Rot. Calcium nitrate, $Ca(NO_3)_2.4H_2O$. It causes the rapid disintegration of mortar.†

Saltpetre, Soda. See CHILE SALTPETRE.

Saltpetre Superphosphate. A fertilizer made by mixing nitre with calcium superphosphate.†

Saltpetre, Wall. See NITROCALCITE.

Salts. See EPSOM SALTS.

Salufer. The sodium salt of hydrofluosilicic acid, Na_2SiF_6. An antiseptic.†

Salumin. (Salumen). Aluminium salicylate.†

Salunol. An aqueous solution of sodium hypochlorite. A disinfectant.†

Salupres. Hydrochlorothiazide, reserpine and potassium chloride. Applications: For the treatment of mild to severe hypertension. (Merck Sharp & Dohme).*

Salurene. Hexamethylene-tetramine salicylate.†

Saluric. Chlorothiazide. Applications: For the treatment of oedema and hypertension. (Merck Sharp & Dohme).*

Saluron. Hydroflumethiazide. Applications: Antihypertensive; diuretic. (Bristol Laboratories, Div of Bristol-Myers Co).

Salvarom. The diethyl ester of phthalic acid. A solvent.†

Sal Vegetable. Potassium tartrate, $K_2C_4H_4O_6.1/2H_2O$.†

Salvex. Compounded plastic material. (Mitsubishi Petrochemical Co).*

Sal Volatile. Commercial ammonium carbonate, $(NH_4)_2CO_3$.†

Salysal. Salicyl-salicylate.†

Salzburg Vitriol. See MIXED VITRIOL.

Samaron. Disperse dyestuffs for synthetic surfaces. (Hoechst UK Ltd).

Samite. A trade name for a carborundum product. An abrasive.†

Samli. A clarified butter from East Africa.†

Samorin. Isometamidium salts. (May & Baker Ltd).

Samson Steel. A proprietary trade name for nickel-chromium steel containing 1.25 per cent nickel and 0.6 per cent chromium.†

Samsonite. An explosive containing nitro-glycerin, collodion cotton, potassium nitrate, wood meal, and ammonium oxalate. Used in coal mines.†

San Nai. See SANNA.

Sanachlor. Hypochlorite for sterilizing dairy equipment. (Ciba-Geigy PLC).

Sanaklenz. Agricultural disinfectant. (The Wellcome Foundation Ltd).

Sanatank. Bulk milk tank sanitizer. (The Wellcome Foundation Ltd).

Sanatogen. Casein sodium glycerophosphate. (Fisons PLC, Pharmaceutical Div).

Sancos. A proprietary preparation of pholcodine, menthol and glycerin. A cough linctus. (Sandoz).†

Sancos Compound. A proprietary preparation of pholcodine, pseudoephedrine hydrochloride and chlorpheniramine maleate. A cough linctus. (Sandoz).†

Sand Acid. Hydrofluosilicic acid, H_2SiF_6.†

Sandalwood. See REDWOODS.

Sandalwood, Red. See REDWOODS.

Sandalwood, White. The wood of *Santalum album*.†

Sandarac Resin. A resin obtained from the N.W. African tree *Callitris quadrivalis*. It is used in the manufacture of spirit varnishes. Pine gum or Australian sandarac is obtained from *Callitris* species in Australia, and resembles the African variety.†

Sandaracha. Synonym for Realgar.†

Sandel Red. The red resinous colouring principle of Sanderswood.†

Sandelwood. See REDWOODS.

Sanderswood. See REDWOODS.

Sander's Blue. Anhydrous basic copper carbonate.†

Sandiver. (Glass Gall). The scum formed on the surface of molten glass. It consists of calcium and sodium sulphates, with about one-tenth of its weight of glass.†

Sandix. See ORANGE LEAD.

Sando-K. A proprietary preparation of potassium chloride. (Sandoz).†

Sandocal. A proprietary preparation of calcium lactate gluconate and sodium and potassium bicarbonate, used in the treatment of osteoporosis. (Sandoz).†

Sandoce. Methyl-saccharin, $C_6H_3(CH_3)$. $CO.SO_2.NH$. A sweetening substance.†

Sandofix. Fixing agents. Applications: Direct and reactive dyes. (Sandoz Products Ltd).

Sandogen. Aromatic sulphonate, levelling agent. Applications: Barry polyamide. (Sandoz Products Ltd).

Sandolube. Softening and lubricating agent. Applications: Improve handling of natural and synthetic fibres. (Sandoz Products Ltd).

Sandomigran. Pizotyline. Applications: Anabolic; antidepressant; serotonin inhibitor. (Sandoz Pharmaceuticals, Div of Sandoz Inc).

Sandopan. Very low foam detergent. Applications: Removal of winding preparations from polyester fibres. (Sandoz Products Ltd).

Sandoptal. Butalbital. Applications: Sedative. (Sandoz Pharmaceuticals, Div of Sandoz Inc).

Sandopur. Washing off assistant; dyestuff complexing and fixing agent. Applications: Printed polyamide fabrics; improvement of wet fastness of metal complex and acid milling dyes on wool. (Sandoz Products Ltd).

Sandotex. Low soiling antistatic agent. Applications: Finishing synthetic fibre carpets. (Sandoz Products Ltd).

Sandoz Effervescent Compound. A proprietary preparation of caffeine, pseudoephedrine hydrochloride and paracetamol. An analgesic. (Sandoz).†

Sandozin. Powerful and economic wetting agent. Applications: Neutral and weakly acid media; many uses in textile processing. (Sandoz Products Ltd).

Sandril. Reserpine. Applications: Antihypertensive. (Eli Lilly & Co).

Sandscale. Impurities formed in the pan during the concentration of brine. It consists of calcium carbonate, $CaCO_3$.†

Sandstone. A stone consisting of grains of sand cemented together by a cementing material, the most common being silica, carbonate of iron, and oxide of iron. It is called calcareous sandstone when the grains are united together with calcium carbonate, argillaceous sandstone when a clayey cement unites the particles, micaceous when scales of mica are present, felspathic when grains of felspar are there, and ferruginous when ferric oxide or hydroxide is present.†

Sanfoin. See ORCHIDEE.

Sangajol. A trade mark for a fraction of Bormeo petroleum distillate boiling at 160-170° C. It contains cyclic hydrocarbons, and is used as a turpentine substitute and resin solvent.†

Sanguial. A preparation of blood and iron, containing 10 parts hemoglobin, 44 parts muscle albumin, and 46 parts blood salts. Prescribed for chlorosis.†

Saniblanket. Urea resin based foams. Applications: Cover for refuse and hazardous waste. (Sanifoam Inc).*

SaniFoam. Urea resin based foams. Applications: Cover for refuse and hazardous waste. (Sanifoam Inc).*

Sani-Soil-Set. Dust laying composition. (Chevron).*

Sanitant. Dairy hygiene detergent sterilizer. (The Wellcome Foundation Ltd).

Sanitas. An aqueous liquid prepared by blowing air through warm oil of turpentine, in contact with water. It contains hydrogen peroxide and thymol, and is used as a disinfectant.†

Sanna. (San Nai, Kapur Kachri, Sitruti, Sheduri). A Chinese drug. It is the dried roots and stems of *Hedychium spicatum*.†

Sanoform. A disinfectant consisting of a mixture of the disinfecting constituents of various tar oils, with calcium chloride and magnesium chloride in a saponified form. The term is also used for the methyl ester of diiodo-salicylic acid, an iodoform substitute.†

Sanoleum. A mixture of crude cresols with hydrocarbons, used in disinfecting urinals.†

Sanoma. A proprietary preparation of carisprodol. A muscle relaxant. (Pfizer International).*

Sanorex. Mazindol. Applications: Anorexic. (Sandoz Pharmaceuticals, Div of Sandoz Inc).

Sanoscent. (Camphortar). Preparations containing camphor as the main ingredient. Disinfectants.†

Sansalid. Uredofos. Applications: Anthelmintic (veterinary). (Beecham Laboratories).

Sanse. The residual cakes obtained from pressed Italian olives. When dried, and the oil extracted with carbon disulphide, it gives the so-called sulphocarbon oil, or sulphur olive oils. Used in the manufacture of green soap for use in the textile industries.†

Sanse Oil. See SANSE.

Sansert. Methysergide maleate. Applications: Vasoconstrictor. (Sandoz Pharmaceuticals, Div of Sandoz Inc).

Sanspor. Fungicide for use on potatoes containing captafol. (ICI PLC).*

Santal. The red colouring matter of Sanderswood.†

Santal Oil. Oil of sandal wood.†

Santalwood. See REDWOODS. It contains 16 per cent of santalin, $C_{15}H_{14}O_5$. The extract is used to colour confectionery and liqueurs.

Santel. Water-based acrylic resins and coatings. Applications: Surface coatings of paper, board, foil and films, gloss enhancers, surface protection, barrier coatings, grease and oil resistance for paper. (ADM Tronics Unlimited Inc).*

Santheose. Theobromine, $C_5H_2(CH_3)_2.N_4O_2$.†

Santiago New Yellows E, K. Dyestuffs. They are fustic preparations.†

Santiciser. A proprietary trade name for vinyl plasticisers coded as follows:
M-17. Methyl phthalyl ethyl glycollate.
1-H. N-cyclohexyl paratoluenesulphonamide.
107. Di-2-ethyl hexyl phthalate.
140. Cresyl diphenyl phosphate.
141. Octyl diphenyl phosphate.
160. Butyl benzyl phthalate.†

Santiciser SC. A proprietary trade name for a vinyl plasticizer, the triglycol ester of a vegetable oil fatty acid. (Harwick Standard Chemical Co).†

Santicizer DUP. Plasticizer. (Monsanto PLC).

Santicizer 10. A proprietary trade name for a plasticizer. *o*-Cresyl-*p*-toluene sulphonate.†

Santicizer 141. Alkylaryl phosphate. Applications: Used in PVC films, sheets, extrusions, mouldings, organosols and plastisols, flame retardant plasticizer. (Monsanto Co).*

Santicizer 143. Modified triaryl phosphate. Applications: plasticizer compatible with PVC, vinyl nitrile elastomers, late emulsions and cellulosic materials. (Monsanto Co).*

Santicizer 148. Alkyl diaryl phosphate. Applications: Flame-retardant plasticizer for PVC resins. (Monsanto Co).*

Santicizer 154. Triaryl phosphate. Applications: Flame-retardant plasticizer, compatible with PVC resins and vinyl nitrile rubber, PVA emulsion and cellulosics. (Monsanto Co).*

Santicizer 160. Butyl benzyl phthalate. Applications: General purpose plasticizer for PVC, used in flooring industry. (Monsanto Co).*

Santicizer 160. Benzyl butyl phthalate plasticizer. (Monsanto PLC).

Santicizer 711. Dialkyl adipate. Applications: General purpose plasticizer for PVC resins, outperforms DOP. (Monsanto Co).*

Santicizer 97. Dialkyl phthalate. Applications: plasticizer for PVC film, sheet and coatings; gives low temperature flexibility. (Monsanto Co).*

Santobrite. A proprietary trade name for sodium pentachlorphenate. A preservative used in paints, adhesives, etc.†

Santocel. A proprietary trade name for silica gel (*qv*), a porous form of silica. It is used as a heat insulator, drying agent, etc.†

Santochlor. A proprietary trade name for *p*-dichlorobenzene. Used as a deodorizer, moth preventative, etc.†

Santocure. N-Cyclohexyl-2-benzothiazolesulphenamide. Applications: vulcanization accelerator. (Monsanto Co).*

Santocure MOR/MOR90. 4-(benzothiazole-2-sulphenyl morpholine rubber accelerators. (Monsanto PLC).

Santoflex. Antidegradant, antioxidant, antiozonant. Applications: Rubber processing chemical. (Monsanto Co).*

Santoflex A. A proprietary trade name for a mixed ketone-amine and diphenyl-*p*-phenylene-diamine. A rubber anti-oxidant.†

Santoflex AW. Rubber antiozonant. (Monsanto PLC).

Santoflex B. A proprietary trade name for the condensation product of acetone and *p*-amino-diphenyl. A rubber antioxidant.†

Santoflex BX. A proprietary trade name for a constant composition blend of Santoflex B and diphenylparapheny lenediamine. A rubber antioxidant.†

Santoflex DD. Rubber antioxidant. (Monsanto PLC).

Santoflex DPA. Rubber antioxidant. (Monsanto PLC).

Santoflex 1P. Antiozonant/antioxidant. (Monsanto PLC).

Santoflex 13. Antiozonant. (Monsanto PLC).

Santoflex 77. Antiozonant. (Monsanto PLC).

Santogard PVI. N-(Cyclohexylthio)phthalimide. Applications: Prevulcanization inhibitor for natural and synthetic rubber. (Monsanto Co).*

Santolite. A proprietary trade name for synthetic resins of the sulphonamide aldehyde type, eg. toluene sulphonamide-formaldehyde, for use in lacquers, etc.†

Santomerse. A proprietary trade name for an alkylated aryl sulphonate. Used as a wetting agent.†

Santonox. 4,4'-Thiobis(6-tert-butyl-M-cresol). Applications: Antioxidants for polyethylene and other plastic resins. (Monsanto Co).*

Santophen 20. A proprietary trade name for pentachlorphenol. A preservative for paints, wood, adhesives, etc.†

Santoprene. Themoplastics rubber. Applications: For engineered industrial rubber applications including mechanical rubber goods, hose, automotive products, wire and cable and sheeting. (Monsanto Co).*

Santoquin. Ethoxyquin. Applications: Feed preservative. (Monsanto Co).*

Santo-Res. Wet strength paper resins. Applications: Cationic retention aid for paper. (Monsanto Co).*

Santoresin. A proprietary trade name for a synthetic resin.†

Santorin Earth. A volcanic ash found in the island of Santorin. Used to convert lime into hydraulic lime.†

Santosite. A proprietary trade name for anhydrous sodium sulphite. A reducing agent.†

Santotan KR. A proprietary trade name for a basic chromium sulphate, $Cr_2(SO_4)_3.(OH)_2$. A tanning agent.†

Santotrac. Synthetic hydrocarbons. Applications: High temperature lubricant. (Monsanto Co).*

Santovac. Polyphenyl ether. Applications: Vacuum diffusion pump fluid. (Monsanto Co).*

Santovar. 2,5-Di(tert-amyl)hydroquinone. Applications: Antioxidant. (Monsanto Co).*

Santovar A. Antioxidant for unvulcanized rubber. (Monsanto PLC).

Santovar 0. See SANTOVAR A.

Santovar-O. Insoluble sulphur 60.†

Santowax. Mixed isomeric terphenyls. Applications: High melting hydrocarbons. (Monsanto Co).*

Santoweb. Treated cellulosic short fibre. Applications: Rubber processing special material. (Monsanto Co).*

Santowhite. 4,4'-Butylidenebis(6-tert-butyl-m-cresol). Applications: Antioxidant for polypropylene, polyethylene, nylon moulding powders and other polymer resins. (Monsanto Co).*

Sanyan. A silk from a wild silkworm of Nigeria.†

Sap Green. (Buckthorn Green, Vegetable Green, Bladder Green, Chinese Green, Lokas, Iris Green). The colouring matter obtained by evaporating to dryness a mixture of lime, or sometimes a little alum, and indigo carmine with the juice of the berries of buckthorn. It is employed in China for giving green shades on silk.†

Sap Yellow. A vegetable dyestuff obtained from the half-ripe berries of various species of *Rhamnus*. Used in the form of a lake in painting.†

Sapamine. A proprietary trade name for diethylaminoethyloleylamino acetate and similar compounds. Used in conjunction with dyes.†

Sapamine. Finishing agent. (Ciba-Geigy PLC).

Sapan Wood. See REDWOODS.

Sapecron. Organophosphorus insecticide. (Ciba-Geigy PLC).

Sapene. A liquid soap, a vehicle for medicaments.†

Saphire. Lube oil. (Chevron).*

Sapin. A mixture of Japan wax with heavy mineral oil (soft or liquid paraffin). A superfatting agent for soaps.†

Sapocarbol. See LYSOL.

Sapoform. A product containing oleic acid, alcohol, potassium hydroxide, formalin, and distilled water.†

Sapogenat. Alkylphenol polyglycol detergent base. (Hoechst UK Ltd).

Sapogenat T. Range of nonionic surfactants of the tri-butyl phenol ethoxylate type in liquid, paste or wax form. Application: Auxiliaries in textile and paper manufacturing, domestic and industrial cleaning agents, emulsifiers and plant protection agents. (Hoechst UK Ltd).

Sapogenin. A decomposition product ($C_{14}H_{22}O_2$) of saponin.†

Saponification Olein. See OLEIN OF SAPONIFICATION.

Saponine. A name usually applied to the active constituent of Panama bark, which is used instead of soap, for washing and producing a lather. The term is also used for a boring and cutting oil.†

Saponite. See STEATITE.

Saprol. (Disinfection Oil). A mixture of crude cresols, hydrocarbons, and pyridine bases. Used for disinfecting lavatories.†

Saquadil. Sulphaquinoxaline/diaveridine. (May & Baker Ltd).

Saran. A range of polyvinylidene chloride plastics. (Dow Chemical Co Ltd).*

Saranex. Coextruded multilayered films. (Dow Chemical Co Ltd).*

Saratoga Steel. A proprietary trade name for a non-shrinking steel containing small quantities of manganese, chromium, tungsten, and carbon.†

Sarcine. (Sarkine). Hypoxanthine, $C_5H_4N_4O$.†

Sarco. A material made from elaterite (a mineral rubber). Used in rubber mixings.†

Sarcocoll. A gum resin from *Penoea sarcocolla*, of Africa.†

Sarcosine. See METHYL-GLYCOCOLL.

Sarenin. Saralasin Acetate. Applications: Antihypertensive. (Norwich Eaton Pharmaceuticals Inc).

Saridone. A proprietary preparation of phenacetin with caffeine and phenyl-dimethyl-isopropyl-pyrazolone. Also, a proprietary preparation of isopropyl-antipyrine, phenacetin and caffeine. An analgesic. (Roche Products Ltd).*

Sarkalyt. A solution of the sodium salt of adenosin-phosphoric acid.†

Sarkine. See SARCINE.

Sarkosyl NL30 and O. Anionic surfactants of the N-acyl sarcosine type, in the form of a colourless solution. NL30 is sodium lauroyl sarcosinate; O is oleoyl sarcosine in acid form. Properties include corrosion inhibition, enzyme inhibition, bacteristatic activity and surfactant properties which are often enhanced by combination with other surfactants eg lauryl sulphate. Applications: Cosmetics, toilet goods, pharmaceuticals, particularly useful in dentifrices, hair, carpet and upholstery shampoos; speciality and alkaline detergents; window cleaners, hand dishwashing formulations; fine fabric detergents; synthetic toilet soap; emulsion polymerization; metal processing; food products. (Ciba-Geigy PLC).

Saroten. A proprietary preparation of amitriptyline hydrochloride. An anti-depressant. (Warner).†

Sarpol. A preparation of crude phenol (carbolic acid). A disinfectant.†

Sarsaparilla. The dried root of *Smilax officinalis* and other species. A tonic.†

Sasetone. Acetone. Applications: Solvent in paints, varnishes, lacquer thinners, printing inks, nail polish removers, acetylene in filling of cylinders, bituminous paints, polyester resins, PVC cloth manufacture, explosives, adhesives. Raw material for manufacture of methyl isobutyl ketone, diacetone alcohol, hexylene glycol and fine chemicals. (Sasolchem).*

Sasolwaks. Hard, high melting point, crystalline paraffin wax. Average molecular formula $C_{50}H102x$. Applications: Constituent in polishes, plastics to enhance gloss, colour dispersion etc., insulating components in electric cables, paper conversion, chewing gums, carbon paper backing, printing inks, paints, hot melt adhesives, lubricant in rubber, laundry machines and plastic moulding. (Sasolchem).*

Sassafras. The dried bark of the root of *Sassafras variifolium*. The extract is used as a carminative.†

Sassolin. Tuscan boric acid.†

Sassy Bark. (Saucy Bark). The bark of *Erythrophleum guineanse*.†

Satco Metal. A proprietary trade name for a lead-base bearing alloy modified by the addition of tin, calcium, magnesium, mercury, aluminium, potassium lithium, all in very small quantities except tin which may rise to 1 per cent.†

Satessa. A highly dispersed pyrogenic silica preparation. Applications: In the textile industry as a lubricating additive; to increase strength of yarn; to increase fibre friction; to increase nonslip characteristics. (Degussa).*

Satin Green. See CHROME GREENS.

Satin Rouge. A variety of lamp-black used for polishing.†

Satin White. A pigment. It consists of gypsum mixed with alumina. A mixture of calcium sulphate with aluminium sulphate is also known under this name.†

Satin-gloss Black. See GAS BLACK.

Satinite. See GYPSUM.

Sativic Acid. Trihydroxy-stearic acid, $C_{17}H_{31}(OH)_3.COOH$.†

Satrapol. A photographic developer containing monomethyl *p*-amino-phenolsulphate.†

Satric. Metronidazole. Applications: Antiprotozoal. (Savage Laboratories, Div of Byk-Gulden Inc).

Satulan. Hydrogenated lanolin. (Croda Chemicals Ltd).

Saturn Glace. Polymers, floral carbons composition. Applications: Applied to new and used vehicles. (Adasco-Inc).*

Saturn Red. See RED LEAD and ORANGE LEAD.

Sauconite. A clay containing zinc.†

Saucy Bark. See SASSY BARK.

Sauflon PW. Lidofilcon B. The material contains 79% of water. Applications: Contact lens material. (Visiontech Inc).

Sauflon 70. Lidofilcon A. The material contains 70% of water. Applications: Contact lens material. (Visiontech Inc).

Saunderswood, Red. See REDWOODS.

Saurol. A distillation product of Meride shale. It resembles ichthyol in its therapeutic properties.†

Savacort. Prednisolone Acetate. Applications: Glucocorticoid. (Savage Laboratories, Div of Byk-Gulden Inc).

Savall. Insecticide, containing quinalphos. (ICI PLC).*

Saventrine. A proprietary preparation of isoprenaline hydrochloride, used as a cardiac stimulant. (Pharmax).†

Savinase. A dust-free granulate - a liquid preparation containing a proteolytic enzyme. Applications: Used in the detergent industry as an additive to powder detergents to improve the detergency towards protein containing stains and used in the detergent industry, primarily as an additive to non-built liquid detergents. (Novo Industri A/S).*

Savloclens. Broad spectrum antiseptic with added detergent properties containing chlorhexidine. (ICI PLC).*

Savlodil. Broad spectrum antiseptic with added detergent properties containing chlorhexidine. (ICI PLC).*

Savlon. A proprietary preparation of CHLORHEXIDINE and Cetrimide, used for a range of antiseptics. (Same sold in UK by Care Laboratories Ltd). (ICI PLC).*

Savlon Babycare. A range of products containing chlorhexidine, principally antiseptics, for use by babies and infants. (ICI PLC).*

Savol. A medicated soap. It contains salol (phenyl salicylate) with perfumes.†

Savonade. (Texapon, Texalin, Hydralin). Liquid hexalin and methyl-hexalin soaps.†

Savonette Oil. A mixture of vegetable fatty acids and resin acids, a by-product of paper manufacture. It is recommended as a substitute for oleic acid in soap manufacture.†

Saxifragin. An explosive mixture containing 76 per cent barium nitrate, 2 per cent potassium nitrate, and 22 per cent charcoal.†

Saxin. A proprietary preparation of saccharin.†

Saxin. Artificial sweetening agent. (The Wellcome Foundation Ltd).

Saxol. See PARAFFIN, LIQUID.

Saxoline. See OZOKERINE.

Saxon. Lotion for men (woodspice and musk), aftershave skin conditioner. (Richardson-Vicks Inc).*

Saxon Blue. The term is usually applied to a solution of indigo in sulphuric acid, but it is also a synonym for smalt.†

Saxon Green. A pigment. It is a green earth found in Saxony.†

Saxon Verdigris. A pigment which approaches Brunswick green in composition, and is prepared by precipitating a mixture of copper sulphate and sodium chloride with milk of lime. It also contains gypsum or Vienna white.†

Saxonite. An explosive similar to Samsonite in composition. The term is also used for a mineral which is a mixture of olivine and enstatite.†

Saxony Blue. See SMALT.

SB-VAC. SB-1 strain, frozen chicken herpes virus marek's vaccine. Applications: immunization of poultry. (Intervet America Inc).*

SB-VAC Plus Marexine-CA. For combination use. Applications: immunization of poultry. (Intervet America Inc).*

Scabene Lotion. Lindane. Applications: Pediculicide; scabicide. (Stiefel Laboratories Inc).

Scadoplast RA3L, RA350. A proprietary trade name for an adipic acid polyester vinyl plasticizer. (DSM Kunstharze Gmbh).†

Scadoplast RS 20, RS 150. A proprietary trade name for a sebacic acid polyester vinyl plasticizer. (DSM Kunstharze Gmbh).†

Scagliola. A stone manufactured from Keene's cement mixed with colouring matter, to which is added water containing dissolved glue or isinglass.†

Scale. Crude Scotch paraffin wax is known in commerce by this name.†

Scale Cleen. Dry acid descaler. (Dearborn Chemicals Ltd).

Scalol. A photographic developer containing methyl-*p*-amino-phenol, $C_6H_4(OH)(NH.CH_3)$, as the active constituent.†

Scandia. Scandium oxide, Sc_2O_3.†

Scarab. Urea formaldehyde compounds. Applications: Moulding. (BIP Chemicals Ltd).*

Scarat. A proprietary trade name for a synthetic resin of the urea type.†

Scarlet B. See BIEBRICH SCARLET.

Scarlet EC. See BIEBRICH SCARLET.

Scarlet F. See NEW COCCINE.

Scarlet for Cotton. A dyestuff. It is a mixture of Chrysoidine with Safranine.†

Scarlet for Silk. See FAST RED B.

Scarlet G. See PONCEAU R.

Scarlet GR. (Scarlet R, Orange N, Orange L, Brilliant Orange R, Xylidine Orange, Naphtharene Orange R, Scarlet RL). A dyestuff. It is the sodium salt of xyleneazo-β-naphthol monosulphonic acid, $C_{18}H_{15}N_2Na.O_4S$. Dyes wool yellowish-red from an acid bath.†

Scarlet GT. A dyestuff obtained from *p*-toluidine and β- naphtholsulphonic acid.†

Scarlet J, JJ. See EOSIN BN.

Scarlet Lake. A pigment. It is a carmine lake with a scarlet hue, which is imparted to it by mixture with vermilion.†

Scarlet Ochre. See INDIAN RED.

Scarlet Phosphorus. (Schenk's Phosphorus). Obtained by heating 70 parts phosphorus with 30 parts phosphorus tribromide, PBr_3.†

Scarlet R. See SCARLET GR.

Scarlet Red. See IODINE RED and BIEBRICH SCARLET.

Scarlet RL. A dyestuff. It is a British equivalent of Scarlet GR.†

Scarlet S. See FAST PONCEAU 2B.

Scarlet Spirit. (Bowl Spirit). A Tin Spirit (*qv*); used by wool dyers for producing cochineal scarlet.†

Scarlet Vermillon. (Extract of Vermilion, Chinese Vermilion, Orange Vermilion, Field's Orange Vermilion). Varieties of vermilion, see CINNABAR.†

Scarlet 000. See CROCEINE 3BX.

Scarlet 2R, 2RL, R, 3R. British equivalents of Ponceau.†

Scarlet 3B, 3R 4R. Varieties of Biebrich Scarlet (*qv*).†

Scarlet 5R. See PONCEAU 5R.

Scarlet 6R. (Ponceau 6R, Amaranth). A dyestuff. It is the sodium salt of *p*-sulphonaphthaleneazo-β-naphthol-trisulphonic acid, $C_{20}H_{10}N_2Na_4]O_{13}S_4$. Dyes wool red from an acid bath.†

Scatole. (Skatole).*N*-Methylindole.†

Scent Sticks. Flammable, scented, oil saturated, compressed wood pulp on a sandalwood stick (incense). Applications: Used as a scented air freshener (incense), novelty item. (Ambrosia Scents).*

Schaeffer's Acid. (Baum's Acid, Armstrong Acid). α-Naphthol-2-sulphonic acid, $C_{10}H_6(OH)(SO_3H)$, also β-naphthol-6-sulphonic acid.†

Schaeffer's Salt. The sodium salt of β-naphthol-6-sulphonic acid.†

Schallerite. An arseno-silicate found in the Franklin furnace. It approximates to the formula, $9MnSiO_3.Mn_3As_2O_8.7H_2O$.†

Scheeletine. (Scheelinite). A tungstate of lead, $PbWO_4$.†

Scheele's Acid. A 4 per cent solution of hydrocyanic acid, HCN.†

Scheele's Green. (Mineral Green, Swedish Green). A pigment consisting of copper arsenite, $CuHAsO_3$.†

Scheelinite. See SCHEELETINE.

Scheerite. A mineral wax resembling ozokerite.†

Scheiber Oil. The glyceride of dehydrated ricinoleic acid. It is suitable for varnishes.†

Scheibler's Reagent. Sodium-phosphotungstate, obtained by dissolving 100 grams sodium tungstate and 70 grams sodium phosphate in 500 cc water, and acidifying with nitric acid. Used as a testing reagent for alkaloids.†

Scheiderite. A mixture of trinitro naphthalene and ammonium nitrate. An explosive.†

Schellan Solution. A colloidal solution of a synthetic resin made from urea and formaldehyde, and kept from gelatinising by means of sodium acetate. Used as a dressing material for textiles.†

Schenk's Phosphorus. See SCARLET PHOSPHORUS.

Schercassist AC. Quaternary. Applications: Dye-levelling agent for acrylics. (Scher Chemicals Inc).*

Schercemol BE. Behenyl erucate. Applications: Emollient wax. (Scher Chemicals Inc).*

Schercemol CM. Cetyl myristate. Applications: Solid emollient, lubricant and body builder. (Scher Chemicals Inc).*

Schercemol CO. Cetyl octanoate. Applications: Good solvency properties, useful for makeup removers. (Scher Chemicals Inc).*

Schercemol CP. Cetyl palmitate. Applications: Synthetic spermaceti wax. (Scher Chemicals Inc).*

Schercemol CS. Cetyl stearate. Applications: Waxy emollient for creams and lotions, thickener and body builder. (Scher Chemicals Inc).*

Schercemol DEGMS. PEG-2 stearate. Applications: Primary emulsifier in creams and lotions. (Scher Chemicals Inc).*

Schercemol DEIS. Decyl isostearate. Applications: Emollient, lubricant and penetrant with unusual pigment-dispersing properties. (Scher Chemicals Inc).*

Schercemol DIA. Diiosopropyl adipate. Applications: Penetrating emollient and solvent for creams and lotions. (Scher Chemicals Inc).*

Schercemol DICA. Diisocetyl adipate. Applications: Low viscosity emollient, useful in skin and hair preparations. (Scher Chemicals Inc).*

Schercemol DID. Diisopropyl dimerate. Applications: Emollient, excellent ingredient for lipstick and lipgloss preparations. (Scher Chemicals Inc).*

Schercemol DIS. Diisopropyl sebacate. Applications: Emollient, solubilizer and coupling agent in creams, lotions and bath oils. (Scher Chemicals Inc).*

Schercemol DISD. Diisostearyl dimerate. Applications: Heavy moisturizing emollient. Suitable for rich night creams, lipsticks and makeup formulations. (Scher Chemicals Inc).*

Schercemol DO. Decyl oleate. Applications: Emollient, lubricant and penetrant. (Scher Chemicals Inc).*

Schercemol EE. Erucyl erucate. Applications: Emollient ester for use in skin, hair and suntanning preparations. (Scher Chemicals Inc).*

Schercemol EGMS. Glycol stearate. Applications: Emulsifier, opacifier and pearling agent in hair and skin preparations. (Scher Chemicals Inc).*

Schercemol GMIS. Glycerol isostearate. Applications: Emulsifier and emollient for creams and lotions. (Scher Chemicals Inc).*

Schercemol GMS. Glycerol stearate. Applications: Primary emulsifier for creams and lotions. (Scher Chemicals Inc).*

Schercemol ICS. Isocetyl stearate. Applications: Light liquid emollient. (Scher Chemicals Inc).*

Schercemol ISE. Isostearyl erucate. Applications: Lubricating emollient for skin and bath preparations. (Scher Chemicals Inc).*

Schercemol MEL-3. Myreth-3 laurate. Applications: Emollient, solubilizer and coupling agent in creams, lotions and bath oils. (Scher Chemicals Inc).*

Schercemol MEM-3. Myreth-3 myristate. Applications: Emollient, solubilizer and coupling agent in creams, lotions and bath oils. (Scher Chemicals Inc).*

Schercemol MEP-3. Myreth-3 palmitate. Applications: Emollient, solubilizer and coupling agent in creams, lotions and bath oils. (Scher Chemicals Inc).*

Schercemol MM. Myristyl myristate. Applications: Solid emollient for creams and lotions. Viscosity builder. (Scher Chemicals Inc).*

Schercemol MP. Myristyl propionate. Applications: Liquid emollient for anti-perspirants, body oils, creams and lotions. (Scher Chemicals Inc).*

Schercemol MS. Myristyl stearate. Applications: Waxy emollient for creams and lotions. (Scher Chemicals Inc).*

Schercemol NGDC. Neopentyl glycol dicaprate. Applications: Good solvency properties, used in make-up removers. (Scher Chemicals Inc).*

Schercemol OLO. Oleyl oleate. Applications: Emollient, cosolvent and solubilizer in cosmetic preparations. (Scher Chemicals Inc).*

Schercemol OP. 2-Ethyl hexyl palmitate. Applications: Anti-tack agent for antiperspirants, creams and lotions. (Scher Chemicals Inc).*

Schercemol OPG. 2-Ethyl hexyl pelargonate. Applications: Emollient and binder for cosmetic preparations. (Scher Chemicals Inc).*

Schercemol PGDP. Propylene glycol dipelargonate. Applications: Emollient and cosolvent for cosmetics. (Scher Chemicals Inc).*

Schercemol PGML. Propylene glycol laurate. Applications: Emollient and solvent in lotions and lipsticks. (Scher Chemicals Inc).*

Schercemol PGMS. Propylene glycol stearate. Applications: Primary emulsifier for lotions and low viscosity creams. (Scher Chemicals Inc).*

Schercemol SE. Stearyl erucate. Applications: Emollient wax with the look and feel of real cocoa butter. (Scher Chemicals Inc).*

Schercemol TIST. Triisostearyl trimerate. Applications: Has emolliency, shine, viscosity and good binding properties. (Scher Chemicals Inc).*

Schercemol TT. Triisopropyl trimerate. Applications: Binder for pigmented products. Imparts gloss and sheen in makeup and hair preparations. (Scher Chemicals Inc).*

Schercemol 1688. Cetearyl octanoate. Applications: Emollient for use in bath and skin preparations where a silky, water-resistant barrier is required. (Scher Chemicals Inc).*

Schercemol 1818. Isostearyl isostearate. Applications: Emollient in creams and lotions. Cosolvent and solubilizer in perfumes. (Scher Chemicals Inc).*

Schercemol 185. Isostearyl neopentanoate. Applications: Emollient for bath oils, creams and lotions that impart freeze-thaw stability. Binder for pigment systems in make-up preparations. (Scher Chemicals Inc).*

Schercemol 318. Isopropyl isostearate. Applications: Low cloud point emollient for creams. (Scher Chemicals Inc).*

Scherco Finish AL. Resin dispersion. Applications: Lubricant for sewing, cutting, napping and softening all textiles. (Scher Chemicals Inc).*

Scherco Softener £1. Quaternary. Applications: Softener and finishing agent for orlon and acrilan. (Scher Chemicals Inc).*

Scherco Softener £2. Quaternary. Applications: Softening agent for acrylics and synthetics. (Scher Chemicals Inc).*

Schercoat OE-44. Fatty oil, edible grade. Applications: Coating for liquor and food glass containers. (Scher Chemicals Inc).*

Schercoat OE-44K. High purity fatty oil. Applications: Protective coating for glass food containers. (Scher Chemicals Inc).*

Schercoat P-110. Modified polyethylene emulsion. Applications: Protective coatings for glass bottles. (Scher Chemicals Inc).*

Schercoat PC-550. Substantive poly emulsion. Applications: Lubricant for glass containers. (Scher Chemicals Inc).*

Schercoat S-220. Modified vinyl resin ionomer emulsion. Applications: Protective coatings for glass containers. (Scher Chemicals Inc).*

Schercoat S-330. Stabilized vinyl resin ionomer emulsion. Applications: Roller coating of glass bottles to impart scuff resistance. (Scher Chemicals Inc).*

Schercolene SB. Detergent. Applications: Heavy duty textile scouring agent. (Scher Chemicals Inc).*

Schercolube 707. Fatty ester ethoxylate. Applications: Softener for knit goods and lubricant for nylon separator threads in sweater bodies. (Scher Chemicals Inc).*

Schercomid EAC. Modified coco amide. Applications: Wool scouring and fulling agent effective at low temperature. (Scher Chemicals Inc).*

Schercomid EACS-100. Mixed fatty amide. Applications: Cold water textile detergent and wool fulling agent. (Scher Chemicals Inc).*

Schercomid 304. Modified coco amide. Applications: Dry-cleaning detergent. (Scher Chemicals Inc).*

Schercopon 2WD. Ethoxylated sulphosuccinate. Applications: Water-white, dry-cleaning detergent. (Scher Chemicals Inc).*

Schercoquat DAS. Quaternium-61. Applications: Liquid quaternary for cosmetic preparations. (Scher Chemicals Inc).*

Schercoquat IAS. Isostearamidopropyl ethyl dimonium ethosulphate. Applications: Contributes body, compatibility and antistatic properties to shampoos. (Scher Chemicals Inc).*

Schercoquat IB. Isostearamidopropyl alkonium chloride. Applications: Liquid quaternary, possessing some bactericidal activity, used in conditioners, hair rinses and skin lotions. (Scher Chemicals Inc).*

Schercoquat IEP. Quaternium 62. Applications: Specialty quaternary. (Scher Chemicals Inc).*

Schercoquat IIB. Isostearyl benzyl imidonium chloride. Applications: Quaternary possessing bactericidal activity. (Scher Chemicals Inc).*

Schercoquat IIS. Isostearyl ethyl imidonium ethosulphate. Applications: Concentrated quaternary for cosmetic preparations. (Scher Chemicals Inc).*

Schercoquat SOAB. Soyamidopropyl benzyldimonium chloride. Applications: Conditioning agent used in hair preparations. (Scher Chemicals Inc).*

Schercoquat SOAS. Soyamidopropylethyldimonium ethosulphate. Applications: Used in hair conditioners. (Scher Chemicals Inc).*

Schercosol DS. Chlorinated solvent. Applications: Dry side rapid stain remover. (Scher Chemicals Inc).*

Schercosol NL. Modified coco amide. Applications: Wet side spotter and fibre lubricant. (Scher Chemicals Inc).*

Schercosol P. Sulphated amide. Applications: Protein stain remover. (Scher Chemicals Inc).*

Schercosol T. Acid stable detergent. Applications: Tanin stain remover. (Scher Chemicals Inc).*

Schercotarder. Fatty amide. Applications: Low temperature wool scouring and fulling agent. Post-scouring agent for dyed or printed goods. (Scher Chemicals Inc).*

Schercowet DOS-70. Sulphosuccinate. Applications: Wetting and rewetting agent for all wet processing. (Scher Chemicals Inc).*

Schercozoline C. Cocoyl imidazoline. Applications: Detergent, wetting agent and antistat. (Scher Chemicals Inc).*

Schercozoline I. Isostearyl imidazoline. Applications: Surfactant, softener and antistatic agent. (Scher Chemicals Inc).*

Schercozoline L. Lauryl imidazoline. Applications: Detergent, wetting agent and antistat. (Scher Chemicals Inc).*

Schercozoline O. Oleyl imidazoline. Applications: W/o emulsifier and corrosion inhibitor. (Scher Chemicals Inc).*

Schercozoline S. Stearyl imidazoline. Applications: Surfactant, softener, and antistatic agent. (Scher Chemicals Inc).*

Schericur. A proprietary preparation of hydrocortisone and clemizole- hexachlorophane for dermatological use (Schering Chemicals Ltd).†

Schering I and II. See MINERAL TABLETS.

Scheriproct. A proprietary preparation of PREDNISOLONE. cinchocaine hexachlorophane and CLEMIZOLE undecylenate used in the treatment of haemorrhoids. (Schering Chemicals Ltd).†

Scheroba Oil. Isostearyl-erucyl erucate. Applications: Similar properties to jojoba oil, but has advantages of low price, product consistency and availability. (Scher Chemicals Inc).*

Scherpol LSB. Ethoxylated alcohol. Applications: Non-foaming, jet-dyeing assistant for polyester. minimizes subsequent smoke formation. (Scher Chemicals Inc).*

Schersoftoil P. Ester of natural oils. Applications: Winding lubricant applied in a package dye machine. (Scher Chemicals Inc).*

Schiff's Reagents. (1) Consists of a solution of rosaniline hydrochloride, decolourized by sulphur dioxide, and is used to test for aldehydes. (2) Furfuraldehyde and hydrochloric acid, employed for testing for urea. (3) Concentrated sulphuric acid, followed by ammonia, a test for cholesterol.†

Schimose. See LYDDITE.

Schlempe. Beet sugar waste. It is the thick brown liquor remaining after the extraction of all possible sugar. It is also called Vinasse.†

Schlippe's Salt. Sodium-thio-antimonate, $Na_3SbS_4.9H_2O$.†

Schneiderite. An explosive. It contains 88 per cent ammonium nitrate, 11 per cent dinitronaphthalene, and 1 per cent resin.†

Schnitzer's Green. A pigment similar to Arnaudon's green, except that crystallized sodium phosphate is used instead of ammonium phosphate in its preparation. Chromium phosphate is also known by this name.†

Schoellkopf's Acids. 1-Naphthol-4, 8-disulphonic acid, and 1-naphthylamine-8-sulphonic acid.†

Schoenanthe. Oil of lemon grass.†

Schoenite. (Schonite). Potassium-magnesium sulphate, $K_2SO_4.MgSO_4.6H_2O$.†

Scholine. A proprietary preparation of suxamethonium chloride. Short- acting muscle relaxant. (Allen & Hanbury).†

Schollpopf Acid. See SCHOELLKOPF'S ACIDS.

Schonberg's Alloy. A die-casting alloy containing 87 per cent zinc, 10 per cent tin, and 3 per cent copper.†

Schonite. See SCHOENITE.

Schorl. (Shorle). A black tourmaline.†

Schorl Rock. An aggregate of black tourmaline and quartz.†

Schou Oil. (Paalsguard Oil). An emulsifier made from soya-bean oil.†

Schraufite. A fossil resin found in Carpathian sandstone.†

Schreibersite. (Dyslytite). An iron nickel phosphide found in meteorites. A chromium sulphide has also been called Schreibersite.†

Schultze's Reagents. (1) Phospho-antimonic acid, made from sodium phosphate and antimony pentachloride, an alkaloidal reagent. (2) Consists of 25 parts dry zinc chloride, 8 parts potassium iodide, 8 1/2 parts water, and iodine. It gives a blue colour with cellulose.†

Schultze's Smokeless Powder. Consists of 62 per cent nitro-lignin, 2 per cent potassium nitrate, 26 per cent barium nitrate, 5 per cent petroleum jelly, and 3f per cent starch.†

Schultze's Stain. A microscopic stain. It consists of equal parts of a 2 per cent solution of β-naphthol sodium and a 2 per cent solution of dimethyl-*p*-phenylenediamine hydrochloride. The solutions are mixed and filtered.†

Schungite. A mineral. It is carbon in an amorphous form.†

Schutzenberger's Salt. Sodium hydrosulphite, $NaHSO_2$.†

Schweinfurth Green. See EMERALD GREEN.

Schweitzer's Reagent. A solution of copper hydroxide, $Cu(OH)_2$, in strong ammonia. A solvent for cellulose.†

Schwelkohle. A brown coal of Germany. It is light brown in colour.†

Scian Turpentine. See CHIAN TURPENTINE.

Scillin. See SCILLIPICRIN.

Scillipicrin. (Scillotoxin, Scillin). Commercial names for pharmaceutical preparations of squill, the fleshy bulb of *Urginea scilla*.†

Scillotoxin. See SCILLIPICRIN.

Scintillase. Papain. (ABM Chemicals Ltd).*

Scintran. Reagents for scintillation counting. (BDH Chemicals Ltd).

Sclair. Polyethylene resins. (Du Pont (UK) Ltd).

Sclerolac. A suggested name for hard lac resin (*qv*).†

Scleron Alloys. See SELERON.

Sclevoveine. A solution of pure sodium salicylate.†

Scolaban. Bunamidine hydrochloride. (The Wellcome Foundation Ltd).

Scopacron. A proprietary trade name for thermosetting acrylic resins modifiable by means of epoxy resins. (Styrene Co-Polymers Ltd).†

Scopacron 50, 75 and 80. A proprietary trade name for a thermo-setting acrylic resin capable of cross-linking with amino and epoxy compounds. Primarily intended for use with melamine formaldehyde resin for motor car top coats. (Styrene Co-Polymers Ltd).†

Scopacryl. A proprietary trade name for thermoplastic acrylic resin solutions used for wall paints and road marking applications. (Styrene Co-Polymers Ltd).†

Scopasol 550. A proprietary trade name for a water-dilutable thermosetting acrylic resin. Used for high performance white gloss coatings to be applied by electrophoresis techniques. (Styrene Co-Polymers Ltd).†

Scopol 58M, 58SP. A proprietary trade name for a vinyl toluene modified alkyd resin. (Styrene Co-Polymers Ltd).†

Scopol 85X. A propetary trade name for a styrene modified alkyd resin. Used for quick drying coatings with exceptional adhesion properties. (Styrene Co-Polymers Ltd).†

Scopolamine. Hyoscine, $C_{17}H_{21}NO_4$, an alkaloid.†

Scopolux 221SP. A proprietary trade name for a medium oil alkyd based on linseed oil. (Styrene Co-Polymers Ltd).†

Scorbital. Tablets of phenobarbitone with ascorbic acid. (British Drug Houses).†

Scorchex. Magnesium oxide products. Applications: Rubber goods. (Croxton and Garry Ltd).*

Scotch Cement. A cement prepared from feebly hydraulic limes, by the addition of 5 per cent plaster of Paris, and grinding.†

Scotch Foundry Pig. A pig iron made for foundry purposes from Scotch clay-band or black-band ores. It usually contains from 0.7-1 per cent phosphorus, and 2.5 per cent silicon.†

Scotch Gin. See SPIRIT OF SWEET NITRE.

Scotch Soda. Impure sodium carbonate, Na_2CO_3.†

Scotch Topaz. Golden topaz, a yellow variety of quartz.†

Scotphos. Fertilizers containing phosphate. (Scottish Agricultural Industries PLC).

Scouring Slag. A slag produced in making spiegel. It is black in colour and contains up to 8 per cent of oxide of iron.†

Scram. Dog and cat repellent. (Chevron).*

Scrap Rubber. Formed by the drying of the latex on the bark at the tapping cut. It is variable in quality and colour.†

Scratch-Guard. Abrasion resistant coating. Applications: Protective coating for various films i.e. polyester, polycarbonate etc. (Custom Coating and Laminating Corporation).*

Screen. Seed protectant. (Monsanto Co).*

Screen Plate Brass. See BRASS.

Screen Star Photo Emulsion. Direct screen making emulsion for professional screen printing. Applications: Textiles, electronic circuits and paper stock. (Bond Adhesives Co).*

Screte. Sulphur concrete. (Chevron).*

Screw Brass. See BRASS.

Screw Bronze. An alloy of 93.5 per cent copper, 5 per cent zinc, 1 per cent tin, and 0.5 per cent lead.†

Scripset. Styrene-maleic anhydride copolymers. Applications: Paper coating and specialty coating resins. (Monsanto Co).*

Scurane V. Polyurethane varnishes. (Rhone-Poulenc NV/CdF Chimie AZF).*

SD-1. Super dispersible rheological additive. Applications: Paint. (NL Industries Inc).*

SD-2. Super dispersible rheological additive. Applications: Paint. (NL Industries Inc).*

SE Wax. A Montan Wax ester containing an emulsifier used in the preparation of non-ionic self polishing emulsions for floors. (Bush Beach Ltd).†

SE-458. A proprietary silicone rubber compound used for bonding to unprimed surfaces during the curing process. (General Electric).†

Sea-Legs. Meclozine hydrochloride tablets-travel sickness remedy. (British Drug Houses).†

Sea-water Bronze. (Sheathing Bronze). An alloy of 32.5 per cent nickel, 45 per cent copper, 5.5 per cent zinc, 16 per cent tin, and 1 per cent bismuth. It resists sea water.†

SeaCure. A solution of copper sulphate and citric acid for treatment of protozoan parasites of marine fishes. Applications: Treatment of marine aquariums. (Aquarium Systems Inc).*

SeaGarden. Soluble nutrients for algae, particularly in marine aquariums. Applications: Aquarium water supplement. (Aquarium Systems Inc).*

Seair. Non-toxic, concentrated solution of neutralized resin. Applications: An admixture for concrete to increase workability, reduce bleeding of the mixing water, provide a more uniform concrete mix and reduce frost damage and scaling. (Secure Inc).*

eal and Heal. Fungicide and pruning paint. (May & Baker Ltd).

ealac. Sealers. (The Scottish Adhesives Co Ltd).*

ealite. A liquid containing glucose, corn starch, glycerol, calcium chloride, and glue. Used to prevent evaporation from oil storage tanks.†

ealum. Mastic rubber tape. Applications: Tape sealant for metal buildings. (Chemseco).*

eaTest. A series of colorimetric tests for analysing seawater. Applications: Aquarium water testing and natural seawater testing. (Aquarium Systems Inc).*

ebacil-Emulsion. For control of all ectoparasites, especially mange mites of domestic animals. Applications: Veterinary medicine. (Bayer & Co).*

ebastine. A dynamite explosive.†

ebizon. Sulphacetamide Sodium. Applications: Antibacterial. (Schering-Plough Corp).

ebkanite. A crude potassium chloride obtained by the evaporation and crystallization of the water of the salt lake in Tunis.†

ebond. Modified acrylic emulsion. Applications: Used to bond new concrete to either new or old concrete. For patching and resurfacing precast architectural panels, industrial concrete floors, highway and bridge deck repair. (Secure Inc).*

Sebrite. A clear, transparent, penetrating liquid sealer. Applications: For protecting and beautifying mechanically textured concrete, exposed aggregate and stone surfaces. (Secure Inc).*

Secadrex. A proprietary preparation of acebutolol plus hydrochlorothiazide. Applications: Treatment of hypertension. (May & Baker Ltd).*

Secaline. Trimethylamine.†

Seclomycin. A proprietary preparation containing streptomycin, benzylpenicillin sodium and procaine penicillin. An antibiotic. (Glaxo Pharmaceuticals Ltd).†

Secolan S-1, BA-1, BA-1G. Soluble animal collagen. (RITA Corporation).*

Secolat. Disodium alkyl sulphosuccinamate as a clear yellow liquid. Applications: Anionic surfactant used in latex foams. (KWR Chemicals Ltd).

Seconal. Secobarbital. Applications: Hypnotic; sedative. (Eli Lilly & Co).

Seconal Sodium. Secobarbital Sodium. Applications: Hypnotic; sedative. (Eli Lilly & Co).

Secondary Vermilion. A pigment. It is vermilion mixed with heavy spar.†

Seconesin. A proprietary preparation of quinalbarbitone and mephenesin. A hypnotic. (Crookes Laboratories).†

Secosol AL 959. Disodium monolauryl sulphosuccinate as a white paste. Applications: Anionic surfactant for shampoos; foam baths; creams. (KWR Chemicals Ltd).

Secosol ALL/40. Disodium monolauryl ether sulphosuccinate as a water white liquid. Applications: Anionic surfactant used in foam baths; shampoos; liquid soaps. (KWR Chemicals Ltd).

Secosol AL/MG 50. Anionic surfactant in which the anion is mono- lauryl sulphosuccinate and the cations are sodium and magnesium. Supplied as a white paste. Applications: Shampoos; foam baths; creams. (KWR Chemicals Ltd).

Secosol DOS/70. Sodium dioctyl sulphosuccinate in liquid form. Applications: Wetting and emulsifying agent. (KWR Chemicals Ltd).

Secosol EA/40. Disodium monoalkyl ethanolamide sulphosuccinate as a clear yellow liquid. Applications: Anionic surfactant for special shampoos; foam baths; liquid soaps. (KWR Chemicals Ltd).

Secosov. Emulsifiers. Applications: Mineral oils; solvents; dispersant for phytosanitaires; cosmetic creams and milks; insecticides. (KWR Chemicals Ltd).

Secosyl. Sodium lauroyl sarcisinate as a clear yellow liquid. Applications:

Anionic surfactant used in shampoos and foam baths. (KWR Chemicals Ltd).

Secretan. An alloy of from 91-95 per cent copper, 5-9 per cent aluminium, 1.5 per cent magnesium, and 0.5 per cent phosphorus.†

Secretol. A fat-splitting material similar and equal to Twitchell's reagent in its action.†

Secrodyl. Tablets dimethisterone with ethinyloestradiol - in gynaecological disorders. (British Drug Houses).†

Sectral. A proprietary preparation of ACEBUTOLOL hydrochloride used in the treatment of cardiac arrhythmias. (May & Baker Ltd).*

Sectral. Acebutolol. Applications: Anti-adrenergic. (Ives Laboratories Inc).

Secure. Sprayable liquid resins - concrete curing compound. Applications: For application to freshly placed concrete following the finishing. For use in preventing cracking and crazing caused by rapid moisture loss due to hot, windy weather. (Secure Inc).*

Securite. A safety explosive for mines. It is a mixture of 26 per cent *m*-dinitrobenzene and 74 per cent ammonium nitrate. It sometimes contains dinitronaphthalene and potassium nitrate.†

Securitol. A sodium silicate used to hasten the setting of cements.†

Securopen. A broad-spectrum penicillin especially active against Pseudomonas aeruginosa. (Bayer & Co).*

Sedacol. A proprietary preparation of clioquinol and phanquone. Antidiarrhoeal. (Zyma (UK) Ltd).†

Sedan Blue. See SOLUBLE BLUE.

Sedatin. See ANTIPYRINE.

Sedative Salt. Boric acid, H_3BO_3.†

Sedatussin. A proprietary preparation of cephaeline hydrochloride, sodium benzoate, syrup of squill and syrup of tolu. A cough linctus. (Eli Lilly & Co).†

Sedeff. An effervescent preparation containing opium, bismuth, and digestive ferments.†

Sedestran. A proprietary preparation of stilboestrol and PHENOBARBITONE used in the treatment of menopausal disorders. (Ciba Geigy PLC).†

Sedex. Ceramic foam filters to prevent no[n]-metallic inclusions in iron castings. (Foseco (FS) Ltd).*

Sedifloc Flocculant Aids. Organic polymer polyacrylamide, water soluble, polymers, cationic, anionic, nonionic. Applications: For dewatering, settling and flotating municipal and industrial solids found in their waste water treatment plant. (Benzsay & Harrison Inc).*

Sednine. A proprietary preparation of pholcodine and pseudoephedrine hydrochloride. A cough linctus. (Allen & Hanbury).†

Seed, Kola. See KOLA NUT.

Seed, Kombé. See KOMBÉ ARROW POISON.

Seed-lac. Stick-lac, after washing free fro[m] the colouring matter soluble in water. See LAC and LAC-DYE.†

Seekay Pitch. A registered trade name for chlorinated naphthalene products available in various grades.†

SEF. Modacrylic fibres. (Monsanto Co).*

Segetan. A silver cyanide with a copper complex. A seed preservative.†

Segoldus. A pigment consisting of 90-92 per cent zinc oxide and 6 per cent lead.

Sehta. Indian jeweller's name for cobaltite.†

Seiba Gum. See TUNO GUM.

Seidlitz Powder. (Effervescent Tartrated Soda Powder). Consists of 3 parts rochelle salt with 1 part sodium bicarbonate in the blue paper, and 1 part tartaric acid in the white paper.†

Seidlitz Powder, Double. Contains a doubl[e] dose of rochelle salt.†

Seidlitz Salt. A name applied to magnesiu[m] sulphate, $MgSO_4.7H_2O$ (Epsom salts), found in the mineral waters of Seidlitz.

Seidschütz Salt. Native magnesium sulphate, $MgSO_4.7H_2O$.†

Seifert Solder. An alloy of 73 per cent tin, 21 per cent zinc, 5 per cent lead, 0.5 pe[r] cent phosphorus and 0.5 per cent tin.†

Seignette's Salt. See ROCHELLE SALT.

Sekawrap. Polypropylene with various plasticizers and stabilizers. Applications: Shrink film for presentation applications. (S Kempner Ltd).*

Sel de Barnit. Zinc tannate.†

Sel de Sagesse. (Sel de Science). See SALT OF ALEMBROTH.

Sel d'Angleterre. Magnesium sulphate, $MgSO_4$.†

Sel-oxone. Selective weedkiller. (ICI PLC, Plant Protection Div).

Selacryn. Ticrynafen. Applications: Diuretic; uricosuric; antihypertensive. (Smith Kline & French Laboratories).

Seladon Green. See BOHEMIAN EARTH.

Selar. Barrier resin. (Du Pont (UK) Ltd).

Selastin EL-10, EL-30, SE EM 95. Hydrolyzed animal elastin. (RITA Corporation).*

Selazate. A proprietary accelerator. Selenium diethyl dithiocarbamate. (Naugatuck (US Rubber)).†

Selbax. Synthetic lubricant for textiles. (Crosfield Chemicals).*

Select-A-Sorb. Hydrous magnesium silicate - industrial talc. Applications: Filler, extender and reinforcing agent for rubber, paper (pitch control), plastics. (Vanderbilt Chemical Corporation).*

Seleen. Selenium Sulphide. Applications: Antifungal; antiseborrheic. (Abbott Laboratories).

Selek. Sealants to prevent metal penetration between ingot moulds and bottom plates. (Foseco (FS) Ltd).*

Selektan. A proprietary preparation of 2-hydroxy-5-iodopyridine.†

Selenac. A proprietary trade name for selenium diethyldithiocarbamate.†

Seleniol. Colloidal selenium.†

Selenite. Crystals of gypsum, $CaSO_4.2H_2O$. Used in optical instruments.†

Selenoxene. Dimethylselenophene, C_6H_8Se.†

Seleron. (Aeron). A group of aluminium alloys containing 85 per cent aluminium, with copper, nickel, zinc, manganese, silicon, and lithium, as the other ingredients. They are claimed to be useful for electrical apparatus.†

Selex. See CRISCO.

Self-hardening Steel. See MUSHET STEEL.

Seliwanoff's Reagent. A solution of 0.05 gram resorcinol in 100 cc dilute (1 : 2) hydrochloric acid. It gives a red colour with fructose.†

Seljut. Special processing emulsifying agent. (Crosfield Chemicals).*

Sella Acid. Acid dyes for leather. (Ciba-Geigy PLC).

Sella Fast. Dyes for leather. (Ciba-Geigy PLC).

Sellacron. Dyes for leather. (Ciba-Geigy PLC).

Sellaflor. Dyes for leather. (Ciba-Geigy PLC).

Sellasol. Synthetic tanning agents for leather. (Ciba-Geigy PLC).

Selora. A proprietary preparation of potassium chloride used as a substitute for table salt. (Winthrop Laboratories).*

Selsun. A proprietary preparation of selenium sulphide and a detergent used as a treatment for dandruff. (Abbott Laboratories).*

Selsun Blue. Selenium Sulphide. Applications: Antifungal; antiseborrheic. (Abbott Laboratories).

Seltzers. Usually consist of 25 parts sodium carbonate, 5 parts sodium chloride, 6 parts sodium sulphate, and 1,000 parts water.†

Selvigon. A proprietary preparation of pipazethate hydrochloride, a cough linctus. (Smith Kline and French Laboratories Ltd).*

Selwynite. See YELLOW OCHRE.

Semap. Penfluridol. Applications: Antipsychotic. (McNeil Pharmaceuticals, McNEILAB Inc).

Sembonit/Erostabil. Several compositions of natural caoutchouc, synthetic caoutchouc, filling material, synthetic resins depending on resistance demands. Applications: Surface protection against corrosion and erosion of vessels, tanks, tubes and industrial equipment. Special corrosion protection for flue-gas-desulphurization plants. (Schaumstoff und Kunststoff GmbH).*

Semeron. Triazine herbicide. (Ciba-Geigy PLC).

Semicoke. A fuel made from coal by low temperature carbonization. It is a smokeless fuel with a low ash.†

Semilente Iletin. Insulin zinc, prompt. Applications: Antidiabetic. (Eli Lilly & Co).

Semilente Insulin. Insulin zinc, prompt. Applications: Antidiabetic. (E R Squibb & Sons Inc).

Seminose. Mannose, $C_6H_{12}O_6$.†

Semi-opal. An impure opaque opal. See OPAL.†

Semirit. Electrically fused corundum, semi-friable grade. Applications: Production of abrasives, abrasive paper, discs and cloth. (Dynamit Nobel Wien GmbH).*

Semi-steel. A metal having properties between cast iron and cast steel. Used for filter-press plates. The term is applied to grey cast irons of low carbon content.†

Sempatap. Insulating material for walls, ceilings and floors - glass fibre non-woven with SBR-latex backing. Applications: Interior acoustic and thermal insulation. (Ebnother Group).*

Sempollan. Polyurethane Casted. Applications: For components required to display high strength, lasting resilience, high wear and oil resistance and a maximum useful life. (Schaumstoff und Kunststoff GmbH).*

Senate. Lube oil. (Chevron).*

Sencorex WG. Water dispersible granular formulation containing 70% w/w metribuzin. Applications: To control annual weeds in early and maincrop potatoes. (Bayer & Co).*

Sendoxan. SEE ENDOXAN. (Degussa).*

Sendust. A proprietary trade name for an iron-silicon-aluminium alloy.†

Seneca Oil. A name given to American petroleum, used in medicine.†

Senegal Gum. (West African Gum) A gum arabic ranking second to Khordofan gum. It is derived from *Acacia senegal* and other species of *Acacia*. It gives a good adhesive mucilage.†

Senegin. See STRUTHIIN.

Seneprolin. See PARAFFIN, LIQUID.

Sengite. An American explosive. It has a guncotton base and is similar to Tonite (*qv*), except that sodium nitrate replaces barium nitrate.†

Senna. The dried leaflets of *Cassia acutifolia*.†

Sennaar Gum. See SUAKIN GUM.

Sensitizer. Diazo photosensitizer. (ABM Chemicals Ltd).*

Sensitol, Red and Green. The German Pinacyanol and Pinaverdol (*qv*).†

Sensorcaine. Bupivacaine hydrochloride. Applications: Anaesthetic. (Astra Pharmaceutical Products Inc).

Sentry Cyclomethicone. Cyclomethicone. Applications: Pharmaceutic aid. (Union Carbide Corp).

Sentry Dimethicone. Dimethicone. Applications: Prosthetic aid. (Union Carbide Corp).

Sentry Simethicone. Simethicone. Applications: Antiflatulent. (Union Carbide Corp).

Seominal. A proprietary preparation of reserpine, phenobarbitone and theobromine. (Bayer & Co).*

Sep 6. Isolating solution for waxes. Applications: Dental preparation. (Bayer & Co).*

Separan. Polyacrylamides used as flocculants. (Dow Chemical Co Ltd).*

Separit. Parting powder. (Foseco (FS) Ltd).*

Separol. Liquid parting medium. (Foseco (FS) Ltd).*

Sepia. A brownish-black pigment derived from the ink-bag of the cuttle-fish. Used as water colour.†

Sepramar. Reagents for amino acids analysis. (BDH Chemicals Ltd).

Sepratek. Fluid emulsion parting agent. (Foseco (FS) Ltd).*

Septal. Fungicide. (FBC Ltd).

Septex No. 1. A proprietary skin cream containing boric acid. zinc oleate and zinc oxide. (Norton, H N Co).†

eptex No. 2. A proprietary skin cream containing boric acid, zinc oxide, zinc oleate and sulphathiazole. (Norton, H N Co).†

eptra. Tablets, suspension and IV infusion. Proprietary formulations of trimethorprim and sulphamethoxazole. Applications: Treatment of urinary tract infections, acute otitis media, acute exacerbations of chronic bronchitis, shigellosis and *Pneumocystis carinii* pneumonitis. (The Wellcome Foundation Ltd).*

eptrin. A proprietary preparation of trimethoprim and sulphamethoxazole. An antibiotic. (The Wellcome Foundation Ltd).*

equens. A proprietary preparation of mestranol (15 white tablets) and mestranol and chlormadinone acetate (5 peach tablets). Oral contraceptive. (Eli Lilly & Co).†

equest-All. A dry non-toxic potable water treatment. It is used to control minerals in water, to prevent 'red water, scale, build-up and corrosion in the distribution system. Applications: Municipal water systems, irrigation systems, cooling towers, boilers, apartments and hotels. (SPER Chemical Corporation).*

equestrene. Iron chelates-food additives foliar feeds. (Ciba-Geigy PLC).

eracelle. A proprietary cellulose acetate packing material.†

eractide. A corticotrophic peptide. Ala^{26}-Gly^{27}- Ser^{31}-α^{1-39}-corticotrophin.†

eradix. Plant hormone. (May & Baker Ltd).

erax. Oxazepam. Applications: Tranquilizer. (Wyeth Laboratories, Div of American Home Products Corp).

erc. Betahistine. Applications: Counteracts disorders of the inner ear, Menière's disease. (Duphar BV).*

Serdet DCK. Sodium alkyl ether sulphate based on a natural alcohol (C12-C14) in paste or liquid form. Applications: Anionic surfactant used in shampoos and bubble bath formulations and dishwashing. (Chemische Fabriek Servo BV).

Serdet DFK. Sodium alkyl sulphate, based on a natural alcohol (C12/C14), in liquid or paste form. Applications: Detergent and emulsifier for shampoos and bubble baths; toothpaste; dishwashing; emulsion polymerization. (Chemische Fabriek Servo BV).

Serdet DFL, DFM and DFN. Anionic surfactants in liquid form. Applications: Foaming agents for shampoos and bubble baths. (Chemische Fabriek Servo BV).

Serdet DM and DMK. Dodecylbenzene sulphonate in acid form or as sodium salt. Applications: Biodegradable primary emulsifiers used in scouring powders, liquid detergents and emulsion polymerization. (Chemische Fabriek Servo BV).

Serdet DML. Triethanolamine dodecylbenzene sulphonate in liquid form. Applications: Biodegradable anionic surfactant used in shampoos and bubble baths. (Chemische Fabriek Servo BV).

Serdet DNK. Sodium nonylphenol 4EO-sulphate in liquid form. Applications: Detergent base for liquid detergent formulations; emulsifier in emulsion polymerization. (Chemische Fabriek Servo BV).

Serdet DPK. Sodium alkyl ether sulphate based on a synthetic alcohol (C12-C15) in liquid or paste form. Applications: Anionic surfactant used in shampoo and bubble bath formulations and dishwashing. (Chemische Fabriek Servo BV).

Serdet DSK. Sodium 2-ethyl-hexyl sulphate in liquid form. Applications: Alkali stable wetting agent used for latex stabilization. (Chemische Fabriek Servo BV).

Serdolamide. Foam stabilizer; re-fatting agent; viscosity modifier and improver. Applications: Detergent, shampoo and bubble bath formulations. (Chem-Y, Fabriek van Chemische Producten BV).

Serdox. Emulsifier; antistatic agent. Applications: Crude and vegetable oils; plastics; textiles processing; cosmetic emulsions. (Chem-Y, Fabriek van Chemische Producten BV).

Serdox NNPQ 7/11. Nonionic surfactant of the alkylphenol ethoxylate type in liquid form. Applications: Low foaming wetting agent. (Chemische Fabriek Servo BV).

Serdox NNP10 and NNP12. Nonylphenol ethoxylate nonionic surfactant in liquid form. Applications: Scouring of textiles; soaking assistant for leather; household and industrial detergents; emulsifier for insecticides and herbicides; plasticizer for mortar and concrete; paper manufacture. (Chemische Fabriek Servo BV).

Serdox NNP15, NNP20 and NNP25. Nonylphenol ethoxylate nonionic surfactant in liquid or solid form. Applications: Detergents and wetting agents for use at high temperatures and electrolyte concentrations; emulsifier for fatty acids and waxes; NNP20 is used as a stabilizer for synthetic latices; NNP25 is an emulsifier for emulsion polymerization. (Chemische Fabriek Servo BV).

Serdox NNP30 and NNP30/70. Nonylphenol ethoxylate nonionic surfactant in solid or liquid form. Applications: Dyeing assistant; lime soap dispersing agent; emulsifier and stabilizer for emulsion polymerization. (Chemische Fabriek Servo BV).

Serdox NNP4. Nonylphenol ethoxylate nonionic surfactant in liquid form. Applications: Oil soluble detergents; emulsifier for insecticides and herbicides; dispersing agent. (Chemische Fabriek Servo BV).

Serdox NNP5 and NNP6. Nonylphenol ethoxylate nonionic surfactant in liquid form. Applications: Emulsifier for insecticides and herbicides; oil soluble detergents. (Chemische Fabriek Servo BV).

Serdox NNP7, NNP8.5 and NNP9. Nonylphenol ethoxylate nonionic surfactant in liquid form. Applications: Scouring of textiles; soaking assistant for leather; household and industrial detergents; emulsifier for insecticides and herbicides; plasticizer for mortar and concrete. (Chemische Fabriek Servo BV).

Serdox NOP 30/70. Octylphenol ethoxylate nonionic surfactant in liqui[d] form. Applications: Emulsifier and stabilizer used in emulsion polymerization. (Chemische Fabriek Servo BV).

Serdox NOP9. Octylphenol ethoxylate nonionic surfactant in liquid form. Applications: Scouring of textiles; soaking assistant for leather; household and industrial detergents; emulsifier fo[r] insecticides and herbicides; plasticizer for mortar and concrete; paper manufacture. (Chemische Fabriek Servo BV).

Sereen. Travel sickness tablets. (The Boot[s] Company PLC).

Serenace. A proprietary preparation of haloperidol. A sedative. (Searle, G D & Co Ltd).†

Serenid Forte. A proprietary preparation o[f] OXAZEPAM. A tranquillizer. (Wyeth).†

Serenid-D. A proprietary preparation of oxazepam. A sedative. (Wyeth).†

Seretin. Carbon tetrachloride, CCl_4.†

Serfene. Poly vinylidene chloride dispersio[n] coatings. (Williams (Hounslow) Ltd).

Serge Blue. A lower quality of Soluble Blu[e] (*qv*).†

Sericine. (Silk Size, Silk Rubber). The gu[m] surrounding the silk from the silk spinner. See CUITE.†

Sericose. Cellulose acetate, used for makin[g] artificial silk and dope.†

Serilan. A range of disperse and acid dye mixtures. Applications: Dyeing of polyester/wool blended fibres. (Yorkshire Chemicals Plc).*

Serilene. A range of disperse dyes. Applications: Dyeing of polyester fibres. (Yorkshire Chemicals Plc).*

Serine. α-Amino-β-hydroxy-propionic acid, $CH_3(OH)CH(NH_2)COOH$.†

Serinyl. A range of disperse dyes. Applications: Dyeing of polyamide fibres. (Yorkshire Chemicals Plc).*

Serisol. A range of disperse dyes. Applications: Dyeing of cellulose acetate fibres. (Yorkshire Chemicals Plc).*

ritox. Selective weedkiller. (May & Baker Ltd).

rizyme. A proprietary trade name for an enzyme used in desizing acetate fabrics and similar materials which contain protein.†

rmag. Liquid magnesium oxide oil fuel additive. (Steetley Refractories Ltd).

rmix. Alkaloid animal feed additive. (Ciba-Geigy PLC).

rmul. Emulsifier, biodegradable. Applications: Mineral oils; pesticide formulations; white spirit and turpentine; vegetable and animal oils. (Chem-Y, Fabriek van Chemische Producten BV).

rmul EA 88. Calcium dodecylbenzene sulphonate in liquid form. Applications: Biodegradable emulsifier for pesticide formulations. (Chemische Fabriek Servo BV).

ermul EA129. Ammonium lauryl sulphate in liquid form. Applications: Emulsifier in emulsion polymerization. (Chemische Fabriek Servo BV).

ermul EA150. Sodium lauryl sulphate in paste form. Applications: Emulsifier in emulsion polymerization. (Chemische Fabriek Servo BV).

ermul EA176. Sodium mono nonylphenol 10-EO sulphosuccinate in liquid from. Applications: Emulsifier for emulsion polymerization. (Chemische Fabriek Servo BV).

ermul EA188, EA136 and EA205. Anionic surfactants of the phosphate ester type. Supplied as liquids in acid form. Applications: Emulsifiers for emulsion polymerization. (Chemische Fabriek Servo BV).

ermul EA54, EA151 and EA146. Anionic surfactants of the ether sulphate type in liquid form. Applications: Emulsifiers for emulsion polymerization. (Chemische Fabriek Servo BV).

ernylan. Phencyclidine Hydrochloride. Applications: Anaesthetic. (Parke-Davis, Div of Warner-Lambert Co).

eromycin. Cycloserine. Applications: Antibacterial. (Eli Lilly & Co).

erosine. Bromanilid.†

Serosteron. A proprietary trade name for Dimethisterone. (British Drug Houses).†

Serpasil. Reserpine. Applications: Antihypertensive. (Ciba-Geigy Corp).

Serpasil-esidrex. Anti-hypertensive/diuretic. (Ciba-Geigy PLC).

Serpatonil. A proprietary preparation of reserpine, and methyl phenidate hydrochloride. (Ciba Geigy PLC).†

Serpiloid. Reserpine. Applications: Antihypertensive. (Riker Laboratories Inc, Subsidiary of 3M Company).

Serseal. Heat conserving compound for use in metal pre-treatment. (ICI PLC).*

Serum-casein. See GLOBULIN.

Servamine KAC 422. N-coco N-N-dimethyl-N-benzalkonium chloride in liquid form. Application: Cationic surfactant used as a bactericide, fungicide, sanitizer and germicide. (Chemische Fabriek Servo BV).

Servamine KEP 4527. N(palmityl amido propyl)N-N-N-trimethyl ammonium chloride in liquid form. Application: Cationic surfactant emulsifier with bactericide properties. (Chemische Fabriek Servo BV).

Servamine KET 350. N-(tall oil amido propyl)N-N-dimethyl amine based on tall oil. Supplied as a liquid. Application: Cationic surfactant used as an adhesion agent and corrosion inhibitor in bitumen. (Chemische Fabriek Servo BV).

Servamine KET 4542. N-(alkylamido propyl)N-ethyl N-N-dimethyl ammonium ethosulphate, based on tall oil, in liquid form. Application: Cationic emulsifier. (Chemische Fabriek Servo BV).

Servamine KOO 330. Amino ethyl oleyl imidazoline in liquid form. Application: Cationic adhesion agent and corrosion inhibitor for bitumen. (Chemische Fabriek Servo BV).

Servamine KOO 330B. Oleylamido ethyl oleyl imidazoline in liquid form. Application: Basic material in the manufacture of quaternary imidazolines. (Chemische Fabriek Servo BV).

Servamine KOO 360. Hydroxy ethyl oleyl imidazoline in liquid form. Application: Cationic adhesion agent and corrosion inhibitor for bitumen. (Chemische Fabriek Servo BV).

Servo Ampholyt (B) JA110. Modified imidazoline in liquid form. Applications: Non eye-irritating for shampoo formulations. (Chemische Fabriek Servo BV).

Servo Ampholyt (B) JA140. Modified imidazoline in liquid form. Applications: Non eye-irritating and non skin-irritating for hair shampoos. (Chemische Fabriek Servo BV).

Servo Ampholyt (B) JB130. Betaine structure, liquid form. Applications: Mild shampoo raw material with hair stimulating properties, for baby shampoos. (Chemische Fabriek Servo BV).

Servo Brilliant Oil B AZ 75. Sodium castor oil sulphonate in liquid form. Applications: Anionic surfactant used in softener and finishing oils, and pasting oil for dyestuffs. (Chemische Fabriek Servo BV).

Servoxyl VLA 2170. Sodium di-2-ethyl hexylsulphosuccinate in liquid form. Applications: Wetting and rewetting agent. (Chemische Fabriek Servo BV).

Servoxyl VLB 1123. Sodium monoalkyl polyglycol ether sulphosuccinate in liquid form. Applications: Raw material for high quality baby shampoos and mild hair shampoos; cleaning agent. (Chemische Fabriek Servo BV).

Servoxyl VLE 1159. Sodium mono nonylphenol 10-EO sulphosuccinate in liquid form. Applications: Emulsifier for emulsion polymerization. (Chemische Fabriek Servo BV).

Servoxyl VP. Range of anionic surfactants of the phosphate ester type, supplied mainly as liquids in acid form. Those based on alcohol and alkyl polyglycol ethers are biodegradable. Applications: Detergents and emulsifiers used in dry cleaning; formulation of metal cleaners; emulsion polymerization; pesticide formulation; cosmetic preparations. (Chemische Fabriek Servo BV).

Seseal. A thermoplastic acrylic sealer. Applications: It seals, hardens and dustproofs concrete. (Secure Inc).*

Seseal 8. Cure, seal, hardener and dustproofer for concrete. Applications Prevents mortar and concrete dropping from bonding to floors, reducing clean-up cost, base for mastic adhesives. (Secure Inc).*

Sestrip. Form release agent. Applications For application to all types of concrete forms prior to concrete placement to ensure release of the forms and to minimize form clean up. (Secure Inc).

Setac. Waxes for dental laboratory technology in bead form and block for Applications: Casting and modelling waxes blue, green, sticky wax red. (Bayer & Co).*

Setacure. Multi functional acrylic monomers and prepolymers. Applications: UV, EBC, radiation curing applications. (Synthese BV).*

Setafix. Acrylic and polyester resins. Applications: Photocopy toners. (Synthese BV).*

Setair. A clear or coloured solution of PV polymer. Applications: Coating zinc sprayed, shot or sand blasted steel, asbestos, concrete, brick, plywood, softboard, strawboard, chipboard and hardboard. (Llewellyn Ryland Ltd).*

Setal. Saturated polyester and alkyd resin Applications: Decoration and industria paints. (Synthese BV).*

Setalana. A proprietary name for a natura nest silk produced by worms of the gen *Anaphe*, introduced into Germany fro Africa. The fibre resembles tussah silk, but is stated to be not so strong.†

Setalin. Modified phenolic resins - modifie hydrocarbon resins. Waterthinnable acrylic resins - acrylic dispersions. Alkyd resins - modifies phenolic resins modified hydrocarbon resins - varnishes - acrylic resins. Applications Rotogravure inks, packaging inks and offset inks. (Synthese BV).*

Setalux. Acrylic resins. Applications: Airdrying, thermosetting, isocyanate curing resins for industrial paint (automotive, refinishing and general industry). (Synthese BV).*

etamine US. Melamine-formaldehyde resins. Applications: Ovendrying industrial paints (automotive and general industry). (Synthese BV).*

etamol. Textile dispersing agents protective colloids. (BASF United Kingdom Ltd).

etarol. Unsaturated polyester resins. Applications: Fibre reinforced polyesters. (Synthese BV).*

etatack A. Acrylic polyols. Applications: Pressure sensitive adhesives. (Synthese BV).*

etatack AF. Thermoplastic/thermosetting acrylics. Applications: Hotmelt adhesives. (Synthese BV).*

etatack LP. Linear polyesterpolyols. Applications: PUR elastomers and prepolymers. (Synthese BV).*

etatack P. Polyesterpolyols. Applications: 2 pack PUR adhesives. (Synthese BV).*

etatack T. Modified rosin esters. Applications: Lamination/paint lamination adhesives. (Synthese BV).*

ethotope. Selenomethionine Se 75. Applications: Diagnostic aid. (E R Squibb & Sons Inc).

etilon. Fatty alcohol compound with emulsifiers, nonionic. Applications: Softener with scrooping effect especially developed for absorbent cotton, silky finish of cotton fabrics. Softening agent for the raising of synthetic goods. (Henkel Chemicals Ltd).

etilose. A French cellulose acetate artificial silk.†

etocyanine. (Brilliant Glacier Blue, Acronol Brilliant Blue). A dyestuff. It is the hydrochloride of diethyl-diamino-*o*-chlorophenylditolyl-carbinol, $C_{25}H_{28}N_2Cl$. Dyes silk and tannined cotton greenish-blue.†

etoglaucine. See VICTORIA GREEN 3B.

etopaline. A triphenyl-methane dyestuff closely related to Erioglaucine A.†

etreat. High solids linseed oil based penetrating concrete sealer. Applications: For treating rough finished, porous concrete to penetrate and seal the surfaces to prevent the absorption of water, salts and other contaminants harmful to concrete. (Secure Inc).*

Setsit. Dithiocarbamate blends - latex. Applications: Primary and ultra accelerators for latex compounds. (Vanderbilt Chemical Corporation).*

Setyrene. Styrenated acrylated alkyds. Applications: Quick drying industrial paints. (Synthese BV).*

Sevamine KOV 4342B. Cationic surfactant composed of quaternary imidazoline in liquid form. Application: Raw material in the preparation of laundry softeners. (Chemische Fabriek Servo BV).

Sevelyte K. Potassium lauryl amino propionate, in orange yellow liquid form. Applications: Pigment dispersant for paints and inks. (KWR Chemicals Ltd).

Sevin. 1-naphthyl N-methylcarbamate. A proprietary preparation of carbaryl used as a veterinary insecticide.†

Sextate. Cyclohexanol (hexahydro-phenol) acetate. It boils at 170- 195° C., has a specific gravity of 0.94-0.96, and a flash point of 155° F. Also stated to be methyl cyclohexyl acetate. (Laporte Industries Ltd).*

Sextol Z. Dimethyl cycohexanol. (Laporte Industries Ltd).*

Sextone. Cyclohexanone. (Laporte UK Trading).

Sextone B. Methylcyclohexanone. A solvent.†

Seymourite. A proprietary trade name for an alloy of 64 per cent copper, 18 per cent nickel, and 18 per cent zinc.†

SH 420. A proprietary preparation of NORETHISTERONE acetate used in the treatment of breast cancer. (Schering Chemicals Ltd).†

Shadeacrete. Dry powdered colourants, iron oxides, ochres, umbers and composite pigments. Applications: Colours for mortars, concrete roofing tiles, floor tiles, sand-lime bricks, concrete blocks, reconstructed stone, split blocks, paving slabs and cement sheets. (W Hawley & Son Ltd).*

Shadocol. Sodium tetraiodophenol-phthalein.†

Shadow. Clay. Applications: Sun reflector for protection against sunburn on agronomic and ornamental crops. (Drexel Chemical Company).*

Shaku-do. A Japanese alloy. It usually contains 94-96 per cent copper, 3.76-4.16 per cent gold, and 0.08-1.55 per cent silver.†

Shale. A dark-grey or black mineral containing 73-80 per cent mineral matter and 20-27 per cent organic matter. It is a source of oil for lubricating purposes.†

Shale Oil. The tarry oil obtained by the distillation of certain bituminous shales. It contains unsaturated hydro-carbons.†

Shanghai Oil. A variety of Colza Oil (*qv*).†

Sharps. See BRAN.

Shawinigan Black. Acetylene black. (Chevron).*

Shawinigan's Black. See ACETYLENE BLACK.

Shawplas. Abrasive compounds and polishes for plastic articles. Applications: Fabricated components, turned parts, buttons, buckles, spectacle frames in acrylic, cellulose acetate, casein, polyester. (Shawplas Ltd).*

Shé-Chuang-Tzu. A Chinese drug. It is the fruit of *Selinum monnieri*, and contains an essential oil which contains *l*-pinene, camphene, and bormyl-iso-valerate.†

Shea Butter. (Bambuk Butter). The fat obtained from the seeds of *Butyrospernum parkii* or *Bassio parkii.*†

Shea Gutta. See GUTTA-SHEA.

Sheathing Bronze. See SEA-WATER BRONZE.

Sheduri. See SANNA.

Sheet Brass. See BRASS.

Sheet-lac. See LAC.

Shellac. See LAC.

Shellac, Arizona. See SONORA GUM.

Shellackose. An alcohol-soluble phenol-formaldehyde resin. Used in the preparation of lacquers.†

Shellac Substitute. See IDITOL.

Shellflex Process and Extender Oils. These are hydrocarbon solvents which generally have lower aromatic content than the Dutrex grades, covering a wid viscosity range. They are used as extender and process oils for natural a synthetic rubber. (Shell Chemicals).†

Shell-head Brass. See BRASS.

Shellite. An explosive. It is a mixture of ammonium perchlorate and paraffin wax.†

Shell-lac. (Shellac). See LAC.†

Shell Limestone. A variety of calcium carbonate in massive form.†

Shellsol D40, D60, D70. These highly refined solvents have a very low aroma content and a very slight sweet odour. Their principal applications are similar to white spirits, in low odour paints an metal cleaning products. (Shell Chemicals).†

Shellsol E, A, AB, R. They are high-boilin aromatic hydrocarbons. These versatil solvents are used in paints, varnishes and in the preparation of agricultural chemical formulations. (Shell Chemicals).†

Shellsol T. A high-boiling, isoparaffinic, aliphatic solvent with high flash-point. It is virtually without odour and is therefore useful in odourless paints, household aerosols, fragrant polishes and cosmetic creams. It is also used as a catalyst carrier in polymerization reactions. (Shell Chemicals).†

Shellswim - 11T. A polymeric additive dissolved in toluene to facilitate its effective distribution in crude oil. It enables waxy crude oils to be transported at temperatures below thei pour point. It also facilitates restarting of a pipeline following a shut-down. (Shell Chemicals).†

Shellswim - 5X. A polymeric additive dissolved in xylene to facilitate its homogeneous distribution in fuel and crude oils. It can be added to waxy residual fuels and waxy crude oils to permit storage, handling and pumping below their natural pour point. (Shell Chemicals).†

Shellvis 50 (SAP 150). It is normally supplied in the form of bales which require shredding before dissolving in lubricating oil, although it is available as a concentrate or in a crumb form.

Shellvis 50 is a styrene-based hydrocarbon viscosity index improver for engine lubricants. (Shell Chemicals).†

elspra. See MEXPHALTE.

erpa. Insecticide. (May & Baker Ltd).

erwood Oil. See PETROLEUM ETHER.

ibu-ichi. A Japanese alloy containing 51-67 per cent copper, 32-49 per cent silver, and traces of gold and iron.†

ield. Selective herbicide. (Murphy Chemical Ltd).

ikimole. Safrole, $C_{10}H_{10}O_2$, the chief constituent of oil of sassafras.†

ikon. The dried roots of *Lithospermum erythrorhizon*.†

ilajatu. An Indian mineral gum.†

imose. See LYDDITE.

imosite. A Japanese explosive, the chief constituent of which is picric acid.†

hinnamu. A vegetable dye obtained from a species of maple found in Korea.†

nio Liao. A Chinese cement for marble, porcelain, etc., made from 54 per cent, slaked lime, 6 per cent alum, and 40 per cent blood.†

hipley's Solutions. Solutions of pyro-gallol and caustic soda in water, usually 10 cc of 1 : 1 caustic soda solution, 1 and 4 cc water, and 2 and 10 grams pyrogallol. Used for the absorption of oxygen.†

hoddy. The recovered and broken up wool of old cloth.†

hoe Nail Brass. See BRASS.

hoemaker's Black. Ferrous sulphate, $FeSO_4$.†

hoemaker's Paste. A paste made by allowing the gluten from flour to putrefy, rolling it out thin, and making it into a paste. Used for securing leather to leather, paper, or other material.†

horle. See SCHORL.

hort Oil Varnishes. See LONG OIL VARNISHES.

hostakovsky Balsam. Polyvinox. A A proprietary preparation of synthetic vinyl butyl ether. (Leopold Charles & Co).†

hot Lead. See SHOT METAL.

Shot Metal. (Shot Lead, Bullet Metal). An alloy of lead with not more than 3 per cent arsenic. One alloy contains 99.8 per cent lead and 0.2 per cent arsenic.†

Sialonite. A mineral, Be_8Se_3.†

Siapton. Liquid organic foliar feed. (ICI PLC).*

Sibley Alloy. An alloy of 67 per cent aluminium and 33 per cent zinc.†

Sibor. A proprietary safety glass.†

Sical. An alloy of from 22-29 per cent aluminium, 50-51 per cent silicon, 2-4 per cent titanium, 1 per cent calcium, 0.2-0.3 per cent carbon, and the remainder iron.†

Sicalite. See GALLATITE.

Siccative. Manganese borate, MnB_4O_7. Used as a siccative mixed with linseed oil and resin, for impregnating leather.†

Siccolam. Compound titanium dioxide paste. Desiccant for exudatory dermatoses. (British Drug Houses).†

Sicily Oil. Inferior olive oil.†

Siclor. Tetrachloroisophthalonitrile. Applications: Fungicide for the preparation of antifouling marine paints, aqueous paints, wood primers, adhesives. (Caffaro Industrial).*

Sicoflex MBS. A thermoplastic material based on methyl methacrylate, butadiene and styrene. (Mazzucchelli Celluloide Spa).†

Sicoflex 80. A proprietary A.B.S. terpolymer possessing very high flow properties. (Mazzucchelli Celluloide Spa).†

Sicoflex 85. A proprietary A.B.S. terpolymer possessing high flow properties. (Mazzucchelli Celluloide Spa).†

Sicoflex 90. A proprietary A.B.S. terpolymer having high impact strength. (Mazzucchelli Celluloide Spa).†

Sicoflex 93. A proprietary general purpose A.B.S. terpolymer. (Mazzucchelli Celluloide Spa).†

Sicoflex 95. A proprietary A.B.S. terpolymer possessing high tensile strength. (Mazzucchelli Celluloide Spa).†

Sicoflex 99. A proprietary A.B.S. terpolymer offering high resistance to heat. (Mazzucchelli Celluloide Spa).†

Sicromo Steel. A proprietary trade name for a chromium-silicon-molybdenum steel containing from 2.25-2.75 per cent chromium, 0.5-1.0 per cent silicon, 0.4-0.6 per cent molybdenum and up to 0.15 per cent carbon.†

Sidanyl. Polyamide film. Applications: Incorporation into laminates, well suited for thermoforming. (UCB nv Film Sector).*

Sident. Precipitated silica. Applications: Abrasive and thickening agent for the production of transparent, translucent and opaque toothpastes. (Degussa).*

Sideraphthite. An alloy resembling silver. It contains 64.5 per cent iron. 22.5 per cent nickel, 4.5 per cent each aluminium and copper, and 4 per cent tungsten. It is stated to be non-oxidizable.†

Siderine Yellow. A basic chromate of iron, used to a small extent as a water colour, and mixed with waterglass as a paint.†

Sidero Cement. A cement in which iron ores are wholly or partly substituted for the clay.†

Sidot's Blende. A phosphorescent zinc sulphide.†

Sidros. A proprietary preparation containing ferrous gluconate and ascorbic acid. A haematinic. (Horlicks).†

Siemensite. A refractory material produced by fusing a mixture of chromite, bauxite, magnesite, and a reducing agent in the arc furnace to obtain a slag containing from 20-40 per cent. Cr_2O_3, 25-45 per cent. Al_2O_3, 18-30 per cent. MgO, and 8-14 per cent other constituents.†

Sienna. A pigment. It consists of hydrated oxide of iron, mixed with a little manganese, and clay. It contains from 50-70 per cent. Fe_2O_3, 8-12 per cent. SiO_2, 2-8 per cent. Al_2O_3, 2-5 per cent. $CaSO_4$ or CaO, and water.†

Sienna, American. See BURNT SIENNA and INDIAN RED.

Sierra Leone Butter. See LAMY BUTTER.

Sifbronze. A proprietary trade name for brass containing some ferromanganese and tin.†

Siflox. Highly dispersed precipitated silic Applications: Applied in shoe soling materials, hoses, cable sheeting, profil etc. (Chemische Fabriken Oker und Braunschweig AG).*

Siflural. A trade name for a solution of aluminium fluosilicate. A disinfectant.

Sigal. A proprietary alloy of 10 per cent. Si and 90 per cent. Al. A pigment.†

Sigma. Lube oil. (Chevron).*

Sigmalium. A proprietary trade name for an alloy of aluminium containing 1 pe cent silicon, 4 per cent copper, and 0.7 per cent magnesium.†

Sigmamycin. A proprietary preparation o tetracycline and troleandomycin or oleandomycin phosphate. An antibiot (Pfizer International).*

Sigmathane. A proprietary single pack moisture cured polyurethane coating. (Sigma Coatings).*

Signal Red. A dyestuff. It is a British equivalent of Lithol red B.†

Silain. Simethicone. Applications: Antiflatulent. (A H Robins Co Inc).

Silajit. A preparation containing benzoic and hippuric acids, gums, albuminoids resin, and fatty acids.†

Sil-al. Aluminium hydro-silicate.†

Silal. A proprietary trade name for a grey iron with 5 per cent silicon and 2.5 pe cent total carbon. It is stated to resist oxidation, growth, and scaling up to 7: C.†

Silanca. A stainless silver with a high silv content.†

Silane. (Silicane). Silicon tetrahydride, SiH_4.†

Silantox. A proprietary colloidal silicon dioxide.†

Silaonite. A mineral. It is a mixture of bismuth and bismuth trisulphide.†

Silargel. A German product. It is a silver chloride-silica gel preparation, a white odourless powder containing 5 per cent silver. It is an adsorbent and disinfecta for the external treatment of burns.†

Silastic. Silicone rubbers, sealants, RTVs. (Dow Corning Ltd).

Silastomer. Silicone rubbers. (Midland Silicones).†

Silastoseal. Room temperature curing silicone rubber sealants. (Midland Silicones).†

Silbamine. Silver fluoride, AgF.†

Silberit. A jewellery alloy. It contains aluminium, nickel, and silver.†

Silcar. A proprietary trade name for a pigment comprising a mixture of silicon dioxide and silicon carbide.†

Silcasil S. An inorganic filler. Applications: Uses include carrier material for insecticides, auxiliary for improving the free-flowing and grinding properties of powders, thickener for liquids. (Bayer & Co).*

Silchrome. A heat-resisting alloy containing 86 per cent iron, 9.5 per cent chromium, 4 per cent silicon, and 0.5 per cent carbon.†

Silchrome R.A. A proprietary trade name for a steel containing 16 per cent chromium, 1 per cent silicon, 1 per cent copper, and 0.12 per cent carbon.†

Silchrome Wire. An alloy of iron with 18 per cent chromium, 3 per cent silicon, 3 per cent tungsten, and 0.3 per cent carbon.†

Silchrome 46M. A proprietary trade name for a chromium steel containing 4-6 per cent chromium, 0.5 per cent molybdenum, and 0.2 per cent carbon.†

Silcolapse. Textile auxiliary chemicals. (ICI PLC).*

Silcolease. Silicone coatings. (ICI PLC).*

Silcon. Silafilcon. Applications: Contact lens material. (Dow Corning Ophthalmics Inc).

Silcoset. Rubber curing agents. (ICI PLC).*

Silcron. Registered trade mark for a fine-particle silica. (SCM Corp, Cleveland, Ohio).†

Silderm. A proprietary preparation of triamcinolone acetonide, neomycin sulphate and undecanoic acid. A steroid skin cream. (Lederle Laboratories).*

Sildura. Registered trade mark for a range of silicone-rubber compositions. curable by the application of heat. (General Electric).†

Silene. A proprietary trade name for a precipitated calcium silicate. Used in rubber mixes to give wear-resistance.†

Silent Spirit. See SPIRIT OF WINE.

Silesia Powder. An explosive. It is a mixture of 75 per cent potassium chlorate, with pure or nitrated resin, and a little castor oil.†

Silesite. A tin silicate with 55 per cent tin, found in the Bolivian tin deposits.†

Silester. Ethyl silicate. (Monsanto PLC).

Silex. A name applied to silica (SiO_2). It is used also for tripoli employed as a filler in paints. A ground flint is also known as silex.†

Silexon. See CARBORA.

Silfbergite. A mineral. It is a variety of dannemorite.†

Sil-fos. A proprietary trade name for a phosphor-silver brazing solder containing 80 per cent copper, 15 per cent silver, and 5 per cent phosphorus.†

Silfrax. (Silicized Carbon). A product obtained by the action of silicon on carbon and consisting of carbon with a coating of silicon carbide and carbon. It is stated to be tougher and stronger than carborundum, and is used as a refractory material in the manufacture of pyrometer tubes for electrical heating elements.†

Silica Gel. The name applied to a colloidal form of silica, prepared by treating sodium silicate with acetic or hydrochloric acid, washing the gelatinous silica, and drying. It is highly absorbent, and is used to absorb water vapours.†

Silica Glass. See VITREOSIL.

Silicam. Silicon imido-nitride, Si_2N_3H, formed when silicon diimide is heated to 900° C. in an atmosphere of dry nitrogen.†

Silicane. See SALINE.

Silicargol. A colloidal silver preparation for wound treatment.†

Silicate Cotton. See SLAG WOOL.

Silicate of Carbon. See GAS BLACK.

Silicated Soap. A soap to which water glass (sodium silicate) has been added. A detergent.†

Silicex. Silicone products. Applications: Antifoams, release agents, moulding rubbers and sealants. (Siliconas Hispania SA).*

Silicised Carbon. See SILFRAX.

Silicium. A proprietary trade name for silicon used as a pigment in the Atephen system (*qv*). A chemically resistant coating.†

Silicoderm F. A preparation for the protection and the care of the skin. (Bayer & Co).*

Silicol. Ferro-silicon, usually containing 84 per cent silicon, for use in the preparation of hydrogen by the action of caustic soda.†

Silicolloid. A natural siliceous material free from iron. Suitable for use in paper manufacture, cleansers, and tooth pastes.†

Silico-manganese. An alloy of silicon and manganese made in the electric arc type of furnace. It contains 60-75 per cent manganese, 20-25 per cent silicon, and the rest iron.†

Silicon Brass. An alloy of 81 per cent copper, 14 per cent zinc, and 3 per cent silicon.†

Silicon Bronze. (Silicum Bronze). An alloy of 97.37 per cent copper, 1.32 per cent tin, 1.24 per cent zinc, and 0.7 per cent silicon.†

Silicon Copper. Alloys of copper with small amounts of silicon. Used for the manufacture of telephone and telegraph wires. An alloy with 10 per cent silicon is also called silicon-copper.†

Silicon Nickel Brass. An alloy of 81 per cent copper, 14 per cent zinc, 3 per cent silicon, and 2 per cent nickel.†

Silicon Pig. See SILICON-EISEN.

Silicon Steel. A steel made by melting steel and ferro-silicon in crucibles. It is used for making sheets, springs, and acid-resisting plants.†

Silicon-Eisen. (Silicon Pig). A pig iron containing from 5-15 per cent silicon.†

Silicones. A generic name for compounds prepared with consistencies varying from greases to tough solids, in which silicon atoms are linked together in lon chains by alternate oxygen atoms and i which alkyl or aryl groups fill the third and fourth valencies of the silicon aton The compounds are unique in that thei physical properties are almost independent of temperature.

Silicone compounds of which details have been disclosed include the following:

E300 and *E301*: Polydimethyl siloxane gums; *E302* and *E303*: Polymethyl-vinyl siloxane gums; *E350* and *E351*: Polymethyl phenyl vinyl siloxane gums; *E367*: A partially-filled silicone gum.†

Silicon, Ferro. See FERRO-SILICON.

Silicosehl. Room temperature vulcanizing two-part silicone rubber systems. Applications: Casting and potting requiring exact surface detail, flexibilit stability and excellent electrical characteristics over wide temperature range. Chemical and weather resistant Non-toxic. (Solochart Ltd).*

Silico-spiegel. An alloy of 20 per cent manganese, 12 per cent silicon, and the rest iron.†

Silico-superphosphate. A preparation made by mixing superphosphate with kieselguhr or precipitated silicic acid. It is stated to give better results on mediu and light soils.†

Silico-titanium. A titanium-silicon alloy used in the steel industry.†

Silicum Bronze. Silicon bronze.†

Siligaz. Simethicone. Applications: Antiflatulent. (Menley & James Laboratories).

Siligen. Textile finishing auxiliaries. (BAS United Kingdom Ltd).

Siligran. A range of inorganic smelted products and their mixtures. Applications: Used as antiscale and fluxing compounds in steel production. (Bayer & Co).*

Silipact. Elastic sealants. Applications: Fo the construction industry and formed-in-place gaskets. (Lonza Limited).*

Siliporite. Molecular sieves. Applications: Drying of liquids and gases, desulphuration, separation of gases and isomers, in the chemical and petrochemical industries. Drying agent for the double glazing industry and for the polyurethane formulations. (British Ceca Co Ltd).*

Siliset. Foundry chemical hardeners. (Foseco (FS) Ltd).*

Silistren. Silicic acid tetra-glycollic ester.†

Silit. A material made by exposing mixed silicon, silicon carbide, and carbon, to the action Of carbon monoxide at 1500° C. It is made in three qualities. (1) A material for resistance subjected to permanent losses, (2) for electric heating work up to 1400° C., and (3) a fireproof material capable of withstanding violent changes of temperature.†

Silitonite. See FRANKONITE.

Silk Blue. See SOLUBLE BLUE.

Silk Blue 0. A dyestuff. It is a British equivalent of Soluble blue.†

Silk Grass. A term applied to pineapple fibre, obtained from the pineapple plant. Used for making cloth in the Phihppine Islands.†

Silk Green. See CHROME GREENS.

Silk Grey. An azine dyestuff obtained by oxidizing the product of the interaction of dimethyl- or diethyl-pheno-safranine and formaldehyde. Dyes silk from an acid bath.†

Silk Gum. See CUITE and SERICINE.

Silk Rubber. A rubber from an African tree, *Funtumia elastica*. See CUITE and SERICINE.†

Silk Scarlet S. A dyestuff. It is a British equivalent of Fast red B.†

Silk Size. See CUITE and SERICINE.

Silk Wadding. The waste from the spinning of silk.†

Silkin. A proprietary cellulose nitrate silk.†

Silkiol. Additive. Applications: Adhesive for carded worsted and cotton spinning. (Henkel Chemicals Ltd).

Silk, Anaphe. The silk obtained from a caterpillar in German East Africa. It has a specific gravity of 1.282.†

Silk, Basinetto. See GALETTAME SILK.

Silk, Cuprate. See CUPRAMMONIUM SILK.

Silk, Gelatin. See VANDURA SILK.

Silk, Neri. See GALETTAME SILK.

Silk, Ricotti. See GALETTAME SILK.

Silk, Sea. See BYSSUS SILK.

Silk, Tasar. See TUSSAR SILK.

Silk, Tubulated. See LUFTSEIDE.

Sillimanith. See PRESSOLITH.

Sillitin - Aktisil. Quartz - kaolinite. Applications: Filler for rubber and paint, soft abrasive for polishing agent. (Hoffmann Mineral).*

Sillitin N. A general purpose filler for rubber. It is a natural, finely divided mixed product of silicic acid and kaolin with a particle size of less than 20 μ.†

Sillman Bronze. An alloy of 86 per cent copper, 10 per cent aluminium, and 4 per cent iron.†

Silm. Photographic fixer stain remover. (May & Baker Ltd).

Silman Steel. A proprietary trade name for a silicon steel containing 2.1 per cent silicon, 0.85 per cent manganese, 0.3 per cent vanadium, 0.25 per cent chromium, and 0.55 per cent carbon.†

Sil-o-cel. A brand of kieselguhr, also a heat insulator made from kieselguhr.†

Siloid. A registered trade mark for micron-sized silica gels.†

Silopren. Hot air and room temperature vulcanizing silicone rubber. Applications: For seals, gaskets, dampening components, hoses, profiled strip, electrical insulating material, production of sealants, casting and coating compounds, fabric proofings and impression coatings. (Bayer & Co).*

Silopren HV. A basic compound of silicone rubber enriched with crosslinking agents, pigments etc., ready for processing. Applications: Technical moulded and extruded articles which have to meet high requirements in terms of resistance to hot air and weathering as well as to low temperature flexibility. (Rhein-Chemie Rheinau).*

Siloxicon. A fireproof material and a resistant to the action of acids and alkalis. It is made by heating powdered silica with a small quantity of carbon in the electric furnace, the composition approximating to Si_2C_2O. It is produced with silicon carbide in the carborundum furnace. Employed alone, or with binding materials, for making crucibles or muffles.†

Siloxide. A mixture of silica with a little titanium, or zirconium oxide.†

Silsoft. Elastofilcon A. Applications: Contact lens material. (Dow Corning Ophthalmics Inc).

Silteg. A range of highly and medium-active aluminium silicates. Applications: For shoe sole materials and technical rubber articles. (Degussa).*

Silumin. (Alpax). Proprietary alloys of aluminium and silicon containing 12 per cent silicon. They have a specific gravity of 2.63-2.65. Also see ALUDUR, ALUMINAC, WILMIL and SIGAL.†

Silumin-Y. A proprietary aluminium-silicon alloy with small additions of manganese and magnesium. It has high corrosion resistance.†

Siluminite. An electric insulator consisting of 75 per cent of mineral matter (asbestos, calcium silicate, and aluminium silicate), with pitch as the binding material.†

Silundum. A product similar to carborundum. Articles, such as crucibles and tubes, are made by shaping pieces of graphite, embedding them in carborundum, and subjecting them to the action of silicon vapour at high temperatures in the electric furnace. It has a high electrical resistance, and is used for making electrodes.†

Silva. A name used mainly in Germany for a type of artificial silk.†

Silvatol. Pretreatment agent. (Ciba-Geigy PLC).

Silvaz. A proprietary trade name for an alloy used for the manufacture of steel. It contains iron with 40-45 per cent silicon, 6.0-6.5 per cent vanadium, 6.0-6.5 per cent aluminium, and 6.0-6.5 per cent zirconium.†

Silvel. A proprietary trade name for an alloy containing 67.9 per cent copper, 16 per cent zinc, 6.5 per cent nickel, 0.5 per cent lead, 2.2 per cent iron, and 6.8 per cent manganese.†

Silver Alum. An aluminium-silver sulphate, $Al_2(SO_4)_3\ Ag_2SO_4\ 24H_2O$.†

Silver Amalgam. An alloy of mercury and silver. It occurs as a mineral, but is also prepared artificially.†

Silver Bell Metal. An alloy of 40-42 per cent copper and 58-60 per cent tin.†

Silver Bronze. An alloy of 64 per cent copper, 17 per cent manganese, 13 per cent zinc, 5 per cent silicon, and 1 per cent aluminium. An electrical resistance alloy.†

Silver Foil. An alloy of from 90-97 per cent tin, 0-2.5 per cent copper, and 0.10 per cent zinc, is known by this name.†

Silver Grain. The cochineal insect killed in an oven at three months old is called silver grain.†

Silver Grey. A Bohme dyestuff. It contains extracts of logwood and redwood, together with a chrome and iron mordant. Also see SLATE GREY and ZINC GREY.†

Silver Ink. A mixture of gum arabic and ground white mica. Employed for inlaying buttons.†

Silver Leaf. An alloy of 91 per cent tin, 8 per cent zinc, 0.35 per cent lead, and 0.2 per cent iron. Another alloy contains 91 per cent tin, 8.25 per cent zinc, and 0.4 per cent antimony.†

Silver Metal. An alloy of 66.5 per cent zinc and 33.5 per cent silver. Also see ALUMINIUM SILVER.†

Silver Methylene Blue. The silver salt of methylene blue. A germicide.†

Silver Quinaseptolate. Silver-oxychino-line-sulphonate.†

Silver Saltpetre. A name which has been applied to silver nitrate, $AgNO_3$.†

Silver Sand. Quartz sand.†

Silver Solder. Variable alloys. Usually they contain silver, copper, and zinc. A soft silver solder contains 67 per cent silver and 33 per cent brass, and is suitable for sheet. A hard silver solder consists of 80 per cent silver and 20 per cent copper. Some alloys contain smaller amounts of silver, and an ordinary one of this type is composed of 47 per cent copper, 47 per cent zinc and 6 per cent silver. Sometimes a little tin is present.†

Silver Ultramarine. See YELLOW ULTRAMARINE.

Silver White. See FLAKE WHITE.

Silver-salt. Sodium anthraquinone monosulphonate, obtained in alizarin manufacture.†

Silverine. An alloy of 77 per cent copper, 17 per cent nickel, 2 per cent iron, 2 per cent zinc, and 2 per cent cobalt.†

Silvering Solutions. Usually consist of solutions of silver cyanide and ammonium cyanide in water. Used for the electro-deposition of silver.†

Silverite. See NICKEL SILVERS.

Silveroid. An alloy of 45 per cent nickel, 54 per cent copper, and 1 per cent manganese.†

Silverstone. Non-stick finishes. (Du Pont (UK) Ltd).

Silver, Blue. See NIELLO SILVER.

Silver, China. See ARGYROLITH.

Silver, Crede's. See COLLARGOL.

Silver, Frosted. See DEAD SILVER.

Silver, German. See NICKEL SILVERS.

Silver, Nevada. See NICKEL SILVERS.

Silver, oxidized. Silver covered with a thin film of sulphide, by immersion in a solution obtained by boiling sulphur with potash.†

Silver, Peru. See CHINESE SILVER.

Silver, Potosi. See NICKEL SILVERS.

Silver, Refined. See STANDARD SILVER.

Silver, Sterling. See STANDARD SILVER.

Silver, Victoria. See NICKEL SILVERS.

Silver, Virginia. See NICKEL SILVERS.

Silvestrite. A mineral. It is siderazote.†

Silvet. Aluminium pigments for plastics. Applications: All plastics, automobiles, toys, bottles etc. (Silberline Mfg Co Inc).*

Silvex. Aluminium pigments for plastics. Applications: All plastics, automobiles, toys, bottles etc. (Silberline Mfg Co Inc).*

Silvital. Fertilizer for vitalizing damaged forest. (Süd-Chemie AG).*

Silvoline. Aluminium paint. Applications: For marine and general purposes. (Llewellyn Ryland Ltd).*

Silzin Bronze. An alloy of copper with 10-20 per cent zinc and 4.5-5.5 per cent silicon.†

Simadex. Selective and total herbicide. (FBC Ltd).

Simanex. Active ingredient: simazine; 2-chloro-4,6-bis(ethylamino)-1,3,5-triazine. Applications: Pre-emergence herbicide for control of weeds in a variety of crops as well as a soil sterilant. (Agan Chemical Manufacturers Ltd).†

Simax. Silicon carbide. Applications: Used in metallurgical applications. (Lonza Limited).*

Simazol. Active ingredients: azolan plus simanex. Applications: Multipurpose herbicidal mixture which eradicates a wide spectrum of established weeds, while preventing further weed germination for extended periods. (Agan Chemical Manufacturers Ltd).†

Simeco. A proprietary preparation of aluminium hydroxide, sucrose and siroethicone used as a gastro-intestinal sedative. (Wyeth).†

Simetite. Sicilian amber of wine-red to garnet-red colour.†

Simflex. Flexible circuitry. (Carl Freudenberg).*

Simflow. Simazine formulated as a flowable liquid. Applications: It is a soil residual herbicide suitable for selective weed control in shrubberies and may also be used as a total weedkiller to keep pathways, bare ground and industrial installations free of weeds and grasses. (Burts & Harvey).*

Simflow Plus. A liquid composition containing aminotriazole and simazine. Applications: It can be used in situations where total weed control is required, including industrial sites, paths, kerbs and channels, drives and hard tennis courts, hardstanding and storage areas. (Burts & Harvey).*

Similor. A rich-coloured brass. It usually contains from 80-89 per cent copper, 9-20 per cent zinc, and 0-7 per cent tin.†

Simmering. Rotating shaft seal. (Carl Freudenberg).*

Simplex Steel. A proprietaryry trade name for a nickel-chromium steel containing 1.25 per cent nickel and 0.6 per cent chromium.†

Simplotan. A proprietary preparation of tinidazole. An antiprotozoal. (Pfizer International).*

Simrax. Face seals. (Carl Freudenberg).*

Simrit. Seals and packing rings. (Carl Freudenberg).*

Sin Red. Potassium permanganate, $KMnO_4$.†

Sinapoline. Diallyl-urea, $(C_3H_5NH)_2CO$.†

Sinatron. A range of polyester resins which include orthophthalic, isophthalic, bisphenolic, neopentilic and self extinguishing resins. (Lonza Limited).*

Sinaxar. A proprietary preparation of STYRAMATE used in the treatment of muscle spasm. (Armour Pharmaceutical Co).†

Sinazine. Herbicide. (Murphy Chemical Ltd).

Sinbar. Terbacil weedkiller. (Du Pont (UK) Ltd).

Sin-Chu. (Japanese Brass). An alloy of 66.5 per cent copper, 33.4 per cent zinc, and 0.1 per cent iron.†

Sindanyo. A trade name for proprietary asbestos products.†

Sinecain. Quinine hydrochloride and antipyrine for hypodermic use.†

Sinemet. Levodopa and carbidopa. Applications: For the treatment of Parkinsonism. (Merck Sharp & Dohme).*

Sinemet-Plus. Levodopa and carbidopa. Applications: For the treatment of Parkinsonism. (Merck Sharp & Dohme).*

Sinequan. A proprietary preparation of doxepin. An antidepressant. (Pfizer International).*

Single Muriate of Tin. An acid solution of stannous chloride, used as a mordant.†

Single Nickel Salt,. Nickel sulphate, $NiSO_4.7H_2O$, used in the plating trade.†

Singlet. Used primarily in the treatment of coughs, colds and upper respiratory conditions. (Dow Chemical Co Ltd).*

Singoserp. A proprietary trade name for Syrosingopine.†

Sinigrin. Potassium myronate, $KC_{10}H_{16}NS_2O_9$. A constituent of black mustard seed.†

Sinodor. A basic magnesium acetate, containing an excess of magnesium hydrate. Used for disinfecting purposes.†

Sinopis. See INDIAN RED.

Sinter-corundum. A proprietary preparation. It is a ceramic material produced from pure alumina at a temperature of about 1800° C. The thermal conductivity at 16° C. is about twenty times as high as that of porcelain, and it is stated to be not attacked by hydrofluoric acid or hot alkali.†

Sinterit. A proprietary trade name for a form of sponge iron (*qv*) used for coupling packings.†

Sinterloy. A proprietary trade name for a steel powder.†

Sinthrome. A proprietary preparation of nicoumalone. An anticoagulant. (Ciba Geigy PLC).†

Sinvabond. Flexographic printing inks. Applications: Printing laminated film structures. (Allied Signal Sinclair and Valentine).*

Sinvaset. Web heatset offset printing inks. Applications: Publication printing. (Allied Signal Sinclair and Valentine).*

Siogel. Approximately 99.7% silcon dioxide. Applications: For dehydration of air and other gases. (Chemische Fabriken Oker und Braunschweig AG).*

Sionon. A propetary sugar substitute. It is sorbitol.†

Siopel. A preparation of DIMETHICONE and cetrimide. A barrier skin cream. (Sold in UK by Care Laboratories Ltd). (ICI PLC).*

Sioplas. Polyethylene cross linking technology. (Dow Corning Ltd).

Sipalin AOC. A proprietary solvent for cellulose nitrate. It is dicyclohexyl adipate.†

Sipalin AOM. A proprietary solvent for cellulose nitrate and plasticizer for rubber. It is dimethylcyclohexyl adipate.

It boils at from 225-232° C., has a specific gravity of 1.011, and a flash-point of 189° C.†

Sipalin MOM. A proprietary solvent and plasticizer. It is dimethylcyclo-hexyl β-methyladipate. It boils from 216-224° C., has a specific gravity of 1.009, and flashes at 195° C.†

Sipeira. Bebeeru bark.†

Sipernat. Range of spray dried hydrophillic precipitated silicas. Applications: Free flow/anti-caking aids. As carrier substances for production of highly concentrated pulverulent formulations of liquid or paste-like active substances. (Degussa).*

Sipex DS. Anionic surfactant in acid form, supplied as a viscous liquid. Applications: Basic material for general detergents eg in dishwashers, and in other formulations requiring a cheap source of anionic active matter. (Henkel Chemicals Ltd).

Sipex 30. Anionic surfactant as a liquid paste. Applications: General detergent; emulsifier for emulsion polymerization. (Henkel Chemicals Ltd).

Sipilite. See BAKELITE.

Sipon. A registered trade mark currently awaiting re-allocation by its proprietors. (Casella AG).†

Sirene. Dodecylbenzene.†

Sirflex. Alkylates for textiles.†

Sirit. Urea resins.†

Sirius. A range of direct dyestuffs. Applications: For dyeing cellulosics. (Bayer & Co).*

Sirius Sirius Supra Sirius Supra LL. Selected direct dyes. (Bayer UK Ltd).

Sirius Yellow G. A dyestuff. It is 1 : 2-benzanthraquinone.†

Sirlene. Feed grade propylene glycol, an emulsifying agent and general purpose food additive. Applications: As a conditioner in animal feed, a preservative, humectant, energy source, lubricant, extender, palatability improver and fines reducer. In dairy cattle, it is used for the prevention and treatment of acetonemia (ketosis). (Dow Chemical Co Ltd).*

Sirpol. Acetovinylic resins.†

Sirtene. L-D polyethylene.†

Sisacan. Dronabinol. Applications: Anti-emetic. (PARS Pharmaceutical Laboratories Inc).

Siseptin. Sisomicin Sulphate. Applications: Antibacterial. (Schering-Plough Corp).

Siserskite. See IRIDOSMINE.

Sitara Fast Red. See HELIO FAST RED RL.

Sitilan,. The methylcyclohexyl ester of adipic acid. A solvent for cellulose and rubber.†

Sitol. A proprietary trade name for the sodium salt of *m*-nitrobenzenesulphonic acid. Used in dyeing.†

Sit-ruti. See SANNA.

Sivex. Ceramic foam filters for removal of non-metallic inclusions from aluminium alloys. (Foseco (FS) Ltd).*

Sivex F. Ceramic foam filters used for the production of aluminium and copper based alloys. (Foseco (FS) Ltd).*

Size. Textile sizing agents. (BASF United Kingdom Ltd).

Size. Usually consists of a starch solution containing small amounts of tallow or oil, and China clay or French chalk.†

Size, Silk. See CUITE and SERICINE.

SKA. A butadiene polymer of Soviet origin, derived from petroleum.†

SK-Amitriptyline. Amitriptyline hydrochloride. Applications: Antidepressant. (Smith Kline & French Laboratories).

SK-Ampicillin. Ampicillin. Applications: Antibacterial. (Smith Kline & French Laboratories).

SKB. A butadiene polymer of Soviet origin, derived from alcohol.†

SK-Bamate. Meprobamate. Applications: Sedative. (Smith Kline & French Laboratories).

SK-Doxycycline. Doxycycline hyclate. Applications: Antibacterial. (Smith Kline & French Laboratories).

SK-Erythromycin. Erythromycin stearate. Applications: Antibacterial. (Smith Kline & French Laboratories).

SK-Estrogens. Oestrogens, esterified. Applications: Oestrogen. (Smith Kline & French Laboratories).

SK-Furosemide. Furosemide. Applications: Diuretic. (Smith Kline & French Laboratories).

SK-Hydrchlorothiazide. Hydrochlorothiazide. Applications: Diuretic. (Smith Kline & French Laboratories).

SK-Lygen. Chlordiazepoxide hydrochloride. Applications: Sedative. (Smith Kline & French Laboratories).

SK-Metronidazole. Metronidazole. Applications: Antiprotozoal. (Smith Kline & French Laboratories).

SK-Penicillin G. Penicillin G Potassium. Applications: Antibacterial. (Smith Kline & French Laboratories).

SK-Penicillin VK. Penicillin V Potassium. Applications: Antibacterial. (Smith Kline & French Laboratories).

SK-Phenobarbital. Phenobarbital. Applications: Anticonvulsant; hypnotic; sedative. (Smith Kline & French Laboratories).

SK-Potassium Chloride. Potassium Chloride. Applications: Replenisher. (Smith Kline & French Laboratories).

SK-Pramine. Imipramine hydrochloride. Applications: Antidepressant. (Smith Kline & French Laboratories).

SK-Prednisone. Prednisone. Applications: Glucocorticoid. (Smith Kline & French Laboratories).

SK-Probenecid. Probenecid. Applications: Uricosuric. (Smith Kline & French Laboratories).

SK-Propantheline Bromide. Propantheline Bromide. Applications: Anticholinergic. (Smith Kline & French Laboratories).

SK-Quinidein Sulfate. Quinidine Sulphate. Applications: Cardiac depressant. (Smith Kline & French Laboratories).

SK-Reserpine. Reserpine. Applications: Antihypertensive. (Smith Kline & French Laboratories).

SK-Soxazole. Sulphisoxazole. Applications: Antibacterial. (Smith Kline & French Laboratories).

SK-Tetracycline. Tetracycline Hydrochloride. Applications: Anti-amebic; antibacterial; antirickettsial. (Smith Kline & French Laboratories).

SK-Thioridazine HC1. Thioridazine Hydrochloride. Applications: Antipsychotic; sedative. (Smith Kline French Laboratories).

SK-Tolbutamide. Tolbutamide. Applications: Antidiabetic. (Smith Kline & French Laboratories).

SK-Triamcinolone. Triamcinolone. Applications: Glucocorticoid. (Smith Kline & French Laboratories).

SK-65. A proprietary preparation of DEXTROPROPOXYPHENE. An analgesic. (Smith Kline and French Laboratories Ltd).*

Skane. Mildewcide. Applications: Coatings. (Rohm and Haas Company).*

Skatole. See SCATOLE.

Skefron. A proprietary preparation of dichlorodifluoromethane and trichlorofluoromethane. An analgesic spray. (Smith Kline and French Laboratories Ltd).*

Skeladin. A proprietary trade name for Metaxalone.†

Skelleftea. A variety of Stockholm tar.†

Skellysolve. A proprietary trade name for a series of petroleum solvents.†

Skiargan. A colloidal silver solution.†

Skin Wiz. Proprietary, polymeric film forming coating to protect moulds and moulded parts which releases easily when required. Applications: Fibreglas tools, injection moulds and moulded or fabricated plastic or metal parts. (Axel Plastics Research Laboratories Inc).*

Skiodan Sodium. Methiodal sodium. Applications: Diagnostic aid. (Sterling Drug Inc).

Skleron. A proprietary trade name for an aluminium alloy containing 12 per cent zinc, 3 per cent copper, 0.6 per. cent manganese, 0.25 per cent silicon, and a small amount of nickel.†

Skliro. Woolwax fatty acids. (Croda Chemicals Ltd).

Sky Blue. See IMMEDIAL PURE BLUE, WILLOW BLUE, and COERULEUM.

Skybond. Polyamide resin varnish. Applications: Designed for structural, electrical and specialty applications where extended exposure to high temperature is required. (Monsanto Co).*

kydrol. Fire resistant hydraulic fluid. Applications: Distributed to commercial airlines. (Monsanto Co).*

kyllex. Dialkyl dimethyl ammonium chloride. Supplied as a mixture of water and isopropanol suspension forming a stiff paste. Application: Cationic surfactant used in fabric softeners and as a bactericide. (Efkay Chemicals Ltd).

K65. Propoxyphene Hydrochloride. Applications: Analgesic. (Smith Kline & French Laboratories).

lack Wax. A soft parsffin wax from the pressing of paraffin distillate.†

lag A. A British chemical standard. It is a basic slag containing 44.5 per cent. CaO, 16.15 per cent. SiO_2, 12.93 per cent. P_2O_5., 8.97 per cent. Fe, and 6.9 per cent. MgO.†

Slag Sand. Blast furnace slag is run out of the furnace to fall into a running stream of water, when it is broken up into a fine sand.†

Slag Wool. (Mineral Cotton, Mineral Wool). Blast furnace slag (essentially a glass composed of silicates of aluminium and calcium), which has had air blown through it. It resembles spun glass, and is used for packing steam pipes.†

Slagbestos. A similar product to slag wool (blast furnace slag).†

Slaked Lime. Calcium hydroxide, $Ca(OH)_2$.†

Slate Black. See MINERAL BLACK.

Slate Dust. (Slate Filler). A ground slate used as a filler in rubber mixings. It usually has a specific gravity of 2.7-2.8.†

Slate Grey. (Stone Grey, Silver Grey, Mineral Grey). Grey pigments obtained by grinding and levigating special kinds of grey slate, which occur in Germany. Used as priming paint, and for the preparation of putty. They are imitated by mixtures of white clay, blacks, ochres, and ultramarine.†

Slate Lime. A mixture of 60 per cent lime with 40 per cent of calcined slate powder used in the manufacture of porous concrete.†

Slax. Slag coagulants. (Foseco (FS) Ltd).*

Slaymor. Rodenticide. (Ciba-Geigy PLC).

S-lec. Polyvinyl butyral resin. Applications: Interlayer film for safety glass. (Sekisui Chemical Co Ltd).*

Slick. A sulphurized processing aid for glass manufacturing. (Specialty Products Co).*

Slicker Solder. See PLUMBER'S SOLDER.

Slip-Ayd. Dispersed slipping agents. Applications: Paints, inks etc. (Cornelius Chemical Co Ltd).*

Slix. Heat resisting refractory cement. (Foseco (FS) Ltd).*

Sloeline. See INDULINE, SPIRIT SOLUBLE.

Sloeline RS, BS. See INDULINE, SOLUBLE.

Slop Wax. The wax present in the heavier wax distillates obtained in the refining of petroleum waxes. It is commonly considered unpressable and therefore different from the paraffin wax pressed from lighter wax distillates.†

Slow-Fe. A proprietary preparation of ferrous sulphate in a slow release base. Iron supplement. (Ciba Geigy PLC).†

Slow-FE Folic. Iron replacement. (Ciba-Geigy PLC).

Slow-K. Potassium Chloride. Applications: Replenisher. (Ciba-Geigy Corp).

Slow-Sodium. A proprietary preparation of sodium chloride in sustained release form. (Ciba Geigy PLC).†

Slow Trasicor. Beta-blocker. (Ciba-Geigy PLC).

Sludge Acid. Sulphuric acid which has been used in the refining of petroleum.†

Slug. Slug killers. (Murphy Chemical Ltd).

Slug-Geta. Slug and snail bait. (Chevron).*

Slugit Liquid. Metaldehyde slug killer. (Murphy Chemical Ltd).

Slugoids. 3% metaldehyde slug killer pellets (containing animal repellant). Applications: Slug/snail control. (Doff Portland Ltd).*

Slug Pellets. Metaldehyde slug killer. (Murphy Chemical Ltd).

Slug Snail Killer. Molluscididide pellets. (Fisons PLC, Horticulture Div).

SMA. A proprietary milk feed for babies. (Wyeth).†

SMA 17352 A. A proprietary copolymer of styrene and maleic anhydride, of low molecular weight, used as a levelling agent in polishes. (Arco Chemical Europe Inc).†

SMA 2625 A. A proprietary styrene maleic anhydride copolymer of low molecular weight, used as a levelling resin in floor polishes. (Arco Chemical Europe Inc).†

SMA 3840. A proprietary styrene-maleic anhydride copolymer of low molecular weight used for coating cans and drums. (Arco Chemical Europe Inc).†

SMA 5500. A proprietary copolymer of Styrene and maleic anhydride, partially esterified and of low molecular weight, used as a vehicle for thermosetting electro-deposited coatings. (Arco Chemical Europe Inc).†

Smalt. (Saxony Blue, Saxon Blue, King's Blue, Royal Blue, Zaffer, Zaffre, Bleu D'Azure, Bleu de Saxe, Azure Blue). A potash glass containing oxide of cobalt. It is prepared by mixing zaffre (cobalt oxide) with powdered quartz and potassium carbonate and heating. The resulting product is a double silicate of cobalt and potassium, containing about 6 per cent cobalt oxide. Used as a pigment. Ash blue and pale smalt are finer qualities of smalt. Also see COBALT BLUE, ESCHEL, and STREWING SMALT.†

Smalt, Pale. See SMALT.

Smaragdgreen. See BRILLIANT GREEN.

Smaragdine. A trade name for a solidified alcohol consisting of alcohol and gun-cotton, coloured with malachite green.†

Smelite. Synonym for Kaolinite.†

Smithite. A mineral. It is a silver sulpharsenite.†

Smithsonite. A mineral. See CALAMINE.†

Smitter-Lénian. An alloy containing 72 per cent copper, 12.75 per cent nickel, 9.75 per cent zinc, 2.3 per cent iron, 2.25 per cent tin, and 1 per cent bismuth.†

Smoke. Uncut disperse and solvent dyestuffs. Applications: Colouration of smoke grenades mainly for use by army. (Holliday Dyes & Chemicals Ltd).*

Smoke Black. A carbon black used as a pigment. It contains 99.75 per cent carbon.†

Smoke Blue. A Bohme dyestuff. It contains logwood extract and a chrome mordant.†

Smokeless Diamond Powder. A 33 grain powder consisting of insoluble nitro-cellulose, with 15 per cent metallic nitrates, 6 per cent charcoal, and 3 per cent petroleum jelly.†

Smoking Deterrent. A proprietary preparation of magnesium carbonate, LOBELIN sulphate and tribasic calcium phosphate. (Campana Corp, Batavia, Ill).†

Smoking Salts. Impure hydrochloric acid.†

Smoky Quartz. A quartz containing organic matter or hydrocarbons. It is usually brown in colour.†

S Monel. A Monel metal with 3.75 per cent silicon used in valves, etc., which are subject to corrosion.†

Smooth-On. Iron and foundry cements, epoxy adhesives and cements. Other formulations based on epoxy, polysulphide and polyurethane polymers. Applications: Maintenance and repair, structural bonding and plastic tooling applications (flexible moulds, cast and laminated plastic tools). (Smooth-On Inc).*

Smoothex. A range of glycerol esters. Applications: Used as emulsifiers, stabilizers and antifoam agents in the food and cosmetic industries. (Harcros Industrial Chemicals).*

SM945. Lightly milled monoclinic zirconium dioxide with a zirconia content (including hafnia) of 94.5%. Applications: Manufacture of ceramic pigments and welding fluxes. (Ferro Corporation).*

SM975. Lightly milled monoclinic zirconium dioxide with a zirconia content (including hafnia) of 97.5%.

Applications: Manufacture of ceramic pigments and welding fluxes. (Ferro Corporation).*

SM987. Lightly milled monoclinic zirconium dioxide with a zirconia content (including hafnia) of 98.7%. Applications: Manufacture of ceramic pigments. (Ferro Corporation).*

SM992. Lightly milled monoclinic zirconium dioxide with a zirconia content (including hafnia) of 99.2%. Applications: Manufacture of lead-zirconate-titanate piezo electric ceramics, zirconates, zirconia technical ceramics and oxygen sensors. (Ferro Corporation).*

SM994. Lightly milled monoclinic zirconium dioxide with a zirconia content (including hafnia) of 99.4%. Applications: Manufacture of lead-zirconate-titanate piezo electric ceramics, zirconates, zirconia technical ceramics and oxygen sensors. (Ferro Corporation).*

Snake Root. Senega root.†

Sniafil. An Italian synthetic wool substitute. It is a product obtained in a similar way to Viscose silk, but differs from it in the treatment of the viscose solution.†

Sniafoam. Glass fibre reinforced polyester foam. (Lonza Limited).*

Sniamid. Nylon granules. Applications: Engineering plastic. (SNIA (UK) Ltd).†

Snomelt. Calcium chloride pellets. Applications: Snow and ice melting, dust control and tyre weighting. (Standard Tar Products Company, Inc).*

Snow White. See ZINC WHITE, and OIL WHITE.

Snowtack. Adhesive tackifier emulsion. (Tenneco Malros Ltd).

S-nyl. Polyvinyl acetate homopolymer. Applications: Chewing gum base. (Sekisui Chemical Co Ltd).*

So-luminum. An aluminium solder. It is an alloy of 55 per cent tin, 33 per cent zinc, 11 per cent aluminium, and 1 per cent copper.†

Soa. A proprietary trade name for sucrose octa-acetate. It is used as a plasticizer.†

Soap Balsam. Soap liniment.†

Soap Bark. The bark of *Quillaia saponaria*, of Chile. The commercial product consists of the layer of bast with the dead bark removed. The active principle is saponin. It is used to clean clothes.†

Soap, Clay. See BENTONITE.

Soap, Mountain. See STEATITE.

Soap Root. Ordinary soap root is composed of the stems and root of the soap wort, *Saponaria officinalis*. The white soap root is the root of species of *Gypsophila*.†

Soap, Stearin. Curd soap.†

Soapstone. See STEATITE.

Soap, Tallow. Curd soap.†

Sobee. A proprietary baby feed based on soya, used in cases when an infant is intolerant of milk. (Bristol-Myers Co Ltd).†

Soborol. Methyl *p*-hydroxybenzoate.†

Sobralite. Synonym for Pyroxmargite.†

Socal. Precipitated calcium carbonate. (Laporte UK Trading).

Socaloin. (Zanaloin). Aloin from Socotrine or Zanzibar aloes.†

Sochamine A 271. Coconut and lauric carboxy/sulphate. Applications: Non-eye-stinging shampoos. (Witco Chemical Ltd).

Sochamine A 7525. Coconut dicarboxylate in liquid form. Applications: Hair and baby shampoo. (Witco Chemical Ltd).

Sochamine A 7527. Coconut dicarboxylate in liquid form. Applications: Hard surface cleaners, hair shampoos, textile treatment, strongly acid or alkaline cleaners. (Witco Chemical Ltd).

Sochamine A 8955. Alkyl imidazoline dicarboxylate in liquid form. Applications: Low foam detergents. (Witco Chemical Ltd).

Soda. Sodium carbonate and bicarbonate are both known by this term.†

Soda Ash. Practically anhydrous sodium carbonate, Na_2CO_3. A commercial variety of soda ash used for softening boiler feed water is known as 58 per cent soda ash and contains 58 per cent. Na_2O.†

Soda Ash Blocks. Supplementary fluxes used in the melting of cast iron. (Foseco (FS) Ltd).*

Soda Blue. (Gas Blue). Impure Prussian blues prepared by using sodium ferrocyanide instead of the potassium salt.†

Soda Bordeaux Mixture. Made with 6 lb copper sulphate, 2 lb caustic soda, and 50 gallons water. See BORDEAUX MIXTURE.†

Sodacopperas. Synonym for Natro jarasite.†

Soda Glass. See SODA-LIME GLASS.

Soda Glass, Soluble. See SOLUBLE GLASS.

Soda Greens. Pigments. They are arsenic greens, obtained by neutralizing the mother liquor containing white arsenic, acetic acid, and dissolved emerald green, which is produced in the preparation of the latter with sodium carbonate. If the neutralization is carried out with milk of lime, " lime arsenic greens " are produced.†

Sodaheterosite. Synonym for Heterosite.†

Soda-lime Glass. A glass usually containing from 71-78 per cent. SiO_2, 12-17 per cent. Na_2O, 5-15 per cent. CaO, 1-4 per cent. Al_2O_3 and Fe_2O_3, and 0-2 per cent. K_2O.†

Soda-lime, Sofnol. See SOFNOL SODALIME G.

Sodalumite. A soda alum in cubic form.†

Soda Lye Obtained by boiling a solution of sodium carbonate with slaked lime.†

Soda Nitre. See CHILE SALTPETRE.

Soda-olein. A sulphonated castor oil.†

Soda Pulp. Wood pulp obtained by means of caustic soda.†

Soda Saltpetre. See CHILE SALTPETRE.

Sodasorb. Soda Lime. Applications: Carbon dioxide absorbant. (W R Grace & Co).

Soda Tar. The name applied to an alkaline solution which has been used to purify petroleum oils after they have been treated with sulphuric acid.†

Soda, Tartrated. See ROCHELLE SALT.

Sodatol. An agricultural explosive. It is a mixture of sodium nitrate and trinitrotoluene.†

Soda Ultramarine. See ULTRAMARINE

Soda, Vitriolated. Sodium sulphate, Na_2SO_4.†

Soda Water Glass. A mixture of sodium silicates.†

Soderseine. Colloidal bismuth.†

Sodiformasal. See FORMASAL.

Sodital. Pentobarbital Sodium. Applications: Hypnotic; sedative. (American Critical Care, Div of American Hospital Supply Corp).

Sodium Aeroflot Promoter. Sodium diethyl dithiophosphate. (Cyanamid BV).*

Sodium Alum. Aluminium-sodium sulphate, $Al_2(SO_4)_3$ Na_2SO_4 $24H_2O$.†

Sodium Amytal. Sodium isoamyl-ethyl barbiturate. A proprietary preparation of araylobarbitone sodium. A hypnotic. (Eli Lilly & Co).†

Sodium Anoxynaphthonate. Sodium 4' - anilino-8-hydroxy-1,1'-azo-naphthalene - 3,6,5' - trisulphonate. Coomassie Blue.†

Sodium Citrotartrate. A mixture of sodium citrate and tartrate.†

Sodium Diuril. Chlorothiazide sodium. Applications: Diuretic; antihypertensive. (Merck Sharp & Dohme, Div of Merck & Co Inc).

Sodium Morrhuate. The sodium salt of the fatty acids of cod-liver oil.†

Sodium Muriate. See MURIATE OF SODA.

Sodium para-aminohippurate. Applications: For intravenous use to measure effective renal plasma flow and tubular secretory capacity. (Merck Sharp & Dohme).*

Sodium Permutite. A sodium zeolite, made artificially.†

Sodium Phosphate, Effervescent. A mixture of sodium phosphate and bicarbonate and citric and tartaric acids.†

Sodium Versenate. Edetate disodium. Applications: Chelating agent; pharmaceutic aid. (Riker Laboratories Inc, Subsidiary of 3M Company).

Sodos. A mixture of sodium dihydrogen phosphate and sodium bicarbonate. Used in medicine.†

Sofanate. A fungicide for fruit storage. (Plant Protection (Subsidiary of ICI)).†

Sofibex. Sodium hypophosphite. Applications: Surface treatment, electroless nickel plating. (British Ceca Co Ltd).*

Soflens. Polymacon. Applications: Contact lens material. (Bausch & Lomb, Professional Products Div).

Sofnol Soda-lime G. (Sofnolite). A proprietary form of soda-lime containing a little manganic acid. It is stated to absorb much more carbon dioxide than ordinary soda-lime, and to change colour as the degree of saturation is approached.†

Sofnolite. See SOFNOL SODALIME G.

Soft Amber. See GEDANITE.

Soft Copal. A name applied to varieties of Australian sandarac resin.†

Soft Platinum. Commercially pure platinum, containing about 1 per cent iridium.†

Soft Solder. See PLUMBER'S SOLDER.

Softcon. Vifilcon. Applications: Contact lens material (hydrophilic). (Parke-Davis, Div of Warner-Lambert Co).

Softenol. Fatty acid esters. Applications: For technical applications. (Dynamit Nobel Wien GmbH).*

Softex. Proprietary trade name for pure red oxide (*qv*).†

Softex. Stearoyl lactylates. Applications: Used as dough conditioners and anti staling additives in bread and baked goods. (Harcros Industrial Chemicals).*

Softigen. Emulsifier, superfatting or re-fatting agent. Applications: Skin protection creams, bath preparations, lotions, shaving preparations, soaps etc. (Dynamit Nobel Wien GmbH).*

Softisan. Ointment and cream bases. Applications: For creams, ointments, emulsions and lipsticks. (Dynamit Nobel Wien GmbH).*

SoftMate DW. Hefilcon A. 2-Hydroxyethyl methacrylate polymer with 1-vinyl-2-pyrrolidinone and ethylene dimethacrylate. Applications: Contact lens material. (Barnes-Hind Inc).

Softrite. A proprietary rubber softener. It has a zinc laurate base.†

Soie Nouvelle. See LUFTSEIDE.

Soil Pests Killer. Granular insecticide. (Fisons PLC, Horticulture Div).

S-Oils. Sulphur-containing oils obtained by the distillation of crude petroleum oil in the presence of sulphur. They have a strong antiseptic action against wood-destroying fungi.†

Soil TRIGGRR. A liquid containing cytokinin, a plant growth regulator, used to increase crop yields and quality. Applications: Applied to the soil. Used for a wide variety of crops including corn, peanuts, sorghum, soybeans, fruits and vegetables. (Westbridge Research Group).*

Soilime. A lime residue from cyanamide manufacture. It contains 50 per cent of lime.†

Soja Bean Oil. See SOYA BEAN OIL.

Sokoff. Industrial grease solvent. (The Wellcome Foundation Ltd).

Sol Rubber. The portion of rubber which enters solution when unmilled raw rubber is treated with a solvent.†

Solacen. Tybamate. Applications: Tranquilizer (minor). (Wallace Laboratories, Div of Carter-Wallace Inc).

Solactol. A proprietary trade name for ethyl lactate.†

Soladox. Chlorosulphonated polyethylene in a solvent base, with dense pigmentation, for use as a corrosion preventative. Applications: Offshore oil, all marine, roofing, any corrosive environment where complete protection is needed. (Liquid Plastics Ltd).*

Soladox 112. Chlorosulphonated polyethylene in a solvent base, with dense pigmentation - for use as weatherproofing compound. Applications: Reflects infra red radiation - for use in defense related environments. (Liquid Plastics Ltd).*

Solaesthin. Methylene chloride, CH_2Cl_2.†

Solan. Water soluble lanoline derivative; emollient; solubilizer; wetting agents. Applications: Shampoos and foam baths; hair sprays lacquers and wave sets; nail varnish remover. (Croda Chemicals Ltd).

Solan E. Polyethoxylated lanolin. (Croda Chemicals Ltd).

Solane. Sulphur dyes. (ICI PLC).*

Solanthrene. Vat dyes. (ICI PLC).*

Solar Oil. The name given to various hydrocarbons obtained as by-products in the treatment of brown coal tar in paraffin works.†

Solar Salt. Salt (sodium chloride) obtained by the evaporation of sea-water.†

Solar Stearin. Lard stearin.†

Solar Steel. A proprietary trade name for a silicon steel containing 1 per cent silicon, 0.5 per cent molybdenum, 0.4 per cent manganese, and 0.5 per cent carbon.†

Solasol. Vat dyes. (ICI PLC).*

Solasulphone. Solapsone.†

Solatene. Beta carotene. Applications: Ultra violet screen. (Hoffmann-LaRoche Inc).

Solatol. A preparation of crude phenol (carbolic acid). A disinfectant.†

Solbrol. *p*-Hydroxybenzoic acid ester. Applications: For the preservation of pharmaceuticals, cosmetics and foodstuffs as well as technical goods. (Bayer & Co).*

Solcod. A proprietary trade name for sulphonated cod oil.†

Solcornol. A proprietary trade name for a sulphonated corn oil.†

Solder. The various alloys or mixtures which constitute solder are usually classified as hard or soft, according to their melting-point. Hard solder includes brazing solder, silver solder, and gold solder, whilst the soft solders usually consist of tin and lead, and melt below 300° C. The addition of cadmium and bismuth tends to lower the melting point, and antimony to raise it.†

Solder, Coarse. See PLUMBER'S SOLDER.

Solder, Fine. See PLUMBER'S SOLDER and TINSMITH'S SOLDER.

Soldering Acid. Hydrochloric acid, HCl.†

Soldering Salt. Ammonium and zinc chlorides.†

Soldering Solution. A solution of zinc chloride. Also see GAUDUIN'S FLUID and MULLER'S FLUID.†

Solder, Slicker. See PLUMBER'S SOLDER.

Solder, Soft. See PLUMBER'S SOLDER.

Soldis. A disinfectant containing phenolic and cresylic bodies. It is miscible with water.†

Soldo. A flux used for tinning metals. It is mixed with powdered tin.†

Soledon. Solubilised vat dyes. (ICI PLC).*

Solef. Polyvinylidene fluoride (PVDF) homopolymers and copolymers. Applications: High purity and corrosion resistant applications in the chemical processing and semiconductor manufacturing industries, including pipe and fittings, valves, pumps and vessels. Protective coatings. Wire and cable jacketing and fibre optics buffer tubing. (Soltex Polymer Corporation).*

Solene. See PETROLEUM ETHER.

Solenhofen Stone. A porous limestone containing clay.†

Solenite. An explosive. It is an Italian smokeless powder, and contains 30 per cent nitro-glycerin, 40 per cent. "insoluble," and 30 per cent. "soluble" nitro-cellulose.†

Solester. Vat dyes. (ICI PLC).*

Solfa. Fungicide for mildew control in a wide range of crops. (ICI PLC).*

Solferino. See MAGENTA.

Solfoton. Phenobarbital. Applications: Anticonvulsant; hypnotic; sedative. (Poythress Laboratories Inc).

Solganal. Aurothioglucose. Applications: Antirheumatic. (Schering-Plough Corp).

Solgen. Industrial detergent. (Crosfield Chemicals).*

Solicum. A material made from waste rubber and oil.†

Solidago. The dried herb of *Solidago odora*. It is used medicinally as a stimulant, carminative, and diuretic.†

Solid Alcohol. A soapy mass containing about 20 per cent water, 20 per cent sodium stearate, and 60 per cent alcohol.†

Solid Blue. 2R, B. See INDULINE, SOLUBLE.†

Solidex. Photographic developer. (May & Baker Ltd).

Solid Green. J, TTO. See BRILLIANT GREEN.†

Solid Green. See DINITROSORESORCIN and MALACHITE GREEN.

Solid Green Crystals. See MALACHITE GREEN.

Solid Green G. See GALLANILIC GREEN.

Solid Green O. See MALACHITE GREEN.

Solid Hydrogen Peroxide. See HYPEROL.

Solidified Alcohol. A name applied to a solution of nitrocellulose in ethyl alcohol for use in heaters, etc., as a fuel.†

Solidified Linseed Oil. (Oxidized Linseed Oil, Linoxyn). A flexible solid mass obtained when linseed oil is exposed to oxidation.†

Solidite. A proprietary trade name for a range of moulded products made from shellac, bitumen, and fillers. Electrical insulation.†

Solid Storax. See RED STORAX.

Solid Violet. See GALLOCYANINE DH and BS.

Solid Yellow S. See ACID YELLOW.

Solidogen LT-13. Resin (cationic). Applications: Improves wet fastness of direct and developed dyeings on cellulosic fibres. On suede leathers for garments or gloves to increase fastness to washing and drycleaning and promote level drying. (GAF Corporation).*

Soligen. A proprietary trade name for certain metallic naphtenates used as paint driers.†

Solinure. Range of soluble fertilizers. (Fisons PLC, Horticulture Div).

Solkote Hi/Sorb-II. Proprietary. Applications: Selective optical coating specifically formulated for solar applications. High and low temperature air and liquid absorbers. Trombe walls. Photographic applications. High temperature applications. (Solec - Solar Energy Corporation).*

Sollacaro's Aluminium Solder. An alloy of 64 per cent zinc, 30 per cent tin, and 6 per cent lead.†

Solochrome. After-chrome dyes. (ICI PLC).*

Solochrome Black F. A dyestuff. It is a British equivalent of Diamond black F.†

Solochrome Black 6B. A dyestuff. It is equivalent to Chrome fast cyanine G.†

Solochrome Yellow Y. A dyestuff. It is a British brand of Milling yellow.†

Soloform. A proprietary product. It is a triiodophenol preparation.†

Solophenyl. Direct dyes. (Ciba-Geigy PLC).

Solosil. Foundry binder for the CO_2 process. (Foseco (FS) Ltd).*

Solox. A proprietary trade name for an alcohol-type solvent, a fuel for alcohol lamps, blow torches, portable stoves, etc. It contains ethyl alcohol mainly, with small quantities of ethyl acetate and petrol.†

Solozone. A proprietary trade name for a sodium peroxide containing 20.5 per cent available oxygen.†

Solpadeine. A proprietary preparation of paracetamol, codeine phosphate and caffeine. An analgesic. (Sterling Research Laboratories).†

Solpolac. Chloropolyethylene. Applications: Resin for the preparation of paints and varnishes. (Caffaro Industrial).*

Solpolac. Chlorinated polyethylene. (SNIA (UK) Ltd).†

Solprene. A therzaoplastic elastomer.†

Solprin. A proprietary preparation of aspirin, calciunl carbonate, citric acid and saccharin. An analgesic. (Reckitts).†

Soltair. Residual total herbicide. (ICI PLC, Plant Protection Div).

Soltercin. A proprietary preparation of soluble aspirin, phenacetin and buto barbitone. (Cox, A H & Co Ltd, Medical Specialities Divn).†

Solu-Cortril. A proprietary preparation of hydrocortisone sodium hemisuccinate. A corticosteroid. (Pfizer International).*

Solu-Delta-Cortef. Prednisolone Sodium Succinate. Applications: Glucocorticoid. (The Upjohn Co).

Solu-Medrol. Methylprednisolone sodium succinate. Applications: Glucocorticoid. (The Upjohn Co).

Solu-Predalone. Prednisolone Sodium Phosphate. Applications: Glucocorticoid. (O'Neal, Jones & Feldman Pharmaceuticals).

Soluble Algin. Alginate of soda, obtained from seaweed.†

Soluble Blue. (Water Blue, Water Blue 6B Extra, Navy Blue, Serge Blue, Lyons Blue, Marine Blue, Pure Blue, Blackley Blue, Sedan Blue, Silk Blue, Guernsey Blue, China Blue, London Blue Extra, Cotton Blue, Blue, Acid Blue, Soluble Blue B, Conc. NS, L, 2R, Silk Blue O, Pure Soluble Blue, Water Blue B, R, Ink Blue, Ink Blue 8671, 7567). Dyestuffs. They consist of the ammonium, sodium, or calcium salts of the trisulphonic acid (with some disulphonic acid), of triphenyl-rosaniline, and triphenyl-pararosaniline. Free acids, $C_{38}H_{31}N_3O_9S$, and $C_{37}H_{29}N_3O_9S$. They dye silk and mordanted cotton blue. Also see ALKALI BLUE and METHYL BLUE.†

Soluble Blue B, Conc. NS, L, 2R. British equivalents of Soluble blue.†

Soluble Blue XG. See ALKALI BLUE XG.

Soluble Blue XL. See METHYL BLUE.

Soluble Blue 8B, 10B. See METHYL BLUE.

Soluble Calomel. See CALOMELOL.

Soluble Castor Oils. See BLOWN OILS.

Soluble Cream of Tartar. Cream of Tartar (potassium bitartrate, $C_4H_5O_6K$) dissolved in a solution of boric acid or borax.†

Soluble Glass. (Soluble Soda-Glass, Water Glass, Glass Liquor). A syrupy solution containing 50 per cent of sodium silicate, Na_4SiO_4, and Na_3SiO_3. It is used to impregnate articles to render them fire-resistant, as an adhesive for glass and porcelain, and as adulterant in soap, in dyeing, and for egg-preserving.†

Soluble Indigo. See INDIGO CARMINE.

Soluble Oil. See TURKEY RED OILS.

Soluble Phenyle. A fluid containing coal-tar creosote, rosin oil, potassium oleate, and caustic soda. It gives an emulsion with water, and is used as a sheep dip.

Soluble Pitch. See PIK SOLUBILIS.

Soluble Potash Glass. Potassium silicate, K_2SiO_3.†

Soluble Primrose. See ERYTHROSIN.

Soluble Prussian Blue. A blue pigment. It is potassium ferrous ferrocyanide, $FeK.Fe(CN)_6$.†

Soluble Regina Purple. A dyestuff obtaine by the sulphonation of Spirit violet.†

Soluble Salumin. Aluminium ammonium salicylate, $Al_2(C_6H_4(ONH_4)CO_2)_62H_2O$. An astringent.†

Soluble Soda Glass. See SOLUBLE GLASS.

Soluble Starch. (Amylodextrin). Obtained by heating starch with glycerin and adding alcohol. An emulsifying agent. A soluble starch is also made by boilin starch in water and adding a little caustic soda to clear it.†

Soluble Tartar. Potassium tartrate, $(CHOH)_2(COOK)_2$.†

Solublon. Packaging materials and plastic film of polyvinylalcol which dissolves either in cold or hot water. Applicatio Packaging for toxic, skin-irritating or strongly coloured materials used in aqueous solution, to protect the handli personnel. Water-soluble laundry bags for packing the contaminated lines in the hospital before laundering in hot water. (Aicello Chemical Co Ltd).*

Solubor. A highly soluble form of sodium borate $Na_2B_8O_{13}.4H_2O$. It is used to correct boron deficiency in plants by applying either as a foliar spray, in nutrient feeds or with herbicides. (Bora Consolidated Ltd).*

Solu-Cortef. Hydrocortisone sodium succinate. Applications: Glucocorticoid. (The Upjohn Co).

Soluene 100 and 350. 0.5 N quaternary ammonium hydroxides in toluene. Application: Cationic surfactants with the ability to solubilise a wide variety o biological samples. (Packard Bekker BV).

Solufeed. Soluble fertilizer. (ICI PLC).*

Solulan. Lanolin alcohols. Applications: Cosmetics, toiletries, pharmaceuticals. (D F Anstead Ltd).

Solulan C-24. Ethoxylated (24 mol) complex of cholesterol-lanolin alcohol fraction and related fatty alcohols. Applications: Source of water-soluble cholesterol for hair conditioning. plasticizer for water-alcohol soluble resins. (Amerchol Corporation).*

Solulan C-25. Ethoxylated (24 mol) water-soluble complex of lanolin alcohols and related fatty alcohols. Applications: Nonionic o/w emulsifier and solubilizer. (Amerchol Corporation).*

Solulan PB-10. Propoxylated (10 mol) lanolin alcohols. Applications: Spreading agent, pigment wetting agent, glosser and hydrophobic emollient. (Amerchol Corporation).*

Solulan PB-2. Propoxylated (2 mol) lanolin alcohols. Applications: Hydrophobic emollient, adds gloss, body and tack to a variety of products. Wetting agent for pigmented makeup items. (Amerchol Corporation).*

Solulan PB-20. Propoxylated (20 mol) lanolin alcohol. Applications: plasticizer, spreading agent and conditioner. (Amerchol Corporation).*

Solulan PB-5. Propoxylated (5 mol) lanolin alcohols. Applications: Water-resistant conditioner for skin and hair. Pigment wetter and glosser. (Amerchol Corporation).*

Solulan 16. Ethoxylated (16 mol) complex of lanolin alcohols and related fatty alcohols. Applications: Nonionic solubilizer, wetting agent and o/w emulsifier. Excellent foam stabilizer and conditioning agent in shampoos. (Amerchol Corporation).*

Solulan 5. Ethoxylated (5 mol) complex of lanolin alcohols and related fatty alcohols. Applications: Nonionic w/o emulsifier, stabilizer for o/w systems. Non-tacky lubricant, emollient and moisturizer. Wetting and dispersing aid for cosmetic pigments. (Amerchol Corporation).*

Solulan 75. Polyoxyethylene (75 mol) ether of whole lanolin. Applications: Water-soluble conditioning additive for detergent systems and toilet soap, improves afterfeel and lather. (Amerchol Corporation).*

Solulan 97. Acetylated complex of Polysorbate 80 and Acetulan. Applications: plasticizer for anhydrous aerosols and resins. Conditioner in detergent systems. (Amerchol Corporation).*

Solulan 98. Partially acetylated complex of Polysorbate 80 and Acetulan. Applications: Nonionic o/w emulsifier, solubilizer and pigment wetting agent. Conditioner for shampoos. (Amerchol Corporation).*

Solumedrone. A proprietary preparation Of METHYLPREONISOLONE used in the treatment of shock. (Upjohn Ltd).†

Solumin. Anionic surfactant. (ABM Chemicals Ltd).*

Solumin F. Large range of sodium sulphated alkylphenol ethoxylates as aqueous solutions. Applications: Wetting, foaming and detergency, dispersion and emulsification, for example in emulsion polymerization; pigment dispersion; cleaning formulations. (ABM Chemicals Ltd).

Solumin PFN. Range of anionic surfactants consisting of phosphate esters of ethoxylated alkylphenols, as viscous liquids. Application: Emulsion hydrotroping, corrosion inhibition and conductivity improvement agents used in emulsion polymerization and as a conductivity additive. (ABM Chemicals Ltd).

Solumin PV27. Phosphate ester of ethoxylated alcohol in yellow liquid form. Application: Anionic surfactant with good stability, detergency and corrosion resistance, used in industrial cleaners, hydrotroping and lubricants. (ABM Chemicals Ltd).

Solumin T45S. Sodium sulphated synthetic alcohol ethoxylate in aqueous solution. Applications: Foaming,wetting and dispersing agent used in shampoos, bubble baths and surgical scrubs. (ABM Chemicals Ltd).

Solumin V27SD. Sodium sulphated synthetic alcohol ethoxylate in aqueous solution. Applications: Foaming, wetting, dispersing and emulsification agent. (ABM Chemicals Ltd).

Solutene. Textile auxiliary chemicals. (ICI PLC).*

Solution SBR. See UNIDENE.

Solutol. An alkaline solution of sodium cresol in an excess of cresol, obtained by treating cresol with caustic soda. A disinfectant.†

Solux. A proprietary trade name for a rubber antioxidant. It is *p*-hydroxy phenylmorpholine.†

Solvatone. A mixture of approximately 80 per cent acetone, 10 per cent isopropyl alcohol, and 10 per cent toluene. A solvent for lacquers.†

Solvay Soda. A registered trade name for a sodium carbonate for water softening.†

Solvene. A proprietary solvent. It is a heavy grade coal-tar naphtha. It is a solvent for ester gum, pitches, etc.†

Solvenol. A proprietary product obtained from the steam distillation of waste wood and stumps of the Southern yellow pine. It is stated to be a terpene with properties similar to turpentine but superior in its solvent power. It is a volatile thinner.†

Solvenol 1. Terpene liquids. Applications: Used as thinners and antiskinning agents in paints, and are reclaiming agents for natural and synthetic rubbers. (Hercules Inc).*

Solvenol 2. Terpene liquids. Applications: Used as thinners and antiskinning agents in paints, and are reclaiming agents for natural and synthetic rubbers. (Hercules Inc).*

Solvenol 226. Terpene liquids. Applications: Used as thinners and antiskinning agents in paints, and are reclaiming agents for natural and synthetic rubbers. (Hercules Inc).*

Solvent Naphtha. A fraction of coal tar distillation of specific gravity 0.875. Also the wood naphtha recovered from grey acetate of lime, prepared from the distillation of wood.†

Solvent 401. Mixture of esters, alcohols, ketones, hydrocarbons. Applications: Paint thinner for car refinishing. (Solrec Ltd).*

Solvent 78. Mixture of esters, alcohols, ketones, hydrocarbons. Applications: Paint thinners, wash solvent. (Solrec Ltd).*

Solventol. See TERPURILE.

Solveol. Cresols made soluble in water by the addition of sodium cresotinate. A disinfectant and substitute for guaiacol and creosote.†

Solvesso. Aromatic solvents of high purity. (Exxon Chemical International Inc).*

Solvetek. Pigment dispersions. Applications: Colouration of solvent-based coatings. (Pacific Dispersions Inc).*

Solvethane. Trichloroethane. (Laporte Industries Ltd).

Solvic. PVC resins. (Laporte Industries Ltd).

Solvifog (N.R.I.). Diluent and carrier for thermal fogging of pesticides. (Makhteshim Chemical Works Ltd).†

Solvigran. Insecticide, containing disulfoton. (ICI PLC).*

Solvochin. A proprietary preparation of basic quinine, 25 per cent solution.†

Solvoclarin. Combination of special detergents, finishing agents antistats and deodorants. Applications: Detergent for chlorinated and fluorinated hydrocarbons. (Henkel Chemicals Ltd).

Solvol. A proprietary solvent. It is tetrahydro-naphthol acetate.†

Solvtext. Solvent publications. (Exxon Chemical Ltd).*

Solway Blue. See ALIZARIN SAPPHIROL B.

Solway Purple. See ALIZARIN IRISOL.

Soma. Carisoprodol. Applications: Relaxant. (Wallace Laboratories, Div of Carter-Wallace Inc).

Somacount. Stabilized and treated blood cells in a milk-like medium. Applications: Reference controls for automated milk cell analysers. (Coulter Electronics Ltd).*

Somafix. Formaldehyde-based cell fixative. Applications: To allow the automated counting of somatic cells in milk. (Coulter Electronics Ltd).*

Somali Gum. An acacia gum from *Acacia glaucophylla* and *Acacia abyssinica*.†

Somaton. Alcoholic saline diluent. Applications: To allow counting of somatic cells in milk by automated analysers. (Coulter Electronics Ltd).*

Somatonorm. Sterile lyophilized powder of genetically produced human somatotropin for the treatment of short stature due to deficiency of growth hormone. (KabiVitrum AB).*

Sombulex. Hexobarbital. Applications: Sedative. (Riker Laboratories Inc, Subsidiary of 3M Company).

Somnos. Chloral hydrate. Applications: An effective non-barbiturate sedative and hypnotic. (Merck Sharp & Dohme).*

Somophyllin. Aminophylline. Applications: Relaxant. (Fisons Corp).

Sonacide. Glutaral. Applications: Disinfectant. (Ayerst Laboratories, Div of American Home Products Corp).

Sonalgin. Butobarbitone/codeine/paracetamol. (May & Baker Ltd).

Sonergan. Butobarbitone/promethazine hydrochloride tablets. (May & Baker Ltd).

Soneryl. Butobarbitone. (May & Baker Ltd).

Sonilyl. A proprietary preparation of sulphchlorpyridazine. An antibiotic. (Mallinckrodt Inc).*

Sonnenschein's Reagent. An alkaloidal reagent prepared by adding phosphoric acid to a warm solution of ammonium molybdate in nitric acid, boiling the precipitate produced in *agua regia*, evaporating to dryness, and dissolving in 10 per cent nitric acid.†

Sonora Gum. (Arizona Shellac). A variety of shellac obtained from *Larrea Mexicana*.†

Soothe. Tetrahydrozoline Hydrochloride. Applications: Adrenergic. (Alcon Laboratories Inc).

Soothease. Throat drops. (Richardson-Vicks Inc).*

Sopanox. Soap antioxidant. (Monsanto PLC).

Soprodac. Food preservative. (BP Chemicals Ltd).*

Soprofor. Surfactant (brand). Applications: PVC viscosity depressant, emulsifier for latexes (acrylic, vynilic, copolymers) and baby-shampoo thickener. (Geronazzo S.p.A).*

Sorane. Two component polyurethane systems. (Avalon Chemical Co Ltd).

Sorban. A strong solution of sorbitol, $HO.CH_2.(CH.OH)_4.CH_2.OH$.†

Sorbichew. Chewable tablets containing isosorbide dinitrate to prevent or abort acute attacks of angina pectoris. (ICI PLC).*

Sorbid. Isosorbide dinitrate tablets for protection against angina pectoris. (ICI PLC).*

Sorbidel. A proprietary preparation containing sorbitol. A laxative. (Rona Laboratories).†

Sorbimacrogol Laurate 300. Polysorbate 20. (See under other Polysorbates for equivalent Sorbimacrogol esters.)†

Sorbin. See SORBINOSE.

Sorbin Red. (Azogrenadine S, Lana-fuchsine SB). A dyestuff prepared from acetyl-*p*-phenylenediamine and *α*-naphthol-3, 6-disulphonic acid. Lana-fuchsine SB has similar properties, but gives a yellow shade of red.†

Sorbinose. (Sorbin). Sorbose, $C_6H_{12}O_6$, a sugar.†

Sorbismal. A German preparation of finely divided bismuth in oil.†

Sorbistat. Sorbic acid. (Pfizer Ltd).

Sorbistat K. Potassium sorbate. (Pfizer Ltd).

Sorbite. A constituent of iron, which is formed in the transformation of austenite, the stage following trootsite and osmondite, and preceding pearlite.†

Sorbitol (EGIC). A proprietary preparation of sorbitol used in intravenous nutrition. (Servier).†

Sorbitrate. Isosorbide dinitrate tablets for protection against angina pectors. (ICI PLC).*

Sorbo. Sorbitol syrups. (ICI PLC).*

Sorbolene. The registered trade mark for a fat liquor for the leather trade. It is used in the tanning and dyeing process to give greater elasticity to the leather, and enables fuller shades to be obtained in dyeing.†

Sorbonorit. Highly activated granular carbon. Applications: Solvent recovery. (Norit).*

Sorbonorit (B). Solvent recovery agents. Applications: Recovery of solvents (waste streams, rotogravure, chemical industry etc). (Norit).*

Sorbosil. Thickening and polishing agents for toothpastes. (Crosfield Chemicals).*

Sorbsil. Dessicant silica gels. (Crosfield Chemicals).*

Soreflon. A proprietary range of PTFE polymers. (Montedison UK Ltd).*

Sorel Cement. (Xylolite Xylolith, Magnesia Cement, Lito-silo). A magnesium oxychloride cement, made from magnesite (magnesium carbonate, $MgCO_3$) and magnesium chloride, $MgCl_2$.†

Sorel's Gutta-percha Substitutes. Substitutes containing rosin, pitch, rosin oil, slaked lime, and gutta-percha. Some are filled with china clay, and in others coal tar is used.†

Sorensen's Salt. Sodium phosphate, $Na_2HPO_4.2H_2O$.†

Soricin. Sodium ricinoleate.†

Sorlate. Polysorbate 80. Applications: Pharmaceutic aid. (Abbott Laboratories).

Sornyl. Benzoyl-benzyl-succinic ester.†

Soromin. Textile spin finishes softening agents. (BASF United Kingdom Ltd).

Soromine AT. Complex fatty amido compound. Applications: Imparts an excellent hand, good body and draping qualities, and lubricity to synthetic, cellulosic and animal fibres and leathers. (GAF Corporation).*

Sorrel Salt. See SALT OF SORREL.

Sorrel's Alloy. An alloy of 98 per cent zinc, 1 per cent iron, and 1 per cent copper. Another alloy contains 80 per cent zinc, 10 per cent iron, and 10 per cent copper.†

Sostenil. A proprietary preparation of vincamine. A cerebral vasodilator. (Pfizer International).*

Sotacor. A proprietary preparation of SOTALOL hydrochloride used in the treatment of angina pectoris. (Bristol-Myers Co Ltd).†

Soubieran's Ammonical Salt. Mercury ammonium-nitrate, $(NH_2.Hg_2O)NO_3$.†

Soucol. Industrial chemical. (Crosfield Chemicals).*

Souesite. A nickel-iron alloy which occurs naturally.†

Soulan's Cement. Consists of 7 parts resin, 10 parts ether, 15 parts collodion, and aniline red. A semi-transparent varnish, used for sealing corks into bottles.†

Southalite. A phenol-formaldehyde condensation product, with a filler of paper, used for insulating purposes.†

Sovatex C1. Alkyl-aryl sulphonate in liquid form. Applications: Anionic scouring and milling agent. (Standard Chemical Company).

Sovatex EP 5288. Coconut imidazoline amphoteric in liquid form. Applications: Antistats; lubricants; corrosion inhibitors, detergents. (Standard Chemical Company).

Sovatex IM12H. Hydroxyethyl imidazoline of coconut fatty acid in the form of a semi-liquid. Applications: Corrosion inhibitor; lubricant; antistatic agent. Used as a base for cationic surface active agents. (Standard Chemical Company).

Sovatex IM12N. Aminoethyl imidazoline of coconut fatty acid in the form of a semi-liquid. Applications: Corrosion inhibitor; lubricant; antistatic agent. Used as a base for cationic surface active agents. (Standard Chemical Company).

Sovatex IM17H. Hydroxyethyl imidazoline of oleic acid in liquid form. Applications: Corrosion inhibitor; lubricant; antistatic agent. Used as a base for cationic surface active agents. (Standard Chemical Company).

Sovatex IM17N. Aminoethyl imidazoline of oleic acid in liquid form. Applications: Corrosion inhibitor; lubricant; antistatic agent. Used as a base for cationic surface active agents. (Standard Chemical Company).

Sovatex MP/1. Oleyl imidazoline amphoteric in liquid form.

Applications: Antistats; lubricants; corrosion inhibitors; detergents. (Standard Chemical Company).

Sovatex WA. Sulphosuccinate surfactant in liquid form. Applications: Concentrated anionic wetting agent. (Standard Chemical Company).

Sovprene. A Russian chloroprene synthetic rubber obtained by the polymerization of acetylene to form divinyl-acetylene and then the formation of chloroprene by treating with hydrogen chloride, followed by polymerization.†

Soxhlet's Solution. A modified Fehling's solution. It consists of (1) a solution of 34.639 grams copper sulphate in 500 cc water and (2) 50 grams caustic soda and 173 grams potassium sodium tartrate in 500 cc water. Used for the determination of sugars.†

Soxomide. Sulphisoxazole. Applications: Antibacterial. (The Upjohn Co).

Soy-che. Zinc, iron, copper, manganese, sulphur. Applications: Chelated micronutrient for soybeans. (Drexel Chemical Company).*

Soya Bean Oil. An oil obtained from the seeds of *Soja hispida* by expression or extraction with a solvent. It is used as an edible oil, and is also employed in soap-making, paints and varnishes, and in the linoleum industry. It is also known as Chinese bean oil.†

Spalerite. Zinc blende, ZnS.†

Span. Sorbitan fatty acid esters. (ICI PLC).*

Span 20. Sorbitan Monolaurate. Applications: Pharmaceutic aid - surfactant. (ICI Americas Inc).

Span 20. A proprietary trade name for Sorbitan Monolaurate.†

Span 40. Sorbitan Monopalmitate. Applications: Pharmaceutic aid - surfactant.(ICI Americas Inc).

Span 60. Sorbitan Monostearate. Applications: Pharmaceutic aid - surfactant. (ICI Americas Inc).

Span 65. Sorbitan Tristearate. Applications: Pharmaceutic aid - surfactant). (ICI Americas Inc).

Span 80. Sorbitan Monooleate. Applications: Pharmaceutic aid - surfactant. (ICI Americas Inc).

Span 85. Sorbitan Trioleate. Applications: Pharmaceutic aid - surfactant. (ICI Americas Inc).

Spandofoam. Rigid polyurethane block foam. (Baxenden Chemical Co Ltd).*

Spandra Transparent Dressing. Fabric supported tecoflex polyurethane film. Moisture vapour permeable, hypoallergenic dressing. Applications: Transparent intravenous dressing. (Thermedics Inc).*

Spaneph Spansules. A proprietary preparation of ephedrine sulphate in a sustained release form, used in the treatment of bronchospasm. (Smith Kline and French Laboratories Ltd).*

Spangite. This is phillipsite, a zeolite containing potassium.†

Spanish Ochre. Burnt Roman ochre. See YELLOW OCHRE.†

Spanish Oil. Inferior olive oil.†

Spanish Oxide. A natural red pigment. It is a red oxide of iron, and contains over 80 per cent. Fe_2O_3. Also see INDIAN RED.†

Spanish Soap. An olive oil soap.†

Spanish Turpentine. See TURPENTINE.

Spanish White. See BISMUTH WHITE.

Spanish White, Fard's. See BISMUTH WHITE.

Spanish Yellow. See CHINESE YELLOW.

Sparine. Promazine Hydrochloride. Applications: Anticholinergic. (Wyeth Laboratories, Div of American Home Products Corp).

Spar, Derbyshire. See FLUORSPAR.

Spar, Dolomite. Dolomite, $MgCa\ (CO_3)_2$.†

Spar, Green. See MALACHITE.

Sparkaloy. A proprietary trade name for silicon-manganese-nickel alloy used for spark-plug wire.†

Sparkle Silver. Non-leafing aluminium pigments. Applications: Used for metallic colours (aesthetics), automobiles, trucks, bicycles, furniture etc. (Silberline Mfg Co Inc).*

Sparkle Silvet. Aluminium pigments for plastics. Applications: All plastics, automobiles, toys, bottles etc. (Silberline Mfg Co Inc).*

Sparkle Silvex. Aluminium pigments for plastics. Applications: All plastics, automobiles, toys, bottles etc. (Silberline Mfg Co Inc).*

Sparkolac. High gloss nitrocellulose lacquers. (The Scottish Adhesives Co Ltd).*

Spar, Light. See GYPSUM.

Sparmite. A bleached barium sulphate pigment. (Pfizer International).*

Spar, Tabular. See WOLLASTONITE.

Spartakon. A proprietary preparation containing levamisole hydrochloride. Applications: Veterinary antihelmintic (pigeons). (Janssen Pharmaceutical Ltd).*

Spartase. Potassium Aspartate and Magnesium Aspartate. Applications: Nutrient. (Wyeth Laboratories, Div of American Home Products Corp).

Spartrix. A proprietary preparation containing carnidazole. Applications: Veterinary antihelmintic (pigeons). (Janssen Pharmaceutical Ltd).*

Spar, Zinc. See CALAMINE.

Spasmo-Dolviran. An antispasmodic analgesic. (Bayer & Co).*

Spasmonal. A proprietary preparation of alverine citrate used in the treatment of colonic disorders. (Norgine).†

Spasor. A non-residual herbicide containing 360g/litre glyphosphate for the control of annual and perennial broad-leaved weeds and grasses. Applications: Clearing ground prior to planting, weed control in all hard surfaces; control of floating and emergent aquatic weeds. (Burts & Harvey).*

Spastipax. A proprietary preparation of hyoscyamine sulphate. atropine sulphate, hyoscine hydrobromide and amylobarbitone. (Nicholas).†

Spathic Zinc Ore. See CALAMINE.

Spauldite. A proprietary trade name for a phenol-formaldehyde synthetic resin laminated product.†

SPD. A rubber vulcanization accelerator. It is sodium-pentamethylene dithiocarbonate.†

Spear Pyrites. See MARCASITE.

Special Extender. Talc, lamellar structure. Applications: Extender for paints and as a general purpose filler. (Bromhead & Denison Ltd).*

Special Grey R. See NEW GREY.

Specpure. Spectrographically standardized metals and chemicals. (Johnson Matthey Chemicals Ltd).*

Spectam. Spectinomycin Hydrochloride. Applications: Antibacterial. (Abbott Laboratories).

Spectazole. Econazole nitrate. Applications: Antifungal. (Ortho Pharmaceutical Corp).

Spectinomycin. An antibiotic produced by Streptomyces spectabilis. Trobicin.†

Spectra-Sorb UV 24. Dioxybenzone. Applications: Ultraviolet screen. (American Cyanamid Co).

Spectra-Sorb UV 284. Sulisobenzone. Applications: Ultraviolet screen. (American Cyanamid Co).

Spectra-Sorb UV 531. Octabenzone. Applications: Ultraviolet screen. (American Cyanamid Co).

Spectra-Sorb UV 5411. Octrizole. Applications: Ultraviolet screen. (American Cyanamid Co).

Spectra-Sorb UV 9. Oxybenzone. Applications: Ultraviolet screen. (American Cyanamid Co).

Spectraban. A proprietary preparation of isoamyl-p-N, N-dimethylaminobenzoate in ethanol. A lotion used to protect skin from ultra-violet light. (Stiefel Laboratories (UK) Ltd).*

Spectrathene. Colour and additive concentrates for polyethylene and other plastics. Applications: Colourants, additives: antistat, slip, antiblock, UV inhibitors, processing aids and flame retardants. (USI Chemicals).*

Spectrim. Multi-component systems used in the manufacture of polyurethane products via reaction injection moulding. (Dow Chemical Co Ltd).*

Spectrobid. A proprietary preparation of bacampicillin hydrochloride. An antibiotic. (Pfizer International).*

Spectroflux. Buffer mixtures for spectrographic analysis. (Johnson Matthey Chemicals Ltd).*

Spectromel. Powder mixtures for spectographic analysis of relatively pure materials. (Johnson Matthey Chemicals Ltd).*

Spectron. Selective herbicide. (FBC Ltd).

Spectrosol. Materials for spectroscopy. (BDH Chemicals Ltd).

Specularite. (Specular Haematite). Iron oxide, Fe_2O_3, with a bright metallic lustre.†

Speculum Metal. An alloy of 66 per cent copper and 34 per cent tin, with a little arsenic. An alloy of 64 per cent copper, 32 per cent tin, and 4 per cent nickel. It has a specific gravity of 8.6 and a melting-point of 750° C. Used for making mirrors of reflecting telescopes.†

Speed X Accelerator. A proprietary rubber vulcanization accelerator containing 60 per cent diphenylguanidine and 40 per cent zinc oxide.†

Speedway. Residual total herbicide. (ICI PLC, Plant Protection Div).

Speetan SB60. Special 58/60 basic chrometan powder. (Lancashire Chemical Works Ltd).

Spelter. Zinc used in galvanizing. The term is also used for hard solder.†

Spenbond. Waterborne adhesives. Applications: Film, foil and paper lamination. (NL Industries Inc).*

Spence Metal. A material obtained by melting ferrous sulphide with sulphur. Used as a jointing material.†

Spenkel. Aromatic urethane resins. Applications: Paint, adhesives and textile. (NL Industries Inc).*

Spenlite. Aliphatic urethane resins. Applications: Paint, adhesives and textile. (NL Industries Inc).*

Spensol. Water dispersible resins. Applications: Paint, adhesives and textile. (NL Industries Inc).*

Spermaceti. A wax obtained from the head of the sperm whale. The crude product is obtained by chilling the head and blubber oils. It consists principally of cetyl palmitate, $C_{16}H_{33}\ OCOC_{15}H_{31}$.†

Spermaceti, Vegetable. See CHINESE WAX.

Spermolin. A linseed oil product. It is a proprietary binder for sands used as cores in metal casting.†

Spermoline Oil. A compound spindle oil used for lubrication.†

Spersol. Sodium and ammonium polyacrylates. Applications: Used as anti-redeposition agents in laundry detergents, as detergents, as water reducing agents in slurries and as dispersants in boiler water. (Harcros Industrial Chemicals).*

SPG Gelatine. Gelatine powder. Applications: Used for granulation of fine powders to improve free flowing and dispersion characteristics. (Rousselot Ltd).*

Sphagni. An insulating material prepared from the white moss found on the Swedish peat moors.†

Spidax. Specialty polyolefin resin. Applications: Moulding compounds. (Sumitomo Bakelite, Japan).*

Spiegeleisen. A ferro-manganese alloy containing from 10-35 per cent manganese, 60-85 per cent iron, 1 per cent silicon, and 4-5 per cent carbon.†

Spiegler Jolle's Reagent. A solution of 2 grams mercuric chloride, 4 grams succinic acid, and 4 grams sodium chloride in 100 cc water. A reagent for albumin in urine.†

Spiegler's Reagent. This consists of 40 grams mercuric chloride and 20 grams tartaric acid dissolved in 500 cc water. To this solution is added 100 grams glycerol and 50 grams, sodium chloride, and the whole made up to 1,000 cc It is used for proteins.†

Spiller's Resin. The oxidation products of rubber are sometimes called by this name.†

Spinflam. A proprietary range of flame retardants for polymers. (Montedison UK Ltd).*

Spinning Brass. See BRASS.

Spinuvex. A proprietary range of hindered amine light stabilisers for polymers. (Montedison UK Ltd).*

Spiragas. Dimethyl silicone membrane. Applications: Gas separations, primarily oxygen enrichment. (Allied-Signal Fluid Systems).*

Spiramycin. An antibiotic produced by Streptomyces ambofaciens. Rovamycin.†

Spirit Black. SEE INDULINE, SPIRIT SOLUBLE.†

Spirit Blue O. See ANILINE BLUE, SPIRIT SOLUBLE.

Spirit, Green Wood. See ACETONE ALCOHOL.

Spirit, Manhatton. See ACETONE ALCOHOL.

Spirit Eosin. (Ethyl Eosin, Eosin S, Eosin BB, Rose JB, Spirit Primrose, Primrose). A dyestuff. It is the potassium salt of tetrabromofluorescein ethyl ether, $C_{22}H_{11}Br_4O_5K$. It dyes wool yellowish-red with slight fluorescence. Also see ERYTHRIN.†

Spirit of Alum. Sulphuric acid, H_2SO_4.†

Spirit of Hartshorn. A solution of ammonia. See VOLATILE ALKALI.†

Spirit of Nitre. Spirit of nitrous ether. (See SPIRIT OF SWEET NITRE.) The term is also applied to nitric acid.†

Spirit of Red Lavender. Compound tincture of lavender.†

Spirit of Salt. Strong impure hydrochloric acid.†

Spirit of Sulphur. Sulphurous acid, H_2SO_3.†

Spirit of Sweet Nitre. Spirit of nitrous ether, consisting of a solution of 1.52-2.66 per cent of ethyl nitrite in alcohol.†

Spirit of Sweet Wine. Ethyl chloride, C_2H_5C1, a local anaesthetic.†

Spirit of Tar. See OIL OF TAR.

Spirit of Tin. Stannic chloride, $SnC1_4$.†

Spirit of Turpentine. See OIL OF TURPENTINE.

Spirit of Vitriol. Sulphuric acid, H_2SO_4.†

Spirit of Vitriol, Sweet. Spirit of ether.†

Spirit of Wood. Methyl alcohol, CH_3OH.†

Spirit Oil. The first fraction from the distillation of Yorkshire Grease (*qv*). Used for making black varnish.†

Spirit, Potato. See POTATO OIL.

Spirit Primrose. See SPIRIT EOSIN.

Spirit Purple. (Spirit Violet, Opal Violet, Regina Spirit Purple). Dyestuffs which consist of diphenylated rosanilines.†

Spirit Red III. See SUDAN IV.

Spirit, Resin. See ROSIN SPIRIT.

Spirit, Silent. See SPIRITS OF WINE.

Spirit, Sky Blue. See DIPHENYLAMINE BLUE, SPIRIT SOLUBLE.

Spirit Soluble Blue. See ANILINE BLUE, SPIRIT SOLUBLE.

Spirit Soluble Eosin. See ERYTHRIN.

Spirit Soluble Quinoline Yellow. A quinoline dyestuff prepared from quinaldine, phthalic anhydride, and zinc chloride.†

Spirit, Standard Wood. See ACETONE ALCOHOL.

Spirit Varnishes. Prepared by mixing resin with such solvents as methylated spirit or turpentine.†

Spirit Vinegar. Made from potato or grain spirit. It contains up to 12 per cent of acetic acid.†

Spirit Violet. See SPIRIT PURPLE.

Spirit Yellow. See ANILINE YELLOW.

Spirit Yellow I. A dyestuff. It is a British brand of Sudan I.†

Spirit Yellow R. (Yellow Fat Colour). A dyestuff prepared from *o*-toluidine, $C_{14}H_{15}N_3$.†

Spirits of Wine. (Ethanol, Silent Spirit). Ethyl alcohol, C_2H_5OH. Commercial spirits of wine contain 84 per cent by weight of alcohol.†

Spirittine. Soft wood tar creosote. A wood preservative.†

Spiro-32. Spirogermanium Hydrochloride. Applications: Antineoplastic. (Unimed Inc).

Spirolite. Plastic pipe. (Chevron).*

Spirolone. Spironolactone. Applications: For the treatment of congestive cardiac failure, essential hypertension, hepatic cirrhosis, malignant ascites, idiopathic oedema and nephrotic syndrome. (Berk Pharmaceuticals Ltd).*

Spironal. Sodium citro-bismuthate.†

Spiropitan. Spiperone. Applications: Antipsychotic. (Janssen Pharmaceutica).

Split Nut. See ORDEAL BEAN.

Spodium, Black. See ANIMAL CHARCOAL.

Spodium, White. See BONE ASH.

Sponge Iron. A finely porous form of iron obtained by reducing iron oxide at a temperature where no sintering or fusion takes place. A reagent for the precipitation of copper, lead, and other metals from solution.†

Spoon Metal. See NICKEL SILVERS.

Sporocide. A wood preservative mainly consisting of potassium-*o*-dinitro-cresylate.†

Sportak. Fungicide. (FBC Ltd).

Sportak Alpha. Fungicide. (FBC Ltd).

Sporting Ballistite. A smokeless powder consisting of 37.6 per cent nitroglycerin and 62.3 per cent nitro cotton.†

Spramex. See MEXPHALTE.

Spray Guard. Splash and spray suppressing rain flaps. (Monsanto Co).*

Spray-Add 77. AG Chem spreader-sticker. (Chevron).*

Spreading Agent. Complex alkyl ether. (Croda Chemicals Ltd).

Spreitan. Blend of fatty acid esters and emulsifiers, nonionic. Applications: Universal coning oil for mono and multifilament and for textures synthetic yarns. (Henkel Chemicals Ltd).

Sprengel's Explosives. Cakes of potassium chlorate which have absorbed combustible liquids.†

Sprengsalpeter. An explosive consisting of 75 per cent sodium nitrate, 15 per cent brown coal, and 10 per cent sulphur.†

Sprills. Pelleted pesticide. (Chevron).*

Spring Brass. See BRASS.

Springbok. Bakery phosphates. (Albright & Wilson Ltd, Phosphates Div).

Springclene 2. Selective herbicide. (FBC Ltd).

Sprodco. Textile yarn lubricant processing aid. (Specialty Products Co).*

Sprödglaserz. Synonym for Polybasite.†

Spruce Ochre. See YELLOW OCHRE.

Spuncote. Coatings for centrifugal dies used in the spinning of cast iron pipes. (Foseco (FS) Ltd).*

Sputamin. A German preparation. It is a powder containing 80 per cent chloramine. An antiseptic.†

Squaw Root. Caulophyllum.†

Squill. Bulb of *Urginea maritima*.†

SRA Dyestuffs. A series of colours used for dyeing of acetyl silk. They are of British manufacture. The letters "SRA" are the initial letters of the words "sulpho-ricinoleic acid," and the dyestuffs are treated with this acid to render them in suitable condition (colloidal dispersion) for use in dyeing acetyl silk.†

SSF. Sodium silico-fluoride, Na_2SiF_6.†

SS Nitrocellulose. Nitrocellulose. Applications: Used in flexographic inks where an alcohol rich solvent is desirable and is used in heat sealing coatings. (Hercules Inc).*

ST 137. A solution of hexyl-resorcinol in glycerin and water. A germicide.†

St Joseph. Aspirin. Applications: Analgesic; antipyretic; antirheumatic. (Plough Inc).

St Joseph Cough Syrup. Dextromethorphan hydrobromide. Applications: Antitussive. (Plough Inc).

ST 52. SEE HONVAN. (Degussa).*

ST-Size. Modified rosin emulsion size. (Hercules Inc).*

Stabgel. Soil consolidation agents. (ICI PLC, Petrochemicals & Plastics Div).

Stabifix. Dyeing auxiliary. Applications: Fixing agent for direct dyes to improve water and wash fastness. (Henkel Chemicals Ltd).

Stabil-9. Sodium aluminium phophate. Applications: Leavening agent for self-raising flour, corn meal, prepared mixes. (Monsanto Co).*

Stabilarsan. A preparation of salvarsan. It is a combination of salvarsan with glucose, and is used in the treatment of syphilis.†

Stabilator A.R. A proprietary anti-oxidant for rubber. It is phenyl-β-naphthylamine. It melts at 108° C., and is recommended for white mixings.†

Stabilite Alba. A proprietary rubber vulcanization accelerator. It is di-*o*-tolyl-ethylene-diamine.†

Stabilizer. Potassium hydroquinone monosulphate. (ABM Chemicals Ltd).*

Stabilizer No. 1. A proprietary trade name for 1 : 3 : 5-isopropyl-cresol.†

Stabillin. Penicillin preparations. (The Boots Company PLC).

Stabillin V-K. A proprietary preparation containing phenoxymethylpenicillin Potassium. An antibiotic. (Boots Company PLC).*

Stabiloid. Colourant dispersions. Applications: Colouring of paper coating and saturation compositions, textile inks and latex paints. (Nueodex Inc).*

Stabilor. A range of precious metal alloys. Applications: For dentistry and dental engineering. (Degussa).*

Stabinol. A proprietary trade name for Isobuzole.†

Stabiram. Amines and derivatives. Primary, secondary, tertiary fatty mono-, di- and polyamines from C8 to C22. Quaternary ammonium salts. Amine oxides. Amine salts. Ethoxylated amines and polyamines. Amino-acids. Special nonionic derivatives. Additives. Applications: Surface active agents used as auxiliaries in road, mining, textile and fertilizer industries, in metallurgy, as lubricating agents, additives for fuel oils and soil stabilization. (British Ceca Co Ltd).*

Stabismol. A proprietary preparation. It is a solution of α-carbonyl- cyclohexanyl acetate in olive oil.†

Stablex. A stabilized bitumen used for protective coatings to resist acids.†

Stabochlor. A proprietary chloride of lime specially prepared.†

Stacol. A complex sodium borophosphate, an inorganic water-soluble resin stable to acids and alkalis.†

Stadis. Fuel oil antistatic additive. (Du Pont (UK) Ltd).

Stadol. Butorphanol tartrate. Applications: Analgesic; antitussive. (Bristol Laboratories, Div of Bristol-Myers Co).

Staff. A mixture of plaster-of-Paris and tow. Used for mouldings.†

Staffelite. A mineral. It contains calcium phosphate with calcium chloride or fluoride.†

Staffordshire All Mine Pig. A pig iron made in Staffordshire from ore. It contains about from 0.5-0.75 per cent of phosphorus.†

Staflene Ho. Polyethylene.†

Staflex CP. A mixed alkyl phthalate. A vinyl plasticizer.†

Stafoxil. A proprietary preparation of flucloxacillin. Applications: An antibiotic. (Brocades).*

Stagnine. An astyptic obtained from the spleen.†

Stahl's Sulphur Salt. Potassium sulphite, $K_2SO_3.2H_2O$.†

Stainless Invar. A Japanese alloy of 54 per cent cobalt, 36.5 per cent iron, and 9.5 per cent chromium. It has a low coefficient of expansion and with stands corrosion well.†

Stainless Iron. This is really stainless steel, and usually contains from 0.1-0.2 per cent carbon, 12-27 per cent chromium, and up to 0.5 per cent silicon.†

Stainless Silver. This is usually an alloy of 92.5 per cent silver with copper and antimony, and is used for table ware.†

Stainless Steel. (Rustless Steel). A chromium-steel alloy containing 12-15 per cent chromium, and not more than 0.45 per cent carbon. Used for cutlery, acid pumps, turbine blades, and exhaust valves for engines. Some alloys contain 12-18 per cent chromium, 8-12 per cent nickel, 74-76 per cent iron, sometimes with a little tungsten and carbon. Also see CHROME STEELS.†

Stalactites. Deposits of calcium carbonate in the form of icicles, formed when water containing calcium carbonate drips from the roofs of caves.†

Stalagmites. Similar deposits to stalactites, except that they are formed on the floors of caves.†

Stalloy. A proprietary trade name for an alloy containing 3.5-4.0 per cent silicon and 0.1-0.2 per cent aluminium. It has a specific resistance of about 55 michroms. cm. Its magnetic hysteresis is much lower than that of pure iron. It is used in the construction of cores for field and armature magnets.†

Stamford Powder. An explosive containing from 68-72 per cent ammonium nitrate,

21-23 per cent sodium nitrate, 3-4 per cent trinitrotoluene, and 3 1/2-4 1/2 per cent ammonium chloride.†

Stamglan. A registered trade mark for L.D. polyethylene. (AKU Holland).†

Stamping Brass. See BRASS.

Stamylan HD. High density polyethylene. Applications: Used in the plastics processing industry for production of crates, household articles, bottles, containers, tubes and pipes, cables, nets, packaging film, toys etc. (DSM NV).*

Stamylan LD. Low density polyethylene. Applications: Used in the plastics processing industry for production of packaging film, heavy-duty bags, extrusion-coated cardboard and paper, tubes and pipes, cable sheathing, household articles, toys, agricultural film, foamed board etc. (DSM NV).*

Stamylan P. Polypropylene. Applications: Used in the plastic processing industry for a wide variety of applications: injection moulding of car components (bumpers, accumulator cases, boot linings, housings), electronic equipment, furniture and thin-walled containers (dairy products); for extrusion of film, fibres and non-woven fabrics, tape and belts; for blow-moulding of bottles and containers. (DSM NV).*

Stamylex PE. Special linear low density and medium density polyethylene. Applications: Used for very special high-performance applications e.g. packaging articles such as tanks, containers, covers, special food packaging films, leisure-time products, cables, monofilaments, fasteners, caps and stoppers and special technical applications. (DSM NV).*

Stanaprin. A proprietary preparation of choline salicylate. An analgesic. (Lloyd's Pharmaceuticals).†

Stanclere. Organotin stabilizers for PVC. (Akzo Chemie GmbH, Düren).†

Stand Oil. (Standöl Varnish, Dicköl Varnish, Lithographer's varnish). Linseed oil boiled strongly, and allowed to burn until it has the desired thickness.†

Standacol. United Kingdom foodstuffs colours. (Williams Div of Morton Thiokol Ltd). *

Standalloy. Dental silver alloys and mercury for mixing amalgams. Applications: For dentistry and dental engineering. (Degussa).*

Standard Benzine. Light petroleum spirit of specific gravity 0.695-0.705 at 15° C., of which 95 per cent boils between 65° and 95° C. Used for the determination of asphalt in oils.†

Standard Gold. (Sterling Gold). It is 22-carat gold containing 91.6 per cent gold with 8.4 per cent other metals, usually copper, to render it harder. American standard gold contains 90 per cent gold and 10 per cent copper. This latter alloy has a specific gravity of 17.7 and melts at 940° C.†

Standard Silver. (Sterling Silver). Silver, 92.5 per cent. with another metal, usually copper, to harden it. American standard silver contains 90 per cent silver and 10 per cent copper.†

Standard Wood Spirit. See ACETONE ALCOHOL.

Standöl Varnish. See STAND OIL.

Standup. Preparation containing 40% chlormaquat. Applications: Growth regulator for cereals. (L W Vass (Agricultural) Ltd).*

Stan-Fast. Chemical additive to a colouring bath for metals such as anodized aluminium. Applications: Used in anodizing aluminium. (Reynolds Metal Co).*

Staniform. A proprietary preparation of methyl stannic iodide. A remedy for boils, carbuncles, small wounds, and injuries.†

Stanley Red. See CLAYTON CLOTH RED.

Stanleys Crow Repellant. Active ingredients: Refined coal tar and creosote oil. Applications: Seed protectant to prevent sprout pulling by birds in newly planted corn. (Borderland Products Inc).*

Stanmine. Acid inhibitor. (ABM Chemicals Ltd).

Stannekite. A resinous hydrocarbon, $C_{20}H_{22}O_3$, found in coal deposits in Bohemia.†

Stannex. Covering and cleansing fluxes for tin and tin-lead alloys. (Foseco (FS) Ltd).*

Stannicide. Fungicides, bactericides and algicides. (Akzo Chemie GmbH, Düren).†

Stannicide. Formulated organotin compounds. Applications: Fungicides, biocides and algaecidal application in water based coatings and adhesives and in the water treatment industry. (Thomas Swan & Co Ltd).*

Stannine. Acid inhibitor. (ABM Chemicals Ltd).*

Stanniol. An alloy of 96.2 per cent tin, 2.4 per cent lead, 1 per cent copper, 0.3 per cent nickel, and 0.1 per cent iron.†

Stannite. A mineral. It is tin pyrites.†

Stannoxyl Liquid. A solution of tin chloride, $SnCl_2$, in glycerin. Used as a lotion in the treatment of boils.†

Stannum. Tin, Sn. A proprietary trade name for a bearing metal. A tin-lead alloy.†

Stanolax. See PARAFFIN, LIQUID.

Stanolind. See PARAFFIN, LIQUID.

Stanolone. 17β-Hydroxy-5α-androstan-3-one. Anabolex.†

Stanozolol. 17β-Hydroxy-17α-methyl-5α-androstano[3,2- *c*]-pyrazole. Stromba.†

Stantienite. A brown resin found with Prussian amber.†

Stanvis. Pyroxylin embedding solution for microscopy. (BDH Chemicals Ltd).

Stanyl. An engineering plastic with excellent impact strength and high temperature resistance. Applications: Used in the electrical and automotive industries and as material for technical yarns. (DSM NV).*

Stanza. Fungicide. (FBC Ltd).

Stanzaite. A mineral. It is a variety of Andalusite.†

Stapenor. An antistaphylococcal penicillin. (Bayer & Co).*

Staphcillin. A proprietary preparation of sodium methicillin. An antibiotic. (Bristol-Myers Co Ltd).†

Staple Artificial Silk. See STAPLE FIBRE.

Staple Fibre. (Staple Artificial Silk, Artificial Wool, Artificial Chappe). This fibre consists of artificial threads of cellulose or cellulose compounds possessing a definite medium length. It is worked up by ordinary spinning machinery and is suitable for mixing with cotton or wool.†

Star Antimony. Pure antimony, Sb.†

Staralox. A trade mark for abrasive good made essentially of alumina.†

Starane. A line of herbicides based primarily on fluroxypyr. (Dow Chemic Co Ltd).*

Starane 2. Selective post-emergence herbicide. (Murphy Chemical Ltd).

Star Bowls. Antimony metal obtained by refining with iron. The metal containin about 91 per cent antimony with abou 7 per cent iron is mixed with crude antimony and salt and heated. The product is known as star bowls. It contains about 99.5 per cent antimony

Starch Cellulose. See FARINOSE.

Starch Glazes. Made by adding borax, powdered stearic acid, or paraffin to potato starch.†

Starch Glue. Prepared by adding 3 pints water and 1/2 lb nitric acid to 2 1/2 lb starch, warming, then heating.†

Starch Gum. See BRITISH GUM.

Starch Paste. Mucilage of starch.†

Starch Sugar. Glucose.†

Starch Syrups. Glucose mixed with dextrine. They are used in the place of sugar for various purposes.†

Starglo. Blend of alcohols, aldehydes and non-ionic wetters. Applications: Tin electroplating additive. (Taskem Inc).*

Starim. RIM nylon (Reaction Injection Moulding), prepared by in-mould polymerization of caprolactam (feedstock for nylon) to a nylon-6 block polymer with utilization of a catalyst 2.0 agents. Applications: Industry in general, in agriculture and automotive (body parts, for instance). (DSM NV).

Starkey's Soap. Turpentine soap.†

Starlite. A proprietary synthetic resin.†

Starpass. A proprietary urea - formaldehyde synthetic resin.†

Stasite. A mineral. It is a hydrated phosphate of uranium and lead,

$8UO_3.4PbO.3P_2O_5.12H_2O$, found in Katanga.†

Stassanised Milk. Milk which has been heated in cylindrical tubes 1 mm diameter, so that a great surface is exposed, thereby killing all germs.†

Stassfurt Salts. See ABRAUM SALTS.

Staszicite. A mineral from Meidzianka, containing 39 per cent. As_2O_5, 26.5 per cent. CuO, 20.8 per cent. CaO, and 7.3 per cent. ZnO.†

Statexan. Antistatic agents. Applications: For the textile industry. (Bayer & Co).*

Statexan KI. An antistatic agent. Applications: For rigid PVC, polystyrene, ABS and other plastics. (Bayer & Co).*

Statil. Aldose reductase inhibitor. (ICI PLC).*

Statuary Bronze. A variable alloy. It usually contains from 75-95 per cent copper, 1-10 per cent tin, 0-5 per cent zinc, 0-6 per cent lead, 0.12-0.34 per cent phosphorus, and 0.19-0.7 per cent nickel.†

Statuary Marble. Marble, $CaCO_3$, with a crystalline or saccharoid structure.†

Statyl. A proprietary preparation of METHYL BENZOQUATE. A veterinary anti-protozoan.†

Staufen. PVC film. (ICI PLC).*

Staurolite. A mineral. It is a basic aluminium ferrous iron silicate, $HFeAl_5Si_2O_{13}$.†

Staurotide. See STAUROLITE.

Staybelite. Hydrogenated rosin. Applications: Used as a modifier for wax-elastomer ethylene adhesive compositions, used in electrical cable paper saturants, in ceramic ink vehicles, in metal resinates and soldering fluxes. (Hercules Inc).*

Staybrite. A proprietary trade name for stainless steels containing chromium and nickel. They usually contain 18 per cent chromium, 8 per cent nickel, 74 per cent iron, sometimes with molybdenum and occasionslly with titanium and tungsten. They possess extreme malleability and are very resistant to corrosion. See also ANKA STEEL.†

Stcherbokov's Solder. An aluminium solder. It contaims 49 per cent zinc, 46 per cent tin, and 1.5 per cent aluminium.†

Steadite. Iron-phosphorus eutectic, consisting of about 61 per cent iron-phosphide, Fe_3P, with iron, a constituent of cast iron. The same name has been applied to a basic calcium-silico phosphate, $3(CaO.P_2O_5).2CaO(2CaO.SiO_2)$, found in the basic slag of the Thomas-Gilchrist process for the dephosphorization of iron.†

Stead's Reagent. A reagent consisting of 100 cc methyl alcohol, 18 cc water, 2 cc concentrated hydrochloric acid, 1 gram copper chloride ($CuCl_2.2H_2O$), 4 grams magnesium chloride ($MgCl_2.6H_2O$). An etching reagent used in the examination of steels.†

Steam Black. See LOGWOOD.

Steam Glue. (Russian Steam Glue). A preparation of glue made by treating glue with nitric acid.†

Steam Orange, Green, and Olive. Dyestuffs prepared from Persian berries, and used in calico printing. See GAMBINE Y.†

Steamed Bone Meal. A fertilizer. It consists of crushed bones, which have been treated with superheated steam and benzene, to remove fat and glue. It contains about 1 per cent nitrogen.†

Steapsin. Lipase, a lipolytic ferment.†

Stearex. A trade mark for a standardized stearic acid. It is a commercially pure, free fatty acid prepared for rubber manufacture. There are two grades: (*a*) Double pressed stearic acid, and (*b*) single pressed stearic acid. The (*b*) quality contains more oleic acid than (*a*).†

Steargillite. A variety of the mineral Montmorillonite.†

Stearin Pitch. (Candle Pitch, Candle Tar). A pitch obtained in the sulphuric acid treatment of fats. After distillation in steam of the washed acids (stearic, palmitic, and oleic), stearin pitch remains to the extent of 2 per cent.†

Stearite. A proprietary trade name for synthetic stearic acid produced by hydrogenation of certain oils.†

Stearodine. Calcium iodostearate.†

Stearopodis. Magnesium stearate, used in the preparation of soap and face creams.†

Stearosan. A compound of santalol and stearic acid.†

Steclin. Tetracycline Hydrochloride. Applications: Anti-amebic; antibacterial; antirickettsial. (E R Squibb & Sons Inc).

Steel. Iron containing combined carbon up to 1.5 per cent. High carbon steels contain from 0.5-1.5 per cent carbon, and mild steels from a trace to 0.5 per cent carbon. Steel containing 1 per cent carbon has a specific gravity of 7.8, and melts at 1430° C. The specific heat of a steel containing 0.004 per cent carbon is 0.107 cal./gm./° C. at 20° C. and 0.117 cal./gm./° C. at 100° C.†

Steel A2. A British chemical standard. It is a carbon steel containing 0.037 per cent carbon, 0.034 per cent silicon, 0.020 per cent sulphur, 0.008 per cent phosphorus, 0.043 per cent manganese, 0.031 per cent arsenic, 0.059 per cent nickel, 0.013 per cent chromium, 0.067 per cent copper, 0.04 per cent oxygen, and 99.72 per cent iron.†

Steel Bronze. See UCHATIUS BRONZE.

Steel B4. A British chemical standard. It is a carbon steel containing 0.400 per cent carbon, 0.026 per cent silicon, 0.046 per cent sulphur, 0.103 per cent phosphorus, 0.735 per cent manganese, and 0.140 per cent arsenic.†

Steel C. A British chemical standard. It is a carbon steel containing 0.093 per cent carbon.†

Steel E. A standard steel containing 0.115 per cent carbon and 0.491 per cent manganese. It is used as the colorimetric standard for the determination of carbon in steels containing more than 0.100 per cent carbon.†

Steel F. A German steel containing 0.67-1.1 per cent silicon and 0.1-0.14 per cent carbon.†

Steel Guard. Petroleum base. Applications: Rust inhibitors for steel, industrial, commercial and vehicle applications. (Adasco-Inc).*

Steel H. A British chemical standard. It i a carbon steel, and contains 0.428 per cent carbon, 0.047 per cent sulphur, a 0.035 per cent phosphorus.†

Steel I. A carbon steel containing 0.521 p cent carbon and 0.726 per cent manganese. It is a British chemical standard.†

Steel M. A British chemical standard. It i a carbon steel containing 0.228 per ce carbon and 0.057 per cent silicon.†

Steel N. A carbon steel containing 0.17 p cent carbon, 0.117 per cent silicon, 0.0 per cent sulphur, 0.037 per cent phosphorus, 0.432 per cent manganese, and 0.029 per cent arsenic. It is a Brit chemical standard.†

Steel N1. A carbon steel containing 0.153 per cent carbon, 0.176 per cent silicon, 0.050 per cent sulphur, 0.036 per cent phosphorus, 0.527 per cent manganese, 0.030 per cent arsenic, 0.260 per cent nickel, and 0.04 per cent copper. It is a British chemical standard.†

Steel Ore. A variety of cinnabar containin 75 per cent mercury.†

Steel P. A high silicon and phosphorus steel. It is a British chemical standard.

Steel R. A British chemical standard. It is a carbon steel containing 0.786 per cen carbon, 0.053 per cent sulphur, and 0.914 per cent manganese.†

Steel S1. A British chemical standard. It i a carbon steel containing 0.921 per cen carbon and 0.051 per cent phosphorus.

Steel T. A nickel steel containing 3.367 pe cent nickel. It is a British chemical standard.†

Steel U. A carbon steel containing 1.203 per cent carbon, 0.472 per cent manganese, and 0.608 per cent nickel. It is a British chemical standard.†

Steel V. A British chemical standard. It is an alloy steel containing 0.548 per cent carbon, 0.161 per cent silicon, 0.063 pe cent sulphur, 0.024 per cent phosphoru 0.542 per cent manganese, 0.861 per cent chromium, and 0.273 per cent vanadium.†

Steel V2A. A rustless steel containing iron with 20 per cent chromium, 7 per cent nickel, and 0.2 per cent carbon.†

eel W. A British chemical standard. It is an alloy steel containing 0.695 per cent carbon, 0.187 per cent silicon, 0.075 per cent sulphur, 0.028 per cent phosphorus, 0.101 per cent manganese, 0.44 per cent nickel, 3.01 per cent chromium, 0.791 per cent vanadium. 4.76 per cent cobalt, and 16.21 per cent tungsten.†

teel W2. A British chemical standard high-speed alloy steel. It contains 0.17 per cent carbon, 0.14 per cent silicon, 0.051 per cent sulphur, 0.220 per cent manganese, 3.29 per cent chromium, 0.82 per cent vanadium, 16.12 per cent tungsten, 4.35 per cent cobalt, 0.43 per cent nickel, and 0.55 per cent molybdenum.†

teel 0. A British chemical standard. It is a nickel steel containing 0.325 per cent carbon, 0.590 per cent manganese, and 3.985 per cent nickel.†

teel 01. A carbon steel containing 0.333 per cent carbon, 0.162 per cent silicon, 0.032 per cent sulphur, 0.031 per cent phosphorus, 0.617 per cent manganese, 0.024 per cent arsenic, 0.162 per cent nickel, 0.017 per cent chromium, and 0.037 per cent copper. It is a British chemical standard.†

teelite. An explosive consisting of potassium chlorate, mixed with oxidized resin, and a little castor oil.†

teel, Blue. See PRUSSIAN BLUE.

teel, Electro-granodised. See ELECTROGRANODISED IRON and STEEL.

teel, Japanese. See MAGNET STEEL.

teel, Open Hearth. See MARTIN STEEL.

teel, Rustless. See CHROME STEELS.

teel, Self-hardening. See MUSHET STEEL.

teel, Stainless. See CHROME STEELS.

teel, V2A. See ANKA STEEL; also STAYBRITE.

teels, Aircraft Construction. (Ternary and Quaternary). These are usually iron with from 0.25-1 per cent carbon, 0.15-3.25 per cent nickel, and 0.45-1.25 per cent chromium.†

teinazid SBU 185. Undecylenic acid alkylolamide sulphosuccinate in liquid form. Applications: Anti-dandruff agent, fungicidal and bacteriostatic additive for shampoos, hair lotions, foam baths, etc. (REWO Chemicals Ltd).

Steinbuhl Yellow. (Gelbin Yellow Ultra marine). A pigment. A chromate of calcium is sold under these names, but Barium yellow also frequently passes under the same terms.†

Stelabid Tablets. Combination of isopropamide iodide and trifluoperazine hydrochloride. Applications: Gastro-intestinal disorders in which hypersecretion and/or painful spasms are a problem with condition complicated by emotional factors. (Smith Kline and French Laboratories Ltd).*

Steladex. A proprietary preparation of trifluoperazine dihydrochloride and dexamphetamine sulphate. (Smith Kline and French Laboratories Ltd).*

Stelazine. Trifluoperazine Hydrochloride. Applications: Antipsychotic; sedative. (Smith Kline & French Laboratories).

Stelazine Tablets, Spansule Capsules, Syrup, Injection and Concentrate. Trifluoperazine hydrochloride (phenothiazine tranquillizer). Applications: Schizophrenia and other psychotic states. In low dosage for treatment of anxiety states. (Smith Kline and French Laboratories Ltd).*

Stellachrome Black L757. A dyestuff. It is equivalent to Palatine chrome black.†

Stellak. Chemicals for boiler water treatment. (Steetley Chemicals Ltd).

Stellited Metal. Metals treated with an alloy consisting chiefly of chromium, tungsten, and cobalt (stellite). The metals treated are usually steel, cast iron, malleable iron, and semi-steel. It is an economical method for these treated metals are rendered suitable for wear-resisting parts of machinery.†

Stellos. Calcium hypochlorite. Applications: Used for chlorination of water. (Stella Meta Filters).*

Stellox. Cereal herbicide. (Ciba-Geigy PLC).

Stelogen. Flux for degassing steel. (Foseco (FS) Ltd).*

Stelopack. Uphill teeming flux for killed steel ingots. (Foseco (FS) Ltd).*

Stelorit. Covering and cleansing fluxes for steels. (Foseco (FS) Ltd).*

Stelotol. Powder flux for uphill teeming. (Foseco (FS) Ltd).*

Stemetil. A proprietary preparation of prochlorperazine dimaleate or mesylate. An anti-emetic and sedative. (May & Baker Ltd).*

Stemex. Paramethasone Acetate. Applications: Glucocorticoid. (Syntex Laboratories Inc).

Stempor. Fungicide containing carbendazim for use on cereals or fruit. (ICI PLC).*

Stenol. A proprietary trade name for technical stearyl alcohol.†

Stenorol. Halofuginone hydrobromide. Applications: Antiprotozoal. (Roussel-UCLAF).

Stenosin. Disodium monomethyl arsonate.†

Stental. Phenobarbital. Applications: Anticonvulsant; hypnotic; sedative. (A H Robins Co Inc).

Stentor Steel. A proprietary trade name for a non-shrinking steel containing 1.6 per cent manganese, 0.25 per cent silicon, and 0.9 per cent carbon.†

Steol. Alkyl ether sulphates. Applications: Detergents, emulsifiers, foaming agents. (Stepan Company).*

Steol CA-460 and KA-460. Ammonium fatty ether sulphate in liquid form. Application: Shampoo, bubble baths, dish detergents and degreasers. (KWR Chemicals Ltd).

Steol CS-460 and KS-460. Sodium fatty ether sulphate in liquid form. Application: Shampoo, bubble baths, dish detergents and degreasers. (KWR Chemicals Ltd).

Steol CS-760 and 7N. Sodium fatty ether sulphate as a pale yellow liquid. Application: Very mild base for shampoos and bubble baths. (KWR Chemicals Ltd).

Steol FA. Ammonium fatty ether sulphate as a pale yellow liquid. Application: General detergent uses and manufacture of gypsum board. (KWR Chemicals Ltd).

Steol 3 OS. Sodium lauryl ether sulphate as a clear yellow viscous liquid. Application: Anionic surfactant used in shampoos, foam baths and liquid detergents. (KWR Chemicals Ltd).

Steol 4N. Sodium fatty ether sulphate as nearly water white liquid. Application: Shampooo, bubble bath, liquid cleaner (KWR Chemicals Ltd).

Steol 7T. Triethanolamine fatty ether sulphate as a pale yellow liquid. Application: Very mild base for shampoos and bubble baths. (KWR Chemicals Ltd).

Stepan. Emollient. Applications: Bath oil anti-perspirants etc. (KWR Chemicals Ltd).

Stepanate. Hydrotropes. Applications: Coupling agent, cloud point depressant (Stepan Company).*

Stepanflo. Surfactant. Applications: Enhanced oil recovery. (Stepan Company).*

Stepanflote. Various surfactants. Applications: Ore flotation reagent. (Stepan Company).*

Stepanform. Anionic - nonionic surfactant blend. Applications: Foaming agent, emulsifier. (Stepan Company).*

Stepanol. Alkyl sulphates. Applications: Mild detergent - foaming agent. (Stepan Company).*

Stepanol AM. Ammonium fatty alcohol sulphate as a pale yellow liquid. Applications: Shampoo; bubble bath; liquid detergents. (KWR Chemicals Ltd).

Stepanol DEA. Diethanolamine fatty alcohol sulphate as a pale yellow liquid Applications: Shampoo. (KWR Chemicals Ltd).

Stepanol ME. Sodium fatty alcohol sulphate as a white powder. Applications: Powdered detergents. (KWR Chemicals Ltd).

Stepanol Mg. Magnesium fatty alcohol sulphate as a pale yellow liquid. Applications: Rug and upholstery shampoos. (KWR Chemicals Ltd).

Stepanol SPT. Triethanolamine lauryl sulphate as a clear yellow viscous liquid Applications: Shampoo; foam baths;

liquid detergents. (KWR Chemicals Ltd).

Stepanol WA-100. Sodium fatty alcohol sulphate as a white powder. Applications: Anionic surfactant used as a dentifrice, and in the pharmaceutical industry. (KWR Chemicals Ltd).

Stepanol WAT. Triethanolamine fatty alcohol sulphate as a nearly water white liquid. Applications: Shampoos. (KWR Chemicals Ltd).

Stepanol WA, WAC, WAQ. Ionic surfactants from the Stepanol WA range. Applications: Shampoos, bubble baths, liquid and paste detergents. (KWR Chemicals Ltd).

Stepantan. Alkyl sulphonate. Applications: Dispersant. (Stepan Company).*

Stepantan A. Anionic surfactant in powder form. Applications: Dispersant, tanning agent. (KWR Chemicals Ltd).

Stepantan NP 80. Anionic surfactant in powder form. Applications: Dispersant for phyto-sanitary products and wettable powders. (KWR Chemicals Ltd).

Stepantex Q90B. Dialkyl methoxysulphate as an amber viscous liquid. Applications: Cationic surfactant used in textile softeners. (KWR Chemicals Ltd).

Stepenor/Stepenor-Retard. Veterinary preparation. Applications: For prophylaxis and treatment of bacterial infections of the udder. (Bayer & Co).*

Steposol. Alkyl ether sulphate. Applications: Foaming agent. (Stepan Company).*

Ster-Zac. A proprietary preparation of HEXACHLOROPOANE. A topical antiseptic. (Hough, Hogeason).†

Sterane. Prednisolone. Applications: Glucocorticoid. (Pfizer Inc).

Sterane IM and IA. Prednisolone Acetate. Applications: Glucocorticoid. (Pfizer Inc).

Sterbon. See CARBORA.

Stercorite. Sodium ammonium hydrogen phosphate. It occurs in guano.†

Sterculia Gum. Indian tragacanth.†

Sterculia Kernals. See OLIVES OF JAVA.

Stereon. Styrene/butadiene SBR block copolymers. Applications: Thermoformed products, injection moulding and adhesives. (Firestone Synthetic).*

Stereosine Grey. A dyestuff which dyes wool and mixed goods bluish-brown from a neutral bath.†

Stereotype Plate. An alloy of 85 per cent lead and 14 per cent antimony, sometimes with the addition of a little tin. See TYPE METAL.†

Steresol. An antiseptic varnish made by dissolving 270 parts purified shellac, 10 parts benzoin, 10 parts balsam of tolu, 100 parts phenol, 6 parts oil of cinnamon, and 6 parts saccharine in alcohol, to make 1,000 parts.†

Sterethox. Steriliser containing dichlorodifluoro methane. (ICI PLC).*

Steribath. An antiseptic solution containing an iodophore. (ICI PLC).*

Steridex. Fungicidal water-based elastomeric protective coating, applied by brush or spray. Applications: For totally eradicating mould growth and bacteria in all hygiene sensitive environments - hospitals, food factories, breweries. (Liquid Plastics Ltd).*

Steriflux. A range of sterile non pyrogenic intravenous infusions. (The Boots Company PLC).

Sterilite. Disinfectants. (Tenneco Organics Ltd).

Sterillium. Synthetic phenolic germicides in a detergent base. Applications: A disinfectant for laundry use. (William Pearson).*

Sterisil. A proprietary trade name for Hexetidine.†

Sterisol. Hexedine. Applications: Antibacterial. (Parke-Davis, Div of Warner-Lambert Co).

Sterline. An alloy of 68 per cent copper, 17-18 per cent nickel, 13-14 per cent zinc, 0.75-0.8 per cent iron, and 0-0.8 per cent lead. It is a nickel silver (German silver).†

Sterling Brass. See BRASS.

Sterling Gold. See STANDARD GOLD.

Sterling Silver. See STANDARD SILVER.

Sterling Solder. An alloy of 61.6 per cent tin, 15.2 per cent zinc, 11.2 per cent aluminium, 8.3 per cent lead, 2.5 per cent copper, and 1.2 per cent antimony.†

Sterlite. A proprietary trade name for a nickel brass containing 25 per cent nickel, 20 per cent zinc, and small amounts of iron, manganese, silicon, and carbon.†

Sterlith. A trade mark for materials of the refractory and abrasive type. They consist essentially of crystalline alumina.†

Sternite. Phenol formaldehyde resin. Applications: Moulding powders. (Manchem Ltd).*

Sternite. Phenolic and polystyrene moulding materials. SPF 5092 is mineral filled. (Sterling Moulding Materials).*

Sterotabs. Water treatment tablets. (The Boots Company PLC).

Sterovum. Ethynerone. Applications: Anti-fertility agent. (Merck Sharp & Dohme).*

Sterox. Dodecylphenol-ethylene oxide condensate (alkylaryl polyoxyethylene ether). Applications: Nonionic surface active agent. (Monsanto Co).*

Sterox DF, DJ. Anionic surface active agent. (Monsanto PLC).

Steroxin-Hydrocortisone. A proprietary preparation of chlorquinaldol and hydrocortisone used in dermatology as an antibacterial agent. (Ciba Geigy PLC).†

Steroxol. Chlorinated detergent. (ABM Chemicals Ltd).*

Sterpon. A proprietary polyester laminating resin. (Convert (Ets G)).†

Sterro Metal. See AICH METAL.

Sthenosised Cotton. Cotton which has been treated with formaldehyde. It becomes resistant to alkalis and cannot be mercerized. It has a greatly decreased affinity for direct dyestuffs.†

Stibiated Tartar. See TARTAR EMETIC.

Stibium. Antimony, Sb.†

Stibocaptate. Antimony (111) sodium meso-2,3-dimercaptosuccinate.†

Stick-lac. See LAC and LAC-DYE.

Stickstoffoxydbaryt. Barium nitrite, $Ba(NO_2).H_2O$.†

Stiedex. An oily cream available in three grades: 0.25%, LP 0.05% and LPN 0.05%. All grades contain desoxymethasone in an oily cream base. The LPN 0.05% also contains neomycin. Applications: For the treatment of a wide range of acute inflammatory and allergic conditions and for chronic skin disorders. (Stiefel Laboratories (UK) Ltd).*

Stil de Grain. See BROWN PINK.

Stilbene Yellow G, 4G, 6G, 8G. Dyestuffs. They are alkaline condensation products of dinitrodibenzyldisulphonic acid and dinitrosostilbenedisulphonic acid. Dyes cotton greenish-yellow from a salt or sodium sulphate bath.†

Stilbene Yellow 2G, 3G, 8G. Dyestuffs. They are British brands of Mikado yellow.†

Stillingia Oil. An oil obtained by crushing the kernel of *Stillingia sebifera.* Alss see CHINESE TALLOW.†

Stilphostrol. Diethylstilbestrol diphosphate. Applications: Oestrogen. (Miles Pharmaceuticals, Div of Miles Laboratories Inc).

Stimate Injection. Desmopressin acetate. Applications: Antidiuretic. (Armour Pharmaceutical Co).

Stimplete. A proprietary preparation of phenobarbitone dexamphetamine sulphate, thiamine hydrochloride, riboflavine, pyridoxine hydrochloride, nicotinamide and alcohol. (Wyeth).†

Stimufol. Soluble fertilizer. (ICI PLC).*

Sting. Herbicide. (Monsanto PLC).

Stink Quartz. (Foetid Quartz). A quartz which has a bad odour, due to organic matter.†

Stink-stone. (Oil-stone). A bituminous schist found in the Tyrol. A source of ichthyol.†

Stirling Metal. See STERLING BRASS.

Stirlingite. Synonym for Roepperite.†

Stirling's Gentian Violet. A microscopic stain. It contains 5 grams gentian violet, 10 cc 95 per cent alcohol, 2 cc aniline, and 88 cc water.†

Stockalite. A proprietary product. It is a very highly refined china clay used as a filler for tyres, cables, and high grade mixes.†

Stockholm. Tar (Pine Tar). A tar obtained principally from pine-wood distillation. It is obtained from *Pinus sylvestris* and other species of *Pinus*. It is used as a preservative paint for ships and roofing and as a rubber softener. It has anti-oxidant properties.†

Stockholm Petroleum. See PETROLEUM, STOCKHOLM.

Stockholm Pitch. Pine-wood tar pitch. It is soluble in alkalies, and is used in the preparation of varnishes, in the rubber and gutta-percha trades, and in the preparation of impervious cements.†

Stoco. A proprietary bituminous plastic.†

Stoddard Solvent. A proprietary trade name for a refined petroleum product for dry cleaning.†

Stoffertite. A calcium phosphate, $CaHPO_4.5H_2O$. It occurs in guano.†

Stoic Metal. An alloy similar in composition to Invar.†

Stoke's Reagent. A reducing agent prepared by dissolving 30 grams ferrous sulphate and 20 grams tartaric acid in 1 litre of water. When required for use, strong ammonia is added until the precipitate first formed is redissolved.†

Stomahesive. A proprietary preparation containing gelatin pectin, carboxy-methyl cellulose and polyisobutylene on a protective film, used for the protection of skin around surgical stomata. (Squibb, E R & Sons Ltd).*

Stomosan. Ethylamine phosphate, used in medicine.†

Stone Black. Animal charcoal.†

Stone, Coal. Anthracite.†

Stone, Green. See BOHEMIAN EARTH.

Stone, Grey. See SLATE GREY.

Stone, Honey. See MELLITE.

Stone, Mercury. Mercuric chloride, $HgCl_2$, in lumps.†

Stone, Oil. See STINK-STONE.

Stone Red. See INDIAN RED.

Stone Root. Collinsonia.†

Stone, Tin. See STREAM TIN.

Stone, Touch. See LYDIAN STONE.

Stone Wax. A name applied to carnauba wax.†

Stone Yellow. See YELLOW OCHRE.

Stone's Bronze. An alloy of 87 per cent copper, 11 per cent tin, and 2 per cent phosphor-copper.†

Stonite. An explosive consisting of 68 per cent nitro-glycerin, 20 per cent kieselguhr, 8 per cent potassium nitrate, and 4 per cent wood meal.†

Stoodite. A proprietary trade name for a high manganese steel.†

Stop. Stannous Fluoride. Applications: Dental caries prophylactic. (Oral-B Laboratories Inc).

Stora. A proprietary trade name for a Swedish charcoal iron used for making malleable iron.†

Storalon. See CARBORA.

Storax. (Styrax). An oleo-resin, the product of the tree *Liquidambar orientalis*. The crude material contains 20-30 per cent water and fragments of bark, etc. Prepared storax is used as a drug.†

Storax Calamita. The powdered bark of *Liquidambar styracflua*, most of the resin being first extracted. The product has no connection with storax.†

Storax, Liquid. See LIQUID STORAX and LIQUIDAMBAR.

Storax, Solid. See RED STORAX.

Stortex. Malt extract. (ABM Chemicals Ltd).*

Stovarsol. Acetarsol. (May & Baker Ltd).

Stowite. An explosive containing 58-61 per cent nitro-glycerin, 4.5-5 per cent nitro-cotton, 18-20 per cent potassium nitrate, 6-7 per cent wood meal, and 11-15 per cent ammonium oxalate.†

Stowmarket Powder. A 33-grain powder.†

Stoxil. Idoxuridine. Applications: Antiviral. (Smith Kline & French Laboratories).

Strandex. Chemical compositions for use in industry as additives in polymer and plastics processing - all containing metal compounds. (Associated Lead Manufacturers Ltd).*

Strandol. Mould lubricant for continuous casting of steel billets. (Foseco (FS) Ltd).*

Strass. (Paste). A kind of glass used to imitate precious stones. It is made from 100 parts sand, 40 parts minium, 24 parts potassium carbonate, 20 parts borax, and 12 parts potassium nitrate. This gives a colourless product, and various oxides are added to colour it.†

Strassburg Turpentine. The oleo-resin from the silver fir, *Pinus picea*.†

Strasser Solder. An alloy of 62 per cent tin, 12 per cent zinc, 4 per cent aluminium, 8 per cent lead, 5 per cent copper, 5 per cent bismuth, and 4 per cent cadmium.†

Strata-Fire. A fuel additive used to reduce engine wear, improve engine performance, increase fuel economy and reduce emission of air pollutants. Applications: Used for all types of internal combustion engines, both diesel and gasoline. The current rate of addition is one ounce to forty gallons of fuel (50 ml per 300 litres of fuel). (SN Corp/Appropriate Technology Ltd).*

Stratyl. A proprietary polyester laminating compound. (Pchiney-St Gobain, France).†

Straus Metal. See CAMITE.

Strawlink. Mineral/vitamin animal feed supplement for straw. (ICI PLC).*

Strelax. Road nosing compounds. (ICI PLC).*

Strenes Metal. A proprietary trade name for a nickel-chromium-molybdenum cast iron.†

Strepolin. A proprietary preparation containing streptoroycin sulphate. An antibiotic. (Glaxo Pharmaceuticals Ltd).†

Strepsils. Antiseptic throat lozenges. (The Boots Company PLC).

Streptets. A proprietary preparation of zinc bacitracin, polymixin B sulphate, and neomycin sulphate. Antibiotic lozenges. (Wyeth).†

Streptohydrazid. Streptonicozid. Applications: Antibacterial. (Pfizer Inc).

Streptokinase. An enzyme obtained from cultures of various strains of Streptococcus haemolyticus and capable of changing plasminogen into plasmin. Kabikinase.†

Streptonivin. A proprietary trade name for Novobiocin.†

Streptorex. A proprietary preparation of streptomycin sulphate. An antibiotic. (Pfizer International).*

Streptotriad. A proprietary preparation of streptomycin, sulphadiazine, sulphathiazole and sulphadimidine, used in the treatment of dysentery. (May & Baker Ltd).*

Stresnil. A proprietary preparation containing azaperone. Applications: Veterinary sedative (pigs). (Janssen Pharmaceutical Ltd).*

Stretonex. A proprietary preparation of streptomycin sulphate. An antibiotic. (Pfizer International).*

Strewing Smalt. The coarsest powdered Smalt (*qv*).†

Strim. Rimming agent for steel ingots. (Foseco (FS) Ltd).*

Stripcote. Liquid parting agents and release agents. (Foseco (FS) Ltd).*

Stripping Salt. See ABRAUM SALTS.

Strobane. A trade mark for an insecticide and acaricide. It is based on polychlorinated terpine and contains 66% chlorine.†

Stromba. A proprietary preparation of stanozolol. An anabolic agent. (Bayer & Co).*

Stronscan-85. Strontium Chloride Sr 85. Applications: Radioactive agent. (Abbott Laboratories).

Strontia. Strontium oxide, SrO.†

Strontian White. A pigment. It consists of strontium sulphate, $SrSO_4$.†

Strontian Yellow. (Yellow Ultramarine). A pigment. Originally it was a chromate of strontium, $SrCrO_4$, but a more durable pigment is now sold under this name.†

Strotope. Strontium Nitrate Sr 85. Applications: Radioactive agent. (E R Squibb & Sons Inc).

Struthiin. (Githagin, Polygalin, Polygallic Acid, Senegin). Saponin, $C_{19}H_{30}O_{10}$, a glucoside found mainly in the common soapwort.†

trycin. Streptomycin Sulphate. Applications: Antibacterial. (E R Squibb & Sons Inc).

tuart's Granolithic Stone. A stone similar to Ward's stone.†

tucco. A specially hard plaster which can be polished. There are two kinds: (1) made from plaster-of-Paris; and (2) made from lime. They are usually mixed with size.†

tugeron. A proprietary preparation of Cinnarizine. An anti-nauseant. (Janssen Pharmaceutical Ltd).*

tugeron Forte. A proprietary preparation containing cinnarizine. Applications: Peripheral arterial disease. (Janssen Pharmaceutical Ltd).*

tuk. Adhesives for footwear industry. (Avalon Chemical Co Ltd).

tupp. A mercurial soot condensed in the chambers during the treatment of mercury ores. It contains about 20 per cent mercury as metal, and sulphate.†

turcal. Ultrafine precipitated calcium carbonate. Applications: Dentifrice, pharmaceutical, food, confectionery, plastics, paint and fermentation. (Sturge Lifford).*

turcarb. Whiting. (John & E Sturge Ltd, Lifford).

typhen I. A proprietary anti-oxidant. It is a mixture of styrenated phenols. (Corning Glass Works - Zircoa Products).†

typhnic Acid. Trinitroresorcinol (2 : 4-dihydroxy-1 : 3 : 5-trinitro-benzene), $C_6H_3O_8N_3$. Used in explosives.†

typticine. Cotarnine hydrochloride, used medicinally as a styptic.†

tyquin. Butamisole hydrochloride. Applications: Anthelmintic (American Cyanamid Co).

Styracin. Cinnamyl cinnamate, $C_6H_5.CH : CH.CH_2.O.CO.CH : CH.C_6H_5$.†

Styrafil. A registered trade name for flame-retardant polystyrene.†

Styraloy 22, 22A. A proprietary trade name for an elastomeric styrene derivative.†

Styramate. β-Hydroxyphenylethyl carbamate.†

Styramic H.T. and M.T. Proprietary trade names for polystyrene thermo-plastics possessing a higher softening point than usual. They are stated to be polydichlorstyrenes.†

Styrocell. Expanded and expansible polystyrene. (Shell Chemicals UK Ltd).†

Styrodur. Extruded cellular polystyrene insulation board. Applications: Thermal insulation of roofs, walls and floors in buildings. (BASF United Kingdom Ltd).*

Styrofan D. Polystyrene dispersion. (BASF United Kingdom Ltd).*

Styrofill. Loose-fill expanded polystyrene packaging material. Applications: Packaging. (BASF United Kingdom Ltd).*

Styroflex. A proprietary trade name for a synthetic resin said to be a flexible polymer of styrene.†

Styrofoam. Brand plastic foam used as insulation in roof construction, in walls and ceilings of homes, general buildings and cold storage structures. (Dow Chemical Co Ltd).*

Styrogallol. Dihydroxyanthracoumarin, $C_{16}H_8O_5$, a yellow dyestuff.†

Styrol. (Styrolene, Cinnamene, Cinnamol). Styrene, $C_6H_5.CH : CH_2$. Styrol is also the name for a colloidal silver preparation.†

Styrolene. See STYROL.

Styrolux. Plastic granules for extrusion and injection moulding. Styrene-butadiene block copolymer. Applications: Clear packaging - extrusion. Various transparent injection moulded parts. (BASF United Kingdom Ltd).*

Styrolyl Alcohol. (Styryl Alcohol). Phenyl-glycol, $C_6H_5.CHOH.CH_2OH$. Used in perfumery.†

Styromol. Refractory coatings for expanded polystyrene patterns used in Replicast FM process. (Foseco (FS) Ltd).*

Styron. General purpose and high impact polystyrene resins used in packaging, housewares, toys, medical and electronics. (Dow Chemical Co Ltd).*

Styrone. Cinnamyl alcohol, $C_6H_5.CH : CH.CH_2OH$.†

Styropor. Expandable polystyrene bead containing pentane. Applications: Packaging and building insulation. (BASF United Kingdom Ltd).*

Styropor. Expandable polystyrene. (Mitsubishi Petrochemical Co).*

Styxol. See UBA.

Suakin Gum. (Talca Gum, Talka Gum, Sennaar Gum). A brittle variety of gum acacia from *Acacia fistula.* It gives a ropy mucilage.†

Subacetate of Lead. Monobasic lead acetate, $(C_2H_3O_2)_2Pb+PbO+H_2O$.†

Subeston. A preparation of Estone (see LENICET). It is a double basic acetate of aluminium.†

Subitol. See ICHTHYOL.

Sublaprint. Uncut disperse dyestuffs. Applications: For the production of inks for printing heat transfer papers. (Holliday Dyes & Chemicals Ltd).*

Sublimaze. A proprietary preparation of fentanyl. An analgesic. (Janssen Pharmaceutical Ltd).*

Sublimed Blue Lead. A pigment produced by heating mixed ores of zinc and lead in a furnace with an air blast. It usually contains 50 per cent lead sulphate, 20 per cent lead oxide, PbO, 11 per cent lead sulphide, PbS, 8 per cent lead sulphite, and 3 per cent zinc oxide.†

Sublimed Calomel. See CALOMEL.

Sublimed White Lead. A white lead manufactured from mixed ores of galena and zinc blende. They are roasted in the presence of an air blast, and the lead sulphate, lead oxide, and zinc oxide formed is collected in large chambers. The average composition is 75 per cent lead sulphate, 20 per cent lead oxide, and 5 per cent zinc oxide. A pigment.†

Sublimo-phenol. Chloro-phenolate of mercury. Used in antiseptic surgery.†

Sublimoform. A mercury-formaldehyde preparation. A seed preservative.†

Subox. A protective coating paint consisting of a suspension of colloidal lead in linseed oil.†

Substitute Gutta-percha. See SOREL'S SUBSTITUTES.

Substitute of Tartar. See SUPERARGO[L]

Substitute Saffron. See VICTORIA YELLOW.

Sub-Vitralen. High molecular weight hig[h] density polyethylene sheet. Applications: Orthopaedic splints. (Stanley Smith & Co Plastics Ltd).*

Sucaryl. Saccharin Sodium. Applications Sweetener. (Abbott Laboratories).

Succinellite. Succinic acid obtained from amber.†

Succinite. Baltic amber which contains succinic acid. Also the name for a mineral, a lime-aluminium garnet.†

Succinol. An oil obtained by the distillati[on] of amber.†

Succinoxate. See ALPHOGEN.

Suchar. A decolourizing carbon used for sugar juices. It is prepared from waste sulphite-cellulose liquors.†

Sucker Plucker Concentrate. Fatty alcoho[l] mixture. Applications: Contact tobacc[o] sucker control. (Drexel Chemical Company).*

Sucker Stuff. Potassium salt of maleic hydrazide. Applications: Systemic control of tobacco suckers. (Drexel Chemical Company).*

Sucostrin. Succinylcholine Chloride. Applications: Blocking agent. (E R Squibb & Sons Inc).

Sucramine. (Lyons Sugar). The ammoniu[m] salt of saccharin, used in France as a sweetening substance.†

Sucrase. (Saccharase, Invertin). Invertase, an enzyme which decomposes saccharose into glucose and levulose.†

Sucrate of Hydrocarbonate of Lime. See SUCRO-CARBONATE OF LIME.

Sucrene. See DULCINE.

Sucro-carbonate of Lime. (Sucrate of Hydrocarbonate of Lime). A complex compound of lime, calcium sucrate, an[d] calcium carbonate formed in the production of sugar from the beet, whe[n] carbon dioxide gas is passed into a solution of sucrate of lime.†

Sucro-levulose. See LEVULOSE.

Sucrol. See DULCINE.

Sucrose. (Cane Sugar, Beet Sugar). Saccharose, $C_{12}H_{22}O_{11}$.†

Suction Gum. (Suction Powder). Powdered gum tragacanth.†

Sucuaryl. A proprietary trade name for sodium cyclamate.†

Sudafed. Pseudoephedrine Hydrochloride. Applications: Adrenergic. (Burroughs Wellcome Co).

Sudan G. (Carminaph J, Cerasine Orange G, Oil Orange O, Oil Yellow). A dyestuff. It is dihydroxyazobenzene, $C_{12}H_{10}N_2O_2$.†

Sudan Glycerin. The dyestuff Sudan III (0.01 gram) is dissolved in 5 cc 90 per cent alcohol, and 5 cc glycerin added. Used as a microscopic stain.†

Sudan I. (Carminaph, Fast Oil Orange I, Oil Orange, E, Spirit Yellow I). A dyestuff. It is benzeneazo-β-naphthol, $C_{16}H_{12}N_2O$. Used for colouring oils and varnishes.†

Sudan II. (Red B, Scarlet G, Fast Oil Orange II, Oil Scarlet L, Y). A dyestuff. It is xyleneazo-β-naphthol, $C_{18}H_{16}N_2O$. Used for colouring oils and varnishes.†

Sudan III. (Red C, Cerasine Red, Fast Oil Scarlet III, Oil Scarlet AS, Spirit Red III). A dyestuff. It is benzeneazo-benzeneazo-β-naphthol, $C_{22}H_{16}N_4O$. Used for colouring oils and varnishes.†

Sudan IV. (Fat Ponceau, Biebrich Scarlet R Medicinal, Spirit Red III). A dyestuff. It is *o*-tolueneazo-*o*-toluene-azo-β-naphthol.†

Sudan R. (Oil Vermilion). A dyestuff similar to Sudan III.†

Südflock. Inorganic precipitants, flocculants and adsorbents for the purification of industrial and domestic effluents. (Süd-Chemie AG).*

Sufafed Products. Proprietary formulations of pseudoephedrine hydrochloride. Applications: Temporary relief of symptoms associated with the common cold, hay fever or other upper respiratory allergies. (The Wellcome Foundation Ltd).*

Sufatone SCS/B. Concentrated cationic surfactant in paste form. Applications: Softening and antistatic agent used for all fabrics, particularly synthetics including acrylics. (Standard Chemical Company).

Sufatone SCS/CL. Cationic surfactant in liquid form. Applications: General purpose mild softening and antistatic agent used for all fibres. (Standard Chemical Company).

Sufatone SC/L and SMC/L. Cationic surfactant in liquid form. Applications: Softening agent for most fibres particularly wool and acrylics. (Standard Chemical Company).

Sufatone SMC/W. Concentrated cationic surfactant in the form of a semi-liquid. Applications: Mild softening agent for most fibres, particularly wool and chlorinated wool. (Standard Chemical Company).

Sufenta. A proprietary preparation containing sufentanil citrate. Applications: Analgesic supplement to anaesthesia. (Janssen Pharmaceutical Ltd).*

Suffa. Sulphur. Applications: Fungicide for fruit and vegetable crops. (Drexel Chemical Company).*

Sufrexal. A proprietary preparation containing ketanserin tartrate. Applications: Antihypertensive, serotonin antagonist. (Janssen Pharmaceutical Ltd).*

Sugamo. A Japanese seaweed suggested for use in paper-making.†

Sugar, Beechwood. See WOOD SUGAR.

Sugar, Beet. See SUCROSE.

Sugar Cane Wax. A wax obtained from the dried filter press cake from sugar mills by benzine extraction. The African cake contains 14-17 per cent of wax, and the Java cake 4 per cent. The wax obtained is not a pure product, but is a mixture of wax and fatty material with 7 per cent glycerin. The pure wax is obtained by crystallization and distillation. The crude wax contains about 7 per cent glycerin, 61 per cent free and combined acids, and 28 per cent unsaponifiable matter. The wax has been used in the manufacture of polishes.†

Sugar Charcoal. (Lampblack). Amorphous carbon.†

Sugar, Diabetic. See GLUCOSE.

Sugar, Fruit. See LEVULOSE.

Sugar House Black. A bone black pigment. It is a by-product of the sugar mills.†

Sugar of Gelatin. Glycine.†

Sugar of Lead. Normal lead acetate, $(CH_3COO)_2Pb+3H_2O$. Used as a mordant in dyeing and printing, and for the preparation of lead salts and paints.†

Sugar of Milk. Lactose, $C_{12}H_{22}O_{11}+H_20$.†

Sugar of Saturn. See SALT OF SATURN.

Sugar Sand. See MAPLE SUGAR SAND.

Sugar Vinegar. See GLUCOSE VINEGAR.

Sugracillin. Penicillin G Potassium. Applications: Antibacterial. (The Upjohn Co).

Suhler White Copper. An alloy of 40 per cent copper, 32 per cent nickel, 25 per cent zinc, 2.6 per cent tin, and 0.6 per cent cobalt. It is a nickel silver.†

Suicalm. Azaperone. Applications: Antipsycotic. (Janssen Pharmaceutica).

Suladrin. Sulphisoxazole Diolamine. Applications: Antibacterial. (Alcon Laboratories Inc).

Sulamyd Sodium. Sulphacetamide Sodium. Applications: Antibacterial. (Schering-Plough Corp).

Sulf-10. Sulphacetamide Sodium. Applications: Antibacterial. (CooperVision Inc).

Sulfacet-R. Sulphacetamide Sodium. Applications: Antibacterial. (Dermik Laboratories Inc).

Sulfacide. Acid dyes. (ICI PLC).*

Sulfactol. Sodium Thiosulphate. Applications: Antidote to cyanide poisoning. (Sterling Drug Inc).

Sulfads. Essentially dipentamethylene thiuram tetrasulphide. Primary and secondary accelerator. vulcanizing agent. (Vanderbilt Chemical Corporation).*

Sulfalar. Sulphisoxazole. Applications: Antibacterial. (Parke-Davis, Div of Warner-Lambert Co).

Sulfamin. Anionic surfactant in liquid form. Applications: Liquid cleaners. (Berol Kemi (UK) Ltd).

Sulfamylon. A proprietary preparation of mafenine acetate in a cream, used in the treatment of infected burns. (Winthrop Laboratories).*

Sulfanol. Sulphur dyes. (ICI PLC).*

Sulfarine. A mixture of magnesium sulphate with 15 per cent sulphuric ac[id]. It is used against potato scab.†

Sulfasan. 4,4'-Dithiodimorpholine. Applications: vulcanizing agent for natural and synthetic rubbers. (Monsanto Co).*

Sulfasan R. Vulcanizing agent. (Monsant[o] PLC).

Sulfasuxidine. Succinylsulphathiazole. Applications: Antibacterial. (Merck Sharp & Dohme).*

Sulfathalidine. Phthalylsulphathiazole. Applications: Antibacterial. (Merck Sharp & Dohme).*

Sulfato de Cobre Valles. Crystallized copper sulphate. Applications: Manufacture of agricultural fungicides and many industrial products. (Industrias Quimicas Del Valles SA).*

Sulfatol E3. Sodium lauryl ether sulphate (coconut/palm kernel C12-C14 alcohol) as a clear almost colourless lo[w] viscosity liquid. Applications: Production of all types of liquid and lotion shampoos and bubble baths. Als[o] a raw material for light duty liquid detergents, dishwashing detergents and auto shampoos. (Efkay Chemicals Ltd).

Sulfatryl. Trisulphapyrimidines (oral suspension). Mixture of Sulphadiazine Sulphamerazine and Sulphamethazine. Applications: Antibacterial. (Wallace Laboratories, Div of Carter-Wallace Inc).

Sulfidal. Colloidal sulphur powder. Applications: Used in powders, lotions, and ointments for treatment of skin diseases. (Nueodex Inc).*

Sulfil. A registered trade name for flame retardant polysulphone.†

Sulfoderm. Silicic acid with 1 per cent colloidal sulphur.†

Sulfogenol. A crude mineral oil obtained from bituminous shale, is saturated wit[h] sulphur, and sulphonated. Sulfogenol is the ammonium salt of the sulphonated product. It has similar properties to ichthyol.†

Sulfomyl. A proprietary preparation of mafenide propionate, used in the form of anti-infective eyedrops. (Winthrop Laboratories).*

ulfonsol. Trisulphapyrimidines (oral suspension). Mixture of Sulphadiazine, Sulphamerazine and Sulphamethazine. Applications: Antibacterial. (Merrell Dow Pharmaceuticals Inc, Subsidiary of Dow Chemical Co).

ulfopon LS. Sodium lauryl sulphate (C12-C18) in liquid/paste form. Applications: Cream shampoos and bubble baths. (Henkel Chemicals Ltd).

ulfopone. See SULPHOPONE.

ulfose. Trisulphapyrimidines (oral suspension). Mixture of Sulphadiazine, Sulphamerazine and Sulphamethazine. Applications: Antibacterial. (Wyeth Laboratories, Div of American Home Products Corp).

ulfosept Oil. The next higher fraction to thiosept oil (*qv*). Used as an insecticide.†

ulfosoft. Anionic surfactant in acid form. Applications: Detergent production. (Berol Kemi (UK) Ltd).

ulframin. Surfactant for cosmetics, toiletries, pharmaceutical, processing, agricultural and other industries. (Baxenden Chemical Co Ltd).*

ulframin 1250. Sodium alkylaryl sulphonate in liquid form. Applications: Anionic surfactant. (Witco Chemical Ltd).

ulframin 14-16 AOS. Sodium olefine sulphonate (C14-C16) as a clear amber liquid. Applications: Foaming agent and detergent for cosmetic and household uses. (Witco Chemical Ltd).

ulframin 33. Sodium olefine sulphonate and sodium ether sulphate as a clear amber liquid. Applications: Foaming agent and detergent for industrial and household liquid detergents. (Witco Chemical Ltd).

ulla. Sulphameter. Applications: Antibacterial. (A H Robins Co Inc).

ulmet. Industrial detergent. (Crosfield Chemicals).*

ulphamagna. A proprietary preparation of streptomycin sulphate, phthalyl sulphathiazole, sulphadiazine and activated attapulgite. (Wyeth).†

ulphamezathine. A proprietary preparation containing sulphadimidine sodium. (ICI PLC).*

Sulphammonium. A solution of sulphur in liquid ammonia to form a purple solution.†

Sulphanil Black. See SULPHUR-BLACK T.

Sulphanil Brown. A dyestuff obtained by converting 2, 4-dinitro-4-amino diphenylamine into its sulphonic acid, and heating this product with sodium sulphide and sulphur.†

Sulphanilic Acid. Aniline-*p*-sulphonic acid, $C_6H_4NH_2SO_3H$.†

Sulpharsenol. See SULFARSENOL.

Sulphate Pulp. Wood pulp obtained by the treatment of wood with alkali liquors containing sodium sulphate.†

Sulphate Resin. See LIQUID RESINS.

Sulphate Ultramarine. Artificial ultramarine in which sodium sulphate is used as a constituent. See ULTRAMARINE.†

Sulphated Oils. See TURKEY RED OILS.

Sulphatine. A fungicide. It is a mixture of 73 per cent sulphur, 20 per cent lime, and 7 per cent copper sulphate. It is used against black rot.†

Sulphatol A and TL/B. Anionic surfactant with a blend of cations and a sulphated fatty alcohol as anion. Paste or liquid form. Applications: Detergent for industrial applications including textiles and leather. (Standard Chemical Company).

Sulphatol B6. Ammonium/triethanolamine lauryl sulphate, derived from the C12-C14 fraction of coconut/palm kernel fatty alcohol. Clear, golden yellow viscous liquid. Applications: Raw material for all types of liquid shampoos and bubble baths. (Efkay Chemicals Ltd).

Sulphatol CL. Sodium sulphated fatty alcohol in paste form. Applications: Anionic surfactant which is stable to hard water and disperses lime soaps, used for various industrial applications including textiles and leather. (Standard Chemical Company).

Sulphatol LS3. Combined sulphated fatty alcohol and nonionic surfactant in liquid form. Applications: Detergent and washing off liquid for textiles etc. (Standard Chemical Company).

Sulphatol LX/B. Sodium sulphated fatty alcohol in paste form. Applications: Detergent, wetting, levelling and softening agent for industrial applications including textiles and leather. (Standard Chemical Company).

Sulphatol PD/B. Potassium sulphated fatty alcohol in paste form. Applications: Detergent with good water solubility, for industrial applications including textiles and leather. (Standard Chemical Company).

Sulphatol 33. Sodium lauryl sulphate, derived from the C12-C14 fraction of coconut/palm kernel fatty alcohol. Clear liquid or thin paste. Applications: Raw material for liquid cream and egg shampoos; emulsifier for cosmetic products. (Efkay Chemicals Ltd).

Sulphatol 33 MO. Monoethanolamine lauryl sulphate, derived from the C12-C14 fraction of coconut/palm kernel fatty alcohol. Clear pale yellow liquid. Applications: Raw material for clear, oil and other liquid shampoos. (Efkay Chemicals Ltd).

Sulphatriad. A proprietary preparation of sulphathiazole, sulphadiazine and sulphamerazine. An antibiotic. (May & Baker Ltd).*

Sulphesatyd. A substance stated to be 3-thiooxindole.†

Sulphex. A proprietary preparation of sulphathiazole and hydroxyamphetamine hydrobromide. (Smith Kline and French Laboratories Ltd).*

Sulphide Dyestuffs. A class of dyestuffs prepared by the fusion of organic amines and other substances with sulphur and sodium sulphide. They are used for cotton dyeing, and are usually fixed by oxidizing agents.†

Sulphide of Arsenic, Yellow. See ORPIMENT.

Sulphide White,. See LITHOPONE.

Sulphiformin. Formaldehyde-sulphurous acid, $HO.CH_2.SO_3H$. An antiseptic. A 1 per cent solution has been used for spraying vines.†

Sulphine. See PRIMULINE.

Sulphine Brown. (Cattu Italiano). A dyestuff obtained by the action of sodium polysulphides upon oils, fats, c fatty acids. Dyes cotton dark brown, changing to reddish-brown by oxidation.†

Sulphite Carbon. A decolourizing carbon used for sugar juices. It is prepared fr sulphite-cellulose liquors.†

Sulphite Pulp. Wood pulp obtained by means of calcium bisulphide. It is ma by digesting the disintegrated wood under pressure with the calcium bisulphite, which gives a mass of cellulose fibres free from lignocellulos amounting to about 45 per cent of the wood.†

Sulphite Turpentine. A by-product obtain from the pulping of spruce by the sulphite process. The main constituent is *p*-cymene.†

Sulphite Turpentine Oil. (Cellulose Turpentine Oil). A by-product obtaine in the manufacture of cellulose. When decolourized it resembles turpentine oil.†

Sulpho Blacks. See CROSS DYE BLACKS.

Sulphocarbon Oil. See SANSE.

Sulphocol. See THIOCOLL.

Sulphoform. Triphenylstibine sulphide, $(C_6H_5)_3SbS$.†

Sulphol. Sulphur dyestuffs. (James Robinson & Co Ltd).

Sulphonated Oils. See TURKEY RED OILS.

Sulphone Acid Blue B. A tetrazo dyestuff. It dyes wool from a bath containing sodium sulphate and acetic acid.†

Sulphonol. A range of acid dyes. Applications: Dyeing of wool and similar fibres. (Yorkshire Chemicals Plc).*

Sulphophone. (Sulfopone). A trade mark for a mixture of zinc sulphide and calcium sulphate. It is an analogous product to lithopone.†

Sulphosol. Solubilized sulphur dyestuffs. (James Robinson & Co Ltd).

Sulphourea. Thiourea, CH_4N_2S.†

Sulphramin B and TPB. Alkylaryl sulphonic acid in liquid form. Applications: Emulsifier for powder and liquid detergents. (Witco Chemical Ltd).

Sulphurated Antimony. (Antimony Crocus, Saffron of Antimony). A mixture of antimony pentasulphide, Sb_2S_5, with a little oxide, Sb_4O_6, and some free sulphur. Formerly used in making tartar emetic.†

Sulphurated Oil. See BALSAM OF SULPHUR.

Sulphurated Potash. (Liver of Sulphur). A mixture of sulphides, mainly $K_2S_2O_3$, and K_4S_3, obtained by heating potassium carbonate with one-half its weight of sulphur. When fresh and carefully prepared it is the colour of liver, and was called liver of sulphur. It is sometimes used in the form of ointment. Calcium sulphide, CaS, is also called liver of sulphur, and is used in the leather industry as a depilating agent.†

Sulphur Auratum. Antimony sulphide.†

Sulphur Black T. (Thional Black, Katigene Black, Pyrrol Black, Thiogene Blue, Sulphanil Black, Pyrogene Blacks and Blues, Melanogen Black). Dyestuffs of the same class as Immedial black V. They are prepared by the action of sodium polysulphides upon various aminooxyderivatives of diphenylamine. They dye from a sulphide bath.†

Sulphur Black T Extra. See DINITROPHENOL BLACK.

Sulphur Blue B. (Sulphur Blue L Extra; Sulphur Brown G, 2G; Sulphur Cutch R, G; Sulphur Corinth B; Sulphur Indigo B). Dyestuffs of a similar type to sulphur black T (*qv*).†

Sulphur Blue L Extra. See SULPHUR BLUE B.

Sulphur Brown G, 2G. See SULPHUR BLUE B.

Sulphur Corinth B. See SULPHUR BLUE B.

Sulphur Cutch R, G. See SULPHUR BLUE B.

Sulphuretted Hydrogen. Hydrogen monosulphide, H_2S.†

Sulphur, Flour. See FLOUR OF SULPHUR.

Sulphur Gold. Antimony pentasulphide, Sb_2S_5. Used for vulcanizing and imparting a red colour to rubber.†

Sulphur Hypochlorite. A mixture of sulphur and sulphur chloride. Used in rubber vulcanizing.†

Sulphuric Ether. (Phosphoric Ether). Diethyl ether.†

Sulphur Indigo B. See SULPHUR BLUE B.

Sulphurion. Colloidal sulphur.†

Sulphurite. A name applied to a sulphur from Java, which contained 29 per cent arsenic.†

Sulphur, Liver of. See SULPHURATED POTASH.

Sulphur, Nacreous. See MOTHER OF PEARL SULPHUR.

Sulphur Olive Green. See THIOCHEM SULPHUR GREEN.

Sulphur Olive Oils. A name for the oil dissolved out from residual olive oil cake by means of carbon disulphide. It is also called sulphocarbon oil. It is rich in stearin. Also see SANSE.†

Sulphur, Ruby. See REALGAR.

Sulphur Soap. Usually a yellow medicated soap to which has been added about 10 per cent powdered sulphur.†

Sulphur Waste. The residue from the distillation of iron pyrites.†

Sulphur Yellow S. See NAPHTHOL YELLOW S.

Sulsol. A proprietary trade name for a colloidal sulphur preparation for horticultural purposes.†

Sultan Red 4B. See BENZOPURPURIN 4B.

Sumac. See SUMACH.

Sumacel. A diatomaceous earth containing 80 per cent. SiO_2, 5-3 per cent. Fe_2O_3 and Al_2O_3, 2.02 per cent. CaO, and 8.16 per cent. H_2O. It is stated to be suitable as a filtering medium for sugars.†

Sumach. (Sumac). The dried and finely powdered leaves and shoots of species of *Rhus*. Used for tanning leather, also for dyeing and printing, Sicilian sumach consists of the leaves of *Rhus coriaria*. The material is often imported in the form of powder containing from 25-28 per cent tannin. It is often adulterated with leaves of *Pistacia lentiscus.* Venetian or Turkish sumach consists of the leaves of *Rhus cotinus* (the wood of this tree gives young fustic), and contains 17 per cent tannin. American sumach is obtained from varieties of *Rhus,* chiefly *Rhus glabra,* and usually contains about 25 per cent tannin. Virginian sumach consists of the leaves of *Rhus typhina.* French sumach is obtained from *Coriari myrtifolia,* and contains 15.6 per cent tannin. Cape sumach is from the leaves of *Colpoon compressum,* and contains 23 per cent tannin, and Russian sumach comes from *Arctostaphylos uva-ursi,* and contains 14 per cent tannin.†

Sumach Wax. See JAPAN TALLOW.

Sumalban. Alban obtained from Sumatra gutta-percha.†

Sumaphos. A mixture of diatomaceous earth and acid phosphate, containing 36.22 per cent. P_2O_5.†

Sumatra Wax. See JAVA WAX.

Sumet Processed Lead. An alloy of 70-80 per cent copper and 15-30 per cent lead.†

Sumibond PA. Phenolic resin. Applications: Adhesives. (Sumitomo Bakelite, Japan).*

Sumicool. EVA resin. Applications: Mats. (Sumitomo Bakelite, Japan).*

Sumiflex. Polyvinyl chloride family. Applications: Moulding compounds. (Sumitomo Bakelite, Japan).*

Sumikon. A proprietary range of phenolic moulding materials. (Sumitomo Bakelite, Japan).*

Sumikon AM. Diallyl phthalate resin. Applications: Moulding compounds. (Sumitomo Bakelite, Japan).*

Sumikon EM, EME. Epoxy resin. Applications: Moulding compounds. (Sumitomo Bakelite, Japan).*

Sumikon IM. Polyimide resin. Applications: Moulding compounds. (Sumitomo Bakelite, Japan).*

Sumikon PM. Phenolic resin. Applications: Moulding compounds. (Sumitomo Bakelite, Japan).*

Sumikon TM. Polyester resin. Applications: Moulding compounds. (Sumitomo Bakelite, Japan).*

Sumikon VM. Polyvinyl chloride resin. Applications: Moulding compounds. (Sumitomo Bakelite, Japan).*

Sumilac PC. Phenolic resin. Applications: Industrial resins. (Sumitomo Bakelite, Japan).*

Sumilite CEL. Composite resin. Applications: Sheets. (Sumitomo Bakelite, Japan).*

Sumilite EI. Epoxy resin. Applications: Laminate materials. (Sumitomo Bakelite, Japan).*

Sumilite EL. Epoxy resin. Applications: Laminated sheets. (Sumitomo Bakelite, Japan).*

Sumilite ELC. Epoxy resin. Applications: Copper clad laminates. (Sumitomo Bakelite, Japan).*

Sumilite FS. PES. PEI. resin. Applications: Sheets. (Sumitomo Bakelite, Japan).*

Sumilite IL. Polyimide resin. Applications: Laminated sheets. (Sumitomo Bakelite, Japan).*

Sumilite ILC. Polyimide resin. Applications: Copper clad laminates. (Sumitomo Bakelite, Japan).*

Sumilite ILI. Polyimide resin. Applications: Laminate materials. (Sumitomo Bakelite, Japan).*

Sumilite NS. Polypropylene resin. Applications: Sheets. (Sumitomo Bakelite, Japan).*

Sumilite PL. Phenolic resin. Applications: Laminated sheets. (Sumitomo Bakelite, Japan).*

Sumilite PLC. Phenolic resin. Applications: Copper clad laminates. (Sumitomo Bakelite, Japan).*

Sumilite Resin PR. Resorcinol resin. Applications: Adhesives. (Sumitomo Bakelite, Japan).*

Sumilite Resin PR. Phenolic resin. Applications: Industrial resins. (Sumitomo Bakelite, Japan).*

Sumilite Resin PR. Epoxy resin. Applications: Industrial resins. (Sumitomo Bakelite, Japan).*

umilite STS. Polystyrene resin. Applications: Sheets. (Sumitomo Bakelite, Japan).*

umilite TFC. Polyimide and other resins. Applications: Copper clad laminates. (Sumitomo Bakelite, Japan).*

umilite TFP. Polyimide and other resins. Applications: Flexible printed circuit boards. (Sumitomo Bakelite, Japan).*

umilite VSL. Polyvinyl chloride and metal foil. Applications: Sheets. (Sumitomo Bakelite, Japan).*

umilite VSS. Polyvinyl chloride resin. Applications: Sheets. (Sumitomo Bakelite, Japan).*

uminet. Polyethylene resin. Applications: Materials for land improvements. (Sumitomo Bakelite, Japan).*

umipipe. Polyethylene resin. Applications: Materials for land improvement. (Sumitomo Bakelite, Japan).*

umitac EA. Epoxy resin. Applications: Adhesives. (Sumitomo Bakelite, Japan).*

umitac GA. Polyurethane resin. Applications: Adhesives. (Sumitomo Bakelite, Japan).*

Sumitac VA. Polyvinyl chloride resin. Applications: Adhesives. (Sumitomo Bakelite, Japan).*

Sumner's Reagents. For glucose determination. 3 , 5 - Dinitro - salicylic acid (10 grams) are dissolved in 500 cc warm water in a 1,000-cc flask and made up to 1,000 cc Sodium hydroxide (13.5 grams) is dissolved in 300 cc cold water in a 2-litre beaker, then 880 cc of the salicylic acid solution are added and mixed; 225 grams of potassium sodium tartrate (Rochelle salt) are added and dissolved by stirring, and the whole transferred to a bottle. Standard iron-alum solution is made by weighing 345 mg violet ferric ammonium sulphate and transferring to a 1,000-cc flask, adding 1 gram 3 : 5-dinitro-salicylic acid and then 500 cc water. The colour of the solution is equivalent to a glucose solution containing 1 gram treated by the Sumner method.†

Sumycin. A proprietary preparation of tetracycline base buffered with potassium metaphosphate. An antibiotic. (Squibb, E R & Sons Ltd).*

Sun Bronze. An alloy of from 40-60 per cent copper, 30-40 per cent tin, and 10 per cent aluminium. Used in jeweller's work. The name is also used for an alloy of from 50-60 per cent cobalt, 30-40 per cent copper, and 10 per cent aluminium.†

Sun Gold. See HELIOCHRYSIN.

Sunaptic Acids. High molecular weight naphthenic acids. Applications: Corrosion inhibitor, oil well drilling mud formulations, emulsifiers, foundry binders. (Sun Refining & Marketing Co).*

Sunaptol. Textile auxiliary chemicals. (ICI PLC).*

Sunaptol NP100. Nonylphenol ethoxylate nonionic surfactant in liquid form. Applications: Solubilization of essential oils and perfumes. (Pechiney Ugine Kuhlmann Ltd).

Sunaptol NP140. Nonylphenol ethoxylate nonionic surfactant in the form of a white waxy solid. Applications: Iodine complexing for iodofor sterilizers; production of self emulsifiable oleines. (Pechiney Ugine Kuhlmann Ltd).

Sunaptol NP350. Nonylphenol ethoxylate nonionic surfactant in the form of a white solid. Applications: Ready moulding into solid block for detergent blocks and tablets; emulsifier and stabilizer for vinyl acetate and acrylic polymer emulsions. (Pechiney Ugine Kuhlmann Ltd).

Sunaptol NP55. Nonionic surfactant of the nonylphenol ethoxylate type in liquid form. Applications: Emulsifier and intermediate used in natural waxes, mineral oils; ethoxysulphate manufacture; emulsifiable solvents for metal degreasing; general cleaning applications. (Pechiney Ugine Kuhlmann Ltd).

Sunaptol NP65 and NP70. Nonionic surfactants of the nonylphenol ethoxylate type in liquid form. Applications: Low temperature wool scouring; emulsifier for silicone and mineral oils; emulsifier and stabilizer for kerosine based hand cleaning gels. (Pechiney Ugine Kuhlmann Ltd).

Sunaptol NP80 and NP95. Nonylphenol ethoxylate nonionic surfactant in liquid form. Applications: Detergency, wetting and emulsifying agent used in wool scouring; metal cleansing; mineral oils, usually in combination with a hydrophobic surfactant. (Pechiney Ugine Kuhlmann Ltd).

Sundora. A proprietary trade name for cellulose acetate.†

Sungard. Sulisobenzone. Applications: Ultraviolet screen. (Miles Pharmaceuticals, Div of Miles Laboratories Inc).

Sunimac ECR. Epoxy resin. Applications: Industrial resins. (Sumitomo Bakelite, Japan).*

Sunimac GCR. Polyurethane resin. Applications: Industrial resins. (Sumitomo Bakelite, Japan).*

Sunnol. A range of alkyl aryl sulphonates. Applications: Used as scouring agents, dyeing assistants and in emulsion polymerization. (Harcros Industrial Chemicals).*

Sunolith. A proprietary trade name for a pigment containing 71 per cent barium sulphate and 29 per cent zinc sulphide. See also PONOLITH.†

Sunoxol. See QUINOSOL.

Sunproof. Blend of waxes for all types of stock to inhibit static atmospheric cracking and frosting. (Uniroyal).*

Sunshine. Resistant decorating colours for porcelain, bone china and earthenware. (Degussa).*

Suntei Tallow. A white sweetish fat expressed from the seeds of *Palaquium oleosum*.†

Suparen. Fermentation - derived rennet. (Pfizer Ltd).

Super-A. Chemical products for general industrial use. Applications: Organometallic chemicals, organic chemicals, catalysts, styrene butadiene polymers, urethane and epoxy polymers, urethane and epoxy curatives and activated carbons. (Anderson Development Company).*

Super A. Vitamin A. Applications: Vitam (anti-xerophthalmic). (The Upjohn Co).

Super AD-IT. Di(phenylmercuric) dodecenyl succinate. Applications: Preservative and fungicide for aqueous coating compositions. (Nueodex Inc).*

Superam. A fertilizer obtained by neutralizing the acids of ordinary supe phosphate with ammonia gas.†

Superargol. (Tartar Cake, Substitute of Tartar). Preparations containing simply acid sodium sulphate. Others contain oxalates, and a few tartaric acid and sulphuric acid.†

Super-ascoloy. A ferrous alloy containing 8 per cent nickel and 18 per cent chromium.†

Superba. A proprietary trade name for a carbon black.†

Superbasique Metal. A modification of cas iron. It is resistant to alkalis.†

Superbeckacite. A proprietary trade name for pure phenolic varnish and lacquer resins.†

Superbrillantoline. Hydromethylabietate. Applications: Raw material for cosmetics. (Laserson & Sabetay).*

Super Bronze. An alloy of from 57-69 per cent copper, 1.2-5.1 per cent aluminium 1.3-2 per cent iron, 21-37 per cent zinc, and 3-3.2 per cent manganese.†

Super Cat. Metal salts of organic acids. Applications: Fuel oil additives. (Nueodex Inc).*

Super Cement. An ordinary Portland cement to which has been added a waterproofing material.†

Super-cliffite. Explosives. No. 1 contains 10 per cent nitro-glycerin, 1 per cent collodion cotton, 60 per cent ammonium nitrate, 16 per cent sodium chloride, 11 per cent ammonium oxalate, and 6 per cent wood meal. No. 2 has the sodium chloride increased to 20 per cent. and the ammonium oxalate reduced to 6 per cent.†

Super D. Cod liver oil. Applications: Sourc of vitamins A and D. (The Upjohn Co).

Super Die. A proprietary trade name for a tool steel containing 10.5 per cent chromium, 1 per cent tungsten, and 1 per cent silicon.†

Super-excellite. An explosive containing 73.5-77 per cent ammonium nitrate, 6.5-8 per cent potassium nitrate, 2-4 per cent wood meal, 3.5-5 per cent nitro-glycerin, and 9-11 per cent ammonium oxalate.†

Superfast Power Pack. Liquid epoxy resin and hardener. Applications: Fast setting system - general purpose adhesive. (Wessex Resins & Adhesives Ltd).*

Superfiltchar. A proprietary product. It is an active decolourizing carbon made from sawdust.†

Superfloc. High MW polymers based on acrylamide in powder, solution or emulsion form. Nonionic, anionic, cationic. (Cyanamid BV).*

Superfloc C507. Melamine formaldehyde resin. (Cyanamid BV).*

Superfloc C521. Monomethylamine-epichlorohydrin condensation products. (Cyanamid BV).*

Superfloc C573-C577-C567. Demethylamine-epichlorohydrine condensation products. (Cyanamid BV).*

Superforcite. A Belgian gelatin dynamite containing 64 per cent nitro-glycerin.†

Super Glue. Cyanoacrylate adhesive. Applications: Bonding two similar or dis-similar materials in seconds. (Novest Inc).*

Supergreen. Lawn fertilizer. (May & Baker Ltd).

Super Green. Various granular fertilizer blends. Applications: fertilizers for lawns, gardens and flowers. (Horn's Crop Service Center).*

Super Hartolan. Distilled lanolin alcohols BP/DAB. (Croda Chemicals Ltd).

Superinone. Tyloxapol. Applications: Detergent. (Sterling Drug Inc).

Superior Alloy. A heat-resisting alloy containing 78 per cent nickel, 19.5 per cent chromium, 2 per cent manganese, and 0.5 per cent iron.†

Superite. An explosive consisting of 80-84 per cent ammonium nitrate, 9-11 per cent potassium nitrate, 2-5 per cent starch, and 3.5-4.5 per cent nitro-glycerin.†

Superjet. A carbon black pigment. (Pfizer International).*

Super-karma. An alloy wire containing 80 per cent nickel and 20 per cent chromium.†

Super-kolax No. 2. An explosive containing nitro-glycerin, collodion cotton, potassium nitrate, barium nitrate, wood meal, starch, and ammonium oxalate.†

Super-ligdynite. A coal mine explosive containing from 15-17 per cent nitro-glycerin, 15-17 per cent ammonium nitrate, 23-25 per cent sodium nitrate, 10-12 per cent flour, 19-21 per cent wood pulp, and 9-11 per cent sodium chloride.†

Superlit. A proprietary synthetic resin.†

Superlite. Tin (IV) oxide. (Keeling & Walker Ltd).

Superlith. See PENOLITH.

Superlock. Anaerobic. Applications: Seal and lock parts in place. (Devcon Corporation).*

Superloid. High viscosity ammonium alginates. Applications: Latex creaming, tyre sealant, ceramic binder and mould release. (Kelco, Div of Merck & Co Inc).*

Super Lubracon. A water based food industry lubricant, with cleaning and anti-microbial properties. Applications: Bottle conveyor lubricant, can conveyor lubricant, keg conveyor lubricant, crate conveyor lubricant. (Harshaw Chemicals Ltd).*

Super Moss Killer & Lawn Fungicide. Dichloropren fungicide/moss killer. (Murphy Chemical Ltd).

Super Mosstox. A liquid formulation containing 34% dichlorophen. Applications: Controls moss in fine turf, footpaths, hard tennis courts, playgrounds, roof and other affected hard surfaces. (Burts & Harvey).*

Super Nickel. A proprietary trade name for alloys of 20-30 per cent nickel with 70-80 per cent copper. They are corrosion resisting.†

Superneutral Metal. A silicon-iron alloy. Used for nitric acid plants.†

Superpalite. (Diphosgene, Green Cross Gas) . Trichloromethyl chloroformate, $ClCOOCCl_3$. A military poison gas.†

Superphosphate. (Mineral Superphosphate, Superphosphate of Lime). A fertilizer. It consists of mono-calcium phosphate, $CaH_4(PO_4)_2$, mixed with calcium sulphate, and contains 25-28 per cent soluble phosphate.†

Superphosphate, Ammonium. See NITRO-PHOSPHATE.

Superphosphate, Mineral. See SUPER-PHOSPHATE.

Superphosphate of Lime. See SUPER-PHOSPHATE.

Superprill. Urea (prilled). Applications: fertilizer. (Columbia Nitrogen Corporation).*

Super-Quench. Metal working oil. (Chevron).*

Super-rippite. A smokeless powder containing from 51-53 parts nitroglycerin, 2-4 parts nitro-cotton, 13.5-15.5 parts potassium nitrate, 15.5-17.5 parts dried borax, and 7-9 parts potassium chloride.†

Super-rippite No. 2. An explosive for coal mines containing 51 per cent nitro glycerin, 3 per cent nitro-cotton, 11 per cent potassium perchlorate, 24 per cent borax, and 10 per cent potassium chloride.†

Supersat. Hydrogenated lanolin. Applications: Odourless, white, highly stable, of particular value in products where a light fragrance or colour is needed. Recommended for a wide range of products - clear gels, makeup removers, polish removers, creams, lotions, shaving creams, ointments and suppositories. (RITA Corporation).*

Supersevtox. Selective herbicide. (FBC Ltd).

Super Solvitax. Cod liver oil BP veterinary. Applications: Dietary supplement with vitamins A and D. (Marfleet Refining Co).*

Super-Sorb "C" Water Absorbant. Copolymer acrylamide sodium acrylate. Applications: Increases water holding capacity of soils and horticultural media. Used in greenhouses, nurseries and landscaping. (Aquatrols Corporation of America).*

Super-Sorb "F" Water Absorbant. Copolymer acrylamide sodium acrylate. Applications: Holds water as a gel around plant roots. Used for transplanting and tranporting of bare root plant material, in reforestation, landscaping and crop production. (Aquatrols Corporation of America).*

Supersorbon. Moulded activated carbon for the recovery of solvents. (Degussa).*

Supersoy. A high grade soya bean flour.†

Superspray. Chlorhexidine teat spray. (Ciba-Geigy PLC).

Super Sterol Ester. C10-C39 carboxylic acid ester of lanolin sterols. (Croda Chemicals Ltd).

Superstyrex. See POLYSTYRENE.

Super-sulphur. Thiuram disulphide, a vulcanization accelerator.†

Super Sulphur No. 1. A proprietary rubber vulcanization accelerator. It is the oxidized zinc salt of dimethyl dithiocarbamic acid.†

Super Sulphur No. 2. A proprietary rubber vulcanization accelerator. It is lead dimethyldithiocarbamate.†

Superthin. Tetrafilcon A. Applications: Contact lens material. (American Optical Corp).

Supertox. Selective weedkiller. (May & Baker Ltd).

Superturpentine. Spirits of turpentine specially rectified *in vacuo*. It boils at 155° C., and distils completely below 160° C.†

Super Verdone. Garden selective herbicide. (ICI PLC, Plant Protection Div).

Super Weedex. Total weedkiller. (Murphy Chemical Ltd).

Superwipes. Chlorhexidine/cetrimide udder wipes. (Ciba-Geigy PLC).

Suprac. Decolourizing and absorptive activated carbons. Applications: Dry cleaning and chemical purification. (Lancashire Chemical Works Ltd).*

Supracen. A range of acid wool dyes with outstanding levelling power and very good lightfastness. (Bayer & Co).*

Supracet. Disperse dyestuffs. Applications: Dyeing of acetate, triacetate and nylon fibres and blends. (Holliday Dyes & Chemicals Ltd).*

Supracide. Organophosphorus insecticide. (Ciba-Geigy PLC).

Supradyn. Proprietary multivitamin preparation containing the B complex vitamins, vitamins A, C, D and E, biotin, nicotinamide, folic acid, (in addition to minerals and trace elements Ca, Fe, Mg, Mn, Cu, Zn, Mo and P). (Roche Products Ltd).*

Suprafix Paste. Vat dyes. (Bayer UK Ltd).

Suprafrax. A clay with a high percentage of alumina. It is used as a furnace lining.†

Supramica. Same as MYKROY/MYCALEX except manufactured with synthetic mica (man made) which does not contain the impurities or Hydroxel ion's (water). Applications: Same as MYKROY/MYCALEX except it is used in many high frequency and high temperature applications which require low outgassing properties. (Mykroy/Mycalex).*

Supramin. A range of acid wool dyestuffs with superior fastness to water, washing and perspiration. (Bayer & Co).*

Supranol. Acid wool dyes. Applications: Used in dyeings which are fast to washing, water and seaweed. (Bayer & Co).*

Suprarenin. Epinephrine bitartrate. Applications: Adrenergic. (Sterling Drug Inc).

Supraresen. The residue obtained when dammar is prepared for use in lacquers and is soluble in hydrocarbons. Used in the varnish industry.†

Suprasec. Isocyanates for general application. (ICI PLC).*

Suprathion. Active ingredient: methidathion; S-2,3-dihydro-5-methoxy-2-oxo-1,3,4-thiadiazol-3-ylmethyl-0,0-dimethyl phosphorodithioate. Applications: Organophosphorous insecticide with high degree of insecticidal activity; also efficient in controlling scales. (Makhteshim Chemical Works Ltd).†

Suprefact. Buserelin acetate; luteinizing hormone-releasing factor. Applications: Gonad-stimulating principle. (Hoechst-Roussel Pharmaceuticals Inc).

Suprex. See CATALPO.

Suprex White. A highly purified precipitated calcium carbonate for use as a rubber filler in the place of blanc fixe. It has a specific gravity of 2.7.†

Suprexcel. Fast-to-light direct cotton dyestuffs. (Holliday Dyes & Chemicals Ltd).

Suprilent. Isoxsuprine. Applications: Promotes blood flow rate. (Duphar BV).*

Suprofix. High speed photographic fixer. (May & Baker Ltd).

Suprol. Photographic developer. (May & Baker Ltd).

Supronal Preparations. Sulphonamide mixture. Applications: For the treatment of bacterial infections. (Bayer & Co).*

Supronic. Non-ionic surfactant. (ABM Chemicals Ltd).*

Supronic B10, B25, B5O, B75 and B1OO. A proprietary range of low-foam surface-active agents. They are polyoxy alkylated polyalkyleoe glycols. (Glover (Chemicals) Ltd).†

Supronic E800. A solid, non-ionic surface-active agent. It is a polyoxy-ethylene polyoxypropylene condensate.†

Supronics. Low foam surface active agents. (ABM Chemicals Ltd).

Surbex T. A proprietary multi-vitamin preparation. (Abbott Laboratories).*

Sure-Curd (Suparen). A standardized solution of fermentation derived milk clotting enzyme elaborated by *Endothia parasitica*. Applications: Used in the manufacture of cheese, especially Swiss and Italian varieties. (Pfizer International).*

Suresperse. Deposit control agents for air washer systems. (Ashland Chemical Company).*

Surexin. Pyrinoline. Applications: Cardiac depressant. (McNeil Pharmaceuticals, McNEILAB Inc).

Surf Ac 820. Alkylaryl polyethoxyethanol and n-butanol. Applications: Non-ionic biodegradable surfactant. (Drexel Chemical Company).*

Surfacaine. Cyclomethycaine sulphate. Applications: Anaesthetic. (Eli Lilly & Co).

Surfactant N-42. Nonylphenol ethoxylate nonionic surfactant in the form of a clear oil. Applications: General surfactant. (Rohm & Haas (UK) Ltd).

Surfactant XQS20. Phophate ester in free acid form. Aqueous solution. Applications: Wetting agent and detergent; textile lubricant; antistatic agent; industrial emulsifiers. (Rohm & Haas (UK) Ltd).

Surfageen. Anionic surfactant of the alkyl ether phosphate type in solid form. Applications: Emulsifier for emulsion polymerization; deinking of waste paper; rust inhibitor. (Chem-Y, Fabriek van Chemische Producten BV).

Surfageen S30. Fatty alcohol ether sulphosuccinate as a clear liquid. Applications: Detergent, foaming, wetting and emulsifying agent used as a mild raw material for shampoos and foam baths. (Chem-Y, Fabriek van Chemische Producten BV).

Surfak. Docusate calcium. Applications: Stool softener. (Hoechst-Roussel Pharmaceuticals Inc).

Surf-A-Seis. A suface explosive. Applications: It is used for surface energy sources in portable seismic operations. (Hercules Inc).*

Surfonic. A series of p-nonylphenol ethoxylates. Applications: Surfactants in agricultural chemicals, industrial cleaners, heavy-duty detergents, paper industry. (Texaco Chemical Co).*

Surfonic N. Series of nonionic surfactants of the nonylphenol ethoxylate type. Applications: All types of detergency and cleaning; ceramics and concrete; dust control; paper; wallpaper removal; photographic film developing; fire fighting; emulsion polymerization; cutting oils; drilling muds; rubber latex as stabilizers. (Texaco Ltd).

Surfynol. A range of acetylenic glycols. Applications: Used as wetting agents with very low foam for paints. (Harcros Industrial Chemicals).*

Surgam. Tiaprofenic acid. Applications: Rheumatoid arthritis, osteo-arthritis. (Roussel Laboratories Ltd).*

Surgical Solution. See PHYSIOLOGICAL SALT SOLUTION.

Surgicel. Oxidized regenerated cellulose. Applications: Haemostatic. (Johnson & Johnson Products Inc).

Surlyn. A registered trade mark for a range of ionomer resins. Constituants of the range include the following:- *Surlyn 1555.* Possesses good flow properties. Used in injection moulding.

Surlyn 1558. A 25-mesh resin used in rotational moulding.

Surlyn 1559. Used in injection moulding.

Surlyn 1560. Used in injection moulding where maximum clarity is required.

Surlyn 1603. Used in film extrusion when good slip is required.

Surlyn 1605. Used as an extruded coating on paper.

Surlyn 1652. Used as an extruded coating on foil.

Surlyn 1707. Used in the extrusion of high clarity sheet and for blow moulding.

Surlyn 1800. Used as a tough coating for wires and cables. (Du Pont (UK) Ltd).†

Surmabond 'Lining'. Pigmented solvent-free epoxy systems. Applications: Seamless flooring and tank lining composition. (Surmak Products Ltd).*

Surmabond 'Roadway'. Tar modified solvent-free epoxy system with special aggregate. Applications: Light weight screed for footbridges, concrete floors. (Surmak Products Ltd).*

Surmabond 'Screeding'. Pigmented solvent-free epoxy system with special aggregates. Applications: Chemical resistant heavy duty, light weight screed and concrete repair material. (Surmak Products Ltd).*

Surmafil. Bituminous mixture with liquid resins and aggregates. Applications: Instant repair of roads and parking areas, cold applied. (Surmak Products Ltd).*

Surmaglaze U12. Moisture cured polyurethane. Applications: Clear or coloured dust-proofing sealer for concrete, stone and timber floors. (Surmak Products Ltd).*

Surmaplast. Solvent-free epoxy knifing systems. Applications: Gap filling of blowholes in castings, repair of petrol, oil and water tanks. Quick setting. (Surmak Products Ltd).*

Surmaseal 101. Chlorinated rubber paint. Applications: Acid resistant paint for steelwork, concrete floors and swimming pools. (Surmak Products Ltd).*

Surmaseal 102. Solvent-based pigmented epoxy paint, polyamide cured. Applications: Chemical and wear resistant floor coating, also for corrosion protection of steel. (Surmak Products Ltd).*

Surmatar. Solvent-based epoxy pitch coating. Applications: Heavy duty anti-corrosive coating. (Surmak Products Ltd).*

Surmontil. A proprietary preparation of trimipramine. An antidepressive drug. (May & Baker Ltd).*

Surolan. A proprietary preparation containing miconazole nitrate, prednisolone. Applications: Veterinary ear and skin infections (cats and dogs). (Janssen Pharmaceutical Ltd).*

Surophosphate. (Dasag). A fertilizer made from sewage, other waste material, and peat. It is of German origin.†

Surrogate, Saffron. See VICTORIA YELLOW.

Susadrin. Nitroglycerin. Applications: Vasodilator. (Merrell Dow Pharmaceuticals Inc, Subsidiary of Dow Chemical Co).

Suscardia. A proprietary preparation of isoprenaline hydrochloride, used as a cardiac stimulant. (Pharmax).†

Suscon Blue. Insecticide containing chlorpyrifos. Applications: For the long term control of soil insect pests in sugar cane and pineapples. (Incitec International).*

Suscovax. Pig samonellosis vaccine. (The Wellcome Foundation Ltd).

Susini. An alloy of aluminium containing from 1.5-4.5 per cent copper, 0.5-1.5 per cent zinc, and 1-8 per cent manganese.†

Suspendex. Expanded polystyrene foam loose-fill used for cushioning and packaging applications. (Dow Chemical Co Ltd).*

Sus-Phrine. Epinephrine. Applications: Adrenergic. (Berlex Laboratories Inc, Subsidiary of Schering AG).

Sustac. A proprietary preparation of glyceryl trinitrate. A vasodilator used in angina pectoris. (Pharmax).†

Sustamycin. A proprietary preparation of TETRACYCLINE hydrochloride. An antibiotic. (MCP Pharmaceuticals).†

Sustane 1-F. Flaked butylated hydroxyanisole, FCC grade. Used as an antioxidant. Applications: stabilization of fats and oils for food use. (UOP Inc)*

Sustilan N. Fibre preserving agent. Applications: Dyeing and printing auxilliary. (Bayer & Co).*

Sutermeister's Stain. For paper (a) Contains 1.3 grams iodine and 1.8 grams potassium iodide in 100 cc water, and (b) consists of a clear saturated solution of calcium chloride.†

SVC. Acetarsol vaginal compound. (May & Baker Ltd).

SW 5063. Racephenicol. Applications: Antibacterial. (Sterwin Chemicals Inc).

Swale Powder. An explosive containing potassium perchlorate, nitro-glycerin, collodion cotton, ammonium oxalate, wood meal, and a little nitro toluene.†

Swalite. An explosive for coal mines, similar to Swale powder (*qv*).†

Swan. A paper based grade of TUFNOL industrial laminates. (Tufnol).*

Swedelec. A Swedish charcoal iron. It has a high magnetic permeability.†

Swedish Factory Tar. A tar obtained from waste wood in charcoal kilns, as a by-product in charcoal burning.†

Swedish Green. See SCHEELE'S GREEN.

Swedish Liquid Resin. See TALLOEL.

Swedish Turpentine. See TURPENTINE.

Sweeta. Saccharin. Applications: Pharmaceutical aid. (E R Squibb & Sons Inc).

Sweet Bark. (Sweet Wood Bark, Eleuthera Bark). Cascarilla, used for extracting cascarilla oil and as an ingredient in insecticides, etc.†

Sweetex. Low calorie sweetener. (The Boots Company PLC).

Sweetex Plus. Low calorie sweetener. (The Boots Company PLC).

Sweet Nitre. See SPIRIT OF SWEET NITRE.

Sweet Wine Spirit. See SPIRIT OF SWEET WINE.

Sweet-water. Consists of glycerin and water, obtained in the distillation of crude glycerol.†

Sweetzyme. An immobilized glucose isomerase produced from a selected strain of Bacillus coagulans. Applications: Sweetzyme Type Q developed specially for long-term use in a continuous fixed-bed column process for production of fructose syrup. Sweetzyme Type A is used for batch operation which however is characterized by much higher residence times and enzyme cost compared to the fixed-bed operation. (Novo Industri A/S).*

Swipe. Cereal herbicide. (Ciba-Geigy PLC).

Swiss Polyamid Grilon. Copolyamide. Applications: Tape weaving, filter fabric and filter manufacturing process, string, embroidery etc. (EMS-Chemie AG).*

Swiss Polyamid Grilon. Polyamide 6. Applications: Textile floor coverings, clothes, non wovens, technical fabrics and paper felts. (EMS-Chemie AG).*

Swiss Polyester Grilene. Polyester. Applications: Woven and knitted fabrics, home textiles and sewing threads, technical applications, various non-wovens and fibre-fill. (EMS-Chemie AG).*

Syanthrose. Levulin, $C_6H_{10}O_5$.†

Sybol. Garden insecticide. (ICI PLC).*

Sycorin. See SACCHARIN.

Sycose. See SACCHARIN.

Sykose. See SACCHARIN.

Syl. A proprietary preparation of dimethicone 350, benzalkonium solution and nitrocellulose used in derroatology as an antibacterial agent. (Lloyd, Hamol).†

Sylade. Silage preservative. (ICI PLC).*

Sylgard. Silicone elastomer potting components. (Dow Corning Ltd).

Syllact. Psyllium Husk. Applications: Laxative. (G D Searle & Co).

Syl-off. Silicone release coatings for paper and films. (Dow Corning Ltd).

Syloid 72. A proprietary trade name for a silica gel for addition (2 per cent.) to plasticized vinyls to prevent plate out.†

Sylopal. A proprietary preparation of DIMETHICONE, magnesium oxide and aluminium hydroxide. A gastro-intestinal sedative. (Norton, H N Co).†

Sylphane S. PVC shrink film. Applications: Sales display and bundling. (UCB nv Film Sector).*

Sylphrap. A proprietary trade name for a regenerated cellulose transparent sheet.†

Syltherm. Silicone heat transfer fluids. (Dow Corning Ltd).

Sylvan. α-Methylfuran, C_5H_6O. A constituent of wood tar.†

Sylvania Cellophane. A proprietary trade name for regenerated cellulose.†

Sylvic Acid. Impure abietic acid.†

Sylvid. A registered trade mark for a range of silica fillers for plastics processing. (Grace, W R & Co).†

Symax. Laminates of nomex, presspaper, leatheroid, paper, melinex, mylar and kapton. Applications: Slot liner and closure material used in the manufacture and repair of electrical motors, transformers and other electrical equipment. (Fothergill Tygaflor Ltd).*

Symcor. Tiamenidien Hydrochloride. Applications: Antihypertensive. (Hoechst-Roussel Pharmaceuticals Inc).

Symel. Extruded silicone elastomer sleeving, fusible silicone rubber tapes and fabrics. Applications: High temperature electrical applications such as lead out wires, marker sleeves and peristaltic pumps. (Fothergill Tygaflor Ltd).*

Symmetrel. Amantadine hydrochloride. Applications: Antiviral. (Endo

Laboratories Inc, Subsidiary of E I du Pont de Nemours & Co).

Symmetrel. A proprietary preparation Of AMANTADINE hydrochloride, used in the treatment of parkinsonism. (Ciba Geigy PLC).†

Sympatol. A proprietary preparation of oxedrine tartrate used in the treatment of cardiac disorders. (Lewis Laboratories).†

Syn Lube. Water soluble, biodegradable lubricants. Applications: Fibre lubricants, textile softeners, yarn processing aids. (Milliken & Company).*

Synacril. Acrylic dyes. (ICI PLC).*

Synacthen. A proprietary preparation of tetracosactrin. (Ciba Geigy PLC).†

Synacthen Depot. Adrenocortical. (Ciba-Geigy PLC).

Synacto. Emulsifiers and corrosion preventives. (Exxon Chemical Ltd).*

Synadryn. A proprietary preparation of prenylamine lactate. Treatment of angina pectoris. (Hoechst Chemicals (UK) Ltd).†

Synalar. A proprietary preparation of fluocinolone acetonide with chinoform or neomycin sulphate for dermatological use. (ICI PLC).*

Synalar-HP. Fluocinolone acetonide. Applications: Glucocorticoid. (Syntex Laboratories Inc).

Synandone. A proprietary trade name for preparations of the acetonide of Fluocinolone. (ICI PLC).*

Synandrets. Methyltestosterone. Applications: Androgen. (Pfizer Inc).

Synandrol. Testosterone Propionate. Applications: Androgen. (Pfizer Inc).

Synandrol F. Testosterone. Applications: Androgen. (Pfizer Inc).

Synandrotabs. Methyltestosterone. Applications: Androgen. (Pfizer Inc).

Synanthic. Oxfendazole. Applications: Anthelmintic. (Syntex Laboratories Inc).

Synaqua. Water soluble resins. (Cray Valley Products Ltd).

Synasol. A proprietary trade name for a denatured ethyl alcohol.†

Synchrocept. Prostalene. Applications: Prostaglandin. (Syntex Laboratories Inc).

Synciliin. A proprietary preparation of potassium phenethicillin. An antibiotic. (Bristol-Myers Co Ltd).†

Synclyst. FCC catalysts. (Crosfield Chemicals).*

Syncrolube. Fatty acid esters. (Croda Chemicals Ltd).

Syncrowax. Synthetic waxes. (Croda Chemicals Ltd).

Syndite. An explosive consisting of 10-22 per cent nitro-glycerin, 0.1-0.3 per cent collodion cotton, 45-49 per cent ammonium nitrate, 7-9 per cent sodium nitrate, 2-5 per cent glycerin, 2-5 per cent starch, and 26-28 per cent sodium chloride.†

Syndraw. Water reducible lubricants containing no mineral oil. Additives include surfactants, lubricity additives, fatty acid soaps, synthetic corrosion inhibitors, biocides and polymers. (Franklin Oil Corporation (Ohio)).*

Synektan. Leather dyes. (ICI PLC).*

Synemol. Fluocinolone acetonide. Applications: Glucocorticoid. (Syntex Laboratories Inc).

Synerone. Testosterone Propionate. Applications: Androgen. (Merrell Dow Pharmaceuticals Inc, Subsidiary of Dow Chemical Co).

Synfluid. Synthetic base stocks for lubes and hydraulic fluids. (Chevron).*

Syngesterone. Progesterone. Applications: Progestin. (Pfizer Inc).

Syngestrets. Progesterone. Applications: Progestin. (Pfizer Inc).

Synkavit. Vitamin K analogue preparations. (Roche Products Ltd).

Synkayvite. Menadiol sodium diphosphate. Applications: Vitamin. (Hoffmann-LaRoche Inc).

Synmold. Phenol formaldehyde resin. Applications: Moulding powders. (Manchem Ltd).*

Synmold. Phenolic moulding powders. (Sterling Moulding Materials).*

Synocryl. Thermoplastic and thermosetting acrylic resins. (Cray Valley Products Ltd).

Synocryl 8205 and 821S. Proprietary trade names for hydroxyacrylics curing at 120° C. with melamine. Used for thermosetting flow enamels for the car industry. (Cray Valley Products Ltd).†

Synocure. Crosslinking acrylic resins: radiation curable prepolymers. (Cray Valley Products Ltd).

Synocure 867S. A proprietary acrylic resin used for coating metals, particularly aluminium. It is hydroxyl-functional. (Cray Valley Products Ltd).†

Synocure 868S. A proprietary flexible acrylic resin used for coating rigid surfaces. It is hydroxyl-functional. (Cray Valley Products Ltd).†

Synocure 869S. A proprietary fast drying acrylic resin used as a commercial wood finishing. (Cray Valley Products Ltd).†

Synogist. A proprietary preparation of sodium sulphosuccinate and undecylenic monoalkyl amide, used as a shampoo for dandruff. (Maltown).†

Synolac. Alkyds: unsaturated polyester resins: epoxy esters. (Cray Valley Products Ltd).

Synolide. Polyamide resins. (Cray Valley Products Ltd).

Synoplast. Mixed coal tar products with different viscosity. Applications: Synoplast-plastix used as a corrosion inhibitor. (Caramba Chemie GmbH).*

Synouryn. Dehydrated castor oil. (Akzo Chemie UK Ltd).

Synova. Mixed coal tar products with different viscosity with and without fillers. Applications: Synova-protective paint, Synova-roof paint, Synova-injection paint, Synova-agents. (Caramba Chemie GmbH).*

Synperonic. Non-ionic surface active agents. (ICI PLC).*

Synperonic NP10 and NP12. Nonylphenol eyhoxylat nonionic surfactants in liquid form. Application: Water-soluble detergents, detergent additives, solubilisers, dispersants and stabilisers. (ICI PLC, Petrochemicals & Plastics Div).

Synperonic NP13 and NP15. Nonylphenol ethoxylate nonionic surfactants in liquid or paste form. Application: Used in conjunction with an oil-soluble anionic surfactant, they are good emulsifiers for a range of solvents, agrochemical pesticides and herbicides. (ICI PLC, Petrochemicals & Plastics Div).

Synperonic NP20 and NP30. Nonylphenol ethoxylate nonionic surfactants in liquid or solid form. Application: Solubilising agents and emulsifiers or co-emulsifiers for highly polar substrates. (ICI PLC, Petrochemicals & Plastics Div).

Synperonic NP4, NP5 and NP6. Nonylphenol ethoxylate nonionic surfactants in liquid form. Application: Oil-soluble detergents and emulsifiers, used as intermediates for sulphation and phosphorylation to give anionic detergents, lubricants and antistatic agents. Emulsifying agents for wide range of oils, waxes and solids, compatible with all other surfactants. (ICI PLC, Petrochemicals & Plastics Div).

Synperonic NP8 and NP9. Nonylphenol ethoxylate nonionic surfactants in liquid form. Application: Water-soluble high performance detergents and wetting agents used e.g. in textile scouring. Emulsifiers for medium polarity oils and solvents. (ICI PLC, Petrochemicals & Plastics Div).

Synperonic N, NX, NXP and NDB. Nonylphenol ethoxylate nonionic surfactants in liquid form. Application: General purpose detergents and wetting agents for textile processing, metal treatment, dust suppression and general cleaning applications. Emulsification of medium polarity oils and solvents. (ICI PLC, Petrochemicals & Plastics Div).

Synperonic OP. Range of nonionic surfactants of the octylphenol ethoxylate type in liquid form. Application: Water-soluble general purpose detergents, wetting agents and emulsifiers with good solution properties in the presence of alkalis and at higher temperatures. (ICI PLC, Petrochemicals & Plastics Div).

Synperonic 3S27 and 3S60S. Sodium 3EO sulphate of Synprol in liquid form. Applications: Emulsifier, detergent, wetting and foaming agent, stable in high electrolyte concentrations and hard water, but tends to hydrolyse in acid solution. Used in liquid household detergents; industrial and domestic cleaning formulations; emulsifying systems; shampoos and bubble baths. (ICI PLC, Petrochemicals & Plastics Div).

Synperonic 3S60A. Ammonium 3EO sulphate of Synprol in liquid form. Applications: Emulsifier, detergent, wetting and foaming agent, stable in high electrolyte concentrations and hard water, but tends to hydrolyse in acid solution. Used in liquid household detergents; industrial and domestic cleaning formulations; emulsifying systems; shampoos and bubble baths. (ICI PLC, Petrochemicals & Plastics Div).

Synpro. Metallic soaps of naturally occurring fatty acid. The metals in order of importance are calcium, zinc, magnesium, aluminium, barium and cadmium. The fatty acids in order of importance are stearic, palmitic and lauric. Applications: Plastic lubricants and stabilizers for thermoplastics and thermosets; the major markets are in PVC, ABS, polystyrene, polyolefins and phenolics. Also used as lubricants in powdered metals and cosmetics. The soaps find uses as thixotropic agents and bodying materials in greases, oils and oil well drilling muds, as well as paints. (Synthetic Products Company).*

Synprol. Detergent alcohol. (ICI PLC).*

Synprol Sulphate. Anionic surfactant as a cream viscous liquid. Applications: Shampoos, bubble baths, liquid detergents and emulsifying systems. (ICI PLC, Petrochemicals & Plastics Div).

Synprolam. Synthetic fatty amines and derivative. (ICI PLC).*

Synprolam 35. Synthetic (C13/C15) alkyl primary amine in liquid form. Applications: Surfactant intermediate; corrosion inhibitor; flotation agent; fertilizer anti-caking agent. (ICI PLC, Petrochemicals & Plastics Div).

Synprolam 35 BQC. Benzyl quaternary ammonium chloride of a synthetic (C13/C15) dimethyl tertiary amine, in liquid form. Applications: Emulsifier; general sanitizer; biocide; corrosion inhibitor; textile dyeing auxilliary; timber preservative. (ICI PLC, Petrochemicals & Plastics Div).

Synprolam 35 DM. Synthetic (C13/C15) dimethyl tertiary amine in liquid form. Applications: Cationic surfactant intermediate. (ICI PLC, Petrochemicals & Plastics Div).

Synprolam 35 DMA. Acetic acid salt of a synthetic (C13/C15) dimethyl tertiary amine. Applications: Emulsifier; biocide; timber preservative. (ICI PLC, Petrochemicals & Plastics Div).

Synprolam 35A. Acetic acid salt of a synthetic (C13/C15) alkyl primary amine in solid form. Applications: Emulsifier; fertilizer anti-caking agent; mineral flotation. (ICI PLC, Petrochemicals & Plastics Div).

Synprolam 35N3. N-(C13/C15) alkyl-1,3-propane diamine in liquid form. Applications: Corrosion inhibitor; bitumen adhesion agent/emulsifier. (ICI PLC, Petrochemicals & Plastics Div).

Synpron. Proprietary mixtures of various metallic soaps and salts of organic acid, antioxidants, organophosphites and lubricants supplied as solids or liquids. Applications: Heat and light stabilizers for PVC flexible and rigid compounds. (Synthetic Products Company).*

Synpron 1032 and 1033. A proprietary range of liquid-antimony mercaptides used as heat stabilisers. (Dart Industries Inc).†

Synpro-Ware. Dispersions of rubber or plastic chemicals in elastomers, silicones, pastes, wetted powder or pellet form. Applications: Chemical dispersion for ease of handling, incorporation and safety for use in rubber, wire and cable and plastic compounding. (Synthetic Products Company).*

Synresin RD 461. A proprietary blocked, one-component polyurethane resin. It is thermosetting and is used as a rubber flock adhesive. (Synres International NV, Holland).†

SYNSOFT. Polymacon. Applications: Contact lens material. (Syntex Ophthalmics Inc).

Synsolve. General name for range of proprietary cleaning and maintenance products. Applications: Many areas of routine and maintenance cleaning. (Synthite Ltd).*

Synstryp. General name for a range of proprietary paint removers and surface coating products. Applications: Removal of paint-surface coatings. (Synthite Ltd).*

Syntamol V. An auxiliary for tanning and dyeing chrome leather. It is a mixture of the neutral salts of aromatic sulphonic acids.†

Syntan. See NERADOL.

Syntetrin. A proprietary preparation of rolitetracycline. An antibiotic. (Bristol-Myers Co Ltd).†

Syntex. A proprietary trade name for an oil modified alkyd resin (*qv*).†

Synteze. Proprietary mixture. Applications: Penetrating oil. (Synthite Ltd).*

Synthacalk. Polysulphide Sealant. Applications: Used for general caulking and sealing in the construction industry, vertical joints and perimeters of doors and windows, etc. (Pecora Corporation).*

Synthacryl. Thermoplastic and thermosetting acrylic resins. Applications: Automotive stoving finishes, industrial stoving finishes and pressure sensitive adhesives. (Resinous Chemicals Ltd).*

Synthamel. Air drying and stoving finishes. (ICI PLC).*

Synthamica. S_1O_2 AL_2O_3 M_gO K_2O F + 3% impurities. Applications: Additive to oil, paint, grease, plastics, glass and other inorganic materials which require a high dielectric or high heat resistant material. A coating for welding rods requiring special atmospheric controls. A thermal barrier and high electrical resistivity. (Mykroy/Mycalex).*

Synthane. A proprietary trade name for phenol-formaldehyde synthetic resin laminated products and other plastics.†

Synthappret. Special products for the non-felting finishing of wool. Applications: For the textile industry. (Bayer & Co).

Synthaprufe. Pitch rubber emulsion for damp proofing. (Thomas Ness Ltd).

Syntharesin. Products for obtaining nonslip effects. Applications: Used by the textile industry. (Bayer & Co).*

Synthasil. Colourless silicone water repellant for masonry brick etc. (Thomas Ness Ltd).

Synthawax. Hydrogenated castor oil. (Unilever).†

Synthecite. A rubber softener. It is a distillate from vulcanized rubber containing vegetable oils and waxes.†

Synthe-plastic. A reaction product of a terpene base. It is a rubber plastic, and is stated to contain no pitches or waxes.

Synthetic Bone Ash. Calcium hydroxyapatite - tricalcium phosphate. Applications: Used to coat moulds when casting molten purified copper and copper aloys. (Murlin Chemical Inc).*

Synthetic Peru Balsam,. See PERUGEN.

Synthetic Rutile. 94% titanium dioxide. (Kerr-McGee Chemical Corp).*

Synthetic Tannin. See NERADOL.

Synthin. A product obtained by heating Synthol (*qv*), at 400° C. in an autoclave. A liquid results which contains saturated hydrocarbons and sulphuric acid. A liquid fuel.†

Synthite. A proprietary trade name for formaldehydes. (Synthite Ltd).*

Synthocarbone. A specially prepared charcoal for use as a fuel.†

Synthol. A liquid fuel containing hydro carbons, acids, alcohols, aldehydes, and esters. It is obtained by reducing carbon monoxide in water gas at high temperatures and under pressure, using iron borings coated with potassium carbonate as contact material.†

Syntholvar. A proprietary trade name for extruded polyvinyl chloride.†

Synthroid. Levothyroxine sodium. Applications: Thyroid hormone. (Flint Laboratories, Div of Travenol Laboratories Inc).

Syntocinon. A proprietary preparation containing oxytocin. Used for promoting uterine contraction during and following labour. (Sandoz).†

yntol K77 and N77. Sodium lauryl ether sulphate (synthetic C12- C15 alcohol) as a pourable gelled paste or a low viscosity liquid. Applications: Production of all types of liquid and lotion shampoos and bubble baths. Also a raw material for light duty liquid detergents, dishwashing detergents and auto shampoos. (Efkay Chemicals Ltd).

yntometrin. A proprietary preparation of ERGOMETRINE maleate and Oxytocin, used in the treatment of post partum haemorrhage. (Sandoz).†

yntopon A,B,C and D. Nonylphenol ethoxylate in liquid form. Applications: Nonionic surfactant. (Witco Chemical Ltd).

yntopon F,G and N. Nonylphenol ethoxylate in solid form. Applications: Nonionic surfactant. (Witco Chemical Ltd).

yntopon 8. Series of nonionic surfactants of the octylphenol ethoxylate type in liquid form. Applications: Surfactants with range of properties and uses. (Witco Chemical Ltd).

yntopressin. A proprietary preparation of lypressin. (Sandoz).†

Syntroil. A series of polyol ester base fluid oils with various organic compounds. Applications: Various OEM applications, primarily small electric motors. (Ultrachem Inc).*

Syntron. Phenol formaldehyde resin. Applications: Moulding powders. (Manchem Ltd).*

Synvaren. A proprietary trade name for a phenol formaldehyde resin adhesive.†

ynvarol. A proprietary trade name for an urea formaldehyde resin adhesive.†

Syrian Asphalt. A natural asphalt containing about 100 per cent bituminous matter. It has a specific gravity of about 1.06, a melting-point of about 100° C., and practically no mineral matter.†

Syringa Vulcanine. An organic rubber vulcanization accelerator.†

Syrtussar. A proprietary preparation of dextromethorphan hydrobromide, phenylpropanolamine hydrochloride, sodium citrate, citric acid and chloroform. A cough linctus. (Armour Pharmaceutical Co).†

Syrup, Mineral. See PARAFFIN, LIQUID.

Systamex. Oxfendazole-based veterinary anthelmintic. (The Wellcome Foundation Ltd).

Systemic Funigicide. Garden fungicide. (Murphy Chemical Ltd).

Systemic Insecticide. Dimethoate insecticide. (Murphy Chemical Ltd).

Systhane. Light yellow solid - common name myclobutanil - chemical name -α-butyl-α-(4-chlorophenyl-1*H*-1,2,4-triazole-1-propanenitrile. Applications: Broad spectrum systemic fungicide. (Rohm and Haas Company).*

Systogen. *p* - Hydroxy - phenyl - ethyl-amine.†

Systral. Chlorphenoxamine. Tablets, cream and jelly. Applications: Antihistaminic and antiallergic. (Degussa).*

Sytam. Systemic organo phosphorous insecticide. (Murphy Chemical Co).†

Sytobex. Cyanocobalamin. Applications: Vitamin. (Parke-Davis, Div of Warner-Lambert Co).

Syton. Silica sol. (Monsanto PLC).

Sytron. A proprietary preparation of sodium iron edetate. A haematinic. (Parke-Davis).†

T

Tabasan. A proprietary preparation of ephedrine, theobromine and salicylamide. Applications: An antiasthmatic. (Ayrton Saunders plc).*

Tabbyite. See WURTZILLITE.

Tabloid. A proprietary formulation of thioguanine. Applications: Treatment of aute nonlymphocytic leukaemias. (The Wellcome Foundation Ltd).*

Tab Rybar Co. A proprietary preparation of isoprenaline sulphate, methylephedrine hydrochloride, butethamate citrate, and theophylline. A bronchial antispasmodic. (Rybar).†

Tacamahac Oil. See LAUREL NUT OIL.

Tacamahac Resin. (West Indian Anime Resin). A resin obtained from various plants, usually from *Calophyllum* species.†

Tacaryl. Methdilazine. Applications: Antipruritic. (Westwood Pharmaceuticals Inc, Subsidiary of Bristol-Myers Co).

TACE. Chlorotrianisene. Applications: Oestrogen. (Merrell Dow Pharmaceuticals Inc, Subsidiary of Dow Chemical Co).

Tachiol. (Tachyol). Silver fluoride, AgF.†

Tachostyptan. A proprietary preparation of THROMBOPLASTIN used to control bleeding. (Consolidated Chemicals).†

Tachyiite. A dark volcanic glass.†

Tachyol. See TACHIOL.

Tacitin. A proprietary preparation of BENZOCTAMINE hydrochloride. A tranquillizer. (Ciba Geigy PLC).†

Tack. Carbon black pastes. Applications: Used for simple and dust-free dyeing of paints, lacquers, paper, cardboard, plastics, synthetic fibres, printing inks and mineral binders. (Degussa).*

Tackidex. Dextrinified starch. Applications: A wide range of applications in the adhesives and food industries according to viscosity requirements. (Roquette (UK) Ltd).*

Tackol. A mixture of oils and resins used as a rubber plasticizer.†

Tactel. Polyamide textile fibres. (ICI PLC).*

Tactix. Epoxy resins used for matrix resin for the fabrication of aerospace components. (Dow Chemical Co Ltd).*

Taffy. A residue from the neutralization of the mixed organic acids produced by the fermentation of kelp-seaweed in the production of acetone. It consists chiefly of calcium propionate.†

Taflite 900. A proprietary range of weather resistant high-impact poly styrenes made from an EPDM graft polymerized with styrenes and dispersed as spherical microgels in polystyrene phases. (Mitsui Toatsu, Japan).†

Tag. Aqueous emulsion of natural and synthetic waxes. Applications: Fruit coating wax. (Makhteshim Chemical Works Ltd).†

Tagamet Tablets, Syrup, Injection and Infusion. Cimetidine. Applications: H_2-receptor antagonist for treatment of peptic ulcer, oesophageal reflux, dispeptic symptons, Mendelson's syndrome and Zollinger-Ellison syndrome. (Smith Kline and French Laboratories Ltd).*

agat. Polyoxyethylene glycerol fatty acid esters. Applications: Solubilizers for water insoluble substances such as flavours, perfumes, vitamin oils; dispersing and antistatic agents for technical purposes. (Th Goldschmidt Ltd).

a-Hong. A lead glass containing ferric oxide. Used by the Chinese as a red enamel on porcelain.†

ailor's Chalk. This material consists of French chalk (magnesium silicate) mixed with a little China clay.†

ak. Mould sealing compound. (Foseco (FS) Ltd).*

aka-diastase. An enzyme from the fungus *Eurotium orzoe,* grown on rice. A proprietary preparation containing aspergillus oryzae enzymes. An antacid. (Parke-Davis).†

akatol. *p*-Aminophenol, $C_6H_4OH.NH_2$.†

akazyma. A proprietary preparation containing takadiastase (*qv*) magnesium carbonate, bismuth carbonate, ginger and calcium carbonate. An antacid. (Parke-Davis).†

'akizolit. A red micro-crystalline kaolin found in Japan, and having the composition, $2Al_2O_3.7SiO_2.7H_2O$. It also contains appreciable amounts of rare earth oxides.†

'a-Kong. A lead glass containing ferric oxide and used by the Chinese as a red enamel.†

'aktene 1252. High cis-1,4-polybutadiene rubber containing 37.5 parts per hundred of a highly aromatic oil.†

'aktic. Animal health insecticide. (FBC Ltd).

'aktic. See AMITRAZ.

'albor's Powder. Cinchona bark in powder form.†

'alca Gum. See SUAKIN GUM.

'alcid. An antacid preparation. (Bayer & Co).*

'alc, Laminated. See MICA.

'alent. Herbicide. (May & Baker Ltd).

'alide. Tungsten carbide material.†

Talipot. (Raw Palmira Root Flour). A starch obtained from a palm, *Corypha umbraculifera.*†

Talisman. Contains 500g/litre chlortoluron. Applications: Cereal herbicide. (Farmers Crop Chemicals Ltd).*

Talite. A siliceous earth containing 84 per cent silica with small quantities of oxides of iron and aluminium. It is used as a rubber filler. See also TALITOL.†

Talitol. (Talite). An alcohol, $CH_2OH(CHOH)_4CH_2OH$.†

Talka Gum. See SUAKIN GUM.

Talloel. A Swedish liquid resin obtained as a by-product in the production of cellulose from Swedish fir by the soda process. It is stated to consist mainly of resin acids, and is closely related to rosin. It contains 87.5 per cent resin acids, 8 per cent unsaponifiable matter, and 3 per cent oxy-acids. It has an acid value of 171. Also see LIQUID RESINS.†

Tall Oil. A by-product of sulphate pulp manufacture. It contains 2.2 per cent of material soluble in petroleum ether, 12.4 per cent unsaponifiable matter, 30.4 per cent resin acid, and 54.9 per cent fatty acids. The resin acid consists of abietic acid, and the fatty acids contain oleic, linoleic, and linolenic acids.†

Tallow. The solid fat of oxen (beef tallow), and sheep (mutton tallow). It consists of tristearin, tripalmitin, and triolein.†

Tallow Clays. Clays containing varying proportions of zinc silicate.†

Tallow, Mineral. See BITUMEN.

Tallow, Mountain. See BITUMEN.

Tallow, Rock. See BITUMEN.

Tallow Seed Oil. Stillingia oil, obtained from the seeds of *Stillingia sebifera.*†

Talmi Gold. See ABYSSINIAN GOLD.

Talon. Rodenticide containing brodifacoum. (ICI PLC).*

Talotalo Gum. (Kau Drega). A gum somewhat resembling gutta-percha, from Fiji.†

Ta-Lou. The Chinese term for a glass flux used for enamelling on porcelain. It is mainly a silicate of lead with a little copper.†

Talpen. A proprietary preparation of talampicillin. Applications: An antibiotic. (Beecham Research Laboratories).*

Talpex. Titanium-aluminium organic complex. Applications: Thixotropic agent for latex paints. (Manchem Ltd).*

Talpheno. Phenobarbital. Applications: Anticonvulsant; hypnotic, sedative. (Merrell Dow Pharmaceuticals Inc, Subsidiary of Dow Chemical Co).

Talusin. A proprietary trade name for Proscillaridin.†

Talwaan. A tanning material. It is the root of *Elephantorrhiza burchelli.*†

Talwin. Pentazocine Hydrochloride. Applications: Analgesic. (Sterling Drug Inc).

Tamarac. The dried bark of *Larix larcina,* an American larch. The extract is used as an astringent and stimulant.†

Tambac. See TOMBAC.

Tambocor Injection. Each ampoule contains 15 ml of a solution of flecainide acetate. (Riker Laboratories/3M Health Care).†

Tambocor Tablets. Each tablet contains flecainide acetate 100 mg (Riker Laboratories/3M Health Care).†

Tambookie Grass. The product of *Hyperrhenice glauca.* It is stated to be suitable for paper-making.†

Tamclad 7200. A proprietary PVC or ganosol. (Tamite Industries Inc, Miami, Fla).†

Tamguard 840, 840H and 840S. A proprietary range of PVC plastisols used for coating electroplating racks. Their Shore hardnesses are A90, D35 and A70 respectively. (Tamite Industries Inc, Miami, Fla).†

Tamol. A combination of formaldehyde and naphthalenesulphonic acid. Used as a precipitant for the production of lakes, from basic dyestuffs.†

Tampicin. A resin, obtained from *Ipomoea simulans.*†

Tamtam. An alloy of 78 per cent copper and 22 per cent tin.†

Tanacetone. Thujone, $C_{10}H_{16}O$.†

Tanacetyl. Tannin acetic ester.†

Tanal. Chromium/aluminium complexes. Applications: Tanning. (Lancashire Chemical Works Ltd).*

Tanalith. Copper/chrome/arsenate waterborne wood preservative to preve fungal decay and insect attack. Applications: Pressure treated timber for construction, fencing, agriculture and any application where timber requires protection. (Hickson & Welc Ltd).*

Tanatol. See UBA.

Tanbase. Magnesium oxide for leather tanning. (Steetley Refractories Ltd).

Tancolin. A prorietary preparation of dextromethorpan hydrobromide, theophylline, sodium citrate, citric aci ascorbic acid and glycerin. A cough linctus. (Maws Pharmacy Suppliers).*

Tandacote. A proprietary preparation Of OXYPHENBUTAZONE. An anti-inflammatory drug. (Ciba Geigy PLC).†

Tandaigesic. A proprietary preparation of OXYPHENBUTAZONE and PARACETAMOL. An anti-inflammatory and analgesic drug. (Cib Geigy PLC).†

Tandearil. Oxyphenbutazone. Applications: Anti-inflammatory; antirheumatic. (Ciba-Geigy Corp).

Tandem. A line of herbicides based primarily on tridiphane. (Dow Chemic Co Ltd).*

Tanderil. A proprietary preparation of oxyphenbutazone. (Ciba Geigy PLC).†

Tanekaha. The bark of *Phyllocladus trichomanoides.* Used in tanning leather.†

Tanigan. Range of tanning materials comprising advanced syntans which make the tanning process safe and economical and improve the quality of the leather. (Bayer & Co).*

Tanked Oil. Linseed oil from which the moisture and other matter has settled out. It has a higher value than the ordinary oil.†

Tannacetin. See ACETANNIN.

Tannal. (Tannalum). A basic aluminium tannate, $Al(OH)_2.(C_{14}H_9O_9)+5H_2O$. An astringent used as a dusting powder.†

Tannaline Fiims. Gelatin films hardened b formaldehyde, used for photographic purposes.†

Tannalum. See TANNAL.

Tanner's Wool. (Glover's Wool). A wool pulled from the carcases of slaughtered sheep with the assistance of lime. It does not dye well.†

Tannesco. Tanning agents for leather. (Ciba-Geigy PLC).

Tannic Acid. See GALLOTANNIC ACID.

Tannic Indigo. See GALLANILIC BLUE.

Tannin Orange R. A dyestuff. It is dimethylaminotolueneazo-β naphthol, $C_{19}H_{19}N_3O$. Dyes leather and tannined cotton orange.†

Tanninphenolmethane. A combination of tannin, formaldehyde, and phenol.†

Tannin, Synthetic. See NERADOL.

Tannismuth. Bismuth bitannate.†

Tanolin. A proprietary trade name for a basic chromium chloride for use in chrome tanning baths.†

Tanret's Reagent. To a solution of 1.35 grams mercuric chloride in 25 cc water is added a solution of 3.32 grams potassium iodide in 25 cc water. This is made up to 60 cc with water and 20 cc glacial acetic acid.†

Tansel. A specially prepared salt for curing hides.†

Tansul. Clarifying agent. Applications: Beer. (NL Industries Inc).*

Tantcopper. A copper alloy analogous to tantiron.†

Tantiron. An alloy of 84 per cent iron, 15 per cent silicon, and 1 per cent carbon. It has a specific gravity of 6.8 and is acid-resisting.†

Tantnickel. A nickel alloy analogous to tantiron.†

Tao. A proprietary preparation of troleandomycin. An antibiotic. (Pfizer International).*

Tap Aid. ASTM S-215 Oil, 1.1.1. trichlorethane, mask odour No 3. Applications: Small hole, drilling, reaming and tapping. Excellent for wire drawing and grinding. (Doyle Specialties).*

Tap Cinder. The basic silicate of iron constituting the slag, and flowing through the tap-hole of the puddling furnace.†

Tapar. Acetaminophen. Applications: Analgesic; antipyretic. (Parke-Davis, Div of Warner-Lambert Co).

Tapazole. Methimazole. Applications: Thyroid inhibitor. (Eli Lilly & Co).

Tara. The tannin from the pods of *Coesalpinia tinctoria*.†

Taractan. Proprietary preparation of chlorprothixene. Major tranquillizer of thioxanthene group. (Roche Products Ltd).*

Tarband. A proprietary bandage impregnated with zinc oxide and coal tar paste, used in the treatment of eczema. (Seton Products Ltd).*

Tar, Bone. See BONE OIL.

Tar Camphor. Naphthalene, $C_{10}H_8$.†

Tar, Candle. See STEARIN PITCH.

Tarcortin. A proprietary preparation of coal tar and hydrocortisone. Applications: For dermatological use - ezcema. (Stafford-Miller).*

Tardex. A range of brominated - organics. Applications: Flame retardant additives. (ISC Chemicals Ltd).*

Tardocillin 1200. A long acting penicillin. (Bayer & Co).*

Tardrox. A proprietary preparation of chlorhydroxyquinolone and tar for dermatological use. (Carlton Laboratories).†

Tari. See WHITE TAN.

Tarmac. A proprietary preparation of blast furnace slag, refined tar, and other ingredients. Used for road dressing.†

Tarmex. A proprietary name for a combination of prepared tar and Mexphalte. Used for road dressing.†

Tarnovicite. See TARNOWITZITE.

Tarnowitzite. (Plumbocalcite). A mineral, (Ca.Pb) CO_3.†

Tar Oil. See OIL OF TAR.

Tarola. A coal-tar product used as a sheep dip.†

Tar, Regenerated. A mixture of pitch and anthracene oil which does not crystallize.†

Tar, Skelleftea. See SKELLEFTEA.

Tarslag. A proprietary preparation of cold blast slag which has been treated with a bituminous compound. Used as a road dressing.†

Tar Spirit. Benzene. See OIL OF TAR.†

Tartar. See ARGOL.

Tartar Cake. See SUPERARGOL.

Tartar Emetic. (Tartrated Antimony).Potassium-antimonyl-tartrate, $C_4H_4O_6(SbO)K+1/2H_2O$. Used in medicine as an emetic, and in dyeing as a mordant.†

Tartar Emetic Powder. (Tartar Emetic Substitute, Antimony Mordant). Mixtures of tartar emetic and zinc sulphate.†

Tartar Emetic Substitute. See TARTAR EMETIC POWDER.

Tartar Substitute. See SUPERARGOL.

Tartar Yellow FS, FS Conc. Dyestuffs. They are British equivalents of Tartrazine.†

Tartarised Borax. Potassium boro-tartrate.†

Tartarline. Potassium bisulphate, used as a substitute for tartaric acid for industrial purposes.†

Tartars. Raw materials which contain more than 40 per cent tartaric acid are termed tartars.†

Tartar, Chalybeated. See TARTRATED IRON.

Tartar, Crude. See ARGOL.

Tartar, Salt of. See POTASH.

Tartar, Stibiated. See TARTAR EMETIC.

Tartar, Vitriolated. Potassium sulphate, K_2SO_4.†

Tar Tea. Tar water.†

Tartrachromin G.G. See ALIZARIN YELLOW 5G.

Tartrate Lime. See LIMO.

Tartrated Antimony. See TARTAR EMETIC.

Tartrated Iron. Iron and potassium tartrate. Used medicinally.†

Tartrated Soda. See ROCHELLE SALT.

Tartrated Soda Powder, Effervescent. See SEIDLITZ POWDER.

Tartratol Yellow L. A British brand of Tartrazine.†

Tartrazine. (Acid Yellow 79210, Hydroxine Yellow G. L. L Conc, Tartar Yellow FS, FS Conc, Tartratol Yellow L, Tartrine Yellow O). A dyestuff, $C_{16}H_{10}O_{10}N_4S_2Na_4$. Dyes silk and wool.†

Tarvia. A proprietary trade name for a specially refined coal-tar.†

Tarwar. The bark of *Cassia auriculate.* A tanning material.†

Tasar Silk. See TUSSAR SILK.

Tasmaderm. Motretinide. Applications: Keratolytic. (Hoffmann-LaRoche Inc).

Tasprin. .-L proprietary preparation of soluble aspirin. (Unichem).†

Tasteless Salts. Sodium phosphate, HNa_2PO_4.†

Taurin. Aminoethanesulphonic acid.†

Tavist. Clemastine fumarate. Applications: Antihistaminic. (Dorsey Pharmaceuticals, Pharmaceutical Div of Sandoz Inc).

Taxafor. Non-ionic surfactant. (ABM Chemicals Ltd).

Taxol. A proprietary preparation of pancreatin, bile salts, aloes and AGAR used in the treatment of constipation. (Cox, A H & Co Ltd, Medical Specialities Divn).†

Taylor. A proprietary trade name for a phenol-formaldehyde synthetic resin laminated product.†

Taylorite. See BENTONITE.

Taylor Oil. A patented binding material obtained by boiling raw linseed oil with driers (litharge), then forcing air through the oil when heated to 300° F., and finally heating it for some time at 500- 600° F.†

Taylor Solder. An alloy of 60 per cent tin, 12 per cent lead, 12 per cent silver, 8 per cent zinc, 4 per cent aluminium, and 4 per cent copper.†

Tazoline. A proprietary preparation of antazoline hydrochloride, octaphonium chloride, titanium dioxide and calaroine. (Rybar).†

Tc 99m Lungaggregate. Technetium Tc 99m Aggregated Albumin. Applications: Diagnostic aid; radioactive agent. (Medi-Physics Inc).

TCA. A rubber vulcanization accelerator. It is thiocarbanilide.†

TCC. 3,4,4'-trichlorocarbanilide. Applications: Bacteriostatic agents for bar soaps. (Monsanto Co).*

Tcha-Lau. A blue powder containing copper. Used by the Chinese for obtaining a blue colour on porcelain.†

TCP. (Plastic X, Plastol X). Trade names for tricresyl phosphate, a plasticizer for cellulose lacquers, and poly-vinylchloride. It has a specific gravity of 1.185-1.189, a boiling range of 430-440° C., and a flash- point of 215° C. The term TCP appears to be also applied to an aqueous solution of trichloro-phenyl-iodo-methyl-salicyl, an antiseptic and germicide.†

Teaberry Oil. Methyl salicylate, $C_6H_4OH.COO.CH_3$.†

Teatcote Plus. A proprietary preparation of polyhexanide. A veterinary antibacterial.†

Tea, Jesuit's. See MATÉ.

Tea-Lead. An alloy of from 97-99 per cent lead and 1-3 per cent zinc. Also an alloy of lead with 2 per cent tin used for wrapping tea.†

Teal Oil. See GINGELLY OIL.

Tea Oil. See OIL OF TEA.

Tear-Efrin. Phenylephrine Hydrochloride. Applications: Adrenergic. (CooperVision Inc).

Tearisol. Hydroxypropyl methylcellulose. Applications: Pharmaceutic aid. (CooperVision Inc).

Tea, St. Bartholomew. See MATÉ.

Tec. A proprietary trade name for cellulose acetate varnish resins.†

Tecagg. Very low density mineral aggregates comprising foamed waste products or clays. Density 0.3 - 0.8. Applications: Insulating building products, refractory insulation. (Filtec Ltd).*

Tecali Onyx. See ONYX OF TECALI.

Tecane. Selective herbicide. (FBC Ltd).

Teccel. Hollow glass microspheres, expanded minerals. White, mono or multicellular lightweight fillers of density 0.15 - 0.6 gm/cc. Applications: Explosives, deep submergence buoyancy, paints, cultured marble. (Filtec Ltd).*

Tec-Char. A proprietary trade name for a granular charcoal.†

Tecfil - frequently shortened to 'T' with a suffix e.g. T.300. Name of a range of lightweight mineral fillers in particular cenospheres - hollow ceramic microspheres derived from fly-ash. Composed primarily of silica and alumina, size 5 - 300 u, density 0.5 to 0.8 gm/cc. Applications: Low density cementing, auto noise attenuation, auto underbody coating, low density plasters, refractory insulation, thermoset resin extender. (Filtec Ltd).*

Techmate. Resilient moulded foam used for cushion packaging applications. (Dow Chemical Co Ltd).*

TechneColl. Technetium Tc 99m Sulphur Colloid. Applications: Radioactive agent. (Mallinckrodt Inc).

TechneScan MAA. Albumin, aggregated. Applications: Diagnostic aid. (Mallinckrodt Inc).

TechneScan PYP. Stannous Pyrophosphate. Applications: Diagnostic aid. (Mallinckrodt Inc).

TechneScan SSC. Stannous Sulphur Colloid. Applications: Diagnostic aid. (Mallinckrodt Inc).

Techroline. Gasoline additive. (Chevron).*

Techron. Gasoline additive. (Chevron).*

Tecnocin. A proprietary range of curing agents for fluorocarbon rubbers. (Montedison UK Ltd).*

Tecnoflon.. A proprietary range of fluorocarbon based rubbers. (Montedison UK Ltd).*

Tecnoprene. A proprietary range of filled polymer moulding granules. (Montedison UK Ltd).*

Tecoflex Polyurethane. Linear segmented aliphatic polyether polyurethane, medical grade elastomer. Applications: Medical products, tubing. (Thermedics Inc).*

Tecpril. High quality foamed clay aggregates, in form of white, regular pellets. Density 0.1 - 0.6 gm/cc. Applications: High tech insulation products, fireproof composites, aerospace/hydrospace composites. (Filtec Ltd).*

Tecquinol. A proprietary anti-oxidant. Bydroquinone. (Eastman Chemical Products).†

Tectilon. Acid dyes. (Ciba-Geigy PLC).

Tedimon. Isocyanates. (Montedison UK Ltd).*

Tedion V-18. Tetradifon. Applications: Selective acaricide for use against mite infestation in orchards, citrus fruit plantations, hop fields, groundnut plantations, vegetable plots, cotton fields and on ornamental plants. (Duphar BV).*

Tedlar. A proprietary trade name for a clear or pigmented polyvinyl fluoride film. It has high resistance to weathering and is generally chemically inert. (Du Pont (UK) Ltd).†

Tedral. A proprietary preparation of theophylline, ephedrine hydrochloride, and phenobarbitone. A bronchial antispasmodic. (Warner).†

Teebrix. Preformed inserts for extending bottom plate lives. (Foseco (FS) Ltd).*

Teefroth. Polyglycol ethers. (ICI PLC, Petrochemicals & Plastics Div).

Teejel. A proprietary preparation of choline salicylate and cetalkonium chloride. An analgesic gel for teething pain. (Napp Laboratories Ltd).*

Teel Oil. See GINGELLY OIL.

Teepol CM44. An aqueous solution of sodium salt of C_9-C_{13} alkyl benzene sulphonate. It is mainly used as a constituent for other detergent blends. (Shell Chemicals).†

Teepol FC5. A high foam detergent for repackers, being a solution of a primary alcohol ethoxysulphate. (Shell Chemicals).†

Teepol GD53. A highly biodegradable, clear, pale amber liquid containing formalin as a preservative. It is specially formulated for dishwashing and general cleaning. (Shell Chemicals).†

Teepol HB6. A highly biodegradable aqueous solution of the sodium salts of C_9-C_{13} primary alcohol sulphate. It contains formalin as a preservative. The solution is used as a solubilizer and emulsifier component of hard surface cleaners and germicidal cleaners. (Shell Chemicals).†

Teepol PB. Sodium primary alcohol ethoxy sulphate as a clear aqueous solution. Application: Purpose developed foaming agent used in gypsum wallboard manufacture. (Shell Chemicals UK Ltd).

Teerlack. Coal tar pitch.†

Teevax. A proprietary preparation of CROTAMITON and HALOPYRAMINE used in the treatment of pruritis. (Ciba Geigy PLC).†

Teflon. A proprietary polytetrafluoroethylene (P.T.F.E.) plastic material having good resistance to high temperatures. (Du Pont (UK) Ltd).†

Teflon F.E.P. A proprietary range of hexafluoropropylene copolymers. (Du Pont (UK) Ltd).†

Tefzel 200. A proprietary ETFE fluoropolymer resin extruded for use as wire insulation. (Du Pont (UK) Ltd).†

Tegda. Triethylene glycol diacetate. Applications: plasticizer for the manufacture of cigarette filter tips, plasticizer for paint and varnish systems. (Bayer & Co).*

Tegin. A patented preparation. It is a neutral ester closely related to the natural fats. Recommended as a salve base.†

Tegison. Etretinate. Applications: Antipsoriatic. (Hoffmann-LaRoche Inc).

Teglac. A proprietary trade name for an alkyd synthetic varnish and lacquer resin.†

Tegmer. Specialty plasticizers. Applications: Rubber and vinyl. (The C P Hall Company).*

Tegmin. An emulsion made from 1 part yellow wax, 2 parts acacia, and 3 parts water. It also contains 5 per cent zinc oxide, and a little wool fat. Used as a surgical dressing.†

Tego Films. Proprietary products. They are tough films produced by impregnating very thin transparent paper with adhesive material along with cresol and formaldehyde. The papers are placed between layers of wood and subjected to heat and pressure.†

Tego-Betain HS. Fatty amidopropyl dimethylaminoacetic acid betaine (C11 - C17), as a clear low viscosity liquid. Applications: Baby shampoos. (Th Goldschmidt Ltd).

Tego-Betain L7 and L10. Fatty amidopropyl dimethylaminoacetic acid betaine (C11 - C17), as a clear low viscosity liquid. Applications: Forms stable emulsions with fatty alcohols over wide pH range. Used in shampoos, bath products, personal hygiene preparations and baby care products. (Th Goldschmidt Ltd).

Tego-Betain T. Fatty amidopropyl dimethylaminoacetic acid betaine (C11 - C17), as a clear low viscosity liquid. Applications: Cleansing agents eg hand-washing gel, household cleanser, floor cleaner, carwash. (Th Goldschmidt Ltd).

Tegofan. See ALLOPRENE.

Tegoglätte. A litharge having smaller particles than the ordinary type. Used in rubber mixings.†

Tegopen. A proprietary preparation of cloxacillin. An antibiotic. (Bristol-Myers Co Ltd).†

Tegretol. A proprietary preparation of carbamezepine. An analgesic. (Ciba Geigy PLC).†

Tegul. A proprietary sulphur jointing compound for bell and spigot pipes. It contains sulphur and sand.†

Tegula. Bitumen sheet. Applications: Underneath stretcher strip. (Vedag GmbH).*

Teka Oil. A proprietary trade name for an extract from stand oil (*qv*) from which bases and acids have been removed.†

Teknol. Photographic developer. (May & Baker Ltd).

Tekpak-Tekbent. Reactor part of anhydrite/bentonite mixture. (Foseco (FS) Ltd).

Tekpak-Tekcem. Alumina cement for pump packing. (Foseco (FS) Ltd).

Telconax. A patented insulating compound made from selected bitumen, waxes, and rubber.†

Telconite. A proprietary insulating material made in various colours.†

Telconstan. A non-magnetic nickel-copper alloy prepared in induction furnaces and having exceptional purity and very low temperature coefficient of resistance. It is used for resistances where standard of resistance with temperature is important.†

Telcothene. A registered trade name for polythene powder, tube and sheet. (Telcon Plastics Ltd).†

Telcovin. A registered trade name for polyvinyl chloride tube and sheet. (Telcon Plastics Ltd).†

Teldrin. Chlorpheniramine maleate. Applications: Antihistaminic. (Menley & James Laboratories).

Teleblock. A proprietary range of thermoplastic rubbers. (Phillips Chemical Co).†

Telegraph Bronze. (Telegraph Metal, Electric Metal). An alloy of 80 per cent copper, 7.5 per cent lead, 7.5 per cent zinc, and 5 per cent tin.†

Telepaque. Iopanoic acid. Applications: Diagnostic aid. (Sterling Drug Inc).

Telloy. A proprietary product. It is stated to be a form of elementary tellurium specially pulverised and purified for use as a rubber vulcanizing agent. It has a specific gravity of 6.27.†

Telluretted Hydrogen. Hydrogen telluride, H_2Te.†

Tellurit. Chill producing mould dressing containing tellurium. (Foseco (FS) Ltd).*

Tellurium Lead. An alloy containing 0.05 per cent tellurium with lead. It resists sulphuric acid.†

Tellurium Tubes. Copper tubes containing pre-determined quantities of tellurium. (Foseco (FS) Ltd).*

Telmin. A proprietary preparation containing mebendazole. Applications: Veterinary antihelmintic (horses). (Janssen Pharmaceutical Ltd).*

Telogen. 1;2 metal complex disperse dyes for fast dyeing of polyamide fibres. (Bayer UK Ltd).

Telon. Selected acid dyestuffs. Applications: For the dyeing of polyamide fibres and wool/polyamide blends. (Bayer & Co).*

Telone. Soil fumigants. Agricultural products containing 1,3-dichloropropene as the active ingredient. They are applied to soil prior to planting to control soil pests such as nematodes which feed on the roots of plants and reduce yields. (Dow Chemical Co Ltd).*

Telopar. Oxantel pamoate. Applications: Anthelmintic. (Pfizer Inc).

Telsit. A gelatin explosive containing from 10-15 per cent dinitrotoluene or liquid trinitrotoluene.†

Teluran. Acrylonitrile-butadiene-styrene polymers and related products. (BASF United Kingdom Ltd).*

Tem-Tuf. Aluminious metal sheet. (Reynolds Metal Co).*

Temadex. Veterinary skin dressing. (The Wellcome Foundation Ltd).

Temaril. Trimeprazine Tartrate. Applications: Antipruritic. (Smith Kline & French Laboratories).

Temasept IV. Tribromsalan. Applications: Disinfectant. (Hexcel Chemical Products).

Temetex. Proprietary topical corticosteroid containing diflucortolone as active ingredient. (Roche Products Ltd).*

Temlock. A proprietary trade name for a board made from wood fibres impregnated with resin and subjected to pressure.†

Tempaloy. A patented alloy of approximate composition, 95 per cent copper, 4 per cent nickel, and 1 per cent silicon.†

Temper. Alloys of arsenic and lead or copper, and tin. Used as hardening materials for shot or pewter.†

Tempered Lead. See NOHEET METAL.

Temperite Alloys. A proprietary trade name for alloys of lead, tin, and cadmium.†

Tempo. Proprietary cellulose esters.†

Tempo. Herbicide, containing linuron and terbatryne. (ICI PLC).*

Tempo. Soft antacid, for relief of acid indigestion. (Richardson-Vicks Inc).*

Tempra. Acetaminophen. Applications: Analgesic; antipyretic. (Mead Johnson & Co).

Tempro. See BELPRO. (ICI PLC).*

Tenacite. See BAKELITE.

Tenacity. Fluxes for silver alloy brazing. (Johnson Matthey Chemicals Ltd).

Tenamine 1. A proprietary anti-oxidant. N butylated-*p*-aminophenol. (Eastman Chemical Products).†

Tenamine 2. A proprietary anti-oxidant. N,N[1]-di-*sec*-butyl-*p*-phenylene diamine. (Eastman Chemical Products).†

Tenamine 3. A proprietary anti-oxidant. 2,6-di-*tert*-butyl-*p*-cresol. (Eastman Chemical Products).†

Tenasco. A proprietary trade name for synthetic fibre resembling Nylon.†

Tenasco Fibre. A proprietary trade name for a fibre obtained by stretching viscose fibre when in a plastic condition.†

Tenatine. Thermoplastic polyester moulding compounds. (Ciba-Geigy PLC).

Tenavoid. A proprietary preparation of bendrofluazide and meprobamate. Prophylaxis for premenstrual syndrome. (Leo Laboratories).†

Tenax Metal. A zinc alloy containing from 0.35-2.56 per cent copper, 0.2-4.42 per cent aluminium, 0-0.35 per cent iron, and up to 1.2 per cent lead. Used for the manufacture of guide rings.†

Tenaxatex VA 632. A proprietary trade name for a high molecular weight vinyl acetate homopolar water emulsion containing 55 per cent. solids. Used as an adhesive base. (H A Smith).†

Tenaxatex VA 956. A proprietary trade name for a vinyl acetate/acrylate copolymer emulsion containing 55 per cent. solids. A medium viscosity adhesive base. VA 957 and VA 959 are as above but are low viscosity materials (H A Smith).†

Tenazit. A proprietary trade name for laminated bakelite or similar synthetic resin.†

Tenebryl. Diiodomethane sodium sulphonate.†

'enephrol. A proprietary preparation of lithium iodide (31 per cent solution).†

'enite 1. A proprietary cellulose acetate moulding compound.†

'enite 2. A proprietary trade name for a cellulose acetate-butyrate moulding compound.†

'enite 7 D.R.D. A proprietary polyethylene terephthalate resin used for injection moulding. (Eastman Chemical International AG).†

Tennal. A proprietary trade name for certain aluminium alloys for casting purposes.†

Tennant's Salt. Chlorinated lime.†

Tenncol. Saponified rosin size. (Tenneco Malros Ltd).

Tennessee Phosphates. Mineral phosphates containing from 60-70 per cent calcium phosphate. Fertilisers.†

Tenoban. Arecoline-acetarsol. (The Wellcome Foundation Ltd).

Tenoran. Substituted urea herbicide. (Ciba-Geigy PLC).

Tenoret. Anti-hypertensive containing atenolol and chlorthiadone as a diuretic. (ICI PLC).*

Tenoretic. Combined anti-hypertensive and diuretic containing atenolol and chlorthalidone. (ICI PLC).*

Tenormin. Anti hypertensive containing atenolol. (ICI PLC).*

Tenrez. Rosin esters. (Tenneco Malros Ltd).

Tensabit. Emulsifier; alkaline emulsion. Applications: Bitumen. (Tensia SA).

Tensactol. Self emulsifying wax. Applications: Cosmetics and pharmaceuticals. (Tensia SA).

Tensadal. Reviving and softening agent. Applications: Cotton and cellulose fibres. (Tensia SA).

Tensagex. Liquid blend. Applications: Baby shampoos. (Tensia SA).

Tensagex BV. Anionic surfactant in the form of a low viscosity liquid. Applications: Liquid detergents; liquid shampoos; bubble baths. (Tensia SA).

Tensagex DMY. Sodium alkyl ether sulphate as a low viscosity liquid. Applications: Low irritant anionic surfactant used in baby shampoos and foam baths. (Tensia SA).

Tensagex DP24. Sodium alkylphenol ether sulphate in liquid form. Applications: Wetting and dispersing agent for pigments and metal degreasing. (Tensia SA).

Tensagex EOC. Sodium lauryl alcohol/2.5EO sulphate in liquid or gel form. Applications: High foaming anionic surfactant used in shampoos, bubble baths, liquid detergents and emulsion polymerization. (Tensia SA).

Tensagex SPDL. Sodium alcohol ether sulphate (C12-C15) in liquid form. Applications: High foaming anionic surfactant used in shampoos, bubble baths, liquid detergents and emulsion polymerization. (Tensia SA).

Tensamina. Adhesion improver. Applications: Bitumen on wet stones. (Tensia SA).

Tensamine C, O, S and SH. Cationic surfactants in the form of primary amines, with the alkyl portion being coconut, oleic, tallow and hydrogenated tallow respectively. Liquid or solid form. Application: Emulsification, pigment dispersion, synthesis intermediate. (Tensia SA).

Tensaminox. Foaming agents; dispersing agents; thickening agents. Applications: Shampoos; liquid detergents; pigments, textiles. (Tensia SA).

Tensanyl. A proprietary preparation of bendrofluazide, reserpine, and potassium chloride. Antihypertensive. (Leo Laboratories).†

Tensarane SBTE. Anionic surfactant in which the cation is triethanolamine and the anion is straight chain dodecylbenzene sulphonate. Supplied as a liquid. Applications: Wetting agent and dyeing auxiliary for cottons. (Tensia SA).

Tensaryl. Low foam; foaming detergent base. Applications: Light duty detergents. (Tensia SA).

Tensaryl DF90. Sodium dodecylbenzene sulphonate Applications: Foaming agent for fine fabrics washing. (Tensia SA).

Tensaryl DX54Sp. and DX62. Sodium dodecylbenzene sulphonate DX54Sp also contains perborate. Applications: Anionic surfactant for heavy duty products. (Tensia SA).

Tensaryl KD. Tetrapropylene benzene sulphonate as a viscous liquid. Applications: Intermediate for liquid and powder detergents when biodegradability is not requested. (Tensia SA).

Tensaryl L48. Sodium dodecylbenzene sulphonate. Applications: Low foaming anionic surfactant for fine fabrics washing. (Tensia SA).

Tensaryl SB. Straight chain dodecylbenzene sulphonate in acid form. Supplied as a viscous liquid. Applications: Raw material for the manufacture of liquid, pasty or solid surfactants. (Tensia SA).

Tensaryl SB Ca. Calcium dodecylbenzene sulphonate in powder form. Applications: Water insoluble anionic surfactant used as a co- emulsifier in organic systems. (Tensia SA).

Tensaryl SBD. Triethanolamine dodecylbenzene sulphonate in liquid form. Applications: Liquid detergent for dishwashing, textiles and car shampoos. (Tensia SA).

Tensaryl SB85P. Sodium dodecylbenzene sulphonate in paste form. Applications: Wetting and dispersing agent with low salt content, used in emulsion polymerization. (Tensia SA).

Tensaryl S30P and S70P. Sodium tetrapropylene benzene sulphonate in the form of a liquid or paste with low salt content. Applications: Anionic surfactant used for emulsion polymerization. (Tensia SA).

Tensaryl 40CC, 50B, 80B and 82F. Sodium tetrapropylene benzene sulphonate as powder, beads or flakes. Applications: Detergent, anti-caking agent and wetting agent, for emulsion polymerization, wettable powders and various industrial uses. (Tensia SA).

Tensatil DA120. Ammonium octyl/decyl-sulphate in liquid form. Applications: Anionic surfactant used as a wetting and foaming agent in froth flotation. (Tensia SA).

Tensatil DB120. Short chain alcohol sulphate in liquid form. Applicatio Anionic surfactant used as a wettin agent in electrolyte solutions. (Tens SA).

Tensatil DEH120. Sodium 2-ethyl-hex sulphate in liquid form. Applicatio Anionic surfactant used as a wettin agent in metal cleaning preparation styrene polymerization. (Tensia SA

Tensatil D100. Sodium octylsulphate i liquid form. Applications: Anionic surfactant used as a wetting agent i polymerization. (Tensia SA).

Tensiamix. Range of low foaming dete base. Applications: Heavy duty detergents. (Tensia SA).

Tensianol. Sensitive skin special washir composition. Applications: Cosmeti synthetic toilet bars. (Tensia SA).

Tensibet 50. N-Alkylbetaine in liquid f Applications: Foam booster and detergent with antistatic effect, usec cosmetics and baby shampoos. (Ten SA).

Tensibet 55. Alkylamidobetaine in liqu form. Applications: Foam stabilizer thickening agent which is mild to th skin, used in lauryl ether sulphate formulations. (Tensia SA).

Tensid. An antihypertensive preparatio (Bayer & Co).*

Tensidef. Anti-foaming agents. Applications: Paper industry. (Tensi SA).

Tensidye. Nonionic liquid. Wetting age Applications: Cosmetic. (Tensia SA

Tensilac 39. A proprietary rubber vulcanization accelerator. It is a soft form of ethylidene-aniline.†

Tensilac 40. A proprietary rubber vulcanization accelerator. It is a resinous condensation product.†

Tensilac 41. A proprietary rubber vulcanization accelerator. It is a har form of ethylidene-aniline.†

Tensilite. An aluminium bronze. It cont from 64-67 per cent copper, 3.1-4.4 cent aluminium, 0-1.2 per cent iron, 3.8 per cent manganese, and 24-29 p cent zinc.†

ensilon. A proprietary preparation of edrophonium chloride used in the diagnosis of myasthenia gravis. (Roche Products Ltd).*

ensimul. Emulsifier. Applications: Oils, solvents, natural and synthetic waxes. (Tensia SA).

ensiofix. Range of emulsifiers. Applications: Pesticide formulations. (Tensia SA).

ensioquat C50. Benzalkonium chloride in aqueous solution. Application: Cationic surfactant used in disinfection and emulsification. (Tensia SA).

ensioquat C75. Benzalkonium chloride in liquid IPA form. Application: Cationic surfactant used in the paint industry. (Tensia SA).

ensiorex. Conditioning agent; concentrated pearling agent. Applications: After shampoo; shampoos. (Tensia SA).

ensiostat. Antistatic agent. Applications: Cellulosic fibres and polymers. (Tensia SA).

'ensipar. Anti-foaming agent. Applications: Industrial uses including food. (Tensia SA).

'ensitex. Wetting agent. Applications: Caustic iye and mercerising. (Tensia SA).

'ensloy. A proprietary trade name for an alloy of iron with approximately 1.5 per cent nickel and 0.5 per cent chromium.†

'ensocide. Antiseptic. Applications: Paper Mill. (Tensia SA).

Tensol. (2) Cements for vinyl and acrylic sheets. (ICI PLC).*

Tensol. Photographic activator/stabilizer chemicals. (May & Baker Ltd).

Tensol. (1) A proprietary trade name for a dispersing and emulsifying agent containing a sulphonated ether.†

Tensoleate. Pigment grinding aid; dispersing agent. Applications: Paint industry. (Tensia SA).

Tensoline. Emulsifier; levelling and dispersing agent. Applications: Oiling and fulling of wool; dyeing of woollen and acrylic fibres. (Tensia SA).

Tensomel. Superamide; foam stabilizer; additive; anti-corrosion. Applications: Shampoos; liquid detergents; bubble baths; cutting oils. (Tensia SA).

Tensomin. Levelling agent. Applications: Dyeing woollen, acrylic, polyamide and polyester fibres; dyeing with acid and metallizing dyes. (Tensia SA).

Tensopac. Wetting and dispersing agents. Applications: Paper mill. (Tensia SA).

Tensopane D. Series of nonionic surfactants of the octylphenol ethoxylate type. Applications: Emulsification; detergent compounding; industrial cleaning; metal pickling; stabilizer in emulsion polymerization. (Tensia SA).

Tensophene D12, D15 and D18. Nonionic surfactants of the nonylphenol ethoxylate type. Applications: Foam control; dispersion; oil emulsifier; coupling agent. (Tensia SA).

Tensophene H10, I10, DT, D36, D42EC, D45, D60 and D90. Nonionic surfactants of the nonylphenol ethoxylate type. Applications: All types of detergent; metal cleaning; emulsion polymerization. (Tensia SA).

Tensophene 2D30. Dinonylphenol ethoxylate nonionic surfactant. Applications: Emulsifier; detergent; metal cleaning; emulsion polymerization. (Tensia SA).

Tensopol LT. Triethanolamine lauryl sulphate in liquid form. Applications: Anionic surfactant used in liquid shampoos, bubble baths and hair lotions. (Tensia SA).

Tensopol ACL and PCL. Sodium lauryl sulphate in the form of needles or powder. Applications: Detergency, foaming and wetting agent for shampoos, emulsion polymerization and pigment dispersion. (Tensia SA).

Tensopol AG and MG. Magnesium lauryl sulphate in the form of needles or powder. Applications: Anionic surfactant used in shampoos and toothpastes. (Tensia SA).

Tensopol A.7 and USP. Sodium lauryl sulphate in the form of needles or powder. Applications: Detergency and foaming agent for toothpastes, shampoos, pharmaceuticals, emulsion polymerization and pigment dispersion. (Tensia SA).

Tensopol DX85 and FL. Sodium alcohol sulphate in liquid or paste form. Applications: Foaming agent for liquid shampoos, including carpet types; latex rug backing. (Tensia SA).

Tensopol N. Ammonium lauryl sulphate in liquid form. Applications: Anionic surfactant used in liquid shampoos and bubble baths. (Tensia SA).

Tensopol SPK. Potassium lauryl sulphate in powder form. Applications: Anionic surfactant used as a base in synthetic toilet bars. (Tensia SA).

Tensopol VAL. Sodium lauryl sulphate in liquid form. Applications: Liquid detergents, shampoos, bubble baths and carpet shampoos. (Tensia SA).

Tensopol 12A and 12P. Sodium dodecyl sulphate in needle or powder form. Applications: Anionic surfactant used in pharmaceuticals, toothpastes and shampoos. (Tensia SA).

Tensopol 30E and LDS. Sodium dodecylbenzene sulphonate LDS also contains optical dye. Applications: Foaming detergents. (Tensia SA).

Tensoprene. Low foaming surfactant. Rinse aid. Detergent. Applications: Dishwashing formulations; industrial cleaners. (Tensia SA).

Tensostat. Bactericide and fungicides. Applications: Paper mill. (Tensia SA).

Tensovax. Emulsifier, solubilizer; wetting agent. Applications: Oils, perfumes, vitamins; metal working; textile specialities. (Tensia SA).

Tensovyl. Dispersing agent. Applications: Post-tinctorial washing of polyester fibres. (Tensia SA).

Tensuccin D8. Sodium di-octyl sulphosuccinate in liquid form. Applications: Anionic surfactant used in textile desizing; dry cleaning; cosmetics. (Tensia SA).

Tensuccin HS40. Disodium stearic hemi-ester sulphosuccinate in the form of a paste. Applications: Anionic surfactant used as a dispersing agent. (Tensia SA).

Tensuccin H724 and H925. Disodium alkyl ether hemi-ester sulphosuccinate in water solution. Applications: Anionic surfactant used as a polymerization stabilizer. (Tensia SA).

Tensuccin ML, MO and MS. Anionic surfactants of the sulphosuccinama type, in liquid or soft paste form. Applications: Foaming agent used i latex for carpet backing, and foam insulation. (Tensia SA).

Tensyl 30. Sodium lauroyl sarcosinate liquid form. Applications: Anionic surfactant used in cosmetics, toothpastes and baby shampoos. (T Ltd).

Tensynvac. 1133 strain viral arthritis. Applications: immunization of poul (Intervet America Inc).*

Tentor. A proprietary preparation of phenylbutazone. (RP Drugs).†

Tenuate. A proprietary preparation of diethylpropion hydrochloride. An a obesity agent. (Richardson-Vicks Ir

Tenuate Dospan. A proprietary prepara of diethylpropion hydrochloride in a slow release base. (Richardson-Vick Inc).*

Tenvate. Diethylpropion hydrochloride. Applications: Anorexic. (Merrell D Pharmaceuticals Inc, Subsidiary of Chemical Co).

Tepanil. Diethylpropion hydrochloride. Applications: Anorexic. (Riker Laboratories Inc, Subsidiary of 3M Company).

Tephal. A wetting agent and detergent Applications: Used in the textile industry for pretreatment, desizing dyeing. (Degussa).*

Tepperite. A proprietary polystyrene.†

Teralan. Dyes for transfer polyester/w (Ciba-Geigy PLC).

Teraprint. Dyes for transfer printing. (Ciba-Geigy PLC).

Terasil. Disperse dyes. (Ciba-Geigy PL

Terate 101. Balsamic resins derived fro petroleum aromatic hydrocarbons. Applications: Used in adhesives and sealants, in alkyd coatings, in mould goods and in rubber compounding. (Hercules Inc).*

Terate 131. Balsamic resins derived fro petroleum aromatic hydrocarbons. Applications: Used in adhesives and sealants, in alkyd coatings, in mould goods and in rubber compounding. (Hercules Inc).*

Terate 202. Thermoplastic resins. Aromatic polyester polyols derived from polycarbomethoxy-substituted diphenyls, polyphenyls and benzyl esters of the toluate family. Applications: Used to extend reactive polyurethane and polyisocyanurate urethane polyols in the manufacture of rigid urethane foam. (Hercules Inc).*

Terate 203. Thermoplastic resins. Aromatic polyester polyols derived from polycarbomethoxy-substituted diphenyls, polyphenyls and benzyl esters of the toluate family. Applications: Used to extend reactive polyurethane and polyisocyanurate urethane polyols in the manufacture of rigid urethane foam. (Hercules Inc).*

Terate 204. Thermoplastic resins. Aromatic polyester polyols derived from polycarbomethoxy-substituted diphenyls, polyphenyls and benzyl esters of the toluate family. Applications: Used to extend reactive polyurethane and polyisocyanurate urethane polyols in the manufacture of rigid urethane foam. (Hercules Inc).*

Terathane. Polyether glycol. (Du Pont (UK) Ltd).

Terbalin. Active ingredients: terbutrex plus triflurex. Applications: Selective pre-emergence herbicidal mixture. (Agan Chemical Manufacturers Ltd).†

Terbufos. A pesticide. S-*tert*-butylthiomethyl-o,o-diethyl phosphorodithioate.†

Terbutol. Para-tert-butylphenol. (ICI PLC, Petrochemicals & Plastics Div).

Terbutrex. Active ingredient: terbutryne; 2-tert-butylamino-4-ethylamino-6-methylthio-1,3,5-triazine. Applications: Pre-emergence and post-emergence weed control. (Agan Chemical Manufacturers Ltd).†

Terbytex. Softening and antistatic. Applications: Natural and synthetic fibres. (Tensia SA).

Tercin. A proprietary preparation of aspirin, phenacetin and butobarbitone. An analgesic. (Cox, A H & Co Ltd, Medical Specialities Divn).†

Tercoton. Dyes for polyester/cotton. (Ciba-Geigy PLC).

Terebene. Acid-isomerized turpentine. It consists of a mixture of dipentene and other hydrocarbons.†

Terebenthene. Pinene, $C_{10}H_{16}$.†

Terebine. (Liquid Drier, Japan Drier). Made by heating oxides of lead and manganese with linseed oil or rosin, or mixtures of the oil and rosin, and thinning with turpentine or turpentine substitute. A drier for paints. It is not to be confused with Terebene.†

Terephane. A proprietary name for polyethylene terephthalate film. (French Origin).†

Terephthal Brilliant Green. A dyestuff, $C_{48}H_{60}N_4Cl_2.3ZnCl_2$, prepared from phthalyl chloride, diethylaniline, and zinc chloride. Dyes wool and silk yellowish-green.†

Terephthal Green. A dyestuff, $C_{40}H_{44}N_4Cl_2.3ZnCl_2$, prepared from phthalyl chloride, dimethylaniline, and zinc chloride. Dyes wool and silk yellowish-green.†

Terfenol. Rosin derivatives. Applications: Paints and inks. (Resinas Sinteticas SA).*

Terfluzin. Trifluoroperazine. (May & Baker Ltd).

Terfonyl. Trisulphapyrimidines (oral suspension). Mixture of Sulphadiazine, Sulphamerazine and Sulphamethazine. Applications: Antibacterial. (E R Squibb & Sons Inc).

Tergitex KW. Polyether solvent blend. Applications: Oil removal from garnetted wool and knitted goods. (Scher Chemicals Inc).*

Tergitol S. A proprietary trade name for a series of biodegradable non- ionic intermediates comprising ethoxylates and ethoxysulphates of linear secondary alcohols. Used in the production of biodegradable detergents. (Union Carbide (UK) Ltd).†

Tergitols. A proprietary trade name for wetting agents consisting of the sodium salts of the sulphates of higher alcohols.†

Tergraf. Rosin derivatives. Applications: Paints and inks. (Resinas Sinteticas SA).*

Tergum. Rosin derivatives. Applications: Paints, inks, adhesives, chewing gum. (Resinas Sinteticas SA).*

Terinda. Polyester yarns. (ICI PLC).*

Terlac. Rosin derivatives. Applications: Paints and inks. (Resinas Sinteticas SA).*

Terlan. Polyamide resins. Applications: Adhesive/moulding applications. (The Terrell Corporation).*

Terluran. ABS colour compounds. (Norsk Hydro Polymers Ltd).

Terluran 846 L. A proprietary ABS of medium rigidity and toughness used for injection moulding, extrusion and thermoforming. (BASF United Kingdom Ltd).*

Terluran 8760 Galvano. A special grade of ABS used for electroplating. (BASF United Kingdom Ltd).*

Terluran 886. A tough grade of ABS. (BASF United Kingdom Ltd).*

Term-X (Various Herbicides/Insecticides). Varies with each product. Applications: Do-it-yourself products for household and commercial use and application. (Puma Chemical Co Inc).*

Termamyl. A liquid enzyme preparation containing an outstandingly heat-stable alpha-amylase produced by a selected strain of Bacillus licheniformis. Applications: Used in the following industries: starch, alcohol, brewing, sugar and textile. (Novo Industri A/S).*

Termex. Laminates of nomex, presspaper, leatheroid, paper, melinex, mylar and kapton. Applications: Slot liner and closure material used in the manufacture and repair of electrical motors, transformers and other electrical equipment. (Fothergill Tygaflor Ltd).*

Terminate. *Bacillus thuringiensis* wettable powder. Applications: Applied by spray to control larvae of *Lepidopteran* insects. (Westbridge Research Group).*

Ternary Steels. Alloy steels containing one special element in addition to the iron and carbon.†

Terne Metal. An alloy of 80 per cent lead, 18 per cent tin, and 2 per cent antimony.†

Terne Plate. An alloy of lead and tin, coa on iron plate, and intended for use in roofing.†

Terolut. Dydrogesterone. Applications: Progestative, counteracts complaints caused by hormonal disorders in wome (Duphar BV).*

Terpalin. A proprietary preparation of EUCALYPTOL, menthol TERPINE hydrate and codeine phosphate. A coug syrup. (Norton, H N Co).†

Terpenato. Rosin derivatives. Application Paints and inks. (Resinas Sinteticas SA).*

Terpestrol. A powder containing lactose with 5 per cent oil of turpentine.†

Terpex D,K-3,S. A proprietary trade name for terpene vinyl plasticizers. (Glidden Co).†

Terpigol. A proprietary trade name for terpinyl monoethylene glycol ether. (BTP Cocker Chemicals Ltd).†

Terpilenol. See LILACIN.

Terpine. A turpentine substitute. It is a product of the distillation of petroleum.

Terpinol. A mixture of terpenes containing terpinene, dipentene, also terpineol and cineol.†

Terpoin. A proprietary preparation of eucalyptol, terpin hydrate, codeine phosphate, menthol and guaiphenesin. A cough linctus. (Hough, Hogeason).†

Terposol No. 3. A proprietary trade name for a solvent consisting of terpene meth ethers.†

Terposol No. 8. A proprietary trade name for a solvent consisting of terpene glycol ethers.†

Terpurile. (Cycloran, Solventol). Wetting-out agents consisting of soaps with organic solvents.†

Terr-o-gas. Methylbromide with chloropicrin. (Great Lakes Chemical (Europe) Ltd).

Terra Alba. See GYPSUM.

Terra-Bron. A proprietary preparation of oxytetracycline, ipecacuanha and ephedrine hydrochloride. (Pfizer International).*

Terra Cariosa. See ROTTEN STONE.

Terra Catechu. See CUTCH.

Terra-Cortril. A proprietary preparation of oxytetracycline hydrochloride, hydrocortisone. An anti-infective and anti-inflammatory. (Pfizer International).*

Terracote. Coatings for dies, moulds, chills etc. (Foseco (FS) Ltd).*

Terra-Cotta. A building material made from clay.†

Terra-Cotta F. A dyestuff. It is the sodium salt of primulineazophenylenediamineazonaphthalenesulphonic acid. Dyes cotton brown from a neutral or alkaline bath.†

Terra-Cotta G. See INGRAIN BROWN.

Terra-Cotta R. See ALIZARIN YELLOW R.

Terra di Sienna. See INDIAN RED and YELLOW OCHRE.

Terradust. Brass foundry 'caster's flour'. (Foseco (FS) Ltd).*

Terra Fullonica. Fuller's earth.†

Terra Japonica. See CUTCH.

Terraklene. Herbicide. (ICI PLC).*

Terraline. See PARAFFIN, LIQUID.

Terram. Non-woven civil engineering fabric. (ICI PLC).*

Terra Merita. See TURMERIC.

Terramycin. A proprietary preparation of oxytetracycline. An antibiotic. (Pfizer International).*

Terramycin Hydrochloride. Oxytetracycline hydrochloride. Applications: Antibacterial; antirickettsial. (Pfizer Inc).

Terramycin Pediatric Drops. Oxytetracycline calcium. Applications: Antibacterial. (Pfizer Inc).

Terraneb SP Turf Fungicide. 65% chloroneb wettable powder. Applications: For control of snow mould (*Typhula*) and pythium blight. (Kincaid Enterprises Inc).*

Terrapaint. Coatings for foundry moulds and cores. (Foseco (FS) Ltd).*

Terra Ponderosa. Barium sulphate, $BaSO_4$.†

Terrapowder. Ferrous mould and core dressings. (Foseco (FS) Ltd).*

Terrar. A preparation from earthy zirconia, in Brazil. Used as an opacifying agent in enamels and glazes.†

Terra-Systam. Systemic organophosphorus insecticides. (Murphy Chemical Co).†

Terra Verte. A pigment. It is a green earthy material found in the Mendip Hills. It consists of a species of ochre, and is essentially silica with oxide of iron and small quantities of other oxides. See also BOHEMIAN EARTH.†

Terravest 801. A stereospecific, low molecular weight polybutadiene. Applications: It consolidates soil against erosion and binds dust, e.g. on heaps of all sorts. (Huls AG).†

Terroline. See PARAFFIN, LIQUID.

Terry's Stain. A microscopic stain. It contains 20 cc of a 1 per cent aqueous solution of methylene blue, 20 cc of a 1 per cent solution of potassium carbonate, and 60 cc of water. This is boiled, cooled, and 10 cc of a 10 per cent solution of acetic acid added, and the whole made up with water to 100 cc.†

Tertroxin. A proprietary preparation of liothyronine sodium. Thyroid hormone preparation. (Glaxo Pharmaceuticals Ltd).†

Terylene. A trade mark for synthetic polyester textile fibre, resistant to most dry cleaning solvents, possesses good wear resistance. It is polyethylene terephthalate produced from dimethyl terephthalate and ethylene glycol. (ICI PLC).*

Teslac. Testolactone. Applications: Antineoplastic. (E R Squibb & Sons Inc).

Tessalon. Benzonatate. Applications: Antitussive. (Endo Laboratories Inc, Subsidiary of E I du Pont de Nemours & Co).

Tessalon. Benzonatate.†

Testalin. An aluminium soap, made by treating ordinary soap with aluminium sulphate. Used for the cementing together of sandstone to form a solid block.†

Testate. Testosterone Enanthate. Applications: Androgen. (Savage Laboratories, Div of Byk-Gulden Inc).

Testifas Oil. A fraction of petroleum distillation. Used as a burning oil.†

Testred. Methyltestosterone. Applications: Androgen. (ICN).

Tesuloid. Technetium Tc 99m Sulphur Colloid. Applications: Radioactive agent. (E R Squibb & Sons Inc).

Tetanol. A proprietary preparation of calcium hevulinate.†

Tetiothalein. Sodium tetraiodophenolphthalein.†

Tetjamer. An aluminium bronze. It contains from 86-93 per cent copper, 5-10 per cent aluminium, 1-3 per cent silicon, and 0.72-0.98 per cent iron.†

Tetmosol. A proprietary preparation of tetraethylthiuram monosulphide, used as an insecticide in soap or solution form, for either human or veterinary use. (ICI PLC).*

Tetra. See TETRAPHOSPHATE.

Tetra-Base-Paper. See TETRA-PAPER.

Tetracarnit. A mixture of pyridine and its homologues with Turkey red oil or similar substances. It is used as a wetting-out agent to assist the penetration of textiles by liquids. Also see NEKAL and AVIVAN.†

Tetrachel. A proprietary preparation of tetracycline hydrochloride. An antibiotic. (Berk Pharmaceuticals Ltd).*

Tetracyn P. A proprietary preparation containing tetracycline hydrochloride, sodium hexametaphosphate. An antibiotic. (Pfizer International).*

Tetracyn S.F. A proprietary preparation of tetractcline with vitamin supplements. An antibiotic. (Pfizer International).*

Tetradine. See LYCETOL.

Tetraflon. A proprietary polytetrafluoroethylene (P.T.F.E.). (Nitto Chemical, Japan).†

Tetraform. A specially pure carbon tetrachloride.†

Tetralin. Tetrahydronaphthalene, $C_{10}H_{12}$. It is a solvent for gums, oils, waxes, and resins, and is used as a substitute for turpentine.†

Tetralin Extra. A mixture of Tetralin (*qv*) and Dekalin (*qv*).†

Tetraline. An old name for tetrachloroethane.†

Tetralitbenzol. A mixed fuel for internal combustion engines. It contains 50 per cent benzol, 25 per cent tetralin, and 2 per cent of 95 per cent alcohol.†

Tetralite. See TETRYL.

Tetralol. Tetrahydro-β-naphthol. An antiseptic.†

α-Tetralone.. A synonym for α-keto-tetrahydronaphthalene.†

Tetramet-125. Thyroxine I 125. Applications: Radioactive agent. (Abbott Laboratories).

Tetramethyl Base. Tetramethyldiaminodiphenylmethane, $(CH_3)_4.N_2.(C_6H_4)_2.CH_2$.†

Tetranitrin. See TETRA-NITROL.

Tetra-nitrol. (Tetranitrin, Butane Tetrol). Erythrol-tetranitrate, $C_4H_6(NO_3)_4$. Has been used in angina.†

Tetranyl. An explosive. It is 2, 3, 4, 6-tetranitro-aniline.†

Tetra-paper. (Tetra-base-paper). Paper which has been treated with dimethyl or tetramethyl-*p*-phenylene-diamine. It is used in testing for ozone.†

Tetraphosphate. (Tetra). Produced by mixing natural phosphate rock powder with 6 per cent of a powder containing equal parts of the carbonates of calcium, sodium, and magnesium, with a little sulphate of soda. The mixture is roasted at from 600-800° C then treated with cold phosphoric acid.†

Tetrapol. See AVIVAN.

Tetrathal. Tetrachlorophthalic anhydride. Applications: Flame retardant for polyester resins and polyols. (Monsanto Co).*

Tetrazets. Bacitracin, tyrotricin, neomycin and benzocaine. Applications: For the relief of minor mouth and throat infections. (Merck Sharp & Dohme).*

Tetrazine. (Hydrazine Yellow). A dyestuff, $C_{16}H_{10}N_4Na_2S_2O_9$. Dyes wool and silk yellow from an acid bath.†

'etrex. Tetracycline Phosphate Complex. Applications: Antibacterial. (Bristol Laboratories, Div of Bristol-Myers Co).

'etrex Bidcaps. A proprietary preparation containing Tetracycline Phosphate Complex. An antibiotic. (Bristol-Myers Co Ltd).†

'etrex PMT. A proprietary trade name for Rolitetracycline.†

'etron. Tetra sodium pyrophosphate. (Albright & Wilson Ltd, Phosphates Div).

'etronal. (Ethyl sulphonal). Diethyl-sulphodiethylmethane, $(C_2H_5)_2.C.(SO_2.C_2H_5)_2$. A hypnotic allied to Sulphonal.†

'etrone. Rubber accelerator. (Du Pont (UK) Ltd).

'etrosan 3,4 D. Alkyl dimethyl 3:4 dichlorobenzyl ammonium chloride in liquid form. Applications: Disinfectant, deodorant and germicide with high biocidal activity. Veterinary, pharmaceutical and agricultural uses. (Millmaster-Onyx UK).

Tetroxone. Selective weedkiller containing bromoxynil, dichlorprop, ioxynil and MCPA as potassium salts. (ICI PLC).*

Tetryl. (Tetralite). A trade name for trinitrophenylmethylnitramine, $C_6H_2(NO_2)_3.NCH_3.NO_2$, used in explosives.

A detonator known as "tetryl" contains 0.4 gram tetranitro-phenylmethylnitramine, and 0.3 gram of a mixture of 87.5 per cent mercury fulminate and 12.5 per cent potassium chlorate.†

Texaco BQ. An insecticide having a petroleum base. It is used for killing the boll weevil.†

Texalin. See SAVONADE.

Texalys. Modified starch. Applications: Textile industry in precoating compounds and also in laminating double backs. (Roquette (UK) Ltd).*

Texanol. A trade mark. 2,2,4-trimethyl-1,3-pentanediolmonoisobutyrate. An intermediate for the manufacture of plasticisers, surfactants, urethanes and pesticides. (Eastman Chemical Products).†

Texapon. See SAVONADE.

Texapon ALS and ALS/S. Ammonium lauryl sulphate in liquid/paste form. Applications: Basic material for liquid shampoos. (Henkel Chemicals Ltd).

Texapon ASV. Mixture of special fatty alcohol ether sulphates in liquid form. Applications: Basic material for liquid shampoos, especially baby shampoos. (Henkel Chemicals Ltd).

Texapon ESI/S, N25 and N40. Sodium lauryl ether sulphate in liquid form. Applications: Basic material for liquid shampoos and bubble bath preparations. (Henkel Chemicals Ltd).

Texapon EVR and EST. Sodium lauryl ether sulphate with special additives, in the form of a liquid emulsion. Applications: Basic material for emulsion shampoos and pearly sheen shampoos. (Henkel Chemicals Ltd).

Texapon IES. Alkanol amine salt of fatty alcohol ether sulphate with fatty acid alkanol amide in liquid form. Applications: Basic material for liquid shampoos and bubble bath preparations. (Henkel Chemicals Ltd).

Texapon L20C. Ammonium amine lauryl sulphate in liquid form. Applications: Basic material for liquid shampoos. (Henkel Chemicals Ltd).

Texapon L20M and L20M/S. Anionic surfactant in liquid form. Applications: Basic material for liquid shampoos. (Henkel Chemicals Ltd).

Texapon L230 and T42. Triethanolamine lauryl sulphate in liquid form. Applications: Basic material for liquid shampoos and bubble bath preparations. (Henkel Chemicals Ltd).

Texapon NSF. Sodium lauryl ether sulphate in liquid form, with extremely low salt contents. Applications: Basic material for cosmetic preparations such as shampoos, bubble bath preparations etc. (Henkel Chemicals Ltd).

Texapret. Textile finishing agents. (BASF United Kingdom Ltd).

Texicote. Emulsion polymers and copolymers based on vinyl acetate. Powder coatings. Applications: Paint, textiles, adhesives. (Scott Bader Co Ltd).*

Texicryl. Emulsion polymers and copolymers based on acrylic acid esters. Applications: Paint, textiles, adhesives, paper coating. (Scott Bader Co Ltd).*

Texigel. Emulsion or solution polymers based on water soluble monomers. Applications: Thickeners, dispersing agents. (Scott Bader Co Ltd).*

Texileather. A proprietary trade name for pyroxylin-coated leather cloth.†

Texilose Yarn. See XYLOLIN YARN.

Texin. A proprietary polyester-based polyurethane. (Mobay Chemical Corp).*

Texipol. Polymeric materials in the form of solids, liquids, emulsions or gels. Applications: Paper, textiles, adhesives, paints, thickening agents and dispersants. (Scott Bader Co Ltd).*

Texoderm. A cellulose product. It is an imitation leather.†

Texofor. Non-ionic surfactant. (ABM Chemicals Ltd).*

Texofor A and B. A proprietary range of higher fatty alcohol-based polyoxy-alkylene condensates used as non-ionic surfactants. (Glover (Chemicals) Ltd).†

Texofor C. A proprietary range of unsaturated fatty acid-based polyoxy-alkylene condensates used as non-ionic surfactants. (Glover (Chemicals) Ltd).†

Texofor D. A proprietary range of glyceride oil-based polyoxyalkylene condensates used as non-ionic surfactants. (Glover (Chemicals) Ltd).†

Texofor E and ED. A proprietary range of saturated fatty acid-based polyoxy-alkylone condensates used as non-ionic surfactants. (Glover (Chemicals) Ltd).†

Texofor FN and FP. A proprietary range of alkyl phenol-based polyoxy- alkylene condensates used as non-ionic surfactants. (Glover (Chemicals) Ltd).†

Texofor FN, FP and FX. Range of nonionic surfactants of the alkylphenol ethoxylate type. Application: Industrial and household cleaners; emulsion polymerization; agriculture etc. (ABM Chemicals Ltd).

Texofor G. A proprietary unsaturated fatty acid-based polyoxyalkylene condensate used as a non-ionic surfactant. (Glover (Chemicals) Ltd).†

Texofor J4. A proprietary bio-degradable non-ionic emulsifier. It is a linear fatty alcohol ethoxylate. (Glover (Chemicals) Ltd).†

Texofor M. A proprietary range of unsaturated fatty acid-based polyoxy-alkylene condensates used as non-ionic surfactants. (Glover (Chemicals) Ltd).

Texofor N. A proprietary range of fatty alcohol-based polyoxyalkylene condensates used as non-ionic surfactants. (Glover (Chemicals) Ltd).

Texofor P. A proprietary range of comple amide-based polyoxyalkylene condensates used as non-ionic surfactants. (Glover (Chemicals) Ltd).

Texofor T. A proprietary range of higher fatty alcohol-based polyoxy- alkylene condensates used as non-ionic surfactants. (Glover (Chemicals) Ltd).

Texogent. Blend of surfactants and solvents. Applications: Degreasing, cleaning, anti-spotting in textile and engineering industries. (ABM Chemicals Ltd).

Texowax. Water soluble waxy coatings - release lubricant; softening. Applications: Paper coating; plastic; textiles; adhesives. (ABM Chemicals Ltd).

Texsolve. Aliphatic hydrocarbon solvents. (Texaco Chemical Co).*

Textase. A diastase preparation.†

Textile Wax. Textile preparation and finishing agent. (BASF United Kingdom Ltd).

Textulite. A proprietary trade name for phenol formaldehyde laminated synthetic resin and moulded compounds.†

TFC. Thin film composite spiral wound membrane. Applications: Reverse osmosis water treatment. (Allied-Signal Fluid Systems).*

Tfol. An argillaceous earth containing free gelatinous silica. Can be used as a soap

TG-8. Tri-ethylene glycol di-caprylate. (Ruco Divn).†

T-gas. The commercial mixture of ethylene oxide and carbon dioxide. Used as an insecticide.†

Thaio Green. No. 1. A halogenated copper phthalocyanine green. Extremely light and heat-fast. (Reckitts).†

Thalamonal. A proprietary preparation of fentanyl, and droperidol, used as a premedication in anaesthesia. (Janssen Pharmaceutical Ltd).*

Thalazole. A proprietary preparation containing phthalylsulphathiazole. (May & Baker Ltd).*

Thallium. Thallous Chloride TI 201. Applications: Diagnostic aid; radioactive agent. (Amersham Corp).

Thallium Alum. A double sulphate of thallium and aluminium, $Tl_2SO_4.Al_2(SO_4)_3.24H_2O$.†

Thalo Blue No. 1. An alpha, solvent sensitive, red shade phthalocyanine blue pigment. Used for solventless printing inks. (Reckitts).†

Thalo Blue No. 2. A beta, solvent stable, green shade phthalocyanine blue pigment. (Reckitts).†

THAM. Tromethamine. Applications: Alkalizer. (Abbott Laboratories).

Thancat. A series of amine-type urethane catalysts. Applications: Catalysts for production of flexible and rigid urethane foams, elastomers and sealants. (Texaco Chemical Co).*

Thanol. A series of urethane polyols. Applications: Intermediates for production of flexible and rigid urethane foams, elastomers and sealants. (Texaco Chemical Co).*

Thao. A gelatinous preparation made in Cochin China from seaweed. It has frequently appeared in England under the names of Japanese or Chinese isinglass, and is used for the same purposes as isinglass.†

Thapsia Resin. The resin of *Thapsia garganica* root. It contains caprylic acid and thapsic acid.†

Thawpit. A preparation of carbon tetrachloride used for cleaning materials.†

Theelin. Estrone. Applications: Oestrogen. (Parke-Davis, Div of Warner-Lambert Co).

Theic. A proprietary preparation of tris-(2-hydroxyethyl)isocyanurate used in the manufacture of heat-resistant wire lacquers. (BASF United Kingdom Ltd).*

Theine. (Guaranine). Caffeine.†

Theobroma Oil. See COCOA BUTTER.

Theobromine. Dimethylxanthine, $C_7H_8O_2N_4$. The active principle of the cocoa bean.†

Theobromose. The lithium compound of theobromine, $C_7H_7N_4O_2Li$.†

Theocal. A double salt of calcium theobromine and calcium lactate. A diuretic.†

Theocalcin. A proprietary preparation of theobromine and calcium salicylate.†

Theocyl. Theobromine acetylsalicylate.†

Theodrex. Each tablet contains aminophylline BP 195 mg and dried aluminium hydroxide gel BP 260 mg (Riker Laboratories/3M Health Care).†

Theogardenal. Theobromine and phenobarbitone. (May & Baker Ltd).

Theograd. A proprietary preparation of theophylline. A bronchodilator. (Abbott Laboratories).*

Theominal. A proprietary preparation of phenobarbitone and theobromine. An antihypertensive. (Bayer & Co).*

Theonar. A proprietary preparation of THEOPHYLLINE and NOSCAPINE. A bronchodilator. (MCP Pharmaceuticals).†

Theophen. A proprietary preparation of butethamate citrate, amylobarbitone, ephedrine hydrochloride and theophylline. A bronchial antispasmodic. (Rybar).†

Thephorin. A proprietary preparation of phenindamine hydrogen tartrate. (Roche Products Ltd).*

Therabloat. Poloxalene. Liquid nonionic surfactant polymer of the polyethylene-polypropylene glycol type. Applications: Pharmaceutic aid. (Norden Laboratories Inc).

Theralax. Bisacodyl. Applications: Laxative. (Beecham Laboratories).

Therban. Hydrogenated NBR. (Bayer UK Ltd).

Theriodide-131. Sodium Iodide I 131. Applications: Antineoplastic; diagnostic aid; radioactive agent. (Abbott Laboratories).

Therlo. An electrical resistance alloy containing 85 per cent copper, 13 per cent manganese, and 2 per cent aluminium.†

Therm-Chek. Applications: For heat and light stabilizers for vinyl halide compositions, in Int Class 1. (Ferro Corporation).*

Thermaflo. PVC compounds. Applications: A wide range of mouldings and extrusions including applications in footwear, building, cable, automotive etc.(Evode Plastics Ltd).*

Thermalene. An intimate mixture of acetylene and vaporised oils. It is used for the production of high temperatures, in the cutting and welding metals.†

Thermalloy. A patented form of thermit containing 50 per cent iron oxide, 27 per cent aluminium, and 23 per cent sulphur. The name appears to be applied also to an alloy containing 66.5 per cent nickel, 30 per cent copper, and 2 per cent iron. It has a magnetic permeability which decreases at higher temperature. An alloy containing 75-85 per cent iron, 10-20 per cent chromium, 2-6 per cent silicon, 0.5-1 per cent manganese, 0.5-1 per cent, tungsten, and 0.2-2 per cent carbon is also known as Thermalloy.†

Thermalloy A. A proprietary trade name for an alloy containing 67.5 per cent nickel, 0.15 per cent carbon, 0.15 per cent silicon and 30 per cent copper.†

Thermalloy B. A proprietary trade name for an alloy containing 57.8 per cent nickel, 0.15 per cent carbon, 0.15 per cent silicon, and 40 per cent copper.†

Thermatomic Carbon. A fine carbon produced by "cracking" natural gas into carbon and hydrogen by passing the gas over heated brickwork. Used as a rubber filler.†

Thermax. Carbon black. Applications: Reinforcer for rubber and plstics. Pigment for paint and plastics. Used also in metallurgy. (Vanderbilt Chemical Corporation).*

Thermazote. A trade mark for an expande thermosetting plastic, manufactured in densities between 7 and 30 lb per cubic foot. It is non-inflammable and odourless and withstands temperatures as high as 300° C. It has a low thermal conductivity and is used in building construction, etc.†

Thermex. Heat transfer media. (ICI PLC).*

Thermexo. Highly exothermic metal producing compounds. (Foseco (FS) Ltd).*

Thermica. S_1O_2 AL_2O_3 M_gO K_2O F + 3% impurities. Applications: Additive to oil, paint, grease, plastics, glass and other inorganic materials which require a high dielectric or high heat resistant material. A coating for welding rods requiring special atmospheric controls. A thermal barrier and high electrical resistivity. (Mykroy/Mycalex).*

Thermine. Tetrahydro-β-naphthyl-amine hydrochloride.†

Therminol. Heat transfer fluids. (Monsanto Co).*

Thermisilio. A proprietary trade name for a chemical-resisting iron- silicon alloy in which the brittleness has been diminished.†

Thermisilio Extra. Similar to the above but has a higher silicon content, is more resistant to acids and has greater hardness and brittleness.†

Thermisilizid. A Swedish iron-silicon alloy of the acid-resisting type.†

Thermit. A world-wide registered trade mark for (a) Alumino-thermic mixtures consisting essentially of nearly equal parts of powdered aluminium and metal oxides, usually iron or manganese oxides. These mixtures burn with a high temperature, and are used for welding metals. They are also used as an ingredient in incendiary bombs. (b) A bearing metal containing lead, antimony (20 per cent.) and small quantities of tin, nickel and copper.†

Thermit Manganese. Manganese metal made by the Thermit reduction method. It contains approximately 98 per cent manganese.†

Thermit Metal. A German bearing metal containing: 14-16 per cent antimony, 5-7 per cent tin, 0.8-1.2 per cent copper, 0.7-1.5 per cent nickel, 0.3-0.8 per cent arsenic, 0.7-1.5 per cent cadmium and 72-78.5 per cent lead.†

Thermlo F. A proprietary rubber vulcanization accelerator. It is an organic polysulphide.†

Thermocast. A proprietary trade name for an ethyl cellulose composition.†

Thermodek. Dry roof screed system. (Vencel Resil Ltd).

Thermoflex. A proprietary trade name for di-para-methoxy- diphenylamine. An antioxidant. (Imperial Chemical Industries), (Du Pont (UK) Ltd).†

Thermoflex A. A proprietary trade name for an antioxidant. It contains 50 parts of phenyl-beta-naphthylamine, 25 parts of methoxy-diphenylamine and 25 parts of diphenyl-para-phenylene diamine. (Imperial Chemical Industries), (Du Pont (UK) Ltd).†

Thermolastic. A registered trade name for extrudable thermoplastic rubber-like materials based upon styrene butadiene copolymers not requiring vulcanization. (Shell Chemicals).†

Thermonit. (Kerarnonit). A refractory cement made in the electric furnace. It is stated to be used as a paint or mortar, and to resist high temperatures. Keramonit is the cement reinforced with metal mesh.†

Thermoplaste. Plastics dyes. (ICI PLC).*

Thermoprene. Products obtained by heating rubber with either an organic sulphonyl chloride or an organic sulphonic acid at 125-135° C. for several hours. *p*-Toluene-sulphonyl chloride and *p*-toluene- sulphonic acid are suitable reagents. One product is a protective paint which is resistant to acids, alkalis, and corrosive gases, and has a low permeability to water. Products resembling gutta-percha, balata, and shellac are also obtained. See also CYCLORUBBERS.†

Thermorun. Thermoplastic elastomer. (Mitsubishi Petrochemical Co).*

Thermoseal. Heat seal lacquers. (The Scottish Adhesives Co Ltd).*

Thermotex. Thermit bottom plate patching compound. (Foseco (FS) Ltd).*

Thetmex. Heat transfer medium. (ICI PLC, Petrochemicals & Plastics Div).

Thial. Hexamethylenetetramine hydroxymethylsulphonate. An antiseptic.†

Thiamin Yellow. A direct cotton dyestuff produced by the action of formaldehyde upon primuline.†

Thiate. Trimethylthiourea, 1,3-diethylthiourea, 1,3-dibutylthiourea. Rubber accelerators. Applications: Used in a wide range of rubber acceleration applications. (Vanderbilt Chemical Corporation).*

Thiate E. A proprietary accelerator. Trimethyl thiourea. (K & K Greef Chemicals Ltd).†

Thiazamide. Sulphathiazole. (May & Baker Ltd).

Thiazina. A proprietary preparation containing thiacetazone, isoniazid. An antituberculous agent. (Smith and Nephew).†

Thiazine Brown G. A dyestuff similar to the R mark.†

Thiazine Brown R. A direct cotton dyestuff, dyeing shades of red-brown.†

Thiazine Red G. An azo dyestuff prepared from dehydrothiotoluidinesulphonic acid. A direct cotton colour.†

Thiazine Red R. See ROSOPHENINE 10 B.

Thiazol Yellow. See CLAYTON YELLOW.

Thiazol Yellow G. A dyestuff. It is equivalent to Clayton yellow.†

Thiazol Yellow R. A dyestuff. It is equivalent to Nitrophenin.†

Thiazoline. (Thiazylamine). Aminothiazole, $C_3H_4N_2S$.†

Thiazylamine. See THIAZOLINE.

Thibenzole. A proprietary trade name for Thiabendazole.†

Thickened Mineral Oils. Mineral oils which have been thickened by dissolving soap, usually aluminium soap, in them.†

Thickened Oils. See BLOWN OILS.

Thickener. See OIL-PULP.

Thi-Di-Mer. Trisulphapyrimidines (oral suspension). Mixture of Sulphadiazine, Sulphamerazine and Sulphamethazine. Applications: Antibacterial. (Merrell Dow Pharmaceuticals Inc, Subsidiary of Dow Chemical Co).

Thiel-Stoll Solution. A saturated solution of lead chlorate, $Pb(ClO_4)_2$. It has a density of 2.6 and is used for the determination of the specific gravity of minerals.†

Thiersch's Antiseptic Solution. A solution containing salicylic acid and boric acid.†

Thiery's Solution. A solution of picric acid for the treatment of burns.†

Thiet-sie. A resinous substance used as a varnish by the Burmese.†

Thilaven. Ammonium ichthosulphonate.†

Thimerosal. A proprietary trade name for Thiomersal.†

Thinner No. 22. An industrial solvent containing approximately 55 per cent terpene hydrocarbon and 45 per cent gasoline.†

Thinners. Formulated thinners for use with spirit based coatings. (Foseco (FS) Ltd).*

Thinoline. (Vulcanized Oils). vulcanized linseed oil, used in rubber mixings.†

Thiocamf. A liquid formed by exposing camphor to the action of sulphur dioxide. Used as a disinfectant as it evolves sulphur dioxide on exposure to air.†

Thiocarmine R. A dyestuff. It is the sodium salt of diethyldibenzyldiamino phenazthioniumdisulphonic acid, $C_{30}H_{28}N_3O_6S_3Na$. Dyes wool and silk indigo-carmine shades from an acid bath.†

Thiocatechine. (Thiocatechine S). A dyestuff obtained by the fusion of *p*-diamines or acetylnitramines with sodium polysulphide. The S mark is the sulphite compound. Dyes cotton brown.†

Thiocatechine S. See THIOCATECHINE.

Thiochem Sulphur Green. (Sulphur Olive Green). A dyestuff produced by the fusion of benzeneazophenol with a copper salt at 180-200° C.†

Thiochromogen. See PRIMULINE.

Thiochrysine. A gold-sodium thiosulphate.†

Thio Cotton Black. A dyestuff obtained by the fusion of a mixture of dinitrophenol and *p*-aminophenolsulphonic acid with sodium polysulphide. Dyes cotton black.†

Thioctacid. α- Liponacid. Tablets and ampoules. Applications: Liver protection. (Degussa).*

Thiocyanosin. An eosin dyestuff. It is tetrabromothiodichlorofluoresceinmethyl ether.†

Thiodet. Residual thiosulphate test kit. (May & Baker Ltd).

Thiodigo Red BG. A dyestuff. It is 5, 5'-dichlorothioindigo.†

Thiofide. 2,2'-Dithiobis(benzothiazole, benzothiazyl disulphide). Applications: vulcanization accelerator. (Monsanto Co).*

Thioflavine S. (Chromine G). A dyestuff. It is the sodium salt of methylated primuline. Dyes cotton, silk, and half-silk goods greenish-yellow from an alkaline bath.†

Thioflavine T. (Acronol Yellow T.). A dyestuff. It is dimethyldehydrothio-toluidine methylochloride, $C_{17}H_{19}Cl.N_2S$. Dyes tannined cotton greenish yellow, and silk yellow with green fluorescence.†

Thiofurfuran. Thiophene, C_4H_4S.†

Thiogene Black. (Thiogene Purple ; Thiogene Dark Red G, R ; Thiogene Rubine O). Sulphide dyestuffs obtained by the action of sulphur upon aminohydroxyphenazines.†

Thiogene Blue. See SULPHUR BLACK T.

Thiogene Dark Red, G, R. See THIOGENE BLACK.

Thiogene Purple. See THIOGENE BLACK.

Thiogene Rubine O. SEE THIOGENE BLACK.†

Thiogene Violet V. A sulphide dyestuff manufactured by heating together sulphur and phenosafranine, then heating the product with sodium sulphide. It is further heated with sodium polysulphide.†

Thioguanine. 2-Aminopurine-6-thiol. LANVIS.†

Thioindigo B. See THIOINDIGO RED B.

Thioindigo Grey B. A dyestuff. It is 7, 7'-diaminothioindigo.†

Thioindigo Orange R. A dyestuff. It is 6, 6'-diethoxythioindigo.†

Thioindigo Pink BN. A dyestuff. It is 6, 6'-dibromodimethylthioindigo.†

Thioindigo Red B. (Vat Dye B, Thioindigo B, Durindone Red B). An indigo dyestuff, $C_{16}H_8O_2S_2$, prepared from phenylthioglycol-*o*-carboxylic acid, by boiling it with alkalis, heating the product with acids, and oxidizing the thioindoxyl produced. It is a yellow-red dyestuff.†

Thioindigo Red 3B. A dyestuff. It is 5, 5'-dichloro-6, 6-dimethylthioindigo.†

Thioindigo Scarlet. An indigo dyestuff, $C_{16}H_9O_2NS$, prepared by condensing thioindoxyl with isatin. A yellow vat dye.†

Thioindigo Scarlet S. A dyestuff. It is 6 : 6'-dithioxylthioindigo.†

Thioindigo Violet 2B. A dyestuff. It is dichloro-dimethyl-dimethoxy-thioindigo.†

Thiokol F.A. A proprietary ethylene dichloride and dichloroethyl formal condensed with sodium sulphide. THIOKOL ZR300.†

Thiokol LP. Liquid polysulphide polymers with SH-Terminals. Applications: Thiokol LP is used as a basis polymer for the production of sealants for insulating glass, building joints, caulking ship decks etc. (Thiokol Gesellschaft mbH).*

Thiokol RD. A proprietary trade name for an oil-resistant synthetic rubber. It is an interpolymer of butadiene and acrylonitrile with a third unspecified component. It contains 4 per cent nitrogen.†

Thiokol TP-90-B. and TP-95. See TP-90.

Thiokol ZR 300. See THIOKOL F.A.

Thiol. (Tumenol, Petrosulfol). Names given to artificial substitutes for ammonium ichtho-sulphonates. Thiol is a product similar to ichthyol (*qv*), and is obtained by sulphonating gas oil (brown coal tar oil), with sulphur, treating the product with sulphuric acid, and pouring the whole into water. Used medicinally in the same way as ichthyol.†

Thiolim. Hypo eliminator. (May & Baker Ltd).

Thiolin. See ICHTHYOL.

Thiolite. An insulator prepared from formaldehyde, cresol, and sulphur chloride.†

Thion Black. A dyestuff prepared by heating sodium tetrasulphide and sodium dinitro-phenoxide at 140-180° C. for from 2-3 hours.†

Thion Blue B. A sulphide dyestuff derived from *p*-nitro- *o*-amino-*p*-hydroxy diphenylamine.†

Thion Green. A sulphide dyestuff obtained by the action of sodium hydroxide upon *p*-hydroxyphenylthiocarbamide.†

Thion Yellow. A dyestuff obtained by heating thio-*m*-tolylenediamine with sodium sulphide solution.†

Thional Black. See SULPHUR BLACK T.

Thional Bronze. A sulphide dyestuff obtained by the fusion of β-hydroxynaphthoquinoneanilide with sodium polysulphides.†

Thional Brown R. A sulphide dyestuff which dyes unmordanted cotton.†

Thional Green. A sulphide dyestuff.†

Thionalide. A commercial name for thioglycollic acid, β-amino-naphthalide, $HS.CH_2CO.NH.C_{10}H_7$. An analytical reagent.†

Thionex. Tetramethyl-thiuram-monosulphide. An ultra-accelerator for rubber vulcanization.†

Thionex. Active ingredient: endosulfan; 6,7,8,9,10,10-hexachloro-1,5,5a,6,9,9a-hexahydro-6,9-methano-2,4,3-benzo e dioxathiepin-3-oxide. Applications: A chlorinated cyclic sulphurous acid ester having broad-spectrum insecticidal activity of long-lasting effect. (Makhteshim Chemical Works Ltd).†

Thionhydrol. Colloidal sulphur.†

Thionine. See LAUTH'S VIOLET.

Thionine Blue G, O Extra. A dyestuff. It is the zinc double chloride of trimethylethyldiaminophenazthionium chloride, $(C_{17}H_{23}N_2SCl)_2ZnCl_2$. Dyes tannined cotton blue.†

Thionine Red-brown B. A British brand of Immedial maroon B.†

Thionite A. A synthetic rubber. It is a polymerized compound of ethylene diglycoside and sodium tetrasulphide.†

Thionol. A registered trade name for certain dyestuffs.†

Thionol Black. See SULPHUR BLACK T.

Thionoline. Hydroxyaminoiminodiphenylsulphide, $C_{12}H_8N_2OS$.†

Thiophen Green. An analogue of malachite green.†

Thiophenol. Phenyl mercaptan, C_6H_5SH.†

Thiophenol Black T Extra. A dyestuff prepared by the fusion of dinitrophenol with sodium polysulphides. (Dinitrophenol Black.)†

Thiophloxin. An eosin dyestuff. It is tetrabromothiodichlorofluoresceine.†

Thiophor. A registered trade mark currently awaiting re-allocation by its proprietors to cover a range of dyestuffs. (Casella AG).†

Thiophor Bronze 5G. A dyestuff obtained by the fusion of *p*-phenylene-diamine and *p*-aminoacetanilide with sulphur.†

Thiophor Indigo. A dyestuff obtained by heating the indophenol derivative from α-naphthol and *p*-aminodimethyl aniline with sodium sulphide and sulphur.†

Thiophosgene. Thiocarbonyl chloride, $CSCl_2$.†

Thiophosphine J. See CHLORAMINE YELLOW.

Thioprene-48. An elastomeric mercaptan-terminated polymer used for sealing glass. A registered trade mark. (Polymeric Systems Inc, Valley Forge, Pa).†

Thiorubin. An azo dyestuff. It is thio-*p*-toluidine diazotized and combined with β-naphtholdisulphonic acid.†

Thiosaccharine. A compound, $C_6H_5(CS)SO_2.NH$, obtained by heating a mixture of saccharin and phosphorus pentasulphide at 220° C., and extracting with hot benzene.†

Thiosal. Disodium dithiosalicylate.†

Thiosept. A product containing sulphur obtained from oil shale. It is used in salves.†

Thiosept Oil. A distillation product of shale oil. It contains sulphur and has a boiling range of 100-350° C.†

Thiostab. A proprietary pure sodium thiosulphate in ampoules.†

Thiostop E. Sodium diethyl dithiocarbamate (25-30% aqueous solution). An ultra-accelerator for NR and SBR latices. An activator for guanidine type accelerators. (Uniroyal).*

Thiostop N. Sodium dimethyldithiocarbamate (40% aqueous solution). A non-staining and non-discolouring polymerization short-stop for SBR and similar rubbers. (Uniroyal).*

Thiostop N. Sodium dimethyl dithiocarbamate (40% aqueous solution). An ultra-accelerator for NR and SBR latices. (Uniroyal).*

Thiosulfil. Sulphamethizole. Applications: Antibacterial. (Ayerst Laboratories, Div of American Home Products Corp).

Thiotan. Reserving agent. Applications: Polyamide/elastomer fibres. (Sandoz Products Ltd).

Thiotax. 2-Mercaptobenzothiazole. Applications: vulcanization accelerator. (Monsanto Co).*

Thiotepa. Injection containing triethylenethiophosphoramide. Applications: Antineoplastic agent. (Lederle Laboratories).*

Thioxine Blacks. Dyestuffs for cotton.†

Thioxine Orange. (Thioxine Yellow G). Yellow sulphide dyestuffs.†

Thioxine Yellow G. See THIOXINE ORANGE.

Thioxydant Lumière. Ammonium persulphate, $(NH_4)_2S_2O_8$.†

Thiozin. Ammonium ichthosulphonate.†

Thisol. Bitumen emulsions. Applications: Industrial. (Vedag GmbH).*

Thissirol. An aqueous solution of about 57 per cent castor oil soap and 29 per cent chloroxylenol mixture. A bactericide.†

Thitsi. (Burma Black Varnish). A natural lacquer. It is the sap of the black varnish tree, *Melanorrhoea visitata*.†

Thiurad. Bis(dimethyl-thiocarbamyl)disulphide; tetramethylthiuram disulphide. Applications: vulcanization accelerator. (Monsanto Co).*

Thiuretic. Hydrochlorothiazide. Applications: Diuretic. (Parke-Davis, Div of Warner-Lambert Co).

Thixatrol. An organic derivative of castor oil. A rheological additive designed to impart thixotropy, viscosity and anti-settling properties. Applications: Solvent systems - paints, inks, caulks, mastics, plastisols. (NL Chemicals (UK) Ltd)*

Thixcin. An organic derivative of castor oil. A rheological additive designed to impart thixotropy, viscosity and anti-settling properties. Applications: Solvent systems, paints, solvent-free epoxies, mastics, inks, cosmetics. (NL Chemicals (UK) Ltd)*

Thixolan. Viscosity modifiers. (Lankro Chemicals Ltd).

Thixomen. Thixotropic additive. (ICI PLC).*

Thixseal. Inorganically modified castor oil derivative. Imparting sag and slump control, good flow and ease of application. Applications: Solvent based sealants, caulks and thick film coatings. (NL Chemicals (UK) Ltd).*

Thomas Meal. Ground slag obtained from the Thomas process for iron. Used as a fertilizer. See BASIC SLAG.†

Thomas Phosphate. See BASIC SLAG.

Thomas Slag. See BASIC SLAG.

Thomasite. A compound, $6CaO.P_2O_5Fe_2SiO_4$. It is a constituent of the basic slag of the Thomas-Gilchrist process for the dephosphorization of iron.†

Thonzide. Thonzonium Bromide. Applications: Detergent. (Parke-Davis, Div of Warner-Lambert Co).

Thoracin. A proprietary preparation of phenylethyl nicotinate, guaiacol furoate, tetrahydrofurfuryl salicylate, camphor and eucalyptol. (Lloyd, Hamol).†

Thoragol. A proprietary preparation of bibenzonium bromide. A cough linctus. (Lloyd, Hamol).†

Thoran. An alloy of 96 per cent tungsten with 4 per cent carbon.†

Thorazine. Chlorpromazine. Applications: Anti-emetic; antipsycotic. (Smith Kline & French Laboratories).

Thoren. A technical diamond substitute. It is an alloy made from tungsten and tungsten carbide.†

Thoria. Thorium dioxide, ThO_2.†

Thoroclear. A silicone-based water repellent coating for limestone. A trade mark. (Standard Dry Wall Products Inc, Miami, Fla).†

Thoron. Radon.†

Thorosheen. An acrylic paint for masonry. A registered trade mark. (Standard Dry Wall Products Inc, Miami, Fla).†

Thorotrast. A proprietary colloidal thorium dioxide preparation.†

Thoulet's Solution. A concentrated solution of potassium and mercury iodides in water. Used to determine the density of minerals.†

Thovaline. A proprietary preparation of talc, kaolin, zinc oxide and cod-liver oil, used in dermatology. (Ilon Laboratories).*

Thowless Solder. An alloy of tin and zinc with small amounts of aluminium and silver.†

Three Elephant Boric Acid. 99.8% minimum H_3BO_3. Two technical granular grades (granular and fine granular), and powdered grade. (Kerr-McGee Chemical Corp).*

Three Elephant Pyrobor Dehydrated Borax. 99% minimum $Na_2B_4O_7$. Standard (-20 to +200 US mesh) and fine (-100 to +325 US mesh) technical grades. (Kerr-McGee Chemical Corp).*

Three Elephant V-Bor Refined Pentahydrate Borax. 99.8% minimum $Na_2B_4O_7.5H_2O$. Standard grade.. (Kerr-McGee Chemical Corp).*

Thresh's Reagent. Potassium bismuth iodide. Used for testing alkaloids.†

Thrombase. A clotting enzyme. It coagulates blood.†

Thronothane Hydrochloride. Pramoxine Hydrochloride. Applications: Anaesthetic. (Abbott Laboratories).

Throsil. A proprietary preparation of CETALKONIUM CHLORIDE and amethocaine hydrochloride. An antiseptic mouth lozenge. (Cox, A H & Co Ltd, Medical Specialities Divn).†

Thsing-Hoa-Liao. The Chinese name for a cobaltiferous aluminic silicate. Used in the manufacture of porcelain.†

Thurmalox. A line of silicon-based, heat and corrosion resistant coatings for protection of metal structures or vessels subjected to high temperatures up to 1600°F (870°C). Applications: Stacks, breechings, furnaces, heat exchangers, exhaust manifolds, kilns, chemical process equipment, prevention of stress-corrosion cracking of stainless steel, wood stoves, barbecue grills, solar collector panels. (Dampney Company Inc.)*

Thurston's Alloy. An alloy of 80 per cent zinc, 14 per cent tin, and 6 per cent copper.†

Thurston's Brass. See BRASS.

Thus, Gum. See GUM THUS.

Thwaites' Solution. A mixture of alcohol, creosote, and chalk in water. Used to preserve animal tissues.†

Thylogen Maleate. Pyrilamine Maleate. Applications: Antihistaminic. (William H Rorer Inc).

Thyme Camphor. (Thymic acid). Thymol, $C_6H_3(OH).CH_3.C_3H_7$.†

Thymene. The residual oils obtained from the preparation of thymol. Used as a cheap perfume for soaps.†

Thymic Acid. See THYME CAMPHOR.

Thymine. 5-Methyluracil, $C_5H_6N_2O_2$, obtained by the hydrolysis of nucleic acids.†

Thymol Blue. Thymolsulphonphthalein. An indicator.†

Thyodene. Analytical reagent CCCN 3819 for iodine and iodometry. Applications: As a white water soluble powder it is superior to starch solution, it is stable and used direct from bottle to solutions to be titrated: for iodine and iodometry. (Campbell Williams & Co).*

Thyol. A substitute for Ichthyol (*qv*) obtained by treating tar oils with sulphur.†

Thyractin. Thyroglobulin. Applications: Thyroid hormone. (Sterling Drug Inc).

Thyrar. Thyroid. Applications: Thyroid hormone. (USV Pharmaceutical Corp).

Thyrocalcitonin. See CALCITONIN.

Thyroidectin. Dried antithyroid serum.†

Thyrolar. Liotrix. A mixture of liothyronin sodium and levothyroxine sodium in a ratio of 1:1 in terms of biological activity, or in a ratio of 1:4 in terms of weight. Applications: Thyroid hormone (USV Pharmaceutical Corp).

Thyroprotein. Thyroglobulin. Applications: Thyroid hormone. (Parke-Davis, Div of Warner-Lambert Co).

Thytropar. A proprietary preparation of thyrotropin. (Armour Pharmaceutical Co).†

Tiazyme. Mixtures of the enzymes Trypsin and Chymotrypsin. Applications: Pharmaceutical. (Hans Rahn & Co).*

Tiberal Roche. Proprietary preparation of ornidazole. Anti-infective agent. (Roche Products Ltd).*

Tibirox. Proprietary combination antibacterial product containing tetroxoprim and sulphadiazine. (Roche Products Ltd).*

Tibricol. A proprietary preparation of nifedipine for the treatment of angina and hypertension. (Pfizer International).*

Ticar. A proprietary preparation of ticarcillin. Applications: An antibiotic. (Beecham Research Laboratories).*

Ticelgesic. A proprietary preparation of PARACETAMOL. An analgesic. (Unichem).†

Ticevite. A proprietary preparation of Vitamins A, D, E and B complex. (Unichem).†

Ticillin V.K. A proprietary preparation of Penicillin V. An antibiotic. (Unichem).†

Ticipect. A proprietary preparation of DIPHENHYDRAMINE hydrochloride, ammonium chloride, Sodium citrate, menthol and chloroform. A cough syrup. (Unichem).†

Tico. An electrical resistance alloy containing 67.5 per cent iron, 30.5 per cent nickel, and small quantities of manganese and copper.†

Tidolith. See CRYPTONE.

Tiers Argent. An alloy containing 66.6 per cent aluminium and 33.3 per cent silver.†

Tiff. Barytes, $BaSO_4$. A local name for the mineral barite.†

Tifolic. A proprietary preparation of ferrous fumarate and folic acid, used in the treatment of anaemia in pregnancy. (Unichem).†

T-I-Gammagee. Tetanus Immune Globulin. Applications: immunizing agent. (Merck Sharp & Dohme, Div of Merck & Co Inc).

Tigan. Trimethobenzamide Hydrochloride. Applications: Anti- emetic. (Beecham Laboratories).

Tigason. Proprietary retinoid preparation containing etretinate as active ingredient. Anti-psoriatic. (Roche Products Ltd).*

Tiglyssin. A proprietary trade name for the hydrobromide of TIGLOIDINE, used in the treatment of muscular spasm. (Duncan, Flockhart).†

Tiguvon. Veterinary preparation. Applications: For use on domestic animals against warble infestation and lice. (Bayer & Co).*

Til. Compounds of titanium. (Tioxide UK Ltd, Til Div).

Tilcom. Organic compounds of titanium and zirconium. Applications: Catalysts, cross-linkers, thixatropes for emulsion paints and adhesion promoters - especially for printing inks onto plastic film. (Tioxide UK Ltd).*

Tile Ore. An earthy variety of native cuprous oxide.†

Tilite. Self sinking aluminium grain refiner. (Foseco (FS) Ltd).*

Tillantin B. A copper-arsenic compound used in a 0.2 per cent solution against smut of cereals.†

Tilly Drops. Dutch drops.†

Til Oil. See GINGELLY OIL.

Tilt. Cereal fungicide. (Ciba-Geigy PLC).

Tilt Turbo. Cereal fungicide. (Ciba-Geigy PLC).

Timails. A range of transparent and coloured enamels. Applications: For use in the surface coating of metals, particularly steel and aluminium. (Bayer & Co).*

Timang Steel. A proprietary trade name for a high manganese steel.†

Timbo. The root rind of a variety of *Conchocarpus*. A narcotic.†

Timbor. A specially soluble form of di-sodium octaborate tetrahydrate ($Na_2B_8O_{13}.4H_2O$). It is highly toxic to wood destroying insects and decay fungi and is used for the preservation of building timber by diffusion impregnation at the sawmill. (Borax Consolidated Ltd).*

Timborised. Preserved timber. (Borax Consolidated Ltd).

Timbrelle. Carpet yarns. (ICI PLC).*

Time Bomb. Trade name for total release fogger insecticide. Applications: Household insecticide controls flying insects, roaches, fleas etc. (Colonial Products Inc).*

Timica. Titanium dioxide coated mica to meet cosmetic standards. (Cornelius Chemical Co Ltd).*

Timodine. A proprietary preparation of NYSTATIN, HYDROCORTISONE, BENZALKONIUM CHLORIDE and DIMETHICONE used in the treatment of infected eczema. (Lloyd, Hamol).†

Timolate. Timolol Maleate. Applications: Anti-adrenergic. (Merck Sharp & Dohme, Div of Merck & Co Inc).

Timonox. Antimony oxide. (Anzon Ltd).

Timonox Blue Star. A proprietary preparation of pure antimony oxide. The arsenic amounts to 0.0018 per cent.†

Timoptic. Timolol Maleate. Applications: Anti-adrenergic. (Merck Sharp & Dohme, Div of Merck & Co Inc).

Timoptol. Timolol Maleate. Applications: Anti-adrenergic. (Merck Sharp & Dohme, Div of Merck & Co Inc).

Tin-copper Green. See GENTELES GREEN.

Tinactin. Tolnaftate. Applications: Antifungal. (Schering-Plough Corp).

Tinaderm. A proprietary preparation of tolnaftate. A skin fungicide. (Glaxo Pharmaceuticals Ltd).†

Tin Amalgam. A tin-mercury alloy containing 44-51 per cent tin. It is prepared by electrolysis.†

Tinamul. Partial glycerides. Applications: Special emulsifier to be used in the manufacture of tahina and halva. (Dynamit Nobel Wien GmbH).*

Tin Ash. Stannic oxide, SnO_2. A polishing powder.†

Tin Brilliants. See FAHLUN DIAMONDS.

Tin Bronze. An alloy of 89 per cent copper and 11 per cent tin.†

Tincal. (Tinkal). An impure borax.†

Tin Crystals. See TIN SALTS.

Tindal. Acetophenazine maleate. Applications: Antipsychotic. (Schering-Plough Corp).

Tinder, German. See AMADOU.

Tin, Dropped. See GRAIN TIN.

Tineafax. Fungicide ointment. (The Wellcome Foundation Ltd).

Tinegal. Dyeing and printing assistant. (Ciba-Geigy PLC).

Tinegal AC. A cationic dyeing agent for acrylics.†

Tin Green. See GENTELES GREEN.

Ting-yu. See TSÉ-IÉOU.

Tinkal. See TINCAL.

Tinkalzit. Synonym for Ulexite.†

Tinnevelly Senna. Indian senna.†

Tinoclarite. Bleaching stabilizers. (Ciba-Geigy PLC).

Tinofil. Dispersed pigments. (Ciba-Geigy PLC).

Tinol. A proprietary preparation of PARACETAMOL and DIPHENHYDRAMINE hydrochloride. An analgesic. (Unichem).†

Tinopal. Fluorescent whitening agents for paper and detergents. (Ciba-Geigy PLC).

Tin Ore. A mineral. It is tinstone, SnO_2.†

Tinorex. Proofing agents. (Ciba-Geigy PLC).

Tinosol. Dyeing and printing assistants. (Ciba-Geigy PLC).

Tinovetin. Biodegradable detergent; highly effective wetting and scouring agent. Applications: All textile processing; scouring, wetting and emulsifying greases. (Ciba-Geigy PLC).

Tin, Phosphor. See PHOSPHOR TIN.

Tin Prepare Liquor. Sodium stannate, $Na_2SnO_3.3H_2O$. Used as a mordant for dyes in calico, printing.†

Tin Salts. (Tin Crystals). Stannous chloride, $SnCl_2$. Used as a wool mordant for dyeing cochineal scarlet, for dyeing blacks on silk, for weighting silk, and for calico printing.†

Tinset. A proprietary preparation containing oxatomide. Applications: Allergic conjunctivitis and rhinitis, other antihistamine indications, food allergy. (Janssen Pharmaceutical Ltd).*

Tin, Stone. See STREAM TIN.

Tin Tacks. Tinned iron tacks.†

Tintacrete. Dry powdered colourants, iron oxides, ochres, umbers and composite pigments sold in small packages for DIY trade. Applications: Colours for mortars, concrete roofing tiles, floor tiles, sand-lime bricks, concrete blocks, reconstructed stone, split blocks, paving slabs and cement sheets. (W Hawley & Son Ltd).*

Tint-Ayd. Pigment dispersions. Applications: Paints, inks etc. (Cornelius Chemical Co Ltd).*

Tinuvin. A trade mark for ultraviolet light absorbers for incorporation in plastics materials. (Ciba-Geigy PLC).†

Tinuvin P. A substituted benzotriazole derivative having a peak absorption at 340 mμ. It is recommended for PVC, polystyrene and acrylics. (Ciba Geigy PLC).†

Tinuvin 770. A proprietary ultra-violet-light stabilizer used in the making of polyolefin plastics. It is a modified hindered amine. (Ciba-Geigy PLC).†

Tin White. A pigment. It is stannic hydroxide, $Sn(OH)_4$. Used in enamel and glass-making.†

Tin, Wood. See STREAM TIN.

Tiona. Range of titanium dioxide pigments including both anatase and rutile cryst[al] forms, with surface treated grades for enhanced performance. Applications: Opacifying and whitening of all paint systems, plastics and floorcoverings, paper, textiles, inks, ceramics, rubber and vitreous enamels. (SCM Chemicals Ltd).*

Tiox. Tioxidazole. Applications: Anthelmintic. (Schering-Plough Corp).

Tioxide. Titanium dioxide pigment. Applications: Decorative and industrial paints, plastics, paper, printing inks, ceramics and man-made fibres. (Tioxide Group Plc).†

TIP. A proprietary preparation. It is tetraiodophenolphthalein for use in cholecystography.†

Ti-pure. Titanium pigments. (Du Pont (UK) Ltd).

Tirucalli Gum. A product of an Indian plant of the *Euphorbia* species. It somewhat resembles gutta-percha.†

Tisco Steel. A proprietary trade name for a high manganese steel containing up to 15 per cent manganese.†

Tised. A proprietary preparation of MEPROBAMATE. A sedative. (Unichem).†

Tissalys. Modified starch. Applications: Textile industry for sizing natural, artificial and synthetic fibres. (Roquette (UK) Ltd).*

Tissier's Metal. An alloy of 97 per cent copper, 2 per cent zinc, and 1 per cent arsenic.†

TIS-U-SOL. Pentalyte. A combination of Sodium chloride, potassium chloride, magnesium sulphate, sodium phosphate dibasic, potassium phosphate monobasic. Applications: Electrolyte combination used to prepare a physiologic irrigation solution intended for use on wounds and open tissue surfaces. (Travenol Laboratories Inc).

Titan. Design for titanium dioxide pigments. Applications: Paint, paper, ink, plastics, ceramics and glass. (NL Industries Inc).*

Titan Blue 3B, R. A direct cotton dyestuff.†

Titan Brown 0, Y, R. Direct cotton dyestuffs. The O brand gives a yellowish, and Y and R reddish shades.†

Titan Cements. Cements obtained by fusing a mixture of titaniferous iron ore, limestone and coke. It consists essentially of calcium titanate ($CaTiO_3$) with small amounts of ferrites, aluminates and calcium silicate, together with from 2-10 per cent ferric oxide.†

Titan Como G, R, S. Direct cotton dyestuffs, giving blue shades.†

Titan Design. Design for titanium dioxide pigments. Applications: Paint, plastics, ink, paper, ceramics and glass. (NL Industries Inc).*

Titaneisstein. (Iserin). A titaniferous iron sand from Iserwiesl and Riesengbirge.†

Titanellow. Titanium oxalate, $Ti_2(C_2O_4)_3.10H_2O$. A mordant.†

Titan Grey. A direct cotton dyestuff.†

Titaniferous Iron. See ILMENITE.

Titaniferous Iron Ore. See ILMENITE.

Titanital. The trade name for a proprietary titanium white. The golden seal brand contains from 95-98 per cent titanium oxide. TiO_2. and the silver seal grade is a mixture of 80 per cent titanium dioxide with 20 per cent zinc oxide.†

Titanite. See SPHENE. Also a proprietary aluminium-manganese alloy containing titanium.

Titanite No. 1. An explosive consisting of 85-88 per cent ammonium nitrate, 6-8 per cent trinitro-toluene, and 4.5-6.5 per cent charcoal.†

Titanium Alloy. A ferro-titanium is called by this name.†

Titanium Green. Titanium ferro-cyanide.†

Titanium Lithopone. See TI-TONE.

Titanium Putty. Titanium reinforced epoxy resin. Applications: Repairing worn or gouged parts, rebuilding wear surfaces and reseating worn or oversized bearings. (Devcon Corporation).*

Titanium White. A pigment with varying amounts of titanium dioxide and barium sulphate, but usually consisting of 25 per cent titanium oxide with 75 per cent barium sulphate. One quality known in trade as Extra X contains 70 per cent titanium oxide, 10 per cent barium sulphate, and 20 per cent calcium phosphate.†

Titanium-calcium Pigment. A pigment similar to titanox, but the titanium oxide is precipitated on calcium sulphate instead of barium sulphate.†

Titan Navy. A dyestuff giving bluish and reddish-navy blue on cotton.†

Titanoferrite. A variety of the mineral ilmenite.†

Titanolith. See CRYPTONE.

Titanox. Titanium dioxide pigments. Applications: Paint, paper, plastics, ink, ceramics and glass. (NL Industries Inc).*

Titanox Design. Design for titanium dioxide pigments. Applications: Paint, plastics, paper, glass and ceramics. (NL Industries Inc).*

Titanox RA-39. A proprietary trade name for a stearate coated titanium dioxide pigment easily dispersible in polystyrene and polyolefines. (Laporte Industries Ltd).*

Titanox-B-30. A pigment containing 30 per cent titanium dioxide instead of 25 per cent in " B.'†

Titan Pink 3B. A bluish-pink dyestuff for cotton.†

Titan Red. See GERANINE.

Titan Red for Wool. A dyestuff for cotton or wool.†

Titan Rose. See GERANINE.

Titan Scarlet C, S. Direct cotton dyestuffs. The brand C gives yellowish, and S bluish-scarlet shades.†

Titanspinel. Synonym for Ulvospinel.†

Titanweiss (C, Extra T, Standard T, Standard A). Trade names for titanium dioxide pigments extended with calcium or barium sulphates.†

Titan Yellow. A dyestuff. It is equivalent to Clayton yellow.†

Titan Yellow GG, G, R, Y. Direct cotton dyestuffs. The brand GG gives a greenish-yellow, and R a dull reddish yellow shade.†

Titite. A proprietary rubber cement, partly made from rubber. It is waterproof, and is used for mending cloth, paper, rubber, leather, and wood.†

Ti-tone. A titanium lithopone containing 15 per cent titanium dioxide, 25 per cent zinc oxide, and 60 per cent barium sulphate. Its specific gravity is 4.25, a it is stated to have a covering power 6 per cent greater than ordinary lithopone.†

Titralac. Each tablet contains calcium carbonate BP 420 mg and glycine 18(mg (Riker Laboratories/3M Health Care).†

Tixo K100. A cyanoacrylate adhesive.†

Tixogel. Anti-settling and thickening agents for paints, varnishes, lubricant adhesives, coatings, putties and cosmetics. (Süd-Chemie AG).*

Tixoton. Bentonites with high swelling properties for the drilling and buildin industry. (Süd-Chemie AG).*

Tixylix. Paediatric cough linctus. (May & Baker Ltd).

Tizit. An alloy of 40-80 per cent tungste 4-15 per cent titanium, 4 per cent chromium, 2-4 per cent carbon, 1-5 pe cent cerium, and 3-40 per cent iron.†

T-lim. Modified rosins in aqueous emulsions. Applications: Sizing agent for paper and paperboard. (Hercules Inc).*

T Metal. An alloy of 95 per cent alumini 4 per cent magnesium, 0.5 per cent silicon, 0.5 per cent iron, and 0.1 per cent copper.†

TMP. Amine salts of organic acids, aromatic acid, aromatic and aliphatic petroleum distillate. Applications: Fo use with propanil herbicide in spraying rice to control evaporation, prevent crystalization of propanil and control drift. (Stull Chemical Company).*

TNA. Tetranitroaniline.†

TNB. Trinitrobenzene.†

TNT. See TROTYL.

TNX. Tetranitroxylene.†

Toad's Eye. See STREAM TIN.

Tobias Acid. 2-Naphthylamine-1-sulphoni acid.†

Tobin Bronze. Alloys of 59-83 per cent copper, 3-48 per cent zinc, 0.9-12.4 pe cent tin, 0.31-2.14 per cent lead, and 0.1-0.8 per cent iron. One alloy contai 58.79 per cent copper, 40.43 per cent zinc, and 0.88 per cent tin.†

obrex. Tobramycin. Applications: Antibacterial. (Alcon Laboratories Inc).

ochlorine. A proprietary preparation. It is chloramine-T.†

ocopherex. Vitamin E. Applications: Vitamin E supplement. (E R Squibb & Sons Inc).

ocopherol. See VITAMIN E.

offix. Special hard fat. Applications: Manufacture of caramels and chewing sweets. (Dynamit Nobel Wien GmbH).*

ofranil. A proprietary preparation of imipramine hydrochloride. An anti-depressant. (Ciba Geigy PLC).†

ogocoll. 1- and 2-component polyurethane and others. Applications: Sealers, adhesives and primers, underbody coatings, windshield-adhesives and corrosion preventative coatings (waxes) for automotive industries. (EMS-Chemie AG).*

oile Micanite. See MICANITE CLOTH.

oisin's Solution. A microscopic stain used for staining white blood corpuscles. Based on methyl violet.†

-ol. Dimethylthioanthrene. See SINTOL. A plasticiser.

Tolamine. See CHLORAMINE T.

Tolanase. A proprietary preparation of tolazamide. An oral hypoglycaemic agent. (Upjohn Ltd).†

Tolane. Diphenyl-acetylene, $C(C_6H_5)C(C_6H_5)$.†

Tolane Red. An azo dyestuff. Dyes wool a brilliant red from an acid bath.†

Tolectin. Tolmetin Sodium. Applications: Anti-inflammatory. (McNeil Pharmaceuticals, McNEILAB Inc).

Toledo Blue V. A polyazo dyestuff. It dyes cotton direct from a sodium sulphate and sodium carbonate bath, or a soap bath.†

Toleron. Ferrous fumarate. Applications: Haematinic. (Mallincrkodt Inc).

Tolgard. Flame retardants. (Tenneco Organics Ltd).

Tolinase. Tolazamide. Applications: Antidiabetic. (The Upjohn Co).

Tolite. See TROTYL.

Tolkan. Herbicide. (May & Baker Ltd).

Tollen's Reagent. A solution of ammoniacal silver nitrate containing free caustic soda. It is prepared when required by mixing (1) 10 per cent caustic soda with (2) ammoniacal silver nitrate, obtained by dissolving sufficient silver nitrate to yield a 10 per cent solution, in a mixture of equal volumes of concentrated ammonia and distilled water. It is used to test for aldehydes and other reducing substances.†

Tolnate. A proprietary preparation of prothipendyl hydrochloride. A sedative. (Smith Kline and French Laboratories Ltd).*

Tolochrome. Photographic colour developer. (May & Baker Ltd).

Toloy 45. An alloy of 45 per cent nickel and 20 per cent chromium with other materials. Used where stress corrosion resistance is required. The material conforms to BS.1648 Grade H.†

Tolplaz. Speciality plasticizers. (Tenneco Organics Ltd).

Tolu Balsam. The oleo-resin of *Myroxylon toluifera*, of South America.†

Toluidine Blue 0. (Methylene Blue T50). A dyestuff. Dyes tannined cotton blue.†

Toluol. Commercial toluene, $C_6H_5CH_3$. It contains traces of benzene, xylene, paraffin, and thiophenes.†

Tolurex. Active ingredient: chlortoluron; 3-(3-chloro-p-tolyl)-1,1-dimethylurea. Applications: Selective pre- and post-emergence herbicide in winter cereals for control of annual grasses and broad leaved weeds. (Agan Chemical Manufacturers Ltd).†

Tolusafranine. See SAFRANINE RAE.

Toluylene. Stilbene, $C_6H_5.CH : CH.C_6H_5$.†

Toluylene Blue. (Witt's Toluylene Blue). $C_{15}H_{18}N_4.HC1$. An indamine dyestuff obtained by the oxidation of dimethyl-*p*-phenylenediamine and *m*-toluene-diamine, or by the combination of nitrosodimethylaniline hydrochloride and *m*-toluenediamine. Dyes cotton.†

Toluylene Brown G. A dyestuff. It is the sodium salt of sulphotoluene-disazo-*m*-phenylenediamine, $C_{13}H_{11}N_6\ SO_3Na$. Dyes cotton yellowish-brown.†

Toluylene Brown R. A dyestuff. It is the sodium salt of sulphotoluene-disazobi-*m*-phenylenediamineazo-naphthalene-sulphonic acid, $C_{39}H_{29}N_{12}O_9S_3Na_3$. Dyes cotton brown from a soap bath.†

Toluylene Orange G. (Kanthosine J). A dyestuff.†

Toluylene Orange R. (Kanthosine R). A dyestuff. It is the sodium salt of ditolyldisazobi-*m*-toluene-diaminesulphonic acid, $C_{28}H_{28}N_8O_6S_2Na_2$. Dyes cotton reddish-orange.†

Toluylene Yellow. A dyestuff. It is sulphotoluenedisazobinitro-*m*-phenylenediamine, $C_{19}H_{17}N_{10}O_7SNa$. Dyes cotton yellow.†

Tolycaine. Methyl 2-diethylaminoaceta-mido-*m*-toluate. Baycain is the hydrochloride.†

Tolyl Peri Acid. 1-Tolylnaphthylamine-8-sulphonic acid.†

Tomahawk. An emulsifiable concentrate containing flucythrinate. Applications: A pyrethroid insecticide for the control of aphids, whitefly, caterpillars and red spider mite. (Fisons PLC).*

Tomaset. Active ingredient: N-m-t; N-m-tolylphthalamic acid. Applications: A flower and fruit plant growth regulator. (Agan Chemical Manufacturers Ltd).†

Tombac. An alloy usually containing 89 per cent copper, 5.5 per cent zinc, and 5.5 per cent tin.†

Tombac, Red. (Red Brass, Tambac). A brass containing less than 18 per cent zinc. It usually consists of 90 per cent copper and 10 per cent zinc, and is used for rolling into leaf.†

Tombac, White. (White Copper). An alloy containing 75 per cent copper and 25 per cent tin.†

Tombasil. A proprietary trade name for an alloy consisting mainly of tombac metal with silicon.†

Tomophan. A proprietary viscose packing material.†

Tomorite. Liquid fertilizer. (Fisons PLC, Horticulture Div).

Toncan. A corrosion-resisting alloy containing pure iron, copper, and molybdenum.†

Toncas Metal. An alloy of 29 per cent nickel, 36 per cent copper, 7.1 per cent iron, 7.1 per cent zinc, 7.1 per cent lead 7.1 per cent tin. and 7.1 per cent antimony. Used for ornamental work.

Tonite. (Potentite). An explosive. It consists of mixtures of granulated gun cotton and barium nitrate. No. 1 contains 50-52 parts guncotton and 40-47 parts barium nitrate. No. 2 contains charcoal also. No. 3 consists of 18-20 parts guncotton, 67-70 parts barium nitrate, and 11-31 parts dinitro- benzene. Also a name for chloracetone, $CH_3COCH_2.C1$.†

Tonka-bean Camphor. Coumarin, found in species of *Dipteryx*.†

Tonocard. Tocainide. Applications: Cardiac depressant (anti-arrhythmic). (Merck Sharp & Dohme, Div of Merck & Co Inc).

Tonophosphan. A proprietary preparation of toldimfos sodium. A source of phosphorus used for veterinary purposes.†

Tonox. p,p'-Diaminodiphenyl-methane. Curing agent for epoxy resins, imparts high heat distortion point. Accelerator for neoprene. Anti-frosting agent in vulcanized rubber. Lessens tendency fc sagging and distortion of unvulcanized natural, SBR and butyl stock. Anti-reversion agent in sulphur-cured butyl rubber tyre curing bags. (Uniroyal).*

Tonquinol. Butyltrinitrotoluene.†

Tonsil. For the adsorbtive decolourization and purification of oils and fats, hydrocarbons, waxes and other liquid intermediate products. (Süd-Chemie AG).*

Tonsil. See Frankonite.

Tool Life. Water based synthetic cutting fluid (extreme pressure). Applications: For all metal removing machining operations where physical and chemical extreme pressure assistance is required to assist functional cooling properties on all metals, except of course, magnesium (Sumner Oil Industries).*

Toolife. A broad line of industrial fluids, both oil and synthetic in nature, water soluble and straight. (Specialty Products Co).*

)onu Gum. See TUNO GUM.

)p 7 Mosaic and Pebble. Expanded polystyrene veneer 7mm thick. Applications: Decorative veneers for domestic ceilings. (Vencel Resil Ltd).*

)pal. A reflux suppressant and antacid. (ICI PLC).*

opanol. A proprietary trade name for a range of antioxidants. (ICI PLC).*

opas. Fungicide. (Ciba-Geigy PLC).

opclip Dridress. Animal health organophosphorus wound dressing. (Ciba-Geigy PLC).

opclip Fly and Scab Dip. Organophosphorus/phenols sheep dip. (Ciba-Geigy PLC).

opclip Foot Rot Aerosol. Antiseptic sheep treatment. (Ciba-Geigy PLC).

opclip Formalin. Formaldehyde for sheep foot rot control. (Ciba-Geigy PLC).

opclip Gold Shield. Scab approved organophosphorus sheep dip. (Ciba-Geigy PLC).

opclip Marker Aerosols. Sheep marker dyes. (Ciba-Geigy PLC).

opclip Marker Fluid. Sheep marker fluid. (Ciba-Geigy PLC).

opclip Parasol. Cypermethrin pour-on for sheep and cattle. (Ciba-Geigy PLC).

'opclip Scab Dip. Organophosphorus/ phenols sheep dips. (Ciba-Geigy PLC)

'opclip Sheep Dip. Organophosphorus sheep dip. (Ciba-Geigy PLC).

'opclip Vaccines. Various sheep vaccines. (Ciba-Geigy PLC).

'opclip Wormer. Organophosphorus anthelmintic for sheep and pigs. (Ciba-Geigy PLC).

'opex. Buffered acne medication, 10% benzoyl peroxide. Applications: To help clear pimples without overdrying the skin. (Richardson-Vicks Inc).*

'opexane. Antibacterial face wash to help clear acne blemishes. (Richardson-Vicks Inc).*

'öpfer's Reagent. Dimethylamino-azobenzene (0.5 gram) in 100 cc of 95 per cent alcohol. It is used to test acidity in stomach contents.†

Tophet. An electrical resistance alloy containing 61 per cent nickel, 10 per cent chromium, 26 per cent iron, and 3 per cent manganese.†

Tophet A. A proprietary trade name for 80 per cent nickel, 20 per cent chrome resistance wire.†

Tophet C. A proprietary trade name for nickel chrome iron resistance wire.†

Topicort. Desoximetasone. Applications: Anti-inflammatory. (Hoechst-Roussel Pharmaceuticals Inc).

Topicycline. Tetracycline Hydrochloride. Applications: Anti- amebic; antibacterial; antirickettsial. (Norwich Eaton Pharmaceuticals Inc).

Topostasin. Proprietary preparation of thrombin. Local haemostatic. (Roche Products Ltd).*

Toppel. Insecticide, containing cypermethrin. (ICI PLC).*

Topsol. Carding and combing oil. (Crosfield Chemicals).*

Topsyn. Fluocinonide. Applications: Glucocorticoid. (Syntex Laboratories Inc).

Torbanite. A variety of cannel coal.†

Torbutrol. Butorphanol tartrate. Applications: Analgesic; antitussive. (Bristol Laboratories, Div of Bristol-Myers Co).

Tordon. Herbicides based primarily on picloram. Broadleaf and brush killers. Applications: Forestry, grain and corn. (Dow Chemical Co Ltd).*

Torecan. A proprietary trade name for Thiethylperazine or its salts. An anti-emetic. (Sandoz).†

Torelle. Fospirate. Applications: Anthelmintic. (Dow Chemical Company Ltd).

Toric Contact Lens. Hefilcon B. 2-Hydroxyethyl methacrylate with 1-vinyl-2-pyrrolidinone and ethylene dimethacrylate. Applications: Contact lens material. (Bausch & Lomb, Professional Products Div).

Torlon 4203L. Polyamide-amide engineering resin essentially unfilled. Applications: Connectors, switches, relays, thrust washers, spline liners, valve seats, poppets, mechanical linkages, brushings, wear rings, insulators, cams, picker fingers, ball bearings, rollers and thermal insulators. (Amoco Chemicals Co).*

Torlon 4275. Polyamide-imide engineering resins filed with 20% graphite powder. Applications: Bearings, thrust washers, wear pads, strips, piston rings, seals, vanes and valve seats. (Amoco Chemicals Co).*

Torlon 4301. Polyamide-imide engineering resins filled with 12% graphite powder, 3% fluorocarbon. Applications: Bearings, thrust washers, wear pads, strips, piston rings, seals, vanes, valves, seats. (Amoco Chemicals Co).*

Torlon 4347. Polyamide-imide engineering resins filled with 12% graphite powder, 8% fluorocarbon. Applications: Bearings, thrust washers, wear pads, strips, piston rings and seals. (Amoco Chemicals Co).*

Torlon 5030. Polyamide-imide engineering resins filled with 30% glass fibre, 1% fluorocarbon. Applications: Burn-in sockets, gears, valve plates, fairings, tube clamps, impellers, rotors housing, back-up rings, terminal strips, insulators and brackets. (Amoco Chemicals Co).*

Torlon 7130. Polyamide-imide engineering resins filled with 30% graphite fibre, 1% fluorocarbon. Applications: Metal replacements, housings, mechanical linkages, gears, fasteners, spline linears, cargo rollers, brackets, valves, labyrinth seals, fairings, tube clamps, standoffs, impellers shrouds, potential use for EMI shielding. (Amoco Chemicals Co).*

Tormentil. The dried rhizome of *Poten-tilla tormentilla*. Used for tanning and as an astringent.†

Tormol. A nickel-chromium-molybdenum steel highly resistant to shock and fatigue.†

Tormosyl. Fluproquazone. Applications: Analgesic. (Sandoz Pharmaceuticals, Div of Sandoz Inc).

Tornalate. Bitolterol mesylate. Applications: Bronchodilator. (Sterling Drug Inc).

Tornesit. A trade name for a protective coating base prepared by the chlorination of rubber.†

Tornusil. A proprietary 2 pack moisture cured inorganic zinc silicate primer. (Sigma Coatings).*

Toron. A sulphur-terpene compound prepared by heating turpentine with sulphur. It is a black viscid liquid or semi-solid. Used for waterproofing cloth, preparing rubberised cloth, and for attaching or coating metal surface with rubber.†

Torqseal. Anaerobic thread locking fluid. Applications: Suitable for thread locking and securing studs and bearin Easily undone with normal hand tools (Hermetite Products Ltd).*

Torrax. Malt flour. (ABM Chemicals Ltd).*

Tosmilen. A proprietary preparation of DEMECARIUM BROMIDE as eye-drops, used in the treatment of glaucoma. (Astra Chemicals Ltd).†

Totacillin. Ampicillin. Applications: Antibacterial. (Beecham Laboratories)

Tota-col. Paraquat based herbicide. (ICI PLC, Plant Protection Div).

Totocillin. Penicillinase-resistant broad-spectrum preparation. Applications: For the treatment of mastitis in cattle, sheep and goats. (Bayer & Co).*

Totocillin. A broad-spectrum penicillin combination with an extended range o action. Also known as Cervantal. (Bayer & Co).*

Totolin. A proprietary preparation of PHENYLPROPANOLAMINE hydrochloride and GUAIPHENESIN. A cough linctus. (Galen).†

Totomycin. A proprietary preparation containing tetracycline hydrochloride. An antibiotic. (Boots Company PLC).*

Totril. Herbicide. (May & Baker Ltd).

Toucas Metal. An alloy of 35.75 per cent copper, 28.56 per cent nickel, 7.1 per cent zinc, 7.2 per cent tin, 7.1 per cent lead, 7.2 per cent antimony, and 7.1 pe cent iron.†

Touch & Go. A proprietary preparation o camphor, ether, cajuput oil and tolu balsam. Applications: A toothache solution. (Ayrton Saunders plc).*

Touchpong Gum. See POUCKPONG GUM.

Touchstone. See LYDIAN STONE.

Tough Copper. Commercial copper, containing impurities such as arsenic.†

Toughened Caustic. Consists of 95 per cent silver nitrate and 5 per cent potassium nitrate, fused together.†

Tournant Oil. A commercial brand of olive oil obtained from fermented marc of expressed olives. It contains free fatty acids, and is used as a Turkey- red oil.†

Tournay's Metal. An alloy of 82.5 per cent copper and 17.5 per cent zinc.†

Tournesol. See LITMUS.

Tournesol en Drapeaux. A blue colouring matter allied to litmus, manufactured from *Croton tinctorium*. It is used for colouring cheese wrappers to detect ripeness in them. When lactic acid is formed the wrapping changes to a red colour.†

Tous-les-mois Starch. (Queensland Arrowroot). The starch from the rhizomes of *Canna edulis*.†

Tova. A proprietary preparation of cyclical hormones consisting of 16 white tablets of ethinyl oestradiol and 5 pink tablets of ethinyl oestradiol and dimethisterone. Used for control of menstruation. (British Drug Houses).†

Tower Brick. Sodium phosphates, wetting agents and corrosion inhibitors. Applications: Used in open recirculating cooling water systems to prevent lime scale and corrosion from fouling up the system. (Delaware Chemical Corp).*

Tower Treat. Liquid algaecide/biocide. Applications: Cooling water treatment. (Schaefer Chemical Products Company).*

Toximul. Sulphonate/nonionic blend. Applications: Agricultural blender. (Stepan Company).*

TPG. Triphenylguanidine.†

Trabuk. An alloy containing 87.5 per cent tin, 5.5 per cent nickel, 5 per cent antimony, and 2 per cent bismuth. It resists vegetable acids.†

Tracervial-131. Sodium Iodide I 131. Applications: Antineoplastic; diagnostic aid; radioactive agent. (Abbott Laboratories).

Trachine. Fowl laryngotracheitis. Applications: immunization of poultry. (Intervet America Inc).*

Trachyte. A volcanic rock composed of felspar with some hornblende and mica.†

Tracilon. Triamcinolone Diacetate. Applications: Glucocorticoid. (Savage Laboratories, Div of Byk-Gulden Inc).

Tracrium Injection. A proprietary formulation of atracurium besylate. Applications: An adjunct to general anaesthesia, to facilitate endo-tracheal intubation and to provide skeletal muscle relaxation during surgery or mechanical ventilation. (The Wellcome Foundation Ltd).*

Tractium. Atracurium besylate. Applications: Relaxant. (Burroughs Wellcome Co).

Tradenal. Proscillaridin. Applications: Cardiotonic. (Knoll Pharmaceutical Co).

Tragasol. A gum obtained by steeping locust-bean kernels in water. Used as a binding material. See also INDUSTRIAL GUM.†

Train Oil. Whale Oil.†

Tramacin. Triamcinolone Acetonide. Applications: Glucocorticoid. (Johnson & Johnson Products Inc).

Tramisol. Levamisole hydrochloride. Applications: Anthelmintic. (American Cyanamid Co).

Trancopal. A proprietary preparation of chlormezanone. A sedative. (Bayer & Co).*

Trancoprin. Tablets containing chlormezanone and aspirin. Applications: An analgesic with a mild tranquillizer/muscle relaxant. (Winthrop Laboratories).*

Tran-cor. A proprietary trade name for high silicon steel used in transformers.†

Trandate. Labetalol hydrochloride. Applications: Anti- adrenergic; anti-adrenergic. (Glaxo Inc).

Tranquo - Adamon. The registered trade name for an antispasmodic and tranquillizer.†

Transderm-Nitro. Nitroglycerin. Applications: Vasodilator. (Ciba-Geigy Corp).

Trans Gard. Blend of petroleum additives and petroleum distillates. Applications: Automatic transmission stop leak and fluid conditioner. (Gard Corporation).*

Transite Board. (Transite is a trade mark.) See PETRUSAL.†

Transpafill. Precipitated aluminium silicate. Applications: Used as filler for printing inks. (Degussa).*

Transpar. Lactic acid and buffered lactic acid mixtures.†

Transparent Gold Ochre. See YELLOW OCHRE.

Transparent Oxide of Chromium. Sesquioxide of chromium, a green pigment.†

Transpex 1. A proprietary trade name for an unplasticized polymethylmethacrylate.†

Transpex 2. A proprietary trade name for unplasticized polystyrene.†

Transpulmin Balsam. Quninine base, eucalyptus oil, menthol and camphor. Applications: Expectorant, antitussive, secretolytic and secretomotoric. Balsam. (Degussa).*

Transpulmin Syrup. Thiophenylpyridylamine-10-carboxylic acid piperidino-ethoxyethylester HCL (pipazetat), peppermint oil, anise oil, eucalyptus oil, liquorice extract, guaiacol glycerin ether, polyoxyethylene hexadecyl ether N-dimethylamino-isopropyl-thiophenyl-pyridylamine HCI (Isothypendyl). Applications: Expectorant, antitussive, secretolytic and secretomotoric. Syrup. (Degussa).*

Transvasin. A proprietary preparation of tetrahydrofurfuryl salicylate, ethyl nicotinate, N-hexyl nicotinate and ethyl *p*- aminobenzoate. (Lloyd, Hamol).†

Tranxene. A proprietary preparation of potassium clorazepate. A sedative. (Boehringer Ingelheim Ltd).†

Trasicor. A proprietary preparation of OXPRENOLOL hydrochloride used in the treatment of angina pectoris and hypertension. (Ciba Geigy PLC).†

Trasidrex. Beta-blocker/diuretic. (Ciba-Geigy PLC).

Trass. A volcanic material found on the bank of the Rhine. It is used in Holland as an addition to lime, to convert it into hydraulic lime.†

Trasulphane. A name for ammonium sulpho-ichthiolicum.†

Trasylol. A proteinase inhibitor. (Bayer & Co).*

Travamin. Protein Hydrolysate. Applications: Replenisher. (Travenol Laboratories Inc).

Travase. Sutilains. Applications: Enzyme. (Flint Laboratories, Div of Travenol Laboratories Inc).

Travertine. (Calc Sinter, Calcareous Tufa). A limestone deposited by calcareous springs.†

Treacles. See GOLDEN SYRUP.

Tread-Brite. Embossed aluminium plate. (Reynolds Metal Co).*

Trebizond Opium. Persian opium.†

Treble Superphosphate. This consists of mono-calcium phosphate, containing 48-49 per cent. P_2O_5 (41-42 per cent water soluble P_2O_5). A fertilizer.†

Trecator-SC. Ethionamide. Applications: Antibacterial. (Ives Laboratories Inc).

Tree Bug-Lok Adhesive. Polyisobutylene. Applications: For the control of Gypsy Moth caterpillars. Non-toxic, ecologically safe, holds forever, traps any crawling insect including ants and cankerworms. (TACC International Corp).*

Tree Copal. A name applied to white Zanzibar copal.†

Tree Gum. (Wood Gum). Xylan, $C_6H_{10}O_5$.†

Tree Wax. See CHINESE WAX.

Trefol. A name applied to amyl salicylate.†

Tre-Hold. Growth regulator. (A H Marks & Co Ltd).

Trelit. A proprietary pyroxylin plastic.†

Tremin. Trihexyphenidyl Hydrochloride. Applications: Anticholinergic; antiparkinsonian. (Schering-Plough Corp).

Tremvac. Calnek strain avian encephalomyelitis. Applications: immunization of poultry. (Intervet America Inc).*

Tremvac-FP. Calnek strain avian encephalomyelitis-fowl pox.

Applications: immunization of poultry. (Intervet America Inc).*

rench's Flameless Explosive. An explosive containing ammonium nitrate.†

renimon. A proprietary preparation of triaziquone. An antimitotic. (FBA Pharmaceuticals).†

rentadil. A proprietary preparation of bamifylline hydrochloride. (Armour Pharmaceutical Co).†

rental. A proprietary preparation of oxypentifylline used in the treatment of peripheral vascular disease. (Hoechst Chemicals (UK) Ltd).†

rényline. A proprietary trade name for rubber vulcanization accelerator. It is triphenylguanidine.†

rescatyl. Ethionamide. (May & Baker Ltd).

ret-o-Lite. A patented preparation for the destruction of petroleum emulsions. It consists of 83 per cent sodium oleate, 5.5 per cent sodium resinate, 5.0 per cent sodium silicate, 4.0 per cent phenol, and 1.5 per cent paraffin wax.†

revintex. Prothionamide. (May & Baker Ltd).

revira. Polyester fibre. (Hoechst UK Ltd).

RH-Roche. Proprietary preparation of protirelin. Thyroid function test. (Roche Products Ltd).*

Tri. An abbreviation for trichloroethylene, C_2HCl_3, a solvent.†

Triacetin. Glycerol triacetate. Applications: Used in the manufacture of cigarette filter tips, plasticizer for paint and varnish systems, fixative for use in perfumes, auxiliary for use in foundries. (Bayer & Co).*

Tri-Ad. Gasoline additive. (Chevron).*

Tri-Adcortyl. A proprietary preparation of triamcinolone acetonide, neomycin sulphate, gramicidin and nystatin used in dermatology as an anti-bacterial agent. (Squibb, E R & Sons Ltd).*

Triadcortyl Cream. Triamcinolone acetonide, nystatin, gramicidin and neomycin sulphate in cream base. (E R Squibb & Sons Ltd).

Triadcortyl Ointment. Triamcinolone acetonide, nystatin, gramicidin and neomycin sulphate in ointment base. (E R Squibb & Sons Ltd).

Triadimefon. A pesticide. 1-(4-Chlorophenoxy)-3,3-dimethyl-1-(1, 2, 4-triazol-1-yl) butan-2-one.†

Triafol. Film based on cellulose acetate butyrate and triacetate. Applications: Particularly suitable for coil insulation. (Bayer & Co).*

Triamolone 40. Triamcinolone Diacetate. Applications: Glucocorticoid. (O'Neal, Jones & Feldman Pharmaceuticals).

Triamonide 40. Triamcinolone Acetonide. Applications: Glucocorticoid. (O'Neal, Jones & Feldman Pharmaceuticals).

Trianisoline. See RHODAMINE 6G.

Triatox. See AMITRAZ.

Triatrix. See AMITRAZ.

Tribiotic Spray. Topical antibacterial in aerosol form containing neomycin sulphate BP 500,000 units, bacitracin zinc BP 10,000 units, polymyxin B sulphate BP 150,000 units. (Riker Laboratories/3M Health Care).†

Tribonol. Powder coating for direct application to sand moulds and cores. (Foseco (FS) Ltd).*

Tri-Borne. Paint dip. (ICI PLC).*

Tribovax. Combined cattle vaccine. (The Wellcome Foundation Ltd).

Tribrissen. Trimethoprim/sulphadiazine. (The Wellcome Foundation Ltd).

Tribunil. A wettable powder containing 70% w/w methabenzthiazuron. Applications: To control broadleaved weeds and meadow-grasses in autumn sown wheat, barley and oats, in winter rye, winter triticale, perennial ryegrass (leys and seed crops) and spring barley. For light to moderate infestations of black-grass in autumn sown wheat, barley and oats. (Bayer & Co).*

Tricaderm. A proprietary preparation of triamcinolone acetonide, salicylic acid and benzalkonium chloride used in the treatment of eczema. (Squibb, E R & Sons Ltd).*

Tricap. Triethyleneglycol dicaprylate/caprate. (Croda Chemicals Ltd).

Trichlormethine. Trimustine.†

Trichorad. Acinitrazole.†

Tri-clene. A proprietary trade name for trichlorethylene used in dry-cleaning. See also TRIKLONE.†

Triclofos. 2,2,2-Trichloroethyl dihydrogen phosphate. Tricloryl is the monosodium salt.†

Tricloryl. A proprietary preparation of triclofos sodium. (The monosodium salt of trichlorethyl phosphate.). A sedative. (Glaxo Pharmaceuticals Ltd).†

Triclos. Triclofos Sodium. Applications: Hypnotic; sedative. (Merrell Dow Pharmaceuticals Inc, Subsidiary of Dow Chemical Co).

Tricoid. Cine film cement. (May & Baker Ltd).*

Tricresol. A purified mixture of the three cresols. It contains about 35 per cent ortho, 20 per cent meta, and 25 per cent para-cresol.†

Tridesilon. Desonide. Applications: Anti-inflammatory. (Miles Pharmaceuticals, Div of Miles Laboratories Inc).

Tridia. A proprietary preparation of neomycin sulphate, clioquinol and kaolin. An antidiarrhoeal. (Crookes Laboratories).†

Tridil. Nitroglycerin. Applications: Vasodilator. (American Critical Care, Div of American Hospital Supply Corp).

Tridione. Trimethadione. Applications: Anticonvulsant. (Abbott Laboratories).

Trieline. A term for trichloroethylene, C_2HCl_3.†

Triethanolamine. Tri-β-hydroxy-ethylamine. The industrial material contains from 20-25 per cent of secondary, 75-80 per cent of tertiary, and 0.5 per cent primary amines.†

Tri-Farmon. Herbicide, containing linuron and trifluralin. (ICI PLC).*

Triflurex. Active ingredient: trifluralin; 2,6-dinitro-N,N-dipropyl-4-trifluromethylaniline. Applications: Selective pre-emergence herbicide. (Agan Chemical Manufacturers Ltd).†

Triformin. Glyceryl formate.†

Trigard. Copper corrosion inhibitors. (Ciba-Geigy PLC).

Trigger. Herbicide. (May & Baker Ltd).

Trigonal. UV catalyst for polyester resins. (Akzo Chemie UK Ltd).

Trigonox. Organic peroxides. (Akzo Chemie UK Ltd).

Trihyde. Alumina trihydrate fire retardant. Applications: PVC, rubber, polyester, epoxy resins. (Croxton and Garry Ltd)

Triketol. See NINHYDRIN.

Triklone. Trichloroethylene solvents. (ICI PLC).*

Trikresol. See TRICRESOL.

Trilactine. A preparation of lactic acid bacilli.†

Trilafon. Perphenazine. Applications: Antipsychotic. (Schering-Plough Corp).

Trilaurin. Glyceryl tri-laurate, $C_3H_5(OOC.C_{11}H_{23})_3$.†

Trilene. A trade mark for trichlorethyllene, a general anaesthetic. (ICI PLC).*

Trilisate. Choline magnesium trisalicylate. Applications: Arthritis. (Napp Laboratories Ltd).*

Trilite. See TROTYL.

Trillat's Reagent. An acetic acid solution of tetramethyldiaminodiphenyl-methane, $[(CH_3)_2N.C_6H_4].CH_2$. The reagent is prepared by dissolving 5 grams of the base in 100 cc of 10 per cent acid. Lead dioxide gives a blue colour with the reagent, but the same colour is given by manganese dioxide and other oxidizing agents.†

Trilombrin. A proprietary preparation of pyrantel pamoate. An antihelmintic. (Pfizer International).*

Trilon. Complexing agents, water softeners. (BASF United Kingdom Ltd).

Trim. Quintozene in dry granular form. Applications: Controlling soil borne diseases in tulips and flower crops. (Wheatley Chemical Co Ltd).*

Trimax. Magnesium trisilicate. Applications: Antacid. (Sterling Drug Inc).

Trimellitic Acid. 1, 2, 4-Benzene - tricarboxylic acid, $C_6H_3(COOH)_3$.†

Trimene. A proprietary rubber vulcanization accelerator. It is the stearic acid salt of the condensation product of ethylamine and formaldehyde (trimene base). It is sold in the concentrated form as trimene base, dissolved in latex, as latene, and in a paste form as trimene. (Naugatuck (US Rubber)).†

Trimene Base. Ethyl chloride, ammonia and formaldehyde reaction product. High temperature accelerator with a medium to long curing range. Prevents sagging of stock in early stages of air curing. It is a latex foam stabilizer which prevents foam collapse by causing gelling to take place at a higher pH. Used in natural and SBR rubbers and latices. (Uniroyal).*

Trimesitinic Acid. Trimesic acid, (1, 3, 5-benzene-tricarboxylic acid), $C_6H_3(COOH)_3$.†

Trimetso. Industrial detergent. (Crosfield Chemicals).*

Trimmit. Grass growth regulator. (ICI PLC).*

Trimogal. White uncoated tablets containing 100mg or 200mg trimethoprim BP. Applications: For the treatment of acute urinary tract infections and long term prophylaxis of recurrent urinary tract infections. Respiratory tract infections, in particular acute and chronic bronchopneumonia and pneumonia caused by organisms sensitive to trimethoprim. Particularly useful for patients sensitive to sulphonamides. (Lagap Pharmaceuticals Ltd).*

Trimopan. Trimethoprim. Applications: Urinary tract infections, respiratory tract infections. (Berk Pharmaceuticals Ltd).*

Trimox. Amoxicillin. Applications: Antibacterial. (E R Squibb & Sons Inc).

Trimpex. Trimethoprim. Applications: Antibacterial. (Hoffmann-LaRoche Inc).

Trinamide. Triple sulphonamide association. (May & Baker Ltd).

Trinidad Asphalt. A natural asphalt obtained from the Trinidad pitch lake. The crude pitch contains from 40-46 per cent bitumen, 24-30 per cent mineral matter (clay), and 21-29 per cent water.†

Trinitrol. Erythrol tetranitrate, $C_4H_6(NO_3)_4$.†

Trinol. A proprietary preparation of benzhexol, used for Parkinsonism. (279). See TROTYL.†

Triodine. Iodophor low foaming sanitizer. (Ciba-Geigy PLC).

Triolam. A proprietary preparation of tinidazole. An antiprotozoal. (Pfizer International).*

Triolein. See OLEIN.

Trioleotope. Triolein I 131. Applications: Radioactive agent. (E R Squibb & Sons Inc).

Trione. Ready to use ninhydrin reagent solution containing organic modifier and buffer. Applications: Amino acid analysis, amine analysis. (Pickering Laboratories Inc).*

Triostam. Trivalent sodium stibogluconate injection. (The Wellcome Foundation Ltd).

Triox. Vegetation killer. (Chevron).*

Trioxitol. Ethyltrigol. (Shell Chemicals UK Ltd).

Trioxone. Selective weedkiller. (ICI PLC).*

Trip. Ferric oxide, Fe_2O_3.†

Triperidol. White uncoated tablets containing 0.5mg trifluperidol and 1mg trifluperidol. Applications: Used in the treatment of acute and chronic schizophrenia. Particularly useful in the main phase of the disease. (Lagap Pharmaceuticals Ltd).*

Triperidol. A proprietary preparation of trifluperidol. A sedative. (Janssen Pharmaceutical Ltd).*

Triphan. See IRIPHAN.

Triplastic. An explosive. It is prepared by mixing trinitro-toluene, together with some lead nitrate and chlorate, with a gelatin made from dinitrotoluene and nitrocellulose.†

Triplastite. An explosive consisting of a mixture of di- and trinitrotoluene 70 parts, guncotton 1.2 parts, and lead nitrate 28.8 parts.†

Triple Salts. (Trisalytes). Used in the electro-deposition of metals. They consist of the cyanide of the metal to be deposited, potassium cyanide, and potassium sulphite.†

Triple Sulfas. Trisulphapyrimidines (oral suspension). Mixture of Sulphadiazine, Sulphamerazine and Sulphamethazine. Applications: Antibacterial. (Lederle Laboratories, Div of American Cyanamid Co).

Triplematic. Metering, proportioning, mixing and dispensing machines for multicomponent reactive resin systems. (Hardman Inc).*

Triplevac. B_1 type, B_1 strain, Newcastle with Massachuttes and Connecticut types. Applications: immunization of poultry. (Intervet America Inc).*

Triplopen. A proprietary preparation containing Benethamine Penicillin, Procaine Penicillin and Benzylpenicillin Sodium. An antibiotic. (Glaxo Pharmaceuticals Ltd).†

Tripoli. (Tripoli Powder, Rotten Stone). A mineral. It consists mainly of silica associated with small quantities of alumina and iron oxide, but the composition varies. The variety of tripoli powder found in Derbyshire, is called Rotten stone. Used as an abrasive.†

Tripoli Powder. See TRIPOLI.

Tripolite. See INFUSORIAL EARTH.

Tripsa. Tribasic phosphate of soda. Used for the prevention of incrustation on boilers.†

Triptafen. A proprietary preparation of amitriptyline hydrochloride and perphenazine. An antidepressant. (Allen & Hanbury).†

Triptil. Portiptyline Hydrochloride. Applications: Antidepressant. (Merck Sharp & Dohme, Div of Merck & Co Inc).

Tripwite. Hydrogen peroxide for tripe dressing. (Interox Chemicals Ltd).

Trisalytes. See TRIPLE SALTS.

Triscal. A proprietary preparation of calcium and magnesium carbonates. An antacid. (Nicholas).†

Trisec. Drying additives. (ICI PLC).*

Trisem. Trisulphapyrimidines (oral suspension). Mixture of Sulphadiazine, Sulphamerazine and Sulphamethazine. Applications: Antibacterial. (Beecham Laboratories).

Trisophone. Non-ionic surfactant. (ABM Chemicals Ltd).*

Trisoralen. Trixsalen. Applications: Pigmentation agent. (Elder Pharmaceuticals Inc).

Tri-Star Antifoam £27. Synergistic blend of organic chemicals, 100% active, water dispersible. Applications: Rapid knockout of existing foam and the prevention of foam formation in systems containing detergents, in latex emulsions, in industrial processes, paints, glues, paper coating formulations, water disposal systems, sewage plants, dye baths, textile printing, paper manufacturing, oil well drilling muds, refineries and petrochemical plants. (Tri-Star Chemical Co Inc).*

Tri-Star Padding Compounds. Modified polyvinyl acetate emulsions. Applications: Fast setting padding compounds. (Tri-Star Chemical Co Inc).*

Tri-Star White Glues. Modified polyvinyl acetate emulsions. Applications: Used for packaging, bookbinding, labelling and woodworking, etc. (Tri-Star Chemical Co Inc).*

Tri-Sulfameth. Trisulphapyrimidines (oral suspension). Mixture of Sulphadiazine, Sulphamerazine and Sulphamethazine. Applications: Antibacterial. (USV Pharmaceutical Corp).

Tritane. Triphenyl-methane, $(C_6H_5)_3CH$.†

Trithac. Fungicide. (Murphy Chemical Ltd).

Tritheon. Acinitrazole.†

Trithion. Insecticide seed treatment. (Murphy Chemical Ltd).

Tritiotope. Tritiated Water. Applications: Radioactive agent. (E R Squibb & Sons Inc).

Tritole. See TROTYL.

Tritolo. See TROTYL.

Triton. See TROTYL.

Triton. Range of alkylaryl polyether alcohol surfactants. Applications: Cosmetics, household products and industrial cleaners. (Rohm and Haas Company).*

iton A-20. Tyloxapol. Applications: Detergent. (Rohm & Haas Co).

iton CF-10. Nonionic surfactant of the alkylphenol ethoxylate type in which the hydrophobe is a benzyl ether of octylaryl- polyether. Light amber liquid. Applications: Very low foam surfactant with good detergent properties used as a de-foamer for food soils and a rinse aid. (Rohm & Haas (UK) Ltd).

riton CF-21. Nonionic surfactant of the alkylphenol ethoxylate type in which the hydrophobe is alkylaryl polyether. Clear amber liquid. Applications: Low foam surfactant and wetting agent used as a textile detergent and rinse aid. (Rohm & Haas (UK) Ltd).

riton CF-87. Nonionic surfactant of the alkylphenol ethoxylate type in which the hydrophobe is a terminated form of alkylaryl polyethoxy. Straw coloured liquid. Applications: Low foam surfactant used as a rinse aid in mechanical dishwashing and low foam detergents. (Rohm & Haas (UK) Ltd).

riton GR-5M. Sodium di-octyl sulphosuccinate in equal parts of isopropanol and water. Slightly viscous pale yellow liquid. Applications: Emulsifier, dispersant and wetting agent, readily water soluble. Used for emulsion polymerization, solubilization of water in organic liquids, improved wetting of textiles, formation of emulsions. (Rohm & Haas (UK) Ltd).

riton GR-7M. Sodium di-octyl sulphosuccinate in a light petroleum distillate. Slightly viscous light amber liquid, insoluble in water, readily forms an emulsion. Applications: Emulsifier and dispersant used in dry cleaning and dewatering. (Rohm & Haas (UK) Ltd).

riton H-55. Potassium phosphate ester as a clear light amber liquid. Applications: Hydrotrope, useful at high builder levels, for nonionic and anionic surfactants in built liquid concentrates. (Rohm & Haas (UK) Ltd).

riton H-66. Potassium phosphate ester as a clear light yellow liquid. Applications: Solubilization of low foam nonionics in built liquid concentrates without increasing the foaming tendency of the system. (Rohm & Haas (UK) Ltd).

Triton N-40, N-57, N-60, N-101, N-111 and N-150. Nonylphenol ethoxylate nonionic surfactant in liquid/oil form. Applications: General surfactant properties plus stabilization, anti-fogging, anti-static, dispersion and corrosion inhibition. N-57 used as detergent emulsifier in solvent type cleaners; N-101 used in hard surface and laundry detergents and metal cleaning. (Rohm & Haas (UK) Ltd).

Triton QS-15. An oxyethylated sodium salt, in the form of an amber liquid. Applications: Surfactant ingredient in highly alkaline built cleaners. (Rohm & Haas (UK) Ltd).

Triton QS9, QS30 and QS44. Phosphate ester in free acid form. Liquid or aqueous solution. Applications: Wetting agents and detergents; primary emulsifiers in emulsion polymerization; textile lubricants for fibres; antistatic agents for fibres, plastics and films. QS30 is used for metal cleaning and QS44 is used as a hydrotrope. (Rohm & Haas (UK) Ltd).

Triton X-100, X-102, X-114, X-120 and X-165. Octylphenol ethoxylate nonionic surfactants in liquid, aqueous solution or powder form. Applications: Emulsifiers in aqueous systems with aromatic solvents; acid/alkaline cleaning; wettable powders; cosmetics and toiletries. (Rohm & Haas (UK) Ltd).

Triton X-155. Nonionic surfactant of the alkylphenol ethoxylate type in which the hydrophobe is alkylaryl polyether. Clear amber liquid. Applications: Emulsifier for aromatics, pesticides; wetting agent for wettable powders and detergents. (Rohm & Haas (UK) Ltd).

Triton X-15, X-35 and X-45. Octylphenol ethoxylate nonionic surfactants in liquid form. Applications: Emulsifiers and emulsion stabilizers in oil systems of the more hydrophobic compounds. They improve the detergency of organic solvents in dry cleaning. (Rohm & Haas (UK) Ltd).

Triton X-200 and X-202. Sodium alkyl-aryl polyether sulphonate as a viscous, creamy aqueous dispersion. Application: Detergent and emulsification agent used in latex stabilization, shampoo formulation (X-200 only), and post stabilization of synthetic emulsions. (Rohm & Haas (UK) Ltd).

Triton X-207. Nonionic surfactant of the alkylphenol ethoxylate type in which the hydrophobe is alkylaryl polyether. Oil-soluble clear amber liquid. Applications: Emulsifier and lubricant for synthetic fibre finish formulations. (Rohm & Haas (UK) Ltd).

Triton X-301. Sodium alkyl-aryl polyether sulphate as a viscous, creamy dispersion. Application: Wetting, detergency and emulsification agent used in household and industrial liquid detergents and the preparation of emulsion polymers. (Rohm & Haas (UK) Ltd).

Triton X-305, X-405 and X-705. Octylphenol ethoxylate nonionic surfactants in aqueous solution or solid form. Applications: Emulsion polymerization; agricultural emulsion concentrates; wettable powders. (Rohm & Haas (UK) Ltd).

Triton 770 and W30. Sodium alkyl-aryl polyether sulphate as a clear amber liquid. Application: Wetting, detergency and emulsification agent used, for example, in the post stabilization of emulsions. (Rohm & Haas (UK) Ltd).

Tritox. An aqueous concentrate containing MCPA, mecoprop and dicamba. Applications: A selective herbicide for the control of broad-leaved weeds in turf. (Fisons PLC).*

Trivax. A proprietary vaccine used to give protection against diphtheria, tetanus and pertussis. (The Wellcome Foundation Ltd).*

Trivexin. Braxy-blackleg-pulpy kidney vaccine. (The Wellcome Foundation Ltd).

Trixene. Isocyanates and polyisocyanates for use in coatings and adhesives. (The Baxenden Chemical Co Ltd).

Trixene. Moisture-curing urethane prepolymers for surface coatings, adhesives, mastics and sealants. Fully reacted urethane polymers in solution. Blocked isocyanates for heat-activated systems. Moisture scavengers. Solvent and water-based acrylic polymers. (Baxenden Chemical Co Ltd).*

Trixidin. An emulsion of antimony trioxide, containing 30 per cent. Sb_2O_3.†

Trobicin. Spectinomycin Hydrochloride. Applications: Antibacterial. (The Upjohn Co).

Trocinate. Thiphenamil Hydrochloride. Applications: Relaxant. (Poythress Laboratories Inc).

Trocor. Mechanically resistant material (corundum). Aluminium oxide. Applications: Admixtures for the building industry. (Dynamit Nobel Wien GmbH).*

Trodax. Nitroxynil. (May & Baker Ltd).

Trogamid T. Special polyamide, thermoplastic moulding compound. Applications: Injection, blow and extrusion moulding compounds used in various branches of industry. (Dynamit Nobel Wien GmbH).*

Trogamid T G35. Special polyamide, glass fibre reinforced. Applications: Injection moulding compounds for electrical engineering, electronics, telecommunication, mechanical and apparatus engineering, precision mechanics. (Dynamit Nobel Wien GmbH).*

Troilite. Haidinger's name for the ferrous sulphide which occurs in meteorites.†

Troisdorf Powder. An explosive. It is a gelatinized nitro-cellulose flake powder.†

Trojan SC. Selective herbicide. (FBC Ltd)

Trolene. Ronnel. Applications: Insecticide. (Dow Chemical Company Ltd).

Trolit F. A proprietary pyroxylin product.†

Trolit S and Special. Proprietary phenol-formaldehyde resin moulding compounds.†

Trolit W. A proprietary cellulose acetate product.†

Trolite. A synthetic resin of the phenol formaldehyde type. It is a term also applied to trinitro-toluene. See TROTYL.†

rolon. A proprietary phenol-formaldehyde synthetic resin. See BAKELITE.†

rolon. Phenolic resins. Applications: Main fields of application are moulded laminated materials, wood working, metal casting, abrasives, friction linings, moulded plastics. (Dynamit Nobel Wien GmbH).*

'rolone. Sulthiame. Applications: Anticonvulsant. (Riker Laboratories Inc, Subsidiary of 3M Company).

'rolovol. A basic drug for the treatment of rheumatoid arthritis. (Bayer & Co).*

'rolovol (Pendramine, Depen). D-Penicillamin. Tablets. Applications: Antirheumatic. (Degussa).*

Troluoil. A proprietary trade name for a petroleum solvent.†

Tromexan. A proprietary preparation of ethyl bicoumacetate. An anticoagulant. (Ciba Geigy PLC).†

Trona Anhydrous Sodium Sulphate. Minimum purity 99% Na_2SO_4 in fine, standard, coarse and special coarse granulations. (Kerr-McGee Chemical Corp).*

Trona Boron Tribromide. 99.8% minimum BBr_3. (Kerr-McGee Chemical Corp).*

Trona Boron Trichloride. 99.9% minimum BCl_3. (Kerr-McGee Chemical Corp).*

Tronacarb Sodium Bicarbonate. White granular solid, industrial and animal feed grades. (Kerr-McGee Chemical Corp).*

Trona Elemental Boron. 'Amorphous" type, standard grade, fine, dark brown powder meeting specifications PA-PD-451, OS11608 and MIL-B-51092 (ORD). Boron content is 90-92%. (Kerr-McGee Chemical Corp).*

Tronalight Light Soda Ash. Na_2CO_3, extracted from the brines of Searles Lake, CA. (Kerr-McGee Chemical Corp).*

Tronalight Light Soda Ash. Grades: Dense-99.7% min. Na_2CO_3; Granular-97.7% min. Na_2CO_3; Light-98.3% min. Na_2CO_3 (dry basis). (Kerr-McGee Chemical Corp).*

Tronamang Electrolytic Manganese Metal. Chip form. Grades: Low-Hy (0.005%H_2), Extra Low-Hy (0.001%H_2), and Nitor-6 nitrided (6%N_2). (Kerr-McGee Chemical Corp).*

Tronamang-75 Manganese Aluminium Briquettes. 75% manganese and 25% aluminium in briquette form for aluminium alloying. (Kerr-McGee Chemical Corp).*

Trona Muriate of Potash. White agricultural grade. 60.5% minimum K_2O in coarse, standard and fine grades. (Kerr-McGee Chemical Corp).*

Trona Potassium Chloride. High purity white industrial grade. 96.8% KCI (61.8% K_2O equivalent). (Kerr-McGee Chemical Corp).*

Trona Potassium Sulphate. High purity white industrial grade. 50% K_2O minimum in granular and standard grades. (Kerr-McGee Chemical Corp).*

Trona Red. See ST. DENIS RED.

Trona Salt Cake. Minimum purity 98% Na_2SO_4. (Kerr-McGee Chemical Corp).*

Trona Soda Ash. Grades: Dense-99.7% min. Na_2CO_3; Granular-97.7% min. Na_2CO_3; Light-98.3% min. Na_2CO_3 (dry basis). (Kerr-McGee Chemical Corp).*

Trona Sulphate of Potash. Standard white agricultural grade with 50% minimum K_2O. (Kerr-McGee Chemical Corp).*

Tronolane. Pramoxine Hydrochloride. Applications: Anaesthetic. (Abbott Laboratories).

Tronox Titanium Dioxide Pigments, Chloride Process. Nine grades for paint, plastics, printing inks and paper applications. (Kerr-McGee Chemical Corp).*

Troostite. A mineral, $2(Zn.Mn)O.SiO_2$. It is also the name for a constituent of steel tempered at a high temperature. It occurs in the transformation of austenite, the stage following marten site, and preceding sorbite.†

Troosto-Sorbite. A constituent of steel. It is similar to Troostite.†

Tropaeoline D. See ORANGE III.

Tropaeoline G. See METANIL YELLOW.

Tropaeoline OO. See ORANGE IV.

Tropaeoline OOOO. A dyestuff. It is benzeneazo-α-naphtholsulphonic acid, $C_6H_5.N_2.C_{10}H_5.HSO_3.OH$.†

Tropaeoline R. See TROPAEOLINE O.

Tropaeoline Y. See TROPAEOLINE O.

Tropaeoline 0. (Tropaeoline R, Tropaeoline Y, Acid Yellow RS, Acme Yellow, Yellow T, Chrysoine, Resorcin Yellow, Chrysoline, Chryseoline Yellow, Chryseoline, Gold Yellow, Golden Yellow, Chrysoine Extra, Resorcin Yellow O Extra). A dyestuff. It is the sodium salt of dihydroxy-benzeneazobenzene-*p*-sulphonic acid, $C_6H_3(OH)_2.N_2.C_6H_4.\ SO_3Na$. Dyes wool reddish-yellow from an acid bath.†

Tropaeoline 000 No. 2. See ORANGE II.

Trophysan. A proprietary preparation of aminoacids, minerals and vitamins in SORBITOL used for intravenous feeding. (Servier).†

Tropium. A proprietary preparation of CHLORDIAZEPOXIDE. A tranquillizer. (DDSA).†

Tropotox. Herbicide. (May & Baker Ltd).

Trosiplast. Rigid and plasticized PVC compounds. Applications: Injection, blow and extrusion moulding compounds used in various branches of industry. (Dynamit Nobel Wien GmbH).*

Trosiplast M. Compound PVC. Applications: Hollow body blow moulding, injection moulding, profile extruding, calendering. (Dynamit Nobel Wien GmbH).*

Trosiplast S. Suspension PVC. Applications: Further processing into injection moulding and extrusion moulding compounds. (Dynamit Nobel Wien GmbH).*

Trosyd. A proprietary preparation of ticonazole. An antifungal. (Pfizer International).*

Trotter Oil. Neatsfoot oil.†

Trotyl. (Trolite, Trilite, Tritolo, Trinol, Tolite, Triton, Tritole, T.N.T.). Trinitrotoluene, $CH_3.C_6H_2(NO_2)_3$. An explosive constituent.†

Troyes White. A white pigment consisting of calcium carbonate.†

TRS Rubber. A proprietary air dried fast-curing rubber. (Mitsui Co Ltd).†

Tru-Color. Colour anodized aluminium. (Reynolds Metal Co).*

Trubin. Macrolide antibiotic. Applications: Especially useful against CRD in poultry and enzootic pneumonia in pigs. Growth promoter. (Bayer & Co).*

Trucal. Calcium carbonate. Applications: Foodstuffs for animals, additives, included in Class 31 for use in animal foodstuffs but not including cereals or cereal products. (Tilcon Ltd).†

Trucarb. Calcium carbonate. Applications: Industrial limestone powders and granules for use in the manufacture of carpets, paints, glues, glass, PVC products, floor covering, mastics, agrochemicals, ceramics, roofing felt, rubber, resins, pigments and pharmaceuticals. (Tilcon Ltd).†

Trufree. A range of special flours for dietary use. Applications: Gluten free, wheat free, low salt diets. (Larkhall Laboratories Ltd).*

Trulime. Hydrated lime $Ca(OH)_2$. Applications: Horticulture/agriculture, alkalinity, building industry, civil engineering (soil stabilization and soil modification), leather processing, organic and inorganic chemicals, petrochemicals, plasterwork, sewage treatment and water treatment. (Tilcon Ltd).†

Truozine. Trisulphapyrimidines (oral suspension). Mixture of Sulphadiazine, Sulphamerazine and Sulphamethazine. Applications: Antibacterial. (Abbott Laboratories).

Truzone. Hydrogen peroxide for the hairdressing trade. (Interox Chemicals Ltd).

Trycite. Polystyrene film used for packaging. (Dow Chemical Co Ltd).*

Trydil. A safe gel heavy duty handcleanser. Applications: Handcleansing. (Borax Consolidated Ltd).*

Trylose S. Methylcellulose.†

Trymer. Rigid polyisocyanurate bunstock used in the manufacture of insulation. (Dow Chemical Co Ltd).*

Trymex Cream and Ointment. Triamcinolone acetonide USP. Applications: Indicated for the relief of the inflammatory and pruritic manifestations of corticosteroid responsive dermatoses. (Altana Inc).*

Trypan Blue. An azo dyestuff derived from tolidine and naphthalene. It destroys the trypanosome of the cattle disease "piroplasmosis.'†

Tryparosan. Chlorinated *p*-fuchsine.†

Tryptanol. Amitriptyline. Applications: Antidepressant with sedative properties. (Merck Sharp & Dohme).*

Tryptizol. Amitriptyline. Applications: Antidepressant with sedative properties. (Merck Sharp & Dohme).*

Trypure Novo. A proprietary preparation of trypsin. Used to promote wound healing and as an inhalation to reduce sputulo viscosity. (British Drug Houses).†

Trysul. Triple sulfa vaginal cream. Applications: Indicated for the treatment of haemophilus vaginalis vaginitis. (Altana Inc).*

Tsé-Hong. A mixture of white lead, aluminia, ferric oxide, and silica, used by the Chlnese for painting on porcelain.†

Tsé-Iéou. (Ting-yu). An oil expressed from Chinese tallow seeds (seeds of *Sapium sebiferum*).†

T-Siloxide. Product of silica fused with 0.1-2 per cent titania. A silica glass.†

Tsing-Lieu. A red pigment used in porcelain painting. It consists of a mixture of stannic and plumbic silicates, with copper oxide or cobalt, and gold.†

T-Size. Rosin emulsion size. (Hercules Inc).*

Tuads. Tetrabutylthiuram disulphide, tetramethyl thiuram disulphide and 60:40 blend methyl and ethyl. Rubber accelerators. Applications: Sulphur donors, accelerators, vulcanizing agents. (Vanderbilt Chemical Corporation).*

Tuasol 100. Tribromsalan. Applications: Disinfectant. (Merrell Dow Pharmaceuticals Inc, Subsidiary of Dow Chemical Co).

Tubania. Jeweller's alloys of varying composition, usually containing copper or brass, antimony, tin, and bismuth. The English alloy contains 12 parts brass, 12 parts tin, 12 parts antimony and 12 parts bismuth. German Tubania consists of 4 parts copper, 3 1/4 parts tin, and 42 parts antimony.†

Tubarine. Tubocurarine Chloride. Applications: Blocking agent (neuromuscular). (Burroughs Wellcome Co).

Tubazole. TBZ and iodophor as a foggable solution. Applications: Controlling various diseases in stored potatoes. (Wheatley Chemical Co Ltd).*

Tubazole M. TBZ and iodophor as a sprayable solution. Applications: Controlling various diseases in stored potatoes. (Wheatley Chemical Co Ltd).*

Tube Brass. See BRASS.

Tuberculin Old Tine Test. Stainless steel disc with four prongs that have been dipped in Old Tuberculin. Applications: Intradermal test for tuberculosis. (Lederle Laboratories).*

Tubergran. Quintozene in dry granular form. Applications: Controlling common scab and rhizoctonia in growing potatoes. (Wheatley Chemical Co Ltd).*

Tubotin. Fungicide. (May & Baker Ltd).

Tubotox. Dinoseb. (May & Baker Ltd).

Tubulated Silk. See LUFTSEIDE.

Tuc-tur Metal. A nickel silver. It contains from 59-61 per cent copper, 21-28 per cent zinc, 12-18 per cent nickel, and 0.3 per cent iron.†

Tuex. Tetramethylthiuram disulphide. Short curing range in natural with normal to high sulphur. Fast curing SBR and is flat curing. Accelerator with sulphur for nitrile, butyl and EPDM rubbers. (Uniroyal).*

Tuex. A proprietary trade name for a rubber vulcanization accelerator. It is tetramethyl-thiuram-disulphide. (Naugatuck (US Rubber)).†

Tufcote. Acrylic emulsion finish. (Du Pont (UK) Ltd).

Tuf-Draw. Lubricant containing mineral oil. May contain additives such as emulsifiers, corrosion inhibitors, biocides, surfactants and lubricating additives. (Franklin Oil Corporation (Ohio)).*

Tuff Stuff. Two-component flexible adhesive. (Hardman Inc).*

Tuf-Lube. Fluorocarbon dry surface lubricant, releasant. (Specialty Products Co).*

Tufnol. Laminated plastics materials bonded with synthetic resins incorporating fillers such as cotton fabric, paper and asbestos fabric. The materials are used for electrical and mechanical components in most manufacturing industries. (Tufnol).*

Tufseal. A trade mark for a range of polymerisable mixtures of asphalt, polyols and isocyanates used as adhesives. (Robertson Co, Pittsburgh, Pa).†

Tufset. A rigid polyurethane used for engineering purposes. (Tufnol).*

Tuf Stuf. Two part epoxy putty. Applications: Filling and bonding most materials. When set (rock hard) can be drilled, tapped, filed, sanded, contoured, painted and polished. Resistant to oil, water, fuel, bleach and dilute acids. (Hermetite Products Ltd).*

Tuftane. High performance polyester and polyether based urethane films. Applications: High performance applications such as fabric lamination, belting, protective covers and similar uses where the durability of urethane is required. (Lord Corporation (UK) Ltd).*

Tugon. A proprietary preparation of metriphonate. An insecticide.†

Tuinal. A proprietary preparation of quinalbarbitone sodium and amylobarbitone sodium. A hypnotic. (Eli Lilly & Co).†

Tula Metal. An alloy of silver, copper, and lead.†

Tumbleblite. Systemic fungicide. (Murphy Chemical Ltd).

Tumblebug. Garden insecticide. (Murphy Chemical Ltd).

Tumblemoss. Moss killer preventer. (Murphy Chemical Ltd).

Tumbleslug. Slug and snail killer. (Murphy Chemical Ltd).

Tumbleweed Gel. Weedkiller for spot application. (Murphy Chemical Ltd).

Tumenol. See THIOL.

Tumenol Powder. Tumenol-sulphonic acid. See THIOL.†

Tumeson. A proprietary preparation of prednisolone, sulphonated distillate of shale oils and titanium dioxide for dermatological use. (Hoechst Chemicals (UK) Ltd).†

Tuncast. Gunning refractory for lining tundishes used in continuous casting of steel. (Foseco (FS) Ltd).*

Tundak. Preformed insulating refractory cones for lining nozzle wells in tundishes. (Williams Div of Morton Thiokol Ltd).*

Tung Oil. (Chinese Wood Oil). The oil obtained by pressure from the seeds of *Aleurites cordata* and *Aleurites fordii*, of China and Japan. The seeds contain from 40-53 per cent of oil.†

Tunga Resin. A neutral glycerol-rosin ester, made with the aid of tung oil as esterifying catalyst.†

Tungophen B. Rosin-free, tung oil reactive phenolic resin. Applications: For improving the hardness and through-drying of alkyd paints and varnishes. (Bayer & Co).*

Tungsten Blue. A colloidal solution of the blue oxide of tungsten (ditungsten pentoxide). It may be used for dyeing silk.†

Tungsten Brass. (Wolfram Brass). An alloy of 60 per cent copper, 22 per cent zinc, 14 per cent nickel, and 4 per cent tungsten. An alloy containing 60 per cent copper, 34 per cent zinc, 2.8 per cent aluminium, 2 per cent tungsten, 0.7 per cent manganese and 0.15 per cent tin is also known by this name.†

Tungsten Bronze. (Wolfram Bronze). An alloy made by fusing potassium tungstate with pure tin. Used for decorative purposes.

It is also the name for an alloy of 95 per cent copper, 3 per cent tin, and 2 per cent tungsten. An alloy containing

90 per cent copper and 10 per cent tungsten is also known by this term.

The term is also applied to sodium ditungstate and tungsten dioxide, $Na_2W_3O_7$-($Na_2W_2O_5$+WO_2), in which tungsten amounts to from 70-85 per cent.†

Tungsten Iron. An alloy of iron and tungsten.†

Tungsten Steel. A very hard alloy of steel and tungsten. It usually contains from 5-8 per cent tungsten, often 4 per cent chromium, and 1.25 per cent carbon. Used for armour plates, projectiles, firearms, and high speed tools. Tool steels contain 1-4 per cent tungsten, and a rifle-barrel steel contains 3-6 per cent tungsten.†

Tungsten Steel, High. These steels usually contain from 80-85 per cent iron with more than 14 per cent tungsten. Some alloys contain from 77- 81 per cent iron, 15-18 per cent tungsten, 3-4 per cent chromium, and 0.15- 0.35 per cent silicon.†

Tungsten Steel, Low. These alloys usually contain about 96 per cent iron with 1.5-2 per cent tungsten, 0.5-1 per cent chromium, and 0.15-0.35 per cent silicon.†

Tungsten Yellow. A pigment. It consists of tungstic acid, and is prepared by decomposing wolframite with sodium carbonate, reacting upon the alkali tungstate with calcium chloride, then adding the calcium tungstate to warm hydrochloric or nitric acid, and washing the precipitated tungstic acid.†

Tuno Gum. (Seiba Gum, Toonu Gum, Tunu Gum). A gum obtained from a tree in Nicaragua. The coagulated gum, or latex, is a sticky product, and is mixed with balata for belting. See also CHICLE.†

Tunu Gum. See TUNO GUM.

Turacine. A red colouring matter contained in the feathers of the turaco birds of Africa. The colouring matter contains 8 per cent copper.†

Turbadium Bronze. An alloy of 46 per cent copper, 44 per cent zinc, 5 per cent lead, 2 per cent nickel, 1.5 per cent manganese, and small quantities of tin and aluminium. Used for propeller castings.†

Turbex. Non-ionic surfactant. (ABM Chemicals Ltd).*

Turbine Brass. See BRASS.

Turbiston Bronze. An alloy containing 55 per cent copper, 41 per cent zinc, 2 per cent nickel, 1 per cent aluminium, 0.84 per cent iron and 0.16 per cent manganese. It resists sea water.†

Turbith. Mineral. See TURPETH MINERAL.

Turbo-Grass . Mineral and plant extracts in a water base containing cytokinin, B-vitamin, morphogenic and porphyrin activity to aid in increased plant metabolism and yield. Applications: For all agricultural, horticultural and forestry products. See also Agrispon and Agro-Vita. (SN Corp/Appropriate Technology Ltd).*

Turboclean. A blend of detergents and surfactants with inhibitors. Applications: A cleaning fluid for compressors of gas turbine engines. (The Kent Chemical Company).*

Turbonit. A German synthetic varnish paper product used for electrical insulation.†

Turfclear. A suspension concentrate containing carbendazim. Applications: Dual purpose treatment for turf disease and worm cast control. (Fisons PLC).*

Turgoids,. A name applied to substances such as textile fibres, hide, tissue, leather, and wood fibres, which swell in water but do not dissolve.†

Turin Yellow. See TURNER'S YELLOW.

Turkey Blue. A dyestuff of the same group as Chrome violet, obtained by the condensation of tetramethyl-aminobenzhydrol with *p*-nitrotoluene.†

Turkey Gum. See KHORDOFAN GUM.

Turkey Red. See ALIZARIN and INDIAN RED.

Turkey Red Oils. (Sulphonated Oils, Monopol Oil, Soluble Oil, Sulphated Oil, Red Oil, Oleine). Viscid, transparent liquids, used in the preparation of cotton fibre for printing Turkey red (Alizarin red).

Concentrated sulphuric acid is run

into castor oil, stirred, the product washed with water, and finally, ammonia or soda added to give an emulsion. Turkey red oil F is completely soluble in water, and Turkey red oil S is partially soluble in water.†

Turkey Red Oxide. See INDIAN RED.

Turkey Rhubarb. Rhubarb root.†

Turkey Umber. See UMBER.

Turkish Geranium Oil. See ROSÉ OIL.

Turmeric. (Indian Saffron, Terra Merita, Curcuma). A natural dyestuff obtained from the underground stems of rhizome of *Curcuma longa* and *C. rotunda*. The dyeing principle is curcumin, $C_{21}H_{20}O_6$. It dyes cotton greenish-yellow, and is also a colouring matter for wool, silk, oil, butter, cheese, curry powder, wood, and wax.†

Turmerine. See CLAYTON YELLOW.

Turnbull's Blue. (Gmelin's Blue). Ferrous ferricyanide, $Fe_3[Fe(CN)_6]_2$.†

Turner's Black. Animal charcoal.†

Turner's Yellow. (Patent Yellow, Cassel Yellow, Verona Yellow, Montpelier Yellow, Mineral Yellow, Veronese Yellow, Turin Yellow, Paris Yellow, English Yellow). Pigments which consist of oxychlorides of lead, usually $3PbO.PbCl_2$.†

Turnsole. See LITMUS.

Turpenteen. See TURPENTYNE.

Turpentine. The exudation from incisions made in certain varieties of pine, fir, and larch. The terms American, French, German, Mexican, Portuguese, and Spanish Turpentine are usually used for the balsam turpentine; German, Finnish, Polish, Russian, and Swedish Turpentine often refer to wood turpentine (*qv*), but German, Finnish, or Swedish Oil can refer to refined sulphite turpentine oil.†

Turpentine Oil, Cellulose. See SULPHITE TURPENTINE OIL.

Turpentine Spirit. See OIL OF TURPENTINE.

Turpentine, Artificial. Camphene, $C_{10}H_{16}$.†

Turpentine, Bordeaux. See FRENCH TURPENTINE.

Turpentine, Chio. See CHIAN TURPENTINE.

Turpentine, Chios. See CHIAN TURPENTINE.

Turpentine, Cyprian. See CHIAN TURPENTINE.

Turpentine, Essence. See OIL OF TURPENTINE.

Turpentine, Finnish. See TURPENTINE.

Turpentine, German. See TURPENTINE.

Turpentine, Larch. See VENICE TURPENTINE.

Turpentine, Mexican. See TURPENTINE.

Turpentine, Polish. See TURPENTINE.

Turpentine, Portuguese. See TURPENTINE.

Turpentine, Scian. See CHIAN TURPENTINE.

Turpentine, Spanish. See TURPENTINE.

Turpentine, Swedish. See TURPENTINE.

Turpentyne. (Turpenteen). A turpentine substitute composed of rosin spirit, shale spirit, petroleum spirit, and coal tar naphtha.†

Turpeth Mineral. (Turbith Mineral, Queen's Yellow). A yellow basic sulphate of mercury, $HgSO_4.2HgO$.†

Turpex. Finishing agents. (Ciba-Geigy PLC).

Turps. See OIL OF TURPENTINE.

Turquoise Blue. A dyestuff prepared by the condensation of tetramethyl-diaminobenzhydrol and *p*-nitrotoluene. Used in calico printing, and in the manufacture of lakes.†

Turquoise Green. A colour used chiefly in porcelain painting. It is usually prepared by heating a mixture of aluminium hydroxide, chromium hydroxide, and cobalt carbonate.†

Tursione. Biocidal preparation. (BDH Chemicals Ltd).

Tusadin. A proprietary product. It is an agent for protection against frost in motor engines.†

Tusana. A proprietary preparation of liquid extracts of cocillana, ipecacuanha, squill, senega and senna and glycerin and dextromethorphan hydrobromide. A cough linctus. (Boots Company PLC).*

Tuscan Red. See MADDER INDIAN RED.

Tussar Silk. (Tasar Silk). The product of the caterpillar of *Antheraca paphia*, of India.†

Tussend. Used primarily in the treatment of coughs, colds and upper respiratory conditions. (Dow Chemical Co Ltd).*

Tussiex. A proprietary preparation of ammonium chloride, sodium citrate, ephedrine hydrochloride, pholcodine and menthol. A cough linctus. (Crookes Laboratories).†

Tussifan S. A proprietary preparation of belladonna extract, potassium citrate, ipecacuanha, squill, anise oil and chloroform spirit. A cough linctus. (Norton, H N Co).†

Tussiplegyl. An antitussive preparation. (Bayer & Co).*

Tussol. A registered trade mark currently awaiting re-allocation by its proprietors to cover a range of pharmaceuticals. (Casella AG).†

Tutania. Alloys. An English one contains 91 per cent tin, 8 per cent lead, 0.7 per cent copper, and 0.3 per cent zinc; and another 80 per cent tin, 16 per cent antimony, 2.7 per cent copper, and 1.3 per cent zinc. A German alloy contains 92 per cent antimony, 7 per cent tin, and 1 per cent copper; and another consists of 62 per cent antimony, 31 per cent copper and 7 per cent tin.†

Tutenag. (Tutenague, Tutenay). A nickel silver. It consists of from 44- 46 per cent copper, 16-40 per cent zinc, and 15-40 per cent nickel.†

Tutenague. See TUTENAG.

Tutia. See TUTTY POWDER.

Tutol. A registered trade name for certain explosives.†

Tutol No. 2. An explosive similar to Rexite (*qv*). It contains sodium nitrate instead of potassium nitrate, and 12 per cent of the explosive base is replaced by sodium chloride.†

Tutty Powder. (Tutia). An impure oxide of zinc, formed during the smelting of lead ores containing zinc.

Sometimes a mixture of blue clay and copper filings is sold under this name.†

Tween. Surfactant: polyexyethylene sorbitan fatty acid esters. (ICI PLC).*

Tween 20. Polysorbate 20. Applications: Pharmaceutic aid. (ICI Americas Inc).

Tween 20, 40, 60, 65, 80, 85. A proprietary trade name for Polysorbate 20, 40, 60, 65, 80, and 85.†

Tween 40. Polysorbate 40. Applications: Pharmaceutic aid. (ICI Americas Inc).

Tween 60. Polysorbate 60. Applications: Pharmaceutic aid. (ICI Americas Inc).

Tween 80. Polysorbate 80. Applications: Pharmaceutic aid. (ICI Americas Inc).

Twin-Tak. Herbicide. (May & Baker Ltd).

Twitchells Reagent. Benzenestearo-sulphonic acid. Used in the decomposition of fats.†

Two Cubed Eight. A gamma ferric oxide for magnetic media use. (Pfizer International).*

Twosward. Fertilizers. (ICI PLC).*

Tycel. Urethane laminating adhesives. Applications: Lamination of many polymeric materials to rigid and flexible substrates. (Lord Corporation (UK) Ltd).*

Tygacell. Composite moulding with carbon, glass, aramid and ceramic reinforcement. Applications: Structural components for racing cars and power boats, components for civil aircraft, space defence, medical x-ray equipment and high performance applications. (Fothergill Tygaflor Ltd).*

Tygadure. High performance insulated wires and cables and optical fibres. Applications: Avionics and electronics, industrial applications, short haul data communications, secure and hazardous environments. (Fothergill Tygaflor Ltd).*

Tygaflon. Fluorocarbons with fibre, metal and other fillers. Applications: Moulded and machined custom components for all industries. High performance electrical applications, expansion joints and shaft covers. (Fothergill Tygaflor Ltd).*

Tygaflor. Fluoronated coated glass and aramid fabrics. Applications: Process conveying belt systems used in the baking and packaging industries and other industrial uses, lightweight membrane roofs and radomes. (Fothergill Tygaflor Ltd).*

Tygafluor. A proprietary trade name for an aqueous dispersion of PTFE (*qv*) with a curing temperature of 90-140° C.†

Tygalam. Composite moulding with carbon, glass, aramid and ceramic reinforcement. Applications: Structural components for racing cars and power boats, components for civil aircraft, space defence, medical x-ray equipment and high performance applications. (Fothergill Tygaflor Ltd).*

Tygan. Polyvinylidene chloride coated fabric. Applications: Filtration, insect screening, glare-reducing blinds. (Fothergill Tygaflor Ltd).*

Tygatape. Engineered, high performance PTFE and silicone tapes. Applications: Masking applications in light/heavy engineering, electrical and electronics, aerospace. (Fothergill Tygaflor Ltd).*

Tygavac. Materials for vacuum bag moulding of composite components, TFE aerosol release sprays. Applications: Aerospace, automotive, medical engineering, industrial applications. (Fothergill Tygaflor Ltd).*

Tyglas. Glass fibre woven fabrics. Applications: Electrical insulation, filtration, reinforced plastics, thermal insulations, industrial plant applications. (Fothergill Tygaflor Ltd).*

Tygon F. A proprietary trade name for a furan resin.†

Tylac. A proprietary range of butadiene copolymer rubbers. (Revertex, Harlow, Essex).†

Tylenol. Acetaminophen. Applications: Analgesic; antipyretic. (McNeil Consumer Products Co).

Tylnatrin. Sodium acetylsalicylate, $C_2H_3O.OC_6H_4.COONa$.†

Tylose. Water soluble cellulose ethers. (Hoechst UK Ltd).

Tylosin. An antibiotic derived from an actinomycete resembling *Streptomyces fradioe*. TYLAN.†

Tymahist. A proprietary preparation of CHLORPROPHENPYRIDAMINE. (Mason Pharm, Inc, Sacramento, California).†

Tymtran. Ceruletide diethylamine. Applications: Stimulant. (Adria Laboratories Inc).

Type Metal. A variable alloy of lead and antimony, frequently with the addition of tin, and sometimes copper or bismuth. The lead is present to the extent of from 50-93 per cent. the antimony 4-30 per cent. the tin 2-40 per cent. copper 0-5 per cent. and bismuth 0-29 per cent. A German type metal contains 60 per cent lead, 12 per cent tin, 18 per cent antimony, 4.7 per cent copper, 4.7 per cent nickel, and 1 per cent bismuth. Founder's type usually contains from 20-25 per cent tin, 25 per cent antimony and the rest lead, for hand type. Alloys poorer in tin are used for linotype and often contain from 2-5 per cent tin, 10-12 per cent antimony, and the rest lead. Monotype metal contains from 6-10 per cent tin, 15 per cent antimony, and the rest lead.†

Typewriter Metal. An alloy of 57 per cent copper, 20 per cent nickel, 20 per cent zinc, and 3 per cent aluminium.†

Tyraline. See MAUVEINE.

Tyramine. (Phenomime). *p*-Hydroxyphenylethylamine. The base is a constituent of ergot.†

Tyrenka. A proprietary trade name for a synthetic fibre resembling Nylon.†

Tyrian Purple. The colouring matter obtained from the shell-fish, *Murex trecuculus*, and from *M. brandaris*. That from *M. trecuculus* contains indigo.†

Tyril. Styrene-acrylonitrile resin (SAN) used in injection moulding, blow moulding, and extrusion with major applications in automotive, medical, housewares. (Dow Chemical Co Ltd).*

Tyrimide. A proprietary preparation containing isopropamide iodine. An antispasmodic. (Smith Kline and French Laboratories Ltd).*

Tyrin Brand. Chlorinated polyethylene resins and elastomers used in a wide variety of applications including roofing membranes, wire and cable, automotive tubing, ignition wire. (Dow Chemical Co Ltd).*

Tyrite. Urethane structural adhesives. Applications: Structural bonds for wood, plastics, metals, foams, fabrics and elastomers. (Lord Corporation (UK) Ltd).*

Tyrol-2, 32B, 6, CEP. Flame retardant materials for plastics. (Stauffer Chemical).†

Tyrolean Earth. See BOHEMIAN EARTH.

Tyrolite. (Cupriferous Calamine, Copper Froth). A basic copper arsenate of green colour, found in the Tyrol. It has the formula, $Cu_3As_2O_8.2Cu(OH)_2.7H_2O$.†

Tyrosinase. An oxidizing enzyme.†

Tyrosine. α-Amino-β-hydroxy-phenylpropionic acid, $HO.C_6H_4.CH_2.CH(NH_2)COOH$. A protein aminoacid.†

Tyrothricin. An antibiotic produced by a strain of Bacillus brevis.†

Tyrozets. Tyroethrocin and benzocaine. Applications: For minor mouth and throat infections. (Merck Sharp & Dohme).*

Tysonite. (1) A mineral. It is a fluoride of cerium metallic elements with thorium.

(2) A blend of Gibsonite and vulcanized vegetable oils. (R T Vanderbilt Co Inc).†

Tyvek. Spunbonded polyolefin. (Du Pont (UK) Ltd).

Tyzine. A proprietary preparation of tetrahydrozoline hydrochloride. A nasal decongestant. (Pfizer International).*

Tyzor. Organic titanate. (Du Pont (UK) Ltd).

U

Uba. (Styxol, Nicoschwab, Tanatol). Preparations containing sodium fluosilicate, Na_2SiF_6, as the main ingredient.†

Ubretid. A proprietary trade name for distigmine bromide, used in the treatment of urinary retention. (Berk Pharmaceuticals Ltd).*

UBS. Dry cleaning soap and paint remover. (S & D Chemicals Ltd).†

Uchatius Bronze. (Steel Bronze). An alloy containing 92 per cent copper and 8 per cent tin.†

Ucicline. Boiler and cooling water treatment. (Laporte Industries Ltd).

Ucinite. A proprietary trade name for a phenol-formaldehyde resin laminated product.†

Ucipol. Boiler and cooling water treatment. (Laporte Industries Ltd).

Ucon. Synthetic lubricants and fluids. (Union Carbide (UK) Ltd).†

Ucrete. A proprietary cement-modified polyurethane resin used for flooring. (ICI PLC).*

Ucuhuba Fat. A fat obtained from the seeds of *Myristica bicuhyba*. It contains 92 per cent fatty acids.†

Udel. A proprietary polysulphone. A high-performance, high-temperature thermoplastic resin used for injection moulding and extrusion. (Union Carbide (UK) Ltd).†

Udet 950. Sodium alkyl aryl sulphonate. Applications: Fast dissolving, high active sulphonate for detergent powders. (DeSoto Inc).*

Udikral. ABS polymers. Applications: Injection moulding, sheet extrusions. (Borg Warner Chemicals).*

Udilo Oil. See LAUREL NUT OIL.

Ufacid. Linear alkylbenzene sulphonic acid, branched dodecyl benzene sulphonic acid and branched tridecylbenzene sulphonic acid. Applications: Active ingredient in detergent powders and liquids. Used as emulsifier in herbicide and pesticide systems. Air entrainment agent in cement. (Unger Fabrikker AS).*

Ufanon. Alkanolamide. Applications: Additive to shampoos, bath products and liquid detergents to boost and stabilize foam and aid viscosity adjustment. (Unger Fabrikker AS).*

Ufaryl. Sodium alkylbenzene sulphonate powder - drum and spray dried, active ingredient in detergent powders. Applications: Used as emulsifier in herbicide and pesticide systems. Air entrainment agent in cement. (Unger Fabrikker AS).*

Uffelmann's Reagent. A lactic acid reagent, prepared by adding a ferric chloride solution to a 2 per cent phenol solution until it is of a violet colour. The colour of the reagent is changed to deep yellow by the addition of lactic acid.†

Uformite. A proprietary trade name for an urea-formaldehyde synthetic resin.†

U-Gencin. Gentamicin sulphate. A complex antibiotic substance, produced by *micromonospora purpurea* nsp. It has three components, sulphates of gentamicin C_1, gentamicin C_2, and gentamicin C_{1A}. Applications: Antibacterial. (The Upjohn Co).

Ugikral RA, RB and SN. Proprietary trade names for ABS terpolymers.†

Ugurol. An antifibrinolytic. (Bayer & Co).*

Uhligite. A mineral. It is a titanate of zirconium, calcium, and aluminium, found in East Africa.†

Uintahite. See GILSONITE.

Uintaite. See GILSONITE.

Ulcatite. A proprietary trade name for a rubber vulcanization accelerator containing hexamethylenetetramine, benzthiazyl disulphide and diphenylguanidine.†

Ulcedal. A proprietary preparation of deglycyrrhizinized liquorice extract used in the treatment of peptic ulcers. (Boehringer Ingelheim Ltd).†

Ulcerban. Sucralfate. Applications: Anti-ulcerative. (Marion Laboratories Inc).

Ulco. A metal used as a substitute for Babbitt metal. It contains 98-99 per cent lead, the rest being barium and calcium.†

Ulcony. An alloy of 65 per cent copper and 35 per cent lead.†

Ulexine. Cytisine, $C_{10}H_{14}N_2O$, an alkaloid found in laburnum and furze.†

Ulmal. A proprietary trade name for an alloy containing aluminium with 10 per cent magnesium, 1 per cent silicon and 0.5 per cent manganese.†

Ulmin Brown. See VANDYCK BROWN.

Ulmite. A name proposed for the dark-coloured fibre covering the grains of sandstone found on the coast of New South Wales. In its properties it resembles those of humus, obtained from brown peat.

Ulon. A proprietary polyurethane elastomer. (Unitex Ltd).*

Ultandren. A proprietary preparation of fluoxymesterone. (Ciba Geigy PLC).†

Ultem. Polyetherimide. Applications: Plastic components for automotive, electrical, electronics, lighting, medical, packaging, audio etc. (GE Plastics Ltd).*

Ulto Accelerator. A proprietary rubber vulcanization accelerator. It is a zinc salt of a complex dithiocarbamate.†

Ultra 206. Organic brightener system. Applications: Bright nickel electroplating (high levelling, fast brightening). (Harshaw Chemicals Ltd).*

Ultracef. Cefadroxil. Applications: Antibacterial. (Bristol Laboratories, Div of Bristol-Myers Co).

Ultracene. A proprietary rubber vulcanization accelerator. It is a guanidine derivative.†

Ultrachem Assembly Fluid 1. A tacky polymer used in assembly of o-rings. Applications: Assembly of o-rings used in helicopter transmissions, jet turbines, pumps, etc. (Ultrachem Inc).*

Ultracortenol. A proprietary preparation of prednisolone pivalate. (Ciba Geigy PLC).†

Ultracut. Water reducible lubricants containing mineral oil, but when diluted in water forms a micro-emulsion which is translucent to transparent. Additives include emulsifiers, corrosion inhibitors, surfactants, biocides and lubricating additives. (Franklin Oil Corporation (Ohio)).*

Ultradil. A proprietary preparation of fluocortolone pivalate and hexanoate used in the treatment of eczema. (Schering Chemicals Ltd).†

Ultra-DMC. A proprietary trade name for a vulcanization accelerator. It is dimethylamine dimethyldithiocarbamate.†

Ultradur. Polybutylene terephthalate granules. Applications: Injection moulding - electrical engineering components, key buttons. (BASF United Kingdom Ltd).*

Ultraferran. A colloidal iron.†

Ultra-Flat. Aluminium alloy sheet. (Reynolds Metal Co).*

Ultrafloc. Tabletted carrageenan. (ABM Chemicals Ltd).*

Ultraform. Polyacetal granules. Applications: Injection moulding - mechanical engineering, cassette hubs. (BASF United Kingdom Ltd).*

Ultralente Iletin. Insulin zinc, extended. Applications: Antidiabetic. (Eli Lilly & Co).

Ultralente Insulin. Insulin zinc, extended. Applications: Antidiabetic. (E R Squibb & Sons Inc).

Ultra-light Alloys. Alloys having a specific gravity below 2 are known by this name. Magnesium-aluminium-zinc and magnesium-copper are alloys of this type.†

Ultralin. A rosin-fatty acid reaction product.†

Ultralog. Grade of chemical with 99.8% to 100% purity. (Chemical Dynamics Corp).*

Ultralumin. A jeweller's alloy. It contains more than 90 per cent aluminium, with nickel, copper, and some rare earth metals of the thorium group. It is specially resistant to sea water.†

Ultramarine. (Lapis-Lazuli Blue, Oriental Blue, Brilliant Ultramarine, French Blue, New Blue, Permanent Blue, French Ultramarine, Soda Ultramarine). A blue colouring matter formerly prepared from the rare mineral lapis-lazuli, by powdering and washing. It is now prepared artificially by fusing together kaolin, sulphur, with soda, or with a mixture of sodium sulphate and charcoal.†

Ultramarine Ash. In obtaining ultramarine from lapis-lazuli, a blue product is first yielded, then a pale blue, and finally a pale bluish-grey material, which is called Ultramarine ash.†

Ultramarine Blue. See WILLOW BLUE, ULTRAMARINE, and GREEN ULTRAMARINE.

Ultramarine Yellow. See BARIUM YELLOW, STEINBUHL YELLOW, STRONTIAN YELLOW and ZINC YELLOW.

Ultramarine, Acid-proof. An ultramarine which resists the action of alum, due to an excess of silica in its composition. It does not resist true acids, but because alum in solution gives an acid reaction it is called acid-proof, and can be used for colouring paper, fabric, and soap where alum is also used.†

Ultramarine, Artificial. See ULTRAMARINE.

Ultramarine, Brilliant. See ULTRAMARINE.

Ultramarine, Cobalt. See COBALT BLUE.

Ultramarine, French. See ULTRAMARINE.

Ultramarine, Gahn's. See COBALT BLUE.

Ultramarine, Green. See CHROME GREENS and GREEN ULTRAMARINE.

Ultramarine, Silver. See YELLOW ULTRAMARINE.

Ultramarine, Soda. See ULTRAMARINE.

Ultramarine, Sulphate. See SULPHATE ULTRAMARINE and ULTRAMARINE.

Ultramid. A trade mark for a wide range of Nylons. Grades of ULTRAMID include the following:

Ultramid A. Nylon 66.

A3K and *A3W.* stabilized 6.6 nylon copolymers. *A3WG5.* A stabilized 6.6 nylon copolymer, 25 per cent glass-loaded.

A3WC6. As above, but 30 per cent glass-loaded.

A3WC7. As above, but 35 per cent glass-fibre-loaded.

A3WC10. As above, but 50 per cent glass-fibre-loaded.

Ultramid A4. A stabilized 6.6 nylon copolymer.†

A4H. As above, but pigmented brown.

AEHG5. A4H 25 per cent loaded with glass fibre.

Ultramid B. Nylon 6.

B3. A general-purpose nylon 6 copolymer with low viscosity.

B3K. A stabilized version of B3 above.

B3W G5. A nylon 6 copolymer 25 per cent loaded with glass fibre.

B3W G6. As above, but stabilized and 30 per cent loaded with glass fibre.

B3W G7. As above, but stabilized and 35 per cent loaded.

B3W G10. As above, but stabilized and 50 per cent loaded.

B4. A nylon 6 copolymer with medium viscosity.

B4K. A stabilized grade of *B4*.
B5. A nylon 6 copolymer with high melt viscosity.
B6. As above, but with very high melt viscosity.
B35. As above, but with low-to-medium viscosity.
B35W. A stabilized grade of *B35*.
Ultramid S. Nylon 6.10.
S3. A general-purpose nylon 6.10 copolymer with medium viscosity.
S3K. A stabilized grade of *S3*.
S4. A modified stabilized 6.10 nylon copolymer. See also GRILON; MARANYL.†

Ultramid A + B. Polyamide 6 + 6.6 granules. Applications: Injection moulding - engineering components. Extrusion - rod, film, monofilament. (BASF United Kingdom Ltd).*

Ultramoll. Plasticizer. (Bayer UK Ltd).

Ultramoll I, II, III. A range of polyadipates. Applications: Polymeric plasticizers. (Bayer & Co).*

Ultramoll M. A polymeric plasticizer. Applications: For use in the formulation of flame retardant pastes for polyester based Moltopren. (Bayer & Co).*

Ultranox. Processing aids for thermoplastics. (Borg Warner Chemicals).

Ultrapas. Melamine resin moulding compounds. Applications: Manufacture of tableware, bathroom and publicity items, screw tops. (Dynamit Nobel Wien GmbH).*

Ultrapen. A proprietary preparation containing propicillin (as potassium salt). An antibiotic. (Pfizer International).*

Ultraphan. Acetate film. (Lonza Limited).*

Ultraproct. A proprietary suppository containing fluocortolone pivalate and hexanoate, cinchocaine hydrochloride, CLEMOZOLE undecanoate and HEXACHLOROPHANE used in the treatment of haemorrhoids. (Schering Chemicals Ltd).†

Ultrasil. A range of precipitated silicas and silicates. Applications: For use as highly dispersed reinforcing fillers with various degrees of activity for heavy duty bright rubber articles. (Degussa).*

Ultrasol. Solvents doubly distilled in glass. (British Drug Houses).†

Ultrastyr. A proprietary range of high impact polystyrene resins. (Montedison UK Ltd).*

Ultrasul. Sulphamethizole. Applications: Antibacterial. (Alcon Laboratories Inc).

Ultratard. Insulin zinc, extended. Applications: Antidiabetic. (E R Squibb & Sons Inc).

Ultratex. Finishing agent. (Ciba-Geigy PLC).

Ultrathane. Cast polyurethane products based on toluene diisocyanate, naphthalene diisocyanate and methylene diisocyanate combined with various polyols. Applications: Solid tyres, mining parts, oil scraper cups and tooling products. (Watts Urethane Products).*

Ultrathene. Ethylene-vinyl acetate copolymer resins. Applications: Blown and cast film, injection moulding, blow moulding, extrusion coating and sheet and profile extrusion. (USI Chemicals).*

Ultra Touch. Water, glycerine, stearic - TEA, butyl stearate, parasepts, fragrance. Applications: Moisturizing hand and body lotion. (Virkler Chemical Co).*

Ultra Violet Dyestuffs. Quinhydrones, obtained by the condensation of a leuco-gallocyanine with a gallocyanine.†

Ultravon. Pretreatment agent. (Ciba-Geigy PLC).

Ultravon AN. A proprietary trade name for a fatty acid amide derivative. It emulsifies oils and fats more effectively and maintains the handle and colour of wool better than conventional detergents. A special detergent primarily for wool scouring. (Ciba Geigy PLC).†

Ultrawet. High foaming versatile nonionic surfactant. Applications: Liquid detergent formulations. (Cornelius Chemical Co Ltd).

Ultrawet DS. Anionic surfactant supplied as cream coloured flakes. Applications: Detergency, wetting, sudsing, dispersing and emulsifying agent for speciality cleaning and industrial processing. (Cornelius Chemical Co Ltd).

Ultrawet K and AOK. Anionic surfactant supplied as cream coloured flakes. Applications: Various industrial and heavy-duty household detergents. (Cornelius Chemical Co Ltd).

Ultra Zinc DMC. A proprietary trade name for a rubber vulcanization accelerator. It is zincdimethyldithiocarbamate.†

Ultrazym. A purified pectolytic enzyme preparation produced from a selected strain of *Aspergillus niger*. Applications: Can be used in any case where the aim is breaking down of soluble and insoluble pectins with varying degrees of esterification for reduction of viscosity, clarification, maceration of plant tissue and depectinization. (Novo Industri A/S).*

Ultroil. A proprietary trade name for a wetting agent for textiles. It is a sulphonated vegetable oil.†

Ultryl 6010. A proprietary non-stabilized PVC resin of the suspension type.†

Ultryl 6500. A proprietary PVC polymer of the suspension type, used in the production of glass-clear film, tube, etc. (Philips Petroleum International).†

Ultryl 6800. A proprietary plasticizer free PVC resin used in the manufacture of pipe and profiles. (Philips Petroleum International).†

Ultryl 7100. A proprietary PVC resin of the suspension type with easy processing properties. (Philips Petroleum International).†

Ultryl 7150. A proprietary PVC resin of the suspension type containing additives to give high clarity. (Philips Petroleum International).†

Ulvio Cocoa. A German proprietary food material prepared by exposing cocoa to ultra-violet radiation.†

Umber. (Umber Brown, Mineral Brown, Velvet Brown, Chestnut Brown, Manganese Velvet Brown, Burnt Umber). Mineral varieties of umber. They are ochres coloured brown by oxides of manganese, and containing varying amounts of clay. The best is obtained from Cyprus, and is known as Raw Turkey Umber. Umber contains MnO_2 or Mn_3O_4, to the extent of from 6-12 per cent., Fe_2O_3 25-40 per cent. and SiO_2, 16-32 per cent. The term umber has also been applied to brown earthy products which contain lignite as the chief constituent. The so-called "Cologne earth', Coal brown, and Cassel brown belong to this class. Also see VANDYCK BROWN. Burnt umber is the calcined product of umber. Also See BURNT SIENNA.†

Umber Brown. See UMBER.

Umber, Mineral. See UMBER.

Umbrathor. A solution of thorium dioxide.†

Umbrenal. A 25 per cent solution of lithium iodide in ampoules.†

Umbrite A. An explosive containing 49 per cent nitroguanidine, 38 per cent ammonium nitrate, and 13 per cent silicon.†

Umbrite B. An explosive containing 37.5 per cent nitroguanidine, 49.5 per cent ammonium nitrate, and 13 per cent silicon.†

Umburana Seed. The product of *Amburana claudii*. Used in Brazil for perfuming tobacco.†

Umea Tar. A pale Swedish pine-wood tar. It is a good variety of Stockholm tar produced in the Umea district.†

UN-28 and UN-32. Fertilizer solutions containing 28% and 32% nitrogen respectively. Applications: Designed for direct agricultural use as a three-way source of nitrogen. (Hercules Inc).*

Unads. Tetramethylthiuram monosulphide - rubber accelerator. Applications: For NR and synthetic rubbers especially in neoprene. (Vanderbilt Chemical Corporation).*

Unakalm. Ketazolam. Applications: Tranquilizer. (The Upjohn Co).

Unal. A photographic developer. It is Rodinal in a solid form, containing, besides *p*-aminophenol, the ingredients necessary for solidification.†

Unburn. A proprietary anti-burn cream containing benzocaine, HEXACHLOROPHANE, orthophenyl phenol, menthol and lanolin. (Leeming/Pacquin, NY).†

ngerol. Sodium fatty alcohol sulphate and sodium fatty alcohol ether sulphate. Applications: Active ingredient in shampoos, bath products and liquid detergents, emulsifying agent in polymerization. (Unger Fabrikker AS).*

nguentum. A proprietary preparation of silicilic acid, liquid paraffin, soft paraffin, cetostearyl alcohol, polysorbate-40, glycerol, oil, sorbic acid and propylene glycol. A protective skin cream. (E Merck, Darmstadt).†

ni-Cal 66. Colourant dispersions. Applications: Colouring of non-aqueous industrial and maintenance coating compositions. (Nueodex Inc).*

Jnicell. Lightweight waterproof mortars for high build and overhead applications. Applications: Repairs to concrete and stonework. (Ronacrete Ltd).*

Jnicor. Corrosion inhibitors. (Universal-Matthey Products Ltd).

Jnicrylic. Solvent and acrylic sealant. Applications: Used for glazing of windows, panels and general caulking. (Pecora Corporation).*

Jnidem. Demulsifiers. (Universal-Matthey Products Ltd).

Jnidiarea. A proprietary preparation of NEOMYCIN sulphate, CLIOQUINOL and attapulgite. An anti-diarrhoeal. (Unigreg).†

Jniflood. Refinery and oilfield chemicals. (Universal-Matthey Products Ltd).

Jniflu. A proprietary preparation of DIPHENHYDRAMINE hydrochloride, PARACETAMOL, caffeine, PHENYLEPHRINE hydrochloride and codeine phosphate. A remedy for colds. (Unigreg).†

Jnifog. Insecticidal formulation. (Mitchell Cotts Chemicals Ltd).

Jni G. Motor oil. (Chevron).*

Jnigel. A nitroglycerin dynamite. Applications: Construction and building industry, explosives, mining, petroleum and related industries. (Hercules Inc).*

Unigest. A proprietary preparation of DIMETHICONE, aluminium hydroxide and magnesium hydroxide and carbonate. An antacid. (Unigreg).†

Unihepa. A proprietary preparation of dimethionine, choline, inositol, thiamine, pyridoxine, biotin, vitamin E, cyanocobalamin, panthenol and folic acid, used to counteract senil muscular degeneration. (Unigreg).†

Unihib. Organic phosphates. Applications: For use as corrosion inhibitors in cooling towers and boilers. (Lonza Limited).*

Unilab Surgibone. Surgibone. Applications: Prosthetic aid. (Unilab Inc).

Unilink 450. 4,4' bis (sec-butylamino) diphenyl methane. Used as a chain extender. Applications: Production of specialty polyurethane plastics. (UOP Inc)*

Uniloy Chrome Steels. A proprietary trade name for alloys containing 4-6 per cent chromium, 0.1-0.25 per cent carbon, up to 0.6 per cent manganese and 0.4-0.6 per cent molybdenum and 1.0-1.25 per cent tungsten.†

Unimite. Ammonia dynamite. Applications: It is used for wet hole blasting in hard rock for operations such as quarrying and coal stripping. (Hercules Inc).*

Unimoll. Phthalates plasticizers. (Bayer UK Ltd).

Unimoll BB. Benzyl butyl phthalate. Applications: Monomeric plasticizer. (Bayer & Co).*

Unimoll DB. Dibutyl phthalate. Applications: Monomeric plasticizer. (Bayer & Co).*

Unimoll DM. Dimethyl phthalate. Applications: Fixative for perfumes, plasticizer for cellulose acetate and acetate butyrate. (Bayer & Co).*

Unimoll 66 and 66M. Dicyclohexyl phthalate. Applications: Monomeric plasticizer. (Bayer & Co).*

Unimycin. A proprietary preparation of OXYTETRACYCLINE hydrochloride. An antibiotic. (Unigreg).†

Union Black B, R. Union dyestuffs which dye from a bath containing sodium sulphate. The brand B gives a deep black, and R a reddish-black.†

Union Black B, 2B, S. Direct cotton or union dyestuffs.†

Union Fast Claret. See ORCHIL RED A.

Unipel. Pelleted fertilizer. (Chevron).*

Unipen. A proprietary preparation of carindacillin sodium. An antibiotic used in the treatment of genito-urinary tract infections. (Pfizer International).*

Uniperol. Textile dyeing auxiliaries. (BASF United Kingdom Ltd).

Uniphyllin. Theophylline in a controlled release tablet. Applications: Asthma. (Napp Laboratories Ltd).*

Uniplast. A proprietary trade name for phenol-formaldehyde moulding compound.†

Unipor. Pour point depressants. (Universal-Matthey Products Ltd).

Uniquat. Alkyl imidazoline benzyl quaternary ammonium compounds. Applications: Germicidal applications in hospitals, institutions and industrial water treatment. (Lonza Limited).*

Unique. A smokeless powder. Applications: It is designed for use in light through heavy shotshell loads. It can also be used in handgun loads. (Hercules Inc).*

Uniroid. A proprietary preparation of NEOMYCIN sulphate, POLYMIXIN B sulphate, HYDROCORTISONE and cinchocaine hydrochloride, used in the treatment of haemorrhoids. (Unigreg).†

Unisol. Textile auxiliary chemicals. (ICI PLC).*

Unisom. Doxylamine succinate. Applications: Antihistaminic. (Pfizer Inc).

Unisperse-E. Pigment pastes for emulsion paints. (Ciba-Geigy PLC).

Unisperse-P. Aqueous pigment dispersions for wallpaper printing mass colouration of paper and paper coating. (Ciba-Geigy PLC).

Unithane. Urethane polymers. (Cray Valley Products Ltd).

Unithane 640 W and 641 W. A proprietary range of urethane oils used in the manufacture of tough chemical resistant coatings such as floor finishes, etc. (Cray Valley Products Ltd).†

Unitop. Cuprimyxin. Applications: Antibacterial; antifungal. (Hoffmann-LaRoche Inc).

Unitreat. Refinery and oilfield chemicals. (Universal-Matthey Products Ltd).

Univadine. Dyeing and printing assistant. (Ciba-Geigy PLC).

Univan. A nickel-vanadium steel.†

Universal Balsam. Consists of 1 part camphor, 6 parts lead acetate, 16 parts beeswax, and 48 parts rape oil.†

Univest. A stereospecific, low-molecular weight polybutadiene. Applications: A universal binder for almost all types of sand materials - neither cement nor water needed. (Huls AG).†

Univol U304. A proprietary trade name for a mixture of distilled C20/22 acids. (UOP Chemicals).†

Univol U308. A proprietary trade name for 90 per cent. caprylic acid. (UOP Chemicals).†

Univol U310. A proprietary trade name for 90 per cent. capric acid. (UOP Chemicals).†

Univol U312. A proprietary trade name for a mixture of caprylic and capric acids. (UOP Chemicals).†

Univol U314/a/b. A proprietary trade name for 90/98 per cent. lauric acids. (UOP Chemicals).†

Univol U320. A proprietary trade name for 90 per cent. myristic acid. (UOP Chemicals).†

Univol U332. A proprietary trade name for 90 per cent. palmitic acid. (UOP Chemicals).†

Univol U334. A proprietary trade narne for 90 per cent. stearic acid. (UOP Chemicals).†

Univol U342. A proprietary trade name for 85/90 per cent. erucic acid. (UOP Chemicals).†

Univol U344. A proprietary trade name for 85/90 per cent. behenic acid. (UOP Chemicals).†

Unna's Stain. A microscopic stain. It contains 0.15 gram methyl green, 0.5 gram pyronin, 5 cc 95 per cent alcohol,

20 cc glycerin, and the whole made up to 100 cc with 2 per cent carbolic acid solution.†

Unna's Zinc Paste. A paste made from gelatin, zinc oxide, glycerin, and water.†

U/o. Chlophedianol hydrochloride. Applications: Antitussive. (Riker Laboratories Inc, Subsidiary of 3M Company).

UOP. Antioxidants. (Universal-Matthey Products Ltd).

UOP. Catalysts. (Universal-Matthey Products Ltd).

UOP 88. A proprietary anti-oxidant. N-N[1] dioctyl-*p*- phenylene diamine. (UOP Chemicals).†

UOP 288. A proprietary anti-oxidant. N,N[1]-bis-(1- methyl/peptyl)-*p*-phenylene diamine. (UOP Chemicals).†

Up-Start. Plant starter. (Chevron).*

Upas. An arrow poison obtained from the *Upas antjar* and *U. radja*. Used in the East Indies.†

Upixon. An anthelmintic also known as Uvilon. (Bayer & Co).*

Uplees Powder. An explosive containing 62-65 per cent ammonium nitrate, 12 1/2-14 1/2 per cent sodium nitrate, 4-6 per cent trinitro-toluene, 13 1/2 per cent ammonium chloride, and 2-4 per cent starch.†

Urac. A proprietary trade name for urea-formaldehyde adhesives.†

Uracil. 2, 6-Dioxypyrimidine. A hydrolytic product of nucleic acid.†

Uracil Mustard. A proprietary preparation ot URAMUSTINE used in the treatment of leukaemia. (Upjohn Ltd).†

Uracryl. A trade mark for a range of acrylic synthetic resins in emulsion form. (Unilever).†

Uradal. See ADALIN.

Uradex. Active ingredients: diurex plus uragan. Applications: Selective pre-emergence herbicide mixture. (Agan Chemical Manufacturers Ltd).†

Uradil. A proprietary range of resins dispersible in water. (963). *Uradil 580/585* are used for air-drying and storing. *Uradil 587/588* are acrylic resins cross-linked by water-thinnable amino resins. *Uradil 503* and *415* are non-oxidizing oil-free polyesters.†

Uragan. Active ingredient: bromacil; 5-bromo-3-sec-butyl-6-methyluracil. Applications: Versatile herbicide for control of established annual and perennial broadleaf weeds and grasses and brush. (Agan Chemical Manufacturers Ltd).†

Uraline. Chloral-urethane, $CCl_3.CH(OH).NH.CO_2.C_2H_5$. A hypnotic.†

Uralite (2). A proprietary trade name for urea-formaldehyde.†

Uramil. Amidomalonylurea, $C_4H_5N_3O_3$.†

Uramon. A proprietary trade name for a fertilizer containing 43 per cent nitrogen in the form of urea or similar compounds.†

Uranine. (Fluoresceine). The sodium or potassium salt of fluoresceine, $C_{20}H_{10}O_5Na_2$. Dyes silk and wool yellow.†

Uranium Red. Prepared by passing sulphuretted hydrogen into a solution of uranium nitrate, containing caustic potash. The orange precipitate formed is treated with potassium carbonate, when it is converted into uranium red.†

Uranium Yellow. Hydrated sodium uranate, $Na_2U_2O.6H_2O$. Used for painting and staining glass and porcelain, and for making fluorescent uranium glass. Hydrated ammonium uranate is also known by this name.†

Uranium, Yellow Sesquioxide. See ORANGE OXIDE OF URANIUM.

Uranox. A high nickel stainless steel.†

Urantoin. A proprietary preparation of NITROFURANTOIN. An antibiotic. (DDSA).†

Urea and Thiourea Resins. Resins obtained by the reaction between urea or thiourea and formaldehyde. Trade names for resins of this class are:- Bonnyware, Ciba Formica, Pertinit Pollopas, Prystaline, Rainbow Ware, Roanoid, Starpass.†

Urea Glue. A glue formed from the condensate of urea and formaldehyde. It is used in conjunction with a hardener.†

Urea-Bromine. A combination of urea and calcium bromide, $4CO(NH_2)_2.CaBr_2$, and containing 36 per cent bromine.†

Ureaphil. A proprietary preparation of urea used as an osmotic diuretic and as an abortifacient. (Abbott Laboratories).*

Ureaphos. A fertilizer containing phosphate of ammonia and urea.†

Urecholine. Bethanechol chloride. Applications: Cholinergic. (Merck Sharp & Dohme, Div of Merck & Co Inc).

Urecoll. Urea-formaldehyde condensation products. (BASF United Kingdom Ltd).*

Ureit. A German name for urea-formaldehyde resins.†

Ureka. A mixture of diphenylguanidine and mercaptobenzothiazole. A rubber vulcanization accelerator. Also a constant composition blend of diphenyl guanidine and 2, 4-dinitrophenylthio-benzothiazole.†

Ureka B. A blend similar to Ureka with a portion of D.P.G. replaced by Guantal. (General Electric).†

Ureka C. Benzothiazyl-thiobenzoate. Rubber vulcanization accelerator. (General Electric).†

Ureol. Polyurethane tooling resins. (Ciba-Geigy PLC).

Urepan. Polyurethane rubber to be crosslinked with Desmodur II or peroxides. Applications: It can be processed on machinery employed in the rubber industry. (Bayer & Co).*

Urethane. Ethyl carbamate, $NH_2.COOC_2H_5$.†

Urethon. Plastic film for sunblinds. (May & Baker Ltd).*

Uretix. A proprietary urea formaldehyde moulding material. (Nisshin Boseki).†

Uretrol. Sodium dimethylamino-azobenzene-*m*-sulphonate.†

Urex. Methenamine hippurate. Applications: Antibacterial. (Riker Laboratories Inc, Subsidiary of 3M Company).

Urexpan. Polyurethane sealant. Applications: Self-levelling sealant for caulking dead-level horizontal joints subject to heavy foot and vehicular traffic. (Pecora Corporation).*

Uricedin. Lithium succinate.†

Urifluine. Lithium succinate.†

Urisol. See HEXAMINE.

Urispas. Favoxate hydrochloride. Applications: Relaxant. (Smith Kline & French Laboratories).

Uristix. A proprietary test strip impregnated with a citrate buffer, tetra bromophenol blue, glucose oxidase, peroxidase and *o*-toluidine, used for the detection of protein and glucose in urine (Ames).†

Uritone. See HEXAMINE.

Uro-Hexoids. A proprietary preparation of hexamine and lithium benzoate tablets. Product now discontinued. (British Drug Houses).†

Urobac. A proprietary preparation of carindacillin sodium. An antibiotic used in the treatment of genito-urinary tract infections. (Pfizer International).*

Urobilistix. A proprietary test strip impregnated with *p*- dimethylamino - benzaldehyde in an acid buffer, used to detect urobilinogen in urine. (Ames).†

Urobiotic. A proprietary preparation of oxytetracycline, sulphamethiazole, phenazopyridine for therapy of genito-urinary infections. (Pfizer International).*

Uroformine. Hexamine (hexamethylene-tetramine), $C_6H_{12}N_4$.†

Urokinase. A plasminogen activator isolated from human urine.†

Urolucosil. A proprietary preparation containing sulphamethizole. A urinary antiseptic. (Warner).†

Uromat PE. A proprietary trade name for an aqueous dispersion based on titanium dioxide used for delustring synthetic fibre fabrics. (Ciba Geigy PLC).†

Urometin. See HEXAMINE

Uromide. A proprietary preparation of sulphacarbamide and PHENAZOPYRIDINE. A urinary anti-spasmodic. (Consolidated Chemicals).†

Uromiro, also Uromiron. Iodamide. Applications: Diagnostic aid. (Bracco Industria Chimica SPA).

Uromitexan. Mesna. Ampoule form. Applications: Uroprotector. (Degussa).*

Uropen. Hetacillin potassium. Applications: Antibacterial. (Bristol Laboratories, Div of Bristol-Myers Co).

Uroplas. A range of urea moulding compounds. Applications: Electrical field, sanitary field, electrotechnical industry and pottery. (SPREA Spa).*

Uropol. A proprietary preparation of TETRACYCLINE phosphate complex, SULPHAMETHIZOLE and PHENAZOPYRIDINE hydrochloride. A urinary antibiotic. (Bristol-Myers Co Ltd).†

Urotropine. (Urometin). See HEXAMINE.

Urovist. Diatrizoate meglumine. Applications: Diagnostic aid. (Berlex Laboratories Inc, Subsidiary of Schering AG).

Urovist Sodium 300. Diatrizoate sodium. Applications: Diagnostic aid. (Berlex Laboratories Inc, Subsidiary of Schering AG).

Ursol D. (Ursol P, Ursol DD). Dyestuffs. They consist of the hydrochlorides of *p*-phenylenediamine, *p*-aminophenol, and diaminodiphenylamine respectively. Used for dyeing fur, feathers, and hair, brown to black.†

Ursol DD. See URSOL D.

Ursol P. See URSOL D.

Urtal. An acrylonitrile-butadiene-styrene resin.†

Urtenol. A combination of oils for the textile trades. It possesses penetration and detergent properties.†

Usacert. High strength water soluble powder food colours certified by F.D.A., U.S.A. Applications: Colouring of foodstuffs and pharmaceuticals. (Williams Div of Morton Thiokol Ltd). *

Usalake. Water insoluble, aluminium lake food colours. Applications: Colouring of foodstuffs and pharmaceuticals. (Williams Div of Morton Thiokol Ltd). *

Usébe Green. See ALDEHYDE GREEN.

Usol Copper Green. Wood preservative formulated from copper naphthenate for use in solvent or water reducible bases. Applications: Brushing, dipping, soaking or mopping of liquid to various wood species. (Standard Tar Products Company, Inc).*

Usoline. See PARAFFIN, LIQUID.

Usol Organiclear. Proprietary wood preservatives to prevent wood rot, decay, mould and termite attack. Applications: Brushing, dipping, soaking or mopping of liquid to various wood species. (Standard Tar Products Company, Inc).*

Usol Zinclear. Zinc naphthenate wood preservative formulated in solvent or water reducible bases. Applications: Brushing, dipping, soaking or mopping of liquid to various wood species. (Standard Tar Products Company, Inc).*

Uspulun. A material containing sodium sulphate, sodium hydroxide, aniline, and mercury-chloro-phenol. A fungicide.†

U-T-C. Colourant dispersions. Applications: Colouring of non-aqueous coating compositions. (Nueodex Inc).*

Uteplex. A proprietary preparation of uridine-5-triphosphoric acid used to relieve muscle spasm. (Rona Laboratories).†

Utica Steel. A proprietary trade name for a die steel. It contains 1.4 per cent tungsten, 1.25 per cent carbon, 0.4 per cent chromium and 0.2 per cent vanadium.†

Uticillin. A proprietary preparation of carfecillin. Applications: An antibiotic. (Beecham Research Laboratories).*

Uticillin VK. Penicillin V Potassium. Applications: Antibacterial. (The Upjohn Co).

Uticort. Betamethasone benzoate. Applications: Glucocorticoid. (Parke-Davis, Div of Warner-Lambert Co).

Utimox. Amoxicillin. Applications: Antibacterial. (Parke-Davis, Div of Warner-Lambert Co).

Utinor. Norfloxacin. Applications: Urinary antiseptic. (Merck Sharp & Dohme).*

Utocyl. Antibiotic sulphonamide veterinary ethical. (Ciba-Geigy PLC).

Utopar. Ritodrine. Applications: Uterospasmolytic, prevents premature birth. (Duphar BV).*

Uval. Sulisobenzone. Applications: Ultraviolet screen. (Dorsey Laboratories, Div of Sandoz Inc).

UV-Chek. Applications: For ultraviolet stabilizers and absorbers for polyolefins and related polymers, in Int Class 1. (Ferro Corporation).*

Uvilon. An anthelmintic also known as Upixon. (Bayer & Co).*

Uvi-Nox 1494. Hindered phenolic compound. Applications: Primary antioxidant particularly effective in polyolefins as a thermal stabilizer and as an inhibitor of monomers to prevent prepolymerization. (GAF Corporation).*

Uvistat. (1). A proprietary preparation of mexenone, used in dermatology. (Ward, Blenkinsop & Co Ltd).†

Uvistat 12, 24, 247, 2211. A proprietary trade name for a series of additives for protecting plastics against ultra violet light. They have the general formula R_1,C_6H_5-CO-C_6H_4OH-OR. Uvistat 247 is particularly effective for the stabilization of polyolefines. (Ward, Blenkinsop & Co Ltd).†

Uvistat 247. A proprietary trade name for 2 - hydroxy - 4 - *n* -heptoxybenzophenone. A virtually non-toxic ultra violet light absorber. (Ward, Blenkinsop & Co Ltd).†

Uvitex. A proprietary trade name for a series of fluorescent brighteners for incorporation in soap-based and synthetic detergents as follows: Uvitex SFC is a stilbenic derivative giving high intensity whites on cellulosic fibres. Uvitex SK is a benzoazole derivative effective on a wide variety of fibres, stable to hypochlorite and chlorisocyanate. Uvitex SOF is also a benzoxazole derivative. Uvitex ERN CONC P is a benzoxazole derivative-a fluorescent brightener for polyester fibres. (Ciba Geigy PLC).†

Uvitex MA. A proprietary trade name for an imidazole derivative fluorescent brightener for acrylic fibres. It is applied in the dope before spinning. (Ciba Geigy PLC).†

Uvitex MP. A proprietary trade name for a heterocyclic-stilbene type fluorescent brightener for polyamid fibres. (Ciba Geigy PLC).†

V

V2A Steel. See ANKA STEEL.

Vacancin Blue. A basic dyestuff giving a dark blue shade on tannined cotton.†

Vacancin Scarlet. An azo red produced on the fibre.†

Vac Blue. See DOINE BLUE R.

Vacsol. Organic solvent wood preservative. Applications: Pressure treatment for joinery and carcassing timber used above ground contact. (Hickson & Welch Ltd).*

Vactran. High grade chemicals for vacuum evaporation. (British Drug Houses).†

Vacuum Salt. A pure salt, NaCl, obtained by boiling brine under a vacuum.†

Vacuum Silicon Iron. Alloys containing about 0.15 or 3.4 per cent silicon, made by melting *in vacuo*. They are annealed at 1100° C., and contain about 0.01 per cent carbon. They have remarkable magnetic properties.†

Vaderm. Alclometasone dipropionate. Applications: Anti-inflammatory. (Schering-Plough Corp).

Valadol. Acetaminophen. Applications: Analgesic; antipyretic. (E R Squibb & Sons Inc).

Valclene. Drycleaning fluid. (Du Pont (UK) Ltd).

Valearin. Valeryltrimenthylammonium chloride.†

Valerian. The dried rhizome of *Valeriana officinalis*.†

Valerone. Diisobutyl ketone, $C_4H_9.CO.C_4H_9$.†

Valfor. Sodium aluminosilicates. Synthetic zeolites (molecular sieves) produced in varying SiO_2:AlO_2 ratios, pore sizes, and crystal structure. Applications: Catalysts for petroleum refining, builder for household detergents (as phosphate replacement), desiccant for drying natural and manufactured gases, for insulating windows etc. (The PQ Corporation).*

Valide. See VALYL.

Valine. α-Aminoisovaleric acid, $(CH_3)_2CH.-CH(NH_2)COOH$. A protein aminoacid.†

Valium. A proprietary preparation of diazepam. A sedative. (Roche Products Ltd).*

Valium Roche. Anti-anxiety agent. (Roche Products Ltd).

Valledrine. Trimeprazine tartrate cough linctus. (May & Baker Ltd).

Vallergan. A proprietary preparation of trimeprazine tartrate. (May & Baker Ltd).*

Vallex. Cough linctus. (May & Baker Ltd).

Valmid. Ethinamate. Applications: Sedative. (Eli Lilly & Co).

Valmidate. A proprietary preparation of ethinamate. A sedative. (Eli Lilly & Co).†

Valoid. Cyclizine hydrochloride. (The Wellcome Foundation Ltd).

Valonia. The acorn cups of *Quercus oegilops*. They contain about 35 per cent of tannin, and are used in the leather industry.†

Valox. Polybutylene terephthalate. Applications: Plastic components for automotive, electrical, electronics, lighting, medical, packaging, audio etc. (GE Plastics Ltd).*

Valpin 50. Anisotropine methylbromide. Applications: Anticholinergic. (Endo Laboratories Inc, Subsidiary of E I du Pont de Nemours & Co).

Valray Alloy 1. A trade mark for an alloy of 20 per cent. chromium and the balance nickel with controlled manganese, carbon and silicon. (Wiggin Alloys Ltd).†

Valve Brass. See Brass.

Valve Bronze. An alloy of from 83-89 per cent copper, 4-5 per cent tin, 3-7 per cent zinc, and 3-6 per cent lead.†

Valvoline. A registered trade name for lubricating oils. They are American petroleum products.†

Valyl. A registered trade mark currently awaiting re-allocation by its proprietors. (Casella AG).†

Vam. A proprietary preparation of vinyl and ethyl ethers. General anaesthetic. (May & Baker Ltd).*

Vamac. Ethylene acrylic elastomer. (Du Pont (UK) Ltd).

Vamin Series. Parenteral solutions. Applications: Crystalline amino acids for intravenous nutrition. (KabiVitrum AB).*

Vamitox. Herbicide. (May & Baker Ltd).

Vanadium Alum. An ammonium-vanadium sulphate, $(NH_4)_2SO_4 \cdot V_2(SO_4)_3 \cdot 24H_2O$.†

Vanadium Brass. An alloy of 70 per cent copper, 29.5 per cent zinc, and 0.5 per cent vanadium.†

Vanadium Bronze. Metavanadic acid, HVO_3. Used as a pigment in the place of gold bronze. It is also the name for an alloy of 61 per cent copper, 38.5 per cent zinc, and 0.5 per cent vanadium.†

Vanadium Steel. An alloy of steel with vanadium.†

Vanadium-Manganese Brass. An alloy containing 58.56 per cent copper, 38.54 per cent zinc, 1.48 per cent aluminium, 1 per cent iron, 0.48 per cent manganese, and 0.03 per cent vanadium.†

Vanadium-Molybdenum Steels. Alloys containing from 0.1-1.0 per cent carbon, 0.52-6.0 per cent molybdenum and 0.1-1.0 per cent vanadium.†

Vanadium-tin Yellow. A pigment. It is a mixture of vanadium pentoxide and tin oxide, and is used in the manufacture of yellow glass.†

Vanair. A proprietary preparation of benzoyl peroxide and sulphur used in the treatment of acne. (Carter-Wallace).†

Van-Amid. Epoxy curing agents. Applications: Used as curing agents for vanoxy resins. (Vanderbilt Chemical Corporation).*

Vanax. Broad range of sulphenamide, dithiocarbamate, thiourea, thiadiazine, isophthalate, guanidine, aldehyde-amine accelerators. Applications: Used in natural, synthetic and latex rubbers as both primary and secondary accelerators. (Vanderbilt Chemical Corporation).*

Van Bac. Paper mill effluents. Applications: Paper processing. (Vanderbilt Chemical Corporation).*

Vanbeenol. Ethyl vanillin. (Bush Boake Allen).*

Vancenase. Beclomethasone dipropionate. Applications: Glucocorticoid. (Schering-Plough Corp).

Vanceril. Beclomethasone dipropionate. Applications: Glucocorticoid. (Schering-Plough Corp).

Vanchem. Corrosion inhibitor and chemical intermediates. Applications: Chemical intermediates - petroleum and other industries. (Vanderbilt Chemical Corporation).*

Vanchem HM-4346. Aromatic polyisocyanate in toluene. Applications: Adhesion promotion or primer or crosslinking agent. (Vanderbilt Chemical Corporation).*

Vanchem HM50. Aromatic polyisocyanate in monochlorobenzene (50). Applications: Adhesion promotion or primer or crosslinking agent. (Vanderbilt Chemical Corporation).*

Vancide. Fungicides. Applications: Used as fungicides for paint, paper, ceramics, plastics, household products, agriculture etc. (Vanderbilt Chemical Corporation).*

Vanclay. Kaolin. Applications: Adsorbant. (R T Vanderbilt Co Inc).

Vancocin Hydrochloride. Vancomycin Hydrochloride. Applications: Antibacterial. (Eli Lilly & Co).

Vancomycin. An antibiotic produced by *Streptomyces orientalis.* Vancocin is the hydrochloride.†

Vancure D.A.A. A proprietary accelerator. N-cyclohexyl-2-benzthiazyl sulphonamide. (K & K Greef Chemicals Ltd).†

Vandex. A proprietary selenium compound used in the rubber industry for imparting ageing properties and high abrasion.†

Vandike P.360. A proprietary trade name for a vinyl acetate-dioctyl maleate copolymer emulsion used in water-based adhesives. (BOC International Ltd).†

Vandike 7085. A proprietary trade name for a vinyl acetate-butyl acrylate copolymer emulsion used in emulsion paints. (BOC International Ltd).†

Vandike 7086. A proprietary trade name for a vinyl acetate-butyl acrylate copolymer emulsion. Used in the manufacture of "non-drip" emulsion paints. (BOC International Ltd).†

Vandura Silk. (Gelatin Silk). An artificial silk prepared from gelatin and formaldehyde. It is also the name for a silk made from casein and formaldehyde.†

Vandyck Red. See FLORENTINE BROWN and INDIAN RED.

Vandyke Brown. (Rubens Brown, Ulmin Brown, Cologne Brown). Brown pigments of vegetable origin prepared from peat, cotton, or soot. The modern vandyke brown is a purified bituminous ochre. Genuine vandyke brown is a natural earth containing a large proportion of organic matter and with an ash usually amounting to 10 per cent. The artificial product generally consists of a carbon black with red oxide and possibly yellow ochre.†

Van Ermengem's Stain. A microscopic stain. Solution (A) contains 1 gram of osmic acid, 20 grams tannin, 150 cc distilled water, and 8 drops glacial acetic acid. Solution (B) is a 0.25-0.5 per cent silver nitrate solution. Solution C contains 6 grams tannin, 1 gram gallic acid, 20 grams sodium acetate, and 700 cc water.†

Vanex. Blended organic and inorganic acids in combination with surfactants, wetting agents and inhibitors. Applications: Cleaner for brick in new construction that may contain vanadium or metals. (Nova Chemical Inc).*

Van Gel. Processed magnesium aluminium silicate - smectite. Applications: Used as viscosity stabilizer, dispersion adjuster and mineral filler for rubber, paint, household products, ceramics and plastics. (Vanderbilt Chemical Corporation).*

Vanilla. The cured unripe fruit of *Vanilla planifolia.* Used as an aromatic and flavouring material.†

Vanillin. 3-Hydroxy-4-methoxybenzaldehyde. A flavourant.†

Vanitox. Selective weed killer. (May & Baker Ltd).*

Vankalite. A proprietary trade name for a beryllium-copper alloy used for setting diamonds in drills.†

Vanlube. A full line of antioxidants, antiwear, extra fine pressure additives, metal deactivators and friction reducers for petroleum lubricants. Applications: Antioxidants, antiwear and extreme pressure additives, metal deactivators and friction reducers for petroleum lubricants. (Vanderbilt Chemical Corporation).*

Vanobid. Candicidin. Applications: Antifungal. (Merrell Dow Pharmaceuticals Inc, Subsidiary of Dow Chemical Co).

Vanox. A full line of amine and phenol rubber antioxidants both primary and secondary. Applications: Used in a variety of rubers, also with other antioxidants. Also used in plastics as an antioxidant and mineral deactivator. (Vanderbilt Chemical Corporation).*

Vanoxy. Epoxy resins. Applications: Uses include protective coatings, laminates, adhesives, castings, tooling, flooring, surfacing, potting and encapsulating. (Vanderbilt Chemical Corporation).*

Vanplast. Epoxy plasticizers. Applications: Premium quality epoxy plasticizers.

Some have FDA approval. (Vanderbilt Chemical Corporation).*

Vanplast 201. Barium dinonylnaphthalene sulphonate. Applications: Corrosion inhibitor used in automotive rubber protective coatings. (Vanderbilt Chemical Corporation).*

Vanquin. A proprietary preparation containing viprynium carbonate. An anthelmintic. (Parke-Davis).†

Vanquish. Bactericidal detergent. (ICI PLC).*

Vansil. A proprietary preparation of oxamniquine. An antihelmintic. (Pfizer International).*

Vansil. Calcium meta silicate - wollastonite. Applications: Filler, reinforcing agent, bright colour used in ceramics, rubber, cosmetics, plastics and paint etc. (Vanderbilt Chemical Corporation).*

Vanstay. A full line of vinyl stabilizers. Applications: Premium quality epoxy plasticizers. Some have FDA approval. (Vanderbilt Chemical Corporation).*

Van Swieten's Solution. See LIQUEUR DE VAN SWIETEN.

Vantac. Acrylic emulsion pressure sensitive adhesives. (Bevaloid Ltd).

Vantalc. Hydrous magnesium silicate - industrial talc. Applications: Used primarily in paint - also used in rubber and plastics. (Vanderbilt Chemical Corporation).*

Vantoc. Industrial disinfectants and bactericides. (ICI PLC).*

Vantoc AL. Aqueous blend of higher alkyl tri-methyl ammonium bromide. Pale, straw coloured liquid. Applications: Bactericide in the brewing and food processing industries. (ICI PLC, Organics Div).

Vantoc CL. Lauryl dimethyl benzyl ammonium chloride in aqueous solution. Applications: General disinfection of plants and equipment in brewing, soft drinks and foodstuffs industries. (ICI PLC, Organics Div).

Vantocil. Biocides and bactericides. (ICI PLC).*

Vantropol. Detergent/sanitizer. (ICI PLC).*

Van Wax. Protective, petroleum and wax blends. Applications: Used as sunlight and ozone checkers for rubber. (Vanderbilt Chemical Corporation).*

Vanzak. Wire life extender. Applications: Used in paper manufacturing. (Vanderbilt Chemical Corporation).*

Vanzyme. Starch converting enzymes. Applications: Used in paper manufacturing. (Vanderbilt Chemical Corporation).*

Vapo-Iso. Isoproterenol hydrochloride. Applications: Adrenergic. (Fisons Corp).

Vapona. A proprietary preparation of dichlorvos. An insecticide.†

Vaporole. A proprietary formulation of amyl nitrite. Applications: Indicated for rapid relief of angina pectoris due to coronary artery disease. (The Wellcome Foundation Ltd).*

Vaporole. A proprietary formulation of strong ammonia solution. Applications: A respiratory stimulant for inhalation to prevent or treat fainting. (The Wellcome Foundation Ltd).*

Varac. See KELP.

Varbian. Cardiac stimulant. (Ciba-Geigy PLC).

Varech. See VRAIC.

Variclene. Green aqueous gel containing 0.5% w/w brilliant green BP, 0.5% w/w lactic acid BP. Applications: An aid in the topical treatment of venous and other types of skin ulcers. (Dermal Laboratories Ltd).*

Vari-Cut. Fiberod product approximately 1/8" diameter cut to lengths between 1/4" and 4 inches. Applications: Thermoplastic moulding compounds for compression, transfer and injection moulding. Application areas include automotive, appliances, equipment, sporting and others. (Polymer Composites Inc).*

Varidase. A proprietary preparation streptokinase and streptodornase. A fibrinolytic drug. (Lederle Laboratories).*

Variotin. A proprietary trade name for PECILOCIN, used in the treatment of fungal skin infections. (Leo Laboratories).†

Variton. Diphemanil methylsulphate. Applications: Anticholinergic. (Schering-Plough Corp).

Varitox. Sodium trichloroacetate. (May & Baker Ltd).*

Varnish. A mixture of boiled oil with various gum resins, and oil of turpentine.†

Varnish, Burma Black. See THITSI.

Varnish, Lithographer's. See STAND OIL.

Varnish, Short Oil. See LONG OIL VARNISHES.

Varnish, Standöl. See STAND OIL.

Varnish, Victoria. See ZAPON VARNISH.

Varnodag. A trade mark for a varnish made from phenol-formaldehyde synthetic resin with colloidal graphite.†

Varnoline. A petroleum distillate used as a lubricant.†

Var Oil. See AIX OIL.

Varox. Peroxides. Applications: vulcanizing chemicals for a wide range of crosslinking agents. (Vanderbilt Chemical Corporation).*

Varsol. General purpose hydrocarbon solvents. (Exxon Chemical International Inc).*

Varsol. White spirit and aliphatic solvents. (Exxon Chemical Ltd).*

Vascardin. A proprietary preparation of sorbide nitrate. A vasodilator used for angina pectoris. (Nicholas).†

Vascoloy-Ramet D. A proprietary trade name for a corrosion-resisting alloy of 80 per cent tantalum carbide, 20 per cent tungsten and nickel.†

Vasconite BT. A proprietary anti-corrosion agent added to poor-quality boiler fuels. It is a suspension of magnesium compounds and combustion catalysts. (Gamlen Chemical Co (UK) Ltd).†

Vascuals. Vitamin E. Applications: Vitamin E supplement. (USV Pharmaceutical Corp).

Vasculit. The sulphate of 2-*n*-Butyl-amino - 1 - (4 - hydroxyphenyl)ethanol (bamethan sulphate). (Boehringer Ingelheim Ltd).†

Vasoclear. Naphazoline hydrochloride. Applications: Adrenergic. (CooperVision Inc).

Vasocon. Naphazoline hydrochloride. Applications: Adrenergic. (CooperVision Inc).

Vasocort. A proprietary preparation of hydrocortisone, hydroxyamphetamine hydrobromide and phenylephrine hydrochloride, used as a nasal spray in cases of allergic rhinitis. (Smith Kline and French Laboratories Ltd).*

Vasodilan. Isoxsuprine. Applications: Promotes blood flow rate. (Duphar BV).*

Vasogen. (1) A proprietary preparation of dimethicone, zinc oxide and calamine used as a protective cream for the skin. (330).

(2) A trade mark for an ointment vehicle - an oxygenated petroleum. PAROGEN.†

Vasolastine. A proprietary preparation of lipoxydase citrogenase, amino acid oxydase and tyrosinase complex. (FAIR Laboratories).†

Vasoliment. See PAROGEN.

Vasomotal. Betahistine. Applications: Counteracts disorders of the inner ear, Menière's disease. (Duphar BV).*

Vasotran. A proprietary preparation of ISOXSUPRINE hydrochloride used in the treatment of peripheral vascular disease. (Bristol-Myers Co Ltd).†

Vasoxine. Injection of methoxamine hydrochloride. (The Wellcome Foundation Ltd).

Vasoxyl. Methoxamine hydrochloride. Applications: Adrenergic. (Burroughs Wellcome Co).

Vasoxyl Injection. A proprietary formulation of methoxamine hydrochloride. Applications: For supporting, restoring or maintaining blood pressure during anaesthesia. (The Wellcome Foundation Ltd).*

Vassy Cement. A cement similar to Roman Cement.†

Vasylox. Methoxamine hydrochloride. (The Wellcome Foundation Ltd).

Vat Dye B. See THIOINDIGO RED B.

Vat Dyes. Dyestuffs insoluble in water which must be treated with a reducing agent, and dissolved in an alkaline solution for dyeing. The colour is produced by oxidation. Indigo is a type of these colours.†

Vat Indigo. See INDIGO WHITE.

Vat Red Paste. An acid dyestuff used to replace barwood as the ground colour for indigo vat blue.†

Vatensol. A proprietary preparation of guanoclor sulphate. An antihypertensive drug. (Pfizer International).*

Vaucher's Alloy. An alloy of 75 per cent zinc, 18 per cent tin, 4.5 per cent lead, and 2.5 per cent antimony.†

Vauquelin's Salt. A compound obtained by treating palladium chloride with ammonia, $[Pd(NH_3)_4]Cl_2.PdCl_2$.†

Vazo. Vinyl polymerization catalyst. (Du Pont (UK) Ltd).

V-Cil-K. A proprietary preparation containing Phenoxymethylpenicillin Potassium. An antibiotic. (Eli Lilly & Co).†

V-Cil-K Sulpha. A proprietary preparation of phenoxymethylpenicillin potassium, and sulphadimidine. An antibiotic. (Eli Lilly & Co).†

V-Cillin. Penicillin V. Applications: Antibacterial. (Eli Lilly & Co).

V-Cillin K. Penicillin V Potassium. Applications: Antibacterial. (Eli Lilly & Co).

Vebonol. Anabolic steroid veterinary ethical. (Ciba-Geigy PLC).

Vec. A proprietary trade name for a vinylidene chloride synthetic resin.†

Vecopyrin. Phenazone-iron chloride.†

Vecortenol. Steroid veterinary ethical. (Ciba-Geigy PLC).

Vecortenol-Vioform. Steroid quinoline topical veterinary ethicals. (Ciba-Geigy PLC).

Vectal. Selective and total herbicide. (FBC Ltd).

Vectra. Liquid crystal polymer - wholly aromatic polyester. Applications: Fibreoptics, electrical/electronics, chemical and hostile environments, mechanical bearing and wear. (Celanese Limited).†

Vedacoll. Bitumen mastix. (Vedag GmbH).*

Vedacolor. Lacquer with an aqueous solvent and artificial resin base. Applications: Roof surface. (Vedag GmbH).*

Vedafix. Skirting rails. (Vedag GmbH).*

Vedaflex. Modified bitumen membrane. Applications: For sealing flat roofs and sloped roofs without additional surface protection. (Vedag GmbH).*

Vedaform. Bitumen shingles in different shapes and colours. Applications: Steep roofing. (Vedag GmbH).*

Vedag BM. Emulsion for the improvement of cement mortar. (Vedag GmbH).*

Vedagit. Bitumen plus filler. Applications: Pebble-bedding compound for cold application. (Vedag GmbH).*

Vedagolan. Bitumen emulsions. (Vedag GmbH).*

Vedagully System. PU-integrated hard foam. Applications: Roof inlets. (Vedag GmbH).*

Vedagum. Bituminous crevice filler, approved for underlays. Applications: Hot application. (Vedag GmbH).*

Vedalith Facade System. Glass fibre reinforced cement. Applications: Bracket-mounted, respiratory, heat insulating system facade, shock-resistant, non-combustible. (Vedag GmbH).*

Vedaphalt. Bitumen emulsions. Applications: Road surfaces treatment. (Vedag GmbH).*

Vedaphon. Anti-drone materials. Applications: Motor cars. (Vedag GmbH).*

Vedapor. Hard foam together with insulation strips. Applications: Roll form insulation. (Vedag GmbH).*

Vedapurit. PS or PU hard foam, sandwiched or non-sandwiched. Applications: Hard foam insulating boards. (Vedag GmbH).*

Vedasin. Universal sealing compound. Applications: Cellars, subgrade purposes. (Vedag GmbH).*

edastar. Domelight. (Vedag GmbH).*

edatect. Bitumen roof sheets with various inserts. (Vedag GmbH).*

edatex. Bitumen solvent glue. Applications: Steam barriers and heat insulating material. (Vedag GmbH).*

edathene. Self-adhesive sheet. Applications: For waterproofing systems. (Vedag GmbH).*

edatherm. Shingle thermal insulating board. Applications: For insulating sloped roofs. EPS with chipboard (V 100 G). (Vedag GmbH).*

'edril. Polymethacrylates. (Montedison UK Ltd).*

'eegum. Magnesium aluminium silicate. Applications: Pharmaceutic aid. (R T Vanderbilt Co Inc).

'eepa Oil. (Veppam Oil, Neem Oil). Margosa oil, obtained from the seeds of *Melia azadirachta*.†

'eetids. Penicillin V Potassium. Applications: Antibacterial. (E R Squibb & Sons Inc).

/ega Brown R. See ACID BROWN R.

/ega Red S. A dyestuff. It is a British brand of Brilliant sulphone red B.†

/egetable Alkali. Potassium hydroxide, KOH.†

/egetable Black. A very light lamp-black containing 99 per cent carbon.†

Vegetable Butter. (Lactine, Vegetaline, Cocoaline, Laureol, Nucoline, Albene, Palmine, Cocose). Names for an edible fat prepared from coco-nut oil and palm-nut oil. Used in chocolate manufacture as a substitute for cocoa-butter.†

Vegetable Calomel. The resin of *Podophyllum*.†

Vegetable Casein. Legumin, found in leguminous seeds.†

Vegetable Ethiops. A form of charcoal obtained by the incineration of *Fuci*.†

Vegetable Fibre. See VULCANIZED FIBRE.

Vegetable Gelatin. Agar-agar.†

Vegetable Glue. (Aparatine). A glue obtained by treating starch with alkali. An adhesive. Also see AGAR-AGAR.†

Vegetable Green. See SAP GREEN.

Vegetable Gum. See BRITISH GUM.

Vegetable Ivory. (Corajo). Tagua nut, the fruit of *Phytelephas macrocarpa*, of South America.†

Vegetable Jelly. Pectin (*qv*), found in vegetable juices.†

Vegetable Rouge. Carthamin, the colouring matter of *Carthamus tinctorius* mixed with French chalk. Used as a cosmetic. See SAFFLOWER.†

Vegetable Salt. Potassium tartrate.†

Vegetable Soda. The general name for the ash of soda plants (land plants).†

Vegetable Spermaceti. See CHINESE WAX.

Vegetable Sulphur. Lycopodium.†

Vegetable Tallow. The name applied to vegetable fats similar to tallow, such as Chinese tallow and Malabar tallow. See CHINESE TALLOW.†

Vegetable Wax. See JAPAN TALLOW.

Vegetable Wool. A product obtained from green pine and fir cones by processes of fermentation, washing, and disintegration. It is mixed with cotton for the production of yarns.†

Vegetaline. The name given to a preparation of lactic acid, used in tanning processes for the removal of lime. It is obtained from the drainage water of preserve manufacture by evaporation, and contains from 8.6-9.6 per cent lactic acid.†

Vegetoil. Vegetable oil plus emulsifiers. Applications: Maximize performance of pesticides. (Drexel Chemical Company).*

Vegolysen. Hexamethonium bromide. (May & Baker Ltd).

Vegolysin T. A proprietary preparation of hexamethoniura tartrate. (May & Baker Ltd).*

Velampishin. See WOOD-APPLE GUM.

Velan. A proprietary product. It is a complex organic compound soluble in water which renders fabric fibres water repellant.†

Velbacil. A proprietary preparation of bacampicillin hydrochloride. An antibiotic. (Pfizer International).*

Velban. Vinblastine Sulphate. Applications: Antineoplastic. (Eli Lilly & Co).

Velbe. A proprietary trade name for the sulphate of VINBLASTINE, used in the treatment of malignant diseases. (Eli Lilly & Co).†

Velcorin. An organic preservative. Applications: For use in the cold sterilization of acid beverages. (Bayer & Co).*

Velicren. Acrylic fibre. Applications: Sewing threads, carpets, fur fabrics. (SNIA (UK) Ltd).†

Velocite. A rubber vulcanization accelerator. It is thiocarbanilide.†

Velon. A proprietary trade name for a vinylidene chloride synthetic resin.†

Velosan. A proprietary rubber vulcanization accelerator. It is aldehyde ammonia.†

Velosef. A proprietary preparation of cephraorine. An antibiotic. (Squibb, E R & Sons Ltd).*

Veloset. Catalysts for sodium silicate bonded sands. (Foseco (FS) Ltd).*

Velpar. Weedkiller. (Du Pont (UK) Ltd).

Velpeau's Caustic Powder. A caustic consisting of burnt alum and powdered savin tops.†

Veltol. Maltol. (Pfizer Ltd).

Veltol-Plus. Ethyl maltol. (Pfizer Ltd).

Velva Coat. Foundry core and mould coatings implying refractories and alcohol. Applications: Foundry core or mould coating utilized to improve the surface finish of cast metals. (Ashland Chemical Company).*

Velva Dri. Foundry core and mould coatings implying refractories and chlorinated solvents. Applications: Foundry core or mould coating utilized to improve the surface finish of cast metals. (Ashland Chemical Company).*

Velvalite. Foundry core and mould coatings implying refractories and alcohol. Applications: Foundry core or mould coating utilized to improve the surface finish of cast metals. (Ashland Chemical Company).*

Velvaplast. Foundry core and mould coatings implying refractories and water solvent. Applications: Foundry core or mould coating utilized to improve the surface finish of cast metals. (Ashland Chemical Company).*

Velva Wash. Foundry core and mould coatings implying refractories and water. Applications: Foundry core and mould coatings utilized to improve surface finish of cast metals. (Ashland Chemical Company).*

Velvet Black. A variety of gas carbon black.†

Velvet Brown. See UMBER.

Velveteen. An imitation velvet made from cotton.†

Velvetex. A proprietary carbon black (thermatomic carbon) in a soft form used in rubber mixings.†

Velvet Red. A pigment. It is a reddish brown powder consisting of ferric oxide coloured by a mixture of spirit soluble rosaniline blue and fuchsine.†

Vendril. Polymethyl methacrylate.†

Venetian Bole. A pigment. It is Indian red (*qv*).†

Venetian Lake. See CRIMSON LAKE.

Venetian Red. See INDIAN RED.

Venetian White. A white lead pigment. It is white lead with barium sulphate, and is sold as No. I, II, or III, containing respectively 20, 40, and 60 per cent barium sulphate.†

Venice Soap. An olive oil soap.†

Venice Turpentine. The oleo-resin of the larch, *Pinus larix*. It contains 20-22 per cent essential oil and 74-80 per cent rosin.A substance is often sold under this name consisting of a mixture of rosin oil, rosin, and turpentine. Used in the varnish industry.†

Venice White. See VENETIAN WHITE.

Venite. A proprietary pyroxylin product.†

Ventilago Madraspanta. (Ouralpatti, Pitti, Lokandi). An Indian dyestuff obtained from a climbing shrub.†

Ventolin. Albuterol sulphate. Applications: Bronchodilator. (Glaxo Inc).

Venzar. Lenacil weedkiller. (Du Pont (UK) Ltd).

VeoVa 10 Monomer. A vinyl monomer containing the tertiary versatic structure can be copolymerized with vinyl acetate and used in all types of emulsion paints. (Shell Chemicals).†

Vepesid. Etoposide. Applications: Antineoplastic. (Bristol Laboratories, Div of Bristol-Myers Co).

Veppam Oil. See VEEPA OIL.

Veracolate. A proprietary preparation of cascara, bile salts and phenol-phthalein. A laxative. (Warner).†

Veractil. A proprietary preparation of methotriroeprazine maleate. A tranquillizer. (May & Baker Ltd).*

Veracur. A proprietary preparation of formaldehyde used to treat warts. (Typharm).†

Verafil. Thermoplastic moulding compounds. (Ciba-Geigy PLC).

Veratrine. A mixture of various alkaloids obtained from the seeds of *Veratrum sabadilla*. It causes sneezing and irritation when inhaled.†

Veratrole. 1,3-Dimethoxybenzene.†

Verazinc. Zinc Sulphate. Applications: Astringent (ophthalmic). (O'Neal, Jones & Feldman Pharmaceuticals).

Verbena Oil. See OIL OF VERBENA.

Vercazol. Furfuramide, an accelerator used in rubber vulcanization.†

Vercyte. Pipobroman. Applications: Antineoplastic. (Abbott Laboratories).

Verdict. A line of herbicides based primarily on haloxyfop. (Dow Chemical Co Ltd).*

Verdigris. Basic copper acetate. It is usually a mixture of mono-, di- , and tri-acetates of copper. Green verdigris consists chiefly of the basic acetate, $2(C_2H_3O_2)_2Cu_2O$. Blue verdigris consists mainly of the basic acetate, $(C_2H_3O_2)_2Cu_2O$. The various forms of verdigris are used in dyeing and calico printing, and for the preparation of oil and water colours.†

Verdigris Green. See EMERALD GREEN.

Verdigris Green. See VERDIGRIS.

Verditer Blue. (Verditer Green, Bremen Green, Mineral Blue, Bremen Blue). An anhydrous basic copper carbonate, produced by the addition of sodium carbonate to a hot solution of copper sulphate or nitrate. Verditer green is an intermediate product. Used for paper-staining. Copper hydrate and copper carbonate are both sold under these names.†

Verditer Green. See VERDITER BLUE.

Verditers. Highly basic copper carbonates.†

Verdiviton. A proprietary preparation of sodium, calcium, potassium and manganese glycerophosphates, and vitamin B complex. (Squibb, E R & Sons Ltd).*

Verdiviton Elixir. Multi vitamin and mineral mix in aqueous solution. (E R Squibb & Sons Ltd).

Verdley. Range of composts and soil conditioners based on peat or bark. (ICI PLC).*

Verdone. Selective weedkiller. (ICI PLC).*

Vergum. Magnesium aluminium silicate - semotite. Applications: Thixotropic agents, filler, extenders for paint, ceramics, cosmetics, pharmaceuticals, agriculture, paper etc. (Vanderbilt Chemical Corporation).*

Veridian. See CHROMIUM GREEN.

Verilite. An alloy of 96 per cent aluminium, 2.5 per cent copper, 0.7 per cent nickel, 0.4 per cent silicon, and 0.3 per cent manganese.†

Veritas. Lube oil. (Chevron).*

Verjuice. The old name for the very sour juice of unripe green grapes, and of crabapples. It contains tartaric, racemic, and malic acids.†

Vermiculite. A mineral, $3MgO.(Fe.Al)_2O_3 3SiO_2$. Used for thermal insulation.†

Vermilion. See CINNABAR.

Vermilion, American. See CHROME RED.

Vermilion, Chinese. See SCARLET VERMILION.

Vermilionettes. Red pigments. They are combinations of white lead or zinc white, and eosin.†

Vermilion, Field's Orange. See SCARLET VERMILION.

Vermilion, Orange. See SCARLET VERMILION.

Vermox. A proprietary preparation containing mebendazole. Applications: Antihelmintic. (Janssen Pharmaceutical Ltd).*

Vernafine. Organic pigment pastes. Applications: Paints. (Colour-Chem Limited).*

Vernalin. Speciality dyestuffs for dyeing all types of leathers. Applications: Leather. (Colour-Chem Limited).*

Vernamine Binders. Binders based on synthetic resin dispersions based on acrylic and other monomers. Applications: Leather. (Colour-Chem Limited).*

Vernaminol Liquors. Synthetic oil derived from wax and hydrocarbons and their derivatives, free from fatty acids. Applications: Leather. (Colour-Chem Limited).*

Vernasein. Extremely fine dispersions of organic pigments in suitable aqueous medium. Applications: Leather. (Colour-Chem Limited).*

Vernasol. Disperse dyes. Applications: Polyester and polyester component in blended fibres and fabrics. (Colour-Chem Limited).*

Vernatan. Synthetic tanning agents, comprising replacement syntans, resin tanning agents and acrylic syntans. Applications: Leather. (Colour-Chem Limited).*

Vernisol Z60. A filling compound with a specific gravity of 1.028. Brand Z61 has a gravity of 1.04.†

Vernol Liquors. Fat liquors derived from natural vegetable oils, animal fats and synthetic esters. Applications: Leather. (Colour-Chem Limited).*

Vernonite. (Lucitone). Proprietary trade names for acrylic synthetic resins for denture bases, etc.†

Verofix. A range of reactive dyestuffs. Applications: For the dyeing of wool and polyamide. (Bayer & Co).*

Verona Brown. A pigment obtained by calcining a ferruginous earth.†

Verona Yellow. See TURNER'S YELLOW.

Veronese Earth. See GREEN EARTH.

Veronese Green. See GREEN EARTH.

Veronese Yellow. See TURNER'S YELLOW

Versabacs. A polycellular carpet-backing system. (Dow Chemical Co Ltd).*

Versalon 1140. A registered trade name for a polyamide resin used as an adhesive between plasticized vinyl resins and metal.†

Versamid. A trade mark for polyamide curing agents for epoxy resins. (114). See also EPICURE.†

Versamid (R). Polyamide resins, solid thermoplastic and liquid reactable. (Cray Valley Products Ltd).

Versapen. Hetacillin. Applications: Antibacterial. (Bristol Laboratories, Div of Bristol-Myers Co).

Versapen K. Hetacillin potassium. Applications: Antibacterial. (Bristol Laboratories, Div of Bristol-Myers Co).

Versatic. Saturated tertiary monocarboxylic acid. (Shell Chemicals UK Ltd).

Versatint Fugitive tints. Water soluble, biodegradable colourants. Applications: Tinting for fibre and yarn identification. (Milliken & Company).*

Versed. Misonidazole. Applications: Antiprotozoal. (Hoffmann-LaRoche Inc).

Versene Acid. Edetic acid. Applications: Pharmaceutic aid. (Dow Chemical Company Ltd).

Versene AG. Brand chelated micronutrients used by growers to provide their crops with these necessary nutrients: zinc, managanese, iron, copper and magnesium. These elements stimulate growth hormones and contribute to plant-crop health and yields. (Dow Chemical Co Ltd).*

Versene CA. Edetate calcium disodium. Applications: Chelating agent. (Dow Chemical Company Ltd).

Versenol AG. Brand chelated micronutrients used by growers to provide their crops with these necessary nutrients - zinc, managanese, iron, copper and magnesium. These elements stimulate growth hormones and contribute to plant-crop health and yields. (Dow Chemical Co Ltd).*

Versicaine. Lignocaine hydrochloride. (May & Baker Ltd).

Versiflex. A proprietary trade name for a transparent vinyl chloride acetate.†

Versilan. Strong soil penetrating action, detergency and emulsification. Applications: Hard surface cleaners; metal degreasing; laundry powders; raw wool scouring. (Diamond Shamrock Process Chemicals Ltd).

Versilan. Alkyl aryl sulphonates, modified non-ionics, phosphate esters. Applications: Used as hydrotropes, wetters in electrolyte solutions. (Lankro Chemicals Ltd).†

Versilyt. Formulated surfactants. (Lankro Chemicals Ltd).

Verstarktes Chromammonit. An explosive containing 70 parts ammonium nitrate, 10 parts potassium nitrate, 12.5 parts trinitro-toluene, 7 parts chromium-ammonium alum, and 6 parts petroleum jelly.†

Verstran. Prazepam. Applications: Sedative. (Parke-Davis, Div of Warner-Lambert Co).

Vert D'eau. A jeweller's alloy containing 60 per cent gold and 40 per cent silver.†

Vert Diamant. See MALACHITE GREEN.

Vert d'Usebe. See ALDEHYDE GREEN.

Vertan. Chelating agents. Vertan 600 controls the hardness of boiler feedwater. Other Vertan chelating agents are used for removing water hardness deposits and metal oxide scale from industrial process equipment. (Dow Chemical Co Ltd).*

Vertifume. Fumigant contains carbon tetrachloride and carbon disulphide plus a fire inhibitor. It is used for controlling insects infesting stored grains. It may be applied as a liquid to the grain mass or via a closed recirculating system. Treated grain must be aired thoroughly before use. (Dow Chemical Co Ltd).*

Vertigon. A proprietary preparation of prochlorper-azine maleate. An anti-emetic. (Smith Kline and French Laboratories Ltd).*

Vertigon Spansule Capsule. Prochlorperazine maleate. Applications: Sustained release preparation for use as tranquillizer, anti-emetic and vestibular sedative. (Smith Kline and French Laboratories Ltd).*

Vert Paul Veronese Green. See EMERALD GREEN.

Vert Sulpho J. See ACID GREEN.

Verton. Long fibre reinforced thermoplastic compounds. (ICI PLC).*

Verv. Calcium stearyl-2-lactylate. A dough conditioner for yeast-leavened bakery products.†

Vesaloin. See HEXAMINE.

Vesalvine. See HEXAMINE.

Vespel. A proprietary trade name for polyimide in the form of prefabricated parts. (Du Pont (UK) Ltd).†

Vesprin. Triflupromazine hydrochloride. Applications: Antipsychotic. (E R Squibb & Sons Inc).

Vesta Phosphate. See RHENANIA PHOSPHATE.

Vestalin. See FERROZOID.

Vestamelt. A range of thermoplastic copolyesters. Applications: Hot melt adhesives for bonding textiles. (Huls AG).†

Vestamid. A large range of polyamides and copolyamides. Applications: Suitable for injection moulding, extrusion, thermoforming, blow moulding, rotational moulding and fluidized bed coating. (Huls AG).†

Vestenamer. Polyoctylene. Applications: Suitable for the manufacture of rubber blends for injection mouldings, extruded products (profiles, hoses), calendered articles and tyres. (Huls AG).†

Vestiform. High molecular polymethacrylate for processing polyvinyl chloride. Applications: Particularly for improving the deep-draw properties of films and sheets. (Huls AG).†

Vestinol. A range of alkyl phthalates. Applications: Primary plasticizers for polyvinyl chloride and paints etc. (Huls AG).†

Vestinol AH. A proprietary trade name for a vinyl plasticizer. Dioctyl phthalate. (Chemische Werke Hüls).†

Vestodur. A range of polybutylene terephthalate with and without various additives. Applications: Vestodur without fillers is suitable for injection moulding and extrusion and Vestodur containing fillers is suitable for injection moulding. (Huls AG).†

Vestolen A. A range of high density polyethylenes. Applications: Suitable for extrusion, compression moulding, injection moulding and rotational moulding. (Huls AG).†

Vestolen AS. A trade mark for a flame resistant high density polyethylene building material. (Chemische Werke Hüls).†

Vestolen EM. A range of elastomer-modified polypropylene. Applications: It has a wide range of possible uses in the automobile industry: bumper coverings, bumper corners and instrument panels. (Huls AG).†

Vestolen P. A range of polypropylenes. Applications: Suitable for woven tapes, twine, monofils, thermoformed film, medical products and laminating film. (Huls AG).†

Vestolit HI. High impact polyvinyl chloride. Applications: Used for window sections, films, profiles and sheets. (Huls AG).†

Vestolit S. Suspension polyvinyl chloride. Applications: Useful for profiles, cable sheathings and wire coverings, tubes, films, sheets, footwear and moulded articles. (Huls AG).†

Vestopal. A range of unsaturated polyester resins. Applications: Used for hand lay-up and compression moulding, industrial mouldings, for polymer-concrete and in fibre spraying processes. (Huls AG).†

Vestoplast. A range of predominantly amorphous olefin copolymers. Applications: Used in hot melts, adhesives, anti-corrosion strips, putties, sealing compounds and road marking compounds. (Huls AG).†

Vestopren. Thermoplastic rubber. Applications: A rubber concentrate for improving the impact stength of polyolefins, especially polypropylene. (Huls AG).†

Vestoran. Styrene acrylonitrile. (Chemische Werke Hüls).†

Vestorian Blue. See EGYPTIAN BLUE.

Vestowax. A range of Fischer-Tropsch waxes. Applications: Suitable for use in hot melts, printing inks, lacquers and as aids for processing rubber and plastics. (Huls AG).†

Vesturit. A range of saturated polyesters for lacquers. Applications: Used as binder components for stoving finishes containing solvents, for 'medium' and 'high' solids paints, for water soluble stoving finishes and as lacquer resins for highly flexible coatings. (Huls AG).†

Vestypor. A range of expandable polystyrene. Applications: Suitable for a wide range of moulding applications. (Huls AG).†

Vestyron. A range of polystyrenes. Applications: Suitable for a wide range of moulding applications. (Huls AG).†

Vestyron X984 and X1260AK. Trade marks. Polystyrenes with high impact resistance. (Chemische Werke Hüls).†

Vestyron 550. A trade mark. Polystyrene containing low residual monomer. A packaging material. (Chemische Werke Hüls).†

Vestyron 551. A trade mark. Polystyrene packaging materials with low residual monomer and exceptional stress cracking resistance in contact with oils. (Chemische Werke Hüls).†

Vesulong. Sulphonamide veterinary ethicals. (Ciba-Geigy PLC).

Vesuvine. See BISMARCK BROWN.

Vesuvine B. See MANCHESTER BROWN EE.

Vetalar. Ketamine hydrochloride. Applications: Anaesthetic. (Parke-Davis, Div of Warner-Lambert Co).

Vetalog. Triamcinolone Acetonide. Applications: Glucocorticoid. (E R Squibb & Sons Inc).

Vetalog Injection. Trimacinolone acetonide in aqueous vehicle. (Ciba-Geigy PLC).

Vetalog Plus Cream. Triamcinolone acetonide and halquinol in cream base. (Ciba-Geigy PLC).

'etanabol. Anabolic steroid veterinary ethical. (Ciba-Geigy PLC).

'etibenzamine. Antihistamines. (Ciba-Geigy PLC).

'etidrex. Diuretic veterinary ethicals. (Ciba-Geigy PLC).

'etol. A proprietary pure vegetable oil palm product.†

'iacutan. A proprietary preparation of methargen in a cream base. Antiseptic skin cream. (Ward, Blenkinsop & Co Ltd).†

'ialon Fast. Dyestuffs for polyamide fibres. (BASF United Kingdom Ltd).

'i-Alpha. Vitamin A. Applications: Vitamin (anti-xerophthalmic). (Lederle Laboratories, Div of American Cyanamid Co).

'ia Rasa. An insecticide. It is the calcium salt of *p*-toluenechlorosulphonamide.†

'ibalt. Cyanocobalamin. Applications: Vitamin. (J B Roerig, Div of Pfizer Laboratories).

'ibatex. Finishing agent. (Ciba-Geigy PLC).

'ibazine. Buclizine hydrochloride. Applications: Antinauseant. (Pfizer Inc).

'ibra-Tabs. Doxycycline hyclate. Applications: Antibacterial. (Pfizer Inc).

Vibrabond. Acrylic polymer cement flooring for heavy duty and chemical resistance. (Prodorite).*

Vibrac. A nickel chrome steel.†

Vibrac Steel. A nickel - chromium - molybdenurn steel.†

Vibramycin. A proprietary preparation of doxycycline. An antibiotic. (Pfizer International).*

Vibrathane. A range of polyether-based and polyester-based prepolymers. They offer high abrasion resistance, chemical resistance and electrical properties. (Uniroyal).*

Vibratussal. A proprietary preparation of doxycycline hyclate and codeine. An antibiotic and antitussive. (Pfizer International).*

Vibratussan. A proprietary preparation of doxycycline hyclate and codeine. An antibiotic and antitussive. (Pfizer International).*

Vibriomune. A proprietary cholera vaccine. (Duncan, Flockhart).†

Vibrocil. A proprietary preparation of dimethindine hydrogen maleate, phenylephrine hydrochloride and neomycin Sulphate. (Zyma (UK) Ltd).†

VIC Coatings. A liquid polyester or acrylic polyol mixed with a crosslinking resin which cures in minutes after being exposed to amine vapour via conventional air-supplied spray gun equipment. The catalyst vapour is created by a VIC generator which delivers a pre-determined concentration of amine catalyst on the air stream to a conventional spray gun. Applications: High performance coatings for primers and topcoats on plastics, metals and wood used in the automotive, general industrial and maintenance markets. (Ashland Chemical Company).*

Vicalloy. A proprietary trade name for a high permeability alloy containing iron with 36-62 per cent cobalt and 6-16 per cent vanadium.†

Vichy Salt. Sodium hydrogen carbonate, $NaHCO_3$.†

Vicks Medinite. Night time colds medicine, for relief of symptons. (Richardson-Vicks Inc).*

Viclan. Polyvinylidene chloride copolymer resins and latices. (ICI PLC, Petrochemicals & Plastics Div).

Vicmos Powder. A smokeless 33-grain powder.†

Vicron. Highly refined calcite. (Pfizer International).*

Vicryl. Polyglactin. Applications: surgical aid - surgical suture material, absorbable. (Ethicon Inc).

Victor Bronze. An alloy of 58.5 per cent copper, 38.5 per cent zinc, 1.5 per cent aluminium, 1 per cent iron, and 0.03 per cent vanadium.†

Victor Metal. An alloy of 50 per cent copper, 34.3 per cent zinc, 15.4 per cent nickel, 0.28 per cent iron, and 0.11 per cent aluminium. Used for sand castings and marine work.†

Victor Powder. A trade mark for a smokeless powder containing nitro-

ammonium nitrate, wood meal, and potassium chloride.†

Victoria Aluminium. See PARTINIUM.

Victoria Black G,. 5G. Dyestuffs†

Victoria Blue B. A dyestuff, $C_{33}H_{32}Cl.N_3$. Dyes silk and wool blue from an acid bath, and cotton mordanted with tannin.†

Victoria Blue 4R. A dyestuff, $C_{34}H_{34}Cl.N_3$. Dyes silk and wool reddish-blue.†

Victoria Green. 3B (Setoglaucine, New Fast Green 3B, New Solid Green 3B). A dyestuff. Dyes silk and wool, and cotton mordanted with tannin and tartar emetic, bluish green.†

Victoria Green. See MALACHITE GREEN.

Victoria Orange. See VICTORIA YELLOW.

Victoria Red. See BENZOPURPURIN 4B and CHROME RED.

Victoria Rubine. See FAST RED D.

Victoria Silver. See NICKEL SILVERS.

Victoria Varnish. See ZAPON VARNISH.

Victoria Violet 4BS. (Ethyl Acid Violet S4B, Coomassie Violet AV). A dyestuff. It is the sodium salt of *p*-amino benzeneazo-1, 8-dihydroxynaphthalene-disulphonic acid, $C_{16}H_{11}N_3S_2O_8Na_2$. Dyes wool bluish-violet from an acid bath.†

Victoria Violet 8BS. A dyestuff of the same class as Victoria Violet 4BS.†

Victoria Yellow. (Gold Yellow, Saffron Surrogate, English Yellow, Victoria Orange, Saffron Substitute, Aniline Orange). A dyestuff. It is a mixture of the potassium or ammonium salts of dinitro-*o*-cresol, and dinitro-*p*-cresol, $C_7H_5KN_2O_5$. Dyes wool and silk orange. Has been used for spraying trees infected with the caterpillar of *Liparis monacha*. It is explosive when dry.†

Victoria Yellow Extra Conc. See METANIL YELLOW.

Victorium. A proprietary trade name for lignin thermoplastic materials.†

Victors. Cough drops. (Richardson-Vicks Inc).*

Victrex. Polyaromatic resins. (ICI PLC).*

Victron. A proprietary trade name for polystyrene and vinylite resins.†

Viczsal. An ammoniacal solution of copper and zinc phenolates. A wood preservative.†

Vidal Black. (Vidal Black S). A dyestuff obtained by the fusion of *p*-aminophenol (or of *p*-aminophenol and other compounds) with sodium polysulphide. Vidal Black S is the bisulphite compound. Cotton is dyed greenish- to bluish-black.†

Vidal Black S. See VIDAL BLACK.

Vidal's Caustic Powder. A caustic consisting of burnt alum and powdered savin tops.†

Vi-Daylin. A proprietary preparation containing Vitamins A, D, C, thiamine, riboflavine, nicotinamide and pyridoxine. (Abbott Laboratories).*

Videne Disinfectant Solution. Pre-operative skin antiseptic containing povidone-iodine USP. (Riker Laboratories/3M Health Care).†

Videne Disinfectant Tincture. Pre-operative skin antiseptic containing povidone-iodine USP. (Riker Laboratories/3M Health Care).†

Videne Powder. Topical antiseptic containing povidone-iodine USP. (Riker Laboratories/3M Health Care).†

Videne Surgical Scrub. Surgical scrub containing povidone-iodine USP. (Riker Laboratories/3M Health Care).†

Videobil. Iopronic acid. Applications: Diagnostic aid. (Bracco Industria Chimica SPA).

Videocolangio. Iodoxamic acid. Applications: Diagnostic aid. (Bracco Industria Chimica SPA).

Vi-Dom-A. Vitamin A. Applications: Vitamin (anti-xerophthalmic). (Miles Pharmaceuticals, Div of Miles Laboratories Inc).

Vidopen. A proprietary preparation of ampicillin. An antibiotic. (Berk Pharmaceuticals Ltd).*

Vidox. Zinc oxide. Applications: Rubber, paint, ceramics. (Manchem Ltd).*

Vielle Powder. See POUDRE B.

Vienna Blue. See COBALT BLUE.

ienna Caustic. Potassium hydroxide with lime. See VIENNA PASTE.†

ienna Cement. A metallic cement made from 86 per cent copper and 14 per cent mercury. An imitation gold.†

ienna Green. See EMERALD GREEN.

ienna Lake. See FLORENCE LAKE.

ienna Paste. A mixture of lime and potash.†

ienna Red. See CHROME RED.

iennese Tombac. An alloy of 97 per cent copper and 2.8 per cent zinc.†

igantol. Vitamin D_3 in oily solution. Applications: Veterinary preparation to prevent rickets and osteomalacia. (Bayer & Co).*

igantol-E-Comp. Vitamins A, D_3 and H. Applications: Used against vitamin deficiency diseases in veterinary medicine. (Bayer & Co).*

igazoo. Sublenox. Applications: Growth stimulant. (American Cyanamid Co).

igil. Broad spectrum fungicide containing diclobutrazol. (ICI PLC).*

igorite. See BAKELITE.

igorite. A safety explosive for mines. It consists of 30 per cent nitroglycerin, 49 per cent potassium chlorate, 7 per cent potassium nitrate, 9 per cent wood pulp and 5 per cent magnesium carbonate.†

ikane. Fumigant based primarily on sulphuryl fluoride is used specifically for the control of drywood termites and wood boring beetles infesting wood in structures, furniture and lumber. The fumigant is odourless, colourless, non-corrosive and does not react to produce malodours. (Dow Chemical Co Ltd).*

Viking. Tyres. (Chevron).*

Vikro. Proprietary alloys containing from 63-65 per cent nickel, 13-23 per cent chromium, 0.5-1 per cent silicon, up to 1 per cent manganese and carbon, and the balance iron.†

Vileda. Household cloths. (Carl Freudenberg).*

Viledon Compact. Non-woven table cover. (Carl Freudenberg).*

Viledon Filter. Non-woven air filter. (Carl Freudenberg).*

Vilene. Freudenberg non-wovens. (Carl Freudenberg).*

Vilit. A range of soluble vinyl chloride copolymers. Applications: Used as binders for paints, heat sealable lacquers for aluminium foils, for coatings on metal, concrete and cardboard. (Huls AG).†

Villescon. A proprietary preparation of prolintane with vitamins. (Boehringer Ingelheim Ltd).†

Viluite. Synonym for Grossular.†

Vimlite. A proprietary trade name for a cellulose acetate plastic.†

Vimopyrine. *p*-Phenetidine tartrate.†

Vinaccia Tartar. Tartar obtained from the manufacture of wines.†

Vinaconic Acid. Propene-3,3-dicarboxylic acid (vinyl-malonic acid) $CH_2{:}CH.CH.(COOH)_2$.†

Vinacron. Polyvinyl chloride plastisol dispersion. Applications: Protective and decorative coatings, rotomoulding, spread coating and dip moulding. (Loes Enterprises Inc).*

Vinacryl R3929, R3940. Proprietary trade names for a 55 per cent. concentrated vinyl-acrylic copolymer emulsions used for non-fray carpet backings. (Vinyl Products, Carshalton).†

Vinacryl 4001/B. A proprietary trade name for a 50 per cent. concentrated acrylic copolymer emulsion used as a cement additive. (Vinyl Products, Carshalton).†

Vinacryl 4005. A proprietary acrylic copolymer emulsion soluble in alkali. (Vinyl Products, Carshalton).†

Vinacryl 4152, 4500/X, 4501/X. Proprietary trade names for vinyl acrylic copolymer emulsions used as adhesives. (Vinyl Products, Carshalton).†

Vinacryl 4160. A proprietary trade name for an acrylic copolymer emulsion used as a 46.5 per cent. concentrate for a binder in paper board manufacture. (Vinyl Products, Carshalton).†

Vinacryl 4260. A proprietary emulsion. Poly-2-ethoxyethyl methacrylate. (Vinyl Products, Carshalton).†

Vinacryl 4290. A proprietary polybutyl methacrylate emulsion. (Vinyl Products, Carshalton).†

Vinacryl 4320. A proprietary self-reactive vinyl acrylic copolymer emulsion used in the finishing of textiles. (Vinyl Products, Carshalton).†

Vinacryl 4322. A proprietary self-reactive vinyl acrylic copolymer emulsion. (Vinyl Products, Carshalton).†

Vinacryl 4450. A proprietary trade name for a vinyl-acrylic copolymer. Used in crack filling compounds. (Vinyl Products, Carshalton).†

Vinacryl 4512. A proprietary acrylic polymer emulsion. (Cray Valley Products Ltd).†

Vinacryl 7170, 7172 and 7175. A proprietary range of styrene acrylic copolymer emulsions. (Vinyl Products, Carshalton).†

Vinal. A proprietary trade name for a synthetic vinyl resin.†

Vinalak 5150. A proprietary self-reactive acrylic polymer solution in isopropyl acetate. (Vinyl Products, Carshalton).†

Vinamul 3240. A proprietary vinyl acetate-ethylene copolymer emulsion. (Vinyl Products, Carshalton).†

Vinamul 3250. A proprietary vinyl acetate-ethylene copolymer emulsion containing a non-ionic emulsifying system. (Vinyl Products, Carshalton).†

Vinamul 6000. A proprietary vinyl acetate emulsion of the unsaturated acid copolymer type, soluble in alkali. (Vinyl Products, Carshalton).†

Vinamul 6050. A proprietary vinyl acetate-vinyl caprate-unsaturated acid terpolymer emulsion, internally plasticized. (Vinyl Products, Carshalton).†

Vinamul 6208. A proprietary trade name for a 50 per cent concentrated vinyl-acrylic copolymer emulsion used for wallpaper grounding, printing and overcoating. (Vinyl Products, Carshalton).†

Vinamul 6275. A proprietary trade name for a 55 per cent. concentrated vinyl-acrylic copolymer emulsion used in the production of emulsion paints. (Vinyl Products, Carshalton).†

Vinamul 6705. A proprietary ethylene grafted vinyl acetate copolymer emulsion. (Vinyl Products, Carshalton).†

Vinamul 6888. A proprietary modified vinyl acetate-acrylate copolymer emulsion. (Vinyl Products, Carshalton).†

Vinamul 6930. A proprietary trade name for 52 per cent. concentrated vinyl acetate - Veo Va 911 copolymer emulsion used for emulsion paints. (Vinyl Products, Carshalton).†

Vinamul 7700. A proprietary polystyrene emulsion. (Vinyl Products, Carshalton).†

Vinamul 7715. A proprietary polystyrene emulsion containing 15 per cent. dibutyl phthalate in a non-volatile plasticizer. (Vinyl Products, Carshalton).†

Vinamul 8400. A proprietary trade name for a 50 per cent. emulsion of polyvinyl acetate. Used for adhesives. (Vinyl Products, Carshalton).†

Vinamul 8430. A proprietary polyvinyl acetate emulsion. (Vinyl Products, Carshalton).†

Vinamul 8460 and 9000. Proprietary polyvinyl acetate emulsions. (Vinyl Products, Carshalton).†

Vinapol R3626. A proprietary trade name for a water-dispersible alkali soluble vinyl acetate powder. (Vinyl Products, Carshalton).†

Vinapol R.3800, R.3863, R10, 030. A proprietary trade name for a water dispersible polyvinyl acetate powder. (Vinyl Products, Carshalton).†

Vinapol 1000. A proprietary polyvinyl acetate powder dispersible in water. (Vinyl Products, Carshalton).†

Vinapol 1030. A proprietary polyvinyl acetate powder plasticized with 10 per cent. dibutyl phthalate and dispersible in water. (Vinyl Products, Carshalton).†

Vinapol 1070. A proprietary finely-divided polyvinyl acetate powder. (Vinyl Products, Carshalton).†

Vinapol 1088. A proprietary acrylic processing and used in the processing of rigid PVC compounds. (Vinyl Products, Carshalton).†

Vinatex. PVC plastisols. (Norsk Hydro Polymers Ltd).

Vinavil. Vinyl acetate, homopolymers and copolymers. (Montedison UK Ltd).*

Vinblastine. An alkaloid extracted from *Vinca rosea.* An antitumour agent.†

Vincapront. A proprietary preparation of vincamine. A cerebral vasodilator. (Pfizer International).*

Vincaron. A proprietary preparation of vincamine. A cerebral vasodilator. (Pfizer International).*

Vincennite. A poison gas. It was hydrocyanic acid mixed with stannic chloride.†

Vinchel 11. A proprietary metal-free liquid organic complex with a stabilizer system of barium and cadmium soaps, used as a chelating agent in PVC compounds. (Vinyl Products, Carshalton).†

Vinchel 20. A proprietary metal-free liquid organic complex with a stabilizer system of barium and cadmium soaps and tribasic barium sulphate, used as a chelating agent in PVC compounds. (Vinyl Products, Carshalton).†

Vinchel 22. A proprietary zinc-based chelating agent for PVC compounds. A basic lead carbonate stabilizing system is employed. (Vinyl Products, Carshalton).†

Vinchel 35. A proprietary zinc-based chelating agent for PVC compounds. A stabilizer system of barium and cadmium soaps, and barium and cadmium liquids is employed. (Vinyl Products, Carshalton).†

Vinco A183. A proprietary liquid complex used as a stabilizer and initiator in the production of expanded PVC. (Victor Wolf N L Ltd).†

Vinco A33. A proprietary liquid patassium/zinc complex used as a stabilizer and initiator in the production of expanded PVC. (Victor Wolf N L Ltd).†

Vinco 248. A proprietary stabilizer of the liquid-barium and cadmium- complex type, used with PVC polymers sensitive to zinc. (Victor Wolf N L Ltd).†

Vinco 249C. A proprietary stabilizer for PVC with a liquid barium, cadmium and zinc base. (Victor Wolf N L Ltd).†

Vinco 265. A proprietary stabilizer for PVC pastes with a liquid barium, cadmium and zinc base. (Victor Wolf N L Ltd).†

Vinco 99A. A proprietary stabilizer for PVC and PVA of the liquid-barium/cadmium-zinc complex type. (Victor Wolf N L Ltd).†

Vinco 99G. A stabilizer similar ti VINCO 99A used with paste-grade resins in rotational moulding. (Victor Wolf N L Ltd).†

Vincristine. An alkaloid obtained from *Vinca rosea.* An antitumour agent.†

Vinegar Naphtha. See ACETIC ESTERS.

Vinegar Salts. Calcium acetate. $Ca(C_2H_3O_2)_2.H_2O$.†

Vinegar, Martial. Ferric acetate.†

Vinegar, Sugar. See GLUCOSE OR SUGAR VINEGAR.

Vinescol 23. A fluorinated synthetic rubber.†

Vinic Ether. Ethyl ether, $C_2H_5O.C_2H_5$.†

Vinidur. Vinyl chloride homopolymer Gs grafted on acrylic rubber. Applications: Extrusion: window profiles and cladding. Injection moulding: fittings. (BASF United Kingdom Ltd).*

Vinisil. Povidone. Applications: Pharmaceutic aid (dispersing and suspending agent). (Abbott Laboratories).

Vinnapas. Homopolymer or copolymer, polyvinyl acetate solutions, dispersions, solid resins. Applications: For lacquers and adhesives, paints. (Wacker Chemie GmbH).*

Vinnathen. A proprietary trade name for a vinyl acetate/ethylene copolymer which can be cross-linked with peroxides. It can be used for cable jackets.†

Vinnol. A proprietary grade of poly-vinyl chloride. (Wacker Chemie GmbH).*

Vinoflex. Vinyl chloride homopolymer Gs, vinyl chloride homopolymer Ge. Applications: Extrusion: pipes, profiles and cables. Kalandering: plasticized and UPVC - films. Fittings, injection moulding, blowmoulding: bottles. (BASF United Kingdom Ltd).*

Vinoflex 377. A proprietary PVC-emulsion homopolymer used in the production of

PVC film. (BASF United Kingdom Ltd).*

Vinoflex 516. A proprietary PVC suspension-type homopolymer. It is an easily-flowing powder used in the extrusion of rigid PVC. (BASF United Kingdom Ltd).*

Vinoflex 526. A proprietary PVC homopolymer similar to VINOFLEX 516. (BASF United Kingdom Ltd).*

Vinoflex 534. A proprietary PVC suspension-type homopolymer used in the extrusion of high-quality cables. (Anic Agricoltura Spa).†

Vinoflex 535. A proprietary PVC suspension-type homopolymer in the form of an easy-flowing powder with porous particles, used in the making of soft, calendered products. (BASF United Kingdom Ltd).*

Vinoflex 719. A proprietary PVC suspension-type polymer used in the making of rigid, tough, weather-resistant products. (BASF United Kingdom Ltd).*

Vinous Alcohol. Ethyl alcohol, C_2H_5OH.†

Vinsalyn. A range of thermoplastic resins. Applications: They are used as binders and/or binder extenders in industrial adhesives and mastics. (Hercules Inc).*

Vinsol. Dark pine resin. (Hercules Ltd).

Vinsol. A proprietary trade name for the black residue from the extraction of rosin with solvents. Used as an insulating varnish.†

Vinsol Emulsion. An emulsion of aliphatic hydrocarbon insoluble resin. Applications: Used as a modifier of water-based adhesives and coatings. (Hercules Inc).*

Vinsol Resin. A proprietary trade name for a resin produced by the distillation of wood.†

Vinuran. αMS/AN / acrylates, MABS / modified acrylates, MBS. Applications: Modifiers mainly for UPVC, processing aids, impact modifiers. Impact modifiers for increased vicat softening temperature. (BASF United Kingdom Ltd).*

Vinychlon. A registered trade mark for a series of Japanese vinylchloride polymers. (Mitsui Co Ltd).†

Vinyl Acetate Resins. See GELVA.

Vinylite. A proprietary trade name for polyvinyl acetate, polyvinyl chloride-acetate and polyvinyl chloride synthetic resins.†

Vinylite V. A proprietary name for an inte polymer of PVC and P.V.A. (*qv*).†

Vinylite X. A proprietary trade name for polyvinyl chloride-acetate.†

Vinyloid. A proprietary trade name for a polyvinyl acetate resin.†

Vinyl Resins. See VYDON, VINNAPAS, RESOVIN, KORON, POLYVINYL-CHLORIDE.

Vinylseal. A proprietary trade name for vinyl acetate resin adhesives.†

Vinyon. A proprietary trade name for viny resins for textile fibres.†

Vinyon Fiber. A proprietary trade name fo a material manufactured from polyviny chloride-acetate.†

Vinyzene. A fungicide and bactericide. bromchlorenone.†

Viocin. A proprietary preparation containing viomycin sulphate. An anti-tuberculous agent. (Pfizer International).*

Vioflor. A proprietary preparation of volatile hydrocarbons. Used to deodorize turpentine substitutes.†

Vioform Powder. Treatment of skin condition. (Ciba-Geigy PLC).

Vioform-Hydrocortisone. A proprietary preparation of clioquinol and hydrocortisone used in dermatology as an antibacterial agent. (Ciba Geigy PLC).†

Vioform.. Clioquinol. Applications: Anti-amebic; anti-infective, topical. (Ciba-Geigy Corp).

Vioglaze. Ultra-violet curing coatings. Applications: Gloss varnish for roller coating paper and board and roller coating for flooring. (Coates Industrial Finishes Ltd).*

Violamine B. See FAST ACID VOILET B.

Violamine G. See ACID ROSAMINE A.

Violamine R. See FAST ACID VIOLET A2R.

Violamine 3B. See FAST ACID BLUE R.

Violaniline. A blue colouring matter belonging to the induline class. See INDULINE, SPIRIT SOLUBLE.†

Violanthrene. A dyestuff prepared from anthranol by heating it with glycerin.†

Violeine. See MAUVEINE.

Violet Bronze. See MAGENTA BRONZE and SAFFRON BRONZE.

Violet C. See CRYSTAL VIOLET.

Violet Carmine. A pigment from the root of *Anchusa tinctoria.*†

Violet Copper. (Red Copper). Reduced copper, prepared by reducing copper oxide.†

Violet Imperial Rouge. See REGINA PURPLE.

Violet Moderne. See BLUE 1900.

Violet Phenylique. See REGINA PURPLE.

Violet Phosphorus. The coarse-grained red variety is metallic or violet phosphorus.†

Violet Powder. Perfumed starch powder.†

Violet Root. Orris root.†

Violet R, RR, 5R. See HOFMANN'S VIOLET.

Violet Smelling Salts. See ENGLISH SALT.

Violet Tungsten. Potassium tritungstate, $K_2W_3O_9.W_2O_5$. A pigment.†

Violet Ultramarine. A pigment obtained from blue ultramarine by the action of dry hydrochloric acid gas and oxygen at 150-180° C. See ULTRAMARINE.†

Violet 3B Extra. See METHYL VIOLET B.

Violet 4RN. See HOFMANN'S VIOLET.

Violet 5B. See BENZYL VIOLET.

Violet 6B. See BENTYL VIOLET.

Violet 7B Extra. See CRYSTAL VIOLET.

Violettol. A mixture of 10 per cent ionone and 90 per cent salicyl aldehyde. Used to strengthen natural violet perfume.†

Violetton. (Rhodione). Trade names for the commercial ionones. Used as perfumes.†

Violuric Acid. Isonitroso-barbituric acid, $C_4H_3N_3O_4$.†

Viomycin. An antibiotic produced by certain strains of *Sstreptomyces griseus* var. *purpureus.*†

Viomycin Sulphate. A proprietary preparation containing viomycin sulphate. An antituberculous agent. (Parke-Davis).†

Vionactane. A proprietary preparation of viomycin pantothenate and sulphate. An antituberculous agent. (Ciba Geigy PLC).†

Vionate Powder. Multi vitamin and mineral mix for veterinary use. (Ciba-Geigy PLC).

Vipla. Emulsion polyvinyl chloride homopolymers.†

Vipophan PVC. Shrink film. (Lonza Limited).*

Vira-A. Vidarabine. Applications: Antiviral. (Parke-Davis, Div of Warner-Lambert Co).

Virazole. Ribavirin. Applications: Antiviral. (ICN Nutritional Biochemicals Corp).

Virco. Various. Applications: Textile dyeing auxiliaries. (Virkler Chemical Co).*

Virexen. Idoxuridine in DMSO solution. Applications: Pharmaceutical speciality: Antiviral for the treatment of herpes simplex and herpes zoster. (Laboratorios Viñas SA).*

Virginia Silver. See NICKEL SILVERS.

Virginiamycin. An antibiotic produced by *Streptomyces virginioe.*†

Virgo Fibre Half-stuff. A fibre produced from the short fibres or so-called seed-lint by treatment with caustic soda.†

Viridian. (French Veronese Green). A green pigment. It is hydrated sesquioxide of chromium, $Cr_2O_3.2H_2O$. See CHROMIUM GREEN.†

Viridine. See ALKALI GREEN.

Vironex. Folpet 40%, cymoxanil 4%. Applications: Wettable powder used as protective fungicide for foliage application to ornamental and crop plants. (Industrias Quimicas Del Valles SA).*

Viroptic Ophthalmic Solution. A proprietary formulation of trifluridine. Applications: Treatment of primary keratoconjunctivitis and recurrent epithelia keratitis due to Herpes simplex virus types 1 and 2. (The Wellcome Foundation Ltd).*

Virormone. Injectable testosterone propionate. Applications: May be used as replacement therapy in castrated adults and in those who are hypogonadal due to either pituitary or testicular disease. Also for the control of carcinoma in post menopausal women. (Paines & Byrne).*

Visacor. Cardiovascular preparations. (ICI PLC).*

Viscalex EP 30. A proprietary acrylic copolymer emulsion used as a thickener in water-based paints. (Allied Colloids Ltd).†

Viscanite. A vegetable glue.†

Vischem. A series of diester and polyolester synthetic based greases with various thickeners. Applications: High temperature greases for all kinds of bearings, worm gears, slides, etc. (Ultrachem Inc).*

Viscoid. A mixture of viscose and clay with powdered horn, or zinc oxide.†

Viscolane. An ointment base prepared from the bark of the mistletoe.†

Viscolith. The hard mass obtained when a viscose solution coagulates.†

Viscoloid. (Nixenoid, Fiberloid). A proprietary trade name for pyroxylin plastics.†

Viscom. See OIL PULP.

Viscoplex. Solutions in mineral oil of long chain fatty alcohol/acrylic/methacrylic acid esters. Applications: Pour point depressants and viscosity index improvers for mineral oils. (Cornelius Chemical Co Ltd).*

Viscose. The sodium salt of cellulose xanthate, obtained by the action of carbon disulphide and alkali upon cellulose. In thin sheets, it is used as a substitute for glass and celluloid. It is also employed as a thickening and dressing substance, as a partial substitute for resin glue in paper manufacture, and in the production of artificial silk. Other synonyms are: Cellushi, Clar-Apel, Crystex, New-wrap, Sidac, Tomophan, Zellwonet.†

Viscose Silk. An artificial silk produced when Viscose (*qv*) is forced through narrow orifices into ammonium chloride solution.†

Viscosin. A proprietary refined oil tar.†

Viscosine. See VALVOLINE.

Vi-Siblin. A proprietary preparation of ispaghula husks and THIAMINE. A laxative. (Parke-Davis).†

Visken. A proprietary preparation of PINDOLOL used in the treatment angina pectoris. (Sandoz).†

Visqueen. Polyethylene film. (ICI PLC).*

Vistaflex. Thermoplastic elastomers. (Exxon Chemical Ltd).*

Vistaflex. Thermoplastic elastomer. (Exxon Chemical International Inc).*

Vistal. Polyethylene films in widths up to 12 metres. Films based on polyethylene LD, MD, LLD in thicknesses of 10 to 800 microns. Applications: Shrink and stretch films for packaging, bundling and palletising, coextruded films, embossed and repellent films for industrial applications, films for horticulture and agriculture, peelable films for easy opening packaging, special films for incorporating into laminates, for balloons for space research and for building and road construction. (UCB nv Film Sector).*

Vistalon. Polyolefin. (Exxon Chemical International Inc).*

Vistalon. Ethylene propylene copolymer. (Exxon Chemical Ltd).*

Vista-Marc. Etafilcon A. Applications: Contact lens material. (Vistakon Inc).

Vistanex. Isobutylene polymers. (Exxon Chemical Ltd).*

Vistanex. Polyisobutylene. (Exxon Chemical International Inc).*

Vistanex. A proprietary trade name for polyisobutylene. A synthetic rubber. See also OPPANOL, ISOLENE, P.I.B. Hydrocarbon resin.†

Vistaril. A proprietary preparation of hydroxyzine. An ataractic. (Pfizer International).*

Vistaril Pamoate. Hydroxyzine pamoate. Applications: Tranquillizer. (Pfizer Inc).

Vistone. Mild E P Oiliness agent. (Exxon Chemical Ltd).*

istra. An artificial silk which is made in a similar way to viscose silk.†

istron. See POLYSTYRENE and other trade names under the latter.

ita-E. A proprietary preparation of D-α-tocopherol used in the treatment of vascular disorders. (Bioglan Laboratories).*

ita-E Gelucaps. A proprietary preparation of D-α -tocopherol acetate. Applications: Used in the treatment of vascular disorders. (Bioglan Laboratories).*

italum. A trade mark for materials of the abrasive class and consisting essentially of alumina.†

itamalt. A proprietary preparation containing vitamins A, B, C, and D.†

itamin B$_1$. Thiamine, aneurin. Deficiency in the diet in man produces Beri-beri and peripheral neuritis.†

itamin B$_{12}$. Cobalamin, cyanocobalamin. Deficiency in man causes pernicious anaemia and neural degeneration.†

itamin B$_2$. Riboflavin. Deficiency in man produces minor symptoms only. Also known as lactoflavin.†

itamin B$_6$. Pyridoxine.†

itamin C. Ascorbic acid. Deficiency in man causes scurvy. It probably plays a part in the production of collagen in the tissues.†

itamin D. A fat soluble vitamin, deficiency of which causes rickets, in children and osteomalacia in adults.†

itamin D$_2$. Calciferol. A synthetic vitamin produced by the irradiation of ergosterol with ultra-violet light. Available as Sterogyl-15, and in many multivitamin preparations.†

itamin D$_3$. Naturally occuring Vitamin D. Derived from fish oils.†

itamin D$_4$. A synthetic vitamin derived from 22-dihydroergosterol by irridation with ultra-violet light. It has less activity than D$_2$ and D$_3$.†

itamin E. Tocopherol. A fat soluble vitamin. Deficiency in rats produces sterility.†

itamin G. See RIBOFLAVIN.

itamin K$_1$. Phytomenadione. A fat soluble vitamin. 2-methyl-3-phytyl-1, 4-naphthaquinone. Deficiency gives rise to haemorrhage.†

Vitamin K$_2$. A fat soluble vitamin. 2-methyl-3-difarnesyl-1,4-naptha-quinone. Synthesized in the gut by bacteria.†

Vitamin M. Folic acid.†

Vitamin P. QUERCITIN.†

Vitamin U. A vitamin extracted from cabbage.†

Vitaminets. Proprietary multivitamin preparation containing the B complex vitamins, vitamins A, C, D and E, biotin, nicotinamide and various minerals or trace elements (Ca, Fe, Mg, Mn and P). (Roche Products Ltd).*

Vitapet. Veterinary oil blend for pets. Applications: Dietary supplement with vitamins A and D. (Marfleet Refining Co).*

Vitapyrena. Medicated hot lemon drink powder, for symptomatic relief of colds. (Richardson-Vicks Inc).*

Vita Zinc. Low purity zinc oxide. Applications: Animal feed supplement. (Manchem Ltd).*

Viten. Vital wheat gluten. Applications: Flour fortification (bread making). (Roquette (UK) Ltd).*

Vitoline Yellow 5G. See PHOSPHINE.

Viton. Fluoroelastomer. (Du Pont (UK) Ltd).

Viton A. A proprietary vinylidene fluoride-hexafluoropropylene of resistance to heat, oil and solvents. its Mooney viscosity at 100° C. is 67. (Du Pont (UK) Ltd).†

Viton AHV. A copolymer similar to VITON A but possessing greater strength at high temperatures. Its Mooney viscosity at 100° C. is 180. (Du Pont (UK) Ltd).†

Viton A35. A copolymer similar to VITON A but possessing easier processing qualities. Its Mooney viscosity at 100° C. is 35. (Du Pont (UK) Ltd).†

Viton E-430. A proprietary fluoroelastomer with good processing and storage properties. (Du Pont (UK) Ltd).†

Viton LN. A waxy semi-solid fluoroelastomer of the VITON type used as a plasticizer to improve moulding and extrusion characteristics. (Du Pont (UK) Ltd).†

Vitradur. Ultra high molecular weight high density polyethylene. Applications: Paper and textile accessories, bunker linings and skating rinks. (Stanley Smith & Co Plastics Ltd).*

Vitrafix. Co-ordination compound for glass laminates. (ICI PLC).*

Vitrahose. Polyvinyl chloride. Applications: Tubes and hoses for many applications. (Stanley Smith & Co Plastics Ltd).*

Vitralen. Ultra high molecular weight high density polyethylene sheet. Applications: Orthopaedic splints. (Stanley Smith & Co Plastics Ltd).*

Vitralene. Polypropylene sheet and block. Applications: Anti-acid fabrications, cutting boards. (Stanley Smith & Co Plastics Ltd).*

Vitralex. Acrylonitrile/butadiene/styrene (ABS), polyvinylidene fluoride (PVDF) sheet and block. Applications: Anti-acid fabrications. (Stanley Smith & Co Plastics Ltd).*

Vitrapad. Polyvinylchloride or polyolefine based sheet and block. Applications: Cutting boards for all industries. (Stanley Smith & Co Plastics Ltd).*

Vitraplas. Polyolefines. Applications: General purpose plastics applications. (Stanley Smith & Co Plastics Ltd).*

Vitrathene. Polyethylene sheet, rod, block and massive castings. Applications: Radiation protection, chemically resistant plant, electronic and radar insulation, orthopaedics, textile and paper trade accessories, bunker linings, cutting boards and tabletops. (Stanley Smith & Co Plastics Ltd).*

Vitre-colloid. A proprietary cellulose acetate.†

Vitreo-colloid. A proprietary trade name for cellulose acetate plastic.†

Vitreon. A porcelain used in the condensing plant for nitric acid.†

Vitreosil. (Silica Glass). Fused Silica, SiO_2.†

Vitriolated Magnesia. Magnesium sulphate, $MgSO_4$.†

Vitriolated Soda. Sodium sulphate, Na_2SO_4.†

Vitriolated Tartar. Potassium sulphate, K_2SO_4.†

Vitriol, English. Ferrous sulphate, $FeSO_4$.

Vitriol, Green. See IRON VITRIOL.

Vitriolic Acid. See OIL OF VITRIOL.

Vitriolized Bones. Bones which have been treated for a long time with sulphuric acid to obtain the phosphates in a more soluble form. A fertilizer.†

Vitriol Oil. See OIL OF VITRIOL

Vitriol, Salt of. Zinc sulphate, $ZnSO_4$.†

Vitriol, Salzburg. See MIXED VITRIOL.

Vitriol Stone. Impure ferric sulphate obtained by the oxidation of pyrites. Used in the manufacture of fuming sulphuric acid.†

Vitrite. Multivitamin syrup. Applications Dietary supplement. (Marfleet Refining Co).*

Vitrite. Polyolefine based sheet and block. Applications: Nuclear shielding and radiation protection. (Stanley Smith & Co Plastics Ltd).*

Vitromail. A range of inorganic decoration colours. Applications: Available in the form of silk screen pastes, thermoplastics and powder for use in the enamelling and glass industries. (Bayer & Co).*

Vitrone. Polyvinyl chloride sheet. Applications: Thermoformed packaging, chemically resistant fabrications. (Stanley Smith & Co Plastics Ltd).*

Vivactil. Protriptyline Hydrochloride. Applications: Antidepressant. (Merck Sharp & Dohme, Div of Merck & Co Inc).

Vivalan. A proprietary preparation of VILOXAZINE hydrochloride. An anti-depressant. (ICI PLC).*

Vivol. Cloth and textile oils. (Crosfield Chemicals).*

Vizor. Weedicide, containing lenacil. (ICI PLC).*

Vlem-Dome. Sulphurated lime. A solution of lime, sublimed sulphur and water. Applications: Scabicide. (Miles

Pharmaceuticals, Div of Miles Laboratories Inc).

Vlemasque. Sulphurated lime. A solution of lime, sublimed sulphur and water. Applications: Scabicide. (Dermik Laboratories Inc).

Vlieseline. Non-wovens for the apparel industry. (Carl Freudenberg).*

V.M. and P. Naphtha. Varnish-maker's and painter's naphtha, a deodorized petroleum product which is practically a gasoline (benzine) of 100-160° C. boiling range and specific gravity 0.730.†

Vocol. Zinc-0,0-di-n-butylphosphorodithioate. Applications: vulcanization accelerator. (Monsanto Co).*

Vogan. A vitamin A concentrate.†

Vogel's Alloy. An alloy containing 8 parts copper, 1 part zinc, 2 parts tin, and 1 part lead. Used for polishing steel.†

Voidform. Cut polystyrene shapes and moulded cylinders. Applications: Formation of voids within reinforced concrete structures and for providing profiles with cast concrete. (Vencel Resil Ltd).*

Voidmaster. Moulded expanded polystyrene panels. Applications: Used between pre-stressed concrete beams to provide floor insulation. (Vencel Resil Ltd).*

Voidox 100 per cent. A registered trade name for a food grade antioxidant. It is a modified fatty acid derivative of a substituted phenol. (The Guardian Chemical Corporation).†

Vol. Ammonium carbonate, $(NH_4)_2CO_3$.†

Volatile Alkali. (Alkaline Air, Spirit of Hartshorn). Ammonia, NH_3.†

Volatile Liniment. Liniment of ammonia.†

Volatile Oil of Bitter Almonds. Benzaldehyde, $C_6H_5.CHO$.†

Volatile Salt. Ammonium carbonate, $(NH_4)_2CO_3$.†

Volborthite, Turkestan. See TANGEITE.

Volcanite. A mixture of selenium and sulphur found in the Lipari Islands.†

Volck. Oil insecticide. (Chevron).*

Volckmann's Solution. A solution containing thymol, alcohol, glycerin, and water.†

Vole. A cotton fabric grade of TUFNOL industrial laminates. (Tufnol).*

Volenite. A rubber substitute made from rosin, oil, and some fibrous material.†

Volidan and Volidan 21. Megestrol acetate with ethinyloestradiol tablets - oral contraceptives. (British Drug Houses).†

Volital. A proprietary preparation of pemoline. Central nervous stimulant. (LAB Ltd).*

Volkite. A moulded rubber product used for electrical insulation.†

Volomite. An abrasive consisting of tungsten carbide.†

Volpar. Spermicidal comtraceptive containing phenylercuric acetate. Gels, paste and foaming tablets. (British Drug Houses).†

Volpo. Polyethoxylated fatty alcohols. (Croda Chemicals Ltd).

Voltalef. PCTFE. Applications: Moulded parts. (Atochem Inc).*

Voltaren. Diclofenac sodium. Applications: Anti-inflammatory. (Ciba-Geigy Corp).

Voltarol. Anti-inflammatory/analgesic. (Ciba-Geigy PLC).

Voltarol Retard. Anti-inflammatory /analgesic. (Ciba-Geigy PLC).

Voltoids. A registered trade name for compressed tablets of ammonium chloride used in the preparation of voltaic cells.†

Volucon. Standard volumetric concentrates. (May & Baker Ltd).

Volusol. Standard volumetric solutions. (May & Baker Ltd).

Vomiting Salt. Zinc sulphate, $ZnSO_4$.†

Von Forster Powder. A gelatinized nitrocellulose flake powder, with a little calcium carbonate.†

Von Vetter's Solution. An aqueous solution of glycerin, sugar, and potassium nitrate. Used to preserve anatomical specimens.†

Vonges Dynamite. An explosive containing 75 per cent nitro-glycerin, 20.8 per cent randanite (decomposed felspar), 3.8 per cent quartz, and 0.4 per cent magnesium carbonate.†

Vontrol. Diphenidol hydrochloride. Applications: Anti- emetic. (Smith Kline & French Laboratories).

Voran. A proprietary polyether-based polyurethane elastomer cross-linked with diamine. (Dow Chemical Co Ltd).*

Voranate. Toluene diisocyanate-based products used in the manufacture of polyurethane products. (Dow Chemical Co Ltd).*

Voranol. Polyether polyols. Applications: Used with isocyanates in the formation of urethanes. (Dow Chemical Co Ltd).*

Vortel. A proprietary preparation of chlorprenaline, ethomoxane and methapyrilene. (Eli Lilly & Co).†

VPC Coatings. A liquid polyester polyol mixed with a crosslinking resin which cures in seconds after being exposed to a tertiary amine catalyst which is contained in a VPC curing chamber to form a tough polyurethane finish. Applications: utilized for high speed, continuous coating applications such as vinyl laminates, paper and wood. (Ashland Chemical Company).*

Vraic. (Varech). French names for Kelp (*qv*).†

VR Coving. Expanded polystyrene. Applications: Coving for domestic properties. (Vencel Resil Ltd).*

Vresamin. Etherified urea/formaldehyde resins. Applications: Stoving finishes and acid curing lacquers. (Resinous Chemicals Ltd).*

V-thane. Solid Polyurethane elastomers. Applications: Engineering based. (Hallam Polymer Engineering Ltd).*

Vuelite. A proprietary trade name for a cellulose acetate plastic. Vuelite-reinforced is a transparent cellulose acetate sheet reinforced with wire mesh.†

Vuepak. A proprietary trade name for an acetate wrapping material.†

Vulcabest. Asbestos faced cladding panels. (Vulcan Plastics Ltd).*

Vulcaboard. Vandal resistant panel. (Vulcan Plastics Ltd).*

Vulcabrite. Stainless steel faced cladding panels. (Vulcan Plastics Ltd).*

Vulcaflex. An anti-oxidant which offers protection against flex-cracking in rubber materials. A complex substituted secondary amine of the dimethoxy-diphenylamine type. (Vulnax International Ltd).*

Vulcaid. A proprietary litharge of small particle size.†

Vulcaid DPG. A proprietary trade name for a rubber vulcanization accelerator. It is diphenyl-guanidine.†

Vulcaid LP. A proprietary vulcanization accelerator. It is lead-penta- methylene dithio-carbamate.†

Vulcaid P. A rubber vulcanization accelerator. It is piperidinepenta-methylenedithiocarbamate. It melts at 172° C., and is soluble in water, alcohol and benzene.†

Vulcaid ZP. A proprietary rubber vulcanization accelerator. It is zinc pentamethylenedithiocarbamate.†

Vulcaid 111. A proprietary trade name for a rubber vulcanization accelerator. It is butyraldehyde-aniline.†

Vulcaid 222. A proprietary trade name for a rubber vulcanization accelerator. It is tetramethyl-thiuram monosulphide.†

Vulcaid 27. A proprietary trade name for a rubber vulcanization accelerator. It is zinc butyl xanthate.†

Vulcaid 28. A proprietary trade name for a rubber vulcanization accelerator. It is dibenzylamine.†

Vulcaid 33. A proprietary trade name for an antioxidant. A liquid amine condensation product.†

Vulcaid 44. A proprietary trade name for an antioxidant. It is a naphthol-amine reaction product.†

Vulcaid 444B. A proprietary trade name for a rubber vulcanization accelerator. It is heptaldehyde-aniline.†

Vulcaid 55. A proprietary trade name for an antioxidant. It is an acetaldehyde-aniline condensation product.†

Vulcalap. GRP shiplap weatherboard. (Vulcan Plastics Ltd).*

Vulcalon. GRP faced cladding panels. (Vulcan Plastics Ltd).*

Vulcalucent. GRP translucent sheets and domelights. (Vulcan Plastics Ltd).*

Vulcamel. A proprietary rubber vulcanization accelerator. It is butyraldehyde-ammonia.†

Vulcamin. Aluminium faced cladding panels. (Vulcan Plastics Ltd).*

Vulcan. Herbicide, containing clopyralid and bromoxynil. (ICI PLC).*

Vulcan Bronze. A proprietary trade name for a bearing bronze containing 1.0 per cent silicon with iron and nickel.†

Vulcan Powder. An explosive containing 30 per cent nitro-glycerin, 52.5 per cent sodium nitrate, 7 per cent sulphur, and 10.5 per cent charcoal.†

Vulcan Red. A pigment for rubber. It is stated to be a dye deposited upon a base.†

Vulcan Red MO. A pigment for rubber. It is said to be a highly dispersed iron oxide, of German origin.†

Vulcanex. A proprietary rubber vulcanization accelerator. It is a schiff's base.†

Vulcaniline. A proprietary trade name for a rubber vulcanization accelerator. It is paranitroso-dimethylaniline.†

Vulcanine. A patented material made from rubber, asbestos, litharge, lime, sulphur, and zinc oxide.†

Vulcanized Oils. See THINOLINE.

Vulcanized Paper. See VULCANIZED FIBRE.

Vulcanite. See EBONITE. It is also the name for a nitro-glycerin explosive.

Vulcanized Fibre. A material made by treating sheets of paper with zinc chloride solution and subjecting the gelatinized sheets to pressure. The paper is sometimes mixed with glycerin and vulcanized oils. Used for making valve discs, brake blocks, and tubes. Other names for this and similar products are Hard Fibre, Red Fibre, Grey Fibre, Vegetable Fibre, Whalebone Fibre, Egyptian Fibre, Fiberoid and Horn Fibre.†

Vulcanol. A proprietary rubber vulcanization accelerator.†

Vulcanox Crack and Joint Sealant. Polyurethane. Applications: Poured or extruded into cracks and joints as a moisture-proof sealant. Cures to a solid with rubber band elasticity. (Metalcrete Mfg Co).*

Vulcaperl. Special physical form of Vulcafor flowable and dust free. Applications: Rubber industry. (Vulnax International Ltd).*

Vulcaplas. A proprietary trade name for an organic polysulphide synthetic elastic material.†

Vulcaplast. GRP faced, aluminium framed cladding panels. (Vulcan Plastics Ltd).*

Vulcapont. A proprietary rubber vulcanization accelerator. It consists of equal parts of thionex and vulcanol.†

Vulcasbeston. A mixture of rubber and asbestos. A heat and electrical insulator.†

Vulcase. A proprietary preparation of colloidal sulphur.†

Vulcasteel. Steel faced cladding panels. (Vulcan Plastics Ltd).*

Vulcastop. Short stoppers used in water emulsion polymerization processes. Applications: Synthetic elastomers production. (Vulnax International Ltd).*

Vulcatex Colours. A proprietary trade name for rubber colours containing the vulcafor colours with rubber and stearic acid.†

Vulcatuf. Bandit resistant panel. (Vulcan Plastics Ltd).*

Vulcazol. Furfuramide, a vulcanization accelerator.†

Vulco-asbestos. A similar material to amianite, and used for electrical insulation.†

Vulcoferran. A trade mark for linings principally of rubber or ebonite for chemical apparatus.†

Vulcogene. A proprietary trade name for a rubber vulcanization accelerator. It is thiocarbanilide.†

Vulcogene ND. A proprietary trade name for a rubber vulcanization accelerator. It is diphenylguanidine.†

Vulcoid. A proprietary trade name for a phenolic resin impregnated vulcanized fibre.†

Vulconex. A proprietary trade name for a rubber vulcanization accelerator. It is ethylidineaniline.†

Vul-Cup. a,a'-Bis(t-butylperoxy)-diisopropylbenzene. Applications:

Cross-linking agent for rubber and plastics. (Hercules Inc).*

Vul-Cup 40KE and R. Peroxide catalysts and vulcanizing agents comprising di-(t-butylperoxyisopropyl)benzene. Applications: Used as vulcanizing and polymerization agents. (Hercules Inc).*

Vulkacit. Range of accelerators. (Bayer UK Ltd).

Vulkacit A. Aldehyde-ammonia, a vulcanization accelerator.†

Vulkacit BP. A rubber vulcanization accelerator. It is a paste, and consists of a mixture of bases.†

Vulkacit CA. A rubber vulcanization accelerator. It is thio-carbanilide. Also see VULCAFOR IV.†

Vulkacit CT. A proprietary trade name for a rubber vulcanization accelerator.†

Vulkacit D. Diphenyl-guanidine, a vulcanization accelerator.†

Vulkacit DM. A proprietary trade name for a rubber vulcanization accelerator. It is benzthiazyl disulphide.†

Vulkacit FP. A proprietary trade name for a rubber vulcanization accelerator. It is methylene-*p*-toluidine.†

Vulkacit H. Hexamethylenetetramine, a vulcanization accelerator.†

Vulkacit M. A proprietary rubber vulcanization accelerator. It is mercapto-benzothiazole.†

Vulkacit Mercapto. A proprietary trade name for a semi-ultra rubber vulcanization accelerator. It is mercaptobenzothiazole.†

Vulkacit NP. 5-Methylhexahydro-1,3,5-thriazine-2-thione.†

Vulkacit P. Piperidinepiperidyldithio-formate, a vulcanization accelerator.†

Vulkacit P Extra. The zinc salt of ethylphenyldithiocarbaminic acid, a rubber vulcanization accelerator.†

Vulkacit Thiuram. A rubber vulcanization accelerator. It is tetra- methylthiuram-disulphide.†

Vulkacit TR. A rubber vulcanization accelerator. It is a mixture of the free bases of polyamines of ethylene.†

Vulkacit 1000. A rubber vulcanization accelerator. It is *o*- tolylbiguanide.†

Vulkacit 470. A rubber vulcanization accelerator. It is a condensation produ[ct] of homologous acroleins with aromatic bases.†

Vulkacit 576. A condensation product of homologous acroleins aromatic bases. [A] rubber vulcanization accelerator most suitable for regenerated rubber.†

Vulkacit 774. A proprietary trade name f[or] an ultra-accelerator for rubber. It is th[e] dithiocarbamate of cyclohexyl ethylamine.†

Vulkadur. Reinforcing resin. (Bayer UK Ltd).

Vulkalent. Retarders. (Bayer UK Ltd).

Vulkanol. Synthetic plasticizer. (Bayer UK Ltd).

Vulkanox. Antioxidants. (Bayer UK Ltd).

Vulkasil. Silica fillers. (Bayer UK Ltd).

Vulklor. Tetrachloro-p-benzoquinone. Combines with R6 to form an effective bonding system for compounds featuring steel cord reinforcement. Used to activate GMF. It also function[s] as a vulcanizing agent without sulphur. Used in natural, SBR, nitrile, butyl and chlorobutyl rubbers. (Uniroyal).*

Vulkollan. Polyols for the Vulkollan system (casting type non-cellular polyurethane elastomers). Applications: Used in the manufacturing of technical goods. (Bayer & Co).*

Vulnopol KM. Solution of potassium dimethyldithiocarbamate. Applications: polymerization short stops in the copolymerization of styrene and butadiene. (Alco Chemical Corporation).*

Vulpinite. A variety of anhydrite mixed with silica.†

Vultex. A trade mark for vulcanized rubbe[r] latex preserved with ammonia.†

Vulvan. Testosterone Propionate. Applications: Androgen. (American Critical Care, Div of American Hospital Supply Corp).

Vydate. Insecticide/nematicide. (Du Pont (UK) Ltd).

Vydax. Fluorocarbon telomers. (Du Pont (UK) Ltd).

Vydon. A proprietary vinyl resin.†

ydyne. Nylon and nylon copolymer resins. Applications: Family of 66, 66/6 copolymers, 69, designed for moulding and extrusion applications. (Monsanto Co).*

yflex NT80S. A proprietary PVC powder coating material. Containing no toxic metals and supporting no micro-biological growth, it can be safely brought into contact with drinking water. (Plastic Coatings Ltd).*

ygen. A trade mark for PVC resins.†

Vykamol 83G. Sorbitan ester/polysorbate blend. (Croda Chemicals Ltd).

Vyloc. Oven wall catalysts. (Du Pont (UK) Ltd).

Vynamon. Pigments for polyvinyl chloride. (ICI PLC).*

Vynathene. High vinyl acetate-ethylene copolymer resins. Applications: Adhesive base polymers, rubber gum stock, lacquer coatings, impact modifiers and processing aids. (USI Chemicals).*

W

W2 Beta. A proprietary name for a special Angelo shellac. (Zinsser, NV).†

Wackenroder's Solution. A solution obtained by passing sulphuretted hydrogen through an aqueous solution of sulphurous acid. The solution contains $H_2S_4O_6$, $H_2S_3O_6$, $H_2S_5O_6$, and colloidal sulphur.†

Wackerschellak. An artificial shellac obtained by the condensation and polymerization of acetaldehyde.†

Wad. (Bog Manganese). An earthy variety of hydrated manganese oxide, MnO_2.†

Wagner's Reagent. A solution of 2 grams potassium iodide in 100 cc water.†

Wahnerit. A German synthetic varnish paper product used for electrical insulation.†

Wahoo Bark. The root bark of *Euonymus atropurpureus*.†

Wakefield Grease. See YORKSHIRE GREASE.

Walchowite. See RETINITE.

Walkerite. A clay of the Fuller's earth type.†

Walker's Earth. Fuller's earth.†

Wall Saltpetre. See NITROCALCITE.

Walsrode. Powder. A proprietary smokeless powder containing 98.6 per cent nitro-cotton and 1.4 per cent volatile matter.†

Wando Steel. A proprietary non-shrinking steel containing 1.05 per cent manganese, 0.5 per cent chromium, 0.5 per cent tungsten, and 0.95 per cent carbon.†

W.A. Powder. An American smokeless powder. It is a guncotton-nitro-glycerin powder, with barium and potassium nitrates.†

Waras. (Wars, Warrus). A resinous powd[er] which covers the seed pods of *Flemingi[a] congesta*, of India. It is used in Arabia as a dye.†

Warburg's Tincture. A quinine preparatio[n] containing aloes, rhubarb, gentian, camphor, and oils. Used in India for th[e] treatment of malaria.†

Warcodet D. Alkylphenol ethoxylate nonionic surfactant as a colourless viscous liquid. Applications: Detergent and emulsifier with wetting, penetrating and soil suspending properties used for textiles. (Warwick Chemical Ltd).

Warcodet K54. Sodium alkyl-aryl sulphonate. Anionic surfactant in the form of a clear golden brown liquid. Applications: Detergent with wetting, penetrating and soil suspension properties. Used in general purpose detergents and in the textile industry eg for scouring. (Warwick Chemical Ltd).

Warcodet V. Phosphate ester in the form of sodium/potassium salts. Applications: Anionic surfactant for wetting, scouring and washing off. (Warwick Chemical Ltd).

Warcodye CLP. Low foam cationic surfactant. Application: Levelling wool and polyamide with acid reactive dyes. (Warwick Chemical Ltd).

Warcodye RWL. Ethoxylated amine based amphoteric. Applications: Leveller used in reactive dyes on wool. (Warwick Chemical Ltd).

Warcosoft WSC. Cationic surfactant in liquid form. Application: Permanent softener for use with acrylics. (Warwick Chemical Ltd).

Warcowet O. Sodium di-octyl sulphosuccinate in liquid form. Applications: Wetting agent with good penetration, stable to boiling, most effective at 30-60°C. Used in textiles eg in scouring, bleaching, package dyeing, piece dyeing and printing. (Warwick Chemical Ltd).

Wardol. Non-ionic emulsifiers. (Courtaulds Chemicals & Plastics Leek Chemicals Group).

Ward's Stone. A concrete composed of limestone and Portland cement.†

Warefog. Chlorpropham as a foggable solution. Applications: Controlling sprouting in ware potatoes. (Wheatley Chemical Co Ltd).*

Warm Sepia. Sepia, warmed by mixing it with a redder brown. A pigment.†

Warmaline. Expanded polystyrene veneer 2mm thick. Applications: Lining for domestic walls for use beneath wallpapers. (Vencel Resil Ltd).*

Warne's Metal. An alloy of 26 per cent nickel, 37 per cent tin, 26 per cent bismuth, and 11 per cent cobalt.†

Warrengas. LPG. (Chevron).*

Warrior. Residual herbicide. (ICI PLC, Plant Protection Div).

Warrus. See WARAS.

Wars. See WARAS.

Wasa Tar. A wood tar of a similar type to Stockholm tar, and sometimes sold under this name.†

Wasc. Nonionic surfactant consisting of nonylphenol ethoxylate in liquid form. Application: Liquid cleaners for hard water. (Berol Kemi (UK) Ltd).

Wash. The fermented wort of the distilleries.†

Washer Brass. See BRASS.

Washington Bleach. See CHLORAMINE T.

Wash-saver. Laundry bleach. (Interox Chemicals Ltd).

Wasp Destroyer. Carbaryl insecticide. (Murphy Chemical Ltd).

Waspend. Pestkiller for flying or crawling pests. (ICI PLC).*

Watchmaker's Alloy. An alloy of 59 per cent copper, 40 per cent zinc, and 1.2 per cent lead.†

Water Blue. See NICHOLSON'S BLUE and SOLUBLE BLUE.

Water Blue B, R. Dyestuffs. They are British equivalents of Soluble blue.†

Water Blue 6B Extra. See SOLUBLE BLUE.

Water Brite. Sodium Hexametaphosphate. Applications: Water treatment. (Flexibulk Ltd).*

Water-dag. See AQUADAG.

Water-eosines. Dyestuffs. They are the alkali salts of eosin and are water soluble. Used in the paper-staining industry.†

Water Gas. The general name for a mixture of gases obtained by the decomposition of steam by incandescent carbon. It usually contains from 43-44 per cent carbon monoxide, 48-49 per cent of hydrogen, 3-4 per cent of carbon dioxide, and 3-4 per cent of nitrogen. Formerly used for heating and lighting.†

Water-glass. See SOLUBLE GLASS.

Water-glass, Sodium. See SOLUBLE GLASS.

Water Mica. Clear transparent Muscovite (potash mica).†

Water Nigrosine Crystals W. A British equivalent of Water-soluble nigrosine.†

Water of Ammonia. A solution of ammonia, NH_4OH.†

Water of Saturn. A dilute solution of lead subacetate.†

Water Paints. Paints which contain saponified oil and colouring matter.†

Waterproof Paints. See MOUNTFORD'S PAINT.

Water Soluble Eosin. See EOSIN.

Water Varnishes. Varnishes made by dissolving gums or glue in water.†

Watsonite. A proprietary material used as a mica substitute. It is made from scrap mica with a binding agent.†

Wattle Bark. (Mimosa). A tanning material obtained from species of *Acacia*. The amount of tannin varies from 12-49 per cent.†

Watt's and Li's Solution. A solution used for the electro-deposition of iron. It contains 150 grams ferrous sulphate, 75 grams ferrous chloride, 120 grams ammonium sulphate, and 1,000 cc water.†

Wax Butter. (Wax Oil). A thick oil obtained by the distillation of beeswax. It consists mainly of cerotene $C_{27}H_{54}$: melissine $C_{30}H_{60}$, and palmitic acid, formerly used medicinally externally and internally.†

Wax C. A high melting point amide wax. Recommended as an internal lubricant in processing ABC, PVC and polystyrene. It is N,N'-distearyl ethylene diamine. (Hoechst Chemicals (UK) Ltd).†

Wax, Earth. See OZOKERITE.

Waxemul. Wax emulsion. (Tenneco Malros Ltd).

Waxene. A proprietary rubber vulcanization accelerator.†

Wax, Fossil. See OZOKERITE.

Waxigel. Modified maize starch (pregelatinized). (Roquette (UK) Ltd).*

Wax, Getah. See JAVA WAX.

Waxilys. Maize starch. Applications: Foods, textiles, cardboard and paper. (Roquette (UK) Ltd).*

Wax, Insect. See CHINESE WAX.

Waxit. A wetting agent used by gold- and silversmiths and in the jewellery industry. (Degussa).*

Wax, Japanese. See CHINESE WAX.

Wax, Mineral. See OZOKERITE, MINERAL WAX, and LIGNITE WAX.

Wax, Montana. See IRISH PEAT WAX.

Wax, Montanin. See IRISH PEAT WAX.

Wax, Oil. See WAX BUTTER.

Waxolan P-5. Propoxylated (5 mol) lanolin wax. Applications: Enhances lubricity and rigidity of lipsticks. (Amerchol Corporation).*

Waxolene. A trade mark for oil and wax soluble dyes. (ICI PLC).*

Waxsol. A proprietary preparation of dioctyl sodium sulphosuccinate. Ear drops for wax removal. (Norgine).†

Wax Spirit. A colourless watery distillate from beeswax. It contains acetic and propionic acids.†

Wax Tailings. The remaining petroleum distillate after the paraffin wax has bee removed.†

Wax, Tree. See CHINESE WAX.

WB 200SL. A single layer microcracked chromium process. (Hanshaw Chemicals).†

Wear-Dated. Textile products. (Monsanto Co).*

Webas. Bitument emulsion for road surfac treatment. (Vedag GmbH).*

Webert Alloy. A proprietary copper silicon alloy containing small amounts of manganese.†

Webnerite. A mineral. It is a variety of andorite.†

Wedel's Oil. Consists of 1 part bergamot, 4 parts camphor, and 32 parts oil of almonds.†

Wedl's Stain. A microscopic stain containing 1 gram orseille, 20 cc absolute alcohol, 5 cc 60 per cent acetic acid, and 40 cc water.†

Weedazin. Total herbicide. (FBC Ltd).

Weedazol-TL. A liquid containing 225g of aminotriazole and ammonium thiocyanate per litre. Applications: To control common couch, docks, thistles and other weeds in fallows, autumn stubbles and headlands before direct drilling of winter wheat, volunteer potatoes in barley stubble, as a pre-planting spring treatment and in established apple and pear orchards. (Bayer).*

Weed Be Gone 45. Emulsifiable concentrate of ethyl ester of 2,4-D acid. Applications: Broadleaf herbicide for corn, sugar cane and wheat crops and pasture land. (Invequimica & CIA SCA).*

Weed Be Gone 50. Amine salt of 2,4-D acid. Applications: Broadleaf herbicide for sugar cane, corn, wheat, barley crops. (Invequimica & CIA SCA).*

Weed-B-Gon. Weed killer. (Chevron).*

Weed Ender. Cacodylic acid. Applications: A post emergence contact weed killer for use on non planted areas and the elimination of unwanted vegetation around plants, trees, patios, fences, driveways, etc. (Lawn & Garden Products Inc).*

Weedex. Herbicide. (Murphy Chemical Ltd).

Weed Gun. Ready for use herbicide spray. (ICI PLC).*

Weed Hoe. MSMA. Applications: Post emergence on turf to control established crabgrass. (Lawn & Garden Products Inc).*

Weedol. Weedkiller containing paraquat and diquat for gardeners. (ICI PLC).*

Weichhaltungsmittel PA. *o*-Phthalic ester of ethyl-glycol. A softening agent for cellulose esters.†

Weichhaltungsmittel PM. The *o*-phthalic ester of methyl-glycol. A softening agent for cellulose esters.†

Weichharz 398A. A proprietary trade name for a non-drying alkyd made from adipic acid and trimethylene glycol. (Chemische Werke, Albert).†

Weichmacher T. Dimethylthioanthrene. See SINTOL.†

Weichmacher 238S. A proprietary trade name for an ester formed from adipic acid and C_4-C_9 synthetic fatty acids with pentaerythritol. (Chemische Werke, Albert).†

Weichmacher 333A. As for 238 above except that it has a lower fatty acid content. (Chemische Werke, Albert).†

Weichmacher 90. A polymer from acetylene, glycerine and ethylene oxide.†

Weigert's Stain. A microscopic stain. It is made by dissolving fuchsine and resorcinol in ferric chloride solution.†

Weighted Silk. Silk impregnated with various inorganic and organic substances in order to increase the weight. Tannin and metallic salts are often used.†

Weisalloy. A proprietary sheet aluminium alloy.†

Weiss-Kupfer. See NICKEL SILVERS.

Weld. (Luteolin). A yellow dyestuff obtained from the dried stalks and leaves of the herbaceous plant known as *Reseda luteola*. Luteolin is the colouring principle. It dyes wool and silk. Weld green is produced on silk by adding indigo carmine and sulphuric acid to the weld dye-bath.†

Weld Extract. An extract of Weld (*qv*). The dyeing principle is luteolin (tetrahydroxyflavone), $C_{15}H_{10}O_6$. It is employed to a small extent for dyeing silk and wool mordanted with tin or alumina.†

Weld Green. See WELD.

Weldox. Adhesives compositions. (Hardman Inc).*

Welgum. Alginates for ceramics, electrodes and water treatment. (Alginate Industries Ltd).†

Welladyne. Iodophor detergent germicide for cleaning. (Ciba-Geigy PLC).

Wellbutrin. Bupropion hydrochloride. Applications: Antidepressant. (Burroughs Wellcome Co).

Wellcome Brand Athropine Sulphate Injection. A proprietary formulation of atropine sulphate. Applications: For preanaesthetic medications to inhibit secretory glands and to suppress both vagal activity associated with the use of halogenated hydrocarbons during inhalation anaesthesia and reflex excitation arising from mechanical stimulation during surgery. (The Wellcome Foundation Ltd).*

Wellcome Brand Scopolamine Hydrobromide Injection. A proprietary formulation of scopolamine hydrobromide. Applications: For preanaesthetic sedation and in conjunction with analgesic agents for obstetric amnesia or for calming delirium. (The Wellcome Foundation Ltd).*

Wellcovorin. Tablets and injection. Proprietary formulations of leucovorin calcium. Applications: For the prophylaxis and treatment of undesired hematopoietic effects of folic acid antagonists. (The Wellcome Foundation Ltd).*

Welldorm. A proprietary preparation of dichloralphenazone. A hypnotic. (Smith and Nephew).†

Wellferon. Interferon. Protein formed by the interaction of animal cells with viruses capable of conferring on animal cells resistance to virus infection. Applications: Antineoplastic; antiviral. (Burroughs Wellcome Co).

Welmet. A proprietary chromium nickel-molybdenum steel.†

Welvic. A trade-mark for a range of plasticized and unplasticized polyvinyl chlorides used in the manufacture of cables, flooring, pipes, etc. (ICI PLC).*

Wepco. Caulking, waterproofing, cementitious materials. (Weatherguard/Marbleloid Products Inc).*

Werderol. A formic acid preservative used for fruit preparations.†

WesBio. A liquid, biodegradable microbiocide containing sodium salts of dimethyl and ethylene dithiocarbamates. Applications: Used to prevent fermentation in drilling fluid and corrosion in producing wells. (Westbridge Research Group).*

Wescodyne. Iodophor disinfectants. (Ciba-Geigy PLC).

WesLoTemp. A liquid, sodium polyacrylate polymer-based drilling fluid additive used as a dispersant/deflocculant. Applications: Used in clay-based fresh water drilling fluid systems subject to low temperatures. (Westbridge Research Group).*

Wessalith. Sodium aluminium silicate. Applications: Phosphate replacement in laundry detergents. (Degussa).*

Wessalon. A range of spray-dried silicas. Applications: Used as carrier substances and grinding aids for pesticides. (Degussa).*

WesScaleStop. A liquid, acrylic homopolymer scale inhibitor. Applications: Used in fresh water-based drilling fluid systems. (Westbridge Research Group).*

Wessell's Silver. An alloy of 51-65 per cent copper, 19-32 per cent nickel, 12-17 per cent zinc, and 2 per cent silver. It is a nickel silver.†

WesSperse. A dry, blended sodium polyacrylate/chrome free lignosulphonate drilling fluid additive used as a dispersant/deflocculant. Applications: Used in clay-based drilling fluid systems subject to calcium, magnesium and chloride contamination and high temperatures. (Westbridge Research Group).*

West African Copaiba. Illurin balsam, an oleo-resin, is known by this name. It is used as a substitute for balsam of copaiba.†

West African Gum. A gum arabic resembling Senegal gum, obtained from *Acacia nilotica*.†

West Indian Anime Resin. See TACAMAHAC RESIN.

West System Brand Products. Liquid epoxy resin, hardeners and accessories. Applications: Wood encapsulation, bonding, filling and fairing. (Wessex Resins & Adhesives Ltd).*

WesTemp. A liquid, sodium polyacrylate polymer-based drilling fluid additive used as a dispersant/deflocculant. Applications: Used in clay-based fresh water drilling fluid systems subject to high temperatures. (Westbridge Research Group).*

WesTemp K+. A liquid, potassium polyacrylate polymer-based drilling fluid additive used as a dispersant/deflocculant. Applications: Used in clay-based fresh water drilling fluid systems subject to high temperatures. (Westbridge Research Group).*

Westfalite No. 3. An explosive consisting of 58-61 per cent ammonium nitrate, 13-15 per cent potassium nitrate, 4-6 per cent trinitro-toluene, and 20-22 per cent ammonium chloride.†

Westhin. A liquid, sodium polyacrylate polymer-based drilling fluid additive used as a dispersant/deflocculant. Applications: Used in clay-based fresh water drilling fluid systems. (Westbridge Research Group).*

WesThin K+. A liquid, potassium polyacrylate polymer-based drilling fluid additive used as a dispersant/deflocculant. Applications: Used in clay-based fresh water drilling fluid systems. (Westbridge Research Group).*

Westo-Flocs. Polymer flocculants. Applications: Wastewater. (Western Chemical Co).*

Weston. Processing aids for thermoplastics. (Borg Warner Chemicals).

Weston Additives. Various chemicals. Applications: Injection moulding, sheet extrusions. (Borg Warner Chemicals).*

Westoran. A registered trade name for a cleaning agent for cotton. It contains emulsified hydrocarbons, and is also used as an insecticide.†

Westphalian Essence. See ESSENCE OF SMOKE.

Westphalite I. A safety explosive for mines, consisting of 95 per cent ammonium nitrate and 5 per cent resin.†

Westphalite II. (Westphalite, Improved). A safety explosive for mines, containing 92 per cent ammonium nitrate, 3 per cent potassium nitrate, and 5 per cent resin.†

Westphalite, Improved. See WESTPHALITE II.

Westrol. A registered trade name for a cleaning liquid for cotton. It contains oils with a solvent. Soaps containing trichlorethylene. A degreasing agent.†

Westropol. A registered trade name for a cleaning and degreasing agent.†

Westrosol. A registered trade name for a preparation of trichlorethylene, $CHCl : CCl_2$.†

WesVis. A liquid, ammonium polyacrylate invert emulsion drilling fluid additive used as a viscosifier, bentonite extender, selective flocculant and hole sweep. Applications: Used in water-based drilling fluid systems. (Westbridge Research Group).*

Wet 6. Dental preparation. Applications: Wetting agent for waxes. (Bayer & Co).*

Weta Material. A porcelain substitute consisting of fine, uniformly distributed carborundum particles with silicates and metals of the iron series, cobalt and nickel, and sinters after firing at 1400° C.†

Wetanol. A proprietary trade name for a wetting agent for textiles, etc. It is a modified sulphated fatty acid ester.†

Wetfix. Range of cationic surfactants comprised of amino groups and selected aliphatic hydrocarbon chains. Liquid form, heat stable. Application: Promotes and retains adhesion between asphalt and aggregate. Incorporated into asphalt which will eventually be used for surface dressing, asphaltic macadams and asphaltic concrete. (Thomas Swan & Co Ltd).

Wetherillite. Synonym for Hetaerolite.†

Wetter-Dynamite. A safety explosive for mines, consisting of 53 per cent nitroglycerin, 14 per cent kiesel-guhr, and 33 per cent magnesium sulphate.†

Wetter-Dynammon. An Austrian explosive containing 94 per cent ammonium nitrate, 2 per cent potassium nitrate, and 4 per cent charcoal.†

Wetteren Powder. A guncotton powder, containing a little calcium carbonate, gelatinized with amyl acetate.†

Wetter-Fulminite. An explosive containing ammonium nitrate.†

Wetter Nobelit B. A gelatinous permitted explosive (group P1). (Dynamit Nobel Wien GmbH).*

Wetz. Miscellaneous. Applications: Surfactant with antifoam. (Drexel Chemical Company).*

Weyl and Zeitler's Solution. A solution used to absorb oxygen. It consists of pyrogallol in sodium hydroxide solution.†

Whale. A cotton fabric grade of TUFNOL industrial laminates. (Tufnol).*

Whalebone Fibre. See VULCANIZED FIBRE.

Wheat Germ Oil. A vitamin E concentrate.†

Wheat Oil. See OIL OF WHEAT.

Wheel Brass. See BRASS.

Wheeler's Solution. A mixture of pyrogallol in potassium hydroxide for the absorption of oxygen.†

Whetstone. (Oilstone, Honestone). Hard rocks, usually siliceous in character, used for sharpening tools. Suitable rocks include hornstone, sandstone, slate, lydian stone, schist, etc.†

Whipcide. PHTHALOVYNE. It is mono (1 - ethyl - 1 - methyl - 2 - propynyl) phthalate. (Pitman-Moore, Englewood NJ).†

White Acid. A mixture of hydrofluoric acid and ammonium fluoride. Used for etching glass†

White Alkali. Refined sodium carbonate from the Le Blanc soda process.†

White Alloy. An alloy of 10 per cent cast iron, 10 per cent copper, and 80 per cent zinc. This name is also applied to alloys containing 49-53 per cent copper, 23-24 per cent zinc, 22-24 per cent nickel, and 2 per cent iron. They are nickel silvers.†

White Argol. See ARGOL.

White Bole. See CHINA CLAY.

White Brass. Variable alloys containing from 2-45 per cent copper, 33-80 per cent zinc, 0-81 per cent tin, 0-13 per cent lead, and 0-11 per cent aluminium.†

White Button Alloy. A nickel-silver containing from 49-53 per cent copper, 23-24.5 per cent zinc, 22-24 per cent nickel, and 2-2.5 per cent iron.†

White Cast Iron. A good variety of cast iron. It usually contains 97 per cent iron and 3 per cent carbon, mainly in the uncombined state.†

White Caustic. Colourless sodium hydroxide.†

White Cerate. Spermaceti ointment.†

White Clay. See CHINA CLAY.

White Copper. A nickel silver usually containing 70 per cent copper, 18 per cent zinc, and 12 per cent nickel. See NICKEL SILVERS.†

White Copperas. (Cinquinolite. A mineral. It is ferric sulphate, $Fe_2(SO_4)_3.9H_2O$. The name White copperas is also used for zinc sulphate.†

White Cosmetic. A trade name for basic nitrate or mixture of basic nitrates obtained by adding water to bismuth nitrate.†

White Dammar. Manila copal resin, obtained from *Vateria indica*, is known by this name.†

White Drying Oil. Linseed oil which has been bleached.†

White Fish-bone. See OS SEPIAE.

White Gold. Water soluble polymer for oilfield use. (BP Chemicals Ltd).*

White Gold. Various alloys are known by this term. A jeweller's alloy of gold whitened by means of silver, is called white gold. An alloy of 90 per cent gold, and 10 per cent palladium, and a platinum substitute, containing 59 per cent nickel, and 41 per cent gold, are both known under this name. Other alloys consisting of from 70-85 per cent gold, 8-10 per cent nickel, and 2-9 per cent zinc, are also sold under this term.†

White Gunpowder. A mixture of 2 parts potassium chlorate and 1 part each potassium ferrocyanide and sugar. An ingredient of explosives.†

White House Cement Paint. Coloured cement. Applications: Decorative finish for stonework, masonry, etc. (Calder Colours (Ashby) Ltd).*

White Indigo. See INDIGO WHITE.

White Insect Wax. (White Lac.). Arjun wax of India produced by the insect *Ceroplastes ceriferus*.†

White Iron Pyrites. See MARCASITE.

White Lac. Shellac which has been bleached. Also See WHITE INSECT WAX.†

White Lead. (Ceruse). A pigment. It is a basic carbonate of lead, the composition of which is variable. Also see KREMSER WHITE and FLAKE WHITE.†

White Lead Colours. Mixtures of lead chloride and sulphate. Pigments.†

White Lead, Freeman's. See FREEMAN'S NON-POISONOUS WHITE LEAD.

White Metal. An alloy of 54 per cent copper, 24 per cent nickel, and 22 per cent zinc. It is a nickel silver. (See NICKEL SILVERS.) It is also the name applied to bearing metals. (See ANTI-FRICTION METALS and BABBITT'S METALS.) The term is also used for Matte Copper, consisting mainly of copper sulphide.†

White Metal A. A British standard alloy. It contains 82.58 per cent lead, 12.05 per cent antimony, 4.64 per cent tin, 0.34 per cent copper, 0.08 per cent zinc,

0.07 per cent iron, 0.06 per cent arsenic, and 0.03 per cent bismuth.†

White Ochre. Ordinary clay is known by this name.†

White Oils. (Egg Oils). A liniment usually containing turpentine, acetic acid, and eggs. Sometimes ammonia and camphor are added.†

White Paste. Copper sulphocyanide, $Cu_2(CNS)_2$.†

White Pine Resin. See PINE GUM.

White Poppyseed Oil. The oil obtained from poppy seeds pressed cold.†

White Portland Cement. A Portland cement in which iron compounds are absent.†

White Precipitate. (Ammoniated Mercury, Lemery's White Precipitate). Mercury ammonium chloride, NH_2HgCl. Used for the preparation of cinnabar, and in medicine.†

White Precipitate, Fusible. See MERCURAMMONIUM CHLORIDE.

White Precipitate, Lemery's. See WHITE PRECIPITATE.

White Pyrites. (Efflorescent Pyrites). A variety of iron pyrites, FeS.†

White Ramie. See GLASS GRASS.

White Rouge. See PLATE POWDER.

White Sennaar Gum. See PICKED TURKEY GUM.

White Solder. An alloy of 10 per cent nickel, 45 per cent copper, and 45 per cent zinc. A soldering alloy. Also see BUTTON SOLDER.†

White Spirit. A turpentine substitute. It is usually a petroleum product, having flash-point and degree of evaporation similar to turpentine.†

White Swan. Lanolin BP. (Croda Chemicals Ltd).

White Tan. (Tari, Teri). *Coesalpinia digyna*, containing 30-50 per cent tannin.†

White Tar. Naphthalene.†

White Tellurium. (Krennerite, Bunsenine). A mineral. It contains from 25-29 per cent gold, 2.7-14.6 per cent silver, and 2.5-19.5 per cent lead, as tellurides.†

White Tung Oil. Tung oil obtained by the cold pressing method.†

White Ultramarine. Obtained by heating aluminium silicate, sodium carbonate, sulphur, and carbon, in the absence of air.†

White Vitriol. Zinc sulphate, $ZnSO_4.7H_2O$.†

Whitewash. A mixture of lime and water.†

White Wash. A dilute solution of lead subacetate. It is also called Goulard's lotion and Goulard's water.†

White-water. The technical name given to the waste from pulp-paper mills. It contains fine particles of cellulose.†

White Wax. White beeswax.†

Whitewood Bark. See CANELLA.

Whitex. A filler for plastics. Calcined clay.†

White Zinc. Zinc carbonate. $ZnCO_3$.†

Whiting. (Chalk). Calcium carbonate.†

Whitworth's Steel. Steel which has been subjected to high pressures to eliminate blow-holes.†

Whole Latex Rubber. Rubber which contains all the solid constituents of latex except any which may be volatile with water vapour. It is obtained by evaporating the water from latex.†

Wiborg Phosphate. A German fertilizer made by heating mineral phosphate with soda. It consists mainly of a tetraphosphate.†

Wichmann's Substitute. A material made from casein and albumen.†

Wickenol 101. Isopropyl myristate. Applications: Pharmaceutic aid. (Wickhen Products Inc).

Wickenol 111. Isopropyl palmitate. Applications: Pharmaceutic aid. (Wickhen Products Inc).

Wickenol 303. Aluminium chlorohydrate. Applications: Anhydrotic. (Wickhen Products Inc).

Wickenol 308. Aluminium sesquichlorohydrate. Applications: Anhydrotic. (Wickhen Products Inc).

Wickenol 363 D. Aluminium chlorohydrex: co-ordination complex of basic aluminium chloride and propylene glycol or polyethylene glycol. Applications: Astringent. (Wickhen Products Inc).

Wiegold Alloy. A dental alloy. It is a brass containing aluminium, and resembles gold in appearance. It is said to consist of 67.73 per cent copper, 32 per cent zinc, and 0.27 per cent aluminium, but some analysts state that it contains 0.25-0.5 per cent lead.†

Wij's Solution. Iodine trichloride (9.4 grams) and iodine (7.2 grams) are dissolved separately in glacial acetic acid, and the solutions added together. Used for the determination of the iodine value of fats and oils.†

Wilcoloy. A proprietary tungsten carbide material.†

Wild Ginger Oil. The oil of *Asarum canadense*.†

Wiles. Fertilizers. (BritAg Ltd).

Wilhelmit. A German explosive containing potassium chlorate and mineral oil.†

Wilkinite. (Jelly Rock). A colloidal clay. It is suggested as a substitute for china clay as a paper filter.†

Willesden Fabrics. Vegetable textiles are passed through a solution of cuprous hydrate in concentrated ammonia (Cuprammonium or Willesden solution). They become coated with a film of gelatinized cellulose containing copper oxide, as the solution dissolves cellulose. It renders paper or other vegetable textiles waterproof and antiseptic.†

Williamson's Blue or Violet. A pigment, $KFe[Fe(CN)_6] + H_2O$, produced from Everitt's salt by treatment with dilute nitric acid, and warming.†

Willow Blue. (Mazarine Blue, Ultramarine Blue, Celeste, Sky Blue) Preparations of cobalt blue used for colouring pottery.†

Wills Metallic 'O' Rings. Metallic sealing systems. Applications: Extreme pressure and temperature sealing for fluid and vacuum service. (Fothergill Tygaflor Ltd).*

Wilmil. Alloys similar in composition to silumin.†

Wilmot's Aluminium Solder. An alloy of 86 per cent tin and 14 per cent bismuth.†

Wilouite. Synonym for Grossular.†

Wilpo. Phentermine Hydrochloride. Applications: Appetite suppressant. (Dorsey Laboratories, Div of Sandoz Inc).

Wilson's Ointment. Ung. Zinci.†

Win-Kinase. Urokinase. Applications: Plasminogen activator. (Sterling Drug Inc).

Wingstay S. A proprietary anti-oxidant. It is styrenated phenol. (Goodyear Tyre and Rubber).†

Wingstay T. A proprietary anti-oxidant. It is a blend of substituted phenols. (Goodyear Tyre and Rubber).†

Wingstay 100. A proprietary anti-oxidant. It is an alkyl aryl amine. (Goodyear Tyre and Rubber).†

Winnofil. A trade name for precipitated calcium carbonate surface treated with calcium stearate. (ICI PLC).*

Winstrol. Stanozolol. Applications: Androgen. (Sterling Drug Inc).

Winter Oils. Lubricating oils which remain liquid at low temperatures.†

Wintergreen Oil. See OIL OF WINTERGREEN.

Winter's Bark. Pepper bark.†

Wintomylon. Nalidixic acid. Applications: Antibacterial. (Sterling Drug Inc).

Wipla Metal. V2A steel (see ANKA STEEL). Used for dental purposes.†

Wire Brass. See BRASS.

Wischnewite. Synonym for Vishnevite.†

Wisdom, Salt of. See SALT OF ALEMBROTH.

Witafrol. Antifoam agents. Applications: Used in mining industry, food industry, water engineering, paper industry. (Dynamit Nobel Wien GmbH).*

Witamol. Plasticisers. Applications: Used in PVC processing, imitation leather, heat resistant cables and sheets, special products with high resistance to cold and weather, application to lacquer. (Dynamit Nobel Wien GmbH).*

Witcamide. Surfactant for cosmetics, toiletries, pharmaceutical, processing, agricultural and other industries. (Baxenden Chemical Co Ltd).*

Witcamine AL42-12. Cationic surfactant in the form of tall oil imidazoline.

Application: Antistatic and corrosion inhibitor for use in carwash and wax formulations. (Witco Chemical Ltd).

Witcamine E-607. N(Lauroyl colamino formyl methyl) pyridinium chloride. Application: Cationic surfactant used in deodorants, after shave and hair rinses. (Witco Chemical Ltd).

Witcamine E-607S. N(Stearoyl colamine formyl methyl) pyridinium chloride. Application: Cationic surfactant used in hair conditioners and as a non-irritant emollient. (Witco Chemical Ltd).

Witcamine 209 and 211. Cationic surfactants in the form of complex imidazolines. Application: Corrosion inhibitors; intermediates. (Witco Chemical Ltd).

Witcamine 210. Cationic surfactant in the form of an alkyl amidoamine. Application: Corrosion inhibitor; intermediate. (Witco Chemical Ltd).

Witcizer. Proprietary trade names :
100. Butyl oleate.
312. Dioctyl phthalate.
313. Diisooctyl phthalate.
412. Dioctyl adipate.†

Witco TX. Modified toluene sulphonic acid in liquid form. Applications: Anionic surfactant used as a catalyst. (Witco Chemical Ltd).

Witcobond. Solvent based polyurethane adhesives for laminating fabrics and foam. Solvent bared polyurethanes for laminating films and boards. Water based polyurethane coatings for textiles, leather, glass fibre sizing, paints, lacquers and others. (Baxenden Chemical Co Ltd).*

Witcodet 100 and P280. Alkylaryl sulphonate in liquid form. Applications: Anionic surfactant used for liquid detergents eg windscreen washer. (Witco Chemical Ltd).

Witcoflex. Solvent based polyurethane textile coatings. (Baxenden Chemical Co Ltd).*

Witcolate AE-3S. Anionic surfactant in liquid form. Cation: sodium. Anion: synthetic alcohol 3EO sulphate. Application: Household and industrial liquid detergents. (Witco Chemical Ltd).

Witcolate D5-10. Sodium 2-ethyl-hexyl sulphate in liquid form. Applications: Wetting agent for emulsion polymerization. (Witco Chemical Ltd).

Witcolate SE-5. Sodium alcohol ether sulphate in liquid form. Application: Hair shampoos, foam baths, liquid detergents. (Witco Chemical Ltd).

Witconate D24-25. Calcium alkylaryl sulphonate. Applications: Oil soluble emulsifier. (Witco Chemical Ltd).

Witconate PTSA. Toluene sulphonic acid in crystal form. Applications: Anionic surfactant with hydrotrope properties, used as a catalyst. (Witco Chemical Ltd).

Witconate P10-45 and P10-59. Anionic surfactant in liquid form. Applications: Emulsifier for dry cleaning. (Witco Chemical Ltd).

Witconate P10-49. Anionic surfactant in liquid form. Applications: Liquid detergents. (Witco Chemical Ltd).

Witconate SCS. Sodium cumene sulphonate in liquid form. Applications: Anionic surfactant. (Witco Chemical Ltd).

Witconate STS. Sodium toluene sulphonate in liquid form. Applications: Anionic surfactant. (Witco Chemical Ltd).

Witconate SXS. Sodium xylene sulphonate in liquid form. Applications: Anionic surfactant. (Witco Chemical Ltd).

Witconol. Emollient. Applications: Bath oils, hair preparations; cosmetic preparations. (Witco Chemical Ltd).

Witepsol. Neutral hard fats based on mixtures of triglycerides. Applications: Preparation of suppositories. (Dynamit Nobel Wien GmbH).*

Withnell Powder. An explosive containing 88-92 per cent ammonium nitrate 4-5 per cent trinitro-toluene, and 4-6 per cent flour.†

Witocan. Special solid fats. Applications: Used in the chocolate and confectionery industry for the manufacture of substitute chocolate tablets, compound coatings, bars and moulded articles. (Dynamit Nobel Wien GmbH).*

Wittenburg Weather Dynamite. An explosive, consisting of 25 per cent nitroglycerin, 34 per cent potassium nitrate, 38.5 per cent rye meal, 1 per cent wood meal, 1 per cent barium nitrate, and 0.5 per cent sodium bicarbonate.†

Wittol Wax. A wax possessing similar properties to beeswax. It is a proprietary material, and is suitable for acid-proof linings.†

Witt's Phenylene Blue. An indamine dyestuff, $C_{14}H_{16}N_4HC1$. It is the dimethyl derivative of Witt's phenylene violet.†

Witt's Phenylene Violet. An indamine dyestuff, $C_{12}H_{12}N_4HC1$, obtained by the oxidation of *p*- phenylene diamine with *m*-phenylenediamine.†

Witt's Toluylene Blue. See TOLUYLENE BLUE.

Woad. A dark, clay-like preparation made from the leaves of the woad plant, *Isatis tinctoria.* It is used for the purpose of exciting fermentation in the indigo vat.†

Wolf N Lamid IG. A proprietary polyamide resin soluble in alcohol, used as a base for varnishes. (Victor Wolf N L Ltd).†

Wolfaid. Saturated and unsaturated polyesters (liquid and solid). Applications: Processing aids for PVC flooring, pigment dispersing aid for unsaturated polyester. (NL Victor Wolf Ltd).*

Wolfamid. Non-reactive polyamide resins. Applications: Packaging inks, cold seal release lacquers, thermographic systems, thixotropic alkyd resins. (NL Victor Wolf Ltd).*

Wolfert. A rubber substitute consisting of felt which has been impregnated with a vulcanized oil.†

Wolfin 18. Thermoplastic, bitumen proof insulating foil for structures. Applications: For flat roofs, foundation insulation, protection against leakage oil, protection against seepage and pressure water. (Degussa).*

Wolfkur. Reactive polyamide resins, polyamino amides and amine adducts. Applications: Epoxy curing agents for coatings and adhesives. (NL Victor Wolf Ltd).*

Wolflex. Saturated polyesters. Applications: Polymeric plasticizers for PVC plastics. (NL Victor Wolf Ltd).*

Wolfol. Saturated polyesters with hydroxyl groups. Applications: Polyester polyols for polyurethanes used in elastomers and adhesives. (NL Victor Wolf Ltd).*

Wolfram Brass. See TUNGSTEN BRASS.

Wolfram Bronze. See TUNGSTEN BRONZE.

Wolframium. An alloy of 98 per cent aluminium, 1.4 per cent antimony, 0.4 per cent copper, 0.1 per cent tin, and 0.04 per cent tungsten.†

Wolfram Ochre. See WOLFRAMINE.

Wolfram White. A pigment. It is barium tugstate.†

Wollastokup. Chemically coupled wollastonite, high aspect ratio and fine particle sizes: CaO, SiO_2, Fe_2O_3, Al_2O_3 MnO, MgO, TiO_2. Applications: Polymer composites, high performance coatings, adhesives, elastomers and friction products. (NYCO).*

Wollastonite. Calcium metasilicate. Applications: Inert filler. (Cornelius Chemical Co Ltd).*

Wollaston's Cement. Consists of 1 part beeswax, 4 parts resin, and 5 parts plaster of Paris. Used for fossils.†

Wolle. An abbreviation for Collodium-wolle nitrocellulose in various viscosities.†

Wongshy. (Wongsky). Chinese names for the pods of *Gardenia grandiflora*, which contain large quantities of crocin for saffron. Dyes silk and wool yellow.†

Wongsky. See WONGSHY.

Wood, Alligator. See ALLIGATOR WOOD.

Wood-apple Gum. (Katbél-ki-gond, Velampishin, Kapithamia Piscum). The gum of *Feronia elephantum*.†

Wood, Bitter. Quassia.†

Wood-cloth. Strips of wood treated with sulphurous acid or alkaline bisulphite, making the fibre stronger.†

Wood, Cuba. See FUSTIC.

Wood Ether. Dimethyl ether, CH_3OCH_3.†

Wood, Fernambuco. See REDWOODS.

Wood Flour. (Wood Meal). Finely powdered wood, usually white pine.

Used as a rubber, linoleum, or soap filler.†

Wood, Jamaica. See LOGWOOD.

Wood, Kambe. See REDWOODS.

Wood, Lima. See REDWOODS.

Wood, Nicaragua. See REDWOODS.

Wood Oil. The final fractions obtained in the distillation of wood spirit, containing high boiling ketones. Also see CHINESE WOOD OIL and GURJUN BALSAM or OIL.†

Wood, Peach. See REDWOODS.

Wood, Pernambuco. See REDWOODS.

Wood Potash. Potash salts obtained from the ash of certain woods.†

Wood, Red. See REDWOODS.

Wood Rosin. Rosin obtained from the stumps and top wood of felled trees which are useless as timber. It is usually of yellow pine.†

Wood, Sandal. See REDWOODS.

Wood, Sapan. See REDWOODS.

Wood Spirits, Green. See ACETONE ALCOHOL.

Wood Spirit, Standard. See ACETONE ALCOHOL.

Wood Stone. See XYLOLITE.

Wood Sugar. Xylose, $C_6H_{10}O_5$.†

Wood, Yellow. See FUSTIC.

Wood's Alloys. Low-melting alloys of bismuth, tin and lead, usually containing cadmium. One alloy contains 50 per cent bismuth, 27 per cent lead, 13 per cent tin and 10 per cent cadmium while another contains 50 per cent bismuth, 25 per cent lead, and 25 per cent tin. A third contains 50 per cent bismuth, 25 per cent lead, 12.5 per cent tin and 12.5 per cent cadmium.†

Wool, Artificial. See STAPLE FIBRE

Wool Blue S. A dyestuff. It is a mixture of Acid violet 7B with Blue green S.†

Wool Fat, Hydrous. See LANOLIN.

Wool, Glover's. See TANNER'S WOOL.

Wool Green S. (Lissamine Green B). A dyestuff. Dyes silk and wool sea green shades.†

Wool Grey B, G, and R. A dyestuff obtained by the action of aniline (or *p*-toluidine) upon the condensation product from nitrosodimethylaniline and β-naphtholsulphonic acid S. Dyes wool grey.†

Wool Milk. An emulsion, obtained from the treatment of wool fat with caustic soda, and dilution. Lanolin is obtained from this wool milk.†

Wool, Mineral. See SLAG WOOL.

Wool Oil. An impure oleic acid used for oiling wool, and for making lubricants and soaps.†

Wool Pitch. A pitchy material obtained as a residue after the distillation of wool grease (Yorkshire grease).†

Wool Red Extra. See FAST RED D.

Wool Scarlet R. A dyestuff. It is the sodium salt of xyleneazo-α- naphthol disulphonic acid, $C_{18}H_{14}N_2O_7S_2Na_2$. Dyes wool red from an acid bath.†

Wool, Vegetable. See LANELLA.

Wool Violet S. A dyestuff. It is the sodium salt of dinitrobenzeneazo diethylmetasulphanilic acid, $C_{16}H_{16}N_5SO_7Na$. Dyes wool reddish-violet from an acid bath.†

Wool Yellow. See PATENT FUSTIN.

Wormseed. The flower-heads of *Arte-misia maritima*, used in medicine.†

Wormseed, American. The fruit of *Chenopodium ambrosoides*.†

Wormwood. Absinthium, the dried leaves and flowering-tops of *Artemisia absinthium*.†

Wormwood, Salt of. See SALT OF WORMWOOD.

Wort. Malt is crushed and heated with water until the starch is converted into sugar by the diastase in the malt. The resulting liquid is known as wort.†

Wovco SP. Trade mark for a range of polyethylene plastics reinforces with carbon fibre. (Worcester Valve Co, Haywards Heath).†

WRA Epoxy Resin Underwater Series. Liquid and putty epoxies. Applications: Specifically formulated for underwater use. General bonding, filling and fairing. (Wessex Resins & Adhesives Ltd).*

WRA System 100 Laminating Composition. Liquid epoxy resin and hardener,

Applications: Laminating resin specifically for use with glass cloth, carbon fibre, aramid and hybrids. (Wessex Resins & Adhesives Ltd).*

WRA System 17. Thixotropic resin and hardener. Applications: General purpose adhesive (bonding, laminating), bonding wood, concrete, most metals, stone, china, GRP, unglazed ceramics, rubber. (Wessex Resins & Adhesives Ltd).*

WRA System 80. PVA cross linking liquid. Applications: General purpose wood glue-waterproof-complies with DIN68602 Section B and BS4071. (Wessex Resins & Adhesives Ltd).*

WRA1000 Varnish. Liquid polyurethane resin and hardener. Applications: Used for varnishing directly onto wood or on top of epoxy coatings. Has a u/v inhibitor. (Wessex Resins & Adhesives Ltd).*

Wresinate. Metal resinates. Applications: Marine antifoulings. Improve glass and drying in alkyd finishes. (Resinous Chemicals Ltd).*

Wright's Stain. A microscopic stain for white blood corpuscles. It consists of 1 gram methylene blue eosin mixture in 600 cc methyl alcohol.†

Wrought Iron. See MALLEABLE IRON.

Wurster's Blue. An oxidation product of tetramethyl-*p*-phenylenediamine. An indicator.†

Wurster's Red. An oxidation product of *p*-aminodimethylaniline.†

Wurtzillite. (Tabbyite, Aegenite, Aeonite). An asphaltic mineral, soluble hot in water.†

WW (Wet Wax). Release agent for epoxy compounds, car wax. (ADC Resins).*

Wyamine Sulfate. Mephentermine sulphate. Applications: Adrenergic. (Wyeth Laboratories, Div of American Home Products Corp).

Wyamycin E. Erythromycin ethylsuccinate. Applications: Antibacterial. (Wyeth Laboratories, Div of American Home Products Corp).

Wyamycin S. Erythromycin stearate. Applications: Antibacterial. (Wyeth Laboratories, Div of American Home Products Corp).

Wycillin. Penicillin G Procaine. Applications: Antibacterial. (Wyeth Laboratories, Div of American Home Products Corp).

Wydase. Hyaluronidase. Applications: Spreading agent. (Wyeth Laboratories, Div of American Home Products Corp).

Wymox. Amoxicillin. Applications: Antibacterial. (Wyeth Laboratories, Div of American Home Products Corp).

Wyovin. A proprietary preparation containing dicyclomine hydrochloride. An antispasmodic. (Wyeth).†

Wytensin. Guanabenz acetate. Applications: Antihypertensive. (Wyeth Laboratories, Div of American Home Products Corp).

Wytox ADP. A registered trade name for alkylated diphenylamine. An anti-oxidant for rubber for protection against heat ageing and flex cracking. (National Polychemicals, Wilmington, Mass).†

Wytox BHT. A registered trade name for alkylated *p*-cresol. A non-staining antioxidant for plastics. (National Polychemicals, Wilmington, Mass).†

Wytox LT. A registered trade name for dilauryl thiodipropionate. An anti-oxidant suitable for use in plastics of the polyolefine and ABS type in contact with food. (National Polychemicals, Wilmington, Mass).†

Wytox 312. A registered trade name for tris-nonylphenyl phosphite, a non staining, non-discolouring, low volatility antioxidant for polyolefins, vinyl chloride polymers, high impact polystyrenes, etc. (National Polychemicals, Wilmington, Mass).†

Wytox 335. A registered trade name for a modified polymeric phosphite stabilizer for emulsion type styrene-butadiene polymers. It has outstanding suppression of gel build-up. It is exceptionally resistant to hydrolysis. (National Polychemicals, Wilmington, Mass).†

X

X2B. A proprietary hard rubber.†

Xala. A name for borax.†

Xametrin. See HEXAMINE.

Xanax. Alprazolam. Applications: Sedative. (The Upjohn Co).

Xantalgin. Alginate impression material. Applications: Dental preparation. (Bayer & Co).*

Xanthano. Impression plaster. Applications: Dentistry. (Bayer & Co).*

Xanthine. See PHOSPHINE.

Xanthophyll. The yellow pigment of leaves. It has the formula, $C_{40}H_{56}O_2$.†

Xanthopicrin. A yellow colouring matter from the bark of *Xanthoxylum caribgum*.†

Xanthopicrite. A yellow resin from *Xanthoxylum* species.†

Xanthopone. Zinc ethylxanthate. A rubber vulcanization accelerator.†

Xanthopurpurin. See PURPUROXANTHIN.

Xantogum. Addition curing elastomeric high precision impression material for the single-phase technique (monophase), 2-paste system. (Bayer & Co).*

Xantopren. Precision impression material on an elastomer basis in two different consistencies. Applications: Dentistry. (Bayer & Co).*

Xantopren Function. Special impression material for functional impressions. Applications: Dentistry. (Bayer & Co).*

Xantopren Plus. Organic silicone polymer for precision impressions. Applications: Dentistry. (Bayer & Co).*

Xantygen. Thermoplastic impression material in rods and plates. Applications: Dentistry. (Bayer & Co).*

Xenacryl. Polyacrylete copolymer solutions and emulsions for paint and adhesives. (The Baxenden Chemical Co Ltd).

Xeneisol 133. A proprietary trade name for the 133 isotope of XENON. (Mallinckrodt Inc).*

Xenith. Blend of polymers, amine and solvents. Applications: Zinc electroplating additive. (Taskem Inc).*

Xenomatic. Xenon Xe 133. Applications: Radioactive agent. (Mallinckrodt Inc).

Xenon. A heavy inert gaseous element present in minute quantities in the atmosphere.†

Xenon XE 133. See XENEISOL 133.

Xenon Xe 133-VSS. Xenon Xe 133. Applications: Radioactive agent. (Medi-Physics Inc).

Xenoy. Polymer alloys. Applications: Plastic components for automotive, electrical, electronics, lighting, medical, packaging, audio etc. (GE Plastics Ltd).*

Xeroderm S100, L67. Silicone-based water repellent agent, especially for chrome leathers and chrome suede. Applications: Leather industry. (Bayer & Co).*

Xerol. A proprietary trade name for glyceryl mono-stearate.†

X-Ite. A proprietary alloy containing 37-39 per cent nickel and 17-19 per cent chromium with iron.†

X.F.L.X. A substituted ethylene mixed with secondary aromatic amines. (C P Hall (Akron)).†

X.L. High boiling tar acids. (Coalite Fuels & Chemicals Ltd).†

XL Carmoisine 6R. A dyestuff. It is a British equivalent of Chromotrope 2R.†

XL Opal Blue. See DIPHENYLAMINE BLUE, SPIRIT SOLUBLE.

X.L.O. A rubber vulcanization accelerator consisting of magnesia and diphenyl guanidine.†

XL Soluble Blue. See METHYL BLUE.

X-Prep. Sennocides A and B. Applications: Laxative/bowel evacuant. (Napp Laboratories Ltd).*

Xuprin. Isoxsuprine. Applications: Promotes blood flow rate. (Duphar BV).*

XX 601. A proprietary brand of zinc oxide used in rubber compounding.†

Xylan 1052. A proprietary lubricant based on PTFE, for use under extreme pressures and in extremes of temperature. (Whitfield Plastics, Runcorn, Cheshire).†

Xylan 330. A proprietary aerosol form of PTFE used as a mould-releasing agent in plastics processing. (Whitfield Plastics, Runcorn, Cheshire).†

Xylene Musk. An artificial musk perfume. It is trinitrotertiarybutyl-*m*-xylene.†

Xylenol Blue. 1, 4-Dimethyl-5-hydroxybenzenesulphonphthalein. An indicator used in biochemistry.†

Xylidine Orange. See SCARLET GR.

Xylidine Red. See PONCEAU R.

Xylidine Scarlet. See PONCEAU R and AZOCOCCIN 2R.

Xylite. A proprietary rubber vulcanization accelerator. It is a tarry diphenylguanidine.†

Xylocaine. Lidocaine. Applications: Anaesthetic. (Astra Pharmaceutical Products Inc).

Xylocard. A proprietary preparation of LIONOCAINE hydrochloride used in the treatment of cardiac arrhythroias. (Astra Chemicals Ltd).†

Xylock 225. A proprietary condensation product of phenols with an aryl alkyl ether, used as a high-performance, heat-stable moulding resin. (Albright & Wilson Ltd).*

Xyloidine. See NITRO-STARCH.

Xylok. High performance resins. (Albright & Wilson Ltd, Phosphates Div).

Xylol. Commercial xylene. It consists of a mixture of about 60 per cent *m*-xylene, 10-25 per cent *o*- and *p*-xylene, ethylbenzene, and small quantities of trimethyl-benzene, paraffin, and thioxene.†

Xylolite. (Xylolith). A cement composed of sawdust mixed with Sorel cement (*qv*).†

Xylolith. See XYLOLITE and SOREL CEMENT.

Xylon. Wood cellulose.†

Xylon FR. A proprietary nylon containing a flame-retarding additive. (Dart Industries Inc).†

Xylonite. See CELLULOID.

Xylopal. Wood opal.†

Xyloproct. A proprietary preparation of LIGNOCAINE, aluminium acetate, zinc oxide and HYDROCORTISONE acetate used in the treatment of haemorrhoids. (Astra Chemicals Ltd).†

Xyloquinone. Dimethylbenzoquinone, $C_6H_2(CH_3)_2O_2$.†

Xylorcinol. Dimethylorcinol, $C_6H_2(CH_3)_2.(OH)_2$.†

Xylose. A proprietary benzyl cellulose.†

Y

Yaba Bark. The bark of *Andira excelsa*.†

Y Alloy. An aluminium alloy containing 4 per cent copper, 2 per cent nickel, and 1.5 per cent magnesium with small amounts of iron and silicon. It has a specific gravity of 2.8. It is used for die-cast pistons, etc.†

Yaltox. A free flowing granule containing 5 per cent w/w carbofuran. Applications: To control a wide range of soil and seedling pests including cabbage root fly, cabbage stem weevil, flea beetle, cabbage stem flea beetle, early aphids in brassicas, turnip root fly, frit fly, millipedes, symphilids, beet leaf miner, springtails, wireworms, free living nematodes, potato cyst eelworm, carrot fly and carrot willow aphid. (Bayer & Co).*

Yama-Mai Silk. A silk produced by the Japanese oak caterpillar of Japan, China, and India.†

Yara-Yara. See NEROLIN.

Yarmor. See PINE OIL.

Yarmor. Pine oil terpene liquids. Applications: Anti-skinning agents in protective coatings, for pigment grinding aids and as an antifoam agent. (Hercules Inc).*

Yarn, Chemical. Artificial silk.†

Yarn, Filastic. See FILASTIC.

Yarrow. The dried leaves of *Achillea millefolium*. A tonic.†

Yeast, Beer. *Foex medicinalis*.†

Yellow Acid. 1,3-Dihydroxynaphthalene-5, 7-disulphonic acid.†

Yellow Bark. The bark of *Cinchona calisaya*.†

Yellow Basilicon. Resin ointment.†

Yellow Berries. See PERSIAN BERRIES.

Yellow Brass. See BRASS.

Yellow Carmine. (Italian Pink, Yellow Lake). Pigments prepared by precipitating the glucoside quercitrin with alumina.†

Yellow Catechu. See CUTCH.

Yellow, Chinese. See CHINESE YELLOW and OCHRE.

Yellow, Cobalt. See AUREOLIN.

Yellow, Cologne. See CHROME YELLOW.

Yellow Cross Gas. See MUSTARD GAS.

Yellow Earth. See OCHRE.

Yellow Fast to Soap. A dyestuff. It is the sodium salt of *m*-carboxybenzeneazodiphenylamine, $C_{19}H_{14}N_3O_2Na$. Dyes cotton orange with a chrome mordant.†

Yellow Fat Colour. See SPIRIT YELLOW R.

Yellow GA. See METANIL YELLOW.

Yellow Gold. An alloy of 53 per cent gold, 25 per cent silver, and 22 per cent copper.†

Yellow Lake. See YELLOW CARMINE.

Yellow Liquors. The drainage from alkali waste heaps.†

Yellow Ochre. (Roman Ochre, Transparent Gold Ochre, Brown Ochre, Terra di Sienna, Stone Yellow, Roman Yellow, Mineral Yellow, Oxford Yellow, Golden Ochre). Yellow pigments. They are native earths consisting chiefly of silica and alumina coloured by hydrated ferric oxide. Some ochres are adulterated with calcium carbonate and barium sulphate, and occasionally chrome yellow or a yellow dyestuff is used to improve colour.†

Yellow Oil. A material containing higher alcohols, such as hexyl alcohol, obtained during the production of butanol from corn.†

Yellow OO. See FAST YELLOW N.

Yellow Precipitate. Yellow mercury oxide, HgO. Ammonium phosphomolybdate is also given this name.†

Yellow Prussiate of Potash. (Ferro prussiate of potassium). Potassium ferrocyanide, $K_4Fe(CN)_6$.†

Yellow Sesquioxide of Uranium. See ORANGE OXIDE OF URANIUM.

Yellow Soda Ash. A soda ash (sodium carbonate) containing traces of iron oxide.†

Yellowstone. Wyoming Sodium Bentonite. Applications: Steel production and foundries. (Bromhead & Denison Ltd).*

Yellow Sulphide of Arsenic. See ORPIMENT.

Yellow T. See TROPAEOLINE O.

Yellow Ultramarine. (Silver Ultramarine, Barium Yellow). A pigment prepared by replacing the sodium constituent of ultramarine with silver. A yellow ultramarine is also prepared by treatment of red ultramarine with hydrochloric acid above 360° C.†

Yellow W. See FAST YELLOW R.

Yellow Wax. A viscous, semi-solid, difficultly volatile substance obtained by the distillation of the still residues of petroleum. It contains anthracene and other hydrocarbons.†

Yellow Wood. See FUSTIC.

Yellow WR. See BRILLIANT YELLOW S.

Yenshee. The dregs and carbonized opium which remains after smoking. It contains from 1-10 per cent morphine.†

Yeoman. Lanolin BP. (Croda Chemicals Ltd).

Yerba Mate. Paraguay tea.†

Ylang-Ylang Oil. Orchid oil.†

Yodoxin. Iodoquinol. Applications: Anti-amebic. (Glenwood Inc).

Yolk Powder. See LECITHIN.

Yoloy. A proprietary alloy. It is a steel containing 1 per cent copper, 2 per cent nickel, and up to 0.2 per cent carbon.†

Yomesan. Veterinary preparation. Applications: Used against tapeworm infestation in dogs and cats. (Bayer & Co).*

Yomesan. A proprietary preparation of niclosamide. An anthelmintic. (Bayer & Co).*

Yonckite. A Belgian explosive consisting of ammonium perchlorate, ammonium nitrate, sodium nitrate, and trinitrotoluene or nitronaphthalene.†

Yoracryl. A range of modified basic dyes. Applications: Dyeing of acrylic fibres. (Yorkshire Chemicals Plc).*

Yorkshire Grease. (Wakefield Grease). The recovered fatty acids from wool grease.†

Young Fustic. (Cotinin). A natural dyestuff the dyeing principle of which is fisetin, $C_{15}H_{10}O_6$. It has a limited use for dyeing wool orange or scarlet (chrome or tin mordant), and for dyeing leather.†

Yperite. See MUSTARD GAS.

Y-Tack. Polyglycol ether and polyacrylic copolymerides in aqueous solution. Applications: Warp and weft treatment chemicals which provide a combination of film forming yarn cover, intra fibre bonding, lubrication and static control. (Thomas Swan & Co Ltd).*

Ytterbium Yb-169 DTPA. Pentetate Calcium Trisodium Yb 169. Applications: Radioactive agent. (Minnesota Mining & Mfg Co).

Yukalon. Trademark for a proprietary grade of polyethylene. (Mitsubishi Petrochemical Co).*

Yutopar. Ritodrine. Applications: Uterospasmolytic, prevents premature birth. (Duphar BV).*

Z

Zaccatila. See COCHINEAL.

Zaditen. Ketotifen fumarate. Applications: Anti-asthmatic. (Sandoz Pharmaceuticals, Div of Sandoz Inc).

Zadstat. Metronidazole. Applications: Antibiotic. (Lederle Laboratories).*

Zaffer. See SMALT.

Zaffre. See SMALT.

Zakin Rubber. A rubber-like substance prepared from glue or similar material.†

Zala. Borax.†

Zam Metal. A proprietary alloy of zinc with aluminium and magnesium.†

Zam-Buk. Antiseptic ointment. (Fisons PLC, Pharmaceutical Div).

Zamak Alloys. Proprietary alloys of zinc with aluminium and sometimes small amounts of copper and magnesium. Copper-aluminium-zinc alloys, suitable for die castings contain : 3.9-4.3 per cent aluminium, 0.9-2.9 per cent copper, 0.003-0.06 per cent magnesium, remainder zinc. They have a specific gravity of 6.64-6.7.†

Zambesi Blue BX, B, RX, R. Direct cotton colours which are developed with β-naphthol.†

Zambesi Brown G, GG. Direct cotton colours which are diazotized and developed. The GG mark gives yellowish, and the G mark dark-brown shades.†

Zambesi Grey B. A direct cotton colour. It dyes cotton or wool bluish grey, and is diazotized and developed on the fibre.†

Zambesi Indigo Blue. A tetrazo dyestuff, which dyes cotton direct, and when diazotized on the fibre and developed with β-naphthol, gives a reddish-blue colour.†

Zanaloin. See SOCALOIN.

Zanchol. A proprietary trade name for florantyrone.†

Zanil. A proprietary preparation of oxyclozanide. A veterinary anthelmintic.†

Zanosar. Streptozocin. Applications: Antineoplastic. (The Upjohn Co).

Zapoglobin. Reagent which destroys red blood cells to leave white blood cells in suspension for analysis. Applications: White blood cell and haemoglobin determination. (Coulter Electronics Ltd).*

Zapon Varnish. (Brassoline, Cristalline, Victoria Varnish). Celluloid varnishes.†

Zaponin. Reagent which destroys red blood cells to leave white blood cells in suspension for analysis. Applications: White blood cell determination using semi-automated cell counters. (Coulter Electronics Ltd).*

Zapoto Gum. See CHICLE.

Zarontin. Ethosuximide. Applications: Anticonvulsant. (Parke-Davis, Div of Warner-Lambert Co).

Zaroxolyn. A proprietary preparation of metolazone. An anti-hypertensive. (Pennwalt Corp).*

Zauberin. A material containing Chloramine T. A detergent and bleaching agent. Also see CHLORAMINE T, Mannolit, Gansil, Glekosa, Purus, and Washington Bleach are names for other washing and bleaching agents, the active principle of which is Chloramine T.†

ZBX. See ACCELERATOR ZBX.

ZB2335. Zinc borate for flame retardancy. (Borax Consolidated Ltd).

Zeasorb. A proprietary preparation of microporous cellulose hexachlorophane, chloroxylenol, aluminium dihydroxy-allantoinate and purified talc, used in dermatology. (Stiefel Laboratories (UK) Ltd).*

Zedox. Zirconium oxide. (Anzon Ltd).

Zeese. Honey substitute for diabetics containing sorbitol, fructose, water, citric acid, permitted colour E150 and flavouring. (LAB Ltd).*

Zeiodelite. A mixture obtained by stirring 24 parts powdered glass into 20 parts melted sulphur. Used as a cement, and for taking casts.†

Zeise's Salt. A salt, $[Pt(C_2H_4)Cl_3]K$, formed when potassium chloride is added to a solution of platinous chloride saturated with ethylene.†

Zelco Metal. An alloy of 83 per cent zinc, 15 per cent aluminium, and 2 per cent copper.†

Zelcon. Fabric conditioner. (Du Pont (UK) Ltd).

Zelec. Anti-static agent. (Du Pont (UK) Ltd).

Zeller's Ointment. Ammoniated mercury ointment.†

Zellner's Paper. Fluoresceine paper.†

Zellwonet. A proprietary viscose packing material.†

Zelulone. A proprietany artificial yarn made from wood pulp.†

Zenadrid. Prednisone. Applications: Glucocorticoid. (Syntex Laboratories Inc).

Zendium. Sodium Fluoride. Applications: Dental caries prophylactic. (Oral-B Laboratories Inc).

Zenith. Organic brightener system. Applications: Bright nickel electroplating (high levelling, fast brightening). (Harshaw Chemicals Ltd).*

Zenker's Fluid. A solution containing 2.5 grams potassium chromate, 1 gram sodium sulphate, 5 grams mercuric chloride, 5 cc glacial acetic acid, and 100 cc water.†

Zentralin. Dimethyldiphenylurea, $(CH_3)_2(C_6H_5)_2CON_2$. Used for explosives.†

Zentralit I. Diethyldiphenylurea.†

Zentralit II. Dimethyldiphenylurea.†

Zeolex. Synthetic sodium aluminium silicate. Applications: Flow promoter, pigment extender. (Cornelius Chemical Co Ltd).*

Zeolites. Hydrated aluminium silicates containing alkali or alkaline earth metals. They occur naturally. Used in ion-exchangers.†

Zeothix. A ground silica. Applications: Thickening agent. (Cornelius Chemical Co Ltd).*

Zeotokol. A coarse dolerite (an igneous rock composed essentially of labradorite and anorthite, with augite and sometimes olivine) ground up and used as a fertilizer.†

Zepel. Fabric fluoridizer. (Du Pont (UK) Ltd).

Zephiran Chloride. Benzalkonium chloride. Applications: Pharmaceutic aid. (Sterling Drug Inc).

Zephirol. A preparation for the disinfection of hands and instruments. (Bayer & Co).*

Zephyr. Inks used in the screen printing process. Applications: Billboards, signs and displays. (Allied Signal Sinclair and Valentine).*

Zerex. A proprietary trade name for a polyvinyl alcohol antifreeze compound.†

Zerofil. A proprietary rock wool (*qv*) which has been treated with asphalt.†

Zerol. Refrigeration fluid. (Chevron).*

Zerone. A proprietary trade name for a methanol and polyvinyl alcohol anti-freeze product.†

Zerotherm. Self drying coatings for sand moulds and cores. (Foseco (FS) Ltd).*

Zerox. Hydrazine solutions. (FBC Ltd).

Zetabon. Plastic coated steel and aluminium used in wire and cable for armour, corrosion and lightening protection. (Dow Chemical Co Ltd).*

Zetar. Coal tar. Applications: Anti-eczematic. (Dermik Laboratories Inc).

Zetax. 2-Mercaptobenzothiazole - primary rubber accelerator with zinc. Applications: Primary accelerator for both natural and synthetic rubbers. (Vanderbilt Chemical Corporation).*

Zettnow's Stain. A microscopic stain. Solution (A) contains 10 grams tannic acid to which has been added 30 cc of a 5 per cent solution of tartar emetic. Solution (B) contains 1 gram silver sulphate in 250 cc water. Take 50 cc and add ethylamine until precipitate redissolves.†

Zettyn. Cetalkonium chloride. Applications: Anti-infective, topical. (Sterling Drug Inc).

Zeus. An alloy of 20 per cent silver and 80 per cent copper. Used for fuse wire.†

Zewa Powder. Sodium lignin sulphonate obtained by evaporation of waste sulphite lyes. It has detergent and water softening properties.†

Zewaphosphate. A phosphate fertilizer.†

Ziehl's Stain. See CARBOLFUCHSINE.

Zienam. Imipenem and cilastatin. Applications: Broad-spectrum beta-lactam antibiotic. (Merck Sharp & Dohme).*

Zigueline. A red oxide of copper.†

Zilloy. A proprietary zinc alloy containing zinc with 1 per cent copper, 0.01 per cent magnesium, and lead and cadmium in addition. The rolled sheets are suitable for building purposes.†

Zimalium. Alloys containing 74-93.5 per cent aluminium, 2.8-14.8 per cent zinc, and 3.7-11.2 per cent magnesium.†

Zimate. Zinc diamyl, dubutyl, diethyl and dimethyldithiocarbamates. Rubber accelerators. Applications: Full range of accelerator needs. Both solid and liquid. For both NR of all types and latex. (Vanderbilt Chemical Corporation).*

Zimco. Vanillin. Applications: Pharmaceutic aid (flavour). (Sterwin Chemicals Inc).

Zinamide. Pyrazinamide. Applications: Antituberculous agent (Merck Sharp & Dohme).*

Zincaband. A proprietary zinc paste bandage. Applications: Subacute and chronic eczema and lichenification. (Seton Products Ltd).*

Zincalium. See ZIMALIUM.

Zinc, Aluminium. See ALZEN and ZISCON.

Zinc Anhydride. A variety of lithopone, also known as Zinc Barytes. It consists of a mixture of calcium and barium sulphates and zinc oxide.†

Zincazol. A proprietary rubber vulcanization accelerator. It is zinc-α-phenylbiguanide.†

Zinc-Baryta White. See LITHOPONE.

Zinc Barytes. See ZINC ANHYDRIDE.

Zinc Blende. (Black Jack, Blende, Rosinjack, Pseudo-galena). A mineral. It is zinc sulphide, ZnS. A pigment.†

Zinc Borate 2335. $2ZnO.3B_2O_3$ $3.5H_2O$. Applications: A specialty flame retardant additive to plasticized PVC and other polymers to reduce afterglow and smoke. (Borax Consolidated Ltd).*

Zinc Bronze. See ADMIRALTY GUN METAL.

Zinc Chrome. (Citron Yellow). A pigment. It is zinc chromate, $ZnCrO_4$. Also see ZINC YELLOW.†

Zinc Chrome Yellow. See ZINC YELLOW.

Zinc Dust. See BLUE POWDER.

Zinced Iron. See GALVANIZED IRON.

Zinc Flowers. See PHILOSOPHER'S WOOL.

Zinc Formosul. A basic zinc-formaldehyde-sulphoxylate. It is employed in fat-splitting.†

Zinc Fume. See BLUE POWDER.

Zinc Greens. The name is now applied exclusively to mixtures of Zinc yellow and Prussian blue, that is Zinc-yellow greens. The palest varieties are known as Parrot greens. The term was formerly applied to Cobalt green (*qv*).†

Zinc Grey. The name originally used for zinc dust employed for painting on iron. The term is now used for finely ground zinc blende. A mixture of zinc oxide with finely divided charcoal is sold under this name. It is produced in the manufacture of zinc. Other names for this product are Diamond grey, Silver grey, and Platinum grey.†

Zincocalcite. A calcite containing zinc carbonate.†

Zincofol. Fungicide. (Chevron).*

Zincolith. See LITHOPONE.

Zincocalcite. A calcite containing zinc carbonate.†

Zincofol. Fungicide. (Chevron).*

Zincolith. See LITHOPONE.

Zinc Omadine. Pyrithione Zinc. Applications: Antibacterial; antifungal; antiseborrheic. (E R Squibb & Sons Inc).

Zincon Dandruff Shampoo. Pyrithione Zinc. Applications: Antibacterial; antifungal; antiseborrheic. (Lederle Laboratories, Div of American Cyanamid Co).

Zinconal. See EKTOGAN.

Zincore, Flinty. See FLINTY ZINC ORE.

Zinc Ore, Spathic. See CALAMINE.

Zinc Perhydrol. See EKTOGAN.

Zinc Powder. Zinc oxide, ZnO.†

Zincrex. Flux for zinc and alloys. (Foseco (FS) Ltd).*

Zinc Spar. See CALAMINE.

Zinc Sulphide Grey. (Calamine White). A pigment. It is a dense zinc oxide used for painting iron, and is artificially made by tinting lithopone with ochres and charcoal.†

Zinc Sulphide White. See LITHOPONE.

Zinctrace. Zinc Chloride. Applications: Astringent; dentin desensitizer. (Armour Pharmaceutical Co).

Zinc Vitriol. Zinc sulphate, $ZnSO_4$.†

Zinc White. (Chinese White, Permanent White, Snow White). A pigment. It is zinc oxide, ZnO. Chinese white is a very dense oxide, and Snow White a very pure one. The following brands of zinc oxide are sold : White seal, Green seal, Red seal, Yellow seal, and Grey Seal. White seal is the purest mark, and contains 99 per cent zinc oxide. The Grey seal contains metallic zinc.†

Zinc Yellow. (Zinc Chrome Yellow, Buttercup Yellow, Lemon Yellow). A pigment consisting of zinc chromate, $ZnCrO_4$.†

Zinc Yellow Greens. See ZINC GREENS.

Zineb. Zinc dithiocarbamate. A fungicide.†

Zinkan. A proprietary combination of aluminium coated with zinc. It is obtained by rolling at elevated temperatures.†

Zinkgrau. Cheap, off colour zinc oxide pigment.†

Zinkhausmannit. Synonym for Hetaerolite.†

Zinkoyd Aktiv. Technical grade oxides and zeolites. Applications: For use as an additive in the paint and rubber industries and as a catalyst in the chemical industry. (Bayer & Co).*

Zinkweiss Weissiegel. Zinc oxide, white seal, best quality zinc white.†

Zinnal. A proprietary dual metal consisting of aluminium sheet coated on both sides with tin.†

Zinol. Powder flux for treating dross on hot-dip galvanizing baths. (Foseco (FS) Ltd).*

Zinox. A paint pigment. It is a hydrated zinc oxide.†

Zinsser's Insulating Wax. (1) Consists of beeswax ; (2) consists of shellac, rosin, and oxide of iron.†

Zintox. A proprietary trade name for an agricultural spray containing basic zinc arsenate.†

Zip Grip. Cyanoacrylate. Applications: Bonding closely mated surfaces and maintenance, production and prototype bonding. (Devcon Corporation).*

Zipan. Promethazine Hydrochloride. Applications: Anti-emetic; antihistaminic. (Savage Laboratories).

Zipcillin. Veterinary intramammary procaine penicillin. (The Wellcome Foundation Ltd).

Ziploc. Brand plastic storage bags. (Dow Chemical Co Ltd).*

Zippo. A trade name for an aluminium solder for joining aluminium to itself, to copper, zinc, tin, or brass.†

Zircomplex. Ammoniacal zirconium complex. Applications: Thixotrope in emulsion paints. (Manchem Ltd).*

Zirconia. Zirconium dioxide, ZrO_2.†

Zircosil. Zirconium silicates. (Anzon Ltd).

Zircosol P. Recycled paper additive. (Magnesium Elektron Ltd).

Zirgel. Thixotropic gelling agent. (Magnesium Elektron Ltd).

Zirmax. Zirconium hardener. (Magnesium Elektron Ltd).

Ziscon. (Ziskon). An alloy of 60 per cent aluminium and 40 per cent zinc.†

Zisium. An alloy of from 82-83 per cent aluminium, 1-3 per cent copper, 15 per cent zinc, and 0-1 per cent tin.†

Ziskon. See ZISCON.

Z-M-L. A proprietary rubber vulcanizing accelerator. It is the zinc salt of mercaptobenzenethiazole with laurex.†

Zoamix. Coccidiostat. Often used on an exchange programme with COYDEN. The main purpose in rotating Zoamix coccidiostat with Coyden coccidiostat is to prevent poultry from developing a resistance to the latter product. (Dow Chemical Co Ltd).*

Zoaquin. Di-iodohydroxyquinoline. (May & Baker Ltd).

Zodiac. Organic brightener system. Applications: Bright nickel electroplating. (Harshaw Chemicals Ltd).*

Zoladex. Anti-cancer preparations. (ICI PLC).*

Zolone. Insecticide. (May & Baker Ltd).

Zolyse. Chymotrypsin. Applications: Enzyme. (Alcon Laboratories Inc).

Zomax. Zomepirac Sodium. Applications: Analgesic; anti-inflammatory. (McNeil Pharmaceuticals, McNEILAB Inc).

Zonarez. Polyterpene resins of alpha-pinene, beta-pinene and dipentene. Applications: Tackifying resins for adhesives and sealants, polyolefin film additives, components of investment casting waxes and wax coatings for paper and paperboard, concrete parting and curing agent components, components of chewing gum base. (Arizona Chemical Company).*

Zonatac. Modified polyterpene resins. Applications: Tackifying resins for adhesives and sealants. (Arizona Chemical Company).*

Zonester 25, 40, 55, 65, 75, 85, 100 Resin Esters. Three glycerol esters, two pentaerythritol esters and two diethylene glycerol esters. Softening point range 25°C-95°C, colours as light as 6 Gardener colour. Available in bulk, solid or beaded form. Applications: Gum base, tackifier resin used in a wide variety of adhesive formulations, printing inks and coatings. (Arizona Chemical Company).*

Zonester 85-D60 Resin Ester Dispersion. A 60 per cent solids aqueous dispersion of an 85°C softening point resin ester. Applications: Tackifier rosin ester used in water based adhesive formulations, particularly SBR and acrylic systems. (Arizona Chemical Company).*

Zonite. A proprietary trade name for sodium hypochlorite solution for disinfecting and antiseptic uses.†

Zonolite. A proprietary registered trade mark for a mineral product made by the heat treatment of vermiculite, a mineral which is similar to a crude mica. It is used in the manufacture of building materials and for use as high temperature insulation.†

Zonyl. Fluorochemical surfactant. (Du Pont (UK) Ltd).

Zootic Acid. Hydrocyanic acid, HCN.†

Zootinsalz. Synonym for Nitratine.†

Zopaque. A proprietary form of titanium oxide for rubber mixing.†

Zorite. A proprietary alloy containing 35 per cent nickel, 15 per cent chromium, 1.75 per cent manganese, and 0.5 per cent carbon with iron.†

Zoroxin. Norfloxacin. Applications: Urinary antiseptic. (Merck Sharp & Dohme).*

Zovirax Products. Proprietary formulations of acyclovir or acyclovir sodium. Applications: For oral and intravenous treatment of initial episodes and the management of recurrent episodes of herpes genitalis and in topical treatment in limited nonlife-threatening mucocutaneous Herpes simplex virus infections in immunocompromised patients, depending on the formulation. (The Wellcome Foundation Ltd).*

Z-Siloxide. A trade name for a compound obtained by fusing silica with 0.1-2 per cent zirconia. It is a silica glass.†

Z Span Spansule Capsule. Zinc sulphate monohydrate. Applications: Sustained release preparation for use when inadequate diet calls for supplementary zinc and treatment of zinc deficiency where indicated. (Smith Kline and French Laboratories Ltd).*

Zulite. A bituminous paint similar to Melanoid but for use as a preservative for wood.†

Zumisite. Yeast food. (ABM Chemicals Ltd).*

Zwickau Yellow. See CHROME YELLOW.

Zyklon. A proprietary hydrogen cyanide fumigant for use against insect and vermin pests.†

Zyloprim Tablets. A proprietary formulation of allopurinol. Applications: The management of primary and secondary gout, recurrent calcium oxalate calculi and leukaemia and in patients with lymphoma and other malignancies who are receiving therapy which causes elevations in serum and urinary uric acid levels. (The Wellcome Foundation Ltd).*

Zyloric. A proprietary preparation of Allopurinol. Used in gout. (The Wellcome Foundation Ltd).*

Zymase. An alcohol-producing enzyme secreted by yeast cells. It decomposes grape sugar.†

Zymin. (Permanent Yeast). A product obtained by partially drying ordinary yeast, immersing it in acetone for fifteen minutes, which kills yeast, drying on filter paper, and washing with ether. It produces alcohol from grape sugar.†

Zymocasein. A phospho-protein obtained from yeast. It is similar to Caseinogen.†

Zymogen. A commercial product. It consists of a nitrogenous substance, to provide food for yeast in fermentation.†

Zytel. A registered trade-mark for a range of nylon resins coded as follows:- *42.* A grade used for the extrusion of tubes, with high melting- point, good resistance to abrasion and impact and a high degree of stiffness.

58 HS-L. A flexible grade used for general-purpose jacketing.

63. A grade soluble in alcohol.

70G. Nylon 66 reinforced with short glass fibres to give good dimensional stability.

70 GHR. Similar to *70G*, but stabilized against hydrolysis.

71 G. A series of modified Nylon 66 reinforced with short glass fibres.

77 G. A series of Nylon 6.12 reinforced with glass fibres.

91 HS. A heat-stabilized and plasticized grade used to make flexible tubing.

101. A nylon moulding compound.

105 BK-10. A black nylon composition with good resistance to weathering.

109 L. A chemically-modified nylon used in the moulding of heavy sections.

122. Zytel 101 modified to give enhanced resistance to hydrolysis.

131 L. Zytel 101 modified to permit fast moulding cycles.

141. A nylon with extra toughness and strength.

151 L. A grade of Nylon 6.12 with low absorption of moisture and good dimensional stability.

153 HS-L. A grade of Nylon 6.12 modified to provide heat stability.

158 L. A grade of Nylon 6.12 possessing higher melt viscosity and extra toughness.

408. A modified natural Nylon 66, tough but easily processed.

408 HS. A heat-stabilized grade of Nylon 66.

410 BK-10. A black Nylon 66 with good resistance to weathering.

3606. A grade of nylon stabilized against heat and weather.

ST 801. A modified 66 polyamide claiming outstanding toughness. (Du Pont (UK) Ltd).†

Z.P.D. The zinc salt of pentamethylene-dithiocarbamic acid. An ultra-rubber vulcanization accelerator.†

1900. Ultra high molecular weight polymers. Applications: Aerospace and aviation, machinery, mining, sporting goods and toys, textiles. (Hercules Inc).*

A

Abatron Inc,

141 Center Drive, Industrial District,
Gilberts, Ill 60136
USA

Abocast, Abocrete, Abocure, Aboseal.

Abbott Laboratories Limited,

Queensborough,
Kent ME11 5EL
UK

Abbott Laboratories,
Chemical & Agricultural Products Division,
14th Sheridan Road,
North Chicago, IL 60064
USA

Ross Laboratories,
Div of Abbott Laboratories,
625 Cleveland Ave,
Columbus, Ohio 43216
USA

A-hydroCort, A-methaPred, Abbaflox, Abbalgesic, Abbenclamide, Abbloraz, Abbocillin-DC, Abbokinase, Abbolactone, Abbopramide, Abbopurin, Abboxapam, Abboxide, Aerotrol, Amical Biocides, Amidate, Aminosol, Ancyte, Butesin Picrate, Cartrol, Clear Eyes, Collokit, Cordilox, Cylert, Cystemme, Cystorelin, Depakene, Depakote, Diasone Sodium Enterab, EES, Endrate, Enduron, Erthro, Ery-Ped, Ery-Tab, EryDerm, Erythrocin, Erythrocin Lactobionate-IV, Erythromid, Erythroped, Ethrane, Eutonyl, Fero-Gradumet, Ferrograd, Ferrograd C, Ferrograd Folic, Forane, Gemonil, Harmogen, Heparin Lock Flush, Hexamic Acid, Holocaine Hydrochloride, Irofol C, Janimine, K-Lor, Karaya Paste, Kelfizina, K.Tab, Mammol, Mono-Kay, Murine Ear Drops, Murine Plus, Narcan, Natopherol, Nembutal, Nembutal Sodium, Nitropress, Norisodrine Aerotrol, Norisodrine Sulfate,, Ogen, Oretic, Oriodide-131, Panwarfin, Paradione, Pediaflor, Pediamycin, Penthrane, Pentothal, Pentothal Sodium,

Abbott Laboratories Limited,

Pertscan-99m, Phenurone, Placidyl, Quelicin, Rachromate-51, Radiocaps-131, Raolein 131, Rondec, Rondec DM, Seleen, Selsun, Selsun Blue, Sorlate, Spectam, Stronscan-85, Sucaryl, Surbex T, Tetramet-125, THAM, Theograd, Theriodide-131, Thronothane Hydrochloride, Tracervial-131, Tridione, Tronolane, Truozine, Ureaphil, Vercyte, Vi-Daylin, Vinisil.

ABM Chemicals Ltd,

Poleacre Lane, Woodley,
Stockport, Cheshire, SK6 1PQ
England

Ambazyme, Ambiteric, Ambiteric D, Amfaid, Amphionic, Amphionic 25B, Amylozyme, Anonaid, Anonaid TH, Assaf, Aziplex, Bacterase, Bitran, Bitran, Catafor, Cataid, Clarifloc, Clortol, Collone, Coupler, Diamalt, Diamex, Duoteric, Effesay, Ferriplex, Ferriplus, Fomescol, Gelatase, Glocure, Glofoam, Glokem, Glokill, Glokill PQ, Glokill 77, Glomeen, Glopol LS6 and L6, Glopol 461, Gloquat, Gloquat 1032, Hydan, Hyflux, Hyflux M, Intrex, Intrex Asa, Intrex DW81, Intrex HA70, Lucilite, Miravon, Morpan, Morpan BC, Nerco, Nercol, Nercolan, Nercosol, Nerfinol, Nerloate, Nervan, Nervan CP, Nervanaid, Nervanase, Nonaid, Oxzone, Panazyme, Pentrone, Pentrone ON, Pentrone S, Pentrosan, Petrone A4 and A6C, Pielanase, Pistol, Proteinase Pulluzyme, Ranotex, Resistone, Resistone QD, Scintillase, Sensitizer, Solumin, Solumin F, Solumin PFN, Solumin PV27, Solumin T45S, Solumin V27SD, Stabilizer, Stanmine Stannine, Steroxol, Stortex, Supronic, Supronics, Taxafor, Texofor, Texofor FN, FP and FX, Texogent, Texowax, Torrax, Trisophone, Turbex, Ultrafloc, Zumisite.

Abril Industrial Waxes,

78-79 Long Lane,
London EC1A 9ET
UK

Abril, Cerabrit, Cosmic.

Acheson Colloids Company,
Acheson Industries (Europe) Limited,

Prince Rock, Plymouth,
Devon PL4 0SP
UK

Aquadag, Dag, Electrodag, Emralon, Molydag, Oildag.

Active Organics,

7715 Densmore Avenue,
Van Nuys, Ca 91406
USA

Acticulum, Actigen, ActiMoist, Actiphyte, Actiplex.

Adasco-Inc,
Industrial & Automotive Division,

10609 Briggs Avenue,
Cleveland, Ohio 44111
USA

Aqua Magic, Saturn Glace, Steel Guard.

ADC Resins,

2410 Peninsula Road,
Oxnard CA 93030,
USA

ADC, WW (Wet Wax).

Addagrip Surface Treatments UK Limited,

Bird-in-Eye Hill,
Uckfield, East Sussex TN22 5HA
UK

Addabond, Addacoat, Addaflex, Addaflor, Addagrout, Addalevel, Addamortar, Addapitch, Addaprime, Addaseal, Addasure.

ADM Tronics Unlimited Inc,

153 Ludlow Avenue,
Northvale, New Jersey 07647
USA

Acrilester, Aquaforte, Polaqua, Santel.

Adria Laboratories Inc,

PO Box 16529,
Columbus, Ohio 43216
USA

Chymex, Endyne, Epsilan-M, Evac-Q-Mag, Evac-Q-Tabs, Fluidil, Ilopan, Ilozyme, Jaon, Kaon-Cl, Modane Soft, Neosar, Tymtran.

Adshead Ratcliffe & Company Limited,

Derby Road,
Belper, Derbyshire DE5 1WY
UK

Adshead Ratcliffe & Company Limited,

Arbo, Arbocaulk, Arbocrylic, Arboflex, Arbofoam, Arbokol, Arbolite, Arbomast, Arboseal, Arbosil, Arbostrip.

Agfa-Gevaert,

Septestraat 27,
B 2510 Mortsel,
Belgium

Agfa-Gevaert.

Agrichem Limited,

Padholme Road,
Peterborough PE1 5XL
UK

Hytrol.

A H Marks & Company Ltd
see Marks, A H & Company Ltd

Aicello Chemical Company Limited,

45 Koshikawa, Ishimaki Honmachi,
Toyohashi City, Aichi Pref 441 11
Japan

Aicello, Boselon, Solublon.

Akzo Chemie UK Ltd,

1-5 Queens Road, Hersham,
Waltham-on-Thames, Surrey KT12 5NL
England

Akzo Chemie Nederland BV,
PO Box 975,
3800 AZ Amersfoort,
The Netherlands

Ameen, Armac, Armeen, Armeen DM Series, Armid, Armofilm, Armoflo, Armogard, Armogel, Armogloss, Armohib, Armohib 25 and 28, Armostat, Armoteric LB, Armoteric SB, Aromox, Arquad, Butanox, Cleroxide, Cyclonox, Diadem Chrome, Dissolvine, Distec, Duomac, Duomeen, Duoquad, Elfan A432, Elfan KT550, Elfan NS 243S, Elfan NS 682 KS, Elfan NS242, Elfan NS252 S, Elfan OS 46, Elfan WA Series, Elfan 200, Elfan 240 and 240S, Elfan 240M and 240M/S, Elfan 240T and 240T/S, Elfan 280, Elfan 680, Elfanol 510, Elfanol 616, Elfanol 850, Elfanol 883, Elfapur N50, Elfapur N70, Elfapur N90, N120 and N150, Epilink, Epilok, Ethofat, Interstab,

Intraval Sodium, Kurade, Lauridit, Laurydol, Lucidol, Lucipal, Novadelox, Nutramin, Perkadox, Phosclere, Purochem, Stanclere, Stannicide, Synouryn, Trigonal, Trigonox.

lbright & Wilson Ltd,

Phosphates Division,
Albright & Wilson House, PO Box 3,
Hagley Road West, Oldbury B86 0NN
England

Albright & Wilson Ltd,
Detergents Division (Marchon),
Marchon Works, Whitehaven,
Cumbria CA28 9QQ
England

Albright & Wilson Ltd,
Detergents Division,
PO Box 15, Whitehaven,
Cumbria CA28 9QQ
England

Tenneco Malros Ltd,
Albright & Wilson Resins & Organics Division,
Rockingham Works, Avonmouth,
Bristol BS11 0YT
England

Albright & Wilson Limited,
1 Knightsbridge Green,
London SW1X 7QD
UK

Syntex Ophthalmics Inc,
1100 E Bell Road, PO Box 39600,
Phoenix, Ariz 85069-9600
USA

Accomet, Accomet C, Albrichrome, Albricide, Albrifloc, Albrightex, Albrilan, Albrilene, Albrilon, Albrilube, Albrinol, Albrinyl, Albriquest, Albriscour, Albrisolve, Albrisperse, Albritone, Albrivap, Alecra, Alexis, Amcide, Amgard, Ansa, Antelope, Aquarite, Aspon, Astralex, Banox, Bewoid, Bewopac, Bex, Briphos, Briquest, Bu-White, Bumal, Burez, Calbrite, Calgon, Calgonite, Caliment, Calipharm, Clearsol, Cresolox, Eltesol, Eltesol ACS 60, Eltesol CA65 and CA96, Eltesol PSA, Eltesol PX, Eltesol ST90, ST Pellets and PT90, Eltesol SX30, Eltesol SX93 and SX Pellets, Eltesol TA, TA65 and TA96, Eltesol TPA, Eltesol XA, XA65 and XA90, Eltesol 4009 and 4018, Eltesol 4402, 4403 and FDA 55/8, Empicol, Empicol AL30/T, Empicol DLS, Empicol EAB, Empicol EGB, Empicol EL, Empicol ESB and ESC, Empicol ETB, Empicol LM and LMV, Empicol LQ, Empicol LX and LXV, Empicol LY28/S, Empicol LZP, Empicol LZ, LZV,LZG and LZGV, Empicol LZ/E, LZV/E, LZ/D and LZV/D, Empicol MD, Empicol ML26, Empicol SCC, SDD, SFF, SGG and STT, Empicol TC30/T and TCR/T, Empicol TDL, Empicol TL40/T, TLP/T and TLR/T, Empicol 0045, Empicryl, Empigen, Empigen AB, AH, AM and AY, Empigen AS and AT, Empigen BAC, Empigen BB, Empigen BT, Empigen CDR, Empigen CM, Empigen

Albright & Wilson Ltd,

XDR, Empilan, Empilan BD, Empilan NP9, Empimin, Empimin KSN, Empimin LAM, Empimin LR28, Empimin LSM, Empimin LS30, Empimin MA, Empimin MHH, Empimin MKK, Empimin MSS, Empimin MTT, Empimin OT, Empiphos, Empiquat, Empiwax, Flo-Con, Gazelle, Hagafilm, Hagatreat, Hagevap, Hydros, Ibex, Kalipol, Kalipol 18, Kanigen, Keydime, Laurex, Malros, Measac, Metagon, Micromet, Nansa, Nansa AS40, Nansa BMC, Nansa HS, Nansa LES42, Nansa SL and SS, Nansa SSA, Nansa TS60, Nansa UCA/S and UCP/S, Nansa YS94, Nansa 1042, Nansa 1042/P, Nansa 1169/P, Nubex, Octorez, Octosol, Phosbrite, Phosclene, Phostin, Plusbrite, Polygon, Proban, Puromix, Puron, Pyrobrite, Snowtack, Springbok, Sterilite, Tenncol, Tenrez, Tetron, Tolgard, Tolplaz, Waxemul, Xylock 225, Xylok.

Alco Chemical Corporation,

P O Box 5401, 909 Mueller Drive,
Chattanooga, TN 37406
USA

Alcodrill HPD-D, Alcodrill HPD-L, Alcogum AN 10, Alcogum L-11, Alcogum L-27, Alcogum L-36, Alcogum L-52, Alcogum L-60, Alcogum 296-W, Alcogum 310, Alcogum 9635, Alcogum 9661, Alcogum 9710, Alcosperse 104, Alcosperse 107-D, Alcosperse 144, Alcosperse 149-C, Alcosperse 169, Alcosperse 175, Alcosperse 249, Alcosperse 602, Alcotreat PC 95, Alcotreat 182, Alstromed A 18 LV, Aquamet M, Aquatreat AR-225-D, Aquatreat AR-232, Aquatreat AR-626, Aquatreat AR-648, Aquatreat AR-7-H, Aquatreat AR-900, Aquatreat DNM-30, Astrowet 0-75, Vulnopol KM.

Alcon Laboratories Inc,

PO Box 1959,
Ft Worth, Texas 76101
USA

Balanced Salt Solution, Dendrid, Isopto Alkaline, Isopto Atropine, Isopto Carbachol, Isopto Carpine, Isopto Eserine, Isopto Frin, Isopto Homatropine, Isopto Hydrocortisone, Isopto Hyoscine, Isopto P-ES, Isopto Plain, Isoptocetamide, Lyophrin, Maxidex, Maxitrol.

Aldo Products Company Inc,

18005 Lisa Lane,
Brookfield, WI 53005
USA

Aldobond, Aldocoat.

Allergan Pharmaceuticals Inc,

2525 Dupont Drive,
Irvine, Calif 92715
USA

Bleph-10 Liquifilm, Bleph-10 SOP, Chloroptic, Exsel, FML Liquifilm, Genoptic Liquifilm, also Genoptic SOP, Herplex Liquifilm, HMS Liquifilm, Ophthetic, P V Carpine Liquifilm, Pilofrin, Pred Forte, Pred Mild, Prefin Liquifilm, Propine.

Allied Chemical Corporation,

Morristown,
New Jersey 07960
USA

Aclar, Aclon PCTFG, Capron, Capron Alpha 8200 C, 8202 C and 8203 C, Capron 8200, Capron 8202, Capron 8206 S, Capron 8230, Capron 8231 and 8233, Capron 8250, 8251 and 8253, Capron 8270, Caprun, Halar, Halar E-CTFE, Halon, Petra.

Allied Colloids Ltd,

PO Box 38, Low Moor, Bradford,
West Yorkshire BD12 0JZ
England

Alcolec 532, Alcolube CL, Alcopol AH New, Alcopol FA, Alcopol O, Alcopol OB, Alcopol OD, Alcopol OS, Alcopol T, Alcosperse, Antioxygene A, Antioxygene AFL, Antioxygene AN, Antioxygene BN, Antioxygene CAS, Antioxygene INC, Antioxygene MC, Antioxygene RES, Antioxygene RM, Antioxygene RO, Antioxygene STN, Antioxygene WBC, Colcar D, Dispex A40 and N40, Dispex G40 and GA40, Glascol HA2, Glascol HA4, Glascol HN2, Glascol HN4, Glascol PA6, Glascol PA8, Glascol PN 8, Viscalex EP 30.

Allied Corporation,

PO Box 1139R,
Morristown, NJ 07960
USA

Allied Corporation,
Water Treatment Division,
PO Box 1139R,
Morristown, NJ 07960
USA

Allied-Signal Fluid Systems,
10124 Old Grove Road,
San Diego, CA 92131
USA

Allied Signal,
Sinclair and Valentine Division,
2520 Pilot Knob Road,
St Paul, MN 55120
USA

Allied Corporation,

Allied Signal,
Planarization & Diffusion Products Division,
1090 S Milpitas Blvd.,
Milpitas, CA 95035
USA

Norplex,
PO Box 1448,
La Crosse, WI 54601
USA

UOP Inc,
PO Box 5017,
Des Plaines, IL 60017-5017
USA

Accuglass, Accuspin, AID, Clarifloc, Comtek, Genesolv A Solvent, Genesolv D Solvent, Genetron Dry Refrigerants, Magnum, Norplex laminates, Polygloss, Roga, Rotoval, Sinvabond, Sinvaset, Spiragas, Sustane 1-F, TFC, Unilink 450, Zephyr.

Altana Inc,
Savage Laboratories Division,

60 Baylis Road,
Melville, NY 11747
USA

Alphatrex, Betatrex, Brexin, Chromagen, Chromagen, Dilor Elixir, Dilor-G, Ditate-DS, Ethiodol, Mytrex, Mytrex F, Nystex, Trymex Cream and Ointment, Trysul.

Altex Chemical Co Ltd,

Clayfield Works, Slaithwaite,
Huddersfield, West Yorkshire BD12 0JZ
England

Alstat, Altolube.

Amax Specialty Metals Corporation,
Amax Mineral and Energy Division,

1 Greenwich Plaza,
Greenwich, CT 06836
USA

Amax - XLP, Amcron, Amsil, Amsulf, Amtel, Amzirc, OFHC Copper.

Ambrosia Scents,

524 E Ohio,
Princeton, IN 47670
USA

Scent Sticks.

Amerchol Corporation,

PO Box 351, Talmadge Road,
Edison, N Jersey 08817
USA

Acetulan, Amerchol BL, Amerchol C, Amerchol CAB, Amerchol H-9, Amerchol L-101, Amerchol L-99, Amerchol RC, Amerchol 400, Amerlate LFA, Amerlate P, Amerlate W, Isopropylan 33, Isopropylan 50, Lanamine, Lanocerin, Lanogel 21, Lanogel 31, Lanogel 41, Lanogel 61, Lanogene, Modulan, Ohlan, Solulan C-24, Solulan C-25, Solulan PB-10, Solulan PB-2, Solulan PB-20, Solulan PB-5, Solulan 16, Solulan 5, Solulan 75, Solulan 97, Solulan 98, Waxolan P-5.

**American Critical Care,
Div of Hospital Supply Corp,**

1600 Waukegan Road,
McGaw Park, Ill 60085
USA

American McGaw,
American Hospital Supply Corp,
2525 McGaw Ave,
Irvine, Calif 92714-5895
USA

Americaine, Bretylol, Brevibloc, Calciparine, Hespan, Intropin, Liqui-Cee, Lotioblanc, Metro IV, Resectisol, Sodital, Tridil, Vulvan.

**American Cyanamid Co,
Fine Chemicals,**

Wayne,
NJ 07470
USA

Abequito, Cyasorb 5411, Myprozine, Spectra-Sorb UV 24, Spectra-Sorb UV 284, Spectra-Sorb UV 531, Spectra-Sorb UV 5411, Spectra-Sorb UV 9, Styquin, Tramisol, Vigazoo.

American Hoechst Corp,

American Hoechst Corp,

Route 202-206 North,
Somerville, NJ 08876
USA

Locron P.

American Laboratories Inc,

4410 South 102nd Street,
Omaha Ne 68127
USA

Microbiotone, Pharmatone, Platone.

American McGaw
see American Critical Care

American Norit Company Inc,

420 Agmac Avenue,
Jacksonville, Florida 32205
USA

Hydrodarco.

American Optical Corp,
Soft Contact Lens Business,

55 New York Ave,
Framingham, Mass 01701
USA

Aosoft, Superthin.

Amersham Corp,

2636 S Clearbrook Dr,
Arlington Heights, Ill 60005
USA

Amerscan MDP Kit, EHIDA Kit, Ibrin, Thallium.

Amoco Chemicals Company,
Engineering Resins Department,

200 E Randolph Drive,
Chicago, Illinois 60601
USA

Torlon 4203L, Torlon 4275, Torlon 4301, Torlon 4347, Torlon 5030, Torlon 7130.

Ampacet Corporation,

250 South Terrace Avenue,
Mount Vernon, NY 10550
USA

Kromaplast, Mastercolor, Polygrade.

Anaquest,
Div of BOC International,

2005 W Beltline Highway,
Madison, Wis 53713
USA

Fluoromar, Idoklon, Vandike P.360, Vandike 7085, Vandike 7086.

Anchor Chemical Company
see Pacific Chemical Industries Pty Limited

Anderson Development Company,

1415 E Michigan Street,
Adrian, MI 49221-3499
USA

Andersil, Andrez 8000, Andur, Curene, Her, Q-Therm, Super-A.

Anstead, D F Ltd,

Radford Way, Billericay,
Essex CM12 0DE
England

Amerchol Polysorbate, Amerscreen, Monafax, Monamate CPA, Monamate OPA, Monamine, Monaterge, Monateric CA-35%, Monateric CAB, Monateric CDX38, Monateric CEM-38%, Monateric CSH 32, Monateric Cy Na-50%, Monateric ISA-35%, Monateric LF, Monateric 1000, Monateric 811, Monateric 85, Monawet MB-45, Monawet MM-80, Monawet MO, Monawet MT, Monawet SNO-35, Monawet TD-30, Monazoline CY, C, O and T, Solulan.

Antec International Limited,

Windham Road, Chilton Industrial Estate,
Sudbury, Suffolk CO10 6XD
UK

Antec Farm Fluid S, Antec Longlife 250 S, Antec OO-Cide.

Anzon Ltd,

Cookson House, Willington Quay,
Wallsend, Tyne & Wear NE28 6UQ
England

Anzon, Oncor 75, Ongard, Timonox, Zedox, Zircosil.

Aquarium Systems Inc.,

8141 Tyler Boulevard,
Mentor, Ohio 44060
USA

Instant Ocean, SeaCure, SeaGarden, SeaTest.

Aquatrols Corp of America,

1432 Union Avenue,
Pennsauken, NJ 08110
USA

Aqua Gro 'G' Granular, Aqua-Gro 'L' Liquid, Aqua-Gro 'S' Spreadable, Asbesto-Wet, Folicote Transpiration Minimizer, Super-Sorb 'C' Water Absorbant, Super-Sorb 'F' Water Absorbant.

Arcmann-Denmark A/S,

Strandparken 15,
DK 8000 Aarhus C
Denmark

Arctite Injection Mortar, Arctite Slurry 200 B, Arctite Tanking Mortar 500.

Arizona Chemical Company,
Tall Oil Fatty Acids & Rosin Products Division,

200 South Sudduth Place,
Panama City, FL 32404
USA

Acintene, Acintol D40LR, D30LR, D25LR, Acintol FA-1, FA-1 Special, FA-2, FA-3, EPG, 746, Acintol R Type SFS, Acintol R Type SM4, Acintol R Type S, R Type SB, R Type 3A, R Type L03A, Tall Oil Rosin, Acintol, Liquaros, Arizole Anethole Extra, Arizole Pine Oil, Arizona DRS-40, DRS-42, DRS-43, DRS-50, DRS-51E Disproportionated Tall Oil Rosin Soaps, Arizona DR22 Disproportionated Tall Oil Rosin, Arizona DR24 Disproportionated Tall Oil Rosin, Arizona DR25 Disproportionated Tall Oil Rosin, Arizona 208 Tall Oil Fatty Acid Ester, Arizona 258 Tall Oil Fatty Acid Ester, Demix 7730, 7740, 7750 Emulsifiers, Zonarez, Zonatac, Zonester 25, 40, 55, 65, 75, 85, 100 Resin Esters, Zonester 85-D60 Resin Ester Dispersion.

Arlington Mills,

1430 East Davis Street,
Arlington Heights, IL 60005
USA

Arlinflex.

Armour Pharmaceutical Co,

303 S Broadway,
Tarrytown, NY 10591
USA

Acthar, Acthar Gel, Albuminar, Amfac, Amisyn, Aquasol A, Aquasol E, Arneel DN, Arneel HF, Arneel S, Arneel TOD, A.A.A. Spray, Barotrast, Biopar Forte, Calcitare, Calsynar, Chrometrace, Chymacort, Chymar, Chymar Ointment, Chymocyclar, Chymoral, Clysodrast, Coppertrace, Dialume, Esophotrast, Factorate, Gammar, Gastalar, H P Acthar Gel, Mangatrace, Montothene G50, Oratrast, Panar, Prothar, Sinaxar, Stimate Injection, Syrtussar, Thytropar, Trentadil, Zinctrace.

A/S Cheminova
see Cheminova, A/S

Ashe Chemicals,

Ashetree Works, Kingston Road,
Leatherhead, Surrey
UK

Amplex.

Ashe Laboratories Limited,

Ashetree Works, Kingston Road,
Leatherhead, Surrey KT22 7JZ
UK

Tancolin.

Ashland Chemical Company,

Ashland Chemical Company,

PO Box 2219,
Columbus, Ohio 43216
USA

Ashland Chemical Company,
Drew Industrial Division,
1 Drew Plaza,
Boonton, N J 07005
USA

Ashland Chemical Company,
Specialty Polymers & Adhesives Division,
1745 Cottage Street,
Ashland, Ohio 44805
USA

Ashland Chemical Company,
Resin & Chemicals Division,
2620 Royal Windsor Drive,
Mississauga, Ontario L5J 4ET
Canada

Advantage, AME 4000, Amercor, Amerfloc, Amerfloc Plus, Amergel, Amergize, Amergy, Ameroyal, Amerplex, Amersite, Amerstat, Amerstat 252, Amertrol, Amerzine, Arofene, Aroflat, Arofoam, Arolon, Aroplaz, Aropol, Aropol Phase Alpha, Aropol Phase II, Aropol WEP, Aroset, Arotap (Phenolic), Arotech, Ashland Hi-Sol 10, Ashland Hi-Sol 15, Ashland Kwik-Dri, Ashland Lacolene, Biosperse, Chem-Rez, Drewclean, Drewcor, Drewfax, Drewfax 400 Series, Drewfax 600 Series, Drewfax 800 Series, Drewfloc, Drewplex, Drewplus, Drewtrol, Envirez, Hetron, Hi Sol, Isocure, Isopaste, Isoset, Isoset WD3-A322 Emulsion Resin, Isoset WD3-CM402 Emulsion Resin, Kwik Dri, Lacolene, LinoCure, Maxigard, Pep Set, Piliogrip Adhesive System for Styructural Bonding, Plio-Caulk, Plio-Nail, Plio-Seam, Plio-Tac, Plio-Tac 38, Pliobond, Suresperse, Velva Coat, Velva Dri, Velva Wash, Velvalite, Velvaplast, VIC Coatings, VPC Coatings.

Ashley Polymers,

5114 Fort Hamilton Parkway,
Brooklyn, NY 11219
USA

Ashlene.

Associated Lead Manufacturers Limited,

Crescent House,
Newcastle upon Tyne NE99 1GE
UK

Almstab, Caldiox, Cooksons, Negex, Strandex.

Astor Chemical Limited,

Tavistock Road,
West Drayton, Middlesex UB7 7RA
USA

Okerin.

Astra Pharmaceutical Products Inc,

50 Otis St,
Westboro, Mass 01581
USA

Aptine, Betaloc, Bricanyl, Bricanyl Expectorant, Citanest, Duranest, Dyclone, Nesacaine, Nobecutane, Polocaine, Sensorcaine, Tosmilen, Xylocaine, Xylocard, Xyloproct.

Astra-werke Degussa-Pharma-Gruppe,

Daimlerstr 25,
D-6000 Frankfurt 1
Germany

Degussa Limited,
Paul Ungerer House, Earl Road,
Stanley Green, Handforth, Wilmslow,
Cheshire SK9 3RL
UK

Degussa AG,
Postfach 11 05 33,
D-6000 Frankfurt 11
Federal Republic of Germany

Degussa,
Weissfrauenstrasse 9,
Frankfurt am Main,
Germany

Acturin, Aerosil, Aerosil 130, 150, 200, 300, 380, Allergospasmin, Aquaphoril, Arbyl, Arbylen, Becosal, Briosil, Bronchodil, Bronchospasmin (Bronchodil), Calsil, Carbosal, Caroat, Ceracolor, Cofill, Colcolor, Comelian, Contex, Corasole, Cormelian (Comelian, Labitan), Cupalit, Cytoxan, Degadur, Degalan, Degalex, Degament, Degapas, Degaplast, Deglas, Degubond, Degucast, Degudent, Degulor, Degusorb, Depen, Derussole, Desimpal, Deva, Dicosal, Diurex, Diurexan (Diurex, Aquaphoril), Duallor, Duralloy, Durferrit, Durosil, Efweko, Egalisal, Elcema, Endoxan (Cytoxan, Genoxal, Sendoxan, Endoxana), Endoxana, Extrusil, Ferri-Darotin, Fixegal, Flammastik, Fleur, Formac 40, Fosfostilben, Genoxal, Granuform, Holoxan (Mitoxana, Ifomide), Honvan (Honvol, ST 52, Fosfostilben), Honvol, Ifomide, Ildamen, Indio, Kamillosan, a Katadolon, Labitan, Lamalgin, Lamefin, Lamefix, Lamegum, Lamephan, Lamepon, Lameprint, Lepandin, Luxalloy, Metasal, Mitoxana, Neacid, Neocosal, Orchidee, Osimol, Palliag, Paraglas, Pasilex, Pendramine, Percarbamid, Permador, Pontallor,

Astra-werke Degussa-Pharma-Gruppe,

Printogen, Proventin 7, Quab 151, Quab 188, Satessa, Sendoxan, Sident, Sident, Silteg, Sipernat, ST 52, Stabilor, Standalloy, Sunshine, Supersorbon, Systral, Tack, Tephal, Thioctacid, Transpafill, Transpulmin Balsam, Transpulmin Syrup, Trolovol (Pendramine, Depen), Ultrasil, Uromitexan, Waxit, Wessalith, Wessalon, Wolfin 18.

Astro Industries Inc,

PO Box 2559, 114 Industrial Boulevard,
Morganton, NC 28655
USA

Aricel, Astro Floctite, Astro Mel.

Atlas Chemical Industries (UK) Ltd,

Cleeve Road, Leatherhead,
Surrey KT22 7SW
England

Atpet, G-3300.

Atochem Incorporated,

Chemical Division,
266 Harristown Road, PO Box 607,
Glen Rock, NJ 07452
USA

Atochem Incorporated,
Polymers Division,
266 Harristown Road,
Glen Rock NJ 07452
USA

Altene DG, Baltane CF, Evatane, Foraflon, Pebax, Platamid, Plataril, Rilsan, Rilsan 11 and 12, Voltalef.

Australian Synthetic Rubber Co Limited,

Maidstone Street, PO Box 33,
Altona 3018
Australia

Austrapol.

Avalon Chemical Co Ltd,

Hitchen Lane, Shepton Mallet,
Somerset BA4 5TZ
England

Avabond, Europolymer, Purez, Sorane, Stuk.

Avicon Inc,

6201 S Freeway, PO Box 85,
Ft Worth, Texas 76101
USA

Aviamide-6, Aviester, Avoilefin.

Avitrol Corporation,

7644 E 46th Street,
Tulsa, OK 74145
USA

Avitrol.

Avoca Pharmaceuticals,

Old Sawmills Road,
Faringdon, Oxon SN7 7DS
UK

Avoca.

Axel Plastics Research Laboratories Inc,

Box 855,
Woodside, NY 11377
USA

Axelglo, Clean Wiz-9, Melax, Mold Wiz Ext, Mold Wiz Int, Skin Wiz.

Ayerst Laboratories
see Wyeth Laboratories

Ayrton Saunders plc,

34 Hanover Street,
Liverpool, Merseyside L1 4LN
UK

Almagel, Analgesic Balm (AS), Astingol, Ayrtol, Beehive Balsam, Bikini Cream, Coterpin, Cremosan, Dispello, Halaurant, Heart Shape Indigestion Tablets, Iglodine, Inhalit, Insect Bite Cream (AS), Kastor, Linituss, Tabasan, Touch & Go.

B

Barnes-Hind Inc

895 Kifer Road,
Sunnyvale, Calif 94086
USA

Alpha Chymar, Barseb, Barseb HC, Comfort Eye Drops, Degest-2, Eppy/N, Ful-Glo, GP-II, Komed HC, Pro-cort, SoftMate DW.

BASF United Kingdom Ltd,

PO Box 4, Earl Road,
Cheadle Hulme, Cheshire, SK8 6QG
England

BASF UK Limited,
Plastics and Fibre Raw Materials Division,
PO Box 4, Earl Road,
Cheadle Hulme, Ches SK8 6QC
UK

BASF UK Limited,
Polyolefines & PVC Division
PO Box 4, Earl Road,
Cheadle Hulme, Ches SK8 6QC
UK

BASF UK Limited,
Technical Plastics Division,
PO Box 4, Earl Road,
Cheadle Hulme, Ches SK8 6QC
UK

BASF Wyandotte Corp,
100 Cherry Hill Road,
Parsippany, NJ 07054
USA

Acronal 14D, Acronal 160D, Acronal 21D, 27D and 30D, Acronal 350D, Acryl, Akaustan, Anthraquinone, Antistatin, Basacryl Salt, Basacryl/Bafixan, Bascal, Basilen, Basolan DC, Basopal, Basosoft, Blankit, Brilliant Indigo, Butofan D, Cellestren, Celliton, Condensol, Cottestren, Cyclanon, Decrolin, Dekol, Diofan, Diofan D, Emulan, Emulan PO, Flixapret, Glizarin Binder, Glyezin, Helizarin, Iporka, Kieralon, Kurofan, Kurofan D, Lanestren, Latensol AP8, Leophen, Leukotrop W, Lubasin, Lucobit, Ludigol, Lufibrol, Lupolen, Lupolen V-2524EX and V-3510K, Lupolen 1800 H/M/S, Lupolen 1810E, Lupolen 1810H, Lupolen 1812D and 1812EH, Lupolen 1814E, Lupolen 1852E/H, Lupolen 2040EX and 2410DX, Lupolen 2410S, Lupolen 2424H and 2425K, Lupolen 2430H, Lupolen 2452 E, Lupolen 3010 S, Lupolen 3020 D, Lupolen 3020 KX and 3025 KX, Lupolen 4261 AX, Lupolen 5011 K, Lupolen 5052 C, Lupolen 6011 K, Lupolen 804H and 1814H, Luprimol, Luprintan ATP, Luprintol, Luran, Luran KR 2517, Luran S, Luran 378P, Luran 757R, Luran 776S, Lurantin Supra, Lurotex A25, Lusynton, Lutensit A-ES, Lutensit A-LBA, Lutensit A-PS, Lutensit An 10, Lutensit AS, Lutensit K-LC, Lutensol, Lutensol AP, Lutexal, Lutofan, Lutofan D, Lutonal, Lutonal D, Lutonal LC, Luvican M170, Nekal BX, Nekal SBS, Nekanil, Neoplen, Novolen, Oppanol, Oppanol D, Ortol, Ortolan, Palamoll 632, Palamoll 644 and 646, Palamoll 645 and 647, Palamoll 855, Palanil, Palanil Carrier, Palatal, Palatal P5, Palatal P8, P50T and P52TL, Palatal S333, Palatin Fast, Palatinol AH, Palatinol DN, Palatinol D10, Palatinol K, Palatinol Z, Palusol, Perapret, Persistol, Plastamol, Plasticizer 13, Plastomoll BMB, Plastomoll DMA, Plastomoll DOA, Plastomoll NA, Plastomoll TAH, Plastomoll WH, Plastomoll 34, Plastopal 11, Pluronic F-68, Pluronic F68LF, Pluronic L-101, Pluronic L-62LF, Pluronic L62D, Poligen MMV, Poligen PE, Polystyrol, Polystyrol KR 253 and KR 2538, Polystyrol KR 2536, Polystyrol 143E, Polystyrol 165H, Polystyrol 168N, Polystyrol 427M, Polystyrol 432F, Polystyrol 466 I, Polystyrol 472 D, Polystyrol 473 E, Polystyrol 475 K, Prestogen, Primasol, Primazin, Propiofan, Propiofan D, Ramasit, Rongalit, Setamol, Siligen, Size, Soromin, Styrodur, Styrofan D, Styrofill, Styrolux, Styropor, Teluran, Terluran 846 L, Terluran 8760 Galvano, Terluran 886, Texapret, Textile Wax, Theic, Trilon, Ultradur, Ultraform, Ultramid A + B, Uniperol, Urecoll, Vialon Fast, Vinidur, Vinoflex, Vinoflex 377, Vinoflex 516, Vinoflex 526, Vinoflex 535, Vinoflex 719, Vinuran.

Bausch & Lomb,
Professional Products Div,

1400 N Goodman St,
Rochester, NY 14692
USA

B & L 70, CW 79, Soflens, Toric Contact Lens.

The Baxenden Chemical Co Ltd,

Paragon Works, Baxenden,
Accrington, Lancs BB5 2SL
England

Castomer, Castomer, Cyclo, Emcol, Flexocel, Florafoam, Fomrez, Formrez, Futura Flex, Futura Thane, Isofoam, Isofoam, Neopon, Spandofoam, Sulframin, Trixene, Trixene, Witcamide, Witcobond, Witcoflex, Xenacryl.

Bayer UK Ltd,

Bayer UK Ltd,

Bayer House, Strawberry Hill,
Newbury, Berkshire RG13 1JA
England

Bayer UK,
Agrochem Division,
Eastern Way,
Bury St Edmunds, Suffolk IP32 7AH
UK

Bayer Antwerpen NV,
Kanaaldok B1,
B-2040 Antwerp
Belgium

Bayer AG,
Bayerwerk,
D5000 Leverkusen
W Germany

Miles Pharmaceuticals,
Div of Miles Laboratories Inc,
400 Morgan Lane,
West Haven, Conn 06516
USA

Abasin, Acaprin, Aciderm, Acidol-Pepsin, Acilan, Acraconc, Acrafloc, Acralen A, Acramin, Acramin, Actal, Adalat, Adaptinol, Adimoll BO, Adimoll DN, Adimoll DO, Adiro, Agestan 68, Alevaire, Alizarin, Alkydal, Alkynol, Alumails, Anti-Oxidant AH, Anti-Oxidant AP, Anti-Oxidant DDA, Anti-Oxidant DNP, Anti-Oxidant DOD, Anti-Oxidant EM, Anti-oxidant MB, Anti-Oxidant PAN, Anti-Oxidant RR 10 N, Anti-Oxidant SP, Antifoam ET, Antifoam T, Antiozonant AFD, Antracol, Aolept, Apernyl, Apolomine, Aqua Ivy, AP, Aricyl, Arubren, Ascinin P, Ascinin R, Ascinin Special, Aspiquinol, Aspirol, Astra, Astraflex, Astragal, Astrazon, Asuntol, Atosil, Avertin, Avolan, Azlin, A.T. 10, Bacdip, Bayblend, Baybond, Baycast, Baychrom A, Baycillin, Baycoll, Baycryl, Baycuten, Bayderm A, Bayderm KF, Bayderm Lacquers Auxiliaries, Bayer Base Plates Glass-Clear, Bayer CM, Bayertitan, Bayer's Tonic, Bayferon, Bayferrox, Bayfidan, Bayfill, Bayfit, Bayflex, Bayfolan, Baygal, Baygen, Baygen Lacquers and Auxiliaries, Baygenal, Bayhibit AM, Baykanol AK, HLX, SL, Baykanol Liquor TN, Bayleton, Bayleton BM, Bayleton CF, Bayleton 5, Baylith, Baylon, Baylube, Baymer, Baymicin, Baymid, Baymidur, Baymin, Baymix, Baymol A and D, Baynat, Bayo-n-ox, Bayolin, Baypen, Bayplast, Baypreg, Baypren, Baypress, Baysical, Baysilone, Baysin, Baysport, Baystal, Baysynthol, Baytan, Baytec, Baytherm, Bayticol, Bendizon, Benzamin, Benzo, Benzo Cuprol, Bilarcil, Bilevon-Solution, Biltricide, Binotal, Blancorol AC, Blankophor, Blattanex, Bolfo, Bronchilator, Broncho-Binotal, Buna CB, Cadmopur, Cafaspin, Campolon, Campovit, Caprinol, Carbo-Pulbit, Carboraffin, Catosal, Cellidor, Cellit, Ceres, Cervantal, Chinaspin, Chromosal B, Cismollan BH, Citarin-L, Clont, Clophen, Cohedur, Colfarit, Combelen, Comital/Comital L, Compolon Forte, Compralgyl, Concurat-L, Contrastol W, Corephen 10, Cort-Dome, Cotolan Fast, Crasnitin, Crelan, Cyren-A, DC Cristobalite, Debenal, Decholin, Decopress, Delegol, Delegol-T, Delicron, Delta-Dome, Demodur RF, Depot-Impletol, Depot-Padutin, Desavin, Desmalkyd, Desmocap, Desmocoll, Desmocoll, Desmoderm, Desmoderm Foil, Desmodur, Desmodur R, Desmodur 1L, Desmodurs, Desmoflex,

Desmolac, Desmopan, Desmophens, Desmorapid, Detigon, Diadavin, Dibotin, Dichlofuanide, Dichlor-Stapenor, Dipterex 80, Dirame, Disflamoll DPK, Disflamoll DPO, Disflamoll TCA, Disflamoll TKP, Disflamoll TOF, Disflamoll TP, Disyston FE-10, Dolviran, Domeboro, Dontalol, Doroma, Draza, Droncit, Drossa, DTIC-Dome, Durethan, Durethan B, Durethan BKV, Durethane, Duroterm, Ectimar, Edolan, Egalon Colours, Egalon Auxiliaries, Egalon Thinners, Ektebin, Elaol VI, Eleudron-Solution, Elityran, Elvaron, Emulvin, Endojodin, Enusin Colours, Epontol, Ercusol, Erkantol, Escorpal, Euderm, Eukanol, Eukanol Colours, Eulan WA, Eulan 33, Eusin, Evidorm, Evipan Sodium, Fabahistin, Faringets, Fenoil, Fergon, Fluisil S55K, Fluorfolpet, Folimat, Folpet, Fouadin, Franol, Geostone, Germanin, Gestinal, Getosedine, Gina, Goltix, Gusathion MS, Helio, Helio and Helio Fast, Levanox, Heliofil, Hemrids, Herrifex DS, Herrisol, Hexopal, Hydraffin, IBR/IVP/P13, Impletol, Impranil, Incidal, Indanthren, Integrin, Irisol, Irisol Fast, Isoderm, Isolan, Isolan K, Isonal, Iversal/Iversal-A cum anaesthetico, Kalleonicit, Kannasyn, Katonium, Lampit, Lasonil, Latibon, Legupren, Leguval, Lekutherm, Lenium, Leromoll, Levafil, Levafix, Levaflex, Levaform, Levalan N, Levalin, Levapon, Levapren, Levasint, Levegal, Levepox, Levesol, Levo-Chrome-Cobalt-Assortment, Levogen, Levogen LF, Levophed, Levopress, Levoxin, Lewasorb, Lewatit, Lida-Mantle, Lingraine, Lobak, Lumicon, Lumicon Non Gamma 2, Lumicon Silver Amalgam/Powder, Luminal, Luminalettes, Macro-lex, Macrolex, Magnaphoscal, Makrofol, Makrolon, Mansonil, Maretin, Marfanil-Prontalbin, Masoten, Masticillin C/Masticillin M, Mefarol, Megaphen, Mersolat, Mesamoll, Mesamoll-Verdingrin, Mesitol, Metasystox, Mezlin, Microcillin, Migrane-Dolviran, Mintacol Solubile, Moldabaster/Moldabaster S, Moldano, Moldaroc, Morestan, Motopren, Murnil, Mycelex, Mycospor, Myocardol, Nadavin, Naganol, Naphthopone E, Naphtopon E, Negram, Neguvon, Neo Duroterm L, Neo Duroterm 3, 5 and 7, Neoteben, Neurocil, Nibren, Niclocide, Nico Padutin, Nigrosin Bases, Nimotop, Norodur, Novodur, Odylen, Optocillin, Optosil Liquid, Optosil Plus, Optosil/Optosil Hard, Ospolot/Ospolot Mite, Padutin, Padutin 100, Panadeine CO, Panadol, Panasorb, Pecutrin, Pekafix, Pekatop, Pekatray, Pergut, Perlit, Persiderm, Persiderm Black, Persoftal, Persoftal PE, Persoftal PE Special, Petramin, Phanodorm, Phenodorm, Phtalofix FN, Phtalogen, Phtalogen K/Phtalogen N1, Phtalotrop B, Pipanol, Plaquenil, Plastifix, Pocan, Porofor, Porofor, Poskydal, Predne-Dome, Prent, Prepon, Presinol, Presinol Mite, Preventol A2, Preventol CL 5, Primaquine Bayer, Pro-Diaban, Prolan, Prolongal, Prominal, Pyratex, Racumin, Rapidogen, Rapidogen and Rapidogen N, Rapidosept, Recoil, Redul, Renacit 4, Resiren, Resistherm, Resistopen, Resithren, Resochin, Resocoton, Resolamin, Resolin, Resolin P, Resonium-A, Resotren, Respumit, Retaminol, Retingan, Retingan R6, R7, R4B, Rhenoblend, Rhenocure, Rhenodin, Rhenofit, Rhenomag, Rhenopor, Rhenosorb, Rintal, Rintal Plus, Roccal, Rompun, Sali-Presinol, Sebacil-Emulsion, Securopen, Sencorex WG, Seominal, Sep 6, Setac, Silcasil S, Silicoderm F, Siligran, Silopren, Sirius, Sirius Sirius Supra Sirius Supra LL, Solbrol, Spasmo-Dolviran, Stapenor, Statexan, Statexan KI, Stepenor/Stepenor-Retard, Stilphostrol, Stromba, Sungard, Supracen, Suprafix Paste, Supramin, Supranol, Supronal Preparations, Sustilan N, Synthappret, Syntharesin, Talcid, Tanigan, Tardocillin 1200, Tegda, Telogen, Telon, Tensid, Theominal, Therban, Tiguvon, Timails, Totocillin, Totocillin, Trancopal, Trasylol, Triacetin, Triafol, Tribunil, Tridesilon, Trolovol, Trubin, Tungophen B, Tussiplegyl, Ugurol, Ultramoll, Ultramoll I, II, III, Ultramoll M, Unimoll, Unimoll BB, Unimoll DB, Unimoll DM, Unimoll 66 and 66M, Upixon, Urepan, Uvilon, Velcorin, Verofix, Vi-Dom-A, Vigantol, Vigantol-E-Comp, Vitromail, Vlem-Dome, Vulkacit, Vulkadur, Vulkalent, Vulkanol, Vulkanox, Vulkasil, Vulkollan, Weedazol-TL, Wet 6, Xantalgin, Xanthano, Xantogum, Xantopren, Xantopren Function, Xantopren Plus, Xantygen, Xeroderm S100, L67, Yaltox, Yomesan, Yomesan, Yomesan, Zephirol, Zinkoyd Aktiv.

BBU,

A-9020 Klagenfurt, Radetzkystrasse 2,
Postfach 95
Austria

BBU,

Austrostab, Austrox, Optox.

BDH Chemicals Ltd,

Broom Road, Poole,
Dorset BH12 4NN
England

Airbron, Algistat, Almacarb, Anahaemin, Analar, Ancofen, Ancolan, Ancoloxin, Ancovert, Anorvit, Aquaclene, Aristar, Azoene, Bisoxyl, Bromelains, Calcichrome, Carbosorb, Certistain, Chlorotex, Clelands Reagent, Colliron, Collubarb, Convol, Corangil, Coscopin, Coscopin Paediatric, Crystran, Dehydrocholin, Dilosyn, Dindevan, Disamide, D.F. 118, Electran, Emdite, Entacyl, Entair Expectorant, Entair-A, Estigyn, Ethiodan, Extil, Extil Compound Linctus, Falapen, Ferbelan, Ferlucon, Fluoderm, Gechophen, Gurr, Hipersolv, Hippuryl Amide, Ipexon, Lastil, Locan, Magtran, Marcain, Marevan, Mepilin, Mersalyl BDH, Mollifex, Mucomycin, Multivite, Myanesin, Mycil, M.F.C, Numorphan Oral, Nuvacon, Oestroform, Onadox-118, Optran, Panabath, Panacide, Panaclean, Paramol-118, Parvol, Pectamol, Pernivit, Peroheme 40, Phylatol, Phytoforol, Prokayvit Oral, Radiostol, Rapitard, Rondase, Scintran, Scorbital, Sea-Legs, Secrodyl, Sepramar, Serosteron, Siccolam, Spectrosol, Stanvis, Tova, Trypure Novo, Tursione, Ultrasol, Uro-Hexoids, Vactran, Volidan and Volidan 21, Volpar.

Beacon Chemical Company Inc

125 MacQuesten Parkway South,
Mount Vernon, NY 10550
USA

Magna Tac, Magnacryl.

Beecham Pharmaceuticals,

Clarendon Road, Worthing,
West Sussex BN14 8QH
England

Beecham Research Laboratories,
Brentford,
Middlesex TW8 9BD
England

Beecham Laboratories,
501 Fifth St,
Bristol, Tenn 37620
USA

Amoxil, Ampiclox, Augmentin, Bactocill, Broxli, Celbenin, Cerazole, Cloxapen, Codiazine, Diphentoin, Dycill, Fastin, Floxapen, Homagenets Aoral, Hycal, Klebcil,

Larotid, Magnapen, Maxolon, Maxolon, Menest, Migen, Nacton, Napental, Orbenin, Orovite, Paramax, Parentrovite, Penbritin, Penisem, Phe-Mer-Nite, Pollinex, Pyopen, Sansalid, Talpen, Theralax, Ticar, Tigan, Totacillin, Trisem, Uticillin.

Belzak Corporation,

850 Bloomfield Avenue,
Clifton, NJ 07012
USA

Belzak AC, Belzak BL-50.

Bengue & Co Limited,

Syntex House, St Ives Road,
Maidenhead, Berks SL6 1RD
UK

Bengue's Balsam, Cortenema, Ethyl Chloride BP, Metanium Ointment, Nestosyl Ointment, Opobyl (Pills), Pulmo Bailly.

Benzsay & Harrison Inc,
Specialty Chemicals Division,

938 Pearse Road,
Schenectady, NY 12309
USA

Benzofloc, Moore Floc, Sedifloc Flocculant Aids.

Berger Elastomers,

Portland Road,
Newcastle-upon-Tyne NE2 1BL
UK

Berpak, Flextron.

Berger Jenson & Nicholson Limited,
Berger Decorative Paints Division,

Petherton Road, Hengrove,
Bristol BS99 7JA
UK

Berger Colorizer - Full Gloss/Vinyl Matt/Vinyl Silk, Berger Cuprinol Woodpaints and Woodstains, Brolac Dualcote Acrylic Primer/Undercoat, Brolac Eggshell Low Odour, Brolac Full Gloss, Brolac PEP Vinyl Matt & Vinyl Silk Emulsions, Brolac Primers, Sealers and Surface Preparation Products, Brolac Specialist Coatings, Brolac Superflat Emulsion, Brolac Tartaruga, Brolac Undercoat, Brolac Varnishes, Brolac Weathercoat

Berger Jenson & Nicholson Limited,
Berger Decorative Paints Division,

No. 1 - finely textured, Brolac Weathercoat No. 2 - smooth, Brolac Weathercoat No. 3, Magicote Masonry Paint, Magicote Non Drip and Liquid Gloss, Magicote Solid Emulsion, Magicote Vinyl Matt, Magicote Vinyl Silk.

Bergvik Sales Limited,

Glen House, Stag Place, Victoria,
London SW1E 5AG
UK

Bevacid, Beviros, Bevitack Resins.

Berk Pharmaceuticals Limited,

St Leonards House, St Leonards Road,
Eastbourne, Sussex BN21 3YG
UK

Arvin, Asilone, Asilone Paediatric, Atensine, Berkatens, Berkazide, Berkmycen, Berkolol, Berkozide, Biogastrone, Bioral, Caplenal, Ceplac, Cremalgin, Domical, Dopamet, Dryptal, Duogastrone, Frumil, Imbrilon, Kest, Ladropen, Mucodyne, Nitrados, Pacitron, Pevidine, Ponoxylan, Pramidex, Primperan, Spirolone, Tetrachel, Trimopan, Ubretid, Vidopen.

Berk Spencer Acids Ltd,

Canning Road, Stratford,
London E15 3NX
England

BAA, Efflox.

Berlex Laboratories Inc,
Subsidiary Schering AG, W Germany

110 E Hanover Ave,
Cedar Knolls, NJ 07927
USA

Angiovist 282, Bilivist, Kay Ciel, Quinaglute, Sus-Phrine, Urovist, Urovist Sodium 300.

Berol Kemi (UK) Ltd,

55/57 Clarendon Road, Watford,
Herts WD1 1SP
England

Berol, Berol 259, Berol 26 and 02, Berol 267 and 09, Berol 269, Berol 272 and 716, Berol 278, 281, 282, 291 and 292, Berol 452 and 475, Berol 472, Berol 474, Berol 480,

Berol 484, Berol 490, Berol 496, Berol 513 and 525, Berol 518, Berol 563, Berol 594, Berol 733, 521 and 522, Berol 822, Fintex 572, Sulfamin, Sulfosoft, Wasc.

Bevaloid Ltd,

PO Box 3, Flemingate, Beverley,
North Humberside HU17 0NW
England

Bevalaloid 35 and 36, Bevaloid, Bevaloid, Bevaloid DA 6805, Bevaloid 111, Bevaloid 1299, Bevaloid 211, Bevaloid 35 and 36, Bevaloid 6423, Bevaloid 6522, Bevaloid 6703, Bevaloid 6744, Gensil, Vantac.

B F Goodrich Chemical Group
see Goodrich, B F, Chemical Group

BioContacts Inc,

111 Sutter St, Suite 600,
San Francisco, Calif 94104
USA

Revlen.

Bioglan Laboratories Limited,

Bridge Road,
Letchworth, Herts SG6 4ET
UK

Pronel Capsules, Vita-E, Vita-E Gelucaps.

BIP Chemicals Ltd,

PO Box 6, Popes Lane, Oldbury,
Warley, W Midlands B69 4PD
England

Beetle, Beetle Resin BT 333, Beetle Resin BT 334, Beetle Resin W69, Filon, Melmex, Mouldrite, Scarab.

Blagden Chemicals Limited,

Amp House, Dingwall Road,
Croydon, Surrey CR9 3QU
UK

Blagden Resins, Blagdenite, Crexathix.

Boehringer Ingelheim Pharmaceuticals Inc,

Boehringer Ingelheim Pharmaceuticals Inc,

90 East Ridge,
Ridgefield, Conn 06877
USA

Alupent, Alupent Expectorant, Alupent Obstetric, Alupent-Sed. A, Atrovent, Berotec, Bisoivomycin, Bisolvon, Buscopan, Camyna, Catapres, Catapres-TTS, Dexa-Rhinaspray, Dulcodos, Dulcolax, Finalgon, Lendorm, Persantin, Persantine, Preludin, Tranxene, Ulcedal, Vasculit, Villescon.

Bond Adhesives Company,

301 Frelinghuysen Avenue,
Newark, NJ 07114
USA

Screen Star Photo Emulsion.

The Boots Company PLC,

Nottingham,
NG2 3AA
England

Abicol, Allisan, Anestan, Anodesyn, Aprinox, Ardinex, Bistabillin, Blandlax, Blandthax, Bronopol-Boots, Bronotabs, Brufen, Burnol Acriflavine Cream, B.A.L, Clearine, Cobastab, Compimide, Coppesan, Cornox, Cortistab, Cream 45, Creamaffin, Delax, Deltastab, Diaflex, Difusor, Dijex, Dillex, Dytransin, D.F.P, Entamide, Eptoin, E45 Cream, Famel, Febrilix, Fenox, Flowfusor, Froben, Furamide, Harvesan, Hydrea Capsules, Hydrenox-M, Hydromycin-D, Hyperdol, Iso-Cornox, Isophane, Ivax, Kaodene, Karvol, Katorin, K285, Ley-Cornox, Lignostab, Meloids, Melsed, Melsedin, Monophane, Moorland Tablets, Mustine, Myacide AS, Mycivin, Mycota, Mylol, Nirolex Expectorant Linctus, Nivebaxin, Nivemycin, Nurofen, Ocusol, Opsan, Optrex, Otox, Panets, Penetrol, Pertusa, Piprelix, Plurivite, Polyfusor Solutions, PR Spray, PR Tablets, P.R. Spray, Quicksol, Sereen, Stabillin, Stabillin V-K, Steriflux, Sterotabs, Strepsils, Sweetex, Sweetex Plus, Totomycin, Tusana.

Borax Consolidated Ltd,

Cox Lane, Chessington,
Surrey KT9 1SJ
England

Borax Research Limited,
Cox Lane,
Chessington, Surrey KT9 1SJ
UK

Borascu, Borateem, Boraxo, Borester, Borocil, Boroxo, BR Destral, Braxo, Calbor, Dehybor, Destral, Destral BR, Firebrake, Lembor, Neobor, Pentabor, Polybor, Rasorite, Solubor, Timbor, Timborised, Trydil, ZB2335, Zinc Borate 2335.

Borden (UK) Ltd,

North Baddesley, Southampton,
Hants SO5 9ZB
England

Borden, Casco, Casco-resin, Cascophen, Epophen, Geoseal.

Borderland Products Inc,

560 Fulton Street, PO Box 1005,
Buffalo, NY 14240
USA

Borderland Black, Crow Chex, Stanleys Crow Repellant.

Borg Warner Chemicals,

20 Coventry Road, Cubbington,
Leamington Spa, Warks CV32 7JW
UK

Beldex, Blendex Modifiers, Cyclolac, Cycolac, Maldene, Marbon Latex, Marbon Resins, Marphos, Prevex, Udikral, Ultranox, Weston, Weston Additives.

Bostik Limited,

Ulverscroft Road,
Leicester LE4 6BW
UK

Bostik.

BP Chemicals Ltd,

Belgrave House, 76 Buckingham Palace Road,
London SW1W 0SU
England

Add F, Add-F, Add-H, Add-M, Aquaprint, Bakelite, Bio-add, Biocide, Bisoflex, Bisoflex DNA, Bisoflex L79, Bisoflex L911, Bisoflex ODN, Bisoflex 100, Bisoflex 1002, Bisoflex 1007, Bisoflex 104, Bisoflex 106, Bisoflex 130, Bisoflex 610, Bisoflex 619, Bisoflex 79A, Bisoflex 791, Bisoflex 799, Bisoflex 8N, Bisoflex 81, Bisoflex 819, Bisoflex 82, Bisoflex 88, Bisol, Bisolene, Bisolite, Bisolube, Bisomer, Bisomer DALP, Bisomer DAM, Bisomer DBF, Bisomer DBM, Bisomer DNM, Bisomer DOM, Bisomer D10M, Bisomer 2HEA, Bisomer 2HEMA, Bisomer 2HPMA, Bisoprufe, BP LDPE, BP Mycocide, BP Polystyrene, Breon, Breon GA 301A, Breon GA 302A, Breon GA 314A, Breox, Butaclor, Calprona K, Cellobond, Epok, Eprylac, Formodac, Gemex, Gemex, Hibosol, Hysa, Hyvis, Lobosol, Napelec, Napgel, Napryl, Naptel, Napvis, Nonex, Polyblack, Polyblends, Polyurax, Propcorn, Rigidex, Rigidex X4RR, Rigidex 3, Rigidex 9, Rigidoil, Rigipore, Soprodac, White Gold.

Bracco Industria Chimica SPA,

Bracco Industria Chimica SPA,

PO Box 12064, Via E Folli 50,
20134 Milano,
Italy

Bilimiro, also Bilimiron, Endobil, Endomirabil, Jodomiron, Uromiro, also Uromiron, Videobil, Videocolangio.

Bright Enterprises, R F Limited,

London Road, West Kingsdown,
Sevenoaks, Kent TN15 6AP
UK

Chemset.

Bristol-Myers Co,

345 Park Ave,
New York, NY 10022
USA

Bristol Laboratories,
Div of Bristol-Myers Co,
PO Box 4755,
Syracuse, NY 13221
USA

Westwood Pharmaceuticals Inc,
Subsidiary of Bristol-Myers Co,
468 Dewitt St,
Buffalo, NY 14213
USA

Alminate, Amikin, Balnetar Liquid, Betapen-VK, BiCNU, Blenoxane, Bristagen, Bristamycin, Bristocycline, Bufferin, CeeNU, Cefa-Lake, Cefadyl, Cilloral, Datril, Defencin, Dicloxin, Dricol, Dynapen, Entericin, Flo-Cillin, H-K Mastitis, Hetacin-K, Kanfotrex, Kantrex, Kantrexil, Kantrim, Ketaject, Ketaset, Ketavet, Lofenalac, Lysodren, Mexate, Mutamycin, Nafcil, Nulogyl, Nutramigen, Palapent, Pentrexyl, Platinol, Polycillin, Polycillin-N, Polycycline Hydrochloride, Polymox, Portagen, Precef, Prosobee, Prostaphlin, Questran, Saluron, Sobee, Sotacor, Stadol, Staphcillin, Synciliin, Syntetrin, Tacaryl, Tegopen, Tetrex, Tetrex Bidcaps, Torbutrol, Ultracef, Uropen, Uropol, Vasotran, Vepesid, Versapen, Versapen K.

BritAg Ltd,

Skeldergate Bridge,
York YO1 1DR
England

Wiles.

British Ceca Company Limited,

Rowan Court, 56 High Street,
Wimbledon Village, London SW19 5EE
UK

Acticarbone, Amphoram, Bactiram, Cataflot, Cecagel, Cecaperl, Cecarbon, Cecasil, Clar-O-Cel, Clarcel, Clarcel Flo, Dinoram, Fluidiram, Inipol, Noram, Noramac, Noramium, Noramox, Norust, Polyram, Porosil-Clarcel, Prochinor, Siliporite, Sofibex, Stabiram.

British Chrome & Chemicals Ltd,

Urlay Nook, Eaglescliffe,
Stockton-on-Tees, Cleveland TS16 0QG
England

Accrotan, Accrox, Chrometan.

Brocades (GB) Limited,

Brocades House, Pyrford Road
West Byfleet, Surrey KT14 6RA
UK

Amfipen, Antraderm, Bicillin, Brocadopa, Cyclospasmol, De-Nol, Depocillin, Locobase Cream and Ointment, Meditar, Pimafucin, Stafoxil.

Broemmel Pharmaceuticals,
Riker Laboratories Inc,

19901 Nordhoff St,
Northridge, Calif 91324
USA

Nizin, Op-Sulfa 30.

Bromhead & Denison Limited,

7 Stonebank,
Welwyn Garden City, Herts AL8 6NQ
UK

BentoPharm, Carrisorb, Crimidesa, K-Bond, K-White, Lubestine, Percolaye, Special Extender, Yellowstone.

Buckeye Cellulose Corp,

Buckeye Cellulose Corp,

PO Box 8407,
Memphis, Tenn 38108
USA

CLD 2.

Burmah-Castrol (UK) Limited,

Burmah House, Pipers Way,
Swindon, Wilts SN3 1RE
UK

Agricastrol, Castrol GTX, Castrol Turbomax, Petromor, Primor, Ravoien, Ravolen 11(T).

Burroughs Wellcome Co
see Wellcome Foundation Limited

Burts & Harvey,

Crabtree Manorway North,
Belvedere, Kent DA17 6BQ
UK

Atraflow Plus, Bras-sicol, Burtolin, Carbon 4E, Dormone, Drat, Keytrol, Krenite, Mowchem, Prefix D, Simflow, Simflow Plus, Spasor, Super Mosstox.

Bush Boake Allen Ltd,

Blackhorse Lane, London,
E17 5QP
England

Abbarome, Abbavert, Abracol, Aldones, Alpha Daphnone, Alphamint, Amborate, Amborol, Amboryl Acetate, Aromex, Benatol, Bentalol, Benteine, Bucarpolate, Butex, Cephreine, Cephrol, Curodex, Emulsene, Fermenticide, Fortex, Hydronal, Iphaneine, Lemol, Lindenol, Mephaneine, Meraneine, Olex, Pentoxyl M, Phanteine, Phantol, Phyllol, Vanbeenol.

BV Nekami
see Nekami, BV

BXL Plastics Ltd,

Buchanan House, 3 St James's Square,
London SW1Y 4JS
England

BXL Plastics Ltd,

Evazote, Plastazote.

C

The C P Hall Company
see Hall, The C P Company

Cabot Plastics Limited,

Gate Street,
Dukinfield, Cheshire SK16 4RU
UK

Cabelec compounds, Plasblak masterbatches, Plaswite masterbatches.

Caffaro SpA,
Industrial Division,

Via privata Vasto No 1,
1-20121 Milan
Italy

Cloparin, Cloparten, Clortex, Lipoclor, Prolit, Siclor, Solpolac.

Calbiochem-Behring Corp,

10933 N Torrey Pines Rd,
La Jolla, Calif 92307
USA

Asellacrin.

Calder Colours (Ashby) Limited,

Ashby de la Zouch, Leics,
UK

Addacol, White House Cement Paint.

Callery Chemical Company, Division, Mine Safety Appliances Company

PO Box 242,
Pittsburgh, PA 15230
USA

First Choice Electroless Palladium.

Campbell Williams & Company,

14 St Neots Road,
Abbotsley, Huntingdon, Cambs
UK

Thyodene.

Caramba Chemie GmbH,

Wanheimer Strasse 334/6,
D-4100 Duisburg 1
Germany

ARO, Caramba, Caramba Felgenglanz, Caramba Felgenneu, Caramba Lackkrone, Caramba Perlglanz, Pagid, Synoplast, Synova.

Carboxyl Chemicals Ltd,

Oriel Street, Vauxhall Road,
Liverpool L3 6DU
England

Flexade Regular.

Cargo Fleet Chemical Co Ltd,

Eaglescliffe Industrial Estate,
Eaglescliffe, Stockton, Cleveland
England

Caflon, Caflon MIS, Caflon MS33, Caflon NAS 25, Caflon SA and SNA, Caflon SS28.

Carl Freudenberg,

Postfach 1369,
D-6940 Weinheim (Bergstr)
Germany

Carl Freudenberg,

Fibral, Fliselina, Frelen, Freudenberg Megulastik, Glitzi, Kluberlubrication, Laysa, Laysa Plan, Lutrabond, Lutradur, Nora, Noraflor, Noralastic, Noralen, Norament, Noramid, Noraplan, Simflex, Simmering, Simrax, Simrit, Vileda, Viledon Compact, Viledon Filter, Vilene, Vlieseline.

Carless Solvents Ltd,

St James's House, Eastern Road,
Romford, Essex RM1 3NL
England

Caromax, Clairsol, Presol.

Carnrick, G W Co Limited,

Acres Down, Furze Hill,
Shipton on Stour, Warks CV36 4EP
UK

Hormonin, Midrid.

Celanese Limited,

78-80 St Albans Road,
Watford, Herts WD2 4AP
UK

Aquacoat, Cartolac, Cryoseal, Curolac, Cytrel, Foilcote, Foilgrip, Gravulac, Masterbond, Mirrolac, Omnilac, Permabond, Resistolac, Sealac, Sparkolac, Thermoseal.

Centerchem Products Inc,

475 Park Ave South,
New York, NY 10016
USA

Alphamine.

Chem-Trend,

32-5 E Grand River,
Howell, MI 48843
USA

Mono-Coat, Mono-Lube, Safety-Cool.

Chem-Y, Fabriek van Chemische Producten BV,

PO Box 50,
2410 AB Bodegraven,
The Netherlands

Akypoquat 129 and 130, Akypoquat 131 and 8188R, Akyporox NP105, Akyporox NP150, NP200 and NP300, Akyporox NP40 , Akyporox NP475, NP500, NP1000, NP1200 and NP1500, Akyporox OP250 and OP400V, Akyporox OP40 and OP115, Akyposal ALS, Akyposal BD and NPS, Akyposal DE, DEG, LFS and LFS/G, Akyposal DS and 23ST, Akyposal EO and RLM, Akyposal MGLS, Akyposal MLES, Akyposal MLS, Akyposal MS, PM and RO/E, Akyposal NLS and SDS, Akyposal TLS, Serdolamide, Serdox, Sermul, Surfageen, Surfageen S30.

Chemetall GmbH,

Reuterweg 14, Postfach 10 15 01,
D-6000 Frankfurt a M 1
Germany

Megum, Mould Release Agent N 32, Naftocit, Naftolen, Naftonox, Naftopast, Naftozin.

Chemical Combine,

6005 Stara Zagora - SHK
Bulgaria

Agridin 60, Agriphlan 24.

Chemical Dynamics Corp,

3001 Hadley Road,
South Plainfield, NJ 07080
USA

Acculog, Chemalog, Ultralog.

The Chemical & Insulating Company Ltd,

West Auckland Road, Darlington,
Co Durham DL3 0UR
England

Dalfratex, Magrods, Paraform.

Cheminova, A/S

PO Box 9,
DK-7620 Lemvig
Denmark

Depat, Dimethoate, EP-1, EP-2, Ethion, Ethyl Parathion, Fyfanon, Methyl Parathion, MP-1, MP-2, MPEM, Novathion, NPP-1.

Chemische Fabriek Servo BV,

POB 1,
7490 AA Delden
The Netherlands

Serdet DCK, Serdet DFK, Serdet DFL, DFM and DFN, Serdet DM and DMK, Serde DML, Serdet DNK, Serdet DPK, Serdet DSK, Serdox NNPQ 7/11, Serdox NNP10 and NNP12, Serdox NNP15, NNP20 and NNP25, Serdox NNP30 and NNP30/70, Serdox NNP4, Serdox NNP5 and NNP6, Serdox NNP7, NNP8.5 and NNP9, Serdox NOP 30/70, Serdox NOP9, Sermul EA 88, Sermul EA129, Sermul EA150, Sermul EA176, Sermul EA188, EA136 and EA205, Sermul EA54, EA151 and EA146, Servamine KAC 422, Servamine KEP 4527, Servamine KET 350, Servamine KET 4542, Servamine KOO 330, Servamine KOO 330B, Servamine KOO 360, Servo Ampholyt (B) JA110, Servo Ampholyt (B) JA140, Servo Ampholyt (B) JB130, Servo Brilliant Oil B AZ 75, Servoxyl VLA 2170, Servoxyl VLB 1123, Servoxyl VLE 1159, Servoxyl VP, Sevamine KOV 4342B.

Chemische Fabrik Weyl GmbH,
Organic Intermediates Division,

Sandhofer Strasse 96,
D-6800 Mannheim 31,
Germany

Impra, Impra-biolan, Impra-color, Impra-elan, Impraleum, Impralit.

Chemische Fabriken Oker und
Braunschweig AG,

3380 Goslar 1, Im Schleeke 77,
Postfach 1328
W Germany

Siflox, Siogel.

Chemoxy International Limited,

All Saints Refinery, Cargo Fleet Road,
Middlesbrough, Cleveland TS3 6AF
UK

Estasol, Hypax, Norfroth.

Chemseco,

4800 Blue Parkway,
Kansas City, MO 64130
USA

Astick, Press N Seal, Sealum.

Chevron Research Company,

PO Box 7141,
San Francisco, California CA 94120-7141
USA

Bitumuls, Bitusize, Brush-B-Gon, Bu-Gas, Bug-Geta, Contax, Crown, Cruisemaster, Delo, Dibrom, Dichevrol, Dieselect, Dieselmotive, Dieselube, Difolatan, Driftol, Eskimo, F-310, Flea-B-Gon, Flotox, Gilsonite and Design, Good Gulf, Greenol, Gulf Lite, Gulf Lubcote, Gulf No-Rust, Gulfad-C, Gulfco, Gulfcrest, Gulfcrown, Gulfcut, Gulfgem, Gulfknit, Gulfleet, Gulflex, Gulflube, Gulfpride, Gulfspin, Gulftene, Gulftex, Gulftow, Gulftronic, Gulfwax, Harmony, Hyjet, ICR, Isocracking, Isotox, Kerolite, Kleenup, Legion, Life, Monitor, No-Nox, OFA, OGA, OLOA, ORA, Oronite, Orthene, Orthenex, Ortho, Ortho-Gro, Ortho-Klor, Orthocide, Orthomatic, Orthorix, Orthotrol, Pacer, Paralux, Paramid, Paramount, Paraquat + Plus, Pearl, Pest-B-Gon, Phaltan, Poly-Eth, Poly-Eth Hi-D, Poly-Pro, Polypel, PRC, Pro-Gas (Gas Disclaimed), Ra-Pid-Gro, Rheniforming, RPM, Sani-Soil-Set, Saphire, Scram, Screte, Senate, Shawinigan Black, Sigma, Slug-Geta, Spirolite, Spray-Add 77, Sprills, Super-Quench, Synfluid, Techroline, Techron, Tri-Ad, Triox, Uni G, Unipel, Up-Start, Veritas, Viking, Volck, Warrengas, Weed-B-Gon, Zerol, Zincofol.

Ciba-Geigy PLC,

30 Buckingham Gate,
London SW1E 6LH
England

Ciba-Geigy Agrochemicals,
Whittlesford, Cambridge,
CB2 4QT
England

Ciba-Geigy Dyestuffs & Chemicals,
Ashton New Road, Clayton,
Manchester M11 4AR
England

Ciba-Geigy PLC,

Ciba-Geigy Industrial Chemicals,
Tenax Road, Trafford Park,
Manchester M17 1WT
England

Ciba-Geigy Pharmaceuticals,
Wimblehurst Road, Horsham,
W Sussex RH12 4AB
England

Ciba-Geigy Pigments,
Ashton New Road, Clayton,
Manchester M11 4AR
England

Ciba-Geigy Plastics,
Duxford, Cambridge,
CB2 4QA
England

Ciba-Geigy Corp,
556 Morris Ave,
Summit, NJ 07901
USA

Adelphane, Aerocol, Aerodux, Aerolite, Aerophen, Albatex, Albatex OR, Albegal, Albegal CL, Aldocorten, Alfacron, Alfadex, Alromin Ru 1000, Ambilhar, Amine 0, Anafranil, Andursil, Antistin, Antistitin-Privine, Antrenyl Duplex, Anturan, Anturane, Apresoline, Apresoline Hydrochloride, Aracast, Araldite, Ardux, Arigal PMP, Aturbane, Avilon, Avivan, Basudin, Belclene, Belgard, Belite, Bellasol, Bellauxine, Belloid, Beltherm, Bradophen, Bradosol, Brasoran 50WP, Butacote, Butazolidin, Butazolidin Alka, Butazolidin with Xylocaine, Calcium Chel 330, CG-80, Chel, Ciba, Ciba 1906, Cibacet, Cibacron, Cibalith-S, Cibamin, Cibanone, Cibaphasol 6042, Cibatex PA, Cibatex 248, Cinquasia, Circadet, Circaline MK 11, Clarite, Clarosan, Coprantex, Crastine, Cromophtal C-20, Cromophtals, Cromphytal M-20, Cuprophenyl, Cytadren, Dairy Fly Spray, Demavet, Descale, Desferal, Desogen, Dianabol, Diazitol Liquid, Dibistin, Dicron 45Sc, Dicrylan, Dicrylan 270, Dicurane, Dicurane Duo, Diphasol, Diphenyl, Do Do, Doriden, Dosulphin, Dyne, Dynemate 200, Elocril, Erio, Eriochrome, Erioclarite, Erional, Erionyl, Eriopon, Ertilen, Esidrex, Esidrex-K, Esidrix, Eurax-Hydrocortisone, Fabrol, Famid, Fasinex, Fenitrothion EC, Fibrredux, Filamid, Filester, Filofin, Flor Sherry, Flovan, Forhistal Maleate, Fornax, Forociben Premix, Fubol, Fumexol, Fungitex 656, Gesagard, Gesaprim, Gesatop, Glyvenol, Hispor, Horna, Hydrobol, Hydrophobol, Hygroton, Hygroton K, Hypertensin, Hytane, Ilcocillin, Industrial Dyne, Industrial Dynemate, INH, Insidon, Invaderm, Invaderm C9B, Invalon, Invasol, Iosan CCT, Iosan D, Iosan Super Dip, Iosan Teat Dip, Iosan Udder Cream, Iosan 4, Iragcet, Irgaclarol, Irgacure, Irgaderm, Irgaferm BC Champagne, Irgafin, Irgafiner, Irgafos, Irgalan, Irgalevone, Irgalite, Irgalite Blue GST, Irgalite C-20, Irgalite Dispersed, Irgalite M-20, Irgalite MPS, Irgalite PDS, Irgalite PR, Irgalite Yellow BGW, Irgalite Yellow F4G, Irgalon, Irganol, Irganox, Irgapadol, Irgaphor, Irgaplastol M-20, Irgapyrol, Irgarol, Irgasan, Irgasan DP 300, Irgasol, Irgasperse-s, Irgastab, Irgatan, Irgatron, Irgawax, Irgazin, Irgazin C-20, Irgazin M-20, Irgoferm CM Montrachet, Ismelin, Ismelin, Ismelin-Navidrex K, Knittex, Lamprene, Lithobid, Locorten, Locorten-N, Lopresor, Lopresoretic, Lopressor, Lopressor SR, Ludiomil, Lyofix, Lyofix F, Lyofix 363, Lypsyl, Maloran, Mantin, Maxilon, Medomin, Melafix DM, Melolam, Melopas, Metandren, Metopirone, Micoren, Micracet, Microlith-A,

Migafar, Migafar AL, Migatex, Mirvale, Mitin, Mofix, Monaspor, Mucron, Navidrex, Navidrex-K, Neoamfo, Neocidol Veterinary Powder, Neolan, Neovadine, Neuro-Transentin, Nitraline, Nogos, Nupercainal, Nupercaine, Nupercaine Hydrochloride, Nuvan, Nuvan Fly Spray, Nuvan Top Aerosol, Nuvanol, Nuvanol N, Ocusert, Oligan, Ophthaine Solution, Opogard, Opticorten, Oracet, Orasol, Oresmasin, Orimeten, Orisulf, Ostrilan, Otrivin Hydrochloride, Otrivine, Otrivine-Antisitin, Oxytracyl, Panolog, Parazolidin, Parsolin, Patoran, Percorten Acetate, Percorten M Crystules, Percorten Pivalate, Percorten 'M' Crystules, Pergacid, Pergantine, Pergaprint, Pergascript, Pergasol, Pergopak, Pertofran, Phenolene and Phenolene Supra, Phobol, Phobotex FTN, Phobotone, Polar, Prebane, Pretolone, Primatol AA, Primatol AD, Primatol AP, Primatol SE, Priscol, Priscoline Hydrochloride, Privine Hydrochloride, Probimer, Proflex, Pyribenzamine Citrate, Pyribenzamine Hydrochloride, Pyrovatex, Quixalud, Reginal, Regitine Mesylate, Rengasil, Reochlor (LF and 54), Reoflam, Reofos, Reolube, Reolube FAD, Reomet, Reomol, Reomol BCF, Reomol P, Reomol PBPS, Reomol 4PG, Reoplast, Reoplex, Reoplex 200, 220, 300, Reoplex 901, Reoplex 902, Resicart, Resin WP, Respirot, Revona, Ridomil, Rimactane, Rimactane, Rimactazid, Ritalin Hydrochloride, Rogitine, Roseline, Sanachlor, Sapamine, Sapecron, Sarkosyl NL30 and O, Sedestran, Sella Acid, Sella Fast, Sellacron, Sellaflor, Sellasol, Semeron, Sequestrene, Sermix, Serpasil, Serpasil-esidrex, Serpatonil, Silvatol, Sinthrome, Slaymor, Slow Sodium, Slow Trasicor, Slow-Fe, Slow-FE Folic, Slow-K, Slow-Sodium, Solophenyl, Stellox, Steroxin-Hydrocortisone, Superspray, Superwipes, Supracide, Swipe, Symmetrel, Synacthen, Synacthen Depot, Tacitin, Tandacote, Tandaigesic, Tandearil, Tanderil, Tannesco, Tectilon, Teevax, Tegretol, Tenatine, Tenoran, Teralan, Teraprint, Terasil, Tercoton, Tilt, Tilt Turbo, Tinegal, Tinoclarite, Tinofil, Tinopal, Tinorex, Tinosol, Tinovetin, Tinuvin, Tinuvin P, Tinuvin 770, Tofranil, Topas, Topclip Dridress, Topclip Fly and Scab Dip, Topclip Foot Rot Aerosol, Topclip Formalin, Topclip Gold Shield, Topclip Marker Aerosols, Topclip Marker Fluid, Topclip Parasol, Topclip Scab Dip, Topclip Sheep Dip, Topclip Vaccines, Topclip Wormer, Transderm-Nitro, Trasicor, Trasidrex, Trigard, Triodine, Tromexan, Turpex, Ultandren, Ultracortenol, Ultratex, Ultravon, Ultravon AN, Unisperse-E, Unisperse-P, Univadine, Ureol, Uromat PE, Utocyl, Uvitex, Uvitex MA, Uvitex MP, Varbian, Vebonol, Vecortenol, Vecortenol-Vioform, Verafil, Vesulong, Vetalog Injection, Vetalog Plus Cream, Vetanabol, Vetibenzamine, Vetidrex, Vibatex, Vioform Powder, Vioform-Hydrocortisone, Vioform., Vionactane, Vionate Powder, Voltaren, Voltarol, Voltarol Retard, Welladyne, Wescodyne.

Cindu Chemicals BV,

Postbus 9,
1420 AA Uithoorn
Netherlands

Cindumix.

Clintwood Chemical Company,

4342 South Wolcott Avenue,
Chicago, IL 60609
USA

Clindrol, Clindrol EGDS, Clindrol SDG, Clindrol SEG, Clindrol SEG-S, Rodea.

Coalite Fuels & Chemicals Ltd,

Coalite Fuels & Chemicals Ltd,

Central Refinery, PO Box 21,
Chesterfield, Derbyshire S44 6AB
England

Coalite N.T.P, Coaltec, Cotane, D.X.L, R.X.X.L, X.L.

Coates Industrial Finishes Ltd,

Station Lane,
Witney, Oxon OX8 6XZ
UK

Delta, Vioglaze.

Col Polymers Incorporated,

2115 Gaylord Street,
Long Beach, CA 90813
USA

Calthane.

Colloids Limited,

Dennis Road,
Widnes, Cheshire WA8 0SN
UK

Alkasperse 25, Mastercarb, Masterwood, Microsperse, Plastosperse, Plastosperse 40.

Colonial Products Inc,

1830 10th Avenue North,
Lake Worth, FL 33461
USA

Ant Flip, Flea Flip, Rat Flip, Time Bomb.

Colorcon Ltd,

Murray Road, St Paul's Cray,
Orpington, Kent BR5 3QY
UK

Opacode, Opacolor, Opadry, Opalux, Opaspray, Opatint.

Colour-Chem Limited,

Ravindra Annexe, Dinshaw Vachha Road,
194 Churchgate Reclamation, Bombay 400 020
India

Colour-Chem, Dermafill, Dermalac, Lustre, Vernafine, Vernalin, Vernamine Binders, Vernaminol Liquors, Vernasein, Vernasol, Vernatan, Vernol Liquors.

Colourex Limited,

11 Wimbledon Avenue,
Brandon, Suffolk IP27 0NZ
UK

Colex 1000 FR.

Columbia Nitrogen Corporation,

PO Box 1483 (13),
Augusta, GA 30913
USA

Nipro (i), Nipro (ii), Old Plantation, Superprill.

Conap Inc,

1405 Buffalo Street,
Olean, NY 14760
USA

Conacure, Conapoxy, Conathane.

Concept Pharmaceuticals Limited,

The Old Coach House, Amersham Hill,
High Wycombe, Bucks HP13 6NQ
UK

Eludril Mouthwash, Eludril Spray, Paleva.

Continental Carbon Australia Pty Limited,

Private Mail Bag, Cronulla,
Sydney, NSW 2230
Australia

Continex Carbon Black.

Continental Pharma SA,

Continental Pharma SA,

Ave Louise 135,
B 1050 Brussels
Belgium

Mervan.

Cook-Waite Laboratories Inc,

90 Park Ave,
New York, NY 10016
USA

Neo-Cobefrin, Ravocaine Hydrochloride.

The Cool-Amp Conducto-Lube Company,

8603 SW 17th Avenue,
Portland, Or 97219
USA

Conducto-Lube, Cool-Amp.

CooperVision Inc,
Optics Div,

2801 Orchard Pkwy,
San Jose, Calif 95134
USA

Atropisol, Catarase, Funduscein, Goniosol, Inflamase, Permalens, Pilomiotin, Sulf-10, Tear-Efrin, Tearisol, Vasoclear, Vasocon.

CooperVision Pharmaceuticals Inc,

San German,
Puerto Rico 00753
USA

Glybrom, Glyrol, Miochol.

Cornelius Chemical Co Ltd

Ibex House, Minories,
London EC3N 1HY
England

Cornelius Chemical Co Limited,
St James's House, 27-43 Eastern Road,
Romford, Essex RM1 3NN
UK

BWF, Constab, Disperse-Ayd, Finntitan, Flat-Ayd, Geracryl, Kemira Phlogopite Mica, Leneta, Mearl Film, Mearlin, Methylon, Min-U-Sil, Plexigum, Pleximon, Plexisol, Plextol, Premix, Rohafloc, Slip-Ayd, Timica, Tint-Ayd, Ultrawet, Ultrawet DS, Ultrawet K and AOK, Viscoplex, Wollastonite, Zeolex, Zeothix.

Corporacion de Desarrollo Tecnologico CA,
Catalysis Division,

Urb Industrial el Bosque-Avda,
Hans Neumann-Parcela 13-Valencia,
Estado Carabobo
Venezuela

Cordetec 100.

Coulter Electronics Limited,

Northwell Drive,
Luton, Beds LU3 3RH
UK

Hemoterge, Isoterge, Isoton, Lyse S, Somacount, Somafix, Somaton, Zapoglobin, Zaponin, 4C.

Courtaulds Chemicals & Plastics,

Sulphur Chemicals Group,
Barton Dock Road, Stretford,
Manchester M32 0TD
England

Courtaulds Chemicals & Plastics,
Leek Chemicals Group,
Bridge End Works, Leek,
Staffs ST13 8LG
England

Courtaulds Chemicals & Plastics,
Speciality Chemicals Group,
PO Box 5, Spondon,
Derby DE2 7BP
England

Courtaulds Chemicals & Plastics,

Courtaulds Chemicals & Plastics,
Water Soluble Polymers Group,
PO Box 5, Spondon,
Derby DE2 7BP
England

Celacol, Courcel, Courlose, Wardol.

Cray Valley Products Ltd,

Farnborough,
Kent BR6 7EA
England

Antisettle, Craymer, Crayvallac, Gelkyd, Genamid (R), Synaqua, Synocryl, Synocryl 8205 and 821S, Synocure, Synocure 867S, Synocure 868S, Synocure 869S, Synolac, Synolide, Unithane, Unithane 640 W and 641 W, Versamid (R), Vinacryl 4512.

Croda Chemicals Ltd,

Cowick Hall, Snaith,
Goole, N Humberside DN14 9AA
England

Acylan, Acylan, Adinol, Adinol CT, Adinol T, Agnowax, Amino Gluten MG, Aminofoam C, Cholesterol, Cithrol, Cithrol GMS A/S, Collasol, Conditioner Base, Corona, Coronet, Cosmowax, Cremba, Crestalan, Crester KZ, Creto, Crill, Crillet, Crillon, Crocell, Croda Bath Oil Disperant, Croda Fluid, Crodacol, Crodafos, Crodalan, Crodalan AWS, Crodalan C24, Crodalan IPL, Crodalan LA, Crodalan 1PL, Crodamet, Crodamet, Crodamine 1, Crodamine 2.C, 2.S and 2.HT, Crodamine 3A, Crodamine 3ABD, Crodamine 3AED, Crodamine 3AHRD and 3ARD, Crodamine 3AOD, Crodamol, Crodapearl, Crodasinic, Crodasinic L and LS35, Crodaterics, Crodax, Crodesta, Crodet, Crodex, Crodinhib, Crodol, Croduret, Crodyne BY 19, Crolactil, Crolan, Crolastin, Cromeen, Cropeptone, Cropeptone, Cropol 60, Croquat, Crosilk, Crotein A,C,, Crotein CAA, Crotein HKP, Crotein HWE, Crotein Q, Dastar, Datagel, Datem, Dicrodamine, Etocas, Fluilan, Fluinlan, Gant, Glycerox, Grodex, Hartolan, Hartolite, Hydrosoy 2000, Incomate IDL, Incromate CDL, Incromate CDP, Incromate ODL, Incromate SDL, Incromectant AMEA-70, Incromectant LMEA-70, Incromine BB, Incromine CB, Incromine IB, Incromine OPB, Incromine Oxide B, Incromine Oxide C, Incromine Oxide I, Incromine Oxide L, Incromine Oxide M, Incromine Oxide O, Incromine Oxide OD-50, Incromine Oxide S, Incromine SB, Incronam B-40, Incronam OP-30, Incronam 1-30, Incronam 30, Incroquat S-85 and SDQ-25, Incrosoft CF1-75, Incrosoft S-75 and S-90, Incrosoft T-75 and T-90, Incrosperse, Isocreme, Kathro, Kester, Lanexol, Lanoiac, Lanolic, Lanolic Acid, Lanosol, Lanosterol, Lanpol, Larnol, Liquid Bases, Lowerite, Metacon, Microlan, Novol, Penta G.P. 79, Pentalan, Polychol, Procetyl AWS, Proil, Propstearyl, Reticusol, Rocsol, Satulan, Skliro, Solan, Solan E, Spreading Agent, Super Hartolan, Super Sterol Ester, Syncrolube, Syncrowax, Tricap, Volpo, Vykamol 83G, White Swan, Yeoma.

Croda Resins Ltd,

Crabtree Manorway, Belvedere,
Kent DA17 6BA
England

Hytex, Hythane, Lamex 173/FR, Lamex 185, Lamex 186, Laquanol, Okstan XO, Plastamid, Plastokyd, Plastokyd SC, Plastokyd 310, Plastoprene, Plastoprene No2/LV, Plastyrol, Plastyrol E6X, Plastyrol S-77X, Plastyrol S88X.

Croda Universal Ltd,

Cowick Hall, Snaith,
Goole, N Humberside DN14 9AA
England

Crodamine.

Crosfield Chemicals,

PO Box 26,
Warrington, Cheshire WA5 1AB
UK

Acsil, Alusil, Alusil ET, Alvex, Besconus, Claysil, Cormix, Crosanol, Croscolor, Croscour, Crosdurn, Crosfield, Crosfield EP, Crosfield HP, Crosfield SP, Crosil, Croslube, Crosoft, Crostat, Crystal, Doucil, Estol, Flucsil, Gasil, Gasil EBC and EBN, Hexo, Lanaire, Metrax, Microcal, Microcal ET, Microsil, Neosyl, Nicat, Polystat, Pyramid, Pyramol, Pyranet, Quatum, Rezex, Selbax, Seljut, Solgen, Sorbosil, Sorbsil, Soucol, Sulmet, Synclyst, Topsol, Trimetso, Vivol.

Croxton & Garry Limited,

Curtis Road,
Dorking, Surrey RH4 1XA
UK

Fortrex (2), Garoflam, Garomix, Garosorb, Garospers, Garozinc, Protexulate, Scorchex, Trihyde.

Custom Coating & Laminating Corporation,

715 Plantation Street,
Worcester MA 01605
USA

Conducto-Wrap, Intimate Contact, Scratch-Guard.

Cutter Laboratories,
Miles Laboratories Inc,

4th & Parker, PO Box 1986,
Berkeley, Calif 94701
USA

Cutter Laboratories,
Miles Laboratories Inc,

Gamastan, Gamimune, Hyper-Tet, Hyperab, HyperHep, Hypertussis, Koate, Konyne, Plasbumin.

Cyanamid Fothergill Limited,

Abenbury Way, Wrexham Industrial Estate,
Wrexham LL13 9UF
UK

Fothergill Composite & Polymer Technologies Limited,
Dunball Park,
Dunball, Bridgwater, Somerset TA6 4TP
UK

Fothergill Engineered Surfaces Limited,
Godiva Place,
Coventry CV1 5PN
UK

H D Symons & Company Limited,
Horace Road,
Kingston upon Thames, Surrey KT1 2SN
UK

Fothergill Engineered Surfaces Limited,
Long Causeway,
Leeds LS9 0NY
UK

Fothergill Engineered Fabrics Limited,
PO Box 1,
Littleborough, Lancashire OL15 9QP
UK

Fothergill Tygaflor Limited,
PO Box 2, Summit,
Littleborough, Lancashire OL5 0LT
UK

Fothergill Cables Limited,
PO Box 3, Church Street,
Littleborough, Lancashire OL15 8HG
UK

Fothergill Composite & Polymer Technologies Limited,
PO Box 7, Tweed Road,
Clevedon, Avon BS11 6ST
UK

Armourcote, Carboform, Crenette, Cycom, Duracore, Niflor, Nuglas, Symax, Symel, Termex, Tygacell, Tygadure, Tygaflon, Tygaflor, Tygalam, Tygan, Tygatape, Tygavac, Tyglas, Wills Metallic 'O' Rings.

Cyanamid GB Ltd

Bowling Park Drive,
Bradford, BD4 7TT
England

Cyanamid BV,
Postbus 1523,
NL 3000 BM Rotterdam
The Netherlands

Accosize, Accostrength, Accostrength 72, Accurac, Accurac 33/35/41, Accurac 39, Acrylamide, Acrylite H, Acrylite M, Aero 301 Xanthate, Aero 303 Xanthate, Aero 317 Xanthate, Aero 343 Xanthate, Aero 3477 Promoter, Aero 350 Xanthate, Aero 3501 Promoter, Aerodri 100, Aerodri 104, Aerodri 200, Aerofloat 208 Promoter, Aerofloat 211 Promoter, Aerofloat 238 Promoter, Aerofroth 65, Aerofroth 76, Aerofroth 88, Aerofroth 99, Aerophine 3418A, Aerosol A-102, Aerosol A-103, Aerosol A-196, Aerosol A-268, Aerosol AY, Aerosol C-61, Aerosol C61, Aerosol IB45, Aerosol MA-80, Aerosol OS, Aerosol OT and GPG, Aerosol TR-70, Aerosol 18, Aerosol 200, Aerosol 22, Anti-oxidant 2246, Anti-oxidant 425, Aquastore, Cyanamer P35 - P70, Cyfloc 6000, Cypan, Cyrez 963, Cyrez 963/4 Powders, Glyoxal 40%, Hydroblok, Isobu-M-AMD, M.B.A, NM-AMD, Parez Resins, RMD, RMR, Sodium Aeroflot Promoter, Superfloc, Superfloc C507, Superfloc C521, Superfloc C573-C577-C567.

Cyclo Corporation,

7500 N W 66th Street,
Miami, Florida 33166
USA

Citrest, Cyclochem, Cyclofor, Cyclomide, Cyclonette, Cyclophos, Cyclopol, Cycloryl, Cycloteric, Cycloton.

Cyro Industries,

155 Tice Blvd, PO Box 8588,
Woodcliff Lake, NJ 07675
USA

Acrylite.

D

Dainippon Pharmaceutical Co,

3-25 Dosho-machi, Higashi-ku,
Osaka 541
Japan

Furanace.

Dampney Company Inc.,

85 Paris Street
Everett, Ma. 02149
USA

Apexior, Dymacryl, Elastoid 1300, Endcor, Epodur, Thurmalox.

Darmex Corporation,

71 Jane Street,
Roslyn NY 11577
USA

Darmex, Darmex Plus.

Datac Adhesives Limited,

Globe Lane Industrial Estate,
Dukinfield, Cheshire SK16 4XE
UK

Datac.

Davis & Geck,

One Casper St,
Danbury, Conn 06813
USA

Dexon, Maxon, Novafil.

Dawood Hercules Chemicals Limited,

Corporate Office, 35A Shahrahe Abdul Hameed Bin Baadees, PO Box 1294,
Lahore
Pakistan

Bubber Shet.

DCS Color & Supply Company Inc,

2011 South Allis Street,
Milwaukee, WI 53207
USA

Blackox, Foundrox, No Vein Compound, Rainbow Custom Coloured Mortars.

Deanshanger Oxides Ltd,

Deanshanger, Milton Keynes,
MK19 6HA
England

Deanox.

Dearborn Chemicals Ltd,

Foundry Lane, Widnes,
Cheshire WA8 8UD
England

Aquafloc, Biomate, Poly Check, Polymate, Scale Cleen.

Degussa Limited
see Astra-werke Degussa-Pharma-Gruppe

Delandale Laboratories Limited,

Delandale House, 37 Old Dover Road,
Canterbury, Kent CT1 3JF
UK

Dicynene, Priadel.

Delaware Chemical Corporation,

PO Box 126,
Daleville IN 47334
USA

Delaware Chemical Corporation,

Algae Treat, Caust X, Limex, Limex 'G', Tower Brick.

Dermal Laboratories Limited,

Tatmore Place, Gosmore,
Hitchin, Herts SG4 7QR
UK

Anhydrol Forte, Callusolve, Capitol, Dioderm, Dithocream, Dithrolan, Emulsiderm, Exolan Cream, Exterol Ear Drops, Glutarol, Psoriderm Bath Emulsion, Psoriderm Cream, Psoriderm Scalp Lotion, Salactol, Variclene.

Dermik Laboratories Inc,

1777 Walton Rd,
Blue Bell, PA 19422
USA

Anthra-Derm, Durrax, Hytone, Sulfacet-R, Vlemasque, Zetar.

DeSoto Inc,

Chemical Specialties Division,
P O Box 2199, Fort Worth,
Tx 76113
USA

Armix 146, Armix 176, Armul 17, Armul 22, Armul 44, Armul 66, Armul 88, DeSonate AOS, DeSonate SA, DeSonate SA-H, DeSonate 50-S, DeSonate 60-S, DeSonic DA-4, DeSonic DA-6, DeSonic N Series (ie. 4N, 9N etc), DeSonic S Series, DeSonic 30C, DeSonol A, DeSonol AE, DeSonol S, DeSonol SE, DeSonol SE-2, DeSonol T, DeSotan SMO, DeSotan SMO-20, DeSotan SMT, DeSotan SMT-20, Flo-Mo DEL, Flo-Mo DEH, Flo-Mo Lowfoam, Flo-Mo Suspend, Flo-Mo 1082, Flo-Mo 1093, Flo-Mo 5BMP, Flo-Mo 80/20, Morwet EFW, Petrac CP-11, Petrac CP-12, Petrac ZN-44, Petro AG Special, Petro BAF, Petro P, Petro S, Petro ULF, Petro WP, Petro 11, Petro 22, Udet 950.

Devcon Corporation,

30 Endicott Street,
Danvers MA 01923
USA

Flexane, MVP, Plastic Steel, Pump Repair Putty, Superlock, Titanium Putty, Zip Grip.

D F Anstead Ltd
see Anstead, D F Ltd

Diaflex Limited,

Chemeonics, Cannon Lane,
Tonbridge, Kent TN9 1PP
UK

Bromox, Dialpha, Diavite, Florox.

Diamond Shamrock Chemicals Company,

351 Phelps Court,
Irving, TX 75015
USA

Dymsol 38C.

Diamond Shamrock Process Chemicals Ltd,

PO Box 1, Eccles,
Manchester M30 0BH
England

Diamond Shamrock Europe Corp,
Ave Reine Astrid 7,
B 1430 Wautheir-Braine, Brussels
Belgium

Ampholan B171, Arylan CA, Arylan PWS, Arylan S, Arylan SBC, Arylan SC, Arylan SNS, Arylan SP, Arylan SX, Arylan TE/C, Ethylan BAB 20, Ethylan BCP and KEO, Ethylan BV, Ethylan BZA, Ethylan DP, Ethylan ENTX, Ethylan GMF, Ethylan HA, Ethylan HP, Ethylan NP1, Ethylan N92, Ethylan PQ, Ethylan 20, Ethylan 44, Ethylan 77 and TU, Foamaster, Hyonic, Lankropol KO Special, Lankropol KO2, Lankropol KSB 22, Lankropol KSG 72, Lankropol ODS, Lankrosol SXS, Levelan P148, Levelan P208, Levelan P307, Levelan P357, Monosulph, Nopalcol, Nopco Foamaster, Nopco Worsted Oil 12, Nopcocastor, Nopcochex, Nopcofloc, Nopcogen, Nopcolene, Nopcolube, Nopcone, Nopcosant, Nopcosize, Nopcosulph, Nopcotan, Nopcote, Nopcotex, Nopcowax, Perlankrol ADP-3, Perlankrol ATL, Perlankrol DAF, Perlankrol DSA, Perlankrol ESD, Perlankrol ESK-29, Perlankrol O, Perlankrol PA, FF and FD, Perlankrol SN and RN, Perlankrol TM1, Phospholan ALF-5, Phospholan KPE-4, Phospholan PDB-3, Phospholan PNP-9, Polyclear, Polymul, Quadrilan AT, Quadrilan BC, Quadrilan MY 211, Quadrilan OD-75, Versilan.

Dista Products Ltd,

Fleming Road, Speke,
Liverpool L24 9LN
England

Dista Products Limited,
Kingsclere Road,
Basingstoke, Hants RG21 2XA
UK

Dista Products Ltd,

Allegron, Capastat, Capreomycin, Distaclor, Distalgesic, Distamine, Distaquaine V-K, Fenopron, Haelan, Haelan Tape, Haelan-C, Haelan, Haelan-X, Illosome, Ilosone, Kefadol, Napsalgesic.

The Distillers Company (Carbon Dioxide) Ltd,

Cedar House, 39 London Road,
Reigate, Surrey RH2 9QE
England

Cardice.

D L Forster Limited
see Forster, D L Limited

Doak Pharmacal Co,

700 Shames Drive,
Westbury, NY 11590
USA

Buro-sol Concentrate, Formula 405.

Doff Portland Limited,

Bolsover Street,
Hucknall, Nottingham NG15 7TY
UK

Doff, Slugoids.

Dorsey
see Sandoz

Dover Chemicals Corporation,

PO Box 40,
Dover, Ohio 44622
USA

Chlorez, Paroil.

Dow Chemical Company Ltd,

Stana Place, Fairfield Avenue,
Staines, Middlesex TW18 4SX
England

The Dow Chemical Company,
PO Box 2166, Midland,
Michigan 48641-2166
USA

Merrell Dow Pharmaceuticals Inc,
Subsidiary of Dow Chemical Co,
Cincinnati,
Ohio 45215
USA

Abavit B, Abavit S, Acrex, Aerothene, Alar, Aldrin Dust, Alysine, Ambiflo, Ambitrol, Analoam, Ascorbin, Bentyl, Bexton, Brussels System, Calibre, Campaign, Cantil, Castethane, Chlorothene, Chlorothene (VG), Chronulac, Clomid, Combined Seed Dressing, Corban, Coverite, Coyden, CPE, Crusader, Curithane, C.P.R, Dalpad, De De Tane, Decapryn Succinate, Defolia, Delinal, Derakane, Derakane (470-45), Derakane (510-40), Deraspan, Diazinon Liquid, Dinamene, Dow DBR, Dow Plasticizer No. 5, Dow Plasticizer No. 55, Dow V9, Dow 276-V2, Dowanol, Dowclene, Dowco (179), Dowetch Deadline, Dowex, Dowex Monosphere, Dowfax, Dowfax 2A1, Dowfax 9N12, 9N14/15 and 9N12W, Dowfax 9N2, 9N3 and 9N4, Dowfax 9N5, 9N6 and 9N7, Dowfax 9N8, 9N9 and 9N10, Dowfrost, Dowfroth, Dowgard, Dowicide, Dowicil, Dowlex, Downright, Dowper, Dowpon, Dowtherm 209, Dufox, Dursban, Dursban, DV, D.E.H, D.E.N, D.E.R, Ethafoam, Ethocel, E.C.A, Fentro, Filmite, Folia-Feed, Forlay, FR-1360, FR-2406, Fungex, Gallant, Gamma-BHC Dust, Gamma-HCH Dust, Garlon, Generon, Grazon 90, Handi-Wrap, Hedulin, Hormone Rooting Powder, Isonate, Isonol, Jefron, Jeunite, Kaydox, Kevadon, Korlan, Kuron, Lawn Weedkiller, Lightguard, Liquid Copper Fungicide, Liquidow, Lontrel, Lorelco, Lorsban, Magnum, Malathion Dust, Malathion Liquid, Mecpa, Metahydrin, Methocel, Methocel A, Methocel E,F,K, Mini Slugit Pellets, Mist-o-Matic, Mist-o-matic Ferrax, Mocap 10G, Morto, Mouse Killer, Murald, Murcurite, Murdiel, Murfite, Murfixtan, Murfly, Murfotox, Murfume, Murphex, Murphicol, Murphos, N-Serve, New Murbetex, Nicorette, Nicotine 40% Shreds, Norpramin, Novafed, Novahistine, Opticite, Ostamer, Overnite, Oxytril P, Papi, Parabis, Paradow, Parencillin, Parenzyme, Pazo, Pelaspan, Pelaspan Mold-a-Pac, Pelaspan 333FR, Pelaspan-Pac, Pellethane, Pentac Aquaflow, Pexid, Phemox,, Phenmerzyl Nitrate, Phenoxine, Phosphomort, Plictran, Polyfilm, Premerge, Primacor, Pulse, Purifloc, Pusher, Quatrex, Quide, Quinamin, Reldan 50, Rifadin, Rifamate, Rimthane, Roachban, Root Guard, Rovel, Saran, Saranex, Separan, Shield, Sinazine, Singlet, Sirlene, Slug, Slug Pellets, Slugit Liquid, Spectrim, Starane, Starane 2, Styrofoam, Styron, Sulfonsol, Super Moss Killer & Lawn Fungicide, Super Weedex, Susadrin, Suspendex, Synerone, Systemic Funigicide, Systemic Insecticide, Sytam, TACE, Tactix, Talpheno, Tandem, Techmate, Telone, Tenvate, Terra-Systam, Thi-Di-Mer, Tordon, Torelle, Triclos, Trithac, Trithion, Trolene, Trycite, Trymer, Tuasol 100, Tumbleblite, Tumblebug, Tumblemoss, Tumbleslug, Tumbleweed Gel, Tussend, Tyril, Tyrin Brand, Vanobid, Verdict, Versabacs, Versene Acid, Versene AG, Versene CA, Versenol AG, Vertan, Vertifume, Vikane, Voran, Voranate, Voranol, Wasp Destroyer, Weedex, Zetabon, Ziploc, Zoamix, 2-G.

Dow Corning Ltd,

Avco House, Castle Street,
Reading, Berkshire RG1 7DZ
England

Dow Corning Ltd,

Dow Corning Ophthalmics Inc,
Midland,
Mich 48640
USA

Gelflex, Lubolid, Molykote, Releasil, Silastic, Silcon, Silsoft, Sioplas, Syl-off, Sylgard, Syltherm.

Doyle Specialties,

9800 Cozycroft Avenue,
Chatsworth, CA 91311
USA

Cut Aid, Tap Aid.

Draxel Chemical Company,

PO Box 9306,
Memphis, TN 38109
USA

Add It To Oil, Ancrack, Antak, Croak, Damoil, Defol, Drexar 530, Dynamyte, Genche, Mix, Peptoil, Retard, Shadow, Soy-che, Sucker Plucker Concentrate, Sucker Stuff, Suffa, Surf Ac 820, Vegetoil, Wetz, Zinche.

DSM NV,

Postbus 65,
Heerlen N1 6400 AB
The Netherlands

Kelburon, Kelprox, Kelrinal, Keltan, Keltan TP, Marvylan, Nyrim, Plexar, Ronfalin, Ronfaloy E, Ronfaloy V, Stamylan HD, Stamylan LD, Stamylan P, Stamylex PE, Stanyl, Starim.

du Pont de Nemours, E I & Co,

Wilmington,
Del 19898
USA

Endo Laboratories Inc,
Subsidiary of E I du Pont de Nemours & Co,
1000 Stewart Ave,
Garden City, NY 11530
USA

Coumadin, Moban, Nubain, Numorphan, Remsed, Symmetrel, Tessalon, Valpin 50.

Du Pont (UK) Ltd,

Wedgwood Way, Stevenage,
Herts SG1 4QN
England

Du Pont (UK) Ltd,
Polymer Products Department,
Maylands Avenue, Hemel Hempstead,
Herts HP2 7DP
England

Adiprene, Akroflex DAZ, Alathon, Alcryn, Alkanol, Ally, Anton N, Antox N, Avamid, Avitex, Avitone, Bexloy, Birox, Butacite, Bynel, Caytur 21 & 22, Caytur 4, Centari, Certi-fired, Clysar, Corlar, Curzate, Dacron, DCI-3, Delrin, Delsene, Dexlar, Diak, Duponol, Dymel, Dytel, Elvacite, Elvaloy, Elvamide, Elvanol, Elvax, Elvax D, Finesse, Formon, Freon, F.E.P, Ganicin, Glean, HVA-2, Hylene, Hypalon, Hytrel, Hyvar, Hyvar X, Imlar, Imron, Ionomer Resins, Kalrez, Kapton, Karmex, Keldax, Kevlar, Kevlar 49, Krytox, Lannate, Lucite, Ludox, Manzate, Merpol, Merpol DSR, Merpol SH, Methacrol, Minlon, Nafion, Nalan, Nalorex, NBC, Neozone B, Neozone E, Nomex, Nordel, Nordel 2744, Nucrel, Ortholeum, Penta-erythritol Tetrastearate (PET), Permalux, Petrowet, Pyralin, Pyre-ML, Rynite, Sclair, Selar, Silverstone, Sinbar, Stadis, Surlyn, Tedlar, Teflon, Teflon F.E.P, Tefzel 200, Terathane, Tetrone, Thermoflex, Thermoflex A, Ti-pure, Tufcote, Tyvek, Tyzor, Valclene, Vamac, Vazo, Velpar, Venzar, Vespel, Viton, Viton A, Viton AHV, Viton A35, Viton E-430, Viton LN, Vydate, Vydax, Vyloc, Zelcon, Zelec, Zepel, Zonyl, Zytel.

Duphar BV,

PO Box 900,
1380 DA Weesp
Netherlands

Betaserc, Bifiteral, Cardilan, Casoron G, Cephulac, Chronulac, Colofac, Dimilin, Du-Ter, Duphalac, Duphaston, Duspatal, Duspatalin, Duvadilan, Duvadilan Retard, Fevarin 50, Fibervorm, Fibravorma, Flammacerium, Flammazine, Floxifral, Fydulan, Fydumas, Fydusit, Influvac, Portalac, Pre-Par, Serc, Suprilent, Tedion V-18, Terolut, Utopar, Vasodilan, Vasomotal, Xuprin, Yutopar.

Dupont-NEN Medical Products,

331 Treble Code Rd,
North Billerica, Mass 01862
USA

Hepatolite, Pulmolite.

Durham Chemicals Ltd,

Birtley, Chester-le-Street,
Co Durham DH3 1QX
England

Durham Chemicals Ltd,

Activox, Activox B, Chemlok, Decelox, Durocide, Duroseal, Durostabe, Electrox, Entrox, Extrox, Microx, Novor, Nuocure 28, Nuodex 321 Extra, Nuodex 87.

Dylon Industries Inc,

120 First Avenue,
Berea OH 44017
USA

Dylon, Dylonite, Leak Detector.

Dymax Engineering Adhesives,

51 Green Woods Road,
Torrington, CT 06791
USA

Dymax Multi-Care Structural Adhesives, Might-Weld Multi-Care Structural Adhesives.

Dynamit Nobel (UK) Ltd,

Gateway House, 302-308 High Street,
Slough, Berkshire SL1 1HF
England

Dynamit Nobel Wien GmbH,
A 1015 Wien,
Postfach 74,
Austria

Dynamit Nobel Aktiengesellschaft,
Chemicals Division,
Postfach 1261,
5210 Troisdorf
West Germany

Ammonit C (Anfo-explosives), Bergauf, Bikorit, Desodora, Dirubin, Donarit 1, Donarit 2, Donarit 3, Dycron, Dyflor 2000, Dynacal, Dynacast, Dynacerin, Dynacet, Dynacoll, Dynaflock, Dynagrout, Dynagunit, Dynamag, Dynamullit, Dynapol H, Dynapol L, Dynapol LH, Dynapol P, Dynapol S, Dynapor, Dynasan, Dynasil, Dynaspinell, Dynasylan, Dynasylan BSM, Dynatherm, Dynazirkon, Gelatine Donarit S, Gelatine Donarit 1, Gelatine Donarit 2, Gelatine Donarit 3, Gelatine Donarit 2 E, Icdal, Imwitor, Kendurit, Knauerit S, Knauerit 2, Lambrex, Lambrit (Anfo-explosives), Lawinit 100, Massa Estarinum, Miglyol, Pelonit D, Phamosan, Plasvita, Pollopas, Polydur, Redurit, Rewagit, Rhenoflex, Semirit, Softenol, Softigen, Softisan, Tinamul, Toffix, Trocor, Trogamid T, Trogamid T G35, Trolon, Trosiplast, Trosiplast M, Trosiplast S, Ultrapas, Wetter Nobelit B, Witafrol, Witamol, Witepsol, Witocan.

Dynamold Incorporated,

2905 Shamrock Avenue,
Fort Worth, Texas 76107
USA

DM-2, MS-26, DMS-4-828, R-502, MR-502K, MR-502Y, MR-502P.

E

Eastman Chemical Products Inc,
Subsidiary of Eastman Kodak Co,

PO Box 511,
Kingsport, Tenn 37662
USA

C-A-P.

Ebnother Group,

CH-6203 Sempach-Station
Switzerland

Elotex, Sempatap.

Efkay Chemicals Ltd,

375 Regents Park Road,
London N3 1DG
England

Skyllex, Sulfatol E3, Sulphatol B6, Sulphatol 33, Sulphatol 33 MO, Syntol K77 and N77.

E I du Pont de Nemours & Co
see du Pont de Nemours, E I & Co

Elder Pharmaceuticals Inc,

3300 Hyland Ave,
Costa Mesa, Calif 92626
USA

Benoquin, Elaqua XX, Eldecort, Eldopaque, Eldoquin, Eloxyl, Fototar, Oxsoralen, Pabanol, Psoranide, RVPaba Lipstick, Trisoralen.

Eli Lilly & Co,

Lilly Corporate Center,
Indianapolis, Ind 46285
USA

Alphalin, Amesec, Amytal, Amytal Sodium, Anhydron, ASA, Atasorb, Atasorb-N, Aventyl, Aventyl Hydrochloride, Betalin 12 Crystalline, Brevital Sodium, Brietal, C-Quens, Capastat Sulfate, Ceclor, Cevalin, Cinobac, Co-Elorine Pulvules, Co-Pyronil, Coban, Coco-Diazine, Coco-Quinine, Cologel, Cordran, Crystodigin, Darvon, Darvon-N, Deltalin, Dexytal, Digiglusin, Dimelor, Dobutrex, Dolasan, Dolophine Hydrochloride, Doloxene, Doloxene Compound-65, Doloxytal, Drenison, Drolban, Duracillin, Dymelor, Eldisine, Eprolin, Ergotrate Maleate, Glucagon, Gravidox, Haldrate, Haldrone, Hepicebrin, Hexa-Betalin, Histadyl E.C, Histalog, Humulin, Ilosone, Ilotycin, Kamoran, Keflex, Keflin, Kefzol, Lente Iletin, Lextron, Mandol, Merthioiate, Metubine Iodide, Monotheamin and Amytal Pulvules, Monotheamin Pulvules, Monteban, Moxam, Nalfon, Nebcin, Neotrizine, NPH Iletin, Nu-Seals Aspirin, Nu-Seals Potassium Chloride, Nu-Seals Sodium Salicylate, Pantholin, Para-Thor-Mone, Penicillin V Pulvules, Penicillin V Sulpha, Pyronil, Rumensin, Sandril, Seconal, Seconal Sodium, Sedatussin, Semilente Iletin, Sequens, Seromycin, Sodium Amytal, Surfacaine, Tapazole, Tuinal, Ultralente Iletin, V-Cil-K, V-Cil-K Sulpha, V-Cillin, V-Cillin K, Valmid, Valmidate, Vancocin Hydrochloride, Velban, Velbe, Vortel.

Elkem Chemicals Inc,

Parkwest Office Center,
Pittsburgh, PA 15275
USA

Elkem Microsilica, EMS 209, Emsac Concrete Additive.

E/M Corporation,
Sub Great Lakes Chemical Corporation,

Highway 52 NW,
West Lafayette, IN 47906
USA

Everlube, Formkote, Krona-Syn, Kronagold, Kronaplate, Microseal, No-Swab, Perma-Slik.

Emerson & Cuming (UK) Limited,

866-868 Uxbridge Road,
Hayes, Middlesex UB4 0RR
UK

Emerson & Cuming (UK) Limited,

Eccobond Paste 99, Eccobond SF40, Eccobond 114, Eccocoat SJB, Eccofloat, Eccofloat EG35, Eccofloat Encapsulant 1421, Eccofloat HG452, Eccofloat PC61, Eccofloat PP22 and 24, Eccofloat SP 12, 20, Eccofloat SS40, Eccofloat UG 36, Eccofloat US 35, Eccofoam PP, Eccosorb Coating 268E, Eccosorb MF, Eccosorb 269E, Eccospheres.

EMS-Grilon SA,

CH-7013
Domat/Ems
Switzerland

EMS-Grilon (UK) Limited
Drummond Road, Astonfields Industrial Estate,
Stafford, Staffs ST16 3EL
UK

EMS-Chemie AG,
Selnaustrasse 16,
8039 Zürich
Switzerland

Emsodur, Grilamid, Grilamid TR, Grilbond, Grilene Swiss Polyester, Grilesta, Grilon, Grilon BT, Grilon C, Grilonit, Grilpet, Griltex, Swiss Polyamid Grilon, Swiss Polyamid Grilon, Swiss Polyester Grilene, Togocoll.

Emser Industries,

PO Box 1717,
Sumter, SC 29151-1717
USA

Emsodur, Grilamid.

Endo Laboratories Inc
see du Pont de Nemours, E I & Co

Engelhard Corporation,
Specialty Chemicals Division,

Menlo Park, CN 28,
Edison, New Jersey 08818
USA

Attaclay, Attacote, Attaflow, Attagel, Attapulgus, Attasorb, Emcor, Pharmasorb.

Enso-Gutzeit OY,
Chemical Products Division,

Imatrankoski Plant,
SF 55 100 Imatra 10
Finland

Enso DTO 10 - 30, Enso Rosin, Ensol 2.

Envhy Limited,

Padholme Road,
Peterborough PE1 5XL
UK

GRO-HY.

E R Squibb & Sons Inc
see Squibb, E R & Sons Inc

Ethicon Inc,

Somerville,
NJ 08876
USA

Ethibond, Ethiflex, Mersilene, Nurolon, PDS, Prolene, Vicryl.

Evans,

318 High Street North,
Dunstable, Beds LU6 1BE
UK

Acriflex, Adexolin, Anethaine, Dequacaine, Dequadin, Fluvirin, Haliborange, Mevilin-L, Minadex, Mycil Ointment, Mycil Powder, Nylax, Optbas.

Evode Plastics Limited,

Wanlip Road, Syston,
Leicester LE7 8PD
UK

Evo-stik 873 Super, Evoprene, Thermaflo.

Exxon Chemical Ltd,

Exxon Chemical Ltd,

Arundel Towers, Portland Terrace,
Southampton SO9 2GW
England

Exxon Chemical International Inc,
363 Mechelsesteenweg,
B-1950 Kraainem
Belgium

Actrel, Barabar, Breaxit, Breaxit, Corexit, Corexit, Covar, Eca, Escaid, Escane, Escopol Escorene, Escorez, Escoweld, Exxate, Exxsol, Isopar, Jayflex, Norpar, Parabar, Paradyne, Paradyne, Paraflow, Paraflow, Paramins, Paranox, Paranox, Parapak, Parapoid, Parapoid, Paratac, Paratone, Paratone, Rustban, Solvesso, Solvtext, Synacto, Varsol, Vistaflex, Vistalon, Vistanex, Vistone.

F

Fairfax Biological Laboratory Inc

PO Box 242, Electronic Road,
Clinton Corners, NY 12514
USA

Doom, Japidermic.

Farley Health Products,

Torr Lane,
Plymouth, Devon PL3 5UA
UK

Glucodin.

Farmers Crop Chemicals Limited,

County Mills,
Worcester WR1 3NU
UK

Maximate, Talisman.

Farmitalia Carlo Erba,

Via Carlo Imbonati 24,
20159 Milano
Italy

Farmitalia (Farmaceutici Italia) Spa,
Viale E Bezzi 24,
20146 Milano
Italy

Adriblastina, Deflamene, Diprofarn, Fluderma, Kelfizine W, Levius.

Fatsco,

Fatsco,

251 North Fair Avenue,
Benton Harbor, Michigan 49022
USA

Fatsco.

FBC Ltd,

Hauxton, Cambridge,
CB2 5HU
England

Asset, Banlene Plus, Bardew, Benazalox, Betanal E, Bronox, Clearon, Cornox Plus, Cornoxynil, Docklene, Ethidium, FBC CMPP, FBC Fly Dip, FBC MCPA, FBC Pirimicarb 50, FBC Protectant Fungicide, FBC Slug Destroyer, FBC Winter Dip, Fi-Clor, Ficam, Focal, Fyzol 11E, Garvox 3G, Gramoxone 100, Iso-Cornox 57, Legumex Extra, Louse Powder, Mayclene, Mitac 20, Nortron, Phenoxylene 50, Pilot, Post-Kite, Pre-Empt, Pre-Kite, Remtal SC, Rizolex, Rogor E, Septal, Simadex, Spectron, Sportak, Sportak Alpha, Springclene 2, Stanza, Supersevtox, Taktic, Tecane, Trojan SC, Vectal, Weedazin, Zerox.

F E Knight Inc
see Knight, F E Inc

Ferguson & Menzies Limited,

312 Broomloan Road,
Glasgow G51 2JW
UK

Fergapol, Fergatac, Ferquatac.

Ferro Corporation,

Bedford Chemical Division,
7050 Krick Road, Bedford,
Ohio 44146
USA

Ferro Corporation,
Zirconia Operation Division,
Dow Road,
Bow, NH 03301
USA

Cata-Chek, DZ910, Ferro-Cure, Hiotrol, Micro-Chek, Ottasept, Oxi-Chek, PCMX, Permyl B-100, Plas-Chek, Poly-Chek, Pyro-Chek, SM945, SM975, SM987, SM992, SM994, S945, S975, S987, S992, S994, Therm-Chek, UV-Chek.

Fillite USA, Inc,

State Rt 2 & Industrial Lane,
P O Box 3074, Huntingdon, WV 25702
USA

Fillite Hollow Microspheres, Fillite Solid Microspheres - PFA.

Filtec Limited,

39 Keswick Drive,
Frodsham, Cheshire WA6 7LT
UK

Tecagg, Teccel, Tecfil - frequently shortened to 'T' with a suffix e.g. T.300, Tecpril.

The Firestone Tire Co,
Firestone Synthetic Division,

381 W Wilbeth Road,
Akron, OH 44301
USA

Diene, Duradiene, Stereon.

Fischer Instrumentation (GB) Ltd,

Arnhem Road, Bone Lane Industrial Estate,
Newbury, Berks RG14 5RU
UK

Couloscope (Coulometric coating thickness gauge).

Fisons PLC,

Fison House, Princes Street,
Ipswich IP1 1QH
England

Fisons plc,
Horticultural Division,
Paper Mill Lane, Bramford,
Ipswich, Suffolk IP8 4BZ
UK

Fisons PLC,
Pharmaceutical Division,
12 Derby Road, Loughborough,
Leicestershire LE11 0BB
England

Fisons PLC,

Fisons Corp,
2 Preston Court,
Bedford, Mass 01730
USA

Acnil, Albert, Aluphos, Ant Killer, Auralgicin, Autumn Lawn Food, Banwee, Banweed, Banweed-S, Barquinol HC, Basilex, Bilston, Bromodan, Cambilene, Cardophylin, Cojene, Cortipix, Cudgel, Cystopurin, Deep Feed, Delta-Genacort, Dextran, Dextraven, Dimyril, Enterfram, Ergomar, Evergreen, Fennite, Fi-Chlor, Fi-Cryl, Fi-Gard, Fi-Line, Fi-Vi, Ficel, Ficoid, Ficote, Filex, Finestol, Fisons MCPB, Fisons P.C.P, Fisons 18-15, FL7P, Framycort, Framygen, Framyspray, Genacort, Genasprin, Genatosan Skin Bar, Genisol, Genitron, GH5, Glasgro, Gleptosil, Greenfly & Blackfly Killer, Greenkeeper, Greenkeeper Mosskiller, Helarion, Heptomer, Herbazin Plus, Herbazin Special, Herbazin Total, Herbazin 50, Houseplant Long Lasting Feed, Hyalase, Imferon, Imposil, Insect Spray for House Plants, Instoms, Intal Compound, K-Slag, Kil, Kondremul, Lawn Food, Lawn Spot Weeder, Lawn Weeds Killer, Liquinure, Lomodex, Lomudase, Lomupren, Lomusol, Malazide, Medac Cream, Metaquest, MG2/MG4, Mosskil, Nalcrom, Nasalcrom, Neo-Cultol, Neotulle, New Legumex, NFT Fertilizer, Nitro-26, Nomaze, Opticrom, Opticrom, Paracodol, Path Weeds Killer, Pestex, Phenoxylene Plus, Prenomiser, Proferdex, Psorox, PS3/PS4/PS5, Rose Food, Roskens, Rynabond, Rynacrom, Sanatogen, Slug Snail Killer, Soil Pests Killer, Solinure, Somophyllin, Tomahawk, Tomorite, Tritox, Turfclear, Vapo-Iso, Zam-Buk.

Flex-Shield,

PO Box 200,
Gilbert, Arizona 85234
USA

FAR Mark I through X, Flex-Shield.

Flexibulk Limited,

Davidson House, Upper Saint John Street,
Lichfield, Staffs WS14 9DU
UK

Celatom, Water Brite.

Flint Laboratories
see Travenol Laboratories Inc

FMC Corporation,
Food and Pharmaceutical Products Div,

2000 Market St,
Philadelphia, Pa 19103
USA

Ac-Di-Sol, Avicel, Dapon M, Dapon 35.

Forster, D L Limited,

12 The Ongar Road Trading Estate,
Great Dunmow, Essex CM6 1EU
UK

Delf HD Aerosol Adhesive, Delf Silicone Aerosol, Delf 534 Aerosol Adhesive.

Fortafix Limited,

First Drove,
Fengate, Peterborough PE1 5BJ
UK

Fortafix.

Foseco (FS) Ltd,

Drayton Manor, Tamworth,
Staffordshire B78 3TL
England

Foseco Limited,
Metallurgical Division,
Tamworth,
Staffordshire B78 3TL
UK

Adal, Afax, Albral, Ambersil, Aniscol, Askure, Bakfil, Bentokol, Borfax, Brix, Brixil, Brixil, Carbonin, Carset, Carsil, Celtex, Cemset, Ceramol, Chilcote, Chrombral, Corfix, Corseal, Coveral, Cupolloy, Cuprex, Cuprit, Degaser, Deoxidizing Tubes, Dephosphex, Desulfex, Dexil, Dustallay., Dy-Chek, Dycastal, Dycote, Eliminal, Esrakon, Exgraphite, Fabrex, Feedercalc, Feedex, Feedol, Fenotec, Ferad, Ferrogen, Ferrotubes, Ferrux, Firit, Flamco, Flomac, Foset, Foshell, Fosoil, Fostap, Fracton, Freeteem, Furotec, Galag, Garcrete, Garnex, Garpak, Garseal, Gartop, Gartube, Gasbinda, Geolith, Geostop, Hardcote, Herculite, Holcote, Impad, Inertex, Ingotol, Inoculin, Inopak, Inotab, Insural, Intob, IPS, Isomol, I.P.S, Jectoflo, Jectomag, Kalbord, Kalcrete, Kalkor, Kalmex, Kalmin, Kalminex, Kalorex, Kalpack, Kalpad, Kalseal, Kalsert, Kaltek, Kaltop, Kapex, Kompak, Koolkat, Koron, Korpad, Ladelloy, Litefax, Lodol, Logas, Lomag, Lubix, Luscin, Lutron, Magcoke, Magrex, Masol, Metallic Sodium, Miranol, Mixad, Molco, Moldcote, Nametal, Navac, Nodulant, Nokol, Notak, Nozolex, Nucleant, Optemet, Pattrex, Pattrit, Pelamag, Platol II, Platol 11, Plumbral, Plumbral, Plumbrex, Plumbrit, Plus-Gas 'C', Powax, Procal, Profax, Radex, Ramtap, RD10, Recupex, Regenex, Remex, Replicast CS, Replicast FM, Rescon, Sedex, Selek, Separit, Separol, Sepratek, Siliset, Sivex, Sivex F, Slax, Slix, Soda Ash Blocks, Solosil, Spuncote, Stannex, Stelogen, Stelopack, Stelorit, Stelotol, Strandol, Strim, Stripcote, Styromol, Tak, Teebrix, Tekpak-Tekbent, Tekpak-Tekcem, Tellurit, Tellurium Tubes, Terracote, Terradust, Terrapaint, Terrapowder, Thermexo, Thermotex, Thinners, Tilite, Tribonol, Tuncast, Tundak, Zerotherm, Zincrex, Zinol.

Fothergill
see Cyanamid Fothergill Limited

Franklin Mineral Products Co,

Franklin Mineral Products Co,

635 Main Street, Wilmington,
MA 01887
USA

Alsibronz.

Franklin Oil Corporation (Ohio),

40 South Park,
Cleveland, Ohio 44146
USA

Syndraw, Tuf-Draw, Ultracut.

Freeman Chemical Corp,
Subsidiary of H H Robertson Co,

222 E Main St, Box 247,
Port Washington, WI 53074
USA

Acpol, Aquapol.

Fries & Fries of Mallinckrodt Inc,

110 E 70th Street,
Cinti, OH 45216
USA

Mallinckrodt Inc,
675 McDonnell Blvd, PO Box 5840,
St Louis, Mo 63134
USA

Angio-Conray, Barosperse, Cal Plus, Chobile, Cholebrine, Conray 325, also Conray-400, Conray, also Conray 30 and Conray 43, Cysto-Conray, Ferrutope, Hippuran I 131, Ichthymall, IHSA I-125, MD 50, MD 60, Methadose, Phosphocol P32, Prime Flavours, Riamat, Robengatope I-125, Sonilyl, TechneColl, TechneScan MAA, TechneScan PYP, TechneScan SSC, Toleron, Xeneisol 133, Xenomatic.

Frodingham Cement Company Limited,

Brigg Road,
Scunthorpe, Humberside DN16 1AW
UK

Cemsave.

G

GAF (Gt Britain) Ltd, Chemical & Industrial Products

Tilson Road, Wythenshawe,
Manchester M23 9PH
England

GAF Corporation
1361 Alps Road,
Wayne, NJ 07470
USA

Alipal CO-128, Alipal CO-433, Alipal CO-436, Alipal EP-110, EP-115, EP-120, Alipal HF-433, Alipal SE-463, Antara HR-719, Antara LB-400, Antara LE-500, Antara LE-600, Antara LE-700, Antara LF-200, Antara LK-500, Antara LM-400, Antara LM-600, Antara LP-700, Antara LS-500, Antarox BL-214, Antarox BL-225, Antarox BL-236, Antarox BL-240, Antarox BL-330, Antarox CA, Antarox CO-210 and CO-430, Antarox CO-520 and CO-530, Antarox CO-610, CO-630, CO-660, CO-710, CO-720 and CO-730, Antarox CO-850, CO-880 and CO-887, Antarox CO-890, CO-897, CO-970, CO-977, CO-980, CO-987, CO-990 and CO-997, Antarox CTA-639, Antarox DM, Antarox LF-222, Antarox LF-224, Antarox LF-330, Antarox LF-344, Antarox RC-520, Antarox RC-620 and RC-630, Biopal NR-20, Biopal NR-20 W, Biopal VRO-20, Blancol N, Blandofen CAZ, Blandofen CT, Blandofen FA, BLO, Butoxyne 497, Cheelox B-13, Cheelox BF Acid, Cheelox BF-12, Cheelox BF-13, Cheelox BF-78, Cheelox DTPA-14, Cheelox FE-12, Cheelox HE-24, Cheelox NTA-Na3, Cheelox NTA-14, CHP, Diazopon SS-837, Emulphogene BC-420, Emulphogene BC-610, Emulphogene BC-720, Emulphogene BC-840, Emulphogene DA-530, Emulphogene DA-630, Emulphogene DA-639, Emulphogene LM-710, Emulphogene TB-970, Emulphor EL-620, Emulphor EL-719, Emulphor EL-980, Emulphor LA-630, Emulphor ON-870, Emulphor VN-430, Emulphor VT-650, Fenopon AC-78, Fenopon CD, Fenopon CN-42, Fenopon CO, Fenopon EP, Fenopon SE, Fenopon T-33 and T-43, Fenopon T-51, Fenopon T-77, Fenopon TC-42, Fenopon TK32, Fenopon TN-74, Gafac BG-510, Gafac BH-650, Gafac BI-729 and BI-750, Gafac BP-769, Gafac GB-520, Gafac LO-529, Gafac MC-470, Gafac PE-510, Gafac RA-600, Gafac RB-400, Gafac RD-510, Gafac RE, Gafac RE-410, Gafac RE-610, Gafac RE-877, Gafac RE-960, Gafac RK-500, Gafac RL-210, Gafac RM, Gafac RM-410, Gafac RM-510, Gafac RM-710, Gafac RP-710, Gafac RS, Gafac RS-410, Gafac RS-610, Gafac RS-710, Gafamide CDD-518, Gafen LB-400, LE-500 and LS-500, Gafen LE-700, LP-700 and LK-500, Gafen LM-400, Gafen LM-600, Gaffix VC-713, Gafgard 233 and 233E, Gafgard 238, Gafgard 245, Gafgard 277, Gafgard 280, Gafite, Gafite LW, Gaflex,

GAF (Gt Britain) Ltd, Chemical & Industrial Products,

Gafoam AD, Gafquat, Gafstat AD-510 and AE-610, Gafstat AS-610 and AS-710, Gaftuf, Ganex P-904, Ganex V, Gantrez, HEP, Humifen, Humifen BA-77, Humifen BX-78, Igepal CA-210, CA-420 and CA-520, Igepal CA-620 and CA-630, Igepal CA-720, Igepal CA-887, 890 and 897, Igepal CO-210, Igepal CO-430, Igepal CO-520, Igepal CO-530, Igepal CO-610, Igepal CO-620, Igepal CO-630, Igepal CO-660, Igepal CO-710, Igepal CO-720, Igepal CO-730, Igepal CO-850, Igepal CO-880, Igepal CO-887, Igepal CO-890, Igepal CO-970, Igepal CTA-639W, Igepal DM-430, Igepal DM-530, Igepal DM-710, Igepal DM-730, Igepal DM-970, Igepal OD-410, Igepal RC-520, Igepal RC-620, Igepon AC-78, Igepon T-33, Igepon T-43, Igepon T-51, Igepon T-77, Igepon TC-42, Igepon TK-32, Igepon TN-74, Iguafen, Katapol OA-910, Katapol PN-430, Katapol PN-730, Katapol PN-810, Katapol VP-532, Katapone VV-328, Ludigol, Mulgofen, Nekal BA-77, Nekal BX-78, Nekal NF, Nekal WS-25-1 and WS-25, Nekal WT-27, NEP, Peregal O, Peregal OK, Peregal ST, Plasdone, Plasdone C, Plasdone K, Polectron 430, Polyclar, Polyplasdone, Polyplasdone XL, PVP-Iodine, Solidogen LT-13, Soromine AT, Uvi-Nox 1494.

Gard Corporation,

2727 Roe Lane,
Kansas City, KS 66103
USA

Oil Gard, Power Gard, Trans Gard.

G D Searle & Co
see Searle, G D & Co

Genentech Inc,

460 Point San Bruno Blvd,
South San Francisco, Calif 94080
USA

Protropin.

Geoliquids Inc,

3127 W Lake Street,
Chicago, IL 60612
USA

Bromoform, Mi-Gee Brand, Neothene.

Georgia Kaolin Co,

2700 US Highway 22 East,
PO Box 3110, Union, NJ 07083
USA

Altowhite.

G E Plastics Limited,

Birchwood Park,
Risley, Warrington WA3 6DA
UK

Lexan, Noryl, Ultem, Valox, Xenoy.

Gerhardt Pharmaceuticals Limited,

Thornton House, Hook Road,
Surbiton, Surrey KT6 5AR
UK

Danbar, Dethlac, Dethmor, Rinoxin.

Geronazzo S.p.A.,

Via Milano 78,
20021 Ospiate di Bollate,
Milan
Italy

Geronol, Geropon, Soprofor.

Givaudan SA,

CH-1214 Vernier
Switzerland

Givgard DXN, Givsorb UV-2.

Glaxo Inc,

Five Moore Dr,
Research Triangle Park, NC 27709
USA

Beclovent Inhaler, Beconase Nasal Inhaler, Trandate, Ventolin.

Glenwood Laboratories Limited,

19 Wincheap,
Canterbury, Kent CT1 3TB
UK

Glenwood Inc,
83 Summit St,
Tenafly, NJ 07670
USA

Glenwood Laboratories Limited,

Myotonine, Potaba, Yodoxin.

**Glessner Corporation Inc
(GGI Products) DBA,**

1301 Sansome Street,
San Francisco 94111
USA

Glaze 'N Seal Concrete and Masonry Sealer, Glaze 'N Seal Waterbase Clear Concrete and Brick Sealer, Hi-Gloss I.

Goldschmidt, Th Ltd, Chemical Products,

Initial House, 150 Field End Road,
Eastcote, Middlesex, HA5 1SA
England

Aminoxid, Datamuls, Perglanz-Konzen-Trat B48 and B30, Tagat, Tego-Betain HS, Tego-Betain L7 and L10, Tego-Betain T.

Goodrich, B F, Chemical Group,

6100 Oak Tree Blvd,
Cleveland, Ohio 44131
USA

Abson A.B.S, Abson A.B.S. 213, Abson A.B.S. 230, Abson A.B.S. 500, BB Accelerator Carbopol 910, Carbopol 934, Carbopol 934P, Carbopol 940, Carbopol 941, Carboset Resins, Estane, Geon 140X31, Geon 460X6, Geon 590X3, Geon 590X4, Geon 590X6, Good-rite Polyacrylates, Goodrite, Hycar ATBN, Hycar Reactive Liquid Polymer (RLP), Hycar VTBN, Hycar 1203X17, Hycar 1204X5, Hycar 1204X9, Hycar 1205X3, Hycar 1273, Hycar 1402 H82, Hycar 1402 H83, Hycar 1403 H84, Hycar 2100, Hycar 2550H33, Hycar 2550H5, Hycar 2550H55, Hycar 2570H28 and 2570H29, Hycar 2570X5, Hycar 2671H49, Hycar 4021, Hycar 4032, Hycar 4043, Hycar 4201, Novitane P-51.

GP66 Chemical Corporation,

PO Box 8832,
Baltimore, MD 21224
USA

GP66 Miracle Cleaner.

Grace, W R Ltd,

Northdale House, North Circular Road,
London NW10 7UH
England

Grace, W R & Co,
Dewey and Almy Chemical Div,
5225 Phillip Lee Dr,
Atlanta, Ga 30336
USA

Daratac SP 1025, Fabrethane, Hamposyl L, C and O, Sodasorb, Sylvid.

Graesser Laboratories Ltd,

Sandycroft, Deeside,
Clwyd CH5 2PX
England

Micropaque.

Great Lakes Chemical (Europe) Ltd,

28 Bakers Road, Uxbridge,
Middlesex UB8 1QS
England

Aquadrome, Bromicide, Bromo-Gas, Dihalo, Meth-O-Gas, Mycocide, Presol W, Terr-o-gas.

Greef, R W & Co Inc,

1445 E Putnam Ave,
Old Greenwich, Conn 06870
USA

Heliophan.

Greene, Tweed,

Box 305, Kulpsville,
Pennsylvania 19443,
USA

Fluoraz (US, UK, France, Germany only).

Guardian Chemical,
Div of United-Guardian Inc,

PO Box 2500,
Smithtown, NY 11787
USA

Clorpactin WCS-90, Clorpactin XCB, Voidox 100 per cent.

G W Carnrick Co Limited

G W Carnrick Co Limited
see Carnrick, G W Co Limited

Haifa Chemicals Limited,

PO Box 1809,
Haifa
Israel

Magnisal, Polyfeed.

H

Hallam Polymer Engineering Limited

Callywhite Lane, Dronfield,
Sheffield S18 6XR
UK

Betathane, V-thane.

Hall, The C P Company,

7300 S Central,
Chicago, IL 60638
USA

E-Z Mix, Hallcomid, Monoplex, Paraplex, Plasthall, Tegmer.

**Hans Rahn & Company,
Pharma Division,**

Wehntalerstrasse 79,
CH-8057 Zurich
Switzerland

Tiazyme.

Harcros Industrial Chemicals,

3-5 Alan Street,
Rydalmere, NSW 2116
Australia

Armeen, Bioacid, Decol, Decolamide, Duomeen, Ethoduomeen, Ethomeen, Ethomid, Ethoquad, Kemmat, Kemmest, Kemonic, Kemopol, Kemotan, Kemwax, Manoxol OT60, Maypon, Orzan, Panex, Polyfon, Quatramine, Smoothex, Softex, Spersol, Sunnol, Surfynol.

Hardman Incorporated,

Hardman Incorporated,

600 Cortlandt Street,
Belleville, NJ 07109
USA

Acrylweld, Butox, Condux, Cyclo-Rubber, Double/Bubble, DPR, Epocap, Epocrete, Epocure, Epolast, Epomarine, Eposet, Eposolve, Epoweld, Hardset, Isolene, Kalar, Kalene, Kalex, Mix-Kit, Monopoxy, Nu-Set, Nylox, Once, Phenoweld, Polymeric Sealant Gun, Rolox, Triplematic, Tuff Stuff, Weldox.

Harshaw Chemicals Limited,
Industrial Finishing Division,

P.O. Box 4,
Daventry, Northants NN11 4HF
UK

Alka, Alkastar 83, Crystalite, DC 150, DC700, Elektra, Mark 80, Neostar, Novadine, Super Lubracon, Ultra 206, Zenith, Zodiac.

Hawley, W & Son Limited,

Colour Works,
Duffield, Derby DE6 4FG
UK

Shadeacrete, Tintacrete.

Haysite Reinforced Plastics,

5599 New Perry Highway,
Erie PA 16509
USA

Haysite.

Henkel Chemicals Ltd,
Organic Products Division,

Merit House, The Hyde,
Edgeware Road, London, NW9 5AB
England

Adalin, Adipon, Aethoxal, Amphocerin, Aversin, Avivage, Belsoft, Breviol, Cottoclarin, Cutina, Dehymuls, Dehyquart A, Dehyquart C, Dehyquart CDB, Dehyquart DAM, Dehyquart LDB, Dehyquart LT, Dehyquart SP, Dehyton AB-30, Dehyton K, Euperlan, Flexin, Floranit, Loxiol G-70, G-71, G-72 and G-73, Miltopan, Myritol, Olinor, Polyquart H, Product AAS 90, Setilon, Silkiol, Sipex DS, Sipex 30, Solvoclarin, Spreitan, Stabifix, Sulfopon LS, Texapon ALS and ALS/S, Texapon ASV, Texapon

ESI/S, N25 and N40, Texapon EVR and EST, Texapon IES, Texapon L20C, Texapon L20M and L20M/S, Texapon L230 and T42, Texapon NSF.

Herbert Laboratories,
Dermatology Div of Allergan Pharmaceuticals Inc,

2525 Dupont Drive,
Irvine, Calif 92713
USA

Aeroseb-Dex, Aeroseb-HC, Clear by Design, Danex, Fluonid, Fluoroplex, Maxiflor.

Hercules Ltd,

20 Red Lion Street,
London WC1R 4PB
England

Hercules Inc,
910 Market Street,
Wilmington DE 19899
USA

A-Fax, Abalyn, Abitol, Adriamycin, Adrucil, Adtac, Amine D, Aqua Mer, Aqualon, Aquapel, AS, Belro, Bexphane, Blanose, Blue Dot, Bresin 2, Bresin 2E, Brisgo II, Bullseye, BX 310, BXT, Canagel 75, Cascade, Cellolyn, Cellolyn, Cellulose Gum, Chlorofin 42, Clorafin, Coaldet, Delfloc, Delnet, Delweve, Di-Cup, Dipentene No. 122, Dresinate, Dresinol, Dymerex, Echo, Fiber Pare, Flav-O-Lok, Flogel, Foral (AX, 85 and 105), Frimulsion, Frother 4171, Gel Flo, Gel Power, Gelamite D, Gelaprime F, Gelcharg, Genu, Genuzan, Green Dot, HBR, Herchlor (C, C85 and C110), Herclor, Herco-Prills, Hercobind DS, Hercoflat, Hercoflav, Hercoflex 600, Hercoflex 707, Hercoflex 707A, Hercoflex 900, Hercofloc, Hercofroth, Hercol 2, Hercol 2X, Hercolube, Hercolyn, Hercolyn D, Hercomix, Hercon 2, Hercon 2X, Hercon 32, Hercon 40, Hercon 48, Hercoprime, Hercopruf, Hercosett, Hercosett 125, Hercosol TP-S, Hercosplit WR, Hercotac AD, Hercotac LA, Hercotuf, Herculine FR, Herculon, Hercures, HVP, Indalca, Indalca, Inkovar 335, Inkovar 617, Kaochlor, Kaon, Klucel, Kristalex, Krystalex, Kymene, Lewisol 28, Luxor, Magan, Magmet, Magnamite, Magnox, Mangnamite, Melhi N, NS and NLM, Merigraph, Metalyn 582, Millidet, Minflo, Modane, Mydflex, Natrosol, Neolyn, Neuphor 100, Pamak, Pamolyn, Pamolyn 100, Pamolyn 100 FGK, Pamolyn 100FG, Pamolyn 125, Pamolyn 200, Pamolyn 240, Pamolyn 300, Pamolyn 327B, Pamolyn 380, Paracol, Parel 58, Parlon, Parlon P, Pental A, Pental G, Pental X, Pental 28, Pental 8D, Pental 802A, Pentalyn, Pentrex, Permalyn, Petrex 7-75T, Pexalyn, Pexate, Pexite, Pexol, Pexol, Picco 5000, Picco 6000, Piccodiene, Piccolastic, Piccolyte A, Piccolyte C, Piccolyte HM110, Piccolyte S, Piccomer XX, Picconol A100, Picconol A200, Picconol A300, Picconol A400, Picconol A500, Picconol A600, Piccopale, Piccopyn, Piccotac, Piccotex, Piccotoner, Piccovar AB, Piccovar AP, Piccovar L, Pinene, Poly Pale, Poly-Pale, Polyrad, Pro-Fax, Pulpex E and P, Red Dot, Redd Citrus Specialties, Reloder 7, Reten, RS Nitrocellulose, Solvenol 1, Solvenol 2, Solvenol 226, SS Nitrocellulose, ST-Size, Staybelite, Surf-A-Seis, T-lim, T-Size, Terate 101, Terate 131, Terate 202, Terate 203, Terate 204, UN-28 and UN-32, Unigel, Unimite, Unique, Vinsalyn, Vinsol, Vinsol Emulsion, Vul-Cup, Vul-Cup 40KE and R, Vulcup, Yarmor, 1900.

Hermadex Limited,

Hermadex Limited,

832 High Road,
London N12 9RA
UK

Florisil, Min-U-Gel, Refinex.

Hermetite Products Limited,

Tavistock Road,
West Drayton UB7 7RA
UK

Ali-Clean, Golden Hermetite, Instant Gasket, Kwikfill, Lusol, Pro-Etch, Red Hermetite, Torqseal, Tuf Stuf.

Hexcel, Chemical Products,

205 Main St,
Lodi, NJ 07644
USA

Germ-i-Tol, Temasept IV.

Hi Temp Lubricants Inc,

7019 Corporate Way,
Dayton, Ohio 45459
USA

Hi Temp EC-1000, Hi Temp EC-4000, Hi Temp EC-5000.

Hickson & Welch Ltd,

Castleford,
W Yorkshire WF10 2JT
England

Antiblu/Antiboror, Avirol, Bradsyn, Garbritol, Gardinal, Hickstor, Lanbritol, Lanette, Lorol, Photine, Photine, Pyrochlor, Pyrolith, Tanalith, Vacsol.

H Marcel Guest Limited
see Marcel Guest, H Limited

Hoechst UK Ltd,
Hoechst House, Salisbury Road,
Hounslow, Middlesex TW4 6JH
England

Hoechst AG,
Postfach 800320,
D6230 Frankfurt (Main) 80
W Germany

Afalon, Albaphos Dental Na 211, Alberit MF, Alberit MP, Alberit PF, Alberit VP, Alcan, Antistatic 812 and 813, Antistatic 816, Appretan, Arelon, Aresin, Aristoflex, Arkopal, Arkopal N, Arkopon T, Bohrmittell Hoechst, Bozefloc, Brassicol, Butoxyl, Cambison, Ceridust, Chaldegal, Colanyl, Curasol, Daneral, Daneral-SA, Daonil, Delrin, Dentplus Special, Dipar, Dodigen, Dolan, Dolanit, Emulsogen, Exolit, Femipausin, Flexonyl, Fluorescent Red 5B, Frigen, Genamin, Genamin KDM, Genapol, Genapol LRO, Genapol ZRO, Genomoll P, Genotherm, Gingicain, Hordaflex, Hordalub, Hordamer, Hostacain, Hostacor, Hostadur, Hostaflam, Hostaflex, Hostaflon, Hostaflon C2, Hostaflon ET, Hostaflon TF, Hostaform, Hostaform C, Hostalen, Hostalen G, Hostalen GM, Hostalen GP, Hostalen OO, Hostalen PP, Hostalit Z, Hostalub, Hostamid, Hostanox, Hostaperm, Hostaphan, Hostaphane, Hostaphat, Hostapon CAS, Hostapon CT, Hostapon KA, Hostapon KTW new, Hostapon STT, Hostapon TF, Hostapor, Hostaprint, Hostapur OS, Hostapur SAS, Hostatint, Hostatron, Hostavin, Hostyren, Ivorin-Profalon, Ivosit, Jadit, Jadit H, Lasix, Lasix + K, Ledmin LPC, Licomer, Licowet, Lumilux, Medialan KA Conc, Medialan KF, Medialan LD, Mergal, Morocide, Mowilith, Mowiol, Mowital, Nata, Octopirox, Pergalen, Polysolvan O, Prapagen WK, WKL and WKT, Prepagen, Prepagen WK, Pressimmune, Rastinon, Remafin, Remazol, Samaron, Sapogenat, Sapogenat T, Synadryn, Trental, Trevira, Tumeson, Tylose, Wax C.

Hoechst-Roussel Pharmaceuticals Inc
see Roussel Laboratories Limited

Hoffmann Mineral
Franz Hoffmann & Sohne KG,

Munchener Str 75,
8858 Neuburg/Do
F R Germany

Sillitin - Aktisil.

Hoffmann-LaRoche Inc
see Roche Products Ltd

Holliday Dyes & Chemicals Ltd,

PO Box B22, Leeds Road,
Huddersfield HD2 1UH
England

Acetyl, Alizarine, Blancol, Cyanine, Cyanine Fast, Elbasol, Elbelan, Elbelene, Elbenyl, Elbesol, Elbestret, Elite Fast, Fluorescent, Merantine, Panacryl, Plastic, Polycron, Smoke, Sublaprint, Supracet, Suprexcel.

Holt Lloyd Corporation,

Holt Lloyd Corporation,

4647 Hugh Howell Road,
Tucker, GA 30084
USA

LPS Brake Cleaner, LPS Electro Contact Cleaner, LPS Engine Degreaser, LPS Heavy-Duty Silicone Lubricant, LPS Instant Cold Galvanize, LPS Instant Super Cleaner/Degreaser, LPS Paint Remover, LPS Tap-All, LPS 1 Greaseless Lubricant, LPS 2 General Purpose Lubricant, LPS 3 Heavy Duty Rust Inhibitor, LPS 500 Plus.

Hommel Pharmaceuticals,

Industriering 34, CH-8134,
Aduswil, Zurich
Switzerland

Hicoseen.

Horace Cory PLC,

Nathan Way, London,
SE28 0AY
England

Corbrite, Corcert, Corfast, Corgran, Corlake, Cortone.

Horn's Crop Service Center,

P O Box 326,
Bellevue, OH 44811
USA

Green Magic, Horn O' Plenty, Super Green.

Howlett Adhesives Limited,

Horsley Road, Off Kingsthorpe Road,
Northampton NN2 6LL
UK

Adflex, Howstik, Howtex, Pvacote.

Huls (UK) Ltd,

Cedars House, Farnborough Common,
Orpington, Kent BR6 7TE
England

Arova 16, Avistin, Buna AP, Buna EM, Buna SL, Buna Vi, Bunatex, Dionil, Drivanil, Driverit, Driverol MPL, Driverol OMM, Driveron, Drivolan, Duranit, Ilexan E, Ilexan HT, Ilexan P, Ilexan S, Lipinol O, Lipinol T, Lipolan, Litex, Marlamid, Marlazin, Marlican, Marlinat, Marlipal, Marlon, Marlon AMX, Marlon AS3, Marlon A350, Marlon A360, A365 and A375, Marlon A390, A396 and ARL, Marlophen, Marlophen DNP, Marlophen X, Marlophen 80 Series and 800 Series, Marlophor, Marlophor CS, DS and NP, Marlophor FC, Marlophor MD, Marlophor T10, Marlopon, Marlopon ADS50, Marlopon AT50, Marlosol, Marlotherm, Marlowet, Marlox, Marlox, Polyvest C70, Polyvest 25, Terravest 801, Univest, Vestamelt, Vestamid, Vestenamer, Vestiform, Vestinol, Vestodur, Vestolen A, Vestolen EM, Vestolen P, Vestolit HI, Vestolit S, Vestopal, Vestoplast, Vestopren, Vestowax, Vesturit, Vestypor, Vestyron, Vilit.

Humber Fertilisers Plc,

P O Box 27, Stoneferry,
Hull HU8 8DQ
UK

Eclipse, Humber.

Hyland Therapeutics
see Travenol Laboratories Inc

Hynson, Westcott & Dunning,
Div of Becton Dickinson & Co,

Charles & Chase Sts,
Baltimore, Bd 21201
USA

BAL in Oil, Bromsulphalein, Cardio-Green, Indigo Carmine.

Hytak Limited,

Greenhill Industrial Estate,
Leabrooks, Derby DE55 4BR
UK

Hytak.

I

ICN Nucleic Acid Research Institute
ICN Pharmaceuticals Inc,

2727 Campus Drive,
Irvine, Calif 92715
USA

Bendopa, Testred.

ICN Nutritional Biochemicals Corp,

26201 Miles Road,
Cleveland, Ohio 44128
USA

Acillin, Nicocap, Virazole.

Ilon Laboratories (Hamilton) Limited,

Lorne Street, Hamilton,
Lanarkshire, ML3 9AB,
UK

Ilonium, Thovaline.

Imperial Chemical Industries plc,

PO Box 6, Bessemer Road,
Welwyn Garden City, Herts AL7 1HD
UK

ICI PLC,
Agricultural Division,PO Box 1, Billingham,
Cleveland TS23 1LB
England

ICI PLC,
Mond Division,
PO Box 13, The Heath,
Runcorn, Cheshire WA7 4QF
England

ICI PLC,
Organics Division,
PO Box 42, Hexagon House,
Blackley, Manchester M9 3DA
England

ICI PLC,
Petrochemicals & Plastics Division,
PO Box 6, Bessemer Road,
Welwyn Garden City, Hertfordshire AL7 1HD
England

ICI PLC,
Plant Protection Division,
Fernhurst, Haslemere,
Surrey GU27 3JE
England

ICI Americas Inc,
Wilmington,
Del 19897
USA

Stuart Pharmaceuticals,
Div of ICI Americas Inc,
Wilmington,
Del 19897
USA

Abol, Abol G, Acid Yellow OO, Acorga, Acrodel, Actellic, Actellifog, Activex, Activol, Actomol, Advance, Advizor, Afrol, Afrol, Agma, Agral, Agramm, Agrocide, Agrosan, Agrosol, Agrothion, Agroxone, AL terna GEL, Alboleum, Albolineum, Alcian, Alibi, Alistell, Alkathene, Alloprene, Alocrom, Alphanol, Ambush, Amichrome, Antasil, Apatef, Aphox, Aphrogene, Aquabase, Aquaperle, Arbocel, Arbogard, Arcton, Aretan, Arklone, Arlacel, Arlacel 80, Arlacel 83, Arlacel 85, Arlacide, Arlagard, Arlamol, Arlamol E, Arlatone, Arlatone 507, Aromasol, Aromix, Arotex, Arsinette, Asterite, Atiran, Atlac, Atlox, Atmos, Atped 400, Atped 600, Atpeg 300, Atromid-S, Avgard, Avloclor, Avlosulfon, Azotox, Baquacil, Baquatop, Battal, BCF, Belco, Belpro, Benlate, Berelex, Betrox, Bexfilm, Bexfilm 'A', Bexfilm 'O', Bexfilm 'P', Bexfilm 'S', Bexfilm 'T', Biopol, Birlane, Blex, Bolda, Bonzi, Botrilex, Boxolon, Bozzle, Bri-Nylon, Brij, Brij 96, Brij 97, Bronocot, Bug Gun, Butoxone, Calaton, Calbux, Caledon, Calsolene Oil, Cambrelle, Captan, Captan 83 WP, Captan-Col, Captan-50, Carbolan, Carp Brand, Cepton, Cereclor, Ceresol, Cerevax, Cerone, Cetaped, Cetavlex, Cetavlon, Chlorazol, Chloros, Chromastral, Cirrasol, Clean-Up, Clerit, Clerite, Clipper, Cobex, Coomassie, Coptal, Coriacide, Corilene, Coriumine, Corsodyl, Corvic, Cosmocil, Cranco, Crex, Cultar, Cuprodine, Cutlass, Cymag, Cymbush, Cymperator, Daltocel, Daltoflex, Daltogard, Daltolac, Daltoped, Daltorez, Daltorol, Degopol, Degopur, Delan-Col, Demetox, Densil, Deoxidine, Deoxylyte, Depsoline, Derris Dust, Detergyl, Di-Farmon, Diakon, Dialose, Dialose Plus, Diamonine, Diazamine, Diazol, Didi-Col, Didigram,

Didimac, Didimac, Dielan-Col, Diethoxol, Diglyme, Dimacide, Diprivan, Disadine, Dispersol, Dispray, Disulphine Blue, Dormakil, Dosaflo, Dragon, Dragonmat, Drikold, Du-Ter, Dulceta, Dulux, Durazol, Ecosyl, Edicol, Edunine, Effersyllium, Ekatin, Electrodyn, Endanil, Endegal, Epodyl, Esterolane, Ethoxol, Ethrel, Exelderm, Farmacel, Farmaneb, Farmon, Ferna-Col, Fernasan, Fernasul, Fernesta, Fernex, Fernide, Fernimine, Fernoxone, Ferrax, Fettel, Filtram, Fixogene, Floratex, Fluolite, Fluon, Fluoranar, Fluorolux, Fluothane, Folosan, Fomac, Foraperle, Forest Bark, Format, Fortimax, Fortress, Fosferno, Foulon, Fouramine, Francolor, Frenokone, Fulcin, Fumite, Fusarex, Fusilade, Fytospore, Gamma-Col, Gammalex, Gammalin, Gammasan, Gammatrol, Gammexane, Ganocide, Garlon 2, Genclor, Genklene, Glazamine, Grain Store Smoke, Gramazine, Gramixel, Gramonol, Gramoxone, Gramuron, Granodine, Granolube, Granstock, Groundhog, Haloflex, Harrier, Heliane, Hemoxone, Hexacal, Hexafoam, Hexaplas, Hexaplus, Hi-Build, Hibiclens, Hibidil, Hibiscrub, Hibisol, Hibispray, Hibitane, Hobane, Hotspur, Icipen 300, Impact, Imperacin, Imprez, Imsol, Inderal, Inderetic, Inochrome, Inoderme, Iodosorb, Kafil, Kallodoc, Kalten, Karate, Kasof, Kaynitro, Kemick, Kephos, Kericompost, Kerigrow, Keriguards, Keriroot, Kerishine, Kerispikes, Kerispray, Keristicks, Ketrax, Klerat, Lampronol, Lanapex, Lancer, Lapudrine, Lawn Plus, Lawnsman, Lawnsman Spring Feed, Lawnsman Weed and Feed, Lawnsman Winterizer, Leefex, Lenetol, Limbaki, Limbux, Linalux, Lissamine, Lissanol, Lissapol, Lithoform, Lontrel, Lontrel Plus, Lorexane, Lubrol, Lubrol N13, Lubrol N5, Lutetia, Lyracamine, Magspa, Maranyl, Marsipol, Matexil, Matikus, Melacos, Melinar, Melinex, Mephetol, Mergamma, Merolan, Mesgamma, Metallichrome, Methasol, Methic, Methoklone, Methoxone, Metosyn, Mi-Col, Mil-Col, Milcap, Milcurb, Milgo, Milstem, Mitrelle, Mixol, Monastral, Monit, Monnex, Monolite, Moss Gun, Mouser, Mylanta, Mylar, Mylicon, Myrj, Myrj 52, Myrj 53, Mysoline, Nalfleet, Nalfloc, Naseptin, Necol, Neopralac, Neramine, Neutrichrome, Neutrogene, New Verdone, Nilergex, Nimrod, Nimrod T, Nitracc, Nitram, Nitro-chalk, Nolvadex, Nonanol, Nonox, Nonox ZA, Novapel, Novaplaste, Novester, Nufol, Nuram, Nutramon, Nylomine, Occlusin, Oncol, Organol, Ovicide, Paddox, Paludrine, Para-col, Parafil, Paragrid, Paralink, Paraloop, Paratie, Paraweb, Parlay, Patafol, Patafol Plus, Path Gun, Pathclear, Pentol, Perecot, Pereman, Perenox, Perenox, Perklone, Permalose, Perminal, Permobel, Perspex, Petrinex, Phosteem, Phyomone, Picket, Pirimor, Plantvax, Plastpak, Plictran, Plondrel, Plus, Polisax, Politarp, Polymon, Primicid, Primotec, Probe, Procilene, Procinyl, Procion, Propafilm, Propafoil, Propaklone, Propathene, Proxel, Pruteen, Qazul, Quintesse, Radar, Rapid, Ratak, Razoxin, Reglone, Reglox, Regulex, Renektan, Renex, Residuren, Resovyl, Ridoline, Ridosol, Rose Plus, Roseclear, Rotalin, Saffil, Saisan, Salamac, Sanspor, Savall, Savloclens, Savlodil, Savlon, Savlon Babycare, Sel-oxone, Serseal, Siapton, Silcolapse, Silcolease, Silcoset, Siopel, Sofanate, Solane, Solanthrene, Solasol, Soledon, Solester, Solfa, Solochrome, Soltair, Solufeed, Solutene, Solvigran, Sorbichew, Sorbid, Sorbitrate, Sorbo, Span, Span 20, Span 40, Span 60, Span 65, Span 80, Span 85, Speedway, Stabgel, Statil, Staufen, Stempor, Sterethox, Steribath, Stimufol, Strawlink, Strelax, Sulfacide, Sulfanol, Sulphamezathine, Sunaptol, Super Verdone, Suprasec, Sybol, Sylade, Synacril, Synalar, Synandone, Synektan, Synperonic, Synperonic NP10 and NP12, Synperonic NP13 and NP15, Synperonic NP20 and NP30, Synperonic NP4, NP5 and NP6, Synperonic NP8 and NP9, Synperonic N, NX, NXP and NDB, Synperonic OP, Synperonic 3S27 and 3S60S, Synperonic 3S60A, Synprol, Synprol Sulphate, Synprolam, Synprolam 35, Synprolam 35 BQC, Synprolam 35 DM, Synprolam 35 DMA, Synprolam 35A, Synprolam 35N3, Synthamel, Tactel, Talon, Teefroth, Tempo, Tempro, Tenoret, Tenoretic, Tenormin, Tensol, Terbutol, Terinda, Terraklene, Terram, Terylene, Tetmosol, Tetroxone, Thermex, Thermoplaste, Thetmex, Thixomen, Timbrelle, Topal, Topanol, Toppel, Tota-col, Tri-Borne, Tri-Farmon, Triklone, Trilene, Trimmit, Trioxone, Trisec, Tween, Tween 20, Tween 40, Tween 60, Tween 80, Twosward, Ucrete, Unisol, Vanquish, Vantoc, Vantoc AL, Vantoc CL, Vantocil, Vantropol, Verdley, Verdone, Verton, Viclan, Victrex, Vigil, Visacor,

Visqueen, Vitrafix, Vivalan, Vizor, Vulcan, Vynamon, Warrior, Waspend, Waxolene, Weed Gun, Weedol, Welvic, Winnofil, Zoladex.

Incitec International,

PO Box 140, Morningside,
Queensland 4170
Australia

Suscon Blue.

Industrial Adhesives Company,
Bond-Plus Adhesives & Coatings,

2632 W Washington Blvd,
Chicago Ill 60612
USA

Bond-Plus, Bond-Plus HM.

Industrias Quimicas Del Valles SA,

Rafael de Casanovas, 73,
Mollet del Valles, Barcelona,
Spain

Caldo Bordeles Valles, Cupertine, Cupertine Folpet, Cupertine Super, Curenox-50, Sulfato de Cobre Valles, Vironex.

Interferon Sciences Inc,

783 Jersey Avenue,
New Brunswick, NJ 08901
USA

Alferon.

International Coatings,

13929 E 166th Street,
Cerritos, California 90701
USA

Aquasoft.

Interox Chemicals Ltd,

Interox Chemicals Ltd,

PO Box 7, Warrington,
Cheshire WA4 6HB
England

Capa, Chloritane, Interox H48, Ixper, Oxyper, Proxitane, Tripwite, Truzone, Washsaver.

Intervet America Inc.,

P O Box 318,
Millsboro, DE 19966
USA

Breedervac-I, Breedervac-II, Breedervac-III, Breedervac-IV, Bron-Newcavac-M, B1Vac, Clonevac D-78, Clonevac-30, Clonevac-30T, Combovac-30, Duovac-C, Duovac-M, FP-Vac, IB-VAC, IB-VAC-H, IB-VAC-M, Marexine-CA, Mildvac-C, Mildvac-M, Newcavac, Newcavac-T, PP-Vac, RE-VAC II, SB-VAC, SB-VAC Plus Marexine-CA, Tensynvac, Trachine, Tremvac, Tremvac-FP, Triplevac.

Invequimica & CIA. S.C.A.,

PO Box 3227,
Medellin
Columbia

Alcaphos 24, Deoxiphos 600, Invephos 20, Invephos 21C, Inveres EVH, Inveres K-82, Malezafin LV-4, Malezafin 55 Plus, Malezafin 57 LV, Potenzol V, Weed Be Gone 45, Weed Be Gone 50.

Irathane International Limited,

78 Holmethorpe Avenue,
Redhill, Surrey RH1 2PF
UK

Irabond, Irasolve, Irathane.

ISC Alloys Ltd,

Alloys House, PO Box 34, Willenhall Lane,
Bloxwich, Walsall WS3 2XW
England

Delaphos, Delaville.

ISC Chemicals Ltd,

St Andrews Road, Avonmouth,
Bristol BS11 9HP
England

Flutec, Formel - NF, Isceon, Tardex.

Ives Laboratories Inc,

685 Third Ave,
New York, NY 10017
USA

Cerubidine, Isordil, Minihist, Monitan, Piportil, Sectral, Trecator-SC.

J

James Briggs & Sons Limited

Lion Works, Old Market Street,
Blackley, Manchester M9 3DU
UK

Antiquax, Fillpak, Hycote.

James Robinson & Co Ltd,

PO Box B.3, Hillhouse Lane,
Huddersfield HD1 6BU
England

Chromol, Endurol, Sulphol, Sulphosol.

Jan Dekker Bv,
Naarden International Chemicals Division,

PO Box 10,
1520 AA Wormerveer,
The Netherlands

Aldosperse, Rexoteric XCE, Rexoteric XCG, Rexoteric XCO, Rexoteric XJO, Rexoteric XOO, Rexoteric YCB, Rexoteric YCE, Rexoteric YJE, Rexoteric YOB, Rexoteric ZXCO.

Janssen Pharmaceutical Limited,

Grove, Wantage,
Oxon OX12 0DQ
UK

Janssen Pharmaceutical,
40 Kingsbridge Rd,
Piscataway, NJ 08854
USA

A-Vitan, Acnidazil, Anquil, Arret, Brentan, Cinnar, Clinafarm, Clinium, Daktacort, Daktarin, Dermonistat, Dioxatrine, Dipidolor, Droleptan, Equivurm Plus, Flubenol, Flukiver, Fluvermal, Gyno-Daktarin, Haldol, Haldol Decanoate, Hismanal, Hypnodil, Hypnomidate, Hypnomidate Concentrate, Hypnorm, Imaverol, Imodium, Inapsine, Mebatreat, Mebenvet, Motilium, Nizoral, Oftentral, Operidine, Ovitelmin, Rapifen, Ripercol, Spartakon, Spartrix, Spiropitan, Stresnil, Stugeron, Stugeron Forte, Sublimaze, Sufenta, Sufrexal, Suicalm, Surolan, Telmin, Thalamonal, Tinset, Triperidol, Vermox.

J C Thompson & Co (Duron) Ltd
see Thompson, J C & Co (Duron) Ltd

John & E Sturge Ltd (Selby),

Denison Road, Selby,
N Yorkshire YO8 8EF
England

Sturge Lifford,
Lifford Chemical Works, Lifford Lane,
Birmingham B30 3JW
UK

John & E Sturge Ltd (Lifford),
Lifford Lane, Kings Norton,
Birmingham B30 3JW
England

Aeromatt, Calofil, Calofort S, Calopake, Caloxal CLP 45, Caloxol, Caloxol CP2, Citraclean, Glucox, Hydrolact, Sturcal, Sturcarb.

Johnson Matthey Chemicals Ltd,

Orchard Road, Royston,
Herts SG8 5HE
England

AuSub, Easy-FLo, Honeycat, Mattheylec, Puratronic, Reacton, Specpure, Spectroflux, Spectromel, Spectromel, Tenacity.

Johnson & Johnson Products Inc,

501 George St,
New Brunswick, NJ 08903
USA

Johnson & Johnson Products Inc,

Surgicel, Tramacin.

JR Technology Limited,

81 North End, Meldreth,
Royston, Herts SG8 6NU
UK

JR Surfacer, Masterblok.

K

KabiVitrum AB

112 87 Stockholm
Sweden

Aminess, Antithrombin, Aunativ, Cetiprin, Cetiprin Novum, Crescormon, Cyklokapron, Epsikapron, Gammonativ, Intralipid, Kabikinase, Octonativ, Preconativ, Rhesonativ, Somatonorm, Vamin Series.

Keeling & Walker Ltd,

Whielden Road, Stoke-on-Trent,
Staffs ST4 4JA
England

Superlite.

Kelco, Division of Merck & Co Inc,

8355 Aero Drive, PO Box 23576,
San Diego, CA 92123
USA

Amoloid HV, Amoloid LV, Kelcoloid HV, Kelcoloid LV, Kelcosol, Kelgin F, Kelgin HV, Kelgin LV, Kelgin MV, Kelgin QH, Kelgin QL, Kelgin QM, Kelgin XL, Kelgo-Gel HV, Kelgo-Gel LV, Kelmar, Keltex, Keltex S, Keltose, Kelzan, Kelzan D, Kelzan S, Superloid.

Kelco/AIL International Ltd,

22 Henrietta Street, London,
WC2E 8NB
England

Alginade, Collatex, Kelgin, Keltrol, Lacticol, Manucol, Manucol Ester, Manugel, Manutex, Municol.

Kempner, S Limited,

498 Honeypot Lane,
Stanmore, Middlesex HA7 1JZ
UK

Reynolon, Sekawrap.

Kemtron Itl Inc.,

P O Box 2508,
Newark, N J 07114
USA

Diepoxy, Kemflorseal, Kempoxy, Kemtop.

Keno Gard (UK) Ltd,

18a Reading Road, Henley-on-Thames,
Oxon RG9 1AG
England

Amine, Ampholyte SKKP 70, Diamine, Etheramine, Lilaflot, Lilamin, Lilaminox, Querton.

Kenrich Petrochemicals Inc,

East 22nd Street, PO Box 64,
Bayonne, NJ 07002
USA

Ken-React, Kenflex, Kenflex A, Kenmag, Kenplast.

Kensol Corporation,

PO Box 3179,
Allentown, PA 18106
USA

Kensol KM Metal Cleaner, Kensol KV Rust Retarder, Kensol KX Oxide Resistor.

The Kent Chemical Company,

George House, Bridewell Lane,
Tenterden, Kent TN30 6HS
UK

Aviashine, Aviawash, Turboclean.

Kerr-McGee Chemical Corporation,

PO Box 25861,
Oklahoma City, OK 73125
USA

KM Ammonium Metavanadate, KM Fly Ash, KM Muriate of Potash, KM Pebble Lime, KM Phosphate Rock, KM Potassium Chloride, KM Potassium Perchlorate, KM Sodium Chlorate, KM Sodium Perchlorate, KM Vanadium Pentoxide, MK Ammonium Perchlorate, Synthetic Rutile, Three Elephant Boric Acid, Three Elephant Pyrobor Dehydrated Borax, Three Elephant V-Bor Refined Pentahydrate Borax, Trona Anhydrous Sodium Sulphate, Trona Boron Tribromide, Trona Boron Trichloride, Trona Elemental Boron, Trona Muriate of Potash, Trona Potassium Chloride, Trona Potassium Sulphate, Trona Salt Cake, Trona Soda Ash, Trona Sulphate of Potash, Tronacarb Sodium Bicarbonate, Tronalight Light Soda Ash, Tronalight Light Soda Ash, Tronamang Electrolytic Manganese Metal, Tronamang-75 Manganese Aluminium Briquettes, Tronox Titanium Dioxide Pigments, Chloride Process.

Kincaid Enterprises Inc,

Box 671,
Nitro, WV 25143
USA

Chloroneb Systemic Flowable Fungicide, Chloroneb 65W Fungicide, Marlate Methoxychlor Insecticide, Marlate 2-MR Emulsifiable Insecticide, Marlate 300 Flowable, Marlate 400 Flowable Concentrate, Marlate 50 WP, Terraneb SP Turf Fungicide.

Knight, F E Inc,

8 Perry Drive,
Foxboro, MA 02035
USA

Castaldo.

Knoll Pharmaceutical Co,

30 N Jefferson Rd,
Whippany, NJ 07981
USA

Akineton, Akinetone, Dilaudid, Euresol, Isoptin, Tradenal.

Kodak Ltd,

Acornfield Road, Kirkby,
Liverpool L33 7UF
England

Kodak Ltd,

Chromatogram.

KWR Chemicals Ltd,

Kingsley House, 8 Bream's Buildings
London EC4A 1HP
England

Amphosol CA, Amphosol DM and DMA, Bio-terge AS-40 and AS-90F, Biosoft C100, Biosoft D, Biosoft N-300, Biosoft S and D-35X, Biosoft S-100 and JN, Catigene BR 80 B, Catigene DC/100, Catigene SR, Catigene T80, Catigene 4513, Catisol AO 100, Lanthanol LAL, Makon NP6, NP10 and 4,8,12,14 and 30, Makon OP6 and OP9, Nacconol 35SL, Nacconol 90F and 40F, Ninate 401, Ninate 411, Ninate 415, Ninox, Petrostep A-70, Secolat, Secosol AL 959, Secosol ALL/40, Secosol AL/MG 50, Secosol DOS/70, Secosol EA/40, Secosov, Secosyl, Sevelyte K, Steol CA-460 and KA-460, Steol CS-460 and KS-460, Steol CS-760 and 7N, Steol FA, Steol 3 OS, Steol 4N, Steol 7T, Stepan, Stepanol AM, Stepanol DEA, Stepanol ME, Stepanol Mg, Stepanol SPT, Stepanol WA-100, Stepanol WAT, Stepanol WA, WAC, WAQ, Stepantan A, Stepantan NP 80, Stepantex Q90B.

Kyowa Hakko Kogyo Co Ltd,

Ohtemachi Building, 1-6-1 Ohtemachi,
Chiyoda-ku, Tokyo 100,
Japan

Glumal.

L

L & K Fertilisers Ltd

Saxilby,
Lincoln LN1 2LS
England

Ansax.

Laboratories for Applied Biology,

91 Amhurst Park,
London N16 5DR
UK

Cerumol, Duromorph, Emeside, Halycitrol, Koladex, Labiton, Labophylline, Laboprin, Labosept, Micryston, Monphytol, Pernomol, Volital, Zeese.

Laboratorios Viñas SA,

Torrente Vidalet 29,
08012-Barcelona
Spain

Virexen.

Lagap Pharmaceuticals Limited,

Woolmer Way,
Bordon, Hants GU35 9QE
UK

Aloral, Ampilar, Bedranol, Bendogen, Diuresal, Doxylar, Erythrolar, Ibular, Indolar, Labrocol, Laracor, Laractone, Laraflex, Laratrim, Lugacin, Malix, Metrolyl, Parmid, Phasal, Trimogal, Triperidol.

Lakeland Laboratories Ltd,

Peel Lane, Astley Green,
Tyldesley, Manchester M29 7FE
England

Lakeland AMA, Lakeland AMA LF, Lakeland PP.

Lancashire Chemical Works Ltd,

High Street West, Glossop,
Derbyshire SK13 8ES
England

Chromeduol, Cromalit, Cromalit 150, Purac, Speetan SB60, Suprac, Tanal.

Lancashire Tar Distillers Ltd,
Lanstar Industrials Division,

Liverpool Road, Cadishead,
Manchester M30 5DT
England

Arborsan, Lanstar.

Langley Smith & Co Limited,

8-10 Paul Street,
London EC2A 4JH
UK

Achilles Dipentene, Achilles Pine Oil, Achilles Tall Oil Fatty Acid.

Lankro Chemicals Ltd,

PO Box 1, Eccles,
Manchester M30 0BH
England

Agrilan, Agrimul, Agrisol, Ampholan, Arylan, B.A.R, Estolan, Ethylan, Isocon, Lahkrostat, Lancare, Landemul, Lankro Mark, Lankroflex, Lankroflex ED3, Lankroflex ED6, Lankroflex GE, Lankrol, Lankrolan, Lankroline, Lankrolyte, Lankromark, Lankromul, Lankroplast, Lankropol, Lankrosol, Lankrosperse, Lankrostat, Lankrothane, Levelan, Mellite, Monolan, Perlankrol, Phospholan, Photomer, Polystab, Propamine, Propamine D, Propocon, Propylan, Propylan A350, Propylan G600, Propylan RF55, Quadrilan, Quasilan, Quell Oil, Thixolan, Versilan, Versilyt.

Lanstar Chemicals,

Liverpool Road, Cadishead,
Manchester M30 5OT
England

Lanstar NP100/50, Lanstar NP2 and NP4, Lanstar NP40, NP50 and NP100, Lanstar PCH, PC2 and PCO, Lanstar PS, Lanstar PSW.

Laporte Industries Ltd,

Hanover House, 14 Hanover Square,
London W1R 0BE
England

Laporte Industries Limited,
Laporte Inorganics Division,
PO Box 2, Moorfield Road,
Widnes, Cheshire WA8 0JU
UK

Alferric, Aluminoferric, Aquasil, Aridex, Avantine, Bariform, Barkite B, Benvic, Brakol, Brebent, Brebond, Bregel, Briklens, Clovean, Comox, Contradet, Cresavon, Dairos, Dairozon, Dappol, Eltex, Eltex P, Emulsifier L.W, Fulacolor, Fulbent, Fulbond, Fulcat, Fulcat Catalysts, Fullasorb, Fulmont, Fulmont Activated Bleaching Earths, Genoxide, Howflex, Howsorb, Howtol, Ixan, Ixol, Lapofloc, Laponite, Lapotan, Lasilso, Nimox, Opol, Polaris, Resin M.S.2, Runa, Rutiox, Sextate, Sextol Z, Solvethane, Solvic, Titanox RA-39, Ucicline, Ucipol.

Laporte UK Trading,

PO Box 8, Kingsway,
Luton, Beds LU4 8EW
England

Sextone, Socal.

Larkhall Laboratories Limited,

225 Putney Bridge Road,
London SW15 2PY
UK

Cantamega 1000, Cantamega 2000, DLPA 375, Ruthmol, Trufree.

Laserson & Sabetay,

BP 57 91151 Etampes,
Cedex
France

Laserson & Sabetay,

Cetaffine, Cosbiol, Dermaffine, Preservals, Superbrillantoline.

Lawn & Garden Products Inc.,

P O Box 5317,
Fresno CA 93755
USA

Kwik-Green, Monterey Bayleton, Monterey Foliar Nutrient 11-4-6, Monterey Herbicide Helper, Monterey Iron Chelated 10%, Monterey Perc-O-Late Plus, Weed Ender, Weed Hoe.

Lederle Laboratories,

Fareham Road,
Gosport, Hants PO13 0AS
UK

Lederle Laboratories,
Div of American Cyanamid Co,
Pearl River,
NY 10965
USA

Acetazolamide, Achromycin, Amicar, Amstat, Aristocort, Aristocort Acetonide, Aristocort Forte Parenteral, Aristocort Syrup, Aristospan, Artane, Artane Sustets, Audicort, Aureocort, Aureomycin, Avotan, Cinopal, Cisplatin, Cyclocort, Declomycin, Deteclo, Diamox, Diamox, Dolene, Folvite, Gevral, Hydromox, Ledercillin VK, Ledercort, Lederfen, Ledermix, Ledermycin, Lederplex, Lederspan, Leucovorin, Levoprome, Loxitane, Methotrexate, Minocin, Minocyn, Myambutol, Mynah, Neoloid, Neptazane, Noltam, Novantrone, Orimune, Parfenac, Pathilon, Pipracil, Pipril, Silderm, Thiotepa, Triple Sulfas, Tuberculin Old Tine Test, Varidase, Vi-Alpha, Zadstat, Zincon Dandruff Shampoo.

Leuchstoffwerk GmbH,

D 6900 Heidelberg,
Im Klingenbuhl 8
West Germany

Leutalux.

Leverton-Clarke Limited,

Beech Dene, Crawley Ridge,
Camberley, Surrey GU15 2AN
UK

Levcarb.

Lilachim SA,

Rue de la Loi 33,
B-1040 Brussels,
Belgium

Lilaminox.

Lilly, Eli & Co
see Eli Lilly & Co

Lin Pac Polymers,

Moor Lane Trading Estate,
Sherburn in Elmet, North Yorkshire
UK

Crystal Polystyrene.

Lipha SA,

115 ave Lacassagne,
69212 Lyon Cedex 1
France

Cantabiline, Praxilene, Migraleve.

Liquid Crystal Lens Co,

29 Bolinas Rd,
Fairfax, Calif 94930
USA

Licryl-55, Licryl-70.

Liquid Plastics Limited,

PO Box 7, London Road,
Preston, Lancs PR1 4AJ
UK

Decadex, Firecheck, Flexcrete, Isoclad, K154, Monolastex Smooth, Soladox, Soladox 112, Steridex.

Llewellyn Ryland Limited,

Haden Street,
Birmingham B12 9DB
UK

Llewellyn Ryland Limited,

Double Shield, Duralac, Powers Terebine, Priormatt, Rylard, Setair, Silvoline.

Loes Enterprises Inc,

1457 Iglehart Avenue,
St Paul, MN 55104
USA

MRV 1000, Vinacron.

Lonza Limited,

Münchensteinerstrasse 38,
PO Box CH-4002 Basle,
Switzerland

Abradux, Abramant, Abramax, Abrarex, Abrasit, Airex, Alkawet, Amphobac, Amphoterge, Bardac, Bardyne, Barlox, Barquat, Carbogran, Carbogran E, Carbogran UF, Carbomant, Carsilon, Carsonol, Carsonon, Carsoquat, Carsosoft, Compalox, Diadur, Dural, Forex, Hyamine 1622, Hyamine 3500 and 2389, Hydex, Hystar, Lonsicar, Lonzaine, Lonzest, Martifin, Martinal, Martipol, Martisorb, Martoxin, Pokalon, Resilon, Rollit, Silipact, Simax, Sinatron, Sniafoam, Ultraphan, Unihib, Uniquat, Vipophan PVC.

Lord Corporation (UK) Limited,

Stretford Motorway Industrial Estate,
Barton Dock Road, Stretford,
Manchester M32 0ZH
UK

Chemglaze, Flocklok, Photoglaze, Tuftane, Tycel, Tyrite.

Luitpold-Werk,

Hayes Gate House, 27 Uxbridge Road,
Hayes, Middlesex UB4 0JN
UK

Adequan, Anacal, Hirudoid, Movelat, Propain.

Lukens International Corporation,

1117 Jefferson Street, PO Box 1040,
Lynchburg, Virginia 24505
USA

Lukens Bone Wax.

L W Vass (Agricultural) Limited
see Vass (Agricultural), L W Limited

M

3M Health Care

Morley Street,
Loughborough, Leicestershire LE11 1EP
UK

Riker Laboratories Inc,
Subsidiary 3M Company,
19901 Nordhoff St,
Northridge, Calif 91324
USA

3M,
3M Center, St Paul,
Minnesota 55144
USA

Minnesota Mining & Manufacturing Co,
3M Center,
St Paul, Minn 55144
USA

3M United Kingdom plc,
Commercial Chemicals Division,
3M House, PO Box 1,
Bracknell, Berks RG12 1JU
UK

Acupan, Acupan Injection, Acupan Tablets, Alsimag, Alsimag 754, Alsimag 779, Alu-Cap, Asmatane Mist, Blenderm, Calcisorb, Calcium Disodium Versenate, Circanol, Complamin, Conadil, Diafen, Difflam Cream, Difflam Oral Rinse, Disalcid, Disalcid Capsules, Disipal (as HCl), Dorbane, Dorbanex Capsules, Dorbanex Forte, Dorbanex Liquid, Duo-Autohaler, Duromine, Durophet, Dynamar Brand Specialities, Fluorel, Hiprex, Intralgin Gel, Iso-Autohaler, Kel-F 81, Kel-F-Elastomer 3700, Lergoban, Lipo-Hepin, Medihaler Ergotamine, Medihaler Iso, Medihaler-Duo, Medihaler-epi, Medihaler-Ergotamine, Medihaler-iso, Medihaler-iso Forte, Medihaler-Tetracaine, Metraspray, Micropore, Norflex, Norflex Injection, Norflex Tablets, Norgesic, Nuelin Liquid, Nuelin SA-250, Nuelin Tablets, Nuelin-SA, Numotac, Peptard, Pholtex, Pulmadil Auto, Pulmadil Inhaler, Rauwiloid, Rikospray Balsam, Rikospray Silicone,

Salbulin Inhaler, Salbulin Syrup, Salbulin Tablets 2 mg, Salbulin Tablets 4 mg, Serpiloid, Sodium Versenate, Sombulex, Tambocor Injection, Tambocor Tablets, Tepanil, Theodrex, Titralac, Tribiotic Spray, Trolone, Urex, U/o, Videne Disinfectant Solution, Videne Disinfectant Tincture, Videne Powder, Videne Surgical Scrub, Ytterbium Yb-169 DTPA.

Macarthys Laboratories Limited,

Snaygill Industrial Estate, Keighley Road,
Skipton, North Yorkshire BD23 2RW
UK

Budale, Millophyline, Nibiol, Pardale.

Macfarlan Smith Ltd,

Wheatfield Road, Edinburgh,
EH11 2QA
Scotland

Bitrex.

Magnesium Elektron Ltd,

Regal House, London Road,
Twickenham TW1 3QA
England

Bacote, Zircosol P, Zirgel, Zirmax.

Mallinckrodt Inc
see Fries & Fries of Mallinckrodt Inc

Manchem Ltd,

Ashton New Road, Manchester,
M11 4AT
England

Alchemy, Aliso, Alsynates, Alumedia, Alusec, Borester, Borester 7, Cozirc, Fotofax, Liquibor, Liquibor 169, Manalox, Manchem, Manfloc, Mangal, Manoblend, Manobond, Manocat, Manofast, Manofil, Manomet, Manosec, Manosil, Manosperse, Manox, Manoxol, Manoxol MA, Manoxol OT, OT/P and OT/B, Multimet, Pharmakon, Rubox, Sternite, Synmold, Syntron, Talpex, Vidox, Vita Zinc, Zircomplex.

Manox Ltd,

Manox House, Coleshill Street,
Miles Platting, Manchester M10 7AA
England

Easisperse, Manox.

Manro Products Ltd,

Bridge Street, Stalybridge,
Cheshire,
England

Manro, Manro ALS, Manro BA and NA, Manro BA and SBS, Manro BES, Manro D Paste, Manro DL28, Manro DS 35, Manro HA, Manro HCS, Manro KXS, Manro MA 35, Manro ML33, Manro NEC, Manro NP, Manro PTSA, Manro SDBS, Manro SLS28, Manro SLS45, Manro STS, Manro SXS, Manro S.I.O.S, Manro TDBS, Manro TL40, Manro XSA.

Marcel Guest, H Limited,

Riverside Works, Collyhurst Road,
Manchester M10 7RU
UK

Acrythane.

Marfleet Refining Company Limited,

Hedon Road, Marfleet,
Hull, North Humberside HU9 5NJ
UK

Chamotan, Enervite, Fax, Faxola, Femin-9, Mainstay, MaxEPA, Super Solvitax, Vitapet, Vitrite.

Marion Laboratories Inc,

10236 Bunker Ridge Rd,
Kansas City, Mo 64137
USA

Carafate, Ditropan, Nico-400, Pavabid, Ulcerban.

Marks, A H & Company Ltd,

Wyke, Bradford,
W Yorkshire BD12 9EJ
England

Centrifugal Syrup, Certrol, Ethrel-E, Ethrel-R, Tre-Hold.

Marley Adhesives,

Bath Road,
Beenham, Reading, Berks RG7 5PU
UK

Embond 168, Embond 212, Safebond 3.

M A Steinhard Limited
see Steinhard, M A Limited

May & Baker Limited,
Pharmaceutical Division,

Rainham Road South,
Dagenham, Essex RM10 7XS
UK

Acelon, Acetylarsan, Actril, Actrilawn, Agavin, Agritox, Alamask, Aliette, Altan, Amandol, Ametox, Amfix, Ansolysen, Anthical, Anthiomaline, Anthiphen, Anthisan, Ascabiol, Astroplax, Astryl, Asulox, Avisol, Avomine, Axall, Azosan, Baker's Anaesthetic Ether, Banistyl, Brevidil, Brevidil E, Brevidil M, Brittox, Brolene, Brulidine, Buctril, Cabufix, Cabulite, Carbergan, Carbetamex, Caritrol, Cascade, Cercobin, Cerubidin, Cervagem, Cestarsol, Checkmate, Chromalay, Chromolay, Clampdown, Clout, Clovotox, Cobrol, Compak, Compitox, Compron, Conray 420, Cunitex, Cyclonal Sodium, Cyclosan, Cymbilide, Decalex, Deccox, Diaginol, Dibrogan, Dicestal, Dicotox, Doublet, Droxychrome, Duomatic, Embacel, Embacide, Embacoid, Embadot, Embafix, Embafume, Embalith, Embamix, Embanox, Embaphase, Embaspeed, Embatex, Embathion, Embatype, Embazin, Embedyne, Embequin, Embesafe, Embesol, Embrol, Embutox, Emtryl, Entramin, Entramin A, Ethulon, Euthatal, Exprol, Fixaplus, Fixatek, Flagyl, Flagyl Compak, Flaxedil, Formax, Fragaroma, Fungus Fighter, Gardenal, Gardenal Sodium, Genochrome, Gonacrine, Grafix, Graslam, Grodex, Halothane M & B, Hycon, Intraval, Iotect, Iotex, Iso-Planotox, Kilmet, Kilnet, Kival, Largactil, Lenticillin, Leucarsone, Lipiodul Ultra-Fluid, Lysivane, M & B 693, Majeptil, Matalex, Mecufix, Meculon, Merfusan, Mersil, Mos-Tox, Multicoild, Mydochrome, Mydoneg, Mydoprint, Myocrisin, Myspamol, M.B, Nefrolan, Neulactil, Nivaquine, Nivaquine, Nivembine, Novidium, Novofix, Novoline, Novolith, Novomatic, Novotak, Novotriad, Nozinan, Nuvamide, Orbitol, Orudis, Oruvail, Osbil, Otamidyl, Otoryl, Oxytril, Pamergan, Parafix, Penitriad, Penjectin, Pentelex, Perfix, Peroxyl, Pevafix, Pevalon, Phenergan, Phensedyi Linctus, Phensedyl, Phytodermine, Piportil Depot, Planetol, Planidets, Planocaine, Planochrome, Planofix, Planotox, Polycon, Pradone Plus, Premaline, Premalox, Primalan, Prolith, Promicrol, Pronalys, Qaulineg, Qualidot, Qualifix, Qualitol, Rapadex, Rolafix, Ronstar, Rota, Rovamycin, Rovral, Sagatal, Samorin, Saquadil, Seal and Heal, Secadrex, Sectral, Seradix, Seritox, Sherpa, Silm, Solidex, Sonalgin, Sonergan, Soneryl, Stemetil, Stovarsol, Streptotriad, Sulphatriad, Supergreen, Supertox, Suprofix, Suprol, Surmontil, SVC, Talent, Teknol, Tensol, Terfluzin, Thalazole, Theogardenal, Thiazamide, Thiodet, Thiolim, Tixylix, Tolkan, Tolochrome, Totril, Trescatyl, Trevintex, Tricoid, Trigger, Trinamide, Trodax, Tropotox, Tubotin, Tubotox, Twin-Tak, Urethon, Valledrine, Vallergan, Vallex, Vam,

May & Baker Limited,
Pharmaceutical Division

Vamitox, Vanitox, Varitox, Vegolysen, Vegolysin T, Veractil, Versicaine, Volucon, Volusol, Zoaquin, Zolone.

MB Powder Metal Group,

P O Box 19, Elton Park Works, Hadleigh Road,
Ipswich IP2 0HX
UK

Oilite.

McKechnie Chemicals Ltd,

PO Box No 4, Ditton Road,
Widnes, Cheshire WA8 0PG
England

Burcop, Comac, Cutonic, Mackechnie, Macuprax, Powmet.

McNeil Consumer Products Co,

Camp Hill Rd,
Fort Washington, Penna 19304
USA

Aberel, Clistin, Clistin-D, PediaCare 1, Tylenol.

McNeil Pharmaceutical,
McNEILAB Inc,

Spring House,
Penna 19477
USA

Aciquel, Anthelvet, Axiquel, Metha-Meridiazine, Nacton, Orap, Pancrease, Paraflex, Semap, Surexin, Tolectin, Zomax.

Mead Johnson & Co,

2404 W Pennsylvania St,
Evansville, Ind 47721
USA

Buspar, Cogesic, Colace, Deapril-ST, Desyrel, Duricef, Enkade, Estrace, Fer-In-Sol, Ifex, K-Lyte, K-Lyte/C1, Megace, Mucomyst, Peri-Colace, Questran, Tempra.

Mechema Chemicals Limited,

Talbot Wharf Chemical Works,
Port Talbot, West Glamorgan SA13 1RL
UK

Aresenid, Calac, Calcars, Cobac, Cobox, Coclopet, Coclor, Cocloran, Cucar, Cuclat, Cufor, Cupac, Cupar, Cuproid, Cusamon, Cusatrib, Cusyd, Cyanolime, Kypfarin, Leadoxe, Leclo, Ledac, Ledca, Ledfo, Ledni, Lepro, Lethi, Manal, Mancobride Mancanese, Mangoxe, Mansu, Mekure T1, Mekure T2, Microsan, Nicar, Nicfo, Nico, Nicor.

Medi-Physics Inc,

5801 Christie Ave,
Emeryville, Calif 94608
USA

MPI DMSA Kidney Reagent, MPI Indium DTPa In III, MPI Indium Oxine In 111, MPI Krypton Kr 81m Gas Generator, Neoscan, Nephroflow, Perfusamine, Tc 99m Lungaggregate, Xenon Xe 133-VSS.

Menley & James Laboratories,

1500 Spring Garden St,
Philadelphia, Penna 19101
USA

Ecotrin, Feosol, Siligaz, Teldrin.

Mercian Minerals & Colours Limited,

Queens Chambers, 61 Boldmere Road,
Sutton Coldfield B73 5XA
UK

Ferroxide, Gilsonite, Plastorit.

Merck Sharpe & Dohme Limited,

Hertford Road,
Hoddesdon, Herts EN11 9BU
UK

Merck Sharp & Dohme,
Div of Merck & Co Inc,
West Point,
Penna 19486
USA

Merck Sharpe & Dohme Limited

Adju-Fluax, Aldomet, Aldomet Ester Hydrochloride, Aldoretic, Aldoril, Alpha-Cillin, AlphaRedisol, Alpivicin, Amilorin, Amprol, Amuno, Aquamephyton, Aramine, Arpocox, Arthrobid, Attenuvax, Axoridin, Benemid, Biavax II, Blocadren, Blocazide, Blocuretic, Caligesic, Chlotride, Clinoril, Co-Deltra, Co-Hydeltra, Codelcortone, Codelsol, Codelspray, Cogentin, Colbenemid, Concordin, Cortone Acetate, Cosmegen, Cremodex, Cremomycin, Cremostrep, Cremosuxidine, Crystoids, Cuprimine, Cyclaine, Daranide, Deca-Indocid, Decaderm, Decadron, Decadron Duofase, Decadron Shock-Pak, Decadron-LA, Decadronal, Decaspray, Deltacortone, Demser, Dermogesic, Dichlosuric, Diuril, Dolobid, Dugro, Duo-Decadrin, Edecrin, Edecrin Sodium, Elavil, Elspar, Equiben, Eqvalan, Estergel, Flexeril, Floropryl, Fluax, Gammagee, H-B-Vax, Hep-B-Gammagee, Heptuss, Hexyltan, Humorsol, Hydelta, Hydeltra-TBA, Hydeltrasol, Hydrocortone, Hydroderm, Hydrodiuril, Hydromet, Hydrosaluric, Hydrospray, Hydrozets, Indocid, Indocid-R, Indocin, Indoptol, Infavina, Innovace, Inversine, Inversine, Ivomec, Lipivas, Lodosin, Lodosyn, Lyovac Sodium Edecrin, M-M-R, M-M-VAX, M-R-VAX II, Mefoxin, Meningovax-C, Mephyton, Meruvax, Meruvax II, Mevasine, Midamide, Midamor, Mintezol, Moducren, Moduret 25, Moduretic, Mumpsvax, Mustargen, Myochrysine, Noroxin, Novazole, Novulatin, Novulatum, Oxxef, Pepcidine, Periactin, Periactin, Periactin-Vita, Pivatil, Pneumovax, Propadrine, Pulmovax, Ranide, Redisol, Salupres, Saluric, Sinemet, Sinemet-Plus, Sodium Diuril, Somnos, Sterovum, Sulfasuxidine, Sulfathalidine, T-I-Gammagee, Tetrazets, Timolate, Timoptic, Timoptol, Tonocard, Triptil, Tryptanol, Tryptizol, Tyrozets, Urecholine, Utinor, Vivactil, Zienam, Zinamide, Zoroxin.

Merix Chemical Company.

2234 East 75th Street,
Chicago, Il. 60649
USA

Frost-Off, Merix, Parabolix 100.

Merquinsa,

Gran Vial 17, Montmel
Barcelona
Spain

Mertan, Quiacryl, Quidur, Quilastic.

Merrell Dow Pharmaceuticals Inc
see Dow Chemical Company Ltd

Metalcrete Mfg. Company,

10330 Brecksville Road,
Cleveland, OH 44141
USA

Diamite Epoxy Brushkote, Diamite Epoxy Flooring, Lexite Granular Carpet, Lexite 100, Vulcanox Crack and Joint Sealant.

Miles Pharmaceuticals
see Bayer

Milliken & Company, Chemical Division

PO Box 817,
Inman, SC 29349
USA

Blazon, Bullseye, Lubestat, Syn Lube, Versatint Fugitive tints.

Millmaster-Onyx UK,

Marlborough House, 30-2 Yarm Road,
Stockton-on-Tees, Cleveland,
England

Ammonyx, Ammonyx CA, CA Special, 4 and 4B, Ammonyx CETAC, Ammonyx DME, Ammonyx KP, Ammonyx SKD, Ammonyx T, Ammonyx 27, Ammonyx 4002, Ammonyx 4080, Ammonyx 485 and 490, Ammonyx 781, Anionyx 12EO, Anionyx 12S, Aston RC, BTC, Cationic Softener X Concentrate, Maprofix ES-2, Maprofix ESY, Maprofix MG, Maprofix NH and NHL, Maprofix TAS, Maprofix TLS, Maprofix WAC-LA and LCP, Maprofix WA, WAC and WAQ, Maprofix 563 and LK.USP, Maprofix 60S and 60N, Maprosyl 30, Neutronynx S-60, Neutronyx 600 and 656, NSAE Powder, Onamine RO, Onamine 12, Onamine 14, Onamine 16, Onamine 18, Onamine 65, 835 and 1214, Onyxide 172, Onyxide 3300, Onyxide 75, Tetrosan 3,4 D.

Milton Roy Co,

PO Drawer 849,
Sarasota Fla 33578
USA

Naturvue.

Minas de Gador SA,

General Zabala 24, 3o A, Esc Izq,
28002 Madrid
Spain

Clarstabil, Clarvin, Fulbond, Gadorgel, Gadorgel Ocma, Lubrite B33.

Minnesota Mining & Manufacturing Co
see 3M

Mission Pharmacal Co,

1325 E Durango Blvd, PO Box 1676,
San Antonio, Texas 78296
USA

Calcet, Calcibind, Folicet, Homapin, Lithostat, Mission Prenatal, Prulet.

Mitchell Cotts Chemicals Ltd,

PO Box No 6, Steanard Lane,
Mirfield, W Yorkshire WF14 8QB
England

Afrisect, Cyperkill, Cypersect, Glowtein, Heptokill, Methokill, Permasect, Permatag, Pynosect, Unifog.

Mitsubishi Petrochemical Co Limited,

Mitsubishi Bldg., 5-2 Marunouchi 2-chome,
Chiyoda-ku, Tokyo 100
Japan

Dobanol, Ecolo, Ecoro, HFC, Linklon, Linklon-X, Mitsubishi Yuka-ECX, Mitsubishi Yuka-SPX, Modic, Rabalon, Salvex, Styropor, Thermorun, Yukalon.

Mobay Chemical Company,

Mobay Road,
Pittsburgh, PA 15205-9741
USA

Bayblend, Makroblend, Merlon, Nydur, Petlon, Pocan, Texin.

Mobil Plastics Europe,

Zoning de Latour,
6761 Virton
Belgium

Bicor, Mobilrap, Mobilrapper, Oppalyte.

Monsanto PLC,

Monsanto House, Chineham Court, Chineham,
Basingstoke, Hants RG24 0UL
England

Monsanto Co,
800 N Lindbergh Blvd,
St Louis, Mo 63167
USA

Monsanto,
Detergents Division,
Rue Laid Burniat,
1348 Louvain la Neuve
France

A-1 Thiocarbanilide, ACL, ACL 56, ACL 59, ACL 60, ACL 90 Plus, Acrilan, Aflaban, Alimet, Asagran, Astroturf, Avadex, Avadex BW, FAR-GO, Bronco, Butasan Vulcanization Accelerator, Butvar, Cadon, Coolanol, Dequest, EMA, Ema Resins, Emulsi-Phos, Ethavan, Ethylthiurad, Far-Go/Avadex BW, Filtrez, Filtros, Flectol H, Flectol ODP, Flectol Pastilles, Fome-Cor, Formvar, Gelva, Gelvatol, HB-40, H.T, Lasso, Levn-Lite, Limit, Limit 33, Losilphos, Lustran, Lustran ABS, Lustran SAN, Lustran Ultra ABS, Lustrex, Machete, Mersize, Methasan, MHA, Modaflow, Mono Thiurad, Multiflow, NFB, NTA, Nutrifos, Oilfos, Pan-O-Lite, PC-1244, PC-1344, Phos-Chek, Prism, Py-Ran, Ramrod, Resimene, RJ-100, Rodeo, Roundup, Saflex, Santicizer DUP, Santicizer 141, Santicizer 143, Santicizer 148, Santicizer 154, Santicizer 160, Santicizer 160, Santicizer 711, Santicizer 97, Santo-Res, Santocure, Santocure MOR/MOR90, Santoflex, Santoflex AW, Santoflex DD, Santoflex DPA, Santoflex 1P, Santoflex 13, Santoflex 77, Santogard PVI, Santonox, Santoprene, Santoquin, Santotrac, Santovac, Santovar, Santovar A, Santowax, Santoweb, Santowhite, Screen, Scripset, SEF, Silester, Skybond, Skydrol, Sopanox, Spray Guard, Stabil-9, Sterox, Sterox DF, DJ, Sting, Sulfasan, Sulfasan R, Syton, TCC, Tetrathal, Therminol, Thiofide, Thiotax, Thiurad, Vocol, Vydyne, Wear-Dated.

Montedison UK Limited,

7/8 Lygon Place, Ebury Street,
London SW1W 0JR
UK

Algoflon, Cosmegin Lyovac, Dutral-Co, Dutral-Ter, Edistir, Esaflon, Extir, Ferlosa, Fomblin, Gabrosa, Galden, Glendion, Kostil, Linearil, Moplefan, Moplen, Noblen, Peroximon, Pibiter, Retilox, Soreflon, Spinflam, Spinuvex, Tecnocin, Tecnoflon., Tecnoprene, Tedimon, Ultrastyr, Vedril, Vinavi.

Morton Thiokol Inc
see Williams Division of Morton Thiokol Ltd

Multicrom SA,

Alte Brown 778,
1704 Ramos Mejia
Argentina

Acticrom, Multicel, Multicet, Multicrom, Multicuer, Multiluz, Multisil, Multiter, Nylocrom.

Multitherm Corporation,

125 South Front Street,
Colwyn PA 19023
USA

Multitherm IG-2, Multitherm PG-1.

Murlin Chemical Incorporated,

Balligo Road,
West Conshohocken, PA 19428
USA

Murlin Premium Ladle Wash, Natural Bone Ash, BCP 400 and BCP 600, Synthetic Bone Ash.

Mykroy/Mycalex,
Spaulding Fibre Company Inc,

125 Clifton Boulevard,
Clifton, New Jersey 07011
USA

Mykroy/Mycalex, Supramica, Synthamica, Thermica.

N

Napp Laboratories

Cambridge Science Park, Milton Road,
Cambridge CB4 4BH
UK

Akrotherm, Audax, Betadine, Brovon, Carylderm, Comploment, Cradocap, Diumide-K, Esoderm, Ferrocontin, K-Contin, Morhulin, Morsep, MST, Nitrocontin, Phyllocontin, PIB, Plesmet, Prioderm, Teejel, Trilisate, Uniphyllin, X-Prep.

Nekami, BV

Nyverheidstraat 32, PO Box 436,
Gouda 2800 AK
Netherlands

Edelwit.

Nelson Research,

1001 Health Sciences Road West,
Irvine, Calif 92715
USA

Azone.

Neste Oy
see Unifos Kemi AB

Nipa Laboratories Ltd,

Nipa Industrial Estate, Llantwit Fardre,
Mid Glamorgan CF38 2SN
Great Britain

Nipa Laboratories Ltd

Nipabenyzl, Nipabutyl, Nipacombin, Nipagin, Nipagin A, Nipagin M, Nipanox, Nipantiox, Nipasept, Nipasol, Nipasol M, Nipastat, Phenodip, Phenonip, Phenosept, Phenoxetol, Progallin.

Nitto Electric Industrial Co Ltd,

1-2 Shimohozumi 1-chome,
Ibaraki, Osaka 567
Japan

Nitto Nitoflon, Nitto SPV.

NL Chemicals (UK) Limited,

St Ann's House,
Wilmslow, Cheshire SK9 1HG
UK

NL Victor Wolf Limited,
West Sleekburn, Bedlington,
Northumberland NE22 7DH
UK

Admerol, Aroflat, Aroflint, Arolon, Aroplaz, Arothix, Baragel, Barley Bloom, Benathix, Bentone, Bentone Gel, Bentone SD, Cykelin, Drisoy, Esskol, F 31, Kelecin, Kellox, Kelmer, Kelpol, Kelpoxy, Kelsol, Kelthix, Keltrol, Kronds, Kronos, Linaqua, M-P-A, Macaloid, MPA, M50, Nalzin, Nykon, Oncor, Oncor Split Sphere, Polytrope, Post-4, Quickfloc, Rheolate, Rheox, SD-1, SD-2, Spenbond, Spenkel, Spenlite, Spensol, Tansul, Thixatrol, Thixatrol, Thixcin, Thixseal, Titan, Titan Design, Titanox, Titanox Design, Vinco A183, Vinco A33, Vinco 248, Vinco 249C, Vinco 265, Vinco 99A, Vinco 99G, Wolf N Lamid IG, Wolfaid, Wolfamid, Wolfkur, Wolflex, Wolfol.

Norda Inc,

140 Route 10,
East Hanover, NJ 07936
USA

Filtrosol A.

Norden Laboratories Inc,

601 W Cornhusker Highway,
Lincoln, Nebr 68501
USA

Therabloat.

Nordox Industrier AS,

istensjyveien 13,
N-0661 Oslo 6
Norway

Nordox.

Nordson (UK) Limited,

Thames Industrial Estate, Wenman Road,
Thame, Oxon OX9 3SW
UK

'R' Type Solvent.

Norit UK Limited,

Clydesmill Place,
Cambuslang Industrial Estate,
Glasgow G32 8RF
UK

Norit NV,
Postbus nr 105,
3800 PM Amersfoort
Netherlands

Celanese Nylon 6/6, Celanex, Darco, Duraloy, Durel, Kematal, Norit, Norit C, Norit PK, Norit R, Norit RO, Norithene, Sorbonorit, Sorbonorit (B), Vectra.

Norplex
see Allied Corporation

Norsk Hydro AS,

Bygdyy AllBe 2,
0257 Oslo 2
Norway

Hy-Vin, Hydro, Norvinyl, Pallgrip, Pevikon.

Norsk Hydro Polymers Ltd,

Aycliffe Industrial Estate,
Newton Aycliffe,
Co Durham DL5 6EA
England

Arpylene, Terluran, Vinatex.

Norwich Eaton Pharmaceuticals Inc,

17 Eaton Ave,
Norwich, NY 13815
USA

Bezalip, Buprenex, Dantafur, Dantrium, Didronel, Dopar, Duvoid, Furacin, Furadantin, Furamazone, Furoxone, Macrodantin, Nebs, Sarenin, Topicycline.

Nova Chemical Inc,
Pro-Spec Division,

1520 Erie,
North K C, MO 64116
USA

Deck Seal-PD, Epoxidized X-70 and X-75, Glass Guard, Masonry Stain and Seal, Methaplex, New Brick, New Brick (Heavy Duty), Paragon-15, Restoration Cleaner, Restoration Cleaner (Heavy Duty), Restoration Cleaner (Super Heavy Duty), Restoration Rinse, Vanex.

Novest Inc,
Pro Seal Products Division,

6500 Glenway Avenue,
Cincinnati, OH 45211
USA

Pro Seal, Pro Weld, Super Glue.

Novo Industri A/S,

Novo All,
DK 2880 Bagsvaerd
Denmark

Alcalase, Aquazym, Celluclast, Cereflo, Ceremix, Dextrozyme, Esperase, Finizym, Fungamyl, Gamanase, Glucanex, Lactozym, Lecitase, Lipozyme, Neutrase, Novozym, Palatase, Pectinex, Promozyme, Rennilase, Savinase, Sweetzyme, Termamyl, Ultrazym.

Nuodex Inc,

Turner Place, PO Box 365,
Piscataway, New Jersey 08854
USA

Admex, Anderol, Aqua Thix, Aquasperse, Cal-Tint, Chroma-Chem, Cobalt 254, Colortrend, Copac, DLG-10, DLG-20, EP Lead, Exkin, Foamacure, Fungitrol, Fungitrol Tinox, G-P-D, HiGel, Kromosperse, LoGel, MedGel, Nuact, Nuade, Nuocure, Nuodex, Nuodex NA, Nuodex 100, Nuodex 84, Nuolate, Nuophene, Nuoplaz, Nuosept, Nuosperse, Nuostabe, Nuosyn, Nuvis, Nuxtra, PMA 18, PMA 60, Polytrend, Quindex, Stabiloid, Sulfidal, Super AD-IT, Super Cat, U-T-C, Uni-Cal 66.

NYCO,

P O Box 368
Willsboro, NY 12996
USA

Micacoat, Nyad Wollastonite, Nycoat, Nycor Barytes, Nycor Celestite, Nyflake Muscovite Mica, Nyglas, Nyspheres Hollow Glass Spheres, Wollastokup.

O

Oak International Chemical Inc

1160 White Street, P O Box 837,
Sturgis, Michigan 49091
USA

Oak Oils, Oak Syncrolube.

Occidental Chemical Corporation,

Occidental Chemical Center,
360 Rainbow Boulevard South, Box 728,
Niagara Falls, New York 14302
USA

Dechlorane Plus 515, 25, 2520, 1000, Durez, Durez 18783, Ferrophos Pigment, Fluorolubes, Halso AG-125, Halso 99, MPS 500.

Octavius Hunt Limited,
Fumite Smoke Pesticides Division,

Dove Lane,
Redfield, Bristol BS5 9NQ
UK

Fumite Dicloran, Fumite Dicofol, Fumite Lindane, Fumite Permethrin, Fumite Pirimiphos Methyl, Fumite propoxur, Fumite TCNB, Fumite Tecnalin.

Onyx Chemical Co,
Millmaster Onyx Group,

Jersey City,
NJ 07302
USA

Isothan Q-75.

Optech Inc,

7310 S Alton Way,
Englewood, Col 80112
USA

Fre-Flex.

Oral-B Laboratories Inc,

170 S Whisman Rd,
Mountain View, Calif 94041
USA

Checkmate, Fluoral, Fluorinse, Gel II, Pro-Portion, Stop, Zendium.

Organon Inc,

375 Mt Pleasant Ave,
West Orange, NJ 07052
USA

Accelerase, Cortrophin Gel ACTH, Cortrophin Zinc ACTH, Cortrosyn, Cotazym, Deca-Durabolin, Doca Acetate, Durabolin, Hexadrol, Liquaemin sodium, Liquamar, Maxibolin, Neo-Hombreol, Norcuron, Orgatrax, Pavulon, Pregnyl, Regonol.

Orkla Exolon A/S & Co,

PO Box 25,
N-7301 Orkanger
Norway

Carborex.

Ortho Pharmaceutical Corp,

Highway 202,
Raritan, NJ 08869
USA

Cloderm, Conceptrol, Grifulvin V, Gynol, Intercept, Meclan Cream, Micatin, Micronor, Monistat-Derm, Retin-A, Spectazole.

Oulu Oy,
Chemical Division,

POB 196,
SF 90101 Oulu 10,
Finland

Oulu Oy
Chemical Division

Oulumer 70, Oulutac 105, Oulutac 20 EP, Oulutac 30, Oulutac 30 D, Oulutac 80, Oulutac 80 D, Oulutac 90, Oulutac 90 D.

Owen Laboratories Div,
Dermatological Products of Texas,

Alcon Laboratories Inc, Affiliate,
San Antonio, Texas 78296
USA

Adsorbocarpine, Alcaine, Alcon-Efrin, Anestacon, Azo-Standard, Barbivis, Benzac, Cetacourt, Cetamide, Cyclogyl, Cystospaz, Econochlor, Econopred, Epinal, Flexsol 43, Fluorescite, Ionil, Ismotic, Isopto Atropine, Isopto Cetamide, IsoptoCarbachol, Locoid, Maxidex Ointment, Miostat, Mydfrin, Mydriacyl, Naphcon, Natacyn, Nutracort, Op-Thal-Zin, Oratrol, Osmoglyn, Pabagel, Pantopaque, Persadox, Phenitol, Soothe, Suladrin, Tobrex, Ultrasul, Zolyse.

O'Neal, Jones & Feldman Pharmaceuticals,

2510 Metro Blvd,
Maryland Heights, MO 63043
USA

ACTH 40, Adeno, Almora, Andro LA 200, Andro 100, Androgyn LA, Anergan 25, Anergan 50, Antilirium, Banflex, Beesix, Betaprone, Choron, Codroxomin, Col-Evac, Cyomin, Dalcaine, depAndro 100, depAndro 200, depGynogen, Depmedalone 40, also Depmedalone 80, depPredalone, Fellozine, Feostat, Gesterol 50, Gynogen LA 10, Malogen CYP 200, Malogen LA 200, Metra, Nandrobolic, Nandrobolic LA, Niac, Nobese, Obermine Black & Yellow, Oxymycin, Paral, Pentryate 80, Petameth, Predalone TBA, Proklar, Quiess, Rauverid, Solu-Predalone, Triamolone 40, Triamonide 40, Verazinc.

P

Pacific Chemical Industries Pty Limited

6 Grand Avenue,
Camellia, N.S.W. 2142
USA

Anchor Chemical Company,
777 Canterbury Road,
Westlake, OH 44145
USA

Accelerator 2P, Accelerator 4P, Age-Rite, Age-Rite AK, Age-Rite Alba, Age-Rite Gel, Age-Rite Hipar, Age-Rite HP, Age-Rite Resin D, Age-Rite Spar, Age-Rite Stalite, Age-Rite Stalite S, Age-Rite White, Ancaflex, Ancamide 280, 400, Ancamine LO, Ancamine LT, Ancamine MCA, Ancatax, Ancazate BU, Ancazate EPH, Ancazate ET, Ancazate ME, Ancazate Q, Ancazate XX, Ancazide ET, Ancazide IS, Ancazide ME, Anchorlube G-771, Anhydrite, Cyanacryl, Forane, Furbac, Jectothane, Pacwet, Pepton 65.

Pacific Dispersions Inc.,

4615 Ardine Street,
Cudahy, CA 90201
USA

AIT, Bond-A-Tint, Foam Tint, GMD, HiTint, Hydrotek, Kelstar, Plast-E-Tint, Solvetek.

Packard-Bekker BV
see Robinson Brothers Ltd

Paines & Byrne Limited,

Bilton Road, Greenford,
Middlesex UB6 7HG
UK

Paines & Byrne Limited

Benztrone, Ce-Cobalin, Cobalin-H, Dalivit, Di-Sipidin, Folicin, Gestone, Gonadotraphon FSH, Gonadotraphon LH, Hemoplex, Ketovite, Pabrinex, Pancrex, Pancrex V, Virormone.

Parke-Davis,
Div of Warner-Lambert Co,

201 Tabor Rd,
Morris Plains, NJ 07950
USA

Abidec, Acetosulphone, Adrenalin, Adroyd, Agarol, Alophen, Ambodryl, Amcill, Amitril, Amsidyl, Anatola, Anugesic, Anugesic-HC, Anusol, Anusol-HC, Aplisol, Aplitest, Arlef-100, Arquel, Bardase, Bebate, Benacine, Benadryl, Benafed, Benylin DM, Benylin Expectorant, Beta-Air, Biomydrin, Bodryl, Caladryl, Camoquin, Capsolin, Carbrital, Celevac, Celontin, Celontin Kapseals, Centrac, Centrax, Chloretone, Chlormytol, Chloromycetin, Chloromycetin Intramuscular, Chloromycetin Kapseals, Chloromycetin Palmitate Suspension, Chloromycetin Pure, Chloromycetin Succinate, Chloromycetin Suppositories, Choledyl, Citralka, Coly-Mycin M Parenteral, Coly-Mycin S, Con-Fer, Cosylan, Cyclopar, D-S-S, D-S-S Plus, Depronal S.A, Desibyl Kapseals, Dibexin, Digifortis Kapseals, Dilantin, Dopastat, Duraquin, Easprin, Ecomytrin, Elase, Emplets Potassium Chloride, Epanutin, Ergodryl, Ergostat, ERYC, Erypar, Estrovis, Euthroid, Fluogen, Gelusil, Hexavibex, Humagel, Humatin, Hygroton, Immu-Tetanus, Indon, Kaogel, Ketalar, Lavacol, Lentizol, Lipo-Lutin, Loestrin 20, Lopid, Lucofen S A, Mandelamine, Maxicam, Meclomen, Metatone, Midicel, Milontin, Mylocon, Nardil, Nema, Nitrostat, Norlestrin, Norlestrin 21, Norlutate, Norlutin, Norlutin 'A', Opilon, Oraldene, Orlest 28, Oxycel, Oxycel, Pacatal, Paladac, Panteric, Paracort, Paracortol, Paroidin, Parsidol, Penapar VK, Pentovis, Perifenil, Peritrate, Petrin, Phemerol Chloride, Pictressin Tannate, Pitocin, Pitressin, Pitressin Tannate, Ponstan, Ponstel, Portyn Kapseals, Povan, Procan, Prodoxol, Proloid, Promacetin, Promapar, Pyridium, Ranestol, Renoquid, Rolaids, Rubelogen, Saroten, Sernylan, Softcon, Sterisol, Sulfalar, Sytobex, Sytron, Taka-diastase, Takazyma, Tapar, Tedral, Theelin, Thiuretic, Thonzide, Thyroprotein, Urolucosil, Uticort, Utimox, Vanquin, Veracolate, Verstran, Vetalar, Vi-Siblin, Viomycin Sulphate, Vira-A, Zarontin.

PARS Pharmaceutical Laboratories Inc,

763 Concord Ave,
Cambridge, Mass 02138
USA

Sisacan.

Pechiney Ugine Kuhlmann Ltd,
Chemical Division,

Smiths Road, Bolton,
Lancashire BL3 2QT
England

Sunaptol NP100, Sunaptol NP140, Sunaptol NP350, Sunaptol NP55, Sunaptol NP65 and NP70, Sunaptol NP80 and NP95.

Pecora Corporation,

165 Wambold Road,
Harleysville, PA 19438
USA

Dynatred, Dynatrol, Dynaweld, Synthacalk, Unicrylic, Urexpan.

PEI, Precision Elastomers Inc,

95 Rantoul Street,
Beverly, MA 01915
USA

Electrathane.

Penick Corp

1050 Wall St West,
Lyndhurst, NJ 07071
USA

Didrate, Phthalamaquin, Rauserfia.

Pennwalt Corporation,
Fluorochemicals Division,

Pennwalt Building, Three Parkway,
Philadelphia PA 19102
USA

Pennwalt Corp,
Pharmacraft Div,
PO Box 1212,
Rochester, NY 14623
USA

Pennwalt Corp, Prescription Div,
Pennwalt Prescription Products,
PO Box 1710,
Rochester, NY 14603
USA

Adapin, Cholan DH, Ionamin, Kynar SL, Kynar 500, Zaroxolyn.

Pentagon Chemicals Limited,

Northside,
Workington, Cumbria CA14 1JJ
UK

Pentagon Chemicals Limited

Pentamin BDMA etc, Pentonate DB, Pentonium 50, 80 etc.

Pentagon Urethanes Limited,

Northside,
Workington, Cumbria CA14 1JJ
UK

Pentamid CI2, Pentamid KH, Pentonite.

The Permafuse Corporation,

675 Main Street,
Westbury, NY 11590
USA

Permafuse.

Person & Covey Inc,

616 Allen Ave,
Glendale, Calif 91201
USA

Enisyl.

Perstorp Ferguson Limited,

Aycliffe Industrial Estate,
Newton Aycliffe, Co Durham DL5 6EF
UK

Perstorp Phenolic Moulding Compound, Perstorp Urea Moulding Compound.

Pfizer Ltd,
Chemical Division,

10 Dover Road, Sandwich,
Kent CT13 0BN
England

Pfizer International Inc,
235 East 42nd Street,
New York, NY 10017
USA

Pfizer Inc,
Pfizer Laboratories Div,
Pfizer Int Div, Pfipharmecs Div,
235 E 42nd St,
New York, NY 10017
USA

Roerig, J B,
Div of Pfizer Laboratories,
235 E 42nd St,
New York, NY 10017
USA

Albacar, Alcor 7, Alexan, Amino-PF, Amoxicap, Amoxisyrup, Ampen, Analock, Antiminth, Antitrem, Antivert, Atarax, Axetin, Azapen, Bacacil, Bactipront, Banminth, Beloc, Betacortril, Betacortril Forte, Biostat A.1, Bonine, Bonomycin, Bronchopront, Brotopon, Cal-Grid, Carindaden, Cefobid, Cefobine, Cefobis, Ceravase, Cinnaloid, Citroflex A-2, Citroflex A-4, Citroflex A-8, Citroflex A2, Citroflex 2, Citroflex 4, Clarin, Cloxicap, Cloxisyrup, Clozan, Codidoxal, Codiphen, Combantrin, Cor-Tyzine, Cortifoam, Cortril, Cortril Acetate-AS, Cotinazin, Coxistac, Curatin, Dabinese, Daricol, Daricon, Daritran, Daritrax, Daxid, Delta Cortril, Deltacortril, Depixol, Diabiformin, Diabinese, Diabiphage, Diapen, Diastatin, Dilangio, Diogyn, Diogyn E, Diogynets, Diplovax, Drenusil, Duramycin, Dynamyxin, Efuranol, Enadel, Envacar, Equipose, Equivert, Fasigyn, Fecap, Felden, Feldene, Fenocin, Fenocin Forte, Ficortril, Floclean 103, Floclean 106, Floclean 107, Floclean 108, Floclean 303, Floclean 307, Flocon 100 - Antiscalent, Fovane, Fugata, GDL, Geminimycin, Geocillin, Geopen, Glibenese, Guntapite, Helmex, Hopp II, Hypovase, Irribral, Isoject-Streptomycin Injection, Isonex, Isonex Forte, Istin, Kroma Red, Liquamycin, Liquapen, Lithane, Macocyn, Mansil, Marax, Masmoran, Matromycin, Matromycin Tao, Matromycin-T, Mecadox, Medium VS, Medium 10, Medium 7, Mellinese, Methral, Minipress, Minizide, Mithracin, Mono-Line, Morflex P50, Morflex 125, Morflex 210, Morflex 240, Morflex 310, Morflex 325, Morflex 330, Morflex 410, Morflex 510, Morflex 525, Morflex 530, Mucopront, Navane, Navane Hydrochloride, Nebacortril, Nebasulf, Neobiotic, Nephril, Niamid, Niamide, Nigrin, Obinese, Oleandocyn, Orbicin, Orbigastril, Orbinamon, Oxicebral, Paratect Bolus, Parixistil, Pasonex, Pathozone, Pen A, Pen A/N, Peripress, Permapen, Permapen, Pferrico, Pferrisperse, Pferritan, Pferrocal, Pferromet, Pferrox, Pfico2-Hop, Pfico2-Isohop, Pfico2-Redihop, Pfiklor, Pfinodal, Pfizer-E, Pfizer-E, Pfizer-EM, Pfizercycline, Pfizerpen, Pfizerpen A, Pfizerpen A, Pfizerpen Vk, Pfizerpen-AS, Pfizerpen-AS, Pfizerquine, Polypress, Posistac, Procardia, Q-Cast, Q-Crete, Q-Gun, Q-Vibe, Quantril, Renese, Resmax, Rondomycin, Safapryn, Safapryn-Co, Sanoma, Sigmamycin, Simplotan, Sinequan, Solu-Cortril, Sorbistat, Sorbistat K, Sostenil, Sparmite, Spectrobid, Sterane, Sterane IM and IA, Streptohydrazid, Streptorex, Stretonex, Suparen, Superjet, Sure-Curd (Suparen), Synandrets, Synandrol, Synandrol F, Synandrotabs, Syngesterone, Syngestrets, Tao, Telopar, Terra-Bron, Terra-Cortril, Terramycin, Terramycin Hydrochloride, Terramycin Pediatric Drops, Tetracyn P, Tetracyn S.F, Tibricol, Trilombrin, Triolam, Trosyd, Two Cubed Eight, Tyzine, Ultrapen, Unipen, Unisom, Urobac, Urobiotic, Vansil, Vansil, Vatensol, Velbacil, Veltol, Veltol-Plus, Vibalt, Vibazine, Vibra-Tabs, Vibramycin, Vibratussal, Vibratussan, Vicron, Vincapront, Vincaron, Viocin, Vistaril, Vistaril Pamoate.

Pharmaceutical Basics Inc,
Cook Group Inc,

301 South Cherokee Street,
Denver, Colorado 80223
USA

Pharmaceutical Basics Inc
Cook Group Inc

Cogest, LAX-1C, LAX-2C, LAX-3, LAX-4, LAX-42, Meprobase-200, Meprobase-400.

Pharmacia Laboratories,
Div of Pharmacia Inc,

800 Centennial Ave,
Piscataway, NJ 08854
USA

Hyskon, Macrodex, Rheomacrodex.

Phillips 66 Company,
Phillips Plastics Resins Division,

PO Box 792,
Pasadena, TX 77501
USA

K-Resin Polymer KR01, K-Resin Polymer KR03, K-Resin Polymer KR04, K-Resin Polymer KR05, K-Resin Polymer KR10, Marlex BMN TR-880, Marlex BMN 55500, Marlex CL-100, Marlex CL-50, Marlex EHM 6003, Marlex EHM 6003, Marlex EHM 6006, Marlex EHM 6007, Marlex EMN TR-885, Marlex ER9-0002, Marlex ER9-0020, Marlex HGH-050, Marlex HGL-050-01, Marlex HGL-050-01 (Antistatic), Marlex HGL-120-01 (Antistatic), Marlex HGL-200 (Antistatic), Marlex HGL-350 (Antistatic), Marlex HGN-020-01, Marlex HGN-020-01, Marlex HGN-120-01 (Nucleated), Marlex HGN-200 (Nucleated), Marlex HGN-200A, Marlex HGN-350 (Nucleated), Marlex HGX-010, Marlex HGX-030, Marlex HGX-040, Marlex HGX-330 (Controlled Rheology), Marlex HGZ-050-02, Marlex HGZ-120-02, Marlex HGZ-120-04, Marlex HGZ-200, Marlex HGZ-350, Marlex HHM TR-130, Marlex HHM TR-140, Marlex HHM TR-144, Marlex HHM TR-210, Marlex HHM TR-226, Marlex HHM TR-230 Black, Marlex HHM TR-232 Black, Marlex HHM TR-250 Black, Marlex HHM TR-418 (Black, Orange), Marlex HHM 4903, Marlex HHM 4903, Marlex HHM 5202, Marlex HHM 5202, Marlex HHM 5502, Marlex HHM-4515, Marlex HLM-020, Marlex HLN-120-01, Marlex HLN-200 (Antistatic, Nucleated), Marlex HLN-350 (Antistatic, Nucleated), Marlex HMN TR-942, Marlex HMN 5060, Marlex HMN 54140, Marlex HMN 5580, Marlex HMN 6060, Marlex HMN-4550, Marlex HMN-938, Marlex HMX-020-01 (Lubricant), Marlex HXM 50100, Marlex HXM 50100, Marlex RGX-020, Marlex RGX-020, Marlex RGX-020 (Antistat), Marlex RMN-020C, Marlex RMN-020C, Marlex RMX-020, Ryton A-100, Ryton R-4, Ryton R-7.

Piccadilly Products Limited,

199 Piccadilly,
London W1V 9LE
UK

Reflectafoam, Rollofix X100.

Pickering Laboratories Inc,

1951 Colony Street,
Mountain View, CA 94043
USA

Trione.

PJ1 Corporation,

7345 Topanga Canyon Blvd,
Canoga Park, CA 91303
USA

PJ1 Chain Lube, Blue Label, PJ1 Octane Plus, PJ1 Super Cleaner.

Plascoat Systems Limited,
Materials & Equipment Division,
Trading Estate,
Farnham, Surrey HU9 9NY
UK

Duraguard, Plascoat Plasinter, Plascoat PPA.

Plastic Coatings Limited,

Woodbridge Industrial Estate,
Guildford, Surrey GU1 1BG
UK

Deconyl, Plascoat PPA 31 series, Vyflex NT80S.

Plough Inc,

3030 Jackson Ave,
Memphis, Tenn 38151
USA

Aftate, Black and White Bleaching Cream, Duration, Moroline, Nujol, St Joseph, St Joseph Cough Syrup.

Plysolene Limited,

Southwater Business Park, Worthing Road,
Southwater, West Sussex RH13 7HE
UK

Plysolene (PIB).

PMC Specialties Group Inc,

10051 Romandel, Santa Fe Springs,
CA 90650
USA

Cao, Dyphene, Prodox.

Polomyx Industries Inc.,

14 Jewel Drive,
Wilmington Mass 01887
USA

Polomyx.

Polymer Composites Incorporated,

5152 West 6th Street,
Winona, MN 55987
USA

Fiberod, Vari-Cut.

Polysciences Inc,

400 Valley Road,
Warrington PA 18976
USA

Cellufluor, Fluoresbrite, Fungi-Fluor, Immuno-bed, JB-4, Polybead, Poly/Bed 812, Poly/Sep 47.

Polyvinyl Chemie Holland BV,

PO Box 123,
5140 AC Waalwijk,
The Netherlands

NeoCryl, NeoRez, NeoVac.

Poythress Laboratories Inc,

16 N 22nd St, PO Box 26946,
Richmond, Va 23261
USA

Merpectogel, Solfoton, Trocinate.

PPF International Limited,

Bromborough Port,
Wirral, L62 4SU
UK

Admul, Advitacon, Advitagel, Advitamix, Advitaroma, Amerone, Aromaplas, Cablinol, Ceritone, Chlor-Tabs, Comprena, Filpro, Flavocents, Flavotint, Hymono, Jasmacyclene, Jasmolide, Omarsan.

The PQ Corporation,

PO Box 840,
Valley Forge, PA 19482
USA

Britesil, Britesorb, Kasil, Metso, N, Nyacol, Q-Cel, Valfor.

Procter & Gamble Co,

11511 Reed Hartman,
Cincinnati, Ohio 45241
USA

Head and Shoulders.

Prodorite Limited,

Eagle Works,
Wednesbury, West Midlands WS10 7LT
UK

Asplit, Cement Prodor, Chematex, Chemdur, Chemline, CTW, Formula 'S', Furacin, Metro Tiles, Orglas, Prodoraqua, Prodorbond, Prodorcrete GT, Prodorfilm, Prodorflor, Prodorglas, Prodorglaze, Prodorguard, Prodorlac, Prodorshield, Vibrabond.

Protective Rubber Coatings (Limpetite) Limited,

Paynes Shipyard, Coronation Road,
Bristol BS3 1RP
UK

Limpetite.

Puma Chemical Company Inc,

1601 109th St,
Grand Prairie, Tx 75050
USA

Puma Chemical Company Inc

Pool-Chem (Chemical line for swimming pools), Restor-E (Restoration Chemical Products), Term-X (Various Herbicides/Insecticides).

Punati Chemical Corporation,

1100 N Woodward,
Birmingham MI 48011
USA

Aromabator PC-80, Aromabator PC-88, Microbator PC-78.

The Purdue Frederick Co,

100 Connecticut Ave,
Norwalk, Conn 06856
USA

A-Sol, Athrombin-K, MS Contin, Pheny-PAS-Tebamin, Phyllocontin.

Q

Ques Industries

P O Box 02590, 3009 W. 47,
Cleveland, OH 44102
USA

Bioques, Bioques Q, Bioques Z, Quesfloc, Quesfloc F11283-1.

Quinoderm Limited,

Manchester Road, Hollinwood,
Oldham, Lancs OL8 4PB
UK

Ceanel, Eczederm, Gelcotar, Hioxyl, Quinocort, Quinoderm Cream, Quinoderm Cream with Hydrocortisone, Quinoped.

R

Reagent Chemical & Research Inc

1300 Post Oak Blvd Suite 650,
Houston TX 77056
USA

Magnesol.

Reed Lignin Inc,

81 Holly Hill Lane,
Greenwich, CT 06830
USA

Additive-A, Ameri-Bond, Dynasperse, Glutrin/Goulac, Kelig, Lignosol, Marabond, Maracarb, Maracell, Maracon, Marasperse, Norlig, Petrolig.

Regal Chemical Company,

PO Box 900, Alpharetta,
Georgia 30201
USA

Dylon, Regal Crown, Regalstar.

Reheis Chemical Co,

235 Snyder Ave,
Berkeley Heights, NJ 07922
USA

Chlorhydrol, F-1000, Macrospherical 95, Micro Dry, Rehydrol.

Rendell, W J Limited,

Ickleford Manor,
Hitchin, Herts SG5 3XE
UK

Genexol, Rendells.

Resadhesion Limited,

29 Eve Road,
Woking, Surrey GU21 5JS
UK

Resad.

Resart - IHM AG,

Gassnerallee 40,
D-6500 Mainz 1
West Germany

Resarit, Resarix SF, Resart, Resart-PMMA XT, Resartglas GS, Resartherm.

Resinas Sinteticas SA,

Aribau, 185-6a planta,
08021 Barcelona
Spain

Reacrone, Reactal, Reafor, Reafree, Reamul, Reatane, Reater, Resibon, Resilita, Resipol, Terfenol, Tergraf, Tergum, Terlac.

Resinous Chemicals Limited,

Cross Lane,
Dunston, Tyne & Wear NE11 9HQ
UK

Additol, Albertol, Alftalat, Alnovol, Alpex, Alpolit, Alresat, Alresen, Becxopox, Daotan, Duroxyn, Macrynal, Maprenal, Phenodur, Resydrol, Synthacryl, Vresamin, Wresinate.

Restech Industries Inc,

590 S Seneca (PO Box 2747),
Eugene, OR 97402
USA

Restore-X Exterior Paint Remover, Restore-X Weathered Wood Renewer.

REWO Chemicals Ltd

9th Flour, Crown House,
Morden, Surrey, SM4 5DV
England

Melioran F6, Preservative, Rewo-amid, Rewo-LAN/5, Rewocor, Rewoderm S 1333, Rewominoxid, Rewophos EAK 8190, Rewophos TD40, Rewophos TD70 and OP80, Rewopol B2003, Rewopol DLS, Rewopol MLS, Rewopol NL3, Rewopol SB FA/30, Rewopol SB IP, Rewopol SB-DO 70, Rewopol SBL 203, Rewopol SLS, Rewopol SMS, Rewopol TLS, Rewopol TMS and ODS, Rewopon IM-BT, Rewopon IM-OA, Rewoquat CPEM, Rewoquat CR 3099, Rewoquat W7500, Reworyl ACS, Reworyl C, Reworyl KXS, Reworyl NCS, Reworyl NTS, Reworyl NXS, Reworyl T, Reworyl X, Rewoteric AM-B13, Rewoteric AM-B13/T, Rewoteric AM-CA, Rewoteric AM-DML, Rewoteric AM-G30, Rewoteric AM-KSF, Rewoteric AM-LL, Rewoteric AM-V, Rewoteric AM-2C, Rewoteric AM-2L, Steinazid SBU 185.

Reynolds Metal Company,

PO Box 27003 A,
Richmond, Virginia 23261
USA

Alreco, Ladalrod, Paperad, Perma-Leaf, Reyalite, Reyalith, Reycomp, Reydox, Reynolds Wrap, Reynolon, Stan-Fast, Tem-Tuf, Tread-Brite, Tru-Color, Ultra-Flat.

R F Bright Enterprises Limited
see Bright Enterprises, R F Limited

Rhein-Chemie Rheinau GmbH,

6800 Mannheim 18,
Miilheimer Strasse 24-28,
Postfach 810409
West Germany

Aflux, Aktiplast, Antidust F, Antidust 2, Antilux, Antilux AOL, Factice, Megapren C 150, Megapren Si 10, 20, 30 and 60, Megapren U225, Rhenodiv, Rhenogran, Rhenomag, Rhenopren, Rhenosin, Rhenosorb, Silopren HV.

Rñzone-Poulenc Chemie NV,
Fabriek Rieme,

Kuhlmannkaai 1,
9020 Gent
Belgium

Rñzone-Poulenc Industries,
22 Avenue Montaigne,
75360 Paris Cedex 08
France

BI Ammonium Phosphate, Flug{Ae{ne 113, Gobapur Acide Pur, Kerimid 500, Kerimid 501, Kerimid 502, Kerimid 503, Kerimid 601, Kinel 5502, Kinel 5514, Kinel 5517, Mono Ammonium Phosphate, Mono Ammonium Phosphate (Agricultural Grade), M33, MN3, Phosphoric Acid, Primofax, Rhodester, Rhodia-Phos, Rhodopas, Rhodopas AX, Rhodopas X, Rhodoviol, Rhovinal B, Rhovinal F, Scurane V.

Richardson-Vicks Inc,

10 Westport Road,
Wilton, Connecticut 06897
USA

Acta, Airets, Aquasun, AZ, Benzodent, Biactol, Clearasil Adult Care, Clearasil Super Strength, Climacel, Colac, Complete, Debendox, DenClen, Denquel, Eversun, Fasteeth, Fasteeth Extra Hold, Firmadent, Fixodent, H{Be{gor, Infacare, Infasoft, Iron Man, K de Krizia, Keramine H, Kleenite, Kolanticon, Kolantyl, Kolantyl-NV, Kukident, Lemon Delph, Lemon Plus, Licuado Instante, Manucr{Ae{me, Mediker, Merbentyl, Merocets, Mila, Milgard, Mill Creek, Milton, Moncler Derma, Monodral, Mytolac, Napisan, Nethaprin Dospan, Nethaprin Expectorant, Night of Olay Nightcare, Oil of Olay, Olay Beauty Bar, Olay Beauty Cleanser, Pantene, Percogesic, P{Be{trole Hahn, Pexid, Praims, Primus, Protasan, Respiral, Rog{Be{ Cavaill{Ae{s, Sal de Uvas Picot, Saxon, Soothease, Tempo, Tenuate, Tenuate Dopan, Topex, Topexane, Vicks Medinite, Victors, Vitapyrena.

Riker Laboratories Inc
see 3M

RITA Corporation,

PO Box 556, Crystal Lake,
Illinois 60014
USA

Acritamers, Forlan C-24, Forlan Series, Forlan 'LM', Forlan 'L', Grilloten ZT40, ZT80, PSE 141G, LSE 87, LSE 87K, Laneto Series, Ritacetyl, Ritachol, Ritaderm, Ritalafa, Ritalan, Ritalan 'C', Ritasol, Ritawax, Secolan S-1, BA-1, BA-1G, Selastin EL-10, EL-30, SE EM 95, Supersat.

**RJ Manufacturing Inc.,
Sealants Division,**

PO Box 34475,
San Antonio, Texas 78265
USA

Marweld M-17.

Roayle Polymers Limited

Poucher Street, Hilltop,
Kimberworth, Rotherham, S Yorks S61 2ET
UK

Black Grip.

Robinson Bros Limited,

Phoenix Street,
West Bromwich, West Midlands B70 0AH
UK

Arbeflex, Arbeflex 489, Arbeflex 550, Arbestab, Robac, Robacure.

Robinson Brothers Ltd,

Phoenix Street, West Bromwich,
West Midlands B70 0AH
England

Packard-Bekker BV,
PO Box 9403,
9703 LP Groningen
The Netherlands

Soluene 100 and 350.

Robins, A H Co Inc,

1407 Cummings Drive,
Richmond, Va 23220
USA

Adabee, Allbee with C, Asthmatussin, Asthmatussin-T, Dimetane, Dimotane Expectorant, Dimotane Expectorant DC, Dimotapp Elixir, Donnagel, Donnagel-PG, Donnatal, Donnazyme, Dopram, Elanone, Exna, Imavate, Micro-K, Phenaphen, Pondimin, Quinidex, Reglan, Robalate, Robaxin, Robicillin Vk, Robimycin, Robinul, Robitet, Robitussin, Robitussin A.C, Silain, Stental, Sulla.

Robnorganic Systems Limited,

Highworth Road,
South Marston, Swindon, Wilts SN3 4TE
UK

Emdithene.

Roche Products Ltd,

Dalry,
Ayrshire KA24 5JJ
Great Britain

Roche Products Limited,
PO Box 8,
Welwyn Garden City, Herts AL7 3AY
UK

Hoffmann-LaRoche Inc,
340 Kingsland St,
Nutley, NJ 07110
USA

Accutane, Airol Roche, Alcobon, Alfavet, Alloferin, Ancobon, Ancotil, Arfonad, Arovit, Asterol, Avatec (as sodium), Bactrim, Bactrim Roche, Becosym, Beflavine Roche, Beflavit, Benadon, Benerva, Benerva Compound, Benexol Roche, Benical, Bepanthen, Berocca, Bovatec (as sodium), Bumex, Cal-C-Vita, Carophyll, Cipralan, Clonopin, Coactabs, Coactin, Cycloserine Roche, Dalmane, Declinax, Dormicum, Dromoran, Efudex, Efudix, Emcyt, Ephynal, Estracyt, Fanasil, Fansidar, Fanzil, Fluorouracil Roche, Fluprim, Gantanol, Gantrisin, Glutril, Hypnovel, Imadyl, Injacom, Ipropran, Konakion, Lariam, Larocin, Larodopa, Laroxyl, Levo-Dromoran, Lexotan, Librax, Libraxin, Librium, Limbitrol, Lipo Gantrisin, Liquemin, Litrison, Lorfan, Madopar, Madribon, Marcoumar, Marplan, Marsilid, Matulane, Mestinon, Mogadon, Natulan, Nipride, Nisentil, Nitoman, Nobrium, Noludar, Omnopon, Omnopon-Scopolamine, Pantene Grooming Lotion, Pentrium, Pethidine Roche, Pethilorfan, Phenoro, Pondinil Roche, Posicor, Prostigmin, Protovit, Provocholine, Redoxon, Redoxon, Rimadyl, Rimifon, Rivotril, Ro-A-Vit, Roaccutane, Rocaltrol, Rocephin, Rohypnol, Romilar, Roniacol, Ronicol, Ronoxan, Rovigon, Rovimix, Rovisol, Saridone, Solatene, Supradyn, Synkavit, Synkayvite, Taractan, Tasmaderm, Tegison, Temetex, Tensilon, Thephorin, Tiberal Roche, Tibirox, Tigason, Topostasin, TRH-Roche, Trimpex, Unitop, Valium, Valium Roche, Versed, Vitaminets.

Roehm Ltd,

Hexoran Division, Derwent Street,
Belper, Derby DE5 1WQ
England

Hexoran, Hexoran A15, Octoran.

Roerig, J B
see Pfizer Ltd

Rohm & Haas (UK) Ltd,

Lennig House, 2 Mason's Avenue,
Croydon, Surrey CR9 3NB
England

Rohm & Haas (UK) Ltd

Rohm & Haas Co,
Independence Mall West,
Philadelphia, Penna 19105
USA

Rohm and Haas Company,
European Operations Division,
Chesterfield House, Bloomsbury Way,
London WC1A 2TP
UK

Acrysol, Acrysol LMW, Amberlite, Amberlite IRP-64, Amberlite IRP-88, Dithane, Duraplus, Hyamine 10-X, Karamate, Karathane, Kathon, Kathon 886, Kathon 893, Kelthane, Kerb, Korad A, Matrikerb, Oroglas DR, Orotan, Pancil T, Paraloid EXL 2607, Paraloid K120N and KM-228, Paraplex G.62, Plexol, Primal, Primapel, Primax, Primenes, RH Maneb 80, Rocryl, Ropaque, Skane, Surfactant N-42, Surfactant XQS20, Systhane, Triton, Triton A-20, Triton CF-10, Triton CF-21, Triton CF-87, Triton GR-5M, Triton GR-7M, Triton H-55, Triton H-66, Triton N-40, N-57, N-60, N-101, N-111 and N-150, Triton QS-15, Triton QS9, QS30 and QS44, Triton X-100, X-102, X-114, X-120 and X-165, Triton X-155, Triton X-15, X-35 and X-45, Triton X-200 and X-202, Triton X-207, Triton X-301, Triton X-305, X-405 and X-705, Triton 770 and W30.

Ronacrete Limited,

Ronac House, 269 Ilford Lane,
Ilford, Essex IG1 2SD
UK

Monoset, Ronafix, Unicell.

Ronsheim & Moore Ltd,

Castleford,
West Yorkshire WF10 2JT
England

Aremsol A, Lorol DA, Lorol MA and MR, Lorol NH, Lorol TA and TAR, Lorol TN and TNR, Product MB320.

Roquette (UK) Limited,

Pantiles House,2 Nevill Street,
Tunbridge Wells, Kent TN2 5TT
UK

Actisize, Alburex, Anidrisorb, Carbilys, Clearam, Collys, Cremalys, Eurylon, Fluitex, Glucidex, Hi-Cat, Lycasin, Lysamine, Lysase, Maltisorb, Meliose, Neosorb, Nutralys, Roferose, Tackidex, Texalys, Tissalys, Viten, Waxigel, Waxilys.

Ross Laboratories
see Abbott Laboratories Limited

Rostine Manufacturing & Supply Company,

PO Box 8192, 4227C W Church,
Springfield, MO 65801
USA

Casto-Magic.

Roussel Laboratories Limited,

Broadwater Park, North Orbital Road,
Uxbridge, Middlesex UB9 5HP
UK

Hoechst-Roussel Pharmaceuticals Inc,
Route 202-206 North,
Somerville, NJ 08876
USA

Roussel Uclaf, Fine Chemicals,
Tour Roussel Nobel, Cedex No 3,
F-92080 Paris la Defense,
France

Alnovin, Altacaps, Altacite Plus, Arlix, A/T/S, Caloreen, Claforan, Clinifeed Favour, Clinifeed ISO, Clinifeed Protein Rich, Clinifeed 400, Corvaton, Cryptolin, Depovirin, Diabeta, Dormonoct, Doxinate, EsCort, Finaplix, Flavomycin, Haemate-P, Lanitop, Loprox, Merital, Molipaxin, Nutranel, Panacur, Proctofibe, Pymafed, Regu-mate, Stenorol, Suprefact, Surfak, Surgam, Symcor, Topicort.

Rousselot Limited,
Sanofi-Elf Bio-Industries Division,

Mill Reef House, 9-14 Cheap Street,
Newbury, Berks RG14 5DN,
UK

Gat 15, Rousselot Gelatine, S.P.G Gelatine.

Rowell Laboratories Inc,

Baudette,
Minn 56623
USA

Balneol, C-Ron, C-Ron Forte, Cin-Quin, Colrex Compound, Colrex Expectorant, Dermacort, Dexone, Indomed, Lithonate, Lithotabs, Orasone, Proctocort, Quine, Ro-Bile, Ronase.

R T Vanderbilt Co Inc

R T Vanderbilt Co Inc
see Vanderbilt, R T Co Inc

Ruscoe, W J Company,

485 Kenmore Boulevard,
Akron, Ohio 44301
USA

Color-Max, En-Dur-Lon, Flexcote, Hylite Color-Max, Plasticote, Plastitoy, Pliobond Adhesives.

Rustlan Chemical Company,

11 Cypress Street,
Salem NH 03079
USA

Extrusion-Plus, Rust-Tap, Rustlan Oil.

RV Chemicals Ltd,

Lugsdale Road, Widnes,
Cheshire WA8 6ND
England

Formusol, Formusol SA, Hypacel, Rediclear, Reduxol Z.

R W Greef & Co Inc
see Greef, R W & Co Inc

Rynco Scientific Corp,

31 Stewart St, PO Box 270,
Floral Park, NY 11002
USA

RX-56.

Rystan Company Inc,

PO Box 214,
Little Falls, NJ 07424
USA

Chloresium, Derifil.

S

Sandoz Ltd

Calverley Lane, Horsforth,
Leeds LS18 4RP
England

Dorsey Laboratories,
Div of Sandoz Inc,
NE US No 6 & Interstate 80,
Lincoln, Nebr 68501
USA

Sandoz Pharmaceuticals,
Div of Sandoz Inc,
59 Route 10 E,
Hanover, NJ 07936
USA

Dorsey Pharmaceuticals,
Pharmaceutical Div of Sandoz Inc,
59 Route 10,
East Hanover, NJ 07936
USA

Acylanid, Belladenal, Bellergal, Biarsan, Cafergot, Calcibronat, Cedilanid, Cedilanid Ampoules, Chromium Sandoz, Clozaril, Deseril, Diapid, Digilanid, Dilasoft, Dilatin NA Liquid, Dorantamin, Dormethan, Dorsacaine, Dorsital, Ekaline, Elfugin, Exoderil, Felamine, Femergin, Glysennid, Gris-PEG, Hydantil, Hydergine, Imacol, Ipesandrine, Landromil, Lyocol, Lyogen, Mellaril, Mellaril-S, Melleril, Mesantoin, Mesontoin, Metaprel, Methergin, Methergine, Neo-Calglucon, Neutrolactis, Nylofixan, Pamelor, Parlodel, Proresid, Pursennid, Retargal, Revatol, Sancos, Sancos Compound, Sando-K, Sandocal, Sandofix, Sandogen, Sandolube, Sandomigran, Sandopan, Sandoptal, Sandopur, Sandotex, Sandoz Effervescent Compound, Sandozin, Sanorex, Sansert, Syntocinon, Syntometrin, Syntopressin, Tavist, Thiotan, Torecan, Tormosyl, Uval, Visken, Wilpo, Zaditen.

Sanifoam Inc.,

1370 Logan Avenue,
Costa Mesa, California 92626
USA

Sanifoam Inc.

Saniblanket, SaniFoam.

Sas Pharmaceuticals Limited,

Sas Group House, 45 Wycombe End,
Beaconsfield, Bucks HP9 1LZ
UK

Fluor-Amps, Ophthalmadine.

Sasolchem,

PO Box 62177,
Marshalltown 2107
South Africa

Atar Phenol, Ethylol, Meketone, Propylol, Sabutol, Sasetone, Sasolwaks.

**Savage Laboratories,
Div of Byk-Gulden Inc,**

PO Box 2006,
Melville, NY 11747
USA

P 13 N.

Schaefer Chemical Products Company,

3000 Carrollton Road,
Saginaw, MI 48604
USA

Boiler-Aid, Condens-Aid, Cool-Treet, Oil-Treet, Tower Treat.

Schaumstoff und Kunststoff GmbH,

A-4021 Linz-Wegscheid, Ober{
osterreich,
Eduard-Sue-Stra{{{e 19
Austria

Sembonit/Erostabil, Sempollan.

Scher Chemicals Inc.,

Industrial West Corner Styertowne Road,
Allwood, PO Box 1236,
Clifton, N Jersey 07012
USA

Colsol, Dipsal, Katemul IG-70, Katemul IGU-70, Lubrisol, Nyloset Finish, Polysoft CA, Repello DC, Schercassist AC, Schercemol BE, Schercemol CM, Schercemol CO, Schercemol CP, Schercemol CS, Schercemol DEGMS, Schercemol DEIS, Schercemol DIA, Schercemol DICA, Schercemol DID, Schercemol DIS, Schercemol DISD, Schercemol DO, Schercemol EE, Schercemol EGMS, Schercemol GMIS, Schercemol GMS, Schercemol ICS, Schercemol ISE, Schercemol MEL-3, Schercemol MEM-3, Schercemol MEP-3, Schercemol MM, Schercemol MP, Schercemol MS, Schercemol NGDC, Schercemol OLO, Schercemol OP, Schercemol OPG, Schercemol PGDP, Schercemol PGML, Schercemol PGMS, Schercemol SE, Schercemol TIST, Schercemol TT, Schercemol 1688, Schercemol 1818, Schercemol 185, Schercemol 318, Scherco Finish AL, Scherco Softener £1, Scherco Softener £2, Schercoat OE-44, Schercoat OE-44K, Schercoat P-110, Schercoat PC-550, Schercoat S-220, Schercoat S-330, Schercolene SB, Schercolube 707, Schercomid EAC, Schercomid EACS-100, Schercomid 304, Schercopon 2WD, Schercoquat DAS, Schercoquat IAS, Schercoquat IB, Schercoquat IEP, Schercoquat IIB, Schercoquat IIS, Schercoquat SOAB, Schercoquat SOAS, Schercosol DS, Schercosol NL, Schercosol P, Schercosol T, Schercotarder, Schercowet DOS-70, Schercozoline C, Schercozoline I, Schercozoline L, Schercozoline O, Schercozoline S, Scheroba Oil, Scherpol LSB, Schersoftoil P, Tergitex KW.

Schering AG,

1000 Berline 65,
Mullerstrasse 170-178,
W Germany

Wesley-Jessen,
Div of Schering Corp (USA),
400 W Superior St,
Chicago, Ill 60610
USA

Androcur, Anovlar, Biligrafin, Biloptin, Depostat, DuraSoft, Eugynon 30 and 50, Microgynon 30, Microlut, Neogest, Primobolan, Primobolan Depot, Primodos, Primogyn Depot, Primolut Depot, Primolut N, Primoteston Depot, Pro-Viron, Progynova, Schericur, Scheriproct, SH 420, Ultradil, Ultraproct.

Schering-Plough Corp,

Galloping Hill Rd,
Kenilworth, NJ 07033
USA

Afrin, Afrinol, Akrinol, Aquacillin, Banamine, Celestone, Chlor-Trimeton, Cypromin, Diprosone, Estinyl, Fulvicin-P/G, Fulvicin-U/F, Garamycin, Gyne-Lotrimin, Hyperstat, Intron, Lotrimin, Meti-Derm, Meticortelone Acetate, Meticorten, Metreton,

Schering-Plough Corp

Mol-Iron, Naqua, Netromycin, Normodyne, Optimine, Oreton, Oreton Methyl, Oreton Propionate, Ovaban. , Paxipam, Permitil, Polaramine, Prantal, Proglycem, Proventil Inhaler, Regutol, Rela, Sebizon, Siseptin, Solganal, Sulamyd Sodium, Tinactin, Tindal, Tiox, Tremin, Trilafon, Vaderm, Vancenase, Vanceril, Variton.

SCM Chemicals Ltd,

PO Box 26, Grimsby,
S Humberside DN37 8DP
England

Tiona.

ScotFoam Corp,

1500 E Second St, Eddystone,
PA 19013
USA

Aerofonic, Afonic.

Scott Bader Co Limited,

Wolaston,
Wellingborough, Northants NN9 7RL
UK

Crystic, Kollercast, Polidene, Texicote, Texicryl, Texigel, Texipol.

Scottish Agricultural Industries PLC,

25 Ravelston Terrace,
Edinburgh EH4 3ET
Great Britain

Enmag, Firesaife, Hortus, Longlife Turf Foods, Nutraphos, Scotphos.

Searle, G D & Co,

4901 Searle Pkwy,
Skokie, Ill 60077
USA

Aldactide, Aldactone, Aminophyllin, Anavar, Apyrogen, Banthine, Cergem, Conovid, Conovid-E, Cytotec, Dartalan, Demulen, Demulen 50, Depepsen, Diodoquin, Diulo, Dramamine, Enavid 5mg, Enavid-E, Equal, Femulen, Flagyl IV, Floraquin, Flurogestone Acetate, Lomotil, Lomotil with Neomycin, Metamucil, Metruien M. A, Metrulen, Nilevar, Nitrodisc, Norpace, NutraSweet, Ovulen 1 mg, Ovulen 50, Pro-Banthine, Serenace, Syllact.

Secure Incorporated,

PO Box 2099,
Waxahachie, Texas 75165
USA

Aquanol, Color Seal, Esterox, Hard Cure, Perma Shield, Polyseal, Seair, Sebond, Sebrite, Secure, Seseal, Seseal 8, Sestrip, Setreat.

Sekisui Chemical Co Limited,

4-4 Nishitemma 2-chome,
Kita-ku, Osaka
Japan

S-lec, S-nyl.

Seton Products Limited,

Turbiton House, Medlock Street,
Oldham, Lancs OL1 3HS
UK

Calaband, Ichthaband, Quinaband, Tarband, Zincaband.

S.F.C.,

3 Rue des Carrieres,
93800 Epinay s/seine
France

Grafitix (Anti-graffiti), Grafitix (Bâtiment), Grafitix (Ravalement).

Shawplas Limited,
Polishing Products Division,

Alms Close, Stukeley Meadows Industrial Estate,
Huntingdon PE18 6DY
UK

Shawplas.

Shell Chemicals UK Ltd,

1 Northumberland Avenue, Trafalgar Square,
London WC2N 5LA
England

Apiezon, Caradate, Caradol, Cardura, Cardura E, Cariflex, Cariflex Butadiene Rubber (BR), Cariflex Isoprene Rubbers (IR), Cariflex S, Cariflex Styrene-Butadiene Rubbers

(SBR), Cariflex Thermoplastyic Rubbers (TR), Carina, Carinex, Carinex SB41, Carinex SI 73, Carlona LB 157, Carlona LF 456, Carlona LF 459, Carlona P PLZ 532, Carlona P PY 61, Carlona 460, Carlona 462, Carlona 463, Carlona 55-004, Carlona 60-010, Carlona 60-060, Carlona 60-120, Dielmoth, Dioxitol, Dobane (Detergent Alkylate), Dobanic Acids JN and 83, Dobanol Ethoxylates, Dobanol Ethoxysulphates, Dobanols, Dobanox, Dobatex, Dutral, Dutrex, Dutrex Process and Extender Oils, Dutrex 20, 25, Elexar, Epicure, Epikote, Epikote DX-209-B-80, Epikote DX-210-B-80, Epikote DX-231-B-91, Epikure 3400, Epon, Epon 8280, Ionex, Ionol, Ionol CP, Ionol CPA-Feed, Ionol J65, Ionox, Ionox 220, Ionox 330, Ionox 901, Ionox 99, Kraton, Lenka, Lensine, Linevol Phthalates, Linevol 79, Linevol 911, Nonidet, Oxilube, Oxilube, Oxilubes, Oxitex, Oxitex, Oxitol, Oxitol, Planavin, Shellflex Process and Extender Oils, Shellsol D40, D60, D70, Shellsol E, A, AB, R, Shellsol T, Shellswim - 11T, Shellswim - 5X, Shellvis 50 (SAP 150), Styrocell, Teepol CM44, Teepol FC5, Teepol GD53, Teepol HB6, Teepol PB, Thermolastic, Trioxitol, VeoVa 10 Monomer, Versatic.

Sherex Chemical Co Inc,

PO Box 646, Dublin,
OH 43017
USA

Adol 62, Adol 66, Adol 85, Hydrofol Acid 1655.

**Sigma Coatings Allweather Paints Limited,
Protective Coatings Division,**

Petrofina House, Ashley Avenue,
Epson, Surrey KT18 5AD
UK

Colturiet, Emaline, Nucol, Pervon, Sigmathane, Tornusil.

Sihi-Ryaland Pumps Limited,

Broadheath,
Altrincham, Cheshire
UK

Myras, Pacvac, Pioneer, Ryax C, Ryax F, O, Ryflex, Rytherm, Ryvin.

Silberline Mfg Co Inc.,

PO Box A,
Lansford, PA 18232
USA

EternaBrite, Silvet, Silvex, Sparkle Silver, Sparkle Silvet, Sparkle Silvex.

Siliconas Hispania SA,

Balmes 357,
08006 Barcelona
Spain

Acrilpact, Emulsil, Silicex.

S Kempner Limited
see Kempner, S Limited

Smith Kline & French Laboratories Limited,

Welwyn Garden City,
Herts AL7 1EY
UK

Smith Kline & French Laboratories,
1500 Spring Garden St,
Philadelphia, Penna 19101
USA

Amylozine Spansule, Ancef, Anspor, Benzedrex, Cefizox, Cendevax, Compazine, Cytomel, Daprisal, Darbid, Dexedrine Tablets, Dibenyline, Dibenyline Capsules, Dibenzyline, Drinamyl, Dyazide Tablets, Dyrenium, Dytac, Dytide, Edrisal, Ervevax, Eskabarb, Eskacef, Eskacillin, Eskacillin 100, Eskacillin 100 Sulpha, Eskacillin 200, Eskacillin 200 Sulpha, Eskadiazine, Eskalith, Eskamel Cream, Eskornade, Eskornade Spansule Capsules, Expansyl Spansule, Fefol, Fefol Spansule Capsule, Fefol Z Spansule Capsule, Fefol-Vit Spansule Capsule, Fenbid Spansule Capsul, Feospan, Feospan Spansule Capsule, Feospan Z Spansule Capsule, Fesovit, Fesovit Spansule Capsule, Fesovit Z Spansule Capsule, Hispril, Histryl, Histryl Spansule Capsule, Iodex, Kerecid, Liskonum Tablets, Mebryl, Micralax, Micralax Micro-enema, Mio-Pressin, Monocid, Nasoflu, Neuro-Phosphates, Octovit Tablets, Paredrine, Parnate Tablets, Parstelin, Parstelin Tablets, Phenobarbitone Spansule Capsules, Poliomyelitis Vaccine, Pragmatar, Pragmatar Ointment, Redeptin Injection, Redeptin. A, Ridaura, Rimevax Injection, Selacryn, Selvigon, SK-Amitriptyline, SK-Ampicillin, SK-Bamate, SK-Doxycycline, SK-Erythromycin, SK-Estrogens, SK-Furosemide, SK-Hydrchlorothiazide, SK-Lygen, SK-Metronidazole, SK-Penicillin G, SK-Penicillin VK, SK-Phenobarbital, SK-Potassium Chloride, SK-Pramine, SK-Prednisone, SK-Probenecid, SK-Propantheline Bromide, SK-Quinidein Sulfate, SK-Reserpine, SK-Soxazole, SK-Tetracycline, SK-Thioridazine HC1, SK-Tolbutamide, SK-Triamcinolone, SK-65, Skefron, SK65, Spaneph Spansules, Stelabid Tablets, Steladex, Stelazine, Stelazine Tablets, Spansule Capsules, Syrup, Injection and Concentrate, Stoxil, Sulphex, Tagamet Tablets, Syrup, Injection and Infusion, Temaril, Thorazine, Tolnate, Tyrimide, Urispas, Vasocort, Vertigon, Vertigon Spansule Capsule, Vontrol, Z Span Spansule Capsule.

Smooth-On Inc,

1000 Valley Road,
Gillette NJ 07933
USA

Smooth-On Inc

Metalset, Smooth-On.

SNIA (UK) Limited,

25-27 Oxford Street,
London W1A 4AE
UK

Chlorothalonil, Clarfina, Cloparin, Cloparol 50, Cloparten Z, Lilion 6, Lilion 66, Lipoclor S, Oromid, Prolit, Sniamid, Solpolac, Velicren.

Solaver SA,

Rue de l'Invasion,
4800 Verviers,
Belgium

Lalicopharsol, Lalitecsol, Lancosol, Lanpharsol, Lantecsol.

Solec - Solar Energy Corporation,

Box 3065,
Princeton, NJ 08543-3065
USA

LO/MIT-1, Solkote Hi/Sorb-II.

Solochart Limited,

Brookhampton Lane,
Kineton, Warks CV35 0JA
UK

Durosehl, Silicosehl.

Solrec Limited,

Middleton Road,
Morecambe, Lancs LA3 3JW
UK

Printwash, Solvent 401, Solvent 78.

Soltex Polymer Corporation,

3333 Richmond Avenue,
Houston, Texas 77098
USA

Fortiflex, Fortilene, Solef.

Spaulding Fibre Co,

1300 S 7th Street,
DeKalb, IL 60115
USA

Armite.

Specialty Products Company,
15 Exchange Place,
Jersey City, NJ 07303
USA

Biosol, Golden Wax, Honey Wax, Kantmelt, KantStik, Kleenmold, Klenal, Regnis, Resin Release N, Slick, Sprodco, Toolife, Tuf-Lube.

SPER Chemical Corporation,

P O Box 5566,
Clearwater, Florida 33518
USA

Sequest-All.

SPREA Spa,

Via Camperio 9,
20123 - Milano,
Italy

Eponac, Melsprea, Moldesite, Polnac, Uroplas.

Squibb, E R & Sons Inc,

PO Box 191,
New Brunswick, NJ 08903
USA

Squibb, E R & Sons Limited,
Medical Division,
Squibb House, 141/149 Staines Road,
Hounslow, Middlesex TW3 3JA
UK

Actrapid, Actrapid Human, Adcortyl Cream, Adcortyl in Orabase, Adcortyl Injection, Adcortyl Ointment, Adcortyl Spray, Adcortyl with Graneodin Cream, Adcortyl with Graneodin Ointment, Albumotope I-131, Azactam., Capoten, Capoten Tablets, Capozide, Choletec, Cholografin, Cholografin Meglumine, Chromalbin, Chromitope Sodium, Cobatope-60, Corgard, Corgard Tablets, Corgaretic, Crysticillin, Cystografin,

Delalutin, Delatestryl, Delestrogen, Dexacillin, Dirocide, Dolmatil, Dynamutilin, Dynamutilin Aqueous Solution, Dynamutilin Water Soluble Powder, Econacort, Ecostatin, Ecostatin Cream, Ecostatin Lotion, Ecostatin Pessaries, Ecostatin Powder Solution, Equipoise, Ethril, Florinef, Florinef Acetate, Fluzone, Follutein, Fungilin, Fungilin Cream, Fungilin Lozenges, Fungilin Ointment, Fungizone, Fungizone for Infusion, Gastrografin, Graneodin, Graneodin Ointment, Halciderm, Halog, Hipputope, Hydrea, Iodotope I-125, Iodotope I-131, Iodotope Therapeutic, Ipral, Isovue, Kenacort, Kenacort Diacetate Syrup, Kenalog, Kenalog Injection, Kinevac, Lentard, Lenté Insulin, Mint-O-Mag, Modecate, Moditen, Moditen Enanthate, Moditen Tablets, Monotard, Monotard Human, Motipress, Motipress Motival Tablets, Motival, Multilind, Multilind Ointment, Mycostatin, Mysteclin Capsules, Mysteclin Syrup, Mysteclin Tablets, Natritope Chloride, Naturetin, Noctec, Noctec Capsules, Nydrazin, Nystadermal, Nystan, Nystan Cream, Nystavescent, Nystavescent pessaries, O-V Statin, Ophthaine, Opthaine, Ora-Testryl, Orabase, Oragrafin Calcium, Oragrafin Sodium, Orahesive, Oravue, Pentids, Phosphotope, Plastibase, Princillin, Principen, Principen/N, Prolixin, Prolixin Enanthate, Pronestyl, Pronestyl Solution, Pronestyl Tablets, Protophane, Rau-Sed, Raudixin, Raudixin, Rautrax Sine K, Rautrax Tablets, ReCovr (veterinary), Remiderm, Remiderm Cream, Remiderm Ointment, Remiderm Spray, Remotic, Remotic Capsules, Rencal, Reno-M, Renovue-DIP, Renovue-65, Rezifilm, Rheaform, Robengatope I-131, Rubramin PC, Rubratope-57, Semilente Insulin, Sethotope, Steclin, Stomahesive, Strotope, Strycin, Sucostrin, Sumycin, Sweeta, Terfonyl, Teslac, Tesuloid, Tocopherex, Tri-Adcortyl, Triadcortyl Cream, Triadcortyl Ointment, Tricaderm, Trimox, Trioleotope, Tritiotope, Ultralente Insulin, Ultratard, Valadol, Veetids, Velosef, Verdiviton, Verdiviton Elixir, Vesprin, Vetalog, Zinc Omadine.

Stafford-Miller Limited,

32-36 The Common,
Hatfield, Herts AL10 0NZ
UK

Alphosyl, Alphosyl HC., Colifoam, Dexamist Ear Spray, Epifoam, Phazyme, Proctofoam HC, Quellada, Tarcortin.

Standard Chemical Company,

Mill Lane, Cheadle,
Cheshire SK8 2NX
England

Atolex ASL/C, Atolex ASL/C100, Atolex AST/3, Atolex DA/25, Atolex Polythene Emulsions, Atolex QE, Sovatex C1, Sovatex EP 5288, Sovatex IM12H, Sovatex IM12N, Sovatex IM17H, Sovatex IM17N, Sovatex MP/1, Sovatex WA, Sufatone SCS/B, Sufatone SCS/CL, Sufatone SC/L and SMC/L, Sufatone SMC/W, Sulphatol A and TL/B, Sulphatol CL, Sulphatol LS3, Sulphatol LX/B, Sulphatol PD/B.

Standard Tar Products Company Inc,

2456 W Cornell Street,
Milwaukee, WI 53209
USA

Ice Melt, Snomelt, Usol Copper Green, Usol Organiclear, Usol Zinclear.

Stanley Smith & Co Plastics Limited,

Worple Road,
Isleworth, Middlesex TW7 7AU
UK

Ignicide, Lockite, Rigilene, Sub-Vitralen, Vitradur, Vitrahose, Vitralen, Vitralene, Vitralex, Vitrapad, Vitraplas, Vitrathene, Vitrite, Vitrone.

Staveley Chemicals Ltd,

Staveley Works, Chesterfield,
Derbyshire S43 2PB
England

Aseptisil, Diuron.

Steetley Chemicals Ltd,

Berk House, PO Box 56, Basing View,
Basingstoke, Hants RG21 2EG
England

Aripol, Flochel, Florosal, Hymod, Stellak.

Steetley Refractories Ltd,
Magnesia Materials Division,

PO Box 8, Hartlepool,
Cleveland TS24 0BY
England

Anscor, Flamarret, Insulmag, Lycal, Magnaspheres, Sermag, Tanbase.

Steinhard, M A Limited,
Pharmaceutical Division,

32-36 Minerva Road,
London NW10 6HJ
UK

Almazine, Aluline, Alunex, Alupram, Fortunan, Glibadone, Metox, Nidazol, Oxanid, Paxadon, Paxalgesic, Paxalol, Paxane, Paxofen, Paxofenac, Paxolax.

Stella-Meta Filters

Laverstoke Mill,
Whitchurch, Hants RG28 7NR
UK

Metasil AL, Metasil ALAG, Metasil A, A+. Extra A+, B, C, E, Metasil D, Metasil DA, Metasil MQC, Metasil Purasil, Metasil R, Metasil SA, SB, Metasil W/2, Pafra, Escomer, Escor, Escorene, Exxtraflex, Jayflex, Vistaflex, Vistalon, Vistanex, Stellos.

Stepan Company,
Surfactant Department,

22 Frontage Road,
Northfield, IL 60093
USA

Amidox, Bio Soft, Bio Terge, Kessco, Lathanol, Makon, Nacconal, Ninate, Ninol, Ninox, Pestilizer, Petrostep, Polystep, Steol, Stepanate, Stepanflo, Stepanflote, Stepanform, Stepanol, Stepantan, Steposol, Toximul.

Sterling Drug Inc,

90 Park Ave,
New York, NY 10016
USA

Alveograf, Amenide, Amipaque, Aralen Hydrochloride, Aralen Phosphate, Atabrine Hydrochloride, Benylate, Bilopaque, Blockain Hydrochloride, Breokinase, Bronkaid Mist, Bronkephrine, Bronkometer, Bronkosol, Bryrel, Carbocaine, Cariod, Cartose, Chronogyn, Cybis, Danocrine, Demerol, Dexawin, Diaparene, Dilabil, Dinacrin, Drisdol, Ducobee-Hy, Enzeon, Etrynit, Eumydrin, Evipal, Falmonox, Flavaxin, Forit, Hypaque Meglumine, Hytakerol, Inocor, Istizin, Isuprel Hydrochloride, Kayexalate, Lavema, Lotusate, Luminal Sodium, Marcaine, Measurin, Mebaral, Meretestate, Monacrin, Nalutron, NegGram, Neo-Synephrine Hydrochloride, Novaldin, Novocain, Omnipaque, Oxucide, Palohex, Parenamine, Periograf, Pontocaine, Pontocaine Hydrochloride. , Pressonex Bitartrate, Roxadyl, Skiodan Sodium, Sulfactol, Superinone, Suprarenin, Talwin, Telepaque, Thyractin, Tornalate, Trimax, Win-Kinase, Winstrol, Wintomylon, Zephiran Chloride, Zettyn.

Sterling Moulding Materials Limited,

Ashton New Road,
Manchester M11 4AT
UK

Bakelite, Rutaform, Sternite, Synmold.

Sterling-Winthrop Group Limited,
Sterling Research Laboratories,

Onslow Street,
Guildford, Surrey GU1 4YS
UK

Calcium Resonium, Danol, Eradacin, Franol Plus, Glurenorm, Hayphryn, Mictral, Modrenal, Pameton, Phisomed, Pyrogastrone, Selora, Sulfamylon, Sulfomyl, Trancoprin.

Sterwin Chemicals Inc,

90 Park Ave,
New York, NY 10016
USA

SW 5063, Zimco, CSI, POLYCON II, SYNSOFT.

Stiefel Laboratories (UK) Limited,

Holtspur Lane, Woodburn Green,
High Wycombe, Bucks HP10 0AU
UK

Stiefel Laboratories Inc,
Ponce de Leon Blvd,
Coral Gables, Fla 33134
USA

AcetOxyl, Acne-Aid Detergent Soap, Anthranol, Benoxyl, Brasivol, Covermark, Driclor, Duofilm, Ichthyol, Lacticare, Lasan, Oilatum Application, Oilatum Emollient, Oilatum Soap, PanOxyl Aquagel, PanOxyl Wash, Polytar, Scabene Lotion, Spectraban, Stiedex, Zeasorb.

Stuart Pharmaceuticals
see Imperial Chemical Industries plc

Stull Chemical Company,

PO Box 47907,
San Antonio, TX 78265
USA

Bivert, T M P.

Sturge Lifford
see John & E Sturge Ltd

Sud-Chemie AG

Postfach 20 22 40,
8000 Munchen 2
West Germany

Advitrol, Agriben, Clarit, Copisil, Edasil, Fixat, Geko, Montigel, Nettolin, Opazil, Rheocin, Silvital, Sudflock, Tixogel, Tixoton, Tonsil.

Sumitomo Bakelite Co Limited,

2-2, 1-chome, Uchisaiwai-Cho,
Chiyoda-Ku, Tokyo 100
Japan

Daponite Sheet, Deco Board P, Deco Poly, Decoart, Decola, Decola Back Sheet, Decola Excel, Decola F, Decola FG, Decola MA, Decola MF, Decola New Marine, Decola PFC, Encapsulation, Igetaleim MA, Igetaleim UA, Mateflex, Moldex, Nonflammable Decobest DA, Plastic Magnet, Spidax, Sumibond PA, Sumicool, Sumiflex, Sumikon, Sumikon AM, Sumikon EM, EME, Sumikon IM, Sumikon PM, Sumikon TM, Sumikon VM, Sumilac PC, Sumilite CEL, Sumilite EI, Sumilite EL, Sumilite ELC, Sumilite FS, Sumilite IL, Sumilite ILC, Sumilite ILI, Sumilite NS, Sumilite PL, Sumilite PLC, Sumilite Resin PR, Sumilite Resin PR, Sumilite Resin PR, Sumilite STS, Sumilite TFC, Sumilite TFP, Sumilite VSL, Sumilite VSS, Suminet, Sumipipe, Sumitac EA, Sumitac GA, Sumitac VA, Sunimac ECR, Sunimac GCR.

Sumner Oil Industries,

500 Talcott Avenue,
St Louis, MO 63147
USA

Ordnance 204, 500, Tool Life.

Sun Refining & Marketing Company,

1801 Market Street, 25/10PC,
Philadelphia, PA 19103
USA

Sunaptic Acids.

Surmak Products Limited,

93-95 Mabgate,
Leeds LS9 7DR
UK

Surmabond 'Lining', Surmabond 'Roadway', Surmabond 'Screeding', Surmafil, Surmaglaze U12, Surmaplast, Surmaseal 101, Surmaseal 102, Surmatar.

Symons, H D & Company Limited
see Cyanamid Fothergill Limited

Syntex Laboratories Inc,

3401 Hillview Ave,
Palo Alto, Calif 94304
USA

Anadrol, Anapolon, Anaprox, Delmate, Domoso, Equiproxen, Lidex, Masteril, Masterone, Metilar, Monocortin, Naprosyn, Nor-Q D, Norinyl-1, Norinyl-2, Stemex, Synalar-HP, Synanthic, Synchrocept, Synemol, Topsyn, Zenadrid.

Syntex Ophthalmics Inc
see Albright & Wilson Ltd

Synthese BV,

Ringersweg 5,
4612 PR Bergen op Zoom
Netherlands

Setacure, Setafix, Setal, Setalin, Setalux, Setamine US, Setarol, Setatack A, Setatack AF, Setatack LP, Setatack P, Setatack T, Setyrene.

Synthetic Products Company,

16601 St Clair Avenue,
Cleveland, Ohio 44110
USA

Synpro, Synpro-Ware, Synpro-Ware, Synpron.

Synthite Ltd,

Ryders Green Road, West Bromwich,
W Midlands B70 0AX
England

Alcoform, Synsolve, Synstryp, Synstryp, Synteze, Synthite.

T

TACC International Corporation

Air Station Industrial Park,
Rockland, MA 02370
USA

Tree Bug-Lok Adhesive.

TAP Pharmaceuticals,

North Chicago,
Ill 60064
USA

Cefmax, Cefomonil, Lupron.

Taskem Inc

8525 Clinton Road,
Cleveland, Ohio 44144
USA

Criterion, Starglo, Xenith.

Tenneco Malros Ltd
see Albright & Wilson Ltd

Tensia SA,

Avenue des Tilleuls 62,
B-4000 Liege,
Belgium

Tensia Ltd,
North Road Industrial Estate, Bridgend,
Mid Glam CF31 3TR
England

Ditensamine C, O and S, Dumacene C13, NP707, NP7710 and NPX10, Egalex, Emultex, Tensabit, Tensactol, Tensadal, Tensagex, Tensagex BV, Tensagex DMY, Tensagex DP24, Tensagex EOC, Tensagex SPDL, Tensamina, Tensamine C, O, S and SH, Tensaminox, Tensarane SBTE, Tensaryl, Tensaryl DF90, Tensaryl DX54Sp. and DX62, Tensaryl KD, Tensaryl L48, Tensaryl SB, Tensaryl SB Ca, Tensaryl SBD, Tensaryl SB85P, Tensaryl S30P and S70P, Tensaryl 40CC, 50B, 80B and 82F, Tensatil DA120, Tensatil DB120, Tensatil DEH120, Tensatil D100, Tensiamix, Tensianol, Tensibet 50, Tensibet 55, Tensidef, Tensidye, Tensimul, Tensiofix, Tensioquat C50, Tensioquat C75, Tensiorex, Tensiostat, Tensipar, Tensitex, Tensocide, Tensoleate, Tensoline, Tensomel, Tensomin, Tensopac, Tensopane D, Tensophene D12, D15 and D18, Tensophene H10, I10, DT, D36, D42EC, D45, D60 and D90, Tensophene 2D30, Tensopol LT, Tensopol ACL and PCL, Tensopol AG and MG, Tensopol A.7 and USP, Tensopol DX85 and FL, Tensopol N, Tensopol SPK, Tensopol VAL, Tensopol 12A and 12P, Tensopol 30E and LDS, Tensoprene, Tensostat, Tensovax, Tensovyl, Tensuccin D8, Tensuccin HS40, Tensuccin H724 and H925, Tensuccin ML, MO and MS, Tensyl 30, Terbytex.

The Terrell Corporation,

820 Woburn Street,
Wilmington, Mass, 01887
USA

Terlan.

Testworth Laboratories Inc,

401 S Main Street, Columbia City,
Indiana 46725
USA

Curaseal, Paracure, Paralac, Paratol.

Tetrahedron Association Inc,

5060 A Convoy St,
San Diego, CA 92111
USA

Audrey.

Texaco Ltd,

195 Knightsbridge,
London SW7 1RU
England

Texaco Ltd

Texaco Chemical Company,
Oxides & Specialties Division,
4800 Fournace Place, PO Box 430,
Bellaire, TX 77401
USA

Diglycolamine Agent, Jeffamine, Jeffox, Surfonic, Surfonic, Surfonic, Surfonic N, Texsolve, Thancat, Thanol.

Th Goldschmidt Ltd, Chemical Products
see Goldschmidt, Th Ltd, Chemical Products

Thermedics Inc.,

470 Wildwood Street, PO Box 2999,
Woburn, MA 01888-1799
USA

Spandra Transparent Dressing, Tecoflex Polyurethane.

Thiokol Gesellschaft mbH,

Sandhofer Strasse 96,
6800 Mannheim 31,
Germany

Thiokol LP.

Thomas Ness Ltd,

Eastwood Hall, Eastwood,
Nottingham NG16 3EB
England

Adamac, Bi-Tarco, Hevikote, Presomet, Synthaprufe, Synthasil.

Thomas Swan & Co Ltd,

Crookhall, Consett,
Co Durham, CH8 7ND
England

Casabet, Casabet 655, Casahib, Casamer, Casamid, Casamids, Casamine, Casamox, Casaquat, Casateric, Casathane, Catex, Cormul, Cygna, Stannicide, Wetfix, Y-Tack.

Thompson, J C & Co (Duron) Ltd,

Duron Works, Drummond Road,
Bradford BD8 8DX
England

Duralcon, Duroil, Durolube, Duroslip, Durosoft, Durosol, Durotex, Durotint, Durowynd.

Thomspon, Weinman & Co,

PO Box 130, Cartersville,
GA 30120
USA

Atomite.

Tioxide UK Limited,
TIL Division

Haverton Hill Road,
Billingham, Cleveland TS23 1PS
UK

Tioxide Group PLC,
10 Stratton Street,
London W1A 4XP
UK

Til, Tilcom, Tioxide.

Travenol Laboratories Inc,

One Baxter Parkway,
Deerfield, Ill 60015
USA

Flint Laboratories,
Div of Travenol Laboratories Inc,
One Baxter Parkway,
Deerfield, Ill 60015
USA

Hyland Therapeutics,
Div of Travenol Laboratories Inc,
444 W Gelnoaks Blvd,
Glendale, Calif 91202
USA

Amigen, Buminate, Choloxin, Discase, Flint SSD, Gentran 40, Gentran 75, Hemofil, Hu-Tet, Osmitrol, Proplex, Synthroid, TIS-U-SOL, Travamin, Travase.

Tri-Star Chemical Company Inc,

P O Box 38627,
Dallas, TX 75238
USA

Tri-Star Antifoam £27, Tri-Star Padding Compounds, Tri-Star White Glues.

Tufnol Limited,

PO Box 376, Wellhead Lane,
Perry Barr, Birmingham B42 2TB
UK

Adder, Asp, Bear, Carp, Crow, Heron, Kite, Lynx, Swan, Tufnol, Tufset, Vole, Whale.

UCB nv Film Sector
Sidac Films Division,

Ottergemsesteenweg 801,
B-9000 Gent
Belgium

Cellophane, Fresh Pak, Sidanyl, Sylphane S, Vistal.

U

UCO Optics Inc

3000 Winton Rd,
Rochester, NY 14623
USA

Aquaflex, Permathin.

Ulfcar International A/S,

P O Box 1020, Lindenborg,
4000 Roskilde
Denmark

Acrydur.

Ultrachem Inc,

1400 North Walnut Street, P O Box 2053,
Wilmington, DE 19806
USA

Chemlube, Omnilube, Syntroil, Ultrachem Assembly Fluid 1, Vischem.

Unger Fabrikker AS,

PO Box 306,
N-1601 Fredrikstad,
Norway

Ufacid, Ufanon, Ufaryl, Ungerol.

Unichema Chemie BV

Unichema Chemie BV

The Sales Organisation of Unichema International,
PO Box 2,
2800 AA Goude,
The Netherlands

Priamid.

Unifos Kemi AB,

PO Box 44,
S-44401 Stenungsund
Sweden

Neste Oy,
Chemicals Division,
SF-06850 Kulloo
Finland

Neste Polyethylene.

Unilab Inc,

764 Ramsey Ave,
Hillside, NJ 07205
USA

Unilab Surgibone.

Unimed Inc,

35 Columbia Rd,
Somerville, NJ 08876
USA

Marinol, Spiro-32.

Union Carbide Corp,

Old Saw Mill River Rd,
Tarrytown, NY 10591
USA

Carbowax, Carbowax Sentry, Cellosolve, Halowax 4000 B-2, Parylene N, Sentry Cyclomethicone, Sentry Dimethicone, Sentry Simethicone, Tergitol S, Ucon, Udel.

Uniroyal Chemical,
Division of Uniroyal Limited,

Brooklands Farm, Cheltenham Road,
Evesham, Worcs WR11 6LW
UK

Uniroyal Chemical Company Inc,
Elm Street,
Naugatlick CT 06770
USA

Uniroyal Chemical Group,
World Headquarters, Middlebury,
Connecticut 06749
USA

Activator 736, Adiprene, Aminox, Antioxidant 431, Antioxidant 449, Antioxidant 451, Aranox, Arazate, Beutene, BIK, BLE, Bonding Agent M 3, Bonding Agent P 1, Bonding Agent R 6, Butazate, Butazate 50D, BXA, Celogen, CPB, DBA Accelerator, Delac MOR, Delac NS, Delac S, Dibenzo GMF, Ethazate, Ethazate 50D, Ethyl Tuex, Flexamine, Flexzone, GMF, Hepteen Base, JZF, M B T, M B T S, Monex, Naugawhite, Naugex SD-1, Nauguard BHT, Nauguard H, Nauguard K, Nauguard PAN, Nauguard Q, Nauguard SP, Nauguard T, Nauguard 445, Nauguard 475, Nauguard 477, Nauguard 495, O X A F, O X A F 50D, Octamine, Paracril (NBR), Polygard, Polylite, Retarder ESEN, Royalac 133, Royalac 136, Royalac 140, Royalene (EPDM), Sunproof, Thiostop E, Thiostop N, Thiostop N, Tonox, Trimene Base, Tuex, Vibrathane, Vulklor.

Unitex Limited,

Halfpenny Lane,
Knaresborough, N Yorkshire HG5 0PP
UK

Celulon, Ulon.

Universal-Matthey Products Ltd,

Jeffreys Road, Brimsdown,
Enfield, Middlesex EN3 7PN
England

Unicor, Unidem, Uniflood, Unipor, Unitreat, UOP, UOP.

UOP Inc
see Allied Corporation

The Upjohn Co,

7171 Portage Rd,
Kalamazoo, Mich 49002
USA

The Upjohn Co

Acetonyl, Acti-Dione, Albamycin Capsules, Albamycin G.U, Albamycin T, Alphadrol, Ansaid, Arbacet, Baciguent, Berubigen, Biosol, Calderol, Cheque, Cleocin, Colestid, Cordex, Cortaid, Cortef, Cortef Acetate, Cortef Oral Suspension, Cyclo-Prostin, Cytosar, Cytosar-U, DAB, Dalacin C, Delta Cortef, Delta-Cortef, Deltasone, Depo-Medrol, Depo-Penicillin, Depo-Provera, Depomedrone, Deracyn, Didrex, Diurnal-Penicillin, E-Mycin, E-Mycin E, Feminone, Florone, Halcion, Halotestin, Hylorel, Kaomycin, Kaopectate, Lincocin, Linocin, Loniten, Lutalyse, Maolate, Medro - Cordex, Medrol, Medrol ADT Pak, also Medrol Dosepak, Medrol Enpak, Medrol Stabisol, Medrone, Medrone Medules, Medrone Veriderm, Melanate, MGA, Micronase, Mitaban, Motrin, Motrin-A, Mycifradin, Myciguent, Mylosar, Myringacaine Drops, Neo Cortef, Neo-Medrone Acne Lotion, Neomedrone Veriderm, Neomix, Nuprin, Orinase, Orinase Diagnostic, Orthoxicol, Orthoxine, Oxylone, Paminal, Pamine, Panmycin, Panmycin Hydrochloride, Panmycin Syrup, Phenolax, Polykol, Predef, Prostin F2 Alpha, Prostin VR Pediatric, Prostin/15M, Provera, Provera Dosepak, Regamycin, Reserpoid, Solu-Cortef, Solu-Delta-Cortef, Solu-Medrol, Solucortef, Solumedrone, Soxomide, Sugracillin, Super A, Super D, Tolanase, Tolinase, Trobicin, U-Gencin, Unakalm, Uracil Mustard, Uticillin VK, Xanax, Zanosar.

US Ethicals Inc,

37-02 48th Avenue,
Long Island City, NY 11101
USA

Eclabron, Klavikordal, Niong, Nitronet.

USI Chemicals,

11500 Northlake Drive,
Cincinnati, OH 45249
USA

Microthene, Petrothene, Petrothene XL, Punctilious Ethyl Alcohol, Spectrathene, Ultrathene, Vynathene.

USV Pharmaceutical Corp,

303 S Broadway,
Tarrytown, NY 10591
USA

Arlidin, Azolid, Calcimar, Cerespan, Dalnate, DDAVP, Dema, E-Toplex, Levothroid, Lozol, Nicobid, Nicolar, Nitrospan, Oxalid, Panthoderm, Pentritol, Pertofrane, Prednis, Presamine, Thyrar, Thyrolar, Tri-Sulfameth, Vascuals.

V

VAMP srl

Viale Vigliani 19,
20148 Milan
Italy

Masterflam.

Van Dyk & Co Inc,

Main and William Sts,
Belleville, NJ 07109
USA

Escalol 507.

Vanderbilt Chemical Corporation,

Penny Road,
Murray, KY 42071
USA

Activ 8, Agerite, Altax, Amax and Amax No 1, Amyl Ledate, Antozite, Bilt-Plates, Bismate, Black Out, Bondogen, Butyl Namate, Captax, Ceramitalc, Continental Clay, Darvan, Dixie Clay, Durax, Ethyl Cadmate, Ethyl Tellurac, I T Talc, Langford Clay, Leegen, McNamee Clay, Methyl and Ethyl Selenac, Methyl Cumate, Methyl Isobutyl Niclate, Methyl Ledate, Molyvan, Morfax, Nacap, Nytal, Octoate 2, Par Clay, Plastogen, Propyl Zithate, Reogen, Rodo, Rokon, Rotax, Select-A-Sorb, Setsit, Sulfads, Thermax, Thiate, Tuads, Unads, Van Bac, Van Gel, Van Wax, Van-Amid, Vanax, Vanchem, Vanchem HM-4346, Vanchem HM50, Vancide, Vanlube, Vanox, Vanoxy, Vanplast, Vanplast 201, Vansil, Vanstay, Vantalc, Vanzak, Vanzyme, Varox, Vergum, Zetax, Zimate.

Vanderbilt, R T Co Inc,

30 Winfield St,
Norwalk, Conn 06855
USA

Vanderbilt, R T Co Inc

Rio Resin, Tysonite, Vanclay, Veegum.

Vass (Agricultural), L W Limited,

Springfield Farm, Silsoe Road,
Maulden, Bedford MK45 2AX
UK

D S M, Standup.

Vedag GmbH,

Flinschstrasse 10-16,
6000 Frankfurt am Main 60,
Germany

Alytol, Coripact, Emaillit, Kaltas, Tegula, Thisol, Vedacoll, Vedacolor, Vedafix, Vedaflex, Vedaform, Vedag BM, Vedagit, Vedagolan, Vedagully System, Vedagum, Vedalith Facade System, Vedaphalt, Vedaphon, Vedapor, Vedapurit, Vedasin, Vedastar, Vedatect, Vedatex, Vedathene, Vedatherm, Webas.

Velsicol Chemical Corp,

341 East Ohio Street,
Chicago, Ill 60611
USA

Benzoflex.

Vencel Resil Limited,

Arndale House, 18-20 Spital Street,
Dartford, Kent DA12 2HT
UK

Claymaster, Jabclad, Jabdec, Jabdie, Jablina Insulating Panels, Jablite, Jablite Cavity, Jablite Flooring, Jablite Insulation Board, Jablite Thermacel, Jablite Thermoclik, Thermodek, Top 7 Mosaic and Pebble, V R Coving, Voidform, Voidmaster, Warmaline.

Venture Chemical Products Ltd,

Boxgrove House, Little Heath Road, Tilehurst,
Reading, Berks RG3 5TX
England

Miranol CM, Miranol C2M, Miranol C2M-SF, Miranol DM, Miranol JEM, Miranol L2M-SF, Miranol SM.

Victor Wolf Limited
see NL Victor Wolf Limited

Vikwood Botanicals Inc,

1817 N 5th Street,
Sheboygan WIS. 53081
USA

Margosan-O.

Virkler Chemical Company,

1022 Pressley Road,
Charlotte, NC 28210
USA

Ultra Touch, Virco.

Visiontech Inc,

Sauflon Plant, 305 N Oak St,
Inglewood, Calif 90302
USA

Sauflon PW, Sauflon 70.

Vistakon Inc,

1325 San Marco Blvd,
NJ 08512
USA

Hydro-Marc, Vista-Marc.

Vulcan Plastics Limited,

Hosey Hill,
Westerham, Kent TN16 1TB
UK

Vulcabest, Vulcaboard, Vulcabrite, Vulcalap, Vulcalon, Vulcalucent, Vulcamin, Vulcaplast, Vulcasteel, Vulcatuf.

Vulnax International Limited,

PO Box 60, Chesford Grange,
Woolston, Warrington WA1 4SE
UK

Vulnax International Limited

Vulnax International Limited,
321 Bureaux de la Coline,
92213 Saint Claud Cedex
France

Vulcaflex, Vulcaperl, Vulcastop.

Wacker-Chemie GmbH,

Prinzregentenstrasse 22,
8000 München 22
Federal Republic of Germany

Elastosil, Polyviol, Vinnapas, Vinnol.

W

Wallace Laboratories

Div of Carter-Wallace Inc,
Cranbury, NJ 08512
USA

Avazyme, Bactratycin, Bepadin, Dormate, Meprospan, Miltown, Organidin, Ovol, Solacen, Soma, Sulfatryl, Vanair.

Ward Blenkinsop & Co Ltd,

Halebank, Widnes,
Cheshire WA8 8NS
England

Adaprin, Aduvex, Auracet, Conotrane, Cortitrane, Coumalux, Ekammon, Elargol, Gon, G.500, Lithostar, Mekad, Merquinox, Mortha, Octaflex, Pavacol, Pavacol Diabetic Cough Syrup, Penotrane, Placadol, Prolugen, Quantacure, Uvistat, Uvistat 12, 24, 247, 2211, Uvistat 247, Viacutan.

Warwick Chemical Ltd,

Wortley Moor Road,
Leeds LS12 4JE
England

Mykon, Mykon 817, Mykonaid, Warcodet D, Warcodet K54, Warcodet V, Warcodye CLP, Warcodye RWL, Warcosoft WSC, Warcowet O.

Watts Urethane Products,

Church Road, Lydney,
Gloucester GL15 5EN
UK

Ultrathane.

W D Service Company Inc

W D Service Company Inc

780 Creek Road, PO Box 147,
Bellmawr, New Jersey 08031
USA

Pos O Print.

Weatherguard/Marbeloid Products Inc,

2515 Newbold Avenue,
Bronx, NY 10462
USA

Durolastik, Kompolite, Marbleloid, Wepco.

The Wellcome Foundation Limited,

PO Box 129, The Wellcome Building,
183 Euston Road, London NW1 2BP
UK

The Wellcome Foundation Ltd,
Temple Hill, Dartford,
Kent DA1 5AH
England

Burroughs Wellcome Co,
3030 Cornwallis Rd,
Research Triangle Park, NC 27709
USA

Actidil, Actifed, Actifed with Codeine Cough Syrup, Actified Compound Linctus, Activax, Aerosporin Sterile Powder, Alcopar, Alkeran Tablets, Almevax, Ancaris, Angised, Antepan, Antepar, Antepar, Antoban, Arilvax, Banocide, Bercotox, Blanthax, Borofax Ointment, Bovinox, Bretylate, Calkleen, Calpol, Calquat, Calsan, Calstrip, Canopar, Cardilate, Carovax, Chlordispel, Cicatrin, Circacid, Circosan, Contrapar, Coopane, Coopaphene, Coopercote, Coopermatic, Coppertox, Coriban, Cortisporin, Cortisporin, Cortisporin Ointment, Covexin, Cyclimorph, Daraclor, Daraclor, Daraprim Tablets, Darvisul, Diconal, Digibind, Empirin, Empracet Codeine Phosphate, Epivax, Esbatal, Esbatal, Famosan, Fascol, Faunolen, Fedrazil Tablets, Ferromyn, Flammex, Flolan, Franocide, Frantin, Freeflo, Gammatox, Gletvax, Haloxil, Histantin, Hypon, Injex, Iodron, Kemadrin Tablets, Kerol, Lactosan, Lanoline, Lanoxicaps, Lanoxin, Lanoxine-PG, Lanvis, Leptovax-Plux, Leukeran Tablets, Linctifed, Linctified Expectorant, Lubafax, Magnilor, Maloprim, Mantadil Cream, Marezine, Marezine Injection, Marzine, Mastrite, Medicaire, Melimax, Mellavax, Migil, Migril, Monopar, Multispray, Myleran, Myleran Tablets, Neopybuthrin, Neosporin Products, Nix Cream Rinse, Omnivax, Otosporin, Ovigest, Parahypon, Pentostam, Perigen, Physeptone, Physeptone Linctus, Pif-Paf, Pipricide, Polybactrin, Polyfax, Polysporin, Pro-Actidil, Proloprim, Proloprim Tablets, Protogest, Protormone, Puri-Nethol, Purinethol, Pybuthrin, Quamilin, Ratox, Reslin, Reslin S, Rinsan, Rythmatine, Sanaklenz, Sanatank, Sanitant, Saxin, Scolaban, Septra, Septrin, Sokoff, Sudafed, Sufafed Products, Suscovax, Systamex, Tabloid, Temadex, Tenoban, Tineafax, Tracrium

Injection, Tractium, Tribovax, Tribrissen, Triostam, Trivax, Trivexin, Tubarine, Valoid, Vaporole, Vaporole, Vasoxine, Vasoxyl, Vasoxyl Injection, Vasylox, Viroptic Ophthalmic Solution, Wellbutrin, Wellcome Brand Athropine Sulphate Injection, Wellcome Brand Scopolamine Hydrobromide Injection, Wellcovorin, Wellferon, Zipcillin, Zovirax Products, Zyloprim Tablets, Zyloric, Zyloric.

Wesley-Jessen
see Schering AG

Wessex Resins & Adhesives Limited,

189-193 Spring Road,
Sholing, Southampton, Hants SO2 7NY
UK

Carpenters Wood Glue, Cascamite, Cascophen Resorcinol Resin RS216/RXS-8, Epoxy Putty Pack (EP-3/EHP-12), Extra Bond, Superfast Power Pack, West System Brand Products, WRA Epoxy Resin Underwater Series, WRA System 100 Laminating Composition, WRA System 17, WRA System 80, WRA1000 Varnish.

Westbridge Research Group,
Westbridge Oilfields Products Division,

2055 Silber Road, £104,
Houston, TX 77055
USA

Westbridge Research Group,
Westbridge Agricultural Products Division,
9920 Scripps Lake Drive, Suite 103,
San Diego, CA 92131
USA

Dominate, Dry Seed TRIGGRR, Foliar TRIGGRR, Soil TRIGGRR, Terminate, WesBio, WesLoTemp, WesScaleStop, WesSperse, WesTemp, WesTemp K+, Westhin, WesThin K+, WesVis.

Westbrook Lanolin Company,

Argonaut Works, Laisterdyke,
Bradford, West Yorkshire BD4 8AU
UK

Acetadeps, Aqualose, Argobase, Argonol, Argowax, Golden Dawn, Golden Fleece, Lanesta, Nicrolan.

Western Chemical Company,

1345 Taney,
North Kansas City, MO 64116
USA

Western Chemical Company

Corafilm, Coravol, Dry Lightning, Liquid Lightning, MultiGuard, Multionic, MultiSperse, Westo-Flocs.

Westwood Pharmaceuticals Inc
see Bristol-Myers Co

W Hawley & Son Limited
see Hawley, W & Son Limited

Wheatley Chemical Company Limited,

Langthwaite Grange Industrial Estate,
South Kirkby, Pontefract, W Yorks WF9 3AP
UK

Byacin, Byatran, Bygran F, Bygran S, Captan Granular, Nebulin, Nemacin, Trim, Tubazole, Tubazole M, Tubergran, Warefog.

Whitehall Laboratories
see Williams Division of Morton Thiokol Ltd

Whitmoyer Laboratories Inc,

99 S Fairlane Ave, PO Box 288,
Myerstown, PA 17067
USA

Carb-O-Sep.

Wickhen Products Inc,

Big Pond Rd,
Huguenot, NY 12746
USA

Wickenol 101, Wickenol 111, Wickenol 303, Wickenol 308, Wickenol 363 D.

William F Nye Inc,

P O Box G-927,
New Bedford MA 02742
USA

Nyebar, Nyogel.

William H Rorer, Inc,

500 Virginia Drive,
Fort Washington, Penna 19034
USA

Ananase, Camalox, Lidarral, Maalox, Perium, Thylogen Maleate.

William Pearson Limited,

Clough Road,
Hull HU6 7QA
UK

Creolin, Eucopine, Hycol, Hycolin, Pearsol, Sterillium.

Williams Division of Morton Thiokol Ltd,

Greville House, Hibernia Road,
Hounslow, Middlesex TW3 3RX
UK

Morton Thiokol Inc,
Carstab Division,
2000 West Street,
Cincinnati, OH 45215-6323
USA

Whitehall Laboratories,
Div of American Home Products Corp,
685 Third Ave,
New York, NY 10017
USA

Adcote, Advastab, Advawax, Alcovar, Ariabel, Ariagran, Arianor, Ariavit, Arigran, Automate, Canadian Certicol, Certolake, Conlex, Conrex, Hytherm, Linsol, Lucidene, Lustranyl, Luxol, Lytron, Morfast, Oilsol, Perlex, Perlextra, Perox, Polycryl, Serfene, Standacol, Usacert, Usalake, Veloset.

Witco Chemical Ltd,

Union Lane, Droitwich,
Worcs WR9 9BB
England

Witco Chemical Corp,
Organics Div,
520 Madison Ave,
New York, NY 10022
USA

Witco Chemical Ltd

Alconate L-80, Calcium Petronate 25H, 25C and 300, Cyclogol, Cycloryl 580 and 585N, Di-Petronate Series, Emcol CC-55, Emcol CC-9, CC-36 and CC-42, Emcol E-607, Emcol K8300, Emcol 4100M, 4150 and 4161-L, Emcol 4300, Emcol 4350, Emcol 4500, Emcol 4600, Emcol 4776, Emphos, Emulpon, Ethotal, Formrez, Hexaryl D60L, Neopon LAM, Neopon LOA/F, Neopon LOS, LOS/F and LOS/NF, Neopon LOT/F, Neopon LS, Neopon LT, Neopon 33, Petromix 9, Petronate L, HL, K, CR and S, Petronate RP, Pyronate, Sochamine A 271, Sochamine A 7525, Sochamine A 7527, Sochamine A 8955, Sulframin 1250, Sulframin 14-16 AOS, Sulframin 33, Sulphramin B and TPB, Syntopon A,B,C and D, Syntopon F,G and N, Syntopon 8, Witcamine AL42-12, Witcamine E-607, Witcamine E-607S, Witcamine 209 and 211, Witcamine 210, Witco TX, Witcodet 100 and P280, Witcolate AE-3S, Witcolate D5-10, Witcolate SE-5, Witconate D24-25, Witconate PTSA, Witconate P10-45 and P10-59, Witconate P10-49, Witconate SCS, Witconate STS, Witconate SXS, Witconol.

Witco Corporation,
Kendall/Amalie Division,

77 N Kendall Avenue,
Bradford PA 16701
USA

Kendex OCTG, Kendex 0220, Kendex 0834, Kendex 0842, Kendex 0847, Kendex 0866, Kendex 0898, Kensol 10, Kensol 13, Kensol 30, Kensol 48T, Kensol 50T, Kensol 51, Kensol 53, Kensol 61, Kensol 80.

W J Rendell Limited
see Rendell, W J Limited

W J Ruscoe Company
see Ruscoe, W J Company

W R Grace Ltd
see Grace, W R Ltd

Wyeth Laboratories,
Div of American Home Products Corp,

PO Box 8299,
Philadelphia, PA 19101
USA

Ayerst Laboratories,
South Way,
Andover, Hants SP10 5LT
UK

Ayerst Laboratories,
Div of American Home Products Corp,
685 Third Ave,
New York, NY 10017
USA

Advil, Algipan Balm, Almocarpine, Aludrox, Aludrox CO, Aludrox SA, Amphojel, Antabuse, Antepsin, Antromid-S, Apisate, APL, Ativan, Auralgan, Aygestin, BC500 With Iron, Beplete, Beplex, Bicillin L-A, B.C. 500, Calthor Suspension, Calthor Tablets, Cindol, Cyclamycin, Cyclapen-W, Dalgan, Darcil, Davenol, Dermoplast, Diucardin, Dristan Inhaler, Dristan Long Lasting Nasal Mist, Dry and Clear, Dryvax, Durapro, Endrine, Enzactin, Epitrate, Equadiol, Equagesic, Equanil, Equaprin, Equatrate, Esorb, Ethobral, Evadyne, Evramycin, Factrel, Fluor-I-Strip, Gluferate, Grisactin, Havapen, HRF Ayerst Laboratories, Indorm, Isordil Tablets, Isordil Tembids Capsules, Largon, Lentopen, Lodine, Mazanor, Membrettes, Mephine, Micolette Micro-enema, Mucaine, Mysoline, Omnipen, Omnipen-N, Ophthalgan, Ossivite, Ovran, Ovranette, Ovrette, Pathocil, Pen Vee Dural, Pen-Vee, Pen-Vee Drops, Pen-Vee K, Pen-Vee Suspension, Penidural All Purpose, Peptavlon, Phosphaljel, Phospholine Iodide, Plastules with Folic Acid, Plastules with Liver, Plegine, Pregfol, Premarin Vaginal Cream, Prempak, Prempak-C, Prepcort, Primatene Mist, Prondol, Protopam Chloride, Prozine, Riopan, Serax, Serenid Forte, Serenid-D, Simeco, Sonacide, Sparine, Sparine, Spartase, Stimplete, Streptets, Sulfose, Sulphamagna, S.M.A, Thiosulfil, Wyamine Sulfate, Wyamycin E, Wyamycin S, Wycillin, Wydase, Wymox, Wyovin, Wytensin.

Y

Yokkaichi Chemical Company Limited,

5F Nakasho Bldg,
1-5-4 Nihonbashi Bakuro cho,
Chuo ku, Tokyo 103
Japan

Alkanolamine, Catiomaster-C, Glycidyl Ether.

Yorkshire Chemicals plc,
Colours Division,

Kirkstall Road,
Leeds LS3 1LL
UK

Airedale, Benzanil, Diphone, Dyamul, Dyapol, Fernol, Fluisol, Novanyl, Optinol, Paradol, Paralene, Paramel, Parvol, Serilan, Serilene, Serinyl, Serisol, Sulphonol, Yoracryl.